AINSWORTH & BISBY'S

DICTIONARY OF THE FUNGI

CABI is an intergovernmental, not-for-profit organization specializing in scientific publishing, research and communication. We work to bridge the gap between the discovery of scientific knowledge and its application in solving real life problems.

CABI is a leading provider of authoritative scientific information on the applied life sciences; we publish CAB Abstracts, the most comprehensive bibliographic database in the applied life sciences. Covering over 150 countries, and over 50 languages, it gives researchers access to an abundance of information often not available from other databases. We also publish multimedia compendia, books and internet resources, including the serial publications *Index of Fungi* [from which the web resource *Index Fungorum* (www.indexfungorum.org) is derived] and *Bibliography of Systematic Mycology*.

CABI's focus on mycology derives from the very early days of the organization. It all began with the establishment of the Imperial Bureau of Mycology (IBM) in 1920, funded by a consortium of British colonial governments to the princely sum of £2000 per year. Its function was as an information provider and identification service for its member nations. In 1930 the name of the organization was changed to the Imperial Mycological Institute, from where the well-known 'herbarium' and culture collection acronym IMI is derived. Several names later, to reflect changing governance and politics, CABI continues to provide expert services to mycology world-wide. CABI holds a dried fungal reference collection (herb. IMI) with in excess of 400,000 specimens representing about 32,000 different species, and a living collection (incorporating the UK National Collection of Fungus Cultures) holding more than 19,000 living isolates representing about 4,500 species. These resources, coupled with 80 years of experience, enable us to offer a range of microbial diagnostic and consultancy services. We are also now actively screening our living collection for novel molecules of benefit to human health and development.

CABI provides research and consultancy in the following areas:

- systematics and ecology of fungi (including lichens, mushrooms and yeasts), nematodes, plant and soil bacteria
- preservation of organisms using cryogenic techniques
- biochemical, physiological and molecular characterization of strains
- biodiversity inventorying and monitoring
- ecology (especially relating to agroecosystems and invasive species management)
- crop protection (especially integrated pest management)
- soil health
- environmental and industrial microbiology
- food spoilage, public health, biodeterioration and biodegradation

CABI provides an authoritative identification service, especially for microfungi of economic and environmental importance (other than certain human and animal pathogens), and for plant pathogenic bacteria and spoilage yeasts.

AINSWORTH & BISBY'S

DICTIONARY OF THE FUNGI

by

P.M. Kirk, P.F. Cannon, D.W. Minter
and J.A. Stalpers

with the assistance of

T.V. Andrianova, A. Aptroot, G.L. Benny, R. Berndt,
T.W. Kuyper, F. Pando, P.J. Roberts, K. Vánky

and others

Tenth Edition

prepared by CABI Europe – UK

CABI is a trading name of CAB International

CABI Head Office
Nosworthy Way
Wallingford
Oxfordshire OX10 8DE
UK

Tel: +44 (0)1491 832111
Fax: +44 (0)1491 833508
Email: cabi@cabi.org
Web site: www.cabi.org

CABI North American Office
875 Massachusetts Avenue
7th Floor
Cambridge, MA 02139
USA

Tel: +1 617 395 4056
Fax: +1 617 354 6875
Email: cabi-nao@cabi.org

A catalogue record for the hardcover edition of this book is available from the British Library, London, UK.

ISBN-13: 978-0-85199-826-8 (hardback edition)
ISBN-13: 978-1-84593-933-5 (paperback edition)

First Edition 1943
Second Edition 1945
Third Edition 1950
Fourth Edition 1954
Fifth Edition 1961
Sixth Edition 1971
Seventh Edition 1983
Eighth Edition 1995
Ninth Edition 2001
Tenth Edition 2008
Paperback Tenth Edition 2011

Printed and bound by CPI Group (UK) Ltd, Croydon, CR0 4YY

Preface

This *Dictionary*, now in its 65th year, aims to provide an entry point into the sum total of our accumulated knowledge on systematic mycology for all those who work with fungi. All organisms traditionally studied by mycologists are covered, including lichens, mushrooms, slime moulds, water moulds and yeasts.

As more molecular data have become available it has been possible to attain greater certainty about the higher-level relationships of fungi and to see some enigmatic taxa at last find a home. While many of the classes and phyla recognized in the ninth edition of this *Dictionary* are retained here, we are aware that further significant change is likely among the fungi *sensu stricto*, with the proposal of several new high-level taxa in the near future. Likewise, we can expect further significant changes in the chromistan and protozoan fungus-like analogues as sequence data for more taxa become available. It has been our aim to recognize such changes while at the same time maintaining a servicable and comprehensive hierarchy for users.

In preparing the tenth edition, therefore, our efforts have been directed most of all to revision of the classification of higher ranks within the *Fungi*, largely based on the results from the AFTOL (Assembling the Fungal Tree of Life) project to which several of the Editors of this *Dictionary* had inputs. Phylogenetic information gained from multi-gene sequence analyses, culminating in 2006-7 with the results of the first phase of the AFTOL project, have revolutionized our understanding of how this kingdom should be classified. Phylogenetic analyses tend to stimulate recognition of many levels of the systematic hierarchy, and in partial response to this trend we now recognize the rank of subphylum in addition to classes and subclasses for the kingdom *Fungi*.

The second major development area for Edition 10 of the *Dictionary of the Fungi* has been to incorporate taxa at family level into the new classificatory framework, as the AFTOL project focused only on ranks at order and above. Where possible this has been carried out using molecular data, but there still remains a substantial number of fungal families for which sequence information is not available. More information may be found in the *Dictionary*'s new sister publication, *Fungal Families of the World* (CABI, 2007).

Many recent phylogenetic studies have been hypothesis-driven, designed to test the accuracy in evolutionary terms of traditional morphology-based classifications. As anamorph taxa have only recently started to be incorporated fully into holomorphic systems, they are substantially under-represented in molecular phylogenetic studies. Edition 9 of this *Dictionary* was the first to abolish separate classification systems for anamorphs and teleomorphs, though for the overwhelming proportion of genera it was only possible to assign them at subphylum level – i.e. to the filamentous *Ascomycota* or *Basidiomycota*. Recent studies have allowed more accurate placement of many asexual taxa, but today we still cannot place two thirds of the 3000-odd anamorph genera included in Edition 10 even to class level. Now the basic classificatory framework has been established to an acceptable degree of certainty, we hope that attention will be shifted towards insertion of these orphan taxa into their rightful place within the fungal system.

The already large and rapidly increasing body of evidence from molecular studies has also led us to the radical decision that this edition should comprise three parts – a Dictionary of the *Fungi*, a Dictionary of the chromistan/stramenopile fungi-like organisms, and a Dictionary of the protozoan fungi-like organisms. Many people, unfamiliar with classifications which have now been accepted by systematists for many years, still think of fungi as 'plants'. But in reality fungi are a disparate assemblage of organisms from at least three different kingdoms, their unifying characteristic being that they are studied by mycologists. In terms of evolutionary origin, the sister group of the kingdom *Fungi* is *Animalia*: *Fungi* are more closely related to the humans who study them than to green plants which they were previously classified with. But this statement also hides the fact that chromistan fungus-like organisms, of which *Phytophthora infestans* (the causal agent of potato blight) is perhaps the best known example, are only very distantly related to *Fungi*, being instead more allied with the brown seaweeds, among others – a clear indication that the mycelial way of life evolved on at least two separate occasions. Surprisingly, however, protozoan fungus-like organisms are closer to the *Fungi*, being classified in the *Amoebozoa* with other protozoan amoebae. *Fungi*, together with *Animalia* and a few other protozoan groups constitute the *Opisthokonta*; and this group and the *Amoebozoa* form the first major branches at the base of the *Eukaryota*.

In earlier editions, for historical reasons, some biographies and longer entries (i.e. the essay-style accounts of topics relevant to mycologists) seem to have been written from the viewpoint of a native speaker of English and to have treated the fungi as an adjunct of botany. Given that the *Dictionary* is now truly international in character and its theme is clearly not botanical, some effort has been made to adjust these entries so that, in addition to being updated, they are seen from a global and explicitly mycological perspective. One result of this has been a considerable increase in the number of eminent but deceased mycologists commemorated by a biography in these pages, notably from India, Japan and Russia, but also including for the first time scientists native to Argentina, China, Cuba, Pakistan, Portugal, Puerto Rico, Spain and Ukraine. Another has been a sprinkling of new topics covered by long entries, in particular covering the new technologies which have come in, and the gradually developing infrastructure of mycology as a science. Limited resources have meant that the work of updating the essay-style accounts has been incomplete and imperfect, in a few cases to the extent that it has been necessary to flag the entry with a warning note. For this edition, all of the biographies, definitions and other longer entries are located in the first part, even when they might more appropriately belong in one of the other two parts. Resources have also, again, not allowed us to update the keys to families and these continue to be omitted. As higher taxa of fungi are increasingly defined using molecular rather than morphological characteristics, it remains to be seen whether morphology-based keys at this level of the new systematic hierarchy can be made workable.

The overall style of the individual entries in this *Dictionary* remains similar to those of previous editions. References are cited in full throughout the taxonomic name entries. Much bibliographic information is becoming available on the Internet and the tenth edition of this *Dictionary* reflects the increasing availability of information from this source. CABI has been producing the *Bibliography of Systematic Mycology* since 1943 and production was computerized in the late 1980s. This database has been available on the internet since late 1999 and users of this *Dictionary* should visit that web site (www.indexfungorum.org) for up-to-date bibliographic references on the systematics of fungi.

Having been intimately involved in the compilation and proof-stage revisions, we are acutely aware of imperfections and improvements that we would have liked to have made. We can do no more than repeat the comment in the ninth edition that our aspiration is that this edition will at least prove to be the same 'marvellously imperfect work needed by all'.

Do send us your corrections and comment so that the database, and whatever product succeeds this book, will be less imperfect and of even more value to mycologists of all disciplines world-wide.

The tenth edition may well be the last 'ink-on-paper' version of Ainsworth & Bisby's Dictionary of the Fungi – it will certainly be the last for which three of the main editors are at the helm. For like the tenth, in its 65th year, the next edition, if there will be one, will be produced after the retirement from formal, full-time employment of these editors. As such, like so many good things, ...

P.M. Kirk
P.F. Cannon
D.W. Minter
J.A. Stalpers

CABI Europe – UK, Egham and CBS, Utrecht

Contributors

Tatiana Andrianova (M.G. Kholodny Institute of Botany, Kiev), biographies and miscellaneous long entries
André Aptroot (ABL Herbarium, Soest), *Pyrenulales*, *Verrucariales*
Alan Archer (National Herbarium of New South Wales, Sydney), *Ostropales*
Jerry Benny (University of Florida), *Zygomycota*
Reinhard Berndt (Universität Tübingen), *Pucciniomycetes*
Meredith Blackwell (Baton Rouge), biographies
Eric Boa (CABI Europe-UK), edible fungi
Uwe Braun (Martin-Luther-Universität Halle-Wittenberg), *Erysiphales*, *Venturiaceae*
Alan Buddie (CABI Europe-UK), long entries relating to molecular techniques
Pedro Crous (Centraalbureau voor Schimmelcultures, Utrecht), *Botryosphaeriales*, *Capnodiales*
Ove Eriksson (Umeå Universitet), *Lahmiales*, *Neolectales*, *Pleosporales*
Harry Evans (CABI Europe-UK), long entries relating to fungi and insects
João Baptista Ferreira (Lisbon), biographies
Ovidiu Constantinescu (University of Uppsala), *Oomycetes*
Neil Gow (Aberdeen), biographies
Peter Johnston (Landcare Research, Auckland), *Helotiales*, *Rhytismatales*
Jan Kohlmeyer and Brigitte Volkmann-Kohlmeyer (University of North Carolina, Morehead City), *Lulworthiales*, *Microascales*
Clete Kurtzman (USDA Peoria), *Saccharomycetales*
Thom Kuyper (Wageningen University), agaricoid *Basidiomycota*
Thomas Læssøe (Botanical Institute, University of Copenhagen), *Xylariales*
Ronny Larsson (University of Lund), *Microsporidia*
Pete Letcher (University of Alabama),*Chytridiomycota*
Bob Lichtwardt (University of Kansas), *Trichomycetes*
Robert Lücking (Field Museum, Chicago), *Arthoniales*, *Ostropales*
Thorsten Lumbsch (Field Museum, Chicago), *Agyriales*, *Pertusariales*
François Lutzoni and Jolanta Miądlikowska (Duke University, North Carolina), *Acarosporales*, *Peltigerales*
Paco Pando (Madrid), *Mycetozoa*
Don Pfister and Matt Smith (Harvard University Herbaria, Cambridge MS), *Pezizales*
Alan Phillips (Universidade Nova de Lisboa), *Botryosphaeriales*
Martina Réblová (Czech Academy of Sciences, Průhonice), *Calosphaeriales*, *Chaetosphaeriales*, *Trichosphaeriales*
Peter Roberts (Royal Botanic Gardens, Kew), *Tremellales* and similar fungi
Amy Rossman (USDA Beltsville), *Diaporthales*, *Hypocreales*
Matt Ryan (CABI Europe-UK), long entries relating to genetic resource collections
Arthur Schüßler (Ludwig-Maximilians-Universität München) and Chris Walker (Edinburgh), *Glomeromycota*
Joey Spatafora (Oregon State University, Corvallis), *Hypocreales*
Brian Spooner (Royal Botanic Gardens, Kew), *Helotiales*, *Leotiales*
Anders Tehler (Swedish Museum of Natural History, Stockholm), *Arthoniales*
Marco Thines (University of Hohenheim), *Oomycetes*
Wendy Untereiner (Brandon University, Manitoba), *Chaetothyriales*, *Dothideales*
Kálmán Vánky (HUV- Tübingen), *Ustilaginomycetes*
Mats Wedin (Swedish Museum of Natural History, Stockholm), *Agyriales*, *Ostropales*, *Peltigerales*, *Teloschistales*
Merlin White (Boise State University, Boise, ID),*Trichomycetes*
Mike Wingfield, Wilhelm de Beer and Marieka Gryzenhout (FABI, Pretoria), *Cryphonectriaceae*, *Ophiostomatales*

Acknowledgements

Many people, too numerous to mention here, have provided information on corrections or omissions in the ninth edition; we would, however, particularly like to thank Ove Eriksson for discussion on the system adopted for the *Ascomycota* and David Hunt for assistance with the illustrations.

User's Guide

To extract the maximum amount of information from this *Dictionary* with the minimum of effort it is necessary to understand the scope of the compilation and certain conventions.

Content. The longest series of entries are those of the generic names (both accepted names and synonyms) complied to the end of *Index of Fungi* **7**(15) January 2008. Every accepted generic name is referred to a higher group (family, order, class, or phylum) and brief descriptions are given of these higher taxa. The systematic entries are supplemented by a glossary of terms, some English common names, and the names of important fungal antibiotics, toxins, etc. In addition, there are entries on general mycological topics, ecology and distribution, applied mycology, and historical and biographical notes on some well known mycologists and major reference collections.

Names. Every generic name is followed by the name (abbreviated according to Kirk & Ansell, 1992; see Author) of the author(s) who first proposed the genus and the year of publication. The place of publication of a generic name can be found on the CABI database web site at www.indexfungorum.org where additional information on typification is available. A similar layout is adopted for suprageneric names but only those at the rank of family and accepted names above order can be relied upon as well researched and thus likely to be correct. The available Catalogues of names are listed under 'Literature'.

The list of generic names is as complete as possible. Some dates and authorities differ from those that may be found in the literature, many of which have been checked in the original, some names omitted from previous compilations are included, as are some which are not validly published (included as nevertheless present in the mycological literature).

For generic names consigned to synonymy, the authority for the disposition is usually given. For each accepted genus estimates are given for the number of its species and its geographical distribution. Where possible these data are based on recent revisions or the personal knowledge of specialists, but in the majority of cases they have not been updated in the absence of such authorities. In the case of larger genera particularly, we have not revised species numbers upwards even though many may have been described since the last edition, in the absence of modern treatments (see Numbers of fungi). This policy is adopted as critical reassessments in such genera usually result in reductions in species numbers.

The distributions given are approximate, especially for genera not critically revised in recent years, and should be regarded as indicative rather than comprehensive. Whenever possible users should verify the facts for themselves and draw their own conclusions.

Coding. The coding used for anamorphic fungi follows that of the ninth Edition and is explained under that entry. This system, borrowing from that given in the seventh Edition, uses letters or symbols instead of numbers to provide a 'mnemonic' for the conidiomatal and conidial characters. With the removal of traditional morphological groupings of conidial fungi we hope that the new codes will make it easier to gain an idea of the morphological features. Some recently published generic names have not been assessed and are not coded.

Abbreviations. See p. 1.

Validation of names in this Edition

Naumovozyma Kurtzman, *nom. nov.*
≡ *Naumovia* Kurtzman, *FEMS Yeast Res.* **4**: 240 (2003), non *Naumovia* Dobrozr., *Bolezni rastenii* **16**: 197 (1928) ['1927'].

Naumovozyma castellii (Capr.) Kurtzman, comb. nov.
Saccharomyces castellii Capr., *Studi sassar.* III Agr. **14**: 457 (1967) ['1966'].
Naumovia castellii (Capr.) Kurtzman, *FEMS Yeast Res.* **4**: 241 (2003).

Naumovozyma dairenensis (H. Nagan.) Kurtzman, comb. nov.
Saccharomyces dairenensis H. Nagan. (as '*dairensis*'), *Bot. Mag.* Tokyo **31**: 107 (1917).
Naumovia dairenensis (H. Naganishi) Kurtzman, *FEMS Yeast Res.* **4**: 241 (2003).

Helicobasidiaceae P.M. Kirk, fam. nov.
with the characters of the *Helicobasidiales* R. Bauer, Begerow, J.P. Samp., M. Weiss & Oberw. *Mycol. Progr.* **5**: 48 (2006) [q.v. for Latin diagnosis]; type *Helicobasidium* Pat. 1885.

Trappeaceae P.M. Kirk, fam. nov.
with the characters of *Trappea* Castellano, *Mycotaxon* **38**: 2 (1990) [q.v. for Latin diagnosis]; type *Trappea* Castellano 1990.

Gallaceaceae Locq. ex P.M. Kirk, fam. nov.
Gallaceaceae Locq., *De Taxia Fung.* **1A**: 52 (1974), nom. inval., Art. 36.1
with the characters of *Mesophellia scleroderma* Cooke, *Grevillea* **14**(no. 69): 11 (1885) [q.v. for Latin diagnosis, measurements excluded]; type *Gallacea* Lloyd 1905.

Sclerogastraceae Locq. ex P.M. Kirk, fam. nov.
Sclerogastraceae Locq., *De Taxia Fung.* **1A**: 48 (1974), nom. inval., Art. 36.1
with the characters of *Sclerogaster* sensu Saccardo, *Syll. fung.* (Abellini) **11**: 169 (1895) [q.v. for Latin diagnosis]; type *Sclerogaster* R. Hesse 1891.

Contents

Section	Page
Dictionary of the Fungi	1
Dictionary of the chromistan fungal analogues	747
Dictionary of the protozoan fungal analogues	759

This page intentionally left blank

Dictionary of the Fungi

a- (**an-**) (prefix), not having; not; as in acaudate, anaerobe, aniso-.
AAA pathway, alpha-aminoadipic acid pathway for lysine synthesis (cf. DAP pathway).
Aaosphaeria Aptroot (1995), ? Dacampiaceae. Anamorph *Microsphaeropsis*. 1, widespread. See van der Aa (*Stud. Mycol.* **31**: 15, 1989; as *Didymosphaeria*), Aptroot (*Nova Hedwigia* **60**: 325, 1995; posn).
ab- (prefix), position away from.
Abacina Norman (1853) ≡ Diplotomma Flot.
Abaphospora Kirschst. (1939) = Strickeria fide Bose (*Phytopath. Z.* **41**, 1961), Aptroot (*Nova Hedwigia* **66**: 89, 1998).
abaxial (of a basidiospore), the side away from the long axis of the basidium (Corner, 1948); cf. adaxial.
Abbreviations. Abbreviations and signs frequently used in this work are:
adj(ective)
Afr(ica)
Am(erica)
Ann(ales) Myc(ologici)
Auct(ores), authors; used esp. as the authority of a name to indicate frequent (and usually incorrect) usage
Austr(alasia)
bibl(iography)
Biog(raphie)s
B(ulletin Trimestriel de la) S(ociété) M(ycologique de) F(rance)
C(anadian) J(ournal of) B(otany)
C(entral)
(International) Code (of Botanical Nomenclature)
c(irca), approximately
c(on)f(er), compare; make a comparison with
cosmop(olitan), probably in almost all countries
D(ematiaceous) H(yphomycetes) (1971)
E(ast)
Ed(itor)
Ed(itor)s
ed(itio)n
et al(ia), and others
e(xempli) g(ratia), for example
em(ended by)
esp(ecially)
Eur(ope)
Fam(ily, -ilies)
fide, used for 'on the authority of'
Fig(ure)
f(orm) cat(egory)
gen(us, -era)
Hemisph(ere)
hypog(eous)
I(ndex) N(ominum) G(enericorum)
Isl(and, -s)
L(ichen-forming)
Lit(erature)
Mediterr(anean region)
M(ore) D(ematiaceous) H(yphomycetes) (1976)
Mycol(ogia)
Mycol(ogical) Pap(ers)
M(ycological) R(esearch)
n(oun)
N(orth)
nom(en) cons(ervandum), nom(en) rej(iciendum), nom(en) utique rej(iciendum); see Nomenclature
obit(uarie)s
obsol(ete), no longer in use
p(atho)v(ar)
Philipp(ine Islands)
pl(ural)
portr(ait)
pos(itio)n
p(ro) p(arte), in part
Publ(ication)s, principal mycological publications
q(uod) v(ide), which see
R(eview of) A(pplied) M(ycology)
R(eview of) P(lant) P(athology)
S(ystema) A(scomycetum)
s(ensu) l(ato), in the broad sense; widely
s(ensu) str(icto), in the strict sense; narrowly
S(outh)
sp(ecies), spp. (pl.)
syn(onym, -s) (q.v.)
T(axonomic) L(iterature) (edition)-2
T(ransactions of the) B(ritish) M(ycological) S(ociety)
T(ransactions of the) M(ycological) S(ociety of) J(apan)
temp(erate parts)
trop(ics), -(ical)
v(erb)
W(est)
widespr(ead), in a number of countries
O, I, II, III, see *Pucciniales*
=, is heterotypic (taxonomic, facultative) a synonym of
≡ , is homotypic (nomenclatural, obligate) a synonym of
(), sign for 'is the cause of'; e.g. *Ascochyta pinodella* (foot rot of pea)
±, more or less
µm, micron
:, in references precedes page number; in author citations, see Nomenclature.

See also Anamorphic fungi for abbreviations for conidiomatal types (1-9), spore groups (A1, B1, etc.), and conidiogenous events (1-44).

Most abbreviations of names of periodicals, except for those noted above, are taken from the *World List of Scientific Periodicals*, 1952 and 1965-67.

And see Authors' names.
Abeliella Mägd. (1937), Fossil Fungi (mycel.) Fungi. 2 (Cretaceous, Oligocene), Europe.
Abelspora C. Azevedo (1987), Microsporidia. 1. See Azevedo (*J. Parasit.* **49**: 83, 1987).
aberrant, an organism that deviates in one or more ways from the norm.
Abgliophragma R.Y. Roy & Gujarati (1966) ? = Wiesneriomyces fide Roy & Gujarati (*TBMS* **49**: 363, 1966), Pirozynski (*Mycol. Pap.* **129**, 1972).
abhymenial, opposite the spore-producing surface.
abjection, the separating of a spore from a sporophore or sterigma by an act of the fungus.
abjunction, the cutting off of a spore from a hypha by a septum.
Abkultur, see Normkultur.
aboospore, a parthenogenetic oospore.
Abortiporus Murrill (1904), ? Meruliaceae. Anamorph *Sporotrichopsis*. 3, widespread. See Ryvarden (*Syn. Fung.* **5**: 104, 1991).
abraded (of lichen thalli), having the surface worn; eroded.
Abropelta B. Sutton (1986), anamorphic *Pezizomycotina*, Cpt.≡ eH.15. 1, India. See Sutton (*TBMS* **86**: 1, 1986).
Abrothallomyces Cif. & Tomas. (1953) = Dacty-

lospora fide Hawksworth *et al.* (*Dictionary of the Fungi* edn 8, 1995).

Abrothallus De Not. (1845), Pezizomycotina. Anamorph *Vouauxiomyces*. *c.* 41 (on lichens), widespread. See Bellemère *et al.* (*Cryptog. Mycol.* **7**: 47, 1986; ultrastr.), Hafellner (*Herzogia* **13**: 139, 1998), Bernasconi *et al.* (*Aust. Syst. Bot.* **15**: 527, 2002; Patagonia), Suija (*Ann. bot. fenn.* **43**: 193, 2006).

abrupt, as if cut off transversely; truncate.

abscission, separating by disappearance of a joining layer or wall, as of conidia from a conidiogenous cell.

Absconditella Vězda (1965), Stictidaceae (L). *c.* 9, widespread (esp. Europe & N. America). See Vězda & Vivant (*Folia geobot. phytotax.* **10**: 205, 1975; key 5 spp.), Vězda & Pišút (*Nova Hedwigia* **40**: 341, 1984), Nimis (*Lichenologist* **30**: 427, 1998; generic concept), Ceynowa-Gieldon (*Acta Mycologica* Warszawa **38**: 99, 2003; Poland), Grube *et al.* (*MR* **108**: 1111, 2004; phylogeny), Lücking *et al.* (*Mycol.* **96**: 283, 2004), Lumbsch *et al.* (*Mol. Phylogen. Evol.* **31**: 822, 2004; phylogenetic position), Kantvilas (*Muelleria* **21**: 91, 2005).

Absidia Tiegh. (1878), Mucoraceae. 18 (esp. in soil), widespread. See Hesseltine & Ellis (*Mycol.* **56**: 568, 1964; cylindrical-spored spp.), Hesseltine & Ellis (*Mycol.* **57**: 234, 1965; globose-spored spp.), Ellis & Hesseltine (*Sabouraudia* **5**: 59, 1966), Hesseltine & Ellis (*Mycol.* **58**: 761, 1966; ovoid-spored spp.), Zycha *et al.* (*Mucorales*, 1969), Nottebrock *et al.* (*Sabouraudia* **12**: 64, 1974), Váňová (*Česká Mykol.* **37**: 151, 1983), Burmester *et al.* (*Curr. Genet.* **17**: 155, 1990; transformations), Hesseltine *et al.* (*Mycol.* **82**: 523, 1990; key), Schipper (*Persoonia* **14**: 133, 1990; key), Wöstemeyer *et al.* (*Curr. Genet.* **17**: 163, 1990; somatic hybrids), Ginman & Young (*Microbios* **66**: 39, 1991; ultrastr.), Kayser & Wöstemeyer (*Curr. Genet.* **19**: 279, 1991; karyotype), Pajak *et al.* (*Mycopathologia* **118**: 109, 1992; keratinolysis), Wöstemeyer & Burmester (*Microbiol. Res.* **149**: 407, 1994; rDNA), Lopes *et al.* (*Mycopathologia* **130**: 89, 1995; mucormycosis), Mimura *et al.* (*J. Med. Vet. Mycol.* **33**: 137, 1995; mucormycosis), Chen & Zheng (*Mycotaxon* **69**: 173, 1998; thermophile), O'Donnell *et al.* (*Mycol.* **93**: 286, 2001; phylogeny), Voigt & Wöstemeyer (*Gene* **270**: 113, 2001; phylogeny), Kwasna *et al.* (*MR* **110**: 501, 2006; phylogeny soil isolates), White *et al.* (*Mycol.* **98**: 872, 2006; phylogeny), Hoffman *et al.* (*MR* **111**: 1169, 2007; phylogeny, classification).

Absidiaceae Arx (1982) = Mucoraceae.
Lit.: Kirk (*in litt.*).

absorb, to obtain food by taking up water and dissolved substances across a membrane. Cf. ingest.

Abstoma G. Cunn. (1926), Agaricaceae. 2, widespread. See Wright & Suarez (*Cryptog. Bot.* **1**: 372, 1990; key).

abstriction, abjunction and then abscission, esp. by constriction.

Abundisporus Ryvarden (1999), Polyporaceae. 7, widespread. See Ryvarden (*Belg. Jl Bot.* **131**: 154, 1998), Dai *et al.* (*Ann. bot. fenn.* **39**: 169, 2002; China).

Abyssomyces Kohlm. (1970), Pezizomycotina. 1 (marine, on hydrozoans in deep water), S. Atlantic. See Kohlmeyer (*Ber. dt. bot. Ges.* **83**: 505, 1970), Kohlmeyer & Volkmann-Kohlmeyer (*Bot. Mar.* **34**: 1, 1991), Kohlmeyer & Volkmann-Kohlmeyer (*Bot. Mar.* **46**: 285, 2003).

Acallomyces Thaxt. (1903), Laboulbeniaceae. 3 (on staphylinid beetles), widespread. See Tavares (*Mycol.* **65**: 929, 1973), Tavares (*Mycol. Mem.* **9**: 627 pp., 1985; monogr.), Santamaria *et al.* (*Treb. Inst. Bot. Barcelona* **14**: 123 pp., 1991; European spp.).

Acalyptospora Desm. (1848) nom. dub., Plantae. Based on gland-like hairs.

acantha, a sharp pointed process; a spine.

Acantharia Theiss. & Syd. (1918), Venturiaceae. Anamorphs *Fusicladium*, *Stigmina*-like. 7 (on leaves, necrotrophic), widespread. See Bose & Müller (*Indian Phytopath.* **18**: 340, 1965), Sivanesan (*TBMS* **82**: 507, 1984; anamorphs), Barr (*Sydowia* **41**: 25, 1989; N America), Hsieh *et al.* (*MR* **99**: 917, 1995; key).

Acanthellorhiza P. Roberts (1999), anamorphic *Heteroacanthella*. 2 (saprobic on dead wood), widespread. See Roberts (*Rhizoctonia-forming fungi*, 1999).

Acanthobasidium Oberw. (1965), Stereaceae. 3, Europe. See Oberwinkler (*Sydowia* **19**: 45, 1965), Boidin *et al.* (*BSMF* **101**: 345, 1994).

Acanthocystis (Fayod) Kühner (1926) = Hohenbuehelia fide Singer (*Agaric. mod. Tax.* edn 3, 1975).

acanthocyte, spiny cell produced on a short branch from the vegetative mycelium of *Stropharia* spp. (Farr, *Mycotaxon* **11**: 241, 1980).

Acanthoderma Syd. & P. Syd. (1917), anamorphic *Pezizomycotina*, ?.≡ eH.?. 1, Philippines.

Acanthodochium Samuels, J.D. Rogers & Nagas. (1987), anamorphic *Acanthocystis, Collodiscula, Rosellinia*, Hsp.0eH.10. 2 (on dead bamboo culms), widespread (esp. tropical). See Samuels *et al.* (*Mycotaxon* **28**: 453, 1987), Ju & Rogers (*Mycol.* **82**: 342, 1990; *Rosellinia* teleomorph), Ju & Rogers (*Mycotaxon* **73**: 343, 1999), Kang *et al.* (*Fungal Diversity* **2**: 135, 1999).

Acanthofungus Sheng H. Wu, Boidin & C.Y. Chien (2000), Stereaceae. 3, widespread. See Wu *et al.* (*Mycotaxon* **76**: 154, 2000).

Acanthographina (Vain.) Walt. Watson (1929) ≡ Acanthothecis.

Acanthographis (Vain.) Walt. Watson (1929) = Acanthothecis fide Staiger (*Biblthca Mycol.* **85**, 2002).

Acanthogymnomyces Udagawa & Uchiyama (2000), Gymnoascaceae. 2 (from soil etc.), India; Japan. See Udagawa & Uchiyama (*Mycotaxon* **76**: 412, 2000).

acanthohyphidium, see hyphidium.

Acantholichen P.M. Jørg. (1998), Corticiaceae. 1, Costa Rica. See Jørgensen (*Bryologist* **101**: 444, 1998).

Acanthomyces Thaxt. (1892) [non *Akanthomyces* Lebert 1858] ≡ Rhachomyces.

Acanthonitschkea Speg. (1908), Nitschkiaceae. Anamorph *Acremonium*-like. 8 (on wood and lichens), widespread. See Nannfeldt (*Svensk bot. Tidskr.* **69**: 49, 1975), Subramanian & Sekar (*Kavaka* **18**: 19, 1993; Indian spp.), Huhndorf *et al.* (*MR* **108**: 1384, 2004; phylogeny, rel. with *Nitschkia*).

Acanthophiobolus Berl. (1893), Tubeufiaceae. 2 (saprobic on plants), widespread. See Walker (*Mycotaxon* **11**: 1, 1980), Scheuer (*Biblthca Mycol.* **123**: 274 pp., 1988; Austria), Barr (*Mycotaxon* **64**: 149, 1997), Crane *et al.* (*CJB* **76**: 602, 1998; key), Kodsueb *et al.* (*Fungal Diversity* **21**: 105, 2006; phylogeny).

Acanthophysellum Parmasto (1967), Stereaceae. 3, widespread. See Parmasto (*Izv. Akad. Nauk Estonsk.*

SSR Ser. Biol. **16**: 377, 1967), Larsson & Larsson (*Mycol.* **95**: 1037, 2003; phylogeny) Close to *Xylobolus*.

Acanthophysiaceae Boidin, Mugnier & Canales (1998) = Stereaceae.

acanthophysis, see hyphidium.

Acanthophysium (Pilát) G. Cunn. (1963), Stereaceae. *c.* 20, widespread. See Cunningham (*Bull. N.Z. Dept. Sci. Industr. Res., Pl. Dis. Div.* **145**: 150, 1963).

Acanthorhynchus Shear (1907), Hyponectriaceae. 1 (saprobic on leaves of *Vaccinium*), N. America. See Barr (*Mycol.* **68**: 611, 1976), Fallah & Shearer (*Mycol.* **93**: 566, 2001; as *Physalospora*).

Acanthorus Bat. & Cavalc. (1967), anamorphic *Capnodiaceae*, Cpt.0eH.?. 1 (on leaves of *Bertholletia*), Brazil. See Batista & Cavalcanti (*Atas Inst. Micol. Univ. Pernambuco* **4**: 246, 1967).

Acanthosphaeria Kirschst. (1939), Trichosphaeriaceae. 2, Europe. See Petrak (*Annls mycol.* **38**: 198, 1940).

Acanthostigma De Not. (1863), Tubeufiaceae. Anamorphs *Helicomyces*, *Helicosporium*. 8 (saprobic on wood or other fungi), widespread. See Réblová & Barr (*Sydowia* **52**: 258, 2000; monogr.), Kodsueb *et al.* (*Mycol.* **96**: 667, 2004; Hong Kong), Kodsueb *et al.* (*Fungal Diversity* **21**: 105, 2006), Tsui *et al.* (*Mycol.* **98**: 94, 2006; rels with *Tubeufia* and helicosporous anamorphs), Tsui *et al.* (*Mycol.* **99**: 884, 2007; phylogeny, anamorph).

Acanthostigmella Höhn. (1905), Tubeufiaceae. Anamorph *Xenosporium*. 6 (saprobic), widespread. See Barr (*Mycotaxon* **6**: 17, 1977; key), Untereiner (*MR* **99**: 897, 1995), Crane *et al.* (*CJB* **76**: 602, 1998).

Acanthostigmella Rick (1933) = Acanthostigma fide Hawksworth *et al.* (*Dictionary of the Fungi* edn 8, 1995).

Acanthostigmina Höhn. (1909) = Acanthostigma fide Rossman (*Mycol. Pap.* **157**, 1987), Crane *et al.* (*CJB* **76**: 602, 1998), Réblová & Barr (*Sydowia* **52**: 286, 2000; monogr.).

Acanthostoma Theiss. (1912) = Phaeodimeriella Speg. fide Müller & von Arx *in* Ainsworth *et al.* (Eds) (*The Fungi* **4A**: 87, 1973).

Acanthotheca Clem. & Shear (1931) [non *Acanthotheca* DC. 1838, *Compositae*] ≡ Acanthotheciella.

Acanthotheciella Höhn. (1911), Sordariomycetes. Anamorph *Ypsilonia*. 3 (on dead scale insects), Asia (tropical); S. America. See Nag Raj (*CJB* **55**: 1599, 1977), Barr (*Mycotaxon* **39**: 43, 1990; posn).

Acanthotheciopsis Zahlbr. (1923) = Acanthothecis fide Hawksworth *et al.* (*Dictionary of the Fungi* edn 8, 1995), Staiger (*Biblthca Lichenol.* **85**: 526 pp., 2002; revision).

Acanthothecis Clem. (1909), Graphidaceae (L). 25, S. America (primarily tropical). See Staiger & Kalb (*Mycotaxon* **73**: 69, 1999), Staiger (*Biblthca Lichenol.* **85**, 2002), Archer (*Biblthca Lichenol.* **94**, 2006; revision), Archer (*Systematics & Biodiversity* **5**: 9, 2007; Solomon Is), Makhija & Adawadkar (*Lichenologist* **39**: 165, 2007; India, key).

Acanthothecium Speg. (1889) = Ypsilonia See Nag Raj (*CJB* **55**: 1599, 1977).

Acanthothecium Vain. (1890) ≡ Acanthothecis.

Acanthothecomyces Cif. & Tomas. (1953) ≡ Acanthothecis.

Acanthotrema Frisch (2006), Graphidaceae (L). 1, Africa (tropical); S. America. See Frisch (*Biblthca Lichenol.* **92**: 3, 2006), Frisch *et al.* (*Biblthca Lichenol.* **92**: 517, 2006; phylogeny), Staiger *et al.* (*MR* **110**: 765, 2006; inclusion in *Graphidaceae*).

Acarella Syd. (1927), anamorphic *Morenoina*, Cpt.0eH.?. 3 (saprobic on leaves etc.), C. America. See Farr (*Sydowia* **38**: 65, 1985).

Acarellina Bat. & H. Maia (1960), anamorphic *Pezizomycotina*, Cpt.0eH.?. 1 (on leaves of *Psidium*), Brazil. See Batista & Maia (*Publções Inst. Micol. Recife* **246**: 4, 1960).

Acariniola T. Majewski & J. Wiśn. (1978) = Pyxidiophora fide Lundqvist (*Bot. Notiser* **133**: 121, 1980), Blackwell & Malloch (*MR* **94**: 415, 1990; recognition as ascospores), Weir & Blackwell (*Insect-Fungal Associations* Ecology and Evolution: 119, 2005; biology).

Acaroconium Kocourk. & D. Hawksw. (2008), anamorphic *Pezizomycotina*. 1 (lichenicolous), Europe; N. America. See Kokourcová & Hawksworth (*Lichenologist* **40**: 105, 2008).

Acarocybe Syd. (1937), anamorphic *Pezizomycotina*, Hsy.0eP.28. 3, Africa; Brazil. See Ellis (*Mycol. Pap.* **76**, 1960; key), Mena-Portales *et al.* (*MR* **103**: 1032, 1999; comparison with *Acarocybiopsis*).

Acarocybella M.B. Ellis (1960), anamorphic *Pezizomycotina*, Hso.≡ eP.28. 1, pantropical.

Acarocybellina Subram. (1992), anamorphic *Pezizomycotina*, Hso.≡ eP.26. 1, widespread (tropical). See Lanzoni (*Boll. Gruppo Micol. 'G. Bresadola'* **28**: front & inside, 1985; separation from *Sporidesmium*), Mena-Portales *et al.* (*MR* **103**: 1032, 1999; comparison with *Acarocybiopsis*).

Acarocybiopsis J. Mena, A. Hern. Gut. & Mercado (1999), anamorphic *Pezizomycotina*, Hso.e≡ P.?. 1 (saprobic on wood), Cuba. See Mena-Portales *et al.* (*MR* **103**: 1032, 1999; description), Mercado-Sierra *et al.* (*Nova Hedwigia* **75**: 533, 2002; comparison with *Veracruzomyces*).

Acaromyces Boekhout, Scorzetti, Gerson & Sztejnb. (2003), anamorphic *Exobasidiomycetidae*. 1 (from mites), Israel. See Boekhout *et al.* (*Int. J. Syst. Evol. Microbiol.* **53**: 1662, 2003; phylogeny, family placement), Yasuda *et al.* (*Mycoscience* **47**: 36, 2006; phylogeny).

Acaropeltis Petr. (1937), anamorphic *Pezizomycotina*, Cpt.0eH.?. 1 (on living leaves), C. America. See Petrak (*Annls mycol.* **35**: 95, 1937; orig. description).

Acarospora A. Massal. (1852), Acarosporaceae (L). *c.* 128, widespread. See Weber (*Lichenologist* **4**: 16, 1968; sect. *Xanthothallia*), Golubkova & Shapiro (*Nov. Sist. niz. Rast.* **13**: 150, 1976; sect. *Trochia*), Clauzade & Roux (*Bull. Mus. Hist. nat. Marseille* **41**, 1981; key 69 Eur. spp.), Castello & Nimis (*Lichenologist* **26**: 283, 1994; Antarct.), Stenroos & DePriest (*Am. J. Bot.* **85**: 1548, 1998; DNA), Lutzoni *et al.* (*Am. J. Bot.* **91**: 1446, 2004; posn), Reeb *et al.* (*Mol. Phylogen. Evol.* **32**: 1036, 2004; posn, phylogeny), Temina *et al.* (*Nova Hedwigia* **80**: 433, 2005; Israel and vicinity), Wedin *et al.* (*MR* **109**: 159, 2005; position within *Lecanoromycetes*), Crewe *et al.* (*MR* **110**: 521, 2006; molecular phylogeny), Miądlikowska *et al.* (*Mycol.* **98**: 1088, 2006; phylogeny), Knudsen *et al.* (*Opuscula Philolichenum* **5**: 1, 2008; S America).

Acarosporaceae Zahlbr. (1906), Acarosporales (L). 11 gen. (+ 10 syn.), 183 spp.

Lit.: Golubkova (*Lishaĭniki semeĭstva Acarosporaceae Zahlbr. v. SSSR*, 1988; keys 8 gen., 91 spp.), Bellemère (*Bull. Soc. linn. Provence* **45**: 355, 1994), Hafellner (*Cryptog. Bot.* **5**: 99, 1995), David & Coppins (*Lichenologist* **29**: 291, 1997), Kocourková-Horáková (*Czech Mycol.* **50**: 271, 1998), Rambold & Hagedorn (*Lichenologist* **30**: 473, 1998), Seppelt *et al.* (*Lichenologist* **30**: 249, 1998), Stenroos & DePriest (*Am. J. Bot.* **85**: 1548, 1998; DNA), Navarro-Rosinés *et al.* (*CJB* **77**: 835, 1999), Lutzoni *et al.* (*Nature* **411**: 937, 2001; posn), Kauff & Lutzoni (*Mol. Phylogen. Evol.* **25**: 138, 2002; posn), Lutzoni *et al.* (*Am. J. Bot.* **91**: 1446, 2004; posn), Miądlikowska & Lutzoni (*Am. J. Bot.* **91**: 449, 2004; posn), Reeb *et al.* (*Mol. Phylogen. Evol.* **32**: 1036, 2004), Wedin *et al.* (*MR* **109**: 159, 2005), Crewe *et al.* (*MR* **110**: 521, 2006), Miądlikowska *et al.* (*Mycol.* **98**: 1088, 2006; phylogeny), Hofstetter *et al.* (*Mol. Phylogen. Evol.* **44**: 412, 2007; phylogeny).

Acarosporales Reeb, Lutzoni & Cl. Roux (2007). Acarosporomycetidae. 1 fam., 11 gen., 183 spp. Fam.:
Acarosporaceae
For *Lit.* see under fam.

Acarosporina Sherwood (1977), Stictidaceae. Anamorph *Phacidiella*-like. 4, widespread. See Johnston (*Mycotaxon* **24**: 359, 1985; anamorph), Miądlikowska *et al.* (*Mycol.* **98**: 1088, 2006; phylogeny), Schoch *et al.* (*MR* **110**: 257, 2006; phylogeny).

Acarosporium Bubák & Vleugel ex Bubák (1911), anamorphic *Pycnopeziza*, St.1eH-P.39. 4, north temperate. See Korzenok (*Mikol. Fitopatol.* **25**: 107, 1991; Russia).

Acarosporomyces Cif. & Tomas. (1953) = Pleopsidium fide Hawksworth *et al.* (*Dictionary of the Fungi* edn 8, 1995).

Acarosporomycetidae Reeb, Lutzoni & Cl. Roux (2004), Lecanoromycetes. Ord.:
Acarosporales
Lit.: Lumbsch *et al.* (*MR* **111**: 257, 2007; phylogeny), Lutzoni *et al.* (*Am. J. Bot.* **91**: 1446, 2004), Miądlikowska *et al* (*Mycol.* **98**: 1088, 2006), Reeb *et al.* (*Mol. Phylogenet. Evol.* **32**: 1036, 2004), Wedin *et al.* (*MR* **109**: 159, 2005).

Acarothallium Syd. (1937) = Wentiomyces fide Müller & von Arx (*Beitr. Kryptfl. Schweiz* **11** no. 2, 1962).

acaryallagic, see caryallagic.

acaudate, not having a tail.

Acaulium Sopp (1912) ≡ Scopulariopsis fide Raper & Thom (*Manual of the Penicillia*, 1949).

Acaulopage Drechsler (1935), Zoopagaceae. 27, widespread. See Drechsler (*Mycol.* **27**: 185, 1935), Drechsler (*Mycol.* **28**: 363, 1936), Drechsler (*Mycol.* **30**: 137, 1938), Drechsler (*Mycol.* **31**: 128, 1939), Drechsler (*Mycol.* **33**: 248, 1941), Drechsler (*Mycol.* **34**: 274, 1942), Drechsler (*Mycol.* **37**: 1, 1945), Drechsler (*Mycol.* **38**: 120, 1946), Drechsler (*Mycol.* **39**: 253, 1947), Drechsler (*Mycol.* **40**: 85, 1948), Drechsler (*Mycol.* **47**: 364, 1955), Drechsler (*Mycol.* **51**: 787, 1959), Drechsler (*Am. J. Bot.* **49**: 1089, 1962), Saikawa & Kadowaki (*Nova Hedwigia* **74**: 365, 2002; amoeba capture).

Acaulospora Gerd. & Trappe (1974), Acaulosporaceae. 26, widespread. See Mosse (*Arch. Mikrobiol.* **70**: 167, 1970), Mosse (*Arch. Mikrobiol.* **74**: 120, 1970; life cycle, ultrastr.), Mosse (*Arch. Mikrobiol.* **74**: 146, 1970; life cycle, ultrastr.), Schenck *et al.* (*Mycol.* **76**: 685, 1984; key), Berch (*Mycotaxon* **23**: 409, 1985; emend.), Morton (*Mycol.* **78**: 787, 1986; effect of mountants & fixatives on spores), Błaszkowski (*Karstenia* **27**: 32, 1987), Sieverding & Toro (*Angew. Bot.* **61**: 217, 1987), Sieverding (*Angew. Bot.* **62**: 373, 1988), Błaszkowski (*Mycol.* **82**: 794, 1990), Gazey *et al.* (*MR* **96**: 643, 1992; sporulation), Maia & Kimbrough (*MR* **97**: 1183, 1993; spore wall ultrastr.), Błaszkowski (*Mycorrhiza* **4**: 173, 1994), Ingleby *et al.* (*Mycotaxon* **50**: 99, 1994), Błaszkowski (*MR* **99**: 237, 1995), Yao *et al.* (*Kew Bull.* **50**: 349, 1995), Błaszkowski (*Mycotaxon* **61**: 193, 1997), Morton *et al.* (*MR* **101**: 625, 1997; synanamorph), Saikawa *et al.* (*Mycoscience* **39**: 477, 1998; phylogeny & synanamorph), Zhang *et al.* (*Mycosystema* **17**: 15, 1998), Schultz *et al.* (*Mycol.* **91**: 676, 1999), Redecker & Raab (*Mycol.* **98**: 885, 2006; phylogeny), Velazquez *et al.* (*Mycotaxon* **103**: 171, 2008; Argentina).

Acaulosporaceae J.B. Morton & Benny (1990), Diversisporales. 2 gen., 31 spp.
Lit.: Morton & Benny (*Mycotaxon* **37**: 471, 1990), Maia & Kimbrough (*MR* **97**: 1183, 1993), Azcon-Aguilar & Barea (*Mycorrhiza* **6**: 457, 1996), Morton *et al.* (*MR* **101**: 625, 1997), Sawaki *et al.* (*Mycoscience* **39**: 477, 1998), van der Heijden *et al.* (*Nature* Lond. **396**: 69, 1998), Stürmer & Morton (*Mycol.* **91**: 849, 1999), Rodriguez *et al.* (*New Phytol.* **152**: 159, 2001), Schüßler *et al.* (*MR* **105**: 1413, 2001), Fracchia *et al.* (*Nova Hedwigia* **77**: 383, 2003), Pringle *et al.* (*Mycorrhiza* **13**: 227, 2003), Redecker (*Glomeromycota* Arbuscular mycorrhizal fungi and their relative(s). *In* The Tree of Life Web Project, http://tolweb.org: [unpaginated], 2005).

accumbent, resting against anything.

acellular, not divided into cells, e.g. a myxomycete plasmodium.

Acephala Grünig & Sieber (2005), anamorphic *Vibrisseaceae*, sterile. 1 (associated with conifer roots), Europe; N. America. See Grünig & Sieber (*Mycol.* **97**: 634, 2005), Grünig & Sieber (*Mycol.* **97**: 628, 2005; descr., phylogeny), Grünig *et al.* (*Fungal Genetics Biol.* **43**: 410, 2006; population genetics).

Acephalis Badura & Badurowa (1964) = Syncephalis fide Skirgiełło & Zadara (*Beih. Sydowia* **8**: 366, 1979).

acephalous, not having a head.

Acerbia (Sacc.) Sacc. & P. Syd. (1899) ? = Rosenscheldia fide Eriksson & Yue (*SA* **13**: 129, 1994).

Acerbiella Sacc. & D. Sacc. (1905), Sordariomycetes. 2, S. America; Java.

acerose, needle-like and stiff; like a pine needle (Fig. 23.3).

acervate, massed up; heaped; growth in heaps or groups.

Acerviclypeatus Hanlin (1990), anamorphic *Ophiodothella*, St.0fH.?. 1 (on living leaves of *Vaccinium*), USA. See Hanlin (*Mycotaxon* **37**: 379, 1990; descr.), Hanlin (*Mycol.* **95**: 506, 2003; development).

Acervulopsora Thirum. (1945) = Maravalia fide Cummins & Hiratsuka (*Illustr. Gen. Rust Fungi rev. edit.*, 1983).

acervulus (pl. **-i**; adj. **-lar**), a ± saucer-shaped conidioma (embedded in host tissue) in which the hymenium of conidiogenous cells develops on the floor of the cavity from a pseudoparenchymatous stroma beneath an integument of host tissue which ruptures at maturity; acervular conidioma (Fig. 10 O).

Acervus Kanouse (1938), Pyronemataceae. 4, widespread. See Pfister (*Occ. Pap. Farlow Herb. Crypt. Bot.* **8**: 1, 1974; key), Pant (*TBMS* **71**: 326, 1978), Pfister & Bessette (*Mycotaxon* **22**: 435, 1985),

Kimbrough & Curry (*Mycol.* **78**: 735, 1986; ultrastructure), Zhuang & Wang (*Mycotaxon* **69**: 339, 1998; 3 spp. China), Prasad & Pant (*Journal of Mycology and Plant Pathology* **34**: 147, 2004; spore ornamentation), Perry *et al.* (*MR* **111**: 549, 2007; phylogeny, isolated posn within *Pyronemataceae*).

Acetabula (Fr.) Fuckel (1870) = Helvella fide Dissing (*Dansk bot. Ark.* **25** no. 1, 1966).

Acetabularia (Berk.) Massee (1893) [non *Acetabularia* J.V. Lamour. 1812, *Algae*] ≡ Cyphellopus fide Singer (*Agaric. mod. Tax.*, 1951).

acetabuliform, saucer-like in form.

Achaetobotrys Bat. & Cif. (1963), Antennulariellaceae. Anamorph *Antennariella*. 3 (probably saprobic on plant exudates), widespread (primarily tropical). See Hughes (*Mycol.* **68**: 693, 1976), Barr & Rogerson (*Mycotaxon* **71**: 473, 1999; USA).

Achaetomiaceae Mukerji (1978) = Chaetomiaceae.

Achaetomiella Arx (1970) = Chaetomium fide Udagawa (*TMSJ* **21**: 34, 1980), Cannon (*TBMS* **87**: 50, 1986).

Achaetomium J.N. Rai, J.P. Tewari & Mukerji (1964), Chaetomiaceae. 7 (from soil etc.), widespread (pantropical). See von Arx (*Proc. Indian Acad. Sci.* Pl. Sci. **94**: 341, 1985), Cannon (*TBMS* **87**: 50, 1986; key), Sultana *et al.* (*Biologia* Lahore **34**: 257, 1988; Pakistan spp.), von Arx *et al.* (*Beih. Nova Hedwigia* **94**, 1988), Lee & Hanlin (*Mycol.* **91**: 434, 1999; DNA), Rodríguez *et al.* (*Stud. Mycol.* **50**: 77, 2004; key).

Acharius (Erik; 1757-1819; Sweden). Country doctor, Vadstena. A pupil of Linnaeus (q.v.) defending his dissertation in 1776, and correspondent of Fries (q.v.). Laid scientific basis for the study and classification of lichen-forming fungi, and responsible for the terms thallus, podetium, apothecium, perithecium, soredium, cyphella as applied to those organisms. Described many new species, especially from Europe. Main collections in **H**, other material in **BM**, **UPS**, **LINN** (Smith collection). *Publs. Methodus qua Omnes Detectos Lichenes.* (1803); *Lichenographia Universalis* (1810); *Synopsis Methodica Lichenum* (1814). *Biogs, obits etc.* Galloway (*Bulletin of the British Museum of Natural History* Botany **18**: 149, 1988 [influence on British lichenology, specimens in **BM**]); González Bueno & Rico (*Acta Botanica Malacitana* **16**: 141, 1991 [impact on Spanish lichenology]); Grummann (1974: 469); Vitikainen (Introduction, *Lichenographia Universalis*, 1976 [reprint]); Stafleu & Cowan (*TL-2* **1**: 4, 1976); Stafleu & Mennega (*TL-2*, *Suppl.* **1**: 14, 1992); Tibell (*Annales Botanici Fennici* **24**: 257, 1987 [*Caliciales*]).

Achitonium Kunze (1819) = Pactilia fide Hawksworth *et al.* (*Dictionary of the Fungi* edn 8, 1995).

Achlyella Lagerh. (1890), ? Chytridiales. 1, Europe.

Achlyites Mesch. (1902), Fossil Fungi. 1 (Silurian, Tertiary), Atlantic.

Achlyogeton Schenk (1859), Chytridiales. 1, widespread (north temperate). See Blackwell & Powell (*Mycotaxon* **64**: 91, 1997).

Achorella Theiss. & Syd. (1915), ? Dothideomycetes. 10, widespread. Type material is inadequate. See Müller & von Arx (*Beitr. Kryptfl. Schweiz* **11** no. 2, 1962).

Achorion Remak (1845) = Trichophyton fide Hawksworth *et al.* (*Dictionary of the Fungi* edn 8, 1995).

Achorodothis Syd. (1926), Mycosphaerellaceae. 2 (in leaves), Costa Rica. See Müller & von Arx (*Beitr. Kryptfl. Schweiz* **11** no. 2, 1962).

Achoropeltis Syd. (1929), anamorphic *Pezizomycotina*, Cpt.0eH.?. 1 (on living leaves), Costa Rica.

achroic (**achromatic**, **achrous**), having no colour or pigment; see Colour.

Achroomyces Bonord. (1851) nom. dub, Agaricomycotina. See Donk (*Persoonia* **4**: 145, 1966; syn. of *Platygloea*).

Achrotelium Syd. (1928), Chaconiaceae. *c.* 5 (on dicots, 3 on *Asclepiadaceae*), Philippines; USA; India; Zimbabwe.

Acia P. Karst. (1879) [non *Acia* Schreb. 1791, *Rosaceae*] ≡ Mycoacia.

acicular, slender and pointed; needle-shaped (Fig. 23.33).

Aciculariella Arnaud (1954), anamorphic *Pezizomycotina*, Hso.0fP.?. 2, Europe. The two original spp. were not formally described.

Aciculoconidium D.S. King & S.C. Jong (1976), anamorphic *Saccharomycetales*, Hso.0eH.3. 1, USA. See King & Jong (*Mycotaxon* **3**: 407, 1976), Smith *in* Kurtzman & Fell (Eds) (*Yeasts, a taxonomic study* 4th edn: 439, 1998), Kurtzman (*Antonie van Leeuwenhoek* **88**: 121, 2005), Suh *et al.* (*Mycol.* **98**: 1006, 2006; phylogeny).

Aciculopsora Aptroot & Trest (2006), ? Ramalinaceae (L). 1, Costa Rica. See Aptroot *et al.* (*J. Hattori bot. Lab.* **100**: 617, 2006; Costa Rica).

Aciculosporium I. Miyake (1908), Clavicipitaceae. Anamorph *Albomyces*. 2 (in living bamboos), Japan. See Kao & Leu (*Pl. Prot. Bull. Taiwan* **18**: 276, 1976), Tubaki & Ando (*Acta Mycol. Sin.* Suppl. **1**: 426, 1987), Tsuda *et al.* (*Bull. natn. Sci. Mus.* Tokyo, B **23**: 25, 1997; host range), Oguchi (*Mycoscience* **42**: 217, 2001), Tanaka *et al.* (*Mycoscience* **43**: 87, 2002; phylogeny), Pažoutová *et al.* (*MR* **108**: 126, 2004; conidial devel.), Walker (*Australas. Pl. Path.* **33**: 211, 2004; comp. with *Cepsiclava*).

Acid rain. The wet acidic deposition of air pollutants can affect fungi including lichen-forming species. While many show a decline, a small number of generalist species may actually increase in incidence in response to this pollution (Kowalski & Stanczykiewicz, *Phytopathologia Polonica* **19**: 69, 2000). Endophytes possibly implicated in pH regulation within leaves of forest trees (Stephan, *Eur. J. For. Path.* **3**: 112, 1973) may be particularly vulnerable (Ei-Ichiro Asai *et al.*, *MR* **102**: 1316, 1998). Lichen-forming fungi with cyanobacterial partners are strongly affected and have declined dramatically in some parts of Europe (Farmer *et al.*, *in* Bates & Farmer, 1992: 284); nitrogenase activity may be affected (Fritz-Sheridan, *Lichenologist* **17**: 27, 1985). Reductions in many mycorrhizal fungi in Europe have been correlated with acid rain, though it is not often clear whether this is a cause of or a result from damage seen in the trees. The decline in fruiting of *Cantharellus cibarius* has been especially noticeable (Jansen & van Doben, *Ambio* **16**: 211, 1987; Derbsch & Schmitt, *Atlas der Pilze des Saarlandes* **2**, 1987). *Russula mustelina* fruiting has been singled out as a valuable early indicator of acid rain problems in European forests (Felher, *Agric. Ecosyst. Envir.* **28**: 115, 1990). *Rhytisma acerinum* is also strongly affected (Greenhalgh & Bevan, *TBMS* **71**: 491, 1978), perhaps because of damage to the delicate mucilaginous sheaths around ascospores during dispersal in wet weather. Mycorrhizal fungi may mollify the effect of acid rain on trees (Blum *et al.*, *Nature* **417**: 729, 2002). In Europe, with legislation to control acid rain pollution, there has been some amelioration of the problem.

Lit.: Arnolds (*in* Hawksworth (Ed.), *Frontiers in mycology*: 243, 1991), Bates & Farmer (Eds) (*Bryophytes and lichens in a changing environment*, 1992), Pegler *et al.* (Eds) (*Fungi of Europe*, 1993), Richardson (*Pollution monitoring with lichens*, 1992).

See Air pollution, Bioindication.

acid-fast (of bacteria), keeping carbol fuchsin stain after the addition of 25 per cent sulphuric acid (H_2SO_4).

acidiphilous (**acidophilous**, **acidophilic**), growing on or in conditions of low hydrogen ion concentration (q.v.); e.g. *Scytalidium acidophilum* with an optimum pH for growth of 3, with good growth even at pH 1 (Miller *et al.*, *Internat. Biodet.* **20**: 27, 1984); also used of lichens on peaty soils or bark of a pH below 5.

Acidomyces B.J. Baker, M.A. Lutz, S.C. Dawson, P.L. Bond & Banfield (2004), ? Teratosphaeriaceae. 1 (from acid mine drainage), California. See Baker *et al.* (*Appl. Environm. Microbiol.* **70**: 6270, 2004), Hoog *et al.* (*Stud. Mycol.* **51**: 33, 2005), Crous *et al.* (*Stud. Mycol.* **58**: 1, 2007; posn).

Aciella (P. Karst.) P. Karst. (1899) [non *Aciella* Tiegh. 1894, *Loranthaceae*] = Asterodon fide Donk (*Taxon* **5**: 69, 1956).

Aciesia Bat. (1961) nom. dub. ? = Tricharia Fée fide Lücking *et al.* (*Lichenologist* **30**: 121, 1998).

Acinophora Raf. (1808) nom. dub., Agaricales.

Acinula Fr. (1822), anamorphic *Pezizomycotina*, Sc.-.-. 1, Europe. Apparently sterile.

Acitheca Currah (1985), Gymnoascaceae. 1 (on bark), USA. See Currah (*Mycotaxon* **24**: 1, 1985), Currah (*SA* **7**: 1, 1988; key).

Ackermannia Pat. (1902) = Sclerocystis fide Zycha *et al.* (*Mucorales*, 1969).

Acladium Link (1809), Botryobasidiaceae. 20. See Wright (*Cryptog. Bot.* **1**: 26, 1989), Partridge *et al.* (*Mycotaxon* **82**: 41, 2002; key).

Acleistia Bayl. Ell. (1917), anamorphic *Calycina*, Ccu.0eH.15. 1 (saprobic on *Alnus* catkins), Europe. The connexion with *Calycina* is not well established. See Bayliss Elliott (*TBMS* **5**: 417, 1916).

Acleistomyces Bat. (1961) = Sporopodium fide Lücking *et al.* (*Lichenologist* **30**: 121, 1998).

Acmosporium Corda (1839) = Aspergillus fide Hughes (*CJB* **36**: 727, 1958).

Acoliomyces Cif. & Tomas. (1953) = Thelomma fide Hawksworth *et al.* (*Dictionary of the Fungi* edn 8, 1995).

Acolium (Ach.) Gray (1821), Caliciaceae (L). *c.* 5, widespread. See Hawksworth *et al.* (*Dictionary of the Fungi* edn 8, 1995).

Acolium Trevis. (1862) ≡ Pseudacolium.

Acompsomyces Thaxt. (1901), Laboulbeniaceae. 7 (on insect cuticles), widespread. See Benjamin (*Mem. N. Y. bot. Gdn* **49**: 210, 1989; key, ontogeny), Santamaria (*Mycotaxon* **49**: 313, 1993; Spain), Santamaría (*Fl. Mycol. Iberica* **5**, 2003; Iberian spp.).

Acontiopsis Negru (1961) nom. inval., Nectriaceae. 1 (on twigs of *Crataegus*), Europe. See Hawksworth *et al.* (*Dictionary of the Fungi* edn 8, 1995; ? syn. of *Cylindrocladiella*), Crous (*Taxonomy and Pathology of Cylindrocladium (Calonectria) and Allied Genera*: 278 pp., 2002).

Acontium Morgan (1902), anamorphic *Pezizomycotina*, Hso.0eH.?. 4, N. America.

acquired immunity, see immune.

acquired resistance, see resistance.

acrasin, a chemotactically active substance which controls the streaming together of the myxamoebae of *Dictyostelium discoideum* (Bonner, *J. exp. Zool.* **110**: 259, 1949) and other *Acrasiales*.

Acremoniella Sacc. (1886) nom. illegit. = Harziella Costantin & Matr. fide Groves & Skolko (*Can. J. Res.* C **24**: 74, 1946), Holubová-Jechová (*Folia Geobot. Phytotax.* **9**: 315, 1974), Warcup (*MR* **95**: 329, 1991; synonymy of *A. atra* auct. with *Harziella*).

Acremonites Pia (1927), Fossil Fungi. 1 (Oligocene), Europe.

Acremoniula G. Arnaud (1954) nom. inval. ≡ Acremoniula G. Arnaud ex Cif.

Acremoniula G. Arnaud ex Cif. (1962), anamorphic *Pezizomycotina*, Hso.0eP.1. 6 (on sooty moulds, esp. *Schiffnerula* and *Meliola*), pantropical. See Deighton (*Mycol. Pap.* **118**, 1969), Mercado Sierra *et al.* (*Mycotaxon* **55**: 491, 1995; Mexico), Hosagoudar *et al.* (*J. Econ. Taxon. Bot.* **25**: 281, 2001; India).

Acremonium Link (1809), anamorphic *Hypocreales*, Hso.0eH.15. *c.* 117, widespread. Still polyphyletic and perhaps best considered as a basic structural type rather than a genus. Grass endophytes formerly placed here are now considered to be *Neotyphodium* spp. See Gams (*Cephalosporium-artige Schimmelpilze*, 1971; monograph), Gams (*TBMS* **64**: 389, 1975), Samuels (*N.Z. Jl Bot.* **14**: 231, 1976; teleomorphs), Walz (*Biblthca Mycol.* **147**: 1, 1992; *A. chrysogenum*), Lowen (*Mycotaxon* **53**: 81, 1995; lichenicolous spp.), Alfaro-García *et al.* (*Mycol.* **88**: 804, 1996; on *Cucurbitaceae*), Glenn *et al.* (*Mycol.* **88**: 369, 1996; phylogeny), Abad *et al.* (*Diagnosis and Identification of Plant Pathogens. Proceedings of the 4th International Symposium of the European Foundation for Plant Pathology*: 287, 1997; VCGs), Ito *et al.* (*MR* **104**: 77, 2000), Rossman (*Stud. Mycol.* **45**: 27, 2000; spp. with Hypocrealean affinities), Seifert & Gams *in* McLaughlin *et al.* (Eds) (*The Mycota* A Comprehensive Treatise on Fungi as Experimental Systems for Basic and Applied Research **7A**: 307, 2001; polyphyly), Wang *et al.* (*Mycosystema* **21**: 192, 2002; Chinese spp.), Lin *et al.* (*Plant Pathology Bulletin* Taichung **13**: 91, 2004; *A. lactucae*), Hsiao *et al.* (*J. Clin. Microbiol.* **43**: 3760, 2005; identification using arrays), Ma *et al.* (*Life in Ancient Ice*: 159, 2005; in glacial ice), Rakeman *et al.* (*J. Clin. Microbiol.* **43**: 3324, 2005; mol. analysis of clinical spp.).

acro- (combining form), at the end; apical; terminal.

acroauxic (of conidiophores), growth in length restricted to the apical region.

Acrocalymma Alcorn & J.A.G. Irwin (1987), anamorphic *Massarina*, Cpd.0eH.15. 1 (on *Medicago*), Australia. See Alcorn & Irwin (*TBMS* **88**: 163, 1987), Shoemaker *et al.* (*CJB* **69**: 569, 1991; teleomorph), Aptroot (*Nova Hedwigia* **66**: 89, 1998; tax. placement).

acrochroic, see Colour.

Acrocladium Petr. (1949) [non *Acrocladium* Mitt. 1869, *Musci*] = Periconiella fide von Arx (*Persoonia* **11**: 389, 1981).

Acroconidiella J.C. Lindq. & Alippi (1964) ? = Cladosporium fide Lindquist & Alippi (*Darwiniana* **13**: 612, 1964), Dugan *et al.* (*Schlechtendalia* **11**,

2004).

Acroconidiellina M.B. Ellis (1971), anamorphic *Pezizomycotina*, Hso.1eP.26. 3, widespread (tropical). See Ellis (*Mycol. Pap.* **125**: 22, 1971).

Acrocordia A. Massal. (1854), Monoblastiaceae (L). 10, widespread (esp. north temperate). See Coppins & James (*Lichenologist* **10**: 179, 1978; UK spp.), Harris (*More Florida Lichens*, 1995).

Acrocordiaceae Oksner ex M.E. Barr (1987) = Monoblastiaceae.

Acrocordiella O.E. Erikss. (1982) = Requienella fide Boise (*Mycol.* **78**: 37, 1986; synonymy), Eriksson & Hawksworth (*SA* **7**: 59, 1988).

Acrocordiomyces Cif. & Tomas. (1953) = Acrocordia fide Hawksworth *et al.* (*Dictionary of the Fungi* edn 8, 1995).

Acrocordiopsis Borse & K.D. Hyde (1989), Melanommataceae. 2 (marine), widespread. See Borse & Hyde (*Mycotaxon* **34**: 535, 1989), Alias *et al.* (*Fungal Diversity* **2**: 35, 1999).

Acrocorelia R. Doll (1982) nom. nud., ? Dothideales (L).

Acrocylindrium Bonord. (1851), anamorphic *Pezizomycotina*, Hso.0eH.?. 3, Europe. ? = Sarocladium fide Gams (*in litt.*). See Gams & Hawksworth (*Kavaka* **3**: 60, 1976).

Acrodesmis Syd. (1926) = Periconiella fide Ellis (*Mycol. Pap.* **111**, 1967).

Acrodictyella W.A. Baker & Partridge (2001), anamorphic *Pezizomycotina*. 1, Alabama. See Baker *et al.* (*Mycotaxon* **78**: 30, 2001), Baker & Morgan-Jones (*Mycotaxon* **85**: 371, 2003; contrast with *Pseudacrodictys*).

Acrodictyopsis P.M. Kirk (1983), anamorphic *Pezizomycotina*, Hso.#eP.1. 1, British Isles. See Kirk (*Mycotaxon* **18**: 260, 1983), Kendrick (*CJB* **81**: 75, 2003; morphogenesis).

Acrodictys M.B. Ellis (1961), anamorphic *Pezizomycotina*, Hso.#eP.1/19. *c.* 38 (saprobic on wood etc.), widespread. See Ellis (*Dematiaceous Hyphomycetes*, 1971), Ellis (*More Dematiaceous Hyphomycetes*, 1976), Chang (*Bot. Bull. Acad. sin.* Taipei **38**: 197, 1997), Whitton *et al.* (*Fungal Diversity* **4**: 159, 2000; on *Pandanaceae*;), Cai *et al.* (*Nova Hedwigia* **75**: 525, 2002; Philippines), Baker & Morgan-Jones (*Mycotaxon* **85**: 371, 2003; contrast with *Pseudacrodictys*), Kodsueb *et al.* (*Cryptog. Mycol.* **27**: 111, 2006; Thailand).

Acrodontiella U. Braun & Scheuer (1995), anamorphic *Pezizomycotina*, Hso.?.?. 1, Austria. See Braun & Scheuer (*Sydowia* **47**: 146, 1995), Braun (*Monogr. Cercosporella, Ramularia Allied Genera (Phytopath. Hyphom.)* **2**, 1998).

Acrodontium de Hoog (1972), anamorphic *Pezizomycotina*, Hso.0eH.1. 9, widespread. See de Hoog (*Stud. Mycol.* **1**, 1972), Sutton *et al.* (*Guide to Clinically Significant Fungi*, 1998; clinical taxa), van Wyk *et al.* (*S. Afr. J. Sci.* **96**: 580, 2000; conidiogenesis), Czeczuga *et al.* (*Feddes Repert.* **112**: 81, 2001; Czech Republic).

Acrogenospora M.B. Ellis (1971), anamorphic *Farlowiella*, Hso.0eP.19. 6 (saprobic on wood and bark), widespread. See Hughes (*N.Z. Jl Bot.* **16**: 312, 1978), Goh *et al.* (*MR* **102**: 1309, 1998; key), Zhu *et al.* (*Mycotaxon* **92**: 383, 2005; China).

Acrogenotheca Cif. & Bat. (1963), Dothideomycetes. Anamorph *Hiospira*. 2, widespread (tropical). See Hughes (*N.Z. Jl Bot.* **5**: 504, 1967), Hughes (*Mycol.* **68**: 693, 1976).

acrogenous, development at the apex.

Acrogynomyces Thaxt. (1931), Laboulbeniaceae. 6 (on insect exoskeletons), Africa. See Tavares (*Mycol. Mem.* **9**: 627 pp., 1985), Santamaría (*MR* **99**: 1071, 1995).

acronema, extension of flagellum tip containing the two central microtubules but none of the nine peripheral elements.

acropetal (1) describes chains of conidia in which the youngest is at the apex, basifugal; cf. basipetal; (2) a pattern of apical growth.

Acrophialophora Edward (1961), anamorphic *Pezizomycotina*, Hso.0eH.15. 2, widespread. See Samson & Mahmood (*Acta Bot. Neerl.* **19**: 804, 1970; key), Al-Mohsen *et al.* (*J. Clin. Microbiol.* **38**: 4569, 2000; clinical), Kendrick (*CJB* **81**: 75, 2003; morphogenesis).

Acrophragmis Kiffer & Reisinger (1970), anamorphic *Pezizomycotina*, Hso.≡ eP.19. 4, widespread (esp. tropical). See Kiffer & Reisinger (*Rev. Écol. Biol. Sol* **7**: 16, 1970), Mercado Sierra & Mena Portales (*Acta bot.* Szeged **32**: 189, 1986; Cuba), Rao & Hoog (*Stud. Mycol.* **28**, 1986; India), Wu & Zhuang (*Fungal Diversity Res. Ser.* **15**, 2005; China).

Acrophytum, see *Akrophyton*.

acropleurogenous, formed at the end and on the sides.

Acrorixis Trevis. (1860) = Thelenella fide Mayrhofer & Poelt (*Herzogia* **7**: 13, 1985), Hawksworth *et al.* (*Dictionary of the Fungi* edn 8, 1995).

Acroscyphus Lév. (1846), Caliciaceae (L). 1, widespread. See Tibell (*Symb. bot. upsal.* **32** no. 1: 291, 1997; anam), Tibell (*Biblthca Lichenol.* **71**: 107 pp., 1998), Joneson & Glew (*Bryologist* **106**: 443, 2003; N America), Tibell & Thor (*J. Hattori bot. Lab.* **94**: 205, 2003; Japan).

Acrospeira Berk. & Broome (1857), anamorphic *Pezizomycotina*, Hso.#eP.1. 1 (parasitic on *Castanea*), widespread (north temperate). See Wiltshire (*TBMS* **21**: 211, 1938).

Acrospermaceae Fuckel (1870), Acrospermales. 4 gen. (+ 3 syn.), 15 spp.

Lit.: Webster (*TBMS* **39**: 361, 1956), Eriksson (*Ark. Bot.* **6**: 381, 1967), Eriksson (*Mycotaxon* **15**, 1982), Nograsek (*Biblthca Mycol.* **133**: 271 pp., 1990), Barr (*Mycol.*, 1994; included in *Xylariales*), Winka & Eriksson (*Phylogenetic Relationships Within the Ascomycota Based on 18S rDNA Sequences* Akademisk Avhandling [Thesis (PhD), Department of Ecology and Environmental Science, Umeå University]: [17] pp., 2000).

Acrospermales Minter, Peredo & A.T. Watson (2007). Dothideomycetes. 1 fam., 4 gen., 15 spp. Fam.:
Acrospermaceae

Lit.: Minter *et al.* (*Bol. Soc. Argent. Bot.* **42**: 107, 2007).

Acrospermoides J.H. Mill. & G.E. Thomps. (1940), ? Acrospermaceae. 1, USA. See Miller & Thompson (*Mycol.* **32**: 1, 1940; descr.), Barr (*Mycotaxon* **39**: 43, 1990; family placement).

Acrospermum Tode (1790), Acrospermaceae. Anamorph *Gonatophragmium*. 11 (saprobic, esp. on grasses), widespread. See Webster (*TBMS* **39**: 361, 1956; conidia), Eriksson (*Ark. Bot.* ser. 2 **6**: 381, 1967), Sherwood (*Mycotaxon* **5**: 39, 1977; posn), Winka & Eriksson (*Phylogenetic Relationships Within the Ascomycota Based on 18S rDNA Sequences* Akademisk Avhandling [Thesis (PhD), Department of Ecology and Environmental Science, Umeå University]: [17] pp., 2000; phylogeny), Minter *et al.* (*Bol. Soc. Argent. Bot.* **42**: 107, 2007).

Acrosphaeria Corda (1842) = Xylaria Hill ex Schrank fide Læssøe (*SA* **13**: 43, 1994).

Acrospira Mont. (1857), anamorphic *Pezizomycotina*, Hso.?.?. 1, Europe.

acrospore, an apical spore.

Acrosporella Riedl & Ershad (1977) = Cladosporium fide Sutton (*in litt.*).

Acrosporium Bonord. (1851), anamorphic *Pezizomycotina*, Hso.0eH.?. 1, Germany.

Acrosporium Nees (1816) nom. rej. = Oidium Link (1824) fide Hawksworth *et al.* (*Dictionary of the Fungi* edn 8, 1995).

acrosporogenous (of conidial maturation), cells delimited and maturing in sequence from base to apex as the tip of the conidium expands (Luttrell, 1963).

Acrostalagmus Corda (1838), anamorphic *Hypocreales*. 2 (isol. ex soil etc.), widespread. See Zare & Gams (*MR* **108**: 576, 2004), Zare *et al.* (*MR* **108**: 576, 2004; rels with *Verticillium*, connexions), Gams *et al.* (*Taxon* **54**: 179, 2005; nomencl.), Pantou *et al.* (*MR* **109**: 889, 2005; phylogeny).

Acrostaphylus G. Arnaud ex Subram. (1971) = Nodulisporium fide Jong & Rogers (*Tech. Bull. Wash. agric. Exp. Stn* **71**, 1972).

Acrostaurus Deighton & Piroz. (1972), anamorphic *Pezizomycotina*, Hso.0bP.19. 1 (fungicolous), widespread (tropical). See Deighton & Pirozynski (*Mycol. Pap.* **128**: 94, 1972).

Acrostroma Seifert (1987), anamorphic *Batistia*, Hsy.0eH.15. 1, Venezuela. See Seifert (*CJB* **65**: 2197, 1987), Samuels & Rodrigues (*Mycol.* **81**: 52, 1989; connexion).

Acrotamnium Nees (1816) nom. dub. ? = Tomentella Pat. fide Stalpers (*Stud. Mycol.* **24**: 72, 1984).

Acrotellomyces Cif. & Tomas. (1953) ≡ Acrotellum.

Acrotellum Tomas. & Cif. (1952) = Thelidium fide Hawksworth *et al.* (*Dictionary of the Fungi* edn 8, 1995).

Acrothamnium, see *Acrotamnium*.

Acrotheca Fuckel (1860) = Ramularia Unger fide Braun (*Monogr. Cercosporella, Ramularia Allied Genera (Phytopath. Hyphom.)* **2**, 1998).

Acrotheciella Koord. (1907), anamorphic *Pezizomycotina*, Hsp.≡ eP.?24. 1, Java.

Acrothecium (Corda) Preuss (1851), anamorphic *Pezizomycotina*, Hso.≡ eH.?. *c.* 15, widespread.

acroton, a spinule in lichens bearing side branches.

Actidiographium Lar.N. Vassiljeva (2000), ? Hysteriaceae. 1, Eastern Russia. See Vasil'eva (*Mikol. Fitopatol.* **34**: 4, 2000).

actidione, trade name for cycloheximide (q.v.).

Actidium Fr. (1815), Mytilinidiaceae. 9, Europe; N. America. See Zogg (*Ber. schweiz. bot. Ges.* **70**: 195, 1960; key).

Actigea Raf. (1814) = Scleroderma fide Stalpers (*in litt.*).

Actigena, see *Actigea*.

actin and **mycosin** are proteins associated with contraction and relaxation of muscle; also present in several lower eukaryotic organisms and responsible for the periodic reversal of protoplasmic streaming in the plasmodium of *Mycetozoa*.

Actiniceps Berk. & Broome (1876), Pterulaceae. 3, widespread (tropical). See Boedijn (*Persoonia* **1**: 11, 1959), Tanaka & Hongo (*Mycoscience* **42**: 433, 2001; Japan).

Actiniopsis Starbäck (1899) = Trichothelium fide Santesson (*Symb. bot. upsal.* **12** no. 1: 1, 1952), Samuels (*N.Z. Jl Bot.* **14**: 232, 1976), Rossman *et al.* (*Stud. Mycol.* **42**: 248 pp., 1999).

Actinobacteria (Actinomycetes; 'Ray Fungi'). A group of morphologically diverse but usually filamentous Gram positive bacteria which have occasionally been mistaken for conidial fungi. *Actinobacteria* are typically saprobes (esp. in soil) but a few are pathogenic for humans, animals, and plants; some (esp. *Streptomyces*) are important sources of antibiotics (see amphotericin, cycloheximide, nystatin, streptomycin); some form lichen-like associations with green algae (see actinolichen).

Lit.: The literature on *Actinobacteria* is extensive. A hierarchical description has been produced by Stackebrandt *et al.* (*Int. J. Syst. Bacteriol.* **47**: 479, 1997). Generic names are listed by Skerman *et al.* (*Approved lists of bacterial names*, Amended Edn, 1989). See Williams *et al.* (Eds) (*Bergey's manual of systematic bacteriology* **4**, The actinomycetes, 1989), Balows *et al.* (*The procaryotes*, 2nd edn, 1992), Goodfellow *et al.* (Eds) (*Biology of the actinomycetes*, 1984), Ortiz-Ortiz *et al.* (Eds) (*Biological, biochemical, and biomedical aspects of actinomycetes*, 1984), Goodfellow *et al.* (Eds) (*Actinomycetes in biotechnology*), Goodfellow & Williams (*Ann. Rev. Microbiol.* **37**: 189, 1983).

Actinocephalum Saito (1905) = Cunninghamella fide Hesseltine (*Mycol.* **47**: 344, 1955).

Actinochaete Ferro (1907) nom. conf., anamorphic *Pezizomycotina*. = Aspergillus (Trichocom.) p.p. and Septobasidium (Septobasid.) p.p. fide Ellis (*in litt.*).

Actinocladium Ehrenb. (1819), anamorphic *Pezizomycotina*, Hso.0bP.1. 5, widespread. See Wu & Zhuang (*Fungal Diversity Res. Ser.* **15**, 2005; China).

Actinocymbe Höhn. (1911), Chaetothyriaceae. 1 or 2, widespread (tropical). See Verma & Kamal (*Indian Phytopath.* **40**: 410, 1988).

Actinodendron G.F. Orr & Kuehn (1963) = Oncocladium fide Hughes (*CJB* **46**: 939, 1968).

Actinodermium Nees (1816) ≡ Sterbeeckia.

Actinodochium Syd. (1927), anamorphic *Pezizomycotina*, Hsp.0eH.3. 2, C. America; India.

Actinodothidopsis F. Stevens (1925) = Venturia Sacc. fide Müller & von Arx (*Beitr. Kryptfl. Schweiz* **11** no. 2, 1962).

Actinodothis Syd. & P. Syd. (1914) = Amazonia fide Hansford (*Beih. Sydowia* **2**, 1961).

Actinoglyphis Mont. (1856) = Sarcographa fide Hawksworth *et al.* (*Dictionary of the Fungi* edn 8, 1995).

Actinogyra Schol. (1934) = Umbilicaria fide Hawksworth *et al.* (*Dictionary of the Fungi* edn 8, 1995).

actinogyrose (**actinogyr**) (of apothecia), disc gyrose and having no proper margin.

actinolichen, a lichen-like association between a green alga and an actinomycete (e.g. *Chlorella* and *Streptomyces* sp.; Lazo & Klein, *Mycol.* **57**: 804, 1965) occurring in nature and also in mixed laboratory cultures. See Kalakoutskii *et al.* (*Actinomycetes*, n.s. **1**(2): 27, 1990; lab. expts, bibliogr.).

Actinomadura H. Lechev. & M.P. Lechev. (1968), Actinobacteria. q.v.

Actinomma Sacc. (1884) = Atichia fide Hawksworth *et al.* (*Dictionary of the Fungi* edn 8, 1995).

Actinomortierella Chalab. (1968) = Mortierella fide Gams (*Nova Hedwigia* **18**: 30, 1969).

Actinomucor Schostak. (1898), Mucoraceae. 1, widespread. See Benjamin & Hesseltine (*Mycol.* **49**: 240, 1957), Jong & Yuan (*Mycotaxon* **23**: 261, 1985), Voigt & Wöstemeyer (*Gene* **270**: 113, 2001; phylogeny), Zheng & Liu (*Nova Hedwigia* **80**: 419, 2005), Khan *et al.* (*Antonie van Leeuwenhoek* **94**: in press, 2008; zygomycosis, n.sp.).
Actinomyce Meyen (1827) nom. dub., ? Fungi.
Actinomyces Harz (1877), Actinobacteria. q.v.
Actinomycetes, see Actinobacteria.
Actinomycites D. Ellis (1916), Fossil Fungi, Actinobacteria. 1 (Jurassic), British Isles. q.v.
Actinomycodium K.M. Zalessky (1915), Fossil Fungi (anamorphic fungi) or Actinomycetes anamorphic *Pezizomycotina*. 1 (Permo-Carboniferous), former USSR.
Actinomyxa Syd. & P. Syd. (1917), Microthyriaceae. 1, Australia.
Actinonema Fr. (1849) = Spilocaea fide Sutton (*Mycol. Pap.* **141**, 1977).
Actinonema Pers. (1822) nom. dub., anamorphic *Pezizomycotina*. The type contains sterile mycelium, but often used for *Marssonina rosae* (teleomorph *Diplocarpon rosae*) (black spot of rose). See Sutton (*Mycol. Pap.* **141**, 1977).
Actinonemella Höhn. (1916) = Asteroma fide Sutton (*Mycol. Pap.* **141**, 1977).
Actinopelte Sacc. (1913) ≡ Tubakia.
Actinopelte Stizenb. (1861) = Solorinella fide Hawksworth *et al.* (*Dictionary of the Fungi* edn 8, 1995).
Actinopeltella Doidge (1924) = Actinopeltis fide von Arx & Müller (*Stud. Mycol.* **9**, 1975).
Actinopeltis Höhn. (1907), Microthyriaceae. 11, widespread. See Ellis (*TBMS* **68**: 145, 1977), Spooner & Kirk (*MR* **94**: 223, 1990), Geel & Aptroot (*Nova Hedwigia* **82**: 313, 2006; fossil taxa).
Actinophora Merr. (1943) ≡ Acinophora.
Actinoplaca Müll. Arg. (1891), Gomphillaceae (L). 4, widespread (primarily tropical). See Vězda & Poelt (*Folia geobot. phytotax.* **22**: 180, 1987), Lücking (*Biblthca Lichenol.* **65**: 1, 1997; Costa Rica), Aptroot *et al.* (*Mycotaxon* **88**: 41, 2003; Yunnan), Farkas (*Biblthca Lichenol.* **88**: 111, 2004; S Africa), Lücking *et al.* (*Lichenologist* **37**: 123, 2005; phenotype cladistics), Lücking (*Cryptog. Mycol.* **27**: 121, 2006; French Guiana).
Actinoplacomyces Cif. & Tomas. (1954) ≡ Actinoplaca.
Actinoplanes Couch (1950), Actinobacteria. q.v.
Actinopolyspora Gochn., K.G. Johnson & Kushner (1975), Actinobacteria. q.v.
Actinoscypha P. Karst. (1888) = Micropeziza fide Nannfeldt (*Bot. Notiser* **129**: 323, 1976).
Actinosoma Syd. (1930) ? = Actinopeltis fide Spooner & Kirk (*MR* **94**: 223, 1990), Eriksson & Hawksworth (*SA* **9**: 6, 1991; status).
Actinospira Corda (1854) ≡ Myxotrichum.
Actinospora Ingold (1952) [non *Actinospora* Turcz. 1835, *Ranunculaceae*] ≡ Actinosporella.
Actinosporella Descals, Marvanová & J. Webster (1999), anamorphic *Miladina*, Hso.1bH.23. 1 (in water), widespread. See Descals (*TBMS* **67**: 208, 1976), Descals & Webster (*TBMS* **70**: 466, 1978; teleomorph), Descals *et al.* (*CJB* **76**: 1647, 1998), Descals (*MR* **109**: 545, 2005).
Actinostilbe Petch (1925), anamorphic *Lanatonectria*, Hsp.0-1eH.15. 3, widespread. See Sutton (*TBMS* **76**: 97, 1981; synonym of *Sarcopodium*), Samuels & Seifert *in* Sugiyama (Ed.) (*Pleomorphic Fungi: The Diversity and its Taxonomic Implications*: 29, 1987), Rossman *et al.* (*Stud. Mycol.* **42**: 248 pp., 1999).
Actinostroma Klotzsch (1843) = Cymatoderma fide Donk (*Taxon* **6**: 17, 1957).
Actinosynnema T. Haseg., H. Lechev. & M.P. Lechev. (1978), Actinobacteria. q.v.
Actinoteichus Cavalc. & Poroca (1971) = Asterothyrium Müll. Arg. fide Lücking *et al.* (*Lichenologist* **30**: 121, 1998).
Actinotexis Arx (1960), anamorphic *Pezizomycotina*, Cpt.0fH.?. 1, Brazil. See von Arx (*Publções Inst. Micol. Recife* **289**: 4, 1960).
Actinothecium Ces. (1854), anamorphic *Pezizomycotina*, Cpt.0eH.?. 5, widespread.
Actinothecium Flot. (1855) = Verrucaria Schrad. fide Hawksworth (*Bull. Br. Mus. nat. Hist.* Bot. **14**: 43, 1985; placement), Hawksworth *et al.* (*Dictionary of the Fungi* edn 8, 1995).
Actinothyrella Edward, Kr.P. Singh, S.C. Tripathi, M.K. Sinha & Ranade (1974) nom. dub., anamorphic *Pezizomycotina*. See Sutton (*Mycol. Pap.* **141**, 1977).
Actinothyrium Kunze (1823), anamorphic *Pezizomycotina*, Cpt.0fH.?. 10, widespread. See Barnes *et al.* (*Stud. Mycol.* **50**: 551, 2004; links with *Dothistroma*).
Actinotrichum Wallr. [not traced] nom. nud., anamorphic *Pezizomycotina*. See Sutton (*Mycol. Pap.* **141**, 1977).
Actonia C.W. Dodge (1935) nom. dub., Fungi. See Batra *in* Subramanian (Ed.) (*Taxonomy of fungi* **1**: 187, 1978).
Actycus Raf. (1815) nom. dub., Fungi.
aculeate, having narrow spines (Fig. 20.3).
aculeolate, having spine-like processes.
acuminate, gradually narrowing to a point.
Acumispora Matsush. (1980), anamorphic *Pezizomycotina*, Hso.≡ eH-P.1. 3, Taiwan. See Matsushima (*Matsush. Mycol. Mem.* **1**: 2, 1980), Matsushima (*Matsush. Mycol. Mem.* **6**, 1989).
Acurtis Fr. (1849) nom. dub., Physalacriaceae. A sterile form of *Armillaria mellea* s.l. when parasitized by *Entoloma abortivum* (Czederpilz *et al.*, *Mycol.* **93**: 84, 2001), not the opposite (*E. abortivum* as parasitized by *Armillaria* as suggested by Watling (*Bull. Soc. linn. Lyon* **43**(Suppl.): 449, 1970), so technically a hyphal anamorph.
acute (1) pointed (Fig. 23.41); (2) less than a right angle.
Acutocapillitium P. Ponce de León (1976), ? Agaricaceae. 3, America (tropical). See Demoulin (*in litt.*), Calonge *et al.* (*Boll. Gruppo Micol. 'G. Bresadola'* **43**: 51, 2000) ? = Glyptoderma (Lycoperd.) fide.
Adamson's fringe, the downward growing hyphae of a dermatophyte in the region above the bulb of a hair.
adapted race (Magnus), see physiologic race.
adaxial (of a basidiospore), the side next to the long axis of the basidium, usually that with the apiculus (Corner, 1948); cf. abaxial.
Adea Petr. (1928) = Seiridium fide Nag Raj & Kendrick (*Sydowia* **38**: 179, 1986).
Adella Petr. (1936) = Wojnowicia fide Sutton (*Česká Mykol.* **29**: 97, 1975).
Adelococcaceae Triebel (1993), Verrucariales. 2 gen., 13 spp.

Lit.: Triebel (*Biblthca Lichenol.* **35**: 278 pp., 1989), Matzer & Pelzmann (*Nova Hedwigia* **52**: 1, 1991), Triebel (*Sendtnera* **1**: 273, 1993), Hoffmann & Hafellner (*Biblthca Lichenol.* **77**: 181 pp., 2000), Orange (*Mycotaxon* **81**: 265, 2002).

Adelococcus Theiss. & Syd. (1918), Adelococcaceae. 6 (on lichens), Europe. See Matzer & Hafellner (*Biblthca Lichenol.* **37**, 1990), Matzer & Pelzmann (*Nova Hedwigia* **52**: 1, 1991; ascospores), Etayo & Breuss (*Öst. Z. Pilzk.* **7**: 203, 1998).

Adelodiscus Syd. (1931), Helotiales. 1, Philippines.

Adelolecia Hertel & Hafellner (1984), Ramalinaceae (L). 3, Europe; N. America. See Hertel & Rambold (*Biblthca Lichenol.* **57**: 211, 1995), Ekman (*Op. Bot.* **127**, 1996), Lumbsch *et al.* (*Mol. Phylogen. Evol.* **31**: 822, 2004; posn).

Adelomyces Thaxt. (1931) = Phaulomyces fide Tavares (*Mycol. Mem.* **9**, 1985).

Adelomycetes, see Anamorphic fungi (Langeron, *Précis de Mycologie*, edn 1, 1945).

Adelopus Theiss. (1918) = Phaeocryptopus fide von Arx & Müller (*Stud. Mycol.* **9**, 1975).

adelphogamy, pseudomictic copulation of mother and daughter cells, as in some yeasts (Gäumann & Dodge, 1928: 13).

adenose, having glands; gland-like.

Aderkomyces Bat. (1961), Gomphillaceae (L). 25, neotropics. See Lücking *et al.* (*Lichenologist* **30**: 121, 1998; synonymy with *Tricharia*), Lücking *et al.* (*Lichenologist* **37**: 123, 2005; accepted genus), Lücking (*Cryptog. Mycol.* **27**: 121, 2006; French Guiana).

Adermatis Clem. (1909) = Lecania fide Hawksworth *et al.* (*Dictionary of the Fungi* edn 8, 1995).

adherance (of fungicides), the ability of a fungicide (or other crop protectant) to stick to a surface. Cf. retention.

adhesive disc, see holdfast.

adhesorium, the organ developed from a resting zoospore of *Plasmodiophora* for attachment to, and penetration of, the host (Aist & Williams, *CJB* **49**: 2023, 1971).

Adhogamina Subram. & Lodha (1964) = Gilmaniella fide Barron (*The genera of hyphomycetes from soil*, 1968).

adiaspiromycosis, pulmonary infection in animals (particularly soil-burrowing rodents) and rarely humans by *Emmonsia* spp., esp. *E. parva* (syn. *Haplosporangium parvum*) and *E. crescens* (Jellison, *Adiaspiromycosis* (syn. *Haplomycosis*), 1969); haplomycosis. Cf. adiaspore.

adiaspore, a large spherical chlamydospore produced in the lungs of animals by the enlargement of an inhaled conidium of *Emmonsia* spp.; cf. adiaspiromycosis. *Chrysosporium pruinosum* produces similar spores in culture (Carmichael, *CJB* **40**: 1167, 1962).

adjunct (in brewing), any legally permitted substance lacking nutritional properties added to the fermentation.

adnate (of lamellae or tubes), joined to the stipe; if lamellae, proximal end not notched (cf. sinuate); sometimes restricted to lamellae widely joined to the stipe (Fig. 19C) (cf. adnexed); (of pellicle, scales, etc.), tightly fixed to the surface.

adnexed (of lamellae), narrowly joined to the stipe (Fig. 19B) (cf. adnate); an ambiguous term.

Adomia S. Schatz (1985), Sordariomycetes. 1 (marine, on *Avicennia*), Egypt; Australia. Perhaps part of the *Ceriospora* complex, or related to *Urosporellopsis*. See Schatz (*TBMS* **84**: 555, 1985; descr.).

adpressed, see appressed.

adspersed, of wide distribution; scattered.

aduncate, bent; hooked; crooked.

Adustomyces Jülich (1979), ? Pterulaceae. 1, Europe; Africa. See Jülich (*Persoonia* **10**: 325, 1979).

adventitious septum, see septum.

adventive branching (of fruticose lichens), branching not of the normal pattern; e.g. regenerate branches produced after damage to the original branches in *Cladonia*.

Aecidiconium Vuill. (1892), ? Pucciniales. 1 (on *Pinus* (*Pinaceae*)), France.

Aecidiella Ellis & Kelsey (1897) = Pucciniosira fide Arthur (*N. Amer. Fl.* **7**: 126, 1907).

Aecidiolum Unger (1832), anamorphic *Pucciniales*. 12. Anamorph name for (0).

aecidiospore, see *Pucciniales*.

Aecidites Debey & Ettingsh. (1859), Fossil Fungi. 4 (Cretaceous, Tertiary), Europe.

Aecidium Pers. (1796), anamorphic *Pucciniales*. *c.* 600 (on angiosperms), widespread. Anamorph name for (I). The name originally applied to the aecial stage of *Puccinia* but is also widely used for the 'aecioid' aecial stages of other rust families. A number may be 'duplicate' names; some may be species of *Endophyllum* (q.v.). As with other anamorphic fungi, an *Aecidium* name is sometimes used even when there is a named teleomorphic (telial, III) state.

aecidium, see *Pucciniales*.

aeciospore, see *Pucciniales*.

aeciotelium, see *Pucciniales*.

aecium, see *Pucciniales*.

Aeciure Buriticá & J.F. Hennen (1994), anamorphic *Arthuria*. 1 (on *Croton* (*Euphorbiaceae*)), Brazil. Anamorph name for (II).

Aedispora P.J. Kilochitskii (1997), Microsporidia. 8.

Aedycia Raf. (1808) nom. rej. = Mutinus fide Stalpers (*in litt.*).

Aegerita Pers. (1801), anamorphic *Bulbillomyces*. 1, Europe. See Hennebert (*Persoonia* **7**: 191, 1973), Julich (*Persoonia* **8**: 59, 1974).

Aegeritella Bałazy & J. Wiśn. (1974), anamorphic *Pezizomycotina*, Hsp.0eH.1. 4 (on ants), Europe; Brazil. See Bałazy & Wiśniewski (*Prace Komisji Nauk Rolniczych i Komisji Nauk Leśnych* **38**: 13, 1974), Espadaler & Wisniewski (*Butlletí de la Institució Catalana d'Història Natural* secció de Botànica **54**: 31, 1987; Spain), Bałazy *et al.* (*MR* **94**: 273, 1990; Morocco).

Aegeritina Jülich (1984), anamorphic *Subulicystidium*. 1, Europe. See Jülich (*Int. J. Mycol. Lichenol.* **1**: 282, 1984).

Aegeritopsis Höhn. (1903) nom. dub., Fungi.

Aenigmatomyces R.F. Castañeda & W.B. Kendr. (1994), anamorphic *Fungi*, Hso.0eH.1. 1 (on ? *Pythium*), Canada. See Castañeda Ruiz & Kendrick (*Mycol.* **85**: 1023, 1993).

Aenigmatospora R.F. Castañeda Ruiz, Saikawa, Guarro & Calduch (1999), anamorphic *Pezizomycotina*. 1, Cuba. See Castañeda *et al.* (*Cryptog. Mycol.* **20**: 115, 1999).

aequi-hymeniiferous (of hymenial development in agarics), having basidia which mature and shed their spores evenly over the surface of each lamella; the non-*Coprinus* type (Buller, *Researches* **2**: 19, 1922). cf. inaequi-hymeniiferous.

aero-aquatic fungi, fungi that grow under water but produce spores in the air above (van Beverwijk, *TBMS* **34**: 280, 1951). See Aquatic fungi.

aerobe, an organism needing free oxygen for growth; cf. anaerobe.

aerobiological pathway, the process (comprising the source, liberation, dispersion, deposition, and impact on another living organism) by which air-borne microorganisms are dispersed (Edwards, *Aerobiology*, 1979).

aerogenic, describes an organism that produces detectable gas during the breakdown of carbohydrate.

aerole (of lichens), a scale-like area on the thallus delimited by cracks or depressions.

Aerophyton Eschw. (1824) nom. dub., anamorphic *Pezizomycotina*.

Aeruginospora Höhn. (1908), ? Tricholomataceae. 2, Australia; Southeast Asia. See Horak (*N.Z. Jl Bot.* **28**: 255, 1990).

Aessosporon Van der Walt (1970), Sporidiobolaceae R.T. Moore. Anamorphs *Bullera*, *Sporobolomyces*. 2, Netherlands. See van der Walt (*Antonie van Leeuwenhoek Ned. Tijdschr. Hyg.* **36**: 54, 1970).

aethalium (of *Mycetozoa*), a sessile fruit-body made by a massing of all or a part of the plasmodium.

aetiology, the science of the causes of disease; etiology (Amer.).

Aetnensis Lloyd (1910) nom. nud., Fungi.

Aflatoxins. A series of toxic polybutole metabolites (mycotoxins) esp. of *Aspergillus flavus* strains when growing on groundnuts, cereals, etc., particularly in warm and moist conditions; most well known mycotoxin; most developed countries have statutory limits; gene probes available; the cause of **aflatoxicosis** in poultry and cattle and carcinogenic for rats and humans.

Lit.: Abbas (*Aflatoxin and food safety*, 2005), Hesseltine *et al.* (*Bact. Rev.* **30**: 795, 1966), *Aflatoxin bibliography, 1960-67*, 1968), Goldblatt (Ed.) (*Aflatoxin: scientific background, control and implications*, 1969), Racovitza (*J. gen. Microbiol.* **57**: 379, 1969; aflatoxin toxic to the mite *Glyciphagus domesticus*), Heathcote & Hibbert (*Aflatoxins: chemical and biological aspects*, 1978), Eaton & Groopman (*The toxicology of aflatoxins*, 1994), Flannigan (Ed.) (*Internat. Biodet.* **22** (Suppl.), 1986; in cereals and stored products), Williams *et al.* (*Am. J. Clin. Nutrition* **80**: 1106, 2004), Wylie & Morehouse (Eds) (*Mycotoxic fungi, mycotoxins, mycotoxicoses* **1-3**, 1977-8), Mycotoxicoses.

African histoplasmosis, infection of humans or animals by *Histoplasma capsulatum* var. *duboisii*.

African Mycological Association, Founded in 1995; recognized as the Committee for Africa within the International Mycological Association (q.v.); structure comprises individual and corporate members, and an elected executive; organizes Regional Mycology Conferences in Africa. Publications: *Mycoafrica, the AMA Newsletter*. Website: http://194.203.77.69/AfricanMycologicalAssociation.

Afroboletus Pegler & T.W.K. Young (1981), Boletaceae. 7, Africa (tropical). See Pegler & Young (*TBMS* **76**: 130, 1981), Watling & Turnbull (*Edinb. J. Bot.* **49**: 343, 1993; South and East Central Africa), Heinemann & Rammeloo (*Bulletin du Jardin Botanique National de Belgique* **64**: 215, 1995; Burundi).

AFTOL (Assembling the Fungal Tree of Life) is the title of a major project funded by the National Science Foundation of the USA, starting as a proposal in 2002 and in its second stage at the time of this edition going to press. The project has involved more than 100 collaborators in over 20 countries. The objective: to enhance understanding of evolution in the kingdom Fungi, and thereby of life on Earth in general, leading to development of diagnostic tools to aid discovery of the very many fungal species believed to exist but as yet unknown. In its first stage, the project developed broad datasets of molecular and non-molecular (i.e. morphological) characters across the kingdom, leading to the first unified phylogenetic classification system for higher ranks of the Fungi. It also resulted in the first database of fungal subcellular characters and character states, and various informational tools for studying phylogeny. The project has already made a profound impact on fungal systematics, and its findings have been incorporated in this edn of the *Dictionary*. See: Hibbett *et al.* (*MR* **111**: 509, 2007). Website: http://aftol.org.

agamic (**agamous**), asexual.

agar (**agar-agar**), a substance from certain red algae (*Gelidium* (Japan, USA), *Gracilaria* (USA), *Gigartina* (UK), *Pterocladia* (NZ), etc.) used to make culture media into gels which few microorganisms can liquefy. See Chapman (*Seaweeds and their uses*, 1950), Newton (*Seaweed utilization*, 1951), Humm (*Econ. Bot.* **1**: 317, 1947); a possible substitute using granulated tapioca or tapioca pearls (*Manihot esculenta*, cassava) has been proposed for use where agar is unavailable or prohibitively priced (Nene & Sheila, *Indian J. mycol. Pl. Path.* **24**: 159, 1994). Cf. gelatin, Media.

agaric (1) one of the *Agaricales*; **fly -**, *Amanita muscaria*; **honey -**, *Armillaria mellea*; (2) (in early medicine, obsol.), species of *Fomes* or *Polyporus*; **female, white**, or **purging -** (agaricum), *F. officinalis*; **male -**, *Phellinus igniarius* (*F. igniarius*).

Agaricaceae Chevall. (1826), Agaricales. 85 gen. (+ 80 syn.), 1340 spp.

Lit.: Kreisel (*Feddes Repert.* **64**: 89, 1962), Homrich & Wright (*Mycol.* **65**: 779, 1973), Kreisel (*Biblthca Mycol.* **36**, 1973; Germany), Brodie (*The Bird's Nest Fungi*: 199 pp., 1975), Brodie (*Lejeunia* n.s. **112**: 1, 1984; suppl.), Pegler (*Kew Bull.* Addit. Ser. **12**: 519 pp., 1986), Singer (*Agaric. mod. Tax.* 4th ed, 1986), Malloch *et al.* (*Mycol.* **79**: 839, 1987), Pegler & Young (*MR* **98**: 904, 1994), Breitenbach & Kränzlin (*Fungi of Switzerland* **4** Agarics, 2nd part: *Entolomataceae, Pluteaceae, Amanitaceae, Agaricaceae, Coprinaceae, Bolbitiaceae, Strophariaceae*: 368 pp., 1995), Sarasini & Pina (*Riv. Micol.* **38**: 237, 1995), Hibbett *et al.* (*Proc. natn Acad. Sci. U.S.A.* **94**: 12002, 1996), Kreisel & Moreno (*Feddes Repert.* **107**: 83, 1996), Sarasini & Pina (*Riv. Micol.* **39**: 115, 1996), Suárez & Wright (*Mycol.* **88**: 655, 1996), Coetzee *et al.* (*Bothalia* **27**: 117, 1997), Grgurinovic (*Larger Fungi of South Australia*: 725 pp. + 34 [m, 1997), Portman *et al.* (*Mycotaxon* **62**: 435, 1997), Sarasini & Pina (*Riv. Micol.* **40**: 19, 1997), Calonge (*Fl. Mycol. Iberica* **3**: 271 pp., 1998), Kreisel (*Öst. Z. Pilzk.* **7**: 215, 1998), Powell & Blackwell (*Mycotaxon* **68**: 505, 1998), Shinners & Tewari (*Mycol.* **90**: 980, 1998), Xu *et al.* (*Mol. Ecol.* **7**: 19, 1998), Hopple & Vilgalys (*Mol. Phylogen. Evol.* **13**: 1, 1999), Johnson (*Mycol.* **91**: 443, 1999), Mitchell & Bresinsky (*Mycol.* **91**: 811, 1999), Diehl (*Sydowia* **52**: 16, 2000), Krüger *et al.* (*Mycol.* **93**: 947, 2001), Redhead *et al.* (*Taxon* **50**: 203, 2001), Agerer (*Nova Hedwigia* **75**: 367, 2002), Binder & Bresinsky (*Mycol.* **94**: 85, 2002), Moncalvo *et al.* (*Mol. Phylogen. Evol.* **23**: 357, 2002), Baseia (*Mycotaxon* **88**: 107, 2003),

Krüger & Kreisel (*Mycotaxon* **86**: 169, 2003), Vellinga (*Mycol.* **95**: 442, 2003), Geml *et al.* (*Mycol. Progr.* **3**: 157, 2004), Lebel *et al.* (*MR* **108**: 210, 2004), Terashima *et al.* (*Mycoscience* **45**: 251, 2004), Vellinga (*MR* **108**: 354, 2004), Didukh *et al.* (*MR* **109**: 729, 2005), Kerrigan (*Mycol.* **97**: 12, 2005), Miller *et al.* (*Mycol.* **97**: 530, 2005), Stott *et al.* (*MR* **109**: 205, 2005), Walther *et al.* (*MR* **109**: 525, 2005).

Agaricales Underw. (1899). Agaricomycetidae. 33 fam., 413 gen., 13233 spp. Mushrooms and toadstools, Gill fungi, Agarics. Terrestrial, lignicolous, sometimes muscicolous or fungicolous, saprobic, mycorrhizal (ectomycorrhizal, exceptionally orchid mycorrhizal), rarely parasitic on plants or fungi; edible, poisonous and hallucinogenic; cosmopolitan.

The mycelium, which is frequently seen in leaf mould and decaying wood, may be perennial (with ages more than thousand years, Smith *et al.*, *Nature* **256**: 428, 1992); the expanding mycelium frequently forms fairly rings (q.v.); some species form sclerotia, hyphal cords or rhizomorphs.

Classification: Fries (*Syst. mycol.* **1-3**, 1821-1832) put almost all fleshy, lamellate toadstools in the genus *Agaricus*, his tribus being the common genera of today. He subsequently elevated several of these infrageneric groups to generic level, but later authors (Staude, Kummer, Quélet, Gillet, Karsten) made most of the changes. Fries based his genera on macroscopic characters of the basidiocarp and colour of spore print and his system had been widely used as it had the advantage that many genera could be identified on field characters. Microscopic studies of basidiocarp structure, initiated by Fayod and Patouillard, have shown a number of Fries's groupings to be unnatural, and new genera and families have been proposed. Singer's monumental work, *The Agaricales in modern taxonomy* (4th ed., 1986), treated three major groups within the *Agaricales* s. l., viz. *Agaricales* s. str., *Boletales*, and *Russulales*. These groups are still accepted in modern treatments based on molecular characters, as the euagarics clade, bolete clade, and russuloid clade (Hibbett & Thorn, *The Mycota*, **7B**, 2001) and are accepted as separate orders in this edition of the *Dictionary*. Hibbett *et al.* (*Proc. nat. Acad. Sci. USA* **94**: 1202, 1997; see also Hibbett & Thorn, *The Mycota* **7B**, 2001) concluded that the lamellate hymenophore has independently arisen in at least 5 out of the 8 clades of the *Homobasidiomycetes*. The results from the AFTOL project now recognize some 20 orders of the *Agaricomycetes* (Hibbett *et al.* (*Mycol.* **98**: 917, 2006; molecular phylogeny), Hibbett *et al.* (*MR* **111**: 109, 2007). The *Agaricales* s. str. (euagarics clade) also contain fungi of the reduced series (cyphelloid fungi; q.v.), some aphyllophorales (q.v.) and gasteromycetes (q.v.). Consequently, the *Agaricales* and most of its families cannot be characterised in morphological terms and for that reason diagnoses are not provided for many of the families. Fams:

(1) **Agaricaceae**
(2) **Amanitaceae**
(3) **Amylocorticiaceae**
(4) **Bolbitiaceae**
(5) **Broomeiaceae**
(6) **Clavariaceae**
(7) **Cortinariaceae**
(8) **Cyphellaceae**
(9) **Cystostereaceae**
(10) **Entolomataceae**
(11) **Fistulinaceae**
(12) **Gigaspermaceae**
(13) **Hemigasteraceae**
(14) **Hydnangiaceae**
(15) **Hygrophoraceae**
(16) **Inocybaceae**
(17) **Limnoperdaceae**
(18) **Lyophyllaceae**
(19) **Marasmiaceae**
(20) **Mycenaceae**
(21) **Niaceae**
(22) **Phelloriniaceae**
(23) **Physalacriaceae**
(24) **Pleurotaceae**
(25) **Pluteaceae**
(26) **Psathyrellaceae**
(27) **Pterulaceae**
(28) **Schizophyllaceae**
(29) **Stephanosporaceae**
(30) **Strophariaceae**
(31) **Tapinellaceae**
(32) **Tricholomataceae**
(33) **Typhulaceae**

Lit.: Josserand (*La description des champignons supérieurs*, 1952 (revised 1983)), Reijnders (*Les problèmes du développement des carpophores des Agaricales et de quelques groupes voisins*, 1963), Reijnders & Stalpers (*Stud. Mycol.* **34**, 1992), Clémençon (*Anatomie der Hymenomycetes*, 1997), Moore, Pegler & Young (*Beih. Nova Hedwigia* **35**, 1971; spore morphology), Gill & Steglich (*Progr. Chem. Nat. Prod.* **51**, 1987; pigment chemistry), Singer (*The Agaricales in modern taxonomy*, 4th ed., 1986), Kühner (*Les Hyménomycètes agaricoïdes, études génerales et classification*, 1980; classification), Horak (*Synopsis generum Agaricalium*, 1968), Donk (*Beih. Nova Hedwigia* **2**, 1961; nomenclature), Hibbett & Thorn (*The Mycota***7B**, 2001; phylogeny), Moncalvo *et al.* (*Syst. Biol.* **49**: 278, 2000; phylogeny), See Krüget *et al.* (*Mycol.* **93**: 947, 2001; phylogeny. See also under *Basidiomycetes*, Macromycetes and fams.

agaricic acid, a hydroxylated tribasic acid from *Fomes officinalis*; used to control tubercular night sweats (Milner, *Med. Klin.* **62**: 1443, 1967).

agaricicolous, living on agarics.

Agaricites Mesch. (1891), Fossil Fungi. 4 (Tertiary, Quaternary), Europe.

Agarico-carnis Paulet (1793) ≡ Fistulina.

Agaricochaete Eichelb. (1906), ? Pleurotaceae. 4, Africa; Asia. Perhaps *Tricholomataceae*. See Pegler (*Kew Bull.* Addit. Ser. **6**, 1977) Position uncertain, could be *Tricholomataceae*.

Agaricodochium X.J. Liu (1981), anamorphic *Pezizomycotina*, Hsp.0eH.15. 1, China. See Liu (*Acta Microbiol. Sin.* **21**: 160, 1981).

agaricoid, of a form resembling *Agaricus*; with a stipe, cap (pileus) and gills (lamellae).

Agarico-igniarium Paulet (1793) ≡ Fomes.

Agaricomycetes Doweld (2001), Agaricomycotina. 17 ord., 100 fam., 1147 gen., 20951 spp. Ords:

(1) **Agaricales**
(2) **Atheliales**
(3) **Auriculariales**
(4) **Boletales**
(5) **Cantharellales**
(6) **Corticiales**

(7) **Geastrales**
(8) **Gloeophyllales**
(9) **Gomphales**
(10) **Hymenochaetales**
(11) **Hysterangiales**
(12) **Phallales**
(13) **Polyporales**
(14) **Russulales**
(15) **Sebacinales**
(16) **Thelephorales**
(17) **Trechisporales**

Lit. (see also under Macromycetes): **General**: Donk (1951-63), Generic names proposed for Hymenomycetes, I ('Cyphellaceae'), II (Hymenolichenes), III ('Clavariaceae'), IV (Boletaceae), *Reinwardtia* **1**: 199, **2**: 435, **3**: 275, 1951-58, V ('Hydnaceae'), *Taxon* **5**: 69, 95, 1956, VI (Brachybasidiaceae, Cryptobasidiaceae, Exobasidiaceae), *Reinwardtia* **4**: 113, 1956, VII ('Thelephoraceae'), VIII (Auriculariaceae, Septobasidiaceae, Tremellaceae, Dacrymycetaceae), *Taxon* **6**: 17, 68, 106, **7**: 164, 193, 236, 1957-58, IX ('Meruliaceae', *Cantharellus*), *Fungus* **28**: 7, 1958, X ('Polyporaceae'), *Persoonia* **1**: 173, 1960 (additions and corrections, **2**: 201, 1962): XI (Agaricaceae); *Beih. Nova Hedw.* **5**, 1962, XII (Deuteromycetes), XIII (additions and corrections); *Taxon* **11**: 75, **12**: 113, 1962-63. [I-IX, XII, XIII, reprinted as 1 vol., 1966; X reprinted, 1968. In this valuable series of papers many taxonomic points are also discussed.] Donk (1954-62) Notes on resupinate hymenomycetes: I (*Pellicularia*), *Reinwardtia* **2**: 425, 1954; II (Tulasnelloid fungi), **3**: 363, 1956; III, IV, V, *Fungus* **26**: 3, **27**: 1, **28**: 16, 1956-58; VI, *Persoonia* **2**: 217, 1962. Rea (1922), Bourdot & Galzin (1927), Killerman (1928), Eriksson (*Symb. bot. upsal.* **16**(1): 1-172, 1958; N. Sweden), Donk (1954-62; *Reinwardtia* **2**: 425, 1954; **3**: 363, 1956; *Fungus* **26**: 3, **27**: 1, **28**: 16, 1956-58; *Persoonia* **2**: 217, 1962; resupinates), Donk (*Persoonia* **3**: 199, 1964; conspectus of families), Shaffer (*in* Parker, 1982, **1**: 248), Stephanova-Kartavenko ([Aphyllophorous fungi of the Urals], 1967; gen. keys), Parmasto (*The Lachnocladiaceae of the Soviet Union with a key to boreal species*, 1970 [*Scripta mycol.* **2**]), Pegler (*The polypores*, 1973 [*Bull. BMS Suppl.*]; keys world gen., Br. spp.), Strid (*Aphyllophorales of N. Central Scandinavia*, 1975 [*Wahlenbergia* **1**]), Domański (*Mala Flora Grzylów* **1**, *Aphyllophorales*, 1975), Rattan (1977), Stalpers (1978). Clémençon (Ed.) (*The species concept in Hymenomycetes*, 1977). Donk (1966), *Persoonia* **4**: 145, 1966; **8**: 33, 1974; checklists of European heterobasidiomycetes, annotations, ref., index. Lowy, *Taxon* **17**: 118, 1968; (heterobasidiomycete taxonomy); Talbot, *Taxon* **17**: 620, 1968. Kühner (*TBMS* **68**: 1, 1977; nuclear behaviour, review), Moser (*Röhrlinge und Blätterpilze*, 1978), Jülich (*Bibl. Mycol.* **85**, 1992), Jülich & Stalpers (*The resupinate non-poroid Aphyllophorales of the temperate Northern hemisphere*, 1980), Kühner (*Les Hyménomycetètes agaricoïdes (Agaricales, Tricholomatales, Pluteales, Russulales)*, 1980), Parmasto (*Windhalia* **16**: 3, 1986), Corner (*Ad Polyporaceas* **1-7** (*Beih. Nova Hedw.*], 1983-1991), Moser & Jülich (*Farbatlas der Basidiomyceten* **1-12**, 1994), Fell *et al.* (*Int. J. Syst. Evol. Microbiol.* **50**: 1351, 2000; mol. phylogeny basidiomycetous yeasts).

Regional: **America, North**, Shaffer (*Keys to genera of higher fungi*, edn 2, 1968; mostly hymenomycetes), **South**, Singer (*Beih. Nova Hedw.* **29**, 1969; Agaricales, Aphyllophorales, Gasteromycetes). **Europe**, Donk (1966); **Great Britain**, Rea (*British Basidiomycetae*, 1922; Suppl. *TBMS* **12**: 205, **17**: 35, 1927-32, incl. gasteromycetes), Reid & Austwick (*Glasgow Nat.* **18**: 255, 1963; annot. list of Scottish basidiomycetes, incl. gasteromycetes, excl. rusts and smuts). **France**, Bourdot & Galzin (*Hyménomycetes de France, Hetérobasidiés, Homobasidiés gymnocarpes*, 1927). **Portugal**, Da Camara (*Catalogus systematicus fungorum omnia Lusitaniae. I, Basidiomycetes.* Pars 1, *Hymeniales*, 1956; Pars 2, *Gasterales, Phalloidales, Tremelloidales, Uredinales* et *Ustilaginales*, 1958). **former USSR**, Raitviir [Key to Heterobasidiomycetidae of the USSR, 1967].

Agaricomycetidae Parmasto (1986), Agaricomycetes. Ords.:
(1) **Agaricales**
(2) **Atheliales**
(3) **Boletales**

For *Lit.* see fam.

Agaricomycotina Doweld (2001), Basidiomycota. Class.:
(1) **Agaricomycetes**
(2) **Dacrymycetes**
(3) **Tremellomycetes**

For *Lit.* see fam.

Agaricon Tourn. ex Adans. (1763) ≡ Fomitopsis.

Agarico-pulpa Paulet (1793) ≡ Fomitopsis.

Agaricostilbaceae Oberw. & R. Bauer (1989), Agaricostilbales. 3 gen. (+ 1 syn.), 16 spp. Basidiospores produced in a yeast-like manner.

Lit.: Oberwinkler & Bauer (*Sydowia* **41**: 224, 1989), Kendrick & Gong (*Mycotaxon* **54**: 19, 1995), Swann & Taylor (*MR* **99**: 1205, 1995), Frieders & McLaughlin (*CJB* **74**: 1392, 1996), Bandoni & Boekhout *in* Kurtzman & Fell (Eds) (*Yeasts, a taxonomic study* 4th edn: 639, 1998), Scorzetti *et al.* (*FEMS Yeast Res.* **2**: 495, 2002).

Agaricostilbales Oberw. & R. Bauer (1989). Agaricostilbomycetes. 3 fam., 9 gen., 43 spp. Fams:
(1) **Agaricostilbaceae**
(2) **Chionosphaeraceae**
(3) **Kondoaceae**

Lit.: Oberwinkler & Bauer (*Sydowia* **41**: 224, 1989).

Agaricostilbomycetes R. Bauer, Begerow, J.P. Samp., M. Weiss & Oberw. (2006), Pucciniomycotina. 2 ord., 3 fam., 10 gen., 47 spp. Ords:
(1) **Agaricostilbales**
(2) **Spiculogloeales**

Lit.: Bauer *et al.* (*Mycol. Progress* **5**: 41, 2006).

Agaricostilbum J.E. Wright (1970), Agaricostilbaceae. 3, Argentina; Congo-Kinshasa; India. See Wright *et al.* (*Mycol.* **73**: 880, 1981), Brady *et al.* (*TBMS* **83**: 540, 1984; nomencl.), Bauer *et al.* (*Syst. Appl. Microbiol.* **15**: 259, 1992; ultrastr.), Fell *et al.* (*Int. J. Syst. Evol. Microbiol.* **50**: 1351, 2000; mol. phylogeny), Bauer *et al.* (*Mycol. Progr.* **5**: 41, 2006).

Agarico-suber Paulet (1793) ≡ Daedalea.

Agaricum P. Micheli ex Haller (1768) ≡ Fomitopsis fide Donk (*Proc. K. ned. Akad. Wet.* Ser. C, Biol. Med. Sci. **74**: 125, 1971).

Agaricum Paulet (1812) ≡ Agaricon.

Agaricus L. (1753), Agaricaceae. *c.* 200, widespread (esp. temperate). *A. bisporus* (= *A. brunnescens* fide Malloch *et al.*, *Mycol.* **68**: 912, 1976), the cultivated mushroom (see Mushroom cultivation). The name

Agaricus was initially used for a group that more or less coincides with the lamellate *Agaricales*. See Möller (*Friesia* **4**: 1, 1950-52; Danish species, as *Psalliota*), Pilát (*Acta Mus. Nat. Prag.* **7**, 1951; key Europ. spp.), Möller (*Friesia* **4**: 135, 1952; Danish species, as *Psalliota*), Heinemann (*Sydowia* **30**: 6, 1978; key), Freeman (*Mycotaxon* **8**: 50, 1979; key N. Am. spp.), Capelli (*Agaricus L. :Fr. ss. Karsten (Psalliota Fr.)*, 1984; key Europ. spp.), Bunyard *et al.* (*Fungal Genetics Biol.* **20**: 243, 1996; phylogeny), Mitchell & Bresinsky (*Mycol.* **91**: 811, 1999; phylogeny), Robison *et al.* (*Mycol.* **93**: 30, 2001; phylogeny), Redhead *et al.* (*Mycotaxon* **83**: 19, 2002; phylogeny), Challen *et al.* (*Mycol.* **95**: 61, 2003; phylogeny *Agaricus* sect. *Duploannulatae*), Fukuda *et al.* (*Mycoscience* **44**: 431, 2003; genetic variation in *Agaricus blazei*), Geml *et al.* (*Mycol. Progr.* **3**: 157, 2004; molecular evolution), Vellinga (*MR* **108**: 354, 2004; phylogeny), Didukh *et al.* (*MR* **109**: 729, 2005; *Agaricus* section *Duploannulati*), Kerrigan *et al.* (*Mycol.* **97**: 1292, 2005; *Agaricus* section *Xanthodermatei* phylogeny).

Agaricus Murrill (1905) ≡ Daedalea.

Agaricus Raf. (1830) ? = Amanita Pers. fide Stalpers (*in litt.*).

agaritine, an amino acid from *Agaricus bisporus*.

Agarwalia D.P. Tiwari & P.D. Agrawal (1974), anamorphic *Pezizomycotina*, Hso.0eP.3. 1 (from soil), India. See Tiwari & Agrawal (*J. Indian bot. Soc.* **52**: 134, 1973), Kendrick (*CJB* **81**: 75, 2003; morphogenesis).

Agarwalomyces R.K. Verma & Kamal (1987), anamorphic *Pezizomycotina*, Hsy.0eP.3. 1, India. See Verma & Kamal (*TBMS* **89**: 596, 1987).

Agglomerata J.I.R. Larsson & Yan (1988), Microsporidia. 5. See Larsson & Yan (*Arch. Protistenk.* **135**: 271, 1988).

agglutinate, fixed together as if with glue.

agglutinin, see antigen.

aggregate (1) (in taxonomy; '**agg.**' or '**aggr.**'), see species; (2) (in descriptions), near together, crowded.

aggregate plasmodium, see plasmodium.

Aglaocephalum W. Weston (1933) nom. nud. = Pulchromyces fide Pfister *et al.* (*Mycotaxon* **1**: 137, 1974).

Aglaopisma De Not. ex Bagl. (1856) = Caloplaca fide Hawksworth *et al.* (*Dictionary of the Fungi* edn 8, 1995).

Aglaospora De Not. (1844) = Massaria fide Eriksson (*SA* **5**: 116, 1986), Barr (*N. Amer. Fl.* ser. 2 **13**: 129 pp., 1990; separate from *Massaria*).

Aglaothecium Groenh. (1962) nom. rej. = Gyalidea fide Hafellner (*Beih. Nova Hedwigia* **79**: 241, 1984), Lumbsch *et al.* (*Taxon* **40**: 331, 1991; nomencl.).

Agmasoma E.I. Hazard & Oldacre (1975), Microsporidia. 2.

Agmocybe Earle (1909) = Inocybe fide Kauffman (*N. Amer. Fl.* **10**, 1924).

Agonimia Zahlbr. (1909), Verrucariales (L). 10, widespread. See Coppins & James (*Lichenologist* **10**: 179, 1978), Harada (*J. Jap. Bot.* **68**: 166, 1993; Japan), Aptroot *et al.* (*Biblthca Lichenol.* **64**, 1997), Czarnota & Coppins (*Graphis Scripta* **11**: 56, 2000; Poland), Aragón & Sarrión (*Nova Hedwigia* **77**: 169, 2003; Spain), Lumbsch *et al.* (*Mol. Phylogen. Evol.* **31**: 822, 2004; phylogeny), Geiser *et al.* (*Mycol.* **98**: 1053, 2006; phylogeny), Aptroot *et al.* (*Biblthca Lichenol.* **97**, 2008; Costa Rica).

Agonimiella H. Harada (1993) = Agonimia fide Aptroot *et al.* (*Biblthca Lichenol.* **64**, 1997).

Agonium Oerst. (1844) nom. dub., ? Fungi. or Cyanobacteria.

Agonomycetales. True conidia absent, but non-dehiscent propagules (allocysts, bromatia, bulbils, chlamydospores, sclerotia etc.) produced in some genera. *Agonomycetes* may be states of basidiomycetes, ascomycetes or other anamorphic fungi. *Rhizoctonia* and *Sclerotium* include important plant pathogens.

Lit.: Watling (*in* Kendrick (Ed.), *The whole fungus* **2**: 453, 1979; states of basidiomycetes), von Arx (*Genera of fungi sporulating in pure culture*, 1981; keys gen.), Domsch *et al.* (*Compendium of soil fungi*, 1980; identification, refs.).

Agostaea (Sacc.) Theiss. & Syd. (1915) = Anhellia fide von Arx (*Persoonia* **2**: 421, 1963).

Agrabeeja Subram. (1995), anamorphic *Pezizomycotina*, Hso.?.?. 1, Singapore. See Subramanian (*Kavaka* **20/21**: 2, 1992/1993).

Agrestia J.W. Thomson (1961) = Aspicilia fide Weber (*Aquilo* Bot. **6**: 43, 1967).

agroclavine, a clavine alkaloid (an intermediate in the biosynthesis of ergoline alkaloids) which is a major alkaloidal constituent of *Claviceps fusiformis* sclerotia. Cf. ergot.

Agrocybe Fayod (1889), Strophariaceae. *c.* 100, widespread. See Singer (*Sydowia* **30**: 194, 1978; key), Flynn & Miller (*MR* **94**: 1103, 1990; taxonomy), Moncalvo *et al.* (*Syst. Biol.* **49**: 278, 2000; phylogeny), Thomas & Manimohan (*Mycotaxon* **86**: 317, 2003; India), Nauta (*Persoonia* **18**: 429, 2004; Netherlands).

Agrogaster D.A. Reid (1986), Bolbitiaceae. 1, New Zealand. Basiodioma gasteroid. See Reid (*TBMS* **86**: 429, 1986).

Agyriaceae Corda (1838), Agyriales (±L). 6 gen. (+ 7 syn.), 32 spp. See *Agyriales* for descr.

Lit.: Hertel & Rambold (*Biblthca Lichenol.* **38**: 145, 1989), Rambold & Triebel (*Notes R. bot. Gdn Edinb.* **46**: 375, 1990), Bellemère (*Bull. Soc. linn. Provence* **45**: 355, 1994), Brodo (*Biblthca Lichenol.* **57**: 59, 1995), Lunke *et al.* (*Bryologist* **99**: 53, 1996), Moberg & Carlin (*Symb. bot. upsal.* **31** no. 3: 319, 1996), Lumbsch (*J. Hattori bot. Lab.* **83**: 1, 1997), Lumbsch *et al.* (*MR* **105**: 16, 2001), Lumbsch *et al.* (*MR* **105**: 265, 2001), Schmitt *et al.* (*Mycol.* **95**: 827, 2003), Reeb *et al.* (*Mol. Phylogen. Evol.* **32**: 1036, 2004), Wedin *et al.* (*MR* **109**: 159, 2005), Miądlikowska *et al.* (*Mycol.* **98**: 1088, 2006; phylogeny), Hofstetter *et al.* (*Mol. Phylogen. Evol.* **44**: 412, 2007; phylogeny), Lumbsch *et al.* (*MR* **111**: 1133, 2007).

Agyriales Clem. & Shear (1931). Ostropomycetidae. 4 fam., 17 gen., 147 spp. Thallus absent. Ascomata apothecial, sometimes elongated, often domed, hymenium usually gelatinous, not blueing in iodine. Interascal tissue of branched and anastomosing paraphyses, sometimes with a well-developed pigmented epithecial layer. Asci varied in form, opening by eversion through a vertical split, and blueing faintly in iodine. Ascospores small, hyaline, aseptate, without a gelatinous sheath. Anamorphs pycnidial. Saprobic on bark and wood, esp. on conifers.

The *Agyriales* was treated for some years as a sub-order of the *Lecanorales*, but molecular data confirm its placement within the *Ostropomycetidae*. It may be appropriate to place the order in synonymy with the

Pertusariales, but more studies are required. Fams:
(1) **Agyriaceae**
(2) **Anamylopsoraceae**
Lit.: Lumbsch (*J. Hattori Bot. Lab.* **83**: 1, 1997), Lumbsch *et al.* (*MR* **105**: 16, 265, 2001), Lumbsch *et al.* (*MR* **111**: 257, 2007; phylogeny), Lumbsch *et al.* (*MR* **111**: 1133, 2007; phylogeny), Miądlikowska *et al* (*Mycol.* **98**: 1088, 2006), Rambold & Triebel (*Notes R. bot. Gdn, Edin.* **46**: 375, 1990).

Agyriella Ellis & Everh. (1897) ≡ Agyriopsis.

Agyriella Sacc. (1884), anamorphic *Pezizomycotina*, Hsp.0eH-P.15. 2, Europe. See Ellis (*Dematiaceous Hyphomycetes*, 1971).

Agyriellopsis Höhn. (1903), anamorphic *Pezizomycotina*, St.0eH.15. 2, Europe.

Agyrina (Sacc.) Clem. (1909) = Steinia fide Nannfeldt (*Nova Acta R. Soc. Scient. upsal.*, 1932).

Agyriopsis Sacc. & P. Syd. (1899) = Schizoxylon fide Sherwood (*Mycotaxon* **6**: 215, 1977).

Agyrium Fr. (1822), Agyriaceae. 3 (saprobic), widespread (temperate). See Lumbsch (*J. Hattori bot. Lab.* **83**: 1, 1997), Kantvilas (*Muelleria* **16**: 65, 2002; Australia), Zhuang & Yang (*Mycotaxon* **96**: 169, 2006; China).

Agyrona Höhn. (1909) = Molleriella fide von Arx (*Persoonia* **2**: 421, 1963).

Agyronella Höhn. (1909) = Schizothyrium fide von Arx & Müller (*Stud. Mycol.* **9**, 1975).

Agyrophora (Nyl.) Nyl. (1896) = Umbilicaria fide Hawksworth *et al.* (*Dictionary of the Fungi* edn 8, 1995).

Ahlesia Fuckel (1870) = Thelocarpon fide Poelt & Hafellner (*Phyton* Horn **17**: 67, 1975), Rossman *et al.* (*Stud. Mycol.* **42**: 248 pp., 1999).

Ahmad (Sultan; 1910-1983; Pakistan). MSc degree (1932) then BEd (1934) then PhD (1950) then DSc (1957), University of the Punjab, Lahore; academic staff (1947 onwards) then Professor and Head of Department of Botany (to 1970), Government College, Lahore (1970); Professor Emeritus, University of the Punjab, Lahore (1972 onwards). Pioneer in studies of the mycota of Pakistan, collaborating particularly with E. Müller (q.v.) and Petrak (q.v.); founder of the Biological Society of Pakistan, and editor of its journal *Biologia* (1955-1983); Fellow of the Academy of Sciences of Pakistan (1974). His specimens are in the fungal reference collection, Department of Botany, University of the Punjab, Lahore (many duplicates in **BPI** and **IMI**). *Publs*. Fungi of West Pakistan. *Monographs. Biological Society of Pakistan* (1956); Fungi of West Pakistan. Supplement I. *Biologia* Lahore (1969); Ascomycetes of Pakistan Parts I & II. *Monographs. Biological Society of Pakistan* (1978). *Biogs, obits etc.* Ghaffar & Ali (*Pakistan Journal of Botany* **26**: 201, 1994).

Ahmadia Syd. (1939), anamorphic *Pezizomycotina*, Cac.≡ eH.15. 1, Pakistan.

Ahmadiago Vánky (2004), ? Ustilaginaceae. 1 (on *Euphorbia*), India. See Vánky (*Mycotaxon* **89**: 55, 2004), Piątek (*Mycotaxon* **92**: 33, 2005).

Ahmadinula Petr. (1953) = Truncatella fide Sutton (*Mycol. Pap.* **141**, 1977), Shoemaker *et al.* (*Sydowia* **41**: 308, 1989; synonymy).

Ahtia M.J. Lai (1980) ≡ Cetrariopsis.

Ahtiana Goward (1986), Parmeliaceae (L). 3, N. America. See Thell *et al.* (*Bryologist* **98**: 596, 1995; monogr.), Thell (*Folia Cryptog. Estonica* **32**: 113, 1998), Thell *et al.* (*Mycol. Progr.* **1**: 335, 2002; phylogeny), Mattsson & Articus (*Symb. bot. upsal.* **34** no. 1: 237, 2004; phylogeny), Thell *et al.* (*Mycol. Progr.* **3**: 297, 2004; phylogeny).

AIDS, Acquired immunity deficiency syndrome. See Bossche *et al.* (Eds) (*Mycoses in AIDS patents*, 1989; infections by fungi in AIDS patients). See Medical and Veterinary mycology, *Pneumocystis*.

Aigialus Kohlm. & S. Schatz (1986), ? Pleosporales. 6 (marine, on mangroves), widespread. See Borse (*TBMS* **88**: 424, 1987; key 4 spp.), Hawksworth (*SA* **6**: 338, 1987; status), Barr (*N. Amer. Fl.* ser. 2 **13**: 129 pp., 1990; posn), Hyde (*MR* **96**: 1044, 1992), Tam *et al.* (*Bot. Mar.* **46**: 487, 2003; posn).

Ailographium, see *Aulographum*.

Ainoa Lumbsch & I. Schmitt (2001), Baeomycetales (L). 2. See Lumbsch *et al.* (*MR* **105**: 272, 2001), Lumbsch *et al.* (*Mol. Phylogen. Evol.* **31**: 822, 2004; phylogeny), Hermansson (*Graphis Scripta* **17**: 41, 2005; Sweden), Wedin *et al.* (*MR* **109**: 159, 2005; phylogeny), Lumbsch *et al.* (*MR* **111**: 257, 2007; phylogeny), Lumbsch *et al.* (*MR* **111**: 1133, 2007).

Ainsworth (Geoffrey Clough; 1905-1998; England). Assistant Mycologist, Imperial Mycological Institute, Kew (1939-1946); Head of Mycological Department, Wellcome Physiological Research Laboratories, Beckenham (1946-1948); Lecturer / Reader, University of the South West, Exeter (1948-1957); Assistant Editor (1957-1960) then Assistant Director (1961-1964) then Director (1964-1968), Commonwealth Mycological Institute, Kew. A mycological scholar, campaigner and visionary; with Bisby (q.v.) co-founder of this Dictionary, the first edition being prepared at night during fire-watch duty in world-war II during the bombing of London; a founder and Honorary President for Life of the International Mycological Association (q.v. Societies and organizations), he chaired the organizing committee of the first International Mycological Congress (Exeter, 1971). *Publs.* (with Sparrow & Sussman) *The Fungi, an Advanced Treatise* 4 vols (1965-1973); *Introduction to the History of Mycology* (1976); *Introduction to the History of Plant Pathology* (1981); *Introduction to the History of Medical Mycology* (1987). *Biogs, obits etc.* Webster (*Mycol.* **91**: 714, 1999); Hawksworth (*MR* **104**: 110, 2000) [portrait].

Ainsworthia Bat. & Cif. (1962) [non *Ainsworthia* Boiss. 1844, *Umbelliferae*] = Phaeosaccardinula fide von Arx & Müller (*Stud. Mycol.* **9**, 1975).

Aipospila Trevis. (1857) = Lecania fide Hafellner (*Beih. Nova Hedwigia* **79**: 241, 1984).

Air pollution. Human introduction of biological materials, chemicals and particulate matter into the atmosphere can harm fungi. Effects on many foliicolous and stem fungi, and on lichen-forming species on all substrata are well documented.

Lichens are arguably the most sensitive organisms to sulphur dioxide known, some being affected at mean levels of about 30 μg m^{-3}. The algae or cyanobacteria in lichens are particularly sensitive to pollutants such as sulphur dioxide which disrupt membranes leading to chlorophyll breakdown. Nylander (q.v.) suggested lichens could be used to monitor air quality in 1866 and there is now a vast literature on this subject. Fluorides are also highly toxic to lichens but particulate deposits (e.g. smoke), heavy metals, and photochemical smog components have less effect. Differential sensitivity due to physiological, structural, and chemical characters enables zones to estimate pollution levels to be constructed (Hawksworth & Rose, *Nature* **227**: 145, 1970; Gilbert, *New Phytol.* **69**: 629, 1970); recolonization in response to falling sulphur dioxide levels can be dramatic (Hawksworth & McManus, *Bot. J. Linn. Soc.* **100**: 99, 1989; London); statistical and computer assisted approaches are increasingly used (e.g. Nimis *et al.*, *Stud. Geobot.* **11**, 1991).

Erysiphales and *Pucciniales* are amongst the other most sensitive fungi; *Diplocarpon rosae* (Saunders, *Ann. appl. Biol.* **58**: 103, 1966) and *Rhytisma acerinum* (Bevan & Greenhalgh, *Environ. Pollut.* **10**: 271, 1976) can also be used as pollution monitors. Numerous studies of forest decline, often in response to acid rain, have shown that endophyte and saprobic microfungi can be very strongly affected, with typically a small number of resistant (generalist) species increasing in abundance, and most other species declining in numbers (e.g. Asai *et al.*, *MR* **102**: 1316, 1998). Leaf-dwelling yeasts (*Sporobolomyces*, *Tilletiopsis*) can be cultured and the density of sporing has been found to be directly related to acidic air pollution (Dowding, *in* Richardson, *Biological indicators of pollution*: 137, 1987).

Radiation pollution has become more important since the 1986 Chernobyl disaster. In this and other cases, the amount of metal and radionuclides taken up by lichens has been used to map the extent of affected areas (Steinne *et al.*, *J. Environ. Radioact.* **21**: 65, 1993). Certain hypogeous fungi, particularly species of *Elaphomyces* accumulate radionuclides in greater quantities than almost any other living organism. After Chernobyl, radionuclides were found to be transmitted from those fungi, along a food chain via wild boar into the human population (Vilic *et al.*, *J. Environ. Radioact.* **81**: 55, 2005). Increases in lead contents from traffic, and falls since the introduction of unleaded fuel, are documented by Lawrey (*Bryologist* **96**: 339, 1993).

Fungal spores may themselves be a component of air pollution. This can be particularly problematical in modern buildings where, for example, ventilation is insufficient. In those conditions, fungi may trigger various allergic, toxic or other responses, sometimes collectively described as 'sick-building syndrome'.

Lit.: Bates & Farmer (Eds) (*Bryophytes and lichens in a changing environment*, 1992), Coleman (*J. Building Appraisal* **1**: 362, 2005), Ferry *et al.* (Eds) (*Air pollution and lichens*, 1973; incl. reviews effects on all plants and fungi), Hawksworth & Rose (*Lichens as pollution monitors*, 1976), Henderson (*Lichenologist* 1974-; twice-yearly bibl.), Nash & Wirth (Eds) (*Lichens, bryophytes and air quality*, [*Bibl. Lich.* **30**], 1988), Nieboer *et al.* (*in* Mansfield, 1976: 61; review sulphur dioxide toxicity), Purvis *et al.* (Eds) (Lichens in a changing pollution environment. *Environmental pollution* **146**: 291, 2007), Richardson (*Bot. J. Linn. Soc.* **96**: 31, 1988; *Pollution monitoring with lichens*, 1992). See also Acid rain, Allergy, Bioindicators, Ecology, Index of Atmospheric Purity, lichen desert.

Air spora. Airborne particles originating from fungi and other organisms are collectively referred to as the air spora or bioaerosol. Fungal spores are important components of the air spora. Prevalent genera are *Alternaria*, *Aspergillus*, *Aureobasidium*, *Cladosporium*, *Curvularia*, *Epicoccum*, *Fusarium*, *Geotrichum*, *Nigrospora*, *Neurospora*, *Penicillium*, *Phoma* and *Pithomyces*. Probably most originate from saprobes growing in soil or on leaf surfaces (see e.g. Levetin & Dorsey, *Aerobiologia* **22**: 3, 2006), but some may be animal or plant pathogens. Knowledge of their occurrence in air was revolutionized by use of continuously operating volumetric samplers (Hirst, *Ann. appl. Biol.* **39**: 257, 1952) out of doors and a realization of the importance of the sampling and collection efficiencies of different trapping methods in determining what is caught. The Hirst and subsequent Burkard traps have revealed the importance in the air spora of ascospores and basidiospores that were previously underestimated by using exposed horizontal sticky slides and open Petri dishes. Indoors, fungal spores are often abundant when stored products are handled but their sampling and enumeration require different methods from those used out of doors because of their smaller size and greater concentrations (see Cox & Wathes, *Bioaerosols handbook*, 1994; Elbert *et al.*, *Atmospheric Chemistry and Physics Discussions* **6**: 11317, 2006). Molecular and immunological techniques are now applied in studying and identifying air spora (see Lacey & West, 2006).

Out of doors, fungal spores are almost always present in the air but their numbers and types depend on time of day, weather, season, geographical location and the nearness of large local spore sources. Total spore concentrations may range from fewer than 200 to 2 million m^{-3}. Terrestrial fungi most commonly produce wind-dispersed spores which then settle by sedimentation, impaction or rain-wash. Active spore discharge provides a means to avoid local settling, to reach potentially turbulent air currents for more distant dispersal. In many basidiomycete species stipe and gills provide a vertical escape path for the spores. Then even delicate air current can change the gradual fall and divert them into turbulent air. Violent ascospore release is more moisture dependent; when the turgid ascus bursts, the wall contracts and spores are ejected into the air. Spores released passively (e.g. of powdery mildews, rusts and smuts) are also often abundant in the air spora, since these mostly disseminate from diseased plant material above ground.

Spores of different species exhibit characteristic circadian periodicities in their occurrence in the air spora because their method of liberation is correlated with time of day (see Spore discharge and dispersal). Spores with active mechanisms requiring water are usually most numerous in the air at night, following dew formation, or rain; those dependent on drying are most numerous in the early morning as the sun dries their colonies; those released through mechanical disturbance occur during the middle of the day, when temperatures are highest and wind speeds, turbulence and convection are greatest. However, some discomycetes release their spores after sunrise, those with large apothecia being later than those with smaller, perhaps because some drying is needed to increase pressure on the asci. *Cladosporium* is the most numerous daytime spore type throughout most of the world although, in some seasons it may be exceeded by *Alternaria* in warm dry climates or by *Curvularia* or *Drechslera* in humid climates. At night time, ascospores, basidiospores and the ballistospores of *Sporobolomyces* and related 'mirror' yeasts become most numerous. Rain initially causes an increase in spore concentrations through 'tap and puff' (Hirst & Stedman, *J. gen. Microbiol.* **33**: 335, 1963), then washes spores from the air, and, afterwards, stimulates release of ascospores.

After exceeding canopy height, fungal spores can migrate long but measurable distances before settling (Nagarajan & Singh, *Ann. Rev. Phytopathol.* **28**: 139, 1990). Intercontinental dispersal of rust spores has been demonstrated for *Puccinia* (Asai, *Phytopathology* **50**: 535, 1960). Variations in the vertical profile of air spora and in their atmospheric concentrations has been used in prognoses for plant disease and allergy development (Lyon *et al.*, *Grana* **23**: 123, 1984; Wu *et al.*, *Atmospheric Environment* **38**: 4879, 2004; Zoppas *et al.*, *Aerobiologia* **22**: 119, 2006). For many fungi, horizontal spore concentration in air is normally minimal at 100-200 m from the source and the vertical concentration decreases logarithmically with height above ground. Fungal spore viability is important in determining migration capacity: rusts spores remain viable for many days and can carry infections great distances.

Large seasonal differences in spore concentrations occur in temperate regions, with few airborne spores in winter (see Li & Kendrick, *Grana* **34**: 199, 1995). In tropical regions, spores may be numerous all the year round although some types may be particularly favoured by wet or dry seasons (see Ogunlana, *Appl. Microbiol.* **29**: 458 (1975); Troutt & Levetin, *International J. Biometeorology* **45**: 64, 2001). Air is rich in spores of common moulds, rusts, downy and powdery mildews in dry weather, and in short-lived ascospores soon after rain. Growing crops form large sources of spores, especially of phytopathogenic fungi, whose occurrence may be correlated with crop growing seasons (see Lacey, *in* Cole & Kendrick (Eds), *Biology of conidial fungi*: 373, 1981). Sometimes, fungi pathogenic to humans can become airborne in dust in desert areas (e.g., *Coccidioides immitis*) or when deposits of guano beneath bird roosts are disturbed (*Histoplasma capsulatum*) (see also Medical mycology).

Indoors, numbers and types of airborne spores are determined by their source and, with stored products, the conditions in which they have been stored, the degree of disturbance of the substrate and the position and amount of ventilation. Concentrations of fungal spores may exceed 100 million m^{-3} air when mouldy hay and grain are handled, with *Aspergillus* and *Penicillium* spp. predominant. *Aspergillus fumigatus*, an opportunistic pathogen and frequent cause of asthma and mycotic abortion in cattle, may also be abundant. Concentrations of oyster mushroom (*Pleurotus ostreatus*) basidiospores may reach 27 million m^{-3} in growing sheds while up to 14 million m^{-3} *Penicillium* spores can be released when mouldy cork is handled. These concentrations may cause occupational allergies (see Allergy). Sampling of air indoors has shown seasonal variation in fungal spore composition, with *Cladosporium* species in one study predominating during warm periods, and *Penicillium* and *Aspergillus* predominating in winter (Medrela-Kuder, *International biodeterioration & biodegradation* **52**: 203, 2003). Species of *Cladosporium* common in indoor air spora can trigger allergic reactions. In Japan, *Trichosporon* sp. present in indoor air spora has been correlated with development of allergic alveolitis (Summerbell *et al.*, *Journal of Medical and Veterinary Mycology* Suppl. **1**: 279, 1992).

Lit.: Dimmick & Akers (Eds) (*An introduction to experimental aerobiology*, 1969), Edmonds (*Aerobiology, the ecological systems approach*, 1979), Gregory (*Microbiology of the atmosphere*, 2nd edn, 1973), Lacey & West *The Air Spora: A manual for catching and identifying airborne biological particles*, 2006, Samson *et al.* (Eds) (*Introduction to food- and airborne fungi*, edn 7, 2004).

Aithaloderma P. Syd. (1913), ? Capnodiaceae. Anamorph *Ciferrioxyphium*. 15, widespread (tropical). See Hughes (*Mycol.* **68**: 693, 1976), Olejnik & Ingrouille (*MR* **103**: 333, 1999; numerical taxonomy), Reynolds & Gilbert (*Aust. Syst. Bot.* **18**: 265, 2005; Australia).

Aithalomyces Woron. (1926) = Euantennaria fide Hughes (*N.Z. Jl Bot.* **10**: 225, 1972).

Aivenia Svrček (1977), Dermateaceae. 4, former Czechoslovakia. See Svrček (*Česká Mykol.* **43**: 215, 1989).

Ajello (Libero; 1916-2004; USA). Largely self-taught medical mycologists, working on tinea pedis among army recruits, Georgia (1943) then Johns Hopkins University (1944-1945); PhD, Columbia University (1947); Diagnostic Reference & Research Unit, Communicable Disease Centre, eventually as Head of the World Health Organizations Collaborating Center for Mycotic Diseases there, Atlanta (1948-1990). Outstanding medical mycologist of the 20th century, with over 400 publications, playing a pivotal role in the International Society for Human and Animal Mycology, and as an editor of its journal *Medical Mycology*; a great mentor who developed courses for the teaching of medical mycology run within the USA and in many other countries. He also significantly provided editorial support for non-English speaking scientists, particularly from Latin America. *Publs.* The medical mycological iceberg. *HSMHA health rep.* **86**: 437, 1971; (with Arora, Mukerji & Elander) *Handbook of applied mycology* vol. **2**, 1991; (with Hay) *Medical mycology. Topley and Wilson's microbiology and microbial infections*, edn 9, 2002. *Biogs, obits etc.* Goodman & DiSalvo (*Mycopathologia* **157**: 359, 2004), Müller (*Mycoses* **46**: 5, 2003).

Ajellomyces McDonough & A.L. Lewis (1968), Ajellomycetaceae. Anamorphs *Blastomyces*, *Histoplasma*. 3, widespread (esp. tropical). *A. dermatitidis* (anamorph *Blastomyces zymonema* (syn. *B. dermatitidis*); see blastomycosis), *A. capsulata* (anamorph *Histoplasma capsulatum*; see histoplasmosis). See Sigler (*J. Med. Vet. Mycol.* **34**: 303, 1996), Guého *et al.* (*Mycoses* **40**: 69, 1997; phylogeny), Sugiyama *et al.* (*Mycoscience* **40**: 251, 1999; phylogeny), Taylor *et al.* (*Fungal Genetics Biol.* **31**: 21, 2000; species concepts), Berbee (*Physiological and Molecular Plant Pathology* **59**: 165, 2001; phylogeny), Sugiyama *et al.* (*Stud. Mycol.* **47**: 5, 2002; phylogeny), Untereiner *et al.* (*Stud. Mycol.* **47**: 25, 2002; phylogeny), Untereiner *et al.* (*Mycol.* **96**: 812, 2004; fam. Placement), Pujol *et al.* (*Evolutionary Genetics of Fungi*: 149, 2005; population genetics).

Ajellomycetaceae Unter., J.A. Scott & Sigler (2004), Onygenales. 7 gen. (+ 3 syn.), 14 spp.

Lit.: Currah (*Mycotaxon* **24**: 1, 1985), Fukushima *et al.* (*Mycopathologia* **116**: 151, 1991), Sigler (*J. Med. Vet. Mycol.* **34**: 303, 1996), Guého *et al.* (*Mycoses* **40**: 69, 1997), Larone *et al.* (*Manual of Clinical Microbiology*: 1259, 1999), Sano *et al.* (*Mycopathologia* **143**: 165, 1998), Bialek *et al.* (*J. Clin. Microbiol.* **38**: 3190, 2000), San-Blas *et al.* (*Medical

Mycology **40**: 225, 2002), Semighini *et al.* (*Diagn. Microbiol. Infect. Dis.* **44**: 383, 2002), Sugiyama *et al.* (*Stud. Mycol.* **47**: 5, 2002), Untereiner *et al.* (*Stud. Mycol.* **47**: 25, 2002), Feitosa *et al.* (*Fungal Genetics Biol.* **39**: 60, 2003), Sigler (*Mycology Series* **16**: 195, 2003), Ueda *et al.* (*Veter. Pathol.* **94**: 219, 2003), Untereiner *et al.* (*Mycol.* **96**: 812, 2004).

Ajrekarella Kamat & Kalani (1964), anamorphic *Pezizomycotina*, St.0eH.19. 1, India. See Sutton (*Mycopath. Mycol. appl.* **33**: 76, 1967; redescr.).

Akanthomyces Lebert (1858), anamorphic *Cordyceps, Torrubiella*, Hsp.0eH.?. 9 (on insects and spiders), widespread. See Mains (*Mycol.* **42**: 566, 1950), Samson & Evans (*Acta Bot. Neerl.* **23**: 28, 1974), Hywel-Jones (*MR* **100**: 1065, 1996; Thailand), Hsieh *et al.* (*Mycol.* **89**: 319, 1997; Taiwan), Artjariyasripong *et al.* (*Mycoscience* **42**: 503, 2001; phylogeny), Stensrud *et al.* (*MR* **109**: 41, 2005; phylogeny), Sung *et al.* (*Stud. Mycol.* **57**: 1, 2007; phylogeny, biology).

Akaropeltella M.L. Farr (1972), ? Micropeltidaceae. Anamorph *Sporidesmium*-like. 1. See Farr (*Mycol.* **64**: 252, 1972), von Arx & Müller (*Stud. Mycol.* **9**, 1975; connexion), Réblová (*Mycotaxon* **71**: 13, 1999).

Akaropeltis Bat. & J.L. Bezerra (1961) [non *Acaropeltis* Petr. 1937] ≡ Akaropeltella.

Akaropeltopsis Bat. & Peres (1966) ? = Stomiopeltis fide von Arx & Müller (*Stud. Mycol.* **9**, 1975), Smith *et al.* (*Phytophylactica* **17**: 101, 1985).

akaryote (of *Plasmodiophoraceae*), the stage in the nuclear cycle before meiosis in which no or little chromatin is seen in the nucleus.

Akenomyces G. Arnaud (1954) nom. inval. = Akenomyces G. Arnaud ex D. Hornby fide Stalpers (*in litt.*).

Akenomyces G. Arnaud ex D. Hornby (1984), anamorphic *Agaricomycetes*. 1 (with clamp connexions), Europe. See Hornby (*TBMS* **82**: 653, 1984).

akinete (1) a non-motile reproductive structure; (2) a resting cell.

Akrophyton Lebert (1858) = Cordyceps fide Tulasne & Tulasne (*Select. fung. carpol.* **3**: 4, 1865), Sung *et al.* (*Stud. Mycol.* **57**: 1, 2007).

alate, winged.

Alatosessilispora K. Ando & Tubaki (1984), anamorphic *Pezizomycotina*, Hso.1bH.1. 1, Japan. See Ando & Tubaki (*TMSJ* **25**: 24, 1984).

Alatospora Ingold (1942), anamorphic *Leotiaceae*, Hso.1bH.15. 5 (freshwater), widespread. See Marvanová & Descals (*J. Linn. Soc.* Bot. **91**: 1, 1985; key), Gönczöl & Révay (*Fungal Diversity* **12**: 19, 2003; ecology), Belliveau & Bärlocher (*MR* **109**: 1407, 2005; phylogeny), Descals (*MR* **109**: 545, 2005; morphology), Baschien *et al.* (*Nova Hedwigia* **83**: 311, 2006; morphology, phylogeny).

Albatrellaceae Nuss (1980), Russulales. 7 gen. (+ 3 syn.), 45 spp.

Lit.: Fogel (*CJB* **57**: 1718, 1979; as *Leucogastraceae*), Beaton *et al.* (*Kew Bull.* **40**: 827, 1985; as *Leucogastraceae*), Keller (*Mycol. helv.* **2**: 1, 1986), Corner (*Beih. Nova Hedwigia* **96**: 218 pp., 1989), Stalpers (*Persoonia* Suppl. **14**: 537, 1992), Zheng *et al.* (*Acta Mycol. Sin.* **11**: 107, 1992), Valenzuela *et al.* (*Revta Mex. Micol.* **10**: 113, 1994), Agerer *et al.* (*Mycotaxon* **59**: 289, 1996), Ginns (*CJB* **75**: 261, 1997), Bruns *et al.* (*Mol. Ecol.* **7**: 257, 1998), de Hoog *et al. in* Kurtzman & Fell (Eds) (*Yeasts, a taxonomic study* 4th edn: 201, 1998; as *Leucogastraceae*), Dai & Zeng (*Mycosystema* **18**: 226, 1999), Montecchi & Sarasini (*Funghi Ipogei d'Europa*: 714 pp., 2000; as *Leucogastraceae*), Thorn (*Karstenia* **40**: 181, 2000), Binder & Hibbett (*Mol. Phylogen. Evol.* **22**: 76, 2002), Larsson & Larsson (*Mycol.* **95**: 1037, 2003), Ryman *et al.* (*MR* **107**: 1243, 2003), Binder *et al.* (*Systematics and Biodiversity* **3**: 113, 2005), Albee-Scott (*MR* **111**: 653, 2007; as *Leucogastraceae*).

Albatrellopsis Teixeira (1993) ≡ Albatrellus.

Albatrellus Gray (1821), Albatrellaceae. 16 (mycorrhizal), widespread (north temperate). See Donk (*Persoonia* **1**: 173, 1960; as *Scutiger*), Ginns (*CJB* **75**: 261, 1, 1975), Nuss (*Hoppea* **39**: 127, 1980; posn), Zheng (*Mycotaxon* **90**: 291, 2004; China).

Albertiniella Kirschst. (1936), Cephalothecaceae. Anamorph *Acremonium*-like. 1 (on *Ganoderma*), Europe; Japan. See Lundqvist (*Svensk bot. Tidskr.* **86**: 261, 1992), Suh & Blackwell (*Mycol.* **91**: 836, 1999; phylogeny), Huhndorf *et al.* (*Mycol.* **96**: 368, 2004; phylogeny).

Albigo Ehrh. ex Steud. (1824) ? = Sphaerotheca Lév. fide Hawksworth *et al.* (*Dictionary of the Fungi* edn 8, 1995).

Albocrustum Lloyd (1925) = Biscogniauxia See Pouzar (*Česká Mykol.* **33**: 207, 1979), Læssøe (*SA* **13**: 43, 1994).

Alboffia Speg. (1899) = Corynelia fide Fitzpatrick (*Mycol.* **12**: 239, 1920).

Alboffiella Speg. (1898) = Itajahya fide Stalpers (*in litt.*).

Alboleptonia Largent & R.G. Benedict (1970) = Entoloma fide Stalpers (*in litt.*).

Albomyces I. Miyake (1908), anamorphic *Aciculosporium*. 1 (on bamboos), Japan. See Oguchi (*Mycoscience* **42**: 217, 2001; morphology, biology).

Albonectria Rossman & Samuels (1999), Nectriaceae. Anamorph *Fusarium*. 3 (decaying wood and other plant parts), widespread (esp. tropical). See Rossman *et al.* (*Stud. Mycol.* **42**: 248 pp., 1999), Samuels *et al.* (*Tropical Mycology* **2**: 13, 2002; key), Summerbell & Schroers (*J. Clin. Microbiol.* **40**: 2866, 2002; phylogeny).

Albophoma Tak. Kobay., Masuma, Omura & Kyoto Watan. (1994), anamorphic *Pezizomycotina*, Cpd.0eH.19. 1 (from soil), Japan. See Kobayashi *et al.* (*Mycoscience* **35**: 399, 1994), Bills *et al.* (*Mycol. Progr.* **1**: 3, 2002).

Albosynnema E.F. Morris (1967), anamorphic *Bionectriaceae*, Hsy.≡ eP.1. 2, C. America; Caribbean. See Morris (*Mycopath. Mycol. appl.* **33**: 179, 1967), Bills *et al.* (*Sydowia* **46**: 1, 1994), Rossman *et al.* (*Mycol.* **93**: 100, 2001; posn).

Albotricha Raitv. (1970), Hyaloscyphaceae. *c.* 19, widespread (north temperate). See Raitviir (*Scripta Mycol.* **1**: 1, 1970; key), Raitviir (*Folia Cryptog. Estonica* **2**: 13, 1973), Raitviir (*Folia Cryptog. Estonica* **12**: 1, 1981), Zhuang (*Mycotaxon* **69**: 359, 1998), Leenurm *et al.* (*Sydowia* **52**: 30, 2000; ultrastr.), Wu (*Mycotaxon* **88**: 387, 2003; Taiwan).

Alciphila Harmaja (2002), anamorphic *Pezizomycotina*. 1 (on urine-impregnated ground), Scandinavia. See Harmaja (*Karstenia* **42**: 34, 2002).

Aldona Racib. (1900), Parmulariaceae. 3 (on leaves of *Pterocarpus*), pantropical. See Müller & Patil (*TBMS* **60**: 117, 1973; key), Inácio *et al.* (*Mycol. Progr.* **4**: 133, 2005).

Aldonata Sivan. & A.R.P. Sinha (1989), Parmulariaceae. 1 (on leaves of *Pterocarpus*), India. See

Sivanesan & Sinha (*MR* **92**: 246, 1989).

Aldridgea Massee (1892) nom. dub., Agaricomycetes. See Donk (*Taxon* **6**: 18, 1957).

Aldridgiella, see *Aldrigiella*.

Aldrigiella Rick (1934) nom. dub., Fungi. See Donk (*Taxon* **6**: 18, 1957).

ale, see beer.

Alectoria Ach. (1809), Parmeliaceae (L). 8, widespread (montane-boreal and bipolar). See Brodo & Hawksworth (*Op. bot. Soc. bot. Lund* **42**, 1977; key), Mattsson & Wedin (*Lichenologist* **31**: 431, 1999), Peršoh *et al.* (*Mycol. Progr.* **3**: 103, 2004; asci), Thell *et al.* (*Symb. bot. upsal.* **34** no. 1: 429, 2004; biogeography), Miądlikowska *et al.* (*Mycol.* **98**: 1088, 2006; phylogeny).

Alectoria Link (1833) = Usnea fide Hawksworth *et al.* (*Dictionary of the Fungi* edn 8, 1995).

Alectoriaceae Tomas. (1949) = Parmeliaceae.

Alectoriomyces Cif. & Tomas. (1953) ≡ Alectoria Ach.

Alectoriopsis Elenkin (1929) = Ramalina fide Eriksson & Hawksworth (*SA* **6**: 112, 1987).

Alectorolophoides Battarra ex Earle (1909) = Cantharellus fide Stalpers (*in litt.*).

alepidote, having no scales or scurf; smooth.

aleukia disease (**alimentary toxic aleukia**; **ATA**), see trichothecenes.

Aleuria (Fr.) Gillet (1879) ≡ Peziza Fr.

Aleuria Fuckel (1870), Pyronemataceae. 17 (on soil), widespread (north temperate). See Rifai (*Verh. K. ned. Akad. Wet.* tweede sect.: 1, 1968; Australian spp.), Moravec (*Česká Mykol.* **26**: 74, 1972), Kaushal (*Mycol.* **68**: 1021, 1976; Indian spp.), Häffner (*Rheinl.-Pfälz. Pilzj.* **3**: 6, 1993), Spooner & Yao (*MR* **99**: 1515, 1995; excl. spp.), Landvik *et al.* (*Nordic Jl Bot.* **17**: 403, 1997; DNA), Hansen & Pfister (*Mycol.* **98**: 1029, 2006; phylogeny), Perry *et al.* (*MR* **111**: 549, 2007; phylogeny).

Aleuriaceae Le Gal (1947) = Pyronemataceae.

Aleuriella P. Karst. (1871) = Mollisia fide Saccardo (*Syll. fung.* **8**: 1, 1889).

Aleurina (Sacc.) Sacc. & P. Syd. (1902) = Peziza Fr. fide Eckblad (*Nytt Mag. Bot.* **15**: 1, 1968).

Aleurina Massee (1898), Pyronemataceae. 11, widespread. See Zhuang & Korf (*Mycotaxon* **26**: 361, 1986; key), Perry *et al.* (*MR* **111**: 549, 2007; phylogeny).

aleuriospore (obsol.), formerly used for a thick-walled and pigmented but sometimes thin-walled and hyaline conidium developed from the blown-out end of a conidiogenous cell or hyphal branch from which it secedes with difficulty, as in *Aleurisma*, *Mycogone*, *Microsporum*; 'chlamydospore' sensu Hughes (1953); gangliospore. Since introduced by Vuillemin (1911), aleuriospore has been used in various senses, see Mason (1933, 1937) and Barron (1968), and finally rejected as a confused term (Kendrick, *Taxonomy of Fungi imperfecti*, 1971).

Aleurisma Link (1809) = Trichoderma Pers. (1794) fide Hughes (*CJB* **36**: 727, 1958), Carmichael (*CJB* **40**: 1137, 1962; synonym of *Chrysosporium* in sense of Vuillemin (1911)).

Aleurismataceae Vuill. (1911) = Hypocreaceae.

Aleurobotrys Boidin (1986), Stereaceae. 10. See Boidin & Gilles (*BSMF* **102**: 291, 1986) *Aleurodiscus* s.l.

Aleurocorticium P.A. Lemke (1964) = Dendrothele fide Lemke (*CJB* **42**: 723, 1965).

Aleurocystidiellum P.A. Lemke (1964), Russulales. 2, widespread. See Lemke (*CJB* **42**: 277, 1964), Larsson & Larsson (*Mycol.* **95**: 1037, 2003; phylogeny).

Aleurocystis Lloyd ex G. Cunn. (1956), Stereaceae. Anamorph *Matula*. 3, widespread. See Cunninham (*Trans. & Proc. Roy. Soc. New Zealand* **84**: 234, 1956), Rajchenberg & Robledo (*Mycotaxon* **92**: 317, 2005; Argentina).

Aleurodiscaceae Jülich (1982) = Stereaceae.

Aleurodiscus Rabenh. ex J. Schröt. (1888) nom. cons., Stereaceae. 12, widespread. See Lemke (*CJB* **42**: 213, 1964; key 26 amyloid-spored spp.), Núñez & Ryvarden (*Syn. Fung.* **12**, 1997), Wu *et al.* (*Mycol.* **93**: 720, 2001; phylogeny), Larsson & Larsson (*Mycol.* **95**: 1037, 2003; phylogeny).

Aleurodomyces Buchner (1912), anamorphic *Pezizomycotina*. 1 (on *Insecta*), Europe.

Aleuromyces Boidin & Gilles (2002), Stereaceae. 1, Gabon. See Boidin & Gilles (*BSMF* **117**: 176, 2001).

Aleurophora O. Magalh. (1916) ? = Chrysosporium fide Dodge (*Medical Mycology*, 1935).

Aleurosporia Grigoraki (1924) = Trichophyton fide Dodge (*Medical Mycology*, 1935).

Alexopoulos (Constantine John; 1907-1986; USA). University teacher, Michigan, Iowa, Texas. Wrote books on general mycology (see Literature) and *Mycetozoa* (q.v.) which became standard texts; pioneered modern recording of fungi in Greece. *Biogs, obits etc.* Brodie (*Mycol.* **79**: 163, 1986); Blackwell (*TBMS* **90**: 153, 1988) [portrait]; Grummann (1974: 201); Stafleu & Mennega (*TL-2*, *Suppl.* **1**: 67, 1992).

Alfvenia J.I.R. Larsson (1983), Microsporidia. 1.

Algacites Schloth. (1825), Fossil Fungi, Algae.

algae (fungi as parasites and mutualists of), see Kohlmeyer (*Veröff. Inst. Meersforsch. Bremerh., Suppl.* **5**: 339, 1974), Kohlmeyer & Kohlmeyer (*Marine mycology*, 1979), Lichens, mycophycobiosis, photobiont.

algal-layer (of lichen thalli), the photobiont-containing layer (usually between the upper cortex and the medulla) of the thallus.

algicolous, living on algae; **- fungi** see van Donk & Brumsz (*in* Reisser (Ed.), *Algae and symbiosis*: 567, 1992; review), algae.

Algincola Velen. (1939), ? Helotiales. 1, former Czechoslovakia.

Algonquinia R.F. Castañeda & W.B. Kendr. (1991), anamorphic *Pezizomycotina*, Hsp.0eH.12. 1, Canada. See Castañeda Ruiz & Kendrick (*Univ. Waterloo Biol. Ser.* **35**: 4, 1991).

Algorichtera Kuntze (1891) ≡ Scorias.

aliform, wing-like in form.

Alina Racib. (1909), Parodiopsidaceae. Anamorph *Septoidium*. 1, Java.

Alinocarpon Vain. (1928) = Thelocarpon fide Hawksworth *et al.* (*Dictionary of the Fungi* edn 8, 1995).

Aliquandostipitaceae Inderb. (2001), Jahnulales. 6 gen., 25 spp.

Lit.: Hawksworth (*Sydowia* **37**: 43, 1984), Inderbitzin *et al.* (*Am. J. Bot.* **88**: 54, 2001), Pang *et al.* (*MR* **106**: 1031, 2002), Raja *et al.* (*Mycotaxon* **91**: 207, 2005), Raja & Shearer (*Mycol.* **98**: 319, 2006), Campbell *et al.* (*CJB* **85**: 873, 2007; phylogeny).

Aliquandostipite Inderb. (2001), Aliquandostipitaceae. 3 (on wood in freshwater), pantropical. See Inderbitzin *et al.* (*Am. J. Bot.* **88**: 54, 2001), Pang *et al.* (*MR* **106**: 1031, 2002; placement), Raja *et al.* (*Mycotaxon* **91**: 207, 2005), Campbell *et al.* (*CJB* **85**: 873,

2007; phylogeny).

aliquot part, a portion that is contained an exact number of times in the whole; not the equivalent of 'sample' in which the concepts of both uniformity and representation are implicit (Emmons, *Bact. News* 1960: 17).

alkaphilic, used or organisms growing well at high pH values; e.g. *Fusarium* sp. at pH 10 (Hiura & Tanimura, *in* Horrikoshi & Grant (Eds), *Superbugs: microorganisms in extreme environments*: 287, 1991).

allantoid (esp. of spores), slightly curved with rounded ends; sausage-like in form (Fig. 23.8).

Allantomyces M.C. Williams & Lichtw. (1993), Legeriomycetaceae. 2 (in *Ephemeroptera*), Australia; Mexico. See Williams & Lichtwardt (*CJB* **71**: 1109, 1993), Valle *et al.* (*Mycol.* **100**: 149, 2008; Mexico).

Allantonectella, see *Allonectella.*

Allantonectria Earle (1901) = Nectria fide Rossman *et al.* (*Mycol.* **85**: 685, 1993), Rossman *et al.* (*Stud. Mycol.* **42**: 248 pp., 1999).

Allantoparmelia (Vain.) Essl. (1978), Parmeliaceae (L). 1, Arctic. See Esslinger (*Mycotaxon* **7**: 46, 1978), Feuerer (*Recollecting Edvard August Vainio*: 47, 1998), Thell *et al.* (*Symb. bot. upsal.* **34** no. 1: 429, 2004; biogeography).

Allantophoma Kleb. (1933) nom. inval., anamorphic *Pezizomycotina*. See Sutton (*Mycol. Pap.* **141**, 1977).

Allantophomoides S.L. Wei & T.Y. Zhang (2003) ? = Septoria Sacc. fide Wei & Zhang (*Mycosystema* **22**: 9, 2003).

Allantophomopsis Petr. (1925), anamorphic *Phacidium*, St.0eH.15. 7, widespread. See Carris (*CJB* **68**: 2283, 1990; gen. revision).

Allantoporthe Petr. (1921) = Diaporthe. Probably polyphyletic. fide Barr (*Mycol. Mem.* **7**, 1978), Zang (*Acta Mycol. Sin.* Suppl. **1**: 407, 1986; phylogeny).

Allantosphaeriaceae Höhn. (1918) = Diatrypaceae.

Allantospora Wakker (1895) = Cylindrocarpon fide Booth (*Mycol. Pap.* **104**, 1966).

Allantozythia Höhn. (1923) = Phlyctema fide Petrak (*Annls mycol.* **27**: 370, 1929), Sutton (*Mycol. Pap.* **141**, 1977).

Allantozythiella Danilova (1951) = Endothiella fide Sutton (*Mycol. Pap.* **141**, 1977).

Allantula Corner (1952), Pterulaceae. 1, Brazil. See Corner (*Ann. Bot., Lond.* n.s. **16**: 270, 1952).

Allarthonia (Nyl.) Zahlbr. (1903) = Arthonia fide Hawksworth *et al.* (*Dictionary of the Fungi* edn 8, 1995).

Allarthoniomyces E.A. Thomas (1939) nom. inval. ≡ Arthonia.

Allarthotheliomyces Cif. & Tomas. (1953) ≡ Allarthothelium.

Allarthothelium (Vain.) Zahlbr. (1908) = Arthonia fide Hawksworth *et al.* (*Dictionary of the Fungi* edn 8, 1995).

Allelochaeta Petr. (1955) = Seimatosporium fide Sutton (*Mycol. Pap.* **141**, 1977).

Alleppeysporonites Ramanujam & Rao (1979), Fossil Fungi, anamorphic *Pezizomycotina.* 1 (Miocene), India. Fossil *Grallomyces.*

Allergy. An acquired, specific, altered capacity to react. It is acquired by exposure to allergenic particles; the sensitivity acquired from a single exposure is specific to one or a few closely related species, although multiple exposures may result in multiple sensitivities; and subsequent re-exposure results in an altered capacity to react or allergic reaction. The form of that reaction depends on the nature of the allergenic particle, for instance, its size and chemical characteristics, the immunological reactivity of the subject and the circumstances of exposure. The two forms of allergy of most concern in this context are an immediate reaction, characterized by rhinitis and hay fever-like symptoms and a late reaction, characterized by alveolitis or pneumonitis. Fungal spores have been implicated as causative agents of both types of allergic reaction. Rhinitis and asthma are caused by normal everyday exposure to airborne allergens in subjects who are constitutionally predisposed (atopic) and who produce specific IgE antibodies against the allergen. Symptoms occur within a few minutes of exposure and may be provoked by 10^4 spores/m^{-3} air, or fewer, typically of fungi with spores larger than 10 µm. The spores may be components of the normal air spora, including *Alternaria, Cladosporium* and *Didymella*, or they may be associated with work environments, for instance cereal rusts and smuts, and *Verticillium lecanii* spores when harvesting *Agaricus bisporus* and *Boletus edulis* when preparing mushroom soup, and *Aspergillus flavus* and *A. awamori* from surface fermentations. Asthma may also be associated with exposure to fungal enzymes during their production. Allergic alveolitis occurs in non-atopic subjects after intense exposures to spores, typically 10^6-10^{10} spores/m^{-3}. At least 10^8 spores/m^{-3} may be required for sensitization but species differ in their antigenicity. Symptoms occur about 4 h after exposure and persist for 24-36 h if there is no further exposure. They include influenza-like symptoms, feverishness, chills, a dry cough, breathlessness and weight loss. With repeated exposure, breathlessness becomes increasingly severe and eventually permanent lung damage may occur with fibrosis, and the increased load on the heart may lead to death. Specific IgG antibodies develop and may be an aid to diagnosis although implication of a fungus in the disease may require further tests. The disease is typically occupational and associated with poorly stored agricultural products. The classic form is farmer's lung, usually caused by thermophilic actinomycetes but sometimes by fungi, including *Aspergillus flavus, A. versicolor* and *Eurotium rubrum* (syn. *Aspergillus umbrosus*). Other forms of allergic alveolitis include cheese-washer's lung (*Penicillium casei*), malt-worker's lung (*Aspergillus clavatus, A. fumigatus*), maple-bark stripper's lung (*Cryptostroma corticale*), mushroom picker's lung (*Aspergillus fumigatus, Cephalotrichum stemonitis, Pholiota nameko, Pleurotus ostreatus*), sawmill worker's lung (*Rhizopus rhizopodiformis, Penicillium* spp., *Aspergillus fumigatus, Trichoderma viride*), sequoiosis (*Aureobasidium pullulans, Graphium* spp.), suberosis (*Penicillium frequentans*), and allergic alveolitis from citric acid fermentations (*Aspergillus fumigatus, A. niger, Penicillium* spp.). Mouldy lichens have also been reported to cause allergic alveolitis.

Allergic skin reactions may be caused by spores of the *Arthrinium arundinis* state of *Apiospora montagnei* in workers cutting the canes of *Arundo donax* in France, by contact with lichens in wood-cutters and people using lichens in decorations (Richardson, *in* Galun (Ed.), *CRC Handbook of lichenology* **3**: 98, 1988; review), and secondary to dermatophyte infections (see mycid). Allergic reactions are also common in response to certain fungal products, the best known example being allergy to antibiotics such as penicillin.

For further information, see Pepys (*Hypersensitivity diseases of the lungs due to fungi and organic dusts*, 1969), Wilken-Jensen & Gravesen (*Atlas of moulds in Europe causing respiratory allergy*, 1984), Lacey (*in* Hawksworth (Ed.), *Frontiers in mycology*: 157, 1991), Lacey & Crook (*Ann. occup. Hyg.* **32**: 515, 1988), Lacey & Dutkiewicz (*J. Aerosol Sci.*, 1994).

Allescheria R. Hartig (1899) ≡ Hartigiella fide Vuillemin (*Annls mycol.* **3**: 341, 1905).

Allescheria Sacc. & P. Syd. (1899) = Monascus fide Malloch (*Mycol.* **62**: 727, 1970).

Allescheriella Henn. (1897), anamorphic *Botryobasidium*. 2, widespread. See Hughes (*Mycol. Pap.* **41**, 1951), Petrak (*Sydowia* **23**: 265, 1970).

Allescherina Berl. (1902) = Cryptovalsa fide Clements & Shear (*Gen. Fung.*, 1931).

Allewia E.G. Simmons (1990), Pleosporaceae. Anamorph *Embellisia*. 2, Australia. See Eriksson & Hawksworth (*SA* **9**: 2, 1991; synonymy with *Lewia*), Berbee *et al.* (*MR* **107**: 169, 2003; recombination), Schoch *et al.* (*Mycol.* **98**: 1041, 2006; phylogeny).

alliaceous, having a taste or smell of onions or garlic; cepaceous.

alliance, see phytosociology.

Alliospora Pim (1883) ? = Aspergillus fide Bisby (*TBMS* **27**: 101, 1944).

Allocetraria Kurok. & M.J. Lai (1991), Parmeliaceae (L). 11, widespread. See Kärnefelt *et al.* (*Acta Bot. Fenn.* **150**: 79, 1994), Thell *et al. in* Daniels *et al.* (Eds) (*Flechten Follmann* Contributions to Lichenology in Honour of Gerhard Follmann: 353, 1995), Thell *et al.* (*Mycol. Progr.* **1**: 335, 2002; phylogeny), Mattsson & Articus (*Symb. bot. upsal.* **34** no. 1: 237, 2004; phylogeny), Randlane & Saag (*Symb. bot. upsal.* **34** no. 1: 359, 2004; chemistry), Thell *et al.* (*Mycol. Progr.* **3**: 297, 2004; phylogeny), Randlane & Saag (*Central European Lichens*: 75, 2006; key).

allochronic, occurring at different time periods, e.g. contemporary and fossil specimens.

allochrous (allochroous), changing from one colour to another.

allochthonous, transported to the place where found; not indigenous; cf. autochthonous.

Allochytridium Salkin (1970), Endochytriaceae. 2, N. America. See Barr & Désaulniers (*Mycol.* **79**: 193, 1987; morphol., physiol., ultrastr.).

Alloclavaria Dentinger & D.J. McLaughlin (2007), Agaricomycetes. 1, Europe. *Hymenochaetales* or *Agaricales* (*Rickenella* clade). See Dentinger & McLaughlin (*Mycol.* **98**: 757, 2007; syst. posn).

allocyst, a chlamydospore-like structure in *Flammula gummosa* (Kühner, 1946).

Allodium Nyl. (1896) = Chaenotheca fide Hawksworth *et al.* (*Dictionary of the Fungi* edn 8, 1995).

Allodus Arthur (1906) = Puccinia fide Arthur (*Manual Rusts US & Canada*, 1934).

Alloglugea Paperna & Lainson (1995), Microsporidia. 1.

Allographa Chevall. (1824) ? = Graphina fide Hawksworth *et al.* (*Dictionary of the Fungi* edn 8, 1995).

Allomyces E.J. Butler (1911), Blastocladiaceae. 9 (in soil), widespread (esp. tropical). See Emerson (*Lloydia* **4**: 77, 1941; life cycle, taxonomy), Teter (*Mycol.* **36**: 194, 1944; sexuality), Emerson & Wilson (*Mycol.* **46**: 393, 1954; cytogenetics and cyto taxonomy), Taylor *et al.* (*Nature* **367**: 601, 1994; fossil from Devonian), Steciow & Elíades (*Darwiniana* **39**: 15, 2001; Argentina).

Allonecte Syd. (1939), Tubeufiaceae. 1, Ecuador. See Rossman (*Mycotaxon* **8**: 485, 1979), Crane *et al.* (*CJB* **76**: 602, 1998), Kodsueb *et al.* (*Fungal Diversity* **21**: 105, 2006; phylogeny).

Allonectella Petr. (1950), Nectriaceae. 2 (on stromata of *Phyllachora*), S. America. See Rossman (*Mycotaxon* **8**: 485, 1979), Rossman *et al.* (*Stud. Mycol.* **42**: 248 pp., 1999).

Allonema Syd. (1934), anamorphic *Pezizomycotina*, Hso.0eH.?. 1, Europe. See Sartory & Meyer (*Annls mycol.* **33**: 101, 1935).

Alloneottiosporina Nag Raj (1993), anamorphic *Pezizomycotina*, Cpd.≡ eH.19. 2, USA; Australia. See Nag Raj (*Coelomycetous Anamorphs with Appendage-bearing Conidia*: 121, 1993).

allopatric, occurring in different geographical regions. Cf. sympatric.

Allophoron Nádv. (1942), Lecanorales (L). 1, Colombia. See Tibell (*Recollecting Edvard August Vainio*: 95, 1998), Tibell *et al.* (*Mycotaxon* **87**: 3, 2003).

Allophylaria (P. Karst.) P. Karst. (1870), Helotiaceae. *c.* 6, Europe. See Carpenter (*Mem. N. Y. bot. Gdn* **33**: 17, 1981), Arendholz (*Mycotaxon* **36**: 283, 1989; nomencl.), Huhtinen (*Karstenia* **29**: 45, 1989).

Allopsalliota Nauta & Bas (1999), Agaricaceae. 1, Netherlands. See Nauta (*Belg. Jl Bot.* **131**: 189, 1998), Vellinga (*MR* **108**: 354, 2004; phylogeny).

Allopuccinia H.S. Jacks. (1931) = Sorataea fide Cummins & Hiratsuka (*Illustr. Gen. Rust Fungi rev. edit.*, 1983).

Allosoma Syd. (1926), Dothideomycetes. Anamorph *Periconiella*. 1, C. America. See *Acrodesmis*.

Allosphaerium Link (1826) nom. dub., Agaricomycotina. See Saccardo (*Syll. fung.* **15**, 1901; syn. of *Rhizoctonia* s. lat.).

Allotelium Syd. (1939), Raveneliaceae. 1 (on *Calliandra* (*Leguminosae*)), S. America. May include *Diabolidium*.

Allothyriella Bat., Cif. & Nascim. (1959), anamorphic *Pezizomycotina*, Cpt.≡ eP.?. 2, C. America; Africa. See Batista *et al.* (*Mycopath. Mycol. appl.* **11**: 11, 1959).

Allothyrina Bat. & J.L. Bezerra (1964), anamorphic *Pezizomycotina*, Cpt.≡ eH.?. 1, Brazil. See Batista & Bezerra (*Portugaliae Acta Biologica* Série B **7**: 384, 1964).

Allothyriopsis Bat., Cif. & H. Maia (1959), anamorphic *Pezizomycotina*, Cpt.≡ eP.?. 1, Ghana. See Batista *et al.* (*Mycopath. Mycol. appl.* **11**: 14, 1959).

Allothyrium Syd. (1939), Asterinaceae. 1, Ecuador.

Almbornia Essl. (1981), Parmeliaceae (L). 2, S. Africa. See Brusse (*Mycotaxon* **40**: 265, 1991), Thell *et al.* (*Mycol. Progr.* **3**: 297, 2004; phylogeny), Crespo *et al.* (*Mol. Phylogen. Evol.* **44**: 812, 2007; phylogeny).

Almeidaea Cif. & Bat. (1962) [non *Almeidaea* Post & Kuntze 1903, *Rutaceae*] = Chaetothyrium fide von Arx & Müller (*Stud. Mycol.* **9**, 1975), Panwar & Jagtap (*Geobios* New Rep. **9**: 121, 1990).

Alnicola Kühner (1926) = Naucoria fide Reid (*TBMS* **82**: 191, 1984).

Alocospora J.C. Krug (1990) nom. inval., Xylariaceae. 1, Europe. See Krug (*Fourth International Mycological Congress Abstracts*: 30, 1990).

Aloysiella Mattir. & Sacc. (1908) = Antennularia fide von Arx & Müller (*Stud. Mycol.* **9**, 1975).

Alpakesa Subram. & K. Ramakr. (1954), anamorphic *Pezizomycotina*, Hso.0eH.1. 4, India. See Morgan-Jones *et al.* (*CJB* **50**: 877, 1972), Matsushima (*Matsush. Mycol. Mem.* **5**, 1987), Punithalingam (*Stud. Mycol.* **31**: 113, 1989; appendages), Abbas *et al.* (*Pakist. J. Bot.* **35**: 249, 2003).

Alpakesiopsis Abbas, B. Sutton, Ghaffar & A. Abbas (2003), anamorphic *Pezizomycotina*. 1, Pakistan. See Abbas *et al.* (*Pakist. J. Bot.* **35**: 249, 2003).

alpha-spore (**A-spore**, **α-spore**), a fertile, fusoid to oblong, biguttulate spore of an anamorph of the *Valsaceae* (*Phomopsis*). Cf. beta-spore.

Alphitomorpha Wallr. (1819) nom. superf. = Erysiphe fide Fries (*Syst. mycol.* **3**: 234, 1829).

Alphitomorphaceae Corda (1842) = Erysiphaceae.

Alphitomyces Reissek (1856) ? = Isaria fide Samson (*Stud. Mycol.* **6**, 1974).

alpine mycology, see Polar and alpine mycology.

Alpova C.W. Dodge (1931), Paxillaceae. 20, widespread (esp. north temperate). See Trappe (*Beih. Nova Hedwigia* **51**: 279, 1975), Bruns *et al.* (*Mol. Ecol.* **7**: 257, 1998; phylogeny), Grubisha *et al.* (*Mycol.* **93**: 82, 2001; genus probably polyphyletic), Nouhra *et al.* (*Mycol.* **97**: 598, 2005).

Alternaria Nees (1816), anamorphic *Lewia*, Hso.#eP.26. 299, widespread. *A. brassicae* (leaf spot of crucifers), *A. cucumerina* (cucurbit leaf spot), *A. longipes*, and others, on tobacco, *A. solani* (early blight of potato) which produces the highly phytotoxic antibiotic alternaric acid (q.v.). A number are common cosmop. Saprobes. See Neergard (*Danish Species of Alternaria and Stemphylium*, 1945; Denmark), Joly (*Le Genre Alternaria*, 1964; monogr.), Simmons (*Mycol.* **59**: 73, 1967; typification), Rao (*Nova Hedwigia* **17**: 219, 1969; India), Simmons (*Mycol.* **61**: 1, 1969; teleomorphs), Ellis (*Dematiaceous Hyphomycetes*, 1971; descriptions), Ellis (*More Dematiaceous Hyphomycetes*, 1976; descriptions), Ando & Takatori (*Mycopathologia* **100**: 17, 1987; keratomycosis), Samson & Frisvad (*Proc. Jap. Assoc. Mycotoxic.* **32**: 3, 1990; mycotoxins), Simmons (*Mycotaxon* **38**: 251, 1990; teleomorphs), Chełkowski & Visconti (*Alternaria. Biology, Plant Diseases and Metabolites* Topics in Secondary Metabolism vol. **3**, 1992), Simmons (*Alternaria. Biology, Plant Diseases and Metabolites* Topics in Secondary Metabolism vol. **3**, 1992; review), McCartney *et al.* (*Pl. Path.* **42**: 280, 1993; dispersal of conidia), Lopes & Boiteux (*Pl. Dis.* **78**: 1107, 1994; *Ipomoea*), Rotem (*The genus Alternaria*, 1994; biology, epidemiology, pathogenicity), Verma & Saharan (*Technical Bulletin, Saskatoon Research Centre, Research Branch, Agriculture and Agri-Food Canada* **1994-6E**, 1994; *Cruciferae*), Visconti & Sibilia (*Mycotoxins in Grain. Compounds Other Than Aflatoxin*: 315, 1994; toxins), Jasalavich *et al.* (*MR* **99**: 604, 1995; *Cruciferae*), Kusaba & Tsuge (*Curr. Genet.* **28**: 491, 1995; toxigenic spp.), Andersen & Thrane (*Mycotoxin Research* **12**: 54, 1996; metabolites), Zhang & David (*Mycosystema* **8-9**: 109, 1995; *Euphorbiaceae*), Kusaba & Tsuge (*Ann. phytopath. Soc. Japan* **63**: 463, 1997; mt DNA), Mims *et al.* (*CJB* **75**: 252, 1997; ultrastructure), Bottalico & Logrieco *in* Sinha & Bhatnagar (eds), (*Mycotoxins in Agriculture and Food Safety*: 65, 1998; mycotoxins, toxigenic spp.), McKay *et al.* (*Eur. J. Pl. Path.* **105**: 157, 1999; *Linum*), Peever *et al.* (*Phytopathology* **89**: 851, 1999; population biology), Simmons (*Mycotaxon* **70**: 325, 1999; toxigenic spp.), Simmons (*Mycotaxon* **70**: 263, 1999; *Citrus*), Inoue & Nasu (*J. Gen. Pl. Path.* **66**: 18, 2000; *Prunus*), Magan & Evans (*Journal of Stored Products Research* **36**: 319, 2000; volatile metabolites), Morris (*MR* **104**: 286, 2000; *Lycopersicon*), Pryor & Gilbertson (*MR* **104**: 1312, 2000; phylogeny), Simmons (*Mycotaxon* **75**: 1, 2000; *Solanaceae*; 101556; small-spored spp.), Andersen *et al.* (*MR* **105**: 291, 2001; *A. gaisen* and similar spp.), Halaby *et al.* (*J. Clin. Microbiol.* **39**: 1952, 2001; phaeohyphomycosis), Romano *et al.* (*Mycoses* **44**: 73, 2001; onychomycoses), Andersen *et al.* (*MR* **106**: 170, 2002; phylogeny, chemistry), Bock *et al.* (*MR* **106**: 428, 2002; *A. brassicicola*), Chou & Wu (*MR* **106**: 164, 2002; phylogeny, morphology), Dugan & Peever (*Mycotaxon* **83**: 229, 2002; *Gramineae*), Hoog & Horré (*Mycoses* **45**: 259, 2002; clinical strains), Peever *et al.* (*Phytopathology* **92**: 794, 2002; *Citrus*), Pryor & Gilbertson (*Mycol.* **94**: 49, 2002; *A. radicina* group), Serdani *et al.* (*MR* **106**: 561, 2002; *Malus*), Simmons (*Mycotaxon* **83**: 127, 2002; teleomorphs), Simmons (*Mycotaxon* **82**: 1, 2002; *Caryophyllaceae*), Strandberg (*Phytoparasitica* **30**: 269, 2002; selective media), Akimitsu *et al.* (*Molecular Plant Pathology* **4**: 435, 2003; *Citrus*), Berbee *et al.* (*MR* **107**: 169, 2003; recombination), Kang *et al.* (*Pl. Path. J.* **19**: 221, 2003; phylogeny, toxins), Kwasna & Kosiak (*MR* **107**: 371, 2003; *Avena*), Pryor & Bigelow (*Mycol.* **95**: 1141, 2003; phylogeny), Simmons (*Mycotaxon* **88**: 163, 2003; *Malvaceae*), Guo *et al.* (*Fungal Diversity* **16**: 53, 2004; endophytes), Hong & Pryor (*Can. J. Microbiol.* **50**: 461, 2004; selective media), Peever *et al.* (*Mycol.* **96**: 119, 2004; *Citrus*), Pérez Martínez *et al.* (*Eur. J. Pl. Path.* **110**: 399, 2004; *A. solani*), Waals *et al.* (*Pl. Dis.* **88**: 959, 2004; *Solanum* in S Africa), Andersen *et al.* (*Phytopathology* **95**: 1021, 2005; image analysis), Bock *et al.* (*MR* **109**: 227, 2005; recombination in *A. brassicicola*), Dubois *et al.* (*Mycopathologia* **160**: 117, 2005; phaeohyphomycosis), Hong *et al.* (*Fungal Genetics Biol.* **42**: 119, 2005; allergenics), Hong *et al.* (*MR* **109**: 87, 2005; IGS polymorphism), Peever *et al.* (*Phytopathology* **95**: 512, 2005; *Citrus*), Quayyum *et al.* (*CJB* **83**: 1133, 2005; *Panax*), Goetz & Dugan (*Pacific Northwest Fungi* **1**: 1, 2006; *A. malorum*), Hong *et al.* (*MR* **110**: 1290, 2006; *Corylus*, *Juglans*), Mercado Vergnes *et al.* (*Pl. Path.* **55**: 485, 2006; *Triticum*), Schoch *et al.* (*Mycol.* **98**: 1041, 2006; phylogeny), Simmons (*CBS Diversity Ser.* **6**, 2007; revision, nomenclator).

Alternariaceae Earle (1934) = Pleosporaceae.

Alternariaster E.G. Simmons (2007), Pleosporaceae. 1. See Simmons (*Alternaria: an Identification Manual*, 2007).

alternaric acid, a metabolite produced by *Alternaria solani* which inhibits spore germination in some fungi and causes wilting and necrosis in higher plants.

alternate host, one or other of the two unlike hosts of an heteroecious rust. See *Teliomycetes*.

alternation of generations, the succession of gametophyte and sporophyte or sexual and asexual phases in a life cycle: **homologous** when the two generations are like in form; **antithetic** if unlike, when the gametophyte is named the **protophyte** and the sporophyte the **antiphyte** (Celakovsky).

Alutaceodontia (Parmasto) Hjortstam & Ryvarden (2002), Schizoporaceae. 1. See Hjortstam & Ryvarden (*Syn. Fung.* **15**: 8, 2002).

alutaceous, the colour of buff leather.

alveola (1) a small surface cavity or hollow; (2) a pore of a polypore (obsol.).

Alveolaria Lagerh. (1892), Pucciniosiraceae. 2 (on *Cordia* (*Boraginaceae*)), America (tropical). See Buriticá & Hennen (*Fl. Neotrop.* **24**: 22, 1980).

alveolate, marked with ± 6-sided (honey-comb-like) hollows; faveolate.

Alveolinus Raf. (1815) nom. dub., Fungi. No spp. included.

Alveomyces Bubák (1914) = Uromyces fide Nattrass (*First list Cyprus fungi*, 1937).

Alveophoma Alcalde (1952), anamorphic *Pezizomycotina*, Cpd.0eH.10. 1, Spain. See Sutton (*TBMS* **47**: 497, 1964).

Alysia Cavalc. & A.A. Silva (1972) = Vouauxiella fide Sutton (*The Coelomycetes*, 1980), Lücking *et al.* (*Lichenologist* **30**: 121, 1998).

Alysidiella Crous (2006), Pezizomycotina. 1 (on *Eucalyptus* leaves), S. Africa. See Crous (*Fungal Diversity* **23**: 325, 2006).

Alysidiopsis B. Sutton (1973), anamorphic *Pezizomycotina*, Hso.0-1eP.3. 4, widespread. See Sutton (*Mycol. Pap.* **132**: 5, 1973), Currah (*CJB* **65**: 1957, 1987; Mexico), Kendrick (*CJB* **81**: 75, 2003; morphogenesis).

Alysidium Kunze (1817), anamorphic *Botryobasidium*. 4, Europe. See Ellis (*Dematiaceous Hyphomycetes*, 1971), Partridge & Morgan-Jones (*Mycotaxon* **83**: 335, 2002).

Alysisporium Peyronel (1922) = Phragmotrichum fide Sutton & Pirozynski (*TBMS* **48**: 349, 1965).

Alysphaeria Turpin (1827) nom. dub., ? Fungi (L).

Alytosporium Link (1824) nom. dub., Fungi. See Donk (*Taxon* **12**: 156, 1963) See also, Stalpers (*Rev. Mycol.* **39**: 99, 1975).

Alyxoria Gray (1821) = Opegrapha Ach. fide Hawksworth *et al.* (*Dictionary of the Fungi* edn 8, 1995).

AM, arbuscular mycorrhiza; see Mycorrhiza.

amadou, the context of *Fomes fomentarius* or *Phellinus igniarius* after the addition of saltpetre ($NaNO_3$); tinder; touchwood; punk.

Amallospora Penz. (1897), anamorphic *Pezizomycotina*, Hsp.1bH.?. 1, Java. See Ho *et al.* (*Mycol.* **92**: 582, 2000), Descals (*MR* **109**: 545, 2005).

Amandinea M. Choisy (1950) ≡ Amandinea M. Choisy ex Scheid. & M. Mayrhofer.

Amandinea M. Choisy ex Scheid. & M. Mayrhofer (1993), Caliciaceae (L). 34, widespread. See Sheard & May (*Bryologist* **100**: 159, 1997; N. Am.), Grube & Arup (*Lichenologist* **33**: 63, 2001; polyphyly), Nordin & Mattsson (*Lichenologist* **33**: 3, 2001; morphology, phylogeny), Helms *et al.* (*Mycol.* **95**: 1078, 2003; phylogeny), Peršoh *et al.* (*Mycol. Progr.* **3**: 103, 2004; asci), Simon *et al.* (*J. Mol. Evol.* **60**: 434, 2005; introns), Miądlikowska *et al.* (*Mycol.* **98**: 1088, 2006; phylogeny).

Amanita Adans. (1763) nom. dub., Agaricales. See Donk (*Beih. Nova Hedwigia* **5**, 1962).

Amanita Dill. ex Boehm. (1760) nom. rej. ≡ Agaricus L.

amanita factor B, see pantherine; **- - C**, see ibotenic acid.

Amanita Pers. (1797), Amanitaceae. *c.* 500, widespread. Many species ectomycorrhizal, but members of subgen. *Lepidella* partly saprobic. Both edible (e.g. *A. caesarea* and poisinous (e.g. *A. phalloides*) species. See Malençon (*Rev. Mycol.* **20**: 81, 1955; development), Bas (*Persoonia* **5**: 285, 1969; key sect. *Lepidella*), Bas (*Beih. Nova Hedwigia* **51**: 53, 1975; relationship to *Amanita*), Campbell & Petersen (*Mycotaxon* **1**: 239, 1975; culture), Horak (*Mycol.* **84**: 64, 1992), Pegler & Shah-Smith (*Mycotaxon* **61**: 389, 1997; key eastern Africa), Wood (*Aust. Syst. Bot.* **10**: 723, 1997; key Australia), Yang (*Biblthca Mycol.* **170**: 1, 1997; key Southwest China), Weiß *et al.* (*CJB* **76**: 1070, 1998; phylogeny), Drehmel *et al.* (*Mycol.* **91**: 610, 1999; phylogeny), Yang *et al.* (*Amanita – Ectomycorrhizal fungi, key genera in profile*, 1999; ecology), Miller & Lodge (*Mycotaxon* **79**: 289, 2001; Dominican Republica), Tulloss *et al.* (*Mycotaxon* **77**: 455, 2001; Pakistan), Bougher & Lebel (*Aust. Syst. Bot.* **15**: 514, 2002), Moncalvo *et al.* (*Mol. Phylog. Evol.* **23**: 357, 2002; nesting within *Amanita*), Oda *et al.* (*Mycoscience* **43**: 351, 2002; Japan), Simmons *et al.* (*Persoonia* **17**: 563, 2002; Guyana), Bhatt *et al.* (*Mycotaxon* **88**: 249, 2003; India), Neville & Hemmes (*Fungi Europaei* **9**: 1120 pp., 2004; Eur.), Yang (*Frontiers in Basidiomycote Mycology*: 315, 2004; Chile), Tulloss (*Mycotaxon* **93**: 189, 2005; distribution).

Amanitaceae R. Heim ex Pouzar (1983), Agaricales. 3 gen. (+ 23 syn.), 521 spp.

Lit.: Hibbett *et al.* (*Nature* **407**: 506, 2000), Moncalvo *et al.* (*Syst. Biol.* **49**: 278, 2000) not supported by other data; see.

Amanitaria E.-J. Gilbert (1941) = Amanita Pers. fide Singer (*Agaric. mod. Tax.* edn 3, 1975).

Amanitella Earle (1909) = Amanita Pers. fide Singer (*Agaric. mod. Tax.* edn 3, 1975).

Amanitella Maire (1913) = Limacella fide Singer (*Agaric. mod. Tax.* edn 3, 1975).

amanitin, see amatoxins.

Amanitina E.-J. Gilbert (1940) = Amanita Pers. fide Singer (*Agaric. mod. Tax.* edn 3, 1975).

Amanitopsis Roze (1876) nom. cons. = Amanita Pers. fide Singer (*Agaric. mod. Tax.* edn 3, 1975) not conserved against *Amanita*.

Amarenographium O.E. Erikss. (1982), anamorphic *Amarenomyces*, St.#eP.15. 1, Europe. See Nag Raj (*CJB* **67**: 3169, 1989; redescr.).

Amarenomyces O.E. Erikss. (1981), Pleosporales. Anamorph *Amarenographium*. 1 (on *Ammophila*), Europe. See Shoemaker & Babcock (*CJB* **67**: 1500, 1989).

Amarrendia Bougher & T. Lebel (2002) = Amanita. The gastroid form has been previously recognised as a separate genus. fide Kuyper (*in litt.*).

Amastigis Clem. & Shear (1931) ≡ Amastigosporium.

Amastigomycètes Clem. & Shear (1931) ≡ Amastigosporium.

Amastigomycota, the zygo-, asco-, and basidiomycetes (Whittaker, 1969).

Amastigosporium Bond.-Mont. (1921) = Mastigosporium fide Hughes (*Mycol. Pap.* **36**, 1951).

amatoxins, cyclic octopeptides (including α-amanitin and β-amanitin, amanin, and the non-toxic amanillin) toxic to humans from *Amanita phalloides*, etc. See Wieland (*Science* **159**: 951, 1968), Wieland (*Peptides of poisonous Amanita mushrooms*, 1986). Cf. phallotoxins.

Amaurascopsis Guarro, Gené & De Vroey (1992), Gymnoascaceae. 2, Burundi; Honduras. See Guarro et al. (*Mycotaxon* **45**: 171, 1992), Hentic (*BSMF* **116**: 173, 2000; phylogeny), Sugiyama *et al.* (*Stud. Mycol.* **47**: 5, 2002; phylogeny).

Amauroascaceae Arx (1987) ? = Onygenaceae.

Amauroascus J. Schröt. (1893), ? Onygenaceae. Anamorph *Chrysosporium*. 14, widespread. See von Arx (*Persoonia* **6**: 374, 1971), Currah *in* Hawksworth (Ed.) (*Ascomycete Systematics. Problems and Perspectives in the Nineties* NATO ASI Series vol. **269** **269**: 370, 1994), Sugiyama *et al.* (*Mycoscience* **40**: 251, 1999; DNA), Udagawa & Uchiyama (*Mycoscience* **40**: 277, 1999), Udagawa & Uchiyama (*Mycoscience* **40**: 291, 1999), Hentic (*BSMF* **116**: 173, 2000; phylogeny), Sugiyama *et al.* (*Stud. Mycol.* **47**: 5, 2002; phylogeny).

Amauroderma (Pat.) Torrend (1920) = Amauroderma Murrill fide Donk (*Persoonia* **1**: 184, 1960).

Amauroderma Murrill (1905), Ganodermataceae. *c.* 30, widespread (tropical). See Furtado (*Mem. N. Y. bot. Gdn* **34**: 1, 1980), Ryvarden & Johansen (*Prelim. Polyp. Fl. E. Afr.*: 315, 1980; key 11 Afr. spp.), Corner (*Beih. Nova Hedwigia* **75**: 45, 1983; keys S. Am. & Malaysian spp.), Ryvarden (*Syn. Fung.* **18**: 57, 2004), Decock & Herrera Figueroa (*Cryptog. Mycol.* **27**: 3, 2006; neotropical spp.).

Amaurodon J. Schröt. (1888), Thelephoraceae. 9, widespread. See Kõljalg (*Syn. Fung.* **9**: 32, 1996; key), Agerer & Bougher (*Aust. Syst. Bot.* **14**: 599, 2001; blue-spored sp.).

Amaurohydnum Jülich (1978), Meruliaceae. 1, Australia. See Jülich (*Persoonia* **9**: 455, 1978).

Amauromyces Jülich (1978), Meruliaceae. 2, Australia; Japan; Réunion. See Jülich (*Persoonia* **9**: 455, 1978), Chen & Oberwinkler (*Mycol.* **96**: 418, 2004; Japan).

Amazonia Theiss. (1913), Meliolaceae. 29 (from leaves), widespread (pantropical). See Hosagoudar (*Nova Hedwigia* **52**: 81, 1991; ascospore germination), Hosagoudar (*Meliolales of India*: 363 pp., 1996; India), Hu *et al.* (*Flora Fungorum Sinicorum* **4**. Meliolales: 270 pp., 1996; China), Hu *et al.* (*Flora Fungorum Sinicorum* **11**, 1999; China), Hosagoudar (*Zoos' Print Journal* **18**: 1243, 2003; endemism), Hosagoudar (*Sydowia* **55**: 168, 2003; diagnostic formulae).

Amazoniella Bat. & H. Maia (1960) = Amazonia fide Hughes (*in litt.*).

Amazonomyces Bat. & Cavalc. (1964), Arthoniaceae (L). 2, neotropics. See Lücking *et al.* (*Lichenologist* **30**: 121, 1998), Grube & Lücking (*MR* **105**: 1007, 2001; ascogenous hyphae).

Amazonotheca Bat. & H. Maia (1959), Schizothyriaceae. 1, Philippines. See Batista & Maia (*Publções Inst. Micol. Recife* **56**: 408, 1959).

Amazonspora C. Azevedo & E. Matos (2003), Microsporidia. 1. See Azevedo & Matos (*J. Parasit.* **89**: 336, 2003).

Amber. This is an important medium for the study of fossil fungi because soft structures may be retained which are generally lost in rock-preserved fossils. Hyphomycetes and coelomycetes associated with spruce seedlings have been found preserved in baltic amber (Dörfelt & Schmidt, *Bot. J. Linn. Soc.* **155**: 449, 2007), and coelomycetes have been found preserved in Dominican amber (Poinar, *MR* **107**: 117, 2003). Basidiomycetes (including basidiomycete parasites on other basidiomycetes) have been reported from early cretaceous Burmese amber (Poinar & Buckley, *MR* **111**: 503, 2007). For reports of fungi on arthropods in amber, including *Entomophthora* sp. on *c.* 25 million year old winged termite from Oligocene-Miocene (Dominican Republic), and for reports on carnivorous fungi in amber, see Fossil fungi. See Poinar & Thomas (*Mycol.* **74**: 332, 1982; lichens), Rikkinen & Poinar (*MR* **104**: 7, 2000; lichens), Waggoner & Poinar (*J. Protozool.* **39**: 639, 1992; myxomycete). See also Fossil fungi.

ambimobile, systemic fungicides which can move upward in the xylem or downward in the phloem.

ambiregnal (of organisms), ones that can be classified in more than one kingdom according to different systematic viewpoints; esp. of those which can potentially be treated under different *Codes*. See Nomenclature, Corliss (*BioSystems* **28**: 1, 1993), Patterson & Larsen (*Regnum veg.* **123**: 197, 1991).

Ambispora C. Walker, Vestberg & A. Schüssler (2007), Ambisporaceae. 4, widespread. See Walker *et al.* (*MR* **111**: 147, 2007).

Ambisporaceae C. Walker, Vestberg & A. Schüssler (2007), Archaeosporales. 1 gen., 4 spp.

Ambivina Katz (1974), Corticiaceae. 1, USA. See Katz (*Nova Hedwigia* **25**: 811, 1974).

Amblyospora E.I. Hazard & Oldacre (1975), Microsporidia. 19.

Amblyosporiopsis Fairm. (1922) = Oedocephalum fide Clements & Shear (*Gen. Fung.*, 1931).

Amblyosporium Fresen. (1863), anamorphic *Pezizomycotina*, Hso.0eH.40. 3, Europe. *A. botrytis* (on agarics, esp. *Lactarius*). See Nicot & Durand (*BSMF* **81**: 623, 1966), Pirozynski (*CJB* **47**: 325, 1969), Kendrick (*CJB* **81**: 75, 2003; morphogenesis).

Ambrodiscus S.E. Carp. (1988), Helotiales. 1 (bark beetle galleries), USA. See Carpenter (*Mycol.* **80**: 320, 1988).

ambrosia fungi, Fungi, often yeasts (e.g. *Ambrosiozyma*, *Ascoidea* and *Dipodascus* spp., etc.) or yeast-like (conidial *Ophiostomatales*), that grow mutualistically in tunnels of ambrosia beetles (wood-boring *Scolytidae*) and serve as food for larvae and adults; many are specific for the particular insect (Batra, *Trans. Kansas Acad. Sci.* **66**: 213, 1963; *Mycol.* **59**: 981, 1968; key gen.); some are associated with devastating tree diseases.

Lit.: Mueller & Gerardo (*Ann. Rev. Entom.* **36**: 563, 2005), Wingfield *et al.* [Eds] (*Ceratocystis and Ophiostoma: taxonomy, ecology and pathology*, 1993). **- gall**, see gall.

Ambrosiaemyces Trotter (1934), Pezizomycotina. 1 (on wood damaged by ambrosia beetles), Sri Lanka.

Ambrosiella Brader (1964), anamorphic *Ceratocystidaceae*, Hsy.0eH.1/38. 9 (in bark beetle galleries), widespread. Polyphyletic; some species belong to the *Ophiostomataceae*. See Batra (*Mycol.* **59**: 986, 1968; key), Cassar & Blackwell (*Mycol.* **88**: 596, 1996; convergent evolution), Blackwell & Jones (*Biodiv. Cons.* **6**: 689, 1997; biology), Rollins *et al.* (*Mycol.* **93**: 991, 2001; phylogeny), Spatafora (*Cellular Origin and Life in Extreme Habitats* **4**: 591, 2002; symbiosis), Zhang *et al.* (*Mycol.* **98**: 1076, 2006; phylogeny).

Ambrosiozyma Van der Walt (1972), ? Saccharomycetales. 6, widespread. Perhaps allied to the *Saccharomycopsidaceae*. See Goto & Takami (*J. gen. appl. Microbiol.* Tokyo **32**: 271, 1986), Jones & Blackwell (*MR* **102**: 661, 1998), Smith *in* Kurtzman & Fell (Eds) (*Yeasts, a taxonomic study* 4th edn: 129,

1998), Suh *et al.* (*Mycol.* **98**: 1006, 2006; phylogeny).
Ameghiniella Speg. (1888), Helotiaceae. 2, N. & S. America. See also *Ionomidotis*. See Zhuang (*Mycotaxon* **31**: 261, 1988; key), Gamundí (*MR* **95**: 1131, 1991), Gamundí & Romero (*Fl. criptog. Tierra del Fuego* **10**, 1998).
amend, the act and result of making an alteration, not necessarily to correct a fault or error. Cf. emend.
Amepiospora Locq. & Sal.-Cheb. (1980), Fossil Fungi. 5, Cameroon.
Ameris Arthur (1906) = Phragmidium fide Arthur (*Manual Rusts US & Canada*, 1934).
Amerobotryum Subram. & Natarajan (1976) = Agaricostilbum fide Subramanian & Natarajan (*Mycol.* **69**: 1224, 1977).
Amerodiscosiella M.L. Farr (1961), anamorphic *Pezizomycotina*, Cpt.0eH.15. 1, Cambodia; Brazil. See Sutton (*TBMS* **60**: 525, 1973), Nag Raj (*CJB* **53**: 2435, 1975), Farr (*Taxon* **26**: 580, 1977; typification), Patil (*Geobios* New Rep. **9**: 173, 1990; India).
Amerodiscosiellina Bat. & Cavalc. (1966), anamorphic *Pezizomycotina*, Cpt.0eH.?. 1, Brazil. See Batista & Cavalcanti (*Atas Inst. Micol. Univ. Pernambuco* **3**: 185, 1966).
Amerodothis Theiss. & Syd. (1915) = Botryosphaeria fide von Arx & Müller (*Beitr. Kryptfl. Schweiz* **11** no. 1, 1954).
Ameromassaria Hara (1918), Pezizomycotina. 1, Japan.
Ameropeltomyces Bat. & H. Maia (1967) = Arthonia fide Lücking *et al.* (*Lichenologist* **30**: 121, 1998).
amerospore, a 1-celled (i.e. non-septate) spore with a length/width ratio ‹ 15:1 (cf. scolecospore); if elongated, axis single and not curved through more than 180° (cf. helicospore); any protuberances ‹ $^1/_4$ spore body length (cf. staurospore). See Anamorphic fungi.
Amerosporiella Höhn. (1916) nom. illegit., anamorphic *Pezizomycotina*, Hso.0eH/1eP.?. 1, Europe.
Amerosporina (Petr.) Petr. (1965) = Amerosporium fide Sutton (*The Coelomycetes*, 1980).
Amerosporiopsis Petr. (1941), anamorphic *Pezizomycotina*, Cpd.0eH.15. 1, Iran. See Sutton (*The Coelomycetes*, 1980), Nag Raj & DiCosmo (*Univ. Waterloo Biol. Ser.* **20**, 1982).
Amerosporis Clem. & Shear (1931) ≡ Amerosporiella.
Amerosporium Speg. (1882), anamorphic *Zoellneria*, St.0eP.15. 2, widespread. See Sutton (*The Coelomycetes*, 1980), Johnston & Gamundí (*N.Z. Jl Bot.* **38**: 493, 2000).
Amerostege Theiss. (1916) nom. dub., ? Fungi. See von Arx & Müller (*Beitr. Kryptfl. Schweiz* **11** no. 1, 1954).
Amerosympodula Matsush. (1996), anamorphic *Pezizomycotina*, Hso.?.?. 1, Peninsular Malaysia. See Matsushima (*Matsush. Mycol. Mem.* **9**: 1, 1996).
Ameson Sprague (1977), Microsporidia. 4.
Amethicium Hjortstam (1983), Phanerochaetaceae. 1, Tanzania. See Hjortstam (*Mycotaxon* **17**: 557, 1983).
ametoecious, see autoecious (q.v.; de Bary).
Amicodisca Svrček (1987), Hyaloscyphaceae. 5, Europe. See Svrček (*Česká Mykol.* **41**: 16, 1987), Huhtinen (*Karstenia* **29**: 45, 1990), Raitviir (*Czech Mycol.* **52**: 289, 2001; key), Raitviir (*Mycotaxon* **87**: 359, 2003), Raitviir (*Scripta Mycologica* Tartu **20**, 2004).
Amidella E.-J. Gilbert (1940) = Amanita Pers. fide Singer (*Agaric. mod. Tax.* edn 3, 1975).
amixis, see heterothallism.
ammonia fungi, a chemoecological group in which reproductive structures develop after the addition of ammonia, urea, etc. or alkalis to the soil (Sagara, *Contrib. biol. Lab. Kyoto Univ.* **24**: 205, 1975).
Amoebochytrium Zopf (1884), Cladochytriaceae. 1, Europe.
amoeboid, not having a cell wall and changing in form, like an amoeba.
amoeboid cell (of *Ameobidiales*), uninucleate cells, formed by protoplasmic cleavage within the fungal thallus, which lack a rigid wall and when released usually encyst, the cysts, in time, producing cystospores.
Amoebomyces Bat. & H. Maia (1965) = Strigula fide Lücking *et al.* (*Lichenologist* **30**: 121, 1998).
Amoebophilus P.A. Dang. (1910), Cochlonemataceae. 4, Europe; N. America. See Drechsler (*Mycol.* **27**: 33, 1935), Drechsler (*Mycol.* **51**: 787, 1959), Barron (*CJB* **61**: 3091, 1983).
Amoenodochium Peláez & R.F. Castañeda (1996), anamorphic *Pezizomycotina*, Hso.?.?. 1, Goa. See Peláez & Castañeda Ruíz (*Mycotaxon* **60**: 258, 1996).
Amoenomyces R.F. Castañeda, Saikawa & Hennebert (1996), anamorphic *Pezizomycotina*, Hso.?.?. 1, Cuba. See Castañeda Ruíz *et al.* (*Mycotaxon* **59**: 453, 1996), Castañeda Ruíz *et al.* (*MR* **104**: 107, 2000; comp. with *Bulbocatenospora*).
Amogaster Castellano (1995), Agaricales. 1, USA. Perhaps *Boletales*. See Castellano (*Mycotaxon* **55**: 186, 1995).
Amorosia Mantle & D. Hawksw. (2006), ? Sporormiaceae. 1 (from intertidal sediment), Bahamas. See Mantle & Hawksworth (*MR* **110**: 1373, 2006).
Amorphomyces Thaxt. (1893), Laboulbeniaceae. 13 (on insect exoskeleton), widespread. See Santamaría (*MR* **104**: 1389, 2000; key), Santamaría (*Fl. Mycol. Iberica* **5**, 2003; Iberian spp.).
Amorphotheca Parbery (1969), Amorphothecaceae. Anamorph *Hormoconis*. 1 (on resin, hydrocarbon products etc.), widespread. *A. resinae* (putative anamorph *Hormoconis resinae*; kerosene fungus, q.v.). See Parbery (*Aust. J. Bot.* **17**: 331, 1969), Sheridan *et al.* (*Tuatara* **19**: 130, 1972), Braun *et al.* (*Mycol. Progr.* **2**: 3, 2003; phylogeny), Seifert *et al.* (*Stud. Mycol.* **58**: 235, 2007; phylogeny, nomenclature).
Amorphothecaceae Parbery (1969), Eurotiomycetidae (inc. sed.). 2 gen. (+ 1 syn.), 2 spp. Possibly allied with *Myxotrichaceae*, but molecular data are contradictory.
Lit.: Parbery (*Aust. J. Bot.* **17**: 331, 1969), Braun *et al.* (*Mycol. Progr.* **2**: 8, 2003), Abliz *et al.* (*FEMS Immunol. Med. Microbiol.* **40**: 41, 2004), Stchigel & Guarro (*MR* **111**: 1100, 2007).
Amparoina Singer (1958), ? Tricholomataceae. 2, S. America. See Singer (*Mycol.* **50**: 110, 1958).
Amparoinaceae Singer (1976) nom. rej. = Tricholomataceae.
Ampelomyces Ces. ex Schltdl. (1852), anamorphic *Phaeosphaeriaceae*, Cpd.0eP.15. 1 or 2 (on *Erysiphales*), widespread. See Foitzik & Triebel (*Arnoldia* **6**: 15, 1993; typification), Kiss (*MR* **101**: 1073, 1997), Kiss & Nakasone (*Curr. Genet.* **33**: 362, 1998), Nischwitz *et al.* (*MR* **109**: 421, 2005; rel. with *Eudarluca*), Szentiványi *et al.* (*MR* **109**: 429, 2005; speciation), Liang *et al.* (*Fungal Diversity* **24**: 225, 2007; phylogeny).
amphi- (prefix), the two (sorts, sides).

Amphiacantha Caullery & Mesnil (1914), Microsporidia. 3.
Amphiamblys Caullery & Mesnil (1914), Microsporidia. 3.
Amphiblistrum Corda (1837) = Oidium Link (1824) fide Linder (*Lloydia* **5**: 165, 1942).
Amphichaeta McAlpine (1904) = Seimatosporium fide Shoemaker (*CJB* **42**: 411, 1964).
Amphichaete Kleb. (1914) ≡ Amphichaetella.
Amphichaetella Höhn. (1916), anamorphic *Pezizomycotina*, Hsp.0eH.?. 1, Europe; Australia. See Morgan-Jones (*CJB* **51**: 1431, 1973), Alcorn (*Australas. Mycol.* **21**: 111, 2002; Australia).
Amphichorda Fr. (1825) = Isaria fide Fries (*Syst. mycol.* **3**: 1, 1832).
Amphiciliella Höhn. (1919) nom. dub., anamorphic *Pezizomycotina*. See Sutton (*Mycol. Pap.* **141**, 1977).
Amphiconium Nees (1816) nom. dub., Algae. Based on algae fide Fries (*Syst. mycol.* **3** (index): 51, 1832).
Amphicypellus Ingold (1944) = Chytriomyces fide Dogma (*Kalikasan* **5**: 136, 1976), Letcher & Powell (*Mycotaxon* **84**: 447, 2002).
Amphicytostroma Petr. (1921), anamorphic *Amphiporthe*, St.0eH.15. 2, Europe.
Amphididymella Petr. (1928) = Acrocordia fide Yue & Eriksson (*Mycotaxon* **24**: 293, 1985).
Amphidium Nyl. (1891) [non *Amphidium* Schimp. 1856, *Musci*] = Epiphloea fide Gyelnik (*Rabenh. Krypt.-Fl.* **9** 2.2, 1940).
Amphiernia Grüss (1926) = Sporobolomyces fide Derx (*Annls mycol.* **28**: 1, 1930).
amphigenous, making growth all round or on two sides.
amphigynous (of *Pythiaceae*), having an antheridium through which the oogonial incept grows.
Amphilogia Gryzenh., H.F. Glen & M.J. Wingf. (2005), Cryphonectriaceae. 2 (on *Elaeocarpus*), Sri Lanka; New Zealand. See Gryzenhout *et al.* (*Taxon* **54**: 1017, 2005), Gryzenhout *et al.* (*FEMS Microbiol. Letters* **258**: 161, 2006).
Amphiloma Körb. (1855) ≡ Gasparrinia.
Amphiloma Nyl. (1855) = Lepraria fide Hawksworth *et al.* (*Dictionary of the Fungi* edn 8, 1995).
Amphilomopsis Jatta (1905) = Chrysothrix fide Hawksworth *et al.* (*Dictionary of the Fungi* edn 8, 1995).
amphimixis, copulation of two cells and nuclei which are not near relations, e.g. egg and sperm; cf. apomixis, automixis and pseudomixis.
Amphimyces Thaxt. (1931), Laboulbeniaceae. 1, W. Africa; Europe. See Hindley (*Wiltshire Archaeological and Natural History Magazine* **79**: 214, 1985; monogr.), Santamaria *et al.* (*Treb. Inst. Bot. Barcelona* **14**: 1, 1991; Europe).
Amphinectria Speg. (1924) nom. dub., ? Tubeufiaceae. See Rossman *et al.* (*Stud. Mycol.* **42**: 248 pp., 1999).
Amphinema P. Karst. (1892), Atheliaceae. 6 (mycorrhizal), widespread. See Sutton & Crous (*MR* **101**: 215, 1997).
Amphinomium Nyl. (1888) = Pannaria fide Galloway & Jørgensen (*Lichenologist* **19**: 345, 1987).
Amphiporthe Petr. (1971), Gnomoniaceae. Anamorph *Amphicytostroma*. 3, Europe; N. America. See Barr (*Mycol. Mem.* **7**, 1978), Zhang & Blackwell (*Mycol.* **93**: 355, 2001; phylogeny).
Amphirosellinia Y.M. Ju, J.D. Rogers, H.M. Hsieh & Vasilyeva (2004), Xylariaceae. 5 (saprobic in bark), north temperate. See Læssøe & Spooner (*Kew Bull.* **49**: 1, 1994; as *Rosellinia*), Ju *et al.* (*Mycol.* **96**: 1393, 2004).
Amphischizonia Mont. (1856) nom. inval. = Cryptodictyon fide Santesson (*Symb. bot. upsal.* **12** no. 1: 1, 1952).
Amphisphaerella (Sacc.) Kirschst. (1934), Xylariales. 8 (from bark), Europe. See Eriksson (*Svensk bot. Tidskr.* **60**: 315, 1966), Kang *et al.* (*Fungal Diversity* **2**: 135, 1999; posn), Wang *et al.* (*Fungal Diversity Res. Ser.* **13**, 2004).
Amphisphaerellula Gucevič (1952), Pezizomycotina. 1, former USSR. See Gucevič (*Bot. Mater. Otd. Sporov. Rast. Bot. Inst. Komarova Akad. Nauk S.S.S.R.* **8**: 142, 1952).
Amphisphaeria Ces. & De Not. (1863) nom. cons., Amphisphaeriaceae. Anamorph *Bleptosporium*. 85 (from wood and bark), widespread. See Kang *et al.* (*Fungal Diversity* **1**: 147, 1998; DNA), Kang *et al.* (*MR* **103**: 53, 1999), Kang *et al.* (*Mycotaxon* **81**: 321, 2002; phylogeny), Jeewon *et al.* (*MR* **107**: 1392, 2003; posn).
Amphisphaeriaceae G. Winter (1885), Xylariales. 32 gen. (+ 47 syn.), 499 spp.

Lit.: Samuels *et al.* (*Mycotaxon* **28**: 473, 1987; anamorphs), Barr (*Mycotaxon* **51**: 191, 1994; family rels), Nag Raj & Mel'nik (*Mycotaxon* **50**: 435, 1994), Okane *et al.* (*CJB* **74**: 1338, 1996), Goh & Hyde (*MR* **101**: 85, 1997), Hyde (*MR* **101**: 609, 1997), Graniti (*Ann. Rev. Phytopath.* **36**: 91, 1998), Kang *et al.* (*Fungal Diversity* **1**: 147, 1998; DNA), Kang *et al.* (*Fungal Diversity* **2**: 135, 1999; excluded genera), Kang *et al.* (*MR* **103**: 53, 1999; genera), Strobel *et al.* (*Syst. Appl. Microbiol.* **22**: 432, 1999), Jeewon *et al.* (*Mol. Phylogen. Evol.* **25**: 378, 2002), Jeewon *et al.* (*Fungal Diversity* **17**: 39, 2004).

Amphisphaerina Höhn. (1919), Pezizomycotina. 3, Europe; N. America.
amphispore, a second, special type of urediniospore; see *Pucciniales*.
amphithallism, see homothallism.
amphithecium, the thalline margin of an apothecium (L).
Amphitiarospora Agnihothr. (1963) = Dinemasporium fide Sutton (*Mycol. Pap.* **141**, 1977).
amphitrichous (amphitrichiate), having one flagellum at each pole.
Amphitrichum T. Nees (1818) nom. dub., Pezizomycotina. See Hughes (*CJB* **36**: 727, 1958).
Amphobotrys Hennebert (1973), anamorphic *Botryotinia*, Hso.0eH/1eP.7. 1, USA. See Hennebert (*Persoonia* **7**: 192, 1973), Holcomb *et al.* (*Pl. Dis.* **73**: 74, 1989), Hong *et al.* (*Pl. Path. J.* **17**: 357, 2001), Kendrick (*CJB* **81**: 75, 2003; morphogenesis).
Amphophialis R.F. Castañeda, W.B. Kendr. & Guarro (1998), anamorphic *Pezizomycotina*, Hso.?.?. 1, Cuba. See Castañeda Ruíz *et al.* (*Mycotaxon* **68**: 12, 1998).
Amphopsis (Nyl.) Hue (1892) = Pyrenopsis fide Hawksworth *et al.* (*Dictionary of the Fungi* edn 8, 1995).
Amphoridium A. Massal. (1852) = Verrucaria Schrad. fide Hawksworth *et al.* (*Dictionary of the Fungi* edn 8, 1995).
Amphoridium, see *Amphoridium A. Massal.*
Amphoroblastia Servít (1953) = Polyblastia. p.p., Thelidium (Verrucar.) p.p. and Verrucaria (Verrucar.) p.p. fide Hawksworth *et al.* (*Dictionary of the Fungi* edn 8, 1995).

Amphoromorpha Thaxt. (1914) = Basidiobolus fide Blackwell & Malloch (*Mycol.* **81**: 735, 1989).

Amphoropsis Speg. (1918) ? = Pyxidiophora fide Blackwell & Malloch (*Mycol.* **81**: 735, 1989), Blackwell (*Mycol.* **86**: 1, 1994).

Amphoropycnium Bat. (1963), anamorphic *Pezizomycotina*, Cpd.0eH.15. 2, Brazil; Philippines. See Batista (*Quad. Lab. crittogam., Pavia* **31**: 19, 1963).

Amphorothecium P.M. McCarthy, Kantvilas & Elix (2001), ? Myeloconidiaceae (L). 1, Australia. See McCarthy *et al.* (*Lichenologist* **33**: 292, 2001).

Amphorula Grove (1922) = Chaetoconis fide Petrak (*Sydowia* **13**: 180, 1959), Sutton (*CJB* **46**: 183, 1968).

Amphorulopsis Petr. (1959), Pezizomycotina. 1, former Yugoslavia. See Petrak (*Sydowia* **13**: 181, 1959).

amphotericin, **A** and **B**, polyene antibiotics from actinomycetes (*Streptomyces* spp.); antifungal; **- B** (fungizone) is used in the therapy of systemic mycoses of humans.

Amplariella E.-J. Gilbert (1940) = Amanita Pers. fide Singer (*Agaric. mod. Tax.* edn 3, 1975).

amplectant, covering; embracing.

ampliate, made greater; enlarged.

Ampliotrema Kalb (2004) ≡ Ampliotrema Kalb ex Kalb fide Kalb (*Biblthca Lichenol.* **88**: 302, 2004).

Ampliotrema Kalb ex Kalb (2006), Thelotremataceae (L). 5, pantropical. See Frisch (*Biblthca Lichenol.* **92**: 3, 2006), Frisch *et al.* (*Biblthca Lichenol.* **92**: 517, 2006; phylogeny, links with *Ocellularia*).

ampoule effect, Corner's (*New Phytol.* **47**: 48, 1948) term for the normal working of a basidium which is compared to an ampoule from which the contents are discharged into the basidiospores by the enlargement of a basal vesicle.

ampoule hypha, see hypha.

ampulla (1) the swollen tip of a conidiogenous cell which produces synchronous blastic conidia (as in *Gonatobotrytum*); (2) a conidiophore which develops a number of short branches or discrete conidiogenous cells (as in *Aspergillus*).

Ampullaria A.L. Sm. (1903) [non *Ampullaria* Couch 1963, *Actinomycetes*] = Melanospora Corda fide Cannon & Hawksworth (*J. Linn. Soc.* Bot. **84**: 115, 1982).

Ampullariella Couch (1964), Actinobacteria. q.v.

Ampullifera Deighton (1960), anamorphic *Pezizomycotina*, Hso.0eH.4. 7 (on foliicolous lichens), pantropical. See Hawksworth (*Bull. Br. Mus. nat. hist.* Bot. **6**: 183, 1979), Hawksworth & Cole (*Mycosystema* **22**: 359, 2003; China).

Ampulliferella Bat. & Cavalc. (1964) = Ampullifera fide Hawksworth (*Bull. Br. Mus. nat. hist.* Bot. **6**: 183, 1979).

Ampulliferina B. Sutton (1969), anamorphic *Pezizomycotina*, Hso.1eP.38. 2, Canada; British Isles. See Sutton (*CJB* **47**: 609, 1969).

Ampulliferinites Kalgutkar & Sigler (1995), Fossil Fungi, anamorphic *Pezizomycotina*. 1 (Eocene), Canada. See Kalgutkar & Sigler (*MR* **99**: 515, 1995).

Ampulliferopsis Bat. & Cavalc. (1964) = Ampullifera fide Hawksworth (*Bull. Br. Mus. nat. hist.* Bot. **6**: 183, 1979).

ampulliform, flask-like in form (Fig. 23.30).

Ampullina Quél. (1875) = Leptosphaeria fide von Arx & Müller (*Stud. Mycol.* **9**, 1975).

Ampulloclitocybe Redhead, Lutzoni, Moncalvo & Vilgalys (2002), Hygrophoraceae. 3, widespread. See Redhead *et al.* (*Mycotaxon* **83**: 36, 2002), Harmaja (*Ann. bot. fenn.* **40**: 213, 2003).

Amygdalaria Norman (1852) ? = Porpidia fide Inoue (*J. Hattori bot. Lab.* **56**: 321, 1984; key), Brodo & Hertel (*Herzogia* **7**: 493, 1987; key 8 spp.), Esnault & Roux (*An. Jard. bot. Madr.* **44**: 211, 1987), Purvis *et al.* (*Lichen Flora of Great Britain and Ireland*, 1992), Buschbom & Mueller (*Mol. Phylogen. Evol.* **32**: 66, 2004; phylogeny), Fryday (*Lichenologist* **37**: 1, 2005; placement).

Amylaria Corner (1955), Bondarzewiaceae. 1, Bhutan. See Stalpers (*Stud. Mycol.* **40**: 48, 1996).

Amylariaceae Corner (1970) = Bondarzewiaceae.

Amylascus Trappe (1971), Pezizaceae. 2 (hypogeous), Australasia. See Trappe (*TBMS* **65**: 496, 1975; key), Hansen *et al.* (*Mycol.* **93**: 958, 2001; phylogeny), Hansen *et al.* (*Mol. Phylogen. Evol.* **36**: 1, 2005; phylogeny), Læssøe & Hansen (*MR* **111**: 1075, 2007; phylogeny).

Amylirosa Speg. (1920) nom. dub., Dothideales. See von Arx & Müller (*Stud. Mycol.* **9**, 1975).

Amylis Speg. (1922), Pezizomycotina. 1, S. America.

Amylo process (Amylomyces process). A method for the commercial production of alcohol by the saccharification of starchy materials by *Amylomyces rouxii* or *Rhizopus* spp. The amylo process is used in preparation of ragi, sufu and tempeh (see Fermented food and drinks). Ragi and ragi-like products from different countries of Asia contain a more or less stable mycota of *Amylomyces*, *Mucor* and *Rhizopus* species as well as various yeasts and bacteria (Hesseltine *et al.*, *Mycopathologia* **101**: 141, 1988). *Amylomyces rouxii*, used in Asia to ferment cassava and rice, has the enzyme glucoamylase which occurs in only one form (Wang *et al.*, *Journal of Food Science* **49**: 1210-1211, 1984). *Rhizopus formosaensis* is a powerful glucoamylase-producing fungus, with one strain suitable for fermentation of a highly concentrated starchy broth (Ling *et al.*, [*Hok Fermentation Engineering Magazine*, Society for Bioscience and Engineering, Japan] **49**: 101, 1971). See Erb & Hildebrandt (*Industr. engin. Chem.* **38**: 792, 1946), Hesseltine (*Mycol.* **57**: 149, 1965; 1991; *Mycologist* **5**: 166, 1991), Johnson (*Ann. Rev. Microbiology* **1**: 159, 1947), Panda (*The complete technology book on starch and its derivatives*, 540 pp., 2004).

Amyloathelia Hjortstam & Ryvarden (1979), ? Amylocorticiaceae. 3, Europe; S. America. See Hjortstam & Ryvarden (*Mycotaxon* **10**: 201, 1979).

Amylobasidium Ginns (1988), Corticiaceae. 1, USA. See Ginns (*Mycol.* **80**: 63, 1988), Ginns (*Mycol.* **90**: 1, 1997).

Amylocarpus Curr. (1859), ? Leotiomycetes. 1 (on wood, marine), Europe. Affinities are unclear. See Crumlish & Curran (*Mycologist* **8**: 83, 1994), Landvik *et al.* (*Mycoscience* **37**: 237, 1996; phylogeny), Landvik *et al.* (*Mycoscience* **39**: 49, 1998; phylogeny), Læssøe & Hansen (*MR* **111**: 1075, 2007; phylogeny).

Amylocorticiaceae Jülich (1982), Agaricales. 10 gen., 45 spp.

Amylocorticiellum Spirin & Zmitr. (2002), Amylocorticiaceae. 4, widespread. See Zmitrovich & Spirin (*Mikol. Fitopatol.* **36**: 22, 2002).

Amylocorticium Pouzar (1959), Amylocorticiaceae. 11, widespread. See Zmitrovich (*Novosti Sistematiki Nizshikh Nov. sist. Niz. Rast.* **36**: 31, 2002; Russian spp.), Gilbertson & Hemmes (*Mem. N. Y. bot. Gdn*

89: 81, 2004; Hawaii).

Amylocystis Bondartsev & Singer (1944), Fomitopsidaceae. 1, Europe.

Amylodontia M.I. Nikol. (1967) = Dentipellis fide Stalpers (*Stud. Mycol.* **40**: 54, 1996; key).

Amyloflagellula Singer (1966), Marasmiaceae. 4, America (tropical); Asia. See Singer (*Darwiniana* **14**: 14, 1966), Antonín (*Czech Mycol.* **54**: 235, 2003), Bodensteiner *et al.* (*Mol. Phylogen. Evol.* **33**: 501, 2004; phylogeny).

Amylofungus Sheng H. Wu (1996), ? Peniophoraceae. 2, New Zealand; Japan. See Wu (*Mycol.* **87**: 886, 1995).

Amylohyphus Ryvarden (1978), Stereaceae. 1, Rwanda. See Ryvarden (*Bulletin du Jardin Botanique National de Belgique* **48**: 81, 1978).

amyloid (of asci, spores, etc.), stained blue by iodine (see Iodine, Stains); cf. dextrinoid. See Dodd & McCracken (*Mycol.* **64**: 1341, 1972; nature of fungal starch), amylomycan.

Amylolepiota Harmaja (2002), Agaricaceae. 1, Europe. See Harmaja (*Karstenia* **42**: 39, 2002).

amylomycan, a name proposed for the I+ blue or red compounds associated with asci (Common, *Mycotaxon* **41**: 67, 1991).

Amylomyces Calmette (1892), Mucoraceae. 1, Asia. See Ellis *et al.* (*Mycol.* **68**: 131, 1976), Voigt & Wöstemeyer (*Gene* **270**: 113, 2001; phylogeny), Abe *et al.* (*Biosc., Biotechn., Biochem.* **70**: 2387, 2006; phylogeny).

Amylonotus Ryvarden (1975), Auriscalpiaceae. 3, widespread (tropical). See Ryvarden (*Norw. Jl Bot.* **22**: 26, 1975) = *Wrightoporia* fide, Stalpers (*Stud. Mycol.* **40**: 129, 1996).

Amylophagus Scherff. (1925), Monad. q.v.

Amyloporia Singer (1944), Polyporaceae. 5, widespread. See Vampola & Pouzar (*Česká Mykol.* **46**: 213, 1993).

Amyloporiella A. David & Tortič (1984), Polyporaceae. 5, Europe; N. America. See David & Tortíc (*TBMS* **83**: 659, 1984; key).

Amylora Rambold (1994), ? Trapeliaceae (L). 1, widespread. See Rambold (*Bull. Soc. linn. Provence* **45**: 344, 1994), Lumbsch & Heibel (*Lichenologist* **30**: 95, 1998), Rambold & Hagedorn (*Lichenologist* **30**: 473, 1998), Lumbsch *et al.* (*MR* **111**: 1133, 2007).

Amylosporaceae Jülich (1982) = Bondarzewiaceae.

Amylosporomyces S.S. Rattan (1977), Stereaceae. 2, widespread. See Rattan (*Biblthca Mycol.* **60**: 244, 1977).

Amylosporus Ryvarden (1973), Bondarzewiaceae. 6, widespread (tropical). See Stalpers (*Stud. Mycol.* **40**: 129, 1996; key).

Amylostereaceae Boidin, Mugnier & Canales (1998), Russulales. 1 gen. (+ 2 syn.), 4 spp.

Amylostereum Boidin (1958), Amylostereaceae. 4, widespread. See Boidin (*Revue Mycol.* Paris **23**: 345, 1958), Legon & Pegler (*Mycologist* **16**: 124, 2002; *Amylostereum areolatum*), Larsson & Larsson (*Mycol.* **95**: 1037, 2003; phylogeny), Slippers *et al.* (*South African Journal of Science* **99**: 70, 2003; association with woodwasps).

Amyloxenasma (Oberw.) Hjortstam & Ryvarden (2005), Amylocorticiaceae. 5, widespread. See Hjortstam & Ryvarden (*Syn. Fung.* **20**: 34, 2005).

an-, see a-.

anaerobe, an organism able to grow without free oxygen. An **obligate -** grows only without free oxygen; a **facultative -** grows with or without free oxygen. See Zehnder (Ed.) (*Biology of anaerobic microorganisms*, 1988).

Anaerobic fungi. Most fungi grow only aerobically (obligate aerobes), some prefer oxygen, but can grow anaerobically and others are oxygen indifferent (facultative anaerobes) (Emerson & Held, *Amer. J. Bot.* **56**: 1103, 1969). Anaerobic fungi occur widely in association with large herbivores, in both the foregut of ruminant-like animals and the hindgut of hindgut fermenters. A well-illustrated account of these fungi is provided by Mountfort (*Anaerobic Fungi (Mycology Series)* **12**: 1, 1994). Rumen fungi specifically colonise and grow on plant vascular tissues, produce active cellulases and xylanases (Bauchop, *Biosystems* **23**: 53, 1989). The flagellate gut fungi (Neocallimastigales) are the sole group which lack mitochondria and grow only without oxygen (obligate anaerobes), although they are tolerant of oxygen during transfer between hosts. They use diverse substrata and produce formate, acetate, lactate, ethanol, succinate, CO_2 and H_2. See Li & Heath (*Can. J. Microbiol.* **39**: 1003, 1993), Trinci *et al.* (*MR* **98**: 129, 1994; review, bibliogr.). Tetronasin and cycloheximide can reduce populations of anaerobic fungi in the rumen of sheep (Gordon & Phillips, 1993).

Lit.: Bauchop (*Biosystems* **23**: 53, 1989), Gordon & Phillips (*Letters in Applied Microbiology* **17**: 220, 1993), Mountfort *Anaerobic Fungi* Mycology Series, vol. 12. CRC, 1994).

Anaeromyces Breton, Bernalier, Dusser, Fonty, B. Gaillard & J. Guillot (1990), Neocallimastigaceae. 2, France; Australia. See Breton *et al.* (*FEMS Microbiol. Lett.* **70**: 181, 1990).

analogous, showing a resemblance in form, structure, or function which is not considered to be evidence of evolutionary relatedness; cf. homologous.

Anamika K.A. Thomas, Peintner, M.M. Moser & Manim. (2002), Cortinariaceae. 1, China; India; Japan. See Thomas *et al.* (*MR* **106**: 246, 2002), Yang *et al.* (*MR* **109**: 1259, 2005).

anamorph (1) (of shapes), a deformed figure appearing in proportion when correctly viewed; (2) (of fungi), see States of fungi.

Anamorphic fungi (Deuteromycotina, Deuteromycetes, Fungi Imperfecti, asexual fungi, conidial fungi, mitosporic fungi) (a few L). These are fungi that are disseminated by propagules not formed from cells where (by inference from a small number of studied examples) meiosis has occurred. Most of these propagules can be referred to as conidia (q.v.) but some are derived from unspecialized vegetative mycelium. Many are correlated with fungal states that produce spores derived from cells where meiosis has, or is inferred to have, occurred (i.e. the teleomorph). These are, where known, members of the ascomycetes or basidiomycetes however, in many cases, they are still undescribed, unrecognized ('unconnected') or poorly known. Some anamorphs have appeared to have lost sexuality and its functions are sometimes replaced by such mechanisms as the parasexual cycle. These fungi have taken independent evolutionary paths from the related holomorphs (holomorphic anamorphs of Hennebert, 1993). See Kendrick (*Sydowia* **41**: 6, 1989), Sutton (*in* Reynolds & Taylor, *The fungal holomorph*: 27, 1993), Hennebert (*in* Reynolds & Taylor, *The fungal holomorph*: 283, 1993).

TABLE 1. Mitosporic fungi coding for conidiomata and conidia (for conidiogenous events see text).

Conidiomata

Hyphomycetes (H)	**Coelomycetes** (C)	**Other**
Hso solitary (hyphal)	**Cpd** pycnidial	**St** stromatic
Hsy synnematal	**Cpt** pycnothyrial	**Sc** sclerotial
Hsp sporodochial	**Cac** acervular	
	Ccu cupulate	

Conidial shape and septation

shape	septation		H conidia hyaline or bright (hyalo-)	P conidia pigmented or dark (phaeo-)
e ellipsoid	**0** aseptate	amerosporae	hyalosporae	phaeosporae
f filiform	**1** 1-septate	didymosporae	hyalodidymae	phaeodidymae
h helical	**≡** 2-multiseptate	phragmosporae	hyalophragmae	phaeophragmae
b branched	**#** muriform	dictyosporae	hyalodictyae	phaeodictyae
		scolecosporae		
		helicosporae		
		staurosporae		

Although more teleomorph/anamorph state connexions are being established, a permanent residue of unconnected conidial fungi is likely to remain. DNA sequencing makes it possible now to place these remaining taxa within the groups of teleomorphic fungi from which they are or were once derived. On morphological grounds this has already been done for some groups. It is traditional to treat anamorphs of the zygomycetes, *Erysiphales*, and *Pucciniales*, for example, in association with their teleomorphic states. The *Code* (see Nomenclature) provides for the use of separate names for the different states of pleomorphic fungi, but rules that the name of the holomorph (the whole fungus in all its correlated states) is that of the teleomorph. The *Code* also recommends that new names for anamorphs are not introduced when the telemorphic connection is firmly established and there is no practical need for separate names. Anamorphic fungi are some of the most frequently encountered fungi and many of them are of considerable economic significance.

Three morphological groups have been recognized that have in the past been named as classes:

(1) **Hyphomycetes** - mycelial forms which bear conidia on separate hyphae or aggregations of hyphae (as synnematous or sporodochial conidiomata) but not inside discrete conidiomata.

(2) **Agonomycetes** - mycelial forms which are sterile, but may produce chlamydospores, sclerotia and/or related vegetative structures.

(3) **Coelomycetes** - forms producing conidia in pycnidial, pycnothyrial, acervular, cupulate or stromatic conidiomata.

To recognize or delimit a taxonomic entity for the anamorphic fungi, such as subdivision Deuteromycotina, while convenient for practical purposes, is meaningless in terms of natural or phylogenetic classification. Therefore entries for anamorphic genera in this *Dictionary* assign them to the appropriate known level in the teleomorphic hierarchy. Informally, well-known groups of anamorphic genera, e.g. 'hyphomycetes' and 'coelomycetes', are likely to continue to be used but their adoption as formal taxa should be avoided. Integrated systems for Mitosporic fungi as a whole were suggested by Höhnel (1923) and Sutton (1980); see also Luttrell *in* Kendrick (1977). Arrangement of correlated anamorphs with ascomycete systematics has been reviewed by Kendrick & DiCosmo (*in* Kendrick (Ed.), *The whole fungus*: 283, 1979) and Sutton & Hennebert (*in* Hawksworth (Ed.), *Ascomycete systematics*: 77, 1994). For more information on the various approaches to the classification of anamorphic fungi see Sutton (*in* Sutton (Ed.), *A Century of Mycology*: 135, 1996).

Coding system in entries for anamorphic genera. Three categories of information are coded:

(i) **Conidiomatal types** listed in Table 1, e.g. Hso, indicates hyphal, Hsy, synnematal etc.

(ii) **Saccardo's spore groups**. Saccardo arranged 'imperfect' fungi (and also many ascomycetes, particularly those of the Sphaeriales) according to the septation or form of the spores and their colour – whether dark or hyaline – and the coined Latin names for these different groupings are set out in Table 1,

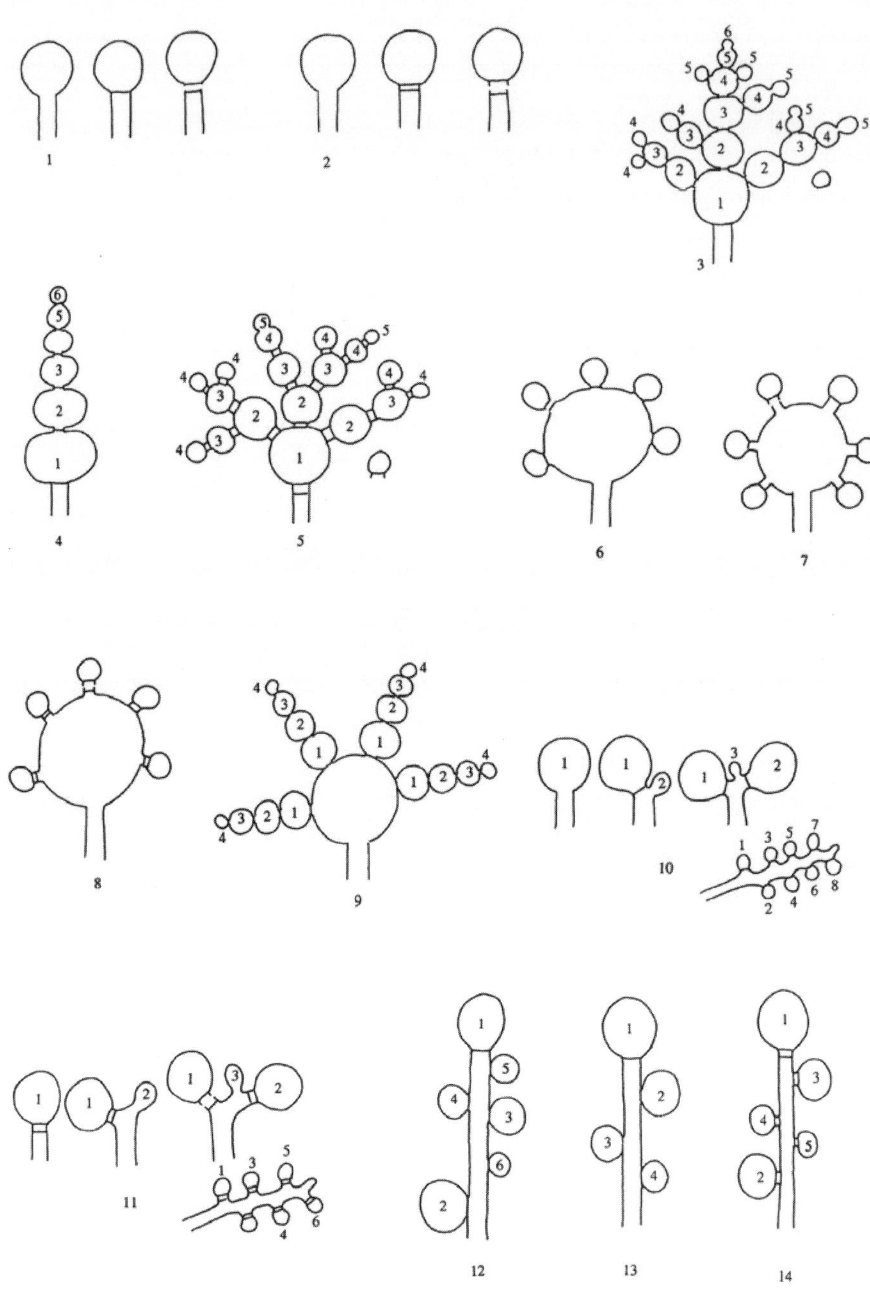

Fig. 1. Conidiogenous events (cc - conidiogenous cell). 1, conidial ontogeny holoblastic, 1 locus per cc, solitary conidia, delimited by 1 septum, maturation by diffuse wall-building, secession schizolytic, no proliferation of cc; 2, conidial ontogeny holoblastic, 1 locus per cc, solitary conidia, delimitation by 2 septa (or a separating cell), secession rhexolytic or by fracture of the cc, maturation by diffuse wall-building, no proliferation of cc; 3, conidial ontogeny holoblastic, apical wall-building random at more than one locus per cc and conidia becoming conidiogenous to form connected branched chains, each conidium delimited by 1 septum, maturation by diffuse wall-building, secession schizolytic, no cc proliferation; 4, conidial ontogeny holoblastic, apical wall-building at 1 locus per cc and each conidium with 1 locus to form a connnected unbranched chain, each conidium delimited by 1 septum, maturation by diffuse wall-building, secession schizolytic, no proliferation of cc; 5, conidial ontogeny holoblastic, apical wall-building randomly at more than 1 locus per cc and conidia becoming conidiogenous to form connected branched chains, each conidium delimited by 2 septa (or a separating cell), secession rhexolytic or by fracture of the cc, maturation by diffuse wall-building, no cc proliferation; 6, conidial ontogeny holoblastic, with localized apical wall-building simultaneously at different loci over the whole cc, each locus forming 1 conidium, delimited by 1 septum, maturation by diffuse wall-building, secession schizolytic, no cc proliferation; 7, conidial ontogeny holoblastic, with localized apical wall-building simultaneously at different loci on denticles over the whole cc, each locus forming 1 conidium, delimited by 1 septum, maturation by diffuse wall-building, secession by rupture of denticle, no cc proliferation; 8, conidial ontogeny holoblastic, with localized apical wall-building simultaneously at different loci over the whole cc, each conidium delimited by 2 septa (or a separating cell), secession rhexolytic or by fracture of the cc, each locus forming 1 conidium, maturation by diffuse wall-building, no cc proliferation; 9, conidial ontogeny holoblastic, apical wall-building simultaneously at several loci per cc and conidia becoming conidiogenous to form connected branched chains, each conidium delimited by 1 septum, maturation by diffuse wall-building, secession schizolytic, no cc proliferation; 10, conidial ontogeny holoblastic, regularly alternating with holoblastic sympodial cc proliferation, maturation by diffuse wall-building, each conidium delimitated by 1 septum, secession schizolytic; 11, conidial ontogeny holoblastic, regularly alternating with holoblastic sympodial cc proliferation, maturation by diffuse wall-building, each conidium delimited by 2 septa (or a separating cell), secession rhexolytic or by fracture of the cc; 12, conidial ontogeny holoblastic, each from apical or lateral loci, delimited by 1 septum, secession schizolytic, holoblastic cc proliferation sympodial or irregular, maturation by diffuse wall-building; 13, conidial ontogeny holoblastic, first from an apical locus, delimited by 1 septum, secession schizolytic, other conidia from lateral loci proceeding down the cc, maturation by diffuse wall-building; 14, conidial ontogeny holoblastic, first from an apical locus, each conidium delimited by 2 septa (or a separating cell), secession rhexolytic or by fracture of the cc, other conidia from lateral loci proceeding down the cc, maturation by diffuse wall-building.

e.g. e≡ H, indicates multiseptate hyaline conidia, hP, helical brown etc.

(iii) **Conidiogenous events**. The matrix system used is based on Minter *et al.* (*TBMS* **79**: 75, 1982; *TBMS* **80**: 38, 1983; *TBMS* **81**: 109, 1983) who showed a continuum of developmental processes associated with conidial production, including ontogeny, delimitation and secession of conidia and proliferation and regeneration of the cells bearing them (see conidiogensis). For the 43 combinations of events so far recognized see Figs 24-26, e.g. 15, indicates a succession of holoblastic conidial ontogeny, delimitation by a transverse septum, schizolytic secession, percurrent enteroblastic conidiogenous cell proliferation followed by holoblastic conidial ontogeny, successive conidia seceding at the same level.

Use of '?' means that insufficient information is available for the feature to be coded, and '-', that the feature is absent, e.g. 'Sc.-.-.' indicates presence of sclerotia but no conidia, and 'Cpd.e1P.?', that pycnidial conidiomata produce 1-septate brown conidia but their genesis is not known.

Lit.: General works on the anamorphic fungi include: Saccardo (*Syll. Fung.* **3**, **4**, **10**, **11**, **14**, **16**, **18**, **22**, **25**, **26**, 1884-1972), Lindau (*Naturlichen Pflanzenfam.*, 1900), Jaczewski (*Key to Fungi* **2**, *Fungi Imperfecti*, 1917), v. Höhnel (*Mykol. Unters.* **3**: 301-369, 1923), Clements & Shear (1931), Kendrick (Ed.) (*Taxonomy of Fungi Imperfecti*, 1971), Barnett & Hunter (*Illustrated genera of imperfect fungi*, 3 edn, 1972), Ainsworth *et al.* (Eds) (*The Fungi* **4**, 1973), Cole & Kendrick (*Biology of conidial fungi*, 1981), Minter *et al.* (*TBMS* **79**: 75, 1982; **80**: 39, 1983; **81**: 109, 1983), Stewart *et al.* (*Deuteromycotina and selected Ascomycotina from wood and wood products*, 1988; bibliogr. and guide to taxonomic lit.), Wilken-Jensen & Gravesen (*Atlas of moulds in Europe causing respiratory allergy*, 1984), Matsumoto & Ajello (*Handb. Appl. Mycol.: Humans, animals & insects* **2**: 117, 1991; dematiaceous fungi pathogenic to humans and lower animals), Campbell (*Handb. Appl. Mycol.: Humans, animals & insects* **2**: 395, 1991; conidiogenesis in fungi pathogenic to man and animals), McGinnis *et al.* (*Jl Med. Vet. Mycol.* **30**(Suppl. 1): 261, 1992), Howard (Ed.) (*Fungi pathogenic for humans and animals* A, 1993), Reynolds & Taylor (Eds), *The fungal holomorph*, 1993), Kiffer & Morelet (*The Deuteromycetes*, 2000), Seifert & Gams (*in* MacLaughlin *et al.* (Eds), *The Mycota* **VIIA**: 307, 2001). See also under *Coelomycetes* and *Hyphomycetes*.

Anamylopsora Timdal (1991), Anamylopsoraceae (L). 1, widespread. See Timdal (*Mycotaxon* **42**: 250, 1991), Lumbsch *et al.* (*Pl. Syst. Evol.* **198**: 275, 1995; fam.), Döring & Lumbsch (*Lichenologist* **30**: 489, 1998; ontogeny), Lumbsch *et al.* (*MR* **105**: 16, 2001; phylogeny), Lumbsch *et al.* (*MR* **105**: 265, 2001; asci), Peršoh *et al.* (*Mycol. Progr.* **3**: 103, 2004; asci).

Anamylopsoraceae Lumbsch & Lunke (1995), Ostropomycetidae (inc. sed.) (L). 1 gen., 1 spp.

Lit.: Timdal (*Mycotaxon* **42**: 250, 1991), Huneck & Elix (*Herzogia* **9**: 647, 1993), Lumbsch *et al.* (*Pl. Syst. Evol.* **198**: 275, 1995), Lumbsch (*J. Hattori bot. Lab.* **83**: 1, 1997), Döring & Lumbsch (*Lichenologist* **30**: 489, 1998), Lumbsch *et al.* (*MR* **105**: 16, 2001), Lumbsch *et al.* (*MR* **105**: 265, 2001), Lumbsch *et al.* (*MR* **111**: 1133, 2007).

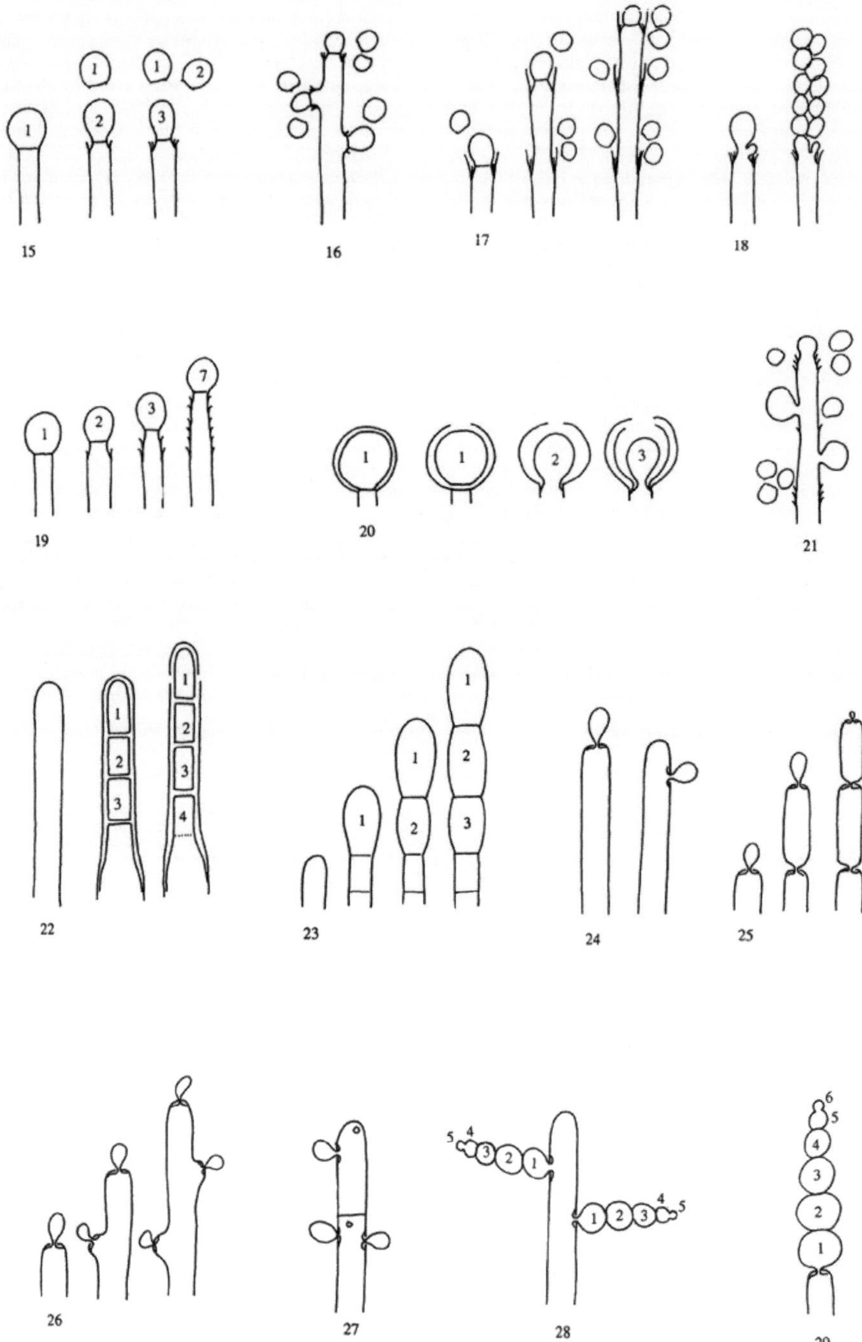
1
1
2
1
2
3
15
16
17
18
1
2
3
7
19
1
1
2
3
20
21
1
2
3
1
2
3
4
22
1
1
2
1
2
3
23
24
25
26
27
5
4
3
2
1
1
2
3
4
5
28
6
5
4
3
2
1
29

Fig. 2. Conidiogenous events (cc - conidiogenous cell). 15, conidial ontogeny holoblastic, delimitation by 1 septum, schizolytic secession, maturation by diffuse wall-building, percurrent enteroblastic cc proliferation followed by conidial ontogeny by replacement apical wall-building, successive conidia seceding at the same level, sometimes in unconnected chains, collarette variable; 16, same as 15 but with several random or irregular conidiogenous loci to each cc; 17, conidial ontogeny holoblastic, delimitation by 1 septum, schizolytic secession, maturation by diffuse wall-building, percurrent enteroblastic cc proliferation followed by conidial ontogeny by replacement apical wall-building, successive conidia seceding at the same level, collarette variable, conidiogenous activity interspersed periodically with percurrent vegetative proliferation; 18, conidial ontogeny holoblastic, delimitation by 1 septum, schizolytic secession, maturation by diffuse wall-building, percurrent and sympodial enteroblastic cc proliferation followed by conidial ontogeny by replacement apical wall-building, successive conidia seceding at the same level, collarette variable; 19, conidial ontogeny holoblastic, delimitation by 1 septum, schizolytic secession, maturation by diffuse wall-building, percurrent enteroblastic cc proliferation followed by conidial ontogeny by replacement apical wall-building, successive conidia seceding at progressively higher levels, sometimes in unconnected chains, collarette variable; 20, conidial ontogeny enteroblastic, delimitation by 1 septum, schizolytic secession, maturation by diffuse wall-building, outer wall of the cc remaining as a conspicuous collarette, percurrent enteroblastic cc proliferation followed by conidial enteroblastic ontogeny by replacement apical wall-building, successive conidia seceding at the same level, a succession of collarettes formed; 21, combination of 10, 12 and 19, where the sequences occur at random, irregularly or interchangeably; 22, conidial ontogeny holoblastic with new inner walls constituting the conidia laid down retrogressively by diffuse wall-building, delimitation retrogressive, loss of apical wall-building followed by replacment ring wall-building at the base of the cc adding more retrogressively delimited conidia, the outer (original) cc wall breaks as a connected chain of conidia is formed, collarette variable, 1 locus per cc, secession schizolytic; 23, conidial ontogeny holoblastic, 1 locus per cc, first conidium delimited by 1 septum, maturation by diffuse wall-building, loss of apical wall-building, replaced by ring wall-building below the delimiting septum which produces conidia in a connected unbranched chain, secession schizolytic, no proliferation of cc; 24, conidial ontogeny holoblastic, simultaneous with minimal enteroblastic percurrent proliferation at the preformed pore in the outer cc wall, conidia solitary, delimited by 1 septum, secession schizolytic, maturation by diffuse wall-buiilding, 1 locus per cc; 25, conidial ontogeny holoblastic, simultaneous with minimal enteroblastic percurrent proliferation at the preformed pore in the outer cc wall, conidia solitary, delimited by 1 septum, secession schizolytic, maturation by diffuse wall-buiilding, after one conidium formed extensive enteroblastic percurrent proliferation by apical wall-building occurs until the next apical locus is formed; 26, same as 24 but with holoblastic sympodial proliferation of the cc with conidiogenesis occurring between loci; 27, same as 24 but with several conidiogenous loci produced in the apical cc and laterally below septa in other ccs constituting the conidiophore; 28, same as 24 but several loci to each cc and first and subsequent conidia becoming conidiogenous by apical wall-building to form unbranched connected chains; more than one locus to a conidium will produce branched chains; 29, same as 24 but first conidium becoming conidiogenous by apical wall-building to form an unbranched connected chain.

anaphylaxis, manifestation of a change (immediate hypersensitivity) in a living animal from the uniting of an antibody with its antigen which may result in the death of the animal; cf. allergy.

anaphysis, a thread-like conidiophore persisting in apothecia of *Ephebe*.

Anaphysmene Bubák (1906), anamorphic *Pezizomycotina*, Cac.1eH.19. 2, Europe; Guatemala. See Sutton (*TBMS* **59**: 285, 1972), Sutton & Hodges (*Mycol.* **82**: 313, 1990), Mel'nik (*Opredelitel' Gribov Rossii* Klass Coelomycetes Byp. **1**. Redkie i Maloizvestnye Rody, 1997).

Anaptychia Körb. (1848), Physciaceae (L). *c*. 11, widespread. See also *Heterodermia*. See Kurokawa (*Beih. Nova Hedwigia* **6**, 1962), Poelt (*Nova Hedwigia* **9**: 21, 1965), Kurokawa (*J. Hattori bot. Lab.* **37**: 563, 1973), Swinscow & Krog (*Lichenologist* **8**: 103, 1976; Africa), Kashiwadani *et al.* (*Bull. natn. Sci. Mus.* Tokyo, B **16**: 147, 1990; chemistry, 23 spp., Peru), Heibel *et al.* (*Schriftenreihe der Landesanstalt für Ökologie, Bodenordnung und Forsten/Landesamt für Agrarordnung* **17**: 225, 1999; conservation, Germany), Lohtander *et al.* (*Mycol.* **92**: 728, 2000; Fennoscandia), Dahlkild *et al.* (*Bryologist* **104**: 527, 2001; photobionts), Grube & Arup (*Lichenologist* **33**: 63, 2001; phylogeny), Nordin & Mattsson (*Lichenologist* **33**: 3, 2001; phylogeny), Scheidegger *et al.* (*Lichenologist* **33**: 25, 2001; evolution), Helms *et al.* (*Mycol.* **95**: 1078, 2003; phylogeny), Peršoh *et al.* (*Mycol. Progr.* **3**: 103, 2004; asci), Miądlikowska *et al.* (*Mycol.* **98**: 1088, 2006; phylogeny), Esslinger (*Bryologist* **110**: 788, 2007; N America), Honegger & Zippler (*MR* **111**: 424, 2007; mating systems), Lohtander *et al.* (*Ann. bot. fenn.* **45**: 55, 2008; phylogeny).

Anaptychiaceae Körb. (1859) = Physciaceae.

Anaptychiomyces E.A. Thomas (1939) nom. inval. ≡ Anaptychia.

Anapyrenium Müll. Arg. (1880) nom. conf. = Buellia. p.p. fide Eriksson (*Op. Bot.* **60**, 1981).

Anarhyma M.H. Pei & Z.W. Yuan (1986), anamorphic *Pezizomycotina*, St.#eP.1. 1, China. See Pei & Yuan (*Bull. bot. Res.* Harbin **6**: 119, 1986).

Anariste Syd. (1927), Asterinaceae. 1, C. America. See Hosagoudar *et al.* (*Journal of Mycopathological Research* **39**: 61, 2001).

Anastomaria Raf. (1820) nom. rej. = Gyrodon fide Kuyper (*in litt.*).

anastomosing, joining irregularly to give a vein-like network.

anastomosis (pl. **anastomoses**), the fusion between branches of the same or different hyphae (or other structures) to make a network.

Anastomyces W.P. Wu, B. Sutton & Gange (1997), anamorphic *Basidiomycota*. 1 (fungicolous), China. See Wu *et al.* (*MR* **101**: 1318, 1997).

Anastrophella E. Horak & Desjardin (1994), Marasmiaceae. 3, New Zealand; Hawaii; Japan. See Horak & Desjardin (*Aust. Syst. Bot.* **7**: 162, 1994), Tanaka & Hongo (*Mycoscience* **42**: 433, 2001).

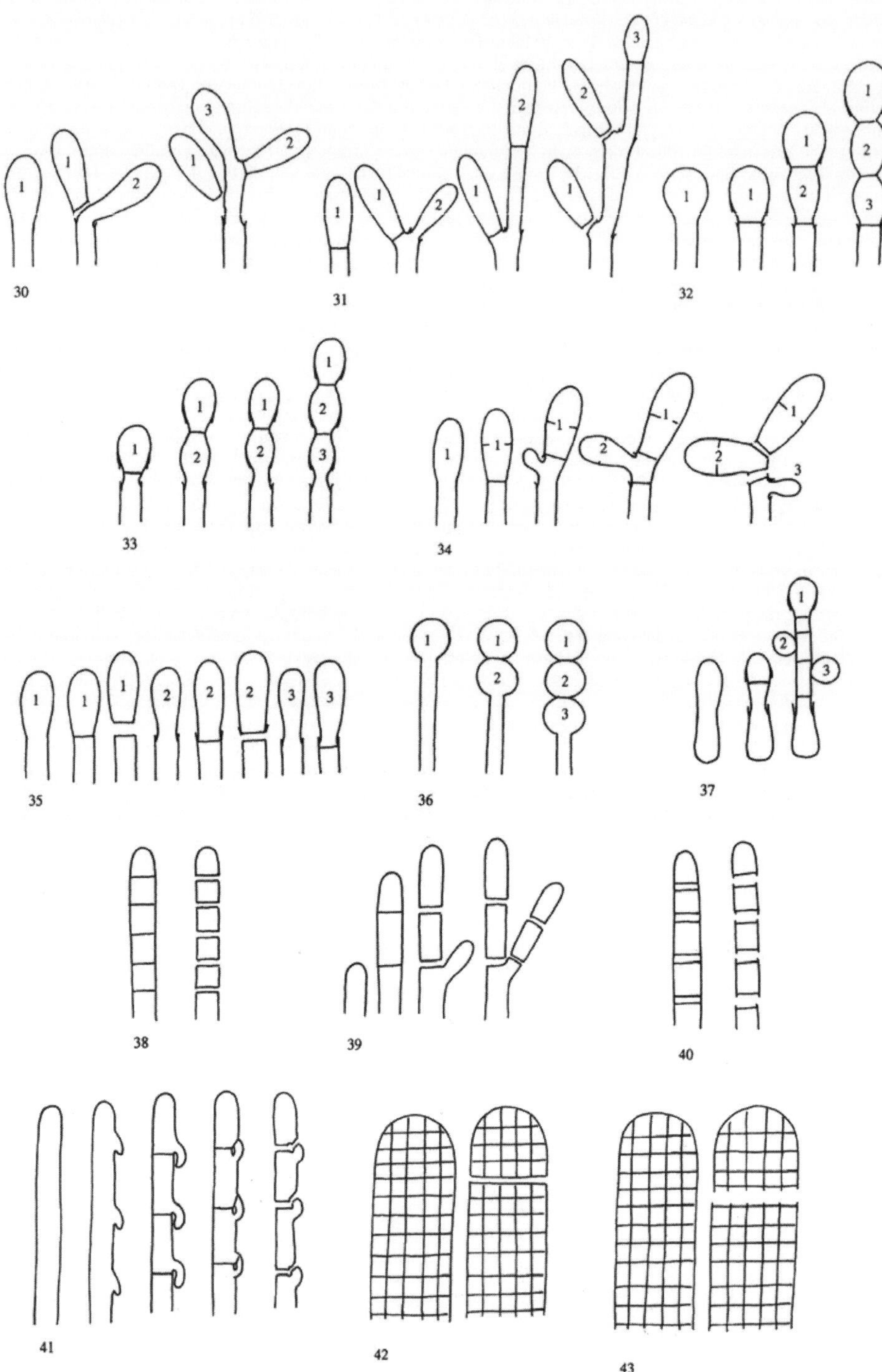
1
1
2
3
1
2
30
1
1
2
1
2
1
2
3
31
1
1
1
2
1
2
3
32
1
1
2
1
2
1
2
3
33
1
1
1
2
1
2
1
2
3
34
1
1
1
2
2
2
3
3
35
1
1
2
1
2
3
36
1
2
3
37
38
39
40
41
42
43

Fig. 3. Conidiogenous events (cc - conidiogenous cell). 30, conidal ontogeny holoblastic, delimitation by 1 septum, maturation by apical and diffuse wall-building, secession schizolytic and coincident with enteroblastic sympodial cc proliferation below the previous locus; subsequent conidia formed similarly but with holoblastic sympodial cc proliferation; 31, conidial ontogeny holoblastic, delimitation by 1 septum, maturation by apical and diffuse wall-building, secession schizolytic and coincident with enteroblastic sympodial cc proliferation below the previous conidiogenous locus, the sequence giving geniculate conidiophores; 32, conidial ontogeny holoblastic, with new inner walls continuous with all conidia laid down by diffuse wall-building, delimitation by 1 septum, loss of apical wall building followed by replacement continuous ring wall-building immediately below delimiting septum, the outer cc wall breaks between the first conidium and the cc to produce a variable collarette, followed by alternation of holoblastic conidial ontogeny by ring wall-building giving connected chains of conidia, maturation by diffuse wall-building, retrogressive delimitation, secession schizolytic; 33, conidial ontogeny holoblastic with new inner walls laid down by diffuse wall-building, delimitation by 1 septum, loss of apical wall-building followed by replacement ring wall-building immediately below delimiting septum, the outer cc wall breaks between the first conidium and the cc to produce a variable collarette, subsequent conidia formed by new inner walls for each conidium by ring wall- building giving connected chains of conidia, maturation by diffuse wall-building, retrogressive delimitation, secession schizolytic; 34, conidial ontogeny holoblastic, delimitation by 1 septum, secession schizolytic, enteroblastic sympodial cc proliferation below the previous locus and delimiting septum, the second and subsequent conidia formed from proliferations and delimited retrogressively, cc reduced in length with each conidium formed; 35, conidial ontogeny holoblastic, maturation by diffuse wall-building, delimitation by 1 septum, secession schizolytic, enteroblastic percurrent cc prolferation with retrogressive delimitation of next conidium, producing unconnected chains of conidia, the cc reduced in length with each conidium formed; 36, conidial ontogeny holoblastic, delimitation by 1 septum with loss of apical wall-building but replaced by diffuse wall-building below the previous conidium to form the next conidium which is retrogressively delimited giving an unconnected chain of conidia, secession schizolytic, cc reduced in length with each conidium formed; 37, conidial ontogeny holoblastic, delimitation by 1 septum with loss of apical wall-building, replaced by ring wall-building below the delimiting septum, outer wall of first conidium and cc breaks, followed by enteroblastic percurrent proliferation by ring wall-building, succeeding conidia holoblastic, delimited laterally and retrogressively, secession schizolytic, several loci per cc; 38, conidial ontogeny holothallic, ccs formed by apical wall-building coincident with conidial ontogeny, random delimitation by 1 septum at each end, no maturation during conidiogenesis, secession randomly schizolytic; 39, conidial ontogeny holothallic, ccs formed by apical wall-building coincident with conidial ontogeny, random delimitation by 1 septum at each end, no maturation during conidiogenesis, secession randomly schizolytic, cc proliferation holoblastic, irregular or sympodial, constituent cells conidiogenous; 40, same as 38 but conidial delimitaiton by 2 septa or separating cells at each end, secession rhexolytic; 41, conidal ontogeny holothallic, ccs formed in association with clamp connexions, random delimitation by septa in cc and the backwardly directed branch in the clamp connexion, maturation by diffuse and localized apical wall-building, secession randomly schizolytic, individual conidia comprised of part of the preceding and following clamp connexions; 42, conidial ontogeny holoblastic by simultaneous apical wall-building in adjacent cells, delimitation by septa in each of these cells, maturation by diffuse wall-building, secession simultaneous, multicellular, schizolytic, no cc proliferation; 43, conidial ontogeny holoblastic by simultaneous apical wall-building in adjacent cells, delimitation by septa in each of these cells, maturation by diffuse wall-building, followed by replacement apical wall-building in conidia to form additional conidia in connected chains, secession simultaneous, multicellular, rhexolytic, no cc proliferation

Anatexis Syd. (1928) = Englerula fide Müller & von Arx (*Beitr. Kryptfl. Schweiz* **11** no. 2, 1962).

Anatolinites Elsik, V.S. Ediger & Bati (1990), Fossil Fungi. 7 (Eocene – Holocene), widespread. See Elsik *et al.* (*Palynology* **14**: 92, 1990).

Anavirga B. Sutton (1975), anamorphic *Vibrissea*, Hso.0bP.1/10. 3, Europe. See Hamad & Webster (*Sydowia* **40**: 60, 1988), Descals (*MR* **109**: 545, 2005; conidia).

anbury, see club root.

Ancistrosporella G. Thor (1995), Roccellaceae (L). 3, Australia. See Thor (*Op. Bot.* **103**, 1990; as *Ancistrospora*), Egea *et al.* (*Mycotaxon* **59**: 47, 1996; New Guinea), Grube (*Bryologist* **101**: 377, 1998; phylogeny), Komposch *et al.* (*Lichenologist* **34**: 223, 2002; Venezuela, orthography).

Ancistrospora G. Thor (1991) [non *Ancistrospora* C.A. Menéndez & Azcuy 1972, fossil sporae-dispersae] ≡ Ancistroporella.

Anconomyces Cavalc. & A.A. Silva (1972) = Lyromma fide Lücking *et al.* (*Lichenologist* **30**: 121, 1998).

Ancoraspora Mig. Rodr. (1982), anamorphic *Pezizomycotina*, Hso.≡ eP.1. 1, Cuba. See Rodríguez Hernández (*Revta Jardín bot. Nac.* Univ. Habana **2**: 20, 1981), Mena Portales *et al.* (*MR* **102**: 736, 1998).

Ancorasporella J. Mena, Mercado & Heredia (1998), anamorphic *Pezizomycotina*, Hso.?.?. 1, Mexico. See Mena Portales *et al.* (*MR* **102**: 736, 1998).

Ancylistaceae J. Schröt. (1893), Entomophthorales. 3 gen. (+ 2 syn.), 45 spp.

Lit.: Wolf (*Nova Hedwigia* **46**: 121, 1988), Humber (*Mycotaxon* **34**: 441, 1989; emend.), Voigt *et al.* (*J. Clin. Microbiol.* **37**: 3957, 1999), Tanabe *et al.* (*Mol. Phylogen. Evol.* **30**: 438, 2004), Keller & Petrini (*Sydowia* **57**: 23, 2005), Tadano *et al.* (*Revta Soc. Bras. Med. Trop.* **38**: 188, 2005), Kędra & Boguś (*J. Invert. Path.* **91**: 50, 2006).

Ancylistales J. Schröt. (1893) = Entomophthorales.

Ancylistes Pfitzer (1872), Ancylistaceae. 5 (on *Closterium*), widespread (north temperate). See Berdan (*Mycol.* **30**: 396, 1938), Sparrow (*Aquatic Phycomycetes* Edn 2: 1065, 1960; key), Tucker (*Mycotaxon* **13**: 481, 1981; key).

Ancylospora Sawada (1944) = Pseudocercospora fide Deighton (*Mycol. Pap.* **140**, 1976), Crous & Braun

(*CBS Diversity Ser.* **1**: 571 pp., 2003).
Andebbia Trappe, Castellano & Amar. (1996), Mesophelliaceae. 1, Australia. See Trappe *et al.* (*Aust. Syst. Bot.* **9**: 808, 1996).
Andreaea Palm & Jochems (1923) [non *Andreaea* Hedw. 1801, *Musci*] ≡ Andreaeana.
Andreaeana Palm & Jochems (1924) = Acremonium. *fide* Gams (*in litt.*).
Andreanszkya Tóth (1968) = Podospora fide Lundqvist (*Symb. bot. upsal.* **20** no. 1, 1972).
androgynous, having the antheridium and its oogonium on one hypha; in de Bary's original sense (*Bot. Zeit.* **46**: 597, 1888) covers hypogynous, etc. Cf. monoclinous.
androphore, a branch forming antheridia, as in *Pyronema*.
Androsaceus (Pers.) Pat. (1887) = Marasmius fide Saccardo (*Syll. fung.* **5**: 1, 1887).
Anekabeeja Udaiyan & V.S. Hosag. (1992) ? = Pycnidiophora fide Eriksson & Hawksworth (*SA* **12**: 24, 1993), Korf (*Mycotaxon* **54**: 413, 1995; nomencl.).
Anellaria P. Karst. (1879) = Panaeolus fide Dennis *et al.* (*TBMS* **43**, 1960).
Anema Nyl. ex Forssell (1885) nom. cons., Lichinaceae (L). 13, widespread. See Moreno & Egea (*Acta Bot. Barcinon.* **91**: 1, 1992; key), McCune *et al.* (*Conservation and Management of Native Plants and Fungi* Proceedings of an Oregon Conference. Corvallis, Oregon, November 15-17, 1995: 234, 1997; conservation, Oregon), Schultz & Büdel (*Lichenologist* **34**: 39, 2002; key).
Anematidium Gronchi (1931) = Zasmidium fide Ciferri & Montemartini (*Atti Ist. bot. Univ. Lab. crittog. Pavia* sér. 5 **17**: 274, 1959).
anemophilous (of spores), taken about by air currents.
aneuploid, having a chromosome number which is not a multiple of the haploid set.
Angatia Syd. (1914), Saccardiaceae. 4 or 5, widespread (tropical).
Angelina Fr. (1849), Dermateaceae. 1, N. America. See Durand (*J. Mycol.* **8**: 108, 1906).
angio- (of a sporocarp), closed at least till the spores are mature. Cf. endo-, gymno-, hemi-angiocarpous, and cleistocarp.
angiocarpous (of a basidiome), hymenial surface at first exposed but later covered by an incurving pileus margin and/or excrescences from the stipe (Singer, 1975: 26); also used in a parallel way for *Ascomycota*.
Angiococcus E. Jahn (1924) nom. dub., ? Fungi. See Peterson & McDonald (*Mycol.* **58**: 962, 1967).
Angiophaeum Sacc. (1898) ≡ Phaeangium Pat.
Angiopoma Lév. (1841) nom. rej. = Drechslera fide Sutton (*Mycotaxon* **3**: 377, 1976).
Angiopomopsis Höhn. (1912), anamorphic *Pezizomycotina*, Cpd.≡ eP.19. 1, Java. See Sutton (*Česká Mykol.* **29**: 97, 1975), Farr *et al.* (*Mycol.* **90**: 290, 1998).
Angiopsora Mains (1934) = Phakopsora fide Ono *et al.* (*MR* **96**: 825, 1992) See.
Angiosorus Thirum. & M.J. O'Brien (1974) = Thecaphora fide Mordue (*Mycopathologia* **103**: 177, 1988).
Angiotheca Syd. (1939) = Dictyonella fide von Arx (*Persoonia* **2**: 421, 1963).
-angium (**-ange**, suffix), a structure having no opening; a cavity.
ang-kak (red rice), an Oriental food colouring obtained by growing *Monascus purpureus* on polished rice; see Fermented food and drinks.
Anguillomyces Marvanová & Bärl. (2000), anamorphic *Basidiomycota*. 1 (freshwater), Canada. See Marvanová & Bärlocher (*Mycotaxon* **75**: 411, 2000).
Anguillospora Ingold (1942), anamorphic *Pleosporales*, Hso.≡ eH.2. 11 (aquatic), widespread. See Petersen (*Mycol.* **54**: 117, 1962; key), Jooste & van der Merwe (*S. Afr. J. Bot.* **56**: 319, 1990; ultrastr.), Marvanová (*Tropical Mycology*: 169, 1997; tropical spp.), Kendrick (*CJB* **81**: 75, 2003; morphogenesis), Belliveau & Bärlocher (*MR* **109**: 1407, 2005; phylogeny), Descals (*MR* **109**: 545, 2005; diagnostic characters), Baschien *et al.* (*Nova Hedwigia* **83**: 311, 2006; phylogeny, morphology).
Anguillosporella U. Braun (1995), anamorphic *Mycosphaerellaceae*, Hso.?.?. 2 (on lving leaves), USA. See Redhead & White (*CJB* **63**: 1429, 1985; as *Anguillospora*), Braun (*Monogr. Cercosporella, Ramularia Allied Genera (Phytopath. Hyphom.)* **1**: 233, 1995).
anguilluliform, worm-like or eel-like in form.
angular septum, see septum.
Angulimaya Subram. & Lodha (1964), anamorphic *Bombardioidea*, Hso.0eH.19. 1 (coprophilous), India. See Subramanian & Lodha (*Antonie van Leeuwenhoek Ned. Tijdschr. Hyg.* **30**: 329, 1964), Krug & Scott (*CJB* **72**: 1302, 1994; connexion).
Angulospora Sv. Nilsson (1962), anamorphic *Pezizomycotina*, Hso.0fH.2. 1 (aquatic), Venezuela. See Nilsson (*Svensk bot. Tidskr.* **56**: 354, 1962), Goh (*Biodiversity of Tropical Microfungi*: 189, 1997), Marvanová (*Tropical Mycology*: 169, 1997).
Angusia G.F. Laundon (1964) = Maravalia fide Ono (*Mycol.* **76**: 892, 1984).
angustate, narrowed.
anheliophilous, preferring diffuse light. Cf. heliophilous.
Anhellia Racib. (1900), ? Myriangiaceae. 7, widespread (tropical). See von Arx (*Persoonia* **2**: 421, 1963), Barreto & Evans (*MR* **98**: 1107, 1994), Inácio & Dianese (*MR* **102**: 695, 1998), Pereira & Barreto (*Fungal Diversity* **12**: 155, 2003).
Animal mycophagists. Fungi, particularly basidiomycetes and larger ascomycetes, can form an important part of the diet of various mammals, including deer, pigs, rabbits, squirrels and various other rodents (Buller, *TBMS* **6**: 355, 1920; *Researches* **2**: 195, 1922; Hastings & Mottram, *TBMS* **5**: 364, 1916; Minter, *IMI Descriptions of Fungi and Bacteria*, Set 172, 2007). In the case of hypogeous fungi, this has evolved as mutualism, the feeding animal benefiting the fungus by dispersing its spores; the resulting digging and soil aeration carried out by mycophagist mammals in search of fruitbodies can contribute significantly to the dynamics of woodland and forest soils. Animal mycophagists and fungi may also have a role as mutualists in seed dispersal (Pirozynski & Malloch, *in* Pirozynski & Hawksworth (Eds), 1988: 227). Conservation studies in North Am. on the northern spotted owl demonstrated that fungi form a key element in the food chain supporting that highly endangered bird (Minter, *IMI Descriptions of Fungi and Bacteria*, Set 172, 2007). Some fungi accumulate radioactive pollutants sufficiently strongly to impact on the food chains they support (Hughman & Huchschlag, *European J. of Wildlife Res.* **51**: 263,

2005; Iceland moss). Lichens may form an important component of food for reindeer (see Reindeer lichen). Fungi are also consumed by invertebrates, particularly slugs (Elliott, *TBMS* **8**: 84, 1922), snails (*Polygyra thyroides*) (Wolf & Wolf, *Bull. Torrey bot. Cl.* **66**: 1, 1939) and arthropods (see Ambrosia fungi, Insects and fungi, Termite fungi). See also Coevolution; Fungi and radiation; Hypogeous fungi; Iceland moss.

Aniptodera Shearer & M.A. Mill. (1977), Halosphaeriaceae. 9 (aquatic and marine), widespread. See Shearer (*Mycol.* **81**: 139, 1989), Volkmann-Kohlmeyer & Kohlmeyer (*Bot. Mar.* **37**: 109, 1994; table chars 9 spp.), Chen *et al.* (*Mycol.* **91**: 84, 1999; DNA), Hyde *et al.* (*Mycoscience* **40**: 165, 1999), Kong *et al.* (*MR* **104**: 35, 2000; DNA), Hyde (*Cryptog. Mycol.* **23**: 5, 2002), Zhang *et al.* (*Mycol.* **98**: 1076, 2006; phylogeny).

aniso- (prefix), unequal.

Anisochora Theiss. & Syd. (1915) = Apiosphaeria fide Hawksworth *et al.* (*Dictionary of the Fungi* edn 8, 1995).

Anisochytridiales Karling (1943) = Hyphochytriales.

anisogamy, the copulation of gametes of unlike form or physiology, i.e. of **-gametes**; heterogamy; cf. isogamy.

Anisogramma Theiss. & Syd. (1917), Valsaceae. 3 (from bark), Europe; N. America. See Osterbauer *et al.* (*Phytopathology* **84**: 1150, 1994; DNA).

anisokont, having flagella of unequal length; heterokont.

Anisomeridium (Müll. Arg.) M. Choisy (1928) nom. cons., Monoblastiaceae (L). *c.* 100, widespread (esp. tropical). See Harris (*More Florida Lichens*, 1995; key 75 spp.), Harada (*Hikobia* **13**: 411, 2001), Komposch (*Lichenologist* **37**: 519, 2005), Aptroot *et al.* (*Biblthca Lichenol.* **97**, 2008; Costa Rica).

Anisomyces Pilát (1940) = Gloeophyllum fide Donk (*Persoonia* **1**: 173, 1960).

Anisomyces Theiss. & Syd. (1914), ? Valsaceae. 1, America (tropical). See Cannon (*Fungal Diversity* **7**: 17, 2001).

Anisomycopsis I. Hino & Katum. (1964), Diaporthales. 1, Japan. See Hino & Katumoto (*J. Jap. Bot.* **39**: 325, 1964).

anisospory, having spores of more than one kind.

Anisostagma K.R.L. Petersen & Jørg. Koch (1996), Halosphaeriaceae. 1 (marine), Denmark. See Petersen & Koch (*MR* **100**: 209, 1996).

Anisostomula Höhn. (1919) = Hyponectria fide Barr (*Mycol.* **68**: 611, 1976).

anisotomic dichotomic branching, branching where one dichotomy becomes stouter and forms a main stem so that the other branch of the dichotomy appears to be lateral, as in *Alectoria ochroleuca* ; cf. isotomic dichotomic branching.

Anixia Fr. (1819) nom. dub., Agaricomycetidae. ? 'gasteromycetes' fide Demoulin (*in. litt.*).

Anixia H. Hoffm. (1862) = Orbicula fide Hughes (*Mycol. Pap.* **42**, 1951).

Anixiella Saito & Minoura ex Cain (1961) = Neurospora fide von Arx (*Persoonia* **7**: 367, 1973), García *et al.* (*MR* **108**: 1119, 2004; phylogeny).

Anixiopsis E.C. Hansen (1897) = Aphanoascus fide Vries (*Mykosen* **12**: 111, 1969), Guého & de Vroey (*CJB* **64**: 2207, 1986; SEM ascospores), Cano & Guarro (*MR* **94**, 1990).

Ankistrocladium Perrott (1960) = Casaresia fide Ellis (*Dematiaceous Hyphomycetes*, 1971).

Ankultur, see Normkultur.

Annajenkinsia Thirum. & Naras. (1955) = Puttemansia fide Pirozynski (*Kew Bull.* **31**: 595, 1977).

Anncaliia I.V. Issi, S.V. Krylova & V.M. Nikolaeva (1993), Microsporidia. 4.

Annella S.K. Srivast. (1976), Fossil Fungi. 2 (Jurassic), British Isles.

annellate (of asci), ones with a thickened apical pore (e.g. *Leotiales*); see ascus; **annellations**; see annellidic.

annellidic (of conidiogenesis), holoblastic conidiogenesis in which the conidiogenous cell (**annellide**, annellophore) by repeated enteroblastic percurrent proliferation produces a basipetal sequence of conidia (**annelloconidia**, annellospores) leaving the distal end marked by transverse bands (**annellations**). See Conidial nomenclature.

Annellodentimyces Matsush. (1985), anamorphic *Pezizomycotina*, Hso.≡ eP.19. 1, Japan. See Matsushima (*Matsush. Mycol. Mem.* **4**: 2, 1985), Ho *et al.* (*Mycol.* **97**: 238, 2005).

Annellodochium Deighton (1969), anamorphic *Pezizomycotina*, Cpd.1eP.19. 1 (on *Diatrype*), Sierra Leone. See Deighton (*Mycol. Pap.* **118**: 28, 1969).

Annellolacinia B. Sutton (1964), anamorphic *Pezizomycotina*, Cac.0eP.19. 2, widespread (tropical). See Frölich *et al.* (*MR* **97**: 1433, 1993).

Annellophora S. Hughes (1952), anamorphic *Pezizomycotina*, Hso.≡ eP.19. 11, widespread (tropical). See Ellis (*Mycol. Pap.* **70**, 1958; key), Manoharachary *et al.* (*Indian Phytopath.* **58**: 454, 2005), Castañeda Ruíz *et al.* (*Mycotaxon* **96**: 151, 2006).

annellophore, see annellidic.

Annellophorella Subram. (1962), anamorphic *Pezizomycotina*, Hso.#eP.19. 1, S. Africa. See Subramanian (*Proc. Indian Acad. Sci.* series B **55**: 6, 1962).

Annellophragmia Subram. (1963), anamorphic *Pezizomycotina*, Hsy.≡ eP.19. 1, India. See Subramanian (*Proc. Indian Acad. Sci.* series B **58**: 349, 1963).

Annellospermosporella P.R. Johnst. (1999), anamorphic *Pezizomycotina*, Hso.?.?. 1, New Zealand. Probably synonymous with *Spermosporella*. See Johnston (*N.Z. Jl Bot.* **37**: 290, 1999).

Annellosympodia McTaggart, R.G. Shivas & U. Braun (2007), Pezizomycotina. 1, Australia. See McTaggart *et al.* (*Australas. Pl. Path.* **36**: 573, 2007).

annular, ring-like; ring-like arrangement.

Annularia (Schulzer) Gillet (1876) [non *Annularia* Sternb. 1825, fossil *Pteridophyta*] = Chamaeota fide Stalpers (*in litt.*).

Annularia Raf. (1815) nom. dub., Fungi. No spp. included.

Annularius Roussel (1806) = Coprinus fide Redhead *et al.* (*Mycotaxon* **50**: 203, 2001).

Annulatascaceae S.W. Wong, K.D. Hyde & E.B.G. Jones (1998), Sordariomycetidae (inc. sed.). 21 gen. (+ 5 syn.), 75 spp.

Lit.: Wong *et al.* (*SA* **16**: 17, 1998), Ho *et al.* (*Mycol.* **91**: 885, 1999), Ho *et al.* (*Fungal Diversity* **3**: 87, 1999), Ranghoo *et al.* (*Fungal Diversity* **2**: 159, 1999), Ho & Hyde (*Fungal Diversity* **4**: 21, 2000), Inderbitzin (*Mycoscience* **41**: 167, 2000), Campbell & Shearer (*Mycol.* **96**: 822, 2004), Réblová (*Mycol.* **98**: 68, 2006).

Annulatascus K.D. Hyde (1992), Annulatascaceae. 14 (wood, aquatic), Australia. See Hyde (*Aust. Syst. Bot.* **5**: 117, 1992), Wong *et al.* (*SA* **16**: 17, 1998), Wong *et al.* (*MR* **103**: 561, 1999; ultrastr.), Tsui *et al.* (*Mycoscience* **43**: 383, 2002), Campbell & Shearer (*Mycol.* **96**: 822, 2004), Huhndorf *et al.* (*Mycol.* **96**: 368, 2004; phylogeny).

Annulohypoxylon Y.M. Ju, J.D. Rogers & H.M. Hsieh (2005), Xylariaceae. 27, widespread. See Ju & Rogers (*Mycol. Mem.* **20**: 365 pp., 1996; as *Hypoxylon* sect. *Annulata*), Ju *et al.* (*Mycol.* **97**: 855, 2005), Bitzer *et al.* (*MR* **112**: 251, 2008; phylogeny, chemistry).

annulus (1) (of basidiomata), a ring-like partial veil, or part of it, round the stipe after expansion of the pileus (Fig. 4C); hymenial veil; apical veil; ring; an - near the top of the stipe is **superior** (an **armilla**, fide Gäumann & Dodge, 1928: 453), one lower down, **inferior**; (2) (in *Papulospora*), the ring of cells around a bulbil; (3) (of asci), the apical ring; anneau apicale; (4) (in *Alternaria*), thickening in apices of conidiogenous cells, fide Campbell (*Arch. Mikrobiol.* **69**: 60, 1970).

Annulusmagnus J. Campb. & Shearer (2004), Annulatascaceae. 1 (on submerged wood), Australia; N. America; Venezuela. See Campbell & Shearer (*Mycol.* **96**: 826, 2004), Zhang *et al.* (*Mycol.* **98**: 1076, 2006; phylogeny).

anoderm, having no skin.

Anodotrichum (Corda) Rabenh. (1844) = Blastotrichum fide Saccardo (*Syll. fung.* **4**: 1, 1886).

Anomalemma Sivan. (1983), ? Melanommataceae. Anamorph *Exosporiella*. 1, Europe. See Sivanesan (*TBMS* **81**: 313, 1983).

Anomalographis Kalb (1992), Graphidaceae (L). 1, Madeira. See Kalb & Hafellner (*Herzogia* **9**: 49, 1992), Staiger (*Biblthca Lichenol.* **85**, 2002).

Anomalomyces Vánky, M. Lutz & R.G. Shivas (2006), ? Ustilaginaceae. 1 (on *Panicum trachyrhachis* (*Poaceae*)), Australia. See Vánky *et al.* (*Mycol. Balcanica* **3**: 120, 2006).

Anomoloma Niemelä & K.H. Larss. (2007), Fomitopsidaceae. 4, widespread. See Niemelä & Larsson (*Mycotaxon* **100**: 312, 2007).

Anomomorpha Nyl. ex Hue (1891), Graphidaceae (L). 5, pantropical. See Hawksworth *et al.* (*Dictionary of the Fungi* edn 8, 1995), Archer (*Systematics & Biodiversity* **5**: 9, 2007; Solomon Is).

Anomomyces Höhn. (1928) nom. dub., anamorphic *Pezizomycotina*. See Sutton (*Mycol. Pap.* **138**, 1975).

Anomoporia Pouzar (1966), ? Fomitopsidaceae. 8, north temperate. See Pouzar (*Česká Mykol.* **20**: 172, 1966).

Anomothallus F. Stevens (1925) nom. dub., Fungi. See Petrak (*Sydowia* **5**: 328, 1951).

Anopeltis Bat. & Peres (1960), ? Capnodiaceae. 1, Venezuela. See Batista & Peres (*Nova Hedwigia* **2**: 472, 1960).

Anopodium N. Lundq. (1964), Lasiosphaeriaceae. 2, Europe (northern). See Mirza & Cain (*CJB* **47**: 1999, 1969; ? = *Podospora*).

Ansatospora A.G. Newhall (1944) nom. inval. = Mycocentrospora fide Deighton (*Taxon* **21**: 716, 1972).

Anserina Velen. (1934) [non *Anserina* Dumort. 1827, *Chenopodiaceae*] = Ascobolus fide Eckblad (*Nytt Mag. Bot.* **15**: 1, 1968).

antabuse, tetraethylthiuramdisulphate (disulfiram); after ingestion reacts with alcohol to give unpleasant symptoms; used in the treatment of chronic alcoholism; see coprine.

antagonism, a general name for associations of organisms damaging to one or more of the associates (cf. antibiosis, symbiosis). Though parasitism is an example of antagonism, the term is used esp. for the effects of toxic metabolic products (see Staling substances) or of undetermined causes on fungi and bacteria in competition. Much experimental work has been done on the antagonism between bacteria, bacteria and fungi, and fungi; and esp. on the competition between microorganisms in the soil; for example, on the effect of saprobic soil fungi on pathogenic species, e.g. *Trichoderma viride* on *Rhizoctonia*, *Pythium*, and other damping-off fungi.

Lit.: Waksman (*Soil Sci.* **43**: 51, 1937; *Bact. Rev.* **5**: 231, 1941); Porter & Carter, and Weindling (*Bot. Rev.* **4**: 165, 475, 1938) give long reference lists, and Hawksworth (*in* Cole & Kendrick, *Biology of conidial fungi* **1**: 171, 1981) more recent ones; Moreau & Moreau (*BSMF* **72**: 250, 1956) (types of association and antagonism). Cf. antibiotic substances.

antarctic mycology, see Polar and alpine mycology.

Antarctomia D.C. Linds. (1975) = Placynthium fide Henssen (*Lichenologist* **13**: 307, 1981).

Antarctomyces Stchigel & Guarro (2001), Thelebolaceae. Anamorph *Sporothrix*-like. 1, Antarctica. See Stchigel *et al.* (*MR* **105**: 377, 2001), Hoog *et al.* (*Stud. Mycol.* **51**: 33, 2005).

Antenaglium F.C. Albuq. (1969) = Gliocephalotrichum fide Carmichael *et al.* (*Genera of Hyphomycetes*, 1980).

Antennaria Link (1809) [non *Antennaria* Gaertn. 1791, *Compositae*] = Antennularia fide Hawksworth *et al.* (*Dictionary of the Fungi* edn 8, 1995).

Antennariella Bat. & Cif. (1963), anamorphic *Antennulariella*, Cpd.0eH.?. *c.* 5, widespread (tropical). See Hughes (*Mycol.* **68**: 693, 1976), Sutton (*Mycol. Pap.* **141**, 1977), Hughes (*CJB* **78**: 1215, 2000).

Antennataria Rchb. (1841) ≡ Antennularia.

Antennatula Fr. ex F. Strauss (1850), anamorphic *Euantennaria*, Hso.≡ eP.1. 8, widespread. See Hughes (*N.Z. Jl Bot.* **12**: 299, 1974), Hughes & Arnold (*Mem. N. Y. bot. Gdn* **49**: 198, 1989).

Antennella Theiss. & Syd. (1918) = Scorias fide von Arx & Müller (*Stud. Mycol.* **9**, 1975).

Antennellina J.M. Mend. (1925) ? = Scorias fide von Arx & Müller (*Stud. Mycol.* **9**, 1975).

Antennellopsis J.M. Mend. (1930) = Phragmocapnias fide Reynolds (*Mycotaxon* **8**: 917, 1979).

Antennina Fr. (1849) ≡ Antennularia.

Antennopsis R. Heim (1952), anamorphic *Pezizomycotina*, Hso.≡ eP.1. 1 (on termites), Europe (southern); Florida. See *Gloeohaustoriales*. See Rossi & Blackwell (*Mycol.* **82**: 138, 1990).

Antennospora Meyers (1957), Halosphaeriaceae. 2 (marine), widespread. See Jones *et al.* (*Bot. Mar.* **27**: 129, 1984), Yusoff *et al.* (*MR* **98**: 997, 1994; ultrastr.).

Antennula, see *Antennatula*.

Antennularia Rchb. (1828), Venturiaceae. *c.* 30, widespread. See also *Protoventuria*. See Müller & von Arx (*Beitr. Kryptfl. Schweiz* **11** no. 2, 1962), Hughes (*N.Z. Jl Bot.* **8**: 156, 1970; as nom. dub.), Hughes & Seifert (*Sydowia* **50**: 192, 1998).

Antennulariaceae Locq. (1984) = Venturiaceae.

Antennulariella Woron. (1915), Antennulariellaceae. Anamorphs *Antennariella*, *Capnodendron*. 4, widespread. See Hughes (*Mycol.* **68**: 693, 1976), Reynolds (*Mycotaxon* **27**: 377, 1986), Hughes (*CJB* **78**: 1215, 2000), Hughes (*Mycol.* **99**: 628, 2007).

Antennulariellaceae Woron. (1925), Capnodiales. 6 gen. (+ 3 syn.), 27 spp.
Lit.: Hughes (*Mycol.* **68**: 693, 1976; gen. names, anamorphs), Reynolds (*Mycotaxon* **27**: 377, 1986; status), Reynolds (*CJB* **76**: 2125, 1998; phylogeny), Barr & Rogerson (*Mycotaxon* **71**: 473, 1999), Hughes (*CJB* **78**: 1215, 2000).

anterior (1) at or in the direction of the front; (2) (of lamellae), the end at the edge of the pileus.

Anthasthoopa Subram. & K. Ramakr. (1956) = Coniella fide Sutton (*CJB* **47**: 603, 1969).

antheridiol, a sex hormone (sterol) of *Achlya bisexualis* which induces antheridial formation in male strains of *Achlya* (McMorris & Barksdale, *Nature* **215**: 320, 1967; Barksdale, *Science* **166**: 831, 1969).

antheridium (pl. **-a**, **antherid**), the male gametangium, either formed from a haplophase thallus, or in which meiosis occurs after delimitation.

antherozoid, a motile male cell; a sperm.

Anthina Fr. (1832), anamorphic *Pezizomycotina*, sterile. 5, widespread (temperate). *A. citri* and *A. brunnea* ('leaf felt' in *Citrus*). See Treu & Rambold (*Mycotaxon* **45**: 71, 1992; possible link with *Cordyceps*).

Anthoblastomyces Verona & Zardetta (1954) nom. inval., anamorphic *Pezizomycotina*.

Anthomyces Dietel (1899), Raveneliaceae. 1 (on *Leguminosae*), Brazil. See Araujo *et al.* (*Fitopatol. Brasil* **30**: 510, 2005; Brazil).

Anthomyces Grüss (1918) = Metschnikowia fide von Arx *et al.* (*Stud. Mycol.* **14**: 1, 1977).

Anthomycetella Syd. & P. Syd. (1916), ? Raveneliaceae. 1 (on *Canarium* (*Burseraceae*)), Philippines.

Anthopeziza Wettst. (1885) = Microstoma Bernstein fide Eckblad (*Nytt Mag. Bot.* **15**: 1, 1968).

Anthopsis Fil. March., A. Fontana & Luppi Mosca (1977), anamorphic *Pezizomycotina*, Hso.0eP.15. 3, Europe; Japan. See Bonfante-Fasolo & Marchisio (*Allionia* **23**: 13, 1970; ultrastr. phialide), Ando & Tubaki (*TMSJ* **26**: 151, 1985; Japan).

Anthoseptobasidium Rick (1943) nom. dub., Agaricomycotina.

Anthostoma Nitschke (1867) = Cryptosphaeria Ces. & De Not. fide Eriksson (*Svensk bot. Tidskr.* **60**: 315, 1966), Rappaz (*Mycol. Helv.* **5**: 21, 1992), Læssøe & Spooner (*Kew Bull.* **49**: 1, 1994).

Anthostomaria (Sacc.) Theiss. & Syd. (1918), Pezizomycotina. 1 (on *Umbilicaria*), former USSR.

Anthostomella Sacc. (1875), ? Xylariaceae. 133, widespread. See Eriksson (*Svensk bot. Tidskr.* **60**: 315, 1966), Francis (*Mycol. Pap.* **139**, 1975; key 30 Eur. spp.), Rappaz (*Mycol. Helv.* **7**: 99, 1995; on hardwoods, Eur., N. Am.), Hyde (*Nova Hedwigia* **62**: 273, 1996; on palms), Lu *et al.* (*Fungal Diversity* **3**: 99, 1999; Australia), Lu & Hyde (*Mycotaxon* **74**: 379, 2000; Portugal), Lu & Hyde (*Mycoscience* **41**: 223, 2000; Brunei), Lu & Hyde (*Fungal Diversity Res. Ser.* **4**, 2000; monogr.), Lu *et al.* (*MR* **104**: 742, 2000; S. Afr.), Davis *et al.* (*Am. J. Bot.* **90**: 1661, 2003; endophytes), Lee & Crous (*MR* **107**: 360, 2003; S Africa), Zhang *et al.* (*Mycol.* **98**: 1076, 2006; phylogeny).

Anthostomellina L.A. Kantsch. (1928), Pezizomycotina. 1, former USSR.

anthracnose, a plant disease having characteristic limited lesions, necrosis, and hypoplasia, generally caused by one of the acervular coelomycetes. See Jenkins (*Phytopathology* **23**: 389, 1933); **spot -**, a disease caused by *Elsinoë* or its anamorph *Sphaceloma* (Jenkins; see *RAM* **26**: 255, 1947).

Anthracobia Boud. (1885), Pyronemataceae. Anamorph *Scytalidium*-like. *c.* 15, widespread (north temperate). See Delattre-Durand & Parguey-Leduc (*BSMF* **95**: 355, 1979; ontogeny), Hohmeyer & Schnacketz (*Beitr. Kenntn. Pilze Mitteleur.* **3**: 427, 1987; key 9 spp.), Yao & Spooner (*MR* **99**: 1519, 1995; Brit. spp.), Yao *et al.* (*Mycologist* **12**: 32, 1998; key Brit. spp.), Hansen & Pfister (*Mycol.* **98**: 1029, 2006; phylogeny), Perry *et al.* (*MR* **111**: 549, 2007; phylogeny).

anthracobiontic, obligately inhabiting burnt areas; **anthracophilous**, sporulation favoured by burnt areas (*see* Pyrophilous fungi); **anthracophobic**, sporulation suppressed or checked on burnt areas; **anthracoxenous**, incidence and growth not affected by burnt areas (Moser, 1949).

Anthracocarpon Breuss (1996), Verrucariaceae (L). 2, Europe. See Breuss (*Annln naturh. Mus. Wien* Ser. B, Bot. Zool. **98**: 40, 1996).

Anthracocystis Bref. (1912) = Sporisorium fide Vánky (*in litt.*).

Anthracoderma Speg. (1888), anamorphic *Pezizomycotina*, St.0eH.?. 3, S. America. See Petrak & Sydow (*Annls mycol.* **33**: 188, 1935).

Anthracoidea Bref. (1895), Anthracoideaceae. Anamorph *Crotalia*. *c.* 75 (in seeds of *Cyperaceae*), widespread (esp. northern hemisphere). See Kukkonen (*Ann. bot. Soc. Zool.-Bot. Fenn. Vanamo* **34** no. 3, 1963), Kukkonen (*Ann. bot. fenn.* **1**: 161, 1964; keys), Kukkonen (*TBMS* **47**: 273, 1964; spore germination), Kukkonen (*Ann. bot. fenn.* **1**: 257, 1964; homothallism), Braun & Hirsch (*Feddes Repert.* **89**: 43, 1978; keys), Nannfeldt (*Symb. bot. upsal.* **22** no. 3: 1, 1979; 34 Nordic spp.), Vánky (*Bot. Notiser* **132**: 221, 1979; species concepts, 1987), Ingold (*MR* **92**: 245, 1989; spore germination, posn), Salo & Sen (*CJB* **71**: 1406, 1993; isoenzyme analysis), Hendrichs *et al.* (*MR* **109**: 31, 2005; molecular phylogenetic approach).

Anthracoideaceae Denchev (1997), Ustilaginales. 20 gen. (+ 7 syn.), 198 spp.
Lit.: Vánky (*TBMS* **89**: 61, 1987), Vánky (*Cryptog. Stud.* **1**: 159 pp., 1987), Ingold (*MR* **92**: 245, 1989), Vánky (*Europ. Smut Fungi*: 570 pp., 1994), Vánky & Oberwinkler (*Nova Hedwigia* Beih. **107**: 96 pp., 1994), Ingold (*MR* **99**: 140, 1995), Piepenbring (*CJB* **73**: 1089, 1995), Vánky (*Mycotaxon* **54**: 215, 1995), Vánky & Websdane (*Mycotaxon* **56**: 217, 1995), Bauer *et al.* (*CJB* **75**: 1273, 1997), Denchev (*Mycotaxon* **65**: 411, 1997), Vánky (*Mycotaxon* **63**: 143, 1997), Begerow *et al.* (*CJB* **75**: 2045, 1998), Ingold (*MR* **103**: 1071, 1999), Piepenbring *et al.* (*Mycol.* **91**: 485, 1999), Vánky (*Mycotaxon* **70**: 17, 1999), Piepenbring (*Nova Hedwigia* **70**: 289, 2000), Vánky (*Mycotaxon* **74**: 343, 2000), Piepenbring (*Bot. Jb.* **24**: 241, 2003), Begerow *et al.* (*MR* **108**: 1257, 2004), Vánky (*Mycol. Balcanica* **1**: 175, 2004), Hendrichs *et al.* (*MR* **109**: 31, 2005), Stoll *et al.* (*MR* **109**: 342, 2005).

Anthracomyces Renault (1898), Fossil Fungi (mycel.) Fungi. 2 (Carboniferous), France.

Anthracophlous Mattir. ex Lloyd (1913) = Rhizopogon fide Stalpers (*in litt.*).

Anthracophyllum Ces. (1879), Marasmiaceae. 10, widespread (tropical). See Pegler & Young (*MR* **93**: 352, 1989; key).

Anthracostroma Petr. (1954), Dothideomycetes. Anamorph *Camarosporula*. 1, Australia. See Petrak (*Sydowia* **8**: 96, 1954).

Anthracothecium Hampe ex A. Massal. (1860), Pyrenulaceae (L). *c.* 29, widespread (esp. tropical). See Johnson (*Ann. Mo. bot. Gdn* **27**: 1, 1940), Singh (*Feddes Repert.* **93**: 67, 1982), Singh & Raychaudhury (*New Botanist* **9**: 32, 1983; India), Singh (*Geophytology* **14**: 69, 1984), Singh (*Geophytology* **15**: 98, 1985), Harris (*Mem. N. Y. bot. Gdn* **49**: 74, 1989; key 5 N. Am. spp.), Aptroot (*Australasian Lichenology* **60**: 34, 2007; key Australian spp.), Aptroot *et al.* (*Biblthca Lichenol.* **97**, 2008; Costa Rica).

Anthracothecomyces Cif. & Tomas. (1953) = Pyrenula Ach. (1814) fide Harris (*Mem. N. Y. bot. Gdn* **49**, 1989).

Anthropomorphus Seger (1745) nom. inval. = Geastrum fide Stalpers (*in litt.*) Used by Lloyd but see, Donk (*Reinwardtia* **1**: 205, 1951).

anthropophilic (of dermatophytes, etc.), preferentially pathogenic for man. Cf. zoophilic.

Anthurus Kalchbr. & MacOwan (1880) = Clathrus fide Dring (*Kew Bull.* **35**: 1, 1980).

anti- (in combination), against.

antiamoebin, an antibiotic from *Emericellopsis poonensis*, *E. synnematicola*, and *'Cephalosporium' pimprinum*; anti-protozoa and helminths (*Hindustan Antibiot. Bull.* **11**: 27, 1968).

antibiosis, antagonism (q.v.) between two organisms resulting in one overcoming the other.

antibiotic (1) (adj.) damaging to life; esp. of substances produced by microorganisms which are damaging to other microorganisms; (2) (n.) any antibiotic substance, esp. one used as a therapeutant, cf. toxin. See Waksman (*Mycol.* **39**: 565, 1947) for a discussion on the use of this term. **- substances** are produced by fungi (esp. *Penicillium* and *Aspergillus*), actinomycetes (esp. *Streptomyces*; see amphotericin, blasticidin, cycloheximide, streptomycin), and other microorganisms.

Lit.: Grayon (Ed.) (*Antibiotics, chemotherapeutics and antibacterial agents for disease control*, 1982), Chadwick & Whelan (Eds) (*Secondary metabolites: their function and evolution*, 1992), Demain *et al.* (Eds) (*Novel microbial products for medicine and agriculture*, 1989), Jong *et al.* (Eds) (*ATCC names of industrial fungi*, 1994).

Antibiotics. Substances antagonistic to and inhibiting growth of fungi, bacteria and other micro-organisms, even at high dilutions. Fleming (q.v.) is usually credited with their discovery, but several people (e.g. Duchesne, q.v.) made similar observations earlier. Penicillin, discovered by Fleming (q.v.) and exploited by Chain (q.v.), Florey (q.v.) and others, is a fungal product, and many fungi when grown under appropriate conditions are now known to produce antibiotics; see the reviews by Brian (*Bot. Rev.* **17**: 357, 1951) and Broadbent (*PANS* **B 14**: 120, 1968). Important or interesting antibiotics from fungi include antiamoebin, alternaric acid, calvacin, cephalosporins, dendrochin, flammulin, fumigillin, fumigatin, fusidic acid, gliotoxin, griseofulvin, helenin, lepiochlorin, patulin, penatin, penicillic acid, penicillin, phomin, poricin, proliferin, sparassol, statolin, trichomycin, trichothecin, trypacidin, ustilagic acids, variecolin, viridin, wortmannin (q.v.).

The market for antibiotic drugs has been estimated as exceeding US$25 billion annually. In addition to their use in human health, antibiotics are very widely and sometimes indiscriminately used in animal feeds (see Mellon *et al.*, *Hogging it! Estimates of antimicrobial abuse in livestock*, 2001). Misuse of antibiotics has caused a rise in numbers of strains resistant to them.

Fungicolous fungi (e.g. *Trichoderma*) produce a complex range of antibiotics including peptaibols and isonitriles. See Howell (in Harman & Kubicek, *Trichoderma* and *Gliocladium* **2**: 173, 1998).

Some lichen products (q.v.) are antibiotics. In general they are most effective against gram-positive bacteria. Usnic acid is used commercially ('Usno', 'Binan', 'Usniplant') and strongly inhibits *Mycobacterium*. Sodium usnate is effective against tomato canker (*Corynebacterium michiganense*) and several lichen acids are active against *Trichosporon*. Usnic acid inhibits *Neurospora crassa* and this and lichen extracts inhibit wood-rotting fungi (Henningsson & Lundström, *Mater. Organ.* **5**: 19, 1970). Hale (*Biology of lichens*, 1967; edn 2, 1974; review), Virtanen *et al.* (*Suomen Kem.* **B27-B30**, 1954-7; many papers on 'Usno'), Vartia (*in* Ahmadjian & Hale (Eds), *The lichens*: 547, 1974; review), Lowe & Elander (*Mycol.* **75**: 361, 1983; antibiotic industry in USA).

antibody, see antigen.

anticlinal, perpendicular to the surface; cf. periclinal.

antigen, a substance which when introduced into the tissues of a living animal induces the development in the blood serum (see **-serum**) of another substance (see Drouhet *et al.* (Eds), *Fungal antigens*, 1988). (the **-body**) with which it reacts specifically; antibodies may be classified according to whether they cause lysis (**lysins**), agglutination (**agglutinins**), or precipitation (**precipitins**) of the antigen; see anaphylaxis, complement-fixation, ELISA, Serology.

Antilyssa Haller ex M. Choisy (1929) = Peltigera fide Hawksworth *et al.* (*Dictionary of the Fungi* edn 8, 1995).

Antimanoa Syd. (1930), Pezizomycotina. 1, S. America.

Antimanopsis Petr. (1948) = Monostichella fide von Arx (*Verh. K. ned. Akad. Wet. Amst.* C **51**: 1, 1957).

antimetabolite, a substance which resembles in chemical structure some naturally occurring compound essential in a living process and which specifically antagonizes the biological action of such an essential compound. See Woolley (*Science, NY* **129**: 615, 1959; review).

Antinoa Velen. (1934) ? = Pezizella Fuckel fide Lizoň (*Mycotaxon* **45**: 1, 1992).

antiphyte, see alternation of generations.

Antipodium Piroz. (1974), anamorphic *Ophionectria*, Hso.≡ eH.15. 1, C. America. See Pirozynski (*CJB* **52**: 1143, 1974), Samuels (*Mycol.* **81**: 347, 1989), Bartoshevich *et al.* (*Journal of Basic Microbiology* **30**: 313, 1990), Castañeda Ruiz *et al.* (*Mycotaxon* **100**: 327, 2007).

antiserum, blood serum (the fluid fraction of coagulated blood) containing antibodies to one or more antigens (q.v.).

antithetic, see alternation of generations.

Antlea P.A. Dang. (1890) nom. dub., ? Fungi. or Protozoa.

Antonospora I. Fries, R.J. Paxton, J. Tengö, J.A. da Silva, S.B. Slemenda, N.J. Pieniazek (1999), Microsporidia. 2. See Fries *et al.* (*Eur. J. Protist.* **35**: 183, 1999).

Antrocarpon A. Massal. (1856) = Ocellularia. p.p. and Thelotrema (Thelotremat.) p.p. fide Hale (*Bull. Br. Mus. nat. hist.* Bot. **8**: 227, 1981).

Antrocarpum G. Mey. (1825) ≡ Thelotrema.

Antrodia P. Karst. (1879), Fomitopsidaceae. 46, Europe; N. America. See Donk (*Persoonia* **4**: 339, 1966), Niemelä & Ryvarden (*TBMS* **65**: 427, 1975; typification), Lombard (*Mycol.* **82**: 185, 1990; culture).

Antrodiella Ryvarden & I. Johans. (1980), Phanerochaetaceae. *c.* 50, USA. See Niemelä (*Karstenia* **22**: 11, 1982), Gilbertson & Ryvarden (*Europ. Polyp.* **1**: 147, 1993), Kim *et al.* (*Antonie van Leeuwenhoek* **83**: 81, 2003; phylogeny), Spirin & Zmitrovich (*Karstenia* **43**: 67, 2003; Russia), Dai (*Mycotaxon* **89**: 389, 2004; China).

Antromyces Fresen. (1850), anamorphic *Pezizomycotina*, Hsy.0eH.3/39. 2 (fimicolous), Europe; S. America. See Seifert *et al.* (*Univ. Waterloo Biol. Ser.* **27**, 1983).

Antromycopsis Pat. & Trab. (1897), anamorphic *Pleurotus*. 3, widespread. See Pollack & Miller (*Mem. N. Y. bot. Gdn* **28**: 174, 1976; teleomorph), Moore (*CJB* **55**: 1251, 1977), Moore (*TBMS* **82**: 377, 1984), Stalpers *et al.* (*CJB* **69**: 6, 1991; gen. revision, key), Capelari & Fungaro (*MR* **107**: 1050, 2003; RAPD).

antrorse, directed upwards or forwards.

Anulohypha Cif. (1962), anamorphic *Pezizomycotina*, Hso.-.-. 1, Dominican Republic. See Ciferri (*Atti Ist. bot. Univ. Lab. crittog. Pavia* sér. 5 **19**: 88, 1962).

Anulomyces Bydgosz (1932) nom. dub., Fungi.

Anulosporium Sherb. (1933) nom. dub., Fungi. See Drechsler (*Mycol.* **26**: 135, 1934), Rubner (*Stud. Mycol.* **39**, 1996; = *Arthrobotrys* or *Monacrosporium* (*Orbiliaceae*)).

Anungitea B. Sutton (1973), anamorphic *Venturiaceae*, Hso.1eP.3/9. 15, widespread. See Sutton (*Mycol. Pap.* **132**: 10, 1973), Crous *et al.* (*CJB* **73**: 224, 1995; S Africa), Castañeda Ruíz *et al.* (*Mycotaxon* **65**: 93, 1997; Cuba), Crous *et al.* (*Stud. Mycol.* **58**: 185, 2007; phylogeny).

Anungitopsis R.F. Castañeda & W.B. Kendr. (1990), anamorphic *Venturiaceae*, Hso.≡ eP.?28. 7, widespread. See Castañeda Ruiz & Kendrick (*Univ. Waterloo Biol. Ser.* **33**: 6, 1990), Castañeda Ruíz *et al.* (*Mycotaxon* **59**: 203, 1996; Cuba), Jørgensen (*Symb. bot. upsal.* **32** no. 1: 113, 1997; S Africa), Ho *et al.* (*Mycotaxon* **72**: 115, 1999; key), Crous *et al.* (*Stud. Mycol.* **58**: 185, 2007; phylogeny).

Anzia Garov. (1868) ≡ Lichenothelia.

Anzia Stizenb. (1861) nom. cons., Parmeliaceae (L). 35, widespread. See Culberson (*Brittonia* **13**: 381, 1961), Kurokawa & Jinzenji (*Bull. natn. Sci. Mus. Tokyo*, B **8**: 369, 1965), Yoshimura & Elix (*J. Hattori bot. Lab.* **74**: 287, 1993), Yoshimura *et al.* (*Biblthca Lichenol.* **58**: 439, 1995; New Guinea), Calvelo (*Mycotaxon* **58**: 147, 1996; S. Am.), Yoshimura *et al.* (*J. Hattori bot. Lab.* **82**: 343, 1997; Indian spp.), Kärnefelt *et al.* (*Nova Hedwigia* **67**: 71, 1998), Yoshimura *in* Marcelli & Seaward (Eds) (*Lichenology in Latin America. History, Current Knowledge and Applications* [Proceedings of GLAL-3, Terceiro Encontro do Grupo Latino-Americano de Liquenólogos, São Paulo, Brazil, 24-28 September, 1997]: 117, 1998; Am.), Rikkinen & Poinar (*MR* **106**: 984, 2002; fossil taxa), Thell *et al.* (*Mycol. Progr.* **3**: 297, 2004; phylogeny), Arup *et al.* (*Mycol.* **99**: 42, 2007; phylogeny), Crespo *et al.* (*Mol. Phylogen. Evol.* **44**: 812, 2007; morphology and phylogeny).

Anziaceae M. Satô (1939) = Parmeliaceae.

Anziella Gyeln. (1940) = Placynthium fide Hawksworth *et al.* (*Dictionary of the Fungi* edn 8, 1995).

Anzina Scheid. (1982), ? Arthrorhaphidaceae (L). 1, Europe. See Scheidegger (*Nova Hedwigia* **41**: 191, 1985), Lumbsch (*J. Hattori bot. Lab.* **83**: 1, 1997), Lumbsch *et al.* (*MR* **105**: 265, 2001; asci), Lumbsch *et al.* (*Mol. Phylogen. Evol.* **31**: 822, 2004; phylogeny), Wedin *et al.* (*MR* **109**: 159, 2005; phylogeny), Lumbsch *et al.* (*MR* **111**: 1133, 2007).

Aorate Syd. (1929) = Titaea fide Boedijn (*Sydowia* **5**: 211, 1951).

Aoria Cif. (1962), anamorphic *Pezizomycotina*, St.0eH.10. 1, Dominican Republic. See Ciferri (*Atti Ist. bot. Univ. Lab. crittog. Pavia* sér. 5 **19**: 89, 1962), Nag Raj & DiCosmo (*Univ. Waterloo Biol. Ser.* **20**, 1982).

apandrous, forming oospores when no antheridia are present.

Aparaphysaria Speg. (1922), Pyronemataceae. 2, India; Tierra del Fuego. See Kimbrough (*Mem. N. Y. bot. Gdn* **49**: 326, 1989).

Apatelomyces Thaxt. (1931), Laboulbeniaceae. 1, W. Africa. See Nannfeldt (*Svensk bot. Tidskr.* **43**: 468, 1949).

Apatomyces Thaxt. (1931), Laboulbeniaceae. 1, Philippines. See Tavares (*Mycol. Mem.* **9**: 627 pp., 1985), Santamaría (*MR* **99**: 1071, 1995).

Apatoplaca Poelt & Hafellner (1980), Teloschistaceae (L). 1, N. America. See Bellemère *et al.* (*Cryptog. Bryol.-Lichénol.* **7**: 189, 1986; ultrastr.), Kantvilas & McCarthy (*Lichenologist* **35**: 397, 2003).

Aphanandromyces W. Rossi (1982), Laboulbeniaceae. 1, Europe. See Rossi (*Mycol.* **74**: 520, 1982), Tavares (*Mycol. Mem.* **9**: 627 pp., 1985), Santamaria *et al.* (*Treb. Inst. Bot. Barcelona* **14**: 1, 1991; Europe), Santamaría (*Fl. Mycol. Iberica* **5**, 2003; Iberian peninsula).

Aphanistis Sorokīn (1889), ? Chytridiales. 1 or 2, former USSR.

Aphanoascus Zukal (1890), Onygenaceae. Anamorph *Chrysosporium*. 12, widespread. See Cano & Guarro (*MR* **94**: 455, 1990; key), Sugiyama *et al.* (*Mycoscience* **40**: 251, 1999; DNA), Cano *et al.* (*Stud. Mycol.* **47**: 153, 2002; phylogeny), Pivkin & Khudyakova (*Mycotaxon* **81**: 7, 2002), Sugiyama *et al.* (*Stud. Mycol.* **47**: 5, 2002).

Aphanobasidium Jülich (1979), Pterulaceae. 15, widespread. See Jülich (*Persoonia* **10**: 326, 1979), Boidin *et al.* (*BSMF* **119**: 333, 2003; subgen. *Aphanobasidium*).

Aphanocladium W. Gams (1971), anamorphic *Nectriaceae*, Hso.0eH.15. 2 (on myxomycetes), widespread. Several species are now placed in *Lecanicillium*. See Gams *et al.* (*CJB* **76**: 1570, 1998), Sung *et al.* (*Nova Hedwigia* **72**: 311, 2001; phylogeny), Zare & Gams (*Rostaniha* Supplement 3, 2004).

Aphanofalx B. Sutton (1986), anamorphic *Pezizomycotina*, St.0eH.1. 2, Zambia; Pakistan. See Sutton & Abbas (*TBMS* **87**: 640, 1987).

Aphanopeltis Syd. (1927), Asterinaceae. Anamorph *Elachopeltis*. 7, America (tropical); Indonesia. See Hosagoudar *et al.* (*Journal of Mycopathological Research* **39**: 61, 2001).

aphanoplasmodium, see plasmodium.

Aphanopsidaceae Printzen & Rambold (1995), Lecanorales (L). 2 gen. (+ 2 syn.), 3 spp.

Lit.: Eriksson (*SA* **9**: 24, 1990) places it outside of the *Lecanoromycetidae*, Printzen & Rambold (*Lichenologist* **27**: 99, 1995), Kantvilas & McCarthy (*Lichenologist* **31**: 555, 1999).

Aphanopsis Nyl. ex P. Syd. (1887), Aphanopsidaceae (L). 1, Europe. See Coppins & James (*Lichenologist* **16**: 241, 1984), Printzen & Rambold (*Lichenologist* **27**: 91, 1995).

Aphanostigme Syd. (1926), ? Pseudoperisporiaceae. *c.* 12, widespread. See Hansford (*Mycol. Pap.* **15**, 1946), Müller (*Sydowia* **18**: 86, 1965), Rossman (*Mycol. Pap.* **157**, 1987), Verma & Kamal (*Indian Phytopath.* **42**: 561, 1990).

Aphanotria Döbbeler (2007), Bionectriaceae. 1, S. America (tropical). See Döbbeler (*MR* **111**: 1406, 2007).

Apharia Bonord. (1864), Pezizomycotina. 1, Europe.

Aphelaria Corner (1950), Aphelariaceae. 20, widespread. See Roberts (*Kew Bull.* **54**: 517, 1999; Cameroon).

Aphelariaceae Corner (1970), Cantharellales. 3 gen., 22 spp. Basidioma ramarioid.

Lit.: Corner (*Ann. Bot. Mem.* [A monograph of Clavaria and allied genera] **1**: 1, 1950), Corner (*TBMS* **49**: 205, 1966), Petersen & Zang (*Acta Bot. Yunn.* **8**: 281, 1986), Roberts (*Kew Bull.* **54**: 517, 1999).

Aphelariopsis Jülich (1982), ? Septobasidiaceae. 2, Sarawak; S. America. See Jülich (*Persoonia* **11**: 402, 1982).

Aphelidium Zopf (1885) nom. dub., Fungi. Protozoa or fungi in algal cells).

Aphidomyces Brain (1923), ? Saccharomycetales. 5 (in *Insecta*), widespread.

Aphotistus Humb. (1793) = Rhizomorpha Roth fide Mussat (*Syll. fung.* **15**, 1901) nom. dub. fide, Donk (*Taxon* **11**: 79, 1962).

Aphragmia Trevis. (1880) [non *Aphragmia* Nees 1836, *Acanthaceae*] = Ionaspis fide Hawksworth *et al.* (*Dictionary of the Fungi* edn 8, 1995).

Aphyllophorales. Order proposed by Rea (after Patouillard) for basidiomycetes having macroscopic basidiocarps in which the hymenophore is flattened (*Thelephoraceae*), club-like (*Clavariaceae*), tooth-like (*Hydnaceae*) or has the hymenium lining tubes (*Polyporaceae*) or sometimes on lamellae, the poroid or lamellate hymenophores being tough and not fleshy as in the *Agaricales*. Traditionally the order has had a core of 4 fam. (as indicated above) based on hymenophore shape but detailed microscopic studies of basidiocarp structure and molecular evidence has shown these groupings to be unnatural. Keys to 550 spp. in culture are given by Stalpers (*Stud. mycol.* **16**, 1978).

Aphyllotus Singer (1974), ? Marasmiaceae. 1, Colombia. See Singer (*Sydowia* Beih. **7**: 29, 1974).

Aphysa Theiss. & Syd. (1917) = Coleroa fide Müller & von Arx (*Beitr. Kryptfl. Schweiz* **11** no. 2, 1962).

Aphysiostroma Barrasa, A.T. Martínez & G. Moreno (1986), Hypocreaceae. Anamorph *Verticillium*-like. 1 (coprophilous), Spain. See Barrasa *et al.* (*CJB* **63**: 2439, 1985), Spatafora & Blackwell (*Mycol.* **85**: 912, 1993; DNA), Rossman *et al.* (*Stud. Mycol.* **42**: 248 pp., 1999), Suh & Blackwell (*Mycol.* **91**: 836, 1999; phylogeny), Sung *et al.* (*Nova Hedwigia* **72**: 311, 2001; phylogeny), Sung *et al.* (*Stud. Mycol.* **57**: 1, 2007).

apical, at the end (or **apex**); **- granule**, a deeply staining granule at the hyphal apex, esp. in *Basidiomycetes*; the 'Spitzenkorper' of Brunswik (1924); **- veil**, see annulus; **- wall building**, see wall building.

apiculate, having an apiculus.

apiculus (of a spore), a short projection at one end; a projection by which it was fixed to the sterigma (Josserand); apicule; hilar appendage.

apileate, having no pileus; resupinate.

Apinisia La Touche (1968), Onygenales. Anamorph *Chrysosporium*. 2 or 3, Europe; Australia. See Guarro *et al.* (*Mycotaxon* **42**: 193, 1991), Sugiyama *et al.* (*Mycoscience* **40**: 251, 1999; DNA), Sugiyama *et al.* (*Stud. Mycol.* **47**: 5, 2002; phylogeny).

Apiocamarops Samuels & J.D. Rogers (1987), Boliniaceae. 3, C. & S. America. See Samuels & Rogers (*Mycotaxon* **28**: 54, 1987), Rogers & Samuels (*Mycol.* **80**: 738, 1988), Rogers & Ju (*Sydowia* **55**: 359, 2003).

Apiocarpella Syd. & P. Syd. (1919), anamorphic *Pezizomycotina*, Cpd.1eH.1. 8, widespread. See Mel'nik (*Nov. Sist. niz. Rast.* **13**: 93, 1976), Punithalingam (*Mycol. Pap.* **142**, 1979; synonym of *Ascochyta*), Vanev & Sofia (*Fitologiya* **29**: 39, 1985; key).

Apioclypea K.D. Hyde (1994), ? Clypeosphaeriaceae. 1 (saprobic on palms), Papua New Guinea. See Hyde *et al.* (*Sydowia* **50**: 21, 1998), Kang *et al.* (*Mycoscience* **40**: 151, 1999), Smith *et al.* (*Fungal Diversity* **13**: 175, 2003; rel. to *Apiospora*), Taylor & Hyde (*Fungal Diversity Res. Ser.* **12**, 2003).

Apiocrea Syd. & P. Syd. (1921) = Hypomyces fide Rogerson & Samuels (*Mycol.* **81**: 413, 1989), Rossman *et al.* (*Stud. Mycol.* **42**: 248 pp., 1999).

Apiodiscus Petr. (1940), ? Rhytismatales. 1, Iran.

Apiodothina Petr. & Cif. (1932) = Coccoidea fide Müller & von Arx (*Beitr. Kryptfl. Schweiz* **11** no. 2, 1962).

Apiognomonia Höhn. (1917), Gnomoniaceae. Anamorphs *Discula*, *Gloeosporidina*. 10 (from stems and leaves), Europe; N. America. *A. erythrostoma* (cherry leaf scorch), *A. quercina* (oak anthracnose). See von Arx (*Antonie van Leeuwenhoek* **17**: 259, 1951), Barr (*Mycol. Mem.* **7**, 1978; key), Monod (*Sydowia* **37**: 222, 1984), Barr (*Mycotaxon* **41**: 287, 1991; N Am. spp.), Haemmerli *et al.* (*Molecular Plant-Microbe Interactions* **5**: 479, 1992; DNA), Viret & Petrini (*MR* **98**: 423, 1994), Butin & Kehr (*Eur. J. For. Path.* **28**: 297, 1998; anam.), Castlebury *et al.* (*Mycol.* **94**: 1017, 2002), Castlebury *et al.* (*Mycoscience* **44**: 203, 2003), Sogonov *et al.* (*Sydowia* **57**: 102, 2005; typification), Sogonov *et al.* (*MR* **111**: 693, 2007; revision).

Apioplagiostoma M.E. Barr (1978), Gnomoniaceae. 3, Europe; N. America. See Mouchacca (*Cryptog. Mycol.* **8**: 141, 1987), Fröhlich & Hyde (*MR* **99**: 727, 1995), Zhang & Blackwell (*Mycol.* **93**: 355, 2001; phylogeny).

Apioporthe Höhn. (1917) = Anisogramma fide Müller & von Arx *in* Ainsworth *et al.* (Eds) (*The Fungi* **4A**: 87, 1973).

Apioporthella Petr. (1929), ? Valsaceae. 1 (from stems etc.), Europe; N. America. See Barr (*Mycotaxon* **41**: 287, 1991).

Apiorhynchostoma Petr. (1923), Clypeosphaeriaceae. 5 (saprobic on wood), Europe. See Sivanesan (*TBMS* **65**: 19, 1975), Rogers *et al.* (*Mycol.* **86**: 700, 1994), Waldner (*Beitr. Kenntn. Pilze Mitteleur.* **11**: 67, 1997), Hyde *et al.* (*Sydowia* **50**: 21, 1998), Réblová (*Sydowia* **50**: 229, 1998), Kang *et al.* (*Mycoscience*

40: 151, 1999; posn).

Apiosordaria Arx & W. Gams (1967), Lasiosphaeriaceae. Anamorph *Cladorrhinum*. 11, widespread. See Krug *et al.* (*Mycotaxon* **17**: 553, 1983), Guarro & Cano (*TBMS* **91**: 587, 1988), Mouchacca & Gams (*Mycotaxon* **48**: 415, 1993; anamorphs), Hyde *et al.* (*Mycoscience* **38**: 437, 1997), Stchigel *et al.* (*Mycol.* **92**: 1206, 2000), Stchigel *et al.* (*Mycol.* **95**: 1218, 2003), Huhndorf *et al.* (*Mycol.* **96**: 368, 2004; phylogeny), Miller & Huhndorf (*Mol. Phylogen. Evol.* **35**: 60, 2005; phylogeny), Zhang *et al.* (*Mycol.* **98**: 1076, 2006; phylogeny).

Apiosphaeria Höhn. (1909), Phyllachoraceae. Anamorph *Oswaldina*. 5 (from living leaves), widespread (neotropics). See Dianese *et al.* (*Sydowia* **46**: 233, 1994; anamorph), Hyde *et al.* (*Sydowia* **50**: 21, 1998), Hyde & Cannon (*Mycol. Pap.* **175**: 114, 1999; spp. on palms).

Apiospora Sacc. (1875), Apiosporaceae. Anamorphs *Arthrinium*, *Cordella*, *Pteroconium*. 7 (on *Palmae*, grasses etc.), widespread. See Samuels *et al.* (*N.Z. Jl Bot.* **19**: 137, 1981), Müller (*Boln Soc. argent. Bot.* **28**: 201, 1992; key), Hyde *et al.* (*Sydowia* **50**: 21, 1998), Smith *et al.* (*Fungal Diversity* **13**: 175, 2003; phylogeny), Huhndorf *et al.* (*Mycol.* **96**: 368, 2004; phylogeny), Zhang *et al.* (*Mycol.* **98**: 1076, 2006; phylogeny).

Apiosporaceae K.D. Hyde, J. Fröhl., Joanne E. Taylor & M.E. Barr (1998), Sordariomycetidae (inc. sed.). 6 gen. (+ 16 syn.), 47 spp.

Lit.: Samuels *et al.* (*N.Z. Jl Bot.* **19**: 137, 1981), Müller (*Boln Soc. argent. Bot.* **28**: 201, 1992), Hyde *et al.* (*Sydowia* **50**: 21, 1998), Wang & Hyde (*Fungal Diversity* **3**: 159, 1999), Huhndorf *et al.* (*Mycol.* **96**: 368, 2004).

Apiosporella Höhn. ex Theiss. (1917) = Pseudomassaria fide Barr (*Mycol.* **68**, 1976).

Apiosporella Speg. (1910) ≡ Apiocarpella.

Apiosporella Speg. (1912) = Aplosporidium fide Hawksworth *et al.* (*Dictionary of the Fungi* edn 8, 1995).

Apiosporina Höhn. (1910) = Venturia Sacc. See also *Dibotryon*. fide Barr (*Sydowia* **41**: 25, 1989), Crous *et al.* (*Stud. Mycol.* **58**: 185, 2007; phylogeny), Winton *et al.* (*Mycol.* **99**: 240, 2007; phylogeny).

Apiosporina Petr. (1925) ≡ Pseudomassaria fide Müller & von Arx (*Beitr. Kryptfl. Schweiz* **11** no. 2, 1962).

Apiosporium Kunze (1817), anamorphic *Capnodium*, St.0eH.?. 2. See Kunze (*Mykologische Hefte* Leipzig **1**, 1817).

Apiosporopsis (Traverso) Mariani (1911), ? Melanconidaceae. 1, Europe. See Reid & Dowsett (*CJB* **68**: 2398, 1990).

apiosporous (of two-celled spores), where one cell is markedly smaller then the other.

Apiothecium Lar.N. Vassiljeva (1987) = Apioporthella fide Barr (*Mycotaxon* **41**: 287, 1991).

Apiothyrium Petr. (1947), Hyponectriaceae. 1, Finland. See Wang & Hyde (*Fungal Diversity* **3**: 159, 1999).

Apiotrabutia Petr. (1929) = Munkiella fide Müller & von Arx (*Beitr. Kryptfl. Schweiz* **11** no. 2, 1962).

Apiotrichum Stautz (1931) = Trichosporon fide Middelhoven *et al.* (*FEMS Yeast Res.* **1**: 15, 2001; taxonomy).

Apiotypa Petr. (1925), Pezizomycotina. 1, Philippines. Type material is missing. See Hyde *et al.* (*Sydowia* **50**: 21, 1998), Hyde & Cannon (*Mycol. Pap.* **175**, 1999).

Aplacodina Ruhland (1900) = Pseudomassaria fide Barr (*Mycol.* **68**, 1976).

aplanetism, the condition of having non-motile spores in place of zoospores.

Aplanocalenia Lücking, Sérus. & Vězda (2005), Gomphillaceae (L). 1. See Lücking *et al.* (*Lichenologist* **37**: 163, 2005).

aplanogamete, a non-motile gamete.

aplanospore (1) a naked, amoeboid or non-amoeboid mobile cell; (2) a sporangiospore.

Aplanosporites R.K. Kar (1979), Fossil Fungi. 1, India. See Kar (*Palaeobotanist* **26**: 35, 1977).

Aplectosoma Drechsler (1951), Cochlonemataceae. 1, USA. See Drechsler (*Mycol.* **43**: 173, 1951).

aplerotic, of an oospore which occupies ‹ 60% of the oogonial volume (Shahzad *et al.*, *Bot. J. Linn. Soc.* **108**: 143, 1992).

Aplopsora Mains (1921), Chaconiaceae. *c.* 6 (on dicots), N. America; Brazil; Russian far east; China; Japan. See Buriticá (*Revista de la Academia Columbiana de Ciencias Exactas, Físicas y Naturales* **22** no. 84: 325, 1998; neotrop. spp).

Aplosporella Speg. (1880), anamorphic *Botryosphaeriaceae*, St.0eP.1. 66, widespread (esp. tropical). See Petrak (*Sydowia* **6**: 336, 1952), Tilak & Ramchandra Rao (*Mycopath. Mycol. appl.* **24**: 362, 1964), Ramchandra Rao (*Mycopath. Mycol. appl.* **28**: 45, 1966; Indian spp.), Ramchandra Rao (*Mycopath. Mycol. appl.* **28**: 68, 1966; Indian spp.), Pandey (*Perspectives in mycological research* **2**: 77, 1990; review), Pande & Rao (*Nova Hedwigia* **60**: 79, 1995; key to 44 spp.), Damm *et al.* (*Fungal Diversity* **27**: 35, 2007; posn).

Aplosporidium Speg. (1912) = Asteromella fide Sutton (*Mycol. Pap.* **141**, 1977).

Aplotomma A. Massal. ex Beltr. (1858) ? = Buellia fide Hawksworth *et al.* (*Dictionary of the Fungi* edn 8, 1995).

Apoa Syd. (1931) = Pachypatella fide von Arx & Müller (*Stud. Mycol.* **9**, 1975).

apobasidiomycete, a gasteromycete having apobasidia.

apobasidium, see basidium.

Apocoryneum B. Sutton (1975) = Massariothea fide Sutton (*Mycol. Pap.* **141**, 1977).

apocyte, multinucleate cell in which the multinucleate condition is accidental, transitory or secondary. See coenocyte.

Apocytospora Höhn. (1924) = Plectophomella fide Petrak (*Annls mycol.* **27**: 368, 1929).

apodial, having no stalk; sessile.

Apodospora Cain & J.H. Mirza (1970), Lasiosphaeriaceae. 4 (coprophilous), N. America; Europe. See Lundqvist (*Symb. bot. upsal.* **20** no. 1, 1972), Barr (*Mycotaxon* **39**: 43, 1990; posn).

Apodothina Petr. (1970), Phyllachoraceae. 1 (on living leaves of *Yucca*), USA. See Petrak (*Sydowia* **23**: 276, 1969).

Apodus Malloch & Cain (1971), ? Lasiosphaeriaceae. 1 (coprophilous), N. America. See Malloch & Cain (*CJB* **49**: 869, 1971), Cai *et al.* (*MR* **110**: 137, 2006; phylogeny), Cai *et al.* (*MR* **110**: 359, 2006; polyphyly).

Apogaeumannomyces Matsush. (2003) nom. inval., ? Chaetosphaeriales. Anamorph *Cercosporula*. 1 (on palm leaf), Peru. See Matsushima (*Matsush. Mycol. Mem.* **10**: 152, 2001).

apogamy, the apomictic development of diploid cells.

Apogloeum Petr. (1954), anamorphic *Pezizomycotina*, St.0eH.?. 1, Tasmania. See Petrak (*Sydowia* **8**: 57, 1954).

Apoharknessia Crous & S.J. Lee (2004), anamorphic *Diaporthales*. 1, pantropical. See Lee *et al.* (*Stud. Mycol.* **50**: 239, 2004).

Apomelasmia Grove (1937), anamorphic *Diaporthales*, St.0eH.15. 2, Europe. See Mel'nik (*Nov. Sist. niz. Rast.* **28**: 69, 1992).

Apomella Syd. (1937) = Botryosphaeria fide Sutton (*in litt.*).

apomixis (adj. **apomictic**), the development of sexual cells into spores, etc., without being fertilized. Cf. amphimixis, automixis, and pseudomixis.

Aponectria (Sacc.) Sacc. (1883) = Nectria fide Rossman *et al.* (*Stud. Mycol.* **42**: 248 pp., 1999).

apophysis, a swelling or a swollen filament, e.g. at the end of a sporangiophore below the sporangium in Mucorales (cf. columella) or on the stem of some species of *Geastrum*; (in basidiomycetes), the swelling at the tip of a sterigma from which the basidiospore develops and which becomes the hilar appendage (q.v.).

Apophysomyces P.C. Misra (1979), Radiomycetaceae. 1, India. See Misra *et al.* (*Mycotaxon* **8**: 377, 1979), Ellis & Ajello (*Mycol.* **74**: 144, 1982), Lakshmi *et al.* (*J. Clin. Microbiol.* **31**: 1368, 1993; zygomycosis), Eaton *et al.* (*J. Clin. Microbiol.* **32**: 2827, 1994; mucormycosis), Meis *et al.* (*J. Clin. Microbiol.* **32**: 3078, 1994; osteomyelitis), Voigt & Wöstemeyer (*Gene* **270**: 113, 2001; phylogeny), Liang *et al.* (*J. Clin. Microbiol.* **44**: 892, 2006; rhino-orbitocerebral mucormycosis).

apoplasmodial (of *Acrasiales*), having non-fusion of the myxamoebae.

apoplastic, movement of substances via the cell walls, not entering the living cell; cf. symplastic.

Aporella Syd. (1939) [non *Aporella* Podp. 1916, *Musci*] ≡ Aporellula fide Sutton (*Mycotaxon* **3**: 377, 1976).

Aporellula B. Sutton (1986), anamorphic *Pezizomycotina*, St.1-≡ eH.15. 1, Ecuador. See Sutton (*Sydowia* **38**: 324, 1985).

Aporhytisma Höhn. (1917) = Diaporthe fide von Arx & Müller (*Beitr. Kryptfl. Schweiz* **11** no. 1, 1954), Petrak (*Sydowia* **24**: 249, 1971), Castlebury *et al.* (*Mycoscience* **44**: 203, 2003).

Aporia Duby (1862) = Lophodermium fide Hawksworth *et al.* (*Dictionary of the Fungi* edn 8, 1995).

Aporidicellaesporites Frunzescu & Bacaran (1990), Fossil Fungi. 1. See Frunzescu & Bacaran (*Revue roum. Géol., Géophys. Géogr.* Géol. **34**: table 1, 1990).

Aporimonocellasporites Frunzescu & Bacaran (1990), Fossil Fungi. 1. See Frunzescu & Bacaran (*Revue roum. Géol., Géophys. Géogr.* Géol. **34**: 12, 1990).

Aporimonodicellaesporites Frunzescu & Bacaran (1990), Fossil Fungi. 1. See Frunzescu & Bacaran (*Revue roum. Géol., Géophys. Géogr.* Géol. **34**: table 1 + pl. 1, fig. 8, 1990).

Aporimulticellaesporites Frunzescu & Bacaran (1990), Fossil Fungi. 1. See Frunzescu & Bacaran (*Revue roum. Géol., Géophys. Géogr.* Géol. **34**: 24, 1990).

Aporitetracellaesporites Frunzescu & Bacaran (1990), Fossil Fungi. 1. See Frunzescu & Bacaran (*Revue roum. Géol., Géophys. Géogr.* Géol. **34**: table 1, 1990).

Aporitricellaesporites Frunzescu & Bacaran (1990), Fossil Fungi. 1. See Frunzescu & Bacaran (*Revue roum. Géol., Géophys. Géogr.* Géol. **34**: table 1, 1990).

Aporomyces Thaxt. (1931), Laboulbeniaceae. 8 (on *Limnichideae* and *Strophylinidae*), widespread. See Benjamin (*Aliso* **12**: 335, 1989; key), Kaur & Mukerji (*Mycoscience* **37**: 61, 1996), Santamaría (*Fl. Mycol. Iberica* **5**, 2003; Europe).

Aporophallus Möller (1895), Phallaceae. 1, Brazil.

Aporothielavia Malloch & Cain (1973) ? = Chaetomidium fide Malloch & Cain (*Mycol.* **65**: 1055, 1973), Suh & Blackwell (*Mycol.* **91**: 836, 1999; phylogeny), Untereiner *et al.* (*CJB* **79**: 321, 2001; phylogeny, genus concept), Cai *et al.* (*MR* **110**: 359, 2006; phylogeny), Cai *et al.* (*MR* **110**: 137, 2006), Greif & Currah (*MR* **111**: 70, 2007; ontogeny).

Aporpiaceae Bondartsev & Bondartseva (1960) = Auriculariaceae.

Aporpium Bondartsev & Singer (1944) = Protomerulius fide Núñez (*Folia cryptog. Estonica* **33**: 99, 1998).

Aposphaeria Berk. (1860) nom. rej., anamorphic *Pezizomycotina*. See Sutton (*Mycol. Pap.* **141**, 1977).

Aposphaeria Sacc. (1880) nom. cons., anamorphic *Melanomma*, Cpd.0eH.15. 101, widespread. See Chesters (*TBMS* **22**: 116, 1938), Heiny *et al.* (*Mycotaxon* **44**: 137, 1992).

Aposphaeriella Died. (1912) = Zignoëlla fide Höhnel (*Sber. Akad. Wiss. Wien* Math.-naturw. Kl., Abt. 1 **126**: 283, 1917).

Aposphaeriopsis Died. (1913) = Cephalotheca fide Chesters (*TBMS* **19**: 261, 1935).

Aposporella Thaxt. (1920), anamorphic *Pezizomycotina*, Hso.0eP.38. 1 (on *Insecta*), Africa.

apospory, direct incorporation in a spore of an oogonial or antheridial diploid nucleus with cytoplasm uninfluenced by any meiosis at the time of spore wall formation (Dick, 1972).

Apostemidium P. Karst. (1871) ≡ Apostemium.

Apostemium (P. Karst.) P. Karst. (1870) = Vibrissea fide Graddon (*TBMS* **48**: 639, 1965; key), Sánchez & Korf (*Mycol.* **58**: 733, 1966).

Apostrasseria Nag Raj (1983), anamorphic *Phacidium*, St.0eH.15. 4, New Zealand; N. America. See Kramer (*Stud. Mycol.* **30**: 151, 1987).

Apotemnoum Corda (1833) = Clasterosporium fide Saccardo (*Syll. fung.* **4**: 382, 1886).

apothecium (pl. **apothecia**), a cup-like or saucer-like ascoma in which the hymenium is exposed at maturity, sessile or stipitate, the stipes sometimes lichenized (podetium; q.v.). See the following for terminology of anatomical structures of apothecia: Degelius (*Sym. bot. upsal.* **13** (2), 1954; tabulation of terms), Korf (*Sci. Rep. Yokohama nat. Univ.* II **7**: 7, 1958; *in* Ainsworth *et al.* (Eds), *The Fungi* **4A**: 249, 1973), Letrouit-Galinou (*Bryologist* **71**: 297, 1969), Maas Geesteranus (*Blumea* **6**: 41, 1947), Sheard (*Lichenologist* **3**: 328, 1967).

Apoxona Donk (1969) = Hexagonia Fr. fide Bondartsev & Singer (*Polyporaceae of the European part of the U.S.S.R. and Caucasus*: 1106 pp., 1953).

Appelia (Sacc.) Trotter (1931) = Trichoconis fide Hawksworth *et al.* (*Dictionary of the Fungi* edn 8, 1995).

appendage, a process (outgrowth) of any sort. For coelomycete conidial appendage terminology see

Nag Raj (*Coelomycetous anamorphs*, 1993).

Appendichordella R.G. Johnson, E.B.G. Jones & S.T. Moss (1987), Halosphaeriaceae. 1 (marine), Europe; N. America. See Johnson *et al.* (*CJB* **65**: 931, 1987), Kohlmeyer & Volkmann-Kohlmeyer (*Bot. Mar.* **34**: 1, 1991).

Appendicispora Spain, Oehl & Sieverd. (2006) = Ambispora fide Walker & Schüssler (*MR* **112**: 297, 2008).

Appendicisporaceae C. Walker, Vestberg & A. Schüssler (2007) nom. illegit. = Ambisporaceae.
Lit.: Walker *et al.* (*MR* **111**: [253], 2007), Walker *et al.* (*MR* **111**: 137, 2007).

Appendicisporonites R.K. Saxena & S. Khare (1991), Fossil Fungi. 1, India. See Saxena & Khare (*Geophytology* **21**: 40, 1991).

Appendicospora K.D. Hyde (1995), ? Xylariales. 2 (dead palm fronds), widespread (tropical). See Hyde (*Sydowia* **47**: 31, 1995), Yanna *et al.* (*Mycoscience* **38**: 395, 1997), Hyde *et al.* (*Sydowia* **50**: 21, 1998; posn).

Appendicularia Peck (1885) [non *Appendicularia* DC. 1828, *Melastomataceae*] ≡ Appendiculina.

appendiculate (1) (of an agaric basidioma), having the edge of the expanded pileus fringed with tooth-like remains of the veil, as in *Psathyrella candolleana*; (2) (of a spore), having one or more setulae.

Appendiculella Höhn. (1919), Meliolaceae. 250 (from leaves), widespread (tropical). See Hughes (*Mycol. Pap.* **166**, 1993), Song (*Mycosystema* **17**: 214, 1998; China), Song *et al.* (*Mycosystema* **21**: 177, 2002; China), Hosagoudar (*Sydowia* **55**: 162, 2003; placement), Rodríguez & Piepenbring (*Mycol.* **99**: 544, 2007; Panama).

Appendiculina Berl. (1889) = Stigmatomyces fide Thaxter (*Proc. Amer. Acad. Arts & Sci.* **25**: 8, 1890).

Appendispora K.D. Hyde (1994), Didymosphaeriaceae. 2, Brunei. See Hyde (*Sydowia* **46**: 29, 1994), Hyde *et al.* (*Nova Hedwigia* **69**: 449, 1999).

Appendixia B.S. Lu & K.D. Hyde (2000), Xylariaceae. 1, USA. Questionably distinct from *Anthostomella*. See Lu & Hyde (*Fungal Diversity Res. Ser.* **4**: 224, 2000).

Appianoporites S.Y. Sm., Currah & Stockey (2004), Fossil Fungi. 1, Canada. See Smith *et al.* (*Mycol.* **96**: 181, 2004).

applanate, flattened.

apple canker, disease caused by *Nectria galligena*.

appressed (**adpressed**), closely flattened down.

appressorium, a swelling on a germ-tube or hypha, esp. for attachment in an early stage of infection, as in certain *Pucciniales* and in *Colletotrichum*; the '. expression of the genotype during the final phase of germination', whether or not morphologically differentiated from vegetative hyphae, as long as the structure adheres to and penetrates the host (Emmett & Parbery, *Ann. Rev. Phytopath.* **13**: 146, 1975); the term hyphopodium (q.v.) is probably best treated as a synonym.

Apra J.F. Hennen & F.O. Freire (1979), Raveneliaceae. 1 (on *Mimosa* (*Leguminosae*)), Brazil. See Hennen & Freire (*Mycol.* **71**: 1053, 1979).

Apterivorax S. Keller (2005), Neozygitaceae. 2, widespread. See Keller & Petrini (*Sydowia* **57**: 47, 2005), Keller & Petrini (*Sydowia* **57**: 23, 2005; key), Keller (*Sydowia* **58**: 75, 2006; validation of *A. acaricida*).

Aptrootia Lücking & Sipman (2007), Trypetheliaceae (L). 1, Costa Rica; Papua New Guinea. See Lücking *et al.* (*Lichenologist* **39**: 187, 2007), Aptroot *et al.* (*Biblthca Lichenol.* **97**, 2008; Costa Rica).

apud, in; sometimes used to indicate a name published by one author in the work of another; cf. ex.

Apus Gray (1821) ≡ Schizophyllum.

Apyrenium Fr. (1849) nom. dub., anamorphic *Hypocreales*. See Donk (*Taxon* **7**: 164, 1958).

Aquadiscula Shearer & J.L. Crane (1985), Helotiaceae. 2 (aquatic), USA. See Shearer & Crane (*Mycol.* **77**: 441, 1985), Fallah & Shearer (*Mycol.* **93**: 566, 2001).

Aquadulciospora Fallah & Shearer (2001), Hyponectriaceae. 1, USA.

Aqualignicola V.M. Ranghoo, K.M. Tsui & K.D. Hyde (2001), Annulatascaceae. 1, Hong Kong. See Ranghoo *et al.* (*MR* **105**: 628, 2001).

Aquamarina Kohlm., Volkm.-Kohlm. & O.E. Erikss. (1996), ? Dothideomycetes. 1, North Carolina. See Kohlmeyer *et al.* (*MR* **100**: 393, 1996).

Aquamortierella Embree & Indoh (1967), Mortierellaceae. 1, New Zealand; Japan. See Embree & Indoh (*Bull. Torrey bot. Club* **94**: 464, 1967), Indoh (*TMSJ* **8**: 28, 1967).

Aquaphila Goh, K.D. Hyde & W.H. Ho (1998), anamorphic *Tubeufia*, Hso.?.?. 2, Australia. See Goh *et al.* (*MR* **102**: 588, 1998), Tsui *et al.* (*Mycol.* **99**: 884, 2007; phylogeny, anamorph).

Aquapoterium Raja & Shearer (2008), Helotiales. 1 (from fresh water), USA. See Raja *et al.* (*Mycol.* **100**: 141, 2008).

Aquascypha D.A. Reid (1965), Meruliaceae. 1, C. & S. America. See Reid (*Nova Hedwigia* Beih. **18**: 51, 1965), Ryvarden (*Syn. Fung.* **18**: 76, 2004).

Aquasphaeria K.D. Hyde (1995), ? Annulatascaceae. 1 (submerged wood), Queensland. See Hyde (*Nova Hedwigia* **61**: 119, 1995).

Aquathanatephorus C.C. Tu & Kimbr. (1978) = Thanatephorus fide Stalpers & Anderson *in* Sneh *et al.* (Eds) (*Rhizoctonia Species* Taxonomy, Molecular Biology, Ecology, Pathology and Disease Control: 58, 1996).

Aquatic fungi. Living in water. Over 3000 species of Fungi and almost 150 chromistans have been recorded from freshwater, brackish and marine environments (Shearer *et al.*, *Biodiversity and Conservation* **16**: 49, 2007). Here the term is restricted to freshwater in contrast to Marine fungi (q.v.). The chief zoosporic fungi of freshwater are *Chytridiomycota* and chromistans, esp. *Chytridiales* and *Saprolegniales*: Sparrow (*Aquatic phycomycetes*, 1943 [edn 2, 1960]; *Mycol.* **50**: 797, 1959, phylogeny), Emerson (*Mycol.* **50**: 589, 1959; culture), Fuller & Jaworski (*Zoosporic fungi in teaching and research*, 1987). Many are fish parasites; some parasitize freshwater plankton: Canter & Lund (*Ann. Bot., Lond.*, n.s. **14-15**, 1950-51; *New Phytol.* **47**: 238, 1948; *TBMS* **36**: 13, **37**: 111, 1953-54), Cook (*Am. J. Bot.* **50**: 580, 1943, on desmids), Khulbe (*Manual of aquatic fungi (Chytridiomycetes and Oomycetes)*, 2001), Paterson (*Mycol.* **50**: 85, 483, 1958).

'Hyphomycetes' of freshwater have received much attention (Ingold, *TBMS* **25**: 339, 1942). These fungi frequently have branched or sigmoid spores as an adaptation (typically of convergent evolution, see Tsui & Berbee, *Molecular Phylogenetics and Evolution* **39**: 587, 2006) to life on decaying leaves in fast running water (Ingold, *Mycol.* **58**: 43, 1966), but may also show other forms of adaptation, for example empty cells acting as float chambers in the genus *Ru-*

bikia. Over 100 anamorph gen. and 300 spp. have been recorded (Ingold, *Am. J. Bot.* **66**: 218, 1979; *An illustrated guide to aquatic and water-borne hyphomycetes* [*Publs Freshwater biol. Assn* **30**], 1975, keys, illustr.; *Biol. J. Linn. Soc.* **7**: 1, 1975, convergent evolution), Nilsson (*Symb. bot. upsal.* **18** (2), 1964), Webster & Descals (*in* Cole & Kendrick, *Biology of conidial fungi* **1**: 295, 1981). **Ecology**: Bärlocher (Ed.) (*The ecology of aquatic hyphomycetes*, 1992). **Teleomorphs**: Webster (*in* Bärlocher, *The ecology of aquatic hyphomycetes*: 99, 1992). **Regional surveys**. **China**: Zhu & Yu (*Acta Mycol. Sin.* **11**: 43, 1992). **Cuba**: Marvanová & Marvan (*Česká Myk.* **23**: 135, 1969). **Ghana**: Dixon (*TBMS* **42**: 174, 1959). **Hawaii**: Ranzoni (*Mycol.* **71**: 786, 1979). **Iceland**: Johnson (*J. Elisha Mitch. sci. Soc.* **84**: 179, 1968). **Jamaica**: Hudson & Ingold (*TBMS* **43**: 469, 1960). **Japan**: Tubaki (*Bull. Nat. Sci. Mus. Tokyo* **41**: 149, 1957). **Malaysia**: Nawawi (*Malayan Nature Journal* **39**: 75, 1985). **New Zealand**. Aimer & Segedin (*N.Z. J. Bot.* **23**: 273, 1985). **Nigeria**: Ingold (*TBMS* **39**: 108, **42**: 479, 1956-59). **N. Am.**: Peterson (*Mycol.* **54**: 117, **55**: 18, 570, 1962-63; gen. key). **Norway**: Bråthen (*Nord. J. Bot.* **4**: 375, 1984). **Puerto Rico**: Santos-Flores & Betancourt-López (*Caribbean J. Sci.* Special Publ. **2**: 1, 1997), Nieves-Rivera & Santos-Flores (*J. Agric. Univ. Puerto Rico* **89**: 97, 2005). **Sierra Leone**: Le'John (*TBMS* **48**: 261, 1965). **S. Am.**: Schoenlein-Crusius & Grandi (*Brazilian J. Microbiol.* **34**: 183, 2003). **Uganda**: Ingold (*TBMS* **41**: 109). **Ukraine**: Dudka ([*Aquatic hyphomycetes of the Ukraine*], 1974). **Venezuela**: Nilsson (*Svensk bot. Tidskr.* **56**: 351, 1962). **Zimbabwe**: Ingold (*TBMS* **41**: 109). See also aero-aquatic fungi.

Over 200 ascomycetes have also been recorded from freshwater habitats (Shearer, *Nova Hedw.* **56**: 1, 1993) and the tropics are now proving extremely rich in novel ascomycete genera (e.g. Hyde, *MR* **98**: 719, 1994).

Some saxicolous lichens, mainly of the *Lichinaceae* and the gen. *Dermatocarpon*, *Hymenelia*, *Placynthium*, *Polyblastia*, *Staurothele*, *Verrucaria* (q.v.), occur in freshwater; some may be always submerged (e.g. *Collema fluviatile*, *Hydrothria venosa*). They can form zones on river and lake margins related to the frequency of submersion (Rosentreter, *Northwest Sci.* **58**: 108, 1984; Santesson, *Medd. Lunds Univ. Limnol. Inst.* **1**, 1939, Sweden; Scott, *Lichenologist* **3**: 368, 1967, Zimbabwe), and can be used in the determination of river channel capacity (Gregory, *Earth Surface Processes* **1**: 273, 1976; Australia); a 'lichen-line' on trees can also indicate highwater levels (Hale, *Bryologist* **87**: 261, 1984).

A small number of smuts are associated with aquatic plants and may show some adaptation themselves to a freshwater environment (Piatek, *Polish Bot. J.* **51**: 173, 2006). In addition to plant debris saprobes and animal parasites, various other substrata in freshwater have been investigated for fungi (Czeczuga *et al.*, *Polish J. Environmental Sci.* **13**: 21, 2004). Yeasts are also known from aquatic environments, and may contribute to water self-purification (Dynowska *et al.*, *Int. J. Ecohydrology and Hydrobiol.* **5**: 147, 2005).

At least some aquatic fungi also occupy dry land habitats, for example as endophytes (Sati *et al.*, *Nat. Acad. Sci. Letters* **29**: 351, 2006). The land environment adjacent to fresh water can markedly affect the aquatic mycota. Introduced forest trees, for example, may result in a change in the range of aquatic fungi colonizing fallen leaves (Ferreira *et al.*, *Archiv für Mikrobiol.* **166**: 467, 2006). Diverse fungi are found in polluted water and sewage: Cooke (*Sydowia, Beih.* **1**: 136, 1957, list; *A laboratory guide to fungi in polluted waters, sewage and sewage treatment systems*, 1963; *Our mouldy earth*, 1970 [reprints and summarizes his studies in this field]). There have been many studies of aquatic fungi in relation to pollution (Krauss *et al.*, Aquatic fungi in heavy metal and organically polluted habitats, in Deshmukh & Rai (Eds) *Biodiversity of fungi, their role in human life*, 2005). Some attention has been given to possibilities of using aquatic fungi in bioremediation of oil pollution (Etim & Antai, *Global J. Env. Sci.* **6**: 33, 2007).

Aquaticheirospora Kodsueb & W.H. Ho (2007), anamorphic *Pleosporales*, H?.?.?. 1, Thailand. See Kodsueb *et al.* (*Botanical Journal of the Linnean Society* **155**: 283, 2007; descr.).

Aquaticola W.H. Ho, K.M. Tsui, Hodgkiss & K.D. Hyde (1999), Annulatascaceae. 5, Australia; Hong Kong. See Ho *et al.* (*Fungal Diversity* **3**: 87, 1999), Fallah & Shearer (*Mycol.* **93**: 566, 2001), Tsui *et al.* (*Nova Hedwigia* **77**: 161, 2003), Campbell & Shearer (*Mycol.* **96**: 822, 2004; phylogeny).

Arachniaceae Coker & Couch (1928) = Lycoperdaceae.

Arachnion Schwein. (1822), Agaricaceae. 6, widespread (subtropical). See Demoulin (*Nova Hedwigia* **21**: 641, 1972), Quadraccia (*Mycotaxon* **58**: 331, 1996; Italy).

Arachniopsis Long (1917) [non *Arachniopsis* Spruce 1882, *Hepaticae*] = Arachnion fide Demoulin (*Nova Hedwigia* **21**: 641, 1972).

Arachniotus J. Schröt. (1893), Gymnoascaceae. 3, Poland. See Orr *et al.* (*Mycol.* **69**: 126, 1977), Currah (*Mycotaxon* **24**: 1, 1985), Udagawa & Uchiyama (*Mycoscience* **41**: 303, 2000), Sugiyama & Mikawa (*Mycoscience* **42**: 413, 2001), Solé *et al.* (*Stud. Mycol.* **47**: 141, 2002; synonymy with *Gymnascella*).

Arachnocrea Z. Moravec (1956), Hypocreaceae. Anamorph *Verticillium*-like. 1 (on old polypores and plant tissues), widespread. See Rossman *et al.* (*Stud. Mycol.* **42**: 248 pp., 1999), Põldmaa (*Stud. Mycol.* **45**: 83, 2000), Samuels *et al.* (*CBS Diversity Ser.* **4**, 2006; USA).

arachnoid, covered with, or formed of, delicate hairs or fibres; araneose.

Arachnomycelium Grüss (1931), Fossil Fungi. 1.

Arachnomyces Massee & E.S. Salmon (1902), Arachnomycetaceae. Anamorph *Onychocola*. 10, Europe; America. See Malloch & Cain (*CJB* **48**: 839, 1970), Currah (*Mycotaxon* **24**: 1, 1985), Gibas *et al.* (*Medical Mycology* **40**: 573, 2002; anam.), Gibas *et al.* (*Stud. Mycol.* **47**: 131, 2002; phylogeny, links with *Eurotiales*), Sugiyama *et al.* (*Stud. Mycol.* **47**: 5, 2002; phylogeny), Gibas *et al.* (*Stud. Mycol.* **50**: 525, 2004), Geiser *et al.* (*Mycol.* **98**: 1053, 2006; phylogeny).

Arachnomycetaceae Gibas, Sigler & Currah (2002), Arachnomycetales. 2 gen., 11 spp.

Lit.: Gibas *et al.* (*Stud. Mycol.* **47**: 131, 2002), Geiser *et al.* (*Mycol.* **98**: 1053, 2006).

Arachnomycetales Gibas, Sigler & Currah (2002). Eurotiomycetidae. 1 fam., 2 gen., 11 spp. Fam.:
Arachnomycetaceae

For *Lit.* see under fam.

Arachnopeziza Fuckel (1870), ? Hyaloscyphaceae. 15, widespread (north temperate). See Korf (*Lloydia* **14**: 129, 1951), Huhtinen (*Mycotaxon* **30**: 9, 1987), Cantrell & Hanlin (*Mycol.* **89**: 745, 1997; DNA), Yu & Zhuang (*Nova Hedwigia* **74**: 415, 2002; China).

Arachnopezizella Kirschst. (1938) = Arachnopeziza fide Korf (*Lloydia* **14**: 129, 1951).

Arachnophora Hennebert (1963), anamorphic *Pezizomycotina*, Hso.≡ eP.1. 4, widespread (esp. north temperate). See Hughes (*N.Z. Jl Bot.* **17**: 139, 1979; descr.), Castañeda Ruíz *et al.* (*Nova Hedwigia* **64**: 473, 1997), Kendrick (*CJB* **81**: 75, 2003; morphogenesis).

Arachnoscypha Boud. (1885) = Arachnopeziza fide Korf (*Lloydia* **14**: 129, 1951), Svrček (*Česká Mykol.* **41**: 193, 1987).

Arachnospora R.F. Castañeda, Minter & Camino (2003), anamorphic *Pezizomycotina*, H?.?.?. 1 (on decaying leaves), Cuba. See Castañeda Ruíz *et al.* (*Mycotaxon* **87**: 386, 2003).

Arachnotheca Arx (1971), Onygenales. Anamorph *Chrysosporium*. 3, widespread. See Currah (*Mycotaxon* **24**: 1, 1985), Uchiyama *et al.* (*Mycoscience* **36**: 211, 1995), Sugiyama & Mikawa (*Mycoscience* **42**: 413, 2001), Sugiyama *et al.* (*Stud. Mycol.* **47**: 5, 2002; phylogeny).

Arachnula Cienk. (1876), Biomyxida. q.v.

Araeocoryne Corner (1950), Gomphaceae. 1, Malaysia. See Corner (*Ann. Bot. Mem.* [A monograph of Clavaria and allied genera] **1**: 194, 1950).

Araneomyces Höhn. (1909), anamorphic *Paranectriella*, Hso.1bH.1. 1 (mycoparasitic), Brazil. See Sutton (*TBMS* **83**: 399, 1984), Rossman (*Mycol. Pap.* **157**: 71 pp., 1987), Wu *et al.* (*MR* **101**: 1318, 1997).

Araneosa Long (1941), Agaricaceae. 1, USA. Basidioma gasteroid. See Long (*Mycol.* **33**: 351, 1941).

araneose (**araneous**), see arachnoid.

Arberia Nieuwl. (1916) [non *Arberia* C.D. White 1908, fossil ? *Pteridophyta*] ≡ Asteridium.

Arborella Zebrowski (1936), Fossil Fungi ? Chytridiomycetes. 2 (Cambrian to ? Recent), Australia.

arboricolous, growing on trees.

Arborillus Munt.-Cvetk. & Gómez-Bolea (1998), anamorphic *Pezizomycotina*, Hso.?.?. 1 (lichenicolous), Spain. See Muntañola-Cvetkovic & Gómez-Bolea (*Mycotaxon* **68**: 152, 1998).

Arborispora K. Ando (1986), anamorphic *Pezizomycotina*, Hso.1bH.1/10. 3 (aquatic), Japan. See Ando (*TMSJ* **27**: 120, 1986), Gönczöl & Révay (*Fungal Diversity* **12**: 19, 2003; ecology).

arbuscle (**arbuscule**), see mycorrhiza.

Arbuscula Bat. & Peres (1965) [non *Arbuscula* H.A. Crum, Steere & L.E. Anderson 1964, *Musci*] ≡ Neoarbuscula.

Arbusculidium B. Sutton (1982) [non *Arbusculidium* J. Deunff 1968, fossil *Acritarcha*] ≡ Neoarbuscula.

Arbusculina Marvanová & Descals (1987), anamorphic *Pezizomycotina*, Hso.1bH.19. 2 (aquatic), widespread. See Marvanová & Descals (*TBMS* **89**: 499, 1987), Marvanová (*TBMS* **90**: 607, 1988), Descals (*MR* **109**: 545, 2005; propagules).

Arbusculites Paradkar (1976), Fossil Fungi. 1 (Cretaceous), India. See Paradkar (*Journal of Palynology* **10**: 120, 1974).

Arcangelia Sacc. (1890) = Didymella fide von Arx & Müller (*Stud. Mycol.* **9**, 1975).

Arcangeliella Cavara (1900) = Lactarius fide Miller *et al.* (*Mycol.* **93**: 344, 2001).

Archaea (**archaebacteria**), an heterogeneous group of prokaryotic organisms belonging to the Domain Archaea. See bacteria.

archaeascus, see ascus.

Archaeoglomus N. Sharma, R.K. Kar, A. Agarwal & R. Kar (2005), Fossil Fungi, Glomeraceae. 1.

Archaeomarasmius Hibbett, D. Grimaldi & Donoghue (1997), Fossil Fungi. 1, New Jersey. See Hibbett *et al.* (*Am. J. Bot.* **84**: 982, 1997).

Archaeospora Morton & Redecker (2001), Archaeosporaceae. 1, USA. See Morton & Redecker (*Mycol.* **93**: 183, 2001), Redecker & Raab (*Mycol.* **98**: 885, 2006; phylogeny).

Archaeosporaceae J.B. Morton & D. Redecker (2001), Archaeosporales. 1 gen., 1 spp.

Lit.: Azcon-Aguilar & Barea (*Mycorrhiza* **6**: 457, 1996), van der Heijden *et al.* (*Nature* Lond. **396**: 69, 1998), Morton & Redecker (*Mycol.* **93**: 183, 2001), Schüßler *et al.* (*MR* **105**: 1413, 2001), Spain (*Mycotaxon* **87**: 109, 2003), Hafeel (*Mycorrhiza* **14**: 213, 2004), Redecker (*Glomeromycota* Arbuscular mycorrhizal fungi and their relative(s). Version 01 July 2005. http://tolweb.org/Glomeromycota/28715/2005.07.01 in The Tree of Life Web Project, http://tolweb.org: [unpaginated], 2005), Walker *et al.* (*MR* **111**: 137, 2007).

Archaeosporales C. Walker & A. Schüssler (2001). Glomeromycetes. 3 fam., 3 gen., 6 spp. Fams:
(1) **Ambisporaceae**
(2) **Archaeosporaceae**
(3) **Geosiphonaceae**

For *Lit.* see under fam.

Archagaricon A. Hancock & Atthey (1869), Fossil Fungi (mycel.) Fungi. 5 (Carboniferous), British Isles.

Archecribraria Locq. (1983), Fossil Fungi. 2, Sahara.

Archemycota. Name in the rank of phylum including the groups treated in this *Dictionary* as *Chytridiomycota* and *Zygomycota* (incl. *Trichomycetes*); see Cavalier-Smith (*in* Rayner *et al.* (Eds), *Evolutionary biology of the fungi*: 339, 1987; *in* Osawa & Honjo (Eds), *Evolution of life*: 271, 1991).

Archeomycelites Bystrov (1959), Fossil Fungi (mycel.) Fungi. 1 (Devonian), former USSR.

Archephoma Watanabe, H. Nishida & Kobayashi (1999), Fossil Fungi. 1, Japan. See Watanabe *et al.* (*Int. J. Pl. Sci.* **160**: 436, 1999).

Archeplax Locq. (1985), Fossil Fungi. 1, Sahara.

Archeterobasidium Koeniguer & Locq. (1979), Fossil Fungi, Agaricomycetes. 1 (Miocene), Libya.

Archiascomycetes = Taphrinomycetes. Class of *Ascomycota* provisionally proposed by Nishida & Sugiyama (*Mycoscience* **35**: 361, 1994) for *Pneumocystis*, *Protomyces*, *Saitoella*, *Schizosaccharomyces* and *Taphrina* based on 18S rRNA sequences; considered by the authors to perhaps not be monophyletic but to have originated before *Euascomycetes* and *Hemiascomycetes*.

archicarp (of ascomycetes), the cell, hypha, or coil which later becomes the ascoma or part of it.

Archilichens, lichens in which the algae are bright green (obsol.).

Archimycetes (obsol.). Name used rarely for *Plasmodiophoromycota* and *Chytridiomycota*. *Myxochytridiales*.

Architrypethelium Aptroot (1991), Trypetheliaceae

(L). 3, widespread (tropical). See Aptroot (*Biblthca Lichenol.* **44**: 120, 1991), Aptroot *et al.* (*Biblthca Lichenol.* **97**, 2008; Costa Rica).

archontosome, an electron-dense body occurring near nuclei at all stages from crozier formation to the development of young ascospores in *Xylaria polymorpha*. See Beckett & Crawford (*J. gen. Microbiol.* **63**: 269, 1970).

Arcispora Marvanová & Bärl. (1998), anamorphic *Basidiomycota*. 1 (aquatic), Canada. See Marvanová & Bärlocher (*Mycol.* **90**: 531, 1998).

arctic mycology, see Polar and alpine mycology.

Arcticomyces Savile (1959) = Exobasidium fide Donk (*Persoonia* **4**: 287, 1966).

Arctocetraria Kärnefelt & A. Thell (1993), Parmeliaceae (L). 2, Europe. See Kärnefelt *et al.* (*Bryologist* **96**: 394, 1993), Thell *et al.* (*Mycol. Progr.* **1**: 335, 2002; phylogeny), Mattsson & Articus (*Symb. bot. upsal.* **34** no. 1: 237, 2004; phylogeny), Randlane & Saag (*Central European Lichens*: 75, 2006; key).

Arctoheppia Lynge (1938) ≡ Thelignya fide Jørgensen & Henssen (*Taxon* **39**: 343, 1990).

Arctomia Th. Fr. (1860), Arctomiaceae (L). 5, Europe; N. America. See Henssen (*Svensk bot. Tidskr.* **63**: 126, 1969), Jørgensen (*Lichenologist* **35**: 287, 2003; China), Wedin *et al.* (*MR* **109**: 159, 2005; phylogeny, link with *Hymeneliaceae*).

Arctomiaceae Th. Fr. (1860), Ostropomycetidae (inc. sed.) (L). 3 gen., 7 spp.
Lit.: Jørgensen (*Lichenologist* **35**: 287, 2003), Lumbsch *et al.* (*Lichenologist* **37**: 291, 2005), Jørgensen (*Nordic Lichen Flora* **3**: Cyanolichens: 9, 2007), Lumbsch *et al.* (*MR* **111**: 257, 2007; phylogeny).

Arctoparmelia Hale (1986), Parmeliaceae. 5, widespread. See Thell *et al.* (*Symb. bot. upsal.* **34** no. 1: 429, 2004; biogeography), Blanco *et al.* (*Mol. Phylogen. Evol.* **39**: 52, 2006; phylogeny).

Arctopeltis Poelt (1983), Lecanoraceae (L). 1, Arctic. See Arup & Grube (*Lichenologist* **30**: 415, 1998; DNA), Feige & Lumbsch (*Cryptog. Bryol.-Lichénol.* **19**: 147, 1998; ontogeny), Grube *et al.* (*MR* **108**: 506, 2004; phylogeny).

Arctosporidium Thor (1930) nom. dub., ? Fungi. 1, Svalbard.

Arcuadendron Sigler & J.W. Carmich. (1976), anamorphic *Pezizomycotina*, Hso.0eH.40. 2, India; former Yugoslavia. See Sigler & Carmichael (*Mycotaxon* **4**: 355, 1976).

arcuate, arc-like.

ardella, a small spot-like apothecium, as in the lichen *Arthonia*.

Ardhachandra Subram. & Sudha (1978), anamorphic *Pezizomycotina*, H. 4, widespread. Placed into synonymy with *Rhinocladiella* by some authors. See Onofri & Castagnola (*Mycotaxon* **18**: 337, 1983), Keates & Carris (*Cryptog. Bot.* **4**: 336, 1994; from *Vaccinium macrocarpon*), Chen & Tzean (*MR* **99**: 364, 1995; key to 4 spp.).

ardosiaceous (ardesiaceous), slate-coloured.

Arecacicola Joanne E. Taylor, J. Fröhl. & K.D. Hyde (2001), Sordariomycetes. 1 (on palm trunk), Indonesia. See Taylor *et al.* (*Mycoscience* **42**: 370, 2001).

Arecomyces K.D. Hyde (1996), Hyponectriaceae. 9 (saprobic on palms), widespread. See Hyde (*Sydowia* **48**: 224, 1996).

Arecophila K.D. Hyde (1996), Xylariales. 13 (on *Palmae*), S.E. Asia. Placed in the *Cainiaceae* by some authors, but relationships are obscure. See Hyde (*Nova Hedwigia* **63**: 81, 1996), Jeewon *et al.* (*MR* **107**: 1392, 2003).

Aregma Fr. (1815) = Phragmidium fide Vánky (*in litt.*).

Arenaea Penz. & Sacc. (1901) = Lachnum fide Korf in Ainsworth *et al* (*The Fungi* **4A**, 1973).

Arenariomyces Höhnk (1954), Halosphaeriaceae. 4 (marine), widespread. See Jones *et al.* (*J. Linn. Soc.* Bot. **87**: 193, 1983), Kohlmeyer & Volkmann-Kohlmeyer (*MR* **92**: 413, 1989; key), Jones *et al.* (*CJB* **74** Suppl. 1: S790, 1995; ultrastr.).

Arenicola Velen. (1947) = Entoloma fide Kuyper (*in litt.*).

Areolaria Kalchbr. (1884) = Phellorinia fide Stalpers (*in litt.*).

areolate, having division by cracks into small areas.

Areolospora S.C. Jong & E.E. Davis (1974), Xylariaceae. 1, widespread. Treated as *Phaeosporis* by Hawksworth (*SA* **13**: 1, 1994). See Hawksworth (*Norw. Jl Bot.* **27**: 97, 1980), Krug *et al.* (*Mycol.* **86**: 581, 1994).

arescent, becoming crustose on drying.

Argentinomyces N.I. Peña & Aramb. (1997), ? Sordariomycetes. 1 (on driftwood), Argentina. See Peña & Arambarri (*Mycotaxon* **65**: 333, 1997).

Argomyces Arthur (1912) ≡ Argotelium.

Argomycetella Syd. (1922) = Maravalia fide Mains (*Bull. Torrey bot. Club* **66**: 173, 1939).

Argopericonia B. Sutton & Pascoe (1987), anamorphic *Pezizomycotina*, Hso.0eH.6/10. 2, Australia; India. See Sutton & Pascoe (*TBMS* **88**: 41, 1987), D'Souza *et al.* (*Mycotaxon* **82**: 133, 2002; Andaman Is).

Argopsis Th. Fr. (1857), ? Brigantiaeaceae (L). 3, widespread (sub-Antarctica). See Lamb (*J. Hattori bot. Lab.* **38**: 447, 1974).

Argotelium Arthur (1906) = Puccinia fide Arthur (*Am. J. Bot.* **5**: 485, 1918).

Argylium Wallr. (1833) = Melanogaster fide Stalpers (*in litt.*).

Argynna Morgan (1895), Argynnaceae. 1, N. America. See Shearer & Crane (*TBMS* **75**: 193, 1980).

Argynnaceae Shearer & J.L. Crane (1980), ? Dothideomycetes (inc. sed.). 2 gen., 2 spp.
Lit.: Shearer & Crane (*TBMS* **75**: 193, 1980), Hawksworth (*SA* **6**: 153, 1987).

arid, dry.

Ariefia Jacz. (1922) = Zopfiella fide Cannon (*in litt.*).

Ariella E.-J. Gilbert (1941) = Amanita Pers. fide Singer (*Agaric. mod. Tax.* edn 3, 1975).

Aristadiplodia Shirai (1919) nom. dub., anamorphic *Pezizomycotina*. 1, Japan. See Sutton (*Mycol. Pap.* **141**, 1977).

Aristastoma Tehon (1933), anamorphic *Pezizomycotina*, Cpd.≡ eH.1. 5, widespread. *A. oeconomicum* (zonate leaf spot of cowpea, *Vigna*). See Sutton (*Mycol. Pap.* **97**, 1964; key), Hyde & Philemon (*MR* **95**: 1151, 1991).

Arkoola J. Walker & Stovold (1986), Venturiaceae. 1, Australia. See Walker & Stovold (*TBMS* **87**: 23, 1986).

Armata W. Yamam. (1958), ? Micropeltidaceae. 1, Japan. See Yamamoto (*Science Reports of the Hyogo University of Agriculture* Series Agricultural Biology **3**: 89, 1958).

Armatella Theiss. & Syd. (1915), Meliolaceae. 12 (from leaves), widespread (tropical). See von Arx (*Fungus* Wageningen **28**: 1, 1958), Hosagoudar (*J.*

Econ. Taxon. Bot. **15**: 195, 1991; India), Hosagoudar & Abraham (*J. Econ. Taxon. Bot.* **25**: 560, 2001), Hosagoudar (*Sydowia* **55**: 162, 2003; family placement).

Armatellaceae Hosag. (2003) = Meliolaceae.
Lit.: Hosagoudar (*Sydowia* **55**: 162, 2007).

armilla, see annulus.

Armillaria (Fr.) Staude (1857), Physalacriaceae. 35, widespread. Most species cause serious root diseases in woody plants; some form orchid mycorrhizas. Application of a biological species concept has led to recognition of a larger number of biological species, which cannot always be recognised morphologically. See Ullrich & Anderson (*Exp. Mycol.* **2**: 119, 1978; karyology), Shaw & Kile (*Armillaria root disease*, 1991), Anderson & Stasovski (*Mycol.* **84**: 506, 1992; phylogeny), Smith *et al.* (*Nature* **256**: 428, 1992; population biology), Piercey-Normore *et al.* (*Mol. Phylogen. Evol.* **10**: 49, 1998; phylogeny), Fox (*Armillaria root rot: biology and control of honey fungus*, 2000).

Armillariella (P. Karst.) P. Karst. (1881) = Armillaria fide Dennis *et al.* (*TBMS* **37**: 33, 1954).

armillate, edged; fringed; frilled.

Arnaudia Bat. (1960) = Acantharia fide Müller & von Arx (*Beitr. Kryptfl. Schweiz* **11** no. 2, 1962).

Arnaudiella Petr. (1927), Microthyriaceae. Anamorph *Xenogliocladiopsis*. 3 or 6, widespread. See Crous & Kendrick (*CJB* **72**: 59, 1994).

Arnaudina Trotter (1931), anamorphic *Pezizomycotina*, Hsp.≡ eP.?. 1, Brazil. See Carmichael *et al.* (*Genera of Hyphomycetes*, 1980).

Arnaudovia Valkanov (1963) = Polyphagus fide Karling (*Chytriomyc. Iconogr.*, 1977).

Arniella Jeng & J.C. Krug (1977), Lasiosphaeriaceae. 2 (coprophilous), USA; Venezuela. See Jeng & Krug (*Mycol.* **69**: 73, 1977), Huhndorf *et al.* (*Mycol.* **96**: 368, 2004).

Arnium Nitschke ex G. Winter (1873), Lasiosphaeriaceae. 12 (mostly coprophilous), widespread. See Krug & Cain (*CJB* **50**: 367, 1972; key), Krug & Cain (*CJB* **55**: 83, 1977), Lorenzo & Havrylenko (*Mycol.* **93**: 1221, 2001; Argentina), Miller (*Sydowia* **55**: 267, 2003; ascomata), Huhndorf *et al.* (*Mycol.* **96**: 368, 2004).

Arnoldia A. Massal. (1856) [non *Arnoldia* Cass. 1824, *Compositae*] = Lempholemma fide Hawksworth *et al.* (*Dictionary of the Fungi* edn 8, 1995).

Arnoldia D.J. Gray & Morgan-Jones (1980) ≡ Arnoldiomyces.

Arnoldiella R.F. Castañeda (1984), anamorphic *Pezizomycotina*, Hso.0-≡ hP.10. 1, Cuba. See Castañeda (*Revta Jardin bot. Nac.* Univ. Habana **5**: 58, 1984), Goos (*Mycol.* **79**: 1, 1987).

Arnoldiomyces Morgan-Jones (1980), anamorphic *Hypomyces*, Hso.≡ eH.10. 2, Americas. See Morgan-Jones (*Mycotaxon* **11**: 446, 1980), Samuels & Seifert *in* Sugiyama (Ed.) (*Pleomorphic Fungi: The Diversity and its Taxonomic Implications*: 29, 1987).

Arongylium, see *Strongylium (Ach.) Gray*.

Aropsiclus Kohlm. & Volkm.-Kohlm. (1994), Xylariales. 1, USA. Perhaps related to *Phomatospora*. See Kohlmeyer & Volkmann-Kohlmeyer (*SA* **11**: 95, 1993), Kohlmeyer & Volkmann-Kohlmeyer (*SA* **13**: 24, 1994).

Aroramyces Castellano & Verbeken (2000) = Hysterangium fide Hosaka *et al.* (*Mycol.* **98**: 949, 2006; systematic position, nested in *Hysterangium*).

Arpinia Berthet (1974), Pyronemataceae. 4, Europe. See Hohmeyer (*Mycol. Helv.* **3**: 221, 1988; key), Häffner (*Mitt. Arbeitsg. Pilzk. Niederrhein* **7**: 132, 1989; key), Perry *et al.* (*MR* **111**: 549, 2007; phylogeny).

arrect, stiffly upright.

Arrhenia Fr. (1849), Tricholomataceae. *c.* 25, widespread (temperate). Recognition of *Arrhenia* makes *Omphalina* paraphyletic. See Redhead (*CJB* **62**: 865, 1984), Lutzoni & Vilgalys (*CJB* **73** Suppl. 1: S649, 1995), Lutzoni (*Syst. Biol.* **46**: 373, 1997), Redhead *et al.* (*Mycotaxon* **83**: 19, 2002; phylogeny), Barrasa & Rico (*Mycol.* **95**: 700, 2003; Iberian Peninsula).

Arrhenosphaera Stejskal (1974), Ascosphaeraceae. 1 (from bee hives), Venezuela. See Stejskal (*J. Apicult. Res.* **13**: 39, 1974).

Arrhytidia Berk. & M.A. Curtis (1849) = Dacrymyces fide Donk (*Persoonia* **4**: 269, 1966).

arsenic detection, see *Scopulariopsis*.

Artallendea Bat. & H. Maia (1960) = Armatella fide Katumoto (*Bull. Fac. Agr. Yamag. Univ.* **13**: 291, 1962).

Artheliopsis Vain. (1896) = Echinoplaca fide Hawksworth *et al.* (*Dictionary of the Fungi* edn 8, 1995).

Arthonaria Fr. (1825) ? = Enterographa fide Hawksworth *et al.* (*Dictionary of the Fungi* edn 8, 1995).

Arthonia Ach. (1806) nom. cons., Arthoniaceae (±L). Anamorph *Septocyta*. *c.* 491 (partly on lichens), widespread. See Santesson (*Symb. bot. upsal.* **12** no. 1: 1, 1952; foliicolous spp.), Coppins (*Lichenologist* **21**: 195, 1989; Brit. Isl.), Tehler (*CJB* **68**: 2458, 1990; cladistics), Grube *et al.* (*Lichenologist* **27**: 25, 1995; key 9 spp. on lichens), Lücking (*Lichenologist* **27**: 127, 1995; Costa Rica), Myllys *et al.* (*Bryologist* **101**: 70, 1998; phylogeny), Sundin & Tehler (*Lichenologist* **30**: 381, 1998; phylogeny), Wedin & Hafellner (*Lichenologist* **30**: 59, 1998; lichenicolous spp.), Grube & Lücking (*MR* **105**: 1007, 2001; ascogenous hyphae), Follmann & Werner (*J. Hattori bot. Lab.* **94**: 261, 2003; on *Roccellaceae*).

Arthoniaceae Rchb. (1841), Arthoniales (±L). 12 gen. (+ 48 syn.), 603 spp.
Lit.: Santesson (*Symb. bot. upsal.* **12** no. 1: 1, 1952; foliicolous spp.), Coppins (*Lichenologist* **21**: 195, 1989), Diederich (*Flechten Follmann* Contributions to Lichenology in Honour of Gerhard Follmann: 179, 1995), Lücking (*Lichenologist* **27**: 127, 1995; 25 foliicolous spp. Costa Rica), Ferraro & Lücking (*Phyton* **37**: 61, 1997; foliicolous spp.), Grube & Matzer (*Biblthca Lichenol.* **68**: 1, 1997), Thor (*Symb. bot. upsal.* **32** no. 1: 267, 1997), Grube (*Bryologist* **101**: 377, 1998), Makhija & Patwardhan (*Mycotaxon* **67**: 287, 1998), Myllys *et al.* (*Bryologist* **101**: 70, 1998), Sundin & Tehler (*Lichenologist* **30**: 381, 1998), Wedin & Hafellner (*Lichenologist* **30**: 59, 1998), Lutzoni (*Am. J. Bot.* **91**: 1446, 2004).

Arthoniactis (Vain.) Clem. (1909) = Lecanactis Körb. fide Hawksworth *et al.* (*Dictionary of the Fungi* edn 8, 1995).

Arthoniales Henssen ex D. Hawksw. & O.E. Erikss. (1986), (±L). Arthoniomycetes. 3 fam., 74 gen., 1538 spp. Sister-group of the *Dothideomycetes*.

Thallus varied, crustose but sometimes very poorly developed or absent; ascomata usually apothecial but sometimes with a poroid opening, often elongated and branched, ascomatal wall poorly to well-developed. Interascal tissue composed of branched paraphysoids in a gel matrix. Asci thick-walled, ±

fissitunicate, usually with a large apical dome, blueing in iodine; ascospores simple or septate, sometimes becoming brown and ornamented, without sheath. Anamorph pycnidial. Forming crustose lichens with green photobionts (esp. trentepohlioid), lichenicolous or saprobes; on a wide range of substrata, incl. many trop. foliicolous and corticolous spp. Fams:
(1) **Arthoniaceae**
(2) **Chrysothricaceae**
(3) **Roccellaceae** (syn. *Opegraphaceae*)

Lit.: Grube (*Bryologist* **101**: 377, 1998; phylogeny), Henssen & Thor (in Hawksworth (ed.), *Ascomycete Systematics*: *Problems and Perspectives in the Nineties*: 43, 1994), Letrouit-Galinou *et al.* (*Bull. Soc. linn. Provence* **45**: 389, 1994; ultrastr. asci), Myllys *et al.* (*Bryologist* **101**: 70, 1998; phylogeny), Renobales & Barreno (*Anales jard. bot. Madrid* **46**: 263, 1989; asci), Spatafora et al. (*Mycol.* **98**: 1018, 2006), Tehler (*CJB* **68**: 2458, 1990, phylogeny; *Crypt. Bot.* **5**: 82, 1995, molec. & morph. phylogeny).

Arthoniomyces E.A. Thomas ex Cif. & Tomas. (1953) = Arthonia.

Arthoniomycetes O.E. Erikss. & Winka (1997), Pezizomycotina. 1 ord., 4 fam., 78 gen., 1608 spp. Ord.: **Arthoniales**

For *Lit.* see ord. and fam.

Arthoniomycetidae, see *Arthoniomycetes*.

Arthoniopsis Müll. Arg. (1890) = Arthonia fide Santesson (*Symb. bot. upsal.* **12** no. 1: 1, 1952).

Arthophacopsis Hafellner (1998), Arthoniales (±L). 1, widespread. See Grube (*Bryologist* **101**: 377, 1998; phylogeny), Hafellner (*Cryptog. Bryol.-Lichénol.* **19**: 159, 1998), Diederich (*Herzogia* **16**: 41, 2003; USA).

Arthopyrenia A. Massal. (1852), Arthopyreniaceae (±L). *c.* 117, widespread. See Swinscow (*Lichenologist* **3**: 55, 1965), Harris (*Mich. Bot.* **12**: 1, 1973), Riedl (*Sydowia* **29**: 115, 1977; *A. punctiformis*-group), Coppins (*Lichenologist* **20**: 305, 1988; Brit. Isl.), Foucard (*Graphis Scripta* **4**: 49, 1992; key 13 spp. on bark, Sweden), Gams (*Taxon* **41**: 99, 1992; nomencl.), Harris (*More Florida Lichens*, 1995), Aptroot *et al.* (*Biblthca Lichenol.* **64**: 220 pp., 1997; New Guinea), Kainz *et al.* (*Nova Hedwigia* **72**: 209, 2001; Namibia), Mohr *et al.* (*MR* **108**: 515, 2004), Del Prado *et al.* (*MR* **110**: 511, 2006; phylogeny), Aptroot *et al.* (*Biblthca Lichenol.* **97**, 2008; Costa Rica).

Arthopyreniaceae Walt. Watson (1929), Pleosporales (±L). 3 gen. (+ 11 syn.), *c.* 162 spp.

Lit.: Coppins (*Lichenologist* **20**: 305, 1988), Upreti & Pant (*Bryologist* **96**: 226, 1993), Harris (*More Florida Lichens*, 1995), Aptroot (*Nova Hedwigia* **64**: 169, 1997), Sérusiaux & Aptroot (*Bryologist* **101**: 144, 1998), Del Prado *et al.* (*MR* **110**: 511, 2006; phylogeny).

Arthopyreniella J. Steiner (1911) = Mycoglaena fide Harris (*in litt.*).

Arthopyreniomyces Cif. & Tomas. (1953) ≡ Pyrenyllium fide Aguirre-Hudson (*Bull. Br. Mus. nat. hist.* Bot. **21**: 85, 1991).

Arthotheliomyces Cif. & Tomas. (1953) = Arthothelium fide Hawksworth *et al.* (*Dictionary of the Fungi* edn 8, 1995).

Arthotheliopsidomyces Cif. & Tomas. (1953) ≡ Arthotheliopsis.

Arthotheliopsis Vain. (1896), Gomphillaceae (L). 4, neotropics. See Hawksworth *et al.* (*Dictionary of the Fungi* edn 8, 1995).

Arthothelium A. Massal. (1852), Arthoniales (±L). *c.* 121, widespread. See Santesson (*Symb. bot. upsal.* **12** no. 1: 1, 1952), Coppins & James (*Lichenologist* **11**: 27, 1979; key Brit. spp.), Tehler (*CJB* **68**: 2451, 1990; posn), Makhija & Patwardhan (*J. Hattori bot. Lab.* **78**: 189, 1995; India), Makhija & Patwardhan (*Trop. Bryol.* **10**: 205, 1995; nomencl.), Grube & Giralt (*Lichenologist* **28**: 15, 1996; Mediterranean), Grube (*Bryologist* **101**: 377, 1998; phylogeny).

arthric (of conidiogenesis), thallic conidiogenesis characterized by the conversion of a pre-existing, determinate hyphal element into a conidium (**arthroconidium**, thallic-arthroconidium, arthrospore), as in *Geotrichum*. See arthrocatenate.

Arthriniaceae Nann. (1934) = Apiosporaceae.

Arthriniites Babajan & Tasl. (1977), Fossil Fungi. 1 (Tertiary), former USSR.

Arthrinium Kunze (1817), anamorphic *Apiospora*, Hso.0eP.37. *c.* 31, widespread (temperate). See Ellis (*Mycol. Pap.* **103**, 1965; key), Minter (*Proc. Indian Acad. Sci.* Pl. Sci. **94**: 281, 1985; relationships with other anamorphic fungi), Rai (*Mycoses* **32**: 472, 1989; *A. phaeospermum* var. *indicum* in humans), Scheuer (*Öst. Z. Pilzk.* **5**: 1, 1996), Kang *et al.* (*Mycotaxon* **81**: 321, 2002; phylogeny).

arthro- (prefix), jointed.

Arthroascus Arx (1972) = Saccharomycopsis Schiønning See von Arx (*Fungus* Wageningen **28**: 1, 1958), Smith *et al.* (*Antonie van Leeuwenhoek* **58**: 249, 1990; DNA), Hosagoudar (*J. Econ. Taxon. Bot.* **15**: 195, 1991; India), Kurtzman & Robnett (*Antonie van Leeuwenhoek* **73**: 331, 1998), Hosagoudar & Abraham (*J. Econ. Taxon. Bot.* **25**: 560, 2001), Hosagoudar (*Sydowia* **55**: 162, 2003; family placement), Naumov *et al.* (*J. gen. appl. Microbiol.* Tokyo **49**: 267, 2003; reinstatement of genus), Naumov & Kondrat'eva (*Doklady Biological Sciences* **403**: 298, 2005; Japan), Naumov *et al.* (*Int. J. Syst. Evol. Microbiol.* **56**: 1997, 2006; phylogeny).

Arthrobotryaceae Corda (1842) = Orbiliaceae.

Arthrobotryella Sibilia (1928) ? = Cordana fide Hughes (*N.Z. Jl Bot.* **16**: 326, 1978).

Arthrobotryomyces Bat. & J.L. Bezerra (1961) nom. dub., anamorphic *Pezizomycotina*, Hsy.≡ eH.? (L). 1, Brazil. See Lücking *et al.* (*Lichenologist* **30**: 121, 1998).

Arthrobotrys Corda (1839), anamorphic *Orbilia*, Hsy.1eH.6. 63 (nematophagous), widespread. See also Predacious fungi. See Haard (*Mycol.* **60**: 1140, 1969; key), Jarowaja (*Acta Mycologica* Warszawa **6**: 337, 1970; key), van Oorschot (*Stud. Mycol.* **26**: 61, 1985; key), Werthmann-Cliemas & Lysek (*TBMS* **87**: 656, 1987; synnema formation), Rubner (*Stud. Mycol.* **39**, 1996), Liou & Tzean (*Mycol.* **89**: 876, 1997; phylogeny), Ahrén *et al.* (*FEMS Microbiol. Lett.* **158**: 179, 1998; phylogeny), Hagedorn & Scholler (*Sydowia* **51**: 27, 1999; phylogeny), Scholler *et al.* (*Sydowia* **51**: 89, 1999; gen. concept, comb. Novs), Zhang *et al.* (*Mycosystema* **20**: 51, 2001; morphogenesis), Mo *et al.* (*Fungal Diversity* **18**: 107, 2005; synanamorphs), Li *et al.* (*Mycol.* **97**: 1034, 2005; phylogeny).

Arthrobotryum Ces. (1854), anamorphic *Pezizomycotina*, Hsy.≡ eP.19. 4, widespread. See Hughes (*Naturalist* Hull: 171, 1951), Illman & White (*CJB* **63**: 423, 1985).

Arthrobotryum O. Rostr. (1916) ≡ Gonyella.

arthrocatenate (of thalloconidia), formed in chains by the simultaneous or random fragmentation of a hypha.

Arthrocladia Golovin (1956) [non *Arthrocladia* Duby 1830, *Algae*] ≡ Arthrocladiella.

Arthrocladiella Vassilkov (1960), Erysiphaceae. Anamorph *Oidium* subgen. *Graciloidium*. 3, widespread. See Vassilkov (*Botanicheskiĭ Zhurnal* **45**: 1368, 1960), Braun *et al.* (*The Powdery Mildews* A Comprehensive Treatise: 13, 2001; review), Cunnington *et al.* (*Australas. Pl. Path.* **32**: 421, 2003; diagnosis), Takamatsu (*Mycoscience* **45**: 147, 2004; phylogeny), Cook *et al.* (*MR* **110**: 672, 2006; on *Catalpa*), Wang *et al.* (*Mycol.* **98**: 1065, 2006; phylogeny).

Arthrocladium Papendorf (1969), anamorphic *Pezizomycotina*, Hso.≡ eP.1/10. 1 (from soil), S. Africa. See Papendorf (*TBMS* **52**: 483, 1969).

arthroconidium, see arthric.

Arthrocristula Sigler, M.T. Dunn & J.W. Carmich. (1982), anamorphic *Pezizomycotina*, Hso.0eP.40. 1, USA; Sri Lanka. See Sigler *et al.* (*Mycotaxon* **15**: 409, 1982).

Arthroderma Curr. (1860), Arthrodermataceae. Anamorphs *Microsporum*, *Trichophyton*. 25 (on skin etc.), widespread. See Padhye & Carmichael (*CJB* **49**: 1525, 1971; key 13 spp.), Currah (*Mycotaxon* **24**: 1, 1985), Takashio *et al.* (*Mycol.* **77**: 166, 1985; ontogeny), Weitzman *et al.* (*Mycotaxon* **15**: 505, 1986), Kawasaki *et al.* (*Mycopathologia* **118**: 95, 1992), Ito *et al.* (*Mycoses* **41**: 133, 1998; ultrastr.), Kano *et al.* (*Curr. Microbiol.* **37**: 236, 1998; chitin synthase phylogeny), Makimura *et al.* (*J. Clin. Microbiol.* **36**: 2629, 1998; phylogeny), Harmsen *et al.* (*Mycoses* **42**: 67, 1999; DNA), Kano *et al.* (*Mycoses* **42**: 71, 1999; primers), Blanz *et al.* (*Mycoses* **43** Suppl. 1: 11, 2000; diagnosis), Gräser *et al.* (*Medical Mycology* **38**: 143, 2000; phylogeny), Kuraishi *et al.* (*Antonie van Leeuwenhoek* **77**: 179, 2000; ubiquinones), Simpanya (*Revta Iberoamer. Micol.* **17** [Special]: 1, 2000; ecology), Kim *et al.* (*Mycoses* **44**: 157, 2001; populations), Gupta *et al.* (*Stud. Mycol.* **47**: 87, 2002; diagnosis), Kano *et al.* (*Mycoses* **45**: 277, 2002; Japan), Kano *et al.* (*Stud. Mycol.* **47**: 49, 2002; chitin synthase genes), Sugiyama *et al.* (*Stud. Mycol.* **47**: 5, 2002; phylogeny), Summerbell *et al.* (*Stud. Mycol.* **47**: 75, 2002; biological species), Takahashi *et al.* (*Jap. J. Med. Mycol.* **44**: 31, 2003; epidemiology), Bedard *et al.* (*MR* **110**: 86, 2006; clonal spp.), Geiser *et al.* (*Mycol.* **98**: 1053, 2006; phylogeny).

Arthrodermataceae Locq. ex Currah (1985), Onygenales. 5 gen. (+ 29 syn.), 65 spp.

Lit.: Currah (*Mycotaxon* **24**: 1, 1985), Weitzman *et al.* (*Mycotaxon* **25**: 505, 1986), Currah (*SA* **7**: 1, 1988), Amer *et al.* (*Int. J. Dermat.* **32**: 97, 1993), Guillamón *et al.* (*Antonie van Leeuwenhoek* **69**: 223, 1996), Chandler (*Topley & Wilson's Microbiology and Microbial Infections* Edn 9. Vol. 4 Medical Mycology: 111, 1998), Hoog *et al.* (*Medical Mycology* **36** Suppl. 1: 52, 1998), Bastert *et al.* (*Mycoses* **42**: 525, 1999), Gräser *et al.* (*Medical Mycol.* **37**: 105, 1999; phylogeny), Harmsen *et al.* (*Mycoses* **42**: 67, 1999), Kano *et al.* (*Mycoses* **42**: 71, 1999), Makimura *et al.* (*J. Clin. Microbiol.* **37**: 807, 1999; phylogeny), Sugiyama *et al.* (*Mycoscience* **40**: 251, 1999), Gräser *et al.* (*Medical Mycology* **38**: 143, 2000), Simpanya (*Revta Iberoamer. Micol.* **17** [Special]: 1, 2000), Summerbell (*Revta Iberoamer. Micol.* **17**: 30, 2000).

Arthrodochium R.F. Castañeda & W.B. Kendr. (1990), anamorphic *Agaricomycetes*, Hsp.0eH.38. 1 (with clamp connexions), Cuba. See Castañeda & Kendrick (*Univ. Waterloo Biol. Ser.* **32**: 6, 1990).

Arthrographis G. Cochet ex Sigler & J.W. Carmich. (1976), anamorphic *Eremomyces*, Hso.0eH.38. 8, widespread. See Sigler & Carmichael (*Mycotaxon* **18**: 495, 1983), Ayer & Nozawa (*Can. J. Microbiol.* **36**: 83, 1990; inhibitory metabolite), Sigler *et al.* (*Can. J. Microbiol.* **36**: 77, 1990; n.sp.), Uchida *et al.* (*TMSJ* **34**: 275, 1993; *A. cuboidea*), Chin-Hong *et al.* (*J. Clin. Microbiol.* **39**: 804, 2001; epidemiology).

Arthrographium Ces. (1854) = Arthrobotryum Ces. fide Mussat (*Syll. fung.* **15**, 1901).

Arthromitus Leidy (1849) nom. dub., Bacteria. 1, USA. Bacteria occurring as trichomes in intestines of millipedes, cockroaches and toads; formerly incorrectly placed in *Trichomycetes*.

Arthromyces T.J. Baroni & Lodge (2007), Tricholomataceae. 2, C. America. See Baroni *et al.* (*MR* **111**: 572, 2007).

Arthroon Renault (1894), Fossil Fungi ? Fungi. 1 (Carboniferous), France.

Arthropsis Sigler, M.T. Dunn & J.W. Carmich. (1982), anamorphic *Onygenales*, Hso.0eH/1eP.38. 4, widespread. See Ulfig *et al.* (*Mycotaxon* **54**: 281, 1995).

Arthropycnis Constant. (1992), anamorphic *Rhynchostoma*, Cpd.0eP.39. 1, widespread. See Constantinescu & Tibell (*Nova Hedwigia* **55**: 174, 1992).

Arthrorhaphidaceae Poelt & Hafellner (1976), Ostropomycetidae (inc. sed.) (±L). 2 gen. (+ 4 syn.), 12 spp.

Lit.: Galloway & Bartlett (*N.Z. Jl Bot.* **24**: 393, 1986), Obermayer (*Nova Hedwigia* **58**: 275, 1994), Santesson & Tønsberg (*Lichenologist* **26**: 295, 1994), Hansen & Obermayer (*Bryologist* **102**: 104, 1999), Wedin *et al.* (*MR* **109**: 159, 2005), Miądlikowska *et al.* (*Mycol.* **98**: 1088, 2006; phylogeny), Lumbsch *et al.* (*MR* **111**: 257, 2007; phylogeny).

Arthrorhaphis Th. Fr. (1860) nom. cons., Arthrorhaphidaceae (±L). 11, widespread (temperate; montane). See Galloway & Bartlett (*N.Z. Jl Bot.* **24**: 393, 1986; NZ spp.), Obermayer (*Nova Hedwigia* **58**: 275, 1994; key 5 Eur. spp.), Hafellner & Obermayer (*Cryptog. Bryol.-Lichénol.* **1 6**: 177, 1995; key fungi on), Obermayer (*J. Hattori bot. Lab.* **80**: 331, 1996; Himalaya), Hansen & Obermayer (*Bryologist* **102**: 104, 1999; Greenland), Miądlikowska *et al.* (*Mycol.* **98**: 1088, 2006; phylogeny).

Arthrorhynchus Kolen. (1857), Laboulbeniaceae. 3 (on insects), widespread. See Benjamin *in* Thaxter (*Mem. Am. Acad. Arts Sci. 1896-1931* **12-16**, 1971), Blackwell (*Mycol.* **72**: 159, 1980; morphology), Santamaria (*Nova Hedwigia* **82**: 349, 2006).

Arthrospora Th. Fr. (1861) ≡ Arthrosporum.

arthrospore (1) see arthric; (2) a specialized uninucleate cell functioning as a spore and derived from the disarticulation of cells of a formerly vegetative branch (*Asellariales*).

Arthrosporella Singer (1970), ? Tricholomataceae. Anamorph *Nothoclavulina*. 1, S. America. See Singer (*Fl. Neotrop.* Monogr. **3**: 17, 1970).

Arthrosporia Grigoraki (1925) = Trichophyton fide Hawksworth *et al.* (*Dictionary of the Fungi* edn 8, 1995).

Arthrosporium Sacc. (1880), anamorphic *Pezizomycotina*, Hsy.≡ eH.10. 2, Italy; N. America. See Wang

(*Mycol.* **64**: 1175, 1972), Kendrick (*CJB* **81**: 75, 2003; morphogenesis).

Arthrosporum A. Massal. (1853), Ramalinaceae (L). 1, Europe. See Timdal (*Op. Bot.* **110**, 1991), Ekman (*MR* **105**: 783, 2001; phylogeny).

arthrosterigma (of lichens), a septate conidiophore (spermatiophore) (obsol.).

Arthrowallemia R.F. Castañeda, D. García & Guarro (1998), anamorphic *Pezizomycotina*, Hso.?.?. 1, Cuba. See Castañeda Ruíz *et al.* (*MR* **102**: 17, 1998).

Arthroxylaria Seifert & W. Gams (2002), anamorphic *Xylariaceae*. 1 (on dung), USA. See Seifert *et al.* (*Czech Mycol.* **53**: 299, 2002).

Arthur (Joseph Charles; 1850-1942; USA). Botanist, Agricultural Experiment Station, Geneva, NY (1884-1887); Purdue University (1887-1915). Noted for his work on *Uredinales*: the 'Arthur Herbarium' at Purdue is one of the most important collections of rusts (75,000 specimens; see Baxter & Kern, *Proceedings of the Indiana Academy of Science* **71**: 228, 1962). *Publs.* Pucciniales *North American Flora* **7** (1907-1927); *Plant Rusts* (1929); *Manual of the Rusts in United States and Canada* (1934). *Biogs, obits etc.* Cummins (*Annual Review of Phytopathology* **16**: 19, 1978); Kern (*Phytopathology* **32**: 833, 1942); Mains (*Mycol.* **34**: 601, 1942); Stafleu & Cowan, (*TL-2* **1**: 70, 1976); Stafleu & Mennega (*TL-2, Suppl.* **1**: 173, 1992); Urban (*Česká Mykologie* **25**: 185, 1971).

Arthurella Zebrowski (1936), Fossil Fungi ? Chytridiomycetes. 1 (Cambrian to ? Recent), Australia.

Arthuria H.S. Jacks. (1931), Phakopsoraceae. Anamorph *Aeciure*. 6 (on *Euphorbiaceae*, 1 on *Asclepiadaceae*), S. America; Carribean; India. See Buriticá (*Revista de la Academia Columbiana de Ciencias Exactas, Físicas y Naturales* **22**: 325, 1998; neotrop. sp.).

Arthuriomyces Cummins & Y. Hirats. (1983), Phragmidiaceae. 3 (on *Rubus* (*Rosaceae*)), N. America; Russia; China; Japan. See Cummins & Hiratsuka (*Illustr. Gen. Rust Fungi rev. edit.*: 114, 1983), Cummins & Hiratsuka (*Illustr. Gen. Rust Fungi* edn 3: 225 pp., 2003; syn. of *Gymnoconia*).

Articularia Höhn. (1909), anamorphic *Pezizomycotina*, Hsy.0eH.?. 1, N. America. See Charles (*Mycol.* **27**: 74, 1935).

Articulariella Höhn. (1909) = Microstoma Bernstein fide von Arx (*Gen. Fungi Sporul. Cult.*, 1970).

articulated, jointed.

Articulis Clem. & Shear (1931) ≡ Articulariella.

Articulophora C.J.K. Wang & B. Sutton (1982), anamorphic *Pezizomycotina*, Hso.#eP.11. 1, USA. See Wang & Sutton (*Mycol.* **74**: 489, 1982).

Articulospora Ingold (1942), anamorphic *Hymenoscyphus*, Hso.0-≡ bH.10. 6 (aquatic), widespread. See Petersen (*Mycol.* **54**: 143, 1962; key), Jooste *et al.* (*MR* **94**: 947, 1990; S Africa), Sivichai *et al.* (*Mycol.* **95**: 340, 2003; Thailand), Belliveau & Bärlocher (*MR* **109**: 1407, 2005; phylogeny), Descals (*MR* **109**: 545, 2005; propagules).

Artocarpomyces Subram. (1996), anamorphic *Pezizomycotina*. 1, S.E. Asia. See Subramanian (*Kavaka* **22/23**: 52, 1994).

Artocreas Berk. & Broome (1875) nom. ambig., Fungi.

Artolenzites Falck (1909) = Trametes fide Donk (*Verh. K. ned. Akad. Wet.* tweede sect. **62**: 1, 1974).

Artomyces Jülich (1982) = Clavicorona fide Stalpers (*in litt.*).

Artymenium Berk. ex E. Fisch. (1933) nom. inval. = Secotium fide Stalpers (*in litt.*).

Arualis Katz (1980), anamorphic *Agaricomycetes*, Hso.0eH-P.1. 1 (with clamp connexions), USA. See Katz (*Mycotaxon* **11**: 230, 1980).

Arwidssonia B. Erikss. (1974), Hyponectriaceae. 2 (on *Empetrum* and *Loiseleuria*), Europe. See Eriksson (*Svensk bot. Tidskr.* **68**: 199, 1974), Holm *et al.* (*Karstenia* **39**: 59, 1999), Wang & Hyde (*Fungal Diversity* **3**: 159, 1999).

Arx (Joseph Adolf, von; 1922-1988; Switzerland, later Netherlands). Student of Gäumann (q.v.) [and contemporary of E. Müller (q.v.) with whom he collaborated for many years], Zürich (1942-1948); Phytopathologist, Willie Commelin Scholten Phytopathological Laboratories, Baarn (1949); Director of **CBS** (1963-1987). An outstanding general mycologist, noted for his review of the fungal kingdom (1970, expanded in 1974 and 1987). He was able to recognize and carry out necessary taxonomic rationalization, leading to publications synonymizing many redundant names, particularly in *Gloeosporium*. *Publs.* Beiträge zur Kenntnis der Gattung *Mycosphaerella*. *Sydowia* (1949); (with Müller) Die Gattungen der amerosporen Pyrenomyceten. *Beiträge zur Kryptogamenflora der Schweiz* (1954); Revision der zu *Gloeosporium* gestellten Pilze. *Verhandelingen der Koninklijke Nederlandse Akademie van Wetenschappen afd. Natuurkunde* Tweede Sectie (1957); (with Müller) Die Gattungen der didymosporen Pyrenomyceten. *Beiträge zur Kryptogamenflora der Schweiz* (1962); (with Müller) Pyrenomycetes: Meliolales, Coronophorales, Sphaeriales. In Ainsworth, Sparrow & Sussman [eds] *The Fungi. An Advanced Treatise* **4A** (1973); *The Genera of Fungi Sporulating in Pure Culture* (1981) [edn 3]. *Biogs, obits etc.* van der Aa *et al.* (*Studies in Mycology* **31**, 1989) [bibliography, portrait]; Arnold (*Boletus* **13**: 24, 1989) [portrait]; Müller (*Sydowia* **41**: 1, 1989) [portrait].

Arxiella Papendorf (1967), anamorphic *Pezizomycotina*, Hso.1eH.9. 2, widespread. See Papendorf (*TBMS* **50**: 73, 1967), Matsushima (*Matsush. Mycol. Mem.* **6**, 1989), Mercado-Sierra *et al.* (*Mycotaxon* **63**: 369, 1997; Cuba).

Arxiomyces P.F. Cannon & D. Hawksw. (1983), ? Ceratostomataceae. 2 (from wood etc.), Europe; Japan. See Horie *et al.* (*Mycotaxon* **25**: 229, 1986).

Arxiozyma Van der Walt & Yarrow (1984) = Kazachstania fide Augustyn *et al.* (*Syst. Appl. Microbiol.* **13**: 44, 1990; fatty acids), Kurtzman *et al.* (*J. Clin. Microbiol.* **43**: 101, 2005; phylogeny).

Arxula Van der Walt, M.T. Sm. & Y. Yamada (1989) = Blastobotrys fide Yamada & Nogawa (*J. gen. appl. Microbiol.* Tokyo **36**: 425, 1990; molecular phylogeny), Kunze & Kunze (*Microbiol. Eur.* **2**: 24, 1994; comparative morphology), Kunze & Kunze (*Antonie van Leeuwenhoek* **65**: 29, 1994; DNA fingerprinting of *A. adeninivorans*), Smith *in* Kurtzman & Fell (Eds) (*Yeasts, a taxonomic study* 4th edn: 441, 1998), Kurtzman & Robnett (*FEMS Yeast Res.* **7**: 141, 2007).

Asahina (Yasuhiko; 1881-1975; Japan). Assistant, University for Pharmacognosy, Tokyo (*c.* 1907); Professor, University of Tokyo (1912-1941). Natural product chemist and later lichenologist; established use of chemotaxonomy for lichen-forming fungi and introduced use of PD (see Metabolic products) in 1934 and microcrystal tests in 1936-1940 (*Journal of Japanese Botany* **12** 516, 1936). *Publs. Lichens of*

Japan 3 vols (1950-1956); (with Shibata) *Chemistry of Lichen Substances* (1954) [reprint 1971]; *Atlas of Japanese Cladoniae* (1971). *Biogs, obits etc.* Culberson & Culberson (*Bryologist* **79**: 258, 1976) [portrait]; Grummann (1974: 585); Kurokawa (*Lichenologist* **8**: 93, 1976) [portrait]; Lichenological Society of Japan *Dr Yasuhiko Asahina's Lichenological Bibliography* 1980 [281 titles]); Stafleu & Cowan (*TL-2* **1**: 72, 1976); Stafleu & Mennega (*TL-2*, *Suppl.* **1**: 184, 1992).

Asahinea W.L. Culb. & C.F. Culb. (1965), Parmeliaceae (L). 3, widespread (circumpolar). See Gao (*Nordic Jl Bot.* **11**: 483, 1991), Saag (*Dissertationes Biologicae Universitatis Tartuensis* **34**, 1998; evolution), Thell *et al.* (*Mycol. Progr.* **1**: 335, 2002; phylogeny), Thell *et al.* (*Symb. bot. upsal.* **34** no. 1: 429, 2004; Scandinavia), Miądlikowska *et al.* (*Mycol.* **98**: 1088, 2006; phylogeny), Randlane & Saag (*Central European Lichens*: 75, 2006; key).

Asaphomyces Thaxt. (1931), Laboulbeniaceae. 4, widespread. See Rossi & Máca (*Sydowia* **58**: 110, 2006).

Asbolisia Bat. & Cif. (1963), anamorphic *Aithaloderma*, Cpd.0eH.?. 6, widespread. See Batista & Ciferri (*Quaderno Ist. Bot. Univ. Pavia* **31**: 37, 1963), Reynolds & Gilbert (*Cryptog. Mycol.* **27**: 249, 2006; Panama).

Asbolisia Speg. (1918) nom. dub., anamorphic *Pezizomycotina*. See Sutton (*Mycol. Pap.* **141**, 1977).

Asbolisiomyces Bat. & H. Maia (1961) nom. dub., anamorphic *Pezizomycotina*, Cpd.0eH.? (L). 1, Brazil. See Lücking *et al.* (*Lichenologist* **30**: 121, 1998).

Ascagilis K.D. Hyde (1992) = Jahnula fide Hyde (*Aust. Syst. Bot.* **5**: 109, 1992), Hyde & Wong (*Nova Hedwigia* **68**: 489, 1999).

ascending (**ascendent**) (of an annulus), having the free edge above attached, cf. descending; (of conidiophores), curving up, cf. erect; (of lamellae), on a cone-like or an unexpanded pileus.

Aschersonia Endl. (1842) nom. rej. ≡ Laschia Jungh.

Aschersonia Mont. (1848) nom. cons., anamorphic *Hypocrella*, St.0-1eH.15. *c.* 21 (on whiteflies (*Aleyrodidae*) and scale insects (*Coccidae*)), widespread (subtropical). See Petch (*Ann. R. bot. Gdns Peradeniya* **7**: 167, 1921), Mains (*Lloydia* **22**: 215, 1960), Hywel-Jones & Evans (*MR* **97**: 871, 1993; ecology), Evans (*MR* **98**: 165, 1994; spore germination), Obornik *et al.* (*Pl. Protection Science* **35**: 1, 1999; molecular characterization and phylogeny), Evans (*Mycology Series* **19**: 517, 2003; biocontrol), Liu *et al.* (*Mycol.* **97**: 246, 2005), Liu *et al.* (*MR* **110**: 537, 2006; *A. aleyrodis* group), Chaverri *et al.* (*Stud. Mycol.* **60**, 2008; phylogeny, monogr. Neotropics).

Aschersoniopsis Henn. (1902) = Munkia fide Höhnel (*Sber. Akad. Wiss. Wien* Math.-naturw. Kl., Abt. 1 **126**: 283, 1917).

Aschion Wallr. (1833) ≡ Tuber.

Aschizotrichum Rieuf (1962) = Wiesneriomyces fide Carmichael *et al.* (*Genera of Hyphomycetes*, 1980).

Ascidiophora Rchb. [not traced] ? = Mucor Fresen. fide Mussat (*Syll. fung.* **15**, 1901).

Ascidium Fée (1824) nom. rej. = Ocellularia fide Hale (*Bull. Br. Mus. nat. hist.* Bot. **8**: 227, 1981).

Ascidium Tode (1782) nom. dub., Fungi. Based on insect eggs fide Fries (*Syst. mycol.* **3**, Index: 52, 1832).

ascigerous, having asci.

ascigerous centrum, the special tissue which produces the asci and hamathecium.

Ascitendus J. Campb. & Shearer (2004), Annulatascaceae. 1 (on wood in freshwater), Austria. See Campbell & Shearer (*Mycol.* **96**: 829, 2004).

Ascluella DiCosmo, Nag Raj & W.B. Kendr. (1983), Dermateaceae. 1 (from living leaves), India. See DiCosmo *et al.* (*Mycotaxon* **21**: 1, 1984).

asco- (prefix), pertaining to an ascus.

Ascoblastomycetes, see *Blastomycota*.

Ascobolaceae Boud. ex Sacc. (1884), Pezizales. 6 gen. (+ 9 syn.), 129 spp.

Lit.: van Brummelen (*Persoonia* Suppl. **1**: 1, 1967; monogr.), Dissing (*Op. bot.* **100**: 43, 1989), van Brummelen (*Persoonia* **14**: 203, 1990), Gargas (*Fungal Genetics Newsl.* Suppl. **38**: 26, 1991), Wu & Kimbrough (*Taiwania* **41**: 7, 1996; devel.), Jahn *et al.* (*Z. Mykol.* **63**: 133, 1997), Landvik *et al.* (*Nordic Jl Bot.* **17**: 403, 1997), Prokhorov (*Mikol. Fitopatol.* **31**: 27, 1997), Landvik *et al.* (*Mycoscience* **39**: 49, 1998; DNA), Ranalli & Mercuri (*Mycotaxon* **67**: 505, 1998), van Brummelen (*Persoonia* **16**: 425, 1998), Wu & Kimbrough (*Int. J. Pl. Sci.* **162**: 91, 2001), Hansen & Pfister (*Mycol.* **98**: 1029, 2006; phylogeny).

Ascobolus Pers. (1792), Ascobolaceae. Anamorph *Rhizostilbella*. 61 (mainly coprophilous), widespread. See van Brummelen (*Persoonia* Suppl. **1**: 1, 1967; key), Wells (*Univ. Calif. Publs Bot.* **62**, 1972; ontogeny), Paulsen & Dissing (*Bot. Tidsskr.* **74**: 67, 1979; key 20 spp.), Kaushal & Thind (*J. Indian bot. Soc.* **62**: 16, 1983; W. Himalayas, key 12 spp.), Parrettini (*Boll. Gruppo Micol. 'G. Bresadola'* **28**: 140, 1985; col. pls), Dissing (*Op. Bot.* **100**: 43, 1989), Kempken (*Biblthca Mycol.* **128**, 1989; extrachromosomal DNA), Prokhorov (*Mikol. Fitopatol.* **28**: 17, 1994; key to Russian spp.), Landvik *et al.* (*Nordic Jl Bot.* **17**: 403, 1997), Landvik *et al.* (*Mycoscience* **39**: 49, 1998; DNA), Antonín & Moravec (*Czech Mycol.* **52**: 295, 2001; variation), Wu & Kimbrough (*Int. J. Pl. Sci.* **162**: 91, 2001; ultrastructure), Dokmetzian *et al.* (*Mycotaxon* **92**: 295, 2005; isozymes), Hansen & Pfister (*Mycol.* **98**: 1029, 2006; phylogeny).

Ascobotryozyma J. Kerrigan, M.T. Sm. & J.D. Rogers (2001), Saccharomycetales. Anamorph *Botryozyma*. 2 (associated with nematodes), Italy; USA. See Kerrigan *et al.* (*Antonie van Leeuwenhoek* **79**: 15, 2001), Kerrigan *et al.* (*MR* **107**: 1110, 2003), Suh *et al.* (*Mycol.* **98**: 1006, 2006; phylogeny).

Ascocalathium Eidam ex J. Schröt. (1893), ? Pyronemataceae. 1, Europe.

Ascocalvatia Malloch & Cain (1971), Onygenaceae. 1 (coprophilous), Canada. See Sugiyama *et al.* (*Mycoscience* **40**: 251, 1999; DNA).

Ascocalyx Naumov (1926), Helotiaceae. Anamorph *Bothrodiscus*. 6, widespread. See Groves (*CJB* **46**: 1273, 1968; key), Schlapfer-Bernard (*Sydowia* **22**: 1, 1969), Müller & Dorworth (*Sydowia* **36**: 193, 1983; key, 6 spp.), Petrini *et al.* (*CJB* **67**: 2805, 1989), Bernier *et al.* (*Appl. Environm. Microbiol.* **60**: 1279, 1994; DNA), Wang (*MR* **101**: 1195, 1997; genetic variation), Wang *et al.* (*CJB* **75**: 1460, 1997; population structure, as *Gremmeniella*), Hamelin *et al.* (*Phytopathology* **88**: 582, 1998; N. Am. introd., as *Gremmeniella*).

ascocarp, see ascoma.

Ascocephalophora K. Matsush. & Matsush. (1995), ? Endomycetaceae. Anamorphs *Fusidium*, *Trichosporiella*. 1, Japan. See Matsushima & Matsushima (*Matsush. Mycol. Mem.* **8**: 45, 1995).

Ascochalara Réblová (1999), Chaetosphaeriaceae. Anamorph *Chalara*-like. 1 (from coniferous wood), Czech Republic. See Réblová (*Sydowia* **51**: 210, 1999).

Ascochyta Lib. (1830), anamorphic *Didymella*, Cpd.1eH.15. 388, widespread. *A. fabae* (on *Vicia*), *A. phaseolorum* (on *Phaseolus*), *A. pisi* (on pea), *A. rabiei* (on chickpea), and others. The genus is in need of redefinition using molecular data. See Armstrong *et al.* (*Can. J. Pl. Path.* **23**: 110, 2001; genetic diversity), Barve *et al.* (*Fungal Genetics Biol.* **39**: 151, 2003; mating types), Fatehi *et al.* (*Mycopathologia* **156**: 317, 2003; genetic diversity), Chongo *et al.* (*Pl. Dis.* **88**: 4, 2004; genetic diversity), Peever *et al.* (*Mol. Ecol.* **13**: 291, 2004; population structure), Lichtenzveig *et al.* (*Eur. J. Pl. Path.* **113**: 15, 2005; mating types), Priest (*Fungi of Australia: Septoria* **0**, 2006; Australian spp.), Schoch *et al.* (*Mycol.* **98**: 1041, 2006; phylogeny), Henson (*Eur. J. Pl. Path.* **119**: 141 pp., 2007; special issue on leume-associated spp.), Peever *et al.* (*Mycol.* **99**: 59, 2007; on legumes).

Ascochytella Tassi (1902) = Ascochyta fide Buchanan (*Mycol. Pap.* **156**, 1987).

Ascochytites Babajan & Tasl. (1973), Fossil Fungi. 1 (Tertiary), former USSR.

Ascochytites Barlinge & Paradkar (1982), Fossil Fungi, anamorphic *Pezizomycotina*. 1 (Cretaceous), India.

Ascochytopsis Henn. (1905), anamorphic *Pezizomycotina*, Ccu.0fH.15. 5 (on *Leguminosae*), widespread. See Sutton (*The Coelomycetes*, 1980), Matsushima (*Matsush. Mycol. Mem.* **10**, 2001).

Ascochytula (Potebnia) Died. (1912) = Ascochyta fide Buchanan (*Mycol. Pap.* **156**, 1987).

Ascochytulina Petr. (1922), anamorphic *Pezizomycotina*, Cpd.1eP.15. 3, Europe. See Buchanan (*Mycol. Pap.* **156**, 1987), Mel'nik (*Opredelitel' Gribov Rossii* Klass Coelomycetes Byp. **1**. Redkie i Maloizvestnye Rody, 1997).

Ascoclavulina Y. Otani (1974), Helotiaceae. Anamorph *Gliomastix*-like. 1, Japan. See Otani (*TMSJ* **15**: 5, 1974).

Ascocodinaea Samuels, Cand. & Magni (1997), Hypocreomycetidae. Anamorph *Dictyochaeta*-like. 2 (on old polypores), USA. See Samuels *et al.* (*Mycol.* **89**: 156, 1997), Réblová *et al.* (*Sydowia* **51**: 49, 1999), Huhndorf *et al.* (*Mycol.* **96**: 368, 2004; phylogeny).

Ascocoma H.J. Swart (1987), ? Phacidiaceae. Anamorph *Coma*. 1, Australia. See Swart (*TBMS* **87**: 603, 1987), Beilharz & Pascoe (*Mycotaxon* **91**: 273, 2005; microconidial state).

ascoconidiophore, the phialide bearing an ascoconidium in *Ascoconidium* (Seaver, *Mycol.* **34**: 412, 1942).

Ascoconidium Seaver (1942), anamorphic *Sageria*, Hsp.≡ eH.15. 2, widespread (north temperate). See Funk (*CJB* **44**: 39, 1966), Nag Raj & Kendrick (*Monogr. Chalara Allied Genera*, 1975).

ascoconidium, a conidium formed directly from an ascospore, esp. when still within the ascus (e.g. *Claussenomyces*).

Ascocorticiaceae J. Schröt. (1893), ? Helotiales. 1 gen., 2 spp.

Lit.: Vellinga & Vries (*Coolia* **30**: 50, 1987).

Ascocorticiellum Jülich & B. de Vries (1982), Pezizomycotina. 1, Europe. See Jülich & de Vries (*Persoonia* **11**: 410, 1982), Vries (*Coolia* **39**: 18, 1996).

Ascocorticium Bref. (1891), Ascocorticiaceae. 2, widespread (temperate). See Cooke (*Ohio J. Sci.* **68**: 161, 1968), Eriksson *et al.* (*Göteborgs Svampkl. Årsskr.*: 1, 1981).

Ascocoryne J.W. Groves & D.E. Wilson (1967), Helotiales. Anamorph *Coryne*. *c*. 8, widespread (north & south temperate). See Christiansen (*Friesia* **7**: 75, 1963; anamorph, Danish spp.), Roll-Hansen & Roll-Hansen (*Norw. Jl Bot.* **26**: 193, 1979), Verkley (*MR* **99**: 187, 1995; asci), Gamundí & Romero (*Fl. criptog. Tierra del Fuego* **10**, 1998), Wang *et al.* (*Mycol.* **98**: 1065, 2006; phylogeny).

Ascocorynium S. Ito & S. Imai ex S. Imai (1934) = Neolecta fide Korf (*Phytologia* **21**: 201, 1971).

Ascocratera Kohlm. (1986), ? Lophiostomataceae. 1 (marine), Belize. Possibly belongs to the *Trypetheliaceae* fide Erikssson (*in litt.*). See Kohlmeyer (*CJB* **64**: 3036, 1986), Harris *in* Aptroot (Ed.) (*Biblthca Lichenol.* **44**, 1991), Kohlmeyer & Volkmann-Kohlmeyer (*Bot. Mar.* **34**: 1, 1991).

Ascocybe D.E. Wells (1954) = Cephaloascus fide von Arx (*Antonie van Leeuwenhoek* **38**: 289, 1972).

Ascodesmidaceae J. Schröt. (1893), Pezizales. 3 gen. (+ 2 syn.), 21 spp. Clusters within *Pyronemataceae* in a recent molecular study.

Lit.: Currah (*Mycol.* **78**: 198, 1986), Brummelen (*Persoonia* **14**: 1, 1989), Kimbrough (*Mem. N. Y. bot. Gdn* **49**: 326, 1989; fam. limits), van Brummelen (*Persoonia* **14**: 1, 1989; ascus ultrastr.), Landvik *et al.* (*Nordic Jl Bot.* **17**: 403, 1997; DNA), Landvik *et al.* (*Mycoscience* **39**: 49, 1998), Hansen & Pfister (*Mycol.* **98**: 1029, 2006; phylogeny), Hansen *et al.* (*Mycol.* **97**: 1023, 2005), Perry *et al.* (*MR* **111**: 549, 2007; phylogeny).

Ascodesmis Tiegh. (1876), Ascodesmidaceae. 6, widespread. See Obrist (*CJB* **39**: 943, 1961; key), Delattre-Durand & Janex-Favre (*BSMF* **95**: 49, 1979; ontogeny), van Brummelen (*Persoonia* **11**: 377, 1981), Patil & Ghadge (*Indian Phytopath.* **40**: 30, 1987; 5 spp.), van Brummelen (*Persoonia* **14**: 1, 1989), van Brummelen (*Stud. Mycol.* **31**: 41, 1989; ultrastr.), Landvik *et al.* (*Nordic Jl Bot.* **17**: 403, 1997), Landvik *et al.* (*Mycoscience* **39**: 49, 1998; DNA), Hansen & Pfister (*Mycol.* **98**: 1029, 2006; phylogeny), Perry *et al.* (*MR* **111**: 549, 2007; phylogeny, paraphyly of *Pyronemataceae*).

Ascodesmisites Trivedi, Chaturv. & C.L. Verma (1973), Fossil Fungi. 1 (Eocene), Malaysia. See Korf (*Mycotaxon* **6**: 193, 1977).

Ascodichaena Butin (1977), Ascodichaenaceae. Anamorph *Polymorphum*. 2, widespread (esp. temperate). See Hawksworth (*Taxon* **32**: 212, 1983; nomencl.), Butin & Marmolejo (*Sydowia* **42**: 8, 1990).

Ascodichaenaceae D. Hawksw. & Sherwood (1982), Rhytismatales. 5 gen. (+ 10 syn.), 11 spp.

Lit.: Hawksworth & Sherwood (*Mycotaxon* **16**: 262, 1982), Butin & Marmolejo (*Sydowia* **42**: 8, 1990), Minter (*Shoot and Foliage Diseases in Forest Trees* Proceedings of a Joint Meeting of the Working Parties: Canker & Shoot Blight of Conifers, Foliage Diseases: 65, 1995), Yuan *et al.* (*Australas. Pl. Path.* **29**: 215, 2000).

Ascofascicula Matsush. (2003), ? Pezizales. 1, Japan. See Matsushima (*Matsush. Mycol. Mem.* **10**: 190, 2001).

ascogenous (ascogenic), ascus-producing or ascussupporting.

ascogonium, the cell or group of cells in Ascomycotina fertilized by a sexual act.

Ascographa Velen. (1934), ? Helotiales. 1, former Czechoslovakia.

Ascohansfordiellopsis D. Hawksw. (1979) = Koordersiella fide Hawksworth (*Bull. Br. Mus. nat. hist.* Bot. **6**, 1979).

Ascohymeniales Nannf. (1932). *Ascomycota* having asci (and paraphyses) developing as a hymenium and not in a pre-formed stroma, as in *Pyrenomycetes* and *Discomycetes* (Nannfeldt, 1932); *Hymenoascomycetes*. Cf. *Ascoloculares*.

Ascoidea Bref. (1891), Ascoideaceae. 4, widespread. See Batra & Francke-Grossman (*Mycol.* **56**: 632, 1964; key), von Arx & Müller (*Sydowia* **37**: 6, 1984), de Hoog *in* Kurtzman & Fell (Eds) (*Yeasts, a taxonomic study* 4th edn: 136, 1998), Suh *et al.* (*Mycol.* **98**: 1006, 2006; phylogeny).

Ascoideaceae J. Schröt. (1894), Saccharomycetales. 1 gen., 4 spp.

Lit.: von Arx & Müller (*Sydowia* **37**: 6, 1984), Batra (*Stud. Mycol.* **30**: 415, 1987), Hoog *in* Kurtzman & Fell (Eds) (*Yeasts, a taxonomic study* 4th edn: 136, 1998), Kurtzman & Blanz *in* Kurtzman & Fell (Eds) (*Yeasts, a taxonomic study* 4th edn: 69, 1998), Kurtzman & Robnett (*Antonie van Leeuwenhoek* **73**: 331, 1998), Suh *et al.* (*Mycol.* **98**: 1006, 2006; phylogeny).

Ascoideales = Saccharomycetales.

Ascolacicola Ranghoo & K.D. Hyde (1998), Sordariales. Anamorph *Trichocladium*. 1 (on wood in freshwater), Austria; Hong Kong. See Ranghoo & Hyde (*Mycol.* **90**: 1055, 1998), Ranghoo *et al.* (*Fungal Diversity* **2**: 159, 1999; DNA), Réblová & Winka (*Mycol.* **93**: 478, 2001), Campbell & Shearer (*Mycol.* **96**: 822, 2004).

Ascolanthanus Cailleux (1967) = Pyxidiophora fide Lundqvist (*Bot. Notiser* **133**: 121, 1980).

Ascolectus Samuels & Rogerson (1990), Saccardiaceae. 1, Brazil. See Samuels & Rogerson (*Mem. N. Y. bot. Gdn* **64**: 177, 1990).

Ascoloculares Nannf. (1932). *Ascomycota* having asci (and paraphyses) developing in cavities in a preformed stroma, as in *Loculoascomycetes* (Nannfeldt, 1932). Cf. *Ascohymeniales*.

ascoma (pl. **ascomata**), an ascus-containing structure, ascocarp.

Ascomauritiana Ranghoo & K.D. Hyde (1999), Pezizomycotina. 1, Mauritius. See Ranghoo & Hyde (*MR* **103**: 938, 1999).

Ascominuta Ranghoo & K.D. Hyde (2000), ? Dothideomycetes. 1 (in freshwater), Hong Kong. See Ranghoo & Hyde (*Mycoscience* **41**: 1, 2000).

Ascomyces Mont. & Desm. (1848) = Taphrina. Sometimes used for *Ginanniella* (Tillet.) anamorphs. fide Mussat (*Syll. Fung.* **15**: 51, 1901).

ascomycete, one of the *Ascomycota*.

Ascomycetella Peck (1881) = Cookella fide Hawksworth *et al.* (*Dictionary of the Fungi* edn 8, 1995).

Ascomycetella Sacc. (1886) ≡ Myriangiopsis.

Ascomycetes. Originally introduced by Berkeley (*Intr. Crypt. Bot.*: 270, 1857) but without a clear indication of rank; see Whittaker (*Quart. Rev. Biol.* **34**: 210, 1959) and Hibbett *et al.* (*MR* **111**: 509, 2007). Commonly used in an equivalent sense to *Ascomycota* and/or *Pezizomycotina* in this *Dictionary*.

Ascomycota Caval.-Sm. (1998), Fungi. 15 class., 68 ord., 327 fam., 6355 gen., 64163 spp. (Ascolichenes, Ascomycoptera, Thecamycetes); sac fungi; ascomycetes. Saprobes, parasites (esp. of plants), or lichen-forming; cosmop. The largest group of *Fungi*, for which the ascus (q.v.) is the diagnostic character. The presence of lamellate hyphal walls with a thin electron-dense outer layer and a relatively thick electron-transparent inner layer also appears diagnostic; this enables anamorphic fungi to be recognized as ascomycetes even in the absence of asci. In the past they have often been grouped on fruit-body type and ascus arrangement (e.g. *Hemiascomycetes*, *Plectomycetes*, *Pyrenomycetes*, *Discomycetes*, *Loculoascomycetes*; q.v.). In recent decades the development of the ascomata, especially the structure and discharge method of the asci, were considered important, but in the last 5-10 years molecular sequence data (especially of the ribosomal genome) have come to the fore.

The size of the group makes it difficult to embrace the enormous range of structures in the group, and to determine which morphological features should be stressed in the recognition of higher categories in addition to sequence data. In many instances, molecular and morphological data are congruent, but integration of these data have proved to be intractable in some cases. Further problems have been encountered with the need to assign families and orders to higher taxa where molecular data are not available. The desire of many systematic and applied mycologists to begin the process of amalgamating anamorph genera into the overall ascomycete system has become rapidly more volubly expressed (see for example Seifert *et al.*, *Stud. Mycol.* **45**, 2000) and in response to this all genera of anamorphic fungi in this *Dictionary* with ascomycetous affinities have been provisionally assigned at least to a higher taxon of *Ascomycota*.

The classification in this edition of the *Dictionary* is based on a series of major phylogenetic studies of fungi under the umbrella of the 'Deep Hypha' and AFTOL ('Assembling the Fungal Tree of Life') projects, as well as other resources including *Myconet*. See esp. Blackwell *et al.* (*Mycol* **98**: 829, 2006), Hibbett *et al.* (*MR* : **111**: 509, 2007), James et al. (*Nature* **443**: 818, 2006) and Lutzoni et al. (*Am. J. Bot.* **91**: 1446, 2004).

Three Subphyla are accepted here. However, many accepted families are not referred to any specific order or class within these subphyla, and over 3200 genera could not be assigned with confidence to any family.

Subphyl.:
(1) **Pezizomycotina** (syn. *Ascomycotina*)
(2) **Saccharomycotina**
(3) **Taphrinomycotina**

A significant number of orders and many families have yet to have any members in them sequenced, and this lack of molecular data means that any current phylogenetic framework contains many 'holes'. As in previous editions, we have attempted to place non-sequenced taxa within the overall classification structure, but many further changes are to be expected. The *Fungi* are treated in more detail at family level in a companion publication (Cannon & Kirk, *Fungal Families of the World*, 2007).

Lit.: **General**: von Arx (*in* Kendrick (Ed.), *The whole fungus* **1**: 201, 1979, classif., anamorphs; *Genera of fungi sporulating in pure culture*, edn 3, 1981; keys gen., lit.; *Plant Pathogenic Fungi*, 1987), von

Arx & Müller (*Beitr. Krypt.-Fl. Schweiz* **11** (1), 1954; gen. amerospored pyrenom.; *Stud. mycol.* **9**, 1975; bitunicate gen., keys), Barr (*Mem. N.Y. bot. Gdn* **28** (1): 1, 1976, classif.; *Prodromus to class Loculoascomycetes*, 1987, keys gen.; *Mycotaxon* **39**: 43, 1990, keys pyren. gen.; *in* Parker, 1982, **1**: 201), Benny & Kimbrough (*Mycotaxon* **12**: 1, 1980; plectomycete gen.), Berbee (*Mol. Biol. Evol.* **13**: 462, 1996, loculoascomycete evolution; *Physiol. Mol. Pl. Path.* **59**: 165, 2001, phylogeny of pathogens), Clements & Shear (*Genera of Fungi*, 1931), Eriksson (*Opera Bot.* **60**, 1981; bitunicate fams, *Myconet*, 1997-; annual system), Eriksson & Hawksworth (*SA* **12**: 51, 1993; outline classif., orders, fam., gen.), Eriksson & Winka (*Myconet* **1**: 1, 1997; supraordinal taxa), Hafellner (*in* Galun, *CRC Handbook of lichenology* **3**: 41, 1988; lichenized gps), Hanlin (*Illustrated genera of ascomycetes*, 1990), Hansford (*Mycol. Pap.* **15**, 1946; foliicolous spp.), Hawksworth (*Proc. Indian Natn Acad. Sci., Pl. Sci.* **94**: 319, 1985, development classif. systems; (Ed.), *Ascomycete systematics: problems and perspectives in the nineties* [NATO ASI Ser. A **269**], 1994), Henssen & Jahns (*Lichenes*, ['1974'] 1973), Kohlmeyer & Kohlmeyer (*Marine mycology*, 1979; *Bot. Mar.* **34**: 1, 1991, keys 255 spp.), Korf (*The Fungi* **4A**: 249, 1973; keys discomycete gen.), Lipscomb *et al.* (*Cladistics* **14**: 303, fungi and eukaryote phylogeny), Liu *et al.* (*Mol. Biol. Evol.* **16**: 1799, 1999; phylogeny based on *RPB2* sequences), Liu & Hall (*Proc. Nat. Acad. Sci.* USA **101**: 4507, 2004; phylogeny related to ascoma structure), Lopandic *et al.* (*Mycol. Progr.* **4**: 205, 2005; rDNA phylogeny, chemotaxonomy), Luttrell (*Univ. Miss. Stud.* **24** (3), 1951 [reprint 1969]), Lutzoni *et al.* (*Nature* **411**: 937, 2001; lichen ancestry), McLaughlin *et al.* (*The Mycota* **7A**, 2001), Müller & von Arx (*Beitr. Krypt.-Fl. Schweiz* **11** (2), 1962, didymospored gen.; *The Fungi* **4A**: 87, 1973; keys pyrenomycete gen.), Munk (*Dansk bot. Arkiv.* **15** (2), 1953; system), Nag Raj (*Coelomycetous Anamorphs with Appendage-Bearing Conidia*, 1993), Nannfeldt (*Nova Acta Reg. Soc. Sci. upsal.*, iii, **8** (2), 1932; inoperculate discom.), Nishida & Sugiyama (*Mycoscience* **35**: 361, 1994; archiascomycetes), Padovan *et al.* (*J. Mol. Evol.* **60**: 726, 2005; molecular clocks), Poelt (*in* Ahmadjian & Hale, *The lichens*: 599, ['1973'] 1974; fams), Reynolds (Ed.) (*Ascomycete systematics*, 1981; ascus, centrum types), Robbertse *et al.* (*Fungal Genet. Biol.* **43**: 715, 2006; phylogenomic analysis), Santesson (*Symb. bot. upsal.* **12** (1), 1952; foliicolous L), Seifert *et al.* (*Stud. Mycol.* **45**, 2000; anamorph integration), Sivanesan (*The bitunicate ascomycetes and their anamorphs*, 1985; keys), Spatafora *et al.* (*Mycol.* **98**: 1018, 2006; overview of *Pezizomycotina* phylogeny), Sugiyama (*Mycoscience* **39**: 487, 1998; phylogeny), Tehler *et al.* (*MR* **107**: 901, 2003; rDNA phylogeny), Walker (in Grgurinovic & Mallett (Eds), *Fungi of Australia* **1A**, 1996; system, key to orders), Wehmeyer (*The pyrenomycetous fungi*, 1975), Zahlbruckner (*Nat. Pflanzenfam.* **8**: 61, 1926; L gen.).

Regional: **Australia**, Walker (in Grgurinovic & Mallett (Eds), *Fungi of Australia* **1A**, 1996; system, key to orders), McCarthy (ed.), *Flora of Australia* **54**, 1992 et seq.; lichenized groups). **British Isles**, Cannon *et al.* (*The British Ascomycotina: an annotated checklist*, 1985; 5100 spp.), Dennis (*British Ascomycetes*, 1968; edn 2, 1978), Ellis & Ellis (*Microfungi on land plants*, 1985; *Microfungi on miscellaneous substrates*, 1988). **Brazil**, Da Silva & Minter (*Mycol. Pap.* **169**, 1995; Batista & co-workers collns). **Caribbean**, Minter *et al.*, *Fungi of the Caribbean*, 2001, checklist. **Denmark**, Munk (*Dansk bot. Arkiv* **17**(1), 1957). **Germany**, Schmidt & Schimdt (*Ascomyceten im Bild* **1-**, 1990 on). **Hungary**, Bánhegyi *et al.* (*Magyaroszág* **1-3**, 1985-87; keys). **Nordic countries**, Hansen & Knudsen, *Nordic Macromycetes* **1**. *Ascomycota*, 2000). **North America**, Brodo *et al.*, *Lichens of North America*, 2001; Ellis & Everhart (*North American Pyrenomycetes*, 1892). **Pakistan**: Ahmad (*Monogr. Biol. Soc. Pakistan* **7-8**, 1978; keys). **Romania**, Sandu-Ville (*Ciuperci Pyrenomycetes - Sphaeriales din România*, 1971). **Spain**, López (*Aportacion al conocimento de los ascomicetes (Ascomycotina) de Cataluña*, **1**, 1987). **Sweden**, Eriksson (*The non-lichenized pyrenomycetes of Sweden*, 1992; 1524 spp.). **Switzerland**, Breitenbach & Kränzlin (*Pilze der Schweiz* **1**, 1981). **U.S.A.**, Farr *et al.* (*Fungi on plants and plant products in the United States*, 1989; checklist). **Venezuela**, Dennis (*Fungus flora of Venezuela and adjacent countries*, 1970).

See also under *Discomycetes*, Geographical distribution, Lichens, *Loculoascomycetes*, Plant pathogenic fungi, and Yeasts.

Ascomycotina. = *Pezizomycotina*.

ascoparaphysis, see paraphysis.

Ascophanella Faurel & Schotter (1965) = Thecotheus fide Korf *in* Ainsworth *et al.* (Eds) (*The Fungi* **4A**: 249, 1973).

Ascophanopsis Faurel & Schotter (1965) = Thecotheus fide Krug & Khan (*Mycol.* **79**: 200, 1987).

Ascophanus Boud. (1869), ? Ascobolaceae. 20 (mostly on dung), widespread (temperate). See Kimbrough (*CJB* **44**: 697, 1966), Moravec (*Česká Mykol.* **25**: 150, 1971), Pouzar & Svrček (*Česká Mykol.* **26**: 25, 1972; typification), van Brummelen *in* Hawksworth (Ed.) (*Ascomycete Systematics. Problems and Perspectives in the Nineties* NATO ASI Series vol. **269** **269**: 398, 1994; posn), Prokhorov (*Mikol. Fitopatol.* **31**: 27, 1997), Huhtinen & Spooner (*Kew Bull.* **58**: 749, 2003).

Ascophora Tode (1790) nom. rej. = Mucor Fresen. fide Hesseltine (*Mycol.* **47**: 344, 1955).

ascophore (1) an ascus-producing hypha, esp. the stalk-like hyphae supporting asci in *Cephaloascus*; (2) apothecium (obsol.).

ascophyte, hypothetical autotrophic ancestor of the *Ascomycota* (Cain, 1972), see Phylogeny, cf. basidiophyte.

ascoplasm, epiplasm (q.v.).

Ascopolyporus Möller (1901), Cordycipitaceae. 5, C. & S. America. See Heim (*BSMF* **69**: 417, 1954), Doi *et al.* (*Bull. natn. Sci. Mus.* Tokyo, B **3**: 22, 1977), Bischoff *et al.* (*Mycol.* **97**: 710, 2005; ecology, phylogeny).

Ascoporia Samuels & A.I. Romero (1993) = Pseudosolidum fide Samuels & Romero (*Bolm Mus. paraense 'Emílio Goeldi'* sér. bot. **7**: 263, 1991), Kutorga & Hawksworth (*SA* **15**: 1, 1997), Rossman *et al.* (*Stud. Mycol.* **42**: 248 pp., 1999).

Ascoporiaceae Kutorga & D. Hawksw. (1997), ? Dothideomycetes (inc. sed.). 1 gen. (+ 1 syn.), 1 spp.

Lit.: Samuels & Romero (*Bolm Mus. paraense 'Emílio Goeldi'* sér. bot. **7**: 263, 1991), Kutorga & Hawksworth (*SA* **15**: 1, 1997).

TABLE 2. Classification of the Ascomycota from the 9th Edition and as adopted in the 10th Edition

Dictionary 2001 (9th Edition)	Dictionary 2008 (10th Edition)
Agyriales (Lecanoromycetidae)	Acarosporales (Acarosporomycetidae)
Arthoniales (Arthoniomycetidae)	Acrospermales (Dothideomycetes)
Boliniales (Sordariomycetidae)	Agyriales (Ostropomycetidae)
Calosphaeriales (Sordariomycetidae)	Arachnomycetales (Eurotiomycetidae)
Capnodiales (Dothideomycetidae)	Arthoniales (Arthoniomycetes)
Chaetothyriales (Chaetothyriomycetidae)	Ascosphaerales (Eurotiomycetidae)
Coryneliales (Dothideomycetidae)	Baeomycetales (Ostropomycetidae)
Diaporthales (Sordariomycetidae)	Boliniales (Sordariomycetidae)
Dothideales (Dothideomycetidae)	Botryosphaeriales (Dothideomycetes)
Elaphomycetales (Eurotiomycetidae)	Calosphaeriales (Sordariomycetidae)
Erysiphales (Erysiphomycetidae)	Candelariales (Lecanoromycetes)
Eurotiales (Eurotiomycetidae)	Capnodiales (Dothideomycetidae)
Gyalectales (Lecanoromycetidae)	Chaetosphaeriales (Sordariomycetidae)
Halosphaeriales (Sordariomycetidae)	Chaetothyriales (Chaetothyriomycetidae)
Helotiales (Leotiomycetidae)	Coniochaetales (Sordariomycetidae)
Hypocreales (Sordariomycetidae)	Coronophorales (Hypocreomycetidae)
Hysteriales (Dothideomycetidae)	Coryneliales (Eurotiomycetidae)
Laboulbeniales (Laboulbeniomycetidae)	Cyttariales (Leotiomycetes)
Lahmiales (Dothideomycetidae)	Diaporthales (Sordariomycetidae)
Lecanorales (Lecanoromycetidae)	Dothideales (Dothideomycetidae)
Lichinales (Lecanoromycetidae)	Erysiphales (Leotiomycetidae)
Lulworthiales (Sordariomycetidae)	Eurotiales (Eurotiomycetidae)
Medeolariales (Leotiomycetidae)	Helotiales (Leotiomycetes)
Meliolales (Meliolomycetidae)	Hypocreales (Sordariomycetidae)
Microascales (Sordariomycetidae)	Hysteriales (Dothideomycetes)
Microthyriales (Dothideomycetidae)	Jahnulales (Dothideomycetes)
Mycocaliciales (Incertae sedis)	Laboulbeniales (Laboulbeniomycetidae)
Mycosphaerellales (Dothideomycetidae)	Lahmiales (Pezizomycotina)
Myriangiales (Dothideomycetidae)	Lecanorales (Lecanoromycetidae)
Neolectales (Neolectomycetidae)	Lecideales (Lecanoromycetidae)
Onygenales (Eurotiomycetidae)	Leotiales (Leotiomycetidae)
Ophiostomatales (Sordariomycetidae)	Lichinales (Lichinomycetes)
Ostropales (Incertae sedis)	Lulworthiales (Spathulosporomycetidae)
Patellariales (Dothideomycetidae)	Medeolariales (Pezizomycotina)
Peltigerales (Lecanoromycetidae)	Melanosporales (Hypocreomycetidae)
Pertusariales (Lecanoromycetidae)	Meliolales (Meliolomycetidae)
Pezizales (Pezizomycetidae)	Microascales (Hypocreomycetidae)
Phyllachorales (Sordariomycetidae)	Microthyriales (Dothideomycetes)
Pleosporales (Dothideomycetidae)	Mycocaliciales (Mycocaliciomycetidae)
Pneumocystidales (Pneumocystidomycetidae)	Myriangiales (Dothideomycetidae)
Pyrenulales (Dothideomycetidae)	Neolectales (Neolectomycetidae)
Pyxidiophorales (Laboulbeniomycetidae)	Onygenales (Eurotiomycetidae)
Rhytismatales (Leotiomycetidae)	Ophiostomatales (Sordariomycetidae)
Saccharomycetales (Saccharomycetidae)	Orbiliales (Orbiliomycetes)
Schizosaccharomycetales (Schizosaccharomycetidae)	Ostropales (Ostropomycetidae)
Sordariales (Sordariomycetidae)	Patellariales (Dothideomycetes)
Spathulosporales (Spathulosporomycetidae)	Peltigerales (Lecanoromycetidae)
Taphrinales (Taphrinomycetidae)	Pertusariales (Ostropomycetidae)
Teloschistales (Lecanoromycetidae)	Pezizales (Pezizomycetidae)
Thelebolales (Leotiomycetidae)	Phyllachorales (Sordariomycetes)
Triblidiales (Incertae sedis)	Pleosporales (Pleosporomycetidae)
Trichosphaeriales (Sordariomycetidae)	Pneumocystidales (Pneumocystidomycetidae)
Trichotheliales (Incertae sedis)	Pyrenulales (Chaetothyriomycetidae)
Verrucariales (Incertae sedis)	Pyxidiophorales (Laboulbeniomycetidae)
Xylariales (Sordariomycetidae)	Rhizocarpales (Lecanoromycetidae)
	Rhytismatales (Leotiomycetes)
	Saccharomycetales (Saccharomycetidae)
	Schizosaccharomycetales (Schizosaccharomycetidae)
	Sordariales (Sordariomycetidae)
	Taphrinales (Taphrinomycetidae)
	Teloschistales (Lecanoromycetidae)
	Thelebolales (Leotiomycetes)
	Triblidiales (Pezizomycotina)
	Trichosphaeriales (Sordariomycetes)
	Trypetheliales (Dothideomycetes)
	Umbilicariales (Lecanoromycetes)
	Verrucariales (Chaetothyriomycetidae)
	Xylariales (Xylariomycetidae)

Ascorhiza Lecht.-Trinka (1931), Pezizomycotina. 1, Europe. See Benny & Kimbrough (*Mycotaxon* **12**: 1, 1980; referral to *Ascosphaerales*).

Ascorhizoctonia Chin S. Yang & Korf (1985), anamorphic *Tricharina*, Sc.-.-. 7, widespread (cool temperate). See Yang & Korf (*Mycotaxon* **23**: 468, 1985), Yang & Kristiansen (*Mycotaxon* **35**: 313, 1989), Barrera & Romero (*Mycotaxon* **77**: 31, 2001).

Ascorhombispora L. Cai & K.D. Hyde (2007), Pleosporales. 1, China. See Cai & Hyde (*Cryptog. Mycol.* **28**: 291, 2007).

Ascoronospora Matsush. (2003), ? Pleosporales. Anamorph *Coronospora*. 1, Japan. See Matsushima (*Matsush. Mycol. Mem.* **10**: 179, 2001).

Ascosacculus J. Campb., J.L. Anderson & Shearer (2003), Halosphaeriaceae. 2, Australia. See Campbell *et al.* (*Mycol.* **95**: 545, 2003).

Ascosalsum J. Campb., J.L. Anderson & Shearer (2003), Halosphaeriaceae. 3, France; USA. See Campbell *et al.* (*Mycol.* **95**: 546, 2003), Pang & Jones (*Nova Hedwigia* **78**: 269, 2004).

Ascoscleroderma Clémencet (1932) = Elaphomyces fide Trappe (*Mycotaxon* **9**: 247, 1979).

Ascosorus Henn. & Ruhland (1900), Pezizomycotina. 1, N. America.

Ascosparassis Kobayasi (1960), Pyronemataceae. 1, widespread (tropics). See Korf (*Lloydia* **26**: 23, 1963), Pfister & Halling (*Mycotaxon* **35**: 283, 1989), Wang & Chou (*Fungal Science* Taipei **11**: 45, 1996; Taiwan).

Ascospermum Schulzer (1863) nom. dub., Fungi. Based on sterile mycelium.

Ascosphaera L.S. Olive & Spiltoir (1955), Ascosphaeraceae. Anamorph *Chrysosporium*-like. 17 (associated with bees), widespread (north temperate; esp. Europe). *A. apis* on larvae of honey bees causing chalk brood. See McManus & Youssof (*Mycol.* **76**: 830, 1984), Rose *et al.* (*Mycotaxon* **19**: 41, 1984; key 7 spp., N. Am.), Bisset (*CJB* **66**: 2541, 1988; key), Skou (*Mycotaxon* **31**: 173, 1988; 7 spp. nov. Japan), Bisset *et al.* (*Mycol.* **88**: 797, 1996; key), Anderson & Gibson (*Aust. Syst. Bot.* **11**: 53, 1998; Australia), Anderson *et al.* (*MR* **102**: 541, 1998; phylogeny), James & Skinner (*J. Invert. Path.* **90**: 98, 2005; PCR diagnosis), Aronstein *et al.* (*Mycol.* **99**: 553, 2007; mating type genes).

Ascosphaeraceae L.S. Olive & Spiltoir (1955), Ascosphaerales. 3 gen. (+ 1 syn.), 19 spp.

Lit.: Skou (*Friesia* **10**: 1, 1972; monogr.), Brady (*IMI Descr. Fungi Bact.* **62**: [1, 1979), Skou (*Mycotaxon* **15**: 487, 1982; emended concept), Kowalska (*Polskie Arch. Wet.* **24**: 7, 1984; biochem. syst.), Skou (*Aust. J. Bot.* **32**: 225, 1984; spp.), Bissett (*CJB* **66**: 2541, 1988), Kish *et al.* (*Mycol.* **80**: 312, 1988), Skou (*Mycotaxon* **31**: 191, 1988; rank), Berbee & Taylor (*BioSystems* **28**: 117, 1992), Laṇdvik *et al.* (*Mycoscience* **37**: 237, 1996), Anderson & Gibson (*Aust. Syst. Bot.* **11**: 53, 1998), Anderson *et al.* (*MR* **102**: 541, 1998), Geiser & LoBuglio *in* McLaughlin *et al.* (Eds) (*The Mycota* A Comprehensive Treatise on Fungi as Experimental Systems for Basic and Applied Research **7A**: 201, 2001).

Ascosphaerales Gäum. ex Benny & Kimbr. (1980). Eurotiomycetidae. 1 fam., 3 gen., 19 spp. Fam.:
Ascosphaeraceae

Apparently nests within *Onygenales*, but further molecular studies are needed and the limits of that order also require research.

For *Lit.* see under fam.

Ascosphaeromycetes = Eurotiomycetes. Used by Skou (*Mycotaxon* **31**: 191, 1988) to accommodate the single order (and family) *Ascosphaerales* (*Ascosphaeraceae*).

Ascospora Fr. (1825) = Mycosphaerella fide Hawksworth *et al.* (*Dictionary of the Fungi* edn 8, 1995).

Ascospora Mont. (1849) nom. conf., anamorphic *Pezizomycotina*. See Sutton (*Mycol. Pap.* **141**, 1977).

Ascosporaceae Bonord. (1851) = Mycosphaerellaceae.

ascospore, a spore produced in an ascus by 'free cell formation'; the ascospore wall is multilayered, it consists of an outer **perispore**, an intermediary layer, the proper wall (**epispore**) and sometimes an internal **endospore**; major differences in which layers are thickened, folded or pigmented can give rise to considerable variation even in a single family (e.g. *Lasiosphaeriaceae*); see Bellemère (*in* Hawksworth (Ed.), *Ascomycete systematics*: 111, 1994), basidiospore, spore wall.

Ascosporium Berk. (1860) = Taphrina fide Saccardo (*Syll. fung.* **8**: 817, 1889).

ascostome, a pore in the apex of an ascus (obsol.).

Ascostratum Syd. & P. Syd. (1912), ? Dothideomycetes. 1, S. Africa.

Ascostroma Bonord. (1851) = Kretzschmaria fide Læssøe (*SA* **13**: 43, 1994).

ascostroma, a stroma in or on which asci are produced, usually restricted to groups with ascolocular ontogeny.

Ascosubramania Rajendran (1997) = Microascus Zukal fide Rajendran (*J. Med. Vet. Mycol.* **35**: 336, 1997), Guarro (*Medical Mycology* **36**: 349, 1998; synonymy).

Ascotaiwania Sivan. & H.S. Chang (1992), ? Sordariales. Anamorphs *Brachysporiella*, *Trichocladium*, *Monotosporella*. 12 (mostly from wood), widespread (tropical). See Wong *et al.* (*SA* **16**: 17, 1998), Ranghoo *et al.* (*Fungal Diversity* **2**: 159, 1999; DNA), Chang (*Fungal Science* Taipei **16**: 35, 2001; anamorph), Réblová & Winka (*Mycol.* **93**: 478, 2001), Wong & Hyde (*Cryptog. Mycol.* **22**: 19, 2001), Campbell & Shearer (*Mycol.* **96**: 822, 2004; phylogeny).

Ascotremella Seaver (1930), Helotiaceae. 2, widespread (north & south temperate). See Gamundí & Dennis (*Darwiniana* **15**: 14, 1969; status), Gamundí & Romero (*Fl. criptog. Tierra del Fuego* **10**, 1998).

Ascotremellopsis Teng & S.H. Ou ex S.H. Ou (1936) = Myriodiscus fide Liu & Guo (*Acta Mycol. Sin.* Suppl. **1**: 97, 1988).

Ascotricha Berk. (1838), Xylariaceae. Anamorph *Dicyma*. 15, widespread. See Hawksworth (*Mycol. Pap.* **126**, 1971; key), Horie *et al.* (*TMSJ* **34**: 123, 1993), Læssøe (*SA* **13**: 43, 1994; posn), Udagawa *et al.* (*Mycotaxon* **52**: 215, 1994), Lee & Hanlin (*Mycol.* **91**: 434, 1999; DNA), Stchigel *et al.* (*Mycol.* **92**: 805, 2000).

Ascotrichella Valldos. & Guarro (1988), ? Coniochaetaceae. Anamorph *Humicola*-like. 1, Chile. See Valldosera & Guarro (*TBMS* **90**: 601, 1988), Læssøe (*SA* **13**: 43, 1994; posn).

Ascovaginospora Fallah, Shearer & W.D. Chen (1997), ? Hyponectriaceae. 1 (dead submerged stems), USA. See Fallah *et al.* (*Mycol.* **89**: 812, 1997; DNA), Chen *et al.* (*Mycol.* **91**: 84, 1999; DNA), Huhndorf *et al.* (*Mycol.* **96**: 368, 2004; phylogeny).

Ascoverticillata Kamat, Subhedar & V.G. Rao (1979)

= Crocicreas fide Eriksson (*SA* **5**: 119, 1986).

Ascovirgaria J.D. Rogers & Y.M. Ju (2002), Xylariaceae. Anamorph *Virgaria*. 1 (on wood), Hawaii. See Rogers & Ju (*CJB* **80**: 478, 2002).

Ascoxyta Lib. (1830) nom. dub., Pezizomycotina. See Holm (*Taxon* **24**: 475, 1975).

Ascoyunnania L. Cai & K.D. Hyde (2005), ? Sordariomycetes. 1 (in submerged bamboo stems), China. Possibly linked with *Ustilaginoidea*. See Cai *et al.* (*Fungal Diversity* **18**: 2, 2005).

Ascozonus (Renny) E.C. Hansen (1877), Thelebolaceae. *c.* 6, widespread (north temperate). See Kimbrough (*CJB* **44**: 693, 1966), Prokhorov (*Mikol. Fitopatol.* **31**: 27, 1997), Landvik *et al.* (*Mycoscience* **39**: 49, 1998; DNA), van Brummelen (*Persoonia* **16**: 425, 1998; ultrastr.), Brummelen & Richardson (*Persoonia* **17**: 487, 2000), Hoog *et al.* (*Stud. Mycol.* **51**: 33, 2005).

ascus (pl. **asci**), term introduced by Nees (*Syst. Pilze*: 164, 1817) for the typically sac-like cell (first figured in *Pertusaria* by Micheli in 1729; q.v.) characteristic of *Ascomycota* (q.v.), in which (after karyogamy and meiosis) ascospores (generally 8) are produced by 'free cell formation' (Fig. 11). Asci vary considerably in structure, and work in the last two decades has shown previous separation into only 2-3 categories (e.g. **bitunicate**, **prototunicate**, **unitunicate**) to be an over simplification. Sherwood (1981) illustrated 9 main types distinguishable by light microscopy (reproduced on p. 36 of edn 7 of this *Dictionary*): prototunivate, bitunicate, astropalean, annellate, hypodermataceous, pseudoperculate, operculate, lecanoralean, and verrucariod). Eriksson (1981) distinguished 7 types of dehiscence in bitunicate asci with an ectotunica and distinct endotunica (see p. 37 of edn 7). These classifications mask a much wider range of variation; Bellemère (1994) recognized 3 predehiscence types and 11 dehiscence categories (Fig. 1). The details of the asci are stressed in ascomycete systematics, esp. in lichen-forming orders where reactions with iodine are emphasized (q.v.) (Hafellner, 1984).

Bitunicate asci with two functional wall layers; those splitting at discharge (**fissitunicate**; 'jack-in-the-box') had been correlated with an ascolocular ontogeny by Luttrell (1951). Reynolds (1989) critically examined this paradigm and found the term to be applied to different ascus types and that an exclusive link to ascostromatic fungi could not be upheld; he also introduced the term **extenditunicate** for asci which extend without any splitting of the wall layers (Reynolds, *Cryptog. Mycol.* **10**: 305, 1989).

Much variation depends on the modifications in the various wall layers, especially the thickness of the walls and the *c* and *d* layers, and the details of apical differentiation (Bellemère, 1994) (Fig. 2). Caution is needed in comparing ascus staining reactions (see iodine) and structures in the absence of ultrastructural data. For terms used to describe the various structures see Fig. 2.

Also encountered are - **crown** (annular thickenings in *Phyllachora*), and - **plug** (thickening in the apex through which the spores are forcibly discharged).

Lit.: Bellemère (*Ann. Sci. nat., Bot.* **12**: 429, 1971; *Rev. Mycol.* **41**: 233, 1977, bitunicate discom.; *in* Hawksworth, 1994: 111, review), Bellemère & Letrouit-Galinou (*Bibl. Lich.* **25**: 137, 1987; ultrastr.), van Brummelen (*Persoonia* **10**: 113, 1978; operculate), Chadefaud (*Rev. Mycol.* **7**: 57, 1942; **9**: 3, 1944; apical apparatus), Chadefaud *et al.* (*Mém. Soc. bot. Fr.* **79**, 1968; lichen asci), Eriksson (*Opera bot.* **60**, 1981; bitunicate types), Griffiths (*TBMS* **60**: 261, 1973; unitunicate pyrenom.), Hafellner (*Beih Nova-Hedw.* **79**: 24, 1984), Hawksworth (*J. Hattori bot. Lab.* **52**: 323, 1982; evolution types; (Ed.), *Ascomycete systematics: problems and perspectives in the nineties*, 1994), Holm (*Symb. bot. upsal.* **30**(3): 21, 1995; history of term), Honneger (*Lichenologist* **10**: 47, 1978; lecanoralean, *Peltigera*; **12**: 157, 1980; *Rhizocarpon*; **14**: 205, 1982, *Pertusaria*; **15**: 57, 1983, *Baeomyces*, *Cladonia*, *Leotia* etc.; *J. Hattori bot. Lab.* **52**: 417, 1982, review lecanoralean types), Janex-Favre (*Revue bryol. Lichén.* **37**: 421, 1971; lich. pyrenom.), Letrouit-Galinou (*Bryologist* **76**: 30, 1973; archaeasceous), Parguey-Leduc (*Ann. Sci. nat., Bot.* XII, **7**: 33, 1966; ascoloc.; *Rev. Mycol.* **41**: 281, 1977; pyrenom.), Parguey-Leduc & Janex-Favre (*Cryptogamie Mycol.* **5**: 171, 1984; ultrastr. 'unitunicate' types), Reynolds (Ed.) (*Ascomycete systematics: the Luttrellian concept*, 1981; *Bot. Rev.* **55**: 1, 1989; bitunicate paradigm), Sherwood (*Bot. J. Linn. Soc.* **82**: 15, 1981; main types), Ziegenspeck (*Bot. Arch. Koenigsberg* **13**: 341, 1926).

ascus plug, thickening in the apex through which the spores are forcibly discharged.

Aseimotrichum Corda (1831) nom. dub., ? Fungi.

Asellaria R.A. Poiss. (1932), Asellariaceae. 9 (in *Isopoda*), widespread. See Lichtwardt (*Mycol.* **65**: 1, 1973; morphology), Manier (*C.R. Hebd. Séanc. Acad. Sci. Paris* **276**: 3429, 1973; ultrastr.), Lichtwardt (*The Trichomycetes. Fungal associates of arthropods*, 1986; revision, key), Valle (*Fungal Diversity* **21**: 167, 2006; Spain), White *et al.* (*Mycol.* **98**: 872, 2006; phylogeny), Valle & Cafaro (*Mycol.* **100**: 122, 2008; zygospores).

Asellariaceae Manier ex Manier & Lichtw. (1968), Asellariales. 3 gen. (+ 1 syn.), 14 spp.

Lit.: Moss (*TBMS* **65**: 115, 1975), Moss & Young (*Mycol.* **70**: 944, 1978), Lichtwardt (*The Trichomycetes. Fungal associates of arthropods*: 343 pp., 1986), Cafaro (*Mycol.* **91**: 517, 1999), Benny *in* McLaughlin *et al.* (Eds) (*The Mycota* A Comprehensive Treatise on Fungi as Experimental Systems for Basic and Applied Research **7A**: 147, 2001), Lichtwardt (*Cellular Origin and Life in Extreme Habitats* **4**: 577, 2002).

Asellariales Manier ex Manier & Lichtw. (1978). Kickxellomycotina. 1 fam., 3 gen., 14 spp. Fam.: **Asellariaceae**

Lit.: Manier (*Ann. Sci. nat., Bot.* sér. 12 **10**: 565, 1969; taxonomy), Scheer (*Z. binnenfischerei DDR* **19**: 369, 1972; taxonomy), Lichtwardt & Manier (*Mycotaxon* **7**: 441, 1978; taxonomy), Moss & Young (*Mycol.* **70**: 944, 1978; phylogeny), Moss (*in* Batra (Ed.), *Insect-fungus symbiosis*: 175, 1979), Lichtwardt (1986; taxonomy, key), White *et al.* (*Mycol.* **98**: 860, 2006; molecular phylogeny), Hibbett *et al.* (*MR* **111**: 109, 2007), Valle & Cafaro (*Mycol.* **100**: 122; zygospores), and see under Family.

aseptate, having no cross walls.

aseptic, free from damaging microorganisms.

Aseroë Labill. (1800), Phallaceae. 2, widespread (tropical). See Spooner (*Mycologist* **8**: 153, 1994), Baseia & Calonge (*Mycotaxon* **92**: 169, 2005; Brazil).

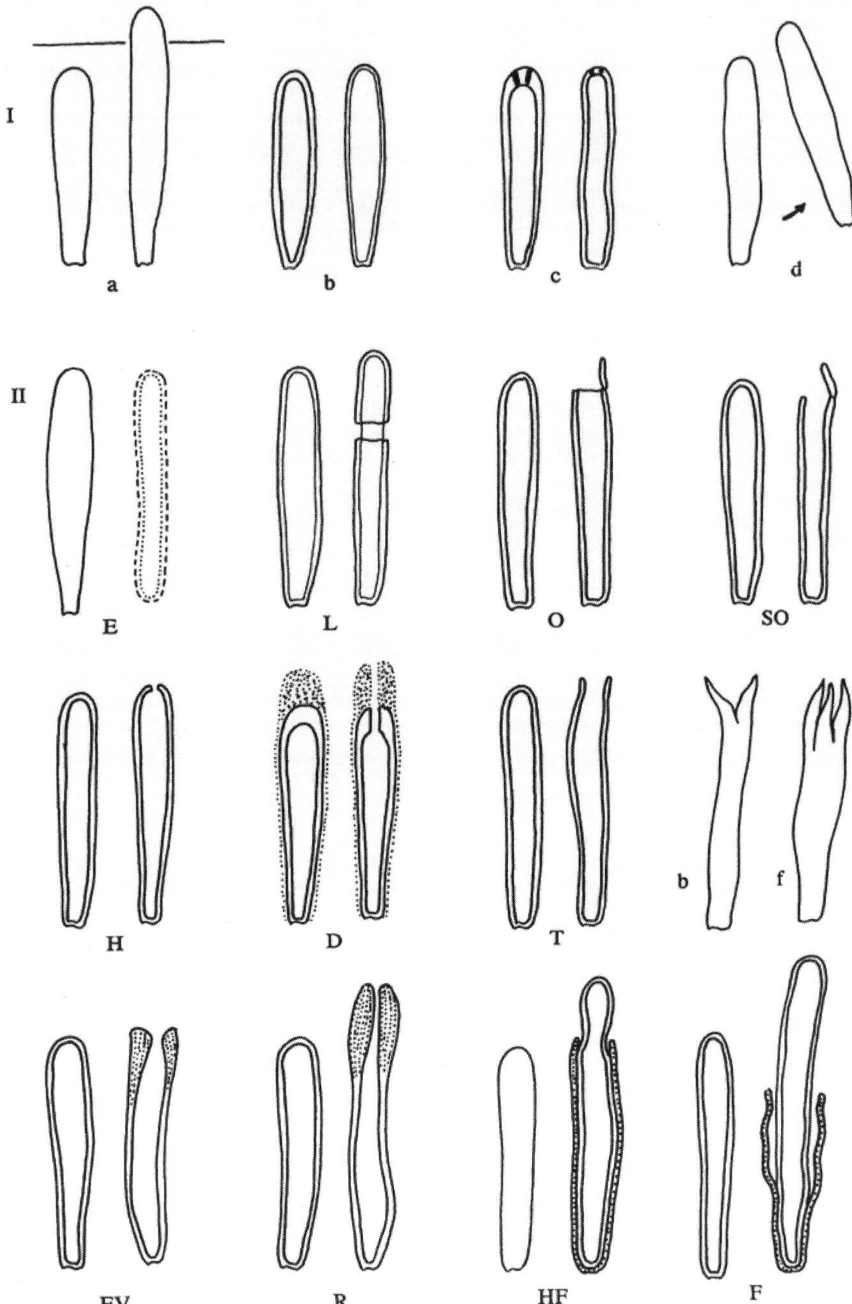

Fig. 4. I. Predehiscence stage of asci. a = protruding ascus; b = ascus wall becoming thinner; c = change in apical structure; d = ascus liberation. II. Dehiscence stage of asci; evanescent ascus (E); rupture of lateral wall (L); subapical rupture (O, operculate, and SO, suboperculate dehiscence); rupture by apical wall without extrusion (H, pore-like dehiscence); D, *Dactylospora*-type; T, *Teloschistes*-type = extenditunicate (b = bivalve, f = fissurate variants); rupture with extrusion (EV, eversion; R, rostrate; HF, hemifissitunicate; F, fissitunicate). After Belleméré, *in* Hawksworth (Ed.) (*Ascomycete Systematics*: 111, 1994).

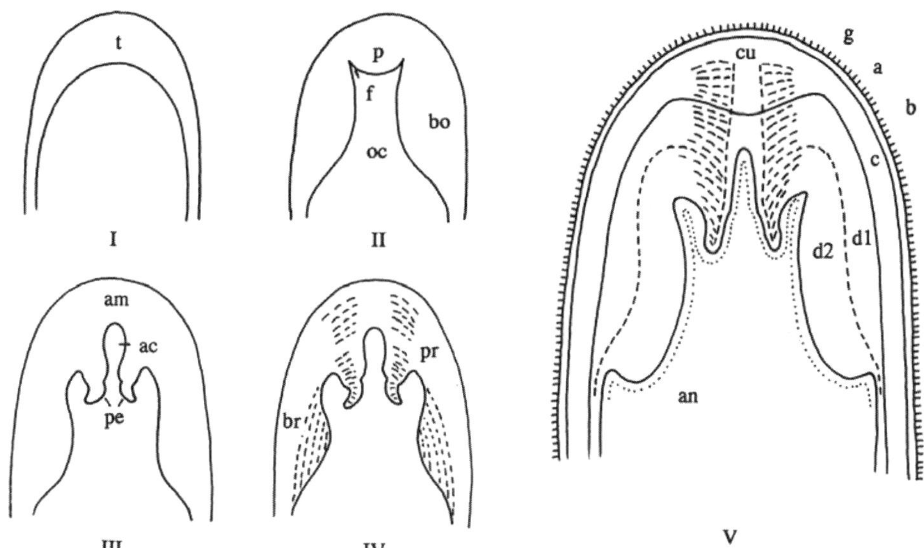

Fig. 5. I-V. Ascus apex components. ac = axial canal; am = axial mass; bo = bourrelet; br = ring in bourrelet; f = furrow; oc = ocular chamber; p = plug; pe = pendant; pr = rings in the plug and pendant; t = tholus; V, ascus apex structure. a = a layer; an = apical nasse; b = b layer; c = c layer; cu = cushion; d1 and d2 = sublayers of the d layer. After Bellemére (*in* Hawksworth (Ed.), *Ascomycete Systematics*: 111, 1994).

Aserophallus Mont. & Lepr. (1845) = Clathrus fide Dring (*Kew Bull.* **35**: 1, 1980).

asexual, without sex organs or sex spores; vegetative.

Ashbia Cif. & Gonz. Frag. (1928) ≡ Ashbya.

Ashbya Guilliermond. (1928) = Eremothecium fide Batra (*USDA Tech. Bull.* **1469**, 1973), Kurtzman (*J. Industr. Microbiol.* **14**: 523, 1995), Prillinger *et al.* (*Yeast* Chichester **13**: 945, 1997), de Hoog *et al. in* Kurtzman & Fell (Eds) (*Yeasts, a taxonomic study* 4th edn: 201, 1998; synonymy with *Eremothecium*), Kroken *et al.* (*Proc. natn Acad. Sci. U.S.A.* **100**: 15670, 2003; polyketide synthase genes), Kurtzman (*FEMS Yeast Research* **4**: 233, 2003; synonymy), Dietrich *et al.* (*Science* N.Y. **304** no. 5668: 304, 2004; genomic studies), Kohn (*Ann. Rev. Phytopath.* **43**: 279, 2005; speciation), Brachat *et al.* (*Topics in Current Genetics* **15**: 197, 2006; genome).

Ashtaangam Subram. (1995), anamorphic *Pezizomycotina*, Hso.0bP.1. 1, Malaysia. See Subramanian (*Korean J. Mycol.* **20**: 281, 1992), Subramanian (*Kavaka* **20/21**: 58, 1992).

Asirosiphon Nyl. (1873) = Spilonema fide Henssen (*Symb. bot. upsal.* **18** no. 1, 1963).

Asociación Latino-Americana de Micología. Founded in 1990; recognized as the Committee for Latin America within the International Mycological Association (q.v.); structure comprises individual members, an elected executive, and national representatives from Latin American and other countries; organizes Latin American Mycological Congress every three or four years. Website: www.almic.org/principal.php.

Asordaria Arx, Guarro & Aa (1987) = Sordaria fide Eriksson & Hawksworth (*SA* **7**: 61, 1988), Cai *et al.* (*MR* **110**: 137, 2006; phylogeny).

asperate, rough with projections or points.

Aspergillaceae Link (1826) = Trichocomaceae.

Aspergillales = Eurotiales.

aspergilliform (of a sporulating structure), resembling that of an *Aspergillus* conidiophore.

aspergillin (1) a black, water-insoluble pigment of *Aspergillus niger* spores (Linossier, 1891); (2) various antibiotics produced by *Aspergillus* spp. See Tobie (*Nature* **158**: 709, 1946).

Aspergillites Trivedi & C.L. Verma (1969), Fossil Fungi. 1 (Tertiary), Malaysia.

Aspergilloides Dierckx (1901) = Penicillium Link fide Raper & Thom (*Manual of the Penicillia*, 1949).

aspergilloma, a 'fungus ball' composed principally of hyphae of *Aspergillus*, found in a pre-existing cavity (esp. in an upper lobe of the lung) or a bronchus, which usually has a relatively benign or asymptomatic effect; cf. aspergillosis.

Aspergillopsis Sopp (1912) nom. dub., anamorphic *Pezizomycotina*. See Raper & Thom (*Manual of the Penicillia*, 1949).

Aspergillopsis Speg. (1910) = Aspergillus fide Hawksworth *et al.* (*Dictionary of the Fungi* edn 8, 1995).

Aspergillosis, Any disease in humans or animals caused by *Aspergillus* (esp. *A. fumigatus*); esp. common in birds; in humans usually respiratory and taking one of four forms: invasive (usually only in immuno-compromised patients, but with a high mortality rate), non-invasive, chronic pulmonary and aspergilloma, or severe asthma with fungal sensitisation (see Chute *et al.* (1971), Ainsworth & Austwick (1973) under Medical and veterinary mycology). See Austwick, (*in* Raper & Fennell, Eds, *The genus As-*

pergillus, 1965), Bossche *et al.* (Eds) (*Aspergillus and aspergillosis*, 1988).

Aspergillus P. Micheli ex Link (1809), anamorphic *Emericella, Eurotium, Neosartorya*, Hsy.0eH/1eP.32. 266, widespread. *A. niger* ('smut' of fig and date; black mould of cotton bolls, fruits, vegetables, etc., environmental contaminant and model organism), *A. fumigatus* (principal causal organism of aspergillosis). See aflatoxins, aspergillin, aspergillosis, Industrial Mycology. See Thom & Church (*The aspergilli*, 1926), Thom & Raper (*Manual of the Aspergilli*, 1945), Benjamin (*Mycol.* **47**: 669, 1955), Raper & Fennell (*The genus Aspergillus*, 1965), Locci (*Rivta Patol. Veget.*, 1972; SEM of teleomorphs), Kozakiewicz (*TBMS* **70**: 175, 1978; SEM of conidia), Samson (*Stud. Mycol.* **18**, 1979; compilation spp. described since 1965), Bennett *in* Demain & Solomon (Eds) (*Biology of industrial microorganisms*: 359, 1985; taxonomy and biology), Currah (*Mycotaxon* **24**: 1, 1985), Kurtzman *et al.* (*Mycol.* **78**: 955, 1986; DNA relatedness), Sekhon *et al.* (*Diagn. immunol.* **4**: 112, 1986; exoantigen grouping), Anon. *in* Bossche *et al.* (Ed.) (*Aspergillus and aspergillosis*, 1988), Bennett & Papa (*Adv. Pl. Path.* **6**: 263, 1988; aflatoxigenic spp.), Bilai & Koval (*Aspergilly*: 1, 1988; spp. from former USSR), Bojovic-Cvetic & Vujicic (*TBMS* **91**: 619, 1988; polysaccharide cytochemistry), Klich & Pitt (*TBMS* **91**: 99, 1988; differentiation of *A. flavus* from *A. parasiticus*), Anon. *in* Samson & Pitt (Eds) (*Modern concepts in Penicillium and Aspergillus classification* [NATO ASI Series A: Life Sciences], 1989), Bezjak (*Mycoses* **32**: 187, 1989; abnormal conidial structures), Hull *et al.* (*Mol. Microbiol.* **3**: 553, 1989; L-proline catabolism gene cluster in *A. nidulans*), Klich & Mullaney (*Mycol.* **81**: 159, 1989; differentiation of *A. parasiticus* from *A. sojae*), Kozakiewicz (*Mycol. Pap.* **161**: 1, 1989; spp. on stored products), Pitt (*J. Appl. Bact.* **67** Suppl. Symp. Ser. 18: 375, 1989; recent developments in systematics), Weidenbörner *et al.* (*J. Phytopath.* **126**: 1, 1989; preparation for SEM), Wirsel *et al.* (*Mol. Microbiol.* **3**: 3, 1989; amylase genes in *A. oryzae*), Clutterbuck (*Fungal Genetics Newsl.* **37**: 80, 1990; bibliography of *A. nidulans*), Moody & Tyler (*Appl. Environm. Microbiol.* **56**: 2453, 1990; restriction enzyme analysis of mtDNA, and DNA RLFPs of *A. flavus* group), Moody & Tyler (*Appl. Environm. Microbiol.* **56**: 2441, 1990; restriction enzyme analysis of mtDNA, and DNA RLFPs of *A. flavus* group), Novak & Kohn (*Exp. Mycol.* **14**: 339, 1990; developmental proteins), Samson & Frisvad (*Proc. Jap. Assoc. Mycotoxic.* **32**: 3, 1990; species concepts and mycotoxins), Tzean *et al.* (*Aspergillus and related teleomorphs from Taiwan* Mycological Monograph of the Food Industry Research & Development Institute, 1990), Chang *et al.* (*J. gen. appl. Microbiol.* Tokyo **37**: 289, 1991; phylogeny), Fragner (*Česká Mykol.* **45**: 113, 1991; spp. from humans and animals), Kozakiewicz *et al.* (*Taxon* **41**: 109, 1992; spp. nom. cons. prop.), Klich (*Mycol.* **85**: 100, 1993; sect. *Versicolores*), Pitt & Samson (*Regnum veg.* **128**: 13, 1993), Tiedt (*MR* **97**: 1459, 1993; ultrastr. conidiogenesis in *A. niger*), Anon. *in* Smith (Ed.) (*Aspergillus* [*Biotechnology Handbooks* 7], 1994), Chang *et al.* (*Appl. Environm. Microbiol.* **61**: 40, 1995; aflatoxin genes), Horn & Greene (*Mycol.* **87**: 324, 1995; VCGs), Ismail *et al.* (*Mycotaxon* **53**: 391, 1995; synoptic key), Peterson (*MR* **99**: 1349, 1995; phylogeny of sects *Wentii* and *Cremei*), Rinyu *et al.* (*J. Clin. Microbiol.* **33**: 2567, 1995; variability in *A. fumigatus*), Verwei *et al.* (*J. Mycol. Médic.* **5**: 194, 1995; ELISA test), Woloshuk *et al.* (*Appl. Environm. Microbiol.* **61**: 3019, 1995; *A. flavus*), Kevei *et al.* (*Antonie van Leeuwenhoek* **70**: 59, 1996; *A. carbonarius*), Kumeda & Asao (*Appl. Environm. Microbiol.* **62**: 2947, 1996; SSCP analysis in sect. *Flavi*), Rodriguez *et al.* (*J. Clin. Microbiol.* **34**: 2559, 1996; clinical strains of *A. fumigatus*), Kaufman *et al.* (*J. Clin. Microbiol.* **35**: 2206, 1997; immunological diagnostics), Law *et al.* (*Mycoses* **39**: 433, 1996; pyrolysis MS), Nemec *et al.* (*FEMS Microbiol. Lett.* **149**: 201, 1997; sterols and fatty acids), Parenicová *et al.* (*MR* **101**: 810, 1997; RFLP of black aspergilli), Bart-Delabesse *et al.* (*J. Clin. Microbiol.* **36**: 2413, 1998; microsatellite markers), Birch *et al.* (*Medical Mycology* **36**: 127, 1998; polar lipids), Fletcher *et al.* (*Journal of Clinical Pathology* **51**: 617, 1998; detection of *A. fumigatus*), Geiser *et al.* (*Proc. natn Acad. Sci. U.S.A.* **95**: 388, 1998; cryptic speciation), Geiser *et al.* (*Nature* Lond. **394** no. 6689: 137, 1998; pathogen of sea fans), Geiser *et al.* (*Mycol.* **90**: 831, 1998; phylogeny of sect. *Fumigati*), Katz *et al.* (*FEMS Immunol. Med. Microbiol.* **20**: 283, 1998; DNA sequence variation), McAlpin *et al.* (*Pl. Dis.* **82**: 1132, 1998; genetic diversity of *A. parasiticus*), Nikkuni *et al.* (*J. gen. appl. Microbiol.* Tokyo **44**: 225, 1998; *A. oryzae*), Brenier-Pinchart *et al.* (*J. Mycol. Médic.* **9**: 16, 1999; molecular diagnostics), Latgé (*Clin. Microbiol. Rev.* **12**: 310, 1999; aspergillosis), Samson (*Contributions to Microbiology* **2**: 5, 1999; *A. fumigatus* group), Sigler & Kennedy (*Manual of Clinical Microbiology*: 1212, 1999; review), Frisvad & Samson (*Stud. Mycol.* **45**: 201, 2000; subgenus *Circumdati*), Geiser *et al.* (*Integration of Modern Taxonomic Methods for Penicillium and Aspergillus Classification*: 381, 2000; molecular analytical tools), Geiser *et al.* (*Fungal Genetics Biol.* **31**: 169, 2000; phylogeny of *A. flavus* group), Hanazawa *et al.* (*Journal of Medical Microbiology* **49**: 285, 2000; detection of *A. fumigatus*), Klich & Cleveland (*Integration of Modern Taxonomic Methods for Penicillium and Aspergillus Classification*: 425, 2000; mycotoxin biosynthesis), Okuda *et al.* (*Integration of Modern Taxonomic Methods for Penicillium and Aspergillus Classification*: 83, 2000; morphological variation), Peterson (*Integration of Modern Taxonomic Methods for Penicillium and Aspergillus Classification*: 323, 2000; phylogeny), Pitt & Samson (*Integration of Modern Taxonomic Methods for Penicillium and Aspergillus Classification*: 51, 2000; typification), Pitt *et al.* (*Integration of Modern Taxonomic Methods for Penicillium and Aspergillus Classification*: 9, 2000; accepted species), Seifert (*Integration of Modern Taxonomic Methods for Penicillium and Aspergillus Classification*: 139, 2000; synoptic key), Tamura *et al.* (*Integration of Modern Taxonomic Methods for Penicillium and Aspergillus Classification*: 357, 2000; phylogeny), Tran-Dinh *et al.* (*Integration of Modern Taxonomic Methods for Penicillium and Aspergillus Classification*: 435, 2000; toxigenicity), Varga *et al.* (*Integration of Modern Taxonomic Methods for Penicillium and Aspergillus Classification*: 397, 2000; black aspergilli), Varga *et al.* (*Can. J. Microbiol.* **46**: 593, 2000; *A. ochraceus* group), Moraes *et al.* (*Mycotaxon* **78**: 413, 2001; from mosquitoes), Rath (*Mycoses* **44**: 65, 2001; reference strains), Yokoyama *et al.* (*FEMS Microbiol. Lett.* **200**: 241, 2001; phylogeny of sect. *Nigri*), Zhao *et al.* (*J. Clin. Microbiol.* **39**: 2261, 2001; identification using nested PCR), Montiel *et al.* (*MR* **107**: 1427, 2003; AFLPs in sect. *Flavi*), Varga *et al.* (*Antonie van Leeuwenhoek* **83**: 191, 2003; sect. *Clavati*), Aguirre *et al.* (*J. Clin. Microbiol.* **42**: 3495, 2004; rapid diagnostics), Frisvad *et al.* (*Stud. Mycol.* **50**: 23, 2004; ochratoxigenic species), Raghukumar *et al.* (*Deep Sea Research* Part I: Oceanographic Research Papers **51**: 1759, 2004; deep sea sediments), Samson *et al.* (*Stud. Mycol.* **50**: 45, 2004; sect. *Nigri*), Sugita *et al.* (*Medical Mycology* **42**: 433, 2004; PCR identification), Varga *et al.* (*Eur. J. Pl. Path.* **110**: 627, 2004; agriculturally important species), Cary *et al.* (*Mycol.* **97**: 425, 2005; aflatoxigenic species), Dörfelt & Schmidt (*MR* **109**: 956, 2005; fossil in amber), Dyer & Paoletti (*Medical Mycology* **43** Suppl. 1: S7, 2005; sexuality in *A. fumigatus*), Galagan (*Nature* Lond. **438** no. 7071: 1105, 2005; sequencing of *A. nidulans*), Halliday *et al.* (*J. Clin. Microbiol.* **43**: 5366, 2005; real-time PCR), Hong *et al.* (*Mycol.* **97**: 1316, 2006; polyphasic taxonomy of *A. fumigatus* group), Leinberger *et al.* (*J. Clin. Microbiol.* **43**: 4943, 2005; microarrays), Nierman (*Nature* Lond. **438** no. 7071: 1151, 2005; genome of *A. fumigatus*), Pringle *et al.* (*Evolution* Lancaster, Pa. **59**: 1886, 2005; cryptic speciation in *A. fumigatus*), Varga *et al.* (*Antonie van Leeuwenhoek* **88**: 141, 2005; *A. terreus* group), Geiser *et al.* (*Mycol.* **98**: 1053, 2006; phylogeny), Hong *et al.* (*Mycol.* **97**: 1316, 2005; polyphasic taxonomy of *A. fumigatus* group), Balajee *et al.* (*Stud. Mycol.* **59**: 39, 2007; clinical identification), Frisvad *et al.* (*Stud. Mycol.* **59**: 31, 2007; chemistry, species recognition), Geiser *et al.* (*Stud. Mycol.* **59**: 1, 2007; review, identification methods), Houbraken *et al.* (*Stud. Mycol.* **59**: 107, 2007; section *Usti*), Klaasen & Osherov (*Stud. Mycol.* **59**: 47, 2007; strain typing), Pál *et al.* (*Stud. Mycol.* **59**: 19, 2007; mating type and VC genes), Perrone *et al.* (*Stud. Mycol.* **59**: 53, 2007; in agricultural products), Pitt & Samson (*Stud. Mycol.* **59**: 67, 2007; nomencl.), Rokas *et al.* (*Stud. Mycol.* **59**: 11, 2007; comparative genomics), Samson *et al.* (*Stud. Mycol.* **59**: 71, 2007; species concepts), Samson *et al.* (*Stud. Mycol.* **59**: 129, 2007; black-spored spp.), Samson *et al.* (*Stud. Mycol.* **59**: 147, 2007; section *Fumigati*), Varga *et al.* (*Stud. Mycol.* **59**: 75, 2007; section *Candidi*), Varga *et al.* (*Stud. Mycol.* **59**: 89, 2007; section *Clavati*), Peterson (*Mycol.* **100**: 205, 2008; 4-locus phylogeny).

Asperisporium Maubl. (1913), anamorphic *Pezizomycotina*, Hsp.1eP.10. 12, America. *A. caricae* (*Carica papaya* leaf spot). See Baker *et al.* (*Mycotaxon* **76**: 247, 2000), Schubert & Braun (*Fungal Diversity* **20**: 187, 2005).

Asperopilum Spooner (1987), Hyaloscyphaceae. 1 (on *Juncus*), Australasia. See Spooner (*Biblthca Mycol.* **116**, 1987).

Asperotrichum, see *Asporothrichum*.

asperulate, delicately asperate.

Aspicilia A. Massal. (1852) nom. cons., Megasporaceae (L). *c.* 230, widespread. See Clauzade & Roux (*Bull. Soc. bot. Centre-Ouest* Nouv. sér. **15**: 127, 1984; Eur., gen. concept.), Laundon & Hawksworth (*Taxon* **37**: 478, 1988; nomencl.), Rosentreter *in* Glenn *et al.* (Eds) (*Lichenogr. Thomsoniana*: 163, 1998), Wedin *et al.* (*MR* **109**: 159, 2005; posn), Miądlikowska *et al.* (*Mycol.* **98**: 1088, 2006; phylogeny), Schmitt *et al.* (*J. Hattori bot. Lab.* **100**: 753, 2006; phylogeny), Nordin *et al.* (*Biblthca Lichenol.* **96**: 247, 2007; phylogeny, chemistry), Nordin *et al.* (*Lichenologist* **40**: 127, 2008; phylogeny).

Aspiciliella M. Choisy (1932) = Aspicilia fide Purvis *et al.* (*Lichen Flora of Great Britain and Ireland*: 710 pp., 1992).

aspicilioid (of lecanorine apothecia), more or less immersed in the thallus, at least when young.

Aspiciliomyces Cif. & Tomas. (1953) ≡ Pachyospora.

Aspiciliopsis (Müll. Arg.) M. Choisy (1929), Trapeliaceae (L). 2. See Schmitt *et al.* (*Mycol.* **95**: 827, 2003; phylogeny), Lumbsch *et al.* (*MR* **111**: 1133, 2007).

Aspidelia Stirt. (1900) = Parmelia fide Culberson (*Bryologist* **69**: 113, 1966), Hale (*Bull. Br. Mus. nat. hist.* Bot. **8**: 227, 1981).

Aspidella E.-J. Gilbert (1941) = Amanita Pers. fide Singer (*Agaric. mod. Tax.* edn 3, 1975).

Aspidopyrenis Clem. & Shear (1931) ≡ Aspidopyrenium.

Aspidopyrenium Vain. (1890) = Aspidothelium fide Hawksworth *et al.* (*Dictionary of the Fungi* edn 8, 1995), Lücking (*Willdenowia* **29**: 299, 1999).

Aspidothea Syd. (1927) = Inocyclus fide Müller & von Arx (*Beitr. Kryptfl. Schweiz* **11** no. 2, 1962).

Aspidotheliaceae Räsänen ex J.C. David & D. Hawksw. (1991), Pezizomycotina (inc. sed.) (L). 1 gen. (+ 6 syn.), 13 spp.

Lit.: Aptroot & Sipman (*Lichenologist* **25**: 121, 1993), Malcolm & Vězda (*Australas. Lichenol. Newsl.* **37**: 13, 1995), Lücking & Sérusiaux (*Nordic Jl Bot.* **16**: 661, 1996), Lücking (*Willdenowia* **29**: 299, 1999), Arie *et al.* (*J. gen. appl. Microbiol.* Tokyo: 257, 2000).

Aspidotheliomyces Cif. & Tomas. (1953) ≡ Aspidothelium.

Aspidothelium Vain. (1890), Aspidotheliaceae (L). 13, widespread (tropical). See Santesson (*Symb. bot. upsal.* **12** no. 1: 1, 1952), Lücking (*Trop. Bryol.* **15**: 45, 1998; Guyana), Lücking (*Willdenowia* **29**: 299, 1999; Ecuador), McCarthy (*Flora of Australia* **58** A: 242 pp., 2001), Flakus & Wilk (*J. Hattori bot. Lab.* **99**: 307, 2006; Bolivia), Aptroot *et al.* (*Biblthca Lichenol.* **97**, 2008; Costa Rica).

Aspilaima Bat. & H. Maia (1961), anamorphic *Pezizomycotina*, Cpt.0eP.?. 1, Brazil. See Batista & Maia (*Publções Inst. Micol. Recife* **338**: 5, 1961).

Aspilidea Hafellner (2001), Lecanorales (L). 1. See Hafellner & Türk (*Stapfia* **76**: 149, 2001).

A-spore, see alpha-spore.

asporogenic (**asporogenous**), not forming spores.

asporogenic yeasts, see Yeasts.

Asporomyces Chaborski (1918) = Torulopsis Berl. fide Mrak *et al.* (*Mycol.* **34**: 139, 1942), Barnett *et al.* (*Yeasts: Characteristics and Identification* 3rd edn, 2000).

Asporothrichum Link (1809) nom. dub., anamorphic *Agaricomycetidae*. Based on mycelium fide Fries (*Syst. mycol.* **3**, index: 1832). = Sporotrichum (Agaricomycetidae, inc. sed.) fide Streinz (*Nom. fung.*, 1862).

Asproinocybe R. Heim (1970), ? Tricholomataceae. 5, Africa (tropical). See Heinemann & Thoen (*Fl. Illustr. Champ. Afr. centr.* **5**: 102, 1977), Guzmán *et al.* (*Docums Mycol.* **33** no. 131: 23, 2004).

Aspropaxillus Kühner & Maire (1934) = Leucopaxillus fide Singer (*Agaric. mod. Tax.* edn 3, 1975).

assimilative (1) taking in; (2) (of hyphae) having to do with the growth phase before reproduction; non-reproductive; vegetative.

Assoa Urries (1944), Pezizomycotina. 1, Spain.

association, see phytosociology.

astatocoenocytic (of nuclear behaviour in basidiomycetes), haplont mycelium cells coenocytic, diplont binucleate but coenocytic and without clamps when aeration insufficient, basidioma binucleate; in contrast to **holocoenocytic** (haplont and diplont coenocytic, only developing basidium binucleate), **heterocytic** (haplont regularly coenocytic), and the **normal** condition when the haplont is uninucleate, the diplont binucleate (Boidin, *in* Petersen (Ed.), *Evolution in the higher basidiomycetes*: 129, 1971).

Astelechia Cif. (1962), anamorphic *Pezizomycotina*, Hso.≡ eP.?. 2, Dominican Republic. See Ciferri (*Atti Ist. bot. Univ. Lab. crittog. Pavia* sér. 5 **19**: 90, 1962).

Asterella (Sacc.) Sacc. (1891) [non *Asterella* P. Beauv. 1805, *Hepaticae*] = Asterina fide Müller & von Arx (*Beitr. Kryptfl. Schweiz* **11** no. 2, 1962).

Asterella Hara (1936) [non *Asterella* P. Beauv. 1805, *Hepaticae*] = Astrosphaeriella fide Hawksworth (*J. Linn. Soc.* Bot. **82**: 35, 1981).

Asteridiella McAlpine (1897), Meliolaceae. 300, widespread (tropical). See Hansford (*Sydowia* **10**: 41, 1956), Hughes (*Mycol. Pap.* **166**, 1993), Song *et al.* (*Acta Mycol. Sin.* **15**: 247, 1996; China), Dianese & Furlanetto (*Progress in Microbial Ecology* Proceedings of the Seventh International Symposium on Microbial Ecology. Santos, São Paulo, Brazil 1995: 207, 1997; Brazil), Hosagoudar *et al.* (*The Meliolineae* A Supplement: 201 pp., 1997; India), Mibey & Hawksworth (*Mycol. Pap.* **174**: 108 pp., 1997; Kenya), Hsieh *et al.* (*Taiwan Ascomycetes* Pyrenomycetes and Loculoascomycetes, 2000; Taiwan), Song & Li (*Mycotaxon* **89**: 201, 2004; China).

Asteridiellina Seaver & Toro (1926) = Actinopeltis fide von Arx & Müller (*Stud. Mycol.* **9**, 1975).

Asteridium (Sacc.) Speg. ex Sacc. (1891) = Meliola fide Höhnel (*Sber. Akad. Wiss. Wien* Math.-naturw. Kl., Abt. 1 **119**: 414, 1910), Hosagoudar (*Sydowia* **55**: 162, 2003).

Asterina Lév. (1845), Asterinaceae. Anamorphs *Asterostomella*, *Clasterosporium*-like. *c.* 336 (on leaves), widespread (esp. tropical). See Doidge (*Bothalia* **4**: 273, 1942), Hansford (*Mycol. Pap.* **15**, 1946), Reynolds (*Cryptog. Mycol.* **8**: 251, 1987; asci), Rahayu & Parbery (*MR* **95**: 731, 1991; Australia), Mibey & Hawksworth (*Mycol. Pap.* **174**, 1997; Kenya), Hosagoudar & Abraham (*J. Econ. Taxon. Bot.* **24**: 557, 2000; nomenclator), Hosagoudar *et al.* (*Journal of Mycopathological Research* **39**: 61, 2001; posn), Song & Li (*Mycotaxon* **84**: 407, 2002; China), Hosagoudar (*Zoos' Print Journal* **18**: 1280, 2003; India), Hofmann & Piepenbring (*Mycol. Progr.* **7**: 87, 2008; Panama).

Asterinaceae Hansf. (1946), ? Capnodiales. 46 gen. (+ 39 syn.), 653 spp.

Lit.: Doidge (*Bothalia* **4**: 273, 1942; S. Afr.), Farr (*Mycol.* **78**: 269, 1986), Swart (*TBMS* **87**: 81, 1986), Farr (*Mycol.* **79**: 97, 1987), Reynolds (*Cryptog. Mycol.* **8**: 251, 1987), Rahayu & Parbery (*MR* **95**: 731, 1991), Hosagoudar & Goos (*Mycotaxon* **52**: 467, 1994), Hosagoudar *et al.* (*Mycotaxon* **58**: 489, 1996), Mibey & Hawksworth (*Mycol. Pap.* **174**: 108 pp., 1997), Mibey & Hawksworth (*Mycol. Pap.* **174**, 1997; Kenya), Goos (*Mycotaxon* **73**: 455, 1999).

Asterinales M.E. Barr ex D. Hawksw. & O.E. Erikss. (1986) = *Capnodiales*. Perhaps synonymous with *Capnodiales*, but very few molecular data are available. See *Asterinaceae*.

Asterinella Theiss. (1912), Microthyriaceae. Anamorphs *Asterostomula*, *Asteromella*-like. *c.* 6, widespread (subtropical). See Hosagoudar *et al.* (*Mycotaxon* **58**: 489, 1996), Hosagoudar & Abraham (*MR* **102**: 184, 1998).

Asterinema Bat. & Gayão (1953), Microthyriaceae. Anamorph *Eriothyrium*. 1, Brazil. See Farr (*Mycol.* **75**: 1036, 1983).

Asterinites Doub. & D. Pons ex Kalgutkar & Janson. (2000), Fossil Fungi. 1, Colombia. See Kalgutkar & Jansonius (*AASP Contributions Series* **39**: 31, 2000).

Asterinites Doub. & D. Pons (1973), Fossil Fungi. 2 (Paleocene), Colombia.

Asterinites Krassilov (1967), Fossil Fungi. 2 (Cretaceous), former USSR.

Asterinopeltis Bat. & H. Maia (1958) = Platypeltella fide von Arx & Müller (*Stud. Mycol.* **9**, 1975).

Asterinotheca Bat. & H. Maia (1958) ? = Asterina fide Müller & von Arx (*Beitr. Kryptfl. Schweiz* **11** no. 2, 1962).

Asterinothyriella Bat. & Cif. (1959), anamorphic *Pezizomycotina*, Cpt.≡ eH.?. 1, Uganda. See Batista & Ciferri (*Atti Ist. bot. Univ. Lab. crittog. Pavia* sér. 5 **16**: 85, 1959).

Asterinothyrium Bat., Cif. & H. Maia (1959), anamorphic *Pezizomycotina*, Cpt.0eH.?. 1, S. Africa. See Batista *et al.* (*Mycopath. Mycol. appl.* **11**: 27, 1959).

Asterinula Ellis & Everh. (1889) = Leptothyrella fide Saccardo (*Syll. fung.* **10**: 1, 1892).

Asterisca G. Mey. (1825) = Sarcographa fide Hawksworth *et al.* (*Dictionary of the Fungi* edn 8, 1995).

Asteristion Leight. (1870) = Chapsa fide Hale (*Bull. Br. Mus. nat. hist.* Bot. **8**: 227, 1981).

Asteristium Clem. (1909) ≡ Asteristion.

Asteritea Bat. & R. Garnier (1961), ? Microthyriaceae. 1, Brazil. See Batista & Garnier (*Broteria* ser. bot. **30**: 41, 1961).

Asterobolus Redhead & P.W. Perrin (1972) = Valdensia fide Redhead & Perrin (*CJB* **50**: 2083, 1972).

Asterocalyx Höhn. (1912), Sclerotiniaceae. 2, widespread. See Dumont & Carpenter (*Mycol.* **70**: 68, 1978), Spooner (*Biblthca Mycol.* **116**, 1987).

Asterochaete (Pat.) Bondartsev & Singer (1941) [non *Asterochaete* Nees 1834, *Cyperaceae*] ≡ Echinochaete.

Asteroconium Syd. & P. Syd. (1903), anamorphic *Pezizomycotina*, Cac.1bH.1/10. 2, C. America; India; China. See Sutton (*The Coelomycetes*, 1980).

Asterocyphella W.B. Cooke (1961), Cyphellaceae. 3, widespread. See Cooke (*Beih. Sydowia* **4**: 118, 1961; key).

Asterocystis De Wild. (1893) = Olpidium fide Sampson (*TBMS* **23**: 199, 1939).

Asterodon Pat. (1894), Hymenochaetaceae. 1, widespread (north temperate). See Corner (*TBMS* **31**: 234, 1948), Müller *et al.* (*MR* **104**: 1485, 2000).

Asterodontaceae Parmasto (2001) = Hymenochaetaceae.

Asterodothis Theiss. (1912), Asterinaceae. Anamorph *Asterostromina*. 1, Africa. See Hosagoudar *et al.* (*Journal of Mycopathological Research* **39**: 61, 2001).

Asterogastraceae R. Heim (1934) = Russulaceae.

asteroid body, a stellate cell of *Sporothrix schenckii* (more rarely *Aspergillus* or other pathogens) in animal tissues resulting from an antigen-antibody complex precipitate deposited on the cell wall (Lurie & Snell, *Sabouraudia* **7**: 64, 1969).

Asteroides Puntoni & Léon (1940) nom. dub., Fungi.

Asterolibertia G. Arnaud (1918), Asterinaceae. *c.* 18, widespread (subtropical). See Hosagoudar & Abraham (*Journal of Mycopathological Research* **35**: 55, 1997; India).

Asteroma DC. (1815), anamorphic *Gnomoniella, Plagiostoma*, Cac.0eH.15. 14, widespread (esp. north temperate). See Sutton (*The Coelomycetes*, 1980).

Asteromassaria Höhn. (1917), Pleomassariaceae. Anamorph *Scolicosporium*. 11, Europe; N. America. See Barr (*Mycotaxon* **15**: 349, 1982), Spooner & Kirk (*TBMS* **78**: 247, 1982; anamorph), Sivanesan (*TBMS* **91**: 317, 1988; key 9 spp.), Mehrotra & Sivanesan (*MR* **93**: 557, 1989), Barr (*Mycotaxon* **49**: 129, 1993; key 8 N. Am. spp.), Tanaka *et al.* (*Mycoscience* **46**: 248, 2005; Japan).

Asteromella Pass. & Thüm. (1880), anamorphic *Dothideomycetes*, Cpd.0eH.15. 234, widespread. Almost certainly polyphyletic. See Batista *et al.* (*Saccardoa* **1**: 17, 1960), Sutton (*The Coelomycetes*, 1980), Vanev & van der Aa (*Persoonia* **17**: 47, 1998; annotated list).

Asteromellopsis H.E. Hess & E. Müll. (1951), anamorphic *Dothidea*, St.0eH.15. 1, Switzerland. See Hess & Müller (*Ber. schweiz. bot. Ges.* **61**: 18, 1951), Goodwin & Zismann (*Mycol.* **93**: 934, 2001; phylogeny).

Asteromidium Speg. (1888), anamorphic *Pezizomycotina*, Cac.≡ eH.10. 3, Brazil. See Petrak & Sydow (*Annls mycol.* **34**: 14, 1936), Ferreira & Muchovej (*Mycotaxon* **30**: 97, 1987; addit. spp.), Pomella *et al.* (*Mycotaxon* **64**: 83, 1997).

Asteromites Poinar (2003), Fossil Fungi. 1, Chiapas. See Poinar (*MR* **107**: 121, 2003).

Asteromyces Moreau & M. Moreau ex Hennebert (1962), anamorphic *Pezizomycotina*, Hso.0eP.11/14. 1, France. See Hennebert (*CJB* **40**: 1211, 1962), Kohlmeyer & Volkmann-Kohlmeyer (*Bot. Mar.* **34**: 1, 1991).

Asteromyxa Theiss. & Syd. (1918) ≡ Dimeriella.

Asteronaevia Petr. (1929) = Diplonaevia fide Hein (*Nova Hedwigia* **38**: 669, 1983).

Asteronectrioidea Cant. (1949), anamorphic *Pezizomycotina*, St.1eH.15. 1, Africa.

Asteronema Trevis. (1845) nom. dub., ? Fungi.

Asteronia (Sacc.) Henn. (1895), Microthyriaceae. 2, Brazil. See Sutton (*Mycol. Pap.* **141**, 1977).

Asteropeltis Henn. (1904) = Trichothelium fide Hawksworth *et al.* (*Dictionary of the Fungi* edn 8, 1995).

Asterophlyctis H.E. Petersen (1903) = Diplophlyctis fide Dogma (*Nova Hedwigia* **25**: 121, 1974).

Asterophoma D. Hawksw. (1981), anamorphic *Chaenothecopsis*, Cpd.0eH.15. 1 (on *Calicium*), widespread. See Tibell (*CJB* **69**: 2427, 1991; ultrastr.).

Asterophora Ditmar (1809), Lyophyllaceae. Anamorph *Ugola*. 3 (2 on other agarics, esp. *Russula*), widespread (temperate). On basidioma of *Russula* and *Lactarius*. See Redhead & Seifert (*Taxon* **50**: 243, 2001; nomencl.), Walther *et al.* (*MR* **109**: 525, 2005; conidiogenesis).

asterophysis, see seta.

Asteroporomyces Cif. & Tomas. (1953) ≡ Asteroporum.

Asteroporum Müll. Arg. (1884), Arthoniales (±L). 3, widespread (tropical). See McCarthy (*Flora of Australia* **58** A: 242 pp., 2001).

Asteropsis Gonz. Frag. (1917), anamorphic *Pezizomycotina*, Cpd.0eP.?. 1, Spain.

Asteroscutula Petr. (1948), anamorphic *Pezizomycotina*, Cpt.0eP.?. 1, Ecuador.

asteroseta (1) see cystidium; (2) see seta.

Asterosphaeria (Höhn.) Syd. (1913) = Astrosphaeriella fide Hawksworth (*J. Linn. Soc.* Bot. **82**: 35, 1981).

Asterosporales = Russulales.

Asterosporium Kunze (1819), anamorphic *Pezizomycotina*, Cac.0bP.1. 5, widespread (temperate). See Murvanishvili & Dekanoidze (*Mikol. Fitopatol.* **26**: 27, 1992; key).

Asterostomella Speg. (1886), anamorphic *Asterina*, Cpt.0eP.?. 39, widespread (tropical). See Batista & Ciferri (*Mycopathologia* **11**: 44, 1959), Hosagoudar & Goos (*Mycotaxon* **52**: 467, 1994; India), Hofmann & Piepenbring (*Mycol. Progr.* **7**: 87, 2008; Panama).

Asterostomidium Lindau (1900) ≡ Asteromidium.

Asterostomopora Bat. & H. Maia (1960), anamorphic *Pezizomycotina*, Cpt.0eP.?. 1, Jamaica. See Batista & Maia (*Publções Inst. Micol. Recife* **221**: 5, 1960).

Asterostomopsis Bat., Cif. & H. Maia (1959), anamorphic *Pezizomycotina*, Hsp.0eP.?. 1, Ghana. See Batista *et al.* (*Mycopath. Mycol. appl.* **11**: 56, 1959).

Asterostomula Theiss. (1916), anamorphic *Pezizomycotina*, Cpt.0eP.?. 4, widespread (tropical).

Asterostomulina Bat., J.L. Bezerra & H. Maia (1964), anamorphic *Pezizomycotina*, Cpt.0eH.?. 1, Brazil. See Batista *et al.* (*Portugaliae Acta Biologica* Série B **7**: 385, 1964).

Asterostroma Massee (1889), Lachnocladiaceae. 14, widespread. See Boidin (*BSMF* **113**: 269, 1997; key), Wagner (*Mycotaxon* **79**: 235, 2001; phylogenetics).

Asterostromataceae Pouzar (1983) = Hymenochaetaceae.

Asterostromella Höhn. & Litsch. (1907) = Vararia fide Burt (*Ann. Mo. bot. Gdn* **9**: 1, 1922).

Asterostromina Bat. & A.F. Vital (1957), anamorphic *Asterodothis*, Cpt.0eH.?. 1, S. Africa. See Batista & Vital (*Revta Biol.* Lisb. **1**: 116, 1957).

Asterotexis Arx (1958), Asterinaceae. 2, America (tropical); Nepal. See von Arx (*Fungus* Wageningen **28**: 6, 1958), Hosagoudar *et al.* (*Journal of Mycopathological Research* **39**: 61, 2001; posn).

Asterotheca I. Hino (1938) [non *Asterotheca* C. Presl 1846, fossil *Pteridophyta*] ≡ Astrotheca.

Asterothecium Wallr. (1836) ≡ Stephanoma.

Asterothelium, see *Astrothelium*.

Asterothrix Kutz. (1843) [non *Asterothrix* Cass. 1827, *Compositae*] ≡ Asteronema.

Asterothyriaceae Walt. Watson ex R. Sant. (1952), Ostropales (L). 3 gen. (+ 12 syn.), 59 spp.

Lit.: Santesson (*Symb. bot. upsal.* **12** no. 1: 1, 1952), Sérusiaux & de Sloover (*Veröff. geobot. Inst., Zürich* **91**: 260, 1986; hypophores), Sérusiaux & Sloover (*Veröff. geobot. Inst., Zürich* **91**: 260, 1986), Hansen *et al.* (*Herzogia* **7**: 367, 1987), Vězda & Poelt (*Phyton* Horn **30**: 47, 1990), Vězda & Poelt (*Nova Hedwigia* Beih. **53**: 99, 1991), Etayo & Vezda (*Lichenologist* **26**: 333, 1994), Boom & Vězda (*Mycotaxon* **54**: 421, 1995), Lücking (*Cryptog. Mycol.* **20**: 193, 1999; phylogeny, inclusion of *Solorinellaceae*), Lücking (*Willdenowia* **29**: 299, 1999), Henssen & Lücking (*Ann. bot. fenn.* **39**: 273, 2002),

Lücking *et al.* (*Mycol.* **96**: 283, 2004).

Asterothyriomyces Cif. & Tomas. (1953) ≡ Asterothyrium Müll. Arg.

Asterothyrites Cookson (1947), Fossil Fungi. 4 (Tertiary), widespread. = Phragmothyrites (Fossil fungi) fide Selkirk.

Asterothyrium Henn. (1904) ≡ Septothyrella.

Asterothyrium Müll. Arg. (1890), Asterothyriaceae (L). 24, widespread (tropical). See Santesson (*Symb. bot. upsal.* **12** no. 1: 1, 1952), Vězda & Poelt (*Phyton* Horn **30**: 47, 1990; posn), Henssen & Lücking (*Ann. bot. fenn.* **39**: 273, 2002; anatomy, ontogeny), Lücking *et al.* (*Mycol.* **96**: 283, 2004; phylogeny, links with *Gomphillaceae*).

Asterotrema Müll. Arg. (1884) ? = Arthonia fide Aptroot (*Biblthca Lichenol.* **44**, 1991).

Asterotremella Prillinger, Lopandic & Sugita (2007), anamorphic *Trichosporonaceae*. 5, widespread. See Prillinger *et al.* (*J. gen. appl. Microbiol.* **53**: 167, 2007).

Asterotremellaceae Prillinger, Lopandic & Sugita (2007) = Trichosporonaceae.

Asterotrichum Bonord. (1851) = Asterophora fide Saccardo (*Syll. fung.* **4**: 1, 1886).

Asterotus Singer (1943) = Resupinatus fide Thorn *et al.* (*Mycol.* **97**: 1140, 2005).

Asterula (Sacc.) Sacc. (1891) = Venturia Sacc. fide von Arx & Müller (*Stud. Mycol.* **9**, 1975).

Astiothyrium Bat. (1964) = Eudimeriolum fide von Arx & Müller (*Stud. Mycol.* **9**, 1975).

Astoma Gray (1821) = Sclerotium fide Rabenhorst (*Deutsch. Krypt. Fl.* **1**: 1, 1844).

astomate (astomous), lacking an ostiole.

Astomella Thirum. (1947), Pezizomycotina. 1, India.

Astrabomyces Bat. (1961) nom. dub., anamorphic *Pezizomycotina*, Hso.≡ eP.? (L). 1, Brazil. See Lücking *et al.* (*Lichenologist* **30**: 121, 1998).

Astraeaceae Zeller ex Jülich (1982) = Diplocystidiaceae.

Astraeus Morgan (1889), Diplocystidiaceae. 2, widespread. *A. hygrometricus*, a mycorrhizal earth-star common in dry places. See Phosri *et al.* (*Mycotaxon* **89**: 453, 2004; Thailand), Sarasini (*Gasteromiceti Epigei*: 406 pp., 2005), Phosri *et al.* (*MR* **111**: 275, 2007).

Astragoxyphium Bat., Nascim. & Cif. (1963) = Leptoxyphium fide Hughes (*Mycol.* **68**: 693, 1976), Inácio & Dianese (*MR* **102**: 695, 1998).

Astrocitum Raf. (1806) ≡ Astrycum.

Astrocystidaceae Hara (1913) = Xylariaceae.

Astrocystis Berk. & Broome (1873), Xylariaceae. Anamorph *Acanthodochium*. 15 (esp. on bamboo and palms), widespread (tropical). See Diehl (*Mycol.* **17**: 185, 1925; morphology), Ju & Rogers (*Mycol.* **82**: 342, 1990; as *Rosellinia*), Læssøe & Spooner (*Kew Bull.* **49**: 1, 1994; key, monograph), Dulymamode *et al.* (*MR* **102**: 1325, 1998), Smith & Hyde (*Fungal Diversity* **7**: 89, 2001), Petrini (*N.Z. Jl Bot.* **41**: 71, 2003; New Zealand), Bahl *et al.* (*Mycol.* **97**: 1102, 2005; phylogeny).

Astrodochium Ellis & Everh. (1897), anamorphic *Pezizomycotina*, Hsp.0eP.?. 1, N. America. See Harrison (*Pl. Dis.* **77**: 1263, 1993), Carris (*Sydowia* **47**: 150, 1995).

astrogastraceous fungi, gasteroid members of the *Russulales*. See also *Hymenogastrales*, *Podaxales*.

Astronatelia Bat. & H. Maia (1962), anamorphic *Pezizomycotina*, Cpt.0eH.?. 1, USA. See Batista & Maia (*Publções Inst. Micol. Recife* **209**: 8, 1962).

Astroplaca Bagl. (1858) = Placolecis fide Hafellner (*Beih. Nova Hedwigia* **79**: 241, 1984).

Astrosphaeriella Syd. & P. Syd. (1913), ? Melanommataceae. Anamorph *Pleurophomopsis*. 51, widespread. See Hawksworth (*J. Linn. Soc.* Bot. **82**: 35, 1981), Hawksworth & Boise (*Sydowia* **38**: 114, 1986; key 10 spp.), Hyde & Fröhlich (*Sydowia* **50**: 81, 1998; spp. on palms), San Martín *et al.* (*Acta Bot. Mexicana* **46**: 19, 1999), Hyde *et al.* (*Nova Hedwigia* **70**: 143, 2000), Zhou *et al.* (*Cryptog. Mycol.* **24**: 191, 2003), Chen & Hsieh (*Bot. Bull. Acad. sin.* Taipei **45**: 171, 2004).

Astrosporina J. Schröt. (1889) = Inocybe fide Kauffman (*N. Amer. Fl.* **10**, 1924) See, Horak (*Persoonia* **10**: 157, 1979; key to 30 spp. from Indomalaya, Australasia).

Astrotheca I. Hino (1938) = Astrosphaeriella fide Hawksworth (*J. Linn. Soc.* Bot. **82**: 35, 1981).

Astrotheliaceae Zahlbr. (1898) = Trypetheliaceae.

Astrothelium Eschw. (1824), Trypetheliaceae (L). *c.* 40, widespread (tropical). See Harris (*Acta Amazon.* Supl. **14**: 55, 1984; key 13 spp. Brazil), Makhija & Patwardhan (*Biovigyanam* **15**: 61, 1989; fam. status), McCarthy (*Flora of Australia* **58** A: 242 pp., 2001; Australia), Aptroot *et al.* (*Biblthca Lichenol.* **97**, 2008; Costa Rica).

Astrycum Raf. (1809) ? = Geastrum fide Stalpers (*in litt.*).

Astylospora Fayod (1889) = Psathyrella fide Singer (*Agaric. mod. Tax.* edn 3, 1975).

asymmetric (of spores), having one side flattened or concave.

Asymmetricospora J. Fröhl. & K.D. Hyde (1998), Melanommataceae. 1, Australia. See Fröhlich & Hyde (*Sydowia* **50**: 183, 1998).

Asyregraamspora Locq. & Sal.-Cheb. (1980), Fossil Fungi. 1, Cameroon.

ATBI (All-Taxon Biodiversity Inventory), a record of the total diversity of living organisms present in one area. See Cannon (*Inoculum* **46**(4): 1, 1995), Inventorying.

ATCC, American Type Culture Collection (Rockville, Md, USA); a not-for-profit service collection founded in 1925; see *American Type Culture Collection profile* (1992).

Ateleothylax M. Ota & Langeron (1923), anamorphic *Pezizomycotina*, Hso.?.?. 1, Sweden.

Atelocauda Arthur & Cummins (1933), Pileolariaceae. 2 (on *Leguminosae*: *Faboideae*), C. America; Australia. See Walker (*Australasian Mycologist* **20**: 3, 2001; monogr., emend., key).

Atelosaccharomyces Beurm. & Gougerot (1909) = Cryptococcus Vuill. fide von Arx *et al.* (*Stud. Mycol.* **14**: 1, 1977).

Atelosaccharomycetaceae Guilliern. (1928) = Filobasidiaceae.

Atestia Trevis. (1861) = Oropogon fide Hawksworth *et al.* (*Dictionary of the Fungi* edn 8, 1995).

Athecaria Nyl. (1897) ? = Aspicilia fide Santesson (*ING* **1**: 155, 1979; typification).

Athelia Pers. (1822), Atheliaceae. Anamorph *Fibularhizoctonia*. 28, widespread. See Jülich (*Willdenowia* Beih. **7**, 1972), Jülich (*Persoonia* **10**: 149, 1978; key 'lichenized' spp.) *A. arachnoidea* is an important pathogen of lichens, esp. *Lecanora conizaeoides*, also epiphytic green algae (causing brownish white lesions in its colonies; see, Arvidsson

(*Svensk bot. Tidskr.* **72**: 285, 1979), Okabe & Matsumoto (*MR* **107**: 164, 2003; phylogeny *Athelia rolfsii*) See also *Sclerotium*.

Atheliaceae Jülich (1982), Atheliales. 22 gen. (+ 3 syn.), 106 spp.

Lit.: Gilbertson & Lindsey (*Mem. N. Y. bot. Gdn* **49**: 138, 1989), Nakasone (*Mycol. Mem.* **15**: 412 pp., 1990), Ryvarden (*Syn. Fung.* **5**: 363 pp., 1991), Harlton *et al.* (*Phytopathology* **85**: 1269, 1995), Adams & Kropp (*Mycol.* **88**: 464, 1996), Stalpers & Andersen *in* Sneh *et al.* (Eds) (*Rhizoctonia Species* Taxonomy, Molecular Biology, Ecology, Pathology and Disease Control: 58, 1996), Boidin *et al.* (*Mycotaxon* **66**: 445, 1998), Ginns (*Mycol.* **90**: 19, 1998), Kirschner & Oberwinkler (*Mycoscience* **40**: 345, 1999), Hibbett *et al.* (*Nature* Lond. **407**: 506, 2000), Larsson *et al.* (*MR* **108**: 983, 2004), Binder *et al.* (*Systematics and Biodiversity* **3**: 113, 2005).

Atheliales Jülich (1981). Agaricomycetidae. 1 fam., 22 gen., 106 spp. Fam.:

Atheliaceae

For *Lit.* see under fam.

Athelicium K.H. Larss. & Hjortstam (1986), Atheliaceae. 2, Europe. See Larsson & Hjortstam (*Windahlia* **15**: 49, 1986).

Athelidium Oberw. (1966), Stephanosporaceae. 1, Europe. See Oberwinkler (*Sydowia* **19**: 62, 1965).

Athelium Nyl. (1886) = Thelocarpon fide Hawksworth *et al.* (*Dictionary of the Fungi* edn 8, 1995).

Atheloderma Parmasto (1968), Hymenochaetales. 2, Europe; Asia. See Parmasto (*Consp. System. Corticiac.*: 73, 1968).

Athelopsis Oberw. ex Parmasto (1968), Atheliaceae. 10, widespread. Polyphyletic. See Hjortstam (*Mycotaxon* **42**: 149, 1991), Kotiranta & Saarenoksa (*Ann. bot. fenn.* **42**: 335, 2005; Finland).

Athrismidium Trevis. (1860) ? = Tomasellia fide Harris (*More Florida Lichens*, 1995).

Atichia Flot. (1850), anamorphic *Seuratia*, Hsy/Ccu.0bH-P.1. 6, widespread. See Meeker (*CJB* **53**: 2483, 1975), Parbery & Brown (*Microbiology of the Phyllosphere*: 101, 1986), Kendrick (*CJB* **81**: 75, 2003; morphogenesis).

Atichiaceae Racib. (1900) = Seuratiaceae.

Atichiopsis R. Wagner (1900) = Seuratia fide Meeker (*CJB* **53**: 2462, 1975).

Atkinson (George Francis; 1854-1918; USA). Professor of Botany, Cornell University (1896-1918). His work did much to stimulate interest in the *Agaricaceae* in the USA. *Publs. Mushrooms Edible and Poisonous* (1901) [edn 2]; Phylogeny and relationships in the ascomycetes. *Annals of the Missouri Botanical Garden* (1915); also other papers on the *Agaricaceae*, phylogeny, and plant diseases. *Biogs, obits etc.* Farlow *et al.* (*American Journal of Botany* **6**: 301, 1919); Stafleu & Cowan (*TL-2* **1**: 78, 1976); Stafleu & Mennega (*TL-2, Suppl.* **1**: 200, 1992).

Atkinsonella Diehl (1950), Clavicipitaceae. Anamorphs *Ephelis*, *Sphacelia*. 2, widespread (north temperate). See Leutchmann & Clay (*Mycol.* **81**: 692, 1989), Morgan-Jones & White (*Mycotaxon* **35**: 455, 1989), Schardl *et al.* (*Pl. Syst. Evol.* **178**: 27, 1991; phylogeny), Morgan-Jones & White (*Mycotaxon* **44**: 89, 1992; culture), Leutchmann & Clay (*Am. J. Bot.* **83**: 1144, 1996; isozymes), Reddy *et al.* (*Mycol.* **90**: 108, 1998; DNA).

Atkinsonia Lloyd (1916) [non *Atkinsonia* F. Muell. 1865, *Loranthaceae*] = Sebacina fide Donk (*Persoonia* **4**: 305, 1966).

atlantic, confined to the Atlantic seaboard. For classification of different types of atlantic distribution in Europe see Ratcliffe (*New Phytol.* **67**: 365, 1968).

Atmospheric pollution, see Air pollution.

atomate, having a powdered surface.

Atopospora Petr. (1925), ? Venturiaceae. Anamorph *Didymochora*. 2, widespread (north temperate). See Sivanesan (*Bitunicate Ascomycetes and their Anamorphs*, 1984), Barr (*Sydowia* **41**: 25, 1989).

Atractiella Sacc. (1886), Phleogenaceae. 6, widespread. See Donk (*Persoonia* **4**: 209, 1966), Bandoni & Inderbitzin (*Czech Mycol.* **53**: 265, 2002; n.sp.), Aime *et al.* (*Mycol.* **98**: 896, 2006; phylogeny).

Atractiellaceae R.T. Moore (1996) = Phleogenaceae.

Atractiellales Oberw. & Bandoni (1982). Atractiellomycetes. 3 fam., 10 gen., 34 spp. Fams:

(1) **Atractogloeaceae**

(2) **Mycogelidiaceae**

(3) **Phleogenaceae** (syn. *Atractiellaceae*, *Hoehnelomycetaceae*)

Lit.: Oberwinkler & Bauer (*Sydowia* **41**: 224, 1989), Bauer *et al.* (*Mycol. Progress* **5**: 41, 2006).

Atractiellomycetes R. Bauer, Begerow, J.P. Samp., M. Weiss & Oberw. (2007). Pucciniomycotina. 1 ord., 3 fam., 10 gen., 34 spp. Ord.:

Atractiellales

Lit.: Bauer *et al.* (*Mycol. Progress* **5**: 41, 2006).

Atractilina Dearn. & Barthol. (1924), anamorphic *Pezizomycotina*, Hso/Hsy.≡ eH-P.10. 2 (on leaf ascomycetes), widespread (tropical). See Deighton & Pirozynski (*Mycol. Pap.* **128**, 1972).

Atractina Höhn. (1904) = Sterigmatobotrys fide Hughes (*CJB* **36**: 727, 1958).

Atractium Link (1809), anamorphic *Pezizomycotina*, Hsy.≡ eH.?. 5, widespread.

Atractobasidium G.W. Martin (1935) = Patouillardina Bres. fide Rogers (*Mycol.* **28**: 398, 1936).

Atractobolus Tode (1790), Pezizomycotina. 1, Europe. See Spooner (*Biblthca Mycol.* **116**: 1, 1987).

Atractocolax R. Kirschner, R. Bauer & Oberw. (1999), Microbotryomycetes. 1 (associated with bark beetles), Europe. See Kirschner *et al.* (*Mycol.* **91**: 542, 1999).

Atractodorus Klotzsch (1832) nom. dub., Fungi.

Atractogloea Oberw. & Bandoni (1982), Atractogloeaceae. 1, USA. See Oberwinkler & Bandoni (*Mycol.* **74**: 634, 1982).

Atractogloeaceae Oberw. & R. Bauer (1989), Pucciniomycotina (inc. sed.). 1 gen., 1 spp.

Lit.: Oberwinkler & Bandoni (*Mycol.* **74**: 634, 1982), Oberwinkler (*Stud. Mycol.* **30**: 61, 1987), Oberwinkler & Bauer (*Sydowia* **41**: 224, 1989).

Atrichophyton Castell. & Chalm. (1919) = Chrysosporium fide Carmichael *in* Ainsworth *et al.* (Eds) (*The Fungi* **4A**: 390, 1973).

Atricordyceps Samuels (1983) = Podocrella fide Samuels (*N.Z. Jl Bot.* **21**: 171, 1983), Samson *et al.* (*Atlas of Entomopathogenic Fungi*, 1988), Shimazu & Glockling (*MR* **101**: 1371, 1997), Chaverri *et al.* (*Mycol.* **97**: 433, 2005).

Atrocybe Velen. (1947), ? Helotiales. 1, former Czechoslovakia.

Atropellis Zeller & Goodd. (1930), Dermateaceae. 4, N. America. *A. pinicola* (pine canker). See Reid & Funk (*Mycol.* **58**: 428, 1966; key), Smith *et al.* (*Quarantine Pests for Europe* Edn 2, 1997).

Atroporus Ryvarden (1973) = Polyporus P. Micheli ex

Adans. fide Reid (*Mem. N. Y. bot. Gdn* **28**: 197, 1976).

Atrosetaphiale Matsush. (1995), anamorphic *Pezizomycotina*, Hso.0fH.15. 1, Peru. See Matsushima (*Matsush. Mycol. Mem.* **8**: 14, 1995), Mel'nik *et al.* (*Mycol. Progr.* **3**: 19, 2004).

Atrotorquata Kohlm. & Volkm.-Kohlm. (1993), Cainiaceae. 1, USA. Very similar and possibly related to *Cainia*. See Kohlmeyer & Volkmann-Kohlmeyer (*SA* **12**: 7, 1993), Kang *et al.* (*MR* **103**: 1621, 1999).

Attamyces Kreisel (1972), anamorphic *Leucoagaricus*. 1 (in ants nests), Cuba. See Singer (*Nova Hedwigia* **26**: 435, 1975).

attenuate (1) narrowed; (2) (of a pathogen), having lowered pathogenicity or virulence.

Atylospora, see *Astylospora*.

atypical, not normal.

Auerswaldia Rabenh. (1857) = Melanospora Corda fide Hawksworth *et al.* (*Dictionary of the Fungi* edn 8, 1995).

Auerswaldia Sacc. (1883), ? Dothideaceae. 30, widespread (pantropical). See von Arx & Müller (*Stud. Mycol.* **9**, 1975), Hyde & Cannon (*Mycol. Pap.* **175**, 1999).

Auerswaldiella Theiss. & Syd. (1914), ? Botryosphaeriaceae. 4, widespread (tropical). See Sivanesan & Hsieh (*MR* **93**: 340, 1989; key), Eriksson & Hawksworth (*SA* **14**: 45, 1995; posn).

Auerswaldiopsis Henn. (1904) ? = Patouillardiella fide Höhnel (*Sber. Akad. Wiss. Wien* Math.-naturw. Kl., Abt. 1 **119**: 432, 1910).

Aulacographa Leight. (1854) = Graphis fide Hawksworth *et al.* (*Dictionary of the Fungi* edn 8, 1995).

Aulacostroma Syd. & P. Syd. (1914), Parmulariaceae. 5, pantropical. See Luttrell & Muthappa (*Mycol.* **66**: 563, 1974), Magnes (*Biblthca Mycol.* **165**, 1997), Inácio & Cannon (*MR* **107**: 82, 2003).

Aulaxina Fée (1825), Gomphillaceae (L). 12, widespread (tropical). See Santesson (*Symb. bot. upsal.* **12** no. 1: 1, 1952), Vězda & Poelt (*Folia geobot. phytotax.* **22**: 179, 1987).

Aulaxinomyces Cif. & Tomas. (1953) = Arthonia fide Lücking & Hawksworth (*Taxon* **56**: 1274, 2007).

auleate (of gasteromycete basidiomata), a closed basidioma in which pleated plates of trama project into the glebal cavity from top and sides. See Dring (1973); after Kreisel (1969).

Aulographaceae Luttr. ex P.M. Kirk, P.F. Cannon & J.C. David (2001), Microthyriales. 2 gen., 31 spp.

Lit.: Batista (*Publções Inst. Micol. Recife* **56**, 1959), McKenzie & Foggo (*N.Z. Jl Bot.* **27**: 91, 1989), Petrini *et al.* (*Mycol. helv.* **3**: 263, 1989).

Aulographella Höhn. (1917) = Morenoina fide Müller & von Arx (*Beitr. Kryptfl. Schweiz* **11** no. 2, 1962).

Aulographina Arx & E. Müll. (1960), ? Asterinaceae. Anamorph *Thyrinula*. 3, widespread. See Wall & Keane (*TBMS* **82**: 257, 1984).

Aulographopsis Petr. (1938) nom. nud., anamorphic *Pezizomycotina*.

Aulographum Lib. (1834), Aulographaceae. 30, widespread. See Petrini *et al.* (*Mycol. Helv.* **3**: 263, 1989).

Aulospora Speg. (1909), Pezizomycotina. 1, Argentina. See Eriksson & Hawksworth (*SA* **5**: 120, 1986).

Aurantiochytrium R. Yokoy. & D. Honda (2007), Thraustochytriaceae. (marine). See Yokoyama, R.; Honda, D. (*Mycoscience* **48**: 199, 2007).

Aurantiosacculus Dyko & B. Sutton (1979), anamorphic *Pezizomycotina*, St.0fH.1. 1 (on *Eucalyptus*), Australia. See Dyko & Sutton (*Mycol.* **71**: 922, 1979).

Aurantiosporium M. Piepenbr., Vánky & Oberw. (1996), Ustilentylomataceae. 4 (on *Cyperaceae*), widespread (tropical). See Piepenbring *et al.* (*Pl. Syst. Evol.* **199**: 62, 1996), Aime *et al.* (*Mycol.* **98**: 896, 2006; phylogeny).

Aurantiporellus Murrill (1905) = Pycnoporellus fide Pegler (*The polypores [Bull. BMS Suppl.]*, 1973).

Aurantiporus Murrill (1905), Polyporaceae. 5, north temperate. See Ryvarden (*Syn. Fung.* **5**: 363 pp., 1991).

Aurapex Gryzenh. & M.J. Wingf. (2006), anamorphic *Cryphonectriaceae*. 1, Colombia. See Gryzenhout *et al.* (*Mycol.* **98**: 112, 2006), Gryzenhout *et al.* (*FEMS Microbiol. Lett.* **258**: 161, 2006).

Auraspora J. Weiser & Purrini (1980), Microsporidia. 1.

Aureobasidiaceae Cif. (1958) = Dothioraceae.

Aureobasidium Viala & G. Boyer (1891), anamorphic *Discosphaerina*, Hsy.0eP.16. 7, widespread. *A. pullulans*, a variable sp. with many syn. See Cooke (*Mycopathologia* **17**: 1, 1962), Joly (*BSMF* **81**: 402, 1965; 149 refs), Pugh & Buckley (*TBMS* **57**: 227, 1971; endophytic in trees), Hermanides-Nijhof (*Stud. Mycol.* **15**: 141, 1977; 14 spp., distinction from *Hormonema* and *Sarcinomyces*), Park (*TBMS* **78**: 385, 1982; Y-M dimorphism), Yoshikawa & Yokoyama (*Ann. phytopath. Soc. Japan* **53**: 606, 1987; *A. microstictum* on *Hemerocallis*), Elinov *et al.* (*Mikol. Fitopatol.* **23**: 425, 1989; physiology and biochemistry), Mokrousov & Bulast (*Genetika* **28**: 31, 1992; DNA unhybridizable UP-PCR patterns in *A. pullulans*), Untereiner & Naveau (*Mycol.* **91**: 67, 1999; phylogeny), Urzì *et al.* (*J. Microbiol. Meth.* **36**: 95, 1999; infraspecific diversity), Yurlova *et al.* (*Stud. Mycol.* **43**: 63, 1999), Nelson *et al.* (*BioTechniques* **29**: 874, 2000; image analysis), Bolignano & Criseo (*J. Clin. Microbiol.* **41**: 4483, 2003; clinical strain), Abliz *et al.* (*FEMS Immunol. Med. Microbiol.* **40**: 41, 2004; molecular diagnostics), Hsiao *et al.* (*J. Clin. Microbiol.* **43**: 3760, 2005; arrays), Rakeman *et al.* (*J. Clin. Microbiol.* **43**: 3324, 2005; molecular diagnostics), Ruibal *et al.* (*Mycol. Progr.* **4**: 23, 2005; rock-inhabiting fungi), Schoch *et al.* (*Mycol.* **98**: 1041, 2006; phylogeny), Sterflinger (*The Yeast Handbook* **[1]**: 501, 2006; ecology).

Aureobasis Clem. & Shear (1931) ≡ Aureobasidium.

Aureoboletus Pouzar (1957), Boletaceae. 5, widespread. See Wolf (*Biblthca Mycol.* **69**, 1980), Watling *et al.* (*British Fungus Flora. Agarics and Boleti* Rev. & Enl. Edn **1**: 173 pp., 2005; Brit. spp.).

Aureofungus Hibbett, Binder & Wang (2003), Agaricales. 1, Dominican Republic. See Hibbett *et al.* (*Mycol.* **95**: 685, 2003).

Aureohyphozyma Hosoya & Y. Otani (1995) = Gelatinipulvinella fide Hawksworth *et al.* (*Dictionary of the Fungi* edn 8, 1995).

Aureomyces Ruokola & Salonen (1970) = Cephaloascus fide von Arx *et al.* (*Antonie van Leeuwenhoek* **38**: 289, 1972).

Auricula Battarra ex Kuntze (1891) [non *Auricula* Castrac. 1873) nom. cons., *Algae*] ≡ Auricularia.

Auricula Lloyd (1922) = Punctularia fide Donk (*Taxon* **6**: 21, 1957).

Auricularia Bull. ex Juss. (1789), Auriculariaceae. *c.* 8, widespread. The edible *A. polytricha* is cultured on poles of *Quercus* in China; *A. auricula-judae*, Jew's ear fungus, is sometimes parasitic, esp. on *Sambucus*. See Lowy (*Mycol.* **43**: 351, 1951; key), Lowy (*Mycol.* **44**: 656, 1952), Donk (*Taxon* **7**: 168, 1958), Donk (*Persoonia* **4**: 154, 1966; nomencl.), McLaughlin (*Am. J. Bot.* **67**: 1225, 1980; meta basidium ultrastr.), Yan *et al.* (*Mycosystema* **21**: 47, 2002; RAPD), Cao & Pan (*Mycosystema* **24**: 53, 2005; ERIC).

Auriculariaceae Fr. (1838), Auriculariales. 7 gen. (+ 12 syn.), 112 spp.
Lit.: Wong & Wells (*Mycol.* **79**: 847, 1987), Corner (*Beih. Nova Hedwigia* **96**: 218 pp., 1989; as *Aporpiaceae*), Lü & McLaughlin (*Mycol.* **83**: 322, 1991), Ryvarden (*Syn. Fung.* **5**: 363 pp., 1991), Reid (*Persoonia* Suppl. **14**: 465, 1992), Roberts (*MR* **97**: 473, 1993), Lü & McLaughlin (*CJB* **73**: 315, 1995), Núñez (*Mycotaxon* **61**: 177, 1997), Begerow *et al.* (*CJB* **75**: 2045, 1998), Núñez (*Folia cryptog. Estonica* **33**: 99, 1998), Roberts (*Mycotaxon* **69**: 209, 1998; key), Yan *et al.* (*Mycosystema* **18**: 206, 1999), Weiss & Oberwinkler (*MR* **105**: 403, 2001), Larsson *et al.* (*MR* **108**: 983, 2004), Weiss *et al.* (*MR* **108**: 1003, 2004), Wells *et al.* (*Frontiers in Basidiomycote Mycology*: 237, 2004).

Auriculariales J. Schröt. (1887). Agaricomycetes. 1 fam., 32 gen., 198 spp. Basidiocarps hemiangiocarpous and sessile; metabasidium cylindrical and horizontally septate, 1-4 cells each bearing a sterigma and basidiospore; hyphae with septal dolipores. Fam.: **Auriculariaceae** (syn. *Exidiaceae*)
Lit.: Donk (1951-63) VIII; (1966: 208), Bandoni (*Trans. mycol. Soc. Japan* **25**: 521, 1984).

Auriculariella (Sacc.) Clem. (1909) = Auricularia fide Donk (*Persoonia* **4**: 158, 1966).

Auriculariopsidaceae Jülich (1982) = Schizophyllaceae.

Auriculariopsis Maire (1902), Schizophyllaceae. 1, widespread. See Donk (*Persoonia* **1**: 76, 1959).

Auriculibuller Samp. & Fonseca (2004), Tremellaceae. Anamorph *Bullera*. 1, Portugal. See Sampaio *et al.* (*Int. J. Syst. Evol. Microbiol.* **54**: 988, 2004).

Auriculora Kalb (1988), Lecanorales (L). 1, S. America. See Henssen & Titze (*Bot. Acta* **101**: 131, 1990), Ekman (*Op. bot.* **127**: 148 pp., 1996).

Auriculoscypha D.A. Reid & Manim. (1985), Septobasidiaceae. 1, India. See Lalitha *et al.* (*MR* **98**: 64, 1994; basidiosp. germin.), Kumar *et al.* (*MR* **111**: 268, 2007; phylogeny).

Aurificaria D.A. Reid (1963), Hymenochaetaceae. 1, widespread. See Reid (*Kew Bull.* **17**: 278, 1963).

Auriporia Ryvarden (1973), Fomitopsidaceae. 3, north temperate. See Parmasto (*Mycotaxon* **11**: 173, 1980; key), Coelho (*Mycol.* **97**: 263, 2005; Brazil spp.).

Auriscalpiaceae Maas Geest. (1963), Russulales. 6 gen. (+ 4 syn.), 38 spp.
Lit.: Donk (*Taxon*: 245, 1951-63) See also *Lit.* under *Hydnaceae*, Berbee & Wells (*Mycol.* **81**: 20, 1989), Wu & Petersen (*Mycosystema* Suppl. **4**: 33, 1991), Petersen & Cifuentes (*MR* **98**: 1427, 1994), Hibbett & Donoghue (*CJB* **73**: S853, 1995), Stalpers (*Stud. Mycol.* **40**: 185 pp., 1996), Ginns (*Mycol.* **90**: 19, 1998), Pine *et al.* (*Mycol.* **91**: 944, 1999), Miller & Methven (*Mycol.* **92**: 792, 2000), Desjardin & Ryvarden (*Sydowia* **55**: 153, 2003), Larsson & Larsson (*Mycol.* **95**: 1037, 2003), Lickey *et al.* (*Sydowia* **55**: 181, 2003), Binder *et al.* (*Systematics and Biodiversity* **3**: 113, 2005).

Auriscalpium Gray (1821), Auriscalpiaceae. 8, widespread. See Maas Geesteranus (*Persoonia* **9**: 493, 1978; key), Stalpers (*Stud. Mycol.* **35**: 29, 1996; key), Ryvarden (*Harvard Pap. Bot.* **6**: 193, 2001; monogr.).

Auritella Matheny & Bougher (2006), Inocybaceae. 7, Australia. See Matheny & Bougher (*Mycotaxon* **97**: 232, 2006), Matheny & Bougher (*Mycol. Progr.* **5**: 2, 2006).

Aurophora Rifai (1968), Sarcoscyphaceae. 1, widespread (pantropical). See Cabello (*Boln Soc. argent. Bot.* **25**: 395, 1988; numerical taxonomy), Zhuang & Wang (*Mycotaxon* **69**: 339, 1998; China).

Australasian Mycological Association. Founded in 1995; recognized as the Committee for Australasia within the International Mycological Association (q.v.); structure comprises individual members, and an elected executive; organizes occasional conferences. Publications: *Australasian Mycologist*. Website: http://bugs.bio.usyd.edu.au/AustMycolSoc/Home/ams.html.

Australiaena Matzer, H. Mayrhofer & Elix (1997), Caliciaceae (L). 1, Australia; N. America. See Matzer *et al.* (*Lichenologist* **29**: 35, 1997), Sheard & May (*Bryologist* **100**: 159, 1997; N. Am.), Scheidegger *et al.* (*Lichenologist* **33**: 25, 2001; evolution).

Australiasca Sivan. & Alcorn (2002), Chaetosphaeriaceae. Anamorph *Dischloridium*. 1, Australia. See Sivanesan & Alcorn (*Aust. Syst. Bot.* **15**: 741, 2002).

Australicium Hjortstam & Ryvarden (2002), Phanerochaetaceae. 2, widespread. See Hjortstam & Ryvarden (*Syn. Fung.* **15**: 19, 2002), Hjortstam *et al.* (*Syn. Fung.* **20**: 42, 2005; Venezuela).

Australohydnum Jülich (1978), ? Phanerochaetaceae. 1, Australia; Europe. See Jülich (*Persoonia* **10**: 138, 1978), Melo & Hjortstam (*Nova Hedwigia* **74**: 527, 2002; Europe).

Australoporus P.K. Buchanan & Ryvarden (1988), Polyporaceae. 1, Australia. See Buchanan & Ryvarden (*Mycotaxon* **31**: 5, 1988).

Austrella P.M. Jørg. (2004), Pannariaceae (L). 1, Australasia. See Jørgensen (*Biblthca Lichenol.* **88**: 230, 2004).

Austrobasidium Palfner (2006), Exobasidiaceae. 1 (causing galls on *Hydrangea serratifolia*), Chile. See Palfner (*Aust. Syst. Bot.* **19**: 431, 2006).

Austroblastenia Sipman (1983), ? Megalosporaceae (L). 2, Australasia. See Kantvilas (*Lichenologist* **26**: 349, 1994).

Austroboletus (Corner) Wolfe (1980), Boletaceae. *c.* 30, America; Australia. See Wolfe (*Biblthca Mycol.* **69**, 1980), Watling (*Aust. Syst. Bot.* **14**: 407, 2001; diversity and possible origins).

Austrocenangium Gamundí (1997), Helotiaceae. Anamorph *Endomelanconium*. 2, S. America. See Gamundí (*Mycotaxon* **63**: 261, 1997), Gamundí & Romero (*Fl. criptog. Tierra del Fuego* **10**, 1998).

Austroclitocybe Raithelh. (1972), ? Tricholomataceae. 1, S. America (temperate). See Raithelhuber (*Metrodiana* **3**: xxvii, 1972).

Austrogaster Singer (1962), Paxillaceae. 3, S. America (temperate); New Zealand. Basidioma gasteroid. See Singer (*Boln Soc. argent. Bot.* **10**: 57, 1962).

Austrogautieria E.L. Stewart & Trappe (1985), Gomphaceae. 6, Australia. See Stewart & Trappe (*Mycol.* **77**: 674, 1985; key).

Austrolecia Hertel (1984), Catillariaceae (L). 1, Antarctica. See Hertel (*Nova Hedwigia* Beih. **79**: 452, 1984), Rambold (*Biblthca Lichenol.* **34**: 345 pp., 1989).

Austrolentinus Ryvarden (1991), Polyporaceae. 1, Australia; Solomon Islands. See Ryvarden (*Syn. Fung.* **5**: 115, 1991).

Austroomphaliaster Garrido (1988), Tricholomataceae. 1, S. America (temperate). See Garrido (*Biblthca Mycol.* **120**: 199, 1988).

Austropaxillus Bresinsky & Jarosch (1999), Serpulaceae. 9, widespread (southern temperate). See Bresinsky *et al.* (*Pl. Biol.* **1**: 327, 1999; phylogeny).

Austropeltum Henssen, Döring & Kantvilas (1992), Sphaerophoraceae (L). 1, Australasia. See Wedin & Döring (*MR* **103**: 1131, 1999; phylogeny), Wedin *et al.* (*Lichenologist* **32**: 171, 2000; phylogeny), Lin *et al.* (*Plant Pathology Bulletin* Taichung **13**: 91, 2004; ascus evolution).

Austropezia Spooner (1987), Hyaloscyphaceae. 1, New Zealand. See Spooner (*Biblthca Mycol.* **116**, 1987).

Austrosmittium Lichtw. & M.C. Williams (1990), Legeriomycetaceae. 4 (in *Diptera*), Australia; New Zealand. See Williams & Lichtwardt (*CJB* **68**: 1045, 1990), Lichtwardt & Williams (*Mycol.* **84**: 384, 1992), White (*MR* **110**: 1011, 2006; phylogeny).

autecology, ecological studies on a single species and its relationship to the biological and physiochemical aspects of its environment.

aut-eu-form, an autoecious rust having all the spore stages.

authentic (of specimens, cultures, etc.), identified by the author of the name of the taxon to which they are referred.

author citations, see Nomenclature.

Authors' names. It is customary to cite Authors' names as authorities for the scientific names of taxa, to provide a clue where the name was published. There is frequently much variation and ambiguity in the ways such names are cited by different writers and uniformity in usage is desirable. For the fungi, Kirk & Ansell (Authors of fungal names. *Index of Fungi Supplement*, 1992) provided a list of over 9,000 authors of scientific names of fungi with recommended forms of their names, including abbreviations. This source, also available on-line in an updated form (see Internet), is now generally accepted as providing the standard. The format adopted by Kirk & Ansell for an author is the surname, or an abbreviation of it, or rarely a contraction of it, with or without initials or other distinguishing appendages. Among the more important criteria used in determining a standard form are: (1) names are in Roman characters; (2) every standard form must be unique to one person; (3) the same surname (i.e. identical spelling) must always be given in the same form, unless it is part of a compound name, and different surnames must not be given the same form; (4) all abbreviations and contractions are terminated by a full-stop but the full-stop does not make a standard form different from the same spelling without a full-stop; (5) the standard forms recommended in *TL-2* (see Literature) are retained in most cases, one of a few exceptions being conflict with particularly well established abbreviations used elsewhere; (6) names are never abbreviated before a consonant; (7) names are usually not abbreviated unless more than two letters are eliminated and replaced by a full-stop.

The above cited list was produced in collaboration with a similar scheme for botanists (Brummitt & Powell (Eds), *Authors of plant names*, 1992) and covers names of authors of all fungal taxa whose nomenclature is governed by the international code of nomenclature used for fungi (see Nomenclature).

The following list of deceased authors for which there are **biographical notices** in this *Dictionary* provides a representative series of examples of author abbreviations. Letters after the authors' dates refer to *Index Herbariorum* codes for the major location of the collections.

Ach(arius, E. 1757-1819); **H** (**BM**, **LD**, **PH**, **UPS**)
S. Ahmad (1910-1983); **BPI**, **IMI**
Ainsw(orth, G.C. 1905-1998); **IMI**
Ajello (L. 1916-2004)
Alexop(oulos, C.J. 1907-1986); **P** (**BPI**)
Arthur (J.C. 1850-1942); **PUR**
Arx (J.A. von 1922-1988); **CBS**
Asahina (Y. 1881-1975); **TNS**
G.F. Atk(inson, 1854-1918); **CUP**

M.E. Barr (Bigelow 1923-2008); **NY** (**DAOM**)
Bat(ista, A.C. 1916-1967); **URM**
Berk(eley, M.J. 1803-1889); **K** (**E**)
Berl(ese, A.N. 1864-1903); **PAD**
E.A. Bessey (1877-1957); **MSC**, **NEB**
Bilgrami (K.S. 1933-1996); **IMI**
Bisby (G.R. 1889-1958); **DAOM**, **IMI**, **WIN**
Bolton (J. 1750-1799)
Bondartsev (A.S. 1877-1968); **LE**
Boud(ier, J.L.É. 1828-1920); **PC**
Bourdot (H. 1861-1937); **PC**
Bref(eld, J.O. 1839-1925); **B**
Bres(àdola, G. 1847-1929); **S** (**BPI**, **L**, **TO**)
W. Br(own, 1888-1975)
Buller (A.H.R. 1874-1944); **WIN**
Bull(iard, J.B.F. 1752-1793); **PC**
J.H. Burnett (1922-2007)
Burt (E.A. 1859-1939); **BPI**, **FH**
E.J. Butler (1874-1943); **HCIO**

Chardón (C.E. 1897-1965); **BPI**, **RPPR**
Cif(erri, R. 1897-1964); **BPI** (**PAV**)
Cooke (M.C. 1825-1914); **K** (**E**, **PAV**, **PC**)
Corda (A.K.J. 1809-1849); **PR** (**K**)
Corner (E.J.H. 1906-1997); **E**
Costantin (J.N. 1857-1936)
G.(H.) Cunn(ingham 1892-1962); **IMI**, **K**, **PDD**
M.A. Curtis (1808-1872); **FH** (**BPI**, **BRU**, **K**, **NEB**, **NYS**)

P.(C.)A. Dangeard (1862-1947); **PC**
Dearn(ess, J. 1852-1954); **DAOM** (**BPI**, **CAN**, **CUP**, **IAC**, **NY**)
de Bary (H.A. 1831-1888); **BM**, **STR**
Deighton (1903-1992); **IMI**
Dennis (1910-2003); **K**
De Not(aris, G. 1805-1877); **RO** (**BM**, **GE**, **PAD**, **PC**, **TO**)
Desm(azières, J.B.H.J. 1786-1862); **BR**, **PC**
Dietel (P. 1860-1947); **B**, **K**, **S**
Dill(enius, J.J. 1684-1747); **OXF**
Dodge (C.W. 1895-1988); **FH**
Doidge (E.M. 1887-1965); **PRE**
Donk (M.A. 1908-1972); **L** (**BO**)

Ellis (J.B. 1829-1905); **NY** (**BPI**, **FH**)
M.B. Ellis (1911-1996); **IMI**
Erikss(on, J. 1848-1931); **S**

Farl(ow, W.G. 1844-1919); **FH**
Fée (A.L.A. 1789-1874); **BM**, **FI**, **PC**, **STR**
E. Fisch(er, 1861-1939); **BERN** (**B**, **BAS**, **KIEL**, **PC**)
Fitzp(atrick, H.M. 1886-1949); **CUP** (**FH**, **IAC**, **NY**)
Friedmann (E.I. 1921-2007)
Fr(ies, E.M. 1794-1878); **UPS** (**B**, **LD**)
Th.(M.) Fr(ies, 1832-1913); **UPS** (**LD**)
Fuckel (K.W.G.L. 1821-1876); **G**

Gäum(ann, E.A. 1893-1963); **BERN**
Golovin (P.N. 1897-1968); **LE**
Gonz(ález) Frag(oso, R. 1862-1928)
Gorlenko (M.V. 1908-1994)
Grev(ille, R.K. 1794-1866); **E** (**GL**)
Grove (W.B. 1848-1938); **K**
J.W. Groves (1906-1970); **DAOM**
Gruby (D. 1810-1898)
Guillierm(ond, M.A.A. 1876-1945); **PC**
Gyeln(ik, V.K. 1906-1945) ; **BP**

Hale (M.E. 1928-1990); **US**
Hansf(ord, C.G. 1900-1966); **EA** (**IMI**, **K**)
E.C. Hansen (1842-1909); **C** (**K**)
(H.J.A.)R. Hartig (1839-1901)
R. Heim (1900-1979); **PC**
Henn(ings, P.C. 1841-1908); **B** (**HBG**, **K**, **KIEL**, **L**, **S**, **W**)
Hirats(uka) f. (Naohide 1903-2000); **TMI** (**PUR**)
Höhn(el, F.X.R. von 1852-1920); **FH** (**K**)

Jacz(ewski, A.L.A. 1863-1932); **LE**

P.(A.) Karsten (1834-1917); **H** (**BPI**, **UPS**)
Kauffman (C.H. 1869-1931); **MICH** (**NY**)
Kniep (K.J.H. 1881-1930)
Körb(er, G.W. 1817-1885); **L** (**G**, **W**, **WRSL**)
J.G. Kühn (1825-1910)
Kusano (S. 1874-1962); **B**, **NY**

I.M. Lamb (1911-1990); **FH**
J.E. Lange (1864-1941); **C**
Langeron (M.C.P. 1874-1950); **PC**
Lév(eillé, J.-H. 1796-1870); **K** (**E**, **G**, **L**, **PC**)
Lindau (G. 1866-1923); **B** (**C**, **L**)
Linds(ay, W.L. 1829-1880); **E** (**BM**)
Link (J.H.F. 1767-1851); **B** (**L**)
L(innaeus, C. 1709-1778); **LINN** (**S**)
Liro (J.I. 1872-1943); **H** (**IMI**)
Lister (A. 1830-1908); **BM**
G. Lister (1860-1949); **BM**
Lloyd (C.G. 1859-1926); **BPI**
Luttr(ell, E.S. 1916-1988)

McAl(pine, D. 1849-1932); **VPRI**
(A.)H. Magn(usson 1885-1964); **UPS**
Maire (R.C.J.E. 1878-1949); **AL** (**MPU**)
G.W. Martin (1886-1971); **BPI**, **IA**
E.W. Mason (1890-1975); **IMI**
Massal(ongo, A.B. 1824-1862); **VER** (**PAD**)
Massee (G.E. 1850-1917); **K**, **NY**
P.(A.) Micheli (1679-1737); **FI**
Millardet (P.M.A. 1838-1902)
Mont(agne, J.P.F.C. 1784-1866); **PC** (**BM**, **L**, **UPS**)
Morochk(ovsky, S.F. 1897-1962); **KW**
M.M. Moser (1924-2002); **IB**
S.T. Moss (1943-2001)
Müll(er) Arg(oviensis, J. 1828-1896); **G** (**BM**)
E. Müll(er 1920-2008); **ZT**
Mundk(ur, B.B. 1896-1952)
Murrill (W.A. 1869-1957); **NY**

Nann(enga)-Bremek(amp, N.E. 1916-1996)
Nannf(eldt, J.A. 1904-1985); **UPS**
Naumov (N.A. 1888-1959); **LEP**
Nees (von Esenbeck, C.G.D. 1776-1858); **STR** (**L**, **UPS**)
T.(F.L.) Nees (von Esenbeck 1787-1837); **GZU**, **STR**
Nevod(ovsky, G.S. 1874-1952); **AA** (**LE**)
Niessl (von Meyendorf, G. 1839-1919); **M**
Nyl(ander, W. 1822-1899); **H** (**BM**, **NY**, **STR**, **UPS**)

Oudem(ans, C.A.J.A. 1825-1906); **L**

Pasteur (L. 1822-1895)
Pat(ouillard, N.T. 1854-1926); **FH** (**PC**)
Peck (C.H. 1833-1917); **NYS**
Pers(oon, C.H. 1761-1836); **L** (**G**, **GOET**, **PC**, **STR**, **TO**)
Petch (T. 1870-1948); **K**
Petr(ak, F. 1886-1973); **W** (**S**)
Pilát (A. 1903-1974); **PRM**
Poelt (J. 1924-1995); **W**
Potebnia (A.A. 1870-1919); **CWU**
Preuss (C.G.T. 1795-1855); **B**

Quélet (L. 1832-1899); **PC**

Rabenh(orst, G.L. 1806-1881); **B**
Racib(orski, M. 1863-1917); **KRAM** (**FH**, **ZT**)
Ramsb(ottom, J. 1885-1974)
Rehm (H. 1826-1916); **S** (**B**)
Rick (J.E. 1869-1946); **PACA** (**B**, **BPI**, **CUP**, **FH**, **IAC**, **IACM**, **K**, **MICH**, **R**, **RB**, **S**, **SFPA** and **SI**)
Mig(uel) Rodr(íguez Hernández 1949-2003); **HAJB**
Rogerson (C.T. 1918-2001); **NY**
Rostaf(iński, J.T. 1850-1928)
Rostrup (E. 1831-1907); **C**, **CP**

Sabour(aud, R. 1864-1938)
Sacc(ardo, P.A. 1845-1920); **PAD**
Savile (D.B.O. 1909-2000)
Săvul(escu, T. 1889-1963); **BUCM**
Schwein(itz, L.D. von 1780-1830); **PH** (**BPI**, **FH**, **K**)
Schwend(ener, S. 1829-1919)
Seaver (F.J. 1877-1970); **NY**
Shear (C.L. 1865-1956); **BPI**
Shvarzman (S.R. 1912-1975); **AA**
Singer (R. 1906-1994); **F**
A.H. Sm(ith 1904-1986); **MICH**
A.L. Sm(ith 1854-1937); **BM** (**K**)
E.F. Sm(ith 1854-1927); **BPI**
W.G. Sm(ith 1935-1917)
Sorauer (P.C.M. 1839-1916)
Sowerby (J. 1757-1822); **BM**, **K** (**LINN**)
Sparrow (F.K. 1903-1977); **MICH**
Speg(azzini, C.L. 1858-1926); **LPS**
Stakman (E.C. 1885-1979); **MPPD**
Syd(ow, H. 1879-1946); **S** (**B**)
P. Syd(ow 1851-1925); **S** (**B**, **DAR**)

Teng (S.-c. 1902-1970)
Teterevn(ikova-Babayan D.N 1904-1988); **ERE**
Thaxter (R. 1858-1932); **FH**
Theiss(en, F. 1877-1919); **W (FH)**
Thind (K.S. 1917-1991); **PAN**
Thom (C. 1872-1956)
Tode (H.J. 1733-1797); fungal reference collection and herbarium destroyed
Tomilin (B.A. 1928-2008); **LE**
Tranzschel (W.A. 1868-1942); **LE (CWU)**
Trevis(an, V. 1818-1887); **PAD**
Tubaki (K. 1924-2005)
Tuck(erman, E. 1817-1886); **FH (US**
Tul(asne, L.R. 1815-1885); **PC**
C. Tul(asne 1816-84); **PC**

Uljan(ishchev, V.I. 1898-1996) **BAK**
Unger (F. 1800-1870); **W**

Vain(io, E.A. 1853-1929); **TUR (BM, BR, C, STE, US)**
Vele(novský, J. 1858-1949) **PRM**
Viégas (A.R. 1906-1986)
Vuill(emin, P. 1861-1932); **PAD, PAV**

Wakefield (E.M. 1886-1976); **K**
H.M. Ward (1854-1906)
Westerd(ijk, J. 1883-1961)
Weston (W.H. 1890-1978); **FH**
Whetzel (H.H. 1877-1944); **CUP**
(H.)G. Winter (1848-1887); **B**
Wormald (H. 1879-1953)
Woronin (M.S. 1838-1903)
J.E. Wright (1922-2005); **BAFC**

Zahlbr(uckner, A. 1860-1938); **W (PAD, STE, US)**
Zopf (W. 1846-1909); **B**

For further information on particular authors see also History (Literature), Internet, Literature (Bibliographies), Medical and veterinary mycology, Reference Collections. Currently active mycologists are listed in society membership lists and regional compilations (e.g. Anon, *Revista Iberoamer. Micol.* **10**: ix, 1993 [Latin Am.]; Bakloushinskaya & Minter *Vorontsov's Who's Who in Biodiversity Sciences*, 2001 [countries formerly in the Soviet Union]; Buyck & Hennebert, *Directory of African Mycology*, 1993).

auto- (prefix), self-inducing, -producing, etc.

autobasidium, see basidium.

autochthonous (1) indigenous; cf. allochthonous; (2) (of soil organisms), continuously active, as opposed to **zymogenous** organisms which become active when a suitable substrate becomes available (Winogradsky, 1924); cf. exochthonous (Park, 1957).

autodeliquescent (of lamellae and pileus of *Coprinus*), becoming liquid by **-digestion**.

autoecious, completing the life cycle on one host (esp. of rusts; cf. heteroecious); ametoecious (de Bary).

autogamy, the fusion of nuclei in pairs within the female organ, without cell fusion having taken place.

Autoicomyces Thaxt. (1908), Ceratomycetaceae. 27, widespread. See Majewski (*Acta Mycologica* Warszawa **34**: 7, 1999; Poland), Santamaría (*Fl. Mycol. Iberica* **5**, 2003; Iberian peninsula), Ye & Shen (*Mycosystema* **22**: 2, 2003; China).

autolysis, self digestion of a cell or tissue by endogenous enzymes.

automictic sexual reproduction, karyogamy between daughter nuclei of different meioses in the same gametangium (Dick, 1972).

automixis, self-fertilization by the fusion of two closely related sexual cells or nuclei; cf. amphimixis, apomixis, pseudomixis.

Autophagomyces Thaxt. (1912), Laboulbeniaceae. 24, widespread. See Tavares (*Mycol. Mem.* **9**: 627 pp., 1985), Benjamin (*Aliso* **19**: 99, 2000).

autotroph (adj. **autotrophic**) (of a living organism), one not using organic compounds as primary sources of energy, i.e. using energy from light or inorganic reactions as do green plants, lichen-forming fungi, and the photosynthetic iron and sulphur bacteria. See Fry & Peel (Eds) (*Autotrophic micro-organisms*, 1954), Lees (*Biochemistry of autotrophic bacteria*, 1955); cf. heterotrophic.

auxanogram, the differential growth of a yeast in Petri dishes prepared by the auxanographic method of Beijerinck (as modified by Lodder, *Die anaskosporogenen Hefen*, 1934, and Langeron, 1952: 430) for determining the carbon and nitrogen requirements of the organism. See also Lodder & van Rij (1952), Pontecorvo (*J. gen. Microbiol.* **3**: 122, 1949; auxanographic techniques in biochemical genetics).

Auxarthron G.F. Orr & Kuehn (1963), Onygenaceae. Anamorph *Malbranchea*-like. 15, widespread. See Samson (*Acta Bot. Neerl.* **21**: 517, 1972), Sugiyama *et al.* (*Mycoscience* **40**: 251, 1999; DNA), Kuraishi *et al.* (*Antonie van Leeuwenhoek* **77**: 179, 2000; ubiquinones), Sugiyama & Mikawa (*Mycoscience* **42**: 413, 2001; phylogeny), Sigler *et al.* (*Stud. Mycol.* **47**: 111, 2002; anamorphs), Solé *et al.* (*MR* **106**: 388, 2002; phylogeny), Solé *et al.* (*Stud. Mycol.* **47**: 103, 2002), Skinner *et al.* (*Mycol.* **98**: 447, 2006; ontogeny).

auxiliary zoospore, first-formed zoospore, formed and flagellate within the sporangium, in a species with dimorphic zoospores (Dick, 1973); flagellar insertion apical or sub-apical.

auxotroph, a biochemical mutant which will only grow on the minimal medium (q.v.) after the addition of one or more specific substances.

avenacein, see enniatin.

avenacin, a fungus inhibitor from oats (*Avena*) (Turner, *Nature* **186**: 325, 1960).

aversion, the inhibition of growth at the adjacent edges of colonies of microorganisms, esp. in a culture of one species. Cf. antagonism; barrage.

Avesicladiella W.P. Wu, B. Sutton & Gange (1997), anamorphic *Pezizomycotina*, Hso.?.?. 2, Europe; China. See Wu *et al.* (*Mycoscience* **38**: 11, 1997).

Avettaea Petr. & Syd. (1927), anamorphic *Pezizomycotina*, Cpd.0eP.15. 3, widespread. See Abbas & Sutton (*TBMS* **90**: 491, 1988).

Avrainvillea Decne. (1842), Algae. Algae.

Awasthia Essl. (1978), Physciaceae (L). 1, India. See Esslinger (*Bryologist* **81**: 445, 1978).

Awasthiella Kr.P. Singh (1980), Verrucariaceae (L). 1, India. See Singh (*Norw. Jl Bot.* **27**: 34, 1980), Singh & Sinha (*Lichen Flora of Nagaland*, 1994; India).

axenic (of cultures), consisting of one organism; uncontaminated; a pure culture. Cf. gnotobiotic.

axeny, inhospitality; 'passive' as opposed to 'active' resistance of a plant to a pathogen (Gäumann, 1946).

axial canal (**- mass**), see ascus.

Axisporonites Kalgutkar & Janson. (2000), Fossil Fungi. 1, India. See Kalgutkar & Jansonius (*AASP*

Contributions Series **39**: 36, 2000).

axoneme, the main core of a flagellum composed of 2 central microtubules surrounded by 9 double microtubules.

Aylographum, see *Aulographum*.

Ayria Fryar & K.D. Hyde (2004), ? Annulatascaceae. 1, Brunei. See Fryar & Hyde (*Cryptog. Mycol.* **25**: 248, 2004), Vijaykrishna *et al.* (*Fungal Diversity* **23**: 351, 2006).

Azbukinia Lar.N. Vassiljeva (1989), Pezizomycotina. 1, Europe; Russia. Possibly belongs to the Thyridiaceae, but identity of the type is uncertain. See Vasil'eva (*Mycotaxon* **35**: 395, 1989), Eriksson & Hawksworth (*SA* Reprint of Volumes 1-4 (1982-1985) **8**: 97, 1990; typific.).

Azosma Corda (1831) = Cladosporium fide Fries (*Syst. mycol.* **3** Index: 55, 1832).

azotodesmic, nitrogen-fixing (Pike & Carroll, *in* Alexopoulos & Mims, *Introductory mycology*, edn 3, 1980).

Azureothecium Matsush. (1989), ? Eurotiomycetes. 1 (from soil), Australia. See Matsushima (*Mycol. Mem.* **6**, 1989).

Azygites Fr. (1832) = Syzygites fide Hesseltine (*Mycol.* **47**: 344, 1955).

azygospore, a parthenogenetic zygospore; characteristic of some *Mucorales*. See Benjamin (*Aliso* **5**: 235, 1963; list).

Azygozygum Chesters (1933) = Mortierella fide Plaats-Niterink *et al.* (*Persoonia* **9**: 85, 1976).

Azymocandida E.K. Novák & Zsolt (1961) = Candida fide Lodder (*Yeasts, a taxonomic study* 2nd edn, 1970).

Azymohansenula E.K. Novák & Zsolt (1961) = Pichia fide Hawksworth *et al.* (*Dictionary of the Fungi* edn 8, 1995).

Azymomyces E.K. Novák & Zsolt (1961) = Torulaspora fide Lodder (*Yeasts, a taxonomic study* 2nd edn, 1970), Batra *in* Subramanian (Ed.) (*Taxonomy of fungi* **1**: 187, 1978).

Azymoprocandida E.K. Novák & Zsolt (1961) = Candida fide Lodder (*Yeasts, a taxonomic study* 2nd edn, 1970).

B, Botanischer Garten und Botanisches Museum Berlin-Dahlem (Berlin, Germany); founded 1815; from 1995 part of the Free University of Berlin; see Kohlmeyer (*Willdenowia* **3**: 63, 1962).

Babjevia Van der Walt & M.T. Sm. (1995) = Dipodascopsis fide Smith *et al.* (*Antonie van Leeuwenhoek* **67**: 177, 1995), Smith *in* Kurtzman & Fell (Eds) (*Yeasts, a taxonomic study* 4th edn: 141, 1998), Suh *et al.* (*Mycol.* **98**: 1006, 2006; phylogeny), Kurtzman *et al.* (*FEMS Yeast Res.* **7**: 1027, 2007; phylogeny).

baccate, soft throughout like a berry.

baccatin, a wilt toxin from *Gibberella baccata* (Gäumann *et al.*, *Phytopath. Z.* **36**: 114, 1959); antibacterial.

Bachmannia Zschacke (1934) [non *Bachmannia* Pax 1897, *Capparaceae*] = Verrucaria Schrad. fide Swinscow (*Lichenologist* **4**: 34, 1968).

Bachmanniomyces D. Hawksw. (1981), anamorphic *Pezizomycotina*, Cpd.0eH.19. 1 (on lichens, *Cladonia*), Europe; N. America. See Hawksworth (*Bull. Br. Mus. nat. Hist.* Bot. **9**: 10, 1981), Zhurbenko & Alstrup (*Symb. bot. upsal.* **34** no. 1: 477, 2004).

Bacidia De Not. (1846), Ramalinaceae (±L). *c.* 229, widespread. Molecular data suggest placement within *Ramalinaceae*. See Vězda (*Čas. Slez. Mus.* Ser. A, Hist. Nat. **10**: 103, 1961), Vězda (*Folia geobot. phytotax.* **15**: 75, 1980; key foliicolous spp.), Awasthi & Mathur (*Proc. Indian Acad. Sci.* Pl. Sci. **97**: 481, 1987; key 8 spp. India), Sérusiaux (*Nordic Jl Bot.* **13**: 447, 1993; differentiation 9 segr. gen.), Ekman (*Op. Bot.* **127**, 1996; N. Am.), Farkas *et al.* (*Sauteria* **9**: 133, 1998; Hungary), Aptroot & Herk (*Lichenologist* **31**: 121, 1999), Llop & Gómez-Bolea (*Mycotaxon* **72**: 79, 1999), Ekman & Wedin (*Pl. Biol.* **2**: 350, 2000), Kalb *et al.* (*Mycotaxon* **75**: 281, 2000; segregate), Ekman (*MR* **105**: 783, 2001; phylogeny), Lücking *et al.* (*Lichenologist* **33**: 189, 2001), Haug (*Mycol. Progr.* **1**: 167, 2002; key to Iberian spp.), Aragón & Martínez (*Bryologist* **106**: 143, 2003), Miądlikowska *et al.* (*Mycol.* **98**: 1088, 2006; phylogeny), Llop *et al.* (*Nova Hedwigia* **85**: 445, 2007; *B. rubella* group).

Bacidiaceae Walt. Watson (1929) = Ramalinaceae.

Lit.: Bellemère & Letrouit-Galinou (*Biblthca Lichenol.* **25**: 137, 1987), Brako (*Mycotaxon* **35**: 1, 1989), Brako (*Fl. Neotrop.* Monogr.: 67 pp., 1991), Boom (*Nova Hedwigia* **54**: 229, 1992), Rambold (*Sendtnera* **1**: 281, 1993), Kalb & Elix (*Biblthca Lichenol.* **57**: 265, 1995), Kantvilas & Elix (*Biblthca Lichenol.* **58**: 199, 1995), Ekman (*Op. bot.* **127**, 1996), Ekman (*Op. bot.* **127**: 148 pp., 1996), Ekman (*Symb. bot. upsal.* **32**: 17, 1997), Feige *et al.* (*Biblthca Lichenol.* **67**: 25, 1997), Aptroot & Herk (*Lichenologist* **31**: 121, 1999), Ekman & Wedin (*Pl. Biol.* **2**: 350, 2000), Kalb *et al.* (*Mycotaxon* **75**: 281, 2000), Ekman (*MR* **105**: 783, 2001), Lumbsch *et al.* (*Mol. Phylogen. Evol.* **31**: 822, 2004).

Bacidiactis M. Choisy (1931) ? = Lecanactis Körb. fide Hawksworth *et al.* (*Dictionary of the Fungi* edn 8, 1995).

Bacidina Vězda (1991), Ramalinaceae (L). 11, widespread. Probably congeneric with *Woessia*, but *Bacidina* is the more familiar name. See Santesson (*Lichens and lichenicolous fungi of Sweden and Norway*, 1993), Sérusiaux (*Biblthca Lichenol.* **58**: 411, 1995), Ekman (*Op. Bot.* **127**, 1996), Ekman (*MR* **105**: 783, 2001; phylogeny, reln with *Woessia*), Miądlikowska *et al.* (*Mycol.* **98**: 1088, 2006; phylogeny).

Bacidiomyces Cif. & Tomas. (1953) = Bacidia.

Bacidiopsis Bagl. (1861) = Pachyphiale fide Hawksworth *et al.* (*Dictionary of the Fungi* edn 8, 1995).

Bacidiopsora Kalb (1988), Ramalinaceae (L). 1, S. America. See Ekman (*Op. Bot.* **127**, 1996), Aptroot *et al.* (*J. Hattori bot. Lab.* **100**: 617, 2006; Costa Rica).

bacillar (**bacilliform**), rod-like in form (Fig. 23.4*a*).

Bacillaria Mont. (1840) [non *Bacillaria* J.F. Gmel. 1791, *Angiospermae*] nom. nud. ≡ Camillea.

Bacillidium Janda (1928), Microsporidia. 5.

Bacillina Nyl. (1897) = Toninia fide Hawksworth *et al.* (*Dictionary of the Fungi* edn 8, 1995).

Bacillispora Sv. Nilsson (1962) = Cylindrocarpon fide Marvanová & Descals (*TBMS* **89**: 499, 1987).

Bacillopeltis Bat. (1957), anamorphic *Pezizomycotina*, Cpt.0eH.?. 1, Brazil. See Batista (*An. Soc. Biol. Pernambuco* **15**: 400, 1957).

Bacillopsis Petsch. (1908) nom. dub., ? Saccharomycetales.

Backusella Hesselt. & J.J. Ellis (1969), Mucoraceae. 3, widespread. See Benny & Benjamin (*Aliso* **8**: 301, 1975; key), Stalpers & Schipper (*Persoonia* **11**: 39, 1980), Roux *et al.* (*Proc. Microsc. Soc. S. Afr.* **27**: 56,

1997; ultrastr.), O'Donnell *et al.* (*Mycol.* **93**: 286, 2001; phylogeny), Voigt & Wöstemeyer (*Gene* **270**: 113, 2001; phylogeny), Nyilasi *et al.* (*Clin. Microbiol. Infect.* **14**: 393, 2008; molecular identification).

Backusia Thirum., M.D. Whitehead & P.N. Mathur (1965) = Monascus fide Cole & Kendrick (*CJB* **46**: 987, 1968).

Bacteria. Heterogeneous group of (usually) unicellular prokaryotic organisms. In the three-domain classification of life, two prokaryotic domains were proposed the *Archaea* (formerly *Archaebacteria*) and *Bacteria* (formerly *Eubacteria*) (Woese *et al.*, *Proc. Natl. Acad. Sci.* **87**: 4576, 1990). The informal name 'bacteria' is occasionally used loosely in the literature to refer to all of the prokayrotes, and care should be taken to interpret its meaning in any particular context. See Woese (*Proc. Natl. Acad. Sci.* **97**: 8392, 2000). Some bacteria are pathogenic for plants, a few are pathogenic for fungi: *Bacillus polymyxa* (bacterial pit) of cultivated mushroom; *Pseudomonas agarici* (drippy gill) of cultivated mushroom (see Geels *et al.*, *J. Phytopathol.* **140**: 249, 1994); *Pseudomonas tolaasi* (brown blotch) of cultivated mushroom (see Cole & Skellerup, *TBMS* **87**: 314, 1986); *Pantoea agglomerans* pv. *uredovora* (syn. *Erwinia uredovora*) a parasite of rust uredinia. Bacteria, in particular *Actinomycetes*, *Bacillus* spp. and *Pseudomonas* spp. produce a variety of biologically active molecules (antibiotics, enzymes etc.) and have been used as biological control agents against fungal plant diseases.

bacteriostatic, of a substance, or a concentration of a bactericide, which will not let growth of bacteria take place but which is not bactericidal.

bactivory, bacteria-feeding; known in fungi only amongst certain marine *Thraustochytriales* (Raghukumar, *Mar. Biol.* **113**: 165, 1992).

bactobiont, see photobiont.

Bactrexcipula Höhn. (1918) ? = Rhizothyrium fide Petrak (*Sydowia* **15**: 185, 1962).

Bactridiopsis Gonz. Frag. & Cif. (1927) = Phillipsiella fide Rossman *et al.* (*Sydowia* **46**: 66, 1994).

Bactridium Kunze (1817), anamorphic *Pezizomycotina*, Hsp.≡ eH.1. 5, widespread. See Hughes (*N.Z. Jl Bot.* **4**: 522, 1966), Berch (*Mycotaxon* **14**: 227, 1982), Yuan *et al.* (*Mycosystema* **17**: 376, 1998; China), Kendrick (*CJB* **81**: 75, 2003; morphogenesis).

Bactroboletus Clem. (1909) ≡ Filoboletus.

Bactrodesmiastrum Hol.-Jech. (1984) ? = Janetia.

Bactrodesmiella M.B. Ellis (1959), anamorphic *Pezizomycotina*, Hsp.≡ eP.19/40. 2, British Isles; NZ. See Ellis (*Mycol. Pap.* **72**: 14, 1959), Hughes (*N.Z. Jl Bot.* **27**: 449, 1989), Dianese & Câmara (*Sydowia* **46**: 225, 1994).

Bactrodesmiites Babajan & Tasl. (1977), Fossil Fungi. 1 (Tertiary), former USSR.

Bactrodesmium Cooke (1883), anamorphic *Stuartella*, Hsp.≡ eP.1. 30, widespread. See Ellis (*Mycol. Pap.* **72**, 1959; key), Hughes & White (*Fungi Canad.*, 1983), Rao & Hoog (*Stud. Mycol.* **28**, 1986; India), Matsushima (*Matsush. Mycol. Mem.* **5**, 1987), Zucconi & Lunghini (*Mycotaxon* **63**: 323, 1997).

Bactropycnis Höhn. (1920) = Coleophoma fide Sutton (*Mycol. Pap.* **141**, 1977).

Bactrosphaeria Penz. & Sacc. (1897), Pezizomycotina. 1, Java.

Bactrospora A. Massal. (1852), Roccellaceae (L). 20, widespread (esp. temperate). See Egea & Torrente (*Lichenologist* **25**: 211, 1993; key), Letrouit-Galinou *et al.* (*Bull. Soc. linn. Provence* **45**: 389, 1994; ultrastr.), Nordin (*Lichenologist* **28**: 287, 1996; Scandinavia), Kantvilas (*Symb. bot. upsal.* **34** no. 1: 183, 2004), Ponzetti & McCune (*Bryologist* **109**: 85, 2006; N Am.), Sparrius *et al.* (*Lichenologist* **38**: 27, 2006; SE Asia).

Bactrosporaceae Rabenh. (1870) = Roccellaceae.

Bactrosporomyces Cif. & Tomas. (1953) ≡ Bactrospora.

baculate (1) (**baculiform**) (of spores), rod-shaped; (2) (of surface ornamentation), rod-shaped (Fig. 20.8).

Baculea Loubès & Akbarieh (1978), Microsporidia. 1.

Baculifera Marbach & Kalb (2000), Caliciaceae (L). 14, pantropical. See Marbach (*Biblthca Lichenol.* **74**: 113, 2000).

Baculospora Zukal (1887), Pezizomycotina. 1, Europe. See Lundqvist (*SA* **7**: 62, 1988).

Badarisama Kunwar, J.B. Manandhar & J.B. Sinclair (1986), anamorphic *Pezizomycotina*, Hso.#eP.1. 1 (bulbil-forming), USA. See Kunwar *et al.* (*Mycotaxon* **27**: 120, 1986).

Badimia Vězda (1986) nom. cons., Ectolechiaceae (L). 11 (on leaves), widespread (tropical). See Lücking *et al.* (*Bot. Acta* **107**: 393, 1994), Lücking & Vězda (*Taxon* **44**: 227, 1995), Aptroot *et al.* (*Biblthca Lichenol.* **64**: 220 pp., 1997; New Guinea), Lücking (*Trop. Bryol.* **15**: 45, 1998; Guyana), Lücking (*Cryptog. Mycol.* **27**: 121, 2006; French Guiana).

Badimiella Malcolm & Vězda (1994), Ectolechiaceae (L). 2, south temperate. See Malcolm & Vězda (*Nova Hedwigia* **59**: 518, 1994), Garnock-Jones & Malcolm (*Biblthca Lichenol.* **78**: 65, 2001), Lücking *et al.* (*Global Ecology and Biogeography* **12**: 21, 2003; ecology).

Baeoderma Vain. (1922) = Sphaerophorus fide Hawksworth *et al.* (*Dictionary of the Fungi* edn 8, 1995).

Baeodromus Arthur (1905), Pucciniosiraceae. 8 (on *Compositae*, *Urticaceae*, *Ranunculaceae* (0, III)), America; Russian Far East. Probably heterogeneous. See Buriticá (*Rev. Acad. Colomb. Cienc.* **18**: 131, 1991), Hernández (*Mycotaxon* **76**: 329, 2000).

Baeomyces Pers. (1794), Baeomycetaceae (L). *c.* 9, widespread. See Thomson (*Bryologist* **70**: 285, 1967), Jahns (*Herzogia* **2**: 133, 1971; ontogeny), Honegger (*Lichenologist* **15**: 57, 1983; asci), Sérusiaux (*Taxon* **32**: 646, 1983; gen. nomencl.), Kumar (*Geophytology* **15**: 159, 1985; key 4 spp. India), Gierl & Kalb (*Herzogia* **9**: 593, 1993; concept), Rambold *et al.* (*Biblthca Lichenol.* **53**, 1993), Stenroos & DePriest (*Am. J. Bot.* **85**: 1548, 1998), Platt & Spatafora (*Lichenologist* **31**: 409, 1999), Stenroos *et al.* (*Mycol. Progr.* **1**: 267, 2002; phylogeny), Peršoh *et al.* (*Mycol. Progr.* **3**: 103, 2004; asci), Miądlikowska *et al.* (*Mycol.* **98**: 1088, 2006; phylogeny), Lumbsch *et al.* (*MR* **111**: 257, 2007; phylogeny).

Baeomycetaceae Dumort. (1829), Baeomycetales (L). 2 gen. (+ 6 syn.), 14 spp.

Lit.: Christensen & Alstrup (*Nova Hedwigia* **51**: 469, 1990), Gierl & Kalb (*Herzogia* **9**: 593, 1993), Rambold *et al.* (*Biblthca Lichenol.* **53**, 1993), Ihlen (*Nova Hedwigia* **64**: 137, 1997), Stenroos & DePriest (*Am. J. Bot.* **85**: 1548, 1998), Platt & Spatafora (*Lichenologist* **31**: 409, 1999), Galloway (*Lichenologist* **32**: 294, 2000), Kauff & Lutzoni (*Mol. Phylogen. Evol.* **25**: 138, 2002), Lumbsch *et al.* (*Mol. Phylogen. Evol.* **31**: 822, 2004), Peršoh *et al.* (*Mycol. Progr.* **3**:

103, 2004), Wedin *et al.* (*MR* **109**: 159, 2005), Miądlikowska *et al.* (*Mycol.* **98**: 1088, 2006; phylogeny), Hofstetter *et al.* (*Mol. Phylogen. Evol.* **44**: 412, 2007; phylogeny).

Baeomycetales Lumbsch, Huhndorf & Lutzoni (2007). Ostropomycetidae. 2 fam., 3 gen., 16 spp. Ascomata sessile or stalked, apothecial, discoid, the true exciple hyaline or pigmented, annular or cupulate. Interascal tissue composed of branched paraphyses, the hymenium not blueing in iodine. Asci thin-walled, with a non-amyloid or slightly amyloid apical portion, 8-spored. Ascospores hyaline, septate or aseptate, sometimes with a sheath. Fam.:
(1) **Baeomycetaceae**
(2) **Trapeliaceae**
Lit.: Hibbett *et al.* (*MR* **111**: 509, 2007); Lumbsch *et al.* (*MR* **111**: 1133, 2007).

Baeomycetomyces E.A. Thomas ex Cif. & Tomas. (1953) ≡ Dibaeis.

Baeomycomyces, see *Baeomycetomyces*.

Baeopodium Trevis. (1857) = Gomphillus fide Hawksworth *et al.* (*Dictionary of the Fungi* edn 8, 1995).

Baeospora Singer (1938), ? Marasmiaceae. 10, widespread (north temperate; tropical). See Singer (*Revue Mycol.* Paris **3**: 193, 1938), Clémençon (*Persoonia* **18**: 411, 2004; basidiome development).

Baeostratoporus Bondartsev & Singer (1944) = Flaviporus fide Donk (*Persoonia* **1**: 173, 1960).

Baetimyces L.G. Valle & Santam. (2002), Legeriomycetaceae. 1 (in *Baetis*), Spain. See Valle & Santamaria (*Mycol.* **94**: 321, 2002).

Baeumleria Petr. & Syd. (1927), ? Phyllachoraceae. 2, widespread. See Clements & Shear (*Gen. Fung.*, 1931), Sutton (*Mycol. Pap.* **141**, 1977), von Arx (*SA* **6**: 213, 1987), Johnston (*N.Z. Jl Bot.* **37**: 703, 1999).

Bagcheea E. Müll. & R. Menon (1954), Valsaceae. 3, India; Taiwan. See Hsieh & Li (*MR* **95**: 895, 1991).

Baggea Auersw. (1866), ? Patellariaceae. 1, Europe. See Kutorga & Hawksworth (*SA* **15**: 1, 1997), Magnes *et al.* (*Mycotaxon* **68**: 321, 1998).

Bagliettoa A. Massal. (1853), Verrucariaceae (L). 1, widespread (esp. temperate). Perhaps synonymous with *Verrucaria*. See Santesson (*Lichens and lichenicolous fungi of Sweden and Norway*, 1993), Halda (*Acta Musei Richnoviensis* Sect. Natur. **10**, 2003), Gueidan & Roux (*Bull. Soc. linn. Provence* **58**: 181, 2007), Gueidan *et al.* (*MR* **111**: 1145, 2007; phylogeny).

Bagliettoaceae Servít (1954) = Verrucariaceae.

Bagnisiella Speg. (1880), ? Dothioraceae. 20, S. America; India. See Patil & Patil (*Indian J. Mycol. Pl. Path.* **13**: 169, 1985), Rajak *et al.* (*J. Econ. Taxon. Bot.* **18**: 546, 1995), Hyde & Cannon (*Mycol. Pap.* **175**, 1999), Barr (*Harvard Pap. Bot.* **6**: 25, 2001).

Bagnisimitrula S. Imai (1942), ? Helotiales. 1, Japan. Position uncertain; previously placed in the *Geoglossaceae*.

Bagnisiopsis Theiss. & Syd. (1915), Phyllachoraceae. 10 (on living leaves of *Melastomataceae*), S. America. See Hyde & Cannon (*Mycol. Pap.* **175**, 1999).

Bahianora Kalb (1984), Lecideaceae (L). 1, Brazil. See Kalb (*Lichenes Neotropici* Fascicle **VIII** (nos 301-350) **8**: 4, 1984).

Bahuchashaka Subram. (1978), anamorphic *Pezizomycotina*, Hso.≡ eP.1. 1, Japan. See Subramanian (*Kavaka* **5**: 97, 1977).

Bahugada K.A. Reddy & V.G. Rao (1984) ? = Monodictys.

Bahukalasa Subram. & Chandrash. (1979), anamorphic *Pezizomycotina*, Hso.0eH.10/12. 1, India. See Subramanian & Chandrashekara (*Kavaka* **6**: 42, 1978).

Bahupaathra Subram. & Lodha (1964) = Cladorrhinum fide Mouchacca & Gams (*Mycotaxon* **48**: 415, 1993).

Bahusaganda Subram. (1994) nom. inval., anamorphic *Pezizomycotina*, Hso.≡ eP.3. 2, India; Panama. See Subramanian & Srivastava (*Proc. Indian natn Sci. Acad.* Part B. Biol. Sci. **60**: 277, 1994), Kirschner & Piepenbring (*Fungal Diversity* **21**: 93, 2006).

Bahusakala Subram. (1958), anamorphic *Dothideomycetes*, Hso.≡ eP.39/40. 7, widespread (esp. tropical). See Sigler & Carmichael (*Mycotaxon* **4**: 349, 1976), Matsushima (*Matsush. Mycol. Mem.* **6**, 1989).

Bahusandhika Subram. (1956), anamorphic *Pezizomycotina*, Hso.≡ eP.3. 2, India. See Subramanian (*J. Indian bot. Soc.* **35**: 469, 1956).

Bahusutrabeeja Subram. & Bhat (1977), anamorphic *Pezizomycotina*, Hso.0eH.15. 5, Australia; India. See Subramanian & Bhat (*CJB* **55**: 2204, 1977), McKenzie (*Mycotaxon* **61**: 303, 1997).

Bainieria Arnaud (1952) nom. inval., anamorphic *Pezizomycotina*, Hso.0eH.19. 1, Europe.

Bakanae disease, of rice (*Gibberella fujikuroi*); see gibberellin.

Bakeromyces Syd. & P. Syd. (1917) = Trichosphaeria fide Höhnel (*Annls mycol.* **16**: 77, 1918).

Bakerophoma Died. (1916) nom. dub., anamorphic *Pezizomycotina*. See Sutton (*Mycol. Pap.* **141**, 1977).

Balaniopsis P.M. Kirk (1985), anamorphic *Pezizomycotina*, Hso.0eP.40. 3, paleotropical. See Kirk (*Mycotaxon* **23**: 308, 1985), Whitton *et al.* (*Mycoscience* **43**: 67, 2002).

Balanium Wallr. (1833), anamorphic *Pezizomycotina*, Hso.1eP.8. 1, Europe. See Hughes (*CJB* **39**: 1505, 1961), Whitton *et al.* (*Mycoscience* **43**: 67, 2002), Kendrick (*CJB* **81**: 75, 2003; morphogenesis).

balanoid, acorn-shaped.

Balansia Speg. (1885), Clavicipitaceae. Anamorph *Ephelis*. 27 (on grasses), widespread (subtropical). See Diehl (*USDA agric. Monogr.* **4**, 1950), Morgan-Jones & White (*Mycotaxon* **44**: 89, 1992; culture), Phelps & Morgan-Jones (*Mycotaxon* **50**: 61, 1994; anamorph), White *et al.* (*Mycol.* **87**: 172, 1995; development), White *et al.* (*Mycol.* **88**: 89, 1996; endophytes), Reddy *et al.* (*Mycol.* **90**: 108, 1998; DNA), White *et al.* (*Stud. Mycol.* **45**: 95, 2000), Lewis *et al.* (*Mycol.* **94**: 1066, 2002; Mexico), Lewis *et al.* (*Mycology Series* **19**: 151, 2003), Zhang *et al.* (*Mycol.* **98**: 1076, 2006; phylogeny).

Balansiella Henn. (1904) = Claviceps fide Rogerson (*Mycol.* **62**: 865, 1970).

Balansina G. Arnaud (1918) = Dothidasteromella fide Müller & von Arx (*Beitr. Kryptfl. Schweiz* **11** no. 2, 1962).

Balansiopsis Höhn. (1910), Clavicipitaceae. 3 (on grasses), America. See Diehl (*USDA agric. Monogr.* **4**, 1950), Reddy *et al.* (*Mycol.* **90**: 108, 1998; DNA).

Balazucia R.K. Benj. (1968), Laboulbeniaceae. 2, Mexico; Japan. See Terada (*TMSJ* **21**: 193, 1980), Tavares (*Mycol. Mem.* **9**: 627 pp., 1985).

Balladyna Racib. (1900), Parodiopsidaceae. Anamorph *Tretospora*. *c.* 16 (on *Rubiaceae*), widespread (tropical). See Müller & von Arx (*Beitr. Kryptfl. Schweiz* **11** no. 2, 1962), Eboh & Cain (*CJB* **51**: 61, 1973).

Balladynastrum Hansf. (1941) = Balladynopsis fide

Sivanesan (*Mycol. Pap.* **146**, 1981).

Balladynella Theiss. & Syd. (1918) = Dysrhynchis fide Müller & von Arx (*Beitr. Kryptfl. Schweiz* **11** no. 2, 1962).

Balladynocallia Bat. (1965), Parodiopsidaceae. 3, pantropical. See Sivanesan (*Mycol. Pap.* **146**, 1981; key), Hosagoudar & Abraham (*New Botanist* **20**: 109, 1993).

Balladynopsis Theiss. & Syd. (1918), Parodiopsidaceae. Anamorph *Tretospora*. 13, widespread (tropical). See Sivanesan (*Mycol. Pap.* **146**, 1981; monogr., key), von Arx & Müller (*Sydowia* **37**: 6, 1984; synonym of *Balladyna*).

ballistospore, a forcibly abjected basidiospore. See Nyland (*Mycol.* **41**: 688, 1950); **- discharge**, see Buller (*Researches* **2-6**, 1922-34), Olive (*Science, N.Y.* **146**: 524, 1964), Ingold (*Friesia* **9**: 66, 1969; review). See also Buller's drop, *Itersonilia*.

Ballistosporomyces Nakase, G. Okada & Sugiy. (1989), anamorphic *Sporidiobolales*. 2 (ballistosporic yeasts), Japan. See Nakase *et al.* (*J. gen. appl. Microbiol.* Tokyo **35**: 289, 1989).

Ballocephala Drechsler (1951), Meristacraceae. 3 (on tardigrades), widespread. See Richardson (*TBMS* **55**: 307, 1970), Pohlad & Bernard (*Mycol.* **70**: 130, 1978), Tucker (*Mycotaxon* **13**: 481, 1981; key), Saikawa & Oyama (*TMSJ* **33**: 305, 1992; EM of infection of tardigrades), Saikawa & Sakuramata (*TMSJ* **33**: 237, 1992; zygospores).

Balsamia Vittad. (1831), Helvellaceae. 6, widespread. See Gilkey (*N. Amer. Fl.* **2**, 1954; key N. Am. spp.), Hawker (*Phil. Trans. R. Soc. Lond.* **237**: 429, 1954; key Eur. spp.), Pegler *et al.* (*British truffles*, 1993), O'Donnell *et al.* (*Mycol.* **89**: 48, 1997; phylogeny), Harrington *et al.* (*Mycol.* **91**: 41, 1999; phylogeny), Percudani *et al.* (*Mol. Phylogen. Evol.* **13**: 169, 1999; phylogeny), Agerer (*Mycol. Progr.* **5**: 67, 2006; mycorrhizas), Hansen & Pfister (*Mycol.* **98**: 1029, 2006; phylogeny).

Balsamiaceae E. Fisch. (1897) = Helvellaceae.

Baltomyces Cafaro (1999), ? Asellariaceae. 1 (in *Isopoda*), USA. See Cafaro (*Mycol.* **91**: 517, 1999), White *et al.* (*Mycol.* **98**: 872, 2006; phylogeny).

Balzania Speg. (1899), Thyridiaceae. 1 (from dead wood), S. America. See Rossman *et al.* (*Stud. Mycol.* **42**: 248 pp., 1999).

Bangia Lyngb. (1819), Algae. Algae.

Banhegyia Zeller & Tóth (1960), Patellariaceae. 1, Europe. See Kohlmeyer & Kohlmeyer (*Icones fung. maris*, 1968), Kutorga & Hawksworth (*SA* **15**: 1, 1997; posn).

Bankera Coker & Beers ex Pouzar (1955), Bankeraceae. 2 (mycorrhizal), widespread. See Stalpers (*Stud. Mycol.* **35**: 29, 1993; key), Dickson (*Field Mycology* **1**: 99, 2000; key British spp.), Arnolds (*Coolia* **46** 3, Suppl.: 96 pp., 2003; Netherlands and Belgium).

Bankeraceae Donk (1961), Thelephorales. 6 gen. (+ 2 syn.), 98 spp.

Lit.: Donk (*Persoonia* **3**: 199, 1964) see also under *Hydnaceae*, Baird (*Biblthca Mycol.* **104**: 156 pp., 1986), Baird & Khan (*Brittonia* **30**: 171, 1986), Harrison & Grund (*Mycotaxon* **28**: 419, 1987), Arnolds (*Atti del IV Convegno Internazionale di Micologia, Borgo val di Taro – I, Funghi Atque Loci Natura (Funghi ed Ambiente)*: 163, 1989), Agerer (*Nova Hedwigia* **55**: 501, 1992), Stalpers (*Stud. Mycol.* **35**: 168 pp., 1993), Hibbett *et al.* (*Nature* Lond. **407**: 506, 2000), Kõljalg & Renvall (*Karstenia* **40**: 71, 2000), Hibbett & Binder (*Proc. R. Soc. Lond.* B. Biol. Sci. **269**: 1963, 2002), Ainsworth (*English Nature Research Reports* **600**: 115 pp., 2004; conservation status in UK), Larsson *et al.* (*MR* **108**: 983, 2004), Hrouda (*Czech Mycol.* **57**: 57, 2005; Europe), Hrouda (*Czech Mycol.* **57**: 279, 2005; Europe).

Banksiamyces G.W. Beaton (1982), ? Helotiaceae. 4 (on *Banksia* cones), Australia. See Zhuang (*Mycotaxon* **32**: 97, 1988; ? = *Encoelia* (Sclerotin.)), Fuhrer & May (*Victorian Nat.* **110**: 73, 1993; host specificity).

Bannoa Hamam. (2002), anamorphic *Erythrobasidiales*. 1, Japan. See Hamamoto *et al.* (*Int. J. Syst. Evol. Microbiol.* **52**: 1027, 2002).

Bapalmuia Sérus. (1993), Pilocarpaceae (L). 17, widespread (tropical). See Ekman (*Op. bot.* **127**: 148 pp., 1996), Kalb *et al.* (*Lichenologist* **75**: 281, 2000; key).

Barbariella Middelh. (1949) = Asaphomyces fide Benjamin (*Mycol.* **47**, 1955).

Barbarosporina Ķirulis (1942), anamorphic *Pezizomycotina*, Cac.0eH.?. 1 (on *Rhytisma*), former USSR.

barbate, having one or more groups of hairs; bearded.

Barbatosphaeria Réblová (2008), Sordariomycetes. Anamorphs *Sporothrix*-like, *Ramichloridium*-like. 1, Europe; N. America. See Réblová (*Mycol.* **99**: 723, 2007).

Barbatospora M.M. White, Siri & Lichtw. (2006), Legeriomycetaceae. 1 (in *Simuliidae*), USA. See White *et al.* (*Mycol.* **98**: 335, 2006).

Barcheria T. Lebel (2004), Agaricaceae. 1, Australia. See Lebel *et al.* (*MR* **108**: 210, 2004).

Barclayella Dietel (1890) = Chrysomyxa fide Dietel (*Nat. Pflanzenfam.* **6**, 1928).

Bargellinia Borzí (1888) = Wallemia fide von Arx (*Gen. Fungi Sporul. Cult.* Edn 3, 1981).

Barklayella Sacc. (1892) [non *Barclayella* Dietel 1890] = Neobarclaya fide Nag Raj (*CJB* **56**: 706, 1978).

Barlaea Sacc. (1889) [non *Barlaea* Rchb. 1876, *Orchidaceae*] ≡ Barlaeina.

Barlaeina Sacc. & P. Syd. (1899) = Lamprospora fide Eckblad (*Nytt Mag. Bot.* **15**: 1, 1968).

barm (1) the froth on the surface of fermenting malt liquids; (2) baker's yeast.

Barnettella D. Rao & P. Rag. Rao (1964), anamorphic *Pezizomycotina*, St.#eP.43. 3, India. See Satyanarayana & Rao (*Mycopath. Mycol. appl.* **45**: 267, 1971), Rao & Rao (*Indian Phytopath.* **26**: 233, 1973; key).

Barnettia Bat. & J.L. Bezerra (1962) = Microcallis fide von Arx & Müller (*Stud. Mycol.* **9**, 1975).

Barr Bigelow (Margaret Elizabeth; 1923-2008; Canada, later USA). BS (1950 then MS (1952), University of British Columbia; PhD under Wehmeyer, University of Michigan (1957); Instructor (1957) then Professor and, eventually, Ray Ethan Torrey Professor (1986-1989), University of Massachusetts, Amherst. An outstanding systematist and expert on the bitunicate ascomycetes, with enormous experience of the diversity of that group; contributed to activities of the Mycological Society of America with time and money, as Editor-in-Chief of *Mycol.* (1976-1980), Vice President (1979-1980) and President (1980-1981). Specimens are mostly in the fungal reference collection of New York Botanic Garden (**NY**) with some in **DAOM**. *Publs.* The Diaporthales in North America with emphasis on *Gnomonia* and its segregates. *Mycologia Memoir* (1978); *Prodromus to*

Class Loculoascomycetes (1987); Some dictyosporous genera and species of Pleosporales in North America. *Memoirs of the New York Botanical Garden* (1990); Prodromus to nonlichenized, pyrenomycetous members of class Hymenoascomycetes. *Mycotaxon* (1990); Melanommatales (Loculoascomycetes). *North American Flora* Series II (1990).

barrage, the space between two mycelia which have an aversion for one another (Vandendries & Brodie, 1933). Cf. zone lines.

Barrella Ahn & Shearer (1995) = Pseudoyuconia fide Eriksson & Hawksworth (*SA* **15**: 139, 1997).

Barria Z.Q. Yuan (1994), ? Phaeosphaeriaceae. 1, China. See Yuan (*Mycotaxon* **51**: 313, 1994), Jaklitsch *et al.* (*Öst. Z. Pilzk.* **11**: 93, 2002).

Barrina A.W. Ramaley (1997), Coniochaetaceae. Anamorph *Phialophora*-like. 1, USA. See Ramaley (*Mycol.* **89**: 962, 1997), Huhndorf *et al.* (*Mycol.* **96**: 368, 2004), García *et al.* (*MR* **110**: 1271, 2006; phylogeny).

Barrmaelia Rappaz (1995), Xylariaceae. Anamorph *Libertella*-like. 6, Europe; N. America. See Rappaz (*Mycol. Helv.* **7**: 99, 1995; monogr.), Winka & Eriksson (*Mycoscience* **41**: 97, 2000; DNA).

Barssia Gilkey (1925), Helvellaceae. 2, widespread. Close to *Balsamia*. See Ławrynowicz & Skirgiełło (*Acta Mycologica* Warszawa **20**: 277, 1986), Kimbrough *et al.* (*Mycol.* **88**: 38, 1996; ultrastr.), O'Donnell *et al.* (*Mycol.* **89**: 48, 1997; phylogeny), Landvik *et al.* (*Mycol.* **91**: 278, 1999), Percudani *et al.* (*Mol. Phylogen. Evol.* **13**: 169, 1999; phylogeny), Hansen & Pfister (*Mycol.* **98**: 1029, 2006; phylogeny).

Bartalinia Tassi (1900), anamorphic *Amphisphaeriaceae*, Hso.≡ eH.19. 16, widespread. See Morgan-Jones *et al.* (*CJB* **50**: 877, 1972), Roux & Van Warmelo (*MR* **94**: 109, 1990; conidioma ontogeny), Jeewon *et al.* (*MR* **107**: 1392, 2003; phylogeny), Wong *et al.* (*CJB* **81**: 1083, 2003; ultrastr.).

Bartaliniopsis S.S. Singh (1974) = Doliomyces fide Sutton (*Mycol. Pap.* **141**, 1977).

Bartheletia Arnaud (1954) nom. inval., anamorphic *Pezizomycotina*, Hsp.?.?. 1, France.

Bartlettiella D.J. Galloway & P.M. Jørg. (1990), Lecanorales (L). 1, New Zealand. See Galloway & Jørgensen (*N.Z. Jl Bot.* **28**: 5, 1990).

Barubria Vězda (1986), Ectolechiaceae (L). 1 (on leaves), Africa; C. America. See Vězda (*Folia geobot. phytotax.* **21**: 207, 1986), Lücking (*Phyton* Horn **39**: 131, 1999).

Barya Fuckel (1864) [non *Barya* Klotzsch 1854, *Begoniaceae*] ≡ Neobarya.

Baryeidamia H. Karst. (1888) = Papulaspora fide Saccardo (*Syll. fung.* **9**: 339, 1891).

Baryella Rauschert (1988) ≡ Neobarya.

basal body (1) (of *Blastocladiaceae*), the part of the thallus fixed to the substratum by rhizoids at the lower end (Indoh, 1940); (2), see blepharoplast.

basal frill (of a spore), the apical part of a conidiogenous cell, or basal part of a cell which is carried away with the detached conidium following rhexolytic secession (Sutton, *CJB* **45**: 1251, 1967). Cf. marginal frill.

Basauxia Subram. (1995), anamorphic *Pezizomycotina*, Hso.≡ eP.1/21. 1, Malaysia. See Subramanian (*Kavaka* **20/21**: 58, 1992/1993).

basauxic (of conidiophores), elongating by a basal growing point (Hughes, *CJB* **31**: 650, 1953). See meristem arthrospore.

Basavamyces V.B. Hosag. (2005), Meliolaceae. 1, India. See Biju *et al.* (*Nova Hedwigia* **80**: 480, 2005).

base ratio, see Molecular biology.

Basiascella Bubák (1914) ? = Piggotia fide Sutton (*Mycol. Pap.* **141**, 1977).

Basiascum Cavara (1888) = Spilocaea fide Hughes (*CJB* **36**: 727, 1958).

Basidydyma Cif. (1962), anamorphic *Pezizomycotina*, Hso.0eH.10. 1, Dominican Republic. See Ciferri (*Atti Ist. bot. Univ. Lab. crittog. Pavia* sér. 5 **19**: 94, 1962).

Basidiella Cooke (1878) = Aspergillus fide Subramanian (*Hyphomycetes*, 1971) See, Seifert (*TBMS* **85**: 123, 1985).

Basidioascus Matsush. (2003), Dipodascaceae. Anamorph *Geotrichum*-like. 1, Australia. See Matsushima (*Matsush. Mycol. Mem.* **10**: 98, 2001).

Basidioblastomycetes. Class used by Moore (*Bot. Mar.* **23**: 361, 1980; key 13 gen.) for *Malasseziales* and *Sporobolomycetales*, and placed by him in *Blastomycota*.

Basidiobolaceae Engl. & E. Gilg (1924), Basidiobolales. 1 gen. (+ 1 syn.), 4 spp.

Lit.: Dykstra (*Mycol.* **86**: 494, 1994), Dykstra & Bradley-Kerr (*Mycol.* **86**: 336, 1994), Humber (*J. gen. Microbiol.*, 1995), Nagayama *et al.* (*Mycol.* **87**: 203, 1995), Jensen *et al.* (*Fungal Genetics Biol.* **24**: 325, 1998), Tanabe *et al.* (*Mol. Phylogen. Evol.* **30**: 438, 2004).

Basidiobolales Caval.-Sm. (1998). Zygomycota. 1 fam., 1 gen., 4 spp. Fam.:
Basidiobolaceae
The order may belong in the *Chytridiales*.

Lit.: Nagahama *et al.* (*Mycol.* **87**: 203, 1995), Cavalier-Smith (*Biol. Rev.* **73**: 230, 1998), James *et al.* (*Mycol.* **98**: 860, 2006; molecular phylogeny), Hibbett *et al.* (*MR* **111**: 109, 2007).

Basidiobolus Eidam (1886), Basidiobolaceae. 4, widespread. *B. ranarum*, a widespr. saprobe; causal agent of subcutaneous phycomycosis in humans. See Drechsler (*Mycol.* **48**: 655, 1956), Benjamin (*Aliso* **5**: 223, 1962; key), Greer & Friedman (*Sabouraudia* **4**: 231, 1966), Srinivasan & Thirumalachar (*Mycopath. Mycol. Appl.* **33**: 56, 1967), Coremans-Pelseneer (*Acta Zool. Path.* **60**: 1, 1974; biology), Chien (*TMSJ* **28**: 445, 1987), Cochrane *et al.* (*Mycol.* **81**: 504, 1989; isozymes), Nelson *et al.* (*Exp. Mycol.* **14**: 197, 1990; ribosomal DNA), Dykstra (*Mycol.* **86**: 494, 1994; spore formation), Dykstra & Bradley-Kerr (*Mycol.* **86**: 336, 1994; ultrastr.), James *et al.* (*Nature* **443**: 818, 2006; phylogeny), Liu *et al.* (*BMC Evolutionary Biology* **6**: 1, 2006; phylogeny), van den Berk *et al.* (*BMC infectious diseases* **6**: 140, 2006; basidiobolomycosis).

Basidiobotrys Höhn. (1909) = Xylocladium fide Jong & Rogers (*Tech. Bull. Wash. agric. Exp. Stn* **71**, 1972).

basidiocarp, see basidioma.

Basidiodendron Rick (1938), Auriculariales. *c.* 15, widespread. See Wells (*Mycol.* **51**: 541, 1960; key), Luck-Allen (*CJB* **41**: 1033, 1963; key N. temp. spp.), Oberwinkler (*Ber. bayer. bot. Ges.* **36**: 42, 1963; Bavarian spp.), Wells & Raitviir (*Mycol.* **67**: 909, 1975), Weiss & Oberwinkler (*MR* **105**: 403, 2001; phylogeny), Kotiranta & Saarenoksa (*Annales Botanici Fennici* **42**: 11, 2005; Finland).

basidiograph, the straight-line graph obtained by plotting the ratio of the length (l) to the width (w) against the length of the basidia of a species of agaric (Corner, 1947: 214). Cf. sporograph.

basidiole, a basidium-like hymenial element that lacks sterigmata because it is either young or permanently sterile; best restricted to immature basidia fide Singer (1962).

Basidiolum Cienk. (1861), ? Zoopagales. 1. See White (*MR* **107**: 245, 2003).

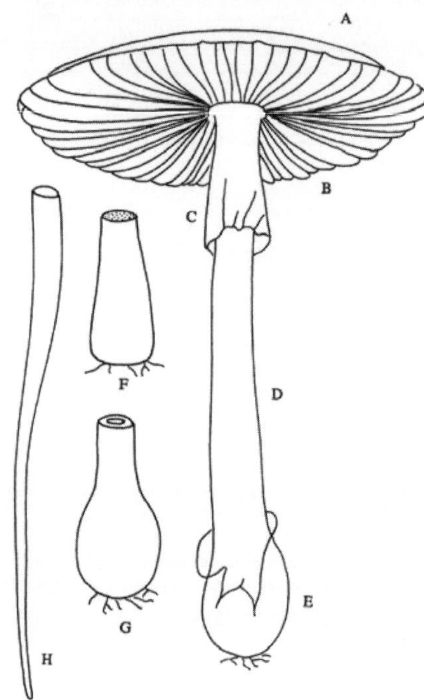

Fig. 6. Basidioma (sporophore) of *Amanita phalloides*. A, pileus (cap); B, lamellae (gills); C, annulus (ring); D, stipe (stalk); E, volva. Other types of stipe: F, stuffed, with base truncate; G, hollow, with base bulbous; H, solid with base radicate.

basidioma (pl. **-ata**), a basidium-producing organ; basidiome (Donk, *Taxon* **18**: 666, 1969); basidiocarp; carpophore; fruit-body; hymenophore; sporophore. See Fig. 4.

basidiomycete, one of the *Basidiomycota*.

Basidiomycetes G. Winter (1880), see *Agaricomycotina.*

Basidiomycota R.T. Moore (1980). Fungi. 16 class., 52 ord., 177 fam., 1589 gen., 31515 spp. (Basidiomycotina, basidiomycetes). 3 subphyla and 6 'unassigned' classes, cosmop. The diagnostic character of this phylum is the presence of a basidium (q.v.) bearing basidiospores. A typical basidium is aseptate and has four 1-celled haploid basidiospores (ballisto- or statismospores) dispersed by air currents but the basidium may be transversely or longitudinally septate and the number of spores (which may be statismospores) are occasionally fewer or more than four. Other diagnostic characters are clamp connexions (Fig. 9), dolipore septa (Fig. 13), and a double-layered wall, lamellate and electron-opaque in electron microscopy. *Basidiomycota* are typically mycelial but some are yeasts (or have a yeast-like state). Such yeasts may be distinguished from ascomycetous yeasts by the morphology of the bud scars, giving a red colour with diazonium blue B, being urease +, having a high GC percentage (see base ratio), and other ultrastructural characters (see Moore, *Bot. Mar.* **23**: 361, 1980; vonArx, 1987) some of which may apply to basidiomycetes in general; molecular sequence data can also clearly place other fungi in the phylum.

The typical life cycle (on which there are variations) involves the germination of the basidiospore to give a septate primary (haploid) mycelium which may produce 'oidia' but anamorphic states (except for yeasts, rusts, and smuts) have been neglected (see Kendrick & Watling, *in* Kendrick (Ed.), *The whole fungus* **2**: 473, 1979). Later by diploidization, the homo- or heterothallic primary mycelium becomes a secondary (dikaryotic) mycelium which frequently has clamp connexions. There is nuclear fusion in the young basidium and meiosis before basidiospore development. The mycelium may be perennial in soil or wood and may form 'fairy rings', sclerotia, rhizomorphs or mycorrhizas. The basidiomata are typically macroscopic and take a variety of forms.

While there is widespread agreement on the use of certain families within the *Basidiomycota* there is less uniformity in grouping the families into orders and esp. supraordinal taxa because of the lack of good differential characters. During recent years the acceptance of supraordinal taxa by specialists has been based on microscopical basidial characters. Talbot (*Taxon* **17**: 620, 1968; 1971), after separating the *Uredinales* and *Ustilaginales* as the *Teliomycetes* grouped the remaining orders as the *Phragmobasidiomycetes* and *Holobasidiomycetes* and distinguished 'gasteromycetes' within the last. Donk (*Proc. K. nederl. Akad. Wet.* C **75**: 365; **76**: 109, 1972-73) separated the *Ustilaginales* as the *Hemibasidiomycetes* and classified the remaining forms as the *Heterobasidiomycetes* (which included the *Uredinales*) and the *Homobasidiomycetes*. Shaffer (*Mycol.* **67**: 1, 1975) advocated no supraordinal groupings of the 14 orders (59 fams.) of *Basidiomycetes* which he recognized. Khan & Kimbrough (*Mycotaxon* **15**: 103, 1982) proposed the adoption of four classes based mainly on septal ultrastructure: *Teliomycetes*, *Hemibasidiomycetes*, *Phragmobasidiomycetes* and *Holobasidiomycetes*.

The classification in this edition of the *Dictionary* is based on a series of major phylogenetic studies of fungi under the umbrella of the 'Deep Hypha' and AFTOL ('Assembling the Fungal Tree of Life') projects. See esp. Blackwell *et al.* (*Mycol* **98**: 829, 2006), Hibbett *et al.* (*MR* **: 111**: 509, 2007), James et al. (*Nature* **443**: 818, 2006). The pragmatic subdivisions in the 9th Edition can no longer be supported. The presence or absence of a macroscopic basidioma, the life-form and life-style (including host specialization) and the traditional 'hymenomycetes' and 'gasteromycetes' can only be regarded as informal and not monophyletic categories; a consequence of this is that some of the older literature cited may be in conflict with the taxon it is cited under.

Three Subphyla are accepted here. However, the *Entorrhizomycetes* and *Wallemiomycetes* are not referred to any of these subphyla, and a number of

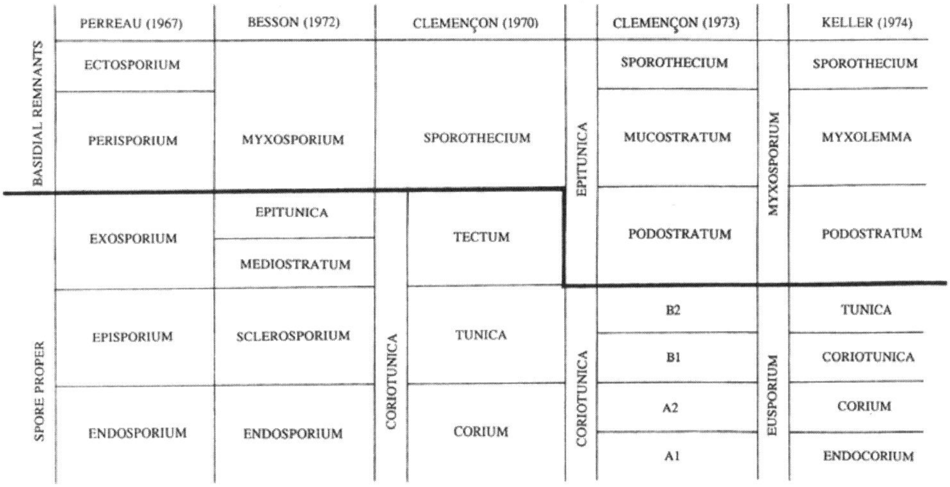

	PERREAU (1967)	BESSON (1972)	CLEMENÇON (1970)		CLEMENÇON (1973)		KELLER (1974)
BASIDIAL REMNANTS	ECTOSPORIUM			EPITUNICA	SPOROTHECIUM	MYXOSPORIUM	SPOROTHECIUM
BASIDIAL REMNANTS	PERISPORIUM	MYXOSPORIUM	SPOROTHECIUM	EPITUNICA	MUCOSTRATUM	MYXOSPORIUM	MYXOLEMMA
SPORE PROPER	EXOSPORIUM	EPITUNICA / MEDIOSTRATUM	TECTUM	EPITUNICA	PODOSTRATUM	MYXOSPORIUM	PODOSTRATUM
SPORE PROPER	EPISPORIUM	SCLEROSPORIUM	CORIOTUNICA: TUNICA	CORIOTUNICA	B2 / B1	EUSPORIUM	TUNICA / CORIOTUNICA
SPORE PROPER	ENDOSPORIUM	ENDOSPORIUM	CORIOTUNICA: CORIUM	CORIOTUNICA	A2 / A1	EUSPORIUM	CORIUM / ENDOCORIUM

Fig. 7. Basidiospore wall-layer terminology, based on transmission electron microscopy (TEM) sections.

genera could not be assigned with confidence to any family.

Subphyl.:

(1) **Agaricomycotina**

(2) **Pucciniomycotina**

(3) **Ustilaginomycotina**

Basidiomycotina Ainsw. (1966), see *Basidiomycota*, *Agaricomycotina*, *Pucciniomycotina*, *Ustilaginomycotina*.

basidiophytes, Cain's (1972) term for hypothetical autotrophic ancestors of basidiomycetes; see Phylogeny. Cf. ascophyte.

Basidiopycnis Oberw., R. Kirschner, R. Bauer, Begerow & Arenal (2006), Phleogenaceae. 1, Germany. See Oberwinkler *et al.* (*Mycol.* **98**: 639, 2006), Hausner *et al.* (*Mycotaxon* **103**: 279, 2008).

Basidioradulum Nobles (1967), Schizoporaceae. 1, Europe; N. America. See Nobles (*Mycol.* **59**: 192, 1967), Legon (*Mycologist* **19**: 81, 2005) Also considered a syn. of *Hyphodontia* s.l.

basidiospore, a propagative cell (typically a ballistospore but in gasteromycetes a statismospore) containing one or two haploid nuclei produced, after meiosis, on a basidium (Fig. 8). The colour, form and ornamentation of the basidiospore are fundamental to basidiomycete classification and an essential part of any specific description. Greater use of electron microscopy has revealed an increasing complexity of the wall layers or teguments (for a comparison of the terminologies applied to these layers see Fig. 5). Also see Spore wall.

Lit.: Perreau (*Ann. Sci. nat., Bot.* sér. 12, **8**: 639, 1967; homobasidiomycetes), Clemençon (*Z. Pilzkde* **36**: 113, 1970; wall ultrastr.), Pegler & Young (*Beih. Nova Hedw.* **35**, 1971; morphology in *Agaricales*), Kühner (*Persoonia* **7**: 217, 1973), Locquin (*Bull. Soc. bot. Fr., Coll. Palyn.*: 135, 1975).

Basidiosporites Elsik (1968), Fossil Fungi. 1 (Paleocene), USA.

basidium (pl. **-ia**), (1) the cell or organ, diagnostic for basidiomycetes, from which, after karyogamy and meiosis, basidiospores (generally 4) are produced externally each on an extension (sterigma, q.v.) of its wall (Fig. 8); (2) a conidiophore or phialide (obsol.).

The confused terminology applied to basidia (sense 1) and their parts has been traced by Clémençon (*Z. Mykol.* **54**: 3, 1988) and is analyzed by Talbot (*TBMS* **61**: 497, 1973) whose recommended usage (basically that of Donk) and synonymy is adopted in the series of definitions which follow (see Fig. 8): **pro-**, the morphological part or developmental stage of the basidium in which karyogamy occurs; primary basidial cell; probasidial cyst; **hypo-** (Martin) p.p.; teliospore of Uredinales. **meta-**, (1) the morphological part or developmental stage in which meiosis occurs; hypo- (Martin) p.p.; **epi-** (Martin) p.p.; promycelium of *Pucciniales*. (When the whole meta- includes pro- remnants the distal and functional part may be distinguished as a **pario-** (Talbot, 1973).) (2) See proto- below. **holo-**, a basidium (e.g. of *Agaricus*) in which the meta- is not divided by primary septa (see septum) but may become adventitiously septate (see septum) (Talbot, *Taxon* **17**: 625, 1968). A holo- may be a **sticho-**, cylindrical, with nuclear spindles longitudinal and at different levels, or a **chiasto-**, clavate, with nuclear spindles across the basidium and at the same level (see Fig. 8A,B). **phragmo-**, a basidium in which the meta- is divided by primary septa, usually cruciate (e.g. *Tremella*) or transverse (e.g. *Auricularia*) (Talbot, 1968).

Among other terms applied to basidia are: **apo-**, one with non-apiculate spores borne symmetrically on the sterigmata and not forcibly discharged (Rogers, *Mycol.* **39**: 558, 1947); **auto-**, one with spores borne asymmetrically and forcibly discharged; **endo-**, one developing within the basidioma, as in gasteromycetes; **epi-**, Martin's term for protosterigma, see sterigma; **hetero-**, a basidium of the *Heterobasidiomycetes*, usually a phragmo-; **homo-**, a

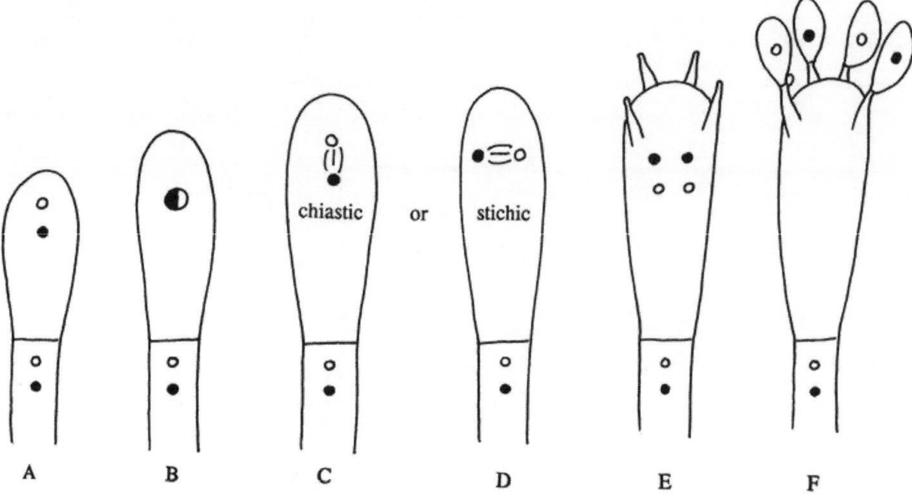

Fig. 8. Basidiospore development (diagrammatic). A-E meiosis (C, stichic, and D, chiastic). E, diploid probasidium; F, basidium (metabasidium) with four basidiospores on sterigmata.

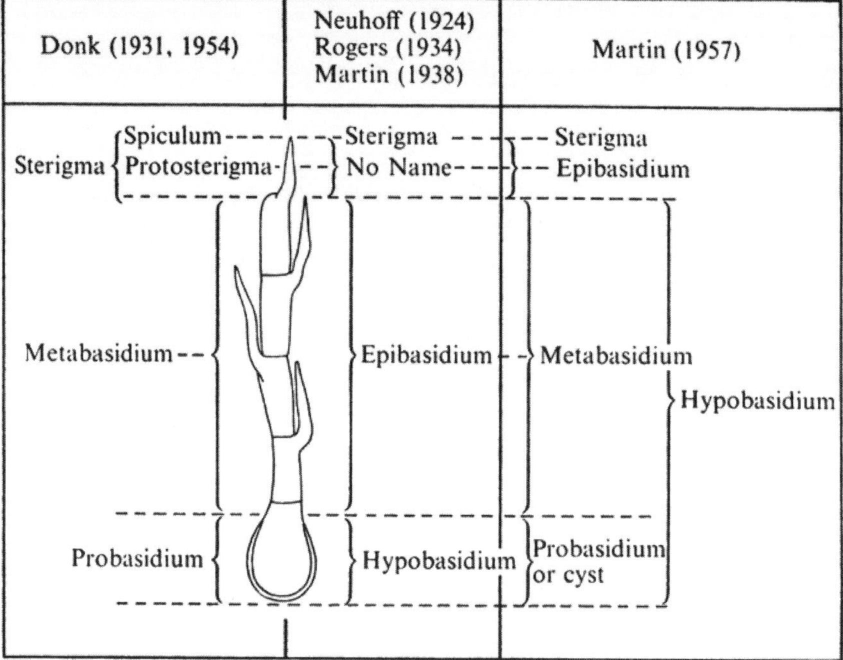

Fig. 9. Basidium terminology, to compare the terminology of different authors, illustrated with reference to the *Septobasidium*-type (Talbot, 1973). Note that in this extreme case the metabasidium of Donk coincides with that of Martin.

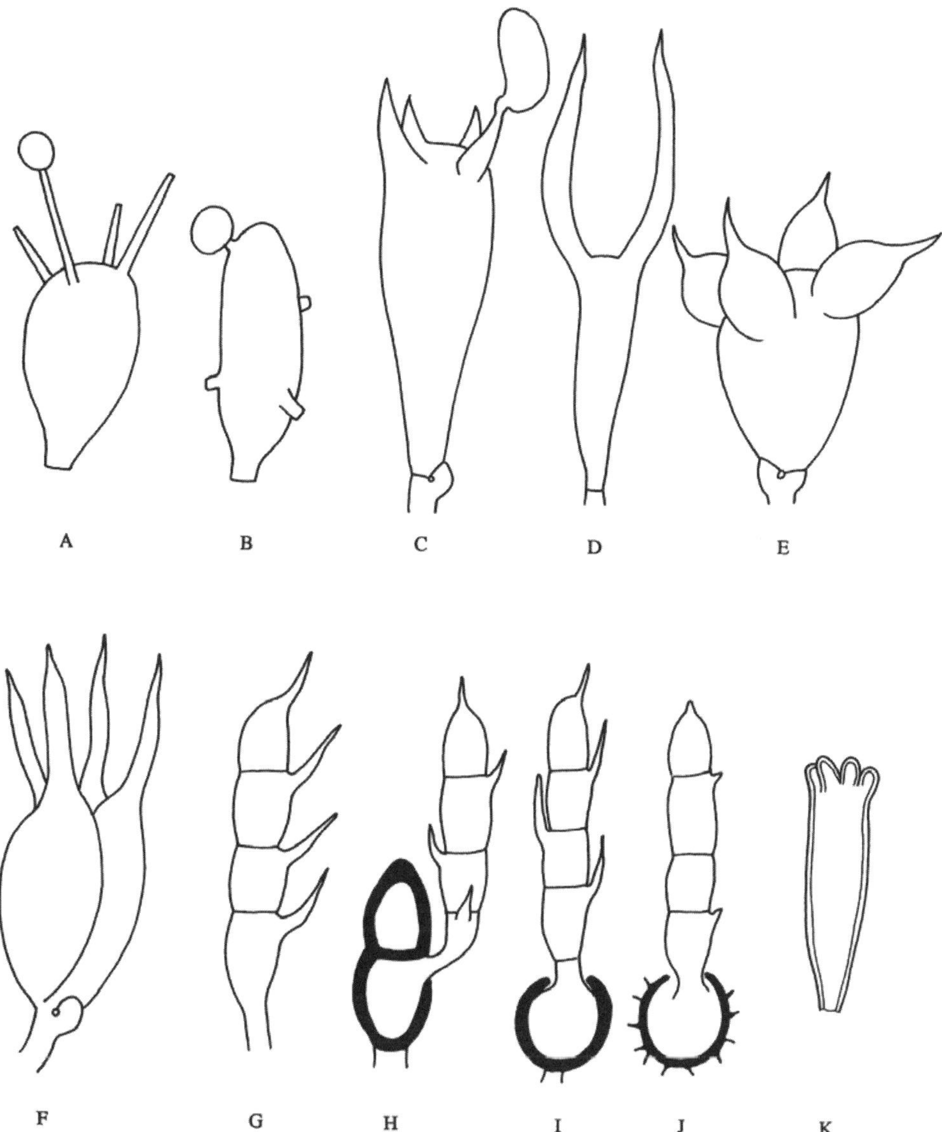

Fig. 10. Basidial types. A-E, holobasidial (A-B, apobasidial; C-E, autobasidial). A, *Lycoperdales*; B, *Tulostomatales*; C, *Agaricales*; D, *Dacrymycetales*; E, *Tulasnellales*. F-K, phragmobasidial (F-G, *Basidiomycetes*; H-I, *Teliomycetes*; J-K, *Ustomycetes*). F, *Tremellales*; G, *Auriculariales*; H, *Uredinales*; I, *Septobasidiales*; J, *Ustilaginales*; K, *Cryptobasidiales*.

basidium of the *Homobasidiomycetes*, usually a holo-; **hypo-**, = pro- (Donk), meta- (Martin); (of *Septobasidium*) = basidium (Martin); **pleuro-**, one relatively broad at the base and with bifurcated spreading 'roots', as in *Pleurobasidium* (Donk); **proto-**, a primitive basidium; the opposite of meta- in the sense of changed or degenerate basidium; **repeto-**, see Chadefaud (*Rev. mycol.* **39**: 173, 1975); **sclero-**, the thick-walled, encysted, gemma-like pro- of the *Pucciniales* (teliospore) and the *Auriculariales* (Janchen,

1923). See also Wells & Wells (*Basidium and basidiocarp evolution, cytology, function and development* 1982).

Basidopus Earle (1909) = Mycena fide Singer (*Agaric. mod. Tax.* edn 3, 1975).

Basifimbria Subram. & Lodha (1968), anamorphic *Xylariales*, Hso.0eH.13. 1 (coprophilous), India. See von Arx (*Sydowia* **35**: 10, 1982; syn. with *Dicyma*).

basifugal, development from the base up, acropetal.

Basilocula Bubák (1914) ? = Ceuthospora Grev. fide Sutton (*Mycol. Pap.* **141**, 1977).

basionym (basinym, basonym) (in nomenclature, q.v.), the name-bringing or epithet-bringing synonym on which a new transfer or new combination is based. Donk (*Bull. Jard. bot. Buitenz.* sér. 3 **18**: 274, 1949) uses **isonym** for a name derived from a basionym. Cf. synisonym, synonym.

basipetal, describes a chain of conidia in which new spores are formed at the base, the oldest at the apex, cf. acropetal.

Basipetospora G.T. Cole & W.B. Kendr. (1968), anamorphic *Monascus*, Hso.0eH.36. 1, widespread. See Cole & Kendrick (*CJB* **46**: 991, 1968), Stchigel *et al.* (*Stud. Mycol.* **50**: 299, 2004).

Basipilus Subram. (1961) = Seimatosporium fide Sutton (*Mycol. Pap.* **88**, 1963), Shoemaker (*CJB* **42**: 411, 1964).

Basisporium Molliard (1902) = Nigrospora fide Mason (*Mycol. Pap.* **3**: 60, 1933).

Basitorula Arnaud (1954) = Gliomastix fide Dickinson (*Mycol. Pap.* **115**, 1968).

basket fungi, *Clathrus* spp.

basocatenate (of conidia), formed in chains with the youngest conidium at the basal or proximal end of the chain.

Basramyces Abdullah, Abdulk. & Goos (1989), anamorphic *Pezizomycotina*, Hso.0-≡ hP.1. 1, Germany; Iraq. See Abdullah *et al.* (*Int. J. Mycol. Lichenol.* **4**: 182, 1989).

Bassi (Agostino; 1773-1856; Italy). A lawyer who later became a farmer. He elucidated the etiology of muscardine disease of silkworms (*Beauveria bassiana*) and was thereby the first to demonstrate by experiment pathogenicity of a fungus for an animal. *Publs. Del Mal del Segno* (1835-1836) [for English translation of part I, 1835, see *Phytopathological Classics* **10**, 1958, portrait]; *Opere do Agostino Bassi*, Pavia (1925) [collected works]. *Biogs, obits etc.* [bicentenary tributes] Verona (*Agostino Bassi nel 200 Anno dalla Nascita*, 1973) [bibliography, portrait]; Porter (*Bacteriological Reviews* **37**: 284, 1973) [portrait].

Bastien treatment, treatment for amanitin poisoning, involving: (1) twice-daily injection of 1g vitamin C, (2) 2 capsules of nifurazide, three times a day, (3) 2 tablets of dihydrostreptomycin, three times a day, (4) penicillin therapy, (5) maintenance of fluid and electrolyte balance.

Batarrea, see *Battarrea*.

Batcheloromyces Marasas, P.S. van Wyk & Knox-Dav. (1975), Teratosphaeriaceae. 4, widespread. See Sutton & Pascoe (*MR* **92**: 210, 1989), Taylor *et al.* (*MR* **103**: 1478, 1999), Taylor *et al.* (*MR* **107**: 653, 2003), Crous *et al.* (*CBS Diversity Ser.* **2**, 2004), Crous *et al.* (*Stud. Mycol.* **58**: 1, 2007; phylogeny).

Bathelium Ach. (1803) nom. rej. = Trypethelium.

Bathelium Trevis. (1861) nom. cons., Trypetheliaceae. *c.* 30. fide Harris (*More Florida Lichens*, 1995), Prado *et al.* (*MR* **110**: 511, 2006; phylogeny), Aptroot *et al.* (*Biblthca Lichenol.* **97**, 2008; Costa Rica).

Bathyascus Kohlm. (1977), Halosphaeriaceae. 5 (marine), widespread. See Ravikumar & Vittal (*MR* **95**: 370, 1991).

Bathystomum Füisting (1868) ? = Massaria fide Saccardo (*Syll. fung.* 326 **19**: 142, 1910).

Batista (Augusto Chaves; 1916-1967; Brazil). Professor of Phytopathology, Escola Superior de Agricultura; founder and Director, Instituto de Micologia da Universidade do Recife [now Universidade Federal de Pernambuco]. A visit to IMI in 1951 (Batista, *Boletim da Secretaria de Agricultura, Industria e Comercio do Estado de Pernambuco* **22**: 29, 1953) influenced his decision to establish the Instituto de Micologia in Recife. Pioneered study of tropical ascomycetes in Brazil. Most collections in Recife (**URM**), some in **IMI**. *Publs.* Author of more than 600 papers and monographs, mainly accounts and revisions of Brazilian ascomycetes, often with various co-workers; most of these works appeared as a numbered series, *Publicações. Instituto de Micologia da Universidade do Recife* [IMUR] (1954 on), some published in a range of journals, others in a long series of offset printed pamphlets published by the IMUR itself [*Publicação* **301** lists the titles of nos 1-300; *Publicação* **674** the titles of nos 301-673]. *Biogs, obits etc.* Carmeiro (*Mycol.* **60**: 1137, 1969) [portrait]; Singer (*Sydowia* **22**: 343, 1969); Aguilar (*Acta Amazonica* **18**: 39, 1988 [list of Batista types in INPA]); Da Silva & Minter (*Mycological Papers* **169**, 1995 [checklist of 3,340 species, with host, substratum and state indexes, and a bibliography); Grummann (1974: 771).

Batistaella Cif. (1962) = Phaeosaccardinula fide von Arx & Müller (*Stud. Mycol.* **9**, 1975).

Batistamnus J.L. Bezerra & Cavalc. (1967) nom. dub., Fungi. See von Arx & Müller (*Stud. Mycol.* **9**, 1975).

Batistia Cif. (1958), Batistiaceae. Anamorph *Acrostroma*. 1, Brazil. See Samuels & Rodrigues (*Mycol.* **81**: 52, 1989), Huhndorf *et al.* (*Mycol.* **96**: 368, 2004; phylogeny).

Batistiaceae Samuels & K.F. Rodrigues (1989), ? Sordariomycetes (inc. sed.). 2 gen., 2 spp.

Lit.: Samuels & Rodrigues (*Mycol.* **81**: 52, 1989), Huhndorf *et al.* (*Mycol.* **96**: 368, 2004), Lumbsch & Huhndorf (*MR* **111**: 1064, 2007).

Batistina Peres (1961), anamorphic *Pezizomycotina*, Cpt.1eH.?. 1, Brazil. See Peres (*Publções Inst. Micol. Recife* **317**: 6, 1961).

Batistinula Arx (1960), Asterinaceae. Anamorph *Triposporium*. 1, Brazil. See von Arx (*Publções Inst. Micol. Recife* **287**: 4, 1960), Hosagoudar *et al.* (*Journal of Mycopathological Research* **39**: 61, 2001).

Batistopsora Dianese, R.B. Medeiros & L.T.P. Santos (1993), Phakopsoraceae. 1 (on *Annona* (*Annonaceae*)), neotropics. Very similar to *Phakopsora*. See Hernández *et al.* (*Sydowia* **57**: 189, 2005; Guyana), Aime (*Mycoscience* **47**: 112, 2006).

Batistospora J.L. Bezerra & M.P. Herrera (1964), Pezizomycotina. 1, Brazil. See von Arx & Müller (*Stud. Mycol.* **9**, 1975).

Batkoa Humber (1989), Entomophthoraceae. 9 (pathogens of *Homoptera*, *Hemiptera* and other *Insecta*), widespread. See Villacarlos & Keller (*Phillip. Ent.* **11**: 81, 1997), Keller & Petrini (*Sydowia* **57**: 23, 2005; key), Keller (*Sydowia* **58**: 38, 2006; Switzerland), Huang *et al.* (*Mycotaxon* **100**: 227, 2007; USA).

Batrachochytrium Longcore, Pessier & D.K. Nichols (1999), Rhizophydiales. 1 (pathogen of amphibian skin), widespread. See Berger *et al.* (*Proc. natn Acad. Sci. U.S.A.* **95**: 9031, 1998), Pessier *et al.* (*Vet. Diagn Invest.* **11**: 194, 1999), Morehouse *et al.* (*Mol. Ecol.* **12**: 395, 2003), Boyle *et al.* (*Diseases of Aquatic Organisms* **60**: 141, 2004; PCR detection), Weldon *et al.* (*Emerging Infectious Diseases* **10**: 2100, 2004; origin).

Batschiella Kirschst. (1938) = Clypeoporthella fide Barr (*Mycol. Mem.* **7**: 232 pp., 1978).

Battareopsis, see *Battarreopsis*.

Battarraea, see *Battarrea*.

Battarraeastrum R. Heim & T. Herrera (1960) ≡ Battarreoides.

Battarraeoides, see *Battarreoides*.

Battarrea Pers. (1801), Agaricaceae. 3, widespread (esp. warmer areas). The volvate stipe to 30 cm. See Maublanc & Malençon (*BSMF* **46**: 43, 1930), Rea (*Mycol.* **34**: 563, 1942), Long (*Mycol.* **35**: 546, 1943), Esqueda *et al.* (*Mycotaxon* **82**: 207, 2002; Mexico).

Battarreaceae Corda (1842) = Agaricaceae.

Battarreoides T. Herrera (1953), Agaricaceae. 1, N. America (deserts). See Petrak (*Sydowia* **9**: 591, 1955).

Battarreopsis Henn. (1902) = Dictyocephalos fide Long & Plunkett (*Mycol.* **32**: 696, 1940).

Battarrina (Sacc.) Clem. & Shear (1931), Hypocreales. 1, Europe. Probably in the *Bionectriaceae*, but the only available material is in poor condition. See Rossman *et al.* (*Stud. Mycol.* **42**: 248 pp., 1999).

Bauch test, A macroscopic test for determining whether monosporidial lines of smuts are compatible (the mixed culture has white aerial mycelium) or incompatible (aerial mycelium absent) (Bauch, *Biol. Zbl.* **42**: 9, 1922).

Baudoinia J.A. Scott & Unter. (2007), Capnodiales. 1, widespread. *B. compniacensis*, causing black growth on walls of distilleries etc. See Ewaze *et al.* (*MR* **111**: 1422, 2007; physiology), Scott *et al.* (*Mycol.* **99**: 592, 2007).

Bauerago Vánky (1999), Microbotryaceae. 8 (in seeds of *Commelinaceae*, *Cyperaceae*, *Juncaceae*), widespread. See Vánky (*Mycotaxon* **70**: 44, 1999; key), Denchev (*Mycotaxon* **87**: 127, 2003), Aime *et al.* (*Mycol.* **98**: 896, 2006; phylogeny).

Bauhinus R.T. Moore (1992) = Microbotryum fide Vánky (*Mycotaxon* **48**: 41, 1993), Bauer & Oberwinkler (*Mycotaxon* **64**: 312, 1997) but see, Denchev *et al.* (*Mycol. Balcan.* **3**: 71, 2006).

Baumanniella Henn. (1897) = Physalacria fide Corner (*Ann. Bot. Mem.* [A monograph of Clavaria and allied genera] **1**, 1950).

Baumiella Henn. (1903) = Leptosphaeria fide von Arx & Müller (*Stud. Mycol.* **9**, 1975).

Bayrhofferia Trevis. (1857) = Lecania fide Hawksworth *et al.* (*Dictionary of the Fungi* edn 8, 1995).

Bdellospora Drechsler (1935), Cochlonemataceae. 1, N. America. See Drechsler (*Mycol.* **27**: 25, 1935).

beaded (of a lamella), having a line of small drops of liquid on the edge.

beak (of ascoma or conidioma), an elongated neck through which the spores are discharged. See rostrum.

beard moss, species of *Alectoria*, *Bryoria*, *Ramalina* and *Usnea*.

Beauveria Vuill. (1912), anamorphic *Cordycipitaceae*, Hso.0eH.10. 9 (on *Insecta*), widespread. One of the most important genera for biocontrol (q.v.) of insects. See McLeod (*CJB* **32**: 818, 1954), de Hoog (*Stud. Mycol.* **1**, 1972), Riba *et al.* (*Fundamental and Applied Aspects of Invertebrate Pathology*: 205, 1986; isoesterase variability), von Arx (*Mycotaxon* **25**: 153, 1986), Shimizu & Aizawa (*J. Invert. Path.* **52**: 348, 1988; serological classification of *B. bassiana*), Mugnai *et al.* (*MR* **92**: 199, 1989; chemotaxonomic evaluation), Bridge *et al.* (*Mycopathologia* **111**: 85, 1990; chemotaxonomy of *B. bassiana*), Kosir *et al.* (*Can. J. Microbiol.* **37**: 534, 1991; RFLPs of virulent and avirulent *B. bassiana*), Rakotonirainy *et al.* (*J. Invert. Path.* **57**: 17, 1991; rRNA sequence comparison with *Tolypocladium*), St Leger *et al.* (*MR* **96**: 1007, 1992; genetic variation), Pfeifer & Khachatourians (*J. Invert. Path.* **61**: 231, 1993; electrophoretic karyotyping), Couteaudier & Viaud (*FEMS Microbiology Ecology* **22**: 175, 1997; population structure on *Melolontha*), Cravanzola *et al.* (*Lett. Appl. Microbiol.* **25**: 289, 1997; genetic polymorphism by RAPD-PCR on *B. brongniartii*), Neuvéglise *et al.* (*Mol. Ecol.* **6**: 373, 1997; rDNA introns), Couteaudier *et al.* (*Molecular Variability of Fungal Pathogens*: 95, 1998; genetic structure), Glare & Inwood (*MR* **102**: 250, 1998; NZ strains), Kück *et al.* (*Molecular Variability of Fungal Pathogens*: 73, 1998; transposons), McCullough *et al.* (*J. Clin. Microbiol.* **36**: 1035, 1998; strains on *Alphitobius*), Piatti *et al.* (*Lett. Appl. Microbiol.* **26**: 317, 1998; molecular characterization), Todorova *et al.* (*MR* **102**: 81, 1998; carbohydrate utilization), Castrillo *et al.* (*J. Invert. Path.* **73**: 269, 1999; population genetics), Enkerli *et al.* (*MR* **105**: 1079, 2001; microsatellite markers), Bidochka *et al.* (*Archs Microbiol.* **178**: 531, 2002; population biology), Coates *et al.* (*Curr. Genet.* **41**: 414, 2002; rDNA introns), Coates *et al.* (*MR* **106**: 40, 2002; rDNA), Gaitan *et al.* (*MR* **106**: 1307, 2002; strains on *Hypothenemus*), Henke *et al.* (*J. Clin. Microbiol.* **40**: 2698, 2002; clinical strain), Huang *et al.* (*Mycotaxon* **81**: 229, 2002; teleomorph of *B. bassiana*), Liu *et al.* (*MR* **106**: 1100, 2002; teleomorph connexions), Padmavathi *et al.* (*MR* **107**: 572, 2003; telomere fingerprinting), Pernfuss *et al.* (*IOBC/WPRS Bulletin* **26**: 121, 2003; species distinction), Wang *et al.* (*Environmental Microbiology* **5**: 908, 2003; population structure), Castrillo *et al.* (*J. Invert. Path.* **86**: 26, 2004; VCGs and recombination), Tucker *et al.* (*J. Clin. Microbiol.* **42**: 5412, 2004; human infection), Uribe & Khachatourians (*MR* **108**: 1070, 2004; mitochondrial RFLPs), Entz *et al.* (*MR* **109**: 1302, 2005; PCR diagnostics), Muro *et al.* (*MR* **109**: 294, 2005; strains on *Eurygaster*), Rehner (*Insect-Fungal Associations* Ecology and Evolution: 3, 2005; phylogeny), Rehner & Buckley (*Mycol.* **97**: 84, 2005; cryptic speciation), Wang *et al.* (*MR* **109**: 1364, 2005; population structure), Gürcan *et al.* (*Mycoses* **49**: 246, 2006; human infection).

Beauveriphora Matsush. (1975), anamorphic *Pezizomycotina*, Hso.0eH.10. 1, Japan. See Matsushima (*Icon. microfung. Matsush. lect.*: 14, 1975), Matsushima (*Matsush. Mycol. Mem.* **8**: 1, 1995).

Beccaria Massee (1892) [non *Beccaria* Müll. Hal. 1872, *Musci*] ≡ Beccariella.

Beccariella Ces. (1879) = Cymatoderma fide Donk (*Taxon* **6**: 17, 1957).

Beccopycnidium F. Stevens (1930), anamorphic *Pezizomycotina*, Cpd.0fH.?. 1, S. America.

Beckhausia Hampe ex Körb. (1865) nom. inval. =

Tomasellia fide Hawksworth *et al.* (*Dictionary of the Fungi* edn 8, 1995).

beef-steak fungus (or **liver fungus**), basidioma of the edible *Fistulina hepatica*; cf. brown oak.

Beejadwaya Subram. (1978), anamorphic *Pezizomycotina*, Hso.0eP.3. 1, Japan. See Subramanian (*Kavaka* **5**: 97, 1977).

Beejasamuha Subram. & Chandrash. (1977) = Beauveria.

Beeli formulae, Numerical designations for coding the ascospore, perithecial, setae, and hyphopodial characteristics of *Meliola* spp. together with their sizes devised by Beeli (*Bull. Jard. bot. Brux.* **8**: 89, 1920) and modified by Stevens (*Ann. Myc.* **25**: 405, 1927) Hansford (*Sydowia* **2**, 1961), Farr (*Mycopath.* **43**: 161, 1971) and Mibey & Hawksworth (*Mycol. Pap.* **174**, 1997); e.g. *Asteridiella westermannii* 3101.5340.

Beelia F. Stevens & R.W. Ryan (1925), Elsinoaceae. 1, Hawaii. See Petrak (*Sydowia* **7**: 321, 1953).

Beenakia D.A. Reid (1956), Clavariadelphaceae. 7, widespread (tropical). See Núñez & Ryvarden (*Sydowia* **46**: 321, 1994; key), Nakasone (*Cryptog. Mycol.* **24**: 131, 2003), Borgarino *et al.* (*BSMF* **121**: 187, 2005).

Beenakiaceae Jülich (1982) = Clavariadelphaceae.

beer, an alcoholic drink obtained by the fermentation of wort. The two main series of beers are the **ales** (produced by top-yeast (*Saccharomyces cerevisiae*) fermentation) and the **lagers** (produced by bottom-yeast (*S. carlsbergensis*) fermentation). See also Brewing, cider, porter, wine, yeast.

behind (of lamellae), the end nearest the stipe.

Belaina Bat. & Peres (1961) = Polynema fide Punithalingam (*Nova Hedwigia* **49**: 297, 1989).

Belainopsis Bat. & H. Maia (1965) nom. dub., anamorphic *Pezizomycotina*. See Punithalingam (*Nova Hedwigia* **49**: 297, 1989).

Belemnospora P.M. Kirk (1981), anamorphic *Pezizomycotina*, Hso.0-1eP.19. 5, widespread. See Sutton *et al.* (*MR* **92**: 354, 1989; key).

Belizeana Kohlm. & Volkm.-Kohlm. (1987), Dothideomycetes. 1 (marine), Belize. See Kohlmeyer & Volkmann-Kohlmeyer (*Bot. Mar.* **30**: 195, 1987), Kohlmeyer & Volkmann-Kohlmeyer (*Bot. Mar.* **34**: 1, 1991).

Bellemerea Hafellner & Cl. Roux (1984), Porpidiaceae (L). 8, esp. montane and polar. See Hafellner & Roux (*Bull. Soc. bot. Centre-Ouest* Nouv. sér. **15**: 129, 1984), Buschbom & Mueller (*Mol. Phylogen. Evol.* **32**: 66, 2004; phylogeny), Peršoh *et al.* (*Mycol. Progr.* **3**: 103, 2004; asci and photobionts).

Bellemerella Nav.-Ros. & Cl. Roux (1997), Verrucariaceae (L). 3 (lichenicolous), France; Spain. See Navarro-Rosinés & Roux (*Mycotaxon* **61**: 443, 1997), Calatayud & Navarro-Rosinés (*Nova Hedwigia* **72**: 473, 2001).

Bellulicauda B. Sutton (1967), anamorphic *Pezizomycotina*, St.0eH.19. 1, Africa. See Sutton (*CJB* **45**: 1254, 1967).

Belonia Körb. (1856), ? Gyalectaceae (L). 12, widespread. See Vězda (*Prirodov. Čas. slezsky* **20**: 241, 1959), Henssen *in* Brown *et al.* (Eds) (*Lichenology: progress and problems*: 107, 1976; ontogeny), Jørgensen *et al.* (*Lichenologist* **15**: 45, 1983; Eur.), Jørgensen *et al.* (*SA* **5**: 121, 1986; posn), Hafellner & Kalb (*Biblthca Lichenol.* **57**: 161, 1995; posn), Malcolm & Coppins (*Australasian Lichenology* **41**: 30, 1997; NZ), McCarthy & Kantvilas (*Lichenologist* **29**: 489, 1997; Tasmania), Navarro-Rosinés & Llimona (*Lichenologist* **29**: 15, 1997; Spain), Coppins & Malcolm (*Lichenologist* **30**: 563, 1998; NZ), Messuti *et al.* (*Bryologist* **102**: 314, 1999; S America), Grube *et al.* (*MR* **108**: 1111, 2004; phylogeny), Kauff & Büdel (*Bryologist* **108**: 272, 2005; ontogeny).

Belonidium Mont. & Durieu (1846) = Lachnum fide Raitviir (*Eesti NSV Tead. Akad. Toim.* Biol. seer **36**: 313, 1987), Spooner (*Biblthca Mycol.* **116**, 1987), Leenurm & Raitviir (*Folia cryptog. Estonica* **36**: 57, 2000; ultrastructure).

Beloniella (Sacc.) Boud. (1885) ≡ Odontura.

Beloniella Rehm (1892) = Calloria fide Hawksworth *et al.* (*Dictionary of the Fungi* edn 8, 1995).

Beloniella Th. Fr. (1877) = Belonia fide Hawksworth *et al.* (*Dictionary of the Fungi* edn 8, 1995).

Beloniomyces Cif. & Tomas. (1953) ≡ Belonia.

Belonioscypha Rehm (1892) = Cyathicula fide Baral & Krieglsteiner (*Beih. Sydowia* **6**, 1985).

Belonioscyphella Höhn. (1918), Helotiaceae. 1 (on mosses), Europe. See Carpenter (*Mem. N. Y. bot. Gdn* **33**, 1981; ? synonymy with *Gloeopeziza*), Döbbeler (*Ber. bayer. bot. Ges.* **57**: 153, 1986).

Belonium Sacc. (1884) = Pyrenopeziza. The genus is in need of revision. fide Baral (*SA* **13**: 113, 1994; synonymy with *Pyrenopeziza*), Nauta & Spooner (*Mycologist* **14**: 21, 2000; UK).

Belonopeziza Höhn. (1917) ? = Calloria fide Hawksworth *et al.* (*Dictionary of the Fungi* edn 8, 1995).

Belonopsis (Sacc.) Rehm (1891), Dermateaceae. Anamorph *Cystodendron*. 7, Europe. See Aebi (*Nova Hedwigia* **23**: 49, 1972), Nannfeldt (*Sydowia* **38**: 194, 1986), Nauta & Spooner (*Mycologist* **14**: 21, 2000; UK), Raitviir & Leenurm (*Folia cryptog. Estonica* **38**: 63, 2001; Estonia), Raitviir (*Mycotaxon* **87**: 359, 2003; Greenland), Nauta (*Coolia* **47**: 8, 2004; Netherlands), Pärtel & Raitviir (*Mycol. Progr.* **4**: 149, 2005; ultrastructure).

Belospora Clem. (1909) ≡ Hymenoscyphus.

Beltraminia Trevis. (1857) ≡ Dimelaena fide Hawksworth *et al.* (*Dictionary of the Fungi* edn 8, 1995).

Beltrania Penz. (1882), anamorphic *Pezizomycotina*, Hso.0eP.10. 13, widespread. See Hughes (*Mycol. Pap.* **47**, 1951), Pirozynski (*Mycol. Pap.* **90**, 1963; key), Pirozynski & Patil (*CJB* **48**: 567, 1970), Ouanyou & Rambelli (*Micol. Ital.* **19**: 33, 1990; substrate and morphology), Rambelli & Pasqualetti (*G. bot. ital.* **124**: 753, 1990; variability), Morelet (*Cryptog. Mycol.* **22**: 29, 2001; key).

Beltraniella Subram. (1952), anamorphic *Pseudomassaria*, Hso.0eP.10. 15, India. See Pirozynski (*Mycol. Pap.* **90**, 1963), Kabi (*Micol. Ital.* **16**: 27, 1987; substrate and morphology), Castañeda Ruíz *et al.* (*Mycotaxon* **58**: 243, 1996; Cuba), Kendrick (*CJB* **81**: 75, 2003; morphogenesis), Gusmão (*Mycol.* **96**: 150, 2004).

Beltraniomyces Manohar., Agarwal & Rao (2003) = Beltrania fide Manoharachary *et al.* (*Indian Phytopath.* **56**: 418, 2003).

Beltraniopsis Bat. & J.L. Bezerra (1960), anamorphic *Pezizomycotina*, Hso.0eP.10. 7, S. America. See Pirozynski (*Mycol. Pap.* **90**, 1963), Gusmão *et al.* (*MR* **104**: 251, 2000; key), Kendrick (*CJB* **81**: 75, 2003; morphogenesis), Castañeda Ruíz *et al.* (*Mycotaxon* **96**: 151, 2006; Cuba).

Benedekiella Negru & Verona (1964) = Physalospora fide Müller & von Arx *in* Ainsworth *et al.* (Eds) (*The*

Fungi **4A**: 87, 1973).

Benekea Bat. & J.L. Bezerra (1960) ? = Geastrumia fide Sutton (*Mycol. Pap.* **141**, 1977).

Benguetia Syd. & P. Syd. (1917), ? Helotiales. 1, Philippines.

Beniowskia Racib. (1900), anamorphic *Pezizomycotina*, Hso.0eP.1. 1, widespread (tropical). See Mason (*Mycol. Pap.* **2**: 26, 1928), Hanlin (*Mycotaxon* **28**: 219, 1987), Kirschner *et al.* (*Mycoscience* **43**: 15, 2002).

Benjaminella I.I. Tav. (1981) = Benjaminiomyces fide Eriksson & Hawksworth (*SA* **5**: 122, 1986; nomencl.), Tavares (*Mycotaxon* **75**: 205, 2000).

Benjaminia Pidopl. & Milko (1971) [non *Benjamina* Vell. 1835, *Rutaceae*] ≡ Benjaminiella.

Benjaminia S. Ahmad (1967), anamorphic *Pezizomycotina*, Cpd.#eP.?. 1, Pakistan. See Ahmad (*Biologia* Lahore **13**: 21, 1967).

Benjaminiella Arx (1981), Mucoraceae. 3, Canary Islands; India; USA. See Cole *et al.* (*Can. J. Microbiol.* **26**: 35, 1980; dimorphism & wall chemistry), Benny *et al.* (*Mycotaxon* **22**: 119, 1985), Forst & Pillinger (*Z. Mykol.* **54**: 139, 1988; dimorphism), Kirk (*Mycotaxon* **35**: 121, 1989), O'Donnell *et al.* (*Mycol.* **93**: 286, 2001; phylogeny), Voigt & Wöstemeyer (*Gene* **270**: 113, 2001; phylogeny).

Benjaminiomyces I.I. Tav. (2000), Laboulbeniaceae. 4 (on insects), N. & S. America. See Tavares (*Mycotaxon* **75**: 205, 2000).

Benjpalia Subram. & Bhat (1989), anamorphic *Pezizomycotina*, Hso.1eP.10/12/13. 1, India. See Subramanian & Bhat (*Kavaka* **15**: 43, 1987).

benomyl ('Benlate'), the first systemic fungicide: one of the benzimidazoles, has very low toxicity to plants and animals, and controls ascomycetous fungal plant pathogens (incl. their anamorphs) by interfering with spindle formation during nuclear division.

Bensingtonia Ingold (1986), anamorphic *Agaricostilbaceae*. 11, widespread. See Nakase *et al.* (*J. gen. appl. Microbiol.* Tokyo **39**: 107, 1993; mol. phylogeny), Takashima *et al.* (*J. gen. appl. Microbiol.* Tokyo **41**: 131, 1995; mol. phylogeny), Barnett *et al.* (*Yeasts: Characteristics and Identification* 3rd edn, 2000), Fell *et al.* (*Int. J. Syst. Evol. Microbiol.* **50**: 1351, 2000; mol. phylogeny), Wang *et al.* (*Int. J. Syst. Evol. Microbiol.* **53**: 2085, 2003), McLaughlin *et al.* (*Am. J. Bot.* **91**: 808, 2004; mitosis).

Berengeria Trevis. (1853) = Rinodina fide Hawksworth *et al.* (*Dictionary of the Fungi* edn 8, 1995).

Berggrenia Cooke (1879), ? Pezizales. 1 or 2, New Zealand. See Buchanan & May (*N.Z. Jl Bot.* **41**: 407, 2003; NZ conservation).

Bergorea Nieuwl. (1916) ≡ Robergea.

Berkelella (Sacc.) Sacc. (1891), Clavicipitaceae. Anamorph *Polycephalomyces*. 2 (on *Myxomycetes*), widespread. See Seifert (*Stud. Mycol.* **27**, 1985; as *Byssostilbe*), Rossman *et al.* (*Stud. Mycol.* **42**: 248 pp., 1999), Bischoff *et al.* (*Mycotaxon* **86**: 433, 2003).

Berkeley (Miles Joseph; 1803-1889; England). Studied at Cambridge University; became a priest, working first in Margate and later Northamptonshire, living at King's Cliffe till 1868, then at Sibbertoft. A great mycologist, once described as 'the last of the old race', Berkeley's chief work was with fungi (including lichen-forming species), though his interests in natural history were very wide: he also wrote about molluscs, mosses, algae, and gardening. He recognized that potato blight was caused by a fungus (1846). He was in touch with Fries, Montagne, and other mycologists, and from 1848 did much work with Broome. Berkeley's early work was done with great care but later, because of the large number of microfungi sent to him, there were more errors. In 1879 his collections were amalgamated with those of the Royal Botanic Gardens, Kew (**K**). *Publs.* His first printed mycological work was *British Fungi* (for which he distributed exsiccati) in Smith's *English Flora* (**5** (2), 1836) [= Hooker's *British Flora*, **2** (2)]; he supplemented this (in collaboration with Broome) by *Notices of British Fungi* (1837-1885) [index in *TBMS* **17**: 308, 1933; reprint, 1967]; in other papers he provided accounts of fungi from the then British colonies, the USA (especially collections of Curtis, q.v.), and other countries, and names for about 6,000 new species; of his 400 papers on mycology, some were short notes, such as those in the *Gardeners' Chronicle* (1845 onwards, including the group of papers on 'vegetable pathology', 1854-1857 [see *Phytopathological Classics* **8**, with biography and portrait]), and others books, e.g. *Introduction to Cryptogamic Botany* (1857) and *Outlines of British Fungology* (1860). *Biogs, obits etc.* Ainsworth (*Mycologist* **21**: 126, 1987) [portrait]; Buczacki (*British Mycological Society Symposium Series* **17**: 1, 1991); Grummann (1974: 368); Stafleu & Cowan (*TL-2* **1**: 192, 1976); Stafleu & Mennega (*TL-2, Suppl.* **2**: 98, 1993).

Berkeleyna Kuntze (1898) ≡ Cephalotrichum Link.

Berkleasmium Zobel (1854), anamorphic *Pleosporales*, Hsp.#eP.1. 21, widespread. See Moore (*Mycol.* **51**: 734, 1961; key), Ellis (*Dematiaceous Hyphomycetes*, 1971), Ellis (*More Dematiaceous Hyphomycetes*, 1976), Yip (*Australas. Pl. Path.* **17**: 31, 1988), Bussaban *et al.* (*Fungal Diversity* **8**: 73, 2001; Thailand), Somrithipol & Jones (*Fungal Diversity* **12**: 169, 2003), Zhao & Zhang (*Mycotaxon* **89**: 241, 2004; China), Pinnoi *et al.* (*Mycol.* **99**: 378, 2007; phylogeny), Wang *et al.* (*MR* **111**: 1268, 2007; phylogeny).

Berlese (Augusto Napoleone; 1864-1903; Italy). Professor of Plant Pathology, School of Agriculture, Milan. Responsible for parts of Saccardo's *Sylloge* and for starting (with his brother, Antonio Berlese, the zoologist) the *Rivista di Patologia Vegetale* (1892 onwards). *Publs. Fungi Moricolae* (1885-1889); *Icones Fungorum ad usum Sylloges Saccardianae Adcommodatae* (1890-1905); papers on *Pleospora* (1888), the *Lophiostomataceae* (1890), *Dematophora* and *Rosellinia* (1892), *Cladosporium* and *Dematium* (1895), and many more. *Biogs, obits etc.* Antonio Berlese (*Rivista di Patologia Vegetale* **10**: 347, 1904); Cavara (*Annales Mycologici* **1**: 178, 1903); Grummann (1974: 513); Stafleu & Cowan (*TL-2* **1**: 197, 1976); Stafleu & Mennega (*TL-2, Suppl.* **2**: 101, 1993).

Berlesiella Sacc. (1888) = Capronia fide Müller *et al.* (*TBMS* **88**: 63, 1987; synonymy), Barr (*Mycotaxon* **41**: 419, 1991), Untereiner & Naveau (*Mycol.* **91**: 67, 1999).

Berteromyces Cif. (1954) = Passalora fide Deighton (*Mycol. Pap.* **112**, 1967), Crous & Braun (*CBS Diversity Ser.* **1**: 571 pp., 2003).

Bertia De Not. (1844), Bertiaceae. 14 (from wood), widespread. See Corlett & Krug (*CJB* **62**: 2561, 1985), Subramanian & Sekar (*Kavaka* **18**: 19, 1993;

Indian spp.), Huhndorf *et al.* (*MR* **108**: 1384, 2004; phylogeny), Zhang *et al.* (*Mycol.* **98**: 1076, 2006; phylogeny).

Bertiaceae Smyk (1981), Coronophorales. 1 gen., 14 spp.

Lit.: Subramanian & Sekar (*Kavaka* **18**: 19, 1993), Huhndorf *et al.* (*MR* **108**: 1384, 2004; phylogeny).

Bertiella (Sacc.) Sacc. & P. Syd. (1899) = Massarina fide Eriksson & Yue (*Mycotaxon* **27**: 247, 1986).

Bertiella Kirschst. (1906) ≡ Kirschsteinia.

Bertossia Cif. & Tomas. (1953) = Mycoglaena fide Hawksworth *et al.* (*Dictionary of the Fungi* edn 8, 1995).

Bertramia Mesnil & Caullery (1897), Chytridiomycota. 1, widespread. See Weiser & McCauley (*Z. Parasitkde* **43**: 299, 1974).

Bertrandia R. Heim (1936) = Hygrocybe fide Kuyper (*in litt.*).

Bertrandiella R. Heim (1966) = Lactocollybia fide Singer (*Agaric. mod. Tax.* edn 3, 1975).

Berwaldia J.I.R. Larsson (1981), Microsporidia. 2. See Larsson (*Parasitology* **83**: 325, 1981).

Bessey (Ernst Athearn; 1877-1957; USA). Professor of Botany and Mycology, Michigan Agricultural College, East Lansing, 1910-1946. Author of the first American textbook of mycology (1935, 1950 [edn 2]) [see Literature]. *Biogs, obits etc.* Barnett (*Mycol.* **50**: 1, 1958) [bibliography, portrait]; Grummann (1974: 204); Stafleu & Cowan (*TL-2* **1**: 219, 1976); Stafleu & Mennega (*TL-2, Suppl.* **2**: 141, 1993).

beta-spore (**B-spore**, **β-spore**), a fertile, usually hamate, spore of an anamorph of the *Valsaceae* (*Phomopsis*). Cf. alpha-spore.

Bettsia Skou (1972), Ascosphaeraceae. Anamorph *Chrysosporium*-like. 1 (in pollen on honeycomb), Europe. See Skou (*Friesia* **11**: 62, 1975), Maghrabi & Kish (*Mycol.* **78**: 676, 1986), Maghrabi & Kish (*Mycol.* **79**: 519, 1987; isozymes), Kuraishi *et al.* (*Antonie van Leeuwenhoek* **77**: 179, 2000; ubiquinones), James & Skinner (*J. Invert. Path.* **90**: 98, 2005; molecular diagnostics).

Betulina Velen. (1947), Hyaloscyphaceae. 2, Europe. See Graddon (*TBMS* **63**: 477, 1974), Huhtinen (*Karstenia* **29**: 45, 1990).

Beverwykella Tubaki (1975), anamorphic *Pezizomycotina*, Hso.#eH.1. 3 (aquatic), widespread. See Tubaki (*TMSJ* **16**: 138, 1975), Nawawi & Kuthubutheen (*TBMS* **90**: 487, 1988), Voglmayr & Delgado-Rodríguez (*MR* **107**: 236, 2003; key).

Bharatheeya D'Souza & Bhat (2002), anamorphic *Pezizomycotina*, H?.?.?. 2, India. See D'Souza & Bhat (*Mycotaxon* **83**: 399, 2002).

Bhargavaella Sarj. Singh & K.S. Srivast. (1980), anamorphic *Pezizomycotina*, Hso.#eP.3. 1, India. See Singh & Srivastava (*J. Indian bot. Soc.* **58**: 391, 1979).

bi- (prefix), twice; having two; two-.

biallelic (of an incompatibility system), having 2 alleles per locus; cf. multiallelic.

Biannularia Beck (1922) = Catathelasma fide Singer (*Agaric. mod. Tax.* edn 3, 1975).

Biannulariaceae Jülich (1982) nom. rej. = Tricholomataceae.

Biatora Ach. (1809) = Stenhammarella fide Hertel (*Beih. Nova Hedwigia* **24**, 1967).

Biatora Fr. (1817), Ramalinaceae (L). *c.* 42, widespread (temperate). See Hafellner (*Herzogia* **8**: 53, 1989), Printzen (*Biblthca Lichenol.* **60**, 1995), Printzen (*Cryptog. bot.* **5**: 105, 1995), Printzen & Tønsberg (*Bryologist* **102**: 692, 1999; N Am.), Printzen *et al.* (*Lichenologist* **31**: 491, 1999; RAPDs), Ekman (*MR* **105**: 783, 2001; phylogeny), Printzen & Tønsberg (*Biblthca Lichenol.* **86**: 133, 2003; pigments), Printzen & Tønsberg (*Symb. bot. upsal.* **34** no. 1: 343, 2004; N Am.), Printzen & Otte (*Graphis Scripta* **17**: 56, 2005; key to European and Macaronesian spp.).

Biatoraceae A. Massal. ex Stizenb. (1862) = Ramalinaceae.

Biatorella De Not. (1846), Biatorellaceae (L). *c.* 31, widespread. See Magnusson (*Rabenh. Krypt.-Fl.* **9** 5.1: 15, 1935), Magnusson (*Annals Cryptog. Exot.* **7**: 115, 1935), Hafellner & Casares-Porcel (*Nova Hedwigia* **55**: 309, 1992; typification), Weber & Nash (*Lichenologist* **24**: 101, 1992; N Am.), Hafellner (*Cryptog. bot.* **5**: 99, 1995).

Biatorellaceae M. Choisy ex Hafellner & Casares (1992), Lecanorales (L). 2 gen. (+ 3 syn.), 33 spp.

Lit.: Hafellner & Casares-Porcel (*Nova Hedwigia* **55**: 309, 1992), Weber & Nash (*Lichenologist* **24**: 101, 1992).

Biatorellina Henn. (1903) ? = Tryblidiopsis fide Hawksworth *et al.* (*Dictionary of the Fungi* edn 8, 1995).

Biatorellopsis C.W. Dodge (1965) = Pleopsidium fide Castello & Nimis (*Lichenologist* **26**: 283, 1994).

Biatoridina Schczedr. (1964) = Epithyrium fide Sutton (*Mycol. Pap.* **141**, 1977).

Biatoridium J. Lahm ex Körb. (1860), Pezizomycotina (L). 3, widespread (temperate). See Hafellner (*Acta Bot. Fenn.* **150**: 39, 1994), Jørgensen (*Taxon* **53**: 521, 2004; nomencl.), Reeb *et al.* (*Mol. Phylogen. Evol.* **32**: 1036, 2004; phylogeny).

Biatorina A. Massal. (1852) nom. rej. = Catinaria fide Jørgensen & Santesson (*Taxon* **42**: 881, 1993).

biatorine (of apothecia), of the lecideine type, pale or more or less coloured and soft in consistency.

Biatorinella Dechatres & Werner (1974) = Fuscidea fide Hawksworth *et al.* (*Dictionary of the Fungi* edn 8, 1995).

Biatorinopsis Müll. Arg. (1881) = Coenogonium Ehrenb. fide Rivas Platas *et al.* (*Fungal Diversity* **23**: 255, 2006).

Biatoropsis Räsänen (1934), Tremellaceae. Anamorph *Hormomyces*-like. 1 (causing galls, 'carpoids', on *Usnea*), widespread. See Diederich & Christiansen (*Lichenologist* **26**: 47, 1994).

Biatriospora K.D. Hyde & Borse (1986), Dothideomycetes. 1 (on mangrove wood), Seychelles. See Hyde & Borse (*Mycotaxon* **26**: 263, 1986), Kohlmeyer & Volkmann-Kohlmeyer (*Bot. Mar.* **34**: 1, 1991).

Bibanasiella R.F. Castañeda & W.B. Kendr. (1991), anamorphic *Pezizomycotina*, Hsp.1bH.12. 1, Cuba. See Castañeda Ruiz & Kendrick (*Univ. Waterloo Biol. Ser.* **35**: 14, 1991).

Bibbya J.H. Willis (1956) = Toninia fide Santesson (*Muelleria* **1**: 91, 1959).

Bibliography of Systematic Mycology. Produced by CABI, and appearing twice annually since 1943, this lists new and recently published work on systematic mycology, arranged in categories which include the major taxonomic groups of fungi, biographies, checklists, obituaries, regional mycotas etc., and fungal generic name indexes. The great thesauri of mycological literature produced by Ciferri, Lindau and Sydow provided coverage of mycological literature up to the year 1930. The *Bibliography of Systematic Mycology* provides a current continuation of these works, from the early 1940s onwards. The years between still remain uncovered by any single publication. See also Literature.

Bibulocystis J. Walker, Beilharz, Pascoe & Priest (2006), Raveneliaceae. 2 (on on *Albizia*, *Daviesia* (*Leguminosae*)), Australia; New Caledonia. See Walker *et al.* (*Australas. Pl. Path.* **35**: 11, 2006).

bibulous (of the surface of a pileus), able to take up water.

bicampanulate, like two bells arranged mouth to mouth (Fig. 23.20).

Biciliopsis Diederich (1997), ? Chaetothyriaceae. 1 (lichenicolous), Papua New Guinea. See Aptroot *et al.* (*Biblthca Lichenol.* **64**, 1997), Hoffmann & Hafellner (*Biblthca Lichenol.* **77**: 1, 2000).

Biciliospora Petr. (1952) = Nitschkia fide von Arx (*Sydowia* **34**: 13, 1981), Subramanian & Sekar (*Kavaka* **18**: 19, 1993).

Biciliosporina Subram. & Sekar (1993), Nitschkiaceae. 1 (from bark), India. See Subramanian & Sekar (*Kavaka* **18**: 19, 1993).

biconic (of spores), like two cones attached base to base (Fig. 23.27).

Biconiosporella Schaumann (1972), Sordariomycetidae. 1 (from wood, marine), Europe. See Kohlmeyer & Volkmann-Kohlmeyer (*Bot. Mar.* **34**: 1, 1991), Huhndorf *et al.* (*Mycol.* **96**: 368, 2004).

Bicornispora Checa, Barrasa, M.N. Blanco & A.T. Martínez (1996), ? Coryneliaceae. Anamorph *Exophiala*. 1 (dead stems), Spain. See Checa *et al.* (*MR* **100**: 500, 1996).

Bicricium Sorokīn (1889) nom. conf., Chytridiales.

Bicrouania Kohlm. & Volkm.-Kohlm. (1990), Melanommataceae. 1 (marine), France. See Kohlmeyer & Volkmann-Kohlmeyer (*MR* **94**: 685, 1990).

Bidenticula Deighton (1972) = Fusarium fide Booth & Sutton (*in litt.*).

Bidonia Adans. (1763) = Hydnum fide Donk (*Taxon* **5**, 1956).

bifarious, in two lines or series; distichous.

bifid, having a crack or division near the middle; forked.

Bifidocarpus Cano, Guarro & R.F. Castañeda (1994), ? Onygenaceae. 2, Cuba. See Cano *et al.* (*Mycotaxon* **52**: 53, 1994), Udagawa & Uchiyama (*Cryptog. Mycol.* **21**: 75, 2000).

Biflagellospora Matsush. (1975), anamorphic *Pezizomycotina*, Hso.0eH.1/19. 1, Japan. See Matsushima (*Icon. microfung. Matsush. lect.*: 16, 1975), Sivichai & Hywel-Jones (*MR* **103**: 908, 1999; Thailand).

Biflagellosporella Matsush. (1993), anamorphic *Pezizomycotina*, Hso.1bH.1. 1, Peru. See Matsushima (*Matsush. Mycol. Mem.* **7**: 44, 1993).

Biflua Jørg. Koch & E.B.G. Jones (1989), Pezizomycotina. 1, Denmark. See Koch & Jones (*CJB* **67**: 1187, 1989), Kohlmeyer & Volkmann-Kohlmeyer (*Bot. Mar.* **34**: 1, 1991).

Bifrontia Norman (1872) = Naetrocymbe Körb. fide Keissler (*Rabenh. Krypt.-Fl.* **9, 1**, 1933).

Bifusella Höhn. (1917), Rhytismataceae. Anamorph *Crandallia*. 10, widespread. *B. faullii* (needle-cast in *Abies*), *B. linearis* (on pine). See Cannon & Minter (*Mycol. Pap.* **155**, 1986), Hansen & Lewis (*Compendium of Conifer Diseases*, 1997), Hou *et al.* (*CJB* **83**: 37, 2005; Canada).

Bifusepta Darker (1963), Rhytismataceae. 1, N. America. See Darker (*Mycol.* **55**: 816, 1963).

Biharia Thirum. & Mishra (1953) = Stenella fide Ellis (*Dematiaceous Hyphomycetes*, 1971), Braun (*Monogr. Cercosporella, Ramularia Allied Genera (Phytopath. Hyphom.)* **2**: 374, 1998), Braun *et al.* (*Mycol. Progr.* **2**: 3, 2003).

bilabiate (1) two-lipped; (2) (of asci), ones in which the ectotunica splits in a lip-like manner to expose the endotunica (e.g. *Pertusaria*).

bilaminate, two-layered.

Bilboque Viégas (1960), anamorphic *Pezizomycotina*, Hso.1eP.?. 1, Brazil. See Viégas (*Bragantia* **19**: 903, 1960).

Bilgrami (Krishna Sahai; 1933-1996; India). Student (1954-1957) then Lecturer (1958-1963), Allahabad University; Reader, Jodhpur University (1963-1970); Professor & Head, Botany Department, Bhagalpur University (1970-1993). Noted for his work on environmental pollution, mycotoxin contamination of animal and human food, and environmental biology, specifically in relation to ecology of the Ganga river and thermal springs of Bihar; compiled the first comprehensive checklist of India's fungi. *Publs.* (with Jamaluddin & Rizwi) *Fungi of India, Lists and References* edn 2 (1991). *Biogs, obits etc.* Tandon (*Indian Phytopathology* **50** (1): 159, 1997).

Bilgramia Panwar, Purohit & Chouhan (1974), anamorphic *Pezizomycotina*, Hso.#eP.3. 1, India. See Carmichael *et al.* (*Genera of Hyphomycetes*, 1980).

Bilimbia De Not. (1846), Lecanorales (L). Probably close to *Lecania*. See Veldkamp (*Lichenologist* **36**: 191, 2004), Llop (*Lichenologist* **38**: 279, 2006), Naesborg *et al.* (*MR* **111**: 581, 2007; phylogeny).

Bilimbiospora Auersw. (1861) nom. rej. = Leptosphaeria fide Holm (*Taxon* **24**: 475, 1975).

Bimeris Petr. (1949), anamorphic *Pezizomycotina*, Cpd.1eH.?. 1, Ecuador.

Bimuria D. Hawksw., Chea & Sheridan (1979), ? Montagnulaceae. 1 (from soil), New Zealand. See Hawksworth (*SA* **6**: 238, 1987; posn), Kruys *et al.* (*MR* **110**: 527, 2006; posn), Schoch *et al.* (*Mycol.* **98**: 1041, 2006; phylogeny).

binate, in two parts.

binding hyphae, see hyphal analysis.

binucleate-phase, the dikaryo-phase.

Binucleospora A.M. Bronnvall & J.I.R. Larsson (1995), Microsporidia. 1. See Bronnvall & Larsson (*Eur. J. Protist.* **31**: 63, 1995).

bio- (prefix), pertaining to life.

biocide, a substance which kills living organisms. Cf. biostat.

Bioconiosporium Bat. & J.L. Bezerra (1964), anamorphic *Pezizomycotina*, Hso.#eP.1. 3, widespread. See Batista & Bezerra (*Publções Inst. Micol. Recife* **417**: 4, 1964), Castañeda Ruiz & Heredia (*Mycotaxon* **76**: 125, 2000).

biocontrol, see Biological control.

bioconversion, the conversion of one material, usually a waste, into a product of increased value (e.g. of lignocellulosic residues for ethanol production); see Saddler (Ed.) (*Bioconversion of forest and agricultural plant residues*, 1993), Biodegradation, Biodeterioration, Biotechnology.

Biodegradation. A term sometimes used synonymously with Biodeterioration (q.v.). However, it is more correctly employed to describe breakdown of materials in a manner beneficial to man, e.g. the removal of and/or use of wastes. It has been defined as 'the harnessing, by man, of the decay abilities of organisms to render a waste material more useful or acceptable' (Allsopp & Seal, *Introduction to biodeterioration*, 1986), and as such covers use of fungi and other organisms (mainly bacteria) in solid or liquid state fermentations to improve digestibilities of, for example, lignocellulosic wastes to ruminants or to produce single cell proteins from wastes. Examples of economically beneficial fungal biodegradation (often in combination with other organisms) include: composting of straw, manure, agricultural waste and bark; mushroom cultivation, including biodegradation of logs by shiitake fungi; production of single cell proteins using alkanes, brewery wastes, molasses and sulphite waste liquids; solid waste treatment of sludge, sewage; pulp and paper mill effluent treatment; waste water treatment, particularly from distilleries and tanneries. Although the term 'biodegradable' is sometimes used to describe decay of natural materials such as forest leaf litter, the concept of biodegradability has come to be used more colloquially in recent years in connexion with detergents, plastics etc. which may break down when discarded into the environment. There is as yet no universally accepted definition of the term **biodegradable**, particularly when applied to these materials. There are outstanding questions regarding whether the term implies complete mineralization and hence removal of the waste from the environment or whether degradation to a small non-toxic molecule will suffice. Several bodies are working to produce a satisfactory definition, e.g. International Biodeterioration Research Group (IBRG), American Society for the Testing of Materials (ASTM).

See Bioconversion, Biodegradation, Bioremediation. See Bennett *et al.* (*Use of fungi in biodegradation* in Hurst [Ed.], *Manual of Environmental Microbiology* edn 2, 2002). See also: Biodeterioration, Bioremediation, Biotransformation.

Biodeterioration. Any undesirable change in the properties of a material caused by the vital activities of organisms (Hueck, 1965). Fungi play an important part in biodeterioration. Some representative examples of fungal damage are: to **animal feeding stuffs**, Snow *et al.* (*Ann. appl. Biol.* **31**: 102, 1944); **building materials**, Batista *et al.* (*Atas Inst. Mycol. Recife* **5**: 311, 1967; fungi, incl. lichens), Martin & Johnson (*Biodet. Abstr.* **6**: 101, 1992; bibl. control lichens), May *et al.* (*Biodet. Abstr.* **7**: 109, 1993; review), Richardson (*The vanishing lichens*, 1975), Singh (*Building mycology*, 1994), see also monuments (below), weathering; **stone**, May *et al.* (*Biodet. abstr.* **7**: 109, 1993); **electrical equipment**, Wasserbauer (*Internat. Biodet. Bull.* **3**: 1, 1967); **food**, see Food spoilage; **fuel**, see kerosene fungus; **glass and optical equipment**, Ohtsuki (*Bot. Mag. Tokyo* **75**: 221, 1962), Nagamuttu (*Internat. Biodet. Bull.* **3**: 25, 1967); **grain**, Christensen & Kauffman (*Grain storage. The role of fungi in quality and loss*, 1969); **gunpowder**, Lacey (*TBMS* **74**: 195, 1980); **leather**, Musgrave (*Ann. appl. Biol.* **34**, 1947), Gordon (*in* O'Flaherty *et al.* (Eds) (*The chemistry and technology of leather*, **4**, 1965); **meat**, Jensen (*Microbiology of meats*, 1942); **monuments**, Nimis *et al.* (*Licheni i conservazione dei monumenti*, 1992), see also building materials (above), weathering; **paint**, Eveleigh (*Ann. appl. Biol.* **49**: 403, 1961; *TBMS* **44**: 573, 1961, esp. *Phoma violacea*), 7th Paint Research Institute Symposium (*J. Coatings Technology* **50**: 35, 1978); **paper**, Sée (*Les maladies du papier piqué*, 1919), Wang (*Tech. Publ. Sta. Univ. Coll. Forestry Syracuse* **87**, 1965; fungi of pulp and paper; cellulolytic fungi); **polyurethane**, Pathirana & Seal (*Internat. Biodet.* **20**: 163, 229, 1984); **structural timber**, see Wood-attacking fungi; **textiles**, Thaysen & Bunker (*The microbiology of cellulose, hemicellulose, pectins and gums*, 1927), Prindle (*Microbiology of textile fibres, Textile Rev.* 1933-6), Morris (*J. Text. Inst.* **18**: T99, 1927), Galloway (*J. Text. Inst.* **26**: T123, 1935); **tobacco**, Papavassiliou *et al.* (*Mycopath.* **44**: 117, 1971; cigarettes); **wood, archaeological**, Blanchette (*Biodet. Abstr.* **9**: 113, 1995). See also Seaward & Brightman (*in* Seaward, *Lichen ecology*: 253, 1977; lichens on man-made substrata, etc.).

The International Biodeterioration and Biodegradation Society (www.biodeterioration.org) covers all aspects of the biodeterioration of materials and bidegradation of wastes and holds regular meetings and international symposia, with published proceedings.

Lit.: Allsopp *et al.* (*Introduction to biodeterioration*, edn 2, 2004; Japanese edn, 1991), Hueck (*Material und Organismen* **1**: 5, 1965), Singh (Ed.) (*Building mycology*, 1994), *Biodeterioration Abstracts* (1987-2001; quarterly), *International Biodeterioration & Biodegradation* (1965-; originally as *International Biodeterioration Bulletin*). See also: Biodegradation, Bioremediation.

Biodiversity (Biological diversity). The variety and value of Life on Earth from the genetic through the organismal to the ecological levels.

The variety and value of fungi at all of those levels is no smaller than for animals or plants: for fungi, the very small proportion which are satisfactorily known (Cannon & Hawksworth, *Adv. Pl. Path.* **11**: 277, 1995), the large numbers of species (see Numbers of fungi), and their importance in ecosystem functioning (Christensen, *Mycol.* **81**: 1, 1989) are all factors arguing for prioritized attention in biodiversity studies. There is a pressing need for inventories (q.v.), which are one of the principal ways of assessing biodiversity, since in many parts of the world, many ecosystems remain almost unexplored for fungi, and biodiversity 'hotspots' for fungi may be different from those for animals or plants.

The UN Convention on Biological Diversity, negotiated in Rio de Janeiro in 1992, became effective at the end of 1993; by June 2008, 191 governments were parties to the convention (but not Andorra, the Holy See, Iraq, Somalia or the USA). The Convention has focused the subject of biodiversity in political circles; many countries have now developed or are developing programmes to survey and value their natural biotic resources and investigate how they can best be conserved and sustainably used. Initiatives such as the IUBS/UNESCO/SCOPE DIVERSITAS programme, the GEF/UNEP Global Biodiversity Assessment and the Species 2000 and Systematics Agenda 2000 projects are prominent elements of such research and development efforts.

Although the convention explicitly states that it covers all forms of biodiversity, it is worded in terms of 'animals, plants and microorganisms', categories into which fungi do not easily fit. Not surprisingly, therefore, many (perhaps most) of the reports and biodiversity strategy and action plans produced by individual countries in response to the convention contain no mention of fungi. To date, the convention has not been subjected to critical evaluation by the mycological community in respect of its impact on conservation of fungal diversity and the sustainable use of fungi.

Lit.: Allsopp *et al.* (Eds) (*Microbial Diversity and Ecosystem Function*, 1995); Aptroot (*Fungal Diversity* **6**: 1, 2001) [200 spp. on one tree]; Galloway (*Biodiv. Conserv.* **1**: 312, 1992) [lichens]; Groombridge (Ed.) (*Global Biodiversity*, 1992); Hawksworth (*MR* **95**: 641, 1991); Hawksworth (Ed.) (*Biodiversity: Measurement and Estimation*, 1995); Heywood (Ed.) (*Global Biodiversity Assessment*, 1995); Isaac *et al.* (Eds) (*Aspects of Tropical Mycology*, 1993); Oberwinkler (*Biodiv. Conserv.* **1**: 293, 1992); Rossman *et al.* (*Protocols for an All Taxa Biodiversity Inventory of Fungi.*, 1998); Schulze & Mooney (Eds) (*Biodiversity and Ecosystem Function*, 1994); Solbrig *et al.* (Eds) (*Biodiversity and Global Change*, edn 2, 1994).

biogenous, living on another living organism; parasitic.

Biographical notices, for list see Authors' names.

Bioindication. The use of organism(s) (**bioindicators**) expressing particular symptoms or responses to indicate changes in some environmental influence; various fungi, including leaf-inhabiting, mycorrhizal and lichen-forming species, are used as bioindicators of acid rain (q.v.), air pollution (q.v.), ammonium eutrophication (Brown, *in* Bates & Farmer, 1992: 259), dating surfaces (see Lichenometry), ecological continuity (see RIEC), fire (Wolseley *et al.*, *Global Ecol. Biogeogr. Lett.* **4**: 116, 1995), heavy metals (Gordon *et al.*, *J. Trop. Ecol.* **11**: 1, 1995; Sosak-Świderska, *Geophysical Res. Abstracts* **10**: EGU2008-A-11527, 2008), radionuclides (see air pollution), and water levels (see Aquatic fungi).

Lit.: Bates & Farmer (Eds) (*Bryophytes and lichens in a changing environment*, 1992), Boddy *et al.* (Eds) (*Fungi and ecological disturbance* [*Proc. R. Soc. Edinb.* B **94**], 1988), Burton (*Biological monitoring of environmental contaminants: plants*, 1986), Ellenberg *et al.* (*Biological monitoring: signals from the environment*, 1991), Fellner (*Agriculture Ecosystems and Environment* **28**, 115, 1989), Hawksworth (*in* Swaminathan & Jana, *Biodiversity: implications for global food security*: 184, 1992), Jeffrey & Madden (Eds) (*Bioindicators and environmental management*, 1991), Richardson (*Pollution monitoring with lichens*, 1992).

biologic form (Marshall Ward) or **race** (Klebahn), see physiologic race.

Biological control (Biocontrol). The use of one or more organisms (agents) to maintain another organism (pest) at a level at which it is no longer a problem. Fungal pathogens, parasites and antagonists are being exploited to control a range of agricultural pests, including arthropods, nematodes, weeds and crop diseases (Association of Applied Biologists, *The exploitation of micro-organisms in applied biology*, 1990; Burge, *Fungi in biological control systems*, 1988; Butt *et al*, *Fungi as biocontrol agents: progress, problems and potential*, 2001); although rarely openly discussed, they have also been used in political and military contexts (see Bioterrorism and fungi). They may be used alone or as part of an integrated pest management scheme (see Maredia *et al.*, *Integrated pest management in the global arena*, 2003). Two distinct approaches can be adopted: classical, involving the release of a coevolved fungal pathogen into an exotic environment where the target pest is an alien or non-indigenous species; inundative, through the application of a mass-produced, typically necrotrophic fungus as a mycopesticide (q.v.). Entomopathogenic fungi in genera such as *Beauveria*, *Metarhizium*, *Paecilomyces* and *Verticillium*, are being used for inundative biological control (see mycopesticides), while *Entomophthora radicans* has been classically introduced into Australia for control of Lucerne aphid (Milner *et al.*, *J. Aust. Ent. Soc.* **21**: 113, 1982).

Nematophagous fungi include the nematode trapping fungi (*Arthrobotrys*, *Dactylella*, *Geniculifera*, *Monacrosporium*), endoparasites (*Hirsutella*, *Catenaria*, *Meria*, *Nematoctonus*, *Nematophthora*) and highly specific egg parasites (*Dactylella oviparasitica*, *Paecilomyces lilacinus*, *Verticillium chlamydosporium*) (Stirling, *in* Burge, 1988). Immediate prospects for the exploitation of nematophagous fungi as biological control agents are uncertain since problems of formulation and application still have to be overcome (Stirling, *Biological control of plant parasitic nematodes*, 1992; Mankau, *J. Nematology* **12**: 244, 1980).

Fungal pathogens for the biological control of weeds have been extensively investigated (Charudattan & Walker (Eds), *Biological control of weeds with plant pathogens*, 1982; TeBeest (Ed.), *Microbial control of weeds*, 1991), using both the classical approach with obligate pathogens such as rusts and smuts, and the inundative approach, with necrotrophic pathogens in the genera *Colletotrichum* and *Phytophthora* (see mycopesticides). Successful control of skeleton weed has been achieved in Australia following release of the European rust *Puccinia chondrillina* (Hasan, *Ann. appl. Biol.* **99**: 119, 1981), and of mistflower in Hawaii with the white smut *Entyloma ageratinae* from the Caribbean (Trujillo, *Proc. VI Int. Symp. Biol. Control Weeds* 25, 1985; Barreto & Evans, *TBMS* **91**: 81, 1988), and of blackberry in Chile with the imported rust *Phragmidium violaceum* (Oehrens, *FAO Pl. Prot. Bull.* **25**: 26, 1977). For risk assessment see Evans (*Aust. Pl. Path.* **29**: 1, 2000).

Mycoparasites and antagonistic fungi, particularly in the genera *Gliocladium*, *Sphaerellopsis*, *Trichoderma* and *Verticillium*, have been evaluated for biological control of crop diseases (Cook & Baker, *The nature and practice of biological control of plant pathogens* 1984; Hornby (Ed.), *Biological control of soil-borne plant pathogens*, 1990), Kubicek & Harman (*Trichoderma and Gliocladium* **1**. Basic biology, taxonomy and genetics, 1998), Harman & Kubicek (*Trichoderma & Gliocladium* **2**. Enzymes, biological control and commercial applications, 1998); see Mycopesticides for more details; bacteria and protozoa as well as other fungi have been tried against *Phytophthora cinnamomi* (Scott, *Adv. Pl. Path.* **11**: 131, 1995).

Biological control is particularly important in forestry where economics rarely permit other options. A classic example is control of the conifer pathogen *Heterobasidion annosum* which colonizes stumps of freshly felled trees, then spreading to adjacent living trees, by stump application of spores of a competitor, *Phlebiopsis gigantea* (Rishbeth, *Ann. Appl. Biol.* **52**:

63, 1963).

Fungal diseases may themselves be subject to biological control by other organisms. *Streptomyces* has been used to suppress fungi causing damping off of alfalfa seedlings (Jones & Samac, *Biological control* 7: 196, 1996) and transmissible hypovirulence using dsRNA viruses (q.v.) has been used with considerable success as a form of biological control for chestnut blight caused by *Cryphonectria parasitica* (see Diamandis & Perlerou, *Advances in hort. sci.* **20**: 50, 2006).

Biological warfare and fungi. see Bioterrorism and fungi.

bioluminescence, see Luminescent fungi.

biomass, the quantity (vol., wt, etc.) of organisms (or living material) in a particular environment (e.g. fungi in soil); sometimes extended to the quantity of organic matter in a material (e.g. domestic refuse). See Boucher & Stone (*in* Carroll & Wicklow, *The fungal community*, edn 2 : 538, 1992; in lichens on trees), Frankland (*Soil Biol. Biochem.* **7**: 339, 1975; **10**: 323, 1978; estimation live biomass), Newell (*in* Carroll & Wicklow, 1992: 521; in litter), Ritz *et al.* (Eds) (*Beyond the biomass*, 1994).

biomass support particles (**ESPs**), large open structures made from knitted stainless steel and crushed into spheres or reticulated polyurethane foam cut into cubes used for immobilization (q.v.) of fungal cells.

Bionectria Speg. (1919), Bionectriaceae. Anamorph *Clonostachys*. 4 (on living and dead leaves, and old woody tissues), widespread. See Rossman *et al.* (*Mycol.* **85**: 685, 1993), Schroers & Samuels (*Z. Mykol.* **63**: 149, 1997), Rossman *et al.* (*Stud. Mycol.* **42**: 248 pp., 1999), Schroers *et al.* (*Mycol.* **91**: 365, 1999), Schroers *et al.* (*Stud. Mycol.* **45**: 63, 2000), Samuels *et al.* (*Tropical Mycology* **2**: 13, 2002; key), Samuels *et al.* (*CBS Diversity Ser.* **4**, 2006; USA), Zhang *et al.* (*Mycol.* **98**: 1076, 2006; phylogeny), Hirooka & Kobayashi (*Mycoscience* **48**: 81, 2007; Japan).

Bionectriaceae Samuels & Rossman (1999), Hypocreales. 35 gen. (+ 24 syn.), 281 spp.

Lit.: Lowen (*A monograph of the Genera Nectriella, Nitschke and Pronectria Clements. With Reference to Charonectria, Cryptonectriella, Hydronectria and Pseudonectria* [Thesis (PhD) Graduate Faculty of Biology, City University, New York]: 331 pp., 1991), Kohlmeyer & Volkmann-Kohlmeyer (*MR* **97**: 753, 1993), Rehner & Samuels (*CJB* **73**: S816, 1995), Rossman *et al.* (*Stud. Mycol.* **42**: 248 pp., 1999), Schroers *et al.* (*Mycol.* **91**: 375, 1999), Okuda *et al.* (*Mycoscience* **41**: 239, 2000), Rossman (*Stud. Mycol.* **45**: 27, 2000), Schroers (*Stud. Mycol.* **45**: 63, 2000), Rossman *et al.* (*Mycol.* **93**: 100, 2001; rDNA phylogeny), Schroers (*Stud. Mycol.* **46**: 1, 2001), Castlebury *et al.* (*MR* **108**: 864, 2004).

BioNET-INTERNATIONAL (BI). Established in June 1993 and with a secretariat hosted by CABI, BioNET is an international not-for-profit initiative dedicated to promoting taxonomy, especially in the biodiversity rich but economically poorer countries of the world. Focused on less industrialised countries and working via ten regional partnerships, BioNET provides a forum for collaboration for all taxonomists and other users of taxonomy. Working with partners locally and internationally, our work contributes to raising awareness of the importance of taxonomy to society, building and sharing of capacity, and meeting taxonomic needs via innovative tools and approaches. BioNET supports implementation of UNEP's Global Taxonomy Initiative having been cited as an actor over forty times in key decisions of the Convention on Biological Diversity. See also: www.bionet-intl.org (website).

biont, a living organism; commonly used as a suffix to a word indicating the nature or position of the biont; see Symbiosis.

biophagous, see biogenous.

biophilous, see biogenous.

Biophomopsis Petr. (1931), anamorphic *Pezizomycotina*, St.0eH.?. 1, West Indies.

Bioporthe Petr. (1929), Diaporthales. 1. See Müller & von Arx *in* Ainsworth *et al.* (Eds) (*The Fungi* **4A**: 87, 1973), Barr (*Mycol. Mem.* **7**, 1978), Cannon (*Fungal Diversity* **7**: 17, 2001).

Bioprospecting. The action of surveying natural ecosystems for economically valuable biotic products. For fungi, such products might include novel edible fungi, valuable enzymes for biotechnology companies, metabolites for pharmaceutical investigation, or new biological control agents. The extent to which the provisions of the Convention on Biological Diversity (1992) will impinge on bioprospecting will depend on national legislation and regulation systems relating to indigenous intellectual property rights (yet to be developed by most countries); for aspects of the Convention relating to microbial groups see Kelley (*in* Allsopp *et al.* (Eds), *Microbial diversity and ecosystem function*: 415, 1995), Sands (*in* Kirsop & Hawksworth (Eds), *The biodiversity of microorganisms and the role of microbial resource centres*: 9, 1994), Reid *et al.* (Eds) (*Biodiversity prospecting*, 1993). Bioprospecting may also relate to use of organisms to detect and even extract valuable minerals. Some macrofungi are known to accumulate gold (Borovička *et al.*, *MR* **109**: 951, 2005). See Bioindication, Patent protection.

Bioremediation. The use of fungi and other organisms to remove, reduce or ameliorate pollution or potentially polluting materials from the environment. This may be brought about by adding suitable nutrients or selected strains or mixtures of organisms to the substratum; they may be naturally occurring or genetically manipulated. There is an extensive literature on this topic. Fungi (particularly white rot fungi) have been used for degradation of lignocellulosic wastes and more recently for xenobiotics. There have also been studies of possible use of fungi for decomposing dyes in industrial effluents (e.g. Muthezhilan *et al.*, *Res. J. Microbiol.* **3**: 204, 2008; see also Marine fungi), for degrading chlorpyrifos (Fang *et al.*, *International Biodet. & Biodeg.* **61**: 294, 2008) and polycyclic aromatic hydrocarbons (Wu *et al.*, *Biodegradation* **19**: 247, 2008) in contaminated soils. The recent discovery that some fungi appear to be attracted by and grow better in the presence of nuclear radiation has aroused considerable interest with further possibilities for fungal bioremediation. See Alexander (*Biodegradation and bioremediation*, 1994), Baker & Herson (Eds) (*Bioremediation*, 1994), Gadd (*MR* **111**: 3, 2007; review), Kerr (*Handbook of bioremediation*, 1994), Lamar *et al.* (*in* Leather (Ed.), *Frontiers in industrial mycology*: 127, 1992), Scheremaker *et al.* (*in* Betts (Eds), *Biodegradation: natural and synthetic materials*: 157, 1992), Zhdanova *et al* (*MR* **108**: 1089, 2004). See also: Biodegradation, Biodeterioration.

bios, a mixture of aneurin (thiamin, vitamin B1), 'biotin', and other substances in yeasts which, on addition to culture media, gives a better growth of yeast (Wildiers, 1901; see Bonner, *Bot. Rev.* **3**: 616, 1937).

Bioscypha Syd. (1927), Helotiaceae. Anamorph *Chalara*-like. 2, C. America. See Carpenter (*Mem. N. Y. bot. Gdn* **33**, 1981), Samuels & Rogerson (*Brittonia* **42**: 105, 1990), Paulin & Harrington (*Stud. Mycol.* **45**: 209, 2000; anamorph).

biostat, a substance which causes living organisms to stop growing. Cf. biocide.

Biostictis Petr. (1950), Stictidaceae. Anamorph *Rhinocladiella*-like. 4, widespread. See Sherwood (*Occ. Pap. Farlow Herb. Crypt. Bot.* **15**: 105, 1980; key).

biosystematics (1) biological systematics (see systematics); (2) (in botany), experimental taxonomy, including genetical, cytological and ecological aspects. The first usage has a wide currency amongst zoologists and is the preferred term for use in a general context.

biotechnology (1) (in mycology and microbiology), 'all lines of work by which products are produced from raw materials with the aid of living organisms' (Ereky, 1919); see Arora (*Fungal biotechnology in agriculture, food and environmental applications*, 2003), Bud (*Nature* **337**: 10, 1989), Coombs (*Macmillan dictionary of biotechnology*, 1986), Hui & Khachatourians (Eds) (*Food biotechnology: microorganisms*, 1995), Tkacz & Lange (*Advances in fungal biotechnology for industry, agriculture and medicine*, 2004), Wainwight (*An introduction to fungal biotechnology*, 1992); (2) technology concerned with machines in relation to human needs (obsol.).

See Genetic engineering, Industrial mycology, Molecular biology.

Bioterrorism and fungi. Biological weapons are devices which disseminate disease-causing organisms or poisons to kill or harm humans, animals or plants. They generally comprise two parts – an agent and a delivery device. In addition to their military use as strategic weapons or on a battlefield, they can be used for assassinations (having a political effect), can cause social disruption (for example, through enforced quarantine), kill or remove from the food-chain livestock or agricultural produce (thereby causing economic losses), or create environmental problems.

Almost any disease-causing organism (such as bacteria, viruses, fungi, prions or rickettsiae) or toxin (poisons derived from animals, plants or microorganisms, or similar substances synthetically produced) can be used in biological weapons. Historical efforts to produce biological weapons have included: aflatoxin; anthrax; botulinum toxin; foot-and-mouth disease; glanders; plague; Q fever; rice blast; ricin; Rocky Mountain spotted fever; smallpox; and tularaemia. The agents can be altered from their natural state to make them more suitable for use as weapons.

Delivery devices can also take any number of different forms. Some more closely resemble weapons than others. Past programmes have constructed missiles, bombs, hand grenades and rockets. A number of programmes also constructed spray-tanks to be fitted to aircraft, cars, trucks, and boats. Efforts have also been documented to develop delivery devices for use in assassination or sabotage missions, including a variety of sprays, brushes, and injection systems as well as contaminated food and clothes.

As well as concerns that these weapons could be developed or used by states, modern technology is making it increasingly likely they could be acquired by private organisations, groups of people or even individuals. Biological weapons have been used in politically-motivated or criminal acts on a number of occasions (United Nations Office at Geneva, www.unog.ch). Fungi thus have known and potential applications as biological weapons. Their use in this way is not new (there are documented examples, both government-sponsored and by private individuals). But such use is now prohibited by the Biological Weapons Convention. Those now involved in weaponizing fungi are thus unsurprisingly secretive, public awareness is generally low, and the topic has received little public debate, although there are many websites devoted to the issue. *Lit.*: Casadevall & Pirofski (*Medical mycol.* **44**: 689, 2006; weapon potential of human pathogenic fungi), Evans (*Phytopathol.* **97**: 1640, 2007; use of plant pathogen by private individuals for political purposes), Klassen-Fischer (*Clinics in Lab. Medicine* **26**: 387, 2006; review of fungi as bioweapons), Madden & Wheelis (*Ann. Rev. Phytopath.* **41**: 155, 2005; review of plant pathogens as weapons), Meselson & Robinson (The yellow rain affair: lessons from a discredited allegation. *In* Clunan *et al*, *Terrorism, war or disease? Unraveling the use of biological weapons*, 2008; alleged use of *Fusarium* toxins in Indo-China), Paterson (*MR* **110**: 1003, 2006; review of fungi and fungal toxins as weapons). See also Biological control, Hallucinogenic fungi, Mycopesticides, Mycotoxicoses.

Biotransformation. Also known as biological or microbial transformation, or more generally bioconversion; the use of fungi and other organisms to modify organic compounds to produce industrially, medically or environmentally important products. These are usually enzymatic reactions where the substrate may be metabolized or co-metabolized. The most useful reactions are quoted as oxidations, reductions, hydrolysis, condensation, isometisation, formation of new c-c bonds and introduction of hetero functions (Crueger & Crueger, 1990). Many transformations have been described but few are used industrially (e.g. transformations of antibiotics, steroids and sterols). *Rhizopus stolonifer* has been used to produce 11α-hydroxyprogesterone from progesterone. *Chaetomium gracile* is widely used in the sugar industry to supply dextranase needed during processing of bagasse (Eggleston & Monge, *Process Biochemistry* **40**: 1881, 2005). In several processes fungal spores are used directly to catalyze biotransformations. See Crueger & Crueger (*Microbial transformations in biotechnology: a textbook of industrial microbiology*, 1990), O'Sullivan (*in* Fogarty & Kelley (Eds), *Microbial enzymes and biotechnology*, edn 2: 295, 1990). See also: Biodegradation.

biotroph (adj. **-trophic**), an obligate parasite (cf. necrotroph, saprotroph), growing on another organism, in intimate association with its cytoplasm.

Biotyle Syd. (1929) = Pseudomeliola fide von Arx (*Acta Bot. Neerl.* **7**: 503, 1958).

biotype (1) (Scheibe) = physiologic race; (2) one individual; a group of individuals having a like genetic make up (Christensen & Rodenhiser, *Bot. Rev.* **6**: 389, 1940; Waterhouse & Watson, *Proc. Linn. Soc. NSW* **66**: 269, 1941).

bipartite, having division into two.

bipolar (1) (of spore), at the two ends (poles); (2) (of an incompatibility system), having 1 locus; unifactorial; cf. tetrapolar; (3) occurring in both Arctic and Antarctic regions;.

Bipolaris Shoemaker (1959), anamorphic *Cochliobolus*, Hso.≡ eP.26. *c.* 73, widespread. See Luttrell (*Revue Mycol.* Paris **41**: 271, 1977), Alcorn (*Mycotaxon* **13**: 339, 1981), Alcorn (*Mycotaxon* **17**: 1, 1, 1983; gen. concepts), Sivanesan (*Mycol. Pap.* **158**, 1987; keys), Alcorn (*Ann. Rev. Phytopath.* **26**: 37, 1988; gen. taxonomy), Muchovej *et al.* (*Fitopatol. Brasil* **13**: 211, 1988; keys), Hanau *et al.* (*Exp. Mycol.* **13**: 337, 1989; conidiogenous cell development), Alcorn (*Mycotaxon* **39**: 361, 1990; additions to genus), Alcorn (*Mycotaxon* **41**: 329, 1991; n. combs and syns), Khasanov (*Opredelitel' Gribov-Vozbul. 'Gelmintosporiozov' Rasten. iz Rodov Bipolaris, Drechslera i Exserohilum*, 1992), Berbee *et al.* (*Mycol.* **91**: 964, 1999; teleomorph phylogeny), Schell *et al. in* Murray *et al.* (Eds) (*Manual of Clinical Microbiology*: 1295, 1999), Chen *et al.* (*Mycotaxon* **76**: 149, 2000; S Afr.), Olivier *et al.* (*Mycol.* **92**: 736, 2000; phylogeny), Emami & Hack (*Curr. Microbiol.* **45**: 303, 2002; xylanase genes), Buzina *et al.* (*J. Clin. Microbiol.* **41**: 4885, 2003; clinical), Chand *et al.* (*Z. PflKrankh. PflPath. PflSchutz* **110**: 27, 2003; variation), Gafur *et al.* (*Mycobiology* **31**: 19, 2003; infraspecific variation), Castelnuovo *et al.* (*Mycoses* **47**: 76, 2004; clinical), Kim *et al.* (*Pl. Path. J.* **20**: 165, 2004; stem rot of cacti), Tsukiboshi *et al.* (*Mycoscience* **46**: 17, 2005; *B heveae* on grass hosts).

Biporipsilonites Kalgutkar & Janson. (2000), Fossil Fungi. 1, widespread. See Kalgutkar & Jansonius (*AASP Contributions Series* **39**: 37, 2000).

Biporispora J.D. Rogers, Y.M. Ju & Cand. (1999), Xylariales. 1 (on wood), France. See Rogers *et al.* (*Nova Hedwigia* **68**: 421, 1999).

Biporisporites Ke & Shi (1978), Fossil Fungi. 1 (Tertiary), China.

birch canker, Siberian chaga fungus, sterile basidiomata of *Inonotus obliquus*; **- fungus**, *Piptoporus betulinus*.

bird's nest fungi, the *Nidulariaceae*.

Bireticulasporis R. Potonié & Sah (1960), Fossil Fungi. 1 (Miocene), India.

Birsiomyces F. Schaarschm. (1966), Fossil Fungi, Ascomycota. 1 (Triassic), Switzerland.

Bisby (Guy Richard; 1889-1958; USA, later Canada, England). Professor of Plant Pathology, Manitoba Agricultural College, Winnipeg (1920-1936); Mycologist, Commonwealth Mycological Institute, Kew (1937-1954). With Ainsworth (q.v.) co-founder of this Dictionary; noted for his ability to make good decisions quickly. *Biogs, obits etc.* Gregory (*TBMS* **42**: 129, 1959) [portrait]; Johnson (*Phytopathology* **49**: 323, 1959) [portrait]; Stafleu & Cowan (*TL-2* **1**: 219, 1976); Stafleu & Mennega (*TL-2, Suppl.* **2**: 174, 1993).

Bisbyella Boedijn (1951) ≡ Agyriopsis.

Bisbyopeltis Bat. & A.F. Vital (1957), anamorphic *Coccodiniaceae*, Cpt.0fH.?. 1, USA; Australia. See Batista & Vital (*An. Soc. Biol. Pernambuco* **15**: 402, 1957), Reynolds & Gilbert (*Aust. Syst. Bot.* **18**: 265, 2005; *Microxyphium* as synanamorph).

Biscladinomyces Cif. & Tomas. (1953) = Cladonia fide Hawksworth *et al.* (*Dictionary of the Fungi* edn 8, 1995).

Biscogniauxia Kuntze (1891), Xylariaceae. Anamorphs *Nodulisporium*, *Periconiella*-like. 25, widespread. See Martin (*J. S. Afr. Bot.* **35**: 267, 1969; as *Nummulariola*), Jong & Benjamin (*Mycol.* **63**: 862, 1971; N America, key, as *Nummularia*), Pouzar (*Česká Mykol.* **33**: 207, 1979), Callan & Rogers (*CJB* **64**: 842, 1986; anamorphs), Petrini & Müller (*Mycol. Helv.* **1**: 501, 1986; key 5 spp. Eur.), Pouzar (*Česká Mykol.* **40**: 1, 1986; Eur. spp.), Granmo *et al.* (*Op. Bot.* **100**: 59, 1989; Nordic country keys), Whalley *et al.* (*MR* **94**: 237, 1990), González & Rogers (*Mycotaxon* **47**: 229, 1993; key 13 spp. Mexico), Rogers *et al.* (*MR* **100**: 669, 1996), Ju *et al.* (*Mycotaxon* **66**: 1, 1998; monogr.), Ju & Rogers (*Mycotaxon* **73**: 343, 1999; Taiwan), Rogers *et al.* (*Nova Hedwigia* **71**: 431, 2000; Venezuela), Sánchez-Ballesteros *et al.* (*Mycol.* **92**: 964, 2000; phylogeny), Ju & Rogers (*MR* **105**: 1123, 2001; global key), Mazzaglia *et al.* (*MR* **105**: 670, 2001; phylogeny), Smith & Hyde (*Fungal Diversity* **7**: 89, 2001; on palms), Stadler *et al.* (*MR* **105**: 1191, 2001; chemistry), Hsieh *et al.* (*Mycol.* **97**: 844, 2005; phylogeny), Luchi *et al.* (*Lett. Appl. Microbiol.* **41**: 61, 2005; RT-PCR).

biseriate (**biserial**), in two series.

Biseucladinomyces Cif. & Tomas. (1953) = Cladonia.

Bispora Corda (1837), anamorphic *Pezizomycotina*, Hso.1eP.4. 7, widespread (temperate). See Sutton (*CJB* **47**: 609, 1969), Wang (*Mem. N. Y. bot. Gdn* **49**: 20, 1989; pleomorphism), Hawksworth & Cole (*Fungal Diversity* **11**: 87, 2002; exclusion of lichenicolous spp.).

Bispora Fuckel (1870) ≡ Bisporella.

Bisporella Sacc. (1884), Helotiales. Anamorph *Bloxamia*. 19, widespread. The taxonomic position is inconclusive in preliminary phylogenetic studies. See Korf & Carpenter (*Mycotaxon* **1**: 57, 1974), Dumont & Korf (*Caldasia* **12**: 339, 1978), Sharma & Korf (*Mycotaxon* **16**: 326, 1982), Korf & Bujakiewicz (*Agarica* **6**: 302, 1985), Seifert & Carpenter (*CJB* **65**: 1262, 1987; anamorph), Johnston (*Mycotaxon* **31**: 345, 1988; anamorph), Lizoň & Korf (*Mycotaxon* **54**: 471, 1995), Gamundí & Romero (*Fl. criptog. Tierra del Fuego* **10**, 1998), Wang *et al.* (*Mol. Phylogen. Evol.* **41**: 295, 2006; phylogeny), Wang *et al.* (*Mycol.* **98**: 1065, 2006; phylogeny).

Bisporomyces J.F.H. Beyma (1940) = Chloridium fide Hughes (*CJB* **36**: 727, 1958).

Bisporostilbella Brandsb. & E.F. Morris (1971), anamorphic *Microascales*, Hsy.1eP.?. 1, USA. See Brandsberg & Morris (*Mycol.* **63**: 1078, 1971), Seifert *et al.* (*Czech Mycol.* **53**: 297, 2002).

Bisseomyces R.F. Castañeda (1985), anamorphic *Pezizomycotina*, Hso.#eP.10. 1, Cuba. See Castañeda (*Deuteromycotina de Cuba* Hyphomycetes **II**: 3, 1985).

Bitancourtia Thirum. & Jenkins (1953) = Elsinoë fide von Arx & Müller (*Stud. Mycol.* **9**, 1975).

Bitrimonospora Sivan., Talde & Tilak (1974) = Monosporascus fide von Arx (*Kavaka* **3**: 33, 1976).

bitunicate (1) having two walls; (2) (of asci), with two functional layers, that may or may not rupture or extend at discharge; see ascus.

Bitunicostilbe M. Morelet (1971) = Spiropes fide Morelet (*Bulletin de la Société des Sciences naturelles et d'Archéologie de Toulon et du Var* **195**: 7, 1971), Deighton (*MR* **94**: 1096, 1990).

Bitzea Mains (1939) = Chaconia fide Thirumalachar & Cummins (*Mycol.* **41**: 523, 1949).

biuncinate, two-hooked.

Bivallum P.R. Johnst. (1991), Rhytismataceae. 6 (on conifers), Australasia; Chile. See Johnston (*Aust. Syst. Bot.* **4**: 355, 1991), Johnston (*Aust. Syst. Bot.* **14**: 377, 2001; Australasia).

bivalvate (1) (of spores), lens-shaped and having a hyaline rim, as in *Arthrinium*; (2) (of asci), see ascus.

Biverpa (Fr.) Boud. (1907) ? = Helvella fide Eckblad (*Nytt Mag. Bot.* **15**: 1, 1968).

biverticillate (of a penicillus), having branching at two levels, i.e. having metulae bearing phialides.

Bivonella (Sacc.) Sacc. (1891) = Thyridium Nitschke fide Cannon (*SA* **8**: 78, 1989).

Bizozzeria, see *Bizzozeria Sacc. & Berl.*

Bizzozeria Sacc. & Berl. (1885) = Lasiosphaeria See Lundqvist (*Symb. bot. upsal.* **20** no. 1, 1972).

Bizzozeria Speg. (1889) ≡ Thaxteria Sacc.

Bizzozeriella Speg. (1888), anamorphic *Pezizomycotina*, Hsp.0eH.?. 1, S. America. See Donk (*Persoonia* **1**: 189, 1960; nomencl.).

Bjerkandera P. Karst. (1879), Meruliaceae. 2, widespread (north temperate). See Ryvarden & Gilbertson (*Europ. Polyp.* **1**: 168, 1993), Anon. (*Mycologist* **18**: 174, 2004; *Bjerkandera adusta*).

Bjerkanderaceae Jülich (1982) = Hapalopilaceae.

black blotch, of clovers (*Cymadothea trifolii*): **- crottle**, see crottle; **- dot** of potato (*Colletotrichum coccodes*); **- jelly fungus**, basidioma of edible *Auricularia* spp.; **- knot** of plum and cherry (*Apiosporina morbosa*); **- leg** of beet (*Phoma betae*, *Pythium*, etc.); of pelargonium (*Pythium* spp.); **- line**, see zone lines; **- mildews**, *Meliolales*; **- piedra**, infection of hair shafts by *Piedraia hortae*; **- pustule** of *Ribes* (*Plowrightia ribesia*); **- root rot** of tobacco and other plants (*Thielaviopsis basicola*); of grapes (*Vitis*) (*Guignardia bidwellii*); **- scurf** of potato (*Thanatephorus cucumeris*, syn. *Corticium solani*); **- slime** of hyacinth (*Sclerotinia bulborum*); **- spot** of apple, see scab, apple; of rose (*Diplocarpon rosae*); **- stem rust** of cereals (*Puccinia graminis*); **- tip** of banana (*Musa*) (*Deightoniella torulosa*); **- tree lichen** (*Bryoria fremontii*) (Turner, *Econ. Bot.* **31**: 461, 1977); **- yeasts**, see yeasts.

blackfellows' bread (or native bread), the sclerotium (*Mylitta australis*) of the Australian *Polyporus mylittae*. See McAlpine (*J. Dep. Agric. Vict.* **2**, 1904), Willis (*Muelleria* **1**: 203, 1967; bibliogr.), Macfarlane *et al.* (*TBMS* **71**: 359, 1978; structure). There are similar sclerotia in India ('little mans' bread') and China.

Blakeslea Thaxt. (1914), Choanephoraceae. 2, widespread (esp. tropical). See Thaxter (*Bot. Gaz.* **58**: 353, 1914), Mehrotra & Baijal (*J. Elisha Mitchell scient. Soc.* **84**: 207, 1968), Kirk (*Mycol. Pap.* **152**: 61 pp., 1984; key), Zheng & Chen (*Acta Mycol. Sin.* Suppl. **1**: 40, 1986), Voigt & Wöstemeyer (*Gene* **270**: 113, 2001; phylogeny), Ho & Chang (*Taiwania* **48**: 232, 2003; Taiwan), Idnurm *et al. in* Heitman *et al.* (Eds.) (*Sex in Fungi*: 407, 2007; mating).

Blarneya D. Hawksw., Coppins & P. James (1980), anamorphic *Pezizomycotina*, Hso.1eH.38 (L). 1, Europe (western). See Hawksworth *et al.* (*J. Linn. Soc.* Bot. **79**: 358, 1979).

Blasdalea Sacc. & P. Syd. (1902), Vizellaceae. Anamorph *Chrysogloeum*. 1, Brazil. See Petrak (*Sydowia* **7**: 343, 1953).

Blasiphalia Redhead (2007), Agaricomycetes. 1, USA. *Hymenochaetales* or *Agaricales* (*Rickenella* clade). See Redhead (*Mycol.* **98**: 934, 2007).

Blastacervulus H.J. Swart (1988), anamorphic *Pezizomycotina*, Cac.0eP.4. 1, Australia. See Swart (*TBMS* **90**: 289, 1988).

Blastenia A. Massal. (1852) = Caloplaca fide Hawksworth *et al.* (*Dictionary of the Fungi* edn 8, 1995).

Blasteniomyces Cif. & Tomas. (1953) ≡ Protoblastenia.

Blasteniospora Trevis. (1853) nom. rej. = Xanthoria fide Hawksworth *et al.* (*Dictionary of the Fungi* edn 8, 1995).

blasteniospore, a polarilocular (q.v.) ascospore.

blastic (of conidiogenesis), one of the two basic sorts of conidiogenesis (cf. thallic), characterized by a marked enlargement of a recognizable conidial initial *before* the initial is delimited by a septum. The conidium is differentiated from *part* of a cell (Kendrick, 1971: 255); **entero-**, when the inner wall (see tretic) or neither wall (see phialidic) of the blastic conidiogenous cell contributes to the formation of the **conidium** (blastic conidium) (cf. holoblastic); **holo-**, when both outer and inner walls of the blastic conidiogenous cell contribute to the formation of the conidium (cf. enteroblastic and see annellidic); **mono-**, when a conidiogenous cell has only one conidiogenous locus; **poly-**, when a conidiogenous cell has several conidiogenous loci.

Blasticomyces I.I. Tav. (1985), Laboulbeniaceae. 2, Asia. See Tavares (*Mycol. Mem.* **9**: 155, 1985), Majewski (*TMSJ* **29**: 249, 1988).

blastidium, a lichen propagule produced by the budding of thalli in a yeast-like manner (Poelt, *Flora, Jena* **169**: 23, 1980). Fig. 22E.

Blastobasidiomycetes, see *Basidiomycota*, *Ustomycetes*.

Blastobotrys Klopotek (1967), anamorphic *Trichomonascus*, Hso.0eH.6. 13, widespread. See Marvanová (*TBMS* **66**: 217, 1976), de Hoog *et al.* (*Antonie van Leeuwenhoek* **51**: 79, 1985; keys), de Hoog *in* Sugiyama (Ed.) (*Pleomorphic Fungi: The Diversity and its Taxonomic Implications*: 221, 1987; developmental cycle), Hoog & Smith *in* Kurtzman & Fell (Eds) (*Yeasts, a taxonomic study* 4th edn: 443, 1998), Suh *et al.* (*Mycol.* **98**: 1006, 2006; phylogeny), Blackwell *et al.* (*British Mycological Society Symposium Series* **25**: 357, 2007; ecology), Kurtzman (*Int. J. Syst. Evol. Microbiol.* **57**: 1154, 2007), Kurtzman & Robnett (*FEMS Yeast Res.* **7**: 141, 2007).

Blastocapnias Cif. & Bat. (1963) = Aithaloderma fide von Arx & Müller (*Stud. Mycol.* **9**, 1975).

Blastocatena Subram. & Bhat (1989), anamorphic *Pezizomycotina*, Hsy.≡ eP.4. 1, India. See Subramanian & Bhat (*Kavaka* **15**: 43, 1987).

blastocatenate (of blastoconidia), formed in chains with the youngest at the apical or distal end of the chain.

Blastocladia Reinsch (1877), Blastocladiaceae. *c.* 15 (saprobes in water), widespread. See Emerson & Cantino (*Am. J. Bot.* **35**: 157, 1948), Das Gupta & John (*Indian Phytopath.* **41**: 521, 1988), Steciow (*Darwiniana* **37**: 335, 1999; Argentina).

Blastocladiaceae H.E. Petersen (1909), Blastocladiales. 5 gen. (+ 4 syn.), 40 spp.

Lit.: Olson (*Op. bot.* **73**: 1, 1984), Remy *et al.* (*Am. J. Bot.* **81**: 690, 1994), Bullerwell *et al.* (*Nucl. Acids Res.* **31**: 1614, 2003), Tanabe *et al.* (*J. gen. appl. Microbiol.* Tokyo **51**: 267, 2005).

Blastocladiales H.E. Peterson (1909). Blastocladiomycetes. 5 fam., 14 gen., 179 spp. Thallus monocentric

or polycentric; zoospores with a prominent nuclear cap of ribosomes, microtubule root consisting of 27 microtubules in groups of 3 extending from proximal end of kinetosome; freshwater or terricolous, saprobic or parasitic (*Coelomomyces* on mosquito larvae); cosmop. Fams:
(1) **Blastocladiaceae**
(2) **Catenariaceae**
(3) **Coelomomycetaceae**
(4) **Physodermataceae**
(5) **Sorochytriaceae**
Lit.: Sparrow (1960: 605; 1973), Fitzpatrick (1930: 130), Emerson & Robertson (*Am. J. Bot.* **61**: 303, 1974), Karling (1977, 1978), Dewel & Dewel (*CJB* **68**: 1968, 1977), Lange & Olson (*TBMS* **74**: 449, 1980), Olson (*Opera Bot.* **73**: 1, 1984; key fams), Dewel (*CJB* **63**: 1525, 1985), James *et al.* (*Mycol.* **98**: 860, 2006; molecular phylogeny), Hibbett *et al.* (*MR* **111**: 109, 2007), and see under Familes.

Blastocladiella V.D. Matthews (1937), Blastocladiaceae. 13, widespread. See Couch & Whiffen (*Am. J. Bot.* **29**: 582, 1942), Sparrow (*Aquatic Phycomycetes* Edn 2: 660, 1960; key), Cantino *in* Meynell & Gooder (Eds) (*Microbial reaction to environment*: 243, 1961; morphogenesis), Karling (*Mycopath. Mycol. appl.* **49**: 169, 1973; subgen.).

Blastocladiomycetes Doweld (2001). Blastocladiomycota. 1 ord., 5 fam., 14 gen., 179 spp. Ord.:
Blastocladiales
Lit.: James *et al.* (*Mycol.* **98**: 860, 2006; molecular phylogeny), and see under Fam. and Ord.

Blastocladiomycota T.W. James (2007), Fungi. 1 class., 1 ord., 5 fam., 14 gen., 179 spp. Class:
Blastocladiomycetes
Lit.: James *et al.* (*Mycol.* **98**: 860, 2006; molecular phylogeny), and see under Classes and Orders.

Blastocladiopsis Sparrow (1950), Blastocladiaceae. 2, USA; Cuba. See Sparrow (*Journal of the Washington Academy of Science* **40**: 52, 1950).

blastoconidium, a blastic (q.v.) conidium.

Blastoconium Cif. (1931), anamorphic *Pezizomycotina*, Hso.#eP.?. 2, widespread (tropical).

Blastodendrion (M. Ota) Cif. & Redaelli (1925) = Candida fide Zobel (*Arch. Hyg. Berlin* **130**: 205, 1943), Bai *et al.* (*Int. J. Syst. Evol. Microbiol.* **50**: 417, 2000).

Blastoderma B. Fisch. & Brebeck (1894) nom. ambig., anamorphic *Pezizomycotina*.

Blastodesmia A. Massal. (1852), Arthopyreniaceae (L). 1 (on *Fraxinus*), Europe. See Keissler (*Rabenh. Krypt.-Fl.* **9**: 384, 1937), Aptroot (*Biblthca Lichenol.* **44**, 1991; posn).

Blastodictys M.B. Ellis (1976), anamorphic *Pezizomycotina*, Hso.#eP.10. 1, Uganda. See Ellis (*More Dematiaceous Hyphomycetes*: 149, 1976).

Blastofusarioides Matsush. (1996), anamorphic *Pezizomycotina*, Hso.?.?. 1, Japan. See Matsushima (*Matsush. Mycol. Mem.* **9**: 2, 1996).

Blastomyces Costantin & Rolland (1888), Ajellomycetaceae. 1, Africa; America. See *Ajellomyces*, blastomycosis. See van Oorschot (*Stud. Mycol.* **20**, 1980), Watts *et al.* (*Am. Jl Clin. Path.* **93**: 575, 1990; giant forms of *B. dermatitidis*), Geber *et al.* (*J. Gen. Microbiol.* **138**: 395, 1992; phylogeny, rRNA sequence and phylogeny in *B. dermatitidis*), Hurst & Kaufman (*J. Clin. Microbiol.* **30**: 3043, 1992; serology), Walsh *et al.* (*Manual of Clinical Microbiology* Edn 6: 749, 1995; review), Guého *et al.* (*Mycoses* **40**: 69, 1997; epidemiology), Di Salvo (*Topley & Wilson's Microbiology and Microbial Infections* Edn 9. Vol. **4** Medical Mycology: 337, 1998; review), Larone *et al.* (*Manual of Clinical Microbiology*: 1259, 1999; review), Huerre *et al.* (*J. Mycol. Médic.* **12**: 5, 2002; morphology), Untereiner *et al.* (*Mycol.* **96**: 812, 2004; phylogeny), Bialek *et al.* (*FEMS Immunol. Med. Microbiol.* **45**: 355, 2005; molecular diagnosis), Pounder *et al.* (*J. Clin. Microbiol.* **44**: 2977, 2006; molecular diagnosis).

Blastomyces Gilchrist & W.R. Stokes (1898) = Zymonema fide Dodge (*Medical Mycology*, 1935).

Blastomycetes. Class often used for anamorphic yeasts (q.v.) and then divided into two Orders, (1) *Cryptococcales* (reproduction by budding, ballistospores absent; ascomycetous affinities, but see under Order), and (2) *Sporobolomycetales* (reproduction by budding and ballistospores; basidiomycetous affinities). This simplistic scheme is no longer tenable (see Kendrick, *The fifth kingdom*, edn 2, 1992). Included as anamorphic fungi in this edition of the *Dictionary*.

blastomycin, an antigen made from *Blastomyces dermatitidis*, esp. for skin testing; **- S**, an antifungal antibiotic from *Streptomyces griseochromogenes* (Fukunage *et al.*, *Bull. agric. chem. Soc. Japan* **19**: 181, 1955) used against rice blast (*Pyricularia oryzae*).

Blastomycoides Castell. (1928) ≡ Zymonema.

blastomycosis (1) a disease in humans caused by *Blastomyces dermatitidis* (teleomorph *Ajellomyces dermatitidis*; see Al-Doory & Di Salvo, *Blastomycosis*, 1992); **N. American -**; Gilchrist's disease; (2) any mycotic disease in humans having budding cells in the parasitized tissues; **cheloidal -**, see lobomycosis; **European -**, see cryptococcosis; **S. American -**, see *Paracoccidioides*.

Blastomycota. Proposed for *Ascoblastomycetes* and *Basidioblastomycetes*, anamorphic yeasts with ascomycetous and basidiomycetous affinities respectively (Moore, *Bot. Mar.* **23**: 361, 1980).

Blastophoma Kleb. (1933) ? = Sclerophoma fide Sutton (*Mycol. Pap.* **141**, 1977).

Blastophorella Boedijn (1937), anamorphic *Pezizomycotina*, Hsy.1eP.6. 1, Java; Sumatra.

Blastophorum Matsush. (1971), anamorphic *Pezizomycotina*, Hso.≡ eH.10/18. 4, Papua New Guinea. See Matsushima (*Microfungi of the Solomon Islands and Papua-New Guinea*: 8, 1971), Matsushima & Matsushima (*Matsush. Mycol. Mem.* **9**: 31, 1996).

Blastophragma Subram. (1995), anamorphic *Pezizomycotina*. 2, S.E. Asia. See Subramanian (*Kavaka* **20/21**: 57, 1995).

Blastoschizomyces Salkin, M.A. Gordon, Sams. & Rieder (1982) = Dipodascus fide Polachek *et al.* (*J. Clin. Microbiol.* **30**: 2318, 1992; taxonomic review), Kurtzman & Fell (*Yeasts, a taxonomic study* 4th edn, 1998), D'Antonio *et al.* (*J. Clin. Microbiol.* **37**: 2927, 1999; onychomycosis).

Blastospora Dietel (1908), Mikronegeriaceae. 3 (on *Rosaceae* or *Cupressaceae* (0, I); on *Smilacaceae* or *Betulaceae* (II, III)), Japan; China; Korea; Nepal. See Mains (*Am. J. Bot.* **25**: 677, 1938), Kaneko & Hiratsuka (*Mycol.* **73**: 577, 1981), Ono *et al.* (*Mycol.* **78**: 253, 1986).

blastospore, a spore formed by marked enlargement of a recognizable conidium initial before the initial is delimited by a septum. The conidium differentiates from part of the cell. See Kendrick (Ed.) (*Taxonomy of fungi imperfecti*, 1971).

Blastosporella T.J. Baroni & Franco-Mol. (2007), Lyophyllaceae. 1, Colombia. See Baroni *et al.* (*MR* **111**: 572, 2007).

Blastosporidium M. Hartmann (1912) ? = Coccidioides fide Sutton (*in litt.*).

Blastostroma C.Z. Wei, Y. Harada & Katum. (1998), anamorphic *Mycodidymella*, Hso.?.?. 1, Japan. See Wei *et al.* (*Mycol.* **90**: 337, 1998).

Blastotrichum Corda (1838) nom. dub., anamorphic *Pezizomycotina*, Hso.≡ eH.?. 5, widespread (esp. Europe). See Gams & Hoozemans (*Persoonia* **6**: 99, 1970).

blematogen (**blematogen layer**), the undifferentiated tissue which becomes the universal veil in agarics (Atkinson, *Am. J. Bot.* **1**: 3, 1914).

Blennorella Kirschst. (1944) ? = Colletotrichum fide Sutton (*Mycol. Pap.* **141**, 1977).

Blennoria Moug. & Fr. (1825), anamorphic *Pezizomycotina*, St.0eH.15. 1, Europe. See Sutton (*Taxon* **21**: 319, 1972).

Blennoriopsis Petr. (1920), anamorphic *Pezizomycotina*, Hso.0eH.?. 1, Europe.

Blennothallia Trevis. (1853) = Collema F.H. Wigg. fide Hawksworth *et al.* (*Dictionary of the Fungi* edn 8, 1995).

Blepharia (Pers.) Ainsw. & Bisby (1943) nom. inval. = Dematium fide Mussat (*Syll. fung.* **15**: 62, 1901).

blepharoplast (of zoospores), the basal body or granule (**kinetosome**) from which arise the longitudinal fibres constituting the axoneme of a flagellum; joined to the nucleus by a **rhizoplast.**

Bleptosporium Steyaert (1961), anamorphic *Amphisphaeriaceae*, Cpd.≡ eP.19. 1, Argentina. See Sutton (*Mycol. Pap.* **88**, 1963).

blewits (blewitt, blue leg, bluette), basidiomata of the edible *Lepista saeva* (syn. *Tricholoma personatum*); **wood -**, *L. nuda* (syn. *T. nudum*).

blight, a common name for a number of different diseases of plants (and for insect attack), esp. when leaf damage is sudden and serious; **potato -**, **late -** (*Phytophthora infestans*); **early -** (*Alternaria solani*).

blister rust (of 5-needled pines), *Cronartium ribicola*.

Blistum B. Sutton (1973), anamorphic *Clavicipitaceae*. 2 (on myxomycetes), widespread. See Seifert (*in litt.*), Bischoff *et al.* (*Mycotaxon* **86**: 433, 2003; phylogeny).

Blitridium De Not. (1863) = Triblidium fide Nannfeldt (*Nova Acta R. Soc. Scient. upsal.*, 1932).

Blodgettia E.P. Wright (1881) ≡ Blodgettia Harv.

Blodgettia Harv. (1858), anamorphic *Pezizomycotina*, Hso.-.-. 2 (mycophycobionts; marine), widespread. See Kohlmeyer & Kohlmeyer (*Marine Mycology*, 1979), Hawksworth (*Notes R. bot. Gdn Edinb.* **44**: 549, 1987), Kohlmeyer & Volkmann-Kohlmeyer (*Bot. Mar.* **34**: 1, 1991).

Blodgettiomyces Feldmann (1939) ≡ Blodgettia Harv. fide Hawksworth (*Notes R. bot. Gdn Edinb.* **44**: 549, 1987).

Blogiascospora Shoemaker, E. Müll. & Morgan-Jones (1966), Amphisphaeriaceae. Anamorph *Seiridium*. 1, Europe. See Shoemaker *et al.* (*CJB* **44**: 247, 1966), Kang *et al.* (*MR* **103**: 53, 1999).

Bloxamia Berk. & Broome (1854), anamorphic *Bisporella*, Hsp.0eP.22. 8, widespread. See Nag Raj & Kendrick (*Monogr. Chalara Allied Genera*, 1975), Johnston (*Mycotaxon* **31**: 345, 1988; teleomorph), Arambarri *et al.* (*Mycotaxon* **43**: 327, 1992; key).

blue cheeses, ripened and flavoured by *Penicillium roqueforti* (Mitosp. fungi), e.g. Roquefort, Stilton, Gorgonzola, Danish Blue etc.

blue stain, blue-grey colouration of wood caused by the growth of brown fungal hyphae in the surface layers.

Blumenavia Möller (1895), Phallaceae. 3, S. America; Africa. See Vargas-Rodriguez & Vázquez-García (*Mycotaxon* **94**: 7, 2005; Mexico).

Blumeria Golovin ex Speer (1975), Erysiphaceae. Anamorph *Oidium* subgen. *Oidium*. 1, widespread. *B. graminis* (cereal and grass mildew). See Braun (*Beih. Nova Hedwigia* **89**, 1987), Caffier *et al.* (*Pl. Path.* **48**: 582, 1999; genetic diversity), Mori *et al.* (*Mycoscience* **41**: 437, 2000; phylogeny), Salari *et al.* (*Iran. Jl agric. Res.* **34**: 353, 2003; physiological races), Wyand & Brown (*Molecular Plant Pathology* **4**: 187, 2003; infraspecific variation), Takamatsu (*Mycoscience* **45**: 147, 2004; phylogeny), Wang *et al.* (*Mycol.* **98**: 1065, 2006; phylogeny), Inuma *et al.* (*Mol. Phylogen. Evol.* **44**: 741, 2008; phylogeny).

Blumeriaceae V.P. Gelyuta (1988) = Erysiphaceae.

Blumeriella Arx (1961), Dermateaceae. Anamorphs *Microgloeum*, *Phloeosporella*. 5 (on *Rosaceae*), N. America; Europe. *B. jaapii* (anamorph *P. padi*), cherry leaf spot. See von Arx (*Phytopath. Z.* **42**: 161, 1961), Jakobsen & Jørgensen (*Tidsskr. Plant.* **90**: 161, 1986), Williamson & Bernard (*CJB* **66**: 2048, 1988; on *Spiraea*), Nauta & Spooner (*Mycologist* **14**: 21, 2000; UK).

blusher, the, basidioma of the edible *Amanita rubescens*.

Blytridium, see *Blitridium*.

Blyttiomyces A.F. Bartsch (1939), Chytridiaceae. *c.* 10, widespread. See Dogma & Sparrow (*Mycol.* **61**: 1149, 1970), Dogma (*Kalikasan* **8**: 237, 1979; key), Letcher *et al.* (*Australasian Mycologist* **22**: 99, 2004; Australia).

BM, The Natural History Museum (London, UK); founded 1753; known by official name The British Museum (Natural History) up to 1989; governed by a Board of Trustees and funded through the Office of Arts and Libraries; most non-lichenized fungi were transferred to **K** in 1969 (Brenan & Ross, *Lichenologist* **4**: 157, 1970); see Anon. (*The history of the collections contained in the Natural History Department of the British Museum*, **1**, 1904), Stearn (*The Natural History Museum at South Kensington*, 1981).

boathook hair, characteristic terminally bifid hair, produced intracellularly in principal-form zoospore of *Saprolegnia*, and which ornaments the cyst formed by this zoospore; principal-form cyst or secondary cyst (see Beakes, 1983).

Bodinia M. Ota & Langeron (1923) = Trichophyton fide Sutton (*in litt.*).

Boedijnopeziza S. Ito & S. Imai (1937) = Cookeina fide Denison (*Mycol.* **59**: 306, 1967), Pfister (*Phytologia* **27**: 55, 1973), Weinstein *et al.* (*Mycol.* **94**: 673, 2002), Iturriaga & Pfister (*Mycotaxon* **95**: 137, 2006).

Boehmia Raddi (1806) nom. rej. = Arrhenia fide Kuyper (*in litt.*) Rejected against *Leptoglossum* but should also be rejected against *Arrhenia*.

Boeomycomyces, see *Baeomycomyces*.

Boerlagella Penz. & Sacc. (1897) [non *Boerlagella* Cogn. 1891, *Sapotaceae*] ≡ Boerlagiomyces.

Boerlagellopsis C. Ramesh (1988) nom. nud., ? Dothideales. 1, India. See Eriksson & Hawksworth (*SA* **8**: 62, 1989).

Boerlagiomyces Butzin (1977), Tubeufiaceae. 6, widespread (tropical). See Crane *et al.* (*CJB* **76**: 602, 1998), Kodsueb *et al.* (*Fungal Diversity* **21**: 105, 2006; phylogeny).

Bogoriella Zahlbr. (1928), ? Verrucariaceae (L). 1, Java.

Bogoriellomyces Cif. & Tomas. (1954) ≡ Bogoriella.

Bohleria Trevis. (1860) = Placidiopsis fide Hawksworth *et al.* (*Dictionary of the Fungi* edn 8, 1995).

Bohuslavia J.I.R. Larsson (1985), Microsporidia. 1.

Boidinia Stalpers & Hjortstam (1982), Russulaceae. 10, widespread. Polyphyletic. See Ginns & Freeman (*Biblthca Mycol.* **157**, 1994), Wu & Buchanan (*Mycotaxon* **67**: 123, 1998).

Bojamyces Longcore (1989), Legeriomycetaceae. 3 (in *Ephemeroptera*), Spain; Mexico; USA. See Longcore (*Mycol.* **81**: 482, 1989), Valle & Santamaría (*Mycol.* **96**: 1386, 2004; Spain), White (*MR* **110**: 1011, 2006; phylogeny), Valle *et al.* (*Mycol.* **100**: 149, 2008; Mexico).

Bolacotricha Berk. & Broome (1851) = Chaetomium fide Hawksworth *et al.* (*Dictionary of the Fungi* edn 8, 1995).

Bolbitiaceae Singer (1948), Agaricales. 17 gen. (+ 16 syn.), 287 spp.

Lit.: Watling & Gregory (*Biblthca Mycol.* **82**, 1981; nomenclator), Pegler (*Kew Bull.* Addit. Ser. **12**: 519 pp., 1986), Singer (*Agaric. mod. Tax.* 4th ed, 1986), Watling & Taylor (*Biblthca Mycol.* **117**: 61 pp., 1987), Singer & Hausknecht (*Pl. Syst. Evol.* **159**: 107, 1988), Watling (*Op. bot.* **100**: 259, 1989), Young (*Aust. Syst. Bot.* **2**: 75, 1989), Singer & Hausknecht (*Pl. Syst. Evol.* **180**: 77, 1992), Fukihara & Hongo (*Mycoscience* **36**: 425, 1995), Gerhardt (*Biblthca Botanica* **147**: 149 pp., 1996), Guidot *et al.* (*Appl. Environm. Microbiol.* **65**: 903, 1999), Aanen *et al.* (*Mycol.* **92**: 269, 2000), Moncalvo *et al.* (*Syst. Biol.* **49**: 278, 2000)

Lit.:, Peintner *et al.* (*Am. J. Bot.* **88**: 2168, 2001), Hallen *et al.* (*MR* **107**: 969, 2003), Walther *et al.* (*MR* **109**: 525, 2005), Walther & Weiss (*Mycol.* **98**: 792, 2006).

Bolbitius Fr. (1838), Bolbitiaceae. *c.* 25, widespread. See Singer (*Sydowia* **30**: 216, 1977; keys world spp.), Watling & Gregory (*Biblthca Mycol.* **82**, 1981), Watling (*British fungus flora* **3**, 1982; key 6 Br. spp.), Arnolds (*Persoonia* **18**: 201, 2003; The Netherlands), Walther & Weiss (*Mycol.* **98**: 792, 2006; anam.).

Boletaceae Chevall. (1826), Boletales. 35 gen. (+ 26 syn.), 787 spp.

Lit.: Pegler & Young (*TBMS* **72**: 353, 1979), Alessio (*Boletus Dill. ex L.*: 712 pp., 1985), Singer (*Agaric. mod. Tax.* 4th ed, 1986), Thiers (*Mem. N. Y. bot. Gdn* **49**: 355, 1989), Watling & Hollands (*Notes R. bot. Gdn Edinb.* **46**: 405, 1990), Singer *et al.* (*Beih. Nova Hedwigia* **102**: 99 pp., 1991), Lannoy & Estadès (*Monographie des Leccinum d'Europe*: 229 pp., 1995), Lakhanpal (*Stud. Cryptog. Bot.* **1**: 170 pp., 1996), Bruns *et al.* (*Mol. Ecol.* **5**: 257, 1998), Kretzer & Bruns (*Mol. Phylogen. Evol.* **13**: 483, 1999), Binder & Besl (*Micologia 2000*: 75, 2000), Watling (*Aust. Syst. Bot.* **14**: 407, 2001), Binder & Bresinsky (*Mycol.* **94**: 85, 2002), Binder & Bresinsky (*Feddes Repert.* **113**: 36, 2002), Lebel & Castellano (*Mycol.* **94**: 327, 2002), Trappe *et al.* (*Mycotaxon* **81**: 195, 2002), Peintner *et al.* (*MR* **107**: 659, 2003), Bakker & Kuyper (*Mycol.* **96**: 102, 2004), Bakker *et al.* (*New Phytol.* **163**: 201, 2004), Mello *et al.* (*J. Biotechn.* **121**: 318, 2006).

Boletales E.-J. Gilbert (1931). Agaricomycetidae. 17 fam., 96 gen., 1316 spp. Terrestrial or lignicolous, saprobic (when lignicolous usually causing brown rot), ectomycorrhizal, sometimes parasitic on ectomycorrhizal members of the Boletales, edible, a few species poisonous, cosmopolitan. Formerly used for the 'boletes', a group of fungi with fleshy fruitbodies and poroid hymenophore, usually growing terrestrially. The poroid hymenophore has arisen at least 5 times in the 8 clades of the Homobasidiomycetes (Hibbett & Thorn, *The Mycota* **7B**, 2001).

Classification: Fries (*Syst. mycol.* **1-3**, 1821-1832) put the fleshy poroid fungi in the genus *Boletus*. A number of his infragenerc groups have subsequently been raised to generic level, and a number of genera have been added. In general, bolete taxonomy reflected its eurocentric bias but Singer (*Agaricales mod. taxon.*, 4th ed., 1986) pointed out that bolete taxonomy can only be understood on the basis of knowledge of tropical species. The *Boletales* as accepted in this edition of the Dictionary also contain lamellate forms, gastroid forms (both epigeous and hypogeous). Consequently, they cannot be characterised in morphological terms, and diagnoses for the various families are not provided. Chemotaxonomic classifications (Gill & Steglich, *Progr. Chem. Nat. Prod.* **51**, 1987) are largely consistent with molecular data; see Binder & Hibbett (*Mycol.* **98**: 971, 2006), Hibbett *et al.* (*MR* **111**: 109, 2007). Fams:

(1) **Boletaceae**
(2) **Boletinellaceae**
(3) **Calostomataceae**
(4) **Coniophoraceae**
(5) **Diplocystidiaceae**
(6) **Gastrosporiaceae**
(7) **Gomphidiaceae**
(8) **Gyroporaceae**
(9) **Hygrophoropsidaceae**
(10) **Paxillaceae**
(11) **Protogastraceae**
(12) **Rhizopogonaceae**
(13) **Sclerodermatacea**
(14) **Serpulaceae**
(15) **Suillaceae**

Lit.: Singer (*Agaricales mod. taxon.*, 4th ed., 1986), Arpin & Kühner (*Bull. Soc. linn. Lyon* **46**: 83-108, 181-208, 1977; classification), Clémençon (*Anatomie der Hymenomyceten*, 1997), Gill & Steglich (*Progr. Chem. Nat. Prod.* **51**, 1987; pigment chemistry, Horak (*Synopsis generum Agaricalium*, 1967), Donk (*Reinwardtia* **3**: 275, 1955; nomenclature), Høiland (*Nord. J. Bot.* **7**: 705, 1987; chemotaxonomy).

Regional: [see also under Macromycetes]. Africa: Heinemann (*Bull. Jard. bot. Etat Brux.* **21**: 223, 1951; **30**: 21, 1960; **34**: 425, 1964; Katanga, Zaire, **Uganda). America, North: Snell & Dick (*The* ***Boleti of Northeastern North America*, 1970), Smith & Thiers (*The Boletes of Michigan*, 1971). Belgium: Heinemann (*Naturalistes belges* **42**: 333, 1961). Europe, Central: Singer (*Die Röhrlinge*, Teil 2, *Die Boletoideae und Strobilomyceteae*, 1967 [*Die Pilze Mitteleuropas*, **6**]). British Isles: Pearson (*Naturalist* Hull, 1946; key), Watling (*British Fungus Flora* **2**, 1970). France: Gilbert (*Les bolets*, 1931), Blum (*Les bolets*, 1963; key 66 spp.; suppl. and revisions, *BSMF* **80**: 297, 1964; **81**: 478, 1965; **84**: 309, 577, 1969; *Revue mycol.* **34**: 249, 1970), Leclair & Essette (*Les *bolets*, 1968; col. pl.). Malaysia: Corner (*Boletus in Malaysia*, 1972). New Zealand: McNabb (*N.Z. Jl Bot.* **6**: 137, 1968). Nova Scotia: Grund & Harrison (*Nova Scotia boletes*, 1976). Poland: Skirgiello (Grzyby (Fungi), Podstawczaki (Basidiomycetes), Barowikwe (Boletales), 1960 [Engl. transl. see *Mycol.* **68**: 1136]). **USA**: Coker & Beers (*The Boletaceae of North Carolina*, 1943), Singer (*The Boletineae of Florida*, 1945-47 [reprinted 1970 from *Farlowia* **2** and *Am. midl. Nat.* **37**]), Thiers (*California mushrooms: a field guide to the boletes*, 1975), Mycogeography in the South Pacific Region (*Austr. J. Bot.* Suppl. Ser. No. 10, 1983), Singer, Araujo & Ivory (*Beih. Nova Hedw.* **77**, 1983), and under fams.

bolete, one of the *Boletales*.

Boletellaceae Jülich (1982) = Boletaceae.

Boletellites P. Briot, Lar.-Coll. & Locq. (1983), Fossil Fungi. 1, Australia.

Boletellus Murrill (1909), Boletaceae. *c.* 50, widespread (esp. subtropical). Separation of *Boletellus* and *Boletus* is still controversial, as spores with longitudinal ridges also occur in *Boletus*. See Singer (*Sydowia* **30**: 221, 1978; key), Singer *et al.* (*Beih. Nova Hedwigia* **105**: 3, 1992; key C. Am. spp.), Binder & Fischer (*Boll. Gruppo Micol. 'G. Bresadola'* **40**: 79, 1997; phylogeny, relationship with *Boletus*).

Boletinellaceae P.M. Kirk, P.F. Cannon & J.C. David (2001), Boletales. 2 gen. (+ 2 syn.), 14 spp.

Lit.: Corner (*Beih. Nova Hedwigia* **33**: 1, 1970), Cotter & Miller (*Mycol.* **77**: 927, 1985), Watling & Meijer (*Edinb. J. Bot.* **54**: 231, 1997), Bruns *et al.* (*Mol. Ecol.* **7**: 257, 1998; phylogeny), Deschamps & Moreno (*Mycotaxon* **72**: 205, 1999), Kretzer & Bruns (*Mol. Phylogen. Evol.* **13**: 483, 1999; phylogeny).

Boletinellus Murrill (1909), Boletinellaceae. 2 (on roots of *Fraxinus* but not forming ectomycorrhiza), N. America; Japan. See Brundrett & Kendrick (*Symbiosis* **3**: 315, 1987; ecology), Gruhn *et al.* (*Mycol.* **84**: 528, 1992; ecology), Nagasawa (*Rep. Tottori Mycol. Inst.* **39**: 1, 2001).

boletinoid (of hymenophores), having a structure intermediate between pores and gills.

Boletinus Kalchbr. (1867) = Suillus Gray fide Smith & Thiers (*The Boletes of Michigan*, 1971) See, Singer (*Sydowia* **30**: 227, 1977; key).

Boletium Clem. (1909) ≡ Volvoboletus.

Boletochaete Singer (1944), Boletaceae. 3, Africa; S.E. Asia. See Singer (*Mycol.* **36**: 359, 1944), Zang & Petersen (*Acta Bot. Yunn.* **26**: 619, 2004; Tibet).

Boletogaster Lohwag (1926) = Boletellus fide Singer (*Farlowia* **2**: 223, 1945).

Boletolichen Juss. (1789) = Helvella fide Mussat (*Syll. fung.* **15**, 1901).

Boletopsidaceae Bondartsev & Singer ex Jülich (1982) = Bankeraceae.

Boletopsis Fayod (1889), Bankeraceae. 5 (mycorrhizal), Europe. See Stalpers (*Stud. Mycol.* **35**: 29, 1993; key), Lohmeyer (*Mycologia Bavarica* **6**: 41, 2003), Bohlin (*MR* **108**: 3, 2004; conservation in Europe).

Boletopsis Henn. (1898) = Suillus Gray fide Singer (*Lilloa* **22**: 654, 1951).

Boletus Fr. (1821), Boletaceae. *c.* 300 (ectomycorrhizal), widespread. Some are edible (see cep). See Corner (*Boletus in Malaysia*, 1972), Singer (*Sydowia* **30**: 227, 1978; key), Lannoy & Estadès (*Docums Mycol.* Mémoire Hors Série **44**: 253, 2001; Europ.), Redeuilh & Simonini (*BSMF* **118**: 139, 2002; nomencl.), Beugelsdijk (*A Taxonomic Review of Boletus, Section Boletus in The Netherlands Using Molecular Tools, Part 1: Report; Part 2: Appendices*: 129 pp., 2004; Netherlands), Šutara (*Czech Mycol.* **57**: 1, 2005; anatomical characters), Watling *et al.* (*British Fungus Flora. Agarics and Boleti* Rev. & Enl. Edn **1**: 173 pp., 2005; Brit.).

Boletus L. (1753) = Phellinus fide Donk (*Persoonia* **1**: 173, 1960).

Boletus Tourn. ex Adans. (1763) = Morchella fide Donk (*Reinwardtia* **3**: 275, 1955).

Bolinia (Nitschke) Sacc. (1882) = Camarops fide Nannfeldt (*Svensk bot. Tidskr.* **66**: 335, 1972).

Boliniaceae Rick (1931), Boliniales. 7 gen. (+ 11 syn.), 40 spp.

Lit.: Nannfeldt (*Svensk bot. Tidskr.* **66**: 335, 1972), Samuels & Rogers (*Mycotaxon* **28**: 45, 1987), Rogers & Samuels (*Mycol.* **80**: 738, 1988), Untereiner (*Mycol. Soc. Amer. Newsl.* **39**: 52, 1988), Andersson *et al.* (*SA* **14**: 1, 1995; posn), Vasilyeva (*Mikol. Fitopatol.* **31**: 5, 1997), Park & Jong (*Mycoscience* **44**: 25, 2003), Huhndorf *et al.* (*Mycol.* **96**: 368, 2004).

Boliniales P.F. Cannon (2001). Sordariomycetidae 2 fam., 8 gen., 41 spp. Stromata immersed to erumpent, crustose or pulvinate, sometimes absent, usually soft-textured, composed of thin-walled hyphal tissue. Ascomata perithecial, long-necked, sometimes vertically elongate, the ostiole periphysate. Interascal tissue of narrow true paraphyses, sometimes thin-walled and evanescent. Asci cylindrical, persistent, thin-walled, not fissitunicate, with a usually small, J- apical ring. Ascospores hyaline or brown, aseptate or transversely septate, sometimes with germ pores. Anamorphs not known. Saprobic in wood and bark, widespr. Affinities of this order are unclear; *Camarops* was traditionally linked to the *Xylariaceae* but molecular data indicates a closer relationship with the *Sordariales*. However, that assemblage has been unacceptably loosely defined in recent years. Fams:

(1) **Boliniaceae**
(2) **Catabotrydaceae**

Lit.: Andersson *et al.* (*SA* **14**: 1, 1995; posn).

Bolosphaera Syd. & P. Syd. (1917) = Phaeostigme fide Hansford (*Mycol. Pap.* **15**, 1946).

Bolton (James; 1750-1799; England). An amateur mycologist (probably in the weaving trade). Produced the first book in the English language dedicated to fungi. *Publs. An History of Fungusses, Growing about Halifax* [Yorkshire], 4 parts (1788-1791) [in German, by Willdenow, as *Geschichte der Merkwurdigsten Pilze*, 1795-1820). *Biogs, obits etc.* Grummann (1974: 257); Laplanche, (*Dictionnaire Iconographique des Champignons Superieurs (Hymenomycetes)*, 1894 [gives Friesian names for figures by Bolton, Bulliard, Paulet, Persoon, Sowerby, and others]); Petersen (*Mycotaxon* **5**: 498, 1977, index); Sartory & Maire (*Interpretation des Planches de J. Bolton An history of fungusses, vols. 1 & 2*, Paris); Shear (*TBMS* **17**: 302, 1933); Stafleu & Cowan (*TL-2* **1**: 264, 1976); Stafleu & Mennega (*TL-2, Suppl.* **2**: 296, 1993); Watling & Seaward (*Archives of Natural History* **10**: 89, 1981, biographical data, bibliography).

Bombardia (Fr.) P. Karst. (1873), Lasiosphaeriaceae. 1 (from wood), Europe; N. America. See Lundqvist (*Symb. bot. upsal.* **20** no. 1, 1972), Jensen (*Mycol.* **77**: 688, 1985; anatomy), Miller (*Sydowia* **55**: 267, 2003; anatomy), Huhndorf *et al.* (*Mycol.* **96**: 368, 2004;

phylogeny), Miller & Huhndorf (*Mol. Phylogen. Evol.* **35**: 60, 2005; anatomy, phylogeny), Zhang *et al.* (*Mycol.* **98**: 1076, 2006; phylogeny).

Bombardiastrum Pat. (1893), Pezizomycotina. 1, S. America.

Bombardiella Höhn. (1909), Sordariales. 1, Java. See Eriksson & Hawksworth (*SA* **6**: 116, 1987).

Bombardioidea C. Moreau ex N. Lundq. (1972), Lasiosphaeriaceae. Anamorph *Angulimaya*. 4 (coprophilous), widespread. See Lundqvist (*Symb. bot. upsal.* **20** no. 1, 1972), Krug & Scott (*CJB* **72**: 1302, 1994; key), Huhndorf *et al.* (*Mycol.* **96**: 368, 2004), Miller & Huhndorf (*Mol. Phylogen. Evol.* **35**: 60, 2005; anatomy, phylogeny).

Bombyliospora De Not. (1852) = Megalospora Meyen fide Hafellner & Bellemère (*Nova Hedwigia* **35**: 207, 1982).

Bombyliosporomyces Cif. & Tomas. (1953) = Megalospora Meyen.

bombysine, like silk.

Bommerella Marchal (1885), Chaetomiaceae. Anamorph *Scopulariopsis*-like. 1 (from soil), widespread. See von Arx *et al.* (*Beih. Nova Hedwigia* **84**, 1986; as *Chaetomium*).

Bomplandiella Speg. (1886), anamorphic *Pezizomycotina*, Hsp.0eP.?. 1, S. America.

Bonanseja Sacc. (1906), ? Rhytismatales. 1, Mexico.

Bonaria Bat. (1959), Micropeltidaceae. 3, widespread (tropical). See Batista (*Publções Inst. Micol. Recife* **56**: 438, 1959).

Bondarcevomyces Parmasto (1999), Tapinellaceae. 1, Asia. See Parmasto & Parmasto (*Mycotaxon* **70**: 219, 1999).

Bondartsev (Apollinaris Semenovich; 1877-1968; Russia). Head of the Phytopathology Department (1913-1931) then Head of Mycology, Department of Cryptogamic Plants (1931-1950), V.L. Komarov Botanical Institute, Leningrad; survived the siege of Leningrad (1941-1943) and awarded Defence of Leningrad medal and Order of Lenin. One of the founders of Russian mycology and plant pathologist; he collaborated with Singer (q.v.) proposing a new taxonomic system for polypores. *Publs.* [*Diseases and Protection of Cultivated Plants*] edn 3 (1931) [in Russian]; [*The Polyporaceae of the European USSR and the Caucasus*] (1953) [in russian; English translation, 1971]; [*Manual for Identification of Domestic Fungi*] (1956) [text in Russian]. *Biogs, obits etc.* Anon. ([*Mikologiya i Fitopatologiya*] **1**: 506, 1967, portrait [in Russian]); Anon. ([*Mikologiya i Fitopatologiya*] **3**: 550, 1969, bibliography, portrait [in Russian]).

Bondarzewia Singer (1940), Bondarzewiaceae. 3, widespread. See Corner (*Beih. Nova Hedwigia* **78**: 205, 1984; key S. E. Asia spp.), See Stalpers (*Stud. Mycol.* **40**: 48, 1996).

Bondarzewiaceae Kotl. & Pouzar (1957), Russulales. 8 gen. (+ 4 syn.), 48 spp.

Lit.: Donk (*Persoonia* **3**: 199, 1964), Korhonen (*Evolutionary Biology of the Fungi* Symposium of the British Mycological Society held at the University of Bristol, April 1986: 301, 1987), Redhead & Norvell (*Mycotaxon* **48**: 371, 1993), Stenlid *et al.* (*MR* **98**: 57, 1994), Stalpers (*Stud. Mycol.* **40**: 185 pp., 1996), Garbelotto *et al.* (*CJB* **76**: 397, 1998), Ginns (*Mycol.* **90**: 19, 1998), Niemelä & Korhonen (*Heterobasidion annosum, Biology, Ecology, Impact and Control*: 27, 1998), Schulze (*J. Phytopath.* **147**: 125, 1999), Ryvarden (*Karstenia* **40**: 153, 2000), Binder & Hibbett (*Mol. Phylogen. Evol.* **22**: 76, 2002), Binder *et al.* (*Systematics and Biodiversity* **3**: 113, 2005).

Bondarzewiales Jülich (1981) = Russulales.

Bondiella Piroz. (1972), ? Mesnieraceae. 1 (*Palmae*), Tanzania. See Pirozynski (*Mycol. Pap.* **129**, 1972), Hyde (*Mycotaxon* **57**: 347, 1996).

Bonia Pat. (1892) [non *Bonia* Balansa 1890, *Gramineae*] ≡ Mycobonia.

Boninogaster Kobayasi (1937), Hysterangiaceae. 1, Bonin Island.

Boninohydnum S. Ito & S. Imai (1940) = Gyrodontium fide Maas Geesteranus (*Persoonia* **3**: 187, 1964).

Bonordenia Schulzer (1866) = Hypomyces fide Rogerson (*Mycol.* **62**: 865, 1970), Rossman *et al.* (*Stud. Mycol.* **42**: 248 pp., 1999).

Bonordeniella Penz. & Sacc. (1901) = Coniosporium fide Ellis (*Dematiaceous Hyphomycetes*, 1971).

booted, see peronate.

Boothiella Lodhi & Mirza (1962), Sordariaceae. 1 (from soil), Asia. See Udagawa & Furuya (*TMSJ* **18**: 302, 1977), von Arx *et al.* (*Beih. Nova Hedwigia* **94**, 1988), Cai *et al.* (*MR* **110**: 137, 2006).

Boothiomyces Letcher (2006), Terramycetaceae. 1 (from soil), New Zealand. See Letcher *et al.* (*MR* **110**: 898, 2006).

boot-lace fungus, the honey agaric, *Armillaria mellea*.

Bordea Maire (1916), Laboulbeniaceae. 14 (on beetles), widespread. See Thaxter (*Memoirs of the American Academy of Arts and Sciences* **16**: 1, 1931), Benjamin (*Aliso* **19**: 99, 2000).

Bordeaux mixture, A spray first used by Millardet in 1883-85 against vine (*Vitis*) mildew (*Plasmopara*), and still in general use for controlling numbers of plant diseases. A common mixture is the '4-4-50': copper sulphate 1.8 kg, quick lime 1.8 kg (or hydrated lime 2.7 kg), water 227 l. When making small amounts the copper sulphate is put in some of the water, the lime in the rest, and the two liquids then mixed. See Ainsworth (*Introduction to the history of plant pathology*, 1981).

Boreoplaca Timdal (1994), Ophioparmaceae (L). 1, Siberia. See Timdal (*Mycotaxon* **51**: 503, 1994), Wedin *et al.* (*MR* **109**: 159, 2005), Miądlikowska *et al.* (*Mycol.* **98**: 1088, 2006; phylogeny).

Boreostereaceae Jülich (1982) = Gloeophyllaceae.

Boreostereum Parmasto (1968), Gloeophyllaceae. 4, widespread (north temperate). See Parmasto (*Consp. System. Corticiac.*: 186, 1968).

Borinquenia F. Stevens (1917) nom. dub., Tubeufiaceae. ? = *Malacaria*, but type material is exhausted. See Rossman (*Mycol. Pap.* **157**: 71 pp., 1987).

Bornetina L. Mangin & Viala (1903), anamorphic *Diacanthodes*. 1. See Donk (*Beih. Nova Hedwigia* **5**, 1962).

Borrera Ach. (1809) [non *Borreria* G. Mey. 1818) nom. cons., *Rubiaceae*] nom. rej. = Teloschistes fide Kurokawa (*Beih. Nova Hedwigia* **6**, 1962).

Bostrichonema Ces. (1867), anamorphic *Pezizomycotina*, Hso.1eH.?. 5, Europe; N. America. See Braun (*Monogr. Cercosporella, Ramularia Allied Genera (Phytopath. Hyphom.)* **2**, 1998), Kendrick (*CJB* **81**: 75, 2003; morphogenesis).

Bostrychia Fr. (1821) = Cytospora fide Fries (*Syst. mycol.* **2**: 1, 1823).

Bostrychonema, see *Bostrichonema*.

Botanamphora Nograsek & Scheuer (1990) = Trematosphaeria fide Nograsek & Scheuer (*SA* **12**: 31, 1993).

Bothia Halling, T.J. Baroni & Binder (2007), Boletaceae. 1, N. America. See Halling *et al.* (*Mycol.* **99**: 310, 2007; descr., phylogeny).

Bothrodiscus Shear (1907), anamorphic *Ascocalyx*, St.≡ eH.10. 3, N. America; former USSR. See Groves (*CJB* **46**: 1273, 1968).

bothrosome (**sagenogen**, **sagenogenetosome**), an invaginated organelle at the cell surface which connects the plasma membrane to the network membranes in *Labyrinthulomycota* (see Porter, 1990).

Botryandromyces I.I. Tav. & T. Majewski (1976), Laboulbeniaceae. 2, widespread. See Tavares & Majewski (*Mycotaxon* **3**: 195, 1976), Tavares (*Mycol. Mem.* **9**: 627 pp., 1985), Santamaría (*Fl. Mycol. Iberica* **5**, 2003).

Botrydiaceae Lindl. (1846) = Sclerotiniaceae.

Botrydiella Badura (1963) = Staphylotrichum fide Ellis (*Dematiaceous Hyphomycetes*, 1971).

Botrydina Bréb. ex Menegh. (1844) nom. utique rej. = Lichenomphalia fide Kuyper (*in litt.*; syn. of *Lichenomphalina* but the name has been used for the anamorph of this genus), Gams (*Öst. bot. Z.* **109**: 376, 1962), Poelt & Jülich (*Herzogia* **1**: 331, 1969), Oberwinkler (*Deutsch. bot Ges.* N.F. **4**: 139, 1970), Redhead & Kuyper (*Arctic & Alpine Mycology* **2**: 319, 1987), Redhead & Kuyper (*Mycotaxon* **31**: 221, 1988), Jørgensen & Ryman (*Taxon* **38**: 305, 1989).

Botrydiplis Clem. & Shear (1931) ≡ Botryodiplodia.

Botrydium Wallr. (1815), Algae. Algae.

Botryella Syd. & P. Syd. (1916) = Sphaerellopsis Cooke fide Sutton (*Mycol. Pap.* **141**, 1977).

botryo-aleuriospore, one of an apical cluster of aleuriospores developed basipetally from the conidiogenous cells.

Botryoascus Arx (1972) = Saccharomycopsis Schiønning fide Kurtzman & Robnett (*CJB* **73** Suppl. 1: S824, 1995), Matsushima (*Matsush. Mycol. Mem.* **9**: 1, 1996).

Botryobasidiaceae Jülich (1982), Cantharellales. 5 gen. (+ 5 syn.), 83 spp.

Lit.: Maekawa (*Trans. Mycol. Soc. Japan* **33**: 317, 1992), Langer (*Biblthca Mycol.* **158**: 459 pp., 1994), Ginns (*Mycol.* **90**: 19, 1998), Langer *et al.* (*MR* **104**: 510, 2000), Langer *et al.* (*Mycoscience* **41**: 201, 2000), Binder & Bresinsky (*Mycol.* **94**: 85, 2002).

Botryobasidium Donk (1931), Botryobasidiaceae. 55, widespread. See Eriksson & Ryvarden (*Cortic. N. Europ.* **2**: 145, 1973; key 11 Eur. spp.), Langer (*Biblthca Mycol.* **158**, 1994; world key).

Botryobasidium Rick (1959), Thelephoraceae. 3, Brazil. See Rick (*Iheringia* Série Botânica **4**: 98, 1959).

botryoblastospore, clusters of conidia borne on the swollen apex (ampulla) of a conidiogenous cell, arising synchronously or asynchronously, either singly or in chains.

Botryobolus Arnaud (1952) = Ballocephala fide Humber (*in litt.*).

Botryochaete Corda (1854) ≡ Phleogena fide Donk (*Persoonia* **4**: 160, 1966).

Botryochaete Rick (1959), Basidiomycota. 3, Brazil. See Rick (*Iheringia* Série Botânica **4**: 122, 1959).

Botryochora Torrend (1914), ? Dothioraceae (?L). 1, Africa. See von Arx & Müller (*Stud. Mycol.* **9**, 1975).

Botryocladium Preuss (1851) = Nematogonum fide Saccardo (*Syll. fung.* **4**: 170, 1886).

Botryoconiaceae Cif. & Vegni (1963) = Cryptobasidiaceae.

Botryoconis Syd. & P. Syd. (1906), Cryptobasidiaceae. 3, Brazil; Japan. See Donk (*Reinwardtia* **4**: 114, 1956), Begerow *et al.* (*Mycol. Progr.* **1**: 187, 2002), Hendrichs *et al.* (*Sydowia* **55**: 33, 2003).

Botryocrea Petr. (1949) = Fusarium fide Samuels *et al.* (*Mycol. Pap.* **164**, 1991).

Botryodeorsum T.P. Devi, N. Mathur, Chowdhry, Jasvir Singh & O. Prakash (2006), anamorphic *Pezizomycotina*. 1, India. See Devi *et al.* (*Indian Phytopath.* **59**: 215, 2006).

Botryoderma Papendorf & H.P. Upadhyay (1969), anamorphic *Pezizomycotina*, Hso.0eH.14. 3, S. Africa; Brazil. See Papendorf & Upadhyay (*TBMS* **52**: 257, 1969), Lopez *et al.* (*Mycotaxon* **55**: 269, 1995).

Botryodiplis Clem. & Shear (1931) ≡ Botryodiplodia.

Botryodiplodia (Sacc.) Sacc. (1884) nom. dub., Diaporthales. See *Lasiodiplodia* for *B. theobromae* and relatives. See Stevens (*Mycol.* **33**: 69, 1941), Zambettakis (*BSMF* **70**: 219, 1954), Crous & Palm (*Sydowia* **51**: 167, 1999).

Botryodiplodina Dias & Sousa da Câmara (1954), anamorphic *Pezizomycotina*, St.1eP.?. 1, Portugal. See Dias & Sousa da Câmara (*Agron. lusit.* **16**: 13, 1954).

Botryodontia (Hjortstam & Ryvarden) Hjortstam (1987), ? Hymenochaetaceae. 6, widespread. Polyphyletic. See Hjortstam (*Mycotaxon* **28**: 20, 1987).

Botryogene Syd. & P. Syd. (1917) nom. dub., anamorphic *Pezizomycotina*. See Sutton (*Mycol. Pap.* **141**, 1977).

Botryohypochnaceae Jülich (1982) = Botryobasidiaceae.

Botryohypochnus Donk (1931) = Botryobasidium Donk fide Langer (*Biblthca Mycol.* **158**, 1994).

Botryohypoxylon Samuels & J.D. Rogers (1986), Dothideomycetes. Anamorph *Iledon*. 1, Brazil. See Samuels & Rogers (*Mycotaxon* **25**: 631, 1986).

Botryola Bat. & J.L. Bezerra (1964), ? Nitschkiaceae. 1, Brazil. See Batista & Bezerra (*Publções Inst. Micol. Recife* **431**: 11, 1964).

Botryolepraria Canals, Hern.-Mar., Gómez-Bolea & Llimona (1997), Lecanorales. 1, Europe. See Canals *et al.* (*Lichenologist* **29**: 339, 1997), Ekman & Tønsberg (*MR* **106**: 1262, 2002; phylogeny), Baruffo *et al.* (*Nova Hedwigia* **83**: 387, 2006; Italy).

Botryomonilia Goos & Piroz. (1975), anamorphic *Pezizomycotina*, Hso.0eH.38. 1, Panama. See Goos & Pirozynski (*CJB* **53**: 2927, 1975), Kendrick (*CJB* **81**: 75, 2003; morphogenesis).

Botryomyces de Hoog & C. Rubio (1982) ≡ Ybotromyces.

Botryomyces Greco (1916), anamorphic *Pezizomycotina*, Hso.?.?. 1 (from man, also in rocks), Argentina. See Benoldi *et al.* (*J. Med. Vet. Mycol.* **29**: 9, 1991; agent of phaeohyphomycosis), Sterflinger (*The Yeast Handbook* **[1]**: 501, 2006; ecology).

Botryonipha Preuss (1852) nom. rej., anamorphic *Pezizomycotina*.

Botryophialophora Linder (1944) = Myrioconium Syd. fide von Arx (*Gen. Fungi Sporul. Cult.*, 1970).

Botryophoma (P. Karst.) Höhn. (1916) = Sclerodothiorella fide Sutton (*Mycol. Pap.* **141**, 1977).

Botryorhiza Whetzel & Olive (1917), Chaconiaceae. 1 (on *Hippocratea* (*Hippocrateaceae*)), Carribean; Brazil.

botryose, racemose; grouped like grapes.

Botryosphaeria Ces. & De Not. (1863), Botryosphaeriaceae. Anamorphs *Diplodia*, *Dothiorella*, *Fusicoccum*, *Lasiodiplodia*, *Sphaeropsis*. *c*. 36, widespread. Many important plant pathogens; molecular species concepts are narrower than traditional morphology-based taxa. See von Arx & Müller (*Beitr. Kryptfl. Schweiz* **11** no. 1, 1954), Barr (*Contr. Univ. Mich. Herb.* **9**: 523, 1972), Jacobs & Rehner (*Mycol.* **90**: 601, 1998; DNA, anams), Denman *et al.* (*Mycol.* **91**: 510, 1999), Silva-Hanlin & Hanlin (*MR* **103**: 153, 1999; DNA), Denman *et al.* (*Stud. Mycol.* **45**: 129, 2000), Smith & Stanosz (*Mycol.* **93**: 505, 2001), Zhou & Stanosz (*Mycol.* **93**: 516, 2001; phylogeny), Ma & Michailides (*Phytopathology* **92**: 519, 2002; California), Phillips (*Phytopath. Mediterr.* **41**: 3, 2002; on grape), Slippers *et al.* (*Mycol.* **96**: 83, 2004; *B. dothidea* s.l.), Slippers *et al.* (*Mycol.* **96**: 1030, 2004; *B. dothidea*, phylogeny), Phillips *et al.* (*Mycopathologia* **159**: 433, 2005; on *Olea*), Crous *et al.* (*Stud. Mycol.* **55**: 235, 2006; phylogeny), Schoch *et al.* (*Mycol.* **98**: 1041, 2006; phylogeny).

Botryosphaeriaceae Theiss. & P. Syd. (1918), Botryosphaeriales. 26 gen. (+ 60 syn.), *c*. 1517 spp.

Lit.: von Arx & Müller (*Beitr. Kryptfl. Schweiz* **11** no. 1, 1954), Sivanesan (*Bitunicate Ascomycetes and their Anamorphs*, 1984), Pennycook & Samuels (*Mycotaxon* **24**: 445, 1985), Bissett (*Mycotaxon* **25**: 519, 1986), Barr (*Prodr. Cl. Loculoasc.*, 1987), Morgan-Jones & White (*Mycotaxon* **30**: 117, 1987), Bissett & Palm (*CJB* **67**: 3378, 1989), Hyde (*Sydowia* **47**: 180, 1995), Jacobs *et al.* (*Phytopathology* **85**: 1206, 1995), Jacobs & Rehner (*Mycol.* **90**: 601, 1998), Denman *et al.* (*Stud. Mycol.* **45**: 129, 2000), Rodrigues *et al.* (*MR* **108**: 45, 2004), Slippers *et al.* (*Mycol.* **96**: 83, 2004), Van Niekerk *et al.* (*Mycol.* **96**: 781, 2004), Barber *et al.* (*MR* **109**: 1347, 2005), Crous *et al.* (*Stud. Mycol.* **55**: 235, 2006), Schoch *et al.* (*Mycol.* **98**: 1041, 2006; phylogeny), Slippers *et al.* (*Pl. Path.* **56**: 128, 2007; on *Rosaceae*), de Wet *et al.* (*Mol. Phylogenet. Evol.* **46**: 116, 2008; phylogeny, host association).

Botryosphaeriales C.L. Schoch, Crous & Shoemaker (2007). Dothideomycetes. 1 fam., 28 gen., 1628 spp. Fam:

Botryosphaeriaceae

For *Lit.* see under fam.

Botryosphaerostroma Petr. (1921) = Botryodiplodia fide Sutton (*Mycol. Pap.* **141**, 1977).

Botryosphaerostroma Petr. & Syd. (1926) ≡ Sphaeropsis Sacc. fide Sutton (*Mycol. Pap.* **141**, 1977).

Botryosporium Corda (1831), anamorphic *Pezizomycotina*, Hso.0eH.6. 4, widespread. See Mason (*Mycol. Pap.* **2**: 27, 1928), Vincent & Blackwell (*Taxon* **36**: 158, 1987; nomencl.), Zhang & Kendrick (*Acta Mycol. Sin.* **9**: 31, 1990; key 4 spp.).

Botryosporium Schwein. (1832) ? = Dictyosporium fide Sutton (*in litt.*).

Botryostroma Höhn. (1911), anamorphic *Venturiaceae*. 2, America (tropical). See Hughes & Seifert (*Sydowia* **50**: 192, 1998).

Botryothecium Syd. (1937) = Rosenscheldiella fide Hansford (*Mycol. Pap.* **15**, 1946).

Botryotinia Whetzel (1945), Sclerotiniaceae. Anamorph *Botrytis*. *c*. 19, widespread. See Hennebert & Groves (*CJB* **41**: 341, 1963), Jarvis (*Can. Dept. Agric. Monogr.* **15**, 1977; taxonomy, physiology, pathogenicity), Giraud *et al.* (*Mol. Biol. Evol.* **14**: 1177, 1997; genetics), Holst-Jensen *et al.* (*Mycol.* **89**: 885, 1997; phylogeny), Giraud *et al.* (*Phytopathology* **89**: 967, 1999; sibling spp.), Holst-Jensen *et al.* (*Nordic Jl Bot.* **18**: 705, 1999; phylogeny), Nielsen & Yohalem (*Mycol.* **93**: 1064, 2001; polyploidy), Nielsen *et al.* (*Pl. Dis.* **86**: 682, 2002; PCR detection), Yohalem *et al.* (*Mycotaxon* **85**: 175, 2003; on onions), Beever & Weeds (*Botrytis: Biology, Pathology and Control*: 29, 2004; review), Fukumori *et al.* (*J. Gen. Pl. Path.* **70**: 256, 2004; spermatia), Staats *et al.* (*Mol. Biol. Evol.* **22**: 333, 2005; phylogeny).

Botryotrichum Sacc. & Marchal (1885), anamorphic *Chaetomium*, Hso.0eP.?. 8, widespread. See Hawksworth (*Persoonia* **8**: 167, 1975), Kushwaha & Agrawal (*TMSJ* **17**: 18, 1976), Santos *et al.* (*Mycotaxon* **48**: 271, 1993; parasitic on nematode eggs), Untereiner *et al.* (*CJB* **79**: 321, 2001; phylogeny).

Botryoxylon Cif. (1962) = Conoplea fide Hughes (*CJB* **36**: 727, 1958).

Botryozyma Shann & M.T. Sm. (1992), anamorphic *Ascobotryozyma*, Hso.0eH.1. 3, widespread. See Smith *et al.* (*Antonie van Leeuwenhoek* **61**: 281, 1992), Kerrigan *et al.* (*Antonie van Leeuwenhoek* **79**: 7, 2001), Kerrigan *et al.* (*MR* **107**: 1110, 2003), Kerrigan *et al.* (*FEMS Yeast Res.* **4**: 849, 2004), Suh *et al.* (*Mycol.* **98**: 1006, 2006; phylogeny).

Botrypes Preuss (1852) nom. dub ? = Ciliciopodium fide Saccardo (*Syll. fung.* **4**: 577, 1886), Seifert (*TBMS* **88**: 123, 1985).

Botrysphaeris Clem. & Shear (1931) = Sphaeropsis Sacc. fide Sutton (*in litt.*).

Botrytis P. Micheli ex Pers. (1794), anamorphic *Botryotinia*, Hso.0eH.6. *c*. 54, widespread. *B. cinerea* (the common grey mould, frequently parasitic); *B. allii* and other spp. (neck rot of onions); *B. paeoniae* (paeony blight); *B. tulipae* (tulip fire). See Hennebert (*Persoonia* **7**: 185, 1973; segregates), Jarvis (*Can. Dept. Agric. Monogr.* **15**, 1977), Coley-Smith *et al.* (*The biology of Botrytis*, 1981), Harrison (*Pl. Path.* **37**: 168, 1987; review of spp. on beans), Shirare *et al.* (*Phytopathology* **79**: 728, 1989; nuclei and mitotic chromosomes in *Botrytis* spp.), Lu & Wu (*Acta Mycol. Sin.* **10**: 27, 1991; *Botryotinia fabae* teleomorph of *Botrytis fabae*, chocolate spot), Verhoeff *et al.* (*Recent advances in Botrytis research*, 1992), van der Vlugt-Bergmans *et al.* (*MR* **97**: 1193, 1993; genetic variation and DNA polymorphisms in *B. cinerea*), van Kan *et al.* (*Neth. Jl Pl. Path.* **99** Suppl. 3: 119, 1993; electrophoretic karyotypes in *B. cinerea*), Giraud et al. (*Mol. Biol. Evol.* **14**: 1177, 1997), Holst-Jensen *et al.* (*Nordic Jl Bot.* **18**: 705, 1998; phylogeny), Förster & Adaskaveg (*Phytopathology* **90**: 171, 2000; detection with DNA primers), Nielsen & Yohalem (*Mycol.* **93**: 1064, 2001; polyploidy), Albertini *et al.* (*MR* **106**: 1171, 2002; speciation), Muñoz *et al.* (*MR* **106**: 594, 2002; Chile), Nielsen *et al.* (*Pl. Dis.* **86**: 682, 2002; on onion), Fournier *et al.* (*Mycol.* **95**: 251, 2003; VCG markers), Moyano *et al.* (*Eur. J. Pl. Path.* **109**: 515, 2003; population genetics), Yohalem *et al.* (*Mycotaxon* **85**: 175, 2003; nomencl.), Beever & Weeds (*Botrytis: Biology, Pathology and Control*: 29, 2004; review), Chilvers *et al.* (*Australas. Pl. Path.* **33**: 29, 2004; on onion in Australia), Fukumori *et al.* (*J. Gen. Pl. Path.* **70**: 256, 2004; spermatia), Fournier *et al.* (*Mycol.* **97**: 1251, 2006; phylogeny), Ma & Michailides (*Pl. Dis.* **89**: 1083, 2005; California), Rehner & Buckley (*My-

col. **97**: 84, 2005; cryptic speciation), Staats *et al.* (*Mol. Biol. Evol.* **22**: 333, 2005; phylogeny).

Botrytites Mesch. (1892), Fossil Fungi. 1 (Oligocene), Europe.

Botrytoides M. Moore & F.P. Almeida (1937) ≡ Rhinocladiella Nannf. fide Schol-Schwarz (*Antonie van Leeuwenhoek* **34**: 140, 1968).

Bottaria A. Massal. (1856) = Mycoporum Flot. ex Nyl. fide Harris (*Evansia* **4**: 28, 1987).

Bottariomyces Cif. & Tomas. (1953) = Pyrenula Ach. (1809) fide Harris (*Mem. N. Y. bot. Gdn* **49**, 1989).

botuliform, cylindrical with rounded ends; sausage-like in form; see allantoid.

Boubovia Svrček (1977), Pezizales. 6, widespread (esp. temperate). See Yao & Spooner (*MR* **100**: 193, 1996), van Brummelen & Kristiansen (*Persoonia* **17**: 265, 1999; Norway), Perry *et al.* (*MR* **111**: 549, 2007; phylogeny).

Boudier (Jean Louis Émile; 1828-1920; France). A pharmacist of Montmorency. An expert on discomycetes, producing some of the finest illustrations of these fungi. Boudier's microscopic measurements are usually *c.* 10% too high (van Brummelen, *Persoonia* **5**: 233, 1969). *Publs. Histoire et Classification des Discomycètes d'Europe* (1907) [reprint 1968]; *Icones Mycologicae ou Iconographie des Champignons de France, Principalement Discomycètes* 5 vols (1905-1910) [reprint 1981-1986]. *Biogs, obits etc.* Mangin (*BSMF* **36**: 181, 1920); Lamy (*BSMF* **100**: CXXXIX, 1984 [correspondence]); Grummann (1974: 268); Stafleu & Cowan (*TL-2* **1**: 290, 1976).

Boudiera Cooke (1877), Pezizaceae. 10, Europe; N. America. See Hirsch (*Wissen. Z. Fried.-Schiller-Univ.* **32**: 1013, 1983), Landvik *et al.* (*Nordic Jl Bot.* **17**: 403, 1997; DNA), Norman & Egger (*Mycol.* **91**: 820, 1999; phylogeny), Hansen *et al.* (*Mol. Phylogen. Evol.* **36**: 1, 2005; phylogeny), Hansen & Pfister (*Mycol.* **98**: 1029, 2006; phylogeny).

Boudiera Lázaro Ibiza (1916) = Phellinus fide Donk (*Verh. K. ned. Akad. Wet.* tweede sect. **62**: 1, 1974).

Boudierella Costantin (1897) ≡ Delacroixia.

Boudierella Sacc. (1895), ? Pyronemataceae. 1, Europe. See Kimbrough & Korf (*Am. J. Bot.* **54**: 9, 1967).

bouillon, meat broth used as a culture medium.

Bourdot (Hubert; 1861-1937; France). Priest, Saint-Priest-en-Murat, Allier (1898-1937). Noted for his studies on *Hymenomycetes*, esp. *Thelephoraceae* and resupinate fungi. Collections in **PC**. *Publs.* (with Galzin) *Hymenomycètes de France* (1928) [reprint 1969]. *Biogs, obits etc.* Gilbert (*BSMF* **55**: 137, 1939) [bibliography, portrait]; Grummann (1974: 268); Stafleu & Cowan (*TL-2* **1**: 294, 1976); Stafleu & Mennega (*TL-2, Suppl.* **2**: 391, 1992).

Bourdotia (Bres.) Bres. & Torrend (1913), Auriculariales. 1, widespread. See Weiss & Oberwinkler (*MR* **105**: 403, 2001; phylogeny).

bourrelet, see ascus.

Bouvetiella Øvstedal (1986), ? Hymeneliaceae (L). 1, Antarctica. See Øvstedal (*Norsk Polarinstitutt Skrifter*: 35, 1986), Timdal *in* Hawksworth (Ed.) (*Ascomycete Systematics. Problems and Perspectives in the Nineties* NATO ASI Series vol. **269 269**: 381, 1994; posn).

Bovetia Onofri & Persiani (1982) = Sarophorum fide Samson & Seifert (*Advances in Penicillium and Aspergillus Systematics* **102**: 397, 1985).

Bovicornua Jørg. Koch & E.B.G. Jones (1993), Halosphaeriaceae. 1 (marine), Europe. See Yusoff *et al.* (*CJB* **71**: 346, 1993), Yusoff *et al.* (*CJB* **72**: 1550, 1994).

Bovilla Sacc. (1883) = Cercophora fide Lundqvist (*Symb. bot. upsal.* **20** no. 1, 1972).

Bovista Pers. (1794), Agaricaceae. *c.* 55, widespread. See Kreisel (*Beih. Nova Hedwigia* **25**, 1967; monogr.), Calonge (*Mycol.* **96**: 1152, 2004; Mexico), Baseia (*Mycotaxon* **91**: 81, 2005; Brazil).

Bovistaria (Fr.) P. Karst. (1889) ≡ Langermannia.

Bovistella Morgan (1892) = Lycoperdon Pers. fide Larsson & Jeppson (*MR* **112**: 4, 2008).

Bovistina Long & Stouffer (1941) = Disciseda fide Ponce (*Fieldiana, Bot.* **38**: 23, 1976).

Bovistoides Lloyd (1919) = Myriostoma fide Suárez & Wright (*Mycotaxon* **71**: 251, 1999).

Boydia A.L. Sm. (1919) = Vialaea fide Redlin (*Sydowia* **41**: 296, 1989).

BPI, US National Fungus Collection (Beltsville, Md, USA); founded 1869, part of the United States Department of Agriculture's (USDA) Agricultural Research Service (ARS): see Cross *et al.* (*Systematic collections of the Agricultural Research Service*, 1977).

BR, National Botanic Garden of Belgium (Meise, Belgium); founded 1870; supported by the Ministry of Agriculture.

Brachiola A. Cali, P.M. Takvorian, S. Lewin, M. Rendel, C.S. Sian, M. Wittner, H.B. Tanowitz, E. Keohane & L. Weiss (1998), Microsporidia. 4. See Cali *et al.* (*Acta Protozool.* **43**: 73, 1998).

Brachiosphaera Nawawi (1976), anamorphic *Aliquandostipitaceae*, Hso.1bH.19. 2 (aquatic), pantropical. See Nawawi (*TBMS* **67**: 213, 1976), Chan *et al.* (*Fungal Diversity* **5**: 89, 2000; Hong Kong), Descals (*MR* **109**: 545, 2005; propagules), Campbell *et al.* (*CJB* **85**: 873, 2007; phylogeny).

brachy- (prefix), short.

Brachyascus Syd. (1917) ≡ Microdiscus Sacc.

Brachybasidiaceae Gäum. (1926), Exobasidiales. 4 gen. (+ 1 syn.), 10 spp.

Lit.: Donk (*Taxon*: 243, 1951-63), Gäumann (*Belg. Jl Bot.*: 358, 1964), Cunningham *et al.* (*Mycol.* **68**: 642, 1976; key), Ingold (*TBMS* **84**: 542, 1985), Oberwinkler (*Stud. Mycol.* **30**: 61, 1987), Barreto & Evans (*TBMS* **91**: 81, 1988), Gruèzo (*Nat. Hist. Bull. Siam Soc.* **38**: 89, 1990), Bauer *et al.* (*CJB* **75**: 1273, 1997), Berndt & Sharma (*MR* **102**: 1484, 1998), Begerow *et al.* (*Mycol. Progr.* **1**: 187, 2002).

Brachybasidiales Donk (1964) = Exobasidiales.

Brachybasidium Gäum. (1922), Brachybasidiaceae. 1 (on *Palmae*), Java. See Gäumann (*Annls Mycol.* **20**: 269, 1922).

Brachycarphium Berk. (1849), Fossil Fungi, anamorphic *Pezizomycotina*. 1 (Oligocene), Baltic.

Brachycladites Mesch. (1892) ≡ Brachycarphium.

Brachycladium Berk. (1848) ≡ Brachycarphium.

Brachycladium Corda (1838), anamorphic *Crivellia*. 1, widespread. See Saccardo (*Syll. fung.* **4**: 489, 1886).

Brachyconidiella R.F. Castañeda & W.B. Kendr. (1990), anamorphic *Dothideomycetes*, Hsy/Hsp.0bP.1. 1, Cuba. See Castañeda Ruiz & Kendrick (*Univ. Waterloo Biol. Ser.* **33**: 11, 1990), Decock *et al.* (*Cryptog. Mycol.* **25**: 137, 2004).

Brachyconidiellopsis Decock, R.F. Castañeda & Adhikari (2004), anamorphic *Microascaceae*. 1 (on dung), Nepal. See Decock *et al.* (*Cryptog. Mycol.* **25**: 140, 2004).

brachycyclic, see *Pucciniales* and Table 5.

Brachydesmiella G. Arnaud ex S. Hughes (1961), anamorphic *Pezizomycotina*, Hso.≡ eP.26. 5, widespread. See Nicot (*BSMF* **86**: 705, 1971), Sivichai *et al.* (*Mycoscience* **39**: 239, 1998), Castañeda Ruiz *et al.* (*Mycotaxon* **85**: 211, 2003; Venezuela), Castañeda Ruíz *et al.* (*Mycotaxon* **95**: 261, 2006; Brazil).

Brachydesmium (Sacc.) Costantin (1888) = Clasterosporium fide Saccardo (*Syll. fung.* **4**: 386, 1886).

brachyform, see *Pucciniales* and Table 5.

Brachyhelicoon Arnaud (1952), anamorphic *Pezizomycotina*, Hso.0-≡ hH.?. 1, Europe. See Arnaud (*BSMF* **68**: 208, 1952), Goos (*Mycol.* **79**: 1, 1987).

brachymeiosis (obsol.), a third division, once claimed to occur in the ascus.

Brachymyces G.L. Barron (1980), Helicocephalidaceae. 1 (from soil), Canada. See Barron (*CJB* **58**: 2450, 1980).

Brachyphoris J. Chen (2007), anamorphic *Hyalorbilia*. 5, widespread. See Chen *et al.* (*Fungal Diversity* **26**: 127, 2007).

Brachysporiella Bat. (1952), anamorphic *Ascotaiwania*, Hso.≡ eP.1. 6, widespread. See Ellis (*Mycol. Pap.* **72**, 1959; key), Ranghoo & Hyde (*Mycol.* **90**: 1055, 1998; key), Fallah *et al.* (*CJB* **77**: 87, 1999), Partridge *et al.* (*Mycotaxon* **73**: 303, 1999), Yanna & Ho (*Cryptog. Mycol.* **25**: 129, 2004).

Brachysporiellina Subram. & Bhat (1989), anamorphic *Pezizomycotina*, Hso.≡ eP.10. 1, India. See Subramanian & Bhat (*Kavaka* **15**: 46, 1987).

Brachysporiopsis Yanna, W.H. Ho & K.D. Hyde (2004), anamorphic *Pezizomycotina*. 1 (on palms), Hong Kong. See Yanna & Ho (*Cryptog. Mycol.* **25**: 130, 2004).

Brachysporisporites R.T. Lange & P.H. Sm. (1971), Fossil Fungi. 2 (Eocene), Australia; Canada.

Brachysporium Sacc. (1886), anamorphic *Cryptadelphia*, Hso.≡ eP.11. 11, widespread. See Ellis (*Dematiaceous Hyphomycetes*, 1971; key), Scheuer & Chlebicki (*Acta Mycologica* Warszawa **32**: 147, 1997), Réblová & Seifert (*Mycol.* **96**: 343, 2004).

brand, a leaf disease caused by a microscopic fungus, esp. a rust or smut (sometimes named the **- fungus**) (obsol.); **- spore**, urediniospore; smut spore.

brandy, see spirits.

Brasiliomyces Viégas (1944), Erysiphaceae. Anamorph *Oidium*. 7, widespread. See Hanlin & Tortolero (*Mycol.* **76**: 439, 1984; key), Zheng (*Mycotaxon* **19**: 281, 1984), Braun (*Beih. Nova Hedwigia* **89**, 1987; key), Mori *et al.* (*Mycol.* **92**: 74, 2000; phylogeny), Braun *et al.* (*The Powdery Mildews* A Comprehensive Treatise: 13, 2001; review), To-Anun *et al.* (*Mycoscience* **44**: 447, 2003), Wang *et al.* (*Mycol.* **98**: 1065, 2006; phylogeny).

Brassia A. Massal. (1860) [non *Brassia* R. Br. 1813, *Orchidaceae*] = Thelotrema fide Hale (*Bull. Br. Mus. nat. hist.* Bot. **8**: 227, 1981).

Braunia Rick (1934) [non *Braunia* Bruch & Schimp. 1846, *Musci*] = Brauniella fide Stalpers (*in litt.*).

Brauniella Rick ex Singer (1955), ? Strophariaceae. 1, S. America. See Singer (*Proc. K. ned. Akad. Wet.* Ser. C, Biol. Med. Sci. **66**: 115, 1963).

Brauniellaceae Singer (1962) = Strophariaceae.

Brauniellula A.H. Sm. & Singer (1959) = Chroogomphus fide Miller (*Mycol.* **95**: 176, 2003).

breathing pore, raised aperture in the upper cortex of *Parmelia exasperata* from the medulla to the exterior.

Brefeld (Julius Oscar; 1839-1925; Germany). Son of a pharmacist in Westphalia; Professor at the Forestry Academy, Eberswaldeln (1876), then Professor of Botany at Münster, and later Breslau. Brefeld is noted for his development of pure culture methods in connexion with work on life-histories and growth of fungi. From the first he saw how necessary it was for culture media and apparatus to be sterile and, 10 years before Koch, was using gelatin to make solid media. Working in turn with different groups of fungi, his writings are still a mine of details on spore germination, and growth and development of fungi. For 25 years he had the use of only one eye and from about 1914 was blind. *Publs. Botanische Untersuchungen über Schimmelpilze* (later *Untersuchungen aus dem Gesammtgebeit der Mykologie*) 15 parts (1872-1912) [in these publications, with the help of beautiful figures, he provided detailed accounts of his investigations]; also responsible for more than 40 papers on mycology and plant diseases. *Biogs, obits etc.* Anon. (*Nature* **116**: 369, 1925); Grummann (1974: 16); Sopp (*Norske Videnskaps-Akademi i Oslo Årbok* **1925**: 83, 1926); Stafleu & Cowan (*TL-2* **1**: 314, 1976); Stafleu & Mennega (*TL-2, Suppl.* **3**: 47, 1995).

Brefeldiella Speg. (1889), Brefeldiellaceae. 2, pantropical. See Müller & von Arx (*Beitr. Kryptfl. Schweiz* **11** no. 2, 1962), Eriksson (*Op. Bot.* **60**, 1981), Reynolds & Gilbert (*Aust. Syst. Bot.* **18**: 265, 2005; Australia), Reynolds & Gilbert (*Cryptog. Mycol.* **27**: 249, 2006; Panamá).

Brefeldiellaceae E. Müll. & Arx (1962), ? Dothideomycetes (inc. sed.). 1 gen., 2 spp.

Lit.: Eriksson (*Op. Bot.* **60**, 1981), Eriksson (*Op. Bot.* **60**: 220 pp., 1981).

Brefeldiellites Dilcher (1965), Fossil Fungi. 2 (Cretaceous, Eocene), Argentina; USA.

Brefeldiopycnis Petr. & Cif. (1932), anamorphic *Pezizomycotina*, ?.0eH.?. 1, West Indies.

Brefeldochium Verkley (2005), anamorphic *Polydesmia*, H?.?.?. 1, widespread. See Verkley (*Nova Hedwigia* **80**: 504, 2005).

Brencklea Petr. (1923) = Scolecosporiella Petr. fide Sutton (*CJB* **46**: 183, 1968).

Brenesiella Syd. (1929), Pezizomycotina. 1, C. America.

Bresàdola (Giacopo; 1847-1929; Italy). An amateur mycologist of Trento. Main collection in Stockholm (**S**). *Publs. Fungi Tridentini Novi et Nondum Delineati, Descripti et Iconibus Illustrati* (1881-1892); *Iconographia Mycologica* 28 vols (1927-60) [mostly *Agaricales*; reprint 1981-1982]. *Biogs, obits etc.* Bresadola (*Iconographia Mycologica* **26**: v, 1933) [portrait]; Zalin & Lazzari (*Carteggio Bresadola-Saccardo*, 1987 [correspondence]); Grummann (1974: 431); Stafleu & Mannega (*TL-2, Suppl.* **3**: 66, 1995).

Bresadolella Höhn. (1903) = Trichosphaerella fide Rossman *et al.* (*Stud. Mycol.* **42**: 248 pp., 1999).

Bresadolia Speg. (1883) = Polyporus P. Micheli ex Adans. fide Donk (*Verh. K. ned. Akad. Wet.* tweede sect. **62**: 1, 1974).

Bresadolina Brinkmann (1909) = Cotylidia fide Donk (*Taxon* **6**: 22, 1957).

Bresadolina Rick (1928), Pezizomycotina. 1, Brazil.

Bretonia C.E. Bertrand & Hovel. (1892), Fossil Fungi (mycel.) Fungi. 1 (Permian), Europe.

Brettanomyces N.H. Claussen ex Custers (1940),

anamorphic *Dekkera*. 10, widespread. See Lee & Jong (*Mycotaxon* **25**: 455, 1986), Hoeben *et al.* (*J. Mol. Evol.* **36**: 263, 1993; mitochondrial genomes), Boekhout *et al.* (*Int. J. Syst. Bacteriol.* **44**: 781, 1994; phylogeny), Cai *et al.* (*Int. J. Syst. Bacteriol.* **46**: 542, 1996; phylogeny), Smith *in* Kurtzman & Fell (Eds) (*Yeasts, a taxonomic study* 4th edn: 450, 1998), Rodrigues *et al.* (*Journal of Applied Microbiology* **90**: 588, 2001; diagnostic media), Cocolin *et al.* (*Appl. Environm. Microbiol.* **70**: 1347, 2004; molecular detection), Suh *et al.* (*Mycol.* **98**: 1006, 2006; phylogeny).

Brevicatenospora R.F. Castañeda, Minter & Saikawa (2006), anamorphic *Pezizomycotina*. 1 (submerged leaves), Cuba. See Castañeda Ruíz *et al.* (*Mycotaxon* **96**: 152, 2006).

Brevicellicium K.H. Larss. & Hjortstam (1978), Hydnodontaceae. 13, widespread. See Larsson & Hjortstam (*Mycotaxon* **7**: 117, 1978), Hjortstam (*Mycotaxon* **79**: 181, 2001; tropical and subtropical species).

brevicollate, short-necked.

Brewing. The process of beer making. Classical brewing comprises a number of stages: (1) **malting**, when water soaked barley grain is allowed to germinate and endogenous enzymes attack the starch and certain proteins of the grain; (2) grinding the malted grain to form **grist**; (3) **mashing**, mixing the grist with water when there is further enzyme action after which (4) the resulting liquid, **wort**, is boiled with female hop (*Humulus lupulus*) flowers and on cooling (5) fermented by *S. cerevisiae* or another yeast. See beer, Wine making, yeast.

Briania D.R. Reynolds (1989), anamorphic *Meliolina*, Hso.0eH.15. 1, Hawaii. See Reynolds (*Pacific Sci.* **43**: 161, 1989).

Briansuttonia R.F. Castañeda, Minter & Saikawa (2004), anamorphic *Pleosporales*. 1, Australia. See Castañeda Ruíz *et al.* (*Mycotaxon* **89**: 304, 2004).

Briardia Sacc. (1885) = Duebenia fide Nannfeldt (*Svensk bot. Tidskr.* **23**: 316, 1929).

Briarea Corda (1831) = Aspergillus fide Hughes (*CJB* **36**: 727, 1958).

Bricookea M.E. Barr (1982) = Lophiostoma fide Shoemaker & Babcock (*Stud. Mycol.* **31**: 165, 1989), Barr (*Mycotaxon* **43**: 371, 1992), Suková (*Czech Mycol.* **56**: 63, 2004; Czech Republic), Eriksson (*Svensk Mykol. Tidskr.* **28**: 38, 2007; synonymy).

Bridgeoporus T.J. Volk, Burds. & Ammirati (1996), Agaricomycetes. 1, USA. *Hymenochaetales* or *Agaricales* (*Oxyporus* clade). See Burdsall *et al.* (*Mycotaxon* **60**: 390, 1996), Redberg *et al.* (*Mycol.* **95**: 836, 2003; phylogeny).

bridging hypha, a branch hypha joining two other hyphae (Buller, *Researches* **4**: 152, 1931); **- species** or **- host**, a plant by which a specialized parasite went, in Marshall Ward's opinion (*Ann. Bot.* **15**: 560, 1902; *Ann. Myc.* **1**: 132, 1903; but see Bean *et al.*, *Ann. Bot.* N.S. **18**: 129, 1954), from a susceptible to a resistant host.

Brigantiaea Trevis. (1853) nom. rej., Brigantiaeaceae (L). 22, widespread. See Hafellner & Bellemère (*Nova Hedwigia* **35**: 237, 1982), Hafellner & Bellemère (*Nova Hedwigia* **38**: 169, 1983; conidia from ascospores), Hafellner (*Acta Univ. Ups. Symb. Bot. Upsal.* **32** no. 1: 35, 1997).

Brigantiaeaceae Hafellner & Bellem. (1982), Lecanorales (L). 2 gen. (+ 2 syn.), 25 spp.

Lit.: Awasthi & Srivastava (*Proc. Indian Acad. Sci.* Pl. Sci. **99**: 165, 1989), Hafellner (*Symb. bot. upsal.* **32** no. 1: 35, 1997).

Brigantiella (Sacc.) Sacc. (1905) ? = Lophiostoma fide Hawksworth *et al.* (*Dictionary of the Fungi* edn 8, 1995).

Briosia Cavara (1888), anamorphic *Pezizomycotina*, Hsy.0eP.38. 2, widespread. See Sutton (*Mycol. Pap.* **132**, 1973), Sigler & Carmichael (*Mycotaxon* **4**: 349, 1976), Nasu *et al.* (*Ann. phytopath. Soc. Japan* **60**: 608, 1994; on grape).

Brobdingnagia K.D. Hyde & P.F. Cannon (1999), Phyllachoraceae. 2 (on living leaves), Africa; Australia. See Hyde & Cannon (*Mycol. Pap.* **175**: 114, 1999), Sivanesan & Shivas (*Fungal Diversity* **11**: 145, 2002).

Brochospora Kirschst. (1944) = Sporormia fide von Arx & Müller (*Stud. Mycol.* **9**, 1975).

Brodoa Goward (1987), ? Parmeliaceae (L). 3, widespread. See Goward (*Bryologist* **89**: 222, 1986), Thell *et al.* (*Symb. bot. upsal.* **34** no. 1: 429, 2004; Scandinavia), Blanco *et al.* (*Mol. Phylogen. Evol.* **39**: 52, 2006; phylogeny).

bromatia, the rounded swellings at the ends of hyphae of ant fungi (see Insects and fungi) which are used by the ants as food.

Bromicolla E.V. Eichw. (1843) = Sclerotium fide Mussat (*Syll. fung.* **15**, 1901) Also considered an alga (*Chlorophyceae* or *Palmellaceae*).

bronchomycosis, see mycosis.

Brooksia Hansf. (1956), Dothideomycetes. Anamorph *Hiospira*. 1, widespread (tropical). See Deighton & Pirozynski (*Mycol. Pap.* **105**, 1966).

broom cells (of agarics), cells bearing apical appendages to give a broom-like appearance on pileus or edge of lamella, as in *Marasmius rotula*; cellules en brosse (Singer, 1962: 62).

Broomeia Berk. (1844), Broomeiaceae. 2, Africa; S. America.

Broomeiaceae Zeller (1948), Agaricales. 1 gen., 2 spp.

Lit.: Bottomley (*Bothalia* **4**: 473, 1948), Ryvarden *et al.* (*An Introduction to the Larger Fungi of South Central Africa*: 200 pp., 1994), Jacobson (*McIlvainea* **12**: 21, 1996), Sharp & Piearce (*Kew Bull.* **54**: 739, 1999).

Broomella Sacc. (1883), Amphisphaeriaceae. Anamorphs *Pestalotia*, *Truncatella*. 7, widespread. See Müller *in* Yuan & Zhao (Eds) (*Sydowia* **44**: 90, 1992; key), Kang *et al.* (*MR* **103**: 53, 1999).

Broomeola Kuntze (1891), anamorphic *Pezizomycotina*, Hsp.1eH.?. 1, British Isles.

broth, a liquid nutrient culture medium, esp. one containing meat extract.

Brown (William; 1888-1975; Scotland, later England). Assistant (1912-1928) then Professor of Plant Pathology (1928-1938) then Professor of Botany (1938-1953), Imperial College of Science & Technology, University of London. Noted for his research on the physiology of parasitism and as a teacher. *Biogs, obits etc.* Garrett (*Biographical Memoirs of Fellows of the Royal Society* **21**: 155, 1975) [bibliography, portrait]; Hawker (*TBMS* **65**: 343, 1975) [portrait].

brown oak, oak wood stained by *Fistulina hepatica* (Cartwright, *TBMS* **21**: 68, 1937).

brown rot fungi, species of *Monilinia* causing fruit rots and other damage to fruit trees. See Wormald (*Tech. Bull. Ministr. Agric.* **3**, 1954), Byrde & Willetts (*The brown rot fungi of fruit*, 1977).

Brucea Rikkinen (2003), Lecanorales. 1 (on conifer resin), USA. See Rikkinen (*Ann. bot. fenn.* **40**: 444, 2003).

Brunaudia (Sacc.) Kuntze (1898) ? = Rhytidhysteron fide Kutorga & Hawksworth (*SA* **15**: 1, 1997).

Brunchorstia Erikss. (1891), anamorphic *Gremmeniella*, St.≡ eH.15. 1, widespread. See Punithalingam & Gibson (*IMI Descr. Fungi Bact.*: no. 369, 1973), Santamaría *et al.* (*Pl. Path.* **54**: 331, 2005; population genetics).

Bruneaudia, see *Brunaudia*.

Brunneiapiospora K.D. Hyde, J. Fröhl. & Joanne E. Taylor (1998), Clypeosphaeriaceae. 6 (saprobic on palms), widespread (tropical). See Hyde *et al.* (*Sydowia* **50**: 21, 1998), Kang *et al.* (*Mycoscience* **40**: 151, 1999), Rogers & Ju (*Sydowia* **55**: 359, 2003).

Brunneocorticium S.H. Wu (2007), Agaricales. 1, China; Taiwan. See Wu *et al.* (*Mycol.* **99**: 302, 2007).

Brunneospora Guarro & Punsola (1987), Onygenaceae. Anamorph *Chrysosporium*. 1, Spain. See Sigler *et al.* (*CJB* **76**: 1624, 1999), Cano *et al.* (*Stud. Mycol.* **47**: 165, 2002).

Brunneosporella V.M. Ranghoo & K.D. Hyde (2001), Annulatascaceae. 1 (on submerged wood), Hong Kong. See Ranghoo *et al.* (*MR* **105**: 625, 2001).

Brunnipila Baral (1985), Hyaloscyphaceae. 1, north temperate. See Raitviir (*Eesti NSV Tead. Akad. Toim.* Biol. seer **36**: 313, 1987), Leenurm *et al.* (*Sydowia* **52**: 30, 2000; ultrastructure), Suková (*Czech Mycol.* **57**: 139, 2005; lignicolous spp.).

Brycekendrickia Nag Raj (1973), anamorphic *Pezizomycotina*, St.0eH.15. 1, India. See Nag Raj (*CJB* **51**: 1337, 1973).

Bryocaulon Kärnefelt (1986), Parmeliaceae (L). 3, widespread (north temperate). See Kärnefelt *et al.* (*Nova Hedwigia* **67**: 71, 1998), Thell *et al.* (*Symb. bot. upsal.* **34** no. 1: 429, 2004; Scandinavia), Randlane & Saag (*Central European Lichens*: 75, 2006; key).

Bryocentria Döbbeler (2004), Bionectriaceae. 3 (on mosses), Europe. See Döbbeler (*Mycol. Progr.* **3**: 247, 2004), Döbbeler (*Mycol.* **97**: 924, 2005; ascospores).

Bryochiton Döbbeler & Poelt (1978), ? Pseudoperisporiaceae. 4 (on *Musci*), Europe. See Döbbeler *in* Laursen *et al.* (Eds) (*Arctic and Alpine Mycology* **2**: 87, 1987), Döbbeler (*Nova Hedwigia* **76**: 1, 2003), Döbbeler (*MR* **111**: 1406, 2007).

Bryochysium Link (1833) nom. dub., Fungi. See Rabenhorst (*Deutsch. Krypt. Fl.* **2** index: 15, 1853; = *Rhizoctonia* s. lat.).

Bryocladium Kunze (1830) = Pisomyxa fide Saccardo (*Syll. fung.* **9**: 374, 1891).

Bryodina Hafellner (2001), Lecanoraceae (L). 1, Europe (central). See Hafellner & Türk (*Stapfia* **76**: 150, 2001).

Bryodiscus B. Hein, E. Müll. & Poelt (1971), Odontotremataceae. 3 (on *Musci*), Europe. See Döbbeler & Poelt (*Svensk bot. Tidskr.* **68**: 369, 1974; key), Döbbeler (*Biodiv. Cons.* **6**: 721, 1997).

Bryoglossum Redhead (1977), Hyaloscyphaceae. 2, Europe; N. America. See also *Mitrula*. See Redhead (*CJB* **55**: 307, 1977), Knudsen & Hansen (*Nordic Jl Bot.* **16**: 211, 1996), Wang *et al.* (*Mol. Phylogen. Evol.* **41**: 295, 2006; phylogeny), Wang *et al.* (*Mycol.* **98**: 1065, 2006; phylogeny).

Bryogomphus Lücking, W.R. Buck, Sérus. & L.I. Ferraro (2005), Pilocarpaceae (L). 1, Caribbean. See Lücking *et al.* (*Bryologist* **108**: 483, 2005).

Bryomyces Döbbeler (1978), ? Pseudoperisporiaceae. 9 (on *Musci*), widespread. See Döbbeler (*Mitt. bot. StSamml.* Münch. **14**: 233, 1978), Döbbeler (*Nova Hedwigia* **76**: 1, 2003).

Bryonectria Döbbeler (1998), Bionectriaceae. 8 (on foliose liverworts and mosses), widespread. See Döbbeler (*Nova Hedwigia* **66**: 334, 1998), Döbbeler (*Sendtnera* **6**: 93, 1999; n. spp.), Rossman *et al.* (*Stud. Mycol.* **42**: 248 pp., 1999), Döbbeler (*Mycol.* **97**: 924, 2005; ascospores).

Bryonora Poelt (1983), Lecanoraceae (L). 11, widespread. See Holtan-Hartwig (*Mycotaxon* **40**: 295, 1991), Poelt & Obermayer (*Nova Hedwigia* **53**: 1, 1991; key), Grube *et al.* (*MR* **108**: 506, 2004; phylogeny).

Bryonosema E.U. Canninga, D. Refardtb, C.R. Vossbrinckc, B. Okamurad & A. Curry (2002), Microsporidia. 2. See Canninga *et al.* (*Eur. J. Protist.* **38**: 247, 2002).

Bryopelta Döbbeler & Poelt (1978), Dothideomycetes. 1 (on *Hepaticae*), Norway; Sweden. See Döbbeler & Poelt (*Mitt. bot. StSamml.* Münch. **14**: 126, 1978).

Bryophagus Nitschke ex Arnold (1862), Ostropales (L). 3, widespread. See Hawksworth *et al.* (*Lichenologist* **12**: 18, 1980; nomencl.), Kantvilas (*Muelleria* **16**: 65, 2002; Australian taxa), Kauff & Lutzoni (*Mol. Phylogen. Evol.* **25**: 138, 2002; phylogeny), Lumbsch *et al.* (*Mol. Phylogen. Evol.* **31**: 822, 2004; phylogeny).

Bryophilous fungi. Fungi growing on *Bryophyta*: muscicolous species on mosses, hepaticolous species on liverworts. A wide range of fungi are restricted to bryophytes (obligate bryophilous fungi). For example, *Synchytrium pyriforme* (*Chytridiomycetes*), *Eocronartium muscicola* and species of *Galerina* (*Strophariaceae*) grow on the gametophytes of mosses, *Epicoccum plagiochilae* (anamorphic *Ascomycota*) on those of hepatics. However, the most such fungi by far belong to the *Pezizomycotina* (*c.* 350 species in 90 genera). They represent an array of taxa of quite different systematic positions (e.g. *Dothideales*, *Hypocreales*, *Leotiales*, *Ostropales*, *Pezizales*), mode of nutrition, infection of host organs and host selection. Several genera are unknown elsewhere (e.g. *Bryodiscus*, *Bryonectria*, *Bryoscyphus*, *Hypobryon*, *Octospora*, *Octosporella*). In other cases mainly non-bryophilous genera contain obligate parasites of mosses and hepatics, for example *Acrospermum adeanum*, *Dactylospora heimerlii*, *Muellerella frullaniae*, *Nectria egens*.

The bryophilous habit has evoked surprising adaptations, e.g. the formation of tiny, frequently gelatinous ascomata which generally prefer those parts of a plant which prevent too rapid loss of moisture and allow at the same time effective spore discharge (leaf axils, border of the ventral leaf side, or perforation of the leaves or perianths in hepatics, interspaces of the photosynthetic leaf lamellae in *Polytrichales*). Some species are necrotrophic (*Belonioscypha hypnorum*, *Bryostroma necans*, *Lizonia emperigonia*, *Nectria muscivora*) with intracellular hyphae, causing necrotic lesions. Necrotic rings in moss cushions in polar regions can be conspicuous (Longton, *The Biology of Polar Bryophytes and Lichens*, pp. 101-104, 1988). Lesions in liverwort colonies are known, too (Hawksworth, *Fld Stud.* **4**: 391, 1976).

Most species represent biotrophic parasites which

do not cause severe damage to their hosts. Hyphae of these species grow over the cell walls or within or between them. Hyphal appendages like appressoria, haustoria or conidiogenous cells are most useful taxonomic characters. Endophytic VA mycorrhizal-like fungi regularly occur inside hepatics; most are sterile and part of a coevolved mutualism (Boullard, 1988; see mycothalli). *Octospora* infects the underground rhizoids of acrocarpic mosses with large appressoria and intracellular haustoria, sometimes causing conspicuous rhizoid galls. *Lizonia* is specific to antheridial cups of *Polytrichum* and related genera. Systematically different species colonizing the spaces betwen the leaf lamellae of *Polytrichales*, a phylogenetically ancient and stable microhabitat, offer striking examples of convergent evolution. Even heavy infections with hundreds of fruit-bodies in a single leaf of *Dawsonia superba* do not induce visible symptoms. Many species have a restricted host range and are specific to a certain host species or group of hosts. The presence of bryophilous fungi in colonies of mosses or hepatics seems to be a very frequent, universal phenomenon. Some hosts apparently never occur without their parasites. These fungi are generally totally neglected despite their number and frequency. *Vezdaea aestivalis* is ± lichenized but intimately associated also with moss leaves. Some lichens can overgrow and kill mosses (Faegri, *Lichenologist* **12**: 248, 1980).

Lit.: Barrio *et al.* (*Boln. Soc. micol. Castellana* **9**: 73, 1985; macromycetes, Spain), Benkert (*Z. Mykol.* **53**: 195, 1987, **64**: 17, 1998, parasitic *Pezizales*), Döbbeler (*Mitt. bot. StSamml., Münch.* **14**: 1, 1978, pyrenocarps; *Nova Hedw.* **31**: 817, 1980, parasitic *Pezizales*; *Mitt. bot. StSamml., Münch.* **17**: 393, 1981, monogr., key 21 spp. on *Dawsonia*; *Arctic Alp. Mycol.* **2**: 87, 1987, ascomycetes on *Polytrichastrum sexangulare*; *Nova Hedw.* **62**: 61, 1996, *Potriphila navicularis* on *Polytrichastrum alpinum*; *Biodiversity and Conservation* **6**: 721, 1997, review; *Nova Hedw.* **66**: 325, 1998, ascomycetes on *Radula falccida*), Duckett *et al.* (*New Phytol.* **118**: 233, 1991; ultrastr.), Felix (*Bot. Helv.* **98**: 239, 1988; review), Kost (*Endocytobiosis & Cell Res.* **5**: 287, 1988, bryophilous basidiomycetes), Poelt (*Sydowia* **38**: 241, 1986; bryophilous lichens), Pocock & Duckett (*Bryol. Times* **31**: 2, 1985; review), Racovitza (*Mém. Mus. natn Hist. Nat., Paris*, n.s. B, **10**, 1959; monogr. including anamorphic fungi and sporophytes as substrata).

Bryophytomyces Cif. (1953), anamorphic *Hymenoscyphus*. 1, Europe; Russia. See Bauch (*Ber. dt. bot. Ges.* **56**: 73, 1938), Redhead & Spicer (*Mycol.* **73**: 940, 1981).

Bryopogon Link (1833) ≡ Alectoria Ach.

Bryopogon Th. Fr. (1860) = Bryoria fide Brodo & Hawksworth (*Op. bot. Soc. bot. Lund* **42**, 1977).

Bryorella Döbbeler (1978), Dothideomycetes. 10 (on *Musci*), widespread. See Döbbeler (*Mitt. bot. StSamml.* Münch. **14**: 128, 1978), Döbbeler (*MR* **111**: 1406, 2007).

Bryoria Brodo & D. Hawksw. (1977), Parmeliaceae (L). 51, widespread (esp. boreal and cool temperate). See Brodo & Hawksworth (*Op. bot. Soc. bot. Lund* **42**, 1977; key 27 N. Am. spp.), Mattsson & Wedin (*Lichenologist* **31**: 431, 1999; DNA), Wedin *et al.* (*MR* **103**: 1152, 1999; phylogeny), Wang & Harada (*Nat. Hist. Res.* **6**: 43, 2001; *B. asiatica* group), Glavich (*Bryologist* **106**: 588, 2003; ecology), Thell *et al.* (*Symb. bot. upsal.* **34** no. 1: 429, 2004; biogeography), Miądlikowska *et al.* (*Mycol.* **98**: 1088, 2006; phylogeny), Wang *et al.* (*J. Hattori bot. Lab.* **100**: 865, 2006; Himalayas).

Bryoscyphus Spooner (1984), Helotiaceae. 7 (on *Bryophyta*, *Cladonia*), widespread. See Verkley *et al.* (*Persoonia* **16**: 383, 1997), Alstrup & Cole (*Bryologist* **10**: 222, 1998).

Bryosphaeria Döbbeler (1978), Dothideomycetes. 9 (on *Musci*), Europe. See Döbbeler (*Mitt. bot. StSamml.* Münch. **14**: 151, 1978).

Bryostigma Poelt & Döbbeler (1979) = Arthonia fide Coppins (*Lichenologist* **21**: 195, 1989).

Bryostroma Döbbeler (1978), Dothideomycetes. 7 (on *Musci*), Europe. See Döbbeler (*Mitt. bot. StSamml.* Münch. **14**: 170, 1978).

Bryothele Döbbeler (1998), Dothideomycetes. 1, Comoros. See Döbbeler (*Nova Hedwigia* **66**: 337, 1998).

B-spore, see beta-spore. Cf. alpha spore.

Bubacia Velen. (1922) nom. nud., Thelephorales. See Donk (*Taxon* **6**: 17, 1957).

Bubakia Arthur (1906) = Phakopsora fide Cummins & Hiratsuka (*Illustr. Gen. Rust Fungi rev. edit.*, 1983).

Bucholtzia Lohwag (1924) = Russula fide Singer & Smith (*Mem. Torrey bot. Club* **21**, 1960; as *Macowanites*).

Buchwaldoboletus Pilát (1969), Boletaceae. 3, Europe; Australia. See Pilát (*Friesia* **9**: 217, 1969), Basezzi & Bottaro (*Rivista di Micologia* **42**: 155, 1999), Watling (*Edinb. J. Bot.* **61**: 41, 2004).

buck-eye rot, a disease of tomato fruits (*Phytophthora nicotianae* var. *parasitica*).

buckle, see clamp-connexion.

budding, a process of multiplication in 1-celled fungi or in spores, in which there is a development of a new cell from a small outgrowth; cf. fission.

Buellia De Not. (1846) nom. cons., Caliciaceae (±L). *c.* 453, widespread. See Sheard (*Lichenologist* **2**: 225, 1964; UK), Schauer (*Mitt. bot. StSamml., München* **5**: 609, 1965), Lamb (*Br. Antarct. Surv. Sci. Rep.* **61**, 1968), Sheard (*Bryologist* **72**: 220, 1969), Hafellner (*Beih. Nova Hedwigia* **62**, 1979; gen. concept), Scheidegger & Ruef (*Nova Hedwigia* **47**: 433, 1988; xanthone spp. Eur.), Scheidegger (*Bot. Chron.* **10**: 211, 1991; distrib. Medit. spp.), Scheidegger (*Lichenologist* **25**: 315, 1993; key 36 spp. on rock, Eur.), Nordin (*Symb. bot. upsal.* **31**: 327, 1996; Nordic spp.), Kalb & Elix (*Mycotaxon* **68**: 465, 1998; chemistry), Moberg *et al.* (*Taxon* **48**: 143, 1999; nomencl.), Nordin (*Bryologist* **102**: 249, 1999; N. Am.), Marbach (*Biblthca Mycol.* **74**: 1, 2000; trop.), Grube & Arup (*Lichenologist* **33**: 63, 2001; polyphyly), Trinkaus *et al.* (*Lichenologist* **33**: 37, 2001; Australia), Molina *et al.* (*Lichenologist* **34**: 509, 2002; phylogeny), Wedin *et al.* (*Taxon* **51**: 655, 2002; phylogeny), Helms *et al.* (*Mycol.* **95**: 1078, 2003; phylogeny), Bungartz & Nash (*Biblthca Lichenol.* **88**: 49, 2004; N America), Bungartz *et al.* (*Bryologist* **107**: 459, 2004; Mexico), Grube *et al.* (*Biblthca Lichenol.* **88**: 163, 2004; Australia), Simon *et al.* (*J. Mol. Evol.* **60**: 434, 2005; introns), Miądlikowska *et al.* (*Mycol.* **98**: 1088, 2006; phylogeny).

Buelliaceae Zahlbr. (1907) = Caliciaceae.

Buelliastrum Zahlbr. (1930), Lecanorales (L). 2, China.

Buelliella Fink (1935), Dothideomycetes. 8 (on lichens), widespread. See Hafellner (*Beih. Nova Hedwigia* **62**, 1979; key), Suija & Alstrup (*Lichenologist* **36**: 203, 2004).

Buelliomyces E.A. Thomas ex Cif. & Tomas. (1953) ≡ Diploicia.

Buelliopsis A. Schneid. (1897) = Buellia fide Hawksworth *et al.* (*Dictionary of the Fungi* edn 8, 1995).

Buergenerula Syd. (1936), Magnaporthaceae. Anamorphs *Passalora*-like, *Nakataea*-like. 3, Europe; N. America. See Barr (*Mycol.* **68**: 611, 1976), McKenzie (*Mycotaxon* **42**: 351, 1991), Kohlmeyer *et al.* (*Bot. Mar.* **40**: 291, 1997).

Buglossoporus Kotl. & Pouzar (1966), Fomitopsidaceae. 10, widespread. See Corner (*Beih. Nova Hedwigia* **78**: 1, 1984).

Buglossus Wahlenb. (1826) = Fistulina fide Stalpers (*in litt.*).

bulbil, a discrete, compact, multicellular, thalloidic propagule initiated in one of several ways but always homogeneous throughout development, with all cells acropetally produced and expanding more or less synchronously to many times (e.g. 4–10×) the diameter of the colourless, thin-walled hyphae from which they arise; pseudoparenchymatous at least at maturity, and lacking internal differentiation (found in certain basidiomycetes such as *Burgoa* and *Minimedusa*, distinguished from sclerotia; see Weresub & LeClair, *CJB* **49**: 2203, 1971); lichenized in *Multiclavula vernalis* (Poelt & Obermayer, *Herzogia* **8**: 289, 1990).

bulbillate (of a stipe), having a small or not clearly marked bulb at the base.

Bulbillomyces Jülich (1974), Meruliaceae. Anamorph *Aegerita*. 1, widespread (northern hemisphere). See Jülich (*Persoonia* **8**: 69, 1974), Legon (*Mycologist* **20**: 81, 2006).

bulbillosis, the condition, in *Agaricales*, in which basidiome sporulation is suppressed and the basidial function taken on by bulbils, as in *Rhacophyllus* (Singer, 1962: 27).

Bulbilopycnis Matsush. (1996), anamorphic *Pezizomycotina*, Hsp.?.?. 1, Transvaal. See Matsushima (*Matsush. Mycol. Mem.* **9**: 3, 1996).

Bulbithecium Udagawa & T. Muroi (1990), ? Bionectriaceae. Anamorph *Acremonium*-like. 1 (coprophilous), Peru. See Udagawa & Muroi (*Bull. natn. Sci. Mus.* Tokyo, B **16**: 13, 1990), Suh & Blackwell (*Mycol.* **91**: 836, 1999; phylogeny), Rossman *et al.* (*Mycol.* **93**: 100, 2001; phylogeny).

Bulbocatenospora R.F. Castañeda & Iturr. (2000), anamorphic *Pezizomycotina*. 1, Venezuela. See Castañeda Ruíz *et al.* (*MR* **104**: 107, 2000).

Bulbomicrosphaera A.Q. Wang (1987) = Erysiphe fide Braun & Takamatsu (*Schlechtendalia* **4**: 3, 2000).

Bulbomollisia Graddon (1984) = Mollisia fide Graddon (*MR* **94**: 231, 1990), Nauta & Spooner (*Mycologist* **13**: 3, 1999).

Bulbopodium Earle (1909) = Cortinarius fide Singer (*Agaric. mod. Tax.* edn 3, 1975).

Bulborrhizina Kurok. (1994), Parmeliaceae (L). 1, Mozambique. See Kurokawa (*Acta Bot. Fenn.* **150**: 105, 1994), Eriksson & Hawksworth (*SA* **13**: 188, 1995).

Bulbothamnidium J. Klein (1870) = Helicostylum fide Upadhyay (*Mycol.* **65**: 735, 1973).

Bulbothricella, see *Bulbotricella*.

Bulbothrix Hale (1974), Parmeliaceae (L). 45, widespread (esp. tropical). See Hale (*Smithson. Contr. bot.* **32**, 1976; monogr.), Lumbsch (*Mycotaxon* **64**: 225, 1997), Blanco *et al.* (*Mol. Phylogen. Evol.* **39**: 52, 2006; phylogeny).

Bulbotricella V. Marcano, Mohali & A. Morales (1996) ? = Bulbothrix

Lit.: Lumbsch (*Mycotaxon* **64**: 225, 1997).

Bulbouncinula R.Y. Zheng & G.Q. Chen (1979) = Erysiphe fide Braun & Takamatsu (*Schlechtendalia* **4**: 3, 2000).

bulbous (1) bulb-like; (2) (of a stipe), having a swelling at the base (Fig. 4G).

Bulgaria Fr. (1822), Bulgariaceae. Anamorph *Endomelanconium*. 1, widespread (north & south temperate). *B. inquinans*, common saprobe on bark of hardwoods after felling, but sometimes parasitic. See Fenwick (*Mycologist* **6**: 177, 1992; culture), Verkley (*The Ascus Apical Apparatus in Leotiales* An Evaluation of Ultrastructural Characters as Phylogenetic Markers in the Families *Sclerotiniaceae*, *Leotiaceae* and *Geoglossaceae*: 209 pp., 1995; ultrastr.), Döring & Triebel (*Cryptog. Bryol.-Lichénol.* **19**: 123, 1998; phylogeny), Gernandt *et al.* (*Mycol.* **93**: 915, 2001; phylogeny), Wang *et al.* (*Mol. Phylogen. Evol.* **41**: 295, 2006; phylogeny), Wang *et al.* (*Mycol.* **98**: 1065, 2006; phylogeny).

Bulgariaceae Fr. (1849), Leotiales. 4 gen. (+ 10 syn.), 7 spp.

Lit.: Gamundí & Arambarri (*Revta Fac. Agron. Univ. nat. La Plata* **59**: 17, 1983), Fenwick (*Mycologist* **6**: 177, 1992), Triebel & Rambold (*Arnoldia* **4**: 15, 1992), Döring & Triebel (*Cryptog. Bryol.-Lichénol.* **19**: 123, 1998), Wang *et al.* (*Mycol.* **98**: 1065, 2006; phylogeny), Wang *et al.* (*Mol. Phylogen. Evol.* **41**: 295, 2006).

Bulgariastrum Syd. & P. Syd. (1913) ? = Dermea fide Korf *in* Ainsworth *et al.* (Eds) (*The Fungi* **4A**: 249, 1973).

Bulgariella P. Karst. (1885), Helotiaceae. 4, Europe; S. America. See Gamundí (*Sydowia* **34**: 89, 1981), Gamundí & Romero (*Fl. criptog. Tierra del Fuego* **10**: 130 pp., 1998; Argentina).

Bulgariopsis Henn. (1902), Helotiaceae. 1 or 2, Brazil.

Bulla Battarra ex Earle (1909) = Agrocybe fide Singer (*Agaric. mod. Tax.* edn 3, 1975).

Bullardia Jungh. (1830) [non *Bulliarda* DC. 1801, *Crassulaceae*] nom. rej. = Melanogaster fide Stalpers (*in litt.*).

Bullaria DC. (1805) = Puccinia fide Arthur (*Manual Rusts US & Canada*, 1934).

Bullaserpens Bat., J.L. Bezerra & Cavalc. (1965), anamorphic *Pezizomycotina*, Cpt.0fH.?. 1, Brazil. See Batista *et al.* (*Atas Inst. Micol. Univ. Pernambuco* **2**: 292, 1965).

bullate (1) having bubble-like or blister-like swellings; (2) (of a pileus), having a rounded projection at the centre (Fig. 19K).

Bullatina Vězda & Poelt (1987), Gomphillaceae (L). 1, widespread (tropical). See Vězda & Poelt (*Folia geobot. phytotax.* **22**: 186, 1987), Brusse (*Bothalia* **22**: 44, 1992; Transkei), Lücking *et al.* (*Lichenologist* **37**: 123, 2005; phylogeny).

Buller (Arthur Henry Reginald; 1874-1944; England, later Canada). Professor of Botany, Manitoba University (1904-1936). A most versatile mycologist, noted for his research on spore discharge and sexuality in basidiomycetes, particularly rusts; also famous for his limerick poems, and an expert billiards player. His library is at the Research Station, Agriculture

Canada, Winnipeg (Dowding, *Mycol.* **50**: 794, 1958; Oliver, *Catalogue of the Buller Memorial Library*, 1965). *Publs.* see Literature. *Biogs, obits etc.* Ainsworth (*Mycologist* **2**: 83, 1988) [portrait]; Bisby (*Nature* **154**: 173, 1944); Brodie & Lowe (*Science* **100**: 305, 1944); Brooks (*Obituary Notices of Fellows of the Royal Society* **5**: 51, 1945) [portrait]; Stafleu & Cowan (*TL-2* **1**: 401, 1976); Stafleu & Mennega (*TL-2, Suppl.* **3**: 217, 1995).

Buller phenomenon. The dikaryotization, in basidiomycetes and ascomycetes, of a homokaryon by a dikaryon (Quintanilha's (1933) term for Buller's discovery; see Buller, 1941); 'di-mon' matings (Papazian, *Bot. Gaz.* **112**: 143, 1950).

Bullera Derx (1930), anamorphic *Tremellaceae*. *c*. 35, widespread. See Ingold & Young (*TBMS* **76**: 165, 1981; spore discharge), Weijman & Rodrigues de Miranda (*Antonie van Leeuwenhoek* **49**: 559, 1983; cell wall composition), Suh *et al.* (*Microbiol.* Reading **141**: 901, 1995), Suh *et al.* (*J. gen. appl. Microbiol.* Tokyo **42**: 501, 1996; mol. phylogeny), Golubev & Nakasa (*FEMS Microbiol. Lett.* **146**: 59, 1997; mycocins), Barnett *et al.* (*Yeasts: Characteristics and Identification* 3rd edn, 2000), Fell *et al.* (*Int. J. Syst. Evol. Microbiol.* **50**: 1351, 2000; mol. phylogeny), Nakase *et al.* (*Journal of General and Applied Microbiology, Tokyo* **48**: 345, 2002; Taiwan n.sp.), Luong *et al.* (*Journal of General and Applied Microbiology, Tokyo* **51**: 335, 2005; Vietnam n.sp.), Fungsin *et al.* (*Journal of General and Applied Microbiology, Tokyo* **52**: 73, 2006; Thailand n.sp.).

Bulleribasidium J.P. Samp., M. Weiss & R. Bauer (2002), Tremellaceae. Anamorph *Bullera*. 1, Germany. See Sampaio *et al.* (*Mycol.* **94**: 874, 2002), Sampaio (*Frontiers in Basidiomycote Mycology*: 49, 2004).

Bulleromyces Boekhout & A. Fonseca (1991), Tremellaceae. Anamorph *Bullera*. 1, widespread (north temperate). See Boekhout & Fonseca (*Antonie van Leeuwenhoek* **59**: 91, 1991), Sampaio (*Frontiers in Basidiomycote Mycology*: 49, 2004).

Buller's drop. A drop of liquid on the hilar appendix of a ballistospore (q.v.) developed immediately before discharge, studied by Buller (*Researches* **2-6**, 1922-34); the drop contains mannitol and hexoses in concentrations enabling water to be taken in from a saturated atmosphere; prior to discharge the Buller's drop expands until it is large enough to come into contact with spore surface; at this point the drop redistributes itself over the spore surface, causing a sudden change in the centre of gravity of the spore sufficient to break the weak apiculus-sterigma connexion linking the spore to the basidium and resulting in discharge (Webster *et al.*, *MR* **99**: 833, 1995). Acceleration has been estimated at 25,000 g (Money, *Mycol.* **90**: 547, 1998). An educational video-film on this topic is commercially available (Webster, *Ballistospore discharge in basidiomycetes*, IWF Wissen und Medien, Göttingen, 1998).

Bulliard (Jean Baptiste François 'Pierre'; 1752-1793; France). Studied at Clairvaux (up to *c.* 1774); lived in Paris (*c.* 1775 onwards). An eminent naturalist; his main mycological work contains excellent early coloured illustrations of fungi; at a time when 'spontaneous generation' was the accepted explanation, he suggested that a fungus caused cereal smut through spore dispersal [a view also expressed, curiously almost simultaneously, in 1794, by the English artist, poet and mystic, William Blake, through his remarkable illustration 'Mildew Blighting the Ears of Corn']. *Publs. Histoire des champignons de la France* (1791-1792) [index: Petersen, *Mycotaxon* **6**: 127, 1977]. *Biogs, obits etc.* Gilbert (*BSMF* **68**: 1, 1952, bibliography, list of biographies, portrait); Gilbert (*Mycopathologia et Mycologia Applicata* **6**: 237, 1952). Grummann (1974: 269); Stafleu & Cowan (*TL-2* **1**: 402, 1976).

Bulliardella (Sacc.) Paoli (1905) = Actidium fide Lohman (*Pap. Mich. Acad. Sci.* **23**: 155, 1938).

Bulliardia, see *Bullardia*.

Bulliardia Lázaro Ibiza (1916) [non *Bulliarda* DC. 1801, *Crassulaceae*] ≡ Cerrena fide Donk (*Verh. K. ned. Akad. Wet.* tweede sect. **62**: 1, 1974).

Bunodea A. Massal. (1855) = Pyrenula Ach. (1809) fide Hawksworth *et al.* (*Dictionary of the Fungi* edn 8, 1995).

Bunodophoron A. Massal. (1861), Sphaerophoraceae (L). 20, widespread (esp. south temperate). See Wedin (*Pl. Syst. Evol.* **187**: 213, 1993), Wedin (*Symb. Bot. Upsal.* **31** no. 1, 1995; key, monograph), Wedin (*Mycotaxon* **53**: 33, 1995), Wedin *et al.* (*Pl. Syst. Evol.* **209**: 75, 1998; phylogeny), Döring & Wedin (*Pl. Biol.* **2**: 361, 2000; anatomy), Wedin *et al.* (*Lichenologist* **32**: 171, 2000; phylogeny), Peršoh *et al.* (*Mycol. Progr.* **3**: 103, 2004; asci).

bunt, a wheat disease (*Tilletia caries* and *T. laevis*, syn. *T. foetida*); stinking smut; **dwarf -**, *T. controversa*. See also *Ustilaginales*.

Burcardia Schmidel (1797) [non *Burchardia* Schreb. 1789, *Liliaceae*] ≡ Sarcosoma.

Burenella Jouvenaz & E.I. Hazard (1978), Microsporidia. 1.

Buerenia M.S. Reddy & C.L. Kramer (1975), Protomycetaceae. 3 (on *Umbelliferae*), widespread. See Döbbeler (*Nova Hedwigia* **60**: 171, 1995).

Burgella Diederich & Lawrey (2007), Cantharellales. 1. See Diederich & Lawrey (*Mycol. Progr.* **6**: 61, 2007; phylogeny).

Burgoa Goid. (1938), anamorphic *Sistotrema*. 4 (propagules bulbils), widespread. See Weresub & LeClair (*CJB* **49**: 2203, 1971).

Burgundy mixture, made as Bordeaux mixture (q.v.) but with sodium carbonate (Na_2CO_3) in place of lime (copper sulphate 1.8 kg, sodium carbonate 2.3 kg, water 227 l).

Burkea Sprague (1977), Microsporidia. 1.

Burnett (John Harrison; 1922-2007; Scotland, later England). Student (1940-1942, 1945-1947) then lecturer (1947-1954), Oxford University; Professor of Botany, St Andrews University (1955-1960); Professor of Botany, University of Newcastle-upon-Tyne (1960-1968); Regius Professor, Glasgow University (1968-1970); Sibthorpian Professor of Rural Economy, Oxford University (1970-1979); Principal and Vice-Chancellor, Edinburgh University (1979-1987). Noted for his work on fungal biology and, in particular, genetics; also a champion of biodiversity conservation, believing strongly that without fungi the rest of the biosphere could not survive; as a commando under Fitzroy Maclean during World War II, his contact with guerilla leader and later president of Yugoslavia, Josep Brod (Tito) was catalytic in establishing postwar national parks in the country; after retirement, he strongly embraced the new digital technology, and well understood its importance for biodiversity conservation; founder and chairman of the International Organization for Plant Information, and secretary of the World Council for the Biosphere (1987-1993); chairman of the UK National Biodiversity Network. Work. *Publs. Fundamentals of Mycology* (1968); *Mycogenetics* (1975); *Fungal Populations and Species* (2003). *Biogs, obits etc.* Anon. (John Harrison Burnett, obituary. *Times* London, 17 August 2007).

Burrillia Setch. (1891), Doassansiaceae. 4 (on aquatic plants), N. C. & S. America; Asia. See Thirumalachar (*Mycol.* **34**: 602, 1947), Vánky (*Sydowia* **34**: 167, 1981).

bursiculate, bag-like.

bursiform, bag-like.

Burt (Edward Angus; 1859-1939; USA). Born into a poor farming family; student, Harvard University (1891-1895); Professor of Natural History, Middlebury College, Vermont (1895-1913); Librarian and Mycologist, Missouri Botanic Garden, and Professor of Botany, Washington University, St Louis (1913-1933). Noted for his work on the *Thelephoraceae*; also, incidentally, a shrewd stock-market investor. *Publs.* The Thelephoraceae of North America. I-XV. *Annals of the Missouri Botanical Garden* (1914-1926). *Biogs, obits etc.* Dodge (*Occasional Papers of the Farlow Herbarium of Cryptogamic Botany* **14** 1, 1979); Lloyd (*Mycological Notes by G.C. Lloyd* **47**: 653, 1917) [portrait].

Buscalionia Sambo (1940), ? Trypetheliaceae (L). 1, Brazil.

Busseëlla Henn. (1902) = Cephaleuros fide Höhnel (*Sber. Akad. Wiss. Wien* Math.-naturw. Kl., Abt. 1 **120**: 411, 1911).

Butler (Edwin John; 1874-1943, Ireland). Imperial Mycologist, India (1900-1919); founding Director of the Imperial Mycological Institute, Kew (1920-1935). *Publs.*the genus *Pythium*. *Memoirs of the Department of Agricirculture in India* Botanical Series 5, **1** (1907); *Fungi and Disease in Plants* (1918); (with H. Sydow & P. Sydow) Fungi Indiae orientalis *Annales Mycologici* **16** (1906); (with Jones) *Plant Pathology* (1950). *Biogs, obits etc.* Güssow (*Phytopathology* **34**: 149, 1944) [portrait]; Johnston (*Review of Tropical Plant Pathology* **7**: 1, 1993); Kulkarni (*Plant Pathology Newsletter* **1**: 5, 1983); Mason (*Obituary Notices of Fellows of the Royal Society* **4**: 455, 1943) [bibliography, portrait]; Stafleu & Mennega (*TL-2*, *Suppl.* **3**: 285, 1995).

Butlerelfia Weresub & Illman (1980), Atheliaceae. 1, Canada; Europe. See Weresub & Illman (*CJB* **58**: 144, 1980).

Butleria Sacc. (1914), Elsinoaceae. 1, India. See Petrak & Sydow (*Annls mycol.* **27**: 87, 1929).

button, a young mushroom (esp. *Agaricus bisporus*) before the pileus has expanded.

Buxetroldia K.R.L. Petersen & Jørg. Koch (1997), Halosphaeriaceae. 1 (marine), Denmark. See Petersen & Koch (*MR* **101**: 1524, 1997), Pang *et al.* (*Nova Hedwigia* **77**: 1, 2003; phylogeny).

Buxtehudea J.I.R. Larsson (1980), Microsporidia. 1.

Byliana Dippen. (1930) = Palawaniella fide Müller & von Arx (*Beitr. Kryptfl. Schweiz* **11** no. 2, 1962).

Byrrha Bat., F. Monnier & J.S. Silveira (1959) = Pichia fide Batra *in* Subramanian (Ed.) (*Taxonomy of fungi* **1**: 187, 1978).

Byrsalis Neck. ex Kremp. (1869) = Peltigera fide Santesson (*Taxon* **3**: 236, 1954).

Byrsomyces Cavalc. (1972) = Microtheliopsis fide Lücking *et al.* (*Lichenologist* **30**: 121, 1998).

Byssiplaca A. Massal. (1860) = Lecanora fide Hawksworth *et al.* (*Dictionary of the Fungi* edn 8, 1995).

byssisede, see byssoid.

Byssitheca Bonord. (1864) ≡ Rosellinia.

Byssoascus Arx (1971), Myxotrichaceae. Anamorph *Oidiodendron*. 1, Canada. See von Arx (*Persoonia* **6**: 377, 1971), Sugiyama *et al.* (*Mycoscience* **40**: 251, 1999; phylogeny), Usuki *et al.* (*Mycoscience* **44**: 97, 2003; mycorrhizas), Wang *et al.* (*Mycol.* **98**: 1065, 2006; phylogeny).

Byssocallis Syd. (1927), Tubeufiaceae. 2 (on *Meliola*), Costa Rica; S. Africa. See Rossman (*Mycotaxon* **8**: 485, 1979), Rossman (*Mycol. Pap.* **157**, 1987; key), Kodsueb *et al.* (*Fungal Diversity* **21**: 105, 2006; phylogeny).

Byssocaulon Mont. (1835), Chrysothricaceae (L). 6, widespread (tropical). See Rogers (*Flora of Australia* **54**: 65, 1992; Australia).

Byssochlamys Westling (1909), Trichocomaceae. Anamorph *Paecilomyces*-like. 4, widespread. *B. fulva* may damage tinned fruits. See Stolk & Samson (*Persoonia* **6**: 341, 1971), Subramanian & Rajendran (*Kavaka* **5**: 83, 1978; ontogeny), Domenech *et al.* (*Microbiology* Reading **145**: 2789, 1999; chemistry, immunology), Samson *et al.* (*Introduction to Food- and Airborne Fungi*: 1, 2002; identification), Luangsa-ard *et al.* (*Mycol.* **96**: 773, 2004; phylogeny), Luangsa-ard *et al.* (*Kasetsart Journal* Natural Sciences **38**: 94, 2004; Thailand), Houbraken *et al.* (*Advances in Experimental Medicine and Biology* **571**: 211, 2006; toxins).

Byssocladiella Gaillon (1833) ≡ Byssocladium.

Byssocladium Link (1815) nom. dub., anamorphic *Pezizomycotina*. See Donk (*Taxon* **11**: 81, 1962).

Byssocorticaceae Jülich (1982) = Atheliaceae.

Byssocorticium Bondartsev & Singer (1944), Atheliaceae. 9 (mycorrhizal), widespread. See Jülich (*Willdenowia* Beih. **7**, 1972), Mervielde (*AMK Mededelingen*: 45, 2003).

Byssocristella M.P. Christ. & J.E.B. Larsen (1970) = Tomentellopsis fide Hjortstam (*Svensk bot. Tidskr.* **68**: 51, 1974).

Byssocystis Riess (1853) = Ampelomyces fide Rogers (*Mycol.* **51**: 96, 1959).

Byssogene Syd. (1922), ? Dothideales. 1, Indonesia.

byssoid, cotton-like; made up of delicate threads; floccose. **- lichens,** see Egea *et al.* (*Lichenologist* **27**: 351, 1995; in *Arthoniales*), Hafellner & Vězda (*Nova Hedw.* **55**: 183, 1992; key 17 gen. thalli).

Byssolecania Vain. (1921), Pilocarpaceae (L). 9, widespread. See Santesson (*Symb. bot. upsal.* **12** no. 1: 1, 1952), Santesson (*SA* **10**: 137, 1991).

Byssoloma Trevis. (1853), Pilocarpaceae (L). Anamorph *Pyriomyces*. 55, widespread. See Vězda (*Folia geobot. phytotax.* **22**: 71, 1987; key 10 Afr. spp.), Kalb & Vězda (*Nova Hedwigia* **51**: 435, 1990; key 11 spp. neotropics), Elix *et al.* (*Biblthca Lichenol.* **58**: 81, 1995; chemistry), Sérusiaux (*Cryptog. Bryol.-Lichénol.* **19**: 197, 1998; Europe), Holien (*Graphis Scripta* **11**: 61, 2000; Scandinavia).

Byssolomataceae Zahlbr. (1926) = Pilocarpaceae.

Byssolophis Clem. (1931), Lophiostomataceae. 1, widespread. See Holm (*Windahlia* **16**: 49, 1986), Barr (*Mycotaxon* **45**: 191, 1992; posn).

Byssomerulius Parmasto (1967), Phanerochaetaceae. 8, widespread. See Parmasto (*Consp. System. Corticiac.*, 1968), Zmitrovich (*Mikol. Fitopatol.* **35**: 9,

2001).

Byssonectria P. Karst. (1881), Pyronemataceae. 10, widespread. See Dennis & Itzerott (*Kew Bull.* **28**: 5, 1973), Benkert (*Gleditschia* **15**: 173, 1987; status), Caillet & Moyne (*Bull. Soc. Hist. nat. Doubs* **84**: 9, 1991; key), Sivertsen (*SA* **9**: 23, 1991; nomencl.), Pfister (*Mycol.* **85**: 952, 1994; development, key 4 spp. N. Am.), Pfister (*Mycotaxon* **53**: 431, 1995), Yao & Spooner (*MR* **100**: 881, 1996; Brit. spp.), Landvik *et al.* (*Nordic Jl Bot.* **17**: 403, 1997; DNA), Kullman (*Mycotaxon* **69**: 199, 1998; sibling spp.), Harrington *et al.* (*Mycol.* **91**: 41, 1999; phylogeny), Hansen *et al.* (*Mol. Phylogen. Evol.* **36**: 1, 2005), Hansen & Pfister (*Mycol.* **98**: 1029, 2006; phylogeny), Perry *et al.* (*MR* **111**: 549, 2007; phylogeny).

Byssoonygena Guarro, Punsola & Cano (1987), Onygenaceae. Anamorph *Malbranchea*. 1, Spain. See Guarro *et al.* (*Mycopathologia* **100**: 159, 1987).

Byssopeltis Bat., J.L. Bezerra & T.T. Barros (1970), Microthyriaceae. 1, Brazil. See Batista *et al.* (*Publções Inst. Micol. Recife* **636**: 5, 1969).

Byssophoropsis (Vain.) Tehler (1993) = Sagenidiopsis fide Egea *et al.* (*Flechten Follmann* Contributions to Lichenology in Honour of Gerhard Follmann: 183, 1995).

Byssophragmia M. Choisy (1931) = Megalospora Meyen fide Sipman (*Biblthca Lichenol.* **18**, 1983).

Byssophytum Mont. (1848), Pezizomycotina (L (sterile)). 2, Java; Tahiti. See Groenhart (*Nederl. Kruid. Arch.* **46**: 774, 1936).

Byssoporia M.J. Larsen & Zak (1978), Atheliaceae. 1 (mycorrhizal), widespread. See Larsen & Zak (*CJB* **56**: 1123, 1978).

Byssopsora A. Massal. (1861) = Bacidia fide Hawksworth *et al.* (*Dictionary of the Fungi* edn 8, 1995).

Byssosphaeria Cooke (1879), Melanommataceae. Anamorph *Pyrenochaeta*. *c.* 12, widespread. See Barr (*Mycotaxon* **20**: 1, 1984; key), Hyde *et al.* (*MR* **103**: 1423, 1999), Chen & Hsieh (*Sydowia* **56**: 24, 2004; Taiwan).

Byssostilbe Petch (1912) = Berkelella fide Rossman *et al.* (*Stud. Mycol.* **42**: 248 pp., 1999).

Byssotheciella Petr. (1923), Pezizomycotina. 1 or 2, Europe; S. America.

Byssothecium Fuckel (1861), Dacampiaceae. Anamorph *Chaetophoma*-like. 5, widespread (temperate). See Boise (*Mycol.* **75**: 666, 1983), Semeniuk (*Mycol.* **75**: 744, 1983; on alfalfa), Crane *et al.* (*Mycol.* **84**: 235, 1992), Lumbsch & Lindemuth (*MR* **105**: 901, 2001; phylogeny), Kruys *et al.* (*MR* **110**: 527, 2006; phylogeny), Schoch *et al.* (*Mycol.* **98**: 1041, 2006; phylogeny).

Byssus L. (1753) nom. rej. prop. ≡ Trentepohlia Formerly used for some filamentous lichenized and other fungus mycelia.

C (1) see Metabolic products. (2) Botanical Museum and Herbarium (Copenhagen, Denmark); founded 1759; part of the University of Copenhagen.

Cabalodontia Piątek (2004), Meruliaceae. 5. See Piątek (*Polish Botanical Journal* **49**: 2, 2004).

CABI, the trading name of CAB INTERNATIONAL (formerly Commonwealth Agricultural Bureaux; founded 1929), an intergovernmental organization established by treaty lodged with the UN, and of which IMI (q.v.) was an Institute; see Scrivenor (*CAB – the first 50 years*, 1980).

Cacahualia Mercado & R.F. Castañeda (1984) = Arachnophora fide Sutton (*in litt.*), Castañeda Ruíz *et al.* (*Nova Hedwigia* **64**: 473, 1997).

Caccobius Kimbr. (1967), Thelebolaceae. 1 (coprophilous), widespread (north temperate). See Landvik *et al.* (*Mycoscience* **39**: 49, 1998; DNA), van Brummelen (*Persoonia* **16**: 425, 1998; ultrastr.).

Cacosphaeria Speg. (1888) = Kacosphaeria fide Hawksworth *et al.* (*Dictionary of the Fungi* edn 8, 1995).

Cacumisporium Preuss (1851), anamorphic *Pezizomycotina*, Hso.≡ eP.19. 9, Europe; N. America. See Goos (*Mycol.* **61**: 52, 1969), Sutton (*Mycol. Pap.* **132**, 1973), Kirk (*Mycotaxon* **43**: 231, 1992), Hyde & Goh (*MR* **102**: 739, 1998), Réblová & Gams (*Czech Mycol.* **51**: 1, 1999; teleomorph), Réblová (*Stud. Mycol.* **45**: 149, 2000; review), Tsui *et al.* (*Mycol.* **93**: 389, 2001; Hong Kong).

cadavericole, an organism living on corpses.

Cadophora Lagerb. & Melin (1927), anamorphic *Helotiales*. 10, widespread. See Conant (*Mycol.* **29**: 597, 1937), Harrington & McNew (*Mycotaxon* **87**: 141, 2003; phylogeny), Bills (*Mem. N. Y. bot. Gdn* **89**: 113, 2004).

caducous (of spores, etc.), falling off readily, deciduous.

Cadyexinis Stach (1957), Fossil Fungi. 3 (Miocene, Carboniferous), Taiwan; Germany.

Caecomyces J.J. Gold (1988), Neocallimastigaceae. 2, British Isles; Canada. See Wubah *et al.* (*Mycol.* **83**: 303, 1991; morphology), Gleason *et al.* (*Australasian Mycologist* **21**: 94, 2002; rhizoid morphology).

Caenomyces E.W. Berry (1916), Fossil Fungi (mycel.) ? Fungi. 7 (Tertiary), Brazil; USA.

Caenothyrium Theiss. & Syd. (1918) ? = Actinopeltis fide Spooner & Kirk (*MR* **94**: 223, 1990).

Caeoma Link (1809), anamorphic *Pucciniales*. *c.* 200, widespread. Anamorph name for aecial anamorphs (I) without bounding structures. See Sato & Sato (*TBMS* **85**: 223, 1985; wider circumscription incl. bounded sori), Crane *et al* (*Mycoscience* **46**: 143, 2005; Japan n.sp., rhododendron rust).

caeoma (pl. **caeomata**), an aecium as in *Caeoma*, i.e. without peridial cells and with or without paraphyses.

caeomatoid (of aecia), resembling caeomata; sometimes, incorrectly, 'caeomoid'.

Caeomurus, see *Coeomurus*.

Caeruleomyces Stalpers (2000), anamorphic *Hymenochaetales*. 1. See Stalpers (*Karstenia* **40**: 177, 2000).

Caerulicium Jülich (1982) = Byssocorticium fide Stalpers (*in litt.*).

Caesar's mushroom, basidioma of the edible *Amanita caesarea*.

caespitose (**cespitose**), in groups or tufts like grass; cf. gregarious.

Caespitotheca S. Takam. & U. Braun (2005), Erysiphaceae. 1. See Takamatsu *et al.* (*MR* **109**: 907, 2005).

caespitulus (pl. **caespituli**), a tuft of spores.

Caesposus Nüesch (1937) = Lyophyllum fide Kuyper (*in litt.*), Cooke (*Spec. Publ. Div. Myc. Dis. Surv. US Department of Agriculture* **3**, 1953; typification).

Cainea S. Hughes (1951) nom. nud. = Apiosordaria fide Hawksworth *et al.* (*Dictionary of the Fungi* edn 8, 1995).

Cainia Arx & E. Müll. (1955), Cainiaceae. 3 (from grasses), widespread (temperate). See Parguey-Leduc (*Revue Mycol.* Paris **28**: 200, 1963; asci, affinities), Krug (*Sydowia* **30**: 122, 1978), Kang *et al.* (*MR* **103**: 1621, 1999), Jeewon *et al.* (*MR* **107**: 1392, 2003;

phylogeny).

Cainiaceae J.C. Krug (1978), Xylariales. 2 gen., 4 spp.
Lit.: Müller & Corbaz (*Sydowia* **10**: 181, 1956), Krug (*Sydowia* **30**: 122, 1978), Kohlmeyer & Volkmann-Kohlmeyer (*SA* **12**: 7, 1993), Kang *et al.* (*MR* **103**: 1621, 1999), Lumbsch *et al.* (*Mycol. Progr.* **1**: 57, 2002).

Cainiella E. Müll. (1957), Hyponectriaceae. 2 (saprobic in dead leaves), Canada; Europe. See Müller (*Sydowia* **10**: 118, 1957).

Cainomyces Thaxt. (1901), ? Laboulbeniales. 1.

Calathaspis I.M. Lamb & W.A. Weber (1972), Cladoniaceae (L). 1, Papua New Guinea. See Stenroos (*Ann. bot. fenn.* **25**: 207, 1988; Melanesia), Stenroos & DePriest (*Am. J. Bot.* **85**: 1548, 1998; DNA), Stenroos *et al.* (*Mycol. Progr.* **1**: 267, 2002; phylogeny).

Calathella D.A. Reid (1964), Marasmiaceae. 9, Europe; N. America. A new name to replace the illegitimate honomyn has not been introduced. See Reid (*Persoonia* **3**: 93, 1964), Singer (*Agar. mod. Tax.* 4th ed.: 387, 1986; syn. on *Flagelloscypha*), Bodensteiner *et al.* (*Mycol.* **93**: 1010, 2001; Bali).

Calathinus Quél. (1886) [non *Calathinus* Rafin. 1836, *Amaryllidaceae*] = Crepidotus fide Kuyper (*in litt.*) Typification is ambiguous, synonymy with *Pleurocybella* has been proposed.

Calathiscus Mont. (1841) ? = Lysurus fide Dring (*Kew Bull.* **35**: 1, 1980).

Calbovista Morse (1935) ≡ Calbovista Morse ex M.T. Seidl.

Calbovista Morse ex M.T. Seidl (1995), Agaricaceae. 1, USA. See Seidl (*Mycotaxon* **54**: 389, 1995).

calcarate, having a projection or spur.

calcareous, containing lime.

Calcarispora Marvanová & Marvan (1963), anamorphic *Pezizomycotina*, Hso.0fH.1. 1 (aquatic), Europe; Canada. See Marvanová & Marvan (*Čas. Slez. Mus.* Ser. A, Hist. Nat. **12**: 109, 1963), Marvanová & Bärlocher (*Czech Mycol.* **53**: 1, 2001; culture).

Calcarisporiella de Hoog (1974), anamorphic *Pezizomycotina*, Hso.0eH.10. 1, British Isles. See de Hoog (*Stud. Mycol.* **7**: 68, 1974).

Calcarisporium Preuss (1851), anamorphic *Hypocreales*, Hso.0eH.10. 2 (on agarics, etc.), Europe. See Barnett (*Mycol.* **50**: 497, 1958), Barnett & Lilly (*Bull. W. Va Exp. Stn* **420T**, 1958), Nicot (*BSMF* **84**: 85, 1968), Rombach & Roberts (*Mycol.* **79**: 153, 1987; symbiosis with *Hirsutella*), Kendrick (*CJB* **81**: 75, 2003; morphogenesis), Hausner & Reid (*CJB* **82**: 752, 2004; phylogeny), Somrithipol & Jones (*Sydowia* **58**: 133, 2006; Thailand).

calceiform, shoe-like in form.

Calceispora Matsush. (1975), anamorphic *Pezizomycotina*, Hso.0eH.15. 2, Japan; Malawi. See Sutton (*Mycol. Pap.* **167**: 11, 1993).

calceolate, see calceiform.

Calceomyces Udagawa & S. Ueda (1988), Xylariaceae. Anamorph *Nodulisporium*. 1, Japan. See Udagawa & Ueda (*Mycotaxon* **32**: 448, 1988).

calcicolous, an organism (**calcicole**) growing on substrates rich in calcium; esp. of spp. on limestone or chalky rocks or soils.

Caldariomyces Woron. (1926), anamorphic *Pezizomycotina*. See Hughes (*Mycol.* **68**: 693, 1976), Reynolds & Faull (*Taxon* **50**: 1183, 2001; conservation proposal), Gams (*Taxon* **54**: 520, 2005; conservation rejected).

Caldesia Rehm (1889) ≡ Holmiella.

Caldesia Trevis. (1869) [non *Caldesia* Parl. 1860, *Alismataceae*] = Arthonia fide Hawksworth *et al.* (*Dictionary of the Fungi* edn 8, 1995).

Caldesiella Sacc. (1877) nom. rej. = Tomentella Pat. fide Larsen (*Taxon* **16**: 510, 1967) See also, Nikolajeva (*Mikol. Fitopatol.* **2**: 198, 1968).

Calenia Müll. Arg. (1890), Gomphillaceae (L). 41, widespread (tropical). See Santesson (*Symb. bot. upsal.* **12** no. 1: 1, 1952), Vězda & Poelt (*Folia geobot. phytotax.* **22**: 179, 1987), Hartmann (*Mycotaxon* **59**: 483, 1996), Lücking (*Biblthca Lichenol.* **65**: 1, 1997), Lücking *et al.* (*Trop. Bryol.* **19**: 55, 2000), Ferraro (*Fungal Diversity* **15**: 153, 2004; hypophores), Lücking *et al.* (*Mycol.* **96**: 283, 2004; phylogeny), Lücking *et al.* (*Lichenologist* **37**: 123, 2005; morphology, cladistics), Papong *et al.* (*Nova Hedwigia* **86**: 201, 2008).

Caleniomyces Cif. & Tomas. (1953) = Caleniopsis.

Caleniopsis Vězda & Poelt (1987), Gomphillaceae (L). 2, widespread (tropical). See Vězda & Poelt (*Folia geobot. phytotax.* **22**: 187, 1987), Lücking *et al.* (*Lichenologist* **37**: 123, 2005; phylogeny), Papong *et al.* (*Nova Hedwigia* **86**: 201, 2008).

Caleutypa Petr. (1934), Pezizomycotina. 1, Europe.

Caliciaceae Chevall. (1826), Teloschistales (±L). 31 gen. (+ 46 syn.), 731 spp.
Lit.: Tibell (*Nova Hedwigia* Beih. **79**: 597, 1984), Tibell (*Symb. bot. upsal.* **27** no. 1: 279 pp., 1987), McCune & Rosentreter (*Bryologist* **95**: 329, 1992), Tibell & Kalb (*Nova Hedwigia* **55**: 11, 1992), Tibell (*Symb. bot. upsal.* **32** no. 1: 291, 1997), Wedin & Tibell (*CJB* **75**: 1236, 1997; phylogeny), Tibell (*Biblthca Lichenol.* **71**, 1998; S Am. spp. s.l.), Tibell (*Biblthca Lichenol.* **71**: 107 pp., 1998), Hladun (*Clementeana* **4**: 48, 1999; key to *Caliciales*), Sarrión *et al.* (*Mycotaxon* **71**: 169, 1999), Tibell (*Nordic Lichen Flora* **1**. Introductory Parts; Calicioid Lichens and Fungi: 20, 1999), Tibell (*Nordic Lichen Flora* **1**. Introductory Parts; Calicioid Lichens and Fungi **1**, 1999; Nordic spp. s.l.), Wedin *et al.* (*CJB* **78**: 246, 2000; rels with *Physciaceae*), Helms *et al.* (*Mycol.* **95**: 1078, 2003; phylogeny), Miądlikowska *et al.* (*Mycol.* **98**: 1088, 2006; phylogeny).

Caliciales Bessey (1907) = Teloschistales.

Calicidium, see *Calycidium*.

Caliciella Vain. (1927) = Calicium fide Hawksworth *et al.* (*Dictionary of the Fungi* edn 8, 1995).

Caliciomyces E.A. Thomas ex Cif. & Tomas. (1953) ≡ Calicium.

Caliciopsis Peck (1880), Coryneliaceae. 28 (on conifers), widespread. See Benny *et al.* (*Bot. Gaz.* **146**: 437, 1985; key), Marmolejo (*Mycotaxon* **72**: 195, 1999), Rikkinen (*Karstenia* **40**: 147, 2000; China), Winka & Eriksson (*Phylogenetic Relationships Within the Ascomycota Based on 18S rDNA Sequences* Akademisk Avhandling [Thesis (PhD), Department of Ecology and Environmental Science, Umeå University]: [17] pp., 2000; phylogeny), Geiser *et al.* (*Mycol.* **98**: 1053, 2006; phylogeny).

Calicium Pers. (1794), Caliciaceae (±L). 30, widespread. See Tibell (*Symb. bot. upsal.* **21** no. 2, 1975), Tibell (*Svensk bot. Tidskr.* **71**: 239, 1977), Tibell (*Beih. Nova Hedwigia* **79**: 597, 1984; gen. concept), Tibell & Kalb (*Nova Hedwigia* **55**: 11, 1992; 9 trop. Am. spp.), Tibell (*Nordic Lichen Flora* **1**. Introductory Parts; Calicioid Lichens and Fungi **1**, 1999; Nordic spp. s.l.), Wedin *et al.* (*Taxon* **51**: 655, 2002; phylogeny), Miądlikowska *et al.* (*Mycol.* **98**: 1088,

2006; phylogeny), Tibell (*J. Hattori bot. Lab.* **100**: 809, 2006; Himalayas).

Calidia Stirt. (1876) = Byssoloma fide James (*Lichenologist* **5**: 175, 1971).

Calidion Syd. & P. Syd. (1919), anamorphic *Uncola*. 2 (on *Lindsaea*, *Polypodium*, *Thelypteris* (*Pteridophyta*)), S. America. Anamorph name for (II).
Lit.: Cummins & Hiratsuka (*Illustr. Gen. Rust Fungi* edn 3: 225 pp., 2003; syn. *Macabuna*, *Physopella* sensu Ono *et al.*, 1992).

Californiomyces U. Braun (1981) = Brasiliomyces fide Zheng (*Mycotaxon* **19**: 281, 1984).

Calkinsia Nieuwl. (1916) ≡ Pterygium.

Callebaea Bat. (1962), ? Capnodiaceae. 1, Uganda. See Batista (*Broteria* ser. bot. **31**: 100, 1962).

Calliderma (Romagn.) Largent (1994) = Entoloma fide Kuyper (*in litt.*).

Callimastix Weissenb. (1912), Coelomomycetaceae. 1 (On *Cyclops*), Germany. See Barr *in* Margulis *et al.* (Eds) (*Handbook of Protoctista*: 454, 1990).

Callimothallus Dilcher ex Janson. & Hills (1977), Fossil Fungi, Microthyriaceae. 7 (Cretaceous, Tertiary), widespread. See Hansen (*Grana* **19**: 67, 1980).

Calliospora Arthur (1905) = Uropyxis fide Arthur (*Manual Rusts US & Canada*, 1934).

Callistodermatium Singer (1981), ? Tricholomataceae. 1, Brazil. See Singer (*Mycol.* **73**: 506, 1981).

Callistospora Petr. (1955), anamorphic *Pezizomycotina*, Hso.≡ eP.19. 1, Australia. See Nag Raj (*CJB* **67**: 3169, 1989; redescr.).

Callistosporium Singer (1944), Tricholomataceae. 13, widespread. See Singer (*Sydowia* **30**: 261, 1978; key), Redhead (*Sydowia* **35**: 223, 1982; *Callistosporium luteo-olivaceum*), Gándara & Guzmán (*Mycotaxon* **96**: 73, 2006).

Callolechia Kremp. (1869) ≡ Collolechia.

Callopis (Müll. Arg.) Gyeln. (1933) = Phyllopsora fide Hawksworth *et al.* (*Dictionary of the Fungi* edn 8, 1995).

Callopisma De Not. (1847) nom. rej. prop. ≡ Caloplaca.

Calloria Fr. (1836), Helotiales. Anamorph *Cylindrocolla*. 4, Europe; N. America. See Hein (*Willdenowia* Beih. **9**, 1976), Spooner (*Biblthca Mycol.* **116**: 711 pp., 1987; Australia), Nauta & Spooner (*Mycologist* **13**: 65, 1999; UK).

Calloriella Höhn. (1918), Helotiales. 1, Europe. See Nannfeldt (*Nova Acta R. Soc. Scient. upsal.* **8**: 193, 1932; aff. with *Orbiliaceae*).

Callorina Korf (1971) ≡ Calloria fide Hein (*Willdenowia* Beih. **9**, 1976).

Calloriopsis Syd. & P. Syd. (1917), ? Helotiaceae. 1 (on *Meliolaceae*), widespread (tropical). See Pfister (*Mycotaxon* **4**: 340, 1976), Lizoň *et al.* (*Mycotaxon* **67**: 73, 1998; posn), Baral & Marson (*Micologia 2000*: 23, 2000).

Callorites Fiore (1932), Fossil Fungi. 1 (Eocene), Italy.

callose, hard or thick and sometimes rough.

Callosisperma Preuss (1855) nom. dub., anamorphic *Pezizomycotina*. See Sutton (*Mycol. Pap.* **141**, 1977).

callosities (of fungi), wall thickenings associated with the penetration of fungicolous parasites (Swart, *TBMS* **64**: 511, 1975). See papillae.

Calocera (Fr.) Fr. (1828), Dacrymycetaceae. *c.* 15, widespread. See McNabb (*N.Z. Jl Bot.* **3**: 31, 1965; key), Kennedy (*CJB* **50**: 413, 1972; basidioma devel.), Reid (*TBMS* **62**: 437, 1974; key Brit. spp.), Mathiesen (*Svampe* **25**: 35, 1992; key Danish spp.).

Caloceraceae Rea (1922) = Dacrymycetaceae.

Calocerás Fr. ex Wallr. (1833) = Calocera fide Kennedy (*Mycol.* **50**: 884, 1958).

Calochaetis Syd. (1935) = Wentiomyces fide Müller & von Arx (*Beitr. Kryptfl. Schweiz* **11** no. 2, 1962).

Calocladia Lév. (1851) [non *Calocladia* Grev. 1836, *Rhodophyta*] ≡ Microsphaera fide Braun (*in litt.*).

Calocline Syd. (1939), anamorphic *Pezizomycotina*, St.0eH.15. 1, Ecuador.

Calocybe Kühner ex Donk (1962), Lyophyllaceae. *c.* 40, widespread. *C. gambosum*, St. George's mushroom. See Singer (*Sydowia* **30**: 264, 1978; key).

Caloderma Petri (1900) = Scleroderma fide Guzmán (*Darwiniana* **16**: 233, 1970).

Calodon P. Karst. (1881) = Hydnellum P. Karst. (1879) fide Donk (*Taxon* **5**: 69, 1956).

Calogloeum Syd. (1924) = Fusamen fide von Arx (*Verh. K. ned. Akad. Wet. Amst.* C **51**: 1, 1957).

Calolepis Syd. (1925) = Pycnoderma fide von Arx (*Persoonia* **2**: 421, 1963).

Calonectria De Not. (1867), Nectriaceae. Anamorph *Cylindrocladium*. *c.* 34, widespread. See Rossman (*Mycotaxon* **8**: 321, 1979), Rossman (*Mycotaxon* **8**: 485, 1979; excl. names), Rossman (*Mycol. Pap.* **150**, 1983), Subramanian & Bhat (*Cryptog. Mycol.* **4**: 269, 1983), Crous & Wingfield (*Mycotaxon* **51**: 341, 1994), Rehner & Samuels (*CJB* **73** Suppl. 1: S816, 1995; phylogeny), Rossman *et al.* (*Stud. Mycol.* **42**: 248 pp., 1999), Schoch *et al.* (*Stud. Mycol.* **45**: 45, 2000), Schoch *et al.* (*Mycol.* **92**: 665, 2000; phylogeny), Crous & Kang (*Mycoscience* **42**: 51, 2001; phylogeny), Kang *et al.* (*Syst. Appl. Microbiol.* **24**: 206, 2001; species concepts), Kang *et al.* (*CJB* **79**: 1241, 2001; connexions), Schoch *et al.* (*MR* **105**: 1045, 2001; phylogeny), Schoch *et al.* (*Pl. Dis.* **85**: 941, 2001; populations), Crous (*Taxonomy and Pathology of Cylindrocladium (Calonectria) and Allied Genera*: 278 pp., 2002; monogr.), Crous *et al.* (*Stud. Mycol.* **50**: 415, 2004; revision), Crous *et al.* (*Stud. Mycol.* **55**: 213, 2006; revision).

Calopactis Syd. & P. Syd. (1912) = Endothiella fide Sutton (*Mycol. Pap.* **141**, 1977).

Calopadia Vězda (1986), Ectolechiaceae (L). 15 (on leaves), widespread (tropical). See Vězda (*Folia geobot. phytotax.* **21**: 208, 215, 1986), Kalb & Vezda (*Folia geobot. phytotax.* **22**: 287, 1987; Brazil), Aptroot *et al.* (*Biblthca Lichenol.* **64**: 220 pp., 1997; New Guinea), Lücking (*Phyton* Horn **39**: 131, 1999; Costa Rica).

Calopadiopsis Lücking & R. Sant. (2002), Ectolechiaceae. 1, Argentina. See Lücking & Santesson (*Bryologist* **105**: 58, 2002).

Calopeltis Syd. (1925) = Cyclotheca fide Müller & von Arx (*Beitr. Kryptfl. Schweiz* **11** no. 2, 1962).

Calopeziza Syd. & P. Syd. (1913) = Dictyonella fide von Arx (*Persoonia* **2**: 421, 1963).

Caloplaca Th. Fr. (1860) nom. cons., Teloschistaceae (L). *c.* 510, widespread. See Wade (*Lichenologist* **3**: 1, 1965; UK), Alon & Galun (*Israel J. Bot.* **20**: 273, 1971; Israel), Nordin (*Caloplaca sect. Gasparrinia i Nordeuropa*, 1972), Wunder (*Biblthca Lichenol.* **3**, 1974; dark apothecia), Hafellner & Poelt (*J. Hattori bot. Lab.* **46**: 1, 1979; key 17 polarilocular spp.), Egea (*Collect. bot.* **15**: 173, 1984; key 59 spp., Spain), Poelt & Pelleter (*Pl. Syst. Evol.* **148**: 51, 1984; key, 10 fructescent spp.), Hansen *et al.* (*Meddr Grønland* Biosc. **25**, 1987; key 43 spp. Greenland), Kärnefelt (*Monogr. Syst. Bot.* Miss. Bot. Gdn **25**:

439, 1988; S. Africa), Søchting (*Op. Bot.* **100**: 241, 1989; key 13 spp.), Nimis (*Not. Soc. Lich. Ital.* **5**: 9, 1992; key 10 spp. Italy), Poelt & Hinteregger (*Biblthca Lichenol.* **50**, 1993; Himalaya), Wetmore (*Mycol.* **86**: 813, 1994; key 17 spp. dark apothecia N. & C. Am.), Arup (*Bryologist* **98**: 129, 1995; key N. Am. maritime spp.), Wetmore (*Bryologist* **99**: 292, 1996; *C. sideritis* group), Wetmore & Kärnefelt (*Bryologist* **101**: 230, 1998; sub fruticose & lobate N. Am. spp.), Wetmore (*Bryologist* **104**: 1, 2001; C. & N. Am.), Gaya *et al.* (*Am. J. Bot.* **90**: 1095, 2003; phylogeny), Kärnefelt (*Biblthca Lichenol.* **86**: 341, 2003; Australia, reproduction), Søchting & Lutzoni (*MR* **107**: 1266, 2003; phylogeny, generic limits), Wetmore (*Bryologist* **106**: 147, 2003; N and C America), Kärnefelt & Kondratyuk (*Biblthca Lichenol.* **88**: 255, 2004; Australia), Wetmore (*Bryologist* **107**: 284, 2004; N and C America), Wetmore (*Bryologist* **107**: 505, 2004; N and C America), Arup (*Lichenologist* **38**: 1, 2006; Scandinavia), Miądlikowska *et al.* (*Mycol.* **98**: 1088, 2006; phylogeny), Kondratyuk *et al.* (*Biblthca Lichenol.* **95**: 341, 2007; Australia), Søchting & Figueras (*Lichenologist* **39**: 7, 2007; despide-producing spp.), Gaya *et al.* (*MR* **112**: 528, 2008; phylogeny), Muggia *et al.* (*MR* **112**: 36, 2008; endolithic spp.).

Caloplacaceae Zahlbr. (1907) = Teloschistaceae.

Caloplacomyces E.A. Thomas (1939) nom. inval. ≡ Caloplaca.

Caloplacopsis (Zahlbr.) B. de Lesd. (1932) = Candelariella fide Hawksworth *et al.* (*Dictionary of the Fungi* edn 8, 1995).

Caloporaceae Bondartseva (1983) = Meruliaceae.

Caloporia P. Karst. (1893) ≡ Caloporus P. Karst.

Caloporus P. Karst. (1881) = Merulius Fr. fide Donk (*Persoonia* **1**: 193, 1960).

Caloporus Quél. (1886) = Albatrellus fide Stalpers (*in litt.*).

Calopposis Lloyd (1925) = Calocera fide McNabb (*N.Z. Jl Bot.* **3**: 31, 1965).

Caloscypha Boud. (1885), Caloscyphaceae. Anamorph *Geniculodendron*. 1, Europe. See Kimbrough & Curry (*Mycol.* **78**: 735, 1986; ultrastr.), Landvik *et al.* (*Nordic Jl Bot.* **17**: 403, 1997; phylogeny), Harmaja (*Karstenia* **42**: 27, 2002; family), Schröder *et al.* (*Forest Pathology* **32**: 225, 2002; anam.), Hansen & Pfister (*Mycol.* **98**: 1029, 2006; phylogeny), Perry *et al.* (*MR* **111**: 549, 2007; phylogeny).

Caloscyphaceae Harmaja (2002), Pezizales. 2 gen. (+ 1 syn.), 2 spp. See Harmaja (*Karstenia* **42**: 27, 2002; family), Hansen & Pfister (*Mycol.* **98**: 1029, 2006; phylogeny).

Calosphaeria Tul. & C. Tul. (1863), Calosphaeriaceae. Anamorph *Calosphaeriophora*. 29, widespread. See Barr (*Mycol.* **77**: 549, 1985), Samuels & Candoussau (*Nova Hedwigia* **62**: 47, 1996), Réblová *et al.* (*Stud. Mycol.* **50**: 533, 2004; phylogeny), Damm *et al.* (*Persoonia* **20**: 39, 2008; asci, relns).

Calosphaeriaceae Munk (1957), Calosphaeriales. 8 gen. (+ 2 syn.), 44 spp.

Lit.: Barr (*Mycol.* **77**: 549, 1985), Romero & Minter (*TBMS* **90**: 457, 1988), Samuels & Candoussau (*Nova Hedwigia* **62**: 47, 1996), Barr (*Cryptog. Bryol.-Lichénol.* **19**: 169, 1998), Mostert *et al.* (*Mycol.* **95**: 646, 2003), Réblová *et al.* (*Stud. Mycol.* **50**: 533, 2004), Damm *et al.* (*Persoonia* **20**: 39, 2008; asci, relns).

Calosphaeriales M.E. Barr (1983). Sordariomycetidae 2 fam., 13 gen., 54 spp. Stromatic tissues almost absent to well-developed, usually pseudostromatic. Ascomata perithecial, immersed, often clustered, with separate or convergent ostioles. Interascal tissue composed of a few elongate paraphyses. Asci formed in fascicles or spicate clusters, croziers sometimes absent, sessile or long-stalked, sometimes polysporous; usually with an inconspicuous J- apical ring. Ascospores hyaline or pale brown, ellipsoidal or allantoid, thin-walled. Saprobes on bark or wood, mainly temp. Fams:

(1) **Calosphaeriaceae**
(2) **Pleurostomataceae**

Lit.: Barr (*Mycol.* **77**: 549, 1985; *Mycotaxon* **39**: 43, 1990), Barr *et al.* (*Mycotaxon* **48**: 529, 1993), Rogers (*in* Hawksworth (Ed.), *Ascomycete systematics*: 321, 1994).

Calosphaeriophora Réblová, L. Mostert, W. Gams & Crous (2004), anamorphic *Calosphaeria*. 1. See Réblová *et al.* (*Stud. Mycol.* **50**: 542, 2004).

Calosphaeriopsis Petr. (1941), Pezizomycotina. 1, Europe.

Calospora Nitschke ex Niessl (1875) = Macrodiaporthe fide Barr (*Mycol. Mem.* **7**: 232 pp., 1978), Holm (*Taxon* **24**: 475, 1978).

Calospora Sacc. (1883) = Prosthecium fide Barr (*Mycol. Mem.* **7**: 232 pp., 1978).

Calosporella J. Schröt. (1897) = Prosthecium fide Barr (*Mycol. Mem.* **7**: 232 pp., 1978).

Calostilbe Sacc. & P. Syd. (1902), Nectriaceae. Anamorph *Calostilbella*. 1 (on rotten wood etc.), pantropical. See Samuels (*CJB* **51**: 1275, 1973), Rossman *et al.* (*Stud. Mycol.* **42**: 248 pp., 1999).

Calostilbella Höhn. (1919), anamorphic *Calostilbe*, Hsy.≡ eP.1. 1, W. Africa; West Indies. See Mason (*Mycol. Pap.* **2**: 29, 1925), Hewings & Crane (*Mycotaxon* **20**: 245, 1984).

Calostoma Desv. (1809), Calostomataceae. 15, widespread. See Boedijn (*Bull. Jard. bot. Buitenz.* ser. 3 **16**: 64, 1938; Indonesia, key), Liu (*J. Shansi Univ., nat. sci. ed.* **1**: 109, 1979; world key [Chinese]), Hibbett *et al.* (*Nature* **407**: 506, 2000; phylogeny), Hughey *et al.* (*Mycol.* **92**: 94, 2000; phylogeny).

Calostomataceae E. Fisch. (1900), Boletales. 1 gen. (+ 3 syn.), 15 spp.

Lit.: Cunningham (*Gast. Austr. N.Z.*: 236 pp., 1942), Hughey *et al.* (*Mycol.* **92**: 94, 2000), Binder & Bresinsky (*Mycol.* **94**: 85, 2002).

Calothricopsis, see *Calotrichopsis*.

Calothyriella Höhn. (1917) = Microthyrium fide Müller & von Arx (*Beitr. Kryptfl. Schweiz* **11** no. 2, 1962).

Calothyriolum Speg. (1919) = Asterina fide Müller & von Arx (*Beitr. Kryptfl. Schweiz* **11** no. 2, 1962).

Calothyriopeltis F. Stevens & R.W. Ryan (1925) nom. dub., Fungi. See Petrak (*Sydowia* **5**: 169, 1951).

Calothyriopsis Höhn. (1919), Microthyriaceae. 2, America (tropical).

Calothyris Clem. & Shear (1931) ≡ Calothyriopeltis.

Calothyrium Theiss. (1912) = Asterinella fide von Arx & Müller (*Stud. Mycol.* **9**, 1975).

Calotrichopsis Vain. (1890), Lichinaceae (L). 3, S. America. See Henssen (*Symb. bot. upsal.* **18** no. 1, 1963), Moreno & Egea (*Acta Bot. Barcinon.* **41**: 1, 1992), Schultz & Büdel (*Lichenologist* **34**: 39, 2002; key).

calvacin, a non-diffusible mucoprotein antibiotic from *Langermannia gigantea*; active against mouse, rat, and hamster tumours (Beneke, *Mycol.* **55**: 257, 1963).

Calvarula Zeller (1939), Phallaceae. 1, USA.

Calvatia Fr. (1849) nom. cons., Agaricaceae. *c.* 40, widespread. See Zeller & Smith (*Lloydia* **27**: 148, 1964; N. Am. spp., keys), Kreisel (*Nova Hedwigia* **48**: 241, 1989; key to segr. *Handkea*, not accepted here), Calonge & Martin (*Boln Soc. Micol. Madrid* **14**: 181, 1990; gen. limits), Demoulin (*Mycotaxon* **46**: 77, 1993), Lange (*Blyttia* **51**: 141, 1993; infr. generic taxa), Coetzee & Van Wyk (*Bothalia* **33**: 156, 2003; *Calvatia* sect. *Macrocalvatia*), Coetzee & Van Wyk (*Taxon* **54**: 541, 2005; nomencl.).

Calvatiella C.H. Chow (1936) = Lycoperdon Pers. fide Kreisel & Calonge (*Mycotaxon* **48**: 13, 1993).

Calvatiopsis Hollós (1929), Agaricaceae. 1, Europe. Perhaps a 'monstrosity'.

calvescent, becoming bare or bald.

Calvitimela Hafellner (2001), ? Tephromelataceae. 5, widespread. See Hafellner & Türk (*Stapfia* **76**: 150, 2001), Arup *et al.* (*Mycol.* **99**: 42, 2007; sister group relations with *Parmeliaceae*).

Calvocephalis Bainier (1882) = Syncephalis fide Benjamin (*Aliso* **4**: 321, 1959).

calvous, naked, bare.

Calycella (Fr.) Boud. (1885) ≡ Calycina.

Calycella (Sacc.) Sacc. (1899), Helotiaceae. 1, Europe. See Korf & Carpenter (*Mycotaxon* **1**: 52, 1974).

Calycella Quél. (1886) = Bisporella fide Hawksworth *et al.* (*Dictionary of the Fungi* edn 8, 1995).

Calycellina Höhn. (1918), Hyaloscyphaceae. Anamorph *Chalara*-like. *c.* 43, widespread. See Thind & Sharma (*J. Indian bot. Soc.* **59**: 350, 1980), Arendholz & Sharma (*Mycotaxon* **20**: 633, 1984), Lowen & Dumont (*Mycol.* **76**: 1003, 1984; key), Baral (*Beitr. Kenntn. Pilze Mitteleur.* **5**: 209, 1989; 4-spored spp.), Huhtinen (*Karstenia* **29**: 45, 1990), Svrček (*Česká Mykol.* **46**: 149, 1993; 5 spp.), Cantrell & Hanlin (*Mycol.* **89**: 745, 1997; DNA), Yu & Zhuang (*Mycosystema* **22**: 42, 2003; phylogeny), Raitviir (*Scripta Mycol.* **20**, 2004).

Calycellinopsis W.Y. Zhuang (1990), Dermateaceae. 1 (on petioles), China. See Zhuang (*Mycotaxon* **38**: 121, 1990).

Calycidiaceae Elenkin (1929), Lecanorales (L). 1 gen. (+ 2 syn.), 1 spp.

Lit.: Tibell (*Symb. bot. upsal.* **27** no. 1: 279 pp., 1987), Wedin (*Taxonomic Studies in Sphaerophoraceae* (Caliciales, Ascomycotina), Acta Universitatis Upsaliensis (Comprehensive Summaries of Uppsala Dissertations from the Faculty of Science and Technology no. 77): 168 pp., 1994), Tibell (*Symb. bot. upsal.* **32** no. 1: 291, 1997), Wedin (*Lichenologist* **34**: 63, 2002).

Calycidiomyces Cif. & Tomas. (1953) ≡ Calycidium.

Calycidium Stirt. (1877), Calycidiaceae (L). 1 (on *Nothofagus*), New Zealand. See Sato (*Miscnea bryol. lichen., Nichinan* **4**: 150, 1968), Tibell (*Symb. bot. upsal.* **27** no. 1, 1987), Tibell (*Symb. bot. upsal.* **32**: 291, 1997; anamorphs), Wedin (*Lichenologist* **34**: 63, 2002; monogr.).

calyciform, cup-like.

Calycina Nees ex Gray (1821), Hyaloscyphaceae. Anamorph *Acleistia*. *c.* 45, widespread. See Baral (*SA* **13**: 113, 1993), Raitviir (*Scripta Mycol.* **20**, 2004), Zhang & Zhuang (*Nova Hedwigia* **78**: 475, 2004; phylogeny).

Calycium DC. (1805) ≡ Calicium.

calycular, cup-like.

Calyculosphaeria Fitzp. (1923) = Nitschkia fide Gaikwad (*Sydowia* **26**: 290, 1974), Nannfeldt (*Svensk bot. Tidskr.* **66**: 49, 1975), Huhndorf *et al.* (*MR* **108**: 1384, 2004).

calyculus, a cup-like or calyx-like structure at the base of the sporangium in *Mycetozoa*.

Calyptella Quél. (1886), Marasmiaceae. 20, widespread. See Singer (*Sydowia* **30**: 270, 1978; key), Young (*Mycologist* **10**: 152, 1996; *Calyptella longipes* in Australia).

Calyptellopsis Svrček (1986), Hyaloscyphaceae. 1, former Czechoslovakia. See Svrček (*Česká Mykol.* **40**: 203, 1986).

Calyptospora J.G. Kühn (1869), Pucciniastraceae. 1, north temperate. Separated from *Pucciniastrum* by the demicyclic life cycle (0, I-III). See Faull (*J. Arnold Arbor.* **20**: 104, 1939), Cummins & Hiratsuka (*Illustr. Gen. Rust Fungi rev. edit.*, 1983; as *Pucciniastrum*), Hiratsuka (*The Rust Flora of Japan*, 1992).

Calyptra Theiss. & Syd. (1918), Dothideomycetes. 2, America.

calyptra, a cap or hood.

Calyptromyces H. Karst. (1849) = Mucor Fresen. fide Sumstine (*Mycol.* **2**: 125, 1910), Hesseltine (*Mycol.* **47**: 344, 1955) See.

Calyptronectria Speg. (1909), Melanommataceae. 1 or 2 (dead branches), S. America. See Rossman *et al.* (*Stud. Mycol.* **42**: 248 pp., 1999).

Calyptrozyma Boekhout & Spaay (1995), Eurotiomycetes. 1, USA. See Boekhout *et al.* (*MR* **99**: 1244, 1995), Schweigkofler *et al.* (*Organ. Divers. Evol.* **2**: 1, 2002; phylogeny).

Camanchaca Follm. & Peine (1999) = Pentagenella fide Follmann & Peine (*J. Hattori bot. Lab.* **87**: 259, 1999), Tehler & Irestedt (*Cladistics* **23**: 432, 2007).

Camaroglobulus Speer (1986), anamorphic *Mytilinidion*, St.0eP.1. 1, Brazil. See Speer (*BSMF* **102**: 100, 1986).

Camarographium Bubák (1916), anamorphic *Pezizomycotina*, St.#eP.15. 4, Europe. See Verkley *et al.* (*Sydowia* **57**: 259, 2005).

Camaropella Lar.N. Vassiljeva (1997) = Camarops fide Vasilyeva (*Mikol. Fitopatol.* **31**: 6, 1997).

Camarophyllopsis Herink (1958), ? Hygrophoraceae. 26, widespread (north temperate). See Printz & Læssøe (*Svampe* **14**: 83, 1986; key), Boertmann (*Biblthca Mycol.* **192**: 168 pp., 2002; monogr.).

Camarophyllus (Fr.) P. Kumm. (1871) = Hygrophorus fide Kuyper (*in litt.*).

Camarops P. Karst. (1873), Boliniaceae. 19, widespread. See Nannfeldt (*Svensk bot. Tidskr.* **66**: 335, 1972), Pouzar (*Česká Mykol.* **40**: 218, 1987; 4 spp. Czech.), Eriksson & Hawksworth (*SA* **7**: 64, 1988; posn), Vassilyeva (*Mikol. Fitopatol.* **22**: 388, 1988), Callan & Rogers (*Sydowia* **41**: 74, 1989; teleomorph in culture), Vasilyeva (*Mikol. Fitopatol.* **31**: 5, 1997; Russian Far East), Huhndorf *et al.* (*Mycol.* **96**: 368, 2004; phylogeny), Catania & Romero (*Sydowia* **57**: 3, 2005; Argentina), Rogers *et al.* (*Sydowia* **58**: 105, 2006; Caribbean), Zhang *et al.* (*Mycol.* **98**: 1076, 2006; phylogeny).

Camaropycnis E.K. Cash (1945), anamorphic *Pezizomycotina*, St.0eH.15. 1, USA.

Camarosporellum Tassi (1902), anamorphic *Pezizomycotina*, Cpd.#eP.1. 3, Europe; USA. See Sutton & Pollack (*Mycopath. Mycol. appl.* **52**: 331, 1974), van Warmelo & Sutton (*Mycol. Pap.* **145**, 1981).

Camarosporiopsis Abbas, B. Sutton & Ghaffar (2000), anamorphic *Pezizomycotina*. 1, Pakistan. See Abbas *et al.* (*Pakist. J. Bot.* **32**: 239, 2000).

Camarosporium Schulzer (1870), anamorphic *Botryosphaeriales*, St.#eP.1/19. *c.* 106, widespread (esp. temperate). A polyphyletic group that is also used for anamorphs of *Cucurbitaria* and related genera. See Butin (*Sydowia* **45**: 161, 1993; pleomorphy), Taylor *et al.* (*Mycotaxon* **78**: 75, 2001; on *Proteaceae*), Crous *et al.* (*Stud. Mycol.* **55**: 235, 2006; phylogeny).

Camarosporula Petr. (1954), anamorphic *Anthracostroma*, St.#eP.1. 1, Australia. See Petrak (*Sydowia* **8**: 99, 1954), Swart (*TBMS* **84**: 733, 1985; conidiogenesis).

Camarosporulum Tassi (1902), anamorphic *Pezizomycotina*, Cpd.#eP.?. 1, widespread.

Camarotella Theiss. & Syd. (1915) = Coccodiella fide Hyde & Cannon (*Mycol. Pap.* **175**, 1999).

Camillea Fr. (1849), Xylariaceae. Anamorph *Xylocladium*. 41, widespread (tropical). See Silveira & Rogers (*Acta Amazon.* Supl. **15**: 7, 1987; Brazil), Læssøe *et al.* (*MR* **93**: 121, 1989; monogr., SEM, key 27 spp.), Rogers *et al.* (*Mycol.* **83**: 274, 1991), González & Rogers (*Mycotaxon* **47**: 229, 1993; key 14 spp., Mexico), Whalley (*Mycologist* **10**: 149, 1996; SEM), Whalley *et al.* (*Kew Bull.* **54**: 715, 1999; SE Asia), Sánchez-Ballesteros *et al.* (*Mycol.* **92**: 964, 2000; phylogeny), Rogers *et al.* (*Sydowia* **54**: 84, 2002; Costa Rica), Triebel *et al.* (*Nova Hedwigia* **80**: 25, 2005; phylogeny).

Campanella Henn. (1895), Marasmiaceae. *c.* 40, widespread (esp. tropical). See Singer (*Nova Hedwigia* **26**: 847, 1976; key), Bougher (*Mycotaxon* **99**: 327, 2007; Australia).

Campanophyllum Cifuentes & R.H. Petersen (2003), Cyphellaceae. 1, Costa Rica. See Cifuentes *et al.* (*Mycol. Progr.* **2**: 287, 2003).

Campanularius Roussel (1806) = Panaeolus fide Kuyper (*in litt.*) This name has not yet been proposed for rejection against *Panaeolus*, nom. cons.

campanulate, bell-like in form (Fig. 23.25).

Campanulospora I.V. Issi, Radischcheva & Dolzhenko (1983), Microsporidia. 1.

Campanulospora Salazar-Yepes, Pardo-Card. & Buriticá (2007), Phragmidiaceae. Anamorph *Gerwasia*. 1, S. America. See Salazar Yepes, M.; Pardo Cardona, V.M.; Buriticá Céspedes, P. (*Caldasia* **29**: 105, 2007).

Campbellia Cooke & Massee (1890) [non *Campbellia* Wight 1849, *Orobanchaceae*] ≡ Rodwaya.

campestroid, agarics having a pileus with a diam. : stipe ratio of 1 or >1. See Freeman (*Mycotaxon* **8**: 1, 1979). Cf. placomycetoid.

Campoa Speg. (1921), Parmulariaceae. 2, S. America; Philippines.

Camposporidium Nawawi & Kuthub. (1988), anamorphic *Pezizomycotina*, Hso.≡ eP.19. 3 (aquatic), tropical. See Nawawi & Kuthubutheen (*Mycotaxon* **32**: 161, 1988), Castañeda-Ruíz & Guarro (*CJB* **76**: 1584, 1998).

Camposporium Harkn. (1884), anamorphic *Pezizomycotina*, Hso.≡ eP.10. 17, widespread. See Hughes (*Mycol. Pap.* **36**, 1951), Peek & Solheim (*Mycol.* **50**: 844, 1959), Watanabe (*TMSJ* **34**: 71, 1993; key 14 spp.), Hyde *et al.* (*S. Afr. J. Bot.* **64**: 151, 1998), Mercado Sierra *et al.* (*Mycotaxon* **67**: 417, 1998), Whitton *et al.* (*Fungal Diversity* **11**: 177, 2002; key).

Campsotrichum Ehrenb. (1819) = Myxotrichum fide Hughes (*CJB* **46**: 939, 1968).

Camptobasidium Marvanová & Suberkr. (1990), Microbotryomycetes. Anamorph *Crucella*. 1, USA. See Marvanová & Suberkropp (*Mycol.* **82**: 209, 1990).

Camptomeris Syd. (1927), anamorphic *Pezizomycotina*, Hsp.≡ eP.10. 9, widespread (tropical). See Hughes (*Mycol. Pap.* **49**, 1952), Bessey (*Mycol.* **45**: 364, 1953; key).

Camptomyces Thaxt. (1894), Laboulbeniaceae. 8, widespread. See Santamaría (*Fl. Mycol. Iberica* **5**, 2003; Iberian peninsula).

Camptosphaeria Fuckel (1870), Lasiosphaeriaceae. 4, widespread. See Krug & Jeng (*Sydowia* **29**: 71, 1977; key), Huhndorf *et al.* (*Mycol.* **96**: 368, 2004).

Camptosporium Link (1818) nom. dub., anamorphic *Pezizomycotina*. See Kirk (*in litt.*), Hughes (*CJB* **36**: 744, 1958; possible synonymy with *Menispora*).

Camptoum Link (1824) ≡ Arthrinium fide Hughes (*CJB* **36**: 727, 1958).

Campylacia A. Massal. ex Beltr. (1858) = Leptorhaphis fide Hawksworth *et al.* (*Dictionary of the Fungi* edn 8, 1995).

campylidium, helmet-shaped conidiomata occurring in various, mainly foliicolous, tropical lichenized genera (e.g. *Badimia*, *Loflammia*, *Sporopodium*); the name *Pyrenotrichum* (Ascomycetes, inc. sed.) has been applied to many of these conidiomata. See Sérusiaux (*Lichenologist* **18**: 1, 1986).

Campylobasidium Lagerh. ex F. Ludw. (1892) nom. rej. = Septobasidium fide Berndt (*in litt.*).

Campylocarpon Halleen, Schroers & Crous (2004), anamorphic *Nectriaceae*. 2 (on *Vitis* roots), S. Africa. See Halleen *et al.* (*Stud. Mycol.* **50**: 448, 2004).

Campylomyces Nakasone (2004), Gloeophyllaceae. 2, Australia; Morocco. See Nakasone (*Sydowia* **56**: 258, 2004).

Campylospora Ranzoni (1953), anamorphic *Pezizomycotina*, Hso.1bH.1. 3 (aquatic), widespread (north temperate). See Ranzoni (*Farlowia* **4**: 373, 1953).

Campylostylus Genev. (1873) nom. dub., ? Fungi.

Campylothecium Ces. (1846) = Cordyceps fide Tulasne & Tulasne (*Select. fung. carpol.* **3**: 18, 1865).

Campylothelium Müll. Arg. (1883), Trypetheliaceae (L). 6, widespread (tropical). See Tucker & Harris (*Bryologist* **83**: 1, 1980), Harris (*Acta Amazon.* Supl. **14**: 55, 1984; Brazil), Prado *et al.* (*MR* **110**: 511, 2006; phylogeny), Aptroot *et al.* (*Biblthca Lichenol.* **97**, 2008; Costa Rica).

Canadian tuckohoe, see stone-fungus.

canal, sometimes applied to the pore connecting the two cells of a polarilocular spore.

canaliculate, having longitudinal grooves (Fig. 20.16).

Canalisporium Nawawi & Kuthub. (1989), anamorphic *Pezizomycotina*, Hsp.#eP.1. 9, widespread. See Nawawi & Kuthubutheen (*Mycotaxon* **34**: 477, 1989), Goh *et al.* (*CJB* **76**: 142, 1998; revision), Goh & Hyde (*Mycol.* **92**: 589, 2000; Australia), Cai *et al.* (*Cryptog. Mycol.* **24**: 3, 2003; China), Ferrer & Shearer (*Mycotaxon* **93**: 179, 2005; Panamá).

Canariomyces Arx (1984), Microascaceae. 1, Canary Islands. See von Arx *et al.* (*Beih. Nova Hedwigia* **94**, 1988; posn).

Canavirgella W. Merr., N.G. Wenner & Dreisbach (1996), Rhytismataceae. 1, USA. See Merrill *et al.* (*CJB* **74**: 1476, 1996).

Cancellaria Brongn. (1825) = Roestelia fide Berndt (*in litt.*) Anamorph name for (I), aecial states of *Gymno-*

sporangium.

cancellate, reticulate; like a network, as the basidioma of *Clathrus*.

Cancellidium Tubaki (1975), anamorphic *Hypocreales*, Hso.#eP.1. 2 (aero-aquatic, conidia hollow), E. Asia. See Tubaki (*TMSJ* **16**: 357, 1975), Shaw (*Mycologist* **8**: 162, 1994; Australia), Yeung *et al.* (*Cryptog. Mycol.* **27**: 295, 2006; phylogeny).

Canceromyces Niessen [not traced] nom. dub., Fungi. Based on a mould from a cancer.

Candelabrella Rifai & R.C. Cooke (1966) = Arthrobotrys fide Cooke (*TBMS* **53**: 475, 1969), Schenck *et al.* (*CJB* **55**: 977, 1977), Rubner (*Stud. Mycol.* **39**, 1996).

Candelabrochaete Boidin (1970), Phanerochaetaceae. 11, widespread. See Hjortstam (*Mycotaxon* **56**: 451, 1995).

Candelabrum Beverw. (1951), anamorphic *Pezizomycotina*, Hso.1bH.1. 7 (aero-aquatic), widespread. See Beverwijk (*Antonie van Leeuwenhoek Ned. Tijdschr. Hyg.* **17**: 283, 1951), Voglmayr (*MR* **102**: 410, 1998; key).

Candelaria A. Massal. (1852), Candelariaceae (L). 11, widespread. See Hillmann (*Rabenh. Krypt.-Fl.* **9** 5.3: 19, 1936), Poelt (*Phyton* Horn **16**: 189, 1974), Jørgensen & Galloway (*Lichenologist* **24**: 407, 1992), Kärnefelt & Westberg (*Mycotaxon* **80**: 456, 2001; southern Africa), LaGreca & Lumbsch (*Biblthca Lichenol.* **78**: 211, 2001; phylogeny), Miądlikowska *et al.* (*Mycol.* **98**: 1088, 2006; phylogeny), Westberg *et al.* (*MR* **111**: 1277, 2007; phylogeny).

Candelariaceae Hakul. (1954), Candelariales (L). 5 gen. (+ 6 syn.), 66 spp.

Lit.: Jørgensen & Galloway (*Lichenologist* **24**: 407, 1992), Castello & Nimis (*Acta Bot. Fenn.* **150**: 5, 1994), LaGreca & Lumbsch (*Biblthca Lichenol.* **78**: 211, 2001), Miądlikowska *et al.* (*Mycol.* **98**: 1088, 2006; phylogeny), Hofstetter *et al.* (*Mol. Phylogen. Evol.* **44**: 412, 2007; phylogeny), Westberg *et al.* (*MR* **111**: 1277, 2007; phylogeny).

Candelariales Miądl., Lutzoni & Lumbsch (2007). Lecanoromycetes. 1 fam., 5 gen., 66 spp. Fam.:

Candelariaceae

For *Lit.* see under fam.

Candelariella Müll. Arg. (1894), Candelariaceae (L). *c.* 48, widespread. See Hakulinen (*Ann. bot. Soc. Zool.-Bot. Fenn. Vanamo* **27** no. 3, 1954; monogr.), Poelt & Reddi (*Ergebn. ForschUnternehmens Nepal Himalaya* **6**: 1, 1969), Laundon (*Lichenologist* **4**: 297, 1970; UK), Harris & Buck (*Mich. Bot.* **17**: 155, 1978; N. Am.), Gilbert *et al.* (*Lichenologist* **13**: 249, 1981; citrine spp.), Castello & Nimis (*Acta Bot. Fenn.* **150**: 5, 1994; Antarctic spp.), LaGreca & Lumbsch (*Biblthca Lichenol.* **78**: 211, 2001; phylogeny), Aragón & Martínez (*Lichenologist* **34**: 81, 2002; Europe), Miądlikowska *et al.* (*Mycol.* **98**: 1088, 2006; phylogeny), Hofstetter *et al.* (*Mol. Phylogen. Evol.* **44**: 412, 2007; phylogeny), Westberg (*Bryologist* **110**: 365, 2007; N America), Westberg (*Bryologist* **110**: 375, 2007; N America), Westberg (*Bryologist* **110**: 391, 2007; N America), Westberg *et al.* (*MR* **111**: 1277, 2007; phylogeny).

Candelariellomyces E.A. Thomas ex Cif. & Tomas. (1953) ≡ Candelariella.

Candelariellopsis Werner (1936) = Candelariella fide Hawksworth *et al.* (*Dictionary of the Fungi* edn 8, 1995).

Candelariopsis (Sambo) Szatala (1959) nom. inval. ? = Caloplaca fide Hawksworth *et al.* (*Dictionary of the Fungi* edn 8, 1995).

Candelina Poelt (1974), Candelariaceae (L). 3, widespread. See Poelt (*Phyton* Horn **16**: 194, 1974), LaGreca & Lumbsch (*Biblthca Lichenol.* **78**: 211, 2001; phylogeny, anatomy), Westberg *et al.* (*MR* **111**: 1277, 2007; phylogeny).

Candelospora Rea & Hawley (1912) = Cylindrocladium fide Boedijn (*in litt.*).

Candelosynnema K.D. Hyde & Seifert (1992), anamorphic *Pezizomycotina*, Hsy.≡ eH.10. 1, Australia. See Hyde & Seifert (*Aust. Syst. Bot.* **5**: 401, 1992).

candicidin, an antibiotic from the actinomycete *Streptomyces griseus*; antibacterial and antifungal (esp. against *Candida albicans*); Lechevalier *et al.* (*Mycol.* **45**: 155, 1953), Kligman & Lewis (*Proc. Soc. exper. Biol. Med.* **82**: 399, 1953).

Candida Berkhout (1923) nom. cons., anamorphic *Saccharomycetales*, Hso.0eH.?. 355, widespread. Pseudomycelium or mycelium present. *C. albicans* (candidiasis, q.v.) and other spp. are pathogenic for humans and animals; *C. utilis*, food yeast. See Shepherd *et al.* (*Ann. Rev. Microbiol.* **39**: 579, 1985; biology, genetics, pathogenicity), Srivastava *et al.* (*Microbial Ecology* **11**: 71, 1985; differentiation of biotypes), Belov & Kamenev (*Mikrobiologiya* **55**: 473, 1986; wall components and morphology), Lehmann *et al.* (*TBMS* **88**: 199, 1987; killer fungi characterize species and biotypes), Magee *et al.* (*J. Bact.* **169**: 1639, 1987; rDNA RFLP), Montrocher & Claisse (*Cellular and Molecular Biology* **33**: 313, 1987; spectrophotometric analysis), Mendling (*Vulvovaginal Candidosis. Theory and Practice*, 1988; review), Viljoen *et al.* (*J. gen. Microbiol.* **134**: 1893, 1988; long chain fatty acids), Weijman & Rodrigues de Miranda (*Antonie van Leeuwenhoek* **54**: 535, 1988; carbohydrate patterns), Weijman *et al.* (*Antonie van Leeuwenhoek* **54**: 545, 1988; redefinition of genus), Hendriks *et al.* (*Syst. Appl. Microbiol.* **12**: 223, 1989; nucleotide sequence of *C.albicans*), Kamiyama *et al.* (*Mycopathologia* **107**: 3, 1989; DNA homology between strains), Kamiyama *et al.* (*J. Med. Vet. Mycol.* **27**: 229, 1989; Adansonian taxonomy of spp.), Merson-Davies & Odds (*J. gen. Microbiol.* **135**: 3143, 1989; morphology index), Montrocher *et al.* (*Yeast* Special Issue **5**: S385, 1989; biochemical analysis), Su & Meyer (*Yeast* Special Issue **5**: S355, 1989; restriction endonuclease analysis of DNA), Boiron (*Bulletin de la Société Française de Mycologie Médicale* **19**: 13, 1990; electrophoretic karyotypes), Iwaguchi *et al.* (*J. gen. Microbiol.* **136**: 2433, 1990; karyotypes), Jensen *et al.* (*Mycoses* **33**: 519, 1990; crossed immunoelectrophoresis to differentiate *C. albicans*), Odds (*Bulletin de la Société Française de Mycologie Médicale* **19**: 5, 1990; molecular biology), Pope (*Phytopathology* **80**: 966, 1990; cellular fatty acids), Rustchenko-Bulgac *et al.* (*J. Bact.* **172**: 1276, 1990; genetic variation in *C. albicans*), Samaranayake & Yaacob (*Oral Candidosis*: 124, 1990; classification of oral candidosis), Scherer & Magee (*Microbiol. Rev.* **54**: 226, 1990; genetics of *C. albicans*), Barns *et al.* (*J. Bact.* **173**: 2250, 1991; evolutionary relationships among pathogenic spp.), Calderone & Braun (*Microbiol. Rev.* **55**: 1, 1991; adherence and receptor relationships in *C. albicans*), Cribb (*Qd Nat.* **31**: 21, 1991; history of delimitation), Hendriks *et al.* (*J. gen. Microbiol.* **137**: 1223, 1991; phylogeny of medical spp.), Lacher & Lehmann (*An-*

nals of Clinical and Laboratory Science **21**: 94, 1991; numerical taxonomy), Meyer *et al.* *in* Kurtzman & Fell (Eds) (*Yeasts, a taxonomic study* 4th edn: 454, 1998; review), Warren & Hazen (*Manual of Clinical Microbiology*: 1184, 1999; review), Hui *et al.* (*Diagn. Microbiol. Infect. Dis.* **38**: 95, 2000; PCR, SSCP analysis), Milde *et al.* (*Veter. Pathol.* **76**: 395, 2000; spp. from tortoises), Peltroche-Llacsahuanga *et al.* (*J. Clin. Microbiol.* **38**: 3696, 2000; FAME analysis), Calderone (*Candida and Candidiasis*, 2001; review), Land (*Trends in Microbiology* **9**: 201, 2001; recombination), Mahmoudabadi *et al.* (*Journal of Applied Microbiology* **93**: 894, 2002; lipids), Nosek *et al.* (*J. Clin. Microbiol.* **40**: 1283, 2002; mt DNA), Suzuki & Nakase (*J. gen. appl. Microbiol.* Tokyo **48**: 55, 2002; phylogeny, ubiquinones), Dodgson *et al.* (*J. Clin. Microbiol.* **41**: 5709, 2003; phylogeny of *C. glabrata*), Himmelreich *et al.* (*Appl. Environm. Microbiol.* **69**: 4566, 2003; NMR spectroscopy), Muñoz *et al.* (*Mycoses* **46**: 85, 2003; lectins), Sampaio *et al.* (*J. Clin. Microbiol.* **41**: 552, 2003; microsatellites), Sandt *et al.* (*J. Clin. Microbiol.* **41**: 954, 2003; FTIR analysis), Starmer *et al.* (*FEMS Yeast Res.* **3**: 441, 2003; yeasts from cacti), Bougnoux *et al.* (*Infect. Genet. Evol.* **4**: 243, 2004; genotyping), Diezmann *et al.* (*J. Clin. Microbiol.* **42**: 5624, 2004; phylogeny), Fundyga *et al.* (*Infect. Genet. Evol.* **4**: 37, 2004; recombination), Graf *et al.* (*Diagn. Microbiol. Infect. Dis.* **48**: 1491, 2004; diagnostics), Jones *et al.* (*Proc. natn Acad. Sci. U.S.A.* **101**: 7329, 2004; genome of *C. albicans*), Pujol *et al.* (*Eukaryotic Cell* **3**: 1015, 2004; genetics), Salomon *et al.* (*Applied Mycology and Biotechnology* **4**: [99], 2004; genomics of *C. albicans*), Dodgson *et al.* (*Fungal Genetics Biol.* **42**: 233, 2005; recombination), Edelmann *et al.* (*J. Clin. Microbiol.* **43**: 6164, 2005; human and animal strain comparison), Foulet *et al.* (*J. Clin. Microbiol.* **43**: 4574, 2005; microsatellites), Goldenberg *et al.* (*J. Clin. Microbiol.* **43**: 5912, 2005; HPLC), Lott *et al.* (*Fungal Genetics Biol.* **42**: 444, 2005; evolutionary origins), Prasad *et al.* (*Int. J. Syst. Evol. Microbiol.* **55**: 967, 2005; from oil sludge), Sampaio *et al.* (*J. Clin. Microbiol.* **43**: 3869, 2005; microsatellite multiplex PCR), Suh & Blackwell (*Mycol.* **97**: 167, 2005; fungus-beetle spp.), Suh *et al.* (*MR* **109**: 1045, 2005; insect associates), Tavanti *et al.* (*J. Clin. Microbiol.* **43**: 284, 2005; taxonomy), White *et al.* (*J. Clin. Microbiol.* **43**: 2181, 2005; detection of invasive strains), Lan & Xu (*Microbiology* Reading **152**: 1539, 2006; population analysis), Lasker *et al.* (*J. Clin. Microbiol.* **44**: 750, 2006; microsatellites), Marot-Leblond *et al.* (*J. Clin. Microbiol.* **44**: 138, 2006; monoclonal antibodies), Nantel (*Fungal Genetics Biol.* **43**: 311, 2006; review, whole-genome sequence), Page *et al.* (*J. Clin. Microbiol.* **44**: 3167, 2006; flow cytometry), Ramos *et al.* (*Antonie van Leeuwenhoek* **89**: 39, 2006; heteroduplex PCR), Romeo *et al.* (*J. Clin. Microbiol.* **44**: 2590, 2006; hyphal wall protein sequences), Suh *et al.* (*Mycol.* **98**: 1006, 2006; phylogeny).

candidiasis, a cosmop. disease of humans (including thrush, mouget, etc.) and animals caused by species of *Candida* (syn. *Monilia* auct.), esp. *C. albicans*; moniliasis; candidosis. Like many other fungal diseases, candidiasis has become more important with the rise in numbers of immunocompromised patients. See Winner & Hurley (*Candida albicans*, 1964; (Eds), *Symposium on Candida infections*, 1966), Odds (*Candida and candidosis*, edn. 2, 1988), Turnbay *et al.* (Eds) (*Candida and candidamycosis*, 1991). Candida Society Website: www.candida-society.org.

candidosis, see candidiasis.

candle-snuff fungus, stromata of *Xylaria hypoxylon*.

canescent, becoming hoary or grey.

caninoid venation, see veins.

canker, a plant disease in which there is sharply-limited necrosis of the cortical tissue, e.g. **apple canker** (*Nectria galligena*).

CANL, Lichen Herbarium, Canadian Museum of Nature (Ottawa, Canada); founded 1882; a government corporation.

Cannanorosporonites Ramanujam & Rao (1979), Fossil Fungi. 1 (Miocene), India.

Canningia J. Weiser, Wegensteiner & Z. Žižka (1995), Microsporidia. 1.

Cannonia Joanne E. Taylor & K.D. Hyde (1999), Xylariaceae. 1, Argentina; Australia. See Taylor & Hyde (*MR* **103**: 1398, 1999).

Canomaculina Elix & Hale (1987) = Parmotrema fide Elix (*Mycotaxon* **65**: 475, 1997), Blanco *et al.* (*Mycol.* **97**: 150, 2005), Blanco *et al.* (*Mol. Phylogen. Evol.* **39**: 52, 2006; phylogeny).

Canoparmelia Elix & Hale (1986), Parmeliaceae (L). *c.* 40, widespread. See Elix *et al.* (*Mycotaxon* **27**: 271, 1986), Adler (*Mycotaxon* **28**: 251, 1987; Argentina), Elix (*Flora of Australia* **55**: 21, 1994), Heiman & Elix (*Mycotaxon* **70**: 163, 1999; N. America), Blanco *et al.* (*Mycol.* **97**: 150, 2005; phylogeny), Blanco *et al.* (*Mol. Phylogen. Evol.* **39**: 52, 2006; phylogeny), Miądlikowska *et al.* (*Mycol.* **98**: 1088, 2006; phylogeny).

Canteria Karling (1971), ? Endochytriaceae. 1, British Isles. See Karling (*Arch. Mikrobiol.* **76**: 129, 1971).

Cantharellaceae J. Schröt. (1888), Cantharellales. 5 gen. (+ 10 syn.), 92 spp.

Lit.: Corner (*Beih. Sydowia* **1**: 266, 1957), Petersen & Mueller (*Boln Soc. argent. Bot.* **28**: 195, 1992), Danell (*Mycorrhiza* **5**: 89, 1994), Feibelman *et al.* (*MR* **101**: 1423, 1997), Redhead *et al.* (*Mycotaxon* **65**: 285, 1997), Watling (*Nature* Lond. **365**: 299, 1997), Li *et al.* (*Acta Sci. nat. Univ. Sunyats.* **38**: 29, 1999), Pine *et al.* (*Mycol.* **91**: 944, 1999), Dahlman *et al.* (*MR* **104**: 388, 2000), Dunham *et al.* (*MR* **107**: 1163, 2003), Dunham *et al.* (*Mol. Ecol.* **12**: 1607, 2003).

Cantharellales Gäum. (1926). Agaricomycetes. 7 fam., 38 gen., 544 spp. Basidioma either funnel-shaped or tubular or stalked and pileate, monomitic, the hymenophore smooth, wrinkled, or folded to form thick gill-like structures; spores smooth, hyaline, non-amyloid; terrestrial, humicolous. Fams:

(1) **Aphelariaceae**
(2) **Botryobasidiaceae**
(3) **Cantharellaceae**
(4) **Ceratobasidiaceae**
(5) **Clavulinaceae**
(6) **Hydnaceae**
(7) **Tulasnellaceae**

Lit.: Donk (1964: 247), Corner (*A monograph of the cantharelloid fungi* [*Ann. Bot. Mem.* **2**], 1966; *New Phytol.* **67**: 219, 1968; *Nova Hedw.* **27**: 325, 1976), Bigelow (*Mycol.* **70**: 707, 1978; New England spp.), Moncalvo *et al.* (*Mycol.* **98**: 937, 2006), Hibbett *et al.* (*MR* **111**: 509, 2007), Weiss *et al.* (*Frontiers in Basidiomycote Mycology*: 7, 2004), Binder *et* al. (*Systematics and Biodiversity* **3**: 113, 2005), and under fams.

Cantharellopsis Kuyper (1986), Tricholomataceae. 1, Europe. See Kuyper (*La Famiglia delle Tricholomataceae* Atti del Convegno Internazionale del 10-15 Settembre 1984, Borgo Val di Taro, Italy: 99, 1986), Redhead *et al.* (*Mycotaxon* **83**: 19, 2002; phylogeny).

Cantharellula Singer (1936), Tricholomataceae. 2, widespread (temperate). See Singer (*Revue Mycol. Paris* **1**: 281, 1936), Redhead *et al.* (*Mycotaxon* **83**: 19, 2002; phylogeny).

Cantharellus Adans. ex Fr. (1821), Cantharellaceae. 65, widespread. *C. cibarius*, the edible chanterelle. See Smith & Morse (*Mycol.* **39**: 497, 1947), Corner (*Beih. Sydowia* **1**, 1957), Heinemann (*Fl. Icon. Champ. Congo* **8**: 154, 1959; keys 17 spp. Congo), Eyssartier & Buyck (*Aust. Syst. Bot.* **14**: 587, 2001; Australia), Redhead *et al.* (*Taxon* **51**: 559, 2002; nomenclature), Gulden (*Sopp og Nyttevekster* **1**: 10, 2005; key nordic spp.).

Cantharocybe H.E. Bigelow & A.H. Sm. (1973), Pleurotaceae. 1, N. America. See Bigelow & Smith (*Mycol.* **65**: 486, 1973).

Cantharomyces Thaxt. (1890), Laboulbeniaceae. 27, widespread. See Weir & Hammond (*Biodiv. Cons.* **6**: 701, 1997; ecology, biodiversity), Rossi & Santamaria (*Mycol.* **92**: 786, 2000; on *Staphylinidae*), Santamaría (*Fl. Mycol. Iberica* **5**, 2003; Iberian peninsula), Shen *et al.* (*Mycosystema* **23**: 303, 2004; China).

Cantharosphaeria Thaxt. (1920) = Eriosphaeria fide Müller & von Arx (*Beitr. Kryptfl. Schweiz* **11** no. 2, 1962).

cap, see pileus.

capillaceous, see capilliform.

Capillaria Pers. (1822), anamorphic *Pezizomycotina*, Hso.?.?. 4, widespread (temperate).

Capillaria Velen. (1947) [non *Capillaria* Roussel 1806, *Algae*] = Lycoperdon Pers. fide Stalpers (*in litt.*).

Capillataspora K.D. Hyde (1989), Dothideomycetes. 1, Brunei. See Hyde (*CJB* **67**: 2522, 1989), Kohlmeyer & Volkmann-Kohlmeyer (*Bot. Mar.* **34**: 1, 1991).

capilliconidium, a secondary conidium produced on a long capillary tube in *Entomophthorales*.

capilliform, hair-like; thread-like; capillaceous.

Capillipes R. Sant. (1956), Helotiaceae. 1, Lapland. See Santesson (*Friesia* **5**: 390, 1956).

Capillistichus Santam. (2004), Laboulbeniaceae. 1, Spain. See Santamaría (*Mycol.* **96**: 763, 2004).

capillitium (of *Mycetozoa* and gasteroid *Agaricomycotina*), a mass of sterile, thread-like elements, tubes or fibres among the spores.

capitate, having a well-formed head (Figs 29.9, 37.18).

capitate-fastigiate (of macrolichens), having a thallus cortex of erect, parallel hyphae terminated by swollen and pigmented apical cells.

capitellum, a little head.

Capitoclavaria Lloyd (1922) ? = Clavaria fide Stalpers (*in litt.*).

Capitorostrum Bat. (1957), anamorphic *Pezizomycotina*, Cpd.0eH.15. 2, Australia; Papua New Guinea. See Hyde & Philemon (*Mycotaxon* **42**: 95, 1991).

Capitotricha (Raitv.) Baral (1985), Hyaloscyphaceae. 3, widespread. See Raitviir (*Eesti NSV Tead. Akad. Toim.* Biol. seer **36**: 313, 1987), Leenurm *et al.* (*Sydowia* **52**: 30, 2000; ultrastr.), Suková (*Czech Mycol.* **57**: 139, 2005; Czech Republic).

Capitularia Flörke (1807) = Cladonia fide Hawksworth *et al.* (*Dictionary of the Fungi* edn 8, 1995).

Capitularia Rabenh. (1851) = Uromyces fide Dietel (*Nat. Pflanzenfam.* **6**, 1928).

capitulum, a stalked globose apical apothecium, as in the *Caliciales*. Cf. mazaedium.

Capnia Vent. (1799) = Dermatocarpon Eschw. fide Hawksworth *et al.* (*Dictionary of the Fungi* edn 8, 1995).

Capniomyces S.W. Peterson & Lichtw. (1983), Legeriomycetaceae. 2 (in *Plectoptera*), Spain; USA. See Lichtwardt (*The Trichomycetes. Fungal associates of arthropods*, 1986), White (*MR* **110**: 1011, 2006; phylogeny), Valle (*Mycol.* **99**: 442, 2007; Spain).

Capnites Theiss. (1916) [non *Capnites* (DC.) Dumort. 1827, *Papaveraceae*] = Phaeosaccardinula fide Hawksworth *et al.* (*Dictionary of the Fungi* edn 8, 1995).

Capnobatista Cif. & F.B. Leal ex Bat. & Cif. (1963) = Trichomerium fide Hawksworth *et al.* (*Dictionary of the Fungi* edn 8, 1995).

Capnobotryella Sugiy. (1987), anamorphic *Capnodiales*, Hso.1eP.1/12. 2, Japan. See Titze & de Hoog (*Antonie van Leeuwenhoek* **58**: 265, 1990; *C. renispora* on roof tile), Sterflinger *et al.* (*Stud. Mycol.* **43**: 5, 1999; ecology), Hambleton *et al.* (*Mycol.* **95**: 959, 2003; morphology), Tsuneda *et al.* (*Rep. Tottori Mycol. Inst.* **41**: 1, 2003; conidiogenesis, phylogeny), Crous *et al.* (*Stud. Mycol.* **58**: 1, 2007), Sert *et al.* (*MR* **111**: 1235, 2007; Turkey).

Capnobotrys S. Hughes (1970), anamorphic *Metacapnodium*, Hso.≡ eP.10. 1, Austria. See Hughes (*N.Z. Jl Bot.* **8**: 205, 1970), Reynolds (*CJB* **76**: 2125, 1998; phylogeny), Tsuneda *et al.* (*Rep. Tottori Mycol. Inst.* **41**: 1, 2003; phylogeny).

Capnocheirides J.L. Crane & S. Hughes (1982), anamorphic *Capnodiales*, Hso.0-≡ eP.38. 1, Europe. See Crane & Hughes (*Mycol.* **74**: 752, 1982).

Capnociferria Bat. (1963) = Antennulariella fide Hughes (*Mycol.* **68**: 693, 1976).

Capnocrinum Bat. & Cif. (1963) = Antennulariella fide Hughes (*Mycol.* **68**: 693, 1976).

Capnocybe S. Hughes (1966), anamorphic *Metacapnodiaceae*, Hso.≡ eP.10. 3, New Zealand; USA. See Hughes (*N.Z. Jl Bot.* **4**: 335, 1966).

Capnodaria (Sacc.) Theiss. & Syd. (1918), Capnodiaceae. 1, Europe.

Capnodendron S. Hughes (1976), anamorphic *Antennulariella*, Hso.≡ eP.3. 3, widespread. See Hughes (*Mycol.* **68**: 750, 1976), Hughes (*CJB* **78**: 1215, 2000), Hughes (*N.Z. Jl Bot.* **41**: 139, 2003).

Capnodenia (Sacc.) Theiss. & Syd. (1917) = Capnodium fide von Arx & Müller (*Stud. Mycol.* **9**, 1975).

Capnodiaceae Höhn. ex Theiss. (1916), Capnodiales. 26 gen. (+ 28 syn.), 117 spp.

Lit.: Batista & Ciferri (*Saccardoa* **2**, 1963), Reynolds (*Taxon* **20**: 759, 1971; hyphal morph.), Reynolds (*Nova Hedwigia* **26**: 179, 1975; growth forms), Hughes (*Mycol.* **68**: 693, 1976; gen. names, anamorphs), Reynolds (*Mycotaxon* **8**: 417, 1979; stalked taxa), Rodríguez Hernández (*Revta Jardín bot. Nac. Univ. Habana* **6**: 53, 1985), Parbery & Brown (*Microbiology of the Phyllosphere*: 101, 1986), Reynolds (*Mycotaxon* **27**: 377, 1986; cladistics), Reynolds *in* Sugiyama (Ed.) (*Pleomorphic Fungi: The Diversity and its Taxonomic Implications*: 157, 1987), Mibey

(*Soft Scale Insects. Their Biology, Natural Enemies and Control World Crop Pests* 7A: 275, 1997), Inácio & Dianese (*MR* **102**: 695, 1998), Reynolds (*CJB* **76**: 2125, 1998; phylogeny), Olejnik & Ingrouille (*MR* **103**: 333, 1999), Reynolds (*CJB* **76**: 2125, 1998), Sterflinger *et al.* (*Stud. Mycol.* **43**: 5, 1999; phylogeny), Reynolds (*Mycopathologia* **148**: 141, 1999), Lindemuth *et al.* (*MR* **105**: 1176, 2001), Lumbsch & Lindemuth (*MR* **105**: 901, 2001), Schoch *et al.* (*Mycol.* **98**: 1041, 2006; phylogeny).

Capnodiales Woron. (1925). Dothideomycetidae. 9 (?+1) fam., 198 gen., 7244 spp. Mycelium superficial, often well-developed, dark, very varied in form, composed of sometimes irregular ± cylindrical or torulose hyphae, sometimes with erect branches, sometimes with mucous coating. Ascomata small, globose or vertically elongated, thin-walled, sometimes covered in a mucous layer, sometimes setose or with hyphal appendages, opening either with a clearly-defined ostiole or a poorly-defined lysigenous pore. Interascal tissue absent or composed of inconspicuous periphysoids. Asci small, ovoid or saccate, fissitunicate, not blueing in iodine or rarely with a J+ exterior layer. Ascospores hyaline to brown, septate, sometimes muriform, rarely ornamented, sheath lacking. Anamorphs very varied. Plant parasites (including many important pathogen genera), saprobes or sooty moulds. Fams.:
(1) **Antennulariellaceae**
(2) **Capnodiaceae**
(3) **Davidiellaceae**
(4) **Euantennariaceae**
(5) **Metacapnodiaceae**
(6) **Mycosphaerellaceae**
(7) **Piedraiaceae**
(8) **Schizothyriaceae**
(9) **Teratosphaeriaceae**

The *Asterinaceae* may also belong here, but molecular data are inadequate.

Lit.: Crous *et al.* (*Stud. Mycol* **58**: 1, 2007; phylogeny), Reynolds (*CJB* **76**: 2125, 1998; phylogeny), Schoch *et al* (*Mycol* **98**: 1041, 2006; phylogeny), Sterflinger *et al.* (*Stud. Mycol.* **43**: 5, 1999; phylogeny).

Capnodiastrum Speg. (1886), anamorphic *Rhytidenglerula*, Cpd.1eP.1. 5, widespread (esp. S. America). See Hughes *in* Sugiyama (Ed.) (*Pleomorphic Fungi: The Diversity and its Taxonomic Implications*: 103, 1987).

Capnodiella (Sacc.) Sacc. (1905) ≡ Sorica.

Capnodina (Sacc.) Sacc. (1926) = Antennulariella fide Hawksworth *et al.* (*Dictionary of the Fungi* edn 8, 1995).

Capnodinula Bat. & Cif. (1963), Capnodiales. 1, Australia. See Batista & Ciferri (*Saccardoa* **2**: 81, 1963).

Capnodinula Speg. (1918) = Wentiomyces fide von Arx & Müller (*Stud. Mycol.* **9**, 1975).

Capnodiopsis Henn. (1902) = Molleriella fide von Arx (*Persoonia* **2**: 421, 1963).

Capnodium Mont. (1848), Capnodiaceae. Anamorphs *Fumagospora*, *Phaeoxyphiella*, *Polychaetella*, *Scolecoxyphium*. c. 15, widespread. See Reynolds (*Bull. Torrey bot. Club* **97**: 253, 1970), Reynolds (*Mycotaxon* **34**: 197, 1989; California), Reynolds (*CJB* **76**: 2125, 1998; phylogeny), Sterflinger *et al.* (*Stud. Mycol.* **43**: 5, 1999; phylogeny), Lumbsch & Lindemuth (*MR* **105**: 901, 2001; phylogeny), Schoch *et al.* (*Mycol.* **98**: 1041, 2006; phylogeny).

Capnofrasera S. Hughes (2003), anamorphic *Antennulariellaceae*. 1, New Zealand. See Hughes (*N.Z. Jl Bot.* **41**: 139, 2003).

Capnogoniella Bat. & Cif. (1963) nom. conf., anamorphic *Pezizomycotina*. See Sutton (*Mycol. Pap.* **141**, 1977).

Capnogonium Bat. & Peres (1961) = Brooksia fide Deighton & Pirozynski (*Mycol. Pap.* **105**, 1966).

Capnokyma S. Hughes (1975), anamorphic *Euantennariaceae*, Hso.≡ eP.1. 2, New Zealand; Venezuela. See Hughes (*N.Z. Jl Bot.* **13**: 638, 1975), Hughes (*Mycol.* **93**: 603, 2001).

Capnophaeum Speg. (1918), ? Capnodiaceae. 2, Asia.

Capnophialophora S. Hughes (1966), anamorphic *Metacapnodium*, Hso.0eH.15. 3, New Zealand. See Hughes (*N.Z. Jl Bot.* **4**: 352, 1966).

Capnosporium S. Hughes (1976), anamorphic *Metacapnodium*, Hso.1-≡ eP.28. 1, New Zealand. See Hughes (*Mycol.* **68**: 752, 1976).

Capnostysanus Speg. (1918) = Stysanus fide Clements & Shear (*Gen. Fung.*, 1931).

Cappellettia Tomas. & Cif. (1952) = Gyalidea fide Hawksworth *et al.* (*Dictionary of the Fungi* edn 8, 1995).

Caprettia Bat. & H. Maia (1965), Monoblastiaceae (L). 6 (foliicolous), pantropical. See Batista & Maia (*Atas Inst. Micol. Univ. Pernambuco* **2**: 377, 1965), Lücking *et al.* (*Lichenologist* **30**: 121, 1998), Sérusiaux & Lücking (*Biblthca Lichenol.* **86**: 161, 2003; revision).

Capricola Velen. (1947), ? Helotiales. 1, Europe.

Capronia Sacc. (1883), Herpotrichiellaceae. Anamorphs *Exophiala*, *Rhinocladiella*, *Cladophialophora*, *Phialophora*. 27, Europe; N. America. See Müller *et al.* (*TBMS* **88**: 63, 1987; key), Barr (*Mycotaxon* **41**: 419, 1991; key N. Am. spp.), Untereiner *et al.* (*MR* **99**: 897, 1995; molec. taxonomy), Au *et al.* (*Mycol.* **91**: 326, 1999; ultrastr.), Haase *et al.* (*Stud. Mycol.* **43**: 80, 1999; phylogeny), Untereiner & Malloch (*Mycol.* **91**: 417, 1999; biochemistry), Untereiner & Naveau (*Mycol.* **91**: 67, 1999; phylogeny), Untereiner (*Stud. Mycol.* **45**: 141, 2000), Lindemuth *et al.* (*MR* **105**: 1176, 2001; phylogeny), Hoog *et al.* (*J. Clin. Microbiol.* **41**: 4767, 2003), Geiser *et al.* (*Mycol.* **98**: 1053, 2006; phylogeny), Sterflinger (*The Yeast Handbook* **[1]**: 501, 2006; ecology).

Caproniella Berl. (1896), Pezizomycotina. 2, Europe. See Holm (*Taxon* **24**: 475, 1975).

Caproniella Berl. (1899) = Capronia fide Müller *et al.* (*TBMS* **88**: 63, 1987).

Caproventuria U. Braun (1998) = Venturia Sacc. fide Braun (*Monogr. Cercosporella, Ramularia Allied Genera (Phytopath. Hyphom.)* **2**: 493, 1998), Braun *et al.* (*Mycol. Progr.* **2**: 3, 2003; phylogeny), Beck *et al.* (*Mycol. Progr.* **4**: 111, 2005; phylogeny), Crous *et al.* (*Stud. Mycol.* **58**: 185, 2007; phylogeny).

Capsicumyces Gamundí, Aramb. & Giaiotti (1979), anamorphic *Pezizomycotina*, Hso.0eH.11. 1, Argentina. See Gamundí *et al.* (*Darwiniana* **22**: 190, 1979).

capsidiol, a phytoalexin (q.v.) from spur pepper (*Capsicum frutescens*).

Capsulasclerotes Malan (1959), Fossil Fungi. 1 (Permian), former Czechoslovakia.

capsule, a hyaline gelatinous sheath surrounding the cell of certain yeasts and bacteria.

Capsulospora K.D. Hyde (1996), Xylariales. 8 (saprobic on palms), S.E. Asia. See Kang *et al.* (*Fungal Di-

versity **1**: 147, 1998; DNA), Kang *et al.* (*Mycoscience* **40**: 151, 1999), Kang *et al.* (*Mycotaxon* **81**: 321, 2002; phylogeny).

Capsulotheca Kamyschko (1960), ? Trichocomaceae. 1, former USSR. See Benny & Kimbrough (*Mycotaxon* **12**, 1980).

Carassea S. Stenroos (2002), Cladoniaceae (L). 1, Brazil. See Stenroos *et al.* (*Mycol. Progr.* **1**: 277, 2002), Zhou *et al.* (*J. Hattori bot. Lab.* **100**: 871, 2006; phylogeny).

Carbacanthographis Staiger & Kalb (2002), Graphidaceae (L). 11, widespread. See Staiger (*Biblthca Lichenol.* **85**: 98, 2002), Nakanishi *et al.* (*Bull. natn. Sci. Mus.* Tokyo, B **29**: 83, 2003; Japan), Archer (*Biblthca Lichenol.* **94**, 2006; Australia), Archer (*Systematics & Biodiversity* **5**: 9, 2007; Solomon Is).

Carbomyces Gilkey (1954), Carbomycetaceae. 3, USA. See Trappe (*TBMS* **57**: 85, 1971), Zak & Whitford (*Mycol.* **78**: 840, 1986; ecology), Trappe & Weber (*Harvard Pap. Bot.* **6**: 209, 2001), Læssøe & Hansen (*MR* **111**: 1075, 2007; phylogeny).

Carbomycetaceae Trappe (1971), Pezizales. 1 gen., 3 spp.

Lit.: Trappe (*Mycotaxon* **9**: 297, 1979), Zak & Whitford (*Mycol.* **78**: 840, 1986), Trappe & Weber (*Harvard Pap. Bot.* **6**: 209, 2001), Læssøe & Hansen (*MR* **111**: 1075, 2007; phylogeny).

carbonaceous, dark-coloured and readily broken; charcoal-like or cinder-like.

Carbonea (Hertel) Hertel (1983), Lecanoraceae (±L). 20 (mainly on lichens), widespread. See Knoph (*Mycotaxon* **72**: 97, 1999), Grube *et al.* (*MR* **108**: 506, 2004; phylogeny).

carbonicolous, living on burnt ground; pyrophilous (q.v.).

Carbosphaerella I. Schmidt (1969), Halosphaeriaceae. 2 (marine), widespread. See Johnson *et al.* (*Bot. Mar.* **27**: 557, 1984), Sundari *et al.* (*Bot. Mar.* **39**: 327, 1996).

Carcinomyces Oberw. & Bandoni (1982), Carcinomycetaceae. 2, north temperate. See Oberwinkler & Bandoni (*Nordic Jl Bot.* **2**: 507, 1982).

Carcinomycetaceae Oberw. & Bandoni (1982), Tremellales. 3 gen. (+ 2 syn.), 25 spp.

Lit.: Hauerslev (*Friesia* **9**: 43, 1969), Gottschalk & Blanz (*Z. Mykol.* **51**: 205, 1985), Ginns (*Mycol.* **78**: 619, 1986), Oberwinkler (*Stud. Mycol.* **30**: 61, 1987), Kotiranta & Larsson (*Windahlia* **18**: 1, 1988), Rath (*Atti Soc. ital. Sci. nat. Mus. Civico Storia nat. Milano* **132**: 13, 1991).

Carestiella Bres. (1897), Stictidaceae (±L). 1, Europe. See Sherwood (*Mycotaxon* **5**: 1, 1977), Wedin *et al.* (*Lichenologist* **37**: 67, 2005; phylogeny), Wedin *et al.* (*MR* **110**: 773, 2006; nests within paraphyletic *Stictis*).

Caribaeomyces Cif. (1962), Microthyriaceae. 1, Dominican Republic. See Ciferri (*Atti Ist. bot. Univ. Lab. crittog. Pavia* sér. 5 **19**: 98, 1962).

carinate, keeled; boat-like.

Carinispora K.D. Hyde (1992), Phaeosphaeriaceae. 1 (on *Nypa*), Brunei. See Hyde (*J. Linn. Soc.* Bot. **110**: 95, 1992).

cariose, decayed.

carioso-cancellate, becoming latticed by decay.

Caripia Kuntze (1898), Marasmiaceae. 1, America (tropical). See Corner (*Ann. Bot. Mem.* [A monograph of Clavaria and allied genera] **1**, 1950), Singer (*Agaric. mod. Tax.* edn 2: 792, 1962).

Carlia Rabenh. (1857) nom. dub., Fungi. See Wakefield (*TBMS* **23**: 215, 1939).

Carlosia G. Arnaud (1954) = Isthmospora fide Kendrick & Carmichae *in* Ainsworth *et al.* (Eds) (*The Fungi* **4A**: 390, 1973).

Carlosia Samp. (1923) = Thelomma fide Hawksworth *et al.* (*Dictionary of the Fungi* edn 8, 1995), Paz-Bermúdez *et al.* (*Taxon* **51**: 771, 2002; typification).

Carmichaelia N.D. Sharma (1980), anamorphic *Pezizomycotina*, Hso.0eP.10. 1, India. See Sharma (*J. Indian bot. Soc.* **59**: 278, 1980), Hambleton *et al.* (*Stud. Mycol.* **53**: 29, 2005).

carminophilic (of basidia), becoming densely granular (= siderophilous (or carminophilous) granulation) after treatment with aceto-carmine stain.

Carneopezizella Svrček (1987), Helotiaceae. 1, former Czechoslovakia. See Svrček (*Česká Mykol.* **41**: 88, 1987).

Carnia Bat. (1960), Pezizomycotina. 1, Brazil. Affinities unknown: the original link suggested with *Pezizaceae* is unlikely.

carnose (carnous), fleshy.

Carnostroma Lloyd (1919) = Xylaria Hill ex Schrank fide von Arx & Müller (*Beitr. Kryptfl. Schweiz* **11** no. 1, 1954), Læssøe (*SA* **13**: 43, 1994; ? synonym of *Xylaria*).

Carnoya Dewèvre (1893) = Mortierella fide Hesseltine (*Mycol.* **47**: 344, 1955).

carnulose, somewhat fleshy.

Caromyxa Mont. (1856) nom. inval. = Mutinus fide Stalpers (*in litt.*).

carotene, a mixture of pigments, chiefly the carotenoid β-carotene, found in various fungi, e.g. *Phycomyces blakesleeanus* (Lilly *et al.*, *Bull. W. Va. agric. Exp. Stn* **441T**, 1960), *Choanephora cucurbitarum* (Chu & Lilly, *Mycol.* **52**: 80, 1961); **carotenoids**, a large group of related polyene compounds, mostly with C_{40}, yellow, red or more rarely colourless. Many have been given trivial names, e.g. torularhodin, neurosporene. See Hesseltine (*Tech. Bull. USDA* **1245**, 1961; *Mucorales*), Arpin (*Bull. mens. Soc. linn. Lyon* **38**, 1968; *BSMF* **84**: 427, 1969; discomycetes), Shibata *et al.* (*List of fungal products*, 1964; 24 refs.), Valadon (*TBMS* **67**: 1, 1976; taxonomic value of carotenoids). See Metabolic products.

Carothecis Clem. (1931) ? = Cephalotheca fide Hawksworth *et al.* (*Dictionary of the Fungi* edn 8, 1995).

Carouxella Manier, Rioux & Whisler (1965), Harpellaceae. 2 (in *Diptera*), France; Argentina. See Lichtwardt (*The Trichomycetes. Fungal associates of arthropods*, 1986), Lichtwardt *et al.* (*Mycol.* **91**: 1060, 1999; Argentina).

Carpenteles Langeron (1922) = Eupenicillium fide Stolk & Scott (*Persoonia* **4**: 391, 1967), Pitt *et al.* (*Integration of Modern Taxonomic Methods for Penicillium and Aspergillus Classification*: 9, 2000).

Carpenterella Tehon & H.A. Harris (1941), Synchytriaceae. 2, USA; India.

Carpobolus P. Micheli ex Paulet (1808) [non *Carpobolus* Schwein. 1822, *Hepaticae*] ≡ Sphaerobolus.

carpogenous, living on fruit.

carpogonium (generally of algae, sometimes of fungi, e.g. *Erysiphaceae*), the female sex organ.

carpoid, see *Biatoropsis*.

Carpoligna F.A. Fernández & Huhndorf (1999), Sordariales. Anamorph *Pleurothecium*. 1, widespread. See Fernández *et al.* (*Mycol.* **91**: 251, 1999), Réblová & Winka (*Mycol.* **93**: 478, 2001).

Carpolithes Brongn. (1822), Fossil Fungi, anamorphic *Fungi*, Hso.?.?. 1.

Carpomycetes, fungi having sporocarps; esp. ascomycetes and basidiomycetes.

carpophore (1) stalk of the sporocarp; (2) sometimes (esp. in France) = basidioma.

carpophoroid, a sterile carpophore-like body, in agarics, of unknown function (Singer, 1962: 22).

Carpophoromyces Thaxt. (1931), Laboulbeniaceae. 1, Sri Lanka. See Nannfeldt (*Svensk bot. Tidskr.* **43**: 468, 1949).

Carpozyma L. Engel (1872) = Hanseniaspora fide Lodder (*Yeasts, a taxonomic study* 2nd edn, 1970).

carrier, an organism harbouring a parasite without itself showing disease (Anon., *TBMS* **33**: 154, 1950).

Carrionia Bric.-Irag. (1938) = Rhinocladiella Nannf. fide Schol-Schwarz (*Antonie van Leeuwenhoek* **34**: 119, 1968).

Carrismyces R.F. Castañ & Heredia (2000), anamorphic *Pezizomycotina*. 1, Mexico. See Castañeda Ruiz & Heredia (*Mycotaxon* **76**: 125, 2000).

cartilaginous, firm and tough but readily bent.

cartilaginous layer, sometimes applied to the sterome in *Cladonia* and the chondroid axis (q.v.) in *Usnea*.

Cartilosoma Kotl. & Pouzar (1958) = Antrodia fide Pegler (*The polypores [Bull. BMS Suppl.]*, 1973), Donk (*Verh. K. ned. Akad. Wet.* tweede sect. **62**: 1, 1974) Accepted by.

caryallagic (of reproduction), having nuclear change; **acaryallagic**, not having nuclear change, as in clone development (Link, *Bot. Gaz.* **88**: 1, 1929).

caryo-, see karyo.

Caryospora De Not. (1855), Zopfiaceae. 6, widespread. See Barr (*Mycotaxon* **9**: 17, 1979), Hawksworth (*TBMS* **79**: 69, 1982), Hyde (*TMSJ* **30**: 333, 1989), Abdel-Wahab & Jones (*Mycoscience* **41**: 379, 2000).

Caryosporella Kohlm. (1985), ? Melanommataceae. 1 (on *Rhizophora*), widespread (pantropical). See Kohlmeyer (*Proc. Indian Acad. Sci.* Pl. Sci. **94**: 355, 1985), Hyde (*TMSJ* **30**: 333, 1989).

Casaresia Gonz. Frag. (1920), anamorphic *Mollisia*, Hso.0bP.1. 1 (aquatic), widespread (north temperate). See Webster *et al.* (*Nova Hedwigia* **57**: 483, 1993; teleomorph connection).

Cashiella Petr. (1951), Dermateaceae. 3, N. America; New Zealand. See Petrak (*Sydowia* **5**: 371, 1951), Gadgil & Dick (*N.Z. Jl For. Sci.* **29**: 440, 1999; New Zealand).

cassideous, helmet-shaped.

Castagnella G. Arnaud (1914) = Rhagadostoma fide Müller & von Arx *in* Ainsworth *et al.* (Eds) (*The Fungi* **4A**: 87, 1973).

Castanedaea W.A. Baker & Partridge (2001), anamorphic *Pezizomycotina*. 1, Cuba. See Partridge *et al.* (*Mycotaxon* **78**: 176, 2001).

Castanedomyces Cano, L.B. Pitarch & Guarro (2002), Onygenaceae. 1, Australia. See Cano *et al.* (*Stud. Mycol.* **47**: 167, 2002).

Castanoporus Ryvarden (1991), Meruliaceae. 1, Japan. See Ryvarden (*Syn. Fung.* **5**: 121, 1991).

Castellania C.W. Dodge (1935) = Candida fide Diddens & Lodder (*Die anaskosporogenen Hefen* **2**, 1942).

Castoreum Cooke & Massee (1887), Mesophelliaceae. 3, Australia. See Cunningham (*Proc. Linn. Soc. N. S. W.* **57**: 313, 1932), Beaton & Weste (*TBMS* **82**: 665, 1984), Trappe & Bougher (*Australasian Mycologist* **21**: 9, 2002; key).

Catabotrydaceae Petr. ex M.E. Barr (1990), Boliniales. 1 gen. (+ 1 syn.), 1 spp.

Lit.: Barr (*Mycotaxon* **39**: 43, 1990), Hyde & Cannon (*Mycol. Pap.* **175**, 1999), Hyde & Cannon (*Mycol. Pap.* **175**: 114 pp., 1999), Rossman *et al.* (*Stud. Mycol.* **42**: 248 pp., 1999), Huhndorf *et al.* (*Mycol.* **96**: 368, 2004).

Catabotrys Theiss. & Syd. (1915), Catabotrydaceae. 1 (on *Palmae*), widespread (tropical). See Seaver & Waterston (*Mycol.* **38**: 180, 1946), Hyde & Cannon (*Mycol. Pap.* **175**, 1999), Huhndorf *et al.* (*Mycol.* **96**: 368, 2004; phylogeny).

Catacauma Theiss. & Syd. (1914) = Phyllachora Nitschke ex Fuckel (1870) fide Petrak (*Annls mycol.* **22**: 1, 1924), Cannon (*Mycol. Pap.* **165**, 1991).

Catacaumella Theiss. & Syd. (1915) = Vestergrenia Rehm fide von Arx & Müller (*Beitr. Kryptfl. Schweiz* **11** no. 1, 1954).

Catachyon (Ehrenb. ex Fr.) Fr. (1832) = Podaxis fide Kuyper (*in litt.*).

catahymenium, see hymenium.

cataphyses, pseudoparaphyses (Groenhart, *Persoonia* **4**: 11, 1965); see hamathecium.

Catapyrenium Flot. (1850), Verrucariaceae (L). 60, widespread (esp. temperate). See Breuss (*Stapfia* **23**: 1, 1990; key 27 spp. Europe), Breuss (*Pl. Syst. Evol.* **185**: 17, 1993; key 13 spp. S. Am.), Breuss (*Nova Hedwigia* **58**: 229, 1994; key 12 N. Afr. spp.), Breuss (*Annln naturh. Mus. Wien* Ser. B, Bot. Zool. **98**: 35, 1996), Breuss (*Annln naturh. Mus. Wien* Ser. B, Bot. Zool. **100**: 671, 1998), Breuss (*Linzer biol. Beitr.* **32**: 1053, 2000; Mexico), Aragón (*An. Jard. bot. Madr.* **60**: 216, 2002; Spain), Heiðmarsson (*MR* **107**: 459, 2003; phylogeny), Gueidan *et al.* (*MR* **111**: 1145, 2007; phylogeny).

Cataractispora K.D. Hyde, S.W. Wong & E.B.G. Jones (1999), Annulatascaceae. 5, S.E. Asia. See Hyde *et al.* (*MR* **103**: 1019, 1999), Ho *et al.* (*Mycol.* **96**: 411, 2004; Hong Kong).

Catarraphia A. Massal. (1860), Arthoniales (L). 1, Malesian-Pacific. See Egea & Torrente (*Cryptog. Bryol.-Lichénol.* **14**: 329, 1993), Grube (*Bryologist* **101**: 377, 1998; phylogeny).

Catarrhospora Brusse (1994), Porpidiaceae (L). 2, S. Africa. See Brusse (*Mycotaxon* **52**: 501, 1994).

cata-species, see *Pucciniales*.

Catastoma Morgan (1892) = Disciseda fide Stalpers (*in litt.*).

catathecium, a flattened ascoma, having the wall more or less radial in structure, and with a basal plate, e.g. *Trichothyrina*; cf. thyriothecium.

Catathelasma Lovejoy (1910), Tricholomataceae. 4, widespread (north temperate). See Singer (*Sydowia* **31**: 193, 1979; key).

Catathelasmataceae Wasser (1985) = Tricholomataceae.

Catatrama Franco-Mol. (1991), Amanitaceae. 1, Costa Rica; India. See Franco-Molano (*Mycol.* **83**: 501, 1991).

Catenaria Sorokīn (1889), Catenariaceae. 11, widespread (temperate). See Birchfield (*Mycopathologia* **13**: 331, 1960), Olson & Reichle (*TBMS* **70**: 423, 1978; meiosis), Singh *et al.* (*MR* **97**: 957, 1993; development).

Catenariaceae Couch (1945), Blastocladiales. 3 gen. (+ 1 syn.), 14 spp.

Lit.: Manier (*Annls Parasit. hum. comp.* **52**: 363,

1977), Tanabe *et al.* (*J. gen. appl. Microbiol.* Tokyo **51**: 267, 2005).

catenate (**catenulate**), in chains or end-to-end series. See arthro-, baso-, blastocatenate.

Catenella Bat. & Peres (1963), anamorphic *Pezizomycotina*, St.0eH.?. 1, Italy. See Batista & Peres (*Publções Inst. Micol. Recife* **222**: 6, 1963).

Catenochytridium Berdan (1939), Endochytriaceae. 6, N. America; Japan.

Catenocuneiphora Matsush. (2003), anamorphic *Pezizomycotina*, Hso.?.?. 1, Japan. See Matsushima (*Matsush. Mycol. Mem.* **10**: 40, 2001).

Catenomyces A.M. Hanson (1944), Catenariaceae. 1, USA.

Catenomycopsis Tibell & Constant. (1991), anamorphic *Chaenothecopsis*, Hso.0eH.3. 1, S. America; Australasia. See Tibell & Constantinescu (*MR* **95**: 556, 1991).

Catenophlyctis Karling (1965), Catenariaceae. 2, widespread. See Golubeva & Stephenson (*N.Z. Jl Bot.* **41**: 319, 2003; from subantarctic).

Catenophora Luttr. (1940), anamorphic *Pezizomycotina*, Cac.0eH.1. 3, USA. See Nag Raj & Kendrick (*CJB* **66**: 898, 1988; key).

Catenophoropsis Nag Raj & W.B. Kendr. (1988), anamorphic *Pezizomycotina*, Cac.0eH.10. 1, Australasia. See Nag Raj & Kendrick (*CJB* **66**: 898, 1988).

catenophysis, a persistent chain of utricular, thin-walled cells formed by the vertical separation of the pseudoparenchyma in the centrum of certain ascomycetes, e.g. some *Halosphaeriaceae* (see Kohlmeyer & Kohlmeyer, *Mycol.* **63**: 857, 1971).

Catenospegazzinia Subram. (1991), anamorphic *Pezizomycotina*, Hso.#eP.37/5. 2, Australia. See Subramanian (*Curr. Sci.* **60**: 657, 1991).

Catenosubulispora Matsush. (1971), anamorphic *Pezizomycotina*, Hso.0fH.3. 1, Guadalcanal. See Matsushima (*Microfungi of the Solomon Islands and Papua-New Guinea*: 10, 1971).

Catenosynnema Kodsueb, E.H.C. McKenzie, W.H. Ho, K.D. Hyde, P. Lumyong & S. Lumyong (2007), anamorphic *Pezizomycotina*. 1, Thailand. See Kodsueb *et al.* (*Cryptog. Mycol.* **28**: 237, 2007).

Catenularia Grove (1886), anamorphic *Chaetosphaeria*, Hso.0eP.17. 10, widespread (temperate). See Hughes (*N.Z. Jl Bot.* **3**: 136, 1965), Réblová (*Stud. Mycol.* **45**: 149, 2000; review), Réblová & Seifert (*Sydowia* **55**: 313, 2003; Thailand).

Catenulaster Bat. & C.A.A. Costa (1959), anamorphic *Pezizomycotina*, Cpt.0eH.?. 1, Brazil. See Batista & Costa (*Mycopath. Mycol. appl.* **11**: 8, 1959).

catenulate, see catenate.

Catenulifera Hosoya (2002), anamorphic *Hyphodiscus*, H?.?.?. 1, north temperate. See Hosoya (*Mycoscience* **43**: 48, 2002), Untereiner *et al.* (*CJB* **84**: 243, 2006; phylogeny, connection).

catenuliform, chain-like.

Catenulopsora Mundk. (1943) = Cerotelium fide Laundon (*Mycotaxon* **3**: 133, 1975).

Catenulostroma Crous & U. Braun (2007), Teratosphaeriaceae. 7, widespread. See Crous *et al.* (*Stud. Mycol.* **58**: 1, 2007; phylogeny, definition).

Catenuloxyphium Bat., Nascim. & Cif. (1963) nom. dub., anamorphic *Pezizomycotina*. See Hughes (*Mycol.* **68**: 693, 1976), Sutton (*Mycol. Pap.* **141**, 1977).

caterpillar fungi, see vegetable caterpillars.

Catharinia (Sacc.) Sacc. (1895) = Julella fide Clements & Shear (*Gen. Fung.*, 1931).

Cathisinia Stirt. (1888) = Sarcogyne Flot. (1851) fide Hawksworth *et al.* (*Dictionary of the Fungi* edn 8, 1995).

Catilla Pat. (1915), Cyphellaceae. 1, Europe.

Catillaria A. Massal. (1852), Catillariaceae (L). *c.* 158, widespread. See Lamb (*Rhodora* **56**: 105, 1956), Vězda (*Folia geobot. phytotax.* **15**: 75, 1980; key foliicolous spp.), Kilias (*Herzogia* **5**: 209, 1981; monogr. Eur. saxic. spp.), Coppins (*Lichenologist* **21**: 217, 1989; UK), Pant & Awasthi (*Proc. Indian Acad. Sci.* Pl. Sci. **99**: 369, 1989; key 10 Indian spp.), Coppins (*Graphis Scripta* **6**: 65, 1994; Sweden), Fryday & Coppins (*Lichenologist* **28**: 507, 1996; Scotland), Tretiach & Hafellner (*Lichenologist* **30**: 221, 1998; Mediterranean), Boom & Etayo (*Lichenologist* **33**: 103, 2001; Iberian peninsula), Andersen & Ekman (*MR* **109**: 21, 2005; phylogeny), Kalb (*Biblthca Lichenol.* **95**: 297, 2007; key).

Catillariaceae Hafellner (1984), Rhizocarpales (±L). 8 gen. (+ 18 syn.), 261 spp.

Lit.: Coppins (*Lichenologist* **21**: 217, 1989), Pant & Awasthi (*Proc. Indian Acad. Sci.* Pl. Sci. **99**: 369, 1989), Fryday & Coppins (*Lichenologist* **28**: 507, 1996), Lumbsch (*J. Hattori bot. Lab.* **83**: 1, 1997), Ekman (*MR* **105**: 783, 2001), Reeb *et al.* (*Mol. Phylogen. Evol.* **32**: 1036, 2004).

Catillariomyces E.A. Thomas ex Cif. & Tomas. (1953) = Cliostomum.

Catillariopsis (Stein) M. Choisy (1950) = Rhizocarpon fide Hawksworth *et al.* (*Dictionary of the Fungi* edn 8, 1995).

Catillochroma Kalb (2007), Megalariaceae (L). 1, north temperate. See Kalb (*Biblthca Lichenol.* **95**: 297, 2007).

Catinaria Vain. (1922) nom. cons., Ramalinaceae (L). 2, widespread (temperate). See Poelt & Vězda (*Biblthca Lichenol.* **16**, 1981), Kalb (*Biblthca Lichenol.* **95**: 297, 2007; key).

Catinariaceae Hale ex Hafellner (1984) = Ramalinaceae.

Catinella Boud. (1907), Dothideomycetes. 1 or 2, widespread. See Spooner & Legon (*Mycologist* **5**: 86, 1991), Nauta & Spooner (*Mycologist* **14**: 21, 2000; UK), Greif *et al.* (*Am. J. Bot.* **94**: 1890, 2007; posn, ontogeny).

Catinella Kirschst. (1924) = Unguicularia fide Raschle (*Sydowia* **29**: 170, 1977).

Catinopeltis Bat. & C.A.A. Costa (1957), anamorphic *Pezizomycotina*, Cpt.≡ eH.?. 1, Brazil. See Batista & Costa (*An. Soc. Biol. Pernambuco* **15**: 405, 1957).

Catinula Lév. (1848), anamorphic *Pezizomycotina*, Cpd.?.?. 10, widespread. See Jae & Sang (*Korean Jl Pl. Path.* **2**: 174, 1986; Korea).

Catocarpon, see *Catocarpus*.

Catocarpus (Körb.) Arnold (1871) = Rhizocarpon fide Hawksworth *et al.* (*Dictionary of the Fungi* edn 8, 1995).

Catolechia Flot. (1850), Rhizocarpaceae (L). 1 (montane), Europe. See Hafellner (*Nova Hedwigia* **30**: 673, 1978), Bellemère & Letrouit-Galinou (*Biblthca Lichenol.* **25**: 137, 1987; ultrastr.), Grube *et al.* (*MR* **108**: 1111, 2004; phylogeny), Miądlikowska *et al.* (*Mycol.* **98**: 1088, 2006; phylogeny).

Catopyrenium Flot. ex Körb. (1855) ≡ Catapyrenium.

Catosphaeropsis Tehon (1939) = Sphaeropsis Sacc. fide Sutton (*The Coelomycetes*, 1980).

catothecium, see catathecium.

cat's ear, basidiome of *Clitopilus passeckerianus*, an

invader of mushroom beds.

Cattanea Garov. (1875) = Dictyosporium fide Damon (*Lloydia* **15**: 118, 1952).

Catulus Malloch & Rogerson (1978), Dothideomycetes. 1 (on *Seuratia*), Canada. See Malloch & Rogerson (*CJB* **56**: 2344, 1978).

cauda, tail; tail-like appendage.

caudate, having a tail.

Caudatispora J. Fröhl. & K.D. Hyde (1995), Sordariomycetidae. 2 (on *Palmae*), America (tropical). See Fröhlich & Hyde (*Sydowia* **47**: 38, 1995), Huhndorf & Fernández (*Sydowia* **50**: 200, 1998).

Caudella Syd. & P. Syd. (1916), Microthyriaceae. 2, America. See Hansford (*Mycol. Pap.* **15**, 1946), Hosagoudar & Goos (*Mycotaxon* **59**: 149, 1996), Pereira & Filardi (*Edinb. J. Bot.* **63**: 263, 2006).

Caudellopeltis Bat. & H. Maia (1960) = Maublancia fide Müller & von Arx (*Beitr. Kryptfl. Schweiz* **11** no. 2, 1962).

Caudomyces Lichtw., Kobayasi & Indoh (1988), Legeriomycetaceae. 2 (in *Antocha*), Canada; China; Japan. See Lichtwardt *et al.* (*TMSJ* **28**: 376, 1987), Strongman & Xu (*Mycol.* **98**: 479, 2006; China), Strongman (*CJB* **85**: 949, 2007; Canada).

Caudophoma B.V. Patil & Thirum. (1968) = Phyllosticta fide van der Aa (*Stud. Mycol.* **5**, 1973).

Caudospora J. Weiser (1946), Microsporidia. 1.

Caudospora Starbäck (1889), Diaporthales. Anamorph *Phomopsis*-like. 1 (on *Quercus*), Europe; N. America. See Barr (*Mycol. Mem.* **7**, 1978; synonymy with *Hercospora*), Rogers (*Mycotaxon* **21**: 475, 1984).

Caudosporella Höhn. (1914) = Harknessia fide Sutton (*Mycol. Pap.* **123**, 1971).

caulescent, having a stem; becoming stemmed.

caulicolous, living on herbaceous stems; **caulicole**, a fungus which does this.

Caullerya Chatton (1907), Microsporidia. 1.

Caulleryetta Dogiel (1922), Microsporidia. 1.

Caulocarpa Gilkey (1947) = Sarcosphaera fide Trappe (*Mycotaxon* **2**: 109, 1975).

Caulochora Petr. (1940) = Glomerella fide von Arx & Müller (*Beitr. Kryptfl. Schweiz* **11** no. 1, 1954).

Caulochytriaceae Subram. (1974), Spizellomycetales. 1 gen., 2 spp.

Lit.: Voos (*Am. J. Bot.* **56**: 898, 1969), Olive (*Am. J. Bot.* **67**: 568, 1980), Barr *in* Margulis *et al.* (Eds) (*Handbook of Protoctista*: 454, 1990).

Caulochytrium Voos & L.S. Olive (1968), Caulochytriaceae. 2 (on *Cladosporium* and *Gloeosporium* conidia), USA. See Voos (*Am. J. Bot.* **56**: 898, 1969).

caulocystidium, see cystidium.

Caulogaster Corda (1831) nom. dub., Fungi. Based on insect eggs fide Tulasne & Tulasne (*Selecta fungorum carpologa* **1**: 125, 1861).

Cauloglossum Grev. (1823) = Podaxis fide Stalpers (*in litt.*).

cauloplane, the stem surface.

Caulorhiza Lennox (1979), Tricholomataceae. 3, USA. See Lennox (*Mycotaxon* **9**: 154, 1979).

Caumadothis Petr. (1971) = Botryosphaeria fide von Arx & Müller (*Stud. Mycol.* **9**, 1975).

Causalis Theiss. (1918) = Coccodiella fide Müller & von Arx *in* Ainsworth *et al.* (Eds) (*The Fungi* **4A**: 87, 1973).

Cautinia Maas Geest. (1967) = Grifola fide Stalpers (*in litt.*).

Cavaraella Speg. (1923), Rhytismatales. 1, Cuba.

Cave fungi. There are many records of fungi from caves and mines in temperate and tropical regions. Although so far no fungi have been reported as peculiar to caves, at least one new species, *Microascus caviariformis*, the 'chicken-sandwich cave fungus', has been described from a cave (Malloch & Hubart, *Can. J. Bot.* **65**: 2384, 1987). Lagarde (*Arch. Zool. exp. gen.* **53**: 277, 1913; **56**: 279, and *Notes et Revue*: 129, 1917; **60**: 593, 1922) recorded *c.* 50 taxa (including most groups of fungi) from caves in France, Spain and N. Africa. Went (*Science* **166**: 385, 1969) and Hasselbring *et al.* (*Mycol.* **67**: 171, 1975) describe fungi associated with stalactites. Cunningham *et al.* (*Environmental geology* **25**: 2, 1995) reported over 90 species of fungi from a cave network in New Mexico. Various yeasts have been reported from caves in Japan (Sugita *et al.*, *Appl. Environ. Microbiol.* **71**: 7626, 2005). Fungi in caves and mines in general reflect the substrata available. Where caves are not too deep, basidiomata of fungi such as *Boletus* spp. or *Hygrocybe* may occasionally be encountered (e.g. Kibby, *Field Mycology* **9**: 66, 2008). Polypores and other wood-destroying fungi are common in mines where the basidiomata are frequently abnormal (Fassatiová, *Česká Myk.* **24**: 162, 1970; mines of Příbram, former Czechoslovakia). Caves may also be a source of dermatophyte fungi (Lurie & May, *Mycol.* **49**: 178, 1957), with many records of miners in South African gold mines with infections of *Sporothrix schenkii* obtained from mine timbers (Doidge, *Bothalia* **5**: 1, 1950); infection by *Histoplasma* can be a serious hazard for speleologists in some parts of the new world (Ashford *et al*, *Am. J. Trop. Med. Hyg.* **60**: 899, 1999). *Fusarium* infections are associated 'white nose syndrome' causing very high mortalities of cave bats (Anon., *New Scientist* **2645**: 6, 2008). Where caves become tourist attractions, resulting environmental changes may increase levels of fungi. This can result in damage to e.g. prehistoric cave paintings (Connor, *Independent* [UK newspaper], 10 May 2006).

cavernose, having hollows or cavities.

cavernula (pl. **-ae**), cavity; esp. the cavities in the lower cortex of *Cavernularia*.

Cavernularia Degel. (1937), Parmeliaceae (L). 2, Europe; N. America. See Ahti & Henssen (*Bryologist* **68**: 85, 1965), Printzen & Ekman (*Lichenologist* **34**: 101, 2002; population biology), Printzen *et al.* (*Mol. Ecol.* **12**: 1473, 2003; phylogeography), Thell *et al.* (*Symb. bot. upsal.* **34** no. 1: 429, 2004; biogeography).

Cavimalum Yoshim. Doi, Dargan & K.S. Thind (1977), Clavicipitaceae. 2, Asia. See Doi *et al.* (*Bull. natn. Sci. Mus.* Tokyo, B **3**: 23, 1977), Bischoff & White (*Mycology Series* **19**: 125, 2003).

Cazia Trappe (1989), Pezizaceae. 1 (hypogeous), USA. See Trappe (*Mem. N. Y. bot. Gdn* **49**: 336, 1989; placement in *Helvellaceae*), Norman & Eggers (*Mycol.* **91**: 820, 1999; phylogeny), Percudani *et al.* (*Mol. Phylogen. Evol.* **13**: 169, 1999; phylogeny), Hansen *et al.* (*Mycol.* **93**: 958, 2001; phylogeny), Hansen & Pfister (*Mycol.* **98**: 1029, 2006; phylogeny), Læssøe & Hansen (*MR* **111**: 1075, 2007; phylogeny).

CBS, Centraalbureau voor Schimmelcultures (Baarn and Delft, Netherlands); founded 1904; an Institute of the Royal Netherlands Academy of Arts and Science; see van Beverwijk (*Ant. v. Leeuwenh.* **25**: 1, 1959), de Hoog (Ed.) (*Centraalbureau voor Schimmelcultures: 75 years culture collection*, 1979).

cecidium, a gall (q.v.); caused by an animal (**zoo-**), esp. an insect; caused by a fungus (**myco-**).

Cecidonia Triebel & Rambold (1988), Lecideaceae. 2 (on lichens), Europe. See Triebel & Rambold (*Nova Hedwigia* **47**: 280, 1988), Buschbom & Mueller (*Mol. Phylogen. Evol.* **32**: 66, 2004; phylogeny).

Ceeveesubramaniomyces Pratibha, K.D. Hyde & Bhat (2005), anamorphic *Basidiomycota*. 1, India. See Pratibha *et al.* (*Kavaka* **32**: 21, 2004).

Cejpia Velen. (1934), Dermateaceae. 2, Europe. See Baral (*SA* **13**: 113, 1994), Nauta & Spooner (*Mycologist* **14**: 21, 2000; UK).

Cejpomyces Svrček & Pouzar (1970) = Thanatephorus fide Langer (*Biblthca Mycol.* **158**, 1994).

Cejpomycetaceae Jülich (1982) = Ceratobasidiaceae.

Celatogloea P. Roberts (2005), Basidiomycota. 1, USA. See Roberts (*Mycologist* **19**: 69, 2005).

Celidiaceae J. Schröt. (1893) = Arthoniaceae.

Celidiopsis A. Massal. (1856) = Arthonia fide Hawksworth *et al.* (*Dictionary of the Fungi* edn 8, 1995).

Celidium Tul. (1852) = Arthonia fide Santesson (*Symb. bot. upsal.* **12** no. 1: 69, 1952), Wedin & Hafellner (*Lichenologist* **30**: 59, 1998), Ertz *et al.* (*Biblthca Lichenol.* **91**: 155 pp., 2005).

cell, a unit of cytoplasm containing one or more nuclei, limited by a membrane (the **cell membrane**), and, in fungi, usually enclosed by a wall (see cell wall chemistry). Adjacent fungal cells do not necessarily contain individual protoplasts (q.v.) because cytoplasm and nuclei may pass from one cell to another.

Cell wall chemistry. See Klis *et al.* (*in* Howard & Gow (Eds), *Biology of the fungal cell*, edn 2, 2007). See also Wessels (*Int. Rev. Cytol.* **104**: 37, 1986; *New Phytologist* **123**: 397, 1993), Bartnicki Garcia (*in* Rayner *et al.* (Eds), *Evolutionary biology of the fungi*: 389, 1987), Gooday (*in* Gow & Gadd, *The growing fungus*: 41, 1995). Polymers found in fungal walls are given in Table 2. The wall (e.g. zygospore wall of *Mucorales*) may also contain melanin (a dark coloured pigment which protects spores from UV light and microbial lysis) and sporopollenin (the most resistant biopolymer known).

cellar fungus, *Coniophora puteana* or *Rhinocladiella ellisii* (sterile mycelium *Zasmidium cellare*; syn. *Racodium cellare*; Hawksworth & Riedl, *Taxon* **26**: 208, 1977).

cellular slime moulds, *Dictyosteliomycota* (q.v.).

Cellularia Bull. (1788) = Trametes fide Donk (*Verh. K. ned. Akad. Wet.* tweede sect. **62**: 1, 1974).

Cellulasclerotes Stach & Pickh. (1957), Fossil Fungi. 3 (Permo-Carboniferous), Europe.

cellulin, a chitan-glucan complex which occurs as granules in the cells and plugs (**- plugs**) at hyphal constrictions in *Leptomitales* (see Lee *et al.*, *Mycol.* **68**: 87, 1976).

cellulolysis adequacy index, an estimate, derived by dividing the rate of cellulolysis by the mycelial growth rate on an agar plate, as to whether the rate of cellulose decomposition by a fungus is adequate to supply its needs for saprobic survival (Garrett, *TBMS* **49**: 59, 1966; Deacon, *TBMS* **72**: 469, 1979).

cellulolytic fungi, fungi able to utilize cellulose-containing materials (incl. plant cellulose, paper, cloth, etc.), e.g. *Chaetomiaceae* (Chahal & Wang, *Mycol.* **70**: 160, 1978). Cellophane or filter paper is often used in the culture of these fungi. See Biodeterioration.

Cellulosporium Peck (1879) nom. dub., anamorphic *Pezizomycotina*. See Sutton (*Mycol. Pap.* **141**, 1977).

Cellypha Donk (1959), ? Tricholomataceae. 10, widespread. See Reid (*Persoonia* **3**: 131, 1964).

Celoporthe Nakab., Gryzenh., Jol. Roux & M.J. Wingf. (2006), Cryphonectriaceae. 1, S. Africa; Indonesia. See Nakabonge *et al.* (*Stud. Mycol.* **55**: 261, 2006).

Celotheliaceae Lücking, Aptroot & Sipman (2008), Pyrenulales. 1 gen., 8 spp.

Celothelium A. Massal. (1860), Celotheliaceae (L). 8, widespread (subtropical and tropical). See Aguirre-Hudson (*Bull. Br. Mus. nat. hist.* Bot. **21**: 85, 1991), Berger & Aptroot (*Herzogia* **13**: 151, 1998; Austria), Prado *et al.* (*MR* **110**: 511, 2006; phylogeny), Aptroot *et al.* (*Biblthca Lichenol.* **97**, 2008; Costa Rica).

Celtidia J.D. Janse (1897), Zopfiaceae. 1, Java. See Hawksworth (*CJB* **57**: 91, 1979).

Celyphus Batten (1973), Fossil Fungi ? Fungi. 1 (late Jurassic – early Cretaceous), British Isles; USA.

CEM, see Societies and organizations.

Cenangella Sacc. (1884) = Dermea fide Nannfeldt (*Nova Acta R. Soc. Scient. upsal.*, 1932).

Cenangiaceae Rehm (1888) = Helotiaceae.

Cenangiella Lambotte (1887) = Scleroderris fide Hawksworth *et al.* (*Dictionary of the Fungi* edn 8, 1995).

Cenangina Höhn. (1909) = Cenangium fide Hawksworth *et al.* (*Dictionary of the Fungi* edn 8, 1995).

Cenangiomyces Dyko & B. Sutton (1979), anamorphic *Agaricomycetes*, Ccu.0eH.41. 1 (with clamp connexions), British Isles. See Dyko & Sutton (*TBMS* **72**: 411, 1979).

Cenangiopsis Rehm (1912), Helotiaceae. 1, Europe.

Cenangiopsis Velen. (1947), ? Helotiales. 1, former Czechoslovakia. See Baral (*SA* **13**: 113, 1994).

Cenangites Mesch. (1892), Fossil Fungi. 1 (Tertiary), Europe.

Cenangium Fr. (1818), Helotiaceae. 25, widespread. *C. ferruginosum* (syn. *C. abietis*) (die-back of pines). See Verkley (*MR* **99**: 187, 1995; asci), Jung *et al.* (*Pl. Path. J.* **17**: 216, 2001; sp. definition).

Cenangiumella J. Fröhl. & K.D. Hyde (2000), Helotiaceae. 1, Brunei. See Fröhlich & Hyde (*Fungal Diversity Res. Ser.* **3**: 240, 2000).

ceno-, see coeno-.

Cenococcum Moug. & Fr. (1829), anamorphic *Dothideomycetes*, Sc.-.-. 1, Europe; N. America. *C. geophilum* is mycorrhizal. See LoBuglio *et al.* (*CJB* **69**: 2331, 1992; variation in rDNA), Massicotte *et al.* (*CJB* **70**: 125, 1992; morphology), Stülten *et al.* (*Mycorrhiza* **5**: 259, 1995; protoplast regeneration), Timonen *et al.* (*New Phytol.* **135**: 313, 1997; ecology), Farmer & Sylvia (*MR* **102**: 859, 1998; ITS variation), LoBuglio (*Ectomycorrhizal Fungi*: 287, 1999; review), Shinohara *et al.* (*Curr. Genet.* **35**: 527, 1999; phylogeography), Portugal *et al.* (*S. Afr. J. Sci.* **97**: 617, 2001; population structure), Brundrett (*New Phytol.* **154**: 275, 2002; coevolution), Jany *et al.* (*New Phytol.* **154**: 651, 2002; genetic diversity), Sakakibara *et al.* (*MR* **106**: 868, 2002; identification techniques), Douhan & Rizzo (*New Phytol.* **166**: 263, 2005; population structure), Douhan *et al.* (*Mycol.* **99**: 812, 2007; population structure).

Cenomyce Ach. (1809) ≡ Cenomyces.

Cenomyces Ach. (1809) = Cladonia fide Hawksworth *et al.* (*Dictionary of the Fungi* edn 8, 1995).

Cenomycetaceae Chevall. (1826) = Cladoniaceae.

Cenozosia A. Massal. (1854) = Ramalina fide Hawksworth *et al.* (*Dictionary of the Fungi* edn 8, 1995).

Centonites Peppers (1964), Fossil Fungi. 1 (Carboniferous), USA.

central body (of ascomycetes), the cell structure (central apparatus) from which astral rays emanate and initiate a cleavage of the cytoplasm (Harper, 1905). See Lindegren *et al.* (*Can. J. Genet. Cytol.* **7**: 37, 1965).

centric (**central**) (of a stipe), at the centre of the pileus; (of oogonium of *Saprolegniaceae*), having one or two layers of fat droplets surrounding the central cytoplasm in contrast to **sub-**, having the cytoplasm surrounded by one layer of droplets on one side, and by two or three layers on the other and **ex-**, having one large drop or a lunate row of droplets on one side (Coker, 1923).

Centridium Chevall. (1826) = Roestelia fide Dietel (*Nat. Pflanzenfam.* **6**, 1928).

centrifugal, from the centre outwards.

centriole, short-cylindrical or barrel-shaped cell organelle 300-500 nm long × 150 nm diam. (kinetosome lacking an axoneme).

centripetal, towards the centre.

Centrolepidosporium R.G. Shivas & Vánky (2007), ? Ustilaginaceae. 1 (on *Centrolepidaceae*), Australia. See Shivas & Vánky (*Mycol. Balcan.* **4**: 1, 2007).

Centrospora Neerg. (1942) [non *Centrospora* Trevis. 1845, *Algae*] ≡ Mycocentrospora.

centrum, the structures within an ascoma, i.e. the asci and interascal tissue (the hamathecium); for types of centrum organization see Reynolds (Ed.) (*Ascomycete systematics*, 1981), and under *Ascomycota*.

cep, basidioma of the edible *Boletus edulis*; **fungi suilli** of Pliny and other classical writers.

cepaceous, see alliaceous.

Cephalacladium, see *Cephalocladium*.

Cephaleuros Kunze (1832), Trentepohliaceae. parasitic on vascular plants. *C. virescens* (syn. *Mycoidea parasitica*) is common on tea (**red-rust**) and other trop. plants; photobiont of *Strigula* (q.v.) and other foliicolous lichens. See Printz (*Nytt Mag. Naturvid.* **80**: 137, 1940; keys), Santesson (*Symb. bot. upsal.* **12**(1), 1952).

Cephaliophora Thaxt. (1903), anamorphic *Pezizales*, Hso.≡ eH.6. 6, widespread (tropical). See Barron *et al.* (*CJB* **68**: 685, 1990; 2 n.spp.), Tanabe *et al.* (*Mycol.* **91**: 830, 1999), Barron (*Biodiversity of Fungi* Inventory and Monitoring Methods: 435, 2004; review).

Cephaloascaceae L.R. Batra (1973), Saccharomycetales. 1 gen. (+ 2 syn.), 1 spp.

Lit.: Batra (*Stud. Mycol.* **30**: 415, 1987), von Arx & van der Walt (*Stud. Mycol.* **30**: 167, 1987), Hausner *et al.* (*Mycol.* **84**: 870, 1992; mol. phylogeny), Spatafora & Blackwell (*Mycol.* **85**: 912, 1993), Kurtzman & Robnett (*CJB* **73** Suppl. 1: S824, 1995; molecular rels), Read & Beckett (*MR* **100**: 1281, 1996), Hoog & Kurtzman *in* Kurtzman & Fell (Eds) (*Yeasts, a taxonomic study* 4th edn: 143, 1998), Malloch & de Hoog *in* Kurtzman & Fell (Eds) (*Yeasts, a taxonomic study* 4th edn: 197, 1998), Suh *et al.* (*Mycol.* **98**: 1006, 2006; phylogeny).

Cephaloascales = Saccharomycetales.

Cephaloascus Hanawa (1920), Cephaloascaceae. 1, Japan; Canada. See Schippers-Lammertse & Heyting (*Antonie van Leeuwenhoek* **28**: 5, 1962), Udagawa *et al.* (*TMSJ* **18**: 399, 1977), Cannon & Minter (*TMSJ* **33**: 51, 1992), de Hoog & Kurtzman *in* Kurtzman & Fell (Eds) (*Yeasts, a taxonomic study* 4th edn: 143, 1998), Spatafora (*Cellular Origin and Life in Extreme Habitats* **4**: 591, 2002; evolution), Suh *et al.* (*Mycol.* **98**: 1006, 2006; phylogeny).

Cephalocladiaceae Corda (1838) = Sclerotiniaceae.

Cephalocladium Rchb. (1828) = Botrytis fide Mussat (*Syll. fung.* **15**: 82, 1901).

Cephalodiplosporium Kamyschko (1961) = Fusarium fide Gams (*in litt.*).

cephalodium (pl. **-ia**), a delimited region within (**internal -**), or a warty, squamulose, or fruticose structure on the surface of, a lichen thallus containing a photobiont different from that characteristic of the rest of the thallus. Generally cephalodia contain cyanobacteria (e.g. *Nostoc*) whilst the rest of the thallus contains a green alga (e.g. *Trebouxia*). *Nostoc* cephalodia fix atmospheric nitrogen (see Millbank & Kershaw, *New Phytol.* **68**: 721, 1969). Known in about 400 lichenized spp. in diverse orders; see cyanotrophy, gall, Lichens, paracephalodium, photomorph, phycotype.

Cephalodochium Bonord. (1851), anamorphic *Pezizomycotina*, Hsp.0eH.?. 1, Europe.

Cephaloedium Kunze (1828) nom. nud. = Exosporium fide Hughes (*CJB* **36**: 744, 1958; nomencl.).

Cephalomyces Bainier (1907) = Cephaliophora fide Ellis (*Dematiaceous Hyphomycetes*, 1971).

Cephalophorum Nees [not traced] nom. dub., anamorphic *Pezizomycotina*. See Benjamin (*Taxon* **17**: 524, 1968).

Cephalophysis (Hertel) H. Kilias (1985), Teloschistaceae (L). 1, Europe. See Kilias (*Herzogia* **7**: 182, 1985), Kärnefelt (*Cryptog. bot.* **1**: 147, 1989; phylogeny).

Cephaloscypha Agerer (1975), Marasmiaceae. 2, widespread. See Agerer (*Sydowia* **27**: 193, 1973-1974).

cephalosporin, one of a series of antibacterial antibiotics from *Acremonium* spp., esp. *A. chrysogenum* (cephalosporin C) and *A. salmosynnematum*); see Roberts (*Mycol.* **44**: 292, 1952), Grosklags & Swift (*Mycol.* **49**: 305, 1957), Kavanagh *et al.* (*Mycol.* **50**: 370, 1958), Sassiver & Lewis (*Adv. appl. Microbiol.* **13**: 163, 1970; structure), Abraham & Newton (*Adv. Chemotherapy* **2**: 23, 1969); cephalosporin N = penicillin N.

Cephalosporiopsis Peyronel (1915), anamorphic *Hypocreales*, Hso.1eH.15. 4, Europe; Africa. fide Carmichael *et al.* (*Genera of Hyphomycetes*, 1980; synonymy with *Acremonium* or *Fusarium*).

Cephalosporium Corda (1839) = Acremonium fide Gams (*Cephalosporium-artige Schimmelpilze*, 1971).

Cephalotelium Syd. (1921) = Ravenelia fide Dietel (*Nat. Pflanzenfam.* **6**, 1928).

Cephalotheca Fuckel (1871), Cephalothecaceae. 6, widespread. See Booth (*Mycol. Pap.* **83**, 1961; key), Suh & Blackwell (*Mycol.* **91**: 836, 1999; phylogeny), Yaguchi *et al.* (*Mycotaxon* **96**: 309, 2006; from human).

Cephalothecaceae Höhn. (1917), Sordariales. 5 gen. (+ 4 syn.), 13 spp.

Lit.: Chesters (*TBMS* **19**: 261, 1935; life history), Lundqvist (*Svensk bot. Tidskr.* **86**: 261, 1992), Fortey *et al.* (*Mycologist* **11**: 132, 1997), Suh & Blackwell (*Mycol.* **91**: 836, 1999; phylogeny), Lumley *et al.* (*Mycotaxon* **74**: 395, 2000), Paulin & Harrington (*Stud. Mycol.* **45**: 209, 2000), Huhndorf *et al.* (*Mycol.* **96**: 368, 2004), Yaguchi *et al.* (*Mycotaxon* **96**: 309,

2006), Stchigel & Guarro (*MR* **111**: 1100, 2007).

Cephalothecium Corda (1838) = Trichothecium fide Hughes (*CJB* **36**: 727, 1958).

cephalothecoid, fragmenting along a series of predefined suture lines, as in ascomata of *Cephalotheca* (q.v.).

Cephalothricum, see *Cephalotrichum Link*.

Cephalotrichum Berk. ex Sacc. (1886) ≡ Trichocephalum.

Cephalotrichum Link (1809), anamorphic *Microascaceae*, Hsy.0eP.19. 8, widespread. See Hughes (*CJB* **36**: 727, 1958), Morton & Smith (*Mycol. Pap.* **86**, 1963).

Cepsiclava J. Walker (2004), Clavicipitaceae. 1 (on living florets of *Phalaris*), Australia. Perhaps synonymous with *Aciculosporium*. See Walker (*Australas. Pl. Path.* **33**: 228, 2004).

Ceracea Cragin (1885), anamorphic *Pezizomycotina*, Hsp.0eH.?. 1, USA. See Martin (*Mycol.* **41**: 77, 1949), Donk (*Persoonia* **4**: 334, 1966).

Ceraceohydnum Jülich (1978) = Mycoaciella fide Hjortstam *et al.* (*Kew Bull.* **45**: 303, 1990; key).

Ceraceomerulius (Parmasto) J. Erikss. & Ryvarden (1973) = Byssomerulius fide Hjortstam & Larsson (*Windahlia* **21**, 1994).

Ceraceomyces Jülich (1972), Amylocorticiaceae. 16, widespread. See Jülich (*Willdenowia* Beih. **7**: 146, 1972), Legon (*Mycologist* **19**: 167, 2005).

Ceraceopsora Kakish., T. Sato & S. Sato (1984), Chaconiaceae. 1 (on *Anemone* (0, I) (*Ranunculaceae*); *Elaeagnus* (II, III) (*Elaeagnaceae*)), Japan. See Kakishima *et al.* (*Mycol.* **76**: 969, 1984).

Ceraceosorales Begerow, Stoll & R. Bauer (2007). Exobasidiomycetes. 1 gen., 1 spp. No families recognized.

Lit. Begerow *et al.* (*Mycol.* **98**: 906, 2006; phylogeny).

Ceraceosorus B.K. Bakshi (1976), Ceraceosorales. 1, India. See Bakshi (*Mycol.* **68**: 649, 1976), Begerow *et al.* (*Mycol.* **98**: 906, 2006; phylogeny).

ceraceous (cereous), wax-like.

Ceraiomyces Thaxt. (1901) = Laboulbenia fide Thaxter (*Proc. Amer. Acad. Arts & Sci.* **51**: 1, 1915).

Ceramoclasteropsis Bat. & Cavalc. (1962), ? Capnodiaceae. 1, Brazil. See Batista & Cavalcanti (*Brotéria* Sér. Ci. Nat. **31**: 101, 1962).

Ceramothyrium Bat. & H. Maia (1956), Chaetothyriaceae. Anamorph *Stanhughesia*. 20, widespread (tropical). See Constantinescu *et al.* (*Stud. Mycol.* **31**: 69, 1989; anamorph), Winka *et al.* (*Mycol.* **90**: 822, 1998; rDNA phylogeny), Haase *et al.* (*Stud. Mycol.* **43**: 80, 1999; phylogeny).

Cerania Gray (1821) nom. rej. = Thamnolia fide Hawksworth *et al.* (*Dictionary of the Fungi* edn 8, 1995).

ceranoid, having horn-like branches.

Cerapora, see *Ceriporia*.

Ceraporia, see *Ceriporia*.

Ceraporus Bondartsev & Singer (1941) nom. nud. ≡ Ceriporia.

Cerasterias Reinsch (1867) nom. dub., ? Algae.

Cerastoma Quél. (1875), (orthographic variant), see *Ceratostoma* Fr.

Cerastomis Clem. (1931) = Ceratostomella fide Hawksworth *et al.* (*Dictionary of the Fungi* edn 8, 1995).

Ceratelium Arthur (1906) ≡ Cerotelium.

Ceratella (Quél.) Bigeard & H. Guill. (1913) [non *Ceratella* Hook. f. 1844, *Compositae*] = Multiclavula.

Ceratella Pat. (1887) [non *Ceratella* Hook. f. 1844, *Compositae*] ≡ Ceratellopsis.

Ceratellopsis Konrad & Maubl. (1937), Gomphaceae. 9, Europe. See Berthier (*Bull. mens. Soc. linn. Lyon* Num. Spéc. **46**: 187, 1976), Shiryaev (*Mikol. Fitopatol.* **38**: 59, 2004; Urals).

Ceratitium Rabenh. (1851) = Gymnosporangium fide Laundon (*Mycol. Pap.* **102**, 1965).

Ceratobasidiaceae G.W. Martin (1948), Cantharellales. 8 gen. (+ 14 syn.), 43 spp. Molecular data suggest most species group together within the *Cantharellales*, except the type, which probably belongs in the *Auriculariales*.

Lit.: Jin & Korpradiskul (*Mycosystema* **17**: 331, 1998), Roberts (*MR* **102**: 1074, 1998), Roberts (*Mycotaxon* **69**: 35, 1998), Roberts (*Sydowia* **50**: 252, 1998), Saunders & Owens (*Mycorrhiza Manual Springer Lab Manual*: 413, 1998), Roberts (*Rhizoctonia-forming fungi*, 1999), Salazar *et al.* (*Mycol.* **91**: 459, 1999), Gonzalez *et al.* (*Mycol.* **93**: 1138, 2001), Pope & Carter (*Mycol.* **93**: 712, 2001), Larsson *et al.* (*MR* **108**: 983, 2004), Otero *et al.* (*Mol. Ecol.* **13**: 2393, 2004), Binder *et al.* (*Systematics and Biodiversity* **3**: 113, 2005), González *et al.* (*Mol. Phylogen. Evol.* **40**: 459, 2006), Sharon *et al.* (*Mycoscience* **47**: 299, 2006).

Ceratobasidiales Jülich (1981) = Cantharellales.

Ceratobasidium D.P. Rogers (1935), Ceratobasidiaceae. Anamorph *Ceratorhiza*. *c*. 18, widespread. See Currah & Zelmer (*Rep. Tottori mycol. Inst.* **30**: 43, 1992), Roberts (*Rhizoctonia-forming fungi*, 1999), Gonzalez *et al.* (*Mycol.* **93**: 1138, 2001; Ribosomal DNA), Shan *et al.* (*Mycol.* **94**: 230, 2002; orchid endophytes), Ma *et al.* (*MR* **107**: 1041, 2003; anamorph phylogeny), Kotiranta & Saarenoksa (*Ann. bot. fenn.* **42**: 237, 2005; Finland).

Ceratocarpia Rolland (1896), Dothideomycetes. 2, Europe; Africa. See Benny & Kimbrough (*Mycotaxon* **12**: 1, 1980).

Ceratochaete Syd. & P. Syd. (1917) = Dysrhynchis fide Müller & von Arx (*Beitr. Kryptfl. Schweiz* **11** no. 2, 1962).

Ceratochaetopsis F. Stevens & Weedon (1927) = Chaetothyrina fide von Arx & Müller (*Stud. Mycol.* **9**, 1975).

Ceratocladia Schwend. (1860) = Alectoria Ach. fide Hawksworth *et al.* (*Dictionary of the Fungi* edn 8, 1995).

Ceratocladium Corda (1839), anamorphic *Pezizomycotina*, Hso.0eH.?6. 1, Europe. See Hughes (*Mycol. Pap.* **47**, 1951), Kendrick (*CJB* **81**: 75, 2003; morphogenesis).

Ceratocladium Pat. (1898) ≡ Xylocladium.

Ceratocoma Buriticá & J.F. Hennen (1991), Pucciniosiraceae. 2 (on *Gompholobium* (*Leguminosae*) or *Xylopia* (*Annonaceae*)), Australia; Africa. See Buriticá & Hennen (*Revista de la Academia Columbiana de Ciencias Exactas, Físicas y Naturales* **18**: 146, 1991).

Ceratocystidaceae Locq. (1972), Microascales. 6 gen. (+ 6 syn.), 79 spp.

Lit.: Malloch & Blackwell (*CJB* **68**: 1712, 1990), Brasier (*Recent Advances in Studies on Oak Decline* Proceedings of an International Congress, Selva di Fasano (Brindisi), Italy, September 13-18, 1992: 241, 1993), Grylls & Seifert *in* Wingfield *et al.* (Eds) (*Ceratocystis* and *Ophiostoma* Taxonomy, Ecology and Pathogenicity: 261, 1993), Hausner *et al.* (*CJB* **71**: 52, 1993), Samuels (*Ceratocystis* and *Ophiostoma* Taxonomy, Ecology and Pathogenicity: 15, 1993), Seifert *et al.* (*Ceratocystis* and *Ophiostoma* Taxonomy, Ecology and Pathogenicity: 269, 1993), Wingfield *et al.* (*Ceratocystis* and *Ophiostoma* Taxonomy, Ecology and Pathogenicity, 1993), Wyk *et al.* (*Ceratocystis* and *Ophiostoma* Taxonomy, Ecology and Pathogenicity: 133, 1993), Beer *et al.* (*Antonie van Leeuwenhoek* **67**: 325, 1995), Harrington & Wingfield (*CJB* **76**: 1446, 1998), Witthuhn *et al.* (*MR* **103**: 743, 1999), Hausner *et al.* (*CJB* **78**: 903, 2000), Okada *et al.* (*Stud. Mycol.* **45**: 169, 2000), Paulin & Harrington (*Stud. Mycol.* **45**: 209, 2000), Baker Engelbrecht & Harrington (*Mycol.* **97**: 57, 2005), Marín *et al.* (*MR* **109**: 1137, 2005), Johnson *et al.* (*Mycol.* **97**: 1067, 2005).

Ceratocystiopsis H.P. Upadhyay & W.B. Kendr. (1975), Ophiostomataceae. Anamorph *Hyalorhinocladiella*. *c.* 11. See Wolfaardt *et al.* (*S. Afr. J. Bot.* **58**: 277, 1992; synoptic key, database), Hausner *et al.* (*MR* **97**: 625, 1993), van Wijk & Wingfield (*CJB* **71**: 1212, 1993; ultrastr.), Benade *et al.* (*CJB* **74**: 891, 1996; anamorph devel.), Hsiau & Harrington (*Mycol.* **89**: 661, 1997; associated with *Coleoptera*), Benade *et al.* (*Mycotaxon* **68**: 251, 1998; conidial devel.), Viljoen *et al.* (*MR* **104**: 365, 2000; taxonomy), Zhou *et al.* (*Fungal Diversity* **15**: 261, 2004; Chile), Zipfel *et al.* (*Stud. Mycol.* **55**: 75, 2006; phylogeny).

Ceratocystis Ellis & Halst. (1890), Ceratocystidaceae. Anamorph *Chalara*. 44, widespread (esp. temperate). See also *Ophiostoma*. See Hunt (*Lloydia* **19**: 1, 1956; monogr., keys), Upadhyay (*monograph of Ceratocystis and Ceratocystiopsis*, 1981; monogr., keys), Kowalski & Butin (*J. Phytopath.* **124**: 236, 1989; 6 spp. on *Quercus*), Wolfaardt *et al.* (*S. Afr. J. Bot.* **58**: 277, 1992; synoptic key, database), Grylls & Seifert *in* Wingfield *et al.* (Eds) (*Ceratocystis* and *Ophiostoma* Taxonomy, Ecology and Pathogenicity: 261, 1993; key), Hausner *et al.* (*CJB* **71**: 52, 1993; posn), Seifert *et al. in* Wingfield *et al.* (Eds) (*Ceratocystis* and *Ophiostoma* Taxonomy, Ecology and Pathogenicity, 1993), Wingfield *et al.* (*Mol. Biol. Evol.* **11**: 376, 1994), Witthuhn *et al.* (*Mycol.* **90**: 96, 1998), Harrington & Wingfield (*CJB* **76**: 1446, 1999; spp. on conifers), Witthuhn *et al.* (*MR* **103**: 743, 1999; phylogeny), Hausner *et al.* (*CJB* **78**: 903, 2000; phylogeny), Paulin & Harrington (*Stud. Mycol.* **45**: 209, 2000), Witthuhn *et al.* (*Mycol.* **92**: 447, 2000; molecular analysis), Barnes *et al.* (*Molecular Plant Pathology* **2**: 319, 2001; microsatellites), Harrington *et al.* (*Pl. Dis.* **86**: 418, 2002; conifer-associated spp.), Paulin-Mahady *et al.* (*Mycol.* **94**: 62, 2002; phylogeny), Baker *et al.* (*Phytopathology* **93**: 1274, 2003; population structure, host specificity), Morin *et al.* (*Phytopathology* **94**: 1323, 2004; Canada), Wyk *et al.* (*Stud. Mycol.* **50**: 365, 2004; Bhutan), Baker Engelbrecht & Harrington (*Mycol.* **97**: 57, 2005; infraspecific taxonomy), Marín *et al.* (*MR* **109**: 1137, 2005; *C. polonica* agg.), Thorpe *et al.* (*Phytopathology* **95**: 316, 2005; on *Araceae*), Wyk *et al.* (*Australas. Pl. Path.* **34**: 587, 2005; Oman), Johnson *et al.* (*Mycol.* **97**: 1067, 2005; N America), Wyk *et al.* (*Fungal Diversity* **21**: 181, 2006; *C. moniliformis* agg.), Zhang *et al.* (*Mycol.* **98**: 1076, 2006; phylogeny), Engelbrecht *et al.* (*Phytopathology* **97**: 1648, 2007; on *Theobroma*).

Ceratogaster Corda (1841) = Elaphomyces fide Mussat (*Syll. fung.* **15**: 83, 1901).

Ceratomyces Thaxt. (1892), Ceratomycetaceae. 20, widespread. See Weir & Rossi (*Mycol.* **93**: 171, 2001; Bolivia), Santamaría (*Fl. Mycol. Iberica* **5**, 2003; Iberian peninsula).

Ceratomycetaceae S. Colla (1934), Laboulbeniales. 12 gen., 85 spp.
Lit.: Tavares (*Mycol. Mem.* **9**: 627 pp., 1985), Bameul (*Nouv. Rev. Entomol.* Nouv. sér. **10**: 19, 1993), Majewski (*Acta Mycologica* Warszawa **34**: 7, 1999), Santamaría (*Nova Hedwigia* **68**: 351, 1999; European spp.), Weir & Hughes (*Mycol.* **94**: 483, 2002).

Ceratonema Pers. (1822) nom. dub., Fungi. Based on mycelium, etc. fide Mussat (*Syll. Fung.* **15**: 83, 1901).

Ceratophacidium J. Reid & Piroz. (1966), Rhytismataceae. 1, USA. See Reid & Pirozynski (*CJB* **44**: 645, 1966).

Ceratophoma Höhn. (1917), anamorphic *Massarina*, St.0eH.?. 1, Europe.

Ceratophora Humb. (1793) nom. rej. = Gloeophyllum fide Donk (*Verh. K. ned. Akad. Wet.* tweede sect. **62**: 1, 1974).

Ceratophorum Sacc. (1880), anamorphic *Pezizomycotina*, Hso.≡ eP.1. 5, widespread. *C. setosum* (lupin leaf spot). See Hughes (*Mycol. Pap.* **36**, 1951), Ellis (*Mycol. Pap.* **70**, 1958), Huseyinov *et al.* (*Mikol. Fitopatol.* **36**: 11, 2002; Turkey).

Ceratophyllum M. Choisy (1951) [non *Ceratophyllum* L. 1753, *Ceratophyllaceae*] = Hypogymnia fide Hawksworth *et al.* (*Dictionary of the Fungi* edn 8, 1995).

Ceratopodium Corda (1837) = Graphium fide Saccardo (*Syll. fung.* **4**: 609, 1886).

Ceratoporia Ryvarden & de Meijer (2002), Ceratobasidiaceae. 1 (saprobic), Brazil. See Ryvarden & Meijer (*Syn. Fung.* **15**: 44, 2002).

Ceratoporthe Petr. (1925), Diaporthales. 1 (on *Sarothamnus*), Europe.

Ceratopycnidium Maubl. (1907) = Byssoloma. Apparently used in two separate senses. fide Bertoni & Cabral (*MR* **95**: 1014, 1991), Bertoni (*Mycotaxon* **52**: 193, 1994; Brazil), Lücking *et al.* (*Lichenologist* **34**: 270, 2002; synonymy).

Ceratopycnis Höhn. (1915), anamorphic *Pezizomycotina*, Cpd.≡ eP.15. 1, Europe. See Morgan-Jones (*Mycotaxon* **2**: 167, 1975), Farr *et al.* (*Mycol.* **90**: 290, 1998).

Ceratopycnium Clem. & Shear (1931) ≡ Ceratopycnidium.

Ceratorhiza R.T. Moore (1987), anamorphic *Ceratobasidium*. *c.* 7, widespread. See Moore (*Antonie van Leeuwenhoek* **55**: 393, 1989), Currah & Zelmer (*Rep. Tottori mycol. Inst.* **30**: 43, 1992), Roberts (*Rhizoctonia-forming fungi*, 1999).

Ceratosebacina P. Roberts (1993), ? Auriculariales. 3, Europe. See Roberts (*MR* **97**: 470, 1993), Weiss & Oberwinkler (*MR* **105**: 403, 2001; phylogeny).

Ceratosperma Speg. (1918) = Nematostoma fide von Arx & Müller (*Stud. Mycol.* **9**, 1975).

Ceratospermopsis Bat. (1951), Meliolaceae. 2 (from leaves), Brazil. See Batista (*Mycopath. Mycol. appl.* **5**: 165, 1951).

Ceratospermum P. Micheli (1729) nom. dub., ? Pezizomycotina.

Ceratosphaeria Niessl (1876), Sordariomycetes. Anamorph *Harpophora*-like. 11, widespread. See Tsuda

& Ueyama (*TMSJ* **18**: 413, 1977), Hyde *et al.* (*Nova Hedwigia* **64**: 185, 1997), Réblová (*Sydowia* **50**: 229, 1998; Czech Republic), Réblová (*Mycol.* **98**: 68, 2006; phylogeny).

Ceratosporella Höhn. (1923), anamorphic *Pezizomycotina*, Hso.2bP.1. 13, widespread. See Hughes (*TBMS* **35**: 243, 1952), Kuthubutheen & Nawawi (*MR* **95**: 159, 1991; key), Castañeda Ruíz *et al.* (*Mycotaxon* **60**: 275, 1996).

Ceratosporiaceae Nann. (1934) = Helminthosphaeriaceae.

Ceratosporium Schwein. (1832), anamorphic *Iodosphaeria*, Hso.0bP.1. 6, widespread. See Hughes (*Mycol. Pap.* **39**, 1951), Hughes (*N.Z. Jl Bot.* **2**: 305, 1965), Kirschner & Chen (*Mycol.* **96**: 917, 2004; Taiwan).

Ceratostoma Fr. (1818) nom. rej. = Melanospora Corda fide Cannon & Hawksworth (*Taxon* **32**: 476, 1983).

Ceratostoma Sacc. (1876) ? = Arxiomyces fide Hawksworth *et al.* (*Dictionary of the Fungi* edn 8, 1995).

Ceratostomataceae G. Winter (1885), Melanosporales. 12 gen. (+ 23 syn.), 63 spp.

Lit.: Doguet (*Botaniste* **39**: 1, 1955; monogr.), Cannon & Hawksworth (*J. Linn. Soc.* Bot. **84**: 115, 1982; key gen.), Horie *et al.* (*Mycotaxon* **25**: 229, 1986), Vakili (*MR* **93**: 67, 1989), Rehner & Samuels (*CJB* **73**: S816, 1995), Goh & Hanlin (*Mycol.* **90**: 655, 1998), Goh *et al.* (*Fungal Science* Taipei **13**: 1, 1998), Jones & Blackwell (*MR* **102**: 661, 1998; DNA), Stchigel *et al.* (*MR* **103**: 1305, 1999), Zhang & Blackwell (*MR* **106**: 148, 2002), Huhndorf *et al.* (*Mycol.* **96**: 368, 2004).

Ceratostomella Sacc. (1878), ? Annulatascaceae. *c.* 18 (on rotten wood etc.), widespread (esp. temperate). See Barr (*Mycotaxon* **46**: 45, 1993), Untereiner (*Mycol.* **85**: 294, 1993), Kang *et al.* (*Mycoscience* **40**: 151, 1999), Inderbitzin (*Mycoscience* **41**: 167, 2000; China), Réblová (*Mycol.* **98**: 68, 2006; phylogeny).

Ceratostomina Hansf. (1946) = Rhynchomeliola fide Müller & von Arx *in* Ainsworth *et al.* (Eds) (*The Fungi* **4A**: 87, 1973).

Ceraunium Wallr. (1833) ≡ Elaphomyces.

Cercidospora Körb. (1865), ? Dothideomycetes. 22 (on lichens), north temperate; Arctic. See Hafellner (*Herzogia* **7**: 353, 1987; key), Eriksson & Yue (*SA* **11**: 166, 1993), Hafellner & Obermayer (*Cryptog. Bryol.-Lichénol.* **16**: 177, 1995; on *Arthrorhaphis*), Zhurbenko (*Mikol. Fitopatol.* **36**: 3, 2002; Russia), Zhurbenko & Triebel (*Biblthca Lichenol.* **86**: 205, 2003; on *Lecidoma*), Ihlen & Wedin (*Lichenologist* **39**: 1, 2007; Fennoscandia, key).

Cercocladospora G.P. Agarwal & S.M. Singh (1974) = Pseudocercospora fide Deighton (*Mycol. Pap.* **140**, 1976), Crous & Braun (*CBS Diversity Ser.* **1**: 571 pp., 2003).

Cercodeuterospora Curzi (1932) = Passalora fide Deighton (*Mycol. Pap.* **137**, 1974), Crous & Braun (*CBS Diversity Ser.* **1**: 571 pp., 2003).

Cercophora Fuckel (1870), Lasiosphaeriaceae. Anamorph *Cladorrhinum*. 42, widespread. See Lundqvist (*Symb. bot. upsal.* **20** no. 1, 1972; coprophilous spp.), Hilber & Hilber (*Z. Mykol.* **45**: 209, 1979; on wood), Ueda (*Mycoscience* **35**: 287, 1994; anamorph), Réblová & Svrcek (*Czech Mycol.* **49**: 207, 1997), Hanlin (*Am. J. Bot.* **86**: 780, 1999; morphology), Miller & Huhndorf (*Sydowia* **53**: 211, 2001; neotropical spp.), Miller (*Sydowia* **55**: 267, 2003; perithecial anatomy), Huhndorf *et al.* (*Mycol.* **96**: 368, 2004; phylogeny), Miller & Huhndorf (*MR* **108**: 26, 2004; phylogeny), Miller & Huhndorf (*Mol. Phylogen. Evol.* **35**: 60, 2005; phylogeny, anatomy), Zhang *et al.* (*Mycol.* **98**: 1076, 2006; phylogeny).

Cercoseptoria Petr. (1925) = Pseudocercospora fide Deighton (*TBMS* **88**: 365, 1987), Crous & Braun (*CBS Diversity Ser.* **1**: 571 pp., 2003).

Cercosperma G. Arnaud ex B. Sutton & Hodges (1983), anamorphic *Pezizomycotina*, Hso.≡ eP.10. 2, pantropical. See Sutton & Hodges (*Nova Hedwigia* **35**: 798, 1981), Dorai & Vittal (*TBMS* **91**: 521, 1988; India).

Cercosphaerella Kleb. (1918) = Mycosphaerella fide von Arx & Müller (*Stud. Mycol.* **9**, 1975).

Cercospora Fresen. (1863) [Aug.] nom. cons. prop., anamorphic *Mycosphaerella*, Hso.0fH.31. *c.* 1416, widespread. Leaf spot of banana (*C. musae*, see *Mycosphaerella*), beet (*C. beticola*), celery (*C. apii*), groundnut (*Arachis*) (*C. arachidicola*, teleomorph *Mycosphaerella arachidis*), tobacco (*C. nicotianae*). See Chupp (*Monograph of the fungus genus Cercospora*, 1954; arranged by host fam.), Sigler *et al.* (*Mycotaxon* **10**: 133, 1979), Ooorschot (*Stud. Mycol.* **20**, 1980; key 22 spp.), Pons *et al.* (*TBMS* **85**: 405, 1985; ultrastructure of *C. beticola*), Assante *et al.* (*Riv. Patol. veg., Pavia* Sér. 4 **22**: 41, 1986; secondary metabolites), Pollack (*Mycol. Mem.* **12**, 1987; annotated compilation of species names), Chabasse (*Bull. Soc. Fr. mycol. Med.* **17**: 373, 1988; key French spp.), Pons & Sutton (*Mycol. Pap.* **160**, 1988; spp. on *Dioscorea*), Stewart & Pflegler (*Journal of the Minnesota Academy of Science* **53**: 34, 1989; synoptic key gen.), Hsieh & Goh (*Cercospora and Similar Fungi from Taiwan*, 1990; Taiwan), Sutton (*Taxon* **40**: 643, 1991; conservation), Skou (*Mycotaxon* **43**: 237, 1992; xerophilic spp., ? related to *Ascosphaeraceae*), Braun (*Cryptog. bot.* **3**: 235, 1993; subgen. *Hyalocercospora*), Braun (*Monogr. Cercosporella, Ramularia Allied Genera (Phytopath. Hyphom.)* **1**: 117 pp., 1995; subg. *Hyalocercospora*), Takizawa *et al.* (*Mycoscience* **35**: 327, 1995; ubiquinone system), Braun (*Sydowia* **48**: 205, 1996), Braun & Mel'nik (*Mikol. Fitopatol.* **30**: 1, 1996; Russian types), Pons & Sutton (*MR* **100**: 815, 1996; on *Heliotropium*), Crous *et al.* (*Mycotaxon* **64**: 405, 1997; Brazil), Crous & Câmara (*Mycotaxon* **68**: 299, 1998; Brazil), Guo (*Mycosystema* **17**: 97, 1998; China), Kobayashi *et al.* (*Mycoscience* **39**: 185, 1998; Japan), Braun (*Schlechtendalia* **2**: 1, 1999), Braun & Sivapalan (*Fungal Diversity* **3**: 1, 1999; Brunei), Crous & Corlett (*CJB* **76**: 1523, 1998; on *Platanus*), Crous *et al.* (*Mycotaxon* **72**: 171, 1999; Brazil), Kim & Shin (*Korean J. Mycol.* **27**: 220, 1999; Korea), Nakashima *et al.* (*Mycoscience* **40**: 269, 1999; Japan), Stewart *et al.* (*MR* **103**: 1491, 1999; phylogeny), Crous *et al.* (*Stud. Mycol.* **45**: 107, 2000; review), Guo & Jiang (*Mycotaxon* **74**: 257, 2000; China), Siboe *et al.* (*J. gen. appl. Microbiol.* Tokyo **46**: 69, 2000; *C. apii* complex), Crous & Braun (*Mycotaxon* **78**: 327, 2001; spp. described by Chupp), Crous *et al.* (*Mycol.* **93**: 1081, 2001; phylogeny), Goodwin *et al.* (*Phytopathology* **91**: 648, 2001; phylogeny), Braun *et al.* (*Feddes Repert.* **113**: 112, 2002; spp. described by Chupp), Crous & Braun (*CBS Diversity Ser.* **1**: 571 pp., 2003; annotated checklist), Guo *et al.* (*Flora Fungorum Sinicorum* **20**, 2003; China), Braun & Crous (*Mycotaxon* **92**: 395, 2005;

nomencl.), Groenewald *et al.* (*Phytopathology* **95**: 951, 2005; phylogeny, sp. concepts), Crous *et al.* (*Stud. Mycol.* **55**: 189, 2006; on maize), Groenewald *et al.* (*Fungal Genetics Biol.* **43**: 813, 2006; mating type genes), Groenewald *et al.* (*Mycol.* **98**: 275, 2006; host range), Schoch *et al.* (*Mycol.* **98**: 1041, 2006; phylogeny).

Cercospora Fuckel (1863) [Apr.] nom. rej. prop. = Pseudocercospora fide Sutton & Pons (*Taxon* **112**: 643, 1991).

Cercosporaceae Nann. (1934) = Mycosphaerellaceae.

Cercosporella Sacc. (1880), anamorphic *Mycosphaerella*, Hsy.0fH.10. 99, widespread. *C. pastinacae* (parsnip (*Pastinaca*) leaf spot). See Deighton (*Mycol. Pap.* **133**, 1973), Braun (*Monogr. Cercosporella, Ramularia Allied Genera (Phytopath. Hyphom.)* **1**: 61, 1995; monogr., keys), Braun (*Monogr. Cercosporella, Ramularia Allied Genera (Phytopath. Hyphom.)* **2**: 401, 1998), Pons & Sutton (*MR* **104**: 1501, 2000; conidiogenesis), Berner *et al.* (*Mycol.* **97**: 1122, 2006; spp. with biocontrol potential).

Cercosporellaceae Nann. (1934) = Mycosphaerellaceae.

Cercosporidium Earle (1901) = Passalora fide Deighton (*Mycol. Pap.* **112**, 1967; key), von Arx (*Proc. K. ned. Akad. Wet.* Ser. C, Biol. Med. Sci. **86**: 15, 1983), Castañeda Ruíz & Braun (*Cryptog. Bot.* **1**: 42, 1989), Liu *et al.* (*Acta phytopath. sin.* **28**: 43, 1998; RAPD), Crous *et al.* (*Mycol.* **93**: 1081, 2001; phylogeny), Crous & Braun (*CBS Diversity Ser.* **1**: 571 pp., 2003; annotated checklist).

Cercosporina Speg. (1911) = Cercospora Fresen. fide Hawksworth *et al.* (*Dictionary of the Fungi* edn 8, 1995).

Cercosporiopsis Miura (1928) nom. illegit. ≡ Pseudocercospora fide Carmichael *et al.* (*Genera of Hyphomycetes*, 1980).

Cercosporites E.S. Salmon (1903), Fossil Fungi. 1 (Miocene), Italy.

Cercosporites Stopes (1913), Fossil Fungi. 1 (Cretaceous), former Czechoslovakia.

Cercosporula Arnaud (1954) nom. inval., anamorphic *Chaetosphaeriales*, Hso.0fH.15. 3, France; neotropics. See Castañeda Ruíz & Arnold (*Revta Jardín bot. Nac.* Univ. Habana **6**: 47, 1985).

Cercostigmina U. Braun (1993) = Pseudocercospora fide Braun (*Cryptog. bot.* **4**: 107, 1993), Crous & Braun (*CBS Diversity Ser.* **1**: 571 pp., 2003; annotated checklist), Taylor *et al.* (*MR* **107**: 653, 2003; phylogeny of spp. on *Proteaceae*).

Cerebella Ces. (1851) = Epicoccum fide Schol-Schwarz (*TBMS* **42**: 149, 1959), Ellis (*Dematiaceous Hyphomycetes*, 1971), Pažoutová & Kolínská (*Czech Mycol.* **54**: 155, 2003).

cerebriform, brain-like; convoluted.

Cereicium Locq. (1979) = Cortinarius fide Kuyper (*in litt.*).

Cereolus (Körb.) Boistel (1903) = Stereocaulon Hoffm. fide Hawksworth *et al.* (*Dictionary of the Fungi* edn 8, 1995).

Cericium Hjortstam (1995), Cystostereaceae. 1, S. America. See Hjortstam (*Mycotaxon* **54**: 184, 1995).

Cerillum Clem. (1931) ≡ Colletomanginia.

Cerinomyces G.W. Martin (1949), Dacrymycetaceae. 12, widespread (temperate). See McNabb (*N.Z. Jl Bot.* **2**: 403, 1964; key), Donk (*Persoonia* **4**: 267, 1966), Huckfeldt & Hechler (*Zeitschrift für Mykologie* **70**: 97, 2004; Germany).

Cerinomycetaceae Jülich (1982) = Dacrymycetaceae. *Lit.*: Maekawa (*CJB* **65**: 583, 1987), Maekawa & Zang (*Mycotaxon* **61**: 343, 1997), Duhem (*Bull. trimest. Soc. mycol. Fr.* **114**: 1, 1998), Larsson *et al.* (*MR* **108**: 983, 2004).

Cerinosterus R.T. Moore (1987), anamorphic *Femsjonia*. 1, widespread. See Moore (*Stud. Mycol.* **30**: 216, 1987).

Ceriomyces Corda (1837) nom. dub. = Ptychogaster fide Patouillard (*Essai taxonomique sur les familles et les genres des Hyménomycètes*, 1900).

Ceriomyces Murrill (1909) = Boletus Fr. fide Singer (*Farlowia* **2**: 223, 1945).

Cerion Massee (1901), Rhytismataceae. 2, Australia; America (tropical). See Sherwood (*Occ. Pap. Farlow Herb. Crypt. Bot.* **15**, 1980), Johnston (*Aust. Syst. Bot.* **14**: 377, 2001; Australia).

Ceriophora Höhn. (1919), Xylariales. 1 (from *Carex* leaves), Europe. Links with the *Cainiaceae* are tentative. See Kang *et al.* (*MR* **103**: 1621, 1999).

Cerioporus Quél. (1886) = Polyporus P. Micheli ex Adans. fide Saccardo (*Syll. fung.* **5**: 1, 1887).

Ceriospora Niessl (1876), Annulatascaceae. Anamorph *Chaetoconis*. 3, widespread (temperate). See Hyde (*Sydowia* **45**: 204, 1993), Campbell *et al.* (*Mycol.* **95**: 41, 2003).

Ceriosporella A.R. Caval. (1966) ≡ Marinospora.

Ceriosporella Berl. (1894) = Lophiostoma fide von Arx & Müller (*Stud. Mycol.* **9**, 1975), Hyde (*Sydowia* **45**: 204, 1993).

Ceriosporopsis Linder (1944), Halosphaeriaceae. 7 (marine), widespread. See Jones & Zainal (*Mycotaxon* **32**: 237, 1988; key), Yusoff *et al.* (*CJB* **72**: 1550, 1994; ultrastr.), Spatafora *et al.* (*Am. J. Bot.* **85**: 1569, 1998; DNA), Chen *et al.* (*Mycol.* **91**: 84, 1999; DNA), Zhang *et al.* (*Mycol.* **98**: 1076, 2006; phylogeny).

Ceriporia Donk (1933), Phanerochaetaceae. 22, widespread. See Pieri & Rivoire (*BSMF* **113**: 193, 1984), Ryvarden & Iturriaga (*Mycol.* **95**: 1066, 2003; Venezuela), Gilbertson & Hemmes (*Mem. N. Y. bot. Gdn* **89**: 81, 2004; Hawai'i).

Ceriporiopsis Domański (1963), Phanerochaetaceae. *c.* 25, widespread. See Domański (*Acta Soc. Bot. Pol.* **32**: 731, 1963), Ryvarden & Iturriaga (*Mycol.* **95**: 1066, 2003; Venezuela).

cernuous, hanging down; drooping; nodding.

Cerocorticium Henn. (1900), Meruliaceae. 2, widespread. See Wassink (*Coolia* **50**: 21, 2007) See also *Radulomyces*.

Cerodothis Muthappa (1969), Dothideomycetes. 1, India. See Muthappa (*Mycol.* **61**: 737, 1969).

Cerophora Raf. (1808) = Lycoperdon Pers. fide Donk (*Taxon* **5**: 73, 1956).

Ceropsora B.K. Bakshi & Suj. Singh (1960), ? Coleosporiaceae. 1 (on *Picea* (*Pinaceae*)), India. See Bakshi & Singh (*CJB* **38**: 260, 1960) May belong to *Chrysomyxa*; similar to *Stilbechrysomyxa*.

Cerotelium Arthur (1906), Phakopsoraceae. *c.* 25, widespread (tropical; subtropical). *C. fici* (fig rust), *C. desmium* (cotton rust). See Ono *et al.* (*MR* **96**: 844, 1992; key), Hüseyın & Selçuk (*Pakist. J. Bot.* **36**: 203, 2004; Turkey), Hernández *et al.* (*Sydowia* **57**: 189, 2005; Guyana), Thaung (*Australasian Mycologist* **24**: 29, 2005; Burma).

Cerradoa J.F. Hennen & Y. Ono (1978) = Edythea fide Cummins & Hiratsuka (*Illustr. Gen. Rust Fungi* edn 3: 225 pp., 2003).

Cerrena Gray (1821), Polyporaceae. 4, widespread. See Westhuizen (*CJB* **41**: 1487, 1963; structure), Pegler (*The polypores [Bull. BMS Suppl.]*, 1973).

Cerrenella Murrill (1905) = Inonotus fide Pegler (*TBMS* **47**: 175, 1964).

Cesatia Rabenh. (1850) [non *Cesatia* Endl. 1838, *Umbelliferae*] = Trullula fide Saccardo (*Syll. fung.* **3**: 1, 1884).

Cesatiella Sacc. (1878), Sordariomycetidae. 1 (from dead wood), Europe. See Kohlmeyer *et al.* (*Bot. Mar.* **40**: 291, 1997), Rossman (*Stud. Mycol.* **42**, 1999).

cespitose, see caespitose.

Ceteraria Ach. (1809) ≡ Cetraria.

Cetradonia J.C. Wei & Ahti (2002), Cladoniaceae (L). 1, USA. See Wei & Ahti (*Lichenologist* **34**: 23, 2002), Miądlikowska *et al.* (*Mycol.* **98**: 1088, 2006; phylogeny), Zhou *et al.* (*J. Hattori bot. Lab.* **100**: 871, 2006; phylogeny).

Cetradoniaceae J.C. Wei & Ahti (2002) = Cladoniaceae.

Lit.: Miądlikowska *et al.* (*Mycol.* **98**: 1088, 2006; phylogeny), Zhou *et al.* (*J. Hattori bot. Lab.* **100**: 871, 2006; phylogeny).

Cetraria Ach. (1803) nom. cons., Parmeliaceae (L). 21, widespread (temperate to Arctic). See Kärnefelt (*Op. Bot.* **46**, 1979; monogr. brown fruticose spp.), Jahns & Schuster (*Beitr. Biol. Pfl.* **55**: 427, 1981; morphogenesis), Awasthi (*Bull. bot. Surv. India* **24**: 1, 1983; key, India), Kärnefelt (*Op. Bot.* **86**: 1, 1986; key, as *Coelocaulon*), Kärnefelt *et al.* (*Pl. Syst. Evol.* **183**: 113, 1992; evol. & phylog.), Kärnefelt *et al.* (*Bryologist* **96**: 394, 1993), Thell (*Cryptog. Bryol.-Lichénol.* **16**: 247, 1995; anamorphs), Thell (*Folia Cryptog. Estonica* **32**: 113, 1998), Thell & Miao (*Ann. bot. fenn.* **35**: 275, 1998; phylogeny), Randlane *et al.* (*Bryologist* **100**: 109, 1999; world list cetrarioid lichens), Thell (*Lichenologist* **31**: 441, 1999; introns), Wedin *et al.* (*MR* **103**: 1152, 1999; phylogeny), Thell *et al.* (*Folia cryptog. Estonica* **36**: 95, 2000; phylogeny), Thell *et al.* (*Mycol. Progr.* **1**: 335, 2002; phylogeny), Thell *et al.* (*Mitteilungen aus dem Institut für Allgemeine Botanik Hamburg* **30-32**: 283, 2002; phylogeny, ecology), Miądlikowska *et al.* (*Mycol.* **98**: 1088, 2006; phylogeny), Randlane & Saag (*Central European Lichens*: 75, 2006; key).

Cetrariaceae Schaer. (1850) = Parmeliaceae.

Cetrariastrum Sipman (1980), Parmeliaceae (L). 5, tropical. See Sipman (*Mycotaxon* **26**: 235, 1986), Kurokawa (*J. Jap. Bot.* **74**: 251, 1999; sorediate spp.), Divakar *et al.* (*Mol. Phylogen. Evol.* **40**: 448, 2006; phylogeny, morphology).

Cetrariella Kärnefelt & A. Thell (1993), Parmeliaceae (L). 2, widespread (northern hemisphere). See Kärnefelt *et al.* (*Bryologist* **96**: 394, 1993), Kärnefelt & Thell (*Biblthca Lichenol.* **75**: 27, 2000; posn), Thell *et al.* (*Mycol. Progr.* **1**: 335, 2002; phylogeny), Thell *et al.* (*Symb. bot. upsal.* **34** no. 1: 429, 2004; biogeography), Randlane & Saag (*Central European Lichens*: 75, 2006; key).

Cetrariomyces E.A. Thomas (1939) nom. inval. ≡ Cetraria.

Cetrariopsis Kurok. (1980), Parmeliaceae (L). 2, widespread. See Kärnefelt *et al.* (*Pl. Syst. Evol.* **183**: 113, 1992), Randlane *et al.* (*Cryptog. Bryol.-Lichénol.* **16**: 35, 1995; status).

Cetrelia W.L. Culb. & C.F. Culb. (1968), Parmeliaceae (L). 17, widespread (esp. Asia). See Culberson & Culberson (*Contr. US natn Herb.* **34**: 449, 1968), Chen (*Acta Mycol. Sin.* Suppl. **1**: 386, 1986; China), Randlane & Saag (*Khemotaksonomicheskoe Izuchenie Sporovykh Rasteniĭ i Gribov, Dostizheniya i Perspektivy Razvitiya*: 193, 1990; chemotaxonomy), Randlane *et al.* (*Ukr. bot. Zh.* **48**: 41, 1991; Ukraine), Randlane & Saag (*Nov. sist. Niz. Rast.* **28**: 118, 1992; Russia), Elix (*Flora of Australia* **55**: 33, 1994; Australia), Barbero *et al.* (*Cryptog. bot.* **5**: 28, 1995; Iberian peninsula), Thell & Miao (*Ann. bot. fenn.* **35**: 275, 1998; phylogeny), Thell *et al.* (*Mycol. Progr.* **1**: 335, 2002; phylogeny), Randlane & Saag (*Central European Lichens*: 75, 2006; key).

Cetreliopsis M.J. Lai (1980), Parmeliaceae (L). 5, Australasia. See Lumbsch (*SA* **7**: 105, 1988), Randlane *et al.* (*Cryptog. Bryol.-Lichénol.* **16**: 35, 1995), Lai & Elix (*Mycotaxon* **84**: 355, 2002; Thailand), Thell *et al.* (*Mycol. Progr.* **1**: 335, 2002; phylogeny).

Ceuthocarpon P. Karst. (1873) = Linospora fide Hawksworth *et al.* (*Dictionary of the Fungi* edn 8, 1995).

Ceuthodiplospora Died. (1912), anamorphic *Splanchnonema*, Cpd.1eH.15. 1, former Czechoslovakia. See Sutton (*Mycol.* **72**: 208, 1980).

Ceuthosira Petr. (1924), anamorphic *Pezizomycotina*, Cpd.0eH.?. 1, Europe.

Ceuthospora Fr. (1825) nom. rej. = Pyrenophora fide Hawksworth *et al.* (*Dictionary of the Fungi* edn 8, 1995).

Ceuthospora Grev. (1826) nom. cons., anamorphic *Phacidium*, St.0eH.15. 106, widespread. See Sutton (*The Coelomycetes*, 1980), DiCosmo *et al.* (*Mycotaxon* **21**: 1, 1984), Ando *et al.* (*Ann. phytopath. Soc. Japan* **55**: 391, 1989; brown zonate leaf blight of tea by *C. lauri*).

Ceuthosporella Höhn. (1929) ≡ Helhonia.

Ceuthosporella Petr. & Syd. (1923), anamorphic *Pezizomycotina*, St.0eH.?. 1, Japan.

CF (CFL), see Nomenclature, Societies and organizations.

Chaconia Juel (1897), Chaconiaceae. 7 (on dicots, esp. *Leguminosae*), widespread (tropical). See Ono & Hennen (*TMSJ* **24**: 369, 1983), Hennen *et al.* (*Catalogue of the Species of Plant Rust Fungi (Uredinales) of Brazil*: 2005, 2005; Brazil spp.).

Chaconiaceae Cummins & Y. Hirats. (1983), Pucciniales. 8 gen. (+ 8 syn.), 75 spp.

Lit.: Dai & Shen (*Mycotaxon* **48**: 193, 1993), Evans (*Mycopathologia* **124**: 163, 1993), Ono & Harada (*Mycoscience* **35**: 179, 1994), Ono *et al.* (*Cryptogams of the Himalayas* **3**. Nepal and Pakistan: 69, 1995), Payak (*Indian Phytopath.* **49**: 307, 1996), Berndt (*Mycol.* **91**: 1045, 1999), Cummins & Hiratsuka (*Illustr. Gen. Rust Fungi* edn 3: 225 pp., 2003), Maier *et al.* (*CJB* **81**: 12, 2003), Wingfield *et al.* (*Australas. Pl. Path.* **33**: 327, 2004), Ritschel (*Biblthca Mycol.* **200**: 132 pp., 2005), Aime (*Mycoscience* **47**: 112, 2006).

Chadefaudia Feldm.-Maz. (1957), Halosphaeriaceae (?L). 6 (marine algae), widespread. See Kohlmeyer & Kohlmeyer (*Marine Mycology*, 1979), Stegena & Kemperman (*Bot. Mar.* **27**: 443, 1984; S. Afr.), Nakagiri & Ito (*Mycol.* **89**: 484, 1997; contrast with *Retrostium*).

Chadefaudiella Faurel & Schotter (1959), Chadefaudiellaceae. 2 (coprophilous), N. Africa. See Faurel & Schotter (*Revue Mycol.* Paris **30**: 330, 1966).

Chadefaudiellaceae Faurel & Schotter ex Benny & Kimbr. (1980), Microascales. 2 gen., 3 spp.

Lit.: Valldosera *et al.* (*Mycotaxon* **30**: 5, 1987), von Arx *et al.* (*Beih. Nova Hedwigia* **94**: 104 pp., 1988), von Arx *et al.* (*Beih. Nova Hedwigia* **94**: 104, 1988).

Chadefaudiomyces Kamat, V.G. Rao, A.S. Patil & Ullasa (1974), Diaporthales. Anamorph *Lasmenia*. 1, India. See Kamat *et al.* (*Revue Mycol.* Paris **38**: 19, 1973).

Chaenocarpus Lév. (1843) nom. dub. = Chaenocarpus Rebent. fide Læssøe (*SA* **13**: 43, 1994).

Chaenocarpus Rebent. (1804) nom. dub., Xylariaceae. 1, Europe. See Dennis (*Revta Biol.* Lisb. **1**: 175, 1958), Læssøe (*SA* **13**: 43, 1994).

Chaenocarpus Spreng. (1831) = Chaenocarpus Rebent. fide Læssøe (*SA* **13**: 43, 1994).

Chaenotheca (Th. Fr.) Th. Fr. (1860), Coniocybaceae (L). 25, widespread (esp. temperate). See Tibell (*Symb. bot. upsal.* **23** no. 1, 1980; key 14 N. Hemisph. spp.), Tibell (*Nova Hedwigia* **79**: 597, 1984; gen. concept), Honegger (*Sydowia* **38**: 146, 1986; asci), Middelborg & Mattsson (*Sommerfeltia* **5**, 1987; Norway), Tibell (*Nordic Jl Bot.* **13**: 441, 1993; anamorphs), Tibell (*Fl. Neotrop.* Monogr. **69**: 78 pp., 1996; neotropical spp.), Tibell (*Symb. bot. upsal.* **32** no. 1: 291, 1997; anamorphs), Tibell (*Biblthca Lichenol.* **71**: 107 pp., 1998; S America), Tibell (*Lichenologist* **33**: 519, 2001; phylogeny, synonymy), Tibell (*Bryologist* **104**: 191, 2001; phylogeny, photobionts), Tibell (*Ann. bot. fenn.* **39**: 73, 2002; morphology, phylogeny), Tibell & Koffman (*Bryologist* **105**: 353, 2002; N America), Aptroot & Tibell (*Australasian Lichenology* **52**: 12, 2003; Papua New Guinea), Rikkinen (*Acta Bot. Fenn.* **175**: 1, 2003; Oregon), Titov *et al.* (*Symb. bot. upsal.* **34** no. 1: 455, 2004; Russia).

Chaenotheciella Räsänen (1943) ? = Chaenothecopsis fide Hawksworth *et al.* (*Dictionary of the Fungi* edn 8, 1995).

Chaenothecomyces Cif. & Tomas. (1953) = Chaenotheca.

Chaenothecopsis Vain. (1927), Mycocaliciaceae. Anamorphs *Asterophoma*, *Catenomycopsis*, *Phialophora*-like. 62 (on lichens, other fungi or wood), Europe. See Schmidt (*Mitt. Inst. Allg. Botanik Hamburg* **13**: 111, 1970), Hawksworth (*TBMS* **74**: 650, 1980), Tibell (*Beih. Nova Hedwigia* **79**: 597, 1984), Tibell (*Symb. bot. upsal.* **27** no. 1, 1987; status, Austral. spp.), Tibell (*CJB* **69**: 2427, 1991; anamorphs), Tibell & Constantinescu (*MR* **95**: 556, 1991; anamorph), Tibell & Ryman (*Nova Hedwigia* **60**: 199, 1995; key 8 spp.), Rikkinen (*Bryologist* **102**: 366, 1999; resinicolous spp.), Rikkinen & Poinar (*MR* **104**: 7, 2000; fossil sp.), Rikkinen (*Mycol.* **95**: 1032, 2003; in amber), Rikkinen (*Mycol.* **95**: 98, 2003; on resin), Tibell & Vinuesa (*Taxon* **54**: 427, 2005; phylogeny), Geiser *et al.* (*Mycol.* **98**: 1053, 2006; phylogeny), Groner (*Lichenologist* **38**: 395, 2006; Switzerland).

Chaetalysis Peyronel (1922) = Acarosporium fide Sutton (*Mycol. Pap.* **141**, 1977).

Chaetantromycopsis H.P. Upadhyay, Cavalc. & A.A. Silva (1986), anamorphic *Pezizomycotina*, Hsy.0eP.15. 1, Brazil. See Upadhyay *et al.* (*Mycol.* **78**: 493, 1986).

Chaetapiospora Petr. (1947) = Pseudomassaria fide Barr (*Mycol.* **68**, 1976).

Chaetarthriomyces Thaxt. (1931), Laboulbeniaceae. 3, widespread. See Scheloske (*Parasitol. Schr. Reihe* **19**: 97, 1969), Majewski (*Acta Mycologica* Warszawa **34**: 7, 1999; Poland), Shen *et al.* (*Mycosystema* **22**: 157, 2003; China).

Chaetasbolisia Speg. (1918), anamorphic *Pezizomycotina*, Cpd.0eH.15. 7, widespread. See Patel *et al.* (*MR* **101**: 335, 1997; India).

Chaetaspis Syd. & P. Syd. (1917) = Rhagadolobium fide Müller & von Arx (*Beitr. Kryptfl. Schweiz* **11** no. 2, 1962).

Chaetasterina Bubák (1909) = Chaetothyrium fide Hawksworth *et al.* (*Dictionary of the Fungi* edn 8, 1995).

Chaetendophragmia Matsush. (1971), anamorphic *Pezizomycotina*, Hso.≡ eP.19. 5, widespread. See Matsushima (*Microfungi of the Solomon Islands and Papua-New Guinea*: 12, 1971), Carris (*MR* **99**: 667, 1995; synanamorph), Wu & Zhuang (*Fungal Diversity Res. Ser.* **15**, 2005).

Chaetendophragmiopsis B. Sutton & Hodges (1978), anamorphic *Pezizomycotina*, Hso.≡ eP.19. 2, pantropical. See Sutton & Hodges (*Nova Hedwigia* **29**: 596, 1978).

Chaethomites, see *Chaetomites*.

Chaetoamphisphaeria Hara (1918), Pezizomycotina. 1, Japan.

Chaetobasidiella Höhn. (1918) nom. dub., anamorphic *Pezizomycotina*. See Sutton (*Mycol. Pap.* **141**, 1977).

Chaetobasis Clem. & Shear (1931) ≡ Chaetobasidiella.

Chaetoblastophorum Morgan-Jones (1977), anamorphic *Pezizomycotina*, Hso.0eH.1. 1, USA. See Morgan-Jones (*Mycotaxon* **5**: 484, 1977).

Chaetobotrys Clem. (1931) ≡ Kusanobotrys.

Chaetocalathus Singer (1943), Marasmiaceae. *c.* 20, widespread (predominantly subtropical). See Singer (*Lilloa* **8**: 441, 1942; monogr.), Singer (*Fl. Neotrop.* **17**: 53, 1976; key 6 neotrop. spp.).

Chaetocarpus P. Karst. (1889) [non *Chaetocarpus* Thwaites 1849) nom. cons., *Euphorbiaceae*] = Veluticeps fide Donk (*Persoonia* **3**: 199, 1964).

Chaetoceratostoma Turconi & Maffei (1912) = Scopinella fide Hawksworth (*TBMS* **64**: 447, 1975).

Chaetoceris Clem. & Shear (1931) ≡ Chaetoceratostoma.

Chaetochalara B. Sutton & Piroz. (1965) = Chalara fide Kirk *in* Kirk & Spooner (*Kew Bull.* **38**: 579, 1984).

Chaetocladiaceae A. Fisch. (1892) = Mucoraceae.
Lit.: Kirk (*in litt.*).

Chaetocladium Fresen. (1863), Mucoraceae. 2 (facultative parasites of *Mucorales*), widespread. See Benny & Benjamin (*Aliso* **8**: 391, 1976), Voigt & Wöstemeyer (*Gene* **270**: 113, 2001; phylogeny), Wöstemeyer & Schimek *in* Heitman *et al.* (Eds.) (*Sex in Fungi*: 431, 2007; trisporic acid & mating).

Chaetoconidium Zukal (1887), anamorphic *Pezizomycotina*, Hso.?.?. 1, Europe.

Chaetoconis Clem. (1909), anamorphic *Ceriospora*, Cpd.≡ eH.15. 2, widespread (north temperate). See Sutton (*CJB* **46**: 183, 1968), Punithalingham (*Nova Hedwigia* **40**: 99, 1984; appendages), Nag Raj (*Coelomycetous Anamorphs with Appendage-bearing Conidia*: 188, 1993).

Chaetocrea Syd. (1927), Tubeufiaceae. 1 (fungicolous), widespread (neotropics). See Rossman *et al.* (*Stud. Mycol.* **42**: 248 pp., 1999).

Chaetocypha Corda (1829) nom. dub., Fungi. See Donk (*Reinwardtia* **1**: 208, 1951).

Chaetocytostroma Petr. (1920), anamorphic *Pezizomycotina*, St.0eH.?. 1, Europe.

Chaetoderma Parmasto (1968), Stereaceae. 2, Europe. See Parmasto (*Consp. System. Corticiac.*: 86, 1968).
Chaetodermataceae Jülich (1982) = Stereaceae.
Chaetodermella Rauschert (1988), Gloeophyllaceae. 1, Europe. See Rauschert (*Haussknechtia* **4**: 52, 1988).
Chaetodimerina Hansf. (1946) = Rizalia fide Pirozynski (*Kew Bull.* **31**: 595, 1977).
Chaetodiplis Clem. (1931), anamorphic *Pezizomycotina*, Cpd.0eP.?. 1, Europe.
Chaetodiplodia P. Karst. (1884), anamorphic *Pezizomycotina*, Cpd.1eP.15. 9, widespread (esp. tropical). See Zambettakis (*BSMF* **70**: 219, 1954).
Chaetodiplodina Speg. (1910), anamorphic *Pezizomycotina*, Cpd.0eP.1. 2, S. America. See Petrak & Sydow (*Annls mycol.* **33**: 181, 1935).
Chaetodiscula Bubák & Kabát (1910) = Hymenopsis fide Sutton (*Mycol. Pap.* **141**, 1977).
Chaetodochis Clem. (1909) ? = Chaetostroma fide Hawksworth *et al.* (*Dictionary of the Fungi* edn 8, 1995).
Chaetodochium Höhn. (1932) = Volutella fide Tulloch (*Mycol. Pap.* **130**, 1972).
Chaetolentomita Maubl. (1915) = Chaetosphaeria fide Müller & von Arx (*Beitr. Kryptfl. Schweiz* **11** no. 2, 1962).
Chaetomastia (Sacc.) Berl. (1891), ? Dacampiaceae. 10, widespread. See Barr (*Mycotaxon* **34**: 507, 1989; key N. Am. spp.), Eriksson & Hawksworth (*SA* **8**: 64, 1989; posn), Barr (*Mycotaxon* **82**: 373, 2002; definition).
Chaetomelanops Petr. (1948) = Pyrenostigme fide von Arx & Müller (*Stud. Mycol.* **9**, 1975).
Chaetomelasmia Danilova (1951) = Diachorella fide Sutton (*Mycol. Pap.* **141**, 1977).
Chaetomeliola (Cif.) Bat., H. Maia & M.L. Farr (1962) = Meliola fide Hughes (*in litt.*).
Chaetomella Fuckel (1870), anamorphic *Leotiomycetes*, Cpd.0eH.15. 5 (esp. in soil), widespread. See Sutton & Sarbhoy (*TBMS* **66**: 297, 1976), Vobis *et al.* (*Boln Soc. argent. Bot.* **28**: 205, 1992; Argentina), Rossman *et al.* (*Mycol. Progr.* **3**: 275, 2004; phylogeny).
Chaetomeris Clem. (1931) ≡ Treubiomyces.
Chaetomiaceae G. Winter (1885), Sordariales. 16 gen. (+ 8 syn.), 205 spp. The family appears to be paraphyletic.

Lit.: Udagawa *in* Kurato & Ueno (Eds) (*Toxigenic Fungi: their toxins and health hazard*: 139, 1984), Cannon (*TBMS* **87**: 45, 1986), von Arx *et al.* (*Beih. Nova Hedwigia* **84**: 162 pp., 1986), von Arx *et al.* (*Beih. Nova Hedwigia* **94**: 104 pp., 1988), von Arx *et al.* (*Beih. Nova Hedwigia* **94**, 1988), Domsch *et al.* (*Compendium of Soil Fungi and Supplement*: 406 pp., 1993), Mouchacca (*Cryptog. Mycol.* **18**: 19, 1997), Silva & Hanlin (*Mycoscience* **37**: 261, 1996), Lee & Hanlin (*Mycol.* **91**: 434, 1999; DNA), Stchigel *et al.* (*MR* **104**: 879, 2000), Untereiner *et al.* (*CJB* **79**: 321, 2001), Stchigel *et al.* (*MR* **106**: 975, 2002), Huhndorf *et al.* (*Mycol.* **96**: 368, 2004), Lumbsch & Huhndorf (*MR* **111**: 1064, 2007).

Chaetomiales = Sordariales.
Chaetomidium (Zopf) Sacc. (1882), Chaetomiaceae. Anamorph *Phialophora*-like. 16 (mainly from soil or coprophilous), widespread. See von Arx (*Stud. Mycol.* **8**, 1975), von Arx (*Beih. Nova Hedwigia* **94**, 1988; keys), Silva & Hanlin (*Mycoscience* **37**: 261, 1997; key), Stchigel *et al.* (*Stud. Mycol.* **50**: 215, 2004).
Chaetomiopsis Mustafa & Abdul-Wahid (1990), Chaetomiaceae. 1 (from soil), Egypt. See Moustafa & Abdul-Wahid (*Mycol.* **82**: 129, 1990).
Chaetomiotricha Peyronel (1914) = Chaetomium fide Hawksworth *et al.* (*Dictionary of the Fungi* edn 8, 1995).
Chaetomites Pampal. (1902), Fossil Fungi. 1 (Tertiary), Italy.
Chaetomium Kunze (1817), Chaetomiaceae. Anamorph *Botryotrichum*. 95 (coprophilous, on straw, wet paper, cloth, cotton fibres (esp. *C. globosum*) many cellulolytic, some thermophilic, some mycotoxic), widespread. See Hawksworth & Wells (*Mycol. Pap.* **134**, 1973; ornamentation terminal hairs), Millner *et al.* (*Mycol.* **69**: 720, 1977; ascospores SEM), Wicklow (*TBMS* **72**: 107, 1979; ecology and hair types), Udagawa *in* Kurato & Ueno (Eds) (*Toxigenic Fungi: their toxins and health hazard*: 139, 1984; toxins), Cannon (*TBMS* **87**: 50, 1986; key spp. with inconspicuous hairs), von Arx *et al.* (*Beih. Nova Hedwigia* **84**, 1986; keys), Guarro & Figueras (*Cryptog. Bot.* **1**: 97, 1989; ontogeny), Koyama (*Proc. Jap. Assoc. Mycotoxic.* **35**: 7, 1992; mycotoxins), Lee & Hanlin (*Mycol.* **91**: 434, 1999; DNA), Untereiner *et al.* (*CJB* **79**: 321, 2001; phylogeny), Hsieh & Hu (*Taiwania* **47**: 264, 2002; Taiwan), Barron *et al.* (*J. Clin. Microbiol.* **41**: 5302, 2003; mycosis), Sun *et al.* (*Mycosystema* **23**: 333, 2004; China), Sun *et al.* (*Mycosystema* **24**: 318, 2005; China), Wang & Zheng (*Nova Hedwigia* **81**: 247, 2005; China), Kubátová (*Czech Mycol.* **58**: 155, 2006; Czech Republic), Zhang *et al.* (*Mycol.* **98**: 1076, 2006; phylogeny), Chang & Wang (*Taiwania* **53**: 85, 2008; key Taiwan spp.).
Chaetomonodorus Bat. & H. Maia (1961) = Microtheliopsis fide Lücking *et al.* (*Lichenologist* **30**: 121, 1998).
Chaetomyces Thaxt. (1893), Laboulbeniaceae. 1, N. & S. America. See Nannfeldt (*Svensk bot. Tidskr.* **43**: 468, 1949).
Chaetonaemosphaera Schwarzman (1968) = Ceratocystis fide Sutton (*Mycol. Pap.* **141**, 1977).
Chaetonaevia Arx (1951), Dermateaceae. 3, Europe. See Dennis (*Kew Bull.* **52**: 451, 1997), Nauta & Spooner (*Mycologist* **13**: 65, 1999; UK).
Chaetonectrioides Matsush. (1996), Dothideomycetes. Anamorph *Mirandina*. 1, Malaysia. See Matsushima (*Matsush. Mycol. Mem.* **9**: 5, 1996).
Chaetopatella I. Hino & Katum. (1958) = Pseudolachnea Ranoj. fide Sutton (*Mycol. Pap.* **141**, 1977).
Chaetopeltaster Katum. (1975), anamorphic *Pezizomycotina*, Cpt.0eH.?. 1, Japan. See Katumoto (*Bulletin of the Faculty of Agriculture, Yamaguchi University* **26**: 98, 1975).
Chaetopeltiopsis Hara (1913) nom. dub., anamorphic *Pezizomycotina*. 1, Japan.
Chaetopeltis Sacc. (1898) [non *Chaetopeltis* Berthold 1878, *Algae*] ≡ Tassia.
Chaetopeltopsis Theiss. (1913) = Chaetothyrina fide von Arx & Müller (*Stud. Mycol.* **9**, 1975).
Chaetophiophoma Speg. (1910), anamorphic *Pezizomycotina*, Cpd.0fH.1. 1, S. America.
Chaetophoma Cooke (1878), anamorphic *Pezizomycotina*, Cpd.0eH.15. 30, widespread. See Hughes (*Mycol.* **68**: 693, 1976).
Chaetophomella Speg. (1918) nom. conf., anamorphic

Pezizomycotina. See Sutton (*Mycol. Pap.* **141**, 1977).

Chaetophorites Pratje (1922), Fossil Fungi (mycel.) ? Fungi. 1 (Silurian, Jurassic, Tertiary), widespread.

Chaetoplaca Syd. & P. Syd. (1917), ? Schizothyriaceae. 1, Philippines.

Chaetoplea (Sacc.) Clem. (1931), Leptosphaeriaceae. Anamorph *Parahendersonia*. 17, widespread (esp. north temperate). See Barr (*Mem. N. Y. bot. Gdn* **62**: 1, 1992; key), Yuan & Barr (*Mycotaxon* **52**: 495, 1994), Eriksson & Hawksworth (*SA* **14**: 48, 1995; posn).

Chaetoporellaceae Jülich (1982) = Schizoporaceae.

Chaetoporellus Bondartsev & Singer (1944) nom. rej. = Hyphodontia fide Langer (*Biblthca Mycol.* **158**, 1994).

Chaetoporus P. Karst. (1890) = Junghuhnia fide Ryvarden (*Persoonia* **7**: 17, 1972).

Chaetopotius Bat. (1951) = Aithaloderma fide von Arx & Müller (*Stud. Mycol.* **9**, 1975).

Chaetopreussia Locq.-Lin. (1977), Sporormiaceae. 1 (coprophilous), Sahara. See Locquin-Linard (*Revue Mycol.* Paris **41**: 185, 1977), Barr (*Mycotaxon* **76**: 105, 2000).

Chaetopsella Höhn. (1930) ≡ Chaetopsis fide Hughes (*TBMS* **34**: 551, 1951).

Chaetopsina Rambelli (1956), anamorphic *Cosmospora*, Hso.0eH.15. 14, widespread. See Kirk & Sutton (*TBMS* **85**: 709, 1985; key), Samuels (*Mycotaxon* **22**: 13, 1985; teleomorphs), Rambelli (*Micol. Ital.* **16**: 7, 1987; bibliogr.), Onofri & Zucconi (*Mycotaxon* **41**: 451, 1991; SEM *C. fulva*), Goh & Hyde (*MR* **101**: 1517, 1997), Okada *et al.* (*Mycoscience* **38**: 409, 1997; phylogeny).

Chaetopsis Grev. (1825), anamorphic *Pezizomycotina*, Hso.1eP.15. 7, widespread. See DiCosmo *et al.* (*Mycol.* **75**: 949, 1983; related genera).

Chaetopyrena Pass. (1881), anamorphic *Pezizomycotina*. 1 or 2, Europe.

Chaetopyrena Sacc. (1883) ≡ Chaetopyrenis.

Chaetopyrenis Clem. & Shear (1931) = Keissleriella fide Hawksworth *et al.* (*Dictionary of the Fungi* edn 8, 1995).

Chaetosaccardinula Bat. (1962) = Strigopodia fide von Arx & Müller (*Stud. Mycol.* **9**, 1975).

Chaetosartorya Subram. (1972), Trichocomaceae. Anamorph *Aspergillus*. 3, Costa Rica. See Subramanian & Rajendran (*Revue Mycol.* Paris **43**: 193, 1979; ontogeny), Kuraishi *et al.* (*NATO ASI Series A: Life Sciences* **185**: 407, 1990; ubiquinones), Chang *et al.* (*J. gen. appl. Microbiol.* Tokyo **37**: 289, 1991; DNA), Peterson (*Integration of Modern Taxonomic Methods for Penicillium and Aspergillus Classification*: 323, 2000; phylogeny), Tamura *et al.* (*Integration of Modern Taxonomic Methods for Penicillium and Aspergillus Classification*: 357, 2000; phylogeny).

Chaetosclerophoma Petr. (1924), anamorphic *Pezizomycotina*, St.0eH.?. 1, Europe.

Chaetoscorias W. Yamam. (1955) = Phragmocapnias fide Hughes (*Mycol.* **68**: 693, 1976).

Chaetoscutula E. Müll. (1959), Dothideomycetes. 1, France; Scotland. See Müller (*Sydowia* **12**: 190, 1958).

Chaetoscypha Syd. (1924) ? = Lachnum fide Hawksworth *et al.* (*Dictionary of the Fungi* edn 8, 1995).

Chaetoseptoria Tehon (1937), anamorphic *Pezizomycotina*, Cpd.0fH.?. 1, N. America. See Sutton (*Mycol. Pap.* **97**, 1964).

Chaetosira Clem. (1931) ≡ Wiesneriomyces.

Chaetospermella Chardón & Toro (1934) ≡ Spermochaetella.

Chaetospermella Naumov (1929) = Chaetospermum fide Sutton (*Mycol. Pap.* **141**, 1977).

Chaetospermopsis Katum. & Y. Harada (1979) = Amphichaetella fide Katumoto & Harada (*TMSJ* **20**: 424, 1979), Alcorn (*Australas. Mycol.* **21**: 111, 2002).

Chaetospermum Sacc. (1892), anamorphic *Pezizomycotina*, St.0eH.1. 5, widespread. See Sutton (*The Coelomycetes*, 1980), Muntañola-Cvetkovic & Gómez-Bolea (*Mycotaxon* **47**: 59, 1993; Iberian peninsula).

Chaetosphaerella E. Müll. & C. Booth (1972), Chaetosphaerellaceae. Anamorphs *Oedemium*, *Veramycina*. 2 (from wood etc.), Europe. See Réblová (*Mycotaxon* **70**: 387, 1999), Huhndorf *et al.* (*Mycol.* **96**: 368, 2004; phylogeny), Zhang *et al.* (*Mycol.* **98**: 1076, 2006; phylogeny).

Chaetosphaerellaceae Huhndorf, A.N. Mill. & F.A. Fernández (2004), Coronophorales. 5 gen. (+ 7 syn.), 8 spp.

Lit.: Réblová (*Mycotaxon* **70**: 387, 1999), Réblová (*Mycotaxon* **71**: 45, 1999), Huhndorf *et al.* (*Mycol.* **96**: 368, 2004), Huhndorf *et al.* (*MR* **108**: 1384, 2004).

Chaetosphaeria Tul. & C. Tul. (1863), Chaetosphaeriaceae. Anamorphs very diverse, including *Cacumisporium*, *Catenularia*, *Chalara*-like, *Chloridium*, *Codinaea*, *Craspedodidymum*, *Cryptophiale*, *Cylindrotrichum*, *Dictyochaeta*, *Exserticlava*, *Fusichalara*, *Gonytrichum*, *Menispora*, *Phaeostalagmus*, *Phialophora*-like and *Zanclospora*. 68, widespread. See Booth (*Mycol. Pap.* **68**, 1957), Gams & Holubová-Jechová (*Stud. Mycol.* **13**, 1976), Barr & Crane (*CJB* **57**: 835, 1979; anamorph), Barr (*Mycotaxon* **39**: 43, 1990; posn), Réblová (*Stud. Mycol.* **45**: 149, 2000), Réblová & Gams (*Mycoscience* **41**: 129, 2000), Réblová & Winka (*Mycol.* **92**: 939, 2000; phylogeny), Huhndorf *et al.* (*Mycol.* **93**: 1072, 2001), Réblová & Seifert (*Sydowia* **55**: 313, 2003; Thailand), Huhndorf *et al.* (*Mycol.* **96**: 368, 2004; phylogeny), Réblová (*Stud. Mycol.* **50**: 171, 2004; New Zealand), Fernández & Huhndorf (*Fungal Diversity* **18**: 15, 2005), Huhndorf & Fernández (*Fungal Diversity* **19**: 23, 2005; anamorph connexions), Fernández *et al.* (*Mycol.* **98**: 121, 2006; morphology, phylogeny), Réblová *et al.* (*MR* **110**: 104, 2006; *Menispora* anam.), Zhang *et al.* (*Mycol.* **98**: 1076, 2006; phylogeny), Atkinson *et al.* (*N.Z. Jl Bot.* **45**: 685, 2007; New Zealand).

Chaetosphaeriaceae Réblová, M.E. Barr & Samuels (1999), Chaetosphaeriales. 28 gen. (+ 33 syn.), 306 spp.

Lit.: Gams & Holubová-Jechová (*Stud. Mycol.* **13**, 1976), Barr (*Mycotaxon* **39**: 43, 1990), Kuthubutheen & Nawawi (*MR* **95**: 1224, 1991), Réblová (*Czech Mycol.* **50**: 73, 1997), Samuels *et al.* (*Mycol.* **89**: 156, 1997), Hyde *et al.* (*MR* **103**: 1432, 1999), Réblová (*Sydowia* **51**: 210, 1999), Réblová (*Mycotaxon* **71**: 45, 1999), Réblová *et al.* (*Sydowia* **51**: 49, 1999), Gams (*Stud. Mycol.* **45**: 192, 2000), Réblová (*Stud. Mycol.* **45**: 149, 2000), Réblová & Winka (*Mycol.* **92**: 939, 2000; phylogeny), Koster *et al.* (*CJB* **81**: 633, 2003), Huhndorf *et al.* (*Mycol.* **96**: 368, 2004), Fernández *et al.* (*Mycol.* **98**: 121, 2006).

Chaetosphaeriales Huhndorf, A.N. Mill. & F.A.

Fernández (2004). Sordariomycetidae. 1 fam., 31 gen., 311 spp. Fam.:
Chaetosphaeriaceae
For *Lit.* see under fam.

Chaetosphaerides Matsush. (2003), ? Trichosphaeriales. 1, Japan. See Matsushima (*Matsush. Mycol. Mem.* **10**: 146, 2001), Réblová (*in litt.*, 2008).

Chaetosphaerites Félix (1894), Fossil Fungi. 3 (Carboniferous, Tertiary), widespread.

Chaetosphaeronema Moesz (1915), anamorphic *Pezizomycotina*, Cpd.1eH.15. 3, widespread. See Punithalingam & Spooner (*Kew Bull.* **57**: 534, 2002; UK).

Chaetosphaeropsis Curzi & Barbaini (1927) = Coniothyriopsis Speg. fide Clements & Shear (*Gen. Fung.*, 1931), Sutton (*Mycol. Pap.* **141**, 1977).

Chaetosphaerulina I. Hino (1938), Tubeufiaceae. Anamorph *Xenosporium*. 5, S. & S.E. Asia. See Sivanesan (*TBMS* **81**: 325, 1983; key, as *Thaxteriellopsis*), Rossman (*Mycol. Pap.* **157**, 1987; synonym of *Herpotrichia*), Crane *et al.* (*CJB* **76**: 602, 1998).

Chaetospora Faurel & Schotter (1965) [non *Chaetospora* R. Br. 1918, *Cyperaceae*] = Neochaetospora fide Sutton & Sankaran (*MR* **95**: 1021, 1991).

Chaetosporium Corda (1833) nom. dub., Fungi. See Saccardo (*Syll. fung.* **4**: 761, 1886).

Chaetosticta Petr. & Syd. (1925), anamorphic *Lasiostemma*, Cpd.≡ eH.15. 3, N. America; Japan. See Crane (*CJB* **49**: 31, 1971), Matsushima (*Matsush. Mycol. Mem.* **10**, 2001).

Chaetostigme Syd. & P. Syd. (1917) = Wentiomyces fide Müller & von Arx (*Beitr. Kryptfl. Schweiz* **11** no. 2, 1962).

Chaetostigmella Syd. & P. Syd. (1917) = Phaeodimeriella Speg. fide Müller & von Arx (*Beitr. Kryptfl. Schweiz* **11** no. 2, 1962).

Chaetostroma Corda (1829) nom. conf., nom. dub., anamorphic *Pezizomycotina*, Hsp.?.?. 10, widespread. See Tulloch (*Mycol. Pap.* **130**, 1972), Holubová-Jechová (*Sydowia* **46**: 238, 1994).

Chaetostromella P. Karst. (1895) nom. conf., anamorphic *Pezizomycotina*. See Petrak (*Sydowia* **7**: 299, 1953).

Chaetostylum Tiegh. & G. Le Monn. (1873) = Helicostylum fide Lythgoe (*TBMS* **41**: 135, 1958), Upadhyay (*Mycol.* **65**: 735, 1973).

Chaetotheca Zukal (1890), ? Trichocomaceae. 1, Europe.

Chaetothiersia B.A. Perry & Pfister (2008), Pyronemataceae. 1, USA. See Perry & Pfister (*Fungal Diversity* **28**: 65, 2008).

Chaetothyriaceae Hansf. ex M.E. Barr (1979), Chaetothyriales. 13 gen. (+ 27 syn.), 98 spp.
Lit.: Batista & Ciferri (*Sydowia* Beih. **3**, 1962), Eriksson & Yue (*Mycotaxon* **22**: 269, 1985), Pohlad (*Mycol.* **80**: 757, 1988), Constantinescu *et al.* (*Stud. Mycol.* **31**: 69, 1989), Panwar & Jagtap (*Geobios* New Rep. **9**: 121, 1990), Winka *et al.* (*Mycol.* **90**: 822, 1998), Haase *et al.* (*Stud. Mycol.* **43**: 80, 1999), Cannon (*IMI Descr. Fungi Bact.* **141**: [22] pp., 1999).

Chaetothyriales M.E. Barr (1987). Chaetothyriomycetidae. 3 fam., 37 gen., 213 spp. Mycelium varied, if external with narrow cylindrical brown hyphae, sometimes with setose appendages. Ascomata erumpent or superficial, sometimes formed beneath a subiculum, spherical or flattened, sometimes setose, often collapsing when dry, the apex ± papillate, with a well-developed periphysate ostiole; peridium thin-walled, composed of compressed pseudoparenchymatous cells, varied in pigmentation; hymenium usually J+; interascal tissue of short apical periphysoids. Asci saccate to clavate, fissitunicate, the inner wall layer often conspicuously thickened in the apical region, sometimes polysporous; ascospores hyaline or pale greyish, transversely septate or muriform. Anamorphs hyphomycetous, sometimes yeast-like (black yeasts, see yeasts). Epiphytic or biotrophic on leaves or saprobic on plants or other fungi, cosmop. Fams.:
(1) **Chaetothyriaceae**
(2) **Coccodiniaceae**
(3) **Herpotrichiellaceae**
For *Lit.* see under fam.

Chaetothyrina Theiss. (1913), Micropeltidaceae. 7, widespread (tropical). *C. musarum* (sooty blotch of banana). See Bitancourt (*Arq. Inst. biol., S. Paulo* **7**: 5, 1936), Panwar *et al.* (*Geobios* New Rep. **10**: 41, 1991), Reynolds & Gilbert (*Aust. Syst. Bot.* **18**: 265, 2005; key).

Chaetothyriolum Speg. (1919) nom. dub., anamorphic *Pezizomycotina*. See Santesson (*Symb. bot. upsal.* **12** no. 1: 1, 1952).

Chaetothyriomycetes O.E. Erikss. & Winka (1997) = Eurotiomycetes. See *Chaetothyriomycetidae*
For *Lit.* see ord. and fam.

Chaetothyriomycetidae Doweld (2001), Eurotiomycetes. Ascomata perithecial, superficial or immersed within a thallus. Asci usually thick-walled and fissitunicate, rarely evanescent, sometimes accompanied by pseudoparaphyses. Ascospores variable in pigmentation and septation. Lichenized, parasitic (esp. on other fungi) or saprobic.
The number of families contained within this subclass continues to increase based on new molecular data. Ord.:
(1) **Chaetothyriales**
(2) **Pyrenulales**
(3) **Verrucariales**
Lit.: Geiser *et al.* (*Mycol.* **98**: 1051, 2006; phylogeny), Liu *et al.* (*Mol. Biol. Evol.* **16**: 1799, 1999; RNA polymerase phylogeny), Sterflinger *et al.* (*Stud. Mycol.* **43**: 5, 1999; phylogeny and ecology), Untereiner *et al.* (*MR* **99**: 897, 1995), Winka *et al.* (*Mycol.* **90**: 822, 1998; rDNA phylogeny).

Chaetothyriopsis F. Stevens & Dorman (1927) ? = Actinopeltis fide Spooner & Kirk (*MR* **94**: 223, 1990), Eriksson & Hawksworth (*SA* **9**: 6, 1991; status).

Chaetothyrium Speg. (1888), Chaetothyriaceae. Anamorph *Merismella*. 26, widespread (esp. tropical). See Hansford (*Mycol. Pap.* **15**, 1946), Panwar & Jagtap (*Geobios* New Rep. **9**: 121, 1990; India), Reynolds & Gilbert (*Aust. Syst. Bot.* **18**: 265, 2005).

Chaetotrichum Rabenh. (1844) nom. dub. ≡ Chaetosporium.

Chaetotrichum Syd. (1927) ≡ Annellophora.

Chaetotyphula Corner (1950), Pterulaceae. 7, tropical. See Dentinger & McLaughlin (*Mycol.* **98**: 746, 2006; phylogeny).

Chaetozythia P. Karst. (1888) nom. dub., Fungi. Based on a mite fide v. Höhnel (*Öst. bot. Z.* **63**: 238, 1913).

Chaetyllis Clem. (1931) ≡ Raciborskiomyces.

chaga fungus, see birch canker.

Chailletia Fuckel (1863) [non *Chailletia* DC. 1811, *Chailletiaceae*] ? = Truncatella fide Sutton (*in litt.*).

Chailletia Jacz. (1913), Valsaceae. 1, Switzerland. See

also *Azbukinia*.

Chailletia P. Karst. (1871) [non *Chailletia* DC. 1811, *Chailletiaceae*] ≡ Karstenia Fr.

Chain (Ernst Boris; 1906-1979; Germany). Friedrich Wilhelm University (1930); Cambridge (1933-1935); Oxford (1935-1946); Istituto Superiore di Sanità, Rome (1946-1964); Imperial College, London (1964-1979). Biochemist who, with Florey (q.v.) organized the first large-scale production of penicillin, the fungal antibiotic discovered by Fleming (q.v.). *Biogs, obits etc.* Mansford (*Nature* **281**: 715, 1979).

Chainia Thirum. (1955), Actinobacteria. q.v.

Chainoderma Massee (1890) = Podaxis fide Cunningham (*Gast. Austr. N.Z.*: 196, 1944).

Chalara (Corda) Rabenh. (1844), anamorphic *Pezizomycotina*, Hso.0-1eH.22. *c*. 103, widespread. *C. quercina* (oak wilt). See Nag Raj & Kendrick (*Monogr. Chalara Allied Genera*, 1975), Kile & Walker (*Aust. J. Bot.* **35**: 1, 1987; *C. australis*, pathogenicity), Nag Raj & Kendrick *in* Wingfield *et al.* (Eds) (*Ceratocystis* and *Ophiostoma* Taxonomy, Ecology and Pathogenicity, 1993; anamorphs ophiostomatoid fungi), Réblová (*Sydowia* **51**: 210, 1999; teleomorph), Paulin & Harrington (*Stud. Mycol.* **45**: 209, 2000; phylogeny), McKenzie *et al.* (*Fungal Diversity* **11**: 129, 2002; key), Paulin-Mahady *et al.* (*Mycol.* **94**: 62, 2002; phylogeny), Wu (*Mycosystema* **23**: 313, 2004; China).

Chalarodendron C.J.K. Wang & B. Sutton (1984), anamorphic *Pezizomycotina*, Hsy.0eH.22. 1, USA. See Wang & Sutton (*Mycol.* **76**: 569, 1984).

Chalarodes McKenzie (1991), anamorphic *Pezizomycotina*, Hso.0eH.15. 2, Pacific Islands. See McKenzie (*Mycotaxon* **42**: 89, 1991).

chalaroplectenchyma, see plectenchyma.

Chalaropsis Peyronel (1916) = Thielaviopsis fide Paulin-Mahady *et al.* (*Mycol.* **94**: 62, 2002; phylogeny).

Chalastospora E.G. Simmons (2007), Pleosporaceae. 1. See Simmons (*Alternaria: an Identification Manual*, 2007).

Chalazion Dissing & Sivertsen (1975), ? Pyronemataceae. 3, Europe. See Kristiansen (*Agarica* **10/11**: 83, 1990; ecology), van Brummelen *in* Hawksworth (Ed.) (*Ascomycete Systematics. Problems and Perspectives in the Nineties* NATO ASI Series vol. **269** **269**: 401, 1994; posn), Francesco *et al.* (*Riv. Micol.* **3**: 203, 1998; key).

Chalciporus Bataille (1908), Boletaceae. 25, widespread. See Singer (*Sydowia* **31**: 196, 1979; key), Šutara (*Czech Mycol.* **57**: 1, 2005; European), Šutara (*Czech Mykol.* **44**: 59, 2005; *Rubinoboletus* in Europe), Klofac & Krisai (*Öst. Z. Pilzk.* **15**: 33, 2006).

Chalcosphaeria Höhn. (1917) = Plagiostoma fide Müller & von Arx *in* Ainsworth *et al.* (Eds) (*The Fungi* **4A**: 87, 1973).

Chalymmota P. Karst. (1879) = Panaeolus fide Singer & Smith (*A monograph on the genus Galerina Earle*, 1964).

Chamaeascus L. Holm, K. Holm & M.E. Barr (1993), Hyponectriaceae. 1, Svalbard. See Holm & Holm (*Blyttia* **51**: 121, 1993), Wang & Hyde (*Fungal Diversity* **3**: 159, 1999).

Chamaeceras Rebent. ex Kuntze (1898) = Marasmius fide Singer (*Agaric. mod. Tax.* edn 3, 1975).

Chamaemyces Earle (1909), Agaricaceae. 1, widespread (north temperate). See Didukh *et al.* (*Mycol. Balcanica* **1**: 89, 2004; Israel), Hausknecht & Pidlich-Aigner (*Öst. Z. Pilzk.* **13**: 1, 2004; Austria), Vellinga (*MR* **108**: 354, 2004; phylogeny).

Chamaenema Kütz. (1833) nom. dub., ? Fungi.

Chamaeota (W.G. Sm.) Earle (1909), Pluteaceae. 9, widespread. See Singer (*Sydowia* **31**: 197, 1979; key), Ying (*Mycotaxon* **54**: 303, 1995; China), Minnis *et al.* (*Mycotaxon* **96**: 31, 2006; USA).

Chamonixia Rolland (1899), Boletaceae. 8, widespread (north temperate). See Bruns *et al.* (*Mol. Ecol.* **7**: 257, 1998; phylogeny), Clémençon (*Persoonia* **18**: 499, 2005; ontogeny).

Chamonixiaceae Jülich (1982) = Boletaceae.

chantarelle, basidioma of the edible *Cantharellus cibarius*.

Chanterel Adans. (1763) = Cantharellus fide Stalpers (*in litt.*).

Chantransiopsis Thaxt. (1914), anamorphic *Pezizomycotina*, Hso.0eH.1. 3 (entomogenous), Java; Europe. See Rossi & Blackwell (*Mycol.* **82**: 138, 1990).

Chaos L. (1753) nom. dub., Fungi. Included protozoa believed to be derived from fungus spores.

Chapeckia M.E. Barr (1978), Sydowiellaceae. 2 (from bark), N. America. See Barr (*Mycol. Mem.* **7**, 1978), Rossman *et al.* (*Mycoscience* **48**: 135, 2007; review).

Chapmanium E.I. Hazard & Oldacre (1975), Microsporidia. 2.

Chapsa A. Massal. (1860), Thelotremataceae (L). 15, pantropical. See Frisch (*Biblthca Lichenol.* **92**: 3, 2006; monogr.), Frisch *et al.* (*Biblthca Lichenol.* **92**: 517, 2006; phylogeny), Mangold *et al.* (*Lichenologist* **40**: 39, 2008; phylogeny).

Characonidia Bat. & Cavalc. (1965), anamorphic *Pezizomycotina*, Cpt.0fH.?. 1, Brazil. See Batista & Cavalcanti (*Atas Inst. Micol. Univ. Pernambuco* **2**: 297, 1965).

Charcotia Hue (1915) = Arthonia fide Hawksworth (*SA* **10**: 127, 1991).

Chardón (Carlos Eugenio; 1897-1965; Puerto Rico). Plant Pathologist, Insular Agricultural Experiment Station, Rio Piedras (1921-1923); Commissioner of Agriculture, San Juan (1923-1930); Chancellor, University of Puerto Rico (1931-1936); Administrator, Puerto Rican Reconstruction Administration (1936); Director, Institute of Tropical Agriculture (1943-1946). Carried out mycological exploration of Colombia, Puerto Rico, Venezuela and the American Virgin Islands in collaboration with Seaver (q.v.), Toro, Whetzel (q.v.) and others. *Publs.* (with Toro) Mycological explorations of Colombia, *Journal of the Department of Agriculture, Porto Rico* (1930); (with Toro) Mycological explorations of Venezuela, *Monographs of the University of Porto Rico* Series B (1932). *Biogs, obits etc.* Kern (*Mycol.* **57**: 839, 1965) [portrait].

Chardonia Cif. (1930), anamorphic *Pezizomycotina*, Hsy.1eH.?. 1, S. America.

Chardoniella F. Kern (1939), Pucciniosiraceae. 4 (on *Compositae* (0, III)), S. America. See Buriticá & Hennen (*Fl. Neotrop.* **24**: 39, 1980), Pardo-Cardona (*Caldasia* **25**: 283, 2003).

Charomyces Seifert (1987), anamorphic *Pezizomycotina*, Hso.0eP.40. 2, Hawaii; Cuba. See Seifert (*CJB* **65**: 230, 1987).

Charonectria Sacc. (1880), Hyponectriaceae. 5 (from dead leaves or lichenicolous), widespread (temperate). See Rossman *et al.* (*Stud. Mycol.* **42**: 248 pp., 1999), Wang & Hyde (*Fungal Diversity* **3**: 159,

1999).

Charrinia Viala & Ravaz (1894) nom. dub., Fungi. See Müller (*SA* **6**: 121, 1987).

chartaceous, paper-like.

chasmothecium, a closed fruitbody having no predefined opening but with the asci arranged in a single basal fascicle, e.g. an ascoma of *Erysiphe* (Braun *et al. in* Belanger *et al.* (Eds). *The Powdery Mildews: A comprehensive Tretise*, 2000). cf. cleistothecium.

Chaudhuria Zahlbr. (1932) = Heterodermia fide Hawksworth *et al.* (*Dictionary of the Fungi* edn 8, 1995).

Chaunopycnis W. Gams (1979), anamorphic *Ophiocordycipitaceae*, Hsp/St.0eH.15. 3 (from soil), widespread. See Gams (*Persoonia* **11**: 75, 1980), Möller *et al.* (*MR* **99**: 681, 1995; population biology), Möller *et al.* (*J. Industr. Microbiol. Biotechnol.* **17**: 359, 1996; chemistry), Bills *et al.* (*Mycol. Progr.* **1**: 3, 2002; phylogeny, chemistry), Matsushima (*Matsush. Mycol. Mem.* **10**, 2001).

Cheese. Fungi are used in the manufacture of many cheeses, some of which have characteristic spp., e.g. camembert (*Penicillium camemberti*), roquefort and other blue cheeses (*P. roqueforti*). See Babel (*Econ. Bot.* **7**: 27, 1953), Marth (*in* Reed, *Prescott & Dunn's Industrial microbiology*, edn 4: 65, 1982), Thom (*Bull. Bur. Anim. Ind. USDA* **82**, 1906); blue cheeses. Fungi known to produce mycotoxins have been detected in some cheeses (Moubasher *et al.*, *Mycopathologia* **66**: 187, 1979).

Cheilaria Lib. (1830), anamorphic *Pezizomycotina*, St.≡ eH.15. 1, widespread. See Sutton (*The Coelomycetes*, 1980).

Cheilariopsis Petr. (1959) ≡ Apomelasmia.

cheilocystidium, see cystidium.

Cheilodonta Boud. (1885) = Orbilia fide Baral (*SA* **13**, 1994).

Cheilophlebium Opiz & Gintl (1856) nom. dub., ? Agaricales. See Donk (*Beih. Nova Hedwigia* **5**: 50, 1962).

Cheilymenia Boud. (1885), Pyronemataceae. 66, widespread (esp. temperate). See Denison (*Mycol.* **56**: 718, 1964; key), Moravec (*Česká Mykol.* **22**: 32, 1968), Moravec (*Česká Mykol.* **38**: 146, 1984), Moravec (*Mycotaxon* **36**: 169, 1989), Wu & Kimbrough (*Bot. Gaz.* **152**: 421, 1992; ascospores), Moravec (*Czech Mycol.* **47**: 7, 1993), Landvik *et al.* (*Nordic Jl Bot.* **17**: 403, 1997; DNA), Moravec (*Czech Mycol.* **50**: 189, 1998), Prokhorov (*Mikol. Fitopatol.* **32**: 14, 1998; key), Moravec (*Czech Mycol.* **55**: 215, 2003; revn), Moravec (*Czech Mycol.* **54**: 135, 2003; revn), Moravec (*Czech Mycol.* **54**: 113, 2003; revn), Hansen & Pfister (*Mycol.* **98**: 1029, 2006; phylogeny), Moravec (*Czech Mycol.* **58**: 149, 2006; anatomy), Perry *et al.* (*MR* **111**: 549, 2007; phylogeny).

Cheimonophyllum Singer (1955), Cyphellaceae. 3, widespread. See Singer (*Sydowia* **9**: 417, 1955).

Cheiroconium Höhn. (1910) = Sirothecium fide Hughes (*CJB* **36**: 727, 1958), Sutton (*Proc. Indian Acad. Sci.* Pl. Sci. **94**: 229, 1985).

cheiroid, see chiroid.

Cheiromoniliophora Tzean & J.L. Chen (1990), anamorphic *Pleosporales*, Hso.#eP.10. 1, widespread. See Tzean & Chen (*MR* **94**: 424, 1990), Castañeda Ruíz *et al.* (*Mycotaxon* **61**: 319, 1997), Ho *et al.* (*Mycol.* **92**: 582, 2000; Hong Kong), Belomesyatseva (*Nov. sist. Niz. Rast.* **35**: 55, 2001; Belorus), Tsui *et al.* (*Fungal Diversity* **21**: 157, 2006; phylogeny), Cai *et al.* (*Persoonia* **20**: 53, 2008; phylogeny).

Cheiromycella Höhn. (1910), anamorphic *Hyaloscypha*, Hsp.1bH.?. 2, Europe; Japan. See Sutton (*Proc. Indian Acad. Sci.* Pl. Sci. **94**: 229, 1985; disposition of names), Kendrick (*CJB* **81**: 75, 2003; morphogenesis).

Cheiromyceopsis Mercado & J. Mena (1988), anamorphic *Pezizomycotina*, Hsp.0bP.1. 1, Cuba. See Mercado & Mena (*Acta Bot. Cubana* **53**: 2, 1988), Delgado-Rodríguez *et al.* (*Cryptog. Mycol.* **23**: 277, 2002), Mercado Sierra *et al.* (*Mycol.* **95**: 860, 2003).

Cheiromyces Berk. & M.A. Curtis (1857), anamorphic *Pezizomycotina*, Hsp.0bP.1. 5, widespread. See Sutton (*Proc. Indian Acad. Sci.* Pl. Sci. **94**: 229, 1985; disposition of names), Ho *et al.* (*Mycol.* **92**: 582, 2000; Hong Kong), Kendrick (*CJB* **81**: 75, 2003; morphogenesis), Promputtha *et al.* (*Nova Hedwigia* **80**: 527, 2005; Thailand).

Cheiromycina B. Sutton (1986), anamorphic *Pezizomycotina*, Hsp.0bP.1 (L). 1, Sweden. See Sutton (*Nordic Jl Bot.* **6**: 831, 1986), Aptroot & Schiefelbein (*MR* **107**: 104, 2003; key), Egorova & Mel'nik (*Mikol. Fitopatol.* **38**: 23, 2004; Russia), Earland-Bennett & Hawksworth (*Lichenologist* **37**: 191, 2005; status), Wang & Zheng (*Nova Hedwigia* **81**: 247, 2005; Austria), Printzen (*Nova Hedwigia* **84**: 261, 2007; key).

Cheiropodium Syd. & P. Syd. (1915) = Clasterosporium fide Ellis (*Dematiaceous Hyphomycetes*, 1971).

Cheiropolyschema Matsush. (1980), anamorphic *Pezizomycotina*, Hso.0bP.28. 1, Taiwan. See Matsushima (*Matsush. Mycol. Mem.* **1**: 15, 1980).

Cheirospora Moug. & Fr. (1825), anamorphic *Pezizomycotina*, Cac.0bP.1. 1, widespread. See Ono & Kobayashi (*Mycoscience* **46**: 352, 2005; Japan).

Cheirosporium L. Cai & K.D. Hyde (2008), anamorphic *Pleosporales*, Hsp.0bP.1. 1, China. See Cai *et al.* (*Persoonia* **20**: 53, 2008; descr.).

Chelisporium Speg. (1910) = Sirothecium fide Sutton (*Proc. Indian Acad. Sci.* Pl. Sci. **94**: 229, 1985).

chemical race, a group of chemically differentiated individuals or populations and not of any particular taxonomic rank (i.e. a chemical race may be a species, var., or chemotype); **- strain**, used as an infraspecific taxonomic rank in lichenized taxa distinguished only by chemical characters (Lamb, *Nature* **168**: 38, 1951).

chemosyndrome, a biogenetically meaningful set of major and minor natural metabolic products produced by a species (Culberson & Culberson, *Syst. Bot.* **1**: 325, 1977).

chemotaxis, a taxis (q.v.) in response to a chemical stimulus. Gooday (*in* Carlile (Ed.), *Primitive sensory and communication systems*, 155, 1975; chemotaxis in fungi), Bonner (*Mycol.* **69**: 443, 1977; in cellular slime moulds).

Chemotaxonomy (biochemical systematics; chemical fungal taxonomy), taxonomy using chemical characteristics, in a broad sense including both primary and secondary metabolites and physiological/chemical tests. Frisvad *et al.* (*MR* **112**: 231, 2008) noted the mycologists and bacteriologists define the term differently. Mycologists limit the term to carbohydrate or lipid-based taxonomy whereas for bacteriologists the term also encompasses the amino acid and nucleic acid complement of the organism. As a result, in mycology, chemotaxonomy declined in importance with the advent of molecular taxonomy (based on PCR technology & DNA sequence analysis), although there have been recent developments (e.g. Frisvad *et al.*, *MR* **112**: 231, 2008; Bitzer *et al.*, *MR* **112**: 251, 2008) which may be enhanced further with developments in metabolomics (see Genomics). Ubiquinone and fatty acid profiles are still used widely in yeast taxonomy (see Barnett *et al.*, *Yeasts: Characteristics and identification* Edn 3, 2000: 18; Deák, *Handbook of food spoilage yeasts*, 2008: 211).

Lit.: **Non-lichen-forming fungi**: Botha & Kock (*Int. J. Food Microbiol.* **19**: 39, 1993; fatty acid profiles, yeasts), Frisvad (*Chemom. Intell. Lab. Syst.* **14**: 253, 1992; secondary metabolites, chemometrics), Frisvad, *in* Hawksworth (Ed.) (*Identification and characterization of pest organisms*: 303, 1994), Frisvad *et al.* (Eds) (*Chemical fungal taxonomy*. 1998; secondary metabolites review, survey), Gómez-Miranda *et al.* (*Exp. Mycol.* **12**: 258, 1988; cell wall composition), Klich & Mullaney (*in* Arora *et al.*, *Handbook of Applied Mycology* **4**: 35, 1992; DNA methods), Meixner (*Chemische Farbreaktionen von Pilzen*, 1975; asco- and basidiomycetes), Micales *et al.* (*in* Arora *et al.*, *Handbook of Applied Mycology* **4**: 57, isozyme analysis), Murray (*J. gen. Microbiol.* **52**: 213, 1968; biochemical tests, pathogenic fungi), Paterson (*in* Frisvad *et al.*, 1998; unsaponifiable lipids review); Rast & Pfyffer (*Bot. J. Linn. Soc.* **99**: 1989; polyols), Rosendahl & Banke (*in*Frisvad *et al.*, 1998; isozyme analysis), Tyrrell (*Bot. Rev.* **35**: 305, 1969; review), Vogel (*Am. Naturalist* **48**: 435, 1964; evolutionary implications of lysine pathways).

Lichens: Culberson & Elix (*Meth. Plant Biochem.* **1**: 509, 1993; review), Feige & Lumbsch (Eds) (*Bibl. Lich.* **53**, 1993; review), Hawksworth (*in* Brown *et al.* (Eds), 1976: 139; review, bibliogr., use), Lumbsch (*in*Frisvad *et al.*, 1998; review), Rogers (*Bot. J. Linn. Soc.* **101**: 229, 1989; chemistry and species concept). See also Metabolic products, Phylogeny.

chemotropism, see tropism.

chemotype, a group of chemically differentiated individuals of a species of unknown or of no taxonomic significance.

Chermomyces Brain (1923) nom. dub., ? Dothideomycetes. See Batra *in* Subramanian (Ed.) (*Taxonomy of fungi* **1**: 187, 1978; applied to yeast-like cells in insect mycetomas).

Cheshunt compound, Copper sulphate ($CuS0_4$) 2 parts, ammonium carbonate ($(NH_4)_2CO_3$) 11 parts; used (at least 24 hr after mixing) at the rate of 56.7 g per 4.5 l of water as a soil disinfectant against damping-off.

chestnut blight, *Cryphonectria parasitica* (syn. *Endothia parasitica*). See Anagnostakis (*Mycol.* **79**: 23, 1987).

Chevalieria G. Arnaud (1920) [non *Chevalieria* Gaudich. ex Beer 1857, *Bromeliaceae*] ≡ Chevalieropsis.

Chevalieropsis G. Arnaud (1923), Parodiopsidaceae. Anamorph *Septoidium*. 1, Africa. See Petrak (*Sydowia* **5**: 346, 1951).

Chiajaea (Sacc.) Höhn. (1920) = Hypomyces fide Rossman *et al.* (*Stud. Mycol.* **42**: 248 pp., 1999).

chiastobasidium, see basidium.

Chiastospora Riess (1852), anamorphic *Pezizomycotina*, Cac.1bH.?. 1, Europe. See Sutton (*Mycol. Pap.* **141**, 1977).

Chiastosporum Dughi (1956) nom. inval. = Collema F.H. Wigg. fide Hawksworth *et al.* (*Dictionary of the Fungi* edn 8, 1995).

Chikaneea B. Sutton (1973), anamorphic *Pezizomycotina*, Hso.≡ eH.10. 1, Canada. See Sutton (*Mycol. Pap.* **132**: 29, 1973), Sinclair *et al.* (*Mycotaxon* **64**: 365, 1997), Marvanová *et al.* (*Nova Hedwigia* **75**: 255, 2002).

Chilemyces Speg. (1910) = Dimerina fide Petrak & Sydow (*Annls mycol.* **32**: 1, 1934).

Chiliospora A. Massal. (1860) = Biatoridium fide Hafellner (*Acta Bot. Fenn.* **150**: 39, 1994), Jørgensen (*Taxon* **53**: 521, 2004; nomencl.).

Chiloella Syd. (1928) = Glomerella fide von Arx & Müller (*Beitr. Kryptfl. Schweiz* **11** no. 1, 1954).

Chilonectria Sacc. (1878) = Nectria fide Rossman *et al.* (*Stud. Mycol.* **42**: 248 pp., 1999).

Chimaeroscypha Raitv. (2004), Hyaloscyphaceae. 1, Tadzhikistan. See Raitviir (*Scripta Mycologica* Tartu **20**: 19, 2004).

chimeroid (of lichen thalli), see Lichens, Phycotype.

Chinese cheese, see sufu.

Chinese mushroom, see straw mushroom.

Chinese rice, see starters.

Chiodecton Ach. (1814), Roccellaceae (L). 24, widespread (esp. tropical). See Tehler (*CJB* **68**: 2458, 1990; cladistics), Thor (*Op. Bot.* **103**, 1990; monogr., key), Henssen & Thor (*Nordic Jl Bot.* **18**: 95, 1998; ontogeny), Cáceres *et al.* (*Lichenologist* **33**: 503, 2001), Thor (*J. Jap. Bot.* **77**: 47, 2002; Japan, Taiwan), Thor (*Biblthca Lichenol.* **95**: 543, 2007; Malaysia).

Chiodectonaceae Zahlbr. (1905) = Roccellaceae.

Chiodectonomyces Cif. & Tomas. (1953) ≡ Enterographa.

Chiographa Leight. (1854) = Phaeographis fide Hawksworth *et al.* (*Dictionary of the Fungi* edn 8, 1995).

Chionaster Wille (1903) nom. dub., Algae. Based on an alga having no chlorophyll.

Chionomyces Deighton & Piroz. (1972), anamorphic *Pezizomycotina*, Hso.≡ eH.10. 4 (on *Meliolaceae*), widespread (tropical). See Deighton & Pirozynski (*Mycol. Pap.* **128**: 74, 1972).

chionophilous, having a preference for growing nearby snow.

Chionosphaera D.E. Cox (1976), Chionosphaeraceae. 4, N. America; Europe. See Kirschner (*MR* **105**: 1403, 2001; new sp. associated with conifer inhabiting bark beetles).

Chionosphaeraceae Oberw. & Bandoni (1982), Agaricostilbales. 4 gen., 18 spp.

Lit.: Oberwinkler & Bandoni (*CJB* **60**: 1726, 1982), Oberwinkler & Bauer (*Sydowia* **41**: 224, 1989), Seifert *et al.* (*Boln Soc. argent. Bot.* **28**: 215, 1992), Diederich (*Biblthca Lichenol.* **61**: 198 pp., 1996), Boekhout *in* Kurtzman & Fell (Eds) (*Yeasts, a taxonomic study* 4th edn: 627, 1998), Kwon-Chung *in* Kurtzman & Fell (Eds) (*Yeasts, a taxonomic study* 4th edn: 643, 1998), Fell *et al.* (*Int. J. Syst. Evol. Microbiol.* **50**: 1351, 2000; mol. phylogeny), Scorzetti *et al.* (*FEMS Yeast Res.* **2**: 495, 2002), Bauer *et al.* (*Mycol. Progr.* **5**: 41, 2006).

Chionyphe Thienem. (1839) = Mucor Fresen. fide Hesseltine (*Mycol.* **47**: 344, 1955).

chiroid (**cheiroid**), shaped like a hand with the fingers together and not divergent, e.g. the conidia of *Cheiromyces*; cf. digitate, palmate.

Chithramia Nag Raj (1988), anamorphic *Pezizomycotina*, St.1eP.1. 1, India. See Nag Raj (*CJB* **66**: 903, 1988).

chitinoclastic, chitin decomposing.

Chitinonectria M. Morelet (1968) = Neonectria fide Rossman *et al.* (*Stud. Mycol.* **42**: 248 pp., 1999).

Chitinozoa, fossil chitinous organisms of uncertain affinity found in the Upper Precambian to the Devonian.

Chitonia (Fr.) P. Karst. (1879) [non *Chitonia* D. Don 1823, *Melastomataceae*] ≡ Clarkeinda.

Chitoniella Henn. (1898) ≡ Clarkeinda fide Singer (*Agaric. mod. Tax.*, 1951).

Chitonis Clem. (1909) ≡ Chitonia.

Chitonomyces Peyr. (1873), Laboulbeniaceae. 98, widespread. See Sugiyama (*TMSJ* **18**: 155, 1977; Japanese spp.), Sugiyama & Nagasawa (*TMSJ* **26**: 3, 1985; Borneo), Santamaria (*Nova Hedwigia* **73**: 339, 2001; Spain).

Chitonospora E. Bommer, M. Rousseau & Sacc. (1891), Xylariales. 1 (from dead grasses), Europe. See Eriksson (*Svensk bot. Tidskr.* **60**: 320, 1966), Kang *et al.* (*Fungal Diversity* **2**: 135, 1999; posn).

chitosan, a partially deacetylated form of chitin characteristic of zygomycetes; **chitases**, enzymes (EC 3.2.1.99) from fungi and bacteria able to hydrolyse chitosan (Monaghan *et al.*, *Nature* (*New Biol.*) **245**: 78, 1973).

chitosome, a small spheroidal structure (40-70 nm diam.), containing chitin synthetase zymogen, found in many fungi (Bracker *et al.*, *Proc. Nat. Acad. Sci. USA* **73**: 4570, 1976).

Chlamydoabsidia Hesselt. & J.J. Ellis (1966), Cunninghamellaceae. 1, India; USA. See Hesseltine & Ellis (*Mycol.* **58**: 761, 1966), Behera & Mukerji (*Norw. Jl Bot.* **21**: 1, 1974), Braun (*Int. J. Mycol. Lichenol.* **3**: 271, 1988).

Chlamydoaleurosporia Grigoraki (1924) = Trichophyton fide Hawksworth *et al.* (*Dictionary of the Fungi* edn 8, 1995).

chlamydocyst, a two-walled resting zoosporangium of *Blastocladiaceae* within a hypha.

Chlamydomucor Bref. (1889) = Mucor Fresen. fide Hesseltine (*Mycol.* **47**: 344, 1955) See, Ellis *et al.* (*Mycol.* **68**: 131, 1976).

Chlamydomyces Bainier (1907), anamorphic *Pezizomycotina*, Hso.1eP.1. 2, widespread. See Mason (*Mycol. Pap.* **2**: 37, 1928), Claudia (*Mycotaxon* **27**: 255, 1986), Hambleton *et al.* (*Stud. Mycol.* **53**: 29, 2005; key).

Chlamydopsis Hol.-Jech. & R.F. Castañeda (1986), anamorphic *Pezizomycotina*, Hso.#bP.1/3. 1, Cuba. See Holubová-Jechová & Castañeda (*Česká Mykol.* **40**: 74, 1986).

Chlamydopus Speg. (1898), Agaricaceae. 1, widespread (desert areas). See Long & Stouffer (*Mycol.* **38**: 619, 1947), Lunghini (*Bollettino dell'Associazione Micologica ed Ecologica Romana* **17**: 22, 2001).

Chlamydorubra K.B. Deshp. & K.S. Deshp. (1966), anamorphic *Pezizomycotina*, Hso.0eP.?. 1, India. See Deshpande & Deshpande (*Mycopath. Mycol. appl.* **29**: 272, 1966).

Chlamydosauromyces Sigler, Hambl. & Paré (2002), Onygenaceae. 1 (from lizard skin), USA. See Sigler *et al.* (*Stud. Mycol.* **47**: 127, 2002).

chlamydospore, an asexual 1-celled spore (primarily for perennation, not dissemination) originating endogenously and singly within part of a pre-existing cell, by the contraction of the protoplast and possessing an inner secondary and often thickened hyaline or brown wall, usually impregnated with hydrophobic material. Originally proposed by de Bary in 1859 for *Asterophora* anamorphs. See Griffiths (*Nova Hedw.* **25**: 503, 1974; origin, structure, function), Hughes (*in* Arai (Ed.), *Filamentous microorganisms*: 1, 1985; definition, occurrence).

Chlamydosporites Paradkar (1975), Fossil Fungi ? Ustilaginales. 1 (Cretaceous), India.

Chlamydosporium Peyronel (1913) = Phoma Sacc. fide Mouchacca & Sutton (*Cryptog. Mycol.* **12**: 251, 1991).

Chlamydotomus Trevis. (1879) nom. dub., ? Algae.

Chlamydozyma Wick. (1964) = Metschnikowia fide Pitt & Miller (*Mycol.* **60**: 663, 1968).

Chlamydozymaceae Wick. (1964) = Metschnikowiaceae.

Chlorangium Link (1849) nom. rej. = Aspicilia fide Laundon & Hawksworth (*Taxon* **37**: 478, 1988; nomencl.), Hawksworth *et al.* (*Dictionary of the Fungi* edn 8, 1995).

Chlorangium Rabenh. (1857) nom. dub., Lecanorales.

Chlorea Nyl. (1855) nom. rej. = Letharia fide Hawksworth *et al.* (*Dictionary of the Fungi* edn 8, 1995), Obermayer (*Progr. Probl. Lichenol. Nineties. Proc. Third Symp. Intern. Assoc. Lichenol.* [*Biblthca Lichenol.* **68**]: 45, 1997).

Chlorencoelia J.R. Dixon (1975), Hemiphacidiaceae. 3, widespread (north temperate). See Zhuang (*Mycotaxon* **31**: 261, 1988), Zhuang *et al.* (*Mycosystema* **19**: 478, 2000; phylogeny), Wang *et al.* (*Mycol.* **98**: 1065, 2006; phylogeny), Wang *et al.* (*Mol. Phylogen. Evol.* **41**: 295, 2006; phylogeny).

Chloridiaceae Nann. (1934) = Chaetosphaeriaceae.

Chloridiella Arnaud (1953) nom. nud. = Idriella fide Nicot & Charpentié (*BSMF* **87**: 621, 1972).

Chloridium Link (1809), anamorphic *Chaetosphaeria*, Hso.0eH.15/18. *c.* 27, widespread. See Gams & Holubová-Jechová (*Stud. Mycol.* **13**, 1976), Wang & Wilcox (*Mycol.* **77**: 951, 1985), Morgan-Jones & Goos (*Mycol.* **84**: 921, 1992; nomencl.), Réblová (*Stud. Mycol.* **45**: 149, 2000; review), Réblová & Gams (*Mycoscience* **41**: 129, 2000), Réblová & Winka (*Mycol.* **92**: 939, 2000; phylogeny), Fernández *et al.* (*Mycol.* **98**: 121, 2006; phylogeny).

Chlorocaulum Clem. (1909) = Stereocaulon Hoffm. fide Hawksworth *et al.* (*Dictionary of the Fungi* edn 8, 1995).

Chlorociboria Seaver ex C.S. Ramamurthi, Korf & L.R. Batra (1958) nom. cons., ? Helotiales. Anamorph *Dothiorina*. 17, widespread. See Tunbridge Ware. See Korf (*Mycol.* **51**: 298, 1951; status), Dixon (*Mycotaxon* **1**: 193, 1975; key), Holst-Jensen *et al.* (*Mycol.* **89**: 885, 1997; DNA), Johnston & Park (*N.Z. Jl Bot.* **43**: 679, 2005; New Zealand), Wang *et al.* (*Mol. Phylogen. Evol.* **41**: 295, 2006; phylogeny), Wang *et al.* (*Mycol.* **98**: 1065, 2006; phylogeny).

Chlorocyphella Speg. (1909) = Pyrenotrichum fide Santesson (*Symb. bot. upsal.* **12** no. 1: 1, 1952), Gams (*Taxon* **54**: 520, 2005; nomencl.).

Chlorodictyon J. Agardh (1870) = Ramalina fide Hawksworth *et al.* (*Dictionary of the Fungi* edn 8, 1995).

Chlorodothis Clem. (1909) = Tomasellia fide Harris (*in litt.*).

Chlorogaster Laessøe & Jalink (2004), Sclerodermataceae. 1, Sabah. See Læssøe & Jalink (*Persoonia* **18**: 421, 2004).

Chlorolepiota Sathe & S.D. Deshp. (1979), Agaricaceae. 1, India. See Sathe & Deshpande (*Curr. Sci.* **48**:

693, 1979).

Chloroneuron Murrill (1911) = Gomphus Pers. fide Petersen (*The genera Gomphus and Gloeocantharellus in North America*, 1972).

Chloropeltigera (Gyeln.) Gyeln. (1934) nom. inval. = Peltigera fide Hawksworth *et al.* (*Dictionary of the Fungi* edn 8, 1995).

Chloropeltis Clem. (1909) = Peltigera fide Hawksworth *et al.* (*Dictionary of the Fungi* edn 8, 1995).

chlorophycophilous (of fungi), lichenized with a green photobiont (Pike & Carroll, *in* Alexopoulos & Mims, *Introductory mycology*, edn 3, 1980).

Chlorophyllum Massee (1898), Agaricaceae. 16, widespread (esp. tropical). Recognition would make *Macrolepiota* non-monophylletic; conservation over *Chlorophyllum* has not yet been proposed. See Heinemann (*Rev. Mycol. nat. Belg.* **31**: 317, 1966; *C. molybdites*, poisonous), Heinemann (*Bull. Jard. bot. nat. Belg.* **38**: 195, 1968), Sundberg (*Madroño* **21**: 15, 1971), Johnson & Vilgalys (*Mycol.* **90**: 971, 1998), Sarasini & Contu (*Rivista di Micologia* **44**: 247, 2001; as *Endoptychum*, Italy), Vellinga (*Mycotaxon* **83**: 415, 2002), Vellinga (*Mycotaxon* **85**: 259, 2003; *Chlorophyllum rachodes* and allies), Hausknecht & Pidlich-Aigner (*Öst. Z. Pilzk.* **13**: 1, 2004; Austria), Vellinga (*MR* **108**: 354, 2004; phylogeny), Ge & Yang (*Mycotaxon* **96**: 181, 2006; China).

Chlorophyllum Murrill (1910) = Gomphus Pers. fide Petersen (*The genera Gomphus and Gloeocantharellus in North America*, 1972).

chloroplast endoplasmic reticulum (CER), a layer of ribosome studded membrane surrounding the plastid.

Chloroscypha Seaver (1931), Helotiales. *c.* 14 (on conifers), widespread (north temperate). See Gremmen (*Nova Hedwigia* **2**: 547, 1963; Netherlands spp.), Kobayashi (*Bull. Govt For. Exp. Stn* **176**, 1965), Butin (*Sydowia* **37**: 15, 1984; S America), Baral (*Z. Mykol.* **53**: 119, 1987; asci), Shoji (*Bulletin of the Forestry and Forest Products Research Institute* Ibaraki, Japan **368**: 23, 1994; ecology), Gernandt *et al.* (*Mycol.* **93**: 915, 2001; phylogeny), Garcia *et al.* (*BSMF* **118**: 125, 2002), Wang *et al.* (*Mycol.* **98**: 1065, 2006; phylogeny), Wang *et al.* (*Mol. Phylogen. Evol.* **41**: 295, 2006; phylogeny).

Chlorosperma Murrill (1922) = Melanophyllum fide Singer (*Agaric. mod. Tax.*, 1951).

Chlorospleniella P. Karst. (1885), ? Helotiales. 1, Brazil. See Groves & Wilson (*Taxon* **16**: 39, 1967).

Chlorosplenium Fr. (1849), Dermateaceae. 5, widespread. See also *Chlorociboria*. See Dixon (*Mycotaxon* **1**: 65, 1974; key), Verkley (*Sydowia* **56**: 343, 2004).

Chlorospora Massee (1898) non *Chlorospora* Speg. 1891 ≡ Melanophyllum.

Chlorostroma A.N. Mill., Lar.N. Vassiljeva & J.D. Rogers (2007), ? Xylariaceae. 1, USA. See Miller *et al.* (*Sydowia* **59**: 138, 2007).

Chlorovibrissea L.M. Kohn (1989), Vibrisseaceae. Anamorph *Phialophora*-like. 4, Australasia. See Kohn (*Mem. N. Y. bot. Gdn* **49**: 112, 1989), Korf (*Mycosystema* **3**: 19, 1991; posn), Buchanan & May (*N.Z. Jl Bot.* **41**: 407, 2003; conservation), Wang *et al.* (*Mycol.* **98**: 1065, 2006; phylogeny), Wang *et al.* (*Mol. Phylogen. Evol.* **41**: 295, 2006; phylogeny).

Chmelia Svob.-Pol. (1966) = Alternaria fide de Hoog & Hermanides-Nijhof (*Stud. Mycol.* **18**, 1976), Hoog & Horré (*Mycoses* **45**: 259, 2002).

Chnoopsora Dietel (1906) = Melampsora fide Cummins & Hiratsuka (*Illustr. Gen. Rust Fungi rev. edit.*, 1983).

Choanatiara DiCosmo (1984), anamorphic *Pezizomycotina*, St.0eH.15. 2, India; Canada. See Vujanovic (*Can. J. Pl. Path.* **20**: 319, 1998).

Choanephora Curr. (1873), Choanephoraceae. 2, widespread (esp. tropical). *C. cucurbitarum* (blossom blight and fruit rot of cucurbits and other plants). See Thaxter (*Rhodora* **5**: 97, 1903), Barnett & Lilly (*Phytopathology* **40**: 80, 1950), Barnett & Lilly (*Mycol.* **47**: 26, 1955; culture), Hesseltine & Benjamin (*Mycol.* **49**: 723, 1957), Higham & Cole (*CJB* **60**: 2313, 1982; sporangiolum ultrastr.), Kirk (*Mycol. Pap.* **152**: 61 pp., 1984; key), Yu & Ko (*MR* **103**: 684, 1999; azygospore formation), Sakai *et al.* (*Am. J. Bot.* **87**: 440, 2000; pollination mutualism), Donnell *et al.* (*Mycol.* **93**: 286, 2001; phylogeny), Voigt & Wöstemeyer (*Gene* **270**: 113, 2001; phylogeny), Papp *et al.* (*Acta Biol. Hung.* **54**: 393, 2003; phylogeny), Feofilova (*App. Biochem. Microbiol.* **42**: 439, 2006; heterothallism), Gareth Jones (*Mycoscience* **47**: 167, 2006; spore appendages).

Choanephoraceae J. Schröt. (1894), Mucorales. 4 gen. (+ 3 syn.), 6 spp.

Lit.: Higham & Cole (*CJB* **60**: 2313, 1982), Kirk (*Mycol. Pap.* **152**: 61 pp., 1984; revision, keys), Benny (*Mycol.* **83**: 150, 1991), Yu & Ko (*MR* **103**: 684, 1999), Sakai *et al.* (*Am. J. Bot.* **87**: 440, 2000), Tanabe *et al.* (*Mol. Phylogen. Evol.* **30**: 438, 2004).

Choanephorella Vuill. (1904) = Choanephora fide Zycha *et al.* (*Mucorales*, 1969).

Choanephoroidea I. Miyake & S. Ito (1935) ? = Choanephora fide Kirk (*in litt.*).

chocolate spot, a disease of *Vicia* and other legumes (*Botrytis cinerea* and *B. fabae*); see Wilson (*Ann. appl. Biol.* **24**: 258, 1937), Harrison (*Pl. Path.* **37**: 168, 1988).

Choeromyces Tul. & C. Tul. (1862) = Choiromyces fide Fischer (*Nat. Pflanzenfam.* **5b**: viii, 1938).

Choiromyces Vittad. (1831), Tuberaceae. 5 (hypogeous), widespread. See Zhang & Minter (*MR* **92**: 91, 1989; cytology), Pegler *et al.* (*British truffles*, 1993), O'Donnell *et al.* (*Mycol.* **89**: 48, 1997; phylogeny), Percudani *et al.* (*Mol. Phylogen. Evol.* **13**: 169, 1999; phylogeny), Ferdman *et al.* (*MR* **109**: 237, 2005; phylogeny), Hansen & Pfister (*Mycol.* **98**: 1029, 2006; phylogeny).

choke, a disease of grasses (*Epichloë typhina*).

Chondroderris Maire (1937), Helotiales. 1, Spain.

Chondrogaster Maire (1925), Mesophelliaceae. 2, Mauritania; Brazil; Europe. See Petrak (*Sydowia* **4**: 373, 1950), Petrak (*Sydowia* **4**: 171, 1950).

chondroid axis, the cartilaginous axis occupying the central portion of the medulla in *Usnea*.

Chondroplea Kleb. (1933) = Discosporium Höhn. fide Sutton (*Mycol. Pap.* **141**, 1977).

Chondropodiella Höhn. (1917), anamorphic *Godronia*, St.0eH.?. 1, USA.

Chondropodiola Petr. & Cif. (1932) nom. dub., anamorphic *Pezizomycotina*. See Sutton (*Mycol. Pap.* **141**, 1977).

Chondropodium Höhn. (1916) = Cornicularielia fide DiCosmo (*CJB* **56**: 1665, 1978).

Chondropsis Nyl. (1879) nom. rej. = Xanthoparmelia fide Rogers (*New Phytol.* **70**: 1069, 1971), Elix & Child (*Brunonia* **9**: 113, 1986), Lumbsch & Kothe (*Lichenologist* **20**: 25, 1988; anatomy), Elix (*Flora of*

Australia **55**: 34, 1994), Blanco *et al.* (*Taxon* **53**: 959, 2004; phylogeny), Gams (*Taxon* **53**: 1067, 2004; nomencl.).

Chondrospora A. Massal. (1860) nom. rej. = Anzia Stizenb. fide Hawksworth *et al.* (*Dictionary of the Fungi* edn 8, 1995).

Chondrostereum Pouzar (1959), Cyphellaceae. 4, widespread. *C. purpureum* (silver leaf of plum, etc.). See Ekramoddoullah *et al.* (*CJB* **15**: 7, 1993; Canadian isolates).

Chondrostroma Syd. (1940) [non *Chondrostroma* Gürich 1906, fossil ? *Algae*] = Phacidiopycnis fide Rupprecht (*Sydowia* **13**: 10, 1959).

Chondrus Stackh. (1797), Algae. Algae.

Chordecystia C.B. Foster (1979), Fossil Fungi. 1.

Chordostylum Tode (1790) nom. dub., Mucoraceae.

Choreospora Constant. & R. Sant. (1987), anamorphic *Pezizomycotina*, Hsp.1bH.15. 1 (on lichens), Australia. See Constantinescu & Santesson (*Lichenologist* **19**: 177, 1987).

Chorioactidaceae Pfister (2008), Pezizales. 3 gen., 7 spp.

Lit.: Pfister *et al.* (*MR* **112**: 513, 2008).

Chorioactis Kupfer (1902), Chorioactidaceae. 1, USA. fide Bellemère *et al.* (*Nova Hedwigia* **58**: 49, 1994; ultrastr.), Rudy & Keller (*Mycologist* **10**: 33, 1996), Harrington *et al.* (*Mycol.* **91**: 41, 1999; phylogeny), Pfister & Kurogi (*Mycotaxon* **89**: 277, 2004; morphology), Hansen & Pfister (*Mycol.* **98**: 1029, 2006; phylogeny), Perry *et al.* (*MR* **111**: 549, 2007; phylogeny), Pfister *et al.* (*MR* **112**: 513, 2008).

Choriphyllum Velen. (1922) ≡ Phaeolus fide Donk (*Persoonia* **8**: 281, 1975).

Chorostate (Nitschke ex Sacc.) Traverso (1906) = Diaporthe fide Barr (*Mycol. Mem.* **7**, 1978).

Chorostella (Sacc.) Clem. & Shear (1931) = Cryptodiaporthe fide Barr (*Mycol. Mem.* **7**, 1978).

Chrismofulvea Marbach (2000), Caliciaceae (L). 4, widespread. See Marbach (*Biblthca Lichenol.* **74**: 151, 2000).

Christiansenia Hauerslev (1969), Carcinomycetaceae. 7, widespread (north temperate). See Hauerslev (*Friesia* **9**: 43, 1969), Rath (*Rivista di Micologia* **30**: 112, 1987).

Christianseniaceae F. Rath (1991) = Carcinomycetaceae.

Christianseniales F. Rath (1991) = Tremellales.

Christiaster Kuntze (1891) ≡ Gonatobotryum.

Chromatium Link (1824) = Dematium fide Mussat (*Syll. fung.* **15**: 86, 1901).

Chromatochlamys Trevis. (1860) = Thelenella fide Mayrhofer (*Biblthca Lichenol.* **26**, 1987; monogr., key), Fryday & Coppins (*Lichenologist* **36**: 89, 2004).

chromatography. Physico-chemical diffusion technique for the identification of chemical products in which a moving *mobile* phase (which may be liquid or gas) is passed over an immobile *stationary* phase (which may be solid or liquid). Phases may be polar or non-polar. Different forms of chromatography may be defined according to the type of matrix used, the physico-chemical principles of separation or even the equipment used; **gas -**, in a gaseous phase with solid support, used in e.g. fatty acid analyses (see Kendrick & Ratledge, *Lipids* **27**: 15, 1992; Stahl 7 Klug, *App. Env. Microbiol.* **62**: 4136, 1996), but is a destructive method: the sample is not eluted/retained for future use; **high-pressure liquid -** (h.p.l.c.), in solvent under pressure, method of choice for screening non-volatile metabolites and has been used widely in mycotoxin screening; often in conjunction with *diode array detector* which allows detection of metabolite maxima/minima across the entire UV range via a polychromator and the diode array (a chip containing a large number of light sensitive diodes) [see Meyer, *Practical high performance liquid chromatography* edn 4, 2004, general text; Frisvad & Thrane, *in* Betina (Ed.) *Chromatography of Mycotoxins*, 253, 1993, use in mycotoxin detection; Smedsgard, *J. Chromatogr.* A **760**: 264, method for screening fungal cultures]; **paper -**, on filter paper, now rarely used (but see Lund, *Lett. Appl. Microbiol.* **21**: 11, 1995 for rapid method with *Penicillium*); **thin-layer -** (t.l.c.), in thin silicate layers on glass, aluminium or plastic plates. Much used in lichenology, e.g. Feige *et al.*, (*J. Chromatogr.* **646**: 417, 1993; h.p.l.c. lichen products), Mietzsch *et al.*, (*Mycotaxon* **47**: 475, 1993; computer progr. lichen products), and for mycotoxin producing fungi, e.g. Abrunhosa *et al.*, (*Lett. Appl. Microbiol* **32**: 240, 2001), (see Chemotaxonomy, Metabolic products); **preparative -** thicker layered plates used for re-extraction of particular metabolites prior to purification and further analysis.

Lit.: **General**: Harborne, (*Phytochemical methods*, 1973), Smith, (*Chromatographic and electrophoretic techniques* **1**, 1960), Frisvad *et al.* (Eds), (*Chemical Fungal Taxonomy*, 1998).

Chromelosporium Corda (1833) = Ostracoderma. *C. fulvum* (peat mould). fide Hughes (*CJB* **36**: 727, 1958; segregation from *Botrytis*), Hughes & Bisalputra (*CJB* **48**: 361, 1970; ultrastructure of conidiogenesis), Hennebert (*Persoonia* **7**: 183, 1973), Hennebert & Korf (*Mycol.* **67**: 214, 1975), Norman & Egger (*Mycol.* **91**: 820, 1999; phylogeny).

Chromendothia Lar.N. Vassiljeva (1993) = Camarops fide Rossman *et al.* (*Stud. Mycol.* **42**: 248 pp., 1999), Castlebury *et al.* (*Mycol.* **94**: 1017, 2002; phylogeny, links to *Endothia*), Zhang *et al.* (*Mycol.* **98**: 1076, 2006; phylogeny).

chromo- (combining form), colour.

Chromocleista Yaguchi & Udagawa (1993), Trichocomaceae. Anamorph *Geosmithia*-like. 2, Japan. See Yaguchi *et al.* (*TMSJ* **34**: 101, 1993), Ogawa *et al.* (*Mycol.* **89**: 756, 1997; DNA), Ogawa & Sugiyama (*Integration of Modern Taxonomic Methods for Penicillium and Aspergillus Classification*: 149, 2000; phylogeny), Kolarík *et al.* (*MR* **108**: 1053, 2004; phylogeny).

Chromocrea Seaver (1910) ≡ Creopus fide Rossman *et al.* (*Stud. Mycol.* **42**: 248 pp., 1999).

Chromocreopsis Seaver (1910) = Thuemenella fide Corlett (*Mycol.* **77**: 272, 1985), Rossman *et al.* (*Stud. Mycol.* **42**: 248 pp., 1999).

Chromocyphella De Toni & Levi (1888), ? Inocybaceae. 5, widespread.

Chromocytospora Speg. (1910) = Phomopsis (Sacc.) Bubák. fide Petrak (*Sydowia* **5**: 335, 1951), Sutton (*Mycol. Pap.* **141**, 1977).

chromogen, a stain-producing organism (Erlich, 1941).

chromogenesis, colour production.

chromogenic (chromogenous), colour-producing.

chromomycosis (**chromoblastomycosis**), a chronic skin disease in humans caused by traumatic inoculation by certain species of various black-yeast genera, including *Exophiala*, *Fonsecaea*, *Phialophora*;

mostly on adult males in the tropics; dermatitis verrucosa; see Al-Doory (*Chromomycosis*, 1972).

chromophilous, deeply staining.

Chromosera Redhead, Ammirati & Norvell (1995), Hygrophoraceae. 1, widespread (north temperate). See Redhead *et al.* (*Beih. Sydowia* **10**: 161, 1995).

chromosome, see Chromosome maps, Chromosome numbers.

Chromosome maps. A chromosome map is a chart of the linear array of genes on a chromosome. The first chromosome map for a fungus was of the sex chromosome of *Neurospora crassa* (Lindegren, *J. Genet.* **32**: 243, 1936), the first for a set of fungal chromosomes was for the 8 chromosomes of *Aspergillus nidulans* (Kafer, *Adv. Genet.* **9**: 105, 1958; for revised maps see Dorn, *Genetics* **56**: 619, 1967). See also Lindegren *et al.* (*Nature* **183**: 800, 1959; *Saccharomyces cerevisiae*). See also Molecular biology.

Chromosome numbers. In fungi, chromosomes numbers are generally low; n=4 is possibly the basic number, although the very small size of fungal chromosomes has probably frequently led to an underestimation of their number in research studies (Wieloch, *J. microbiol. methods* **67**: 1, 2006; visualizing fungal chromosomes). n=6 has been reported for *Cenococcum geophilum* (Portugal *et al.*, *Nucleus* **45**: 14, 2002) and n=5 or n=6 for *Tuber aestivum* (Poma *et al.*, *FEMS microbiol. letters* **167**: 101, 1998). 2- and 3-ploid series are frequent among *Chromista*, and higher polyploids can occur. *Allomyces* has 8-50+ chromosomes (Emerson & Wilson, *Mycol.* **46**: 393, 1954); *Puccinia graminis* n=3 (fide McGinnis, *CJB* **31**: 522, 1953); *Dermatocarpon weberi* n=6 or 8; *Porpidia crustulata* n=2; *Agaricales*, n=2-12 (Ueda, *Trans. mycol. Soc. Japan* **10**: 23, 1969). See also: karyotype.

Rather few fungi have so far been studied in this respect, although pulsed field gel electrophoresis (q.v.) as a technique will accelerate this process and has already been used to show that fungal nuclei contain 3 (*Schizosaccharomyces pombe*) to *c.* 21 (*Ustilago hordei*) chromosomes (see Mills & McCluskey, *Molecular plant-microbe interactions* **3**: 351-357, 1990). (see Cytology). See Altman & Dittmer (*Biological data book*, 1964), Burges (*in* Lousley, *Species studies in the British flora*: 76, 1955), Ornduff (*Index to plant chromosome numbers* [*Regnum veg.* **50**], 1968), Rogers (*Evolution* **27**: 153, 1973; polyploidy in fungi).

Chromosporium Corda (1829), anamorphic *Pezizomycotina*, Hso.0eH.15. 20, widespread.

Chromostylium Giard (1889) ? = Metarhizium fide Tulloch (*TBMS* **66**: 409, 1976).

Chromotorula F.C. Harrison (1927) = Rhodotorula fide Lodder (*Die anaskosporogenen Hefen* **1**, 1934), Fell & Statzell-Tallman *in* Kurtzman & Fell (Eds) (*Yeasts, a taxonomic study* 4th edn: 800, 1998).

Chroocybe Räsänen (1943) nom. inval. ? = Coniocybe fide Hawksworth *et al.* (*Dictionary of the Fungi* edn 8, 1995).

Chroodiscus (Müll. Arg.) Müll. Arg. (1890), Thelotremataceae (L). 30, widespread (tropical). See Lumbsch & Vězda (*Nova Hedwigia* **50**: 245, 1990), Lücking (*Cryptog. Mycol.* **20**: 193, 1999), Kantvilas & Vězda (*Lichenologist* **32**: 325, 2000; Tasmania), Galloway (*Australasian Lichenology* **49**: 16, 2001), Messuti *et al.* (*Lichenologist* **35**: 241, 2003; Argentina), Frisch (*Biblthca Lichenol.* **92**: 3, 2006; Africa), Frisch *et al.* (*Biblthca Lichenol.* **92**: 517, 2006; phylogeny).

Chroogomphus (Singer) O.K. Mill. (1964) nom. cons. prop., Gomphidiaceae. 18, widespread (north temperate). See Singer & Kuthan (*Česká Mykol.* **30**: 81, 1976), Agerer (*Nova Hedwigia* **50**: 1, 1990; ecology), Miller & Aime (*Trichomycetes and Other Fungal Groups, Robert W. Lichwardt Commemoration Volume*: 315, 2001), Miller (*Mycol.* **95**: 176, 2003).

Chrooicia Trevis. (1861) = Pyrenula Ach. (1809) fide Hawksworth *et al.* (*Dictionary of the Fungi* edn 8, 1995).

Chroolepus C. Agardh (1824) = Cystocoleus fide Hawksworth *et al.* (*Dictionary of the Fungi* edn 8, 1995).

Chroostroma Corda (1837) = Pactilia fide Saccardo (*Syll. fung.* **4**: 673, 1886).

Chrysachne Cif. (1938), anamorphic *Pezizomycotina*, Hsp.0eH.?. 1, West Indies.

Chrysalidopsis Steyaert (1961), anamorphic *Pezizomycotina*, Cac.≡ eP.?. 1, Chile. See Steyaert (*Darwiniana* **12**: 171, 1961).

Chryseidea Onofri (1981) ? = Phaeoisaria.

Chrysella Syd. (1926), Pucciniaceae. 1 (on *Mikania* (0, III) (*Compositae*)), C. America.

chryseous, golden yellow.

Chrysobostrychodes G. Kost (1985) ≡ Chrysomphalina.

Chrysocelis Lagerh. & Dietel (1913), ? Mikronegeriaceae. 6 (on dicots, 1 on monocots), widespread (esp. tropical). See Ono & Hennen (*TMSJ* **24**: 369, 1983).

Chrysoconia McCabe & G.A. Escobar (1979), Coniophoraceae. 1, Réunion. See McCabe & Escobar (*Mycotaxon* **9**: 240, 1979).

Chrysoconiaceae Jülich (1982) = Coniophoraceae.

Chrysocyclus Syd. (1925), Pucciniaceae. 3 (on *Compositae*, *Solanaceae*), America (tropical). See Davidson (*Mycol.* **24**: 221, 1932).

chrysocystidium, see cystidium.

Chrysoderma Boidin & Gilles (1991), Meruliaceae. 1, Réunion. See Boidin & Gilles (*Cryptog. Mycol.* **12**: 126, 1991).

Chrysogloeum Petr. (1959), anamorphic *Blasdalea*, Cac.0eH.?. 1, Peru. See Petrak (*Sydowia* **12**: 254, 1958).

Chrysogluten Briosi & Farneti (1904) nom. rej. = Cosmospora fide Rossman & Samuels (*Taxon* **47**: 723, 1998; nomencl.), Rossman *et al.* (*Stud. Mycol.* **42**: 248 pp., 1999).

Chrysoglutenaceae Jatta (1911) = Nectriaceae.

chrysogonidia, photobiont cells of *Trentepohlia* (obsol.).

Chrysomma Acloque (1893) = Caloplaca fide Hawksworth *et al.* (*Dictionary of the Fungi* edn 8, 1995).

Chrysomphalina Clémençon (1982), Hygrophoraceae. 4, Europe. See Lutzoni (*Syst. Biol.* **46**: 373, 1997; phylogeny), Hibbett *et al.* (*Nature* **407**: 506, 2000; suggested placement in *Marasmiaceae*), Moncalvo *et al.* (*Syst. Biol.* **49**: 278, 2000), Redhead *et al.* (*Mycotaxon* **83**: 19, 2002; phylogeny), Barassa *et al.* (*Mycotaxon* **88**: 113, 2003; Spain).

Chrysomyces Theiss. & Syd. (1917) = Perisporiopsis Henn. fide Müller & von Arx (*Beitr. Kryptfl. Schweiz* **11** no. 2, 1962).

Chrysomyxa Unger (1840), Coleosporiaceae. *c.* 23 (on *Picea* (0, I); II (*Caeoma*-like and often (?) with evanescent peridium); III on dicots, or microcyclic III (= *Melampsoropsis*) on *Picea* or *Tsuga* (*Pinaceae*)), widespread (northern hemisphere). See Savile (*Can. J. Res.* C **28**: 318, 1950), Crane (*CJB* **79**: 957, 2001), Crane (*Mycol.* **97**: 534, 2005).

Chrysomyxaceae Gäum. ex Leppik (1972) = Coleosporiaceae.

Chrysonilia Arx (1981), anamorphic *Neurospora*, Hso.0eH.38. 3, widespread. *C. sitophila* (syn. *Monilia sitophila*), red bread mould. See Zeng *et al.* (*FEMS Microbiol. Lett.* **237**: 79, 2004; detection).

Chrysophlyctis Schilb. (1896) = Synchytrium fide Fitzpatrick (*The lower fungi. Phycomycetes*, 1930).

Chrysoporthe Gryzenh. & M.J. Wingf. (2004), Cryphonectriaceae. Anamorph *Chrysoporthella*. 5, widespread. See Hodges (*Mycol.* **72**: 542, 1980; on *Eucalyptus*), van Heerden & Wingfield (*MR* **105**: 94, 2001; genetic diversity of *C. cubensis*), Myburg *et al.* (*Mycoscience* **44**: 187, 2003; on *Myrtaceae*), Roux *et al.* (*Pl. Dis.* **87**: 1329, 2003; on *Eucalyptus*), Roux *et al.* (*Plant Dis.* **87**: 1329, 2003), Gryzenhout *et al.* (*Stud. Mycol.* **50**: 129, 2004), Gryzenhout *et al.* (*Fungal Diversity* **20**: 39, 2005; Ecuador), Rodas *et al.* (*Pl. Path.* **54**: 460, 2005; Colombia), Gryzenhout *et al.* (*MR* **110**: 833, 2006; Colombia), Gryzenhout *et al.* (*Mycol.* **98**: 239, 2006; phylogeny), Gryzenhout *et al.* (*FEMS Microbiol. Letters* **258**: 161, 2006), Heath *et al.* (*Pl. Dis.* **90**: 433, 2006; South Africa), Nakabonge *et al.* (*Plant Dis.* **90**: 734, 2006).

Chrysoporthella Gryzenh. & M.J. Wingf. (2004), anamorphic *Chrysoporthe*. 1. No independent species have been recognized. See Gryzenhout *et al.* (*Stud. Mycol.* **50**: 130, 2004).

Chrysopsora (Vain.) M. Choisy (1951) = Psora Hoffm. (1796) fide Timdal (*Nordic Jl Bot.* **4**: 525, 1984).

Chrysopsora Lagerh. (1892), Pucciniaceae. 1 (on *Gynoxis*, *Mikania* (0, III) (*Compositae*)), America (tropical).

Chrysorhiza T.F. Andersen & Stalpers (1996), anamorphic *Waitea*. 1, widespread. See Stalpers & Andersen *in* Sneh *et al.* (Eds) (*Rhizoctonia Species* Taxonomy, Molecular Biology, Ecology, Pathology and Disease Control: 58, 1996).

Chrysosplenium Allesch. (1898) [non *Chrysosplenium* L. 1753, *Saxifragaceae*] ≡ Chlorosplenium fide Dixon (*Mycotaxon* **1**: 65, 1974).

Chrysospora (orthographic variant), see *Chrysopsora* (Vain.) M. Choisy.

Chrysosporium Corda (1833), anamorphic *Onygenaceae*, Hso.0eH.1. *c.* 62, widespread. Teleomorphs various (also in *Arthrodermataceae*). See Carmichael (*CJB* **40**: 1137, 1962), Boekhout (*Stud. Mycol.* **31**: 29, 1989; septal ultrastr.), Hoog *et al.* (*Medical Mycology* **36** Suppl. 1: 52, 1998; review), Gräser *et al.* (*Medical Mycology* **37**: 105, 1999; phylogeny), Roilides *et al.* (*J. Clin. Microbiol.* **37**: 18, 1999; human disease), Sugiyama *et al.* (*Mycoscience* **40**: 251, 1999; phylogeny), Kushwaha (*Revta Iberoamer. Micol.* **17** [Special]: 66, 2000; physiology, biotechnology), Vidal *et al.* (*Revta Iberoamer. Micol.* **17** [Special]: 22, 2000; phylogeny), Vidal & Guarro (*Stud. Mycol.* **47**: 189, 2002; diagnosis).

Chrysothallus Velen. (1934), Hyaloscyphaceae. 8, former Czechoslovakia. See Korf (*Lloydia* **14**: 129, 1951), Huhtinen (*Karstenia* **29**: 45, 1990), Galán & Raitviir (*Nova Hedwigia* **58**: 453, 1994).

Chrysothricaceae Zahlbr. (1905), Arthoniales (L). 2 gen. (+ 4 syn.), 17 spp.

Lit.: Thor (*Bryologist* **91**: 360, 1988), Tønsberg (*Graphis Scripta* **6**: 31, 1994), Grube (*Bryologist* **101**: 377, 1998).

Chrysothrix Mont. (1852) nom. cons., Chrysothricaceae (L). 11, widespread. See Laundon (*Lichenologist* **13**: 101, 1981; key), Thor (*Bryologist* **91**: 360, 1988; S America), Tønsberg (*Graphis Scripta* **6**: 31, 1994), Grube (*Bryologist* **101**: 377, 1998; phylogeny), Jagadeesh Ram *et al.* (*Lichenologist* **38**: 127, 2006; India), Elix & Kantvilas (*Lichenologist* **39**: 361, 2007; Australia), Harris & Ladd (*Opuscula Philolichenum* **5**: 29, 2008; N America).

Chuppia Deighton (1965), anamorphic *Pezizomycotina*, Hso.#eP.1. 1, Venezuela. See Deighton (*Mycol. Pap.* **101**: 32, 1965).

Chytrella Kirschst. (1941) = Unguicularia fide Raschle (*Sydowia* **29**: 170, 1977).

chytrid, one of the *Chytridiales*. See Canter-Lund & Lund, *Freshwater Algae* (1995) for general account.

Chytridhaema Moniez (1887), ? Olpidiaceae. 1, France.

Chytridiaceae Nowak. (1878), Chytridiales. 33 gen. (+ 5 syn.), 238 spp.

Lit.: Sparrow (*Aquatic Phycomycetes* Edn 2: 1187 pp., 1960), Barr *in* Margulis *et al.* (Eds) (*Handbook of Protoctista*: 454, 1990), James *et al.* (*CJB* **78**: 336, 2000).

Chytridiales Cohn (1879). Chytridiomycetes. 4 fam., 75 gen., 494 spp. Thallus monocentric or polycentric, endogenous or exogenous; zoospores mostly with one conspicuous lipid globule; rhizoids tapering to fine (‹0.5µm) tips; generally aquatic but also in soil; saprobic or parasitic; on algae, microfauna, other fungi, pollen, plant debris, chitin and keratin, rarely terrestrial on higher plants (*Synchytrium*); *Batrachochytrium* pathogen in amphibian skin; cosmop.

The *Spizellomycetales* were removed from *Chytridiales sensu* Sparrow and *Chytridiales* emended using ultrastructural characters of zoospore (Barr, 1980): microtubules extending from one side of kinetosome in a parallel array; non-flagellate centriole parallel and connected to the kinetosome; rumposome often present on surface of lipid globule; ribosomes usually in mass, which is more or less enclosed by endoplasmic reticulum. Families in order are based on type of development, no on phylogeny. Sparrow (1960) divided the order into an operculate and inoperculate series; Karling (1977) used Whiffen's (1944) system based on thallus development types, a system followed by Barr, which is used herein. Fams:

(1) **Chytridiaceae**
(2) **Cladochytriaceae**
(3) **Endochytriaceae**
(4) **Synchytriaceae**

The placement of many genera within the revised *Chytridiales* is still uncertain.

Lit.: Sparrow (1960), Karling (*Chytridiomycetarum iconographia*, 1977; 175 pl.), Barr (*CJB* **58**: 2380, 1980; 1990), Barr & Désaulniers (*CJB* **66**: 869, 1988; precise configuration of organelles in zoospore), Powell & Roychoudhury (*CJB* **70**: 750, 1992; redefinition of chytridialean types of MLC), Longcore (*CJB* **71**: 414, 1993; new type of MLC, *Mycol.* **87**: 25, 1995; discussion of taxonomic characters), James *et al.* (*Mycol.* **98**: 860, 2006; molecular phylogeny), Hibbett *et al.* (*MR* **111**: 109, 2007), and see under Familes.

Chytridioides Tregnoboff (1913) = Chytridiopsis fide

Karling (*Chytriomyc. Iconogr.*: 37, 1977), Canning *in* Margulis *et al.* (Eds) (*Handbook of Protoctista*: 53, 1990; systematic position).

Chytridiomycetes Caval.-Sm. (1998), Chytridiomycota. 3 ord., 10 fam., 98 gen., 678 spp. Ords:
(1) **Chytridiales**
(2) **Rhizophydiales**
(3) **Spizellomycetales**

Lit.: James *et al.* (*Mycol.* **98**: 860, 2006; molecular phylogeny), Hibbett *et al.* (*MR* **111**: 109, 2007), and see under Orders.

Chytridiomycota M.J. Powell (2007), Fungi. 2 class., 4 ord., 14 fam., 105 gen., 706 spp. (Chytridiomycetes, Rumpomycetes). Thallus coenocytic, holocarpic or eucarpic, monocentric, polycentric, or mycelial; cell walls chitinous (at least in hyphal stages); mitochondial cristae flat; zoospores posteriorly (whiplash) monoflagellate or (rarely) polyflagellate, lacking mastigonemes or scales, with unique flagellar 'root' systems. Aquatic saprobes or parasites growing on decaying and living organic material (incl. nematodes, insects, amphibian skin, plant partss, other chytrids and fungi), in freshwater and soils; a few are marine, some are obligate anaerobes on cellulosic substrata in guts of herbivores.

The presence of flagellate zoospores has led to these fungi sometimes being classified as *Mastigomycotina* or being placed with the fungal phyla of the *Chromista* rather than in the *Fungi* (Kendrick, 1992; Kreissel, 1988). Retained in the *Fungi* (Barr, 1992; Bruns *et al.*, 1991; Corliss, 1994; Alexopoulos *et al.* 1996) on ultrastructural (flattened mitochondrial cristae), cell wall chemistry (chitinous) and molecular evidence (see Phylogeny).

Two classes are now recognized (zygospore ultrastructure in *Deckenbachia* and *Thalassochytrium* is sufficiently distinct that it cannot be satisfactority placed):
(1) **Chytridiomycetes**
(2) **Monoblepharidomycetes**

Lit.: Barr (*in* Margulis *et al.* (Eds), *Handbook of Protoctista*: 454, 1990; *Mycol.* **84**: 1, 1992), Bruns *et al.* (*Ann. Rev. Ecol. Syst.* **22**: 525, 1991), Karling (*Sydowia Beih.* **6**, 1966; Indian spp., keys; *Chytridiomycetarum iconographia*, 1977), Powell (*Mycol.* **85**: 1, 1993; roles in environment), Sparrow (*in* Ainsworth *et al.*, *The Fungi* **4B**: 85, 1973; ecology), Longcore (*Mycotaxon* **60**: 149, 1996; taxonomic change since 1960), James *et al.* (*CJB* **78**: 336, 2000; molecular phylogeny), James *et al.* (*Mycol.* **98**: 860, 2006; molecular phylogeny), Hibbett *et al.* (*MR* **111**: 109, 2007), and see under Classes and Orders.

Chytridiopsis W.G. Schneid. (1884), Microsporidia. 3, Europe; S. America. See Canning *in* Margulis *et al.* (Eds) (*Handbook of Protoctista*: 53, 1990; systematic position).

Chytridium A. Braun (1851), Chytridiaceae. *c.* 45, widespread (temperate). See Karling (*Chytriomyc. Iconogr.*, 1977; in algae).

Chytriomyces Karling (1945), Chytridiaceae. 34, widespread. See Johnson (*J. Elisha Mitchell scient. Soc.* **87**: 200, 1971; Icelandic spp.), Karling (*Chytriomyc. Iconogr.*, 1977), Letcher & Powell (*Mycotaxon* **84**: 447, 2002; monogr., key).

Cibalocoryne Hazsl. (1881) = Geoglossum fide Saccardo (*Syll. fung.* **8**: 1, 1889).

Cibdelia Juel (1925), ? Olpidiaceae. 1, Europe. See Juel (*Ark. Bot.* **19**: 9, 1925).

Cibiessia Crous (2007), anamorphic *Teratosphaeriaceae*. 3, Australia; S. Africa. See Crous *et al.* (*Fungal Diversity* **26**: 1, 2007), Crous *et al.* (*Stud. Mycol.* **58**: 1, 2007; phylogeny).

Ciboria Fuckel (1870), Sclerotiniaceae. Anamorph *Myrioconium*. *c.* 21, widespread (temperate). See also *Moellerodiscus*. See Schumacher (*Norw. Jl Bot.* **25**: 145, 1978; Norwegian amenticolous spp.), Spooner (*Biblthca Mycol.* **116**, 1987; 3 spp. Australasia), Holst-Jensen *et al.* (*Mycol.* **89**: 885, 1997; phylogeny), Holst-Jensen *et al.* (*Nordic Jl Bot.* **18**: 705, 1999; phylogeny), Galán & Palmer (*Czech Mycol.* **52**: 277, 2001; Europe), Wang *et al.* (*Mycol.* **98**: 1065, 2006; phylogeny), Wang *et al.* (*Mol. Phylogen. Evol.* **41**: 295, 2006; phylogeny).

Ciboriella Seaver (1951) = Hymenoscyphus fide Schumacher (*Mycotaxon* **38**: 233, 1990).

Ciborinia Whetzel (1945), Sclerotiniaceae. Anamorphs *Myrioconium*, *Sclerotium*. 16, widespread. See Batra (*Am. J. Bot.* **47**: 819, 1960; key), Holst-Jensen *et al.* (*Mycol.* **89**: 885, 1997; phylogeny), Zhuang & Wang (*Mycosystema* **16**: 161, 1997), Holst-Jensen *et al.* (*Nordic Jl Bot.* **18**: 705, 1999; phylogeny), Cook (*Bulletin OEPP* EPPO Bulletin **33**: 257, 2003; diagnostics), Toor *et al.* (*Australas. Pl. Path.* **34**: 319, 2005; genetic diversity), Saito & Kaji (*Mycoscience* **47**: 41, 2006; on *Gentiana*).

Ciboriopsis Dennis (1962) = Moellerodiscus fide Dumont (*Mycol.* **68**: 233, 1976), Johnston (*N.Z. Jl Bot.* **40**: 105, 2002).

Cicadocola Brain (1923), anamorphic *Pezizomycotina*, Hso.0eH.4. 1 (in *Insecta*), S. Africa.

Cicadomyces Šulc (1911), ? Saccharomycetales. 10 (in *Insecta*), widespread.

cicatricose, with longitudinal ridges (Fig. 20.15).

cicatrized (of conidiogenous cells and conidia), having thickened scars.

Cicinnobella Henn. (1904) = Perisporina. Also used for *Dimerium* anamorphs. fide Höhnel (*Sber. Akad. Wiss. Wien* Math.-naturw. Kl., Abt. 1 **120**: 379, 1911).

Cicinobolus Ehrenb. (1853) = Ampelomyces fide Rogers (*Mycol.* **51**: 96, 1959), Donk (*Taxon* **15**: 149, 1966).

Cidaris Fr. (1849), ? Helvellaceae. 1, N. America. Affinities are unclear. See van Brummelen *in* Hawksworth (Ed.) (*Ascomycete Systematics. Problems and Perspectives in the Nineties* NATO ASI Series vol. **269 269**: 399, 1994; posn).

cider, an alcoholic drink obtained by the fermentation of apple juice by various yeasts. Cf. beer.

Ciferri (Raffaele; 1897-1964, Italy, later Dominican Republic). Founder and Director of the Experimental Agricultural Station, Santiago de los Caballeros, Dominican Republic (1925-1932; Vice Director, Italian Cryptogamic Laboratory, Botanical Institute, Pavia (1932-1936); Professor of Botany, Florence University (1936-1942); Professor of Botany, University of Pavia (1942-1964). A most versatile mycologist who, through more than 1,000 publications, made notable contributions to tropical mycology, particularly in Brazil with Batista (q.v.) and the Dominican Republic, with González Fragoso (q.v.); also known for his work with the *Ustilaginales* (q.v.), plant pathogenic fungi (q.v.), medical mycology (q.v.), and the compilation of a mycological bibliography. Founder of the journal *Mycopathologia et Mycologia Applicata*. *Publs. G. Lindau & P. Sydow*

Thesaurus Literaturae Mycologicae et Lichenologicae Supplementum 1911-1930 4 vols (1957-1960) [these volumes are freely available on-line; see Internet: catalogues & thesauri]. *Biogs, obits etc.* Baldacci (*Mycol.* **57**: 198, 1965) [portrait]; Grummann (1974: 516); Stafleu & Cowan (*TL-2* **1**: 503, 1976); Tomaselli (*Atti dell'Istituto Botanico della Università e Laboratorio Crittogamico di Pavia* Serie 5, **21** (Suppl.), 1964; bibliography).

Ciferria Gonz. Frag. (1925), anamorphic *Pezizomycotina*, ?.0fH.?. 1, West Indies.

Ciferriella Petr. (1930), anamorphic *Pezizomycotina*, Cac.≡ eP.19. 1, West Indies.

Ciferrina Petr. (1932), anamorphic *Pezizomycotina*, St.0fH.?. 1, West Indies.

Ciferriolichen Tomas. (1952) = Arthopyrenia fide Hawksworth *et al.* (*Dictionary of the Fungi* edn 8, 1995).

Ciferriomyces Petr. (1932) ? = Pyrenostigme fide Hansford (*Mycol. Pap.* **15**, 1946).

Ciferriopeltis Bat. & H. Maia (1965), anamorphic *Pezizomycotina*, Cpt.0eP.?. 1, Dominican Republic. See Batista & Maia (*Publções Inst. Micol. Recife* **463**: 4, 1965).

Ciferriotheca Bat. & I.H. Lima (1959) = Metathyriella fide Luttrell *in* Ainsworth *et al.* (Eds) (*The Fungi* **4A**: 213, 1973).

Ciferrioxyphium Bat. & H. Maia (1963), anamorphic *Aithaloderma*, Hsy.≡ eH.15. 2, widespread (tropical). See Batista & Maia (*Quad. Lab. crittogam., Pavia* **31**: 65, 1963).

Ciferriusia Bat. (1962) = Yatesula fide von Arx & Müller (*Stud. Mycol.* **9**, 1975), Reynolds & Gilbert (*Cryptog. Mycol.* **27**: 249, 2006; Panamá).

Ciglides Chevall. (1826) = Gymnosporangium fide Sydow & Sydow (*Monographia Uredinearum seu Specierum Omnium ad hunc usque Diem Descriptio et Adumbratio Systematica* **3**, 1915).

Ciglidiaceae Chevall. (1826) = Pucciniaceae.

Ciliaria Quél. (1885) [non *Ciliaria* Stackh. 1809, *Algae*] ≡ Scutellinia fide Clements & Shear (*Gen. Fung.*, 1931).

ciliate, edged with hairs.

Ciliatosporidium I. Foissner & W. Foissner (1995), Microsporidia. 1.

Ciliatula Velen. (1922) = Pezoloma fide Korf (*Phytologia* **21**: 201, 1971).

ciliatulate, thinly ciliate.

Cilicia Fr. (1825), Agaricomycetes (L). 2, Europe. See Donk (*Reinwardtia* **2**: 435, 1954) See, Laundon (*Lichenologist* **13**: 101, 1981).

Ciliciocarpus Corda (1831) = Gautieria fide Pilát (*Fl. ČSR* **B1**: 211, 1958).

Ciliciopodium Corda (1831), anamorphic *Nectria*, Cpt.0eH.?. 5, widespread. See Booth (*Mycol. Pap.* **73**, 1959).

Ciliciopus Clem. & Shear (1931) ≡ Ciliciopodium.

Ciliella Sacc. & P. Syd. (1902), Helotiales. 1, Brazil.

Ciliochora Höhn. (1919), anamorphic *Pezizomycotina*, St.0eH.19. 2, paleotropical. See Nag Raj (*Coelomycetous Anamorphs with Appendage-bearing Conidia*, 1993), Sutton *et al.* (*MR* **100**: 405, 1996; Italy), Moriondo & Menguzzato (*Inftore fitopatol.* **50**: 55, 2000; status as endophyte).

Ciliochorella Syd. (1935), anamorphic *Pezizomycotina*, St.1eP.15. 2, widespread (tropical). See Subramanian & Ramakrishnan (*TBMS* **39**: 314, 1956), Punithalingam (*Stud. Mycol.* **31**: 113, 1989; cytology), Masilamani & Muthumary (*MR* **98**: 857, 1994; ontogeny).

Ciliofusa Clem. & Shear (1931) ≡ Ciliofusarium.

Ciliofusarium Rostr. (1892) = Menispora fide Hughes (*CJB* **36**: 727, 1958).

Ciliofusospora Bat. & J.L. Bezerra (1963), Pezizomycotina. 1, Brazil. See Batista & Bezerra (*Publções Inst. Micol. Recife* **385**: 15, 1963).

Ciliolarina Svrček (1977), Hyaloscyphaceae. Anamorph *Septonema*-like. 6, N. America; Europe (northern). See Huhtinen (*Biblthca Mycol.* **150**: 93, 1993; key), Raitviir (*Scripta Mycologica* Tartu **20**, 2004).

Ciliomyces Höhn. (1906) = Paranectria fide Hawksworth & Pirozynski (*CJB* **55**: 2555, 1977), Rossman *et al.* (*Stud. Mycol.* **42**: 248 pp., 1999).

Ciliophora Petr. (1929), anamorphic *Pezizomycotina*, St.0eH.19. 1, C. America.

Ciliophorella Petr. (1940), anamorphic *Pezizomycotina*, Cpd.0eH.?. 1, Austria.

Cilioplea Munk (1953), Lophiostomataceae. 9, widespread. See Crivelli (*Über die heterogene Ascomycetengattung Pleospora Rbh.*, 1983; key), Barr (*SA* **5**: 125, 1986; posn), Barr (*Mem. N. Y. bot. Gdn* **62**: 92 pp., 1990; North America).

Ciliosculum Kirschst. (1941), ? Hyaloscyphaceae. 1, Germany.

Ciliosira Syd. (1942) = Acarosporium fide Petrak (*Sydowia* **14**: 347, 1960).

Ciliospora Zimm. (1902) = Chaetospermum fide Sutton (*Mycol. Pap.* **141**, 1977), Matsushima (*Matsush. Mycol. Mem.* **5**, 1987).

Ciliosporella Petr. (1927), anamorphic *Pezizomycotina*, Cpd.0eH.15. 2, Austria; Australia. See Yuan & Mohammed (*MR* **101**: 1531, 1997; n.sp. Australia).

cilium (pl. **cilia**), (1) an appendage of animal cells, e.g. protozoa; sometimes used for the flagellum (q.v.) of a zoospore; (2) a hair-like out-growth, e.g. from the edge of an apothecium or lichen thallus.

-cillin (suffix), for penicillins; derivatives of carboxy-6-amino-penicillanic acid (*WHO Chron.* **17**: 400, 1963).

cincinnate (**cincinnal**), rolled round; curled.

Cinereomyces Jülich (1982), Polyporaceae. 1, widespread. See Jülich (*Biblthca Mycol.* **85**: 396, 1981), Spirin (*Karstenia* **45**: 103, 2005; Russia).

cingulate, edged all round.

Cintractia Cornu (1883), Anthracoideaceae. *c.* 15, widespread (tropical to subtropical). The sorus of agglutinated ustilospores is characterized by a sterile stroma; the name has been widely used also for *Anthracoidea* (q.v.). See Ingold (*MR* **99**: 140, 1995; spore germination), Piepenbring *et al.* (*Mycol.* **91**: 485, 1999), Piepenbring (*Nova Hedwigia* **70**: 289, 2000; classification).

Cintractiaceae Vánky (2000) = Anthracoideaceae.

Cintractiella Boedijn (1937), Cintractiellaceae. 2 (induces hypertrophy in *Diplasia* & *Hypolytrum* (*Cyperaceae*)), East Indies; S. America. See Vánky (*Fungal Diversity* **13**: 167, 2003).

Cintractiellaceae Vánky (2003), Ustilaginales. 2 gen., 3 spp.

Lit.: Vánky (*Fungal Diversity* **13**: 167, 2003).

Cintractiomyxa Golovin (1952) = Anthracoidea fide Nannfeldt & Lindeberg (*Svensk bot. Tidskr.* **51**: 503, 1957).

Cionothrix Arthur (1907), Pucciniosiraceae. 6 (on

Compositae, 1 on *Sapindaceae* (0, III)), S. America. See Buriticá & Hennen (*Fl. Neotrop.* **24**: 18, 1980), Buriticá (*Rev. Acad. Colomb. Cienc.* **18** no. 69: 131, 1991), Berndt (*Mycol.* **94**: 523, 2002; Argentina).

Ciposia Marbach (2000), Caliciaceae (L). 1, USA. See Marbach (*Biblthca Lichenol.* **74**: 158, 2000).

circadian, pertaining to a day, e.g. a 24-hr rhythm. Cf. diel, diurnal.

Circinaria Bonord. (1851) = Valsa Fr. fide Hawksworth *et al.* (*Dictionary of the Fungi* edn 8, 1995).

Circinaria Fée (1825) = Coccocarpia fide Hawksworth *et al.* (*Dictionary of the Fungi* edn 8, 1995).

Circinaria Link (1809) nom. rej. = Aspicilia fide Laundon & Hawksworth (*Taxon* **37**: 478, 1988).

Circinaria M. Choisy (1929) = Lecanora fide Hawksworth *et al.* (*Dictionary of the Fungi* edn 8, 1995).

Circinastrum Clem. (1909) ≡ Weinmannodora fide Sutton (*Mycol. Pap.* **141**, 1977).

circinate, twisted round; coiled (Fig. 23.37).

Circinella Tiegh. & G. Le Monn. (1873), Mucoraceae. 9, widespread. See Hesseltine & Fennell (*Mycol.* **47**: 193, 1955; key), Naganishi (*TMSJ* **15**: 175, 1974), Ho (*Fungal Science* Taipei **10**: 23, 1995; ultrastr.), Arambarri & Cabello (*Mycotaxon* **57**: 145, 1996; key to spp.), Roux *et al.* (*Proc. Microsc. Soc. S. Afr.* **28**: 45, 1998; ultrastr.), Voigt & Wöstemeyer (*Gene* **270**: 113, 2001; phylogeny).

Circinoconis Boedijn (1942), anamorphic *Pezizomycotina*, Hso.0-≡ hP.1. 1, East Indies; Australasia. See Aptroot & Iperen (*Nova Hedwigia* **67**: 485, 1998; Papua New Guinea).

Circinomucor Arx (1982) = Mucor Fresen. fide Kirk (*in litt.*).

Circinoniesslia Samuels & M.E. Barr (1998), Niessliaceae. 1 (on dead bark or fungicolous), Puerto Rico. See Samuels & Barr (*CJB* **75**: 2165, 1998).

Circinostoma Gray (1821) ? = Valsa Fr. fide Hawksworth *et al.* (*Dictionary of the Fungi* edn 8, 1995).

Circinotrichum Nees (1816), anamorphic *Pezizomycotina*, Hso.0eH.6. 17, widespread. See Pirozynski (*Mycol. Pap.* **84**, 1962; key), Castañeda Ruíz *et al.* (*Mycotaxon* **63**: 169, 1997).

Circinumbella Tiegh. & G. Le Monn. (1872) = Circinella fide Hesseltine (*Mycol.* **47**: 344, 1955).

Circulocolumella S. Ito & S. Imai (1957), Hysterangiaceae. 1, Bonin Island. See Ito & Imai (*Science Rep. Yokohama Nat. Univ.* Section 2 **6**: 4, 1957).

circum- (prefix), all round; round about.

circumcinct, having a band around the middle.

circumscissile, opening or cracking along a circle.

cirrate (**cirrose**), rolled round (curled) or becoming so.

Cirrenalia Meyers & R.T. Moore (1960), anamorphic *Halosphaeriaceae*, Hso.0-≡ hP.1. 17 (marine and terrestrial), widespread (temperate). See Kohlmeyer (*Ber. dt. bot. Ges.* **79**: 27, 1966), Goos (*Proc. Indian Acad. Sci.* Pl. Sci. **94**: 245, 1985), Ravikumar & Purushothaman (*Curr. Sci.* **57**: 674, 1988; India), Matsushima (*Matsush. Mycol. Mem.* **9**: 1, 1996), Somrithipol *et al.* (*Nova Hedwigia* **75**: 477, 2002; Thailand), Zhao & Liu (*Fungal Diversity* **18**: 201, 2005; review), Tsui & Berbee (*Mol. Phylogen. Evol.* **39**: 587, 2006; phylogeny).

Cirrholus Mart. (1821) nom. dub., Fungi.

Cirrhomyces Höhn. (1903) = Chloridium fide Hughes (*CJB* **36**: 727, 1958).

Cirrosporium S. Hughes (1980), anamorphic *Pezizomycotina*, St.≡ eP.23. 1, New Zealand. See Hughes (*N.Z. Jl Bot.* **18**: 329, 1980).

cirrus (**cirrhus**), a curl-like tuft; a tendril-like mass or 'spore horn' of forced-out spores.

Cirsosia G. Arnaud (1918), Asterinaceae. 7, widespread (tropical). See Batista & Maia (*Rev. brasil. Biol.* **2**: 115, 1960), Hosagoudar & Pillai (*MR* **98**: 127, 1994; India), Reynolds & Gilbert (*Aust. Syst. Bot.* **18**: 265, 2005; Australia), Thaung (*Australas. Mycol.* **25**: 5, 2006; Burma).

Cirsosiella G. Arnaud (1918) = Cirsosia fide Müller & von Arx (*Beitr. Kryptfl. Schweiz* **11** no. 2, 1962).

Cirsosina Bat. & J.L. Bezerra (1960), ? Microthyriaceae. 2 (on *Rhododendron*), Europe; Borneo. See Batista & Bezerra (*Revta Biol.* Lisb. **2**: 132, 1960).

Cirsosiopsis Butin & Speer (1979), Microthyriaceae. 1 (on *Araucaria*), Brazil. See Butin & Speer (*Sydowia* **31**: 10, 1978).

Cissococcomyces Brain (1923), anamorphic *Pezizomycotina*, Hso.?.?. 1 (in *Insecta*), S. Africa.

Cistella Quél. (1886) nom. cons., Hyaloscyphaceae. Anamorph *Phialophora*-like. 38, widespread (temperate). See Dennis (*Mycol. Pap.* **32**, 1949), Raitviir (*Scripta Mycol.* **1**, 1970; keys), Matheis (*Friesia* **11**: 85, 1976), Raitviir (*Scripta Mycol.* **8**: 147, 1978), Helfer (*Libri Botanici* **1**, 1991; anamorph), Cantrell & Hanlin (*Mycol.* **89**: 745, 1997; DNA), Raitviir & Järv (*Proc. Est. Acad. Sci. Biol. Ecol.* **46**: 94, 1997), Gams (*Stud. Mycol.* **45**: 187, 2000; anamorph), Yu & Zhuang (*Mycosystema* **22**: 42, 2003; phylogeny), Raitviir (*Scripta Mycol.* **20**, 2004).

Cistellina Raitv. (1978) = Hyphodiscus fide Svrček (*Česká Mykol.* **41**: 193, 1987), Haines (*Mycotaxon* **35**: 324, 1989), Raitviir (*Scripta Mycol.* **20**, 2004), Zhuang *et al.* (*Flora Fungorum Sinicorum* **21**, 2004; China).

cisternal ring, a ring-like arrangement of the endoplasmic reticulum which appears to bud and give rise to vesicles.

Citeromyces Santa María (1957), Saccharomycetales. 1, widespread. See Kurtzman *in* Kurtzman & Fell (Eds) (*Yeasts, a taxonomic study* 4th edn: 146, 1998), Suzuki & Nakase (*J. gen. appl. Microbiol.* Tokyo **45**: 239, 1999), Nagatsuka *et al.* (*Int. J. Syst. Evol. Microbiol.* **52**: 2315, 2002; Thailand), Suh *et al.* (*Mycol.* **98**: 1006, 2006; phylogeny).

citreoviridin, a polyene toxin of *Penicillium citreoviride*; the cause of cardiac beri-beri in humans.

citrinin, a toxic yellow pigment of *Penicillium citrinum*, *P. viridicatum*, etc.; the cause of nephrotoxicosis in pigs.

Citromyces Wehmer (1893) = Penicillium Link fide Thom (*The Penicillia*, 1930), Peterson & Sigler (*MR* **106**: 1109, 2002; nomencl.).

Civisubramaniania Vittal & Dorai (1986), anamorphic *Pezizomycotina*, Hso.0eP.1. 1, India. See Vittal & Dorai (*TBMS* **87**: 482, 1986).

CK, see Metabolic products.

Cladaria Ritgen (1828) nom. rej. = Ramaria Fr. ex Bonord. fide Corner (*Ann. Bot. Mem.* [A monograph of Clavaria and allied genera, addenda] **17**: 347, 1953).

Cladaspergillus Ritgen (1831) = Aspergillus fide Mussat (*Syll. fung.* **15**: 89, 1901).

clade, a monophyletic group of any magnitude (first used by Huxley, *Nature* **180**: 454, 1957); the recognition and hierarchical arrangement of such taxa constitutes the practice of cladistics (q.v.).

Cladia Nyl. (1870), Cladoniaceae (L). 14, widespread. See Filson (*J. Hattori bot. Lab.* **49**: 1, 1981; monogr., key), Ahti (*Regnum veg.* **128**: 58, 1993; names in use), Stenroos *et al.* (*Pl. Syst. Evol.* **207**: 43, 1997), Wedin *et al.* (*Lichenologist* **32**: 171, 2000; phylogeny), Stenroos *et al.* (*Mycol. Progr.* **1**: 267, 2002; phylogeny), Guo & Kashiwadani (*National Science Museum Monographs* **24**: 207, 2004; phylogeny), Miądlikowska *et al.* (*Mycol.* **98**: 1088, 2006; phylogeny).

Cladiaceae Filson (1981) = Cladoniaceae.

Cladidium Hafellner (1984), Lecanoraceae (L). 2, western USA. See Ryan (*Mycotaxon* **34**: 697, 1989), Ryan & Nash (*Nova Hedwigia* **64**: 393, 1997).

Cladina (Nyl.) Nyl. (1866) = Cladonia fide Ahti (*Ann. bot. Soc. zool.-bot. fenn. Vanamo* **32** no. 1, 1961), Ahti (*Beih. Nova Hedwigia* **79**: 25, 1984), Huovinen & Ahti (*Ann. bot. fenn.* **23**: 93, 1986; chemistry 30 spp.), Wei *et al.* (*Acta Mycol. Sin.* **5**: 240, 1986; 9 spp. China), Rouss & Ahti (*Lichenologist* **21**: 29, 1989), Ahti (*Regnum veg.* **128**: 58, 1993; names in use; accepted gen.), Burgaz & Ahti (*Nova Hedwigia* **59**: 399, 1994), Hammer (*Mycol.* **89**: 461, 1997; morphology), Stenroos *et al.* (*Pl. Syst. Evol.* **207**: 43, 1997; phylogeny), Burgaz & Ahti (*Nova Hedwigia* **66**: 549, 1998; Spain), Ito *et al.* (*Mycopathologia* **144**: 169, 1998; nomencl.), Carbonero *et al.* (*Phytochem.* **61**: 681, 2002; polysaccharides), Guo & Kashiwadani (*National Science Museum Monographs* **24**: 207, 2004; phylogeny).

Cladinomyces Cif. & Tomas. (1953) = Cladonia fide Hawksworth *et al.* (*Dictionary of the Fungi* edn 8, 1995).

Cladistics. See Phylogenetic analysis.

Cladobotryum Nees (1816), anamorphic *Hypomyces*, Hso.≡ eH.15. 35, widespread. See Gams & Hoozemans (*Persoonia* **6**: 96, 1970; key), de Hoog (*Persoonia* **10**: 33, 1978; key 11 spp.), Põldmaa *et al.* (*Sydowia* **49**: 80, 1997; 3 new holomorphic spp.), Põldmaa *et al.* (*CJB* **77**: 1756, 1999; phylogeny).

Cladobyssus Ritgen (1831) = Hypha fide Mussat (*Syll. fung.* **15**, 1901).

Cladochaete Sacc. (1912) [non *Cladochaete* DC. 1838, *Compositae*] = Chaetomium fide Petrak & Sydow (*Reprium nov. Spec. Regni veg., Beih.* **42**: 487, 1927).

Cladochasiella Marvanová (1997), anamorphic *Pezizomycotina*, Hso.?.?. 1, Czech Republic. See Marvanová (*Cryptog. Mycol.* **18**: 285, 1997), Descals (*MR* **109**: 545, 2005; diagnostics).

Cladochytriaceae J. Schröt. (1892), Chytridiales. 10 gen. (+ 1 syn.), 38 spp.

Lit.: Sparrow (*Aquatic Phycomycetes* Edn 2: 1187 pp., 1960), Longcore (*CJB* **71**: 415, 1993), Blackwell *et al.* (*Mycotaxon* **89**: 259, 2004).

Cladochytrium Nowak. (1877), Cladochytriaceae. *c.* 10, widespread. See Sparrow (*Aquatic Phycomycetes* Edn 2: 462, 1960; key).

Cladoconidium Bandoni & Tubaki (1985), anamorphic *Pezizomycotina*, Hso.≡ eH.10. 1, Japan; Canada. See Bandoni & Tubaki (*TMSJ* **26**: 426, 1985), Marvanová & Bärlocher (*Czech Mycol.* **53**: 1, 2001; Canada).

Cladodendron Lázaro Ibiza (1916) ≡ Grifola fide Donk (*Verh. K. ned. Akad. Wet.* tweede sect. **62**: 1, 1974).

Cladoderris Pers. ex Berk. (1842) = Cymatoderma fide Donk (*Taxon* **6**: 17, 1957).

Cladodium (Tuck.) Gyeln. (1934) [non *Cladodium* Brid. 1826, *Musci*] ≡ Cladidium.

cladogram, see Cladistics.

Cladographium Peyronel (1918), anamorphic *Pezizomycotina*, Hsy.0eP.?. 1, Europe.

Cladomeris Quél. (1886) = Grifola fide Pegler (*The polypores [Bull. BMS Suppl.]*, 1973).

Cladona Adans. (1763) nom. rej. prop. = Cladonia fide Hawksworth *et al.* (*Dictionary of the Fungi* edn 8, 1995).

Cladonia P. Browne (1756) nom. cons., Cladoniaceae (L). 518, widespread. See also *Cladina*. See Sandstede (*Rabenh. Krypt.-Fl.* **9** 4.2: 1, 1931; Eur.), Asahina (*Lichens of Japan* **1**, 1950), Thomson (*The lichen genus Cladonia in North America*, 1968), Culberson & Kristinsson (*Bryologist* **72**: 431, 1970), Jahns (*Untersuchungen zur Entwicklungsgeschichte der Cladoniaceen*, 1970), Bird & Marsh (*CJB* **50**: 915, 1972; Alberta), Ahti (*Ann. bot. fenn.* **10**: 163, 1973; *Unciales*), Jahns & Beltman (*Lichenologist* **5**: 349, 1973; ontogeny), Ahti (*Ann. bot. fenn.* **15**: 7, 1978; Eur.), Ahti (*Ann. bot. fenn.* **17**: 195, 1980; *gracilis*-group), Ahti (*Lichenologist* **12**: 125, 1980), Ahti (*J. Hattori bot. Lab.* **52**: 331, 1982; evol. trends), Ahti (*Lichenologist* **14**: 105, 1982; morphology), Honegger (*Lichenologist* **15**: 57, 1983; asci), Archer & Bartlett (*N.Z. Jl Bot.* **24**: 581, 1986; key 44 NZ spp.), Upreti (*Feddes Repert.* **98**: 469, 1987; key 62 spp. India & Nepal), Ahti (*Regnum veg.* **128**: 58, 1993; names in use), Hammer (*Bryologist* **96**: 299, 1993; sect. *Perviae* N. Am.), Ahti & Marcelli (*Biblthca Lichenol.* **52**: 5, 1995; *Cladonia verticillaris*-group), Hammer (*Bryologist* **98**: 1, 1995; 57 NW USA spp.), Hammer (*Mycol.* **87**: 46, 1995; ontogeny), Stenroos *et al.* (*Pl. Syst. Evol.* **207**: 43, 1997), Burgaz & Ahti (*Nova Hedwigia* **66**: 549, 1998; Spain), Hammer (*Lichenologist* **30**: 567, 1998; ontogeny), Stenroos & DePriest (*Am. J. Bot.* **85**: 1548, 1998; DNA), Burgaz *et al.* (*Portugaliae Acta Biologica* Série B **18**: 121, 1999; Portugal), Hammer (*Mycol.* **91**: 334, 1999), Woranovicz-Barreira *et al.* (*FEMS Microbiol. Lett.* **181**: 313, 1999; chemistry), Hammer (*Am. J. Bot.* **87**: 33, 2000; ontogeny), Wedin *et al.* (*Lichenologist* **32**: 171, 2000; phylogeny), Azuaga *et al.* (*Mycotaxon* **79**: 433, 2001; Andorra), Hammer (*Bryologist* **104**: 560, 2001; Australia), Ahti *et al.* (*Lichenologist* **34**: 305, 2002; India), Stenroos *et al.* (*Mycol. Progr.* **1**: 267, 2002; phylogeny), Hammer (*Bryologist* **106**: 410, 2003; New Zealand), Herk & Aptroot (*Biblthca Lichenol.* **86**: 193, 2003; *C. cervicornis* group), Myllys *et al.* (*Mol. Phylogen. Evol.* **27**: 58, 2003; phylogeny), Piercey-Normore (*Mycotaxon* **86**: 233, 2003; Canada), Guo & Kashiwadani (*National Science Museum Monographs* **24**: 207, 2004; phylogeny), Litterski & Ahti (*Symb. bot. upsal.* **34** no. 1: 205, 2004; distribution), Piercey-Normore (*CJB* **82**: 947, 2004; photobionts), Ahti & DePriest (*Taxon* **54**: 183, 2005; nomencl.), Miądlikowska *et al.* (*Mycol.* **98**: 1088, 2006; phylogeny), Ahti (*Biblthca Lichenol.* **96**: 5, 2007; E Asia, N America), Beiggi & Piercey-Normore (*J. Mol. Evol.* **64**: 528, 2007; ITS secondary structure).

Cladoniaceae Zenker (1827), Lecanorales (L). 17 gen. (+ 26 syn.), 570 spp.

Lit.: Jahns (*Nova Hedwigia* **20**: 1, 1970; ontogeny), Ahti (*Lichenologist* **14**: 105, 1982; interpretation of thallus), Galloway & James (*Notes R. bot. Gdn Edinb.* **44**: 561, 1987), Stenroos *et al.* (*Ann. bot. fenn.* **25**: 207, 1988; Melanesia), Stenroos *et al.* (*Fl. criptog. Tierra del Fuego* **43**: 95, 1991), Ahti (*Regnum veg.* **128**: 58, 1993; gen., spp. names in use), Burgaz

& Ahti (*Nova Hedwigia* **59**: 399, 1994), DePriest (*Cryptog. bot.* **5**: 60, 1995), Goward & Ahti (*J. Hattori bot. Lab.* **82**: 143, 1997), Stenroos *et al.* (*Pl. Syst. Evol.* **207**: 43, 1997), Ahti (*Lichenology in Latin America. History, Current Knowledge and Applications* [Proceedings of GLAL-3, Terceiro Encontro do Grupo Latino-Americano de Liquenólogos, São Paulo, Brazil, 24-28 September, 1997]: 109, 1998), Stenroos (*Nova Hedwigia* **66**: 457, 1998), Stenroos & DePriest (*Am. J. Bot.* **85**: 1548, 1998; DNA), Ahti (*Fl. Neotrop.* Monogr. **78**, 2000), Hammer (*Am. J. Bot.* **87**: 33, 2000; thallus morphology), Wedin *et al.* (*Lichenologist* **32**: 171, 2000; phylogeny), Miądlikowska *et al.* (*Mycol.* **98**: 1088, 2006; phylogeny), Hofstetter *et al.* (*Mol. Phylogen. Evol.* **44**: 412, 2007; phylogeny).

Cladoniicola Diederich, van den Boom & Aptroot (2001), anamorphic *Pezizomycotina*. 1, Netherlands. See Diederich *et al.* (*Belg. Jl Bot.* **134**: 127, 2001).

Cladoniomyces E.A. Thomas ex Cif. & Tomas. (1953) ≡ Pyxidium Hill.

Cladoniopsis Zahlbr. (1941) = Baeomyces fide Hawksworth *et al.* (*Dictionary of the Fungi* edn 8, 1995).

Cladophialophora Borelli (1980), anamorphic *Capronia*, Hso.1eP+0eH.15. 4, widespread. See Honbo *et al.* (*Sabouraudia* **22**: 209, 1984; relationship to *Cladosporium carrionii*), Kwon-Chung *et al.* (*J. Med. Vet. Mycol.* **27**: 413, 1989; taxonomy of *C. trichoides*), Yegres & Richard-Yegres *in* Miyaji (Ed.) (*Current problems of opportunistic fungal infections*: 12, 1989; epidemiology of *C. carrionii*), Sekhon *et al.* (*Europ. J. Epidemiol.* **8**: 387, 1992; cerebral phaeohyphomycosis by *C. bantiana*), Kawasaki *et al.* (*Mycopathologia* **124**: 149, 1993; molecular epidemiology of *C. carrionii*), Braun & Feiler (*Microbiol. Res.* **150**: 81, 1995; key, n. sp. and comb.), Braun (*Monogr. Cercosporella, Ramularia Allied Genera (Phytopath. Hyphom.)* **2**: 493, 1998; relationship to *Fusicladium*, *Phaeoramularia* and *Pseudocladosporium*), Haase *et al.* (*Stud. Mycol.* **43**: 80, 1999; phylogenetic posn.), Abliz *et al.* (*J. Clin. Microbiol.* **42**: 404, 2004; molecular diagnostics), Guillot *et al.* (*J. Clin. Microbiol.* **42**: 4901, 2004; from dog), Levin *et al.* (*J. Clin. Microbiol.* **42**: 4374, 2004; review of medical lit.), Crous *et al.* (*Stud. Mycol.* **58**: 185, 2007; phylogeny), Davey & Currah (*MR* **111**: 106, 2007; key, from bryophytes), de Hoog *et al.* (*Stud. Mycol.* **58**: 219, 2007; phylogeny, pathogenicity).

Cladophytum Leidy (1849) nom. dub., ? Fungi. 1, USA.

Cladoporus (Pers.) Chevall. (1826) nom. rej. = Laetiporus fide Donk (*Persoonia* **1**: 173, 1960; based on an abnormal polypore).

Cladopsis Nyl. (1885) = Pyrenopsis fide Henssen (*Lichenologist* **21**: 101, 1989).

Cladopycnidium H. Magn. (1940) = Lecidea fide Hertel (*Ergebn. Forsch Unternehmens Nepal Himalaya* **6**: 145, 1977).

Cladoriella Crous (2006), anamorphic *Dothideomycetes*. 1, S. Africa. See Crous (*Stud. Mycol.* **55**: 56, 2006).

Cladorrhinum Sacc. & Marchal (1885), anamorphic *Apiosordaria, Cercophora*, Hso.0eH.15. 6, widespread. See Domsch *et al.* (*Compendium of soil fungi*, 1980), Mouchacca & Gams (*Mycotaxon* **48**: 415, 1993; key), Chopin *et al.* (*J. Med. Vet. Mycol.* **35**: 53, 1997; keratomycosis).

Cladosarum E. Yuill & J.L. Yuill (1938) ? = Aspergillus fide Raper & Fennell (*The genus Aspergillus*, 1965).

Cladosphaera Dumort. (1822), ? Pezizomycotina. 1, Europe.

Cladosphaeria Nitschke ex Jacz. (1894) = Cryptosphaeria Ces. & De Not. fide Petrak (*Sydowia* **15**: 186, 1962).

Cladosporiaceae Nann. (1934) = Davidiellaceae.

Cladosporiella Deighton (1965), anamorphic *Pezizomycotina*, Hso.≡ eP.3. 1, Malaysia. See Deighton (*Mycol. Pap.* **101**: 34, 1965), Crous & Braun (*CBS Diversity Ser.* **1**: 571 pp., 2003).

Cladosporites Félix (1894), Fossil Fungi. 3 (Tertiary), Europe; USA.

Cladosporium Link (1816), anamorphic *Davidiella*, Hso.≡ eP.3. *c.* 150, widespread. *C. herbarum* is ubiquitous and generally saprobic. Medical species are now included in *Cladophialophora.* See de Vries (*Contribution to the knowledge of the genus Cladosporium Link ex Fr.*, 1952; [reprint 1967]), Yamamoto (*Sci. Rep. Hyogo Univ. Agr.* **4**: 1, 1959; spp. from Japan), Ellis (*DH*, 1971), Ellis (*MDH*, 1976), Latgé *et al.* (*J. Microbiol.* **34**: 1325, 1988; wall ultrastr. *C. cladosporioides*), Morgan-Jones & McKemy (*Mycotaxon* **39**: 185, 1990; taxonomy of *C. uredinicola*), McKemy & Morgan-Jones (*Mycotaxon* **41**: 135, 1991; taxonomy of *C. chlorocephalum*), McKemy & Morgan-Jones (*Mycotaxon* **41**: 397, 1991; taxonomy of *C. oxysporum*), McKemy & Morgan-Jones (*Mycotaxon* **43**: 163, 1992; taxonomy of *C. cucumerinum*), Morgan-Jones & McKemy (*Mycotaxon* **43**: 9, 1992; taxonomy of *C. vignae*), David (*Mycol. Pap.* **172**, 1997; subgen. *Heterosporium*), Ho *et al.* (*Mycotaxon* **72**: 115, 1999; key), Crous *et al.* (*Stud. Mycol.* **45**: 107, 2000; review), Crous *et al.* (*Mycol.* **93**: 1081, 2001; phylogeny), Wirsel *et al.* (*Fungal Genetics Biol.* **35**: 99, 2002; molecular ecology), Braun *et al.* (*Mycol. Progr.* **2**: 3, 2003; phylogeny, teleomorph), Dugan *et al.* (*Schlechtendalia* **11**, 2004; checklist), Park *et al.* (*Mycotaxon* **89**: 441, 2004; indoor environment), Schubert & Braun (*Sydowia* **56**: 296, 2004; morphology), Heuchert *et al.* (*Schlechtendalia* **13**, 2005; fungicolous taxa), Schoch *et al.* (*Mycol.* **98**: 1041, 2006; phylogeny), Crous *et al.* (*Stud. Mycol.* **58**: 33, 2007; phylogeny, definition), Crous *et al.* (*Stud. Mycol.* **58**: 1, 2007; phylogeny), Schubert *et al.* (*Stud. Mycol.* **58**: 105, 2007; morphology, biodiversity), Zalar *et al.* (*Stud. Mycol.* **58**: 157, 2007; *C. sphaerospermum* complex).

Cladosporothyrium Katum. (1984) = Zelopelta fide Sutton (*in litt.*).

Cladosterigma Pat. (1892), ? Exobasidiales. 1, Ecuador. See Petch (*TBMS* **8**: 212, 1923), Seifert (*TBMS* **85**: 123, 1985), Seifert & Bandoni (*Sydowia* **53**: 156, 2001).

Cladotrichum Corda (1831) = Oedemium fide Hughes (*CJB* **36**: 727, 1958).

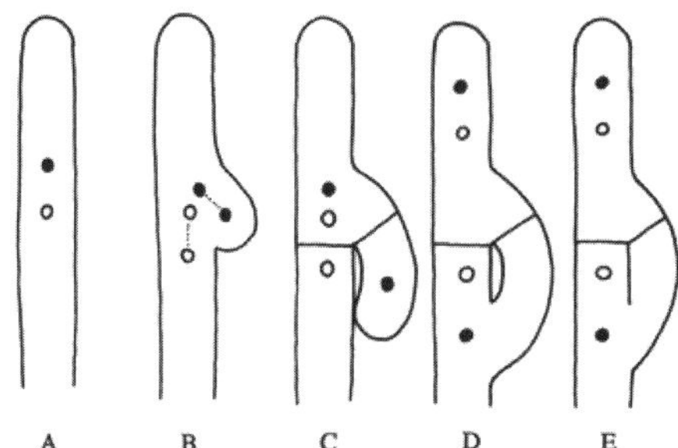

Fig. 11. Clamp connexion formation (diagrammatic). A, dikaryotic hyphal apex; simultaneous nuclear division and formation of a backwardly directed lateral branch into which one of the daughter nuclei passes; C, formation of two cross-walls cutting off an apical cell which contains two compatible nuclei, and a lateral branch with a single nucleus; D, fusion of lateral branch with subapical cell which then becomes dikaryotic; E, final stage.

clamp-connexion (-connection, clamp, clamp-cell) (of basidiomycetes), a hyphal outgrowth which, at cell division, makes a connection between the resulting two cells by fusion with the lower; buckle; nodose septum; by-pass hypha (Fig. 11).

Clarkeinda Kuntze (1891), Agaricaceae. 5, widespread. See Pegler (*J. Linn. Soc.* Bot. **91**: 245, 1985; key).

Classicula R. Bauer, Begerow, Oberw. & Marvanová (2003), Classiculaceae. Anamorph *Naiadella*. 1, Germany. See Bauer *et al.* (*Mycol.* **95**: 757, 2003).

Classiculaceae R. Bauer, Begerow, Oberw. & Marvanová (2003), Classiculales. 2 gen., 2 spp.
Lit.: Bauer *et al.* (*Mycol.* **95**: 757, 2003).

Classiculales R. Bauer, Begerow, Oberw. & Marvanová (2003). Classiculomycetes. 1 fam., 2 gen., 2 spp. Fam.:
Classiculaceae.
For *Lit.* see under fam.

Classiculomycetes R. Bauer, Begerow, J.P. Samp., M. Weiss & Oberw. (2006), Pucciniomycotina. 1 ord., 1 fam., 2 gen., 2 spp. Ord.:
Classiculales
For *Lit.* see ord. and fam.

Classification. The asigning of objects to defined categories; taxonomy. The application of scientific names to the categories into which fungi may be placed and the relative order of those categories is governed by an internationally agreed code (see Nomenclature). The rank of **species** is basic (code Art. 2), but as yet there is no universally applicable definition of species (q.v.). There is fairly general agreement on the ranks to be used for the main infraspecific ranks (although taxa based on pathogenic, physiological, or biochemical characters present difficulties), and on ranks above species from genus to order. Above order, and particularly above class there is more diversity in usage, not least because the code's rules of priority do not apply above family. In the example in Table 3, group-name endings in heavy type are those authorized for fungi in the *Code* (Arts 16-19). The relative order of the various ranks must not be altered (Art. 5) or the names become invalid (Art. 33.5; an exception is made in Art. 33.6 for wrong use of 'tribe' within a genus by Fries in the *Systema mycologicum*, 1821-32). Additional ranks may be intercalated or added, 'provided confusion or error is not thereby introduced' (Art. 4.3).

There is particular instability in the use of ranks above order, and different authors prefer to use or avoid particular categories. 'Division' is peculiar to the code governing fungi, but from 1993 'phylum', familiar to zoologists, was authorized as an alternative which could be used in its place, and this has been used by mycologists (see Kingdoms of fungi). Above the rank of kingdom, additional categories justified by evidence from molecular phylogeny (q.v.) are becoming necessary, 'domain' now being the most widely used.

Different suffixes are available for the terminations of higher taxonomic categories according to whether the code of nomenclature in used, and further under the code governing fungi according to whether the names are considered to be of fungi or algae (i.e. '*-phyta*' not '*-mycota*' as the phylum termination; '*-phyceae*' not '*-mycetes*' for class names; for consistency in this *Dictionary*, and to clearly indicate where phyla are ones traditionally studied by mycologists, the fungal termination has been retained throughout. This also has the practical advantage of keeping phylum names the same regardless of the kingdom in which they are placed.

It is important to note that not all available ranks are used for particular fungi, and there is an increasing tendency for mycologists not to use ranks between the principal ones, or below subspecies.

The rank of special form (q.v.; Art. 4) is available to mycologists wishing to separate morphologically identical fungi by host reactions. Special forms are not regulated by the Code, nor are notations for physiologic races (q.v.) designated by numbers by agreement between interested mycologists.

Any characters judged to represent significant discontinuities, whether biological, morphological, ultrastructural, or molecular, can be used in classifications. Particular emphasis is placed on reproductive structures, ultrastructure, and molecular evidence at levels above family.

Practice as regards the typography of scientific names has varied in different nations. The codes all rule that species names should be differentiated, and *italic* is usually employed, but practice at the family level and above has been inconsistent. The current code governing fungi (Greuter *et al.*, *Regnum Veg.* **138**, 2000) uses *italic* for all scientific names regardless of rank (e.g. *Fungi*, *Ascomycota*, *Lecanorales*, *Russulaceae*).

See Chemotaxonomy, Cladistics, Kingdoms of fungi, Nomenclature, Phylogeny, Species.

Clasterisporium orthographic variant; see *Clasterosporium.*

Clasteropycnis Bat. & Cavalc. (1963), anamorphic *Pezizomycotina*, Cpd.0eP.?. 1, Brazil. See Batista & Cavalcanti (*Quad. Lab. crittogam., Pavia* **31**: 69, 1963)

Clasterosphaeria Sivan. (1984), Magnaporthaceae. Anamorph *Clasterosporium*. 1, Malaysia. See Sivanesan (*TBMS* **83**: 710, 1984).

Clasterosporiaceae Nann. (1934) = Magnaporthaceae.

TABLE 3. Principal, secondary and some other ranks in the nomenclatural hierarchy (botanical).

Rank	Example
Domain	*Eukaryota*
Kingdom	*Fungi*
Subkingdom	†
Phylum	*Basidio**mycota***
Subphylum	† ***-mycotina***
Class	*Telio**mycetes***
Subclass	† ***-mycetidae***
Order	*Uredi**nales***
Suborder	† ***-ineae***
Family	*Puccini**aceae***
Subfamily	† ***-oideae***
Tribe	*Puccini**eae***
Subtribe	† ***-inae***
Genus	*Puccinia*
Subgenus	*Puccinia*
Section	(*Hetero-Puccinia*)
Subsection	†
Series	†
Subseries	†
Species	*Puccinia graminis*
Subspecies	*Puccinia graminis* subsp. *graminis*
Variety	*P. graminis* var. *stackmanii*
Subvariety	†
Form	†
Subform	†
Special form	§*Puccinia graminis* f.sp. *avenae*
Physiologic Race	*P. graminis* f.sp. *avenae* Race 1
Individual	†

† Not necessary for this example.
§ See text.

Clasterosporites Pia (1927), Fossil Fungi. 1 (Eocene), Europe.

Clasterosporium Schwein. (1832), anamorphic *Clasterosphaeria*, Hso.≡ eP.1/19. 10 (mostly on *Cyperaceae*), widespread. Also used for anamorphs of *Asterinaceae*. See Ellis (*Mycol. Pap.* **70**, 1958; key), Sutton *et al.* (*Pl. Path.* **43**: 1066, 1994; *C. flexum* on *Chamaecyparis*), Taylor & Crous (*MR* **104**: 618, 2000), Hosagoudar (*Zoos' Print Journal* **19**: 1437, 2004).

Clathraceae Chevall. (1826) = Phallaceae.
Lit.: Dring (*Kew Bull.* **35**: 1, 1980; monogr.).

clathrate (**clathroid**), like a network; latticed.

Clathrella E. Fisch. (1898) = Clathrus fide Dring (*Kew Bull.* **35**: 1, 1980).

Clathridium, see *Clethridium*.

Clathrina Müll. Arg. (1883) = Cladia fide Hawksworth *et al.* (*Dictionary of the Fungi* edn 8, 1995).

Clathrococcum Höhn. (1911) ? = Epicoccum fide Hawksworth *et al.* (*Dictionary of the Fungi* edn 8, 1995).

Clathroconium Samson & H.C. Evans (1982), anamorphic *Pezizomycotina*, Hso.0-≡ hP.10. 1 (on *Arachnida*), Ghana. See Samson & Evans (*CJB* **60**: 1577, 1982), Samson *et al.* (*Atlas of Entomopathogenic Fungi*: 187 pp., 1988).

Clathrogaster Petri (1900), Hysterangiaceae. 2, Borneo.

Clathroporina Müll. Arg. (1882) = Porina Müll. Arg. fide McCarthy (*Lichenologist* **27**: 321, 1995), McCarthy & Malcolm (*Nova Hedwigia* **62**: 543, 1996), McCarthy & Malcolm (*Lichenologist* **29**: 1, 1997), Lücking & Vezda (*Willdenowia* **28**: 181, 1998; foliicolous spp.).

Clathroporinopsis M. Choisy (1929) ? = Topelia fide Hafellner & Kalb (*Biblthca Lichenol.* **57**: 161, 1995), McCarthy (*Taxon* **45**: 533, 1996; synonymy with *Porina*).

Clathrosphaera Zalewski (1888) nom. conf., anamorphic *Pezizomycotina*. See *Clathrosphaerina*. See van Beverwijk (*TBMS* **34**: 280, 1951).

Clathrosphaerina Beverw. (1951), anamorphic *Hyaloscypha*, Hso.0bP.1. 1 (conidia hollow, net-like), Europe. See Beverwijk (*TBMS* **34**: 289, 1951), Goos (*Mycol.* **79**: 1, 1987; review).

Clathrospora Rabenh. (1857) nom. rej., Pleosporaceae. Anamorph *Alternaria*-like. 5, widespread. See Shoemaker & Babcock (*CJB* **70**: 1617, 1992), Kruys *et al.* (*MR* **110**: 527, 2006; posn).

Clathrosporium Nawawi & Kuthub. (1987), anamorphic *Helotiales*, Hso.#eP.1. 1 (aero-aquatic, conidia hollow, net-like), Malaysia. See Nawawi & Kuthubutheen (*TBMS* **89**: 408, 1987), Hennebert (*CJB* **76**: 1596, 1998), Wang *et al.* (*Mycol.* **98**: 1065, 2006; phylogeny).

Clathrotrichum Pat. (1921) = Beniowskia fide Mason (*Mycol. Pap.* **2**: 27, 1928).

Clathrus P. Micheli ex L. (1753), Phallaceae. *c.* 16, widespread (subtropical; tropical). See Burk (*Mycotaxon* **8**: 463, 1979), Dring (*Kew Bull.* **35**: 1, 1980).

Claudopus Gillet (1876) = Entoloma fide Kuyper (*in litt.*).

Claurouxia D. Hawksw. (1988), ? Candelariaceae (L). 1, Europe. See Hawksworth (*SA* **7**: 65, 1988).

Clausaria Nyl. (1861) = Pertusaria fide Hawksworth *et al.* (*Dictionary of the Fungi* edn 8, 1995).

Claussenomyces Kirschst. (1923), Helotiaceae. Anamorph *Dendrostilbella*. 19, widespread. See Iturriaga (*Mycotaxon* **42**: 327, 1991), Gamundí & Giaiotti (*N.Z. Jl Bot.* **33**: 515, 1995), Verkley (*MR* **99**: 187, 1995; asci), Gamundí & Romero (*Fl. criptog. Tierra del Fuego* **10**, 1998), Medardi (*Czech Mycol.* **59**: 101, 2007; review).

Claustula K.M. Curtis (1926), Claustulaceae. 1, New Zealand. See Buchanan & May (*N.Z. Jl Bot.* **41**: 407, 2003; conservation).

Claustulaceae G. Cunn. (1931), Phallales. 4 gen., 10 spp.

Clautriavia (Pat.) Lloyd (1909) = Phallus fide Stalpers (*in litt.*).

Clauzadea Hafellner & Bellem. (1984), ? Porpidiaceae (L). 4, Europe. See Hafellner & Bellemère (*Nova Hedwigia* Beih. **79**: 319, 1984), Pietschmann (*Nova Hedwigia* **51**: 521, 1990; asci), Meyer (*Sendtnera* **8**: 85, 2002; monogr.), Buschbom & Mueller (*Mol. Phylogen. Evol.* **32**: 66, 2004; phylogeny).

Clauzadeana Cl. Roux (1984), Lecanoraceae (L). 1, widespread. See Roux (*Bull. Soc. linn. Provence* **35**: 101, 1983).

Clauzadella Nav.-Ros. & Cl. Roux (1996), Verrucariaceae (L). 1, France. See Navarro-Rosinés & Roux (*CJB* **74**: 1533, 1996).

Clauzadeomyces Diederich (1994), anamorphic *Pezizomycotina*, Hso.?.?. 1 (lichenicolous), Belgium. See Diederich (*Bull. Soc. linn. Provence* **45**: 418, 1994).

clava, a club-like fruiting structure, e.g. of *Cordyceps*.

clavacin, see patulin.

Clavaria Vaill. ex L. (1753) nom. cons., Clavariaceae. 28, widespread. See Corner (*Ann. Bot. Mem.* [A monograph of Clavaria and allied genera] **1**, 1950), Donk (*Reinwardtia* **2**: 441, 1954), Pine *et al.* (*Mycol.* **91**: 944, 1999; phylogeny), Dentinger & McLaughlin (*Mycol.* **98**: 746, 2006; phylogeny) sensu Linnaeus = *Clavariadelphus* fide.

Clavariaceae Chevall. (1826), Agaricales. 7 gen. (+ 8 syn.), 120 spp.
Lit.: Coker (*The Clavarias of the United States and Canada* [reprint, 1973], 1923; review), Corner (*Ann. Bot. Mem.* [A monograph of Clavaria and allied genera] **1**, 1950), Corner (*Ann. Bot. Mem.* [A monograph of Clavaria and allied genera] **1**: 1, 1950), Donk (*Reinwardtia* **1**, 1951; gen. names), Corner (*Ann. Bot. Mem.* [A monograph of Clavaria and allied genera, addenda] **16**, 1952), Corner (*TBMS* [A monograph of Clavaria and allied genera, addenda] **35**: 285, 1952), Corner (*Ann. Bot. Mem.* [A monograph of Clavaria and allied genera, addenda] **17**, 1953), Pilát (*Acta Mus. Nat. Prag.* **14B**: 129, 1958; European spp.), Thind (*The Clavariaceae of India*, 1961; 15 gen., 92 spp.; keys), Donk (*Persoonia* **3**: 199, 1964), Henry (*Ann. Carnegie Mus.* **39**: 125, 1967; keys 30 spp. W. Penn.), Petersen (*TBMS* **50**: 641, 1967; fam. interrelationships), Petersen (*Sydowia* **21**: 105, 1968), Corner (*Beih. Nova Hedwigia* **33**: 1, 1970), Petersen (*Friesia* **9**: 369, 1971; type studies of *Clavaria* and *Clavulinopsis*), Pilát (*Sborn. národ. muz. Praze* **27B**: 113, 1971; former Czechoslovakia), Berthier (*Biblthca Mycol.* **98**, 1976; world monogr.), Maas Geesteranus (*De fungi van Nederland Die Clavaroide fungi. Auriscalpiaceae, Clavariaceae, Clavulinaceae, Gomphaceae* [*Wetensch. Med. KNVV* **113**], 1976), Pegler & Young (*TBMS* **84**: 207, 1985), Thind & Sharda (*Kavaka* **14**: 9, 1986), Petersen (*Bull. N.Z. Dept. Sci. Industr. Res., Pl. Dis. Div.* [The clavarioid fungi of New Zealand] **263**: 143, 1988), Petersen (*Mycosystema* **2**: 159, 1989), Villarreal & Pérez-Moreno (*Micol. Neotrop. Aplic.* **4**: 119, 1991), Rodríguez-Armas *et al.* (*Docums Mycol.* **22**: 21, 1992), Pine *et al.* (*Mycol.* **91**: 944, 1999), Dentinger & McLaughlin (*Mycol.* **98**: 746, 2006), García-Sandoval *et al.* (*Mycotaxon* **94**: 265, 2005).

Clavariachaeta Lloyd (1922) nom. nud. ≡ Clavariachaete.

Clavariachaete Corner (1950), Hymenochaetaceae. 2, America (tropical). See Corner (*Ann. Bot. Mem.* [A monograph of Clavaria and allied genera] **1**: 268, 1950).

Clavariadelphaceae Corner (1970), Gomphales. 2 gen. (+ 1 syn.), 26 spp.

Clavariadelphus Donk (1933), Clavariadelphaceae. 19, widespread (temperate). See Wells & Kempton (*Mich. Bot.* **7**: 35, 1968; key N. Am. spp.), Methven (*Mycotaxon* **34**: 153, 1989; N. Am. spp.).

Clavariana Nawawi (1976), anamorphic *Pezizomycotina*, Hso.1bH.10. 1 (aquatic), widespread. See Nawawi (*TBMS* **67**: 217, 1976), Marvanová (*Tropical Mycology*: 169, 1997), Descals (*MR* **109**: 545, 2005).

Clavarichaetaceae Jülich (1982) = Hymenochaetaceae.

Clavariella P. Karst. (1881) = Ramaria Fr. ex Bonord. fide Corner (*Ann. Bot. Mem.* [A monograph of Clavaria and allied genera] **1**, 1950).

Clavariopsis De Wild. (1895), anamorphic *Pleosporales*, Hso.1bH.1. 2, widespread. See Petersen (*Mycol.* **55**: 21, 1963; key), Nawawi (*TBMS* **88**: 428, 1987; Malaysia), Belliveau & Bärlocher (*MR* **109**: 1407, 2005; phylogeny).

Clavariopsis Holterm. (1898) ≡ Holtermannia.

Clavascidium Breuss (1996), Verrucariaceae (L). 4, widespread. See Breuss (*Annln naturh. Mus. Wien* Ser. B, Bot. Zool. **98**: 41, 1996), Gueidan *et al.* (*MR* **111**: 1145, 2007; phylogeny).

Clavascina Beneš (1961), Fossil Fungi. 1 (Carboniferous), former Czechoslovakia.

clavate (1) club-like; narrowing in the direction of the base; (2) (of stipes of agarics), narrowing to the apex (Fig. 23.16). Cf. obclavate. See also clava.

clavatin, see patulin.

Clavatisporella K.D. Hyde (1995), ? Hyponectriaceae. 1, Irian Jaya.

Clavatospora Sv. Nilsson ex Marvanová & Sv. Nilsson (1971), anamorphic *Corollospora*, Hso.1bH.1. 3 (aquatic), widespread. See Marvanová & Nilsson (*TBMS* **57**: 531, 1971), Chen (*Fungal Science* Taipei **15**: 47, 2000), Campbell *et al.* (*Fungal Diversity Res. Ser.* **7**: 15, 2002; phylogeny), Descals (*MR* **109**: 545, 2005).

Claviceps Tul. (1853), Clavicipitaceae. Anamorph *Sphacelia*. 41 (mainly on *Poaceae*, rarely *Cyperaceae*), widespread. See Ergot, Poisonous fungi. See Brady (*Lloydia* **25**: 1, 1962; hosts), Loveless (*TBMS* **47**: 205, 1964; identification from honey dew), Tanda (*J. Agric. Sci., Tokyo* **5**: 85, 1981; vars.), Taber *in* Demain & Solomon (Eds) (*Biology of industrial microorganisms*: 449, 1985; review biology), Düvell (*Biblthca Mycol.* **126**, 1989; linear plasmid), Smit (*Biblthca Mycol.* **143**, 1992; transformation), Kuldau *et al.* (*Mycol.* **89**: 431, 1997; DNA), Anon. *in* Kren &

Cvak (Eds) (*Ergot. The Genus Claviceps. Medicinal and Aromatic Plants Industrial Profiles* **6**, 1999), Pažoutová *et al.* (*Pl. Dis.* **84**: 437, 2000; genetic diversity), Scott *et al.* (*Lett. Appl. Microbiol.* **31**: 95, 2000; DNA extraction), Tooley *et al.* (*Phytopathology* **90**: 1126, 2000; population biology), White *et al.* (*Stud. Mycol.* **45**: 95, 2000), Pažoutová (*MR* **105**: 275, 2001; phylogeny), Tooley *et al.* (*Mycol.* **93**: 541, 2001; on sorghum), Tsukiboshi *et al.* (*JARQ* Japan Agricultural Research Quarterly **35**: 221, 2001; on sorghum), Tooley *et al.* (*Pl. Dis.* **86**: 1247, 2002; population biology), Alderman (*Mycology Series* **19**: 195, 2003; speciation), Pažoutová (*Mycology Series* **19**: 329, 2003; evolutionary strategy), Alderman *et al.* (*Pl. Dis.* **88**: 63, 2004; USA), Pažoutová *et al.* (*MR* **108**: 126, 2004; pleomorphy), Walker (*Australas. Pl. Path.* **33**: 211, 2004; Australia), Fisher *et al.* (*Can. J. Pl. Path.* **27**: 389, 2005; diagnostics), Fisher *et al.* (*MR* **109**: 439, 2005; diversity in *C. purpurea*), Pažoutová & Frederickson (*Pl. Path.* **54**: 749, 2005; genetic diversity), Muthusubramanian *et al.* (*MR* **110**: 452, 2006; infraspecific variation), Tooley *et al.* (*MR* **110**: 441, 2006; India), Zhang *et al.* (*Mycol.* **98**: 1076, 2006; phylogeny), van der Linde & Wehner (*Mycol.* **99**: 586, 2007; S Africa).

Clavicipitaceae O.E. Erikss. (1982), Hypocreales. 43 gen. (+ 28 syn.), 321 spp.

Lit.: Doguet (*BSMF* **76**: 171, 1960; status), Kobayasi & Shimizu (*[Iconography of Vegetable Wasps and Plant Worms]*: 280 pp., 1983), Koval' (*Klavitsiptalnyie Grib SSSR*, 1984), Rykard *et al.* (*Mycol.* **76**: 1095, 1984; conidiomata), Jones & Clay (*CJB* **65**: 1027, 1987; ascus ontogeny), Clay (*Ecology* **69**: 10, 1988), Samson *et al.* (*Atlas of Entomopathogenic Fungi*: 61, 1988), Samson *et al.* (*Atlas of Entomopathogenic Fungi*: 187 pp., 1988), Rehner & Samuels (*CJB* **73**: S265, 1991), Schardl *et al.* (*Pl. Syst. Evol.* **178**: 27, 1991; coevolution of endophytes and teleomorphs with hosts), Morgan-Jones *et al.* (*Mycotaxon* **43**: 401, 1992; *Balansieae*), Hywel-Jones & Evans (*MR* **97**: 871, 1993), Spatafora & Blackwell (*Mycol.* **85**: 912, 1993; DNA), Glenn *et al.* (*Mycol.* **88**: 369, 1996), Hodge *et al.* (*Mycol.* **88**: 715, 1996), Bacon & Hill (*Neotyphodium/Grass Interactions*, 1997), Glenn & Bacon (*Neotyphodium/Grass Interactions Proceedings of the Third International Symposium on Acremonium/Grass Interactions, held May 28-31, 1997, in Athens, Georgia*: 53, 1997), Tzean *et al.* (*Atlas of Entomopathogenic Fungi from Taiwan*, 1997), Bacon & White (*Microbial Endophytes*, 2000), Bacon & White (*Microbial Endophytes*: 600 pp., 2000), White *et al.* (*Microbial Endophytes*: 49, 2000), White *et al.* (*Stud. Mycol.* **45**: 95, 2000), Pažoutová *et al.* (*MR* **108**: 126, 2004), Rehner & Buckley (*Mycol.* **97**: 84, 2005), Stensrud *et al.* (*MR* **109**: 41, 2005), Spatafora *et al.* (*Mol. Ecol.* **16**: 1701, 2007), Sung *et al.* (*Stud. Mycol.* **57**: 1, 2007; phylogeny, monogr.), Sung *et al.* (*Mol. Phylogenet. Evol.* **44**: 1204, 2007; phylogeny).

Clavicipitales Nannf. (1932) = Hypocreales.

Clavicorona Doty (1947), Auriscalpiaceae. 10, widespread. See Dodd (*Mycol.* **64**: 746, 1972; key), Lickey *et al.* (*Sydowia* **55**: 181, 2003; phylogeny).

Clavicoronaceae Corner (1970) = Auriscalpiaceae.

Clavicybe Harmaja (2002) ≡ Ampulloclitocybe.

Clavidisculum Kirschst. (1938) = Cistella fide Svrček (*Česká Mykol.* **41**: 193, 1987), Raitviir (*Scripta Mycol.* **20**, 2004).

claviformin, see patulin.

clavine alkaloids, a group of ergoline alkaloids occurring in *Claviceps* sclerotia; also synthesized by *Aspergillus fumigatus* and *Penicillium chermesinum*. See agroclavine, ergoline alkaloids, ergot.

Clavispora Rodr. Mir. (1979), Metschnikowiaceae. Anamorph *Candida*. 2, widespread. See Lachance *et al.* (*Int. J. Syst. Bacteriol.* **36**: 524, 1986), Hendriks *et al.* (*Syst. Appl. Microbiol.* **15**: 98, 1992), Kurtzman *in* Hawksworth (Ed.) (*Ascomycete Systematics. Problems and Perspectives in the Nineties* NATO ASI Series vol. **269 269**: 363, 1994), Lachance & Phaff *in* Kurtzman & Fell (Eds) (*Yeasts, a taxonomic study* 4th edn: 148, 1998), Lachance *et al.* (*FEMS Yeast Res.* **4**: 253, 2003; genetic variation), Arabatzis *et al.* (*Medical Mycology* **42**: 27, 2004; molecular delimitation of clinical strains), Suh *et al.* (*Mycol.* **98**: 1006, 2006; phylogeny).

Clavocephalis Bainier (1882) = Syncephalis fide Kirk (*in litt.*).

Clavochytridium Couch & H.T. Cox (1939) = Blastocladiella fide Couch & Whiffen (*Am. J. Bot.* **29**: 582, 1942).

Clavogaster Henn. (1896), ? Agaricaceae. 1, New Zealand. See Hennings (*Hedwigia* **35**: 303, 1896).

Clavomphalia E. Horak (1987), Tricholomataceae. 1, China. See Horak (*TMSJ* **28**: 176, 1987).

Clavularia P. Karst. (1882) = Cornucopiella fide Seifert (*TBMS* **85**: 123, 1985).

clavulate, somewhat club-like.

Clavulicium Boidin (1957), Clavulinaceae. 4, widespread. See Dentinger & McLaughlin (*Mycol.* **98**: 746, 2006; phylogeny).

Clavulina J. Schröt. (1888), Clavulinaceae. *c.* 40, widespread. See Corner (*Ann. Bot. Mem.* [A monograph of Clavaria and allied genera] **1**, 1950), Petersen (*Mycol.* **59**: 39, 1967), Dentinger & McLaughlin (*Mycol.* **98**: 746, 2006; phylogeny).

Clavulinaceae Donk (1970), Cantharellales. 4 gen. (+ 1 syn.), 59 spp.

Lit.: Coker (*The Clavarias of the United States and Canada* [reprint, 1973], 1923), Corner (*Ann. Bot. Mem.* [A monograph of Clavaria and allied genera] **1**, 1950), Corner (*TBMS* [A monograph of Clavaria and allied genera, addenda] **35**: 285, 1952), Pilát (*Acta Mus. Nat. Prag.* **14B**: 129, 1958; European spp.), Thind (*The Clavariaceae of India*, 1961; 15 gen., 92 spp.; keys), Donk (*Persoonia* **3**: 199, 1964), Henry (*Ann. Carnegie Mus.* **39**: 125, 1967; keys 30 spp. W. Penn.), Petersen (*TBMS* **50**: 641, 1967; fam. interrelationships), Maas Geesteranus (*De fungi van Nederland Die Clavaroide fungi. Auriscalpiaceae, Clavariaceae, Clavulinaceae, Gomphaceae* [*Wetensch. Med. KNVV* **113**], 1976), Petersen (*Mycol.* **77**: 903, 1985), Corner (*Aust. J. Bot.* **34**: 103, 1986), Petersen (*Bull. N.Z. Dept. Sci. Industr. Res., Pl. Dis. Div.* **263**: 143, 1988), Corner (*A monograph of Clavaria and allied genera, addenda* [*Ann. Bot., Lond.* N.S.] **16**, 1989), Corner (*A monograph of Clavaria and allied genera, addenda* [*Ann. Bot., Lond.* N.S.] **17**, 1991), Pine *et al.* (*Mycol.* **91**: 944, 1999), Villegas *et al.* (*Mycotaxon* **70**: 127, 1999), Henkel *et al.* (*Mycol. Progr.* **4**: 343, 2005), Koide *et al.* (*New Phytol.* **165**: 305, 2005).

Clavulinopsis Overeem (1923), Clavariaceae. 33, widespread. See Petersen (*Mycol. Mem.* **2**, 1968; key 10 N. Am. spp.), Petersen (*Mycol.* **70**: 660, 1978), Dentinger & McLaughlin (*Mycol.* **98**: 746, 2006;

phylogeny) = Clavaria (Clavar.) fide.

clavus, the sclerotium of ergot (obsol.).

Cleidiomyces, see *Kleidiomyces*.

Cleistobombardia J.H. Mirza (1968) = Tripterosporella fide Hawksworth *et al.* (*Dictionary of the Fungi* edn 8, 1995).

cleistocarp, see cleistothecium.

Cleistocystis Sousa da Câmara (1931), anamorphic *Pezizomycotina*, Cpd.0eH.?. 1, Portugal.

Cleistoiodophanus J.L. Bezerra & Kimbr. (1976), Pezizaceae. 1, USA.

Cleistonium Speer (1986), anamorphic *Glonium*, St.0eH.?. 1, Brazil. See Speer (*BSMF* **102**: 104, 1986).

Cleistophoma Petr. & Syd. (1927), anamorphic *Pezizomycotina*, Cpd.0eH.?. 2, N. America; Europe. See Sutton (*Mycol. Pap.* **141**, 1977), Aa & Vanev (*A Revision of the Species Described in Phyllosticta*, 2002).

Cleistosoma Harkn. (1884) = Emericella fide Peek & Solheim (*Mycol.* **50**: 844, 1959).

Cleistosphaera Syd. & P. Syd. (1916), Parodiopsidaceae. 1, S. America.

Cleistotheca Zukal (1893) = Pleospora fide von Arx & Müller (*Stud. Mycol.* **9**, 1975).

cleistothecium, a closed fruitbody having no predefined opening; in ascomycota with asci not regularly arranged, e.g. an ascoma of *Thielavia*. cf. chasmothecium.

Cleistothecopsis F. Stevens & E.Y. True (1919) = Pleospora fide Clements & Shear (*Gen. Fung.*, 1931).

Cleistothelebolus Malloch & Cain (1971), Pezizomycotina. 1 (coprophilous), Canada. See van Brummelen *in* Hawksworth (Ed.) (*Ascomycete Systematics. Problems and Perspectives in the Nineties* NATO ASI Series vol. **269 269**: 400, 1994), van Brummelen (*Persoonia* **16**: 425, 1998; posn).

Clelandia Trappe (1979) [non *Clelandia* J.M. Black 1932, *Violaceae*] ≡ Mycoclelandia.

Clematomyces Thaxt. (1900), Laboulbeniaceae. 4 or 5, Asia; west Africa. See Nannfeldt (*Svensk bot. Tidskr.* **43**: 468, 1949).

Cleptomyces Arthur (1918), Pucciniaceae. 1 (on *Leguminosae* or *Verbenaceae*), S. America. See Thirumalachar (*Mycol.* **52**: 688, 1961; syn. of *Stereostratum*), Lindquist (*Royas de la Republica Argentina y zonas limitrofes, Series Coleccion Cientifica no. 20*, 1982).

Clethria, see *Cletria*.

Clethridium (Sacc.) Sacc. (1895) = Discostroma fide Holm (*Taxon* **24**: 475, 1975), Brockmann (*Sydowia* **28**: 275, 1976), Kang *et al.* (*MR* **103**: 53, 1999).

Cletria P. Browne (1756) = Clathrus fide Stalpers (*in litt.*).

Clibanites (P. Karst.) P. Karst. (1871), ? Bionectriaceae. 1 (on rotten wood), Finland. See Rossman *et al.* (*Stud. Mycol.* **42**: 248 pp., 1999).

Climacocystis Kotl. & Pouzar (1958), Fomitopsidaceae. 1, widespread (north temperate). See Ryvarden & Gilbertson (*Europ. Polyp.* **1**: 208, 1993).

Climacodon P. Karst. (1881), Phanerochaetaceae. 6, widespread. See Candoussau (*Bulletin de la Société Mycologique du Béarn* **104**: 3, 1998).

Climacodontaceae Jülich (1982) = Meruliaceae.

Climate change. There is widespread agreement among scientists that Earth's climate is changing as a result of human activity. The impacts of this change, which is towards global warming, cannot yet be predicted accurately, but concern has been expressed about anticipated mass extinctions of animal and plant species. To date, the impact on fungi, their response, and the impact of that response on humanity, have only just begun to be considered: public and political awareness remains very low. If mass extinctions also occur in the fungi as a result of climate change, many will be unrecorded, as so few of the world's fungal species are known (see Numbers of fungi). There is already some evidence that fruiting periods have changed for north-temperate macrofungi. As with animals and plants, the geographical distributions of fungi may move in response to climate change, but at present accessible distributional information is inadequate and there are no agreed criteria for determining whether fungi appearing in new areas are invasives or refugees. The implications for agriculture and forestry, through potential spread of crop diseases, have been largely overlooked. Not enough is known about how fungi will respond to temperature changes, nor about what impact that will have on, for example, the carbon cycle, although Dighton (2003) provides a scholarly introduction to this impending debate. Some habitats are more clearly threatened by global warming than others, and fungi known to occur only in those habitats (nivicolous myxomycetes, other alpine or arctic fungi, specialist mangrove swamp fungi etc.) can confidently be evaluated as under threat. For the rest, the current state of knowledge is rudimentary.

Lit.: Aptroot & Herk (*Environmental pollution* **146**: 293, 2007; effect of global warming on lichen-forming fungi with *Trentepohlia* phycobionts), Blaustein & Dobson (*Nature* **439**: 143, 2006; role of *Batrachochytrium dendrobatis* and climate change in amphibian decline, reviewed in *MR* **110**: 752, 2006), Dighton (*Fungi in ecosystem processes*, 2003), Gange *et al.* (*Science* **316**: 71, 2007; changes in fungal fruiting patterns as evidence of climate change), Mattock *et al.* (*British Wildlife* **18**: 267, 2007; changes in fruiting patterns of spring fungi as evidence of climate change).

Clinoconidium Pat. (1898), Cryptobasidiaceae. 2 (on *Lauraceae*), C. & S. America. See Malençon (*BSMF* **69**: 94, 1953), Begerow *et al.* (*Frontiers in Basidiomycote Mycology*: 81, 2004).

Clinotrichum Cooke (1871) = Acladium fide Kendrick & Carmichae *in* Ainsworth *et al.* (Eds) (*The Fungi* **4A**: 390, 1973).

Clintamra Cordas & Durán (1977), Clintamraceae. 1 (on *Nolina* (*Liliaceae*)), N. America. See Vánky (*Fungal Diversity* **6**: 131, 2001).

Clintamraceae Vánky (2001), Ustilaginales. 1 gen., 1 spp.

Lit.: Cordas & Durán (*Mycol.* **68**: 1239, 1976), Durán (*Ustilaginales of Mexico. Taxonomy, Symptomatology, Spore Germination and Basidial Cytology*: 331 pp., 1987), Vánky (*Cryptog. Stud.* **1**: 159 pp., 1987), Vánky (*Trans. Mycol. Soc. Japan* **32**: 381, 1991), Vánky (*MR* **102**: 513, 1998), Vánky (*Fungal Diversity* **6**: 131, 2001).

Clinterium Fr. (1849) = Topospora fide Groves (*CJB* **43**: 1195, 1965).

Clintoniella (Sacc.) Rehm (1900) = Hypomyces fide Rogerson (*Mycol.* **62**: 865, 1970), Rossman *et al.* (*Stud. Mycol.* **42**: 248 pp., 1999).

Cliostomum Fr. (1825), Ramalinaceae (L). 7, Europe; N. America. See Gowan (*Mycol.* **82**: 766, 1990; key

4 N. Am. spp.), Ekman (*Symb. bot. upsal.* **32** no. 1: 17, 1997), Ekman (*MR* **105**: 783, 2001; phylogeny).

Clisosporium Fr. (1819) nom. rej. = Coniothyrium fide Hawksworth *et al.* (*Dictionary of the Fungi* edn 8, 1995).

Clistosoma Clem. & Shear (1931) ≡ Cleistosoma.

Clithramia Nag Raj (1988), anamorphic *Pezizomycotina*, St.1eP.1. 1 (on *Oryza*), India. See Nag Raj (*CJB* **66**: 903, 1988).

Clithris (Fr.) Bonord. (1851) = Cenangium fide Minter (*in litt.*).

Clitocybe (Fr.) Staude (1857), Tricholomataceae. *c.* 300, widespread (esp. north temperate). *C. augeana* (frequently but incorrectly as *C. dealbata*), a mushroom-bed contaminant. The genus is polyphyletic and several segregate genera have recently been recognised. See Harmaja (*Karstenia* **10**: 5, 1969; key 43 spp. Finland), Singer (*Sydowia* **31**: 199, 1979; key 202 spp.), Bigelow (*Beih. Nova Hedwigia* **72**, 1982), Bigelow (*Beih. Nova Hedwigia* **81**, 1985; N. Am. spp.), Moncalvo *et al.* (*Syst. Biol.* **49**: 278, 2000; probably non-monophyletic), Harmaja (*Ann. bot. fenn.* **40**: 213, 2003), Lachapelle (*Revue du Cercle de Mycologie de Bruxelles* **3**: 17, 2003; key Europ. spp.), Mata & Petersen (*Mycotaxon* **86**: 303, 2003; type studies of neotropical spp.).

Clitocybula (Singer) Singer ex Métrod (1952), Marasmiaceae. *c.* 15, widespread (temperate). See Singer (*Sydowia* **31**: 233, 1979; key).

Clitopilina Arnaud (1952) nom. nud., Xenasmataceae.

Clitopiloidea (Romagn.) Largent (1994) = Entoloma fide Kuyper (*in litt.*).

Clitopilopsis Maire (1937) = Rhodocybe fide Singer (*Am. midl. Nat.* **37**: 527, 1946).

Clitopilus (Fr. ex Rabenh.) P. Kumm. (1871), Entolomataceae. *c.* 30, widespread (north temperate). *C. passeckerianus*, a mushroom-bed contaminant. See Ware (*Gdnrs' Chron.* **97**: 325, 1935), Singer (*Sydowia* **31**: 235, 1979; key), Baroni *et al.* (*Fungal Diversity* **6**: 13, 2001; Thailand), Legon & Roberts (*Mycologist* **16**: 114, 2002; Brit. sp.) See cat's ear.

Clohesyomyces K.D. Hyde (1993), anamorphic *Pezizomycotina*, Cpd.0eH.1. 1 (aquatic), Australia. See Hyde (*Aust. Syst. Bot.* **6**: 170, 1993).

Clohiesia K.D. Hyde (1995), Sordariales. 2, Australia. See Hyde (*Nova Hedwigia* **61**: 119, 1995; similarities with *Clypeosphaeriaceae*), Réblová & Winka (*Mycol.* **93**: 478, 2001), Untereiner *et al.* (*CJB* **79**: 321, 2001; phylogeny).

Clonophoromyces Thaxt. (1931), Laboulbeniaceae. 1, West Indies; Japan. See Terada & Tavares (*TMSJ* **34**: 357, 1993; Japan).

Clonostachyopsis Höhn. (1907) = Clonostachys fide Clements & Shear (*Gen. Fung.*, 1931), Schroers (*Stud. Mycol.* **46**: 1, 2001).

Clonostachys Corda (1839), anamorphic *Bionectriaceae*, Hso.0eH.15. 48, widespread (esp. Europe). See Hawksworth & Punithalingam (*TBMS* **64**: 89, 1975), Schroers *et al.* (*Mycol.* **91**: 365, 1999), Schroers (*Stud. Mycol.* **45**: 63, 2000; phylogeny), Schroers (*Stud. Mycol.* **46**: 1, 2001; monogr.).

Closteroaleurosporia Grigoraki (1924) ≡ Microsporum.

closterospore (in dermatology), a multinucleate phragmospore, as in *Trichophyton* (obsol.).

Closterosporia Grigoraki (1924) = Microsporum fide Hawksworth *et al.* (*Dictionary of the Fungi* edn 8, 1995).

Closterosporium Sacc. (1883), anamorphic *Melanomma*, Cac.≡ eP.?. 1, Europe.

club fungi, the *Clavariaceae*.

club root, a disease of crucifers (*Plasmodiophora brassicae*), see *Plasmodiophora*.

cluster-cup, see aecium.

clypeate, having a clypeus.

Clypeispora A.W. Ramaley (1991), anamorphic *Mycosphaerella*, Cpd.0eH.1. 1, USA. See Ramaley (*Mycotaxon* **40**: 13, 1991).

Clypeocarpus Kirschst. (1941) = Mazzantia fide von Arx & Müller (*Beitr. Kryptfl. Schweiz* **11** no. 1, 1954).

Clypeoceriospora Sousa da Câmara (1946), Pezizomycotina. 1, Europe.

Clypeochorella Petr. (1923), anamorphic *Pezizomycotina*, St.0eH.?. 1, Europe.

Clypeococcum D. Hawksw. (1977), Dacampiaceae. 7 (on lichens), Europe; N. America. See Hawksworth (*J. Linn. Soc.* Bot. **75**: 196, 1977).

Clypeodiplodina F. Stevens (1927) = Ascochytulina fide Clements & Shear (*Gen. Fung.*, 1931).

Clypeolaria Tratt. (1825) nom. dub., Fungi. See Rudolphi (*Linnaea* **4**: 395, 1829; det. as *Insecta*).

Clypeolella Höhn. (1910) = Schiffnerula fide Hughes (*CJB* **62**: 2213, 1984), Hughes *in* Sugiyama (Ed.) (*Pleomorphic Fungi: The Diversity and its Taxonomic Implications*: 103, 1987), Hosagoudar (*Zoos' Print Journal* **18**: 1071, 2003).

Clypeolina Speg. (1923) ≡ Clypeolopsis.

Clypeolina Theiss. (1918), Micropeltidaceae. 1, Brazil.

Clypeolinopsis Bat. (1959) ? = Stomiopeltis fide Müller & von Arx (*Beitr. Kryptfl. Schweiz* **11** no. 2, 1962).

Clypeolopsis F. Stevens & Manter (1925) ? = Stomiopeltis fide Müller & von Arx (*Beitr. Kryptfl. Schweiz* **11** no. 2, 1962).

Clypeolum Speg. (1881), Pezizomycotina. 8, widespread (tropical). See Batista (*Publções Inst. Micol. Recife* **56**, 1959), Müller & von Arx (*Beitr. Kryptfl. Schweiz* **11** no. 2, 1962; synonymy with *Porina*).

Clypeomyces Kirschst. (1935) nom. dub., Fungi. See Petrak (*Sydowia* **1**: 61, 1947).

Clypeopatella Petr. (1942), anamorphic *Pezizomycotina*, St.0eH.?. 1, Rhodes.

Clypeophialophora Bat. & Peres (1962), anamorphic *Pezizomycotina*, St.0eH.?. 1, Brazil. See Batista & Peres (*Publções Inst. Micol. Recife* **358**: 8, 1962).

Clypeophysalospora H.J. Swart (1981), Clypeosphaeriaceae. 1 (from dead leaves), Australia. See Kang *et al.* (*Mycoscience* **40**: 151, 1999).

Clypeoporthe Höhn. (1919), Valsaceae. Anamorph *Phaeocytostroma*. 4 (grass culms), widespread (tropical). See Sivanesan & Waller (*Phytopath. Pap.* **29**, 1986), Barr (*Mycotaxon* **41**: 287, 1991).

Clypeoporthella Petr. (1924), ? Diaporthaceae. Anamorph *Phomopsis*. 2, Europe; N. America. See Barr (*Mycol. Mem.* **7**, 1978).

Clypeopycnis Petr. (1925), anamorphic *Pezizomycotina*, Cpd.0eH.15. 3, widespread. See Sutton & Pascoe (*Stud. Mycol.* **31**: 177, 1989), Priest (*Fungi of Australia: Septoria* **0**, 2006).

Clypeopyrenis Aptroot (1991), Pyrenulaceae (L). 1, C. & S. America. See Aptroot (*Biblthca Lichenol.* **44**: 40, 1991), Aptroot *et al.* (*Biblthca Lichenol.* **97**, 2008; Costa Rica).

Clypeorhynchus Kirschst. (1936) = Diaporthe fide von Arx & Müller (*Beitr. Kryptfl. Schweiz* **11** no. 1,

1954).

Clypeoseptoria F. Stevens & P.A. Young (1925), anamorphic *Pezizomycotina*, St.0fH.?. 3, Hawaii; Brazil. See Ciferri & Batista (*Atti Ist. bot. Univ. Lab. Crittog. Pavia* sér. 5 **14**: 53, 1957).

Clypeosphaeria Fuckel (1870), Clypeosphaeriaceae. 31 (dead wood and bark), widespread. See Barr (*SA* **8**: 1, 1989), Hyde *et al.* (*Sydowia* **50**: 21, 1998), Kang *et al.* (*Mycoscience* **40**: 151, 1999), Tsui *et al.* (*Mycol.* **93**: 1002, 2001; Hong Kong), Jeewon *et al.* (*MR* **107**: 1392, 2003; phylogeny).

Clypeosphaeriaceae G. Winter (1886), Xylariales. 8 gen. (+ 3 syn.), 49 spp.

Lit.: Barr (*SA* **8**: 1, 1989), Barr (*Mycotaxon* **39**: 43, 1990), Hyde *et al.* (*Sydowia* **50**: 21, 1998), Kang *et al.* (*Fungal Diversity* **1**: 147, 1998), Kang *et al.* (*Mycoscience* **40**: 151, 1999; broad concept probably polyphyletic), Jeewon *et al.* (*MR* **107**: 1392, 2003).

Clypeosphaerulina Sousa da Câmara (1939), Pezizomycotina. 1, Portugal.

Clypeostagonospora Punith. (1981), anamorphic *Pezizomycotina*, St.≡ eH.1. 1, USA. See Punithalingam (*Nova Hedwigia* **34**: 67, 1981).

Clypeostigma Höhn. (1919) = Phyllachora Nitschke ex Fuckel (1870) fide von Arx & Müller (*Beitr. Kryptfl. Schweiz* **11** no. 1, 1954).

Clypeostroma Theiss. & Syd. (1914), Dothideomycetes. 2, Australasia. See von Arx & Müller (*Stud. Mycol.* **9**, 1975), Johnston (*N.Z. Jl Bot.* **40**: 265, 2002; revision).

Clypeothecium Petr. (1922) = Exarmidium fide Barr & Boise (*Mycotaxon* **23**: 233, 1985), Aptroot (*Nova Hedwigia* **66**: 89, 1998).

Clypeotrabutia Seaver & Chardón (1926) = Phyllachora Nitschke ex Fuckel (1870) fide Cannon (*Mycol. Pap.* **163**, 1991).

Clypeum Massee (1896) = Parmularia Lév. fide Batista & Vital (*Atas Inst. Micol. Univ. Recife* **1**, 1960).

Clypeus (Britzelm.) Fayod (1889) = Inocybe fide Mussat (*Syll. fung.* **15**, 1901).

clypeus, a shield-like stromatic growth, with or without host tissue, over one or more ascomata or conidiomata.

CMA, see Media.

CMI, see IMI.

Cnazonaria Corda (1829) = Typhula fide Berthier (*Bull. mens. Soc. linn. Lyon* Num. Spéc. **46**, 1976).

co- (prefix), together.

coacervate, massed (heaped), together.

coadnate, united, cohering, connate.

coalescent, joined together.

coarctate, crushed together, crowded, constricted.

Coccidiascus Chatton (1913), ? Eremotheciaceae. 1 (in *Drosophila*), Europe; N. America. See Phaff *in* Kurtzman & Fell (Eds) (*Yeasts, a taxonomic study* 4th edn: 153, 1998), Suh *et al.* (*Mycol.* **98**: 1006, 2006; phylogeny).

Coccidiodictyon Oberw. (1989), Septobasidiaceae. 1, Spain. See Oberwinkler (*Op. bot.* **100**: 188, 1989).

Coccidioidaceae Cif. (1932) ? = Onygenaceae.

Coccidioides G.W. Stiles (1896), anamorphic *Onygenaceae*, ?.0eH.?. 3, widespread (esp. America). *C. immitis* on humans and animals (**coccidiomycosis**, coccidioidal granuloma, San Joaquin Valley Fever). See also *Uncinocarpus*. See Fiese (*Coccidioidomycosis*, 1958), Al Doory (*Mycopathologia* **46**: 113, 1972; bibliogr.), Sigler & Carmichael (*Mycotaxon* **4**: 458, 1976; *Malbranchea* state), Stevens (*Coccidioidomycosis. A text*, 1980), Szaniszlo *et al. in* Howard (Ed.) (*Fungi pathogenic for man and animals* Part A, Biology, 1983; life cycle, dimorphism), Pan *et al.* (*Microbiology* **140**: 1481, 1994; syst. posn), Kaufman *et al.* (*J. Clin. Microbiol.* **36**: 3721, 1998; diagnostics in vivo), Fisher *et al.* (*Mol. Ecol.* **8**: 1082, 1999; population studies), Larone *et al.* (*Manual of Clinical Microbiology*: 1259, 1999; review), Greene *et al.* (*Mycol.* **92**: 406, 2000; molecular detection and identification), Fisher *et al.* (*Mycol.* **94**: 73, 2002; *C. posadasii*), Sugiyama *et al.* (*Stud. Mycol.* **47**: 5, 2002; phylogeny), Millar *et al.* (*J. Clin. Microbiol.* **41**: 5778, 2003; molecular identification), Bialek *et al.* (*J. Clin. Microbiol.* **42**: 778, 2004; molecular identification), Muñoz *et al.* (*J. Clin. Microbiol.* **42**: 1247, 2004; Mexico), Geiser *et al.* (*Mycol.* **98**: 1053, 2006; phylogeny), Pounder *et al.* (*J. Clin. Microbiol.* **44**: 2977, 2006; molecular identification), Umeyama *et al.* (*J. Clin. Microbiol.* **44**: 1859, 2006; molecular identification), Morrow (*Mycol.* **98**: 669, 2006; in Holocene bison).

coccidioidin, an antigen prepared from *Coccidioides immitis*, esp. for skin testing. Cf. spherulin.

coccidiomycosis, see *Coccidioides*; mycosis.

Coccidomyces Buchner (1912), ? Saccharomycetidae. 1 (in *Insecta*), Germany. See Stainhaus (*Insect microbiology*, 1946).

Coccidophthora Syd. & P. Syd. (1913) nom. dub., Fungi. See Petch (*TBMS* **10**: 190, 1925).

Cocciscia Norman (1870), ? Dacampiaceae (?L). 1, Scandinavia.

Coccobolus Wallr. (1833) = Ceuthospora Grev. fide Rabenhorst (*Dtschl. Krypt.-Fl.* **1**: 144, 1844), Sutton (*Mycol. Pap.* **141**, 1977).

Coccobotrys Boud. & Pat. (1900), anamorphic *Lepiota*. 2, Europe; Chile. See Donk (*Taxon* **11**: 82, 1962).

Coccocarpia Pers. (1827), Coccocarpiaceae (L). 24, widespread. See Arvidsson (*Op. Bot.* **67**, 1983; monogr., key), Awasthi (*Kavaka* **12**: 83, 1986; Indian spp.), Marcano *et al.* (*Trop. Bryol.* **10**: 215, 1995; Venezuela), Makhija *et al.* (*Trop. Bryol.* **17**: 47, 1999), Miądlikowska & Lutzoni (*Am. J. Bot.* **91**: 449, 2004; phylogeny), Miądlikowska *et al.* (*Mycol.* **98**: 1088, 2006; phylogeny), Lücking *et al.* (*Biblthca Lichenol.* **95**: 429, 2007; Costa Rica).

Coccocarpiaceae Henssen (1986), Peltigerales (L). 5 gen. (+ 5 syn.), 36 spp.

Lit.: Awasthi (*Kavaka* **13**: 83, 1985), Lumbsch & Kothe (*Mycotaxon* **43**: 277, 1992), Lumbsch *et al.* (*Mycotaxon* **43**: 277, 1992; thallus SEM), Marcano *et al.* (*Trop. Bryol.* **10**: 215, 1995), Wagner (*Bryonora* **15**: 8, 1995), Makhija *et al.* (*Trop. Bryol.* **17**: 47, 1999), Henssen & Tønsberg (*Bryologist* **103**: 108, 2000), Ekman & Jørgensen (*CJB* **80**: 625, 2002), Wedin & Wiklund (*Symb. bot. upsal.* **34** no. 1: 469, 2004), Miądlikowska *et al.* (*Mycol.* **98**: 1088, 2006; phylogeny), Hofstetter *et al.* (*Mol. Phylogen. Evol.* **44**: 412, 2007; phylogeny), Jørgensen (*Nordic Lichen Flora* **3**: Cyanolichens: 12, 2007).

Coccochora Höhn. (1909), Dothideales. 1, Asia. See Bose & Müller (*Indian Phytopath.* **17**: 3, 1964).

Coccochorella Höhn. (1910) ≡ Coccochora.

Coccochorina Hara (1927) = Clypeosphaeria fide Hara (*Byogaichu-hoten*, 1948).

Coccodiella Hara (1910), Phyllachoraceae. 21 (biotrophic on leaves of *Poaceae*), widespread (tropical). See Hyde & Cannon (*Mycol. Pap.* **175**: 114, 1999),

Pearce & Hyde (*Fungal Diversity Res. Ser.* **17**: 308 pp., 2006; Australia), Seixas *et al.* (*Mycol.* **99**: 99, 2007; Brazil).

Coccodiniaceae Höhn. ex O.E. Erikss. (1981), Chaetothyriales. 5 gen. (+ 6 syn.), 18 spp.
Lit.: Eriksson (*Op. Bot.* **60**, 1981), Reynolds (*Mycotaxon* **27**: 377, 1986), Winka *et al.* (*Mycol.* **90**: 822, 1998; phylogeny), Winka (*Phylogenetic Relationships Within the Ascomycota Based on 18S rDNA Sequences, Akademisk Avhandling* [Thesis (PhD), Department of Ecology and Environmental Science, Umeå University]: [91] pp., 2000), Liu & Hall (*Proc. natn Acad. Sci. U.S.A.* **101**: 4507, 2004), Crous *et al.* (*Stud. Mycol.* **58**: 1, 2007; morphology).

Coccodinium A. Massal. (1860), Coccodiniaceae. 2, Europe. See Eriksson (*Op. Bot.* **60**, 1981), Winka *et al.* (*Mycol.* **90**: 822, 1998; phylogeny), Crous *et al.* (*Stud. Mycol.* **58**: 1, 2007; morphology).

Coccodiscus Henn. (1904) = Coccoidea fide Müller & von Arx (*Beitr. Kryptfl. Schweiz* **11** no. 2, 1962).

Coccodothella Theiss. & Syd. (1915) = Coccoidella fide Müller & von Arx (*Beitr. Kryptfl. Schweiz* **11** no. 2, 1962).

Coccodothis Theiss. & Syd. (1914), Parmulariaceae. 2, America. See Petrak (*Sydowia* **8**: 291, 1954), Butin & Marmolejo (*Rev. Mex. Micol.* **4**: 9, 1988), Marmolejo & Butin (*MR* **101**: 1515, 1997).

Coccogloeum Petr. (1955), anamorphic *Pezizomycotina*, Cac.0eH.?. 1, Austria. See Petrak (*Sydowia* **9**: 588, 1955).

Coccoidea Henn. (1900), Coccoideaceae. 2, Japan; India; C. America. See Eriksson (*Op. Bot.* **60**, 1981), Inácio & Cannon (*Fungal Diversity* **9**: 71, 2002).

Coccoideaceae Henn. ex Sacc. & D. Sacc. (1905), ? Dothideales. 3 gen. (+ 4 syn.), 10 spp.
Lit.: Eriksson (*Op. Bot.* **60**, 1981), Sivanesan (*TBMS* **89**: 265, 1987), Yuan *et al.* (*Mycotaxon* **60**: 175, 1996), Inácio & Cannon (*Fungal Diversity* **9**: 71, 2002).

Coccoidella Höhn. (1909), ? Coccoideaceae. Anamorph *Colletogloeum*-like. 7, America (tropical); Australia. See Müller & von Arx (*Beitr. Kryptfl. Schweiz* **11** no. 2, 1962), Sivanesan (*TBMS* **89**: 265, 1987; posn), Yuan *et al.* (*Mycotaxon* **60**: 175, 1996; Australia), Inácio & Cannon (*Fungal Diversity* **9**: 71, 2002).

Coccoidiopsis Hara (1913) ≡ Coccodiella fide Müller & von Arx *in* Ainsworth *et al.* (Eds) (*The Fungi* **4A**: 87, 1973), Cannon (*SA* **11**: 168, 1993).

Coccomycella Höhn. (1917) = Coccomyces fide Sherwood (*Occ. Pap. Farlow Herb. Crypt. Bot.* **15**, 1980).

Coccomyces De Not. (1847), Rhytismataceae. 116, widespread. See Sherwood (*Occ. Pap. Farlow Herb. Crypt. Bot.* **15**, 1980; monogr., key), Cannon & Minter (*Mycol. Pap.* **155**, 1986; India), Johnston (*N.Z. Jl Bot.* **24**: 89, 1986; key 19 spp., NZ), Spooner (*Kew Bull.* **45**: 451, 1990; key 11 spp., Sabah), Johnston (*Mycotaxon* **52**: 221, 1994; ascospores), Johnston (*Aust. Syst. Bot.* **13**: 199, 2000; Australia), Lin *et al.* (*Mycosystema* **20**: 1, 2001; China), Hou *et al.* (*CJB* **83**: 37, 2005; China), Wang *et al.* (*Mycol.* **98**: 1065, 2006; phylogeny).

Coccomycetella Höhn. (1917), Odontotremataceae. 1, Europe. See Sherwood-Pike (*Mycotaxon* **28**: 137, 1987).

Cocconia Sacc. (1889), Parmulariaceae. 6, widespread (tropical). See Hansford (*Mycol. Pap.* **15**, 1946), Batista & Vital (*Atas Inst. Micol. Univ. Recife* **1**: 167, 1960), Inácio & Cannon (*Fungal Diversity* **9**: 71, 2002).

Cocconiopsis G. Arnaud (1918) = Cyclostomella fide Müller & von Arx (*Beitr. Kryptfl. Schweiz* **11** no. 2, 1962).

Coccopeziza Har. & P. Karst. (1890) = Arthonia fide Sherwood (*Mycotaxon* **6**: 215, 1977).

Coccophacidium Rehm (1888) = Therrya fide Nannfeldt (*Nova Acta R. Soc. Scient. upsal.*, 1932).

Coccophysium Link (1833) nom. dub., ? Fungi.

Coccopleum Ehrenb. (1818) = Sclerotium fide Fries (*Syst. Mycol.* **2**: 256, 1823).

Coccospora Kudo (1925), Microsporidia. 1.

Coccosporella P. Karst. (1893) = Mycogone fide Hughes (*CJB* **36**: 727, 1958).

Coccosporium Corda (1831) nom. rej., anamorphic *Pezizomycotina*. See Hughes (*CJB* **36**: 752, 1958), Holubová-Jechová (*Sydowia* **46**: 240, 1994; ? = Corynespora (Corynesporasc.)).

Coccostroma Theiss. & Syd. (1914) = Coccodiella fide Hyde & Cannon (*Mycol. Pap.* **175**, 1999; typification).

Coccostromella Petr. (1968), ? Dothideales. 1, widespread (tropical). See Petrak (*Sydowia* **21**: 267, 1967).

Coccostromopsis Plunkett (1924) = Coccodiella fide Hyde & Cannon (*Mycol. Pap.* **175**, 1999).

Coccotrema Müll. Arg. (1888), Coccotremataceae (L). 19, widespread. See Brodo (*Bryologist* **76**: 260, 1973; N. Am. spp.), Messuti (*N.Z. Jl Bot.* **34**: 57, 1996; S. Am.), Lumbsch *et al.* (*Organ. Divers. Evol.* **1**: 99, 2001; posn), Schmitt *et al.* (*Lichenologist* **33**: 315, 2001; generic limits), Messuti (*Mycotaxon* **82**: 429, 2002; S America), Messuti (*Biblthca Lichenol.* **86**: 129, 2003), Schmitt *et al.* (*J. Hattori bot. Lab.* **100**: 753, 2006; phylogeny).

Coccotremataceae Henssen ex J.C. David & D. Hawksw. (1991), Pertusariales (L). 2 gen. (+ 3 syn.), 20 spp.
Lit.: Brodo (*Bryologist* **76**: 260, 1973), Henssen *in* Brown *et al.* (Eds) (*Lichenology: progress and problems*: 107, 1976; ontogeny), Galloway & Watson-Gandy (*Bryologist* **95**: 227, 1992), Messuti (*N.Z. Jl Bot.* **34**: 57, 1996), Schmitt *et al.* (*Lichenologist* **33**: 315, 2001), Schmitt *et al.* (*J. Hattori bot. Lab.* **100**: 753, 2006; phylogeny).

Coccotrichum Link (1824) = Botrytis fide Saccardo (*Syll. fung.* **4**: 120, 1886), Hughes (*CJB* **36**: 727, 1958; accepted genus).

Coccularia Corda (1829) nom. dub., anamorphic *Pezizomycotina*. Rejected by Fries (*Summ. veg. scand.*: 522, 1849).

coccus, a spherical bacterium.

Cochlearia (Cooke) Lambotte (1888) [non *Cochlearia* L. 1753, *Cruciferae*] = Otidea fide Eckblad (*Nytt Mag. Bot.* **15**: 1, 1968).

cochleariform, spoon-like in form.

cochleate, shell-like in form; twisted like a shell.

cochliobolin, see ophiobolin.

Cochliobolus Drechsler (1934), Pleosporaceae. Anamorphs *Bipolaris*, *Curvularia*. 22, widespread. Important plant pathogens include: *C. heterostrophus* (anamorph *Bipolaris maydis*; leaf spot of maize), *C. miyabeanus* (anamorph *B. oryzae*; brown spot of rice), *C. sativus* (anamorph *B. sorokiniana*, spot blotch and root rot of temperate cereals), *C. victoriae* (anamorph *B. victoriae*; foot rot of cereals); see victorin. See Sivanesan (*Mycol. Pap.* **158**: 261 pp.,

1987), Raguchander *et al.* (*Plant Pathology Newsletter* **6**: 45, 1990; on triticale), Lakshmanan *et al.* (*Z. PflKrankh. PflPath. PflSchutz* **98**: 185, 1991; biocontrol), Panaccione *et al.* (*Phytopathology* **81**: 1156, 1991; *tox* genes), Fetch *et al.* (*Phytopathology* **82**: 1101, 1992; on Hordeum), Simcox *et al.* (*Phytopathology* **82**: 621, 1992; isoenzymes), Tzeng *et al.* (*Genetics* Bethesda **130**: 81, 1992; population structure), Clay *et al.* (*Protoplasma* **178**: 34, 1994; appressoria), Jones & Dunkle (*Molecular Plant-Microbe Interactions* **8**: 476, 1995; virulence genes), Scheffer & Walton (*Helminthosporia Metabolites, Biology, Plant Diseases* Bipolaris, Drechslera, Exserohilum: 61, 1995; toxigenic spp.), Gafur *et al.* (*Mycoscience* **38**: 455, 1997; mating types), Christiansen *et al.* (*MR* **102**: 919, 1998; mating types), Shimizu *et al.* (*J. gen. appl. Microbiol.* Tokyo **44**: 251, 1998; phylogeny), Turgeon & Berbee (*Molecular Genetics of Host-Specific Toxins in Plant Disease*: 153, 1998; pathogenesis), Berbee *et al.* (*Mycol.* **91**: 964, 1999; phylogeny), Zhong & Steffenson (*Phytopathology* **91**: 469, 2001; virulence), Gafur *et al.* (*Mycobiology* **31**: 19, 2003; Indonesia), Schoch *et al.* (*Mycol.* **98**: 1041, 2006; phylogeny).

Cochliomyces Speg. (1912), Euceratomycetaceae. 2, C. & S. America. See Nannfeldt (*Svensk bot. Tidskr.* **43**: 468, 1949).

Cochlonema Drechsler (1935), Cochlonemataceae. 18, widespread. See Drechsler (*Mycol.* **27**: 185, 1935), Drechsler (*Mycol.* **29**: 229, 1937), Drechsler (*Mycol.* **31**: 128, 1939), Drechsler (*Mycol.* **31**: 388, 1939), Drechsler (*Mycol.* **33**: 248, 1941), Drechsler (*Mycol.* **34**: 274, 1942), Drechsler (*Mycol.* **37**: 1, 1945), Drechsler (*Mycol.* **38**: 120, 1946), Drechsler (*Mycol.* **43**: 161, 1951), Drechsler (*Mycol.* **47**: 364, 1955), Drechsler (*Mycol.* **51**: 787, 1959), Jones (*TBMS* **42**: 75, 1959; Kenya), Jones (*TBMS* **45**: 348, 1962; UK), Miura (*J. Jap. Bot.* **47**: 204, 1972; Japan), Dyal (*Sydowia* **27**: 293, 1976; keys parasites of nematodes and amoebae, bibliogr.), Saikawa & Sato (*Mycol.* **83**: 403, 1991; ultrastr.), Michel & Wylezich (*Mikrokosmos* **94**: 75, 2005; biology), Koehsler *et al.* (*Mycol.* **99**: 215, 2007; EM, phylogeny).

Cochlonemataceae Dudd. (1974), Zoopagales. 6 gen., 33 spp.

Lit.: Dyal (*Sydowia* **27**: 293, 1976; keys parasites of nematodes and amoebae, bibliogr.), Saikawa & Saito (*TBMS* **87**: 337, 1986), Saikawa & Sato (*Mycol.* **83**: 403, 1991), Saikawa & Katsurashima (*Mycol.* **85**: 24, 1993), Saikawa & Aoki (*Nova Hedwigia* **60**: 571, 1995), Tanabe *et al.* (*Mol. Phylogen. Evol.* **16**: 253, 2000), Barron (*Biodiversity of Fungi* Inventory and Monitoring Methods: 435, 2004).

Cocoicola K.D. Hyde (1995), Phaeochoraceae. 5 (palm leaves), widespread (tropical). See Hyde & Cannon (*Mycol. Pap.* **175**, 1999).

Codinaea Maire (1937), anamorphic *Chaetosphaeria*. See Gamundí *et al.* (*Darwiniana* **21**: 96, 1977), Kuthubutheen & Nawawi (*MR* **95**: 1224, 1991), Samuels *et al.* (*Mycol.* **89**: 156, 1997; teleomorph).

Codinaeopsis Morgan-Jones (1976), anamorphic *Chaetosphaeriaceae*, Hso.0eH.16. 1, N. America; Japan. See Morgan-Jones (*Mycotaxon* **4**: 166, 1976), Arambarri & Cabello (*Mycotaxon* **38**: 11, 1990), Réblová (*Stud. Mycol.* **45**: 149, 2000; review), Réblová & Winka (*Mycol.* **92**: 939, 2000; phylogeny).

Codonmyces Calat. & Etayo (1999), anamorphic *Pezizomycotina*. 1, Spain. See Calatayud & Etayo (*Lichenologist* **31**: 594, 1999).

Coeloanguillospora Dyko & B. Sutton (1978) = Filosporella fide Dyko & Sutton (*Mycotaxon* **7**: 323, 1978).

Coelocaulon Link (1833) = Cetraria fide Aptroot (*in litt.*), Kantvilas (*Flora of Australia* **55**: 36, 1994; Australia).

Coelographium (Sacc.) Gäum. (1920), anamorphic *Pezizomycotina*, Hsy.0eP.?. 1, Europe.

Coelomomyces Keilin (1921), Coelomomycetaceae. *c.* 70 (in *Insecta*), widespread. See Couch (*J. Elisha Mitchell scient. Soc.* **61**: 124, 1945), Couch & Dodge (*J. Elisha Mitchell scient. Soc.* **63**: 69, 1947), Laird (*Can. J. Zool.* **37**: 781, 1959; key), Sparrow (*Aquatic Phycomycetes* Edn 2: 638, 1960; keys), Laird (*J. Elisha Mitchell scient. Soc.* **78**: 132, 1962), Anon. *in* Couch & Bland (Eds) (*The genus Coelomomyces*, 1988), Scholte *et al.* (*J. Insect Sci.* **4**: 19, 2004).

Coelomomycetaceae Couch ex Couch (1962), Blastocladiales. 2 gen. (+ 1 syn.), 71 spp.

Lit.: Weiser & Žižka (*Česká Mykol.* **28**: 227, 1975) See also *Lit.* under *Coelomomyces*, Whisler *et al.* (*Proc. natn Acad. Sci. U.S.A.* **72**: 693, 1975), Couch & Bland (*The genus Coelomomyces*: 399 pp., 1985), Scholte *et al.* (*J. Insect Sci.* **4**: 1943, 2004).

Coelomorum Paulet (1793) = Helvella fide Hawksworth *et al.* (*Dictionary of the Fungi* edn 8, 1995).

Coelomyces, see *Coelomomyces*.

Coelomycetes (obsol.). Widespr., saprobic or parasitic on higher plants, fungi, lichens, vertebrates, also recovered from the widest range of ecological niches. 1000 gen. (+ 500 syn.), 7000 spp. The term 'coelomycetes' merely indicates that conidia are formed within a cavity lined by fungal or fungal/host tissue. The range of conidiogenous events is more limited than in hyphomycetes (q.v.), but solitary, 'phialidic', and 'annellidic' types predominate, and 'tretic' and 'basauxic' are absent. See Anamorphic fungi. For modern reassessments of coelomycete systematics see Sutton (*in* Ainsworth *et al.* (Eds), *The Fungi* **4A**: 513, 1973), Nag Raj (*in* Cole & Kendrick (Eds), *Biology of conidial fungi* **1**: 43, 1981), Nag Raj (*Coelomycetous anamorphs with appendage-bearing conidia*, 1993).

Lit.: Höhnel (*Mykol. Unters.* **1**: 301-369, 1923; classification), Grove (*British stem- and leaf- fungi* **1** & **2**, 1935, 1937), Petch (*TBMS* **26**: 53-70, 1943; British *Nectrioideae*), Sutton (*Coelomycetes* I - VII, *Mycol. Pap.* 1961-1981; VI, 1977, generic names proposed for *Coelomycetes*), Nag Raj *et al.* (*Icones generum coelomycetum* I-XIII, 1972-1982), Nag Raj *et al.* (*CJB* **49-67**; Genera coelomycetum I-XXVIII), Mathur (*The Coelomycetes of India*, 1979), Kendrick (*The whole fungus*, 1979), Michaelides *et al.* (*Icones generum coelomycetum* suppl.), Sutton (*The Coelomycetes*, 1980), Nag Raj (*in* Cole & Kendrick, *Biology of conidial fungi*: 43-84, 1981). See also under Mitosporic fungi.

This artificial class has been traditionally separated into three orders, Sphaeropsidales, characterized by pycnidial conidiomata, *Melanconiales* with acervular conidiomata, and *Pycnothyriales* with pycnothyrial conidiomata. An alternative system for anamorphic fungi as a whole (now obsol.) was suggested by Sutton (1980) where differences in conidiogenesis were used for separation of taxa at the class, subclass and ordinal levels with conidiomatal structure used at the subordinal level. The traditional separation is:

(1) **Melanconiales** (fams. Melanconiaceae, Stilbosporaceae, Coryneaceae - see Sutton, 1973 for discussion for family names). Mycelium is within the host or substratum. Conidiogenous events are various. Conidiomata are subcuticular, epidermal, subepidermal, peridermal or subperidermal and the conidiogenous layer is formed within the substratum. Dehiscence is by rupture of the overlying tissues and conidial masses may be dry or slimy. Conidiomata become erumpent at maturity and grade into sporodochial conidiomata (tuberculariaceous hyphomycetes). In culture such fungi cannot be distinguished from many hyphomycetes.

Lit.: von Arx (*A revision of the fungi classified as Gloeosporium*, 1970), Guba (*Monograph of Monochaetia and Pestalotia*, 1961), Vassiljevsky & Karakulin (*Fungi imperfecti parasitici*, Pt. II. Melanconiales, 1950).

(2) **Sphaeropsidales** (Phomales, Phyllostictales; fams. Sphaerioidaceae, Nectrioidaceae, Leptostromataceae, Excipulaceae, Discellaceae, Asbolisiaceae etc. – see Sutton, 1973 for discussion of family names). Mycelium may be immersed in the substratum or superficial. Conidia may be dry or slimy and conidiogenous events are various. Conidiomata are superficial, semi-immersed or immersed with the conidiogenous layer lining the walls of the locule(s). Diversity in conidiomatal structure is considerable in terms of tissue composition (see tissue types), number and arrangement of locules, relationship to the substrata and type of dehiscence (see Sutton, 1980; Nag Raj, 1981, for the range of variation).

Lit.: Allescher (*Rabenh. Krypt.-Fl.* **1** (6-7), 1900-03), Diedicke (*Krypt.-Fl. Mk Brandenb.* **9** (7), 1914), Petrak & Sydow (*Beih. Rep. spec. nov. regni veg.* **42**, 1927; *Macrophoma* and its segregates), Bender (*The Fungi Imperfecti: Order Sphaeropsidales*, 1933), Biga *et al.* (*Sydowia* **12**, 1959; *Coniothyrium*), Byzova *et al.* (*Flora Sporovyk rast. Kazak.* **5**, 1970), Punithalingam (*Studies on Sphaeropsidales in culture* I-III, *Mycol. Pap.* 1970-81).

(3) **Pycnothyriales** (fams. Pycnothyriaceae, Microthyriopsidaceae, Peltopycnidiaceae, Actinothyriaceae, Actinopeltaceae, Rhizothyriaceae, Peltasteraceae, see Katumoto, 1975, for discussion of family names). Mycelium may be immersed in the substratum or superficial; when superficial it may bear hyphopodia and/or setae. Conidia are produced in several ways. Conidiomata are superficial or subcuticular, flattened, uni- or multi-locular, sometimes attached to the substratum by a central column of tissue or hypostroma, otherwise attached at the periphery; conidiogenous layer may be restricted to the upper or lower surface or occur on both; tissue structure of the pycnothyrium and nature of marginal cells are important generic criteria. Anamorphs of Dothideales.

Lit.: Naumov (*BSMF* **30**: 423-432, 1915), Arnaud (*Ann. Ecol. Nat. Agr. Montp.* **16**: 1-288, 1918), v. Höhnel (*Mykol. Unters.* **1**: 301-369, 1923), Tehon (*Trans. Ill. St. Acad. Sci.* **33**: 63-65, 1940), Batista & Ciferri (*Mycopath. Mycol. Appl.* **11**: 1-102, 1959), Katumoto (*Bull. Fac. Agr. Yamaguti Univ.* **26**: 45-122, 1975).

Coelomycidium Debais. (1919), Blastocladiales. 1, Europe. See Loubès & Manier (*Prototistologica* **10**: 47, 1974), Weiser & Žižka (*Česká Mykol.* **28**: 159, 1974), Weiser & Žižka (*Česká Mykol.* **28**: 227, 1974, 1974).

Coelopogon Brusse & Kärnefelt (1991), Parmeliaceae (L). 2, south temperate. See Brusse & Kärnefelt (*Mycotaxon* **42**: 35, 1991), Thell *et al.* (*Mitteilungen aus dem Institut für Allgemeine Botanik Hamburg* **30-32**: 283, 2002; phylogeny, ecology), Thell *et al.* (*Mycol. Progr.* **1**: 335, 2002; phylogeny), Mattsson & Articus (*Symb. bot. upsal.* **34** no. 1: 237, 2004; phylogeny).

Coelopus Bataille (1908) = Gyroporus fide Singer (*Am. midl. Nat.* **37**: 527, 1946).

Coelorhopalon Overeem (1925) = Xylaria Hill ex Schrank fide Dennis (*Mycol. Pap.* **62**, 1956), Læssøe (*SA* **13**: 43, 1994; ? synonym of *Xylaria*).

Coelosphaeria Sacc. (1873) ≡ Nitschkia fide Holm (*Taxon* **24**: 275, 1975).

Coelosporidium Mesnil & Marchoux (1897) nom. dub., Fungi. Protozoa or Fungi.

Coelosporium Link (1824), anamorphic *Pezizomycotina*, Hso.?.?. 1, Europe.

Coemansia Tiegh. & G. Le Monn. (1873), Kickxellaceae. 18, widespread. See Linder (*Farlowia* **1**: 49, 1943; key), Benjamin (*Aliso* **4**: 149, 1959), Chein (*Mycol.* **63**: 1046, 1971), Young (*Microbios* **63**: 187, 1990; ultrastr.), Kwansa *et al.* (*MR* **103**: 900, 1999; isolation), Kwansa *et al.* (*MR* **103**: 925, 1999; mycoparasitism), Kurihara & Tokumasu (*Memoirs of the National Science Museum* Tokyo **34**: 205, 2000; Japan), Ho & Hsu (*Taiwania* **50**: 22, 2005; Taiwan), James *et al.* (*Nature* **443**: 818, 2006; phylogeny).

Coemansiella Sacc. (1883) = Kickxella fide Linder (*Farlowia* **1**: 49, 1943).

Coemurus, see *Coeomurus*.

Coenicia Trevis. (1861) ? = Pyrenula Ach. (1809) fide Aguirre-Hudson (*Bull. Br. Mus. nat. hist.* Bot. **21**: 85, 1991).

coeno- (prefix), living together, e.g. multinucleate.

Coenocarpus Fr. (1825) ≡ Chaenocarpus Spreng.

coenocentrum (of oomycetes), a small deeply staining body at the centre of the multinucleate oosphere to which the egg-nucleus goes.

coenocyte (adj. **coenocytic**), a multinucleate mass of protoplasm; (adj., of fungi), non-cellular, in the sense of non-septate; Vuillemin (1912) used **coenocyte** for a cell usually multinucleate and **apocyte** for one temporarily or secondarily multinucleate; **coenocytium** originally used for a structure resulting from nuclear division not followed by cytoplasmic cleavage, in contrast to a **syncytium**, a multinucleate structure resulting from the fusion of several protoplasts.

coenogametes, multinucleate gametangia (q.v.) which, upon fusion, give a **coenozygote**.

Coenogoniaceae Stizenb. (1862), Ostropales. 1 gen. (+ 13 syn.), 87 spp.

Lit.: Davis (*Bryologist* **97**: 186, 1994), Stocker-Worgötter (*Symbiosis* **23**: 117, 1997), Lücking & Kalb (*Bot. Jb.* **122**: 1, 2000), Kauff & Lutzoni (*Mol. Phylogen. Evol.* **25**: 138, 2002), Miądlikowska *et al.* (*Mycol.* **98**: 1088, 2006; phylogeny), Rivas Platas *et al.* (*Fungal Diversity* **23**: 255, 2006).

Coenogoniomycella Cif. & Tomas. (1954) ≡ Coenogonium Ehrenb. fide Hawksworth *et al.* (*Dictionary of the Fungi* edn 8, 1995).

Coenogoniomyces Cif. & Tomas. (1954) ≡ Coenogonium Ehrenb. fide Hawksworth *et al.* (*Dictionary of the Fungi* edn 8, 1995).

Coenogonium Clem. (1909) = Coenogonium Ehrenb. fide Hawksworth *et al.* (*Dictionary of the Fungi* edn 8, 1995).

Coenogonium Ehrenb. (1820), Coenogoniaceae (L). *c.* 87, widespread (esp. tropical). See Santesson (*Symb. bot. upsal.* **12** no. 1: 1, 1952), Uyenco (*Bryologist* **66**: 217, 1964; taxonomy), Uyenco (*Trans. Am. microsc. Soc.* **84**: 1, 1965; *Trentepohlia*), Wang Jang (*Taiwania* **17**: 40, 1972), Xavier Filho *et al.* (*Boln Soc. Bot.* II **56**: 115, 1983; Brazil), Lücking & Kalb (*Bot. Jb.* **122**: 1, 2000; Brazil), Malcolm (*Australasian Lichenology* **54**: 19, 2004; generic limits), Miądlikowska *et al.* (*Mycol.* **98**: 1088, 2006; phylogeny), Rivas Platas *et al.* (*Fungal Diversity* **23**: 255, 2006; Costa Rica).

Coenoicia Trevis. (1861) = Melanotheca fide Hawksworth *et al.* (*Dictionary of the Fungi* edn 8, 1995).

Coenomyces K.N. Deckenb. (1901), Chytridiomycota. 1 (on marine cyanobacteria), Europe; N. America.

Coenomycogonium Cif. & Tomas. (1954) = Coenogonium Ehrenb. fide Hawksworth *et al.* (*Dictionary of the Fungi* edn 8, 1995).

Coenosphaeria Munk (1953) = Keissleriella fide Bose (*Phytopath. Z.* **41**, 1961).

coenozygote, see coenogametes.

Coeomurus Gray (1821) nom. rej. ≡ Uromyces fide Berndt (*in litt.*).

Coevolution. Reciprocal evolutionary adaptations between disparate organisms leading to their interdependence. A process widespread in fungi as about two-thirds of the spp. form intimate relationships with other organisms as commensals, mutualists, or pathogens. Most studied in pathogen-host gene-for-gene (q.v.) 'arms-races'.

Lit.: Cannon (*in* McKey & Sprent (Eds), *Advances in legume systematics* **5**, 1994; in *Phyllachoraceae* and *Leguminosae*), Hedberg (Ed.) (*Symb. bot. upsal.* **22**(4), 1979; parasites as taxonomists), Karatygin (*Koevolutseya gribov i rasteniĭ*, 1993), Kenneth & Palti (*Mycol.* **76**: 705, 1984; in *Compositae*), Pirozynski & Hawksworth (Eds) (*Coevolution of fungi with plants and animals*, 1988), Savile (*Bot. Rev.* **45**: 377, 1979), Stone & Hawksworth (Eds) (*Coevolution and systematics*, 1986), Thompson (*The coevolutionary process*, 1994).

See also ambrosia fungi, Endophytes, Fungicolous fungi, galls, Insects and fungi, Lichens, Mycorrhiza, symbiosis.

Coilomyces Berk. & M.A. Curtis (1853) = Geastrum fide Zeller (*Mycol.* **40**: 649, 1948).

Coinostelium Syd. (1939) = Prospodium fide Thirumalachar & Kern (*Bull. Torrey bot. Club* **82**: 102, 1955).

Cokeromyces Shanor (1950), Mucoraceae. 1, USA. See Shanor *et al.* (*Mycol.* **42**: 271, 1950), Benny & Benjamin (*Aliso* **8**: 391, 1976), McGough *et al.* (*Clin. Microbiol. Newsl.* **12**: 113, 1980; mucormycosis), Jeffries & Young (*Mycol.* **75**: 509, 1983; ultrastr.), Benny *et al.* (*Mycotaxon* **22**: 119, 1985), Axelrod *et al.* (*J. Infect. Dis.* **155**: 1062, 1987; mucormycosis), Murripalli *et al.* (*J. Clin. Microbiol.* **34**: 2601, 1996; mucormycosis), O'Donnell *et al.* (*Mycol.* **93**: 286, 2001; phylogeny), Voigt & Wöstemeyer (*Gene* **270**: 113, 2001; phylogeny), Tanabe *et al.* (*Mol. Phylogen. Evol.* **30**: 438, 2004; phylogeny), White *et al.* (*Mycol.* **98**: 872, 2006; phylogeny).

Colacogloea Oberw. & Bandoni (1991), Heterogastridiaceae. 4, widespread (north temperate). See Fell *et al.* (*Int. J. Syst. Evol. Microbiol.* **50**: 1351, 2000; mol. phylogeny), Bandoni *et al.* (*Czech Mycology* **54**: 31, 2002; Canada).

Colacosiphon R. Kirschner, R. Bauer & Oberw. (2001), anamorphic *Cryptomycocolacaceae*. 1, Germany. See Kirschner *et al.* (*Mycol.* **93**: 643, 2001), Bauer *et al.* (*Mycol.* **95**: 756, 2003; ultrastr.).

Colemaniella Agnihothr. (1974), anamorphic *Pezizomycotina*, Hso.≡ eP.?. 1, India. See Agnihothrudu (*Journal of Coffee Research* **4**: 3, 1974).

Colensoniella Hafellner (1979), ? Dothideales. 1, south temperate. See Hafellner (*Nova Hedwigia* Beih. **62**: 160, 1979), Rosato (*Boln Soc. argent. Bot.* **30**: 163, 1995; Argentina).

Coleocarpon Stubblef., T.N. Taylor, C.E. Mill. & G.T. Cole (1983), Fossil Fungi. 1, USA.

Coleodictyospora Charles (1929), anamorphic *Pezizomycotina*, Hso.#eP.1. 1, West Indies. See Mel'nik (*Mikol. Fitopatol.* **32**: 32, 1998).

Coleodictys Clem. & Shear (1931) ≡ Coleodictyospora.

Coleoma Clem. (1909) ≡ Coleopuccinia.

Coleomyces Moreau & M. Moreau (1937) = Cylindrocarpon fide Booth (*Mycol. Pap.* **104**, 1966).

Coleonaema Höhn. (1924), Pezizomycotina. See Duan *et al.* (*Fungal Diversity* **26**: 187, 2007; revision).

Coleophoma Höhn. (1907), anamorphic *Pezizomycotina*, Cpd.0eH.15. 14, widespread. See Sutton (*The Coelomycetes*, 1980), Masilamani & Muthumary (*Acta Bot. Indica* **22**: 286, 1994; ultrastr.), Masilamani & Muthumary (*MR* **99**: 693, 1995; ontogeny), Wu *et al.* (*MR* **100**: 943, 1996; revision), Bianchinotti & Rajchenberg (*Sydowia* **56**: 217, 2004; on *Proteaceae*), Duan *et al.* (*Fungal Diversity* **26**: 187, 2007; revision).

Coleopteromyces Ferrington, Lichtw. & López-Lastra (1999), Legeriomycetaceae. 1 (in *Scirtidae*), Brazil. See Lichtwardt *et al.* (*Mycol.* **91**: 1064, 1999).

Coleopuccinia Pat. (1889), Pucciniales. 4 (on *Rosaceae* (III)), Japan; China. See Thirumalachar & Whitehead (*Am. J. Bot.* **41**: 120, 1954), Cummins & Hiratsuka (*Illustr. Gen. Rust Fungi rev. edit.*, 1983; syn. of *Gymnosporangium*).

Coleopucciniella Hara ex Hirats. (1937), Pucciniaceae. 2, E. Asia (esp. Japan). See Laundon (*Mycotaxon* **3**: 134, 1975).

Coleoseptoria Petr. (1940), anamorphic *Pezizomycotina*, St.0fH.?. 1, Germany.

Coleosperma Ingold (1954), Dermateaceae. 1 (aquatic), British Isles. See Ingold (*TBMS* **37**: 9, 1954), Nauta & Spooner (*Mycologist* **13**: 65, 1999).

Coleospora Cribb (1959) nom. dub., ? Fungi.

Coleosporiaceae Dietel (1900), Pucciniales. 6 gen. (+ 5 syn.), 131 spp.

Lit.: Berndt (*CJB* **77**: 1469, 1999), Crane (*CJB* **79**: 957, 2001), Ono (*Mycoscience* **43**: 421, 2002), Cummins & Hiratsuka (*Illustr. Gen. Rust Fungi* edn 3: 225 pp., 2003), Hernández & Hennen (*Mycol.* **95**: 728, 2003), Maier *et al.* (*CJB* **81**: 12, 2003), Wingfield *et al.* (*Australas. Pl. Path.* **33**: 327, 2004), Aime (*Mycoscience* **47**: 112, 2006).

Coleosporium Lév. (1847), Coleosporiaceae. *c.* 100 (on *Pinus* (0, I) (*Pinaceae*); on dicots (asp. asterids) and some monocots (*Orchidaceae*) (II *Caeoma*-like, III)), widespread. *C. campanulae* (on *Campanula*), *C. senecionis* (on *Senecio*), *C. solidaginis* (on *Solidago* and *Aster*), 5 microcyclic spp. on *Pinus*. See Kern (*Mycol.* **20**: 60, 1928; key spp. on *Pinus*), Laundon (*Mycotaxon* **3**: 154, 1975; typification), Kaneko (*Rep. Tottori mycol. Inst.* **19**: 1, 1981; keys 28 Jap. spp.), Berndt (*Sydowia* **48**: 263, 1996), Mims &

Richardson (*CJB* **83**: 451, 2005; ultrastr.), Aime (*Mycoscience* **47**: 112, 2006).

Coleroa Rabenh. (1850), Venturiaceae. 8, widespread. See Barr (*Sydowia* **41**: 25, 1989; key 3 N. Am. spp.).

Colispora Marvanová (1988), anamorphic *Pezizomycotina*, Hso.≡ eH.21. 2 (aquatic), widespread. See Marvanová (*TBMS* **90**: 614, 1988), Nawawi & Kuthubutheen (*Mycotaxon* **34**: 497, 1989; Malaysia), Gönczöl & Révay (*Mycotaxon* **59**: 237, 1996), Descals (*MR* **109**: 545, 2005; diagnostics).

collabent, falling in; collapsing.

Collacystis Kunze (1827) nom. dub., anamorphic *Pezizomycotina*. See Sutton (*Mycol. Pap.* **141**, 1977).

collarette, a cup-shaped structure at the apex of a conidiogenous cell.

collariate, having a collar; see collarette.

Collarium Link (1809) nom. dub., anamorphic *Pezizomycotina*.

collarium, the ring of tissue to which the proximal ends of remote lamellae are attached as in *Marasmius* spp., etc.

collateral host, see *Pucciniales*.

Collecephalus J.A. Spencer (1972), anamorphic *Pezizomycotina*, Hso.0eH.6. 1, USA. See Spencer (*Hemerocallis J.* **26**: 15, 1972).

Collection and preservation. Carefully dried specimens of many fungi retain at least their microscopic structure for many years. Their preservation in dried reference collections as research and voucher material is therefore recommended. For most taxa, it is only practicable to retain fruit-bodies in dried collections, though mycelia can be preserved as living cultures. Most fungi with fleshy fruit bodies shrink considerably on drying, and such features as colour are frequently lost. Good field notes on size, shape, colour, texture etc. are thus important, and colour photographs are valuable. Freeze-drying (lyophilization) of fleshy fungi often gives satisfactory results (Kendrick, *Mycol.* **61**: 392, 1969), although the resulting specimens are fragile.

The experienced worker when making a collection gives attention to (1) conservation: ensuring that the gathering is unlikely to affect the continuing presence of the fungus in that location; (2) quality: representative material must be gathered in a range of developmental stages (including those with full-sized spores in or on the cells producing them and appearing ready for release, but also including when possible earlier and later stages); (3) amount: where possible, enough material should be collected for investigation not only by the collector but also later by others; and (4) field notes, ensuring that features which will deteriorate on drying are adequately described, and that as much information as possible is given about the location and timing of the collection (including, where possible, latitude, longitude, altitude in m above sea level and, where appropriate, aspect), abundance (not what is being counted - ascoma, basidioma etc. - the number seen - a logarithmic scale may be appropriate - the time taken to see that quantity, and whether the find was serendipitous or the result of a search - in the case of an unsuccessful search, an abundance of zero may be recorded), associated organisms (e.g. the species of plant on which the fungus occurs), the part of the organism (e.g. leaf) or other substratum, and details of the nature of any association (necrotrophic parasite etc.). Many lichens are very slow-growing, so particular care should be taken to ensure that populations are not unduly denuded by collectors. Field observations may suffice, and necessary chemical tests can sometimes be carried out in situ. It is possible to assess the condition of immersed and stromatic microfungi in the field by cutting transversely across the fruit bodies and examining their contents preferably with a hand-lens; where spores are present the inner surface of the fruit-body is usually shiny, and the contents will swell if a drop of water is added. Collections should be placed in packets or baskets as appropriate, ensuring that fragile specimens are protected from crushing, and that there is an unambiguous link between field notes and specimen. Plastic bags are not recommended due to the uncontrollable build-up of humidity levels. Where feasible, specimens should be collected along with their substratum, to minimize disturbance of their structures. Knives, folding saws or secateurs may be appropriate, and for rock-inhabiting lichens a geological hammer and masonry chisel may be needed. After as short a time as possible, and if necessary after microscopical examination, the material must be dried. Many leaf-fungi are best pressed between drying papers, although delicate hyphomycetes etc. may be damaged by such a process and should be air-dried. Specimens of microfungi on wood and bark, and most macrofungi, such as polypores and some gasteromycetes, should be dried in a current of warm dry air, ideally in a specially constructed dryer. Spore prints (on black paper if the spores are pale) of hymenomycetes are valuable, and can sometimes later be used for culturing.

Once dry, specimens are usually kept loose, to facilitate later observations. Microfungi should be protected by glassine inner bags in folded paper outer packets, which may be fixed onto herbarium sheets or kept in filing cabinets. Fragile specimens, especially myxomycetes and macrofungi, are normally pinned or glued onto small pieces of card and kept in small cardboard boxes. Nowadays, most collection data are stored in computerized databases, but packets must still be labelled with at least basic information in order to facilitate curation. Illustrations, photographs etc. must also be cross-referenced with the specimen. Semi-permanent slide preparations (see Mounting media) should always be kept if possible, either in protective cardboard boxes within the specimen packet or in a separate collection. This minimizes the need for future workers to deplete the specimen further, and is particularly important for type or authentic collections.

See also Genetic Resource Collections, Inventories, Reference collections.

Collema C.A. Browne (1756), ? Pezizomycotina (L). 1, Europe.

Collema Weber ex F.H. Wigg. (1780) nom. cons., Collemataceae (L). 78, widespread. See Degelius (*Symb. bot. upsal.* **13** no. 2: 1, 1954; Europe, keys), Degelius (*Symb. bot. upsal.* **20** no. 2: 1, 1974; extra-Europe, keys), Akhtar & Awasthi (*Biol. Mem.* **5**: 13, 1980; India), Degelius (*Nordic Jl Bot.* **6**: 345, 1986), Swinscow & Krog (*Lichenologist* **18**: 63, 1986; E Africa), Upreti & Singh (*J. Bombay nat. Hist. Soc.* **85**: 234, 1988; India), Degelius (*Nordic Jl Bot.* **9**: 101, 1989; New Zealand), Degelius (*Nordic Jl Bot.* **14**: 229, 1994), Liu & Wei (*Mycotaxon* **86**: 349, 2003; China), Liu & Wei (*Mycosystema* **22**: 531, 2003; China), Miądlikowska & Lutzoni (*Am. J. Bot.* **91**:

449, 2004; phylogeny), Wedin & Wiklund (*Symb. bot. upsal.* **34** no. 1: 469, 2004; phylogeny), Miądlikowska *et al.* (*Mycol.* **98**: 1088, 2006; phylogeny).

Collemataceae Zenker (1827), Peltigerales (L). 8 gen. (+ 35 syn.), 293 spp.

Lit.: Henssen (*Lichenologist* **3**: 29, 1965; simple-spored gen.), Upreti & Singh (*J. Bombay nat. Hist. Soc.* **85**: 234, 1988), Jørgensen & Henssen (*Graphis Scripta* **5**: 12, 1993), Degelius (*Nordic Jl Bot.* **14**: 229, 1994), Verdon & Elix (*Acta Bot. Fenn.* **150**: 209, 1994), Galloway & Jørgensen (*Flechten Follmann* Contributions to Lichenology in Honour of Gerhard Follmann: 227, 1995), Galloway (*Nova Hedwigia* **69**: 317, 1999), Galloway & Knight (*Lichenologist* **31**: 642, 1999), Jørgensen & Tønsberg (*Bryologist* **102**: 412, 1999), Guttová (*Lichenologist* **32**: 291, 2000), Wiklund & Wedin (*Cladistics* **19**: 419, 2003), Miądlikowska & Lutzoni (*Am. J. Bot.* **91**: 449, 2004), Miądlikowska *et al.* (*Mycol.* **98**: 1088, 2006; phylogeny), Hofstetter *et al.* (*Mol. Phylogen. Evol.* **44**: 412, 2007; phylogeny), Jørgensen (*Nordic Lichen Flora* **3**: Cyanolichens: 14, 2007).

Collematomyces E.A. Thomas ex Cif. & Tomas. (1953) = Collema F.H. Wigg.

Collematospora Jeng & Cain (1976), Sordariomycetes. 1 (coprophilous), Venezuela. Position very uncertain. See Jeng & Krug (*CJB* **54**: 2429, 1976), Huhndorf *et al.* (*Sydowia* **51**: 176, 1999).

Collembolispora Marvanová & Pascoal (2003), anamorphic *Pezizomycotina*. 1, Portugal. See Marvanová & Cássio (*Cryptog. Mycol.* **24**: 341, 2003).

Collemis Clem. (1931) ≡ Collema F.H. Wigg.

Collemodes Fink (1918) = Collema F.H. Wigg. fide Hawksworth *et al.* (*Dictionary of the Fungi* edn 8, 1995).

Collemodiopsis (Vain.) B. de Lesd. (1910) = Collema F.H. Wigg. fide Hawksworth *et al.* (*Dictionary of the Fungi* edn 8, 1995).

Collemodium Nyl. (1878) = Leptogium fide Hawksworth *et al.* (*Dictionary of the Fungi* edn 8, 1995).

Collemopsidiomyces Cif. & Tomas. (1953) ≡ Collemopsidium.

Collemopsidium Nyl. (1881), Xanthopyreniaceae (L). 20, Europe; N. Africa. See Henssen (*SA* **5**: 126, 1986), Nordin (*Graphis Scripta* **13**: 39, 2002), Kohlmeyer *et al.* (*Mycol. Progr.* **3**: 51, 2004), Mohr *et al.* (*MR* **108**: 515, 2004; evolution), Paz-Bermúdez *et al.* (*Nova Hedwigia* **80**: 73, 2005; Spain).

Collemopsis Nyl. ex Cromb. (1874) = Psorotichia fide Ellis (*Lichenologist* **13**: 123, 1981).

Collemopsis Trevis. (1880) = Psorotichia fide Moreno & Egea (*Acta Bot. Barcinon.* **41**: 1, 1992).

Colleptogium M. Choisy (1962) ≡ Leptogium.

Colletoconis de Hoog & Aa (1978), anamorphic *Pezizomycotina*, Cac.0eH/1eP.15. 1 (on aecia of *Uredinales*), Argentina. See de Hoog & Aa (*Persoonia* **10**: 48, 1978).

Colletogloeopsis Crous & M.J. Wingf. (1997) = Readeriella fide Crous & Wingfield (*CJB* **75**: 668, 1997), Cortinas *et al.* (*Stud. Mycol.* **55**: 133, 2006), Crous *et al.* (*Stud. Mycol.* **55**: 99, 2006), Hunter *et al.* (*Stud. Mycol.* **55**: 147, 2006), Andjic *et al.* (*MR* **111**: 1184, 2007; phylogeny), Crous *et al.* (*Stud. Mycol.* **58**: 1, 2007; morphology).

Colletogloeum Petr. (1953) = Pseudocercospora fide Sutton (*The Coelomycetes*, 1980), Sutton & Swart (*TBMS* **87**: 93, 1986), Crous *et al.* (*Stud. Mycol.* **45**: 107, 2000), Crous (*in litt.*, 2008).

Colletomanginia Har. & Pat. (1906) ? = Engleromyces fide Lloyd (*Mycol. Writ.*, 1917), Rogers (*Mycol.* **73**: 28, 1981), Læssøe (*SA* **13**: 43, 1994).

Colletosporium Link (1824), anamorphic *Pezizomycotina*. 2, Europe. See Hughes (*CJB* **36**: 727, 1958).

Colletostroma Petr. (1953) = Colletotrichum fide von Arx (*Verh. K. ned. Akad. Wet. Amst.* C **51**: 1, 1957).

Colletotrichella Höhn. (1916) = Kabatia fide Sutton (*Mycol. Pap.* **141**, 1977).

Colletotrichopsis Bubák (1904) ? = Colletotrichum fide von Arx (*Verh. K. ned. Akad. Wet. Amst.* C **51**: 1, 1957).

Colletotrichum Corda (1831), anamorphic *Glomerella*, Cac.0eH.15. 60, widespread. *C. gloeosporioides* (600 syn.; teleomorph *Glomerella cingulata*) (anthracnose of citrus, banana, and many other plants), *C. kahawae* (coffee berry disease; *Phaseolus* (*C. lindemuthianum*; cucumber (*C. orbiculare*, syn. *C. lagenarium*); *C. coccodes* (syn. *C. atramentarium*) (potato black dot and tomato root rot); *C. dematium* f. *circinans* (onion (*Allium*) smudge); *C. lini* (flax (*Linum*) seedling blight). See Sutton (*Mycol. Pap.* **141**, 1977; generic synonymy), Baxter *et al.* (*S. Afr. J. Sci.* **2**: 259, 1983; 11 S. Afr. spp.), Baxter *et al.* (*Phytophylactica* **17**: 15, 1985; lit. review), Park *et al.* (*Korean Jl Pl. Path.* **3**: 85, 1987; spp. separation by electrophoresis), Wang & Li (*Acta Mycol. Sin.* **6**: 211, 1987; key 14 spp. China), Ali *et al.* (*Phytopathology* **79**: 1148, 1989; electrophoresis differentiates *C. graminicola*), Boland & Brochu (*Can. J. Pl. Path.* **11**: 303, 1989; cv. response to races of *C. destructivum*), Koch *et al.* (*Phytophylactica* **21**: 69, 1989; spp. on *Medicago* in S. Afr.), Panaccione *et al.* (*Mycol.* **81**: 876, 1989; conidial dimorphism in *C. graminicola*), Bailey *et al.* (*MR* **94**: 810, 1990; species on *Vigna*), Braithwaite *et al.* (*Aust. Syst. Bot.* **3**: 733, 1990; ribosomal DNA as a taxonomic marker), Braithwaite *et al.* (*MR* **94**: 1129, 1990; RFLPs in *C. gloeosporioides*), Jeffries *et al.* (*Pl. Path.* **39**: 343, 1990; biology and control of tropical fruit crop spp.), Smith (*Pl. Dis.* **74**: 69, 1990; spp. from *Fragaria*), Bonde *et al.* (*Phytopathology* **81**: 1523, 1991; isozyme patterns in spp. on *Fragaria*), Brooker *et al.* (*Phytopathology* **81**: 672, 1991; nitrate non-utilizing mutants and vegetative compatibility), Walker *et al.* (*MR* **95**: 1175, 1991; spp. on *Xanthium*), Yang & Chuang (*Plant Prot. Bull.* **33**: 262, 1991; variation in *C. musae*), Bailey & Jeger (Eds) (*Colletotrichum: biology, pathology and control*, 1992), Gunnell & Gubler (*Mycol.* **84**: 157, 1992; spp. on strawberry), Pain *et al.* (*Physiol. Mol. Plant Pathol.* **41**: 111, 1992; monoclonal antibodies in), Sutton *in* Bailey & Jeger (Eds) (*Colletotrichum: biology, pathology and control*, 1992), Correll *et al.* (*Phytopathology* **83**: 1199, 1993; RAPD analysis of *C. orbiculare*), Freeman *et al.* (*Exp. Mycol.* **17**: 309, 1993; molecular genotyping), Lyanage *et al.* (*Phytopathology* **83**: 113, 1993; curinase in *C. gloeosporioides*), Sherriff *et al.* (*Exp. Mycol.* **18**: 121, 1994; ribosomal DNA analysis and species groups), Sreenivasaprasad *et al.* (*MR* **98**: 186, 1994; nucleotides and identification of *C. acutatum*), Fabre *et al.* (*MR* **99**: 429, 1995; molecular markers and bean isolates), Freeman & Rodriguez (*MR* **99**: 501, 1995; spp. differentiation of strawberry isolates by PCR), Sherriff *et al.* (*MR* **99**: 475, 1995; rDNA sequence differentiates *C. graminicola* and *C. sublineolum*), Bailey *in* Dehne *et al.* (Eds) (*Diagnosis and Identification of Plant Pathogens* Proceedings of the 4th International Symposium of the European Foundation for Plant Pathology: 47, 1997), Poplawski *et al.* (*Aust. J. Bot.* **46**: 143, 1998; DNA markers in a biotype of *C. gloeosporioides*), Browning *et al.* (*Pl. Dis.* **83**: 286, 1999; on *Gramineae*), Buddie *et al.* (*MR* **103**: 385, 1999; on *Fragaria*), Kelemu *et al.* (*Eur. J. Pl. Path.* **105**: 261, 1999; S America, *Stylosanthes*), Lardner *et al.* (*MR* **103**: 275, 1999; *C. acutatum* s.l.), Martín & García-Figueres (*Eur. J. Pl. Path.* **105**: 733, 1999; on *Olea*), Cannon *et al.* (*Colletotrichum* Host Specificity, Pathology, and Host-Pathogen Interaction: 1, 2000; review, species concepts), Correll *et al.* (*Colletotrichum* Host Specificity, Pathology, and Host-Pathogen Interaction: 145, 2000; population structure), Freeman *et al.* (*Phytopathology* **90**: 608, 2000; on *Prunus* etc.), Freeman *et al.* (*Appl. Environm. Microbiol.* **66**: 5267, 2000; on *Anemone*), García Muñoz *et al.* (*Mycol.* **92**: 488, 2000; physiology, biochemistry, on *Fragaria*), Johnston (*Colletotrichum* Host Specificity, Pathology, and Host-Pathogen Interaction: 21, 2000; phylogeny), Gehlot & Purohit (*Indian Phytopath.* **54**: 215, 2001; ultrastr.), Guerber & Correll (*Mycol.* **93**: 216, 2001; teleomorph), Hsiang & Goodwin (*Eur. J. Pl. Path.* **107**: 593, 2001; on *Gramineae*), Shen *et al.* (*MR* **105**: 1340, 2001; on *Nicotiana*), Abang *et al.* (*Pl. Path.* **51**: 63, 2002; on *Dioscorea*), Cullen *et al.* (*Pl. Path.* **51**: 281, 2002; molecular detection), Lim *et al.* (*Pl. Path. J.* **18**: 161, 2002; on *Musa*), Moriwaki *et al.* (*J. Gen. Pl. Path.* **68**: 307, 2002; Japan), Munaut *et al.* (*MR* **106**: 579, 2002; on *Stylosanthes*), Nirenberg *et al.* (*Mycol.* **94**: 307, 2002; on *Lupinus*), Nitzan *et al.* (*Phytopathology* **92**: 827, 2002; on *Solanum*), Varzea *et al.* (*Pl. Path.* **51**: 202, 2002; on *Coffea*), Abang (*Biblthca Mycol.* **197**, 2003; on *Dioscorea*), Guerber *et al.* (*Mycol.* **95**: 872, 2003; *C. acutatum* complex), Mackie *et al.* (*Aust. J. agric. Res.* **54**: 829, 2003; Australia), Moriwaki *et al.* (*Mycoscience* **44**: 47, 2003; Japan), Rodríguez-Guerra *et al.* (*Pl. Path.* **52**: 228, 2003; Mexico), Souza-Paccola *et al.* (*J. Phytopath.* **151**: 329, 2003; genetics), Cano *et al.* (*J. Clin. Microbiol.* **42**: 2450, 2004; clinical strains), Ford *et al.* (*Australas. Pl. Path.* **33**: 559, 2004; on *Lens*), Lu *et al.* (*MR* **108**: 53, 2004; endophytes), Lubbe *et al.* (*Mycol.* **96**: 1268, 2004; on *Proteaceae*), Photita *et al.* (*Fungal Diversity* **18**: 117, 2005; Thailand), Sharma *et al.* (*J. Phytopath.* **153**: 232, 2005; *C. capsici* aggregate), Talhinhas *et al.* (*Appl. Environm. Microbiol.* **71**: 2987, 2005; *C. acutatum* aggregate), Weber *et al.* (*J. Phytopath.* **153**: 318, 2005; mating type genes), Armstrong-Cho & Banniza (*MR* **110**: 951, 2006; teleomorph), Farr *et al.* (*MR* **110**: 1395, 2006; on *Agavaceae*), Liu *et al.* (*Phytopathology* **97**: 1305, 2007; *C. orbiculare* group), Bridge *et al.* (*J. Phytopath.* **156**: 274, 2008; *C. kahawae*), Peres *et al.* (*Phytopathology* **98**: 345, 2008; on *Citrus*).

colliculose (**colliculous**), with rounded swellings; blistered.

Colligerites K.P. Jain & R.K. Kar (1979), Fossil Fungi. 2 (Palaeocene), India.

Collodendrum Clem. (1909) ≡ Tremellodendron.

Collodiscula I. Hino & Katum. (1955), Xylariaceae. Anamorph *Acanthodochium*. 1 (on dead bamboo culms), Japan; Taiwan. See Samuels (*SA* **6**: 121, 1987; status), Samuels *et al.* (*Mycotaxon* **28**: 453, 1987), Ju & Rogers (*Mycol.* **82**: 342, 1990; posn), Læssøe (*SA* **13**: 43, 1994), Læssøe & Spooner (*Kew Bull.* **49**: 1, 1994), Ju & Rogers (*Mycotaxon* **73**: 343, 1999), Kang *et al.* (*Fungal Diversity* **2**: 135, 1999; uncertain posn).

Collodochium Höhn. (1902), anamorphic *Pezizomycotina*, Hsp.0eH.?. 1, Europe.

Collolechia A. Massal. (1854) = Placynthium fide Hawksworth *et al.* (*Dictionary of the Fungi* edn 8, 1995).

Collonaemella Höhn. (1915) = Cornularia fide Clements & Shear (*Gen. Fung.*, 1931), Sutton (*Mycol. Pap.* **141**, 1977).

Collonema Grove (1886) nom. dub., Pezizomycotina. Based on an effete ascoma. See Sutton (*Mycol. Pap.* **141**: 45, 1977).

Collopezis Clem. (1909) ≡ Tjibodasia.

Collopus Earle (1909) = Mycena fide Singer (*Agaric. mod. Tax.*, 1951).

Collostroma Petr. (1947), anamorphic *Pezizomycotina*, St.0eH.?. 1, Austria.

collulum, the neck of a conidiogenous cell (Zaleski, 1927).

Collybia (Fr.) Staude (1857) nom. cons., Tricholomataceae. 3, widespread (north temperate). The majority of taxa formerly classified in *Collybia* have now been transferred to *Gymnopus* and *Rhodocollybia*. See Antonín & Noordeloos (*Libri Botanici* **17**, 1997; European spp.), Antonín *et al.* (*Mycotaxon* **63**: 359, 1997; taxonomy), Moncalvo *et al.* (*Syst. Biol.* **49**: 278, 2000; phylogeny), Hughes *et al.* (*MR* **105**: 164, 2001; phylogeny), Mata & Petersen (*Mycotaxon* **86**: 303, 2003; type studies neotropical species (s.lat.)).

Collybidium Earle (1909) ≡ Flammulina.

Collybiopsis (J. Schröt.) Earle (1909) = Marasmius fide Kuyper (*in litt.*).

Collyria Fr. (1849) nom. dub., Agaricales. See Singer (*Agaric. mod. Tax.* edn 3, 1975).

colon (:) (in author citations), see Nomenclature.

Colonnaria Raf. (1808) = Clathrus fide Dring (*Kew Bull.* **35**: 1, 1980).

Colonomyces R.K. Benj. (1955), Euceratomycetaceae. 1, USA; Europe. See Benjamin (*Aliso* **3**: 186, 1955).

colony (of bacteria and yeasts), a mass of individuals, generally of one species, living together; (of mycelial fungi), a group of hyphae (frequently with spores) which, if from one spore or cell, may be one individual.

Colour. Many fungi can be characterized partly by the colour of their fruitbodies or other organs, or by the colour of pigments exuded into pure culture media. Describing those colours has often been attempted without reference to standards, with unsatisfactory results. Names of colours should therefore be given precision by reference to a colour standard, preferably the Munsell Color System as exemplified in the *Munsell book of color* (various editions available through the internet). This work is expensive and the 'Mycologists' Color Kit' designed for individual ownership and produced by the Munsell Color Co. seems to be no longer available. The *Methuen handbook of colour* by Kornerup, edn 3, 1984 (an English translation of the Danish *Farver i Farver* by Kornerup & Wanscher, edn 2, 1967) is an inexpensive colour chart for which the equivalent Munsell notation is given for each colour; Rayner's *A mycological colour chart*, 1970 (now unavailable), illustrated Dade's nomenclature (see below).

Since its publication in 1912 Ridgway's *Color*

standards and color nomenclature, which covers 1,115 colours, has been much used by biologists but copies of this work must now be referred to with caution due to variations having occurred in some colours depending on the amount of use a copy has received and the care which has been given to its preservation. Tables for Ridgway/Munsell conversions accompany Rayner's chart. The Pantone company produces specialist colour charts which may also be applicable.

Other general colour standards include Oberthur & Dauthenay, *Repertoire de couleurs*, 2 vols., 1905 (for Ridgway equivalents see Snell & Dick, *Glossary of mycology*, rev. edn. 1971); Klinksieck & Vallette, *Code de couleurs*, 1908; Maerz & Paul, *A Dictionary of color*, edn 2, 1950; and Wilson, *Horticultural colour chart*, 2 vols., 1939-42. Kelly & Judd (*The ISCC-NBS method of designating colors and a dictionary of color names*, 1955 [reprint 1963], National Bureau of Standards, Washington Circular 553) list the Inter-Society Color Council-NBS equivalents of 7,500 colour names including those of Ridgway and Maerz & Paul; a supplement provides standard colour patches (ISCC-NRSS, *Centroid color charts*, *Standard sample* No. 2106). Locquin, *Guide des coleurs naturelles* (*Obs. Disp. mycol.* **1** (2), 1975) has about 1,536 colour patches, Pfaff (*Z. Mykol.* **49**: 237, 1983; comparison colour atlases), *RHS Colour Chart*, 2000 (Royal Horticultural Society, UK) inc. 800 samples).

Saccardo (*Chromotaxia*, 1891; edn 3, 1912) gave the Latin, Italian, French, English and German names and synonyms of 50 colours of mycological interest, and Dade (*Mycol. Pap.* **6** (edn 2), 1949) related Saccardo's colours to the Ridgway standard, suggested additional Latin colour terms for mycological use, and gave an extended list of colour names. The colour terminology used by Fries for agarics was discussed by Wharton (*Grevillea* **13**: 25, 1884) whose paper has been reprinted in part by Stearn (*Botanical Latin*, edn 4, 1992) who also reprints Dade's chart and gives additional interesting historical information on colour terminology. Locquin, *Chromotaxis: Code mycologique et pedologique des couleurs*, 1958 (see *Mycol.* **50**: 447, 1958) related a number of colours to a series of colour transparencies. See Tocker *et al.* (*Taxon* **40**: 201, 1991; survey colour charts).

Colour names are omitted from this *Dictionary* although a few general colour terms such as dematiaceous, hyaline, inquinant, etc., and those given below have been compiled.

Corner (*Clavaria*, 1950) proposed the following series of terms for the more precise description of pigmentation of basidiomata and hyphae: **achroic**, without true pigmentation; **euchroic**, having true pigmentation as opposed to **epichroic**, discoloration due to injury; **hysterochroic**, slowly discoloured from base to apex in old age; euchroism may be **acrochroic**, coloured specially in the hyphal tips at the growing point, **metachroic**, changing colour through the appearance of a new pigment in maturer tissue, **ectochroic**, pigment on the outside of the hypha, **mesochroic**, pigment in the hyphal wall, or **endochroic**, pigment inside the cell. The last may be subdivided according to whether the pigment is diffused in the cytoplasm (**cytochroic**), in the cell vacuoles (**cystochroic**) or in oil drops (**lipochroic**). Cf. Pigments.

colour of the reverse, the colour of the underside of a fungus culture in a tube or Petri dish.

Colpoma Wallr. (1833), Rhytismataceae. Anamorph *Conostroma*. 14, widespread. See Twyman (*TBMS* **29**: 234, 1946), Johnston (*N.Z. Jl Bot.* **29**: 405, 1991; NZ), Medardi (*Riv. Micol.* **45**: 239, 2002; Italy), Buchanan & May (*N.Z. Jl Bot.* **41**: 407, 2003; conservation), Hou & Piepenbring (*Forest Pathology* **35**: 359, 2005; China).

Colpomella Höhn. (1926) nom. dub., anamorphic *Pezizomycotina*. See Sutton (*Mycol. Pap.* **141**, 1977).

Coltricia Gray (1821), Hymenochaetaceae. 20, widespread. See Pegler (*The polypores [Bull. BMS Suppl.]*, 1973), Legon (*Mycologist* **17**: 42, 2003; UK).

Coltriciaceae Jülich (1982) = Hymenochaetaceae.

Coltriciella Murrill (1904), Hymenochaetaceae. 7, widespread. See Murrill (*Bull. Torrey bot. Club* **31**: 348, 1904), Wagner & Fischer (*Mycol.* **94**: 998, 2002).

Coltriciopsis Teixeira (1991) = Coltricia fide Ryvarden (*Svensk Bot. Tidskr.* **68**: 276, 1974).

columella, a sterile central axis within a mature fruitbody which may be uni- or multicellular, unbranched or branched, of fungal or host origin; (of gasteromycetes, after Cunningham) **axile -**, when an axis in the gleba; **dendroid -**, having lateral branches, as in *Gymnoglossum*; **percurrent -**, joining the peridium at the apex of gleba; **pseudo-**, embryonic tissue in the mature peridium of *Geastrum*; **simple -**, not branched, as in *Secotium*.

Columnocystis Pouzar (1959) = Veluticeps fide Hjortstam & Tellería (*Mycotaxon* **37**: 53, 1990).

Columnodomus Petr. (1941), anamorphic *Pezizomycotina*, St.0eH.?. 1, Europe.

Columnodontia Jülich (1979), Meruliaceae. 1, S.E. Asia; Australasia. See Jülich (*Persoonia* **10**: 326, 1979).

Columnomyces R.K. Benj. (1955), Laboulbeniaceae. 1, USA. See Benjamin (*Aliso* **3**: 185, 1955), Tavares (*Mycol. Mem.* **9**: 627 pp., 1985), Santamaría (*Mycol.* **96**: 761, 2004).

Columnophora Bubák & Vleugel (1916), anamorphic *Pezizomycotina*, Hso.0eP.1. 1, Europe.

Columnosphaeria Munk (1953), Dothioraceae. Anamorph *Aureobasidium*. 1, widespread. See von Arx & Müller (*Stud. Mycol.* **9**, 1975), Barr (*Harvard Pap. Bot.* **6**: 25, 2001), Schoch *et al.* (*Mycol.* **98**: 1041, 2006; phylogeny).

Columnothyrium Bubák (1916), anamorphic *Pezizomycotina*, Cpt.0eH.?. 1, Europe.

Colus Cavalier & Séchier (1835), Phallaceae. 4, widespread.

Coma Nag Raj & W.B. Kendr. (1972), anamorphic *Ascocoma*, Cac.0bP.19. 1, Australasia. See Sutton (*Nova Hedwigia* **25**: 161, 1974), Swart (*TBMS* **87**: 603, 1987; teleomorph), Beilharz & Pascoe (*Mycotaxon* **91**: 273, 2005).

comate, having hairs; shaggy.

Comatospora Piroz. & Shoemaker (1971), anamorphic *Pezizomycotina*, St.1eH.19. 1, Canada. See Pirozynski & Shoemaker (*CJB* **49**: 539, 1971).

Combea De Not. (1846), Roccellaceae (L). 1, S. Africa; USA. See Tehler (*CJB* **68**: 2458, 1990; cladistics), Myllys *et al.* (*Bryologist* **101**: 70, 1998), Myllys *et al.* (*Lichenologist* **31**: 461, 1999; DNA), Lücking *et al.* (*Mycol.* **96**: 283, 2004; phylogeny), Tehler & Irestedt (*Cladistics* **23**: 432, 2007).

Combodia Fr. (1849) ? = Lasiodiplodia fide Sutton

(*Mycol. Pap.* **141**, 1977).

Comesella Speg. (1923), ? Dothideales. 1 (on *Pteridophyta*), Cuba.

Comesia Sacc. (1884), ? Helotiales. 2 or 3, Europe; N. Africa.

Cometella Schwein. (1835) ≡ Clasterosporium fide Ellis (*Mycol. Pap.* **70**, 1958).

commensalism, a form of symbiosis (q.v.).

Comminutispora A.W. Ramaley (1996), Capnodiales. Anamorph *Hyphospora*. 1, USA. See Ramaley (*Mycol.* **88**: 132, 1996), Hambleton *et al.* (*Mycol.* **95**: 959, 2003; phylogeny).

commissure, a closing join; a seam.

commixt, mixed with; intermingled.

Common names (also vernacular names). For fungi, these should be printed in Roman (not italic) type and decapitalized. They can be divided into two general categories: those derived from the scientific Latin name of a genus or higher taxonomic rank (common in scientific writing), and those of the language of the user, rather than the scientific Latin binomial or trinomial (generally used by field mycologists). Common names used in scientific writing may be either singular or plural, e.g. aspergillus (aspergilli), ascomycete(s), fusarium (fusaria), etc. Such use seems unobjectionable as it enables taxa to be distinguished from the individuals of which they are composed, e.g. phytophthoras = species of *Phytophthora*; see Seifert *et al.*, *Stud. Mycol.* **45**, 2000).

In English, in marked contrast to flowering plants, few fungi (incl. lichen-forming species) have common names which are genuinely ancient; only twenty-five or so fungi (e.g. fly agaric, blewits, morel, puff ball) have achieved recognition by the *Shorter Oxford Dictionary* and *Webster's Dictionary*. Basidiomycetes and other conspicuous fungi have, however, been recognized by common names from earliest times, and many ancient names may be supposed lost along with so much of Britain's folk heritage. There have been numerous attempts with varying success to generate modern common names for larger fungi (and occasionally for microfungi; see Cooke, *Handbook of British fungi*, 1871), frequently by more or less literal English versions of the scientific names, most recently by the British Mycological Society (see Internet). Lichens used by man for dyeing, or other purposes, or eaten by reindeer were also often given common names, e.g. crottle, heather rags, lungwort, manna, old man's beard, orchil, rock tripe. Common names for diseases of plants, man, and animals caused by fungi are also widely used in both technical and popular writing, and a many plant disease names are also applied to the pathogens themselves, e.g. rust, smut, powdery mildew (see British Society for Plant Pathology. Names of British Plant Diseases and their Causes *Phytopathological Papers* **28**, 1984).

Common names for fungi in other languages are often more abundant and diverse than for English, but may not be so well documented. Many examples can be found with Internet search engines (see Internet). See also Lundqvist & Persson (*Svenska Svampnamn*, 1987; 2584) [Swedish names]; Ulvinen *et al.* (*Karstenia* **29** (Suppl.), 1989) [Finnish larger fungi]. See also Classification, Dyeing, Nomenclature.

community, any phytosociological taxon.

Comocephalum Syd. (1939), anamorphic *Pezizomycotina*, Cpd.#eP.?. 1, Ecuador.

Comocheila A. Massal. (1860) = Phlyctis fide Hawksworth *et al.* (*Dictionary of the Fungi* edn 8, 1995).

Comoclathris Clem. (1909), Diademaceae. Anamorph *Alternaria*-like. 21, America; Europe. See Shoemaker & Babcock (*CJB* **70**: 1617, 1992; key), Barr (*SA* **12**: 27, 1993; synonym of *Graphyllium*).

comose, having hairs in groups or tufts.

compaginate, joined tightly together.

companion cell, contiguous donor gametangium of *Olpidiopsidales*.

compatible (of mating types, strains, etc.), able to be cross-mated; cross-fertile; **hemi-**, a homokaryon compatible with 1 of the 2 components of the dikaryon. See Sex.

complanate, flat; smooth.

complement-fixation test, a sensitive test by which antigen-antibody reaction may be detected and quantified. The test depends on the ability of antigens after reacting with their specific antibodies to 'fix' **complement** (a group of proteins normally present in freshly isolated serum) the presence of which is necessary for the lysis of red blood cells by haemolysin (red-cell-immune serum).

Completoria Lohde (1874), Completoriaceae. 1 (on prothalli of *Pteridophyta*), widespread (north temperate). See Atkinson (*Cornell Univ. Agric. Exp. Stn Bull.* **94**: 252, 1895), Tucker (*Mycotaxon* **13**: 481, 1981), Humber (*Mycotaxon* **34**: 441, 1989).

Completoriaceae Humber (1989), Entomophthorales. 1 gen., 1 spp.

Lit.: Atkinson (*Bot. Gaz.* **19**: 467, 1894), Humber (*Mycotaxon* **34**: 441, 1989).

complex, sometimes used to designate a group of closely related species.

Complexipes C. Walker (1979), anamorphic *Tricharina*, Hso.0eP.1. 2, widespread. See Yang & Korf (*Mycotaxon* **23**: 457, 1985), Lee & Ka (*Korean J. Mycol.* **18**: 127, 1990; Korea), Pacioni *et al.* (*Micologia e Vegetazione Mediterranea* **13**: 3, 1998).

complicate, bent upon itself.

compound, made up of a number of parts.

compressed (of a stipe), flattened transversely.

Compsocladium I.M. Lamb (1956), Ramalinaceae (L). 1, New Guinea. See Lamb (*Lloydia* **19**: 157, 1956), Ekman (*Op. bot.* **127**: 148 pp., 1996), Frisch (*Biblthca Lichenol.* **96**: 73, 2007; Ecuador).

Compsomyces Thaxt. (1894), Laboulbeniaceae. 7, widespread. See Santamaría (*Fl. Mycol. Iberica* **5**, 2003; Iberian peninsula).

Compsosporiella Sankaran & B. Sutton (1991) = Anisomeridium fide Sankaran & Sutton (*MR* **95**: 1289, 1991), Harris (*More Florida Lichens*, 1995).

Computing, has numerous applications in mycology, and pertinent papers are cited under appropriate topics. For coding of characters for use in microbial databases see Gams *et al.* (*J. gen. microbiol.* **134**: 1667, 1988) and Rogosa *et al.* (*Coding microbiological data for computers*, 1986). See also Authors, Genetic Resource Collections, Internet, Inventorying, Literature, Neural networks, Numerical taxonomy.

Concamerella W.L. Culb. & C.F. Culb. (1981) = Parmotrema fide Culberson & Culberson (*Bryologist* **84**: 307, 1981), Adler (*Mycotaxon* **38**: 331, 1990; key), Thell *et al.* (*Mycol. Progr.* **3**: 297, 2004; phylogeny), Blanco *et al.* (*Mycol.* **97**: 150, 2005; phylogeny), Blanco *et al.* (*Mol. Phylogen. Evol.* **39**: 52, 2006; phylogeny).

concatenate, in chains; catenulate.

concave (esp. of a pileus), hollowed out; basin-like.
concentric bodies, ultrastructures found in many lichenized fungi and also in some other fungi such as *Rhopographus*, *Sphaerotheca*, *Cercospora*, *Pseudopeziza*, *Sphaceloma*; 'elliptical bodies'. See Griffiths & Greenwood (*Arch. Microbiol.* **87**: 285, 1972), Pons *et al.* (*TBMS* **83**: 181, 1984), Beilharz (*TBMS* **84**: 79, 1985), Meyer (*Mycol.* **79**: 44, 1987; in granules in *Allomyces*).
conceptacle, any hollow structure producing spores or spermatia.
conchate (**conchiform**), like a bivalve shell.
Conchatium Velen. (1934) = Cyathicula fide Baral (*SA* **13**: 113, 1994).
Conchites Paulet (1791) = Auricularia fide Donk (*Taxon* **7**: 174, 1958).
Conchomyces Overeem (1927), Tricholomataceae. 2, Indonesia. See Thorn *et al.* (*Mycol.* **92**: 241, 2000; phylogeny).
Conchyliastrum Zebrowski (1936), Fossil Fungi ? Chytridiomycetes. 2 (Cambrian to ? Recent), Australia.
concolorous, of one colour.
concrescent, becoming joined.
concrete, joined by growth.
conditioning, the process by which fungi enzymically soften up substrata such as dead leaves before detritivorous animals can eat them.
conducting hypha, see hypha.
Condylospora Nawawi (1976), anamorphic *Pezizomycotina*, Hso.0fH.10. 3 (aquatic), pantropical. See Nawawi & Kuthubutheen (*Mycotaxon* **33**: 329, 1988), Santos-Flores *et al.* (*Caribb. J. Sci.* **32**: 116, 1996; Puerto Rico), Chan *et al.* (*Fungal Diversity* **5**: 89, 2000; Hong Kong).
conferted, near together; crowded.
Conferticium Hallenb. (1980), Stereaceae. 4, widespread (north temperate). See Ginns & Freeman (*Biblthca Mycol.* **157**, 1994), Larsson & Larsson (*Mycol.* **95**: 1037, 2003; phylogeny).
Confertobasidium Jülich (1972) = Scytinostromella fide Ginns & Freeman (*Biblthca Mycol.* **157**, 1994).
Confertopeltis Tehon (1933) nom. dub., anamorphic *Pezizomycotina*. See Sutton (*Mycol. Pap.* **141**, 1977).
Conferva L. (1753), Algae. Algae.
confervoid, composed of loose filaments or cells.
Confistulina Stalpers (1983) = Fistulina.
confluent (1) (of sori, etc.), coming together; running into one another; (2) (of the flesh of a stipe), continuous with the trama of the pileus.
congeneric, one of two or more taxa considered to belong to one genus; cf. synonym.
congested, very near together.
conglobate (1) massed into a ball; (2) (of the bases of stipes), together making a fleshy mass.
conglutinate, glued together; esp. of paraphyses in *Lecanorales*.
Conia Vent. (1799) nom. rej. = Lepraria fide Sutton (*in litt.*).
Coniambigua Etayo & Diederich (1995), anamorphic *Pezizomycotina*. 1, Spain. See Etayo & Diederich (*Flechten Follmann* Contributions to Lichenology in Honour of Gerhard Follmann: 207, 1995), Diederich (*Herzogia* **16**: 41, 2003).
Coniangium Fr. (1821) = Arthonia fide Hawksworth *et al.* (*Dictionary of the Fungi* edn 8, 1995), Grube *et al.* (*Mycol.* **96**: 1159, 2004).
Coniarthonia Grube (2001), Arthoniaceae (L). 6, widespread. See Grube (*Lichenologist* **33**: 492, 2001).
Conicomyces R.C. Sinclair, Eicker & Morgan-Jones (1983), anamorphic *Pezizomycotina*, Hsy.0fH.15. 2, widespread. See Illman & White (*CJB* **63**: 419, 1985; addit. spp.), Seifert (*Mycotaxon* **71**: 301, 1999; Canada).
Conicosolen F. Schill. (1927) = Psorotheciopsis fide Hawksworth *et al.* (*Dictionary of the Fungi* edn 8, 1995).
Conida A. Massal. (1856) = Arthonia fide Santesson (*Symb. bot. upsal.* **12** no. 1: 1, 1952), Calatayud *et al.* (*Biblthca Lichenol.* **88**: 67, 2004).
conidange, a small pycnidium (in a lichen thallus) having no stout wall (des Abbayes, 1951) (obsol.).
Conidella Elenkin (1901) ? = Arthonia fide Hawksworth *et al.* (*Dictionary of the Fungi* edn 8, 1995), Calatayud *et al.* (*Biblthca Lichenol.* **88**: 67, 2004).
Conidial nomenclature. The traditional approach to nomenclature of spores of anamorphic fungi is that of Saccardo who differentiated 7 morphological types based on shape and septation (amero- [with no septum], dictyo- [with longitudinal and transverse septa], didymo- [with 1 transverse septum], helico- [with a helix shape, with or without septa], phragmo- [with more than 1 transverse septum], scoleco- [long and thin, with or without septa], and staurospores [with radiating 'arms' or star-shaped, with or without septa]). These terms were further qualified according to whether the spores were pigmented (phaeo-amerospores, etc.) or not (hyalo-amerospores, etc.); see Anamorphic fungi. Kendrick & Nag Raj (*in* Kendrick (Ed.), *The whole fungus* **1**: 43, 1979) discussed Saccardo's categories and offered precise definitions.

Since Hughes (*CJB* **31**: 577, 1953) drew attention to the systematic importance of conidiogenous events (q.v.), a nomenclature for conidia has evolved based on the methods by which the conidia develop, giving rise to terms such as annelloconidia, phialoconidia, tretoconidia etc. (cf.). These attempt to summarize in a single word a complex series of developmental events any of which can vary independently of the others; as a result, they are imprecise, and not universally accepted: a better procedure is to use the word 'conidium' with qualifying adjectives or descriptors. Main contributions to conidial terminology have been by the first Kananaskis Workshop-Conference in 1969 (see Kendrick (Ed.), *Taxonomy of fungi imperfecti*, 1971). Ellis (*Dematiaceous hyphomycetes*, 1971) offered a series of definitions as did Kendrick & Carmichael (*in* Ainsworth *et al.* (Eds), *The Fungi* **4A**: 323, 1973), Cole (*CJB* **53**: 2983, 1975), and Cole & Samson (*Patterns of development in conidial fungi*, 1979). The descriptive terminology in both conidial morphology and conidiogenous events was initiated by Minter *et al.* (*TBMS* **79**: 75, 1982; **80**: 39, 1983; **81**: 109, 1983), developed by Sutton (*in* Reynolds & Taylor (Eds), *The fungal holomorph*: 28, 1993) and Hennebert & Sutton, and Sutton & Hennebert (*in* Hawksworth (Ed.), *Ascomycete systematics*: 65, 77, 1994), and is used in this edn of the *Dictionary*. See Anamorphic fungi.
conidiangium, pycnidium (obsol.).
Conidiascus Holterm. (1898), ? Saccharomycetes. 1, Java.
Conidiobolus Bref. (1884), Ancylistaceae. 38, widespread. See King (*CJB* **54**: 45, 1976), King (*CJB* **54**: 1285, 1976; key), King (*CJB* **55**: 718, 1977), Re-

maudière & Keller (*Mycotaxon* **11**: 323, 1980), Ben-Ze'ev & Kenneth (*Mycotaxon* **14**: 393, 1982; subgen. *Delacrouxia*), Latgé *et al.* (*CJB* **60**: 413, 1982; ultrastr.), Latgé *et al.* (*Exp. Mycol.* **10**: 99, 1986; ultrastr.), Keller (*Sydowia* **40**: 122, 1987), Humber (*Mycotaxon* **34**: 441, 1989), Callaghan *et al.* (*MR* **104**: 1270, 2000; repetitive conidia), Tosi *et al.* (*Mycotaxon* **90**: 343, 2004; n.sp. Antarctica), Tanabe *et al.* (*J. gen. appl. Microbiol.* Tokyo **51**: 267, 2005), Keller (*Sydowia* **58**: 38, 2006), Huang *et al.* (*Mycotaxon* **100**: 227, 2007; USA), Waingankar *et al.* (*Mycopathol.* **165**: 173, 2008; n.sp.).

Conidiocarpus Woron. (1926), anamorphic *Pezizomycotina*, Cpd.0eH.15. 2, widespread. See Hughes (*Mycol.* **68**: 693, 1976), Matsushima (*Matsush. Mycol. Mem.* **10**, 2001; Japan), Reynolds & Gilbert (*Aust. Syst. Bot.* **18**: 265, 2005; Australia), Reynolds & Gilbert (*Cryptog. Mycol.* **27**: 249, 2006; Panamá).

Conidiogenesis. The process of conidium formation. Concepts of conidiogenesis have increasingly been used in anamorphic fungal systematics since Hughes (*CJB* **31**: 577, 1953) classified some hyphomycetes according to the different methods by which conidia develop from conidiophores and the ways in which conidiophores (and conidiogenous cells) grow before, during and after conidia are produced. The historical development of this approach has been covered by Kendrick (Ed.) (*Taxonomy of fungi imperfecti*, 1971), for hyphomycetes, by Sutton (*in* Ainsworth *et al.* (Eds), *The Fungi* **4A**: 513, 1973) for coelomycetes, and by Vobis (*Bibl. Lich.* **14**, 1980) for lichenized pycnidia. Cole & Samson (*Patterns of development in conidial fungi*, 1979) emphasized the contribution of ultrastructural data to developmental concepts. Minter *et al.* (*TBMS* **79**: 75, 1982; **80**: 39, 1983; **81**: 109, 1983) reassessed optical and electron microscopy observations and demonstrated a continuum of developmental processes. See Anamorphic fungi, wall-building.

conidiogenous, producing conidia; **- cell**, any cell from or within which a conidium is directly produced; **- locus**, the place on a conidiogenous cell at which a conidium arises; Kendrick (1971: 258).

conidiole, a small conidium, esp. one on another; a secondary conidium, as in *Empusa*.

conidioma (pl. **-ata**), a specialized multi-hyphal, conidia-bearing structure (Kendrick & Nag Raj *in* Kendrick (Ed.), *The whole fungus* **1**: 51, 1979). See acervulus, pycnidium, sporodochium, synnema (all obsol. nouns, but used adjectivally, e.g. acervular conidioma). See Fig. 10. Cf. conidiophore.

conidiophore, a simple or branched hypha (a **fertile hypha**) bearing or consisting of conidiogenous cells from which conidia are produced; sometimes used when describing reduced structures for the conidiogenous cell.

Conidiosporomyces Vánky (1992), Tilletiaceae. 3 (on *Poaceae*), pantropical. See Vánky & Vánky (*Lidia* **5**: 157, 2002; southern Africa).

Conidiotheca Réblová & L. Mostert (2007), ? Calosphaeriales. 1, USA. See Réblová & Mostert (*MR* **111**: 299, 2007; phylogeny).

Conidioxyphium Bat. & Cif. (1963) = Conidiocarpus fide Hughes (*Mycol.* **68**: 693, 1976).

conidium, a specialized, non-motile (cf. zoospore), asexual spore, usually caducous, not developed by cytoplasmic cleavage (cf. sporangiospore) or free-cell formation (cf. ascospore); in certain *Oomycota* produced through the incomplete development of zoosporangia which fall off and germinate to produce a germination tube. See Sutton (*TBMS* **86**: 1, 1986; derivation etc.); **- initial**, a cell, or part of a cell from which a conidium develops; **- ontogeny**, conidiogenesis (q.v.).

Coniella Höhn. (1918), anamorphic *Schizoparmaceae*, Cpd.0eP.15. 13, widespread. See Sutton (*The Coelomycetes*, 1980; key), Castlebury *et al.* (*Mycol.* **94**: 1017, 2002; phylogeny), Van Niekerk *et al.* (*MR* **108**: 283, 2004; phylogeny), Rossman *et al.* (*Mycoscience* **48**: 135, 2007; review).

Coniobotrys Pouzar (1958) ? = Jaapia Bres. fide Donk (*Persoonia* **3**: 199, 1964).

Coniobrevicolla Réblová (1999), Trichosphaeriaceae. 1 (from wood), Denmark. See Réblová (*Mycotaxon* **70**: 421, 1999).

Coniocarpaceae Rchb. (1841) = Arthoniaceae.

Coniocarpon DC. (1805) nom. rej. = Arthonia fide Hawksworth *et al.* (*Dictionary of the Fungi* edn 8, 1995).

Coniocessia D. García, Stchigel, D. Hawksw. & Guarro (2006), Xylariales. Anamorph *Nodulisporium*-like. 1 (from soil), Jordan. See García *et al.* (*MR* **110**: 1284, 2006).

Coniochaeta (Sacc.) Cooke (1887), Coniochaetaceae. Anamorph *Lecythophora*. 65 (coprophilous or on rotting wood), widespread. Species with *Nodulisporium*-like anamorphs belong to *Coniolaria* (*Xylariales*). See Hawksworth & Yip (*Aust. J. Bot.* **29**: 377, 1981; key 11 spp. in cult.), Mahoney & LaFavre (*Mycol.* **73**: 931, 1981; synopsis), Checa *et al.* (*Cryptog. Mycol.* **9**: 1, 1988; key 15 spp. Spain), Hawksworth *in* Hawksworth (Ed.) (*Ascomycete Systematics. Problems and Perspectives in the Nineties* NATO ASI Series vol. **269 269**: 377, 1994; polyphyly), Kamiya *et al.* (*Mycoscience* **36**: 377, 1995), Lee & Hanlin (*Mycol.* **91**: 434, 1999; DNA), Romero *et al.* (*MR* **103**: 689, 1999; Argentina), Weber (*Nova Hedwigia* **74**: 159, 2002; anam.), Weber *et al.* (*Nova Hedwigia* **74**: 187, 2002; phylogeny), Huhndorf *et al.* (*Mycol.* **96**: 368, 2004; phylogeny), Asgari & Zare (*Nova Hedwigia* **82**: 227, 2006; Iran), García *et al.* (*MR* **110**: 1271, 2006; phylogeny).

Coniochaetaceae Malloch & Cain (1971), Coniochaetales. 5 gen. (+ 8 syn.), 72 spp.

Lit.: Checa *et al.* (*Cryptog. Mycol.* **9**: 1, 1988), Crane & Shearer (*Mycotaxon* **54**: 107, 1995), Kamiya *et al.* (*Mycoscience* **36**: 377, 1995), Guarro *et al.* (*Mycoscience* **38**: 123, 1997), Ramaley (*Mycol.* **89**: 962, 1997), Huhndorf *et al.* (*Sydowia* **51**: 176, 1999), Lee & Hanlin (*Mycol.* **91**: 434, 1999; DNA), Romero *et al.* (*MR* **103**: 689, 1999), Weber *et al.* (*Nova Hedwigia* **74**: 187, 2002), Huhndorf *et al.* (*Mycol.* **96**: 368, 2004), García *et al.* (*MR* **110**: 1271, 2006).

Coniochaetales Huhndorf, A.N. Mill. & F.A. Fernández (2004). Sordariomycetidae. 1 fam., 9 gen., 76 spp. Fam.:

Coniochaetaceae

For *Lit.* see under fam.

Coniochaetidium Malloch & Cain (1971) = Coniochaeta fide von Arx (*Stud. Mycol.* **8**: 25, 1975; key), Udagawa & Tsubouchi (*Mycotaxon* **27**: 63, 1986), Guarroet *al.* (*Mycoscience* **38**: 123, 1997), García *et al.* (*MR* **110**: 1271, 2006; phylogeny), Zhang *et al.* (*Mycol.* **98**: 1076, 2006; phylogeny).

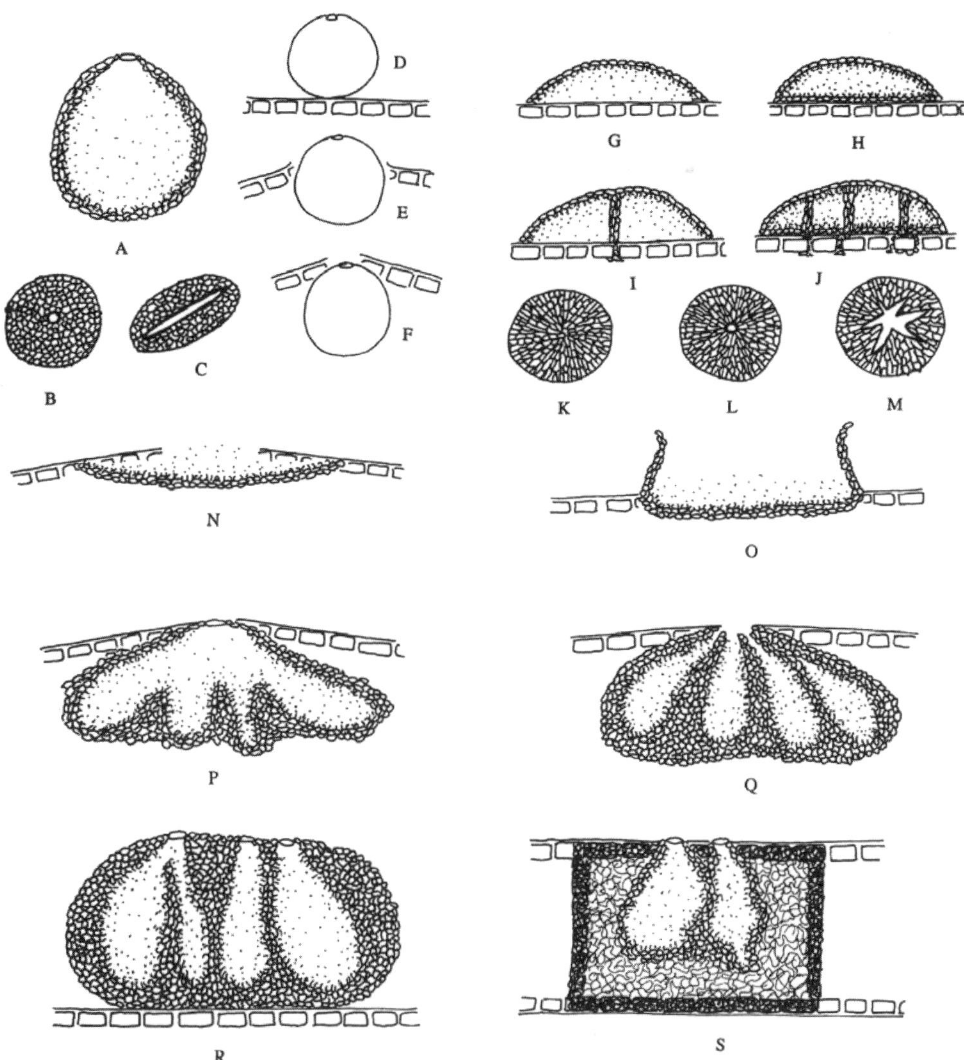

Fig. 12. Conidiomatal types. A-F, pycnidial; B, dehiscence by a central circular ostiole; C, dehiscence by a longitudinal ostiole (raphe); D, superficial; E, semi-immersed; F, immersed; G-M, pycnothyrial; G, with upper wall only; H, with upper and lower walls; I, with a central supporting column; J, multilocular with several supporting columns; K, dehiscence from the margin; L, dehiscence by a central ostiole; M, dehiscence by irregular fissures; N, acervular; O, cupulate; P-R, eustromatic; P, convoluted, immersed; Q, multilocular, immersed; R, multilocular, superficial; S, pseudostromatic.

Coniochila A. Massal. (1860) ? = Ocellularia fide Hawksworth *et al.* (*Dictionary of the Fungi* edn 8, 1995).

Coniocybaceae Rchb. (1837), Lecanoromycetes (inc. sed.) (L). 4 gen. (+ 12 syn.), 56 spp.

Lit.: Honegger (*Lichenologist* **17**: 273, 1985), Honegger (*Sydowia* **38**: 146, 1985), Tibell (*Symb. bot. upsal.* **27** no. 1: 279 pp., 1987), Tibell (*Fl. Neotrop.* Monogr. **69**: 78 pp., 1996), Tibell (*Symb. bot. upsal.* **32** no. 1: 291, 1997), Wedin & Tibell (*CJB* **75**: 1236, 1997), Rikkinen (*Bryologist* **101**: 558, 1998), Tibell (*Biblthca Lichenol.* **71**: 107 pp., 1998), Selva & Tibell (*Bryologist* **102**: 377, 1999), Tibell (*Nordic Lichen Flora* **1**. Introductory Parts; Calicioid Lichens and Fungi: 20, 1999), Tibell (*Bryologist* **104**: 191, 2001), Tibell (*Lichenologist* **33**: 519, 2001), Tibell & Koffman (*Bryologist* **105**: 353, 2002).

Coniocybe Ach. (1816), Coniocybaceae (±L). 24, widespread (esp. temperate). See Tibell (*Symb. bot. upsal.* **21** no. 2: 1, 1975), Tibell (*Svensk bot. Tidskr.* **72**: 171, 1978), Honegger (*Lichenologist* **17**: 273, 1985; anamorph), Redhead (*CJB* **62**: 2514, 1984), Tibell (*Symb. bot. upsal.* **27** no. 1: 279 pp., 1987; Australasia), Selva (*Symb. bot. upsal.* **34** no. 1: 19, 2004).

Coniocybomyces Cif. & Tomas. (1953) ≡ Coniocybe.

Coniocybopsis Vain. (1927) = Microcalicium fide Tibell (*Bot. Notiser* **131**: 229, 1978), Tibell (*Recollecting Edvard August Vainio*: 95, 1998).

Coniodictyum Har. & Pat. (1909), Cryptobasidiaceae. 1, Africa. See Malençon (*BSMF* **69**: 77, 1953), Maier *et al.* (*Stud. Mycol.* **55**: 279, 2006; epidemic on *Zizyphus mucronata* in South Africa).

Coniolaria Seigle-Mur., Guiraud, Steiman & Sage (1995) nom. inval. = Coniolariella fide Seigle-Murandi *et al.* (*Cryptog. Bot.* **5**: 346, 1995), García *et al.* (*MR* **110**: 1271, 2006).

Coniolariella D. García, Stchigel & Guarro (2006), ? Xylariaceae. Anamorph *Rhinocladiella*-like. 1 (from wood), Colorado. See García *et al.* (*MR* **110**: 1285, 2006), García *et al.* (*MR* **110**: 1271, 2006; phylogeny).

Conioloma Flörke (1815) = Arthonia fide Hawksworth *et al.* (*Dictionary of the Fungi* edn 8, 1995).

Coniomela (Sacc.) Kirschst. (1934) = Coniochaeta fide von Arx & Müller (*Beitr. Kryptfl. Schweiz* **11** no. 1, 1954).

Coniomycetes (obsol.). Class for fungi with spores borne in a naked mass, including certain anamorphic fungi, *Pucciniales* and *Ustilaginales* (Fries, *Syst. mycol.* **1**, 1821).

Coniophora DC. (1815), Coniophoraceae. 20, widespread. *C. puteana* (cellar fungus) is a cause of rot in wood. See Lentz (*Mycol.* **49**: 534, 1957; basidial development), Ginns (*CJB* **51**: 249, 1973; type studies), Ginns (*Op. bot.* **61**, 1982; world monogr.), Adaskaveg *et al.* (*Harvard Pap. Bot.* **6**: 15, 2001; isolates from citrus), Huckfeldt & Schmidt (*Mycologist* **20**: 42, 2006; key to strand-forming house-rot fungi).

Coniophoraceae Ulbr. (1928), Boletales. 6 gen. (+ 2 syn.), 29 spp.

Lit.: Cooke (*Mycol.* **49**: 197, 1957; key), Donk (*Taxon*: 254, 1951-63), Hallenberg (*Lachnocladiaceae and Coniophoraceae of North Europe*, 1985), Besl *et al.* (*Z. Mykol.* **52**: 277, 1986; chemosyst.), Hjortstam (*Windahlia* **17**: 55, 1987), Pegler (*Serpula lacrymans. Fundamental Biology and Control Strategies*: 1, 1991), Singh *et al.* (*Mycologist* **7**: 124, 1993), Gilbertson & Hemmes (*Mycotaxon* **65**: 427, 1997), Bigelow *et al.* (*MR* **102**: 257, 1998), Ginns (*Mycol.* **90**: 19, 1998), Thorn *et al.* (*CJB* **76**: 686, 1998), Schmidt & Moreth (*MR* **104**: 69, 2000), Palfreyman *et al.* (*FEMS Microbiol. Lett.* **228**: 281, 2003), Kauserud (*Mycol.* **96**: 232, 2004), Kauserud *et al.* (*MR* **108**: 1264, 2004), Kauserud *et al.* (*Mol. Ecol.* **13**: 3137, 2004), Binder *et al.* (*Systematics and Biodiversity* **3**: 113, 2005), Moreth & Schmidt (*Holzforschung* **59**: 90, 2005), Kauserud *et al.* (*Mol. Ecol.* **15**: 421, 2006).

Coniophorafomes Rick (1934), Stereaceae. 1, S. America.

Coniophorella P. Karst. (1889) = Coniophora fide Ginns (*Op. Bot.* **61**, 1982).

Coniophoropsis Hjortstam & Ryvarden (1986), Boletales. 1, Argentina. See Hjortstam & Ryvarden (*Mycotaxon* **25**: 540, 1986).

Coniophyllum Müll. Arg. (1892) = Calycidium fide Hawksworth *et al.* (*Dictionary of the Fungi* edn 8, 1995).

Conioscypha Höhn. (1904), anamorphic *Conioscyphascus*, Hso.0eP.20. 7, widespread. See Shearer (*Mycol.* **65**: 128, 1973), Kirk (*TBMS* **82**: 177, 1984; n.sp.), Matsushima (*Matsush. Mycol. Mem.* **7**: 47, 1993), Matsushima (*Matsush. Mycol. Mem.* **9**: 7, 1996), Chen & Tzean (*Bot. Bull. Acad. sin.* Taipei **41**: 315, 2000; Taiwan), Réblová & Seifert (*Stud. Mycol.* **50**: 95, 2004; phylogeny).

Conioscyphascus Réblová & Seifert (2004), ? Sordariales. Anamorph *Conioscypha*. 2, Europe. See Réblová & Seifert (*Stud. Mycol.* **50**: 100, 2004).

Conioscyphopsis Goh & K.D. Hyde (1998), anamorphic *Pezizomycotina*, Hso.?.?. 1, Australia. See Goh & Hyde (*MR* **102**: 308, 1998).

Coniosporiella Bat. (1966) = Schiffnerula fide von Arx & Müller (*Stud. Mycol.* **9**, 1975).

Coniosporiopsis Speg. (1918) nom. conf., anamorphic *Pezizomycotina*. See Hughes (*CJB* **36**: 754, 1958).

Coniosporium Link (1809), anamorphic *Chaetothyriales*, Hsp.#eP.23. 7, Europe. See Sterflinger *et al.* (*Antonie van Leeuwenhoek* **72**: 349, 1997), De Leo *et al.* (*Stud. Mycol.* **43**: 70, 1999), Sterflinger *et al.* (*Stud. Mycol.* **43**: 5, 1999; ecology), Sterflinger & Prillinger (*Antonie van Leeuwenhoek* **80**: 275, 2001; phylogeny), Selbmann *et al.* (*Stud. Mycol.* **51**: 1, 2005).

Coniotheciella Speg. (1918) ? = Coniothecium fide Hawksworth *et al.* (*Dictionary of the Fungi* edn 8, 1995).

Coniothecium Corda (1833) nom. dub., anamorphic *Pezizomycotina*. See Hughes (*CJB* **36**: 727, 1958), Hawksworth (*TBMS* **65**: 219, 1975; redisp. lichenicolous spp.).

Coniothele Norman (1868) = Verrucaria Schrad. fide Hawksworth *et al.* (*Dictionary of the Fungi* edn 8, 1995).

Coniothyriaceae W.B. Cooke (1983) = Leptosphaeriaceae.

Coniothyriella Speg. (1889) nom. conf., Pezizomycotina. See Petrak & Sydow (*Beih. Rep. spec. nov. regn. veg.* **42**: 322, 1927).

Coniothyriella Speg. (1910) ≡ Coniothyrina.

Coniothyriites Babajan & Tasl. (1970), Fossil Fungi. 1 (Tertiary), former USSR.

Coniothyrina Syd. (1912), anamorphic *Pezizomycotina*, Cpd.0eP.15. 1 (on *Agave*), widespread (subtropical).

Coniothyrinula Petr. (1923) = Coniothyrium fide Clements & Shear (*Gen. Fung.*, 1931).

Coniothyriopsiella Bender (1932) ? = Cyclothyrium fide Sutton (*Mycol. Pap.* **141**, 1977).

Coniothyriopsis Petr. (1923) = Cyclothyrium fide Sutton (*Mycol. Pap.* **141**, 1977).

Coniothyriopsis Speg. (1910) ? = Microsphaeropsis Höhn. fide Sutton (*Mycol. Pap.* **141**, 1977).

Coniothyris Clem. (1909) ≡ Coniothyriella Speg. (1910).

Coniothyrium Corda (1840) nom. cons., anamorphic *Leptosphaeria*, Cpd.0-1cP.19. 44, widespread. See Westcott (*Mem. Cornell agric. Exp. Stn* **153**, 1934; *C. minitans* (mycoparasite of sclerotia), Biga *et al.* (*Sydowia* **12**: 258, 1959), Sutton (*Mycol. Pap.* **123**, 1971; relationship to *Microsphaeropsis*), Sesan & Crisan (*Stud. Cercet. Biol. Acad. romana* **40**: 71, 1988), Sesan (*Probl. Prot. Plant.* **17**: 29, 1989),

Sandys-Winsch *et al.* (*MR* **97**: 1175, 1993; world distrib.), Wingfield *et al.* (*Mycopathologia* **136**: 139, 1996; on *Eucalyptus*), Swart *et al.* (*S. Afr. J. Bot.* **64**: 137, 1998; on *Proteaceae*), Guarro *et al.* (*Medical Mycology* **37**: 133, 1999; clinical strain), Câmara *et al.* (*MR* **105**: 41, 2001; teleomorphs in *Paraphaeosphaeria*), Muthumeenakshi *et al.* (*MR* **105**: 1065, 2001; phylogeny, population structure), Sterflinger & Prillinger (*Antonie van Leeuwenhoek* **80**: 275, 2001; rock-inhabiting spp.), Taylor & Crous (*Mycoscience* **42**: 265, 2001; on *Proteaceae*), Grendene *et al.* (*MR* **106**: 796, 2002; mycoparasitism), van Zyl *et al.* (*Mycopathologia* **155**: 149, 2002; S Africa), Zyl *et al.* (*MR* **106**: 51, 2002; S Africa, Thailand), Verkley *et al.* (*Stud. Mycol.* **50**: 323, 2004; exclusion of mycoparasitic spp.), Schoch *et al.* (*Mycol.* **98**: 1041, 2006; phylogeny).

Conisphaeria Cooke (1879) nom. dub., Fungi. Based on various pyrenomycetes.

conjugate, joined; in twos; **- nuclei**, two nuclei in one cell which undergo division (**- division**) at the same time, as in basidiomycetes.

conjugation, copulation (q.v.), esp. isogamic copulation; **- tube**, a tube between two copulating cells.

Conjunctospora Udagawa & Uchiy. (1999), anamorphic *Pezizomycotina*, Hso.?.?. 1, Japan. See Udagawa & Uchiyama (*CJB* **76**: 1638, 1998).

conk, the fruit-body (basidioma) of a wood-attacking fungus, esp. of a polypore.

connate, joined by growth.

connective, see disjunctor.

connective hyphae, hyphae of the connective tissue of the context (Fayod, 1889).

Connersia Malloch (1974), Pseudeurotiaceae. 1 (from soil, wood etc.), British Isles; Canada. See Suh & Blackwell (*Mycol.* **91**: 836, 1999; phylogeny), Sogonov *et al.* (*Mycol.* **97**: 695, 2005).

connivent (1) touching but not organically joined; (2) (of a pileus margin), touching the stipe.

Conocybe Fayod (1889) nom. cons., Bolbitiaceae. *c.* 200, widespread. See van Waveren (*Persoonia* **6**: 119, 1970), Watling (*Fl. Illustr. Champ. Afr. centr.* **3**: 57, 1974; key 13 Afr. spp.), Watling (*Revue Mycol. Paris* **40**: 31, 1976), Watling (*British fungus flora* **3**, 1982; key 58 Br. spp.), Arnolds & Hausknecht (*Persoonia* **18**: 239, 2003; Netherlands), Hausknecht (*Öst. Z. Pilzk.* **12**: 41, 2003; sect. *Mixtae*), Hausknecht *et al.* (*Öst. Z. Pilzk.* **13**: 153, 2004; type stud. North American spp.), Hausknecht (*Beitr. Kenntn. Pilze Mitteleur.* **14**: 93, 2005; stirps *Pilosella* in Europe), Hausknecht *et al.* (*Karstenia* **45**: 1, 2005; Finland), Walther & Weiss (*Mycol.* **98**: 792, 2006; Anamorphs).

Conohypha Jülich (1975), Meruliaceae. 2, Europe; USA. See Jülich (*Persoonia* **8**: 303, 1975), Wojewoda (*Acta Mycologica* Warszawa **38**: 3, 2003; Poland).

Conoplea Pers. (1801), anamorphic *Sarcosomataceae*, Hso.0eP.10. 8, widespread. See Hughes (*CJB* **38**: 659, 1960; key), Vittal & Dorai (*Mycotaxon* **51**: 27, 1994; India), Urban *et al.* (*MR* **108**: 749, 2004).

Conopleaceae Chevall. (1826) = Sarcosomataceae.

Conostoma Bat. & J.L. Bezerra (1965), anamorphic *Pezizomycotina*, Hso.0eH.?. 1, Brazil. See Batista & Bezerra (*Riv. Patol. veg., Pavia* Sér. 4 **1**: 42, 1965).

Conostroma Moesz (1921), anamorphic *Colpoma*, St.0eH.10. 2, Europe; India. See Muthumary (*Curr. Sci.* **55**: 1081, 1986; India spp.).

Conotrema Tuck. (1848) = Stictis fide Gilenstam (*Ark. Bot.* ser. 2 **7**: 149, 1969), Sherwood (*Mycotaxon* **5**: 1, 1977), Aptroot *et al.* (*Biblthca Lichenol.* **64**, 1997), Winka *et al.* (*Lichenologist* **30**: 455, 1998), Wedin *et al.* (*New Phytol.* **164**: 459, 2004; ecology, phylogeny), Wedin *et al.* (*Lichenologist* **37**: 67, 2005; phylogeny), Wedin *et al.* (*MR* **110**: 773, 2006; Scandinavia).

Conotrematomyces Cif. & Tomas. (1953) ≡ Conotrema.

Conotremopsis Vězda (1977), Stictidaceae (L). 1, Tasmania. See Vězda (*Folia geobot. phytotax.* **12**: [313], 1977).

Conservation (of fungi). In view of the destruction of many habitats on a global scale and the threats of climate change, there is an increasingly urgent need to attend to conservation of rare and endangered fungi. Many groups of fungi are sensitive to environmental pollution, the most abundant evidence being for lichen-forming species. *In situ* conservation of fungi therefore prioritizes conservation of habitats and regulation of pollution (for *ex situ* fungal conservation, see Genetic resource collections). Fungal conservation is however still in its infancy. Public awareness is very low. Consequently, in many countries there is no explicit legal protection for fungi. Many national biodiversity action plans produced as a result of the Rio Convention (see Biodiversity) fail to consider fungi. The IUCN (The World Conservation Union) currently includes only two committees for fungal conservation in its Species Survival Commission (both listed under plants), one for fungi in general, the other for lichens (there are, in comparison, twenty committees for birds). There are serious practical problems in applying IUCN red-listing criteria to the fungi, and very few fungi have yet been formally evaluated using this system. The broader conservation movement is largely unaware of the need to conserve fungi (frequently only one or two token larger fungi are listed as endangered, and then usually as 'lower' plants), and fungi (for example host-specific species known only on rare endemic plants) may be treated as part of the problem (a threat to the plant) rather than recognized as being themselves in need of protection. Furthermore, the broader conservation movement usually identifies priority habitats on the basis of bird, mammal and flowering plant diversity, with fungi rarely taken into account. This can mean that habitats rich in fungal diversity are being missed by such categorization; host-specific fungi are frequently not found in all places where associated organisms occur; furthermore, like other groups of organisms, fungi are being affected by climate change with, for example, refugee species appearing in areas beyond their traditional geographical limits.

The challenges for fungal conservation are therefore daunting. Efforts to date have emphasized gathering scientific evidence for threats and decline (there have also been successful efforts in the UK to identify 'important fungal areas'), but the political activity necessary to turn that evidence into results remains very small; there is little awareness of the need to maintain a distinction between those responsible for scientific evidence and those involved in the political promotion of conservation, and there is little dialogue between those interested in *in situ* and *ex situ* conservation. Evidence gathering has been directed mainly towards the visually more striking larger fungi, including species forming conspicuous lichens. Basic activities include compilation of distribution and ecological data into regional and national databases from which **Red Lists** are prepared, comprising condensed information on threatened species. National lists have been prepared for most European countries for basidiomycetes and some larger ascomycetes (see Internet) but coverage is patchy. For lichens, there is a preliminary *Red List of Macrolichens in the European Community* (Sérusiaux, 1989) but, although efforts continue, there is still no European level red-list for any other group of fungi. For other parts of the world, red list coverage is usually poor to non-existent, although some impressive fungal conservation work has occurred in the northwest USA as an adjunct to projects conserving the northern spotted owl (e.g. Molina, *MR* **112**: 613, 2008).

On the basis of the Red Lists, it appears that some ecological groups of basidiomycetes are more vulnerable than others: of special concern are (1) wood-inhabiting fungi (esp. bracket spp.) on large logs in virgin and old-growth forests, endangered by intensive forestry (Høiland & Bendiksen, *in* Arnolds & Kreisel, *Conservation of fungi in Europe*: 51, 1993; Kotiranta & Niemelä, *Threatened polypores in Finland*, 1993); (2) fungi of peat bogs, marshland and boggy forests, due to reclamation and drainage (Winterhoff & Krieglsteiner, *Beig. Veroff. Natur. Landschaftspflege Bad. Württ.* **40**, 1984); (3) fungi of sand dunes, due to recreation, digging of sand and afforestation (Winterhoff & Krieglsteiner, 1984); (4) fungi of old, permanent pastures on soils poor in nutrients, due to increased fertilizer application, afforestation or lack of appropriate management (Arnolds, *Opera Bot.* **100**: 7, 1989; Nitare, *Svensk bot. Tidskr.* **82**: 341, 1988); (5) ectomycorrhizal fungi in forests on soils poor in nutrients, due to acidification and/or nitrogen accumulation from air pollution (Arnolds, *Agric. Ecosyst. Environm.* **35**: 209, 1991; Dighton & Jansen, *Environm. Pollut.* **73**: 179, 1991).

There is a wealth of evidence to show that lichens are also greatly affected by many of the above factors, but more particularly deforestation, agricultural practices and a wide variety of atmospheric pollutants (see Air pollution); a review of the principles and priorities of lichen conservation is provided by Seaward (*J. Hattori Bot. lab.* **52**: 401, 1982). Experiments and circumstantial evidence have not demonstrated a strong correlation between collection of mushrooms and their decline (Egli *et al.*, *Mycol. Helv.* **3**: 417, 1990), but rare lichens have disappeared as a result of over-zealous collecting for economic reasons (e.g. as a component of curries, wreaths, decorations) or for reference collections and exsiccata. Maintenance of threatened species and species diversity of fungi in general is often dependent on management practices (Keizer, *in* Pegler *et al.* (Eds), 1993: 251), but in deciding these, fungi are rarely taken into account.

There are very few non-governmental organizations specifically dedicated to fungal conservation. Mycologists in Europe co-operate through the European Council for Conservation of Fungi (ECCF), established by the 9th Congress of European Mycologists in Oslo (1985) and now the conservation wing of the European Mycological Association (see Societies and organizations). The ECCF is composed of national representatives of each country [other persons active in fungal conservation can be placed on the mailing list by request]. The ECCF organizes specialist meetings on fungal conservation between European Congresses, and short business meetings during those Congresses. The Bern Convention's Standing Order Committee failed in 2006 to adopt an ECCF submission presenting the scientific case for including 33 endangered larger fungi in its lists of strictly protected organisms, but in 2007, the Council for Europe adopted guidelines for conservation of macrofungi in Europe prepared by the ECCF. Some other mycological societies also have conservation groups: lichen conservation is catered for by a subcommittee of the International Association for Lichenology (and via its *Newsletter*). There is a Conservation Special Interest Group in the Australasian Mycological Society, and in November 2007, a specialist group for fungal conservation was established in Brazil. Three prototype Specialist Committees have been established for ascomycete conservation, rust & smut conservation, and for conservation of chromistans, chytrids, myxomycetes and zygomycetes (see Internet).

General introductions to fungal conservation are provided by Arnolds and Richardson (*in* Hawksworth (Ed.), *Frontiers in mycology*: 187, 243, 1991); Palm & Chapela (*Mycology in Sustainable Development*, 1997); Moore *et al.* (*Fungal Conservation. Issues and Solutions*, 2001). See also Biodiversity, Climate change.

Consetiella Hol.-Jech. & Mercado (1982), anamorphic *Pezizomycotina*, Hsy.-.-. 1, Cuba. See Holubová-Jechová & Mercado (*Mycotaxon* **14**: 310, 1982), Mercado-Sierra *et al.* (*Mycotaxon* **67**: 417, 1998).

consortium, a form of symbiosis (q.v.) in which two or more organisms live together in an interdependent way (obsol.).

conspecific, of two or more taxa considered to be one species; cf. synonym.

constipate, crowded together.

Constricta R. Heim & Mel.-Howell (1965) nom. inval., Agaricaceae. 1, Ivory Coast. See Heim & Meléndez-Howell (*Revue Mycol.* Paris **30**: 324, 1965).

containment levels, see Safety, Laboratory.

contaminated (1) bearing, or intermixed with, a pathogen, as spores on seeds, fungi in soil (c.f. infection and infested); (2) (of cultures), not pure.

context (of hymenomycetes), the hyphal mass between the superior surface and the subhymenium or the trama of basidiocarps.

contiguous, touching; joining.

contingent, touching.

continuous (1) (of spores, hyphae, etc.), having no septa; (2) (of a stipe), one with the tissue of the pileus or peridium; (3) (of cultures), see culture.

control, to prevent or retard the development of a disease (Anon., *TBMS* **33**: 154, 1950).

Contumyces Redhead, Moncalvo, Vilgalys & Lutzoni (2002), Agaricomycetes. 3, Europe. *Hymenochaetales* or *Agaricales* (*Rickenella* clade). See Redhead *et al.* (*Mycotaxon* **82**: 161, 2002).

convergence, describes two organisms with many characters in common but which are descended from widely separate origins. Difficult to distinguish from parallelism due to intermediate conditions.

convex (of a pileus), equally rounded; broadly obtuse;

convexo-expanded, having the edge bent over; **convexo-plane**, convex when young, flat after expansion.

Cooke (Mordecai Cubitt; 1825-1914, England). Schoolmaster, Lambeth (1851-1860); cataloguer, India Museum, London (1862-1880); cryptogamic botanist, Royal Botanic Gardens, Kew (1880-1892). Like his countryman Berkeley (q.v.), he had wide interests in natural history from which his mycological work was the chief development. He was responsible for a very great number of books, papers, and other writings, producing *Science Gossip* for some years and starting (in 1872) the cryptogamic journal *Grevillea*, most of which was written by him during the twenty years the journal was under his control. He described a great number of new species, and new groupings, esp. of polypores and pyrenomycetes (*Grevillea* 1884-1890). His collection of some 46,000 specimens and 25,000 sheets of drawings, are now in the fungal reference collection of the Royal Botanic Gardens, Kew (**K**). *Publs. Handbook of British Fungi* (1871) [supplemented by 1,300 numbers of *Fungi Britannici Exsiccati*]; *Mycographia* (1875-1878) [pictures and accounts of discomycetes]; *Illustrations of British fungi (Hymenomycetes)* (1881-1891) [with 1,200 coloured plates, see *TBMS* **20**: 33, 1935, for index]; *Handbook of Australian Fungi* (1892); *Fungoid Pests of Cultivated Plants* (1906). *Biogs, obits etc.* Ramsbottom (*TBMS* **5**: 169, 1915); English (*Mordecai Cubitt Cooke*, 1987); Stafleu & Cowan (*TL-2* **1**: 536, 1976).

Cookeina Kuntze (1891), Sarcoscyphaceae. 12, widespread (esp. tropical). See Pfister & Kaushal (*Mycotaxon* **20**: 117, 1984; key), Landvik *et al.* (*Nordic Jl Bot.* **17**: 403, 1997; DNA), Wang (*Mycotaxon* **62**: 289, 1997; China), Harrington *et al.* (*Mycol.* **91**: 41, 1999; phylogeny), Maldonado González (*Revta Jardín bot. Nac.* Univ. Habana **21**: 305, 2000; Cuba), Weinstein *et al.* (*Mycol.* **94**: 673, 2002; phylogeny), Meléndez-Howell *et al.* (*Mycotaxon* **87**: 53, 2003; ultrastr.), Hansen & Pfister (*Mycol.* **98**: 1029, 2006; phylogeny), Iturriaga & Pfister (*Mycotaxon* **95**: 137, 2006; monogr.).

Cookella Sacc. (1878), Cookellaceae. 1 (on *Articularia* and *Microstroma*), widespread. See Tehler (*CJB* **68**: 2458, 1990; relationships, use as outgroup).

Cookellaceae Höhn. ex Sacc. & Trotter (1913), ? Myriangiales. 3 gen. (+ 11 syn.), 13 spp.

Lit.: Eriksson (*Op. Bot.* **60**, 1981), Eriksson & Yue (*Mycotaxon* **38**: 201, 1990), Tehler (*CJB* **68**: 2458, 1990; rels, use as outgroup).

Cooksonomyces H.J. Swart & D.A. Griffiths (1974), anamorphic *Pezizomycotina*, Cac/St.= eP.19. 1, Australia. See Swart & Griffiths (*TBMS* **63**: 152, 1974).

Coolia Huijsman (1943) nom. nud. = Squamanita fide Bas (*Persoonia* **3**, 1965).

coolplate, a temperature-controlled plate on which cultures under a light source may be maintained without an undesirable rise in temperature. See Cooke (*FBPP News* **6**: 37, 1981).

Coonemeria Mouch. (1997), Trichocomaceae. Anamorph *Paecilomyces*-like. 3, widespread. See Mouchacca (*Cryptog. Mycol.* **18**: 1, 1997).

Copelandia Bres. (1912) = Panaeolus fide Kuyper (*in litt.*).

Coppinsia Lumbsch & Heibel (1998), ? Trapeliaceae (L). 1, Europe. See Lumbsch & Heibel (*Lichenologist* **30**: 96, 1998), Heibel & Lumbsch (*Lichenologist* **31**: 203, 1999; Germany), Lumbsch *et al.* (*MR* **111**: 1133, 2007).

Copranophilus Speg. (1909) = Pyxidiophora fide Lundqvist (*Bot. Notiser* **133**: 121, 1980), Rossman *et al.* (*Stud. Mycol.* **42**: 248 pp., 1999).

Coprinaceae Overeem & Weese (1924) = Agaricaceae.

Coprinarius (Fr.) P. Kumm. (1871) nom. rej. = Panaeolus fide Kuyper (*in litt.*).

coprine, a disulfuran-like metabolite of the edible *Coprinus atramentarius* which gives a reaction in humans similar to that of antabuse (q.v.).

Coprinellus P. Karst. (1879), Psathyrellaceae. Anamorph *Hormographiella*. *c.* 100, widespread. See Jørgensen & Stalpers (*Taxon* **50**: 909, 2001; nomencl.), Ko *et al.* (*MR* **105**: 1519, 2001; phylogeography), Redhead *et al.* (*Taxon* **50**: 203, 2001; nomencl.), Gams (*Taxon* **54**: 520, 2005; nomencl.), Cáceres *et al.* (*Antonie van Leeuwenhoek* **89**: 79, 2006; anamorphs).

Coprinites Poinar & Singer (1990), Fossil Fungi. 1, Dominican Republic.

Coprinopsis Beeli (1929) ? = Oudemansiella fide Singer (*Agaric. mod. Tax.* edn 3, 1975).

Coprinopsis P. Karst. (1881), Psathyrellaceae. *c.* 200, widespread. See Jørgensen & Stalpers (*Taxon* **50**: 909, 2001; nomencl.), Redhead *et al.* (*Taxon* **50**: 203, 2001; nomencl.), Keirle *et al.* (*Fungal Diversity* **15**: 33, 2004; Hawaii), Gams (*Taxon* **54**: 520, 2005; nomencl.).

Coprinus Pers. (1797), Agaricaceae. Anamorph *Rhacophyllus*. *c.* 10 (coprophilous, on wood, etc.), widespread. The Ink-Caps. Some are edible; *C. comatus*, the shaggy ink-cap or shaggy-mane, is a common mushroom in waste places. The lamellae and their cystidia 'deliquesce' by autodigestion after spore discharge giving 'ink' that may be used for writing. A phylogenetic reconsideration of *Coprinus* s. lat. has led to a changed generic circumscription. The type species of the genus (and the generitype of the family) belongs to *Agaricaceae*; the other coprinoid taxa belong to *Psathyrellaceae*. See Buller (*Researches* **1-7**, 1909; to 1950), Seaver (*Mycol.* **27**: 83, 1935; lifting power), Lange (*Dansk bot. Ark.* **14** no. 6, 1952; biological species concept), Lange & Smith (*Mycol.* **45**: 747, 1953), Orton (*TBMS* **40**: 263, 1957), Kits van Waveren (*Persoonia* **5**: 131, 1968; 'stercorarius' group), Anderson (*The life history and genetics of Coprinus lagopus*, 1971), Romagnesi (*BSMF* **92**: 189, 1976; *C. micaceus*-group), van der Bogart (*Mycotaxon* **4**: 233, 1976; Western N. Am. Sect. *Coprinus*), Patrick (*Mycotaxon* **6**: 341, 1977; sectional ranks), Orton & Watling (*British Fungus Fl.* **3**, 1979; keys 92 spp.), Reijnders (*Persoonia* **10**: 384, 1979; development), van der Bogart (*Mycotaxon* **8**: 243, 1979; Western N. Am. sect. *Seratuli*), Johnson & Vilgalys (*Mycol.* **90**: 971, 1998), Hopple & Vilgalys (*Mol. Phylogen. Evol.* **13**: 1, 1999), Redhead *et al.* (*Taxon* **50**: 203, 2001; new systematic arrangement).

Coprinusella (Peck) Zerov (1979) = Coprinus fide Kuyper (*in litt.*).

Coprobia Boud. (1885) = Cheilymenia fide Moravec (*Mycotaxon* **38**: 459, 1990), Moravec (*Libri Botanici* **21**, 2005).

Coprobolus Cain & Kimbr. (1970), Thelebolaceae. 1 (coprophilous), Canada. Position is uncertain. See van Brummelen (*Persoonia* **16**: 425, 1998).

coprogen, a growth factor in dung required by *Pilobo-*

lus spp. (Hesseltine *et al.*, *Mycol.* **45**: 7, 1953; Pidacks *et al.*, *J. Am. Chem. Soc.* **75**: 6064, 1953).

Coprolepa Fuckel (1870) = Hypocopra fide Cain (*Univ. Toronto Stud., Biol.* **38**, 1934), Læssøe (*SA* **13**: 43, 1994; ? synonym of *Xylaria*).

coprome, a physically and chemically uniform unit (pellet) of faeces used in experimental studies of coprophilous fungi (Wood & Cooke, *TBMS* **83**: 337, 1984).

Copromyces N. Lundq. (1967), Sordariaceae. 1 (coprophilous), Sweden; Canada. See Lundqvist (*Symb. bot. upsal.* **20** no. 1, 1972), Huhndorf *et al.* (*Mycol.* **96**: 368, 2004; phylogeny), Miller & Huhndorf (*Mol. Phylogen. Evol.* **35**: 60, 2005; phylogeny).

Coprophilous fungi. Fungi living on dung; fimicolous fungi. Fungi on dung include aerial contaminants, non-specialized species which have been eaten and survived passage through the gastro-intestinal tract, symbionts living in the gastro-intestinal tract excreted with the dung, and species specially adapted for living on dung. The last category are the coprophilous fungi, and they comprise many species from a wide range of taxonomic groups, e.g. most *Acrasiales*, some other *Mycetozoa*, *Mucorales*, *Pezizales*, *Sordariales*, *Coprinaceae*, and certain other *Basidiomycota*. Many produce violently discharged spores with sticky sheaths which, particularly for species on herbivorous dung, aid colonization of adjacent vegetation. Some coprophilous fungi are specific to particular types of animal, others seem to be generalists. There is often a distinct succession observable on decaying dung: some later colonizers may produce antibiotics to aid colonization, and among these are species (such as *Poronia punctata*) now endangered possibly partly as a result of antibiotic supplements to domesticated animal feeds.

Lit.: Bell (*Dung fungi: an illustrated guide to the coprophilous fungi in New Zealand*, 1983), Bell (*An illustrated guide to the coprophilous ascomycetes of Australia*, 2005), Cain (*Univ. Toronto Studies, biol. ser.* **38**, 1934 [reprinted 1968]; *Can. J. Res.* **C28**: 566; *CJB* **34**: 675; **35**: 255; **39**: 1633; **40**: 447, 1950-62; coprophilous ascomycetes I-VIII), Eliasson & Lundqvist (*Bot. Notiser* **132**: 551, 1979; coprophilous myxomycetes), Prokhorov (*Mikol. Fitopat.* **28**: 20; key 20 discom. gen.), Richardson (*MR* **105**: 387, 2001; diversity), Richardson & Watling (*Bull. BMS* **2**: 18, 1968; asco., basidio.; **3**: 86, 1969, phyco., keys; edn 2, 1975; edn 3, 1997), Webster (*TBMS* **54**: 161, 1970; review), Dix & Webster (*Fungal ecology*, 1995; review), Wicklow (*in* Carroll & Wicklow, *The fungal community*, edn 2 : 715, 1992). See also ammonia fungi.

Coprotiella Jeng & J.C. Krug (1976), Pezizomycetes. 1, S. America. See van Brummelen (*Persoonia* **16**: 425, 1998; posn).

Coprotinia Whetzel (1944), Sclerotiniaceae. 1 (on dung), widespread (north temperate). See Holst-Jensen *et al.* (*Mycol.* **89**: 885, 1997; relationships), Holst-Jensen *et al.* (*Mycol.* **96**: 135, 2004; phylogeny).

Coprotrichum Bonord. (1851) = Sporendonema fide Mason & Hughes *in* Wood (Ed.) (*Nature* **179**: 328, 1957).

Coprotus Korf & Kimbr. (1967), Thelebolaceae. 23, widespread. See Pfister (*in litt.*), Häffner (*Rheinl.-Pfälz. Pilzj.* **5-6**: 134, 1995), Brummelen (*Persoonia* **16**: 425, 1998; posn), Prokhorov (*Mikol. Fitopatol.* **32**: 40, 1998), Suárez *et al.* (*Mycotaxon* **97**: 257, 2006; isozymes).

copulants, Kniep's name for copulating structures of like form.

copulation, the fusion of sexual elements; conjugation; **gametangial -**, the fusion of two sexual organs; **heterogamic -**, the fusion of gametes morphologically unlike; **isogamic -**, the fusion of gametes morphologically like; conjugation in the narrower sense; **planogamic -**, the fusion of motile gametes to give a motile zygote (a planozygote). Cf. merogamy.

Cora Fr. (1825) = Dictyonema C. Agardh ex Kunth fide Parmasto (*Nova Hedwigia* **29**: 99, 1978).

Coraceae Tomas. ex Tomas. (1950) = Atheliaceae.

Coraemyces Cif. & Tomas. (1954) ≡ Cora.

coral fungi, basidiomata of *Clavariaceae*.

coral spot, a branch disease of shrubs and trees (*Nectria cinnabarina*).

Corallicola Volkm.-Kohlm. & Kohlm. (1992), Halosphaeriaceae. 1 (on coral), Belize. See Kohlmeyer & Volkmann-Kohlmeyer (*Mycotaxon* **44**: 417, 1992).

Corallinopsis Lagarde (1917), anamorphic *Pezizomycotina*, Hsy.0eP.?. 1 (on *Insecta*), France.

Coralliochytrium Domján (1937), Chytridiales. 1, Hungary. See Domján (*Folia Cryptog.* **2**: 22, 1937).

Corallium G. Hahn (1883) = Ramaria Fr. ex Bonord. fide Donk (*Reinwardtia* **2**: 435, 1954).

Corallochytrium Raghuk. (1987), ? Choanozoa. 1, Arabian Sea. See Cavalier-Smith & Allsopp (*Eur. J. Protist.* **32**: 1, 1996), Cavalier-Smith *in* Coombs *et al.* (Eds) (*Evolutionary Relationships Among Protozoa*: 375, 1998).

Corallocytostroma Y.N. Yu & Z.Y. Zhang (1980), anamorphic *Clavicipitaceae*, St.≡ eH.?. 2, China; Australia. See Yu & Zhang (*Acta Microbiol. Sin.* **20**: 232, 1980), Shivas *et al.* (*MR* **101**: 849, 1997), Pažoutová *et al.* (*MR* **108**: 126, 2004; ontogeny, phylogeny).

Corallodendron Jungh. (1838) [non *Corallodendron* Mill. 1754, *Papilionaceae*] ≡ Corallomyces Fr. fide Samson & Seifert *in* Samson & Pitt (Eds) (*Advances in Penicillium and Aspergillus systematics* **102**: 397, 1985; represents a diatom).

Coralloderma D.A. Reid (1965), Meruliaceae. 2, Asia; Australia. See Reid (*Nova Hedwigia* Beih. **18**: 332, 1965).

Corallofungus Kobayasi (1983), Hydnaceae. 1, Japan. See Kobayasi (*J. Jap. Bot.* **56**: 174, 1983).

coralloid, much branched; like coral in form; esp. basidiomata of *Clavaria*. Cf. forate.

Coralloidea Roussel (1806) ≡ Ramaria Fr. ex Bonord.

Coralloides Hoffm. (1789) nom. rej. prop. ? = Stereocaulon Hoffm. fide Hawksworth *et al.* (*Dictionary of the Fungi* edn 8, 1995).

Coralloides Tourn. ex Maratti (1822) nom. rej. prop. = Ramaria Fr. ex Bonord. fide Donk (*Persoonia* **8**: 281, 1975).

Coralloides Wulfen (1776) nom. dub., Fungi (L).

Corallomorpha Opiz (1856) nom. dub., anamorphic *Pezizomycotina*. See Sutton (*Mycol. Pap.* **141**, 1977).

Corallomyces Berk. & M.A. Curtis (1853) = Corallomycetella fide Rogerson (*Mycol.* **62**: 865, 1970), Rossman *et al.* (*Stud. Mycol.* **42**: 248 pp., 1999).

Corallomyces Fr. (1849), anamorphic *Pezizomycotina*, Hsy.0eH.?. 2, Africa; Java.

Corallomycetella Henn. (1904), Nectriaceae. Anamorphs *Fusarium*-like, *Rhizostilbella*. 2 (woody tissues), widespread (pantropical). See Rossman *et al.* (*Stud. Mycol.* **42**: 248 pp., 1999).

corbiculae, protective structures forming a stroma around the telia of certain rusts (Kuhnholtz-Lordat, *Bull. mens. Acad. Sci. Lett. Montpellier* **71**: 91, 1942; see *RAM* **26**: 468, 1947); paraphyses; pseudoparaphyses.

Corbulopsora Cummins (1940), Pucciniaceae. 3 (on *Compositae*), Papua New Guinea; India. See Cummins (*Mycol.* **32**: 364, 1940) Cf. *Miyagia.*

Corda (August Karl Josef; 1809-1849, Czech). A medic; employed in the National Museum, Prague; lost at sea in the West Indies. His collections, previously thought lost with him, were rediscovered and are in the National Museum, Prague (**PRM**), see Pilát, *Acta Musei Naturalis Pragae* **1B**: 139, 1938). *Publs.* Pilze. *Sturm's Deutchlands Flora* (1829-1841); *Icones Fungorum hucusque Cognitorum* 6 parts (1837-1854) [the last part by Zobel]; *Anleitung zum Studium der Mycologie* (1842). *Biogs, obits etc.* Šebek (*Česká Mykologie* **38**: 129, 1984); Šebek (*Mykologický Sborník* **61**: 113, 1984); Stafleu & Cowan (*TL-2* **1**: 546, 1976).

Cordalia Gobi (1885) = Tuberculina fide Saccardo (*Syll. fung.* **4**: 653, 1886).

Cordana Preuss (1851), anamorphic *Pezizomycotina*, Hso.1eP.6. 13, Europe; tropical. *C. musae* (banana leaf spot). See Hughes (*CJB* **33**: 259, 1955), de Hoog *et al.* (*Proc. K. Ned. Akad. Wet.* Ser. C, Biol. Med. Sci. **86**: 197, 1983), Priest (*MR* **94**: 861, 1990; spp. on *Musa*, Australia), Castañeda Ruíz *et al.* (*Mycotaxon* **73**: 1, 1999; Venezuela), Markovskaja (*Mycotaxon* **87**: 179, 2003; Lithuania), Cai *et al.* (*Sydowia* **56**: 222, 2004; China), Soares *et al.* (*Fungal Diversity* **18**: 147, 2005; Brazil).

Cordella Speg. (1886), anamorphic *Apiospora*, Hso.0eP.37. 4, pantropical. See Subramanian (*Proc. Indian natn Sci. Acad.* Part B. Biol. Sci. **55**: 38, 1962), Reynolds & Gilbert (*Aust. Syst. Bot.* **18**: 265, 2005; Australia), Reynolds & Gilbert (*Cryptog. Mycol.* **27**: 249, 2006; Panamá).

Cordierites Mont. (1840), ? Helotiales. 3, widespread. See Samuels & Kohn (*Sydowia* **39**: 202, 1986; NZ), Zhuang (*Mycotaxon* **31**: 261, 1988; key), Wang *et al.* (*Mol. Phylogen. Evol.* **41**: 295, 2006; phylogeny).

Corditubera Henn. (1897), Boletales. 5, Africa (tropical). See Demoulin & Dring (*Bull. Jard. bot. nat. Belg.* **45**: 345, 1975), Malençon (*Cryptog. Mycol.* **4**: 1, 1983), Verbeken & Walleyn (*Boll. Gruppo Micol. 'G. Bresadola'* **46**: 87, 2003).

Cordycepioideus Stifler (1941) = Ophiocordyceps fide Blackwell & Gilbertson (*Mycol.* **76**: 763, 1984), Blackwell & Rossi (*Mycotaxon* **25**: 581, 1986; ecology), Samson *et al.* (*Atlas of Entomopathogenic Fungi*, 1988), Ochiel *et al.* (*Mycologist* **11**: 7, 1997), Suh *et al.* (*Mycol.* **90**: 611, 1998; phylogeny), Sung *et al.* (*Stud. Mycol.* **57**: 1, 2007).

Cordyceps Fr. (1824) nom. cons., Cordycipitaceae. Anamorphs *Beauveria, Isaria, Lecanicillium, Akanthomyces, Mariannaea.* 90 (mostly on insects; caterpillar fungi, vegetable caterpillars), widespread. Recently split into a number of segregates, including *Ophiocordyceps, Metacordyceps* and *Elaphocordyceps*; see also these genera. See Kobayasi (*Bull. Biogeogr. Soc. Japan* **9**: 271, 1939; Japan), Mains (*Mycol.* **50**: 169, 1957; N. Am., key), Willis (*Muelleria* **1**: 2, 1959; Australia, key), Kobayasi & Shimizu (*Bull. natn. Sci. Mus.* Tokyo, B **5**: 69, 1960), Kobayasi & Shimizu (*Bull. natn. Sci. Mus.* Tokyo, B **2**: 133, 1976; New Guinea), Evans & Samson (*TBMS* **79**: 431, 1982), Evans & Samson (*TBMS* **82**: 127, 1984; on ants), Hameed *et al.* (*Revue roum. Biol.* Sér. Biol. Végét. **29**: 147, 1984; *Ophiocordyceps*), Samson *et al.* (*Atlas of Entomopathogenic Fungi*, 1988), Ito & Hirano (*Lett. Appl. Microbiol.* **25**: 239, 1997; DNA), Tzean *et al.* (*Atlas of Entomopathogenic Fungi from Taiwan*, 1997), Chen *et al.* (*Biochemical Genetics* **37**: 201, 1999; DNA), Artjariyasripong *et al.* (*Mycoscience* **42**: 503, 2001; phylogeny), Guzmán *et al.* (*Mycotaxon* **78**: 115, 2001; Mexico), Kinjo & Zang (*Mycoscience* **42**: 567, 2001; *Ophiocordyceps*, China), Liu *et al.* (*J. Invert. Path.* **78**: 178, 2001), Liu *et al.* (*MR* **105**: 827, 2001; *Ophiocordyceps*, anamorph), Nikoh & Fukatsu (*Mol. Biol. Evol.* **18**: 1631, 2001; phylogeny, introns), Park *et al.* (*Mycobiology* **29**: 121, 2001; phylogeny), Huang *et al.* (*Mycotaxon* **81**: 229, 2002; *Beauveria* anamorph), Hywel-Jones (*MR* **106**: 2, 2002; ascospores), Liu *et al.* (*MR* **106**: 1100, 2002; phylogeny, anamorphs), Evans (*Mycology Series* **19**: 517, 2003; biocontrol), Fukatsu & Nikoh (*Mycology Series* **19**: 311, 2003; host relations), Stensrud *et al.* (*MR* **109**: 41, 2005; phylogeny), Yokoyama *et al.* (*FEMS Microbiol. Lett.* **250**: 145, 2005; heterothallism), Torres *et al.* (*Mycotaxon* **94**: 253, 2005), Stensrud *et al.* (*MR* **111**: 409, 2007; evolution in *C. sinensis*), Sung *et al.* (*Stud. Mycol.* **57**: 1, 2007; phylogeny, monogr.).

Cordycipitaceae Kreisel ex G.H. Sung, J.M. Sung, Hywel-Jones & Spatafora (2007), Hypocreales. 15 gen. (+ 18 syn.), 250 spp. See Sung *et al.* (*Stud. Mycol.* **57**: 1, 2007; phylogeny, monogr.), Sung *et al.* (*Mol. Phylogenet. Evol.* **44**: 1204, 2007; phylogeny).

Cordylia Fr. (1818) [non *Cordylia* Pers. 1807, *Leguminosae*] ≡ Cordyceps.

Cordyliceps Fr. (1832) ≡ Cordyceps.

Corella Vain. (1890) = Dictyonema C. Agardh ex Kunth fide Parmasto (*Nova Hedwigia* **29**: 99, 1978).

Coremiaceae Nann. (1934) = Trichocomaceae.

Coremiales, see *Hyphomycetes*.

Coremiella Bubák & K. Krieg. (1912), anamorphic *Pezizomycotina*, Hsy.0eP.40. 1, widespread. See Ellis (*Dematiaceous Hyphomycetes*, 1971), Comerio *et al.* (*Revta Iberoamer. Micol.* **22**: 50, 2005).

Coremiopsis Sizova & Suprun (1957) ? = Isaria fide Samson (*Stud. Mycol.* **6**, 1974).

Coremium Link (1809) = Penicillium Link fide Thom (*Bull. Bur. Anim. Ind. USDA* **118**, 1910), Seifert & Samson (*Advances in Penicillium and Aspergillus Systematics* **102**: 143, 1985).

coremium, see synnema.

Corenohydnum Lloyd (1936) nom. dub., Fungi.

Coreomyces Thaxt. (1903), Laboulbeniaceae. 20, widespread. See Majewski (*Acta Mycologica* Warszawa **9**: 217, 1973; Polish spp.), Tavares (*Mycol. Mem.* **9**: 627 pp., 1985), Majewski (*TMSJ* **29**: 151, 1988; Japan), Santamaría (*Fl. Mycol. Iberica* **5**, 2003; Iberian peninsula).

Coreomycetopsis Thaxt. (1920), ? Laboulbeniaceae. 1 (on termites), West Indies; USA. See Blackwell & Kimbrough (*Mycol.* **68**: 541, 1976; structure and development), Henk *et al.* (*Mycol.* **95**: 561, 2003).

Corethromyces Thaxt. (1892), Laboulbeniaceae. 82 (on beetles), widespread. See Tavares (*Mycol. Mem.* **9**, 1985), Santamaría (*Mycol.* **89**: 325, 1997; on rove beetles), Weir & Rossi (*Mycol.* **93**: 171, 2001; Bolivia), Weir & Hughes (*Mycol.* **94**: 483, 2002; New

Zealand), Rossi & Máca (*Sydowia* **58**: 110, 2006; Czech Republic), Santamaria (*Nova Hedwigia* **82**: 349, 2006; Spain), Rossi & Weir (*Mycol.* **99**: 131, 2007; S America).

Corethropsis Corda (1839), anamorphic *Pezizomycotina*, Hsy.0eH.?. 3, Europe; America.

Corethrostroma Kleb. (1933), anamorphic *Pezizomycotina*, St.0eH.?. 1, Europe.

coriacellate, somewhat coriaceous.

coriaceous, like leather in texture.

Corinophoros A. Massal. (1856) nom. rej. = Peccania fide Hawksworth *et al.* (*Dictionary of the Fungi* edn 8, 1995).

Coriolaceae Singer (1961) = Polyporaceae.

Coriolellus Murrill (1905) = Antrodia fide Donk (*Verh. K. ned. Akad. Wet.* tweede sect. **62**: 1, 1974).

Coriolopsis Murrill (1905), Polyporaceae. 17, widespread. See Pegler (*The polypores [Bull. BMS Suppl.]*, 1973), Ryvarden & Johansen (*Prelim. Polyp. Fl. E. Afr.*: 315, 1980; key 10 Afr. spp.).

Coriolus Quél. (1886) = Trametes fide Singer (*Publções Inst. Micol. Recife* **304**, 1961), Domanski *et al.* (*Polyporaceae* edn 2, 1973) See also.

Coriscium Vain. (1890) nom. utique rej. = Lichenomphalia fide Stalpers (*in litt.*). The name has been used for the anamorph (lichenized squamules) of certain lichenized species of Omphalina and is now listed as nom. utique rej. See Redhead & Kuyper (*Arctic Alpine Mycology* **2**: 319, 1987).

corium, see Spore wall and Fig. 5.

corkir, see cudbear.

Cormothecium A. Massal. (1854) = Rhizocarpon fide Hawksworth *et al.* (*Dictionary of the Fungi* edn 8, 1995).

Corneohydnum Lloyd (1924) nom. nud., ? Agaricomycetes. ? Basidiomycetes.

corneous (1) horn-like in texture; (2) (of a substance), like horn.

Corner (Edred John Henry; 1906-1996; England, later Singapore). Assistant Director, Singapore Botanic Garden (1929-1939, remaining in Singapore until 1945); Principal Field Scientific Officer, UNESCO (1946); Lecturer in Botany (1949), then Reader in Plant Taxonomy (1959), then Professor of Tropical Botany (1966), Cambridge University; Fellow of the Royal Society of London (1955); developed the concept of hyphal systems, using analysis of hyphal development to elucidate phylogeny in basidiomycetes; his taxonomic monographs demonstrated the huge diversity of basidiomycetes in southeast Asia. *Publs.* Ad Polyporaceas I-VII *Beihefte zur Nova Hedwigia* (1983-1989). *Biogs, obits etc.* Watling & Ginns (*Mycol.* **90**: 732, 1998) [portrait]; Watling (*MR* **105**: 1533, 2001) [portrait].

Corneromyces Ginns (1976), Boletales. 1, Sabah. See Ginns (*Mycol.* **68**: 970, 1976).

Corneromycetaceae Jülich (1979) = Coniophoraceae.

Corneroporus Hattori (2001), Bankeraceae. 1, Malaysia. See Hattori (*Mycoscience* **42**: 426, 2001).

Cornicularia (Schreb.) Hoffm. (1792), Parmeliaceae (L). 1, widespread (north temperate). See Kärnefelt (*Op. Bot.* **86**: 1, 1986), Kärnefelt *et al.* (*Pl. Syst. Evol.* **183**: 113, 1992), Thell *et al.* (*Mycol. Progr.* **1**: 335, 2002; phylogeny), Peršoh *et al.* (*Mycol. Progr.* **3**: 103, 2004; asci, photobionts), Thell *et al.* (*Symb. bot. upsal.* **34** no. 1: 429, 2004; biogeography), Randlane & Saag (*Central European Lichens*: 75, 2006; key).

Cornicularia Bonord. (1851) = Clavulinopsis fide Corner (*Ann. Bot. Mem.* [A monograph of Clavaria and allied genera] **1**, 1950).

Cornicularia Schaer. (1850) ≡ Alectoria Ach.

Corniculariaceae Schaer. (1850) = Parmeliaceae.

Corniculariella P. Karst. (1884), anamorphic *Durandiella*, St.0fH.15. 7, widespread. See DiCosmo (*CJB* **56**: 1665, 1978), Illman (*Taxon* **34**: 512, 1985; typification), Verkley (*Nova Hedwigia* **75**: 433, 2002).

corniform, shaped like a horn (Fig. 23.36).

Corniola Gray (1821) [non *Corniola* Adans. 1763, *Leguminosae*] = Arrhenia fide Redhead (*CJB* **62**: 865, 1984).

Cornipulvina Huhndorf, A.N. Mill., F.A. Fernández & Lodge (2005), Boliniaceae. 1, neotropics. See Huhndorf *et al.* (*Fungal Diversity* **20**: 61, 2005).

Cornucopiella Höhn. (1915), anamorphic *Pezizomycotina*, Cpd.0eH.15. 2, Europe. See Seifert (*TBMS* **85**: 123, 1985), Okada & Tubaki (*Sydowia* **39**: 148, 1986).

Cornuella Setch. (1891) [non *Cornuella* Pierre 1891, *Sapotaceae*] ≡ Tracya.

Cornularia Sacc. (1884) = Corniculariella fide DiCosmo (*CJB* **56**: 1665, 1978).

Cornuntum Velen. (1947), ? Helotiales. 1, Europe.

cornute (1) horned; horn-like in form; (2) (of aecia), see roestelia.

Cornutispora Piroz. (1973), anamorphic *Pezizomycotina*, Cpd.1bH.10. 4 (on lichens and *Rhytismatales*), widespread. See Hawksworth (*TBMS* **67**: 151, 1976), Martínez & Hafellner (*Mycotaxon* **69**: 271, 1998), Punithalingam (*MR* **107**: 917, 2003; cytology), Knoph (*Biblthca Lichenol.* **88**: 345, 2004; Japan).

Cornutostilbe Seifert (1990), anamorphic *Pezizomycotina*, Hsy.0eH.1. 1, Indonesia. See Seifert (*Mem. N. Y. bot. Gdn* **59**: 120, 1990).

Cornuvesica C.D. Viljoen, M.J. Wingf. & K. Jacobs (2000), Microascales. Anamorph *Chalara*-like. 1, Europe; N. America. See Viljoen *et al.* (*MR* **104**: 365, 2000), Hausner *et al.* (*CJB* **81**: 40, 2003), Hausner & Reid (*CJB* **82**: 752, 2004; phylogeny).

Corollium Sopp (1912) ≡ Paecilomyces fide Raper & Thom (*Manual of the Penicillia*, 1949).

Corollospora Werderm. (1922), Halosphaeriaceae. Anamorphs *Clavariopsis*-like, *Sigmoidea*, *Varicosporina*. 19 (on wood or sand, marine), widespread. See Nakagiri (*TMSJ* **27**: 197, 1986; anamorphs), Kohlmeyer & Volkmann-Kohlmeyer (*TBMS* **88**: 181, 1987; key 6 spp.), Nakagiri & Tokura (*TMSJ* **28**: 413, 1987; key 13 spp.), Kohlmeyer & Volkmann-Kohlmeyer (*Bot. Mar.* **34**: 1, 1991; key), Read *et al.* (*Bot. Mar.* **35**: 553, 1992; asci), Spatafora *et al.* (*Am. J. Bot.* **85**: 1569, 1998; phylogenetic analysis), Chen *et al.* (*Mycol.* **91**: 84, 1999; DNA), Kong *et al.* (*MR* **104**: 35, 2000; DNA), Campbell *et al.* (*Fungal Diversity Res. Ser.* **7**: 15, 2002; phylogeny, revision), Zhang *et al.* (*Mycol.* **98**: 1076, 2006; phylogeny), Hsieh *et al.* (*Bot. Mar.* **50**: 302, 2007; ontogeny).

Coronasclerotes Stach & Pickh. (1957), Fossil Fungi. 5 (Carboniferous), Germany.

coronate, crowned.

Coronatomyces Dania García, Stchigel & Guarro (2004), ? Sordariaceae. 1, Cuba. Perhaps synonymous with *Boothiella*. See García *et al.* (*Stud. Mycol.* **50**: 144, 2004).

Coronella P. Crouan & H. Crouan (1867) = Kickxella fide Linder (*Farlowia* **1**: 49, 1943).

Coronellaria P. Karst. (1870), Dermateaceae. 4,

Europe. See Scheuer (*Biblthca Mycol.* **123**, 1988), Magnes & Hafellner (*Biblthca Mycol.* **139**, 1991), Raitviir & Leenurm (*Folia cryptog. Estonica* **38**: 63, 2001; Estonia).

Coronicium J. Erikss. & Ryvarden (1975), Pterulaceae. 5, widespread (north temperate); Hawaii. See Eriksson & Ryvarden (*Cortic. N. Europ.* **4**, 1976), Gilbertson & Hemmes (*Mem. N. Y. bot. Gdn* **89**: 81, 2004; Hawaii).

Coronium Bonord. (1864) nom. dub., anamorphic *Pezizomycotina*. See Sutton (*Mycol. Pap.* **141**, 1977).

Coronopapilla Kohlm. & Volkm.-Kohlm. (1990), ? Zopfiaceae. 1 (marine), Belize. See Kohlmeyer & Volkmann-Kohlmeyer (*MR* **94**: 685, 1990), Eriksson & Hawksworth (*SA* **8**: 7, 1991).

Coronophora Fuckel (1864), Nitschkiaceae. 5 (from wood), Europe. See Subramanian & Sekar (*Kavaka* **18**: 19, 1993), Huhndorf *et al.* (*MR* **108**: 1384, 2004).

Coronophoraceae Höhn. (1907) = Nitschkiaceae.
Lit.: Nannfeldt (*Svensk bot. Tidskr.* **69**: 49, 1975), Subramanian & Sekar (*Kavaka* **18**: 19, 1993) but the rank originally intended by Höhnel is uncertain. See.

Coronophorales Nannf. (1932). Hypocreomycetidae. 4 fam., 26 gen., 87 spp. Fams:
(1) **Bertiaceae**
(2) **Chaetosphaerellaceae**
(3) **Nitschkiaceae**
(4) **Scortechiniaceae**
For *Lit.* see under fam.

Coronophorella Höhn. (1909) = Nitschkia fide Müller & von Arx *in* Ainsworth *et al.* (Eds) (*The Fungi* **4A**: 87, 1973).

Coronoplectrum Brusse (1987), Parmeliaceae (L). 1, Namibia. See Brusse (*Mycotaxon* **28**: 131, 1987).

Coronospora M.B. Ellis (1971), anamorphic *Ascoronospora*, Hso.≡ eP.10. 4, S.E. & E. Asia. See Ellis (*Mycol. Pap.* **125**: 16, 1971), Matsushima (*Matsush. Mycol. Mem.* **10**, 2001; connection), Zhang & Zhang (*Mycosystema* **23**: 331, 2004).

Coronotelium Syd. (1921) = Puccinia fide Dietel (*Nat. Pflanzenfam.* **6**, 1928).

correct (of names), see Nomenclature.

correlated species (of *Pucciniales*), a species derived by reduction (of life cycle or morphology) from a parent heteroecious macrocyclic species/or the parent species itself.

Corrugaria Métrod (1949) = Mycena fide Singer (*Agaric. mod. Tax.*, 1951).

corrugate, wrinkled.

cortex, a more or less thick outer covering; **epi-** (q.v.); **corticate**, having a cortex.

corticate, see cortex.

Corticiaceae Herter (1910) nom. cons., Corticiales. 29 gen. (+ 8 syn.), 136 spp. Traditionally this family contained the resupinate homobasidiomycetes, which are now divided over more than 10 orders.
Lit.: Larsson (*MR* **111**: 1040, 2007).

Corticiales K.H. Larss. (2007). Agaricomycetes. 1 fam., 29 gen., 136 spp. Fam.:
Corticiaceae
For *Lit.* see under fam.

Corticifraga D. Hawksw. & R. Sant. (1990), ? Lecanorales. 2 (on lichens, *Peltigerales*), widespread (esp. temperate). See Hawksworth & Santesson (*Biblthca Lichenol.* **38**: 123, 1990), Martínez & Hafellner (*Mycotaxon* **69**: 271, 1998; Sapin, Portugal), Zhurbenko (*Mikol. Fitopatol.* **35**: 34, 2001; Russia), Hafellner *et al.* (*Mycotaxon* **84**: 293, 2002; N America), Hawksworth (*Lichenologist* **35**: 191, 2003; UK, Ireland), Zhurbenko (*Lichenologist* **39**: 221, 2007; review).

corticioid, of a form resembling *Corticium*; specifically, with a smooth hymenium.

Corticioides Lloyd (1908) = Tremella Pers. fide Lloyd (*Mycol. Writ. (Myc. Notes No. 57)* **5**: 816, 1919).

Corticirama Pilát (1957), Corticiaceae. 2, Europe. See Pilát (*Beih. Sydowia* **1**: 128, 1957).

Corticiruptor Wedin & Hafellner (1998), Lecanorales (L). 1, Norway. See Wedin & Hafellner (*Lichenologist* **30**: 86, 1998).

Corticium Fr. (1835) = Phanerochaete fide Donk (*Persoonia* **3**: 199, 1964).

Corticium Pers. (1794), Corticiaceae. 25, widespread. *C. solani* (anamorph *Rhizoctonia solani*), see *Thanatephorus*, and *C. salmonicolor* (pink disease of rubber, tea, and other tropical plants, see *Phanerochaete*. Formerly used for many resupinate basidiomycetes; now restricted to the species previously known as *Laeticorticium*. See Eriksson & Ryvarden (*Cortic. N. Europ.* **4**: 759, 1976; key 5 Eur. spp.), Larsen & Gilbertson (*Norw. Jl Bot.* **24**: 99, 1977), Greslebin & Rajchenberg (*N.Z. Jl Bot.* **41**: 437, 2003; Patagonia), Duhem & Michel (*BSMF* **122**: 145, 2006; key s.str.).

corticolous, living on bark; **corticole**, an organism which does this.

Corticomyces A.I. Romero & S.E. López (1989), anamorphic *Agaricomycetes*. 1 (with clamp-connexions), Argentina. See Romero & López (*Mycotaxon* **34**: 431, 1989).

cortina (of agarics), a partial veil (or part of one), frequently web-like, covering the mature gills.

Cortinaria, see *Cortinarius*.

Cortinariaceae R. Heim ex Pouzar (1983) nom. cons., Agaricales. 12 gen. (+ 28 syn.), 2104 spp.
Lit.: Ammirati *et al.* (*Poisonous Mushrooms of the Northern United States and Canada*: 396 pp., 1986), Bidaud *et al.* (*Atlas des Cortinaires* **8**: 239, 1996), Liu *et al.* (*CJB* **75**: 519, 1997), Brandrud (*Edinb. J. Bot.* **55**: 65, 1998), Hibbett *et al.* (*Nature* **407**: 506, 2000) suggest that *Cortinariaceae* could be the sistergroup to *Hydnangiaceae*, while the inclusion of *Inocybe* is not supported. Several gastroid taxa are also included in the *Cortinariaceae*, Høiland & Holst-Jensen (*Mycol.* **92**: 694, 2000), Moncalvo *et al.* (*Syst. Biol.* **49**: 278, 2000), Bougher & Lebel (*Aust. Syst. Bot.* **14**: 439, 2001), Peintner *et al.* (*Am. J. Bot.* **88**: 2168, 2001), Peintner *et al.* (*Mycol.* **94**: 620, 2002), Garnica *et al.* (*MR* **107**: 1143, 2003), Garnica *et al.* (*Mycol.* **95**: 1155, 2003), Peintner *et al.* (*MR* **107**: 485, 2003), Peintner *et al.* (*Mycol.* **96**: 1042, 2004), Frøslev *et al.* (*Mol. Phylogen. Evol.* **37**: 602, 2005), Matheny (*Mol. Phylogen. Evol.* **35**: 1, 2005), Vesterholt (*Fungi of Northern Europe* **3**: 146 pp., 2005), Garnica *et al.* (*CJB* **83**: 1457, 2005), Moreau *et al.* (*Mol. Phylogen. Evol.* **38**: 794, 2006).

Cortinariales = Agaricales.

cortinarins, toxic fluorescent cyclic decapeptides produced by *Cortinarius* spp. (Laatsch & Matthies, *Mycol.* **83**: 492, 1991). See also mycetismus (Mycetism).

Cortinarius (Pers.) Gray (1821) nom. cons., Cortinariaceae. *c.* 2000, widespread (esp. north temperate). The genus has been subdivided in a number of subgenera, but these usually lack phylogenetic quality or make other groups non-monophyletic. Recognition of *Dermocybe* as a separate genus would make the rest of *Cortinarius* paraphyletic. See Moser (*Die Gattung Phlegmacium*, 1960; key European subgen. *Phlegmacium*), Horak & Moser (*Nova Hedwigia* **10**: 211, 1965; key), Moser (*Z. Pilzk.* **35**: 213, 1969; key European subgen. *Leprocybe*), Moser (*Schweiz. Z. Pilzk.* **50**: 153, 1972; key), Moser & Horak (*Beih. Nova Hedwigia* **52**: 513, 1975; key South American spp.), Brandrud (*Nordic Jl Bot.* **3**: 577, 1983; key European subgen. *Cortinarius*), Høiland (*Op. bot.* **71**: 1, 1984; key European spp. subgen. *Dermocybe*), Beaton *et al.* (*Kew Bull.* **40**: 171, 1985; key Austral.), Gill & Steglich (*Progr. Chem. Nat. Prod.* **51**, 1987; pigment chemistry), Brandrud *et al.* (*Cortinarius Flora Photographica*, 1990; European spp.), Horak & Wood (*Sydowia* **42**: 88, 1990; key Australiasian subgen. *Myxacium* and *Paramyxacium*), Reumaux *et al.* (*Atlas des Cortinaires*, 1990; European spp.), Bendiksen *et al.* (*Sommerfeltia* **19**: 1, 1993; key European spp. subgen. *Myxacium*), Gill (*Aust. Jl Chem.* **48**: 1, 1995; pigments in Australian spp.), Horak (*Beih. Sydowia* **10**: 101, 1995; key Australasian subgen. *Phlegmacium*), Liu *et al.* (*CJB* **75**: 519, 1997; phylogeny), Høiland & Holst-Jensen (*Mycol.* **92**: 694, 2000; phylogeny), Seidl (*Mycol.* **92**: 1091, 2000; phylogeny subgen. *Myxacium*), Soop (*BSMF* **117**: 91, 2001; New Zealand sp.), Peintner *et al.* (*Mycotaxon* **81**: 177, 2002; syn. of *Cortinarius*), Consiglio *et al.* (*Il Genere Cortinarius in Italia* **1**, 2003; Italy), Garnica *et al.* (*MR* **107**: 1143, 2003; South American spp), Consiglio *et al.* (*Il Genere Cortinarius in Italia* **2**, 2004; Italy), Francis & Bougher (*Australasian Mycologist* **23**: 1, 2004; Western Australian spp.), Soop (*Cortinarius in Sweden*, 2004; Sweden), Antonini *et al.* (*Il Genere Cortinarius in Italia*, 2005; keys to subgen.), Bidaud *et al.* (*Atlas des Cortinaires* **15**: 983, 2005), Consiglio *et al.* (*Il Genere Cortinarius in Italia* **3**, 2005; Italy), Consiglio *et al.* (*Rivista di Micologia* **49**: 29, 2006; sect. *Cystidiosi*) Cortinarius poisoning, see Mycetism.

Cortinellus Roze (1876) = Tricholoma fide Singer (*Agaric. mod. Tax.*, 1951).

Cortiniopsis J. Schröt. (1889) ≡ Lacrymaria.

Cortinomyces Bougher & Castellano (1993) ≡ Protoglossum.

Corylomyces Stchigel, M. Calduch & Guarro (2006), Lasiosphaeriaceae. 1, France. See Stchigel *et al.* (*MR* **110**: 1362, 2006).

Corylophomyces R.K. Benj. (1994), Laboulbeniaceae. 5, widespread. See Benjamin (*Aliso* **14**: 42, 1994), Benjamin (*Aliso* **18**: 71, 1999), Santamaría (*Fl. Mycol. Iberica* **5**, 2003).

Corymbomyces Appel & Strunk (1904) = Gliocladium fide Kendrick & Carmichae *in* Ainsworth *et al.* (Eds) (*The Fungi* **4A**: 390, 1973).

corymbose, arranged in clusters.

Corynascella Arx & Hodges (1975), ? Chaetomiaceae. 3, widespread. See von Arx (*Beih. Nova Hedwigia* **94**, 1988; key), Guarro *et al.* (*Mycol.* **89**: 955, 1997), Stchigel *et al.* (*Mycol.* **95**: 1218, 2003).

Corynascus Arx (1973), Chaetomiaceae. Anamorph *Myceliophthora*. 7, widespread. See von Arx (*Stud. Mycol.* **8**: 21, 1975), von Arx (*Beih. Nova Hedwigia* **94**, 1988; key), Mouchacca (*Cryptog. Mycol.* **18**: 19, 1997; thermophilic spp.), Sigler *et al.* (*Mycotaxon* **68**: 185, 1998), Stchigel *et al.* (*MR* **104**: 879, 2000; key).

Coryne Nees (1816), anamorphic *Ascocoryne*, Hsy.0eH.15. 1, widespread (esp. temperate). See Seifert (*Stud. Mycol.* **31**: 157, 1989).

Coryneaceae Corda (1839) = Pseudovalsaceae.

Corynecystis Brusse (1985), Lichinaceae (L). 1, S. Africa. See Brusse (*Bothalia* **15**: 552, 1985).

Corynelia Ach. (1823), Coryneliaceae. 7 (on *Podocarpaceae*), widespread (esp. tropical and south temperate). See Benny *et al.* (*Bot. Gaz.* **146**: 238, 1985), Johnston & Minter (*MR* **92**: 422, 1989), Hawksworth (*SA* **14**: 48, 1995; nomencl.), Winka & Eriksson (*Phylogenetic Relationships Within the Ascomycota Based on 18S rDNA Sequences* Akademisk Avhandling [Thesis (PhD), Department of Ecology and Environmental Science, Umeå University]: [17] pp., 2000; phylogeny), Geiser *et al.* (*Mycol.* **98**: 1053, 2006; phylogeny).

Coryneliaceae Sacc. ex Berl. & Voglino (1886), Coryneliales. 8 gen. (+ 7 syn.), 46 spp.

Lit.: Benny *et al.* (*Bot. Gaz.* **146**: 232, 1985), Benny *et al.* (*Bot. Gaz.* **146**: 437, 1985), Benny *et al.* (*Bot. Gaz.* **146**: 238, 1985), Benny *et al.* (*Bot. Gaz.* **146**: 431, 1985), Johnston & Minter (*MR* **92**: 422, 1989), Marmolejo (*Mycotaxon* **72**: 195, 1999), Rikkinen (*Karstenia* **40**: 147, 2000), Winka & Eriksson (*Phylogenetic Relationships Within the Ascomycota Based on 18S rDNA Sequences* Akademisk Avhandling [Thesis (PhD), Department of Ecology and Environmental Science, Umeå University]: [17] pp., 2000), Inderbitzin *et al.* (*MR* **108**: 737, 2004).

Coryneliales Seaver & Chardón (1926). Eurotiomycetidae. 1 fam., 8 gen., 46 spp. Fam.:

Coryneliaceae

Lit.: Benny *et al.* (*Bot. Gaz.* **146**: 232, 238, 431, 437, 1985), Fitzpatrick (*Mycol.* **12**: 206, 1920; **34**: 464, 1942; keys, monogr.), Geiser *et al.* (*Mycol.* **98**: 1051, 2006), Johnston & Minter (*MR* **92**: 422, 1989; asci).

Coryneliella Har. & P. Karst. (1890), Pezizomycotina. 1, Mauritius. See Fitzpatrick (*Mycol.* **12**: 206, 1920).

Coryneliopsis Butin (1972), ? Coryneliaceae. 2 (with *Cyttaria*), Chile. See Benny *et al.* (*Bot. Gaz.* **146**: 437, 1985; key), Johnston & Minter (*MR* **92**: 422, 1989; asci).

Coryneliospora Fitzp. (1942), Coryneliaceae. 2, widespread. See Benny *et al.* (*Bot. Gaz.* **146**: 437, 1985; key).

Corynelites Babajan & Tasl. (1970), Fossil Fungi. 1 (Tertiary), former USSR.

Corynella Boud. (1885) [non *Corynella* DC. 1825, *Leguminosae*] = Claussenomyces fide Korf & Abawi (*CJB* **49**: 1879, 1971).

Coryneopsis Grove (1933) = Seimatosporium fide Sutton (*Mycol. Pap.* **141**, 1977).

Corynesphaera Dumort. (1822) nom. rej. prop. ≡ Cordyceps.

Corynespora Güssow (1906) nom. cons., anamorphic *Corynesporasca*, Hso.≡ eP.25. 89, widespread. See Wei (*Mycol. Pap.* **34**, 1950), Ellis (*Mycol. Pap.* **65**, 1957; key), Ellis (*Mycol. Pap.* **76**, 1960), Goh *et al.* (*Fungal Diversity* **1**: 85, 1998; rDNA analysis), Gams & Seifert (*Taxon* **48**: 379, 1999; nomencl.), Pereira *et al.* (*Biological Control* **26**: 21, 2003; biocontrol), Silva *et al.* (*MR* **107**: 567, 2003; genetic variation).

Corynesporasca Sivan. (1996), Corynesporascaceae. Anamorph *Corynespora*. 1, Sri Lanka. See Sivanesan (*MR* **100**: 783, 1996).

Corynesporascaceae Sivan. (1996), Pleosporales. 2 gen., 90 spp.

Lit.: Carris (*Mycol. Soc. Amer. Newsl.* **38**: 19,

1987), Morgan-Jones (*Mycotaxon* **31**: 511, 1988), Sivanesan (*MR* **100**: 783, 1996), Goh *et al.* (*Fungal Diversity* **1**: 85, 1998).

Corynesporella Munjal & H.S. Gill (1961), anamorphic *Pezizomycotina*, Hso.≡ eP.?. 4, India. See Munjal & Gill (*Indian Phytopath.* **14**: 7, 1961).

Corynesporina Subram. (1994) nom. inval., anamorphic *Pezizomycotina*, Hso.?.?. 1, Singapore. See Subramanian (*Nova Hedwigia* **59**: 266, 1994), Shoemaker & Hambleton (*CJB* **79**: 592, 2001).

Corynesporopsis P.M. Kirk (1981), anamorphic *Pezizomycotina*, Hso.1-≡ eP.29. 9, widespread. See Sutton (*Sydowia* **41**: 330, 1989), Mercado Sierra *et al.* (*Mycotaxon* **64**: 7, 1997).

Corynetes Hazsl. (1881) = Geoglossum fide Nitare (*Windahlia* **14**: 37, 1984), Spooner (*Biblthca Mycol.* **116**, 1987).

Coryneum Nees (1816), anamorphic *Pseudovalsa*, Cac.≡ eP.19. 21, widespread (esp. temperate). See Sutton (*Mycol. Pap.* **138**, 1975; key), Sutton & Rizwi (*Nova Hedwigia* **32**: 341, 1980), Orsenigo *et al.* (*Mycotaxon* **67**: 257, 1998).

Corynitaceae Kalchbr. (1880) = Phallaceae.

Corynites Berk. & M.A. Curtis (1853) = Mutinus fide Stalpers (*in litt.*).

Corynocladus Leidy (1850) nom. dub., ? Microthyriaceae.

Corynodesmium Wallr. (1828) nom. nud., anamorphic *Pezizomycotina*. See Hughes (*CJB* **36**: 727, 1958).

Corynoides Gray (1821) = Calocera fide McNabb (*N.Z. Jl Bot.* **3**: 31, 1965).

Corynophoron Nyl. ex Müll. Arg. (1894) = Stereocaulon Hoffm. fide Hawksworth *et al.* (*Dictionary of the Fungi* edn 8, 1995).

Corynophorus, see *Corinophoros*.

Coscinaria Ellis & Everh. (1886) = Oomyces fide Rogerson (*Mycol.* **62**: 865, 1970).

Coscinedia A. Massal. (1860) = Myriotrema fide Hale (*Bull. Br. Mus. nat. hist.* Bot. **8**: 227, 1981).

Coscinocladium Kunze (1846), Physciaceae (L). 1, Europe. See Crespo *et al.* (*Taxon* **53**: 405, 2004; phylogeny).

coscinocystidium, a cystidium projecting as a pseudocystidium.

coscinoid, a pitted conducting element in *Linderomyces*.

Coscinopeltella Chardón (1930) = Vestergrenia Rehm fide von Arx & Müller (*Beitr. Kryptfl. Schweiz* **11** no. 1, 1954).

Coscinopeltis Speg. (1909) = Munkiella fide Müller & von Arx (*Beitr. Kryptfl. Schweiz* **11** no. 2, 1962).

Coscinospora Mirza (1963) nom. inval. ≡ Jugulospora.

Cosmariospora Sacc. (1880), anamorphic *Pezizomycotina*, Hsp.1eP.?. 1, Italy.

Cosmospora Rabenh. (1862), Nectriaceae. Anamorphs *Acremonium*-like, *Fusarium*-like, *Chaetopsina*, *Cylindrocladiella*, *Stilbella*, *Volutella*. 47, widespread. Anamorphs varied, the genus may not be monophyletic. See Samuels *et al.* (*Mycol. Pap.* **164**, 1991; keys 40 spp. as *Nectria* subgen. *Dialonectria*), Rossman *et al.* (*Stud. Mycol.* **42**: 248 pp., 1999), Nirenberg & Samuels (*CJB* **78**: 1482, 2000), Samuels *et al.* (*Fusarium* Paul E. Nelson Memorial Symposium: 1, 2001; *Fusarium*-like anamorphs), Samuels *et al.* (*Tropical Mycology* **2**: 13, 2002; key), Summerbell & Schroers (*J. Clin. Microbiol.* **40**: 2866, 2002; phylogeny), Hosoya & Tubaki (*Mycoscience* **45**: 261, 2004), Samuels *et al.* (*CBS Diversity Ser.* **4**, 2006; USA), Zhang & Zhuang (*Mycosystema* **25**: 15, 2006; phylogeny).

Costanetoa Bat. & J.L. Bezerra (1963), anamorphic *Pezizomycotina*, Cpd.1bH.?. 1, Brazil. See Batista & Bezerra (*Quad. Lab. crittogam., Pavia* **31**: 75, 1963).

Costantin (Julien Noël; 1857-1936, France). Maitre de Conférences de Botanique, École Normale Supérieure, Paris (1887-1901); Professor of Culture (1901-1919) then Professor of Organographics (1919 onwards) at the Natural History Museum, Paris. Carried out early work in pure culture with hyphomycetes, particularly species parasitic on other fungi; was an early exponent of the view, later adopted by Vuillemin (q.v.) and Mason (q.v.) that hyphomycetes should be classified by the development of their conidia. *Publs. Les Mucédinées Simples. Histoire, Classification, Culture et Rôle des Champignons Inférieurs dans les Maladies des Végétaux et des Animaux* (1888); *Atlas des Champignons Comestibles et Vénéneux* (1895); (with Durour) *Nouvelle Flore des Champignons. de France* (1891) [many later edns and reprints]. *Biogs, obits etc.* Magrou (*BSMF* **53**: 245, 1937); Stafleu & Cowan (*TL-2* **1**: 555, 1976).

Costantinella Matr. (1892), anamorphic *Morchellaceae*, Hso.0eP.10. 4, Europe; N. America. See Paden (*Persoonia* **6**: 405, 1972), Wong *et al.* (*Fungal Diversity* **8**: 173, 2001; Hong Kong).

Costapeda Falck (1923) ≡ Helvella.

costate, veined or ribbed.

costiferous, see hypha.

cot death (Sudden Infant Death Sydrome; SIDS), hypothesis for involvement of *Scopulariopisis brevicaulis* through biodeteriogenic action on mattresses with subsequent release of toxic gases was proposed by Richardson (*Lancet* **335**: 670, 1990) but not supported by later experiments (Kelley *et al.*, *Human Exp. Toxicol.* **11**: 347, 1992).

Cotylidia P. Karst. (1881), Agaricomycetes. 9, widespread (esp. tropical). *Hymenochaetales* or *Agaricales* (*Rickenella* clade). See Reid (*Beih. Nova Hedwigia* **18**: 56, 1965; key), Boidin *et al.* (*BSMF* **66**: 445, 1998; sub *Phanerochaetaceae*).

cotyliform, plate-like or wheel-like with an upturned edge.

Cougourdella E. Hesse (1935), Microsporidia. 2.

Coulterella Zebrowski (1936), Fossil Fungi ? Chytridiomycetes. 1 (Cambrian to ? Recent), Australia.

Courtoisia L. Marchand (1830) nom. rej. prop. = Rinodina fide Hafellner (*Beih. Nova Hedwigia* **79**: 241, 1984).

Coutinia J.V. Almeida & Sousa da Câmara (1903) = Botryosphaeria fide von Arx & Müller (*Stud. Mycol.* **9**, 1975).

Coutourea Castagne (1845) nom. dub., anamorphic *Pezizomycotina*. See Sutton (*Mycol. Pap.* **141**, 1977).

Cowlesia Nieuwl. (1916) ≡ Macropodia.

cramp balls, the ascomata of *Daldinia concentrica*.

Crandallia Ellis & Sacc. (1897), anamorphic *Duplicaria*, Cpt.0eH.15. 1, USA. See Powell (*Mycol.* **65**: 1362, 1973), Kohlmeyer & Volkmann-Kohlmeyer (*MR* **105**: 500, 2001).

Craneomyces Morgan-Jones, R.C. Sinclair & Eicker (1987), anamorphic *Pezizomycotina*, Hso.0eP.3. 1, S. Africa. See Morgan-Jones *et al.* (*Mycotaxon* **30**: 345, 1987).

Craspedodidymum Hol.-Jech. (1972), anamorphic *Chaetosphaeriaceae*, Hso.0eP.15. 10, widespread. See Holubová-Jechová (*Česká Mykol.* **26**: 70, 1972),

Pinruan *et al.* (*Mycoscience* **45**: 177, 2004; Thailand, on palms), Huhndorf & Fernández (*Fungal Diversity* **19**: 23, 2005; teleomorph).

Craspedon Fée (1825) = Strigula fide Hawksworth *et al.* (*Dictionary of the Fungi* edn 8, 1995).

Crassoascus Checa, Barrasa & A.T. Martínez (1993), Annulatascaceae. 1 (from wood etc.), Spain. See Kang *et al.* (*Mycoscience* **40**: 151, 1999).

Crassochaeta Réblová (1999), Chaetosphaerellaceae. 2 (from wood), widespread. See Réblová (*Mycotaxon* **71**: 45, 1999), Réblová & Winka (*Mycol.* **93**: 478, 2001; phylogeny), Huhndorf *et al.* (*MR* **108**: 1384, 2004; phylogeny).

Cratarellus, see *Craterellus*.

Craterella Pers. (1794) nom. rej. = Cotylidia fide Donk (*Taxon* **6**: 26, 1959).

Craterellaceae Herter (1910) = Cantharellaceae.

Craterellus Pers. (1825), Cantharellaceae. 20, widespread. *C. cornucopioides* ('Horn of Plenty') is edible. See Corner (*Beih. Sydowia* **1**, 1957), Feibelman *et al.* (*MR* **101**: 1423, 1997), Dahlman *et al.* (*MR* **104**: 388, 2000; molecular systematics).

Crateridium Trevis. (1862) = Cyphelium Ach. fide Tibell (*Beih. Nova Hedwigia* **79**: 597, 1984).

crateriform, cup-like or crater-like in form.

Craterocolla Bref. (1888), Sebacinaceae. Anamorph *Ditangium*. 1, Europe. See Donk (*Persoonia* **4**: 164, 1966), Weiss & Oberwinkler (*MR* **105**: 403, 2001; phylogeny).

Craterolechia A. Massal. (1860) ? = Arthonia fide Zahlbruckner (*Catalogus Lichenum Universalis* **2**, 1922).

Crateromyces Corda (1831) nom. dub., Fungi. ? Based on insect eggs.

Cratiria Marbach (2000), Caliciaceae (L). 14, widespread (esp. tropical). See Marbach (*Biblthca Lichenol.* **74**: 160, 2000).

Crauatamyces Viégas (1944), Dothideomycetes. 1, Brazil.

Creangium Petr. (1950) = Saccardia fide von Arx (*Persoonia* **2**: 421, 1963).

Crebrothecium Routien (1949) = Eremothecium fide Hawksworth *et al.* (*Dictionary of the Fungi* edn 8, 1995).

Cremasteria Meyers & R.T. Moore (1960), anamorphic *Pezizomycotina*, Hso.0eP.4. 1 (marine), USA; Mediterranean. See Meyers & Moore (*Am. J. Bot.* **47**: 348, 1960), Kohlmeyer & Volkmann-Kohlmeyer (*Bot. Mar.* **34**: 1, 1991).

Cremeogaster Mattir. (1924) = Leucophleps fide Fogel (*CJB* **57**: 1718, 1979).

Crenasclerotes Stach & Pickh. (1957), Fossil Fungi. 3 (Carboniferous), Germany.

crenate, having the edge toothed with rounded teeth (Fig. 23.43).

crenulate, delicately crenate (Fig. 23.44).

Creodiplodina Petr. (1957), anamorphic *Pezizomycotina*, Hso.1eH.?. 1, Australia. See Petrak (*Sydowia* **10**: 316, 1956).

Creographa A. Massal. (1860) nom. rej. prop. = Phaeographis fide Lücking *et al.* (*Taxon* **56**: 1296, 2007; nomencl.).

Creolophus P. Karst. (1879) = Hericium Pers. fide Larsson & Larsson (*in litt.*).

Creomelanops Höhn. (1920) = Botryosphaeria fide Samuels & Singh (*TBMS* **86**: 295, 1986), Eriksson & Yue (*SA* **6**: 241, 1987).

Creonecte Petr. (1949), anamorphic *Pezizomycotina*, Cpd.0fH.?. 1 (on *Uredo*), S. America.

Creonectria Seaver (1909) = Nectria fide Rogerson (*Mycol.* **62**: 865, 1970), Rossman *et al.* (*Stud. Mycol.* **42**: 248 pp., 1999).

Creopus Link (1833) = Hypocrea fide Rogerson (*Mycol.* **62**: 865, 1970), Rossman *et al.* (*Stud. Mycol.* **42**: 248 pp., 1999), Chaverri & Samuels (*Stud. Mycol.* **48**, 2003).

Creoseptoria Petr. (1937), anamorphic *Pezizomycotina*, St.0fH.?. 1, Caucasus.

creosote fungus, see kerosene fungus.

Creosphaeria Theiss. (1910), ? Xylariaceae. Anamorph *Selenosporopsis*. 3, N. & S. America. See Ju *et al.* (*Mycotaxon* **48**: 219, 1993), Læssøe (*SA* **13**: 43, 1994), Bills & Peláez (*Mycotaxon* **57**: 471, 1996; endophytes), Ju & Rogers (*Mycotaxon* **73**: 343, 1999; Taiwan), Sánchez-Ballesteros *et al.* (*Mycol.* **92**: 964, 2000; phylogeny), Acero *et al.* (*Mycol.* **96**: 249, 2004; phylogeny), Triebel *et al.* (*Nova Hedwigia* **80**: 25, 2005; phylogeny).

Creothyriella Bat. & C.A.A. Costa (1957), anamorphic *Pezizomycotina*, Cpt.0eH.?. 1, India. See Batista & Costa (*Revta Biol.* Lisb. **1**: 97, 1957).

Creothyrium Petr. (1925) = Cylindrocolla fide Sutton (*Mycol. Pap.* **141**, 1977).

Crepidopus (Nees) Gray (1821) nom. rej. = Pleurotus fide Kuyper (*in litt.*).

Crepidotaceae Singer (1951) = Inocybaceae.

Crepidotus (Fr.) Staude (1857), Inocybaceae. *c.* 200, widespread. See Hesler & Smith (*North American species of Crepidotus*, 1965; key N. Am. spp.), Singer (*Beih. Nova Hedwigia* **44**: 341, 1973; key), Senn-Irlet (*Mycotaxon* **52**: 59, 1994; culture), Senn-Irlet (*Persoonia* **16**: 1, 1995; key Eur. spp.), Senn-Irlet & de Meijer (*Mycotaxon* **66**: 165, 1998; key Brazilian spp.), Aime *et al.* (*Am. J. Bot.* **92**: 74, 2005; phylogeny and taxonomy).

Crepidula Simakova, Pankova & I.V. Issi (2003), Microsporidia. 1. See Simakova *et al.* (*Parazitologiya* **37**: 145, 2003).

Crepidulospora Simakova, Pankova & I.V. Issi (2004), Microsporidia. 1. See Simakova *et al.* (*Parazitologiya* **38**: 477, 2004).

Crepinula Kuntze (1891) ≡ Cephalotheca.

crescentic, see lunate.

Cresponea Egea & Torrente (1993), Roccellaceae (L). 12, widespread (esp. tropical). See Egea & Torrente (*Mycotaxon* **48**: 302, 1993), Letrouit-Galinou *et al.* (*Bull. Soc. linn. Provence* **45**: 389, 1994; ultrastr.), Grube (*Bryologist* **101**: 377, 1998; phylogeny), Kantvilas (*Symb. bot. upsal.* **34** no. 1: 183, 2004), Kantvilas (*Australasian Lichenology* **58**: 32, 2006; Australia).

Cresporhaphis M.B. Aguirre (1991), ? Trichosphaeriaceae (±L). 7, widespread (temperate; mediterranean). See Barr (*Mycotaxon* **46**: 64, 1993), Calatayud & Aguirre-Hudson (*MR* **105**: 122, 2001).

Cribbea A.H. Sm. & D.A. Reid (1962), Cortinariaceae. 4, widespread (southern temperate). Basidioma gasteroid. See Smith & Reid (*Mycol.* **54**: 98, 1962), Francis & Bougher (*Australasian Mycologist* **21**: 81, 2002; Australia).

Cribbeaceae Singer, J.E. Wright & E. Horak (1963) = Cortinariaceae. The name *Cribbeaceae* has not yet been proposed for rejection against *Cortinariaceae.*

cribose (**cribriform**), having a network like a sieve.

Cribritaceae Locq., D. Pons & Sal.-Cheb. (1981) = Microthyriaceae.

Cribrites R.T. Lange (1978), Fossil Fungi, Microthyriaceae. 1 (? Eocene), Australia.
Cribropeltis Tehon (1933), anamorphic *Pezizomycotina*, Cpt.0eH.?. 1, USA.
Cribrospora Pacioni & P. Fantini (2000), Agaricales. 1, Europe. See Pacioni & Fantini (*Micologia e Vegetazione Mediterranea* **14**: 171, 1999).
Cricunopus P. Karst. (1881) = Suillus Gray fide Smith & Thiers (*The Boletes of Michigan*, 1971).
Criella (Sacc.) Henn. (1900), Rhytismataceae. 1, widespread (tropical).
Crinigera I. Schmidt (1969), Pezizomycotina. 1 (on *Fucus*, marine), Baltic Sea. See Koch & Jones (*CJB* **67**: 1183, 1989).
Crinipellis Pat. (1889), Marasmiaceae. *c*. 65, widespread. The pathogens of cacao are now placed in *Moniliophthora* (q.v.). See Singer (*Lilloa* **8**: 441, 1942; monogr.), Singer (*Fl. Neotrop.* Monogr. **17**: 9, 1976; key 14 neotrop. spp.), Arruda *et al.* (*Mycol.* **97**: 1348, 2005; new sp.).
Crinitospora B. Sutton & Alcorn (1985), anamorphic *Pezizomycotina*, Cac.1eH.15/19. 1, Australia. See Sutton & Alcorn (*TBMS* **84**: 437, 1985).
Crinium Fr. (1819) ≡ Crinula Fr.
Crinofera Nieuwl. (1916) ≡ Pilophora.
Crinula Fr. (1821), anamorphic *Holwaya*, Hsy.0eP.?. 2, Europe; N. America. See Seifert (*Stud. Mycol.* **27**: 1, 1985), Aronsson (*Svensk bot. Tidskr.* **85**: 9, 1991; Sweden), Roth (*Schweiz. Z. Pilzk.* **69**: 197, 1991; Switzerland).
Crinula Sacc. (1889) ≡ Holwaya.
Criserosphaeria Speg. (1912), ? Helotiales. 1, Argentina. See Petrak & Sydow (*Annls mycol.* **33**: 157, 1935).
crispate, curled and twisted.
crista, tubular, pouch-like or shelf-like inwardly directed fold of the inner membrane of a mitochondrion; site of ATP production during aerobic metabolism.
Cristaspora Fort & Guarro (1984), Trichocomaceae. 1, Spain. See Fort & Guarro (*Mycol.* **76**: 1115, 1984).
cristate, crested.
Cristella Pat. (1887) = Sebacina fide Rogers (*Mycol.* **36**: 70, 1944), Weresub (*Taxon* **16**: 402, 1967), Donk (*Taxon* **17**: 278, 1968) = Trechispora (Hydnodont.) fide, Liberta (*CJB* **51**: 1871, 1973).
Cristelloporia I. Johans. & Ryvarden (1979) = Trechispora fide Larsson (*in litt.*).
Cristidium R. Sant. (1952) nom. nud., anamorphic *Pezizomycotina*. 1 (on *Gyalectidium* in the tropics).
Cristinia Parmasto (1968), Stephanosporaceae. 7, widespread. See Hjortstam & Grosse-Brauckmann (*Mycotaxon* **47**: 405, 1993; key).
Cristiniaceae Jülich (1982) = Atheliaceae.
Cristula Chenant. (1920), anamorphic *Pezizomycotina*, Hso.1bH.1. 1, Europe.
Cristularia (Sacc.) Costantin (1888) = Botrytis fide Saccardo (*Syll. fung.* **4**: 134, 1886).
Cristulariella Höhn. (1916), anamorphic *Nervostroma*, Hso.1bH.1. 4 (on *Acer* etc.), widespread (north temperate). See Redhead (*CJB* **53**: 700, 1975), Niedbalski *et al.* (*Mycol.* **71**: 722, 1979; development in *C. pyramidalis*), Suto & Suyama (*Mycoscience* **46**: 227, 2005), Narumi-Saito *et al.* (*Mycoscience* **47**: 351, 2007).
Cristulospora L.F. Khodzhaeva & I.V. Issi (1989), Microsporidia. 3.
Crivellia Shoemaker & Inderb. (2006), Pleosporaceae. Anamorph *Brachycladium*. 1, widespread. See Inderbitzin *et al.* (*CJB* **84**: 1304, 2006).
Crocicreas Fr. (1849), Helotiaceae. 62, widespread (esp. north temperate). See also *Cyathicula*. See Carpenter (*Mem. N. Y. bot. Gdn* **33**, 1981), Baral & Krieglsteiner (*Beih. Sydowia* **6**, 1985), Galán (*MR* **98**: 1137, 1994), Triebel & Baral (*Sendtnera* **3**: 199, 1996), Iturriaga *et al.* (*MR* **103**: 28, 1999; application).
Crocicreomyces Bat. & Peres (1964) = Calopadia fide Lücking *et al.* (*Lichenologist* **30**: 121, 1998).
Crocodia Link (1833) nom. rej. = Pseudocyphellaria fide Hawksworth *et al.* (*Dictionary of the Fungi* edn 8, 1995).
Crocynia (Ach.) A. Massal. (1860) nom. cons., Crocyniaceae (L). *c*. 2, widespread (tropical). See Hue (*Bull. Soc. bot. Fr.* **71**: 311, 1924; *s.l.* monogr.), Ekman & Tønsberg (*MR* **106**: 1262, 2002; phylogeny), Miądlikowska *et al.* (*Mycol.* **98**: 1088, 2006; phylogeny).
Crocyniaceae M. Choisy ex Hafellner (1984), Lecanorales (L). 1 gen. (+ 1 syn.), *c*. 2 spp.
Lit.: Hafellner (*Nova Hedwigia* Beih. **79**: 241, 1984), Aptroot & Sipman (*Willdenowia* **20**: 221, 1991), Hofstetter *et al.* (*Mol. Phylogen. Evol.* **44**: 412, 2007; phylogeny).
Crocysporium Corda (1837) = Aegerita fide Saccardo (*Syll. fung.* **4**: 662, 1886).
Cronartiaceae Dietel (1900), Pucciniales. 2 gen., 24 spp.
Lit.: Crane *et al.* (*Proceedings of the Fourth IUFRO Rusts of Pines Working Party Conference* Tsukuba: 101, 1995), Hiratsuka (*Proceedings of the Fourth IUFRO Rusts of Pines Working Party Conference* Tsukuba: 1, 1995), Mims *et al.* (*Mycol.* **88**: 47, 1996), Vogler & Bruns (*Mycol.* **90**: 244, 1998), Et-touil *et al.* (*Phytopathology* **89**: 915, 1999), Vogler (*HortTechnol.* **10**: 518, 2000), Hantula *et al.* (*MR* **106**: 203, 2002), Cummins & Hiratsuka (*Illustr. Gen. Rust Fungi* edn 3: 225 pp., 2003), Maier *et al.* (*CJB* **81**: 12, 2003), Kubisiak *et al.* (*Heredity* **92**: 41, 2004), Wingfield *et al.* (*Australas. Pl. Path.* **33**: 327, 2004), Aime (*Mycoscience* **47**: 112, 2006).
Cronartium Fr. (1815), Cronartiaceae. *c*. 20 (on *Pinus* (0, I) (*Pinaceae*); on dicots (II, III)), widespread. *C. ribicola* (blister rust of *Pinus strobus* and other 5-needle pines; II and III on *Ribes*), *C. quercuum* (fusiform rust of pine). See Mielke (*Bull. Sch. Forest. Yale* **52**, 1943; *C. ribicola*), Peterson (*Rep. Tottori mycol. Inst.* **10**: 203, 1973), Hiratsuka & Powell (*Canad. Forest Serv. Tech. Rep.* **4**, 1976), Burdsall & Snow (*Mycol.* **69**: 503, 1977; *C. quercuum* taxonomy), Vogler & Bruns (*Mycol.* **90**: 244, 1998; DNA seq. analysis), Hantula *et al.* (*MR* **106**: 203, 2002; genetics), Hamelin *et al.* (*Phytopathology* **95**: 793, 2005; Mol. epidemiol. white pine blister rust).
Crossopsora Syd. & P. Syd. (1919), Phakopsoraceae. *c*. 17, widespread (tropical). See Peterson (*Rep. Tottori mycol. Inst.* **10**: 203, 1973), Gjærum *et al.* (*Lidia* **5**: 87, 2000; Uganda), Berndt *et al.* (*Mycotaxon* **83**: 265, 2002; Brazil), Engkhaninun *et al.* (*Mycoscience* **46**: 137, 2005; Thailand).
Crotalia Liro (1938), anamorphic *Anthracoidea*. 1, Europe.
Crotone Theiss. & Syd. (1915), Venturiaceae. 1, S. America.
Crotonocarpia Fuckel (1870) = Cucurbitaria fide von Arx & Müller (*Stud. Mycol.* **9**, 1975).

crottle, Scottish term for many lichens (obsol.); often used collectively; **black -**, *Parmelia omphalodes*.

Crouania Fuckel (1870) [non *Crouania* J. Agardh 1842, *Algae*] = Lamprospora fide Eckblad (*Nytt Mag. Bot.* **15**: 1, 1968).

Crouaniella (Sacc.) Lambotte (1888) = Ascobolus fide van Brummelen (*Persoonia* Suppl. **1**: 1, 1967).

crowded (of gills), very close together; conferted.

crown rust (of oats), *Puccinia coronata*.

Crozalsiella Maire (1917) = Ustilago fide Vánky (*Europ. Smut Fungi*: 348, 1994) fide.

crozier, the hook of an ascogenous hypha before ascus-development; ascus crook (Fig. 13).

Crucella Marvanová & Suberkr. (1990), anamorphic *Camptobasidium*. 1, USA. See Marvanová & Suberkropp (*Mycol.* **82**: 212, 1990).

Crucellisporiopsis Nag Raj (1983), anamorphic *Pezizomycotina*, Ccu.1bH.15. 2, Venezuela; New Zealand. See Nag Raj (*CJB* **60**: 2601, 1982).

Crucellisporium M.L. Farr (1968), anamorphic *Pezizomycotina*, Cac/Ccu.1bH.10. 2, USA; Tanzania. See Nag Raj & Kendrick (*CJB* **56**: 713, 1978), Punithalingam (*Nova Hedwigia* **48**: 297, 1989; cytology), Punithalingam (*MR* **107**: 917, 2003).

cruciate (1) in the form of a cross; (2) (of basidial septa), vertical and at right angles.

Crucibulum Tul. & C. Tul. (1844), Agaricaceae. 3, widespread (temperate). See Sarasini & Pina (*Rivista di Micologia* **39**: 115, 1996), Zhou *et al.* (*Fungal Diversity* **17**: 243, 2004; China), Sarasini (*Gasteromiceti Epigei*: 406 pp., 2005).

cruciform (of nuclear division in *Plasmodiophora*), having the chromosomes in a ring around a dumb-bell-shaped nucleolus.

cruciform division, see promitosis.

Cruciger R. Kirschner & Oberw. (1999), anamorphic *Agaricomycetes*. 1, Germany. See Kirschner & Oberwinkler (*Mycoscience* **40**: 345, 1999).

Crucispora E. Horak (1971), Agaricaceae. 2, New Zealand; Asia. See Horak (*N.Z. Jl Bot.* **9**: 489, 1971).

Crumenella P. Karst. (1890), Helotiaceae. 1 (on *Myrica*), Europe.

Crumenula De Not. (1864) = Godronia fide Groves (*CJB* **43**: 1195, 1965), Petrini *et al.* (*CJB* **67**: 2805, 1989).

Crumenula Rehm (1889) ≡ Crumenulopsis.

Crumenulopsis J.W. Groves (1969), Helotiaceae. Anamorph *Digitosporium*. 4, Europe; N. America. See Ennos & Swales (*MR* **95**: 521, 1991; populations), Hanlin *et al.* (*Mycol.* **84**: 650, 1992).

crust, a general term for a hard surface layer, esp. of a sporocarp; crustose.

Crustoderma Parmasto (1968), Meruliaceae. 14, widespread. See Nakasone (*Mycol.* **76**: 40, 1984), Gilbertson & Nakasone (*Mycol.* **95**: 467, 2003; key).

Crustodiplodina Punith. (1988), anamorphic *Pezizomycotina*, St.1eH.19. 1, British Isles. See Punithalingam (*Mycol. Pap.* **159**: 199, 1988).

Crustodontia Hjortstam & Ryvarden (2005), Polyporales. 1, widespread. See Hjortstam & Ryvarden (*Syn. Fung.* **20**: 36, 2005).

Crustomollisia Svrček (1987), Dermateaceae. 1, Europe. See Svrček (*Sydowia* **39**: 219, 1987), Nauta & Spooner (*Mycologist* **13**: 65, 1999).

Crustomyces Jülich (1978), Cystostereaceae. 3, widespread. See Jülich (*Persoonia* **10**: 140, 1978), Legon (*Mycologist* **20**: 118, 2006).

crustose (crustaceous), crust-like; used for lichens having a thallus stretching over and firmly fixed to the substratum by the whole of their lower surface; such thalli generally lack rhizinae and a lower cortex. (Fig. 21B).

Crustospathula Aptroot (1998), Ramalinaceae (L). 1, Papua New Guinea. See Aptroot (*Trop. Bryol.* **14**: 27, 1998).

Crustula Velen. (1934) = Mollisia fide Baral (*SA* **13**: 113, 1994).

Cryocaligula Minter (1986), anamorphic *Ploioderma*, St.1eH.10. 1, C. America. See Minter (*Recent Research on Conifer Needle Diseases* (USDA Forest Service General Technical Report GTR-WO 50): 78, 1986).

Cryomyces Selbmann, de Hoog, Mazzaglia, Friedmann & Onofri (2005), anamorphic *Dothideomycetes*. 2, Antarctica. See Selbmann *et al.* (*Stud. Mycol.* **51**: 1, 2005).

cryopreservation, see Genetic Resource Collections.

Cryphonectria (Sacc.) Sacc. & D. Sacc. (1905), Cryphonectriaceae. Anamorph *Endothiella*. 5, widespread. *C. parasitica* (chestnut (*Castanea*) blight or canker). See also *Endothia*. See Roane in Roane *et al.* (Eds) (*Chestnut blight, other Endothia diseases, and the genus Endothia*: 28, 1986; key 11 spp.), Micales & Stipes (*Phytopathology* **77**: 650, 1987), Milgroom & Lipari (*Mol. Ecol.* **4**: 633, 1995; DNA), Cortesi *et al.* (*Eur. J. For. Path.* **28**: 167, 1998; VCGs), Milgroom & Cortesi (*Proc. natn Acad. Sci. U.S.A.* **96**: 10518, 1999; pop. str.), Myburg *et al.* (*Mycol.* **91**: 243, 1999; DNA), Castlebury *et al.* (*Mycol.* **94**: 1017, 2002; posn), Hoegger *et al.* (*Mycol.* **94**: 105, 2002; *C. radicalis*), Myburg *et al.* (*CJB* **80**: 590, 2002), Marra *et al.* (*Heredity* **93**: 189, 2004; population biology), Myburg *et al.* (*CJB* **82**: 1730, 2004; phylogeny), Myburg *et al.* (*Mycol.* **96**: 990, 2004; phylogeny), Gryzenhout *et al.* (*Taxon* **54**: 539, 2005; nomencl.), Breuillin *et al.* (*MR* **110**: 288, 2006; genetic variation), Gryzenhout *et al.* (*FEMS Microbiol. Lett.* **258**: 161, 2006; *C. parasitica*, taxonomy review), Gryzenhout *et al.* (*Mycol.* **98**: 239, 2006; phylogeny), Liu & Milgroom (*Mycol.* **99**: 279, 2007; E Asia, VCGs).

Cryphonectriaceae Gryzenh. & M.J. Wingf. (2006), Diaporthales. 12 gen. (+ 2 syn.), 26 spp.

Lit.: Barr (*Mycol. Mem.* **7**: 1, 1978), Anagnostakis (*Mycol.* **79**: 23, 1987), Redlin & Rossman (*Mycol.* **83**: 200, 1991), Castlebury *et al.* (*Mycol.* **94**: 1017, 2002), Myburg *et al.* (*Mycoscience* **44**: 187, 2003), Gryzenhout *et al.* (*Stud. Mycol.* **50**: 130, 2004), Myburg *et al.* (*Mycol.* **96**: 990, 2004), Gryzenhout *et al.* (*Mycol.* **98**: 239, 2006), Rossman *et al.* (*Mycoscience* **48**: 135, 2007; phylogeny), Gryzenhout *et al.* (*Taxonomy, phylogeny, and ecology of bark-infecting and tree killing fungi in the Cryphonectriaceae*: in press, 2008; monograph).

crypta, a sleeve-like formation around a tree root (esp. evergreens) in tropics and subtropics developed by certain agarics (Singer, 1962: 20).

Cryptadelphia Réblová & Seifert (2004), Trichosphaeriaceae. Anamorph *Brachysporium*. 6, widespread (temperate). See Réblová & Seifert (*Mycol.* **96**: 348, 2004), Markovskaja & Treigien (*Nova Hedwigia* **84**: 495, 2007; Lithuanian, anamorph).

Cryptandromyces Thaxt. (1912), Laboulbeniaceae. 13 or 19, widespread. See Tavares (*Mycol. Mem.* **9**, 1985), Santamaría (*Fl. Mycol. Iberica* **5**, 2003; Iberian peninsula).

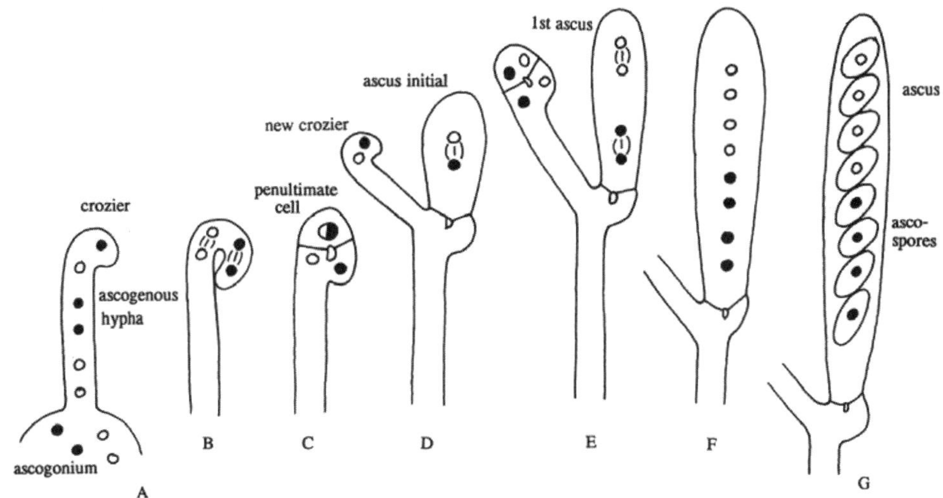

Fig. 13. Ascus and ascospore development (diagrammatic). A, ascogenous hypha with a crozier at the apex developing from an ascogonium; B, conjugate division of the two nuclei in the crozier; C, two septa cut off a binucleate penultimate cell, the nuclei in which fuse to form a diploid fusion nucleus, meanwhile the backwardly directed terminal cell fuses with the ascogenous hypha; D, penultimate cell enlarges to become the ascus within which the fusion nucleus begins to divide meiotically, and a new crozier develops from beneath the ascus and repeats the behaviour of the first; E, second division of meiosis occurs in the young ascus; F, mitotic division of the four haploid nuclei in the ascus; G, ascospores formed.

Cryptascoma Ananthap. (1988), Valsaceae. 1, India. See Ananthapadmanaban (*TBMS* **90**: 479, 1988).

Cryptella Quél. (1875) = Robergea fide Saccardo (*Syll. fung.* **2**: 806, 1883).

Cryptendoxyla Malloch & Cain (1970), ? Cephalothecaceae. Anamorph *Chalara*-like. 1 (from soil etc.), Canada. See von Arx & van der Walt (*Stud. Mycol.* **9**: 167, 1987; posn), Suh & Blackwell (*Mycol.* **91**: 836, 1999; phylogeny).

cryptic, inconspicuous or hidden.

Cryptica R. Hesse (1884) = Pachyphloeus fide Fischer (*Nat. Pflanzenfam.* **5b**: viii, 1938).

Cryptoascus Petri (1909), Pezizomycotina. 2, Europe. *C. oligosporus* on *Olea*, *C. graminis* on wheat.

Cryptobasidiaceae Malençon ex Donk (1956), Exobasidiales. 4 gen. (+ 2 syn.), 9 spp.

Lit.: Ciccarone (*Micol. Ital.* **18**: 29, 1989), Bauer *et al.* (*CJB* **75**: 1273, 1997), Gómez & Kisimova-Horovitz (*Revta Biol. trop.* **45**: 1293, 1998), Begerow *et al.* (*Mycol. Progr.* **1**: 187, 2002), Hendrichs *et al.* (*Sydowia* **55**: 33, 2003).

Cryptobasidiales Jülich (1981) = Exobasidiales.

Cryptobasidium Lendn. (1921) = Botryoconis fide Malençon (*BSMF* **69**: 92, 1953).

Cryptoceuthospora Petr. (1921), anamorphic *Pezizomycotina*, St.0eH.?. 2, Europe.

Cryptochaete P. Karst. (1889) = Peniophora fide Donk (*Persoonia* **3**: 199, 1964).

Cryptocline Petr. (1924), anamorphic *Helotiales*, Cac.0eH.19. 15, widespread (north temperate). See Morgan-Jones (*CJB* **51**: 309, 1973), Sutton (*The Coelomycetes*, 1980), Petrini (*Sydowia* **37**: 238, 1984), Matsushima (*Matsush. Mycol. Mem.* **10**, 2001; Japan).

Cryptococcaceae Kütz. ex Castell. & Chalm. (1919) = Tremellaceae.

Cryptococcales. Order used for *Blastomycetes* (q.v.) with ascomycetous teleomorphs, but *Cryptococcus* itself is now known to be polyphyletic and partly of basidiomycetous affinity (and assiged here to *Tremellaceae*).

cryptococcosis, a disease in humans and animals caused by *Cryptococcus neoformans* (teleomorph *Filobasidiella neoformans*); 'European blastomycosis'; torulosis. See Casadevall & Perfect (*Cryptococcus neoformans*, 1998), Kovacs *et al.* (*Annls Intern. Med.* **1103**: 533, 1985; clinical), Buchanan & Murphy (*Emerg. Infect. Dis.* **4**: 71, 1998; pathogenesis).

Cryptococcus Kütz. (1833) nom. rej. ≡ Cryptococcus Vuill.

Cryptococcus Vuill. (1901) nom. cons., anamorphic *Filobasidiella*. c. 75, widespread. In a wide sense, applied to a range of yeast forming species within the *Tremellomycetes*. See Ellis & Pfeiffer (*J. Clin. Microbiol.* **28**: 1642, 1990; ecology), James & Cherniak (*Infect. Immun.* **60**: 1084, 1992; cell wall biochemistry), Guého *et al.* (*Antonie van Leeuwenhoek* **63**: 175, 1993; mol. phylogeny), Perfect *et al.* (*J. Clin. Microbiol.* **31**: 3305, 1993; PFGE), Currie *et al.* (*J. Clin. Microbiol.* **32**: 1188, 1994; RFLP), Fan *et al.* (*J. med. Vet. Mycol.* **32**: 1163, 1994; mol. phylogeny), Fell *et al.* (*Stud. Mycol.* **38**: 129, 1995; mol. phylogeny), Filonow *et al.* (*Biol. Control* **7**: 212, 1996; biocontrol), Tanaka *et al.* (*J. med. Vet. Mycol.* **34**: 299, 1996; ploidy), Boekhout & Scorzetti (*J. Med. Vet. Mycol.* **35**: 147, 1997; mycocins), Boekhout *et al.*

(*Int. J. Syst. Bacteriol.* **47**: 432, 1997; PFGE, RAPD), Sorrell & Ellis (*Rev. Iberoam. Micol.* **14**: 42, 1997; ecology), Casadevall & Perfect (*Cryptococcus neoformans*, 1998), Krockenberger *et al.* (*Medical Mycology* **39**: 523, 2001; immunohistochemical differentiation of *Cryptococcus neoformans* vars), Hazen & Howell (*Manual of Clinical Microbiology* **2**: 1693, 2003), Lemmer *et al.* (*Medical Mycology* **42**: 135, 2004; PCR fingerprinting compared to serotyping in *Cryptococcus neoformans*), McClelland *et al.* (*Trends in Microbiology* **12**: 208, 2004; mating system in *Cryptococcus neoformans*), Lin *et al.* (*Nature* London **434** no. 7036: 1017, 2005; sexual reproduction in partners of the same mating type in *Cryptococcus neoformans*), Loftus *et al.* (*Science* Washington **307** no. 5713: 1321, 2005; genome map), Lin & Heitman (*Annual Review of Microbiology* **60**: 69, 2006; biology of *Cryptococcus neoformans* species complex), Playford *et al.* (*J. Clin. Microbiol.* **44**: 876, 2006; detection and identification by reverse line blot hybridization).

Cryptocolax R.A. Scott (1956), Fossil Fungi, Eurotiales. 1 (Eocene), USA. See Pirozynski (*Ann. Rev. Phytopath.* **14**: 237, 1976).

Cryptocoryneopsis B. Sutton (1980), anamorphic *Pezizomycotina*, Hso.0bP.1. 1, Australia. See Sutton (*TBMS* **74**: 393, 1980), Ho *et al.* (*Mycol.* **92**: 582, 2000; key), Alcorn (*Australas. Mycol.* **21**: 111, 2002).

Cryptocoryneum Fuckel (1870), anamorphic *Pezizomycotina*, St.0bP.1. 4, widespread (temperate). See Ho *et al.* (*Mycol.* **92**: 582, 2000; key).

Cryptocrea Petr. (1937) = Eudarluca fide Müller & von Arx (*Beitr. Kryptfl. Schweiz* **11** no. 2, 1962).

Cryptoderis Auersw. (1869) ≡ Pleuroceras.

Cryptoderma Imazeki (1943) = Phylloporia fide Wagner & Fischer (*MR* **105**: 781, 2001).

Cryptodesma Leidy (1850) nom. dub., ? Fungi.

Cryptodiaporthe Petr. (1921), Gnomoniaceae. Anamorphs *Diplodina*, *Uniseta*. 24, widespread (north & south temperate). See Barr (*Mycotaxon* **41**: 287, 1991), Redlin & Rossman (*Mycol.* **83**: 200, 1991), Bathgate *et al.* (*MR* **100**: 159, 1996), Castlebury *et al.* (*Mycol.* **94**: 1017, 2002), Castlebury *et al.* (*Mycoscience* **44**: 203, 2003; phylogeny), Sogonov *et al.* (*Sydowia* **57**: 102, 2005; phylogeny), Gryzenhout *et al.* (*Mycol.* **98**: 239, 2006; placement of *C. corni*), Zhang *et al.* (*Mycol.* **98**: 1076, 2006; phylogeny).

Cryptodictyon A. Massal. (1860), Lecideaceae (L). 2, Asia.

Cryptodidymosphaeria (Rehm) Höhn. (1917) = Didymosphaeria fide Aptroot (*Nova Hedwigia* **60**, 1995).

Cryptodidymosphaerites Currah, Stockey & B.A. LePage (1998), Fossil Fungi, Pleosporales. 1. See Currah *et al.* (*Mycol.* **90**: 668, 1998).

Cryptodiscus Corda (1838), ? Stictidaceae. 6, widespread. See Sherwood (*Mycotaxon* **5**: 1, 1977), Sherwood-Pike (*Mycotaxon* **28**: 137, 1987), Wedin *et al.* (*Lichenologist* **37**: 67, 2005; phylogeny).

cryptoendolithic (of organisms, esp. lichens), surviving at low temperatures through modification of the thallus so that it can exist inside rock between rock crystals (Friedmann, *Science, N.Y.* **215**: 1045, 1982). Cf. endolithic.

Cryptogamia (obsol.). Division of the kingdom *Plantae* for the spore-producing plants, i.e. the *Thallophyta* (which traditionally included fungi), *Bryophyta* (mosses and liverworts), and *Pteridophyta* (ferns, etc.); the *Phanerogamia* (*Spermatophyta*) being the division for the flowering (or seed-producing) plants.

Cryptogene Syd. (1939) = Ascochytopsis fide Sutton (*Mycol. Pap.* **141**, 1977).

Cryptogenella Syd. (1939) = Ascochytopsis fide Sutton (*Mycol. Pap.* **141**, 1977).

Cryptohymenium Samuels & L.M. Kohn (1987), Dermateaceae. 1, New Zealand. See Samuels & Kohn (*Sydowia* **39**: 202, 1987).

Cryptolechia A. Massal. (1853), Gyalectaceae (L). 8, widespread (esp. tropical). See Vězda (*Folia geobot. phytotax.* **4**: 443, 1969), Hawksworth & Dibben (*Lichenologist* **14**: 98, 1982; nomencl.), Kauff & Büdel (*Bryologist* **108**: 272, 2005; ontogeny, anatomy), Kalb (*Biblthca Lichenol.* **95**: 297, 2007; key).

Cryptoleptosphaeria Petr. (1923), ? Diaporthales. 1 (on *Leptosphaeria*), Europe. Affinities of this genus remain obscure. See Rossman *et al.* (*Stud. Mycol.* **42**: 248 pp., 1999), Réblová & Seifert (*Stud. Mycol.* **50**: 95, 2004).

Cryptomela Sacc. (1884) = Cryptosporium fide Höhnel (*Sber. Akad. Wiss. Wien* Math.-naturw. Kl., Abt. 1 **125**: 76, 1916).

Cryptomeliola S. Hughes & Piroz. (1997), Meliolaceae. 3, Cuba; Kenya. See Mibey & Hawksworth (*Mycol. Pap.* **174**, 1997), Mibey & Cannon (*Cryptog. Mycol.* **20**: 249, 1999).

Cryptomphalina R. Heim (1966), Polyporaceae. 1, Thailand. See Singer (*Agaric. mod. Tax.* 4th ed: 185, 1986; possibly a teratological form of *Lentinus*).

Cryptomycella Höhn. (1925), anamorphic *Cryptomycina*, St.0eH.15. 1, Europe. See Gabel (*Mycol.* **85**: 861, 1993).

Cryptomyces Grev. (1825), Rhytismataceae. 1, widespread (north temperate). *C. maximus* (on *Salix*). See Alcock (*TBMS* **11**: 161, 1926), Minter & Cannon (*IMI Descr. Fungi Bact.* **148** nos 1471-1480, 2001), Vasil'eva (*Mikol. Fitopatol.* **36**: 17, 2002).

Cryptomycetaceae Höhn. (1917) = Rhytismataceae.

Cryptomycina Höhn. (1917), Pezizomycotina. Anamorph *Cryptomycella*. 3 or 1, widespread (north temperate). *C. pteridis* (common on bracken, *Pteridium aquilinum*). See Bache-Wiig (*Mycol.* **32**: 214, 1940).

Cryptomycocolacaceae Oberw. & R. Bauer (1990), Cryptomycocolacales. 2 gen., 2 spp.

Lit.: Mattsson & Wedin (*Lichenologist* **30**: 463, 1998).

Cryptomycocolacales Oberw. & R. Bauer (1990). Cryptomycocolacomycetes. 1 fam., 2 gen., 2 spp. Fam.:

Cryptomycocolacaceae

For *Lit.* see under fam.

Cryptomycocolacomycetes R. Bauer, Begerow, J.P. Samp., M. Weiss & Oberw. (2006). Pucciniomycotina. 1 ord., 1 fam., 2 gen., 2 spp. Ord.:

Cryptomycocolacales

For *Lit.* see ord. and fam.

Cryptomycocolax Oberw. & R. Bauer (1990), Cryptomycocolacaceae. 1 (parasitic on an ascomycete), Costa Rica. See Oberwinkler & Bauer (*Mycol.* **82**: 672, 1990), Bauer *et al.* (*Mycol. Progr.* **5**: 41, 2006).

Cryptonectriella (Höhn.) Weese (1919) = Nectriella Nitschke ex Fuckel fide Rossman *et al.* (*Stud. Mycol.* **42**: 248 pp., 1999).

Cryptonectriopsis (Höhn.) Weese (1919) = Phomatospora fide Barr (*Mycol. Mem.* **7**: 232 pp., 1978).

Cryptoniesslia Scheuer (1993), Niessliaceae. 1 (from dead leaves), British Isles. See Scheuer (*MR* **97**: 543, 1993), Samuels & Barr (*CJB* **75**: 2165, 1997).

Cryptoparodia Petr. (1950) = Antennularia fide Müller & von Arx (*Beitr. Kryptfl. Schweiz* **11** no. 2, 1962).

Cryptopeltis Rehm (1906) = Porina Müll. Arg. fide Hawksworth *et al.* (*Dictionary of the Fungi* edn 8, 1995).

Cryptopeltosphaeria, see *Cryptoleptosphaeria*.

Cryptopezia Höhn. (1919), Helotiales. 1, Samoa.

Cryptophaeella Höhn. (1917) ? = Microsphaeropsis Höhn. fide Sutton (*Mycol. Pap.* **141**, 1977).

Cryptophallus Peck (1897) = Phallus fide Lloyd (*Mycol. Notes* **26**: 329, 1907).

Cryptophiale Piroz. (1968), anamorphic *Pezizomycotina*, Hso.1eH.15. 21, widespread (tropical). See Sutton *et al.* (*MR* **92**: 354, 1989; key), McKenzie (*Mycotaxon* **49**: 307, 1993; NZ, New Caledonia spp.), Goh & Hyde (*MR* **100**: 999, 1996; key), Umali *et al.* (*Mycoscience* **40**: 189, 1999).

Cryptophialoidea Kuthub. & Nawawi (1987), anamorphic *Pezizomycotina*, Hso.0fH.15. 2, Malaysia. See Kuthubutheen & Nawawi (*TBMS* **89**: 581, 1987), Kuthubutheen & Nawawi (*MR* **98**: 686, 1994), Delgado *et al.* (*Fungal Diversity* **20**: 31, 2005; Cuba).

Cryptoporaceae Jülich (1982) = Polyporaceae.

Cryptoporus (Peck) Shear (1902), Polyporaceae. 2, N. America; S.E. Asia. See Gilbertson & Ryvarden (*N. Amer. Polyp.* **1**: 220, 1986), Wu & Zang (*Mycotaxon* **74**: 415, 2000; China), Giaon *et al.* (*Bollettino dell'Associazione Micologica ed Ecologica Romana* **18-19**: 56, 2002).

Cryptopus Theiss. (1914) [non *Cryptopus* Lindl. 1824, *Orchidaceae*] ≡ Adelopus.

Cryptorhynchella Höhn. (1915) ? = Sphaerographium fide Hawksworth *et al.* (*Dictionary of the Fungi* edn 8, 1995), Verkley (*Nova Hedwigia* **75**: 433, 2002).

Cryptosordaria De Not. ex Sacc. (1891) nom. inval. = Anthostomella See Læssøe (*SA* **13**: 43, 1994; ? synonym of *Xylaria*), Lu & Hyde (*Fungal Diversity Res. Ser.* **4**, 2000).

Cryptosphaerella Sacc. (1882) ? = Coronophora fide Subramanian & Sekar (*Kavaka* **18**: 19, 1993).

Cryptosphaeria Ces. & De Not. (1863) nom. cons., Diatrypaceae. Anamorph *Cytosporina*. 8, widespread (esp. temperate). See Glawe & Rogers (*CJB* **64**: 1493, 1986; anamorph), Glawe & Jacobs (*Mycol.* **79**: 135, 1987), Rappaz (*Mycol. Helv.* **2**: 285, 1987; key), Glawe (*Sydowia* **41**: 122, 1989; anamorph), Rappaz (*Taxon* **38**: 664, 1989; nomencl.), Romero & Carmarán (*Fungal Diversity* **12**: 161, 2003; Argentina), Acero *et al.* (*Mycol.* **96**: 249, 2004; phylogeny, polyphyly).

Cryptosphaeria Grev. (1822) nom. rej. = Diplodia fide Bisby & Mason (*TBMS* **24**: 138, 1940).

Cryptosphaerina Lambotte & Fautrey ex Sacc. & P. Syd. (1902) = Cryptosphaeria Ces. & De Not. fide Pirozynski (*in litt.*).

Cryptospora Tul. & C. Tul. (1863) [non *Cryptospora* Kar. & Kir. 1842, *Cruciferae*] = Cryptosporella fide Holm (*SA* **11**: 29, 1992), Mejía *et al.* (*MR* **112**: 23, 2008; phylogeny).

Cryptosporella Sacc. (1877), Gnomoniaceae. 8, widespread. See Reid & Booth (*CJB* **65**: 1320, 1987), Ananthapadmanaban (*TBMS* **90**: 479, 1988; synonym of *Wuestneia*), Reid & Booth (*CJB* **67**: 879, 1989), Lee *et al.* (*Stud. Mycol.* **50**: 235, 2004), Zhang *et al.* (*Mycol.* **98**: 1076, 2006; phylogeny), Mejía *et al.* (*MR* **112**: 23, 2008; phylogeny, monogr.).

Cryptosporellaceae Arx & E. Müll. (1954) = Gnomoniaceae.

Cryptosporina E.I. Hazard & Oldacre (1975), Microsporidia. 1.

Cryptosporina Höhn. (1905) = Botryosphaeria fide von Arx & Müller (*Beitr. Kryptfl. Schweiz* **11** no. 1, 1954).

Cryptosporiopsis Bubák & Kabát (1912), anamorphic *Pezicula, Ocellaria*, Cac/St.0eH.15. *c.* 29, widespread (esp. temperate). *C. malicorticis* (syn. *Gloeosporium perennans*) (perennial canker of apple). See Dennis (*Kew Bull.* **29**: 157, 1974; key), Johnston & Fullerton (*N.Z. J. Exp. Agr.* **16**: 159, 1988; *C. citri* causing leaf spot of *Citrus*), Gené *et al.* (*MR* **94**: 309, 1990; bud rot of *Corylus*), Dugan *et al.* (*Mycol.* **85**: 551 and 565, 1993; morphology, pathogenicity, cytology etc. of *C. perennans* and *C. curvispora*), Kowalski *et al.* (*MR* **102**: 347, 1998), Verkley (*Stud. Mycol.* **44**: 180 pp., 1999; monogr.), Abeln *et al.* (*Mycol.* **92**: 685, 2000; phylogeny), Sigler *et al.* (*Stud. Mycol.* **53**: 53, 2005; N America).

Cryptosporium Kunze (1817), anamorphic *Pezizomycotina*. 25, widespread (temperate). See Sutton (*Mycol. Pap.* **141**, 1977).

Cryptostictella Grove (1912) = Discosia fide Petrak & Sydow (*Annls mycol.* **23**: 209, 1925), Grove (*British Stem- and Leaf-Fungi (Coelomycetes)* **2**, 1937).

Cryptostictis Fuckel (1866) = Seimatosporium fide Shoemaker (*CJB* **42**: 411, 1964).

Cryptostroma P.H. Greg. & S. Waller (1952), anamorphic *Pezizomycotina*, St.0eP.19. 1, N. America; Europe. *C. corticale* (sooty bark of *Acer*).

Cryptosympodula Verkley (1999), anamorphic *Scleropezicula*, Hso.?.?. 1, Canada. See Verkley (*Stud. Mycol.* **44**: 132, 1999).

Cryptothamnium Wallr. (1842) = Chaenocarpus Spreng. fide Læssøe (*SA* **13**: 55, 1994).

Cryptothecia Stirt. (1876) nom. cons., Arthoniaceae (L). 75, widespread (tropical). See Santesson (*Symb. bot. upsal.* **12** no. 1: 1, 1952), Awasthi & Agarwal (*J. Indian bot. Soc.* **48**: 62, 1969), Makhija & Patwardhan (*Biovigyanam* **11**: 1, 1985; key 24 spp.), Thor (*Bryologist* **94**: 278, 1991), Lücking (*Lichenologist* **27**: 127, 1995; Costa Rica), Thor (*Symb. bot. upsal.* **32**: 267, 1997; Australasia), Grube (*Bryologist* **101**: 377, 1998), Grube & Lücking (*MR* **105**: 1007, 2001; ontogeny), Sipman (*Biblthca Lichenol.* **86**: 177, 2003; Singapore), Sparrius & Saipunkaew (*Lichenologist* **37**: 507, 2005; Thailand), Lücking *et al.* (*Lichenologist* **38**: 235, 2006; *C. candida* complex).

Cryptotheciaceae Bonord. (1864) = Arthoniaceae.

Cryptothecium Penz. & Sacc. (1897) [non *Cryptothecium* Hübener 1851, fossil *Musci*] = Protocreopsis fide Rossman *et al.* (*Stud. Mycol.* **42**: 248 pp., 1999).

Cryptothele Th. Fr. (1866), Lichinaceae (L). 8, widespread. See Henssen & Büdel (*Beih. Nova Hedwigia* **79**: 381, 1984), Schultz & Büdel (*Lichenologist* **34**: 39, 2002; key).

Cryptotheliaceae Walt. Watson (1929) = Trypetheliaceae.

Cryptotheliomyces Cif. & Tomas. (1953) ≡ Cryptothele.

Cryptothelium A. Massal. (1860), Trypetheliaceae (L). 1, tropical. See Eriksson (*Op. Bot.* **60**, 1981), Harris (*Acta Amazon.* Supl. **14**: 55, 1984; key 2 spp. Brazil), Eriksson & Hawksworth (*SA* **11**: 56, 1992;

status), Aptroot *et al.* (*Biblthca Lichenol.* **97**, 2008; Costa Rica).

Cryptotrichosporon I. Okoli & Boekhout (2007), anamorphic *Trichosporonaceae*. 1, Nigeria. See Okoli *et al.* (*FEMS Yeast Res.* **7**: 339, 2007).

Cryptovalsa Ces. & De Not. ex Fuckel (1870), Sordariomycetes. 24, widespread. See Hyde (*MR* **97**: 799, 1993), Inderbitzin *et al.* (*MR* **103**: 1628, 1999; Hong Kong), Mostert *et al.* (*Australas. Pl. Path.* **33**: 295, 2004; on *Vitis*), Vasilyeva & Stephenson (*Fungal Diversity* **19**: 189, 2005; USA).

Cryptovalsaria Lar.N. Vassiljeva & S.L. Stephenson (2007), Sordariomycetes. 1. See Vasiljeva & Stephenson (*Sydowia* **59**: 154, 2007).

Cryptumbellata Udagawa & Uchiy. (1999), anamorphic *Pezizomycotina*, Hso.?.?. 1, China. See Udagawa & Uchiyama (*Mycotaxon* **70**: 186, 1999).

Crystallocystidium (Rick) Rick (1940) nom. dub., Polyporales.

Ctenoderma Syd. & P. Syd. (1920) = Skierka fide Mains (*Mycol.* **31**: 175, 1939).

ctenoid, comb-like.

Ctenomyces Eidam (1880), Arthrodermataceae. Anamorph *Chrysosporium*-like. 1, widespread. See Orr & Kuehn (*Mycopathologia* **21**: 321, 1963), Currah (*Mycotaxon* **24**: 1, 1985), Harmsen *et al.* (*Mycoses* **42**: 67, 1999; DNA), Sugiyama & Mikawa (*Mycoscience* **42**: 413, 2001; phylogeny), Geiser *et al.* (*Mycol.* **98**: 1053, 2006; phylogeny).

Ctenosporites Elsik & Janson. (1974), Fossil Fungi, anamorphic *Pezizomycotina*. 2 (Eocene), widespread. See Lange & Smith (*N. Jb. Geol. Paläont. Mh.* **11**: 649, 1975).

Ctenosporium R. Kirschner (2006), anamorphic *Pezizomycotina*. 1, Panama. Possibly synonymous with the fossil genus *Ctenosporites*. See Kirschner (*Mycol. Progr.* **5**: 136, 2006).

Ctesium Pers. (1827) = Graphis fide Staiger (*Biblthca Lichenol.* **85**, 2002).

Cubamyces Murrill (1905) = Trametes fide Overholts (*Polyporaceae of the United States, Alaska, & Canada*, 1953).

Cubasina R.F. Castañeda (1986), anamorphic *Pezizomycotina*, Hso.#eP.1. 1, Cuba. See Castañeda (*Deuteromycotina de Cuba* Hyphomycetes IV: 6, 1986), Markovskaya & Treigiene (*Mikol. Fitopatol.* **38**: 52, 2004).

Cubonia Sacc. (1889), ? Ascobolaceae. 3, Europe. Most species included in this genus belong in to *Pseudombrophila*, but a revision is needed. See Kimbrough & Korf (*Am. J. Bot.* **54**: 9, 1967), Eckblad (*Nytt Mag. Bot.* **15**: 1, 1968), Durand (*BSMF* **88**: 155, 1973; ontogeny), Pfister (*Mycol.* **76**: 843, 1984).

Cucujomyces Speg. (1917), Laboulbeniaceae. 14, widespread. See Rossi & Weir (*CJB* **74**: 77, 1996; Indonesia), Weir & Rossi (*CJB* **75**: 791, 1997; NZ), Weir & Rossi (*Mycol.* **93**: 171, 2001; Bolivia), Santamaría (*Fl. Mycol. Iberica* **5**, 2003; Iberian peninsula), Hughes *et al.* (*Mycol.* **96**: 1355, 2004; New Zealand).

Cucullaria Corda (1842) = Leotia fide Hawksworth *et al.* (*Dictionary of the Fungi* edn 8, 1995).

cucullate, hood-like or cowl-like in form.

Cucullospora K.D. Hyde & E.B.G. Jones (1986) [non *Cucullispora* Scheuring 1970, fossil-sporae dispersae] ≡ Cucullosporella.

Cucullosporella K.D. Hyde & E.B.G. Jones (1990), Halosphaeriaceae. 1 (marine), Seychelles. See Jones & Hyde (*Mycotaxon* **37**: 197, 1990), Alias *et al.* (*Mycoscience* **42**: 405, 2001; ultrastr.), Pang *et al.* (*Nova Hedwigia* **77**: 1, 2003; phylogeny).

Cucurbidothis Petr. (1921) = Curreya fide von Arx & Müller (*Stud. Mycol.* **9**, 1975), Silva-Hanlin & Hanlin (*MR* **103**: 153, 1999).

Cucurbitaria Gray (1821), Cucurbitariaceae. Anamorphs *Camarosporium*, *Dichomera*, *Pleurostromella*, *Megaloseptoria*, *Pyrenochaeta*, *Diplodia*-like. *c.* 36 (on twigs), widespread. See Mirza (*Nova Hedwigia* **16**: 161, 1968), Silva-Hanlin & Hanlin (*MR* **103**: 153, 1999; DNA), Liew *et al.* (*Mol. Phylogen. Evol.* **16**: 392, 2000; phylogeny), Schoch *et al.* (*Mycol.* **98**: 1041, 2006; phylogeny).

Cucurbitariaceae G. Winter (1885), Pleosporales. 5 gen. (+ 9 syn.), 49 spp.

Lit.: Mirza (*Nova Hedwigia* **16**: 161, 1968), Barr & Boise (*Mem. N. Y. bot. Gdn* **49**: 298, 1989), Barr (*Mem. N. Y. bot. Gdn* **62**: 92 pp., 1990), Berbee (*Mol. Biol. Evol.* **13**: 462, 1996), Ramaley & Barr (*Mycotaxon* **65**: 501, 1997), Silva-Hanlin & Hanlin (*MR* **103**: 153, 1999; DNA).

Cucurbitariaceites R.K. Kar, R.Y. Singh & Sah (1972), Fossil Fungi, Dothideomycetes. 2 (Tertiary), India.

Cucurbitariella Petr. (1917) = Coniochaeta fide von Arx & Müller (*Beitr. Kryptfl. Schweiz* **11** no. 1, 1954).

Cucurbitariopsis C. Massal. (1889) ? = Rhabdospora fide Mussat (*Syll. fung.* **10**: 1, 1892), Sutton (*Mycol. Pap.* **141**, 1977).

Cucurbitariopsis Vassilkov (1960) ≡ Gemmamyces.

Cucurbitopsis Bat. & Cif. (1957), Pezizomycotina. 1, Portugal. See Batista & Ciferri (*Publções Inst. Micol. Recife* **95**: 3, 1957).

Cucurbitula Fuckel (1870) = Coniochaeta fide Petrini (*Sydowia* **44**: 169, 1993).

cudbear (**corkir**), Scottish names for lichens used in making dye, esp. *Ochrolechia tartarea*. See Dyeing.

Cudonia Fr. (1849), Cudoniaceae. 9, widespread (temperate). See Mains (*Am. J. Bot.* **27**: 322, 1940), Nannfeldt (*Ark. Bot.* **30A** no. 4: 1, 1942), Mains (*Mycol.* **48**: 694, 1956; N. Am. spp.), Sharma & Rawla (*Biblthca Mycol.* **91**: 203, 1983; India), Landvik *et al.* (*Mycoscience* **37**: 237, 1997; phylogeny), Döring & Triebel (*Cryptog. Bryol.-Lichénol.* **19**: 123, 1998; phylogeny), Gernandt *et al.* (*Mycol.* **93**: 915, 2001; phylogeny), Wang *et al.* (*Mycol.* **94**: 641, 2002; phylogeny, n. sp.), Wang *et al.* (*Mycol.* **98**: 1065, 2006; phylogeny), Wang *et al.* (*Mol. Phylogen. Evol.* **41**: 295, 2006; phylogeny).

Cudoniaceae P.F. Cannon (2001), Rhytismatales. 2 gen. (+ 3 syn.), 21 spp.

Lit.: Landvik *et al.* (*Mycoscience* **37**: 237, 1996; phylogeny), Döring & Triebel (*Cryptog. Bryol.-Lichénol.* **19**: 123, 1998; phylogeny), Gernandt *et al.* (*Mycol.* **93**: 915, 2001), Wang *et al.* (*Mycol.* **94**: 641, 2002), Wang *et al.* (*Mycol.* **98**: 1065, 2006; phylogeny).

Cudoniella Sacc. (1889), Helotiaceae. Anamorph *Tricladium*. *c.* 30, widespread. See Dennis (*Persoonia* **3**: 72, 1964), Baral & Krieglsteiner (*Beih. Sydowia* **6**, 1985; synonym of *Hymenoscyphus*), Webster *et al.* (*Nova Hedwigia* **60**: 493, 1995; anamorph), Gamundí & Romero (*Fl. criptog. Tierra del Fuego* **10**: 130 pp., 1998; Argentina), Wang *et al.* (*Am. J. Bot.* **92**: 1565, 2005; phylogeny), Wang *et al.* (*Mycol.* **98**: 1065, 2006; phylogeny), Wang *et al.* (*Mol. Phy-

logen. Evol. **41**: 295, 2006; phylogeny).

Cudoniopsis Speg. (1925), Sclerotiniaceae. 1, S. America.

Culbersonia Essl. (2000), Physciaceae. 2, Kenya; USA. See Esslinger (*Bryologist* **103**: 771, 2000).

Culcitalna Meyers & R.T. Moore (1960), anamorphic *Halosphaeriaceae*, Hsp.≡ eP.1. 1 (marine), Newfoundland. See Meyers & Moore (*Am. J. Bot.* **47**: 348, 1960), Hambleton *et al.* (*Stud. Mycol.* **53**: 29, 2005; phylogeny).

Culicicola Nieuwl. (1916) ≡ Lamia.

Culicidospora R.H. Petersen (1960), anamorphic *Pezizomycotina*, Hso.1bH.1. 2 (aquatic), widespread. See Petersen (*Mycol.* **55**: 23, 1963), Descals (*MR* **109**: 545, 2005; diagnostics).

Culicinomyces Couch, Romney & B. Rao (1974), anamorphic *Pezizomycotina*, Hso.0eH.15. 3 (in *Anopheles* and rotifers), USA; Australia. See Inmann & Bland (*CJB* **61**: 2618, 1983; ultrastr. conidiogenesis), Sweeney *et al.* (*J. Invert. Path.* **42**: 224, 1983; ultrastr.), Sigler *et al.* (*Mycol.* **79**: 493, 1987; on *Aedes kochi*).

Culicospora J. Weiser (1977), Microsporidia. 1.

Culicosporella J. Weiser (1977), Microsporidia. 1.

culmicolous, living on stems, esp. those of grasses; caulicolous; **culmicole**, an organism which does this.

culmomarasmin, a wilt toxin of *Fusarium culmorum* (Gäumann *et al.*, *Phytopath. Z.* **36**: 115, 1959).

cultivar, a variety in the horticultural (or agricultural) sense (see Art. 10, *International code of nomenclature for cultivated plants* [*Regnum veg.* **104**], 1980). Snyder *et al.* (*J. Madras Univ.* **27**: 185, 1957) used cultivar for infraspecific taxa of *Fusarium* but this term is only correctly used in mycology for trade varieties of cultivated mushrooms, etc.

cultivated mushroom, basidioma of *A. bisporus* (syn. *A. brunnescens*; name in need of conservation; see Malloch, *Mycol.* **68**: 910, 1976, nomencl.);.

culture, a growth of one organism or of a group of organisms for the purpose of experiment (esp. of fungi and other organisms on laboratory media) or sometimes for trade (e.g. a **mushroom** -); **continuous** -, one in which the culture medium is simultaneously added and withdrawn (harvested) so that the volume remains constant; see Calcott (*Continuous culture of cells*, 2 vols, 1981); **enrichment** -, a culture which favours the growth of the desired organism in a mixed culture or population; **pure** -, a - of one sort of organism; **- medium**, see medium; **type** -, see type.

Culture collections, see Genetic resource collections.

culture methods, see Media.

Cumminsiella Arthur (1933), Pucciniaceae. 8 (on *Berberis*, *Mahonia* (autoecious) (*Berberidaceae*)), Europe; America. See Baxter (*Mycol.* **49**: 864, 1957; key), McCain & Hennen (*Syst. Bot.* **7**: 48, 1982; keys etc.) *C. mirabilissima* is a neomycete in many regions on cultivated *Mahonia*.

Cumminsina Petr. (1955), ? Raveneliaceae. 1 (on *Grewia* (*Tiliaceae*)), Angola. See Petrak (*Sydowia* **9**: 474, 1955).

cumulate, massed together; heaped up.

Cumulospora I. Schmidt (1985), anamorphic *Pezizomycotina*, Hso.#eP.?. 1 (marine), widespread. See Schmidt (*Mycotaxon* **24**: 420, 1985), Peña & Arambarri (*Darwiniana* **35**: 69, 1998; Argentina), Chatmala *et al.* (*Fungal Diversity* **17**: 1, 2004; Thailand).

cuneate, see cuneiform.

cuneiform (**cuneate**), thinner at one end than the other; wedge or axe-blade shaped (Fig. 23.23).

Cuniculitrema J.P. Samp. & R. Kirschner (2001), Cuniculitremaceae. Anamorph *Sterigmatosporidium*. 1, Germany. See Kirschner *et al.* (*Antonie van Leeuwenhoek* **80**: 155, 2001).

Cuniculitremaceae J.P. Samp., R. Kirschner & M. Weiss (2001), Tremellales. 4 gen., 25 spp.

Lit.: Kirschner *et al.* (*Antonie van Leeuwenhoek* **80**: 155, 2001).

Cunningham (Gordon Herriot; 1892-1962; New Zealand). Wounded at Gallipoli (1914); instructor, Horticulture Division, Department of Agriculture, Palmerston North (1917-1920) then Wellington (1920-1928); Head of Mycological Laboratory, Plant Research Station, Palmerston North (1928-1935); Director, Plant Diseases Division, Department of Scientific and Industrial Research, Auckland (1939-1957); Fellow of the Royal Society of London (1950). Pioneer New Zealand plant pathologist and mycologist. *Publs.* The Ustilaginaceae or smuts of New Zealand. *Transactions of the New Zealand Institute* (1924), *Fungus Diseases of Fruit-trees in New Zealand* (1925); *The Rust Fungi of New Zealand* (1931) *Plant Protection by the Aid of Therapeutants* (1935); *Gasteromycetes of Australia an New Zealand* (1944); The Thelephoraceae of Australia and New Zealand. *Bulletin of the New Zealand Department of Scientific and Industrial Research* (1963); Polyporaceae of New Zealand. *Bulletin of the New Zealand Department of Scientific and Industrial Research* (1965). *Biogs, obits etc.* Ramsbottom (*Biographical Memoirs of Fellows of the Royal Society* **10**: 15, 1964) [bibliography, portrait]; Stafleu & Cowan (*TL-2* **1**: 573, 1976).

Cunninghamella Matr. (1903), Cunninghamellaceae. 10, widespread. See Samson (*Proc. K. ned. Akad. Wet.* Ser. C, Biol. Med. Sci. **72**: 322, 1969; key), Hawker *et al.* (*J. gen. Microbiol.* **60**: 181, 1970; sporangiolum ultrastructure), Baijal & Mehrotra (*Sydowia* **33**: 1, 1980; key), Shipton & Lunn (*TBMS* **74**: 483, 1980; taxonomic criteria, key), Lunn & Shipton (*TBMS* **81**: 303, 1983; key), Weitzman (*TBMS* **83**: 527, 1984), Zheng & Chen (*Mycosystema* **5**: 1, 1992), Dermoumi (*Mycoses* **36**: 293, 1993; zygomycoses), Zheng & Chen (*Mycosystema* **7**: 1, 1994), Zheng & Chen (*Mycosystema* **8-9**: 1, 1995-1996), Zheng & Chen (*Mycotaxon* **69**: 187, 1998), Su *et al.* (*MR* **103**: 805, 1999; phylogeny), Voigt & Wöstemeyer (*Gene* **270**: 113, 2001; phylogeny), White *et al.* (*Mycol.* **98**: 872, 2006; phylogeny), Koyama *et al.* (*Respirology* **13**: 309, 2008; mucormycosis).

Cunninghamellaceae Naumov ex R.K. Benj. (1959), Mucorales. 5 gen. (+ 3 syn.), 15 spp.

Lit.: Benny *et al.* (*Mycol.* **84**: 639, 1992; emend.), Dermoumi (*Mycoses* **36**: 293, 1993), Zheng & Chen (*Mycosystema* **7**: 1, 1994), Zheng & Chen (*Mycosystema* **8-9**: 1, 1995-1996), Zheng & Chen (*Mycotaxon* **69**: 187, 1998), Su *et al.* (*MR* **103**: 805, 1999).

Cunninghamia Curr. (1873) [non *Cunninghamia* R. Br. 1826, *Cycadaceae*] ≡ Choanephora.

Cunninghammyces Stalpers (1985), Cyphellaceae. 2, New Zealand; Réunion. See Hjortstam & Larsson (*Windahlia* **21**, 1994; = *Xenasma*), Hjortstam *et al.* (*Syn. Fung.* **20**: 42, 2005).

cup fungus, a discomycete (esp. *Leotiales* or *Pezizales*) ascoma. **- lichen**, a sp. of *Cladonia* having podetia expanded into goblet-like scyphi.

Cuphocybe R. Heim (1951) = Cortinarius fide Peintner *et al.* (*Mycotaxon* **83**: 447, 2002).

Cuphophyllus (Donk) Bon (1985), Hygrophoraceae. *c.* 25, widespread. See Lodge (*Kew Bulletin* **54**: 807, 1999; Lesser Antilles), Bañares & Arnolds (*Persoonia* **18**: 135, 2002; Canary Islands).

cupulate, cup-like in form, as e.g. in conidiomata (Fig. 10 O).

Cupulisporonites Z.C. Song & Liu Cao (1994), Fossil Fungi. 1, Antarctica. See Song & Cao (*Monograph, State Antarctic Committee* China **3**: 38, 1994).

Cupulomyces R.K. Benj. (1992), Laboulbeniaceae. 1, Grenada. See Benjamin (*Aliso* **13**: 355, 1992), Santamaria (*MR* **99**: 1071, 1995).

Curculiospora Arnaud (1954), anamorphic *Pezizomycotina*, Hso.0-≡ hP.1. 1, France.

curling factor, see griseofulvin.

Curreya Sacc. (1883), Cucurbitariaceae. Anamorph *Coniothyrium*-like. 2 (on conifers), widespread (north temperate). See von Arx & van der Aa (*Sydowia* **36**: 1, 1983), Eriksson (*SA* **5**: 127, 1986), Barr (*Mycotaxon* **29**: 501, 1987), Barr (*Mem. N. Y. bot. Gdn* **62**, 1990), Kruys *et al.* (*MR* **110**: 527, 2006; phylogeny).

Curreyella (Sacc.) Lindau (1897) ≡ Discostroma fide Kang *et al.* (*MR* **103**: 53, 1999).

Curreyella Massee (1895) = Plicaria fide Eckblad (*Nytt Mag. Bot.* **15**: 1, 1968).

Curtis (Moses Ashley; 1808-1872; USA). Became priest, Wilmington, North Carolina (1835); teacher, Raleigh, North Carolina (1837-1839); missionary, Hillsboro (1841-1847) then Society Hill (1847-156) then again Hillsboro (1856-1872), North Carolina. Student and collector of American fungi who collaborated with Tuckerman (q.v.) and Berkeley (q.v.), sending them specimens collected personally and received from other field biologists mainly in the USA but also from Cuba. Collections in **FH** and **K**. *Biogs, obits etc.* Berkeley & Berkeley (*A Yankee botanist.*, 1986); Petersen (*Mycotaxon* **9**: 459, 1979); Petersen ('B. & C': the mycological association of M.J. Berkeley and M.A. Curtis. *Bibliotheca Mycologica* **72**: 1, 1980); Shear & Stevens (*Mycol.* **11**: 181, 1919); Snell & Dick (*Mycol.* **45**: 968, 1953); Stafleu & Cowan (*TL-2* **1**: 573, 1976).

Curucispora Matsush. (1981), anamorphic *Pezizomycotina*, Hso.1bH.1. 2, Ponape. See Matsushima (*Matsush. Mycol. Mem.* **2**: 4, 1981).

Curvatispora V.V. Sarma & K.D. Hyde (2001), Clypeosphaeriaceae. 1, Singapore. See Sarma & Hyde (*Nova Hedwigia* **72**: 480, 2001).

Curvibasidium Samp. & Golubev (2004), Microbotryomycetes. 2, Japan; Europe. See Sampaio *et al.* (*Int. J. Syst. Evol. Microbiol.* **54**: 1402, 2004).

Curvicladiella Decock & Crous (2006), anamorphic *Nectriaceae*, Hso.?.?. 1, French Guiana. See Decock & Crous (*Stud. Mycol.* **55**: 225, 2006).

Curvicladium Decock & Crous (1998) ≡ Curvicladiella fide Decock & Crous (*Mycol.* **90**: 276, 1998), Crous *et al.* (*Stud. Mycol.* **55**: 213, 2006).

Curvidigitus Sawada (1943), anamorphic *Pezizomycotina*, Hsy.0-≡ bH.?. 1, Taiwan. See Pirozynski (*Mycol. Pap.* **129**, 1972).

Curvisporium, see *Curvusporium*.

Curvularia Boedijn (1933), anamorphic *Cochliobolus*, Hsy.≡ eP.26. 54, widespread. *C. lunata* common on crop plants, esp. trop. (Lam-Quang-Bach, *Fiches Phytopath. trop.* **15**, 1964). See Kendrick & Cole (*CJB* **46**: 1279, 1968; spore devel.), Sivanesan (*Mycol. Pap.* **158**, 1987; key), Alcorn (*Mycotaxon* **39**: 361, 1990; additions to genus), Alcorn (*Mycotaxon* **41**: 329, 1991), Berbee *et al.* (*Mycol.* **91**: 964, 1999; ITS), Olivier *et al.* (*Mycol.* **92**: 736, 2000; phylogeny), Hosokawa *et al.* (*Mycoscience* **44**: 227, 2003; morphology), Sivanesan *et al.* (*Aust. Syst. Bot.* **16**: 275, 2003; Australia), Sun *et al.* (*Mycoscience* **44**: 239, 2003; species concepts), Carter & Boudreaux (*J. Clin. Microbiol.* **42**: 5419, 2004; clinical), Bhattacharya *et al.* (*BMC Evolutionary Biology* **5**: 58, 2005; introns), Pimentel *et al.* (*J. Clin. Microbiol.* **43**: 4288, 2005; clinical).

Curvulariopsis M.B. Ellis (1961), anamorphic *Pezizomycotina*, Hsy.≡ eP.10. 1, Ecuador. See Ellis (*Mycol. Pap.* **82**: 39, 1961).

Curvusporium Corbetta (1963) = Curvularia fide Sutton (*in litt.*), Hosokawa *et al.* (*Mycoscience* **44**: 227, 2003).

cuspidate (e.g. of a pileus or cystidium), having a well-marked sharp outgrowth or point at the top.

Cuspidatispora Shearer & Bartolata (2006), Lasiosphaeriaceae. 1, USA. See Miller *et al.* (*Mycoscience* **47**: 220, 2006).

Cuspidosporium Cif. (1955) = Exosporium fide Ellis (*Mycol. Pap.* **82**, 1961).

Custingophora Stolk, Hennebert & Klopotek (1968), anamorphic *Sordariomycetes*, Hso.0eP.15. 2, widespread. See Barr & Crane (*CJB* **57**: 835, 1979), Viljoen *et al.* (*MR* **103**: 497, 1999), Réblová & Winka (*Mycol.* **92**: 939, 2000; phylogeny), Pinnoi *et al.* (*Nova Hedwigia* **77**: 213, 2003; Thailand).

Cuticularia Ducomet (1907), anamorphic *Pezizomycotina*, Hso.-.-. 1, Europe.

cutis (**cuticle**) (of basidiomata), the outer layer consisting of compressed hyphae parallel to the surface; the upper and lower layers of the **cutis** are sometimes distinguished as **epi-** and **sub-**. See Shaffer (*Brittonia* **22**: 230, 1970; cuticular terminology in *Russula*), Singer (*Agaricales in Modern Taxonomy* edn 4: 69, 1986); also pellis.

Cutomyces Thüm. (1878) = Puccinia fide Dietel (*Nat. Pflanzenfam.* **6**, 1928).

CYA, see Media.

cyanescent, becoming blue.

Cyanicium Locq. (1979) = Cortinarius fide Kuyper (*in litt.*).

Cyanisticta Gyeln. (1931) = Pseudocyphellaria fide Hawksworth *et al.* (*Dictionary of the Fungi* edn 8, 1995).

Cyanoannulus Raja, J. Campb. & Shearer (2003), Annulatascaceae. 1, USA. See Raja *et al.* (*Mycotaxon* **88**: 8, 2003), Campbell & Shearer (*Mycol.* **96**: 822, 2004).

Cyanobaeis Clem. (1909) = Baeomyces fide Hawksworth *et al.* (*Dictionary of the Fungi* edn 8, 1995).

Cyanobasidium Jülich (1979) = Lindtneria fide Hjortstam (*Mycotaxon* **28**: 19, 1987).

cyanobiont, see photobiont.

Cyanocephalium Zukal (1893) ? = Thelocarpon fide Müller & von Arx (*Beitr. Kryptfl. Schweiz* **11** no. 2, 1962), Rossman *et al.* (*Stud. Mycol.* **42**: 248 pp., 1999).

Cyanochyta Höhn. (1915), anamorphic *Gibberella*, Cpd.1eH.?. 1, Europe. See Sutton (*Mycol. Pap.* **141**, 1977).

Cyanoderma Höhn. (1919) [non *Cyanoderma* Weber Bosse 1887, *Rhodophyceae*] ≡ Cyanodermella.

Cyanodermella O.E. Erikss. (1981), Stictidaceae. 2,

Europe; N. America. See Eriksson (*Ark. Bot.* ser. 2 **6**: 381, 1967; as *Cyanoderma*), Winka *et al.* (*Lichenologist* **30**: 455, 1998; DNA), Rossman *et al.* (*Stud. Mycol.* **42**: 248 pp., 1999), Ekman & Tønsberg (*MR* **106**: 1262, 2002; phylogeny).

Cyanodiscus E. Müll. & M.L. Farr (1971), Saccardiaceae. 1 or 2, USA (subtropical); Cuba. See Müller & Farr (*Mycol.* **63**: 1080, 1971), Reynolds & Gilbert (*Cryptog. Mycol.* **27**: 249, 2006; Panamá).

Cyanodontia Hjortstam (1987), Meruliaceae. 1, E. Africa. See Hjortstam (*Mycotaxon* **28**: 23, 1987).

Cyanohypha Jülich (1982) = Botryobasidium Donk fide Langer (*Biblthca Mycol.* **158**, 1994).

Cyanopatella Petr. (1949), anamorphic *Pezizomycotina*, Cpd.0eH/1eP.?. 1, Iran.

cyanophilous (of spores, etc.), readily absorbing a blue stain such as cotton blue or gentian violet.

Cyanophomella Höhn. (1918), anamorphic *Gibberella*, Cpd.0-1eH.?. 1, Europe.

cyanophycophilous (of fungi), ones lichenized with a cyanobacterium (Pike & Carroll, *in* Alexopoulos & Mims, *Introductory mycology*, edn 3, 1980); see photobiont.

Cyanoporina Groenh. (1951), ? Pyrenothricaceae (L). 1, Java. See Groenhart (*Reinwardtia* **1**: 198, 1951).

Cyanopulvis J. Fröhl. & K.D. Hyde (2000), Xylariaceae. 1, Australia. See Fröhlich & Hyde (*Fungal Diversity Res. Ser.* **3**: 308, 2000).

Cyanopyrenia H. Harada (1995), ? Eurotiomycetes (L). 1 (aquatic), Japan. See Harada (*Lichenologist* **27**: 249, 1995).

Cyanospora Heald & F.A. Wolf (1910) = Robergea fide Nannfeldt (*Nova Acta R. Soc. Scient. upsal.*, 1932).

Cyanosporus McGinty (1909) = Polyporus P. Micheli ex Adans. fide Stevenson & Cash (*Bull. Lloyd Libr. Mus.* **35**: 130, 1936).

Cyanotheca Pascher (1914) nom. dub., ? Fungi.

cyanotrophic (of fungi, esp. lichen-forming spp.), obtaining nutrients (esp. nitrates fixed from the atmosphere) by forming regular connexions to free-living or ± lichenized cyanobacteria; see Poelt & Mayrhofer (*Pl. Syst. Evol.* **158**: 265, 1988); see also cephalodium.

Cyathela Raf. (1819) nom. dub., Fungi. No spp. included.

Cyathella Raf. (1815) nom. dub., Fungi. No spp. included.

Cyathia P. Browne (1756) = Cyathus fide Stalpers (*in litt.*).

Cyathicula De Not. (1863) = Crocicreas fide Carpenter (*Mem. N. Y. bot. Gdn* **33**, 1981), Triebel & Baral (*Sendtnera* **3**: 199, 1996), Gamundí & Romero (*Fl. criptog. Tierra del Fuego* **10**: 130 pp., 1998; Argentina), Iturriaga *et al.* (*MR* **103**: 28, 1999; nomencl.), Gamundí (*Darwiniana* **41**: 29, 2003; Chile).

cyathiform, like a cup, a little wider at the top than at the bottom, and sometimes stalked.

Cyathipodia Boud. (1907) = Helvella fide Dissing (*Dansk bot. Ark.* **25** no. 1, 1966).

Cyathisphaera Dumort. (1822) ≡ Cucurbitaria fide Nannfeldt (*Svensk bot. Tidskr.* **69**: 49, 1975).

Cyathodes P. Micheli ex Kuntze (1891) ≡ Cyathus.

Cyathus Haller (1768), Agaricaceae. *c.* 45 (on soil, wood, etc.), widespread. The Bird's Nest Fungi. See Brodie (*Bot. Notiser* **130**: 453, 1977; world key), Gomez & Pérez-Silva (*Rev. Mex Micol.* **4**: 161, 1988; key Mexican spp.), Zhou *et al.* (*Fungal Diversity* **17**: 243, 2004; China), Sarasini (*Gasteromiceti Epigei*: 406 pp., 2005).

Cybebe Tibell (1984) = Chaenotheca fide Middelborg & Mattson (*Sommerfeltia* **5**, 1987), Sparrius *et al.* (*Mycotaxon* **83**: 357, 2002; Taiwan).

Cyberliber (www.cybertruffle.org.uk/cyberliber). An open source of mycological literature on the internet, providing access to scanned images of all the main mycological catalogues and thesauruses, and of many volumes of mycological books and journals. There are, in addition, lists of published mycological works in alphabetical order by author, and a search facility to find works by any specified mycological author.

Cyclaneusma DiCosmo, Peredo & Minter (1983), Leotiomycetes. 2 (on *Pinus*), widespread. See Cannon & Minter (*Mycol. Pap.* **155**, 1986; India), Choi & Simpson (*Mycotaxon* **54**: 455, 1995; development), Gernandt *et al.* (*Mycol.* **93**: 915, 2001; phylogeny), Minter (*Mycotaxon* **87**: 43, 2003).

Cycledium Wallr. (1833) ≡ Schizoxylon.

Cycledum, see *Cycledium*.

Cyclobium C. Agardh (1821) ≡ Clisosporium.

Cycloconium Castagne (1845) nom. rej. = Spilocaea fide Hughes (*CJB* **36**: 727, 1958), Braun (*Taxon* **54**: 538, 2005; nomencl.).

Cyclocybe Velen. (1939) = Agrocybe fide Kuyper (*in litt.*).

Cyclocytospora Höhn. (1928) = Cytospora fide Hawksworth *et al.* (*Dictionary of the Fungi* edn 8, 1995).

Cycloderma Klotzsch (1832) nom. dub., Agaricales. See Lloyd (*Mycol. Notes* **17**: 181, 1904) probably an unopened *Geastrum*.

Cyclodomella P.N. Mathur, V.V. Bhatt & Thirum. (1959) = Coniella fide Petrak (*Sydowia* **14**: 352, 1960).

Cyclodomus Höhn. (1909), anamorphic *Maculatifrondes*, Cpd.0eH.15. 3, USA; S. America.

Cyclodothis P. Syd. (1913) = Mycosphaerella fide von Arx & Müller (*Stud. Mycol.* **9**, 1975).

Cyclographa Vain. (1921) = Catarraphia fide Egea & Torrente (*Cryptog. Bryol.-Lichénol.* **14**: 329, 1993).

Cyclographina D.D. Awasthi (1979) = Diorygma fide Awasthi (*Norw. Jl Bot.* **26**: 169, 1979), Archer (*Telopea* **10**: 589, 2004; Australia), Kalb *et al.* (*Symb. Bot. Upsal.* **34** no. 1: 140, 2004).

cycloheximide (actidione), an antibiotic from *Streptomyces griseus* (*J. Am. Chem. Soc.* **69**: 174, 1947); antibacterial and antifungal. For use in isolating fungi pathogenic for humans, see Georg *et al.* (*J. Lab. Clin. Med.* **44**: 222, 1954). **- tolerance**, used to distinguish plant pathogenic fungi, e.g. *Ceratocystis* from *Ophiostoma* (Harrington, *Mycol.* **72**: 1123, 1981).

Cyclomarsonina Petr. (1965), anamorphic *Pezizomycotina*, St.1eH/1eP.?. 1, India. See Petrak (*Sydowia* **18**: 391, 1964).

Cyclomyces Kunze ex Fr. (1830) nom. rej. = Hymenochaete fide Wagner & Fischer (*Mycol. Progr.* **1**: 93, 2002).

Cyclomycetella Murrill (1904) = Hymenochaete fide Donk (*Reinwardtia* **1**: 483, 1952; as *Cyclomyces*).

Cyclopeltella Petr. (1953), anamorphic *Cyclopeltis*, Cpt.0eH.?. 1, Philippines. See Petrak (*Sydowia* **7**: 373, 1953).

Cyclopeltis Petr. (1953), Micropeltidaceae. Anamorph *Cyclopeltella*. 1, Philippines. See Petrak (*Sydowia* **7**: 370, 1953).

Cyclophomopsis Höhn. (1920) = Phomopsis (Sacc.) Bubák. fide Sutton (*Mycol. Pap.* **141**, 1977).

Cyclopleurotus Hasselt (1824) = Pleurotus fide Singer (*Agaric. mod. Tax.* 4th ed, 1986).

Cycloporellus Murrill (1907) ≡ Cyclomycetella.

Cycloporus Murrill (1904) ≡ Coltricia fide Pegler (*The polypores [Bull. BMS Suppl.]*, 1973), Ryvarden (*Khumbu Himal.* **6**: 380, 1977).

Cyclopus (Quél.) Barbier (1907) ≡ Agrocybe.

Cycloschizaceae Locq., D. Pons & Sal.-Cheb. (1981) = Parmulariaceae.

Cycloschizella Höhn. (1919) = Cycloschizon fide Müller & von Arx (*Beitr. Kryptfl. Schweiz* **11** no. 2, 1962).

Cycloschizon Henn. (1902), Parmulariaceae. 11, widespread (tropical). See Inácio & Minter (*IMI Descr. Fungi Bact.* **145** nos 1441-1450, 2000).

cyclosis, cytoplasmic streaming; characteristic of eukaryotes.

cyclosporin (**-e**, **cyclosporin A**, Sandimmun), a ring-shaped polypeptide from *Tolypocladium inflatum* first reported by Dreyfuss *et al.* (*Eur. J. appl. Microbiol.* **3**: 125, 1976) which selectively inhibits the immune system in humans, especially affecting T cells; since 1983 approved for general use during kidney, heart, liver, pancreas, and bone marrow transplants, reducing organ rejection rates and increasing patient survival; see Winter (*in* Calhoun, *1986 Yearbook of science and the future*: 160, 1985; review). Also used for selective isolation of basidiomycetes.

Cyclostoma P. Crouan & H. Crouan (1867) ≡ Stictis.

Cyclostomella Pat. (1896), Parmulariaceae. 2, C. America. See Inácio & Minter (*IMI Descr. Fungi Bact.* **145** nos 1441-1450, 2000).

Cyclotheca Theiss. (1914), Microthyriaceae. 9, widespread (tropical). See Hosagoudar & Abraham (*New Botanist* **20**: 109, 1993).

Cyclothyrium Petr. (1923), anamorphic *Thyridaria*, St.0eP.15. 2 (on *Juglans*), British Isles; India. See Sutton (*The Coelomycetes*, 1980).

Cylicogone Emden & Veenb.-Rijks (1974) = Conioscypha fide Sutton (*in litt.*), Réblová & Seifert (*Stud. Mycol.* **50**: 95, 2004).

Cylindrina Pat. (1886) ? = Stictis fide Sherwood (*Mycotaxon* **6**: 215, 1977).

Cylindrium Bonord. (1851) = Fusidium fide Hughes (*CJB* **36**: 727, 1958).

Cylindrobasidiaceae Jülich (1982) = Meruliaceae.

Cylindrobasidium Jülich (1974), Physalacriaceae. 6, widespread. See Jülich (*Persoonia* **8**: 72, 1974).

Cylindrocarpon Wollenw. (1913), anamorphic *Neonectria*, Hsp.≡ eH.15. 58 (esp. in soil; sometimes pathogenic), widespread. See Booth (*Mycol. Pap.* **104**, 1966; key), Samuels (*N.Z. Jl Bot.* **16**: 73, 1978; *Nectria* teleomorphs), Brayford (*TBMS* **89**: 347, 1987; description of *C. bugnicourtii*), Brayford (*Mycopathologia* **100**: 115, 1987; 10 descriptions and illustrations), Brayford & Samuels (*Mycol.* **85**: 612, 1993), Samuels & Brayford (*Sydowia* **45**: 55, 1993), Hennequin *et al.* (*J. Clin. Microbiol.* **37**: 3586, 1999; clinical diagnostics), Iwen *et al.* (*J. Clin. Microbiol.* **38**: 3375, 2000; clinical strain), Mantiri *et al.* (*CJB* **79**: 334, 2001; *Neonectria* teleomorphs), Rossman *et al.* (*Mycol.* **93**: 100, 2001; phylogeny), Watanabe *et al.* (*Mycoscience* **42**: 591, 2001; Japan), Summerbell & Schroers (*J. Clin. Microbiol.* **40**: 2866, 2002; phylogeny), Seifert *et al.* (*Phytopathology* **93**: 1533, 2003; on ginseng), Brayford *et al.* (*Mycol.* **96**: 572, 2004), Halleen *et al.* (*Stud. Mycol.* **50**: 431, 2004), Petit & Gubler (*Pl. Dis.* **89**: 1051, 2005; on *Vitis*), Castlebury *et al.* (*CJB* **84**: 1417, 2006; phylogeny on *Fagus*), Halleen *et al.* (*Stud. Mycol.* **55**: 227, 2006; on *Vitis*), Schroers *et al.* (*MR* **112**: 82, 2008; 3-septate spp.).

Cylindrocarpostylus R. Kirschner & Oberw. (1999), anamorphic *Pezizomycotina*, Hso.?.?. 1 (in bark beetle galleries), Europe. See Kirschner & Oberwinkler (*MR* **103**: 1155, 1999), Kubátová *et al.* (*Czech Mycol.* **53**: 237, 2001; Czech Republic).

Cylindrocephalum Bonord. (1851) = Chalara fide Hughes (*CJB* **36**: 727, 1958).

Cylindrochytridium Karling (1941), Chytridiaceae. 2, USA; British Isles.

Cylindrochytrium, see *Cylindrochytridium*.

Cylindrocladiella Boesew. (1982), anamorphic *Nectricladiella*, Hso.≡ eH.15. 11, widespread. See Crous & Wingfield (*MR* **97**: 433, 1993; key), Victor *et al.* (*MR* **102**: 273, 1998), Schoch *et al.* (*Stud. Mycol.* **45**: 45, 2000; phylogeny), Crous (*Taxonomy and Pathology of Cylindrocladium (Calonectria) and Allied Genera*: 278 pp., 2002; review), Coller *et al.* (*Australas. Pl. Path.* **34**: 489, 2005; on *Vitis*).

Cylindrocladiopsis J.M. Yen (1979) = Cylindrocladium fide Braun (*Monogr. Cercosporella, Ramularia Allied Genera (Phytopath. Hyphom.)* **1**: 216, 1995), Crous & Seifert (*Fungal Diversity* **1**: 53, 1998).

Cylindrocladium Morgan (1892), anamorphic *Calonectria*, Hso.≡ eH.15. 66, widespread. See Wormald (*TBMS* **27**: 71, 1944), Alfenas (*Fitopatol. Brasil* **11**: 275, 1986; key), Stevens *et al.* (*Mycol.* **82**: 436, 1990; aminopepsidase specificity and identification), Crous *et al.* (*S. Afr. For. Jl* **1576**: 69, 1991; spp. in S. Africa forest nurseries), Peerally (*Mycotaxon* **40**: 323, 1991; key and review), Crous & Wingfield (*S. Afr. J. Bot.* **58**: 397, 1992; states of *Calonectria*), Crous *et al.* (*Syst. Appl. Microbiol.* **16**: 266, 1993; techniques for characterization of 3-septate spp.), El-Gholl *et al.* (*CJB* **71**: 466, 1993; *C. ovatum* n.sp. from *Eucalyptus*), Watanabe (*Mycol.* **86**: 151, 1994; 3 spp. on *Phellodendron*), Crous *et al.* (*Sydowia* **50**: 1, 1998), Crous *et al.* (*CJB* **77**: 1813, 1999; phylogeny based on DNA), Schoch *et al.* (*Mycol.* **91**: 286, 1999; population structure), Schoch *et al.* (*Mycol.* **92**: 665, 2000), Kang *et al.* (*Syst. Appl. Microbiol.* **24**: 206, 2001; species concepts), Schoch *et al.* (*Pl. Dis.* **85**: 941, 2001; population genetics), Schoch *et al.* (*MR* **105**: 1045, 2001; phylogeny), Henricot & Culham (*Mycol.* **94**: 980, 2002; on *Buxus*), Crous *et al.* (*Stud. Mycol.* **50**: 415, 2004), Risède & Simoneau (*Eur. J. Pl. Path.* **110**: 139, 2004; on *Musa*), Watanabe & Nakamura (*Mycoscience* **45**: 351, 2004; Japan), Crous *et al.* (*Stud. Mycol.* **55**: 213, 2006).

Cylindrocolla Bonord. (1851), anamorphic *Calloria*, Hsp.0eH.?. 5, widespread (north temperate).

Cylindrodendrum Bonord. (1851), anamorphic *Nectriaceae*, Hso.0eH.15. 3, Europe. *C. album* synanamorph of *Cylindrocarpon*). See Petch (*TBMS* **27**: 81, 1944), Buffina & Hennebert (*Mycotaxon* **19**: 323, 1984), Summerbell *et al.* (*CJB* **67**: 573, 1989).

Cylindrodochium Bonord. (1851) ≡ Cylindrosporium.

Cylindrogloeum Petr. (1941), anamorphic *Pezizomycotina*, Cac.0-1eH.15. 2, Lapland.

Cylindromyces C. Manoharachary, D.K. Agarwal & N.K. Rao (2004), anamorphic *Pezizomycotina*, H?.?.?. 1, India. See Manoharachary *et al.* (*Indian Phytopath.* **57**: 161, 2004).

Cylindronema Schulzer (1866) nom. dub., anamorphic *Pezizomycotina*. See Carmichael *et al.* (*Genera of

Hyphomycetes, 1980).

Cylindrophoma (Berl. & Voglino) Höhn. (1918), anamorphic *Pezizomycotina*, Cpd.0eH.?. 1, Italy.

Cylindrophora Bonord. (1851) nom. dub., anamorphic *Pezizomycotina*. See de Hoog (*Persoonia* **10**: 67, 1978).

Cylindrospora I.V. Issi & Voronin (1986), Microsporidia. 38.

Cylindrosporella Höhn. (1916), anamorphic *Stegophora*. 3. See Sutton (*Mycol. Pap.* **141**, 1977), Kaneko & Kobayashi (*Mycoscience* **43**: 181, 2002; on *Betulaceae*).

Cylindrosporium Grev. (1822), anamorphic *Pyrenopeziza*, Cac.0eH.15. 3, widespread (north temperate). *C. concentricum*, light leaf spot of *Brassica*. See Nag Raj & Kendrick (*CJB* **49**: 2119, 1971), Rawlinson *et al.* (*TBMS* **71**: 425, 1978), Foster *et al.* (*Physiological and Molecular Plant Pathology* **55**: 111, 1999; mating types), Paavolainen *et al.* (*MR* **104**: 611, 2000; on *Betula*), Andrianova & Minter (*IMI Descr. Fungi Bact.* **163** nos 1621-1630, 2005), Priest (*Fungi of Australia: Septoria*, 2006; Australia).

Cylindrosympodium W.B. Kendr. & R.F. Castañeda (1990), anamorphic *Venturiaceae*, Hso.≡ eH.10. 6, widespread. See Castañeda Ruiz & Kendrick (*Univ. Waterloo Biol. Ser.* **32**, 1990), Lücking (*Lichenologist* **35**: 33, 2003), Crous *et al.* (*Stud. Mycol.* **58**: 185, 2007; phylogeny).

Cylindrotaenium Thomé (1867) nom. dub., ? Fungi.

Cylindrotheca Bonord. (1864), Pezizomycotina. 1, Europe.

Cylindrothyrium Maire (1907), anamorphic *Pezizomycotina*, Cpt.0eH.?. 1, France.

Cylindrotrichum Bonord. (1851), anamorphic *Chaetosphaeria*, Hsy.0eH.16. 13, Europe. See Gams & Holubová-Jechová (*Stud. Mycol.* **13**, 1976; key), DiCosmo *et al.* (*Mycol.* **75**: 949, 1983; synonym of *Chaetopsis*), Rambelli & Onofri (*TBMS* **88**: 393, 1987), Réblová & Gams (*Czech Mycol.* **51**: 1, 1999; teleomorph), Réblová (*Stud. Mycol.* **45**: 149, 2000; review), Réblová & Winka (*Mycol.* **92**: 939, 2000; phylogeny), Fernández & Huhndorf (*Fungal Diversity* **18**: 15, 2005).

Cylindroxyphium Bat. & Cif. (1963), anamorphic *Pezizomycotina*, Cpd.0eH.15. 1, USA. See Batista & Ciferri (*Quad. Lab. crittogam., Pavia* **31**: 77, 1963).

Cyllamyces Ozkose, B.J. Thomas, D.R. Davies, G.W. Griff. & Theodorou (2001), Neocallimastigaceae. 1, British Isles. See Ozkose *et al.* (*CJB* **79**: 668, 2001).

Cylomyces Clem. (1931) ≡ Listeromyces.

Cymadothea F.A. Wolf (1935) = Mycosphaerella fide Cannon (*Mycol. Pap.* **163**, 1991; nomencl.).

Cymatella Pat. (1899), Marasmiaceae. 4, Antilles.

Cymatellopsis Parmasto (1985), Marasmiaceae. 1, E. Africa. See Parmasto (*Nova Hedwigia* **40**: 463, 1984).

Cymatoderma Jungh. (1840), Meruliaceae. 9, widespread (tropical). See Welden (*Mycol.* **52**: 856, 1962; Am. spp.), Reid (*Beih. Nova Hedwigia* **18**: 95, 1965; key), Douanla-Meli & Langer (*Mycotaxon* **90**: 323, 2004).

Cymbella Pat. (1886) [non *Cymbella* C. Agardh 1830, *Bacillariophyta*] ≡ Chromocyphella.

cymbiform, boat-shaped; navicular (Fig. 23.32).

Cymbothyrium Petr. (1947), anamorphic *Pezizomycotina*, St.0eP.1. 1, Europe.

Cynema Maas Geest. & E. Horak (1995), Tricholomataceae. 1, Papua New Guinea. See Maas Geesteranus & Horak (*Biblthca Mycol.* **159**: 208, 1995).

Cyniclomyces Van der Walt & D.B. Scott (1971), Saccharomycetales. 1, Europe. See Phaff & Miller *in* Kurtzman & Fell (Eds) (*Yeasts, a taxonomic study* 4th edn: 154, 1998), Suh *et al.* (*Mycol.* **98**: 1006, 2006; phylogeny).

Cynicus Raf. (1815) nom. dub., Fungi. No spp. included.

Cynophallus (Fr.) Corda (1842) nom. rej. ≡ Mutinus.

Cypellomyces, see *Cypheliomyces*.

Cypheliaceae Zahlbr. (1903) = Caliciaceae.

Cypheliomyces E.A. Thomas ex Cif. & Tomas. (1953) = Cyphelium Ach. fide Tibell (*Beih. Nova Hedwigia* **79**: 597, 1984).

Cypheliopsis (Zahlbr.) Vain. (1927) = Thelomma fide Tibell (*Beih. Nova Hedwigia* **79**: 597, 1984).

Cyphelium Ach. (1815), Caliciaceae (L). *c*. 12, widespread (esp. north and south temperate). See Weber (*Bryologist* **70**: 197, 1967), Tibell (*Svensk bot. Tidskr.* **65**: 138, 1971; Eur.), Tibell (*Beih. Nova Hedwigia* **79**: 597, 1984), Middelborg & Mattson (*Sommerfeltia* **5**: 1, 1987; Norway), Tibell (*Symb. bot. upsal.* **27** no. 1: 279 pp., 1987; Australasia), Pant & Awasthi (*Biovigyanam* **15**: 3, 1989; India, Nepal), Tibell (*Fl. Neotrop.* Monogr. **69**: 78 pp., 1996; neotropics), Tibell (*Symb. bot. upsal.* **32** no. 1: 291, 1997; anamorphs), Wedin & Tibell (*CJB* **75**: 1236, 1997; phylogeny), Tibell (*Biblthca Lichenol.* **71**: 107 pp., 1998; S America), Tibell (*Nordic Lichen Flora* **1**. Introductory Parts; Calicioid Lichens and Fungi: 20, 1999; Scandinavia), Wedin *et al.* (*CJB* **78**: 246, 2000; phylogeny), Wedin *et al.* (*Taxon* **51**: 655, 2002; phylogeny), Tibell & Thor (*J. Hattori bot. Lab.* **94**: 205, 2003; Japan), Peršoh *et al.* (*Mycol. Progr.* **3**: 103, 2004; asci), Tibell (*J. Hattori bot. Lab.* **100**: 809, 2006; India).

Cyphelium Chevall. (1826) = Chaenotheca fide Tibell (*Beih. Nova Hedwigia* **79**: 597, 1984).

Cyphelium De Not. (1846), Coniocybaceae. 1, Europe.

Cyphella Fr. (1822), Cyphellaceae. 2, widespread. See Matheny *et al.* (*Mycol.* **98**: 982, 2006; phylogeny).

cyphella (pl. **-ae**), a break in the lower (rarely upper) cortex of a lichen thallus which is roundish or ovate and in section appears as a cup-like structure lined with a layer of loosely connected, frequently globular, cells formed from the medulla, characteristic of *Sticta*; also used for pores open to the halliomedulla in *Oropogon*.

Cyphellaceae Lotsy (1907), Agaricales. 16 gen. (+ 1 syn.), 31 spp. The family name has previously been used for a diversity of fungi. Many of the genera character-ised by a cyphelloid habit have been distributed among several families of the *Agaricales*.

Lit.: Pilát (*Annls mycol.* **22**: 204, 1924), Pilát (*Annls mycol.* **23**: 144, 1925), Cooke (*Mycol.* **49**: 680, 1957; *Porotheleaceae*), Donk (*Persoonia* **1**: 25, 1959), Cooke (*Sydowia* Beih. **4**, 1961; *Porotheleaceae*), Donk (*Persoonia* **2**: 331, 1962), Reid (*Persoonia* **3**: 97, 1964; Michigan), Donk (*Acta Bot. Neerl.* **15**: 95, 1966; reassessment), Donk (*Checklist of European polypores* [*Verh. K. ned. Akad. Wet.* **62**], 1974).

Cyphellathelia Jülich (1972) = Pellidiscus fide Hjortstam (*Windahlia* **15**: 59, 1986).

Cyphellina Rick (1959) nom. dub. = Tomentella Pat. fide Donk (*Persoonia* **3**: 199, 1964).

Cyphellocalathus Agerer (1981), Tricholomataceae. 1, Bolivia. See Agerer (*Mycol.* **73**: 491, 1981) = Lachnella (Tricholomat.) fide, Singer (*Agaric. mod. Tax.* 4th ed: 386, 1986).

Cyphellomyces Speg. (1906) = Phellorinia fide Stalpers (*in litt.*).

Cyphellophora G.A. de Vries (1962), anamorphic *Chaetothyriaceae*, Hso.≡ eP.15. 11 (on human skin etc.), widespread. See Vries *et al.* (*Antonie van Leeuwenhoek* **52**: 141, 1986), Walz & de Hoog (*Antonie van Leeuwenhoek* **53**: 143, 1987; n.sp. and cf. *Annellodentimyces*), Decock *et al.* (*Antonie van Leeuwenhoek* **84**: 209, 2003; phylogeny), Geiser *et al.* (*Mycol.* **98**: 1053, 2006; phylogeny).

Cyphellopsidaceae Jülich (1982) = Marasmiaceae.

Cyphellopsis Donk (1931) = Merismodes fide Kuyper (*in litt.*).

Cyphellopus Fayod (1889) nom. dub., ? Bolbitiaceae. 1, British Isles. See Pearson (*TBMS* **20**: 521, 1935).

Cyphellopycnis Tehon & G.L. Stout (1929) = Phomopsis (Sacc.) Bubák. fide Sutton (*TBMS* **47**: 497, 1964).

Cyphellostereum D.A. Reid (1965), Agaricomycetes. 2, widespread. *Hymenochaetales* or *Agaricales* (*Rickenella* clade). See Reid (*Beih. Nova Hedwigia* **18**: 484 pp., 1965; key), Salcedo *et al.* (*Nova Hedwigia* **82**: 81, 2006).

Cyphidium Magnus (1875) = Olpidium fide Sparrow (*Aquatic Phycomycetes* Edn 2: 1187 pp., 1960).

Cyphina Sacc. (1884) = Sarcopodium fide Sutton (*Mycol. Pap.* **141**, 1977).

Cyphospilea Syd. (1926) = Coleroa fide Müller & von Arx (*Beitr. Kryptfl. Schweiz* **11** no. 2, 1962).

Cyptotrama Singer (1960), Physalacriaceae. 15, S. America. See Singer (*Lilloa* **30**: 375, 1960).

Cyrenella Goch. (1981), anamorphic *Cystobasidiomycetes*. 1 (with clamp connexions), USA. See Gochenaur (*Mycotaxon* **13**: 268, 1981), Bauer *et al.* (*Mycol. Progr.* **5**: 41, 2006).

Cyrta Bat. & H. Maia (1961) [non *Cyrta* Lour. 1790, *Styracaceae*] = Calopadia fide Lücking *et al.* (*Lichenologist* **30**: 121, 1998).

Cyrtidium Vain. (1921), ? Dothideomycetes. 2, Europe.

Cyrtidula Minks (1876), Dothideomycetes. 2, Italy. See Harris (*More Florida Lichens*, 1995).

Cyrtocnon Link ex Rchb. (1828) nom. dub., anamorphic *Pezizomycotina*.

Cyrtographa Müll. Arg. (1894) = Minksia fide Zahlbruckner (*Catalogus Lichenum Universalis* **4**, 1926).

Cyrtopsis Vain. (1921), Dothideomycetes. 1, Europe.

cyst (1) an encysted cell (? the product of meiosis after karyogamy), usually aggregated into a cystosorus which germinates to produce a zoospore (*Plasmodiophorales*); (2) an encysted zoospore (planospore) which becomes a gametangium (*Blastocladiales*); **macro-** (of *Myxomycota*), an encysted aggregate of myxoamoebae; the resting form of a young plasmodium; the alternative to the sorocarp in some cellular slime moulds (dictyostelids) (Nickerson & Raper, *Am. J. Bot.* **60**: 190, 1973); propagule, especially the walled structure of the encysted zoospore (*Peronosporales*); resting spores of chrysomonads etc.; **micro-** (of *Mycetozoa*), an encysted myxamoeba or swarm spore; **spore-**, a cell, hollow organ or sac-like structure enclosing a mass of protoplasm containing spores as in the *Ascosphaeraceae* (Skou, 1982).

Cystangium Singer & A.H. Sm. (1960), Russulaceae. 32, Australia; S. America. Basidioma gasteroid. See Miller *et al.* (*Mycol.* **93**: 344, 2001; basiodioma gasteroid; probably polyphyletic and should be merged with *Russula*), Trappe *et al.* (*Mycotaxon* **81**: 195, 2002; nomenclature), Lebel (*Aust. Syst. Bot.* **16**: 371, 2003; australasian spp.).

cystesium, a cell which differentiates to adhere to a cystidium arising from the opposite hymenium (Horner & Moore, *TBMS* **88**: 488, 1987).

Cystidiella Malan (1943), anamorphic *Pezizomycotina*, Hso.0eH.1. 1, Italy. With secondary conidia.

Cystidiodendron Rick (1943) nom. dub., Hydnaceae. See Donk (*Persoonia* **3**: 199, 1964).

Cystidiodontia Hjortstam (1983), Cystostereaceae. 2, widespread. See Hjortstam (*Mycotaxon* **17**: 571, 1983).

cystidiole (of hymenomycetes), a simple hymenial cell of about the same diameter as the basidia but remaining sterile and protruding beyond the hymenial surface.

Cystidiophorus Bondartsev & Ljub. (1963) nom. inval., Polyporaceae. 1.

cystidium (pl. **-ia**), a sterile body, frequently of distinctive shape, occurring at any surface of a basidioma, particularly the hymenium from which it frequently projects (Fig. 12). Cystidia have been classified and named according to their: (1) *origin*: **hymenial -** (**tramal -**), originating from hymenial (tramal) hyphae; **pseudo-**, derived from a conducting element, filamentous to fusoid, oily contents, embedded not projecting; **coscino-**, see coscinoid; **skeleto-**, the apical part of a skeletal hypha (frequently ± inflated) projecting into or through the hymenium; false seta; **macro-**, arising deep in the trama in Lactario-Russulae; **hypho-**, hypha-like, derived from generative hyphae. (2) *position* (first by Buller, **2**): on the pileus surface (**pileo-**, **dermato-**, Fayod); at the edge (**cheilo-**), side (**pleuro-**), or within (**endo-**) a lamella; on the stipe (**caulo-**). (3) *form*: **lepto-**, smooth, thin-walled; **lampro-**, thick-walled, with or without encrustation (**setiform lampro-**, awl-shaped, wall pigmented; **asteroseta**, a radially branched lampro-; **microsclerid**, a versiform, endolampro-; **lyo-**, cylindrical to conical, very thick-walled, abruptly thin-walled at apex, not encrusted, colourless, as in *Tubulicrinis* (Donk, 1956); **monilioid gloeo-**, (torulose gloeo- (Bourdot & Galzin, 1928); moniliform paraphysis (Burt, 1918); pseudophysis; **schizo-** (Nikolayeve, 1956, 1961)), monilioid, frequently with a beaded apex (as in *Hericiaceae* and *Corticiaceae*). (4) *contents*: **gloeo-**, thin-walled, usually irregular, contents hyaline or yellowish and highly refractile; **chryso-**, like lepto- but with highly staining contents; **hypo-**, (Larsen & Burdsall, *Mem. N.Y. bot. Gdn* **28**: 123, 1976); **oleo-**, having an oily resinous exudate; **pseudo-**, see (1) above. See also hyphidium, seta. Reviews of cystidia include Romagnesi (*Rev. Mycol., Paris* **9** (suppl.): 4, 1944), Talbot (*Bothalia* **6**: 249, 1954), Lentz (*Bot. Rev.* **20**: 135, 1954), Smith (*in* Ainsworth & Sussman (Eds), *The fungi* **2**: 151, 1966), Price (*Nova Hedw.* **24**: 515, 1975; types in polypores).**Cystingophora** Arthur (1907) = Ravenelia fide Arthur (*Manual Rusts US & Canada*, 1934).

Cystoagaricus Singer (1947), Psathyrellaceae. 4, America (subtropical). See Singer (*Mycol.* **39**: 85, 1947), Vellinga (*MR* **108**: 354, 2004; posn.).

Cystobasidiaceae Gäum. (1926), Cystobasidiales. 2 gen., 9 spp.

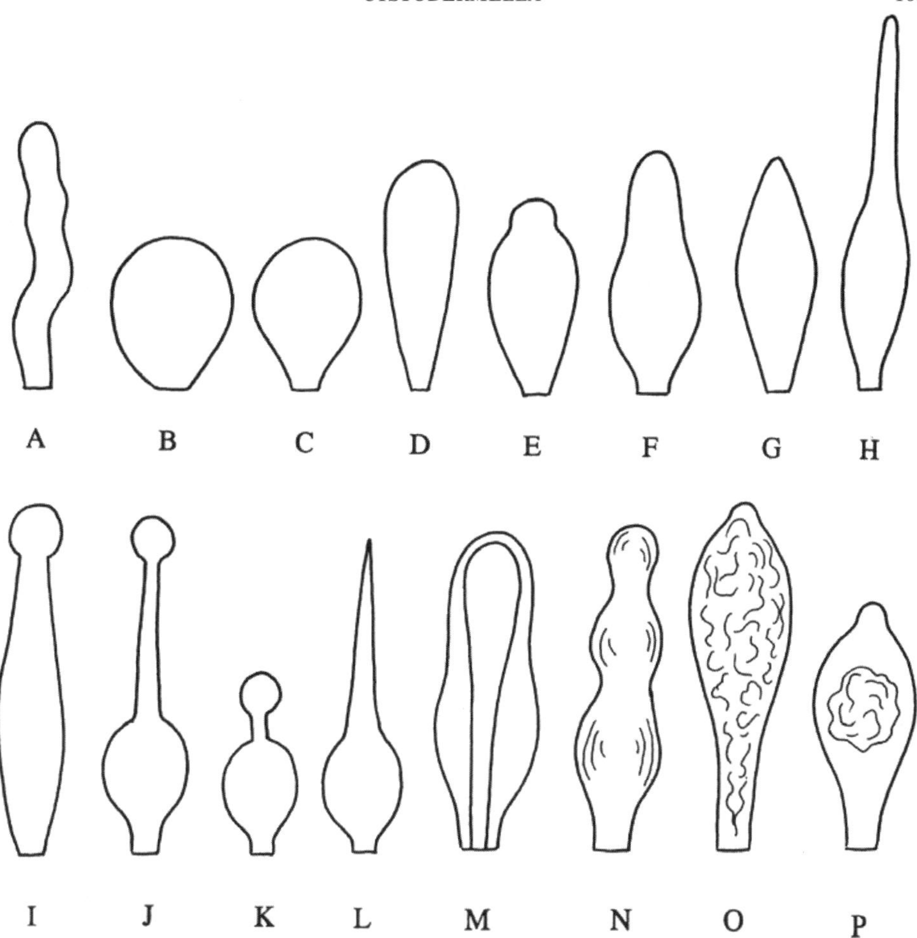

Fig. 14. Cystidia. A, hyphoid (*Collybia*); globose (*Agaricus*); C, pyriform (*Agaricus*); D, clavate (*Inocybe*); E, utriform (*Psathyrella*); F, lageniform (*Pholiota*); G, fusoid (*Psathyrella*); H, lanceolate (*Hypholoma*); I, capitate (*Hyphoderma*); J, tibiiform (*Galerina*); K, lecythiform (*Conocybe*); L, urticoid (*Naucoria*); M, metuloid (*Lentinus*); N, gloeocystidium (*Gloeocystidiellum*); O, macrocystidium (*Russula*); P, chrysocystidium (*Stropharia*). Not to scale.

Cystobasidiales R. Bauer, Begerow, J.P. Samp., M. Weiss & Oberw. (2006). Cystobasidiomycetes. 1 fam., 2 gen., 9 spp. Fam.:
Cystobasidiaceae
For *Lit.* see under fam.

Cystobasidiomycetes R. Bauer, Begerow, J.P. Samp., M. Weiss & Oberw. (2006). Pucciniomycotina. 3 ord., 1 fam., 7 gen., 14 spp. Ord.:
(1) **Cystobasidiales**
(2) **Erythrobasidiales**
(3) **Naohideales**
For *Lit.* see ord. and fam.

Cystobasidium (Lagerh.) Neuhoff (1924), Cystobasidiaceae. *c.* 5, widespread. See Martin (*Mycol.* **31**: 507, 1939), Olive (*Mycol.* **44**: 564, 1952).

cystochroic, see colour.

Cystocoleus Thwaites (1849), anamorphic *Capnodiales*, Hso.-.- (L). 1, widespread. See Muggia *et al.* (*MR* **112**: 50, 2008; phylogeny).

Cystocybe Velen. (1921) nom. dub. ? = Cortinarius fide Singer (*Agaric. mod. Tax.* edn 3, 1975).

Cystodendron Bubák (1914), anamorphic *Mollisia*, Hsp.0eH/1eP.15. 1, Europe. See Gams (*Stud. Mycol.* **45**: 192, 2000), Crous *et al.* (*Eur. J. Pl. Path.* **109**: 845, 2003; phylogeny).

Cystoderma Fayod (1889), Agaricaceae. *c.* 35, widespread (esp. temperate). Probably a separate family is justified. See Justo & Castro (*A Pantorra* **2**: 45, 2002), Saar (*Mycotaxon* **86**: 455, 2003; key temperate Eurasia spp), Jamoni (*Funghi e Ambiente* **94-95**: 51, 2004), Saar & Læssøe (*Mycotaxon* **96**: 123, 2006; species from high Andean Ecuador).

Cystodermella Harmaja (2002), Agaricaceae. 12, widespread. Probably a separate family (with *Cystoderma* is justified. See Harmaja (*Karstenia* **42**: 43, 2002), Saar (*Mycotaxon* **86**: 455, 2003; key).

Cystodium Fée (1837), ? Dothideomycetes. 1, widespread (tropical).

Cystofilobasidiaceae K. Wells & Bandoni (2001), Cystofilobasidiales. 9 gen. (+ 1 syn.), 20 spp.

Lit.: Kwon-Chung (*Stud. Mycol.* **30**: 75, 1987), Wery *et al.* (*Yeast* Chichester **12**: 641, 1996), Fell & Statzell-Tallman *in* Kurtzman & Fell (Eds) (*Yeasts, a taxonomic study* 4th edn: 676, 1998), Kwon-Chung *in* Kurtzman & Fell (Eds) (*Yeasts, a taxonomic study* 4th edn: 646, 1998), Miller & Phaff *in* Kurtzman & Fell (Eds) (*Yeasts, a taxonomic study* 4th edn: 789, 1998), Fell & Blatt (*J. Industr. Microbiol. Biotechnol.* **23**: 677, 1999), Fell *et al.* (*Int. J. Syst. Bacteriol.* **49**: 907, 1999), Diaz & Fell (*Antonie van Leeuwenhoek* **77**: 7, 2000), Sampaio *et al.* (*Int. J. Syst. Evol. Microbiol.* **51**: 221, 2001).

Cystofilobasidiales Fell, Roeijmans & Boekhout (1999). Tremellomycetes. 1 fam., 9 gen., 20 spp. Fam.:

Cystofilobasidiaceae

Lit.: Fell *et al.* (*IJSB* **49**: 911, 1999 (taxonomy), Fell *et al.* (*IJSEM* **50**: 1351, 2000; mol. phylogeny).

Cystofilobasidium Oberw. & Bandoni (1983), Cystofilobasidiaceae. 5 (some marine), widespread. See Oberwinkler *et al.* (*Syst. Appl. Microbiol.* **4**: 114, 1983), Fell *et al.* (*Int. J. Syst. Evol. Microbiol.* **50**: 1351, 2000; mol. phylogeny), Weiss *et al.* (*Frontiers in Basidiomycote Mycology*: 7, 2004).

Cystogloea P. Roberts (2006), Basidiomycota. 1 (on perithecia of *Pseudotrichia*), Sweden. See Roberts (*Acta Mycologica* Warszawa **41**: 25, 2006).

Cystogomphus Singer (1942), Gomphidiaceae. 1, France (introduced).

Cystolepiota Singer (1952), Agaricaceae. *c.* 10, widespread. See Knudsen (*Bot. Tidsskr.* **73**: 124, 1978; key), Johnson & Vilgalys (*Mycol.* **90**: 971, 1998; phylogeny), Johnson (*Mycol.* **91**: 443, 1999), Hausknecht & Pidlich-Aigner (*Öst. Z. Pilzk.* **13**: 1, 2004; Austria), Vellinga (*MR* **108**: 354, 2004; phylogeny).

Cystolobis Clem. (1909) = Knightiella fide Hawksworth *et al.* (*Dictionary of the Fungi* edn 8, 1995).

Cystomyces Syd. (1926), Raveneliaceae. 1 (on *Lonchocarpus* (?) (*Leguminosae*)), C. America. See Walker *et al.* (*Australas. Pl. Pathol.* **35**: 1, 2006).

Cystopage Drechsler (1941), Zoopagaceae. 7, N. America; British Isles. See Drechsler (*Mycol.* **33**: 251, 1941), Drechsler (*Mycol.* **37**: 1, 1945), Drechsler (*Mycol.* **47**: 364, 1955), Drechsler (*Mycol.* **49**: 387, 1957), Drechsler (*Mycol.* **51**: 787, 1959), Dyal (*Sydowia* **27**: 293, 1976; keys parasites of nematodes and amoebae, bibliogr.), Kasim (*Acta Mycologica* Warszawa **33**: 161, 1998), Shimada & Saikawa (*Nippon Kingakukai Kaiho* **47**: 1, 2006; chlamydospore germination).

Cystopezizella Svrček (1983) = Calycina fide Baral (*SA* **13**: 113, 1994).

Cystophora Rabenh. (1844) [non *Cystophora* J. Agardh 1841, *Algae*] ≡ Voglinoana.

Cystopsora E.J. Butler (1910), Pucciniaceae. 2 (on *Antidesma* (0, I, III) (*Euphorbiaceae*); *Olea* (*Oleaceae*)), India; Indonesia. 0, I, III. See Thirumalachar (*Bot. Gaz.* **107**: 74, 1945), Cummins & Hiratsuka (*Illustr. Gen. Rust Fungi rev. edit.*, 1983; syn. of *Zaghouania*).

cystosorus (of *Chytridiales*), a group of united cysts or resting spores.

cystospore (1) an encysted zoospore formed at the exit to the zoosporangium and germinating to produce a new zoospore (planospore) as in *Achlya* (*Oomycetes*), *Achlyogeton* and *Achlyella* (*Chytridiomycetes*); (2) (of *Amoebidiales*) spores released from an encysted amoeboid cell.

Cystosporogenes E.U. Canning, R.J. Barker, J.P. Nicholas & A.M. Page (1984), Microsporidia. 3.

Cystostereaceae Jülich (1982), Agaricales. 6 gen., 16 spp.

Lit.: Chamuris (*Mycol.* **78**: 380, 1986), Chamuris (*Mycol. Mem.* **14**: 247 pp., 1988), Hjortstam (*Mycotaxon* **42**: 149, 1991), Ginns (*Mycol.* **90**: 19, 1998), Larsson *et al.* (*MR* **108**: 983, 2004).

Cystostereum Pouzar (1959), Cystostereaceae. 6, widespread. See Hallenberg & Ryvarden (*Mycotaxon* **2**: 135, 1975).

Cystostiptoporus Dhanda & Ryvarden (1975) = Microporellus fide Ryvarden (*Gen. Polyp.*: 135, 1991).

Cystotelium Syd. (1921) = Ravenelia fide Sydow & Sydow (*Monographia Uredinearum seu Specierum Omnium ad hunc usque Diem Descriptio et Adumbratio Systematica* **3**, 1915), Dietel (*Nat. Pflanzenfam.* **6**, 1928; syn. of *Haploravenelia*).

Cystotheca Berk. & M.A. Curtis (1860), Erysiphaceae. Anamorph *Oidium* subgen. *Setoidium*. 4, Asia; N. America. See Katumoto (*Rep. Tottori mycol. Inst.* **10**: 437, 1973; key), Braun (*Beih. Nova Hedwigia* **89**, 1987; key), Saenz & Taylor (*CJB* **77**: 150, 1999; mol. syst.), Braun & Takamatsu (*Schlechtendalia* **4**: 1, 2000; phylogeny), Takamatsu *et al.* (*MR* **104**: 1304, 2000; phylogeny, host relations), Takamatsu (*Mycoscience* **45**: 147, 2004; evolution).

Cystothyrium Speg. (1888) nom. conf., anamorphic *Pezizomycotina*. See Petrak & Sydow (*Annls mycol.* **33**: 157, 1935).

Cystotricha Berk. & Broome (1850), anamorphic *Helotiales*, St.1eH.1. 1, British Isles. See Wu *et al.* (*MR* **102**: 179, 1998).

Cystotrichiopsis B.A. Abbas, B. Sutton & Ghaffar (2001), anamorphic *Pezizomycotina*, C?.?.?. 1, Pakistan. See Abbas *et al.* (*Pakist. J. Bot.* **33**: 365, 2001).

Cytidia Quél. (1888), Corticiaceae. 5, widespread (esp. N. America). *C. salicina*, a common red fungus on *Salix*. See Cooke (*Mycol.* **43**: 196, 1951; key), Donk (*Persoonia* **1**: 70, 1959), Legg & Roberts (*Field Mycology* **1**: 45, 200; UK).

Cytidiaceae Jülich (1982) = Corticiaceae.

Cytidiella Pouzar (1954) = Auriculariopsis fide Stalpers (*Persoonia* **13**: 495, 1988).

Cytispora Fr. (1823), Pezizomycotina. 1, Europe.

cytochalasin, one of a series of related fungal metabolites (e.g. from *Helminthosporium*, *Metarhizium*, *Phoma*, *Xylariaceae*, *Zygosporium*) which inhibit cytokinesis so that multinucleate cells result (Carter, *Nature* **213**: 261, 1967), Turner (1971: 352); **- B** = phomin.

cytochroic, see colour.

Cytodiplospora Oudem. (1894) = Diplodina fide Sutton (*Mycol. Pap.* **141**, 1977).

Cytodiscula Petr. (1931), anamorphic *Pezizomycotina*, St.0eH.?. 1, Madeira.

Cytogloeum Petr. (1925), anamorphic *Pezizomycotina*, Cac.0eH.15. 1, Europe.

Cytology. The study of cell fungal contents, esp. the nucleus, is usually well covered in mycology textbooks. The nuclear membrane may (*Basidiobolus ranarum*) or may not (*Saccharomyces cerevisiae*) break down during mitosis. Pulse field gel electrophoresis (q.v.) has shown that fungal nuclei contain 3 (*Schizosaccharomyces pombe*) to *c.* 21 (*Ustilago hordei*) chromosomes (see Mills & McCluskey, *Molecular plant-microbe interactions* **3**: 351-357, 1990). Mitotic chromosomes are smaller in fungi than meiotic ones. Usually, a well-defined 'metaphase plate' is absent and disjunction of sister chromatids is asynchronous during mitosis (see review Aist & Morris, *Fungal. Genet. Biol.* **27**: 1-25, 1999). Cyclic changes of condensation/decondensation states of fungal rDNA were observed by fluorescence *in situ* hybridisation (FISH) during mitosis. In the ascomycetes *Cochliobolus heterostrophus* and *Haematonectria haematococca* rDNA is decondensed throughout its entire length at interphase and condensation reaches a maximum at metaphase, remaining in that state through anaphase (Taga *et al.*, *MR* **107**: 1012-1020, 2003). Pulse field gel electrophoresis (PFGE) can aid visualisation of minute chromosomes and reveals that different isolates of the same fungal species frequently have chromosomes differing in size and the presence of supernumerary, conditionally dispensable, chromosomes. Some supernumerary chromosomes carry functional genes, like antibiotic resistance (Miao *et al.*, *Science* **254**: 1773-1776, 1991) or, in *Nectria haematococca*, the ability to cause disease symptoms on host plants (Wasmann & VanEtten, *Mol. Plant-Microb. Interact.* **9**: 793-803, 1996). DNA fingerprinting (Owen, *J. Med. Microbiol.* **30**: 898, 1989), restriction fragment length polymorphisms (RFLP) (Kusters-van-Someren *et al.*, *Curr. Genet.* **19**: 21, 1991) and amplified fragment length polymorphism (AFLP) (*Clin. Microbiol. Rev.* **13**: 332-370, 2000) are used to study fungal genomes, to characterize species and strains of fungi. Single or double stranded DNA or RNA mycoviruses and double stranded DNA or RNA plasmids may be present in fungal cytoplasm. Hyphal tips of extending hyphae contain a very high concentration of macrovesicles (100-250 nm diam.), microvesicles (40-70 nm diam.), and coated vesicles (vesicles surrounded by a basket-like lattice of fibrous protein called clathrin); the microvesicles may correspond to chitosomes (Bracker *et al.*, *Proc. Natl. Acad. Sci. USA* **73**: 4570, 1976); the concentration of vesicles at the hyphal tip can sometimes be observed as a phase-dark body known as the Spitzenkörper (Steinberg, *Eukaryotic Cell* **6**:351-360, 2007). Specialized hyphae, called conidial anastomosis tubes (CATs), are produced by conidia and by conidial germ tubes of some fungi. In contrast to germ tubes, CATs are cellular elements that are morphologically and physiologically distinct: microtubules and nuclei pass through fused CATs, they are thinner, shorter, lack branches, exhibit determinate growth, and home toward each other (Rocka *et al.,Eukaryotic Cell* **4**: 911-919, 2005). The fungal cytoskeleton contains microfilaments (4-9 nm diam.) formed by the polymerization of G-actin, microtubules (30-40 nm diam.) formed by the polymerization of α- and β-tubulin, and filasomes (aggregates of microfilaments coated with vesicles). The cytoplasm contains mitochondria, microbodies, hydrogenosomes (in anaerobic fungi), Golgi bodies (in *Chromista*), smooth and rough endoplasmic reticula, 80S ribosomes, glycogen and lipid globules. Spherical Worinin bodies (e.g. in *Fusarium* spp.) or hexagonal crystals (*Neurospora crassa*) are associated with the septal pores of *Ascomycota*. Nuclei of filamentous fungi can migrate within the cytoplasm along microtubules, driven by the microtubule dependent motor protein dynein. Cellular differentiation, mating, and filamentous growth are regulated in many fungi by environmental and nutritional signals. Signal transduction cascades: protein kinase cascade and cyclic AMP signaling pathway, regulate development, mating and virulence of fungi that has been well studied in *Candida albicans*, *Cryptococcus neoformans*, *Saccharomyces cerevisiae* and *Ustilago maydis* (Lengeler *et al.,Microbiol. Mol. Biol. Rev.* **64**: 746-785, 2000). See also Chromosome numbers.

Lit.: Arora (Ed.) (*Handbook of Fungal Biotechnology*, 2nd ed., 2003), Howard & Gow (Eds) (*Biology of the Fungal Cell*, 2nd ed., 2007), Gull & Oliver *The Fungal Nucleus*, 1981), San-Blas & Calderone (Eds) (*Pathogenic Fungi: Structural Biology and Taxonomy*, 2004), Talbot, N. (Ed.) (*Molecular and Cell Biology of Filamentous Fungi: a Practical Approach*, 2001), Tsuchiya *et al.,Mycol.* **96**: 208-210, 2004).

cytolysis, breaking up or solution of the cell wall.

Cytomelanconis Naumov (1951), ? Melanconidaceae. Anamorph *Cytospora*-like. 1 (from bark), former USSR.

Cytonaema Höhn. (1914), anamorphic *Pezizomycotina*, St.0eH.15. 1, Austria.

Cytophoma Höhn. (1914) = Cytospora fide Défago (*Phytopath. Z.* **14**: 103, 1944).

Cytophyllopsis R. Heim ex R. Heim (1958) = Weraroa fide Singer & Smith (*Bull. Torrey bot. Club* **85**: 324, 1958).

Cytoplacosphaeria Petr. (1920), anamorphic *Pezizomycotina*, St.≡ eH.15. 1, Europe. See Poon & Hyde (*Bot. Mar.* **41**: 141, 1998; Hong Kong).

Cytoplea Bizz. & Sacc. (1885), anamorphic *Roussoella*, St.0eP.15. 5, widespread. See Sutton (*The Coelomycetes*, 1980), Hyde *et al.* (*MR* **100**: 1522, 1996; teleomorph), Hyde (*MR* **101**: 609, 1997), Kang *et al.* (*Fungal Diversity* **1**: 147, 1998).

Cytopleastrum Abbas, Sutton, Ghaffar & A. Abbas (2004), anamorphic *Pezizomycotina*. 1, Pakistan. See Abbas *et al.* (*Pakist. J. Bot.* **36**: 457, 2004).

cytoskeleton, intracellular network of protein filaments that is insoluble in non-ionic detergents.

Cytosphaera Died. (1916), anamorphic *Pezizomycotina*, St.0eH.1. 2, Asia; Australia.

Cytospora Ehrenb. (1818), anamorphic *Valsa*, St.0eH.15. *c.* 110, widespread. See Défago (*Phytopath. Z.* **14**: 103, 1944), Gvritishvili (*Fungi of the genus Cytospora Fr. in the USSR*, 1982), Gille (*Arch. phytopath. Pflanz.* **26**: 237, 1990; spp. on *Prunus*), Old *et al.* (*MR* **95**: 1253, 1991), Pluim *et al.* (*Pl. Dis.* **78**: 551, 1994; infraspecific variation), Adams (*Phytopathology* **85**: 1129, 1995; host/parasite relations), Adams *et al.* (*Mycol.* **94**: 947, 2002; phylogeny).

Cytosporella Sacc. (1884), anamorphic *Pezizomycotina*, St.0eH.15. 32, widespread (temperate). See Aa *et al.* (*N.Z. Jl Bot.* **39**: 543, 2001; New Zealand).

Cytosporina Sacc. (1884) ≡ Dumortieria fide Sutton (*Mycol. Pap.* **141**, 1977).

Cytosporites Babajan & Tasl. (1970), Fossil Fungi. 1 (Tertiary), former USSR.

Cytosporium Sacc. (1884) ≡ Cellulosporium.

Cytosporopsis Höhn. (1918) = Cytospora fide Sutton (*Mycol. Pap.* **141**, 1977).

Cytostaganis Clem. & Shear (1931) ≡ Cytostagonospora.

Cytostagonospora Bubák (1916), anamorphic *Pezizo-*

mycotina, Cpd.0fH.15. 3, Australia; Europe. See Sutton & Swart (*TBMS* **87**: 99, 1986; n. comb.), Abbas *et al.* (*Pakist. J. Bot.* **33**: 229, 2001; Pakistan), Priest (*Fungi of Australia: Septoria*, 2006; Australia).

Cytotriplospora Bayl. Ell. & Chance (1921) = Strasseria fide Sutton (*Mycol. Pap.* **141**, 1977), Parmelee & Cauchon (*CJB* **57**: 1660, 1979).

Cyttaria Berk. (1842), Cyttariaceae. Anamorph *Cyttariella*. 10 (on *Nothofagus*), south temperate. Some species are edible. See White (*TBMS* **37**: 431, 1954), Kobayasi (*TMSJ* **7**: 118, 1966), Gamundí (*Darwiniana* **16**: 461, 1971), Kobayasi (*TMSJ* **19**: 473, 1978), Humphries *et al.* *in* Stone & Hawksworth (Eds) (*Coevolution and systematics*: 55, 1986), Mengoni (*Boln Soc. argent. Bot.* **24**: 393, 1986), Gamundí (*Fl. criptog. Tierra del Fuego* **10** no. 4: 1, 1987), Minter *et al.* (*Mycologist* **1**: 7, 1987), Crisci *et al.* (*Cladistics* **4**: 279, 1988), Gamundí & de Lederkremer (*Ciencia Investig.* **43**: 4, 1989), Mengoni (*Boln Soc. argent. Bot.* **26**: 7, 1989), Gamundí (*SA* **10**: 69, 1991), Landvik & Eriksson *in* Hawksworth (Ed.) (*Ascomycete Systematics. Problems and Perspectives in the Nineties* NATO ASI Series vol. **269** **269**: 225, 1994), Gernandt *et al.* (*Mycol.* **93**: 915, 2001; phylogeny), Gamundí & Minter (*IMI Descr. Fungi Bact.* **160** nos 1591-1600, 2004; descr. 10 spp.), Wang *et al.* (*Mol. Phylogen. Evol.* **41**: 295, 2006; phylogeny), Wang *et al.* (*Mycol.* **98**: 1065, 2006; phylogeny).

Cyttariaceae Speg. (1887), Cyttariales. 2 gen., 11 spp.
Lit.: Humphries *et al.* (*Coevolution and Systematics* **32**: 55, 1986), Minter *et al.* (*Mycologist* **21**: 7, 1987), Crisci *et al.* (*Cladistics* **4**: 279, 1988), Gamundí (*SA* **10**: 69, 1991; review), Landvik & Eriksson (*NATO ASI Series* **269**: 225, 1994), Döring & Triebel (*Cryptog. Bryol.-Lichénol.* **19**: 123, 1998), Landvik *et al.* (*Mycoscience* **39**: 49, 1998; DNA) also see under *Cyttaria*. Placed provisionally within the *Helotiales* but molecular data for many constituent familes is lacking so that order may be polyphyletic, Wang *et al.* (*Mycol.* **98**: 1065, 2006; phylogeny).

Cyttariales Luttr. ex Gamundí (1971). Leotiomycetes. 1 fam., 2 gen., 11 spp. Fam.:
Cyttariaceae
For *Lit.* see under fam.

Cyttariella Palm (1932), anamorphic *Cyttaria*. 1, S. America. See Santesson (*Svensk bot. Tidskr.* **39**: 319, 1945), Gamundí (*Fl. criptog. Tierra del Fuego* **10**: 126 pp., 1986), Mengoni (*Boln Soc. argent. Bot.* **26**: 7, 1989).

Cyttarophyllopsis R. Heim (1968), Bolbitiaceae. 1, India. Basidioma gasteroid. See Heim (*Revue Mycol.* Paris **33**: 211, 1968).

Cyttarophyllum (R. Heim) Singer (1936) = Galeropsis fide Heim (*Revue Mycol.* Paris **15**: 3, 1950).

CZ, see Media.

Dacampia A. Massal. (1853), Dacampiaceae. 3 (on lichens), Europe. See Eriksson (*Op. Bot.* **60**, 1981), Crivelli (*Über die heterogene Ascomycetengattung Pleospora Rbh.*, 1983), Henssen (*Cryptog. Bot.* **5**: 149, 1995), Halici & Hawksworth (*Fungal Diversity* **28**: 49, 2008; key).

Dacampiaceae Körb. (1855), Pleosporales. 16 gen. (+ 7 syn.), 121 spp.
Lit.: Barr (*Prodr. Cl. Loculoasc.*, 1987), Hawksworth & Diederich (*TBMS* **90**: 293, 1988), Alstrup & Hawksworth (*Meddr Grønland* Biosc. **31**: 90 pp., 1990), Wang *et al.* (*MR* **111**: 1268, 2007; phylogeny).

Dacampiosphaeria D. Hawksw. (1980) = Pyrenidium fide Hawksworth (*TBMS* **80**: 547, 1983).

Dacrina Fr. (1825) nom. dub., ? Hydnaceae.

Dacrina Fr. (1832) = Strumella fide Saccardo (*Syll. fung.* **4**: 742, 1886).

Dacrydium Link (1809) nom. conf., anamorphic *Pezizomycotina*. See Sutton (*Mycol. Pap.* **144**, 1977).

Dacrymycella Bizz. (1885), anamorphic *Pezizomycotina*, Hsp.0eH.13. 2, Europe; Java.

Dacrymyces Nees (1816), Dacrymycetaceae. *c.* 39, widespread. See McNabb (*N.Z. Jl Bot.* **11**: 461, 1973; key), Reid (*TBMS* **62**: 449, 1974; key Brit. spp.), Mathiesen (*Svampe* **23**: 46, 1991; Danish spp.), Hahn & Karasch (*Zeitschrift für Mykologie* **68**: 31, 2002), Burdsall & Laursen (*Memoirs of the New York Botanical Garden* **89**: 107, 2004; New Zealand subantarctic islands n.spp.), Shirouzou *et al.* (*Mycoscience* **48**: 388, 2007; phylogeny).

Dacrymycetaceae J. Schröt. (1888), Dacrymycetales. 9 gen. (+ 8 syn.), 101 spp.
Lit.: McNabb (*N.Z. Jl Bot.* **3**: 59, 1965), Reid (*TBMS* **62**: 433, 1974), Maekawa (*CJB* **65**: 583, 1987), Oberwinkler (*Stud. Mycol.* **30**: 61, 1987), Ing (*Mycologist* **4**: 34, 1990), Maekawa & Zang (*Mycotaxon* **61**: 343, 1997), Duhem (*Bull. trimest. Soc. mycol. Fr.* **114**: 1, 1998), Mossebo *et al.* (*Cryptog. Mycol.* **22**: 119, 2001), Weiss & Oberwinkler (*MR* **105**: 403, 2001), Binder & Hibbett (*Mol. Phylogen. Evol.* **22**: 76, 2002), Larsson *et al.* (*MR* **108**: 983, 2004), Shirouzou *et al.* (*Mycoscience* **48**: 388, 2007; phylogeny).

Dacrymycetales Henn. (1898). Dacrymycetes. 1 fam., 9 gen., 101 spp. Fam.:
Dacrymycetaceae
Lit.: Donk (1951-63, VIII; *Proc. Kon. nederl. Akad. Wet.* C **67** (2), 1964; 1966: 264), Kennedy (*Mycol.* **50**: 874, 1959; gen. key), Reid (*TBMS* **62**: 433, 1974; Br. spp.; keys).

Dacrymycetes Doweld (2001), Agaricomycotina. 1 ord., 1 fam., 9 gen., 101 spp. Ord.:
Dacrymycetales
Lit.: Doweld (*Prosyllabus*: LXXVII, 2001), Hibbett *et al.* (*MR* **111**: 509, 2007; phylogeny).

Dacryobasidium Jülich (1982) = Cristinia fide Hjortstam & Grosse-Brauckmann (*Mycotaxon* **47**: 405, 1993).

Dacryobolaceae Jülich (1982) = Meruliaceae.

Dacryobolus Fr. (1849), Fomitopsidaceae. 5, widespread. See Eriksson (*Symb. bot. upsal.* **16** no. 1: 115, 1958), Christiansen (*Dansk bot. Ark.* **19**: 244, 1960), Parmasto (*Consp. System. Corticiac.*: 98, 1968; taxonomy), Manjon *et al.* (*An. Jard. bot. Madr.* **40**: 297, 1984; key Eur. spp.).

Dacryodochium P. Karst. (1896) = Graphiola fide Sutton (*in litt.*).

dacryoid, having one end rounded and the other more or less pointed; pear-like or tear-like in form.

Dacryoma Samuels (1988), anamorphic *Nectria*, St.0-1eH.15. 3, Indonesia; Peru. See Samuels (*Brittonia* **40**: 328, 1988), Matsushima (*Matsush. Mycol. Mem.* **10**, 2001; Peru).

Dacryomitra Tul. & C. Tul. (1872) = Calocera fide McNabb (*N.Z. Jl Bot.* **3**: 31, 1965).

Dacryomyces, see *Dacrymyces*.

Dacryomycetopsis Rick (1958) nom. dub., Basidiomycota. See Rick (*Iheringia* Série Botânica **2**: 53, 1958).

Dacryonaema Nannf. (1947), Dacrymycetaceae. 1,

Europe. See Lisická (*Czech Mycology* **48**: 217, 1995; Slovakia).

Dacryopinax G.W. Martin (1948), Dacrymycetaceae. *c.* 15, widespread (esp. tropical). See McNabb (*N.Z. Jl Bot.* **3**: 59, 1965; key), Lowy (*Mycotaxon* **13**: 428, 1981; key trop. Am. spp.), Sierra & Cifuentes (*Mycotaxon* **92**: 243, 2005; key), Roberts (*Mycotaxon* **96**: 83, 2006; Caribbean).

Dacryopsella Höhn. (1915) = Pistillina fide Donk (*Reinwardtia* **2**: 435, 1954).

Dacryopsida Nees (1816) nom. dub., Agaricomycetes.

Dacryopsis Massee (1891) = Ditiola Fr. fide Kennedy (*Mycol.* **56**: 298, 1964).

Dacryoscyphus R. Kirschner & Zhu L. Yang (2005), anamorphic *Dacrymycetaceae*. 1, China. See Kirschner & Yang (*Antonie van Leeuwenhoek* **87**: 331, 2005).

Dactuliochaeta G.L. Hartm. & J.B. Sinclair (1988), anamorphic *Pleosporales*, Cpd.0eH.15. 1 (on (*Glycine*), Africa. For *Pyrenochaeta* with *Dactuliophora* synanamorph. See Hartman & Sinclair (*Mycol.* **80**: 696, 1988).

Dactuliophora C.L. Leakey (1964), anamorphic *Pleosporales*, Hsp.#eP.42. 4, widespread (esp. Africa). See Datnoff *et al.* (*TBMS* **87**: 297, 1986; as *Pyrenochaeta*), Ramaley (*Mycol.* **91**: 132, 1999).

Dactylaria Sacc. (1880), anamorphic *Helotiales*, Hso.1-≡ eH.10. *c.* 109, widespread. Probably polyphyletic. See Bhatt & Kendrick (*CJB* **46**: 1253, 1968; type sp.), de Hoog & von Arx (*Kavaka* **1**: 55, 1973), de Hoog & van Oorschot (*Proc. K. ned. Akad. Wet.* Ser. C, Biol. Med. Sci. **86**: 55, 1983), de Hoog (*Stud. Mycol.* **26**: 1, 1985; key 41 spp.), Dixon & Salkin (*J. Clin. Microbiol.* **24**: 12, 1986; medical spp.), Sawadogo & Cayrol (*Riv. Scient.*: 27, 1990), Castañeda Ruiz & Kendrick (*Univ. Waterloo Biol. Ser.* **35**: 1, 1991; Cuba), Saikawa & Kaneko (*Mycoscience* **35**: 89, 1994; infection process), Castañeda Ruíz *et al.* (*Mycotaxon* **58**: 253, 1996; Cuba), Rubner (*Stud. Mycol.* **39**, 1996), Goh & Hyde (*MR* **101**: 1265, 1997; revision), Liou & Tzean (*Mycol.* **89**: 876, 1997; phylogeny), Ahrén *et al.* (*FEMS Microbiol. Lett.* **158**: 179, 1998; phylogeny), Paulus *et al.* (*Fungal Diversity* **14**: 143, 2003; Australia).

Dactylariopsis Mekht. (1967) = Arthrobotrys fide Schenck *et al.* (*CJB* **55**: 977, 1977), Scholler *et al.* (*Sydowia* **51**: 89, 1999; synonymy with *Drechlerella*).

Dactylella Grove (1884), anamorphic *Orbilia*, Hso.≡ eH.10. *c.* 62 (mainly on nematodes), widespread (esp. north temperate). See Subramanian (*J. Indian bot. Soc.* **42**: 291, 1963), Zhang *et al.* (*Mycosystema* **7**: 111, 1995; review), Rubner (*Stud. Mycol.* **39**, 1996), Liou & Tzean (*Mycol.* **89**: 876, 1997; phylogeny), Ahrén *et al.* (*FEMS Microbiol. Lett.* **158**: 179, 1998; phylogeny), Webster *et al.* (*MR* **102**: 99, 1998; teleomorph), Hagedorn & Scholler (*Sydowia* **51**: 27, 1999; phylogeny), Zhang *et al.* (*Mycosystema* **20**: 51, 2001; ontogeny), Barron (*Biodiversity of Fungi* Inventory and Monitoring Methods: 435, 2004; review).

Dactylellina M. Morelet (1968) = Monacrosporium fide Scholler *et al.* (*Sydowia* **51**: 89, 1999; gen. concept), Li *et al.* (*Mycol.* **97**: 1034, 2006; phylogeny).

Dactyliaceae Nann. (1934) = Orbiliaceae.

Dactylifera Alcorn (1987), anamorphic *Pezizomycotina*, Hso.0bP.26. 1, Australia. See Alcorn (*Mycotaxon* **28**: 71, 1987).

Dactylina G. Arnaud ex Subram. (1964) ≡ Lactydina.

Dactylina Nyl. (1860), Parmeliaceae (L). *c.* 10, widespread. See Lynge (*Skr. Svalbard Ishavet* **59**, 1933), Follmann *et al.* (*Willdenowia* **5**: 7, 1968), Thomson & Bird (*CJB* **56**: 1602, 1978; N. Am. spp., key), Kärnefelt & Thell (*Nova Hedwigia* **62**: 487, 1996; revision), Thell *et al.* (*Mycol. Progr.* **1**: 335, 2002; phylogeny), Mattsson & Articus (*Symb. bot. upsal.* **34** no. 1: 237, 2004; phylogeny), Miądlikowska *et al.* (*Mycol.* **98**: 1088, 2006; phylogeny), Randlane & Saag (*Central European Lichens*: 75, 2006; key European spp.).

Dactylium Nees (1816) nom. rej. = Monacrosporium fide Gams & Rubner (*Taxon* **46**: 335, 1997).

Dactyloblastus Trevis. (1853) = Phlyctis fide Hawksworth *et al.* (*Dictionary of the Fungi* edn 8, 1995).

dactyloid, finger-like.

Dactylomyces Sopp (1912), Trichocomaceae. Anamorph *Polypaecilum*. 2 (thermophilic), Europe; N. America. See also *Coonemeria*. See Apinis (*TBMS* **50**: 576, 1967), Mouchacca (*Cryptog. Mycol.* **18**: 19, 1997; review).

Dactyloporus Herzer (1893), Fossil Fungi, Agaricomycetes. 1 (Carboniferous), N. America. See Seward (*Fossil plants* **1**: 211, 1898).

Dactylospora Körb. (1855), Dactylosporaceae. *c.* 49 (on lichens, hepatics or wood), widespread. See Hafellner (*Beih. Nova Hedwigia* **62**, 1979), Bellemère & Hafellner (*Cryptog. Mycol.* **3**: 71, 1982; asci), Döbbeler & Triebel (*Bot. Jahrb. Syst.* **107**: 503, 1985; on *Hepaticae*), Au *et al.* (*Mycoscience* **37**: 129, 1996; ultrastr.), Kohlmeyer & Volkmann-Kohlmeyer (*Mycotaxon* **67**: 247, 1998; marine spp.), Sarrión *et al.* (*Lichenologist* **34**: 361, 2002; Spain), Ihlen *et al.* (*Bryologist* **107**: 357, 2004; key).

Dactylosporaceae Bellem. & Hafellner (1982), ? Lecanorales (L). 1 gen. (+ 6 syn.), 49 spp.

Lit.: Döbbeler & Triebel (*Bot. Jahrb. Syst.* **107**: 503, 1985), Au *et al.* (*Mycoscience* **37**: 129, 1996), Kohlmeyer & Volkmann-Kohlmeyer (*Mycotaxon* **67**: 247, 1998), Jones *et al.* (*Mycoscience* **40**: 317, 1999).

Dactylosporangium Thiemann, Pagani & Beretta (1967), Actinobacteria. q.v.

Dactylosporina (Clémençon) Dörfelt (1985), Marasmiaceae. 2, S. America. Or perhaps *Physalacriaceae*. See Dörfelt (*Feddes Repert.* **96**: 236, 1985).

Dactylosporites Paradkar (1976), Fossil Fungi. 1 (Cretaceous), India. See Paradkar (*Journal of Palynology* **10**: 120, 1974).

Dactylosporium Harz (1872), anamorphic *Pezizomycotina*, Hso.#eP.10. 2, Europe; Cuba. See Hughes (*Naturalist* Hull: 63, 1952), Seifert & Hughes (*N.Z. Jl Bot.* **38**: 489, 2000).

Dactylosporium Mekht. (1967) ≡ Dactylellina.

dadih (**dadiah**), a fermented product of buffalo milk popular in Western Sumatra. See Gandjar *et al.* (*IMC3 Abstr.*: 452, 1983; microbiology).

Daedala Hazsl. (1887) = Hypodermina fide Darker (*CJB* **45**: 1399, 1967).

Daedalea Pers. (1801), Fomitopsidaceae. 7, widespread. *D. quercina* (wood rot). See Corner (*Beih. Nova Hedwigia* **86**: 265 pp., 1987), Legon (*Mycologist* **19**: 44, 2005).

Daedaleaceae Jülich (1982) = Fomitopsidaceae.

Daedaleites Mesch. (1892), Fossil Fungi. 2 (Tertiary), Europe.

Daedaleopsis J. Schröt. (1888), Polyporaceae. 6, widespread. See Pegler (*The polypores [Bull. BMS*

Suppl.], 1973), Ko & Jung (*Antonie van Leeuwenhoek* **75**: 191, 1999; molecular phylogeny), Bernicchia *et al.* (*MR* **110**: 14, 2006; neolithic DNA).

Daedaloides Lázaro Ibiza (1916) = Cryptoderma fide Pegler (*The polypores [Bull. BMS Suppl.]*, 1973).

Daldinia Ces. & De Not. (1863) nom. cons., Xylariaceae. Anamorph *Nodulisporium*. *c.* 30, widespread. *D. concentrica* ('calico wood', 'cramp balls' on *Fraxinus*). See Pérez-Silva (*Boln. Soc. mex. Micol.* **7**: 51, 1973; Mexican spp.), Thind & Dargan (*Kavaka* **6**: 15, 1979; Indian spp.), Petrini & Müller (*Mycol. Helv.* **1**: 501, 1986; key 5 spp. Eur.), Læssøe (*SA* **13**: 43, 1994; syn. of *Hypoxylon*), Ju *et al.* (*Mycotaxon* **61**: 243, 1997; monogr.), Rogers *et al.* (*Mycotaxon* **72**: 507, 1999; typification), Johannesson *et al.* (*MR* **104**: 275, 2000; Europ.), Johannesson *et al.* (*Mycol.* **93**: 440, 2001; population structure), Johannesson *et al.* (*Mol. Ecol.* **10**: 1665, 2001; genetic variation), Stadler *et al.* (*Mycotaxon* **80**: 167, 2001; European spp.), Stadler *et al.* (*Mycotaxon* **77**: 379, 2001; biochemistry), Stadler *et al.* (*Mycol. Progr.* **1**: 31, 2002; ascospore ornamentation), Stadler *et al.* (*MR* **108**: 257, 2004; cryptic spp.), Stadler *et al.* (*MR* **108**: 1025, 2004; large-spored spp.), Hsieh *et al.* (*Mycol.* **97**: 844, 2005; phylogeny), Triebel *et al.* (*Nova Hedwigia* **80**: 25, 2005; phylogeny), Bitzer *et al.* (*MR* **112**: 251, 2008; phylogeny, chemistry).

Daleomyces Setch. (1924) = Peziza Fr. fide Korf (*Mycol.* **48**: 711, 1956), Antonín (*Czech Mycol.* **57**: 249, 2005; Czech Republic).

Damnosporium Corda (1842) = Bactridium fide Saccardo (*Syll. fung.* **4**: 691, 1886).

damping-off, a rotting of seedlings at soil level. In **pre-emergent** - the young plant is attacked at germination so that the seedling does not come up. Species of *Fusarium, Phytophthora, Pythium* and *Rhizoctonia* are common **- fungi.** Soil sterilization is frequently the best control measure (see Cheshunt compound).

Danaëa Caneva & Rambelli (1981) [non *Danaëa* Sm. 1793, *Pteridophyta*] ≡ Kiliophora.

Dangeard (Pierre Clement Augustin; 1862-1947; France). On the academic staff, Faculty of Sciences, Caen (1886-1891) then Poitiers (1891-1908) then Paris (1908-1924); Professor of Botany, Sorbonne, Paris (1924 onwards). A distinguished mycologist, protozoologist, algologist, cytologist, and morphologist. He provided data for classification of many fungi, especially those with algal or protozoan affinities. *Publs.* Recherches sur le développement du périthèce chez les Ascomycètes. *Le Botaniste* (1907). *Biogs, obits etc.* Grummann (1974: 313); Heim (*Revue de Mycologie* **12**: 97, 1947) [portrait]; Moreau (*Revue Générale de Botanique* **57**: 193, 1950) [portrait]; Stafleu & Cowan (*TL-2* **1**: 596, 1976).

Dangeardia Schröd. (1898), Chytridiaceae. 7, Europe; N. America. See Batko (*Acta Mycologica* Warszawa **12**: 407, 1970).

Dangeardiana Valkanov ex A. Batko (1970), Chytridiaceae. 4 (on *Eudorina*), Bulgaria. See Batko (*Acta Mycologica* Warszawa **6**: 430, 1970).

Dangeardiella Sacc. & P. Syd. (1899), Melanommataceae. 2 (on *Pteridophyta*), Europe. See Obrist (*Phytopath. Z.* **35**: 379, 1959), Liew *et al.* (*Mol. Phylogen. Evol.* **16**: 392, 2000; phylogeny), Câmara *et al.* (*MR* **107**: 516, 2003; phylogeny).

dangeardien (**dangeardium**), collective term for both asci and basidia; structures where diploid nuclei are formed, undergo meiosis and form haploid spores (Moreau, *Botaniste* **34**: 315, 1949). See van der Walt & Johanssen (*Ant. v. Leeuwenhoek* **40**: 185, 1974; concept in yeast taxonomy).

Dangeardiomycetes, *Basidiomycetes*, *Protobasidiomycetes* and *Ascomycetes* excluding *Periascomycetes* (Moreau, 1953).

DAOM, Canadian National Mycological Herbarium (Ottawa, Canada); founded 1929; part of the Centre for Land and Biological Resources Research, Agriculture Canada; genetic resource collection **CCFC**; see Cody *et al.* (*Systematics in Agriculture Canada at Ottawa 1886-1986*, 1986).

DAP pathway, alpha, beta-diaminopimelic acid pathway for lysine synthesis (cf. AAA pathway).

Dapsilosporium Corda (1837) = Melanconium fide Saccardo (*Syll. fung.* **3**: 1, 1884), Sutton (*Mycol. Pap.* **141**, 1977).

DAR, Plant Pathology Branch Herbarium New South Wales (Rydalmere, NSW, Australia); founded 1890; collections vested in a Trust established by the NSW Parliament from 1983 and supported by the goverment.

Darbishirella Zahlbr. (1898) = Ingaderia See Tehler (*CJB* **68**: 2458, 1990; cladistics), Feige & Lumbsch (*Mycotaxon* **48**: 381, 1993).

Darkera H.S. Whitney, J. Reid & Piroz. (1975), Helotiales. Anamorph *Tiarosporella*. 2, Europe; N. America. See Whitney *et al.* (*CJB* **53**: 3052, 1975), Gernandt *et al.* (*Mycol.* **93**: 915, 2001; phylogeny).

Darluca Castagne (1851) = Sphaerellopsis Cooke fide Sutton (*Mycol. Pap.* **141**, 1977).

Darlucella Höhn. (1919) = Sphaerellopsis Cooke fide Sutton (*Mycol. Pap.* **141**, 1977).

Darlucis Clem. (1931) = Heteropatella fide Sutton (*Mycol. Pap.* **141**, 1977).

Daruvedia Dennis (1988), Dothideomycetes. 1, British Isles. See Barr (*Mycotaxon* **51**: 191, 1994; posn).

Darwiniella Speg. (1887) nom. dub., Pezizomycotina. See Petrak & Sydow (*Annls mycol.* **34**: 22, 1936).

Dasturella Mundk. & Khesw. (1943), Phakopsoraceae. 3 (on *Rubiaceae* (0, I where known); on *Poaceae* (esp., II, III)), India. See Thirumalachar *et al.* (*Bot. Gaz.* **108**: 371, 1947), Cummins & Hiratsuka (*Illustr. Gen. Rust Fungi* edn 3: 225 pp., 2003; syn. of *Kweilingia*), Thaung (*Australasian Mycologist* **24**: 29, 2005; Burma).

Dasybolus Clem. & Shear (1931) ≡ Dasyobolus.

Dasyobolus (Sacc.) Sacc. (1895) = Ascobolus fide van Brummelen (*Persoonia* Suppl. **1**: 1, 1967).

Dasypezis Clem. (1909) = Lachnum fide Nannfeldt (*Nova Acta R. Soc. Scient. upsal.*, 1932).

Dasyphthora Clem. (1909) nom. rej. = Nectriopsis fide Rossman *et al.* (*Stud. Mycol.* **42**: 248 pp., 1999).

Dasypyrena Speg. (1912) = Actinopeltis fide von Arx & Müller (*Stud. Mycol.* **9**, 1975).

Dasyscypha Fuckel (1870) = Neodasyscypha Suková & Spooner.

Dasyscyphella Tranzschel (1898), Hyaloscyphaceae. 23, Asia; N. America. See Dennis (*Kew Bull.* **27**: 273, 1972), Raitviir (*Eesti NSV Tead. Akad. Toim.* Biol. seer **26**: 33, 1977), Baral (*Z. Mykol.* **59**: 3, 1993), Galán & Raitviir (*Nova Hedwigia* **58**: 453, 1994), Raitviir & Järv (*Proc. Est. Acad. Sci. Biol. Ecol.* **46**: 94, 1997), Leenurm *et al.* (*Sydowia* **52**: 30, 2000; ultrastr.), Raitviir (*Polish Botanical Journal* **47**: 227, 2002; revision), Suková (*Czech Mycol.* **57**: 139, 2005; Czech Republic).

Dasyscyphus Nees ex Gray (1821) = Lachnum. Formerly used for many members of the *Hyaloscyphaceae* with distinctive apothecial hairs. fide Holm (*TBMS* **67**: 333, 1976), Korf (*Mycotaxon* **5**: 515, 1977), Holm (*Mycotaxon* **7**: 139, 1978), Dimitrova (*Phytologica Balcanica* **6**: 133, 2000; Bulgaria).

Dasysphaeria Speg. (1912), Pezizomycotina. 1, S. America.

Dasyspora Berk. & M.A. Curtis (1853), Uropyxidaceae. 1 (on *Annonaceae*), America (tropical). See Hennen & Figueredo (*Mycol.* **73**: 350, 1981).

Dasysticta Speg. (1912), anamorphic *Pezizomycotina*, Cpd.0eH.15. 1, S. America.

Dasystictella Höhn. (1919) nom. dub., anamorphic *Pezizomycotina*. See Sutton (*Mycol. Pap.* **141**, 1977).

Datronia Donk (1967), Polyporaceae. 5, north temperate. See Ryvarden & Gilbertson (*Europ. Polyp.* **1**: 230, 1993).

datum (pl. **data**), facts, figures, information and observations. Often used adjectivally, e.g. data bank, data matrix, database.

Davidgallowaya Aptroot (2007), Parmeliaceae (L). 1, Papua New Guinea. See Aptroot (*Biblthca Lichenol.* **95**: 137, 2007).

Davidiella Crous & U. Braun (2003), Davidiellaceae. Anamorph *Cladosporium*. 33, widespread. See Braun *et al.* (*Mycol. Progr.* **2**: 8, 2003), Park *et al.* (*Mycotaxon* **89**: 441, 2004; indoor spp.), Seifert *et al.* (*CJB* **82**: 914, 2004; phylogeny), Crous *et al.* (*MR* **110**: 264, 2006), Schoch *et al.* (*Mycol.* **98**: 1041, 2006; phylogeny), Crous *et al.* (*Stud. Mycol.* **58**: 1, 2007; phylogeny), Schubert *et al.* (*Stud. Mycol.* **58**: 105, 2007; phylogeny).

Davidiellaceae C.L. Schoch, Spatafora, Crous & Shoemaker (2007), Capnodiales. 4 gen. (+ 11 syn.), 185 spp. See Schoch *et al.* (*Mycol.* **98**: 1041, 2006; phylogeny), Crous *et al.* (*Stud. Mycol.* **58**: 1, 2007; phylogeny).

Davincia Penz. & Sacc. (1901) = Crocicreas fide Carpenter (*Mem. N. Y. bot. Gdn* **33**: 1, 1981).

Davinciella (Sacc. & D. Sacc.) Trotter (1928) ≡ Merodontis.

Davisiella Petr. (1924), anamorphic *Pezizomycotina*, Cpd.0eH/#eH.15. 1, USA.

Davisomycella Darker (1967), Rhytismataceae. 10, widespread. See Minter & Ivory (*TBMS* **91**: 171, 1988), Ganley *et al.* (*Proc. natn Acad. Sci. U.S.A.* **101**: 10107, 2004; endophytes), Hou *et al.* (*Nova Hedwigia* **83**: 511, 2006; China).

Davisoniella H.J. Swart (1988), anamorphic *Mycosphaerella*, St.0eP.19. 1, Australia. See Swart (*TBMS* **90**: 289, 1988), Crous *et al.* (*Stud. Mycol.* **55**: 99, 2006; phylogeny).

Dawsicola Döbbeler (1981), Helotiales. 1 (on *Musci*), New Zealand. See Döbbeler (*Sydowia* **38**: 41, 1986), Döbbeler (*Biodiv. Cons.* **6**: 721, 1997).

Dawsomyces Döbbeler (1981), ? Arthoniomycetes. 2 (on *Musci*), New Guinea; Australia. See Döbbeler (*Mitt. bot. StSamml.* Münch. **17**: 426, 1981), Döbbeler (*Biodiv. Cons.* **6**: 721, 1997).

Dawsophila Döbbeler (1981), Dothideomycetes. 2, New Guinea; New Zealand. See Grube (*Bryologist* **101**: 377, 1998).

de Bary (Heinrich Anton; 1831-1888; Germany). Doctor of Medicine, Berlin (1853); practising surgeon, Frankfurt-am-Main, the town of his birth (1853); privat docent in botany, Tübingen (1853-1855); Professor, Freiburg (1855-1867); Professor, Halle (1867-1872); Professor, Strasburg University (1872-1888); editor, *Botanische Zeitung* (1867-1887). His writings on mycology (and on algology, bacteriology and botany) were outstanding, his interests being more biological and physiological than systematic: from his investigations on life histories, parasites and saprobes, *Mycetozoa* (see Martin, *Proceedings of the Iowa Academy of Science* **65**: 20, 1958), the nature of lichens, etc., he made new and important discoveries. Heteroecism in the *Pucciniales* was made clear by his experiments and he gave accounts of development and of sex in a number of 'phycomycetes' and 'ascomycetes'. His teaching (many of his students became noted) and writings had a very great effect on later development of mycology, and it is generally accepted that he was 'the founder of modern mycology'. *Publs. Untersuchungen über die Brandpilze* (1853); *Die Mycetozoen* (1859); *Morphologie und Physiologie der Pilze* (1866) [edn 2, 1884; English edn, 1887]. *Biogs, obits etc.* Reess (*Bericht der Deutschen Botanischen Gesellschaft* **6**: viii, 1888); Smith (*Phytopathology* **1**: 1, 1911) [portrait]; Sparrow (*Mycol.* **70**: 222, 1978); Stafleu & Cowan (*TL-2* **1**: 135, 1976).

de Bary bubbles, air bubbles in ascospores; first described by de Bary (1884: 106). See Dodge (*Bull. Torrey bot. Cl.* **84**: 431, 1957).

de Notaris (Giuseppe; 1805-1877; Italy). Graduate in medicine, Padua (1830); Professor of Botany, University of Genoa (1839-1872); Professor of Botany, Rome (1872-1877). Made early investigations of Italian ascomycetes, including lichen-forming species, and introduced reforms in ascomycete taxonomy; also published extensively on Italian liverworts and mosses. Main collections in Rome (**RO**). *Publs. Micromycetes Italici Novi vel Minus Cogniti* (1839-1855); *Sferiacei Italici* (1863); also other papers, especially on ascomycetes. *Biogs, obits etc.* Cooke (*Grevillea* **5**: 143, 1877); Graniti (*Rendiconti della Accademia Nazionale delle Scienze detta dei XL* Serie V **15**: 9, 1991) [portrait]; Stafleu & Cowan (*TL-2* **1**: 622, 1976).

Dearness (John; 1852-1954; Canada). School inspector, East Middlesex (1874-1899); Vice-Principal (1899-1918) then Principal (1918-1922) London Normal School. Self-taught amateur mycologist; the first Canadian mycologist to receive international recognition. Collections in **DAOM** (Parmelee, *Mycol.* **70**: 509, 1978). *Publs.* New and noteworthy fungi I-VI. *Mycol.* (1916-1929); (with House) New or noteworthy species of fungi I-IV. *Bulletin of the New York State Museum* (1918-1925). *Biogs, obits etc.* Tamblyn (*Mycol.* **47**: 909, 1955) [bibliography, portrait]; Ginns (*Mycotaxon* **26**: 47, 1986); Stafleu & Cowan, *TL-2* **1**: 605, 1976).

Dearnessia Bubák (1916), anamorphic *Pezizomycotina*, Cpd.≡ eH.1. 1, Canada.

death cap, basidiomata of the highly poisonous *Amanita phalloides*.

Debarya Schulzer (1866) = Hypocrea. The type no longer exists, so application is dubious. fide Rossman *et al.* (*Stud. Mycol.* **42**: 248 pp., 1999).

Debaryella Höhn. (1904), ? Sordariomycetes. 2, Europe. The type is lost and affinities are obscure. fide Rossman *et al.* (*Stud. Mycol.* **42**: 248 pp., 1999), Réblová & Seifert (*Stud. Mycol.* **50**: 95, 2004).

Debaryolipomyces Ramírez (1957) nom. nud., ? Saccharomycetes. See Batra *in* Subramanian (Ed.) (*Taxonomy of fungi* **1**: 187, 1978).

Debaryomyces Klöcker (1909) nom. rej. = Torulaspora fide van der Walt & Johanssen (*CSIR Res. Rept.* **325**, 1975).

Debaryomyces Lodder & Kreger-van Rij (1984) nom. cons., Saccharomycetales. 19, widespread. See Hendriks *et al.* (*Syst. Appl. Microbiol.* **15**: 98, 1992; posn), Yamada *et al.* (*J. gen. appl. Microbiol.* Tokyo **38**: 623, 1992; gen. concept), Cai *et al.* (*Int. J. Syst. Bacteriol.* **46**: 542, 1996; rDNA), Nakase *et al. in* Kurtzman & Fell (Eds) (*Yeasts, a taxonomic study* 4th edn: 157, 1998), Ramos *et al.* (*J. gen. appl. Microbiol.* Tokyo **44**: 399, 1998; ITS), Nishikawa *et al.* (*Medical Mycology* **37**: 101, 1999; PCR identification), Corredor *et al.* (*FEMS Microbiol. Lett.* **193**: 171, 2000; DNA probes), Petersen *et al.* (*Int. J. Food Microbiol.* **69**: 11, 2001; DNA typing), Thanh *et al.* (*FEMS Yeast Res.* **2**: 415, 2002; from woodlice), Corredor *et al.* (*Antonie van Leeuwenhoek* **83**: 215, 2003; chromosomes), Petersen & Jespersen (*Journal of Applied Microbiology* **97**: 205, 2004; genetic diversity), Martorell *et al.* (*FEMS Yeast Res.* **5**: 1157, 2005; molecular identification), Romero *et al.* (*FEMS Yeast Res.* **5**: 455, 2005; molecular detection), Quirós *et al.* (*Antonie van Leeuwenhoek* **90**: 211, 2006; IGS analysis), Suh *et al.* (*Mycol.* **98**: 1006, 2006; phylogeny).

Debaryoscyphus Arendh. & R. Sharma (1986) = Hamatocanthoscypha fide Huhtinen (*Karstenia* **29**: 545, 1990).

Debaryozyma Van der Walt & Johannsen (1978) nom. rej. ≡ Debaryomyces Lodder & Kreger ex Kreger.

Decaisnella Fabre (1879), Massariaceae. 11, widespread. See Barr (*Sydowia* **38**: 11, 1986), Barr (*SA* **5**: 127, 1986; status), Barr (*N. Amer. Fl.* ser. 2 **13**: 129 pp., 1990; key, posn), Abdel-Wahab & Jones (*CJB* **81**: 598, 2003; Australia), Checa & Blanco (*Mycotaxon* **91**: 353, 2005; Spain).

Decampia Mudd (1861) ≡ Dacampia.

Decapitatus Redhead & Seifert (2000), anamorphic *Mycena*. 1. See Redhead *et al.* (*Taxon* **49**: 789, 2000).

decay, the destruction of plant or animal material by fungi and other microorganisms.

Deccanodia Singhai (1974), Fossil Fungi, anamorphic *Pezizomycotina*. 1 (Eocene), India.

deciduous (of spores, etc.), falling away at maturity; shed, either with (e.g. teliospores) or without (e.g. urediniospores) a fragment of the pedicel or sporophore; cf. persistent.

Deckenbachia Jacz. (1931) ≡ Coenomyces.

declinate, bent or curved down or forwards.

declivate (**declivous**), sloping.

decolourate, colourless.

Decomposition. In the strict sense, is the breakdown of organic materials through biological activity, although the physical processes of leaching and fragmentation are sometimes considered a part of decomposition. Fungal decomposition, particularly of plant remains, is a key element worldwide in the recycling of nutrients. The products resulting from biological decomposition by fungi are energy for the decomposer, inorganic elements and compounds, and simple organic compounds such as CO_2 or alcohol which result from aerobic and anaeraobic respiration (fermentation), respectively. Products containing complex organic compounds with phenolic rings (humic acids) can result from partial decomposition of lignin, and are an important component of soil organic matter. The release of mineral nutrients through decomposition is known as nutrient mineralization. When fungi decompose organic matter such as wood or straw in which the ratio of carbon to inorganic nutrients (especially nitrogen and phosphorus) is high, the nutrient mineralization phase is often preceded by a nutrient immobilization phase. During nutrient immobilization, the decomposers incorporate mineral nutrients from their organic substrate and sometimes from the surrounding environment into their biomass. See Carroll & Wicklow (Eds) (*The fungal community*, edn 2, 1992), Dix & Webster (*Fungal ecology*, 1995). See also: Biodegradation.

Deconica (W.G. Sm.) P. Karst. (1879) = Psilocybe (Fr.) P. Kumm. fide Singer (*Agaric. mod. Tax.*, 1951).

Decorospora Inderb., Kohlm. & Volkm.-Kohlm. (2002), Pleosporaceae. 1 (marine), widespread. See Inderbitzin *et al.* (*Mycol. Progr.* **94**: 657, 2002).

decorticate, having no cortex.

decumbent, resting on the substratum with the ends turned up.

decurrent (of lamellae), running down the stipe (Fig. 19E).

decurved (of the pileus edge), bent down.

decussate (of lichen thalli), having the surface divided and crossed by dark lines.

Dedalea, see *Daedalea*.

dediploidization (in ascomycetes and basidiomycetes), the making of haploid cells (or hyphae) by a dikaryotic diploid mycelium or cell.

deer balls (**Lycoperdon nuts** or **harts' truffles**), *Elaphomyces* ascomata.

Deflexula Corner (1950) = Pterula fide Kuyper (*in litt.*).

Degelia Arv. & D.J. Galloway (1981), Pannariaceae (L). 11, widespread (temperate). See Jørgensen & James (*Biblthca Lichenol.* **38**: 253, 1990; key), Lumbsch & Kothe (*Mycotaxon* **43**: 277, 1992; thallus SEM), Jørgensen & James (*Lichenologist* **30**: 533, 1998), Jørgensen (*Bryologist* **103**: 670, 2000; N America), Jørgensen *et al.* (*Lichenologist* **32**: 257, 2000; n. spp.), Jørgensen (*Biblthca Lichenol.* **78**: 109, 2001; Australia), Ekman & Jørgensen (*CJB* **80**: 625, 2002; phylogeny), Jørgensen (*N.Z. Jl Bot.* **40**: 327, 2002; New Zealand), Wiklund & Wedin (*Cladistics* **19**: 419, 2003; phylogeny), Jørgensen (*Biblthca Lichenol.* **88**: 229, 2004; S hemisphere), Miądlikowska & Lutzoni (*Am. J. Bot.* **91**: 449, 2004; phylogeny), Wedin & Wiklund (*Symb. bot. upsal.* **34** no. 1: 469, 2004; phylogeny), Jørgensen & Sipman (*J. Hattori bot. Lab.* **100**: 695, 2006; New Guinea), Miądlikowska *et al.* (*Mycol.* **98**: 1088, 2006; phylogeny).

Degeliella P.M. Jørg. (2004), Pannariaceae (L). 2, austral. See Jørgensen (*Biblthca Lichenol.* **88**: 235, 2004).

dehiscence papilla, morphologically and ultrastructurally distinct protuberance on an undischarged sporangium that becomes converted into an exit tube.

dehiscent (**dehiscing**) (of asci or fruit-bodies), opening when mature, by pores or by becoming broken into parts.

Deichmannia Alstrup & D. Hawksw. (1990), anamorphic *Pezizomycotina*, Hsp.#eP.1. 1 (on lichens), Greenland. See Alstrup & Hawksworth (*Meddr Grønland* Biosc. **31**: 26, 1990).

Deighton (Frederick Claude; 1903-1992; England, later Sierra Leone). Plant Pathologist, British Colonial Office (up to 1955); Mycologist, IMI (1955-1973). With Hughes, a pioneer of the mycota of West Africa, in particular Sierra Leone; also noted as an expert on *Cercospora* and related genera. Collections in **IMI**. *Publs*. Seventeen numbers of the series *Mycological Papers*. *Biogs, obits etc*. Ainsworth (*Brief Biographies of British Mycologists* p. 57, 1996).

Deightonia Petr. (1947) = Vanderystiella fide Petrak (*Sydowia* **5**: 328, 1951).

Deightoniella S. Hughes (1952), anamorphic *Mycosphaerellaceae*, Hso.1-≡ eP.19. 10, widespread (esp. tropical). *D. torulosa* (fruit spot ('speckle') of banana. See Ellis (*DH*, 1971), Constantinescu (*Proc. K. Ned. Akad. Wet.* Ser. C, Biol. Med. Sci. **86**: 137, 1983), Ondřej (*Česká Mykol.* **38**: 39, 1984; Czech. spp.), Barr (*Mycol.* **77**: 549, 1985), Crous *et al.* (*Mycosphaerella Leaf Spot Diseases of Bananas: Present Status and Outlook* Proceedings of the 2nd International Workshop on *Mycosphaerella* Leaf Spot Diseases Held in San José, Costa Rica, 20-23 May 2002: 43, 2003).

Deigloria Agerer (1980), Marasmiaceae. 10, widespread (neotropics). See Agerer (*Mycotaxon* **12**: 188, 1980).

Dekkera Van der Walt (1964), ? Pichiaceae. Anamorph *Brettanomyces*. 5, widespread. See Smith *et al.* (*Yeast* Chichester **6**: 299, 1990), Molina *et al.* (*Int. J. Syst. Bacteriol.* **43**: 32, 1993; molec., key), Yamada *et al.* (*Biosc., Biotechn., Biochem.* **58**: 1893, 1994; molec. syst.), Smith *et al. in* Kurtzman & Fell (Eds) (*Yeasts, a taxonomic study* 4th edn: 174, 1998), Cocolin *et al.* (*Appl. Environm. Microbiol.* **70**: 1347, 2004; molecular detection), Martorell *et al.* (*Int. J. Food Microbiol.* **106**: 79, 2006; molecular typing), Suh *et al.* (*Mycol.* **98**: 1006, 2006; phylogeny).

Dekkeromyces Santa María & C. Sánchez-Pinto (1970) = Kluyveromyces fide Batra *in* Subramanian (Ed.) (*Taxonomy of fungi* **1**: 187, 1978).

Dekkeromyces Wick. & K.A. Burton (1956) nom. inval. = Kluyveromyces fide Batra *in* Subramanian (Ed.) (*Taxonomy of fungi* **1**: 187, 1978).

Delacourea Fabre (1879) = Lophiostoma fide Barr (*Sydowia* **38**: 11, 1985), Eriksson & Hawksworth (*SA* **6**: 123, 1987).

Delacroixia Sacc. & P. Syd. (1899) = Conidiobolus fide Tyrrell & McLeod (*J. invert. Path.* **20**: 11, 1972).

Delastreopsis Mattir. (1905) ≡ Lespiaultinia fide Trappe (*TBMS* **65**: 496, 1975).

Delastria Tul. & C. Tul. (1843), Pezizales. 1 (hypogeous), Europe (southern); N. Africa.

Delentaria Corner (1970), Gomphaceae. 1, Brazil. See Corner (*Beih. Nova Hedwigia* **33**: 225, 1970).

Delicatula Fayod (1889), ? Tricholomataceae. 2, widespread (temperate). See Antonín (*Czech Mycol.* **54**: 205, 2003).

deliquescent, becoming liquid, e.g. after maturing.

Delisea Fée (1825) [non *Delisea* J.V. Lamour. 1819, *Algae*] ≡ Plectocarpon.

Delitescor Earle (1909) = Psilocybe (Fr.) P. Kumm. fide Singer (*Agaric. mod. Tax.*, 1951).

Delitschia Auersw. (1866), Delitschiaceae. 51 (coprophilous), widespread. See Luck-Allen & Cain (*CJB* **53**: 1827, 1975; key), Parguey-Leduc (*BSMF* **94**: 409, 1978; ontogeny), Barr (*N. Amer. Fl.* ser. 2 **13**: 129 pp., 1990; posn), Barrasa & Checa (*Rev. Iberoam. Micol.* **7**: 5, 1990; key 8 spp.), Eriksson & Hawksworth (*SA* **10**: 138, 1991), Pelaez *et al.* (*Mycotaxon* **50**: 115, 1994; West Africa), Hyde & Steinke (*Mycoscience* **37**: 99, 1996; aquatic spp.), Richardson (*MR* **102**: 1038, 1998), Winka & Eriksson (*Phylogenetic Relationships Within the Ascomycota Based on 18S rDNA Sequences* Akademisk Avhandling [Thesis (PhD), Department of Ecology and Environmental Science, Umeå University]: [17] pp., 2000; phylogeny), Bell (*CBS Diversity Ser.* **3**, 2005), Kruys *et al.* (*MR* **110**: 527, 2006; phylogeny), Schoch *et al.* (*Mycol.* **98**: 1041, 2006; phylogeny).

Delitschiaceae M.E. Barr (2000), Pleosporales. 3 gen. (+ 2 syn.), 54 spp.

Lit.: Barr (*Mycotaxon* **76**: 105, 2000), Liew *et al.* (*Mol. Phylogen. Evol.* **16**: 392, 2000), Winka & Eriksson (*Phylogenetic Relationships Within the Ascomycota Based on 18S rDNA Sequences* Akademisk Avhandling [Thesis (PhD), Department of Ecology and Environmental Science, Umeå University]: [17] pp., 2000), Bell & Mahoney (*Muelleria* **15**: 3, 2001), Kruys *et al.* (*MR* **110**: 527, 2006).

Delitschiella Sacc. (1905) = Delitschia fide Müller & von Arx (*Beitr. Kryptfl. Schweiz* **11** no. 2, 1962).

Delortia Pat. & Gaillard (1888), anamorphic *Pezizomycotina*, Hsp.0-≡ hH.1. 4 (aquatic), America (tropical); Africa. See Pirozynski (*Mycol. Pap.* **129**, 1972), Goh & Hyde (*MR* **101**: 42, 1997; Australia).

Delphinella (Sacc.) Kuntze (1898), Dothideales. Anamorphs *Dothiorella*, *Sclerophoma*. 6, widespread (esp. north temperate). See Barr (*Contr. Univ. Mich. Herb.* **9**: 523, 1972), Froidevaux (*Nova Hedwigia* **23**: 679, 1973), Barr (*Harvard Pap. Bot.* **6**: 25, 2001; revision), Lumbsch & Lindemuth (*MR* **105**: 901, 2001; phylogeny), Morozova & Vasil'eva (*Mikol. Fitopatol.* **37**: 59, 2003; Siberia), Tsuneda *et al.* (*Mycol.* **96**: 1128, 2004; phylogeny), Schoch *et al.* (*Mycol.* **98**: 1041, 2006; phylogeny).

Delpinoella Sacc. (1899), Pezizomycotina. 1, Africa.

Delpinoina Kuntze (1891), Ascodichaenaceae. Anamorph *Macroallantina*. 2, Europe. See Speer (*BSMF* **103**: 9, 1987).

Delpontia Penz. & Sacc. (1901), ? Stictidaceae. 1 (on *Pteridophyta*), Java. See Sherwood (*Mycotaxon* **5**: 1, 1977).

deltoid, triangular in shape.

Deltosperma W.Y. Zhuang (1988), anamorphic *Unguiculariopsis*, St.0eP.15. 4, N. America; Europe. See Zhuang (*Mycotaxon* **32**: 31, 1988), Zhuang (*MR* **104**: 507, 2000).

DematiaceaeFr. (1832). (obsol.). Having dark-coloured hyphae and/or conidia.

dematiaceous (of mycelium, spores, etc.), pigmented, more or less darkly. Cf. moniliaceous.

Dematiocladium Allegr., Aramb., Cazau & Crous (2005), anamorphic *Nectriaceae*, H?.?.?. 1, Brazil. See Crous *et al.* (*MR* **109**: 836, 2005), Decock *et al.* (*Mycol.* **98**: 488, 2006; phylogeny).

Dematioscypha Svrček (1977), Hyaloscyphaceae. Anamorph *Haplographium*. 3, Europe; New Zealand. See Huhtinen (*Mycotaxon* **30**: 9, 1987; 3 spp.), Spooner (*Biblthca Mycol.* **116**, 1987), Raitviir (*Czech Mycol.* **52**: 289, 2001; key), Raitviir (*Scripta Mycologica* Tartu **20**, 2004).

Dematium Pers. (1801) nom. conf., anamorphic *Pezizomycotina*. See Hughes (*CJB* **36**: 727, 1958), Morgan-Jones & Goos (*Mycol.* **84**: 921, 1992).

Dematoidium Stautz (1931) ? = Aureobasidium fide Hermanides-Nijhof (*Stud. Mycol.* **15**: 143, 1977).

Dematophora R. Hartig (1883), anamorphic

Rosellinia, Hsy.0eP.11. 1, widespread. *D. necatrix* (white root rot of apple and pear). See Watanabe (*Ann. phytopath. Soc. Japan* **58**: 65, 1992; sporulation in vitro), Schnittler & Novozhilov (*Mycotaxon* **71**: 387, 1999; Taiwan), Nakamura *et al.* (*Mycoscience* **41**: 503, 2000; teleomorph), Petrini (*N.Z. Jl Bot.* **41**: 71, 2003; New Zealand), Petrini & Petrini (*MR* **109**: 569, 2005; revision).

-deme (suffix), a neutral term, always used with a prefix, and denoting any group of individuals within a taxon (q.v.; usually a species); first proposed by Gilmour & Gregor (*Nature* **144**: 333, 1939); occasionally used in mycology, e.g. agamodeme (predominantly apomictic), photosymbiodeme (of lichen thalli with different photobionts).

demicyclic, see *Pucciniales* and Table 5.

Dencoeliopsis Korf (1971), Helotiaceae. 2, Europe; USA. See Holm & Holm (*Symb. bot. upsal.* **21** no. 3: 6, 1977; possible synonymy with *Rutstroemia*), Zhuang (*Mycotaxon* **32**: 97, 1988).

Dendrina Fr. (1829) nom. dub., anamorphic *Pezizomycotina*. See Lindau (*Rabenh. Krypt.-Fl.* **1**: 203, 1907).

Dendriscocaulon Nyl. (1888), Lobariaceae (L). 10, widespread. See James & Henssen *in* Brown *et al.* (Eds) (*Lichenology: progress and problems*: 27, 1976), Thomas *et al.* (*Biblthca Lichenol.* **82**: 123, 2001; phylogeny), Tønsberg & Goward (*Bryologist* **104**: 12, 2001; Pacific), Lohtander *et al.* (*MR* **106**: 777, 2002; phylogeny), Takahashi *et al.* (*J. Hattori bot. Lab.* **100**: 783, 2006; photosymbiodemes).

dendritic, irregularly branched; tree-like; dendroid.

Dendrochaete G. Cunn. (1965) = Echinochaete fide Reid (*Kew Bull.* **17**: 278, 1963).

dendrochin, an antifungal antibiotic from *Dendrodochium toxicum* toxic to farm animals (Bilai, *Antibiotic-producing microscopic fungi*: 139, 1963).

Dendrocladium (Pat.) Lloyd (1919) = Ramaria Fr. ex Bonord. fide Corner (*Reinwardtia* **2**: 457, 1954).

Dendroclathra Voglmayr & Delg.-Rodr. (2001), anamorphic *Pezizomycotina*, Hso.?.?. 1 (aero-aquatic), Cuba. See Voglmayr & Delgado-Rodríguez (*CJB* **79**: 995, 2001).

Dendrocollybia R.H. Petersen & Redhead (2001), Tricholomataceae. Anamorph *Tilachlidiopsis*. 1, widespread (temperate). See Hughes *et al.* (*MR* **105**: 164, 2001).

Dendrocorticium M.J. Larsen & Gilb. (1974), Corticiaceae. 7, widespread. See Larsen & Gilbertson (*Norw. Jl Bot.* **24**: 99, 1977).

Dendrocyphella Petch (1922), Agaricales. 1, Sri Lanka. See Kost (*MR* **102**: 505, 1998).

Dendrodochium Bonord. (1851) = Clonostachys fide Tulloch (*Mycol. Pap.* **130**, 1972; status), Schroers *et al.* (*Mycol.* **91**: 365, 1999; syn. of *Clonostachys*), Schroers (*Stud. Mycol.* **46**: 1, 2001), Ramaley (*Mycotaxon* **90**: 181, 2004).

Dendrodomus Bubák (1915), anamorphic *Pezizomycotina*, Cpd.0eH.15. 1, Europe.

Dendrodontia Hjortstam & Ryvarden (1980), Corticiaceae. 1, Africa. See Hjortstam & Ryvarden (*Mycotaxon* **10**: 273, 1980).

Dendroecia Arthur (1906) = Ravenelia fide Sydow & Sydow (*Monographia Uredinearum seu Specierum Omnium ad hunc usque Diem Descriptio et Adumbratio Systematica* **3**, 1915), Dietel (*Nat. Pflanzenfam.* **6**, 1928; syn. of *Haploravenelia*).

Dendrogaster Buchholz (1901) = Hymenogaster fide Smith (*Mycol.* **58**: 100, 1966), Fogel (*Mycol.* **77**: 72, 1985).

Dendrographa Darb. (1895), Roccellaceae (L). 3, N. America; C. America. See Tehler (*CJB* **68**: 2458, 1990; cladistics), Sundin & Tehler (*Bryologist* **99**: 19, 1996), Lohtander *et al.* (*Bryologist* **101**: 404, 1998; species pairs, DNA), Myllys *et al.* (*Lichenologist* **31**: 461, 1999; phylogeny), Lumbsch *et al.* (*Mol. Phylogen. Evol.* **34**: 512, 2005; phylogeny), Tehler & Irestedt (*Cladistics* **23**: 432, 2007).

Dendrographiella Agnihothr. (1972), anamorphic *Pezizomycotina*, Hsy.≡ eP.28. 1 (on latex of *Hevea*), India. See Agnihothrudu (*Proc. Indian Acad. Sci.* series B **75**: 156, 1972).

Dendrographium Massee (1892), anamorphic *Pezizomycotina*, Hsy.≡ eP.27. 2, S. America; India.

dendrohyphidium, see hyphidium.

dendroid, tree-like in form; dendritic.

Dendroleptosphaeria Sousa da Câmara (1932) ? = Leptosphaeria fide Hawksworth *et al.* (*Dictionary of the Fungi* edn 8, 1995).

Dendromyceliates K.P. Jain & R.K. Kar (1979), Fossil Fungi. 1 (Miocene), India. = Cryptophiale (Ascomycetes, inc. sed.) fide Sutton (*in litt.*).

Dendromyces Libosch. (1810) = Battarrea fide Stalpers (*in litt.*).

Dendrophagus Murrill (1905) [non *Dendrophagus* Toumey 1900, *Loranthaceae*] = Ganoderma fide Donk (*Verh. K. ned. Akad. Wet.* tweede sect. **62**: 1, 1974).

Dendrophoma Sacc. (1880) = Dinemasporium fide Sutton (*TBMS* **48**: 611, 1965).

Dendrophora (Parmasto) Chamuris (1987), Peniophoraceae. 3, widespread. See Chamuris (*Mycotaxon* **28**: 543, 1987).

Dendrophysellum Parmasto (1968), Corticiaceae. 1, former USSR. See Parmasto (*Consp. System. Corticiac.*: 146, 1968).

dendrophysis, see hyphidium.

Dendropleella Munk (1953) ? = Hendersonia Berk. fide Barr & Holm (*Taxon* **33**: 109, 1984).

Dendropolyporus (Pouzar) Jülich (1982) = Polyporus P. Micheli ex Adans. fide Ryvarden (*Syn. Fung.* **5**, 1991).

Dendrosarcos, see *Dendrosarcus*.

Dendrosarcus Paulet (1793), Fungi.

Dendroseptoria Alcalde (1948), anamorphic *Pezizomycotina*, Cpd.1bH.1. 1, Europe. See Diederich *et al.* (*Belg. Jl Bot.* **134**: 127, 2001).

Dendrosphaera Pat. (1907), Trichocomaceae. 1, Indo-China; East Indies. See Boedijn (*Bull. Jard. bot. Buitenz.* ser. 3 **13**: 472, 1935), Malloch *in* Samson & Pitt (Eds) (*Advances in Penicillium and Aspergillus systematics* **102**: 365, 1985; posn), Kuraishi *et al.* (*NATO ASI Series A: Life Sciences* **185**: 407, 1990; ubiquinones), Kuraishi *et al.* (*Antonie van Leeuwenhoek* **77**: 179, 2000; ubiquinones).

Dendrosphaeraceae Cif. ex Benny & Kimbr. (1980) = Trichocomaceae.

Dendrospora Ingold (1943), anamorphic *Pezizomycotina*, Hso.1bH.1. 9 (aquatic), British Isles. See Descals & Webster (*TBMS* **74**: 135, 1980), Roldán *et al.* (*Mycotaxon* **29**: 21, 1987; Spain), Sreekala & Bhat (*Frontiers in Microbial Biotechnology and Plant Pathology*: 295, 2002), Kendrick (*CJB* **81**: 75, 2003; morphogenesis), Descals (*MR* **109**: 545, 2005; diagnostics).

Dendrosporium Plakidas & Edgerton ex J.L. Crane (1972), anamorphic *Pezizomycotina*, Hsy.1eH.10. 1,

N. America; India. See Crane (*TBMS* **58**: 423, 1972), Kendrick (*CJB* **81**: 75, 2003; morphogenesis).

Dendrosporomyces Nawawi, J. Webster & R.A. Davey (1977), anamorphic *Agaricomycetes*. 1 (aquatic, with dolipore septa), Malaysia. See Nawawi *et al.* (*TBMS* **68**: 59, 1977).

Dendrostilbe Dearn. (1924) nom. dub., anamorphic *Pezizomycotina*. ? An error for *Dendrostilbella*.

Dendrostilbella Höhn. (1905), anamorphic *Claussenomyces*, Hsp.0eH.15. 7, widespread. See Seifert (*Stud. Mycol.* **27**: 1, 1985).

Dendrothele Höhn. & Litsch. (1907), Corticiaceae. 36, widespread. See Lemke (*CJB* **42**: 723, 1964; as *Aleurocorticium*), Legon *et al.* (*Mycologist* **16**: 114, 2002), Nakasone (*Nova Hedwigia* **83**: 99, 2006).

Dendrotrichoscypha Svrček (1977) = Mollisina fide Sharma (*Portug. Acta Biol.* **15**: 281, 1989; key), Huhtinen (*SA* **11**: 170, 1993).

Dendryphiaceae Corda (1840) ? = Pleosporaceae.

Dendryphiella Bubák & Ranoj. (1914) = Dendryphion fide Hughes (*CJB* **36**: 727, 1958), Ellis (*Dematiaceous Hyphomycetes*, 1971; 2 spp.), Michaelis *et al.* (*Mycol.* **79**: 514, 1987; genetics and systematics), Mohamad *et al.* (*MR* **93**: 400, 1989; separation of spp. by ELISA), Guo & Zhang (*Mycosystema* **18**: 236, 1999).

Dendryphion Wallr. (1833), anamorphic *Pleosporaceae*, Hso.≡ bP.26. 4, widespread. See Siboe *et al.* (*Mycotaxon* **73**: 283, 1999; Kenya), Farr (*Mycol.* **92**: 145, 2000; teleomorph), Farr *et al.* (*Mycol.* **92**: 145, 2000), Inderbitzin *et al.* (*CJB* **84**: 1304, 2006).

Dendryphiopsis S. Hughes (1953), anamorphic *Kirschsteiniothelia*, Hso.≡ eP.24. 5, widespread. See Hughes (*N.Z. Jl Bot.* **16**: 360, 1978), Subramanian & Sekar (*Kavaka* **15**: 87, 1987; teleomorph), Goh *et al.* (*Fungal Diversity* **1**: 85, 1998; ribosomal RNA analysis), Olivier & Loria (*FEMS Microbiol. Lett.* **168**: 235, 1998; PCR primers for detection), Schoch *et al.* (*Mycol.* **98**: 1041, 2006; phylogeny).

Dendryphiosphaera Lunghini & Rambelli (1978), anamorphic *Pezizomycotina*, Hso.≡ eP.1. 3, pantropical. See Lunghini & Rambelli (*G. bot. ital.* **112**: 185, 1978), Nawawi & Kuthubutheen (*Mycotaxon* **32**: 461, 1988), Castañeda Ruíz *et al.* (*Mycotaxon* **67**: 9, 1998).

denigrate, blackened.

Dennis (Richard William George; 1910-2003; England). Mycologist (1944-1951) then Head of Mycology (1951-1976), Royal Botanic Gardens, Kew. A great and prolific mycologist, widely travelled, with very wide experience of arctic, temperate and tropical fungi, but particularly an expert on discomycetes; also known for his interest in Celtic art and the Scottish islands. *Publs.* (with Wakefield q.v.) *Common British Fungi: a guide to the more common larger Basidiomycetes of the British Isles* (1950); Fungus flora of Venezuela and adjacent countries *Kew Bulletin* Additional Series (1970); *British Ascomycetes* (1981) [revised edn with supplement]; *Fungi of the Hebrides* (1986). *Biogs, obits etc.* Spooner & Roberts (*MR* **108**: 1097, 2004) [bibliography, portrait]; Spooner & Roberts (*Mycol.* **96**: 187, 2004).

Dennisiella Bat. & Cif. (1962), Coccodiniaceae. Anamorph *Microxyphium*-like. 7, widespread (esp. tropical). See Hughes (*Mycol.* **68**: 693, 1976), Reynolds & Gilbert (*Aust. Syst. Bot.* **18**: 265, 2005; Australia).

Dennisiodiscus Svrček (1976), Helotiales. 10, Europe; Papua New Guinea. See Baral & Krieglsteiner (*Beih. Sydowia* **6**, 1985), Nauta & Spooner (*Mycologist* **14**: 21, 2000; UK).

Dennisiomyces Singer (1955), Tricholomataceae. 5, S. America. See Singer (*An. Soc. Biol. Pernambuco* **13**: 225, 1955).

Dennisiopsis Subram. & Chandrash. (1977), Pezizales. 2, India. See van Brummelen (*Persoonia* **16**: 425, 1998; comparison with *Coprotus*).

Dennisographium Rifai (1977), anamorphic *Pezizomycotina*, Hsy.0eH.15. 1, Java. See Rifai (*Kew Bull.* **31**: 726, 1977), Seifert (*Mem. N. Y. bot. Gdn* **59**: 109, 1990).

dense body vesicle (**DBV**), cytoplasmic vesicle, associated with phosphorylated glucan metabolism, found in a variety of TEM morphological states ranging from a single (or several) electron-opaque core in an amorphous matrix to a highly structured, myelin-like arrangement of alternating electron-opaque and electron-translucent layers.

Densocarpa Gilkey (1954) = Stephensia fide Trappe (*TBMS* **65**: 496, 1975), Læssøe & Hansen (*MR* **111**: 1075, 2007; phylogeny).

Densospora McGee (1996), Zygomycota. 4, Australia. See McGee (*Aust. Syst. Bot.* **9**: 330, 1996), Gleason & McGee (*Australasian Mycologist* **22**: 73, 2004; ultrastr.).

dentate, toothed (Fig. 23.45). Cf. denticulate.

denticle, a small tooth-like projection esp. one on which a spore is borne.

Denticularia Deighton (1972), anamorphic *Pezizomycotina*, Hso.0eP.3. 5, widespread (esp. tropical). See de Hoog (*Persoonia* **10**: 51, 1978), Chamuris *et al.* (*Mycotaxon* **24**: 319, 1985), Alcorn *et al.* (*Australas. Pl. Path.* **28**: 115, 1999; Australia), Siboe *et al.* (*Mycotaxon* **73**: 283, 1999; Kenya).

denticulate, having small teeth. Cf. dentate.

Dentinaceae Kotl. & Pouzar (1972) = Hydnaceae.

Dentinum Gray (1821) ≡ Hydnum fide Hall & Stuntz (*Mycol.* **63**: 1113, 1971; key).

Dentipellis Donk (1962), Hericiaceae. 4, widespread. See Ginns (*Windahlia* **16**: 35, 1986; key).

Dentipratulum Domański (1965), Auriscalpiaceae. 1, Europe. See Domański (*Acta Mycologica* Warszawa **1**: 6, 1965).

Dentocircinomyces R.F. Castañeda & W.B. Kendr. (1990), anamorphic *Pezizomycotina*, Hso.0eH.10. 1 (on *Eucalyptus*), Cuba. See Castañeda & Kendrick (*Univ. Waterloo Biol. Ser.* **32**: 17, 1990).

Dentocorticium (Parmasto) M.J. Larsen & Gilb. (1974), Polyporaceae. 3, widespread. See Larsen & Gilbertson (*Norw. Jl Bot.* **24**: 99, 1977).

denuded, uncovered or glabrous by loss of scales, etc.

depauperate, poorly developed.

Depazea Fr. (1818) = Asteroma fide Sutton (*Mycol. Pap.* **141**, 1977).

Depazites Geinitz (1855), Fossil Fungi. 16 (Tertiary, Quaternary), Europe.

dependent, hanging down.

Dephilippia Rambelli (1959) = Circinotrichum fide Pirozynski (*Mycol. Pap.* **84**, 1962).

deplanate, flat.

depressed (1) (of a pileus), having the middle lower than the edge (Fig. 19N); (2) (of lamellae), sinuate (q.v.).

depside (**depsidone**), see Metabolic products.

Derexia Naras. (1970), anamorphic *Pezizomycotina*, Hso.0eH.1. 1 (on *Lecanium*), India; Java. See Narasimhan (*Indian Phytopath.* Suppl. Issue **23**: 20,

1970).

derived, of a character that has changed from the form in which it appeared in an ancester.

derm (**dermium**) (of basidiomata), an outer layer in which the hyphae are perpendicular to the surface (Lowag, 1941); cf. cortex;.

Dermapteromyces Thaxt. (1931), Laboulbeniaceae. 3, America (tropical). See Nannfeldt (*Svensk bot. Tidskr.* **43**: 468, 1949).

Dermascia Tehon (1935) ? = Lophodermium fide Darker (*CJB* **45**: 1399, 1967).

Dermatangium Velen. (1926) = Tremella Pers. fide Donk (*Persoonia* **4**: 179, 1966).

Dermatea Fr. (1849) ≡ Dermea.

Dermateaceae Fr. (1849), Helotiales. 33 gen. (+ 31 syn.), 315 spp. Much reduced compared with traditional treatments.

Lit.: Nannfeldt (*Nov. Acta Reg. Soc. Sc. Upsal.* ser. 4 **8** no. 2, 1932), Baral (*Z. Mykol.* **53**: 119, 1987), Gamundí (*Fl. criptog. Tierra del Fuego* **10**: 126 pp., 1986), Dyer *et al.* (*MR* **100**: 1219, 1996), Spiers & Hopcroft (*MR* **102**: 1025, 1998), Nauta & Spooner (*Mycologist* **13**: 3, 1999; keys Br. gen.), Nauta & Spooner (*Mycologist* **13**: 65, 1999; keys Br. gen.), Nauta & Spooner (*Mycologist* **13**: 146, 1999; keys Br. gen.), Spooner & Nauta (*Mycologist* **13**: 98, 1999; keys Br. gen.), Stewart *et al.* (*MR* **103**: 1491, 1999), Verkley (*Stud. Mycol.* **44**: 180 pp., 1999), Abeln (*Mycol.* **92**: 685, 2000; phylogeny of *Peziculoideae*), Abeln *et al.* (*Mycol.* **92**: 685, 2000), Nauta & Spooner (*Mycologist* **14**: 21, 2000; keys Br. gen.), Nauta & Spooner (*Mycologist* **14**: 65, 2000; keys Br. gen.), Dyer *et al.* (*Fungal Genetics Biol.* **33**: 173, 2001), Jong *et al.* (*MR* **105**: 658, 2001), Verkley *et al.* (*MR* **107**: 689, 2003), Cunnington (*Australas. Pl. Path.* **33**: 453, 2004), Pärtel & Raitviir (*Mycol. Progr.* **4**: 149, 2005), Wang *et al.* (*Mycol.* **98**: 1065, 2006; phylogeny), Wang *et al.* (*Mol. Phylogen. Evol.* **41**: 295, 2006).

Dermatella P. Karst. (1871) = Pezicula Tul. & C. Tul. fide Nannfeldt (*Nova Acta R. Soc. Scient. upsal.*, 1932), Verkley (*Stud. Mycol.* **44**, 1999).

Dermateopsis Nannf. (1932), Helotiales. 1, widespread (temperate). See Gamundí (*Fl. criptog. Tierra del Fuego* **10**, 1987).

Dermatina (Almq.) Zahlbr. (1922) = Arthonia. See *Mycoporum*. fide Harris (*Mich. Bot.* **12**: 3, 1973).

Dermatina (Sacc.) Höhn. (1909) = Pezicula Tul. & C. Tul. fide Nannfeldt (*Nova Acta R. Soc. Scient. upsal.*, 1932).

Dermatiscum Nyl. (1867), Caliciaceae (L). 3, S. Africa; N. America. See Brusse (*Mycotaxon* **25**: 161, 1986), Nordin (*Symb. bot. upsal.* **32** no. 1: 195, 1997; ultrastr.), Scheidegger *et al.* (*Lichenologist* **33**: 25, 2001; phylogeny), Helms *et al.* (*Mycol.* **95**: 1078, 2003; phylogeny), Lendemer (*Bryologist* **106**: 311, 2003).

dermatitis verrucosa, see chromomycosis.

Dermatocarpaceae Stizenb. (1862) = Verrucariaceae.

Dermatocarpella H. Harada (1993) = Placidium fide Breuss (*Annln naturh. Mus. Wien* Ser. B, Bot. Zool. **98**: 35, 1996).

Dermatocarpon Eschw. (1824), Verrucariaceae (L). 17 (esp. damp rocks), widespread. See Doppelbauer (*Nova Hedwigia* **2**: 279, 1960; structure), Awasthi & Upreti (*J. Econ. Taxon. Bot.* **7**: 7, 1985; key 4 spp. India), Janex-Favre & Wagner (*BSMF* **102**: 161, 1986; pycnidia), Harada (*Nat. Hist. Res.* **2**: 113, 1993; gen. concept), Breuss (*Öst. Z. Pilzk.* **4**: 137, 1995; section *Polyrhizion*), Heidmarsson (*Bryologist* **99**: 315, 1996; pruina), Heiðmarsson (*Ann. bot. fenn.* **35**: 59, 1998; Scandinavia), Orange (*Lichenologist* **30**: 1, 1998; UK), Breuss (*Biblthca Lichenol.* **86**: 99, 2003; Canada), Heiðmarsson (*MR* **107**: 459, 2003; phylogeny), Schmitt *et al.* (*Mycol.* **97**: 362, 2005; morphology, phylogeny), Amtoft (*Bryologist* **109**: 182, 2006; N America), Geiser *et al.* (*Mycol.* **98**: 1053, 2006; phylogeny), Gueidan *et al.* (*MR* **111**: 1145, 2007; phylogeny), Amtoft *et al.* (*Bryologist* **111**: 1, 2008; N America).

Dermatocarpon W. Mann (1825) = Endocarpon fide Hawksworth *et al.* (*Dictionary of the Fungi* edn 8, 1995).

dermatocyst (**dermatocystidium**), see cystidium (1).

Dermatodea Vent. (1799) = Nephroma fide Hawksworth *et al.* (*Dictionary of the Fungi* edn 8, 1995).

Dermatodothella Viégas (1944), Dothideomycetes. 1, Brazil.

Dermatodothis Racib. ex Theiss. & Syd. (1914), Dothideomycetes. 6, Asia; S. America. See Müller (*Sydowia* **28**: 148, 1976; key).

Dermatoidium Stautz [not traced] nom. dub., anamorphic *Pezizomycotina*.

Dermatomeris Reinsch (1890) = Mastodia fide Hawksworth *et al.* (*Dictionary of the Fungi* edn 8, 1995).

dermatomycosis, see mycosis.

Dermatophilus (Van Saceghem) M.A. Gordon (1964), Actinobacteria. q.v.

Dermatophyte. A fungus parasitizing keratinized tissue (hair, skin, nails) of humans and animals and causing **dermatophytosis** (pl. **-es**) (ringworm, tinea). These fungi, which are typically anamorphic fungi (hyphomycetes) with teleomorphs in the *Arthrodermataceae* (*Onygenales*), have frequently been treated as a special group the 'Dermatophytes' or Ringworm Fungi. Ringworm is cosmop. and dermatophytes, or non- or weakly pathogenic dermatophyte-like fungi, occur widely in soil and other keratin containing substrata such as bird's nests or horns of some mammals. There are 3 main gen. (*Epidermophyton*, *Microsporum*, *Trichophyton* distinguished by characteristic macroconidia), + *c.* 30 syn., *c.* 40 spp. (1,000 names).

Lit.: Sabouraud (*Les Teignes*, 1910), Bruhns & Alexander (*in* Jadassohn, *Handbuch der Haut- und Geschlechtskrankheiten* **2**, 1930), Dawson (*Rev. med. vet. Mycol.* **6**: 223, 1968; ringworm in animals), Emmons (*Arch. Derm. Syph., Chicago* **30**: 337, 1934), Georg (*Animal ringworm in public health*, 1959 [US Dep. Health, Educ., & Welfare]), Gotz (*Die Pilzkrankheiten der Haut durch Dermatophyten*, 1962 [= *Jadassohn Handb., Ergänzungwerk* **4** (3)]), Hironaga (*Jap. J. med. mycol.* **24**: 283, 1983, anamorph-teleomorph connexions), Kushwaha & Guarro (Eds) (*Biology of dermatophytes and other keratinophilic fungi*, 174 pp., 2000), Stockdale (*Biol. Rev.* **28**: 84, 1952; nutrition). See also Lewis *et al.* (1958), Conant *et al.* (1971) under Medical and veterinary mycology; griseofulvin; Wood's light.

dermatophytid, a pustular allergic eruption (**id-reaction**) of the skin at a distance from a primary infection by a dermatophyte.

Dermatosoraceae Vánky (2001) = Anthracoideaceae.

Dermatosorus Sawada ex L. Ling (1949), Anthracoideaceae. 6 (on *Cyperaceae*), Africa; Asia; Australia; S. America. See Langdon (*TBMS* **68**: 447, 1977), Vánky (*TBMS* **89**: 61, 1987), Vánky (*Mycotaxon* **54**:

215, 1995; key), Vánky & Shivas (*Fungal Diversity* **14**: 243, 2003; Australia).

Dermea Fr. (1825), Dermateaceae. Anamorphs *Foveostroma, Micropera, Gelatinosporium, Corniculariella*. 22, widespread (esp. temperate). See Groves (*Mycol.* **38**: 351, 1946), Funk (*CJB* **54**: 2852, 1976), Verkley (*Stud. Mycol.* **44**: 125, 1999), Abeln (*Mycol.* **92**: 685, 2000; phylogeny), Wang *et al.* (*Mol. Phylogen. Evol.* **41**: 295, 2006; phylogeny), Wang *et al.* (*Mycol.* **98**: 1065, 2006; phylogeny).

Derminus (Fr.) Staude (1857) nom. rej. ≡ Pholiota fide Donk (*Bull. Jard. Bot. Buitenzorg* Sér. 3 **18**: 271, 1949).

dermis (of lichens), the limiting layer of a thallus (obsol.).

Dermiscellum Hafellner, H. Mayrhofer & Poelt (1979), Physciaceae (L). 1, N. America. See Hafellner *et al.* (*Herzogia* **5**: 55, 1979), Nordin & Mattsson (*Lichenologist* **33**: 3, 2001; morphology, phylogeny), Scheidegger *et al.* (*Lichenologist* **33**: 25, 2001; evolution), Helms *et al.* (*Mycol.* **95**: 1078, 2003; phylogeny), Lendemer (*Bryologist* **106**: 311, 2003; nomencl.).

Dermocybe (Fr.) Wünsche (1877) = Cortinarius fide Kuyper (*in litt.*; a monophyletic taxon, recognition of which as a genus makes *Cortinarius* non-monophyletic).

Dermoloma J.E. Lange ex Herink (1959), Tricholomataceae. *c.* 15, widespread. See Svrček (*Česká Mykol.* **20**: 256, 1966; key), Arnolds (*Persoonia* **14**: 519, 1992), Arnolds (*Persoonia* **17**: 665, 2002; *Dermoloma magicum*).

Dermolomataceae Bon (1979) nom. rej. = Tricholomataceae.

Dermomycoides Granata (1919), ? Chytridiales. 2, Europe.

Dermosporium Link (1815) = Aegerita fide Saccardo (*Syll. fung.* **4**: 644, 1886).

Descalsia A. Roldán & Honrubia (1989), anamorphic *Pezizomycotina*, Hso.1bH.10. 1 (aquatic), Spain. See Roldán & Honrubia (*MR* **92**: 494, 1989).

Descematia Nieuwl. (1916) ≡ Sphaerocephalum.

descending (descendant) (of an annulus), having the free edge below the attached (cf. ascending).

Descolea Singer (1950), Cortinariaceae. *c.* 15, widespread. See Horak (*Persoonia* **6**: 231, 1971; key), Lago *et al.* (*Mycotaxon* **78**: 37, 2001; *Descolea-Setchelliogaster-Descomyces* complex), Peintner *et al.* (*Am. J. Bot.* **88**: 2168, 2001; phylogeny).

Descomyces Bougher & Castellano (1993), Cortinariaceae. 5, Australasia. Now widespread having spread with *Eucalyptus*. *Descomyces* is not yet conserved against *Hymenangium*. See Francis & Bougher (*Australasian Mycologist* **21**: 81, 2002).

Describing, naming and publishing, see Hawksworth (*Mycologist's handbook*: 48, 1974), Classification, Nomenclature, Species, Systematics.

Desertella Mouch. (1979), anamorphic *Pezizomycotina*, Hso.0eH.10. 1 (from soil), Egypt. See Mouchaca (*Revue Mycol.* Paris **43**: 71, 1979).

Desetangsia Nieuwl. (1916) ≡ Sphaerotheca Lév.

Deshpandiella Kamat & Ullasa (1973), Phyllachoraceae. Anamorph *Mycohypallage*. 1 (biotrophic on leaves), India. See Hawksworth (*SA* **5**: 128, 1986).

Deslandesia Bat. (1962) nom. inval. = Limacinula Höhn. fide Reynolds (*Bull. Torrey bot. Club* **98**: 157, 1971).

Desmaturus (Schltdl.) Kalchbr. (1880) = Lysurus fide Dring (*Kew Bull.* **35**: 1, 1980).

Desmazierella Crié (1878) nom. dub., anamorphic *Pezizomycotina*.

Desmazierella Lib. (1829), Chorioactidaceae. Anamorph *Verticicladium*. 2 (on conifer needles), Europe. See Benkert (*Gleditschia* **19**: 173, 1991), Landvik *et al.* (*Nordic Jl Bot.* **17**: 403, 1997; phylogeny), Meléndez-Howell *et al.* (*Mycotaxon* **68**: 53, 1998; ultrastr.), Harrington *et al.* (*Mycol.* **91**: 41, 1999; phylogeny), Hansen & Pfister (*Mycol.* **98**: 1029, 2006; phylogeny), Pfister *et al.* (*MR* **112**: 513, 2008).

Desmazières (Jean Baptiste Henri Joseph; 1787-1862; France). A merchant of Lille, the 'possesseur d'une belle fortune', and amateur mycologist. *Publs.* The exsiccatum series *Plantes Cryptogames du Nord de la France* (1825-1860), and his papers relating to these collections in *Annales de Science Naturelles* Paris and *Mémoires de la Société Scientifique de Lille*. *Biogs, obits etc.* Anon. (*Bulletin de la Société Royale de Botanique de Belgique* **1**: 102, 1862); Grummann (1974: 274); Stafleu & Cowan (*TL-2* **1**: 630, 1976).

Desmazieria Mont. (1852) [non *Desmazieria* Dumort. 1822, *Poaceae*] ≡ Niebla.

Desmella Syd. & P. Syd. (1919), Pucciniales. 3 (on *Pteridophyta*), America (tropical); Hawaii, Australia. See Cummins (*Annls mycol.* **38**: 335, 1940), Thirumalachar & Cummins (*Mycol.* **40**: 417, 1948), Hennen & Ono (*Mycol.* **70**: 569, 1978), Cummins & Hiratsuka (*Illustr. Gen. Rust Fungi* edn 3: 225 pp., 2003; placed in *Uropyxidaceae*).

Desmellopsis J.M. Yen (1969), Pucciniales. 1 (on *Aframomum* (*Zingiberaceae*)), Gabon. See Yen (*Rev. mycol.* **34**: 20, 1969), Cummins & Hiratsuka (*Illustr. Gen. Rust Fungi rev. edit.*, 1983; syn. of *Puccinia*).

Desmidiospora Thaxt. (1891), anamorphic *Pezizomycotina*, Hso.0bP.1. 2, N. America. See Clark & Prusso (*Mycol.* **78**: 865, 1986; on ants), Evans (*Trichomycetes and Other Fungal Groups* Robert W. Lichwardt Commemoration Volume: 119, 2001; review).

Desmopatella Höhn. (1924) = Phacidiella P. Karst. fide Sutton (*Mycol. Pap.* **141**, 1977).

Desmosorus Ritschel, Oberw. & Berndt (2005), Pucciniales. 1 (on *Cattleya* (*Orchidaceae*)), America (tropical); elsewhere on cultivated orchids. See Ritschel *et al.* (*Mycol. Progr.* **4**: 333, 2005).

Desmotascus F. Stevens (1919) = Botryosphaeria fide von Arx & Müller (*Beitr. Kryptfl. Schweiz* **11** no. 1, 1954).

Desmotelium Syd. (1937) = Chaconia fide Laundon (*Mycotaxon* **3**: 132, 1975).

Desmotrichum Lév. (1843) = Gonatobotrys fide Saccardo (*Syll. fung.* **4**: 169, 1886).

Desportesia I.V. Issi & Voronin (1986), Microsporidia. 1.

destroying angel, the pure white agaric, *Amanita virosa*, ingestion of 1 mg of which can prove fatal; toxins are cyclic polypeptides, esp. amanitins.

Destuntzia Fogel & Trappe (1985), Gomphaceae. 5, N. America. See Fogel & Trappe (*Mycol.* **77**: 732, 1985; key), Albee-Scott (*MR* **111**: 1030, 2007; phylogeny).

determinate (1) clearly marked; definite; (2) (of conidiophores), growth ceasing with the production of terminal conidia.

detersile (of villosity), removable so that the surface becomes bare.

Detonia Sacc. (1889) = Plicaria fide Korf (*Mycol.* **52**: 648, 1960), Moravec (*Mycotaxon* **30**: 473, 1987).
Detonina Kuntze (1891) ≡ Apiospora.
detoxification, the conversion of a toxin (e.g. an inhibitory phytoalexin) to non-toxic (non-inhibitory) products.
deuteroconidium (of dermatophytes), a spore-like cell, the outcome of the division of a hemispore (protoconidium).
deuterogamy, the condition in which other processes replace fusion of gametes, as in some fungi, the macroalgae, and phanerogams; secondary pairing.
Deuterolichenes (obsol.). Introduced by Mameli-Calvino (*Nuovo G. bot. ital.* n.s. **27**: 379, 1930) for *Pyrenotrichum* (q.v.). Also later used for sterile leprose and filamentous lichens as well as lichenized anamorphic fungi.
Deuteromycotina, see Anamorphic fungi, Kendrick (*Sydowia* **41**: 6, 1989; abandonment of term).
Deuterophoma Petri (1929) = Phoma Sacc. fide Kanciaveli & Ghikascvili (*Lav. Ist. prot. piante Georgia* **5**: 1, 1948), Sutton (*Mycol. Pap.* **141**, 1977).
devalidated (of names), names which would have been validly published under the *Code* except for the operation of Art. 13 (see Nomenclature). Such names may be 'revalidated' or 'taken up' by post-starting point authors. Formerly widely used in mycology prior to changes made in this Art. in 1983.
devil's cigar, *Urnula geaster* (see Seaver, *Mycol.* **29**: 60, 1937); **- snuffboxes**, puffballs.
Devriesia Seifert & N.L. Nick. (2004), anamorphic *Teratosphaeriaceae*. 5, widespread (temperate). See Seifert *et al.* (*CJB* **82**: 919, 2004), Crous *et al.* (*Stud. Mycol.* **58**: 1, 2007).
Dexhowardia J.J. Taylor (1970), anamorphic *Pezizomycotina*, Hso.0eH.6. 1, USA. See Taylor (*Mycopathologia* **40**: 306, 1970).
Dexteria F. Stevens (1917) = Hyalosphaera fide Rossman (*Mycol. Pap.* **157**, 1987).
Dextrinocystidium Sheng H. Wu (1996), ? Stereaceae. 2, New Zealand; Ivory Coast. See Wu (*Mycol.* **87**: 888, 1995).
Dextrinocystis Gilb. & M. Blackw. (1988), ? Hydnodontaceae. 1, USA. See Gilbertson & Blackwell (*Mycotaxon* **33**: 376, 1988).
Dextrinodontia Hjortstam & Ryvarden (1980), ? Hydnodontaceae. 1, Tanzania. See Larsson (*in litt.*), Hjortstam & Ryvarden (*Mycotaxon* **12**: 172, 1980) = *Trechispora* (Sistotremat.) fide.
dextrinoid (of spores, etc.), stained yellowish-brown or reddish-brown by Melzer's iodine (see Iodine); pseudoamyloid (Singer); cf. amyloid.
Dextrinosporium Bondartsev (1972) nom. dub. ? = Perenniporia fide Ryvarden (*Syn. Fung.* **5**, 1991).
dhose, a Malaysian fermented food produced my means of yeasts and peas.
Diabole Arthur (1922), Raveneliaceae. 1 (on *Mimosa* (0, III) (*Leguminosae*)), C. America; West Indies; Mexico; Brazil.
Diabolidium Berndt (1995), Raveneliaceae. 1 (on *Calliandra* (*Leguminosae*)), S. America. See Berndt (*Mycotaxon* **54**: 263, 1995), Cummins & Hiratsuka (*Illustr. Gen. Rust Fungi* edn 3: 225 pp., 2003; syn. of *Allotelium*).
Diaboliumbilicus I. Hino & Katum. (1955), Pezizomycotina. 1, Japan. See Hino & Katumoto (*Bulletin of the Faculty of Agriculture, Yamaguchi University* **6**: 40, 1955).
Diacanthodes Singer (1945), Meruliaceae. Anamorph *Bornetina*. 1, widespread. (+ mealy bugs) causes phthiriosis (root disease) of coffee. See Tomaszewski *et al.* (*Int. J. Syst. Evol. Microbiol.* **53**: 1204, 2003).
Diachanthodaceae Jülich (1982) = Meripilaceae.
Diachora Müll. Arg. (1893), Phyllachoraceae. Anamorph *Diachorella*. 5 (on *Leguminosae*), widespread (temperate). See Müller (*Trans. Bot. Soc. Edinb., Suppl.*: 69, 1986; key), Cannon (*Mycol. Pap.* **163**, 1991).
Diachorella Höhn. (1918), anamorphic *Diachora*, St.0fH.15. 5 (on *Leguminosae*), widespread. See Sutton (*The Coelomycetes*, 1980), Wang & Yuan (*Cryptog. bot.* **5**: 360, 1995; China).
Diacrochordon Petr. (1955), Pezizomycotina. 1, Germany. See Petrak (*Sydowia* **9**: 591, 1955).
Diadema Shoemaker & C.E. Babc. (1989), Diademaceae. 8 (esp. on *Poaceae*), widespread. See Shoemaker & Babcock (*CJB* **67**: 1349, 1989).
Diademaceae Shoemaker & C.E. Babc. (1992), Pleosporales. 3 gen. (+ 1 syn.), 30 spp.
Lit.: Shoemaker & Babcock (*CJB* **67**: 1349, 1989), Shoemaker & Babcock (*CJB* **70**: 1617, 1992), Barr (*SA* **12**: 27, 1993), Ahn & Shearer (*CJB* **76**: 258, 1998), Dong *et al.* (*MR* **102**: 151, 1998), Barreto & Torres (*Australas. Pl. Path.* **28**: 103, 1999), Simmons (*Mycotaxon* **75**: 1, 2000), Pryor & Bigelow (*Mycol.* **95**: 1141, 2003).
Diademosa Shoemaker & C.E. Babc. (1992), Diademaceae. 1, USA. See Shoemaker & Babcock (*CJB* **70**: 1617, 1992).
Diademospora B.E. Söderstr. & Bååth (1979), anamorphic *Pezizomycotina*, Hso.≡ eH.1. 1 (from soil), Sweden. See Söderström & Bååth (*TBMS* **72**: 340, 1979).
diageotropism, tendency to horizontal growth in relation to the earth surface.
diagnosis (1) an account, esp. the first (see Nomenclature), of the distinguishing characteristics of a taxonomic group; (2) determining a fungus or disease; **diagnostic**, characteristic; of use for identification.
Dialaceniopsis Bat. (1959), anamorphic *Pezizomycotina*, Cpd.1eH.15. 1, Uganda. See Batista (*An. Soc. Biol. Pernambuco* **16**: 141, 1959).
Dialacenium Syd. (1930) = Rhytidenglerula fide Müller & von Arx (*Beitr. Kryptfl. Schweiz* **11** no. 2, 1962).
Dialhypocrea Speg. (1919), Hypocreaceae. 1 (from dead branches), Brazil. See Rossman *et al.* (*Stud. Mycol.* **42**: 248 pp., 1999).
Dialonectria (Sacc.) Cooke (1884) = Cosmospora fide Rossman *et al.* (*Stud. Mycol.* **42**: 248 pp., 1999).
Dialytes Nitschke (1867) nom. nud. = Diaporthe fide Wehmeyer (*University of Michigan Studies, Science Series* **9**: 1, 1933).
Diamantinia A.N. Mill., Laessøe & Huhndorf (2003), ? Xylariaceae. 1, Brazil. See Miller *et al.* (*Sydowia* **55**: 94, 2003), Huhndorf *et al.* (*Mycol.* **96**: 368, 2004; phylogeny).
Diamphora Mart. (1821) nom. dub., ? Fungi.
Diandromyces Thaxt. (1918), Laboulbeniaceae. 1, Chile. See Nannfeldt (*Svensk bot. Tidskr.* **43**: 468, 1949).
Dianesea Inácio & P.F. Cannon (2002), Coccoideaceae. 1, Costa Rica. See Inácio & Cannon (*Fungal Diversity* **9**: 72, 2002).
Diaphanium Fr. (1836), anamorphic *Pezizomycotina*, Hsp.0eH.?. 2, Europe.

Diaphanopellis P.E. Crane (2005), Coleosporiaceae. 1 (on *Rhododendron* (*Ericaceae*)), Himalaya. See Crane (*Mycol.* **97**: 539, 2005).

diaphanous, transparent or nearly so.

Diaphoromyces Thaxt. (1926), Laboulbeniaceae. 3, widespread. See Weir & Rossi (*CJB* **75**: 791, 1997; NZ), Rossi & Weir (*Mycol.* **90**: 282, 1998).

Diapleella Munk (1953) = Kalmusia fide Eriksson (*SA* **9**: 8, 1990).

Diaporthaceae Höhn. ex Wehm. (1926), Diaporthales. 5 gen. (+ 30 syn.), 335 spp.

Lit.: Uecker (*Mem. N. Y. bot. Gdn* **49**: 38, 1989), Zhang *et al.* (*Phytopathology* **89**: 796, 1999), Mostert *et al.* (*Mycol.* **93**: 146, 2001), Castlebury *et al.* (*Mycol.* **94**: 1017, 2002), Says-Lesage *et al.* (*Phytopathology* **92**: 308, 2002), Castlebury *et al.* (*Mycoscience* **44**: 203, 2003), Myburg *et al.* (*Mycol.* **96**: 990, 2004), Rekab *et al.* (*MR* **108**: 393, 2004), Rossman *et al.* (*Mycoscience* **48**: 135, 2007; phylogeny).

Diaporthales Nannf. (1932). Sordariomycetidae. 10 fam., 144 gen., 1196 spp. Ascomata perithecial, usually aggregated into a pseudostroma, usually long-necked. Interascal tissue absent or of thin-walled unspecialized cells, deliquescing early. Asci usually thick-walled but not fissitunicate, with a conspicuous J- refractive apical ring, detached at maturity. Ascospores varied. Anamorphs varied, coelomycetous. Saprobes and plant parasites, mainly on bark and wood, cosmop. Fams:

(1) **Cryphonectriaceae**
(2) **Diaporthaceae**
(3) **Gnomoniaceae**
(4) **Melanconidaceae**
(5) **Melogrammataceae**
(6) **Pseudovalsaceae**
(7) **Schizoparmaceae**
(8) **Sydowiellaceae**
(9) **Togniniaceae**
(10) **Valsaceae**

Lit.: Barr (*Mycol. Mem.* **7**, 1978), Castlebury *et al.* (*Mycol.* **94**: 1017, 2002), Kobayashi (*Bull. Govt For. Exp. Stn Meguro* **226**, 1972; keys), Malloch (*in* Kendrick (Ed.), *The whole fungus* **1**: 153, 1979; *in* Reynolds (Ed.), *Ascomycete systematics*, 1981; wide concept), Merezhko & Smyk (*Flora Gribov Ukrainy, Diaportal'nye Griby*, 1991; Ukraine, keys), Müller & von Arx (1962, 1973; keys gen.), Rossman *et al* (*Mysoscience* **48**: 135, 2007), Smyk *et al.* (*Ukr. bot. Zh.* **46**: 46, 1989; SEM), Wehmeyer (*The genus Diaporthe*, 1933; *Univ. Michigan Stud.* sci. ser. **14**, 1942), Zhang & Blackwell (*Mycol.* **93**: 355, 2001; molecular phylogeny).

Diaporthe Nitschke (1870), Diaporthaceae. Anamorph *Phomopsis*. 92, widespread (esp. north temperate). Some are parasites, at least in the *Phomopsis* state, e.g. *D. phaseolorum* var. *batatatis* (storage rot of sweet potato, *Ipomoea*), *D. citri* (*Phomopsis* stem-end rot of citrus fruits), *D. perniciosa* (a connexion with die-back of fruit trees), *D. phaseolorum* (pod blight of lima bean, *Phaseolus lunatus*), *D. woodii* (lupinosis of sheep). See Wehmeyer (*The genus Diaporthe*, 1933; revision), Uecker (*Mem. N. Y. bot. Gdn* **49**: 38, 1989; ontogeny), Williamson *et al.* (*MR* **98**: 1364, 1994; toxins), Linders & Van de Aa (*MR* **99**: 1409, 1995; mating types), Zhang *et al.* (*Pl. Dis.* **81**: 1143, 1997; PCR detection), Phillips (*Mycol.* **91**: 1001, 1999; on *Vitis*), Scheper (*MR* **104**: 226, 2000; on *Vitis*), Mostert *et al.* (*Mycol.* **93**: 146, 2001; on *Vitis*), Castlebury *et al.* (*Mycol.* **94**: 1017, 2002; phylogeny), Moleleki *et al.* (*Eur. J. Pl. Path.* **108**: 909, 2002; S Africa), Says-Lesage *et al.* (*Phytopathology* **92**: 308, 2002; on *Helianthus*), Castlebury *et al.* (*Mycoscience* **44**: 203, 2003; phylogeny, generic limits), Pecchia *et al.* (*Mycopathologia* **157**: 317, 2004; on *Helianthus*), Rekab *et al.* (*MR* **108**: 393, 2004; phylogeny, on *Helianthus*), Vergara *et al.* (*Mycopathologia* **158**: 123, 2004; on *Helianthus*), Niekerk *et al.* (*Australas. Pl. Path.* **34**: 27, 2005; on *Vitis*), Janse van Rensburg *et al.* (*Stud. Mycol.* **55**: 65, 2006; on *Aspalathus*), Zhang *et al.* (*Mycol.* **98**: 1076, 2006; phylogeny).

Diaporthella Petr. (1924), ? Valsaceae. 3, Europe; N. America. See Barr (*Mycol. Mem.* **7**, 1978).

diaporthin, a wilt toxin from *Endothia parasitica* (Boller *et al.*, *Helv. chim. Acta* **40**: 875, 1957); antibacterial.

Diaporthopsis Fabre (1883) = Diaporthe fide Castlebury *et al.* (*Mycoscience* **44**: 203, 2003).

Diarimella B. Sutton (1980), anamorphic *Pezizomycotina*, St.0eH.15. 1, India. See Sutton (*The Coelomycetes*: 452, 1980), Vujanovic *et al.* (*CJB* **76**: 2037, 1998).

Diarthonis Clem. (1909) = Arthonia fide Hawksworth *et al.* (*Dictionary of the Fungi* edn 8, 1995).

diaspore (1) any unit of dissemination, e.g. a spore, fragment of mycelium, sclerotium (Sernander, 1927); (2) (of lichens), particularly applied to vegetative propagules. See hormocyst, isidium, soredium, etc. (Fig. 22).

Diathrypton Syd. (1922) = Schiffnerula fide von Arx & Müller (*Stud. Mycol.* **9**, 1975).

Diatractium Syd. & P. Syd. (1921), Diaporthales. 2 (biotrophic on leaves), C. & S. America. See Cannon (*SA* **7**: 23, 1988; possible rel. with *Phyllachoraceae*), Cannon (*MR* **92**: 327, 1989).

Diatrypaceae Nitschke (1869), Xylariales. 13 gen. (+ 20 syn.), 229 spp.

Lit.: Schrantz (*BSMF* **76**: 305, 1961; 27 gen., keys), Glawe & Rogers (*Mycotaxon* **20**: 401, 1984; Pacific NW, keys), Vasil'eva (*Mikol. Fitopatol.* **19**: 3, 1985; gen. relationships), Chlebicki (*TBMS* **86**: 441, 1986), Glawe & Rogers (*CJB* **64**: 1493, 1986), Rappaz (*Mycol. Helv.* **2**: 285, 1987; monogr., keys), Chacón & Medel (*Revta Mex. Micol.* **4**: 323, 1988), Eriksson & Hawksworth (*SA* **7**: 67, 1988; status), Dargan & Bhatia (*Nova Hedwigia* **48**: 405, 1989), Glawe (*Sydowia* **41**: 122, 1989; anamorphs), Ju *et al.* (*Mycotaxon* **41**: 311, 1991), Carmarán & Romero (*Boln Soc. argent. Bot.* **28**: 139, 1992; gen. concepts), Rogers *in* Hawksworth (Ed.) (*Ascomycete Systematics. Problems and Perspectives in the Nineties* NATO ASI Series vol. **269 269**: 321, 1994), DeScenzo *et al.* (*Phytopathology* **89**: 884, 1999), Peros *et al.* (*MR* **103**: 1385, 1999), Acero *et al.* (*Mycol.* **96**: 249, 2004), Trouillas & Gubler (*MR* **108**: 1195, 2004), Lardner *et al.* (*MR* **109**: 799, 2005), Carmarán *et al.* (*Fungal Diversity* **23**: 67, 2006).

Diatrypales Chadef. ex D. Hawksw. & O.E. Erikss. (1986) = Xylariales.

Diatrype Fr. (1849), Diatrypaceae. Anamorphs *Libertella*, *Cytosporina*. 59, widespread. See Janex-Favre (*Revue Mycol.* Paris **42**: 265, 1978; spore formation), Glawe & Rogers (*Mycotaxon* **20**: 401, 1984), Patil & Patil (*Indian J. Mycol. Pl. Path.* **13**: 134, 1985; 22 spp. India), Rappaz (*Mycol. Helv.* **2**:

285, 1987; key), Rappaz (*Mycotaxon* **30**: 209, 1987; *D. stigma* group), Romero & Minter (*TBMS* **90**: 457, 1988; asci), Dargan & Bhatia (*Nova Hedwigia* **48**: 405, 1989; key 11 spp. Himalayas), Chacón (*Docums Mycol.* **32** nos 127-128: 95, 2003; Mexico), Acero *et al.* (*Mycol.* **96**: 249, 2004; phylogeny), Vasilyeva & Stephenson (*Fungal Diversity* **17**: 191, 2004; USA), Chlebicki (*Czech Mycol.* **57**: 117, 2005; Czech Republic), Carmarán *et al.* (*Fungal Diversity* **23**: 67, 2006), Zhang *et al.* (*Mycol.* **98**: 1076, 2006; phylogeny).

Diatrypella (Ces. & De Not.) De Not. (1863), Diatrypaceae. Anamorph *Libertella*. 33, widespread. See Croxall (*TBMS* **33**: 45, 1950), Glawe & Rogers (*Mycotaxon* **20**: 401, 1984), Glawe (*Mem. N. Y. bot. Gdn* **49**: 51, 1989; anamorph), Prášil (*Česká Mykol.* **45**: 37, 1991; stroma), Chacón (*Docums Mycol.* **32** nos 127-128: 95, 2003; Mexico), Acero *et al.* (*Mycol.* **96**: 249, 2004; phylogeny, polyphyly), Vasilyeva & Stephenson (*Fungal Diversity* **19**: 189, 2005; USA).

Diatrypeopsis Speg. (1884) = Camillea fide Læssøe *et al.* (*MR* **93**: 121, 1989), Læssøe (*SA* **13**: 43, 1994; ? synonym of *Xylaria*).

Diatrypoidiella Manoharachary, Kunwar & Agarwal (2005) ? = Jattaea fide Manoharachary *et al.* (*Indian Phytopath.* **58**: 205, 2005).

Dibaeis Clem. (1909), Icmadophilaceae (L). 13, widespread (tropical). See Gierl & Kalb (*Herzogia* **9**: 593, 1993), Platt & Spatafora (*Lichenologist* **31**: 409, 1999; phylogeny), Stenroos *et al.* (*Mycol. Progr.* **1**: 267, 2002; phylogeny), Miądlikowska *et al.* (*Mycol.* **98**: 1088, 2006; phylogeny), Wang *et al.* (*Mol. Phylogen. Evol.* **41**: 295, 2006; phylogeny).

Dibeloniella Nannf. (1932), Dermateaceae. Anamorph *Phialophora*-like. 1, Europe. See Nauta & Spooner (*Mycologist* **14**: 21, 2000; UK).

Dibelonis Clem. (1909) = Leptotrochila fide Nannfeldt (*Nova Acta R. Soc. Scient. upsal.*, 1932).

Dibelonis Clem. & Shear (1931) = Leptotrochila fide Nannfeldt (*Nova Acta R. Soc. Scient. upsal.*, 1932).

Diblastia Trevis. (1857) = Xanthoria fide Hawksworth *et al.* (*Dictionary of the Fungi* edn 8, 1995).

Diblastospermella Speg. (1918) = Cicinnobella fide Petrak (*Sydowia* **5**: 328, 1951), Sutton (*Mycol. Pap.* **141**, 1977).

Diblepharis Lagerh. (1900) = Monoblepharis fide Fitzpatrick (*The lower fungi. Phycomycetes*, 1930).

Dibotryon Theiss. & Syd. (1915), Venturiaceae. Anamorph *Fusicladium*. 1, N. America. *D. morbosum* (syn. *Apiosporina morbosa*; black knot of *Prunus*). See Barr (*Prodr. Cl. Loculoasc.*, 1987), Barr (*Sydowia* **41**: 25, 1989), Fernando *et al.* (*Can. J. Pl. Path.* **27**: 364, 2005; diagnosis), Zhang *et al.* (*Phytopathology* **95**: 859, 2005; population biology), Zhang *et al.* (*Pl. Dis.* **89**: 815, 2005; detection), Winton *et al.* (*Mycol.* **99**: 240, 2007; phylogeny).

Dicaeoma Gray (1821) = Puccinia fide Dietel (*Nat. Pflanzenfam.* **6**, 1928).

Dicantharellaceae Jülich (1982) = Lachnocladiaceae.

Dicarpella Syd. & P. Syd. (1921), ? Melanconidaceae. Anamorph *Tubakia*. 4, N. America. See Reid & Dowsett (*CJB* **68**: 398, 1990), Barr (*Mycotaxon* **41**: 287, 1991), Belisario (*Mycotaxon* **41**: 147, 1991).

Dicarphus Raf. (1808) nom. nud. ? = Hydnum fide Donk (*Taxon* **5**: 75, 1956).

Dicaryomycota Raf. (1808) nom. nud. ? = Hydnum fide Donk (*Taxon* **5**: 75, 1956).

dicaryon, see dikaryon.

Dicellaeporisporites Kalgutkar (1997), Fossil Fungi. 2, Canada; Australia. See Kalgutkar (*Review of Palaeobotany and Palynology* **97**: 210, 1997).

Dicellaesporites Elsik (1968), Fossil Fungi. 22 (Tertiary), widespread.

Dicellispora Sawada (1944), anamorphic *Pezizomycotina*, Hso.0eP.?3. 1, Taiwan.

Dicellomyces L.S. Olive (1945), Brachybasidiaceae. 3 (on *Poaceae*, *Scirpus* and *Calamus*), Europe; India; N. America. See Parmasto (*Consp. System. Corticiac.*, 1968), McNabb & Talbot *in* Ainsworth *et al.* (Eds) (*The Fungi* **4A**: 317, 1973), Reid (*TBMS* **66**: 537, 1976), Berndt & Sharma (*MR* **102**: 1484, 1998; key).

Dicellomycetaceae Parmasto (1968) = Brachybasidiaceae.

Dicephalospora Spooner (1987), Rutstroemiaceae. 4, widespread. See Spooner (*Biblthca Mycol.* **116**, 1987), Zhuang (*Fungal Diversity* **3**: 187, 1999; spp. trop. China), Verkley (*Sydowia* **56**: 343, 2004).

Dichaena Fr. (1849) ≡ Polymorphum fide Hawksworth & Punithalingam (*TBMS* **60**: 501, 1973).

Dichaenaceae Fr. (1849) = Ascodichaenaceae.

Dichaenopsella Petr. (1952) ? = Polymorphum fide Sutton (*Mycol. Pap.* **141**, 1977).

Dichaenopsis Paoli (1905) = Stagonospora fide Butin (*Phytopath. Z.* **100**: 186, 1981).

Dichaetis Clem. (1931) ≡ Wentiomyces.

Dichantharellus Corner (1966), Lachnocladiaceae. 2, Malaysia. See Corner (*Monogr. Cantharelloid Fungi*: 99, 1966).

Dicheirinia Arthur (1907), Raveneliaceae. 12 (on *Leguminosae*), America (tropical); Mauritius; New Caledonia; Canary Islands; Madeira. See Cummins (*Mycol.* **27**: 151, 1935), Cummins (*Bull. Torrey bot. Club* **64**: 39, 1937), Walker *et al.* (*Australas. Pl. Pathol.* **35**: 1, 2006).

Dichelostroma Bat. & Peres (1963), anamorphic *Pezizomycotina*, Cpt.0eH.?. 1, USA. See Batista & Peres (*Publções Inst. Micol. Recife* **395**: 5, 1963).

Dichitonium Berk. & M.A. Curtis (1875) ? = Dendrodochium. See also *Oichitonium*. fide Hawksworth *et al.* (*Dictionary of the Fungi* edn 8, 1995).

Dichlaena Durieu & Mont. (1849), Trichocomaceae. Anamorph *Aspergillus*. 1, N. Africa.

Dichlamys Syd. & P. Syd. (1920) = Uromyces fide Thirumalachar (*Sydowia* **5**: 23, 1951).

Dichobotrys Hennebert (1973), anamorphic *Trichophaea*, Hso.0eH.7. 4, widespread. See Hennebert (*Persoonia* **7**: 193, 1973), Kendrick (*CJB* **81**: 75, 2003; morphogenesis).

Dichochaete Parmasto (2001), Hymenochaetaceae. 2, widespread. See Parmasto (*Folia cryptog. Estonica* **37**: 56, 2000).

Dichocladosporium K. Schub., U. Braun & Crous (2007) = Graphiopsis Trail fide Schubert *et al.* (*Stud. Mycol.* **58**: 95, 2007), Braun *et al.* (*Mycotaxon* **103**: 207, 2008).

Dichodium Nyl. (1888) = Physma fide Hawksworth *et al.* (*Dictionary of the Fungi* edn 8, 1995).

dichohyphidium, see hyphidium.

Dicholobodigitus G.P. White & Illman (1988), anamorphic *Pezizomycotina*, Hso.≡ bP.28. 1, Canada. See White & Illman (*CJB* **66**: 2149, 1988).

Dichomera Cooke (1878), anamorphic *Botryosphaeriaceae*, St.#eP.1/19. 41, widespread. See Butin (*Sydowia* **45**: 161, 1993; pleomorphy), Yuan (*Nova Hedwigia* **70**: 139, 2000; on *Eucalyptus*), Gehlot &

Purohit (*Journal of Mycology and Plant Pathology* **34**: 1, 2004; ultrastr.), Barber *et al.* (*MR* **109**: 1347, 2005; pleomorphy), Burgess *et al.* (*Australas. Pl. Path.* **34**: 557, 2005; Australia), Crous *et al.* (*Stud. Mycol.* **55**: 235, 2006; phylogeny).

Dichomitus D.A. Reid (1965), Polyporaceae. 7, Europe; N. America. See Reid (*Revta Biol.* Lisb. **5**: 149, 1964-5), Ipulet & Ryvarden (*Syn. Fung.* **20**: 87, 2005; Uganda).

Dichomyces Thaxt. (1893) = Peyritschiella fide Tavares (*Mycol. Mem.* **9**, 1985).

Dichonema Blume & T. Nees (1826) = Dictyonema C. Agardh ex Kunth fide Parmasto (*Nova Hedwigia* **29**: 99, 1978).

dichophysis, see hyphidium.

Dichopleuropus D.A. Reid (1965), Lachnocladiaceae. 1, Malaysia. See Reid (*Nova Hedwigia* Beih. **18**: 329, 1965).

Dichoporis Clem. (1909) = Strigula fide Hafellner & Kalb (*Biblthca Lichenol.* **57**: 161, 1995).

Dichosporidium Pat. (1903), Roccellaceae (L). 6, widespread (tropical). See Thor (*Op. Bot.* **103**, 1990), Henssen & Thor (*Nordic Jl Bot.* **18**: 95, 1998; ontogeny), Thor (*J. Jap. Bot.* **77**: 47, 2002; Japan, Taiwan), Thor (*Biblthca Lichenol.* **95**: 543, 2007; Malaysia).

Dichosporium Pat. (1899) ≡ Dichosporidium.

Dichostereaceae Jülich (1982) = Lachnocladiaceae.

Dichostereum Pilát (1926), Lachnocladiaceae. Anamorph *Spiniger*. 11, widespread. See Boidin & Lanquetin (*Mycotaxon* **6**: 277, 1977), Boidin & Lanquetin (*BSMF* **96**: 381, 1980; key), Kotkova (*Mikol. Fitopatol.* **38**: 40, 2004; Russia).

Dichothrix Theiss. (1912) [non *Dichothrix* Zanardini ex Bornet & Flahault. 1886, *Cyanobacteria*] ≡ Schistodes.

Dichotomella Sacc. (1914) = Nigrospora fide Hughes (*CJB* **36**: 727, 1958).

Dichotomocladium Benny & R.K. Benj. (1975), Syncephalastraceae. 5, widespread. See Benny & Benjamin (*Aliso* **8**: 338, 1975), Benny & Benjamin (*Mycol.* **85**: 660, 1993), Voigt & Wöstemeyer (*Gene* **270**: 113, 2001; phylogeny).

Dichotomomyces Saito ex D.B. Scott (1970), Trichocomaceae. Anamorph *Polypaecilum*. 1, Japan. See Malloch (*SA* **6**: 124, 1987; posn), Fort *et al.* (*Boln Soc. Micol. Madrid* **14**: 61, 1990).

Dichotomophthora Mehrl. & Fitzp. ex P.N. Rao (1966), anamorphic *Pezizomycotina*, Hso.≡ eP.26. 2, widespread (esp. tropical). See de Hoog & van Oorschot (*Proc. K. ned. Akad. Wet.* Ser. C, Biol. Med. Sci. **86**: 55, 1983), Hosoe *et al.* (*Phytochem.* **29**: 997, 1990; anthraquinone derivative from *D. lutea*), Eken (*Mycotaxon* **87**: 153, 2003; Turkey), Kendrick (*CJB* **81**: 75, 2003; morphogenesis).

Dichotomophthoropsis M.B. Ellis (1971), anamorphic *Pezizomycotina*, Hso.0-≡ hP.26. 2, India; USA. See Ellis (*Mycol. Pap.* **125**: 20, 1971), Rasheed *et al.* (*Mycol.* **82**: 390, 1990; India).

dichotomous, branching, frequently successively, into two more or less equal arms.

Dicladium Ces. (1852) = Colletotrichum fide Sutton (*CJB* **44**: 887, 1966).

Diclasmia Trevis. (1869) = Sticta fide Hawksworth *et al.* (*Dictionary of the Fungi* edn 8, 1995).

diclinous, having the oogonium and its antheridium on different hyphae. Cf. androgynous.

Diclonomyces Thaxt. (1931), Laboulbeniaceae. 3, pantropical. See Nannfeldt (*Svensk bot. Tidskr.* **43**: 468, 1949).

Dicoccum Corda (1829) nom. ambig., anamorphic *Pezizomycotina*. See Hughes & Pirozynski (*CJB* **50**: 2521, 1972).

Dicollema Clem. (1909) = Collema F.H. Wigg. fide Hawksworth *et al.* (*Dictionary of the Fungi* edn 8, 1995).

Dicrandromyces Thaxt. (1931) = Tetrandromyces fide Tavares (*Mycol. Mem.* **9**, 1985).

Dicranidion Harkn. (1885), anamorphic *Orbilia*, Hsp.1bH.10. 3, America. See Peek & Solheim (*Mycol.* **50**: 844, 1958), Pfister (*Mycol.* **89**: 1, 1997; review), Kendrick (*CJB* **81**: 75, 2003; morphogenesis).

Dicranocladium Sousa da Câmara (1931) nom. provis., anamorphic *Pezizomycotina*.

Dicranophora J. Schröt. (1886), ? Mucoraceae. 1 (on decaying agarics), widespread (north temperate). See Vuillemin (*Annls mycol.* **5**: 33, 1907), Dobbs (*TBMS* **21**: 167, 1938), Volgmayr & Krisai-Greilhuber (*MR* **100**: 583, 1996; development, ultrastr.), Voigt & Wöstemeyer (*Gene* **270**: 113, 2001; phylogeny).

Dicranophoraceae J.H. Mirza (1979) = Mucoraceae.
Lit.: Kirk (*in litt.*).

Dicranotropis Breddin [not traced] nom. dub., Fungi. ? Based on fungi on an insect.

Dicteridium Raf. (1815) nom. dub., Fungi. No spp. included.

dictydine granules, see plasmodic granules.

Dictyoarthrinium S. Hughes (1952), anamorphic *Pezizomycotina*, Hso.#eP.37. 2 or 4, widespread (tropical). See Hughes (*Mycol. Pap.* **48**: 29, 1952), Mena Portales *et al.* (*Revta Iberoamer. Micol.* **12**: 31, 1995; Cuba), Somrithipol (*Mycol.* **99**: 792, 2007; key).

Dictyoarthrinopsis Bat. & Cif. (1958), anamorphic *Pezizomycotina*, Hso.#eH.?. 1, Costa Rica. See Batista & Ciferri (*Atti Ist. bot. Univ. Lab. crittog. Pavia* sér. 5 **15**: 57, 1958).

Dictyoasterina Hansf. (1947), Microthyriaceae. 1, Africa. See Hughes *in* Sugiyama (Ed.) (*Pleomorphic Fungi: The Diversity and its Taxonomic Implications*: 103, 1987).

Dictyobole G.F. Atk. & Long (1902) = Lysurus fide Dring (*Kew Bull.* **35**: 1, 1980).

Dictyocatenulata Finley & E.F. Morris (1967), anamorphic *Pezizomycotina*, Hsy.#eH.23. 1, widespread. See Seifert *et al.* (*Mycol.* **79**: 459, 1987; gen. redescr.).

Dictyocephala A.G. Medeiros (1962) = Pantospora fide Deighton (*Mycol. Pap.* **140**, 1976).

Dictyocephalos Underw. ex V.S. White (1901), Phelloriniaceae. 1, widespread. See Long & Plunkett (*Mycol.* **32**: 696, 1940), Dios *et al.* (*Mycotaxon* **84**: 265, 2002), Fan & Liu (*Mycosystema* **23**: 306, 2004; China), Sarasini (*Gasteromiceti Epigei*: 406 pp., 2005).

Dictyochaeta Speg. (1923), anamorphic *Chaetosphaeria, Ascocodinaea*, Hso/Hsy.0eH/1eH.16. *c.* 65, widespread. See Hughes & Kendrick (*N.Z. Jl Bot.* **6**: 331, 1968), Gamundí *et al.* (*Darwiniana* **21**: 95, 1977), Kuthubutheen & Nawawi (*MR* **95**: 1211, 1991; Malaysian spp.), Kuthubutheen & Nawawi (*MR* **95**: 1220, 1991; Malaysian spp.), Kuthubutheen & Nawawi (*MR* **95**: 1224, 1991; key 59 spp.), Réblová *et al.* (*Sydowia* **51**: 49, 1999; teleomorph), Whitton *et al.* (*Fungal Diversity* **4**: 133, 2000; on *Pandanaceae*), Markovskaja & Treigiene (*Botanica Lithuanica* **7**: 93, 2001; Lithuania), Tsui *et al.* (*Cryp-

tog. Mycol. **22**: 139, 2001; Hong Kong), Fernández *et al.* (*Mycol.* **98**: 121, 2006; phylogeny).

Dictyochaetopsis Aramb. & Cabello (1990), anamorphic *Chaetosphaeriaceae*, Hso.1-≡ eH.16. 12, widespread. See Arambarri & Cabello (*Mycotaxon* **38**: 12, 1990), Whitton *et al.* (*Fungal Diversity* **4**: 133, 2000; on *Pandanaceae*), Calduch *et al.* (*Mycol.* **94**: 1071, 2002; Brazil).

dictyochlamydospore, a non-deciduous multicelled chlamydospore composed of an outer wall separable from the walls of the component cells which are rather easily separated from each other, as in some *Phoma* spp. formerly ascribed to *Peyronellaea* (Luedemann, *Mycol.* **51**: 778, 1961).

Dictyochora Theiss. & Syd. (1914) nom. dub., Fungi.

Dictyochorella Theiss. & Syd. (1915) nom. dub., Fungi. See von Arx & Müller (*Stud. Mycol.* **9**, 1975).

Dictyochorina Chardón (1932) nom. dub., Pezizomycotina. See Petrak (*Sydowia* **5**: 343, 1951).

Dictyocoprotus J.C. Krug & R.S. Khan (1991), Pyronemataceae. 1 (coprophilous), Mexico; USA. See Krug & Khan (*Mycol.* **83**: 103, 1991).

Dictyocyclus Sivan., W.H. Hsieh & Chi Y. Chen (1998), Parmulariaceae (L). 1, Taiwan. See Sivanesan *et al.* (*J. Linn. Soc.* Bot. **126**: 323, 1998).

Dictyodesmium S. Hughes (1951), anamorphic *Pezizomycotina*, Hsp.#eP.1. 3, USA; France. See Hughes (*Mycol. Pap.* **36**: 29, 1951), Zhao & Zhang (*Mycosystema* **24**: 12, 2005; China).

Dictyodochium Sivan. (1984), anamorphic *Gibbera*, Hsp.#eP.1/10. 1, India. See Sivanesan (*TBMS* **82**: 517, 1984).

Dictyodothis Theiss. & Syd. (1915), Dothideaceae. 2, America. See Eriksson (*Myconet* **5**: 2969, 2000).

Dictyographa Darb. (1897) = Darbishirella fide Hawksworth *et al.* (*Dictionary of the Fungi* edn 8, 1995).

Dictyographa Müll. Arg. (1893) = Opegrapha Ach. fide Hawksworth *et al.* (*Dictionary of the Fungi* edn 8, 1995), Grube (*Bryologist* **101**: 377, 1998; phylogeny), Ertz & Diederich (*Lichenologist* **39**: 143, 2007; synonymy).

Dictyolaceae Gäum. (1926) = Tricholomataceae. The *Dictyolaceae* have not yet been proposed as nom. rej. against *Tricholomataceae*.

Dictyolus Quél. (1886) = Arrhenia fide Kuyper (*in litt.*).

Dictyomollisia Rehm (1909) = Uleomyces fide von Arx (*Persoonia* **2**: 421, 1963).

Dictyomorpha Mullins (1961), Chytridiales. 2 (on *Oomycetes*), N. America. See Dick (*Straminipilous Fungi*: 670 pp., 2001).

Dictyonella Höhn. (1909), Saccardiaceae. 7, widespread (tropical). See von Arx (*Persoonia* **2**: 421, 1963), Hsieh *et al.* (*MR* **101**: 897, 1997; Taiwan), Inácio & Dianese (*MR* **102**: 695, 1998; Brazil).

Dictyonema C. Agardh ex Kunth (1822), Agaricomycetidae (L). 7, widespread (esp. tropical). See Parmasto (*Nova Hedwigia* **29**: 99, 1978), Chaves *et al.* (*Bryologist* **107**: 242, 2004; Costa Rica).

Dictyonema Reinsch (1875) nom. dub., ? Fungi.

Dictyonemataceae Tomas. ex Tomas. (1950) = Atheliaceae.

Dictyonematomyces Cif. & Tomas. (1954) = Dictyonema C. Agardh ex Kunth fide Parmasto (*Nova Hedw.* **29**: 108, 1972).

Dictyonia Syd. (1904), Helotiaceae. 2, widespread (tropical).

Dictyopanus Pat. (1900) = Panellus fide Burdsall & Miller (*Beih. Nova Hedwigia* **51**, 1975).

Dictyopeltella Bat. & I.H. Lima (1959), Micropeltidaceae. 2, widespread (tropical). See Batista (*Publções Inst. Micol. Recife* **56**, 1959).

Dictyopeltis Theiss. (1913), Micropeltidaceae. 6, pantropical. See Batista (*Publções Inst. Micol. Recife* **56**, 1959), Müller & von Arx (*Beitr. Kryptfl. Schweiz* **11** no. 2, 1962).

Dictyopeplos Kuhl & Hasselt (1824) = Phallus fide Stalpers (*in litt.*).

Dictyophallus Corda (1842) = Phallus fide Stalpers (*in litt.*).

Dictyophora Desv. (1809) = Phallus fide Dring (*Mycol. Pap.* **98**, 1964).

Dictyophrynella Bat. & Cavalc. (1964), anamorphic *Pezizomycotina*, Hso.≡ eP.?. 1, Brazil. See Batista & Cavalcanti (*Portugaliae Acta Biologica* Série B **7**: 356, 1964).

Dictyoploca Mont. ex Pat. (1890) = Gymnopus (Pers.) Roussel fide Kuyper (*in litt.*).

Dictyopolyschema M.B. Ellis (1976), anamorphic *Pezizomycotina*, Hso.#eP.24. 1, British Isles. See Ellis (*More Dematiaceous Hyphomycetes*: 373, 1976).

dictyoporospore, a deciduous, multicelled porospore the component cells of which are firmly united and not enclosed by an outer wall, as in *Alternaria* (Luedemann, *Mycol.* **51**: 778, 1961).

Dictyoporthe Petr. (1955), Melanconidaceae. Anamorphs *Hendersonula*, *Coryneum*. 3, Pakistan; Taiwan. See Hsieh *et al.* (*MR* **101**: 1092, 1997; Taiwan), Jaklitsch & Barr (*Öst. Z. Pilzk.* **6**: 45, 1997; key).

Dictyopus Quél. (1886) = Boletus Fr. fide Singer (*Farlowia* **2**: 223, 1945).

Dictyorinis Clem. (1909) = Rinodina fide Hawksworth *et al.* (*Dictionary of the Fungi* edn 8, 1995).

Dictyorostrella U. Braun (1999), anamorphic *Pezizomycotina*. 1, Canada. See Braun (*Schlechtendalia* **3**: 35, 1999).

dictyoseptate, having transverse and longitudinal cross walls (septa), like layers of cement between bricks; muriform.

dictyosomes, ± spherical vesicles associated with the edges of the membrane-bound sacs (cisternae) which constitute the golgi apparatus in *Oomycota* and other fungi as shown by electron microscopy.

Dictyospiropes M.B. Ellis (1976), anamorphic *Pezizomycotina*, Hso.#eP.10. 1, India. See Ellis (*More Dematiaceous Hyphomycetes*: 214, 1976).

dictyosporangium, a septate sporangium, as in *Dictyuchus*.

dictyospore, differs from an amerospore (q.v.) by being divided by intersecting septa in more than one plane; muriform spore. See Anamorphic fungi.

Dictyosporites Félix (1894), Fossil Fungi. 2 (Paleocene), widespread.

Dictyosporium Corda (1836), anamorphic *Pleosporales*, Hso.0bP.1. 28, widespread. See Sutton (*Proc. Indian Acad. Sci.* Pl. Sci. **94**: 229, 1985; disposition of names), Tzean & Chen (*MR* **92**: 497, 1989; 2 n.spp.), Chen & Tzean (*Fungal Science* Taipei **14**: 105, 1999; Taiwan), Goh *et al.* (*Fungal Diversity* **2**: 65, 1999), Arambarri *et al.* (*Mycotaxon* **78**: 185, 2001; Argentina), Photita *et al.* (*Mycotaxon* **82**: 415, 2002; Thailand), Cai *et al.* (*Sydowia* **55**: 129, 2003; Yunnan), Cai *et al.* (*Cryptog. Mycol.* **24**: 3, 2003; China, definition), Kodsueb *et al.* (*Cryptog. Mycol.* **27**: 111, 2006; Thailand), Tsui *et al.* (*Fungal Diver-*

sity **21**: 157, 2006; phylogeny), Cai *et al.* (*Persoonia* **20**: 53, 2008; phylogeny).

Dictyostomiopelta Viégas (1944), Micropeltidaceae. 1, Brazil.

Dictyothyriella Rehm (1914) = Micropeltis fide Clements & Shear (*Gen. Fung.*, 1931).

Dictyothyriella Speg. (1924), Micropeltidaceae. 1, Cape Horn.

Dictyothyrina Theiss. (1913), Micropeltidaceae. 2, America.

Dictyothyrium Grove (1932) ? = Mycoporum Flot. ex Nyl. fide Sutton (*Mycol. Pap.* **141**, 1977).

Dictyothyrium Theiss. (1912), Micropeltidaceae. *c.* 15, widespread (tropical).

Dictyotopileos Dilcher (1965), Fossil Fungi, Micropeltidaceae. 1 (Eocene), USA.

Dictyotremella Kobayasi (1971), Tremellaceae. 1, Papua New Guinea. See Kobayasi (*Bull. natn. Sci. Mus.* Tokyo, B **14**: 481, 1971).

Dictyotrichiella Munk (1953) = Capronia fide Müller *et al.* (*TBMS* **88**: 63, 1987), Janex-Favre (*Cryptog. Mycol.* **9**: 133, 1988; ontogeny), Untereiner (*Mycol.* **89**: 120, 1997; morphology), Untereiner & Naveau (*Mycol.* **91**: 67, 1999; phylogeny).

Dicyma Boulanger (1897), anamorphic *Ascotricha*, Hsy.0eP.10. 11, widespread. See Hawksworth (*Mycol. Pap.* **126**, 1971), von Arx (*Proc. k. ned. Akad. Wet.* Ser. C, Biol. Med. Sci. **85**: 21, 1982), Udagawa & Uchiyama (*Mycotaxon* **70**: 177, 1999; teleomorph), Davis *et al.* (*Am. J. Bot.* **90**: 1661, 2003; endophytic taxa), Watanabe *et al.* (*Mycoscience* **44**: 411, 2003; from *Basidiomycota*, Japan), Tavares *et al.* (*Fitopatol. Brasil* **29**: 148, 2004; Brazil).

Didonia Velen. (1934), ? Helotiales. 5, Europe. Perhaps part of the *Hyaloscyphaceae*. See Svrček (*Česká Mykol.* **46**: 41, 1992; status).

Didothis Clem. (1931) ≡ Uleodothis.

Didymaria Corda (1842) = Ramularia Unger fide Hughes (*CJB* **36**: 727, 1958), Braun (*Nova Hedwigia* Beih. **53**: 291, 1991).

Didymariopsis Speg. (1910) = Colletotrichum fide Deighton (*TBMS* **59**: 185, 1972).

Didymascella Maire & Sacc. (1903), ? Hemiphacidiaceae. 3 (on conifers), Europe; N. America. *D. thujina* damages *Thuja*. See Minter (*IMI Descr. Fungi Bact.*: no. 1334, 1998).

Didymascina Höhn. (1905) = Didymosphaeria fide Aptroot (*Nova Hedwigia* **60**, 1995).

Didymascus Sacc. (1896), ? Rhytismatales. 1, Siberia. See Korf (*Mycol.* **54**: 24, 1962).

Didymaster Bat. & H. Maia (1967) = Strigula fide Lücking *et al.* (*Lichenologist* **30**: 121, 1998).

Didymella Sacc. (1880), Pleosporales. Anamorphs *Ascochyta*, *Phoma*. *c.* 88, widespread. *D. applanata* (raspberry spur blight), *D. citrullina* (on cucurbits), *D. lycopersici* (tomato stem and fruit rots). See Corbaz (*Phytopath. Z.* **28**: 375, 1956), Holm (*Taxon* **24**: 475, 1975; nomencl.), Corlett (*CJB* **59**: 2016, 1981), Kaiser *et al.* (*Pl. Dis.* **81**: 809, 1997), Navas-Cortés *et al.* (*Phytoparasitica* **26**: 199, 1998; genetics), Silva-Hanlin & Hanlin (*MR* **103**: 153, 1999; phylogeny), Liew *et al.* (*Mol. Phylogen. Evol.* **16**: 392, 2000; phylogeny), Armstrong *et al.* (*Can. J. Pl. Path.* **23**: 110, 2001; mating types), Somai *et al.* (*Phytopathology* **92**: 997, 2002; molecular variation), Barve *et al.* (*Fungal Genetics Biol.* **39**: 151, 2003; mating types), Fatehi *et al.* (*Mycopathologia* **156**: 317, 2003; *A. pinodes* complex), Kothera *et al.* (*MR* **107**: 297, 2003; AFLP analysis), Lindqvist-Kreuze *et al.* (*Pl. Path.* **52**: 567, 2003; on *Rubus*), Boerema *et al.* (*Phoma Identification Manual* Differentiation of Specific and Infra-Specific Taxa in Culture: 470 pp., 2004), Chongo *et al.* (*Pl. Dis.* **88**: 4, 2004; genetic diversity, Canada), Koch & Utkhede (*Can. J. Pl. Path.* **26**: 291, 2004; molecular detection), Peever *et al.* (*Mol. Ecol.* **13**: 291, 2004; population structure), Lichtenzveig *et al.* (*Eur. J. Pl. Path.* **113**: 15, 2005; Israel), Aptroot (*CBS Diversity Ser.* **5**, 2006), Schoch *et al.* (*Mycol.* **98**: 1041, 2006; phylogeny).

Didymellina Höhn. (1918) = Mycosphaerella fide von Arx & Müller (*Stud. Mycol.* **9**, 1975), Braun *et al.* (*Mycol. Progr.* **2**: 3, 2003).

Didymellopsis (Sacc.) Clem. & Shear (1931), Xanthopyreniaceae. 4 (on lichens), Europe. See Grube & Hafellner (*Nova Hedwigia* **51**: 283, 1990).

Didymobotryaceae Nann. (1934) = Ophiocordycipitaceae.

Didymobotryopsis Henn. (1902) = Hirsutella fide Samson & Evans (*MR* **95**: 887, 1991).

Didymobotrys Clem. & Shear (1931) ≡ Didymobotryopsis.

Didymobotryum Sacc. (1886), anamorphic *Pezizomycotina*, Hsy.1eP.28. 7, N. America; Asia. See Seifert (*Mem. N. Y. bot. Gdn* **59**: 109, 1990; Indonesia), D'Souza & Bhat (*Mycol.* **94**: 535, 2002; India).

Didymochaeta Sacc. & Ellis (1898), anamorphic *Pezizomycotina*, Cpd.1-≡ eH.15. 1, N. America.

Didymochaetina Bat. & J.L. Bezerra (1965), anamorphic *Pezizomycotina*, Cpd.1eH.?. 1, Jamaica. See Batista & Bezerra (*Riv. Patol. veg., Pavia* Sér. 4 **1**: 44, 1965).

Didymochlamys Henn. (1897) [non *Didymochlamys* Hook. 1872, *Rubiaceae*] ≡ Kuntzeomyces.

Didymochora Höhn. (1918), anamorphic *Euryachora*, St.0eH.?. 1, Europe.

Didymocladiaceae Nann. (1934) = Hypocreaceae.

Didymocladium Sacc. (1886) = Cladobotryum fide Hughes (*CJB* **36**: 727, 1958).

Didymocoryne Sacc. & Trotter (1913), Helotiales. 2, widespread (north temperate).

Didymocrater Mart. (1821) nom. dub., ? Fungi.

Didymocrea Kowalski (1965), Pleosporales. 1, India. See Kowalski (*Mycol.* **57**: 404, 1965; devel.), Luttrell (*Am. J. Bot.* **62**: 186, 1975; development), Aptroot (*Stud. Mycol.* **37**, 1995; posn), Rossman *et al.* (*Stud. Mycol.* **42**: 248 pp., 1999), Kruys *et al.* (*MR* **110**: 527, 2006; phylogeny).

Didymocyrtidium Vain. (1921), Dothideomycetes. 2, Europe.

Didymocyrtis Vain. (1921), Dothideomycetes. 1 (on *Caloplaca*), Finland.

Didymolepta Munk (1953), Leptosphaeriaceae. 2, Europe. See Eriksson & Hawksworth (*SA* **7**: 68, 1988), Barr (*Mycotaxon* **43**: 371, 1992).

Didymopeltis Bat. & I.H. Lima (1959) = Schizothyrium fide Müller & von Arx (*Beitr. Kryptfl. Schweiz* **11** no. 2, 1962).

Didymopleella Munk (1953), Dothideomycetes. 4, Europe. See Munk (*Dansk bot. Ark.* **15**: 109, 1953).

Didymoporisporonites Sheffy & Dilcher (1971), Fossil Fungi. 10 (Eocene, Tertiary), China; USA.

Didymopsamma Petr. (1925) = Chaetosphaeria fide Müller & von Arx (*Beitr. Kryptfl. Schweiz* **11** no. 2, 1962).

Didymopsis Sacc. & Marchal (1885), anamorphic *Pezizomycotina*, Hsy.1eH.1. 5, widespread.

Didymopsora Dietel (1899), Pucciniosiraceae. 6 (on dicots (0, III)), S. America; Africa. See Cunningham (*Mycol.* **60**: 769, 1968), Buriticá (*Rev. Acad. Colomb. Cienc.* **18**: 131, 1991).

Didymopsorella Thirum. (1950), Uropyxidaceae. 2 (on *Toddalia* (*Rutaceae*)), Africa; China; India; Sri Lanka. See Thirumalachar (*Sci. Cult.* **16**: 210, 1950).

Didymopycnomyces Cavalc. & A.A. Silva (1972) = Coenogonium Ehrenb. fide Lücking *et al.* (*Lichenologist* **30**: 121, 1998), Rivas Platas *et al.* (*Fungal Diversity* **23**: 255, 2006).

Didymosamarospora T.W. Johnson & H.S. Gold (1957) nom. dub., Fungi. See Kohlmeyer & Kohlmeyer (*Marine Mycology*, 1979).

Didymosamarosporella, see *Didymosamarospora*.

Didymosira Clem. (1909) ≡ Pucciniosira.

Didymosphaerella Cooke (1889), Montagnulaceae. See Aptroot (*Stud. Mycol.* **37**, 1995), Ramaley (*Mycotaxon* **78**: 435, 2001; on *Yucca*).

Didymosphaeria Fuckel (1870) nom. cons., Didymosphaeriaceae. Anamorphs *Fusicladiella*-like, *Phoma*-like. 23, widespread. See Kohlmeyer & Volkmann-Kohlmeyer (*MR* **94**: 685, 1990; redisp. marine spp.), Hyde (*Sydowia* **46**: 29, 1994), Aptroot (*Stud. Mycol.* **37**, 1995; key), Aptroot (*Nova Hedwigia* **60**: 325, 1995; excl. names), Hyde *et al.* (*Nova Hedwigia* **69**: 449, 1999; on palms).

Didymosphaeriaceae Munk (1953), Pleosporales. 4 gen. (+ 6 syn.), 50 spp.

Lit.: Scheinpflug (*Ber. schweiz. bot. Ges.* **68**: 325, 1958), Kohlmeyer & Volkmann-Kohlmeyer (*MR* **94**: 685, 1990), Aptroot (*Nova Hedwigia* **60**: 325, 1995), Aptroot (*Stud. Mycol.* **37**: 160 pp., 1995), Aptroot (*Stud. Mycol.* **37**, 1995), Hyde *et al.* (*Nova Hedwigia* **69**: 449, 1999).

Didymosphaerites Fiore (1932), Fossil Fungi. 1 (Eocene), Italy.

Didymosphaerites Pia (1927), Fossil Fungi. 1.

didymospore, differs from an amerospore (q.v.) in having one transverse septum. See Anamorphic fungi.

Didymosporiella Traverso & Migliardi (1911) nom. dub., anamorphic *Pezizomycotina*. See Sutton (*Mycol. Pap.* **141**, 1977).

Didymosporina Höhn. (1916), anamorphic *Pezizomycotina*, Cac.1eP.19. 1, Europe. See Szabó (*Acta phytopath. entom. Hung.* **32**: 69, 1997).

Didymosporis Clem. & Shear (1931) ≡ Didymosporiella.

Didymosporium Nees (1816) nom. dub., anamorphic *Pezizomycotina*. See Sutton (*Mycol. Pap.* **141**, 1977).

Didymosporium Sacc. (1880) nom. illegit., anamorphic *Pezizomycotina*, Cac.0eP.?. 1, Italy. See Sutton (*Mycol. Pap.* **141**, 1977).

Didymosporonites Sheffy & Dilcher (1971), Fossil Fungi. 4, USA.

Didymostilbe Bres. & Sacc. (1902) nom. dub., Fungi. See Sydow & Sydow (*Annls mycol.* **1**: 176, 1903).

Didymostilbe Henn. (1902), anamorphic *Peethambara*, Hsp.1eH.15. 8, Asia. See Seifert (*Stud. Mycol.* **27**: 130, 1985; 6 spp. Indonesia), Rossman *et al.* (*Mycol.* **93**: 100, 2001; phylogeny), Samuels *et al.* (*Tropical Mycology* **2**: 13, 2002; key).

Didymothozetia Rangel (1915), anamorphic *Pezizomycotina*, Hsp.1-≡ eH.?. 1, Brazil.

Didymothyriella Bat. & I.H. Lima (1959) = Plochmopeltis fide Müller & von Arx (*Beitr. Kryptfl. Schweiz* **11** no. 2, 1962).

Didymotrichella Arnaud (1954), anamorphic *Pezizomycotina*, Hso.1eP.?27. 1, France. See Arnaud (*BSMF* **69**: 284, 1953).

Didymotrichia Berl. (1893) = Neopeckia fide von Arx & Müller (*Stud. Mycol.* **9**, 1975), von Arx & Müller (*Sydowia* **37**: 6, 1984).

Didymotrichiella Munk (1953) = Capronia fide Müller *et al.* (*TBMS* **88**: 63, 1987).

Didymotrichum Bonord. (1851) = Cladosporium fide Hughes (*CJB* **36**: 727, 1958).

Didymotrichum Höhn. (1914) = Dactylaria fide Bhatt & Kendrick (*CJB* **46**, 1968).

Didymozoophaga Soprunov & Galiulina (1951) = Arthrobotrys fide Hawksworth *et al.* (*Dictionary of the Fungi* edn 8, 1995).

Diederichia D. Hawksw. (2003), anamorphic *Dothideomycetes*. 1 (on lichens), Spain. See Etayo & Diederich (*Mycotaxon* **60**: 415, 1996), Hawksworth (*Lichenologist* **35**: 206, 2003).

Diederimyces Etayo (1995), Verrucariaceae. Anamorph *Phaeosporobolus*. 1, Spain. See Etayo (*Nova Hedwigia* **61**: 190, 1995), Calatayud & Barreno (*Lichenologist* **35**: 279, 2003; teleomorph-anamorph connection).

Diedickea Syd. & P. Syd. (1913), anamorphic *Pezizomycotina*, Cpt.0eH.?. 3, widespread (tropical).

Diedickella Petr. (1922) = Stagonospora fide Clements & Shear (*Gen. Fung.*, 1931).

Diehlia Petr. (1951) ? = Phaeangellina fide Korf *in* Ainsworth *et al.* (Eds) (*The Fungi* **4A**: 249, 1973).

Diehliomyces Gilkey (1955), ? Pezizales. 1, widespread. *D. microsporus*; truffle-like fungus invading mushroom beds. See Læssøe & Hansen (*MR* **111**: 1075, 2007; phylogeny).

diel, a 24-hr periodicity. Cf. circadian, diurnal.

Dielsiella Henn. (1903) = Cycloschizon fide Müller & von Arx (*Beitr. Kryptfl. Schweiz* **11** no. 2, 1962).

Dietel (Paul; 1860-1947; Germany). High school teacher, Leipzig. A life-long student of the *Pucciniales*, from 1887 to 1943 he published more than 150 papers on rusts. At the time, his work was generally accepted as the most authoritative on rust classification. *Publs.* Hemibasidii und Uredinales. *Engler & Prantl, Die Natürlichen Pflanzenfamilien* (1898); Unterklasse Hemibasidii (Ustilaginales und Uredinales). *Engler & Prantl, Die Natürlichen Pflanzenfamilien* (1928). *Biogs, obits etc.* Poeverlein (*Sydowia* **4**: 1, 1950) [bibliography, portrait]; Stafleu & Cowan (*TL-2* **1**: 649, 1976).

Dietelia Henn. (1897), Pucciniosiraceae. 12 (on dicots (0, III)), America (± tropical); Africa; Japan; Philippines; New Guinea. See Buriticá (*Rev. Acad. Colomb. Cienc.* **18**: 131, 1991), Gjærum *et al.* (*Lidia* **5**: 18, 2000; Uganda), Evans & Ellison (*Mycol.* **97**: 935, 2005; sp. on *Mikania*), Maier *et al.* (*MR* **111**: 176, 2007).

Dievernia M. Choisy (1931) = Ramalina fide Hawksworth *et al.* (*Dictionary of the Fungi* edn 8, 1995).

differential hosts, the special species or cultivars of host plants the reactions of which are used for determining physiologic races.

diffluent, breaking up in water.

diffract (of a pileus surface), cracked into small areas; areolate.

Diffractella Guarro, P.F. Cannon & Aa (1991), Lasiosphaeriaceae. 1 (from wood), Europe; Japan. See Guarro *et al.* (*SA* **10**: 79, 1991), Huhndorf *et al.* (*Mycol.* **96**: 368, 2004).

diffuse, widely or loosely spreading and having no distinct margin; **- wall building**, see wall building.

digitate, with deep radiating divisions, finger-like.

Digitatispora Doguet (1962), Atheliaceae. 2 (marine), Europe; N. America. See Doguet (*Compte rendu hebdomadaire des Sciences de l'Academie des sciences* Paris **254**: 4338, 1962).

Digitatisporaceae Jülich (1982) = Atheliaceae.

Digitellus Paulet (1791), anamorphic *Lentinus*. 1. See Donk (*Taxon* **11**: 82, 1962).

Digitodesmium P.M. Kirk (1981), anamorphic *Pleosporales*, Hsp.0bP.1. 4, widespread. See Sutton (*Proc. Indian Acad. Sci.* Pl. Sci. **94**: 229, 1985), Ho *et al.* (*Mycol.* **91**: 900, 1999; Hong Kong), Cai *et al.* (*Nova Hedwigia* **75**: 525, 2002; Philippines), Cai *et al.* (*Sydowia* **55**: 129, 2003; Yunnan), Tsui *et al.* (*Fungal Diversity* **21**: 157, 2006; phylogeny), Cai *et al.* (*Persoonia* **20**: 53, 2008; phylogeny).

Digitodochium Tubaki & Kubono (1989), anamorphic *Pezizomycotina*, Hsp.1bH.1. 1, Japan. See Tubaki & Kubono (*Sydowia* **41**: 344, 1989).

Digitomyces Mercado, Calduch & Gené (2003), anamorphic *Pezizomycotina*, H?.?.?. 1, Taiwan. See Mercado Sierra *et al.* (*Mycol.* **95**: 860, 2003).

Digitopodium U. Braun, Heuchert & K. Schub. (2006), anamorphic *Pezizomycotina*, H?.?.?. 1 (on *Hemileia*), Zaire. See Braun *et al.* (*Schlechtendalia* **13**: 66, 2006).

Digitoramispora R.F. Castañeda & W.B. Kendr. (1990), anamorphic *Pezizomycotina*, Hso.0bP.19. 2, widespread. See Castañeda Ruiz & Kendrick (*Univ. Waterloo Biol. Ser.* **33**: 18, 1990), Somrithipol & Jones (*Nova Hedwigia* **77**: 373, 2003; Thailand).

Digitosarcinella S. Hughes (1984), anamorphic *Pezizomycotina*, Hso.1bH.1. 1, Brazil. See Hughes (*CJB* **62**: 2208, 1984).

Digitosporium Gremmen (1953), anamorphic *Crumenulopsis*, St.0bP.1. 1, Finland. See Gremmen (*Acta Bot. Neerl.* **2**: 233, 1953).

Digitothyrea P.P. Moreno & Egea (1992), Lichinaceae (L). 3, Africa; C. America. See Moreno & Egea (*Lichenologist* **24**: 215, 1992), Schultz & Büdel (*Lichenologist* **34**: 39, 2002; key), Schultz & Büdel (*Lichenologist* **35**: 151, 2003; phylogeny).

Digraphis Clem. (1909) = Graphis fide Hawksworth *et al.* (*Dictionary of the Fungi* edn 8, 1995).

Diheterospora Kamyschko ex G.L. Barron & Onions (1966) = Pochonia fide Barron & Onions (*CJB* **44**: 861, 1966), Barron (*CJB* **63**: 211, 1985; 12 spp. from parasitized rotifers), Barron (*CJB* **69**: 494, 1991; segregation of *Rotipherophthora*), Gams & Zare (*Mycology Series* **19**: 17, 2003; review).

Dihyphis Locq. (1985), Fossil Fungi. 1, Estonia.

Dikarya, Fungi. (*Neomycota* Caval.-Sm.). A subkingdom for the (least inclusive clade that contains) *Ascomycota* and *Basidiomycota* (Hibbett *et al.*, *MR* **111**: 509, 2007).

dikaryon (adj. **dikaryotic**), a cell having two genetically distinct haploid nuclei.

dikaryoparaphysis, see hyphidium.

dikaryotization, the conversion of a homokaryon into a dikaryon typically by the fusion of 2 compatible homokaryons, but see Buller phenomenon; **illegitimate -**, the sporadic occurrence of a dikaryon in non-compatible di-mon matings.

dilacerate, torn asunder.

Dillenius (Johann Jacob; 1684-1747; Germany, later England). First Sherardian Professor of Botany, Oxford University (1728). Paid special attention to fungi (incl. lichen-forming species), and mosses. *Publs. Historia Muscorum* (1742) [includes all the then known lichen-forming fungi, many of which are supported by specimens in the fungal reference collection in Oxford (**OXF**)]. *Biogs, obits etc.* Anon. (*Oxford Dictionary of National Biography* concise edn **1**: 796, 1993); Grummann (1974: 9); Jørgensen *et al.* (*Botanical Journal of the Linnean Society* **115**: 261, 1994 [typification of many names of lichen-forming fungi]); Petersen (*Mycotaxon* **5**: 415, 1977 [discussion of mycological work]); Stafleu & Cowan (*TL-2* **1**: 655, 1976).

Dilophia Sacc. (1883) [non *Dilophia* Thomson 1853, *Cruciferae*] ≡ Lidophia.

Dilophospora Desm. (1840), anamorphic *Lidophia*, St.≡ eH.15. 1, widespread. See Walker & Sutton (*TBMS* **62**: 231, 1974; teleomorph), Riley *et al.* (*MR* **102**: 301, 1998; allozyme analysis).

Diluviicola K.D. Hyde, S.W. Wong & E.B.G. Jones (1998), Annulatascaceae. 1 (from freshwater), Brunei. See Hyde *et al.* (*Fungal Diversity* **1**: 141, 1998), Ho & Hyde (*Fungal Diversity* **4**: 21, 2000), Raja *et al.* (*Mycotaxon* **88**: 1, 2003).

Dimargaris Tiegh. (1875), Dimargaritaceae. 7 (mycoparasites of *Mucorales*), widespread. See Benjamin (*Aliso* **6**: 1, 1965; key), Mandelbrot & Erb (*Mycol.* **64**: 1124, 1972; hosts of *D. verticillata*), Saikawa (*J. Jap. Bot.* **52**: 200, 1977; septal ultrastr.), Jeffries & Young (*Ann. Bot.* **47**: 107, 1981; haustoria), Jeffries & Cuthbert (*Protoplasma* **121**: 129, 1984; ultrastr.), Jeffries & Young (*TBMS* **83**: 223, 1984; spore ultrastr.), Kirk & Kirk (*TBMS* **82**: 551, 1984), Tanabe *et al.* (*J. gen. appl. Microbiol.* Tokyo **51**: 267, 2005; phylogeny), James *et al.* (*Nature* **443**: 818, 2006; phylogeny), White *et al.* (*Mycol.* **98**: 872, 2006; phylogeny).

Dimargaritaceae R.K. Benj. (1959), Dimargaritales. 3 gen., 13 spp.

Lit.: Benjamin (*Aliso* **4**: 321, 1959), Benjamin (*Aliso* **5**: 11, 1961), Benjamin (*Aliso* **5**: 273, 1963), Benjamin (*Aliso* **6**: 1, 1965), Beblowska (*Acta Mycologica* Warszawa **27**: 271, 1991), Wrzosek & Gajowniczek (*Acta Mycologica* Warszawa **33**: 265, 1998), Tanabe *et al.* (*Mol. Phylogen. Evol.* **30**: 438, 2004).

Dimargaritales R.K. Benj. (1979). Kickxellomycotina. 1 fam., 4 gen., 14 spp. Fam.:

Dimargaritaceae

Lit.: Benjamin (*Aliso* **4**: 321, **5**: 273, **6**: 1, 1959-1965, *in* Kendrick (Ed.), *The whole fungus* **2**: 573, 1979), Tanabe *et al.* (*Mol. Phylogen. Evol.* **16**: 253, 2000; phylogeny), White *et al.* (*Mycol.* **98**: 860, 2006; molecular phylogeny), Hibbett *et al.* (*MR* **111**: 109, 2007), and see under Family.

Dimastigosporium Faurel & Schotter (1965), anamorphic *Pezizomycotina*, Ccu.0eH.?. 1 (coprophilous), Sahara. See Faurel & Schotter (*Revue Mycol.* Paris **30**: 156, 1964).

Dimaura Norman (1853) = Catolechia fide Hafellner (*Nova Hedwigia* **30**: 673, 1978).

Dimelaena Norman (1853), Caliciaceae (L). 8, widespread. See Sheard (*Bryologist* **80**: 100, 1977; palaeogeography, chemotaxonomy), Leuckert *et al.* (*Nova Hedwigia* **34**: 623, 1981; chemotypes), Mayrhofer *et al.* (*Mycotaxon* **58**: 293, 1996; S. hemisphere), Calatayud & Rico (*Bryologist* **102**: 39, 1999; chemotypes), Nordin & Mattsson (*Lichenologist* **33**:

3, 2001; phylogeny), Elvebakk & Moberg (*Lichenologist* **34**: 311, 2002; Chile), Rico *et al.* (*Lichenologist* **35**: 117, 2003; link with *Buellia*), Obermayer *et al.* (*Symb. bot. upsal.* **34** no. 1: 327, 2004; Tibet), Rico *et al.* (*Cryptog. Mycol.* **27**: 149, 2006; Iberian peninsula).

Dimera Fr. (1825) = Oedemium fide Hughes (*CJB* **36**: 727, 1958).

Dimerella Trevis. (1880) = Coenogonium Ehrenb. fide Santesson (*Symb. bot. upsal.* **12** no. 1: 1, 1952) but see Vězda & Farkas (*Folia geobot. phytotax.* **23**: 187, 1988; key 12 spp. Afr.), Alvarez Andrés & Carballal Durán (*Nova Hedwigia* **73**: 409, 2001; Iberian peninsula), Kauff & Lutzoni (*Mol. Phylogen. Evol.* **25**: 138, 2002; phylogeny), Grube *et al.* (*MR* **108**: 1111, 2004; phylogeny), Malcolm (*Australasian Lichenology* **54**: 19, 2004; New Zealand), Malcolm (*Australasian Lichenology* **56**: 25, 2005; Australasia), Rivas Platas *et al.* (*Fungal Diversity* **23**: 255, 2006; Costa Rica, reln with *Coenogonium*).

Dimeriaceae E. Müll. & Arx ex Arx & E. Müll. (1975) = Pseudoperisporiaceae.

Dimeriella Speg. (1908), Parodiopsidaceae. 2, Brazil. See Farr (*Mycol.* **71**: 243, 1979), Barr (*Prodr. Cl. Loculoasc.*, 1987; posn).

Dimeriellina Chardón (1939) = Auerswaldiella fide von Arx & Müller (*Beitr. Kryptfl. Schweiz* **11** no. 1, 1954).

Dimeriellopsis F. Stevens (1927) = Nematostoma fide Sivanesan (*SA* **6**: 201, 1987).

Dimerina Theiss. (1912), Pseudoperisporiaceae. Anamorph *Ectosticta*. *c.* 10 (on *Meliolaceae*, *Asterinaceae*), widespread (tropical). See Hansford (*Mycol. Pap.* **15**, 1946), Farr *et al.* (*CJB* **63**: 1983, 1985), Reynolds & Gilbert (*Aust. Syst. Bot.* **18**: 265, 2005; Australia).

Dimerinopsis Syd. & P. Syd. (1917) = Dimerina fide von Arx & Müller (*Stud. Mycol.* **9**, 1975).

Dimeriopsis F. Stevens (1917) = Dimerina fide Müller & von Arx (*Beitr. Kryptfl. Schweiz* **11** no. 2, 1962).

Dimerisma Clem. (1909) = Spheconisca fide Hawksworth *et al.* (*Dictionary of the Fungi* edn 8, 1995).

Dimerium (Sacc. & P. Syd.) Sacc. & D. Sacc. (1905) nom. dub., Dothideomycetes. See also *Phaeostigme*. See Hughes (*Mycol. Pap.* **166**, 1993).

Dimerium (Sacc. & P. Syd.) Sacc. & D. Sacc. (1905), Parodiopsidaceae. Anamorph *Cicinnobella*. 1, Chile. See Hughes (*Mycol. Pap.* **166**, 1993).

Dimeromyces Thaxt. (1896), Laboulbeniaceae. 109 (on beetles), widespread. See Santamaria (*Nova Hedwigia* **58**: 177, 1994), Kaur & Mukerji (*Mycoscience* **37**: 61, 1996), Weir & Hammond (*Biodiv. Cons.* **6**: 701, 1997; on beetles), Terada (*Mycoscience* **41**: 39, 2000; Japan), Santamaría (*Fl. Mycol. Iberica* **5**, 2003; Iberian peninsula), Majewski (*Acta Mycologica* Warszawa **41**: 65, 2006; Poland).

Dimerospora Th. Fr. (1860) = Lecania fide Hawksworth *et al.* (*Dictionary of the Fungi* edn 8, 1995).

Dimerosporiella Höhn. (1909) ≡ Dimerosporina.

Dimerosporiella Speg. (1908), Bionectriaceae. Anamorph *Acremonium*-like. 1 (on hyphae of *Meliola*), Brazil. See Rossman *et al.* (*Stud. Mycol.* **42**: 248 pp., 1999).

Dimerosporina Höhn. (1910) = Dysrhynchis fide Müller & von Arx (*Beitr. Kryptfl. Schweiz* **11** no. 2, 1962).

Dimerosporiopsis Henn. (1901) = Antennularia fide Müller & von Arx (*Beitr. Kryptfl. Schweiz* **11** no. 2, 1962).

Dimerosporium Fuckel (1870) = Asterina fide Müller & von Arx (*Beitr. Kryptfl. Schweiz* **11** no. 2, 1962).

dimerous (of basidia), having a constriction between the probasidium and the metabasidium, as in *Brachybasidium*.

dimidiate (1) shield-like; appearing to lack one half, or having one half very much smaller than the other; (2) (of a pileus), without a stalk and semi-circular; (3) (of lamellae), stretching only halfway to the stipe; (4) (of an ascomatal wall), having the outer wall covering only the top part.

dimitic, see hyphal analysis.

dimixis, see heterothallism.

di-mon, see Buller phenomenon.

dimorphic, having two forms; esp. of *Histoplasma*, *Sporothrix*, and other pathogens of humans and animals which have yeast and mycelial habits. See Romano (*in* Ainsworth & Sussman (Eds), *The Fungi* **2**: 181, 1966; review), Szaniszlo (Ed.) (*Fungal dimorphism.*, 1985), San-Blas (*Handb. Appl. Mycol.* **2**. *Humans, animals and insects*: 459, 1991; molec. aspects).

Dimorphocystidaceae Jülich (1982) = Pterulaceae.

Dimorphocystis Corner (1950) = Actiniceps fide Boedijn (*Persoonia* **1**: 11, 1959).

Dimorphomyces Omoifo (1997) nom. illegit., anamorphic *Pezizomycotina*, Hso.?.?. 1, Nigeria. See Omoifo (*Hindustan Antibiot. Bull.* **38**: 8, 1996).

Dimorphomyces Thaxt. (1893), Laboulbeniaceae. 27, widespread. See Weir & Rossi (*Mycol.* **93**: 171, 2001; Bolivia), Santamaría (*Fl. Mycol. Iberica* **5**, 2003; Iberian peninsula).

Dimorphomycetaceae S. Colla (1934) = Laboulbeniaceae.

Dimorphospora Tubaki (1958), anamorphic *Hymenoscyphus*, Hso.0eH.3+15. 1, widespread. See Tubaki (*J. Hattori bot. Lab.* **20**: 156, 1958), Belliveau & Bärlocher (*MR* **109**: 1407, 2005; phylogeny).

Dimorphotricha Spooner (1987), Hyaloscyphaceae. 1, Australia. See Spooner (*Biblthca Mycol.* **116**, 1987).

Dinemasporiella Bubák & Kabát (1912) = Pseudolachnea Ranoj. fide Sutton (*Mycol. Pap.* **141**, 1977).

Dinemasporiella Speg. (1910) nom. dub., anamorphic *Pezizomycotina*. See Sutton (*Mycol. Pap.* **141**, 1977).

Dinemasporiopsis Bubák & Kabát (1914) = Pseudolachnea Ranoj. fide Sutton (*Mycol. Pap.* **141**, 1977).

Dinemasporis Clem. & Shear (1931) ≡ Dinemasporiella Speg.

Dinemasporium Lév. (1846), anamorphic *Phomatospora*, Ccu.0eH.15. 18, widespread. See Nag Raj (*Coelomycetous Anamorphs with Appendage-bearing Conidia*, 1993), Yamaguchi *et al.* (*Mycoscience* **46**: 367, 2005; Japan), Duan *et al.* (*Fungal Diversity* **26**: 205, 2007; key).

Dingleya Trappe (1979), Tuberaceae. 7 (hypogeous), Australia. See Trappe *et al.* (*Aust. Syst. Bot.* **5**: 597, 1992), O'Donnell *et al.* (*Mycol.* **89**: 48, 1997; phylogeny), Percudani *et al.* (*Mol. Phylogen. Evol.* **13**: 169, 1999; DNA), Hansen & Pfister (*Mycol.* **98**: 1029, 2006; phylogeny), Læssøe & Hansen (*MR* **111**: 1075, 2007; phylogeny).

dioecism (adj. **dioecious**), the condition in which the male and female sex structures are on different thalli, e.g. in certain *Laboulbeniales*; also reported in *Lecidea verruca* where the 'male' thalli are mostly smaller (Poelt, *Pl. Syst. Evol.* **135**: 81, 1980). cf. monoecism, heterothallism.

Dioicomyces Thaxt. (1901), Laboulbeniaceae. 32, widespread. See Rossi (*Mycol.* **85**: 125, 1993), Kaur & Mukerji (*Mycoscience* **37**: 61, 1996), Santamaria (*MR* **106**: 615, 2002; revision).

Diomedella Hertel (1984), Lecanoraceae (L). 2, New Zealand; sub-Antarctic islands. See Hertel (*Nova Hedwigia* Beih. **79**: 445, 1984).

Dionysia Arnaud (1952) = Candelabrum fide Bottomley (*TBMS* **37**: 234, 1954).

Diorchidiella J.C. Lindq. (1957), Raveneliaceae. 2 (on *Mimosa* (*Leguminosae*)), S. America. See Lindquist (*Darwiniana* **11**: 416, 1957).

diorchidioid (of teliospores), 2-celled and with septum longitudinal.

Diorchidium Kalchbr. (1882), Raveneliaceae. *c.* 15 (on *Leguminosae*, 1 on *Rubiaceae*), widespread (esp. tropical). See Hennen *et al.* (*Mycol.* **90**: 1079, 1998), Wood (*South African Journal of Botany* **72**: 534, 2006; southern Africa).

Diorygma Eschw. (1824), Graphidaceae (L). 1, widespread. See Hawksworth *et al.* (*Dictionary of the Fungi* edn 8, 1995), Kalb *et al.* (*Symb. bot. upsal.* **34** no. 1: 133, 2004; monogr.), Archer (*Australasian Lichenology* **56**: 10, 2005; Australia), Archer (*Biblthca Lichenol.* **94**, 2006; Australia), Miądlikowska *et al.* (*Mycol.* **98**: 1088, 2006; phylogeny), Watanuki (*Lichenology* **5**: 69, 2006; distribution), Archer (*Systematics & Biodiversity* **5**: 9, 2007; Solomon Is).

Diosporangium, see *Diasporangium*.

Dioszegia Zsolt (1957), anamorphic *Tremellaceae*. 13, widespread. See Wang *et al.* (*Antonie van Leeuwenhoek* **93**: 391, 2008).

Diphaeis Clem. (1909) = Rhizocarpon fide Hawksworth *et al.* (*Dictionary of the Fungi* edn 8, 1995).

Diphaeosticta Clem. (1909) = Pseudocyphellaria fide Hawksworth *et al.* (*Dictionary of the Fungi* edn 8, 1995).

Diphanis Clem. (1909) = Rhizocarpon fide Hawksworth *et al.* (*Dictionary of the Fungi* edn 8, 1995).

Diphanosticta Clem. (1909) = Pseudocyphellaria fide Galloway (*Bull. Br. Mus. nat. hist.* Bot. **17**, 1988).

Diphloeis Clem. (1909) = Toninia fide Hawksworth *et al.* (*Dictionary of the Fungi* edn 8, 1995).

Diphragmium Boedijn (1960) = Diorchidium fide Berndt (*Mycotaxon* **59**: 253, 1996).

Diphratora Trevis. ex Jatta (1900) = Solenopsora fide Hawksworth *et al.* (*Dictionary of the Fungi* edn 8, 1995).

diphycophilous, fungi lichenized with both a green and a blue-green photobiont (Pike & Carroll, *in* Alexopoulos & Mims, *Introductory mycology*, edn 3, 1980).

Diphymyces I.I. Tav. (1985), Laboulbeniaceae. 7, widespread. See Santamaría (*MR* **97**: 791, 1993; Spain), Weir & Rossi (*CJB* **75**: 791, 1997; NZ).

Diplacella Syd. (1930), Diaporthales. 1, America (tropical).

diplanetism (adj. **diplanetic**) (of zoospores of *Oomycota*), a sequence of two motile flagellate phases with an interspersed mobile aplanosporic phase in the zoosporic part of the life-history; the aplanosporic phase as a walled cyst; motile phases may be monomorphic or dimorphic.

diplo- (prefix), two; twice; double.

diplobiontic, see *Diplobionticae*.

Diplobionticae. Subclass introduced by Nannfeldt (1932) for *Ascomycota* in which the life cycle consists of two thalli (bionts), like the gametophyte and sporophyte of many algae, e.g. *Spermophthora* is diplobiontic, because of its two generations, free from each other; cf. *Haplobionticae*.

Diplocarpa Massee (1895), Helotiales. 1, Europe. See Nauta & Spooner (*Mycologist* **14**: 21, 2000; UK).

Diplocarpon F.A. Wolf (1912), Helotiales. Anamorphs *Entomosporium*, *Marssonina*. 6, widespread. *D. rosae* (anamorph *Marssonina rosae*; black spot of rose), *D. earlianum* (anamorph *M. fragariae*; strawberry leaf scorch). See Ali *et al.* (*Proceedings, BCPC Conference* Pests & Diseases 2000: 251, 2000).

Diplocarponella Bat. (1957) = Stomiopeltis fide Müller & von Arx (*Beitr. Kryptfl. Schweiz* **11** no. 2, 1962).

Diploceras (Sacc.) Died. (1915) = Seimatosporium fide Sutton (*TBMS* **64**: 483, 1975).

Diplochora Höhn. (1906) = Pseudophacidium fide Petrak (*Sydowia* **5**: 193, 1951).

Diplochora P. Syd. (1913) ≡ Diplochorella.

Diplochorella P. Syd. (1913), Dothideomycetes. 3, S. Africa; New Zealand. See von Arx & Müller (*Stud. Mycol.* **9**, 1975), Wagner & Wilkinson (*Pl. Dis.* **76**: 212, 1992), Johnston & Cannon (*N.Z. Jl Bot.* **42**: 921, 2004; New Zealand).

Diplochorina Gutner (1933), Dothideomycetes. 1, former USSR.

Diplochytridium Karling (1971) nom. inval., Chytridiaceae. 21. See Desjardin (*Sydowia* **42**: 17, 1990; culture).

Diplochytrium Tomaschek (1878) = Olpidium fide Karling (*Chytriomyc. Iconogr.*, 1977).

Diplocladiaceae Nann. (1934) = Hypocreaceae.

Diplocladiella G. Arnaud ex M.B. Ellis (1976), anamorphic *Pezizomycotina*, Hso.0bP.10. 7, widespread (esp. tropical). See Nawawi (*Mycotaxon* **28**: 297, 1987), Lee *et al.* (*Fungal Diversity* **1**: 165, 1998), Kendrick (*CJB* **81**: 75, 2003; morphogenesis), Cooper (*N.Z. Jl Bot.* **43**: 323, 2005; New Zealand).

Diplocladium Bonord. (1851) = Cladobotryum fide Hughes (*CJB* **36**: 727, 1958), Kirschner & Oberwinkler (*MR* **103**: 1152, 1999; exclusion of *D. gregarium*).

Diplococcium Grove (1885), anamorphic *Helminthosphaeria*, Hso.1eP.28. 20, widespread. See Sinclair *et al.* (*TBMS* **85**: 736, 1985), Subramanian & Sekar (*Kavaka* **15**: 87, 1987; teleomorph), Goh & Hyde (*Fungal Diversity* **1**: 65., 1998; key), Wang & Sutton (*CJB* **76**: 1608, 1998; *Selenosporella* synanamorph), Réblová (*Sydowia* **51**: 223, 1999; teleomorph), Braun *et al.* (*J. Econ. Taxon. Bot.* **25**: 284, 2001; India).

diploconidium, a binucleate conidium.

Diplocryptis Clem. (1909) = Cryptodiscus fide Sherwood (*Mycotaxon* **6**: 215, 1977).

Diplocystidiaceae Kreisel (1974), Boletales. 4 gen. (+ 1 syn.), 5 spp.

Lit.: Kreisel (*Feddes Repert.* **85**: 325, 1974), Miller & Miller (*Gasteromycetes. Morphological and Development Features with Keys to the Orders, Families, and Genera*: 157 pp., 1988), Pegler *et al.* (*British Puffballs, Earthstars and Stinkhorns* An Account of the British Gasteroid Fungi: 255 pp., 1995; as *Astraeaceae*), Hughey *et al.* (*Mycol.* **92**: 94, 2000; as *Astraeaceae*), Binder & Bresinsky (*Mycol.* **94**: 85, 2002; as *Astraeaceae*).

Diplocystis Berk. & M.A. Curtis (1869), Diplocystidiaceae. 1, West Indies. See Kreisel (*Feddes Repert.* **85**: 325, 1974).

Diploderma Link (1816) ? = Astraeus fide Stalpers (*in litt.*).

Diplodermaceae Fr. (1849) = Astraeaceae.

Diplodia Fr. (1834), anamorphic *Botryosphaeriaceae*, Cpd.1eP.1. 960, widespread. See Shear (*Mycol.* **25**: 274, 1933; teleomorphs), Zambettakis (*BSMF* **70**: 219, 1954; species concepts, 24 spp. accepted), Vajna (*Eur. J. For. Path.* **16**: 223, 1986; on *Quercus*), Wang *et al.* (*Mycol.* **78**: 960, 1986; TEM), Luque & Girbal (*Eur. J. For. Path.* **19**: 7, 1989; on *Quercus*), Swart *et al.* (*Phytopathol.* **81**: 489, 1991; variation), Swart *et al.* (*MR* **97**: 832, 1993; biometrics), Smith & Stanosz (*Phytopathol.* **85**: 699, 1995; population biology, USA), Ragazzi *et al.* (*Eur. J. For. Path.* **27**: 391, 1997; VCGs), Jacobs & Rehner (*Mycol.* **90**: 601, 1998; morphology, phylogeny), Crous & Palm (*Sydowia* **51**: 167, 1999; generic limits), Stanosz *et al.* (*MR* **103**: 1193, 1999; DNA, isoenzymes), Denman *et al.* (*Stud. Mycol.* **45**: 129, 2000; teleomorph), Wet *et al.* (*Pl. Dis.* **84**: 151, 2000; biogeography), Burgess *et al.* (*MR* **105**: 1331, 2001; genetic diversity), Zhou & Stanosz (*Mycol.* **93**: 516, 2001; phylogeny), Phillips (*Phytopath. Mediterr.* **41**: 3, 2002; on *Vitis*), Alves *et al.* (*Mycol.* **96**: 598, 2004; on *Quercus*), Burgess *et al.* (*MR* **108**: 1399, 2004; on *Pinus*), Van Niekerk *et al.* (*Mycol.* **96**: 781, 2004; on *Vitis*), Wingfield *et al.* (*Australas. Pl. Path.* **33**: 513, 2004; on *Pinus*), Luchi *et al.* (*J. Phytopath.* **153**: 37, 2005; molecular diagnostics), Alves *et al.* (*Fungal Diversity* **23**: 1, 2006; *D. cupressi*), Crous *et al.* (*Stud. Mycol.* **55**: 235, 2006; phylogeny), Smith & Stanosz (*Pl. Dis.* **90**: 307, 2006; molecular detection), Damm *et al.* (*Mycol.* **99**: 664, 2007; S Africa, on *Prunus*), Phillips *et al.* (*Mycol.* **97**: 513, 2007; genus concept, phylogeny).

Diplodiella (P. Karst.) Sacc. (1884) nom. dub., anamorphic *Pezizomycotina*. See Sutton (*Mycol. Pap.* **141**, 1977).

Diplodiella Petr. (1953) nom. illegit., anamorphic *Pezizomycotina*. See Sutton (*Mycol. Pap.* **141**, 1977).

Diplodina Westend. (1857), anamorphic *Cryptodiaporthe*, Cac.1-≡ eH.15. 5, widespread. See Sutton (*The Coelomycetes*, 1980), Wulf (*Nachricht. Deutsch. Pflanzenschutz.* **42**: 97, 1990; *D. acerina* endophytic and antagonistic to leaf-feeding insects), Bathgate *et al.* (*MR* **100**: 159, 1996; on *Banksia*, Australia).

Diplodinis Clem. (1931), anamorphic *Pezizomycotina*, Cpd.1eH.?. 1, Scandinavia.

Diplodinula Tassi (1902), anamorphic *Pezizomycotina*, Cpd.1eH.?. 68, widespread.

diplodioecious, see heterophytic.

Diplodiopsis Henn. (1904) = Parodiella fide Müller & von Arx (*Beitr. Kryptfl. Schweiz* **11** no. 2, 1962).

diplodiosis, in cattle and sheep, a neuromuscular paretic syndrome caused by *Stenocarpella maydis* (syn. *Diplodia maydis*) infected maize in South Africa.

Diplodites Babajan & Tasl. (1970), Fossil Fungi. 1 (Tertiary), former USSR.

Diplodothiorella Bubák (1916) = Sphaerellopsis Cooke fide Sutton (*Mycol. Pap.* **141**, 1977).

Diplogelasinospora Cain (1961), Sordariales. 2, Asia; N. America. See Udagawa & Horie (*J. Jap. Bot.* **47**: 297, 1972), Udagawa *in* Subramanian (Ed.) (*Taxonomy of fungi* **1**: 225, 1978; anamorph), Novotný (*Czech Mycol.* **53**: 203, 2001; Czech Republic), Cai *et al.* (*MR* **110**: 359, 2006; phylogeny), Cai *et al.* (*MR* **110**: 137, 2006; phylogeny).

Diplogramma Müll. Arg. (1891), Roccellaceae (L). 1. See Staiger (*Biblthca Lichenol.* **85**, 2002), Archer (*Telopea* **10**: 589, 2004; Australia, comparison with *Opegrapha*).

Diplogrammatomyces Cif. & Tomas. (1953) ≡ Diplogramma.

Diplographis Kremp. ex A. Massal. (1860) = Fissurina fide Staiger (*Biblthca Lichenol.* **85**, 2002).

diploheteroecious, see heterophytic.

Diploicia A. Massal. (1852), Caliciaceae (L). *c.* 4, widespread (temperate). See Elix *et al.* (*Mycotaxon* **33**: 457, 1988; chemistry), Matzer *et al.* (*Nordic Jl Bot.* **17**: 433, 1997; South Africa), Nordin (*Symb. bot. upsal.* **32** no. 1: 195, 1997; ultrastr.), Harada *et al.* (*J. Nat. Hist. Mus. Inst.* Chiba **5**: 97, 1999; Japan), Grube & Arup (*Lichenologist* **33**: 63, 2001; evolution), Molina *et al.* (*Lichenologist* **34**: 509, 2002; phylogeny, synonymy with *Diplotomma*), Helms *et al.* (*Mycol.* **95**: 1078, 2003; phylogeny).

diploid (1) (of a nucleus), having the 2*n* number of chromosomes; (2) (of a cell), having the 2*n* number of chromosomes in one (synkaryotic, 2*n*) or two (dikaryotic, *n* + *n*) nuclei; (3) (of a mycelium), made up of dikaryotic diploid cells.

Diploidium G. Arnaud (1923) = Septoidium fide Ellis (*Dematiaceous Hyphomycetes*, 1971).

diploidization, the process by which a haploid cell (or mycelium) becomes a diploid (dikaryotic) cell (mycelium) having conjugate nuclei (Buller, 1941); cf. heterokaryotise.

diplokaryon, see synkaryon.

Diplolabia A. Massal. (1854) orth. var. = Dyplolabia fide Hawksworth *et al.* (*Dictionary of the Fungi* edn 8, 1995).

Diplolaeviopsis Giralt & D. Hawksw. (1991), anamorphic *Pezizomycotina*, Cpd.1eH.19. 1 (on lichens), Spain; USA. See Giralt & Hawksworth (*MR* **95**: 759, 1991), Diederich (*Herzogia* **16**: 41, 2003; USA).

Diplomitoporus Domański (1970), Polyporaceae. 11, widespread. See Gilbertson & Ryvarden (*N. Amer. Polyp.* **1**: 240, 1986).

diplomitotic nuclear cycle, occurrence of two mitotic phases of different ploidy in the nuclear cycle (Dick, 1987); karyogamy - mitosis - meiosis - mitosis - karyogamy -.

diplomonoecious, see homophytic.

Diplomyces Thaxt. (1895), ? Laboulbeniaceae. 3, widespread. See Rossi & Cesari Rossi (*G. bot. ital.* n.s. **112**: 63, 1978), Benjamin (*Aliso* **10**: 345, 1983), Majewski (*Acta Mycologica* Warszawa **19**: 183, 1983; Poland), Tavares (*Mycol. Mem.* **9**: 627 pp., 1985), Santamaría (*Fl. Mycol. Iberica* **5**, 2003; Iberian peninsula).

Diplonaevia Sacc. (1889), Helotiales. 23, widespread (north temperate). See Hein (*Sydowia* **36**: 78, 1983; key), Nannfeldt (*Nordic Jl Bot.* **4**: 791, 1985; key 10 spp. *Juncaceae*), Scheuer (*MR* **95**: 634, 1991), Nauta & Spooner (*Mycologist* **14**: 21, 2000; UK), Raitviir (*Mycotaxon* **87**: 359, 2003; Greenland), Raitviir (*Folia cryptog. Estonica* **40**: 43, 2003), Suková (*Czech Mycol.* **56**: 63, 2004; Czech Republic).

Diplonema P. Karst. (1889) [non *Diplonema* Kjellm. 1855, *Algae*] ≡ Amphinema.

Diploneurospora K.P. Jain & R.C. Gupta (1970), Fossil Fungi, Xylariales. 1 (Miocene), India.

diplont, the thallus of the diploid stage; the sporophyte.

Diploöspora Grove (1916), anamorphic *Pezizomycotina*, Hsy.1eH.?. 3, Europe. See Sreekala & Bhat (*Mycotaxon* **80**: 101, 2001; India).

Diplopeltis Pass. (1889) [non *Diplopeltis* Endl. 1837, *Sapindaceae*] ≡ Pycnoseynesia fide Sutton (*Mycol. Pap.* **141**, 1977).

Diplopeltopsis Henn. ex Höhn. (1911) = Asterothyrium Müll. Arg. fide Hawksworth *et al.* (*Dictionary of the Fungi* edn 8, 1995).

diplophase, the part of a life-history in which the cells are diploid.

Diplophlyctis J. Schröt. (1892), Endochytriaceae. 12, widespread. See Sparrow (*Aquatic Phycomycetes* Edn 2: 386, 1960; key), Dogma (*TBMS* **67**: 255, 1976; key chitinophilic spp.).

Diplophragmia Vain. (1934) = Lecidella fide Santesson (*The lichens and lichenicolous fungi of Sweden and Norway*, 1993).

Diploplacis Clem. & Shear (1931) = Sphaerellopsis Cooke fide Sutton (*Mycol. Pap.* **141**, 1977).

Diploplacosphaeria Petr. (1921) = Sphaerellopsis Cooke fide Sutton (*Mycol. Pap.* **141**, 1977).

Diploplenodomopsis Petr. (1923) = Diplodina fide Clements & Shear (*Gen. Fung.*, 1931).

Diploplenodomus Died. (1912), anamorphic *Pezizomycotina*, St.1eH.?. 2 or 3, Europe. See Boerema *et al.* (*Persoonia* **16**: 141, 1996).

Diplopodomyces W. Rossi & Balazuc (1977), Laboulbeniaceae. 1, Europe. See Rossi & Balazuc (*Revue Mycol.* Paris **41**: 528, 1977), Tavares (*Mycol. Mem.* **9**: 627 pp., 1985).

Diplorhinotrichum Höhn. (1902) = Dactylaria fide Bhatt & Kendrick (*CJB* **46**, 1968).

Diplorhynchus Arnaud (1952) nom. inval., anamorphic *Pezizomycotina*, Hso.0-≡ hH.1. 1, France. See Goos (*Mycol.* **79**: 1, 1987).

Diploschistaceae Zahlbr. (1905) = Thelotremataceae.

Diploschistella Vain. (1926) nom. rej., Gomphillaceae (L). 1, pantropical. See Lumbsch & Hawksworth (*Taxon* **36**: 764, 1987; nomencl.), Lücking *et al.* (*Lichenologist* **37**: 123, 2005; phylogeny), Lücking *et al.* (*Lichenologist* **38**: 131, 2006; key, Costa Rica).

Diploschistes Norman (1853), Thelotremataceae (L). *c.* 43, widespread. See Lumbsch (*J. Hattori bot. Lab.* **66**: 133, 1989; key 14 holarctic spp.), Lumbsch (*Nova Hedwigia* **56**: 227, 1993), Pant & Upreti (*Lichenologist* **25**: 33, 1993; key 14 spp. India & Nepal), Lumbsch & Tehler (*Bryologist* **101**: 398, 1998; cladistic analysis), Winka *et al.* (*Lichenologist* **30**: 455, 1998; DNA), Martín *et al.* (*Pl. Biol.* **2**: 571, 2000; morphology), Kauff & Lutzoni (*Mol. Phylogen. Evol.* **25**: 138, 2002; phylogeny), Lumbsch & Elix (*Biblthca Lichenol.* **86**: 119, 2003; Australia), Martín *et al.* (*Lichenologist* **35**: 27, 2003; phylogeny), Frisch *et al.* (*Biblthca Lichenol.* **92**: 517, 2006; phylogeny), Miądlikowska *et al.* (*Mycol.* **98**: 1088, 2006; phylogeny), Hofstetter *et al.* (*Mol. Phylogen. Evol.* **44**: 412, 2007; phylogeny).

Diploschistomyces Werner (1976) nom. inval. ≡ Diploschistes.

Diplosclerophoma Petr. (1923) = Diplodina fide Sutton (*Mycol. Pap.* **141**, 1977).

Diplosis Clem. (1909) = Toninia fide Hawksworth *et al.* (*Dictionary of the Fungi* edn 8, 1995).

Diplosphaerella Grove (1912) = Delphinella fide von Arx & Müller (*Stud. Mycol.* **9**, 1975).

Diplosporis Clem. (1906) ≡ Geminispora.

Diplosporites Pia (1927), Fossil Fungi. 1 (Oligocene), France.

Diplosporium Link (1824) = Oedemium fide Hughes (*CJB* **36**: 727, 1958).

Diplosporonema Höhn. (1917), anamorphic *Pezizomycotina*, Cac.≡ eH.10. 1, widespread.

diplospory, when a diploid nucleus is incorporated into cytoplasm influenced by adjacent meioses in the coenocytic gametangium, subsequently giving rise to oospores (in *Oomycetes*; Dick, *New Phytol.* **71**: 1151, 1972).

Diplostephanus Langeron (1922) = Emericella fide von Arx (*Gen. Fungi Sporul. Cult.* Edn 3, 1981).

diplostichous, in two lines or groups.

Diplostoma, see *Diploderma*.

diplostromatic, see stroma.

diplosynoecious, see homophytic.

Diplotheca (Zahlbr.) Räsänen (1943) [non *Diplotheca* Hochst. 1846, *Leguminosae*] = Lecidea fide Hawksworth *et al.* (*Dictionary of the Fungi* edn 8, 1995).

Diplotheca Starbäck (1893), Myriangiaceae. 1 (on *Cactaceae*), America.

Diplothrix Vain. (1921) = Calloriopsis fide Santesson (*Svensk bot. Tidskr.* **45**: 300, 1951).

Diplotomma A. Massal. (1852) = Diploicia fide Hawksworth *et al.* (*Dictionary of the Fungi* edn 8, 1995).

Diplotomma Flot. (1849), Caliciaceae (±L). *c.* 29, widespread. See Singh & Awasthi (*Geophytology* **19**: 173, 1990; key 11 Indian spp.), Nordin (*Symb. bot. upsal.* **32** no. 1: 195, 1997; ultrastr.), Molina *et al.* (*Lichenologist* **34**: 509, 2002; phylogeny), Helms *et al.* (*Mycol.* **95**: 1078, 2003; phylogeny), Simon *et al.* (*J. Mol. Evol.* **60**: 434, 2005; introns).

Diplozythia Bubák (1904) nom. conf., anamorphic *Pezizomycotina*. See Sutton (*Mycol. Pap.* **141**, 1977).

Diplozythiella Died. (1916), anamorphic *Pezizomycotina*, St.1eH.15. 1, India.

Dipodascaceae Engl. & E. Gilg (1924), Saccharomycetales. 6 gen. (+ 6 syn.), 71 spp.

Lit.: Hoog *et al.* (*Stud. Mycol.* **29**: 131 pp., 1986), Hoog *et al. in* Kurtzman & Fell (Eds) (*Yeasts, a taxonomic study* 4th edn: 181, 1998), Hoog *et al. in* Kurtzman & Fell (Eds) (*Yeasts, a taxonomic study* 4th edn: 209, 1998), Hoog *et al. in* Kurtzman & Fell (Eds) (*Yeasts, a taxonomic study* 4th edn: 574, 1998), Smith *et al.* (*Antonie van Leeuwenhoek* **77**: 71, 2000), Ueda-Nishimura & Mikata (*Microbiology* Reading **146**: 1045, 2000), Suh *et al.* (*Mycol.* **98**: 1006, 2006; phylogeny).

Dipodascales = Saccharomycetales.

Dipodascopsis L.R. Batra & Millner (1978), Lipomycetaceae. 2, Europe; N. America. See Jansen van Rensburg *et al.* (*Syst. Appl. Microbiol.* **18**: 410, 1995; lipids, DNA), Smith & de Hoog *in* Kurtzman & Fell (Eds) (*Yeasts, a taxonomic study* 4th edn: 178, 1998), Smith *et al.* (*S. Afr. J. Sci.* **96**: 247, 2000; chemistry), Suh & Blackwell (*Insect-Fungal Associations* Ecology and Evolution: 244, 2005; ecology), Suh *et al.* (*Mycol.* **98**: 1006, 2006; phylogeny), Kurtzman *et al.* (*FEMS Yeast Res.* **7**: 1027, 2007; phylogeny).

Dipodascus Lagerh. (1892), Dipodascaceae. Anamorph *Geotrichum*. 16, widespread. See de Hoog *et al.* (*Stud. Mycol.* **29**, 1986; key), de Hoog *et al. in* Kurtzman & Fell (Eds) (*Yeasts, a taxonomic study* 4th edn: 181, 1998), Smith & Poot (*Antonie van Leeuwenhoek* **74**: 229, 1998; DNA), Ueda-Nishimura & Mikata (*Microbiology* Reading **146**: 1045, 2000; rDNA secondary structures), Spatafora (*Cellular

Origin and Life in Extreme Habitats **4**: 591, 2002; ecology), Smith & Poot (*FEMS Yeast Res.* **3**: 301, 2003; phylogeny), Smith *et al.* (*Antonie van Leeuwenhoek* **83**: 317, 2003; chemistry, morphology), Bareetseng *et al.* (*Antonie van Leeuwenhoek* **85**: 187, 2004; ascospores), Hoog & Smith (*Stud. Mycol.* **50**: 489, 2004; phylogeny, key, polyphyly), Suh *et al.* (*MR* **109**: 261, 2005; ecology), Suh & Blackwell (*MR* **110**: 220, 2006; in insect guts), Suh *et al.* (*Mycol.* **98**: 1006, 2006; phylogeny), van Heerden *et al.* (*FEMS Yeast Res.* **7**: 173, 2007; morphology).

Dipodomyces Thaxt. (1931), Laboulbeniaceae. 2, W. Africa; Poland. See Nannfeldt (*Svensk bot. Tidskr.* **43**: 468, 1949).

Diporicellaesporites Elsik (1968), Fossil Fungi. 15 (Paleocene), widespread.

Diporidicellaesporites Frunzescu & Bacaran (1990), Fossil Fungi. 1. See Frunzescu & Bacaran (*Revue roum. Géol., Géophys. Géogr.* Géol. **34**: 12, 1990).

Diporimonocellasporites Frunzescu & Bacaran (1990), Fossil Fungi. 1. See Frunzescu & Bacaran (*Revue roum. Géol., Géophys. Géogr.* Géol. **34**: table 1, 1990).

Diporimonodicellaesporites Frunzescu & Bacaran (1990), Fossil Fungi. 1. See Frunzescu & Bacaran (*Revue roum. Géol., Géophys. Géogr.* Géol. **34**: table 1, 1990).

Diporimulticellaesporites Frunzescu & Bacaran (1990), Fossil Fungi. 1. See Frunzescu & Bacaran (*Revue roum. Géol., Géophys. Géogr.* Géol. **34**: table 1, 1990).

Diporina Clem. (1909) = Strigula fide Harris *in* Hafellner & Kalb (*Biblthca Lichenol.* **57**: 161, 1995).

Diporipollis Kalgutkar & Janson. (2000), Fossil Fungi. 1, USA.

Diporisporites Hammen (1954), Fossil Fungi. 20 (Cretaceous, Tertiary), widespread.

Diporitetracellaesporites Frunzescu & Bacaran (1990), Fossil Fungi. 1. See Frunzescu & Bacaran (*Revue roum. Géol., Géophys. Géogr.* Géol. **34**: table 1, 1990).

Diporitricellaesporites Frunzescu & Bacaran (1990), Fossil Fungi. 1. See Frunzescu & Bacaran (*Revue roum. Géol., Géophys. Géogr.* Géol. **34**: table 1, 1990).

Diporopollis S. Dutta & Sah (1970), Fossil Fungi. 1 (Eocene), India.

Diporotheca C.C. Gordon & C.G. Shaw (1961), Diporothecaceae. 1 (on roots), USA. See Mibey & Hawksworth (*SA* **14**: 25, 1995).

Diporothecaceae Mibey & D. Hawksw. (1995), Pezizomycotina (inc. sed.). 1 gen., 1 spp.
Lit.: Mibey & Hawksworth (*SA* **14**: 25, 1995).

Dipsacomyces R.K. Benj. (1961), Kickxellaceae. 1, Honduras. See Benjamin (*Aliso* **5**: 15, 1961).

Dipyrenis Clem. (1909) nom. dub., Pyrenulaceae. See Hawksworth (*Nova Hedwigia* **43**: 1, 1986).

Dipyrgis Clem. (1909), Dothideomycetes. 1, Australia. See Tibell (*Beih. Nova Hedwigia* **79**: 597, 1984).

Dipyxis Cummins & J.W. Baxter (1967), Uropyxidaceae. 2 (on *Bignoniaceae*), Costa Rica; Mexico; Brazil. See Cummins & Baxter (*Mycol.* **59**: 368, 1967).

direct (of fruit-body development), cell enlargement occurring at the same time as cell division; in **indirect** development cell enlargement mainly occurs after the period of cell division (Corner, 1950).

Dirimosperma Preuss (1855) nom. dub., anamorphic *Pezizomycotina*. See Sutton (*Mycol. Pap.* **141**, 1977).

Dirina Fr. (1825), Roccellaceae (L). 11, widespread. See Tehler (*Op. Bot.* **70**, 1983), Tehler (*Lichenologist* **18**: 295, 1986), Tehler (*Lichenologist* **20**: 398, 1988; anamorph), Tehler (*CJB* **68**: 2458, 1990; phylogeny), Letrouit-Galinou *et al.* (*Bull. Soc. linn. Provence* **45**: 389, 1994; ultrastr.), Tehler *et al.* (*Lichenologist* **27**: 255, 1995; Mexico), Grube (*Bryologist* **101**: 377, 1998; phylogeny), Myllys *et al.* (*Bryologist* **101**: 70, 1998; phylogeny, morphology), Myllys *et al.* (*Lichenologist* **31**: 461, 1999; phylogeny), Follmann (*J. Hattori bot. Lab.* **90**: 251, 2001; S America, key), Follmann (*Mitteilungen aus dem Institut für Allgemeine Botanik Hamburg* **30-32**: 61, 2002; biogeography), Tehler & Irestedt (*Cladistics* **23**: 432, 2007).

Dirinaceae Zahlbr. (1905) = Roccellaceae.

Dirinaria (Tuck.) Clem. (1909), Caliciaceae (L). 29, widespread (esp. tropical). See Awasthi (*Biblthca Lichenol.* **2**, 1975; monogr.), Swinscow & Krog (*Norw. Jl Bot.* **25**: 157, 1978; E. Afr.), Huneck *et al.* (*J. Hattori bot. Lab.* **62**: 331, 1987; chemistry, S America), Risbud & Patwardhan (*Biovigyanam* **15**: 57, 1989; Andaman Is), Scutari (*Darwiniana* **33**: 149, 1995; Argentina), Nordin (*Symb. bot. upsal.* **32** no. 1: 195, 1997; ultrastr.), Czeczuga *et al.* (*Feddes Repert.* **112**: 81, 2001; carotenoids), Nordin & Mattsson (*Lichenologist* **33**: 3, 2001; phylogeny, morphology), Helms *et al.* (*Mycol.* **95**: 1078, 2003; phylogeny), Moberg (*Symb. bot. upsal.* **34** no. 1: 257, 2004; southern Africa), Miądlikowska *et al.* (*Mycol.* **98**: 1088, 2006; phylogeny).

Dirinastromyces Cif. & Tomas. (1953) ≡ Dirinastrum.

Dirinastrum Müll. Arg. (1893) ? = Buellia See Tehler (*Op. Bot.* **70**, 1983).

Dirinella M. Choisy (1931) = Lecanora fide Hawksworth *et al.* (*Dictionary of the Fungi* edn 8, 1995).

Dirinopsis De Not. (1846) = Dirina fide Tehler (*Op. Bot.* **70**, 1983).

Disaeta Bonar (1928) = Seimatosporium fide Shoemaker (*CJB* **42**: 411, 1964).

Disarticulatus G.F. Orr (1977) = Arachniotus fide von Arx (*Persoonia* **9**: 393, 1977).

disc (1) (**disk**) (of discomycetes), the round, plate-like or curved spore-producing part of the ascoma; (2) (of a pileus), the central part of the top surface.

Discales. Used by Le Gal (1953) for discomycetes.

Discaria (Sacc.) Sacc. (1889) [non *Discaria* Hook. 1830, *Rhamnaceae*] ≡ Plicaria.

Discella Berk. & Broome (1850) ? = Rhabdospora fide Sutton (*Mycol. Pap.* **141**, 1977).

Discellaceae Clem. & Shear (1931) = Mycosphaerellaceae.

Dischloridium B. Sutton (1977), anamorphic *Australiasca*, Hso.0eH.15. 9, widespread. See Seifert & Gams (*Mycotaxon* **24**: 459, 1985), Holubová-Jechová (*Česká Mykol.* **41**: 107, 1987), Srivastava (*Sydowia* **39**: 217, 1986).

disciform, round and flat.

Discina (Fr.) Fr. (1849), Discinaceae. 20, widespread (north temperate). See Donadini (*Bull. Soc. linn. Provence* **38**: 161, 1987; 17 spp.), Kimbrough *et al.* (*CJB* **68**: 317, 1990; similarity to *Gyromitra*), O'Donnell *et al.* (*Mycol.* **89**: 48, 1997; phylogeny), Harrington *et al.* (*Mycol.* **91**: 41, 1999; phylogeny), Vizzini (*Boll. Gruppo Micol. 'G. Bresadola'* **46**: 53, 2003; cytology).

Discina Bonord. (1851) nom. dub., Fungi. See Eckblad (*Nytt Mag. Bot.* **15**: 170, 1968).

Discinaceae Benedix (1961), Pezizales. 5 gen. (+ 12

syn.), 58 spp.

Lit.: Gibson & Kimbrough (*CJB* **66**: 1743, 1988), Kimbrough *et al.* (*CJB* **68**: 317, 1990), Kimbrough (*MR* **95**: 421, 1991), Abbott & Currah (*Mycotaxon* **62**: 1, 1997), Landvik *et al.* (*Nordic Jl Bot.* **17**: 403, 1997), O'Donnell *et al.* (*Mycol.* **89**: 48, 1997; phylogeny), Harrington *et al.* (*Mycol.* **91**: 41, 1999; phylogeny), Hansen & Pfister (*Mycol.* **98**: 1029, 2006; phylogeny), Læssøe & Hansen (*MR* **111**: 1075, 2007; phylogeny).

Discinella Boud. (1885), Helotiaceae. *c.* 12, widespread. See Dennis (*Mycol. Pap.* **62**, 1956), Anderson *et al.* (*Mycologist* **15**: 132, 2001).

Discinella P. Karst. (1891), ? Pezizales. 1, Europe.

Disciotis Boud. (1885), Morchellaceae. *c.* 4, widespread (north temperate). See O'Donnell *et al.* (*Mycol.* **89**: 48, 1997; phylogeny), Hansen & Pfister (*Mycol.* **98**: 1029, 2006; phylogeny).

Disciseda Czern. (1845), Agaricaceae. 15, widespread. See Moravec (*Sydowia* **8**: 278, 1954), Grgurinovic (*Larger Fungi of South Australia*: 725 pp. + 34 [m, 1997), Jeppson (*Windahlia* **22**: 33, 1995; peridial morphology), Moreno *et al.* (*Persoonia* **18**: 215, 2003).

Discoascina Beneš (1961), Fossil Fungi. 1 (Carboniferous), former Czechoslovakia.

Discocainia J. Reid & A. Funk (1966), Rhytismataceae. 3, widespread. Some authors place this in the *Helotiaceae*. See Sherwood (*Occ. Pap. Farlow Herb. Crypt. Bot.* **15**, 1980), Livsey & Minter (*CJB* **72**: 549, 1994).

discocarp, an ascoma in which the hymenium is uncovered when the asci and spores are mature; an apothecium.

Discocera A.L. Sm. & Ramsb. (1917) nom. rej. ? = Trapelia fide Hawksworth & David (*Taxon* **38**: 493, 1989), Rambold & Triebel (*Notes R. bot. Gdn Edinb.* **46**: 375, 1990; status).

Discochora Höhn. (1918) = Guignardia fide Bissett (*CJB* **64**: 1720, 1986), Hawksworth *et al.* (*Dictionary of the Fungi* edn 8, 1995).

Discocistella Svrček (1962) ≡ Cistella.

Discocolla Prill. & Delacr. (1894), anamorphic *Pezizomycotina*, Hsp.≡ eH.?. 1 or 2, widespread (north temperate).

Discocurtisia Nannf. (1983), Dermateaceae. 1 (on grasses), N. America. See Nannfeldt (*TBMS* **75**: 292, 1983).

Discocyphella Henn. (1900) = Marasmius fide Kuyper (*in litt.*).

Discodiaporthe Petr. (1921) = Melanconis fide Wehmeyer (*Revision of Melanconis*, 1941).

Discodothis Höhn. (1909) = Rhagadolobium fide Müller & von Arx (*Beitr. Kryptfl. Schweiz* **11** no. 2, 1962).

Discofusarium Petch (1921) = Fusarium fide Wollenweber & Reinking (*Die Fusarien*, 1935).

Discogloeum Petr. (1923), anamorphic *Pezizomycotina*, Cac.0eH.15. 2, Europe. See von Arx (*Biblthca Mycol.* **24**, 1970).

Discohainesia Nannf. (1932), ? Leotiomycetes. Anamorphs *Hainesia*, *Pilidium*. 1, N. America; Europe. The anamorph connection requires further research. See Shear & Dodge (*Mycol.* **13**: 135, 1921), Palm (*Mycol.* **83**: 787, 1991), Rossman *et al.* (*Mycol. Progr.* **3**: 275, 2004; phylogeny, anamorph).

discoid, flat and circular; resembling a disk (Fig. 23.5*a*, *b*).

discolourous, of a different colour, as of the two surfaces of a foliose lichen thallus.

Discomycella Höhn. (1912), ? Helotiales. 1, Java.

discomycete, one of the *Discomycetes*.

Discomycetella Sanwal (1953) ? = Miladina. The type cannot be located. fide Kimbrough (*Bot. Rev.* **36**: 91, 1970), Pfister (*Am. J. Bot.* **60**: 355, 1973).

Discomycetes. **Cup fungi**; class of *Ascomycota* formerly used for taxa with ascomata which are sessile, open, ± saucer-shaped or cup-shaped apothecia, but these may be covered by a membrane at first or be permanently closed and hypogeous, or have the hymenium borne on stipitate convoluted structures; the apothecia are generally ascohymenial in ontogeny, with unitunicate asci (inoperculate or operculate). The *Helotiales*, *Ostropales*, *Pezizales* and *Rhytismatales* are regularly included, but the name has also been applied to some *Dothideales*, *Patellariales* and *Lecanorales*. The class is not accepted in modern classification but 'discomycetes' still has value as a colloquial descriptive term.

Lit.: **General**: Boudier (1907, 1905-10; see Boudier), Nannfeldt (1932), Dennis (1978), Bellèmere (*BSMF* **83**: 393, 753, 1968; inoperc. ontogeny; *Revue mycol.* **41**: 233, 1977; ascus ultrastr.), Kimbrough (*Bot. Rev.* **36**: 91, 1970; classification; fam. key; *in* Parker, 1982, **1**: 232), Korf (1973), Kamaletdinova & Vassilyeva (*Cytology of discomycetes*, 1982; ultrastr.). See also Le Gal (1953), Seaver (1928) cited below, and under *Ascomycota*, ascus, Lichens and Macromycetes.

Regional: **America, North**, Seaver (*North American cup-fungi. (Operculates)*, 1928; suppl. 1942; nomenclatural revision, Pfister, *Occ. Pap. Farlow Herb. Cryptog. Bot.* **17**, 1982); *North American cup-fungi. (Inoperculates)*, 1951). **Argentina**, Gamundí (*Darwiniana* **12**: 386, 1962, *Leotiales*; **13**: 568, 1964, *Pezizales*; keys). **Australasia**, Rifai (*Verh. K. ned. Akad. wet.* ser. 2 **57**(3), 1968; Operculates); Spooner (1987; *Leotiales*) **British Isles**, Ramsbottom & Balfour-Browne (*TBMS* **34**: 38, 1951; checklist), Dennis (1978), Cannon *et al.* (1981). **Canada**, Abbott & Currah (*The larger cup fungi and other ascomycetes of Alberta*, 1989; keys). **former Czechoslovakia**, Velenovský (*Monographia discomycetum Bohemiae*, 2 vols, 1934). **Europe**, see Boudier (1907, 1905-10). **France**, Gillet (*Champignons de France. Les Discomycetes*, 1879-83 [reprint 1979 (*Bull. Soc. bot., C.-o.*; num. spéc. 3)]). **Germany**, Baral & Krieglsteiner (*Z. Mykol., Beih.* **6**, 1985; inoperculates). **India**, Batra & Batra (*Kansas Univ. Sci. Bull.* **44**: 109, 1963; 185 spp. keys), Thind & Singh (*Res. Bull. Punjab Univ.* **22**: 51, 1971). **Japan**, Otani (*Trans. mycol. Soc. Japan* **31**: 117, 1990; keys gen.). **Madagascar**, Le Gal (*Les discomycètes de Madagascar* [*Prodr. Fl. mycol. Madagascar* 40], 1953). **Russia**, Naumov (*Flora gribov Leningradskovoblasti, Vypusk II. Diskomitsety*, 1964; keys). **Switzerland**, Breitenbach & Kränzlin (*Pilzäner Schweiz* **1**, 1984).

Discomycetoidea Matsush. (1993), anamorphic *Pezizomycotina*, Hsy.0eP.22. 1, Ecuador. See Matsushima (*Matsush. Mycol. Mem.* **7**: 49, 1993).

Discomycopsella Henn. (1902) = Phyllachora Nitschke ex Fuckel (1870) fide von Arx & Müller (*Beitr. Kryptfl. Schweiz* **11** no. 1, 1954).

Discomycopsis Müll. Arg. (1893) ? = Euryachora fide Ferdinandsen & Rostrup (*Dansk bot. Ark.* **5** no. 20, 1928).

Discorehmia Kirschst. (1936), ? Helotiaceae. 5, Europe. See Nauta & Spooner (*Mycologist* **13**: 3, 1999).

Discosia Lib. (1837), anamorphic *Amphisphaeriaceae*, St.≡ eP.19. 32, widespread (esp. temperate). See Subramanian & Reddy (*Kavaka* **2**: 57, 1974; types), Reddy *in* Subramanian (Ed.) (*Taxonomy of fungi* **2**: 493, 1984), Nag Raj (*CJB* **69**: 1246, 1991; excluded spp.), Vanev (*Mycotaxon* **41**: 387, 1991; sects.), Nag Raj (*Coelomycetous Anamorphs with Appendage-bearing Conidia*, 1993), Vanev (*Mycotaxon* **49**: 199, 1993), Wu & Sutton (*MR* **100**: 287, 1996), Jeewon *et al.* (*Mol. Phylogen. Evol.* **25**: 378, 2002; phylogeny), Jeewon *et al.* (*MR* **107**: 1392, 2003; phylogeny).

Discosiella Syd. & P. Syd. (1912), anamorphic *Pezizomycotina*. See Subramanian & Reddy (*Proc. Indian natn Sci. Acad.* Part B. Biol. Sci. **75**: 111, 1972), Nag Raj (*CJB* **59**: 2519, 1981), Eriksson & Hawksworth (*SA* **11**: 56, 1992), Kobayashi *et al.* (*Mycoscience* **46**: 78, 2005; Japan).

Discosiellina Subram. & K.R.C. Reddy (1972), anamorphic *Pezizomycotina*, Cpt.0fH.1. 1, India. See Nag Raj (*CJB* **59**: 2531, 1981).

Discosiopsis Edward, Kr.P. Singh, S.C. Tripathi, M.K. Sinha & Ranade (1974) = Pestalotiopsis fide Sutton (*Mycol. Pap.* **141**, 1977).

Discosiospora A.W. Ramaley (1989) = Discosia fide Nag Raj (*Coelomycetous Anamorphs with Appendage-bearing Conidia*, 1993).

Discosphaera Dumort. (1822) nom. rej. prop. = Hypoxylon Bull. See Læssøe (*SA* **13**: 43, 1994; ? synonym of *Xylaria*).

Discosphaerina Höhn. (1917), Dothioraceae. Anamorphs *Kabatia*, *Kabatiella*, *Sarcophoma*, *Selenophoma*, *Hormonema*-like, *Aureobasidium*-like. 11, Europe; N. America. The genus has been confused with *Guignardia*. See Barr (*Contr. Univ. Mich. Herb.* **9**: 523, 1972), Holm *et al.* (*Karstenia* **39**: 59, 1999), Yurlova *et al.* (*Stud. Mycol.* **43**: 63, 1999; DNA), Lumbsch & Lindemuth (*MR* **105**: 901, 2001; phylogeny).

Discospora Arthur (1907) = Pileolaria fide Arthur (*Manual Rusts US & Canada*, 1934).

Discosporella Höhn. (1927) ≡ Conostroma fide Petrak (*Annls mycol.* **27**: 371, 1929).

Discosporiella Petr. (1923) = Cryptosporiopsis fide von Arx (*Verh. K. ned. Akad. Wet. Amst.* C **51**: 22, 1957).

Discosporina Höhn. (1927), anamorphic *Pezizomycotina*, Cac.0eH.15. 4, widespread.

Discosporiopsis Petr. (1921) = Phacidiopycnis fide Sutton (*Mycol. Pap.* **141**, 1977).

Discosporium Höhn. (1915), anamorphic *Cryptodiaporthe*, St.0eH.19. 3, widespread. See Sutton (*The Coelomycetes*, 1980).

Discosporium Sacc. & P. Syd. (1902), anamorphic *Pezizomycotina*, Hsp.0eP.?. 1, Italy.

Discostroma Clem. (1909), Amphisphaeriaceae. Anamorph *Seimatosporium*. 28, widespread. See Brockmann (*Sydowia* **28**: 275, 1976; key), Sivanesan (*TBMS* **81**: 325, 1983), Kang *et al.* (*Fungal Diversity* **1**: 147, 1998; DNA), Kang *et al.* (*MR* **103**: 53, 1999), Jeewon *et al.* (*Mol. Phylogen. Evol.* **25**: 378, 2002; phylogeny), Kang *et al.* (*Mycotaxon* **81**: 321, 2002), Jeewon *et al.* (*MR* **107**: 1392, 2003; phylogeny), Hatakeyama & Harada (*Mycoscience* **45**: 106, 2004; on *Paeonia*), Paulus *et al.* (*Sydowia* **58**: 76, 2006; key).

Discostromella Petr. (1924) = Leptostromella fide Clements & Shear (*Gen. Fung.*, 1931).

Discostromopsis H.J. Swart (1979) = Discostroma fide Sivanesan (*TBMS* **81**: 325, 1983).

Discotheciella Syd. & P. Syd. (1917), anamorphic *Pezizomycotina*, Cpt.1eH.?. 1, Philippines.

discothecium, an ascostroma resembling an apothecium but bearing cylindrical bitunicate asci and differing from a hysterothecium by the weathering away of the covering layer (Korf, *Mycol.* **54**: 25, 1962).

Discothecium Syd. & P. Syd. (1916) ≡ Discotheciella.

Discothecium Zopf (1897) = Endococcus fide Hawksworth (*Bot. Notiser* **132**: 283, 1979).

Discoxylaria J.C. Lindq. & J.E. Wright (1964), Xylariaceae. Anamorph *Hypocreodendron*. 1, Argentina; Mexico. See Rogers *et al.* (*Mycol.* **87**: 41, 1995).

Discozythia Petr. (1922), anamorphic *Pezizomycotina*, ?.0eH.?. 3, Europe.

discrete (1) separate; not joining; (2) (of a conidiogenous cell), not subtended by a conidiophore; cf. integrated.

discrete body, a non-functional cleistothecial initial of a dermatophyte in culture; pseudocleistothecium.

Discula Sacc. (1884), anamorphic *Apiognomonia, Gnomonia, Gnomoniella, Plagiostoma*, Cac.0eH.15. 15, Europe; N. America. See von Arx (*Verh. K. ned. Akad. Wet. Amst.* C **51**: 32, 1957), Petrak (*Sydowia* **15**: 221, 1962), Petrak (*Sydowia* **24**: 270, 1971), Swart *et al.* (*Phytophylactica* **22**: 143, 1990; *D. platani* on plane trees in S. Afr.), Redlin (*Mycol.* **83**: 633, 1991; *Cornus* anthracnose), Haemmerli *et al.* (*Mol. Plant-Microbe Interact.* **5**: 479, 1992; differentiation by RAPDs), Toti *et al.* (*MR* **96**: 420, 1992; morphometry), Pacumbaba & Beyl (*Phytopathology* **83**: 467, 1993; physiology), Stanosz (*Pl. Dis.* **77**: 1022, 1993; on *Acer*), Trigiano *et al.* (*Phytopathology* **83**: 1338, 1993; DNA), McElreath *et al.* (*Curr. Microbiol.* **29**: 57, 1994; double-stranded RNA in *D. destructiva*), Viret & Petrini (*MR* **98**: 423, 1994; colonization), Viret *et al.* (*New Phytol.* **127**: 123, 1994; infection process), Yao *et al.* (*Curr. Microbiol.* **29**: 145, 1994; genetics), Trigiano *et al.* (*Mycol.* **87**: 490, 1995; biogeography), Cohen (*Mycol.* **91**: 917, 1999; isolation), Caetano-Anollés *et al.* (*Curr. Genet.* **39**: 346, 2001; evolution), Rong *et al.* (*Curr. Microbiol.* **42**: 144, 2001; phylogeny), Zhang & Blackwell (*Mycol.* **93**: 355, 2001; phylogeny), Castlebury *et al.* (*Mycol.* **94**: 1017, 2002; phylogeny), Zhang & Blackwell (*Phytopathology* **92**: 1276, 2002; population structure), Green & Castlebury (*MR* **111**: 62, 2007; phylogeny, anamorph).

Disculina Höhn. (1916), anamorphic *Dothideomycetes*, St.0eH.19. 1, Europe.

disinfectant, a substance for the destruction of pathogenic microorganisms.

disjunctor, a cell or projection, sometimes having a short existence, developing through the pores of septal lamellae of adjoining conidia in a chain (e.g. in *Monilinia*); a connective. See Batra (*Mycol.* **80**: 660, 1988).

disk, see disc.

Disparidicellites Kalgutkar & Janson. (2000), Fossil Fungi. 1, China. See Kalgutkar & Jansonius (*AASP Contributions Series* **39**: 94, 2000).

Disperma Theiss. (1916) [non *Disperma* J.F. Gmel. 1792, *Rubiaceae*] ≡ Dicarpella.

dispersal spore, a spore disseminated by wind, water, or other agent; diaspore.

Dispira Tiegh. (1875), Dimargaritaceae. 4 (mycoparasitic on *Mucorales* and *Chaetomium*), widespread. See Benjamin (*Aliso* **5**: 248, 1963; key), Kurtzman (*Mycol.* **60**: 915, 1968; parasitism & culture), Misra & Lata (*Mycotaxon* **8**: 372, 1979), White *et al.* (*Mycol.* **98**: 872, 2006; phylogeny).

dispore, one of the spores of a 2-spored basidium as opposed to a **tetraspore**, one of the spores of a 4-spored basidium (Corner, 1947). Cf. monospore.

Disporotrichum Stalpers (1984), anamorphic *Agaricales*, Hso.0eH.10. 1, Netherlands. See De Hoog & Gerrits (*Mycoses* **41**: 183, 1998; molec. diagn. clinical isol.).

dissepiment, a partition, e.g. that between the pores of a polypore.

Dissitimurus E.G. Simmons, McGinnis & Rinaldi (1987), anamorphic *Pezizomycotina*, Hso.≡ eP.3/10. 1 (from humans), USA. See Simmons *et al.* (*Mycotaxon* **30**: 247, 1987).

Dissoacremoniella Kiril. (1970), anamorphic *Pezizomycotina*, Hso.0eH.6. 1, former USSR. See Kirilenko (*Nov. sist. Niz. Rast.* **7**: 235, 1970).

dissociation, Leonian's name for mutation or saltation.

Dissoconium de Hoog, Oorschot & Hijwegen (1983), anamorphic *Mycosphaerella*, Hso.0-1eH.10. 2, widespread. See de Hoog & Takeo (*Antonie van Leeuwenhoek* **59**: 285, 1991; karyology), Crous *et al.* (*Sydowia* **51**: 155, 1999), Crous *et al.* (*Stud. Mycol.* **45**: 107, 2000; phylogeny), Crous *et al.* (*Mycol.* **93**: 1081, 2001; phylogeny), Crous *et al.* (*Stud. Mycol.* **50**: 195, 2004; on *Eucalyptus*), Hunter *et al.* (*Stud. Mycol.* **55**: 147, 2006; on *Eucalyptus*), Arzanlou *et al.* (*Persoonia* **20**: 24, 2008; phylogeny).

Dissoderma (A.H. Sm. & Singer) Singer (1974) = Squamanita fide Kuyper (*in litt.*).

Dissophora Thaxt. (1914), Mortierellaceae. 3, widespread (north temperate). See Carreiro & Koste (*CJB* **70**: 2177, 1982; growth temperature), Gams & Carreiro (*Stud. Mycol.* **31**: 85, 1985; taxonomy), Benny (*CJB* **73** Suppl. 1: S725, 1995), Voigt & Wöstemeyer (*Gene* **270**: 113, 2001; phylogeny), White *et al.* (*Mycol.* **98**: 872, 2006; phylogeny).

distal, situated away from either the centre of a body or the point of origin; terminal; cf. proximal.

Distichomyces Thaxt. (1905) = Rickia fide Thaxter (*Proc. Amer. Acad. Arts & Sci.* **48**: 363, 1912).

distichous, in two lines.

Distocercospora N. Pons & B. Sutton (1988), anamorphic *Mycosphaerellaceae*, Hso.≡ eP.10. 1, widespread. See Pons & Sutton (*Mycol. Pap.* **160**: 60, 1988), Crous & Braun (*CBS Diversity Ser.* **1**: 571 pp., 2003), Kirschner *et al.* (*Fungal Diversity* **17**: 57, 2004; Taiwan).

Distolomyces Thaxt. (1931), Laboulbeniaceae. 3, Europe; Asia. See Santamaría (*Fl. Mycol. Iberica* **5**, 2003; Iberian peninsula).

Distopyrenis Aptroot (1991), Pyrenulaceae. 8, widespread (tropical). See Harada (*Mycoscience* **41**: 491, 2000).

Distorimula F. San Martín, P. Lavín, Esqueda (1999), Xylariales. 1, Mexico. See San Martín *et al.* (*Mycotaxon* **73**: 263, 1999).

distoseptate (of septation), having the individual cells each surrounded by a sac-like wall distinct from the outer wall, as in *Drechslera* (Luttrell, *Mycol.* **55**: 672, 1963) (Fig. 22A-B). Cf. euseptate.

Distothelia Aptroot (2005), ? Monoblastiaceae (L). 1, Hong Kong. See Seaward & Aptroot (*Bryologist* **108**: 284, 2005).

distribution, see Geographical distribution.

Ditangifibula G.C. Adams (1996), anamorphic *Pezizomycotina*, Hso.?.?. 1, Alaska. See Adams *et al.* (*Mycol.* **87**: 911, 1995), Kirschner & Chen (*Stud. Mycol.* **50**: 337, 2004).

Ditangium P. Karst. (1867), anamorphic *Craterocolla*. 1, Europe. See Donk (*Persoonia* **4**: 165, 1966).

Dithelopsis Clem. (1909) = Thelopsis fide Hawksworth *et al.* (*Dictionary of the Fungi* edn 8, 1995).

dithiocarbamates, organic fungicides; **dimethyl-** (DMDC): thiram, ferbam, ziram; **ethylene-bis-** (EBDC): maneb, mancozeb, zineb.

Dithozetia Clem. & Shear (1931) ≡ Didymothozetia.

Ditiola Fr. (1822), Dacrymycetaceae. *c.* 10, widespread. See McNabb (*N.Z. Jl Bot.* **4**: 546, 1966), Govorova (*Mikologiya i Fitopatologiya* **28**: 1, 1994).

Ditiola P. Browne (1756) ? = Schizophyllum fide Donk (*Beih. Nova Hedwigia* **6**: 89, 1962).

Ditiola Schulzer (1860) ≡ Holwaya.

Ditmaria Lühnem. (1809) nom. dub., ? Fungi.

Ditopella De Not. (1863), Gnomoniaceae. 1 (on *Alnus* bark), Europe. See Reid & Booth (*CJB* **45**: 1479, 1967), Barr (*Mycol. Mem.* **7**: 232 pp., 1978), Castlebury *et al.* (*Mycol.* **94**: 1017, 2002; phylogeny).

Ditopellina J. Reid & C. Booth (1967), Diaporthales. 1, Europe. See Reid & Booth (*CJB* **45**: 1481, 1967), Stoykov & Denchev (*Mycol. Balcanica* **3**: 179, 2006; Bulgaria).

Ditopellopsis J. Reid & C. Booth (1967), Diaporthales. 4, N. America. See Barr (*Mycol. Mem.* **7**, 1978), Barr (*Mycotaxon* **41**: 287, 1991), Monod (*Beih. Sydowia* **9**: 1, 1993).

Ditremis Clem. (1909) nom. rej. = Anisomeridium fide McCarthy (*Aust. Syst. Bot.* **5**: 125, 1992; Australia), McCarthy (*Muelleria* **8**: 1, 1993), Hawksworth *et al.* (*Dictionary of the Fungi* edn 8, 1995).

Ditylis Clem. (1909) = Tylophoron fide Tibell (*Beih. Nova Hedwigia* **79**: 597, 1984).

diurnal, in daylight hours. Cf. circadian, diel.

divaricate, divergent at right angles.

Diversispora C. Walker & Schüssler (2004), Diversisporaceae. 1, USA. See Walker & Schüssler (*MR* **108**: 982, 2004), Redecker & Raab (*Mycol.* **98**: 885, 2006; phylogeny).

Diversisporaceae C. Walker & A. Schüssler (2004), Diversisporales. 1 gen., 1 spp.

Lit.: Azcon-Aguilar & Barea (*Mycorrhiza* **6**: 457, 1996), van der Heijden *et al.* (*Nature* Lond. **396**: 69, 1998), Schüßler *et al.* (*MR* **105**: 1413, 2001), Walker & Schüssler (*MR* **108**: 982, 2004), Redecker (*Glomeromycota* Arbuscular mycorrhizal fungi and their relative(s). Version 01 July 2005. http://tolweb.org/Glomeromycota/28715/2005.07.01 in The Tree of Life Web Project, http://tolweb.org: [unpaginated], 2005).

Diversisporales C. Walker & A. Schüssler (2004). Glomeromycetes. 4 fam., 6 gen., 74 spp. Fams:
(1) **Acaulosporaceae**
(2) **Diversisporaceae**
(3) **Diversisporaceae**
(4) **Pacisporaceae**

For *Lit.* see under fam.

diverticulum, a pocket-like side branch, as on mycelium of *Pythium*.

Divinia Cif. (1955), anamorphic *Pezizomycotina*, Hso.≡ eP.?. 1, Dominican Republic. See Ciferri (*Sydowia* **9**: 326, 1955).

Dixidium R.A. Poiss. (1932) = Smittium fide Manier & Lichtwardt (*Annls Sci. Nat.* Bot., sér. 12 **9**: 519, 1968).

Dixomyces I.I. Tav. (1985), Laboulbeniaceae. 14, widespread. See Tavares (*Mycol. Mem.* **9**: 627 pp., 1985), Terada (*Mycoscience* **39**: 77, 1998; Japan).

Dixophyllum Earle (1909) = Russula fide Singer (*Agaric. mod. Tax.*, 1951).

DNA, see Molecular biology.

DNA fingerprinting. This relies on repetitive sequences which are dispersed through the genome and which have a high variability; these are valuable in the manufacture of genotype-specific probes for species or pathogenic races, e.g. in mycology in *Magnaporthe grisea* (Hamer *et al.*, *Proc. Nat. Acad. Sci., USA* **86**: 9981, 1989). DNA fingerprinting is of particular use in characterization of individual strains and populations, especially in medical mycology (e.g. Girordin *et al.*, *J. Infect. Dis.* **169**: 683, 1994). Most methods use the polymerase chain reaction (PCR, see Molecular Biology). Some more widely-used methods are listed below. In some cases more than one name or acronym has been appropriated to describe essentially the same method but with subtle variations (e.g. AP-PCR [arbitrarily primed PCR] has been used as a synonym for both RAPD [q.v.] and SSR-PCR [q.v.] by separate researchers; strictly speaking it is more accurately applied to the former as SSR-PCR does not involve 'arbitrary' primers).

RAPD (random amplified polymorphic DNA) analysis uses PCR with short primers to produce sufficient amounts of particular parts of a target DNA or RNA sequence that can then be compared in agarose gels; this is especially useful in studies of closely allied fungi where species-specific patterns can be found (Schaad *et al.*, *in* Hawksworth, 1994: 461), and when dealing with small samples (e.g. a single spore or old type collection). Problems with lack of reproducibility have led to this method being largely superseded by other, more consistent, methods (see below).

SSR-PCR / VNTR-PCR (simple sequence repeat PCR / variable number tandem repeat PCR). In general this method may be viewed as a more stringent version of RAPD involving the use of longer oligonucleotide primers and higher annealing temperatures. The primers tend to be made of repeated double, triple or even quadruple units [e.g. (CA)9, (CAG)5, (GACA)4] which are complementary to the flanking regions of variable number tandem repeats of the microsatellite DNA (see Bridge *et al.*, *Lett. Appl. Microbiol.* **24**: 426, 1997).

ISSR-PCR (inter-simple sequence repeat PCR). This method is a modified form of the SSR-PCR method that incorporates higher annealing temperatures and oligonucleotide primers comprising simple sequence repeat units anchored with so-called degenerate ends [e.g 5'-BDB(ACA)5 where 'B' denotes 'not A' and 'D' denotes 'not C'; see Grünig *et al.*, *MR* **105**: 24, 2001; use in dark septate endophytes]. This method is considered superior to many other fingerprinting methods owing to its genome-wide coverage (rather than being localised to particular 'hotspots') and the large number of fragments generated that result in complex banding patterns (see Taylor *et al.*, *Ann. Rev. Phytopathol.* **37**: 197, 1999).

AFLP (amplified fragment length polymorphism) is a technique that combines the resolving power of RAPD (q.v.) with the reproducibility of RFLP analyses (see *Molecular Biology*); first developed by Vos *et al.* (*Nuc. Acids Res.* **23**: 4407, 1995). Has since been used extensively in fungi even from a single spore (e.g. see Rosendahl & Taylor, *Mol. Ecol.* **6**: 821, 1997; use in arbuscular mycorrhizal fungi, Majer *et al.*, *MR* **100**: 1107, 1996; use with *Cladosporium*, Duncan & Cooke, *Mycologist* **16**: 59, 2002; use with *Phytophthora*).

SCARs (sequence-characterised amplified regions) are polymorphic bands (produced normally by one of the methods above and visualised on a polyacrylamide gel for optimal separation) which are found to be diagnostic for a particular population or individual. Such bands are excised from the gel and the DNA extracted and sequenced, allowing new specific primers to be created that may allow detection of the particular strain (e.g. Hermosa *et al.*, *Curr. Gen.* **38**: 343, 2001; SCAR marker in *Trichoderma*).

SSCP (single-strand conformational polymorphism) allows the separation of isolates that produce fragments of identical size from a given amplification. This is achieved by the use of native polyacrylamide gels and, in theory, can allow separation of samples differing in as little as a single base difference (e.g. Bonello *et al.*, *New Phytologist* **138**: 533, 1998; use with *Suillus*).

DGGE (denaturing gradient gel electrophoresis) is a sensitive method that enables the discrimination of similarly sized DNA fragments (obtained by PCR amplification of particular regions but using primers containing GC-rich clamps) on the basis of their actual sequence via the medium of an acrylamide gel containing a gradient of denaturing chemical (usually urea and/or formamide; alternatively temperature gradients may be used [i.e. TGGE]). DGGE has been employed widely in bacterial environmental studies where microbial consortia are present but has also been applied to fungal investigations (e.g. Vainio & Hantula, *MR* **104**: 927, 2000; use with wood-inhabiting fungi).

DNA sequencing. This involves determination of individual consecutive bases within a given region of DNA. Most sequence analysis is undertaken using the Sanger chain termination method (Sanger & Coulson, *J. Mol. Biol.* **93**: 441, 1975) although alternative methods are being developed. In practice the term DNA sequencing encompasses additional aspects of DNA analysis, including comparison of base sequences in particular parts of DNA (or RNA) molecules (for total genome sequencing see Genomics). Ribosomal gene clusters have been most used in phylogenetic studies by mycologists, particularly the genes coding for 5.8S, 18S and 28S rRNA (ribosomal RNA) genes and adjacent non-coding spacers (e.g. internal transcribed spacer [ITS] and intergenic spacer [IGS]) but also functional genes where appropriate (e.g. chitinase for *Trichoderma*, tef for *Fusarium* and *Trichoderma* and β-tubulin for *Botrytis* and *Penicillium*, etc.). The method involves the PCR method (see Molecular biology), and base sequences are determined either manually or by automated sequencing machines. Increasingly DNA sequencing is undertaken by capillary electrophoresis but so-called '(polyacrylamide) slab gel' DNA sequencers are still used. Sequences obtained can be subjected to phylogenetic analyses, including already available sequences accessible from on-line databanks. A basic introduction to the method, which is now pivotal in fungal phylogenetic studies, is provided by Mitchell et al. (*Mycologist* **9**: 67, 1995).

Voucher specimens. In all molecular studies it is advisable to deposit reference material of the fungi actually used in an appropriate Reference Collection (see Collection and preservation).

DNA sequence libraries: *EMBL Nucleotide Sequence Database* (http://www.ebi.ac.uk/embl), maintained at the European Molecular Biology Laboratory's European Bioinformatics Institute; *GenBank* (http://www.ncbi.nlm.nih.gov/Genbank), maintained at the National Center for Biotechnology Information, National Institutes of Health, Bethesda, Md, USA; *DDBJ* (http://www.ddbj.nig.ac.jp), maintained at the DNA Data Bank of Japan. All three databases exchange new sequence data accessions on a daily basis; see also *Nucleic Acid Res.* **36**, 2008; database issue). Recent studies have raised concerns regarding the authenticity/validity of some identifications behind the holdings of these global databases, with suggestions that up to 20% are incorrectly named (see Bridge *et al.*, *New Phytologist* **160**: 43, 2003) reinforcing the need for responsible vouchering of specimens (see Vilgalys, *New Phytologist* **160**: 4, 2003). Validated sequence databases are available commercially which can allow users to make identifications with appropriate levels of confidence if the relevant databases are purchased. DNA sequencing is now being used to facilitate resolution of morphologically similar taxa (e.g. Pérez-Sierra & Henricot, *Mycologist* **16**: 42, 2002).

DNA barcoding. Since the Convention on Biological Diversity (1993) and the resulting Bonn Guidelines (2002), the possibility has been raised of distinguishing taxa mechanically by analogy with the way supermarket products of similar types are distinguished with a hand-held scanner reading a black and white striped product code. The potential for this so-called 'barcoding' of all living organisms is being addressed by the *Consortium for the Barcoding of Life* (CBOL; see http://barcoding.si.edu and www.dnabarcodes.org). For each organism, an initial specimen must be collected, followed by laboratory DNA analysis, database generation and data analysis. The supposition is that there is a single region of DNA in all living organisms characteristic for each individual species. In Animalia a region of the mitochondrial cytochrome oxidase c 1 (*cox I*) gene has been proposed for use. For Fungi this region does not seem to be universally suitable because multiple copies exist in some taxa and there may be large scale length variation (see Rossman, *Inoculum* **58**: 1, 2007). The large volume of work already undertaken with the rDNA gene cluster, especially the ITS regions, has led to it being proposed for use in the Fungi, qualified with further work on an additional appropriate region (as the ITS is not appropriate in all taxa due to multiple non-orthologous copies of the region; e.g. see O'Donnell & Cigelnik, *Mol. Phyl. Evol.* **7**: 103, 1997; for *Fusarium*).

Doassansia Cornu (1883), Doassansiaceae. 12 (on aquatic plants), widespread. Spore ball, immersed in host tissue, has cortex of sterile cells. See Vánky (*Sydowia* **34**: 167, 1981).

Doassansiaceae R.T. Moore ex P.M. Kirk, P.F. Cannon & J.C. David (2001), Doassansiales. 12 gen. (+ 4 syn.), 39 spp.

Lit.: Vánky (*Cryptog. Stud.* **1**: 159 pp., 1987), Begerow *et al.* (*CJB* **75**: 2045, 1998), Piepenbring *et al.* (*Protoplasma* **204**: 155, 1998), Vánky (*MR* **102**: 513, 1998), Begerow *et al.* (*MR* **104**: 53, 2000), Vánky (*Aust. Syst. Bot.* **14**: 385, 2001), Piepenbring (*Bot. Jb.* **24**: 241, 2003).

Doassansiales R. Bauer & Oberw. (1997). Exobasidiomycetes. 3 fam., 14 gen., 42 spp. Fams:
(1) **Doassansiaceae**
(2) **Melaniellaceae**.
(3) **Rhamphosporaceae**
For *Lit.* see under fam.

Doassansiella Zambett. (1970) nom. inval., anamorphic *Doassansiopsis*. 1. See Zambettakis (*Revue Mycol.* Paris **35**: 164, 1970).

Doassansiopsidaceae Begerow, R. Bauer & Oberw. (1998), Urocystidales. 2 gen., 14 spp.

Lit.: Vánky (*Cryptog. Stud.* **1**: 159 pp., 1987), Bauer *et al.* (*CJB* **75**: 1273, 1997), Begerow *et al.* (*CJB* **75**: 2045, 1998), Vánky (*Aust. Syst. Bot.* **14**: 385, 2001), Piepenbring (*Bot. Jb.* **24**: 241, 2003), Begerow *et al.* (*MR* **108**: 1257, 2004).

Doassansiopsis (Setch.) Dietel (1897), Doassansiopsidaceae. Anamorph *Doassansiella*. 13 (on aquatic plants), widespread. See Vánky (*Mycotaxon* **79**: 231, 2001), Vánky (*Mycotaxon* **95**: 1, 2006).

Doassinga Vánky, R. Bauer & Begerow (1998), Doassansiaceae. 1 (aquatic on *Callitrichaceae*), Europe. See Hoog *et al. in* Kurtzman & Fell (Eds) (*Yeasts, a taxonomic study* 4th edn: 574, 1998; review).

Dochmiopus Pat. (1887) = Crepidotus fide Singer & Smith (*A monograph on the genus Galerina Earle*, 1964).

Dochmolopha Cooke (1878) = Seimatosporium fide Shoemaker (*CJB* **42**: 411, 1964).

Dodge (Carroll William; 1895-1988; USA). Student, Middlebury College, Vermont (1912-1916); Fellow, Washington University, St Louis (1915-1918); Sergeant, US Army (1918-1920); Instructor in Botany (1921-1924) then Assistant Professor and Curator, Farlow Herbarium, Harvard University (1924-1931); Professor of Botany, Washington University and Mycologist, Missouri Botanic Garden (1931-1963). Noted for work on hypogeous fungi, medical mycology in Latin America, and on lichen-forming fungi, particularly from Antarctica; he also translated Gäumann's (q.v.) classic work *Comparative Morphology of the Fungi* into English. *Publs. Medical Mycology: Fungous Diseases of Men and other Mammals* (1935); Lichens and lichen parasites. *BANZ Antarctic Research Expedition 1929-1931. Reports* Series B **7** (1948); Some lichens of tropical Africa. I-V. *Beihefte zur Nova Hedwigia* (1953-1971). *Biogs, obits etc.* Randolph (*Mycol.* **82**: 160, 1990) [portrait].

Dodgea Malençon (1939) = Truncocolumella fide Smith & Singer (*Brittonia* **11**: 215, 1959).

Dodgella Zebrowski (1936), Fossil Fungi ? Chytridiomycetes. 3 (Eocene – ? Recent), widespread.

dog lichen, *Peltigera* spp., e.g. *P. canina*; used in folk lore for treatment of bites of a rabid dog.

dog stinkhorn, basidioma of *Mutinus caninus*.

Doguetia Bat. & J.A. Lima (1960) = Trichasterina fide Müller & von Arx (*Beitr. Kryptfl. Schweiz* **11** no. 2, 1962).

Doidge (Ethel Mary; 1887-1965; England, later South Africa). Assistant to Pole Evans, the plant pathologist and mycologist (1908-1912) then Professional Assistant (1912-1929) then Principal Plant Pathologist

(1929-1942), Division of Botany and Mycology, Transvaal Department of Agriculture, Pretoria. Made contributions on South African fungi (especially rusts) and phytopathology, and compiled a monumental checklist of South African fungi. *Publs.* A preliminary study of the South African rust fungi. *Bothalia* (1927); South African rust fungi *Bothalia* (1928); The South African fungi and lichens to the end of 1945. *Bothalia* (1950). *Biogs, obits etc.* Grummann (1974: 396); Gunn (*Bothalia* **9**: 251, 1967, bibliography, portrait).

Dokmaia Promp. (2003) ? = Phaeoisaria fide Promputtha *et al.* (*Sydowia* **55**: 100, 2003).

Dolabra C. Booth & W.P. Ting (1964), Dothideomycetes. 1, Malaysia. See Booth & Ting (*TBMS* **47**: 237, 1964).

dolabrate (**dolabriform**), hatchet-like in form (Fig. 23.24).

dolabriform, having the shape of the head of an axe or cleaver (Fig. 23.24).

dolicho- (in Greek combinations), long.

Dolichoascus Thibaut & Ansel (1970), Saccharomycetales. 1 (in *Insecta*), widespread. See Batra *in* Subramanian (Ed.) (*Taxonomy of fungi* **1**: 187, 1978; typification).

Dolichocarpus R. Sant. (1949), Roccellaceae (L). 1, Chile. See Tehler (*CJB* **68**: 2458, 1990; cladistics), Grube (*Bryologist* **101**: 377, 1998; phylogeny), Follmann (*Mitteilungen aus dem Institut für Allgemeine Botanik Hamburg* **30-32**: 61, 2002; biogeography).

dolichospore, a long spore.

Dolichousnea (Y. Ohmura) Articus (2004), Parmeliaceae (L). 4, widespread. See Articus (*Taxon* **53**: 932, 2004), Arup *et al.* (*Mycol.* **99**: 42, 2007; phylogeny).

doliiform, barrel-like in form (Fig. 23.31).

Doliomyces Steyaert (1961), anamorphic *Amphisphaeriaceae*, Cpd.≡ eP.19. 2, India; S. America. See Nag Raj & Kendrick (*CJB* **50**: 45, 1972).

dolipore septum, a septum of a dikaryotic basidiomycete hypha which flares out in the middle portion forming a barrel-shaped structure with open ends as shown by electron microscopy; see Markham (*MR* **98**: 1089, 1994; review), Moore (*in* Hawksworth (Ed.), *Identification and characterization of pest organisms*: 249, 1994; summary diagr., Fig. 13), Moore & McAlear (*Am. J. Bot.* **49**: 86, 1962); septal pore swelling; cf. parenthesome.

dollar spot, a turf disease caused by *Sclerotinia homeocarpa*.

Domain (Empire, Superkingdom). A category in Classification (q.v.) above that of kingdom; all fungal phyla belong in kingdoms in the domain *Eukaryota*.

Domingoella Petr. & Cif. (1932), anamorphic *Pezizomycotina*, Hsy.0eP.?. 4, widespread (esp. tropical). See Deighton & Pirozynski (*Mycol. Pap.* **128**, 1972).

Domin-scale, A 10-point scale used in ecological tables to indicate the approximate cover of a surface by different taxa: + = single individuals; 1 = a few individuals; 2 = sparsely distributed; 3 = frequent but cover less than 4%; 4 = cover 4-10%; 5 = cover 11-25%; 6 = cover 26-33%; 7 = cover 34-50%; 8 = cover 51-75%; 9 = 76-90%; 10 = cover 91-100%.

Donadinia Bellem. & Mel.-Howell (1990), Sarcosomataceae. 1, Europe; Asia. See Harrington *et al.* (*Mycol.* **91**: 41, 1999; phylogeny), Hansen & Pfister (*Mycol.* **98**: 1029, 2006; phylogeny).

Donk (Marinus Anton; 1908-1972; Indonesia, later Netherlands). Student, later doctoral student, University of Utrecht (1927-1933); Mycologist, Buitenzorg Botanic Garden, Java (1934-1940); interned in prison camp (1942-1945) [saving many lives by culturing a yeast to produce much needed vitamins]; Head of Herbarium Bogoriense, Buitenzorg Botanic Garden, Java (1947-1955); Head of the Mycological Department, Rijksherbarium, Leiden (1956-1972); Corresponding Member (1954-1962) then Full Member (1962-1972), Royal Dutch Academy of Sciences. Outstanding taxonomist (especially of *Agaricomycetes*, q.v.), and nomenclaturalist. *Publs.* The generic names proposed for Hymenomycetes. I-XIII. *Reinwardtia* and *Taxon* various vols (1951-1963); The generic names proposed for *Agaricaceae*. *Beihefte zur Nova Hedwigia* (1962); A conspectus of the families of *Aphyllophorales*. *Persoonia* (1964). *Biogs, obits etc.* Maas Geesteranus (*Persoonia* **7**: 119, 1973) [bibliography, portrait]; Singer (*Mycol.* **65**: 503, 1973) [portrait]; Stafleu & Cowan (*TL-2* **1**: 671, 1976).

Donkella Doty (1950) = Clavulinopsis fide Donk (*Persoonia* **1**: 25, 1959) = Ramariopsis (Clavar.) fide, Petersen (*Mycol.* **70**: 660, 1978).

Donkia Pilát (1937) = Climacodon fide Maas Geesteranus (*Hydnaceous fungi of the eastern old world*, 1971).

Donkioporia Kotl. & Pouzar (1973), Fomitopsidaceae. 1, north temperate. See Kotlaba & Pouzar (*Persoonia* **7**: 214, 1973), Kleist & Seehann (*Z. Mykol.* **65**: 23, 1999; in buildings).

Dontuzia L.D. Gómez (1973), Pezizomycotina. 1, widespread (tropical). See Gómez (*Brenesia* **2**: 21, 1973).

Doratomyces Corda (1829), anamorphic *Microascaceae*. 1, widespread. See von Arx (*Gen. Fungi Sporul. Cult.* Edn 3, 1981; synonym of *Cephalotrichum*), Issakainen *et al.* (*Medical Mycology* **41**: 31, 2003; phylogeny, polyphyly), Zhang *et al.* (*Mycol.* **98**: 1076, 2006; phylogeny).

Doratospora J.M. Mend. (1930) = Rizalia fide Pirozynski (*Kew Bull.* **31**: 595, 1977).

dorsal, back or upper surface; the surface facing away from the axis, cf. ventral; sometimes used for the upper surface of foliose lichens.

Dothichiza Lib. ex Roum. (1880), anamorphic *Dothiora*, St.0eH.?. 15, widespread. See Petrak (*Sydowia* **10**: 201, 1957), Sutton (*Mycol. Pap.* **141**, 1977), Sutton & Livsey (*TBMS* **88**: 271, 1987).

Dothichloë G.F. Atk. (1894) = Balansia fide Diehl (*USDA agric. Monogr.* **4**, 1950), Reddy *et al.* (*Mycol.* **90**: 108, 1998).

Dothiclypeolum Höhn. (1916) = Thyriopsis fide Clements & Shear (*Gen. Fung.*, 1931).

Dothidasteris Clem. & Shear (1931) ≡ Dothidasteromella.

Dothidasteroma Höhn. (1909), Parmulariaceae. Anamorph *Melanoplaca*. 2, Asia; Australasia. See Swart (*TBMS* **91**: 581, 1988).

Dothidasteromella Höhn. (1910), Asterinaceae. 3, widespread. See Swart (*TBMS* **91**: 453, 1988), Hosagoudar *et al.* (*Journal of Mycopathological Research* **39**: 61, 2001).

Dothidea Fr. (1818) nom. cons., Dothideaceae. Anamorph *Asteromellopsis*. 20, widespread (temperate). See Silva-Hanlin & Hanlin (*MR* **103**: 153, 1999; DNA), Liew *et al.* (*Mol. Phylogen. Evol.* **16**: 392, 2000; phylogeny), Lumbsch & Lindemuth (*MR* **105**: 901, 2001; phylogeny), Shoemaker *et al.* (*Taxon* **52**:

623, 2003; nomencl.), Shoemaker & Hambleton (*CJB* **83**: 484, 2005), Schoch *et al.* (*Mycol.* **98**: 1041, 2006; phylogeny).

Dothideaceae Chevall. (1826), Dothideales. 13 gen. (+ 11 syn.), 79 spp.

Lit.: Butin (*Sydowia* **38**: 20, 1985), Hyde & Cannon (*Mycol. Pap.* **175**: 114 pp., 1999), Silva-Hanlin & Hanlin (*MR* **103**: 153, 1999; DNA), Lindemuth *et al.* (*MR* **105**: 1176, 2001), Schoch *et al.* (*Mycol.* **98**: 1041, 2006; phylogeny).

dothideaceous, having the asci in locules in a stroma, as in *Dothidea*.

Dothideales Lindau (1897). Dothideomycetidae. 4 fam., 57 gen., 350 spp. Ascomata perithecial, usually formed as lysigenous locules within stromatic tissue, dark brown. Hymenium not blueing in iodine. Interascal tissue in most cases absent. Asci ± cylindrical, thick-walled, fissitunicate, usually with a clearly defined ocular chamber but lacking other apical structures. Ascospores usually septate, if septate longitudinally asymmetrical, constricted at the primary septum but not always at the others, hyaline or brown, rarely ornamented. Mostly saprobes or necrotrophic parasites of plants. Fams:

(1) **Dothideaceae**

(2) **Dothioraceae**

The *Coccoideaceae* and *Planistromellaceae* are placed here provisionally as molecular data are lacking. *Lit.*: von Arx & Müller (*Stud. Mycol.* **9**, 1975, keys fams, gen.), Barr (*Prodromus to Class Loculoascomycetes*, 1987), Eriksson (*Opera Bot.* **60**, 1981), Schoch *et al* (*Mycol* **98**: 1041, 2006; phylogeny), Shoemaker *et al* (*Taxon* **52**: 623, 2003; nomencl.).

Dothideites Pat. (1893), Fossil Fungi. 1.

Dothidella Speg. (1880), Polystomellaceae. Anamorph *Stictochorella*. 3, widespread (tropical). See Wakefield (*TBMS* **24**: 282, 1940), Müller & von Arx (*Beitr. Kryptfl. Schweiz* **11** no. 2, 1962), Swart (*TBMS* **89**: 483, 1987).

Dothideodiplodia Murashk. (1927), anamorphic *Pezizomycotina*, St.1-≡ eP.19. 1, former USSR.

Dothideomycetes O.E. Erikss. & Winka (1997), Pezizomycotina. 11 ord., 90 fam., 1302 gen., 19010 spp. The largest and most varied class of the *Ascomycota*, including most ascolocular ascomycetes with J-bitunicate asci. Ascomata very varied, apothecial, perithecial or cleistothecial, at least nominally formed as lysigenous locules within stromatic tissue; hymenium sometimes gelatinous but not usually blueing in iodine. Interascal tissue frequently present, usually composed of branched or anastomosed paraphysoids or pseudoparaphyses, at least initially attached at both base and apex. Asci usually clavate or cylindrical, thick-walled, usually fissitunicate, rarely with apical structures. Ascospores almost always septate, longitudinally asymmetrical, constricted at the primary septum but not always at the others, sometimes muriform; hyaline or brown, not often ornamented. Saprobes, parasites, rarely lichen-forming, coprophilous, etc.

Subcl.:

(1) **Dothideomycetidae**

(2) **Pleosporomycetidae**

Several orders belong within the *Dothideomycetes* based on molecular data, but cannot be placed with confidence into a specific Subcl. These include the *Botryosphaeriales*, *Hysteriales*, *Jahnulales* and *Patellariales*. The *Microthyriales* are included here provisionally, as are a further 20 fams and a very large number of gen. without clear affinities, where molecular data are lacking. A number of orders and families placed within the *Dothideomycetidae* sensu *Dictionary* edn 9 are now included within an expanded *Eurotiomycetes*, notably the *Coryneliales* and *Pyrenulales*.

Lit.: von Arx & Müller (*Beitr. Kryptogfl. Schweiz* **11**(1), 1954; *Stud. Mycol.* **9**, 1975, keys fams, gen.), Barr (*Contr. Univ. Mich. Herb.* **9**: 523, 1972, N. Am.; *Mycol.* **71**: 935, 1979; *Prodromus to Class Loculoascomycetes*, 1987; orders, fams), Barr & Huhndorf (*in* McLaughlin *et al.* (Eds), *The Mycota* **7A**: 283, 2001), Benny & Kimbrough (*Mycotaxon* **12**: 1980; non-ostiolate gen.), Berbee (*Mol. Biol. Evol.* **13**: 462, 1996; evolution), Eriksson (*Opera Bot.* **60**, 1981), Hansford (*Mycol. Pap.* **15**, 1946; foliicolous spp.), Harris (*Mich. Bot.* **12**: 3, 1973; L,), Janex (*Revue bryol. lichén.* **37**: 421, 1971; ontogeny, L), Liew *et al.* (*Mol. Phylogenet. Evol.* **16**: 392, 2000; hamathecium, phylogeny), Lindemuth et al. (*MR* **105**: 1176, 2001; phylogeny), Lumbsch & Lindemuth (*MR* **105**: 901, 2001; phylogeny), Luttrell (*Phytopathology* **55**: 828, 1965, *in* Ainsworth *et al.* (Eds), *The Fungi* **4A**: 135, 1973; keys, gen.), Müller & von Arx (*Beitr. Kryptogfl. Schweiz* **11**(2), 1962), Parguey-Leduc (*Ann. Sci. nat.*, Bot. sér. 12, **7**: 33, 1966; ontogeny), Pirozynski (*Kew Bull.* **31**: 595, 1977; trop. fungicolous spp.), Poelt (*in* Ahmadjian & Hale (Eds), *The Lichens*: 599, 1974 ['1973']; fams, L), Reynolds (*Ascomycete Systematics. The Luttrellian Concept*, 1981; centrum types), Schoch et al. (*Mycol.* **98**: 1041, 2006; phylogeny), Theissen & Sydow (*Ann. Myc.* **13**: 149, 1915; gen.), also under ascus, hamathecium.

Dothideomycetidae P.M. Kirk, P.F. Cannon, J.C. David & Stalpers ex C.L. Schoch, Spatafora, Crous & Shoemaker (2007), Dothideomycetes. Ascomata perithecial or formed as lysigenous locules within stromatic tissue, often developing from pigmented external mycelium. Interascal tissue absent. Asci usually clavate, sometimes almost orbicular, thick-walled and fissitunicate. Ascospores almost always septate, longitudinally asymmetrical, constricted at the primary septum but not always at the others, sometimes muriform; hyaline or brown, not often ornamented. Mostly saprobes or parasites of plants. Ords:

(1) **Capnodiales**

(2) **Dothideales**

(3) **Myriangiales**

The *Dothideomycetidae* is much reduced compared with its treatment in the *Dictionary* edn 9, which equates largely to the *Dothideomycetes* of this edn. See Schoch *et al.* (*Mycol.* **98**: 1041, 2006).

Dothideopsella Höhn. (1915) = Leptosphaeria fide von Arx & Müller (*Stud. Mycol.* **9**, 1975).

Dothideovalsa Speg. (1909), Diatrypaceae. 3, widespread. See Petrak (*Sydowia* **5**: 169, 1951), Rappaz (*Mycol. Helv.* **2**: 285, 1987).

Dothidina Theiss. & Syd. (1915) = Coccodiella fide Müller & von Arx *in* Ainsworth *et al.* (Eds) (*The Fungi* **4A**: 87, 1973).

Dothidites Mesch. (1892), Fossil Fungi. 8 (Tertiary), Europe.

Dothidotthia Höhn. (1918), Pleosporales. *c.* 10, widespread (temperate). See Barr (*Mycotaxon* **34**: 517, 1989; key 7 N. Am. spp.), Aptroot (*Nova Hedwigia*

60: 325, 1995), Crous *et al.* (*Stud. Mycol.* **55**: 235, 2006; phylogeny, generic limits), Ramaley (*Mycotaxon* **94**: 127, 2005; anamorph).

Dothiomyces Bat. & J.L. Bezerra (1961) = Byssolecania fide Lücking *et al.* (*Lichenologist* **30**: 121, 1998).

Dothiopeltis E. Müll. (1957), Leptopeltidaceae. Anamorph *Idriella*-like. 1, Switzerland. See Holm & Holm (*Bot. Notiser* **130**: 115, 1977), Crivelli & Müller (*Bot. Helv.* **93**: 33, 1983).

Dothiopsis P. Karst. (1884) nom. dub., anamorphic *Pezizomycotina*. See Sutton (*Mycol. Pap.* **141**, 1977).

Dothiora Fr. (1837) nom. rej. = Melanogramma fide Holm (*Taxon* **24**: 475, 1975).

Dothiora Fr. (1849) nom. cons., Dothioraceae. Anamorphs *Dothichiza*, *Hormonema*. 19, widespread (esp. temperate). See Froidevaux (*Nova Hedwigia* **23**: 679, 1973; key), Yurlova *et al.* (*Stud. Mycol.* **43**: 63, 1999; DNA), Barr (*Harvard Pap. Bot.* **6**: 25, 2001), Schweigkofler *et al.* (*Organ. Divers. Evol.* **2**: 1, 2002; phylogeny).

Dothioraceae Theiss. & P. Syd. (1918), Dothideales. 20 gen. (+ 34 syn.), 153 spp.

Lit.: Froidevaux (*Nova Hedwigia* **23**: 679, 1973), Simon *et al.* (*Can. J. Microbiol.* **41**: 35, 1995), Sivanesan & Hsieh (*MR* **99**: 1295, 1995), Dupont *et al.* (*MR* **102**: 631, 1998), Hoog *et al.* (*Stud. Mycol.* **43**: 31, 1999), Untereiner & Naveau (*Mycol.* **91**: 67, 1999), Urzì *et al.* (*J. Microbiol. Meth.* **36**: 95, 1999), Yurlova *et al.* (*Stud. Mycol.* **43**: 63, 1999; DNA), Verkley *et al.* (*Mycol.* **96**: 558, 2004).

Dothiorales = Dothideales.

Dothiorella Sacc. (1880), anamorphic *Botryosphaeriaceae*. Anamorph *Botryosphaeria*. 56, widespread. See Petrak & Sydow (*Rep. Spec. nov. regni veg.* **42**: 214, 1927), Sutton (*Mycol. Pap.* **141**, 1977; typification), Crous & Palm (*Sydowia* **51**: 167, 1999), Luque *et al.* (*Mycol.* **97**: 1111, 2006; on *Vitis*), Phillips *et al.* (*Mycol.* **97**: 513, 2005; review, phylogeny), Crous *et al.* (*Stud. Mycol.* **55**: 235, 2006; review, phylogeny), Phillips *et al.* (*Persoonia* **21**: in press, 2008; phylogeny).

Dothiorellina Bubák (1911) nom. dub., anamorphic *Pezizomycotina*. See Sutton (*Mycol. Pap.* **141**, 1977).

Dothiorina Höhn. (1911), anamorphic *Chlorociboria*, St.0eH.15. 3, Europe. See Dixon (*Mycotaxon* **1**: 193, 1975; teleomorph), Riedl (*Sydowia* **29**: 146, 1977; key), Sánchez & Bianchinotti (*Mycotaxon* **102**: 395, 2007; ontogeny).

Dothioropsis Riedl (1974), anamorphic *Pezizomycotina*, St.0eH.?38. 1, Europe. See Riedl (*Phyton Horn* **16**: 222, 1974), Mel'nik & Minter (*Mycologist* **20**: 109, 2006; UK).

Dothisphaeropsis Höhn. (1919) = Microsphaeropsis Höhn. fide Sutton (*Mycol. Pap.* **141**, 1977).

Dothistroma Hulbary (1941), anamorphic *Mycosphaerella*, St.0fH.1/10. 1, widespread. *D. septospora* causes defoliation of *Pinus*. See Gibson (*Ann. Rev. Phytopath.* **10**: 51, 1972), Evans (*Mycol. Pap.* **153**, 1984), Wingfield *et al.* (*BioScience* **51**: 134, 2001; biogeography), Barnes *et al.* (*Stud. Mycol.* **50**: 551, 2004; phylogeny, review), Groenewald *et al.* (*Phytopathology* **97**: 825, 2007; mating type genes).

Dothithyrella Höhn. (1918) = Leptopeltis fide Holm & Holm (*Bot. Notiser* **130**: 215, 1977).

Dothivalsaria Petr. (1966), Massariaceae. 1, Europe. See Barr (*N. Amer. Fl.* ser. 2 **13**: 129 pp., 1990).

Dothophaeis Clem. (1931) ≡ Englerodothis.

Dozya P. Karst. (1873) [non *Dozya* Sande Lac. 1866, *Musci*] ≡ Hypocreopsis P. Karst. fide Rossman *et al.* (*Stud. Mycol.* **42**: 248 pp., 1999).

DR curve, a log-probit dosage-response curve (Horsfall, *Principles of fungicidal action*, 1956).

Drechmeria W. Gams & H.-B. Jansson (1985), anamorphic *Clavicipitaceae*, Hso.0eH.15. 2 (from nematodes and protozoans), widespread. See Gernandt & Stone (*Mycol.* **91**: 993, 1999; phylogeny), Gams & Zare (*Mycology Series* **19**: 17, 2003; review).

Drechslera S. Ito (1930) nom. cons., anamorphic *Pyrenophora*, Hso.≡ eP.26. *c.* 29 (graminicolous), widespread. See Shoemaker (*CJB* **37**: 880, 1959), Shoemaker (*CJB* **40**: 809, 1962; key 14 spp.), Pandey & Gupta (*Acta Mycologica* Warszawa **20**: 209, 1986; as a mycoparasite), Sivanesan (*Mycol. Pap.* **158**: 1, 1987; keys), Alcorn (*Ann. Rev. Phytopath.* **26**: 27, 1988; gen. taxonomy), Muchovej *et al.* (*Fitopatol. Brasil* **13**: 211, 1988; keys), Strobel *et al.* (*Phytoparasitica* **16**: 145, 1988; phytotoxins), Ondrej (*Česká Mykol.* **43**: 45, 1989; key 20 Czech spp.), Khazanov (*Opredelitel' Gribov-Vozbul. 'Gel'mintosporiozov' Rast. iz Rodov Bipolaris, Drechslera i Exserohilum*, 1992), Goh *et al.* (*Fungal Diversity* **1**: 85, 1998; ribosomal RNA analysis), Maraite (*Helminthosporium Blights of Wheat: Spot Blotch and Tan Spot* Proceedings of an International Workshop Held at CIMMYT, El Batan, Mexico, 9-14 February 1997: 6, 1998; nomencl.), Turgeon & Berbee (*Molecular Genetics of Host-Specific Toxins in Plant Disease*: 153, 1998; evolution), Jawhar *et al.* (*Cereal Research Communications* **28**: 87, 2000; *D. graminea* group), Mehta (*Fitopatol. Brasil* **26**: 590, 2001; on *Avena*), Zhang & Berbee (*Mycol.* **93**: 1048, 2001; phylogeny), Ondrej & Minariková (*Petria* **12**: 267, 2002; morphology), Friesen *et al.* (*Phytopathology* **95**: 1144, 2005; on *Triticum*), Serenius *et al.* (*MR* **109**: 809, 2005; on *Hordeum*).

Drechslerella Subram. (1964), anamorphic *Orbilia*, Hso.≡ eH.1/10. 1, USA. See Scholler *et al.* (*Sydowia* **51**: 89, 1999; gen. concept & n. combs), Li *et al.* (*Mycol.* **97**: 1034, 2006; phylogeny), Yu *et al.* (*Mycotaxon* **96**: 163, 2006; teleomorph).

Drechsleromyces Subram. (1978), anamorphic *Pezizomycotina*, Hso.0fH.10. 1, USA. See Subramanian (*Kavaka* **5**: 93, 1977), Rubner (*Stud. Mycol.* **39**, 1996).

Dremuspora Sal.-Cheb. & Locq. (1980), Fossil Fungi. 1, Cameroon.

Drepanoconis J. Schröt. & Henn. (1896), Cryptobasidiaceae. 3, S. America. See Linder (*Ann. Mo. bot. Gdn* **16**: 343, 1929), Gómez & Kisimova-Horovitz (*Revta Biol. trop.* **45**: 1293, 1998; Costa Rica), Begerow *et al.* (*Mycol. Progr.* **1**: 187, 2002), Hendrichs (*Sydowia* **55**: 33, 2003; Central & South America).

Drepanomyces Thaxt. (1931), Ceratomycetaceae. 1, Sumatra; Sarawak. See Tavares (*Mycol. Mem.* **9**: 627 pp., 1985), Bameul (*Nouv. Rev. Entomol.* Nouv. sér. **10**: 19, 1993).

Drepanopeziza (Kleb.) Höhn. (1917), Dermateaceae. Anamorphs *Gloeosporidiella*, *Marssonina*, *Monostichella*. 14, widespread (temperate). See Rimpau (*Phytopath. Z.* **43**: 257, 1962; on *Ribes*, *Populus*, *Salix*; keys), Spiers (*Eur. J. For. Path.* **28**: 233, 1998; on *Populus*, New Zealand), Spiers & Hopcroft (*MR* **102**: 1025, 1998), Nauta & Spooner (*Mycologist* **14**: 21, 2000; UK).

Drepanospora Berk. & M.A. Curtis (1875), anamorphic *Pezizomycotina*, Hso.0-≡ hP.1/10. 2, widespread. See Goos (*Mycol.* **81**: 356, 1989).

DRIPs, An assemblage of protistans named from the initial letters of the original constituent organisms (***D**ermocystidium*, **R**osette Agent, ***I**chthyophonus*, ***P**sorospermium*). In phylogenetic trees derived from 18S rRNA, the DRIPs clade appears as a sister group to animals, fungi and choanozoans and usually basal to all three. Members of the group typically possess vesicular mitochondrial cristae and a walled trophic phase. All appear to be parasitic. More recently other taxa have been added: *Rhinosporidium* (See Herr *et al.*, 1999), *Anurofeca* (See Baker *et al.*, 1999) and *Amoebidium parasiticum* (See Ustinova *et al.*, *Protist* **151**: 253, 2000). See also Ragan *et al.* (*Proc. Natl. Acad. Sci. USA***93**: 11907, 1996), Cavalier-Smith (**Biol. Rev**. **73**: 203, 1998).

Drosella Maire (1935) = Chamaemyces fide Singer (*Agaric. mod. Tax.*, 1951).

Drosophila Quél. (1886) = Psathyrella fide Singer (*Agaric. mod. Tax.*, 1951) the name is used by, Romagnesi (*BSMF* **91**: 137, 1975).

Drudeola Kuntze (1891), anamorphic *Pezizomycotina*. 2, USA. See Sutton (*Mycol. Pap.* **141**, 1977).

Drummondia Bat. & H. Maia (1963) [non *Drummondia* Hook. 1828, *Musci*] ≡ Mycousteria fide Farr (*Mycol.* **78**: 280, 1986).

Drumopama Subram. (1957), anamorphic *Pezizomycotina*, Hsy.0eP.10. 1, India; Cuba. See Subramanian (*Proc. Indian Acad. Sci.* series B **46**: 333, 1957), Castañeda Ruiz & Kendrick (*Univ. Waterloo Biol. Ser.* **32**, 1990; Cuba).

Druparia Raf. (1808) nom. dub., Agaricomycetes. ? 'gasteromycetes'.

Drupasia Raf. (1809) ≡ Druparia.

druse, a stellate cluster of large crystals in a lichen thallus.

dry rot fungus (or **house fungus**), *Serpula lacrymans*; see Seehann & Hegarty (*Docs. Internat. Res. Group Wood Preserv.* IRG/WP/1337, 1988; bibliogr.), Watkinson (*Biodet. Abstr.* **8**: 161, 1994; review).

dry spore, a spore that becomes separated without slime from the cell producing it (Mason, 1937); cf. slime spore. See Xerosporae.

Dryadomyces Gebhardt (2005), anamorphic *Ophiostomataceae*, H?.?.?. 1 (from ambrosia beetles), Taiwan. See Gebhardt *et al.* (*MR* **109**: 693, 2005).

dryad's club, basidioma of *Clavaria pistillaris*.

dryad's saddle, basidioma of *Polyporus squamosus*.

Dryinosphaera Dumort. (1822) nom. dub., Pezizomycotina. Used for diverse perithecioid fungi.

Dryodon Quél. ex P. Karst. (1881) = Hericium Pers. fide Donk (*Taxon* **5**: 69, 1956).

Dryophila Quél. (1886) = Pholiota fide Patouillard (*Essai taxonomique sur les familles et les genres des Hyménomycètes*, 1900).

Dryophilum Schwein. (1834) nom. dub., Fungi. ? Based on insect galls.

Dryosphaera Jørg. Koch & E.B.G. Jones (1989), anamorphic *Pezizomycotina*, T. 2 (marine), widespread. See Kohlmeyer & Volkmann-Kohlmeyer (*CJB* **71**: 992, 1993).

dual phenomenon (in anamorphic fungi), the condition in which a fungus is made up of two culturally different elements or individuals (Hansen, *Mycol.* **30**: 442, 1938; Hansen & Snyder, *Am. J. Bot.* **30**: 419, 1943).

dual propagule (in a lichen), one comprising elements of both the fungal and the photosynthetic partner (e.g. isidium, soredium; see Lichens).

Dualomyces Matsush. (1987), anamorphic *Pezizomycotina*, Hso.0fH-P.10/11. 2, Taiwan. See Matsushima (*Matsush. Mycol. Mem.* **5**: 13, 1987).

Dubiocarpon S.A. Hutch. (1955), Fossil Fungi ? Ascomycota. 5 (Carboniferous), Europe; USA. See Baxter (*Paleont. Contrib. Univ. Kansas* **77**, 1975), Pirozynski (*Ann. Rev. Phytopath.* **14**: 237, 1976).

Dubiomyces Lloyd (1921) = Ustilaginoidea fide Diehl (*in litt.*) See, Stevenson & Cash (*Lloyd, The new fungus names*, 1936).

Dubitatio Speg. (1882), Massariaceae. Anamorph *Aplosporella*-like. 1 (on rotten wood), S. America (temperate). See Rossman *et al.* (*Stud. Mycol.* **42**: 248 pp., 1999).

Duboscqia Pérez (1908), Microsporidia. 3.

Dubujiana D.R. Reynolds & G.S. Gilbert (2005), Microthyriaceae. 1, Australia. See Reynolds & Gilbert (*Aust. Syst. Bot.* **18**: 282, 2005).

Duchesne (Ernest; 1874-1912; France). École du Service de Santé Militaire de Lyon (1894). Physician who noted, 32 years before Fleming (q.v.), that certain moulds kill bacteria. *Publs. Contribution à l'étude de la concurrence vitale chez les micro-organismes: antagonisme entre les moisissures et les microbes*, doctoral thesis (1897). *Biogs, obits etc.* Pouillard (*Médecine et armées* **31**: 527, 2003).

Ductifera Lloyd (1917), Auriculariales. *c.* 11, widespread (esp. tropical). See Wells (*Mycol.* **50**: 407, 1958), Weiss & Oberwinkler (*MR* **105**: 403, 2001).

Duddingtonia R.C. Cooke (1969), anamorphic *Orbiliaceae*. See Schenk *et al.* (*CJB* **55**: 977, 1977), Liou & Tzean (*Mycol.* **89**: 876, 1997; phylogeny), Ahrén *et al.* (*FEMS Microbiol. Lett.* **158**: 179, 1998; phylogeny), Ahren *et al.* (*MR* **108**: 1205, 2004; genetic diversity).

Duebenia Fr. (1849), Helotiales. 2, Europe. See Hein (*Willdenowia* Beih. **9**, 1976; key), Nauta & Spooner (*Mycologist* **13**: 65, 1999).

Dufourea Ach. (1809) nom. rej. = Xanthoria fide Hawksworth *et al.* (*Dictionary of the Fungi* edn 8, 1995), Kärnefelt & Thell (*Nova Hedwigia* **62**: 487, 1996; nomencl.).

Dufoureomyces Cif. & Tomas. (1953) = Dactylina Nyl. fide Hawksworth *et al.* (*Dictionary of the Fungi* edn 8, 1995).

Dufouria Trevis. (1861) = Xanthoria fide Hawksworth *et al.* (*Dictionary of the Fungi* edn 8, 1995).

Dumontinia L.M. Kohn (1979), Sclerotiniaceae. 1 (on *Anemone* rhizomes), Europe. See Holst-Jensen *et al.* (*Mycol.* **89**: 885, 1997; phylogeny), Holst-Jensen *et al.* (*Nordic Jl Bot.* **18**: 705, 1999; phylogeny).

Dumortieria Westend. (1857), anamorphic *Pezizomycotina*, St.0eH.?. 2, Belgium. See Sutton (*Mycol. Pap.* **141**, 1977).

Dumoulinia Stein (1883) = Megalospora Meyen fide Sipman (*Biblthca Lichenol.* **18**, 1983).

Duosporium K.S. Thind & Rawla (1961), anamorphic *Pezizomycotina*, Hso.≡ eP.26. 1, India; China. See Thind & Rawla (*Am. J. Bot.* **48**: 862, 1961).

duplex (of the context), in two layers, that adjacent to the lamellae or tubes being harder than the one over it.

Duplicaria Fuckel (1870), Rhytismataceae. Anamorph *Crandallia*. 3, widespread (boreal; alpine). See Powell (*Mycol.* **65**: 1362, 1973), Johnston (*Mycol. Pap.* **176**: 239 pp., 2001).

Duplicariella B. Erikss. (1970), Rhytismataceae. 1,

Scandinavia. See Eriksson (*Symb. bot. upsal.* **19**: 20, 1970).

Duportella Pat. (1915), Peniophoraceae. 12, widespread. See Cunningham (*Trans. roy. Soc. N.Z.* Bot. **85**: 91, 1957), Hjortstam & Ryvarden (*Syn. Fung.* **4**: 19, 1990; key), Boidin *et al.* (*BSMF* **107**: 91, 1991; key trop. spp.) = Peniophora fide, Boidin *et al.* (*Mycotaxon* **56**: 445, 1998).

Duportellaceae Jülich (1982) = Peniophoraceae.

Duradens Samuels & Rogerson (1990), Sordariomycetidae. 1 (from wood), Guyana. See Samuels & Rogerson (*Mem. N. Y. bot. Gdn* **64**: 165, 1990), Kang *et al.* (*Mycoscience* **40**: 151, 1999), Huhndorf *et al.* (*Fungal Diversity* **20**: 59, 2005; phylogeny).

Durandia Rehm (1913) [non *Durandia* Boeck. 1896, *Cyperaceae*] ≡ Durandiella.

Durandiella Seaver (1932), Dermateaceae. Anamorphs *Corniculariella*, *Gelatinosporium*. 10, Europe; N. America. See Groves (*CJB* **32**: 116, 1954), Krieglsteiner (*Z. Mykol.* **44**: 277, 1978; key 5 spp. Europ.).

Durandiomyces Seaver (1928) = Peziza Fr. fide Korf (*Mycol.* **48**: 711, 1956).

Durella Tul. & C. Tul. (1865) = Xylogramma fide Sherwood (*Mycotaxon* **5**: 1, 1977), Magnes (*Biblthca Mycol.* **165**: 177 pp., 1997), Medardi (*Docums Mycol.* **33** no. 131: 29, 2004).

Durietzia (C.W. Dodge) Yoshim. (1998), Lobariaceae (L). 5, S. America. See Yoshimura (*Recollecting Edvard August Vainio*: 90, 1998), Stenroos *et al.* (*CJB* **81**: 232, 2003; photomorphs).

Durietzia Gyeln. (1935) = Ionaspis fide Petterson (*Bot. Notiser*: 100, 1946).

Durispora K.D. Hyde (1994), Diaporthales. 2, Malaysia; Hong Kong. See Hyde (*Sydowia* **46**: 315, 1994), Photita *et al.* (*Nova Hedwigia* **71**: 101, 2000).

Durogaster Lloyd (1919) nom. dub., Fungi. Based on *Helosis brasiliensis* (*Balanophoraceae*) fide Stevenson & Cash (*Bull. Lloyd Libr.* **35**: 178, 1936).

Durosaccum Lloyd (1924) = Pisolithus fide Stalpers (*in litt.*).

Dussiella Pat. (1890), Clavicipitaceae. 2, America. See White (*Am. J. Bot.* **80**: 1465, 1993), Bischoff *et al.* (*Mycol.* **96**: 1088, 2004).

Dutch elm disease, see elm disease.

duvet (of dermatophytes), a soft, thick layer of hyphae like brushed-up cloth.

Dwayaangam Subram. (1978), anamorphic *Orbilia*, Hso.1bH.1. 7, N. America; Malaysia. See Barron (*CJB* **69**: 1402, 1991), Kohlmeyer & Volkmann-Kohlmeyer (*MR* **105**: 500, 2001), Sokolski *et al.* (*Mycol.* **98**: 628, 2006).

Dwayabeeja Subram. (1958), anamorphic *Pezizomycotina*, Hso.≡ fP.1/10. 4, widespread (esp. tropical).

Dwayaloma Subram. (1957), anamorphic *Pezizomycotina*, Hsp.1eP.?. 1, India. See Subramanian (*J. Indian bot. Soc.* **36**: 62, 1957).

Dwayalomella Brisson, Piroz. & Pauzé (1975), anamorphic *Pezizomycotina*, Cac/Ccu.1eH.15. 2, N. America. See Carris (*Mycol.* **81**: 638, 1989).

Dwayamala Subram. (1956) = Dendryphiella fide Kendrick & Carmichae *in* Ainsworth *et al.* (Eds) (*The Fungi* **4A**: 390, 1973).

Dwibahubeeja N. Srivast., A.K. Srivast. & Kamal (1995), anamorphic *Pezizomycotina*, Hso.0bP.1. 1, India. See Srivastava *et al.* (*MR* **99**: 395, 1995).

Dwibeeja Subram. (1995), anamorphic *Pezizomycotina*. 1, Singapore. See Subramanian (*Kavaka* **20/21**: 57, 1992/1993).

Dwiroopa Subram. & Muthumary (1986), anamorphic *Diaporthales*, St.0eP.1. 1, widespread. See Farr & Rossman (*Mycoscience* **44**: 443, 2003; morphology).

Dwiroopella Subram. & Muthumary (1986), anamorphic *Pezizomycotina*, St.1eP.15/19. 1, India. See Subramanian & Muthumary (*Proc. Indian Acad. Sci.* Pl. Sci. **96**: 202, 1986).

Dyadosporites Hammen (1954), Fossil Fungi. 2 (Cretaceous, Paleocene, Tertiary), China; Colombia.

Dyadosporonites Elsik (1968), Fossil Fungi. 11 (Cretaceous, Tertiary), India; N. America.

Dycticia Raf. (1808) ? = Clathrus fide Stalpers (*in litt.*).

Dyctionella, see *Dictyonella*.

Dyctiostomiopelta, see *Dictyostomiopelta*.

Dyctyoblastus Kremp. (1869) = Phlyctis fide Hawksworth *et al.* (*Dictionary of the Fungi* edn 8, 1995).

Dyeing. Some lichens were formerly used (esp. in N. Europe and N. America) as dyes of animal fibres (e.g. wool). **Orchil** (from *Roccella* spp.) is one of the best known giving a reddish purple as did **cudbear** (*Ochrolechia tartarea*), both being direct dyes. These colours, which are fugitive to light, are a complex series of orcein derivatives formed from microaerophilic oxidation of orcinol-type secondary metabolites in the presence of ammonia. Other spp. such as **crottle** (*Parmelia saxatilis* and *P. omphalodes*) give yellowish buff to brown shades and are reactive dyes. These colours are light-fast and are produced by colourless lichen substances which have an aldehyde group such as the depsidone salazinic acid. The fruiting bodies of some macromycetes (e.g. agarics) have been less used and not historically, but they can give a range of pinks, blues, yellows, reds and browns (e.g. *Boletus*, *Cortinarius*, *Hydnellum*, *Hygrocybe* spp.) using mordants such as alum or iron. Lichens in particular should not be collected indiscriminately for dyeing purposes from natural or seminatural habitats as this is contrary to the conservation of biodiversity.

Lit.: Bolton (*Lichens for vegetable dyeing*, 1960; edn 2, 1982), Brough (*Syesis* **17**: 81, 1984; 250 tests on 42 Can. lichens), Cardon (*Guide des teintures naturelles*, 1990), Casselman (*Craft of the dyer: colour from plants and lichens of the Northeast*, 1980), Kok (*Lichenologist* **3**: 248, 1966; orchil dyes), Rice & Beebee (*Mushrooms for color*, edn 2, 1980), Richardson (*The vanishing lichens*, 1975), Solberg (*Acta Chem. Scand.* **10**: 1116, 1956; dyeing wool with lichens and lichen substances).

Dyonisia, see *Dionysia*.

Dyplolabia A. Massal. (1854), Graphidaceae. See Kalb & Staiger (*Hoppea* **61**: 409, 2000; revision), Archer (*Biblthca Lichenol.* **94**, 2006; revision), Staiger *et al.* (*MR* **110**: 765, 2006; phylogeny), Archer (*Systematics & Biodiversity* **5**: 9, 2007; Solomon Is).

Dyrithiopsis L. Cai, R. Jeewon & K.D. Hyde (2003), Amphisphaeriaceae. Anamorph *Monochaetiopsis*. 1, China. See Jeewon *et al.* (*Mycol.* **95**: 912, 2003).

Dyrithium M.E. Barr (1994), Pezizomycotina. 1, N. America; Puerto Rico. See Eriksson & Hawksworth (*SA* **13**: 191, 1995; nomencl.), Kang *et al.* (*Fungal Diversity* **2**: 135, 1999; application of name).

dysgonic (of dermatophytes), growing more slowly in culture, frequently with less aerial mycelium, than a normal, **eugonic**, strain. See Johnstone & La Touche (*TBMS* **39**: 442, 1956).

Dyslachnum Clem. (1909) = Lachnum fide Raitviir (*Scripta Mycol.* **1**: 1, 1970).

Dyslecanis Clem. (1909) = Lecania fide Hawksworth

et al. (Dictionary of the Fungi edn 8, 1995).

Dysrhynchis Clem. (1909), Parodiopsidaceae. 7, widespread (tropical). See Hosagoudar (*Persoonia* **18**: 123, 2002).

Dysticta Clem. (1909) = Sticta fide Hawksworth *et al.* (*Dictionary of the Fungi* edn 8, 1995).

Dystictina Clem. (1909) = Sticta fide Hawksworth *et al.* (*Dictionary of the Fungi* edn 8, 1995).

dystrophic, inadequately nourished. Cf. eutrophic, oligotrophic.

E (1) the ratio of length to width (of spores, basidia, etc.); Corner (*New Phytol.* **46**: 195, 1947). Cf. Q; (2) Royal Botanic Garden, Edinburgh (UK); founded 1670; a direct grant institution of the Department of Agriculture and Fisheries for Scotland; see Fletcher & Brown (*The Royal Botanic Garden, Edinburgh, 1670-1970*, 1970), Hedge & Lamond (*Index of collectors in the Edinburgh herbarium*, 1970), Watling (*Trans. bot. Soc. Edinb.* **45**: 1, 1986).

e- (prefix) (**ex-**), from; out of; without; not having. See ex.

Earlea Arthur (1906) = Phragmidium fide Arthur (*Manual Rusts US & Canada*, 1934).

Earliella Murrill (1905), Polyporaceae. 1, N. America. See Ryvarden (*Mycotaxon* **23**: 169, 1985), Gilbertson & Ryvarden (*N. Amer. Polyp.* **1**: 249, 1986).

earth-balls, basidiomata of *Sclerodermatales*.

earth-stars, basidiomata of *Geastrum*.

earth-tongues, ascomata of *Geoglossum*.

Eballistra R. Bauer, Begerow, A. Nagler & Oberw. (2001), Eballistraceae. 3 (on *Poaceae*), widespread. See Bauer *et al.* (*MR* **416**: 416, 2001; phylogen.), Piepenbring (*Caldasia* **24**: 103, 2002; Colombia), Vánky & Vánky (*Lidia* **5**: 157, 2002; southern Africa), Thaung (*Australasian Mycologist* **24**: 29, 2005; Burma).

Eballistraceae R. Bauer, Begerow, A. Nagler & Oberw. (2001), Georgefischeriales. 1 gen., 3 spp.

Lit.: Bauer *et al.* (*MR* **105**: 423, 2001), Bauer *et al.* (*MR* **109**: 1250, 2005), Vánky (*Mycotaxon* **91**: 217, 2005).

Ebollia Minter & Caine (1980), anamorphic *Pezizomycotina*, St.0fH.1. 1, Chile. See Minter & Caine (*TBMS* **74**: 436, 1980).

Ecchyna Fr. ex Boud. (1885) = Phleogena fide Donk (*Persoonia* **4**: 160, 1966).

Ecchynaceae Rea (1922) = Phleogenaceae.

Eccilia (Fr.) P. Kumm. (1871) = Entoloma fide Kuyper (*in litt.*).

eccrinid, one of the *Eccrinales* or *Amoebidiales*.

Echidnocymbium Brusse (1987), Ramalinaceae (L). 1, S. Africa. See Brusse (*Mycotaxon* **29**: 173, 1987), Ekman (*Op. bot.* **127**: 148 pp., 1996; N America).

Echidnodella Theiss. & Syd. (1918), Asterinaceae. 12, widespread (esp. tropical). See Goos (*Mycotaxon* **73**: 455, 1999), Hosagoudar (*Indian Phytopath.* **58**: 194, 2005; India), Thaung (*Australas. Mycol.* **25**: 5, 2006; Myanmar).

Echidnodes Theiss. & Syd. (1918), Asterinaceae. Anamorph *Asterostomula*. *c.* 17, widespread (tropical). See Hawksworth *et al.* (*Dictionary of the Fungi* edn 8, 1995), Goos (*Mycotaxon* **73**: 455, 1999), Hosagoudar (*Indian Phytopath.* **58**: 194, 2005; India), Thaung (*Australas. Mycol.* **25**: 5, 2006; Myanmar).

echinate (dim. **echinulate**) (of spores, etc.), having sharply pointed spines (Fig. 20.2); spinose.

Echinella Massee (1895) = Pirottaea fide Nannfeldt (*Nova Acta R. Soc. Scient. upsal.*, 1932).

Echinoascotheca Matsush. (1995), ? Phaeotrichaceae. 1, Pakistan. See Matsushima (*Matsush. Mycol. Mem.* **8**, 1995).

Echinobotryaceae Nann. (1934) = Microascaceae.

Echinobotryum Corda (1831), anamorphic *Microascaceae*, Hsy.0eP.10. 1, widespread. Synanamorph *Doratomyces*.

Echinocatena R. Campb. & B. Sutton (1977), anamorphic *Pezizomycotina*, Hso.0eP.3. 1, India. See Campbell & Sutton (*TBMS* **69**: 126, 1977).

Echinochaetaceae Jülich (1982) = Polyporaceae.

Echinochaete D.A. Reid (1963), Polyporaceae. 4, widespread (tropical). See Reid (*Mem. N. Y. bot. Gdn* **28**: 187, 1976), Ryvarden & Johansen (*Prelim. Polyp. Fl. E. Afr.*: 315, 1980), Corner (*Beih. Nova Hedwigia* **78**: 105, 1984), Núñez & Ryvarden (*Syn. Fung.* **10**: 19, 1995), Silveira (*Mycotaxon* **93**: 1, 2005; S. Am.).

Echinochondrium Samson & Aa (1975), anamorphic *Pezizomycotina*, Hso.#eP.1. 1, Sri Lanka. bulbils with spines. See Samson & Aa (*Revue Mycol.* Paris **39**: 103, 1975).

Echinoderma (Locq. ex Bon) Bon (1991) = Lepiota fide Kuyper (*in litt.*).

Echinodia Pat. (1919), anamorphic *Echinoporia*. 1, Singapore.

Echinodiscus Etayo & Diederich (2000), Helotiales. 1 (lichenicolous), Europe. See Etayo & Diederich (*Bull. Soc. Nat. luxemb.* **100**: 64, 2000).

Echinodontiaceae Donk (1961), Russulales. 2 gen. (+ 2 syn.), 7 spp.

Lit.: Donk (*Taxon*: 263, 1951-63), Gross (*Mycopath. Mycol. appl.* **24**: 1, 1964), Brodo (*Bryologist* **76**: 260, 1973), Stalpers (*Taxon* **28**: 414, 1979), Stalpers (*Stud. Mycol.* **40**: 185 pp., 1996), Ginns (*Mycol.* **90**: 19, 1998), Tabata *et al.* (*Mycoscience* **41**: 585, 2000), Binder & Hibbett (*Mol. Phylogen. Evol.* **22**: 76, 2002), Larsson & Larsson (*Mycol.* **95**: 1037, 2003), Maijala *et al.* (*Mycol.* **95**: 209, 2003).

Echinodontium Ellis & Everh. (1900), Echinodontiaceae. 5, N. America; Japan. *E. tinctorium* (Indian paint fungus, conifer wood rot). See Thomas (*Publ. Dep. Agric. Can. For. Bio. Div.* **1041**, 1958), Gross (*Mycopath. Mycol. appl.* **24**: 1, 1964; key), Tabata *et al.* (*Mycoscience* **41**: 585, 2000; molecular phylogeny), Larsson & Larsson (*Mycol.* **95**: 1037, 2003; phylogeny).

Echinodothis G.F. Atk. (1894) ≡ Dussiella fide White (*Am. J. Bot.* **80**: 1465, 1993).

Echinomyces Rappaz (1987), Diatrypaceae. 2, Africa; Australia. See Læssøe (*SA* **13**: 43, 1994; ? synonymy with *Fassia*), Chacón (*Acta Bot. Mexicana* **49**: 15, 1999; Mexico).

Echinophallus Henn. (1898), Phallaceae. 1, East Indies. See Demoulin (*in litt.*), Boedijn (*Bull. Jard. bot. Buitenz.* ser. 3 **12**: 90, 1932) possibly based on *Phallus* 'egg' fide.

Echinoplaca Fée (1824), Gomphillaceae (L). 25, widespread (tropical). See Santesson (*Symb. bot. upsal.* **12** no. 1: 1, 1952), Vězda (*Folia geobot. phytotax.* **14**: 43, 1979), Vězda & Poelt (*Folia geobot. phytotax.* **22**: 179, 1987), Lücking (*Biblthca Lichenol.* **65**: 1, 1997).

Echinopodospora B.M. Robison (1970) = Apiosordaria fide Jong & Davis (*Mycol.* **66**: 467, 1974).

Echinoporia Ryvarden (1980), Schizoporaceae. Anamorph *Echinodia*. 2, widespread (tropical). See Ryvarden & Johansen (*Prelim. Polyp. Fl. E. Afr.*: 325, 1980), Langer (*Biblthca Mycol.* **154**, 1994; syn of *Hyphodontia*).

Echinosphaeria A.N. Mill. & Huhndorf (2004), Helminthosphaeriaceae. Anamorph *Vermiculariopsiella*. 1, widespread. See Miller & Huhndorf (*MR* **108**: 29, 2004), Puja *et al.* (*Cryptog. Mycol.* **27**: 11, 2006).

Echinospora Mirza (1963) nom. inval. ≡ Apiosordaria.

Echinosporangium Malloch (1967) [non *Echinosporangium* Kylin 1956, *Rhodophyta*] ≡ Lobosporangium.

Echinosporella Contu (1992) = Calocybe fide Kuyper (*in litt.*).

Echinosporium Woron. (1913) = Petrakia fide Petrak (*Sydowia* **20**: 186, 1968).

Echinothecium Zopf (1898), ? Capnodiaceae. 2 (on lichens), Europe; N. America. See Navarro-Rosinés & Gómez-Bolea (*Folia Botanica Miscellanea* **6**: 61, 1989; Iberian peninsula), Calatayud *et al.* (*MR* **106**: 1230, 2002; confusion with *Lichenostigma*).

Echinotrema Park.-Rhodes (1955) = Trechispora fide Larsson (*MR* **98**: 1153, 1994).

Echinula Graddon (1977), Hyaloscyphaceae. 1, British Isles. See Graddon (*TBMS* **69**: 255, 1977).

echinulate, see echinate.

Echinus Haller (1768) ≡ Hydnum.

Echusias Hazsl. (1873) = Nitschkia fide Höhnel (*Annls mycol.* **17**: 130, 1919).

eclosion, an explosive series of movements which results in the release of a germinating inner spore from a rigid exosporium, as in *Hypoxylon fragiforme* (Chapela *et al.*, *CJB* **68**: 2571, 1990).

Ecology. Fungal ecology (**myco-**) is the entire field of ecological and coenological study in fungi. See Ainsworth & Sussman (*The Fungi* **3**, 1968; many chs), Allsopp *et al.* (Eds) (*Microbial diversity and ecosystem function*, 1995), Christensen (*Mycol.* **81**: 1, 1989; review, 186 refs.), Cooke (*Ecology* **29**: 376, 1948, lit. fungus sociology, ecology; *Bot. Rev.* **24**: 241, 1958; *Mycopath. Mycol. appl.* **48** (1), 1972, IMC 1 papers), Cooke & Rayner (*Ecology of saprotrophic fungi*, 1984), Dighton *et al.* (*The fungal community: its organization and role in the ecosystem*, edn 3, 2005), Dix & Webster (*Fungal ecology*, 1995) Gadd *et al* (Eds) (*Fungi in the environment. British Mycological Society Symposium Series* **25**, 2007), Gilbert (*La mycologie sur le terrain*, 1928), Grigorova & Norris (Eds) (*Techniques in microbial ecology*, 1990), Harley (*J. Ecol.* **59**: 887, 1971; in ecosystems), Lynch & Hobbie (Eds) (*Micro-organisms in action: concepts and applications in microbial ecology*, 1988), Morten Lange (*Dansk. bot. Arkiv* **13** (1), 1948; agarics of Maglemose), Nuss (*Zur Oekologie der Porlinge*, 1975), Pugh (*TBMS* **74**: 1, 1980; strategies), Ramsbottom (*Mushrooms and toadstools*, 1953), Schultz & Boyle (*MR* **109**: 661, 2005; review), Watling (*TBMS* **90**: 1, 1988; macromycetes), Weir (*Mycol.* **10**: 4, 1918; altitudinal range of forest fungi), Wicklow & Carroll (Eds) (*The fungal community*, 1981; wide ranging surveys, relationship to general ecological concepts; edn 2, 1992), Wilkins *et al.* (*Ann. appl. Biol.* **24**: 703, **25**: 472, **26**: 25, 1937-39; larger fungi).

Lichens are primary colonizers in plant succession, occurring in all pioneer terrestrial habitats. Single species may grow on a wide range of substrata or be restricted to particular types. Whole communities rather than individual species tend to be characteristic of particular substrata growing in a particular climatic region, though a few species are restricted to particular trees (e.g. *Lecanora populicola* on *Populus*), esp. those which are probably not really lichenized (e.g. *Arthopyrenia laburni* on *Laburnum*, *Stenocybe septata* on *Ilex*). On bark and rocks, pH, shade, nutrient enrichment, buffer and ion exchange capacities, air pollution, and disturbance affect the communities developed. Lichens tend to dominate habitats in which competition from other plants is minimal. Vertical zonation of communities occurs on freshwater and maritime rocks (see Aquatic fungi and Marine fungi) and trees (Kershaw, *Lichenologist* **2**: 263, 1964). See Barkman (*Phytosociology and ecology of cryptogamic epiphytes*, 1958), Brodo (*in* Ahmadjian & Hale, 1974: 401 ['1973']; substrate ecology), Seaward (Ed.) (*Lichen ecology*, 1977; comprehensive survey, glossary).

See also Air pollution, Aquatic fungi, Biodiversity, Bioindication, Cave fungi, Coprophilous fungi, Entomogenous fungi, Endophyte, Fungicolous fungi, Insects and fungi, Lichenicolous fungi, Lichenometry, Growth rates, Mycorrhiza, Mycosociology, Plant pathogenic fungi, Phytosociology, Radiation and fungi, RIEC, Sand dune fungi, Soil fungi, Wood-attacking fungi.

ecorticate, having no cortex.

ecotype, part of a population of a species showing morphological, chemical, or physiological characteristics which appear to be genetically determined and correlated with particular ecological conditions, but which are not considered of major taxonomic significance.

ectal, outer; outermost.

ectal excipulum (of ascomata), the outer layers including the subhymenium in a non-lichenized apothecium, sometimes multi-layered; see excipulum.

Ecteinomyces Thaxt. (1903), Laboulbeniaceae. 1, America; Europe. See Benjamin *in* Thaxter (*Mem. Am. Acad. Arts Sci. 1896-1931* **12-16**, 1971), Tavares (*Mycol. Mem.* **9**: 627 pp., 1985), Santamaría (*Fl. Mycol. Iberica* **5**, 2003; Iberian peninsula).

Ectendomeliola Hosag. & D.K. Agarwal (2006), Meliolaceae. 1, India. See Hosagoudar & Agarwal (*Indian Phytopath.* **59**: 98, 2006).

ecto- (prefix), outside.

ectoascus, the outer wall of a fissitunicate (q.v.) ascus, as in *Lecanidion*.

ectochroic, see colour.

Ectochytridium Scherff. (1925) ≡ Zygorhizidium.

Ectographis Trevis. (1853) nom. rej. prop. = Phaeographis fide Lücking *et al.* (*Taxon* **56**: 1296, 2007; nomencl.).

Ectolechia A. Massal. (1853) = Ocellularia fide Hale (*Bull. Br. Mus. nat. hist.* Bot. **8**: 227, 1981).

Ectolechia Trevis. (1853) = Sporopodium fide Hawksworth *et al.* (*Dictionary of the Fungi* edn 8, 1995).

Ectolechiaceae Zahlbr. (1905), Lecanorales (L). 17 gen. (+ 10 syn.), 137 spp.

Lit.: Vězda (*Folia geobot. phytotax.* **21**: 199, 1986), Kalb & Vězda (*Folia geobot. phytotax.* **22**: 286, 1987; 13 spp. Brazil), Lücking *et al.* (*Bot. Acta* **107**: 393, 1994)

Lit.:, Sérusiaux (*Abstracta Botanica* **21**: 145, 1997), Lücking (*Trop. Bryol.* **15**: 45, 1998), Lücking (*Phyton* Horn **39**: 131, 1999), McCarthy *et al.* (*Lichenologist* **32**: 317, 2000).

Ectomyces P. Tate (1927) = Termitaria fide Tate (*Parasitology* **20**: 77, 1968).
ectomycorrhiza, see mycorrhiza.
ectoparasite, a parasite living on the outside of its host.
ectoplacodial, see stroma.
Ectosphaeria Speg. (1921) = Diatrype fide Petrak & Sydow (*Annls mycol.* **32**: 23, 1934).
ectospore (1) an exogenous spore; (2) a basidiospore (obsol.).
Ectosticta Speg. (1912) = Rhizosphaera fide Clements & Shear (*Gen. Fung.*, 1931), Sutton (*Mycol. Pap.* **141**, 1977).
Ectostroma Fr. (1823) nom. dub., anamorphic *Pezizomycotina*. See Carmichael *et al.* (*Genera of Hyphomycetes*, 1980).
ectostroma, see stroma.
ectothecal (of ascomycetes), having the hymenium exposed.
Ectothrix (of ascomycetes), having the hymenium exposed.
ectothrix, living on the surface of hair.
Ectotrichophyton Castell. & Chalm. (1919) = Trichophyton fide Hawksworth *et al.* (*Dictionary of the Fungi* edn 8, 1995).
ectotroph (adj. **ectotrophic**), see mycorrhiza.
ectotropic, curving out.
ectotunica, the outer wall of a bitunicate ascus.
ED, effective dose; **ED_{50}**, the effective dose for a 50% (usually lethal) response (Horsfall, *Principles of fungicidal action*, 1956).
edaphic, pertaining to the soil.
Edenia M.C. González, Anaya, Glenn, Saucedo & Hanlin (2007), Pleosporaceae. 1, Mexico. See González *et al.* (*Mycotaxon* **101**: 251, 2007).
Edhazardia J.J. Becnel, V.Sprague & T. Fukuda (1989), Microsporidia. 1.
Edible fungi. For an extensive review, see Boa (2004). Fungi are an important source of food in many countries, and may be cultivated or gathered from the wild. The size of the gathered wild edible fungus market globally has been estimated as several million tonnes with a value of at least US$2 billion in 2004 (Boa, 2004). Examples of countries where wild edible fungi are gathered for personal consumption, and for sale locally include: Malawi, Tanzania and other countries in an extensive regions of central and southern Africa where Miombo woodland is present; Chile, Guatemala, Mexico and the USA in the Americas; China, Nepal, North Korea, Russia, South Korea, Taiwan and Turkey in Asia; Belarus, the Czech Republic, France, Italy, Poland, Rumania, Russia, Spain, Turkey and Ukraine in Europe (Boa, 2004). Emigrants from countries of the former Soviet Union to Israel, Europe, Asia and Mexico to the USA tend to continue their collecting practices in the country of arrival. Wild edible fungi are gathered extensively for export, most notably matsutake (Tricholoma spp.) to Japan from China, and the Pacific northwest of Canada and the USA. Exports can constitute an important component of local and national economies. Species may be consumed for their nutritional value, which is generally undervalued (de Román *et al.*, 2006), or to enhance the flavour of meals.

The main species gathered or cultivated are fleshy basidiomycetes (esp. of *Agaricus*, *Amanita*, *Boletus*, *Cantharellus*, *Coprinus*, *Lactarius*, *Macrolepiota*, *Pleurotus*, *Russula*, *Termitomyces*, *Tricholoma* etc.), and some larger ascomycetes (e.g. *Cyttaria*, *Morchella*, *Terfezia* and *Tuber*). The range of species consumed often varies geographically (e.g. *Cyttaria* is only consumed in southern South America), but this pattern is becoming eroded with globalization of markets. Fleshy basidiomycetes and subterranean ascomycetes (*Terfezia* and *Tuber*) are more frequent in late summer and autumn; the other larger ascomycetes chiefly occur in spring. The edible part is generally the ascoma or basidioma but in *Laccocephalum mylittae* (blackfellows' bread of Australia), *Poria cocos* (Indian bread or tuckahoe of America; cf. stone-fungus), and some tropical species of *Lentinus* it is the sclerotium (which in *Lentinus* may weigh 10kg or more). Microfungi may also be significant components of food. In Taiwan young *Ustilago esculenta* in *Zizania aquatica* (Canadian rice) is used and in Mexico cobs of maize infected with *U. maydis* ('huitlacoche') is commonly eaten. Yeast (*Saccharomyces cerevisiae*) after autolysis (under the trade names Marmite, Vegex, etc.) is a food for humans and dry yeast such as *Candida utilis* (q.v.) is coming into use, as is mycoprotein (q.v.). Button mushrooms (*Agaricus* spp.), oyster (*Pleurotus* spp.), padi-straw (*Volvariella volvacea*), and shii-take (*Lentinula edodes*) mushrooms, and black truffles (Tuber melansporum) are commonly cultured (see Mushroom cultivation). *Agrocybe aegerita* is sometimes obtained by watering old poplar (*Populus*) wood.

Commercial collection of fungi from the wild is extensive and expanded greatly when countries in Eastern Europe became independent from the former Soviet Union. There was a surge in amounts of chanterelles (*Cantharellus* spp.) gathered and exported to western Europe in the 1990s while Italian companies constantly scout for new sources of *Boletus edulis* from around the world. *Lactarius deliciosus* (niscalos, rovelones) is gathered from the French Alps to western Spain to meet a large demand from Catalans (de Román & Boa, 2006). There are conservation (q.v.) concerns about use of wild edible fungi which are linked to loss of woodland habitat and to excessive harvesting. In some countries, guidelines or regulations concerning wild harvesting have been introduced, see e.g. Ławrynowicz (Conservation of fungi in Poland. *In* Perini (Ed.). *Conservation of fungi*, 25, 1997; regulations in Poland) while licenses must be bought in parts of Italy, a common practice now extended to matsutake areas in the Pacific northwest. The absolute amount gathered is thought to be less important than disturbance to soil when raking for some species (e.g. truffles and matsutake) and compaction from people walking in areas.

There are many ways to prepare edible fungi for the table. If flavour is important, then small pieces of the fungus can be spread throughout the food; if aroma is important, it may be better to place thin slices on top of the food. Various cook books are available which specialize in fungus recipes. Macrofungi may be eaten for ceremonial reasons (see Ethnomycology and Hallucinogenic fungi), and may also provide food for animals (see Animal mycophagists).

Lichens form a very important food for reindeer and caribou (*Bryoria*, *Cladonia*, Iceland moss, q.v.), and for invertebrates. Although not poisonous (except possibly those containing large amounts of vulpinic acid, e.g. *Letharia vulpina*, Wolf's moss, q.v.), lichens are not generally eaten by humans, with only a few exceptions (e.g. rock tripe, shaybah, manna). See Dahl (*Bot. Rev.* **20**: 463, 1954), Llano (*Bot. Rev.* **10**: 1, 1944), Richardson (*The vanishing lichens*, 1975), Seaward (Ed.) (*Lichen ecology*, 1977).

Lit.: Batra (*Biologia*, Lahore **29**: 293, 1983; Afghanistan, Pakistan, N. India), Boa (Wild edible fungi, a global overview of their use and importance to people. *FAO Non-wood Forest Products* **17**: 1, 2004; also available in French and Spanish), de Román & Boa (The marketing of *Lactarius deliciosus* in Spain. *Economic Bot.* **60**: 284, 2006), de Román *et al.* (Wild-gathered fungi for health and rural livelihoods. *Proc. Nutrition Soc.* **65**: 190, 2006), Fidalgo & Prance (*Mycol.* **68**: 201, 1976; Brazilian tribes), Hall *et al.* (*Taming the truffle*, 2007), Huang ([*Edible fungi cyclopedia*], 1993; in Chinese), Mao ([*The macrofungi of China*], 2000; in Chinese), Palm & Chapela (*Mycology in Sustainable Development*, 1997; many refs), Purkayastha & Chandra (*Manual of Indian edible mushrooms*, 1985), Rammeloo & Walleyn (*The edible fungi of Africa south of the Sahara*, 1993; 300 spp.), Singer & Harris (*Mushrooms and truffles. Botany, cultivation and utilization*, 2nd edn, 1987), Villareal and Pérez-Moreno (Los hongos comestibles silvestres de México, un enfoque integral. *Micologia Neotropica Aplicada* **2**: 77, 1989; Mexico), Wasson & Wasson (*Mushrooms, Russia and history*, 1957).

See also Ethnomycology, Hallucinogenic fungi, Industrial mycology, Macromycetes, mycoprotein, Mushroom cultivation, soma.

Edmundmasonia Subram. (1958), anamorphic *Sordariales*. 1, widespread. See Kendrick & Carmichae *in* Ainsworth *et al.* (Eds) (*The Fungi* **4A**: 390, 1973; syn of *Brachysporiella*), Arambarri & Godeas (*Mycotaxon* **52**: 91, 1994; Argentina).

Edrudia W.P. Jord. (1980), Lecanoraceae (L). 1, California. See Poelt & Hafellner (*Mitt. bot. StSamml., München* **16**: 503, 1980), Wright (*Bull. Calif. Lichen Soc.* **7**: 7, 2000).

Edwardiella Henssen (1986), Lichinaceae (L). 1, Marion Island. See Henssen (*Lichenologist* **18**: 51, 1986), Schultz & Büdel (*Lichenologist* **34**: 39, 2002; key).

Edythea H.S. Jacks. (1931), Pucciniales. 3 (on *Berberis* (*Berberidaceae*)), S. America. See Thirumalachar & Cummins (*Mycol.* **40**: 47, 1948), Hennen & Ono (*Mycol.* **70**: 569, 1978), Cummins & Hiratsuka (*Illustr. Gen. Rust Fungi* edn 3: 225 pp., 2003; placed in *Uropyxidaceae*).

Edyuillia Subram. (1972), Trichocomaceae. Anamorph *Aspergillus*. 1, widespread. See Subramanian & Rajendran (*Revue Mycol.* Paris **41**: 223, 1977; ontogeny), Kuraishi *et al.* (*NATO ASI Series A: Life Sciences* **185**: 407, 1990; ubiquinones).

Eeniella M.T. Sm., Bat. Vegte & Scheffers (1981), anamorphic *Saccharomycetaceae*, Hso.0eH.10/19. 1, Sweden. See Boekhout *et al.* (*Int. J. Syst. Bacteriol.* **44**: 781, 1994), Yamada *et al.* (*J. Industr. Microbiol.* **14**: 456, 1995; phylogeny).

effete (1) overmature, exhausted; (2) (of fruiting bodies), empty.

Effetia Bartoli, Maggi & Persiani (1984), ? Sordariales. Anamorph *Virgariella*-like. 1 (from soil), Ivory Coast. See Bartoli *et al.* (*Mycotaxon* **19**: 515, 1984), Réblová (*Mycol.* **95**: 128, 2003).

effigurate (of lichen thalli), obscurely lobed.

efflorescent, bursting out of.

effuse, stretched out flat, esp. as a film-like growth.

effused-reflexed (of *Agaricomycotina*), stretched out over the substratum but turned up at the edge to make a pileus.

Efibula Sheng H. Wu (1990) = Phanerochaete fide Hjortstam & Larsson (*Windahlia* **21**, 1994).

Efibulobasidium K. Wells (1975), Sebacinaceae. 4, widespread. See Wells (*Mycol.* **67**: 148, 1975), Weiss & Oberwinkler (*MR* **105**: 403, 2001; posn, phylogeny), Mahamulkar *et al.* (*Indian Phytopathology* **55**: 464, 2002; n.sp. India).

egg (1) the female gamete; (2) (of phalloids, *Amanita*, etc.), the young basidioma before the volva is broken.

Eichleriella Bres. (1903), Auriculariaceae. *c.* 7, widespread. See Wells & Raitviir (*Mycol.* **72**: 564, 1980), Roberts (*Mycotaxon* **96**: 83, 2006; Jamaica).

Eidamella Matr. & Dassonv. (1901) = Myxotrichum fide Orr *et al.* (*CJB* **41**: 1439, 1963), Currah (*Mycotaxon* **24**: 1, 1985).

Eidamia Lindau (1904) = Harzia fide Carmichael *et al.* (*Genera of Hyphomycetes*, 1980).

Eiglera Hafellner (1984), Hymeneliaceae (L). 2, Europe; N. America (esp. coastal). See Lumbsch (*Bryologist* **100**: 180, 1997).

Eigleraceae Hafellner (1984) = Hymeneliaceae.

Eiona Kohlm. (1968), Pezizomycotina. 1 (marine), Denmark. See Johnson *et al.* (*Bot. Mar.* **34**: 229, 1991; ascospore ontogeny), Kohlmeyer (*SA* **10**: 35, 1991; posn).

Ejectosporus S.W. Peterson, Lichtw. & M.C. Williams (1991), Legeriomycetaceae. 2 (in *Plecoptera*), Canada; USA. See Peterson *et al.* (*Mycol.* **83**: 389, 1991), Strongman (*Mycol.* **97**: 333, 2006).

Eklundia C.W. Dodge (1968) = Candelariella fide Castello & Nimis (*Lichenologist* **26**: 283, 1994).

Ekmanomyces Petr. & Cif. (1932) = Dictyonella fide von Arx (*Persoonia* **2**: 421, 1963).

Elachopeltella Bat. & Cavalc. (1964), anamorphic *Pezizomycotina*, Cpt.0eH.?. 2, Brazil. See Batista & Cavalcanti (*Anais Congr. Soc. Bot. Brasil* **14**: 178, 1963), Farr (*Mycol.* **78**: 269, 1986; Brazi), Reynolds & Gilbert (*Cryptog. Mycol.* **27**: 249, 2006; Panama).

Elachopeltis Syd. (1927), anamorphic *Pezizomycotina*, Cpt.0eH.?. 4, America (tropical). See Farr (*Mycol.* **78**: 269, 1986; Brazi), Reynolds & Gilbert (*Cryptog. Mycol.* **27**: 249, 2006; Panama).

Elachophyma Petr. (1931) = Molleriella fide von Arx (*Persoonia* **2**: 421, 1963).

Eladia G. Sm. (1961) = Penicillium Link fide Smith (*TBMS* **44**: 47, 1961), Pitt *et al.* (*Integration of Modern Taxonomic Methods for Penicillium and Aspergillus Classification*: 9, 2000).

Elaeodema Syd. (1922), anamorphic *Pezizomycotina*, Hso or Cac.0eP.?. 2, China. See Sutton (*Mycol. Pap.* **141**, 1977).

Elaeomyces Kirchn. (1888) nom. dub., ? Dothideomycetes.

Elaphocephala Pouzar (1983), ? Atheliaceae. 1, Europe. See Pouzar (*Česká Mykol.* **37**: 206, 1983), Grosse-Brauckmann & Kummer (*Feddes Repert.* **115**: 90, 2004; Germany).

Elaphocordyceps G.H. Sung & Spatafora (2007), Ophiocordycipitaceae. Anamorph *Verticillium*-like. 21 (mostly parasitic on *Elaphomyces*), widespread. See Sung *et al.* (*Stud. Mycol.* **57**, 2007).

Elaphomyces Nees (1820), Elaphomycetaceae. 25 (ectomycorrhizal, hypogeous), widespread. See Kobayashi (*Nagaoa* **7**: 35, 1960; key 14 Jap. spp.), Trappe (*TMSJ* **17**: 209, 1976; variability), Samuelson *et al.* (*Mycol.* **79**: 571, 1987; ultrastr. ascospore ornam.), Zhang & Minter (*CJB* **67**: 909, 1989; key 5 spp. Canada),), Zhang (*MR* **95**: 973, 1991; key 7 spp. China), Pegler *et al.* (*British truffles*, 1993; key 5 Br. spp.), Landvik *et al.* (*Mycoscience* **37**: 237, 1997; DNA), Xu (*Mycosystema* **18**: 238, 1999; ecology), Miller *et al.* (*MR* **105**: 1268, 2001; phylogeny).

Elaphomycetaceae Tul. ex Paol. (1889), Eurotiales. 2 gen. (+ 6 syn.), 27 spp.

Lit.: Samuelson *et al.* (*Mycol.* **79**: 571, 1987), Zhang & Minter (*CJB* **67**: 909, 1989), Zhang (*MR* **95**: 973, 1991), Landvik & Eriksson (*Ascomycete Systematics. Problems and Perspectives in the Nineties* NATO ASI Series vol. **269 269**: 225, 1994), Landvik *et al.* (*Mycoscience* **37**: 237, 1996), Xu (*Mycosystema* **18**: 238, 1999), Miller *et al.* (*MR* **105**: 1268, 2001), Geiser *et al.* (*Mycol.* **98**: 1053, 2006; phylogeny).

Elaphomycetales Trappe (1979) = Eurotiales.

Elasmomyces Cavara (1898) = Russula fide Miller *et al.* (*Mycol.* **98**: 960, 2006).

Elasmomycetaceae Locq. ex Pegler & T.W.K. Young (1979) = Russulaceae.

elater (1) a free capillitium-thread, e.g. in *Myxogastria* and *Farysia*; (2) a body with spiral or annular markings in the gleba of *Battarrea*.

Elateraecium Thirum., F. Kern & B.V. Patil (1966), anamorphic *Hiratsukamyces*. 3 (on *Salacia* (*Celastraceae*), *Desmodium* (*Leguminosae*)), India; S. Africa. Anamorph name for (I). See Laundon (*Mycotaxon* **3**: 133, 1975), Thirumalachar *et al.* (*Sydowia* **27**: 80, 1973-4; teleomorphs).

Elateromyces Bubák (1912) = Farysia fide Vánky (*in litt.*).

Elattopycnis Bat. & Cavalc. (1964), anamorphic *Pezizomycotina*, Cpt.0eH.?. 1, Brazil. See Batista & Cavalcanti (*Anais Congr. Soc. Bot. Brasil* **16**: 181, 1963).

Elderia McLennan (1961) = Stephensia fide Trappe (*Mycotaxon* **9**: 247, 1979), Læssøe & Hansen (*MR* **111**: 1075, 2007; phylogeny).

Electron microscopy, see Ultrastructure.

Electrophoresis. A method of chromatography (q.v.) in which charged particles (esp. DNA, RNA, proteins) are separated by their differential migration in an inert matrix by an electric field. Agarose provides a matrix of large pores particularly suitable for DNA and RNA electrophoresis, and polyacrylamide (**PAGE**) and starch provide matrices of smaller pore size more suitable for the separation of proteins (or for the single base resolution required for DNA sequencing). DNA and RNA migration is directly related to particle size, but proteins migration may be affected by both size and charge (size only if the dissociating system used). The extraction methods depend on whether proteins or nucleic acids are being examined, and the bands are stained for study. The technique is valuable for studying both total protein patterns and particular enzymes (inter- and extracellular), especially variation in one enzyme between isolates; the variants are **isozymes** (**isoenzymes**) and about 140 can be assayed (e.g. esterases, dehydrogenases, pectinases, phosphatases).

Automated and standardized units are available, which can operate with computer-linked gel-readers and increasingly, slab gels are being replaced by capillary (and other rapid 'cassette'-type) electrophoresis. The techniques have proved especially valuable in characterization of closely related strains of plant pathogenic fungi.

Pulsed field gel - (PFGE), a term used to describe a number of different techniques for separating large pieces of DNA, e.g. chromosomes. Different proprietary systems are available (e.g. CHEF, FIGE, PFGE), but all involve applying electric field to an electrophoresis gel as a series of pulses or variations rather than a single continuous constant field. See Boekhout *et al.* (*Ant. v. Leeuwenhoek* **63**: 157, 1993). Now use is less frequent in mycology, and mainly in bacteriology.

Lit.: Gabriel & Gersten (*Analyt. Biochem.* **45**: 1741, 1992; staining), Loxdale & Den Hollander (Eds) (*Electrophoretic studies on agricultural pests*, 1989), Paterson & Bridge (*Biochemical techniques for filamentous fungi* [*IMI Techn. Handbk* **1**], 1994), Hames (*Gel Electrophoresis of Proteins*, edn 3, 1998); capillary electrophoresis: Price *et al.* (*Med. Mycol.* **41**: 369, 2003; sequencing of fungi), Dresler-Nurmi *et al.* (*J. Microbiol. Methods* **41**: 161, 2000; detection of *Trametes*), Wright *et al.* (*Soil Biol. Biochem.* **30**: 1853, 1998; use in mycorrhizas).

Elegantimyces Goh, K.M. Tsui & K.D. Hyde (1998) = Sporidesmium fide Goh *et al.* (*MR* **102**: 239, 1998).

Elenkinella Woron. (1922) = Molleriella fide von Arx (*Persoonia* **2**: 421, 1963).

Eleutherascus Arx (1971), Ascodesmidaceae. 4, widespread. See Steffens & Jones (*CJB* **61**: 1599, 1983; asci), van Brummelen (*Persoonia* **14**: 1, 1989), van Brummelen (*Stud. Mycol.* **31**: 41, 1989; ultrastr.), Stchigel *et al.* (*MR* **105**: 377, 2001), Hansen & Pfister (*Mycol.* **98**: 1029, 2006; phylogeny).

Eleutheris Clem. & Shear (1931) ≡ Eleutheromycella.

Eleutheromycella Höhn. (1908), anamorphic *Pezizomycotina*, Cpd.0eH.15. 1, Europe. See Seeler (*Farlowia* **1**: 119, 1943).

Eleutheromyces Fuckel (1870), anamorphic *Leotiomycetes*, Hsy.0eH.15. 3 (on macromycetes), widespread (temperate). The genus is highly polyphyletic. See Seeler (*Farlowia* **1**: 119, 1943), Sigler (*Cryptog. Bot.* **1**: 384, 1990; yeast-like synanamorph), Tsuneda *et al.* (*Mycol.* **89**: 867, 1997; black spot disease of *Lentinula edodes*), Maekawa & Tsuneda (*Mycotaxon* **78**: 167, 2001; *Hyphozyma* synanam.).

Eleutherosphaera Grove (1907) ? = Pyxidiophora fide Hawksworth & Webster (*TBMS* **68**: 329, 1977), Rossman *et al.* (*Stud. Mycol.* **42**: 248 pp., 1999).

elf cups, ascomata of *Pezizales*.

Elfvingia P. Karst. (1889) = Ganoderma fide Donk (*Verh. K. ned. Akad. Wet.* tweede sect. **62**: 1, 1974).

Elfvingiella Murrill (1914) ≡ Fomes.

ELISA (enzyme-linked immunosorbent assay). A very widely used method for detecting and quantifying antibodies (see antigen). Species-specific antigens in yeasts (Middelhoven & Notermans, *J. Gen. Appl. Bact.* **34**: 15, 1988).

Elixia Lumbsch (1997), Elixiaceae (L). 1, Europe. See Lumbsch (*J. Hattori bot. Lab.* **83**: 1, 1997), Wedin *et al.* (*MR* **109**: 159, 2005; phylogeny).

Elixiaceae Lumbsch (1997), Umbilicariales (L). 1 gen., 1 spp.

Lit.: Lumbsch *et al.* (*Mol. Phylogen. Evol.* **31**: 822, 2004), Miądlikowska *et al.* (*Mycol.* **98**: 1088, 2006), Lumbsch *et al.* (*MR* **111**: 1133, 2007).

Elletevera Deighton (1969), anamorphic *Mycosphae-*

rellaceae, Hso.≡ eP.10. 1 (on *Phyllachora*), N. America; Africa. See Deighton (*Mycol. Pap.* **118**: 17, 1969), Crous & Braun (*CBS Diversity Ser.* **1**: 571 pp., 2003).

Ellimonia Syd. (1930) = Inocyclus fide Müller & von Arx (*Beitr. Kryptfl. Schweiz* **11** no. 2, 1962).

Elliottinia L.M. Kohn (1979), Sclerotiniaceae. 1, Europe. See Kohn (*Mycotaxon* **9**: 415, 1979), Diamandis (*Micologia 2000*: 143, 2000; conservation).

ellipsoidal (of spores, etc.), elliptical in optical section (Fig. 23.1).

Ellis (Job Bicknell; 1829-1905; USA). Student, Union College, Schenectady, New York (1849-1851) [with Tuckerman (q.v.) and Peck (q.v.)]; school teacher (1851-1864); served on the Union side in the American Civil War (1864-1865); teaching and farming (1865-1875). Pioneering North American mycologist, taking up mycology in 1878, describing more than 4,000 species, mainly in works published in *Bulletin of the Torrey Botanical Club* and *Journal of Mycology*; noted for his work on ascomycetes, particularly pyrenomycetes; with Everhart and Kellerman, founder of *Journal of Mycology*, the forerunner of *Mycol.* His specimens are in the fungal reference collection in New York Botanic Garden (**NY**). *Publs.* (with Everhart) *North American Pyrenomycetes* (1892). *Biogs, obits etc.* Barr, Huhndorf & Rogerson (Pyrenomycetes described by J.B. Ellis. *Memoirs of the New York Botanic Garden* **79**: 1, 1996); Cash (A record of the fungi named by J.B. Ellis. I-III. *USDA Special Publication, Plant Disease Survey* **2**: 1-165, 167-345, 347-518, 1952-1954); Grummann (1974: 187); Kaye (*Mycotaxon* **26**: 29, 1986); Kellerman (*Journal of Mycology* **12**: 41, 1906) [portrait]; Rodrigues (*Mycotaxon* **34**: 577, 1989) [index 202 types with amyloid rings]; Stafleu & Cowan (*TL-2* **1**: 742, 1976); see also the website: Ahn (*A Biography of Job Bicknell Ellis* http://sciweb.nybg.org/science2/hcol/intern/ellis1.asp).

Ellis (Martin Beazor; 1911-1996; Guernsey, later England). Mycologist (1946-1960) then Chief Mycologist (1960-1976), Commonwealth Mycological Institute, Kew. A taxonomist, primarily of pigmented hyphomycetes, the meticulous scholarship and superb artwork of his publications made identification of these organisms possible for the first time by the non-expert. *Publs. Dematiaceous Hyphomycetes* (1971); *More Dematiaceous Hyphomycetes* (1976); (with J.P. Ellis) *Microfungi on Land Plants* (1985); (with J.P. Ellis) *Microfungi on Miscellaneous Substrates* (1988). *Biogs, obits etc.* Sutton & Mordue (*MR* **101**: 510, 1997) [portrait].

Ellisembia Subram. (1992), anamorphic *Sordariomycetes*, Hso.≡ eP.1/19. 17, widespread. See Subramanian (*Proc. Indian natn Sci. Acad.* Part B. Biol. Sci. **58**: 183, 1992), Shenoy *et al.* (*MR* **110**: 916, 2006; phylogeny, polyphyly).

Ellisia Bat. & Peres (1965) [non *Ellisia* L. 1763, *Hydrophyllaceae*] ? = Heteroconium fide Carmichael *et al.* (*Genera of Hyphomycetes*, 1980).

Ellisiella Bat. (1956), anamorphic *Pezizomycotina*. 3, Portugal. See Batista (*An. Soc. Biol. Pernambuco* **14**: 19, 1956).

Ellisiella Sacc. (1881) = Colletotrichum fide Nag Raj (*CJB* **51**: 2463, 1973).

Ellisiellina Sousa da Câmara (1949) = Colletotrichum fide Sutton (*CJB* **44**: 887, 1966).

Ellisiodothis Theiss. (1914) ? = Muyocopron fide Sivanesan (*in litt.*).

Ellisiopsis Bat. (1956) = Beltraniella fide Pirozynski & Patil (*CJB* **48**: 567, 1970).

Ellisomyces Benny & R.K. Benj. (1975), Mucoraceae. 1, USA. See Beakes & Campos-Takaki (*TBMS* **83**: 607, 1984; sporangiolum ultrastr.), Beakes *et al.* (*CJB* **62**: 2677, 1984), Beakes *et al.* (*TBMS* **83**: 593, 1984; thallospore formation and ultrastr.), Voigt & Wöstemeyer (*Gene* **270**: 113, 2001; phylogeny).

Ellula Nag Raj (1980), anamorphic *Agaricomycetes*, St.0eH.2. 1 (with clamp connexions), Brazil. See Nag Raj (*CJB* **58**: 2013, 1980).

Ellurema Nag Raj & W.B. Kendr. (1986), Amphisphaeriaceae. Anamorph *Hyalotiopsis*. 1 (from leaves), India. See Kang *et al.* (*Fungal Diversity* **1**: 147, 1998; DNA), Kang *et al.* (*MR* **103**: 53, 1999), Kang *et al.* (*Mycotaxon* **81**: 321, 2002; phylogeny).

elm disease (**Dutch elm disease**), a vascular disease of elm (*Ulmus*) caused by *Ophiostoma ulmi* or the even more aggressive *O. novo-ulmi*; the last sp. may be related to one from Himalaya; anamorphs in *Pesotum*. See *IMI Descriptions* **361** (1973), *IMI Map* **36**, Holmes & Heybrock (*Phyt. classic* **13**, 1990; transl. early papers), Brasier (*Mycopath.* **115**: 155, 1991; *Nature* **372**: 227, 1994).

Elmeria Bres. (1912) [non *Elmeria* Ridl. 1905, *Saxifragaceae*] ≡ Elmerina.

Elmerina Bres. (1912), Auriculariales. 4, widespread (esp. tropical). See Imazeki (*Revue Mycol.* Paris **20**: 159, 1955; as *Protodaedalea*), Parmasto (*Nova Hedwigia* **39**: 101, 1984; monogr.), Reid (*Persoonia* **14**: 465, 1992), Núñez (*Mycotaxon* **61**: 177, 1997), Núñez (*Folia cryptog. Estonica* **33**: 99, 1998; spp. key), Wang & Liu (*Mycosystema* **20**: 159, 2001; as *Protodaedalea*), Weiss & Oberwinkler (*MR* **105**: 403, 2001), Hattori (*Mycoscience* **44**: 265, 2003).

Elmerinula Syd. (1934), Dothideomycetes. 1, Philippines.

Elmerococcum Theiss. & Syd. (1915) = Plowrightia fide von Arx & Müller (*Stud. Mycol.* **9**, 1975).

Elosia Pers. (1822) = Alternaria fide Hughes (*CJB* **36**: 727, 1958).

Elpidophora Ehrenb. ex Link (1824) = Graphiola fide Mussat (*Syll. fung.* **15**: 134, 1901).

Elsikisporonites P. Kumar (1990), Fossil Fungi. 1, India.

Elsinoaceae Höhn. ex Sacc. & Trotter (1913), Myriangiales. 11 gen. (+ 19 syn.), 126 spp.

Lit.: Eriksson (*Op. Bot.* **60**, 1981), Gardner & Hodges (*Mycol.* **78**: 506, 1986), Gabel & Tiffany (*Mycol.* **79**: 737, 1987), Johnston & Beever (*N.Z. Jl Bot.* **32**: 519, 1994), Palm (*Mycol.* **91**: 1, 1999), Swart *et al.* (*Mycol.* **93**: 366, 2001), Alvarez *et al.* (*Pl. Dis.* **87**: 1322, 2003), Ridley & Ramsfield (*Mycol.* **97**: 1362, 2005), Schoch *et al.* (*Mycol.* **98**: 1041, 2006; phylogeny).

Elsinoë Racib. (1900), Elsinoaceae. Anamorph *Sphaceloma*. 48, widespread (esp. warmer areas). *E. fawcettii* (citrus scab), *E. canavaliae* (scab of *Canavalia*), *E. phaseoli* (scab of Lima bean). See Gabel & Tiffany (*Mycol.* **79**: 737, 1987; development), Alcorn *et al.* (*Australas. Pl. Path.* **28**: 115, 1999; anam. connexion), Swart *et al.* (*Mycol.* **93**: 366, 2001; spp. on *Proteaceae*), Alvarez *et al.* (*Pl. Dis.* **87**: 1322, 2003; genetic diversity, Brazil), Ridley & Ramsfield (*Mycol.* **97**: 1362, 2005; on *Pittosporum*, New Zealand), Schoch *et al.* (*Mycol.* **98**: 1041, 2006; phylogeny).

Elvela, see *Helvella*.

Elytroderma Darker (1932), Rhytismataceae. 2, Europe; N. America. *E. deformans* (witches' brooms on pine). See Minter & Fonseca (*Nova Hedwigia* **37**: 181, 1983), Hansen & Lewis (*Compendium of Conifer Diseases*, 1997), Gernandt *et al.* (*Mycol.* **93**: 915, 2001; phylogeny), Ortiz-García *et al.* (*Mycol.* **95**: 846, 2003; phylogeny, synonymy).

Elytrosporangium Morais, Bat. & Massa (1966) nom. dub., Actinobacteria. q.v.

Emarcea Duong, Jeewon & K.D. Hyde (2004) = Anthostomella fide Duong *et al.* (*Stud. Mycol.* **50**: 255, 2004).

emarginate (1) (of lamellae), see sinuate; (2) (of apothecia), lacking a thalline exciple (excipulum thallinum) (lichens) or a raised proper exciple (excipulum proprium).

Embellisia E.G. Simmons (1971), anamorphic *Allewia*, Hso.≡ eP.26. 19, widespread. See Simmons (*Mycotaxon* **38**: 251, 1990; spp. and teleomorphs), Pryor & Bigelow (*Mycol.* **95**: 1141, 2003; phylogeny, polyphyly).

Emblemia Pers. (1827) ? = Phaeographina fide Hawksworth *et al.* (*Dictionary of the Fungi* edn 8, 1995).

Emblemospora Jeng & J.C. Krug (1976), Lasiosphaeriaceae. 2, S. America. See Jeng & Krug (*CJB* **54**: 1971, 1976).

Embolidium Bat. (1964) = Coniella fide Sutton (*Mycol. Pap.* **141**, 1977).

Embolidium Sacc. (1878) = Calicium fide Schmidt (*Mitt. Inst. Allg. Botanik Hamburg* **13**: 111, 1970).

Embolus Batsch (1783) [non *Embolus* Haller 1768] ? = Chaenotheca fide Hawksworth *et al.* (*Dictionary of the Fungi* edn 8, 1995).

emend, to correct an error. Cf. amend.

Emericella Berk. (1857), Trichocomaceae. Anamorph *Aspergillus*. 36, widespread. See Christensen & Raper (*TBMS* **71**: 177, 1978), Frisvad *in* Samson & Pitt (Eds) (*Advances in Penicillium and Aspergillus systematics* **102**: 437, 1985; chemotaxonomy), Horie *et al.* (*Mycoscience* **37**: 137, 1996; Brazil), Horie *et al.* (*Mycoscience* **37**: 323, 1996; China), Stchigel *et al.* (*MR* **103**: 1057, 1999), Peterson (*Integration of Modern Taxonomic Methods for Penicillium and Aspergillus Classification*: 323, 2000; phylogeny), Zohri (*Folia Microbiol.* Praha **45**: 391, 2000; biochemistry), Klich (*Identification of Common Aspergillus Species*: 116 pp., 2002), Frisvad & Samson (*Syst. Appl. Microbiol.* **27**: 672, 2004; Venezuela), Gugnani *et al.* (*J. Clin. Microbiol.* **42**: 914, 2004; onychomycosis), Cary *et al.* (*Mycol.* **97**: 425, 2005; aflatoxin-producing spp.).

Emericellopsis J.F.H. Beyma (1940), Hypocreales. Anamorph *Acremonium*-like. 11, widespread. See Davidson & Christensen (*TBMS* **57**: 385, 1971; key), Gams (*Cephalosporium-artige Schimmelpilze*: 21, 1971; keys), Belyakova (*Mikol. Fitopatol.* **8**: 385, 1974; key), Wu & Kimbrough (*CJB* **68**: 1877, 1990; ultrastr.), Malloch *in* Hawksworth (Ed.) (*Ascomycete Systematics. Problems and Perspectives in the Nineties* NATO ASI Series vol. **269 269**: 374, 1994; posn), Glenn *et al.* (*Mycol.* **88**: 369, 1996; DNA), Ogawa *et al.* (*Mycol.* **89**: 756, 1997; phylogeny), Rossman *et al.* (*Stud. Mycol.* **42**: 248 pp., 1999; placement in *Bionectriaceae*).

Emilmuelleria Arx (1986), Chaetomiaceae. 1 (coprophilous), USA. See von Arx (*Sydowia* **38**: 6, 1986), Barr (*Mycotaxon* **39**: 43, 1990; posn), Silva & Hanlin (*Mycoscience* **37**: 261, 1997).

Emmia Zmitr., Spirin & Malysheva (2006), Polyporaceae. 2.

Emmonsia Cif. & Montemart. (1959), anamorphic *Ajellomycetaceae*. 1, widespread. See adiaspiromycosis, adiaspore. See Carmichael (*CJB* **40**, 1962), van Oorschot (*Stud. Mycol.* **20**, 1980), Sigler (*J. Med. Vet. Mycol.* **34**: 303, 1996), Drouhet *et al.* (*J. Mycol. Médic.* **8**: 64, 1998), Hoog *et al.* (*Medical Mycology* **36** Suppl. 1: 52, 1998; phylogeny), Peterson & Sigler (*J. Clin. Microbiol.* **36**: 2918, 1998; genetic variation), Wellinghausen *et al.* (*International Jounal of Medical Microbiology* **293**: 441, 2003; in lungs), Untereiner *et al.* (*Mycol.* **96**: 812, 2004; phylogeny, family placement).

Emmonsiella Kwon-Chung (1972) = Ajellomyces fide McGinnis & Katz (*Mycotaxon* **8**: 157, 1979), Fukushima *et al.* (*Mycopathologia* **116**: 151, 1991).

Empusa Cohn (1855) [non *Empusa* Lindl. 1824, *Orchidaceae*] ≡ Entomophthora Fresen.

Empusaceae Clem. & Shear (1931) = Entomophthoraceae.

Lit.: Kirk (*in litt.*).

Enantioptera Descals (1983), anamorphic *Pezizomycotina*, Hso.1bH.10. 1 (from foam), British Isles. See Descals (*TBMS* **80**: 70, 1983), Marvanová (*Czech Mycol.* **56**: 193, 2004; Czech Republic).

Enantiothamnus Pinoy (1911), anamorphic *Pezizomycotina*, Hso.?.?. 1 (on humans), Africa.

Enarthromyces Thaxt. (1896), Laboulbeniaceae. 1, Asia; Africa. See Terada *et al.* (*Bot. Bull. Acad. sin.* Taipei **45**: 165, 2004; Taiwan).

Encephalitozoon Levaditi, Nicolau & Schoen (1923), Microsporidia. 5.

Encephalium Link (1816) = Tremella Pers. fide Donk (*Persoonia* **4**: 178, 1966).

Encephalographa A. Massal. (1854), ? Hysteriaceae (?L). 2 (on lichens), Europe (southern). See Renobales & Aguirre (*SA* Reprint of Volumes 1-4 (1982-1985) **8**: 87, 1990; nomencl., posn), Tretiach & Modenesi (*Nova Hedwigia* **68**: 527, 1999; ecology, poss. links with *Melaspilea*).

Encephalographomyces Cif. & Tomas. (1953) = Poeltinula fide Hawskworth (*SA* **10**: 36, 1991).

Enchnoa Fr. (1849), Nitschkiaceae. 5 (from bark), widespread. Often placed in the *Calosphaeriaceae*. See Barr (*SA* **13**: 192, 1995; posn), Réblová & Svrcek (*Czech Mycol.* **49**: 193, 1997), Mostert *et al.* (*Stud. Mycol.* **54**: 115 pp., 2006).

Enchnosphaeria Fuckel (1870) = Herpotrichia fide Barr (*Mycotaxon* **20**, 1984).

Enchylium A. Massal. (1853) = Pterygiopsis fide Hawksworth *et al.* (*Dictionary of the Fungi* edn 8, 1995).

Enchylium Gray (1821) = Leptogium fide Hawksworth *et al.* (*Dictionary of the Fungi* edn 8, 1995).

Encliopyrenia Trevis. (1860) = Verrucaria Schrad. fide Hawksworth *et al.* (*Dictionary of the Fungi* edn 8, 1995).

Encoelia (Fr.) P. Karst. (1871) nom. cons., Sclerotiniaceae. Anamorph *Myrioconium*. *c.* 15, widespread (north temperate). See Korf & Kohn (*Mem. N. Y. bot. Gdn* **28**: 109, 1976), Arendholz & Sharma (*Mycotaxon* **20**: 633, 1984), Juzwik & Hinds (*CJB* **62**: 1916, 1984; anamorph), Zhuang (*Mycotaxon* **31**: 261, 1988), Verkley (*MR* **99**: 187, 1995; asci), Baral & Richter (*Boletus* **21**: 45, 1997; subgen. *Kirschsteinia*), Holst-Jensen *et al.* (*Mycol.* **89**: 885,

1997), Holst-Jensen *et al.* (*Nordic Jl Bot.* **18**: 705, 1999), Holst-Jensen *et al.* (*Mol. Biol. Evol.* **16**: 114, 1999; posn), Zhuang (*Mycotaxon* **72**: 325, 1999), Zhuang *et al.* (*Mycosystema* **19**: 478, 2000; phylogeny).

Encoeliella Höhn. (1910) = Unguiculariopsis fide Zhuang (*Mycotaxon* **32**: 1, 1988).

Encoeliopsis Nannf. (1932), Helotiaceae. 4, Europe; N. America. See Groves (*CJB* **47**: 1319, 1969), Zhuang (*Mycotaxon* **32**: 97, 1988).

Endacinus Raf. (1814) = Pisolithus fide Saccardo (*Syll. fung.* 7: 148, 1888).

Endaematus Raf. (1814) nom. dub., Fungi.

Endematus (orthographic variant), see *Endaematus*.

endemic, native to one country or geographical region.

endo- (prefix), inside.

endoascospores, spore-like cells produced within ascospores (see Morgan-Jones, *CJB* **51**: 493, 1972).

endoascus, the often extensible inner wall layers of a bitunicate (q.v.) ascus.

endobasidial (of a conidiophore in a lichenized pycnidium), having a secondary sporing branch (obsol.); cf. exobasidial.

Endobasidium Speschnew (1901), Exobasidiaceae. 1, Samarkand (Asia). See Donk (*Reinwardtia* **4**: 116, 1956; basidiomycete affinities doubtful).

endobasidium, see basidium.

endobiotic, making growth inside living organisms.

Endoblastidium Codreanu (1931), ? Blastocladiales. 2 (in *Insecta*), Europe.

Endoblastoderma B. Fisch. & Brebeck (1894) = Candida fide Lodder (*Yeasts, a taxonomic study* 2nd edn, 1970).

Endoblastomyces Odinzowa ex Kudrjanzev (1960) = Issatchenkia fide Kurtzman & Fell (*Yeasts, a taxonomic study* 4th edn: 223, 1998).

Endobotrya Berk. & M.A. Curtis (1874), anamorphic *Pezizomycotina*, Cac.#eP.1. 1, N. America.

Endobotryella Höhn. (1909), anamorphic *Pezizomycotina*, St.#eP.1. 1, Europe.

Endocalyx Berk. & Broome (1876), anamorphic *Pezizomycotina*, Hsp.0eP.10. 9, widespread (tropical). See Montemartini Corte (*Atti Ist. bot. Univ. Lab. Crittog. Pavia* sér. 5 **20**: 260, 1963; key), Okada & Tubaki (*Mycol.* **76**: 300, 1984; cultural behaviour), Mena Portales & Mercado Sierra (*Revta Jardín bot. Nac.* Univ. Habana **5**: 53, 1984; Cuba).

Endocarpaceae Fée ex Zenker (1827) = Verrucariaceae.

Endocarpidium Müll. Arg. (1862) = Placidiopsis fide Hawksworth *et al.* (*Dictionary of the Fungi* edn 8, 1995).

endocarpinoid (of lichenized perithecia), sunk into the tissues of the thallus, as in *Endocarpon* (obsol.).

Endocarpiscum Nyl. (1864) = Heppia fide Hawksworth *et al.* (*Dictionary of the Fungi* edn 8, 1995).

Endocarpomyces E.A. Thomas ex Cif. & Tomas. (1953) ≡ Dermatocarpon Eschw.

Endocarpon Hedw. (1789), Verrucariaceae (L). *c.* 50, widespread (esp. temperate). See Singh & Upreti (*Candollea* **39**: 539, 1984; key 6 spp., India), Wagner (*CJB* **65**: 2441, 1987; ontogeny), Stocker-Wörgötter & Turk (*Pl. Syst. Evol.* **158**: 313, 1988; ultrastr.), Wagner & Letrouit-Galinou (*CJB* **66**: 2118, 1988; squamule development), McCarthy (*Lichenologist* **23**: 27, 1991; 10 spp. Australia), Harada (*Nova Hedwigia* **56**: 335, 1993; key 7 spp. Japan), Harada (*J. Nat. Hist. Mus. Inst.* Chiba **4**: 97, 1997; Micronesia), Breuss (*Öst. Z. Pilzk.* **9**: 147, 2000; Mexico), Geiser *et al.* (*Mycol.* **98**: 1053, 2006; phylogeny), Gueidan *et al.* (*MR* **111**: 1145, 2007; phylogeny), Aptroot *et al.* (*Biblthca Lichenol.* **97**, 2008; Costa Rica).

endocarpous (of gasteromycetes, etc.), having the mature hymenium covered over; angiocarpous.

Endocena Cromb. (1876), Acarosporaceae (L). 1 (sterile), S. America. See Stenroos & DePriest (*Am. J. Bot.* **85**: 1548, 1998; DNA).

Endochaetophora J.F. White & T.N. Taylor (1988), Fossil Fungi, Ascomycota. 1 (Triassic), Antarctica.

endochroic, see colour.

Endochytriaceae Sparrow ex D.J.S. Barr (1980), Chytridiales. 10 gen., 56 spp.

Lit.: Barr *et al.* (*Mycol.* **79**: 587, 1987), Chen & Chien (*Bot. Bull. Acad. sin.* Taipei **36**: 235, 1995), Longcore (*Mycol.* **87**: 25, 1995), Shin *et al.* (*CJB* **79**: 1083, 2001).

Endochytrium Sparrow (1933), Endochytriaceae. 7, widespread (temperate). See Karling (*Mycol.* **33**: 356, 1941).

Endocladis Clem. & Shear (1931) ≡ Endoramularia.

Endococcus Nyl. (1855), Dothideomycetes. 24 (on lichens), widespread (esp. temperate). See Hawksworth (*Bot. Notiser* **132**: 283, 1979), David & Etayo (*Lichenologist* **27**: 314, 1995; on *Collema*), Etayo & Breuss (*Öst. Z. Pilzk.* **10**: 315, 2001).

Endocochlus Drechsler (1935), Cochlonemataceae. 4, N. America; Kenya. See Drechsler (*Mycol.* **27**: 14, 1935), Drechsler (*Mycol.* **28**: 363, 1936), Drechsler (*Mycol.* **41**: 229, 1949), Dyal (*Sydowia* **27**: 293, 1976; keys parasites of nematodes and amoebae, bibliogr.).

Endocoenobium Ingold (1940), Endochytriaceae. 1, British Isles. See Canter (*J. Linn. Soc.* Bot. **91**: 95, 1985).

Endocoleroa Petr. (1969) = Venturia Sacc. fide Sivanesan (*Biblthca Mycol.* **59**, 1977).

Endocolium Syd. (1937), Pezizomycotina. 1, C. America.

endocommensal, an organism living as a commensal (see symbiosis) inside another (e.g. *Trichomycetes* in the gut of *Insecta*).

Endoconia Raf. (1819) nom. dub., ? Agaricomycetes. See Murrill (*Index Rafinesq.*, 1949).

Endoconidioma Tsuneda, Hambleton & Currah (2004), anamorphic *Dothideaceae*. 1, Canada. See Tsuneda *et al.* (*Mycol.* **96**: 1129, 2004).

Endoconidiophora Münch (1907) = Ceratocystis fide Hawksworth *et al.* (*Dictionary of the Fungi* edn 8, 1995).

Endoconidium Prill. & Delacr. (1891), anamorphic *Gloeotinia*, Hsp.0eH.15. 3, widespread (temperate).

endoconidium, a conidium formed inside a hypha, e.g. as in *Thielaviopsis basicola*.

Endoconospora Gjaerum (1971), anamorphic *Pezizomycotina*, Hsp.0eH.15. 2, Norway; India. See Gjaerum (*Norw. Jl Bot.* **18**: 109, 1971), Braun & Hosagoudar (*Mycotaxon* **46**: 259, 1993).

Endocoryneum Petr. (1922), anamorphic *Pezizomycotina*, St.≡ eP.15. 2, Austria; Italy. See Marras *et al. in* Luisi *et al.* (Eds) (*Recent advances in oak decline*: 255, 1993).

Endocreas Samuels & Rogerson (1989) = Valsonectria fide Seifert & Samuels (*Mycol.* **89**: 512, 1997), Rossman *et al.* (*Stud. Mycol.* **42**: 248 pp., 1999).

Endocronartium Y. Hirats. (1969), Cronartiaceae. 4 (on *Pinus* (*Pinaceae*)), Europe; North America; Japan. A heterogeneous group of endo-cyclic rusts related to *Cronartium*. See Imazu & Kakishima (*Proc. 4(th) rust on pines working party* Tsukuba: 27, 1995), Vogler & Bruns (*Mycol.* **90**: 244, 1998; DNA seq. analysis).

endocyanosis, the inclusion of cyanobacteria inside the cells of another organism; in fungi, see *Geosiphon*.

Endocycla Syd. (1927) = Schizothyrium fide Müller & von Arx (*Beitr. Kryptfl. Schweiz* **11** no. 2, 1962).

endocyclic, see *Pucciniales* and Table 5.

endocystidium, see cystidium.

endocystidium-form, see *Pucciniales* and Table 5.

Endodermophyton Castell. (1910) = Trichophyton fide Sutton (*in litt.*).

Endodesmia Berk. & Broome (1871) [non *Endodesmia* Benth. 1862, *Clusiaceae*] ≡ Broomeola.

Endodesmidium Canter (1949), Synchytriaceae. 1 (on desmids), Europe.

Endodothella Theiss. & Syd. (1915) = Phyllachora Nitschke ex Fuckel (1870) fide Cannon (*Mycol. Pap.* **163**: 302 pp., 1991).

Endodothiora Petr. (1929), Dothioraceae. 1 (on *Dothidea*), former USSR.

endoectothrix, making growth in and on a hair.

endogenous (1) living inside; (2) immersed in the substratum; (3) undergoing development within.

Endogloea Höhn. (1915) = Phomopsis (Sacc.) Bubák. fide Sutton (*Mycol. Pap.* **141**, 1977).

Endogonaceae Paol. (1889), Endogonales. 4 gen., 27 spp.

Lit.: Morton & Benny (*Mycotaxon* **37**: 471, 1990; rev.), Yao *et al.* (*Kew Bull.* **50**: 349, 1995), Yao *et al.* (*Genera of Endogonales*: 229 pp., 1996), Jeffries & Dodd (*Appl. Microb. System.*: 73, 2000), Kyde & Gould (*Microbial Endophytes*: 161, 2000), Benny *et al. in* McLaughlin *et al.* (Eds) (*The Mycota* A Comprehensive Treatise on Fungi as Experimental Systems for Basic and Applied Research **7A**: 113, 2001), Gleason & McGee (*Australasian Mycologist* **22**: 73, 2004).

Endogonales Moreau ex R.K. Benj. (1979). Mucoromycotina. 1 fam., 4 gen., 27 spp. Endomycorrhizal and saprobic taxa transferred to *Glomerales* (q.v.) by Morton & Benny (1990). Fam.:

Endogonaceae

Lit.: Morton & Benny (1990), Pegler *et al.* (*British Truffles. A revision of British hypogeous fungi*, 1993), Warcup (*MR* **94**: 173, 1990; Australia), Yao *et al.* (*Kew Bull.* **50**: 349, 1995; gen. names, *Genera of Endogonales*, 1996; review), White *et al.* (*Mycol.* **98**: 860, 2006; molecular phylogeny), Hibbett *et al.* (*MR* **111**: 109, 2007), and see under Family.

Endogone Link (1809), Endogonaceae. *c.* 20 (saprobic or forming ectomycorrhiza), widespread (esp. temperate). See Gerdemann & Trappe (*Mycol. Mem.* **5**, 1974; key NW Am. spp.), Tandy (*Aust. J. Bot.* **23**: 849, 1975; Australia), Bonfante-Fasolo & Scannerini (*Mycopathologia* **59**: 117, 1976; ultrastr.), Trappe & Gerdemann (*Mycol.* **71**: 206, 1979), Berch & Fortin (*Mycol.* **75**: 328, 1983; zygospore germination), Berch & Fortin (*CJB* **61**: 899, 1983; axenic culture), Berch & Fortin (*CJB* **62**: 170, 1984), Berch & Castellano (*Mycol.* **78**: 292, 1986; sporulation in culture), Gibson *et al.* (*Mycol.* **79**: 433, 1987; cytochemistry), Jabaji-Hare *et al.* (*Mycol.* **80**: 54, 1988; fatty acids), Dalpé (*CJB* **68**: 910, 1990; culture), Warcup (*MR* **94**: 173, 1990; ectomycorrhizal association), Yao *et al.* (*Kew Bull.* **50**: 306, 1995), Yao *et al.* (*Genera of Endogonales*: 229 pp., 1996), Błaszkowski (*Mycotaxon* **63**: 131, 1997), Błaszkowski *et al.* (*MR* **102**: 1096, 1998), Gleason & McGee (*Australasian Mycologist* **22**: 73, 2004; ultrastr.), Goto & Maia (*Mycotaxon* **96**: 327, 2006; Brazil), James *et al.* (*Nature* **443**: 818, 2006; phylogeny).

Endogonella Höhn. (1913) = Glaziella fide Zycha *et al.* (*Mucorales*, 1969).

endogonidium, a gonidium having its development inside a receptable or gonidangium (obsol.).

Endogonopsis R. Heim (1966), ? Diplocystidiaceae. 1, S. Asia. See Heim (*Revue Mycol.* Paris **31**: 150, 1966) Possibly based on young *Astraeus* (Astr.), Heim (*Revue Mycol.* Paris **33**: 379, 1968), Heim (*C.R. Acad. Sci. Paris* D **268**: 1489, 1969).

Endohormidium Auersw. & Rabenh. (1869) = Corynelia fide Fitzpatrick (*Mycologia* 7 **34**: 464, 1943).

Endohyalina Marbach (2000), Caliciaceae (L). 2, neotropics. See Marbach (*Biblthca Lichenol.* **74**: 201, 2000).

endohypha, see hypha.

endokapylic, a thallus of a lichenicolous fungus in which no morphologically distinct lichenized structure is formed (Poelt & Vězda, 1984; Rambold & Triebel, *Bibl. Lich.* **48**, 1992).

Endolepiotula Singer (1963), Agaricaceae. 1, Argentina. Basidioma gasteroid.

endolithic, in stone (Kobluk & Kahle, *Bull. Can. Pet. Geol.* **25**: 208, 1977; fungi, bibliogr.); cf. epilithic. See also cryptoendolithic.

Endolpidium De Wild. (1894) = Olpidium fide Sparrow (*Aquatic Phycomycetes* Edn 2: 1187 pp., 1960).

Endomelanconium Petr. (1940), anamorphic *Austrocenangium*, St.0eP.1. 4, widespread. See Verkley & Aa (*Mycol.* **89**: 967, 1997; Papua New Guinea), Yanna *et al.* (*Fungal Diversity* **2**: 199, 1999; Hong Kong).

Endomeliola S. Hughes & Piroz. (1994), Meliolaceae. 1, New Zealand. See Hughes & Pirozynski (*N.Z. Jl Bot.* **32**: 53, 1994).

Endomyces Reess (1870), Endomycetaceae. 3 (on *Armillaria*), widespread (temperate). See Redhead & Malloch (*CJB* **55**: 1701, 1977), von Arx (*Antonie van Leeuwenhoek* **43**: 33, 1977), de Hoog *in* Kurtzman & Fell (Eds) (*Yeasts, a taxonomic study* 4th edn: 194, 1998), Suh *et al.* (*Mycol.* **93**: 317, 2001), Suh *et al.* (*MR* **109**: 261, 2005; ecology), Suh *et al.* (*Mycol.* **98**: 1006, 2006; phylogeny).

Endomycetaceae J. Schröt. (1893), Saccharomycetales. 4 gen. (+ 1 syn.), 6 spp.

Lit.: Weijman (*Antonie van Leeuwenhoek* **42**: 315, 1976), Redhead & Malloch (*CJB* **55**: 1701, 1977), Hoog *in* Kurtzman & Fell (Eds) (*Yeasts, a taxonomic study* 4th edn: 194, 1998), Suh *et al.* (*Mycol.* **93**: 317, 2001), Suh *et al.* (*Mycol.* **98**: 1006, 2006; phylogeny).

Endomycetales Gäum. & C.W. Dodge (1928) = Saccharomycetales.

Endomycetes (Endomycota). Class (or phylum) for sporogenous and asporogenous yeasts, *Taphrinales*, *Exobasidiales*, and *Ustilaginales* (von Arx, *Pilzkunde*, 1967); emended to include *Trichomycetes*, etc. by Kreisel (1969), and emended to exclude *Exobasidiales*, *Taphrinales* and *Ustilaginales* by von Arx (1981); supported by glyceraldehyde-3-phosphate dehydrogenase sequences (Smith, *Proc. natn. Acad. Sci., USA* **86**: 7063, 1989). See *Ustomycota*.

endomycobiont, a fungal biont in a symbiosis completely immersed in the tissues of the host (e.g. the fungal partner in a mycophycobiosis); an inhabitant (see symbiosis); also used of certain mycorrhizas; see also Endophyte.

Endomycodes Delitsch (1943), Saccharomycetales. 2, widespread.

Endomycopsella Boedijn (1960) = Saccharomycopsis Schiønning fide Boedijn (*Mycopathologia* **12**: 163, 1960), Kurtzman & Robnett (*Antonie van Leeuwenh.* **73**: 331, 1998).

Endomycopsis Dekker (1931) ≡ Guilliermondella fide von Arx & Yarrow (*Antonie van Leeuwenhoek* **50**: 799, 1984).

endomycorrhiza, see mycorrhiza.

Endomycota, see *Saccharomycotina*.

Endonema Pascher (1929) ≡ Pascherinema.

Endonevrum Czern. (1845) = Mycenastrum fide Stalpers (*in litt.*).

Endonius Raf. (1819) nom. dub., Fungi. No spp. included.

endo-operculation (of sporangial dehiscence in chytrids), operculum forced off and carried away by the emerging sporogenous contents (cf. exo-operculation).

endoparasite, a parasite living inside its host.

endoperidermal, within the periderm (Lambright & Tucker, *Bryologist* **83**: 170, 1980); endophloeodal + hypophloeodal.

endoperidium, the inner layer of the peridium.

Endoperplexa P. Roberts (1993), Auriculariales. 4, widespread. See Roberts (*MR* **97**: 471, 1993), Weiss *et al.* (*MR* **108**: 1003, 2004; phylogeny).

Endophallus M. Zang & R.H. Petersen (1989), Phallaceae. 1, China. ? A monstrosity from *Phallus*.

Endophis Norman (1852) nom. rej. = Leptorhaphis fide Hawksworth *et al.* (*Dictionary of the Fungi* edn 8, 1995).

Endophlaea Cooke (1888) nom. dub., Pezizomycotina.

endophloeodic (**endophloeodal**, **endophloeic**) (of the thallus of a crustaceous lichen), almost entirely immersed in bark.

Endophragmia Duvernoy & Maire (1920) = Phragmocephala fide Kirk (*in litt.*), Ellis (*Mycol. Pap.* **72**, 1959; key).

Endophragmiella B. Sutton (1973), anamorphic *Echinosphaeria*, Hso.≡ eP.1. *c.* 75, widespread. See Hughes (*N.Z. Jl Bot.* **17**: 139, 1979; key), Castañeda Ruíz *et al.* (*MR* **102**: 548, 1998), Tsui *et al.* (*Cryptog. Mycol.* **22**: 139, 2001; Hong Kong), M[a]noharachary & Agarwal (*Journal of Mycopathological Research* **41**: 117, 2003; India), Miller & Huhndorf (*MR* **108**: 26, 2004; phylogeny).

Endophragmiopsis M.B. Ellis (1966), anamorphic *Pezizomycotina*, Hso.≡ eP.1. 1, India. See Hughes (*CJB* **61**: 1727, 1984).

Endophyllaceae Dietel (1897) = Pucciniaceae.

Endophyllachora Rehm (1913) = Phyllachora Nitschke ex Fuckel (1870) fide Clements & Shear (*Gen. Fung.*, 1931).

Endophylloides Whetzel & Olive (1917) = Dietelia fide Buriticá (*Rev. Acad. Colomb. Cienc.* **18**: 131, 1991).

endophyllous, living within (i.e., below the cuticle) leaves.

Endophyllum Lév. (1825), Pucciniaceae. *c.* 30 (on angiosperms), widespread. *E. sempervivi* (houseleek, *Sempervivum*, rust). The telial are aecidium-like, the spores produce basidia on germination. See Ashworth (*TBMS* **19**: 240, 1935; cytology, life-history), Buriticá (*Rev. Acad. Colomb. Cienc.* **18** no. 69: 131, 1991), Gjærum *et al.* (*Lidia* **5**: 83, 2000; Uganda).

Endophyte. An organism that occurs within a living plant, cf. epiphyte (qv). Endophyte has been used in a variety of ways, giving rise to semantic confusion and ambiguities. In its widest sense the term has been used to include organisms ranging from bacteria to parasitic angiosperms (e.g. mistletoes) that exist completely or partially within any part of any plant-like organism. With respect to fungi, endophyte has been used to refer to those present as mycorrhizas (qv), as pathogenic infections or saprobic colonizations, and as outwardly symptomless infections showing no signs of either disease, decay or fructification. The distinction between latent pathogens that cause disease only under specific environmental circumstances, or latent saprobes that only cause symptoms upon death of the host plant tissues, and those fungi which remain as symptomless infections is not always clear. While there is frequent use of the term in a restricted sense to refer to symptomless, mutualistic (q.v.) fungi within aerial plant parts the term is also used to include all organisms that grow endophytically regardless of their symptoms or disease effects, see Chanway (*CJB* **74**: 321, 1996), Stone *et al.* (*in* Bacon & White (Eds), *Microbial endophytes*: 3, 2000), Wennström (*Oikos* **71**: 535, 1994), Wilson (*Oikos* **73**: 274, 1995). Given its variety of usages, endophyte should be clearly defined when used.

Isolation techniques typically involve plating healthy, surface-sterilized plant tissues on agar media and observing the outgrowth of fungi, or incubation of washed plant tissues under humid conditions and subsequent collection and plating of discharged spores (see Bacon, *in* Labeda (Ed.), *Isolation of Biotechnological Organisms from Nature*: 259, 1990; Bills, in Redlin & Carris (Eds), *Endophytic Fungi in Grasses and Woody Plants*: 31, 1996; Cohen, *Mycologia* **91**: 917, 1999).

Conifer needles, particularly of the *Pinaceae* support a complex mycota of symptomless inhabitants including, notably, species of *Lophodermium* and other members of the *Rhytismataceae*. Several species may inhabit a single needle. Colonization occurs from flushing onwards, with the fungus remaining symptomless in the leaf for several years before fruiting. Incidence varies with needle age and events in the life of the needle can determine which species fruit after needle death (Millar & Richards, *Mitteilungen der Bundesforschungsanstalt für Forst- und Holtzwirtschaft* **108**: 57, 1975). Fungal endophytes in other woody plants have also been investigated (see Redlin & Carris (Eds), *Endophytic Fungi in Grasses and Woody Plants*, 1996; Wilson, *in* Bacon & White (Eds) *Microbial endophytes*: 389, 2000). These species are dispersed by spores and not transmitted by systemic infection of seeds, and while some show a high level of host specificity (e.g. Chapela *et al.*, *Physiol. & Mol. Pl. Path.* **39**: 289, 1991), others have a wide range of host plants. Some plants contain more than one endophyte (e.g. *Castanea*, see Bissegger & Sieber, *Mycologia* **86**: 648, 1994; *Picea*, see Müller & Hallaksela, *MR* **104**: 1139, 2000). In many cases the role of the endophyte is unknown although insect antagonism and anti-herbivory are often suspected. It is possible that many are latent pathogenic or sabrobic infections held in a balanced antagonism with their host (Schulz *et al.*, *MR* **103**: 1275, 1999). There is also some indication that leaf endophytes may influence the onset of leaf senescence and abscission. Fungal endophytes of *Taxus* have the ability to produce taxol, an important anti-cancer drug (Stierle *et al.*, 1995).

Symptomless endophytes of grasses are mainly *Clavicipitaceae* (Glenn *et al.*, *Mycologia* **88**: 369, 1996; White & Reddy, *Mycologia* **90**: 226, 1998), with species of *Epichloë* and related anamorphic *Neotyphodium* spp. (Tredway *et al.*, *MR* **103**: 1593, 1999; Christensen *et al.*, *MR* **104**: 974, 2000) mainly on cool-season grasses, while *Balansia*, *Parepichloë*, and *Myriogenospora* are mainly on warm-season grasses. The fungi occur intercellularly and may fruit on leaves or inflorescences, some do not sporulate and are completely seed-transmitted (White, *Pl. Dis.* **71**: 340, 1987; Sampson, *TBMS* **18**: 337, 1935). While the fungi probably gain enhanced nutrition and greater protection against desiccation, predators and parasites, the host plants although sometimes displaying a degree of sterility often show increased drought tolerance, and are vegetatively invigorated (possibly due to fungally produced auxins, see Porter *et al.*, *J. Nat. Products* **42**: 309, 1979), and exhibit increased resistance to pathogenic fungi and to herbivorous insects, nematodes and mammals. Resistance in many cases is due to the production of toxic secondary metaboilites (qv) by the endophytic fungus. This is of particular importance where forage plants are involved (Lane *et al.*, *in* Bacon & White (Eds), *Microbial endophytes*: 341, 2000), and includes compounds such as pyrrolizidine alkaloids (lolines, active against insects), tremorgenic indole diterpene alkaloids (lolitrems, active against insects and mammals) and ergot alkaloids (mainly ergovaline, affecting mammals and insects). Some of the alkaloids have been implicated in mycotoxicoses (qv), such as lolitrems with staggers in livestock (mainly due to *Neotyphodium lolii* in *Lolium perenne*), or ergot peptides with fescue foot (*N. coenophialum* in *Festuca arundinacea*) and lysergic acid amides with sleepygrass and similar conditions (*Neotyphodium* in *Stipa robusta* and *Achnatherum inebrians*). Other compounds are known but their significance is less clear, including sesquiterpene alcohols which may have antifungal activity, and the widespread guanidium-containing pyrrolpyrazine alkaloid, peramine, which may be active against insects.

Lit.: endophytes of marine algae, Cubit (PhD Thesis, University of Oregon, USA., 1974); of liverworts, Duckett & Read (*New Phytol.* **129**: 439, 1995); of mosses, Petrini (*in* Fokkenna & van den Heuval (Eds), *Microbiology of the Phyllosphere*: 175, 1986); of ferns, Fisher (*New Phytol.***132**: 119, 1996), Swatzell *et al.* (*Int. J. Plant Sci.* **157**: 53, 1996); of tropical palms and other monochots, Rodrigues & Samuels (*MR* **94**: 827, 1990), Dreyfuss & Petrini (*Bot. Helv.* **94**: 33, 1984), Frohlich *et al.* (*MR* **104**: 1202, 2000); of cotton, Gasoni & de Gurfinkel (*MR* **101**: 867, 1997); of maize Yates *et al.* (*MR* **103**: 129, 1999); non-clavicipitaceous grass endophytes, Siegel *et al.* (*Mycologia* **87**: 196, 1995); on Orchidaceae, Bayman *et al.* (*New Phytol.* **135**: 143, 1997); of soya, Walcott *et al.* (*Plant Dis.* **82**: 584, 1998); of radish, Leeman *et al.* (*Eur. J. Plant Path.* **102**: 21, 1996); dark-septate root endophytes, Caldwell *et al.* (*Mycologia* **92**: 230, 2000), Grunig *et al.* (*MR* **105**: 24, 2001); zygomycetous endophytes, Zheng & Jiang (*Mycotaxon* **56**: 455, 1995); of submerged roots, Iqbal *et al.* (*CJB* **73**: 538, 1995); effects on nematodes, Elmi *et al.* (*Grass & Forage Sci.* **55**: 166, 2000); effects on arthropods, Carriere *et al.* (*J. Econ. Entomol.* **91**: 324, 1998), Hata & Futai (*CJB* **73**: 384, 1995), Lewis & Vaughan (*Tests Agrochem. & Cultivars* **18**: 34, 1997), Richmond & Shetlar (*J. Econ. Entomol.* **92**: 1329-1334, 1999); effects on cattle, Cosgrove *et al.* (*Proc. N.Z. Grassland Assoc.* **57**: 43, 1996), Oliver *et al.* (*J. Animal Sci.* **76**: 2853, 1998), Shewmaker *et al.* (*Agronomy J.* **89**: 695, 1997); successional fields, Clay & Holah (*Science* **285**: 1742, 1999); evolution, Moon *et al.* (*Mycologia* **92**: 1103, 2000); taxol production, Stierle *et al.* (*J. Nat. Prod.-Lloydia* **58**: 1315, 1995).

Endoplacodium Petr. (1949), anamorphic *Pezizomycotina*, St.0eH.?. 1, Iran.

endopropagule, a propagule produced inside the body (medical mycology).

Endoptychum Czern. (1845) nom. rej. = Chlorophyllum. Secotioid fungus, formerly recognised as separate genus. fide Vellinga & De Kok (*Taxon* **51**: 563, 2002).

Endopyreniaceae Zahlbr. (1898) = Verrucariaceae.

Endopyreniomyces E.A. Thomas (1939) nom. inval. ≡ Endocarpon.

Endopyrenium Flot. (1855) = Catapyrenium fide Thomson (*Bryologist* **90**: 27, 1987), Hawksworth *et al.* (*Dictionary of the Fungi* edn 8, 1995).

Endoraecium Hodges & D.E. Gardner (1984), ? Raveneliaceae. 2 (on *Acacia* (*Leguminosae*: *Mimosoidea*)), Hawaii. See Scholler & Aime (*Mycoscience* **47**: 159, 2006; emend., incl. *Racospermyces*).

Endoramularia Petr. (1923), anamorphic *Pezizomycotina*, Hsp or Cac.≡ eH.?. 1, Europe.

Endoreticulatus W.M. Brooks, J.J. Becnel & G.G. Kennedy (1988), Microsporidia. 4. See Brooks *et al.* (*J. Protozool.* **35**: 487, 1988).

endosaprophytism, the destruction of an alga by the fungus in a lichen (Elenkin).

endosclerotium, a sclerotium of endogenous origin.

Endoscypha Syd. (1924), Helotiales. 1, New Zealand.

Endospora Scherff. (1925) nom. dub., ? Chytridiomycetes.

endospore (1) the inner wall of a spore (see ascospore, Spore wall); (2) an endogenous spore, e.g. a sporangiospore.

Endosporella Thaxt. (1920) = Pyxidiophora fide Blackwell (*Mycol.* **86**: 1, 1994).

Endosporisorium Vánky (1995) = Macalpinomyces fide Vánky (*Mycotaxon* **62**: 127, 1997).

Endosporoideus W.H. Ho, Yanna, K.D. Hyde & Goh (2005), anamorphic *Pezizomycotina*, H?.?.?. 1, Hong Kong. See Ho *et al.* (*Mycol.* **97**: 239, 2005).

Endosporostilbe Subram. (1958) = Bloxamia fide Nag Raj & Kendrick (*Monogr. Chalara Allied Genera*, 1975).

Endostigme Syd. (1923) = Venturia Sacc. fide von Arx & Müller (*Stud. Mycol.* **9**, 1975).

Endostilbum Malençon (1964), anamorphic *Pezizomycotina*, Hsy.0eH.15. 1, Europe; Morocco. See Korf & Candoussau (*BSMF* **90**: 209, 1974).

endosymbiont, an organism which lives in mutualistic symbiosis within the cells of another organism; the inhabitant.

Endothia Fr. (1849), Cryphonectriaceae. Anamorph

Endothiella. 2, widespread. See also *Cryphonectria*. See Roane *In* Roane *et al.* (Eds) (*Chestnut blight, other Endothia diseases, and the genus Endothia*: 28, 1986; key 11 spp.), Micales & Stipes (*Phytopathology* **77**: 650, 1987), Myburg *et al.* (*Mycol.* **91**: 243, 1999; DNA), Castlebury *et al.* (*Mycol.* **94**: 1017, 2002; phylogeny), Myburg *et al.* (*Mycoscience* **44**: 187, 2003; on *Myrtaceae*), Myburg *et al.* (*Mycol.* **96**: 990, 2004; phylogeny), Myburg *et al.* (*CJB* **82**: 1730, 2004; phylogeny), Gryzenhout *et al.* (*FEMS Microbiol. Letters* **258**: 161, 2006), Gryzenhout *et al.* (*Mycol.* **98**: 239, 2006; phylogeny, n. fam.), Zhang *et al.* (*Mycol.* **98**: 1076, 2006; phylogeny).

Endothiella Sacc. (1906), anamorphic *Cryphonectria*, St.0eH.15. 3, Europe; N. America. See Roane *In* Roane *et al.* (Eds) (*Chestnut blight, other Endothia diseases, and the genus Endothia*: 28, 1986), Castlebury *et al.* (*Mycol.* **94**: 1017, 2002), Myburg *et al.* (*Mycol.* **96**: 990, 2004), Gryzenhout *et al.* (*FEMS Microbiol. Lett.* **258**: 161, 2006).

Endothlaspis Sorokīn (1890), Ustilaginales. 2, central Asia. See Langdon & Fullerton (*Mycotaxon* **6**: 421, 1978), Vánky (*Cryptog. Stud.* **1**: 159 pp., 1987).

Endothrix Sorokīn (1890), Ustilaginales. 2, central Asia. See Langdon & Fullerton (*Mycotaxon* **6**: 421, 1978), Vánky (*Cryptog. Stud.* **1**: 159 pp., 1987).

endothrix, living inside a hair.

Endotrabutia Chardón (1930) = Phyllachora Nitschke ex Fuckel (1870) fide Petrak (*Sydowia* **5**: 336, 1951).

Endotrichum Corda (1838) nom. dub., anamorphic *Pezizomycotina*. See Sutton (*Mycol. Pap.* **141**, 1977).

endotrophic, see mycorrhiza.

Endotryblidium Petr. (1959), ? Patellariaceae. 1, widespread (temperate). See Magnes (*Biblthca Mycol.* **165**, 1997).

endotunica, endoascus (q.v.).

Endoxyla Fuckel (1871), ? Boliniaceae. 9, widespread. See Barr (*Mycol. Mem.* **7**: 232 pp., 1978), Barr (*Mycotaxon* **46**: 45, 1993), Untereiner (*Mycol.* **85**: 294, 1993; key), Réblová (*Mycol.* **98**: 68, 2006; key).

Endoxylina Romell (1892), Diatrypaceae. 5, widespread. See Mhaskar (*Botanique, Nagpur* **3**: 69, 1972), Ju *et al.* (*Mycotaxon* **58**: 419, 1996), Chacón (*Fungal Diversity* **11**: 61, 2002; key).

endozoic, living inside an animal.

Endozythia Petr. (1959), anamorphic *Pezizomycotina*, Cpd.0eH.?. 1 (in *Pleospora* perithecia), former Yugoslavia. See Petrak (*Sydowia* **13**: 116, 1959).

Enduria Norman (1885) nom. dub., ? Dothideomycetes.

Endyllium Clem. (1931) ≡ Magnusiomyces.

Enerthidium Syd. (1939), anamorphic *Pezizomycotina*, Cac.0eP.19. 1, Africa.

Engelhardtiella A. Funk (1973), anamorphic *Pezizomycotina*, Hsp.1bH.15. 1 (on *Botryosphaeria*), N. America. See Funk (*CJB* **51**: 1643, 1973).

Engizostoma Gray (1821) ≡ Valsa Fr.

Englera F. Stevens (1939) = Asterina fide Müller & von Arx (*Beitr. Kryptfl. Schweiz* **11** no. 2, 1962).

Englerodothis Theiss. & Syd. (1915), Parmulariaceae. 2, Africa. See Müller & von Arx (*Beitr. Kryptfl. Schweiz* **11** no. 2, 1962), Dulymamode *et al.* (*MR* **105**: 247, 2001).

Engleromyces Henn. (1900), Xylariaceae. 1, Africa; Asia. See Dennis (*Bull. Jard. bot. Brux.* **31**: 148, 1961), Rogers (*Mycol.* **73**: 28, 1981; posn), Kokwaro (*Bothalia* **14**: 237, 1983; ethnomycol.), Zang (*Acta Bot. Yunn.* **14**: 385, 1992; Asia).

Englerula Henn. (1904), Englerulaceae. Anamorph *Capnodiastrum*. 6, widespread. See Hansford (*Mycol. Pap.* **15**, 1946), Reynolds & Gilbert (*Aust. Syst. Bot.* **18**: 265, 2005; Australia).

Englerulaceae Henn. (1904), ? Dothideomycetes (inc. sed.). 11 gen. (+ 9 syn.), 106 spp.

Lit.: Stevens & Ryan (*Ill. biol. monogr.* **17** no. 2: 1, 1939), Doidge (*Bothalia* **4**: 273, 1942; S. Afr.), Hughes (*CJB* **64**: 1591, 1986), Hughes (*Mycol.* **82**: 657, 1990), Castlebury *et al.* (*Mycotaxon* **54**: 461, 1995).

Englerulaster Höhn. (1910) = Asterina fide Müller & von Arx (*Beitr. Kryptfl. Schweiz* **11** no. 2, 1962).

Englerulella Hansf. (1946) = Rhytidenglerula fide Müller & von Arx (*Beitr. Kryptfl. Schweiz* **11** no. 2, 1962).

Engyodontium de Hoog (1978), anamorphic *Cordycipitaceae*, Hso.0eH.10. 5, widespread. See Gams *et al.* (*Persoonia* **12**: 135, 1984; key), Sekhon *et al.* (*Mycopathologia* **138**: 1, 1997; antigens), Sung *et al.* (*Nova Hedwigia* **72**: 311, 2001; phylogeny), Sung *et al.* (*Stud. Mycol.* **57**: 1, 2007; phylogeny).

enniatin, A (lateratiin) and **B**, peptide antibiotics from *Fusarium orthoceras*; antibacterial (Gäumann *et al.*, *Experientia* **3**: 202, 325, 1947); avenacein, sambucinum.

enokitake (winter mushroom), the edible *Flammulina velutipes*, cultivated in Japan and Taiwan (Chang & Hayes, 1978).

enphytotic, a plant disease of which the damage is constant from year to year; cf. epiphytotic.

Enridescalsia R.F. Castañeda & Guarro (1998), anamorphic *Pezizomycotina*, Hso.?.?. 1, Cuba. See Castañeda Ruíz *et al.* (*MR* **102**: 42, 1998).

Ensaluta Zobel (1854) = Tuber fide Fischer (*Nat. Pflanzenfam.* **5b**: viii, 1938), Læssøe & Hansen (*MR* **111**: 1075, 2007; phylogeny).

ensate (ensiform), narrow and pointed; sword-like in form.

Enslinia Fr. (1835) [non *Enslenia* Raf. 1817, *Acanthaceae*] ≡ Porodiscus Murrill.

Entelexis Van der Walt & Johannsen (1973), ? Saccharomycetales. Anamorph *Torulopsis*. 1, S. Africa. See van der Walt & Johannsen (*Antonie van Leeuwenhoek Ned. Tijdschr. Hyg.* **39**: 646, 1973).

enteroblastic, see blastic.

Enterobotryum Preuss (1853) nom. dub., Fungi. See Lundqvist (*Symb. bot. upsal.* **20** no. 1, 1972).

Enterobrus (orthographic variant), see *Enterobryus*.

Enterocarpus Locq.-Lin. (1977), Microascaceae. 2 (coprophilous), Sahara; Italy. See Locquin-Linard (*Revue Mycol.* Paris **41**: 509, 1977), Doveri *et al.* (*Boll. Gruppo Micol. 'G. Bresadola'* **40**: 187, 1998).

Enterocytozoon Desportes, Le Charpentier, Galian, Bernard, Cochand-Priollet, Lavergne, Ravisse & Modigliani (1985), Microsporidia. 1.

Enterodictyon Müll. Arg. (1892), Roccellaceae (L). 2, East Indies.

Enterodictyonomyces Cif. & Tomas. (1953) ≡ Enterodictyon.

Enterodictyum Clem. & Shear (1931) ≡ Enterodictyon.

Enterographa Fée (1824), Roccellaceae (±L). *c.* 59 (partly on lichens), widespread (esp. tropical). See Santesson (*Symb. bot. upsal.* **12** no. 1: 1, 1952), Coppins & James (*Lichenologist* **11**: 27, 1979; Brit. spp.), Torrente & Egea (*Biblthca Lichenol.* **32**, 1989), Aptroot *et al.* (*Biblthca Lichenol.* **57**: 19, 1995; key 7 foliicolous spp.), Aptroot *et al.* (*Bryologist* **106**: 278, 2003; chemistry), Sparrius (*Biblthca Lichenol.* **89**: 141 pp., 2004; monograph), Ertz *et al.* (*Biblthca Lichenol.* **91**: 155 pp., 2005), Sparrius *et al.* (*Lichenologist* **38**: 27, 2006; SE Asia).

Enteromyxa Ces. (1879) ? = Lycogalopsis fide Stalpers (*in litt.*).

Enteroramus Lichtw., M.M. White, Cafaro & Misra (1999), Pichiaceae. 1, USA. See Lichtwardt *et al.* (*Mycol.* **91**: 697, 1999), Suh *et al.* (*Mycol.* **96**: 756, 2004; phylogeny).

Enterostigma Müll. Arg. (1885) = Thelotrema fide Salisbury (*Lichenologist* **5**: 319, 1972).

Enterostigmatomyces Cif. & Tomas. (1953) ≡ Enterostigma.

Enthallopycnidium F. Stevens (1925), anamorphic *Pezizomycotina*, Cpt.0fH.?. 2, pantropical. See Parbery & Brown (*Microbiology of the Phyllosphere*: 101, 1986).

entheogen, a plant (or fungus) substance used by humans in prehistory associated with religous feelings; see ethnomycology, hallucinogenic fungi, soma.

entire (of edges of lamellae, etc.), not torn; having no teeth.

ento- (prefix), inside.

Entoderma Hanula, Andreadis & M. Blackw. (1991), anamorphic *Pezizomycotina*, Cac.0-1eH.1. 1 (on *Insecta*), USA. See Hanula *et al.* (*J. Invert. Path.* **58**: 328, 1991).

Entodesmium Riess (1854), Lophiostomataceae. 7 (on legumes), widespread (temperate). See Holm (*Symb. bot. upsal.* **14** no. 3: 1, 1957), Shoemaker (*CJB* **62**: 2730, 1984; key 6 spp.), Barr (*Mycotaxon* **43**: 371, 1992; posn), Liew *et al.* (*Mol. Phylogen. Evol.* **16**: 392, 2000; interascal tissue, phylogeny).

Entoleuca Syd. (1922), Xylariaceae. Anamorph *Geniculisporium*. 3, widespread (temperate). See Læssøe (*SA* **13**: 43, 1994), Rogers & Ju (*Mycotaxon* **59**: 441, 1996; monogr.), Mazzaglia *et al.* (*MR* **105**: 670, 2001; phylogeny), Stadler *et al.* (*MR* **105**: 1191, 2001; biochemistry), Petrini (*N.Z. Jl Bot.* **41**: 71, 2003; New Zealand), Ju *et al.* (*Mycol.* **96**: 1393, 2004), Kasanen *et al.* (*MR* **108**: 766, 2004; genetic diversity).

Entoloma (Fr.) P. Kumm. (1871), Entolomataceae. *c.* 1000, widespread (esp. tropical). See Orton (*TBMS* **43**: 328, 1960; key 15 Br. spp.), Largent & Benedict (*Mycol.* **62**: 440, 1970; key subgenera), Pegler & Young (*Kew Bull.* **30**: 19, 1975; key 9 Br. spp.), Horak (*Sydowia* **28**: 171, 1976; cuboid spored spp.), Mazzer (*Biblthca Mycol.* **46**, 1976; monogr.), Horak (*Sydowia* **29**: 289, 1977), Pegler (*Kew Bull.* Addit. Ser. **6**, 1977; key 5 E. Afr. spp.), Horak (*Sydowia* **31**: 58, 1978), Horak (*Sydowia* **30**: 40, 1978; 74 S. Am. spp.), Romagnesi & Gilles (*Beih. Nova Hedwigia* **59**, 1978; Ivory Coast spp.), Horak (*N.Z. Jl Bot.* **17**: 275, 1979; key 9 NZ spp.), Nordeloos (*Persoonia* **10**: 207, 1979; key 11 spp. subgen. *Pouzaromyces*), Nordeloos (*Persoonia* **10**: 427, 1979; subgen. *Nolanea*, key Eur. spp.), Pegler & Young (*Beih. Sydowia* **8**, 1979), Horak (*Beih. Nova Hedwigia* **65**, 1980; Indomalaya, Australasia), Nordeloos (*Persoonia* **11**: 121, 1980; subgen. *Entoloma*, key Eur. spp.), Nordeloos (*Persoonia* **11**: 451, 1980; subgen. *Leptonia*, key Eur. spp.), Baroni (*Beih. Nova Hedwigia* **67**, 1982; revision), Baroni & Petersen (*Mycol.* **79**: 358, 1987), Baroni & Halling (*Mycol.* **84**: 419, 1992; key N. Am. spp.), Nordeloos (*Entoloma s.l.*, 1992; Eur. spp.), Manimohan *et al.* (*MR* **99**: 1083, 1995; key 21 Indian spp.), Bon (*Bull. Trimestr. Féd. Mycol. Dauphiné-Savoie* **41**: 13, 2001; key alpine spp.), Hofstetter *et al.* (*MR* **106**: 1043, 2002; phylogeny), Contu (*Micologia e Vegetazione Mediterranea* **17**: 95, 2002; Italy), Consiglio (*Micologia e Vegetazione Mediterranea* **18**: 143, 2003), Manimohan *et al.* (*Persoonia* **19**: 45, 2006; Indian spp.), Gates & Noordeloos (*Persoonia* **19**: 157, 2007; Tasmania).

Entolomataceae Kotl. & Pouzar (1972), Agaricales. 4 gen. (+ 27 syn.), 1071 spp.

Lit.: Noordeloos (*Beih. Nova Hedwigia* **91**: 419 pp., 1987), Noordeloos & Gulden (*CJB* **67**: 1727, 1989), Baroni & Horak (*Mycol.* **86**: 138, 1994), Manimohan *et al.* (*MR* **99**: 1083, 1995), Baroni (*Kew Bull.* **54**: 777, 1999), Eyssartier *et al.* (*MR* **105**: 1144, 2001), Hofstetter (*MR* **106**: 1043, 2002), Moncalvo *et al.* (*Mol. Phylogen. Evol.* **23**: 357, 2002), Noordeloos & Gulden (*Mem. N. Y. bot. Gdn* **89**: 97, 2004).

Entolomina Arnaud (1952) nom. inval., Thelephoraceae. 1, Europe.

entomo- (prefix), of *Insecta*.

Entomocorticium H.S. Whitney, Bandoni & Oberw. (1987), Peniophoraceae. 1, Canada. See Whitney *et al.* (*CJB* **65**: 96, 1987).

Entomocosma Speg. (1918) ? = Pyxidiophora fide Blackwell (*Mycol.* **86**: 1, 1994).

entomogenous, living in or on insects, esp. as pathogens.

Entomogenous fungi. Range from commensals or mutualists, through ectoparasites which do not seriously affect their arthropod hosts, to pathogens that are lethal and include representatives of all the major groups of fungi. Excluding the laboulbeniomycetes, 750 spp. in 56 gen. known to be pathogens or parasites of arthropod pests. The terms entomopathogens, entomogenous mutualistic symbionts, entomogenous ectoparasites and entomogenous endoparasites have been proposed (Evans, *in* Pirozynski & Hawksworth (Eds), *Coevolution of fungi with plants and animals*: 149, 1988).

Septobasidium has symbiotic associations with scale insects, as has *Stereum sanguinolentum* with *Sirex* (Perkin, *Ann. appl. Biol.* **29**: 268, 1942), and a number of insects make use of fungi in their alimentary systems; e.g. Phaff *et al.* (*Ecology* **37**: 533, 1956; yeasts of *Drosophila*). See Brues (*Insect dietary*, 1946), Steinhaus (*Insect microbiology*, 1946; *Insect pathology*, edn 2, 1963), Shifrine & Phaff (*Mycol.* **48**: 41, 1956; yeasts and bark beetles), Madelin (*Endeavour* **19**: 181, 1960), Batra (Ed.) (*Insect-fungus symbiosis*, 1980).

Trichomycetes and *Laboulbeniales* have little effect on their hosts but the chytrid *Coelomomyces*, the oomycete *Lagenidium giganteum* (both mainly on mosquito larvae), and among zygomycetes the *Entomophthoraceae* (e.g. *Entomophthora*, *Erynia*, *Massospora*, *Neozygites*, *Zoophthora*) are important insect pathogens. Among ascomycetes *Ascosphaera*, *Cordyceps*, *Torrubiella*, and *Hypocrella* and its *Aschersonia* anamorph, frequently cause epizootics as do the hyphomycetes *Beauveria*, *Culicinomyces*, *Hirsutella*, *Metarhizium*, *Nomuraea*, *Paecilomyces*, and *Verticillium*. Several of these genera are being exploited as mycoinsecticides to control a range of arthropod pests (see mycopesticides).

Lit.: **General**: Brady (*Biocontrol News and Information* **2**: 281, 1981), Evans (see Misra & Horn, pp.

119-144; fungi on ants), Evans & Hywel-Jones (*in* Ben-Dov & Hodgson (Eds), *Soft scale insects*: 5, 1994; fungi on coccids), Evans & Prior (*in* Rosen (Ed.), *Armored scale insects*: 3, 1990; fungi on scales), Evans & Samson (*Mycologist* **1**: 152, 1987; fungi on spiders), Hajek (*in* Lumsden & Vaughan (Eds), *Pest Management: biologically based technologies*: 54, 1993; prospects for insect control), Madelin (*Ann. Rev. Entomol.* **11**: 423, 1966; *in* Ainsworth & Sussman (Eds), *The Fungi* **3**: 227, 1968), McCoy (*J. Cell Biochem.* **13A** 156, 1989; review control of pests), Kobayasi & Shimizu (*Iconography of vegetable wasps and plant worms*, 1983; taxonomy), Misra & Horn (*Trichomycetes and other fungal groups*, 2001; ecology, taxonomy), Moore (*Biocontrol News and Information* **9**: 209, 1988; fungi on mealybugs), Müller-Kögler (*Pilzkrankheiten bei Insekten*, 1965), Petch (*TBMS* **7-12**, 1921-27; *Studies* **16-27**, 1931-44, *Notes*), Samson *et al.* (*Atlas of entomopathogenic fungi*, 1988), Scholte *et al.* (*J. insect sci.* **4**: 19, 2004; fungi on mosquitos), Shah & Bell (*Appl. Microbiol. Biotechnol.* **61**: 413, 2003; review on control of pests), Steinhaus (1963), Tavares (*Laboulbeniales, Fungi, Ascomycetes*, 1985), Vega & Blackwell (*Insect-fungal associations: ecology and evolution* 336 pp., 2005), Wheeler & Blackwell (*Fungus-insect relationships*, 1984; compr. review).

Regional: **British Isles**, Petch (*TBMS* **31**: 286, 1948), Leatherdale (*Entomophaga* **15**: 419, 1970; hosts). **Israel**, Kenneth *et al.* (*Israel J. agric. Res.* **21**: 63, 1971), Kenneth & Olmert (*Israel J. Entomol.* **10**: 105, 1975). **N. America**, Charles (*Insect Pest Surv. Bull. US* **21** (Suppl. 9), 1941). **Former USSR**, Koval' (*Key to the entomogenous fungi of the USSR*, 1974), Khachatourians (*Handb. Appl. Mycol.: Humans, Animals & Insects* **2**: 613, 1991; physiology and genetics). **Taiwan**, Tzean *et al.* (*Atlas of entomopathogenic fungi from Taiwan*, 1997). **South Korea**, Sung Jae-Mo (*The insects-born fungus of Korea in color*, 1996).

See ambrosia fungi, Co-evolution, Insects and fungi, *Laboulbeniales*.

Entomopatella Petr. (1927) = Chaetospermum fide Petrak (*Annls mycol.* **32**: 447, 1934).

Entomopeziza Kleb. (1914) = Diplocarpon fide Nannfeldt (*Nova Acta R. Soc. Scient. upsal.*, 1932).

Entomophaga A. Batko (1964), Entomophthoraceae. 17, widespread. *E. grylli* on locusts. See Remaudière & Keller (*Mycotaxon* **11**: 323, 1980), Humber (*Mycotaxon* **21**: 265, 1984; 1989), Keller (*Sydowia* **40**: 122, 1987; key), Murrin & Nolan (*CJB* **65**: 169, 1987; ultrastr.), Nolan (*Can. J. Microbiol.* **33**: 808, 1987; protoplasts), Bidochka *et al.* (*Appl. Environm. Microbiol.* **61**: 556, 1995), Hajek *et al.* (*J. Invert. Path.* **68**: 260, 1996), Bidochka *et al.* (*Mol. Ecol.* **6**: 303, 1997; pathotypes), Hajek *et al.* (*Mycol.* **95**: 262, 2003; PCR-RFLP), Keller & Petrini (*Sydowia* **57**: 23, 2005; key), Keller (*Sydowia* **59**: 75, 2007; Switzerland, n.sp.).

entomophilous (of fungi), having spores distributed by insects.

Entomophthora Fresen. (1856), Entomophthoraceae. *c.* 30 (on *Arthropoda*, *Insecta*), widespread. *E. muscae* on house flies. See Gustafsson (*LantrHogsk Annlr* **31**: 103, 1965; taxonomy), Gustafsson (*LantrHogsk Annlr* **31**: 405, 1966; cultivation, physiology), Remaudière & Keller (*Mycotaxon* **11**: 323, 1980; segregate genera), Waterhouse & Brady (*Bull. BMS* **16**: 113, 1982; keys), Eilenberg *et al.* (*J. Invert. Path.* **48**: 318, 1986; ultrastr.), Keller (*Sydowia* **40**: 122, 1987; key), Humber (*Mycotaxon* **34**: 441, 1989; redescript.), Eilenberg *et al.* (*J. Invert. Path.* **65**: 179, 1995; ultrastr.), Keller (*Sydowia* **54**: 157, 2002; n.sp.), Keller & Petrini (*Sydowia* **57**: 23, 2005; key), Keller (*Sydowia* **58**: 38, 2006; n.sp.), López Lastra *et al.* (*Mycopathologia* **161**: 251, 2006; Argentina).

Entomophthora Krenner (1961) nom. inval. = Entomophthora Fresen. fide Kirk (*in litt.*).

Entomophthoraceae Nowak. (1877), Entomophthorales. 12 gen. (+ 6 syn.), 202 spp.

Lit.: Humber (*Mycotaxon* **34**: 441, 1989), Murrin & Nolan (*CJB* **67**: 754, 1989), Perry & Fleming (*Mycol.* **81**: 154, 1989), Bidochka *et al.* (*Appl. Environm. Microbiol.* **61**: 556, 1995), Bidochka *et al.* (*Mol. Ecol.* **6**: 303, 1997), Jensen *et al.* (*Fungal Genetics Biol.* **24**: 325, 1998), Keller *et al.* (*Sydowia* **51**: 197, 1999), Freimoser *et al.* (*Can. J. Microbiol.* **47**: 1082, 2001), Jensen & Eilenberg (*MR* **105**: 307, 2001), Tymon *et al.* (*MR* **108**: 419, 2004), Keller & Petrini (*Sydowia* **57**: 23, 2005), Tymon & Pell (*MR* **109**: 285, 2005).

Entomophthorales G. Winter (1880). Entomophthoromycotina. 5 fam., 23 gen., 277 spp. Spores forcibly discharged; most are parasites of insects. Fams:

(1) **Ancylistaceae**
(2) **Completoriaceae**
(3) **Entomophthoraceae**
(4) **Neozygitaceae**
(5) **Meristacraceae**

Lit.: Pohlad & Bernard (*Mycol.* **70**: 130, 1978; key spp. on nemat. and tardigr.), Waterhouse (*in* Ainsworth *et al.* (Eds), *The Fungi* **4B**: 219, 1973; key gen.), Lakon (*Nova Hedw.* **5**: 7, 1963; gen. key), Remaudière & Keller (*Mycotaxon* **11**: 323, 1980), Tucker (*Mycotaxon* **13**: 481, 1981; key non-entomogen. spp.), Humber & Ramoska (*in* Samson *et al.*, *Fundamental and applied aspects of invertebrate pathology*: 190, 1986; life cycles), Latgé *et al.* (*in* Samson *et al.*, *Fundamental and applied aspects of invertebrate pathology*: 190, 1986; life cycles), Wolf (*Nova Hedw.* **46**: 121, 1988; parasitism), Humber (*Mycotaxon* **34**: 441, 1989; emend.), Mikawa (*Bull. natn Sci. Mus.* Tokyo B **15**: 49, 1989; Nepal), Papierok (*Ann. Entom. Fenn.* **55**: 63, 1989; Finland), Toriello *et al.* (*J. Inv. Path.* **53**: 358, 1989; immunological separation of gen.), Keller (*Sydowia* **40**: 122, 1987, **43**: 39. 1991; Switzerland), Bałazy (*Flora of Poland* **24**. *Entomophthorales*, 1993), Nagahama *et al.* (*Mycol.* **87**: 203, 1995; phylogeny by 18S RNA), Lacey (Ed.) *(Manual of Techiques in Insect pathology*, 1997), Jensen *et al.* (*Fung. Genet. Biol.* **24**: 325, 1998; phylogeny), Humber (*J. Invert. Pathol.* **98**: 262, 2008; entomopathogenicity).

Entomophthoromycotina G . Winter (1880), see *Entomophthorales*.

Entomospora Sacc. ex Jacz. (1926) = Taphrina fide Eriksson & Hawksworth (*SA* **7**: 70, 1988).

Entomosporium Lév. (1856), anamorphic *Diplocarpon*, Cac.#eH.15. 1 (on *Rosaceae*), widespread (temperate). See Sutton (*The Coelomycetes*, 1980), Muthumary (*Curr. Sci.* **57**: 195, 1987; conidiogenesis), Shin *et al.* (*Korean Jl Pl. Path.* **14**: 732, 1998; Korea).

Entomyclium Wallr. (1833) = Dendryphion fide Hughes (*CJB* **36**: 727, 1958).

Entonaema Möller (1901), Xylariaceae. Anamorph

Nodulisporium. 6, widespread (esp. tropical). Clusters within *Hypoxylon* s.l. See Heim (*BSMF* **76**: 121, 1960; morphol.), Rogers (*Mycol.* **73**: 28, 1981; key, monogr.), Rogers (*Mycotaxon* **15**: 500, 1982; anam.), Sihanonth *et al.* (*MR* **102**: 458, 1998), Stadler *et al.* (*MR* **108**: 239, 2004; biochemistry), Triebel *et al.* (*Nova Hedwigia* **80**: 25, 2005; phylogeny), Bitzer *et al.* (*MR* **112**: 251, 2008; phylogeny, chemistry), Stadler *et al.* (*Mycol. Progr.* **7**: 53, 2008; phylogeny, chemistry).

entoparasitic, parasitic inside the host.

Entopeltacites Selkirk (1972), Fossil Fungi. 5 (Miocene), Australia.

Entopeltis Höhn. (1910) = Vizella fide Hughes (*CJB* **31**: 577, 1953), Swart (*TBMS* **57**: 455, 1971), von Arx & Müller (*Stud. Mycol.* **9**, 1975), Johnston (*N.Z. Jl Bot.* **38**: 629, 2000), Cunnington (*Mycotaxon* **93**: 135, 2005).

Entophlyctidaceae Whiffen (1944) = Endochytriaceae.

Entophlyctis A. Fisch. (1892), Endochytriaceae. *c.* 20, widespread (temperate). See Barr (*CJB* **49**: 2215, 1971), Longcore (*Mycol.* **87**: 25, 1995).

entoplacodial, see stroma.

Entorrhiza C.A. Weber (1884), Entorrhizaceae. 14 (on *Juncaceae* and *Cyperaceae* roots), widespread. See Fineran (*Nova Hedwigia* **29**: 825, 1978), Fineran (*Nova Hedwigia* **30**: 1, 1979), Vánky (*Mycotaxon* **68**: 343, 1998; key).

Entorrhizaceae R. Bauer & Oberw. (1997), Entorrhizales. 2 gen. (+ 1 syn.), 15 spp.

Lit.: Vánky (*Cryptog. Stud.* **1**: 159 pp., 1987), Bauer *et al.* (*CJB* **75**: 1273, 1997), Begerow *et al.* (*CJB* **75**: 2045, 1998), Piepenbring *et al.* (*Bot. Acta* **111**: 444, 1998), Piepenbring *et al.* (*Protoplasma* **204**: 170, 1998), Vánky & McKenzie (*Fungal Diversity Res. Ser.* **8**: 259 pp., 2002), Vánky (*Mycotaxon* **89**: 55, 2004).

Entorrhizales R. Bauer & Oberw. (1997). Entorrhizomycetidae. 1 fam., 2 gen., 15 spp. Fam.:
Entorrhizaceae
For *Lit.* see under fam.

Entorrhizomycetes Begerow, Stoll & R. Bauer (2007), Ustilaginomycotina. 1 ord., 1 fam., 2 gen., 15 spp. Ord.:
Entorrhizales
For *Lit.* see ord. and fam.

Entorrhizomycetidae R. Bauer, Oberw. & Vánky (1997), see *Entorrhizomycetes.*
For *Lit.* see fam.

Entosordaria (Sacc.) Höhn. (1920) = Stereosphaeria fide Hawksworth *et al.* (*Dictionary of the Fungi* edn 8, 1995).

Entosordaria Speg. (1920) nom. nud. = Anthostomella fide Francis (*Mycol. Pap.* **139**, 1975).

Entosthelia (Wallr.) Hue (1915) = Dermatocarpon Eschw. fide Hawksworth *et al.* (*Dictionary of the Fungi* edn 8, 1995).

entostroma, see stroma.

Entropezites Poinar & H.R. Buckley (2007), Fossil Fungi. 1.

Entrophospora R.N. Ames & R.W. Schneid. (1979), Acaulosporaceae. 5, widespread. Unknown phylogenetic affiliation. See Ames & Schneider (*Mycotaxon* **8**: 347, 1979), Schenck *et al.* (*Mycol.* **76**: 685, 1984), Sieverding & Toro (*Mycotaxon* **28**: 209, 1987), Wu *et al.* (*Mycotaxon* **53**: 283, 1995), Wu *et al.* (*Mycol.* **87**: 582, 1995; aeroponic spore ontogeny), Yao *et al.* (*Genera of Endogonales*: 229 pp., 1996), Błaszkowski *et al.* (*Mycotaxon* **68**: 165, 1998), Wang *et al.* (*Mycoscience* **17**: 92, 1998), Fracchia *et al.* (*Nova Hedwigia* **77**: 383, 2003; isolation, culture and host colonization of *Entrophospora schenckii*), Redecker & Raab (*Mycol.* **98**: 885, 2006; phylogeny), Sieverding & Oehl (*J. Appl. Bot. Food Quality* Angew. Botan. **80**: 69, 2006).

Entrophosporaceae Oehl & Sieverd. (2006) = Acaulosporaceae.

Entyloma de Bary (1874), Entylomataceae. Anamorph *Entylomella*. *c.* 180 (on dicots), widespread. Spores hyaline or pale, intercellular. The sori are generally in leaves (causing leaf spots) and there is frequently an anamorphic stage; cf. *Entylomella* and *Cylindrosporium* s.l. *Entyloma dahliae* (on *Dahlia*), *E. australe* (on *Physalis*), *E. calendulae* (on *Calendula*), *E. ellisii* (on *Spinacia*), *E. fuscum* (on *Papaver* and *Glaucium*), *E. petuniae* (on *Petunia*). See Begerow *et al.* (*MR* **106**: 1392, 2002; phylog.), Piątek (*Mycotaxon* **93**: 323, 2005; on *Convolvulaceae*), Spooner & Legon (*Mycologist* **20**: 90, 2006; UK).

Entylomaster Vánky & R.G. Shivas (2006), Doassansiaceae. 2 (on *Araceae*), Australia; Europe (southern). See Vánky & Shivas (*Mycol. Balcanica* **3**: 13, 2006).

Entylomataceae R. Bauer & Oberw. (1997), Entylomatales. 2 gen., 225 spp.

Lit.: Boekhout *et al.* (*Stud. Mycol.* **38**: 175, 1995), Bauer *et al.* (*CJB* **75**: 1273, 1997), Begerow *et al.* (*CJB* **75**: 2045, 1998), Begerow *et al.* (*MR* **104**: 53, 2000), Adejumo *et al.* (*Mycopathologia* **150**: 85, 2001), Begerow *et al.* (*MR* **106**: 1392, 2002), Piepenbring (*Bot. Jb.* **24**: 241, 2003), Jackson (*Evolution* Lancaster, Pa. **58**: 1909, 2004), Zwetko & Blanz (*Biosystem. Ecol. Ser.* **21**: 241 pp., 2004), Boekhout *et al.* (*FEMS Yeast Res.* **6**: 63, 2005).

Entylomatales R. Bauer & Oberw. (1997). Exobasidiomycetes. 1 fam., 2 gen., 225 spp. Fam.:
Entylomataceae
For *Lit.* see under fam.

Entylomella Höhn. (1924), anamorphic *Entyloma*. 45, widespread. See Ciferri (*Omagiu lui Traian Săvulescu cu prilejul împlinirii a 70 de ani*: 175, 1959), Braun & Hill (*Mycol. Progr.* **1**: 19, 2002; anam., New Zealand).

Entylomellaceae Cif. (1959) = Entylomataceae.

Eoagaricus L. Krieg. (1923) ≡ Physalacria.

Eoaleurina Korf & W.Y. Zhuang (1986), Pyronemataceae. 1, C. & S. America. See Zhuang & Korf (*Mycotaxon* **26**: 361, 1986), Korf (*SA* **6**: 127, 1987).

Eocronartiaceae Jülich (1982), Platygloeales. 5 gen. (+ 1 syn.), 9 spp.

Eocronartium G.F. Atk. (1902), Eocronartiaceae. 1 (on *Musci*), Europe; N. America. See Stanley (*Trans. Am. microsc. Soc.* **59**: 407, 1940), Khan & Kimbrough (*CJB* **58**: 642, 1980; ultrastr.), Frieders & McLaughlin (*MR* **105**: 734, 2001; cytol., ultrastr., anam.).

Eoetvoesia Schulzer (1866) nom. dub., anamorphic *Pezizomycotina*.

Eolichen Zukal (1884) ? = Nectria. Affinities are unclear. fide Keissler (*Rabenh. Krypt.-Fl.* **9** 1(2), 1936).

Eolichenomyces Cif. & Tomas. (1954) ≡ Eolichen.

Eomycenella G.F. Atk. (1902) = Mycena fide Kuyper (*in litt.*).

Eomyces F. Ludw. (1894) nom. dub., Algae. Based on an achlorotic alga fide Batra (*in* Subramanian (Ed.), *Taxonomy of fungi* **1**: 187, 1978).

Eomycetopsis J.W. Schopf (1968), Fossil Fungi ? Fungi. 1, Australia.

Eopolyporoides Rigby (1982), Fossil Fungi. 1, Australia. See Playford *et al.* (*Publication* Geological Survey of Queensland **380**: 5, 1982).

Eopyrenula R.C. Harris (1973), Dacampiaceae (L). 3, widespread (north temperate). See Aptroot (*Biblthca Lichenol.* **44**, 1991; posn), Harris (*More Florida Lichens*, 1995), Johansson & Hermansson (*Graphis Scripta* **12**: 19, 2000; Sweden).

Eosphaeria Höhn. (1917), Lasiosphaeriaceae. Anamorph *Phialophora*-like. 1, Europe; N. America. See Petrini *et al.* (*TBMS* **82**: 554, 1984).

Eoterfezia G.F. Atk. (1902), Eoterfeziaceae. 1 or 2, N. America. No recent information is available. See van Brummelen *in* Hawksworth (Ed.) (*Ascomycete Systematics. Problems and Perspectives in the Nineties* NATO ASI Series vol. **269 269**: 400, 1994).

Eoterfeziaceae G.F. Atk. (1902), Pezizomycotina (inc. sed.). 1 gen., 2 spp.
Lit.: Jeng & Cain (*Mycotaxon* **3**: 387, 1976).

Epaphroconidia Calat. & V. Atienza (1995), anamorphic *Pezizomycotina*. 1, Spain; N. America. See Calatayud & Atienza (*MR* **99**: 850, 1995), Diederich (*Herzogia* **16**: 41, 2003; USA).

epapillate, having no papillae.

Ephebaceae Th. Fr. (1860) = Lichinaceae.

Ephebe Fr. (1825), Lichinaceae (L). 12, widespread. See Henssen (*Symb. bot. upsal.* **18** no. 1, 1963), Schultz & Büdel (*Lichenologist* **34**: 39, 2002; key), Schultz & Büdel (*Lichenologist* **35**: 151, 2003; phylogeny).

Ephebeia Nyl. (1875) = Ephebe fide Henssen (*Symb. bot. upsal.* **18** no. 1, 1963).

Ephebella Itzigs. (1857) nom. dub., Fungi. Apparently based on an alga fide Currah (*SA* **5**: 130, 1986).

Ephebomyces Cif. & Tomas. (1953) ≡ Ephebe fide Hawksworth *et al.* (*Dictionary of the Fungi* edn 8, 1995).

Ephedracetes T.C. Huang (1981), Fossil Fungi. 1 (Miocene), Taiwan.

Ephedrosphaera Dumort. (1822) nom. rej. = Nectria fide Rossman *et al.* (*Stud. Mycol.* **42**: 248 pp., 1999).

Ephelidium C.W. Dodge & E.D. Rudolph (1955), anamorphic *Pezizomycotina*, ?.?.? (L). 1, Antarctica. See Dodge & Rudolph (*Ann. Mo. bot. Gdn* **42**: 136, 1955).

Ephelidium Speg. (1920) nom. conf., anamorphic *Pezizomycotina*. See Sutton (*Mycol. Pap.* **141**, 1977).

Ephelina Sacc. (1889) = Leptotrochila fide Schüepp (*Phytopath. Z.* **36**: 236, 1959).

Epheliopsis Henn. (1908) = Eutypa fide Petrak (*Sydowia* **5**: 169, 1951), Ciferri (*Atti Ist. bot. Univ. Lab. Crittog. Pavia* sér. 5 **19**: 105, 1962), Rappaz (*Mycol. helv.* **2**: 285, 1987).

Ephelis Fr. (1849), anamorphic *Balansia, Epichloë*, Hsp.0fH.?. 5 (mostly on grasses), widespread (esp. tropical). See Ullasa (*Mycol.* **61**: 572, 1969), Govindu & Thirumalachar *in* Subramanian (Ed.) (*Taxonomy of fungi* **2**: 328, 1984), Morgan-Jones *et al.* (*Mycotaxon* **43**: 401, 1992; review), Phelps *et al.* (*Mycotaxon* **49**: 91, 1993), Phelps & Morgan-Jones (*Mycotaxon* **50**: 61, 1994), Kuldau *et al.* (*Mycol.* **89**: 431, 1997; phylogeny), Tanaka *et al.* (*MR* **105**: 811, 2001; Asia).

Ephemerellomyces M.M. White & Lichtw. (2004), Harpellales. 1, Norway. See White & Lichtwardt (*Mycol.* **96**: 893, 2004).

Ephemeroascus Emden (1973) = Coniochaeta fide van Emden (*TBMS* **61**: 599, 1973), García *et al.* (*MR* **110**: 1271, 2006; phylogeny, synonymy).

Ephemerocybe Fayod (1889) = Coprinellus fide Kuyper (*in litt.*).

epi- (prefix), upon.

epibasidium, see basidium.

Epibelonium E. Müll. (1963), Saccardiaceae. 1, France. See Müller (*Phytopath. Z.* **47**: 240, 1963).

epibiotic, living on the surface of another organism.

Epibotrys Theiss. & Syd. (1915) = Gilletiella fide von Arx & Müller (*Stud. Mycol.* **9**, 1975).

Epibryon Döbbeler (1978), ? Pseudoperisporiaceae. 30 (on *Musci*), widespread. See Döbbeler (*Sendtnera* **5**: 19, 1998), Wubet *et al.* (*New Phytol.* **161**: 517, 2004; New Zealand), Döbbeler (*Nova Hedwigia* **82**: 257, 2006; Ecuador), Döbbeler (*MR* **111**: 1406, 2007).

epibryophilous, growing over bryophytes.

Epichloë (Fr.) Tul. & C. Tul. (1865), Clavicipitaceae. Anamorph *Neotyphodium*. 18, widespread. *E. typhina* (choke of grasses). See Doguet (*BSMF* **76**: 171, 1960; development), White (*Am. J. Bot.* **84**: 170, 1977; centrum), White (*Mycol.* **85**: 444, 1993; Br. spp.), Bacon & Hill (*Neotyphodium/Grass Interactions*, 1994), Schardl (*Ann. Rev. Phytopath.* **34**: 109, 1996; review), Chung *et al.* (*Phytopathology* **87**: 599, 1997; genetics), Schardl *et al.* (*Mol. Biol. Evol.* **14**: 133, 1997; coevolution with grasses), Bacon & White (*Microbial Endophytes*, 2000), Moon *et al.* (*Mycol.* **92**: 1103, 2000; evolution), White *et al.* (*Stud. Mycol.* **45**: 95, 2000), Craven *et al.* (*Sydowia* **53**: 44, 2001; hybridisation), Craven *et al.* (*Ann. Mo. bot. Gdn* **88**: 14, 2001; phylogeny), Schardl (*Fungal Genetics Biol.* **33**: 69, 2001; on *Festuca*), White *et al.* (*Symbiosis* **31**: 241, 2001; S America), Moon *et al.* (*Mycol.* **94**: 694, 2002; S hemisphere), White *et al.* (*Cellular Origin and Life in Extreme Habitats* **4**: 413, 2002; evolution), Bischoff & White (*Mycology Series* **19**: 125, 2003; review), Brem & Leuchtmann (*Evolution* Lancaster, Pa. **57**: 37, 2003; evolution), Leuchtmann (*Mycology Series* **19**: 169, 2003; biodiversity), Schardl & Moon (*Mycology Series* **19**: 273, 2003; review), Sullivan & Faeth (*Mol. Ecol.* **13**: 649, 2004; gene flow), Gentile *et al.* (*Mol. Phylogen. Evol.* **35**: 196, 2005; Argentina), Spooner & Kemp (*Mycologist* **19**: 82, 2005; UK), Li *et al.* (*Mycol.* **98**: 560, 2006; China), Yanagida *et al.* (*Mycol.* **97**: 1287, 2005; on *Agropyron*), Zhang *et al.* (*Mycol.* **98**: 1076, 2006; phylogeny), Moon *et al.* (*Mycol.* **99**: 895, 2007; on *Stipeae* and *Meliceae*).

Epichloea Giard (1889) nom. dub., Fungi.

epichroic, see colour.

Epichysium Tode (1790) nom. dub., Fungi. ? Based on insect debris fide Fries (*Syst. mycol.* **3**: 293, 1832).

Epicladonia D. Hawksw. (1981), anamorphic *Pezizomycotina*, Cpd.0-1eH.15. 3 (on lichens, esp. *Cladonia*), Europe. See Hawksworth (*Bull. Br. Mus. nat. Hist.* Bot. **9**: 15, 1981).

Epiclinium Fr. (1849), anamorphic *Pezizomycotina*, Ccu.1eH.?. 2, Europe; America.

Epicnaphus Singer (1960), Marasmiaceae. 3, S. America. See Raithelhuber (*Metrodiana* **4**: 52,, 1973).

Epicoccaceae Nann. (1934) = Pleosporaceae.

Epicoccospora Budathoki & S.K. Singh (1995), anamorphic *Pezizomycotina*, Hso.?.?. 1, Nepal. See Budathoki & Singh (*Indian Phytopath.* **48**: 103, 1995).

Epicoccum Link (1815), anamorphic *Pleosporaceae*, Hsp.#eP.1. 4, widespread. *E. nigrum*, *E. andropo-*

gonis (syn. *Cerebella andropogonis*) on *Sphacelia* of *Claviceps* and frequently mistaken as *Ustilaginales*. See Langdon (*Phytopathology* **32**: 613, 1942), Langdon (*Mycol. Pap.* **61**, 1955), Schol-Schwarz (*TBMS* **42**: 149, 1959), Madrigal & Melgarejo (*CJB* **73**: 425, 1995; antibiotic activity), Arenal *et al.* (*MR* **104**: 301, 2000; *Phoma* synanamorph), Arenal *et al.* (*Journal of Applied Microbiology* **93**: 36, 2002; diagnostics), Pažoutová & Kolínská (*Czech Mycol.* **54**: 155, 2003; phylogeny), Arenal *et al.* (*Mycotaxon* **89**: 465, 2004; phylogeny, *Phoma* synanamorph), Wang & Guo (*Mycosystema* **23**: 474, 2004; morphology, phylogeny), Mims & Richardson (*CJB* **83**: 1354, 2005; ultrastr.).

epicortex, a thin polysaccharide-like layer over the surface of the cellular upper cortex in thalli of some *Parmeliaceae* visible by SEM (Hawksworth, *in* Hale, *Smithson. Contr. bot.* **10**: 5, 1973) and which may have regular pores functioning in gas exchange (Hale, *Lichenologist* **13**: 1, 1981). See Hyvärinen (*Lichenologist* **24**: 267, 1992; environmental induction), Lumbsch & Kothe (*Mycotaxon* **43**: 277, 1992; *Coccocarpiaceae, Pannariaceae*).

Epicorticium Velen. (1926) = Phaeomarasmius fide Singer (*Agaric. mod. Tax.* edn 3, 1975).

Epicrea Petr. (1950), ? Clavicipitaceae. 1 (on stromata of *Hypocrella*), Ecuador. See Rossman *et al.* (*Stud. Mycol.* **42**: 248 pp., 1999).

epicutis, see cutis.

Epicymatia Fuckel (1870) = Stigmidium fide Roux & Triebel (*Bull. Soc. linn. Provence* **45**: 451, 1994).

Epicyta Syd. (1926) = Aplosporella fide Clements & Shear (*Gen. Fung.*, 1931).

epidemic (1) (adj.), (of a disease of humans, but used of plants and animals), general and severe in a group for a time; (2) (n.), the disease itself; cf. epiphytotic, epizootic.

epidemiology, the study of disease incidence, distribution and control. See Plant pathogenic fungi.

Epidermella Tehon (1935) ? = Hypoderma De Not. fide Hawksworth *et al.* (*Dictionary of the Fungi* edn 8, 1995).

Epidermidophyton E. Lang (1879) nom. rej., anamorphic *Pezizomycotina*.

Epidermomyces Loeffler (1983) ≡ Epidermophyton Sabour.

Epidermophyton E. Lang (1879) nom. rej. = Epidermophyton Sabour.

Epidermophyton Sabour. (1907) nom. cons., anamorphic *Arthrodermataceae*, Hso.≡ eH.2. 2 (on humans), widespread. Macroconidia pyriform; microconidia absent. *E. floccosum* (syn. *E. inguinale* and *E. cruris*) on humans (glabrous skin) causing tinea cruris and tinea pedis. See Guého *et al.* (*Annales de l'Institut Pasteur* Microbiologie **136B**: 195, 1985; ultrastr.), Sundaram *et al.* (*Curr. Sci.* **55**: 406, 1986; morphology), Cabañes *et al.* (*Revta Iberoamer. Micol.* **10**: 14, 1993; nomencl., morphology), Weitzman *et al.* (*Manual of Clinical Microbiology* Edn 6: 791, 1995; review), Kawasaki *et al.* (*Mycopathologia* **134**: 121, 1996; phylogeny), Gräser *et al.* (*Medical Mycology* **37**: 105, 1999; phylogeny), Kane & Summerbell (*Manual of Clinical Microbiology*: 1275, 1999; review), Kano *et al.* (*Mycopathologia* **146**: 111, 1999; phylogeny), Untereiner *et al.* (*Mycol.* **96**: 812, 2004; phylogeny).

epidermophytosis, see dermatophytosis (esp. tinea cruris and tinea pedis).

Epidochiopsis P. Karst. (1892), anamorphic *Pezizomycotina*, ?.0eH.?. 1, Europe.

Epidochium Fr. (1849) = Tremella Pers. fide Donk (*Taxon* **7**: 193, 1958).

Epidrolithus Raf. (1836) ? = Leptogium fide Merrill (*Index Rafinesq.*, 1949).

epiflora, surface flora; sometimes applied (incorrectly) to the microbiota on seed surfaces; the epibiota.

epigeal (**epigean**, **epigeic**), on the earth.

epigeic (of lichens), not attached to any substrate but blowing about on the surface of the ground; see wandering lichens.

epigenous, growing on the surface.

Epiglia Boud. (1885) = Mniaecia fide Korf *in* Ainsworth *et al.* (Eds) (*The Fungi* **4A**: 249, 1973).

Epigloea Zukal (1889), Epigloeaceae (?L). 13 (on *Algae*), Europe; Antarctica. See Döbbeler (*Beih. Nova Hedwigia* **79**: 203, 1984; key), David (*SA* **6**: 217, 1987), Döbbeler (*Sendtnera* **2**: 277, 1994), Ceynowa-Gieldon (*Acta Mycologica* Warszawa **37**: 3, 2002; Poland), Cykowska & Flakus (*Polish Botanical Journal* **50**: 233, 2005; Poland), Pérez-Ortega & Barreno (*Nova Hedwigia* **83**: 523, 2006; Iberian peninsula).

Epigloeaceae Zahlbr. (1903), Pezizomycotina (inc. sed.) (±L). 1 gen. (+ 2 syn.), 13 spp.

Lit.: David (*SA* **6**: 217, 1987), Döbbeler (*Sendtnera* **2**: 277, 1994).

Epigloeomyces Cif. & Tomas. (1957) ≡ Epigloea.

epigynous, having the antheridium above the oogonium on one hypha.

epihymenium, a thin layer of interwoven hyphae on the surface of the hymenium (Corner, 1950; epithecium).

epikapylic, a thallus of a lichenicolous fungus in which a morphologically distinct lichenized structure is formed (Poelt & Vězda, 1984).

Epilichen Clem. (1909), ? Rhizocarpaceae (±L). 1 (on *Baeomyces*), widespread (north temperate). See Hafellner (*Beih. Nova Hedwigia* **62**, 1979).

Epilithia Nyl. (1853) = Gyalideopsis See Lücking *et al.* (*Lichenologist* **37**: 165, 2005).

epilithic, living on the surface of stones; cf. endolithic.

epinecral layer, see necral layer.

Epinectria Syd. & P. Syd. (1917) = Dimerosporiella Speg. fide Rossman *et al.* (*Stud. Mycol.* **42**: 248 pp., 1999).

Epinyctis Wallr. (1831) = Lepraria fide Hawksworth *et al.* (*Dictionary of the Fungi* edn 8, 1995).

Epipeltis Theiss. (1913) = Schizothyrium fide von Arx & Müller (*Stud. Mycol.* **9**, 1975).

Epiphegia G.H. Otth (1870) ? = Massarina fide Aptroot (*Nova Hedwigia* **66**: 89, 1998).

Epiphloea Trevis. (1880), Lichinaceae (L). 1, Europe. See Gyelnik (*Rabenh. Krypt.-Fl.* **9** 2.2, 1940).

epiphloeodal, living upon bark.

Epiphora Nyl. (1876) [non *Epiphora* Lindl. 1837, *Orchidaceae*] = Plectocarpon fide Santesson (*The lichens and lichenicolous fungi of Sweden and Norway*, 1993).

epiphragm, the membrane over the young fruit-body in the *Nidulariaceae*.

epiphyllous, on the upper surface of a leaf; foliicolous.

Epiphyma Theiss. (1916) = Botryosphaeria fide von Arx & Müller (*Beitr. Kryptfl. Schweiz* **11** no. 1, 1954).

epiphyte, a plant living on another, but not as a parasite.

epiphytic (adj.), frequently = corticolous.
epiphytotic, an epidemic among plants.
epiplasm (of an ascus), the cytoplasm not used up in the 'free cell formation' of ascospores.
Epiploca Kleb. (1918) = Epipolaeum fide Müller & von Arx (*Beitr. Kryptfl. Schweiz* **11** no. 2, 1962).
Epipolaeum Theiss. & P. Syd. (1918), Pseudoperisporiaceae. 17, widespread. See Shoemaker (*CJB* **43**: 631, 1965), Farr (*Mycol.* **71**: 243, 1979).
Episclerotium L.M. Kohn (1984), Helotiaceae. 2, Europe. See Malaval (*Docums Mycol.* **19**: 9, 1989; ascus ultrastr.).
Episeptum J.I.R. Larsson (1986), Microsporidia. 3. See Larsson (*Arch. Protistenk.* **131**: 257, 1986).
Episoma Syd. (1925) = Phaeostigme fide von Arx & Müller (*Stud. Mycol.* **9**, 1975).
Episphaerella Petr. (1924), Pseudoperisporiaceae. 5, America; Africa. See Müller & von Arx (*Beitr. Kryptfl. Schweiz* **11** no. 2, 1962), Farr *et al.* (*CJB* **63**: 1983, 1985), Barr (*Mycotaxon* **64**: 149, 1997).
Episphaeria Donk (1962), Inocybaceae. 1, Europe. Perhaps *Strophariaceae*. See Donk (*Persoonia* **2**: 336, 1962).
epispore, see ascospore, Spore wall.
episporium, see Spore wall.
Episporogoniella U. Braun (1994), anamorphic *Pezizomycotina*, Hso.≡ eP.1. 1 (on *Bryophyta*), Brazil. See Braun (*Mycotaxon* **51**: 41, 1994).
Epistictum Trevis. (1869) = Dermatocarpon Eschw. fide Hawksworth *et al.* (*Dictionary of the Fungi* edn 8, 1995).
Epistigme Syd. (1924), anamorphic *Pezizomycotina*, Cpd.0eP.?. 1, S. Africa.
epistroma, see stroma.
Epitea Fr. (1832) = Phragmidium fide Dietel (*Nat. Pflanzenfam.* **6**, 1928) applying to *Caoma* stage.
epithalline, of a falsely thalline apothecial edge in lichenized fungi.
epithecium, tissue at the surface of an apothecium formed by the branching of the ends of the paraphyses above the asci; cf. epihymenium, pseudoepithecium.
Epithele (Pat.) Pat. (1900), Polyporaceae. 10, widespread. See Boquiren (*Mycol.* **63**: 937, 1971; key), Hjortstam & Ryvarden (*Syn. Fung.* **20**: 23, 2005).
Epitheliaceae Jülich (1982) = Polyporaceae.
epithelium, see cutis.
Epithelopsis Jülich (1976), Polyporaceae. 2, Australia; India; New Zealand. See Jülich (*Persoonia* **8**: 457, 1976), Lepp (*Australasian Mycologist* **23**: 53, 2004).
epithet (1) the second (specific) adjectival part of a Latinized binomial (the 'trivial' name of the zoologist); (2) the third or fourth (varietal, etc.) term.
Epithyrium (Sacc.) Trotter (1931), anamorphic *Sarea*, St.0eP.15. 1 (on conifer resin), widespread. See Hawksworth & Sherwood (*CJB* **59**: 357, 1981).
epitunica, see exosporium; Spore wall.
epitype, see Nomenclature, type.
Epixyla Raf. (1806) nom. dub., Fungi.
epixylic (**epixylous**), living on wood; lignicolous.
Epixylon Füisting (1867) = Hypoxylon Bull. fide Læssøe (*SA* **13**: 43, 1994).
epizoic, living on animals.
epizootic, an epidemic among animals.
epizootic lymphangitis, a disease of horses caused by *Histoplasma farciminosum*.
Epochniella Sacc. (1880) = Stemphylium fide Lindau (*Rabenh. Krypt.-Fl.* **1**: 207, 1907).
Epochnium Link (1809) = Monilia Bonord. fide Hughes (*CJB* **36**: 727, 1958).
epruinose, having no pruina.
Epulorhiza R.T. Moore (1987), anamorphic *Tulasnella*. 3 (mycorrhizal), widespread. See Currah *et al.* (*CJB* **68**: 1171, 1990; mycorrhizal sp.), Currah & Zelmer (*Rep. Tottori mycol. Inst.* **30**: 43, 1992), Roberts (*Rhizoctonia-forming fungi*, 1999), Gleason & McGee (*Australasian Mycologist* **21**: 12, 2002; septal pore cap ultrastructure), Ma *et al.* (*MR* **107**: 1041, 2003; molecular phylogeny).
equal (of a stipe), having the same diameter throughout.
Erannium Bonord. (1860) = Coleosporium fide Saccardo (*Syll. fung.* **18**: 774, 1906).
Erastia Niemelä & Kinnunen (2005), Polyporaceae. 2, widespread. See Niemelä & Kinnunen (*Karstenia* **45**: 76, 2005).
erasure phenomenon (in *Dictyostelium*), the loss by amoebae of the capacity to recapitulate when developing cultures are disaggregated and placed on a growth medium.
Erebonema A. Roem. (1845) nom. dub., anamorphic *Pezizomycotina*.
erect, upright; straight, not curved, up.
Eremascaceae Engl. & E. Gilg (1924), Eurotiomycetidae (inc. sed.). 1 gen., 2 spp.

Lit.: Hocking (*Handbook of Applied Mycology* Vol. 3. Foods and Feeds: 69, 1991), Berbee & Taylor (*Mol. Biol. Evol.* **9**: 278, 1992; DNA), Landvik *et al.* (*Mycoscience* **37**: 237, 1996; DNA), Anderson *et al.* (*MR* **102**: 541, 1998; DNA), Kurtzman & Robnett (*Antonie van Leeuwenhoek* **73**: 331, 1998), Lumbsch *et al.* (*Pl. Biol.* **2**: 525, 2000).

Eremascus Eidam (1883), Eremascaceae. 2, Europe; India. See Harrold (*Ann. Bot.* Lond. **14**: 127, 1950), Berbee & Taylor (*Mol. Biol. Evol.* **9**: 278, 1992; posn), Landvik *et al.* (*Mycoscience* **37**: 237, 1997; DNA), Geiser *et al.* (*Mycol.* **98**: 1053, 2006; phylogeny).
Eremastrella S. Vogel (1955), ? Psoraceae (L). 1, S. Africa; Australia. See Schneider (*Biblthca Lichenol.* **13**, 1980), Pietschman (*Nova Hedwigia* **51**: 521, 1990; posn), Lumbsch & Kothe (*Nova Hedwigia* **57**: 19, 1993; thalli).
Eremiomyces Trappe & Kagan-Zur (2005), Pezizaceae. 1 (hypogeous), S. Africa. See Ferdman *et al.* (*MR* **109**: 244, 2005), Læssøe & Hansen (*MR* **111**: 1075, 2007; phylogeny).
Eremodothis Arx (1976), Sporormiaceae. 1, India; Japan. See Udagawa & Ueda (*J. Jap. Bot.* **56**: 289, 1981), Kruys *et al.* (*MR* **110**: 527, 2006; phylogeny).
Eremomyces Malloch & Cain (1971), Eremomycetaceae. Anamorphs *Arthrographis*-like, *Trichosporiella*-like. 2 (coprophilous), widespread. See Malloch & Sigler (*CJB* **66**: 1929, 1988).
Eremomycetaceae Malloch & Cain (1971), ? Dothideomycetes (inc. sed.). 3 gen. (+ 1 syn.), 11 spp.

Lit.: Malloch & Sigler (*CJB* **66**: 1929, 1988), von Arx *et al.* (*Beih. Nova Hedwigia* **94**: 104 pp., 1988), Lumley *et al.* (*Mycotaxon* **74**: 395, 2000).

Eremotheca Theiss. & Syd. (1917) = Schizothyrium fide Müller & von Arx (*Beitr. Kryptfl. Schweiz* **11** no. 2, 1962).
Eremothecella Syd. & P. Syd. (1917), Arthoniaceae (L). 6, widespread (tropical). See Sérusiaux (*SA* **11**: 39, 1992), Grube (*Bryologist* **101**: 377, 1998; phylogeny), Grube & Lücking (*MR* **105**: 1007, 2001; as-

cogenous hyphae).

Eremotheciaceae Kurtzman (1995), Saccharomycetales. 2 gen. (+ 3 syn.), 3 spp. Some authors treat this as a synonym of *Saccharomycetaceae*.

Lit.: Yamada *et al.* (*Bull. Fac. Agric. Shizuoka Univ.* **44**: 9, 1994), Kurtzman (*J. Industr. Microbiol.* **14**: 523, 1995), Prillinger *et al.* (*Yeast* Chichester **13**: 945, 1997), de Hoog *et al. in* Kurtzman & Fell (Eds) (*Yeasts, a taxonomic study* 4th edn: 201, 1998), Wendland *et al.* (*Curr. Genet.* **35**: 618, 1999), Kurtzman & Robnett (*FEMS Yeast Res.* **3**: 417, 2003), Suh *et al.* (*Mycol.* **98**: 1006, 2006; phylogeny).

Eremothecium Borzí (1888), Eremotheciaceae. 2, widespread. See Batra (*USDA Tech. Bull.* **1469**, 1973; key), Rosing (*Mycol.* **79**: 157, 1987; ultrastr.), Kurtzman (*J. Industr. Microbiol.* **14**: 523, 1995), Prillinger *et al.* (*Yeast* Chichester **13**: 945, 1997; placement in *Saccharomycetaceae*), de Hoog *et al. in* Kurtzman & Fell (Eds) (*Yeasts, a taxonomic study* 4th edn: 201, 1998), Suh *et al.* (*Mycol.* **98**: 1006, 2006; phylogeny).

ergoline alkaloids, from *Claviceps* sclerotia include both lysergic acid derivatives (esp. *C. purpurea*, *C. paspali*) and clavine alkaloids (esp. *C. fusiformis*, *C. gigantea*, *Sphacelia sorghi*). Cf. ergot.

ergometrine (D-lysergic acid propanolamide), an ergot alkaloid from *Claviceps purpurea* sclerotia (esp. Spanish and Portuguese); used in medicine against migraine. Cf. ergotamine.

ergosterol, the commonest sterol of fungi (hence also in lichens) first isolated from *Claviceps purpurea* sclerotia; yeast ergosterol is converted to vitamin D_2 by ultraviolet radiation. Commonly used as an index for living-fungal mass in decaying materials (Newell, Fungal biomass and productivity (*in*Paul (Ed.), *Methods in Microbiology* **30** *Marine Microbiology*) pp. 357-372, 2001).

Ergot (1) the *Claviceps* disease of cereals and other grasses, esp. *Claviceps africana* (on *Sorghum* spp.), *C. fusiformis* (on *Pennisetum*), *C. paspali* (on *Paspalum* spp.), and *C. purpurea* (on *Secale* and other grasses); (2) an ergot fungus; (3) (in trade), the sclerotia of ergot fungi.

Infection (of ovaries) of cereals and grasses is largely limited to outcrossing plants with flowers which open, and occurs by rain- or insect-borne conidia of the *Sphacelia* state which occur suspended in 'honey dew' exuded from the host florets in response to infection. Sclerotial development begins 2-3 weeks after infection. The sclerotia contain a range of alkaloids (of which the most important are lysergic acid derivatives; see Stoll & Hofmann, *in* Manske (Ed.) (*The alkaloids* **8**: 725, 1965); also ergometrine, ergotamine, ergotoxine, including the active principals not only of the poisonous properties of ergot but also of the therapeutic applications (e.g. in migraine, obstetrics, and as hallucinogens). The ergoline ring system is built up from *L*-triptophan and mevalonic acid. *N*-methyl group is derived from methionine.

Ergot of rye contaminating bread causes **ergotism** in humans which is of two main types: the gangrenous (St Anthony's Fire of the Middle Ages) and the spasmodic (see Barger, *Ergot and ergotism*, 1931; Bové, *The story of ergot*, 1970). Human exposure to low levels of ergolines still appears to be widespread and there have been severe outbreaks perhaps in France in 1951 (Fuller, *The day of St. Anthony's fire*, 1968, although mercury poisoning has also been suggested as a cause of this outbreak), and more recently in Ethiopia and India (*WHO Environmental Health Criteria* 105, 1990).

Claviceps paspali causes paspalum staggers in cattle, sheep, and horses (Hopkirk, *NZ Jl Agric.* **53**: 103, 1936) and *C. fusiformis* on *Pennisetum typhoides* has been associated with agalactia in sows resulting in the death of new-borne piglets (Loveless, *TBMS* **50**: 15, 1967).

Ergot of rye was once the only source of the medicinal ergot alkaloids but since the early 1960s semisynthetic alkaloids have been prepared on a large scale from lysergic acid produced by *C. paspali* fermentations. Ergot alkaloids are also produced by other genera (Řeháček & Sajdl, *Ergot alkaloids*, 1990). See also Kren & Cvak (*Ergot: the genus Claviceps*, 518 pp., 1999), Schardl *et al.* (*Alkaloids Chem. Biol.* **63**: 45, 2006), White *et al.* (*Clavicipitalean fungi: evolutionary biology, chemistry, biocontrol and cultural aspects*, 2003). See hallucinogenic fungi.

Ergotaetia E.J. Quekett (1841) = Sphacelia fide Hawksworth *et al.* (*Dictionary of the Fungi* edn 8, 1995).

ergotamine, a cyclic tripeptide derivative of lysergic acid from *Claviceps purpurea* sclerotia; used in medicine against migraine. Cf. ergometrine.

ergotism, ergot poisoning. See ergot.

ergotoxine, a mixture of ergocornine, ergocristine, and ergokryptine; cyclic tripeptide derivatives of lysergic acid from *Claviceps purpurea* sclerotia.

Ericianella Brond. (1828) = Bactridium fide Fries (*Summa veg. Scand.*, 1849).

Eriksson (Jakob; 1848-1931; Sweden). Professor and Director, Department of Plant Physiology and Agricultural Botany, Academy of Agriculture, Stockholm (1885-1913). He discovered physiologic races in rusts (see *Jahrbücher für Wissenschaftliche Botanik* **29**: 499, 1896). *Publs.* (with Hennings) *Die Getreideroste, ihre Geschichte und Natur, sowie Massregeln gegen dieselben* (1896) [first published in Swedish in *Medd. Kongl. Landtbruks Akad. Esper.* **38**, 1894]; Über die Specialisierung des Parasitismus bei den Getreiderostpilzen. *Bericht der Deutschen Botanischen Gesellschaft* (1894); Über die Mykoplasmatheorie. *Biologisches Zentralblatt* (1910); *Die Pilzkrankheiten der landwirtschaftlichen Kulturgegewachse* (1926) [in English, 1930; first printed in Swedish, 1910 and in English, 1912]. *Biogs, obits etc.* Stafleu & Cowan (*TL-2* **1**: 798, 1976).

Erikssonia Penz. & Sacc. (1898), Phyllachoraceae. 5 (biotrophic on leaves), widespread (tropical). See Stevenson (*Mycol.* **35**: 629, 1943).

Erikssonopsis M. Morelet (1971), Helotiaceae. 1, Europe. See Morelet (*Bulletin de la Société des Sciences naturelles et d'Archéologie de Toulon et du Var* **195**: 7, 1971).

Erinaceae Quél. (1886) = Bankeraceae.

Erinacella, see *Ericianella*.

erinaceous, prickly like a hedgehog.

Erinaceus Dill. ex Maratti (1822) = Hydnum fide Donk (*Persoonia* **8**: 279, 1975).

Erinella Quél. (1886) ≡ Lachnum.

Erinella Sacc. (1889) = Lachnum fide Spooner (*Biblthca Mycol.* **116**, 1987).

Erinellina Seaver (1951) = Lachnum fide Hawksworth *et al.* (*Dictionary of the Fungi* edn 8, 1995).

Erineum Pers. (1822) nom. dub., Fungi. Gall on an

outgrowth caused by gall-mites (*Arachnida*; *Eriophydae*).

Eriocaulago Vánky (2005), ? Ustilaginaceae. 2 (on *Eriocaulaceae*), Africa; S. Asia; N. America. See Vánky (*Mycol. Balcanica* **2**: 113, 2005).

Eriocercospora Deighton (1969), anamorphic *Mycosphaerellaceae*, Hso.≡ eP.10. 2 (on *Asterinaceae*), widespread (tropical). See Deighton (*Mycol. Pap.* **118**: 5, 1969), Crous & Braun (*CBS Diversity Ser.* **1**: 571 pp., 2003).

Eriocercosporella Rak. Kumar, A.N. Rai & Kamal ex U. Braun (1998), anamorphic *Mycosphaerellaceae*, Hso.≡ eP.?. 2, India. See Braun (*Monogr. Cercosporella, Ramularia Allied Genera (Phytopath. Hyphom.)* **2**: 398, 1998).

Eriocladus Lév. (1846) nom. rej. ≡ Lachnocladium.

Eriocorys Quél. (1886) ≡ Strobilomyces.

Erioderma Fée (1825), Pannariaceae (L). 40, widespread (esp. tropical). See Elix *et al.* (*Aust. J. Chem.* **40**: 1581, 1987; biochemistry), Jørgensen (*Ann. bot. fenn.* **38**: 259, 2001; E Asia), Jørgensen (*Taxon* **50**: 525, 2001; nomencl.), Jørgensen & Arvidsson (*Nova Hedwigia* **73**: 497, 2001; sorediate spp.), Jørgensen & Arvidsson (*Nordic Jl Bot.* **22**: 87, 2002; S America), Jørgensen & Sipman (*Ann. bot. fenn.* **39**: 201, 2002; SE Asia), Wiklund & Wedin (*Cladistics* **19**: 419, 2003; phylogeny), Jørgensen & Arvidsson (*Symb. bot. upsal.* **34** no. 1: 113, 2004; Ecuador), Wedin & Wiklund (*Symb. bot. upsal.* **34** no. 1: 469, 2004; phylogeny), Miądlikowska *et al.* (*Mycol.* **98**: 1088, 2006; phylogeny).

Eriomene (Sacc.) Clem. & Shear (1931) = Menispora fide Hughes (*CJB* **36**: 727, 1958).

Eriomenella Peyronel (1918) = Menispora fide Hughes (*CJB* **36**: 727, 1958).

Eriomoeszia Vánky (2005), ? Ustilaginaceae. 1 (on *Eriocaulaceae*), S. Asia; N. America. See Vánky (*Mycol. Balcanica* **2**: 105, 2005).

Eriomycopsis Speg. (1910), anamorphic *Pezizomycotina*, Hso.≡ eH.10. *c.* 13 (fungicolous), widespread. See Deighton & Pirozynski (*Mycol. Pap.* **128**, 1972), Braun & Mel'nik (*Mikol. Fitopatol.* **32**: 3, 1998; Russia), Johnston (*N.Z. Jl Bot.* **37**: 289, 1999; New Zealand).

Erionema Maire (1906) = Menispora fide Hughes (*CJB* **36**: 727, 1958).

Eriopezia (Sacc.) Rehm (1892), Hyaloscyphaceae. *c.* 30, Europe; New Zealand. See Korf (*Mycotaxon* **7**: 457, 1978), Spooner (*Biblthca Mycol.* **116**, 1987), Kobler (*Schweiz. Z. Pilzk.* **69**: 30, 1991).

Eriopeziza, see *Eriopezia*.

Erioscypha Kirschst. (1938) = Lachnum fide Korf (*Mycotaxon* **7**: 399, 1978).

Erioscyphella Kirschst. (1938) = Lachnum fide Korf (*Mycotaxon* **7**: 399, 1978), Haines & Dumont (*Mycotaxon* **19**: 1, 1984).

Eriosperma Raf. (1808) nom. dub., ? Agaricales.

Eriosphaera De Toni (1888) [non *Eriosphaera* F. Dietr. 1817, *Compositae*] ≡ Lasiosphaera.

Eriosphaerella Höhn. (1906) = Eriosphaeria fide Müller & von Arx (*Beitr. Kryptfl. Schweiz* **11** no. 2, 1962).

Eriosphaeria Sacc. (1875), Trichosphaeriaceae. 6 (from wood and bark), Asia; Europe. See Réblová (*Czech Mycol.* **50**: 73, 1997), Shenoy *et al.* (*MR* **110**: 916, 2006).

Eriospora Berk. & Broome (1850), anamorphic *Stictidaceae*, Cpd.0fH.10. 5, Europe; S. America. See Petrak (*Sydowia* **1**: 94, 1947).

Eriosporangium Bertero ex Ruschenb. (1831) = Puccinia fide Jackson (*Mycol.* **24**: 62, 1932).

Eriosporella Höhn. (1916), anamorphic *Pezizomycotina*, St.1bH.15. 1, Europe. See Punithalingam (*MR* **107**: 917, 2003; cytology).

Eriosporina Tognini (1894) ? = Sirothecium fide Sutton (*Mycol. Pap.* **141**, 1977).

Eriosporium Vánky (2005), ? Ustilaginaceae. 2 (on *Eriocaulaceae*), Africa. See Vánky (*Mycol. Balcanica* **2**: 113, 2005).

Eriosporopsis Petr. (1947), anamorphic *Pezizomycotina*, Cpd.0fH.?. 1, Europe.

Eriothyrium Speg. (1888), anamorphic *Pezizomycotina*, Cpt.0eH.?. 5, S. America.

Erispora Pat. (1922), Pezizomycotina. 1, Philippines. Type material is lost and affinities are unclear, fide Rossman *et al.* (*Stud. Mycol.* **42**, 1999).

Ernakulamia Subram. (1996), anamorphic *Pezizomycotina*. 1, S.E. Asia. See Subramanian (*Kavaka* **22/23**: 67, 1994).

erogen, a substance controlling the induction and differentiation of sex organs; **erotactin**, a sperm attractant; **erotropin**, a substance inducing a chemotropic response in sex organs (Machlis, *Mycol.* **64**: 238, 1972). See hormones.

Eromitra Lév. (1846) ≡ Mitrophora.

erose (of a lamella, etc.), having delicate tooth-like projections from the edge.

Erostella (Sacc.) Sacc. (1906) = Togninia fide Barr *et al.* (*Mycotaxon* **48**: 529, 1993), Mostert *et al.* (*Mycol.* **95**: 646, 2003).

Erostrotheca G.H. Martin & Charles (1928) = Melanospora Corda fide von Arx & Müller (*Beitr. Kryptfl. Schweiz* **11** no. 1, 1954).

erratic (of lichen thalli), not fixed to the substratum and often blowing around, e.g. *Chondropsis semiviridis*, *Sphaerothallia esculenta* ('manna'); epigaeic; vagrant; wandering lichens.

Erratomyces M. Piepenbr. & R. Bauer (1997), ? Tilletiaceae. 5 (on leaves of *Leguminosae*), Australia; India; S. & N. America. See Castlebury *et al.* (*Mycol.* **97**: 888, 2005; phylog.).

erumpent, bursting through the surface of the substratum. Cf. perrumpent.

Eruptio M.E. Barr (1996), ? Mycosphaerellaceae. Anamorphs *Lecanosticta*, *Dothistroma*. 3, widespread. See Evans (*Mycol. Pap.* **153**, 1984; as *Mycosphaerella*), Barr (*Mycotaxon* **60**: 433, 1996), Crous (*MR* **103**: 607, 1999), Verkley *et al.* (*Mycol.* **96**: 558, 2004; phylogeny).

Erynia (Nowak. ex Batko) Remaud. & Hennebert (1980), Entomophthoraceae. 16 (on *Insecta*), widespread. See Remaudière & Hennebert (*Mycotaxon* **11**: 269, 1980), Humber (*Mycotaxon* **13**: 471, 1981), Humber (*Mycotaxon* **15**: 167, 1982), Butt & Beckett (*Protoplasma* **120**: 72, 1984; ultrastr. mitosis & spindle-pole body), Butt & Beckett (*Protoplasma* **120**: 61, 1984; ultrastr. mitosis & spindle-pole body), Descals & Webster (*TBMS* **83**: 669, 1984; aquatic spp.), Li & Humber (*CJB* **62**: 653, 1984), Ben-Ze'ev (*Mycotaxon* **25**: 1, 1986), Ben-Ze'ev (*Mycotaxon* **27**: 263, 1986), Ben-Ze'ev (*Mycotaxon* **28**: 403, 1987), Humber (*Mycotaxon* **34**: 441, 1989), Perry & Fleming (*Mycol.* **81**: 154, 1989; zygospore germination), Keller (*Sydowia* **43**: 39, 1991), Keller (*Sydowia* **45**: 252, 1993), Robel (*MR* **101**: 573, 1997; polymorphism), Li *et al.* (*Mycosystema* **17**: 91, 1998), Keller

& Petrini (*Sydowia* **57**: 23, 2005; key), Keller (*Sydowia* **58**: 38, 2006; Switzerland), Keller (*Sydowia* **59**: 75, 2007; Switzerland, n.sp.).

Eryniopsis Humber (1984), Entomophthoraceae. 3, Europe; USA. See Humber (*Mycotaxon* **21**: 257, 1984), Keller & Eilenberg (*Sydowia* **45**: 264, 1993), Keller & Petrini (*Sydowia* **57**: 23, 2005; key), Keller (*Sydowia* **59**: 75, 2007; Switzerland, n.sp.).

Eryporus Quél. (1886) = Boletinus fide Murrill (*N. Amer. Fl.* **9**: 133, 1910).

Erysibe Theophr. ex Wallr. (1833) nom. dub., anamorphic *Pucciniales*. Anamorph name mainly for (II).

Erysiphaceae Tul. & C. Tul. (1861), Erysiphales. 19 gen. (+ 40 syn.), 769 spp.

Lit.: Braun (*Beih. Nova Hedwigia* **89**: 700 pp., 1987), Braun (*The Powdery Mildews (Erysiphales) of Europe*: 337 pp., 1995), Cook *et al.* (*MR* **101**: 975, 1997), Saenz (*McIlvainea* **13**: 33, 1998), Takamatsu *et al.* (*Mycoscience* **39**: 441, 1998), Braun (*Schlechtendalia* **3**: 48, 1999), Saenz & Taylor (*CJB* **77**: 150, 1999), Braun & Takamatsu (*Schlechtendalia* **4**: 1, 2000), Mori *et al.* (*Mycol.* **92**: 74, 2000), Takamatsu (*Mycoscience* **45**: 147, 2004), Takamatsu & Matsuda (*Mycoscience* **45**: 340, 2004), Hirose *et al.* (*MR* **109**: 912, 2005), Takamatsu *et al.* (*MR* **110**: 1093, 2006), Wang *et al.* (*Mycol.* **98**: 1065, 2006; phylogeny), Takamatsu *et al.* (*MR* **112**: 299, 2008; phylogeny).

Erysiphales H. Gwynne-Vaughan (1922). Leotiomycetidae. 1 fam., 19 gen., 769 spp. Fam.:
Erysiphaceae

Lit.: **General**: Amano (*Host range and geographical distribution of the powdery mildew fungi*, 1987; compr. host & geogr. distr.), Bélanger *et al.* (Eds) (*The powdery Mildews: A comprehensive Treatise*, 2000), Blumer (*Beitr. Krypt. Flora Schweiz* **7**(1), 1933; *Echte Mehltaupilze (Erysiphaceae)*, 1967; keys; descriptions), Boesewinkel (*Revue mycol.* **41**: 493, 1977; ident. by conidia, key), Braun (*Feddes Repert.* **88**: 655, 1978; taxonomy, *A monograph of the Erysiphales*, 1987 [*Beih. Nova Hedw.* **89**]; keys 435 spp., *Schlechtendahlia* **3**: 48, 1999; generic concepts), Braun & Takamatsu (*Schlechtendalia* **4**: 1, 2000; gen. & fam. concept), Cook *et al.* (*MR* **101**: 975, 1997; anams, SEM), Gelyuta (*Biol. Zh. Armen.* **41**: 351, 1988; fam. concepts), Hirata (*Host range and geographical distribution of the powdery mildews*, 1968; *Sydowia* **25**: 100, 1972), Hirata *et al.* *CJB* **78**: 1521, 2000; ITS, coevolution), Junell (*TBMS* **45**: 539, 1965; nomenclature), Mori *et al.* (*Mycol.* **92**: 74, 2000; mol. syst., *Mycoscience* **41**: 437, 2000; evolution), Saenz (*McIlvainea* **13**: 33, 1998; rDNA), Saenz & Taylor (*CJB* **77**: 150, 1999; mol. syst.), Spencer (Ed.) (*The powdery mildews*, 1978), Takamatsu *et al.* (*Mycoscience* **39**: 441, 1998; DNA; *MR* **104**: 1304, 2000; coevolution), Zheng (*Mycotaxon* **22**: 209, 1985; key),

Regional: **Argentina**: Havrylenko (*Erysiphales de la Region Andino-Patagonica*, 1997; 51 spp.). **Armenia**: Simonyan (*Mikoflora Armenii*, **VII** Erysiphales, 1994; 106 spp.). **Bulgaria**: Fakirova (*Fungi Bulgaricae*, **1** Ordo Erysiphales, 1991; 95 spp.), **Canada**, Parmelee (*CJB* **55**: 1940, 1977; host index). **China**, Zheng & Yu (*Fl. Fung. Sin.* **1**, 1987; 214 spp.), Chen & Gao (*Powdery Mildews in Fujian, Fuzhou, China*, 1993; 90 spp.). **Estonia**, Karis (*Esti jahukastelised (Erysiphaceae)*, 1987). **Europe**, see Blumer (1933, 1967 above), Braun (*The powdery mildews (Erysiphales) of Europe*, 1995). *Greece*, Pantidou (*Annls Inst. Phytopath. Benaki* n.s. **10**: 187, 1971). **India**, Sharma & Patel (*Mycol. Inform.* **4**, 1995; checklist and bibliog.). **Israel**, Chorin & Palti (*Israel J. agric. Res.* **12**: 153, 1963). **Italy**, Ciferri & Camera (*Quaderno Ist. bot. Univ. Pavia* 21, 1962). **Japan**, Homma (*J. Fac. Agric., Hokkaido Imp. Univ. Sapporo* **38**: 183, 1937). **Kazakhstan**, Vasyagina *et al.* ([*Flora sporovikh rastenii Kazakhstana*] **3**, 1961). **Korea**: Shin (*Erysiphaceae of Korea*, 1986; 62 spp.). **Lithuania**: Grigaliūnaitõ (*Mycota Lithuaniae* **III** Erysiphales 1, 1997; 86 spp.). **New Zealand**, Hammett (*N.Z. Jl Bot.* **15**: 687, 1977). **Poland**: Sałata (*Flora Polska Grzyby* 15, 1985), **Portugal**, de Varennes e Mendonca & de Sequeira (*Agron. lusit.* **24**: 87, 1963; **26**: 21, 1965; **33**: 151, 1972). **Romania**, Sandu-Ville (*Ciupercile Erysiphaceae din Romania*, 1967), Eliade (*Lucr. Grăd. Bot. Bucur.* 1990: 105, 1990; 147 spp.). **Russia**, Golovin ([*Powdery mildews parasitizing cultivated and useful plants*], 1960), Bunkina (*Plantae non Vasculares, Fungi et Bryopsida Orientis Extremi Sovietici, Fungi*, **2** Ascomycetes, Erysiphales, Clavicipitales, Helotiales, 1991; 121 spp.). **Slovakia**: Paulech (*Flóra Slovenska* **X/1**, Mycota (Huby), Ascomycetes (Vreckaté), Erysiphales (Múčnatkotvaré), 1995; 108 spp.). **South Africa**, van Jaarsveld (*Phytophylactica* **16**: 155, 1984). **Spain**, Durrieu & Macé (*BSMF* **88**: 175, 1973). **Sweden**, Junell (*Symb. bot. upsal.* **19** (1), 1967). **UK**, Ing (*Mycologist* **4**: 46, 88, 125, 172, 1990, checklist; **5**: 24, 60, 1991, key). **Ukraine**: Gelyuta (*Flora Fungorum RSS Ucrainicae, Ascomycetes, Erysiphales*, 1989; 108 spp.).

Erysiphe R. Hedw. ex DC. (1805), Erysiphaceae. Anamorph *Oidium* subgen. *Pseudoidium*. 326, widespread. See Martin & Gay (*CJB* **61**: 2472, 1983; conidiogenesis), Braun (*Beih. Nova Hedwigia* **89**, 1987; key), Zeller (*Mycol.* **87**: 525, 1995; morphology syst.), Braun (*Schlechtendalia* **2**: 48, 1998; gen. taxonomy), Takamatsu *et al.* (*Mycoscience* **40**: 252, 1999; molec. syst.), Braun & Takamatsu (*Schlechtendalia* **4**: 3, 2000; emend), Mori *et al.* (*Mycoscience* **41**: 437, 2000; phylogeny), Stummer *et al.* (*MR* **104**: 44, 2000; as *Uncinula*, genetic diversity), Havrylenko (*Nova Hedwigia* **72**: 409, 2001; Argentina), Takamatsu *et al.* (*Mycoscience* **43**: 333, 2002; on *Glycine*), Zimmermannová-Pastircáková *et al.* (*Schlechtendalia* **8**: 39, 2002; on *Aesculus*, Europe), Cunnington *et al.* (*Australas. Pl. Path.* **32**: 421, 2003; molecular identification, anamorphs), Nomura *et al.* (*Mycoscience* **44**: 157, 2003; on *Vitis*), Cunnington *et al.* (*Australas. Pl. Path.* **33**: 281, 2004; on *Fabaceae*, Australia), Cunnington *et al.* (*Fungal Diversity* **16**: 1, 2004; genetic diversity), Jackson (*Evolution* Lancaster, Pa. **58**: 1909, 2004; host-jumping), Takamatsu (*Mycoscience* **45**: 147, 2004; phylogeny), Takamatsu & Matsuda (*Mycoscience* **45**: 340, 2004; evolution), Péros *et al.* (*Eur. J. Pl. Path.* **113**: 407, 2005; on *Vitis*, France), Amrani & Corio-Costet (*Pl. Path.* **55**: 505, 2006; pathogenicity on *Vitis*), Braun *et al.* (*Mycol. Progr.* **5**: 139, 2006; on *Carpinus*), Wang *et al.* (*Mycol.* **98**: 1065, 2006; phylogeny), Takamatsu *et al.* (*MR* **111**: 809, 2007; *E. alphitoides* group).

Erysiphella Peck (1876) = Erysiphe fide Salmon (*Mem. Torrey bot. Club* **9**: 1, 1900).

Erysiphites Mesch. (1902), Fossil Fungi. 1 (Tertiary), former USSR.

Erysiphites Pampal. (1902), Fossil Fungi ? Fungi. 1 (Miocene), Sicily. See Salmon (*J. Bot. Lond.* **41**: 127, 1903).

Erysiphomycetidae = Leotiomycetidae. See *Erysiphales*; now placed within the *Leotiomycetes*.

For *Lit.* see fam.

Erysiphopsis Halst. (1899) = Erysiphe fide Salmon (*Mem. Torrey bot. Club* **9**: 1, 1900).

Erysiphopsis Speg. (1910), anamorphic *Pezizomycotina*. 1, S. America. See Petrak & Sydow (*Annls mycol.* **34**: 36, 1936).

Erythricium J. Erikss. & Hjortstam (1970), Corticiaceae. 3, widespread. See Eriksson & Hjortstam (*Svensk bot. Tidskr.* **64**: 165, 1970), Roux & Coetzee (*Pl. Dis.* **89**: 1158, 2005; pink disease in South Africa).

Erythrobasidiales R. Bauer, Begerow, J.P. Samp., M. Weiss & Oberw. (2006). Cystobasidiomycetes. 2 gen., 2 spp. No familes recognized.

Erythrobasidium Hamam., Sugiy. & Komag. (1988), anamorphic *Erythrobasidiales*. 1, Japan. See Hamamoto *et al.* (*J. gen. appl. Microbiol.* Tokyo **37**: 131, 1991; nomencl.), Fell *et al.* (*Int. J. Syst. Evol. Microbiol.* **50**: 1351, 2000; mol. phylogeny), Bauer *et al.* (*Mycol. Progr.* **5**: 41, 2006).

Erythrocarpon Zukal (1885), Chaetomiaceae. 1, Europe. Affinities are obscure, no recent research is available.

Erythrocarpum Sacc. (1891) ≡ Erythrocarpon.

Erythrodecton G. Thor (1991), Roccellaceae (L). 2, widespread. See Henssen & Thor (*Nordic Jl Bot.* **18**: 95, 1998; ontogeny), Thor (*Biblthca Lichenol.* **95**: 543, 2007; Malaysia).

Erythrogloeum Petr. (1953), anamorphic *Pezizomycotina*, St.0eH.15. 1, Costa Rica. See Ferreira *et al.* (*Fitopatol. Brasil* **17**: 106, 1992; anthracnose disease of *Hymenaea* spp.).

Erythrogymnotheca Yaguchi & Udagawa (1994), Trichocomaceae. 1, Japan. See Yaguchi *et al.* (*Mycoscience* **35**: 219, 1994).

Erythromada Huhndorf, A.N. Mill., F.A. Fernández & Lodge (2005), Sordariomycetes. 1, Puerto Rico. See Huhndorf *et al.* (*Fungal Diversity* **20**: 63, 2005).

Erythromyces Hjortstam & Ryvarden (1990), Hymenochaetaceae. 1, widespread. See Hjortstam & Ryvarden (*Mycotaxon* **37**: 55, 1990), Hjortstam *et al.* (*Kew Bull.* **53**: 805, 1998).

Erythrosphaera Sorokīn (1871) = Cephalotheca fide Hawksworth *et al.* (*Dictionary of the Fungi* edn 8, 1995).

Esalque J.F. Hennen, M.B. Figueredo & A.A. Carvalho (2000), ? Raveneliaceae. 1 (on *Caesalpinia* (*Leguminosae*)), Brazil. See Hennen (*Mycol.* **92**: 315, 2000) Cf. *Hennenia*.

Eschatogonia Trevis. (1853), ? Acarosporaceae (L). 2, tropical. See Timdal (*Lichenologist* **40**: 31, 2008; Peru, key).

Escovopsis J.J. Muchovej & Della Lucia (1990), anamorphic *Hypocreales*, Hso.0eH.15. 2 (from attine ant nests), widespread. See Romero *et al.* (*Rev. Mex. Micol.* **3**: 231, 1987; as *Phialocladus*), Seifert *et al.* (*Mycol.* **87**: 407, 1995), Jacobs *et al.* (*Mycol.* **97**: 111, 2005; phylogeny).

esculent, of use as a food; see Edible fungi.

Esdipatilia Phadke (1981), anamorphic *Pezizomycotina*, Hsy.≡ eH.1. 1, India. See Phadke (*TBMS* **77**: 642, 1981).

eseptate, see aseptate.

Esfandiaria Petr. (1955) [non *Esfandiaria* Charif & Aellen 1955, *Chaenopodiaceae*] ≡ Esfandiariomyces.

Esfandiariomyces Ershad (1985), Pezizomycotina. 1, Iran. See Ershad (*Iranian Journal of Plant Pathology* **21**: 8, 1985).

esorediate (**esorediose**), having no soredia.

Esslingeriana Hale & M.J. Lai (1980), Parmeliaceae (L). 1, USA. See Kärnefelt *et al.* (*Pl. Syst. Evol.* **183**: 113, 1992; status), Mattsson & Articus (*Symb. bot. upsal.* **34** no. 1: 237, 2004; phylogeny).

Esteya J.Y. Liou, J.Y. Shih & Tzean (1999), anamorphic *Pezizomycotina*, Hso.?.?. 1, Taiwan; Europe. See Liou *et al.* (*MR* **103**: 243, 1999), Kubátová *et al.* (*Czech Mycol.* **52**: 227, 2000; Czech Republic).

Etheirodon Banker (1902) ≡ Steccherinum fide Banker (*Mycol.* **21**: 145, 1929).

Etheirophora Kohlm. & Volkm.-Kohlm. (1989), Hypocreomycetidae. 3 (marine), Atlantic Ocean; Pacific Ocean. See Kohlmeyer & Volkmann-Kohlmeyer (*MR* **92**: 416, 1989; key), Schoch *et al.* (*MR* **111**: 154, 2007; phylogeny).

Ethnomycology. Mycology as a branch of ethnology, i.e. the attitudes to, lore and use etc. of fungi by different ethnic, racial or national divisions of humanity. To date, most studies in ethnomycology have concentrated on edible and poisonous macrofungi (q.v.), and on the role of fungi, often hallucinogenic, in religion. Studies on other aspects, such as medicinal use of fungi, are only now beginning. In respect of edible and poisonous macrofungi, different human societies exhibit a full range of attitudes from indifference to strong interest, often resolved as either a powerful ignorance and fear of all macrofungi (typical of, for example, the Anglo-Saxon peoples) or an enthusiasm for using wild edible species combined with extensive practical experience of which species are to be avoided (typical of, for example, northern Slav people). Geographical coverage by existing ethnomycological work is patchy, and many human societies in Africa, Asia, the Pacific ocean islands and South America in particular remain completely unstudied or with tantalizing indications of a wealth of cultural experience still undocumented. Pioneering but unpublished work by Teng (q.v.) on the ethnomycology of China was confiscated during the country's 'cultural revolution', and is presumed destroyed. In addition to connexions with established and recognized religions, there are many beliefs linking fungi to the supernatural, for example through fairy rings (q.v.) or as a result of consumption of hallucinogenic fungi (q.v.).

Lit.: Adhikari & Durrieu (*BSMF* **112**: 31, 1996; Nepal), Akpaja *et al.* (*Int. J. Medicinal Mushrooms* **7**: 373, 2005; ethnomycology among the Bini people of Nigeria), Birks (*McIlvainnea* **10**: 89, 1991; fungi in folk medicine), Bulakh (*Int. J. Medicinal Mushrooms* **3**: 125, 2001; medicinal mushrooms in the Russian far east), Buller (*TBMS* 5: 21, 1914; fungus lore or ancient Greeks and Romans), Chamberlain (*Mycologist* **10**: 173, 1996; southwest China), Grzywnowicz (*Int. J. Medicinal Mushrooms* **3**: 154, 2001; medicinal mushrooms in Polish folk medicine), Guzmán (*Int. J. Medicinal Mushrooms* **3**: 95, 2001; medicinal mushrooms in Mexican folk traditions), Härkönen (*Symb. Bot. Ups.* **30**: 145, 1995; Tanzania), Harsh *et al.* (*J. Tropical For.* **9**: 270, 1993; forest fungi and tribal economy in Madhya Pradesh, India), Hobbs (*Medicinal mushrooms: an exploration of tradition,* *healing and culture*, edn 2, 1995), Kalotas (*Fungi of Australia* **18**: 269, 1997; aboriginal knowledge and uses of fungi), Keewaydinoquay (*Puhpohwee for the people: a narrative account of some uses of fungi among the Ahnishinaabeg*, 67 pp, 1998; ethnomycology among some North American tribes), Lowy (*Mycol.* **64**: 816, 1972, Maya codices; **66**: 188, 1974, *Amanita muscaria* and the thunderbolt legend; *Revista Interam. Rev.* **2**: 405, 1972, **5**: 110, 1975, **10**: 94, 1980, mushrooms and religion), Mata (*Rev. Mex. Micol.* **3**: 175, 1987; ethnomycology among the Mayans of Yucatan), Gamundí & Minter (*IMI Descriptions of Fungi and Bacteria* **160**: 1592, 2004; vernacular names and folk uses of *Cyttaria* by now extinct Fuegian natives), Montoya-Esquivel (*McIlvainea* **13**: 6, 1998; ethnomycology of Tlaxcala, Mexico), Morris (*Nyala* **16**: 1, 1992; Malawi), Osemwegie *et al.* (*Int. J. Medicinal Mushrooms* **5**: 313, 2003; ethnomycology among the Igbo people of Nigeria), Oso (*Mycol.* **67**: 311, 1975; ethnomycology of the Yoruba, Nigeria), Piearce (*Bull. Brit. Mycol. Soc.* **15**: 139, 1981; Zambia), Prance (*Advances in economic botany* **1**: 127, 1984; use of fungi by Amazonian Indians), Rai *et al.* (*Mycologist* **7**: 192, 1993; ethnic use of medicinal mushrooms in central India), Redlinger (Ed.) (*The sacred mushroom seeker*, 1990), Saar (*J. Ethnopharmacology* **31**: 175, 1991; fungi in Khanty folk medicine, Siberia), Sillitoe (*Science in New Guinea* **21**: 3, 1995; ethnomycology in Papua New Guinea southern highlands), Vaidya & Rabba (*Mycologist* **7**: 131, 1993; fungi in folk medicine), Wasson & Wasson (*Mushrooms, Russia and history*, 2 vols. 1957). See also: Edible fungi, Hallucinogenic fungi, Soma.

etiology, see aetiology.

eu- (prefix), true; sometimes used, but wrongly for the subgenus or section including the type species of the generic name of which it is an infrageneric taxon.

Euacanthe Theiss. (1917), Scortechiniaceae. 1. See Nannfeldt (*Svensk bot. Tidskr.* **66**: 49, 1975), Subramanian & Sekar (*Kavaka* **18**: 19, 1993), Huhndorf *et al.* (*MR* **108**: 1384, 2004; phylogeny), Schoch *et al.* (*MR* **111**: 154, 2007; phylogeny).

Euactinomyces Langeron (1922) nom. dub., Fungi.

Eualectoria (Th. Fr.) Gyeln. (1934) ≡ Alectoria Ach.

Euantennaria Speg. (1918), Euantennariaceae. Anamorphs *Antennatula, Hormisciomyces*. 6, widespread. See Corlett *et al.* (*N.Z. Jl Bot.* **11**: 213, 1973), Hughes (*N.Z. Jl Bot.* **12**: 299, 1974), Hughes & Arnold (*Mem. N. Y. bot. Gdn* **49**: 198, 1989; anamorph).

Euantennariaceae S. Hughes & Corlett ex S. Hughes (1972), Capnodiales. 9 gen. (+ 6 syn.), 28 spp.

Lit.: Batista & Ciferri (*Saccardoa* **2**, 1963), Reynolds (*Taxon* **20**: 759, 1971; hyphal morph.), Hughes (*N.Z. Jl Bot.* **10**: 225, 1972), Reynolds (*Nova Hedwigia* **26**: 179, 1975; growth forms), Hughes (*Mycol.* **68**: 693, 1976; gen. names, anamorphs), Reynolds (*Mycotaxon* **8**: 417, 1979; stalked taxa), Sugiyama *et al.* (*Studies on Cryptogams in Southern Chile*: 169, 1984), Parbery & Brown (*Microbiology of the Phyllosphere*: 101, 1986), Reynolds (*Mycotaxon* **27**: 377, 1986; status), Hughes & Arnold (*Mem. N. Y. bot. Gdn* **49**: 198, 1989), Hughes & Seifert (*Sydowia* **50**: 192, 1998).

Euascomycetes. Class for pyrenomycetes, discomycetes and laboulbeniomycetes; cf. Hemiascomycetes, Loculoascomycetes.

Euascomycetidae. See *Ascomycota*.

Euaspergillus F. Ludw. (1892) = Aspergillus fide Raper & Fennell (*The genus Aspergillus*, 1965).

Eubelonis Clem. (1909) ? = Calycina fide Arendholz (*Mycotaxon* **36**: 283, 1988).

Eubelonis Höhn. (1926), Helotiaceae. 2, Europe. See Arendholz (*Mycotaxon* **36**: 283, 1988).

Eucantharomyces Thaxt. (1895), Laboulbeniaceae. 26, widespread. See Santamaría (*MR* **98**: 1303, 1994; asci).

eucarpic, developing reproductive structures on limited portions of the thallus; residual nucleate protoplasm remaining and capable of further mitotic growth and regeneration.

Eucaryota (**Eucarya**), see *Eukaryota*.

Eucasphaeria Crous (2007), ? Sordariomycetes. Anamorph *Ascochytopsis*-like. 1. See Crous *et al.* (*Fungal Diversity* **26**: 1, 2007).

Euceramia Bat. & Cif. (1962), Chaetothyriaceae. 1, Brazil.

Euceratomyces Thaxt. (1931), Euceratomycetaceae. 1, N. America. See Santamaría (*Nova Hedwigia* **61**: 65, 1995).

Euceratomycetaceae I.I. Tav. (1980), Laboulbeniales. 5 gen., 7 spp.

Lit.: Tavares (*Mycol. Mem.* **9**: 627 pp., 1985), Santamaria i del Campo (*L'ordre Laboulbenials (Fungi, Ascomycotina) a la Península Ibèrica i Illes Ballears [Thesis]*: 669 pp., 1990), Santamaría (*Nova Hedwigia* **61**: 65, 1995), Santamaría & Rossi (*Pl. Biosystems* **133**: 163, 1999).

euchroic, see colour.

Eucladoniomyces Cif. & Tomas. (1953) = Cladonia fide Hawksworth *et al.* (*Dictionary of the Fungi* edn 8, 1995).

Eucollema (Cromb.) Horw. (1912) ≡ Collema F.H. Wigg.

Eucorethromyces Thaxt. (1900) = Corethromyces fide Thaxter (*Memoirs of the American Academy of Arts and Sciences* **16**: 1, 1931).

eucortex (of lichens), a cortex composed of well-differentiated tissue.

Eucyphelis Clem. (1909) = Chaenotheca fide Hawksworth *et al.* (*Dictionary of the Fungi* edn 8, 1995).

Eudacnus Raf. ex Merr. (1943) ≡ Endacinus.

Eudarluca Speg. (1908), ? Phaeosphaeriaceae. Anamorph *Sphaerellopsis*. 3 or 4 (on rusts), widespread. See Eriksson (*Bot. Notiser* **119**: 33, 1966; biology, systematics), Kranz (*Nova Hedwigia* **24**: 169, 1974; hosts), Yuan *et al.* (*MR* **102**: 866, 1998), Liesebach & Zaspel (*Rust Diseases of Willow and Poplar*: 231, 2005; genetic diversity), Nischwitz *et al.* (*MR* **109**: 421, 2005; evolution, host specialization), Pei & Yuan (*Rust Diseases of Willow and Poplar*: 243, 2005; biocontrol), Bayon *et al.* (*MR* **110**: 1200, 2006; genetic diversity).

Eudimeriolum Speg. (1912), Pseudoperisporiaceae. *c.* 8, widespread (tropical). See Hansford (*Mycol. Pap.* **15**, 1946), Farr (*Mycol.* **71**: 243, 1979), Samuels & Rogerson (*Brittonia* **42**: 105, 1990), Barr (*Mycotaxon* **64**: 149, 1997), Reynolds & Gilbert (*Aust. Syst. Bot.* **18**: 265, 2005; Australia).

Eudimeromyces Thaxt. (1918) = Dimeromyces fide Tavares (*Mycol. Mem.* **9**, 1985).

Euepixylon Füisting (1867), Xylariaceae. 2, Europe; E. Russia. See Læssøe (*SA* **13**: 43, 1994), Vassiljeva (*Nizshie Rasteniya, Griby i Mokhoobraznye Dalnego Vostoka Rossii* Griby **4**: 210, 1998).

euform, see *Pucciniales* and Table 5.

eugonic, see dysgonic.
eugonidium, a bright green lichen photobiont (e.g. *Trebouxia*) (obsol.).
Euhaplomyces Thaxt. (1901), Laboulbeniaceae. 1, British Isles. See Nannfeldt (*Svensk bot. Tidskr.* **43**: 468, 1949).
euhymenium, see hymenium.
Euhypoxylon Füisting (1867) ≡ Hypoxylon Bull.
Eukaryota. (**Eukarya**). The domain (empire, superkingdom) to which all **eukaryotes** belong; i.e. encompassing all organisms with one or more nuclei in their cells bounded by a nuclear membrane and with paired DNA-containing chromosomes (and also other complex organelles, e.g. Golgi bodies, mitochondria). The counterpart of the **Prokaryota** (**Prokarya, prokaryotes**) which is now generally divided into two separate kingdoms, *Archaea* (formerly *Archaebacteria*) and *Bacteria* (formerly *Eubacteria*; including also *Cyanobacteria*), and the viruses that lack the above structures. See Classification, Phylogeny.
eukaryote (adj. **eukaryotic**), one of the *Eukaryota* (q.v.); cf. prokaryote.
Eumela Syd. (1925), Pseudoperisporiaceae. 1, C. America. See Hansford (*Mycol. Pap.* **15**, 1946).
Eumicrocyclus, see *Eumycrocyclus*.
Eumisgomyces Speg. (1912) = Laboulbenia fide Spegazzini (*An. Mus. nac. Hist. nat. B. Aires* **27**: 70, 1915), Tavares (*Mycol. Mem.* **9**, 1985).
Eumitria Stirt. (1881) = Usnea fide Hawksworth *et al.* (*Dictionary of the Fungi* edn 8, 1995).
Eumonoicomyces Thaxt. (1901), Laboulbeniaceae. 2 or 3, widespread.
eumorphic, well-formed.
Eumycetes. (**Eumycota**), true fungi; see *Fungi*.
eumycetoma, see mycetoma.
Eumycota (**Eumycophyta**), see *Eumycetes*.
Eumycrocyclus Hara (1915) ≡ Coccoidella.
Euoidium Y.S. Paul & J.N. Kapoor (1986) = Oidium Link (1824) fide Hawksworth *et al.* (*Dictionary of the Fungi* edn 8, 1995).
Euopsis Nyl. (1875), Lichinaceae (L). 2, widespread. See Hafellner (*Beih. Nova Hedwigia* **79**: 241, 1984), Schultz & Büdel (*Lichenologist* **34**: 39, 2002; key).
Eupelte Syd. (1924), Asterinaceae. Anamorphs *Clasterosporium*, *Septoidium*-like. 3, widespread (tropical).
Eupenicillium F. Ludw. (1892), Trichocomaceae. Anamorph *Penicillium*. 46, widespread. See Udagawa & Horie (*Antonie van Leeuwenhoek* **39**: 313, 1973; ascospores), Pitt (*The genus Penicillium and its teleomorphic states Eupenicillium and Talaromyces*, 1979; keys), Stolk & Samson (*Stud. Mycol.* **23**, 1983), Berbee *et al.* (*Mycol.* **87**: 210, 1995; molec. data suggests affinity with *Aspergillus*), Leal *et al.* (*MR* **101**: 1259, 1997; polysaccharides), Ogawa *et al.* (*Mycol.* **89**: 756, 1997; DNA), Skouboe *et al.* (*MR* **103**: 873, 1999; phylogeny), Ogawa & Sugiyama (*Integration of Modern Taxonomic Methods for Penicillium and Aspergillus Classification*: 149, 2000; phylogeny), Pitt & Samson (*Integration of Modern Taxonomic Methods for Penicillium and Aspergillus Classification*: 51, 2000; types), Pitt *et al.* (*Integration of Modern Taxonomic Methods for Penicillium and Aspergillus Classification*: 9, 2000; accepted names), Prieto *et al.* (*CJB* **80**: 410, 2002; cell wall polysaccharides), Tuthill (*Mycol. Progr.* **3**: 3, 2004; genetic variation), Geiser *et al.* (*Mycol.* **98**: 1053, 2006; phylogeny), Horn & Peterson (*Mycol.* **100**: 12, 2008; on *Aspergillus* heads).
Eupezizella Höhn. (1926) = Hyaloscypha fide Huhtinen (*Karstenia* **29**: 45, 1989).
Euphoriomyces Thaxt. (1931), Laboulbeniaceae. 13, widespread. See Santamaría (*Revta Iberoamer. Micol.* **8**: 43, 1991; key).
Euplotespora Fokin, Di Giuseppe, Erra & Dini (2008), Microsporidia. 1. See Fokin *et al.* (*J. Eukaryot. Microbiol.* **55**: 214, 2008).
Eupropolella Höhn. (1917), Helotiales. 7, Europe. See Défago (*Sydowia* **21**: 1, 1967), Nauta & Spooner (*Mycologist* **14**: 21, 2000; UK).
Eupropolis De Not. (1863) = Phaeotrema fide Sherwood (*Mycotaxon* **5**: 50, 1977).
Eurasina G.R.W. Arnold (1970) = Helminthophora fide de Hoog (*Persoonia* **18**: 33, 1978).
Euricoa Bat. & H. Maia (1955) = Cylindrocarpon fide Kendrick & Carmichae *in* Ainsworth *et al.* (Eds) (*The Fungi* **4A**: 390, 1973).
European Council for Conservation of Fungi, see Conservation.
European mildew, see mildew.
European Mycological Association. Founded in 2003; recognized as the Committee for Europe within the International Mycological Association (q.v.); structure comprises individual and corporate members, an elected executive, with national representatives from European countries; organizes Congress of European Mycologists every three or four years, and various other meetings. Publications: *EMA Newsletter*. Website: www.euromould.org.
Europhium A.K. Parker (1957) = Grosmannia fide Benny & Kimbrough (*Mycotaxon* **12**, 1980), Solheim (*Nordic Jl Bot.* **6**: 199, 1986), von Arx (*SA* **5**: 310, 1986), Hausner *et al.* (*CJB* **78**: 903, 2000), Zipfel *et al.* (*Stud. Mycol.* **55**: 75, 2006).
Eurotiaceae Clem. & Shear (1931) = Trichocomaceae.
Eurotiales G.W. Martin ex Benny & Kimbr. (1980). Eurotiomycetidae. 3 fam., 49 gen., 928 spp. Stromata absent. Ascomata small, cleistothecial, usually solitary, rarely absent; peridium usually thin, membranous, usually brightly coloured, varied in structure and rarely acellular and cyst-like. Interascal tissue absent. Asci clavate or saccate, thin-walled, evanescent, sometimes formed in chains. Ascospores varied, small, aseptate, often ornamented and with equatorial thickening, without a sheath. Anamorphs prominent, many of industrial and medical importance (e.g. *Aspergillus*, *Penicillium*). Saprobic, mainly from soil or decaying plant materials.

Formerly used for ± all ascomycetes with cleistocarpic ascomata, many now placed within groups including ostiolate counterparts from which they have been derived. Fam.:
(1) **Elaphomycetaceae**
(2) **Thermoascaceae**
(3) **Trichocomaceae** (syn. *Aspergillaceae* auct., *Eurotiaceae*)

Lit.: Geiser *et al.* (*Mycol.* **98**: 1051, 2006; phylogeny), Geiser & LoBuglio (*in* McLaughlin *et al.* (Eds), *The Mycota* **7A**: 201, 2001), Guarro *et al.* (*Mycotaxon* **42**: 193, 1991; key 8 spherical-spored gen.), Fennell (*in* Ainsworth *et al.* (Eds), *The Fungi* **4A**: 45, 1973; keys gen.), Malloch (*in* Reynolds (Ed.), *Ascomycete systematics*: 73, 1981).
Eurotiella Lindau (1900) ≡ Allescheria Sacc. & P. Syd.
Eurotiomycetes O.E. Erikss. & Winka (1997), Pezizomycotina. 10 ord., 27 fam., 281 gen., 3401 spp.

Contains a morphologically high heterogenous assemblage of taxa, including *Penicillium* and its relatives, human pathogens, many of the black yeasts, and some lichenized families.

Subcl.:

(1) **Chaetothyriomycetidae**

(2) **Eurotiomycetidae**

(3) **Mycocaliciomycetidae**

Lit.: Ekman & Tønsberg (*MR* **106**: 1262, 2002), Geiser *et al.* (*Mycol* **98**: 1051, 2006), Lutzoni *et al.* (*Am. J. Bot.* **91**: 1446, 2004).

Eurotiomycetidae Doweld (2001), Eurotiomycetes. Stromata absent. Ascomata usually small, cleistothecial, rarely absent, often brightly coloured, varied in structure. Interascal tissue absent. Asci thin-walled, evanescent, sometimes formed in chains. Ascospores varied, small, aseptate, often ornamented and with equatorial thickening, without a sheath. Anamorphs prominent. Ords:

(1) **Coryneliales**

(2) **Eurotiales**

(3) **Onygenales**

For *Lit.* see ord. and fam.

Eurotiopsis Costantin ex Laborde (1897) = Monascus fide Malloch (*Mycol.* **62**: 738, 1970).

Eurotiopsis P. Karst. (1889) nom. dub., anamorphic *Pezizomycotina*.

Eurotites Mesch. (1892), Fossil Fungi. 1 (Oligocene), Europe.

Eurotium Link (1809), Trichocomaceae. Anamorph *Aspergillus*. 25, widespread. See Blaser (*Sydowia* **28**: 1, 1976; key), Pitt *in* Pitt & Samson (Eds) (*Advances in Penicillium and Aspergillus systematics* **102**: 383, 1985; nomencl.), Kozakiewicz (*Mycol. Pap.* **161**, 1989), Chang *et al.* (*J. gen. appl. Microbiol.* Tokyo **37**: 289, 1991; DNA), Pitt & Hocking (*Fungi and Food Spoilage* Edn 2: 593 pp., 1997; review), Samson (*Contributions to Microbiology* **2**: 5, 1999; review), Pitt & Samson (*Integration of Modern Taxonomic Methods for Penicillium and Aspergillus Classification*: 51, 2000; types), Pitt *et al.* (*Integration of Modern Taxonomic Methods for Penicillium and Aspergillus Classification*: 9, 2000; accepted names), Klich (*Identification of Common Aspergillus Species*: 116 pp., 2002; monograph), Zeng *et al.* (*FEMS Microbiol. Lett.* **237**: 79, 2004; DNA probes), Butinar *et al.* (*FEMS Microbiology Ecology* **51**: 155, 2005; hypersaline environments), Peterson (*Mycol.* **100**: 205, 2008; 4-locus phylogeny).

Euryachora Fuckel (1870), Dothideaceae. Anamorph *Placosphaeria*-like. *c.* 5, widespread (temperate). See Obrist (*Phytopath. Z.* **35**: 382, 1959).

Euryancale Drechsler (1939), Cochlonemataceae. 5, Japan; N. America. See Drechsler (*Mycol.* **31**: 410, 1939), Drechsler (*Mycol.* **47**: 364, 1955), Drechsler (*Mycol.* **51**: 787, 1959), Dyal (*Sydowia* **27**: 293, 1976; keys parasites of nematodes and amoebae, bibliogr.), Saikawa & Saito (*TBMS* **87**: 337, 1986; zygospores), Saikawa & Aoki (*TMSJ* **32**: 509, 1991), Saikawa & Katsurashima (*Mycol.* **85**: 24, 1993; ultrastr.), Saikawa & Aoki (*Nova Hedwigia* **60**: 571, 1995).

Euryporus Quél. (1886) = Suillus Gray fide Kuyper (*in litt.*).

Eurytheca De Seynes (1878), ? Myriangiaceae. 2, Europe; West Indies.

euseptate (of conidial septation), having cells separated by multilayered walls of similar structure to lateral walls, as in *Pyricularia* (Luttrell, *Mycol.* **55**: 672, 1963) (Fig. 22C); cf. distoseptate.

Eusordaria Zopf (1883) ? = Sordaria fide Lundqvist (*Symb. bot. upsal.* **20** no. 1: 269, 1972).

Eustegia Fr. (1823) [non *Eustegia* R. Br. 1810, *Asclepiadaceae*] ≡ Stegia.

Eustilbum Arnold (1885) nom. nud. = Dendrostilbella fide Hawksworth *et al.* (*Dictionary of the Fungi* edn 8, 1995).

eustroma, see stroma.

Eusynaptomyces Thaxt. (1931), Ceratomycetaceae. 5, widespread. See Scheloske (*Pl. Syst. Evol.* **126**: 267, 1976), Tavares (*Mycol. Mem.* **9**: 627 pp., 1985), Santamaría (*Fl. Mycol. Iberica* **5**, 2003; Iberian peninsula), Santamaria (*Nova Hedwigia* **82**: 349, 2006; Spain).

euthecium, an ascoma (cleistothecium, perithecium, apothecium) of an euascomycete; cf. pseudothecium.

Euthrypton Theiss. (1916) = Seuratia fide Meeker (*CJB* **53**: 2462, 1975).

euthyplectenchyma, see plectenchyma.

Euthythyrites Cookson (1947), Fossil Fungi, Asterinaceae. 3 (Oligocene, Miocene), Australia; India.

Eutorula H. Will (1916) = Torulopsis Berl. fide Hawksworth *et al.* (*Dictionary of the Fungi* edn 8, 1995).

Eutorulopsis Cif. (1925) ? = Torulopsis Berl. fide Hawksworth *et al.* (*Dictionary of the Fungi* edn 8, 1995).

eutrophic, rich in nutrients; cf. dystrophic, oligotrophic.

eutrophication, nutrient enrichment, usually used when directly or indirectly caused by human influences.

Eutryblidiella (Rehm) Höhn. (1959) = Rhytidhysteron fide Samuels & Müller (*Sydowia* **32**: 277, 1979).

Eutypa Tul. & C. Tul. (1863), Diatrypaceae. Anamorphs *Libertella*, *Cytosporina*. 32, widespread (esp. temperate). See Glawe & Rogers (*Mycotaxon* **20**: 401, 1984), Rappaz (*Mycol. Helv.* **2**: 285, 1987; key), Carter (*Phytopath. Pap.* **32**, 1991; pathology), Ju *et al.* (*Mycotaxon* **41**: 311, 1991; anamorph), Larignon & Dubos (*Eur. J. Pl. Path.* **103**: 147, 1997; on *Vitis*), Peros *et al.* (*Phytopathology* **87**: 799, 1997), DeScenzo *et al.* (*Phytopathology* **89**: 884, 1999; DNA), Peros *et al.* (*MR* **103**: 1385, 1999; pathology), Cortesi & Milgroom (*Journal of Plant Pathology* **83**: 79, 2001; population analysis), Acero *et al.* (*Mycol.* **96**: 249, 2004; phylogeny), Chacón Zapata (*Brenesia* **62**: 41, 2004; Mexico), Trouillas & Gubler (*MR* **108**: 1195, 2004; California, on *Vitis*), Lardner *et al.* (*MR* **109**: 799, 2005; molecular diagnostics), Carmarán *et al.* (*Fungal Diversity* **23**: 67, 2006; phylogeny), Rolshausen *et al.* (*Phytopathology* **96**: 369, 2006; species concepts), Zhang *et al.* (*Mycol.* **98**: 1076, 2006; phylogeny), Lardner *et al.* (*Australas. Pl. Path.* **36**: 149, 2007; Australia, on *Vitis*).

Eutypella (Nitschke) Sacc. (1875) nom. cons., Diatrypaceae. Anamorphs *Cytosporina*, *Libertella*. 78, widespread. See Glawe & Rogers (*Mycotaxon* **20**: 401, 1984; Pacific NW, keys), Rappaz (*Mycol. Helv.* **2**: 285, 1987; key), Ju *et al.* (*Mycotaxon* **41**: 311, 1991; anamorph), Hyde (*MR* **99**: 1462, 1995; Australia), Nordén (*Windahlia* **22**: 65, 1995; Sweden), Acero *et al.* (*Mycol.* **96**: 249, 2004; phylogeny), Chacón Zapata (*Brenesia* **62**: 41, 2004; Mexico), Carmarán *et al.* (*Fungal Diversity* **23**: 67, 2006; phylogeny).

eutypoid, having groups of perithecia in a stroma with

the ostioles vertical and breaking through the surface individually. Cf. valsoid.

Eutypopsis P. Karst. (1878) = Endoxyla fide Barr (*Mycol. Mem.* **7**: 232 pp., 1978).

Euzodiomyces Thaxt. (1900), Euceratomycetaceae. 2, widespread. See Weir & Rossi (*Mycol.* **93**: 171, 2001; Bolivia), Santamaría (*Fl. Mycol. Iberica* **5**, 2003; Iberian peninsula).

evanescent, having a short existence; fugacious.

Evanidomus Caball. (1941), anamorphic *Pezizomycotina*, St.0eP.?. 1, Spain.

Everhartia Sacc. & Ellis (1882), anamorphic *Pezizomycotina*, Hsp.0-≡ hH.?. 3, widespread. See Moore (*Mycol.* **47**: 90, 1955), Yanna & Ho (*J. Linn. Soc.* Bot. **134**: 465, 2000; Hong Kong), Kirschner & Chen (*Stud. Mycol.* **50**: 337, 2004).

Evernia Ach. (1809), Parmeliaceae (L). *c.* 10, widespread. See Ahlner (*Acta phytogeogr. suec.* **22**, 1948), Bird (*CJB* **52**: 2427, 1974; N. Am.), Awasthi (*Bull. bot. Surv. India* **24**: 96, 1983; key 3 spp. India), Tavares (*Mycol. Mem.* **9**: 627 pp., 1985), Golubkova & Shapiro (*Nov. Sist. niz. Rast.* **24**: 144, 1987; 9 spp. former USSR), Feuerer & Marth (*Mitt. Inst. Allg. Botanik Hamburg* **27**: 101, 1997; pseudocyphellae), Kärnefelt *et al.* (*Nova Hedwigia* **67**: 71, 1998; anatomy), Mattsson & Wedin (*Lichenologist* **30**: 463, 1998), Mattsson & Wedin (*Lichenologist* **31**: 431, 1999), Wedin *et al.* (*MR* **103**: 1152, 1999; DNA), Crespo *et al.* (*Taxon* **50**: 807, 2001; genus concepts), Peršoh *et al.* (*Mycol. Progr.* **3**: 103, 2004; asci, photobionts), Thell *et al.* (*Mycol. Progr.* **3**: 297, 2004; phylogeny), Thell *et al.* (*Symb. bot. upsal.* **34** no. 1: 429, 2004; Scandinavia), Piercey-Normore (*New Phytol.* **169**: 331, 2006; photobionts).

Everniaceae Tomas. (1897) = Parmeliaceae.

Everniastrum Hale ex Sipman (1986), Parmeliaceae (L). 33, widespread (esp. tropical). See Sipman (*Mycotaxon* **26**: 235, 1986; key), Eriksson & Hawksworth (*SA* **10**: 37, 1991; nomencl.), Jiang & Wei (*Lichenologist* **25**: 57, 1993; key 9 spp. China), Elix (*Flora of Australia* **55**: 37, 1994), Thell *et al.* (*Mycol. Progr.* **3**: 297, 2004; phylogeny), Blanco *et al.* (*Mol. Phylogen. Evol.* **39**: 52, 2006; phylogeny), Divakar *et al.* (*Mol. Phylogen. Evol.* **40**: 448, 2006; tropical spp.).

Everniicola D. Hawksw. (1982), anamorphic *Pezizomycotina*, Cpd.1eH.15. 1 (on lichens, esp. *Evernia* and *Nephroma*), Europe; N. America. See Alstrup & Hawksworth (*Meddr Grønland* Biosc. **31**, 1990).

Everniomyces Cif. & Tomas. (1953) ≡ Evernia.

Everniopsis Nyl. (1860), Parmeliaceae (L). 2, C. & S. America. See Kärnefelt *et al.* (*Nova Hedwigia* **67**: 71, 1998; anatomy), Thell *et al.* (*J. Hattori bot. Lab.* **100**: 797, 2006).

Eversia J.L. Crane & Schokn. (1977), anamorphic *Pezizomycotina*, Hso.#eP.19. 1, widespread. See Crane & Schoknecht (*Mycol.* **69**: 539, 1977).

everted, turned inside out.

Evicentia Barreno (2000), Biatorellaceae (L). 2, Italy; USA. See Barreno & Schoeninger (*The Fourth IAL Symposium* Progress and Problems in Lichenology at the Turn of the Millennium, Barcelona, 3-8 September 2000. Book of Abstracts: addendum [i], 2000).

Evlachovaea Borisov & Tarasov (1999), anamorphic *Pezizomycotina*, Hso.?.?. 1, Russia. See Borisov & Tarasov (*Mikol. Fitopatol.* **33**: 250, 1999).

Evlachovaia Voronin (1986), Microsporidia. 1.

Evolution, see Coevolution, Fossil fungi, Kingdoms of fungi, Phylogeny.

Evulla Kavina (1939) ? = Neobulgaria fide Korf *in* Ainsworth *et al.* (Eds) (*The Fungi* **4A**: 249, 1973).

ex (1) (in citations, e.g. G. Arnaud ex M.B. Ellis), from; first validly published by the second author(s), see Nomenclature; (2) (prefix), see e-.

ex situ (Lat.), **ex-situ** (Engl.) (of an organism), one taken from its natural habitat; used of living cultures isolated from nature and maintained in Genetic resource collections, and also of non-viable material held in Reference collections; cf. *in situ*.

Exarmidium P. Karst. (1873), ? Hyponectriaceae. 9, widespread. See Barr & Boise (*Mycotaxon* **23**: 233, 1985), Wang & Hyde (*Fungal Diversity* **3**: 159, 1999).

exasperate, roughened with hard projecting points.

excavate, hollow out.

excentric (**eccentric**), (1) one sided; (2) (of a stipe), at one side or not in the centre of the pileus; cf. centric.

Excioconidium Plunkett (1925) = Chalara fide Nag Raj & Kendrick (*Monogr. Chalara Allied Genera*, 1975).

Excioconis Clem. & Shear (1931) ≡ Excioconidium.

exciple, see excipulum.

Excipula Fr. (1823) = Pyrenopeziza fide Nannfeldt (*Nova Acta R. Soc. Scient. upsal.*, 1932), Sutton (*Mycol. Pap.* **141**, 1977).

Excipulaceae Bonord. (1851), Helotiales.

Lit.: Petch (*TBMS* **26**: 53, 1943), Sutton *in* Ainsworth *et al.* (Eds) (*The Fungi* **4A**: 553, 1973), Sutton (*Coelomycetes*, 1980; descr., connexion).

Excipularia Sacc. (1884), anamorphic *Pezizomycotina*, Hsp.≡ eP.10. 2, Europe; Russia. See Petrak (*Sydowia* **16**: 357, 1963), Spooner & Kirk (*TBMS* **78**: 247, 1982), Yurchenko (*Mycena* **1**: 32, 2001; Belarus).

Excipulariopsis P.M. Kirk & Spooner (1982), anamorphic *Pezizomycotina*, Hsp.≡ eP.1. 1, India. See Kirk & Spooner (*TBMS* **78**: 251, 1982).

Excipulella Höhn. (1915) = Heteropatella fide Sutton (*Mycol. Pap.* **141**, 1977).

Excipulina Sacc. (1884) = Heteropatella fide Sutton (*Mycol. Pap.* **141**, 1977).

Excipulites Göpp. (1836), Fossil Fungi. 4 (Cretaceous, Tertiary), Europe.

excipulum (of ascomata), tissue or tissues containing the hymenium in an apothecium, or forming the walls of a perithecium; cf. **ectal -**, **medullary -**. **- proprium**, non-lichenized excipular tissue forming the margins of an apothecium of a lichenized fungus; **- thallinum**, lichenized excipular tissue of a lecanorine apothecium, external to an excipulum proprium (which may be much reduced), usually with a structure like that of the vegetative lichen thallus. See *Lit.* under apothecium, tissue types.

Exesisporites Elsik (1969), Fossil Fungi. 1 (Pleistocene), USA.

exhabitant, see symbiosis.

Exidia Fr. (1822), Auriculariaceae. *c.* 25, widespread. See Donk (*Persoonia* **4**: 166, 1966; syns), Lowy (*Fl. Neotrop.* **6**, 1970), Lowy (*Nova Hedwigia* **19**: 407, 1971; key), Weiss & Oberwinkler (*MR* **105**: 403, 2001; phylogeny), Selosse *et al.* (*New Phytologist* **155**: 183, 2002; ectomycorrhiza), Roberts (*Mycotaxon* **96**: 83, 2006; Jamaica).

Exidiaceae R.T. Moore (1978) = Auriculariaceae.

Lit.: Wells *et al.* (*Frontiers in Basidiomycote Mycology*: 237, 2004).

Exidiopsis (Bref.) Möller (1895), Auriculariaceae. *c.* 30, widespread. See Wells (*Mycol.* **53**: 317, 1962;

key), Wells & Raitviir (*Mycol.* **69**: 987, 1977; former USSR), Roberts (*MR* **97**: 467, 1993; Brit. spp.), Roberts (*Mycotaxon* **87**: 25, 2003; Venezuela n.sp.), Weiss *et al.* (*MR* **108**: 1003, 2004; phylogeny).

exigynous, having the antheridial stalk arising directly from the oogonial cell above the basal septum.

Exiliseptum R.C. Harris (1986), Trypetheliaceae (L). 1, Brazil. See Harris (*Acta Amazon.* Supl. **14**: 55, 1984).

Exilispora Tehon & E.Y. Daniels (1927) = Leptosphaeria fide von Arx & Müller (*Stud. Mycol.* **9**, 1975).

exit tube, extension of the sporangium, produced prior to or during sporangial discharge, which enables sporangial contents to be released outside the host or substrate.

exo- (prefix), outside.

Exoascaceae G. Winter (1884) = Taphrinaceae.

Exoascus Fuckel (1860) = Taphrina fide Mix (*Kansas Univ. Sci. Bull.* **33**: 1, 1949).

Exobasidiaceae J. Schröt. (1888), Exobasidiales. 5 gen. (+ 1 syn.), 56 spp.

Lit.: Bauer *et al.* (*CJB* **75**: 1273, 1997), Begerow *et al.* (*CJB* **75**: 2045, 1998), Begerow *et al.* (*MR* **105**: 809, 2001), Begerow *et al.* (*Mycol. Progr.* **1**: 187, 2002), Crous *et al.* (*Australas. Pl. Path.* **32**: 313, 2003), Nagao *et al.* (*Mycoscience* **45**: 85, 2004).

exobasidial (1) having the basidia uncovered; (2) separated by a wall from the basidium; (3) (of a conidiophore in a lichenized pycnidium sporophore; obsol.), having no secondary branch (Steiner); cf. endobasidial.

Exobasidiales Henn. (1898). Exobasidiomycetes. 4 fam., 17 gen., 83 spp. Gall-forming plant parasites, esp. of *Ericaceae* and *Commelinaceae*. Fams:

(1) **Brachybasidiaceae**
(2) **Cryptobasidiaceae**
(3) **Exobasidiaceae**
(4) **Graphiolaceae**

Lit.: Donk (1951-63, VI: 1966: 280), Blanz (*Zeit. f. Myk.* **44**: 91, 1978; posn), Gäumann (1964: 358), Cunningham *et al.* (*Mycol.* **68**: 642, 1976; key).

Exobasidiellum Donk (1931) = Helicogloea fide Kirk (*in litt.*).

Exobasidiomycetes Begerow, Stoll & R. Bauer (2007), Ustilaginomycotina. 6 ord., 16 fam., 53 gen., 597 spp. Ords:

(1) **Doassansiales**
(2) **Entylomatales**
(3) **Exobasidiales**
(4) **Georgefischeriales**
(5) **Microstromatales**
(6) **Tilletiales**

For *Lit.* see ord. and fam.

Exobasidiomycetidae Jülich (1981), see *Exobasidiomycetes.*

For *Lit.* see fam.

Exobasidiopsis Karak. (1922) = Kabatiella fide Hawksworth *et al.* (*Dictionary of the Fungi* edn 8, 1995).

Exobasidium Woronin (1867), Exobasidiaceae. *c.* 50 (on *Ericaceae*), widespread (esp. north temperate). *E. japonicum*, *Azalea* gall. See Savile (*CJB* **37**: 641, 1959; N. Am.), McNabb (*Trans. roy. Soc. N.Z.* Bot. **1**: 259, 1962; NZ), Nannfeldt (*Symb. bot. upsal.* **23** no. 2: 1, 1981; Eur.), Begerow *et al.* (*Mycol. Progr.* **1**: 187, 2002; phylog.), Boekhout *et al.* (*Int. J. Syst. Evol. Microbiol.* **53**: 1655, 2003; probable paraphyly), Nagao *et al.* (*Mycoscience* **44**: 44, 2003; Japan), Li & Guo (*Mycotaxon* **96**: 323, 2006; China).

Exochalara W. Gams & Hol.-Jech. (1976), anamorphic *Pezizomycotina*, Hso.0eH.32. 2, widespread. See Mercado Sierra *et al.* (*Hifomicetes Demaciáceos de Cuba: Enteroblásticos* Monografie Museo Regionale di Scienze Naturali, Torino **23**, 1997), Ring & Gams (*Mycotaxon* **76**: 451, 2000).

exochthonous (of soil organisms), invaders ill-adapted to live in soil (Park, 1957); cf autochthonous.

exogenization, an hypothetical process whereby endogenously formed spores become exogenously formed: a mechanism proposed to support the evolution of basidiomycetes from the ancestral ascomycetes (Clemençon, *Persoonia* **9**: 363, 1977).

exogenous, undergoing development outside.

Exogone Henn. (1908) = Agyrium fide Höhnel (*Sber. Akad. Wiss. Wien* Math.-naturw. Kl., Abt. 1 **120**: 8, 1911).

exolete (of perithecia, pycnidia, etc.), long over-mature; empty.

Exomassarinula Teng (1940) ? = Melchioria fide Petrak (*Sydowia* **13**: 23, 1959), Müller & von Arx (*Beitr. Kryptfl. Schweiz* **11** no. 2, 1962).

exomycology, mycology of outer space.

exo-operculation (of sporangial dehiscence in chytrids), the operculum is hinged to the rim of the pore; 'true operculation'; cf. endo-operculation.

exoperidium, the outer layer of the peridium.

Exophiala J.W. Carmich. (1966), anamorphic *Capronia*, Hso.0-1eH-P.15. 28, widespread. See McGinnis & Ajello (*Mycol.* **66**: 518, 1974; 3 on fish), de Hoog (*Stud. Mycol.* **15**: 100, 1977), de Bievre *et al.* (*Bull. Soc. Fr. Myc. Med.* **16**: 345, 1987; physiological basis for taxonomy), de Hoog *in* Sugiyama (Ed.) (*Pleomorphic Fungi: The Diversity and its Taxonomic Implications*: 221, 1987; developmental cycle), de Hoog *et al.* (*Proc. 10th ISHAM Congr., Barcelona*: 168, 1988; taxonomy *E. jeanselmei* complex), Pedersen & Langvad (*MR* **92**: 153, 1989; spp. on fish), Kawasaki *et al.* (*Mycopathologia* **110**: 107, 1990; mitochondrial DNA of spp.), Matsumoto *et al.* (*J. Med. Vet. Mycol.* **28**: 437, 1990; synanamorph for *E. dermatitidis* as *Wangiella*), de Hoog *et al.* (*Antonie van Leeuwenhoek* **65**: 143, 1994; pleoanamorphic cycle), Rogers *et al.* (*Stud. Mycol.* **43**: 122, 1999; phylogeny), Untereiner & Naveau (*Mycol.* **91**: 67, 1999), Kano *et al.* (*Veter. Pathol.* **76**: 201, 2000; from dog), Kawasaki *et al.* (*Mycopathologia* **146**: 75, 1999; mitochondrial DNA), Untereiner (*Stud. Mycol.* **45**: 141, 2000; teleomorph), Wang *et al.* (*J. Clin. Microbiol.* **39**: 4462, 2001; phylogeny), García-Martos *et al.* (*Revta Iberoamer. Micol.* **19**: 72, 2002; human infections), Vitale & Hoog (*Medical Mycology* **40**: 545, 2002; phylogeny, antifungal susceptability), Yurlova & Hoog (*Mycoses* **45**: 443, 2002; exopolysaccharides), Chee & Kim (*Mycobiology* **30**: 1, 2003; molecular markers), Hoog *et al.* (*J. Clin. Microbiol.* **41**: 4767, 2003; polymorphism), Matos *et al.* (*Antonie van Leeuwenhoek* **83**: 293, 2003; molecular diversity), Kawasaki *et al.* (*Jap. J. Med. Mycol.* **46**: 261, 2005; genetic variation, Japan), Geiser *et al.* (*Mycol.* **98**: 1053, 2006; phylogeny), Sterflinger (*The Yeast Handbook* **[1]**: 501, 2006; ecology).

Exophoma Weedon (1926), anamorphic *Pezizomycotina*. 1, N. America. See Sutton (*Mycol. Pap.* **141**, 1977).

exopropagule, a propagule formed outside the body (medical mycology).

Exormatostoma Gray (1821) nom. dub., ? Diaporthales.
exospore, see Spore wall.
Exosporella Höhn. (1912), anamorphic *Pezizomycotina*, Hsp.0fH.?. 1, Java.
Exosporiella P. Karst. (1892), anamorphic *Pezizomycotina*, Hsp.≡ eP.19. 1 (on *Corticium*), Europe.
Exosporina G. Arnaud (1921) ≡ Arnaudina.
Exosporina Oudem. (1904) nom. dub., anamorphic *Dothideomycetes*, Hsp.0eP.?. 1, widespread (temperate). See Seifert *et al.* (*Sydowia* **50**: 133, 1998; revision).
Exosporinella Bender (1932) ≡ Arnaudina.
Exosporium Link (1809), anamorphic *Pezizomycotina*, Hso/Hsp.≡ eP.26. 19, widespread. See Ellis (*Dematiaceous Hyphomycetes*, 1971), Ellis (*More Dematiaceous Hyphomycetes*, 1976).
exosporium, see Spore wall.
Exoteliospora R. Bauer, Oberw. & Vánky (1999), Melanotaeniaceae. 1 (on *Osmunda* (*Pteridophyta*)), N. America. See Bauer *et al.* (*Mycol.* **91**: 669, 1999).
exotic (1) (adj.), of another country; not indigenous; (2) (n), an - organism.
Exotrichum Syd. & P. Syd. (1914) ? = Crocicreas fide Höhnel (*Mykol. Unters.* **1**: 359, 1923).
expallant (of a pileus), becoming pale on drying.
expansin, see patulin.
expersate (of oospores in *Saprolegniaceae*), having one large refractive body surrounded by a homogeneous cytoplasm (Howard, *Mycol.* **63**: 684, 1971).
explanate, spread out.
explosive (of asci), see ascus.
Exserohilum K.J. Leonard & Suggs (1974), anamorphic *Setosphaeria*, Hso.≡ eP.26. 34, widespread. See Sivanesan (*Mycol. Pap.* **158**: 1, 1987; keys), Alcorn (*Ann. Rev. Phytopath.* **26**: 37, 1988; gen. taxonomy), Muchovej *et al.* (*Fitopatol. Brasil* **13**: 211, 1988; keys), Khazanov (*Opredelitel' Gribov-Vozbul. 'Gel'mintosporiozov' Rast. iz Rodov Bipolaris, Drechslera i Exserohilum*, 1992), Borchardt *et al.* (*Eur. J. Pl. Path.* **104**: 611, 1998; population analysis), Goh *et al.* (*Fungal Diversity* **1**: 85, 1998; ribosomal RNA analysis), Sun *et al.* (*Mycosystema* **23**: 480, 2004; phylogeny), Sun *et al.* (*Mycotaxon* **92**: 173, 2005; China).
exserted, sticking out; protruding (e.g. a mature ascus of *Ascobolus*).
Exserticlava S. Hughes (1978), anamorphic *Pezizomycotina*, Hso.≡ eP.6. 3, widespread (esp. tropical). See Hughes (*N.Z. Jl Bot.* **16**: 332, 1978), Tsui *et al.* (*Fungal Diversity* **7**: 135, 2001; revision).
exsiccatus (adj.; Latin), dried or dry, e.g. fungus (-i) exsiccatus (-i), planta (-ae) exsiccata (-ae), specimen (specimina) exsiccatum (-a); **exsiccatum** (n.; pl. **-a**), a dried specimen; **exsiccata** (n.; pl. **-ae**; preferred abbreviation, Exs.), a set of dried specimens (fide Jackson, 1928, and Stearn, 1968). Exsiccatae distributed to major reference collections are generally cited in systematic works; those with printed descriptions issued before 1 January 1953 can be the places of valid publication of new taxa (Art. 30.3; see Nomenclature), and separately published labels are acceptable after that date (note that label data were not always cited in some major nomenclators as sources of names, e.g. by Saccardo, 1882-1972).

Lit.: **General**, Sayre (*Mem. N.Y. bot. Gdn* **19**: 1, 1969, general cryptogamic exsiccatae, lichens; **19**: 277, 1975; collectors), Stafleu & Cowen (*TL-2*, **1-7**, 1976-88). **Fungi**, Pfister (*Mycotaxon* **23**: 1, 1985; compr. catalogue), Stevenson (*Beih. Nova Hedw.* **36**, 1971; N. Am.). **Lichens**, Hawksworth (*in* Seaward, 1977: 498; issued 1969-76), Hawksworth & Ahti (*Lichenologist* **22**: 1, 1990; issued 1976-89), Hawksworth & Seaward (*Lichenology in the British Isles 1568-1975*, 1977; UK), Hertel (*Mitt. Bot. StSamml., München* **18**: 297, 1982; labels), Lynge (*Nyt. Mag. Naturvid.* **55-60**, 1915-22; **79**: 233, 1939), Sayre (1969).
extenditunicate, see ascus.
extramatrical (1) living on or near the surface of the matrix or substratum; (2) VAM structures (mycelium, spores) developing outside roots of a phytobiont.
Extrawettsteinina M.E. Barr (1972), Pleosporaceae. 4, Europe; N. America. See Barr (*Prodr. Cl. Loculoasc.*, 1987).
Extrolite, An outwardly directed chemical compound produced during differentiation of a living organism; usually excreted but sometimes accumulated in the cell wall or membrane; being outwardly directed, they are always involved in interactions between the source organisms and adjacent organisms or the abiotic environment (Samson & Frisvad, *Stud. Mycol.* **49**: 1, 2004). The term includes 'secondary metabolites'. See also Chemotaxonomy, secondary metabolites.
extrusome, membrane-bound structure derived from vesicle of the Golgi system and anchored to the cell membrane by proteinaceous particles; contents extruded in respose to stimuli.
Extrusothecium Matsush. (1996) ? = Leptosphaerulina fide Matsushima (*Matsush. Mycol. Mem.* **9**: 10, 1996), Matsushima (*Matsush. Mycol. Mem.* **10**, 2001).
fabiform, see reniform.
Fabisporus Zmitrovich (2001), Polyporaceae. 5. See Zmitrovich (*Mycena* **1**: 93, 2001).
Fabospora Kudrjanzev (1960) = Kluyveromyces fide Batra *in* Subramanian (Ed.) (*Taxonomy of fungi* **1**: 187, 1978), Lachance (*The Yeasts. A Taxonomic Study*: 227, 1998).
Fabosporaceae E.K. Novák & Zsolt (1961) = Saccharomycopsidaceae.
Fabraea Sacc. (1881) = Leptotrochila fide Schüepp (*Phytopath. Z.* **36**, 1959).
Fabrella Kirschst. (1941), Hemiphacidiaceae. 1 (on *Tsuga*), N. America. See Korf (*Mycol.* **54**: 12, 1962), Gernandt *et al.* (*Mycol.* **93**: 915, 2001; phylogeny), Wang *et al.* (*Mol. Phylogen. Evol.* **41**: 295, 2006; phylogeny), Wang *et al.* (*Mycol.* **98**: 1065, 2006; phylogeny).
Fabreola Kuntze (1891) ≡ Urosporella.
facial eczema, see sporidesmin.
facultative (1) sometimes; not necessarily; not obligate (q.v.); (2) (of a parasite), having the power of living as a saprobe; able to be cultured on laboratory media; (3) **- synonym**, see synonym.
Faerberia Pouzar (1981), Polyporaceae. 1, Europe; Mexico. See Pouzar (*Česká Mykol.* **35**: 187, 1981), Estrada-Torres & Cifuentes (*Mycotaxon* **67**: 433, 1998; Mexico).
Faerberiaceae Pouzar (1983) = Polyporaceae. The *Favolaschiaceae* are listed as nom. rej. against *Tricholomataceae*.
Fairmania Sacc. (1906) = Microascus Zukal fide Malloch & Cain (*CJB* **49**: 859, 1971).

Fairmaniella Petr. & Syd. (1927), anamorphic *Pezizomycotina*, Cac.0eP.15. 1, widespread. See Sutton (*Mycol. Pap.* **123**, 1971), swart (*TBMS* **90**: 279, 1988; Australia), Crous *et al.* (*S. African Forestry Jl* **149**: 9, 1989; S Africa), Wingfield *et al.* (*S. African Forestry Jl* **173**: 53, 1995; Chile).

fairy butter, basidiomata of *Tremella albida.*

Fairy rings. Fungus rings, which are generally of basidiomycetes (some 60 recorded species), are very frequent in grass and grassland, and not uncommon in woods. There are three chief types: (1) those in which the development of the sporocarps has no effect on the vegetation, e.g. *Chlorophyllum molybdites*, *Lepista sordida*, myxomycete rings; (2) those in which there is increased growth of the vegetation, e.g. *Calvatia cyathiformis*, *Disciseda subterranea* (*Catastroma subterraneum*), the basidiomata of which are at the outer edge of the ring, *Lycoperdon gemmatum*, *Lepista personata*; (3) those in which the vegetation is damaged, sometimes so badly as to have an effect on its value, e.g. *Agaricus praerimosus* (*A. tabularis*), *Leucopaxillus giganteus*, *Marasmius oreades* (in British Isles), *Calocybe gambosa*. Rings of the third type are frequently made up of outer and inner rings in which the growth of the vegetation is strong with a ring of dead or badly damaged vegetation between. Rings of fungal origin are also sometimes seen as bare circles under trees with the black truffle, *Tuber melanosporum* as a mycorrhizal symbiont. These may occur because of chemicals produced by the fungus which inhibit growth of vegetation.

Rings start when mycelium grows outwards evenly from a central point of origin; the growth being at all times on the outer edge because of the band of decaying mycelium and used-up soil within the ring of active hyphae. The mean growth of a ring of *A. praerimosus* is 12 cm in radius every year (0-30 cm in any one year); that of one of *Calvatia cyathiformis* about 24 cm. From this, the ages of rings of these two fungi in Colorado, 60 and more than 200 m diam., were thought to be 250 and 420 years; parts of *A. praerimosus* rings were possibly 600 years old. A substantial but incomplete fairy ring was observed at the Roman fort of Richborough, near Sandwich, England, with its centre where the defence ditch is now located.

In European cultures, fungus rings have been attributed to supernatural causes (Hall, *Elves in Anglo-Saxon England: matters of belief, health, gender and identity*, 2007), and are often associated with various pagan beliefs (see: Ethnomycology).

Lit.: Shantz & Piemeisel (*J. Agric. Res.* **11**: 191, 1917), Bayliss Elliott (*Ann. appl. Biol.* **13**: 277, 1926), Parker-Rhodes (*TBMS* **38**: 59, 1955), Burnett & Evans (*Nature* **210**: 1368, 1966), Stevenson & Thompson (*J. theor. Biol.* **58**: 143, 1976; kinetics), Gregory (*Bull. BMS* **16**: 161, 1982; 'free' and 'tethered' rings).

fairy-ring champignon, the edible *Marasmius oreades*.

falcarindiol, an antifungal compound produced by carrot roots (Garrod & Lewis, *TBMS* **72**: 515, 1979).

falcate (**falciform**), curved like the blade of a scythe or sickle (Fig. 23.10).

Falcatispora K.L. Pang & E.B.G. Jones (2003) = Ascosalsum fide Pang *et al.* (*Nova Hedwigia* **77**: 13, 2003), Pang & Jones (*Nova Hedwigia* **78**: 269, 2004).

Falciascina Beneš (1961), Fossil Fungi. 1 (Carboniferous), former Czechoslovakia.

Falciformispora K.D. Hyde (1992), Pleosporaceae. 1 (marine), Mexico. See Hyde (*MR* **96**: 26, 1992).

Falcipatella Gucevič (1952) = Heteropatella fide Sutton (*Mycol. Pap.* **141**, 1977).

Falcipatellina Gucevič (1952) = Heteropatella fide Sutton (*Mycol. Pap.* **141**, 1977).

falciphore, see falx.

Falcispora Bubák & Serebrian. (1912) = Selenophoma fide Petrak (*Sydowia* **5**: 328, 1951).

Falcocladium S.F. Silveira, Alfenas, Crous & M.J. Wingf. (1994), anamorphic *Nectriaceae*, Hsy/Hsp.0-1eH.15. 2, Brazil. See Crous *et al.* (*Mycotaxon* **50**: 447, 1994), White & Reddy (*Mycol.* **90**: 226, 1998), Somrithpol *et al.* (*Sydowia* **59**: 148, 2007; review).

false membrane (of a smut), a tissue of sterile fungal cells limiting the sorus, as in *Sphacelotheca*.

false morel, see lorchel.

false truffle, see truffle.

falx, a 'fertile hypha' or conidiophore of *Zygosporium*, having the form of a bill-hook. Falces may be sessile or on special hyphae or **falciphores** (Mason, 1941).

Fanniomyces T. Majewski (1972), Laboulbeniaceae. 2, Europe; N. America. See Balazuc (*Revue Mycol. Paris* **43**: 393, 1979), Tavares (*Mycol. Mem.* **9**: 627 pp., 1985), Santamaria *et al.* (*Treb. Inst. Bot. Barcelona* **14**: 1, 1991; Europe).

farctate (of a stipe), having the centre softer than the outer layer; stuffed.

farinaceous (**farinose**), like meal in form or smell.

Farinaria Sowerby (1803) = Ustilago fide Fries (*Syst. mycol.* **3**, 1829).

Farinodiscus Svrček (1987) = Proliferodiscus fide Baral (*SA* **13**: 113, 1994).

Farlow (William Gilson; 1844-1919; USA). Graduate student, Department of Medicine (1866-1870) then assistant in cryptogamic botany under A. Gray (1870-1872), Harvard University; student of de Bary (q.v.), Strassburg, Germany (*c.* 1872-1874); Assistant Professor (1874-1879) then Professor (1879 onwards), Harvard University. A pioneer plant pathologist in North America, he and Thaxter (q.v.) were responsible for training several people later important in mycology and plant pathology. His writings were on plant diseases, fungi (including lichen-forming species) and bibliography. The fungal reference collection he established in Harvard (**FH**) and its library, both named after him, are among the most extensive in mycology. *Publs. Bibliographical Index of North American Fungi* (1905). *Biogs, obits etc.* Clinton (*Phytopathology* **10**: 1, 1920) [portrait]; Grumman (1974: 188); Pfister (*Bulletin of the Boston Mycological Club* **4**: 6, 1975 [Farlow, *Icones Farlowiana*, 1927]); Stafleu & Cowan (*TL-2* **1**: 813, 1976).

Farlowia Sacc. (1883) [non *Farlowia* J. Agardh 1876, *Algae*] ≡ Farlowiella.

Farlowiella Sacc. (1891), Pleosporales. Anamorph *Acrogenospora*. 2, Europe; Tristan da Cunha. See Zogg (*Beitr. Kryptfl. Schweiz* **11** no. 1, 1962), Goh *et al.* (*MR* **102**: 1309, 1998; anamorph), Lumbsch *et al.* (*Mol. Phylogen. Evol.* **34**: 512, 2005; phylogeny), Zhu *et al.* (*Mycotaxon* **92**: 383, 2005; China), Prado *et al.* (*MR* **110**: 511, 2006; phylogeny), Schoch *et al.* (*Mycol.* **98**: 1041, 2006; phylogeny).

Farnoldia Hertel (1983), Lecanorales (L). 6, widespread. See Pietschmann (*Nova Hedwigia* **51**: 521, 1990; posn), Buschbom & Mueller (*Mol. Phylogen.*

Evol. 32: 66, 2004; phylogeny).

Farringia Stafleu (1979) nom. nud., Fungi.

Farriolla Norman (1885), Pezizomycotina. 1, Norway. See Tibell (*Beih. Nova Hedwigia* **79**: 597, 1984).

Farriollomyces Cif. & Tomas. (1953) ≡ Farriolla.

Farrowia D. Hawksw. (1975) = Chaetomium fide Piepenbring (*Nova Hedwigia* **70**: 289, 2000; phylogeny).

Farysia Racib. (1909), Anthracoideaceae. *c.* 17 (on *Carex*, *Uncinia* (*Cyperaceae*)), widespread. Spore mass interspersed with elater-like fascicles of hyphae. See Vánky & McKenzie (*Fungal Diversity Res. Ser.* **8**: 259 pp., 2002; New Zealand).

Farysiaceae Vánky (2001) = Anthracoideaceae.

Farysporium Vánky (1999), ? Anthracoideaceae. 1 (on *Cyperaceae*), Australasia. See Vánky & McKenzie (*Fungal Diversity Res. Ser.* **8**: 259 pp., 2002; New Zealand).

fasciate (**fasciated**), massed or joined side by side.

Fasciatispora K.D. Hyde (1991), Xylariales. 1 (on *Nypa*), pantropical. See Hyde (*Nova Hedwigia* **61**: 249, 1995), Hyde & Wong (*Fungal Diversity* **2**: 129, 1999; ultrastr.), Kang *et al.* (*Fungal Diversity* **2**: 135, 1999; possibly an outlying member of the *Xylariaceae*), Lu & Hyde (*Mycotaxon* **71**: 393, 1999; USA).

fascicle (1) (esp. of hyphae), a little group or bundle; (2) (of books or exsiccatae), one part, or collection of separate leaves, of a work issued in parts.

fasciculate, having growth in fascicles; **- basidium**, see basidium.

Fassia Dennis (1964), ? Diatrypaceae. 1, Congo. See Cannon (*SA* **5**: 130, 1986), Læssøe (*SA* **13**: 43, 1994), Acero *et al.* (*Mycol.* **96**: 249, 2004).

fastigiate, having parallel, massed, upright branches.

fastigiate cortex (of lichens), made up of parallel hyphae at right angles to the axis of the thallus; cf. fibrous cortex.

Fastigiella Benedix (1969) ≡ Neogyromitra.

fatiscent, cracked or falling apart.

fatty acids, a class of organic compounds, and generally the hydrophobic component of many microbial cellular membrance lipids. Quantitative and qualitative differences in the fatty acid content of microbial cells can be used in classification and identification; widely used in bacteria and also of value for some yeasts (Botha & Kock, *Int. J. Food Microbiol.* **19**: 39, 1993), but yet to be fully explored with filamentous fungi.

Faurelina Locq.-Lin. (1975), ? Chadefaudiellaceae. Anamorph *Arthrographis*-like. 1 (coprophilous), widespread (arid regions). See Parguey-Leduc & Locquin-Linard (*Revue Mycol.* Paris **40**: 161, 1976; ontogeny), Valldosera *et al.* (*Mycotaxon* **30**: 5, 1987), von Arx *et al.* (*Beih. Nova Hedwigia* **94**: 104, 1988).

Fauxtylostoma McGinty (1923) nom. inval., Agaricomycetes.

Favaria Raf. (1815) nom. dub., Fungi. No spp. included.

faveolate (**favose**), honeycombed; alveolate.

favic chandeliers, dichotomously branched, swollen, hyphal tips, growing submerged from the edge of the colony of *Trichophyton schoenleinii*.

Favillea Fr. (1848), Sclerodermataceae. 1, Australia. See Fries (*Fung. Natal.*: 32, 1848).

favoid, like a honeycomb.

Favolaschia (Pat.) Pat. (1892), Mycenaceae. *c.* 50, widespread (esp. tropical). See Singer (*Beih. Nova Hedwigia* **50**, 1974), Parmasto (*Kew Bull.* **54**: 783, 1999).

Favolaschiaceae Singer (1969) = Mycenaceae.

Favolus Fr. (1828) = Polyporus P. Micheli ex Adans. fide Donk (*Persoonia* **1**: 173, 1960).

Favolus P. Beauv. (1805), Polyporaceae. 2, widespread (esp. tropical). See Donk (*Persoonia* **1**: 173, 1960; nomencl.).

Favomicrosporon Benedek (1967) nom. dub., Pezizomycotina. See Ajello (*Sabouraudia* **6**: 153, 1968).

Favostroma B. Sutton & E.M. Davison (1983), anamorphic *Pezizomycotina*, Cpd.0eH.1. 1, Australia. See Sutton & Davison (*TBMS* **81**: 291, 1983).

Favotrichophyton (Castell. & Chalm.) Neveu-Lem. (1921) = Trichophyton fide Hawksworth *et al.* (*Dictionary of the Fungi* edn 8, 1995).

Favraea, see *Fabraea*.

favus, a skin disease in humans (*Trichophyton schoenleinii*).

Fayodia Kühner (1930), Tricholomataceae. *c.* 10, widespread (north temperate). See Bigelow (*Mycotaxon* **9**: 38, 1979), Antonín (*Persoonia* **18**: 341, 2004; type studies European spp.), Antonín (*Mykol. Listy* **90-91**: 20, 2004; key European spp.), Antonín & Noordeloos (*A monograph of the genera Hemimycena,* , 2004), Johnston et al. (N.Z. J. Bot. 44: 65, 2006; New Zealand (*N.Z. J. Bot.* **44**: 65, 2006; New Zealand).

Fayodiaceae Jülich (1982) nom. rej. = Tricholomataceae.

Fechtneria Velen. (1939) = Hymenogaster fide Svrček *in* Pilát (Ed.) (*Fl. ČSR* **B**, **1**: 143, 1958).

federation, see Phytosociology.

Fée (Antoine Laurent Apollinaire; 1789-1874; France). Graduate then Professor of Botany (1832), Strasburg; died in Paris. Much early work on tropical fungi including lichen-forming species; also worked with plants. Main collection in Rio de Janeiro; other material in Paris (**P**), Copenhagen (**C**), Geneva **G**, and London (**BM**). *Publs. Nova Acta Academiae Leopoldino Carolinae Germaniae Naturae Curiosorum* **18**, Suppl. 1 (1841); *Essai sur les Cryptogames des Écorces* (1824-1825) [Suppl. 1837]; *Méthode Lichénographique et Genera* (1825). *Biogs, obits etc.* Anon. (*Journal of Botany* London **12**: 223, 1874); Grummann (1974: 277); Stafleu & Cowan (*TL-2* **1**: 818, 1976).

Feigeana Mies, Lumbsch & Tehler (1995), Roccellaceae (L). 1, Socotra. See Mies *et al.* (*Mycotaxon* **54**: 156, 1995), Grube (*Bryologist* **101**: 377, 1998; phylogeny).

Felisbertia Viégas (1944), Helotiales. 1, S. America.

Felixites Elsik ex Janson. & Hills (1990), Fossil Fungi. 2 (Carboniferous), Europe; N. America.

fellent, bitter like gall.

Fellhanera Vězda (1986), Pilocarpaceae (L). 60, widespread (tropical). See Awasthi & Mathur (*Proc. Indian Acad. Sci.* Pl. Sci. **97**: 481, 1987; key 4 spp. India), Lücking *et al.* (*Bot. Acta* **107**: 393, 1994), Sérusiaux (*Lichenologist* **28**: 197, 1996; key foliicolous spp.), Lücking (*Trop. Bryol.* **13**: 141, 1997), Ferraro & Lücking (*Mycotaxon* **73**: 163, 1999), Sparrius & Aptroot (*Lichenologist* **32**: 515, 2000), Lücking & Santesson (*Lichenologist* **33**: 111, 2001; key sorediate spp.), Sparrius (*Lichenologist* **34**: 86, 2002), Spier *et al.* (*Lichenologist* **34**: 447, 2002; chemistry), Andersen & Ekman (*MR* **109**: 21, 2005; phylogeny).

Fellhaneropsis Sérus. & Coppins (1996), Pilocarpaceae (L). 4, Europe. See Sérusiaux (*Lichenologist* **28**: 198, 1996), Andersen & Ekman (*MR* **109**: 21, 2005; phylogeny).

Fellneria Fuckel (1867) = Colletotrichum fide Duke (*TBMS* **13**: 156, 1928).

Fellomyces Y. Yamada & I. Banno (1984), anamorphic *Cuniculitremaceae*. *c.* 12, widespread. See Yamada *et al.* (*J. gen. appl. Microbiol.* Tokyo **32**: 157, 1986; enzyme systems), Yamada *et al.* (*Agric. Biol. Chem.* **53**: 2993, 1989; phylogeny), Guého *et al.* (*Int. J. Syst. Bacteriol.* **40**: 60, 1990; partial rRNA sequencies), Kurtzman (*Int. J. Syst. Bacteriol.* **40**: 56, 1990; DNA relatedness), Suh *et al.* (*Microbiol. Culture Coll.* **12**: 79, 1996; mol. phylogeny), Prillinger *et al.* (*Syst. Appl. Microbiol.* **20**: 572, 1997; taxonomy), Fell *et al.* (*Int. J. Syst. Evol. Microbiol.* **50**: 1351, 2000; mol. phylogeny), Lopandic *et al.* (*Microbiological Research* **160**: 160, 2005; Mexico).

felt (of citrus), superficial saprobic fungi, such as *Septobasidium pseudopedicellatum*; **leaf -**, *Anthina citri* and other fungi; **root -**, *Helicobasidium mompa* (Japan).

Feltgeniomyces Dieder. (1990), anamorphic *Pezizomycotina*, Hso.1eH.15. 1 (on lichens), widespread. See Diederichs (*Mycotaxon* **37**: 304, 1990), Etayo & Breuss (*Öst. Z. Pilzk.* **7**: 203, 1998; Australia), Calatayud & Etayo (*CJB* **79**: 223, 2001; Spain).

Femsjonia Fr. (1849) = Ditiola Fr. fide Reid (*TBMS* **62**: 474, 1974), Weiss *et al.* (*Frontiers in Basidiomycote Mycology*: 7, 2004).

Fenestella Tul. & C. Tul. (1863), Fenestellaceae. Anamorph *Pleurostromella*. 2 or 2 (on twigs), widespread (temperate). See Barr (*Rept. Kevo Subarct. Res. Stn* **11**: 12, 1974), Huhndorf & Glawe (*Mycol.* **82**: 541, 1990; anamorph).

Fenestellaceae M.E. Barr (1979), Pleosporales. 2 gen. (+ 3 syn.), 12 spp.

Lit.: Barr (*Prodr. Cl. Loculoasc.*, 1987), Vasil'eva (*Mikol. Fitopatol.* **21**: 403, 1987), Barr (*N. Amer. Fl.* ser. 2 **13**: 129 pp., 1990), Huhndorf & Glawe (*Mycol.* **82**: 541, 1990), Yuan & Barr (*Sydowia* **46**: 338, 1994), Lumbsch & Lindemuth (*MR* **105**: 901, 2001).

fenestrate (1) having windows or openings; (2) (of spores), muriform.

Fenestroconidia Calat. & Etayo (1999), anamorphic *Pezizomycotina*. 1 (lichenicolous), Spain. See Calatayud & Etayo (*Lichenologist* **31**: 588, 1999).

Fennellia B.J. Wiley & E.G. Simmons (1973), Trichocomaceae. Anamorph *Aspergillus*. 3, widespread. See Kuraishi *et al.* (*NATO ASI Series A: Life Sciences* **185**: 407, 1990; ubiquinones), Yaguchi *et al.* (*Mycoscience* **35**: 175, 1994), Tamura *et al.* (*Integration of Modern Taxonomic Methods for Penicillium and Aspergillus Classification*: 357, 2000; phylogeny), Peterson (*Mycol.* **100**: 205, 2008; 4-locus phylogeny).

Fennellomyces Benny & R.K. Benj. (1975), Syncephalastraceae. 4, USA; Pakistan; India. See Benny & Benjamin (*Aliso* **8**, 1975), Mirza *et al.* (*Mucorales of Pakistan*, 1979), Misra *et al.* (*Mycotaxon* **10**: 251, 1979), Benny & Schipper (*Mycol.* **84**: 52, 1992), Arambarri & Cabello (*Mycotaxon* **57**: 145, 1996; *Circinella naumovii* transferred to *Fennellomyces*), Voigt & Wöstemeyer (*Gene* **270**: 113, 2001; phylogeny), Kwasna *et al.* (*MR* **110**: 501, 2006; phylogeny).

Feracia Rolland (1905) nom. dub., Pezizomycotina. 1, Europe. Type material is lost and affinities are unclear. See Rossman *et al.* (*Stud. Mycol.* **42**: 248 pp., 1999).

fermentation, chemical changes in organic substrates caused by enzymes, generally those of living micro-organisms. See Fermented food and drinks.

Fermented food and drinks. Fungi (esp. yeasts) and other organisms, particularly bacteria, have been used in the preparation of fermented foods for many centuries, long before their function was recognized or understood. Every country in east Asia has indigenous fermented food, prepared on a scale that ranges from single households to large commercial operations. Often referred to as 'oriental fermentation', the process makes the starting material more digestible or palatable in terms of texture, flavour, aroma, pH and appearance, and furnishes essential nutrients in the form of vitamins, proteins, amino acids and calories.

Fermented foods are the result of the action of specific microbial enzymes. Any enzyme used in the food industry can be manufactured if pure cultures of the appropriate micro-organism are available. Much of our understanding of the microbiology and biochemistry of oriental fermentation stems from work with pure cultures isolated from naturally fermented food products. As a result, most indigenous fermented foods can now be made from raw materials inoculated with cultures which have been obtained through a long selection process. One micro-organism or, more commonly, a combination of two or more micro-organisms work together to produce the final product.

Various substrates are used in fermentations: the most popular are soybeans and rice. Others include cassava, coconut, corn, fish, milk and peanuts. All these foods are staples in the diets of the people who normally consume the fermented foods. A koji or starter is prepared for every fermentation. Portions of the starter culture are then used to inoculate a larger batch of the same or a different substrate. Enzymes and by-products produced by the starter culture accelerate the rate of the next fermentation and provide a better growing environment for the succeeding organism(s). Cultures selected for starters are usually highly proteolytic, lipolytic, and/or amylolytic moulds and predominantly species of *Actinomucor*, *Aspergillus*, *Monascus*, *Mucor*, *Neurospora* or *Rhizopus*. The moulds break down macromolecules to produce amino acids, small fatty acids, vitamins and sugars, which ultimately add to the flavour and digestibility of the product. These smaller fermentation by-products are then used by bacteria to produce organic acids which lower pH and provide a favourable environment for yeasts. The yeasts chosen to complete a fermentation differ according to the desired product. The relatively high alcohol, salt or sugar content of fermented food helps to prolong its shelf life in places where refrigeration is not common.

Asian and Oriental foods and drinks based on fungal fermentations (frequently by *Aspergillus*) include: ang-kak (*Monascus purpureus*), hama-natto, lao-chao, oncom merah, ontojam, sufu, tape, tempeh, and the Japanese sake (rice wine), shoyn (soy sauce), miso (soy cheese), scocho (a distilled spirit) and mizaume (a sugar syrup from rice). Fermented milk drinks involving fungi include kefir (central and eastern Europe, Caucasus), kumiss (Russia), leben (Egypt), mazu (Armenia) and yoghurt (Bulgaria); bread (bakers yeast), *S. cerevisiae*. People who wish to avoid eating meat, often for ethical reasons, are increasing in numbers in developed countries, and the globalized economy has meant that these fermented foods have become popular with them as alternatives to meat (see also Mycoprotein).

Lit.: Batra & Millner (*Mycol.* **66**: 942, 1974; fungi and fermentation in Asian foods and beverages), Hessletine (*Mycologist* **5**: 162, 1991; *Zygomycetes* in food fermentation), Bennett & Kuch (Eds) (*Aspergillus: biology and industrial applications*, 1992), Hesseltine & Wang (*Indigenous fermented food of non-western origin* [*Mycologia Memoir* **11**], 1986), Nout (*in* Carroll & Wicklow, *The fungal community* edn 2, 1992: 817), Powell *et al.* (Eds) (*The genus Aspergillus from taxonomy and genetics to industrial application*, 1994), Samson (*in* Jones (Ed.), *Exploitation of micro-organisms*: 321, 1993), Steinkraus (Ed.) (*Industrialization of indigenous fermented foods*, 1989), Yokotsuka (*in* Arora *et al.* (Eds), *Proteinaceous fermented foods and condiments prepared with koji molds*. [*Handb. Appl. Mycol.*: 329], 1991). See also Brewing, Food and beverage mycology, Mycoprotein, Wine-making.

Fermentotrichon E.K. Novák & Zsolt (1961) = Geotrichum fide von Arx *et al.* (*Stud. Mycol.* **14**: 1, 1977).

Fernaldia Lynge (1937) [non *Fernaldia* Woodson 1932, *Apocynaceae*] = Thelignya fide Jørgensen & Hensen (*Taxon* **39**: 343, 1990).

Ferrarisia Sacc. (1919), Parmulariaceae. 5, widespread (tropical). See Hosagoudar & Goos (*Mycotaxon* **59**: 149, 1996), Hosagoudar & Abraham (*Indian Phytopath.* **51**: 389, 1999).

Ferraroa Lücking, Sérus. & Vězda (2005), Gomphillaceae (L). 1, Costa Rica. See Lücking *et al.* (*Lichenologist* **37**: 164, 2005), Lücking *et al.* (*Lichenologist* **38**: 131, 2006; checklist).

fertile hypha, see conidiophore.

fertilization, the fusion of sex nuclei; **- tube/hypha** hypha developing from an antheridial gametic cell; passing through the antheridium wall and bridging the gap between non-contiguous gametangia to penetrate the oogonium.

Fevansia Trappe & Castellano (2000), ? Rhizopogonaceae. 1, N. America. See Trappe & Castellano (*Mycotaxon* **75**: 155, 2000).

FH, Farlow Reference Library and Herbarium of Cryptogamic Botany, Harvard University (Cambridge, Mass, USA); founded 1919; funded by endowment and the university.

Fibriciellum J. Erikss. & Ryvarden (1975), Hydnodontaceae. 1, Europe. See Eriksson & Ryvarden (*Cortic. N. Europ.* **3**: 373, 1975).

Fibricium J. Erikss. (1958), Hymenochaetales. 5, widespread. Polyphyletic. See Hayashi (*Bull. Govt For. Exp. Stn* **260**, 1974; 4 spp.).

fibril (1) a very small fibre; (2) (in *Usnea*), short simple branches perpendicular to the main branches.

Fibrillanosema G.M. Johanna, S. Galbreath, J.E. Smith, R.S. Terry, J.J. Becnel & A.M. Dunn (2004), Microsporidia. 1. See Johanna *et al.* (*Intern. J. Parasitol.* **34**: 235, 2004).

fibrillar surface coat, fibrous component attached to the flagellar membrane and covering the entire surface of the flagellum.

Fibrillaria Sowerby (1803) nom. dub., Agaricomycetes. Based on sterile hymenomycete mycelium fide Donk (*Taxon* **11**: 84, 1962).

Fibrillithecis A. Frisch (2006), Thelotremataceae (L). 3, pantropical. See Frisch (*Biblthca Lichenol.* **92**: 135, 2006), Frisch *et al.* (*Biblthca Lichenol.* **92**: 517, 2006; phylogeny).

fibrillose, covered with silk-like fibres.

Fibrodontia Parmasto (1968), Hydnodontaceae. 5, widespread. See Langer (*Biblthca Mycol.* **154**, 1994).

Fibropilus (Noordel.) Largent (1994) = Entoloma fide Kuyper (*in litt.*).

Fibroporia Parmasto (1968) = Perenniporia fide Donk (*Verh. K. ned. Akad. Wet.* tweede sect. **62**: 1, 1974).

fibrous cortex (of lichens), made up of loosely woven distinct hyphae parallel with the long axis of the thallus; cf. fastigiate cortex.

Fibulobasidium Bandoni (1979), Sirobasidiaceae. 3, USA. See Bandoni (*CJB* **57**: 264, 1979), Sampaio (*Frontiers in Basidiomycote Mycology*: 49, 2004).

Fibulochlamys A.I. Romero & Cabral (1989), anamorphic *Agaricomycetes*, Hso.0eH.1. 1 (with clamp connexions), Argentina. See Romero & Cabral (*Mycotaxon* **34**: 430, 1989).

Fibulocoela Nag Raj (1978), anamorphic *Agaricomycetes*, St.0eH.1. 1 (with clamp connexions), India; Cuba. See Nag Raj (*CJB* **56**: 1491, 1978).

Fibulomyces Jülich (1972), Atheliaceae. Anamorph *Taeniospora*. 4, widespread (north temperate). See Hjortstam & Larsson (*Windahlia* **21**, 1994).

Fibuloporia Bondartsev & Singer (1944) = Trechispora fide Ryvarden (*Genera of polypores*, 1991).

Fibulorhizoctonia G.C. Adams & Kropp (1996), anamorphic *Athelia*. 3, widespread. See Adams & Kropp (*Mycol.* **88**: 464, 1996).

Fibulosebacea K. Wells & Raitv. (1987), Auriculariaceae. 1, Europe. See Wells & Raitviir (*TBMS* **89**: 344, 1987).

Fibulostilbum Seifert & Oberw. (1992), Chionosphaeraceae. 1, Brazil. See Seifert *et al.* (*Boln Soc. argent. Bot.* **28**: 215, 1992).

Fibulotaeniella Marvanová & Bärl. (1988), anamorphic *Agaricomycetes*. 1 (with clamp connexions, aquatic), Canada. See Marvanová & Bärlocher (*Mycotaxon* **32**: 340, 1988).

field mushroom (common mushroom), basidioma of the edible *Agaricus campestris*.

filamentous (1) thread-like; filamentose; (2) (of lichens), the photobiont forms a filament of cells which is surrounded by hyphae or cells of the mycobiont (e.g. *Cystocoleus*, *Racodium*, *Coenogonium*).

Filariomyces Shanor (1952), Laboulbeniaceae. 1, widespread. See Shanor (*Am. J. Bot.* **39**: 499, 1952), Tavares (*Mycol. Mem.* **9**: 627 pp., 1985), Lee & Na (*Korean J. Mycol.* **26**: 108, 1998; Korea).

Filaspora Preuss (1855) nom. rej. = Rhabdospora fide Hawksworth *et al.* (*Dictionary of the Fungi* edn 8, 1995).

Filicupula Y.J. Yao & Spooner (1996), ? Pezizales. 1 (parasitic on liverworts), Europe. Affinities uncertain; ecology is not typical of the *Pezizales*. See Yao & Spooner (*Kew Bull.* **51**: 193, 1996).

filiform, thread-like (Fig. 23.2).

Filobasidiaceae L.S. Olive (1968), Filobasidiales. 1 gen., 4 spp.

Lit.: Mitchell *et al.* (*J. Med. Vet. Mycol.* **30**: 207, 1992), Fell & Statzell-Tallman *in* Kurtzman & Fell (Eds) (*Yeasts, a taxonomic study* 4th edn: 742, 1998), Kwon-Chung *in* Kurtzman & Fell (Eds) (*Yeasts, a taxonomic study* 4th edn: 663, 1998), Scorzetti *et al.* (*FEMS Yeast Res.* **2**: 495, 2002), Sivakumaran *et al.* (*Mycopathologia* **156**: 157, 2003), Guffogg *et al.*

(*Int. J. Syst. Evol. Microbiol.* **54**: 275, 2004), Suh *et al.* (*MR* **109**: 261, 2005).

Filobasidiales Jülich (1981). Tremellomycetes. 1 fam., 1 gen., 4 spp. Fam.:

Filobasidiaceae

Lit.: Jülich (*Higher taxa Basid.*: 347, 1981), Fell *et al.* (*IJSEM* **50**: 1351, 2000; mol. phylogeny).

Filobasidiella Kwon-Chung (1976), Tremellaceae. Anamorph *Cryptococcus*. 3, widespread. See Kwon-Chung *et al.* (*Stud. Mycol.* **38**, 1995; mol. phylogeny, ultrastr.), Fell *et al.* (*Int. J. Syst. Evol. Microbiol.* **50**: 1351, 2000; mol. phylogeny), Ginns & Malloch (*Mycol. Progr.* **2**: 137, 2003; parasitism of *Verticillium lecanii*), Sivakumaran *et al.* (*Mycopathologia* **156**: 157, 2003; genetics), Weiss *et al.* (*Frontiers in Basidiomycote Mycology*: 7, 2004), Yamada *et al.* (*Mycoses* **47**: 24, 2004; medically important spp.), Loftus *et al.* (*Science, Washington* **307**: 1321, 2005; *Cryptococcus neoformans* genome), Lin & Heitman (*Annual Review of Microbiology* **60**: 69, 2006; *Cryptococcus neoformans* species complex).

Filobasidium L.S. Olive (1968), Filobasidiaceae. Anamorph *Cryptococcus*. 4, Europe; N. America. See Kwon-Chung (*Int. J. Syst. Bacteriol.* **27**: 293, 1977), Kwon-Chung (*Stud. Mycol.* **38**, 1995; mol. phylogeny, ultrastr.), Fell *et al.* (*Int. J. Syst. Evol. Microbiol.* **50**: 1351, 2000; mol. phylogeny), Guffogg *et al.* (*International Journal of Systematic and Evolutionary Microbiology* **54**: 275, 2004; n.sp. Antarctica), Weiss *et al.* (*Frontiers in Basidiomycote Mycology*: 7, 2004), Golubev *et al.* (*MR* **110**: 957, 2006; n.sp. Russia).

Filoboletus Henn. (1900) = Mycena fide Kuyper (*in litt.*).

filoplasmodium, see plasmodium.

filopodium, a slender unbranched process (pseudopodium) from a plasmodium, as in *Schizoplasmodiopsis*. Cf. rhizopodium.

Filosporella Nawawi (1976), anamorphic *Pezizomycotina*, Hso.0fH.19. 6 (aquatic), widespread. See Dyko & Sutton (*Mycotaxon* **7**: 323, 1978), Marvanová *et al.* (*Nova Hedwigia* **54**: 151, 1992; UK), Gulis & Marvanová (*Mycotaxon* **68**: 313, 1998; Belarus), Marvanová & Bärlocher (*MR* **102**: 750, 1998; Canada).

Fimaria Velen. (1934) = Pseudombrophila fide Brummelen (*Persoonia* **13**: 213, 1986), van Brummelen (*Libri Botanici* **14**, 1995; synonymy with *Pseudombrophila*), Prokhorov (*Mikol. Fitopatol.* **32**: 14, 1998), Hansen *et al.* (*Mycol.* **97**: 1023, 2005; phylogeny).

fimbriate, edged; delicately toothed; fringed. Cf. fimbrillate.

fimbrillate, having a very small fringe. Cf. fimbriate.

Fimetaria Griffiths & Seaver (1910) ≡ Sordaria.

Fimetariaceae D.A. Griffiths & Seaver (1910) = Sordariaceae.

Fimetariella N. Lundq. (1964), Lasiosphaeriaceae. 1 (coprophilous), Europe; Canada. See Barr (*Mycotaxon* **39**: 43, 1990; posn), Krug (*CJB* **73**: 1905, 1995), Huhndorf *et al.* (*Mycol.* **96**: 368, 2004).

fimicolous, living on animal droppings. Cf. coprophilous.

fine structure, see ultrastructure.

Finerania C.W. Dodge (1971) ? = Ramonia fide Galloway (*in litt.*).

finger-and-toe, see club root.

Finkia Vain. (1929), Lichinaceae (L). 1, C. America. See Moreno & Egea (*Acta Bot. Barcinon.* **41**: 1, 1992).

Fioriella Sacc. (1905) = Diplodina fide Sutton (*Mycol. Pap.* **141**, 1977).

fireplace fungi, fungi characteristic of burnt ground, etc. See pyrophilous fungi.

Fischer (Edward; 1861-1939; Switzerland). Student of de Bary (q.v.) in Strasburg; Professor of Botany and General Biology at the University and Director of the Botanic Garden and Botanical Institute, Berne (1897-1933). Produced major monographs for central Europe of various ascomycete and basidiomycete groups. *Publs. Die Uredineen der Schweiz* (1904); Tuberineae (1897, 1938) and Gasteromycetes (1900, 1933) in *Engler & Prantl, Die Natürlichen Pflanzenfamilien* (eds 1 and 2); Tuberaceae and Hemiasci in *Rabenhorst's Kryptogamen-Flora* (1897); (with Gäumann) *Biologie der Pflanzenbewohnenden Parasitischen Pilze* (1929). *Biogs, obits etc.* Grummann (1974: 640); Rytz (*Bericht der Schweizerischen Botanischen Gesellschaft* **50**: 793, 1940); Schopper (*Revue de Mycologie* Série 2 **5**: 47, 1940) [portrait]; Stafleu & Cowan (*TL-2* **1**: 834, 1976).

Fischerula Mattir. (1928), Pezizales. 2 (hypogeous), Italy; USA. See also *Leucangium*. See Trappe (*Mycol.* **67**: 934, 1975; key), O'Donnell *et al.* (*Mycol.* **89**: 48, 1997; ; close to *Leucangium* and a sister group of the *Morchellaceae*), Hansen & Pfister (*Mycol.* **98**: 1029, 2006; phylogeny), Venturella *et al.* (*Cryptog. Mycol.* **27**: 201, 2006; Sicily, ecology), Læssøe & Hansen (*MR* **111**: 1075, 2007; phylogeny).

fission (1) becoming two by division of the complete organism; cf. budding; (2) (of conidial liberation), secession by the separation of a double septum; cf. fracture, lysis.

fissitunicate, see ascus.

Fissolimbus E. Horak (1979), ? Marasmiaceae. 1, Papua New Guinea. See Horak (*Beih. Sydowia* **8**: 202, 1979).

Fissuricella Pore, D'Amatao & Ajello (1977), anamorphic *Pezizomycotina*, Hso.#eH.?. 1 (yeast-like), USA. See Pore *et al.* (*Sabouraudia* **15**: 71, 1977).

Fissurina Fée (1824), Graphidaceae (L). 1, widespread. See Hawksworth *et al.* (*Dictionary of the Fungi* edn 8, 1995), Staiger & Kalb (*Mycotaxon* **73**: 69, 1999; morphology), Staiger (*Biblthca Lichenol.* **85**, 2002; monogr.), Nakanishi *et al.* (*Bull. natn. Sci. Mus.* Tokyo, B **29**: 83, 2003; Japan), Aptroot (*Symb. bot. upsal.* **34** no. 1: 31, 2004; Taiwan), Kalb *et al.* (*Symb. bot. upsal.* **34** no. 1: 133, 2004), Aptroot & Rodrigues (*Cryptog. Mycol.* **26**: 273, 2005; Azores), Archer (*Telopea* **11**: 59, 2005; Australia), Archer (*Biblthca Lichenol.* **94**, 2006; Australia), Staiger *et al.* (*MR* **110**: 765, 2006; phylogeny), Archer (*Systematics & Biodiversity* **5**: 9, 2007; Solomon Is), Makhija & Adawadkar (*Lichenologist* **39**: 165, 2007; India, key).

fistular (**fistulose**), hollow, like a pipe.

Fistulariella Bowler & Rundel (1977) = Ramalina See Nimis (*Lichenologist* **30**: 427, 1998; generic concept), Nimis (*Lichenologist* **30**: 427, 1998).

Fistulina Bull. (1791), Fistulinaceae. Anamorph *Confistulina*. 6, widespread. *F. hepatica*, the edible beefsteak fungus, is the cause of brown oak (q.v.).

Fistulinaceae Lotsy (1907), Agaricales. 3 gen. (+ 5 syn.), 8 spp.

Lit.: Donk (*Taxon*: 263, 1964), Niemelä & Kotiranta (*Karstenia* **26**: 57, 1986), Guzmán (*Revta Mex. Micol.* **3**: 29, 1987), Rajchenberg & Greslebin (*Myco-

taxon **56**: 325, 1995), Hibbett *et al.* (*Proc. natn Acad. Sci. U.S.A.* **94**: 12002, 1996), Quanten (*Op. bot. Belg.* **11**: 352 pp., 1997), Binder *et al.* (*Systematics and Biodiversity* **3**: 113, 2005).

Fistulinales = Agaricales.

Fistulinella Henn. (1901), Boletaceae. 15, pantropical. See Singer (*Am. midl. Nat.* **37**: 527, 1946), Guzman (*Boln. Soc. mex. Micol.* **8**: 53, 1974; taxonomy and geography), Pegler & Young (*TBMS* **76**: 103, 1981).

Fitzpatrick (Harry Morton; 1886-1949; USA). Assistant (1908-1911), then Instructor (1911-1913), then Professor (1913-1949), Cornell University. A founder member of the Mycological Society of America, its first Secretary & Treasurer, later President and, finally, Historian. Highly respected as a teacher. *Publs.* Monograph of the Coryneliaceae *Mycol.* (1920); Monograph of the Nitschkiaceae *Mycol.* (1923); *The Lower Fungi. Phycomycetes* (1930). *Biogs, obits etc.* Barrus (*Mycol.* **43**: 249, 1951; bibliography, portrait); Stafleu & Cowan (*TL-2* **1**: 842, 1976).

Fitzpatrickella Benny, Samuelson & Kimbr. (1985), Coryneliaceae. 1 (on *Drimys*), Juan Fernández. See Benny *et al.* (*Bot. Gaz.* **146**: 232, 1985).

Fitzpatrickia Cif. (1928) ? = Acanthonitschkea fide Nannfeldt (*Svensk bot. Tidskr.* **66**: 49, 1975).

Fixatives. Fixation is the process by which biological tissues are preserved from decay while preserving their structure and organization. Fixed cells are killed, thereby terminating any ongoing biochemical activity, the purpose being to preserve the material in as close as possible to its natural state. Chemical fixatives:

Formal-acet-alcohol.
Formalin 13.0 ml
Acetic acid (glacial) 5.0 ml
Ethyl alcohol (50 per cent) 200.0 ml

Flemming's weak solution.
Chromic acid (1 per cent) 25.0 ml
Acetic acid (1 per cent) 10.0 ml
Water 60.0 ml
With addition of Osmic 5.0 ml
acid (2 per cent) before use.

Other fixatives:
Glutaraldehyde
Osmium tetroxide.

Flabellaria Pers. (1818) [non *Flabellaria* Cav. 1790, *Malpighiaceae*] ≡ Schizophyllum.

flabellate (**flabelliform**), like a fan; in the form of a half-circle.

Flabelliforma E.U. Canning, R. Killick-Kendrick & M. Killick-Kendrick (1991), Microsporidia. 3.

Flabellimycena Redhead (1984), ? Mycenaceae. 1, S. America. See Redhead (*CJB* **62**: 886, 1984).

Flabellocladia Nawawi (1985), anamorphic *Pezizomycotina*, Hso.1bH.1. 1 (aquatic), Malaysia. See Nawawi (*TBMS* **85**: 174, 1985).

Flabellomyces Kobayasi (1982) = Coenogonium Ehrenb. fide Kashiwadani (*Bull. natn. Sci. Mus.* Tokyo, B **9**: 159, 1983).

Flabellophora G. Cunn. (1965), Polyporaceae. 18, widespread (pantropical). See Corner (*Beih. Nova Hedwigia* **86**: 18, 1987; key with broad species concept), Ryvarden & Iturriaga (*Mycol.* **95**: 1066, 2003; Venezuela).

Flabellopilus Kotl. & Pouzar (1957) ≡ Meripilus fide Donk (*Verh. K. ned. Akad. Wet.* tweede sect. **62**: 1, 1974).

Flabellospora Alas. (1968), anamorphic *Pezizomycotina*, Hso.1bH.1. 3, widespread (tropical). See Alasoadura (*Nova Hedwigia* **15**: 415, 1968).

flaccid, not stiff; limp; flabby.

Flagelloscypha Donk (1951), Niaceae. *c.* 25, widespread. See Agerer (*Sydowia* **27**: 131, 1975; key), Handa & Harada (*Mycoscience* **46**: 265, 2005; Japan).

Flagellosphaeria Aptroot (1995), Xylariales. 1, Portugal. See Kang *et al.* (*Fungal Diversity* **2**: 135, 1999; uncertain posn).

Flagellospora Ingold (1942), anamorphic *Nectria*, Hso.0fH.15. 3, widespread. See Petersen (*Mycol.* **55**: 570, 1963; key), Jooste & van der Merwe (*S. Afr. J. Bot.* **56**: 319, 1990; ultrastr. *F. penicillioides*), Webster (*Nova Hedwigia* **56**: 455, 1993; telemorph), Scutari & Calvelo (*Ann. bot. fenn.* **32**: 55, 1995; ecology), Belliveau & Bärlocher (*MR* **109**: 1407, 2005; phylogeny).

flagellum (pl. **flagella**), cylindrical extension of a eukaryotic cell, bounded by a plasma membrane and containing an axoneme; two types can be distinguished by electron microscopy, the **whiplash** with a smooth continuous surface (as in *Chytridiomycota*) and the **tinsel**, characteristic of *Hyphochytriomycota*, with the surface covered with hair-like processes (**mastigonemes** or **flimmers**); **-ar apparatus**, complex consisting of one or more basal bodies which may bear flagella, may have microtubular and fibrous roots associated with their bases; **-ar fibrous roots**, roots composed of a bundle of filaments, frequently appearing cross-striated; **-ar hairs**, filamentous appendages usually arranged in one or more rows but not covering the entire surface of a flagellum; **-ar scales**, organic structures of discrete size and shape, often covering the whole surface of the flagellum external to the plasma membrane, usually assembled in the dictyosome. See Barr (*Mycol.* **84**: 1, 1992; terminology flagellar apparatus). See also axoneme, blepharoplast; and cf. cilium.

Flageoletia (Sacc.) Höhn. (1916) = Phomatospora fide Reid & Booth (*CJB* **44**: 445, 1966).

Flahaultia Arnaud (1952) nom. dub., Fungi. See Watling & Kendrick (*Naturalist* Hull **104**: 1, 1979).

Flakea O.E. Erikss. (1992) = Agonimia fide Aptroot *et al.* (*Biblthca Lichenol.* **57**: 19, 1995), Thor & Kashiwadani (*SA* **14**: 87, 1996; chemistry).

Flamingomyces R. Bauer, M. Lutz, Piątek, Vánky & Oberw. (2007), Urocystidaceae. 1 (on *Ruppia maritima* (*Ruppiaceae*)), Europe (southern). See Bauer *et al.* (*MR* **111**: 1199, 2007).

Flaminia Sacc. & P. Syd. (1902), ? Pucciniales. 1 (on *Xanthoxylon* (*Rosaceae*)), Brazil. May apply to *Aecidium* anamorph. See Sherwood (*Mycotaxon* **5**: 1, 1977).

Flammispora Pinruan, Sakay., K.D. Hyde & E.B.G. Jones (2004), Sordariomycetes. 1, Thailand. See Pinruan *et al.* (*Stud. Mycol.* **50**: 384, 2004).

Flammopsis Fayod (1889) = Pholiota fide Singer (*Agaric. mod. Tax.*, 1951).

Flammula (Fr.) P. Kumm. (1871) [non *Flammula* (DC.) Fourr. 1868, *Ranunculaceae*] = Pholiota fide Singer (*Agaric. mod. Tax.*, 1951).

Flammulaster Earle (1909), Inocybaceae. 20, widespread. See Vellinga (*Persoonia* **13**: 1, 1986; NW Europe), Horak & Moreau (*BSMF* **120**: 215, 2004).

flammulin, an anti-tumour antibiotic from *Flammulina velutipes* (Watanabe *et al.*, *Bull. chem. Soc. Japan*

37: 747, 1964).

Flammulina P. Karst. (1891), Physalacriaceae. 10, widespread (temperate). See Ingold (*TBMS* **75**: 107, 1980; *F. velutipes* mycelium, oidia, etc.), Hughes *et al.* (*Mycol.* **91**: 978, 1999; phylogeny), Methven *et al.* (*Mycol.* **92**: 1064, 2000; biogeography) enokitake.

flask fungi, *Ascomycota* with perithecioid ascomata.

Flaviporellus Murrill (1905) = Inonotus fide Ryvarden (*Syn. Fung.* **5**, 1991).

Flaviporus Murrill (1905), Meruliaceae. 13, widespread. See Ginns (*CJB* **58**: 1578, 1980).

Flavobathelium Lücking, Aptroot & G. Thor (1997), Dothideomycetes. 1, Costa Rica. See Lücking *et al.* (*Lichenologist* **29**: 221, 1997).

Flavocetraria Kärnefelt & A. Thell (1994), Parmeliaceae (L). 2, arctic-alpine; boreal. See Kärnefelt *et al.* (*Acta Bot. Fenn.* **150**: 81, 1994), Thell & Miao (*Ann. bot. fenn.* **35**: 275, 1998; phylogeny), Thell *et al.* (*Mycol. Progr.* **1**: 335, 2002; phylogeny), Bjerke & Elvebakk (*N.Z. Jl Bot.* **42**: 647, 2004; biogeography, S America), Thell *et al.* (*Symb. bot. upsal.* **34** no. 1: 429, 2004; biogrography, Scandinavia), Miądlikowska *et al.* (*Mycol.* **98**: 1088, 2006; phylogeny).

Flavodon Ryvarden (1973) = Irpex fide Hjortstam & Larsson (*Windahlia* **21**, 1994).

Flavoparmelia Hale (1986), Parmeliaceae (L). 32, widespread. See Elix & Johnston (*Mycotaxon* **33**: 391, 1988), Elix (*Flora of Australia* **55**: 39, 1994), Thell *et al.* (*Cryptog. bot.* **5**: 120, 1995; asci), Crespo & Cubero (*Lichenologist* **30**: 369, 1998; phylogeny), Crespo *et al.* (*Lichenologist* **31**: 451, 1999; morphology, chemistry, phylogeny), Schumm (*Mitteilungen der Mikroskopischen Arbeitsgemeinschaft Stuttgart* **1999**: 46, 1999; key, Macaronesia), Eliasaro & Adler (*Mitteilungen aus dem Institut für Allgemeine Botanik Hamburg* **30-32**: 25, 2002; Brazil), Thell *et al.* (*Mycol. Progr.* **3**: 297, 2004; phylogeny), Thell *et al.* (*Symb. bot. upsal.* **34** no. 1: 429, 2004; biogeography, Scandinavia), Blanco *et al.* (*Mycol.* **97**: 150, 2005; phylogeny), Blanco *et al.* (*Mol. Phylogen. Evol.* **39**: 52, 2006; phylogeny, chemistry), Miądlikowska *et al.* (*Mycol.* **98**: 1088, 2006; phylogeny).

Flavophlebia (Parmasto) K.H. Larss. & Hjortstam (1977) = Radulomyces fide Jülich (*Persoonia* **10**: 325, 1979).

Flavopunctelia (Krog) Hale (1984), Parmeliaceae (L). 5, widespread (temperate; tropical). See Ferraro (*Phytologia* **61**: 189, 1986; Argentina), Feuerer & Marth (*Mitteilungen aus dem Institut für Allgemeine Botanik Hamburg* **27**: 101, 1997; vegetative reproduction), Crespo & Cubero (*Lichenologist* **30**: 369, 1998; phylogeny), Lumbsch (*Chemical Fungal Taxonomy*: 345, 1998; chemistry), Crespo *et al.* (*Taxon* **50**: 807, 2001; phylogeny), Thell *et al.* (*Folia cryptog. Estonica* **41**: 115, 2005; phylogeography), Blanco *et al.* (*Mol. Phylogen. Evol.* **39**: 52, 2006; phylogeny), Miądlikowska *et al.* (*Mycol.* **98**: 1088, 2006; phylogeny).

Flavoscypha Harmaja (1974) = Otidea fide Dissing (*in litt.*), Liu & Zhuang (*Fungal Diversity* **23**: 181, 2006).

Flegographa A. Massal. (1860) nom. rej. prop. = Phaeographis fide Lücking *et al.* (*Taxon* **56**: 1296, 2007; nomencl.).

Fleischeria Penz. & Sacc. (1901) = Hypocrella fide Rogerson (*Mycol.* **62**: 865, 1970).

Fleischhakia Auersw. (1869) = Preussia fide Clements & Shear (*Gen. Fung.*, 1931).

Fleischhakia Rabenh. (1878) = Psilopezia fide Seaver (*North American Cup Fungi* (Operculates), 1928).

Fleming (Alexander; 1881-1955; Scotland). Student (1901-1906) then Assistant Bacteriologist (1906-1908) then Lecturer (1908-1914 and 1918-1928) then Professor of Bacteriology (1928-1948), St Mary's Hospital, London; Captain, Army Medical Corps, Western Front (1914-1918). Discoverer of penicillin; Fellow of the Royal Society, London 1943; Nobel prize-winner (with Chain & Florey), 1945. *Publs.* On the bacterial action of cultures of a *Penicillium*, with special reference to their use in the isolation of *B. influenzae*. *The British Journal of Experimental Pathology* (1929). *Biogs, obits etc.* Brown, K. (*Penicillin Man: Alexander Fleming and the Antibiotic Revolution*, 2004); A. Maurois (*The Life of Sir Alexander Fleming*, 1959).

flesh, the trama, esp. of the pileus of an agaric or bolete.

fleshy (of sporocarp), soft, not cartilaginous-like or wood-like.

flexuous (**flexuose**), wavy.

flexuous hyphae (of *Pucciniales*), an unbranched or branched haploid hyphal projection from a pycnium, which may be diploidized by a pycniospore of opposite 'sex' (Craigie, see *Nature* **141**: 33, 1938). Cf. receptive body.

flimmergeissel, flagellum bearing two rows of tripartite tubular hairs. Cf straminipilous.

Floccaria Grev. (1827) = Penicillium Link fide Fries (*Syst. mycol.* **3**: 409, 1832).

flocci, cotton like groups or tufts.

Floccomutinus Henn. (1895) = Mutinus fide Demoulin & Dring (*Bull. Jard. bot. nat. Belg.* **45**: 365, 1975).

floccose, cottony; byssoid.

Floccularia Pouzar (1957), Agaricaceae. 6, widespread (north temperate). Probably a separate family is justified. See Pouzar (*Česká Mykol.* **11**: 49, 1957).

flocculent (of a liquid-culture), having small masses of cells throughout or as a deposit.

Flocculina P.D. Orton (1960) = Flammulaster fide Kuyper (*in litt.*).

flocculose, delicately cottony.

flor effect (of yeasts), formation of a pellicle; see yeasts.

flor yeasts, see yeast.

flora (1) the plants of a particular geographical area or habitat; (2) a description, catalogue or list of all or some groups of plants in a particular area. Formerly applied to fungi and lichens (i.e. **fungus -**, **lichen -**), but as fungi are not plants the term mycobiota (q.v.) is preferred.

Florey (Howard, 1898-1968; Australia). University of Adelaide (1917-1921); Oxford (1921-1926); Cambridge (1926-1931); Sheffield (1931-1935); Oxford (1935-1968). Pharmacologist who, with Chain (q.v.) organized the first large-scale production of penicillin, the fungal antibiotic discovered by Fleming (q.v.). *Biogs, obits etc.* Fenne (*Australian Dict. Biography* **14**: 188, 2002).

Floricola Kohlm. & Volkm.-Kohlm. (2000), anamorphic *Mycosphaerellaceae*. 1, USA. See Kohlmeyer & Volkmann-Kohlmeyer (*Bot. Mar.* **43**: 385, 2000).

Flosculomyces B. Sutton (1978), anamorphic *Pezizomycotina*, Hso.#eP.1. 2, widespread. See Sutton (*Mycol.* **70**: 788, 1978).

flowers of tan, the myxomycete *Fuligo septica*.

Fluctua Marbach (2000), Caliciaceae (L). 1, Brazil.

See Marbach (*Biblthca Lichenol.* **74**: 207, 2000).

Fluminicola S.W. Wong, K.D. Hyde & E.B.G. Jones (1999), Sordariales. 1, Philippines. See Wong *et al.* (*Fungal Diversity* **2**: 190, 1999), Ho & Hyde (*Fungal Diversity* **4**: 21, 2000), Raja *et al.* (*Mycotaxon* **88**: 1, 2003; phylogeny).

Fluminispora Ingold (1958) = Dimorphospora fide Ingold (*Guide to aquatic hyphomycetes*, 1975).

fluorescent, giving out light when placed in ultraviolet (or other) radiation.

flush (of fungal growth), the sudden development of a large quantity of mycelium or a periodic surge of basidiomata emergence, esp. in mushroom cultures.

Fluviatispora K.D. Hyde (1994), Halosphaeriaceae. 3 (on submerged fronds of *Livistona*), Papua New Guinea. See Hyde (*MR* **98**: 719, 1994), Fryar & Hyde (*Cryptog. Mycol.* **25**: 245, 2004; key).

Fluviostroma Samuels & E. Müll. (1980), Sordariomycetes. Anamorph *Stromatostilbella*. 1 (on rotten wood), widespread (tropical). See Seifert (*CJB* **65**: 2196, 1987; anamorph), Okada *et al.* (*CJB* **76**: 1495, 1998; rels).

fly agaric (fly fungus, fly mushroom), basidioma of *Amanita muscaria* (see soma).

fly fungus (house fly fungus), *Entomophthora muscae*.

fly-speck fungi, ascomata of *Microthyriaceae*, *Micropeltidaceae* and conidiomata of their anamorphs.

Foetidaria A. St.-Hil. (1835) = Lysurus fide Dring (*Kew Bull.* **35**: 1, 1980).

foliicolous, living on leaves; **- lichens**, see Santesson (*Symb. bot. upsal.* **12** (1), 1952; monogr.), Farkas & Sipman (*Trop. bryol.* **7**: 93, 1993; checklist 482 spp., bibliogr.), 81; bibliogr. 83 papers 1952-85), Lücking (*Beih. Nova Hedw.* **104**: 1, 1992; keys 228 spp. Costa Rica), Ferraro (*Bonplandia* **5**: 191, 1983; S. Am.), Santesson & Tibell (*Austrobaileya* **2**: 529, 1988; 66 spp. Australia), Sérusiaux (*Bot. J. Linn. Soc.* **100**: 87, 1989; review).

foliole, a small leaf-like excrescence on the surface of a foliose lichen.

Foliopollenites Sierotin (1961), Fossil Fungi ? Fungi. 1, widespread.

foliose (1) leaf-like; (2) (of lichens), having a layered (stratose) thallus, usually with a lower cortex, and attached to the substratum either by rhizines or at the base, but not by the whole lower surface (e.g. *Parmelia*, *Peltigera*) (Fig. 21D).

Follmannia C.W. Dodge (1967) = Caloplaca fide Kärnefelt (*Cryptog. Bot.* **1**: 147, 1989).

Follmanniella Peine & B. Werner (1995), Roccellaceae (L). 1, Atacama Desert. See Peine & Werner (*Flechten Follmann* Contributions to Lichenology in Honour of Gerhard Follmann: 289, 1995), Follmann (*J. Hattori bot. Lab.* **90**: 251, 2001; key), Follmann (*Mitteilungen aus dem Institut für Allgemeine Botanik Hamburg* **30-32**: 61, 2002; biogeography).

Fomes (Fr.) Fr. (1849), Polyporaceae. 3, widespread. See Gilbertson & Ryvarden (*N. Amer. Polyp.* **1**: 263, 1986) Traditionally used for perennial species (e.g. Overholts, *The genus Fomes*, 97 pp.); see also *Heterobasidion*, *Fomitopsis*, *Phellinus*, Medical uses of fungi.

Fomesporites T.C. Huang (1981), Fossil Fungi. 1 (Miocene), Taiwan.

Fominia Girz. (1927) = Colletotrichum fide Sutton (*Mycol. Pap.* **141**, 1977).

Fomitaceae Jülich (1982) = Polyporaceae.

Fomitella Murrill (1905) = Fomitopsis fide Ryvarden (*Bull. Jard. Bot. Belg.* **48**: 102, 1978).

Fomites Locq. & Koeniguer (1982), Fossil Fungi. 1, Libya.

Fomitiporella Murrill (1907), Hymenochaetaceae. 5, widespread. See Wagner & Fischer (*Mycol.* **94**: 998, 2002).

Fomitiporia Murrill (1907), Hymenochaetaceae. 11, widespread. See Wagner & Fischer (*Mycol.* **94**: 998, 2002).

Fomitopsidaceae Jülich (1982), Polyporales. 24 gen. (+ 33 syn.), 197 spp.

Lit.: Carranza-Morse & Gilbertson (*Mycotaxon* **25**: 469, 1986), Larsen & Lombard (*Mycotaxon* **26**: 271, 1986), Hjortstam & Ryvarden (*Mycotaxon* **28**: 553, 1987), Blanchette *et al.* (*Mycol.* **84**: 119, 1992), Hibbett & Donoghue (*CJB* **73**: S853, 1995), Rajchenberg (*Nordic Jl Bot.* **15**: 105, 1995), Chang *et al.* (*Mycol. Monogr.* **10**: 126 pp., 1996), Högberg & Stenlid (*Mol. Ecol.* **8**: 703, 1999), Thorn (*Karstenia* **40**: 181, 2000), Parmasto (*Harvard Pap. Bot.* **6**: 179, 2001), Kauserud & Schumacher (*MR* **107**: 155, 2003), Kim *et al.* (*Antonie van Leeuwenhoek* **83**: 81, 2003), Larsson *et al.* (*MR* **108**: 983, 2004), Kim *et al.* (*Mycol.* **97**: 812, 2005).

Fomitopsis P. Karst. (1881), Fomitopsidaceae. 32, widespread. See Pegler (*The polypores [Bull. BMS Suppl.]*, 1973), Kim *et al.* (*Mycol.* **97**: 812, 2005; phylogeny).

Fonsecaea Negroni (1936), anamorphic *Herpotrichiellaceae*. 1 (human pathogen), widespread. See Schol-Schwarz (*Antonie van Leeuwenhoek* **34**: 119, 1968), Ibrahim-Granet *et al.* (*Sabouraudia* **23**: 253, 1985; numerical taxonomy), Wang *et al.* (*Acta Mycol. Sin.* **5**: 14, 1986; conidial ontogeny), Wang *et al.* (*Mycopathologia* **98**: 105, 1987; SEM), Ramos & Borghi (*Bol. Micol.* Valparaíso **4**: 135, 1989; morphology), Ramos & Borghi (*Bol. Micol.* Valparaíso **4**: 141, 1989; biochemistry, ultrastr.), Dennetière *et al.* (*J. Mycol. Médic.* **1**: 302, 1991; dimorphism), Spatafora *et al.* (*J. Clin. Microbiol.* **33**: 1322, 1995; DNA), Attili *et al.* (*Medical Mycology* **36**: 219, 1998; phylogeny), Haase *et al.* (*Stud. Mycol.* **43**: 80, 1999; phylogeny), Untereiner & Naveau (*Mycol.* **91**: 67, 1999; phylogeny), Abliz *et al.* (*J. Clin. Microbiol.* **41**: 873, 2003; molecular identification), Hoog *et al.* (*Medical Mycology* **42**: 405, 2004; molecular ecology), Tanabe *et al.* (*Jap. J. Med. Mycol.* **45**: 105, 2004; strain typing), Surash *et al.* (*Medical Mycology* **43**: 465, 2005; phaeohyphomycosis), Huang *et al.* (*J. Clin. Microbiol.* **44**: 3299, 2006; arrays).

Fontanospora Dyko (1978), anamorphic *Pezizomycotina*, Hso.1bH.1. 2 (aquatic), USA. See Dyko (*TBMS* **70**: 411, 1978), Marvanová *et al.* (*Czech Mycol.* **50**: 3, 1997; UK).

Food and beverage mycology. Fungi are important in the food and drinks industry as both positive and negative factors (see Food spoilage). Filamentous fungi and yeasts are used commercially in the manufacture of a large variety of foods and drinks. Perhaps the most well known use of fungi is in cheese making, notably *Penicillium camembertii* in camembert cheese and *P. roquefortii* for the manufacture of blue-veined cheeses such as roquefort, stilton, and Danish blue. They can also be eaten as themselves (see Edible fungi, Mushroom culture, Truffles), and more recently as mycoprotein (q.v.). However, in S.E. Asia particularly, fungi have been used for centuries for the production of various food fermentations (see Fermented food and drinks). The most well-known of these products is soya sauce, produced by seeding soy beans with *Aspergillus oryzae*.

Yeasts are used for the manufacture of bread, beer and wine (see Brewing, Wine making). In most cases *Saccharomyces cerevisiae* is involved, causing a fermentation of sugar to alcohol (production of wine and beer) and carbon dioxide (in bread, the alcohol evaporates during the baking process). *S. carlsbergensis* is used in the manufacture of lager. Occasionally strongly-flavoured or aromatic fungi are used to produce sauces, such as truffle-oil, or specialist drinks such as truffle-liqueurs. The IUMS has an International Commission on Food Mycology (see Societies and organizations).

Lit.: Beuchat (Ed.) (*Food and beverage mycology*, edn 2, 1987), Pitt & Hocking (*Fungi and food spoilage*, 1985), Querol & Fleet (Eds) (Yeasts in food and beverages. *The yeast handbook* **2**, 2006), Samson *et al.* (*Introduction to food-borne fungi*, edn 4, 1995), Hui & Khachatourians (Eds) (*Food biotechnology*, 1995). See also Mycotoxicoses, Yeasts.

Food spoilage. Can be caused by many fungi which invade and decompose harvested and processed foods and beverages, change their biochemical composition, hyper- (over-) synthesize enzymes, or produce health related effects due to various toxic metabolites. In general, fungi are responsible for a loss of 5-10% of food production in developing countries and up to 40% in some countries. Fungal contamination of food usually results in accumulation of several mycotoxins (e.g. viomellein, zearalenon, aflatoxins, ochratoxins, aurofusarin, fusarin, etc.). Most fungal toxins are not acutely toxic in low amounts, but have severe long-term effects causing cancer, nephro- or hepato-toxicoses (see Mycotoxicoses); some toxic metabolites have immuno-suppressive effects in humans and animals assisting in development of bacterial infections. Legislation exists in some countries to control mycotoxin levels in susceptible imported food commodities, e.g. ochratoxin A in coffee.

Development of food spoilage depends on intrinsic characteristics of the fungus and on external factors such as water, temperature, pH, preservatives, composition of the gaseous atmosphere, etc. Interactions of spoilage organisms with the environment and with each other determine conditions for food safety and quality (Roller, *Journal of Food Microbiology* **50**: 151-153, 1999). In most fresh, moist foods, fungi do not grow well due to competition from bacteria. In foods which have conditions such as lowered water activity (q.v.), pH or refrigerated temperatures, filamentous fungi and yeasts may proliferate. Fungi can contaminate dried foods such as nuts, spices, cereals, dried milk and meat, as well as salted fish, fruit, vegetables, meats, jams, confectionery and dairy products. Mycotoxins of contaminated products can withstand different types of processing, including heat treatment, and can persist, e.g. in flour and bread, even after the fungal source has been killed by milling and baking. Only a limited number of fungal species (spoilage consortia) can inhabit a given food type and cause spoilage there by secondary metabolites (Filtenborg *et al.*, *Int. J. Food Microbiol.* **33**: 85, 1996). It is thus possible to predict the number and identity of mycotoxins that can be found in a particular food product or feed stuff (Andersen & Thrane, *Mycotoxin Res.* **12**: 54, 1996; Frisvad *et al.*, *Chemical Fungal Taxonomy* : 289, 1998; Larsen *et al.*, *Biochem. Syst. Ecol.* **26**: 463, 1998). Common food spoiler fungi include *Absidia*, *Aspergillus*, *Botrytis cinerea*, *Byssochlamys*, *Fusarium*, *Mucor*, *Paecilomyces*, *Penicillium* (and their teleomorphs), *Rhizopus*, *Sclerotinia sclerotiorum*, *Syncephalastrum*, with *Aspergillus* and *Penicillium* dominating. Food-borne fungi which do not produce phytotoxic or herbicidal compounds or mycotoxins may serve as biological control agents. E.g. non-aflatoxigenic *Aspergillus parasiticus* and *A. flavus* strains out compete toxigenic strains of the same species in the field (Dorner & Cole *J. Stored Prod. Res.* **38**: 329, 2002).

Some fungi can resist pasteurization, e.g. *Byssochlamys*, *Neosartorya* and *Talaromyces*, which can be found in canned and bottled fruits, vegetables and juices. Yeasts are particularly preservative resistant and are found in beers, wines, cider, and soft drinks (*Saccharomyces cerevisiae*, *Brettanomyces bruxellensis*, *Zygosaccharomyces bailii*), and pickles and sauces (*Candida krusei*, *Pichia membranaefaciens*). Annual losses to the food industry caused by yeast spoilage are rather high (James *et al.*, *Methods in Biotechnology* **14**: 37, 2000). Techniques which attempt to preserve food and beverages by restricting oxygen entry into these products, which are rich in nutrients, are not promising strategies against yeasts, particularly in respect of growth and gas formation by *Z. bailii* (Rodrigues *et al.*, *Applied and Environmental Microbiology* **67**: 2123, 2001).

Lit.: Beuchat (Ed.) (*Food and beverage mycology*, edn 2, 1987), Blackburn (Ed.) (*Food spoilage microorganisms*, 736 pp., 2006), Cannon (*in* Robinson (Ed.), *Developments in food microbiology* **3**: 141, 1988), Deak & Beuchat (*Handbook of Food Spoilage Yeasts (CRC Series in Contemporary Food Science)*, 1996), Dijksterhuis & Samson (*Food Mycology: A Multifaceted Approach to Fungi and Food. Mycology series* 25: 1, 2007), Fassatiová (*Moulds and filamentous fungi in technical microbiology*, 1986), *Food Science and Technology Abstracts* (1969 on; abstracts), Hocking *et al.* (Eds) (*Advances in Food Mycology*, 2005), Pitt & Hocking (*Fungi and food spoilage*, edn 2, 1997), Samson *et al.* (Eds) (*Introduction to food- and airborne fungi*, edn 7, 2004), Samson *et al.* (Eds) (*Modern Methods in Food Mycology (Developments in Food Science)*,1992), Spencer & Ragout-Spencer (Eds) (*Environmental Microbiology: Methods and Protocols (Methods in Biotechnology)*, 2004). See also Beer, Fermented foods and drinks, Food and beverage mycology, Mycotoxins, Wine making, Yeasts.

foot cell (1) a basal cell supporting the conidiophore in *Aspergillus*; (2) the basal cell of the conidium in *Fusarium*. See Sutton (*TBMS* **86**: 1, 1986; occurrence in anamorphic fungi).

Foraminella S.L.F. Mey. (1982) ≡ Parmeliopsis.

forate (of 'gasteromycete' basidiomata), invagination of the primordial tissue resulting in a series of pits; the type of development generally known as 'coralloid' of which it is the opposite (Dring, 1973).

Forensic mycology. Although a still scarcely developed tool, fungi can be of great value in forensic science. In addition to their importance in cases of poisoning, other examples of use include: determining the minimum interval since death (Hitosugi *et al.*, *Legal medicine* **8**: 240, 2006; Turner & Wiltshire, *Forensic Sci. Internat.* **101**: 113, 1999); or locating clandestine burial sites such as mass graves (Carter & Tibbett, *J. Forensic Sci.* **48**: 4, 2003); determining the presence of illegal hallucinogenic substances in fungi (Nugent & Saville, *Forensic Sci. Internat.* **140**: 147, 2004). As fungal toxins have allegedly been used in biological warfare (Pearce, *New Scientist* **174**: 13, 2002), forensic mycology may also be potentially relevant in international law. See also Mycotoxicoses.

form (pl. **-a**) [**f.**; **ff.**, pl.], the lowest formal taxonomic rank regulated by the Code (see Classification, Nomenclature); **- category**, see Fossil fungi; **- genus**/ **- species**, one used for anamorphs (q.v.); **-a specialis** [**f. sp.**; **-ae speciales**, **f. spp.**, pl.], see special form.

-form (suffix), shape.

fornicate, arched; (of *Geastrum*), having the fibrous and fleshy layers of the fruit body becoming arched over the cup-like mycelial layer.

Forssellia Zahlbr. (1906) = Pterygiopsis fide Henssen (*Bryologist* **73**: 617, 1970).

Fossil fungi (Fungi fossiles). Although fungal fossils are probably not rare, rather little is known about fungi as fossils. The fossil record of Fungi extends back to the early Phanerozoic, and no doubt well into the Proterozoic (see Pirozynski, *Ann. Rev. Phytopath.* **14**: 237, 1976), with chytrid-like forms being recorded from the late Precambrian, and *Glomus*-like forms from the early Paleozoic (see Mycorrhiza). Some, such as the remarkable 9 m high *Prototaxites* have only recently been recognized, tentatively, as fungi (Kibby, *Field Mycology* **9**: 48, 2008); the reported survival of viable fungal spores in glacial ice about 140,000 years old has blurred the edges between fossil and living fungi (see Longevity). Substantial portions of larger soft and fleshy fruitbodies are very rarely found (mainly embedded in amber), and most fungal fossil material comprises spores and other microscopic structures such as hyphae. This has meant that it is not even easy to assess how many extant species are also represented as fossils. Some fossil fungi are assigned extant generic names (or sometimes names of higher taxa) modified by a suffix, usually *-ites* (e.g. *Pleosporites*, *Pleosporonites*, but n.b. names of some genera which are not fossils may also terminate in these letters, e.g. *Muellerites*), or are given new form-generic names. In addition, fossil fungi (typically dispersed spores) may be given terminological names denoting suprageneric form – **form categories** (*f. cat.*) within morphographic classifications developed by palynologists or coal petrologists.

Meschinelli (*in* Saccardo, *Syll. Fung.* **10**: 741, 1892; **11**: 657, 1895) and in his *Fungorum fossilium hucusque cognitorum iconographia* (1902; edn 1, 1898) described numerous species of fossil fungi, as sporocarps, spores and symptoms like those caused by extant parasitic fungi.

Gilled fungi have been reported from amber (Hibbett *et al.*, *Nature* **377**: 487, 1995; Cretaceous *Marasmius*-like basidiome). For basidiomycete fossils, see also: Dennis (*Mycol.* **62**: 578, 1970).

A fungus similar to *Beauveria* was gound overgrowing an ant in Dominican amber (Poinar & Thomas, *Cellular and molecular life sciences* **40**: 578, 1984). A species of *Aspergillus* has been described from Eocene Baltic amber in association with a collembolan insect (Dörfelt & Schmidt, *MR* **109**: 956, 2005). Carnivorous fungi been found from Cretaceous amber (Schmidt *et al.*, *Science* **318**: 1743, 2007) cannot be assigned to known extant genera, and it has been surmised that fungal devices for trapping prey, such as hyphal rings, may therefore have evolved more than once. For ascomycete fossils, see also: Sherwood-Pike & Gray (*Lethaia* **18**: 1, 1985), Stubblefield & Taylor (*Am. J. Bot.* **70**: 387, 1983).

There is little information about fossil lichens, but few who are familiar with these organisms have searched for them as fossils. Because cyanobacteria are well represented in Precambrian biotas and most groups of fungi were highly diversified during the Paleozoic, it seems likely that lichens were also components of the early terrestrial ecosystems. As many have a relatively durable thallus morphology and organization one might expect lichens to be preserved. There is some suggestion that certain Silurian and Devonian and even earlier enigmatic fossils may represent lichens (Retallack, *Palaebiol.* **20**: 523, 1994; Stein, *Am. J. Bot.* **80**: 93, 1993), but reports of Precambrian lichens appear to be based on abiotic features (see *Thuchomyces*). To date the best records are from Tertiary sediments (e.g. Sherwood-Pike, *Lichenologist* **17**: 114, 1985; Richardson & Green, *Lichenologist* **3**: 89, 1965). There are reports of lichens preserved in amber (Mägdefrau, *Ber. dtsch. bot. Ges.* **70**: 433, 1957, Baltic; Poinar *et al.*, *Science* **259**: 222, 1993).

There is even less information about chytrid-like (Taylor *et al.*, *Amer. J. Bot.* **79**: 1233, 1992; *Mycol.* **84**: 901, 1992), zygomycete-like (Pirozynski & Dalpé, *Symbiosis* **7**: 1, 1989) and myxomycete-like (Waggoner & Poinar, *J. Protozool.* **39**: 639, 1992) fossils.

Lit.: *General*: Elsik *et al.* (*Annotated glossary of fungal polynomorphs*, AASP **11**, 1983; terminology), Graham (*J. Palynol.* **36**: 60, 1962; spores), Pia (*Arch. Hydrobiol.* **31**: 264, 1937; microborings in calcareous substrates), Pirozynski (*Ann. Rev. Phytopath.* **14**: 237, 1976 [general review], *Geosci. Canada* **16**: 183, 1989; methods of study), Sherwood-Pike (*Biosystems* **25**: 121, 1991; evolution), Stubblefield & Taylor (*New Phytol.* **108**: 3, 1988; paleomycology), Taylor (*in*: *The Fossil Record* **2**: 9, 1993; geologic ranges). See amber.

Interactions: **Animal Interactions**, White & Taylor (*Mycol.* **81**: 643, 1989), Grahn (*Lethaia* **14**: 135, 1981). **Geological Interactions**, Wright (*Sedimentology* **33**: 831, 1986). **Mutualism**, Stubblefield *et al.* (*Am. J. Bot.* **74**: 1904, 1987). **Parasitism**, Hass *et al.* (*Am. J. Bot.* **81**: 29, 1994), Stidd & Cosentino (*Science* **190**: 1092, 1975), Daghlian (*Palaeontology* **21**: 171, 1978). **Saprophytism**, Stubblefield *et al.* (*Am. J. Bot.* **72**: 1765), Taylor & White (*Am. J. Bot.* **76**: 389, 1989).

Regional: **Africa**, Wolf (*Bull. Torrey bot. Cl.* **93**: 104, 1966). **Antaractica**, Stubblefield & Taylor (*Bot. Gaz.* **147**: 116, 1986). **Argentina**, Singer & Archangelsky (*Am. J. Bot.* **45**: 194, 1958). **Australia**, Selkirk (*Proc. Linn. Soc. NSW* **100**: 70, 1975). **Brazil**, Martill (*Merc. Geol.* **12**: 1, 1989). **Canada**, Currah & Stockey (*Nature* **350**: 698, 1991). **Germany**, Kretzschmar (*Facies* **7**: 237, 1982). **India**, Kalgutkar (*Rev. Palaeobot. Palynol.* **77**: 107, 1993), Kar *et al* (*Curr. Sci.* **89**: 257, 2005). **Mexico**, Magallon-Puebla & Cevallos-Ferriz (*Am. J. Bot.* **80**: 1162, 1993). **N. America**, Dilcher (*Palaeontographica* **B116**: 1, 1965), Wagner & Taylor (*Rev. Palaeobot. Palynol.*

37: 317, 1982). **Russia**, Krassilov (*Lethaia* **14**: 235, 1981). **United Kingdom**, Taylor *et al.* (*Am. J. Bot.* **79**: 1233, 1992, *Mycol.* **87**: 560, 1995; arbuscular fungi), Smith (*Palaeontology* **23**: 205, 1980), Kidston & Lang (*Trans. R. Soc. Edinb.* **52**: 855, 1921). **W. Indies**, Poinar & Singer (*Science* **248**: 1099, 1990).

Fouragea Trevis. (1880) = Opegrapha Ach. fide Santesson (*Symb. bot. upsal.* **12** no. 1: 1, 1952).

foveate, having small holes or cavities; pitted (Fig. 20.11).

Foveodiporites C.P. Varma & Rawat (1963), Fossil Fungi. 1 (Tertiary), India.

foveolate, delicately pitted; dimpled, dim. of foveate.

Foveoletisporonites Ramanujam & Rao (1979), Fossil Fungi. 1 (Miocene), India.

Foveostroma DiCosmo (1978), anamorphic *Pezizomycotina*, St.≡ eH.15. 6, widespread. See Hosagoudar & Balakrishnan (*J. Econ. Taxon. Bot.* **15**: 477, 1989), Abbas *et al.* (*Pakist. J. Bot.* **33**: 1, 2001; Pakistan).

Foxia Castell. (1908) = Exophiala fide de Hoog (*Stud. Mycol.* **15**, 1977), Haase *et al.* (*Stud. Mycol.* **43**: 80, 1999).

Fracchiaea Sacc. (1873), Nitschkiaceae. 1, widespread. See Nannfeldt (*Svensk bot. Tidskr.* **69**: 49, 1975), Nannfeldt (*Svensk bot. Tidskr.* **69**: 289, 1975), Subramanian & Sekar (*Kavaka* **18**: 19, 1990), Huhndorf *et al.* (*MR* **108**: 1384, 2004; phylogeny), Schoch *et al.* (*MR* **111**: 154, 2007; phylogeny).

Fractisporonites R.T. Clarke (1965), Fossil Fungi. 3 (Cretaceous, Eocene), USA.

fracture (of conidial liberation), secession involving the rupture of the wall of an adjacent vegetative or degenerate cell at a point removed from the septum; cf. fission, lysis, rhexolytic.

fragmentation spores, conidia produced by hyphae breaking up into separate cells.

Fragosia Caball. (1928), Saccharomycetales. 1, Europe.

Fragosoa Cif. (1926) ? = Hysterographium fide Clements & Shear (*Gen. Fung.*, 1931).

Fragosoella Petr. & Syd. (1927) ? = Scleropycnium fide Clements & Shear (*Gen. Fung.*, 1931).

Fragosphaeria Shear (1923), Cephalothecaceae. Anamorph *Acremonium*-like. 2 (on rotten wood), widespread. Position uncertain; DNA indicates a possible link with the *Ophiostomatales*. See Malloch & Cain (*CJB* **48**: 1815, 1970), Suh & Blackwell (*Mycol.* **91**: 836, 1999; phylogeny).

Frankia Brunch. (1886) = Frankiella Speschnew fide Hawksworth *et al.* (*Dictionary of the Fungi* edn 8, 1995).

Frankiella Speschnew (1900) = Greeneria fide Sutton (*Mycol. Pap.* **141**, 1977).

Franzpetrakia Thirum. & Pavgi (1957), ? Ustilaginaceae. 2 (on *Poaceae*), S. & E. Asia. See Guo *et al.* (*Mycosystema* **3**: 57, 1990), Vánky (*Mycotaxon* **95**: 1, 2006).

Fraseria Bat. (1962) = Microcallis fide von Arx & Müller (*Stud. Mycol.* **9**, 1975).

Fraseriella Cif. & A.M. Corte (1957), anamorphic *Xeromyces*, Hso.0-1eH.1. 1, widespread. See Ciferri & Corte (*Atti Ist. bot. Univ. Lab. crittog. Pavia* sér. 5 **14**: 109, 1957).

Fraserula Syd. (1938) = Inocyclus fide Müller & von Arx (*Beitr. Kryptfl. Schweiz* **11** no. 2, 1962).

free (of lamellae or tubes), not joined to the stipe (Fig. 19A); cf. remote, seceding.

free cell formation, the process by which the 8 nuclei, each with some adjacent cytoplasm, are cut off by walls in the immature ascus to become ascospores.

freeze drying, see Genetic resource collections, Preservation.

Fremineavia Nieuwl. (1916), Melanconidaceae. 1 (from bark), N. America. See Wehmeyer (*Revision of Melanconis*, 1941).

Fremitomyces P.F. Cannon & H.C. Evans (1999), Phyllachoraceae. 2, E. Africa; Seychelles. See Cannon & Evans (*MR* **103**: 585, 1999).

Fresenia Fuckel (1866) [non *Fresenia* DC. 1836, *Compositae*] ≡ Graphiothecium.

Freynella Kuntze (1891) ≡ Coccosporium.

friable, readily powdered.

Friedman (Emerich Imre; 1921-2007; Hungary, later Israel, USA). Professor, Hebrew University, Jerusalem (1951-1968); Associate Professor then Professor and Director of the Polar Desert Research Center, Florida State University, Tallahassee (1968 onwards). A pioneer in the study of fungi in extreme environments, particularly of hyphomycetes living inside rocks, particularly in Antarctica and Chile; also involved in astrobiology, the search for extraterrestrial life. *Publs. Antarctic Microbiology* (1993). *Biogs, obits etc.* Anon. (*Daily Telegraph* [UK], 22 June 2007).

Friedmanniomyces Onofri (1999), anamorphic *Capnodiales*, Hso.?.?. 2 (cryptoendolithic), Antarctica. See Onofri *et al.* (*Nova Hedwigia* **68**: 176, 1999), Selbmann *et al.* (*Stud. Mycol.* **51**: 1, 2005; revision).

Fries (Elias Magnus; 1794-1878; Sweden). Born in Småland and went to school in Wexiö; an only child, Fries was first interested in flowering plants, following his father, but at age 12 his attention was taken by fungi; within five years he had knowledge of more than 300 species, to which he gave names; studied under Acharius (q.v.); student, University of Lund (1811-1814) [although he was awarded his degree for a thesis on flowering plants, at Lund was able to see the works of Persoon q.v.) and other mycologists and to get current names for some of his fungal specimens]; Associate Professor (1814-1824) then full Professor (1824-1834), University of Lund; Professor of Botany, Uppsala University (1834 onwards). With Persoon, the founder of Mycology. The *Systema Mycologicum*, though produced without the help of a compound microscope, is still one of the most important books on systematic mycology. It is specially important for hymenomycetes, but takes in all groups of fungi (e.g. there are more than 500 pyrenomycetes, in good order), and it advanced taxonomy of most fungi far more than any earlier work. With the exceptions of 'gasteromycetes', lichen-forming fungi, *Mycetozoa*, *Pucciniales* and *Ustilaginales*, names of all fungi mentioned in his *Systema Mycologicum* and *Elenchus Fungorum* have protected santioned status in nomenclature (q.v.). *Publs. Observationes mycologicae* (1815-1818) [index, Petersen, *Mycotaxon* **17**: 87-147, 1983]; *Systema Mycologicum* 3 vols (1821-1832) [reprint 1952]; *Elenchus Fungorum* (1828); *Lichenographia Europaea Reformata* (1831); *Epicrisis Systematis Mycologici* (June 1838) [edn 2 as *Hymenomycetes Europaei*, 1874 (reprinted 1963)]; *Summa Vegetabilium Scandinaviae* (1846-1849); *Monographia Hymenomycetum Sueciae* (1853) [reprint 1963]; in addition, he distributed 450 numbers of his *Scleromyceti Sueciae* exsiccati [pyrenomycetes and coelomycetes; see Holm & Nannfeldt (*Friesia* **7**:

10, 1964) and Pfister (*Mycotaxon* **3**: 185, 1975) for annotated check list], directed production of a series of *Icones Fungorum*, and produced various works on vascular plants. *Biogs, obits etc.* Dudley (*Journal of Mycology* **2**: 91, 1886); Eriksson (*Elias Fries och den Romantiska Biologien*, 1962); T.M. Fries & R. Fries (*Friesia* **5**: 135, 1955) [English translation of Fries' autobiography]; Grummann (1974: 474); Holm (*Symbolae Botanicae Upsalienses* **30** (3): 21, 1995); Krok (*Bibliotheca Botanica Suecicana*, 1925 [bibliography, 171 titles]); Lloyd (*Mycological Writings* **1**: 161, 1904); Lloyd (*Mycological Writings* **3**: 413, 1909); Lundström (*Trans. bot. Soc. Edinb.* **13**: 383, 1879); Nannfeldt (*Symbolae Botanicae Upsalienses* **22** (4): 24, 1979); Stafleu & Cowan (*TL-2* **1**: 878, 1976); Strid (*Catalogue of Fungus Plates Painted under the Supervision of Elias Fries*, 1994).

Fries (Theodore Magnus: 1832-1913; Sweden). Son of E.M. Fries (q.v.); Docent of Botany (1837-1862) then Assistant Professor of Botany and Practical Economy (1862-1877) then Professor of Botany and Practical Economy (1877-1899), University of Uppsala. Edited correspondence of Linnaeus and wrote some 170 publications on many aspects of biology and geography; probably responsible for accelerating acceptance of the concept of Schwendener (q.v.) that lichens are composed of at least two organisms; described many new species of lichen-forming fungi, especially from Europe. Specimens in the fungal reference collection in Uppsala (**UPS**). *Publs. Lichenes Arctoi* (1860); *Lichenographia Scandinavica* (1, 1871; 2, 1874). *Biogs, obits etc.* Grummann (1974: 475); Hemmendorff (*Svensk Botanisk Tidskrift* **8**: 109, 1914); Stafleu & Cowan (*TL-2* **1**: 889, 1976).

Friesia Lázaro Ibiza (1916) [non *Friesia* Spreng. 1818, *Euphorbiaceae*] = Ganoderma fide Donk (*Verh. K. ned. Akad. Wet.* tweede sect. **62**: 1, 1974).

Friesites P. Karst. (1879) = Hericium Pers. fide Donk (*Taxon* **5**: 69, 1956).

Friesula Speg. (1880) nom. dub., Thelephorales. See Singer (*Lilloa* **23**: 152, 1950).

Frigidispora K.D. Hyde & Goh (1999), Pezizomycotina. 1 (aquatic), British Isles. See Hyde & Goh (*MR* **103**: 1564, 1999).

Frigidopyrenia Grube (2005), Xanthopyreniaceae (L). 1, subarctic. See Grube (*Phyton* **45**: 305, 2005).

Fritzea Stein (1879) ? = Psora Hoffm. (1796) fide Hawksworth *et al.* (*Dictionary of the Fungi* edn 8, 1995).

frog cheese, young puff-balls.

Frommea Arthur (1917) = Phragmidium fide Laundon (*Mycotaxon* **3**: 155, 1975).

Frommeëlla Cummins & Y. Hirats. (1983), Phragmidiaceae. 2 (on *Duchesnea*, *Potentilla* (*Rosaceae*)), widespread. See McCain & Hennen (*Mycotaxon* **39**: 249, 1990), Helfer (*Nova Hedwigia* **81**: 325, 2005; Europ. spp.).

Frondicola K.D. Hyde (1992), ? Hyponectriaceae. 1 (on *Nypa*), Brunei. See Hyde (*Sydowia* **45**: 204, 1993; posn).

Frondisphaeria K.D. Hyde (1996), ? Dothideomycetes. 2, Brunei. See Hyde (*Mycoscience* **37**: 169, 1996), Fröhlich & Hyde (*Fungal Diversity Res. Ser.* **3**, 2000).

Frondispora K.D. Hyde (1993), Xylariales. 1 (on *Palmae*), Europe. See Kang *et al.* (*Mycoscience* **40**: 151, 1999).

Frostiella Murrill (1942) nom. nud. = Boletellus fide Singer (*Farlowia* **2**: 223, 1945).

fructicolous, living on fruit.

fructification, see fruit-body.

fruit-body (fructification), a general term for spore-bearing organs in both macrofungi and microfungi. The more precise terms apothecium, ascoma (ascocarp), basidioma (basidiocarp), conidioma, perithecium, sporocarp, etc. are preferred usage.

fruticolous, living on shrubs.

fruticose (1) shrub-like; (2) (of lichens), having an upright or hanging thallus of radiate structure (e.g. *Cladonia*, *Ramalina*, *Usnea*) (Fig. 21E).

fruticulose (of lichens), having a minutely shrubby habit (e.g. *Ephebe*, *Polychidium*).

Frutidella Kalb (1994), Ramalinaceae (L). 1, Europe; Australasia. See Sipman *et al.* (*Australasian Lichenology* **43**: 10, 1998).

Fuckel (Karl Wilhelm Gottlieb Leopold; 1821-1876; Germany). A pharmacist (1836-1852; later of independent means; died of typhus in Vienna. Applied the ideas of the Tulasne brothers (q.v.) regarding asexual and sexual stages of ascomycetes; his collections provided a foundation for a mycota of the Rhine region; also associated with early exploration of arctic fungi. Collections are in l'Herbier Boissier, Geneva (**G**. *Publs. Symbolae Mycologicae, Beiträge zur Kenntniss der Rheinischen Pilze* (1870) [for which exsiccati were distributed; supplements 1871, 1873, and 1875; for publication dates, see Rogers (*Mycol.* **46**: 533, 1954); the types for the new species he made in this work are also in Geneva (**G**), with mass isotypes distributed among the 2,700 colls in his *Funghi Rhenani Exsiccati* (1863-1874)]. *Biogs, obits etc.* Stafleu & Cowan (*TL-2* **1**: 896, 1976).

Fuckelia (Nitschke ex Sacc.) Cooke (1869) ? = Euepixylon See Læssøe (*SA* **13**: 43, 1994; ? synonym of *Xylaria*), Læssøe & Spooner (*Kew Bull.* **49**: 1, 1994).

Fuckelia Bonord. (1864), anamorphic *Godronia*, St.1eH.15. 1, Europe.

Fuckelia Niessl (1875) nom. conf., Fungi. See Holm (*Taxon* **24**: 275, 1975).

Fuckelina Kuntze (1891) ≡ Macropodia.

Fuckelina Sacc. (1875) = Stachybotrys fide Hughes (*CJB* **36**: 727, 1958).

Fucus L. (1753), Algae. Algae.

fugacious, see evanescent.

Fugomyces Sigler = Quambalaria fide Kolařík *et al.* (*Czech Mycol.* **58**: 81, 2006; clinical aspects).

Fujimyces Minter & Caine (1980), anamorphic *Pezizomycotina*, Ccu.0fH.1. 1 (on *Pinus*), British Isles. See Minter & Caine (*TBMS* **74**: 434, 1980).

fulcrum (of lichens), a conidiophore within a pycnidium (obsol.).

Fulgensia A. Massal. & De Not. (1853), Teloschistaceae (L). 8, widespread. See Klement (*Nova Hedwigia* **11**: 495, 1965; key), Poelt (*Mitt. bot. StSamml., München* **5**: 571, 1965), Poelt & Hinteregger (*Biblthca Lichenol.* **50**, 1993; Himalaya), Westberg & Kärnefelt (*Lichenologist* **30**: 515, 1998), Kasalicky *et al.* (*CJB* **78**: 1215, 2000; phylogeny), Gaya *et al.* (*Am. J. Bot.* **90**: 1095, 2003; phylogeny, polyphyly), Gaya *et al.* (*MR* **112**: 528, 2008; phylogeny).

Fulgia Chevall. (1822) = Coniocybe fide Trevisan (*Flora* **45**: 1, 1862).

Fuligomyces Morgan-Jones & Kamal (1984), anamorphic *Pezizomycotina*, Hso.#eP.1. 1, India. See Morgan-Jones & Kamal (*Mycotaxon* **20**: 595, 1984), Khan *et al.* (*Mycotaxon* **49**: 477, 1993).

Fulminaria Gobi (1900) = Harpochytrium fide Atkinson (*Ann. mycol.* **1**: 479, 1903).

Fulminariaceae Gobi (1900) = Harpochytriaceae.

Fulvia Cif. (1954) = Passalora fide von Arx (*Proc. K. ned. Akad. Wet.* Ser. C, Biol. Med. Sci. **86**: 15, 1983), Curtis *et al.* (*Curr. Genet.* **25**: 318, 1994; phylogeny), Okada *et al.* (*FEMS Immunol. Med. Microbiol.* **16**: 39, 1996; ubiquinones), Crous *et al.* (*Mycol.* **93**: 1081, 2001; phylogeny), Crous & Braun (*CBS Diversity Ser.* **1**: 571 pp., 2003; review).

Fulvidula Romagn. (1936) nom. nud. = Gymnopilus fide Singer (*Agaric. mod. Tax.*, 1951).

Fulvifomes Murrill (1914), Hymenochaetaceae. 6, widespread. See Wagner & Fischer (*Mycol.* **94**: 998, 2002).

Fulvisporium Vánky (1997), Ustilentylomataceae. 1 (on *Poaceae*), Australia. See Vánky *et al.* (*Mycotaxon* **64**: 59, 1997), Aime *et al.* (*Mycol.* **98**: 896, 2006; phylogeny).

Fulvoflamma Crous (2006), anamorphic *Rhytismatales*. 1. See Crous (*Stud. Mycol.* **55**: 56, 2006).

fumagillin, an antibiotic (epoxide) from *Aspergillus fumigatus*; an amoebicide (McCowen *et al.*, *Science, N.Y.* **113**: 202, 1951).

Fumago Pers. (1822) nom. conf., anamorphic *Pezizomycotina*. See Friend (*TBMS* **48**: 371, 1965).

Fumagopsis Speg. (1910), anamorphic *Pezizomycotina*. See Kendrick & Carmichae *in* Ainsworth *et al.* (Eds) (*The Fungi* **4A**: 390, 1973), van der Aa & van Oorschot (*Persoonia* **12**: 415, 1985), Wu & Sutton (*MR* **99**: 1450, 1995).

Fumagospora G. Arnaud (1911), anamorphic *Capnodium*, Cpd.#eP.?. 2, Europe. See Hughes (*Mycol.* **68**: 693, 1976).

fumigatin, a benzoquinone antibiotic from *Aspergillus fumigatus*; antibacterial (Anslow & Raistrick, *Biochem. J.* **32**: 687, 1938).

Fumiglobus D.R. Reynolds & G.S. Gilbert (2006), Capnodiaceae. 1, Panama. See Reynolds & Gilbert (*Cryptog. Mycol.* **27**: 252, 2006).

fumitremorgin, a tremorgenic metabolite (indole derivative) of *Aspergillus fumigatus* (Yamazaki *et al.*, *Tetrahedron Lett.* **14**: 1241, 1975).

fumonisin, B1, B2 (**B1, B2**), toxic metabolites of *Fusarium moniliforme* (ear rot of maize). Infected maize eaten by horses causes equine leukocephalomalacia (Plattner *et al.*, *Mycol.* **82**: 698, 1990).

Funalia Pat. (1900) = Trametes fide Ryvarden (*Gen. Polyp.*: 150, 1991).

fungaemia (**fungemia**), fungi in the blood.

fungal (1) (n.), a fungus (obsol.); (2) (adj.), see fungous.

FUNGI. 36 class., 140 ord., 560 fam., 8283 gen. (+ 5101 syn.), 97861 spp. (Carpomycetes, Eumycota, Eumycophyta, Eumycetes, Fungales, Hysterophyta, Inophyta, Mycota, Mycetes, Mycetoideum, Mycetales, Mycetalia, Mycophyta, Mycophytes, Mycophycophytes). Kingdom of *Eukaryota*; the true fungi. 6 phyla. Eukaryotic organisms without plastids, nutrition absorptive (osmotrophic), never phagotrophic, lacking an amoeboid pseudopodial phase; cell walls containing chitin and β-glucans; mitochondria with flattened cristae and peroxiomes nearly always present; Golgi bodies or individual cisternae present; unicellular or filamentous and consisting of multicellular coenocytic haploid hyphae (homo- or heterokaryotic); mostly non-flagellate, flagella when present always lacking mastigonemes; reproducing sexually or asexually, the diploid phase generally short-lived, saprobic, mutualistic, or parasitic.

The organisms studied by mycologists, fungi, are mosly placed here, but other belong to the kingdoms *Chromista* and *Protozoa* (see Kingdoms of fungi, Phylogeny). Some authors use *Eumycota* as the kingdom name which has the advantage of avoiding confusion with 'fungi' (e.g. Barr, 1992). The kingdom name *Fungi* is retained by Cavalier-Smith (1993) and Corliss (1994); as most fungi belong here, and as the name *Fungi* is immediately familiar to most students, we retain that here.

Various arrangements and ranks have been proposed for the main groupings of *Fungi*. The rank of phylum is adopted here as that is currently almost universally used (Barr, 1992; Bruns *et al.*, 1992; Corliss, 1994; Kendrick, 1992) and was already advocated by von Arx (*Sydowia* **37**: 1, 1984); some authors nevertheless prefer lower ranks, e.g. classes (Nishida & Sugiyama, *Molec. Biol. Evol.* **10**: 431, 1993). The 6 phyla accepted here are:

(1) **Ascomycota**
(2) **Basidiomycota**
(3) **Chytridiomycota**
(4) **Glomeromycota**
(5) **Microsporidia**
(5) **Zygomycota**

The *Deuteromycotina* is not accepted as a formal taxonomic category, as in the previous edition of the *Dictionary*; they are not a monophyletic unit, but are fungi which have either lost a sexual phase or which are anamorphs of other phyla (mainly *Ascomycota*; some *Basidiomycota*); with modern molecular or ultrastructural techniques such fungi can be assigned to existing taxa; see under Anamorphic fungi. See Classification, Kingdoms of fungi, Literature, Phylogeny.

Fungi fossiles, see Fossil fungi.

Fungi Imperfecti, see Anamorphic fungi; fungi having no sexual state, presumptively mitotic.

fungi suilli, see cep.

fungi, higher, see higher fungi.

fungi, lower, see lower fungi.

fungicidal, able to kill fungus spores or mycelium.

Fungicides. Substances which are able to kill fungi s.l., esp. if lethal at low concentration. The term is sometimes loosely applied to fungistatic and genestatic substances. **contact -**, those which kill fungi when applied to a substratum surface; **eradicant -**, those applied to a substratum in which a fungus is already present, or used in disease control after infection has been established; **protective -**, those used to protect an organism against infection by a fungal pathogen (*TBMS* **33**: 155, 1950); **systemic -**, substances which are fungicidal (or fungistatic) when taken up systemically by an infected organism, usually a plant (see below). Heat, light, and other radiations have fungicidal properties, see sterilization.

Many chemical substances are fungicidal or inhibit fungal growth. Even pure water sometimes inhibits spore germination (e.g. of *Sclerotinia fructicola*) while conversely very low concentrations of some fungicides may stimulate spore germination or even be a prerequisite for fungal growth by supplying an essential trace element (see Nutrition). Today hundreds of fungicides are available in thousands of formulations for specific uses. Sulphur (elemental or in combination) and inorganic compounds of copper and mercury were the first fungicides employed and these three elements, frequently in organic combination, are still ingredients of important fungicides. Some of the more important fungicides (with indications of their main uses) are:

Sulphur: elemental sulphur (as flowers of sulphur, colloidal sulphur, wettable sulphur, etc.), lime sulphur.

Copper: bordeaux and burgundy mixtures (q.v.), cheshunt compound (q.v.), copper sulphate, basic copper sulphates, chlorides and carbonate, copper oxides (seed treatment), organo-coppers (copper oleate and resinate) (plant protection).

Mercury: mercuric chloride (corrosive sublimate) (wood preservation, soil treatment), mercurous chloride (calomel) (seed and soil treatment). Organomercurials are not now used in agriculture because of non-target effects.

Other inorganic compounds: borax and boric acid (storage decay of fruit), calcium hypochlorite (bleaching powder) (surface sterilization of biological material), potassium iodide (sporotrichosis), potassium dichromate, sodium fluoride, zinc chloride, and arsenic compounds (wood preservation), triphenyl tin salts (crop protection).

Organic compounds: formaldehyde (formalin) (seed and soil disinfection), 5-fluorocytosine (candidiasis, cryptococcosis), guanidines (dodine), dithiocarbamates (thiram, mancozeb, propineb), carboximides (iprodione), imidayols (prochloraz, imazalil), chlorinated hydrocarbons (chlorathalonil, didoran), methyl bromide and cresylic acid (soil sterilization), propionic acid (food preservation), propylene oxide (sterilization of biological materials), salicylanide (Shirlan) (plant protection, textile spoilage); see also antibiotics, copper, mercury, sulphur, systemic fungicides above and below.

Antibiotics: amphotericin (systemic mycoses), kasugamycin (systemic; rice blast), cycloheximide (crop protection and differential isolation of fungi), nystatin (candidiasis).

Systemic fungicides: (against plant diseases) acylalanines (furaxyl, metalaxyl), benzimidazoles (benomyl, carbendazim, thiabendazole), oxathiins (carboxin), pyrimidine derivatives (dimethirimol, ethirimol), triazoles (triadimefon, propinconazole). See Marsh (Ed.) (*Systemic fungicides*, 1972 [edn 2, 1977]).

Use of fungicides should be sparing, no more than necessary, and managed, ideally within the context of an integrated scheme, to avoid unnecessary pollution and the development of resistance. Methyl bromide is now rapidly being phased out of use from most countries because of its toxicity and effect on the ozone layer, although some use continues, notably in the USA. Alternating two different fungicides greatly lowers the risk of resistance development in a crop, as fungal strains which tolerate on fungicide are usually unable to survive the other (see Fungicide Resistance Action Committee, www.frac.info).

Lit.: Martin (*The scientific principles of plant protection* edn 6, 1973), *Guide to the chemicals used in crop protection* (edn 4, 1961; *Canada Dep. Agric., Publ. Res. Branch* **1093**), Horsfall (*Fungicides and their action*, 1945; *Principles of fungicidal action*, 1956), Torgeson (Ed.) (*Fungicides, an advanced treatise*, **1**, *Agricultural and industrial applications. Environmental interactions*, 1967; **2,** *Chemistry and physiology*, edn 2, 1983), Waller & Lenné (*Plant pathologist's pocketbook*, 2001), Smith (Ed.) (*Fungicides in crop protection. 100 years of progress*. 2 vol., 1985 [BCPC Monogr. **31**], Nene & Thapliyal (*Fungicides in plant disease control*. edn 3, 1993), Whitehead (Ed.) (*The UK Pesticide Guide 2001*). For 'Definitions of fungicide terms' and 'Recommended methods' for testing fungicides, see *Phytopath.* (**33**: 624, 1943; **34**: 401, 1944); also Zehr (Ed.) (*Methods for evaluating plant fungicides, nematocides and bactericides*, 1978).

Fungicolous fungi. Fungi growing on other fungi as parasites ('mycoparasites') commensals, or saprobes; fungicoles. Reliable estimates of the total number of fungi involved are not available; Hawksworth (1981) traced records of 1100 spp. of anamorphic fungi alone on 2500 other spp. of fungi. The fungi may be 'necrotrophic' (destructive), or 'biotrophic' (forming balanced relationships). Destructive effects may be mediated at a distance, by contact and coiling via antibiotics or hyphal-wall degrading enzymes, or hyphal interference (contact necrotrophs) or by penetration of the hyphae of the host (invasive necrotrophs). Biotrophic mycoparasitism occurs externally via haustoria (haustorial biotrophs), or via specialized contact cells which may produce colacosomes (Bauer & Oberwinkler, *Bot. Acta* **104**: 53, 1991) and which achieve direct cytoplasmic continuity with the host through fine interhyphal channels (fusion biotroph), or internally by entry of the complete thallus of the mycoparasite into the host (intracellular biotrophs). For a description of these host-parasite interfaces see Jeffries & Young (*Interfungal parasitic relationships*, 1994). Fungicolous fungi are an excellent source of novel biologically active and chemically interesting compounds (Schmidt *et al.*, *J. Nat. Prod.* **70**: 1317, 2007).

Particularly destructive mycoparasites are *Trichoderma* spp. (also of value in biocontrol; see esp. Papavizas (*Ann. Rev. Phytopathol.* **23**: 23, 1985, Kubicek & Harman (*Trichoderma and Gliocladium* **1**. Basic biology, taxonomy and genetics, 1998), Harman & Kubicek (*Trichoderma & Gliocladium* **2**. Enzymes, biological control and commercial applications, 1998); both have very wide host ranges (Domsch *et al.*, 1980). Others are restricted to particular host taxa, usually fams., gen. or spp. Examples (**host**: fungi) are: **Mycetozoa**: *Aphanocladium album*, *Polycephalomyces* spp., *Leucopenicillifer*, *Nectria myxomyceticola* anamorph *Verticillium rexianum*). See Ing (*Bull. BMS* **8**: 25, 1974; key), Rogerson & Stephenson (*Mycol.* **85**: 456, 1993). **Oomycota**: esp. other *Oomycota*. e.g. *Pythium oligandrum* on *Pythium ultimum* (Whipps & Lumsden, *Biocontrol Sci. Technol.* **1**: 75, 1991). **Chytridiomycota**: chytrids esp. other *Chytridiales*, e.g. *Allomyes* by *Catenaria allomycis* (Sykes & Porter, *Mycol.* **72**: 288, 1980; Powell, *Bot. Gaz.* **143**: 176, 1982), *Allomyces* and *Polyphagus* by *Rozella* and *Rozellopsis* (Held, *Bot. Rev.* **47**: 451, 1981; Powell, *Mycol.* **76**: 1039, 1984). **Zygomycota**: *Chaetocladium*, *Piptocephalis*, *Dimargaris* etc. On other *Mucorales* (Benjamin, *Aliso* **4**: 321, 1959; Jeffries, *Bot. J. Linn. Soc.* **91**: 135, 1985); *Stachybotrys chartarum* and *Anguillospora pseudolongissima* on spores of *Glomus* and *Gigaspora* (Paulitz & Menge, *Phytopathol.* **76**: 351, 1986). **Ascomycota**: the zenith for fungicolous fungi. *Ampelomyces quisqualis* on numerous *Erysiphales*; *Cordyceps ophioglossoides* on *Elaphomyces*; *Gona-*

tobotrys simplex; *Hansfordia pulvinata* on *Cercospora* etc.; *Nectria magnusiana* (anamorph *Fusarium epistromum*) on *Diatrypella* spp.; *Nematogonum ferrugineum* on *Nectria coccinea*; *Coniothyrium minitans* on *Botryotinia* and *Sclerotinia* sclerotia; *Pseudofusidium hansfordii* on *Mycovellosiella*; *Sphaerulomyces coralloides* on aquatic hyphomycetes; *Syspastospora parasitica* on *Beauveria*, *Hirsutella*, *Paecilomyces* and *Verticillium* (Tribe, *TBMS* **40**: 489, 1957; Huang, *CJB* **55**: 289, 1977); *Tremella karstenii* on *Colpoma juniperi*; *Tympanosporium parasiticum* on *Nectria cinnabarina*; trop. foliicolous spp. (esp. *Meliolaceae*) support a vast array of fungi (Hansford, *Mycol. Pap.* **15**, 1946; Deighton & Pirozynski, *Mycol. Pap.* **128**, 1972); Pirozynski, *Kew Bull.* **31**: 595, 1977; Hughes, *Mycol. Pap.* **166**, 1993). See also Lichenicolous fungi. **Basidiomycota**: *Eudarluca caricis* (anamorph *Sphaerellopsis filum*) on 226 spp. *Pucciniales* (Kranz, *Nova Hedw.* **24**: 169, 1974); *Tuberculina persicina* on 26 spp. *Uredinales*; *Hypomyces* spp. (anamorphs *Apiocrea*, *Cladobotryum*) esp. common on decaying Agaricales and Boletales (Arnold, *Bibliogr. Mill. Univ. bibliotek Jena* **25**, 1976; bibliogr. host index; Rogerson & Samuels, *Mycol.* **86**: 839, 1994, on agarics); *Mycogone perniciosa* (wet bubble disease) on cult. mushrooms (*CMI Descr.* **499**, 1976); *Asterophora* spp. on *Russula*; *Amblyosporium spongiosum* esp. on *Lactarius* (Nicot & Durand, *BSMF* **81**: 623, 1965); *Helminthosphaeria* spp. on *Clavariaceae*; *Pseudoboletus parasiticus* on *Scleroderma*. See Nicot (*BSMF* **31**: 393, 1967); key spp. on agarics, Tubaki (*Nagaoa* **5**: 11, 1955), de Hoog (*Persoonia* **10**: 33, 1978, keys), *Verticillium biguttatum* on *Rhizoctonia solani* (Boogert *et al.*, *Soil Biol. Biochem.* **24**: 159, 1992) and mushroom culture.

Lit.: (*Ann. Rev. Phytopathol.* **28**: 59, 1990), Barnett & Binder (*Ann. Rev. Phytopath.* **11**: 273, 1973), Buller (**3**), Burge (*Fungi in biological control systems*, 1988), Cooke (*The biology of symbiotic fungi*, 1977), Domsch *et al.* (*Compendium of soil fungi*, 2 vols., 1980), Fletcher *et al.* (*Mushrooms- pest and disease control*, 1986), Gams *et al.* (*in* Mueller *et al.* (Eds), *Biodiversity of fungi: inventory and monitoring methods*, 2004), Hashioka (*Forsch. Geb. Pflanzenkrankh.* **8**: 179, 1973; *Rep. Tottori Mycol. Inst.* **10**: 473, 1973; SEM), Hawksworth (*in* Cole & Kendrick (Eds), *The biology of conidial fungi* **1**: 171, 1981, conidial spp.), Helfer (*Pilze auf Pilzfruchtkörpern. Libri Botanici* **1**: 1, 1991), Jeffries & Young (*Interfungal parasitic relationships*, 1994), Lumsden (*in* Carroll & Wicklow (Eds), *The fungal community*: 275, 1992; review, examples), Madelin (*The Fungi* **3**: 253, 1968; review), Rudakov (*Mycol.* **70**: 150, 1978; physiol. groups), Seeler (*Farlowia* **1**: 119, 1943), Weindling (*Bot. Rev.* **4**: 475, 1938).

See also Ecology, Lichenicolous fungi.

fungiform, mushroom-shaped.

fungistasis (1) see mycostasis; (2) (**fungistatic**) (of a substance, or of a concentration of a fungicide), inhibiting fungus growth but not fungicidal. Cf. genestasis.

Fungites Casp. (1907), Fossil Fungi (mycel.) Fungi. 4 (Oligocene), Baltic Sea; Pacific Ocean.

Fungites Hallier (1865), Fossil Fungi (mycel.) Fungi. 1 (Tertiary), Germany.

fungivorous, using fungi as food; **fungivore**, an organism which does this.

fungizone, trade name for amphotericin B (q.v.).

Fungodaster Haller ex Kuntze (1891) = Leotia fide Hawksworth *et al.* (*Dictionary of the Fungi* edn 8, 1995).

fungoid, fungus-like.

Fungoidaster P. Micheli (1729) nom. inval. = Craterellus fide Pegler (*in litt.*).

Fungoides Tourn. (1719) nom. inval., Fungi. Used for a wide range of fungi (Ascomycota & Basidiomycota) with cup-like fruits, including discomycetes and *Nidulariales*.

fungology, mycology (obsol.).

fungoma, fungus ball formation.

fungophobia, a horror or dread of fungi (Hay, *British Fungi*: 6, 1887) (obsol.).

fungous, of, or having to do with, fungi; fungal.

funguria, the presence of fungi, particularly yeasts, in urine.

fungus (pl. **fungi**), (Lat. **fungus**, orig. *sfungus*, cognate with *spongia* from Gk *sphongis*, a sponge), champignon (Fr.), ciupercă (Rumanian), cogumelo (Port.), fungo (Ital.), gljiva and guba (Croat.), goba (Slovene), gomba (Hung.), grybas (kremblys) (Lith.), grzyb (Pol.), hongo (Span.), houba (Czech.), huba (Slovak), 真菌 (jun, chün) (Chin.), kavak (Hindi), kavaka (Sanskrit), këphurdhe (Albanian), 菌類 (kinrui) (Jap.), μύκης (Gk), makunã (Kikuyu), Pilz (Germ.), poonjalam (Tamil), seen (Estonian), sŋne (Latvian), sieni (Fin.), svamp (Scand.), zwam (Dutch), ГРИБ (Russ.), ГРЫБ ; (Byeloruss.), ГЪЬИВА and ГУБА (Serb.), ГЪБА (Bulgar.); 582.28 (Universal Decimal Classification); MACR.003 (semantic code; Perry & Kent, *Tools for machine literature searching*, 1958), one of the *Fungi*; [in Medicine: an abnormal sponge-like growth].

Fungus Adans. ex Kuntze (1898) = Agaricus L. fide Kirk (*in litt.*).

fungus ball, see aspergilloma.

fungus gnats, the *Mycetophilidae* (*Insecta*); see Mushroom culture.

fungus root, see mycorrhiza.

Fungus Tourn. ex Adans. (1763) ≡ Agaricus L.

fungus, ray, see ray fungi.

funicular, cord-like.

funicular cord (funiculus), the cord of hyphae by which the peridioles in *Nidulariaceae* (e.g. *Cyathus*) are at first fixed to the inner wall of the peridium; see splash cup.

Funicularius K.K. Baker & Zaim (1979), anamorphic *Pezizomycotina*, Hso.0eH.10. 1 (on mosquito), USA. See Baker & Zaim (*J. Invert. Path.* **34**: 200, 1979), Samson *et al.* (*Atlas of Entomopathogenic Fungi*: 187 pp., 1988).

funiculose (of hyphae), aggregated into rope-like strands, plectonematogenous (q.v.).

Funiliomyces Aptroot (2004), Amphisphaeriaceae. 1, Brazil. See Aptroot (*Stud. Mycol.* **50**: 309, 2004).

funoid, composed of rope-like strands or fibres; funicular (q.v.).

Furcaspora Bonar (1965), anamorphic *Pezizomycotina*, St.1bH.10. 3, USA. See Bonar (*Mycol.* **57**: 391, 1965), Punithalingam (*MR* **107**: 917, 2003).

furcate, forked.

Furcouncinula Z.X. Chen (1982) = Erysiphe fide Braun & Takamatsu (*Schlechtendalia* **4**: 3, 2000).

Furculomyces Lichtw. & M.C. Williams (1992), Legeriomycetaceae. 3 (in *Diptera*), Australia; USA. See Lichtwardt & Williams (*CJB* **70**: 1196, 1992),

White (*MR* **110**: 1011, 2006; phylogeny).

furfuraceous, covered with bran-like particles; scurfy.

Furia (A. Batko) Humber (1989), Entomophthoraceae. 17, widespread.

Fusamen (Sacc.) P. Karst. (1890), anamorphic *Pezizomycotina*, Cac.0eH.19+10. 5 (on *Salix*), Europe. See von Arx (*Biblthca Mycol.* **24**, 1970), Sutton (*The Coelomycetes*, 1980).

Fusariaceae Nann. (1934) = Nectriaceae.

fusaric acid (Fusarinsäure, Germ.), a pyridine-carboxylic acid from *Fusarium bulbigenum* var. *lycopersici*, *F. vasinfectum* and other *Hypocreaceae* able to induce wilt symptoms in tomato (Gäumann, *Phytopath.* **47**: 342, 1948, review; **48**: 670, 1958, mechanism of action). Cf. lycomarasmin.

Fusariella Sacc. (1884), anamorphic *Pezizomycotina*, Hso.≡ eP.15. 13, widespread (temperate). See Hughes (*Mycol. Pap.* **28**, 1949), Liu & Zhang (*Mycosystema* **25**: 145, 2006; China).

Fusariellites Babajan & Tasl. (1977), Fossil Fungi. 1 (Tertiary), former USSR.

Fusariopsis Horta (1919), anamorphic *Pezizomycotina*, Hsy.0eH.?. 1 (on humans), France. See Dodge (*Medical Mycology*: 860, 1935).

Fusarium Link (1809), anamorphic *Gibberella, Haematonectria*, Hso.≡ eH.15. *c.* 111 (saprobes and parasites), widespread. This important genus has fusoid, curved, septate macroconidia in slimy masses (sporodochia) on branched conidiophores. Smaller 0- or 1-septate microconidia and chlamydospores are common. A third conidial type, the mesoconidium, is recognized by some authors. The mycelium and spores are generally bright in colour. Important pathogens: *F. avenaceum* and *F. culmorum* (cereal foot rots), *F. bulbigenum* (*Narcissus* basal rot), *F. caeruleum* (dry rot of potato tubers), *F. conglutinans* (cabbage yellows), *F. oxysporum* (potato wilt) and its var. *cubense* (Panama disease of banana), *F. vasinfectum* (cotton wilt), and others. *F solani* is phylogenetically distinct. See Wollenweber & Reinking (*Die Fusarien*, 1935), Toussoun & Nelson (*A Pictorial Guide to the Identification of Fusarium Species*, 1969), Gerlach & Ershad (*Nova Hedwigia* **20**: 725, 1970; Iran), Booth (*The genus Fusarium*, 1971), Joffe (*Mycopath. Mycol. Appl.* **53**: 201, 1974), Sen Gupta (*Nova Hedwigia* **25**: 699, 1974; India), Nirenberg (*Mitt. biol. BundAnst. Ld- u. Forstw.* **169**, 1976; sect. *Liseola*), Booth (*Fusarium. Laboratory guide to the identification of the major species*, 1977), Gerlach *in* Subramanian (Ed.) (*Taxonomy of Fungi* **1**: 115, 1978), Pattin (*Acta Phytotox.* **6**, 1978; toxigenic spp.), Gerlach & Nirenberg (*Mitt. biol. BundAnst. Ld- u. Forstw.* **209**, 1982; revision), Teetor-Barsch & Roberts (*Mycopathologia* **84**: 3, 1983; entomogenous spp.), Gams (*Toxigenic Fungi: their toxins and health hazard*: 129, 1984; synanamorphs), Burgess *et al.* (*Mycol.* **77**: 212, 1985; *F. scirpi*), Marasas *et al.* (*Mycol.* **77**: 971, 1985; S Africa), Puhalla (*CJB* **63**: 179, 1985; *F. oxysporum*, VCGs), Raabe (*Pl. Dis.* **69**: 450, 1985; on *Hebe*), Szécsi & Dobrovolszky (*Mycopathologia* **89**: 89, 1985; DNA reassociation), Ye & Wu (*Acta phytopath. sin.* **15**: 87, 1985; esterases, graminicolous spp.), Burgess & Trimboli (*Mycol.* **78**: 223, 1986; *F. nygamai*), Correll *et al.* (*Phytopathology* **76**: 396, 1986; *F. oxysporum* f.sp. *apii*), Hatai *et al.* (*Journal of Wildlife Diseases* **22**: 570, 1986; in fish), Joffe (*Fusarium species: their biology and toxicology*, 1986), Marasas *et al.* (*Mycol.* **78**: 242, 1986; moniliformin production), Marasas *et al.* (*Mycol.* **78**: 678, 1986; *F. polyphialidicum*), Naiki (*Research Bulletin of the Faculty of Agriculture, Gifu University* **51**: 29, 1986; DNA content), Rubidge (*TBMS* **87**: 463, 1986; DNA plasmids), Tiedt *et al.* (*TBMS* **87**: 237, 1986; ultrastr.), Tiedt *et al.* (*TBMS* **87**: 237, 1986; ultrastr.), Ylimäki & Jamalainen (*Annales Agriculturae Fenniae* **25**: 9, 1986; Finland), Bosland & Williams (*CJB* **65**: 2067, 1987; *F. oxysporum*, *Brassicaceae*), Brayford (*TBMS* **89**: 347, 1987; *F. bugnicourtii*), Corley *et al.* (*Journal of Natural Products* **50**: 897, 1987; trichothecenes), Hocking & Andrews (*TBMS* **89**: 239, 1987; identification methods), Kim & Kim (*Korean Jl Pl. Path.* **3**: 137, 1987; pathogenicity on banana), Logrieco *et al.* (*Abstracts of the European Seminar 'Fusarium - Mycotoxins, Taxonomy, Pathogenicity'* Warsaw, 8-10 September 1987: no. 13, 1987; from potato), Marasas *et al.* (*Bothalia* **17**: 97, 1987; S Africa, bibliography), Okuda *et al.* (*J. Med. Vet. Mycol.* **25**: 177, 1987; clinical strain), Wasfy *et al.* (*Mycopathologia* **99**: 9, 1987; biochemistry, physiology), Wyk *et al.* (*TBMS* **88**: 347, 1987; ontogeny), Baayen & Gams (*Neth. Jl Pl. Path.* **94**: 273, 1988; on carnation), Blumenthal-Yonassi *et al.* (*Proceedings of the Japanese Association of Mycotoxicology* Supplement no. 1: 232, 1988; zearalenone production), Bottalico *et al.* (*Proceedings of the Japanese Association of Mycotoxicology* Supplement no. 1: 228, 1988; mycotoxins), Ellis (*Mycol.* **80**: 255, 1988; sect. *Liseola*), Ellis (*Mycol.* **80**: 734, 1988; sect. *Liseola*), Gerlagh & Blok (*Neth. Jl Pl. Path.* **94**: 17, 1988; on *Cucurbitaceae*), Lamprecht *et al.* (*Bothalia* **18**: 189, 1988; *F. tricinctum*), Logrieco & Bottalico (*TBMS* **90**: 215, 1988; on maize), Nelson *et al.* (*Mycol.* **79**: 884, 1987; *F. beomiforme*), Tiedt & Jooste (*TBMS* **90**: 531, 1988; ultrastr.), Tivoli (*Agronomie* **8**: 211, 1988; on potato, France), Visconti *et al.* (*Proceedings of the Japanese Association of Mycotoxicology* Supplement no. 1: 230, 1988; trichothecenes), Wyk *et al.* (*Phytophylactica* **20**: 73, 1988; on wheat, S Africa), Wyk *et al.* (*TBMS* **91**: 611, 1988; macroconidial development), Wyk *et al.* (*S. Afr. J. Bot.* **54**: 118, 1988; conidial development, cytology), Brayford (*Journal of Applied Bacteriology* Symposium Supplement Series no. 18 **67**: 47S, 1989; review), Burgess *et al.* (*Mycol.* **81**: 818, 1989; Australia), Chelkowski *et al.* (*J. Phytopath.* **124**: 155, 1989; on cereals), Cunfer (*Fusarium. Mycotoxins, Taxonomy and Pathogenicity* [Topics in Secondary Metabolism vol. 2]: 387, 1989; on *Claviceps*), Logrieco *et al.* (*Mycotoxin Research* Supplement: 9, 1989; mycotoxin production), Manka *et al.* (*J. Phytopath.* **124**: 143, 1989; *F. graminearum*), Nirenberg (*Fusarium. Mycotoxins, Taxonomy and Pathogenicity* [Topics in Secondary Metabolism vol. 2]: 179, 1989; Europe), Thrane (*Fusarium. Mycotoxins, Taxonomy and Pathogenicity* [Topics in Secondary Metabolism vol. 2]: 199, 1989; mycotoxins), Thrane & Frisvad (*Mycotoxin Research* Supplement: 21, 1989; mycotoxins), Jacobson & Gordon (*MR* **94**: 734, 1990; population biology), Leslie *et al.* (*Mycol. Soc. Amer. Newsl.* **41**: 24, 1990; sect. *Liseola*), Nelson *et al.* (*Mycol.* **82**: 99, 1990; morphology, physiology), Nirenberg (*Stud. Mycol.* **32**: 91, 1990; review), Ploetz (*Fusarium Wilt of Banana*, 1990), Thrane (*J. Microbiol. Meth.* **12**: 23, 1990; chemistry), Burgess *et al.* (*Australas. Pl. Path.* **20**: 86, 1991; identification media), Correll (*Phytopa-*

thology **81**: 1061, 1991; *F. oxysporum*), Das & Chattopadhyay (*Journal of Mycopathological Research* **29**: 51, 1991; mating populations), Leslie (*Phytopathology* **81**: 1058, 1991; *Gibberella fukjikuroi*), Liddell (*Phytopathology* **81**: 1044, 1991; review), Logrieco *et al.* (*Exp. Mycol.* **15**: 174, 1991; sect. *Sporotrichiella*), Miller *et al.* (*Mycol.* **83**: 121, 1991; trichothecenes), Peterson (*Phytopathology* **81**: 1051, 1991; phylogeny), Samson (*Fungi and Mycotoxins in Stored Products* ACIAR Proceedings no. **36**: 39, 1991; food-borne spp.), Wyk *et al.* (*MR* **95**: 284, 1991; development of microconidia), Gordon & Okamoto (*Exp. Mycol.* **16**: 245, 1992; VCGs), Gordon & Okamoto (*CJB* **70**: 1211, 1992; *F. oxysporum*), Guarro & Gené (*Mycoses* **35**: 109, 1992; mycoses), Nelson (*Mycopathologia* **117**: 29, 1992; *F. moniliforme*), Nelson *et al.* (*Appl. Environm. Microbiol.* **58**: 984, 1992; fumonisin production), Thrane *et al.* (*Modern Methods in Food Mycology* Developments in Food Science vol. **31**: 285, 1990; diagnostic methods), Tiedt & Jooste (*MR* **96**: 187, 1992; ultrastr.), Abramson *et al.* (*Can. J. Pl. Path.* **15**: 147, 1993; trichothecenes, Canada), Bridge *et al.* (*Pl. Path.* **42**: 264, 1993; physiology), Campbell & Leslie (*Phytopathology* **83**: 1413, 1993; genetic diversity), Elias *et al.* (*Molecular Plant-Microbe Interactions* **6**: 565, 1993; on *Lycopersicon*), Nelson *et al.* (*Ann. Rev. Phytopath.* **31**: 233, 1993; fumonisins), O'Donnell (*The Fungal Holomorph: Mitotic, Meiotic and Pleomorphic Speciation in Fungal Systematics*: 225, 1993; review), O'Donnell *et al.* (*Phytopathology* **83**: 1346, 1993; *F. solani*), Parisi *et al.* (*Phytopathology* **83**: 533, 1993; Australia), Summerell *et al.* (*MR* **97**: 1015, 1993; Australia), Wing *et al.* (*MR* **97**: 1441, 1993; toxigenic spp., Australia), Appel & Gordon (*Phytopathology* **84**: 1097, 1994; IGS diversity), Kelly *et al.* (*Phytopathology* **84**: 1293, 1994; RAPDs, *F. oxysporum*), McGinnis *et al.* (*J. Mycol. Médic.* **4**: 45, 1994; endocarditis), Mes *et al.* (*Pl. Path.* **43**: 362, 1994; on *Gladiolus*), Nelson *et al.* (*Clin. Microbiol. Rev.* **7**: 479, 1994; clinical aspects), Edel *et al.* (*Phytopathology* **85**: 579, 1995; molecular methods), Hagen & Hagen (*Mycopathologia* **129**: 143, 1995; morphometrics), Logrieco *et al.* (*Mycopathologia* **129**: 153, 1995; *F. sambucinum*), Thrane & Hansen (*Mycopathologia* **129**: 183, 1995; *F. sambucinum*), Brayford (*Sydowia* **48**: 163, 1996; species concepts), Burgess *et al.* (*Sydowia* **48**: 1, 1996; biodiversity), Leslie (*Sydowia* **48**: 32, 1996; genetics), O'Donnell (*Sydowia* **48**: 57, 1996; phylogeny), Rheeder *et al.* (*Mycol.* **88**: 509, 1996; *F. globosum*), Waalwijk *et al.* (*Sydowia* **48**: 90, 1996; DNA analysis), Gordon & Martyn (*Ann. Rev. Phytopath.* **35**: 111, 1997; evolution), Huang *et al.* (*Pl. Path.* **46**: 871, 1997; *F. moniliforme*), Kistler (*Phytopathology* **87**: 474, 1997; *F. oxysporum*), Leslie & Zeller (*Cereal Research Communications* **25**: 539, 1997; mutant strains), Nirenberg & Aoki (*Mycoscience* **38**: 329, 1997; *F. nisikadoi*), O'Donnell & Cigelnik (*Mol. Phylogen. Evol.* **7**: 103, 1997; rDNA variation), Sands *et al.* (*Pl. Dis.* **81**: 501, 1997; on *Erythroxylum*), Aoki & O'Donnell (*Mycoscience* **39**: 1, 1998; *F. kyushuense*), Hyun & Clark (*MR* **102**: 1259, 1998; *F. lateritium*), Marasas *et al.* (*Mycol.* **90**: 505, 1998; sect. *Arthrosporiella*), Mulè *et al.* (*Bulletin of the Institute for Comprehensive Agricultural Sciences, Kinki University* **6**: 23, 1998; trichothecenes), Nirenberg & O'Donnell (*Mycol.* **90**: 434, 1998), O'Donnell *et al.* (*Fungal Genetics Biol.* **23**: 57, 1998; Quorn fungus), O'Donnell *et al.* (*Mycol.* **90**: 465, 1998; phylogeny), Schütt *et al.* (*Mycotoxin Research* **14**: 35, 1998; moniliformins), Aoki & O'Donnell (*Mycol.* **91**: 597, 1999; *F. pseudograminearum*), Arie *et al.* (*Mycoscience* **40**: 311, 1999; mating types), O'Donnell *et al.* (*J. Phytopath.* **147**: 445, 1999; *F. oxysporum*), Baayen *et al.* (*Phytopathology* **90**: 891, 2000; *F. oxysporum*), O'Donnell *et al.* (*Proc. natn Acad. Sci. U.S.A.* **97**: 7905, 2000; *F. graminearum*), O'Donnell *et al.* (*Mycoscience* **41**: 61, 2000; *G. fujikuroi*), Suga *et al.* (*MR* **104**: 1175, 2000; *F. solani*), Wikler & Gordon (*CJB* **78**: 709, 2000; *F. circinatum*), Aoki *et al.* (*Mycoscience* **42**: 461, 2001; *G. fujikuroi*), Backhouse *et al.* (*Fusarium* Paul E. Nelson Memorial Symposium: 122, 2001; biogeography), Kistler (*Fusarium* Paul E. Nelson Memorial Symposium: 70, 2001; host specificity), Steenkamp *et al.* (*Mycol.* **94**: 1032, 2002; cryptic speciation), Yli-Mattila *et al.* (*MR* **106**: 655, 2002; phylogeny), Skovgaard *et al.* (*Mycol.* **95**: 630, 2003; *F. commune*), Summerell *et al.* (*Pl. Dis.* **87**: 117, 2003; identification), Geiser *et al.* (*Eur. J. Pl. Path.* **110**: 473, 2004; molecular identification), Leslie *et al.* (*Eur. J. Pl. Path.* **110**: 611, 2004; mating populations), O'Donnell *et al.* (*J. Clin. Microbiol.* **42**: 5109, 2004; human pathogens), Geiser *et al.* (*Mycol.* **97**: 191, 2005; *F. xylarioides*), Leslie *et al.* (*Mycol.* **97**: 718, 2005; *F. sacchari*), Akinsanmi *et al.* (*Pl. Path.* **55**: 494, 2006; Australia), Desjardins (*Fusarium Mycotoxins* Chemistry, Genetics and Biology, 2006; mycotoxins), Leslie & Summerell (*Fusarium Labor. Manual*, 2006; identification manual), Scott & Chakraborty (*MR* **110**: 1413, 2006; *F. pseudograminearum*), Bentley *et al.* (*Phytopathology* **98**: 250, 2008; Australia), Suga *et al.* (*Phytopathology* **98**: 159, 2008; Japan).

Fuscidea V. Wirth & Vězda (1972), Fuscideaceae (L). *c.* 34, widespread (esp. temperate). See Inoue (*Hikobia, Suppl.* **1**: 161, 1981), Inoue (*Hikobia, Suppl.* **1**: 177, 1981), Oberhollenzer & Wirth (*Beih. Nova Hedwigia* **79**: 537, 1984; 9 spp.), Oberhollenzer & Wirth (*Stuttg. Beitr. Naturk.*, 1985), Brodo & Wirth *in* Glenn *et al.* (Eds) (*Lichenogr. Thomsoniana*: 149, 1998), Kantvilas (*Biblthca Lichenol.* **78**: 169, 2001; Tasmania), Kantvilas & McCarthy (*Lichenologist* **35**: 397, 2003), Wedin *et al.* (*MR* **109**: 159, 2005; phylogeny), Miądlikowska *et al.* (*Mycol.* **98**: 1088, 2006; phylogeny), Bylin *et al.* (*Biblthca Lichenol.* **96**: 49, 2007; phylogeny).

Fuscideaceae Hafellner (1984), Umbilicariales (±L). 6 gen. (+ 3 syn.), 55 spp.

Lit.: Oberhollenzer & Wirth (*Biblthca Lichenol.* **38**: 363, 1990), Farkas (*Biblthca Lichenol.* **58**: 97, 1995), Ihlen & Tønsberg (*Bryologist* **99**: 32, 1996), Kantvilas & Vězda (*Nordic Jl Bot.* **16**: 325, 1996), Ihlen & Ekman (*Biol. J. Linn. Soc.* **77**: 535, 2002), Reeb *et al.* (*Mol. Phylogen. Evol.* **32**: 1036, 2004), Miądlikowska *et al.* (*Mycol.* **98**: 1088, 2006; phylogeny), Bylin *et al.* (*Biblthca Lichenol.* **96**: 49, 2007; phylogeny).

Fuscoboletinus Pomerl. & A.H. Sm. (1962) = Suillus Gray fide Singer *et al.* (*Mycol.* **55**: 362, 1963).

Fuscocerrena Ryvarden (1982), Polyporaceae. 1, widespread. See Gilbertson & Ryvarden (*N. Amer. Polyp.* **1**: 279, 1987).

Fuscoderma (D.J. Galloway & P.M. Jørg.) P.M. Jørg. & D.J. Galloway (1989), Pannariaceae (L). 3, south temperate. See Jørgensen (*N.Z. Jl Bot.* **37**: 257, 1999), Jørgensen & Sipman (*Lichenologist* **34**: 33, 2002; New Guinea).

Fuscolachnum J.H. Haines (1989), Hyaloscyphaceae. 7 (on leaves), Europe; N. America. See Haines (*Mem. N. Y. bot. Gdn* **49**: 315, 1989).

Fuscopannaria P.M. Jørg. (1994), Pannariaceae (L). 33, widespread (esp. temperate). See Jørgensen (*Op. Bot.* **45**, 1978; Eur. spp., as *Pannaria*), Jørgensen (*N.Z. Jl Bot.* **37**: 257, 1999), Jørgensen (*J. Hattori bot. Lab.* **89**: 247, 2000; E Asia), Jørgensen (*Bryologist* **103**: 670, 2000; N America), Jørgensen (*Bryologist* **103**: 104, 2000; sorediate taxa), Jørgensen (*Biblthca Lichenol.* **78**: 109, 2001; Australia), Ekman & Jørgensen (*CJB* **80**: 625, 2002; phylogeny), Jørgensen (*N.Z. Jl Bot.* **40**: 327, 2002; New Zealand), Jørgensen (*J. Hattori bot. Lab.* **92**: 225, 2002; Asia), Jørgensen & Zhurbenko (*Bryologist* **105**: 465, 2002; Arctic), Jørgensen (*Lichenologist* **36**: 207, 2004; Asia), Jørgensen (*Lichenologist* **37**: 221, 2005; Europe, key), Jørgensen (*Bryologist* **108**: 255, 2005; N America), Jørgensen & Sipman (*J. Hattori bot. Lab.* **100**: 695, 2006; New Guinea), Miądlikowska *et al.* (*Mycol.* **98**: 1088, 2006; phylogeny), Jørgensen (*Lichenologist* **39**: 235, 2007; SE Asia).

Fuscophialis B. Sutton (1977), anamorphic *Pezizomycotina*, Hso.≡ eP.16. 3, widespread (tropical). See Sutton (*Boln Soc. argent. Bot.* **17**: 158, 1977).

Fuscoporella Murrill (1907) = Phellinus fide Pegler (*The polypores [Bull. BMS Suppl.]*, 1973).

Fuscoporia Murrill (1907) = Phellinus fide Donk (*Verh. K. ned. Akad. Wet.* tweede sect. **62**: 1, 1974).

Fuscoscypha Svrček (1987), Hyaloscyphaceae. 1, Europe. See Huhtinen (*Karstenia* **29**: 45, 1990).

fuscous, dusky; too brown for a grey (Corner, 1958).

Fusculina Crous & Summerell (2006), anamorphic *Pleosporales*. 1, Australia. See Crous & Summerell (*Fungal Diversity* **23**: 334, 2006).

fuseau, a fusoid macroconidium of a dermatophyte (e.g. *Microsporum*); a spindle (obsol.).

Fusella Sacc. (1886), anamorphic *Pezizomycotina*, Hso.0eP.?. 3, Europe; Africa.

Fusellites Babajan & Tasl. (1977), Fossil Fungi. 1 (Tertiary), former USSR.

Fusicatena K. Matsush. & Matsush. (1996), anamorphic *Pezizomycotina*, Hso.?.?. 1, Japan. See Matsushima & Matsushima (*Matsush. Mycol. Mem.* **9**: 32, 1996).

Fusichalara S. Hughes & Nag Raj (1973), anamorphic *Pezizomycotina*, Hso.≡ eP.15. 4, widespread. See Hughes & Nag Raj (*N.Z. Jl Bot.* **11**: 662, 1973), Davydkina (*Mikol. Fitopatol.* **25**: 26, 1991; Russia).

Fusicladiella Höhn. (1919), anamorphic *Mycosphaerella*, Hso.1eH.1. 5, widespread (tropical). See Deighton (*Mycol. Pap.* **101**, 1965), Schubert *et al.* (*Schlechtendalia* **9**, 2003).

Fusicladiites Babajan & Tasl. (1973), Fossil Fungi. 1 (Tertiary), former USSR.

Fusicladina Arnaud (1952) nom. inval., anamorphic *Pezizomycotina*, Hso.1eH.?. 1, Europe.

Fusicladiopsis Karak. & Vassiljevsky (1937) ≡ Karakulinia.

Fusicladiopsis Maire (1907) = Stemphylium fide Hughes (*CJB* **36**: 727, 1958).

Fusicladium Bonord. (1851), anamorphic *Acantharia, Venturia*, Hsp.1eP.10. 87, widespread. See Hughes (*CJB* **31**: 560, 1953), Ondřej (*Česká Mykol.* **25**: 165, 1971), Sutton & Pascoe (*Aust. Syst. Bot.* **1**: 79, 1988; on *Parahebe*), Braun *et al.* (*Taxon* **51**: 557, 2002; nomencl.), Braun *et al.* (*Mycol. Progr.* **2**: 3, 2003; phylogeny), Partridge & Morgan-Jones (*Mycotaxon* **85**: 357, 2003), Schubert *et al.* (*Schlechtendalia* **9**, 2003; monograph), Beck *et al.* (*Mycol. Progr.* **4**: 111, 2005; phylogeny), Schubert & Braun (*Mycol. Progr.* **4**: 101, 2005), Ruszkiewicz-Michalska (*Acta Mycologica* Warszawa **41**: 285, 2006; Poland), Crous *et al.* (*Stud. Mycol.* **58**: 185, 2007; phylogeny).

Fusicladosporium Partr. & Morgan-Jones (2003) = Fusicladium fide Partridge & Morgan-Jones (*Mycotaxon* **85**: 360, 2003), Schubert *et al.* (*Schlechtendalia* **9**, 2003).

fusicoccin, a tricarboxylic terpene from *Fusicoccum amygdali* which induces stomatal opening (Turner & Granati, *Nature* **223**: 1070, 1969) and promotes spore germination and cell elongation. See also Chain *et al.* (*Physiol. Pl. Path.* **1**: 495, 1971; toxicity).

Fusicoccum Corda (1829), anamorphic *Botryosphaeria*, St.0eH.1. *c.* 91, widespread. *F. putrefaciens* (cranberry (*Vaccinium*) end rot). See Morgan-Jones & White (*Mycotaxon* **30**: 117, 1987; redescr. anamorph of *Botryosphaeria ribis*), Rayachhetry *et al.* (*Mycol.* **88**: 239, 1996; biocontrol), Gardner (*Mycol.* **89**: 298, 1997; Hawaii), Crous & Palm (*Sydowia* **51**: 167, 1999; review), Denman *et al.* (*Stud. Mycol.* **45**: 129, 2000; review), Phillips (*Mycotaxon* **76**: 135, 2000; on *Populus*), Smith *et al.* (*Mycol.* **93**: 277, 2001; on *Eucalyptus*), Zhou & Stanosz (*MR* **105**: 1033, 2001; phylogeny), Phillips (*Phytopath. Mediterr.* **41**: 3, 2002; on *Vitis*), Slippers *et al.* (*Mycol.* **96**: 1030, 2004; phylogeny), Slippers *et al.* (*Mycol.* **96**: 83, 2004; phylogeny), Crous *et al.* (*Stud. Mycol.* **55**: 235, 2006; phylogeny).

Fusicolla Bonord. (1851), anamorphic *Pezizomycotina*, Hsp.0eH.?. 5, widespread.

Fusicytospora Gutner (1934) = Phomopsis (Sacc.) Bubák. fide Petrak (*Sydowia* **1**: 61, 1947).

fusidic acid, an antibacterial antibiotic from *Fusidium coccineum* (Godfredsen, *Nature* **193**: 987, 1962); = ramycin (from *Micromucor ramanniana*).

Fusidites Mesch. (1902), Fossil Fungi. 1 (Oligocene), Europe.

Fusidium Link (1809) nom. rej., anamorphic *Pezizomycotina*. 1, widespread. See Booth (*Mycol. Pap.* **104**, 1966), Cooper (*N.Z. Jl Bot.* **43**: 323, 2005; New Zealand).

Fusidomus Grove (1929) = Fusarium fide Sutton (*Mycol. Pap.* **141**, 1977).

fusiform, spindle-like; narrowing toward the ends (Fig. 23.1).

fusiform rust (of pine), *Cronartium quercuum* (Czabator, *U.S. For. Serv. Res. Pap.* **65**, 1971).

Fusiformisporites Rouse (1962), Fossil Fungi. 8 (Tertiary), India; N. America.

Fusispora Fayod (1889) = Lepiota fide Singer & Smith (*A monograph on the genus Galerina Earle*, 1964).

Fusisporella Speg. (1910), anamorphic *Pezizomycotina*, Hsp.1eH.?. 1 or 2, America (tropical).

Fusisporium Link (1809) = Fusarium fide Wollenweber & Reinking (*Die Fusarien*: 5, 1935), Hughes (*CJB* **36**: 727, 1958).

Fusitheca Bonord. (1864) nom. dub., ? Fungi.

fusoid, somewhat fusiform.

Fusoidispora Vijaykr., Jeewon & K.D. Hyde (2005), Annulatascaceae. 1, Hong Kong. See Vijaykrishna *et al.* (*Sydowia* **57**: 272, 2005).

Fusoma Corda (1837) nom. dub., anamorphic *Pezizo-*

mycotina. See Hughes (*CJB* **36**: 727, 1958).

Fusticeps J. Webster & R.A. Davey (1980), anamorphic *Pezizomycotina*, Hso.≡ eP.1. 1 (aquatic), Malaysia. See Webster & Davey (*TBMS* **75**: 341, 1980), Hyde & Goh (*Mycoscience* **39**: 199, 1998).

fuzz-ball, see puff-ball.

fuzzy coat, the outer gelatinous coat of an ascus, esp. of one staining blue in iodine; see ascus.

G, Conservatoire et Jardin Botanique de la Ville de Genève (Geneva, Switzerland); founded 1817; funded by the city.

Gabarnaudia Samson & W. Gams (1974), anamorphic *Ceratocystidaceae*, Hso.0eH.15. 5, Europe; Asia. See Samson & Gams (*Stud. Mycol.* **6**: 88, 1974), Hausner & Reid (*CJB* **82**: 752, 2004; phylogeny), Gao *et al.* (*Mycosystema* **24**: 603, 2005; China).

Gabura Adans. (1763) nom. rej. = Collema F.H. Wigg. fide Hawksworth *et al.* (*Dictionary of the Fungi* edn 8, 1995).

Gaertneriomyces D.J.S. Barr (1980), Spizellomycetaceae. 3 (from soil), N. America; Europe. See Barr (*CJB* **58**: 2386, 1980).

Gaeumannia Petr. (1950) = Melchioria fide Müller & von Arx *in* Ainsworth *et al.* (Eds) (*The Fungi* **4A**: 87, 1973).

Gaeumanniella Petr. (1952), Pezizomycotina. 1, USA. See Petrak (*Sydowia* **6**: 162, 1952).

Gaeumannomyces Arx & D.L. Olivier (1952), Magnaporthaceae. Anamorph *Harpophora*. *c*. 11 (on grass roots), widespread. *G. graminis* (with 3 vars.; whiteheads and take-all (q.v.) of cereals). See Simonsen (*Friesia* **9**: 361, 1971; anamorph), Walker (*TBMS* **58**: 427, 1972), Weste (*TBMS* **59**: 133, 1972; root infection), Walker (*Mycotaxon* **11**: 1, 1980; taxonomy), Bryan *et al.* (*Appl. Environm. Microbiol.* **61**: 681, 1995), Fouly *et al.* (*Mycol.* **89**: 590, 1997; restriction analysis), Augustin *et al.* (*J. Phytopath.* **147**: 109, 1999; RAPDs), Bryan *et al.* (*MR* **103**: 319, 1999; DNA, host range), Ward & Bateman (*New Phytol.* **141**: 323, 1999; DNA), Gams (*Stud. mycol.* **45**: 187, 2000; anam.), Rachdawong *et al.* (*Pl. Dis.* **86**: 652, 2002; molecular identification), Wong (*MR* **106**: 857, 2002), Freeman & Ward (*Molecular Plant Pathology* **5**: 235, 2004; review), Saleh & Leslie (*Mycol.* **96**: 1294, 2004; phylogeny).

Gaillardiella Pat. (1895), Nitschkiaceae. 1 (from wood), S. America. See Petrak (*Sydowia* **4**: 158, 1950), Subramanian & Sekar (*Kavaka* **18**: 19, 1993), Huhndorf *et al.* (*MR* **108**: 1384, 2004).

Galactinia (Cooke) Boud. (1885) = Peziza Fr. fide Seaver (*North American Cup Fungi* (Operculates), 1928).

Galactiniaceae Berthet ex Le Gal (1969) = Pezizaceae.

Galactomyces Redhead & Malloch (1977), Dipodascaceae. 6, widespread. See de Hoog *et al.* (*Stud. Mycol.* **29**, 1986; key), de Hoog *et al. in* Kurtzman & Fell (Eds) (*Yeasts, a taxonomic study* 4th edn: 209, 1998), Naumova *et al.* (*Antonie van Leeuwenhoek* **80**: 263, 2001; sibling species), Hoog & Smith (*Stud. Mycol.* **50**: 489, 2004; phylogeny), Pimenta *et al.* (*Int. J. Syst. Evol. Microbiol.* **55**: 497, 2005), Romano *et al.* (*The Yeast Handbook* **2**: 13, 2006; review), Suh & Blackwell (*MR* **110**: 220, 2006; in beetle guts), Suh *et al.* (*Mycol.* **98**: 1006, 2006; phylogeny).

Galactopus Earle (1909) = Mycena fide Singer (*Agaric. mod. Tax.*, 1951).

galeate, hooded; hat-shaped or helmet-shaped.

Galeoscypha Svrček & J. Moravec (1989), Pyronemataceae. 1, Europe. See Svrček & Moravec (*Česká Mykol.* **43**: 210, 1989).

Galera (Fr.) P. Kumm. (1871) [non *Galera* Blume 1825, *Orchidaceae*] = Galerina fide Kuyper (*in litt.*).

Galeraicta Preuss (1852) ? = Rabenhorstia fide Saccardo (*Syll. fung.* **3**: 1, 1884), Sutton (*Mycol. Pap.* **141**, 1977).

Galerella Earle (1909), Bolbitiaceae. 7, widespread. See Hausknecht & Contu (*Öst. Z. Pilzk.* **12**: 31, 2003).

Galerina Earle (1909), Strophariaceae. *c*. 250 (freq. bryophilous), widespread. See Kühner (*Le Genre Galera*, 1935), Smith & Singer (*A monograph on the genus Galerina Earle*, 1964), Barkman (*Coolia* **14**: 49, 1969; key 50 Dutch spp.), Wells & Kempton (*Lloydia* **32**: 369, 1969; Alaskan spp.), Pegler & Young (*Kew Bull.* **27**: 483, 1972; basidiospore form), Kühner (*BSMF* **88**: 41, 1973; alpine spp.), Kühner (*BSMF* **88**: 119, 1973; alpine spp.), Gulden *et al.* (*MR* **105**: 432, 2001; *G. marginata* complex), Wood (*Aust. Syst. Bot.* **14**: 615, 2001; Australia spp.), Gulden (*Mycol.* **97**: 823, 2005; polyphyly).

Galeromycena Velen. (1947) = Macrocystidia fide Pegler (*in litt.*).

Galeropsidaceae Singer (1962) nom. rej. = Bolbitiaceae. The *Galeropsidaceae* are listed as nom. rej. against *Cortinariaceae*.

Galeropsina Velen. (1947) = Psilocybe (Fr.) P. Kumm. fide Singer (*Agaric. mod. Tax.* 4th ed, 1986).

Galeropsis Velen. (1930), Bolbitiaceae. 16 (xerophytic), widespread. Perhaps *Agaricaceae*. See Singer (*Proc. K. ned. Akad. Wet.* Ser. C, Biol. Med. Sci. **66**: 106, 1963; key), Sarasini (*Gasteromiceti Epigei*: 406 pp., 2005).

Galerula P. Karst. (1879) nom. dub. ? = Galerina fide Singer & Smith (*A monograph on the genus Galerina Earle*, 1964).

Galiella Nannf. & Korf (1957), Sarcosomataceae. 8, widespread (north temperate). See Korf (*Mycol.* **49**: 107, 1957), Cabello (*Boln Soc. argent. Bot.* **25**: 395, 1988), Li & Kimbrough (*CJB* **74**: 1651, 1995; spore ontogeny), Zhuang & Wang (*Mycotaxon* **67**: 355, 1998; spp. China), Harrington *et al.* (*Mycol.* **91**: 41, 1999; phylogeny), Pant (*Mycotaxon* **79**: 315, 2001; India), Köpcke *et al.* (*Phytochem.* **60**: 709, 2002; chemistry), Hansen & Pfister (*Mycol.* **98**: 1029, 2006; phylogeny).

Gall. A predictable and consistent plant deformation that occurs in response to the stimulus of a foreign organism, often a swelling or outgrowth; a cecidium.

Fungi producing galls directly on plants include some *Exobasidiales* (Ing, *in* Williams, 1994: 67; key 35 Eur. sp.), *Pucciniales* (Preece & Hicks, *in* Williams (Ed.), 1994: 57; 87 Br. spp.) and *Plasmodiophorales* (Preece, *Mycologist* **16**: 27, 2002). The gall midges (*Cecidomyiidae*) have mycophagous larvae implicated in the gall reaction (Harris, *in* Williams (Ed.), 1994: 201); some are only mycophagous and do not cause galls; **ambrosia -s** are caused by *Macrophoma* and anamorphs of *Botryosphaeria* (Bisset & Borkent, *in* Pirozynski & Hawksworth (Eds), 1988: 203).

Lichens have been interpreted as galls induced by an alga in a fungus (Moreau & Moreau, *BSMF* **34**: 84, 1918; see also cephalodium) but there are significant differences (Hawksworth & Honegger, *in* Williams (Ed.), 1994: 77); **-s on lichens**, may be caused by some lichenicolous fungi (e.g. *Biatoropsis*, *Plec-*

tocarpon, *Polycoccum*, *Thamnogalla*), nematodes (Siddiqi & Hawksworth, *Lichenologist* **14**: 175, 1982), mites (Gerson & Seaward, *in* Seaward (Ed.), *Lichen ecology*: 69, 1977), or be of unknown origin (Grummann, *Bot. Jb.* **80**: 101, 1960).

See Mani (*Ecology of plant galls*, 1964), Redfern & Shirley (*British plant galls*, 2002), Williams (Ed.) (*Plant galls: organisms, interactions, populations*, 1994), Coevolution.

Gallacea Lloyd (1905), Gallaceaceae. 6, Australia; New Zealand. See Trappe & Claridge (*Australasian Mycologist* **22**: 27, 2003).

Gallaceaceae Locq. ex P.M. Kirk (2008), Hysterangiales. 2 gen., 9 spp.

Gallaicolichen Sérus. & Lücking (2007), Pezizomycotina (L). 1, Australasia; Hawaii. See Sérusiaux & Lücking (*Biblthca Lichenol.* **95**: 509, 2007).

Gallowaya Arthur (1906), Coleosporiaceae. 3 (on *Pinus* (0, III, microcyclic) (*Pinaceae*)), N. America; Siberia. *G. crowellii* on *Pinus*. See Dodge (*J. Agric. Res.* **31**: 641, 1925; morphology), Cummins & Hiratsuka (*Illustr. Gen. Rust Fungi rev. edit.*, 1983; syn. of *Coleosporium*).

Galorrheus (Fr.) Fr. (1825) [non *Galorrheus* Haw. 1812, *Euphorbiaceae*] ≡ Lactarius.

Galzinia Bourdot (1922), Corticiaceae. 9, widespread. See Rogers (*Mycol.* **36**: 70, 1944), Olive (*Mycol.* **46**: 794, 1954), Boidin & Gilles (*BSMF* **110**: 185, 1994; key).

Galziniella Parmasto (1968) = Sistotrema Fr. fide Eriksson *et al.* (*Cortic. N. Europ.* **7**, 1984).

Gambleola Massee (1898), Pucciniosiraceae. 1 (on *Berberis* (*Berberidaceae*)), India; Nepal. Sometimes included in *Pucciniosira*. See Buriticá (*Rev. Acad. Colomb. Cienc.* **18** no. 69: 131, 1991; as syn. of *Pucciniosira*).

gametangium (pl. **-ia**) (gametange), cell containing gametes or gametic nuclei; the gametangium may initially be diploid and is the site of meiosis, or haploid. See zygangium.

gamete, naked uninucleate haploid cell with the sole function of fusing with another gamete to produce a zygote; sometimes used for the sex-nuclei of coenogametes.

gametogenesis, the development of gametes.

gametophyte, a haploid or sexual plant; haplont or haplophase. Cf. sporophyte.

gametothallus, a thallus producing gametes; cf. sporothallus.

gamma particle, a DNA-containing cytoplasmic organelle in the zoospore of *Blastocladiella emersonii* (Myers & Cantino, *The gamma particle*, 1974; Barstow & Lovett, *Mycol.* **67**: 518, 1975).

Gamolpidium Vlădescu (1892), ? Chytridiales. 2, Rumania.

Gamonaemella Fairm. (1922) = Gamospora fide Clements & Shear (*Gen. Fung.*, 1931), Sutton (*Mycol. Pap.* **141**, 1977).

Gamosphaera Dumort. (1822) ≡ Nemania.

Gamospora Sacc. (1885) ? = Wiesneriomyces fide Sutton (*Mycol. Pap.* **141**, 1977).

Gamosporella Speg. (1888) nom. dub., anamorphic *Pezizomycotina*. See Sutton (*Mycol. Pap.* **141**, 1977).

Gampsonema Nag Raj (1975), anamorphic *Pezizomycotina*, St.≡ eH.10. 1, widespread. See Nag Raj (*CJB* **53**: 1621, 1975).

Gamsia M. Morelet (1969), anamorphic *Pezizomycotina*, Hso.0eP.19. 2, widespread. See Morelet (*Annales de la Société des Sciences Naturelles et d'Archéologie de Toulon et du Var* **21**: 105, 1969).

Gamsiella (R.K. Benj.) Benny & M. Blackw. (2004), Mortierellaceae. 1, USA. See Benny & Blackwell (*Mycol.* **96**: 147, 2004), White *et al.* (*Mycol.* **98**: 872, 2006; phylogeny).

Gamsylella M. Scholler, Hagedorn & A. Rubner (1999), anamorphic *Orbiliaceae*, Hso.?.?. 6, widespread. See Scholler *et al.* (*Sydowia* **51**: 108, 1999), Li *et al.* (*Mycol.* **97**: 1034, 2005; phylogeny, synonymy with *Dactylellina*).

Gamundia Raithelh. (1979), Tricholomataceae. 6, Europe; S. America (temperate). See Raithelhuber (*Metrodiana* **8**: 34, 1979), Antonín (*Persoonia* **18**: 341, 2004; type studies European spp.), Antonín (*Mykol. Listy* **90-91**: 20, 2004; key European spp.), Antonín & Noordeloos (*A monograph of the genera Hemimycena,* , 2004).

gangliform, having knots; knotted.

Gangliophora Subram. (1992), anamorphic *Pezizomycotina*, Hso.≡ eP.19. 1, Taiwan. See Subramanian (*Proc. Indian natn Sci. Acad.* Part B. Biol. Sci. **58**: 188, 1992).

Gangliophragma Subram. (1978) = Dactylella fide Subramanian (*Kavaka* **5**: 94, 1977), Rubner (*Stud. Mycol.* **39**, 1996).

gangliospore, Subramanian's (*Curr. Sci.* **31**: 410, 1962) term for aleuriospore in the sense of 'holoblastic conidium'.

Gangliostilbe Subram. & Vittal (1976), anamorphic *Pezizomycotina*, Hsy.≡ eP.1. 2, pantropical. See Subramanian & Vittal (*Kavaka* **3**: 70, 1975), Petrini (*Sydowia* **37**: 238, 1984; Ethiopia), Subramanian & Bhat (*Kavaka* **15**: 41, 1987; India), Mercado-Sierra *et al.* (*Nova Hedwigia* **64**: 455, 1997; Costa Rica).

Ganoderma P. Karst. (1881), Ganodermataceae. Anamorph *Thermophymatospora*. *c.* 80, widespread (esp. tropical). *G. applanatum* (wood decay); *G. lucidum* (wood decay); *G. philippii* (syn. *G. pseudoferreum*; root rot of cacao, coffee, rubber, tea, etc.); *G. orbiformum* (syn. *G. boninense*; root rot of palms). See Moncalvo & Ryvarden (*Syn. Fung.* **11**: 114 pp., 1997), Ryvarden (*Mycol.* **92**: 180, 2000), Hong & Jung (*Mycol.* **96**: 742, 2004; phylogeny).

Ganodermataceae Donk (1948), Polyporales. 4 gen. (+ 9 syn.), 117 spp.

Lit.: Steyaert (*Persoonia* **7**: 55, 1972; SE Asian spp.), Steyaert (*Bull. Jard. Bot. Nat. Belg.* **50**: 135, 1980; subgen. classif.), Zhao (*Biblthca Mycol.* **132**: 176 pp., 1989), Moncalvo *et al.* (*Ganoderma: Systematics, Phytopathology and Pharmacology* Proceedings of Contributed Symposium 59A, B, 5th International Mycological Congress, Vancouver, August 14-21, 1994: 31, 1995), Moncalvo *et al.* (*MR* **99**: 1489, 1995), Moncalvo *et al.* (*Mycol.* **87**: 223, 1995), Moncalvo & Ryvarden (*Syn. Fung.* **11**: 114 pp., 1997), Gottlieb *et al.* (*MR* **104**: 1033, 2000), Miller *et al.* (*Ganoderma Diseases of Perennial Crops*: 159, 2000), Moncalvo (*Ganoderma Diseases of Perennial Crops*: 23, 2000), Pilotti *et al.* (*Ganoderma Diseases of Perennial Crops*: 195, 2000), Kim & Jung (*Mycobiology* **29**: 73, 2001), Smith & Sivasithamparam (*Aust. Syst. Bot.* **16**: 487, 2003), Hong & Jung (*Mycol.* **96**: 742, 2004), Ryvarden (*Syn. Fung.* **18**: 57, 2004), Wang & Yao (*Can. J. Microbiol.* **51**: 113, 2005).

Ganodermatales = Polyporales.

Ganodermites A. Fleischm., M. Krings, H. Mayr & R.

Agerer (2007), Fossil Fungi, Ganodermataceae. 1.

Garethjonesia K.D. Hyde (1992), Sordariomycetidae. 1 (aquatic), Australia. See Hyde (*Aust. Syst. Bot.* **5**: 407, 1992).

gari, a fermentation product of manioc, bacteria and *Geotrichum candidum* in west Africa.

Garnaudia Borowska (1977), anamorphic *Pezizomycotina*, Hso.1eP.1. 2, Cuba; Poland. See Borowska (*Acta Mycologica* Warszawa **13**: 169, 1977).

Garovaglia Trevis. (1853) = Polychidium fide Hawksworth *et al.* (*Dictionary of the Fungi* edn 8, 1995).

Garovaglina Trevis. (1880) ≡ Garovaglia.

Gasparrinia Tornab. (1848) [non *Gasparinia* Bertol. 1839, *Umbelliferae*] nom. rej. = Caloplaca fide Hawksworth *et al.* (*Dictionary of the Fungi* edn 8, 1995), Kärnefelt (*Cryptog. Bryol.-Lichénol.* **19**: 93, 1998).

Gassicurtia Fée (1824) nom. rej., Caliciaceae (L). 1, mostly neotropical. See Aptroot (*Taxon* **36**: 474, 1987; nomencl., synonymy with *Buellia*), Kalb & Elix (*Mycotaxon* **68**: 465, 1998; chemistry), Marbach (*Biblthca Lichenol.* **74**, 2000).

Gasterella Zeller & L.B. Walker (1935), Gasterellaceae. 1, USA.

Gasterellaceae Zeller (1948), Boletales. 1 gen., 1 spp.
Lit.: Zeller & Walker (*Mycol.* **27**: 573, 1935), Zeller (*Mycol.* **40**: 639, 1948).

Gasterellopsis Routien (1940), Agaricaceae. 1, USA.

Gasteroagaricoides D.A. Reid (1986), Psathyrellaceae. 1, Australasia. See Reid (*TBMS* **86**: 431, 1986).

gasteroconidium, see gasterospore.

Gasterohymeniales. See *Hymenogastrales* and *Podaxales*.

gasteroid, of a form resembling a member of the gasteromycetes (q.v.); specifically, a fruit body enclosing the hymenium.

Gasteromycetes. Morphological category of *Basidiomycota*, traditionally based on homobasidiomycetes that do not actively discharge their spores (statismospores). The polyphyletic nature of the group has for a long time been known. Epigeous and hypogeous gasteromycetes occur in 4 of the 8 clades of the homobasidiomycetes recognized by Hibbett & Thorn (*The Mycota* **7B**, 2001). Gasteromycetes have been regarded as conspecific (*Gastrosuillus laricinus*), congeneric (*Hydnangium*, *Podohydnangium*), confamilial (*Chamonixia*, *Arcangeliella*) or conordinal (*Sclerodermataceae*, *Lycoperdaceae*) with taxa in the other *Basidiomycota*, while affinities of some groups have not yet been established; for recent information see under these taxa. Integration of gasteromycetes in a natural system of *Basidiomycota* inevitably leads to a taxonomic reduction of many higher taxa; these new placements are not yet universally accepted, particularly at the generic level.

The single-celled basidium (q.v.) is generally 4-spored but the basidiospores are not forcibly discharged and sterigmata may be absent. The basidia and basidiospores mature within the basidioma, which typically has a peridium (sometimes multilayered) covering a ± fleshy mycelial tissue ('gleba') in which the basidia are borne, eventually lining hymenial cavities. The gleba is frequently crossed by a columella and/or veins. After the decay of the gleba, the columella and capillitium (q.v.) may facilitate spore dispersal.

Gasteromycetes are mainly terrestrial and epigeous or hypogeous, partly saprobic and partly ectomycorrhizal; they are cosmopolitan, occurring especially in warm, dry areas.

Lit.: **General**: Reijnders (*MR* **104**: 900, 2000; morphogenesis), Heim (*Evolution of the higher Basidiomycetes*, 505, 1971), Singer (*Agaricales mod. taxon.*, 4th ed., 1986), Dring (*in* Ainsworth *et al.* (Eds), *The Fungi* **4B**: 421, 1973; keys orders, fams., gen.), Fischer (*Nat. Pflanzenfam.* edn 2, **7A**, 1933; unless otherwise stated, synonymies follow this work), Ponce de Léon (*in* Parker, 1982, **1**: 256), Zeller (*Mycol.* **40**: 639, 1948; **41**: 36, 1949; keys orders, fams., gen.), Miller & Miller (*Gasteromycetes. Morphological and developmental features*, 1988).

Regional: **Africa, South**, Bottomley (*Bothalia* **4**: 473, 1948; keys), **East**, Demoulin & Dring (*Bull. Jard. bot. Belg.* **45**: 339, 1975), **West**, Dring (*Mycol. Pap.* **98**, 1964; keys). **Canary Islands**, Beltrán Tejera & Wildpret (*Vieraea* **7**: 49, 1977). **China**, Liu Bo (*The Gasteromycetes of China* [*Beih. Nova Hedw.* **76**], 1984) **Congo**, Dissing & Lange (*Fl. Icon. Champ. Congo* **12-13**, 1963-64; keys). **Zaïre**, Dissing & Lange (*Bull. Jard. bot. Etat. Brux.* **32**: 325, 1962). **America, North**, Burk (*A bibliography of North American gasteromycetes. I Phallales*, 1980), Coker & Couch (*The gasteromycetes of the eastern United States and Canada*, 1928 [reprint 1968]), Smith (*Puffballs and their allies in Michigan*, 1951); **South**, Wright (*Holmbergia* **5**: 45, 1956; key), Fries (*Ark. Bot.* **8** (11), 1909). **Argentina**, Wright (*Lilloa* **21**: 191, 1949). **Brazil**, Homrich (*Rev. Mycol.* **34**: 3, 1969; Rio Grande do Sul). **Mexico**, Herrera (*An. Inst. Biol. Univ. Mex.* **35**: 9, 1965; keys), Guzmán & Herrera (*An. Inst. Biol. Univ. nal. autón. Mex., Bot.* **40**, 1969). **Uruguay**, Lohwag & Swoboda (*Rev. Sudam. Bot.* **7**: 1, 1942). **West Indies**, Dennis (*Kew Bull.* **1953**: 307), Reid (*Kew Bull.* **31**: 657, 1977). **Asia**, **Iran**, Saber (*Iran. J. Plant Pathol.* **22**: 25, 1986). **Israel**, Dring & Rayss (*Israel J. Bot.* **12**: 147, 1964). **Middle East**, Eckblad (*Nytt Mag. Bot.* **17**: 129, 1970; *Iranian J. Bot.* **1**: 65, 1976). **Mongolia**, Kreisel (*Feddes Rep.* **86**: 321, 1975), Dörfelt & Bumžaa (*Nova Hedw.* **43**: 87, 1986). **Nepal**, Kreisel (*Khumbu Himal.* **6**: 25, 1969; *Feddes Rep.* **87**: 83, 1976). **Pakistan**, Ahmad (*Publ. Dept. Bot. Univ. Punjab* **11**, 1952). **Surinam**, Fischer (*Ann. Myc.* **31**: 113, 1933). **Thailand**, Ellingsen (*Nordic Jl Bot.* **2**: 283, 1982), Phanichapol (*Thai For. Bull.* **16**: 233, 1986). **Former USSR**, Sosin (*Opredelitel' Gasteromitsetov SSSR*, 1973), Järva (*Folia crypt. Eston.* **2**: 15, 1973; Estonia), Shvartsman & Filimonova (*Fl. Spor. Rast. Kazakhstana* **6**, 1970; Kazakhstan). **Australasia**, Cunningham (*The Gasteromycetes of Australia and New Zealand*, 1944 [reprint 1979]). **Europe**, **Belgium**, Demoulin (*Les gastéromycètes Belgique*, edn 2, 1975; *Bull. Jard. Bot. nat. Belg.* **38**: 1, 1968). **British Isles**, Palmer (*Nova Hedw.* **25**: 65, 1968; bibliogr.), Demoulin & Marriott (*Bull. BMS* **15**: 37, 1981; key), Pegler *et al.* (*Puffballs, Earthstars & Stinkhorns*, 1995). **Former Czechoslovakia**, Pilát (Ed.) (*Fl. CSR* B **1**, *Gasteromycetes*, 1958). **Germany**, Gross *et al.* (*Beih. Z. Mykol.* **2**, 1980). **Greenland**, Lange (*Meddr Grønl.* **147**(4), 1948). **Hungary**, Hollós (*Die Gasteromyceten Ungarns*, 1904). **Italy**, Petri (*Fl. Ital. Crypt.* **1**(5), 1909). **Netherlands**, Mass Geesteranus (*Coolia* **15**: 49, 1971). **Norway**, Eckblad (*Nytt Mag. Bot.* **4**: 19, 1955; *Astarte* **4**: 7, 1971, Finland). **Portugal**, Almeida (*Port. Acta Biol.* B **11**: 205, 1972). **Spain**, Calonge & Demoulin (*BSMF* **91**: 247, 1975). See also *Lit.* under Macromycetes.

gasterospore (**gasteroconidium**), a thick-walled, globose, chlamydospore of *Ganoderma*; probably apomictic (see Bose, *Mycol.* **25**: 432, 1933; *Sydowia, Beih.* **1**: 176, 1957).

Gastroboletaceae Singer (1962) = Boletaceae.

Gastroboletus Lohwag (1926), Boletaceae. 13, widespread (temperate). See Thiers (*Mem. N. Y. bot. Gdn* **49**: 355, 1989), Nouhra *et al.* (*Mycotaxon* **83**: 409, 2002; key).

Gastrocybe Watling (1968) = Galeropsis fide Moreno *et al.* (*Mycotaxon* **36**: 63, 1989).

Gastrolactarius R. Heim ex J.M. Vidal (2005) = Lactarius fide Kuyper (*in litt.*).

Gastroleccinum Thiers (1989), Boletaceae. 1, N. America. See Thiers (*Mem. N. Y. bot. Gdn* **49**: 357, 1989).

Gastromyces R. Ludw. (1861), Fossil Fungi ? Agaricomycetes. 1 (Carboniferous), former USSR.

Gastropila Homrich & J.E. Wright (1973), Agaricaceae. 4, America. See Demoulin (*in litt.*; ? = *Calvatia*), Ponce de Leon (*Phytologia* **33**: 455, 1976).

Gastrosporiaceae Pilát (1934), Phallales. 1 gen. (+ 1 syn.), 2 spp.

Lit.: Zeller (*Mycol.* **40**: 639, 1948), Domínguez de Toledo & Castellano (*Mycotaxon* **64**: 443, 1997), Iosifidou & Agerer (*Feddes Repert.* **113**: 11, 2002), Kreisel & Hausknecht (*Öst. Z. Pilzk.* **11**: 191, 2002).

Gastrosporium Mattir. (1903), Gastrosporiaceae. 2, widespread. See Monthoux & Röllin (*Candollea* **31**: 119, 1976), Iosifidou & Agerer (*Feddes Repert.* **113**: 11, 2002; systematic position).

Gastrosuillus Thiers (1989) = Suillus Gray. Gastroid form recently derived from *Suillus*. fide Baura *et al.* (*Mycol.* **84**: 592, 1992; recently derived from *Suillus grevillei*), Kretzer & Bruns (*Mycol.* **89**:: 586, 1997).

Gastrotylopilus T.H. Li & Watling (1999) = Fistulinella fide Trappe *et al.* (*Muelleria* **18**: 75, 2003).

Gaubaea Petr. (1942), anamorphic *Pezizomycotina*, St.0eP.19. 2, central Asia.

Gäumann (Ernest Albert; 1893-1963; Switzerland). PhD, Bern (1917); Botanist, Agricultural Experiment Station, Zürich-Oerlikon (1922-1927); Professor of botany, Institut für Spezielle Botanik, Eidgenössische Technische Hochschule, Zürich (1927-1963). Carried out outstanding and extensive research in plant pathology (including antibiotics, defence reactions, immunity and resistance), rust biology and taxonomy of downy mildews. Editor-in-chief, *Phytopathologische Zeitschrift* (1937-1963). *Publs. Vergleichende Morphologie der Pilze* (1926); *Die Pilze* (1949, 1964); *Pflanzliche Infektionslehre* (1946, 1951); *Die Rostpilze Mitteleuropas* (1959). *Biogs, obits etc.* Arx (*Netherlands Journal of Plant Pathology* **70**: 99, 1964); Gardner & Kern (*Mycol.* **57**: 1, 1965) [portrait]; Brown (*Annals of Applied Biology* **53**: 297, 1964); Landholt (*Verhandlungen der Schweizerischen Naturforschenden Gesellschaft* **1963**: 194) [bibliography, portrait]; Minkiavicius ([*Mikologiya i Fitopatologiya*] **3**: 393, 1969, portrait [in Russian]); Stafleu & Cowan (*TL-2* **1**: 903, 1976); Zobrist *et al.* (*Herrn Professor Dr. Ernst Gäumann zum siebzigsten Geburtstag.*, 1963).

Gausapia Fr. (1825) nom. rej. = Septobasidium fide Berndt (*in litt.*).

Gauthieromyces Lichtw. (1983), Legeriomycetaceae. 3 (in *Ephemeroptera*), China; France; India; Mexico. See Lichtwardt (*The Trichomycetes. Fungal associates of arthropods*, 1986), Strongman & Xu (*Mycol.* **98**: 479, 2006; China), Valle *et al.* (*Mycol.* **100**: 149, 2008; Mexico).

Gautieria Vittad. (1831) nom. cons., Gomphaceae. 25, widespread. See Rauschert (*Hercynia* N.F. **12**: 217, 1975), Beaton *et al.* (*Kew Bull.* **40**: 193, 1985; key Austral. spp.).

Gautieriaceae Zeller (1948) = Gomphaceae.

Gautieriales = Gomphales.

GC, ratio; **GC content**; see base ratio, molecular biology.

Geaster P. Micheli ex Fr. (1829) ≡ Geastrum fide Demoulin (*Taxon* **36**: 498, 1984).

Geasterites Pia (1927), Fossil Fungi. 1 (Miocene), USA. But see Tiffney (*TBMS* **76**: 493, 1981).

Geasteroides Long (1917), Geastraceae. 1, USA. See Long (*Mycol.* **37**: 601, 1945).

Geasteropsis Hollós (1903) = Geastrum fide Ponce de Leon (*Fieldiana, Bot.* **31**: 311, 1968).

Geastraceae Corda (1842), Geastrales. 7 gen. (+ 13 syn.), 64 spp.

Lit.: Ponce de Leon (*Fieldiana, Bot.* **31**: 303, 1968; world monogr., broad spp. concept), Ingold (*TBMS* **58**: 179, 1972), Boiffard (*Docums Mycol.* **6** no. 24: 1, 1976; France), Dörfelt (*Die Erdsteine Geastraceae und Asteraceae*: 385 pp., 1985; centr. Eur.), Miller & Miller (*Gasteromycetes. Morphological and Development Features with Keys to the Orders, Families, and Genera*: 157 pp., 1988), Sunhede (*Geastraceae (Basidiomycotina)* [*Synopsis Fungorum* **1**], 1989; monogr. esp. N. Eur.), Domínguez & Castellano (*Mycol.* **88**: 863, 1996), Hibbett *et al.* (*Proc. natn Acad. Sci. U.S.A.* **94**: 12002, 1996), Coetzee *et al.* (*Bothalia* **27**: 117, 1997), Calonge (*Fl. Mycol. Iberica* **3**: 271 pp., 1998), Krüger *et al.* (*Mycol.* **93**: 947, 2001), Binder & Bresinsky (*Mycol.* **94**: 85, 2002), Hibbett & Binder (*Proc. R. Soc. Lond.* B. Biol. Sci. **269**: 1963, 2002), Kreisel & Hausknecht (*Öst. Z. Pilzk.* **11**: 191, 2002), Baseia *et al.* (*Mycotaxon* **85**: 409, 2003), Esqueda *et al.* (*Mycotaxon* **87**: 445, 2003), Estrada-Torres *et al.* (*Mycol.* **97**: 139, 2005), Geml *et al.* (*Mol. Phylogen. Evol.* **35**: 313, 2005), Geml *et al.* (*Mycol.* **97**: 680, 2005).

Geastrales K. Hosaka & Castellano (2007). Phallomycetidae. 1 fam., 7 gen., 64 spp. Fam.:

Geastraceae

For *Lit.* see under fam.

Geastrum Pers. (1794), Geastraceae. *c.* 50, widespread. The earth stars. See Sunhede (*Syn. Fung.* **1**, 1990), Zhou & Yan (*Mycosystema* **21**: 485, 2002; China).

Geastrumia Bat. (1960), anamorphic *Pezizomycotina*, Ccu.1bH.1. 1, widespread. See Pirozynski (*Mycol.* **63**: 897, 1971), Williamson & Sutton (*Pl. Dis.* **84**: 714, 2000).

Geisleria Nitschke (1861) = Strigula fide Ernst (*Herzogia* **9**: 321, 1993), Roux *et al.* (*Biblthca Lichenol.* **90**: 96 pp., 2004).

Geisleriomyces Cif. & Tomas. (1953) ≡ Geisleria.

Geissodea Vent. (1799) = Cetraria fide Hawksworth *et al.* (*Dictionary of the Fungi* edn 8, 1995).

gel tissue, a mixture of gel and hyphae found in members of the *Leotiales* and *Tremellales*; the gel may arise either by direct secretion or by disintegration of hyphae: see Moore (*Mycol.* **57**: 114, 1965; *Am. J. Bot.* **52**: 389, 1965, ontogenesis; *Stain Technol.* **40**: 23, 1965, staining). Cf. gliatope.

Gelasinospora Dowding (1933), Sordariaceae. 24 (from dung, soil etc.), widespread. Placed into synonymy with *Neurospora* by some authors. See Cailleux (*BSMF* **87**: 536, 1972; key), von Arx (*Persoonia* **11**: 443, 1982; key), Khan & Krug (*Mycol.* **81**: 226, 1989), Beatty *et al.* (*MR* **98**: 1309, 1994; genetics), Stchigel *et al.* (*MR* **102**: 1405, 1998; Argentina), Dettman *et al.* (*Fungal Genetics Biol.* **34**: 49, 2001; phylogeny), García *et al.* (*MR* **108**: 1119, 2004; phylogeny, synonymy with *Neurospora*), Miller & Huhndorf (*Mol. Phylogen. Evol.* **35**: 60, 2005; phylogeny, anatomy), Zhang *et al.* (*Mycol.* **98**: 1076, 2006; phylogeny).

Gelasinosporites P. Briot, Lar.-Coll. & Locq. (1983), Fossil Fungi. 1, Australia.

gelatin, product obtained by boiling collagen, soluble in water above *c.* 40°C. Gels of *c.* 4-12% used to test ability of some microorganisms to liquefy or hydrolyse gelatin.

Gelatina Raf. (1806) = Tremella Pers. fide Merrill (*Index Rafinesq.*: 68, 1949).

Gelatinaria Flörke ex Wallr. (1831) nom. inval. ? = Nostoc fide Hawksworth *et al.* (*Dictionary of the Fungi* edn 8, 1995).

Gelatinaria Raf. (1815) nom. dub., Fungi. No spp. included.

Gelatinipulvinella Hosoya & Y. Otani (1995), Leotiaceae. Anamorph *Aureohyphozyma*. 1, Japan. Perhaps better placed in the *Leotiaceae*, but both apothecial anatomy and ascus apex are atypical of this family.

Gelatinocrinis Matsush. (1995), anamorphic *Pezizomycotina*, Hsp.0eH.15. 1, Japan. See Matsushima (*Matsush. Mycol. Mem.* **8**: 20, 1995).

Gelatinodiscaceae S.E. Carp. (1976) = Leotiaceae.

Gelatinodiscus Kanouse & A.H. Sm. (1940), Helotiaceae. 1, USA. See Carpenter (*Mycotaxon* **3**: 209, 1976).

Gelatinopsis Rambold & Triebel (1990), Helotiaceae. 2 (on lichens and other fungi), Europe; N. America. Perhaps better placed in the *Leotiaceae*. See Rambold & Triebel (*Notes R. bot. Gdn Edinb.* **46**: 375, 1990), Aptroot *et al.* (*Nova Hedwigia* **64**: 155, 1997; France), Etayo *et al.* (*Lichenologist* **33**: 473, 2001; Spain).

Gelatinopycnis Dyko & B. Sutton (1979), anamorphic *Pezizomycotina*, St.0fH.15. 1, Germany. See Dyko & Sutton (*CJB* **57**: 375, 1979).

Gelatinosporis Clem. & Shear (1931) ≡ Gelatinosporium.

Gelatinosporium Peck (1873), anamorphic *Durandiella*, St.1eH.15. 1, USA. See DiCosmo (*CJB* **56**: 1665, 1978), Verkley (*Nova Hedwigia* **75**: 433, 2002).

gelatinous, jelly-like; used for the hyphae of tissues which become partly dissolved and glutinous with moisture.

Gelatoporia Niemelä (1985), Meruliaceae. 3, widespread (north temperate). See Niemelä (*Karstenia* **25**: 22, 1985), Spirin & Zmitrovich (*Karstenia* **43**: 67, 2003; Russia).

Gelatosphaera Bat. & H. Maia (1959) = Rhizosphaera fide Sutton (*Mycol. Pap.* **141**, 1977).

Geleenites Dijkstra (1949), Fossil Fungi. 1 (Cretaceous), Canada; Netherlands. See Jansonius *et al.* (*Pollen et Spores* **23**: 557, 1981; ? = Ascotricha (Xylar.)).

Gelidium J.V. Lamour. (1813), Algae. Algae.

Gelimycetes. Class within the *Orthomycotina* (q.v.) including *Auriculariales* p.p., *Dacrymycetales*, *Tremellales* p.p. and *Tulasnellales* (Cavalier-Smith *in* Rayner *et al.* (Eds), *Evolutionary biology of the fungi*: 339, 1987).

Gelineostroma H.J. Swart (1988), Rhytismatales. 2 (on *Arthrotaxis*), Australia. See Swart (*TBMS* **90**: 445, 1988), Johnston (*Aust. Syst. Bot.* **5**: 507, 1992), Johnston (*Aust. Syst. Bot.* **14**: 377, 2001).

Gelona Adans. (1763) nom. rej. = Pleurotus fide Donk (*Beih. Nova Hedwigia* **5**, 1962).

Gelopellaceae Zeller (1939) = Claustulaceae.

Gelopellis Zeller (1939), Claustulaceae. 6, S. America; Japan; Australia. See Homrich (*Revue Mycol.* Paris **34**: 6, 1969), Beaton & Malajczuk (*TBMS*: 478, 1986; Australia).

Geltingia Alstrup & D. Hawksw. (1990), Odontotremataceae. 2 (on lichens), widespread (esp. Arctic). See Rambold & Triebel (*Notes R. bot. Gdn Edinb.* **46**: 375, 1990), Diederich & Etayo (*Lichenologist* **32**: 423, 2000).

Geminaginaceae Vánky (2001), Ustilaginales. 1 gen., 1 spp.

Lit.: Vánky (*Mycoscience* **37**: 173, 1996), Piepenbring *et al.* (*Protoplasma* **204**: 202, 1998), Vánky (*MR* **102**: 513, 1998), Vánky (*Fungal Diversity* **6**: 131, 2001).

Geminago Vánky & R. Bauer (1996), Geminaginaceae. 1 (on *Sterculiaceae*), Africa (tropical). See Vánky & Vánky (*Lidia* **5**: 157, 2002; southern Africa).

Geminella J. Schröt. (1870) [non *Geminella* Turpin 1828, *Algae*] = Schroeteria fide Vánky (*in litt.*).

Geminispora Pat. (1893), ? Phyllachoraceae. 2 (on *Leguminosae*), Africa; S. America. See Cannon (*Mycol. Pap.* **163**, 1991).

Geminoarcus K. Ando (1993), anamorphic *Pezizomycotina*. 2, Japan. See Ando (*TMSJ* **34**: 110, 1993).

gemma (pl. **-ae**), (1) an asexual propagule borne singly or in chains at the ends of hyphae, referred to in older literature as a chlamydospore (*Saprolegniaceae*); (2) another term for oidia in *Agaricomycotina* (Gäumann, *The Fungi*: 449, 1928), rejected for this usage by Kendrick & Watling (*in* Kendrick (Ed.), *The whole fungus* **2**: 477, 1979).

Gemmamyces Casagr. (1969) = Cucurbitaria fide Petrak (*Sydowia* **23**: 265, 1970), Yuan & Wang (*Mycotaxon* **53**: 371, 1995; accepted genus).

Gemmaspora D. Hawksw. & Halıcı (2007), Verrucariales. 1 (lichenicolous), Middle East. See Hawksworth & Halici (*Lichenologist* **39**: 121, 2007).

Gemmina Raitv. (2004), Hyaloscyphaceae. 1, Europe. See Raitviir (*Scripta Mycologica* Tartu **20**: 44, 2004).

Gemmophora Schkorb. (1912) nom. dub., anamorphic *Pezizomycotina*. Probably based on chlamydospores and sterile mycelium.

Gemmularia Raf. ex Steud. (1824) ≡ Pachyma fide Donk (*Taxon* **11**: 85, 1962).

Gemmulina Descals & Marvanová (1999), anamorphic *Pezizomycotina*, Hso.?.?. 1 (aquatic), British Isles. See Descals *et al.* (*CJB* **76**: 1657, 1998).

Genabea Tul. & C. Tul. (1844), Pyronemataceae. 1 (hypogeous, mycorrhizal), Europe; N. America. See Zhang (*MR* **95**: 986, 1991), Pegler *et al.* (*British truffles*, 1993), Smith *et al.* (*Mycol.* **98**: 699, 2006; phylogeny, ecology), Læssøe & Hansen (*MR* **111**: 1075, 2007; phylogeny).

gene probes, see Molecular biology: DNA fingerprinting.

Genea Vittad. (1831), Pyronemataceae. *c.* 32 (hypogeous, mycorrhizal), N. America; Europe. See Pfister (*Mycol.* **76**: 170, 1984; posn), Lazzara & Montecchi (*Revista Micol.* **34**: 44, 1991; 5 spp., Italy), Zhang (*MR* **95**: 986, 1991; concept, key 3 spp. China), Zhang (*SA* **11**: 31, 1992; nuclei), Pegler *et al.* (*British truffles*, 1993), Li & Kimbrough (*Int. J. Pl. Sci.* **155**: 235, 1994; ultrastr.), Gross (*Z. Mykol.* **62**: 176, 1996), Moreno-Arroyo *et al.* (*Boll. Gruppo Micol. 'G. Bresadola'* **44**: 31, 2001; Spain), Hansen & Pfister (*Mycol.* **98**: 1029, 2006; phylogeny), Smith *et al.* (*Mycol.* **98**: 699, 2006; phylogeny, ecology), Læssøe & Hansen (*MR* **111**: 1075, 2007; phylogeny), Perry *et al.* (*MR* **111**: 549, 2007; phylogeny), Erős-Honti *et al.* (*Mycorrhiza* **18**: 133, 2008; mycorrhizas, Hungary).

Geneaceae Trappe (1979) = Pyronemataceae.

gene-for-gene (in host-parasite relationships), the correspondence for each gene determining resistance in the host of a specific and related gene determining virulence in the pathogen; first described by Flor in 1955 for *Melampsora lini* on flax (*Linum usitatissimum*), which has 29 resistance genes each of which the pathogen has avirulent counterparts (Lawrence *et al.*, *Phytopath.* **71**: 12, 1981). See Person (*CJB* **37**: 1101, 1959), Person *et al.* (*Nature* **194**: 561, 1962), Flor (*Ann. Rev. Phytopath.* **9**: 275, 1971), Parlevliet (*in* Pirozynski & Hawksworth (Eds), 1988: 19), Coevolution.

Geneosperma Rifai (1968) = Scutellinia fide Korf & Zhuang (*Acta Mycol. Sin.* Suppl. **1**: 90, 1986; 2 spp.), Moravec (*Mycotaxon* **58**: 233, 1996), Moravec (*Czech Mycol.* **50**: 85, 1997), Wang (*Bull. natn. Mus. Nat. Sci.* Taiwan **11**: 119, 1998; Taiwan).

genera, see genus.

generative hyphae, see hyphal analysis.

genestasis, inhibition of sporulation; **genistat**, a substance preventing or reducing sporulation in fungi without materially affecting vegetative growth; 'antisporulator' (Horsfall, 1947); cf. fungistatic.

Genetic engineering. The insertion or removal of inheritable genetic material from an organism so that its properties are transformed. The topic has a political dimension. Genetic engineering has been used to increase pathogenicity of nematophagous fungi (Ahman *et al.*, *Appl. and environmental microbiol.* **68**: 3408, 2002), to alter *Laccaria bicolor* for use as a biocontrol agent (Bills *et al* (*Mycol.* **91**: 237, 1999). Genes from fungi can also be inserted into plasmid genomes and expressed through *Escherichia coli* or *Saccharomyces cerevisiae*. Genetic engineering of plants to enhance their resistance to fungal diseases is also possible (e.g. transferring genes resistant to *Phytophthora infestans* from *Solanum bulbocastaneum* to *S. tuberosum*, see Song *et al.*, *Proc. nat. acad. sci.* **100**: 9128, 2003). See Bennett & Lasure (Eds) (*Gene manipulations in fungi*, 1985; *More gene manipulation in fungi*, 1991), Fincham (*Microbiol. Rev.* **53**: 148, 1989; review), Kinghorn & Turner (Eds) (*Applied molecular genetics of filamentous fungi*, 1992), Timberlake & Marshall (*Science* **244**: 1313, 1989), Biotechnology.

Genetic resource collections. Genetic resource collections which include fungal cultures are catalogued in Takishima *et al.* (Eds) (*Guide to World Data Center on Microorganisms with a list of culture collections worldwide*, 1989), and Sugawara *et al.* (Eds) (*World directory of collections of cultures of microorganisms*, edn 4, 1993). In 2008, these collections were maintaining some 476,299 fungal strains representing around 25,085 species or subspecies (http://wdcm.nig.ac.jp/statistics.html). Collections are given standard acronyms or abbreviations by the World Data Center, and a selection of those for the larger collections are included in this edn of the *Dictionary*. See also Reference collections. The chief collections of fungal cultures by region are as follows. **Africa**. South Africa (Plant Protection Research Institute, Pretoria); **Asia**. China (University of Hong Kong Culture Collection, Hong Kong); Japan (Institute for Fermentation, Osaka); South Korea (Center for Fungal Genetic Resources, Seoul National University, Seoul). **Australasia**. Australia (Mycology Culture Collection, Women's and Children's Hospital, Adelaide; Plant Pathogen Culture Collection, Indooroopilly, Queensland); New Zealand (International Collection of Microorganisms from Plants, DSIR, Auckland). **Caribbean**. Cuba (INIFAT, Havana). **Europe** Denmark (Culture Collection of Fungi, BioCentrum-DTU, Lyngby); France (Laboratoire de Cryptogamie, Muséum National d'Histoire Naturelle, Paris); Germany (Pilz-Referenz-Zentrum, Friedrich Schiller University, Jena); Italy (Industrial Yeasts Collection, Università di Perugia, Perugia) Netherlands (Centraalbureau voor Schimmelcultures, Baarn; Russia (All-Russian Collection of Microorganisms, Institute of Biochemistry and Physiology of Microorganisms, Pushchino, Moscow oblast'); Slovakia (Culture Collection of Yeasts, Institute of Chemistry, Bratislava); Sweden (Fungal Cultures, University of Goteborg, Goteborg; Culture Collection of Fungi, University of Uppsala, Uppsala); UK (CABI UK Centre, Egham [housing the collection of the former International Mycological institute and the holdings of the UK National Collection of Wood-Rotting Fungi and British Antarctic Survey]; the UK National Collection of Yeast Cultures, Norwich). **North America**. Canada (Canadian Collection of Fungal Cultures, Agriculture and Agri-Food Canada, Ottawa; University of Alberta Microfungus Collection, Edmonton, Alberta); USA (American Type Culture Collection, Washington, DC; ARS Collection of Entomopathogenic Fungi, Ithaca, NY; Pfaff Yeast Culture Collection, University of California, Davis, CA). **South America**. Brazil (Centro Especializado em Micologia Medica, Universidade Federal do Ceara, Ceara; Colecão de Culturas de Fungos, Instituto Oswaldo Cruz, Rio de Janeiro). See World Data Center for Microorganisms (http://wdcm.nig.ac.jp). See also Reference collections.

International collaboration is facilitated by the World Federation for Culture Collections which functions under the auspices of the IUBS and organises meetings of the International Congress of Culture Collections.

Culture **maintenance** is traditionally on slopes of appropriate media stored at laboratory or low temperature but now cryopreservation (see below) when applicable is the method of choice. Lyophilization (freeze-drying) of the fungus spore is an ideal method to facilitate distribution; this method is favoured by service collections (Smith *et al.* in Smith *et al.* Eds, 2001. *The UK National Culture Collection Biological Resource: Properties, maintenance and management.* pp 389. UK National Culture Collection, Egham). The properties of the fungus in culture may be unstable and subject to strain drift through loss of plasmids, spontaneous mutations or genetic recombination due to the presence of heterokaryons, the parasexual cycle or normal sexual events. Therefore conditions of storage should be selected that minimize the risk of such changes. Freezing and storage of fungi at ultra-low temperature such as in or above liquid nitrogen or in -140° to -150°C freezers appears to provide the ideal method. There are, however, several other methods that are used successfully. They range from continuous growth through methods that reduce rates of metabolism to the ideal situation where metabolism is halted, or to a level that for all practical purposes it can be treated as suspended. Microbial resource collections, large or small, set out to maintain organisms in a pure, viable and stable condition to make them available for future use. The method selected may depend upon the requirements of the use of the organism. These vary according to the numbers and range of fungi to be preserved and the facilities available. The cost of materials and labour involved and the desired level of stability and longevity required is also taken into consideration. Preservation methods that allow growth or metabolism can only be used for short-term storage, such methods are subculture, storage under a layer of mineral oil and storage in sterile water. Drying and freeze-drying techniques can be used for long-term storage of fungi but not all will survive. Storage in soil or in or above silica gel produce dry conditions that can allow the desiccated fungus to survive for 8 and up to 20 years. Freeze-drying, the removal of water, dehydration of fungi under reduced pressure by the sublimation of ice is a method widely used but generally only allows the fungus spore or other robust structures, such as sclerotia, to survive. The methods for the above preservation technique have been published widely.

Lit.: **General**, Hawksworth & Kirsop (Eds) (*Living resources for biotechnology: filamentous fungi*, 1988), Kirsop & Kurtzman (Eds) (*Living resources for biotechnology: yeasts*, 1988), Kirsop & Doyle (Eds) (*Maintenance of microorganisms and cultured cells: A manual of laboratory methods*, 1991), Smith & Onions (*The preservation and maintenance of living fungi*, edn 2, 1994), Ryan *et al.* (*World Journal of Microbiology & Biotechnology* **16**: 183, 2000). Ryan & Smith (*MR* **108**: 1351, 2004).

Cryopreservation by frozen storage, Gulya *et al.* (*MR* **97**: 240, 1993), Holden & Smith (*MR* **96**: 473, 1992), Ito (*Inst. Fermen. Osaka Res. Commun.* **15**: 119, 1991), Morris *et al.* (*Cryobiology* **25**: 471, 1988; *J. Gen. Microbiol.* **134**: 2897, 1988), Pearson *et al.* (*Cryoletters* **11**: 205, 1990), Roquebert & Bury (*Wrld J. Myc. Res.* **9**: 651, 1992), Smith (*in* Chang *et al.* (Eds), *Genetics and breeding of edible mushrooms*, 1993; *in* Jennings (Ed.), *Tolerance of fungi*, 1993), Hubalek (*Cryopreservation of microorganisms at ultra-low temperatures*, 1996).

Cryopreservation over liquid nitrogen, Smith (*in* Jennings (Ed.), *Tolerance of fungi*: 145, 1993).

Lyophilization (freeze drying), Ryan & Smith (*in* Day & Stacey (Eds), *Methods Molec. Biol.* **368**: 127, 1995; protocols), Tan *et al.* (*Mycol.* **83**: 654, 1991; *Mycol.* **86**: 281, 1994), Smith (*World Journal of Microbiology & Biotechnology* **14**: 49, 1998).

Mineral oil, Little & Gordon (*Mycol.* **59**: 733, 1967; refs.), Fennell (*Bot. Rev.* **26**: 1, 1960; review).

Anhydrous silica gel, Perkins (*Can. J. Microbiol.* **8**: 591, 1962), Gentles & Scott (*Sabouraudia* **17**: 415, 1979).

Soil, Bakerspigel (*Mycol.* **45**: 596, 1953).

Water, Castellani (*Mycopath.* **20**: 1, 1963).

Comparisons of methods, Onions (*in* Booth, *Methods in microbiology* **4**: 113, 1971), Smith & Onions (*TBMS* **81**: 535, 1983; mineral oil), Smith & Onions (1994), Smith (*TBMS* **79**: 415, 1982; 3,000 fungi preserved up to 13 years), *IMI Culture Collection Catalogue* 10th edn, 1992), Berry & Hennebert (*Mycol.* **83**: 605, 1992; freeze drying).

See also Media, Mites, and Safety.

Genetics. The text of this entry is unrevised and remains the same as for the ninth edition. In general, fungi appear to conform to the well- established patterns of genetical behaviour typical of other groups of organisms and species of *Saccharomyces, Neurospora*, and *Aspergillus* have been widely used in studies on formal genetics. They mostly exhibit haplo- and diplophases in their life-history, and thus undergo meiosis at some stage.

The use of monosporous cultures, and esp. of methods for the isolation of all the spores from one ascus or basidium, has made possible such detailed studies as the transmission of various mycelial characters and segregation of sex. Investigation of the stage of meiosis at which the allelomorphs segregate has been rendered possible in some ascomycetes (e.g. *Neurospora*) because the eventual products, the ascospores, exhibit a linear succession in the ascus, and the frequencies of genetical crossing-over for various characters have been deduced from such meiocyte analyses.

Incompatability mechanisms involving one (*Mucor*) or multiple genes (*Coprinus*) are known (see heterothallism). Mutants induced by X-rays, ultraviolet light, etc., have been shown to be deficient in their capacity to effect particular stages of protein synthesis (*Neurospora*) and fermentation (yeasts).

Lit.: Kniep (*Bibliogr. genet.*, 1929), Fincham, Day & Radford (*Fungal genetics*, edn 4, 1979), Esser & Kuenen (*Genetic der Pilze*, 1965), Sermonti (*Genetics of antibiotic-producing organisms*, 1969), Day (*Genetics of host-parasite interactions*, 1974), Burnett (*Mycogenetics*, 1975), Ullrich & Raper (*Taxon* **26**: 169, 1977; evol. genetic mechanisms, 80 refs), Koltin *et al.* (*Bact. Rev.* **36**: 156, 1972; evol. incomp. factors in higher fungi), Day & Jellis (Eds) (*Genetics and plant pathogens*, 1987), Sidhu (Ed.) (*Genetics of plant pathogenic fungi* [*Adv. Pl. Path.* **6**], 1988), Clutterbuck (*in* Gow & Gadd (Eds), *The growing fungus*: 239, 1995). See also Genetic engineering, Parasexual cycle, Sex, Variation.

Genicularia Rifai & R.C. Cooke (1966) [non *Genicularia* Rouss. 1806, *Algae*] ≡ Geniculifera.

geniculate, bent like a knee.

Geniculifera Rifai (1975) = Arthrobotrys fide Oorschot (*Stud. Mycol.* **26**: 61, 1985), Rubner (*Stud. Mycol.* **39**, 1996), Koppenhöfer *et al.* (*Mycol.* **89**: 220, 1997; ecology), Liou & Tzean (*Mycol.* **89**: 876, 1997).

Geniculisynnema Okane & Nakagiri (2007), anamorphic *Nemania*. 1 (from termite nest), Japan. See Okane, I.; Nakagiri, A. (*Mycoscience* **48**: 240, 2007).

Geniculodendron G.A. Salt (1974), anamorphic *Caloscypha*, Hso.0eH.10. 1, Canada; Europe. See Paden *et al.* (*CJB* **56**: 2375, 1978), Harmaja (*Karstenia* **42**:

27, 2002; teleomorph), Schröder *et al.* (*Forest Pathology* **32**: 225, 2002; Germany).

Geniculospora Sv. Nilsson ex Marvanová & Sv. Nilsson (1971), anamorphic *Leotiomycetes*, Hso.1bH.1. 2 (aquatic), widespread. Probably polyphyletic. A connection with *Hymenoscyphus* has been claimed. See Nolan (*Mycol.* **64**: 1173, 1972), Ingold (*Guide to aquatic hyphomycetes*, 1975; synonymy with *Articulospora*), Belliveau & Bärlocher (*MR* **109**: 1407, 2005; phylogeny).

Geniculosporium Chesters & Greenh. (1964), anamorphic *Nemania, Entoleuca, Rosellinia, Podosordaria*, Hso.0eH.11. 4, widespread. See Rogers *et al.* (*Mycotaxon* **67**: 61, 1998; *Podosordaria* anam.), Ju & Rogers (*Mycotaxon* **73**: 343, 1999).

Geniopila Marvanová & Descals (1985), anamorphic *Pezizomycotina*, Hso.≡ eH.15. 1 (aquatic), former Czechoslovakia; British Isles.

genistat, see genestasis.

Genistella L. Léger & M. Gauthier (1932) [non *Genistella* Ortega 1773, *Papilionaceae*] ≡ Legeriomyces.

Genistellaceae L. Léger & M. Gauthier (1932) = Legeriomycetaceae.

Genistelloides S.W. Peterson, Lichtw. & B.W. Horn (1981), Legeriomycetaceae. 4 (in *Plecoptera*), USA. See Lichtwardt (*The Trichomycetes. Fungal associates of arthropods*, 1986), Williams & Lichtwardt (*Mycol.* **79**: 473, 1987), White & Lichtwardt (*Mycol.* **96**: 891, 2004; Norway), White (*MR* **110**: 1011, 2006; phylogeny).

Genistellospora Lichtw. (1972), Legeriomycetaceae. 5 (in *Diptera*), widespread. See Lichtwardt (*Mycol.* **64**: 167, 1972), Moss & Lichtwardt (*CJB* **54**: 2346, 1976; ultrastr.), Moss & Lichtwardt (*CJB* **55**: 3099, 1977; ultrastr.), Alencar *et al.* (*Memórias do Instituto Oswaldo Cruz* **98**: 799, 2003; Brazil), López Lastra *et al.* (*Mycol.* **97**: 320, 2005; Argentina), White (*MR* **110**: 1011, 2006; phylogeny).

genocentric, see reproductocentric.

genome, the total inheritable genetic material of an organims; a haploid set of chromosomes in eukaryotes. In *Saccharomyces cerevisiae* this is 12 Mgbases coding for some 6,000 genes; in *Neurospora crassa* the genome is much longer, 40 Mgbases; and *Aspergillus nidulans* has some 13,000 genes. See also Chromosome number.

Genomics. Uncertainty still exists concerning the exact meaning of the term, allowing its application to disparate activities. Several categories are now recognised which depend on the initial characteristics of interest: based on system attributes the topic can be divided into structural genomics and functional genomics (q.v.), but further divisions can be recognized when the relationships to other scientific disciplines or even the organisms being studied are considered; see Zhou *et al.* (*Microbial functional genomics*, 2004). A consensus exists, however, that the term encompasses 'the mapping, sequencing, and analysis of genomes' (Zhou *et al.*, ibid.). Techniques, results and discoveries made initially in studies of the molecular biology of bacteria and from ambitious projects such as the Human Genome Project (see The International Human Genome Mapping Consortium *Nature* **409**: 860, 2001) are now being applied increasingly to other living organisms, including the fungi (Prade & Bohnert, *Genomics of plants and fungi*, 2003). The advent of polymerase chain reaction (PCR, q.v.) technology together with automated high throughput capillary DNA sequencing facilities has led to a vast increase in total genome sequencing of fungi. In 1995, when the eighth edn of this *Dictionary* was published, only one entire genome sequence (*Saccharomyces cerevisiae*) had been generated; but by 2006, >40 'complete' fungal genomes were known (see Fitzpatrick *et al.*, *BMC Evol. Biol.* **6**: 99, 2006) with sequencing of many more underway.

Structural genomics, the 'genome-wide structural study of genes, proteins, and other biomolecules' (Zhou *et al.*, ibid.). This has allowed comparative studies (including size of genome; size and number of chromosomes; and number of potential genes) to be made across different completed fungal genomes leading to more detailed phylogenetic analysis (see Wanchanthuek *et al.*, *in* Sunnerhagen & Piskur (Eds) *Comparative genomics using fungi as models*, 2005).

Functional genomics, the elucidation of a 'systems level understanding of the functional aspects of biological systems…using genome-wide approaches' (Zhou *et al.*, ibid.). This involves the combination of whole-genome sequencing and bioinformatic analysis to examine the resultant total RNA, protein and metabolite profiles ('transcriptome', 'proteome' and 'metabolome', respectively). See Castrillo & Oliver, *in* Brown (Ed.) *The Mycota vol. XIII: Fungal genomics*, 2005. This methodology is being used increasingly to try to understand inter-taxon relationships, pathogenicity of fungi and ultimately to develop new targeted pharmaceutical or agrochemical products.

Lit.: Arora & Khachatourians (Eds) (*Fungal genomics. Applied mycology and biotechnology* **4**, 2004), Fitzpatrick *et al.* (*BMC Evol. Biol.* **6**: 99, 2006), Munro *et al.* (*in* Brown (Ed.) *The Mycota vol. XIII: Fungal genomics*, 2005), Thomson & Zhou, (*in* Zhou *et al.*, *Microbial functional genomics*, 2004), Sunnerhagen & Piškur (Comparative genomics using fungi as models. *Topics in current genetics* **15**: 289 pp., 2006).

genotype, the sum of the genetic potential of an organism; in some fungi only part of this is expressed at any given time. See holomorph, teleomorph, anamorph.

genus (pl. **genera**; adj. **generic**), (1) (in taxonomy), one of the principal ranks in the nomenclatural hierarchy (see Classification), the name of which forms the first part of a binomial species name (see species); (2) (more generally), a class of objects or concepts.

As in the case of the species (q.v.), there is no universally applicable criteria by which genera are distinguished, but in general the emphasis is now on there being several discontinuities in fundamental characters, especially concerning the nature of the reproductive structures. In the last century, however, features such as spore colour and septation were accorded a predominant role by some workers (see Anamorphic fungi).

Lit.: Clemençon (Ed.) (*Mycol. Helv.* **6**, 1993; esp. in macromycetes), Hale (*Beih. Nova Hedw.* **79**: 11, 1984; in lichens), Hawksworth (1974), Poelt (*in* Hawksworth (Ed.), *Frontiers in mycology*: 85, 1991; in lichens).

Geocoryne Korf (1978), Leotiaceae. 2, Canary Islands; India. See Lizoň *et al.* (*Mycotaxon* **67**: 73, 1998; posn).

Geodina Denison (1965), Sarcoscyphaceae. 1, Costa Rica. See Cabello (*Boln Soc. argent. Bot.* **25**: 395, 1988; numerical taxonomy).

geofungi, soil fungi (Cooke, 1963).

Geoglossaceae Corda (1838), Pezizomycotina (inc. sed.). 6 gen. (+ 4 syn.), 48 spp. Possibly allied to the *Lichinomycetes*.

Lit.: Imai (*Jl Fac. Agric. Hokkaido Univ.* **45**: 155, 1941; Japanese spp.), Nannfeldt (*Ark. Bot.* **30A** no. 4: 1, 1942), Mains (*Mycol.* **46**: 586, 1954), Maas Geesteranus (*Persoonia* **4**: 19, 1965; keys Indian spp.), Olsen (*Agarica* **7**: 120, 1986), Spooner (*Biblthca Mycol.* **116**: 711 pp., 1987), Verkley (*Persoonia* **15**: 405, 1994; asci), Wang *et al.* (*Mycol.* **98**: 1065, 2006; phylogeny), Wang *et al.* (*Mol. Phylogen. Evol.* **41**: 295, 2006), Hofstetter *et al.* (*Mol. Phylogen. Evol.* **44**: 412, 2007; phylogeny).

Geoglossum Pers. (1794), Geoglossaceae. 24, widespread (temperate). See Nannfeldt (*Ark. Bot.* **30A** no. 4: 1, 1942), Mains (*Mycol.* **46**: 586, 1954; N. Am.), Maas Geesteranus (*Persoonia* **3**: 89, 1964), Raitviir *in* Parmasto (*Zhivaya priroda Dal'nego Vostoka*: 52, 1971; E. former USSR), Benkert (*Mykol. MittBl.* **20**: 47, 1976; Germany), Olsen (*Agarica* **7**: 120, 1986; key 22 spp. Norway), Spooner (*Biblthca Mycol.* **116**, 1987; 9 spp. Australasia, nomencl.), Verkley (*Persoonia* **15**: 405, 1994; asci), Zhuang (*Fl. Fung. Sinicorum* **8**, 1998; 10 spp. China), Wang (*Bull. natn. Mus. Nat. Sci.* Taiwan **13**: 147, 2001; Taiwan), Wang *et al.* (*Mycol.* **98**: 1065, 2006; phylogeny), Wang *et al.* (*Mol. Phylogen. Evol.* **41**: 295, 2006; phylogeny).

Geographical distribution. Knowledge of the geographical distribution of most fungi is inadequate. It is, however, possible to make a few generalizations. Almost every ecological niche has a specialized mycota (see Ecology), and the key factor determining fungal distribution is the substratum, whether it be a particular plant or animal, or some other material able to support their growth. Some fungi are extremely widespread compared with plants. This is especially true for species with no strong substratum preference (e.g. some generalist saprobes, lichen-forming ascomycetes, myxomycetes, polypores, opportunistic moulds, and soil fungi). Many common air-borne and soil moulds are ubiquitous. The thesis that most fungi are everywhere and that the environment selects (Gams, 1993) is probably true only for such species. Other fungi, particularly those with a narrow range of substrata are usually less widespread. Species which only occur in association with particular plants, for example, have distributions within the ranges of their associated organisms.

Some distributions can be related to continental drift, for example *Thamnomyces* spp. in west Africa and South America, *Cyttaria* spp. in Australasia and South America, and the wide distribution of some lichens (Sheard, *Bryologist* **80**: 100, 1977; Kärnefelt, *Bibl. Lich.* **38**: 291, 1990; Tehler, *Opera Bot.* **70**, 1988). Geographical distributions of the *Pucciniales* (Bisby, 1933) and *Erysiphales* (Hirata, 1966) have been stated to conform to the 'age and area' theory of Willis, which postulates that commonest genera are oldest, and the older the genus the wider its distribution. The geographical distribution of fungi with defined substratum preferences can depend not only on the presence of the substratum but also on its accessibility. On some Caribbean islands, for example, the absence of native terrestrial mammals has meant there are few species of dung-inhabiting fungi compared with continental America, even centuries after the introduction of horses (Richardson, pers. comm.; Minter *et al.*, *Fungi of the Caribbean*, 2001). Substratum accessibility may also be a factor in determining fungal biodiversity hotspots. Animal and plant diversity of tropical rain forests may favour generalist fungi over specialized species which have the problem of locating their substratum; the opposite may be the case in ecosystems with lower animal and plant diversity, such as temperate conifer forests. Natural distributions of many fungi have also been affected by acid rain, air pollution (q.v.), forest clearance, and accidental or deliberate disposal by humans (see Conservation).

Some lichen-forming species have exceptionally broad distributions, and wide disjunctions are known; many follow particular forest zones or rock types, and some are even bipolar (Du Rietz, *Acta Phytogeogr. Succ.* **13**: 215, 1940; Lynge, *Naturen* **12**: 367, 1941). Apart from continental drift, the wide distribution of some lichens has also been related to glaciations (Brodo & Hawksworth, *Opera Bot.* **42**, 1977); isolated populations of various species existed by the Cretaceous (Kärnefelt, 1990). In the *Caliciales*, small-spored genera are more likely to be distributed by long-range dispersal than those with large-spores (Tibell, *Bot. J. Linn. Soc.* **116**: 159, 1994).

Because of the importance of the geographical distribution of plant pathogenic fungi for disease control regulations and international movement of plant germ plasm, the distribution of the more important plant pathogenic fungi is comparatively well known. The *IMI Distribution Maps of Plant Diseases* (1943 on) cover over 1300 spp. and are constantly updated, with new editions issued as required.

There has been a renewed interest in mapping distributions in the last two decades, facilitated by developments in computer techniques. Most initiatives are at national level, and concentrate on species that are most easily recorded; dynamically created distribution maps are now also, however, available on-line at global, regional, national and subnational levels, although the information presented on those maps is usually very incomplete (see Internet). Kreisel compiled a bibliography of published distribution maps for non-lichenized fungi (*Feddes Rep.* **82**: 589, 1971, hymeno- and gasteromycetes, 1930-69; **83**: 741, 1973, hemi- and phragmobasidiomycetes; **84**: 619, 1973, basid. suppl.; **85**: 161, 1974, anamorphic fungi and endomycetes, 1941-72; **86**: 329, 1975, phycom.; **87**: 109, 1976, suppl.), and Hawksworth & Ahti (*Lichenologist* **22**: 1, 1990) include publications with maps.

Lit.: Diehl (*J. Wash. Acad. Sci.* **27**: 244, 1937), Bisby (*Am. J. Bot.* **20**: 246, 1933; *Bot. Rev.* **9**: 466, 1943), Gams (*in* Winterhoff (Ed.), *Fungi in vegetation science*: 183, 1992), Hirata (*Host range and geographical distributions of the powdery mildews*, 1966), Lumbsch *et al.* (Eds) (Phylogeography and biogeography of fungi. *MR* **112**, 2008), Pirozynski (*in* Ainsworth & Sussman (Eds), *The Fungi*, **3**: 487, 1968), Pirozynski & Weresub (*in* Kendrick (Ed.), *The whole fungus* **1**: 93, 1979), Pirozynski & Walker (*Aust. J. Bot.* Suppl. **10**, 1983; Pacific).

Major regional studies are cited in this *Dictionary* under systematic entries and also Discomycetes, Inventories, Lichens, Literature, Macromycetes, Plant pathogenic fungi, and Pyrenomycetes. See also Biodiversity, Numbers of fungi, and particular substrata.

Geomorium Speg. (1922) = Underwoodia fide Gamundí (*Darwiniana* **11**: 418, 1957).

Geomyces Traaen (1914), anamorphic *Pseudogymnoascus, Gymnostellatospora*, Hso.0eP.1. 4, widespread (esp. north temperate). See Sigler & Carmichael (*Mycotaxon* **4**: 376, 1976), van Oorschot (*Stud. Mycol.* **20**, 1980), Hocking & Pitt (*Mycol.* **80**: 82, 1988), Kuraishi *et al.* (*Antonie van Leeuwenhoek* **77**: 179, 2000; ubiquinones), Vidal *et al.* (*Revta Iberoamer. Micol.* **17** [Special]: 22, 2000; phylogeny), Gianni *et al.* (*Mycoses* **46**: 430, 2003; skin infection), Jiang & Yao (*Mycotaxon* **94**: 55, 2005; phylogeny, development), Rice & Currah (*Mycol.* **98**: 307, 2006; phylogeny).

Geopetalaceae Jülich (1982) = Pleurotaceae. The diagnosis of the family by Jülich refers to *Geopetalum* as typified by Singer (= *Faerberia*).

Geopetalum Pat. (1887) = Hohenbuehelia fide Kuyper (*in litt.*).

Geopetalum Singer (1951) ≡ Geopetalum Pat.

Geophila Quél. (1886) [non *Geophila* D. Don 1825, *Rubiaceae*] = Psilocybe (Fr.) P. Kumm. fide Kuyper (*in litt.*).

geophilous, earth loving, e.g. of fungi having underground fruit bodies. Cf. terricolous.

Geopora Harkn. (1885), Pyronemataceae. 13, widespread (northern hemisphere). See Burdsall (*Mycol.* **60**: 504, 1968), Senn-Irlet (*Beitr. Kenntn. Pilze Mitteleur.* **5**: 191, 1989; key 8 spp.), Zhang & Yu (*Acta Mycol. Sin.* **11**: 8, 1992; 3 spp. China), Yao & Spooner (*MR* **100**: 72, 1996; key Brit. spp.), Hansen & Pfister (*Mycol.* **98**: 1029, 2006; phylogeny), Læssøe & Hansen (*MR* **111**: 1075, 2007; phylogeny), Perry *et al.* (*MR* **111**: 549, 2007; phylogeny).

Geoporella Soehner (1951) = Hydnotrya fide Trappe (*TBMS* **65**: 496, 1975).

Geopyxis (Pers.) Sacc. (1889), ? Pyronemataceae. 7 (Biotrophic, associated with conifer roots), widespread. See Thind *et al.* (*Acta Bot. Indica* **9**: 115, 1981; key Himalayan spp.), Kimbrough & Gibson (*CJB* **68**: 342, 1990; ultrastr.), Garnoeidner *et al.* (*Z. Mykol.* **52**: 201, 1991), Crous *et al.* (*S. Afr. J. Bot.* **62**: 89, 1996), Landvik *et al.* (*Nordic Jl Bot.* **17**: 403, 1997; DNA), Vrålstad *et al.* (*Mol. Ecol.* **7**: 609, 1998; mycorrhizal status), Zhuang & Liu (*Nova Hedwigia* **83**: 177, 2006).

Georgefischeria Thirum. & Naras. (1963), Georgefischeriaceae. 4 (witches' broom of *Rivea*, *Argyreia* & *Lettsomia* (*Convolvulaceae*)), India. See Bauer *et al.* (*MR* **105**: 416, 2001), Gandhe (*Frontiers in Microbial Biotechnology and Plant Pathology, (Prof. S.M. Reddy Commemoration Volume)*: 69, 2002).

Georgefischeriaceae R. Bauer, Begerow & Oberw. (1997), Georgefischeriales. 2 gen., 27 spp.

Lit.: Bauer *et al.* (*CJB* **75**: 1273, 1997), Begerow *et al.* (*CJB* **75**: 2045, 1998), Bauer *et al.* (*MR* **105**: 423, 2001), Begerow *et al.* (*Mycol. Progr.* **1**: 187, 2002), Begerow *et al.* (*MR* **108**: 1257, 2004).

Georgefischeriales R. Bauer, Begerow & Oberw. (1997). Exobasidiomycetes. 4 fam., 7 gen., 40 spp. Fams:

(1) **Eballistraceae**
(2) **Gjaerumiaceae**
(3) **Georgefischeriaceae**
(4) **Tilletiariaceae**

For *Lit.* see under fam.

Geoscypha (Cooke) Lambotte (1888) = Peziza Fr. fide Eckblad (*Nytt Mag. Bot.* **15**: 1, 1968).

Geosiphon F. Wettst. (1915), Geosiphonaceae. 1 (arbuscular mycorrhiza unknown), Europe. includes cyanobacteria (*Nostoc*) in vesicles; the only known example of endocyanosis (q.v.) in fungi. See Mollenhauer *in* Reisser (Ed.) (*Algae and symbiosis*: 339, 1992), Schüssler *et al.* (*Bot. Acta* **107**: 36, 1994; arbuscular mycorrhizal affinity), Schüssler *et al.* (*Protoplasma* **185**: 131, 1995; endocytobiosis), Gehrig *et al.* (*J. Mol. Evol.* **43**: 71, 1996; phylogeny), Schüssler *et al.* (*Protoplasma* **190**: 53, 1996; ultrastr.), Mollenhauer *et al.* (*Protoplasma* **193**: 3, 1998; nutrition), Schüssler & Kluge *in* McLaughlin *et al.* (Eds) (*The Mycota* A Comprehensive Treatise on Fungi as Experimental Systems for Basic and Applied Research **9**: 151, 2001), Schüssler (*Plant and Soil* **244**: 75, 2002; phylogeny, taxonomy, and evolution), Redecker & Raab (*Mycol.* **98**: 885, 2006; phylogeny).

Geosiphonaceae Engl. & E. Gilg (1924), Archaeosporales. 1 gen. (+ 1 syn.), 1 spp.

Lit.: Azcon-Aguilar & Barea (*Mycorrhiza* **6**: 457, 1996), Gehrig *et al.* (*J. Mol. Evol.* **43**: 71, 1996), van der Heijden *et al.* (*Nature* Lond. **396**: 69, 1998), Schüssler & Kluge *in* McLaughlin *et al.* (Eds) (*The Mycota* A Comprehensive Treatise on Fungi as Experimental Systems for Basic and Applied Research **7A**: 151, 2001), Schüßler *et al.* (*MR* **105**: 1413, 2001), Redecker (*Glomeromycota* Arbuscular mycorrhizal fungi and their relative(s). Version 01 July 2005. http://tolweb.org/Glomeromycota/28715/2005.07.01 in The Tree of Life Web Project, http://tolweb.org: [unpaginated], 2005), Schüßler & Wolf *in* Declerck, Strullu & Fortin (Eds) (*Root-organ culture of mycorrhizal fungi*, 2005).

Geosiphonales Caval.-Sm. (1998) = Archaeosporales.

Geosiphonomyces Cif. & Tomas. (1957) ≡ Geosiphon.

Geosmithia Pitt (1979), anamorphic *Hypocreales*, Hso.0eH.32/33. 10, widespread. Species with *Talaromyces* anamorphs are unrelated. See Ogawa *et al.* (*Mycol.* **89**: 756, 1997; rDNA analysis), Ogawa & Sugiyama (*Integration of Modern Taxonomic Methods for Penicillium and Aspergillus Classification*: 149, 2000; teleomorph), Pitt *et al.* (*Integration of Modern Taxonomic Methods for Penicillium and Aspergillus Classification*: 9, 2000; accepted names), Heredia *et al.* (*Mycol.* **93**: 528, 2001; Mexico), Prieto *et al.* (*CJB* **80**: 410, 2002; cell wall polysaccharides), Kolařík *et al.* (*MR* **108**: 1053, 2004; morphology, phylogeny), Kubátová *et al.* (*Czech Mycol.* **56**: 1, 2004; ecology), Kolařík *et al.* (*MR* **109**: 1323, 2005), Yaguchi *et al.* (*Cryptog. Mycol.* **26**: 133, 2005; thermotolerant sp.), Kolařík *et al.* (*MR* **111**: 1298, 2007; with bark beetles).

Geotrichaceae Cif. ex Subram. (1962) = Dipodascaceae.

Geotrichella Arnaud (1954) nom. dub., anamorphic *Pezizomycotina*, Hso.0eP.40. 2 (on paper), France; N. America. See Sigler & Carmichael (*Mycotaxon* **4**: 349, 1976).

Geotrichites Stubblef., C.E. Mill., T.N. Taylor & G.T. Cole (1985), Fossil Fungi. 1 (on arachnoid in amber), Dominican Republic.

Geotrichoides Langeron & Talice (1932) = Trichosporon fide von Arx *et al.* (*Stud. Mycol.* **14**: 30, 1977).

Geotrichopsis Tzean & Estey (1991), anamorphic *Agaricomycetes*. 1 (with dolipore septa), Canada. See Tzean & Estey (*MR* **95**: 1351, 1991).

geotrichosis, disease in humans or animals caused by

Geotrichum.

Geotrichum Link (1809), anamorphic *Dipodascus, Galactomyces*, Hsy.0eH.39. 33, Europe; America. *G. candidum* (often as *Oospora lactis*) in milk. See Morenz (*Mykol. Schriftenreihe* **1**, 1963), Guého (*Antonie van Leeuwenhoek* **45**: 199, 1979; base comp. and taxonomy), Weijman (*Antonie van Leeuwenhoek* **45**: 119, 1979; carbohydrates and taxonomy), Olesen & Kier (*Nordic Jl Bot.* **4**: 365, 1984; SEM and TEM structure), de Hoog *et al.* (*Stud. Mycol.* **29**: 1, 1986; key), Guého *et al.* (*J. Clin. Microbiol.* **25**: 1191, 1987; DNA relatedness in *G. capitatum*), de Hoog & Amberger (*Antonie van Leeuwenhoek* **58**: 101, 1990; protein patterns in *Geotrichum* and teleomorphs), Jensen *et al.* (*Mycoses* **33**: 519, 1991; crossed immunoelectrophoresis to differentiate *G. candidum*), Watanabe *et al.* (*Mycoscience* **35**: 417, 1994; endoconidia), Pitt & Hocking (*Fungi and Food Spoilage* Edn 2: 593 pp., 1997; food-associated spp.), Hoog *et al. in* Kurtzman & Fell (Eds) (*Yeasts, a taxonomic study* 4th edn: 574, 1998; review), Smith & Poot (*Antonie van Leeuwenhoek* **74**: 229, 1998; genetic characterization), Tsai & Hsieh (*Plant Pathology Bulletin* Taichung **8**: 9, 1999; selective medium), Smith *et al.* (*Antonie van Leeuwenhoek* **77**: 71, 2000; taxonomy), Marcellino *et al.* (*Appl. Environm. Microbiol.* **67**: 4752, 2001; from cheese), Gente *et al.* (*Int. J. Food Microbiol.* **76**: 127, 2002; chromosomes), Hoog & Smith (*Stud. Mycol.* **50**: 489, 2004; phylogeny, review), Suh & Blackwell (*MR* **110**: 220, 2006; on beetles), Suh *et al.* (*Mycol.* **98**: 1006, 2006; phylogeny), Wuczkowski *et al.* (*Int. J. Syst. Evol. Microbiol.* **56**: 301, 2006).

geotropism, see tropisms.

Geotus Pilát & Svrček (1953) = Arrhenia fide Redhead (*CJB* **62**: 865, 1984).

Gerdemannia C. Walker, Błaszk., A. Schüssler & Schwarzott (2004) = Pacispora fide Kirk (*in litt.*).

Gerdemanniaceae C. Walker, Błaszk., A. Schüssler & Schwarzott (2004) = Pacisporaceae.

Gerhardtia Bon (1994) = Lyophyllum fide Kuyper (*in litt.*).

Gerlachia W. Gams & E. Müll. (1980) = Microdochium fide Samuels & Hallett (*TBMS* **81**: 473, 1983).

germ pore, a differentiated, frequently apical area, or hollow, in a spore wall (esp. in rusts) through which a **germ tube** (a germination hyphae) may come out; see Melendez-Howell (*Ann. Sci. nat. Bot.* sér. 12 **8**: 487, 1967; germ pore of basidiospores).

germ slit, a thin area of spore wall usually orientated along the long axis of the spore. In *Bryothele mira* the germ slit is transverse (Dobbeler, *Nova Hedw.* **66**: 337, fig. 3, 5, 1998).

germ tube, a germination hyphae which is formed by a germinating spore.

germicide, a substance causing destruction of microorganisms.

germination by repetition, producing secondary spores in place of germ tubes, as in *Heterobasidiomycetes* and *Sporobolomyces*.

Germslitospora Lodha (1978) = Coniochaeta fide Udagawa & Furuya (*TMSJ* **20**: 5, 1979), von Arx (*Gen. Fungi Sporul. Cult.* Edn 3, 1981), García *et al.* (*Mycol.* **95**: 525, 2003), García *et al.* (*MR* **110**: 1271, 2006; phylogeny).

Gerronema Singer (1951), Marasmiaceae. 13, widespread (esp. subtropical). See Singer (*Nova Hedwigia* **7**: 53, 1964; keys), Singer (*Fl. Neotrop.* **3**: 24, 1970; key 20 neotrop. spp.), Redhead (*CJB* **65**: 1551, 1986; emend. circumscr.), Lutzoni (*Syst. Biol.* **46**: 373, 1997; phylogeny), Redhead *et al.* (*Mycotaxon* **83**: 19, 2002; phylogeny), Bañares *et al.* (*Mycol.* **98**: 455, 2006; Canary Islands).

Gerulajacta Preuss (1855) nom. dub., anamorphic *Pezizomycotina*. See Sutton (*Mycol. Pap.* **141**, 1977).

Gerwasia Racib. (1909), Phragmidiaceae. 9 (on *Rubus, Rosa* (*Rosaceae*)), East Indies; America (tropical); China; Indonesia; Japan; Nepal; Philippines. See Tai (*Farlowia* **3**: 95, 1947), Buriticá (*Rev. I.C.N.E.* **5**: 173, 1994), Berndt (*Frontiers in Basidiomycote Mycology*: 185, 2004; Costa Rica).

Geusia Rühl & Korn (1979), Microsporidia. 1.

ghost fungus, *Pleurotus nidiformis*, an Australian luminous agaric. See Willis (*Muelleria* **1**: 213, 1967).

Giacominia Cif. & Tomas. (1953) = Arthopyrenia fide Hawksworth *et al.* (*Dictionary of the Fungi* edn 8, 1995).

giant puff-ball, *Langermannia gigantea*; see record fungi.

giant stone-fungus, *Polyporus tumulosus*; the pseudosclerotium may exceed $1m^3$.

Gibbago E.G. Simmons (1986), anamorphic *Pleosporaceae*, Hso.#eP.26. 1, N. & S. America. See Simmons (*Mycotaxon* **27**: 108, 1986).

gibber, gibbous (q.v.).

Gibbera Fr. (1825), Venturiaceae. Anamorphs *Dictyodochium*, *Stigmina*-like. *c.* 28, widespread. See Müller (*Sydowia* **8**: 60, 1954), Eriksson (*Svensk bot. Tidskr.* **68**: 192, 1974; key 9 spp.), Sivanesan (*TBMS* **82**: 507, 1984; anamorphs), Samuels *et al.* (*Brittonia* **40**: 392, 1988), Barr (*Sydowia* **41**: 25, 1989; key 4 N. Am. spp.).

Gibberella Sacc. (1877), Nectriaceae. Anamorph *Fusarium*. 23 (saprobes and pathogens), widespread. *G. fujikuroi* (on cotton, maize, rice (Bakanae disease) and other crops in warm areas); *G. zeae* (frequently as *G. saubinetii*) (foot rot and ear blight (scab) of cereals). See also gibberellin. See Nirenberg & O'Donnell (*Mycol.* **90**: 434, 1998), O'Donnell *et al.* (*Mycol.* **90**: 465, 1998; phylogeny), Samuels *et al. in* Summerell (Ed.) (*The genus Fusarium*, 1998), Rossman *et al.* (*Stud. Mycol.* **42**: 248 pp., 1999), O'Donnell *et al.* (*Mycoscience* **41**: 61, 2000), Aoki *et al.* (*Mycoscience* **42**: 461, 2001), Geiser *et al.* (*Mycol.* **93**: 670, 2001), Leslie (*Fusarium* Paul E. Nelson Memorial Symposium: 113, 2001; population genetics), Steenkamp *et al.* (*Molecular Plant Pathology* **2**: 215, 2001; molecular ecology), Britz *et al.* (*Sydowia* **54**: 9, 2002; *G. circinata*), Desjardins (*Ann. Rev. Phytopath.* **41**: 177, 2003; review), Mirete *et al.* (*Int. J. Food Microbiol.* **89**: 213, 2003; fumonisins), Zeller *et al.* (*Mycol.* **95**: 943, 2003; *G. konza*), Láday *et al.* (*Eur. J. Pl. Path.* **110**: 563, 2004; mtDNA), Leslie *et al.* (*Eur. J. Pl. Path.* **110**: 611, 2004; mating populations), Leslie *et al.* (*Appl. Environm. Microbiol.* **70**: 2254, 2004; Kansas), Phan *et al.* (*Stud. Mycol.* **50**: 261, 2004; Australia), Zeller *et al.* (*Mol. Ecol.* **13**: 563, 2004; USA), Adugna *et al.* (*Z. PflKrankh. PflPath. PflSchutz* **112**: 134, 2005; on *Coffea*), Geiser *et al.* (*Mycol.* **97**: 191, 2005; on *Coffea*), Lepoint *et al.* (*Appl. Environm. Microbiol.* **71**: 8466, 2005; on *Coffea*), Leslie *et al.* (*Mycol.* **97**: 718, 2005; *G. sacchari*).

gibberellin, a complex of hormone-like substances from *Gibberella fujikuroi* (anamorph *Fusarium moniliforme*) which causes overgrowth of higher plants, first recognized as the cause of Bakanae disease of rice. Gibberellin **A_1**, **A_2**, **A_3** and other fractions have been distinguished including **gibberellic acid** which has a similar physiological action to gibberellin A_1. Gibberellin is manufactured commercially for use in horticulture. See Stodola (*Source book on gibberellin* 1828-1957, 1958; 632 abstracts), Knapp (Ed.) (*Eigenschaften und Wirkungen der Gibberelline*, 1962), MacMillan & Takahashi (*Nature* **217**: 170, 1968; allocation of trivial names), Jefferys (*Adv. appl. Microbiol.* **13**: 283, 1970).

Gibberellulina Sousa da Câmara (1950) nom. dub., Sordariomycetidae. 1, Europe. See Rossman (*Stud. Mycol.* **42**, 1999).

Gibberidea (Fr.) Kuntze (1898) ≡ Cucurbitaria.

Gibberidea Fuckel (1870), Dothideomycetes. Anamorph *Pleurostomella*. 1, Europe. See Holm (*Svensk bot. Tidskr.* **62**: 217, 1968).

Gibberinula Kuntze (1898) ≡ Gibberidea Fuckel.

gibbous (of a pileus), having a swelling or wide umbo, or having a convex top and a flat underside; gibber, gibbose.

Gibellia Pass. (1886) ≡ Gibellina.

Gibellia Sacc. (1885), ? Melanconidaceae. 1 (from bark), Australia. See von Arx & Müller (*Beitr. Kryptfl. Schweiz* **11** no. 1, 1954).

Gibellina Pass. (1886), ? Magnaporthaceae. 3, Asia; Europe. *G. cerealis* on wheat. See Glynne *et al.* (*TBMS* **84**: 653, 1985).

Gibellula Cavara (1894), anamorphic *Torrubiella*, Hsy.0eH.15/16. 17, widespread. See Petch (*Annls mycol.* **30**: 386, 1932), Samson & Evans (*Mycol.* **84**: 300, 1992; on *Arachnida*), Tzean *et al.* (*Mycol.* **89**: 309, 1997; spp. from Taiwan), Selçuk *et al.* (*Mycol. Balcanica* **1**: 61, 2004; Turkey).

Gibellulopsis Bat. & H. Maia (1959), anamorphic *Plectosphaerellaceae*. 1, widespread. See Zare *et al.* (*Nova Hedwigia* **85**: 463, 2007).

Gibsonia Massee (1909) ? = Melanospora Corda fide Cannon & Hawksworth (*J. Linn. Soc.* Bot. **84**, 1982).

Gigantospora B.S. Lu & K.D. Hyde (2003), Xylariaceae. 1, USA. See Lu & Hyde (*Nova Hedwigia* **76**: 202, 2003).

Gigasperma E. Horak (1971), Gigaspermaceae. 2, Australasia; N. America. The genus is polyphyletic with some members belonging to *Boletales* and others to *Agaricales*. See Beaton & Malajczuk (*TBMS* **87**: 478, 1986), Castellano & Trappe (*Aust. Syst. Bot.* **5**: 613, 1992), Matheny *et al.* (*Mycol.* **98**: 982, 2006; see on-line supplementary material).

Gigaspermaceae Jülich (1982), Agaricales. 1 gen., 2 spp.

Lit.: Castellano & Trappe (*Aust. Syst. Bot.* **5**: 641, 1992), Kropp & Hutchison (*Mycol.* **88**: 662, 1996).

Gigaspora Gerd. & Trappe (1974), Gigasporaceae. 7, widespread. See Sward (*New Phytol.* **87**: 761, 1980), Sward (*New Phytol.* **87**: 661, 1981; spore ultrastr.), Walker & Sanders (*Mycotaxon* **27**: 169, 1986; emend.), Spain *et al.* (*Mycotaxon* **34**: 667, 1989), Tommerup & Sivasithamparam (*MR* **94**: 897, 1991), Maia *et al.* (*Mycol.* **85**: 883, 1993), Maia *et al.* (*Mycol.* **86**: 343, 1994; spore wall ultrastr., germination), Bentivenga & Morton (*Mycol.* **87**: 719, 1995; monogr. & key), Montecchi *et al.* (*Riv. Micol.* **39**: 269, 1996), Yao *et al.* (*Genera of Endogonales*: 229 pp., 1996), Gadhar *et al.* (*Can. J. Microbiol.* **43**: 795, 1997; polymorphism), Bago *et al.* (*New Phytol.* **139**: 581, 1998; molecular analysis), Lanfranco *et al.* (*Mol. Ecol.* **8**: 372, 1999; genetic variation), Redecker & Raab (*Mycol.* **98**: 885, 2006; phylogeny).

Gigasporaceae J.B. Morton & Benny (1990), Diversisporales. 2 gen., 39 spp.

Lit.: Azcon-Aguilar & Barea (*Mycorrhiza* **6**: 457, 1996), Bentivenga & Morton (*Proc. natn Acad. Sci. U.S.A.* **93**: 5659, 1996; fatty acids), van der Heijden *et al.* (*Nature* Lond. **396**: 69, 1998), Lanfranco *et al.* (*Mol. Ecol.* **8**: 37, 1999; sequence variability), Jeffries & Dodd (*Appl. Microb. System.*: 73, 2000), Schüßler *et al.* (*MR* **105**: 1413, 2001), Walker & Schüssler (*MR* **108**: 982, 2004), Redecker (*Glomeromycota* Arbuscular mycorrhizal fungi and their relative(s). Version 01 July 2005. http://tolweb.org/Glomeromycota/28715/2005.07.01 in The Tree of Life Web Project, http://tolweb.org: [unpaginated], 2005), Souza *et al.* (*MR* **109**: 697, 2005).

Gigasporites Carlie J. Phipps & T.N. Taylor (1996), Fossil Fungi. 1, Antarctica. See Phipps & Taylor (*Mycol.* **88**: 709, 1996).

Gilbertella Hesselt. (1960), Choanephoraceae. 1, widespread (tropical). *G. persicaria* pathogenic to peach (*Amygdalus persicae*). See Hesseltine (*Bull. Torrey bot. Club* **87**: 21, 1960), O'Donnell *et al.* (*CJB* **55**: 662, 1977; zygospore ontogeny), Powell *et al.* (*CJB* **59**: 908, 1981; ultrastr. chlamydospore), Powell *et al.* (*Protoplasma* **111**: 87, 1982; ultrastr. membrane), Whitney & Arnott (*Mycol.* **78**: 42, 1986), Whitney & Arnott (*Mycol.* **80**: 707, 1988; calcium oxalate chrystals), Benny (*Mycol.* **83**: 150, 1991), Michaelides *et al.* (*Mycol.* **89**: 609, 1997; zygosporogenesis), Donnell *et al.* (*Mycol.* **93**: 286, 2001; phylogeny), Voigt & Wöstemeyer (*Gene* **270**: 113, 2001; phylogeny), Papp *et al.* (*Acta Biolog. Hung.* **54**: 393, 2003; phylogeny).

Gilbertellaceae Benny (1991) = Choanephoraceae.

Lit.: Kirk (*in litt.*).

Gilbertia Donk (1934) nom. inval. = Amanita Pers. fide Kuyper (*in litt.*).

Gilbertiella R. Heim (1965) [non *Gilbertiella* Boutique 1951, *Annonaceae*] ≡ Gyrodon.

Gilbertina R. Heim (1966) = Gyrodon fide Pegler (*Kew Bull.* Addit. Ser. **9**, 1983).

Gilbertsonia Parmasto (2001), Fomitopsidaceae. 1, USA. See Parmasto (*Harvard Pap. Bot.* **6**: 179, 2001).

Gilchristia Redaelli & Cif. (1934) ≡ Zymonema.

Gilkeya M.E. Sm., Trappe & Rizzo (2007), Pyronemataceae. 1 (hypogeous), N. America. See Smith *et al.* (*Mycol.* **98**: 705, 2006), Læssøe & Hansen (*MR* **111**: 1075, 2007; phylogeny).

gill (of an agaric), commonly used in English for lamella (q.v.) which is to be preferred as a more international term; **- fungi**, members of the *Agaricales*.

Gilletia Torrend (1914) ≡ Telligia.

Gilletiella Sacc. & P. Syd. (1899), Dothideomycetes. Anamorph *Ascochyta*-like. 2 or 2 (on *Chusquea*), America. See Müller & von Arx (*Beitr. Kryptfl. Schweiz* **11** no. 2, 1962), Eriksson (*SA* **7**: 72, 1988).

Gillotia Sacc. & Trotter (1913), Mycosphaerellaceae. Anamorph *Asteromella*-like. *c.* 3, widespread (tropical). See von Arx & Müller (*Stud. Mycol.* **9**, 1975).

Gilmania Bat. & Cif. (1962) = Chaetothyrium fide von Arx & Müller (*Stud. Mycol.* **9**, 1975).

Gilmaniella G.L. Barron (1964), anamorphic *Pezizomycotina*, Hso.0eP.1/10. 7, widespread. See Moustafa & Ezz-Eldin (*MR* **92**: 502, 1989; key),

Umali *et al.* (*MR* **102**: 435, 1998; Hong Kong).
gilvous, pale yellow.
gin, see spirits.
Ginanniella Cif. (1938) = Urocystis fide Nagler (*Z. Mykol.* **53**: 331, 1987), Vánky (*Illustrated genera of smut fungi*, 1987).
ginger beer plant (Californian or American 'bees'), a mixture of a yeast (*Saccharomyces pyriformis*) and a bacterium (*Bacterium vermiforme*) used for fermenting a sugar solution to make a drink (Marshall Ward, *Phil. Trans. R. Soc., Lond.* **B 183**: 125, 1892). See Ramsbottom (*TBMS* **7**: 86, 1921); cf. tibi, tea fungus, teekwass.
Ginzbergerella Zahlbr. (1931) = Gyrocollema fide Henssen (*SA* **5**: 131, 1986).
Giraffachitina Locq. (1985), Fossil Fungi. 1, Estonia.
Girardia Gray (1821) = Bangia fide Henssen (*Symb. bot. upsal.* **18** no. 1, 1963).
Giulia Tassi (1904), anamorphic *Pezizomycotina*, Cpd.0eH.1. 1, Europe. See Pirozynski & Shoemaker (*CJB* **49**: 529, 1971).
Gjaerumia R. Bauer, M. Lutz & Oberw. (2005), Gjaerumiaceae. 3 (on *Liliaceae* s.l.), Europe (northern). See Bauer *et al.* (*MR* **109**: 1250, 2005).
Gjaerumiaceae R. Bauer, M. Lutz & Oberw. (2005), Georgefischeriales. 1 gen., 3 spp.
Lit.: Bauer *et al.* (*MR* **109**: 1250, 2005).
Glabrocyphella W.B. Cooke (1961), Marasmiaceae. 13, widespread. See Cooke (*Beih. Sydowia* **4**: 45, 1961), Redhead *et al.* (*Mycotaxon* **83**: 19, 2002; phylogeny).
Glabrotheca Chardón (1939), Pezizomycotina. 1, Venezuela.
glabrous, smooth; not hairy.
glaireous, slimy.
Glaphyriopsis B. Sutton & Pascoe (1987), anamorphic *Pezizomycotina*, Ccu.≡ eH-P.15. 2, Australia. See Sutton & Pascoe (*TBMS* **88**: 169, 1987).
Glarea Bills & F. Paláez (1999), anamorphic *Helotiales*, Hso.?.?. 1, Spain. See Bills *et al.* (*MR* **103**: 189, 1999).
Glaucinaria Fée ex A. Massal. (1860) = Diorygma fide Hawksworth *et al.* (*Dictionary of the Fungi* edn 8, 1995), Kalb *et al.* (*Symb. bot. upsal.* **34** no. 1: 133, 2004).
Glaucomaria M. Choisy (1929) ? = Lecanora fide Hafellner (*Beih. Nova Hedwigia* **79**: 241, 1984; status).
Glaucospora Rea (1922) = Melanophyllum fide Singer (*Agaric. mod. Tax.*, 1951).
glaucous, having a bluish-grey waxy bloom.
Glaxoa P.F. Cannon (1997), Tubeufiaceae. 1, Great Britain. See Cannon (*SA* **15**: 121, 1997), Kodsueb *et al.* (*Fungal Diversity* **21**: 105, 2006).
Glaziella Berk. (1880), Glaziellaceae. 1, widespread (tropical). See Landvik & Eriksson (*SA* **13**: 13, 1984; posn), Gibson *et al.* (*Mycol.* **78**: 941, 1987), Landvik *et al.* (*Nordic Jl Bot.* **17**: 403, 1997; phylogeny), Harrington *et al.* (*Mycol.* **91**: 41, 1999; phylogeny), Hansen & Pfister (*Mycol.* **98**: 1029, 2006; phylogeny), Læssøe & Hansen (*MR* **111**: 1075, 2007; phylogeny), Perry *et al.* (*MR* **111**: 549, 2007; phylogeny).
Glaziellaceae J.L. Gibson (1986), Pezizales. 1 gen. (+ 1 syn.), 1 spp.
Lit.: Landvik & Eriksson (*SA* **13**: 13, 1984; posn), Gibson *et al.* (*Mycol.* **78**: 941, 1986), Landvik & Eriksson (*SA* **13**: 13, 1994), Landvik *et al.* (*Nordic Jl Bot.* **17**: 403, 1997), Harrington *et al.* (*Mycol.* **91**: 41, 1999), Hansen & Pfister (*Mycol.* **98**: 1029, 2006; phylogeny), Hansen *et al.* (*Mycol.* **97**: 1023, 2005), Perry *et al.* (*MR* **111**: 549, 2007; phylogeny).
Glaziellales J.L. Gibson (1986) = Pezizales.
gleba, the sporing tissue in an angiocarpous sporocarp, esp. of gasteromycetes and hypogeous *Pezizales*; **glebal mass**, the projectile of *Sphaerobolus*.
glebula, a rounded process from a lichen thallus.
Glenospora Berk. & Desm. (1849) nom. rej. = Septobasidium. Used in medical mycology for unrelated fungi. fide Berndt (*in litt.*) Used in medical mycology for unrelated fungi; see, Petch (*TBMS* **12**: 105, 1927).
Glenosporaceae Nann. (1934) = Septobasidiaceae.
Glenosporella Nann. (1931) = Geomyces fide Carmichael (*CJB* **40**, 1962), van Oorschot (*Stud. Mycol.* **20**, 1980).
Glenosporopsis O.M. Fonseca (1943), anamorphic *Pezizomycotina*, Hso.0eP.1/4. 1 (on humans), Brazil. See Taborda *et al.* (*J. Clin. Microbiol.* **37**: 2031, 1999).
gleocystidium, see cystidium.
Gleophyllum, see *Gloeophyllum*.
gleoplerous hyphae (oil hyphae), hyphae with very long cells (or unicellular), with numerous oil drops in the plasma. See Jülich (*in* Gams (Ed.), *Kleine Kryptogamenflora* **II**(7), 1984).
gliatope, a site of heavy gel production (Moore, *Am. J. Bot.* **52**: 391, 1965). See gel tissue.
Glioannellodochium Matsush. (1989), anamorphic *Pezizomycotina*, Hsp.0eH.19. 1, Australia. See Matsushima (*Matsush. Mycol. Mem.* **6**: 19, 1989).
Glioblastocladium Matsush. (1989), anamorphic *Pezizomycotina*, Hso.0eH.10. 1, Australia. See Matsushima (*Matsush. Mycol. Mem.* **6**: 20, 1989).
Gliobotrys Höhn. (1902) = Stachybotrys fide Bisby (*TBMS* **26**: 133, 1943).
Gliocephalis Matr. (1899), anamorphic *Pyxidiophoraceae*, Hsy.0eH.15. 2, Asia; N. America. See Hawksworth (*Bull. Br. Mus. nat. hist.* Bot. **6**: 181, 1979), Jacobs *et al.* (*Mycol.* **97**: 111, 2005; phylogeny, morphology).
Gliocephalotrichum J.J. Ellis & Hesselt. (1962), anamorphic *Leuconectria*, Hso.0eH.15. 5, widespread. See Wiley & Simmons (*Mycol.* **63**: 575, 1971; key), Rossman *et al.* (*Mycol.* **85**: 685, 1993; teleomorph), Schoch *et al.* (*Stud. Mycol.* **45**: 45, 2000; phylogeny), Watanabe & Nakamura (*Mycoscience* **46**: 46, 2005; Japan), Decock *et al.* (*Mycol.* **98**: 488, 2006; French Guiana).
Gliocladiopsis S.B. Saksena (1954), Nectriaceae. 1, pantropical. See Barron (*The genera of hyphomycetes from soil*, 1968), Schoch *et al.* (*Stud. Mycol.* **45**: 45, 2000; phylogeny), Crous (*Taxonomy and Pathology of Cylindrocladium (Calonectria) and Allied Genera*: 278 pp., 2002; revision).
Gliocladium Corda (1840), anamorphic *Sphaerostilbella*, Hsy.0eH.15. 13, widespread. See also *Clonostachys*, *Trichoderma*. See Seifert (*Stud. Mycol.* **27**: 1, 1985; teleomorph), Schroers (*Stud. Mycol.* **46**: 1, 2001), Samuels *et al.* (*CBS Diversity Ser.* **4**, 2006; USA).
Gliocladochium Höhn. (1916) ≡ Periola.
Gliocoryne Maire (1909) = Pistillaria Fr. fide Corner (*Ann. Bot. Mem.* [A monograph of Clavaria and allied genera] **1**, 1950).
Gliodendron Salonen & Ruokola (1969) = Sterigmatobotrys fide Sutton (*Mycol. Pap.* **132**, 1973).
Gliomastix Guég. (1905), anamorphic *Hypocreales*,

Hso.0eP.15. 19, widespread. See Dickinson (*Mycol. Pap.* **115**, 1968; key), Gams (*Cephalosporium-artige Schimmelpilze*, 1971), Hammill (*Mycol.* **73**: 229, 1981; typification), Gams & Boekhout (*Proc. Indian Acad. Sci.* Pl. Sci. **94**: 273, 1985).

Glionectria Crous & S.L. Schoch (2000), Nectriaceae. Anamorph *Gliocladiopsis*. 1, widespread (pantropical). See Schoch *et al.* (*Stud. Mycol.* **45**: 45, 2000), Crous (*Taxonomy and Pathology of Cylindrocladium (Calonectria) and Allied Genera*: 278 pp., 2002; revision).

Gliophorus Herink (1958) = Hygrocybe fide Singer (*Agaric. mod. Tax.* edn 2, 1962).

Gliophragma Subram. & Lodha (1964), anamorphic *Pezizomycotina*, Hsy.1bH.10. 1, India. See Subramanian & Lodha (*CJB* **42**: 1059, 1964).

Gliostroma Corda (1837) nom. dub., anamorphic *Pezizomycotina*. See Holubová-Jechová (*Sydowia* **46**: 242, 1994).

gliotoxin, an antibiotic from *Gliocladium virens* (Webster & Lomas, *TBMS* **47**: 535, 1964), *Aspergillus fumigatus*, *Penicillium cinerascens* (Brian & Hemming, *Ann. appl. Biol.* **32**: 214, 1945; *Biochem. J.* **41**: 570, 1947); antibacterial and antifungal (has been used as a seed dressing). Cf. viridin.

Gliotrichum Eschw. (1822) nom. dub., ? Fungi.

Glischroderma Fuckel (1870), anamorphic *Pezizaceae*, St.0eP.6. 1, Europe. See Hennebert (*Persoonia* **7**: 183, 1973), Norman & Egger (*Mycol.* **91**: 820, 1999; phylogeny), Hansen *et al.* (*Mycol.* **93**: 958, 2001; phylogeny).

Glischrodermataceae Rea (1922) = Pezizaceae.

Globaria Quél. (1873) ≡ Bovista fide Demoulin (*Persoonia* **7**: 152, 1973).

Globifomes Murrill (1904), Polyporaceae. 1, N. America. See Gilbertson & Ryvarden (*N. Amer. Polyp.* **1**: 307, 1987).

Globipilea, see *Globopilea*.

Globoa Bat. & H. Maia (1962), Dothideomycetes. 1, Uganda. See Batista & Maia (*Beih. Sydowia* **3**: 54, 1962).

Globoasclerotes Stach & Pickh. (1957), Fossil Fungi. 1 (Carboniferous), Germany.

globoid (**globose**, **globular**, **globulose**), spherical or almost so (Fig. 23.1).

Globonectria Etayo (2002), Bionectriaceae. 1, Colombia. See Etayo (*Biblthca Lichenol.* **84**: 47, 2002).

Globopilea Beauseign. (1926) = Helvella fide Dissing (*Dansk bot. Ark.* **25** no. 1, 1966).

Globosasclerotes Stach & Pickh. (1957), Fossil Fungi. 1 (Carboniferous, Tertiary), Europe. See Stach & Pickhardt (*Palaeontologische Zeitschrift* **31**: 140, 1957).

Globosomyces Jülich (1980), Polyporales. 1, Borneo. See Jülich (*J. Linn. Soc.* Bot. **81**: 45, 1980).

Globosopyreno Lloyd (1923) nom. dub., ? Fungi.

Globosphaeria D. Hawksw. (1990), Sordariomycetes. 1 (on lichens, esp. *Normandina*), Tasmania. See Hawksworth (*Lichenologist* **22**: 301, 1990), Matzer (*Cryptog. Mycol.* **14**: 11, 1993).

Globuliciopsis Hjortstam & Ryvarden (2004), Polyporales. 2, S. America. See Hjortstam & Ryvarden (*Syn. Fung.* **18**: 22, 2004).

Globulicium Hjortstam (1973), Agaricomycetes. 1, Europe. *Hymenochaetales* or *Agaricales* (*Rickenella* clade). See Hjortstam (*Svensk bot. Tidskr.* **67**: 108, 1973).

Globuligera (Sacc.) Höhn. (1918) = Xylogramma fide Kutorga & Hawksworth (*SA* **15**: 1, 1997).

Globulina Speg. (1888), Dothideomycetes. 1, Brazil. See Rossman (*Mycol. Pap.* **157**, 1987).

Globulina Velen. (1934) [non *Globulina* Link 1820, *Chlorophyta*] = Unguiculella fide Svrček (*Česká Mykol.* **41**: 193, 1987).

Globuliroseum Sullia & K.R. Khan (1984), anamorphic *Pezizomycotina*, Hsp.0eH.1. 1, India. See Sullia & Khan (*Kavaka* **11**: 67, 1983).

glochidiate, covered with barbed bristles.

Gloeandromyces Thaxt. (1931), Laboulbeniaceae. 2, C. & S. America. See Nannfeldt (*Svensk bot. Tidskr.* **43**: 468, 1949).

Gloeoasterostroma Rick (1938) nom. conf., Russulales. See Rick (*Iheringia* Série Botânica **4**: 116, 1959).

Gloeocalyx Massee (1901) = Plectania fide Korf (*Mycol.* **49**: 102, 1957).

Gloeocantharellus Singer (1945), Gomphaceae. 3, N. & S. America. See Petersen (*The genera Gomphus and Gloeocantharellus in North America*, 1972), Vasco-P. & Franco-Molano (*Mycotaxon* **91**: 87, 2005; Colombian Amazonia).

Gloeocercospora D.C. Bain & Edgerton ex Deighton (1971) = Microdochium fide Rawla (*TBMS* **60**: 283, 1973; comparison with *Ramulispora*), Braun (*Monogr. Cercosporella, Ramularia Allied Genera (Phytopath. Hyphom.)* **1**, 1995), Heo *et al.* (*Pl. Path. J.* **15**: 242, 1999; Korea).

Gloeocorticium Hjortstam & Ryvarden (1986), Cyphellaceae. 1, Argentina. See Hjortstam & Ryvarden (*Mycotaxon* **25**: 551, 1986).

Gloeocoryneum Weindlm. (1964) = Leptomelanconium fide Morgan-Jones (*CJB* **49**: 1011, 1971).

Gloeocybe Earle (1909) = Lactarius fide Singer (*Agaric. mod. Tax.*, 1951).

Gloeocystidiellaceae Jülich (1982) = Stereaceae.

Gloeocystidiellum Donk (1931), Stereaceae. 7, widespread. Sensu lato 68 spp. See Donk (*Fungus* Wageningen **26**: 8, 1956), Eriksson & Ryvarden (*Cortic. N. Europ.* **3**: 405, 1975; key 10 Eur. spp.), Stalpers (*Stud. Mycol.* **40**: 59, 1996; key), Larsson & Larsson (*Mycol.* **95**: 1037, 2003; phylogeny).

Gloeocystidiopsis Jülich (1982) = Gloiothele fide Hjortstam & Larsson (*Windahlia* **21**, 1994).

Gloeocystidium P. Karst. (1889) = Dacryobolus fide Donk (*Persoonia* **3**: 199, 1964).

gloeocystidium, see cystidium.

Gloeodes Colby (1920), anamorphic *Pezizomycotina*, Cpt.0eH.?. 1, widespread. *G. pomigena* (sooty blotch of apple and citrus). See Williamson & Sutton (*Pl. Dis.* **84**: 714, 2000), Grabowski (*Phytopathologia Polonica* **34**: 5, 2004; Poland), Williamson *et al.* (*Mycol.* **96**: 885, 2004).

Gloeodiscus Dennis (1961), ? Dothideomycetes. 1, New Zealand. See Dennis (*Kew Bull.* **15**: 319, 1961).

Gloeodontia Boidin (1966), Stereaceae. 5, widespread. See Burdsall & Lombard (*Mem. N. Y. bot. Gdn* **28**: 16, 1976), Stalpers (*Stud. Mycol.* **40**: 59, 1996; key), Larsson & Larsson (*Mycol.* **95**: 1037, 2003; phylogeny).

Gloeoglossum E.J. Durand (1908) = Geoglossum fide Nannfeldt (*Ark. Bot.* **30A** no. 4: 1, 1942).

Gloeohaustoriales. Ordinal name proposed by Heim (*BSMF* **67**: 354, 1951) for *Antennopsis*, *Muiaria*, *Muiogone* and *Chantransiopsis*.

Gloeoheppia Gyeln. (1935), Gloeoheppiaceae (L). 4, widespread (desert areas). See Henssen (*Lichenolo-*

gist **27**: 261, 1995), Schultz & Büdel (*Lichenologist* **34**: 39, 2002), Schultz (*Bryologist* **110**: 286, 2007; USA).

Gloeoheppiaceae Henssen (1995), Lichinales (L). 3 gen., 8 spp.
Lit.: Henssen (*Lichenologist* **27**: 261, 1995).

Gloeohypochnicium (Parmasto) Hjortstam (1987), Russulales. 1, widespread. See Hjortstam (*Mycotaxon* **28**: 19, 1987), Larsson (*Mycol.* **95**: 1037, 2004; phylogeny).

Gloeolecta Lettau (1937) = Bryophagus fide Hawksworth *et al.* (*Lichenologist* **12**: 1, 1980).

Gloeomucro R.H. Petersen (1980), Hydnaceae. 9, widespread. See Petersen (*Mycol.* **72**: 301, 1980).

Gloeomucronaceae Jülich (1982) = Hydnaceae.

Gloeomyces Sheng H. Wu (1996), Stereaceae. 3, Taiwan; Japan. See Wu (*Mycotaxon* **58**: 47, 1996), Larsson & Larsson (*Mycol.* **95**: 1037, 2003; phylogeny).

Gloeopeniophora Höhn. & Litsch. (1907) = Peniophora fide Eriksson *et al.* (*Cortic. N. Europ.* **5**, 1978; as subgen.).

Gloeopeniophorella Rick (1934) nom. dub., Russulales.

Gloeopeziza Zukal (1891), Helotiaceae. 2 or 4 (on *Hepaticae*), widespread. See Döbbeler (*Sydowia* **38**: 41, 1986), Döbbeler (*Sendtnera* **3**: 103, 1996).

Gloeophyllaceae Jülich (1982), Gloeophyllales. 7 gen. (+ 12 syn.), 31 spp.
Lit.: Hof (*Some Wood-Destroying Basidiomycetes* **1**: 55, 1981), Corner (*Beih. Nova Hedwigia* **86**: 265 pp., 1987), Jung (*Biblthca Mycol.* **119**: 260 pp., 1987), Chamuris (*Mycol. Mem.* **14**: 247 pp., 1988), Hjortstam & Ryvarden (*Mycotaxon* **37**: 55, 1990), Nakasone (*Mycol. Mem.* **15**: 412 pp., 1990), Nakasone (*Mycol.* **82**: 622, 1990), Ryvarden & Gilbertson (*Syn. Fung.* **6**: 387 pp., 1993), Ginns (*Mycol.* **90**: 19, 1998), Thorn *et al.* (*Mycol.* **92**: 241, 2000), Hibbett & Donoghue (*Syst. Biol.* **50**: 215, 2001), Adair *et al.* (*FEMS Microbiol. Lett.* **211**: 117, 2002), Binder & Hibbett (*Mol. Phylogen. Evol.* **22**: 76, 2002), Schmidt *et al.* (*Z. Mykol.* **68**: 141, 2002), Yuan *et al.* (*Mycosystema* **23**: 173, 2004), Binder *et al.* (*Systematics and Biodiversity* **3**: 113, 2005), Moreth & Schmidt (*Holzforschung* **59**: 90, 2005).

Gloeophyllales Thorn (2007). Agaricomycetes. 1 fam., 8 gen., 33 spp. Fam.:
Gloeophyllaceae
For *Lit.* see under fam.

Gloeophyllum P. Karst. (1882), Gloeophyllaceae. 13, widespread. See David & Fiasson (*Bull. mens. Soc. linn. Lyon* **46**: 304, 1977; chemotaxonomy), Corner (*Beih. Nova Hedwigia* **86**: 61, 1987; key Malaysia spp.), Schmidt *et al.* (*Z. Mykol.* **68**: 141, 2002; species in buildings), Yuan *et al.* (*Mycosystema* **23**: 173, 2004; China).

Gloeoporus Mont. (1842), Meruliaceae. 26, widespread. See David (*BSMF* **88**: 209, 1972), Coelho *et al.* (*Mycol.* **98**: 821, 2006; Brazil).

Gloeopyrenia Zschacke (1937) = Protothelenella fide Mayrhofer (*Herzogia* **7**: 313, 1987), Hawksworth *et al.* (*Dictionary of the Fungi* edn 8, 1995).

Gloeoradulum Rick (1959) nom. dub., Agaricomycetes.

Gloeosebacina Neuhoff (1924) = Stypella fide Donk (*Persoonia* **4**: 178, 1966).

Gloeosoma Bres. (1920) = Aleurodiscus fide Lemke (*CJB* **42**: 213, 1964).

Gloeosporidiella Petr. (1921), anamorphic *Drepanopeziza*, Cac.0eH.15. 12, widespread. See Rimpau (*Phytopath. Z.* **43**: 257, 1962), Constantinescu *et al.* (*Mycotaxon* **94**: 175, 2005; Sweden).

Gloeosporidiellaceae Melnik (1986) = Dermateaceae.

Gloeosporidina Petr. (1921), anamorphic *Apiognomonia*, Cac.0eH.15. 7, widespread. See Sutton & Pollack (*Mycol.* **65**: 1125, 1973), Kubono (*TMSJ* **34**: 261, 1993), Kubono *et al.* (*Mycoscience* **35**: 279, 1994; *Stromatinia* teleomorph), Butin & Kehr (*Eur. J. For. Path.* **28**: 297, 1998; *Apiognomonia* teleomorph), Yuan *et al.* (*Pl. Dis.* **84**: 510, 2000; on *Eucalyptus*, Australia).

Gloeosporidium Höhn. (1916) = Discula fide von Arx (*Biblthca Mycol.* **24**, 1970).

Gloeosporiella Cavara (1892), anamorphic *Pezizomycotina*, Hsp.1bH.1. 1, Europe. See Sutton (*Nova Hedwigia* **25**: 163, 1974).

Gloeosporina Höhn. (1916) = Asteroma fide Sutton (*Mycol. Pap.* **141**, 1977).

Gloeosporiopsis Speg. (1910) = Colletotrichum fide Petrak & Sydow (*Annls mycol.* **33**: 178, 1935).

Gloeosporium Desm. & Mont. (1849) = Marssonina. Used for many very diverse fungi, v. Arx (1970) lists 735 names and refers 288 to the *Colletotrichum* anamorph of *Glomerella cingulata* and others to 48 other gen. For the plant pathogens *G. album*, see *Phlyctema*; *G. concentricum*, *Cylindrosporium*; *G. musarum*, *Colletotrichum*; *G. perennans*, *Cryptosporiopsis*. fide von Arx (*Biblthca Mycol.* **24**: 203 pp., 1970).

Gloeostereum S. Ito & S. Imai (1933), Cyphellaceae. 1, Japan. See Petersen & Parmasto (*MR* **97**: 1213, 1993).

Gloeosynnema Seifert & G. Okada (1988), anamorphic *Agaricomycetes*, Hsy.0eH.21. 1 (with clamp connexions), Indonesia; Japan. See Seifert & Okada (*Mycotaxon* **32**: 471, 1988).

Gloeotinia M. Wilson, Noble & E.G. Gray (1954), Helotiales. Anamorph *Endoconidium*. 1, widespread. *G. granigena*; blind seed disease of grass seed. See Wilson *et al.* (*TBMS* **37**: 29, 1954), Griffiths (*TBMS* **41**: 461, 1958; sexuality), Hardison (*Mycol.* **54**: 201, 1962; hosts), Schumacher (*Mycotaxon* **8**: 125, 1975; nomencl.), Holst-Jensen *et al.* (*Mycol.* **89**: 885, 1997; nuclear rDNA phylogeny), Alderman (*Mycol.* **90**: 422, 1998; species separation).

Gloeotrochila Petr. (1947) = Cryptocline fide von Arx (*Verh. K. ned. Akad. Wet. Amst.* C **51**: 24, 1957).

Gloeotromera Ervin (1956) = Ductifera fide Wells (*Mycol.* **50**: 407, 1958).

Gloeotulasnella Höhn. & Litsch. (1908) = Tulasnella fide Olive (*Mycol.* **49**: 668, 1957).

Gloiocephala Massee (1892), Physalacriaceae. 30, widespread. See Antonín (*Fungus Flora of tropical Africa* **1**, 2007).

Gloiodon P. Karst. (1879), ? Bondarzewiaceae. 3, Europe. See Maas Geesteranus (*Proc. K. ned. Akad. Wet.* C **66**: 430, 1963), Stalpers (*Stud. Mycol.* **40**: 78, 1996; key), Desjardin & Ryvarden (*Sydowia* **55**: 153, 2003; monogr.).

Gloiosphaera Höhn. (1902), anamorphic *Pezizomycotina*, Hso.0eH.15. 2, Europe; N. America. See Wang (*Mycol.* **63**: 890, 1971), Pollack & McKnight (*Mycol.* **64**: 415, 1972).

Gloiosporae, having slimy spores. Cf. *Xerosporae*. See Wakefield & Bisby (*TBMS* **25**: 50, 1941).

Gloiothelaceae Boidin, Mugnier & Canales (1998) = Stereaceae.

Gloiothele Bres. (1920), Peniophoraceae. 6, widespread. See Ginns & Freeman (*Biblthca Mycol.* **157**, 1994), Legon & Pegler (*Mycologist* **16**: 177, 2002; *Gloiothele lactescens*), Larsson & Larsson (*Mycol.* **95**: 1037, 2003; phylogeny).

Glomeraceae Piroz. & Dalpé (1989), Glomerales. 2 gen. (+ 5 syn.), 86 spp.

Lit.: Pirozynski & Dalpé (*Symbiosis* **7**: 1, 1989), Morton & Benny (*Mycotaxon* **37**: 471, 1990; key), Azcon-Aguilar & Barea (*Mycorrhiza* **6**: 457, 1996), van der Heijden *et al.* (*Nature* Lond. **396**: 69, 1998), Schüßler *et al.* (*MR* **105**: 1413, 2001), Walker & Schüssler (*MR* **108**: 982, 2004), Redecker (*Glomeromycota* Arbuscular mycorrhizal fungi and their relative(s). Version 01 July 2005. http://tolweb.org/Glomeromycota/28715/2005.07.01 in The Tree of Life Web Project, http://tolweb.org: [unpaginated], 2005).

Glomerales J.B. Morton & Benny (1990). Glomeromycetes. 1 fam., 2 gen., 86 spp. Endomycorrhizal on plants. Fam:

Glomeraceae

Morton & Benny (*Mycotaxon* **37**: 471, 1990; rev., keys) recognized two suborders: *Glomineae* for the vesicular-arbuscular mycorrhiza (VAM) forming (1) and (2), and *Gigasporineae* for the arbuscular mycorrhiza forming (3) but inconsistent with molecular phylogeny, see Schwarzott *et. al.* (*Mol. Phylog. Evol.* **21**: 190, 2001).

Lit.: Thaxter (*Proc. Am. Acad. Arts & Sci.* **57**: 291, 1922; sporocarpic spp.), Gerdemann & Trappe (*Mycol. Mem.* **5**, 1974; rev.), Walker (*Mycotaxon* **18**: 443, 1983; spore wall morphology [murograph]), Powell & Bagyaraj (*VA Mycorrhiza*, 1984; review), Burdrett *et al.* (*CJB* **62**: 2128, 1984; arbuscule staining), Berch (*Front. Appl. Microbiol.* **2**: 161, 1986; review), Silvia & Hubbell (*Symbiosis* **1**: 259, 1986; aeroponic culture), Bonfonte-Falso (*Symbiosis* **3**: 249, 1987; review), Stubblefield & Taylor (*New Phytol.* **108**: 3, 1988; palaeomycology), Jabaji-Hare (*Mycol.* **80**: 622, 1988; lipid & fatty acid profiles), Morton (*Mycotaxon* **32**: 267, 1988, checklist 126 spp. and classific., **37**: 493, 1990; evolution, *Mycol.* **82**: 192, 1990; phylogeny, *Mycorrhiza* **2**: 97, 1993; review), Pirozynski & Dalpé (*Symbiosis* **7**: 1, 1989; rev. & palaeomycology), Schenck & Peréz (*Manual for the identification of VA mycorrhizal fungi*, 3rd edn, 1990), Allen (*The evolution of mycorrhizae*, 1991; *Mycorrhizal functioning*, 1992), Bruns *et al.* (*Mol. Phylog. Evol.* **1**: 231, 1992; phylogeny), Millner & Kitt (*Mycorrhiza* **2**: 9, 1992; soil-less culture), Streussy (*Mycorrhiza* **1**: 113, 1992; phylogeny), Read *et al.* (*Mycorrhizas in ecosystems*, 1992), Morton *et al.* (*Mycotaxon* **48**: 491, 1993; INVAM culture collection), Maia *et al.* (*Mycol.* **85**: 323, 1993; fixation & embedding), Simon *et al.* (*Nature* **363**: 67, 1993; phylogeny), Walker & Trappe (*MR* **97**: 339; names & epithets), Sancholle & Dalpé (*Mycotaxon* **49**: 187, 1993; fatty acids), Hass *et al.* (*Am. J. Bot.* **81**: 29, 1994; palaeomycology), Giovannetti & Gianinazzi-Pearson (*MR* **98**: 705, 1994; review), Gianinazzi-Pearson *et al.* (*Mycol.* **86**: 478, 1994; wall chemistry & phylogeny), Morton & Bentivegna (*Plant & Soil* **159**: 47, 1994; taxonomy), Bentivegna & Morton (*in* Pfegler & Linderman (Eds), *Mycorrhizae and plant health*: 283, 1983; systematics), Robson *et al.* (*Management of mycorrhizas in agriculture, horticulture, and forestry*, 1994), Gianinazzi & Schüepp (Eds) (*Impact of arbuscular mycorrhizas on sustainable agriculture and natural ecosystems*, 1994; taxonomy, biodiversity, Eur. culture collection), Mehrotra & Baijal (*in* Devivedi & Pandy (Eds), *Biotechnology In India*, p. 227, 1994), Morton *et al.* (*in* Varma & Hock (Eds), *Mycorrhiza: structure, function, molecular biology and biochemistry*: 669, 1995, *CJB* **73**(suppl. 1): S25, 1995; diversity), Yao *et al.* (*Kew Bull.* **50**: 349, 1995), Clapp (Ed.) (*Species diagnostics protocols* (*Meth. Mol. Biol.* **50**), 1996), Dodd & Rosendahl (*Mycorrhizae* **6**: 275, 1996; identification system), Phipps & Taylor (*Mycol.* **88**: 707, 1996; fossils), Remy *et al.* (*Proc. Nat. Acad. Sci. USA* **91**: 11841, 1994; paleomycology), Sbrana *et al.* (*MR* **99**: 1249, 1995; biochemistry), Sanders *et al.* (*New Phytol.* **133**: 123, 1996; genetic diversity), Simon (*New Phytol.* **133**: 95, 1996; phylogeny), Linder (*in* Carrol & Tudzynski (Eds), *The mycota* **6B**, p. 117, 1997; review), Mehrotra (*Mycorrhiza News* **9**: 1, 1997), Hosny *et al.* (*Fung. Gen. Biol.* **22**: 103, 1997; DNA base composition), Redecker *et al.* (*Appl. Microbiol.* **63**: 1756, 1997; identification, *Fung. Genet. Biol.* **28**: 238, 1999; phylogeny, *Mol. Phylogen. Evol.* **14**: 276, 2000; phylogeny, *Science* **289**: 1884, 2000; phylogeny), Sanders (*Progr. Microbiol. Ecol.*, p. 77, 1997; biodiversity), Smith & Read (*Mycorrhizal Symbiosis*, 2nd ed.), Varma (Ed.) (*Mycorrhiza manual*, 1998), Wu & Lin (*Technical Manual Mycorrhizal Fungi*, 1998), Blackwell (*Science* **289**: 1884, 2000; phylogeny).

Glomerella Spauld. & H. Schrenk (1903), Glomerellaceae. Anamorph *Colletotrichum*. *c.* 11 (necrotrophic on plants), widespread (esp. tropical). *G. cingulata* on many hosts (*CMI Descript.* **315**, 1971); *G. acutata* on many hosts, *G. tucumanensis* on sugarcane. See Sutton *in* Bailey & Jeger (Eds) (*Colletotrichum: Biology, Pathology and Control*: 1, 1992; review), Uecker (*Mycol.* **86**: 82, 1994; ontogeny), Johnston & Jones (*Mycol.* **89**: 420, 1997; DNA), Cannon *et al. in* Prusky *et al.* (Eds) (*Colletotrichum: Host Specificity, Pathology and Host-Pathogen Interaction*: 1, 2000; sp. concepts), Guerber & Correll (*Mycol.* **93**: 216, 2001; *G. acutata*), Rodríguez-Guerra *et al.* (*Mycol.* **97**: 793, 2005; *G. lindemuthianum*), Cook *et al.* (*MR* **110**: 672, 2006; *G. truncata*), Zhang *et al.* (*Mycol.* **98**: 1076, 2006; phylogeny).

Glomerellaceae Locq. ex Seifert & W. Gams (2007), Hypocreomycetidae (inc. sed.). 2 gen. (+ 23 syn.), 71 spp.

Lit.: Sutton (*Colletotrichum: Biology, Pathology and Control*: 1, 1992), Sivanesan & Hsieh (*MR* **97**: 1523, 1993), Uecker (*Mycol.* **86**: 82, 1994), Johnston & Jones (*Mycol.* **89**: 420, 1997), Silva-Hanlin & Hanlin (*MR* **103**: 153, 1999), García Muñoz *et al.* (*Mycol.* **92**: 488, 2000), Prusky *et al.* (*Host specificity, Pathology, and Host–Pathogen Interactionos of Colletotrichum*: 400 pp., 2000), Guerber & Correll (*Mycol.* **93**: 216, 2001), Abang *et al.* (*Pl. Path.* **51**: 63, 2002), Guerber *et al.* (*Mycol.* **95**: 872, 2003), Lu *et al.* (*MR* **108**: 53, 2004), Lubbe *et al.* (*Mycol.* **96**: 1268, 2004), Du *et al.* (*Mycol.* **97**: 641, 2005), Farr *et al.* (*MR* **110**: 1395, 2006), Zhang *et al.* (*Mycol.* **98**: 1076, 2006).

Glomerilla Norman (1869), ? Verrucariaceae (?L). 1, Norway.

Glomerillaceae Norman (1869) = Verrucariaceae.

Glomerobolus Kohlm. & Volkm.-Kohlm. (1996), anamorphic *Stictidaceae*, Hso.?.?. 1, USA. See

Kohlmeyer & Volkmann-Kohlmeyer (*Mycol.* **88**: 329, 1996), Schoch *et al.* (*MR* **110**: 257, 2006; phylogeny, dispersal).

Glomeromycetes Caval.-Sm. (1998), Glomeromycota. 4 ord., 9 fam., 12 gen., 169 spp. Ords:
(1) **Archaeosporales**
(2) **Diversisporales**
(3) **Glomerales**
(4) **Paraglomerales**.
For *Lit.* see ord. and fam.

Glomeromycota C. Walker & A. Schüssler (2001), Glomeromycota. 4 ord., 9 fam., 12 gen., 169 spp. Ords:
(1) **Archaeosporales**
(2) **Diversisporales**
(3) **Glomerales**
(4) **Paraglomerales**.
For *Lit.* see ord. and fam.

glomerospore, an asexual reproductive structure [in *Glomeromycota*] formed at the end of a subtending hyphae, or a bulbous sporogenous cell, or from a sporiferous saccule, in this case with lateral or intercalary position, with or without a germinative component, presenting or not a germinal shield or orb, typically formed by fungi that form an arbuscular mutualistic symbiosis with most terrestrial plants (see Gotoa & Maia, *Mycotaxon* **96**: 129, 2006).

Glomerula Bainier (1903) = Actinomucor fide Hesseltine (*Mycol.* **47**: 344, 1955).

Glomerularia H. Karst. (1849) = Gonatobotrys fide Henderson (*Notes R. bot. Gdn Edinb.* **23**: 497, 1961).

Glomerularia Peck (1879) ≡ Glomopsis.

Glomerulomyces A.I. Romero & S.E. López (1989), anamorphic *Agaricomycetes*, Hso.0eH.6. 1 (with clamp connexions), Argentina. See Romero & López (*Mycotaxon* **34**: 432, 1989).

glomerulus (**glomerule**), a clump or cluster; frequently used for clusters of photobiont cells in lichens.

Glomites T.N. Taylor, W. Remy, Hass & Kerp (1995), Fossil Fungi. 1 (Devonian), Scotland. See Taylor *et al.* (*Mycol.* **87**: 561, 1995).

Glomopsis D.M. Hend. (1961), anamorphic *Herpobasidium*. 2, USA. See Donk (*Persoonia* **4**: 213, 1966), Mel'nik & Sheiko (*Nov. sist. Niz. Rast.* **35**: 90, 2001; Russia).

Glomorphites García Massini (2007), Fossil Fungi. 1. See García Massini, J.L. (*Int. J. Pl. Sci.* **168**: 673, 2007).

Glomospora D.M. Hend. (1961), anamorphic *Platygloeaceae*. 1, British Isles. See Henderson (*Notes R. bot. Gdn Edinb.* **23**: 497, 1961).

Glomosporiaceae Cif. (1963), Urocystidales. 3 gen. (+ 5 syn.), 59 spp.
Lit.: Hughes (*Mycol.* **68**: 693, 1976), Vánky & Berbee (*Mycotaxon* **33**: 281, 1988), Vánky (*Trans. Mycol. Soc. Japan* **32**: 145, 1991), Bauer *et al.* (*CJB* **75**: 1273, 1997), Begerow *et al.* (*CJB* **75**: 2045, 1998), Piepenbring *et al.* (*Protoplasma* **204**: 202, 1998), Vánky (*Mycotaxon* **70**: 35, 1999), Begerow *et al.* (*MR* **104**: 53, 2000), Andrade *et al.* (*Phytopathology* **94**: 875, 2004).

Glomosporium Kochman (1939) = Thecaphora.

Glomus Tul. & C. Tul. (1845), Glomeraceae. *c.* 85, widespread. Genus is non-monophyletic. See Gerdemann & Trappe (*Mycol. Mem.* **5**, 1974), Berch & Fortin (*CJB* **61**: 2608, 1983), Berch & Fortin (*CJB* **62**: 170, 1984), Jabaji-Hare *et al.* (*Mycol.* **76**: 1024, 1984; lipid composition), Morton (*Mycol.* **77**: 192, 1985; morphological variation), Bonfante-Fasolo *et al.* (*Biol. Cell* **57**: 265, 1986; chitin in cell wall), Bonfante-Fasolo & Schubert (*CJB* **65**: 539, 1987; spore wall ultrastr.), Koske & Gemma (*Mycol.* **81**: 935, 1989), Koske & Halvorsol (*Mycol.* **81**: 927, 1989), Almeida & Schenk (*Mycol.* **82**: 703, 1990), Schenck & Peréz (*Manual for the Identification of VA Mycorrhizal Fungi*, 1990), Chabot *et al.* (*Mycol.* **84**: 315, 1992; life cycle in root culture), Meier & Charvat (*Int. J. Pl. Sci.* **153**: 541, 1992; ultrastr. spore germ.), Yao *et al.* (*Mycologist* **6**: 132, 1992; ultrastr.), Wu & Silvia (*Mycol.* **85**: 317, 1993; spore ontogeny), Błaszkowski (*Mycorrhiza* **4**: 201, 1994), Cabello *et al.* (*Mycotaxon* **51**: 123, 1994), Gaspar *et al.* (*Mycotaxon* **51**: 129, 1994; lipid & fatty acid composition), Maia & Kimbrough (*Int. J. Pl. Sci.* **155**: 689, 1994; ultrastr.), Miller & Jeffries (*MR* **98**: 307, 1994; spore wall ultrastr.), Błaszkowski (*Mycol.* **87**: 732, 1995), Lanfranco *et al.* (*Mol. Ecol.* **4**: 61, 1995; identification), Walker *et al.* (*MR* **99**: 1500, 1995), Yao *et al.* (*Kew Bull.* **50**: 349, 1995), Dodd *et al.* (*New Phytol.* **133**: 113, 1996; genetic variation), Lloyd-Macgilp *et al.* (*New Phytol.* **133**: 103, 1996; genetic diversity), Morton (*Mycorrhiza* **6**: 161, 1996), Pfeiffer *et al.* (*Mycotaxon* **59**: 383, 1996), Spain *et al.* (*Mycotaxon* **60**: 137, 1996), Bentivenga *et al.* (*Am. J. Bot.* **84**: 1211, 1997; genetic variation), Błaszkowski (*Mycol.* **89**: 339, 1997), Błaszkowski & Tadych (*Mycol.* **89**: 804, 1997), Morton *et al.* (*MR* **101**: 625, 1997; synanamorph), Strullu *et al.* (*C.R. Acad. Sci. Paris* Sér. III **320**: 41, 1997; life cycle), Stürmer & Morton (*Mycol.* **89**: 72, 1997; development), Zhang *et al.* (*Mycosystema* **16**: 241, 1997), Maia & Kimbrough (*Int. J. Pl. Sci.* **159**: 581, 1998; ultrastr.), Sawaki *et al.* (*Mycoscience* **39**: 477, 1998; phylogeny & synanamorph), Vandenkoornhuyse & Leyval (*Mycol.* **90**: 791, 1998; genetic diversity), Walker & Vestberg (*Ann. Bot.* **82**: 601, 1998; synonymy), Schwarzott *et al.* (*Mol. Phylog. Evol.* **21**: 190, 2001; polyphyletic), Blaszkowski *et al.* (*Mycotaxon* **90**: 447, 2004; n.sp.), Blaszkowski *et al.* (*MR* **110**: 555, 2006; n.sp.), Redecker & Raab (*Mycol.* **98**: 885, 2006; phylogeny).

Gloniella Sacc. (1883), Hysteriaceae. 9, widespread (temperate). See Zogg (*Beitr. Kryptfl. Schweiz* **11** no. 3, 1962), Sivanesan *et al.* (*TBMS* **90**: 665, 1988; Indian spp.), Steinke & Hyde (*Mycoscience* **38**: 7, 1997; S. Africa), Lorenzo & Messuti (*MR* **102**: 1101, 1998; S. America), Checa *et al.* (*Mycol.* **99**: 285, 2007; Costa Rica).

Gloniopsis De Not. (1847), Hysteriaceae. 2 or 3, widespread (temperate). See Zogg (*Beitr. Kryptfl. Schweiz* **11** no. 3, 1962), Lorenzo & Messuti (*MR* **102**: 1101, 1998; S. America), Lee & Crous (*S. Afr. J. Bot.* **69**: 480, 2003; S. Africa), Checa (*Fl. Mycol. Iberica* **6**, 2004; Spain).

Glonium Muhl. (1813), Hysteriaceae. Anamorph *Cleistonium*. *c.* 13, widespread (temperate). See Speer (*BSMF* **102**: 101, 1986), Lorenzo & Messuti (*MR* **102**: 1101, 1998; S America), Vasil'eva (*Mikol. Fitopatol.* **34**: 3, 2000; Russian Far East), Lee & Crous (*S. Afr. J. Bot.* **69**: 480, 2003; S Africa), Messuti & Lorenzo (*Nova Hedwigia* **84**: 521, 2007; S America).

Glossifungites Lomnicki (1886), Fossil Fungi. 1 (Cretaceous), Poland.

Glossodium Nyl. (1855) = Icmadophila fide Rambold *et al.* (*Biblthca Lichenol.* **53**: 217, 1993).

glossoid cell, elongate (tongue-shaped) cell containing an elaborate extrusome (*Haptoglossa*).

Glotzia M. Gauthier ex Manier & Lichtw. (1969), Legeriomycetaceae. 6 (in *Ephemeroptera* and *Plectoptera*), widespread. See Lichtwardt (*The Trichomycetes. Fungal associates of arthropods*, 1986), Lichtwardt & Williams (*CJB* **68**: 1057, 1990), Williams & Lichtwardt (*CJB* **68**: 1045, 1990), White & Lichtwardt (*Mycol.* **96**: 891, 2004; Norway).

glucans, one of the main constituents of fungal walls. **R-glucans** are alkali insoluble; **S-glucans** are soluble.

Glugea Thélohan (1891), Microsporidia. 24.

Glugoides J.I.R. Larsson, D. Ebert, J. Vavra & Voronin (1996), Microsporidia. 1. See Larsson *et al.* (*Eur. J. Protist.* **32**: 251, 1996).

Glukomyces Beij. [not traced] nom. dub., ? Fungi.

gluten (1) a substance on the surface of some agarics, etc., which is sticky when wet; (2) spore mass in *Phallus*.

Glutinaster Earle (1909) = Tricholoma fide Singer (*Agaric. mod. Tax.*, 1951).

Glutinisporidium Thor (1930) nom. dub., ? Fungi. 1, Svalbard.

Glutinium Fr. (1849), anamorphic *Pezizomycotina*, Hso.?.?. 3, Europe; N. America.

Glutinoagger Sivan. & Watling (1980), anamorphic *Agaricomycetes*. 1 (with clamp connexions), Seychelles. See Sivanesan & Watling (*TBMS* **74**: 424, 1980).

glutinous, sticky; made up of, or covered with, gluten.

glyceollin, a phytoalexin (q.v.) from soybean (*Glycine max*).

Glycydiderma Paulet (1808) ≡ Geastrum.

Glycyphila Mont. (1851), anamorphic *Pezizomycotina*, Hso.?.?. 2, Europe.

Glyphidaceae Stizenb. (1862) = Graphidaceae.

Glyphidium A. Massal. (1860) ? = Arthonia fide Hawksworth *et al.* (*Dictionary of the Fungi* edn 8, 1995).

Glyphis Ach. (1814), Graphidaceae (L). 5, widespread (esp. tropical). See Pant (*Geophytology* **20**: 48, 1991; 2 spp. India), Archer (*Telopea* **10**: 589, 2004; Australia), Archer (*Telopea* **11**: 59, 2005; Australia), Staiger *et al.* (*MR* **110**: 765, 2006; phylogeny).

Glyphium Nitschke ex F. Lehm. (1886), Mytilinidiaceae. Anamorph *Peyronelia*. 4 or 6, widespread. See Sutton (*TBMS* **54**: 255, 1970), Goree (*CJB* **52**: 1265, 1974), Vasil'eva (*Mikol. Fitopatol.* **35**: 15, 2001; Russian Far East), Lumbsch *et al.* (*Symb. bot. upsal.* **34** no. 1: 9, 2004; phylogeny), Lorenzo & Messuti (*Boln Soc. argent. Bot.* **40**: 13, 2005; Argentina), Wedin *et al.* (*MR* **109**: 159, 2005; phylogeny), Geiser *et al.* (*Mycol.* **98**: 1053, 2006; phylogeny).

Glypholecia Nyl. (1853), Acarosporaceae (L). 5, Europe; N. America. See Hafellner (*Cryptog. bot.* **5**: 99, 1995), Lutzoni *et al.* (*Am. J. Bot.* **91**: 1446, 2004; phylogeny), Reeb *et al.* (*Mol. Phylogen. Evol.* **32**: 1036, 2004; phylogeny), Wedin *et al.* (*MR* **109**: 159, 2005; phylogeny), Miądlikowska *et al.* (*Mycol.* **98**: 1088, 2006; phylogeny).

glypholecine, having particularly labyrinth-like lirella as in *Glypholecia*.

Glypholeciomyces Cif. & Tomas. (1953) = Glypholecia.

Glyphomyces Cif. & Tomas. (1953) ≡ Glyphis.

Glyphopeltis Brusse (1985), ? Psoraceae (L). 1, Europe; S. Africa. See Timdal (*Mycotaxon* **31**: 101, 1988), Timdal (*Cryptog. Bryol.-Lichénol.* **15**: 171, 1994; asci), Egea *et al.* (*Flechten Follmann* Contributions to Lichenology in Honour of Gerhard Follmann: 183, 1995).

Glyptoderma R. Heim & Perr.-Bertr. (1971), Agaricaceae. 1, America (tropical). See Heim & Perreau-Bertrand (*Revue Mycol.* Paris **36**: 90, 1971).

Glyptospora Fayod (1889) = Lacrymaria fide Kuyper (*in litt.*).

Gnaphalomyces Opiz = Lanosa fide Streinz (*Nom. fung.*: 302, 1862).

Gnomonia Ces. & De Not. (1863), Gnomoniaceae. Anamorphs *Asteroma*, *Cylindrosporella*, *Discula*, *Leptothyrium*, *Zythia*. 60, widespread. *G. platani* (plane (*Platanus*) scorch). See Bolay (*Ber. schweiz. bot. Ges.* **81**: 398, 1972), Barr (*Mycol. Mem.* **7**, 1978; key 35 spp.), Monod (*Beih. Sydowia* **9**: 1, 1983), Bolay (*Mycotaxon* **41**: 287, 1991; N. Am. spp.), Lappalainen & Yli-Mattila (*MR* **103**: 328, 1999; RAPDs), Castlebury *et al.* (*Mycol.* **94**: 1017, 2002), Sogonov *et al.* (*Sydowia* **57**: 102, 2005; type species), Moročko *et al.* (*Eur. J. Pl. Path.* **114**: 235, 2006; on *Fragaria*), Zhang *et al.* (*Mycol.* **98**: 1076, 2006; phylogeny), Green & Castlebury (*MR* **111**: 62, 2007; phylogeny, anamorph), Moročko & Fatehi (*MR* **111**: 603, 2007; on strawberry).

Gnomoniaceae G. Winter (1886) nom. cons., Diaporthales. 18 gen. (+ 26 syn.), 204 spp.

Lit.: Barr (*Mycol. Mem.* **7**: 1, 1978), Monod (*Beih. Sydowia* **9**: 1, 1983), Noordeloos *et al.* (*Persoonia* **14**: 47, 1989), Redlin & Rossman (*Mycol.* **83**: 200, 1991), Lappalainen & Yli-Mattila (*MR* **103**: 328, 1999), Zhang & Blackwell (*Mycol.* **93**: 355, 2001), Castlebury *et al.* (*Mycol.* **94**: 1017, 2002), Sogonov *et al.* (*Sydowia* **57**: 102, 2005), Rossman *et al.* (*Mycoscience* **48**: 135, 2007; phylogeny), Mejía *et al.* (*MR* **112**: 23, 2008; phylogeny).

Gnomoniella Sacc. (1881), Gnomoniaceae. Anamorphs *Asteroma*, *Discula*. 13, widespread. See Barr (*Mycol. Mem.* **7**, 1978), Monod (*Beih. Sydowia* **9**: 1, 1983), Redlin & Stack (*Mycotaxon* **32**: 175, 1988; on *Fraxinus*), Barr (*Mycotaxon* **41**: 287, 1991; N. Am. spp.), Farr *et al.* (*Sydowia* **53**: 185, 2001; phylogeny).

Gnomonina Höhn. (1917) ≡ Laestadia.

Gnomoniopsis Berl. (1892) = Gnomonia fide Bolay (*Ber. schweiz. bot. Ges.* **81**, 1972).

Gnomoniopsis Stoneman (1898) ≡ Glomerella.

gnotobiotic (of cultures), ones in which all the living components are known. Cf. axenic.

Godal Adans. (1763) nom. dub., anamorphic *Pezizomycotina*.

Godfrinia Maire (1902) = Hygrocybe fide Singer (*Agaric. mod. Tax.*, 1951).

Godronia Moug. & Lév. (1846), Helotiaceae. Anamorphs *Sporonema*, *Topospora*. 27, widespread (north temperate). See Groves (*CJB* **43**: 1195, 1965; key), Holm & Holm (*Nordic Jl Bot.* **11**: 675, 1991), Verkley (*Nova Hedwigia* **75**: 433, 2002; anamorph).

Godroniella P. Karst. (1884) = Myxormia fide Clements & Shear (*Gen. Fung.*, 1931), Tulloch (*Mycol. Pap.* **130**, 1972).

Godroniopsis Diehl & E.K. Cash (1929), Helotiaceae. Anamorphs *Micropera*, *Sphaeronaema*, *Dichaenopsella*. 2, N. America. See Petrak (*Sydowia* **6**: 336, 1952).

Goidanichia G. Arnaud (1954) ≡ Goidanichiella G. Arnaud ex G.L. Barron.

Goidanichia Tomas. & Cif. (1952) = Staurothele fide Hawksworth *et al.* (*Dictionary of the Fungi* edn 8, 1995).

Goidanichiella G. Arnaud ex G.L. Barron (1968) ≡ Haplographium fide Gams *et al.* (*Mycotaxon* **38**: 149, 1990).

Goidanichiella G.L. Barron ex W. Gams (1990), anamorphic *Pezizomycotina*, Hso.0eH.15. 2, widespread. See Gams *et al.* (*Mycotaxon* **38**: 149, 1990), Hyde *et al.* (*Fungal Diversity* **11**: 119, 2002; Brunei).

Goidanichiomyces Cif. & Tomas. (1953) ≡ Goidanichia Tomas. & Cif.

Golbergia J. Weiser (1977), Microsporidia. 1.

Golgi body, Dictyosome with large numbers of cisternae which may be visible using light microscopy after staining.

Golovin (Petr Nikolaevich; 1897-1968; Russia, later Uzbekistan). Instructor, Salsk District Station of Plant Protection (1925); student, Institute of Plant Pathology, Leningrad (1925-*c.*1933); Plant Pathologist, Central-Asian Institute of Plant Protection, Tashkent (1930-1933); Senior Lecturer, Plant Anatomy & Morphology and Head of the Mycology Laboratory (1934-1943) then Professor of Non-flowering Plants, Central-Asian State University (1943-1951); Senior Scientist, V.L. Komarov Botanical Institute, Leningrad (1951-1962). Pioneer mycologist in central Asia, exploring the mycota and collecting in the central and western Tien-Shan mountains, the deserts of Kzyl-Kum and Kara-Kum, and the Pamirs (1934-1950); also significant work on the *Erisyphaceae*. Specimens are in the fungal reference collection, V.L. Komarov Botanical Institute, St Petersburg (*LE*). *Publs.* [*Fungus Flora in Central Asia. Parasitic Fungi in Central Asia*] (1949) [in Russian]; [Monographic survey on the genus *Leveillula* Arnaud (Parasitic fungi ? fam. Erysiphaceae)] and [Material for a monograph of the parasitic fungi (family Erysiphaceae) in the USSR]. *Notulae Systematicae e Sectione Cryptogamica Instuti Botanicae Nomine V.L. Komarovii Academiae Scientiarum, USSR* (1956) [in Russian]. *Biogs, obits etc.* Gorlenko & Dunin ([*Mikologiya i Fitopatologiya*] **3**: 94, 1969) [portrait [in Russian].

Golovinia Mekht. (1967) = Drechslerella fide Hawksworth *et al.* (*Dictionary of the Fungi* edn 8, 1995).

Golovinomyces (U. Braun) V.P. Gelyuta (1988), Erysiphaceae. Anamorph *Oidium* subgen. *Reticoloidium*. 32, widespread. *G. cichoracearum* (mildew of composites). See Braun (*Beih. Nova Hedwigia* **89**, 1987; key; as *Erysiphe* sect. *Golovinomyces*), Bardin *et al.* (*Pl. Path.* **48**: 531, 1999; genetic variation, as *Erysiphe*), Braun (*Schlechtendalia* **3**: 48, 1999; generic taxonomy), Cherepanov (*Mikol. Fitopatol.* **35**: 71, 2001; Russia, key), Vakalounakis & Klironomou (*Mycotaxon* **80**: 489, 2001; on cucurbits), Matsuda & Takamatsu (*Mol. Phylogen. Evol.* **27**: 314, 2003; host-parasite relations), Takamatsu & Matsuda (*Mycoscience* **45**: 340, 2004; evolution), Cunnington *et al.* (*Australas. Pl. Path.* **34**: 51, 2005; on *Solanaceae*, Australia), Takamatsu *et al.* (*MR* **110**: 1093, 2006; evolution), Voytyuk *et al.* (*Mycol. Balcanica* **3**: 131, 2006; Israel), Takamatsu *et al.* (*MR* **112**: 299, 2008; phylogeny).

Gomezina Chardón & Toro (1934) = Aphanostigme fide von Arx & Müller (*Stud. Mycol.* **9**, 1975).

Gomphaceae Donk (1961), Gomphales. 13 gen. (+ 17 syn.), 287 spp.

Lit.: Corner (*Ann. Bot. Mem.* [A monograph of Clavaria and allied genera] **1**: 1, 1950), Donk (*Persoonia* **3**: 199, 1964), Corner (*Beih. Nova Hedwigia* **33**: 1, 1970), Marr & Stuntz (*Biblthca Mycol.* **38**: 232 pp., 1973), Petersen (*Biblthca Mycol.* **43**: 161 pp., 1975), Petersen (*Biblthca Mycol.* **79**: 261 pp., 1981), Parmasto & Ryvarden (*Windahlia* **18**: 35, 1988), Núñez & Ryvarden (*Sydowia* **46**: 321, 1994), Pine *et al.* (*Mycol.* **91**: 944, 1999), Villegas *et al.* (*Mycotaxon* **70**: 127, 1999), Humpert *et al.* (*Mycol.* **93**: 465, 2001), Hibbett & Binder (*Proc. R. Soc. Lond.* B. Biol. Sci. **269**: 1963, 2002), Binder *et al.* (*Systematics and Biodiversity* **3**: 113, 2005), Nouhra *et al.* (*Mycorrhiza* **15**: 55, 2005).

Gomphales Jülich (1981). Phallomycetidae. 3 fam., 18 gen., 336 spp. Fams:

(1) **Clavariadelphaceae**
(2) **Gomphaceae**
(3) **Lentariaceae**

Lit.: Villegas *et al.* (*Fungal Diversity* **18**: 157, 2005; spore characters).

Gomphidiaceae Maire ex Jülich (1982), Boletales. 4 gen. (+ 3 syn.), 30 spp. Associated with ectomycorrhizas of members of *Suillaceae* and *Rhizopogonaceae*, always in association with *Pinaceae* (north temperate). Autonomy of families in the suilloid clade (*Gomphidiceae*, *Rhizopogonaceae*, *Suillaceae*) is fairly weak, and the three families are maintained in the present edition for convenience only.

Lit.: Miller (*Mycol.* **63**: 1129, 1971), Agerer (*Nova Hedwigia* Beih. **53**: 127, 1991), Greselin (*Boll. Gruppo Micol. 'G. Bresadola'* **34**: 56, 1991), Besl & Bresinsky (*Pl. Syst. Evol.* **206**: 223, 1997), Miller & Aime (*Trichomycetes and Other Fungal Groups, Robert W. Lichwardt Commemoration Volume*: 315, 2001), Miller *et al.* (*Mycol.* **94**: 1044, 2002), Miller (*Mycol.* **95**: 176, 2003), Watling (*Edinb. J. Bot.* **61**: 41, 2004).

Gomphidius Fr. (1836), Gomphidiaceae. 10, widespread (esp. north temperate). See Watling (*British Fungus Flora* **1**: 77, 1970; key), Miller (*Mycol.* **63**: 1129, 1971; key), Singer (*Taxon* **22**: 445, 1973; typification), Agerer (*Beih. Nova Hedwigia* **53**: 127, 1991; ecology), Miller & Aime (*Mycol.* **94**: 1044, 2002; new spp.), Miller (*Mycol.* **95**: 176, 2003).

Gomphillaceae Walt. Watson ex Hafellner (1984), Ostropales (L). 23 gen. (+ 21 syn.), 318 spp.

Lit.: Vězda & Poelt (*Folia geobot. phytotax.* **22**: 179, 1987), Kalb & Vězda (*Biblthca Lichenol.* **29**: 80 pp., 1988; 43 spp. neotrop.), Lücking (*Biblthca Lichenol.* **65**: 109, 1997; Costa Rica), Dennetière & Péroni (*Cryptog. Bryol.-Lichénol.* **19**: 105, 1998; phylogeny), Lücking *et al.* (*Mycol.* **96**: 283, 2004; phylogeny), Lücking *et al.* (*Bryologist* **110**: 622, 2007; N America).

Gomphillus Nyl. (1855), Gomphillaceae (L). 4, widespread. See Kalb & Vězda (*Biblthca Lichenol.* **29**, 1988), Lücking (*Biblthca Lichenol.* **65**: 109, 1997; Costa Rica), Buck *in* Glenn *et al.* (Eds) (*Lichenogr. Thomsoniana*: 71, 1998), Dennetière & Péroni (*Cryptog. Bryol.-Lichénol.* **19**: 105, 1998; phylogeny), Lücking *et al.* (*Mycol.* **96**: 283, 2004; phylogeny), Ferraro & Lücking (*Bryologist* **108**: 491, 2005; Americas), Lücking *et al.* (*Lichenologist* **37**: 123, 2005; phylogeny), Lücking *et al.* (*Bryologist* **110**: 622, 2007; N America).

Gomphinaria Preuss (1851) = Acrotheca fide Saccardo (*Syll. fung.* **4**: 277, 1886).

Gomphogaster O.K. Mill. (1973), Gomphidiaceae. 1, USA. See Miller (*Mycol.* **65**: 227, 1973).

Gomphora Fr. (1825) ≡ Gomphus Pers. fide Petersen

(*The genera Gomphus and Gloeocantharellus in North America*, 1972).

Gomphos Kuntze (1891) ≡ Cortinarius.

Gomphospora A. Massal. (1852) ? = Schismatomma fide Hawksworth *et al.* (*Dictionary of the Fungi* edn 8, 1995).

Gomphus (Fr.) Weinm. (1826) ≡ Gomphidius.

Gomphus Pers. (1797), Gomphaceae. 10, widespread (temperate). See Petersen (*The genera Gomphus and Gloeocantharellus in North America*, 1972).

Gonapodya A. Fisch. (1892), Gonapodyaceae. 2, widespread (temperate). See Karling (*Chytriomyc. Iconogr.*, 1977), Noyes Mollione & Longcore (*Mycol.* **91**: 727, 1999; ultrastr.), Gandhe & Kurne (*Zoos' Print Journal* **20**: 2059, 2006; reproduction).

Gonapodyaceae H.E. Petersen ex P.M. Kirk, P.F. Cannon & J.C. David (2001), Monoblepharidales. 2 gen., 7 spp.

Lit.: Mollicone & Longcore (*Mycol.* **91**: 727, 1999), Noyes Mollicone & Longcore (*Mycol.* **91**: 727, 1999), Steciow *et al.* (*Boln Soc. argent. Bot.* **36**: 203, 2001), Bullerwell *et al.* (*Nucl. Acids Res.* **31**: 1614, 2003).

Gonatobotrydiaceae Nann. (1934) = Ceratostomataceae.

Gonatobotrys Corda (1839), anamorphic *Melanospora*, Hso.0-1eH.8. 2 (fungicolous), widespread. See Whaley & Barnett (*Mycol.* **55**: 199, 1963; *G. simplex* on *Alternaria* and *Cladosporium*), Walker & Minter (*TBMS* **77**: 299, 1981; key), Vakili (*MR* **93**: 67, 1989; teleomorph).

Gonatobotrytites Pia (1927), Fossil Fungi. 1 (Eocene), Europe.

Gonatobotryum Sacc. (1880), anamorphic *Pezizomycotina*, Hso.0eH/1eP.5. 4 (fungicolous), widespread. See Kendrick *et al.* (*CJB* **45**: 591, 1968; conidiogenesis), Walker & Minter (*TBMS* **77**: 299, 1981; key).

Gonatophragmiella Rak. Kumar, A.N. Rai & Kamal (1994), anamorphic *Pezizomycotina*, Hso.≡ eP.?. 1, India. See Kumar *et al.* (*Indian Phytopath.* **47**: 129, 1994).

Gonatophragmium Deighton (1969), anamorphic *Acrospermum*, Hso.≡ eP.10. 2, widespread (tropical). See Takahashi & Teramine (*Ann. phytopath. Soc. Japan* **52**: 404, 1986; teleomorph), Rai (*MR* **100**: 1263, 1996; India), Siboe *et al.* (*Mycotaxon* **73**: 283, 1999; Kenya).

Gonatopyricularia Z.D. Jiang & P.K. Chi (1989) ≡ Pyriculariopsis.

Gonatorhodis Clem. & Shear (1931) ≡ Gonatorrhodiella.

Gonatorrhodiella Thaxt. (1891) = Gonatobotryum fide Walker & Minter (*TBMS* **77**: 299, 1981).

Gonatorrhodum Corda (1839), anamorphic *Pezizomycotina*, Hso.0eP.?3. 1 or 2, Europe.

Gonatosporium Corda (1839) = Arthrinium fide Hughes (*CJB* **36**: 727, 1958).

Gonatotrichum Corda (1842) ≡ Gonytrichum.

Gondwanamyces G.J. Marais & M.J. Wingf. (1998), Ceratocystidaceae. Anamorph *Knoxdaviesia*. 2 (on *Protea*), S. Africa. See Marais *et al.* (*Mycol.* **90**: 136, 1998), Viljoen *et al.* (*MR* **103**: 497, 1999; phylogeny), Wingfield *et al.* (*MR* **103**: 1616, 1999; phylogeny), Gibb & Hausner (*MR* **107**: 1442, 2003; introns), Roets *et al.* (*Stud. Mycol.* **55**: 199, 2006; phylogeny), Roets *et al.* (*CJB* **84**: 989, 2006; PCR detection).

Gongromeriza Preuss (1851) = Chloridium fide Hughes (*Friesia* **9**: 61, 1969).

Gongronella Ribaldi (1952), Cunninghamellaceae. 2, widespread. See Hessletine & Ellis (*Mycol.* **53**: 406, 1961), Hessletine & Ellis (*Mycol.* **56**: 568, 1964; key), Upadhyay (*Nova Hedwigia* **17**: 65, 1969; key), Kirk (*IMI Descr. Fungi Bact.* **131**, 1997), Voigt & Wöstemeyer (*Gene* **270**: 113, 2001; phylogeny).

Gongylia Körb. (1855) = Arthrorhaphis fide Hawksworth *et al.* (*Lichenologist* **12**: 13, 1980), Jørgensen & Santesson (*Taxon* **42**: 881, 1993), Fryday (*Bryologist* **107**: 231, 2004).

gongylidius (pl. **gongylidia**), a bulbous structure developed by fungi cultivated by termites.

Gongylocladium Wallr. (1833) ≡ Oedemium.

gonidial layer, photobiont layer in a lichen thallus (obsol.).

gonidimium, a hymenial alga (obsol.).

Gonidiomyces Vain. (1921) nom. dub., Pezizomycotina. 1, Philippines.

gonidium, photobiont (obsol.).

gonimium, a cyanobacterial cell in a lichen thallus (obsol.).

goniocyst (**goniocystula**), a group of algal cells derived from a single cell surrounded by a hyphal envelope forming a roundish structure which is not a soralium (e.g. the vegetative thallus '*Botrydina vulgaris*'; q.v.), (Fig. 22H). See Sérusiaux (*Lichenologist* **17**: 1, 1985).

goniocystangium, cup-like structure bearing goniocysts (q.v.) on foliicolous species of *Catillaria* (Vězda, 1980) and *Opegrapha* (Santesson, *Svensk Naturv.* 1968: 176, 1968; Sérusiaux, 1985).

Gonionema Nyl. (1855) = Thermutis fide Hawksworth *et al.* (*Dictionary of the Fungi* edn 8, 1995).

Goniopila Marvanová & Descals (1985), anamorphic *Pezizomycotina*, Hso.0eH.21. 1 (aquatic), USA; Europe. See Marvanová & Descals (*J. Linn. Soc.* Bot. **91**: 14, 1985), Belliveau & Bärlocher (*MR* **109**: 1407, 2005; placement in *Helotiales*), Campbell *et al.* (*MR* **110**: 1025, 2006; placement in *Pleosporales*).

Goniosporium Link (1824) = Arthrinium fide Höhnel (*Mitt. bot. Inst. tech. Hochsch. Wien* ser. 2 **2**: 9, 1925).

goniosporous, having angled spores.

Gonohymenia J. Steiner (1902) = Lichinella fide Moreno & Egea (*Cryptog. Bryol.-Lichénol.* **13**: 237, 1992), Schultz & Büdel (*Lichenologist* **34**: 39, 2002).

Gonohymeniomyces Cif. & Tomas. (1953) ? = Psorotichia.

Gonolecania Zahlbr. (1924) = Byssolecania fide Santesson (*Svensk bot. Tidskr.* **43**: 547, 1949), Santesson (*SA* **10**: 137, 1991).

gonoplasm (of *Peronosporales*), the protoplasm, at the centre of an antheridium, which later undergoes fusion with the oosphere.

gonosphere, a zoospore of the *Chytridiales* (obsol.).

Gonothecis Clem. (1909) = Gyalectidium fide Hawksworth *et al.* (*Dictionary of the Fungi* edn 8, 1995).

Gonothecium (Vain.) Clem. & Shear (1931) = Gyalectidium fide Santesson (*Symb. bot. upsal.* **12** no. 1: 1, 1952).

gonotocont, the organ in which meiosis takes place.

Gonsala (orthographic variant), see *Gonzala*.

Gonyella Syd. & P. Syd. (1919), anamorphic *Pezizomycotina*, Hso.1eP.?. 1, Europe.

Gonytrichella Emoto & Tubaki (1971) = Dicyma fide von Arx (*Proc. k. ned. Akad. Wet.* Ser. C, Biol. Med. Sci. **85**: 21, 1982).

Gonytrichum Nees & T. Nees (1818), anamorphic *Chaetosphaeria, Melanopsammella*, Hsy.0eP.15. 7, widespread. See Gams & Holubová-Jechová (*Stud. Mycol.* **13**, 1976; key), Persiani & Maggi (*Mycotaxon* **39**: 465, 1990; ontogeny), Réblová (*Stud. Mycol.* **45**: 149, 2000; review), Réblová & Winka (*Mycol.* **92**: 939, 2000; phylogeny), Fernández *et al.* (*Mycol.* **98**: 121, 2006; phylogeny, morphology).

Gonzala Adans. ex Léman (1821) = Peziza Fr. fide Streinz (*Nom. fung.*, 1862).

González Fragoso (Romualdo; 1862-1928; Spain). Student of medicine, Seville (*c.* 1879-1882); Keeper of Pacific Collections, Museo de Ciencias Naturales, Madrid (1884-1885); practising doctor, Seville (1885-1911) [providing expert support during the 1888 cholera epidemic in Toledo]; mycologist with state support (1911 onwards); President (1920) then Honorary Member (1921-1928), Real Sociedad Española de Historia Natural, Madrid. Pioneer of mycology in Spain, establishing a national fungal reference collection, and (with Ciferri, q.v.) making significant contributions to the mycota of the Dominican Republic. *Publs.* Contribución a la flora micológica del Guadarrama. Uredinales. *Trabajos del Museo Nacional de Ciencias Naturales. Madrid* vols 3-5 (1914); (with Ciferri) Hongos parásitos y saprofitos de la República Dominicana. *Estación Agronómica de Moca* (1928). *Biogs, obits etc.* Ciferri (*Estación Agronómica de Moca* Serie B, Botánica **12**: 1, 1928) [portrait].

Gooday (Graham William; 1942-2002; England). PhD Bristol (1968); Research Fellow, Leeds University (1967-1969) and with J.H. Burnett (q.v.), Oxford University (1969-1972); Lecturer to full Professor of Microbiology, Aberdeen (1972-2002). First recipient of Society for General Microbiology's Fleming Medal, 1976; FRSE, 1989; President, British Mycological Society, 1993. Over 200 papers, including studies of fungal sex hormones, hyphal growth, basidiomycete fruit body formation, chitin biosynthesis and degradation in a wide range of organisms (including fungi, parasites, fish, plants and man), algal cytology, and algal-invertebrate symbiosis; notable research achievements were autoradiographic confirmation of the apex as the site of wall synthesis in fungi, characterisation of the action of trisporic acid sex pheromone of *Mucor*, and biochemical characterisation of fungal chitin synthases and chitinases; an influential teacher and mentor to many other mycologists. *Publs*: Functions of trisporic acid. *Phil. Trans. Roy. Soc.* B **284**: 509, 1978; An autoradiographic study of hyphal growth of some fungi. *J. Gen. Microbiol.* **67**: 125 1971; The First Fleming Lecture. Biosynthesis of the fungal wall: Mechanisms and implications. *J. Gen. Microbiol.* **99**: 1, 1977.

Goosia B. Song (2003), Englerulaceae. 1, China. See Song (*Mycotaxon* **87**: 413, 2003).

Goosiella Morgan-Jones, Kamal & R.K. Verma (1986), anamorphic *Pezizomycotina*, Hso.0-≡ hH.2.10. 1, India. See Morgan-Jones *et al.* (*Mycol.* **78**: 496, 1986).

Goosiomyces N.K. Rao & Manohar. (1989), anamorphic *Pezizomycotina*, Hso.0bP.12. 1, India. See Rao & Manoharachary (*MR* **92**: 250, 1989).

Goossensia Heinem. (1958), Cantharellaceae. 1, Congo. See Heinemann (*Bull. Jard. Bot. État* **28**: 424, 1958).

Goplana Racib. (1900), Chaconiaceae. 12 (on dicots, (II, III)), widespread (tropical). See Ono & Hennen (*TMSJ* **24**: 369, 1983), Berndt (*Mycol.* **91**: 1045, 1999).

Gorgadesia Tav. (1964), Roccellaceae (L). 1, Cape Verde. See Tavares (*Revta Biol.* Lisb. **4**: 131, 1964), Follmann & Mies (*Courier Forschungsinstitut Senckenberg* **105**: 57, 1988), Tehler (*CJB* **68**: 2458, 1990; cladistics), Myllys *et al.* (*Bryologist* **101**: 70, 1998; phylogeny).

Gorgomyces M. Gönczöl & Révay (1985), anamorphic *Pezizomycotina*, Hso.0fH.1/10. 1 (nematophagous, aquatic), Hungary; Spain. See Gönczöl & Révay (*Nova Hedwigia* **41**: 453, 1985), Roldán (*Mycotaxon* **34**: 381, 1989).

Gorgoniceps (P. Karst.) P. Karst. (1871), Helotiaceae. 6, widespread. See Seaver (*Mycol.* **38**: 548, 1946), Korf (*Mycol.* **58**: 724, 1966), Huhtinen & Iturriaga (*Mycotaxon* **29**: 189, 1987; culture).

Gorlenko (Mikail Vladimirovich; 1908-1994; Russia). Graduated, Voronezh State University (1930); Phytopathologist then Director, Voronezh Plant Protection Station (1929-1941); Head of Laboratory then Director, Moscow Plant Protection Station (1941-1955); Head, Mycology and Algology Department, Lomonosov Moscow State University (1955-1994). He provided an academic infrastructure for mycology, through which his teams could work on physiology and evolution of parasitism and specialization, variability and genetics of plant pathogenic fungi, their cytology, soil fungi as antagonists of plant pathogens, antibiotics of fungal origin for plant protection, mushrooms cultivation and other topics; organizer and first Editor-in-Chief, *Mikologiya i Fitopatologiya* (1967-1994). *Publs.* [*Plant diseases and environment*] (1950) [in Russian, Chinese edn 1954]; *Response of Plant Immunity to Infectious Diseases* (1973) [in Russian]; (ed.) *Plant Life: Myxomycetes and Fungi* edn 2 vol. 2 (1991) [in Russian]. *Biogs, obits etc.* Anon. ([*Mikologiya i Fitopatologiya*] **22**: 273, 1988) [portrait, in Russian]); Anon. ([*Mikologiya i Fitopatologiya*] **29**: 66, 1995) [portrait, in Russian]; Litvinov, Novotelnova & Khokhryakov ([*Mikologiya i Fitopatologiya*] **3**: 287, 1969) [portrait, in Russian].

Gorodkoviella Vassilkov (1969) = Pachyella fide Dissing (*SA* **6**: 129, 1987).

gossamers, fine, floating mycelial nets produced by fungi on media lacking added carbon (Wainwright, *Mycologist* **1**: 182, 1987).

Goupilia Mérat (1834) ? = Scleroderma fide Stalpers (*in litt.*).

Govindua Bat. & H. Maia (1960), Microthyriaceae. 1, India. See Batista & Maia (*Revta Biol.* Lisb. **2**: 95, 1960).

Graamspora Locq. & Sal.-Cheb. (1980), Fossil Fungi. 1, Cameroon.

Gracilistilbella Seifert (2000), anamorphic *Stilbocrea*. 4, widespread. See Seifert & Samuels (*Stud. Mycol.* **45**: 18, 2000).

Graddonia Dennis (1955), Helotiales. 1, Europe. See Dennis (*Kew Bull.* **1955**: 359, 1955), Nauta & Spooner (*Mycologist* **14**: 65, 2000; UK).

Graddonidiscus Raitv. & R. Galán (1992), Hyaloscyphaceae. 3, Europe; N. America. See Galán *et al.* (*MR* **98**: 1137, 1994), Raitviir (*Scripta Mycologica* Tartu **20**, 2004).

Grahamiella Spooner (1981), Helotiaceae. 3, Europe. See Spooner (*TBMS* **76**: 265, 1981).

Grallomyces F. Stevens (1918), anamorphic *Pezizomycotina*, Hsy.0bP.19. 1, widespread (tropical). See Deighton & Pirozynski (*Mycol. Pap.* **105**, 1966), Heredia Abarca & Mercado Sierra (*Mycotaxon* **68**: 137, 1998; Mexico).

Gram, + or - (of bacteria), staining or not staining by Gram's method (see Methods).

Gramincola Velen. (1947) nom. dub., Agaricales.

Graminella L. Léger & M. Gauthier ex Manier (1962), Legeriomycetaceae. 3 (in *Ephemeroptera*), widespread. See Lichtwardt & Moss (*TBMS* **76**: 311, 1981), Lichtwardt (*The Trichomycetes. Fungal associates of arthropods*, 1986; key), Valle (*Mycol.* **99**: 442, 2007; Spain), Valle *et al.* (*Mycol.* **100**: 149, 2008; Mexico).

Graminelloides Lichtw. (1998), Legeriomycetaceae. 1 (in *Diptera*), Costa Rica. See Lichtwardt (*Rev. Biol. Trop.* **45**: 1349, 1997), Nelder *et al.* (*Fungal Diversity* **22**: 121, 2006; ecology and taxonomy).

graminicolous, living on *Gramineae*.

Grammothele Berk. & M.A. Curtis (1868), Polyporaceae. 8, widespread (tropical). See Ryvarden (*TBMS* **73**: 1, 1979), Ipulet & Ryvarden (*Syn. Fung.* **20**: 87, 2005; Uganda).

Grammotheleaceae Berk. & M.A. Curtis (1868), Polyporaceae. 8, widespread (tropical). See Ryvarden (*TBMS* **73**: 1, 1979), Ipulet & Ryvarden (*Syn. Fung.* **20**: 87, 2005; Uganda).

Grammothelopsis Jülich (1982), Polyporaceae. 4, Africa; S. America. See David & Rajchenberg (*Mycotaxon* **22**: 299, 1985), Ryvarden & de Meijer (*Syn. Fung.* **15**: 34, 2002; Brazil).

Granatisporites Elsik & Janson. (1974), Fossil Fungi. 5 (Paleocene), Canada. = Brachysporisporites (Fossil fungi) fide Kremp (*BMR Bull.* **192**: 76, 1978).

Grandigallia M.E. Barr, Hanlin, Cedeño, Parra & R. Hern. (1987), Dothideomycetes. 1, Venezuela. See Barr *et al.* (*Mycotaxon* **29**: 196, 1987).

Grandinia Fr. (1838) nom. rej., Agaricomycetidae. See Donk (*Taxon* **5**: 77, 1956) but see, Jülich (*Int. J. Mycol. Lichenol.* **1**: 27, 1982).

Grandiniella P. Karst. (1895) = Phanerochaete fide Eriksson *et al.* (*Taxon* **27**: 51, 1978).

Grandiniochaete Rick (1940) nom. dub., Agaricomycetidae. See Donk (*Taxon* **5**: 79, 1956).

Grandinioides Banker (1906) = Mycobonia fide Donk (*Taxon* **6**: 17, 1957).

Granmamyces J. Mena & Mercado (1988) = Weufia fide Mercado (*AnaNet* **10**: 8, 1990).

Granodiporites C.P. Varma & Rawat (1963), Fossil Fungi. 2, India.

granular (**granulate**, **granulose**) (of a surface), covered with very small particles. Fig. 20.4.

Granularia Roth (1791) = Nidularia Fr. fide Palmer (*Taxon* **10**: 54, 1961).

Granularia Sacc. & Ellis ex Sacc. (1882), anamorphic *Pezizomycotina*, Hsp.0eH.?. 1, N. America.

Granularia Sowerby (1815) = Urocystis fide Vánky (*in litt.*).

Granulina Velen. (1947) nom. dub., Agaricomycetes. 'gasteromycetes'.

Granulobasidium Jülich (1979), Cyphellaceae. 1, N. America. See Jülich (*Persoonia* **10**: 328, 1979), De Meulder (*AMK Mededelingen*: 30, 2005).

Granulocystis Hjortstam (1986) [non *Granulocystis* Hindák 1977, *Algae*] ≡ Leifia.

Granulodiplodia Zambett. ex M. Morelet (1973) = Sphaeropsis Sacc. fide Sutton (*Mycol. Pap.* **141**, 1977).

granuloma, a nodule of firm tissue formed as a reaction to chronic irritation.

Granulomanus de Hoog & Samson (1978), anamorphic *Torrubiella*, Hso.0fH.15. 1 (on *Arachnida*), British Isles. See Samson *et al.* (*Atlas of Entomopathogenic Fungi*: 187 pp., 1988).

Granulopyrenis Aptroot (1991), Requienellaceae (±L). 6, widespread (tropical). See Aptroot (*Biblthca Lichenol.* **44**: 91, 1991), Geiser *et al.* (*Mycol.* **98**: 1053, 2006; phylogeny).

Graphidaceae Dumort. (1822), Ostropales (L). 25 gen. (+ 59 syn.), 897 spp.

Lit.: Wirth & Hale (*Contr. US natn Herb.* **36**: 63, 1963; Mexico), Wirth & Hale (*Smithson. Contr. bot.* **40**, 1978; Dominican Republic), Nakanishi (*J. Sci. Hiroshima Univ.* B(2) **25**: 16, 1987), Harris (*Some Florida lichens*, 1990; gen. concepts), López de Silanes (*Clementeana* **3**: 25, 1997; key Iberian spp.), Winka *et al.* (*Lichenologist* **30**: 455, 1998), Archer (*Telopea* **8**: 273, 1999), Nakanishi & Harada (*Nat. Hist. Res.* **5**: 63, 1999), Staiger & Kalb (*Mycotaxon* **73**: 69, 1999), Archer (*Telopea* **8**: 461, 2000), Kauff & Lutzoni (*Mol. Phylogen. Evol.* **25**: 138, 2002), Kalb *et al.* (*Symb. bot. upsal.* **34** no. 1: 133, 2004), Miądlikowska *et al.* (*Mycol.* **98**: 1088, 2006; phylogeny), Staiger *et al.* (*MR* **110**: 765, 2006), Archer (*Systematics & Biodiversity* **5**: 9, 2007; Solomon Is).

Graphidales Bessey (1907) = Ostropales.

Graphidastra (Redinger) G. Thor (1990), Roccellaceae (L). 2, paleotropical. See Thor (*Op. bot.* **103**: 80, 1991), Grube (*Bryologist* **101**: 377, 1998; phylogeny), Thor (*J. Jap. Bot.* **77**: 47, 2002; Japan, Taiwan), Sparrius *et al.* (*Lichenologist* **38**: 27, 2006; Thailand, Vietnam).

Graphidium Lindau (1909) ? = Paecilomyces fide Kendrick & Carmichae *in* Ainsworth *et al.* (Eds) (*The Fungi* **4A**: 390, 1973).

Graphidomyces E.A. Thomas ex Cif. & Tomas. (1953) = Graphis fide Hawksworth *et al.* (*Dictionary of the Fungi* edn 8, 1995).

Graphidula Norman (1853) = Verrucaria Schrad. fide Staiger (*Biblthca Lichenol.* **85**, 2002).

Graphilbum H.P. Upadhyay & W.B. Kendr. (1975) = Pesotum fide Upadhyay & Kendrick (*Mycol.* **67**: 800, 1975), Mouton *et al.* (*S. Afr. J. Sci.* **90**: 293, 1994; review), Okada *et al.* (*CJB* **76**: 1502, 1998), Harrington *et al.* (*Mycol.* **93**: 111, 2001; phylogeny), Kim *et al.* (*FEMS Microbiol. Lett.* **222**: 187, 2003).

Graphina Müll. Arg. (1880) = Graphis fide Dharne & Raychoudhury (*Bull. bot. Surv. India* **10**: 267, 1969), Archer (*Telopea* **8**: 273, 1999), Archer (*Mycotaxon* **77**: 153, 2001; Australia), Archer (*Mycotaxon* **83**: 361, 2002; Solomon Is), Archer (*Mycotaxon* **88**: 143, 2003; Australia, Solomon Is), Archer (*Mycotaxon* **86**: 31, 2003), Makhija & Adawadkar (*Mycotaxon* **91**: 347, 2005; India), Staiger *et al.* (*MR* **110**: 765, 2006; phylogeny; 107842), Lücking *et al.* (*Taxon* **56**: 1296, 2007; nomencl.).

Graphinella Zahlbr. (1923) = Spirographa fide Santesson (*The lichens and lichenicolous fungi of Sweden and Norway*, 1993).

Graphinellaceae Walt. Watson (1929) = Odontotremataceae.

Graphinomyces Cif. & Tomas. (1953) ≡ Thalloloma fide Lücking & Hawksworth (*Taxon* **56**: 1274, 2007).

Graphiocladiella H.P. Upadhyay (1981) = Leptographium fide Tsuneda & Hiratsuka (*CJB* **62**: 2618,

1984; SEM conidiogenesis), Six *et al.* (*Mycol.* **95**: 781, 2003).

Graphiola Poit. (1824), Graphiolaceae. *c.* 5, widespread (tropical and glasshouses). *G. phoenicis* on *Phoenix*. See Cole (*Mycol.* **75**: 93, 1983), Begerow *et al.* (*Mycol. Progr.* **1**: 187, 2002; phylog.).

Graphiolaceae Clem. & Shear (1931), Exobasidiales. 2 gen. (+ 4 syn.), 6 spp.

Lit.: Oberwinkler *et al.* (*Pl. Syst. Evol.* **140**: 251, 1982), Cole (*Mycol.* **75**: 93, 1983), Blanz & Gottschalk (*Syst. Appl. Microbiol.* **8**: 121, 1986; affin. with *Exobasidium* and some graminicolous smuts from 5S ribosomal RNA data), Bauer *et al.* (*CJB* **75**: 1273, 1997), Begerow *et al.* (*CJB* **75**: 2045, 1998), Sjamsuridzal & Sugiyama (*J. gen. appl. Microbiol.* Tokyo **44**: 355, 1998), Begerow *et al.* (*Mycol. Progr.* **1**: 187, 2002).

Graphiolales = Exobasidiales.

Graphiolites Fritel (1910), Fossil Fungi. 1 (Eocene), France.

Graphiopsis Bainier (1907) = Phaeoisaria fide Mason (*Annotated Acct of Fungi rec'd I.M.I.*, 1937).

Graphiopsis Trail (1889), anamorphic *Davidiellaceae*. 1 (on *Paeonia*), widespread. See Braun *et al.* (*Mycotaxon* **103**: 207, 2008).

Graphiothecium Fuckel (1870), anamorphic *Pezizomycotina*, Hso.0eP.?. 6, widespread (north temperate). See Braun (*Sydowia* **45**: 81, 1993).

Graphis Adans. (1763), Graphidaceae (L). *c.* 386, widespread (esp. tropical). See Winka *et al.* (*Lichenologist* **30**: 455, 1998; DNA), Archer (*Telopea* **8**: 273, 1999; Australia), Nakanishi & Harada (*Natural History Research* **5**: 63, 1999; Micronesia), Archer (*Aust. Syst. Bot.* **14**: 245, 2001; Australia), Nakanishi *et al.* (*Bull. natn. Sci. Mus.* Tokyo, B **27**: 47, 2001; Thailand), Nakanishi *et al.* (*Bull. natn. Sci. Mus.* Tokyo, B **28**: 107, 2002; Vanuatu), Staiger (*Biblthca Lichenol.* **85**, 2002; revision), Bock & Hauck (*Lichenologist* **37**: 105, 2005; Africa), Adawadkar & Makhija (*Mycotaxon* **96**: 51, 2006; India), Archer (*Biblthca Lichenol.* **94**, 2006; Australia), Miądlikowska *et al.* (*Mycol.* **98**: 1088, 2006; phylogeny), Staiger *et al.* (*MR* **110**: 765, 2006; phylogeny), Adawadkhar & Makhija (*Mycotaxon* **99**: 303, 2007; India), Archer (*Systematics & Biodiversity* **5**: 9, 2007; Solomon Is).

Graphium Corda (1837), anamorphic *Microascus, Pseudallescheria*, Hsy.0eH.19. *c.* 30, widespread. See also *Pesotum*, *Ophiostoma*. See Sutton & Laut (*Bi-mon. Res. Nts Can. For.* **26**: 25, 1970; conidiogenesis), Ellis (*Dematiaceous Hyphomycetes*, 1971; key), Wingfield *et al.* (*MR* **95**: 1328, 1991; gen. synonymy), Seifert & Okada *in* Wingfield *et al.* (Eds) (*Ceratocystis* and *Ophiostoma* Taxonomy, Ecology and Pathogenicity: 27, 1993; anamorphs of *Ophiostoma*), Issakainen *et al.* (*J. Med. Vet. Mycol.* **35**: 389, 1997), Issakainen *et al.* (*MR* **103**: 1179, 1999), Okada *et al.* (*CJB* **76**: 1495, 1999; 18S RNA analysis), Okada *et al.* (*Stud. Mycol.* **45**: 169, 2000; typification, phylogeny), Rainer *et al.* (*J. Clin. Microbiol.* **38**: 3267, 2000; genetic variation), Jacobs *et al.* (*Mycol.* **95**: 714, 2003), Zhang *et al.* (*Mycol.* **98**: 1076, 2006; phylogeny).

graphium (pl. **-ia**), the synnema of *Graphium*.

Graphostroma Piroz. (1974), Graphostromataceae. Anamorph *Nodulisporium*-like. 1, N. America. See Glawe & Rogers (*CJB* **64**: 1493, 1986; anamorph), Barr *et al.* (*Mycotaxon* **48**: 529, 1993; posn), Vassiljeva & Stephenson (*Fungal Diversity* **17**: 191, 2004; synonym of *Diatrype*), Zhang *et al.* (*Mycol.* **98**: 1076, 2006; phylogeny).

Graphostromataceae M.E. Barr, J.D. Rogers & Y.M. Ju (1993), Xylariales. 1 gen., 1 spp.

Lit.: Pirozynski (*CJB* **52**: 2129, 1974), Barr (*Mycol.* **77**: 549, 1985), Glawe & Rogers (*CJB* **64**: 1493, 1986), Barr *et al.* (*Mycotaxon* **48**: 529, 1993).

Graphyllium Clem. (1901), Hysteriaceae. 10, N. America. See Shoemaker & Babcock (*CJB* **70**: 1617, 1992; posn), Barr (*SA* **12**: 27, 1993), Checa *et al.* (*Mycol.* **99**: 285, 2007; Costa Rica).

Greeneria Scribn. & Viala (1887), anamorphic *Diaporthales*, Cac.0eP.15. 1, widespread. See Sutton (*The Coelomycetes*, 1980), Kao *et al.* (*Pl. Prot. Bull. Taiwan* **32**: 256, 1990; pathology of *G. uvicola*), Farr *et al.* (*Sydowia* **53**: 185, 2001; phylogeny, review).

gregarious, in companies or groups but not joined together.

Gregorella Lumbsch (2005), Arctomiaceae (L). 1, Europe. See Lumbsch *et al.* (*Lichenologist* **37**: 300, 2005).

Greletia Donadini (1980) = Smardaea fide Korf & Zhuang (*Mycotaxon* **40**: 413, 1991), Hansen *et al.* (*Mol. Phylogen. Evol.* **36**: 1, 2005; phylogeny).

Gremlia Nieuwl. (1916) ≡ Polycephalum.

Gremmenia Korf (1962) = Phacidium fide Reid & Pirozynski (*Mycol.* **60**: 526, 1968), Stone & Gernandt (*Mycotaxon* **91**: 115, 2005).

Gremmeniella M. Morelet (1969), Helotiales. Anamorph *Brunchorstia*. *c.* 3 (conifer canker), widespread (northern hemisphere). See also *Ascocalyx*. See Petrini *et al.* (*CJB* **67**: 2805, 1989), Bernier *et al.* (*Appl. Environm. Microbiol.* **60**: 1279, 1994; DNA), Hamelin & Rail (*CJB* **75**: 693, 1997; DNA), Hamelin *et al.* (*Phytopathology* **88**: 582, 1998; biogeography), Kaitera *et al.* (*MR* **102**: 199, 1998; Finland, Russia), Hamelin *et al.* (*MR* **104**: 527, 2000; PCR detection), Uotila *et al.* (*Forest Pathology* **30**: 211, 2000; hybridisation), Dusabenyagasani *et al.* (*CJB* **80**: 1151, 2002; molecular variation), Hantula & Tuomivirta (*CJB* **81**: 1213, 2003; Europe), Santamaría *et al.* (*Forest Pathology* **34**: 395, 2004; Spain), Wang *et al.* (*Mol. Phylogen. Evol.* **41**: 295, 2006; phylogeny), Wang *et al.* (*Mycol.* **98**: 1065, 2006; phylogeny).

Greville (Robert Kaye; 1794-1866; England, later Scotland). Trained in medicine, London and Edinburgh, then settling in Edinburgh. An ardent collector, he influenced Berkeley (q.v.) to study fungi rather than algae and mosses, and introduced him to W.J. Hooker. His fungal specimens are mostly in the fungal reference collection at Edinburgh Botanic Garden (**E**). His interests also included campaigning for the emancipation of slaves and, in natural history, diatoms, flowering plants and insects. *Publs. Scottish Cryptogamic Flora* (1823-1828); *Flora Edinensis* (1824). *Biogs, obits etc.* Balfour (*Transactions of the Botanical Society of Edinburgh* **8**: 463, 1866); Grummann (1974: 371); Stafleu & Cowan (*TL-2* **1**: 1009, 1976).

grex (1) see plasmodium; (2) (of cultivated plants) progeny of an artificial cross between known parents.

Grifola Gray (1821), Meripilaceae. 5, widespread (temperate). See Corner (*Beih. Nova Hedwigia* **96**: 63, 1989), Shen *et al.* (*Mycol.* **94**: 472, 2002; phylogeny).

Grifolaceae Jülich (1982) = Meripilaceae.

Griggsia F. Stevens & Dalbey (1919), ? Dothideomycetes. 1, West Indies.

Grilletia Renault & C.E. Bertrand (1885), Fossil Fungi, Chytridiomycota. 1 (Carboniferous), France.

Grimmicola Döbbeler & Hertel (1983), Helotiaceae. 1 (on *Grimmia*), Marion Island. See Döbbeler & Hertel (*Sydowia* **36**: 34, 1983).

Griphosphaerella Petr. (1927) = Monographella fide Müller (*Revue Mycol.* Paris **41**: 129, 1977).

Griphosphaeria Höhn. (1918) = Discostroma fide Brockmann (*Sydowia* **28**, 1976).

Griphosphaerioma Höhn. (1918), Amphisphaeriaceae. Anamorphs *Labridella*, *Sarcostroma*. 2, N. America; Japan. See Shoemaker (*CJB* **41**: 1419, 1963), Kang *et al.* (*MR* **103**: 53, 1999), Ono & Kobayashi (*Mycoscience* **44**: 109, 2003).

griseofulvin ('fulvicin', 'grifulvin', 'grisactin'), a Cl-containing antibiotic from *Penicillium griseofulvum*, *P. nigricans*; described and named by Raistrick *et al.* in 1936, and independently as 'curling factor' by Brian *et al.* in 1946. See Brian (*Ann. Bot. Lond.* N.S. **13**: 59, 1949; *TBMS* **43**: 1, 1960; reviews); antifungal. Has been used as a systemic fungicide against plant pathogens (Rhodes *et al.*, *Ann. appl. Biol.* **45**: 215, 1957) and orally against dermatophyte infections in animals and humans (Gentles, *Nature* **182**: 476, 1958; Lauder & O'Sullivan, *Vet. Rec.* **70**: 949, 1958; Davies, *Antifungal chemotherapy*, 1980: 180).

Griseoporia Ginns (1984) = Gloeophyllum fide Ryvarden (*Syn. Fung.* **5**, 1991).

grisette, basidiomata of the edible *Amanita vaginata*.

grist, see Brewing.

Groenhiella Jørg. Koch, E.B.G. Jones & S.T. Moss (1983), Nitschkiaceae. 1 (on rotten wood, marine), Denmark. See Petersen (*Bot. Mar.* **40**: 71, 1997; ultrastr.).

Grosmannia Goid. (1936), Ophiostomataceae. Anamorphs *Leptographium*, *Pesotum*-like. 1 (associated with bark beetles), widespread. See de Hoog (*Stud. Mycol.* **7**, 1974), Zipfel *et al.* (*Stud. Mycol.* **55**: 75, 2006).

Grove (William Bywater; 1848-1938; England). Headmaster of Birmingham High School for Boys (1887-1900); Lecturer in Horticulture and Chemistry, Studley Horticultural College (1900-1905); Lecturer in Botany, Birmingham Municipal Technical School (1905-1927). A great collector whose field experience resulted in accounts of rusts and of coelomycetes which became standard texts. His specimens are now in the fungal reference collection, Royal Botanic Gardens, Kew (**K**). *Publs. British Rust Fungi* (1913); *British Stem- and Leaf- Fungi (Coelomycetes)* (1935-1937); English translation of Tulasne's *Selecta Fungorum Carpologia* (1931). *Biogs, obits etc.* Buller (*North-west Naturalist* **13**: 30, 1938) [portrait]; Mason (*Journal of Botany* London **76**: 86, 1938); Stafleu & Cowan (*TL-2* **1**: 1015, 1976).

Groveola Syd. (1921) = Uromyces fide Dietel (*Nat. Pflanzenfam.* **6**, 1928).

Groveolopsis Boedijn (1951), anamorphic *Pezizomycotina*, Cpd.0fH.19. 1, Java. See Boedijn (*Sydowia* **5**: 351, 1951), Nag Raj & DiCosmo (*Univ. Waterloo Biol. Ser.* **21**, 1980).

Groves (James Walton; 1906-1970; Canada). Mycologist (1936-1951) then Chief of Mycology Section (1951-1970), Canada Agriculture, Ottawa. Noted for research on discomycetes taxonomy (esp. *Dermateaceae* and *Sclerotiniaceae*), and on seed-borne fungi (some with Skolko). *Publs.* The genus *Tympanis*. *Canadian Journal of Botany* (1952); The genus *Durandiella*. *Canadian Journal of Botany* (1954); The genus *Godronia*. *Canadian Journal of Botany* (1965). *Biogs, obits etc.* Shoemaker (*Mycol.* **63**: 1, 1971) [portrait]; Thomson (*Canadian Field Naturalist* **86**: 177, 1972) [portrait].

Grovesia Dennis (1960), Helotiaceae. 1, Venezuela. See Dennis (*Kew Bull.* **14**: 444, 1960).

Grovesiella B. Erikss. (1969) ≡ Erikssonopsis.

Grovesiella M. Morelet (1969), Helotiaceae. Anamorph *Pittostroma*. 3, USA; Europe. See Seiber & Kowalski (*Mycol.* **85**: 653, 1993; anamorph).

Grovesinia M.N. Cline, J.L. Crane & S.D. Cline (1983), Sclerotiniaceae. Anamorph *Hinomyces*. 1, widespread (north temperate). See Harada & Noro (*TMSJ* **29**: 85, 1988), Holst-Jensen *et al.* (*Mycol.* **89**: 885, 1997; phylogeny), Holst-Jensen *et al.* (*Nordic Jl Bot.* **18**: 705, 1999; phylogeny), Holst-Jensen *et al.* (*Mycol.* **96**: 135, 2004; phylogeny), Narumi-Saito *et al.* (*Mycoscience* **47**: 351, 2006; anamorph).

growth form, see habit.

Growth rates. In fungi growth may be unrestricted or restricted.

(1) **Unrestricted growth** occurs when a fungus is grown in batch culture in a medium containing an excess of all nutrients. Under these conditions, growth is exponential and proceeds at the organism's maximum specific growth rate for the conditions (type of nutrients, temperature, pH etc.). Thus, $^{dM}{}_{dt}$ = $\mu_{max}M$ where M = fungal biomass, t = time and μ_{max} = maximum specific growth rate. The fastest μ_{max} recorded for a fungus in batch culture on a glucose-mineral salts medium is 0.61 h^{-1} (doubling time of 1.1 h) for *Geotrichum candidum*. During unrestricted growth, the total hyphal length and the number of tips of a mycelium increase at the same μ_{max}, and consequently the ratio (G, the hyphal growth unit) between these parameters is a constant (Trinci, *J. Gen. Microbiol.* **81**: 225, 1974). Thus, unrestricted growth of a mycelium involves the duplication of a physiological unit of growth (G) consisting of a hyphal tip and a certain length of hypha. G is species- and strain-specific (Trinci, *in* Jennings & Rayner (Eds), *Ecology and physiology of the fungal mycelium*: 23, 1984) varying from *c.* 48 µm (*Penicillium chrysogenum*) to *c.* 682 µm (*Fusarium vaucerium*).

(2) **Restricted growth** of a fungus occurs when not all nutrients (including O_2) are present in excess [e.g. in a chemostat culture (Righelato, *in* Smith & Berry (Eds), *The filamentous fungi* **1**: 79, 1975)] or when factors such as nutrient concentration, pH or mycelial morphology are altered sufficiently to affect μ_{max} [e.g. in submerged batch culture in which growth is decelerating because of nutrient depletion or in which mycelial pellets (spherical colonies) are formed (Trinci, *Arch. Mikrob.* **73**: 353, 1970)]. Batch growth of a fungus on a solid medium eventually results in the establishment of conditions (e.g. nutrient depletion, change in pH etc.) below the centre of the colony which are less favourable for growth than was initially the case. Consequently, mature colonies of fungi increase in radius at a linear rate (K_r). Thus, R = $R_0 + K_r(t_1 - t_0$ where t_0 = time at onset of linear expansion of colony and R_0 = colony radius at t_0. For colonies expanding in radius at a linear rate, growth in a peripheral ring of biomass (w, the peripheral growth zone) occurs at approximately μ_{max} but growth proceeds at below μ_{max} elsewhere in the col-

ony, often falling to zero or near zero at the colony centre, i.e. a colony is heterogeneous and growth of most of its biomass is restricted. Expansion in radius of a mature colony is described by the following equation (Trinci, *J. Gen. Microbiol.* **67**: 325, 1971) $K_r = w\mu$. It follows from this equation that K_r cannot be used to study the effect of an environmental variable on growth if the variable alters w.

Lichens are amongst the slowest growing organisms known; measured rates vary from 0.01-90 mm marginal growth p.a. (most in range 1.0-6.0 mm). Minutely crustose species are generally the slowest, and fruticose ones (e.g. *Ramalina menziesii*) the fastest. Some species have different growth rates in different geographical regions, and competition may limit growth of adjacent thalli (Pentecost, *Lichenologist* **12**: 135, 1980). Growth rates of some species are reported to be accelerated by chemicals (see Barashkova, *Problems of the North* **7**: 149, 1964). A few attempts at modelling lichen growth have been made (Hill, *Lichenologist* **13**: 265, 1981; Topham (*in* Seaward, *Lichen ecology*: 35, 1977; review).

For reviews see Gow & Gadd (Eds) (*The growing fungus*, 1995), Wessels (*in* Hawksworth (Ed.), *Frontiers in mycology*: 27, 1991; molec. aspect).

See also Ecology, Lichenometry, Longevity, Physiology.

Gruby (David; 1810-1898; Serbia, later France). A Hungarian Jew, born in Kis-Kér [then Hungary, now Backo Dobro Polje, Serbia]; studied medicine at Vienna (1828-1839); teacher and researcher, Hospital St Louis and the Foundling Asylum, Paris (*c.* 1840); medical doctor, Paris (1844 onwards) [with patients including Frédéric Chopin and Franz Liszt; experimenting in anaesthesia in 1847-1848]. During 1841-1844, i.e. before Pasteur (q.v.), he published the first descriptions of the fungi causing favus (*Trichophyton schoenleinii*), thrush (*Candida albicans*), and ringworm of the beard (*T. mentagrophytes*) and of the scalp (*Microsporum audouinii*) in humans. *Publs.* Mémoire sur une vegétation qui constitue la vraie teigne. *Comptes Rendus Hebdomadaire des Séances de l'Académie des Sciences, Paris* (1841); Recherches anatomiques sun une plante cryptogame qui constitue le vrai muguet des enfants. *Comptes Rendus Hebdomadaire des Séances de l'Académie des Sciences, Paris* (1842); Sur une espèce de mentagre contagieuse résultant du développement d'un nouveau cryptogame dans la racine des poils de la barbe de l'homme. *Comptes Rendus Hebdomadaire des Séances de l'Académie des Sciences, Paris* (1842); Recherches sur la nature, le siège et le développement du *Porrigo decalvans* ou phytoalopécie. *Comptes Rendus Hebdomadaire des Séances de l'Académie des Sciences, Paris* (1843); Recherches sur les cryptogames qui constituent la maladie contagieuse du cuir chevelu sous le nom de Teigne (Mahon). *Herpes tonsurans* (Cazenave). *Comptes Rendus Hebdomadaire des Séances de l'Académie des Sciences, Paris* (1844). *Biogs, obits etc.* Beeson (David Gruby, M.D. (1810-1898). *Archives of Dermatology and Syphilology* Chicago **23**: 141, 1931); Zakon & Benedek (*Bulletin of the History of Medicine* **16**: 155, 1944, portrait [also English translation of five papers]).

Grubyella M. Ota & Langeron (1923) = Trichophyton fide Hawksworth *et al.* (*Dictionary of the Fungi* edn 8, 1995).

Guanomyces M.C. González, Hanlin & Ulloa (2000), Sordariales. 1, Mexico; Trinidad. See González *et al.* (*Mycol.* **92**: 1138, 2000), Stchigel *et al.* (*MR* **110**: 1361, 2006; phylogeny).

Guceviczia Glezer (1959) = Wojnowicia fide Sutton (*Česká Mykol.* **29**: 97, 1975).

Gudelia Henssen (1995), Gloeoheppiaceae (L). 1, Mexico. See Henssen (*Lichenologist* **27**: 261, 1995).

Guedea Rambelli & Bartoli (1978), anamorphic *Pezizomycotina*, Hso.≡ eP.1. 2, widespread. See Rambelli & Bartoli (*TBMS* **71**: 342, 1978).

Gueguenia Bainier (1907) = Amblyosporium fide Pirozynski (*CJB* **47**: 325, 1969).

Guehomyces Fell & Scorzetti (2004), anamorphic *Cystofilobasidiaceae*. 1. See Fell & Scorzetti (*Int. J. Syst. Evol. Microbiol.* **54**: 997, 2004).

Guelichia Speg. (1886), anamorphic *Puttemansia*, Hsp.0eH.?. 1, S. America. See Petrak & Sydow (*Annls mycol.* **34**: 38, 1936).

Guepinella Bagl. (1870) = Heppia fide Hawksworth *et al.* (*Dictionary of the Fungi* edn 8, 1995).

Guepinia Fr. (1825), Auriculariales. 1, widespread (north temperate). See Donk (*Persoonia* **4**: 185, 1966).

Guepinia Hepp (1864) [non *Guepinia* Bastard 1821, *Cruciferae*] = Heppia fide Hawksworth *et al.* (*Dictionary of the Fungi* edn 8, 1995).

Guepiniopsis Pat. (1883), Dacrymycetaceae. *c.* 7, widespread. See Kennedy (*Mycol.* **50**: 874, 1959), McNabb (*N.Z. Jl Bot.* **3**: 159, 1965; monotypic), Roberts (*Mycotaxon* **96**: 83, 2006; Caribbean).

Guestia G.J.D. Smith & K.D. Hyde (2001), Xylariaceae. 1, Ecuador. See Smith & Hyde (*Fungal Diversity* **7**: 107, 2001), Petrini (*N.Z. Jl Bot.* **41**: 71, 2003; key).

Guignardia Viala & Ravaz (1892) nom. cons., Botryosphaeriaceae. Anamorphs *Phyllosticta*, *Leptodothiorella*. *c.* 65, widespread. *G. aesculi* (leaf blotch of horse chestnut, *Aesculus*), *G. bidwellii* (black rot of grapes, *Vitis*), *G. camelliae* (copper blight of tea, *Camellia*), *G. citricarpa* (on *Citrus*), *G. vaccinii* (scald or blast of cranberry, *Vaccinium*). See van der Aa (*Stud. Mycol.* **5**, 1973; *Phyllosticta* anamorphs), Punithalingam (*Mycol. Pap.* **136**, 1974; culture), Eriksson & Hawksworth (*SA* **5**: 132, 1986), Hyde (*Sydowia* **47**: 180, 1995; spp. on palms), Janex-Favre *et al.* (*MR* **100**: 875, 1996; ontogeny), Silva-Hanlin & Hanlin (*MR* **103**: 153, 1999; DNA), Hoffmann & Hafellner (*Biblthca Lichenol.* **77**, 2000; excl. lichenicolous spp.), Meyer *et al.* (*S. Afr. J. Sci.* **97**: 191, 2001; on *Citrus*, S Africa), Okane *et al.* (*CJB* **79**: 101, 2001; on *Ericaceae*), Zhou & Stanosz (*Mycol.* **93**: 516, 2001), Baayen *et al.* (*Phytopathology* **92**: 464, 2002; *G. citricarpa*), Bonants *et al.* (*Eur. J. Pl. Path.* **109**: 503, 2003; PCR detection), Okane *et al.* (*Mycoscience* **44**: 353, 2003; endophytes), Pandey *et al.* (*MR* **107**: 439, 2003; endophytes), Rodrigues *et al.* (*MR* **108**: 45, 2004; *G. mangiferae*), Paul *et al.* (*Crop Protection* **24**: 297, 2005; biogeography), Crous *et al.* (*Stud. Mycol.* **55**: 235, 2006; phylogeny), Meyer *et al.* (*Pl. Dis.* **90**: 97, 2006; diagnostics), Schoch *et al.* (*Mycol.* **98**: 1041, 2006; phylogeny), Baldassari *et al.* (*Eur. J. Pl. Path.* **120**: 103, 2007; on *Citrus*).

Guignardiella Sacc. & P. Syd. (1902) ≡ Vestergrenia Rehm.

Guilliermond (Marie Antoine Alexandre; 1876-1945; France). Worked at the University of Lyons and from

1921 in Paris; Professor of Botany, the Sorbonne, Paris (1927); Membre de l'Institut (1935). Noted for his studies on cytology, sexuality, and phylogeny of yeasts. Founder of *Revue de Cytologie et de Cytophysiologie Vegetales* (1935 onwards). *Publs. Les Levures* (1912) [English translation, Tanner, 1920]; *Cytoplasm of the Plant Cell* (1941) [gives special attention to fungi and has a long list of Guilliermond's publications]. *Biogs, obits etc.* Verona (*Mycopathologia et Mycologia Applicata* **4**: 124, 1948); Heim (Notice sur la vie et les travaux de Alexandre Guilliermond (1876-1945) déposée en la séance du 21 Avril 1947. *Institut de France Académie des Sciences*).

Guilliermondella Nadson & Krassiln. (1928) = Saccharomycopsis Schiønning fide Kurtzman & Robnett (*CJB* **73** Suppl. 1: S824, 1995).

Guilliermondia Boud. (1904), ? Sordariales. 1, Europe.

Guilliermondia Nadson & Konok. (1911) ≡ Nadsonia.

Guizhounema X. Mu (1977), Fossil Fungi. 1, China.

Gummiglobus Trappe, Castellano & Amar. (1996), Mesophelliaceae. 2, Australia. See Trappe & Bougher (*Australasian Mycologist* **21**: 9, 2002; key).

Gummivena Trappe & Bougher (2002), Mesophelliaceae. 1, Australia. See Trappe & Bougher (*Australas. Mycol.* **21**: 9, 2002).

gummosis, a plant disease having secretion of 'gum' as a well-marked symptom; of cucumber (*Cladosporium cucumerinum*).

Gurleya Doflein (1898), Microsporidia. 7.

Gurleyides Voronin (1986), Microsporidia. 1.

Gussonea Tornab. (1848) = Pleopsidium fide Hafellner (*Nova Hedwigia* **56**: 281, 1993).

Gut fungi, Fungi enter animal guts with food, both non-flagellate (*Trichomycetes*) and flagellate (*Neocallimastigales*) gut fungi, which naturally grow and reproduce only there. Both groups may confer some benefits on their hosts, but are primarily commensals exploiting highly specialized environments. Many *Trichomycetes* attach to the gut lining, especially the hind gut, of diverse larval and adult arthropods. These guts may be somewhat anaerobic, but the non-flagellate gut fungi contain mitochondria and grow best aerobically (Lichtwardt, *in* Howard & Miller, *The Mycota. VI Human and Animal Relationships*: 315, 1996, review, bibliogr.). Flagellate *Neocallimastigales* inhabit the foregut or hindgut fermentation chambers of most herbivores. All lack mitochondria and are obligate anaerobes which colonize ingested plant material and aid its digestion (Li & Heath, *Can. J. Microbiol.* **39**: 1003, 1993; Trinci *et al.*, *MR* **98**: 129, 1994, reviews, bibliogr.). Unidentified chytrids also occur in sea urchin guts (Thorsen, *Mar. Biol.* **133**: 353, 1999) and may thus be more widespread.

guttate (1) having tear-like drops; (2) (of a pileus), marked as if by drops of liquid.

Guttularia W. Oberm. (1913) ? = Melanospora Corda fide Hawksworth *et al.* (*Dictionary of the Fungi* edn 8, 1995).

guttulate (of spores), having one or more oil-like drops (**guttules**) inside.

Gutturomyces Rivolta (1884) ? = Aspergillus fide Sutton (*in litt.*).

Gyalecta Ach. (1808), Gyalectaceae (L). 34, widespread (esp. temperate). See Vězda (*Sborn. Vys. Zemšd. Lesn. Brno* **1**: 21, 1958), Vězda (*Folia geobot. phytotax.* **4**: 443, 1969), Stenroos & DePriest (*Am. J. Bot.* **85**: 1548, 1998; DNA), Alvarez Andrés & López de Silanes (*Nova Hedwigia* **74**: 257, 2002; Iberian peninsula), Kauff & Lutzoni (*Mol. Phylogen. Evol.* **25**: 138, 2002; phylogeny), Peršoh *et al.* (*Mycol. Progr.* **3**: 103, 2004; asci), Kauff & Büdel (*Bryologist* **108**: 272, 2005; ontogeny), Miądlikowska *et al.* (*Mycol.* **98**: 1088, 2006; phylogeny).

Gyalecta Eaton (1829) [non *Gyalecta* Ach. 1808] = Diploschistes fide Hawksworth *et al.* (*Dictionary of the Fungi* edn 8, 1995).

Gyalectaceae Stizenb. (1862), Ostropales (L). 7 gen. (+ 11 syn.), 80 spp.

Lit.: Santesson (*Symb. bot. upsal.* **12** no. 1: 1, 1952), Vězda (*Folia geobot. phytotax.* **1**: 311, 1966), Vězda (*Folia geobot. phytotax.* **4**: 443, 1969), Hansen *et al.* (*Herzogia* **7**: 367, 1987; Greenland), Coppins *et al.* (*Graphis Scripta* **6**: 89, 1994), Lücking (*Trop. Bryol.* **15**: 45, 1998), Stenroos & DePriest (*Am. J. Bot.* **85**: 1548, 1998; DNA), Lücking (*Lichenologist* **31**: 359, 1999), Messuti *et al.* (*Bryologist* **102**: 314, 1999), Kauff & Lutzoni (*Mol. Phylogen. Evol.* **25**: 138, 2002), Wiklund & Wedin (*Cladistics* **19**: 419, 2003), Lumbsch *et al.* (*Mol. Phylogen. Evol.* **31**: 822, 2004), Reeb *et al.* (*Mol. Phylogen. Evol.* **32**: 1036, 2004), Miądlikowska *et al.* (*Mycol.* **98**: 1088, 2006; phylogeny).

Gyalectales Henssen ex D. Hawksw. & O.E. Erikss. (1986) = Ostropales.

Gyalectella J. Lahm (1883) = Dimerella fide Hawksworth *et al.* (*Dictionary of the Fungi* edn 8, 1995).

Gyalectidium Müll. Arg. (1881), Gomphillaceae (L). 43, widespread (tropical). See Santesson (*Symb. bot. upsal.* **12** no. 1: 1, 1952), Vězda (*Folia geobot. phytotax.* **14**: 43, 1979), Vězda & Poelt (*Folia geobot. phytotax.* **22**: 179, 1987), Ferraro *et al.* (*J. Linn. Soc.* Bot. **137**: 311, 2001; monograph), Ferraro & Lücking (*Acta Bot. Brasilica* **17**: 619, 2003; Brazil), Lücking *et al.* (*Lichenologist* **37**: 123, 2005; phylogeny), Safranek & Lücking (*Bryologist* **108**: 295, 2005; USA).

Gyalectina Vězda (1970) ≡ Cryptolechia.

Gyalectomyces E.A. Thomas ex Cif. & Tomas. (1953) = Gyalecta Ach. fide Hawksworth *et al.* (*Dictionary of the Fungi* edn 8, 1995).

Gyalidea Lettau ex Vězda (1966) nom. cons., Asterothyriaceae (L). 30, widespread. See Vězda (*Folia geobot. phytotax.* **1**: 311, 1966), Vězda (*Folia geobot. phytotax.* **14**: 43, 1979; key), Vězda & Poelt (*Phyton* Horn **30**: 47, 1990; 21 spp.), Vězda & Poelt (*Nova Hedwigia* **53**: 99, 1991; key), Etayo & Vezda (*Lichenologist* **26**: 333, 1994; Europe), Malcolm & Vezda (*Nova Hedwigia* **61**: 457, 1995; New Zealand), Foucard & Thor (*Lichenologist* **28**: 101, 1996; Sweden), Harada & Vezda (*Bryologist* **99**: 193, 1996; Japan), Harada & Vežda (*Nat. Hist. Res.* **5**: 57, 1999; Japan), Aptroot & Lücking (*Taxon* **51**: 565, 2002; nomencl.), Henssen & Lücking (*Ann. bot. fenn.* **39**: 273, 2002; anatomy, ontogeny), Aptroot & Lücking (*Biblthca Lichenol.* **86**: 53, 2003; phenotype phylogeny).

Gyalideopsis Vězda (1972) nom. cons., Gomphillaceae (L). 70, widespread. See Vězda (*Folia geobot. phytotax.* **14**: 43, 1979; key), Kalb & Vězda (*Nova Hedwigia* **58**: 511, 1994; trop.), Lücking (*Biblthca Lichenol.* **65**: 1, 1997), Lücking & Sérusiaux (*Lichenologist* **30**: 543, 1998), Sérusiaux (*Nova Hedwigia* **67**: 381, 1998; W Indies), Boom & Vezda (*Öst. Z. Pilzk.* **9**: 27, 2000; Europe), Etayo & Diederich (*Bryologist* **104**: 130, 2001; USA), Ferraro

(*Fungal Diversity* **15**: 153, 2004; hypophores), Lendemer & Lücking (*Bryologist* **107**: 234, 2004; N America), Lücking *et al.* (*Mycol.* **96**: 283, 2004; phylogeny), Lücking *et al.* (*Lichenologist* **37**: 123, 2005; phylogeny), Lücking *et al.* (*Lichenologist* **38**: 131, 2006; Costa Rica), Lücking *et al.* (*Bryologist* **110**: 622, 2007; N America), Vězda (*Biblthca Lichenol.* **96**: 305, 2007; key foliicolous spp.).

Gyalolechia A. Massal. (1852) = Fulgensia fide Kärnefelt (*Cryptog. Bot.* **1**: 147, 1989).

Gyelnik (Vilmos Köfaragó; 1906-1945; Hungary). PhD, Budapest University (1929); worked at the Budapest National Museum during the German occupation; died in Austria as a result of a bomb attack. Published about 100 papers on lichen-forming fungi (1926-1942) proposing new names mainly in *Alectoria*, *Nephroma*, *Parmelia*, and *Peltigera*. His work has been much criticized. Collections now in Budapest (**BP**) [? some missing; see Verseghy (*Typen-Verzeichnis der Flechtensammlung* (1964), catalogue of types in **BP**]. *Biogs, obits etc.* Grummann (1974: 451); Hillmann (*Feddes Repertorium* **46**: 132, 1939); Kušan (*Annales Mycologici* **32**: 57, 1934, criticism); Sjödin (*Acta Horti Gothoburgensis* **19**: 113, 1954; bibliography, catalogue of new names); Stafleu & Cowan (*TL-2* **1**: 1027, 1976).

Gymnascella Peck (1884), Gymnoascaceae. 14, widespread. See Currah (*Mycotaxon* **24**: 1, 1985), Sugiyama *et al.* (*Mycoscience* **40**: 251, 1999; DNA), Iwen *et al.* (*J. Clin. Microbiol.* **38**: 375, 2000; pulmonary infection), Kuraishi *et al.* (*Antonie van Leeuwenhoek* **77**: 179, 2000; ubiquinones), Solé *et al.* (*Stud. Mycol.* **47**: 141, 2002; phylogeny), Sugiyama *et al.* (*Stud. Mycol.* **47**: 5, 2002; phylogeny, links with *Arachniotus*).

Gymnoascaceae Baran. (1872), Onygenales. 12 gen. (+ 10 syn.), 38 spp.

Lit.: Malloch (*Phytophylactica*, 1981)

Lit.:, von Arx (*Gen. Fungi Sporul. Cult.* Edn 3, 1981)Currah (*Mycotaxon* **24**: 1, 1985; monogr.), Ghosh (*Proc. Indian Acad. Sci.* (Pl. Sci.) **94**: 197, 1986; physio-ecology), Ghosh (*Kavaka* **12**: 1, 1986; phylogeny), Currah (*SA* **7**: 1, 1988; key gen.), Dalpé (*New Phytol.* **113**: 523, 1989), Bowman *et al.* (*Mol. Phylogen. Evol.* **6**: 89, 1996), Sugiyama *et al.* (*Mycoscience* **40**: 251, 1999; phylogeny), Iwen *et al.* (*J. Clin. Microbiol.* **38**: 375, 2000), Kuraishi *et al.* (*Antonie van Leeuwenhoek* **77**: 179, 2000), Sugiyama & Mikawa (*Mycoscience* **42**: 413, 2001), Sugiyama *et al.* (*Stud. Mycol.* **47**: 5, 2002), Untereiner *et al.* (*Stud. Mycol.* **47**: 25, 2002), Stchigel & Guarro (*MR* **111**: 1100, 2007).

Gymnoascales G. Winter (1884) = Onygenales.

Gymnoascoideus G.F. Orr, K. Roy & G.R. Ghosh (1977), Gymnoascaceae. Anamorph *Malbranchea*. 1, widespread. See Currah (*Mycotaxon* **24**: 1, 1985), Currah *in* Hawksworth (Ed.) (*Ascomycete Systematics. Problems and Perspectives in the Nineties* NATO ASI Series vol. **269 269**: 291, 1994; posn), Sugiyama *et al.* (*Mycoscience* **40**: 251, 1999; phylogeny), Solé *et al.* (*Stud. Mycol.* **47**: 141, 2002; phylogeny).

Gymnoascopsis C. Moreau & M. Moreau (1959) ? = Ascosorus fide von Arx & Müller (*Stud. Mycol.* **9**, 1975).

Gymnoascus Baran. (1872), Gymnoascaceae. Anamorph *Malbranchea*. 8, widespread (north temperate). See Orr (*Mycotaxon* **5**: 470, 1977; key), von Arx (*Persoonia* **13**: 173, 1986; key 14 spp. s.l.), Sugiyama *et al.* (*Mycoscience* **40**: 251, 1999; DNA), Kuraishi *et al.* (*Antonie van Leeuwenhoek* **77**: 179, 2000; ubiquinones), Sugiyama & Mikawa (*Mycoscience* **42**: 413, 2001; phylogeny), Solé *et al.* (*Stud. Mycol.* **47**: 141, 2002; phylogeny), Sugiyama *et al.* (*Stud. Mycol.* **47**: 5, 2002; phylogeny), Geiser *et al.* (*Mycol.* **98**: 1053, 2006; phylogeny).

gymnocarpous (of a sporocarp), open, with the hymenium appearing and developing to maturity exposed and not enclosed. Cf. angiocarpous.

Gymnocaulon P.A. Duvign. (1956) = Stereocaulon Hoffm. fide Hawksworth *et al.* (*Dictionary of the Fungi* edn 8, 1995).

Gymnochilus Clem. (1896) [non *Gymnochilus* Blume 1859, *Orchidaceae*] = Psathyrella fide Singer (*Agaric. mod. Tax.*, 1951).

Gymnocintractia M. Piepenbr., Begerow & Oberw. (1999) = Ustanciosporium fide Vánky (*in litt.*).

Gymnoconia Lagerh. (1894), Phragmidiaceae. 1 (on *Rubus* (0, I (=$I^{(III)}$, endocyclic pathway)) (*Rosaceae*)), widespread. *G. nitens* (orange rust of *Rubus*). See Laundon (*Mycotaxon* **3**: 133, 1975), Helfer (*Nova Hedwigia* **81**: 325, 2005; Europ. spp.).

Gymnocybe P. Karst. (1879) [non *Gymnocybe* Fr. 1825) nom. rej., *Musci*] = Pholiota fide Kuyper (*in litt.*).

Gymnoderma Humb. (1793) nom. rej., Thelephoraceae.

Gymnoderma Nyl. (1860) nom. cons., Cladoniaceae (L). 3, Asia; N. America. See Yoshimura & Sharp (*Am. J. Bot.* **55**: 635, 1968), Ahti (*Regnum veg.* **128**: 58, 1993), Wei & Ahti (*Lichenologist* **34**: 19, 2002; separation of *Cetradonia*), Zhou *et al.* (*J. Hattori bot. Lab.* **100**: 871, 2006; phylogeny).

Gymnodermatomyces Cif. & Tomas. (1953) ≡ Gymnoderma Nyl.

Gymnodiscus Zukal (1887) [non *Gymnodiscus* Less. 1831, *Compositae*] ≡ Zukalina.

Gymnodochium Massee & E.S. Salmon (1902), anamorphic *Pezizomycotina*, Hsp.1eH.?. 1, Europe.

Gymnoeurotium Malloch & Cain (1973) = Edyuillia fide Benny & Kimbrough (*Mycotaxon* **12**: 1, 1980).

Gymnogaster J.W. Cribb (1956), Agaricaceae. 1, Australia. See Cribb (*Pap. Dept. Bot. (formerly Biol.) Univ. Qd.* **3**: 109, 1956).

Gymnoglossum Massee (1891), Bolbitiaceae. 1, Australia. See Smith (*Mycol.* **58**: 122, 1966).

Gymnogomphus Fayod (1889) nom. dub., Boletales. The name is possibly a synonym of *Boletus*. See Singer (*Agaric. mod. Tax.* 4th ed: 758, 1986).

Gymnographa Müll. Arg. (1887) = Phaeographis fide Staiger (*Biblthca Lichenol.* **85**: 526 pp., 2002; monograph), Archer (*Telopea* **10**: 589, 2004), Archer (*Biblthca Lichenol.* **94**, 2006), Lücking (*Biblthca Lichenol.* **96**: 185, 2007).

Gymnographoidea Fink (1930) = Mazosia fide Aptroot (*in litt.*).

Gymnographomyces Cif. & Tomas. (1953) ≡ Gymnographa.

Gymnographopsis C.W. Dodge (1967), Graphidaceae (L). 3, widespread (southern hemisphere temperate). See Dodge (*Nova Hedwigia* **12**: 307, 1966), Egea & Torrente (*Cryptog. Bryol.-Lichénol.* **17**: 305, 1996; S Africa), Staiger (*Biblthca Lichenol.* **85**, 2002).

Gymnohydnotrya B.C. Zhang & Minter (1989), Discinaceae. 3, Australia. See Zhang & Minter (*MR* **92**: 192, 1989), Læssøe & Hansen (*MR* **111**: 1075, 2007; phylogeny).

Gymnomitrula S. Imai (1941) = Heyderia fide Maas Geesteranus (*Persoonia* **3**: 81, 1964).

Gymnomyces Massee & Rodway (1898) = Russula fide Miller *et al.* (*Mycol.* **98**: 960, 2006) Gastroid forms are polyphyletic and have formerly been morphologically recognised as a separate genus.

Gymnopaxillus E. Horak (1966), Serpulaceae. 4, S. America (temperate); Australia. See Horak (*Nova Hedwigia* **10**: 335, 1966), Claridge *et al.* (*Aust. Syst. Bot.* **14**: 273, 2001; Australia).

Gymnopeltis F. Stevens (1924) = Lecideopsella fide Eriksson & Hawksworth (*SA* **9**: 15, 1991).

Gymnopilus P. Karst. (1879), ? Strophariaceae. *c.* 200, widespread. See Hesler (*Mycol. Mem.* **3**, 1969; 78 N. Am. spp., keys), Guzmán-Dávalos & Ovrebo (*Mycol.* **93**: 398, 2001; Costa Rica and Panama), Fausto-Guerra *et al.* (*Mycotaxon* **84**: 429, 2002; cultural studies), Rees *et al.* (*Mycotaxon* **84**: 93, 2002; phylogeny), Guzmán-Dávalos (*Mycol.* **95**: 1204, 2003; infrageneric classification), Holec (*Mykol. Listy* **93**: 10, 2005; key Czech Republic spp.), Holec (*Sborník Národního Musea v Praze* **61**: 52 pp., 2005; Czech Republic).

gymnoplast, see protoplast.

Gymnopuccinia K. Ramakr. (1951) = Didymopsorella fide Cummins & Hiratsuka (*Illustr. Gen. Rust Fungi rev. edit.*, 1983).

Gymnopus (Pers.) Roussel (1806) nom. rej., Marasmiaceae. *c.* 300, widespread. Segregate of *Collybia*, containing most of its species, as the genus *Collybia* was previously polyphyletic. See Antonín & Noordeloos (*Libri Botanici* **17**, 1997; European spp.), Antonín *et al.* (*Mycotaxon* **63**: 359, 1997; taxonomy), Moncalvo *et al.* (*Syst. Biol.* **49**: 278, 2000; phylogeny), Ortega *et al.* (*Mycotaxon* **85**: 67, 2003; thermophilic taxa), Mata *et al.* (*Fungal Diversity* **16**: 113, 2004; Costa Rica; mating system), Wilson *et al.* (*Sydowia* **56**: 137, 2004; Java and Bali), Wilson & Desjardin (*Mycol.* **97**: 667, 2005; phylogeny), Mata *et al.* (*Sydowia* **58**: 191, 2007).

Gymnopus (Quél.) Quél. ex Moug. & Ferry (1887) ≡ Rostkovites.

Gymnosphaera Tassi (1902) [non *Gymnosphaera* Blume 1828, *Cyatheaceae*] = Stagonospora fide Saccardo & Saccardo (*Syll. fung.* **18**: 361, 1906).

Gymnosporangium R. Hedw. ex DC. (1805), Pucciniaceae. Anamorph *Roestelia*. *c.* 57 (on *Rosaceae* (0, I); on *Cupressaceae* (III)), widespread. *G. gaeumannii*, *G. nootkatense* and *G. paraphysatum* have II; *G. bermudianum* is autoecious (I, III on *Juniperus*); *G. juniperi-virginianae* (apple (*Malus*)-cedar (*Juniperus*) rust of USA); *G. fuscum* (pear (*Pyrus*)-juniper (*Juniperus*) rust); *G. clavipes* (quince (*Cydonia*) rust). See Kern (*Revised taxonomic account of Gymnosporangium*, 1972; key, 400 refs.), Yun *et al.* (*Plant Pathology Journal* **21**: 310, 2005; aecial hosts).

Gymnosporium Corda (1833) nom. dub., anamorphic *Pezizomycotina*. See Hughes (*CJB* **36**: 727, 1958).

Gymnostellatospora Udagawa, Uchiy. & Kamiya (1993), Myxotrichaceae. 1, Japan. See Sigler *et al.* (*Mycoscience* **41**: 495, 2000; key), Tsuneda & Currah (*Mycol.* **96**: 627, 2004; morphogenesis), Rice & Currah (*Mycol.* **98**: 307, 2006; phylogeny, comparison with *Pseudogymnoascus*).

Gymnotelium Syd. (1921) = Gymnosporangium fide Kern (*Revised taxonomic account of Gymnosporangium*, 1972).

gymnothecium, an ascoma in which the peridium is a loose hyphal network, typical of *Gymnoascaceae*.

Gymnotrema Nyl. (1858) = Glyphis fide Staiger (*Biblthca Lichenol.* **85**, 2002).

Gymnoxyphium Cif., Bat. & I.J. Araujo (1963), anamorphic *Pezizomycotina*, St.1eH.?. 2, USA; Brazil. See Ciferri *et al.* (*Quad. Lab. crittogam., Pavia* **31**: 85, 1963).

gynophore (of *Pyronemataceae*), the multinucleate female structure undergoing development.

Gyoerffyella Kol (1928), anamorphic *Pezizomycotina*, Hso.0-≡ hH.?1. 9 (aquatic), Europe; N. America. See Marvanová *et al.* (*Persoonia* **5**: 29, 1967; key), Marvanová (*TBMS* **65**: 555, 1975), Dudka & Mel'nik (*Mikol. Fitopatol.* **24**: 13, 1990; USSR spp.), Mulenko (*Polish Bot. Stud.* **5**: 79, 1993; Poland), Descals (*MR* **109**: 545, 2005; diagnostic features).

Gypsoplaca Timdal (1990), Gypsoplacaceae (L). 1, widespread. See Timdal (*Biblthca Lichenol.* **38**: 419, 1990), Arup *et al.* (*Mycol.* **99**: 42, 2007; sister group relations with *Parmeliaceae*).

Gypsoplacaceae Timdal (1990), Lecanorales (L). 1 gen., 1 spp.

Lit.: Timdal (*Biblthca Lichenol.* **38**: 419, 1990), Goward *et al.* (*Bryologist* **99**: 439, 1996), Arup *et al.* (*Mycol.* **99**: 42, 2007; sister group relations with *Parmeliaceae*).

Gyraria Nees (1816) = Tremella Pers. fide Donk (*Persoonia* **4**: 179, 1966).

gyrate (**gyrose**), curved to the back and to the front in turn; folded and wavy; convoluted like a brain; (of an apothecioid ascoma), concentrically folded, e.g. *Umbilicaria*.

Gyratylium Preuss (1855) ? = Sphaeropsis Sacc. fide Sutton (*Mycol. Pap.* **141**, 1977).

Gyrocephalus Bref. (1888) = Guepinia Fr. fide Donk (*Persoonia* **4**: 145, 1966).

Gyrocephalus Pers. (1824) nom. rej. = Gyromitra fide Hawksworth *et al.* (*Dictionary of the Fungi* edn 8, 1995).

Gyrocerus Corda (1837) ? = Circinotrichum fide Hawksworth *et al.* (*Dictionary of the Fungi* edn 8, 1995).

Gyrocollema Vain. (1929), Lichinaceae (L). 1, C. America. See Moreno & Egea (*Acta Bot. Barcinon.* **41**: 1, 1992).

Gyrocratera Henn. (1899) = Hydnotrya fide Trappe (*TBMS* **65**: 496, 1975).

Gyrodon Opat. (1836), Paxillaceae. 10, widespread. See Bruns *et al.* (*Mol. Ecol.* **7**: 257, 1998; phylogeny), Kretzer & Bruns (*Mol. Phylogen. Evol.* **13**: 483, 1999; phylogeny), Nagasawa (*Rep. Tottori Mycol. Inst.* **39**: 1, 2001; Japan).

Gyrodontaceae Heinem. (1951) = Paxillaceae.

Gyrodontium Pat. (1900), Coniophoraceae. 3, widespread. See Maas Geesteranus (*Persoonia* **3**: 187, 1964), Carlier *et al.* (*Cryptog. Mycol.* **25**: 261, 2004; phylogeny).

Gyroflexus Raith. (1981), Hymenochaetales. 1, widespread. *Hymenochaetales* or *Agaricales* (*Rickenella* clade). See Raithelhuber (*Die Gattung Clitocybe* **1**: 17, 1981).

Gyrolophium Kunze ex Krombh. (1831) = Dictyonema C. Agardh ex Kunth fide Parmasto (*Nova Hedwigia* **29**: 99, 1978).

Gyromitra Fr. (1849) nom. cons., Discinaceae. *c.* 18, widespread (esp. north temperate). *G. esculenta* (the lorchel) has a heat-labile toxin, **gyromitrin** (q.v.), and is carcinogenic. See Azema (*Docums Mycol.* **10**: 1, 1979), Gibson & Kimbrough (*CJB* **66**: 1743, 1988; ascosporogenesis), Kimbrough (*MR* **95**: 421, 1991; septal structure), Crous *et al.* (*S. Afr. J. Bot.* **62**: 89, 1996), Abbott & Currah (*Mycotaxon* **62**: 1, 1997; N. Am.), Landvik *et al.* (*Nordic Jl Bot.* **17**: 403, 1997; phylogeny), O'Donnell *et al.* (*Mycol.* **89**: 48, 1997; phylogeny), Percudani *et al.* (*Mol. Phylogen. Evol.* **13**: 169, 1999; phylogeny), Huhtinen & Ruotsalainen (*Karstenia* **44**: 25, 2004; Finland), Medel (*Mycotaxon* **94**: 103, 2005; Mexico).

gyromitrin (N-formylhydrazone), heat-labile, carcinogenic, cellular toxin produced by *Gyromitra esculenta* (the lorchel; false morel); breaks down to monomethylhydrazine (MMH), which is also extremely toxic.

Gyromitrodes Vassilkov (1942) = Pseudorhizina fide Pouzar (*Česká Mykol.* **15**: 42, 1961).

Gyromium Wahlenb. (1812) ≡ Umbilicaria.

Gyromyces Göpp. (1844), Fossil Fungi ? Fungi. 1 (Carboniferous), Germany.

Gyrophana Pat. (1897) = Serpula fide Cooke (*Mycol.* **49**: 197, 1957).

Gyrophanopsis Jülich (1979), Meruliaceae. 2, widespread. See Jülich (*Persoonia* **10**: 329, 1979), Stalpers & Buchanan (*N.Z. Jl Bot.* **29**: 333, 1991).

Gyrophila Quél. (1886) = Tricholoma fide Singer & Smith (*A monograph on the genus Galerina Earle*, 1964).

Gyrophora Ach. (1803) ≡ Umbilicaria.

Gyrophora Pat. (1887) ≡ Gyrophana.

Gyrophoraceae Zenker (1827) = Umbilicariaceae.

Gyrophoromyces E.A. Thomas ex Cif. & Tomas. (1953) ≡ Umbilicaria.

Gyrophoropsis Elenkin & Savicz (1910) = Umbilicaria fide Wei (*An Enumeration of Lichens in China*, 1991), Hawksworth *et al.* (*Dictionary of the Fungi* edn 8, 1995).

Gyrophragmium Mont. (1843), Agaricaceae. 1, widespread (subtropical). See Kreisel (*Feddes Repert.* **83**: 577, 1973), Migliozzi & Camboni (*Micol. Ital.* **30**: 36, 2001), Vellinga (*MR* **108**: 354, 2004; phylogeny; some species belong in *Agaricus*).

Gyrophthorus Hafellner & Sancho (1990), Pezizomycotina. 2 (lichenicolous), Europe. See Hoffmann & Hafellner (*Biblthca Lichenol.* **77**: 1, 2000).

Gyropodium E. Hitchc. (1825) = Calostoma fide Fischer (*Nat. Pflanzenfam.* **5b**: viii, 1938).

Gyroporaceae Binder & Bresinsky (2002), Boletales. 1 gen. (+ 4 syn.), 10 spp.

Lit.: Bruns & Gardes (*Mol. Ecol.* **2**: 233, 1993), Castro & Freire (*Persoonia* **16**: 123, 1995), Bruns *et al.* (*Mol. Ecol.* **7**: 257, 1998), Hughey *et al.* (*Mycol.* **92**: 94, 2000), Nagasawa (*Rep. Tottori Mycol. Inst.* **39**: 1, 2001), Watling (*Aust. Syst. Bot.* **14**: 407, 2001), Binder & Bresinsky (*Mycol.* **94**: 85, 2002), Buchanan & May (*N.Z. Jl Bot.* **41**: 407, 2003), Bruns & Shefferson (*CJB* **82**: 1122, 2004).

Gyroporus Quél. (1886), Gyroporaceae. 10, widespread. See Nagasawa (*Rep. Tottori Mycol. Inst.* **39**: 1, 2001; Japan), Li *et al.* (*Fungal Diversity* **12**: 123, 2003; China), Watling *et al.* (*British Fungus Flora. Agarics and Boleti* Rev. & Enl. Edn **1**: 173 pp., 2005; UK).

Gyrostomomyces Cif. & Tomas. (1953) ≡ Gyrostomum.

Gyrostomum Fr. (1825) = Glyphis fide Hale (*Bull. Br. Mus. nat. hist.* Bot. **8**: 227, 1981), Staiger (*Biblthca Lichenol.* **85**: 526 pp., 2002; monograph).

Gyrostroma Naumov (1914), anamorphic *Hypocreales*, St.0eH.?. 3, America; former USSR. See Samuels *et al.* (*CBS Diversity Ser.* **4**, 2006).

Gyrothecium Nyl. (1855) = Sporastatia fide Hawksworth *et al.* (*Dictionary of the Fungi* edn 8, 1995).

Gyrothrix (Corda) Corda (1842), anamorphic *Pezizomycotina*, Hso.0eH.6. 20, widespread (esp. tropical). See Cunningham (*Mycol.* **66**: 127, 1974; key), Arambarri *et al.* (*MR* **101**: 1529, 1997).

Gyrothyrium Arx (1950) = Schizothyrium fide Müller & von Arx (*Beitr. Kryptfl. Schweiz* **11** no. 2, 1962).

Gyrotrema A. Frisch (2006), Thelotremataceae (L). 2, S. America (tropical). See Frisch (*Biblthca Lichenol.* **92**: 379, 2006).

Gyrotrichum Spreng. (1827) ≡ Circinotrichum.

H, Botanical Museum, University of Helsinki (Finland); founded *c.* 1750.

H bodies, pairs of sporidia of *Tilletia* fused in pairs while still attached to the promycelium.

Haasiella Kotl. & Pouzar (1966), Tricholomataceae. 2, Europe. See Kotlaba & Pouzar (*Česká Mykol.* **20**: 135, 1966), Redhead *et al.* (*Mycotaxon* **83**: 19, 2002).

habitat, natural place of occurrence of an organism.

Habrostictis Clem. (1909) nom. dub., anamorphic *Pezizomycotina*. See Hein (*Willdenowia* Beih. **9**, 1976).

Habrostictis Fuckel (1870) = Orbilia fide Baral (*SA* **13**: 113, 1994).

Haddowia Steyaert (1972), Ganodermataceae. 3, widespread (pantropical). See Steyaert (*Persoonia* **7**: 108, 1972), Ryvarden (*Syn. Fung.* **19**: 19, 2004).

Haddowiaceae Jülich (1982) = Ganodermataceae.

Hadotia Maire (1906) = Lophodermium fide von Arx & Müller (*Stud. Mycol.* **9**, 1975).

hadromycosis, term coined by Pethybridge (*Sci. Proc. Roy. Dublin Soc.* **15**: 63, 1916) for a plant disease in which the pathogen is confined to the xylem; tracheomycosis. 'Vascular wilt' is the preferred term when wilting is a symptom. Cf. wilt.

Hadronema Syd. & P. Syd. (1909), anamorphic *Pezizomycotina*, Hso.1eP.?. 3, widespread.

Hadrospora Boise (1989), Phaeosphaeriaceae. 2, widespread. See Boise (*Mem. N. Y. bot. Gdn* **49**: 308, 1989), Webster (*Nova Hedwigia* **57**: 141, 1993), Tanaka & Harada (*Mycoscience* **44**: 245, 2003; Japan).

Hadrosporium Syd. (1938), anamorphic *Pezizomycotina*, Hsp.≡ eP.19. 1, Australia.

Hadrotrichum Fuckel (1865), anamorphic *Xylariales*, Hsp.0eP.10. 5, widespread (temperate).

Haematomma A. Massal. (1852) nom. cons., Haematommataceae (L). 53, widespread. See Rogers (*Lichenologist* **14**: 115, 1982; Australia), Rogers & Bartlett (*Lichenologist* **18**: 247, 1986; NZ), Kalb *et al.* (*Biblthca Lichenol.* **59**: 199, 1995), Staiger & Kalb (*Biblthca Lichenol.* **59**: 3, 1995), Lumbsch *et al.* (*Nova Hedwigia* **86**: 105, 2008; phylogeny, chemistry).

Haematommataceae Hafellner (1984), Lecanorales (L). 1 gen. (+ 1 syn.), 53 spp.

Lit.: Rogers (*Lichenologist* **17**: 307, 1985), Honegger (*New Phytol.* **103**: 785, 1986), Rogers & Bartlett (*Lichenologist* **18**: 247, 1986), Brodo & Culberson (*Bryologist* **89**: 203, 1986), Rogers & Hafellner (*Lichenologist* **20**: 167, 1988), Kalb *et al.* (*Biblthca Lichenol.* **59**: 199, 1995), Staiger & Kalb (*Biblthca Lichenol.* **59**: 3, 1995), Lumbsch *et al.* (*Nova Hedwigia* **86**: 105, 2008; phylogeny, chemistry).

Haematommatomyces Cif. & Tomas. (1953) ≡ Ophioparma.

Haematomyces Berk. & Broome (1875) nom. dub., Fungi. Based on resin.

Haematomyxa Sacc. (1884), Pezizomycotina. 2, N. America. See Pirozynski (*SA* **6**: 130, 1987), Kutorga & Hawksworth (*SA* **15**: 1, 1997).

Haematonectria Samuels & Nirenberg (1999), Nectriaceae. Anamorph *Fusarium*-like. 5 (saprobes and pathogens), widespread. Molecular data suggest that *Neocosmospora* is at least closely related. See Samuels & Brayford (*Sydowia* **46**: 75, 1994), Spatafora & Blackwell (*Mycol.* **85**: 912, 1994), O'Donnell (*Sydowia* **48**: 57, 1996), Rossman *et al.* (*Stud. Mycol.* **42**: 248 pp., 1999), O'Donnell (*Mycol.* **92**: 919, 2000; phylogeny, as *Nectria*), Summerbell & Schroers (*J. Clin. Microbiol.* **40**: 2866, 2002; phylogeny, medical strains), Leslie & Summerell (*Fusarium Labor. Manual*, 2006; review), Samuels *et al.* (*CBS Diversity Ser.* **4**, 2006; USA), Zhang *et al.* (*Mycol.* **98**: 1076, 2006; phylogeny), Dupont *et al.* (*Mycol.* **99**: 526, 2007; colonization of cave paintings).

Haematostereum Pouzar (1959) = Stereum fide Donk (*Persoonia* **3**: 199, 1964).

haerangium, the sporulating organ of certain ascomycetes (e.g. *Fugascus*, *Ceratostomella*), classed by Falck as *Herangiomycetes*, in which the 8 ascospores developed from the **octophore** are contained by a membrane and surrounded by a circle of hairs (the **tentacle**) around the ostiole of the perithecium (Falck, 1947).

Hafellia Kalb, H. Mayrhofer & Scheid. (1986), Caliciaceae (L). 5, widespread. See Sheard (*Bryologist* **95**: 79, 1992; key 5 spp., N. Am.), Posswald *et al.* (*Muelleria* **8**: 133, 1994; Tasmania), Nordin (*Symb. bot. upsal.* **32** no. 1: 195, 1997; ultrastr.), Nordin & Mattsson (*Lichenologist* **33**: 3, 2001; phylogeny), Scheidegger *et al.* (*Lichenologist* **33**: 25, 2001; evolution), Etayo & Marbach (*Lichenologist* **35**: 369, 2003; key corticolous spp., Canary Is), Helms *et al.* (*Mycol.* **95**: 1078, 2003; phylogeny), Miądlikowska *et al.* (*Mycol.* **98**: 1088, 2006; phylogeny).

Hafellnera Houmeau & Cl. Roux (1984) = Schaereria fide Houmeau & Roux (*Bull. Soc. bot. Centre-Ouest* Nouv. sér. **15**: 142, 1984), Thor (*Symb. bot. upsal.* **32** no. 1: 267, 1997), Tehler & Irestedt (*Cladistics* **23**: 432, 2007).

Hagenia Eschw. (1824) [non *Hagenia* J.F. Gmel. 1791, *Angiospermae*] = Anaptychia fide Kurokawa (*Beih. Nova Hedwigia* **6**, 1962).

Haglundia Nannf. (1932), Dermateaceae. 5, Europe. See Gamundí (*Fl. criptog. Tierra del Fuego* **10**, 1987), Dougoud (*Schweiz. Z. Pilzk.* **70**: 136, 1992), Nauta & Spooner (*Mycologist* **13**: 146, 1999; UK).

Hainesia Ellis & Sacc. (1884), anamorphic *Discohainesia*, Ccu.0eH.15. 5, widespread. Synanamorph *Pilidium*. See Sutton & Gibson (*IMI Descr. Fungi Bact.*, 1977), Palm (*Mycol.* **83**: 787, 1992; synanamorph of *Pilidium*), Punithalingam & Spooner (*MR* **101**: 1228, 1997; sp.n. from England).

hair (in *Agaricales*), one of the hair-shaped epicuticular elements forming a pilose covering or down under a lens and not homologous with a cystidium, pseudoparaphysis, or seta, e.g. in *Lachnella*, *Crinipellis* (as restricted by Singer, 1962: 61).

Halbania Racib. (1889), Asterinaceae. 1, Java.

Halbaniella Theiss. (1917) = Actinopeltis fide von Arx & Müller (*Stud. Mycol.* **9**, 1975).

Halbanina G. Arnaud (1918) = Cirsosia fide von Arx & Müller (*Stud. Mycol.* **9**, 1975).

Hale (Mason Ellsworth jr, 1928-1990; USA). Born in Connecticut; influenced by A.W. Evans (1868-1959) while at Yale University; PhD, University of Wisconsin (1953); learnt microchemical techniques from Asahina (q.v.); lichenologist at the Smithsonian Institution (1957-1990). Made a unique contribution to understanding the *Parmeliaceae*, preparing monographs of numerous segregate genera and also later of lichen-forming Ostropales; his keys and textbooks contributed to a revival of interest in lichen-forming fungi of North America; made some 80,000 collections from the arctic, antarctic, and esp. the tropics as well as the USA; these are in the Smithsonian Institution (USA); also a collector and user of old printing machines. *Publs. Lichen Handbook* (1957); *How to Know the Lichens* (1979) [edn 2]; *Biology of Lichens* (1983) [edn 3]; (with V. Ahmadjian) [eds] *The Lichens* (1973, published 1974]). *Biogs, obits etc.* Ahmadjian (*Endocytobiosis and Cell Research* **7**: 1, 1990) [portrait]; Culberson (*Bryologist* **94**: 90, 1991) [portrait]; Grummann (1974: 189); Lawrey (*Lichenologist* **22**: 405, 1990) [portrait]; Sipman & Seaward (*International Lichenological Newsletter* **23**(2): 42, 1990).

Halecania M. Mayrhofer (1987), Catillariaceae (L). 12, Europe; N. America. See Mayrhofer (*Herzogia* **7**: 383, 1987), Coppins (*Lichenologist* **21**: 217, 1989; UK), Boom & Elix (*Lichenologist* **37**: 237, 2005; Asia).

Haleomyces D. Hawksw. & Essl. (1993), ? Verrucariaceae. 1 (on lichens, esp. *Oropogon*), widespread (neotropical). See Hawksworth & Esslinger (*Bryologist* **96**: 349, 1993).

Haligena Kohlm. (1961), Halosphaeriaceae. 1 (marine), widespread. See Johnson *et al.* (*CJB* **65**: 931, 1987), Sakayaroj *et al.* (*Mycol.* **97**: 804, 2005; phylogeny, morphology).

haline, found near the sea shore (obsol.).

Halisaria Giard (1889) nom. dub., ? Fungi.

Hallingea Castellano (1996), Gallaceaceae. 3, S. America. See Castellano & Muchovej (*Mycotaxon* **57**: 339, 1996).

Hallucinogenic fungi. Various fungi contain chemicals which, when consumed by humans and other animals can cause hallucinations. **Teonanácatl**, basidiomata of *Psilocybe* (*P. mexicana*, etc.; see Singer and Singer & Smith, *Mycol.* **50**: 239, 1958), *Stropharia*, *Paneolus*, *Lycoperdon cruciatum*, *L. mixtecorum*, etc., eaten by Mexican Indians during magical ceremonies to induce cerebral effects; sclerotia ('ergots') of species of the ascomycete genus *Claviceps* were similarly consumed in ancient Greece as part of the mystic ceremonies at Eleusis; the strong association of toadstools, particularly the fly agaric, with elves, fairies, gnomes and other imaginary little people may have its origins in the hallucinatory effects of consuming that species. Various crystalline active principles have been isolated from these fungi, e.g **psilocybin** and, most famously, the hallucinogenic drug LSD: at least some of these active principles are may be toxins. Possible uses have been extensively and controversially explored by the military. In many countries people who consume hallucinogenic fungi or hallucinogenic drugs derived from fungi are censured and there is frequently legislation to prevent an activity which is viewed as harmful to society in general and the consumers themselves in particular. In response to this, a substantial underground culture has developed, with many internet sites offering advice on how to cultivate or otherwise obtain so called 'magic' mushrooms, and a tourist industry for devotees to visit countries where consumption is permitted. Certain species of *Copelandia*, *Panaeolus* and *Psilocybe* are commonly used and a range of vernacular names, such as 'blue meanies' or 'shrooms', has evolved for these fungi. Much psychedelic art, literature and music has been inspired through fungal-induced hallucinations, particularly since the 1960s, and there is some evidence that native Mexican art was similarly influenced.

Lit.: Castañeda (*The teachings of Don Juan*, 1968), Heim, Wasson *et al.* (*Les champignons hallucinogènes du Mexique*, 1958; *Nouvelles investigations sur les champignons hallucinogènes*, 1967 [reprinted from *Archiv. Mus. Nat. d'Hist. Nat.* sér. 7 **6**, **9**]), Leary *et al.* (*The psychedelic experience*, 1964), Menser (*Magic mushroom handbook*, 1984), Shelton (*A Mexican odyssey*, 1975). *Discography*: Beatles (*Revolver*, 1966), Cream (*Disraeli Gears*, 1967), Gong (*Camembert Electrique*, 1971), Hendrix (*Are You Experienced?*, 1967), Jefferson Airplane (*Surrealistic Pillow*, 1967). See Ethnomycology, Soma.

halmophagous (of ectotrophic mycorrhiza), having a mantle and a Hartig net (Burgeff, 1943).

Haloaleurodiscus N. Maek., Suhara & K. Kinjo (2005), Russulales. 1, Japan. See Maekawa *et al.* (*MR* **109**: 826, 2005).

Halobyssus Zukal (1893) = Monilia Bonord. fide Clements & Shear (*Gen. Fung.*, 1931).

Halocyphina Kohlm. & E. Kohlm. (1965), Niaceae. 1 (marine), USA. See Ginns & Malloch (*Mycol.* **69**: 53, 1977).

Halographis Kohlm. & Volkm.-Kohlm. (1988), Roccellaceae (?L). 1 (marine), Australia; Belize. See Kohlmeyer & Volkmann-Kohlmeyer (*CJB* **66**: 1138, 1988), Kohlmeyer & Volkmann-Kohlmeyer (*Cryptog. Bot.* **2**: 367, 1992).

Haloguignardia Cribb & J.W. Cribb (1956), Lulworthiales. 6 (forming galls on marine *Fucales*), Australasia. See Kohlmeyer & Kohlmeyer (*Marine Mycology*, 1979), Inderbitzin *et al.* (*MR* **108**: 737, 2004; phylogeny), Campbell *et al.* (*MR* **109**: 556, 2005; phylogeny).

halonate (1) (of a leaf-spot), having concentric rings; one of the 'frog-eye' type; (2) (of a spore), having a transparent coat around it.

Halonectria E.B.G. Jones (1965), Hypocreales. 1 (on intertidal wood, marine), British Isles. See Rossman *et al.* (*Stud. Mycol.* **42**: 248 pp., 1999).

Halonia Fr. (1849) nom. dub., Fungi. See von Arx & Müller (*Beitr. Kryptfl. Schweiz* **11** no. 1, 1954).

halophilic, tolerating salt; living in salt water; *Dendryphiella salina* is esp. well-studied and there appears to be no salt accumulation in its vacuoles (see Jennings, *in* Rodriguez-Valera (Ed.), *General and applied aspects of halophilic microorganisms*: 107, 1991; review, bibliogr.). See Marine fungi.

Halophiobolus Linder (1944) = Lulworthia fide Kohlmeyer & Kohlmeyer (*Marine Mycology*, 1979).

Halorosellinia Whalley, E.B.G. Jones, K.D. Hyde & Laessøe (2000), Xylariaceae. Anamorph *Geniculosporium*-like. 1 (mangrove roots), widespread (pantropical). See Whalley *et al.* (*MR* **104**: 368, 2000), Petrini (*N.Z. Jl Bot.* **41**: 71, 2003; key).

Halosarpheia Kohlm. & E. Kohlm. (1977), Halosphaeriaceae. 22 (on wood, marine or freshwater), widespread. See Jones (*CJB* **74** Suppl. 1: S790, 1995; ultrastr.), Hyde *et al.* (*Mycoscience* **40**: 165, 1999), Kong *et al.* (*MR* **104**: 35, 2000; DNA), Abdel-Wahab *et al.* (*Mycoscience* **42**: 255, 2001), Anderson *et al.* (*Mycol.* **93**: 897, 2001; phylogeny), Baker *et al.* (*CJB* **79**: 1307, 2001; ultrastr.), Anderson & Shearer (*Mycotaxon* **82**: 115, 2002; anamorph), Campbell *et al.* (*Fungal Diversity Res. Ser.* **7**: 15, 2002; phylogeny), Pang (*Fungal Diversity Res. Ser.* **7**: 35, 2002; morphology), Campbell *et al.* (*Mycol.* **95**: 530, 2003; phylogeny), Pang & Jones (*Nova Hedwigia* **78**: 269, 2004), Pang *et al.* (*Nova Hedwigia* **83**: 207, 2006; phylogeny).

Halosphaeria Linder (1944), Halosphaeriaceae. Anamorphs *Periconia*-like, *Trichocladium*. 4 (marine wood), widespread. See Shearer & Crane (*Mycol.* **69**: 1218, 1978; anamorph), Kohlmeyer & Kohlmeyer (*Marine Mycology*, 1979), Jones *et al.* (*Bot. Mar.* **27**: 129, 1984), Hyde *et al.* (*Bot. Mar.* **37**: 51, 1994; ultrastr.), Spatafora *et al.* (*Am. J. Bot.* **85**: 1569, 1998; phylogenetic analysis), Kong *et al.* (*MR* **104**: 35, 2000; DNA), Pang (*Fungal Diversity Res. Ser.* **7**: 35, 2002; morphology), Campbell *et al.* (*Mycol.* **95**: 530, 2003), Pang *et al.* (*J. Linn. Soc.* Bot. **146**: 223, 2004; separation of *Okeanomyces*), Zhang *et al.* (*Mycol.* **98**: 1076, 2006; phylogeny).

Halosphaeriaceae E. Müll. & Arx ex Kohlm. (1972), Microascales. 61 gen. (+ 5 syn.), 177 spp.

Lit.: Kohlmeyer & Kohlmeyer (*Marine Mycology*: 511, 1979), Jones *et al.* (*J. Linn. Soc.* Bot. **87**: 193, 1983; *Corollospora* s.l.), Jones *et al.* (*Bot. Mar.* **27**: 129, 1984; *Halosphaeria* s.l.), Nakagiri & Tubaki (*Bot. Mar.* **28**: 485, 1985; anamorphs), Farrant *in* Moss (Ed.) (*Biology of marine fungi*: 231, 1986; asci), Jones & Moss (*SA* **6**: 179, 1987; key 21 gen.), Nakagiri & Tubaki *in* Sugiyama (Ed.) (*Pleomorphic Fungi: The Diversity and its Taxonomic Implications*: 79, 1987), Hyde & Jones (*Bot. Mar.* **32**: 205, 1989), Jones *et al.* (*CJB* **74**: S342, 1989), Kohlmeyer & Volkmann-Kohlmeyer (*Bot. Mar.* **34**: 1, 1991; key), Yusoff *et al.* (*CJB* **72**: 1550, 1994), Jones (*CJB* **73**: S790, 1995), Spatafora *et al.* (*Am. J. Bot.* **85**: 1569, 1998; phylogenetic analysis), Abdel-Wahab *et al.* (*MR* **103**: 1500, 1999), Chen *et al.* (*Mycol.* **91**: 84, 1999), Chen *et al.* (*Mycol.* **91**: 84, 1999; DNA), Hyde *et al.* (*Mycoscience* **40**: 165, 1999), Kong *et al.* (*MR* **104**: 35, 2000; DNA), Anderson *et al.* (*Mycol.* **93**: 897, 2001), Campbell *et al.* (*Mycol.* **95**: 530, 2003), Pang *et al.* (*Nova Hedwigia* **77**: 1, 2003), Sakayaroj *et al.* (*Mycol.* **97**: 804, 2005), Schoch *et al.* (*MR* **111**: 154, 2007; as *Halosphaeriales*).

Halosphaeriales Kohlm. (1986) = Microascales.

Halosphaeriopsis T.W. Johnson (1958), Halosphaeriaceae. 1 (marine), widespread. See Jones *et al.* (*Bot. Mar.* **27**: 129, 1984), Spatafora *et al.* (*Am. J. Bot.* **85**: 1569, 1998; phylogenetic analysis), Kong *et al.* (*MR* **104**: 35, 2000; DNA), Campbell *et al.* (*Mycol.* **95**: 530, 2003; phylogeny).

Halospora, see *Holospora*.

Halotthia Kohlm. (1963), ? Zopfiaceae. 1 (marine), Europe. See Kohlmeyer (*Nova Hedwigia* **6**: 9, 1963).

Halstedia F. Stevens (1920) = Phyllachora Nitschke ex Fuckel (1870) fide Petrak (*Annls mycol.* **32**: 317, 1934).

Halteromyces Shipton & Schipper (1975), Cunninghamellaceae. 1, Australia. See Shipton & Schipper (*Antonie van Leeuwenhoek* **41**: 337, 1975), Voigt & Wöstemeyer (*Gene* **270**: 113, 2001; phylogeny).

Halysiomyces E.G. Simmons (1981), anamorphic *Pezizomycotina*, Hso.1-≡ eH.4. 1, USA. See Simmons (*Mycotaxon* **13**: 408, 1981).

Halysium Corda (1837) nom. conf., anamorphic *Pezizomycotina*. See Hughes (*CJB* **36**: 727, 1958).

hamanatto, an edible oriental product obtained by the fermentation of soybeans with *Aspergillus oryzae*; tao-cho (Malaysia); tao-si (Philippines).

hama-natto, see Fermented food and drinks.

Hamaspora Körn. (1877), Phragmidiaceae. 14 (on *Rubus* (*Rosaceae*)), Africa; Asia; Australasia. See Monoson (*Mycopathologia* **37**: 263, 1969), Gjærum *et al.* (*Lidia* **5**: 74, 2000; Uganda).

Hamasporella Höhn. (1912) = Hamaspora fide Dietel (*Nat. Pflanzenfam.* **6**, 1928).

hamate (**hamose, hamous**), hooked (Fig. 23.35); unciate. Cf. hamulate.

hamathecium (Eriksson, *Opera Bot.* **60**: 15, 1981), a neutral term for all kinds of hyphae or other tissues between asci, or projecting into the locule or ostiole of ascomata; usually of carpocentral origin; interascal tissues. Eriksson recognized seven categories (see Fig. 14A-F):

(A) **Interascal pseudoparenchyma**, carpocentral tissues unchanged or compressed between developing asci; e.g. *Wettsteinina*.

(B) **Paraphyses**, hyphae originating from the base of the cavity, usually unbranched and not anastomosed; e.g. *Pyrenula*, *Xylaria*.

(C) **Paraphysoids** (trabecular pseudoparaphyses; tinophyses), interascal or pre-ascal tissue stretching and coming to resemble pseudoparaphyses; often only remotely septate, anastomosing and very narrow (see Barr, *Mycol.* **71**: 935, 1979); e.g. *Patellaria*, *Melanomma*.

(D) **Pseudoparaphyses** (cellular pseudoparaphyses; cataphyses), hyphae originating above the level of the asci and growing downwards between the developing asci, finally becoming attached to the base of the cavity and often also then free in the upper part; often regularly septate, branched and anastomosing and broader; e.g. *Pleospora*.

(E) **Periphysoids**, short hyphae originating above the level of the developing asci but not reaching the base of the cavity; e.g. *Nectria*, *Metacapnodium*.

(F) **Periphyses**, hyphae confined to the ostiolar canal; unbranched and not anastomosing; can occur in conjunction with (B), (D) or (E); e.g. *Gibberella*, *Pyrenula*.

(G) **Hamathecial tissue absent** (not figured), e.g. *Dothidea*.

Hamatocanthoscypha Svrček (1977), Hyaloscyphaceae. *c.* 13, widespread (temperate). See Huhtinen (*Karstenia* **29**: 45, 1990; key), Raitviir (*Scripta Mycol.* **20**, 2004).

Hamigera Stolk & Samson (1971), Trichocomaceae. Anamorph *Merimbla*. 1. See Pitt (*The genus Penicillium and its teleomorphic states Eupenicillium and Talaromyces*, 1979), Ogawa & Sugiyama (*Integration of Modern Taxonomic Methods for Penicillium and Aspergillus Classification*: 149, 2000), Pitt *et al.* (*Integration of Modern Taxonomic Methods for Penicillium and Aspergillus Classification*: 9, 2000), Bills *et al.* (*MR* **105**: 1273, 2001; anamorph), Geiser *et al.* (*Mycol.* **98**: 1053, 2006; phylogeny), Peterson (*Mycol.* **100**: 205, 2008; 4-locus phylogeny).

hamulate (**hamulose**), having little hooks. Cf. hamate.

Handkea Kreisel (1989) = Lycoperdon Pers. fide Larsson & Jeppson (*MR* **112**: 4, 2008).

hanging drop, see van Tieghem cell.

Hansen (Emil Christian; 1842-1909; Denmark). Microbiologist and Director of the Physiological Department of the Carlsberg Laboratories, Copenhagen. Specialized in yeasts and coprophilous fungi. *Publs.* De Danske Gjøningssvampe (Fungi fimicoli Danici) *Videnskabelige Meddelelser fra Dansk Naturhistorisk Forening i Kjøbenhavn* (1876, published 1877); *Practical Studies in Fermentation: being Contributions to the Life-history of Micro-organisms* (1896) [translation by A.K. Miller]. *Biogs, obits etc.* Stafleu & Cowan (*TL-2* **2**: 46, 1979).

Hansenia Lindner (1905) = Hanseniaspora fide von Arx (*Gen. Fungi Sporul. Cult.* Edn 3, 1981).

Hansenia P. Karst. (1879) [non *Hansenia* Turcz. 1884, *Umbelliferae*] = Trametes fide Donk (*Persoonia* **1**: 173, 1960).

Hansenia Zikes (1911) = Hanseniaspora fide von Arx (*Gen. Fungi Sporul. Cult.* Edn 3, 1981).

Hansenia Zopf (1883) = Strattonia fide Lundqvist (*Symb. bot. upsal.* **20** no. 1, 1972).

Hanseniaspora Zikes (1911), Saccharomycodaceae. Anamorph *Kloeckera*. 10, widespread. See Meyer *et al.* (*Antonie van Leeuwenhoek* **44**: 79, 1978), Boekhout *et al.* (*Int. J. Syst. Bacteriol.* **44**: 781, 1994), Smith *in* Kurtzman & Fell (Eds) (*Yeasts, a taxonomic study* 4th edn: 214, 1998), Esteve-Zarzoso *et al.* (*Antonie van Leeuwenhoek* **80**: 85, 2001; phylogeny), Cadez *et al.* (*FEMS Yeast Res.* **1**: 279, 2002; genetic diversity), Cadez *et al.* (*Int. J. Syst. Evol. Microbiol.* **53**: 1671, 2003; n. spp.), Capece *et al.* (*Journal of Applied Microbiology* **98**: 136, 2005; strain diversity), Cadez *et al.* (*Int. J. Syst. Evol. Microbiol.* **56**: 1157, 2006; phylogeny), Pramateftaki *et al.* (*FEMS Yeast Res.* **6**: 77, 2006; mitochondrial genome), Suh *et al.* (*Mycol.* **98**: 1006, 2006; phylogeny).

Hansenula Syd. & P. Syd. (1919) = Pichia. Still used in applied environments. fide Kreger-van Rij (Ed.) (*Yeasts, a taxonomic study* 3rd edn, 1984), Kurtzman (*Antonie van Leeuwenhoek* **50**: 209, 1984), Barnett *et al.* (*Yeasts: Characteristics and Identification* 2nd edn, 1990), Kurtzman *in* Kurtzman & Fell (Eds) (*Yeasts, a taxonomic study* 4th edn: 273, 1998), Middelhoven (*Hansenula polymorpha* Biology and Applications: 1, 2002), Boekhout (Ed.) (*FEMS Yeast Res.* **7**: 1081, 2007; special journal issue).

Hansenulaceae E.K. Novák & Zsolt (1961) = Pichiaceae.

Hansford (Clifford George; 1900-1966; England, later Jamaica, Uganda, Sri Lanka, Australia, Botswana). Microbiologist, Department of Agriculture, Jamaica; Government Mycologist, Department of Agriculture, Uganda (1926-1946); Plant Pathologist, Tea Research Institute, Sri Lanka (1946-1957); mycologist, Waite Institute, Adelaide; cotton research, Botswana. A pioneer of the mycota of East Africa, in particular Uganda; also noted as an expert on *Meliola* and related genera. Collections in **EA** (**IMI**, **K**). *Publs*. The Meliolineae. A monograph. *Sydowia* Beiheft (1961); Iconographia Meliolinearum. *Sydowia* Beiheft (1963). *Biogs, obits etc.* Ainsworth (*Brief Biographies of British Mycologists* p. 85, 1996).

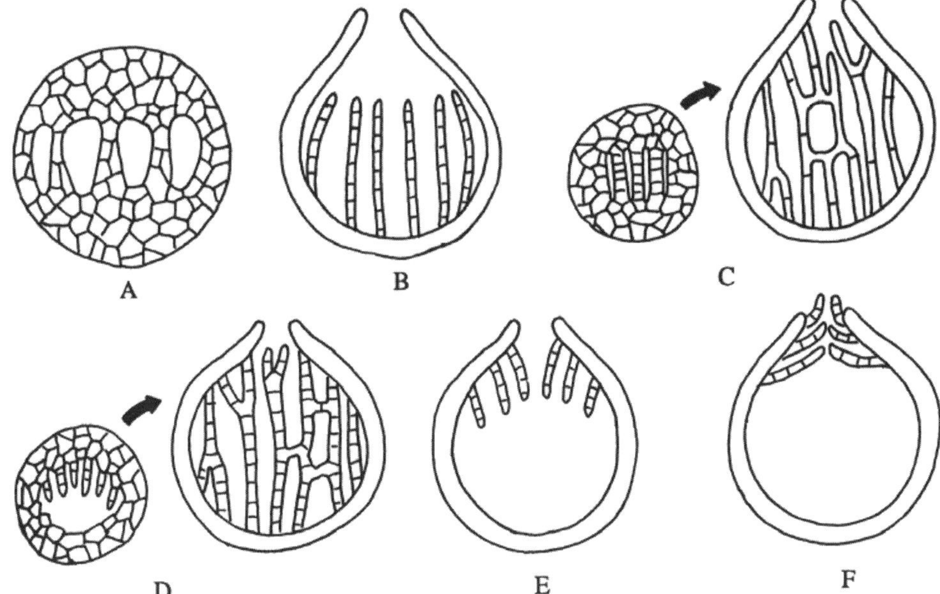

Fig. 15. Hamathecium terminology, following Eriksson (1981). See text for explanation.

Hansfordia S. Hughes (1951), anamorphic *Pezizomycotina*, Hso.0eP.11. 12, widespread. See Deighton (*TBMS* **59**: 531, 1972), von Arx (*Sydowia* **35**: 10, 1982; syn. with *Dicyma*), Gené *et al.* (*MR* **104**: 1404, 2000; Spain).

Hansfordiella S. Hughes (1951), anamorphic *Trichothyrium*, Hso.≡ eP.1. 5, Africa; Philippines. See Subramanian (*Proc. Indian natn Sci. Acad.* Part B. Biol. Sci. **45**: 282, 1957; key spp. on *Meliola* etc.).

Hansfordiellopsis Deighton (1960), anamorphic *Koordersiella*, Hso.#eP.1. 5 (on lichens), Africa; S. America. See Hawksworth (*Bull. Br. Mus. nat. hist.* Bot. **6**: 181, 1979; key).

Hansfordina Bat. (1962) = Microcallis fide von Arx & Müller (*Stud. Mycol.* **9**, 1975).

Hansfordiopeltis Bat. & C.A.A. Costa (1956), anamorphic *Pezizomycotina*, Cpt.0eH.?. 2, Congo. See Batista & Costa (*An. Soc. Biol. Pernambuco* **14**: 40, 1956), Farr (*Mycol.* **78**: 269, 1986).

Hansfordiopeltopsis M.L. Farr (1986), anamorphic *Pezizomycotina*, Cpt.1bH.?. 1, Brazil. See Farr (*Mycol.* **78**: 274, 1986), Reynolds & Gilbert (*Cryptog. Mycol.* **27**: 249, 2006).

Hansfordiopsis Bat. (1959), Micropeltidaceae. 1, Australia. See Batista (*Publções Inst. Micol. Recife* **56**: 407, 1959).

Hansfordiula E.F. Morris (1963) = Phaeoisaria fide de Hoog & Papendorf (*Persoonia* **8**: 407, 1976).

Hantzschia Auersw. (1862) [non *Hantzschia* Grunov 1877, *Bacillariophyceae*] nom. rej. = Kendrickiella fide Kendrick (*CJB* **42**: 1119, 1964), Jacobs *et al.* (*CJB* **79**: 110, 2001).

Hapalocystis Auersw. ex Fuckel (1863), Sydowiellaceae. Anamorph *Phoma*-like. 4 (from bark), Europe; N. America. See Glawe (*Mycol.* **77**: 880, 1985; anamorph), Barr (*Mycotaxon* **41**: 287, 1991), Jaklitsch & Voglmayr (*Stud. Mycol.* **50**: 229, 2004; key), Rossman *et al.* (*Mycoscience* **48**: 135, 2007; review).

Hapalopera Fott (1942) = Phlyctidium (A. Braun) Rabenh. fide Sparrow (*Aquatic Phycomycetes* Edn 2: 1187 pp., 1960) but see Batko (1975); = *Rhizophydium* fide, Karling (*Chytriomyc. Iconogr.*, 1977).

Hapalophragmiopsis Thirum. (1950) = Hapalophragmium fide Monoson (*Mycol.* **69**: 21, 1977), Cummins & Hiratsuka (*Illustr. Gen. Rust Fungi rev. edit.*, 1983), Lohsomboon *et al.* (*MR* **96**: 461, 1992).

Hapalophragmites Ramanujam & Ramachar (1980), Fossil Fungi, Phragmidiaceae. 1, India. See Ramanujam & Ramachar (*Records of the Geological Survey of India* **113**: 82, 1980).

Hapalophragmium Syd. & P. Syd. (1901), Raveneliaceae. 14 (on *Leguminosae*), Africa (tropical). See Monoson (*Mycol.* **69**: 22, 1977; key), Lohsomboon *et al.* (*MR* **96**: 461, 1992), Thaung (*Australasian Mycologist* **24**: 29, 2005; Burma).

Hapalopilaceae Jülich (1982) = Polyporaceae.

Hapalopilus P. Karst. (1881), Polyporaceae. 5, widespread. See Ko *et al.* (*Mycol.* **93**: 270, 2001).

Hapalosphaeria Syd. (1908), anamorphic *Pezizomycotina*, Cpd.0eH.15. 1 (in anthers), Europe.

Haplaria Link (1809) = Botrytis fide Hennebert (*Persoonia* **7**: 188, 1973).

Haplariella Sacc. (1931) = Haplariella Syd. & P. Syd. fide Carmichael *et al.* (*Genera of Hyphomycetes*, 1980).

Haplariella Syd. & P. Syd. (1908), anamorphic *Pezizomycotina*, Cpd.0eH.?. 1, S. America.

Haplariopsis Henn. (1908) ≡ Haplariella Syd. & P. Syd.

Haplariopsis Oudem. (1903), anamorphic *Pezizomycotina*, Hsy.1eH.?. 1, Europe.

haplo- (prefix), one only; single.

Haplobasidion Erikss. (1889), anamorphic *Pezizomycotina*, Hso.0eP.3. 3, Europe; Asia. See Ellis (*Mycol.*

Pap. **67**, 1957; key).

haplobiontic, see Haplobionticae; cf. diplobiontic.

Haplobionticae. Subclass of *Ascomycota* in which the life cycles consists of one thallus (biont); cf. Diplobionticae. Introduced by Nannfeldt (1932).

Haploblastia Trevis. (1860) = Strigula fide Hawksworth *et al.* (*Dictionary of the Fungi* edn 8, 1995).

Haplocarpon M. Choisy (1936) = Porpidia fide Hawksworth *et al.* (*Dictionary of the Fungi* edn 8, 1995), Fryday (*Lichenologist* **37**: 1, 2005).

Haplochalara Linder (1933) = Catenularia fide Mason (*Mycol. Pap.* **5**: 121, 1941).

haploconidium (of *Tremellales*), a uninucleate conidium.

Haplocybe Clem. (1909) ? = Ombrophila Fr. fide Nannfeldt (*Nova Acta R. Soc. Scient. upsal.*, 1932).

Haplocystis Sorokīn (1874), ? Chytridiales. 1, Italy.

Haplodina Zahlbr. (1930), Roccellaceae (L). 3, China.

haplodioicious, see heterothallism.

Haplodothella Werderm. (1923) = Vestergrenia Rehm fide von Arx & Müller (*Beitr. Kryptfl. Schweiz* **11** no. 1, 1954).

Haplodothis Höhn. (1911) = Mycosphaerella fide von Arx & Müller (*Stud. Mycol.* **9**, 1975).

haplogonidia (**haplogonimia**), gonidia (gonimia) in ones, not in groups (obsol.).

Haplographa Anzi (1860) = Lithographa fide Hawksworth *et al.* (*Dictionary of the Fungi* edn 8, 1995).

Haplographiaceae E. Castell. & Chalm. (1919) = Hyaloscyphaceae.

Haplographites Félix (1894), Fossil Fungi. 2 (Tertiary), Europe.

Haplographium Berk. & Broome (1859), anamorphic *Dematioscypha*, Hso.0eH.10. 15, widespread (esp. temperate). See Zucconi & Pagano (*Mycotaxon* **46**: 11, 1993; gen. limits), Kendrick (*CJB* **81**: 75, 2003; morphogenesis).

haploheteroecious, see heterothallism.

haploid (1) (of a nucleus), having the *n* number of chromosomes; (2) (of a cell), having 1 haploid nucleus; (3) (of a mycelium), made up of haploid cells;.

Haplolepis Syd. (1925), anamorphic *Pezizomycotina*. 2, Europe. See Sutton (*Mycol. Pap.* **141**, 1977).

Haploloma Trevis. (1857), ? Lecanorales. 1 (on lichens), Europe.

Haplomela Syd. (1925) = Melanconium fide Clements & Shear (*Gen. Fung.*, 1931).

haplomitotic A nuclear cycle, occurrence of one mitotic haploid phase in the nuclear cycle (Dick, 1987); karyogamy - meiosis - mitosis - karyogamy -.

haplomitotic B nuclear cycle, occurrence of one mitotic diploid phase in the nuclear cycle (Dick, 1987); karogamy - mitosis - meiosis - karyogamy -.

haplomonoecious, see homothallism.

Haplomyces Thaxt. (1893), Laboulbeniaceae. 3, N. America; Europe. See Santamaría (*Fl. Mycol. Iberica* **5**, 2003; Iberian peninsula), De Kesel & Krastina-De Kesel (*Acta Mycologica* Warszawa **41**: 55, 2006; Latvia).

haplomycosis, see adiaspiromycosis.

haplont, the thallus of the haplophase; the gametophyte.

Haplopeltheca Bat., J.L. Bezerra & Cavalc. (1963), Micropeltidaceae. 1, Brazil. See Batista *et al.* (*Publções Inst. Micol. Recife* **388**: 5, 1963).

Haplopeltis Theiss. (1914) = Muyocopron fide von Arx & Müller (*Stud. Mycol.* **9**, 1975).

haplophase, the part of the life history in which the cells are haploid.

Haplophoma Riedl & Ershad (1977) = Phomopsis (Sacc.) Bubák. fide Sutton (*in litt.*).

Haplophyse Theiss. (1916), ? Rhytismatales. 1, Hawaii.

Haploporaceae Jülich (1982) = Polyporaceae.

Haploporus Bondartsev (1953) ≡ Haploporus Bondartsev & Singer.

Haploporus Bondartsev & Singer (1944), Polyporaceae. 1, Europe. See Eriksson (*Symb. bot. upsal.* **16** no. 1: 160, 1958), Niemalä (*Ann. bot. fenn.* **8**: 237, 1971), Zeng & Bai (*Acta Mycol. Sin.* **12**: 12, 1993; China), Piątek (*Polish Botanical Journal* **48**: 81, 2003; Belarus), Zmitrovich *et al.* (*International Journal of Medicinal Mushrooms* **7**: 489, 2005; medicinal properties *Haploporus suaveolens*).

Haplopyrenula Müll. Arg. (1883) = Vizella fide Santesson (*Svensk bot. Tidskr.* **43**: 547, 1949).

Haplopyrenulomyces Cif. & Tomas. (1953) ≡ Haplopyrenula.

Haplopyxis Syd. & P. Syd. (1920) = Uromyces fide Baxter (*Mycol.* **55**: 73, 1963).

Haploravenelia Syd. (1921) = Ravenelia fide Berndt (*in litt.*) Not used by, Arthur (*Manual Rusts US & Canada*, 1934) or, Cummins (*Rust fungi on legumes and composites in North America*, 1978).

Haplospora Räsänen (1943) nom. inval. ≡ Haplopyrenula.

Haplosporangium Thaxt. (1914) = Mortierella fide Gams (*Persoonia* **9**: 381, 1977).

Haplosporella, see *Aplosporella*.

Haplosporidium Caullery & Mesnil (1899) nom. dub., ? Fungi.

Haplosporidium Speg. (1912), anamorphic *Pezizomycotina*, Cpd.0eH.?. 1, Argentina. See Trotter (*Syll. Fung.* **25**: 178, 1925; ? = *Chaetophoma*), Clements & Shear (*Gen. Fung.*: 357, 1931; ? = *Asteroma*).

Haplosporium Mont. (1843) nom. dub., Fungi. See Petrak & Sydow (*Annls mycol.* **27**: 114, 1929).

Haplostroma Syd. & P. Syd. (1916) = Coccodiella fide Müller & von Arx *in* Ainsworth *et al.* (Eds) (*The Fungi* **4A**: 87, 1973).

haplostromatic, see stroma.

Haplostromella Höhn. (1917) nom. nud. = Strasseria fide Sutton (*Mycol. Pap.* **141**, 1977).

haplosynoecious, see homothallism.

Haplotelium Syd. (1922) = Uromyces fide Dietel (*Nat. Pflanzenfam.* **6**, 1928).

Haplotheciella Höhn. (1918) = Didymella fide von Arx & Müller (*Stud. Mycol.* **9**, 1975).

Haplothecium Theiss. & Syd. (1915) = Glomerella fide von Arx & Müller (*Beitr. Kryptfl. Schweiz* **11** no. 1, 1954).

Haplothelopsis Vain. (1921) = Thelopsis fide Hawksworth *et al.* (*Dictionary of the Fungi* edn 8, 1995).

Haplotrichella Arnaud (1954) = Gliomastix fide Carmichael *et al.* (*Genera of Hyphomycetes*, 1980).

Haplotrichum Eschw. (1824) nom. dub., Pezizomycotina.

Haplotrichum Link (1824), anamorphic *Botryobasidium*, Hso.0eH.3. See Hughes (*CJB* **36**: 727, 1958; = *Acladium*), Holubová-Jechová (*Česká Mykol.* **30**: 3, 1976), Partridge *et al.* (*Mycotaxon* **77**: 201, 2001), Partridge *et al.* (*Mycotaxon* **82**: 41, 2002; key).

Haplovalsaria Höhn. (1919) = Didymosphaeria fide Aptroot (*Nova Hedwigia* **60**, 1995).

Hapsidascus Kohlm. & Volkm.-Kohlm. (1991), Pezizomycotina. 1 (marine), Belize. See Kohlmeyer & Volkmann-Kohlmeyer (*SA* **10**: 113, 1991).

Hapsidomyces J.C. Krug & Jeng (1984), Pezizaceae. 1 (coprophilous), Venezuela. See Krug & Jeng (*Mycol.* **76**: 748, 1984), Hansen *et al.* (*Mol. Phylogen. Evol.* **36**: 1, 2005; phylogeny).

Hapsidospora Malloch & Cain (1970), Hypocreales. Anamorph *Acremonium*-like. 1 (from compost), Canada. See Suh & Blackwell (*Mycol.* **91**: 836, 1999; phylogeny), Rossman *et al.* (*Mycol.* **93**: 100, 2001; ? placement in *Bionectriaceae*).

hapteron (1) an aerial organ of attachment of some fruticose lichens (e.g. *Alectoria sarmentosa* subsp. *vexillifera*) formed by a secondary branch which becomes attached to the substratum; (2) attachment organ at base of a funicular cord in *Nidulariaceae*.

Haptocara Drechsler (1975), anamorphic *Pezizomycotina*, Hso.≡ eH.6. 1 (on nematodes), USA. See Drechsler (*Am. J. Bot.* **62**: 1072, 1975).

Haptocillium W. Gams & Zare (2001), anamorphic *Ophiocordycipitaceae*. 12 (on nematodes), widespread. See Gams & Zare (*Nova Hedwigia* **72**: 334, 2001), Glockling & Holbrook (*Mycologist* **19**: 2, 2005), Sung *et al.* (*Stud. Mycol.* **57**, 2007; phylogeny).

haptonema, filamentous appendage (usually coiled) consisting of the plasma membrane, a sheath of endoplasmic reticulum, and a core of several microtubules anchored near the kinetosome.

Haptospora G.L. Barron (1991), anamorphic *Hypocreales*, Hso.1bH.15. 3 (endoparasitic in rotifers), West Indies; New Zealand. See Barron (*CJB* **69**: 503, 1991).

Haradaea Denchev (2006) = Microbotryum fide Vánky (*in litt.*) but see, Denchev *et al.* (*Mycol. Balcanica* **3**: 71, 2006).

Haraea Sacc. & P. Syd. (1913), Meliolaceae. 1 or 2 (from leaves), Japan; West Indies.

Haraella Hara & I. Hino (1955) [non *Haraella* Kudô 1930, *Orchidaceae*] ≡ Hinoa.

Harikrishnaella D.V. Singh & A.K. Sarbhoy (1972) = Chaetomella fide Sutton & Sarbhoy (*TBMS* **66**: 297, 1976).

Hariotia P. Karst. (1889) [non *Hariotia* Adans. 1763, *Cactaceae*] ≡ Delphinella.

Hariotula G. Arnaud (1917) = Cyclotheca fide Müller & von Arx (*Beitr. Kryptfl. Schweiz* **11** no. 2, 1962).

Harknessia Cooke (1881), anamorphic *Wuestneia*, St.0eP.19. 44, widespread. Probably referable to the *Cryptosporellaceae*, but more research is needed. See Sutton (*Mycol. Pap.* **123**, 1971; key), Nag Raj & DiCosmo (*Biblthca Mycol.* **80**, 1981), Sutton & Pascoe (*MR* **92**: 431, 1989; development of ostiole), Crous *et al.* (*Mycol.* **85**: 108, 1993; S. Afr. spp.), Nag Raj (*Coelomycetous Anamorphs with Appendage-bearing Conidia*, 1993), Castlebury *et al.* (*Mycol.* **94**: 1017, 2002), Lee *et al.* (*Stud. Mycol.* **50**: 235, 2004; phylogeny, fam. placement), Rossman *et al.* (*Mycoscience* **48**: 142, 2007).

Harknessiella Sacc. (1889), Dothideomycetes. 1, N. America.

Harmandiana B. de Lesd. (1914) nom. dub., ? Dothideomycetes.

Harmoniella V.N. Boriss. (1981), anamorphic *Pezizomycotina*, Hso.0eP.10. 1, British Islesraine. See Borissova (*Mikol. Fitopatol.* **15**: 89, 1981).

Harpagomyces Wilcz. (1911), anamorphic *Pezizomycotina*, Hso.0bP.1. 1, Europe.

Harpella L. Léger & Duboscq (1929), Harpellaceae. 5 (in *Diptera*), widespread. See Lichtwardt (*Mycol.* **59**: 482, 1967), Reichle & Lichtwardt (*Mycol.* **81**: 103, 1972; ultrastr.), Moss & Lichtwardt (*CJB* **58**: 1035, 1980; ultrastr., taxonomy), Lichtwardt (*The Trichomycetes. Fungal associates of arthropods*, 1986; key), Alencar *et al.* (*Memórias do Instituto Oswaldo Cruz* **98**: 799, 2003; Brazil), White (*MR* **110**: 1011, 2006; phylogeny).

Harpellaceae L. Léger & Duboscq ex P.M. Kirk & P.F. Cannon (2007), Harpellales. 5 gen., 40 spp.

Lit.: Moss (*TBMS* **65**: 115, 1975), Moss & Young (*Mycol.* **70**: 944, 1978), Lichtwardt (*The Trichomycetes. Fungal associates of arthropods*: 343 pp., 1986), Benny *in* McLaughlin *et al.* (Eds) (*The Mycota* A Comprehensive Treatise on Fungi as Experimental Systems for Basic and Applied Research **7A**: 147, 2001), Lichtwardt (*Cellular Origin and Life in Extreme Habitats* **4**: 577, 2002).

Harpellales Lichtw. & Manier (1978). Kickxellomycotina. 2 fam., 38 gen., 200 spp. Thallus simple or branched, septate, with basal holdfast; asexual reproduction by lateral, monosporous, elongate sporangia (trichospores), which upon release exhibit one or more basally attached appendages; sexual reproduction by biconical zygospores; endocommensals or parasites of freshwater arthropods (normally *Insecta*). Fams:

(1) **Harpellaceae**
(2) **Legeriomycetaceae**

Lit.: Lichtwardt (1986; fam. key, *Am. J. Bot.* **51**: 836, 1964; culture), Lichtwardt & Manier (*Mycotaxon* **7**: 441, 1978; taxonomy), Manier (1969; taxonomy), Moss & Lichtwardt (*CJB* **54**: 2346, 1976, **55**: 3099, 1977; ultrastr.), Moss & Young (*Mycol.* **70**: 944, 1978; phylogeny), Moss *et al.* (*Mycol.* **67**: 120, 1975; sexual reproduction), Moss (1979), Peterson & Lichtwardt (*TBMS* **88**: 189, 1987; antigenic variation), White *et al.* (*Mycol.* **98**: 860, 2006; molecular phylogeny), Hibbett *et al.* (*MR* **111**: 109, 2007), and see under Familes.

Harpellomyces Lichtw. & S.T. Moss (1984), Harpellaceae. 3 (in *Diptera*), widespread. See Lichtwardt (*The Trichomycetes. Fungal associates of arthropods*, 1986), Santamaria & Girbal (*An. Jard. bot. Madr.* **55**: 219, 1997; Spain), Lichtwardt *et al.* (*Mycol.* **93**: 764, 2001; Canada), White (*MR* **110**: 1011, 2006; phylogeny), White *et al.* (*Mycol.* **98**: 872, 2006; USA).

Harpezomyces Malloch & Cain (1973) ≡ Chaetosartorya.

Harpidiaceae Vězda ex Hafellner (1984) = Lichinaceae.

Harpidium Körb. (1855), Lichinaceae (L). 2, Europe; N. America. See Sancho & Crespo (*Lazaroa* **5**: 265, 1983), Henssen *et al.* (*Bot. Acta* **101**: 49, 1987; posn), Schultz *et al.* (*Bryologist* **103**: 802, 2000; N America), Schultz & Büdel (*Lichenologist* **34**: 39, 2002; key).

Harpocephalum G.F. Atk. (1897) = Periconia fide Ellis (*Dematiaceous Hyphomycetes*, 1971).

Harpochytriaceae Wille (1900), Monoblepharidales. 1 gen. (+ 2 syn.), 6 spp.

Lit.: Gaurilof *et al.* (*CJB* **58**: 2098, 1980), Gaurilof *et al.* (*CJB* **58**: 2090, 1980), Paquin *et al.* (*Curr. Genet.* **31**: 380, 1997), Einax & Voigt (*Organ. Divers. Evol.* **3**: 185, 2003).

Harpochytriales R. Emers. & Whisler (1968) = Chytridiales.

Harpochytrium Lagerh. (1890), Harpochytriaceae. 6, widespread (north temperate). See Jane (*J. Linn. Soc. Lond.* **53**: 28, 1946), Gaurilof *et al.* (*CJB* **58**: 2090, 1980; ultrastr.), Paquin *et al.* (*Curr. Genet.* **31**: 380, 1997; within *Monoblepharidales*), James *et al.* (*Mycol.* **98**: 860, 2006; phylogeny).

Harpographium Sacc. (1880), anamorphic *Pezizomycotina*, Hsy.0eP.10. 10, widespread. See Morris (*Am. midl. Nat.* **68**: 319, 1962), Deighton (*TBMS* **85**: 738, 1985).

Harpophora W. Gams (2000), anamorphic *Gaeumannomyces*. 4, widespread. See Gams (*Stud. Mycol.* **45**: 192, 2000), Freeman & Ward (*Molecular Plant Pathology* **5**: 235, 2004; review), Saleh & Leslie (*Mycol.* **96**: 1294, 2004; on *Zea*).

Harposporella Höhn. (1925) nom. dub., anamorphic *Pezizomycotina*. See Sutton (*Mycol. Pap.* **141**, 1977).

Harposporium Lohde (1874), anamorphic *Podocrella*, Hso.0fH.15. *c.* 29, widespread (temperate). See Drechsler (*Mycol.* **38**: 1, 1946), Saikawa & Endo (*TMSJ* **27**: 341, 1986; TEM nematode infection), Barron & Szuarto (*CJB* **69**: 1284, 1991), Hodge *et al.* (*MR* **101**: 1377, 1997; *Hirsutella* synanamorph), Glockling (*MR* **102**: 891, 1998), Gams & Zare (*Mycology Series* **19**: 17, 2003), Chaverri *et al.* (*Mycol.* **97**: 433, 2005).

Harpostroma Höhn. (1928), anamorphic *Leptosillia*, St.0fH.?. 1, Europe.

Hartiella Massee (1910) = Calostilbella fide Mason (*Mycol. Pap.* **2**: 29, 1925).

Hartig (Heinrich Julius Adolph Robert; 1839-1901; Germany). Teacher at Forestry Academy Eberswalde, 1866-1878, then Professor of Botany, Munich. Noted for his work on forest tree pathogens and symbionts. *Publs. Wichtige Krankheiten der Waldbaume* (1874) [English translation *Phytopathological Classics* **12**, 1975]; *Die Zersetzungserscheinungen des Holzes der Nadelholzbäume und der Eiche* (1878); *Lehrbuch der Baumkrankheiten* (1882) [in French, 1891; in English, 1894; as *Lehrbuch der Pflanzenkrankheiten*, 1900]. *Biogs, obits etc.* Meinecke (*Phytopathology* **5**: 1, 1915) [portrait]; Merrill *et al.* (*Phytopathological Classics* **12**, 1975) [bibliography, portrait]; Ostrofsky & Ostrofsky (*Review of Tropical Plant Pathology* **7**: 237, 1993).

Hartig net, the intercellular hyphal network formed by an ectomycorrhizal fungus upon the surface of a root.

Hartigiella P. Syd. (1900) = Meria fide Vuillemin (*Annls mycol.* **3**: 340, 1905).

Harzia Costantin (1888), anamorphic *Melanospora*, Hso.0eP.1. 3, widespread. See Holubová-Jechová (*Folia geobot. phytotax.* **9**: 315, 1974).

Harziella Costantin & Matr. (1899), anamorphic *Pezizomycotina*, Hso.0eH.15. 1 (on *Lepista nuda*), Europe. See Fontana (*Allionia* **6**: 35, 1960).

Harziella Kuntze (1891) ≡ Trichocladium.

Hasegawaea Y. Yamada & I. Banno (1987), Schizosaccharomycetaceae. 1, Japan. See Yamada & Banno (*Yeast* Special Issue **5**: S393, 1989), Jeffery *et al.* (*Antonie van Leeuwenhoek* **72**: 327, 1997; lipid composition).

Hassallia Trevis. (1848), Algae. Algae.

Hassea Zahlbr. (1902), Dothideomycetes (L). 1, USA.

Hassiella T.N. Taylor, Krings & Kerp (2006), Fossil Fungi. 1, British Isles.

Hasskarlinda Kuntze (1891) ≡ Corallomyces Fr.

hastate, like a spear-head or arrow-head in form.

Hastifera D. Hawksw. & Poelt (1986) = Micarea Fr. (1825) [nom. cons.] fide Hafellner (*Herzogia* **9**: 167, 1992).

Hauerslevia P. Roberts (1998), Auriculariales. 1, Europe. See Roberts (*Cryptog. Mycol.* **19**: 277, 1998).

haustorial cap, an electron-dense, cap-like mass at the end of a lobe of the haustorial apparatus of *Exobasidium camelliae* (Mims, *Mycol.* **74**: 188, 1982).

haustorium, a special hyphal branch, esp. one within a living cell of the host, for absorption of food (see Karling, *Am. J. Bot.* **19**: 41, 1932). Honegger (*New Phytol.* **103**: 785, 1986) distinguishes three main types of fungus-plant cell interactions: (1) **wall-to-wall apposition** with no penetration; (2) **intracellular haustoria** where the fungus penetrates into the plant cell, with or without the formation of special sheath, neckband, or collar; (3) **intraparietal haustoria** where penetration is restricted to the wall layers (common in some groups of lichens).

Haustoria of *Phytophthora* (Blackwell, *TBMS* **36**: 138, 1953), of *Peronosporales* (Fraymouth, *TBMS* **39**: 79, 1956); in lichens (Honegger, 1986); evolution in rusts (Rajendren, *Bull. Torrey bot. Cl.* **99**: 84, 1972); ultrastructure (Beckett *et al.*, *Atlas of fungal ultrastructure*, 1974); cf. rhizoid. Used in addition, by de Bary, for organs of attachment (appressoria).

Hawksworthia Manohar., N.K. Rao, D.K. Agarwal & Kunwar (2004), anamorphic *Pezizomycotina*. 1, India. See Manoharachary *et al.* (*Indian Phytopath.* **57**: 499, 2004).

Hawksworthiana U. Braun (1988), anamorphic *Pezizomycotina*, Hso.0-1eH.10. 1 (on lichens), north temperate. See Braun (*Monogr. Cercosporella, Ramularia Allied Genera (Phytopath. Hyphom.)* **1**, 1995), Hafellner *et al.* (*Mycotaxon* **84**: 293, 2002; N America).

Haynaldia Schulzer (1866) [non *Haynaldia* Schur 1866, *Gramineae*] = Helicostylum fide Hesseltine (*Mycol.* **47**: 344, 1955).

hazard groups (of fungi), see Safety, Laboratory.

Hazardia J. Weiser (1977), Microsporidia. 1.

Hazlinszkya Körb. (1861) ? = Melaspilea fide Hawksworth *et al.* (*Dictionary of the Fungi* edn 8, 1995).

heart rot, decay of the inner wood of trees, caused by basidiomycetes.

heather rags, common name for *Hypogymnia physodes*.

Hebeloma (Fr.) P. Kumm. (1871), Strophariaceae. *c.* 150, widespread (esp. north temperate). See Bruchet (*Bull. Soc. linn. Lyon* **39** Suppl. 6: 1, 1970; key European spp.), Smith *et al.* (*The veiled species of Hebeloma in the western United States*, 1983; key N. Amer. spp.), Vesterholt (*Nordic Jl Bot.* **9**: 289, 1989; key Europ. spp. sect. *Hebeloma*), Vesterholt (*Symb. bot. Upsal.* **30** no. 3: 129, 1995; key Europ. spp *H. crustuliniforme*-complex), Aanen & Kuyper (*Mycol.* **91**: 783, 1999; taxonomy), Marmeisse *et al.* (*Hebeloma – Ectomycorrhizal fungi, key genera in profile*, 1999; ecology), Aanen *et al.* (*Mycol.* **92**: 269, 2000; phylogeny), Moncalvo *et al.* (*Syst. Biol.* **49**: 278, 2000; phylogeny), Bon (*Docums Mycol.* **31** no. 123: 3, 2002; key European spp.), Aanen & Kuyper (*Persoonia* **18**: 285, 2004; species concept), Vesterholt (*Fungi of Northern Europe* **3**: 146 pp., 2005; N. Europ. spp.), Boyle *et al.* (*MR* **110**: 369, 2006; molecular phylogeny Europ. spp.).

Hebelomataceae Locq. (1977) nom. rej. = Strophariaceae.

Hebelomatis Earle (1909) = Hebeloma fide Singer (*Agaric. mod. Tax.*, 1951).

Hebelomina Maire (1925), Strophariaceae. 4, N. Africa; Europe. The genus is polyphyletic and contains albino forms of both *Hebeloma* (*H. domardiana*) and *Gymnopilus* (*H. neerlandica*). See Huijsman (*Persoonia* **9**: 485, 1978; distinction from *Hebeloma*), Gennari (*Rivista di Micologia* **45**: 311, 2002; key), Fraiture & Hayova (*Acta Mycologica* Warszawa **41**: 177, 2006).

Hectocerus Raf. (1806) nom. nud., Fungi. See Merrill (*Index Rafinesq.*, 1949).

Hegermila Raitv. (1995), Hyaloscyphaceae. 4, S. America. See Raitviir & Järv (*Eesti NSV Tead. Akad. Toim.* Biol. seer **44**: 23, 1995).

Heim (Roger; 1900-1979; France). Préparateur (1927-1929) then assistant (1929-1933) then Deputy Director (1933-1945), then Professor (1945-1973) and Director (1951-1965), Muséum National d'Histoire Naturelle, Paris; a member of the French resistance during World War II, arrested by the Gestapo, interned in Buchenwald and subsequently awarded the Croix de Guerre. Widely travelled; main interests: larger basidiomycetes, mycetism, termite fungi, tropical mycology and ethnomycology (q.v.) with Wasson (q.v.). *Publs.* Le genre *Inocybe*. *Encyclopédie Mycologique* (1938); *Les Champignons d'Europe* (1969) [edn 2]; *Les Champignons Toxiques et Hallucinogènes* (1973); *Termites et Champignons* (1977). *Biogs, obits etc.* Ainsworth (*TBMS* **76**: 177, 1981) [portrait]; Batra (*Mycol.* **72**: 1063, 1981) [portrait]; Dorst (*Comptes Rendu de l'Académie des Sciences, Paris* **290**: 120, 1980); Grummann (1974: 326); Romagnesi (*BSMF* **96**: 117, 1980) [portrait]; Stafleu & Cowan (*TL-2* **2**: 137, 1979).

Heimatomyces Peyr. (1873) = Chitonomyces fide Thaxter (*Memoirs of the American Academy of Arts and Sciences* **12**: 187, 1896).

Heimiella Boedijn (1951) [non *Heimiella* Lohmann 1913, *Haptophyceae*] ≡ Heimioporus.

Heimiella Racov. (1959) nom. illegit., anamorphic *Pezizomycotina*, Cpd.1eH.?. 1, France; Rumania. See Racovitza (*Mémoires du Museum National d'Histoire Naturelle* Paris **10**: 193, 1959).

Heimiodiplodia Zambett. (1955) = Botryodiplodia fide Petrak (*Sydowia* **16**: 353, 1963).

Heimiodora Nicot (1960), anamorphic *Pezizomycotina*, Hso.0eH.10. 1 (from soil), Thailand. See Benjamin (*Aliso* **5**: 278, 1963; posn), Rao & Pande (*Perspectives in Mycological Research* **2** (Prof. G.P. Agarwal Festschrift): 17, 1990).

Heimiomyces Singer (1942) = Xeromphalina fide Kuyper (*in litt.*).

Heimioporus E. Horak (2004), Boletaceae. *c.* 15, widespread (mainly (sub)tropical). See Singer (*Am. midl. Nat.* **37**: 527, 1946; as *Heimiella*), Corner (*Beih. Nova Hedwigia*: 1, 1970; as *Heimiella*), Pegler & Young (*TBMS* **76**, 1981; as *Heimiella*), Horak (*Sydowia* **56**: 237, 2004).

Heinemannomyces Watling (1999), Agaricaceae. 1, Peninsular Malaysia. See Watling (*Belg. Jl Bot.* **131**: 133, 1998), Vellinga (*MR* **108**: 354, 2004; phylogeny).

Heleiosa Kohlm., Volkm.-Kohlm. & O.E. Erikss. (1996), Dothideomycetes. 1, USA. See Kohlmeyer *et al.* (*CJB* **74**: 1830, 1996).

helenin (**helenine**), an antiviral antibiotic from *Penicillium funiculosum*; considered to be RNA of viral origin (Banks *et al.*, *Nature* **218**: 545, 1968).

Heleococcum C.A. Jørg. (1922), Bionectriaceae. 4 (in soil and sediments etc.), widespread. See Tubaki (*TMSJ* **8**: 5, 1967), Rehner & Samuels (*CJB* **73** Suppl. 1: S816, 1995; phylogeny), Udagawa *et al.* (*Mycoscience* **36**: 37, 1995), Rossman *et al.* (*Stud. Mycol.* **42**: 248 pp., 1999), Rossman *et al.* (*Mycol.* **93**: 100, 2001; rDNA phylogeny), Bilanenko *et al.* (*Mycotaxon* **91**: 497, 2005; Mongolia).

Helgardia Crous & W. Gams (2003), anamorphic *Oculimacula*, H?.?.?. 4, widespread. See Crous *et al.* (*Eur. J. Pl. Path.* **109**: 845, 2003).

Helhonia B. Sutton (1980), anamorphic *Pezizomycotina*, St.1eH.15. 1, Europe. See Sutton (*The Coelomycetes*: 600, 1980).

Heliastrum Petr. (1931), Pezizomycotina. 1, Philippines.

Helicascus Kohlm. (1969), Pleosporales. 1 (marine), widespread. See Kohlmeyer (*CJB* **47**: 1471, 1969), Hyde (*Bot. Mar.* **34**: 311, 1991), Tam *et al.* (*Bot. Mar.* **46**: 487, 2003).

Helicia Dearn. & House (1925) [non *Helicia* Lour. 1790, *Proteaceae*] = Cylindrocolla fide Sutton (*Mycol. Pap.* **141**, 1977).

Helicobasidiaceae P.M. Kirk (2008), Helicobasidiales. 3 gen. (+ 4 syn.), 17 spp.

Helicobasidiales R. Bauer, Begerow, J.P. Samp., M. Weiss & Oberw. (2006). Pucciniomycetes. 1 fam., 3 gen., 17 spp. Fam.:

Helicobasidiaceae

For *Lit.* see under gen.

Helicobasidium Pat. (1885), Helicobasidiaceae. Anamorphs *Thanatophytum*, *Tuberculina*. *c.* 6, widespread. *H. purpureum* (syn. *H. brebissonii* sensu Donk) (anamorph *Thanatophytum crocorum*) is the cause of violet root rot in a number of plants. See Donk (*Persoonia* **4**: 156, 1966; synonymy), Aimi *et al.* (*Curr. Microbiol.* **44**: 148, 2002; cytology), Lutz *et al.* (*MR* **108**: 227, 2004), Lutz *et al.* (*Mycol.* **96**: 1316, 2004), Nakamura *et al.* (*MR* **108**: 641, 2004; violet root rot).

Helicobasis Clem. & Shear (1931) ≡ Helicobasidium.

Helicobolomyces Matzer (1995), anamorphic *Arthonia*, Hso.0-≡ hH.1. 1 (on lichens), neotropics. See Grube *et al.* (*Lichenologist* **27**: 28, 1995), Lücking *et al.* (*Global Ecology and Biogeography* **12**: 21, 2003).

Helicobolus Wallr. (1833) = Phloeospora fide Sutton (*Mycol. Pap.* **141**, 1977).

Helicocephalidaceae Boedijn (1959), Zoopagales. 3 gen., 12 spp.

Lit.: Boedijn (*Sydowia* **12**: 355, 1958), Borowska (*Acta Mycologica* Warszawa **32**: 129, 1997), Benny *et al. in* McLaughlin *et al.* (Eds) (*The Mycota* A Comprehensive Treatise on Fungi as Experimental Systems for Basic and Applied Research **7A**: 113, 2001).

Helicocephalum Thaxt. (1891), Helicocephalidaceae. 4 (obligate parasites of nematode eggs), widespread. See Thaxter (*Bot. Gaz.* **16**: 201, 1891), Drechsler (*Mycol.* **26**: 33, 1934), Drechsler (*Mycol.* **35**: 134, 1943), Arnaud (*BSMF* **68**: 181, 1952; possible anamorphic Ascomycota), Barron (*TBMS* **65**: 309, 1975), Watanabe & Koizumi (*TMSJ* **17**: 1, 1976), Kitz & Embree (*Mycol.* **81**: 164, 1989), Roux & Botha (*Proc. Elect. Microsc. Soc. S. Afr.* **25**: 40, 1994; SEM), Roux (*S. Afr. J. Bot.* **62**: 104, 1996), Roux (*S. Afr. J. Bot.* **62**: 285, 1996), Borowska (*Acta Mycologica* Warszawa **32**: 129, 1997; possible anamorphic *Ascomycota*).

Helicoceras Linder (1931), anamorphic *Pezizomycotina*, Hsy.≡ eP.1. 4, widespread (north temperate). See Moore (*Mycol.* **47**: 90, 1955; key).

Helicocoryne Corda (1854) = Helicoma fide Hughes (*CJB* **36**: 727, 1958).

Helicodendron Peyronel (1918), anamorphic *Lambertella*, Hso.0-≡ hP.3. *c.* 27 (aero-aquatic), widespread. See Glen-Bott (*TBMS* **38**: 17, 1955), Moore (*Mycol.* **47**: 90, 1955; key), Abdullah (*TBMS* **81**: 638, 1983), Goos *et al.* (*TBMS* **84**: 423, 1985; keys), Abdullah (*Nova Hedwigia* **44**: 339, 1987), Abdullah *et al.* (*Nova Hedwigia* **63**: 425, 1996; Spain), Voglmayr (*Pl. Syst. Evol.* **205**: 185, 1997; Austria), Voglmayr & Fisher (*MR* **101**: 1122, 1997), Tsui & Berbee (*Mol. Phylogen. Evol.* **39**: 587, 2006; phylogeny), Wang *et al.* (*Mycol.* **98**: 1065, 2006; phylogeny).

Helicodesmus Linder (1925) = Helicodendron fide Linder (*Ann. Mo. bot. Gdn* **16**: 329, 1929).

Helicofilia Matsush. (1983), anamorphic *Pezizomycotina*, Hso.0-≡ h-bP.1. 1, India. See Matsushima (*Matsush. Mycol. Mem.* **3**: 11, 1983), Goos (*Mycol.* **79**: 1, 1987).

Helicogermslita Lodha & D. Hawksw. (1983), Xylariaceae. 4, widespread. See Læssøe & Spooner (*Kew Bull.* **49**: 1, 1994), Rappaz (*Mycol. Helv.* **7**: 99, 1995), Lee & Crous (*Sydowia* **55**: 109, 2003; South Africa), Petrini (*N.Z. Jl Bot.* **41**: 71, 2003; key, New Zealand), Ju *et al.* (*Mycol.* **96**: 1393, 2004; key).

Helicogloea Pat. (1892), Phleogenaceae. *c.* 20, widespread. See Baker (*Mycol.* **38**: 630, 1946), Donk (*Persoonia* **4**: 157, 1966), Roberts (*Mycotaxon* **87**: 187, 2003; Caribbean), Kirschner (*Frontiers in Basidiomycote Mycology*: 165, 2004; anam.).

Helicogonium W.L. White (1942), Endomycetaceae. 1 (on *Corticium*), Canada. See Malloch & Hoog *in* Kurtzman & Fell (Eds) (*Yeasts, a taxonomic study* 4th edn: 197, 1998), Baral (*Nova Hedwigia* **69**: 1, 1999), Suh *et al.* (*Mycol.* **98**: 1006, 2006; phylogeny).

Helicogoosia Hol.-Jech. (1991), anamorphic *Pezizomycotina*, Hso.0-≡ hP.1. 1, Czech Republic. See Holubová-Jechová (*Mycotaxon* **41**: 445, 1991).

Helicoma Corda (1837), anamorphic *Tubeufia*, Hso.0-≡ hH.10. 45, widespread. See Goos (*Mycol.* **78**: 744, 1986; key), Castañeda Ruíz *et al.* (*MR* **102**: 58, 1998; Cuba), Tsui & Berbee (*Mol. Phylogen. Evol.* **39**: 587, 2006; phylogeny), Tsui *et al.* (*Mycol.* **98**: 94, 2006; phylogeny).

Helicomina L.S. Olive (1948) = Pseudocercospora fide Deighton (*Mycol. Pap.* **140**, 1976), Crous & Braun (*CBS Diversity Ser.* **1**: 571 pp., 2003).

Helicominites Barlinge & Paradkar (1982), Fossil Fungi, anamorphic *Pezizomycotina*. 1 (Cretaceous), India.

Helicominopsis Deighton (1960), anamorphic *Pezizomycotina*, Hso.0-≡ hH.10. 1, Sierra Leone. See Deighton (*Mycol. Pap.* **78**: 20, 1960), Goos (*Mycol.* **79**: 1, 1987).

Helicomyces Link (1809), anamorphic *Acanthostigma*, Hso.0-≡ hH.10. 12 (aero-aquatic), widespread. See Goos (*Mycol.* **77**: 606, 1985; keys), Réblová & Barr (*Sydowia* **52**: 258, 2000; teleomorph), Kodsueb *et al.* (*Mycol.* **96**: 667, 2004; Hong Kong), Schoch *et al.* (*Mycol.* **98**: 1041, 2006; phylogeny), Tsui *et al.* (*Mycol.* **98**: 94, 2006; phylogeny).

Helicomycetaceae Nann. (1934) = Tubeufiaceae.

Helicomyxa R. Kirschner & Chee J. Chen (2004), anamorphic *Hyaloriaceae*. 1, Taiwan. See Kirschner & Chen (*Stud. Mycol.* **50**: 338, 2004).

Helicoön Morgan (1892), anamorphic *Tubeufiaceae*, Hso.0-≡ hH.10. 19, N. America; Europe. See Moore (*Mycol.* **47**: 90, 1955; key), Goos (*TBMS* **87**: 115, 1986; key 8 spp.), van der Aa & Samson (*MR* **98**: 74, 1994; key amendment), Goh & Hyde (*MR* **100**: 1485, 1996; key), Voglmayr (*MR* **101**: 337, 1997), Chang (*Bot. Bull. Acad. sin.* Taipei **42**: 149, 2001; Taiwan), Tsui & Berbee (*Mol. Phylogen. Evol.* **39**: 587, 2006; phylogeny).

Helicoönites Kalgutkar & Sigler (1995), Fossil Fungi, anamorphic *Pezizomycotina*. 1 (Eocene), Canada. See Kalgutkar & Sigler (*MR* **99**: 519, 1995).

Helicopsis P. Karst. (1889) = Helicoma fide Linder (*Ann. Mo. bot. Gdn* **16**: 302, 1929).

Helicorhoidion S. Hughes (1958), anamorphic *Pezizomycotina*, Hso.0-≡ hH-P.10. 5, widespread. See Sutton (*Sydowia* **41**: 336, 1989; relationships), Hyde *et al.* (*MR* **103**: 1409, 1999; on palms).

Helicosingula P.S. van Wyk, Marasas, Baard & Knox-Dav. (1985), anamorphic *Pezizomycotina*, Hso.0-≡ hP.1. 1, S. Africa. See van Wyk *et al.* (*TBMS* **85**: 183, 1985), Crous *et al.* (*CBS Diversity Ser.* **2**, 2004).

Helicosporangium H. Karst. (1865) ? = Papulospora fide Kendrick & Carmichae *in* Ainsworth *et al.* (Eds) (*The Fungi* **4A**: 390, 1973).

helicospore, a non-septate or septate spore, with a single (usually elongated) axis curved through at least 180° but may describe one or more complete rotations, in two or three dimensions (cf. amerospore, scolecospore); any protuberances, other than setulae, ‹$^1/_4$ spore body length (cf. staurospore). See Goos (*Mycol.* **79**: 4, 1987; terminology). See also Anamorphic fungi.

Helicosporella Arnaud (1954) nom. inval., anamorphic *Pezizomycotina*, Hso.0-≡ hP.1. 1, Europe.

Helicosporiaceae Nann. (1934) = Tubeufiaceae.

Helicosporiates Kalgutkar & Sigler (1995), Fossil Fungi, anamorphic *Pezizomycotina*. 1 (Eocene), Canada. See Kalgutkar & Sigler (*MR* **99**: 520, 1995).

Helicosporidium Keilin (1921), Chlorococcaceae. 2 (in *Insecta* and *Acari*), widespread (tropical). See Wesier (*J. Protozool.* **17**: 436, 1970; ? *Ascomycota*), Lindegren & Hoffman (*J. Invert. Path.* **27**: 105, 1976; ? *Protozoa*), Avery & Undeen (*J. Invert. Path.* **49**: 246, 1987; effect on mosquitoes), Eriksson & Hawksworth (*SA* **8**: 68, 1989; posn), Tartar *et al.* (*Int. J. Syst. Evol. Microbiol.* **52**: 273, 2002; member of *Chlorophyta*).

Helicosporina Arnaud (1954) nom. inval., anamorphic *Pezizomycotina*.

Helicosporina G. Arnaud ex Rambelli (1960), anamorphic *Pezizomycotina*, Hso.0-≡ hP.10. 1, Italy. See Sutton (*in litt.*; ? Synonym of *Troposporella*).

Helicosporium Nees (1816), anamorphic *Tubeufia*, Hso.0-≡ hP.10. 20, widespread. See Goos (*Mycol.* **81**: 356, 1989; key), Schoch *et al.* (*Mycol.* **98**: 1041, 2006; phylogeny), Tsui & Berbee (*Mol. Phylogen. Evol.* **39**: 587, 2006; phylogeny), Tsui *et al.* (*Mycol.* **98**: 94, 2006; phylogeny).

Helicostilbe Höhn. (1902), anamorphic *Pezizomycotina*, Hsy.0-≡ hP.10. 1, Austria.

Helicostilbe Linder (1929) = Trochophora fide Moore (*Mycol.* **47**: 90, 1955).

Helicostylum Corda (1842), Mucoraceae. 2, widespread. See Upadhyay (*Mycol.* **65**: 735, 1973; key), Benny (*Mycol.* **87**: 253, 1995; key), Voigt & Wöstemeyer (*Gene* **270**: 113, 2001; phylogeny), Schmidt *et al.* (*Microbial Ecology* **56**: in press, 2008; coenocytic 'snow molds').

Helicothyrium I. Hino & Katum. (1961), anamorphic *Pezizomycotina*, Cpt.0-≡ hH.?. 1, Japan. See Hino & Katumoto (*Bulletin of the Faculty of Agriculture, Yamaguchi University* **11**: 11, 1961).

Helicotrichum Nees & T. Nees (1818) = Helicosporium fide Hughes (*CJB* **36**: 727, 1958).

Helicoubisia Lunghini & Rambelli (1979), anamorphic *Pezizomycotina*, Hso.0-≡ hP.10. 1, Ivory Coast. See Lunghini & Rambelli (*Micol. Ital.* **8**: 21, 1979), Pinnoi *et al.* (*Sydowia* **56**: 72, 2004).

Heliocephala V.G. Rao, K.A. Reddy & de Hoog (1984), anamorphic *Pezizomycotina*, Hso.≡ eP.1. 1, India; S. Africa. See Kuyper (*in litt.*), Decock *et al.* (*Mycol.* **90**: 330, 1998; S Africa), White *et al.* (*Stud. Mycol.* **45**: 95, 2000; India).

Heliocybe Redhead & Ginns (1985) = Neolentinus fide Rune (*MR* **98**: 543, 1994).

Heliomyces Lév. (1844) = Marasmius fide Singer (*Agaric. mod. Tax.*, 1951).

heliophilous, preferring direct sunlight. Cf. anheliophilous.

heliozooid, amoeba-like, but having well-marked ray-like pseudopodia.

Heliscella Marvanová (1980), anamorphic *Pezizomycotina*, Hso.1bH.15. 2 (aquatic), widespread. See Marvanová (*TBMS* **75**: 224, 1980).

Heliscina Marvanová (1980), anamorphic *Pezizomycotina*, Hso.1eH/1bH.19. 2 (aquatic), Czech Republic. See Marvanová (*TBMS* **75**: 227, 1980).

Heliscus Sacc. (1880), anamorphic *Nectria*, Hsp.≡ eH.15. 4 (aquatic), widespread. See Petersen (*Mycol.* **55**: 570, 1963; key), Baschien *et al.* (*Nova Hedwigia* **83**: 311, 2006; phylogeny).

Helmichia J.I.R. Larsson (1982), Microsporidia. 2. See Larsson (*Protistologica* **18**: 353, 1982).

Helminthascus Tranzschel (1898), ? Clavicipitaceae. 1 (on *Insecta*), former USSR.

Helminthocarpon Fée (1837), Graphidaceae (L). 2, tropical. See Awasthi & Joshi (*Norw. Jl Bot.* **29**: 165, 1979), Aptroot (*Myconet* **2**: 15, 1999).

Helminthocarponomyces Cif. & Tomas. (1953) nom. dub., ? Graphidaceae. See Lücking & Hawksworth (*Taxon* **56**: 1274, 2007).

helminthoid, worm-like in form; vermiform.

Helminthopeltis Sousa da Câmara (1950), ? Microthyriaceae. 1, Europe. See Sousa da Câmara (*Agron. lusit.* **12**: 102, 1950).

Helminthophana Peyr. (1873) = Arthrorhynchus fide Thaxter (*Proc. Amer. Acad. Arts & Sci.* **36**: 395, 1901).

Helminthophora Bonord. (1851), anamorphic *Pezizomycotina*, Hso.≡ eH.15. 1 (fungicolous), widespread (temperate). See de Hoog (*Persoonia* **10**: 55, 1978).

Helminthosphaeria Fuckel (1870), Helminthosphaeriaceae. Anamorph *Diplococcium*. 2 or 8 (on *Clavulina*, *Peniophora* and wood), widespread (north temperate). See Samuels *et al.* (*Mycol.* **89**: 141, 1997), Goh & Hyde (*Fungal Diversity* **1**: 65, 1998; anam.), Réblová (*Mycotaxon* **70**: 387, 1999), Réblová (*Sydowia* **51**: 223, 1999), Miller & Huhndorf (*MR* **108**: 26, 2004; phylogeny).

Helminthosphaeriaceae Samuels, Cand. & Magni (1997), Sordariomycetidae (inc. sed.). 8 gen. (+ 6 syn.), 163 spp.

Lit.: Samuels *et al.* (*Mycol.* **89**: 141, 1997), Goh & Hyde (*Fungal Diversity* **1**: 65, 1998), Goh & Hyde (*CJB* **76**: 1698, 1998), Réblová (*Sydowia* **51**: 223, 1999), Réblová (*Mycotaxon* **70**: 387, 1999), Wang & Sutton (*CJB* **76**: 1608, 1998), Huhndorf *et al.* (*Mycol.* **96**: 368, 2004), Miller & Huhndorf (*MR* **108**: 26, 2004).

helminthosporal, a terpenoid mycotoxin from the *Drechslera* anamorph of *Cochliobolus sativus* toxic to wheat and barley.

Helminthosporiaceae Corda (1837) = Massarinaceae.

Helminthosporiopsis Speg. (1880) nom. dub., anamorphic *Pezizomycotina*. See Seifert (*Sydowia* **45**: 103, 1993).

Helminthosporites Chitaley & M.T. Sheikh (1971), Fossil Fungi. 1 (Palaeocene), India.

Helminthosporites Pia (1927), Fossil Fungi. 1 (Carboniferous), Europe.

Helminthosporium Link (1809) nom. cons., anamorphic *Massarinaceae*, Hso.≡ eP.27. *c.* 35 (lignicolous), widespread. *Bipolaris*, *Drechslera*, *Exserohilum* were once placed within this genus. Molecular data suggests placement in the *Massarinaceae*; the link with *Splanchnonema* (*Pleomass.*) needs further study. See Ellis (*Mycol. Pap.* **82**, 1961; key 10 spp.), Luttrell (*Mycol.* **55**: 643, 1963; spore development; terminology for characters), Subramanian & Sekar (*Kavaka* **15**: 87, 1987; teleomorph), Olivier *et al.* (*Mycol.* **92**: 736, 2000; phylogeny), Errampalli *et al.* (*J. Microbiol. Meth.* **44**: 59, 2001; PCR detection), Eriksson & Hawksworth (*Mycol.* **95**: 426, 2003; phylogeny).

helminthosporoside, a host-specific toxin produced by *Drechslera sacchari* in sugarcane (Strobel & Steiner, *Physiol. Pl. Pathol.* **2**: 129, 1972).

Helmisporium Link (1809) nom. rej. prop. = Helminthosporium fide Hawksworth *et al.* (*Dictionary of the Fungi* edn 8, 1995).

Helocarpaceae Hafellner (1984), Lecanorales. 1 gen., 4 spp.

Lit.: Hafellner (*Beih. Nova Hedwigia* **79**: 241, 1984).

Helocarpon Th. Fr. (1860), Helocarpaceae (L). 4, widespread. See Borghesi & Fantini (*Bollettino del Circolo Micologico 'Giovanni Carini'* **41**: 29, 2001; Pacific Northwest), Andersen & Ekman (*MR* **109**: 21, 2005; phylogeny).

Helochora Sherwood (1979), ? Phyllachoraceae. 1 (from leaves of *Puya*), Chile. See Cannon (*Stud. Mycol.* **31**: 49, 1989), Pearce & Hyde (*Fungal Diversity* **6**: 83, 2001).

Helodiomyces F. Picard (1913), Ceratomycetaceae. 1, Europe; N. Africa. See Santamaría (*Fl. Mycol. Iberica* **5**, 2003; Iberian peninsula).

Helolachnum Torrend (1910) = Lachnum fide Nannfeldt (*Nova Acta R. Soc. Scient. upsal.*, 1932).

Helopodium Ach. ex Michx. (1803) = Cladonia fide Hawksworth *et al.* (*Dictionary of the Fungi* edn 8, 1995).

Helostroma Pat. (1902) = Microstroma fide Saccardo (*Michelia* **1**: 273, 1878).

Helote Hazsl. (1881) ≡ Microglossum Gillet fide Durand (*Annls mycol.* **6**: 387, 1908).

Helotiaceae Rehm (1886) nom. cons., Helotiales. 117 gen. (+ 73 syn.), 826 spp.

Lit.: Dennis (*Mycol. Pap.* **62**, 1956; UK), Korf *in* Ainsworth *et al.* (Eds) (*The Fungi* **4A**: 249, 1973), Korf (*Acta Mycol. Sin.* Suppl., 1973) The *Helotiaceae* as recognized here is still very varied in form despite removal of the *Leotiaceae* and transfer of *Encoelia* to the *Sclerotiniaceae*, and it is likely that it will be further divided in future.

Lit.:, Baral (*Z. Mykol.* **53**: 119, 1987), Zhuang (*Mycotaxon* **32**: 97, 1988; key 19 gen. *Encoelioideae*) See also *Leotiaceae*, Petrini *et al.* (*CJB* **67**: 2805, 1989), Verkley (*Persoonia* **15**: 405, 1994; asci), Verkley (*The Ascus Apical Apparatus in Leotiales* An Evaluation of Ultrastructural Characters as Phylogenetic Markers in the Families *Sclerotiniaceae*, *Leotiaceae* and *Geoglossaceae*: 209 pp., 1995), Korf *et al.* (*Taxon* **45**: 683, 1996; nomencl.), Triebel & Baral (*Sendtnera* **3**: 199, 1996), Döbbeler (*Biodiv. Cons.* **6**: 721, 1997), Gamundí & Romero (*Fl. criptog. Tierra del Fuego* **10**: 130 pp., 1998), Gamundí & Romero (*Fl. criptog. Tierra del Fuego* **10** no. 5, 1998), Monreal *et al.* (*CJB* **77**: 1580, 1999), Verkley (*Stud. Mycol.* **44**: 180 pp., 1999), Baral & Marson (*Micologia 2000*: 23, 2000), Vrålstad *et al.* (*New Phytol.* **155**: 131, 2002), Wang *et al.* (*Mycol.* **98**: 1065, 2006; phylogeny), Wang *et al.* (*Mol. Phylogen. Evol.* **41**: 295, 2006).

Helotiales Nannf. ex Korf & Lizoň (2000). Leotiomycetes. 10 fam., 501 gen., 3881 spp. Stromata usually absent, if present sclerotial. Ascomata apothecial, usually small, often brightly coloured, sessile or stipitate, cupulate or discoid, rarely convex, sometimes surrounded by conspicuous hairs; wall tissues often separated into 2-3 distinct layers with differing cell types. Interascal tissue of simple paraphyses, variously shaped, the apices sometimes swollen. Asci usually small, thin-walled, without separable wall layers, with an apical pore surrounded by a J+ or J- ring which is rather varied in form (14 types in Verkley, 1995). Ascospores usually small, simple or transversely septate, mostly hyaline, usually not quite longitudinally symmetrical, usually smooth. Anamorphs varied, not known from many, hyphomycetous or coelomycetous. Saprobes and plant parasites, few lichenized or lichenicolous.

Inoperculate discomycetes; cup fungi. The taxonomy of the order is unsettled, and molecular data are inadequate to elucidate relationships in many cases. Fams:

(1) **Ascocorticiaceae**
(2) **Dermateaceae**
(3) **Helotiaceae**
(4) **Hemiphacidiaceae**
(5) **Hyaloscyphaceae**
(6) **Loramycetaceae**
(7) **Phacidiaceae**
(8) **Rutstroemiaceae**
(9) **Sclerotiniaceae**
(10) **Vibrisseaceae**

Lit.: Arendholz (*Blattbewohnenden Ascomyceten aus der Ordnung der Helotiales*, 1979), Baral (*Z. Mykol.* **53**: 119, 1987; ascus ultrastr., *Mycotaxon* **44**: 333, 1992; microscopic preparation), Carpenter (*Mycol.* **80**: 127, 1988; ordinal nomencl.), Gamundí (*Fl. Cript. Tierra de Fuego* **10**(4), 1987), Gamundí & Romero (*Fl. Cript. Tierra de Fuego* **10**(5), 1998), Hennebert & Bellemère (1979; anamorphs), Huhtinen (*in* Hawksworth (Ed.), 1994: 295), Korf (*in* Ainsworth *et al.* (Eds), *The Fungi* **4A**: 249, 1973; keys gen.), Korf & Lizoň (*Mycotaxon* **75**: 501, 2000; nomencl.), Landvik *et al.* (*Mycoscience* **37**: 237, 1996; *Mycoscience* **39**: 49, 1998), Nannfeldt (1932), Pfister & Kimbrough (in McLaughlin *et al.* (Eds), *The Mycota* **7A**: 257, 2001), Spooner (*Bibl. Mycol.* **116**, 1987), Verkley (*Persoonia* **15**: 303, 1993; **15**: 405, 1994; *Hymenoscyphoideae* asci; *The ascus apical apparatus in Leotiales*, 1995, ultrastr.), Weber (*Bibl. Mycol.* **140**, 1992; reprod. system), see also under *Ascomycota*, *Discomycetes*, Macromycetes.

Helotidium Sacc. (1884) = Allophylaria fide Nannfeldt (*Nova Acta R. Soc. Scient. upsal.*, 1932).

Helotiella Sacc. (1884) nom. dub., Helotiales. 1, Europe. See Galán (*Riv. Micol.* **36**: 149, 1993).

Helotiopsis Höhn. (1910) = Pithyella fide Korf & Zhuang (*Mycotaxon* **29**: 1, 1987).

helotism, the physiologic relation of alga to fungus in a lichen (obsol.).

Helotium Pers. (1801) = Cudoniella fide Dennis (*Persoonia* **3**: 29, 1964; redisp. spp.), Holm (*TBMS* **67**: 333, 1976; nomencl.).

Helotium Tode (1790) nom. utique rej. = Hemimycena fide Kuyper (*in litt.*).

Helvella L. (1753), Helvellaceae. *c.* 52 (saddle fungi, false morels), widespread (north temperate). See Dissing (*Dansk bot. Ark.* **25** no. 1: 1, 1966), Harmaja (*Karstenia* **14**: 102, 1974; gen. concept), Weber (*Beih. Nova Hedwigia* **51**, 1975; W.N. Am.), Harmaja (*Karstenia* **17**: 45, 1977), Harmaja (*Karstenia* **19**: 33, 1979; cupulate spp.), Liu *et al.* (*Acta Mycol. Sin.* **4**: 208, 1985; key 16 Chinese spp.), Häffner (*Z. Mykol. Beih.*: 7, 1987; key 43 spp., illustr.), Abbott & Currah (*Mycotaxon* **33**: 229, 1988; key 16 spp. Alberta), Gibson & Kimbrough (*CJB* **66**: 771, 1988; ascosporogenesis), Calonge & Arroya (*Mycotaxon* **39**: 203, 1990; key 21 spp. Spain), Abbott & Currah (*Mycotaxon* **62**: 1, 1997; 5 subgen.), Landvik *et al.* (*Nordic Jl Bot.* **17**: 403, 1997), Landvik *et al.* (*Mycol.* **91**: 278, 1999; polyphyly), Wang & Chen (*Fungal Science* Taipei **17**: 11, 2002; Taiwan), Zhuang (*Mycotaxon* **90**: 35, 2004; China), Hansen & Pfister (*Mycol.* **98**: 1029, 2006; phylogeny).

Helvellaceae Fr. (1823), Pezizales. 6 gen. (+ 20 syn.), 63 spp.

Lit.: Dissing (*Dansk Bot. Ark.* **25**, 1966), Donadini (*BSMF* **102**: 373, 1986; characterization of mitochondrial DNA), Donadini (*Bull. trimest. Soc. mycol. Fr.* **102**: 373, 1986), Gibson & Kimbrough (*CJB* **66**: 771, 1988), Calonge & Arroyo (*Mycotaxon* **39**: 203, 1990), Kimbrough & Gibson (*Mycol.* **81**: 914, 1989), Kimbrough (*MR* **95**: 421, 1991), Zhang (*Mycotaxon* **42**: 155, 1991), Pegler *et al.* (*British truffles*, 1993), Kimbrough *et al.* (*Mycol.* **88**: 38, 1996), Abbott & Currah (*Mycotaxon* **62**: 1, 1997), Landvik *et al.* (*Nordic Jl Bot.* **17**: 403, 1997), O'Donnell *et al.* (*Mycol.* **89**: 48, 1997; phylogeny), Harrington *et al.* (*Mycol.* **91**: 41, 1999; phylogeny), Landvik *et al.* (*Mycol.* **91**: 278, 1999), Percudani *et al.* (*Mol. Phylogen. Evol.* **13**: 169, 1999; phylogeny), Hansen & Pfister (*Mycol.* **98**: 1029, 2006; phylogeny), Læssøe & Hansen (*MR* **111**: 1075, 2007; phylogeny).

Helvellella S. Imai (1932) = Pseudorhizina fide Eckblad (*Nytt Mag. Bot.* **15**: 1, 1968).

hemi- (prefix), half; in part. Cf. semi-.

Hemialysidium Hol.-Jech. (1992) nom. inval., anamorphic *Pezizomycotina*, Hso.?.?. 1, Europe. See Holubova-Jechova (*Abstracts, XI Congress of European Mycologists* Kew, England, 7-11 September 1992: 19, 1992).

hemiamyloidity, The red colouration when iodine solutions are applied to the hymenium of most *Ascomycetes* (Baral, *Mycotaxon* **29**: 399, 1987).

hemiangiocarpous (of a sporocarp), opening before quite mature.

Hemiasci (sensu Varitchak, *Botaniste* **25**: 370, 1933), *Dipodascus* + *Ascoidea*.

Hemiascomycetes. *Ascomycota* in which the asci are not produced in ascomata. In addition the thallus usually comprises poorly developed mycelium or is represented by separate cells; mainly included in *Saccharomycetales* here.

hemiascospore, ascospore of a hemiascus.

Hemiascosporiaceae L.R. Batra (1973) = Ascodesmidaceae.

Hemiascosporium L.R. Batra (1973) = Eleutherascus fide von Arx (*Persoonia* **9**: 393, 1977), Batra *in* Subramanian (Ed.) (*Taxonomy of fungi* **1**: 187, 1978).

hemiascus, the atypical multispored ascus of the *Hemiasci* (q.v.).

Hemibasidiomycetes. See *Pucciniales* and *Ustilaginales*.

Hemibeltrania Piroz. (1963), anamorphic *Pezizomycotina*, Hsy.0eP.10. 8, widespread (esp. tropical). See Castañeda Ruíz *et al.* (*MR* **102**: 930, 1998), Shin & Braun (*Mycotaxon* **67**: 317, 1998), Cooper (*N.Z. Jl Bot.* **43**: 323, 2005; New Zealand).

Hemicarpenteles A.K. Sarbhoy & Elphick (1968), Trichocomaceae. Anamorph *Aspergillus*. 2, widespread. See Kuraishi *et al.* (*NATO ASI Series A: Life Sciences* **185**: 407, 1990; ubiquinones), Chang *et al.* (*J. gen. appl. Microbiol.* Tokyo **37**: 289, 1991; DNA), Pitt *et al.* (*Integration of Modern Taxonomic Methods for Penicillium and Aspergillus Classification*: 9, 2000; accepted names), Tamura *et al.* (*Integration of Modern Taxonomic Methods for Penicillium and Aspergillus Classification*: 357, 2000; phylogeny), Udagawa & Uchiyama (*Mycoscience* **43**: 3, 2002; separation of *H. acanthosporus*).

hemicompatible, see compatible.

Hemicorynespora M.B. Ellis (1972), anamorphic *Chaetosphaeria*, Hso.0-1eH-P.25. 9, widespread (tropical). See Ellis (*Mycol. Pap.* **131**: 19, 1972), Mercado Sierra *et al.* (*Mycotaxon* **63**: 155, 1997; Mexico), Sivanesan & Chang (*MR* **101**: 845, 1997; key, teleomorph).

Hemicorynesporella Subram. (1992), anamorphic *Pezizomycotina*, Hso.≡ eP.24. 1, Solomon Islands. See Subramanian (*Proc. Indian natn Sci. Acad.* Part B. Biol. Sci. **58**: 189, 1992).

Hemicybe P. Karst. (1879) = Lentinellus fide Singer (*Agaric. mod. Tax.*, 1951).

Hemicyphe Corda (1831) nom. dub., ? Fungi.

Hemidiscia Lázaro Ibiza (1916) = Postia fide Donk (*Verh. K. ned. Akad. Wet.* tweede sect. **62**: 1, 1974).

Hemidothis Syd. & P. Syd. (1916), anamorphic *Pezizomycotina*, St.0fH.?. 1 or 2, America (tropical). See Petrak (*Annls mycol.* **27**: 374, 1929), Sydow (*Annls mycol.* **28**: 193, 1930).

hemifissitunicate, see ascus.

hemiform, see *Pucciniales* and Table 5.

Hemigaster Juel (1895), Hemigasteraceae. 1, Sweden. See Singer (*Agaric. mod. Tax.* 4th ed: 845, 1986).

Hemigasteraceae Gäum. & C.W. Dodge (1928), Agaricales. 1 gen., 1 spp.

Lit.: Singer (*Agaric. mod. Tax.* 4th ed, 1986), Reijnders (*MR* **104**: 900, 2000).

Hemiglossum Pat. (1890), Helotiales. 2, China; Japan. See Imai (*J. Fac. Agric. Hokkaido Imp. Univ.* **45**: 155, 1941), Zhuang (*Mycotaxon* **32**: 97, 1988; posn).

Hemigrapha (Müll. Arg.) R. Sant. ex D. Hawksw. (1975), ? Parmulariaceae. 9 (on lichens), widespread. See Matzer (*Mycol. Pap.* **171**, 1996), Diederich & Wedin (*Nordic Jl Bot.* **20**: 203, 2000; on *Peltigerales*).

Hemigyalecta, see *Semigyalecta*.

Hemihysteriaceae Sacc. & Traverso (1907) = Asterinaceae.

Hemileccinum Šutara (2008) = Boletus Fr. fide Kuyper (*in litt.*).

Hemileia Berk. & Broome (1869), Pucciniales. *c.* 40 (on angiosperms (II, III only)), widespread (esp. tropical). *H. vastatrix* (coffee rust). See Gopalkrishnan (*Mycol.* **43**: 271, 1951; morphology), Stevenson & Bean (*Spec. Publ.* Div. Mycol. Dis. Survey, USDA **4**, 1953; annotated bibliogr.), Rajendren (*Mycol.* **59**: 918, 1967; telial urediniospores ($II^{(III)}$) in which meiosis occurs in *H. vastatrix*), Rajendren (*Mycopath. Mycol. Appl.* **47**: 81, 1972; morphology), Fulton (Ed.) (*Coffee rust in the Americas*, 1984), Anon. (*IMI Descr. Fungi Bact.* **162**, 2004), Gouveia *et al.* (*Mycol.* **97**: 396, 2005; RAPD), Ritschel (*Biblthca Mycol.* **200**: 132 pp., 2005; monogr.).

Hemileiopsis Racib. (1900) = Hemileia fide Dietel (*Nat. Pflanzenfam.* **6**, 1928) but see, Ono *et al.* (*Mycol.* **78**: 253, 1986; suggests this might be the nomenclaturally correct name for *Hemileia*).

Hemilichenes, Lichens of uncertain systematic position because sporocarps are unknown (obsol.); usually = *Phycolichenes*, *Deuterolichenes*.

Hemimycena Singer (1938), Mycenaceae. *c.* 50, widespread. Perhaps *Marasmiaceae*. See Courtecuisse (*Agarica* **6**: 108, 1985; France), Gröger (*Boletus* **12**: 59, 1988; *Hemimycena cucullata* in Germany), Desjardin (*Mycotaxon* **42**: 187, 1991; USA), Lepp (*Australasian Mycologist* **23**: 105, 2004; *Hemimycena tortuosa* in Australia), Moreau *et al.* (*Mycotaxon* **91**: 323, 2005; new cistophilous sp.).

Hemimyriangium J. Reid & Piroz. (1966) = Molleriella fide von Arx & Müller (*Stud. Mycol.* **9**, 1975).

hemiparasite, a facultative parasite.

Hemiphacidiaceae Korf (1962), Helotiales. 9 gen. (+ 6 syn.), 26 spp.

Lit.: Korf (*Mycol.* **54**: 481, 1962), Reid & Cain (*Mycol.* **54**: 481, 1962), Sherwood-Pike *et al.* (*CJB* **64**: 1849, 1986), Minter (*Shoot and Foliage Diseases in Forest Trees* Proceedings of a Joint Meeting of the Working Parties: Canker & Shoot Blight of Conifers, Foliage Diseases: 65, 1995), Gernandt *et al.* (*Mycol.* **89**: 735, 1997), Gernandt *et al.* (*Mycol.* **93**: 915, 2001), Stone & Gernandt (*Mycotaxon* **91**: 115, 2005), Wang *et al.* (*Mycol.* **98**: 1065, 2006; phylogeny).

Hemiphacidium Korf (1962) = Sarcotrochila fide Reid & Cain (*Mycol.* **54**: 481, 1962), Stone & Gernandt (*Mycotaxon* **91**: 115, 2005; phylogeny, placement), Wang *et al.* (*Mycol.* **98**: 1065, 2006; phylogeny).

Hemipholiota (Singer) Bon (1986) = Pholiota fide Kuyper (*in litt.*).

Hemisartorya J.N. Rai & H.J. Chowdhery (1976) = Neosartorya fide von Arx (*Gen. Fungi Sporul. Cult.* Edn 3, 1981).

Hemiscyphe, see *Hemicyphe*.

Hemisphaeria Klotzsch (1843) nom. rej. prop. ≡ Daldinia.

Hemisphaeriales Theiss. (1913). Obsol., a poorly defined assemblage including *Microthyriales* and several other groups with flattened ascomata.

Hemisphaeropsis Petr. (1947), anamorphic *Pezizomy-*

cotina, Cpt.1eP.?. 1, USA.

Hemispora Vuill. (1906) = Wallemia fide Carmichael *et al.* (*Genera of Hyphomycetes*, 1980).

Hemisporaceae M. Ota (1928) = Wallemiaceae.

hemispore (esp. of dermatophytes), (1) a cell at the end of a filament, which later becomes by division deuteroconidia; protoconidium (after Vuillemin); (2) one of the two cells produced by a primary trans-septum in an ascospore (Eriksson, *Ark. Bot.* II **6**: 339, 1967), see septum.

Hemistropharia Jacobsson & E. Larss. (2007), Cortinariaceae. 1, USA. See Jacobsson & Larsson (*Mycotaxon* **102**: 235, 2007).

Hemisynnema Subram. (1995), anamorphic *Pezizomycotina*, Hsy.≡ eP.1. 1, Malaysia. See Subramanian (*Kavaka* **20/21**: 58, 1992).

Hemithecium Trevis. (1853), Graphidaceae (L). 1, pantropical. See Nakanishi *et al.* (*Bull. natn. Sci. Mus.* Tokyo, B **29**: 83, 2003; Japan), Adawadkar & Makhija (*Mycotaxon* **92**: 387, 2005; India), Archer (*Telopea* **11**: 59, 2005; Australia), Makhija *et al.* (*Mycotaxon* **93**: 365, 2005; India), Archer (*Biblthca Lichenol.* **94**, 2006; revision), Archer (*Systematics & Biodiversity* **5**: 9, 2007; Solomon Is).

Hemmesomyces Gilb. & Nakasone (2003), Corticiaceae. 1, Hawaii. See Gilbertson & Nakasone (*Mycol.* **95**: 467, 2003).

Hendersonia Berk. (1841) nom. rej. = Stagonospora fide Swart & Walker (*TBMS* **90**: 633, 1988; redisposition spp. on *Eucalyptus*).

Hendersonia Sacc. (1884) ≡ Hendersonia Berk.

Hendersoniella (Sacc.) Sacc. (1902) ≡ Hendersoniella Tassi fide Sutton (*Mycol. Pap.* **141**, 1977).

Hendersoniella Tassi (1900), anamorphic *Pezizomycotina*, Cpd.?.?. 1, Italy. See Sutton (*Mycol. Pap.* **141**, 1977).

Hendersonina E.J. Butler (1913), anamorphic *Pezizomycotina*, St.0-1eP.15. 1, Asia.

Hendersoniopsis Höhn. (1918), anamorphic *Pezizomycotina*, St.≡ eP.15. 1, widespread (temperate).

Hendersoniopsis Woron. (1922) = Stenocarpella fide Sutton (*Mycol. Pap.* **141**, 1977).

Hendersonula Speg. (1880), anamorphic *Botryosphaeriales*, St.≡ eP.19. 5 (on fungi), widespread (tropical). See Sutton & Dyko (*MR* **93**: 466, 1989; revision, key).

Hendersonulina Petr. (1951), anamorphic *Pezizomycotina*, St.≡ eP.?. 6, widespread. See Petrak (*Sydowia* **5**: 421, 1951).

Hendersonulina Tassi (1902), anamorphic *Pezizomycotina*, Cpd.≡ eP.?. 43, widespread. See Sutton (*Mycol. Pap.* **141**, 1977).

Hendrickxia P.A. Duvign. (1942) = Everniopsis fide Santesson (*Svensk bot. Tidskr.* **43**: 547, 1949).

Henicospora P.M. Kirk & B. Sutton (1980), anamorphic *Pezizomycotina*, Hso.≡ eP.2. 5, widespread. See Kirk & Sutton (*TBMS* **75**: 249, 1980), Kuthubutheen & Nawawi (*MR* **98**: 677, 1994; Malaysia).

Hennebertia M. Morelet (1969), anamorphic *Pezizomycotina*, Hso.0eP.10+19. 1, Europe. See Hennebert (*TBMS* **51**: 749, 1968).

Hennenia Buriticá (1995), Pucciniales. 1 (on *Annona* (*Annonaceae*)), Colombia. See Buriticá C. (*Revta Acad. colomb. cienc. exact. fis. nat.* **19**: 465, 1995) Cf. *Esqualia*.

Hennings (Paul Christoph; 1841-1908; Germany). Self-taught, beginning as a postal worker; worked at Kiel Botanic Garden (1860-1864, 1874-1879); assistant (1880-1891) then Kustos (1891-1902) then Professor (1902-1908), Berlin Botanic Garden; Royal Professor (1902-1908). Described over 120 fungal genera and many species mainly from tropical material (particularly Africa, but also the Amazon basin, India, Japan and New Guinea), sent to Berlin by contemporary collectors; also an amateur poet. Collections mostly in Berlin (**B**); Batista *et al.*, Fungi Paraenses, *Publicaçoes. Instituto de Micologia da Universidade do Recife* **506**, 1966, listed Hennings' specimens in the Emílio Goeldi Museum, Pará. *Publs.* Fungi. In O. Warburg [ed.] *Monsunia. Beiträge zur Kenntnis der Vegetation des Süd- und Östasiatischen Monsungebietes* (1899, published 1900); also numerous papers published in *Engler's Botanisches Jahrbücher* and *Hedwigia* from *c.* 1891 onwards. (*Biogs, obits etc.* Grummann (1974: 18); Lindau (*Hedwigia* **48**: 1, 1909; bibliography); Perkins (*Botanical Gazette* **47**: 239, 1909); Stafleu & Cowans (*TL-2* **2**: 157, 1979).

Henningsia Möller (1895), Meripilaceae. 1, widespread (neotropical). See Ryvarden (*Gen. Polyp.*: 158, 1991), Corner (*Mycologist* **9**: 127, 1995; close to *Rigidoporus*).

Henningsiella Rehm (1895), Schizothyriaceae. 2, S. America.

Henningsina Möller (1901) = Phylacia fide Dennis (*Kew Bull.* **12**: 297, 1959), Müller & von Arx *in* Ainsworth *et al.* (Eds) (*The Fungi* **4A**: 87, 1973), Læssøe (*SA* **13**: 43, 1994; ? synonym of *Xylaria*).

Henningsomyces Kuntze (1898), Marasmiaceae. *c.* 20, widespread. See Reid (*Persoonia* **3**: 118, 1964), Agerer (*Persoonia* **7**: 389, 1973).

Henningsomyces Sacc. (1905) [non *Henningsomyces* Kuntze 1898] ≡ Dysrhynchis.

Henrica B. de Lesd. (1921), Verrucariaceae (L). 3, Europe; N. America. See Navarro-Rosinés & Hladun (*Cryptog. Bryol.-Lichénol.* **13**: 125, 1992), Breuss (*Bryologist* **105**: 398, 2002; N America).

Henriquesia Pass. & Thüm. (1879) [non *Henriquesia* Spruce ex Benth. 1854, *Rubiaceae*] ≡ Delpinoina.

Hepataria Raf. (1808) ? = Tremella Pers. fide Merrill (*Index Rafinesq.*: 69, 1949).

hepatic (1) concerning the liver; (2) a liverwort.

hepaticolous, growing on liverworts (*Hepaticae*); see Bryophilous fungi.

Heppia Nägeli ex A. Massal. (1854), Lichinaceae (L). 7, widespread (esp. desert or dry areas). See Wetmore (*Ann. Mo. bot. Gdn* **57**: 158, 1971; N. Am.), Swinscow & Krog (*Norw. Jl Bot.* **26**: 213, 1979; E. Afr.), Filson (*Muelleria* **6**: 495, 1988; Australia), Egea (*Biblthca Lichenol.* **31**, 1989; W. Eur. & N. Afr.), Henssen (*Acta Bot. Fenn.* **15**: 57, 1994; key), Schultz & Büdel (*Lichenologist* **34**: 39, 2002; key), Schultz & Büdel (*Lichenologist* **35**: 151, 2003; posn).

Heppiaceae Zahlbr. (1906) = Lichinaceae.

Lit.: Büdel (*Biblthca Lichenol.* **23**: 105 pp., 1987), Upreti & Büdel (*J. Hattori bot. Lab.* **68**: 279, 1990), Henssen (*Acta Bot. Fenn.* **150**: 57, 1994), Schultz *et al.* (*Pl. Biol.* **3**: 116, 2001), Schultz & Büdel (*Lichenologist* **35**: 151, 2003; phylogeny), Jørgensen (*Nordic Lichen Flora* **3**: Cyanolichens: 43, 2007).

Heppiomyces Cif. & Tomas. (1953) ≡ Heppia.

Heppsora D.D. Awasthi & Kr.P. Singh (1977), Tephromelataceae (L). 1, India; Middle East. See Hafellner (*Beih. Nova Hedwigia* **79**: 241, 1989), Poelt & Grube (*Nova Hedwigia* **57**: 1, 1993; status), Mies & Schultz (*Biblthca Lichenol.* **88**: 433, 2004;

Yemen).

Heptameria Rehm & Thüm. (1879), Dothideomycetes. 2, Europe. See Lucas & Sutton (*TBMS* **57**: 283, 1971).

Heptasporium Bref. (1908) = Sistotrema Fr. fide Rogers (*Mycol.* **36**: 70, 1944).

Heptaster Cif., Bat. & Nascim. (1956), anamorphic *Capnodiales*, Hsy.0bP.?. 2, Brazil. See Hughes (*Mycol.* **68**: 810, 1976; ? = *Tripospermum*).

Heraldoa Bat. (1959) = Lembosia fide Müller & von Arx (*Beitr. Kryptfl. Schweiz* **11** no. 2, 1962).

Herbampulla Scheuer & Nograsek (1993), ? Magnaporthaceae. 1 (on grasses), Austria. See Scheuer & Nograsek (*Mycotaxon* **47**: 415, 1993).

herbarium (pl. **-ia**), (1) a collection of dried plants; (2) the place in which such a collection is stored. Often also used for dried Reference collections (q.v.) of fungi, especially when curated along with plant specimens.

herbarium beetle, *Stegobium paniceum*, may eat the spores of certain fungi, esp. *Lycoperdon* and smuts. See Gordon (*TBMS* **21**: 193, 1938), Bridson & Forman (*The herbarium handbook*, rev. edn, 1992; control), Pinniger (*Biodet. Abstr.* **5**: 125, 1991; control).

herbicolous, living on herbs.

Hercospora Fr. (1825), Diaporthales. Anamorph *Rabenhorstia*. 2 (from bark), Europe. See Petrak (*Annls mycol.* **36**: 44, 1938), Müller & von Arx (*Beitr. Kryptfl. Schweiz* **11** no. 2, 1962).

Herculea Fr. (1823) = Podaxis fide Kuyper (*in litt.*).

Hericiaceae Donk (1964), Russulales. 3 gen. (+ 9 syn.), 12 spp.

Lit.: Donk (*Persoonia* **3**: 199, 1964), Ginns (*Windahlia* **16**: 35, 1986), Koski-Kotiranta & Niemelä (*Karstenia* **27**: 43, 1987), Stalpers (*Stud. Mycol.* **40**: 185 pp., 1996), Hibbett & Donoghue (*Syst. Biol.* **50**: 215, 2001), Larsson & Larsson (*Mycol.* **95**: 1037, 2003).

Hericiales = Russulales.

Hericium Fr. (1825) ≡ Hericium Pers. fide Hall & Stuntz (*Mycol.* **63**: 1103, 1971; subgen. key).

Hericium Pers. (1794), Hericiaceae. 5, widespread (north temperate). See Stalpers (*Stud. Mycol.* **40**: 79, 1996; key), Arnolds (*Coolia* **46** 3, Suppl.: 96 pp., 2003; The Netherlands and Belgium), Bohlin (*MR* **108**: 3, 2004; conservation status).

Hericius Juss. (1789) ≡ Hericium Pers.

Heringia Schwein. ex Berk. & Curtis (1853) nom. nud., ? Fungi.

Hermanniasporidium Thor (1930) nom. dub., ? Fungi. 1, Svalbard.

Hermatomyces Speg. (1910), anamorphic *Pezizomycotina*, Hsp.#eP.1. 3, widespread (tropical). See Hughes (*Mycol. Pap.* **50**: 100, 1953).

Herminia R. Hilber (1979) = Eosphaeria fide Petrini *et al.* (*TBMS* **82**: 554, 1984), Barr (*Mycotaxon* **39**: 43, 1990; status), Hilber & Hilber (*The genus Lasiosphaeria and Allied Taxa*, 2002).

Herpobasidium Lind (1908), Eocronartiaceae. Anamorph *Glomopsis*. 4, widespread. See Donk (*Persoonia* **4**: 158, 1966), Donk (*Persoonia* **4**: 214, 1966), Oberwinkler & Bandoni (*TBMS* **83**: 639, 1984), Bauer & Oberwinkler (*CJB* **72**: 1229, 1993; ultrastr., posn).

Herpocladiella J. Schröt. (1894) nom. dub., ? Mucorales. See Hesseltine (*Mycol.* **47**: 344, 1955).

Herpocladium J. Schröt. (1886) [non *Herpocladium* Mitt. 1873, *Hepaticae*] ≡ Herpocladiella.

Herpomyces Thaxt. (1903), Herpomycetaceae. 25 (on cockroaches), widespread. See Tavares (*Mycol.* **57**: 104, 1965; development), Tavares (*Mycotaxon* **11**: 485, 1980), Santamaría (*Fl. Mycol. Iberica* **5**, 2003; Iberian peninsula).

Herpomycetaceae I.I. Tav. (1981), Laboulbeniales. 1 gen., 25 spp.

Lit.: Majewski & Sugiyama (*Trans. Mycol. Soc. Japan* **26**: 295, 1985), Tavares (*Mycol. Mem.* **9**: 627 pp., 1985), Majewski (*Trans. Mycol. Soc. Japan* **29**: 33, 1988), Santamaria *et al.* (*Treb. Inst. Bot. Barcelona* **14**: 1, 1991).

Herposira Syd. (1938), anamorphic *Pezizomycotina*, Hsy/Cpd.0eP.?. 1, Australia.

Herpothallaceae Tomas. ex Tomas. (1950) = Arthoniaceae.

Herpothallon Tobler (1937) = Cryptothecia fide Thor (*Bryologist* **94**: 278, 1991).

Herpothallonomyces Cif. & Tomas. (1954) ≡ Herpothallon.

Herpothrix Clem. (1909) ? = Chaetomastia fide Hawksworth *et al.* (*Dictionary of the Fungi* edn 8, 1995).

Herpotrichia Fuckel (1868), Pleosporales. Anamorph *Pyrenochaeta*. 28, widespread. *H. juniperi* (brown felt blight, snow moulds, of conifers in cold regions). See Sivanesan (*Mycol. Pap.* **127**, 1972; key), Barr (*Mycotaxon* **20**: 1, 1984; key 4 spp., gen. concept), von Arx & Müller (*Sydowia* **37**: 6, 1984), Hyde & Aptroot (*Nova Hedwigia* **66**: 247, 1998), Silva-Hanlin & Hanlin (*MR* **103**: 153, 1999; phylogeny), Chen & Hsieh (*Sydowia* **56**: 24, 2004; Taiwan), Schoch *et al.* (*Mycol.* **98**: 1041, 2006; phylogeny).

Herpotrichiella Petr. (1914) = Capronia fide Müller *et al.* (*TBMS* **88**: 63, 1987).

Herpotrichiellaceae Munk (1953), Chaetothyriales. 13 gen. (+ 22 syn.), 84 spp.

Lit.: Barr (*Rhodora* **78**: 53, 1976), Müller *et al.* (*TBMS* **88**: 63, 1987), Janex-Favre (*Cryptog. Mycol.* **9**: 133, 1988; ontogeny), Braun & Feiler (*Microbiol. Res.* **150**: 81, 1995), Untereiner *et al.* (*MR* **99**: 897, 1995; molec. taxonomy), Untereiner *et al.* (*MR* **99**: 897, 1995), Haase *et al.* (*J. Clin. Microbiol.* **34**: 2049, 1996), Uijthof & Hoog (*Culture Collections to Improve the Quality of Life, Proceedings of the Eighth International Congress for Culture Collections, Veldhoven, The Netherlands, 25-29 August 1996*: 389, 1996), Untereiner (*Mycol.* **89**: 120, 1997), Hoog *et al.* (*Medical Mycology* **36** Suppl. 1: 52, 1998), Au *et al.* (*Mycol.* **91**: 326, 1999), de Hoog (*Stud. Mycol.* **43**, 1999; ecology, phylogeny), Gerrits van den Ende & Hoog (*Stud. Mycol.* **43**: 151, 1999), Haase *et al.* (*Stud. Mycol.* **43**: 80, 1999), Hoog *et al.* (*Stud. Mycol.* **43**: 133, 1999), McKemy *et al.* (*Mycol.* **91**: 200, 1999), Rogers *et al.* (*Stud. Mycol.* **43**: 122, 1999), Untereiner & Naveau (*Mycol.* **91**: 67, 1999), Gams (*Stud. Mycol.* **45**: 192, 2000), Untereiner (*Stud. Mycol.* **45**: 141, 2000), Abliz *et al.* (*FEMS Immunol. Med. Microbiol.* **40**: 41, 2004).

Herpotrichiopsis Höhn. (1914) = Pyrenochaeta fide Clements & Shear (*Gen. Fung.*, 1931).

Herpotrichum Fr. [not traced] nom. dub. = Protonema fide Fries (*Syst. orb. veg.*, 1825) based on mycelium fide, Fries (*Summa veg. Scand.*, 1849).

Herreromyces R.F. Castañeda & W.B. Kendr. (1991), anamorphic *Pezizomycotina*, Hso.0eH.15. 1, Cuba. See Castañeda Ruiz & Kendrick (*Univ. Waterloo Biol. Ser.* **35**: 60, 1991).

Herteliana P. James (1980), Ramalinaceae (L). 4,

widespread. See James (*Lichenologist* **12**: 106, 1980).

Hertelidea Printzen & Kantvilas (2004), ? Stereocaulaceae (L). 4, widespread. See Printzen & Kantvilas (*Biblthca Lichenol.* **88**: 541, 2004), Kantvilas & Elix (*Australasian Lichenology* **59**: 30, 2006; Australia).

Hertella Henssen (1985), Placynthiaceae (L). 2, southern hemisphere. See Henssen (*Mycotaxon* **22**: 382, 1985), Henssen (*Biblthca Lichenol.* **88**: 195, 2004).

Hesperomyces Thaxt. (1891), Laboulbeniaceae. 5, widespread. See Weir & Beakes (*Mycol.* **88**: 677, 1996; ontogeny), Weir & Blackwell (*Mycol.* **93**: 802, 2001; DNA extraction), Santamaría (*Fl. Mycol. Iberica* **5**, 2003; Iberian peninsula).

Hessea Ormières & Sprague (1973), Microsporidia. 1.

Hesseltinella H.P. Upadhyay (1970), Cunninghamellaceae. 1, Brazil. See Upadhyay (*Persoonia* **6**: 111, 1970), Stuart & Young (*TBMS* **89**: 392, 1987; morphology), Benny & Khan (*Scanning Microsc.* **2**, 1988), Benny & Samson (*Mem. N. Y. bot. Gdn* **49**: 11, 1989), Benny & Benjamin (*Mycol.* **83**: 713, 1991), Voigt & Wöstemeyer (*Gene* **270**: 113, 2001; phylogeny), White *et al.* (*Mycol.* **98**: 872, 2006; phylogeny).

Heterina Nyl. (1858) = Heppia fide Hawksworth *et al.* (*Dictionary of the Fungi* edn 8, 1995).

hetero- (prefix), other; not normal; different.

Heteroacanthella Oberw. (1990), Auriculariales. Anamorph *Acanthellorhiza*. 2, Taiwan; USA; British Isles. See Duhem & Trichiès (*Bulletin, Société Mycologique du Limousin* **121**: 119, 2005; France).

Heteroacanthellaceae P. Roberts (1998) = Ceratobasidiaceae.

Heterobasidiaceae Jülich (1982) = Bondarzewiaceae.

Heterobasidiomycetes R.T. Moore (1980), see *Basidiomycota*.

Heterobasidion Bref. (1888), Bondarzewiaceae. Anamorph *Spiniger*. 6, widespread (temperate). *H. annosum* (syn. *Fomes annosus*), serious root rot of conifers. See Stalpers (*Stud. Mycol.* **40**: 79, 1996; key), Bahnweg *et al.* (*J. Phytopath.* **150**: 382, 2002; PCR detection), Dai *et al.* (*MR* **106**: 1435, 2002; sexuality and intersterility in *Heterobasidion insulare* complex), Johannesson & Stenlid (*Mol. Phylogen. Evol.* **29**: 94, 2003; phylogeography), Ota *et al.* (*Mycol.* **98**: 717, 2006; phylogeny Japanese species).

Heterobasidium Massee (1889) nom. conf., Agaricomycetes.

heterobasidium, see basidium.

Heterobotrys Sacc. (1880) = Seuratia fide Meeker (*CJB* **53**: 2462, 1975).

Heterocarpon Müll. Arg. (1885), Verrucariaceae. 1, USA. See Harada (*SA* **10**: 1, 1991).

Heterocephalacria Berthier (1980) = Syzygospora fide Ginns (*Mycol.* **78**: 619, 1986).

Heterocephalum Thaxt. (1903), anamorphic *Pezizomycotina*, Hsy.0eH.15. 1, widespread (tropical). See Onofri *et al.* (*TBMS* **87**: 551, 1987; SEM), Persiani & Maggi (*TBMS* **87**: 631, 1986; Cote d'Ivoire), Sharma *et al.* (*Journal of Mycology and Plant Pathology* **34**: 69, 2004; India).

Heteroceras Sacc. (1915) ≡ Neoheteroceras fide Nag Raj (*Coelomycetous Anamorphs with Appendage-bearing Conidia*, 1993).

Heterochaete Pat. (1892), Auriculariaceae. *c.* 40, widespread (esp. tropical). See Bodman (*Lloydia* **15**: 193, 1952), Roberts (*Mycotaxon* **87**: 25, 2003; Venezuela), Roberts (*Mycotaxon* **96**: 83, 2006; Jamaica).

Heterochaetella (Bourdot) Bourdot & Galzin (1928), Auriculariales. *c.* 2, widespread. See Luck-Allen (*CJB* **38**: 559, 1960), Roberts (*Mycotaxon* **69**: 209, 1998), Weiss & Oberwinkler (*MR* **105**: 403, 2001; phylogeny).

Heterochlamys Pat. (1895) [non *Heterochlamys* Turcz. 1843, *Euphorbiaceae*] ≡ Gilletiella.

Heteroconidium Sawada (1944), anamorphic *Pezizomycotina*, Hsy.≡ eH+0eH.?. 1 (dimorphic), Taiwan.

Heteroconium Petr. (1949), anamorphic *Antennulariella*, Hso.≡ eP.3. 11, S. America; S. Africa. See Ellis (*More Dematiaceous Hyphomycetes*, 1976), Morgan-Jones (*Mycotaxon* **4**: 498, 1976), Chaudhary *et al.* (*MR* **95**: 1070, 1991; India), Castañeda Ruíz *et al.* (*Mycotaxon* **71**: 295, 1999), Hughes & Crane (*Mycol.* **98**: 141, 2006), Hughes (*Mycol.* **99**: 628, 2007), Narisawa *et al.* (*Mycoscience* **48**: 274, 2007; root endophyte), Usuki & Narisawa (*Mycol.* **99**: 175, 2007; symbiosis with *Cruciferae*).

Heterocyphelium Vain. (1927), Lecanorales (L). 1, widespread. See Tibell (*Beih. Nova Hedwigia* **79**: 597, 1984), Tibell (*Symb. bot. upsal.* **27** no. 1, 1987), Tibell (*Fl. Neotrop.* Monogr. **69**: 78 pp., 1996).

heterocytic, see astatocoenocytic.

Heterodea Nyl. (1868), Cladoniaceae (L). 2, Australia. See Filson (*Lichenologist* **10**: 13, 1978; key), Wedin *et al.* (*Lichenologist* **32**: 171, 2000; phylogeny), Miądlikowska *et al.* (*Mycol.* **98**: 1088, 2006; phylogeny), Zhou *et al.* (*J. Hattori bot. Lab.* **100**: 871, 2006; phylogeny).

Heterodeaceae Filson (1978) = Cladoniaceae.
Lit.: Rambold (*Cryptog. Bot.* **5**: 111, 1995), Wedin *et al.* (*Lichenologist* **32**: 171, 2000).

Heterodermia Trevis. (1868), Physciaceae (L). *c.* 80, widespread (esp. tropical). See Swinscow & Krog (*Lichenologist* **8**: 103, 1976; E. Afr. spp.), Kashiwadani *et al.* (*Bull. natn. Sci. Mus.* Tokyo, B **16**: 147, 1990; Peru), Trass (*Folia Cryptog. Estonica* **29**: 2, 1992; tab. key), Esslinger & Bratt *in* Glenn *et al.* (Eds) (*Lichenogr. Thomsoniana*: 25, 1998; N. Am., distr.), Kurokawa (*Folia Cryptog. Estonica* **32**: 21, 1998; checklist), Moberg & Nash (*Bryologist* **102**: 1, 1999; Sonoran Desert), Lohtander *et al.* (*Mycol.* **92**: 728, 2000; Scandinavia), Chen & Wang (*Mycotaxon* **77**: 107, 2001; China), Grube & Arup (*Lichenologist* **33**: 63, 2001; morphology, phylogeny), Trass (*Folia cryptog. Estonica* **37**: 93, 2000; Russia etc.), Helms *et al.* (*Mycol.* **95**: 1078, 2003; phylogeny), Moberg (*Biblthca Lichenol.* **88**: 453, 2004; Europe), Miądlikowska *et al.* (*Mycol.* **98**: 1088, 2006; phylogeny), Lücking *et al.* (*Systematics & Biodiversity* **6**: 31, 2008; Costa Rica).

Heterodoassansia Vánky (1993), Doassansiaceae. 7 (on aquatic plants), widespread. See Vánky (*Mycotaxon* **48**: 28, 1993).

Heterodothis Syd. & P. Syd. (1914) = Strigula fide Hawksworth *et al.* (*Dictionary of the Fungi* edn 8, 1995).

heteroecious (n., **heteroecism**), undergoing different parasitic stages on two unlike hosts, as in the *Pucciniales* (q.v.).

Heteroepichloë E. Tanaka, C. Tanaka, Gafur & Tsuda (2002), Clavicipitaceae. 2, Japan; Java. See Tanaka *et al.* (*Mycoscience* **43**: 92, 2002).

heterogametes, gametes of different form.

heterogamy, the copulation of heterogametes; cf. isogamy.

Heterogastridiaceae Oberw. & R. Bauer (1990), Heterogastridiales. 4 gen., 7 spp.

Lit.: Seeler (*Farlowia* **1**: 119, 1943), Bandoni & Oberwinkler (*CJB* **59**: 1613, 1981), Oberwinkler *et al.* (*Mycol.* **82**: 48, 1990), Swann *et al.* (*Mycol.* **91**: 51, 1999), Scorzetti *et al.* (*FEMS Yeast Res.* **2**: 495, 2002).

Heterogastridiales Oberw. & R. Bauer (1990). Microbotryomycetes. 1 fam., 4 gen., 7 spp. Fam.:

Heterogastridiaceae

For *Lit.* see under fam.

Heterogastridium Oberw. & R. Bauer (1990), Heterogastridiaceae. Anamorph *Hyalopycnis*. 1, widespread (north temperate). See Oberwinkler & Bauer (*Mycol.* **82**: 57, 1990), Bauer *et al.* (*Mycol. Progr.* **5**: 41, 2006).

Heterographa Fée (1824) = Polymorphum fide Hawksworth & Punithalingam (*TBMS* **60**: 501, 1973).

heterokaryosis (adj. **heterokaryotic**), (1) (of cells), the condition of having two or more genetically different nuclei, sometimes as a result of anastomosis, cf. dikaryotic; (2) (of mycelia), being made up of heterokaryotic cells; see Parmeter *et al.* (*Ann. Rev. Phytopath.* **1**: 51, 1963; in plant pathogenic fungi), Caten & Jinks (*TBMS* **49**: 81, 1966; occurrence in nature).

heterokaryotic (1) having two or more slightly (‹ 5%) genetically different nuclei in common cytoplasm (fungi); (2) showing nuclear dimorphism (protists).

heterokaryotise (of rusts and pyrenomycetes), fusion of haploid structures of opposite sex which does not give a conjugate arrangement of the nuclei, cf. diploidization.

heterokont (1) condition whereby a flagellum possesses two rows of tripartite tubular hairs (heterokont flagellum, syn. flimmergeissel, tinsel flagellum); (2) of an organism possessing flagella in pairs, the members of which differ in length; type of motion; external appendages. Usually with kinetosomes mutually attached at a wide angle. Cf. isokont.

heteromerous (1) (of a lichen thallus), having the mycobiont and photobiont in well-marked layers, usually between the medulla and the upper cortex; (2) (of trama in *Russulaceae*), having sphaerocyst nests among filamentous hyphae; cf. homoiomerous.

heteromixis, see heterothallism.

heteromorphic (**heteromorphous**), (1) having variation from normal structure; (2) having organs of different length; (3) (of agaric lamella edge), sterile due to the pressure of cystidia; cf. homomorphous.

Heteromyces L.S. Olive (1957) ≡ Oliveonia.

Heteromyces Müll. Arg. (1889), Cladoniaceae (L). 1, Brazil. See Stenroos & DePriest (*Am. J. Bot.* **85**: 1548, 1998; DNA).

Heteromycophaga P. Roberts (1997), anamorphic *Tremellomycetes*. 2, Great Britain. See Roberts (*Mycotaxon* **63**: 210, 1997).

Heteronectria Penz. & Sacc. (1898) = Cercophora fide Lundqvist (*Symb. bot. upsal.* **20** no. 1, 1972).

Heteropatella Fuckel (1873), anamorphic *Heterosphaeria*, St.≡ eH.10. 5, widespread (temperate). *H. antirrhini* (*Antirrhinum* shot hole). See Leuchtmann (*Mycotaxon* **28**: 261, 1987), Nag Raj (*Coelomycetous Anamorphs with Appendage-bearing Conidia*, 1993; key).

Heteropera Theiss. (1917) = Phomatospora fide Petrak (*Sydowia* **14**: 347, 1960).

Heterophracta (Sacc. & D. Sacc.) Theiss. & Syd. (1918) = Merismatium fide Hawksworth *et al.* (*Dictionary of the Fungi* edn 8, 1995).

heterophytic, the equivalent in the sporophyte generation of dioecious in the gametophyte generation (Blakeslee, *Bot. Gaz.* **42**: 161, 1906); cf. homophytic.

Heteroplacidium Breuss (1996), Verrucariaceae (L). 9, widespread. See Breuss (*Annln naturh. Mus. Wien* Ser. B, Bot. Zool. **98**: 40, 1996), Mies & Schultz (*Biblthca Lichenol.* **88**: 433, 2004; Socotra), Gueidan *et al.* (*MR* **111**: 1145, 2007; phylogeny).

heteroplastic, see heterokaryotic.

Heteroplegma Clem. (1903) = Peziza Fr. fide Eckblad (*Nytt Mag. Bot.* **15**: 1, 1968).

Heteroporimyces M.K. Elias (1966), Fossil Fungi. 1, New Zealand.

Heteroporus Lázaro Ibiza (1916) ≡ Abortiporus fide Fidalgo (*Rickia* **4**: 99, 1969).

Heteroradulum Lloyd (1917) nom. dub., Fungi.

Heterorepetobasidium Chee J. Chen & Oberw. (2002), Auriculariales. 2, Taiwan. See Chen *et al.* (*Mycol.* **94**: 515, 2002).

Heteroscypha Oberw. & Agerer (1979), Auriculariales. 1, S. Africa. See Oberwinkler & Agerer (*Beih. Sydowia* **8**: 31, 1979).

Heteroscyphaceae Jülich (1979) = Auriculariaceae.

Heteroseptata E.F. Morris (1972), anamorphic *Pezizomycotina*, Hsy.≡ eP.15. 1, Costa Rica. See Morris (*Mycol.* **64**: 889, 1972).

Heterospermales. Le Gal's (1953) term for inoperculate discomycetes; cf. *Homospermales*.

Heterosphaeria Grev. (1824), Helotiaceae. Anamorph *Heteropatella*. 6, Europe. See Leuchtmann (*Mycotaxon* **28**: 261, 1987; key 8 spp.).

Heterosphaeriaceae Rehm (1888) = Helotiaceae.

Heterosphaeriopsis Hafellner (1979), Dothideomycetes. 1, Ecuador. See Hafellner (*Nova Hedwigia* Beih. **62**: 175, 1979).

Heterosporiopsis Petr. (1950), anamorphic *Pezizomycotina*, Hsy.≡ eP.?. 1, S. America. See Petrak (*Sydowia* **4**: 521, 1950), Hughes & Seifert (*Sydowia* **50**: 192, 1998; revision).

Heterosporis Schubert (1969), Microsporidia. 3.

Heterosporium Klotzsch ex Cooke (1877) = Cladosporium fide Hughes (*CJB* **36**: 727, 1958), David (*Mycol. Pap.* **172**: 157 pp., 1997; revision), Braun *et al.* (*Mycol. Progr.* **2**: 3, 2003; phylogeny).

Heterosporula (Singer) Kühner (1980) = Gamundia fide Kuyper (*in litt.*).

heterospory (1) having asexually produced spores of more than one kind (de Bary, 1887: 496) (obsol.); (2) having spores which differ in the mating type (+ or -) in heterothallic fungi (e.g. in *Mucorales*) (Blakeslee, *Bot. Gaz.* **42**: 161, 1906); (3) polymorphism of basidiospores in *Agaricales* associated with extreme conditions (Heim, *Rev. Mycol.* **8**: 32, 1943).

Heterostomum Fr. [not traced] nom. dub., Pezizomycotina.

Heterotextus Lloyd (1922) = Guepiniopsis fide Reid (*TBMS* **62**: 474, 1974).

Heterothallism. Condition of sexual reproduction in which 'conjugation is possible only through the interaction of different thalli' (Blakeslee, 1904). Heterothallism and homothallism (q.v.) were first applied by Blakeslee to the methods of zygospore formation in the Mucorales, and he considered the terms to correspond to dioecism and monoecism for higher plants. Heterothallism has, however, been used as the equivalent of haplodioecism or dioecism (as in *Dictyuchus monosporus*, where male and female organs are produced on different individuals), and self-sterility or self-incompatibility (as in *Ascobolus magnificus*, where male and female organs are developed on one individual). Whitehouse (*Biol. Rev.* **24**: 411, 1949) distinguished the first type (haplodioecism) as '**morphological -**', and the second (haploid incompatibility) as '**physiological -**'. Physiological heterothallism may be determined either by two allelomorphs at one locus or by multiple allelomorphs at one or two loci (Whitehouse, *New Phytol.* **48**: 212, 1949). Multiple-allelomorph physiological heterothallism is characteristic of the hymenomycetes and gasteromycetes, among which Whitehouse has estimated 35% are heterothallic and bipolar (with one locus), 55% are heterothallic and tetrapolar (with two loci). Pontecorvo (*Adv. Genetics* **5**: 194, 1953) designated as **relative -** the formation of crossed asci in excess of 50% by the combination of certain homothallic strains of *Aspergillus nidulans*. Drayton & Groves (*Mycol.* **44**: 132, 1952) proposed to restrict heterothallism to morphological heterothallism, Korf (*Nature* **170**: 534, 1952) proposed to use heterothallism in a wide physiological sense, distinguishing a morphologically heterothallic organism as being both haplo-dioecious (dioecious) and heterothallic.

Burnett (*New Phytol.* **55**: 50, 1956) proposed the following terminology for mating systems of fungi: **heteromixis** (adj. heteromictic) for fusion of genetically different nuclei which includes **dimixis** when 2 types of nuclei (= heterothallism sensu Blakeslee, morphological, and 2-allelomorph physiological heterothallism), **diaphoromixis** when more than 2 types of nuclei (= multiple allelomorph physiological heterothallism), and **homoheteromixis** = secondary homothallism, amphithallism; **homomixis** (adj. homomictic) = homothallism; **amixis** (adj. amictic), apomixis in haploid organisms.

Manipulation of the mating locus MAT which governs heterothallism in many ascomycetes has been used experimentally to change the reproductive strategy of one species (*Gibberella zeae* from homothallic to heterothallic, making possible simple procedures to obtain sexual recombinants of this species of value in genetic analysis of pathogenicity and the ability to produce mycotoxins (Lee *et al.*, *Molecular microbiology* **50**: 145, 2003). Heterothallism has also been observed in *Peronospora* and *Pythium* and is thus not confined to true Fungi (Danielsen, *J. basic microbiol.* **41**: 305, 2001). See Bistis (*Fungal genetics and biology* **23**: 213, 1998; review in Euascomycetes), Feofilova (*Prikl. biokhim. Mikrobiol.* **42**, 2006; review for Mucoraceae of biological implications and uses in biotechnology).

Heterothecium Flot. (1850) ≡ Megalospora Meyen fide Santesson (*Symb. bot. upsal.* **12** no. 1: 1, 1952).

Heterothecium Mont. (1852) nom. illegit. ≡ Lopadium.

Heterotolyposporium Vánky (1997), Anthracoideaceae. 2 (on *Cyperaceae*, *Juncaceae*), Australiasia; S. Africa. Has two kinds of spores. See Vánky (*Mycotaxon* **63**: 143, 1997), Vánky & McKenzie (*Fungal Diversity Res. Ser.* **8**: 259 pp., 2002; New Zealand).

heterotroph (adj. **heterotrophic**) (of living organisms), using organic compounds as primary sources of energy, cf. autotrophic.

Heterovesicula C.E. Lange, C.M. Macvean, J.E. Henry, D.A. Streett (1995), Microsporidia. 1.

heteroxenous, having more than one host.

heterozygous, having heterokaryosis resulting from the fusion of gametes.

Heuflera Bail (1860), Pezizomycotina. 1, Europe.

Heufleria Auersw. (1869), Rhytismatales. 1 (on *Elyna*), Europe.

Heufleria Trevis. (1853) ≡ Astrothelium.

Heufleria Trevis. (1861) [non *Heufleria* Trevisan 1853; fide Trevisan 1861] = Cryptothelium fide Hawksworth *et al.* (*Dictionary of the Fungi* edn 8, 1995).

Heufleridium Müll. Arg. (1883), Pyrenulales (L). 2, widespread (tropical; New Zealand).

Hexacladium D.L. Olivier (1983), anamorphic *Pezizomycotina*, Hso.1bH.1/10. 1 (on pollen grains), S. Africa. See Olivier (*TBMS* **80**: 237, 1983).

Hexagona, see *Hexagonia Fr.*

Hexagonella F. Stevens & Guba ex F. Stevens (1925), Schizothyriaceae. 1, Hawaii.

Hexagonia Fr. (1838), Polyporaceae. 16, widespread (esp. tropical). See Bondartzev & Singer (*Polyporaceae of the European part of the U.S.S.R. and Caucasus*: 1106 pp., 1953) = Polyporus (Polypor.) sensu, Fidalgo (*Mem. N. Y. bot. Gdn* **17**: 35, 1968), Fidalgo (*Taxon* **17**: 37, 1968), Pegler (*The polypores [Bull. BMS Suppl.]*, 1973), Ryvarden & Johansen (*Prelim. Polyp. Fl. E. Afr.*: 366, 1980; key 11 Afr. spp.), Jülich (*Persoonia* **12**: 107, 1984; nomencl.) = Apoxona (Coriol.) sensu.

Hexagonia Pollini (1816) nom. rej. = Polyporus P. Micheli ex Adans. fide Ryvarden (*Syn. Fung.* **5**, 1991).

Hexajuga Fayod (1889) ≡ Clitopilus.

Heydenia Fresen. (1852), anamorphic *Pezizomycotina*. 4, widespread (north temperate). See Carmichael *et al.* (*Genera of Hyphomycetes*, 1980), Hairaud & Moreau (*Bulletin Mycologique et Botanique Dauphiné-Savoie* **42** no. 166: 47, 2002; France).

Heydeniopsis Naumov (1915) = Chaenotheca fide Seifert & Brodo (*Sydowia* **45**: 101, 1993).

Heyderia Link (1833), ? Hemiphacidiaceae. 6, widespread. See Maas Geesteranus (*Persoonia* **3**: 87, 1964), Knudsen (*Bot. Tidsskr.* **69**: 248, 1978; Denmark), Benkert (*Gleditschia* **10**: 141, 1983; key 5 spp.), Siepe (*Beitr. Kenntn. Pilze Mitteleur.* **2**: 193, 1986; key 3 spp. Eur.), Malaval (*Docums Mycol.* **19**: 9, 1989; ascus ultrastr.), Wang *et al.* (*Am. J. Bot.* **92**: 1565, 2005; phylogeny), Wang *et al.* (*Mycol.* **98**: 1065, 2006; phylogeny).

hiascent, becoming wide open.

Hiatula (Fr.) Mont. (1854) ? = Mycena fide Singer (*Agaric. mod. Tax.* 4th ed: 846, 1986).

Hiatulopsis Singer & Grinling (1967), Agaricaceae. 2, Brazil; Congo. See Singer & Grinling (*Persoonia* **4**: 364, 1967), Singer (*Fieldiana, New Series* **21**: 1, 1989).

Hidakaea I. Hino & Katum. (1955), ? Microthyriaceae. 1, Japan. See Hino & Katumoto (*Bulletin of the Faculty of Agriculture, Yamaguchi University* **6**: 38, 1955).

Hiemsia Svrček (1969), Pyronemataceae. 1 or 2, Europe. See Hohmeyer (*Mitt. Arbeitsg. Pilzk.Niederrhein* **6**: 11, 1988).

Higginsia Nannf. (1932) [non *Higginsia* Pers. 1805, *Rubiaceae*] ≡ Blumeriella.

higher fungi, *Ascomycota*, *Basidiomycota*, Anamorphic fungi.

hilar appendage (of basidiospores), the small wart-like or cone-like projection which connects the spore with the sterigma; sterigmatal appendage (Smith); apicule (Josserand). Cf. apophysis.

Hilberina Huhndorf & A.N. Mill. (2004), Sordariomycetes. 1. See Miller & Huhndorf (*MR* **108**: 31, 2004).

Hildebrandiella Naumov (1917) ? = Syzygites fide Hesseltine (*Mycol.* **47**: 344, 1955).

Hildenbrandia Nardo (1834), Algae. Algae.

Hilidicellites Kalgutkar & Janson. (2000), Fossil Fungi, anamorphic *Ascomycota*. 18, widespread. See Kalgutkar & Jansonius (*AASP Contributions Series* **39**: 133, 2000).

Hilitzeria Dyr (1941) nom. nud. ? = Dichotomocladium fide Benny (*A taxonomic revision of the Thamnidiaceae (Mucorales)*, 1973).

Hillia, see *Sillia*.

hilum, a mark or scar, esp. that on a spore at the point of attachment to a conidiogenous cell or sterigma.

Himanthites, see *Himantites*.

Himantia (Fr.) Zoll. (1844) = Cylindrobasidium fide Donk (*Taxon* **12**: 161, 1963; nomencl.).

Himantia Pers. (1801), anamorphic *Fungi*. *c.* 4 (sterile mycelium), Europe. See Donk (*Taxon* **11**: 85, 1962; refs).

himantioid (of mycelium), in spreading fan-like cords, as in *Himantia*.

Himantites Debey & Ettingsh. (1859), Fossil Fungi. 1, Europe.

Himantormia I.M. Lamb (1964), Parmeliaceae (L). 1, Antarctica. See Kärnefelt *et al.* (*Nova Hedwigia* **67**: 71, 1998), Fryday (*Lichenologist* **37**: 313, 2005), Thell *et al.* (*Biblthca Lichenol.* **95**: 531, 2007; phylogeny).

Hindersonia Lév. (1846) ≡ Hendersonia Berk. fide Holm (*Taxon* **24**: 275, 1975).

Hindersonia Moug. & Nestl. ex J. Schröt. (1897) ≡ Ceriospora.

Hinoa Hara & I. Hino (1961), anamorphic *Pezizomycotina*. 2, Japan. See Hara & Hino (*Icones Fungorum Bamb. Jap.*: 67, 1961).

Hinomyces Narumi & Y. Harada (2006), anamorphic *Grovesinia*. 2, Japan. See also *Cristulariella*. See Narumi-Saito *et al.* (*Mycoscience* **47**: 351, 2007).

Hiospira R.T. Moore (1962), anamorphic *Brooksia*, Hso.0-≡ hP.1. 1, widespread (tropical). See Moore (*TBMS* **45**: 145, 1962).

Hippocrepidea Sérus. (1997), Gomphillaceae (L). 1, Papua New Guinea. See Aptroot *et al.* (*Biblthca Lichenol.* **64**: 68, 1997).

Hippocrepidium Sacc. (1874) = Hirudinaria fide Hughes (*Mycol. Pap.* **39**, 1951).

Hippoperdon Mont. (1842) nom. rej. = Calvatia fide Stalpers (*in litt.*).

Hiratsuka (Naohide; 1903-2000; Japan). Undergraduate (1923-1926) then graduate (1926-1928) then lecturer (1928-1929), Department of Agricultural Biology, Hokkaido Imperial University; Professor of Botany and Plant Pathology, Tottori Agricultural College (1929-1946); Professor of Mycology and Plant Pathology, College of Agricultural Education, Tokyo (1946-1949); Professor of Mycology and Plant Pathology, Tokyo University of Education, 1949-1967); Director, Totori Mycological Institute (1967-1994). Celebrated for his studies on rusts and widely recognized as the foremost authority on taxonomy and biology of east Asian rust fungi; he named more than 250 species in 31 different rust genera. *Publs. A Monograph of the Pucciniastreae* (1936); *Uredinological Studies* (1955); *Revision Monograph of the Pucciniastreae* (1958); *Rust Flora of Japan* (1992). Azbukina & Kakishima ([*Mikologiya i Fitopatologiya*] **33**: 61, 1999) [portrait, in Russian]; Bandoni (*Mycoscience* **41**: 651, 2000) [portrait].

Hiratsukaia Hara (1948), Pucciniales. 1, Japan. See Cummins & Hiratsuka (*Illustr. Gen. Rust Fungi rev. edit.*, 1983; Excluded from *Pucciniales*).

Hiratsukamyces Thirum., F. Kern & B.V. Patil (1975), Pucciniales. 2 (on *Salacia* (*Celastraceae*)), India. See Thirumalachar *et al.* (*Sydowia* **27**: 80, 1973-4).

Hirneola Fr. (1825) nom. rej. ≡ Mycobonia.

Hirneola Fr. (1848) nom. cons. = Auricularia fide Lowy (*Mycol.* **44**: 656, 1952).

Hirneola Velen. (1939) [non *Hirneola* Fr. 1825, nec *Hirneola* Fr. 1848, nom. cons.] ≡ Clitopilopsis.

Hirneolina (Pat.) Bres. (1905) = Heterochaete fide Lowy (*Nova Hedwigia* **19**: 407, 1971), Donk (*Persoonia* **8**: 33, 1974; ≡ *Eichleriella*).

Hirschioporus Donk (1933) = Trichaptum fide Ryvarden (*Syn. Fung.* **5**, 1991).

hirsute, having long hairs.

Hirsutella Pat. (1892), anamorphic *Ophiocordyceps*, Hso/Hsy.0eH.15. *c.* 46 (on insects), widespread. See Macleod (*CJB* **37**: 695, 1959), Macleod (*CJB* **37**: 819, 1959; nutrition), Minter & Brady (*TBMS* **74**: 271, 1980), Minter *et al.* (*TBMS* **81**: 455, 1983; on eriophyid mites), Samson *et al.* (*Persoonia* **12**: 123, 1984; troglobiotic hyphomycetes), Evans & Samson (*CJB* **64**: 2098, 1986; with *Tetracrium* synanamorph), Fernández-Garcia *et al.* (*MR* **94**: 1111, 1990; on mealybug), Strongman *et al.* (*J. Invert. Path.* **55**: 11, 1990; spp. on spruce budworm), Samson & Evans (*MR* **95**: 887, 1991; *Didymobotryopsis*, a synonym), Strongman & MacKay (*Mycol.* **85**: 65, 1993; RAPDs used for varietal discrimination), Tedford *et al.* (*MR* **98**: 1127, 1994; RAPDs analysis of variation in *H. rhossiliensis*), Hodge *et al.* (*MR* **101**: 1377, 1997; *Harposporium* synanamorphs), Hywel-Jones (*MR* **101**: 1202, 1997; Thailand), Guzmán *et al.* (*Mycotaxon* **78**: 115, 2001; Mexico), Liu *et al.* (*MR* **105**: 827, 2001; teleomorph), Liu *et al.* (*MR* **106**: 1100, 2002; phylogeny), Stensrud *et al.* (*MR* **109**: 41, 2005; phylogeny), Sung *et al.* (*Stud. Mycol.* **57**, 2007; revision).

Hirsutosporos Batson (1983), Microsporidia. 1.

hirtose (**hirtous**), having hairs; hirsute.

Hirudinaria Ces. (1856), anamorphic *Pezizomycotina*, Hso.0bP.1. 1, Europe; N. America. See Hughes (*Mycol. Pap.* **39**, 1951).

hispid, having hairs or bristles.

Hispidicarpomyces Nakagiri (1993), Hispidicarpomycetaceae. 1 (on marine red alga, *Galaxaura*), Japan. See Nakagiri (*Mycol.* **85**: 638, 1993), Nakagiri & Ito (*Mycol.* **89**: 484, 1997), Inderbitzin *et al.* (*MR* **108**: 737, 2004; phylogeny).

Hispidicarpomycetaceae Nakagiri (1993), Lulworthiales. 1 gen., 1 spp.

Lit.: Nakagiri (*Mycol.* **85**: 638, 1993), Nakagiri & Ito (*Mycol.* **89**: 484, 1997).

Hispidocalyptella E. Horak & Desjardin (1994), Marasmiaceae. 1, Australia. See Horak & Desjardin (*Aust. Syst. Bot.* **7**: 165, 1994).

Hispidula P.R. Johnst. (2003), Hyaloscyphaceae. 4, Australasia. See Johnston (*N.Z. Jl Bot.* **41**: 687, 2003).

hispidulous, somewhat, or delicately; cf. hispid.

Histeridomyces Thaxt. (1931), Laboulbeniaceae. 6,

widespread. See Rossi (*Mycol.* **72**: 430, 1980), Weir (*Mycotaxon* **79**: 81, 2001; New Zealand).

histogenous (1) produced from tissue; (2) (of spores), produced from hyphae or cells, without conidiogenous cells.

histolysis, the disappearance or solution of a wall or tissue.

Histoplasma Darling (1906), anamorphic *Ajellomyces, Emmonsiella*, Hso.0eH.1. 2, widespread. *H. capsulatum* on humans and animals (**histoplasmosis**). See Sweany (Ed.) (*Histoplasmosis*, 1960), Cooke (*Mycopathologia* **39**: 1, 1969; 2,300 ref.), Ajello *et al.* (Eds) (*Histoplasmosis symposium*, 1971), Kwong-Chung (*Science* N.Y. **177**: 368, 1972; teleomorphs), Vincent *et al.* (*J. Bact.* **165**: 813, 1986; RFLP and classification), Maresca & Kobayashi (*Microbiol. Rev.* **5 3**: 186, 1989; dimorphism and cell differentiation in *H. capsulatum*), Fukushima *et al.* (*Mycopathologia* **116**: 151, 1991; re-evaluation of teleomorph by ubiquinone), Carter *et al.* (*J. Clin. Microbiol.* **34**: 2577, 1996; recombination), Guého *et al.* (*Mycoses* **40**: 69, 1997; phylogeny, epidemiology), Lasker *et al.* (*Medical Mycology* **36**: 205, 1998; introns), Okeke *et al.* (*Mycoses* **41**: 355, 1998; rDNA analysis), Poonwan *et al.* (*J. Clin. Microbiol.* **36**: 3073, 1998; Thailand), Kasuga *et al.* (*J. Clin. Microbiol.* **37**: 653, 1999; phylogeny), Reyes-Montes *et al.* (*J. Clin. Microbiol.* **37**: 1404, 1999; Mexico), Jiang *et al.* (*J. Clin. Microbiol.* **38**: 241, 2000; strain typing), Taylor *et al.* (*Fungal Genetics Biol.* **30**: 207, 2000; Mexico), Muniz *et al.* (*J. Clin. Microbiol.* **39**: 4487, 2001; genetic diversity), Tamura *et al.* (*Jap. J. Med. Mycol.* **43**: 11, 2002; phylogeny), Bracca *et al.* (*J. Clin. Microbiol.* **41**: 1753, 2003; molecular detection), Kasuga *et al.* (*Mol. Ecol.* **12**: 3383, 2003; phylogeography), Canteros *et al.* (*FEMS Immunol. Med. Microbiol.* **45**: 423, 2005; karyotypes), Komori *et al.* (*Jap. J. Med. Mycol.* **46**: 291, 2005; phylogeny), Zancopé-Oliveira *et al.* (*FEMS Immunol. Med. Microbiol.* **45**: 443, 2005; Brazil), Geiser *et al.* (*Mycol.* **98**: 1053, 2006; phylogeny).

Histoplasmataceae Redaelli & Cif. (1934) = Ajellomycetaceae.

histoplasmin, an antigen prepared from *Histoplasma capsulatum*, esp. for skin testing.

histoplasmosis, see *Histoplasma*.

History. Basidiomycetes and larger ascomycetes, including larger lichen-forming species have been used by man as food and in other ways since early times. There are references to fungi, including lichen-forming species in Greek and Roman literature (see Buller, *TBMS* **5**: 21, 1915) and illustrated accounts can be found in the European printed herbals of the sixteenth and seventeenth centuries (e.g. those of Bauhin, Dodoens, Parkinson) but their systematic investigation stems mainly from the studies of Tournefort, Dillenius, and, particularly, Micheli in the eighteenth century. Many of the earliest studies include a wide range of groups, but as the nature of lichens was not accepted until the end of the nineteenth century lichenology tended to be a rather independent study; integration of the two specialities has accelerated only since the 1970s, and the first well-integrated classifications are even later (see *Ascomycota*).

The first phase of mycology was mainly systematic. The chief contribution of Linnaeus was binomial nomenclature, but before 1800 advances were made by Tode and others. Outstanding among later workers before 1900 were Persoon, Fries, Berkeley, Corda, the Tulasne brothers, de Bary, Brefeld, and Saccardo – for all of whom (and selected others) there are short biographical notices (see Authors' names). The twentieth century saw major developments in the knowledge of cytology, sex, genetics, physiologic specialization, etc. in fungi and a great expansion in medical and industrial mycology and mycological aspects of plant pathology. There was also a growing awareness that fungi need protection and conservation no less than other organisms. In the early twenty-first century, molecular and digital technologies are being applied to mycology, the one revolutionizing classifications, and the other enabling huge amounts of information to be processed and disseminated.

At the beginning of the nineteenth century Acharius put lichenology on a sound scientific basis and many terms and currently accepted genera were introduced by him. With improvements in the microscope by the mid-century spore characters were first used in lichen taxonomy, particularly by de Notaris, Massalongo, Trevisan, and Körber; and the dual nature of lichens was recognized by de Bary and Schwendener. During the latter part of the nineteenth century important regional and monographic studies were made by Nylander, Müller-Argoviensis, Hue, and Crombie. From the 1950s particularly metabolic products (q.v.) came to play an increasing role in lichen systematics, and physiological (see Physiology) and ultrastructural (see Ultrastructure) studies have increased the understanding of their symbiotic nature.

Literature: **General**: Ainsworth (*Introduction to the history of mycology*, 1976). **Mycology**: Bo & Johnson (*Mycol.* **73**: 1098, 1981; mycology in China), Krieger (*Mycol.* **14**: 311, 1922; mycological illustrations), Lazzari (*Storia della micologia italiana*, 1973), Lutjeharms (*Meded. nederl. mycol. Vereen.* **23**, 1936; 18th century), Machol (*Mycologist* **4**: 129, 184, 1990; **5**: 28, 1991; early mushroom books), Ramsbottom (*Proc. Linn. Soc. Lond.* **151**: 280, 1941; mycology since Linnaeus), Raychaudhuri *et al.* (*History of plant pathology of south east Asia with special reference to India*, 1996), Rea (*TBMS* **5**: 211, 1916), Rogers (*A brief history of mycology in North America*, edn 2, 1981), Singer (*Mycologists and other taxa*, 1984). **Lichenology**: Krempelhuber (*Geschichte und Literatur der Lichenologie* **1**, 1867; **2**, 1869-72), Smith (*Lichens*, 1921 [reprint 1975]), Grummann (*Biographisch-bibliographisches handbuch der Lichenologie*, 1974), Hawksworth & Seaward (*Lichenology in the British Isles 1568-1975*, 1977).

See also Authors' names, Ethnomycology, Literature, Medical and Veterinary mycology, Plant pathogenic fungi.

Hjortstamia Boidin & Gilles (2003), Phanerochaetaceae. 12, widespread. See Boidin & Gilles (*BSMF* **118**: 99, 2002), Hjortstam & Ryvarden (*Syn. Fung.* **20**: 33, 2005).

HMAS, Mycology Herbarium, Systematic Mycology and Lichenology Laboratory, Institute of Microbiology (Beijing, China); founded 1953; a specialist Institute of Academia Sinica; the Laboratory was established in 1985; genetic resource collection **CGMCC** (Centre for General Microbiological Culture Collections).

hoary (esp. of a pileus or stipe), covered thickly with silk-like hairs; canescent.

Hobsonia Berk. ex Massee (1891), anamorphic *Atractiellales*, Hsp.0-≡ hH.1. 1, widespread. See Sikaroodi *et al.* (*MR* **105**: 453, 2001; phylogeny), Aime *et al.* (*Mycol.* **98**: 896, 2006; phylogeny).

Hobsoniopsis D. Hawksw. (2001), anamorphic *Pezizomycotina*. 1, Sweden. See Sikaroodi *et al.* (*MR* **105**: 457, 2001).

Hochkultur, see Normkultur.

Hodophilus R. Heim (1965) = Camarophyllopsis fide Kuyper (*in litt.*).

Hoehneliella Bres. & Sacc. (1902), anamorphic *Pezizomycotina*, Ccu.1eP.15. 1, Austria. See Vasant Rao & Sutton (*Kavaka* **3**: 21, 1976).

Hoehnelogaster Lohwag (1926) nom. dub., ? Boletales.

Hoehnelomyces Weese (1920) = Atractiella fide Oberwinkler & Bauer (*Sydowia* **41**: 224, 1989).

Hoehnelomycetaceae Jülich (1982) = Phleogenaceae.

Hohenbuehelia Schulzer (1866) nom. rej., Pleurotaceae. Anamorph *Nematoctonus*. *c.* 50, widespread. See Barron & Dierkes (*CJB* **55**: 3054, 1977; ecology, anamorph), Thorn & Barron (*Mycotaxon* **25**: 321, 1986; N. Am. spp.), Thorn *et al.* (*Mycol.* **92**: 241, 2000; phylogeny), Fazio & Albertó (*Mycotaxon* **77**: 117, 2001; Argentina).

Höhnel (Franz Xavier Rudolf von; 1852-1920; Austria). PhD, Strassburg (1877); Professor of Botany, Vienna Technical College, 1884-1920. Proposed 250 new genera and 500 species of fungi; his taxonomic research on coelomycetes being particularly noted. His most important work was disseminated through two extensive but dispersed groups of papers, both aptly named *Fragmente*. See also (by Weese) v. Höhnel's mykologischen Nachlass-Schriften 1-150. *Mitteilungen der Botanischen Institut der Technischen Hochschule, Wien* **1-12** (1924-1935) [(index **12**: 33]. Most of his collections are in Harvard (**FH**). *Publs*. Fragmente zur Mykologie. *Sitzungsberichten der Kaiserliche Akademie der Wissenschaften in Wien* Mathematische-Naturwissenschaftliche Klasse Abt. I (1902-1923) [index for fragmente 1-1000 printed separately 1916; index to fragmente 1001-1225, see *Mitteilungen der Botanischen Institut der Technischen Hochschule, Wien* **9**: 66, 1932;. see also the facsimile reprint by J. Cramer, *Fragmente zur Mykologie i-xxiv* (1967)]. Mykologische Fragmente *Annales Mycologici* **1-18** (1903-1920) [reprint 1967]. *Biogs, obits etc.* Stafleu & Cowan (*TL-2* **2**: 229, 1979); Weese (*Bericht der Deutschen Botanischen Gesellschaft* **38**: (103), 1921).

Holcomyces Lindau (1904) = Diplodia fide Höhnel (*Annls mycol.* **3**: 187, 1905).

holdfast, a process from the thallus for attachment, e.g. appressorium, hyphopodium, stigmatopodium, and stomatopodium; cf. rhizoid, hapteron, haustorium.

Holleya Y. Yamada (1986) = Eremothecium fide Yamada & Nagahama (*J. gen. appl. Microbiol.* Tokyo **37**: 199, 1991), Kurtzman (*J. Industr. Microbiol.* **14**: 523, 1995), Prillinger *et al.* (*Yeast* Chichester **13**: 945, 1997), de Hoog *et al. in* Kurtzman & Fell (Eds) (*Yeasts, a taxonomic study* 4th edn: 201, 1998).

Hollosia Gyeln. (1939) = Scutula fide Santesson (*Svensk bot. Tidskr.* **43**: 141, 1949).

Holmiella Petrini, Samuels & E. Müll. (1979), Patellariaceae. Anamorph *Cornicularielia*. 1 (on *Juniperus*), widespread. See Kutorga & Hawksworth (*SA* **15**: 1, 1997).

Holmiodiscus Svrček (1992), Helotiaceae. 1, Sweden. See Holm & Nannfeldt (*Thunbergia* **16**: 19, 1992).

holo- (prefix), all; whole; entire.

Holobasidiomycetes, see *Basidiomycota*.

Holobasidiomycetidae, see *Basidiomycota*.

holobasidium, see basidium.

Holobispora Voronin (1986), Microsporidia. 1.

holoblastic, see blastic.

holocarpic, having all the thallus used for the fruitbody.

holocarpous (of lichen thalli), ones formed by colonies of a free-living photobiont being invaded by a mycobiont and developing directly into fruiting bodies (Henssen, *Lichenologist* **18**: 51, 1986).

Holocoenis Clem. (1909) = Coenogonium Ehrenb. fide Hawksworth *et al.* (*Dictionary of the Fungi* edn 8, 1995) but see Rivas Platas *et al.* (*Fungal Diversity* **23**: 255, 2006; Costa Rica).

holocoenocytic, see astatocoenocytic.

Holocoryne (Fr.) Bonord. (1851) = Clavaria fide Corner (*Ann. Bot. Mem.* [A monograph of Clavaria and allied genera] **1**, 1950).

Holocotylon Lloyd (1906), Agaricaceae. 3, Mexico; N. America (subtropical). See Zeller (*Mycol.* **39**: 282, 1947).

Holocryphia Gryzenh. & M.J. Wingf. (2006), Cryphonectriaceae. 1 (on *Eucalyptus* and *Tibouchina*), widespread. See Venter *et al.* (*Sydowia* **54**: 98, 2002), Myburg *et al.* (*Mycol.* **96**: 990, 2004; phylogeny), Gryzenhout *et al.* (*FEMS Microbiol. Letters* **258**: 161, 2006; phylogeny), Gryzenhout *et al.* (*Stud. Mycol.* **55**: 48, 2006), Heath *et al.* (*Australas. Pl. Path.* **36**: 560, 2006; *Tibouchina* as host), Nakabonge *et al.* (*Australas. Pl. Path.* **37**: 154, 2008; population biology).

Holocyphis Clem. (1909) = Thelomma fide Hawksworth *et al.* (*Dictionary of the Fungi* edn 8, 1995).

hologamy, the condition in which all the thallus becomes a gametangium, i.e. there is fusion between two mature individuals as in *Polyphagus*.

Hologloea Pat. (1900) = Favolaschia fide Kuyper (*in litt.*).

Hologymnia Nyl. (1900) = Evernia fide Hawksworth *et al.* (*Dictionary of the Fungi* edn 8, 1995).

holomorph, see States of fungi.

holomorphum, two or more accepted species, each comprising teleomorph and anamorph, but scarcely distinguishable in one of the morphs (Tribe, *Bull. BMS* **17**: 94, 1983).

holophyte, a physiologically self-supporting green plant.

holosaprophyte, a true saprophyte (Johow).

Holospora Tomas. & Cif. (1952) = Polyblastia fide Hawksworth *et al.* (*Dictionary of the Fungi* edn 8, 1995).

Holosporomyces Cif. & Tomas. (1953) ≡ Holospora.

holosporous (of conidial maturation), the conidium approaches its final size and shape before delimiting cells and maturing as a whole (Luttrell, 1963).

Holothelis Clem. (1909) = Thelopsis fide Hawksworth *et al.* (*Dictionary of the Fungi* edn 8, 1995).

holotype, see type.

Holstiella Henn. (1895) = Trypethelium fide von Arx & Müller (*Stud. Mycol.* **9**, 1975).

Holtermannia Sacc. & Traverso (1910), Tremellaceae. 7, S.E. Asia; Brazil. See Kobayasi (*Sci. Rep. Tokyo Bunrika Daig.* B **50**: 75, 1937), Mahamulkar *et al.* (*Indian Phytopathology* **55**: 464, 2002; India n.sp.).

Holttumia Lloyd (1924), Xylariaceae. 1, E. & S.E.

Asia. See Læssøe (*SA* **13**: 43, 1994), Rogers & Ju (*Mycotaxon* **68**: 358, 1998; synonym of *Kretzschmaria*).

Holubovaea Mercado (1983), anamorphic *Pezizomycotina*, Hso.≡ eP.19. 1, Cuba. See Mercado (*Acta Bot. Cubana* **15**: 5, 1983), Mercado Sierra *et al.* (*Monografie* Museo Regionale di Scienze Naturali, Torino **23**, 1997).

Holubovaniella R.F. Castañeda (1985), anamorphic *Pezizomycotina*, Hso.≡ eP.1. 2, Cuba; Kenya. See Castañeda (*Deuteromycotina de Cuba* Hyphomycetes **III**: 14, 1985).

Holwaya Sacc. (1889), ? Bulgariaceae. Anamorph *Crinula*. 1, widespread (north temperate). See Korf & Abawi (*CJB* **49**: 1879, 1971), Krieglsteiner & Häffner (*Z. Mykol.* **51**: 131, 1985), Aronsson (*Svensk bot. Tidskr.* **85**: 9, 1991), Roth (*Schweiz. Z. Pilzk.* **69**: 197, 1991; anamorph), Wang *et al.* (*Mol. Phylogen. Evol.* **41**: 295, 2006; phylogeny), Wang *et al.* (*Mycol.* **98**: 1065, 2006; phylogeny).

Holwayella H.S. Jacks. (1926) = Chrysocyclus fide Cummins & Hiratsuka (*Illustr. Gen. Rust Fungi rev. edit.*, 1983).

Homalopeltis Bat. & Valle (1961), anamorphic *Pezizomycotina*, Cpt.1eP.?. 1, Brazil. See Batista & Valle (*Publções Inst. Micol. Recife* **337**: 5, 1961).

Homaromyces R.K. Benj. (1955), Laboulbeniaceae. 1, USA; Europe. See Benjamin (*Aliso* **3**: 183, 1955), Tavares (*Mycol. Mem.* **9**: 627 pp., 1985).

homeostasis, the maintenance of constant chemical and physical conditions within a living organism. See Whittenbury *et al.* (Eds) (*Homeostatic mechanisms in micro-organisms*, 1987).

homo- (prefix), one and the same.

Homobasidiomycetes R.T. Moore (1971), see *Basidiomycota*.

homobasidium, see basidium.

homobium, a self-supporting association of a fungus and an alga, as in lichens.

Homodium Nyl. ex H. Olivier (1903) = Leptogium fide Hawksworth *et al.* (*Dictionary of the Fungi* edn 8, 1995).

homohetromixis, see heterothallism.

homohylic vesicle, sporangial vesicle, the wall of which is continuous with, and of the same material as the wall layer, or one of the wall layers of the sporangium. See also plasmamembranic vesicle, precipitative vesicle.

homoiomerous (1) (of a lichen thallus), having the mycobiont and phycobiont evenly intermixed throughout the thallus, as in *Collema*; (2) (of trama in agarics), composed of hyphal tissue only; cf. heteromerous.

homokaryotic, having genetically identical nuclei, as in a line of isolates without variation (Brierley, *Ann. appl. Biol.* **18**: 429, 1931).

homokaryotic diplospory, incorporation of one or more homogenetic diploid nuclei in an oospore formed in response to abortive meioses in the oogonium (Dick, 1972).

homologous, showing a resemblance in form or structure, but not necessarily function, which is considered to be evidence of evolutionary relatedness; cf. analogous. See Poelt (*in* Hawksworth (Ed.), *Frontiers in mycology*: 85, 1991; homologous characters in lichens). See also alternation of generations, Cladistics.

homomixis, haplo-monoecism or monoecism. Whitehouse (*Biol. Rev.* **24**: 428, 1949) recognized two types: **primary homothallism** such as occurs in a homokaryotic individual and **secondary homothallism** (= pseudo heterothallism or facultative heterothallism (Dodge), **amphithallism**, Lange, 1952) as in a thallus derived from one heterokaryotic spore containing nuclei of compatible mating types; cf. heterothallism.

homomorphous (of an agaric lamella edge), hymenium on edge not differentiated from that on the faces; cf. heteromorphous.

Homonemeae, the *Algae* and *Fungi* (obsol.).

homonym, a name which (under the *Code*) must not be used because of an earlier name in a different sense, i.e. the names are the same but the types different; Donk (*Bull. bot. Gard. Buitenzorg* ser. 3 **18**: 282, 1949) in addition to heterotypic homonyms recognizes homotypic, non-synonymous, synonymous, and monadelphous homonyms, the last being derived from a common source, particularly from one devalidated name.

Homophron (Britzelm.) W.B. Cooke (1953) = Psathyrella fide Kuyper (*in litt.*).

homophytic, the equivalent in the sporophyte generation of monoecious in the gametophyte generation (Blakeslee, *Bot. Gaz.* **42**: 161, 1906); cf heterophytic.

homoplasmic, see karyotic.

Homopsella Nyl. (1887) = Porocyphus fide Henssen (*Symb. bot. upsal.* **18** no. 1, 1963).

Homopsellomyces Cif. & Tomas. (1953) ≡ Homopsella.

Homospermales. Le Gal's (1953) term for operculate discomycetes; cf. *Heterospermales*.

homospory (1) having asexually produced spores of only one kind (= isospory); (2) having spores which are not differentiated according to mating type (c.f. heterospory).

Homostegia Fuckel (1870), Dothideomycetes. 1 or 5 (on lichens), Europe. See Hawksworth *et al.* (*Biblthca Lichenol.* **88**: 187, 2004).

homothallism (adj. **homothallic**), the condition in which sexual reproduction can occur without the interaction of two differing thalli.

Homothecium A. Massal. (1853), Collemataceae (L). 3, widespread (temperate; S. America). See Henssen (*Lichenologist* **3**: 29, 1965).

homozygous, with identical alleles at the same locus on homologous chromosomes.

honey agaric (**honey fungus**), *Armillaria mellea*; shoestring or boot-lace fungus.

honey-dew (1) a secretion, attractive to insects, associated with the *Sphacelia* phase of *Claviceps* (Mower & Hancock, *CJB* **53**: 2826, 1975); (2) a secretion by aphids.

Honoratia Cif., Vegni & Montemart. (1963) = Preussia fide von Arx & Storm (*Persoonia* **4**: 407, 1967).

Hoornsmania Crous (2007), anamorphic *Davidiellaceae*. 1, Europe. See Crous (*Fungal Planet* **11**, 2007).

Horakia Oberw. (1976) ≡ Verrucospora.

Horakiella Castellano & Trappe (1992), Sclerodermataceae. 1, Australia. See Castellano & Trappe (*Aust. Syst. Bot.* **5**: 641, 1992).

Horakomyces Raithelh. (1983) ? = Melanomphalia fide Kirk (*in litt.*).

Hormiactella Sacc. (1886), anamorphic *Pezizomycotina*, Hso.1eP.3. 4, Europe; West Indies. See Holubová-Jechová (*Folia geobot. phytotax.* **13**: 433, 1978), Borisova (*Mikol. Fitopatol.* **34**: 26, 2000).

Hormiactina Bubák (1916) ? = Hormiactis fide Carmichael *et al.* (*Genera of Hyphomycetes*, 1980).
Hormiactis Preuss (1851), anamorphic *Pezizomycotina*, Hso.1eH.3. 5, Europe; Sri Lanka.
Hormiokrypsis Bat. & Nascim. (1957), anamorphic *Ophiocapnocoma*, Hso.0bP.28. 1, USA. See Hughes (*N.Z. Jl Bot.* **5**: 117, 1967), Hughes (*N.Z. Jl Bot.* **10**: 225, 1972).
Hormisciella Bat. (1956) = Antennatula fide Hughes (*N.Z. Jl Bot.* **8**: 153, 1970).
Hormiscioideus M. Blackw. & Kimbr. (1979), anamorphic *Pezizomycotina*, Hso.?.?. 1 (on *Insecta*), USA. See Blackwell & Kimbrough (*Mycol.* **70**: 1275, 1978).
Hormisciomyces Bat. & Nascim. (1957), anamorphic *Euantennaria*, Hso.0eP.15. 1, Cuba. See Hughes (*N.Z. Jl Bot.* **12**: 299, 1974), Hughes & Arnold (*Mem. N. Y. bot. Gdn* **49**: 198, 1989).
Hormisciopsis Sumst. (1914), anamorphic *Pezizomycotina*, Hso.0eP.3. 1, N. America.
Hormiscium Kunze (1817) = Torula fide Hughes (*CJB* **36**: 727, 1958), Crane & Schoknecht (*Mycol.* **78**: 86, 1986).
Hormoascus Arx (1972) = Ambrosiozyma fide van der Walt & von Arx (*Syst. Appl. Microbiol.* **6**: 90, 1985), Goto & Takami (*J. gen. appl. Microbiol.* Tokyo **32**: 271, 1986), Kurtzman & Robnett (*CJB* **73** Suppl. 1: S824, 1995), Smith *in* Kurtzman & Fell (Eds) (*Yeasts, a taxonomic study* 4th edn: 129, 1998).
Hormocephalum Syd. (1939), anamorphic *Pezizomycotina*, Hso.≡ eP.4. 1, Ecuador.
Hormocladium Höhn. (1923), anamorphic *Pezizomycotina*, Hso.0-1eP.?. 1, Japan. See Braun & Melnik (*Trudy Botanicheskogo Instituta im. V.L. Komarova* **20**: 18, 1997).
Hormococcus Preuss (1852) nom. dub., anamorphic *Pezizomycotina*. See Sutton (*Mycol. Pap.* **141**, 1977).
Hormococcus Robak (1956), anamorphic *Pezizomycotina*, Cpd.0eH.15. 1, Norway. See Robak (*Friesia* **5**: 379, 1956).
Hormoconis Arx & G.A. de Vries (1973), Amorphothecaceae. 1. See Partridge & Morgan-Jones (*Mycotaxon* **83**: 335, 2002), Braun *et al.* (*Mycol. Progr.* **2**: 8, 2003), Seifert *et al.* (*Stud. Mycol.* **58**: 235, 2007; phylogeny, nomenclature).
hormocyst, a propagule or diaspore produced in a special **hormocystangium** comprising a few cyanobacterial cells and fungal hyphae (Fig. 22I); produced by a few gelatinous lichens, e.g. *Lempholemma cladodes*, *L. vesiculiferum*. See Degelius (*Svensk bot. Tidsk.* **39**: 419, 1945), Henssen (*Lichenologist* **4**: 99, 1969).
hormocystangium, see hormocyst.
Hormodendraceae Nann. (1934) = Davidiellaceae.
Hormodendroides M. Moore & F.P. Almeida (1937) = Rhinocladiella Nannf. fide Hawksworth *et al.* (*Dictionary of the Fungi* edn 8, 1995).
Hormodendron (orthographic variant), see *Hormodendrum*.
Hormodendrum Bonord. (1851) = Cladosporium fide de Vries (*Contribution to the knowledge of the genus Cladosporium Link ex Fr.*, 1952), David (*Mycol. Pap.* **172**, 1997).
Hormodochis Clem. (1909) = Trullula fide Sutton (*Mycol. Pap.* **141**, 1977).
Hormodochium (Sacc.) Höhn. (1911) = Trullula fide Sutton (*Mycol. Pap.* **141**, 1977).
Hormographiella Guarro & Gené (1992), anamorphic *Coprinellus*, Hso.0eH.38/39. 3 (from humans, coprophilous), Europe. See Guarro *et al.* (*Mycotaxon* **45**: 179, 1992), Cáceres *et al.* (*Antonie van Leeuwenhoek* **89**: 79, 2006; anamorphs of *Coprinellus domesticus*).
Hormographis Guarro, Punsola & Arx (1986), anamorphic *Pezizomycotina*, Hsy.0eH.40. 1 (from soil), Spain. See Guarro *et al.* (*Mycol.* **78**: 969, 1986).
Hormomitaria Corner (1950) = Physalacria fide Kuyper (*in litt.*).
Hormomyces Bonord. (1851), anamorphic *Tremella*, Hsp.0eH.3. 3, widespread (temperate). See Tubaki (*TMSJ* **17**: 243, 1976; culture), Roberts (*Mycotaxon* **63**: 195, 1997; UK), Zang & Wang (*Acta Bot. Yunn.* **19**: 324, 1997; China).
Hormonema Lagerb. & Melin (1927), anamorphic *Dothiora*, Hso.0eH.1/10. 5, widespread. Apparently polyphyletic. See Hermanides-Nijhof (*Stud. Mycol.* **15**: 166, 1977), Funk *et al.* (*CJB* **63**: 1579, 1985), Shin *et al.* (*J. Clin. Microbiol.* **36**: 2157, 1998; in humans), Yurlova *et al.* (*Stud. Mycol.* **43**: 63, 1999), Peláez *et al.* (*Syst. Appl. Microbiol.* **23**: 333, 2000; endophyte), Bills *et al.* (*Stud. Mycol.* **50**: 149, 2004; Spain).
hormones (sex or sexual), of fungi, see antheridiol, progamones, sirenin, trisporic acids; also erogen, etc. Machlis (*Mycol.* **64**: 235, 1972), Gooday (*Ann. Rev. Biochem.* **43**: 35, 1974), reviews. Cf. pheromone.
Hormopeltis Speg. (1912) = Micropeltis fide Clements & Shear (*Gen. Fung.*, 1931).
Hormosperma Penz. & Sacc. (1897) = Lasiosphaeria fide Hawksworth *et al.* (*Dictionary of the Fungi* edn 8, 1995).
Hormosphaeria Lév. (1863) nom. dub., ? Arthoniales. 1, S. America. See Santesson (*Symb. bot. upsal.* **12** no. 1: 1, 1952).
Hormospora De Not. (1844) [non *Hormospora* Bréb. 1839, *Algae*] = Bombardioidea fide Hawksworth *et al.* (*Dictionary of the Fungi* edn 8, 1995).
Hormosporites Grüss (1928), Fossil Fungi ? Fungi. 1 (Devonian), Europe.
Hormotheca Bonord. (1864) = Coleroa fide Corlett & Barr (*Mycotaxon* **25**: 255, 1986), Müller (*SA* **5**: 310, 1986).
Hormyllium Clem. (1909) ≡ Hormococcus Robak fide Sutton (*Mycol. Pap.* **141**, 1977).
horn of plenty, basidioma of the edible *Craterellus cornucopioides*.
Hornodermoporus Teixeira (1993) = Perenniporia fide Decock (*in litt.*).
horse mushroom, basidiomata of the edible *Agaricus arvensis*.
horse-hair blight fungi, the rhizomorphic mycelia of tropical species of *Marasmius*, e.g. *M. equicrinis* (E. trop.) and *M. sarmentosus* (W. Indies). See Petch (*Ann. R. bot. Gdns Peradeniya* **6**: 43, 1915). Cf. thread blight.
horse-tail lichen, see rock hair.
Hortaea Nishim. & Miyaji (1984), anamorphic *Teratosphaeriaceae*, Hso.1eP.21. 1, widespread. See Mittag (*Mycoses* **36**: 242, 1993; fine structure), Uijthof *et al.* (*Mycoses* **37**: 307, 1994; genotypes of *H. werneckii*), Piontelli L. *et al.* (*Bol. Micol.* Valparaíso **12**: 89, 1997; saline environments, Chile), Rogers *et al.* (*Stud. Mycol.* **43**: 122, 1999; phylogeny), Haritani *et al.* (*Jap. J. Med. Mycol.* **43**: 175, 2002; from guinea pig), Hölker *et al.* (*Antonie van Leeuwenhoek* **86**: 287, 2004; from lignite), Ng *et al.* (*Mycopathologia* **159**: 495, 2005; from blood), Ster-

flinger (*The Yeast Handbook* [1]: 501, 2006; review), Crous *et al.* (*Stud. Mycol.* 58: 1, 2007).

Hosseusia Gyeln. (1940), ? Pannariaceae (L). 1, Argentina.

host (1) a living organism harbouring a parasite; frequently in the sense of 'suscept' (q.v.); sometimes, as in 'host index', in a general sense covering certain substrata; (2) a commercial computer operator keeping databases other than their own on its machines and making them available to others for on-line searching for a fee (see Literature).

house fungus (or **dry rot fungus**), *Serpula lacrimans*.

Hrabyeia J. Lom & I. Dykova (1990), Microsporidia. 1. See Lom & Dykova (*Eur. J. Protist.* **25**: 243, 1990).

hülle cells, terminal or intercalary thick-walled cells which occur in large numbers in association with the ascomata of, e.g. *Aspergillus nidulans*.

Huangshania O.E. Erikss. (1992), Triblidiaceae. 2, China. See Magnes (*Biblthca Mycol.* **165**, 1997).

Hubbsia W.A. Weber (1965), Roccellaceae (L). 3, new world. See Follmann (*Nova Hedwigia* **31**: 285, 1979), Tehler (*CJB* **68**: 2458, 1990; cladistics), Tehler *et al.* (*Symb. bot. upsal.* **32** no. 1: 255, 1997), Myllys *et al.* (*Bryologist* **101**: 70, 1998; phylogeny), Myllys *et al.* (*Lichenologist* **31**: 461, 1999; phylogeny), Follmann (*J. Hattori bot. Lab.* **90**: 251, 2001; key), Follmann (*Mitteilungen aus dem Institut für Allgemeine Botanik Hamburg* **30-32**: 61, 2002; biogeography), Tehler & Irestedt (*Cladistics* **23**: 432, 2007).

Huea C.W. Dodge & G.E. Baker (1938) = Caloplaca fide Hawksworth *et al.* (*Dictionary of the Fungi* edn 8, 1995).

Hueella Zahlbr. (1926) nom. rej. prop. = Fuscopannaria fide Jørgensen (*Taxon* **49**: 812, 2000; nomencl.).

Hueidea Kantvilas & P.M. McCarthy (2003), Fuscideaceae (L). 1, Australia. See Kantvilas & McCarthy (*Lichenologist* **35**: 398, 2003).

Hughesiella Bat. & A.F. Vital (1956) = Chalara fide Nag Raj & Kendrick (*Monogr. Chalara Allied Genera*, 1975).

Hughesinia J.C. Lindq. & Gamundí (1970), anamorphic *Pezizomycotina*, Hso.0bP.28. 1, Chile; Cuba. See Lindquist & Gamundí (*Boln Soc. argent. Bot.* **13**: 54, 1970), Delgado *et al.* (*Fungal Diversity* **20**: 31, 2005; Cuba).

Hugueninia J.L. Bezerra & T.T. Barros (1970), ? Microthyriaceae. 1, New Caledonia. See Hyde (*Sydowia* **49**: 1, 1997), Mouchacca (*Mycotaxon* **67**: 99, 1998).

Huilia Zahlbr. (1930) = Porpidia fide Hawksworth *et al.* (*Dictionary of the Fungi* edn 8, 1995).

Humaria (Fr.) Boud. (1885) ≡ Octospora.

Humaria Fuckel (1870), Pyronemataceae. *c.* 16, widespread (north temperate). See Kanouse (*Mycol.* **39**: 635, 1947), Denison (*Mycol.* **51**: 612, 1961), Eriksson & Hawksworth (*SA* **11**: 179, 1993; nomencl.), Hansen & Pfister (*Mycol.* **98**: 1029, 2006; phylogeny), Perry *et al.* (*MR* **111**: 549, 2007; phylogeny), Erős-Honti *et al.* (*Mycorrhiza* **18**: 133, 2008; mycorrhizas, Hungary).

Humariaceae Velen. (1934) = Pyronemataceae.

Humariella J. Schröt. (1893) ≡ Scutellinia.

Humarina Seaver (1927) ≡ Octospora.

Humboldtina Chardón & Toro (1934) = Leptosphaeria fide Barr (*SA* **5**: 134, 1986).

Humicola Traaen (1914), anamorphic *Chaetomiaceae*, Hso.0eP.1. 6, widespread. See Fassatiová (*Česká Mykol.* **18**: 102, 1964; key), Nicoli & Russo (*Nova Hedwigia* **25**: 737, 1974; 8 spp.), de Bertoldi (*CJB* **54**: 2755, 1976; 13 spp., physiol. analysis), Rodrigues *et al.* (*Revista de Microbiologia* **21**: 232, 1990; cytology), Rodrigues *et al.* (*MR* **95**: 169, 1991; cytology, biochemistry), Weinstein *et al.* (*Mycol.* **89**: 706, 1997; ecology), Hambleton *et al.* (*Stud. Mycol.* **53**: 29, 2005; phylogeny), Singh *et al.* (*Curr. Sci.* **89**: 1745, 2005; in mushroom compost).

Humicolopsis Cabral & S. Marchand (1976), anamorphic *Pezizomycotina*, Hso.0eH.?. 1, Argentina. See Cabral & Marchand (*Boln Soc. argent. Bot.* **17**: 69, 1976).

Humicolopsis Verona (1977), anamorphic *Pezizomycotina*, Hso.0eH.10. 1, Italy. See Verona (*G. bot. ital.* **111**: 88, 1977).

humicolous, living in or on decaying organic matter, soil.

Humidicutis (Singer) Singer (1959), Hygrophoraceae. 10, widespread (north temperate). See Moncalvo *et al.* (*Syst. Biol.* **49**: 278, 2000; molec.; related to *Pluteaceae*).

Humphreya Steyaert (1972), Ganodermataceae. 4, widespread (tropical). See Ryvarden & Johansen (*Prelim. Polyp. Fl. E. Afr.*: 94, 1980; key 4 Afr. spp.), Kotlaba & Pouzar (*Czech Mycol.* **55**: 7, 2003; Cuba).

Husseia Berk. (1847) = Calostoma fide Stalpers (*in litt.*).

Hyalacrotes (Korf & L.M. Kohn) Raitv. (1991), Hyaloscyphaceae. 2, Europe. See Azbukina (*Nizshie Rasteniya, Griby i Mokhoobraznye Sovetskogo Dal'nego Vostoka* Griby. Tom **2**. Askomitsety. Erizifal'nye, Klavitsipital'nye, Gelotsial'nye: 337, 1991), Raitviir (*Scripta Mycologica* Tartu **20**, 2004).

Hyalasterina Speg. (1919) nom. dub., Fungi.

Hyalina, see *Hyalinia*.

hyaline, transparent or nearly so; translucent; frequently used in the sense of colourless.

Hyalinia Boud. (1885) [non *Hyalinia* Stackh. 1809, *Algae*] = Orbilia fide Baral (*SA* **13**: 113, 1994).

Hyalinocysta E.I. Hazard & Oldacre (1975), Microsporidia. 2.

hyalo- (prefix) (of spores), hyaline or brightly coloured, esp. for groups of Anamorphic fungi (q.v.) (**hyalodidymae**, etc.).

Hyalobelemnospora Matsush. (1993), anamorphic *Subbaromyces*, Hso.0eH.19. 1, Peru. See Matsushima (*Matsush. Mycol. Mem.* **7**: 54, 1993).

Hyalobotrys Pidopl. (1948) = Stachybotrys fide Kendrick & Carmichae *in* Ainsworth *et al.* (Eds) (*The Fungi* **4A**: 390, 1973).

Hyalocamposporium Révay & J. Gönczöl (2007), anamorphic *Pezizomycotina*. 4, widespread. See Révay & Gönczöl (*Fungal Diversity* **25**: 197, 2007).

Hyalocapnias Bat. & Cif. (1963) = Scorias fide Hawksworth *et al.* (*Dictionary of the Fungi* edn 8, 1995).

Hyaloceras Durieu & Mont. (1849) = Seiridium fide Höhnel (*Sber. Akad. Wiss. Wien* Math.-naturw. Kl., Abt. 1 **125**: 27, 1916), Sutton (*Mycol. Pap.* **141**, 1977).

Hyalochlorella Poyton (1970) nom. dub., ? Algae. See Alderman (*Veröff. Inst. Meeresf. Bremerhaven* Sonderband **5**: 251, 1974).

Hyalocladium Mustafa (1977), anamorphic *Pezizomycotina*, Hso.#eH.10. 1, Kuwait. See Mustafa (*TBMS* **67**: 537, 1976).

Hyalococcus J. Schröt. (1889) ? = Trichosporon fide Barnett *et al.* (*Yeasts: Characteristics and Identification* 3rd edn, 2000).

Hyalocrea Syd. & P. Syd. (1917), Dothideomycetes. 4, widespread (tropical). See Pirozynski (*Kew Bull.* **31**: 595, 1977), Rossman (*Mycol. Pap.* **157**: 71 pp., 1987).

Hyalocurreya Theiss. & Syd. (1915) = Uleomyces fide von Arx (*Persoonia* **2**: 421, 1963).

Hyalocylindrophora J.L. Crane & Dumont (1978) = Dischloridium fide Bhat & Sutton (*TBMS* **84**: 723, 1985), Seifert & Gams (*Mycotaxon* **24**: 459, 1985).

Hyalodema Magnus (1910) = Coniodictyum fide Donk (*Reinwardtia* **4**: 115, 1956).

Hyalodendriella Crous (2007), anamorphic *Helotiales*. 1, Netherlands. See Crous *et al.* (*Stud. Mycol.* **58**: 33, 2007).

Hyalodendron Diddens (1934) = Trichosporon fide Fell & Scorzetti (*Int. J. Syst. Evol. Microbiol.* **54**: 995, 2004).

Hyaloderma Speg. (1884), Pezizomycotina. 1, S. America. See Pirozynski (*Kew Bull.* **31**: 595, 1977), Rossman (*Mycol. Pap.* **157**: 71 pp., 1987).

Hyalodermella Speg. (1918), Pezizomycotina. 1, S. America.

Hyalodictys Subram. (1962) = Miuraea fide Deighton (*Mycol. Pap.* **133**, 1973).

Hyalodictyum Woron. (1916), anamorphic *Pezizomycotina*, Cac.#eH.1. 1, former USSR. See Mel'nik (*Opredelitel' Gribov Rossii* Klass Coelomycetes Byp. **1**. Redkie i Maloizvestnye Rody, 1997).

Hyalodothis Pat. & Har. (1893) nom. dub. ? = Melanodothis fide Arnold (*Mycol.* **59**: 246, 1967).

Hyaloflorea Bat. & H. Maia (1955) = Cylindrocarpon fide Carmichael *et al.* (*Genera of Hyphomycetes*, 1980).

Hyalohelicomina T. Yokoy. (1974), anamorphic *Pezizomycotina*, Hso.0-≡ hH.1. 1, Japan. See Yokoyama (*TMSJ* **15**: 159, 1974).

hyalohyphomycosis, a mycotic infection of humans or animals caused by a non-dematiaceous fungus. Cf. phaeohyphomycosis.

Hyalomelanconis Naumov (1954) = Melanconis fide Hawksworth *et al.* (*Dictionary of the Fungi* edn 8, 1995).

Hyalomeliolina F. Stevens (1924), Pseudoperisporiaceae. 2, C. & S. America. See Sivanesan (*SA* **6**: 201, 1987), Hughes (*Mycol. Pap.* **166**, 1993).

Hyalopesotum H.P. Upadhyay & W.B. Kendr. (1975) = Pesotum fide Upadhyay & Kendrick (*Mycol.* **67**: 801, 1975), Harrington *et al.* (*Mycol.* **93**: 111, 2001).

Hyalopeziza Fuckel (1870), Hyaloscyphaceae. Anamorph *Phialophora*-like. 20, widespread. See Raschle (*Sydowia* **29**: 170, 1977; key, anam.), Korf & Kohn (*Mycotaxon* **10**: 503, 1980), Huhtinen (*Karstenia* **29**: 45, 1990), Raitviir & Huhtinen (*Mycotaxon* **62**: 445, 1997), Baral & Galán (*Beitr. Kenntn. Pilze Mitteleur.* **12**: 133, 1999), Raitviir (*Scripta Mycol.* **20**, 2004).

Hyalopleiochaeta R.F. Castañeda, Guarro & Cano (1996), anamorphic *Pezizomycotina*, Hso.?.?. 1, Cuba. See Castañeda Ruíz *et al.* (*Mycotaxon* **57**: 458, 1996).

Hyalopsora Magnus (1902), Pucciniastraceae. *c.* 11 (on *Abies* (0, I; where known) (*Pinaceae*); on *Pteridophyta* (II, III)), widespread (esp. north temperate). See Pady (*Ann. Bot.* **49**: 71, 1985; cytology, morphology).

Hyalopus Corda (1838) ? = Acremonium fide Kendrick & Carmichae *in* Ainsworth *et al.* (Eds) (*The Fungi* **4A**: 390, 1973).

Hyalopycnis Höhn. (1918), anamorphic *Heterogastridium*. 1, Europe; N. America. See Seeler (*Farlowia* **1**: 119, 1943), Scorzetti *et al.* (*FEMS Yeast Res.* **2**: 495, 2002; phylogeny).

Hyalopyrenia H. Harada (1996), ? Verrucariaceae (L). 1, Japan. See Harada (*Lichenologist* **28**: 415, 1996).

Hyaloraphidium Korshikov (1931), ? Chytridiomycetes. 1. See Ustinova *et al.* (*Protist* **151**: 253, 2000), Forget *et al.* (*Mol. Biol. Evol.* **19**: 310, 2002; mol. biol.).

Hyalorbilia Baral & G. Marson (2000), Orbiliaceae. Anamorph *Dactylella*. 1, widespread. See Baral & Marson (*Micologia 2000*: 44, 2000), Liu *et al.* (*Nova Hedwigia* **81**: 145, 2005; China), Wu *et al.* (*Fungal Diversity* **25**: 233, 2007).

Hyalorhinocladiella H.P. Upadhyay & W.B. Kendr. (1975), anamorphic *Ceratocystiopsis, Ophiostoma*, Hso.0eH.10. 1, widespread. See Upadhyay & Kendrick (*Mycol.* **67**: 800, 1975), Benade *et al.* (*Mycol.* **87**: 298, 1995), Benade *et al.* (*CJB* **74**: 891, 1996), Zipfel *et al.* (*Stud. Mycol.* **55**: 75, 2006; phylogeny).

Hyaloria Möller (1895), Hyaloriaceae. 1, Brazil. See Wells (*Mycol.* **61**: 77, 1969), Weiss & Oberwinkler (*MR* **105**: 403, 2001), Weiss *et al.* (*Frontiers in Basidiomycote Mycology*: 7, 2004).

Hyaloriaceae Lindau (1897), Auriculariales. 3 gen., 8 spp.

Lit.: Wells (*Mycol.* **61**: 77, 1969), Bandoni (*Trans. Mycol. Soc. Japan* **25**: 489, 1984), Bononi & Capelari (*Rickia* **11**: 109, 1984), Roberts (*Mycotaxon* **69**: 209, 1998), Weiss & Oberwinkler (*MR* **105**: 403, 2001), Kirschner & Chen (*Stud. Mycol.* **50**: 337, 2004).

Hyaloscolecostroma Bat. & J. Oliveira (1967), ? Capnodiaceae. 1, Brazil. See Batista & Oliveira (*Atas Inst. Micol. Univ. Pernambuco* **5**: 448, 1967).

Hyaloscypha Boud. (1885) nom. cons., Hyaloscyphaceae. Anamorphs *Clathrosphaerina, Pseudaegerita, Cheiromycella*. 38, widespread. See Svrček (*Česká Mykol.* **39**: 205, 1985; spp. descr. by Velenovský), Huhtinen (*Karstenia* **29**: 45, 1990; key), Cantrell & Hanlin (*Mycol.* **89**: 745, 1997; DNA), Hosoya & Huhtinen (*Mycoscience* **43**: 405, 2002; Japan), Raitviir (*Scripta Mycol.* **20**, 2004), Wang *et al.* (*Mol. Phylogen. Evol.* **41**: 295, 2006; phylogeny).

Hyaloscyphaceae Nannf. (1932), Helotiales. 74 gen. (+ 61 syn.), *c.* 933 spp.

Lit.: Dennis (*Mycol. Pap.* **32**, 1949; UK), Korf (*Lloydia* **14**: 129, 1951; *Arachnopezizae*), Dennis (*Kew Bull.*: 289, 1954; trop. Am.), Dennis (*Kew Bull.* **13**: 32, 1958; Australia), Dennis (*Kew Bull.* **14**: 418, 1960; Venezuela), Dennis (*Kew Bull.* **17**: 319, 1963; redispos.), Dennis (*Persoonia* **2**: 171, 1972; *Lachneae*), Hein (*Nova Hedwigia* **32**: 31, 1980; SEM hairs), Korf & Kohn (*Mycotaxon* **10**: 503, 1980; spp. glassy hairs), Sharma (*Nova Hedwigia* **43**: 381, 1986; 76 spp. India), Baral (*Z. Mykol.* **53**: 119, 1987), Huhtinen (*Mycotaxon* **29**: 267, 1987; hairs, gen. limits), Raitviir (*Mikol. Fitopatol.* **21**: 200, 1987; subfams.), Spooner (*Biblthca Mycol.* **116**: 711 pp., 1987; Australasia), Svrček (*Česká Mykol.* **41**: 193, 1987; key 50 gen. Eur.), Baral (*Beitr. Kenntn. Pilze Mitteleur.* **5**: 209, 1989), Huhtinen (*Karstenia* **29**: 45, 1989), Haines (*Nova Hedwigia* **54**: 97, 1992), Galán *et al.* (*MR* **98**: 1137, 1994), Verkley (*Nova Hedwigia* **63**: 215, 1996), Cantrell & Hanlin (*Mycol.* **89**: 745, 1997; phylogeny), Hosoya & Otani (*Mycoscience* **38**: 187, 1997; Jap. spp.), Zhuang (*Mycotaxon* **69**: 359, 1998),

Leenurm *et al.* (*Sydowia* **52**: 30, 2000; ultrastr.), Raitviir (*Scripta Mycol.* **20**, 2004), Zhang & Zhuang (*Nova Hedwigia* **78**: 475, 2004), Wang *et al.* (*Mycol.* **98**: 1065, 2006; phylogeny), Wang *et al.* (*Mol. Phylogen. Evol.* **41**: 295, 2006).

Hyaloseta A.W. Ramaley (2001), Niessliaceae. Anamorph *Monocillium*. 1, USA. See Ramaley (*Mycotaxon* **79**: 269, 2001).

Hyalosphaera F. Stevens (1917), Dothideomycetes. 3, C. & S. America. See Rossman (*Mycol. Pap.* **157**, 1987; key).

Hyalospora Nieuwl. (1916) ≡ Catharinia.

Hyalostachybotrys Sriniv. (1958) = Stachybotrys fide Barron (*Mycol.* **56**: 313, 1964).

Hyalosynnema Matsush. (1975), anamorphic *Pezizomycotina*, Hsy.≡ eH.10. 1, Japan; Taiwan. See Matsushima (*Icon. microfung. Matsush. lect.*: 85, 1975), Kirschner *et al.* (*Fungal Science* Taipei **16**: 47, 2001).

Hyalotexis Syd. (1925) = Hyalosphaera fide Petrak (*Annls mycol.* **26**: 385, 1928).

Hyalotheles Speg. (1908), Elsinoaceae. 1, S. America.

Hyalothyridium Tassi (1900), anamorphic *Pezizomycotina*, Cpd.#eH.1. 3, widespread. See Latterell & Rossi (*Mycol.* **76**: 506, 1984; reinstatement of genus), Nag Raj (*CJB* **67**: 3169, 1989).

Hyalothyris Clem. (1909) ≡ Hyalothyridium.

Hyalotia Guba (1961) = Bartalinia fide Nag Raj (*CJB* **53**: 1615, 1975).

Hyalotiastrum Abbas, B. Sutton, Ghaffar & Abbas (2003), anamorphic *Pezizomycotina*. 1, Pakistan. See Abbas *et al.* (*Pakist. J. Bot.* **35**: 449, 2003).

Hyalotiella Papendorf (1967), anamorphic *Pezizomycotina*, St.≡ eP.10. 2, widespread (tropical to subtropical; S. Africa). See Papendorf (*TBMS* **50**: 69, 1967).

Hyalotiopsis Punith. (1970), anamorphic *Ellurema*, St.≡ eP.19. 1, India. See Punithalingam (*Mycol. Pap.* **119**: 12, 1969).

Hyalotricha Dennis (1949) = Hyalopeziza fide Raschle (*Sydowia* **29**: 170, 1977).

Hyalotrochophora Finley & E.F. Morris (1967) ? = Delortia fide Kendrick & Carmichae *in* Ainsworth *et al.* (Eds) (*The Fungi* **4A**: 390, 1973).

Hybogaster Singer (1964), Hybogasteraceae. 1, Chile. See Singer (*Sydowia* **17**: 13, 1964).

Hybogasteraceae Jülich (1982), Russulales. 1 gen., 1 spp.

Lit.: Singer (*Sydowia* **17**: 13, 1964).

hybrid, the result of a cross between organisms belong to different taxa which yield viable progeny. It is unclear how common the process is in nature with fungi; as with plants, hybridization can occur at levels from genus down; e.g. *Saccharomycopsis fibuligera* × *Yarrowia lipolytica* (Nga *et al.*, *J. gen. Microbiol.* **138**: 223, 1992), *Cladonia grayi* × *C. merochlorophaea* (Culberson *et al.*, *Am. J. Bot.* **75**: 1135, 1988); a **sexual -**. A **mechanical -** (in lichens) is formed by the growth together of propagules from different genera, species or genotypes to form a single thallus not involving sexual crossing (Hawksworth, *in* Street (Ed.), *Essays in plant taxonomy*: 211, 1978; see Lichens). The existing rules of nomenclature for hybrids and grafts are available to name both kinds of progeny (Hawksworth, *Internat. Lich. Newsl.* **21**: 59, 1988).

Hydnaceae Chevall. (1826), Cantharellales. 9 gen. (+ 12 syn.), 190 spp.

Lit.: Coker & Beers (*Stipitate hydnums of the Eastern United States*, 1951), Maas Geesteranus (*Fungus* (*The stipitate hydnums of the Netherlands*, I) **26**, 1956), Maas Geesteranus (*Fungus* (*The stipitate hydnums of the Netherlands*, II) **26**, 1956), Maas Geesteranus (*Fungus* (*The stipitate hydnums of the Netherlands*, III) **27**, 1957), Maas Geesteranus (*Persoonia* (*The stipitate hydnums of the Netherlands*, IV) **1**: 115, 1959), Harrison (*The stipitate hydnums of Nova Scotia* [*Can. Dep. Agric. Res. Branch Publ.* **1099**], 1961), Nikolaeva (*Fl. Pl. Crypt. URSS* **6** Fungi **2**, 1961), Donk (*Persoonia* **3**: 199, 1964), Maas Geesteranus (*Hydnaceous fungi of the eastern old world*, 1971), Harrison (*Mycol.* **65**: 277, 1973), Maas Geesteranus (*Verh. K. Ned. Akad. Wet., Natur.* **60**, 1975; Eur. spp.), Maas Geesteranus (*Die terrestrichen Stachelpilze Europas: the terrestrial hydnums of Europe*, 1975), Grand & van Dyke (*J. Elisha Mitchell scient. Soc.* **92**: 114, 1976; SEM spores), Harrison & Grund (*Mycotaxon* **28**: 427, 1987), Harrison & Grund (*Mycotaxon* **28**: 419, 1987), Wilkinson (*CJB* **65**: 150, 1987; as *Sistotremataceae*), Ryvarden (*Syn. Fung.* **5**: 363 pp., 1991; as *Sistotremataceae*), Gulden & Hanssen (*Sommerfeltia* **13**: 58 pp., 1992), Larsson (*MR* **98**: 1153, 1994; as *Sistotremataceae*), Agerer *et al.* (*Nova Hedwigia* **63**: 183, 1996), Larsson (*Nordic Jl Bot.* **16**: 83, 1996; as *Sistotremataceae*), Larsson (*Nordic Jl Bot.* **16**: 73, 1996; as *Sistotremataceae*), Ginns (*Mycol.* **90**: 19, 1998; as *Sistotremataceae*), Pine *et al.* (*Mycol.* **91**: 944, 1999), Greslebin (*MR* **105**: 1392, 2001; as *Sistotremataceae*), Binder & Hibbett (*Mol. Phylogen. Evol.* **22**: 76, 2002), Hibbett & Binder (*Proc. R. Soc. Lond.* B. Biol. Sci. **269**: 1963, 2002; as *Sistotremataceae*), Larsson *et al.* (*MR* **108**: 983, 2004), Binder *et al.* (*Systematics and Biodiversity* **3**: 113, 2005; as *Sistotremataceae*), Küffer & Senn-Irlet (*Mycol. Progr.* **4**: 77, 2005; as *Sistotremataceae*).

Hydnangiaceae Gäum. & C.W. Dodge (1928), Agaricales. 2 gen. 76 spp.

Lit.: Bougher *et al.* (*MR* **97**: 613, 1993; generic delimitation), Kropp & Mueller (*Laccaria – Ectomycorrhizal fungi, key genera in profile*, 1999; generic classification), Hibbett *et al.* (*Nature* **407**: 506, 2000), Moncalvo *et al.* (*Syst. Biol.* **49**: 278, 2000).

Hydnangium Wallr. (1839) = Laccaria fide Kuyper (*in litt.*); originally treated as a sequestrate genus.

Hydnellum P. Karst. (1879), Bankeraceae. 38, widespread. See Coker & Beers (*The Stipitate Hydnums of the Eastern United States*: 56, 1951), Maas Geesteranus (*Persoonia* **2**: 388, 1962), Stalpers (*Stud. Mycol.* **35**: 168 pp., 1993; key), Dickson (*Field Mycology* **1**: 99, 2000; key British spp.), Arnolds (*Coolia* **46** 3, Suppl.: 96 pp., 2003; Netherlands and Belgium), Parfitt *et al.* (*MR* **111**: 761, 2007; molecel. phylog. British spp.).

Hydnellum P. Karst. (1896) nom. dub., Basidiomycota. See Stalpers (*in litt.*).

Hydnites Mesch. (1892), Fossil Fungi. 2 (Tertiary), Europe.

Hydnobolites Tul. & C. Tul. (1843), Pezizaceae. 3 (hypogeous), N. America; Europe. See Kimbrough *et al.* (*Bot. Gaz.* **152**: 408, 1991; posn), Hansen *et al.* (*Mycol.* **93**: 958, 2001; phylogeny), Læssøe & Hansen (*MR* **111**: 1075, 2007; phylogeny).

Hydnocaryon Wallr. (1833) = Genea fide Trappe (*TBMS* **65**: 496, 1975), Læssøe & Hansen (*MR* **111**: 1075, 2007; phylogeny).

Hydnochaete Bres. (1896), Hymenochaetaceae. 8, widespread (esp. tropical). See Ryvarden (*Mycotaxon* **15**: 425, 1982), Parmasto (*Mycotaxon* **91**: 137, 2005).
Hydnochaete Peck (1897) ≡ Hydnochaetella.
Hydnochaetella Sacc. (1898) ≡ Asterodon fide Donk (*Taxon* **5**: 96, 1956).
Hydnocristella R.H. Petersen (1971), Lentariaceae. 1, N. America. See Petersen (*Česká Mykol.* **25**: 130, 1971).
Hydnocystis Tul. (1844), Pyronemataceae. 2 (hypogeous), Europe; Japan. See Burdsall (*Mycol.* **60**: 503, 1968), van Brummelen *in* Hawksworth (Ed.) (*Ascomycete Systematics. Problems and Perspectives in the Nineties* NATO ASI Series vol. **269 269**: 398, 1994; posn), Læssøe & Hansen (*MR* **111**: 1075, 2007; phylogeny).
Hydnodon Banker (1913) = Trechispora fide Ryvarden (*Syn. Fung.* **15**: 31, 2002).
Hydnodontaceae Jülich (1982), Trechisporales. 15 gen. (+ 7 syn.), 105 spp.
Lit.: Larsson (*MR* **111**: 1040, 2007; phylogeny).
Hydnofomes Henn. (1900) = Echinodontium fide Banker (*Mycol.* **5**: 293, 1914).
Hydnogloea Curr. (1871) = Pseudohydnum P. Karst. fide Donk (*Persoonia* **4**: 173, 1966).
hydnoid, of a form resembling *Hydnum*; specifically, with a teeth-like hymenium.
Hydnomerulius Jarosch & Besl (2001), Paxillaceae. 1. Perhaps *Boletaceae*. See Jarosch & Besl (*Pl. Biol.* **3**: 447, 2001).
Hydnophlebia Parmasto (1967), Meruliaceae. 1, N. America. See Parmasto (*Consp. Syst. Cort.*: 384, 1967).
Hydnophysa Clem. (1909) ≡ Hydnofomes.
Hydnoplicata Gilkey (1955), Pezizaceae. 1 (hypogeous), Australia. See Trappe (*Mycotaxon* **2**: 109, 1975), Trappe & Claridge (*Australas. Mycol.* **25**: 33, 2006).
Hydnopolyporus D.A. Reid (1962), Meripilaceae. 3, widespread. See Fildago (*Mycol.* **55**: 713, 1963).
Hydnoporia Murrill (1907) = Hydnochaete Bres. fide Banker (*Mycol.* **5**: 293, 1914).
Hydnopsis (J. Schröt.) Rea (1909) = Tomentella Pat. fide Donk (*Taxon* **5**: 69, 1956).
Hydnopsis Tul. & C. Tul. (1865) nom. dub., anamorphic *Pezizomycotina*. See Sutton (*Mycol. Pap.* **141**, 1977).
Hydnospongos Wallr. (1839) = Gautieria fide Stalpers (*in litt.*).
Hydnotrema Link (1833) ≡ Sistotrema Fr.
Hydnotrya Berk. & Broome (1846), Discinaceae. *c.* 15 (hypogeous), widespread. See Gilkey (*N. Amer. Fl.* **2**, 1954; key), Zhang (*Mycotaxon* **42**: 155, 1991), Pegler *et al.* (*British truffles*, 1993; key 4 Br. spp.), Abbott & Currah (*Mycotaxon* **62**: 1, 1997; N. Am.), O'Donnell *et al.* (*Mycol.* **89**: 48, 1997; phylogeny), Harrington *et al.* (*Mycol.* **91**: 41, 1999; phylogeny), Hansen & Pfister (*Mycol.* **98**: 1029, 2006; phylogeny), Tedersoo *et al.* (*New Phytol.* **170**: 581, 2006; ectomycorrhiza), Læssøe & Hansen (*MR* **111**: 1075, 2007; phylogeny).
Hydnotryaceae M. Lange (1956) = Discinaceae.
Hydnotryopsis Gilkey (1916), Pezizaceae. 2 (hypogeous), N. America. See Gilkey (*N. Amer. Fl.* **2**, 1954; key, as *Choiromyces*), Trappe (*Mycotaxon* **2**: 115, 1975), Hansen *et al.* (*Mol. Phylogen. Evol.* **36**: 1, 2005; phylogeny), Tedersoo *et al.* (*New Phytol.* **170**: 581, 2006; ectomycorrhiza), Læssøe & Hansen (*MR* **111**: 1075, 2007; phylogeny).
Hydnum L. (1753) nom. cons., Hydnaceae. 120, widespread. Some spp. are a cause of heartwood-rot in living trees. See Petersen (*Taxon* **26**: 144, 1977; typification).
Hydrabasidium Park.-Rhodes ex J. Erikss. & Ryvarden (1978) = Scotomyces fide Jülich (*Persoonia* **10**: 334, 1979).
Hydraeomyces Thaxt. (1896), Laboulbeniaceae. 1, widespread. See Siemaszko & Siemaszko (*Polsk. Pismo Entom.* **12**: 125, 1933), Tavares (*Mycol. Mem.* **9**: 627 pp., 1985), Santamaría (*Fl. Mycol. Iberica* **5**, 2003).
Hydrocina Scheuer (1991), Helotiales. Anamorph *Tricladium*. 1 (aquatic), British Isles. See Webster *et al.* (*Nova Hedwigia* **52**: 65, 1991), Wang *et al.* (*Am. J. Bot.* **92**: 1565, 2005; phylogeny), Wang *et al.* (*Mycol.* **98**: 1065, 2006; phylogeny).
Hydrocybe (Fr. ex Rabenh.) Wünsche (1877) = Cortinarius fide Kauffman (*N. Amer. Fl.* **10**, 1932).
Hydrocybium Earle (1909) = Cortinarius fide Kauffman (*N. Amer. Fl.* **10**, 1932).
hydrofungi, aquatic fungi (Cooke, 1963).
Hydrogen-ion concentration (pH). Like other organisms, fungi are sensitive to the pH of their environment, and there is a large literature on this topic, exemplified by the following: in plants, fungal cell wall-degrading enzymes are influenced by pH (Akimitsu *et al.*, *Physiological and molecular plant pathology* **65**: 271, 2004); pathogenicity of some conifer needle endophytes has been related to their ability to change host pH, and host resistance to innate needle buffer capacity (Scholz & Stephan, *Eur. J. Forest Pathol.* **4**: 118, 1974); soil pH can determine populations of keratinophilic fungi (Böhme & Ziegler, *Mycopathologia* **38**: 247, 1969); gene expression can be pH regulated in *Aspergillus nidulans* (Denison, *Fungal genetics and biology* **29**: 61, 2000; general review of pH regulation of gene expression in fungi); ambient pH affects gene expression in *Metarrhizium anisopliae* but can itself be altered by the fungus (St Leger *et al.*, *Microbiology***145**: 2691, 1999). In maintaining pure cultures of fungi, the medium pH is also important. Most fungi grow best at approximately pH 7 but tolerate a wide range from pH 3-10 (or even 11). This property is made use of for freeing fungal cultures from bacteria or inhibiting bacterial growth when making isolates from soil by using an acid medium (pH 4). See Webb (*Ann. Mo. bot. Gdn* **6**: 201, 1919), MacInnes (*Phytopath.* **12**: 290, 1922), acidiphilous, alkaphilic.
Hydrogera F.H. Wigg. ex Kuntze (1891) ≡ Pilobolus.
Hydrometrospora J. Gönczöl & Révay (1985), anamorphic *Pezizomycotina*, Hso.1bH.1. 1 (aquatic), Hungary. See Gönczöl & Révay (*Nova Hedwigia* **40**: 199, 1984).
Hydromycus Raf. (1808) = Dacrymyces fide Merrill (*Index Rafinesq.*: 69, 1949).
Hydromyxales. An order of '*Myxomycetes*' fide Jahn (*Nat. PflFam.* Aufl. 2, **2**, 1928) but excluded from this *Dictionary* as protozoans not now studied by 'mycologists'. Cf. monads.
Hydronectria Kirschst. (1925), ? Nitschkiaceae (L). 1 (algicolous), Europe. Marine species are now placed in *Kallichroma*. See Kohlmeyer & Volkmann-Kohlmeyer (*MR* **97**: 753, 1993), Harris *in* Rossman *et al.* (Eds) (*Stud. Mycol.* **42**, 1999).
Hydronectriaceae Riedl (1987) = Nitschkiaceae.

Hydrophilomyces Thaxt. (1908), Laboulbeniaceae. 12, widespread. See Santamaria (*Nova Hedwigia* **82**: 349, 2006; Spain).

Hydrophora Tode (1791) nom. rej. prop. = Mucor Fresen. fide Hesseltine (*Mycol.* **47**: 344, 1955).

Hydrophorus Battarra ex Earle (1909) = Hygrocybe fide Kuyper (*in litt.*).

Hydropisphaera Dumort. (1822) nom. rej., Bionectriaceae. Anamorph *Acremonium*-like. 18 (on dead plant material and old fungi), widespread (esp. tropical). See Gamundí (*Darwiniana* **18**: 548, 1974), Samuels (*N.Z. Jl Bot.* **14**: 231, 1976), Rossman *et al.* (*Mycol.* **85**: 685, 1993), Rossman *et al.* (*Stud. Mycol.* **42**: 248 pp., 1999), Rossman *et al.* (*Mycol.* **93**: 100, 2001; phylogeny), Samuels *et al.* (*CBS Diversity Ser.* **4**, 2006; USA), Zhang & Zhuang (*Mycosystema* **25**: 15, 2006; phylogeny), Zhang *et al.* (*Mycol.* **98**: 1076, 2006; phylogeny).

Hydropus Kühner ex Singer (1948), Marasmiaceae. *c.* 100, widespread (esp. tropical). See Singer (*Fl. Neotrop.* **32**, 1982; key), Robich (*Rivista di Micologia* **35**: 155, 1992), Hausknecht *et al.* (*Öst. Z. Pilzk.* **6**: 181, 1997; Austrian spp.), Pegler & Legon (*Mycologist* **15**: 60, 2001), Moreau & Courtecuisse (*Mycotaxon* **89**: 331, 2004; *Hydropus kauffmanii* from Europe).

Hydrotelamonia Rob. Henry (1957) nom. inval. = Cortinarius fide Kuyper (*in litt.*) used as a genus but only validly published at subgenus rank.

Hydrothyria J.L. Russell (1856) = Peltigera fide Feige *et al.* (*Herzogia* **8**: 69, 1989), Miądlikowska & Lutzoni (*Int. J. Pl. Sci.* **161**: 925, 2000; phylogeny, morphology, chemistry).

hydrotropism, see tropism.

Hygramaricium Locq. (1979) = Cortinarius fide Kuyper (*in litt.*).

Hygroaster Singer (1955), Tricholomataceae. 2, America (tropical). See Singer (*Sydowia* **9**: 370, 1955), Ludwig (*Z. Mykol.* **63**: 155, 1997), Lechner *et al.* (*Mycotaxon* **91**: 9, 2005; Argentina).

Hygrochroma DC. [not traced] ? nom. dub. = Phloeospora fide Mussat (*Syll. fung.* **15**: 170, 1901).

Hygrocrocis C. Agardh (1824) ≡ Typhoderma.

Hygrocybe (Fr.) P. Kumm. (1871), Hygrophoraceae. *c.* 150, widespread. See Orton (*TBMS* **43**: 248, 1960; key), Hesler & Smith (*North American species of Hygrophorus*, 1963; key), Heinemann (*Fl. Icon. Champ. Congo* **15**: 280, 1966; keys 13 spp. Congo), Singer (*Agaric. mod. Tax.* edn 3, 1975), Bon (*Docums Mycol.* **7** no. 25: 1, 1976; keys), Boertmann (*The genus Hygrocybe. Fungi of Northern Europe* **1**, 1995; N. Europ. spp., key), Boertmann (*Biblthca Mycol.* **192**: 168 pp., 2002; monogr.).

Hygromitra Nees (1816) = Leotia fide Fries (*Syst. mycol.* **1**: 1, 1821).

Hygromyxacium Locq. (1979) = Cortinarius fide Kuyper (*in litt.*).

hygrophanous, having a water-soaked appearance when wet.

hygrophilous, preferring a moist habitat.

Hygrophoraceae Lotsy (1907), Agaricales. 9 gen. (+ 14 syn.), 325 spp. Possibly polyphyletic but recent data are consistent with the monophyly of the taxon; not yet listed as nom. rej. against *Tricholomataceae*.

Lit.: Hesler & Smith (*North American species of Hygrophorus*, 1963; key), Chandrasrikul *et al.* (*Thai J. Agric. Sci.* **18**: 287, 1985), Arnolds (*Persoonia* **14**: 43, 1989), Boertmann (*Nordic Jl Bot.* **10**: 311, 1990), Horak (*N.Z. Jl Bot.* **28**: 255, 1990), Boertmann (*Fungi of Northern Europe* **1**: 184 pp., 1995), Bougher & Young (*Mycotaxon* **63**: 25, 1997), Young (*Australas. Mycol.* **18**: 63, 1999), Hibbett (*Nature* **407**: 506, 2000), Moncalvo *et al.* (*Syst. Biol.* **49**: 278, 2000; phylogeny), Beisenherz (*Regensb. Mykol. Schr.* **10**: 3, 2002), Boertmann (*Biblthca Mycol.* **192**: 168 pp., 2002), Moncalvo *et al.* (*Mol. Phylogen. Evol.* **23**: 357, 2002), Young & Mills (*Muelleria* **16**: 3, 2002), Young & Mills (*Muelleria* **16**: 3, 2002; Tasmania), Young (*Australas. Mycol.* **21**: 114, 2002), Cantrell & Lodge (*MR* **108**: 1301, 2004), Larsson & Jacobsson (*MR* **108**: 781, 2004), Young (*Fungi of Australia: Hygrophoraceae*: 188 pp., 2005), Matheny *et al.* (*Mycol.* **98**: 982, 2006; phylogeny).

Hygrophoropsidaceae Kühner (1980), Boletales. 2 gen., 18 spp.

Lit.: Šutara (*Česká Mykol.* **46**: 50, 1992), Binder *et al.* (*Z. Mykol.* **63**: 189, 1997), Bresinsky *et al.* (*Pl. Biol.* **1**: 327, 1999), Watling (*MR* **105**: 1440, 2001), Tuthill & Frisvad (*Mycol.* **94**: 240, 2002).

Hygrophoropsis (J. Schröt.) Maire ex Martin-Sans (1929), Hygrophoropsidaceae. 5, widespread (esp. temperate). See Singer *et al.* (*Beih. Nova Hedwigia* **98**: 6, 1990; key C. Am. spp.).

Hygrophorus Fr. (1836), Hygrophoraceae. *c.* 100 (usually mycorrhizal), widespread (esp. north temperate). See Heinemann (*Bull. Jard. Bot. État* **33**: 421, 1963; keys Centr. Afr. spp.), Hesler & Smith (*North American species of Hygrophorus*, 1963; keys), Bon (*Docums Mycol.* **7** no. 27-28: 1, 1977; keys), Bird & Grund (*Proc. Nova Scotia Int. Sci.* **29**: 1, 1979; keys 53 Can. spp.), Boertmann (*Biblthca Mycol.* **192**: 168 pp., 2002; monogr.).

Hygrophyllum, see *Hypophyllum Paulet*.

hygroscopic (1) becoming soft in wet air, hard in dry; (2) (of a sporocarp), opening and discharging spores in dry air.

Hygrotrama Singer (1959) = Camarophyllopsis fide Kuyper (*in litt.*).

Hylophila Quél. (1886) [non *Hylophila* Lindl. 1833, *Orchidaceae*] ≡ Hebeloma.

Hylostoma Pers. (1822), Pezizales. 1, Europe.

Hymenagaricus Heinem. (1981), Agaricaceae. 10, widespread (esp. tropical). See Heinemann & Sister Little Flower (*Bull. Jard. bot. nat. Belg.* **54**: 151, 1984), Reid & Eicker (*South African Journal of Botany* **64**: 356, 1998; key South African spp.).

Hymenangiaceae Corda (1842) = Cortinariaceae.

Hymenangium Klotzsch (1839) = Descomyces. *Descomyces* is not yet conserved against *Hymenangium.* fide Kuyper (*in litt.*).

Hymenelia Kremp. (1852), Hymeneliaceae (L). 26, widespread (north temperate). See Lutzoni & Brodo (*Syst. Bot.* **20**: 224, 1995), Lumbsch (*Bryologist* **100**: 180, 1997), Wedin *et al.* (*MR* **109**: 159, 2005; phylogeny), Miądlikowska *et al.* (*Mycol.* **98**: 1088, 2006; phylogeny).

Hymeneliaceae Körb. (1855), Ostropomycetidae (inc. sed.) (L). 5 gen. (+ 4 syn.), 37 spp. *Aspicilia* is now excluded and referred to the *Megasporaceae*.

Lit.: Janex-Favre (*Cryptog. Bryol.-Lichénol.* **6**: 25, 1985; ontogeny), Lutzoni & Brodo (*Syst. Bot.* **20**: 224, 1995), Lumbsch (*Bryologist* **100**: 180, 1997), Buschbom & Mueller (*Mol. Phylogen. Evol.* **32**: 66, 2004), Reeb *et al.* (*Mol. Phylogen. Evol.* **32**: 1036, 2004), Miądlikowska *et al.* (*Mycol.* **98**: 1088, 2006; phylogeny), Schmitt *et al.* (*J. Hattori bot. Lab.* **100**:

753, 2006), Hofstetter *et al.* (*Mol. Phylogen. Evol.* **44**: 412, 2007; phylogeny).

Hymenella Fr. (1822), anamorphic *Pezizomycotina*, Cpd.0eH.?. 20, widespread. See Tulloch (*Mycol. Pap.* **130**, 1972; typification).

hymenial algae (or gonidia), algal cells in the hymenium of a lichenized ascomycete, e.g. *Endocarpon*, *Staurothele*, *Thelendia*; **- cystidium**, see cystidium; **- veil**, see annulus.

hymeniderm (of basidiomata), an outer layer composed of an unstratified layer of single cells or hyphal tips; see derm.

Hymeniopeltis Bat. (1959), anamorphic *Pezizomycotina*, Cpt.0eH.?. 1, Brazil. See Batista (*An. Soc. Biol. Pernambuco* **16**: 147, 1959), Farr (*Mycol.* **78**: 269, 1986).

hymenium, the spore-bearing layer of a fruit-body; (of basidiomycetes) **euhymenium** (Donk, *Persoonia* **3**: 210, 1964), a hymenium in which the basidia and their sterile homologues are the first elements to be formed, as a palisade; in a **static** (non-thickening) **euhymenium** the exhausted basidia are replaced at the same level by intercalation; in a **thickening euhymenium** the tramal hyphae grow between the exhausted basidia to form a new hymenium above the old, as in the Cantharellaceae; **catahymenium** (Lemke, *CJB* **42**: 218, 1964), a hymenium in which hyphidia are the first-formed elements and the basidia embedded at various levels elongate to reach the surface and do not form a palisade. Cf. thecium, apothecium.

Hymenoascomycetes. A class proposed by Kimbrough (*in* Wheeler & Blackwell (Eds), *Fungus-insect relationships*: 184, 1984) for operculate and inoperculate discomycetes and pyrenomycetes; see *Ascohymeniales*.

Hymenobactron (Sacc.) Höhn. (1916), anamorphic *Pezizomycotina*, Hsp.0eH.?. 1 or 2, Europe.

Hymenobia Nyl. (1854), Pezizomycotina. 1 (on lichens), Europe. See Triebel (*Biblthca Lichenol.* **35**, 1989; sub *Hymenobiella*).

Hymenobiella Triebel (1989) ≡ Hymenobia fide Eriksson & Hawksworth (*SA* **9**: 13, 1990).

Hymenobolus Durieu & Mont. (1845), Helotiales. 1 (on *Agave*), N. Africa. See Rieuf (*Al Awamia* **4**: 127, 1962).

Hymenochaetaceae Imazeki & Toki (1954), Hymenochaetales. 27 gen. (+ 47 syn.), 487 spp.

Lit.: Donk (*Persoonia* **3**: 199, 1964), Fiasson & Niemelä (*Karstenia* **24**: 14, 1984; emend.), Dai *et al.* (*Mycotaxon* **65**: 273, 1997), Müller *et al.* (*MR* **104**: 1485, 2000), Parmasto (*Mycotaxon* **79**: 107, 2001), Wagner (*Mycotaxon* **79**: 235, 2001), Wagner & Fischer (*MR* **105**: 773, 2001), Germain *et al.* (*Can. J. Pl. Path.* **24**: 194, 2002), Góes-Neto *et al.* (*Mycotaxon* **84**: 337, 2002), Gottlieb *et al.* (*Mycol. Progr.* **1**: 299, 2002), Kauserud & Schumacher (*CJB* **80**: 597, 2002), Wagner & Fischer (*Mycol. Progr.* **1**: 100, 2002), Wagner & Ryvarden (*Mycol. Progr.* **1**: 105, 2002), Nam *et al.* (*Mycobiology* **31**: 133, 2003), Fischer & Binder (*Mycol.* **96**: 799, 2004), Gatica *et al.* (*Phytopath. Mediterr.* **43**: 59, 2004), Binder *et al.* (*Systematics and Biodiversity* **3**: 113, 2005), Küffer & Senn-Irlet (*Mycol. Progr.* **4**: 77, 2005), Parmasto (*Mycotaxon* **91**: 137, 2005).

Hymenochaetales Oberw. (1977). Agaricomycetes. 2 fam., 48 gen., 610 spp. Fams:

(1) **Hymenochaetaceae**

(2) **Schizoporaceae**

For *Lit.* see under fam.

Hymenochaete Lév. (1846) nom. cons., Hymenochaetaceae. 110, widespread (esp. tropical). *H. agglutinans* (canker of young hardwoods). See Job (*Mycol. helv.* **5**: 1, 1990; temp. S. Hemisph. spp.), Léger (*Biblthca Mycol.* **171**: 1, 19XX; world monograph).

Hymenochaetella P. Karst. (1889) = Hymenochaete fide Donk (*Persoonia* **3**: 199, 1964).

Hymenoconidium Zukal (1888) ? = Marasmius fide Donk (*Beih. Nova Hedwigia* **5**, 1962).

Hymenodecton Leight. (1854) nom. rej. prop. = Phaeographis fide Lücking *et al.* (*Taxon* **56**: 1296, 2007; nomencl.).

Hymenogaster Vittad. (1831), Strophariaceae. *c.* 100, widespread (temperate). See Smith (*Mycol.* **58**: 100, 1966; keys N. Am.), Bougher & Castellano (*Mycol.* **85**: 273, 1993), Peintner *et al.* (*Am. J. Bot.* **88**: 2168, 2001), Binder & Bresinsky (*Mycol.* **7**: 85, 2002; phylogeny).

Hymenogasteraceae, see *Hymenogastraceae*.

Hymenogastraceae Vittad. (1831) = Strophariaceae.

Hymenogastrales = Agaricales. See also *Lit.* under *Gasteromycetes*.

Hymenogloea Pat. (1900), Marasmiaceae. 1, America (tropical).

Hymenogrammaceae Jülich (1982) = Grammotheleaceae.

Hymenogramme Mont. & Berk. (1844), Polyporaceae. 1, S.E. Asia. See Ryvarden (*TBMS* **73**: 9, 1979).

Hymenomarasmius Overeem (1927) ? = Marasmius fide Kuyper (*in litt.*).

Hymenomycetes Fr. (1821), see *Basidiomycota*.

Hymenophallus Nees (1816) = Phallus fide Stalpers (*in litt.*).

hymenophore, a spore-bearing structure, esp. a basidioma, or that part of it bearing the hymenium. Cf. sporophore., incl.

Hymenopleella Munk (1953) = Lepteutypa fide Shoemaker & Müller (*CJB* **43**: 1457, 1965), Müller & von Arx *in* Ainsworth *et al.* (Eds) (*The Fungi* **4A**: 87, 1973).

Hymenopodium Corda (1837) = Clasterosporium fide Hughes (*TBMS* **34**: 577, 1951).

hymenopodium (hymenopode), tissue under the hymenium; subhymenium or hypothecium.

Hymenopsis Sacc. (1886), anamorphic *Pezizomycotina*, Ccu.0eP.15/19. 13, widespread. See Sutton (*The Coelomycetes*, 1980), Nag Raj (*Coelomycetous Anamorphs with Appendage-bearing Conidia*, 1993), Kohlmeyer & Volkmann-Kohlmeyer (*MR* **105**: 500, 2001).

Hymenoscypha (Fr.) W. Phillips (1887) ≡ Hymenoscyphus.

Hymenoscyphaceae Bellem. (1978) = Helotiaceae.

Hymenoscyphus Gray (1821), Helotiaceae. Anamorphs *Idriella*, *Geniculospora*, *Dimorphospora*, *Helicodendron*, *Tricladium*, *Articulospora*, *Scytalidium*-like. *c.* 155, widespread. See Dennis (*Persoonia* **3**: 29, 1964), Kimbrough & Atkinson (*Am. J. Bot.* **59**: 165, 1972), Thind & Sharma (*Nova Hedwigia* **32**: 121, 1980; Indian spp.), Dumont (*Mycotaxon* **12**: 313, 1981; neotrop. spp.), Dumont (*Mycotaxon* **13**, 1981; temp. spp.), Descals *et al.* (*TBMS* **83**: 541, 1984), Lizoň (*Mycotaxon* **45**: 1, 1992), Verkley (*Persoonia* **15**: 303, 1993; asci), Baral (*SA* **13**: 113, 1994; gen. concept), Hengstmengel (*Persoonia* **16**:

191, 1996), Chambers *et al.* (*MR* **103**: 286, 1999; sp. on liverworts), McLean *et al.* (*New Phytol.* **144**: 351, 1999; DNA, ericoid mycorrhizas), Monreal *et al.* (*CJB* **77**: 1580, 1999; DNA, ericoid mycorrhizas), Gernandt *et al.* (*Mycol.* **93**: 915, 2001; phylogeny), Zhang & Zhuang (*Mycosystema* **21**: 493, 2002; China), Sivichai *et al.* (*Mycol.* **95**: 340, 2003; Thailand), Athipunyakom *et al.* (*Kasetsart Journal* Natural Sciences **38**: 216, 2004; Spain), Baral *et al.* (*Sydowia* **58**: 145, 2006; phylogeny, Spain), Wang *et al.* (*Mycol.* **98**: 1065, 2006; phylogeny), Wang *et al.* (*Mol. Phylogen. Evol.* **41**: 295, 2006; phylogeny).

Hymenostilbe Petch (1931), anamorphic *Ophiocordyceps*, Hsy.0eH.?. 15, widespread. See Samson & Evans (*Proc. K. ned. Akad. Wet.* Ser. C, Biol. Med. Sci. **78**: 73, 1975; key), Hywel-Jones (*MR* **99**: 154, 1995; Thailand), Hywel-Jones (*MR* **99**: 1201, 1995; on cockroaches), Guzmán *et al.* (*Mycotaxon* **78**: 115, 2001; Mexico), Stensrud *et al.* (*MR* **109**: 41, 2005; phylogeny), Sung *et al.* (*Stud. Mycol.* **57**, 2007; review, phylogeny).

Hymenula Fr. (1825) ≡ Hymenella.

Hypasteridium Speg. (1923) nom. dub., ? Meliolaceae. See Eriksson & Hawksworth (*SA* **7**: 74, 1988).

hyper- (prefix), above.

Hyperdermium J.F. White, R. Sullivan, Bills & Hywel-Jones (2000), Cordycipitaceae. 2, C. & S. America. See White *et al.* (*Mycol.* **92**: 908, 2000), Sung *et al.* (*Stud. Mycol.* **57**, 2007; review, phylogeny), Chaverri *et al.* (*Stud. Mycol.* **60**, 2008; phylogeny, neotropics).

hyperepiphyllous, growing on epiphyllous or foliicolous lichens or bryophytes, esp. in tropical rain forests.

Hyperomyxa Corda (1839) = Cheirospora fide Hughes (*CJB* **36**: 727, 1958).

hyperparasite, a parasite growing on another parasite (obsol.); Fungicolous fungi (q.v.) is a preferable general term where parasitism has not been established.

Hyperphyscia Müll. Arg. (1894), Physciaceae (L). 1 or 9, widespread (esp. tropical). See Hafellner *et al.* (*Herzogia* **5**: 39, 1979), Kashiwadani (*Bull. natn. Sci. Mus.* Tokyo, B **11**: 91, 1985), Moberg (*Nordic Jl Bot.* **7**: 719, 1987; key 7 spp., E. Afr.), Scutari (*Mycotaxon* **61**: 87, 1997), Lohtander *et al.* (*Mycol.* **92**: 728, 2000; phylogeny), Nordin & Mattsson (*Lichenologist* **33**: 3, 2001; phylogeny), Scheidegger *et al.* (*Lichenologist* **33**: 25, 2001; evolution), Helms *et al.* (*Mycol.* **95**: 1078, 2003; phylogeny), Cubero *et al.* (*MR* **108**: 498, 2004; phylogeny).

hyperplasia, over-development of some sort (e.g. swellings, galls, witches' brooms) as a reaction to a disease-producing agent; cf. hypoplasia.

Hyperrhiza Spreng. (1827) ≡ Uperhiza.

hypersaprophyte, a saprophyte only found on substrates invaded by other saprophytes, e.g. *Herpotrichiellaceae*, *Lasiosphaeria*, *Nectria sanguinea*, etc. (Munk, *Sydowia, Beih.* **1**, 1957).

hypersensitivity, increased sensitivity, as in the condition in which there is death of the host tissue at the point of attack by a pathogen, so that the infection does not spread; esp. of reaction to rusts for which the word was first used by Stakman; intolerance; (and see Allergy, Medical fungi).

hypertonic (of culture media), having an osmotic pressure higher than that of the organism cultured; cf. hypotonic.

hypertrophy (of organs, etc.), the state of having growth greater than normal.

hypertrophyte, a parasitic fungus causing hyperplasia in a plant (Wakker).

Hypha Pers. (1822) nom. dub., anamorphic *Fungi*. See Donk (*Taxon* **11**: 86, 1962).

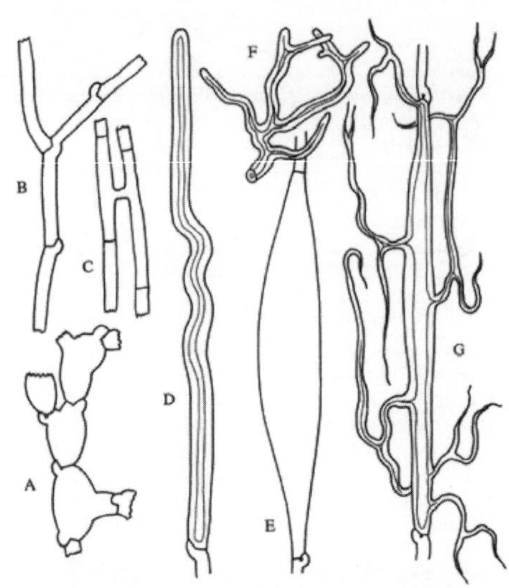

Fig. 16. Hyphal types. A, inflated generative hyphae; B, non-inflated generative hyphae with clamp connexions; C, generative hyphae without clamp connexions; D, unbranched skeletal hypha; E, sarco-hypha; F, highly branched ligative (binding) hypha; G, skeleto-ligative hypha. Not to scale

hypha (pl. **hyphae**), one of the filaments of a mycelium; Vuillemin restricted the term to septate filaments; cf. siphon; **ampoule -**, a swollen hypha as in certain basidiomycetes; **arboriform -**, much branched skeletal - of *Ganoderma*; **ascogenous -**, a dikaryotic hypha from which an ascus develops; **binding -**, see Hyphal analysis; **costiferous -**, transverse ribs (costae) on the inner surface of the hyphal wall of the gill trama in *Paxillus involutus* and *P. filamentosus*, non-amyloid, stained by Congo red and calcofluor but not toluidine blue; **endo-** (intrahyphal -), vegetative or fertile element initiated by the differentiation within a - from the innermost wall layer (Cole & Samson, 1979); **inflated -**, one in which cells behind the growing apex enlarge and cause the apparent rapid rate of growth characteristic of most agaric and gasteromycete basidiocarps; in an **uninflated -** no change of cell size occurs as in most polypores; **mediate -** and **mycelial -**, see Corner (*TBMS* **17**: 54, 1932); **oleiferous -**, do not carry latex (cf. lactifer) but frequently resinous substances (Singer, 1960: 34); **oiliferous -**, a submerged hypha of an endolithic lichen having torulose, guttulate cells; see des Abbayes (*Traité de Lichenologie*, 1951); **racquet -**, one of racket cells (q.v.); **skeletal -**, see Hyphal analysis; **stuffing -**, see Corner (*TBMS* **17**: 54, 1932). See flexuous hypha, hyphal analysis, hyphal peg, textura, Woronin's hypha.

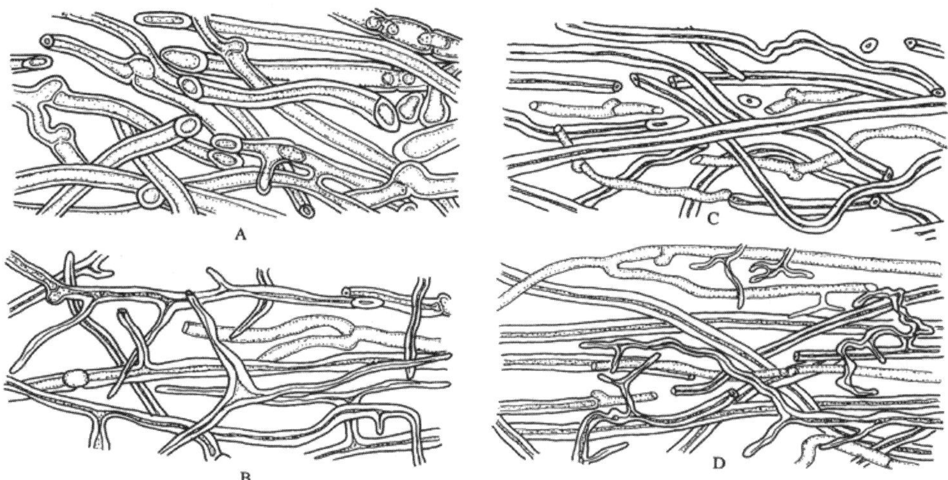

Fig. 17. Hyphal systems. See Pegler (*Bull. BMS* 7(suppl.), 1973). A, monomitic hyphal system, with thick-walled generative hyphae; B, dimitic hyphal system, with generative and ligative (binding) hyphae; C, dimitic hyphal system, with generative and skeletal hyphae; D, trimitic hyphal system, with generative, skeletal and ligative hyphae.

Hyphal analysis. A procedure by which the development and structure of basidiomata can be investigated, providing important taxonomic criteria. Three main types of hyphal systems of increasing complexity were recognized by Corner (*TBMS* **17**: 51, 1932) (Figs 16, 17): **monomitic**, having hyphae of one kind (generative hyphae which are branched, septate, with or without clamp-connexions, thin- to thick-walled, and of unlimited length; they give rise both to other hyphal types and to the hymenium) (Teixeira, *Mycol.* **52**: 30, 1961; gen. hyphae of polypores); **dimitic**, having hyphae of two kinds (generative and skeletal hyphae which are thick-walled, aseptate, and of limited length, with thin-walled apices, generally unbranched but when terminal they can develop arboriform branching or taper) or generative and binding (see below); **trimitic**, having hyphae of three kinds (generative, skeletal and binding (or ligative) hyphae which are aseptate, thick-walled, much branched, either *Bovista*-type with tapering branches or coralloid; they bind the skeletal and generative hyphae together). In *Polyporaceae* and *Lentinaceae*, intercalary skeletal hyphae can give rise to ligative branching, the entire element being termed a skeleto-binding cell (Corner, 1981) or skeleto-ligative hypha (Pegler, 1983). Corner also recognized **sarco-dimitic** (in which the skeletal hyphae are replaced by thick-walled, long, inflating fusiform elements) and **sarco-trimitic** (in which the generative hyphae give both thick-walled inflated elements similar to binding hyphae but septate) types.

Most soft and fleshy basidiomata are monomitic, with hyphae which are generally inflated (most agaricoid and clavarioid fungi). Hard and tough basidiomata may be monomitic with the generative hyphae developing thickened walls, dimitic, with skeletal hyphae (e.g. *Phellinus* spp.) or (esp. when perennial) trimitic (e.g. *Fomes*, *Ganoderma*, *Microporus xanthopus*). Every species has a well-defined and constant construction, which is maintained regardless of changes in the external morphology of the basidioma due to environmental conditions, hence the importance of hyphal analysis in taxonomy.

Lit.: Corner (*Ann. Bot.* **46**: 71, 1932, *Phytomorphology* **3**: 152, 1953, *Beih. Nova Hedw.* **75**: 13, 1983, **78**: 13, 1984), Cunningham (*N. Z. Jl Sci. Tech.* **28**(A): 238, 1946, *TBMS* **37**: 44, 1954), Lentz (*Bot. Rev.* **20**: 135, 1954), Talbot (*Bothalia* **6**: 1, 1951).

hyphal fusions, see anastomosis. Vegetative hyphae of a mycelium may fuse, forming an interconnected network. Fusion between hyphae of different mycelia is controlled by genetic systems which determine sexual (see sex) or vegetative compatibility (q.v.).

hyphal net (‘**Hyphenfilz**’), organ of attachment in some squamulose (placodioid) lichens (e.g. *Psora decipiens*) where a delicately branched reticulate net penetrates the substrate (see Poelt & Baumgärtner, *Öst. bot. Z.* **111**: 1, 1964; rhizinose strand).

hyphal peg (of basidiomata), a bunch of somewhat interwoven hyphae extending from the trama (where it originates) to the hymenium from which it may project (Singer, 1960: 47); (of hyphae), projection from a hypha for fusion (Buller, **5**), peg-hypha.

hyphal rhizoid, a hypha acting as a rhizoid.

Hyphal tip growth, Hyphal elongation is mainly by apex-located synthesis, first confirmed by Gooday (q.v.) and expansion of the hyphal tip (*contra* intercalary growth). Plasma membrane (PM) and cell wall are synthesized in a steep tip-high gradient by fusion of precursor vesicles liberating their contents to be incorporated into the wall as they incorporate their membranes into the PM. Fibrillar wall components (e.g. chitin) are synthesized by tip-located PM-inserted enzymes. The wall is plastic when newly formed but as further wall material is built and the tip extends taking the active growing point away, it gradually becomes inextensible and less plastic. The declining gradient of cell surface extensibility yields

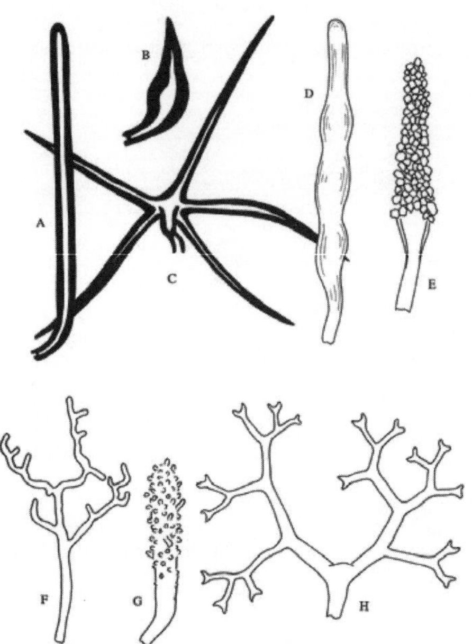

Fig. 18. Hyphidia. A, setal hypha (*Phellinus*); B, seta (*Inonotus*); C, asteroseta (*Asterostroma*); D, gloeohypha (*Gloeocystidiellum*); E, encrusted (*Peniophora*); F, dendrohyphidium (*Cytidia*); G, acanthohyphidium (*Aleurodiscus*); H, dendrohyphidim (*Vararia*). Not to scale

to turgor pressure and is very finely regulated to produce the constant-diameter hypha. Various models have been proposed to help interpret hyphal tip growth. Two models address regulation of extensibility; a) cross-linking of wall polymers, with or without balanced concomitant wall hydrolytic enzyme activity, regulating wall extensibility (Wessels, *in* Heath, *Tip Growth in Plant and Fungal Cells*: 1, 1990) or b) an actin-rich membrane skeleton attached to the inside of the PM regulates extension of the membrane upon which the wall is assembled (Heath, *CJB* **73**(Suppl. 1): S131, 1995). In addition, the gradient of vesicle fusions is postulated to be generated by a 'vesicle supply centre' whose location is coincident with the Spitzenkörper, a labile vesicle- and actin-rich structure present in growing tips of most, but not all, fungi, whose behaviour correlates with tip growth rate and direction (Bartnicki-Garcia, *in* Heath, *Tip Growth in Plant and Fungal Cells*: 211, 1990). Another model attempts to explain the role of calcium gradients in the hyphal tip (Regalado, *Microbiology* **144**: 2771, 1998), while a further model interpret hyphal tip growth in terms of microtubule-based transport (Sugden *et al.*, *Physical Rev. E*, 2007). Current data do not differentiate the relative contributions of these different components, but cellular regulation, possibly involving cytoplasmic Ca^{2+} (Jackson & Heath, *Microbiol. Rev.* **57**: 367, 1993), of each could generate the known diversity of sizes, shapes, branching patterns and directions of growth characteristic of all hyphal structures (see also wall building). See also spitzenkörper).

Hyphasma Rebent. ex Nocca & Balb. (1821) nom. dub., anamorphic *Fungi*. See Donk (*Taxon* **11**: 86, 1962).

Hyphaster Henn. (1903) = Asterostomella fide Höhnel (*Sber. Akad. Wiss. Wien* Math.-naturw. Kl., Abt. 1 **119**: 21, 1910).

Hyphelia Fr. (1825) = Trichothecium. Also used for anamorphs of *Corticium*. fide Hughes (*CJB* **36**: 727, 1958), Donk (*Taxon* **11**: 86, 1962), Hennebert (*Persoonia* **7**: 195, 1973).

hyphidium (pl. **-ia**), (paraphysis, pseudoparaphysis, paraphysoid, dikaryoparaphysis, and pseudophysis sensu Singer (1962) are syn. or near syn.), a little, or strongly, modified terminal hypha in the hymenium of hymenomycetes (Fig. 18). Donk (*Persoonia* **3**: 229, 1964) distinguished; **haplo-** (simple -), unmodified, unbranched or little branched; **dendro-** (dendrophysis), irregularly strongly branched; **dicho-** (dichophysis), repeatedly dichotomously branched; **acantho-** (acanthophysis; bottle-brush paraphysis (Burt, 1918)), having pin-like outgrowths near the apex; in *Corticiaceae* may be botryose, clavate, coralloid, or cylindrical. Cf. cystidium.

Hyphobasidiofera K. Matsush. & Matsush. (1996), Agaricomycetes. 1, Malaysia. See Matsushima & Matsushima (*Matsush. Mycol. Mem.* **9**: 33, 1996).

Hyphochlaena Cif. (1962), anamorphic *Pezizomycotina*, Hso.?.?. 1 (sterile mycelium), Dominican Republic. See Ciferri (*Atti Ist. bot. Univ. Lab. crittog. Pavia* sér. 5 **19**: 110, 1962).

hyphocystidium, see cystidium.

Hyphoderma Fr. (1849) ≡ Hyphelia fide Donk (*Taxon* **11**: 88, 1962).

Hyphoderma Wallr. (1833), Meruliaceae. *c.* 95, widespread. s. str. 35 species. See Donk (*Fungus* Wageningen **27**: 13, 1957), Donk (*Persoonia* **2**: 220, 1962), Eriksson & Ryvarden (*Cortic. N. Europ.* **3**: 448, 1975; key 27 Eur. spp.), Wu (*Acta Bot. Fenn.* **142**: 64, 1990; key Taiwan spp.), Gilbertson & Hemmes (*Mem. N. Y. bot. Gdn* **89**: 81, 2004; Hawaii), Larsson (*MR* **111**: 186, 2007; phylogeny).

Hyphodermataceae Jülich (1982) = Meruliaceae.

Hyphodermella J. Erikss. & Ryvarden (1976), Phanerochaetaceae. 3, widespread. See Melo & Hjortstam (*Nova Hedwigia* **77**: 351, 2003; Portugal).

Hyphodermopsis Jülich (1982) = Gyrophanopsis fide Stalpers & Buchanan (*N.Z. Jl Bot.* **29**: 333, 1991).

Hyphodictyon Millardet (1866) = Atichia fide von Arx & Müller (*Stud. Mycol.* **9**, 1975).

Hyphodiscosia Lodha & K.R.C. Reddy (1974), anamorphic *Pezizomycotina*, Hso.1eH.1/10. 4, India; Europe; Japan. See Lodha & Reddy (*TBMS* **62**: 419, 1974), Watanabe (*Mycol.* **84**: 113, 1992).

Hyphodiscosioides Matsush. (1993), anamorphic *Pezizomycotina*, Hso.0eH.15/19. 1, Peru. See Matsushima (*Matsush. Mycol. Mem.* **7**: 54, 1993).

Hyphodiscus Kirschst. (1906), Helotiales. Anamorph *Catenulifera*. 6, north temperate. See Zhuang (*Mycotaxon* **31**: 411, 1988), Baral (*Z. Mykol.* **59**: 3, 1993), Raitviir & Galán (*SA* **13**: 159, 1995), Hosoya (*Mycoscience* **43**: 47, 2002; Japan), Raitviir (*Scripta Mycol.* **20**, 2004), Untereiner *et al.* (*CJB* **84**: 243, 2006; phylogeny).

Hyphodontia J. Erikss. (1958), Schizoporaceae. 64, widespread. See Weresub (*CJB* **39**: 1475, 1961),

Eriksson & Ryvarden (*Cortic. N. Europ.* **4**: 583, 1976; key 23 Eur. spp.), Wu (*Acta Bot. Fenn.* **142**: 85, 1990; key Taiwan spp.), Langer (*Biblthca Mycol.* **154**, 1994; world key), Wu (*Mycol.* **93**: 1019, 2001; new spp.).

Hyphodontiastra Hjortstam (1999), Meruliaceae. 1, Brazil. See Hjortstam (*Kew Bull.* **54**: 755, 1999).

Hyphodontiella Å. Strid (1975), Clavariaceae. 2, Nordic. See Strid (*Kongel. Norske Vidensk. Selsk. Skr.* 1975 **4**: 19, 1975).

hyphoid (1) like hyphae; cobwebby; (2) (of aecia of *Dasyspora*), having aeciospores on hyphal projections from stomata (Arthur).

Hypholoma (Fr.) P. Kumm. (1871) = Psilocybe (Fr.) P. Kumm. fide Kuyper (*in litt.*).

Hypholomopsis Earle (1909) = Psathyrella fide Singer (*Agaric. mod. Tax.*, 1951).

Hyphomucor Schipper & Lunn (1986), Mucoraceae. 1, paleotropical. See Schipper (*Mycotaxon* **27**: 83, 1986), Voigt & Wöstemeyer (*Gene* **270**: 113, 2001; phylogeny).

Hyphomyces C.H. Bridges & C.W. Emmons (1961) nom. inval., Fungi.

Hyphomycetes, Anamorphic fungi; widespr. in most ecological niches. This artificial class is traditionally separated into three (or four, if agonomycetes are included) orders, based on the presence or absence of conidia and the degree of aggregation of the conidiophores into more complex structures (conidiomata). An alternative system of classification (now obsol.) for the anamorphic fungi as a whole (including the coelomycetes, q.v.) was advanced by Sutton (1980). The traditional separation is:

(1) **Agonomycetales** (Agonomycetaceae) (Mycelia sterilia, q.v.). No conidia generally produced, but in some genera propagules which are liberated by multicellular secession are formed.

(2) **Hyphomycetales** (= Moniliaceae + Dematiaceae). This group comprises the main body of the hyphomycetes; conidiophores are separate, not organized on synnematal or sporodochial conidiomata. The full range of conidiogenous events occurs.

(3) **Stilbellales** (Coremiales, Synnematomycetes, Stilbellaceae). Conidiophores are aggregated as synnemata, and conidiogenous events are various but the 'basauxic' type is absent. See Morris (*Western Ill. Univ. Ser. biol. Sci.* **3**, 1963; key 54 synnem. gen.), Benjamin (*Taxon* **17**: 521, 1968; typification), Seifert (*Stud. Mycol.* **27**, 1985; *Stilbella* and related genera).

(4) **Tuberculariales** (Tuberculariaceae). Conidiophores are aggregated on sporodochial conidiomata and a wide range of conidiogenous events occurs.

Lit.: Costantin (*Les mucedinées simples*, 1888; *Rabenh. Krypt.-Fl.* **1** (8-9), 1907-10), Wakefield & Bisby (*TBMS* **25**: 49, 427, 1941; Brit. list), Hughes (*CJB* **36**: 727, 1958; annotated list 400 gen. names), Delitsch (*Systematik der Schimmelpilze*, 1961), Kendrick & Carmichael (*in* Ainsowrth *et al.*, *The Fungi* **4**: 323, 1973; synoptic keys, illustr., gen., gen. names), Litvinov ([Keys for the Identification of microscopic soil fungi. *Moniliales*], 1967), Barron (*The genera of hyphomycetes from soil*, 1968), Onions *et al.* (*Smith's Introduction to Industrial Mycology* edn 7, 1981), Ellis (*Dematiaceous Hyphomycetes*, 1971; *More Dematiaceous Hyphomyetes*, 1976), Subramanian (*Hyphomycetes. An account of Indian species, except Cercosporae*, 1971), Carmichael *et al.* (*Genera of Hyphomycetes*, 1980), Cole & Kendrick (Eds) (*Biology of conidial fungi*, 2 vols, 1981), Subramanian (*Hyphomycetes: taxonomy and biology*, 1983), Borowska (*Flora Polska, Grzyby (Mycota)*) **16** *Deuterom., Dematiaceae, Phialoconidiae*, 1986; keys to spp. in major genera), Domsch *et al.* (*Compendium of soil fungi*, 2 vols, 1980). See also under Anamorphic fungi.

Hyphonectria (Sacc.) Petch (1937) ≡ Hydropisphaera fide Rossman *et al.* (*Stud. Mycol.* **42**: 248 pp., 1999).

hyphophore, erect stalked peltate asexual sporophores in the *Asterothyriaceae* (e.g. *Echinoplaca*, *Gyalideopsis*, *Tricharia*). See Sérusiaux & De Sloover (*Veröff. Geobot. Inst. ETH, Rübel* **91**: 260, 1986; types), Vězda (*Čas. slez. Muz. Silesiae* A **22**: 67, 1973; *Folia geobot. phytotax., Praha* **14**: 43, 1979).

Hyphopichia Arx & Van der Walt (1976), Saccharomycetales. Anamorph *Candida*. 1 (insect-associated), widespread. See Kurtzman *in* Hawksworth (Ed.) (*Ascomycete Systematics. Problems and Perspectives in the Nineties* NATO ASI Series vol. **269 269**: 361, 1994), Kurtzman (*Antonie van Leeuwenhoek* **88**: 121, 2005), Suh *et al.* (*Mycol.* **98**: 1006, 2006; phylogeny).

hyphopodium, a short branch of one or two cells on epiphytic mycelium of *Meliolales*, etc.; in a **capitate** - the end is rounded, = appressorium (fide Mibey & Hawksworth, *SA* **14**: 25, 1995); **mucronate** - = conidiogenous cell (fide Hughes, *CJB* **59**: 1514, 1981). A **stigmatopodium** (**stigmopodium**) is a hyphopodium in which the end cell or **stigmatocyst** has a haustorium (Arnaud). A stigmatocyst in a hypha is a **node cell**; cf. Doidge (*Bothalia* **4**: 273, 1942). See also Walker (*Mycotaxon* **11**: 1, 1980).

Hyphopolynema Nag Raj (1977), anamorphic *Pezizomycotina*, Hsp.0-1eH.15. 3, Colombia. See Nag Raj (*CJB* **55**: 760, 1977), Kohlmeyer & Volkmann-Kohlmeyer (*Mycotaxon* **70**: 489, 1999).

Hyphoradulum Pouzar (1987), Cyphellaceae. 1, Europe. See Pouzar (*Česká Mykol.* **41**: 26, 1987).

Hyphoscypha Bres. (1903) = Lachnum fide Huhtinen (*SA* **6**: 131, 1987).

Hyphoscypha Velen. (1934), ? Helotiales. 1, Europe.

Hyphosoma Syd. (1924) nom. conf., anamorphic *Pezizomycotina*. See Hughes (*N.Z. Jl Bot.* **8**: 153, 1970).

Hyphospora A.W. Ramaley (1996), anamorphic *Comminutispora*, Hso.?.?. 1, USA. See Ramaley (*Mycol.* **88**: 133, 1996), Hambleton *et al.* (*Mycol.* **95**: 959, 2003; phylogeny), Tsuneda *et al.* (*Mycol.* **96**: 1128, 2004; morphology).

Hyphostereum Pat. (1892), anamorphic *Pezizomycotina*, Hsp/Ccu.0eH.?. 1, S. America.

Hyphothyrium B. Sutton & Pascoe (1989), anamorphic *Pezizomycotina*, Hsp.0-1eP.19. 1, Australia. See Sutton & Pascoe (*MR* **92**: 219, 1989).

Hyphozyma de Hoog & M.T. Sm. (1981), anamorphic *Leotiomycetes*, Hso.0eH.3/4. 4 (on other fungi), Europe; N. America. The genus is highly polyphyletic. See also *Eleutheromyces*. See de Hoog & Smith (*Antonie van Leeuwenhoek* **52**: 39, 1986; key), Hutchison *et al.* (*MR* **97**: 1409, 1993; key), Tsuneda *et al.* (*Mycol.* **89**: 867, 1997; on *Lentinula*), Maekawa & Tsuneda (*Mycotaxon* **78**: 167, 2001; on *Trametes*), Lopandic *et al.* (*Mycol. Progr.* **4**: 205, 2005; phylogeny).

hypnocyst, an *Alternaria*-like group of cells (Chippindale, *TBMS* **14**: 203, 1929; Griffiths, *Nova Hedw.* **25**: 511, 1974).

hypnospora, a resting spore.

Hypnotheca Tommerup (1970), ? Helotiales. Anamorph *Monochaetiellopsis*. 1, Australia. See von Arx & Müller (*Stud. Mycol.* **9**: 11, 1975).
hypo- (prefix), under.
hypobasidium, see basidium.
Hypoblema Lloyd (1902) = Calvatia fide Zeller & Smith (*Lloydia* **27**: 167, 1964).
Hypobryon Döbbeler (1983), Dothideomycetes. 6 (on hepatics), Europe. See Döbbeler (*Nova Hedwigia* **37**: 3, 1983), Döbbeler (*Mycol. Progr.* **5**: 32, 2006).
Hypocapnodium Speg. (1918) = Aithaloderma fide von Arx & Müller (*Stud. Mycol.* **9**, 1975).
Hypocelis Petr. (1929) = Vizella fide von Arx & Müller (*Stud. Mycol.* **9**, 1975).
Hypocenia Berk. & M.A. Curtis (1874) = Topospora fide Groves (*CJB* **43**: 1195, 1965).
Hypocenomyce M. Choisy (1951), Ophioparmaceae (L). 14, widespread. See Timdal (*Nordic Jl Bot.* **4**: 83, 1984; key), Dirig (*Mycotaxon* **37**: 441, 1990; ecology, N America), Timdal (*Mycotaxon* **77**: 445, 2001; n. spp.), Wedin *et al.* (*MR* **109**: 159, 2005; phylogeny), Miądlikowska *et al.* (*Mycol.* **98**: 1088, 2006; phylogeny), Lumbsch *et al.* (*MR* **111**: 257, 2007; phylogeny), Lumbsch *et al.* (*MR* **111**: 257, 2007; phylogeny).
Hypochanum Kalchbr. (1876) ≡ Macowanites.
Hypochnaceae J. Schröt. (1888) = Arthoniaceae.
Hypochnella J. Schröt. (1888), ? Atheliaceae. 1, Europe. See Wakefield (*TBMS* **5**: 127, 1915), Hagara (*Mykol. Listy* **80**: 4, 2002).
Hypochnellaceae Jülich (1982) = Atheliaceae.
Hypochniciellum Hjortstam & Ryvarden (1980), Atheliaceae. 4, widespread. See Legon (*Mycologist* **17**: 41, 2003).
Hypochnicium J. Erikss. (1958), Meruliaceae. 20, widespread. See Eriksson (*Symb. bot. upsal.* **16** no. 1: 100, 1958), Legon (*Mycologist* **17**: 129, 2003; *Hypochnicium analogum*), Nilsson & Hallenberg (*Mycol.* **95**: 54, 2003; phylogeny *Hypochnicium punctulatum* complex).
Hypochnites Mesch. (1898), Fossil Fungi. 1 (Oligocene), Baltic.
hypochnoid, having effused, resupinate, dry, rather loosely intertwined hyphae, as in *Tomentella* (formerly *Hypochnus*).
Hypochnopsis P. Karst. (1889) = Amaurodon fide Kõljalg (*Syn. Fung.* **9**: 32, 1996).
Hypochnus Fr. (1818) ? = Tomentella Pat. fide Donk (*Taxon* **6**: 75, 1957).
Hypochnus Fr. ex Ehrenb. (1820) = Cryptothecia fide Thor (*Bryologist* **94**: 278, 1991).
Hypocline Syd. (1939), anamorphic *Pezizomycotina*, St.0eH.10. 1, Africa.
Hypocopra (Fr.) J. Kickx f. (1867), Xylariaceae. 30 (esp. coprophilous), widespread. See Krug & Cain (*CJB* **52**: 809, 1974; key).
Hypocrea Fr. (1825), Hypocreaceae. Anamorph *Trichoderma*. 171 (on rotten wood, often associated with other fungi), widespread. See Rifai & Webster (*TBMS* **49**: 289, 1964), Rifai & Webster (*TBMS* **49**: 297, 1964), Doi (*Bull. natn. Sci. Mus.* Tokyo, B **12**: 693, 1969), Doi (*Bull. natn. Sci. Mus.* Tokyo, B **15**: 649, 1972), Patil & Patil (*Indian Phytopath.* **36**: 635, 1983; India), Doi & Yamatoya (*Mem. N. Y. bot. Gdn* **49**: 233, 1989; *H. pallida* group), Samuels & Lodge (*Mycol.* **88**: 302, 1996), Lieckfeldt *et al.* (*CJB* **76**: 1507, 1998), Samuels *et al.* (*Stud. Mycol.* **41**, 1998), Rossman *et al.* (*Stud. Mycol.* **42**: 248 pp., 1999), Dodd *et al.* (*MR* **104**: 23, 2000; anamorph phylogeny), Põldmaa (*Stud. Mycol.* **45**: 83, 2000), Chaverri *et al.* (*Mycol.* **93**: 758, 2001), Chaverri *et al.* (*Mycol.* **93**: 1113, 2001), Doi (*Memoirs of the National Science Museum* Tokyo **37**: 113, 2001; Japan), Dodd *et al.* (*Mycol. Progr.* **1**: 409, 2002), Liu *et al.* (*Mycotaxon* **82**: 463, 2002; China), Bezerra *et al.* (*Fitopatol. Brasil* **28**: 408, 2003), Chaverri & Samuels (*Stud. Mycol.* **48**, 2003; spp. with green ascospores), Chaverri *et al.* (*Mol. Phylogen. Evol.* **27**: 302, 2003), Chaverri *et al.* (*Mycol.* **95**: 1100, 2003), Lu & Samuels (*Sydowia* **55**: 255, 2003), Chamberlain *et al.* (*Karstenia* **44**: 1, 2004; stipitate spp.), Chaverri *et al.* (*Mycol. Progr.* **3**: 29, 2004), Druzhinina *et al.* (*Stud. Mycol.* **50**: 401, 2004), Lu *et al.* (*Mycol.* **96**: 310, 2004), Põldmaa & Samuels (*Sydowia* **56**: 79, 2004; Thailand), Jaklitsch *et al.* (*Stud. Mycol.* **56**: 135, 2006), Jaklitsch *et al.* (*Mycol.* **98**: 499, 2006), Overton *et al.* (*Stud. Mycol.* **56**: 39, 2006), Overton *et al.* (*Stud. Mycol.* **56**: 1, 2006; *H. citrina* group), Samuels *et al.* (*CBS Diversity Ser.* **4**, 2006), Zhang *et al.* (*Mycol.* **98**: 1076, 2006; phylogeny).
Hypocreaceae De Not. (1844), Hypocreales. 22 gen. (+ 27 syn.), 454 spp.

Lit.: Bissett (*CJB* **69**: 2357, 1991), Bissett (*CJB* **69**: 2373, 1991), Samuels (*MR* **100**: 923, 1996), Grondona *et al.* (*Appl. Environm. Microbiol.* **63**: 3189, 1997), Kuhls *et al.* (*Mycol.* **89**: 442, 1997), Bulat *et al.* (*MR* **102**: 933, 1998), Castle *et al.* (*Appl. Environm. Microbiol.* **64**: 133, 1998), Gams & Bissett (*Trichoderma and Gliocladium* Vol. **1**. Basic Biology, Taxonomy and Genetics: 3, 1998), Harman *et al.* (*Trichoderma and Gliocladium* Vol. **1**. Basic Biology, Taxonomy and Genetics: 243, 1998), Kindermann *et al.* (*Fungal Genetics Biol.* **24**: 298, 1998), Lieckfeldt *et al.* (*Trichoderma and Gliocladium* Vol. **1**. Basic Biology, Taxonomy and Genetics: 35, 1998), Chen *et al.* (*Appl. Microbiol. Biotechn.* **52**: 246, 1999), Lieckfeldt *et al.* (*CJB* **76**: 1507, 1998), McKay *et al.* (*Appl. Environm. Microbiol.* **65**: 606, 1999), Põldmaa *et al.* (*CJB* **77**: 1765, 1999), Rossman *et al.* (*Stud. Mycol.* **42**: 248 pp., 1999; cell constituent polyamines). Much less widely circumscribed than in previous editions of this *Dictionary*.

Lit.:, Sahr *et al.* (*Mycol.* **91**: 935, 1999), Hermosa *et al.* (*Appl. Environm. Microbiol.* **66**: 1890, 2000), Lieckfeldt *et al.* (*Microbiol. Res.* **155**: 7, 2000), Põldmaa (*Stud. Mycol.* **45**: 83, 2000), Rossman (*Stud. Mycol.* **45**: 27, 2000), Douhan & Rizzo (*MR* **107**: 1342, 2003), Lu *et al.* (*Mycol.* **96**: 310, 2004), Druzhinina *et al.* (*Mycoscience* **47**: 55, 2006), Gams (*Stud. Mycol.* **56**: 177 pp., 2006).
hypocreacous, fleshy and brightly coloured, like *Hypocrea*.
Hypocreales Lindau (1897). Sordariomycetidae. 7 fam., 237 gen., 2647 spp. Ascomata perithecial, rarely cleistothecial, sometimes either in or on a stroma, ± globose, sometimes ornamented, rarely setose, the ostiole periphysate; peridium and stromatal tissues fleshy, usually brightly coloured; interascal tissue of apical paraphyses, often evanescent; asci ± cylindrical, thin-walled, sometimes with a small apical ring or a conspicuous apical cap, not blueing in iodine; ascospores varied, hyaline or pale brown, usually septate, sometimes muriform, sometimes elongate and fragmenting, without a sheath. Anamorphs prominent, hyphomycetous. Saprobes or parasites of plants, often fungicolous or licheni-

colous, rarely coprophilous, cosmop. Fams:
(1) **Bionectriaceae**
(2) **Clavicipitaceae**
(3) **Cordycipitaceae**
(4) **Hypocreaceae**
(5) **Nectriaceae**
(6) **Niessliaceae**
(7) **Ophiocordycipitaceae**

Lit.: Boedijn (*Persoonia* **3**: 1, 1964), Dingley (*Trans. Proc. R. Soc. N.Z.* **79**: 55, 177, 323, 403, **81**: 329, 489, **83**: 643, **84**: 467, 1951-57; NZ), Doi (*Bull. natn Sci. Mus. Tokyo* B **1**: 1, **2**: 119, 1975-76; S. Am.), Müller & von Arx (1962, 1973), Petch (*TBMS* **21**: 243, 1938; UK), Rogerson (*Mycol.* **62**: 865, 1970; gen. names, key gen.), Rossman (*Mycol. Pap.* **150**, 1983; key 52 phragmosporous spp.; *Stud. Mycol.* **45**: 27, 2000; gen. concepts), Rossman *et al.* (*Mycol.* **85**: 685, 1993; key 8 gen. simple spores; *Stud. Mycol.* **42**, 1999; monogr.), Samuels & Rossman (*Whole fungus* **1**: 167, 1979; anamorphs, gen. concepts), Samuels & Seifert (*in* Sugiyama (Ed.), *Pleomorphic fungi*: 29, 1987; anamorphs), Samuels & Seifert (*Sydowia* **43**: 249, 1991; key 20 synnematous spp.), Samuels *et al.* (*Mem. N.Y. bot. Gdn* **59**: 6, 1990; 75 spp., Indonesia), Schoch *et al.* (*MR* **111**: 154, 2007; phylogeny). See also under *Nectria* and anamorph gen. names.

Hypocrella Sacc. (1878), Clavicipitaceae. Anamorph *Aschersonia*. 37 (on *Insecta*), widespread (tropical). See Hywel-Jones & Evans (*MR* **97**: 871, 1993), Hywel-Jones & Samuels (*Mycol.* **90**: 36, 1998), Rossman *et al.* (*Stud. Mycol.* **42**: 248 pp., 1999), Liu *et al.* (*Mycotaxon* **78**: 67, 2001; China), Chaverri *et al.* (*MR* **109**: 1268, 2005), Liu & Hodge (*MR* **109**: 818, 2005), Liu *et al.* (*MR* **110**: 537, 2006; anamorph), Sung *et al.* (*Stud. Mycol.* **57**, 2007), Torres *et al.* (*MR* **111**: 317, 2007; Panama), Chaverri *et al.* (*Stud. Mycol.* **60**, 2008; phylogeny, neotropics).

Hypocreodendron Henn. (1897), anamorphic *Discoxylaria*, Hsy.0eH.19. 1, S. America. See Rogers *et al.* (*Mycol.* **87**: 41, 1995).

Hypocreomycetidae O.E. Erikss. & Winka (1997), Sordariomycetes. Ords:
(1) **Coronophorales**
(2) **Hypocreales**
(3) **Melanosporales**
(4) **Microascales**

Lit.: see under fams, also Schoch *et al.* (*MR* **111**: 154, 2007; marine lineages).

Hypocreophis Speg. (1919) = Hypocrella fide Clements & Shear (*Gen. Fung.*, 1931).

Hypocreopsis G. Winter (1875) ≡ Selinia fide Rossman *et al.* (*Stud. Mycol.* **42**: 248 pp., 1999).

Hypocreopsis P. Karst. (1873), Hypocreaceae. Anamorph *Stromatocrea*. 5 (decaying wood, often associated with resupinate basidiomycetes), widespread (north temperate). See Niemelä & Nordin (*Karstenia* **25**: 75, 1985; Eur.), Candy & Webster (*Mycologist* **2**: 18, 1988), Candoussau (*Mycologist* **4**: 170, 1990), Rossman *et al.* (*Stud. Mycol.* **42**: 248 pp., 1999), Ainsworth (*English Nature Research Reports* **600**, 2004; conservation).

hypocystidium, see cystidium.

Hypodendrum Paulet ex Earle (1909) = Pholiota fide Singer (*Agaric. mod. Tax.*, 1951).

Hypoderma DC. (1805) nom. rej. = Lophodermium See Cannon & Minter (*Taxon* **32**: 572, 1983; gen. nomencl.).

Hypoderma De Not. (1847) nom. cons., Rhytismataceae. 54, widespread (esp. temperate). See Cannon & Minter (*Taxon* **32**: 572, 1983; gen. nomencl.), Johnston (*N.Z. Jl Bot.* **28**: 159, 1990; 12 spp. NZ, gen. concept), Johnston (*Mycotaxon* **52**: 221, 1994; ascospores), Johnston (*Mycol. Pap.* **176**: 239 pp., 2001; on grasses), Hou & Piepenbring (*Nova Hedwigia* **82**: 91, 2006; China).

Hypodermataceae Rehm (1887) = Rhytismataceae.

hypodermataceous (of asci), ones which are essentially unitunicate, lack any apical thickening, and which discharge the spores through a narrow pore; see ascus.

Hypodermella Tubeuf (1895), Rhytismataceae. 3, widespread. See Cannon & Minter (*Mycol. Pap.* **155**: 123 pp., 1986; Indian subcontinent), Hansen & Lewis (*Compendium of Conifer Diseases*, 1997), Johnston (*Biodiversity of Tropical Microfungi*: 241, 1997).

Hypodermellina Höhn. (1917), Rhytismatales. 1, Europe.

Hypodermina Höhn. (1916), anamorphic *Diaporthales*, St.0eH.?. 1, Europe. See Castlebury *et al.* (*Mycol.* **94**: 1017, 2002; as *Mazzantiella*).

Hypodermium Link (1816) ≡ Caeoma fide Laundon (*Mycol. Pap.* **102**, 1965).

Hypodermium Link (1825) nom. dub., anamorphic *Pezizomycotina*. See Sutton (*Mycol. Pap.* **141**, 1977).

Hypodermopsis Earle (1902) = Hysterium Pers. fide Nannfeldt (*Nova Acta R. Soc. Scient. upsal.*, 1932).

Hypodermopsis Kuntze (1898) ≡ Hypoderma De Not.

Hypodiscus Lloyd (1923) [non *Hypodiscus* Nees 1836, *Restionaceae*] ≡ Neohypodiscus.

Hypodrys Pers. (1825) ≡ Fistulina.

Hypoflavia Marbach (2000), Caliciaceae (L). 2, S. America. See Marbach (*Biblthca Lichenol.* **74**: 291, 2000).

Hypogaea E. Horak (1964), Agaricaceae. 1, Argentina. Basidioma gasteroid. See Horak (*Sydowia* **17**: 299, 1964).

Hypogaeaceae E. Horak (1964) = Agaricaceae.

Hypogaeum Pers. (1797) = Elaphomyces fide Fries (*Syst. mycol.* **3**: 57, 1829).

hypogean (**hypogeal**, **hypogeic**, **hypogeous**), in the earth; (of fungi) see Hypogeous fungi.

hypogenous, produced lower down.

Hypogeous fungi. Fungi having subterranean sporocarps comprise many taxonomically unrelated species showing remarkable features of convergent evolution as a result of adaptation to this specialized habitat. They include the truffles (q.v.; esp. *Elaphomycetales*, *Pezizales*), false truffles (*Rhizopogon*, *Hymenogaster*, *Gautieria*), various 'gasteromycetes' (*Radiigera*, *Pyrenogaster*) and a few *Zygomycetes* (e.g. *Endogone*). Most are mycorrhizal; many mutualistic relationships have also evolved between hypogeous fungi animals, particularly insects and some mammals; many show considerable variability within a single fruitbody in numbers of spores produced per ascus or basidium, and their fruitbodies are often involuted or convoluted with a strongly reduced hymenium and thought to be derived from epigeous forms; their importance of some species in the food chain of endangered mammals has meant more is known about their conservation needs than for most fungi (e.g. Loeb *et al.*, *Am. Midl. Nat.* **144**: 286, 2000), see Animal mycophagists; some, particularly species of *Elaphomyces* accumulate radionuclides in exceptional quantities (see Air pollution). Their unusual nature has stimulated much folk lore and many legends about hypogeous fungi including, particularly, a belief that some are aphrodisiacs. See Tulasne (*Fungi hypogaei*, 1851 [reprint 1970]), Hawker (*Biol. Rev.* **30**: 127, 1955), Trappe (*Mycotaxon* **9**: 297, 1979). **Australasia** Trappe *et al.* (*Aust. Syst. Bot.* **5**: 597 & 617, 1992; **9**: 773 & 803, 1996); **British Isles** Hawker (*Phil. Trans.* **B237**: 429, 1954; *TBMS* **63**: 67, 1974, list), Pegler *et al.* (*British truffles. A revision of British hypogeous fungi*, 1993); **Europe** Montecci & Sarasini (*Funghi Ipogei d'Europa*, 2000); **Germany** Hess (*Die Hypogaeen Deutschsland*, 1891-4 [reprint 1971]); **Hungary** (Carpathian Basin), Szemere (*Die unteririschen Pilze des Karpatenbeckens*, 1965); **Malaysia** Corner & Hawker (*TBMS* **36**: 125, 1953); **Mexico** Trappe & Guzmán (*Mycol.* **63**: 317, 1971); **Spain** Calonge *et al.* (*An. Inst. Bot. Cavanilles* **34**: 15, 1977), Moreno Arroyo *et al.* (*Tesoros de nuestros montes. Trufas de Andalucía*, 2005); **USA** Trappe (1979), Castellano *et al.* (*Key to spores of the genera of hypogeous fungi of north temperate forests with special reference to animal mycophagy*, 1989). See Truffle. See also: www.mykoweb.com/biblio/hypo_bib.pdf (bibliography of hypogeous and secodioid basidiomycetes).

Hypogloeum Petr. (1923), anamorphic *Pezizomycotina*, Cac.0eH.?. 1, Europe.

Hypogymnia (Nyl.) Nyl. (1896), Parmeliaceae (L). *c.* 65, widespread. See Bitter (*Hedwigia* **40**: 171, 1901), Krog (*Skr. norsk. Polarinst.* **144**, 1968), Krog (*Lichenologist* **6**: 135, 1974), Beltman (*Biblthca Lichenol.* **11**, 1978), Elix (*Brunonia* **2**: 175, 1980; Australasia), Awasthi (*Kavaka* **12**: 87, 1986; key 11 spp. India), Luo (*Bull. bot. Res.* Harbin **6**: 155, 1986; China), Wei (*Acta Mycol. Sin.* Suppl. **1**: 379, 1987; isidiate spp. China), Elix & Jenkins (*Mycotaxon* **35**: 489, 1989; key 18 Australian spp.), Hyvärinen (*Lichenologist* **24**: 267, 1992; thallus structure), Goward & McCune (*Bryologist* **96**: 450, 1993; N America), Zeybek *et al.* (*Cryptog. bot.* **3**: 260, 1993; chemistry), Crespo & Cubero (*Lichenologist* **30**: 369, 1998; phylogeny), Wedin *et al.* (*MR* **103**: 1152, 1999; phylogeny), McCune & Obermayer (*Mycotaxon* **79**: 23, 2001; typification), McCune & Tchabanenko (*Bryologist* **104**: 146, 2001; E Asia), Yoshida & Kashiwadani (*Bull. natn. Sci. Mus.* Tokyo, B **27**: 35, 2001; typification, E Asia), McCune *et al.* (*Bryologist* **106**: 226, 2003; Himalayas), Thell *et al.* (*Symb. bot. upsal.* **34** no. 1: 429, 2004; Scandinavia), McCune (*Bull. Calif. Lichen Soc.* **13**: 42, 2006; conservation), McCune *et al.* (*Bryologist* **109**: 80, 2006; N America), Miądlikowska *et al.* (*Mycol.* **98**: 1088, 2006; phylogeny), Wei & Wei (*Mycotaxon* **94**: 155, 2005; China).

Hypogymniaceae Poelt ex Elix (1980) = Parmeliaceae.

hypogyny (adj. **hypogynous**), the condition of having the antheridium under the oogonium and on the same hypha.

Hypohelion P.R. Johnst. (1990), Rhytismataceae. 2, Europe; N. America. See Johnston (*Mycotaxon* **39**: 219, 1990), Johnston (*Mycol. Pap.* **176**: 239 pp., 2001).

Hypolepia Raf. (1808) nom. nud., Fungi. See Merrill (*Index Rafinesq.*, 1949).

hypolithic, see endolithic.

Hypolyssus Pers. (1825) nom. conf., Agaricomycetes. sensu Berk. = *Caripia* (Podoscyph.).

Hypomnema Britzelm. (1883) nom. dub., Agaricales.

Hypomyces (Fr.) Tul. (1860), Hypocreaceae. Anamorphs *Cladobotryum*, *Sibirina*, *Verticillium*. 53 (mostly on larger fungi), widespread. See Arnold (*Nova Hedwigia* **21**: 529, 1972; classif.), Tubaki (*Rep. Tottori mycol. Inst.* **12**: 161, 1975; Japanese spp.), Arnold (*Bibliogr. Mitteil. Univ.-bibl. Jena* **25**, 1976; bibliogr.), Samuels (*Mem. N. Y. bot. Gdn* **26**, 1976), Rogerson & Samuels (*Mycol.* **77**: 763, 1985; spp. on discomycetes), Rogerson & Samuels (*Mycol.* **81**: 413, 1989; key 10 spp. on *Boletales*), Rogerson & Samuels (*Mycol.* **85**: 231, 1993; spp. on polypores), Rogerson & Samuels (*Mycol.* **86**: 839, 1994; key 13 spp. on agarics), Rossman *et al.* (*Stud. Mycol.* **42**: 248 pp., 1999), Põldmaa (*Stud. Mycol.* **45**: 83, 2000), Douhan & Rizzo (*MR* **107**: 1342, 2003; host-parasite rels), Põldmaa (*Mycol.* **95**: 921, 2003; on *Stereaceae*), Põldmaa & Samuels (*Sydowia* **56**: 79, 2004; Thailand), Tokiwa & Okuda (*Mycoscience* **46**: 294, 2005; Japan).

Hypomycetaceae Earle (1901) = Hypocreaceae.

Hypomycetales = Hypocreales.

Hypomycopsis Henn. (1904) = Mycosphaerella fide von Arx & Müller (*Stud. Mycol.* **9**, 1975).

hyponecral, see necral layer.

Hyponectria Sacc. (1878), Hyponectriaceae. 16 (on dead leaves etc.), widespread. See Barr (*Mycol.* **69**: 952, 1977; key), Rossman *et al.* (*Stud. Mycol.* **42**: 248 pp., 1999), Wang & Hyde (*Fungal Diversity* **3**: 159, 1999), Winka & Eriksson (*Mycoscience* **41**: 97, 2000; DNA), Sivanesan & Shivas (*Fungal Diversity* **9**: 169, 2002).

Hyponectriaceae Petr. (1923), Xylariales. 22 gen. (+ 15 syn.), 128 spp.

Lit.: Samuels *et al.* (*Mycotaxon* **28**: 473, 1987), Barr (*Indian Journal of Mycology and Plant Pathology*, 1994). The affinities are still uncertain pending further molecular analysis, see also Hyde *et al.* (*Sydowia* **50**: 21, 1998), Mahuku *et al.* (*MR* **102**: 559, 1998), Wang & Hyde (*Fungal Diversity* **3**: 159, 1999), Winka & Eriksson (*Mycoscience* **41**: 97, 2000), Jeewon *et al.* (*MR* **107**: 1392, 2003).

Hyponevris Earle (1909) ≡ Schizophyllum.

Hyponevris Paulet (1808) = Merulius Haller ex Boehm. fide Donk (*Taxon* **12**: 167, 1963).

hyponym, a name only; one having no description or reference to a specimen.

hypoparasite, a hidden parasite; a pathogen dispersed along with another pathogen, such as a mycovirus in an *Ophiostoma ulmi* population.

Hypophloeda K.D. Hyde & E.B.G. Jones (1989), Diaporthales. 1 (mangrove roots), Brunei; Seychelles. See Hyde & Jones (*TMSJ* **30**: 61, 1989).

hypophloeodal, under the periderm or bark; endophloeodal; subcutical; within the bark.

hypophyllous, on the under surface of a leaf.

Hypophyllum Earle (1909) = Lactarius fide Singer (*Agaric. mod. Tax.*, 1951).

Hypophyllum Paulet (1808) ≡ Agaricus L.

hypoplasia, the state of having growth less than normal, cf. hyperplasia.

Hypoplasta Preuss (1855) = Cytospora fide Saccardo (*Syll. fung.* **3**: 1, 1884).

Hypoplegma Theiss. & Syd. (1917) = Perisporiopsis Henn. fide Müller & von Arx (*Beitr. Kryptfl. Schweiz* **11** no. 2, 1962).

Hypopteris Berk. (1854) = Apiospora fide Theissen & Sydow (*Annls mycol.* **13**: 419, 1915).

Hyporrhodius (Fr.) Staude (1857) = Pluteus fide Kuyper (*in litt.*).

Hypospila Fr. (1825) nom. dub., Pezizomycotina. See von Arx (*Antonie van Leeuwenhoek* **17**: 257, 1951).

Hypospilina (Sacc.) Traverso (1913), Diaporthales. Anamorph *Asteroma*. 2, Europe. See Barr (*Mycol. Mem.* **7**, 1978), Monod (*Beih. Sydowia* **9**: 1, 1983).

Hypostegium Theiss. (1916) = Glomerella fide von Arx & Müller (*Beitr. Kryptfl. Schweiz* **11** no. 1, 1954).

Hypostigme Syd. (1925) = Parastigmatea fide von Arx & Müller (*Beitr. Kryptfl. Schweiz* **11** no. 1, 1954).

Hypostomum Vuill. (1896) nom. dub., Fungi.

hypostroma, see stroma.

Hypotarzetta Donadini (1985), Pyronemataceae. 1, Europe. See Donadini (*Docums Mycol.* **15** no. 60: 47, 1985), Fouchier & Neville (*Bulletin Semestriel de la Fédération des Associations Mycologiques Méditerranéennes* **13**: 32, 1998).

hypothallus (1) (of lichens), the first hyphae of the thallus to grow, usually used of a crustaceous lichen which has no photobiont cells or cortex; = prothallus (protothallus), fide Maas Geesteranus (*Blumea* **6**: 47, 1947) who restricts hypothallus to the spongy tissue on the underside of the thallus in *Anzia*, *Pannaria* and *Pannoparmelia*, but see spongiostratum; (2) (of *Mycetozoa*), the thin layer on the surface of the substratum not used up in sporangial development; Ross (*Mycol.* **65**: 477, 1973) distinguished epi- and subhypothallic development.

hypothecium, medullary excipulum; the hyphal layer under the subhymenium in an apothecium; sometimes used indiscriminately for all tissues below the hymenium (including the subhymenium).

Hypothele Paulet (1812) nom. inval. ≡ Hydnum.

hypotonic (of culture media), having an osmotic pressure lower than that of the organism cultured, cf. hypertonic.

Hypotrachyna (Vain.) Hale (1974), Parmeliaceae (L). *c.* 198, widespread (esp. subtropics and montane tropics). See Hale (*Smithson. Contr. bot.* **25**, 1975), Krog & Swinscow (*Norw. Jl Bot.* **26**: 11, 1979; E. Afr.), Eliasaro *et al.* (*Mycotaxon* **69**: 255, 1998; Brazil), Kurokawa & Lai (*Mycotaxon* **77**: 225, 2001; Taiwan), Divakar & Upreti (*Mycotaxon* **86**: 67, 2003), Thell *et al.* (*Mycol. Progr.* **3**: 297, 2004; phylogeny), Elix *et al.* (*Lichenologist* **37**: 101, 2005; saxicolous spp.), Masson (*Cryptog. Mycol.* **26**: 205, 2005; France), Blanco *et al.* (*Mol. Phylogen. Evol.* **39**: 52, 2006; phylogeny), Divakar *et al.* (*Mol. Phylogen. Evol.* **40**: 448, 2006; phylogeny, polyphyly), Miądlikowska *et al.* (*Mycol.* **98**: 1088, 2006; phylogeny).

Hypotrachynicola Etayo (2002), Sordariomycetes. 1 (on lichens), Colombia. See Etayo (*Biblthca Lichenol.* **84**: 50, 2002).

Hypoxylaceae DC. (1805) = Xylariaceae.

Hypoxylina Starbäck (1905) = Hypoxylon Bull. fide Læssøe (*SA* **8**: 25, 1989), Læssøe (*SA* **13**: 43, 1994; ? synonym of *Xylaria*).

Hypoxylites Kirschst. (1925), Fossil Fungi. 2 (Neolithic), Germany.

Hypoxylon Adans. (1763) nom. rej. = Xylaria Hill ex Schrank See Læssøe (*SA* **13**: 43, 1994; ? synonym of *Xylaria*), Hawksworth *et al.* (*Dictionary of the Fungi* edn 8, 1995).

Hypoxylon Bull. (1791) nom. cons., Xylariaceae. Anamorphs *Nodulisporium*, *Virgariella*. 130, widespread. See also *Annulohypoxylon*, *Biscogniauxia*, *Entonaema*, *Nemania*. See Miller (*Monogr. World spec. Hypoxylon*, 1961; keys *Hypoxylon* s.l.), Dennis (*Bull. Jard. bot. Brux.* **33**: 317, 1963; Congo), French *et al.* (*CJB* **47**: 223, 1969), Jong & Rogers (*Tech. Bull. Wash. agric. Exp. Stn* **71**, 1972; anamorphs), Abe (*TMSJ* **25**: 399, 1985; tissue types), Ju & Tzean (*Trans. Mycol. Soc. Rep. China* **1**: 13, 1985; key 13 spp. Taiwan), Petrini & Müller (*Mycol. Helv.* **1**: 501, 1986; key 25 spp. C Eur.), Cherepanov (*Nov. Sist. niz. Rast.* **25**: 109, 1988; key 34 spp. former USSR), Granmo *et al.* (*Op. Bot.* **100**: 59, 1989; Nordic spp.), van der Gucht & van der Veken (*Mycotaxon* **44**: 275, 1992; key 18 Papua New Guinea spp.), Læssøe (*SA* **13**: 43, 1994; gen. concept, nomencl.), Ju & Rogers (*Mycol. Mem.* **20**, 1996; monogr.), Granmo (*Sommerfeltia* **26**, 1999; Norway, key), Sánchez-Ballesteros *et al.* (*Mycol.* **92**: 964, 2000; molecular phylogeny), Mazzaglia *et al.* (*MR* **105**: 670, 2001; phylogeny), Mühlbauer *et al.* (*Mycol. Progr.* **1**: 235, 2002; chemistry), Quang *et al.* (*Phytochem.* **65**: 469, 2004; chemistry), Hsieh *et al.* (*Mycol.* **97**: 844, 2005; phylogeny), Quang *et al.* (*Phytochem.* **66**: 797, 2005; chemistry), Triebel *et al.* (*Nova Hedwigia* **80**: 25, 2005; phylogeny, anamorphs), Stadler & Fournier (*Revta Iberoamer. Micol.* **23**: 160, 2006; chemistry, phylogeny), Suwannasai *et al.* (*Mycotaxon* **94**: 303, 2005; Thailand).

Hypoxylonites Elsik (1990), Fossil Fungi. *c.* 20 (Eocene – Oligocene), USA.

Hypoxylonopsis Henn. (1904) = Valsaria fide Saccardo (*Syll. fung.* **24**: 538, 1926).

Hypoxylonsporites P. Kumar (1990), Fossil Fungi. 2, India.

Hypoxylum Juss. (1789) nom. dub. = Cordyceps fide Hawksworth *et al.* (*Dictionary of the Fungi* edn 8, 1995).

Hypsilophora Berk. (1879) nom. dub., ? Fungi.

Hypsizygus Singer (1947), Lyophyllaceae. 3, widespread (north temperate). See Ponzi (*Rivista di Micologia* **36**: 245, 1993), Dielen (*AMK Mededelingen*: 44, 1999).

Hypsolophora, see *Hypsilophora*.

Hypsostroma Huhndorf (1992), Hypsostromataceae. Anamorph *Pleurophomopsis*-like. 2, C. & S. America. See Huhndorf (*Mycol.* **84**: 750, 1992), Huhndorf (*Mycol.* **86**: 266, 1994; posn).

Hypsostromataceae Huhndorf (1994), ? Dothideomycetes (inc. sed.). 2 gen., 4 spp.

Lit.: Huhndorf (*Mycol.* **84**: 750, 1992), Huhndorf (*Mycol.* **86**: 266, 1994).

Hypsotheca Ellis & Everh. (1885) = Caliciopsis fide Fitzpatrick (*Mycol.* **34**: 464, 1942).

Hysterangiaceae E. Fisch. (1899), Hysterangiales. 4 gen. (+ 1 syn.), 54 spp.

Lit.: Curry & Kimbrough (*Mycol.* **75**: 781, 1983), Beaton *et al.* (*Kew Bull.* **40**: 435, 1985), Castellano & Beever (*N.Z. Jl Bot.* **32**: 305, 1994; NZ spp.), Castellano & Muchovej (*Mycotaxon* **57**: 339, 1996), Calonge (*Fl. Mycol. Iberica* **3**: 271 pp., 1998), Castellano (*Ectomycorrhizal Fungi*: 311, 1999), Lebel & Castellano (*Aust. Syst. Bot.* **12**: 803, 1999), Humpert *et al.* (*Mycol.* **93**: 465, 2001), Binder & Bresinsky (*Mycol.* **94**: 85, 2002).

Hysterangiales K. Hosaka & Castellano (2007). Phallomycetidae. 5 fam., 18 gen., 114 spp. Fams:

(1) **Gallaceaceae**
(2) **Hysterangiaceae**
(3) **Mesophelliaceae**
(4) **Phallogastraceae**

(5) **Trappeaceae**
For *Lit.* see under fam.

Hysterangium Vittad. (1831), Hysterangiaceae. *c.* 50, widespread (esp. temperate). See Zeller & Dodge (*Ann. Mo. bot. Gdn* **16**: 83, 1929; key N. Am. spp.), Soehner (*Sydowia* **6**: 246, 1952; Bayern), Schwarzel (*Schweiz. Z. Pilzk.* **10**: 154, 1979; key 18 spp.), Beaton *et al.* (*Kew Bull.* **40**: 435, 1985; key Austral. spp.), Castellano (*Hysterangium – Ectomycorrhizal fungi, key genera in profile*, 1999; ecology), Castellano *et al.* (*Karstenia* **40**: 12, 2000; as *Hysterangium*), Verbeken & Walleyn (*Boll. Gruppo Micol. 'G. Bresadola'* **46**: 87, 2003; African sp. (as *Hysterangium*)).

Hysteriaceae Chevall. (1826), Hysteriales. 14 gen. (+ 7 syn.), 69 spp.
Lit.: Zogg (*Beitr. Kryptfl. Schweiz* **11** no. 3, 1962; Eur.), Bellemère (*Annls Sci. Nat.* Bot., sér. 12 **12**: 429, 1971), Amano (*TMSJ* **24**: 283, 1983; Japan), Larios & Honrubia (*Revta Ibér. Micol.* **5**: 111, 1988), Sivanesan *et al.* (*TBMS* **90**: 665, 1988), Renobales & Aguirre (*SA* Reprint of Volumes 1-4 (1982-1985) **8**: 87, 1990), van der Linde (*S. Afr. J. Bot.* **58**: 491, 1992; key 5 gen., 12 spp., S. Afr.), Checa (*Mycotaxon* **62**: 349, 1997; Iberian spp.), Goh *et al.* (*MR* **102**: 1309, 1998), Lorenzo & Messuti (*MR* **102**: 1101, 1998; S Am. spp.), Tretiach & Modenesi (*Nova Hedwigia* **68**: 527, 1999), Vasil'eva (*Mikol. Fitopatol.* **33**: 297, 1999), Diederich & Wedin (*Nordic Jl Bot.* **20**: 203, 2000), Liew *et al.* (*Mol. Phylogen. Evol.* **16**: 392, 2000; DNA), Schoch *et al.* (*Mycol.* **98**: 1041, 2006; phylogeny).

hysteriaceous (**hysterioid**, **hysteriiform**), long and cleft, like the **hysterothecium** (ascoma) of the *Hysteriaceae*; lirellate.

Hysteriales Lindau (1896). Dothideomycetes. 1 fam., 14 gen., 69 spp. Ascomata erumpent or superficial, often aggregated, elongated, sometimes branched, opening by a longitudinal split or radial slits; peridium black, very thick-walled, carbonaceous, composed of small pseudoparenchymatous cells or intertwined hyphal tissue. Interascal tissue of narrow cellular or trabeculate pseudoparaphyses. Asci cylindrical, fissitunicate, with a distinct ocular chamber. Ascospores hyaline to brown, variously septate, sometimes with a mucous sheath. Anamorphs varied. Mostly saprobic, cosmop. Fam.:
Hysteriaceae
Lit.: Liew *et al.* (*Mol. Phylog. Evol.* **16**: 392, 2000), Schoch *et al.* (*Mycol* **98**: 1041, 2006).

Hysteridium P. Karst. (1905), anamorphic *Pezizomycotina*, St.0eH.?. 1, Europe. See Sutton (*Mycol. Pap.* **141**, 1977).

Hysterina (Ach.) Gray (1821) = Opegrapha Ach. fide Hawksworth *et al.* (*Dictionary of the Fungi* edn 8, 1995).

Hysteriopsis Geyl. (1887), Fossil Fungi. 1 (Tertiary), Indonesia. = Hysterites (Fossil fungi) fide Meschinelli (1892).

Hysteriopsis Speg. (1906) ? = Hysterographium fide Zogg (*Beitr. Kryptfl. Schweiz* **11** no. 3, 1962).

Hysterites Göpp. (1836), Fossil Fungi. 15 (Tertiary), Europe.

Hysterites Unger (1841), Fossil Fungi. 1 (Tertiary), former Yugoslavia.

Hysterium Pers. (1797), Hysteriaceae. Anamorphs *Coniosporium*, *Hysteropycnis*. 10, widespread (esp. temperate). See Zogg (*Beitr. Kryptfl. Schweiz* **11** no. 3, 1962), Vasil'eva (*Mikol. Fitopatol.* **33**: 225, 1999; Russian Far East), Liew *et al.* (*Mol. Phylogen. Evol.* **16**: 392, 2000; phylogeny), Lee & Crous (*S. Afr. J. Bot.* **69**: 480, 2003; S Africa), Schoch *et al.* (*Mycol.* **98**: 1041, 2006; phylogeny), Checa *et al.* (*Mycol.* **99**: 285, 2007; Costa Rica).

Hysterium Tode (1791) = Colpoma See Cannon & Minter (*Taxon* **32**: 572, 1983; gen. nomencl.).

Hysterocarina H. Zogg (1949), Hysteriaceae. 1, Brazil.

hysterochroic, see colour.

Hysterodiscula Petr. (1942), anamorphic *Rhytismataceae*, St.0eH.1. 2, widespread (north temperate). Reports suggesting genetic links with *Duplicaria* need confirmation. See Darker (*CJB* **45**: 1427, 1967), Eriksson (*Symb. Bot. Upsal.* **19** no. 4: 19, 1970).

Hysterodothis Höhn. (1909) = Sphaerodothis fide von Arx & Müller (*Beitr. Kryptfl. Schweiz* **11** no. 1, 1954).

Hysterogaster Zeller & C.W. Dodge (1928) = Hymenogaster fide Cunningham (*Gast. Austr. N.Z.*: 47, 1944).

Hysteroglonium Rehm ex Lindau (1896), ? Hysteriaceae. 2 or 3, N. America; Europe.

Hysterographium Corda (1842), Hysteriaceae. Anamorph *Hysteropycnis*. 6, widespread. See Zogg (*Phytopath. Z.* **14**: 310, 1944), Zogg (*Beitr. Kryptfl. Schweiz* **11** no. 3, 1962), Larios & Honrubia (*Revta Ibér. Micol.* **5**: 111, 1988; Spain), Vasil'eva (*Mikol. Fitopatol.* **33**: 297, 1999; E Russia), Messuti & Lorenzo (*Nova Hedwigia* **76**: 451, 2003; S America), Checa *et al.* (*Mycol.* **99**: 285, 2007; Costa Rica).

Hysteromyces Vittad. (1844) = Rhizopogon fide Fischer (*Nat. Pflanzenfam.* **5b**: viii, 1938).

Hysteromyxa Sacc. & Ellis (1882) nom. dub., Fungi. See Petrak (*Sydowia* **5**: 196, 1951).

Hysteronaevia Nannf. (1984), Helotiales. 11, widespread (north temperate). See Scheuer (*Biblthca Mycol.* **123**: 274, 1988), Nauta & Spooner (*Mycologist* **14**: 65, 2000; UK), Raitviir (*Folia cryptog. Estonica* **40**: 43, 2003; Estonia), Suková (*Czech Mycol.* **56**: 63, 2004; Czech Republic).

Hysteropatella Rehm (1890), Hysteriaceae. 3, Europe; N. America. See Sherwood-Pike (*Sydowia* **38**: 267, 1986), Schoch *et al.* (*Mycol.* **98**: 1041, 2006; phylogeny).

Hysteropeltella Petr. (1923), Dothideomycetes. 1, Europe.

Hysteropeziza Rabenh. (1874) = Pyrenopeziza fide Nannfeldt (*Nova Acta R. Soc. Scient. upsal.*, 1932).

Hysteropezizella Höhn. (1917), Helotiales. Anamorph *Neottiospora*. *c.* 19, widespread. See Défago (*Sydowia* **21**: 1, 1967), Hein (*Nova Hedwigia* **34**: 449, 1981; paraphyses ornamentation), Scheuer (*Biblthca Mycol.* **123**: 274, 1988).

hysterophyte, a saprophyte (obsol.).

Hysteropsis Rehm (1887), Dothideomycetes. 1, Europe.

Hysteropycnis Hilitzer (1929), anamorphic *Pezizomycotina*, St.0eH.?. 5, Europe; N. America.

Hysterostegiella Höhn. (1917), Helotiales. 10, widespread. See Hein (*Nova Hedwigia* **38**: 669, 1983; key), Scheuer (*Biblthca Mycol.* **123**: 274, 1988).

Hysterostoma Theiss. (1913) = Dothidasteromella fide Müller & von Arx (*Beitr. Kryptfl. Schweiz* **11** no. 2, 1962).

Hysterostomella Speg. (1885), Parmulariaceae. 17, widespread (tropical). See Hansford (*Mycol. Pap.* **15**,

1946), Farr (*Sydowia* **38**: 65, 1986), Inácio & Cannon (*Fungal Diversity* **9**: 71, 2002).

Hysterostomina Theiss. & Syd. (1915) = Hysterostomella fide Hansford (*Mycol. Pap.* **15**, 1946), Batista & Vital (*Atas do IMUR* **1**: 53, 1960), von Arx & Müller (*Stud. Mycol.* **9**, 1975).

hysterothecium, an elongated ascoma like that of the *Hysteriaceae* with a slit-like line of dehiscence.

Hystrichosphaeridium Deflandre (1937), Fossil Fungi, Ascomycota. 1, Europe. See Deflandre (*Annales de Paléontologie* **26**: 68, 1937).

Hystricula Cooke (1884) ? = Winterella (Sacc.) Kuntze fide Saccardo (*Syll. fung.* **1**: 471, 1882).

Hystrix Alstrup & Olech (1993) [non *Hystrix* Moench 1794, *Poaceae*] = Acanthonitschkea fide Eriksson & Santesson (*SA* **14**: 54, 1995).

I, see iodine.

IAL, see Societies and organizations.

Ialomitzia Gruia (1964), anamorphic *Pezizomycotina*, Hso.0eP.?. 1, Rumania. Alternatively a cyanobacterium.

IAP, see Index of Atmospheric Purity.

ibotenic acid, a metabolite of *Amanita muscaria*, etc., toxic to humans and flies (*Musca* spp.); amanita factor C.

ICBN, International Code of Botanical Nomenclature; see Nomenclature.

Iceland moss. *Cetraria islandica*, habitually eaten by reindeer and caribou, and used as a substitute for flour during hard times in Scandinavia and Iceland, is rich in digestible carbohydrates; used in soups and pastilles.After the Chernobyl nuclear accident in April 1986, this species accumulated radioactive Caesium in levels harmful for reindeer populations. See Dahl (*Bot. Rev.* **20**: 463, 1954); Machart *et al.* (*J. Environmental Radioactivity* **97**: 70, 2007); Richardson & Young (*in* Seaward (Ed.), *Lichen Ecology*: 121, 1977); Animal mycophagists; Medical uses of fungi.

Icerymyces Brain (1923) nom. conf., anamorphic *Pezizomycotina*, St.0eH.?. 1 (in insects). See Batra *in* Subramanian (Ed.) (*Taxonomy of fungi* **1**: 187, 1978).

Ichthyochytrium Plehn (1920), Chytridiales. 1 (in *Cyprinus*), Germany.

Ichthyosporidium Caullery & Mesnil (1905), Microsporidia. 2, Europe; N. America. *I. giganteum*, causal agent of fish disease. See Canning *in* Margulis *& al.* (Eds) (*Handbook of Protoctista*: 53, 1990; systematic position).

Icmadophila Trevis. (1853) nom. cons., Icmadophilaceae (L). 6, widespread (northern hemisphere). See Frey (*Rabenh. Krypt.-Fl.* **9** 4.1: 819, 1933), Honegger (*Lichenologist* **15**: 57, 1983; asci), Rambold *et al.* (*Biblthca Lichenol.* **53**, 1993), Stenroos & DePriest (*Am. J. Bot.* **85**: 1548, 1998), Platt & Spatafora (*Lichenologist* **31**: 409, 1999), Galloway (*Lichenologist* **32**: 294, 2000; synonym of *Knightiella*), Miądlikowska *et al.* (*Mycol.* **98**: 1088, 2006; phylogeny).

Icmadophilaceae Triebel (1993), Pertusariales (L). 6 gen. (+ 11 syn.), 58 spp.

Lit.: Rambold *et al.* (*Biblthca Lichenol.* **53**, 1993), Kantvilas (*Herzogia* **12**: 7, 1996), Stenroos & DePriest (*Am. J. Bot.* **85**: 1548, 1998), Platt & Spatafora (*Lichenologist* **31**: 409, 1999), Galloway (*Lichenologist* **32**: 294, 2000), Platt & Spatafora (*Mycol.* **92**: 475, 2000), Lumbsch *et al.* (*Mol. Phylogen. Evol.* **31**: 822, 2004), Reeb *et al.* (*Mol. Phylogen. Evol.* **32**: 1036, 2004), Wedin *et al.* (*MR* **109**: 159, 2005), Miądlikowska *et al.* (*Mycol.* **98**: 1088, 2006; phylogeny), Hofstetter *et al.* (*Mol. Phylogen. Evol.* **44**: 412, 2007; phylogeny).

Icmadophilomyces E.A. Thomas ex Cif. & Tomas. (1953) nom. illegit. ≡ Icmadophila.

icones (Latin), pictures; Figures; plates.

ICSU, see Societies and organizations.

ICTF, see Societies and organizations.

ICZN, International Code of Zoological Nomenclature.

Idiocercus B. Sutton (1967), anamorphic *Pezizomycotina*, Cpd.0eH.19. 3, widespread. See Nag Raj (*Coelomycetous Anamorphs with Appendage-bearing Conidia*, 1993).

Idiomyces Thaxt. (1893), Laboulbeniaceae. 1, Europe; Japan. See Benjamin (*Aliso* **10**: 345, 1983), Tavares (*Mycol. Mem.* **9**: 627 pp., 1985), Hafellner (*Linzer biol. Beitr.* **31**: 507, 1999; Iberian peninsula).

id-reaction, see dermatophytid.

Idriella P.E. Nelson & S. Wilh. (1956), anamorphic *Hymenoscyphus*, Hso.0eP.10. *c.* 30, widespread. See von Arx (*Sydowia* **34**: 30, 1981; key), Castañeda Ruiz & Kendrick (*Univ. Waterloo Biol. Ser.* **35**: 1, 1991; 4 n.spp.), Rodrigues & Samuels (*Mycotaxon* **43**: 271, 1992; endophytic spp. on palms), Lascaris & Deacon (*MR* **98**: 1200, 1994; growth, sporulation).

IFO, Institute for Fermentation (Osaka, Japan); founded 1944 as Kōkū-Hakkō Kenyūsho and supported by the Japanese government and Takeda Chemical Industries Ltd, with a separate borad of trustees; see Anon. (*Res. Comm. IFO 17*: 1, 1995; 50th anniv. issue).

Igneocumulus A.W. Ramaley (2003), ? Hypocreales. Anamorph *Lecythophora*-like. 1, USA. See Ramaley (*Mycotaxon* **88**: 158, 2003).

Ijuhya Starbäck (1899), Bionectriaceae. Anamorph *Acremonium*-like. 7 (on wood and herbaceous debris), widespread. See Samuels (*Mem. N. Y. bot. Gdn* **48**, 1988), Rossman *et al.* (*Stud. Mycol.* **42**: 248 pp., 1999).

ikatake ('ika-take'), the basidiome of *Aseroë arachnoidea*.

IKI, see iodine.

Iledon Samuels & J.D. Rogers (1986), anamorphic *Botryohypoxylon*, St.≡ eP.15. 1, Venezuela. See Samuels & Rogers (*Mycotaxon* **25**: 633, 1986).

Ileodictyon Tul. ex M. Raoul (1844), Phallaceae. 2, widespread (esp. southern hemisphere). See Ka *et al.* (*Mycobiology* **32**: 54, 2004; Korea).

illegitimate, see Nomenclature. Cf. legitimate.

Illosporiopsis D. Hawksw. (2001), anamorphic *Hypocreales*. 1 (lichenicolous), Italy. See Sikaroodi *et al.* (*MR* **105**: 457, 2001).

Illosporium Mart. (1817), anamorphic *Hypocreales*, Cpd.0eH.?. 1 (on lichens, *Peltigera*), widespread (north temperate). See also *Marchandiomyces*. See Hawksworth (*Bull. Br. Mus. nat. hist.* Bot. **6**: 181, 1979), Sikaroodi *et al.* (*MR* **105**: 453, 2001; phylogenetic posn.).

Ilyomyces F. Picard (1917), Laboulbeniaceae. 2, Europe. See Weir (*MR* **99**: 789, 1995), Santamaría (*Fl. Mycol. Iberica* **5**, 2003; Iberian peninsula).

Ilytheomyces Thaxt. (1917), Laboulbeniaceae. 15, widespread. See Rossi (*Mycol.* **90**: 1047, 1998; Bolivia).

IMA, see Societies and organizations.

IMA Committee for Asia. Founded in 1977; recognized as the Committee for Asia within the International Mycological Association (q.v.); structure comprises an executive, with national representatives from Asian countries. Publications: *IMA Committee for Asia Newsletter*. Website: http://web.hku.hk/%7Ekdhyde/imaca.

Imazekia Tak. Kobay. & Y. Kawabe (1992), ? Phyllachoraceae. 1 (from living leaves), Japan. See Kobayashi & Kawabe (*Jap. Jour. Trop. Agr.* **36**: 195, 1992).

Imbricaria (Schreb.) Michx. (1803) [non *Imbricaria* Juss. 1789, *Sapotaceae*] = Anaptychia fide Hawksworth *et al.* (*Dictionary of the Fungi* edn 8, 1995).

Imbricariaceae Chevall. (1826) = Physciaceae.

imbricate (of pilei, scales, squamules, etc.), partly covering one another like the tiles on a roof.

IMC, see International Mycological Congresses.

IMI (Imperial Bureau of Mycology, 1920-29; Imperial Mycological Institute, 1930-47; Commonwealth Mycological Institute, 1948-85; CAB International Mycological Institute, 1986-90; International Mycological Institute, 1990-98). A former Institute of CABI (q.v.), originally based in Kew, then relocated to Egham, Surrey, UK in 1992; see Aitchison & Hawksworth (*IMI: retrospect and prospect*, 1993). The acronym remains in use for the dried reference collection.

Imicles Shoemaker & Hambl. (2001), anamorphic *Pezizomycotina*. 6, widespread. See Shoemaker & Hambleton (*CJB* **79**: 598, 2001).

Imimyces A. Hern. Gut. & B. Sutton (1997) = Polydesmus fide Hawksworth *et al.* (*Dictionary of the Fungi* edn 8, 1995).

immaculate, not spotted.

immarginate, having no well-defined edge.

Immersaria Rambold & Pietschm. (1989), Porpidiaceae (L). 1, widespread. See Calatayud & Rambold (*Lichenologist* **30**: 231, 1998), Buschbom & Mueller (*Mol. Phylogen. Evol.* **32**: 66, 2004; phylogeny).

immersed, embedded in the substratum.

Immersiella A.N. Mill. & Huhndorf (2004), Lasiosphaeriaceae. 2. See Miller & Huhndorf (*MR* **108**: 31, 2004), Zhang *et al.* (*Mycol.* **98**: 1076, 2006; phylogeny).

Immersisphaeria Jaklitsch (2007), Sordariomycetes. 1 (fungicolous), Poland. See Jaklitsch (*Mycotaxon* **101**: 17, 2007).

immobolisation, the controlled, intentional, attachment of fungal cells in fermentation technology (Webb, *Mycologist* **3**: 163, 1989). Cf. biomass support particles.

Immotthia M.E. Barr (1987), Pleosporales. Anamorph *Coniothyrium*-like. 1, Europe; N. America. See Barr (*Mycotaxon* **29**: 501, 1987), Jaklitsch *et al.* (*Öst. Z. Pilzk.* **11**: 93, 2002).

immune, the condition of having qualities which do not allow, or not having qualitites which allow, the development of a disease to take place; **natural immunity** is based on qualities natural to the organism, **acquired immunity**, on the development of such qualities in the course of its life-time, generally as a result of taking the disease naturally or experimentally (not certainly present in plants).

immune (from), exempt from infection (*TBMS* **33**: 155, 1950); having immunity.

immunization, the process of increasing the resistance of, or of giving resistance to, a living organism.

immunosuppressant, a substance such as the fungal metabolite cyclosporin (produced by *Tolypocladium inflatum*), which partly or completely suppresses the immune system; used to prevent rejection of transplanted organs.

imperfect state, see States of fungi; Anamorphic fungi.

imperforate, having no opening.

Imprimospora G. Norris (1986), Fossil Fungi. 1 (Eocene), Canada.

impriorable, illegitimate (obsol.).

Impudentia Vujanović (2003), anamorphic *Pezizomycotina*, H?.?.?. 2 (on insects), Canada. See Vujanovic *et al.* (*Mycotaxon* **88**: 234, 2003).

Imshaugia S.L.F. Mey. (1985), Parmeliaceae (L). 3, widespread (northern hemisphere). See Hinds (*Mycotaxon* **72**: 271, 1999; N. Am.), Elix (*Mycotaxon* **90**: 337, 2004; S America), Thell *et al.* (*Symb. bot. upsal.* **34** no. 1: 429, 2004; biogeography), Miądlikowska *et al.* (*Mycol.* **98**: 1088, 2006; phylogeny).

in situ (Lat.), **in-situ** (Engl.) (of an organism), one living in its natural habitat; cf. *ex situ*.

inaequi-hymeniiferous (of hymenial development in agarics), having basidia which mature and shed their spores in zones; the coprinus type (Buller, *Researches* **2**: 19, 1922). cf. aequi-hymeniiferous.

Inapertisporites Hammen ex Rouse (1959), Fossil Fungi. 28 (Jurassic, Cretaceous, Eocene, Tertiary), widespread.

incertae sedis (inc. sed.), of uncertain taxonomic position.

Incertisporites Hammen (1954), Fossil Fungi. 1 (Cretaceous, Tertiary), Colombia.

Inciliaria Fr. (1825) nom. dub., Fungi (L).

incised, as if cut into; esp. of a pileus margin or lobes of a foliose lichen thallus.

Incolaria Herzer (1893), Fossil Fungi ? Fungi. 1 (Carboniferous), N. America.

incompatible (1) (of sex), unable to be cross-mated due to mating type or fertility barriers; (2) (of vegetative mycelia), unable to form a stable heterokaryon due to genetic differences at one or more vegetative compatibility (vc, het) loci. See vegetative compatibility.

incrassate, made thick.

Incrucipulum Baral (1985), Hyaloscyphaceae. *c.* 4, widespread (temperate). See Baral (*Z. Mykol.* **53**: 119, 1987).

Incrupila Raitv. (1970), Hyaloscyphaceae. *c.* 10, Europe; N. America. See Raitviir & Galán (*MR* **98**: 1137, 1994), Raitviir (*Micologia 2000*: 457, 2000), Raitviir (*Scripta Mycol.* **20**, 2004).

Incrupilella Svrček (1986) = Hyphodiscus fide Baral (*SA* **13**: 113, 1994).

incrusted (of hyphae), having matter excreted on the walls (Corner, 1950).

Incrustocalyptella Agerer (1983), Cyphellaceae. 3, Colombia; Papua New Guinea; Hawaiian Is.; Thailand. See Agerer (*Z. Mykol.* **49**: 160, 1983), Desjardin *et al.* (*Fungal Diversity* **4**: 75, 2000; Thailand).

Incrustoporia Domański (1963) = Skeletocutis fide David (*Naturaliste can.* **109**: 235, 1982).

Incrustoporiaceae Jülich (1982) = Polyporaceae.

incubation period, the time between inoculation and the development of visible symptoms.

indefinite, not sharply limited.

indehiscent (of sporocarps, sporangia, etc.), not opening, or with no special method of opening.

indeterminate (1) having the edge not well-defined, esp. of fruit-bodies and leaf-spots; (2) (of conidiophores), continuing growth indefinitely.

Index of Atmospheric Purity (IAP). A numerical estimate of the purity of the air on the basis of the lichens present on trees (LeBlanc & DeSloover, *CJB* **48**: 1485, 1970). See Air pollution, Bioindication.
Index of Ecological Continuity, see RIEC.
Index of Fungi. Produced by CABI, and appearing twice annually since 1940, this lists new and recently published nomenclatural novelties for fungi s.l. This publication provides a present-day continuation of the great mycological catalogues produced by Petrak, Saccardo, and Zahlbruckner. See Literature.
IndexFungorum (www.indexfungorum.org/Names/IndexFungorum Partnership.htm). The *de facto* world mycological nomenclator. Freely available on-line, this resource provides information about scientific names of fungi at species level and below, including, where available, authors and place of publication, with hyperlinks to a scanned image of the main catalogue entry for each name and, where available, the protologue page.
Indian bread, see tuckahoe.
Indian paint fungus, *Echinodontium tinctorium* (q.v.).
Indiella Brumpt (1906) = Madurella fide Ciferri & Redaelli (*Mycopathologia* **3**: 182, 1941).
indigenous, natural to a country or region; native.
indigenous property, see Bioprospecting, Patent protection.
indirect (of fruit-body development), see direct.
individualism in fungi, mechanisms may exist in nature to 'define' individuals involving co-operative (hyphal fusions, heterokaryosis) and individualistic methods. See Todd & Rayner (*Sci. Progr.* **66**: 331, 1980), incompatible.
indumentum, a covering, such as hairs, etc.
indurated, made hard.
Induratia Samuels, E. Müll. & Petrini (1987), Xylariaceae. Anamorph *Nodulisporium*. 1, New Zealand. See Samuels & Rossman (*Mycol.* **84**: 26, 1992).
indusium (1) cover; (2) (of phalloids), a net-like structure hanging from the top of the stipe under the pileus.
Industrial mycology. Fungi are used in many industrial processes. Because of the ease with which many can be grown in pure culture, and the convenience of their single-cell growth, use of yeasts has dominated, but increasing numbers of other fungi are now being used for new processes.

Some of the more important substances produced, frequently from different forms of carbohydrates and starch, are: alcohol (by *Saccharomyces cerevisiae* from sugar or, after hydrolysis, starch, e.g. cereals, potatoes, or cellulose, e.g. wood, waste sulphite liquor); citric acid (*Aspergillus*, *Penicillium*, *Mucor*); enzyme mixtures such as takadiastase (*Aspergillus*); fats (*Penicillium*); fumaric (*Rhizopus*) and gluconic (*Aspergillus*) acids; glycerol (*S. cerevisiae* var. *ellipsoideus* by the sulphite process); riboflavin (various lactose- fermenting yeasts); itaconic acid (*A. terreus*, etc.; used as a copolymer with acrylic resins); kojic acid (*A. flavus* group); lactic acid (*Rhizopus*); lipoids ('fat') (*Endomycopsis vernalis*; also *Geotrichum candidum*); mycoprotein (*Fusarium venenatum*); rennin (*Rhizomucor pusillus*, *Rhizopus oligosporus*, etc.; O'Leary & Fox, *J. Dairy Res.* **41**: 381, 1974); riboflavin (*Eremothecium*).

Steroid transformations are effected by hyphomycetes and other fungi (Peterson, *in* Rainbow & Rose (Eds), *Biochemistry of industrial micro-organisms*: 537, 1963). Ensilage is another fermentation process and the retting of flax, hemp, and other fibres is dependent on pectin-attacking bacteria and fungi, see Thaysen & Bunker (1927). Fungi play a major role in production of many pharmaceutical commodities, including antibiotics, immunosuppressives, statins and vitamins (). More recent developments include 'mycelial paper' (the addition of mycelium of *Phytophthora cinnamomi* and other phycomycetes to wood pulp); see Johnson & Carlson (*Biotechnol. Bioeng.* **20**: 1063, 1978), fuels from biomass (Jefferies *et al.*, *in* Smith *et al.*, 1980), and use of fungal chitin in applications to heal wounds (Hamlyn & Schmidt, *Mycologist* **8**: 147, 1994).

Lit.: An Zhiquiang [ed.] (*Fungal biotechnology: a handbook of industrial mycology*, 2004), Arora (*Handbook of fungal biotechnology*, 2003), Beuchat (*Food and beverage mycology*, 1978), Bracken (*The chemistry of micro-organisms*, 1955), Brian (*TBMS* **58**: 359, 1972; economic value of fungi), Emmons *et al.* (*Henrici's molds, yeasts and actinomycetes*, edn 2, 1947), Foster (*Chemical activities of fungi*, 1949), Fulmer & Werkman (*An index to the chemical action of micro-organisms on the non-nitrogenous compounds*, 1930), Galloway (*Applied mycology and bacteriology*, edn 3, 1950), Gray (*The relation of fungi to human affairs*, 1959), Hansen (*Jørgensen's Micro-organisms and fermentation*, 1948), Lafar (*Technische Mycologie*, 2 vols, 1896-1907 [English transl. 1889-1910], *Handbuch der technischen Mykologie*, 5 vols, 1904-14), Lemke (*J. industrial microbiol. and biotechnol.* **14**: 355, 1995; industrial mycology and genetics), Leong & Berka (*Molecular industrial mycology: systems and applications for filamentous fungi*, 1991), Onions *et al* (*Smith's Introduction to industrial mycology*, edn 7, 1981), Prescott & Dunn (*Industrial microbiology*, 1940; edn 4, 1982), Ramsbottom (*Rep. Br. Ass.* **1936**: 189; uses of fungi), Smith & Berry (Eds) (*The filamentous fungi*, **1**, *Industrial mycology*, 1975), Smith *et al.* (Eds) (*Fungal biotechnology*, 1980), Thaysen & Bunker (*The microbiology of cellulose, hemicellulose, pectins and gums*, 1927).

See also Antibiotics, Biodegradation, Biodeterioration, Biotechnology, Brewing, Cheese, Edible fungi, Fermented food and drinks, Genetic engineering, Lichens (Economics), Metabolic products, Mushroom cultivation, Nutrition, Pigments, Starters, Wine making.
Inermisia Rifai (1968) = Byssonectria fide Pfister (*Mycol.* **85**: 952, 1994).
inermous, having no spines or prickles.
Inesiosporium R.F. Castañeda & W. Gams (1997), anamorphic *Pezizomycotina*, Hso.?.?. 2, Cuba; Hawaii. See Castañeda Ruiz & Gams (*Nova Hedwigia* **64**: 485, 1997).
infarctate, solid; turgid.
infect (of a pathogen), to enter and establish a pathogenic relationship with an organism; to enter and persist in a carrier (*TBMS* **33**: 155, 1950); to make an attack on an organism; (of an agent), to make infection of an organism take place; **-ed** (of an organism), attacked by a pathogen, cf. contaminated; **-ion**, the act of infecting; **-ious** (of diseases), resulting from infection; sometimes used in the sense of able to be handed on by touch (contagious) or by inoculum; **-ive** (of a pathogen), able to make an attack on a living

organism; (of a vector, medium, etc.), having the power of effecting the transmission of a pathogen.

inferior (of an annulus), low down on the stipe.

infested, attacked by animals, esp. insects; sometimes used of fungi in soil or other substrata in the sense of 'contaminated'.

infissitunicate (Dughi, *C. r. hebd. Séanc. Acad. Sci., Paris* **243**: 750, 1956), see ascus.

inflated hypha, see hypha.

Inflatostereum D.A. Reid (1965), Phanerochaetaceae. 2, S. America; Asia. See Reid (*Nova Hedwigia* Beih. **18**: 143, 1965), Welden (*Revta Biol. trop.* **44** Suppl. 4: 91, 1996).

inflexed (of pileus margin), turned down (Fig. 19F).

infra- (prefix), below; **-generic** (of ranks), all those below that of genus; **-specific** (of ranks), all those below that of species; used in a parallel way for other ranks.

Infrafungus Cif. (1951), anamorphic *Pezizomycotina*, Hso.1-≡ eH.?. 1 (on *Cladosporium*), Philippines. See Ciferri (*Mycopath. Mycol. appl.* **6**: 26, 1951).

Infundibulicybe Harmaja (2003), Tricholomataceae. 13, widespread. See Harmaja (*Ann. bot. fenn.* **40**: 215, 2003).

infundibuliform, funnel-like in form.

Infundibulomyces Plaingam, Somrith. & E.B.G. Jones (2003), anamorphic *Pezizomycotina*. 1, Thailand. See Plaingam *et al.* (*CJB* **81**: 732, 2003).

Infundibulum Velen. (1934) = Peziza Fr. fide Eckblad (*Nytt Mag. Bot.* **15**: 1, 1968), Svrček (*Acta Mus. Nat. Prag.* **32B**: 115, 1976).

Infundibura Nag Raj & W.B. Kendr. (1981), anamorphic *Saccoblastiaceae*. 1, New Zealand; British Isles. See Nag Raj & Kendrick (*CJB* **59**: 544, 1981).

Ingaderia Darb. (1897), Roccellaceae (L). 4, widespread. See Tehler (*CJB* **68**: 2458, 1990; cladistics), Feige & Lumbsch (*Mycotaxon* **48**: 381, 1993), Myllys *et al.* (*Bryologist* **101**: 70, 1998), Follmann (*J. Hattori bot. Lab.* **90**: 251, 2001; S America), Tehler & Irestedt (*Cladistics* **23**: 432, 2007).

ingest, to obtain food by engulfing it; see phagotrophic. Cf. absorb.

Ingoldia R.H. Petersen (1962) = Gyoerffyella fide Marvanová *et al.* (*Persoonia* **5**: 29, 1967), Marvanová & Descals (*TBMS* **89**: 499, 1987).

Ingoldiella D.E. Shaw (1972), anamorphic *Sistotrema*. 3 (with clamp connexions), Australia. See Nawawi & Webster (*TBMS* **78**: 287, 1982).

Ingoldiomyces Vánky (1996), Tilletiaceae. 1 (in seeds of *Poaceae*), N. & S. America. See Castlebury *et al.* (*Mycol.* **97**: 888, 2005; phylog.).

Ingvariella Guderley & Lumbsch (1997), Thelotremataceae (L). 1, widespread (esp. mediterranean). See Guderley *et al.* (*Nova Hedwigia* **64**: 152, 1997).

inhabitant, see symbiosis.

inhibitory substances, see staling substances.

Inifatiella R.F. Castañeda (1985), anamorphic *Pezizomycotina*, Hsy.≡ eP.1. 1, Cuba. See Castañeda (*Deuteromycotina de Cuba* Hyphomycetes **III**: 20, 1985).

ink-caps, basidiomata of *Coprinus*.

innate, bedded in; immersed.

Innatospora J.F.H. Beyma (1929) = Arthrinium fide Ellis (*Mycol. Pap.* **103**, 1965).

Inocephalus (Noordel.) P.D. Orton (1991) = Entoloma fide Kuyper (*in litt.*).

Inocibium Earle (1909) = Inocybe fide Kauffman (*N. Amer. Fl.* **10**, 1924).

inoculate, to put a microorganism, or a substance containing one, into an organism or a substratum.

inoculation, the act of inoculating [of (an organism or substratum)] *with* (the inoculum); *of* (the inoculum) *into* (an organism or substratum); by (an agent or method).

inoculum, the substance, generally a pathogen, used for inoculating.

inoculum potential (of a fungus or other microorganism), the energy of growth available for colonization of a substratum at the surface of the substratum to be colonized (Garrett, 1956).

Inocutis Fiasson & Niemelä (1984), Hymenochaetaceae. 8. See Wagner & Fischer (*Mycol.* **94**: 998, 2002).

Inocybaceae Jülich (1982), Agaricales. 13 gen. (+ 22 syn.), 821 spp.

Lit.: Kupyer (*Persoonia* Suppl. **3**: 1, 1986), Singer (*Agaric. mod. Tax.* 4th ed, 1986), Nordstein (*Syn. Fung.* **2**: 115 pp., 1990), Senn-Irlet (*Persoonia* **16**: 1, 1995), Kropp & Matheny (*Mycol.* **96**: 295, 2004), Aime *et al.* (*Am. J. Bot.* **92**: 74, 2005), Binder *et al.* (*Systematics and Biodiversity* **3**: 113, 2005), Matheny (*Mol. Phylogen. Evol.* **35**: 1, 2005), Matheny & Bougher (*Mycol. Progr.* **5**: 2, 2006).

Inocybe (Fr.) Fr. (1863), Inocybaceae. *c.* 500, widespread (esp. temperate). See Pegler & Young (*Kew Bull.* **26**: 499, 1972; basidiospores, key Br. spp.), Horak (*N.Z. Jl Bot.* **15**: 713, 1977; key 24 NZ spp.), Alessio (*Bres. Icon. Mycol.* **29** Suppl. 3: 1, 1980; col. illustr. & key Eur. spp.), Stangl & Veselský (*Česká Mykol.* **34**: 45, 1980; rough-spored spp.), Kuyper (*Persoonia* Suppl. **3**: 1, 1986).

Inocybella Zerova (1974) = Inocybe fide Kuyper (*in litt.*).

Inocyclus Theiss. & Syd. (1915), Parmulariaceae. 7 (esp. on *Pteridophyta*), widespread (tropical).

Inoderma (Ach.) Gray (1821) = Thrombium fide Hawksworth *et al.* (*Dictionary of the Fungi* edn 8, 1995).

Inoderma Berk. (1881) ≡ Mesophellia.

Inoderma P. Karst. (1879) ≡ Inodermus.

Inodermus Quél. (1886) = Inonotus fide Donk (*Persoonia* **1**: 173, 1960).

Inodosporus Overstreet & Weidner (1974), Microsporidia. 2. See Overstreet & Weidner (*Z. ParasitKde* **44**: 169, 1974).

Inoloma (Fr.) Wünsche (1877) ≡ Cortinarius.

Inonotaceae Fiasson & Niemelä (1984) = Hymenochaetaceae.

Inonotopsis Parmasto (1973), Hymenochaetaceae. 1, widespread (north temperate). See Wagner & Fischer (*Mycol.* **94**: 998, 2002).

Inonotus P. Karst. (1879), Hymenochaetaceae. *c.* 80, widespread. s. str. 30 species. See Pegler (*TBMS* **47**: 175, 1964; key), Gilbertson (*Mem. N. Y. bot. Gdn* **28**: 67, 1976), Wagner & Fischer (*MR* **105**: 773, 2001), Ryvarden (*Syn. Fung.* **15**: 70, 2002; neotropical species), Ryvarden (*Syn. Fung.* **21**: 149 pp., 2005; 101 species).

inoperculate (of an ascus or sporangium), opening by an irregular apical split to discharge the spores, see ascus; cf. operculate.

Inopilus (Romagn.) Pegler (1983) = Entoloma fide Kuyper (*in litt.*).

inordinate, in no order.

inquinant, stained; blackened; dirty (obsol.).

Insecticola Mains (1950) = Akanthomyces fide Samson & Evans (*Acta Bot. Neerl.* **23**: 28, 1974).

Insects (and other invertebrates) and fungi. The relationships between insects and fungi are many and complex: for reviews, see Vega & Blackwell (*Insect-fungal associations*, 2005; ecology, evolution) and Wilding *et al.* (Eds) (*Insect-fungus interactions*, 1989).

Termites, ants (Uphof, *Bot. Rev.* **8**: 563, 1942; Lüscher, *Nature* **167**: 34, 1951; Weber, *Ecology* **38**: 480, 1957; *Lepiota* cultured by *Cyphomyrma*) and ambrosia beetles make 'cultures' of fungi (Bakshi, *TBMS* **33**: 111, 1950). This 'insect agriculture' is highly evolved, often complex and may also involve further species, for example bacteria. The insect may then feed on the resulting cultures directly, or on the fungal substratum after the fungus has made it more digestible (see Termites and fungi). Ambrosia beetles have evolved special structures called *mycangia* which are adaptations for the collection and transport of fungal hyphae or spores (similar structures are also seen in some other beetles and certain mites); each species of ambrosia beetle may have its own specific fungus. 'Ambrosia galls' on plants caused by a mutualistic association between *Cecidomyiidae* flies and anamorphs of *Botryosphaeria* (see galls). Other insects use fungus spores as food (e.g. herbarium beetles, q.v.), and the transmission of fungus spores (esp. slime spores), in addition to pycniospores and other 'diploidizing agents', is frequently effected by insects (see Smell). See Leach (*Insect transmission of plant diseases*, 1940), Carter (*Insects in relation to plant diseases*, 1962; edn 2, 1973); Evans (*Ann. Appl. Biol.* **75**: 331, 1973); also stigmatomycosis. Molluscs such as slugs have also been observed feeding on fungal spores (Buller, *TBMS* **7**: 270, 1922), and some fungi may have evolved mechanisms for deterring this feeding (Wood *et al.*, *Biochemical systematics and ecology* **29**: 531, 2001; Wood & Lefevre, *Biochemical systematics and ecology* **35**: 634, 2007).

Many invertebrates, particularly mites and molluscs, graze on lichens. Mites and ants may be important in the dispersal of soredia and ascospores ingested by rotifers may be viable after excretion. Some lichens have been reported from shells of land snails and some from marine limpets. Some foliose species have been found on *Coleoptera* (see Gressit, *Entom. News* **80**: 1, 1969). Many moths and other insects mimic lichens in larval or adult stages as well as using them for food. Lichen acids do little to deter grazing molluscs and may pass through the gut unchanged (Zopf, *Biol. Zbl.* **16**: 593, 1896). See Lawrey (*Biology of lichenized fungi*, 1984), Richardson (*The vanishing lichens*, 1975), Seaward & Gerson (*in* Seaward (Ed.), *Lichen ecology*: 69, 1977; detailed review), Smith (*Lichens*, 1921 [reprint 1975]); Lichens, soredia.

Certain beetles are frequently associated with myxomycetes, in at least some cases involved with spore dispersal, and there is a growing literature on these associations (e.g. Kotelenets, [*Mikologiya i Fitopatologiya*] **37**: 50, 2003) [in Russian]. Insects, other arthropods, and other invertebrates, including molluscs have been observed to feed on myxomycetes. See reviews by Ing (*Proc. S. Lond. Ent. Nat. Hist. Soc.* **1967**: 18, 1967) and the reports of Keller & Snell (*Mycol.* **94**: 757, 2002) and Waggoner & Poinar (*J. eucaryotic microbiology* **39**: 639, 1992; fossil records of myxomycetes and associated beetles).

See Biological control, Co-evolution, Entomogenous fungi, Symbiosis.

Insiticia Earle (1909) = Mycena fide Singer (*Agaric. mod. Tax.*, 1951).

insititious, of inserted nature, introduced from without.

Insolibasidium Oberw. & Bandoni (1984), Platygloeaceae. 1, N. America; Australia. See Cunnington & Pascoe (*Australas. Pl. Path.* **32**: 433, 2003; Australia).

inspissate, made thick.

Institale Fr. (1829) = Hypoxylon Bull. fide Læssøe (*SA* **13**: 43, 1994).

integrated (of conidiogenous cells), incorporated in the main axis or branches of the conidiophore; cf. discrete.

intellectual property, see Bioprospecting, Patent protection.

inter- (prefix), between; among.

interascal tissue, see hamathecium.

interascicular parenchyma, the paraphysis-like hyphae or paraphysoidal interthecial fibres (Stevens) (obsol.).

interbiotic, living as a parasite on or near one or more living organisms, as certain rhizoidal chytrids.

Intercalarispora J.L. Crane & Schokn. (1983), anamorphic *Pezizomycotina*, Hso.#eP.1. 1 (aquatic), USA. See Crane & Schoknecht (*CJB* **61**: 2243, 1983), Mercado-Sierra *et al.* (*Mycotaxon* **67**: 417, 1998).

intercalary (1) (of growth), between the apex and the base; (2) (of cells, spores, etc.), between two cells.

intercellular, between cells.

International Mycological Association. Founded in 1971; recognized as the Section for General Mycology within the International Union of Biological Societies; objective is to encourage all aspects of mycology; structure comprises individual and corporate members, an elected executive, with regional committees for Africa (see African Mycological Association), Asia (see IMA Committee for Asia), Australasia (see Australasian Mycological Association), Europe (see European Mycological Association), Latin America (see Asociación Latino-americana de Micología) and North America (see Mycological Society of America); organizes International Mycological Congresses (q.v.). Website: www.ima-mycology.org.

International Mycological Congresses (arranged by the International Mycological Association, see Societies and organizations). (1) Exeter, UK (1971; *TBMS* **58**(2), Suppl.: 1-40, 1972); (2) Tampa, USA (1977; *Mycol.* **70**: 253-265, 1978); (3) Tokyo, Japan (1983); (4) Regensberg, Germany (1990); (5) Vancouver, Canada (1994); (6), Jerusalem, Israel (1998); (7), Oslo, Norway (2002); (8), Cairns, Australia (2006). The next congress is scheduled for Edinburgh, UK in 2010.

Internet. Since the previous edn of this *Dictionary*, the internet has become a huge resource for mycological information. Resources available on the internet are developing very rapidly, but individual websites are notoriously ephemeral. As a result, search engines such as Baidu (www.baidu.com), Google (www.google.com), Guruji (www.guruji.com), Naver (www.naver.com) and Yahoo (www.yahoo.com) are essential tools, and collaborative projects such as Hoodong (www.hoodong.com) and Wikipedia (www.wikipedia.org) provide ingress to a vast amount of information, some truly authoritative.

Most (but far from all) mycological information on the internet is in the English language. There is a strong movement among mycologists and other life scientists to make information about scientific names of fungi, distributional (biodiversity) information about living organisms, and taxonomic publications freely and openly available on-line. A selection of non-commercial English language websites which provide free and open information specifically about fungi, are listed below (those also functioning in other languages are marked with an asterisk).

General
Cybertruffle* (www.cybertruffle.org.uk)
IndexFungorum Partnership (www.indexfungorum.org/Names/IndexFungorumPartnership.htm)
Landcare (http://nzfungi.landcareresearch.co.nz/html/mycology.asp) [mainly New Zealand]
Mycobank (www.mycobank.org)
Mycology.net (www.mycology.net)
WWW Virtual Library: Mycology (http://mycology.cornell.edu)
USDA Systematic Mycology and Microbiology Laboratory* (http://nt.ars-grin.gov/fungaldatabases)

Authors of fungal names and other mycologists
Authors of Fungal Names (www.indexfungorum.org/Names/AuthorsOf Fugal-Names.asp)
Cybertruffle's Fungal Valhalla (www.cybertruffle.org.uk/valhalla)
Lichenologist directory (www.botany.hawaii.edu/lichen)
Mycologists on-line (www.fungi.sav.sk/myco)

Biodiversity information (when, where and with what fungi occur)
Cybertruffle's Robigalia* (www.cybertruffle.org.uk/robigalia)
IMI on-line (http://194.203.77.76/herbIMI/IMINumber.asp)
Index of checklists of lichens and lichenicolous fungi (www.biologie.uni-hamburg.de/checklists/lichens/portalpages/index_index.htm)
Landcare (http://nzfungi.landcareresearch.co.nz/html/mycology.asp) [mainly New Zealand]
Mycotaxon regional checklists in downloadable format (www.mycotaxon.com/resources/weblists.html)
USDA Systematic Mycology and Microbiology Laboratory* (http://nt.ars-grin.gov/fungaldatabases)

Conservation
Ascomycete conservation (www.cybertruffle.org.uk/ascos)
IUCN Species Survival Commission Fungi Specialist Group (www.rbg.vic.gov.au/iucnsscfungi/welcome_to_iucn)
Mildew, mould and myxomycete conservation (www.cybertruffle.org.uk/moulds)
Rust and smut conservation (www.cybertruffle.org.uk/rustsmut)

Distribution maps
Cybertruffle's Robigalia* (www.cybertruffle.org.uk/robigalia)
Landcare (http://nzfungi.landcareresearch.co.nz/html/mycology.asp) [New Zealand]

Mycological literature (including catalogues and thesauri)
Bibliography of Systematic Mycology (www.indexfungorum.org/BSM/bsm.asp)
Cyberliber (www.cybertruffle.org.uk/cyberliber)
LibriFungorum (http://194.203.77.76/LibriFungorum/Index.htm)
Mycobank (www.mycobank.org/MycoBiblio.aspx)
USDA Systematic Mycology and Microbiology Laboratory* (http://nt.ars-grin.gov/fungaldatabases/literature/literature.cfm)

Mycological societies
African Mycological Association (http://194.203.77.69/AfricanMycologicalAssociation)
Australasian Mycological Association (http://bugs.bio.usyd.edu.au/AustMycolSoc/Home/ams.html)
European Mycological Association (www.euromould.org)
International Association for Lichenology (www.lichenology.org)
International Mycological Association (www.ima-mycology.org)
International Mycorrhiza Society (www.mycorrhizas.org)
International Society for Human and Animal Mycology (www.isham.org)
International Society for Plant Pathology (www.isppweb.org)
Latin-American Mycological Association* (www.almic.org) [Caribbean, Central America, Mexico and South America]
Mycological Society of America (www.msafungi.org) [North America]
The International Mycological Directory (www.cybertruffle.org.uk/imd) maintains an extensive and updated list of these and other mycological societies.

Fungal reference collections
International Mycological Directory (www.cybertruffle.org.uk/imd)

Photographs of fungi
Lichen-forming fungi (www.nhm.uio.no/botanisk/lavherb.htm)

Scientific names of fungi and their associated organisms [none of these lists are exhaustive]
Cybernome* (www.cybertruffle.org.uk/cybernome)
Dictionary of the Fungi Hierarchy (www.indexfungorum.org/Names/fundic.asp)
Fungus Family Names Database (www.indexfungorum.org/Names/Families.asp)
IndexFungorum (www.indexfungorum.org/Names/Names.asp) [the *de facto* world standard for fungal names]
Mycobank (www.mycobank.org/MycoTaxo.aspx)
Nomen.eumycetozoa.com (www.eumycetozoa.com) ['myxomycete' nomenclator]
USDA Systematic Mycology and Microbiology Laboratory* (http://nt.ars-grin.gov/fungaldatabases/nomen/Nomenclature.cfm)

Scientific names of other organisms associated with fungi [none of these lists are exhaustive]
Cybernome* (www.cybertruffle.org.uk/cybernome)
International Plant Names Index (www.ipni.org) [names of seed-producing plants]
Species2000 (www.sp2000.org) [names of all groups of organisms]

More general English language websites providing information of value to mycologists include:
Biodiversity Heritage Library
(www.biodiversitylibrary.org)
Global Biodiversity Information Facility
(www.gbif.org)
Genbank
(www.ncbi.nlm.nih.gov/sites/entrez?db=taxonomy)
Index Herbariorum
(http://sciweb.nybg.org/science2/IndexHerbariorum.asp)
Text of international code governing nomenclature of fungi (http://ibot.sav.sk/icbn/main.htm).

interspace (of a pileus), the space between the lamellae.

interthecial, between asci.

intervenose (of a pileus), veined in the interspaces.

Intexta J.I.R. Larsson, Steiner & Bjørnson (1997), Microsporidia. 1. See Larsson *et al.* (*Acta Protozool.* **36**: 295, 1997).

Intextomyces J. Erikss. & Ryvarden (1976), Agaricomycetes. 4, widespread. See Eriksson & Ryvarden (*Cortic. N. Europ.* **4**: 735, 1976).

intra- (prefix), within; inside.

intracellular, within the cell.

intrahyphal, see hypha.

Intralichen D. Hawksw. & M.S. Cole (2002), Pezizomycotina. 3, widespread. See Hawksworth & Cole (*Fungal Diversity* **11**: 88, 2002).

intramatrical, living in the matrix or substratum.

intramatrical spores, an alternative name for vesicles produced in host roots by most endomycorrhizal fungi.

intraparietal, within a wall or walls (e.g. of crystals amongst the tissues of an exciple).

Intrapes J.F. Hennen & Figueiredo (1979), anamorphic *Pucciniales*. 1 (on *Couepia* (*Chrysobalanaceae*)), Brazil. Anamorph name for (II). See Cummins & Hiratsuka (*Illustr. Gen. Rust Fungi rev. edit.*, 1983).

Intrapredatorus W.J. Chen, T.L. Kuo, S.T. Wu (1998), Microsporidia. 1. See Chen *et al.* (*Parasitol. Internat.* **47**: 183, 1998).

Intraspora Oehl & Sieverd. (2006) = Archaeospora fide Schüssler & Walker (*in litt.*) but see, Oehl & Sieverding (*J. Appl. Bot. Food Quality* Angew. Botan. **80**: 77, 2006).

Intrasporangium Kalakout., Kirillova & Krassiln. (1967), Actinobacteria. q.v.

intricate cortex (of lichen thalli), made up of hyphae twisted together; cf. textura intricata.

introrse, in the direction of the central axis; inwards.

intumescence, a swelling.

invaginated, covered by a sheath.

Inventories. Assessments of Biodiversity (q.v.) within specific areas, with the aim of cataloguing what is there and valuing its presence. The most prominent inventorying activity to date is the proposed All-Taxa Biodiversity Inventory (ATBI) of the Guanacaste Conservation Area, Costa Rica (Janzen & Hallwachs, *Draft report of an NSF Workshop, Philadelphia*, 1993). Inventories of fungi have been made for various small areas on the initiative of various largely unsupported individuals, classic examples being Esher Common and Slapton Ley, both in the UK, and both with thousands of fungi recorded from a few square km. After many years study of those sites, there is no evidence that the numbers of species new to each site are tailing off. To make an inventory of fungi, even in a small area, thus clearly requires long-term financial investment at a level few if any governments are as yet prepared to contemplate. See Hawksworth *et al.* (*in* Janardhanan *et al.* (Eds), *Tropical mycology*, 1998), Mueller *et al.* (*Biodiversity of fungi - inventory and monitoring methods*, 2004), Rossman (*in* Peng & Chou (Eds), *Biodiversity and terrestrial ecosystems*: 169, 1994), Rossman *et al.* (*Protocols for an All Taxa Biodiversity Inventory of Fungi.*, 1998). See also: Biodiversity, Collection.

involucrellum, tissue forming the upper part of a perithecioid ascoma surrounding the true exciple, not involving host or substrate materials (cf. clypeus) and generally dimidiate (e.g. *Verrucaria*).

Involucrocarpon Servít (1953) = Catapyrenium fide Santesson (*Lichens and lichenicolous fungi of Sweden and Norway*, 1993).

Involucropyrenium Breuss (1996), Verrucariaceae (L). 4, widespread (temperate). See Breuss (*Annln naturh. Mus. Wien* Ser. B, Bot. Zool. **98**: 37, 1996).

Involucroscypha Raitv. (2002), Helotiales. 1. See Raitviir (*Mycotaxon* **81**: 46, 2002).

Involucrothele Servít (1953) = Thelidium fide Hawksworth *et al.* (*Dictionary of the Fungi* edn 8, 1995).

involute, rolled in (Fig. 19I).

Involutisporonites R.T. Clarke (1965), Fossil Fungi. 4 (Cretaceous, Paleocene, Tertiary), China; USA.

Inzengaea Borzí (1885) = Emericella fide von Arx (*Gen. Fungi Sporul. Cult.* Edn 3, 1981).

iodine (**I**, **J**), used as Lugol's solution (I 0.5g, KI 1.5 g, water 100 ml), in potassium iodide (**IKI**; I 1%, KI 3%), and formerly often Melzer's reagent (see Stains) giving blue, red, lavender or violet colours seen best after pre-treatment with potassium hydroxide in spores, asci, hymenial tissues, etc. Reactions can vary according to the kind of iodine solution used, its concentration, the age of the material, and the nature of any pretreatment; all need to be reported when referring to such tests. Extensively used in the systematics of lichenized and non-lichenized fungi since Nylander (*Flora, Jena* **48**: 465, 1865). See Baral (*Mycotaxon* **29**: 399, 1987; caution in use), Common (*Mycotaxon* **41**: 67, 1991; review I+ materials), Kohn & Korf (*Mycotaxon* **3**: 165, 1975; pre-treatment), Nannfeldt (*TBMS* **67**: 283, 1976; ascus plugs), amylomycan, amyloid, dextrinoid.

Iodophanaceae Prokhorov (1993) = Pezizaceae.

Iodophanus Korf (1967), Pezizaceae. Anamorph *Oedocephalum*. 16 (coprophilous), widespread. See Kimbrough *et al.* (*Am. J. Bot.* **56**: 1187, 1969), Thind & Kaushal (*Indian Phytopath.* **31**: 343, 1979), Valadon *et al.* (*TBMS* **74**: 187, 1980; carotenoids, taxonomy), Kimbrough & Curry (*Mycol.* **77**: 219, 1985; posn), de Cachi & Ranalli (*Nova Hedwigia* **49**: 59, 1989; ontogeny), Kimbrough *et al.* *in* Hawksworth (Ed.) (*Ascomycete Systematics. Problems and Perspectives in the Nineties* NATO ASI Series vol. **269** **269**: 398, 1994), Landvik *et al.* (*Nordic Jl Bot.* **17**: 403, 1997; DNA), Prokhorov (*Mikol. Fitopatol.* **31**: 27, 1997; key), Hansen *et al.* (*Mycol.* **93**: 958, 2001; phylogeny), Hansen & Pfister (*Mycol.* **98**: 1029, 2006; phylogeny).

Iodosphaeria Samuels, E. Müll. & Petrini (1987), Iodosphaeriaceae. Anamorphs *Ceratosporium*, *Selenosporella*-like. 2 (from leaves), widespread. See Samuels *et al.* (*Mycotaxon* **28**: 473, 1987), Barr (*Mycotaxon* **39**: 43, 1990; posn), Kang *et al.* (*Fungal Diversity* **2**: 135, 1999; posn), Taylor & Hyde (*Sydowia*

51: 127, 1999; on palms), Hilber & Hilber (*The genus Lasiosphaeria and Allied Taxa*, 2002), Huhndorf *et al.* (*Mycol.* **96**: 368, 2004).

Iodosphaeriaceae O. Hilber (2002), Xylariales. 1 gen., 2 spp.

Lit.: Hilber & Hilber (*Genus Lasiosphaeria and Allied Taxa*, 2002).

Iodowynnea Medel, Guzmán & S. Chacón (1996), Pezizaceae. 1, widespread (tropical). See Medel *et al.* (*Mycotaxon* **59**: 127, 1996), Hansen *et al.* (*Mol. Phylogen. Evol.* **36**: 1, 2005; phylogeny), Hansen & Pfister (*Mycol.* **98**: 1029, 2006; phylogeny), Van Vooren & Moyne (*BSMF* **122**: 137, 2006; France).

Iola, see *Jola*.

Ionaspis Th. Fr. (1871), Hymeneliaceae (L). 7, widespread. See Magnusson (*Acta horti gotoburg.* **8**: 1, 1933), Jørgensen (*Graphis Scripta* **2**: 118, 1989; key 11 spp. Scandinavia), Lutzoni & Brodo (*Syst. Bot.* **20**: 224, 1995).

ionomidotic reaction, release of a dark pigment into aqueous potassium hydroxide (KOH) mounts; an important taxonomic criterion in some dark coloured *Leotiales* apothecia (e.g. *Claussenomyces*; Oullette & Korf, *Mycotaxon* **10**: 255, 1979).

Ionomidotis E.J. Durand ex Thaxt. (1923) ? = Ameghiniella fide Zhuang (*Mycotaxon* **31**: 261, 1988; key), Gamundí (*MR* **95**: 1131, 1991), Gamundí & Romero (*Fl. criptog. Tierra del Fuego* **10**, 1998).

Ionophragmium Peres (1961), anamorphic *Pezizomycotina*, Cpt.≡ eP.?. 1, Brazil. See Peres (*Publções Inst. Micol. Recife* **317**: 11, 1961).

Ioplaca Poelt (1977), Teloschistaceae (L). 2, Nepal. See Poelt & Hinteregger (*Biblthca Lichenol.* **50**, 1993; Himalaya).

Iotidea Clem. (1909) = Peziza Fr. fide Eckblad (*Nytt Mag. Bot.* **15**: 1, 1968).

ipomoearone, a phytoalexin (q.v.) from sweet potato (*Ipomoea batatas*).

Iraniella Petr. (1949), Pezizomycotina. 1, Iran.

Irene Theiss. & P. Syd. (1917) = Asteridiella fide Hawksworth *et al.* (*Dictionary of the Fungi* edn 8, 1995).

Irenina F. Stevens (1927) = Appendiculella fide Hansford (*Beih. Sydowia* **2**, 1961).

Irenopsis F. Stevens (1927), Meliolaceae. 70, widespread (tropical). See Hosagoudar (*Meliolales of India*: 363 pp., 1996; India), Mibey & Hawksworth (*Mycol. Pap.* **174**: 108 pp., 1997; Kenya).

Iridinea Velen. (1934), ? Helotiales. 2, former Czechoslovakia.

Iridionia Sacc. (1902) ≡ Irydyonia.

Irpex Fr. (1828), Meruliaceae. 6, Europe; N. America. See Maas Geesteranus (*Persoonia* **7**: 443, 1974), Buzina *et al.* (*J. Clin. Microbiol.* **43**: 2009, 2005; clinical).

Irpicaceae Spirin & Zmitr. (2003) = Steccherinaceae.

Irpiciochaete, see *Irpicochaete*.

Irpiciporus Murrill (1905) = Spongipellis fide Pegler (*The polypores [Bull. BMS Suppl.]*, 1973).

Irpicium Bref. (1912) nom. dub. = Abortiporus fide Donk (*Verh. K. ned. Akad. Wet.* tweede sect. **62**: 1, 1974).

Irpicochaete Rick (1940) nom. dub., ? Polyporales.

Irpicodon Pouzar (1966), Amylocorticiaceae. 1, Europe. See Wojewoda (*Acta Mycologica* Warszawa **38**: 3, 2003; Poland).

irpicoid, having teeth, or becoming toothed, as in *Irpex*.

Irpicomyces Deighton (1969), anamorphic *Pezizomycotina*, Hso.≡ eP.1. 2 (on *Schiffnerula*), Malaysia; Japan. See Deighton (*Mycol. Pap.* **118**: 24, 1969), Watanabe & Narita (*Mycoscience* **35**: 105, 1994).

Irydyonia Racib. (1900), ? Rhytismatales. 1, Java.

Isaria Pers. (1794), anamorphic *Cordycipitaceae*, Hsy.0eH.10. 7 (entomogenous), widespread. See Petch (*TBMS* **19**: 34, 1934; spp. on insects), de Hoog (*Stud. Mycol.* **1**, 1972; descr. spp., nomencl.), Luangsa-ard *et al.* (*Mycol.* **96**: 773, 2004; phylogeny), Gams *et al.* (*Taxon* **54**: 537, 2005; nomencl.), Hodge *et al.* (*Taxon* **54**: 485, 2005; typification), Luangsa-ard *et al.* (*MR* **109**: 581, 2005; phylogeny), Sung *et al.* (*Stud. Mycol.* **57**, 2007).

Isariaceae Link (1826) = Cordycipitaceae.

Isariella Henn. (1908), anamorphic *Pezizomycotina*, Hsp.0eH.?. 2, C. & S. America. See Seifert (*Mem. N. Y. bot. Gdn* **49**: 202, 1989).

Isariopsella Höhn. (1929) = Ramularia Unger fide Braun (*Monogr. Cercosporella, Ramularia Allied Genera (Phytopath. Hyphom.)* **2**: 316, 1998).

Isariopsis Fresen. (1863) ≡ Phacellium fide Braun (*Nova Hedwigia* **47**: 335, 1989; reassessment of spp.).

Ischnochaeta Sawada (1959) = Erysiphe fide Zheng (*Mycotaxon* **22**: 209, 1985).

Ischnoderma P. Karst. (1879), Fomitopsidaceae. 10, widespread (temperate). See Pouzar (*Česká Mykol.* **25**: 15, 1971), Jahn (*Westf. Pilzber.* **9**: 99, 1973), Corner (*Beih. Nova Hedwigia* **97**: 77, 1989), Roberts & Ryvarden (*Kew Bull.* **61**: 55, 2006; Cameroon).

Ischnodermataceae Jülich (1982) = Hapalopilaceae.

Ischnostroma Syd. & P. Syd. (1914), anamorphic *Pezizomycotina*, Cpt.0fH.?. 1, Philippines.

ISHAM, see Societies and organizations.

Ishwaramyces Hosag. (2004), Asterinaceae. 1, India. See Hosagoudar *et al.* (*J. Econ. Taxon. Bot.* **28**, 2004).

Isia D. Hawksw. & Manohar. (1978), Sordariomycetes. 2, India; Nepal; New Caledonia. See Hawksworth & Manoharachary (*TBMS* **71**: 332, 1978).

Isidiaceae Rchb. (1841) = Pertusariaceae.

isidiate, having isidia.

isidiiferous, of a lichen thallus bearing isidia.

Isidium (Ach.) Ach. (1803) nom. rej. = Pertusaria fide Hawksworth *et al.* (*Dictionary of the Fungi* edn 8, 1995).

isidium (pl. **isidia**), a photobiont-containing protuberance of the cortex in lichens which may be warty, cylindrical, clavate, scale-like, coralloid, simple, or branched (Fig. 22D); occurring directly on the thallus (e.g. *Peltigera praetextata*, *Pseudevernia furfuracea*); may become sorediate (e.g. *Lobaria pulmonaria*). See Bailey (*in* Brown *et al. Lichenology: progress and problems*: 214, 1976; review), Du Rietz (*Svensk bot. Tidskr.* **18**: 141, 1924; classification), Kershaw & Millbank (*Lichenologist* **4**: 214, 1970; rôle), Puymaly (*Botaniste* **48**: 237, 1965).

Isipinga Doidge (1921) = Symphaster fide Müller & von Arx (*Beitr. Kryptfl. Schweiz* **11** no. 2, 1962).

islandicin, toxins of *Penicillium islandicum*; a cause of hepatitis (yellow rice disease) in humans. See also luteoskyrin.

islanditoxin, see islandicin.

iso- (prefix), equal.

isoenzyme, see isozyme.

isogamete, one of two sex cells like in form.

isogamy, the conjugation of isogametes.

isohaplont, a haplont of cells having genotypically like nuclei (Kniep, 1928), cf. miktohaplont.
isokont (**isokontous**) (of cilia and flagella), of equal length. Cf. heterokont.
isolate (v.), (1) to make an isolation; (n), (2) the culture itself, esp. the first 1-spore or pure isolation of a fungus from any place; Lotsy's 'species' (Brierley, *Ann. appl. Biol.* **18**: 420, 1931).
isolation (1) the process of getting a fungus or other organism into pure culture; (2) the culture itself.
Isolation methods. *Aquatic and marine fungi*: Couch (*J. Elisha Mitch. Scientific. Soc.* **55**: 208, 1939; chytrids), Fuller *et al.* (*Mycol.* **56**: 745, 1964; marine 'phycomycetes'), Fuller & Jaworski (*Zoosporic fungi in teaching and research*, 1987), Jones (Ed.) (*Recent advances in aquatic mycology*, 1976), Sparrow (*Aquatic phycomycetes*, edn 2, 1960). *Single spore isolation*: See Hildebrand (*Bot. Rev.* **4**: 627, 1938; **16**: 181, 1950), Badry (*Micromanipulators and micromanipulation*, 1963).

Soil fungi: Gams (*in* Winterhoff (Ed.), *Fungi in vegetation science*: 183, 1992; review); see also Soil fungi.

Spore trapping: See spora.

Use of antibiotics: Georg *et al.* (*Science* **114**: 387, 1951) advocate penicillin 20 units/ml + streptomycin 40 units/ml to suppress unwanted bacteria when isolating fungi (see also cycloheximide); use of rose bengal in isolation media (Ottow, *Mycol.* **64**: 304, 1972).

Isolation media: When isolating fungi, use of translucent media enables developing colonies to be clearly visible. On standard nutrient-rich media e.g. Malt Agar or Potato Dextrose Agar, fungal growth can be rapid. To facilitate separation of colonies, growth rate can be reduced by preparing standard media at quarter or half-strength. Tap Water Agar is an effective isolation medium since it restricts development of fast-growing fungi and discourages bacterial growth. Antibiotics added to media to eliminate bacterial contamination include penicillin G, streptomycin and chloramphenicol (see Media). Bacteria can also be eliminated by lowering the pH of the medium. Selective media are used to obtain particular fungi and exclude others. Many recipes for selective media have been published. See Singleton *et al.* (Eds) (*Methods for research on soilborne phytopathogenic fungi*, 1992), Davet & Rouxel (*Detection and isolation of soil fungi*, 2000), Tuite (*Plant pathological methods. Fungi & bacteria*, 1969). Selective methods often involve the addition of chemical inhibitors to isolation media. See Vaartja (*Phytopathology* **50**: 870, 1960), Edgington (*Phytopathology* **61**: 42, 1971). Broad spectrum growth inhibitors include Cyclosporin A, Cycloheximide, Dichloran and Rose Bengal. See Hocking & Pitt (*Applied and Environmental Microbiology* **39**: 488, 1980), Bills & Polishook (*CJB* **69**: 1477, 1991). Pentachloronitrobenzene (PCNB) can be used to select for basidiomycetes and *Fusarium* (Singleton *et al.*, 1992). Antibiotics of the polyene group, nystatin and pimaricin are also used in selective media for isolating *Pythium* and *Phytophthora*. See Eckert. & Tsao (*Phytopathology* **52**: 771, 1962). Methylbezimidazone carbamate (Benomyl) is selective for some *Oomycota* and *Zygomycota*. At low concentrations, it is also used to select for basidiomycetes. See Hunt & Cobb (*CJB* **49**: 2064, 1971). Selective media for conidial fungi is discussed by Jong (*in* Cole & Kendrick (Eds), *Biology of conidial fungi* **2**, 1981). For selective media used in food-borne fungi, see Samson *et al.* (*Introduction to food-borne fungi*edn 5, 1996).

See also entries on particular habitats, and Culture, and Preservation.
isolichenin, see lichenin.
isomorphic, like in form but unlike in structure.
Isomunkia Theiss. & Syd. (1915), Fungi. 1, Ecuador. See Petrak (*Sydowia* **5**: 337, 1951).
Isomyces Clem. (1931) ≡ Debaryomyces Lodder & Kreger ex Kreger.
isonym, a homotypic synonym; see basionym.
isoplanogamete, one of two motile sex cells like in form.
Isosoma Svrček (1989) = Cudoniella fide Baral (*SA* **13**: 113, 1994).
isospory, see homospory.
Isotexis Syd. (1931) = Elsinoë fide von Arx & Müller (*Stud. Mycol.* **9**, 1975).
Isothea Fr. (1849), Phyllachoraceae. 4 (from living leaves), widespread (arctic-alpine). See Cannon (*MR* **100**: 1409, 1996), Chlebicki & Suková (*Mycotaxon* **90**: 153, 2004; Carpathians).
isotomic dichotomous branching, branching in which both dichotomies are about the same thickness and length so that the dichotomic pattern is visible even in older parts of the thallus, as in *Cladonia evansii*, cf. anisotomic dichotomic branching.
isotype, see type.
isozyme (**isoenzyme**), one of a family of enzymes with different molecular weights and electron charges and so separating in electrophoresis (q.v.).
Issatchenkia Kudrjanzev (1960) = Pichia fide Kurtzman *et al.* (*Int. J. Syst. Bacteriol.* **30**: 503, 1980), Peterson & Kurtzman (*Antonie van Leeuwenhoek* **58**: 235, 1990; molec. phylogeny), Yamada *et al.* (*Biosc., Biotechn., Biochem.* **61**: 577, 1996; phylogeny), Kurtzman *et al. in* Kurtzman & Fell (Eds) (*Yeasts, a taxonomic study* 4th edn: 221, 1998), Thanh *et al.* (*FEMS Yeast Res.* **4**: 113, 2003; Vietnam), Sugiyama *et al.* (*Mycol.* **98**: 996, 2006).
Issia J. Weiser (1977), Microsporidia. 3.
Isthmiella Darker (1967), Rhytismataceae. 4 (on conifers), N. America; Asia. See Darker (*CJB* **45**: 1419, 1967), Hou *et al.* (*CJB* **83**: 37, 2005; China).
Isthmolongispora Matsush. (1971), anamorphic *Pezizomycotina*, Hso.≡ eH.10. 5, Papua New Guinea. See de Hoog & Hennebert (*Proc. K. ned. Akad. Wet.* Ser. C, Biol. Med. Sci. **86**: 343, 1983).
Isthmophragmospora Kuthub. & Nawawi (1992), anamorphic *Pezizomycotina*, Hso.≡ eP.10. 1 (aquatic), Malaysia. See Kuthubutheen & Nawawi (*CJB* **70**: 101, 1992).
Isthmospora F. Stevens (1918), anamorphic *Trichothyrium*, Hso.0bP.1. 1, widespread (tropical).
isthmospore, a spore comprising two or more cells interconnected by a much narrower region, as in the ascospores of *Vialaea* (Cannon, *MR* **99**: 367, 1995); a conidium composed of four more thick-walled cells separated by thin-walled cells as in *Isthmospora* (Hughes, *Mycol. Pap.* **50**, 1953).
Isthmosporella Shearer & J.L. Crane (1999), Phaeosphaeriaceae. 1 (freshwater), USA. See Shearer & Crane (*Mycol.* **91**: 141, 1999).
Isthmotricladia Matsush. (1971), anamorphic *Pezizomycotina*, Hso.1bH.10. 1, Papua New Guinea. See Matsushima (*Bull. natn. Sci. Mus.* Tokyo, B **14**: 478,

1971), Roldan (*Mycotaxon* **42**: 297, 1991; culture).

isthmus (1) the narrower or thinner-walled portion of an isthmospore (q.v.); (2) the thickened medial perforated septum of a polarilocular ascospore.

isthmus disarticulation, protoplasmic retraction during spore formation, with the secretion of an endosporic wall membrane.

Isuasphaera H.D. Pflug (1978), Fossil Fungi, anamorphic *Pezizomycotina*. 1 (*c*. 3800 million yr old Isua quartzite), Greenland.

Itajahya Möller (1895), Phallaceae. 1, widespread (tropical; subtropical). See Ahmad (*Lloydia* **8**: 238, 1945), Malençon (*BSMF* **100**: Atlas pl. 238, 1984).

italicization (of scientific names), see Classification.

Itersonia Rippel & Flehmig [not traced] nom. dub., ? Myxobacteria.

Itersonilia Derx (1948), anamorphic *Cystofilobasidiaceae*. 2, widespread (temperate). See Sowell & Korf (*Mycol.* **52**: 934, 1960), Channon (*Ann. appl. Biol.* **51**: 1, 1963; leaf spot and canker of parsnip by *I. pastinacae*), Webster *et al.* (*TBMS* **82**: 13, 1984; ballistospore discharge), Yamada & Konda (*J. gen. appl. Microbiol.* Tokyo **30**: 313, 1984; significance of coenzyme Q system in classification), Webster *et al.* (*TBMS* **91**: 193, 1988; ballistospore discharge), Boekhout (*MR* **95**: 135, 1991; phenetic systematics), Boekhout & Jille (*Syst. Appl. Microbiol.* **14**: 117, 1991; mitosis and DNA content in *I. perplexans*), Boekhout *et al.* (*Can. J. Microbiol.* **37**: 188, 1991; genomic characterization), Ingold (*Mycologist* **5**: 35, 1991; development), Horita & Yasuoka (*Journal of General Plant Pathology* **68**: 277, 2002; *Itersonilia perplexans* in Japan).

Ithyphallus Gray (1821) nom. rej. ≡ Mutinus.

ITS, see Molecular biology.

Ityorhoptrum P.M. Kirk (1986), anamorphic *Pezizomycotina*, Hso.#eP.19. 1, British Isles. See Kirk (*TBMS* **86**: 417, 1986).

IUBS, see Societies and organizations.

IUCN, International Union for the Conservation of Nature and Natural Resources; the World Conservation Union. See Conservation.

IUMS, see Societies and organizations.

iwatake, see rock tripe.

Iwilsoniella E.B.G. Jones (1991), Halosphaeriaceae. 1 (cooling tower water), British Isles. See Jones (*SA* **10**: 7, 1991).

Ixechinaceae Guzmán (1974) = Boletaceae.

Ixechinus R. Heim (1968) = Fistulinella fide Guzmán (*Boln. Soc. mex. Micol.* **8**: 53, 1974).

ixo- (prefix), sticky.

Ixocomus Quél. (1888) ≡ Suillus Gray.

ixocutis (of a pileus), a slimy cuticle.

Ixodopsis P. Karst. (1870) ≡ Sordaria.

ixotrichoderm (**ixotrichodermium**) (of a pileus), a trichodermium composed of gelatinized hyphae (Snell, 1939; for a discussion, see Shaffer, *Mycol.* **58**: 486, 1966).

Iyengarina Subram. (1958), anamorphic *Pezizomycotina*, Hso.0bP.1. 1, India. See Subramanian (*J. Indian bot. Soc.* **37**: 406, 1958).

J, see iodine.

Jaapia Bres. (1911), Boletales. 2, widespread. See Nannfeldt & Eriksson (*Svensk bot. Tidskr.* **47**: 177, 1953).

Jaapia Kirschst. (1938) ≡ Keisslerina.

Jack o'lantern, the basidioma of the luminous *Clitocybe illudens*.

jack-in-the-box discharge, see ascus.

Jacksonia J.C. Lindq. (1970) [non *Jacksonia* R. Br. ex Aiton 1811, *Papilionaceae*] ≡ Dietelia fide Lindquist (*Rev. Fac. Agron., Univ. Nac. La Plata* **47**: 303, 1972).

Jacksoniella J.C. Lindq. (1972) = Dietelia fide Sathe (*Indian Phytopath.* **27**: 617, 1974), Cummins & Hiratsuka (*Illustr. Gen. Rust Fungi rev. edit.*, 1983) See.

Jacksoniella Kamat & Sathe (1972) = Dietelia fide Cummins & Hiratsuka (*Illustr. Gen. Rust Fungi rev. edit.*, 1983).

Jacksonomyces Jülich (1979), Meruliaceae. 10, widespread. See Wu & Chen (*Bull. natn. Sci. Mus.* Tokyo, B **3**: 260, 1977).

Jackya Bubák (1902) [non *Jackia* Wall. 1823, *Rubiaceae*] = Puccinia fide Dietel (*Nat. Pflanzenfam.* **6**, 1928).

Jacobaschella Kuntze (1891) ≡ Diplosporium.

Jacobia Arnaud (1952) nom. inval., anamorphic *Agaricomycetes*, Hso.0eH.?. 1 (with clamp connexions), Europe.

Jacobia Contu (1998) ≡ Contumyces fide Kuyper (*in litt.*).

Jacobsonia Boedijn (1935), Helotiaceae. 1, Sumatra.

Jaculispora H.J. Huds. & Ingold (1960), anamorphic *Classiculaceae*, Hso.0eH.1. 1 (aquatic), Jamaica. See Hudson & Ingold (*TBMS* **43**: 475, 1960).

Jaczewski (Arthur Louis Arthurovič de; 1863-1932; Russia). Studied under Fischer (q.v.), Berne (1889-1894); founded the laboratory of mycology and plant pathology (1896) then the Central Laboratory for Plant Pathology (1901) then the Bureau for Mycology and Phytopathology (1907), St Petersburg Botanic Garden; first Director, Bureau of Mycology and Phytopathology [later renamed the Jaczewski Institute], St Petersburg (1907-1932). A pioneer of mycology in Russia and, perhaps, the greatest Russian mycologist, with a prolific scientific output; his outstanding managerial skills enabled him to realize remarkable results during very difficult years of revolution and war. *Publs.* [*Key to the Fungi*] (1913-1917), [*Fundamentals of mycology*] (1933) [in Russian]; also numerous other works on mycology and plant pathology. *Biogs, obits etc.* Grummann (1974: 548); Jones (*Phytopathology* **23**: 111, 1933) [portrait]; Stafleu & Cowan (*TL-2* **2**: 413, 1979).

Jaczewskia Mattir. (1912) = Phallus fide Vasil'kov (*Bot. Zh. SSSR* **40**: 596, 1955).

Jaczewskiella Murashk. (1926) = Stigmina fide Sutton (*Mycol. Pap.* **141**, 1977).

Jaffuela Speg. (1921), ? Dothioraceae. 1 (on *Puya*), Chile.

Jafnea Korf (1960), Pyronemataceae. 2, widespread. See Berthet & Korf (*Nat. can.* **96**: 247, 1969), Schumacher (*Mycotaxon* **33**: 149, 1988), Benkert & Klofac (*Öst. Z. Pilzk.* **13**: 55, 2004; Europe), Perry *et al.* (*MR* **111**: 549, 2007; phylogeny).

Jafneadelphus Rifai (1968) = Aleurina Massee fide Zhuang & Korf (*Mycotaxon* **26**: 361, 1986).

Jahniella Petr. (1921), anamorphic *Pezizomycotina*, Cpd.0fH.1. 3, former Czechoslovakia.

Jahnoporus Nuss (1980), Albatrellaceae. 2, widespread (north temperate). See Nuss (*Hoppea* **39**: 176, 1980).

Jahnula Kirschst. (1936), Aliquandostipitaceae. 10 (on submerged wood), widespread. See Hawksworth (*Sydowia* **37**: 43, 1984), Hyde & Wong (*Nova Hedwigia* **68**: 489, 1999; Australia), Pang *et al.* (*MR* **106**:

1031, 2002), Raja & Shearer (*Mycol.* **98**: 319, 2006), Campbell *et al.* (*CJB* **85**: 873, 2007; phylogeny).

Jahnulales K.L. Pang, Abdel-Wahab, El-Shar., E.B.G. Jones & Sivichai (2002). Dothideomycetes. 1 fam., 6 gen., 25 spp. Fam.:
Aliquandostipitaceae
For *Lit.* see under fam.

Jainesia Gonz. Frag. & Cif. (1925), anamorphic *Pezizomycotina*, Hso.≡ eP.?. 1, West Indies.

Jamesdicksonia Thirum., Pavgi & Payak (1961), Georgefischeriaceae. 23 (on *Cyperaceae*, *Poaceae*), S. Asia; N.; C. & S. America; Australia. See Bauer *et al.* (*MR* **105**: 416, 2001), Vánky & McKenzie (*Fungal Diversity Res. Ser.* **8**: 259 pp., 2002; New Zealand).

Jamesiella Lücking, Sérus. & Vězda (2005), Gomphillaceae (L). 1, widespread. See Lücking *et al.* (*Lichenologist* **37**: 165, 2005), Lücking *et al.* (*Lichenologist* **38**: 131, 2006; Costa Rica).

Janacekia J.I.R. Larsson (1983), Microsporidia. 3.

Janannfeldtia Subram. & Sekar (1993), Nitschkiaceae. 2, India. See Subramanian & Sekar (*Kavaka* **18**: 19, 1993), Huhndorf *et al.* (*MR* **108**: 1384, 2004).

Janauaria Singer (1986), Agaricaceae. 1, Brazil. See Singer (*Agaric. mod. Tax.* 4th ed: 495, 1986).

Janetia M.B. Ellis (1976), anamorphic *Pezizomycotina*, Hso.≡ eP.10. 17, widespread. See Sutton & Pascoe (*Aust. Syst. Bot.* **1**: 127, 1988), Sivanesan (*MR* **94**: 566, 1990; synnematous spp.), Calduch *et al.* (*Mycol.* **94**: 355, 2002; Spain), Xu & Guo (*Nova Hedwigia* **75**: 201, 2002; China).

Janospora (Starbäck) Höhn. (1923) = Stilbospora fide Sutton (*Mycol. Pap.* **141**, 1977).

Janseella Henn. (1899) = Phaeotrema fide Hawksworth *et al.* (*Dictionary of the Fungi* edn 8, 1995).

Jansia Penz. (1899) = Mutinus fide Dring *in* Ainsworth *et al.* (Eds) (*The Fungi* **4B**: 458, 1973).

Jansoniisporites Kalgutkar (1997), Fossil Fungi. 1, Canada. See Kalgutkar (*Review of Palaeobotany and Palynology* **97**: 216, 1997).

Japanese mushroom, see matsu-take.

Japewia Tønsberg (1990), Ramalinaceae (L). 3, widespread (north temperate). See Tønsberg (*Lichenologist* **22**: 205, 1990), Printzen (*Bryologist* **102**: 714, 1999).

Japewiella Printzen (1999), Ramalinaceae (L). 4, Mexico. See Printzen (*Bryologist* **102**: 714, 1999).

Japonia Höhn. (1909), anamorphic *Yoshinagaia*, Cpd.1eH.1. 1, Japan.

Japonogaster Kobayasi (1989), Agaricaceae. 1, Japan. ? A monstrosity of *Lycoperdon*.

Jarmania Kantvilas (1996), Ramalinaceae (L). 1, Australia. See Kantvilas (*Lichenologist* **28**: 230, 1996).

Jarxia D. Hawksw. (1989), Naetrocymbaceae. 1, C. America; Florida. See Hawksworth (*Stud. Mycol.* **31**: 93, 1989), Harris (*More Florida Lichens*, 1995).

Jattaea Berl. (1900), Calosphaeriaceae. Anamorphs *Stachybotrys*-like, *Phialophora*-like. 8, widespread. See Barr (*Mycol.* **77**: 549, 1985), Romero & Samuels (*Beih. Sydowia* **43**: 228, 1991), Mostert *et al.* (*Stud. Mycol.* **54**: 115 pp., 2006), Damm *et al.* (*Persoonia* **20**: 39, 2008; asci, relns).

Jattaeolichen Tomas. & Cif. (1952) = Arthopyrenia fide Hawksworth *et al.* (*Dictionary of the Fungi* edn 8, 1995).

Jattaeomyces Cif. & Tomas. (1953) ≡ Jattaeolichen.

Javaria Boise (1986), ? Melanommataceae. 2 (on *Palmae*), N. & S. America. See Boise (*Acta Amazon.* Supl. **14**: 49, 1986), Hyde & Fröhlich (*Sydowia* **50**: 81, 1998; synonym of *Astrosphaeriella*).

Javonarxia Subram. (1995), anamorphic *Pezizomycotina*, Hso.≡ eP.19. 2, Malaysia. See Subramanian (*Kavaka* **20/21**: 57, 1995).

Jeaneliomyces Lepesme (1945) = Dimeromyces fide Hawksworth *et al.* (*Dictionary of the Fungi* edn 8, 1995).

jelly fungi, a term sometimes applied to the *Tremellales* s.l.

Jenmania W. Wächt. (1897), Lichinaceae (L). 2, S. America. See Henssen (*Lichenologist* **5**: 444, 1973; key), Schultz & Büdel (*Lichenologist* **35**: 151, 2003; phylogeny).

Jenmaniomyces Cif. & Tomas. (1953) ≡ Jenmania.

Jerainum Nawawi & Kuthub. (1992), anamorphic *Pezizomycotina*, Hso.#eP.1. 1 (aquatic), Malaysia. See Nawawi & Kuthubutheen (*Mycotaxon* **45**: 409, 1992).

Jew's ear, the basidioma of *Auricularia auricula-judae*.

Jirovecia J. Weiser (1977), Microsporidia. 3.

Jiroveciana J.I.R. Larsson (1981), Microsporidia. 1.

Jobellisia M.E. Barr (1993), ? Diaporthales. 2 (from wood), widespread. See Barr (*Mycotaxon* **46**: 45, 1993), Huhndorf *et al.* (*Sydowia* **51**: 183, 1999; DNA, posn), Kang *et al.* (*Mycoscience* **40**: 151, 1999; posn), Réblová *et al.* (*Stud. Mycol.* **50**: 533, 2004), Mostert *et al.* (*Stud. Mycol.* **54**: 115 pp., 2006).

Joerstadia Gjaerum & Cummins (1982), Phragmidiaceae. 4 (on *Alchemilla* (*Rosaceae*)), Africa; Madagascar. See Gjaerum & Cummins (*Mycotaxon* **15**: 420, 1982).

Johansonia Sacc. (1889), Saccardiaceae. *c.* 11, widespread (tropical).

Johansoniella Bat., J.L. Bezerra & Cavalc. (1966) = Johansonia fide von Arx & Müller (*Stud. Mycol.* **9**, 1975).

Johenrea C.E. Langeb, J.J. Becnel, E. Razafindratiana, J. Przybyszewski & H. Razafindrafara (1996), Microsporidia. 1. See Langeb *et al.* (*J. Invert. Path.* **68**: 28, 1996).

Johncouchia S. Hughes & Cavalc. (1983), anamorphic *Septobasidium*, Hso.0bP.1. 1, widespread. See Hughes & Cavalcanti (*CJB* **61**: 2226, 1983).

Johnkarlingia Pavgi & S.L. Singh (1979), Synchytriaceae. 1 (on *Brassica* roots), India. See Pavgi & Singh (*Mycopathologia* **69**: 53, 1979).

Johnstonia M.B. Ellis (1971) [non *Johnstonia* Walkom 1925, fossil] ≡ Neojohnstonia.

Jola Möller (1895), Eocronartiaceae. 1 (on *Musci*), widespread (tropical). See Martin (*Mycol.* **31**: 239, 1939), Frieders & McLaughlin (*MR* **105**: 734, 2001; cytol., ultrastr., anam.).

Jonaspis, see *Ionaspis*.

Jongiella M. Morelet (1971) = Camillea fide Læssøe *et al.* (*MR* **93**: 121, 1989), Læssøe (*SA* **13**: 43, 1994; ? synonym of *Xylaria*).

Jopex, see *Irpex*.

jordanon, Brierley's (*Ann. app. Biol.* **18**: 420, 1931) application of Lotsy's term for a group of isolates constituting a race or strain, with similar cultural characters and behaviour (obsol.).

Josefpoeltia S.Y. Kondr. & Kärnefelt (1997), Teloschistaceae (L). 3, S. America. See Kondratyuk & Kärnefelt (*Progr. Probl. Lichenol. Nineties. Proc.*

Third Symp. Intern. Assoc. Lichenol. [*Biblthca Lichenol.* **68**]: 22, 1997), Frödén & Lindblom (*Bryologist* **106**: 447, 2003).

Jove's beard, the basidioma of *Odontia barba-jovis*.

Jrpex, see *Irpex*.

Jubispora B. Sutton & H.J. Swart (1986), anamorphic *Pezizomycotina*, St.≡ eP.19. 1, Africa. See Sutton & Swart (*TBMS* **87**: 97, 1986).

Jugglerandia Lloyd (1923) ≡ Holwaya.

Jugulospora N. Lundq. (1972), Lasiosphaeriaceae. 1 (on burnt soil), Europe; N. America. See Lundqvist (*Symb. bot. upsal.* **20** no. 1, 1972), Barr (*Mycotaxon* **39**: 43, 1990; posn), Huhndorf *et al.* (*Mycol.* **96**: 368, 2004; phylogeny), Miller & Huhndorf (*Mol. Phylogen. Evol.* **35**: 60, 2005; phylogeny).

Julella Fabre (1879), Thelenellaceae (±L). 3 or 18, widespread. See Barr (*Sydowia* **38**: 11, 1986), Aptroot & van den Boom (*Mycotaxon* **56**: 1, 1995), Kohlmeyer *et al.* (*Bot. Mar.* **40**: 291, 1997), Tam *et al.* (*Bot. Mar.* **46**: 487, 2003; phylogeny), Aptroot *et al.* (*Biblthca Lichenol.* **97**, 2008; Costa Rica).

Juliohirschhornia Hirschh. (1986), Ustilaginaceae. 1 (on *Paspalum*), S. America. See Vánky (*in litt.*; ? syn. of *Ustilago*), Hirschhorn (*Las Ustilaginales de la Flora Argentina*, 1986).

Jumillera J.D. Rogers, Y.M. Ju & F. San Martín (1997), Xylariaceae. Anamorphs *Geniculosporium*-like, *Libertella*-like. 7, widespread. See Rogers *et al.* (*Mycotaxon* **64**: 39, 1997), Rogers & Ju (*CJB* **80**: 478, 2002; Hawaii).

Juncigena Kohlm., Volkm.-Kohlm. & O.E. Erikss. (1997), Hypocreomycetidae. Anamorph *Cirrenalia*. 1 (stems of *Juncus*), N. America. See Kohlmeyer *et al.* (*Bot. Mar.* **40**: 291, 1997), Schoch *et al.* (*MR* **111**: 154, 2007; phylogeny).

Junctospora Minter & Hol.-Jech. (1981), anamorphic *Pezizomycotina*, Hso.0eH.4. 1, former Czechoslovakia. See Minter & Holubová-Jechová (*Folia geobot. phytotax.* **16**: 202, 1981).

Junewangia W.A. Baker & Morgan-Jones (2002), anamorphic *Pezizomycotina*, H?.?.?. 1, widespread. See Baker *et al.* (*Mycotaxon* **81**: 307, 2002).

Junghuhnia Corda (1842), Meruliaceae. *c.* 20, widespread. See Ryvarden (*Persoonia* **7**: 17, 1972), Pouzar (*Czech Mycol.* **55**: 1, 2003; Cuba), Ipulet & Ryvarden (*Syn. Fung.* **20**: 87, 2005; Uganda).

Jungkultur, see Normkultur.

Junia Dumort. (1822) [non *Junia* Adans. 1763, *Clethraceae*] = Phallus fide Stalpers (*in litt.*).

juvenescence, the process of maturing at a stage of development normally immature.

K (1) see Metabolic products; (2) Royal Botanic Gardens (Kew, Surrey, UK); founded 1841; managed by a Board of Trustees and mainly funded by the UK Ministry of Agriculture, Fisheries and Food (MAFF); most lichen material transferred to **BM** (q.v.); see Blunt (*In for a penny*, 1978), Hepper (Ed.) (*Royal Botanic Gardens, Kew: gardens for science and pleasure*, 1982); (3) see Sörensen coefficient.

Kabataia J. Lom, I. Dyková & K. Tonguthai (1999), Microsporidia. 1. See Lom *et al.* (*Dis. Aq. Organ.* **38**: 39, 1999).

Kabatana J. Lom, I. Dyková & K. Tonguthai (2000), Microsporidia. 3. See Lom *et al.* (*J. Parasit.* **93**: 655, 2000).

Kabathia Nieuwl. (1916) = Sphaerellopsis Cooke fide Sutton (*Mycol. Pap.* **141**, 1977).

Kabatia Bubák (1904), anamorphic *Discosphaerina*, St.0-1eH.1. 9, widespread. See Sutton (*The Coelomycetes*, 1980).

Kabatiella Bubák (1907), anamorphic *Discosphaerina*, Cpd.0eH.?. 17, widespread (esp. north temperate). *K. caulivora* (clover (*Trifolium*) scorch). See Sampson (*TBMS* **13**: 103, 1928), von Arx (*Revision of fungi classified as* Gloeosporium, 1970), Hoog *et al.* (*Stud. Mycol.* **43**: 31, 1999), Yurlova *et al.* (*Stud. Mycol.* **43**: 63, 1999; teleom.), Crous *et al.* (*Australas. Pl. Path.* **29**: 267, 2000; Australia).

Kabatina R. Schneid. & Arx (1966), anamorphic *Dothioraceae*, Cac.0eH.15. 5, Europe; N. America. See Ramaley (*Mycotaxon* **43**: 437, 1992; on *Mahonia*), Butin & Pehl (*MR* **97**: 1340, 1993), Hsiang *et al.* (*Can. J. Pl. Path.* **22**: 79, 2000; Canada).

Kacosphaeria Speg. (1887), ? Calosphaeriaceae. 1, Tierra del Fuego. See Petrak (*Sydowia* **5**: 328, 1951).

Kaernefeltia A. Thell & Goward (1996), Parmeliaceae (L). 2, N. America. See Thell & Goward (*Bryologist* **99**: 125, 1996), Thell *et al.* (*Mycol. Progr.* **1**: 335, 2002; phylogeny), Mattsson & Articus (*Symb. bot. upsal.* **34** no. 1: 237, 2004).

Kafiaddinia Mekht. (1978) = Dactylellina fide Scholler *et al.* (*Sydowia* **51**: 89, 1999).

Kainomyces Thaxt. (1901), Laboulbeniaceae. 3, W. Africa; Asia. See Terada (*TMSJ* **19**: 55, 1978), Tavares (*Mycol. Mem.* **9**: 627 pp., 1985).

Kakabekia Bargh. (1965), Fossil Fungi. 2, Europe; N. America.

Kalaallia Alstrup & D. Hawksw. (1990), Dacampiaceae. 1 (on lichens), Greenland. See Alstrup & Hawksworth (*Meddr Grønland* Biosc. **31**, 1990), Orange (*Mycotaxon* **81**: 265, 2002).

Kalaharituber Trappe & Kagan-Zur (2005), Pezizaceae. 1, Namibia. See Ferdman *et al.* (*MR* **109**: 242, 2005), Læssøe & Hansen (*MR* **111**: 1075, 2007; phylogeny).

Kalbiana Henssen (1988), ? Verrucariaceae (L). 1, Brazil. See Henssen (*Lichenes Neotropici* Fascicle **X** (nos 401-450): 9, 1988).

Kalbographa Lücking (2007), Graphidaceae (L). 1, Caribbean. See Lücking (*Biblthca Lichenol.* **96**: 185, 2007).

Kalchbrennera Berk. (1876) = Lysurus fide Dring (*Kew Bull.* **35**: 68, 1980).

Kalchbrenneriella Diederich & M.S. Christ. (2002), anamorphic *Pezizomycotina*, H?.?.?. 1 (on lichens). See Diederich (*Bryologist* **105**: 411, 2002).

Kaleidosporium Van Warmelo & B. Sutton (1981), anamorphic *Pezizomycotina*, Cac.#eP.15. 1, N. America. See van Warmelo & Sutton (*Mycol. Pap.* **145**: 23, 1981).

Kallichroma Kohlm. & Volkm.-Kohlm. (1993), Bionectriaceae. 2 (submerged wood, marine), widespread. See Kohlmeyer & Volkmann-Kohlmeyer (*MR* **97**: 753, 1993), Rossman *et al.* (*Stud. Mycol.* **42**: 248 pp., 1999), Rossman *et al.* (*Mycol.* **93**: 100, 2001).

Kalmusia Niessl (1872), ? Montagnulaceae. *c.* 12, widespread. See Barr (*Prodr. Cl. Loculoasc.*: 168 pp., 1987), Tanaka *et al.* (*Mycoscience* **46**: 110, 2005; Japan), Eriksson (*Mycotaxon* **95**: 67, 2006; on *Bambusa*).

Kamatella Anahosur (1969), anamorphic *Pezizomycotina*, Cpd.1eP.1. 1, S. Africa; India. See Anahosur (*Bull. Torrey bot. Club* **96**: 207, 1969).

Kamatia V.G. Rao & Subhedar (1976) = Pseudodictyosporium fide Kendrick *in* Carmichael *et al.* (*Gen-

era of Hyphomycetes, 1980), Tsui *et al.* (*Fungal Diversity* **21**: 157, 2006).

Kamatomyces Sathe (1966) = Masseeëlla fide Berndt (*in litt.*).

kamé, truffle (Arabic); **black -**, **brown -**, the edible *Terfezia boudieri* and *T. claveryi*, respectively, of the Middle East (Awameh & Alsheikh, *Mycol.* **72**: 50, 494, 1980).

kames, see terfas.

Kameshwaromyces Kamal, R.K. Verma & Morgan-Jones (1986), anamorphic *Pezizomycotina*, Hso.#eP.1. 1, India. See Kamal *et al.* (*Mycotaxon* **25**: 247, 1986).

Kananascus Nag Raj (1984), Sordariomycetes. Anamorph *Koorchaloma*. 2, widespread. See Nag Raj (*Mycotaxon* **19**: 167, 1984), Treigiene (*Mycotaxon* **96**: 173, 2006; Lithuania).

Kanousea Bat. & Cif. (1962) = Microcallis fide von Arx & Müller (*Stud. Mycol.* **9**, 1975).

Kantvilasia P.M. McCarthy, Elix & Sérus. (2000), Ectolechiaceae. 1, Tasmania; S. America. See McCarthy *et al.* (*Lichenologist* **32**: 317, 2000), Lücking (*Cryptog. Mycol.* **27**: 121, 2006; French Guiana).

Kapooria J. Reid & C. Booth (1989), Diaporthales. 1, India. See Reid & Booth (*CJB* **67**: 879, 1989).

Kappamyces Letcher & M.J. Powell (2005), Kappamycetaceae. 1, USA. See Letcher & Powell (*Nova Hedwigia* **80**: 115, 2005).

Kappamycetaceae Letcher (2006), Rhizophydiales. 1 gen., 1 spp.

Karakulinia N.P. Golovina (1964) = Fusicladium fide Deighton (*Mycol. Pap.* **112**, 1967), Schubert *et al.* (*Schlechtendalia* **9**, 2003).

Karlingia A.E. Johanson (1944) nom. illegit. ≡ Rhizophlyctis fide Blackwell & Powell (*Mycotaxon* **70**: 213, 1999; nomencl.).

Karlingiomyces Sparrow (1960), Chytridiaceae. 5, widespread. See Karling (*Sydowia* Beih. **6**: 58, 1966), Dogma (*Nova Hedwigia* **24**: 393, 1973), Blackwell *et al.* (*Mycotaxon* **89**: 259, 2004; monogr., key) = Karlingia (Chytrid.) fide.

Karoowia Hale (1989), Parmeliaceae (L). 18, widespread. See Elix (*Flora of Australia* **55**: 62, 1994), Elix (*Australasian Lichenology* **46**: 18, 2000; Australia), Blanco *et al.* (*Taxon* **53**: 959, 2004; phylogeny), Thell *et al.* (*Mycol. Progr.* **3**: 297, 2004; phylogeny), Blanco *et al.* (*Mol. Phylogen. Evol.* **39**: 52, 2006; phylogeny).

Karschia Körb. (1865), Dothideomycetes. 4 (on lichens), Europe. See Hafellner (*Beih. Nova Hedwigia* **62**, 1979; key, disp. excl. spp.).

Karsten (Petter Adolf; 1834-1917; Finland). PhD, Helsinki (1859); schoolteacher, Wasa Gymnasium (1862-1864); Lecturer in botany (1864-1900) then Professor (1900-1908), Agricultural and Dairy Institute, Mustiala. Pioneer in Finnish mycology, describing many new species. Collections mostly in Helsinki (**H**). *Publs. Mycologia Fennica* (1871-1878) [reprint 1967]; *Symbolae ad Mycologiam Fennicam* 33 parts (1870-1895) [reprint 1967]; *Rysslands, Finlands och den Skandinaviska Halföns Hattsvampar* (1879-82); *Kritisch Öfversigt af Finlands Basidsvampar (Basidiomycetes; Gastero- et Hymenomycetes)* (1889); also collected mycological papers (1859-1909) [reprint 4 vols, 1973]. *Biogs, obits etc.* Grummann (1974: 609); Hintikka (*Karstenia* **1**: 5, 1950, portrait) [in Finnish]; Lowe (*Mycol.* **48**: 99, 1956) [polypore types]; Stafleu & Cowan (*TL-2* **2**: 505, 1979).

Karstenella Harmaja (1969), Karstenellaceae. 1, Finland. See Harmaja (*Karstenia* **9**: 20, 1969), Harmaja (*Karstenia* **14**: 109, 1974).

Karstenellaceae Harmaja (1974), Pezizales. 1 gen., 1 spp.

Lit.: Harmaja (*Karstenia* **9**: 20, 1969), Harmaja (*Karstenia* **14**: 109, 1974), Kimbrough (*Mem. N. Y. bot. Gdn* **49**: 326, 1989).

Karstenia Britzelm. (1897) [non *Karstenia* Göpp. 1836, fossil *Polypodiaceae*] ≡ Prillieuxia.

Karstenia Fr. (1885) nom. cons., ? Rhytismatales. 9, Europe; America. See Sherwood (*Mycotaxon* **5**: 1, 1977), Sherwood (*Occ. Pap. Farlow Herb. Crypt. Bot.* **15**, 1980), Wilberforce (*Mycologist* **13**: 149, 1999).

Karsteniomyces D. Hawksw. (1980), anamorphic *Pezizomycotina*, Cpd.1eH.1. 1 (on lichens, esp. *Peltigera*), Europe. See Hawksworth (*Bull. Br. Mus. nat. hist.* Bot. **9**: 22, 1981), Boqueras & Diederich (*Mycotaxon* **47**: 425, 1993).

Karstenula Speg. (1879), Melanommataceae. Anamorph *Microdiplodia*. 16, widespread (temperate). See Munk (*Dansk bot. Ark.* **17**, 1957), Eriksson & Hawksworth (*SA* **10**: 140, 1991), Constantinescu (*MR* **97**: 377, 1993; anamorph), Yuan & Mohammed (*Mycotaxon* **63**: 9, 1997).

karyochorisis, somatic nuclear division resulting from a constriction of the nuclear membrane (Moore, *Z. Zellforsch.* **63**: 921, 1964).

karyogamy, the fusion of two sex nuclei after cell fusion, i.e. after plasmogamy.

karyotype, the size and number of chromosomes in an organism. Generally determined by deduction from mating studies, microscopically or from specialized electrophoresis methods. See McCluskey & Mills (*Mol. Pl.-Microbe Interactions* **3**: 366, 1990), pulsed field gel-electrophoresis.

Kaskaskia Born & J.L. Crane (1972) = Gyrostroma fide Bedker & Wingfield (*TBMS* **81**: 179, 1983).

Kathistaceae Malloch & M. Blackw. (1990), Sordariomycetes (inc. sed.). 4 gen. (+ 1 syn.), 8 spp.

Lit.: Malloch & Blackwell (*CJB* **68**: 1712, 1990), Blackwell & Jones (*Biodiv. Cons.* **6**: 689, 1997), Blackwell *et al.* (*Mycol.* **95**: 987, 2003).

Kathistes Malloch & M. Blackw. (1990), Kathistaceae. 3 (coprophilous), widespread (north temperate). See Malloch & Blackwell (*CJB* **68**: 1712, 1990), Blackwell *et al.* (*Mycol.* **95**: 987, 2003; phylogeny).

katothecium, see catathecium.

katsuobushi, a Japanese food obtained by fermenting cooked bonito fish (*Sarda sarda*) with *Aspergillus* spp.

Katumotoa Kaz. Tanaka & Y. Harada (2005), Phaeosphaeriaceae. 1, Japan. See Tanaka & Harada (*Mycoscience* **46**: 313, 2005).

Kauffman (Calvin Henry; 1869-1931; USA). School teacher (1895-1901). Student, University of Wisconsin (1901-1902); Assistant, Cornell University, New York (1902-1904); Instructor (1904-1912) then Assistant Professor (1912-1920 then Associate Professor (1920-1923) then Professor (1923-1931) and Director of Herbarium (1921-1931), Michigan University. Contributed to the knowledge of the *Agaricaceae* particularly in Michigan, the Rocky Mountains and the Pacific Northwest. *Publs. Agaricaceae of Michigan* (1918); also other books and papers on the *Agaricaceae* (q.v.). *Biogs, obits etc.* Mains (*Phytopa-*

thology **22**: 271, 1932); Mains (*Mycol.* **24**: 265, 1932) [portrait]; Stafleu & Cowan (*TL-2* **2**: 507, 1979).

Kaufmannwolfia Galgoczy & E.K. Novák (1962) = Trichophyton fide Hawksworth *et al.* (*Dictionary of the Fungi* edn 8, 1995).

Kavinia Pilát (1938), Lentariaceae. 5, Europe. See Eriksson & Ryvarden (*Cortic. N. Europ.* **4**: 752, 1976), Legon (*Mycologist* **17**: 42, 2003; UK).

Kawasakia Y. Yamada & Nogawa (1995) = Lipomyces fide Yamada & Nogawa (*Bull. Fac. Agric. Shizuoka Univ.* **45**: 31, 1995), Kurtzman *et al.* (*FEMS Yeast Res.* **7**: 1027, 2007; phylogeny).

Kazachstania Zubcova (1971), Saccharomycetaceae. 28, widespread. See von Arx *et al.* (*Stud. Mycol.* **14**: 1, 1977), Kurtzman *in* Kurtzman & Fell (Eds) (*Yeasts, a taxonomic study* 4th edn: 134, 1998; as *Arxiozyma*), Kurtzman & Robnett (*FEMS Yeast Res.* **3**: 417, 2003), Lu *et al.* (*Int. J. Syst. Evol. Microbiol.* **54**: 2431, 2004), Wu & Bai (*Int. J. Syst. Evol. Microbiol.* **55**: 2219, 2005), Suh *et al.* (*Mycol.* **98**: 1006, 2006; phylogeny), Imanishi *et al.* (*FEMS Yeast Res.* **7**: 330, 2007).

Kazulia Nag Raj (1977), anamorphic *Chaetothyriaceae*, Hsp.1bH.10. 1, Brazil. See Nag Raj (*CJB* **55**: 1621, 1977).

kb (kilobases), an abbreviation for 1000 base pairs of DNA.

KC, see Metabolic products.

kefiran, a water soluble polysaccharide from kefir grains (Kooiman, *Carboh. Res.* **7**: 200, 1968).

Keinstirschia J. Reid & C. Booth (1989), Diaporthales. 1, Germany. See Reid & Booth (*CJB* **67**: 897, 1989).

Keisslerellum Werner (1944) nom. inval. = Mycoporellum fide Hawksworth *et al.* (*Dictionary of the Fungi* edn 8, 1995).

Keissleria Höhn. (1918) = Broomella fide Höhnel (*Annls mycol.* **18**: 71, 1920).

Keissleriella Höhn. (1919), ? Massarinaceae. Anamorph *Dendrophoma*-like. *c.* 25, widespread. See Bose (*Phytopath. Z.* **41**: 179, 1961; key 11 spp.), Sivanesan (*Bitunicate Ascomycetes and their Anamorphs*, 1984), Kohlmeyer *et al.* (*MR* **100**: 393, 1996).

Keisslerina Petr. (1920) = Dothiora Fr. (1849) fide Froidevaux (*Nova Hedwigia* **23**: 679, 1973).

Keissleriomyces D. Hawksw. (1981), anamorphic *Pezizomycotina*, Cpd.≡ eH.15. 1 (on *Cladonia*), Europe. See Hawksworth (*Bull. Br. Mus. nat. Hist.* Bot. **9**: 25, 1981).

Keithia Sacc. (1892) [non *Keithia* Spreng. 1822, ? *Capparaceae*] = Didymascella fide Hawksworth *et al.* (*Dictionary of the Fungi* edn 8, 1995).

Kelleria Tomin (1926) [non *Kelleria* Endl. 1848, *Thymelaeceae*] ? = Thelocarpon fide Salisbury (*Lichenologist* **3**: 175, 1966).

Kellermania Ellis & Everh. (1885), anamorphic *Planistromella*, Cpd.0eH/≡ eH.1. 13, USA. See Nag Raj (*Coelomycetous Anamorphs with Appendage-bearing Conidia*, 1993), Ramaley (*Mycotaxon* **47**: 259, 1993; teleomorphs), Ramaley (*Mycotaxon* **55**: 255, 1995; key to 7 spp.), Ramaley (*Mycotaxon* **66**: 509, 1998).

Kellermaniites Babajan & Tasl. (1977), Fossil Fungi. 1 (Tertiary), former USSR.

Kellermanniopsis Edward, Kr.P. Singh, S.C. Tripathi, M.K. Sinha & Ranade (1974) ? = Bleptosporium fide Sutton (*Mycol. Pap.* **88**, 1963).

Kemmleria Körb. (1861) ? = Buellia fide Hawksworth *et al.* (*Dictionary of the Fungi* edn 8, 1995).

Kendrickiella K. Jacobs & M.J. Wingf. (2001), anamorphic *Eurotiales*. 1. See Jacobs *et al.* (*CJB* **79**: 113, 2001), Grünig *et al.* (*CJB* **80**: 1239, 2002; phylogeny).

Kendrickomyces B. Sutton, V.G. Rao & Mhaskar (1976), anamorphic *Pezizomycotina*, St.0eH.15. 1, India; Australia. See Sutton *et al.* (*TBMS* **67**: 243, 1976).

Kensinjia J. Reid & C. Booth (1989), Diaporthales. 1, USA. See Barr (*Mycotaxon* **41**: 287, 1991).

Kentingia Sivan. & W.H. Hsieh (1989), Dothideomycetes. Anamorph ? *Excipulariopsis*. 1, Taiwan; India. See Sivanesan & Hsieh (*MR* **93**: 83, 1989).

Kentrosporium Wallr. (1844) ≡ Claviceps.

kephir, see Fermented food and drinks.

keratin, protein which is the main component of skin, hair, nails, feathers and horns.

Keratinomyces Vanbreus. (1952) = Trichophyton fide Ajello (*Sabouraudia* **6**: 153, 1968), Guillamón *et al.* (*Antonie van Leeuwenhoek* **69**: 223, 1996), Gräser *et al.* (*Medical Mycology* **37**: 105, 1999).

keratinophylic (1) capable of decomposing keratin; (2) of many fungi causing superficial mycoses in humans; see dermatophyte, ringworm.

Keratinophyton H.S. Randhawa & R.S. Sandhu (1964) = Aphanoascus fide Cano & Guarro (*MR* **94**, 1990).

keratomycosis, a fungus infection of the cornea of the eye.

Keratosphaera H.B.P. Upadhyay (1964) ? = Koordersiella fide Eriksson & Hawksworth (*SA* **6**: 133, 1987).

kerion, an inflammatory form of ringworm of the scalp; tinea kerion.

Kermincola Šulc (1907) nom. dub., anamorphic *Pezizomycotina*.

Kernella Thirum. (1949), Pucciniaceae. 1 (on *Litsea* (*Lauraceae*)), India; Nepal; China. May belong to *Puccinia*.

Kernia Nieuwl. (1916), Microascaceae. Anamorphs *Graphium*, *Scopulariopsis*. 5 (coprophilous), widespread. See Malloch & Cain (*CJB* **49**: 855, 1971; key), Locquin-Linard (*Revue Mycol.* Paris **41**: 509, 1977), Lee & Hanlin (*Mycol.* **91**: 434, 1999; DNA), Issakainen *et al.* (*Medical Mycology* **41**: 31, 2003; phylogeny).

Kernia Thirum. (1947) ≡ Kernella.

Kerniomyces Toro (1939), Schizothyriaceae. 1, S. America.

Kernkampella Rajendren (1970), Raveneliaceae. 10 (on *Euphorbiaceae*), widespread (esp. tropical). See Rajendren (*Mycol.* **62**: 1112, 1970; morphology), Laundon (*Mycotaxon* **3**: 142, 1975), Bagyanarayana *et al.* (*Journal of Mycology and Plant Pathology* **33**: 76, 2003; India).

kerosene fungus (**creosote fungus**), *Amorphotheca resinae* (anamorph *Hormoconis resinae*); grows on creosoted wood and petrochemical fuels, also isolated from soil; utilizes n-alkanes with chain lengths between C_9 and C_{19} grows esp. well at fuel/water interfaces forming hyphal mats that can break loose and cause blockages in aircraft wing and other fuel storage tanks, often associated also with accelerated corrosion; a particular problem during development of the airliner Concorde where fuel tanks were used as a heat-sink during supersonic flight, resulting in a warmer than usual environment which promoted growth of the fungus. See David & Kelley (*Mycopath.* **129**: 159, 1995 [*IMI Descr.* **1230**]), Hendley (*TBMS* **47**: 467, 1964), Hill (*Aircraft engineering and aerospace technology* **75**: 497, 2003), Parbery (*Austr. J. Bot.* **17**: 331, 1969; natural occurrence), Ratledge (*in* Wilkinson (Ed.), *Developments in biodegradation of hydrocarbons*: 1, 1987; hydrocarbon uptake), Shennon (*in* Houghton *et al.* (Eds), *Biodeterioration* 7, 248, 1988; biocides).

ketjap, shoyu sauce (q.v.) as made in Indonesia.

Ketubakia Kamat, Varghese & V.G. Rao (1987), anamorphic *Pezizomycotina*, Hsp.≡ eP.1. 1, India. See Kamat *et al.* (*Nova Hedwigia* **44**: 519, 1987).

Keys, construction of. See Farr (*Taxon* **55**: 589, 2006; discussion of on-line keys), Hawksworth (*Mycologist's handbook*, 1974), Pankhurst (*Biological identification*, 1978), Parker-Rhodes (*Bull. BMS* **8**: 68, 1974), Rayner (*Bull. BMS* **10**: 31, 1976), Tilling (*J. Biol. Educ.* **18**: 293, 1984), Korf (*Mycol.* **64**: 937, 1972); also Numerical taxonomy.

Khekia Petr. (1921) = Herpotrichia fide von Arx & Müller (*Sydowia* **37**: 6, 1984).

Khuskia H.J. Huds. (1963), Trichosphaeriales. Anamorph *Nigrospora*. 1, widespread. See Sivanesan & Holliday (*IMI Descr. Fungi Bact.* **311**, 1971).

Kickxella Coem. (1862), Kickxellaceae. 1, Europe; N. America. See Linder (*Farlowia* **1**: 49, 1943), Benjamin (*Aliso* **4**: 149, 1958), Young (*Ann. Bot.* **38**: 873, 1974; spore ultrastr.).

Kickxellaceae Linder (1943), Kickxellales. 12 gen. (+ 2 syn.), 32 spp.

Lit.: Benjamin (*Aliso* **4**: 321, 1959), Benjamin (*Aliso* **5**: 11, 1961), Benjamin (*Aliso* **5**: 273, 1963), Moss & Young (*Mycol.* **70**: 944, 1978), Saikawa (*CJB* **67**: 2484, 1989), Tanabe *et al.* (*Mol. Phylogen. Evol.* **30**: 438, 2004).

Kickxellales Kreisel ex R.K. Benj. (1979). Kickxellomycotina. 1 fam., 12 gen., 32 spp. Fam.:

Kickxellaceae

Lit.: Benjamin (*Aliso* **4**: 149, 321; **5**: 11, 273, 1958-1963; *The whole fungus* **2**: 573, 1979), Young (*New Phytol.* **67**: 823, 1968; spore ultrastr.), Moss & Young (*Mycol.* **70**: 944, 1978; phylogeny), White *et al.* (*Mycol.* **98**: 860, 2006; molecular phylogeny), Hibbett *et al.* (*MR* **111**: 109, 2007), and see under Family.

Kickxellomycotina Benny (2007). 5 fam., 59 gen., 264 spp. Fams:

(1) **Asellariaceae**
(2) **Dimargaritaceae**
(3) **Harpellaceae**
(4) **Kickxellaceae**
(5) **Legeriomycetaceae**

Lit.: Benjamin (*Aliso* **4**: 149, 321; **5**: 11, 273, 1958-1963; *The whole fungus* **2**: 573, 1979), Young (*New Phytol.* **67**: 823, 1968; spore ultrastr.), Moss & Young (*Mycol.* **70**: 944, 1978; phylogeny), White *et al.* (*Mycol.* **98**: 860, 2006; molecular phylogeny), Hibbett *et al.* (*MR* **111**: 109, 2007), and see under Family.

Kiehlia Viégas (1944), Parmulariaceae. 1 or 2, Brazil; Panama. See Farr (*Mycol.* **60**: 924, 1968), Inácio *et al.* (*Mycol. Progr.* **4**: 133, 2005).

kievitone, an isoflavonoid phytoalexin produced by the French bean (*Phaseolus vulgaris*).

Kiliasia Hafellner (1984) = Toninia fide Timdal (*Op. Bot.* **110**, 1991).

Kilikiostroma Bat. & J.L. Bezerra (1961) = Strigula fide Lücking *et al.* (*Lichenologist* **30**: 121, 1998).

Kiliophora Kuthub. & Nawawi (1993), anamorphic *Pezizomycotina*, Hso.0eP.16. 2, Malaysia; Ivory Coast. See Kuthubutheen & Nawawi (*Mycotaxon* **48**: 239, 1993).

Kimbropezia Korf & W.Y. Zhuang (1991), Pezizaceae. 1, Canary Islands. See Korf & Zhuang (*Mycotaxon* **40**: 269, 1991), van Brummelen (*Cryptog. Bryol.-Lichénol.* **19**: 257, 1998; ultrastr.), Norman & Egger (*Mycol.* **91**: 820, 1999; phylogeny), Hansen *et al.* (*Mycol.* **93**: 958, 2001; phylogeny).

Kimuromyces Dianese, L.T.P. Santos, R.B. Medeiros & Furlan. (1995), ? Uropyxidaceae. 1 (on *Astronium* (*Anacardiaceae*)), Brazil. See Dianese *et al.* (*Fitopatol. Brasil* **20**: 251, 1995).

Kineosporia Pagani & Parenti (1978), Actinobacteria. q.v.

kinetid, flagellar apparatus including kinetosomes and their associated tubules and fibres.

kinetosome, intracellular, non-membrane bound organelles; microtubular cylinders (0.2 µm diam.) organized in the 9 (A, B C- microtubules) + 0 pattern and with axoneme extensions in the 9 (A, B- microtubules) + 2 pattern.

KINGDOMS OF FUNGI, In the seventh edition of this *Dictionary*, the fungi were accepted as a single kingdom along with bacteria (*Monera*), plants (*Plantae*), animals (*Animalia*) and protists (*Protista*) according to the five-kingdom scheme of Whittaker (*Science* **163**: 160, 1969). However, with advances in ultrastructural, biochemical, and especially molecular biology, the treatment of the fungi as one of the five kingdoms of life has become increasingly untenable (see Phylogeny). The organisms studied by mycologists are now established as polyphyletic (i.e. with different phylogenies) and have to be referred to at least three different kingdoms as previously adopted in the 8th and 9th editions of this *Dictionary*.

This situation has arisen because of similarities in biology and structure. Essentially, 'fungi', pragmatically defined as organisms studied by mycologists, are eukaryotic, heterotrophic, develop branching filaments (or are more rarely single-celled), and reproduce by spores. Where it is useful to speak of fungi in this polyphyletic sense, the name can be used non-italicized and not capitalized to differentiate this from the kingdom *Fungi*; it is also possible to use other informal names for 'fungi', e.g. eumycotan or protoctistan fungi if considered desirable (Kendrick, 1992), but their replacement by scientific names that could imply either a common ancestry (cf. *Panomycetes*) or more formal rank (e.g. union; Barr, 1992) is not advocated here. Retention of the fungal phylum termination **-mycota** regardless of the kingdom to which the 'fungal' phyla are refered fulfills the same purpose. However, for this edition of the *Dictionary* zoological orthography is adopted for most of the suprafamilial ranks of the 'protozoan fungi'.

The three kingdoms of 'fungi' accepted in this edition of the *Dictionary* are:

(1) **Chromista**
(2) **Fungi**
(3) **Protozoa**

Their distinguishing features and names are discussed under the entries for each of the kingdoms. In recognition of their distinct evolutionary origins, organisms from these three kingdoms which are traditionally studied by mycologists, are now listed separately in this *Dictionary*. Some authors unite the *Chromista* and *Protozoa* into a single highly polyphylectic kingdom *Protoctista* (syn. *Protista*), but that conclusion is not supported by molecular, biochemical, and other evidence (see under kingdom entries). Cavalier-Smith (1993) and Corliss (1994) both retain *Chromista* and *Protozoa* as separate kingdoms. Dick (2001) segregates the *Straminipila* as a kingdom separate from the *Chromista* and including the fungal phyla of that kingdom.

Lit.: Ainsworth *et al.* (Eds) (*The Fungi* **4A-4B**, 1973), von Arx (*The genera of fungi sporulating in pure culture*, edn 3, 1981), Barr (*Mycol.* **84**: 1, 1992), Bessey (*Morphology and taxonomy of fungi*, 1950), Cavalier-Smith (*in* Osawa & Honjo (Eds), *Evolution of life*: 271, 1991; *Microbiol. Rev.* **37**: 953, 1993), Corliss (*Acta Protozool.* **33**: 1, 1994), Hawksworth (*MR* **95**: 641, 1991), Kendrick (*The fifth kingdom*, edn 2, 1992), Kriesel (*Grundzüge eines natürlichen Systems det Pilze*, 1969; *Biol. Rundsch.* **26**: 65, 1988), Leedale (*Taxon* **23**: 261, 1974; review early systems), Margulis (*Symbiosis in cell evolution*, edn 2, 1993), Moore (*Recent advances in microbiology*: 49, 1971; *Bot. Mar.* **23**: 361, 1980; *in* Hawksworth (Ed.), *Identification and characterization of pest organisms*, 1994).

See also Classification, Phylogeny

Kinorhynchospora A.V. Adrianov & A.V. Rybakov (1991), Microsporidia. 1.

Kionocephala P.M. Kirk (1986), anamorphic *Pezizomycotina*, Hso.≡ eP.4. 1, British Isles. See Kirk (*TBMS* **86**: 419, 1986).

Kionochaeta P.M. Kirk & B. Sutton (1986), anamorphic *Pezizomycotina*, Hso.0eH.15. 11, widespread (tropical). See Kirk & Sutton (*TBMS* **85**: 712, 1985), Crous *et al.* (*Mycol.* **86**: 447, 1994; S Africa), Goh & Hyde (*MR* **101**: 1517, 1997), Okada *et al.* (*Mycoscience* **38**: 409, 1997; phylogeny).

Kirchbaumia Schulzer (1866) = Phallus fide Stalpers (*in litt.*).

Kirkia Benny (1995) [non *Kirkia* Oliv. 1868, *Simaroubaceae*] ≡ Kirkomyces.

Kirkomyces Benny (1996), Mucoraceae. 1, India. See Benny (*Mycol.* **87**: 922, 1995), Voigt & Wöstemeyer (*Gene* **270**: 113, 2001; phylogeny).

Kirramyces J. Walker, B. Sutton & Pascoe (1992), anamorphic *Mycosphaerellaceae*, Cpd.≡ eP.19. 7, widespread (esp. Australia). See Walker *et al.* (*MR* **96**: 919, 1992), Crous *et al.* (*S. Afr. J. Bot.* **63**: 111, 1997), Andjic *et al.* (*MR* **111**: 1184, 2007; phylogeny).

Kirschsteinia Syd. (1906) = Nitschkia fide Müller & von Arx (*Beitr. Kryptfl. Schweiz* **11** no. 2, 1962).

Kirschsteiniella Petr. (1923) = Cyclothyrium fide Hawksworth (*J. Linn. Soc.* Bot. **82**: 35, 1981).

Kirschsteiniothelia D. Hawksw. (1985), Dothideomycetes. Anamorph *Dendryphiopsis*. 7, widespread. See Barr (*SA* **12**: 27, 1993), Barr (*Mycotaxon* **49**: 129, 1993; posn), Shearer (*Mycol.* **85**: 963, 1994; asci), Chen & Hsieh (*Sydowia* **56**: 229, 2004; type sp.), Huang *et al.* (*Mycotaxon* **98**: 153, 2006; Taiwan), Schoch *et al.* (*Mycol.* **98**: 1041, 2006; phylogeny), Wang *et al.* (*MR* **111**: 1268, 2007; phylogeny).

Kjeldsenia Colgan, Castellano & Bougher (1995), Claustulaceae. 1, USA. Basidioma gasteroid.

Klasterskya Petr. (1940), ? Ophiostomataceae. Anamorph *Hyalorhinocladiella*-like. 1 (from pine needles), Europe. See Minter (*TBMS* **80**: 162, 1983; posn), Valldosera & Guarro (*MR* **92**: 113, 1989), Malloch & Blackwell (*CJB* **68**: 1712, 1990; posn).

Klastopsora Dietel (1904) = Pucciniostele fide Dietel (*Nat. Pflanzenfam.* **6**, 1928).

Klebahnia Arthur (1906) = Uromyces fide Arthur (*Manual Rusts US & Canada*, 1934).

Klebahnopycnis Kirschst. (1939) = Hoehneliella fide Petrak (*Annls mycol.* **38**: 199, 1940).

Kleidiomyces Thaxt. (1908), Laboulbeniaceae. 3, America. See Nannfeldt (*Svensk bot. Tidskr.* **43**: 468, 1949).

Klöckeria, see *Kloeckera*.

Kloeckera Janke (1923), anamorphic *Hanseniaspora*, Hso.0eH.?. 6, widespread. See Meyer *et al.* (*Antonie van Leeuwenhoek* **44**: 79, 1978), Barnett *et al.* (*Yeasts: Characteristics and Identification* 2nd edn, 1990; 1 sp. accepted), Smith *in* Kurtzman & Fell (Eds) (*Yeasts, a taxonomic study* 4th edn: 580, 1998), Cadez *et al.* (*Int. J. Syst. Evol. Microbiol.* **56**: 1157, 2006; phylogeny), Suh *et al.* (*Mycol.* **98**: 1006, 2006; phylogeny).

Kloeckeraspora Niehaus (1932) = Hanseniaspora fide Lodder (*Yeasts, a taxonomic study* 2nd edn, 1970), Esteve-Zarzoso *et al.* (*Antonie van Leeuwenhoek* **80**: 85, 2001).

Kluyveromyces Van der Walt (1956), Saccharomycetaceae. 20, widespread. See Sidenberg & Lachance (*Int. J. Syst. Bacteriol.* **36**: 94, 1986; isozymes, DNA homology), Fuson *et al.* (*Int. J. Syst. Bacteriol.* **37**: 371, 1987; DNA relatedness), Martini & Martini (*Int. J. Syst. Bacteriol.* **37**: 380, 1987), Kock *et al.* (*Syst. Appl. Microbiol.* **10**: 293, 1988), Martini & Rosini (*Mycol.* **81**: 317, 1989; killer relationships), Kämper (*Biblthca Mycol.* **136**, 1990; DNA killer plasmid), Molnár *et al.* (*Antonie van Leeuwenhoek* **70**: 67, 1996; coenzyme Q system), Belloch *et al.* (*Syst. Appl. Microbiol.* **20**: 397, 1997; mtDNA), Belloch *et al.* (*Syst. Appl. Microbiol.* **21**: 266, 1998; rDNA), Belloch *et al.* (*Yeast* Chichester **14**: 1341, 1998; chromosomes), Lachance *in* Kurtzman & Fell (Eds) (*Yeasts, a taxonomic study* 4th edn: 227, 1998), Belloch *et al.* (*Int. J. Syst. Evol. Microbiol.* **50**: 405, 2000; phylogeny), Kurtzman *et al.* (*Taxon* **50**: 907, 2001; nomencl.), Kurtzman (*FEMS Yeast Res.* **4**: 233, 2003; generic limits), Naumova *et al.* (*FEMS Yeast Res.* **5**: 263, 2004; *K. lactis*), Zivanovic *et al.* (*FEMS Yeast Res.* **5**: 315, 2005; mtDNA sequence), Suh *et al.* (*Mycol.* **98**: 1006, 2006; phylogeny), Lachance (*FEMS Yeast Res.* **7**: 642, 2007; review).

Kmetia Bres. & Sacc. (1902), anamorphic *Pezizomycotina*, Hsp.0fH.?. 1, Europe.

Kmetiopsis Bat. & Peres (1960), anamorphic *Pezizomycotina*, Hsp.0fH.15. 1, Brazil. See Batista & Peres (*Publções Inst. Micol. Recife* **245**: 4, 1960).

Kneallhazia Sokolova & Fuxa (2008), Microsporidia. 1. See Sokolova & Fuxa (*Parasitology* **135**: 903, 2008).

Kneiffia Fr. (1836) [non *Kneiffia* Spach 1835, *Oenotheraceae*] ≡ Kneiffiella Underw. ≡ Kneiffiella Henn. (1898), ≡ Neokneiffia Sacc. (Aug. 1898) (q.v.), ≡ Pycnodon Underw. (Dec. 1898), = Hyphoderma (Cortic.) fide Jülich (*Persoonia* **8**: 59, 1972).

Kneiffiella P. Karst. (1889), Schizoporaceae. *c.* 25 (widespread). See Larsson *et al.* (*Mycol.* **98**: 926, 2007).

Kneiffiella Underw. (1897) ≡ Neokneiffia.

Knemiothyrium Bat. & J.L. Bezerra (1960), anamorphic *Pezizomycotina*, Cpt.1-≡ eP.?. 1, Jamaica. See Batista & Bezerra (*Publções Inst. Micol. Recife* **280**: 5, 1960).

Kniep (Karl Johannes Hans; 1881-1930; Germany). PhD, Jena (1904); Professor, Würzburg (1914); Professor of Plant Physiology (1924-1930) and Director of the Institute of Plant Physiology, Berlin-Dahlem. Noted for his work on the genetics of fungi. *Publs. Die Sexualität der Niederen Pflanzen* (1928); Vererbungserscheinungen bie Pilzen. *Bibliographica Genetica* (1929). *Biogs, obits etc.* Grummann (1974: 97); Harder (*Bericht der Deutschen Botanischen Gesellschaft* **48**: (164), 1930).

Knightiella Müll. Arg. (1886) = Icmadophila fide Galloway (*N.Z. Jl Bot.* **18**: 481, 1980), Stenroos & DePriest (*Am. J. Bot.* **85**: 1548, 1998), Galloway (*Lichenologist* **32**: 294, 2000).

Knoxdaviesia M.J. Wingf., P.S. van Wyk & Marasas (1988), anamorphic *Gondwanamyces*, Hso.0eH.15. 1, S. Africa. See Wingfield *et al.* (*Mycol.* **80**: 26, 1988), Mouton *et al.* (*Mycotaxon* **46**: 363, 1993; ontogeny), Marais *et al.* (*Mycol.* **90**: 136, 1998; teleomorph), Hausner *et al.* (*CJB* **78**: 903, 2000; phylogeny).

Knufia L.J. Hutchison & Unter. (1996), anamorphic *Pezizomycotina*, Hso.?.?. 1, Canada. See Hutchison *et al.* (*Mycol.* **87**: 903, 1995), Tsuneda & Currah (*CJB* **83**: 510, 2005; conidiogenesis), Tsuneda & Currah (*Rep. Tottori Mycol. Inst.* **42**: 1, 2004).

Knyaria Kuntze (1891) ≡ Tubercularia Tode.

Kobayasia S. Imai & A. Kawam. (1958), Phallaceae. 1, Japan. See Imai & Kawamura (*Science Rep. Yokohama Nat. Univ.* Section 2 **7**: 5, 1958).

Kochiomyces D.J.S. Barr (1980), Spizellomycetaceae. 1, USA. See Barr (*CJB* **58**: 2386, 1980).

Kochmania Piątek (2005) = Thecaphora fide Vánky *et al.* (*Mycol. Progr.*: [in press], 2008).

Koch's postulates, Criteria for proving the pathogenicity of an organism. Steps include: (1) the suspected causal organism must be constantly associated with the disease; (2) it must be isolated and grown in pure culture; (3) when a healthy plant is inoculated with it the original disease must be reproduced; (4) the same organism must be reiosolated from the experimentally infected plant. See Evans (*Yale Jl biol. Medic.* **49**: 175, 1976).

Kockovaella Nakase, I. Banno & Y. Yamada (1991), anamorphic *Cuniculitremaceae*, Hso.0eH.?. 11 (ballistosporic yeast), S.E. Asia. See Nakase *et al.* (*J. gen. appl. Microbiol.* Tokyo **37**: 175, 1991; taxonomy), Nakase *et al.* (*J. gen. appl. Microbiol.* Tokyo **39**: 107, 1993; mol. phylogeny), Fell *et al.* (*Int. J. Syst. Evol. Microbiol.* **50**: 1351, 2000; mol. phylogeny), Fungsin *et al.* (*International Journal of Systematic and Evolutionary Microbiology* **52**: 281, 2002; Thailand n.sp.).

Kodamaea Y. Yamada, Tom. Suzuki, M. Matsuda & Mikata (1995), Saccharomycetales. 4, USA; Australia. See Lachance *et al.* (*Can. J. Microbiol.* **45**: 172, 1999), Rosa *et al.* (*Int. J. Syst. Bacteriol.* **49**: 309, 1999), Suh *et al.* (*Mycol.* **98**: 1006, 2006; phylogeny).

Kodonospora K. Ando (1993), anamorphic *Pezizomycotina*, Hso.0bP.15. 1, Japan. See Ando (*MR* **97**: 506, 1993).

Koerberia A. Massal. (1854), Placynthiaceae (L). 2, widespread. See Henssen (*CJB* **41**: 1347, 1963), Burgaz & Martínez (*Nova Hedwigia* **73**: 381, 2001; Iberian peninsula).

Koerberiella Stein (1879), Porpidiaceae (L). 2, Europe. See Rambold *et al.* (*Lichenologist* **22**: 225, 1990), Navarro-Rosinés & Hafellner (*Biblthca Lichenol.* **53**: 179, 1993), Brodo (*Bryologist* **98**: 609, 1995; N. Am.), Buschbom & Mueller (*Mol. Phylogen. Evol.* **32**: 66, 2004; phylogeny).

Koerberiellaceae Hafellner (1984) = Porpidiaceae.

Kohlmeyera S. Schatz (1980) = Mastodia fide Eriksson (*Op. Bot.* **60**, 1981).

Kohlmeyeriella E.B.G. Jones, R.G. Johnson & S.T. Moss (1983), Lulworthiaceae. 1 (marine), widespread (northern hemisphere). See Jones & Moss (*SA* **6**: 179, 1987), Campbell *et al.* (*Fungal Diversity Res. Ser.* **7**: 15, 2002; posn), Campbell *et al.* (*MR* **109**: 556, 2005; phylogeny).

Kohninia Holst-Jensen, Vrålstad & T. Schumach. (2004), Sclerotiniaceae. 1, Norway. See Holst-Jensen *et al.* (*Mycol.* **96**: 139, 2004).

koji mould, *Aspergillus oryzae* and allied spp. used as starter for various fermented Japanese foods (saké, miso, shoyu, mirin, amazaké) by inoculating rice. See Murakami (*J. gen. Microbiol.* **17**: 281, 1971), mould rice, miso, Industrial mycology.

kojic acid, characteristic metabolic product of *Aspergillus flavus, oryzae* and *tamarii* groups (Birkinshaw *et al.*, *Phil. Trans.* **B220**: 127, 1931). It gives a blood-red colour with ferric chloride ($FeCl_3$).

Kokkalera Ponnappa (1970) = Podosphaera fide Braun & Takamatsu (*Schlechtendalia* **4**: 3, 2000).

Koleroga Donk (1958) = Ceratobasidium fide Talbot (*Persoonia* **3**: 371, 1965) but see, Muthappa (*TBMS* **73**: 159, 1979).

Kolletes Kohlm. & Volkm.-Kohlm. (2005), anamorphic *Pezizomycotina*. 1 (halotolerant), USA. See Kohlmeyer & Volkmann-Kohlmeyer (*Bot. Mar.* **48**: 316, 2005).

Kolman Adans. (1763) nom. rej. = Collema F.H. Wigg. fide Hawksworth *et al.* (*Dictionary of the Fungi* edn 8, 1995).

Komagataea Y. Yamada, M. Matsuda, K. Maeda, Sakak. & Mikata (1994), Phaffomycetaceae. 1, Asia; Europe. See James *et al.* (*Int. J. Syst. Bacteriol.* **48**: 591, 1998; phylogeny), Kurtzman *in* Kurtzman & Fell (Eds) (*Yeasts, a taxonomic study* 4th edn: 413, 1998; synonymy with *Williopsis*), Yamada *et al.* (*Biosc., Biotechn., Biochem.* **63**: 827, 1999), Naumova *et al.* (*Syst. Appl. Microbiol.* **27**: 192, 2004).

Komagataella Y. Yamada, M. Matsuda, K. Maeda & Mikata (1995), Saccharomycetales. 1, widespread. See Kurtzman *in* Kurtzman & Fell (Eds) (*Yeasts, a taxonomic study* 4th edn: 273, 1998; synonymy with *Pichia*), Kurtzman (*Int. J. Syst. Evol. Microbiol.* **55**: 973, 2005; phylogeny, definition), Suh *et al.* (*Mycol.* **98**: 1006, 2006; phylogeny).

Kommamyce Nieuwl. (1916) ≡ Biscogniauxia.

Kompsoscypha Pfister (1989), Sarcoscyphaceae. 4 or 4, widespread (tropical). See Pfister (*Mem. N. Y. bot. Gdn* **49**: 339, 1989), Harrington *et al.* (*Mycol.* **91**: 41, 1999; phylogeny).

Kondoa Y. Yamada, Nakagawa & I. Banno (1989), Kondoaceae. 2, Antarctic Ocean; Portugal. See Fell *et al.* (*Int. J. Syst. Evol. Microbiol.* **50**: 1351, 2000; mol. phylogeny), Fonseca *et al.* (*Antonie van Leeuwenhoek* **77**: 293, 2000).

Kondoaceae R. Bauer, Begerow, J.P. Samp., M. Weiss & Oberw. (2006), Agaricostilbales. 1 gen., 2 spp.
Lit.: Bauer *et al.* (*Mycol. Progr.* **5**: 41, 2006).

Konenia Hara (1913), Pezizomycotina. 1, Japan; USA. See Petrak (*Sydowia* **6**: 358, 1952).

Konradia Racib. (1900), Clavicipitaceae. 1, Asia. See Boedijn (*Annls mycol.* **33**: 229, 1935), Bischoff & White (*Mycology Series* **19**: 125, 2003).

Kontospora A. Roldán, Honrubia & Marvanová (1990), anamorphic *Pezizomycotina*, Hso.≡ eH.19. 1 (aquatic), Spain. See Roldán *et al.* (*MR* **94**: 243, 1990).

Koorchaloma Subram. (1953), anamorphic *Kananascus*, Hsp.0eH.15. 10, pantropical. See Nag Raj (*Mycotaxon* **19**: 167, 1984), Sarma *et al.* (*Bot. Mar.* **44**: 321, 2001), Treigiene (*Mycotaxon* **96**: 173, 2006).

Koorchalomella Chona, Munjal & J.N. Kapoor (1958), anamorphic *Pezizomycotina*, Hsp.0eH.15. 1, India; USA. See Nag Raj (*Mycotaxon* **19**: 167, 1984).

Koordersiella Höhn. (1909), Dothideomycetes. Anamorph *Hansfordiellopsis*. 3 (on lichens), widespread (tropical). See Eriksson & Hawksworth (*SA* **6**: 133, 1987).

Koralionastes Kohlm. & Volkm.-Kohlm. (1987), Koralionastetaceae. 5 (marine), widespread (tropics). See Kohlmeyer & Volkmann-Kohlmeyer (*CJB* **68**: 1554, 1990), Volkmann-Kohlmeyer & Kohlmeyer (*Mycotaxon* **44**: 417, 1992).

Koralionastetaceae Kohlm. & Volkm.-Kohlm. (1987), Pezizomycotina (inc. sed.). 1 gen., 5 spp.

Lit.: Kohlmeyer & Volkmann-Kohlmeyer (*Mycol.* **79**: 764, 1987), Kohlmeyer & Volkmann-Kohlmeyer (*CJB* **68**: 1554, 1990).

Körber (Gustav Wilhelm; 1817-1885; Germany). Born in Schlesien; studied in Breslau (now Wrocław, Poland) and Berlin; lecturer and later Professor at the University of Wrocław. An outstanding lichenologist who made major contributions to understanding the nature of lichen-forming fungi and to the systematics of crustose genera using microscopic features. Although his work was often eclipsed by the more prolific Nylander (q.v.), many genera Körber recognized in the 1850s-1860s came into use again in the 1980s (e.g. *Lecidella*, *Porpidia*, *Pyrrhospora*). Main collections in Leiden (**L**; incl. 'Typenherbar'); other material in Wrocław (**WRSL**). *Publs. De Gonidiis Lichenum* (1840); *Systema Lichenum Germaniae* (1854-55); *Parerga Lichenologica* (1859-65). *Biogs, obits etc.* Grummann (1974: 23); Stafleu & Cowan (*TL-2* **2**: 603, 1979).

Kordyana Racib. (1900), Brachybasidiaceae. 5 (parasites esp. of *Commelinaceae*), widespread (tropical). See Gäumann (*Annls mycol.* **20**: 257, 1922), Donk (*Reinwardtia* **4**: 117, 1956).

Kordyanella Höhn. (1904) = Tremellidium fide Rogers (*Mycol.* **49**: 902, 1957).

Korfia J. Reid & Cain (1963), Hemiphacidiaceae. 1, USA. See Reid & Cain (*Mycol.* **55**: 783, 1963).

Korfiella D.C. Pant & V.P. Tewari (1970), Sarcosomataceae. 1, India. See Pant & Tewari (*TBMS* **54**: 493, 1970), Prasad & Pant (*Journal of Mycology and Plant Pathology* **34**: 147, 2004; spore ornamentation).

Korfiomyces Iturr. & D. Hawksw. (2004), Lecanoromycetes. 1, Venezuela. See Iturriaga & Hawksworth (*Mycol.* **96**: 1155, 2004).

Korkir Adans. (1763) ? = Ochrolechia fide Hawksworth *et al.* (*Dictionary of the Fungi* edn 8, 1995).

Korunomyces Hodges & F.A. Ferreira (1981), anamorphic *Pezizomycotina*, Hso.1bH.1. 2, Brazil; Cuba. See Hodges & Ferreira (*Mycol.* **73**: 335, 1981), Seixas *et al.* (*Mycol.* **99**: 99, 2007; Brazil).

Korupella Hjortstam & P. Roberts (2000), Agaricomycetidae. 1, Cameroon. See Roberts (*Kew Bull.* **55**: 817, 2000).

Kostermansinda Rifai (1968), anamorphic *Pezizomycotina*, Hsy.#eP.1. 3, pantropical. See Rifai (*Reinwardtia* **7**: 376, 1968), Arambarri *et al.* (*Mycotaxon* **29**: 29, 1987), Catania (*Lilloa* **40**: 173, 2001).

Kostermansindiopsis R.F. Castañeda (1986), anamorphic *Pezizomycotina*, Hsy.#eP.1. 1, Cuba. See Castañeda (*Deuteromycotina de Cuba* Hyphomycetes IV: 8, 1986), Yanna & Hyde (*Aust. Syst. Bot.* **15**: 755, 2002; key).

Kotlabaea Svrček (1969), Pyronemataceae. 2, Europe. See Benkert & Kristiansen (*Z. Mykol.* **65**: 33, 1999; Norway), Khare (*Nova Hedwigia* **77**: 445, 2003).

Kramabeeja G.V. Rao & K.A. Reddy (1981), anamorphic *Pezizomycotina*, Hso.≡ eP.1. 1, India.

Kramasamuha Subram. & Vittal (1973), anamorphic *Pezizomycotina*, Hso.≡ eP.11. 1, widespread. See Subramanian & Vittal (*CJB* **51**: 1128, 1973).

Kraurogymnocarpa Udagawa & Uchiy. (1999), Gymnoascaceae. 2, Japan. See Udagawa & Uchiyama (*Mycoscience* **40**: 277, 1999).

Kravtzevia Shvartsman (1961), Pezizomycotina. 1, former USSR. See Shvartsman (*Trudy Inst. Bot., Alma-Ata* **9**: 75, 1961).

Kregervanrija Kurtzman (2006), Pichiaceae. 1, widespread. See Suh *et al.* (*Mycol.* **98**: 1006, 2006; phylogeny).

Kreiseliella Braun (1991), anamorphic *Pezizomycotina*, Cac.0-1eH.4/3/10. 1, former USSR. See Braun (*Boletus* **15**: 39, 1991).

Krempelhuberia A. Massal. (1854) nom. rej. = Pseudographis fide Hawksworth *et al.* (*Dictionary of the Fungi* edn 8, 1995).

Kretzschmaria Fr. (1849), Xylariaceae. Anamorph *Geniculosporium*-like. 28, widespread. See Wilkins (*TBMS* **18**: 320, 1934; pathogenicity), Dennis (*Kew Bull.*: 297, 1957), Dennis (*Bull. Jard. bot. Brux.* **31**: 144, 1961), Silveira & Rogers (*Acta Amazon.* Supl. **15**: 7, 1985; key 4 spp. Brazil), Narula & Rawla (*Nova Hedwigia* **40**: 241, 1986; key 3 spp., Himalayas), San Martín & Rogers (*Mycotaxon* **48**: 179, 1993; Mexico), Rogers & Ju (*Mycotaxon* **68**: 345, 1998; monogr.), Hladki & Romero (*Mycotaxon* **79**: 481, 2001; Argentina), Rogers & Ju (*Mycol. Progr.* **3**: 37, 2004; revision), Ju *et al.* (*Mycol.* **99**: 612, 2007; phylogeny).

Kretzschmariella Viégas (1944), Xylariaceae. Anamorph *Mirandina*-like. 1, Brazil. See Ju & Rogers (*Mycotaxon* **51**: 241, 1994; anam.), Læssøe & Spooner (*Kew Bull.* **49**: 1, 1994), Ju & Rogers (*Nova Hedwigia* **74**: 75, 2002).

Kriegeria Bres. (1891), Microbotryomycetes. Anamorph *Zymoxenogloea*. 1, Europe; N. America. See Kao (*Mycol.* **48**: 288, 1956).

Kriegeria G. Winter ex Höhn. (1914) [non *Kriegeria* Bres. 1891] = Rutstroemia fide Dennis (*Mycol. Pap.* **62**, 1956).

Kriegeriella Höhn. (1918), Pleosporaceae. 4, Europe; N. America. See Barr (*Mycotaxon* **2**: 104, 1975), von Arx & Müller (*Stud. Mycol.* **9**, 1975), Barr (*Mycotaxon* **29**: 501, 1987; posn).

Kriegeriellaceae M.E. Barr (1987) = Pleosporaceae.

Krieglsteinera Pouzar (1987), Heterogastridiaceae. 1 (on *Lasiosphaeria*), Europe. See Pouzar (*Beitr. Kenntn. Pilze Mitteleur.* **3**: 404, 1987), Miller *et al.*

(*Mycologist* **17**: 12, 2003), Rödel (*Boletus* **28**: 31, 2005).

Krishnamyces Hosag. (2003), anamorphic *Pezizomycotina*. 1, India. See Hosagoudar (*Zoos' Print Journal* **18**: 1159, 2003).

Krishtalia P.J. Kilochitskii (1997), Microsporidia. 1.

Krispiromyces T.N. Taylor, Hass & W. Remy (1993), Fossil Fungi. 1 (Devonian), British Isles. See Taylor *et al.* (*Mycol.* **84**: 906, 1992).

Krogia Timdal (2002), ? Ramalinaceae (L). 1, Mauritius. See Timdal (*Lichenologist* **34**: 293, 2002), Llop (*Lichenologist* **38**: 519, 2006).

Krombholzia P. Karst. (1881) [non *Krombholzia* Rupr. ex E. Fourn. 1876, *Gramineae*] ≡ Leccinum.

Krombholziella Maire (1937) = Leccinum fide Singer (*Farlowia* **2**: 223, 1945).

Kroswia P.M. Jørg. (2002), Pannariaceae (L). 2, widespread. See Jørgensen (*Lichenologist* **34**: 297, 2002), Jørgensen & Sipman (*J. Hattori bot. Lab.* **100**: 695, 2006).

Kruphaiomyces Thaxt. (1931), Laboulbeniaceae. 1, Fiji. See Nannfeldt (*Svensk bot. Tidskr.* **43**: 468, 1949).

Kryptastrina Oberw. (1990), Pucciniomycotina. 1 (in corticioid fungi), Colombia. See Oberwinkler (*Rep. Tottori Mycol. Inst.* **28**: 118, 1990).

K-selection, adaptation to the long-term colonization of habitats already occupied by other species; in fungi generally involving limited numbers of sexually produced and often long-lived propagules, e.g. large thick-walled ascospores; cf. r-selection (q.v. for *Lit.*).

Kubickia Svrček (1957) ? = Ombrophila Fr. fide Korf *in* Ainsworth *et al.* (Eds) (*The Fungi* **4A**: 249, 1973).

Kubinyia Schulzer (1866) = Mamiania fide Saccardo (*Syll. fung.* **16**: 1263, 1902).

Kuehneola Magnus (1898), Phragmidiaceae. 9 (on *Rosaceae*), widespread. *K. uredinis* (syn. *K. albida*) (blackberry (*Rubus*) stem (or yellow) rust). See Parmelee & Carteret (*Fungi Canadenses* **307**, 1986), Helfer (*Nova Hedwigia* **81**: 325, 2005; Europ. spp.) Species of *Kuehneola* reported from *Anacardiaceae*, *Malvaceae*, *Celastraceae* and *Verbenaceae* may belong to other phakopsoraceous genera.

Kuehneromyces Singer & A.H. Sm. (1946) = Pholiota fide Kuyper (*in litt.*) See Mykoholz.

Kuehniella G.F. Orr (1976), Onygenaceae. 1, N. America. See Currah (*SA* **7**: 1, 1988).

Kuettlingeria Trevis. (1857) = Caloplaca fide Hawksworth *et al.* (*Dictionary of the Fungi* edn 8, 1995), Galloway (*Australasian Lichenology* **55**: 21, 2004).

Kühn (Julius Gotthelf; 1825-1910; Germany). Self-educated farm estate manager (1848-1855, 1857-1862); student, Agricultural Academy, Bonn-Poppelsdorf (1855-1856); Professor of Agriculture and founder of the Agricultural Institute, Halle University (1862 onwards). Demonstrated that ergot sclerotia were a state of *Claviceps*; noted for his important book on plant diseases which resulted in the first widespread practical application of scientific principles of plant pathology. *Publs. Die Krankheiten der Kulturgewächse, ihre Ursachen und ihre Verhütung* (1858). *Biogs, obits etc.* Grummann (1974: 101); Stafleu & Cowan (*TL-2* **2**: 682, 1979); Whetzel (*An outline of the History of Phytopathology* p. 122, 1918) [portrait]; Wilhelm & Tietz (*Annual Review of Phytopathology* **16**: 343, 1978) [portrait].

Kuhneria P.A. Dang. (1933) nom. dub., Fungi. ? 'phycomycetes'.

Kuklospora Oehl & Sieverd. (2006) = Entrophospora fide Walker & Schüssler (*in litt.*) but see, Oehl & Sieverding (*J. Appl. Bot. Food Quality* Angew. Botan. **80**: 74, 2006).

Kulkarniella Gokhale & Patel (1952) = Monosporidium fide Cummins & Hiratsuka (*Illustr. Gen. Rust Fungi rev. edit.*, 1983).

Kullhemia P. Karst. (1878), Dothideomycetes. 1, Europe.

Kumanasamuha P. Rag. Rao & D. Rao (1964), anamorphic *Pezizomycotina*, Hso.0eP.10. 4, widespread. See Rao & Rao (*Mycopathologia* **22**: 333, 1964), Kirschner *et al.* (*Fungal Science* Taipei **16**: 47, 2001; Taiwan).

Kumarisporites Kalgutkar & Janson. (2000), Fossil Fungi, anamorphic *Ascomycota*. 1, India. See Kalgutkar & Jansonius (*AASP Contributions Series* **39**: 157, 2000).

Kumbhamaya M. Jacob & D.J. Bhat (2000) = Cyphellophora fide Jacob & Bhat (*Cryptog. Mycol.* **21**: 82, 2000), Sreekala & Bhat (*Mycotaxon* **84**: 65, 2002), Decock *et al.* (*Antonie van Leeuwenhoek* **84**: 209, 2003).

kumiss, see Fermented food and drinks.

Kunkelia Arthur (1917), Phragmidiaceae. 1, N. America. See Laundon (*Mycotaxon* **3**: 135, 1975; ≡ Gymnoconia), Cummins & Hiratsuka (*Illustr. Gen. Rust Fungi rev. edit.*, 1983), Buriticá (*Rev. Acad. Colomb. Cienc.* **18** no. 69: 131, 1991).

Kuntzeomyces Henn. ex Sacc. & P. Syd. (1899), ? Anthracoideaceae. 2 (on *Rhynchospora*), S. America. See Ling & Stevenson (*Mycol.* **41**: 87, 1949), Molina-Valero (*Caldasia* **13**: 49, 1980), Piepenbring (*MR* **105**: 757, 2001), Piepenbring (*Fl. Neotrop.* Monogr. **86**, 2003).

Kupsura Lloyd (1924) = Lysurus fide Zeller (*Mycol.* **40**: 646, 1948).

Kuraishia Y. Yamada, K. Maeda & Mikata (1994), Saccharomycetaceae. 2 (from rotten wood and insect frass), widespread. See Kurtzman *in* Kurtzman & Fell (Eds) (*Yeasts, a taxonomic study* 4th edn: 273, 1998), Péter *et al.* (*Antonie van Leeuwenhoek* **88**: 241, 2005; phylogeny, anamorph), Suh *et al.* (*Mycol.* **98**: 1006, 2006; phylogeny).

Kurosawaia Hara (1954) = Sphaceloma fide Sutton (*Mycol. Pap.* **141**, 1977).

Kurssanovia Kravtzev (1955), Pezizomycotina. 1, former USSR. See Kravtzev (*Trudy Inst. Bot., Alma-Ata* **2**: 145, 1955).

Kurssanovia Pidopl. (1948) = Fusariella fide Sutton (*in litt.*).

Kurtzmaniella M.A. Lachance & W.T. Starmer (2008), Saccharomycetaceae. 1 (on cactus flowers, associated with beetles), USA. See Lachance & Starmer (*Int. J. Syst. Evol. Microbiol.* **58**: 520, 2008).

Kurtzmanomyces Y. Yamada, Itoh, H. Kawas., I. Banno & Nakase (1989), anamorphic *Chionosphaeraceae*. 3, Europe. See Yamada *et al.* (*Agric. Biol. Chem.* **53**: 2993, 1989; phylogeny), Boekhout & Phaff (*Yeasts in Food* Beneficial and Detrimental Aspects, 2003).

Kusano (Shunsuke; 1874-1962; Japan). DPhil, University of Rigakushi; Academic staff (1896-1908) then Assistant Professor of Botany (1908) then Professor (1927-1934), University of Tokyo; Botanical Laboratory, College of Agriculture, Komaba, Tokyo. An Japanese mycologist noted for his work on the biology of chytrid fungi. *Publs.* On the life-history and cytology of a new species of *Olpidium*, with special reference to the copulation of motile isogametes. *Journal of the College of Agriculture* Tokyo Imperial University (1912); The life-history and physiology of *Synchytrium fulgens* Schroet., with special reference to its sexuality. *Japanese Journal of Botany* (1930, 1936). *Biogs, obits etc.* Grummann (1974: 593).

Kusanoa Henn. (1900) = Uleomyces fide von Arx (*Persoonia* **2**: 421, 1963).

Kusanobotrys Henn. (1904), Dothideomycetes. 1, Japan. See Eriksson & Hawksworth (*SA* **13**: 141, 1994; posn), Hosagoudar (*Persoonia* **18**: 123, 2002).

Kusanoopsis F. Stevens & Weedon (1923) = Uleomyces fide Thirumalachar & Jenkins (*Mycol.* **45**: 781, 1953).

Kusanotheca Bat. & Cif. (1963) = Dysrhynchis fide von Arx & Müller (*Stud. Mycol.* **9**, 1975).

Kutchiathyrites R.K. Kar (1979), Fossil Fungi. 1 (Oligocene), India.

Kutilakesa Subram. (1956) = Sarcopodium fide Sutton (*TBMS* **76**: 97, 1981).

Kutilakesopsis Agnihothr. & G.C.S. Barua (1957) = Sarcopodium fide Sutton (*TBMS* **76**: 97, 1981).

Kuzuhaea R.K. Benj. (1985), Piptocephalidaceae. 1, Japan. See Benjamin (*J. Linn. Soc.* Bot. **91**: 117, 1985), Tanabe *et al.* (*Mol. Phylogen. Evol.* **16**: 253, 2000; phylogeny), Tanabe *et al.* (*J. gen. appl. Microbiol.* Tokyo **51**: 267, 2005; phylogeny).

Kweilingia Teng (1940), Phakopsoraceae. 2 (on *Bambusa* (*Poaceae*)), China. See Vánky (*Illustrated genera of smut fungi*, 1987), Buriticá (*Rev. Acad. Colomb. Cienc.* **22**: 325, 1998; may include *Tunicopsora* and *Dasturella* (on *Poaceae*, *Costaceae*)), Carvalho *et al.* (*Summa Phytopathologica* **27**: 260, 2001).

Kylindria DiCosmo, S.M. Berch & W.B. Kendr. (1983), anamorphic *Chaetosphaeriaceae*, Hso.≡ eH.15. 9, widespread. See Rambelli & Onofri (*TBMS* **88**: 393, 1987; emend.), Fernández & Huhndorf (*Fungal Diversity* **18**: 15, 2005).

Kymadiscus Kohlm. & E. Kohlm. (1971) = Dactylospora fide Kohlmeyer (*in litt.*).

Kyphomyces I.I. Tav. (1985), Laboulbeniaceae. 14, widespread. See Tavares (*Mycol. Mem.* **9**: 242, 1985).

Kyphophora B. Sutton (1991), anamorphic *Pezizomycotina*, St.0eH.38. 1 (on mangroves), Australia. See Sutton (*Sydowia* **43**: 268, 1991).

L, Onderzoekinstituut Rijksherbarium/Hortus Botanicus (Leiden, Netherlands); founded 1575; part of the University of Leiden; see Anon. (*Meded. Rijks Herb.* **62-69**, 1931), Kalkman & Smit (*Blumea* **25**: 1, 1979).

Laaseomyces Ruhland (1900) ? = Scopinella fide Benny & Kimbrough (*Mycotaxon* **12**: 1, 1980).

labiate, having lips; lip-like.

Labidulomyces, see *Labiduromyces*.

Labiduromyces Ishik. (1941) = Filariomyces fide Tavares (*Mycol. Mem.* **9**, 1985).

labium, a lip.

Laboratory safety, see Safety, Laboratory.

Laboulbenia Mont. & C.P. Robin (1853), Laboulbeniaceae. *c.* 593 (on *Insecta*, *Arachnida*), widespread. See Kaur & Mukerji (*Mycoscience* **36**: 311, 1995; India), Weir & Beakes *in* Jolivet & Cox (Eds) (*Chrysomelidae Biology* **2**: 117, 1996), Rossi & Weir (*MR* **101**: 129, 1997), De Kesel (*Sterbeeckia* **18**: 13, 1998; Belgium), Santamaría (*Fl. Mycol. Iberica* **4**, 1998; Iberian spp.), Terada (*Mycoscience* **42**: 1, 2001), Hughes *et al.* (*Mycol.* **96**: 1355, 2004; New Zealand), Terada (*Mycoscience* **45**: 324, 2004; morphological variation), Terada *et al.* (*Bot. Bull. Acad. sin.* Taipei **45**: 165, 2004; on carabids, Taiwan), Terada (*Mycoscience* **46**: 215, 2005; on carabid beetles, Asia), Rossi & Santamaria (*Nova Hedwigia* **82**: 189, 2006).

Laboulbeniaceae G. Winter (1886), Laboulbeniales. 126 gen. (+ 32 syn.), 1931 spp.

Lit.: Bánhegyi *et al.* (*Magyarország Mikroszkopikus Gombáinak Határozókönyve* 2. Kötet Eumycotina (Ascomycetes: A Discomycetestöl, Basidiomycetes, Deuteromycetes): 635 pp., 1985), Huldén (*Karstenia* **25**: 1, 1985), Tavares (*Mycol. Mem.* **9**: 627 pp., 1985), Lee (*Korean J. Pl. Taxon.* **16**: 1, 1986), Santamaría (*Nova Hedwigia* **63**: 63, 1996), Weir & Beakes (*Mycol.* **88**: 677, 1996), Weir & Hammond (*Biodiv. Cons.* **6**: 701, 1997), Santamaría & Rossi (*Pl. Biosystems* **133**: 163, 1999), Rossi & Santamaria (*Mycol.* **92**: 786, 2000), Weir & Hughes (*Mycol.* **94**: 483, 2002), Hughes *et al.* (*Mycol.* **96**: 834, 2004).

Laboulbeniales Lindau (1898). Laboulbeniomycetidae. 4 fam., 146 gen., 2050 spp. Stromata usually present, composed of a basal black haustorium and a dark cellular thallus, formed under tight developmental control. Ascomata perithecial, often surrounded by complex appendages, translucent, ovoid, thin-walled. Interascal tissue absent. Asci few, clavate, thin-walled, evanescent, usually 4-spored. Ascospores hyaline, elongate, 1-septate. Anamorph hyphomycetous, spermatial, fleeting. Ectocommensalists of *Insecta* (few on *Arachnida* and *Diplopoda*), cosmop.

The 'host' is only rarely seriously damaged by the fungus, which generally does not extend below the chitin. Tavares (*Mycol. Mem.* **9**, 1985; key gens., illustr.) is largely followed for generic and family concepts here. Fams:

(1) **Ceratomycetaceae**
(2) **Euceratomycetaceae**
(3) **Herpomycetaceae**
(4) **Laboulbeniaceae**

Lit.: Benjamin (*in* Thaxter reprint, 1971; *in* Ainsworth *et al.* (Eds), *The Fungi* **4A**: 223, 1973), Balazuc (*Bull. mens. Soc. linn. Lyon* **40**: 134, 1971, bibliogr.; **42** (9)- **43** (9), 1973-1974; France), Blackwell (*Mycol.* **86**: 1, 1994; molec. relationships), Blackwell & Rossi (*Mycotaxon* **25**: 581, 1986; on termites), Lee (*Korean J. Pl. Taxon.* **16**: 89, 1986; fams.), Huldén (*Karstenia* **23**: 31, 1983, Finland; **25**: 1, 1985, palearctic), Majewski (*Polish Bot. Stud.* **7**, 1994; 179 spp. Poland, 116 pl.), Rossi & Cesari Rossi (*Giorn. Bot. ital.* **112**: 63, 1978, Italy; *CJB* **57**: 993, 1979, W. Africa), Rossi & Santamaria (*Mycol.* **92**: 786, 2000; spp. on *Staphylinidae*), Santamaria (*L'ordre Laboulbeniales. a la Península Ibèrica*, 1990; Spain), Santamaria *et al.* (*Trebalb d'Inst. Bot. Barcelona* **14**: 1-123, 1991; checklist, host list, and distribution European spp.), Scheloske (*Parasitol SchrReihe* **19**, 1969; biology, ecology, systematics), Shanor (*Mycol.* **47**: 1, 1955; general review), Sugiyama (*Ginkgoana* **2**, 1973; Japan), Tavares (*in* Batra (Ed.), *Insect-fungus symbiosis*, 1979; genera on host groups, *Mycologia Mem.* **9**, 1985), Terada (*Trans. mycol. Soc. Japan* **19**: 55, 1978; Taiwan), Thaxter (*Mem. Am. Acad. Arts Sci.* **12-16**, 1896-1931 [reprint 1971]; monogr.), Weir & Rossi (*Mycol.* **93**: 171, 2001; Bolivian spp.).

Laboulbeniella Speg. (1912) = Laboulbenia fide Thaxter (*Proc. Amer. Acad. Arts & Sci.* **50**: 15, 1914).

Laboulbeniomycetes A. Engler (1897), Pezizomycotina. 2 ord., 5 fam., 151 gen., 2072 spp. Subcl.:
(1) **Laboulbeniomycetidae**
Lit.: Benjamin (*in* Ainsworth *et al.* (Eds), *The Fungi* **4A**: 223, 1973), Balazuc (*Bull. mens. Soc. linn. Lyon* **40**: 134, 1971, bibliogr.; **42** (9)- **43** (9), 1973-1974; France), Blackwell (*Mycol.* **86**: 1, 1994; molec. relationships), Thaxter (*Mem. Am. Acad. Arts Sci.* **12-16**, 1896-1931 [reprint 1971]; monogr.).

Laboulbeniomycetidae Alexop. (1962), Laboulbeniomycetes. Ords:
(1) **Laboulbeniales**
(2) **Pyxidiophorales**
Lit.: Thaxter (*Mem. Amer. Acad. Arts & Sci.* **12-16**, 1896-1931 [reprint 1971]; monogr.), and see fam.

Laboulbeniopsis Thaxt. (1920), Laboulbeniales. 1 (on termites), West Indies. See Kimbrough & Gouger (*J. Invert. Path.* **16**: 205, 1970), Kimbrough & Gouger (*Mycol.* **68**: 541, 1976; ultrastr.), Henk *et al.* (*Mycol.* **95**: 561, 2003; posn).

Labrella Fr. (1828) nom. dub., anamorphic *Pezizomycotina*. See Sutton (*Mycol. Pap.* **141**, 1977).

Labridella Brenckle (1929), anamorphic *Griphosphaerioma*, Cpd.≡ eP.1. 1, N. America. See Jeewon *et al.* (*Mol. Phylogen. Evol.* **25**: 378, 2002).

Labridium Vestergr. (1897) ? = Seimatosporium fide Sutton (*Mycol. Pap.* **141**, 1977).

labriform, lip-shaped; frequently used for terminal soralia of lichens having this shape.

Labyrintha Malcolm, Elix & Owe-Larss. (1995), Porpidiaceae (L). 1, New Zealand. See Malcolm *et al.* (*Lichenologist* **27**: 241, 1995), Fryday (*Biblthca Lichenol.* **88**: 127, 2004).

Labyrinthomyces Boedijn (1939), Tuberaceae. 7 (hypogeous), Australia. See Zhang & Minter (*SA* **7**: 45, 1988), Trappe *et al.* (*Aust. Syst. Bot.* **5**: 597, 1992), O'Donnell *et al.* (*Mycol.* **89**: 48, 1997; phylogeny), Bougher & Lebel (*Aust. Syst. Bot.* **14**: 439, 2001), Hansen & Pfister (*Mycol.* **98**: 1029, 2006; phylogeny), Læssøe & Hansen (*MR* **111**: 1075, 2007; phylogeny).

Lacazia Taborda, V.A. Taborda & McGinnis (1999), anamorphic *Onygenales*, Hso.?.?. 1, S. America; N. America; Europe. *L. loboi* (lobomycopsis (q.v.), diseases in humans and dolphins). See Scholler *et al.* (*Sydowia* **51**: 89, 1999; synonym of *Dactyliellina*), Taborda *et al.* (*J. Clin. Microbiol.* **37**: 2031, 1999), Haubold *et al.* (*Medical Mycology* **38**: 9, 2000; in dolphins), Herr *et al.* (*J. Clin. Microbiol.* **39**: 309, 2001; phylogeny), Leandro *et al.* (*Phytopathology* **91**: 659, 2001; relationship to *Cladosporium*), Mendoza *et al.* (*Revta Iberoamer. Micol.* **22**: 213, 2005), Vilela *et al.* (*J. Clin. Microbiol.* **43**: 3657, 2005).

Laccaria Berk. & Broome (1883), Hydnangiaceae. *c.* 75, widespread. See Fries & Mueller (*Mycol.* **76**: 633, 1984; taxonomy), Mueller (*Fieldiana, Bot.* **30**: 1, 1992; key N. Am. spp.), Mueller *et al.* (*Mycol.* **85**: 850, 1993; spores), Kropp & Mueller (*Laccaria – Ectomycorrhizal fungi, key genera in profile*, 1999; ecology, phylogeny), Contu (*Boll. Gruppo Micol. 'G. Bresadola'* **46**: 5, 2003; Italy), Osmundson (*Mycol.* **97**: 949, 2005; USA).

Laccariaceae Jülich (1982) = Hydnangiaceae. The *Laccariaceae* are listed as nom. rej. against *Tricholomataceae*.

laccate, polished; varnished; shining.

Laccocephalum McAlpine & Tepper (1895), Polyporaceae. 5, Australia; Japan. See Núñez & Ryvarden (*Syn. Fung.* **10**: 29, 1995).

Lacellina Sacc. (1913), anamorphic *Pezizomycotina*, Hso.0eP.13. 4, widespread (tropical). See Ellis (*Mycol. Pap.* **67**, 1957).

Lacellinopsis Subram. (1953), anamorphic *Pezizomycotina*, Hso.0eP.3. 4, widespread (tropical). See Ellis (*Mycol. Pap.* **67**, 1957).

lacerate, as if roughly cut or torn.

Lachancea Kurtzman (2003), Saccharomycetaceae. 6, widespread. See Kurtzman (*FEMS Yeast Res.* **4**: 239, 2003), Naumova *et al.* (*Doklady Biological Sciences* **405**: 469, 2005; speciation), Suh *et al.* (*Mycol.* **98**: 1006, 2006; phylogeny).

Lachnaceae Raitv. (2004) = Hyaloscyphaceae.

Lachnaster Höhn. (1917) = Lachnum fide Korf *in* Ainsworth *et al.* (Eds) (*The Fungi* **4A**: 249, 1973).

Lachnea (Fr.) Gillet (1879) [non *Lachnea* L. 1753, *Thymeleaceae*] ≡ Scutellinia fide Schumacher (*Op. Bot.* **101**, 1990).

Lachnea Boud. (1885) ≡ Humaria Fuckel.

Lachneaceae Velen. (1934) = Pyronemataceae.

Lachnella Boud. (1885) = Lachnum fide Nannfeldt (*Nova Acta R. Soc. Scient. upsal.*, 1932).

Lachnella Fr. (1836), Niaceae. 6, widespread. See Singer (*Fl. Neotrop.* **17**: 58, 1976; key), Bodensteiner *et al.* (*Mol. Phylogen. Evol.* **33**: 501, 2004; phylogeny), Agerer & Treu (*Z. Mykol.* **72**: 115, 2006; Papua New Guinea).

Lachnellaceae Boud. (1907), nom. inval., see *Hyaloscyphaceae*.

Lachnellula P. Karst. (1884), Hyaloscyphaceae. Anamorph *Naemospora*. 40 (on conifers), widespread (temperate). *L. willkommii* (larch (*Larix*) canker). See Manners (*TBMS* **36**: 362, 1953), Manners (*TBMS* **40**: 500, 1957; as *Trichoscyphella*), Dennis (*Persoonia* **2**: 171, 1962), Dharne (*Phytopath. Z.* **53**: 101, 1965; key), Baral (*Beitr. Kenntn. Pilze Mitteleur.* **1**: 143, 1986; key 22 spp., Europ.), Cantrell & Hanlin (*Mycol.* **89**: 745, 1997; DNA), Baral & Matheis (*Z. Mykol.* **66**: 45, 2000; white-haired spp.), Leenurm & Raitviir (*Folia cryptog. Estonica* **38**: 41, 2001; ultrastr.), Minter (*IMI Descr. Fungi Bact.* **165** nos 1641-1650, 2005).

Lachnidium Giard (1891) = Fusarium fide Mussat (*Syll. fung.* **15**: 184, 1901), Summerbell & Schroers (*J. Clin. Microbiol.* **40**: 2866, 2002; anamorph of *Haematonectria*).

Lachnobelonium Höhn. (1926) = Lachnum fide Nannfeldt (*Nova Acta R. Soc. Scient. upsal.*, 1932).

Lachnocaulon Clem. & Shear (1931) [non *Lachnocaulon* Kunth 1841, *Eriocaulaceae*] ? = Stereocaulon Hoffm. fide Hawksworth *et al.* (*Dictionary of the Fungi* edn 8, 1995).

Lachnocaulum (orthographic variant), see *Lachnocaulon*.

Lachnocladiaceae D.A. Reid (1965), Russulales. 8 gen. (+ 3 syn.), 124 spp.
Lit.: Hallenberg (*Lachnocladiaceae and Coniophoraceae of North Europe*, 1985), Hallenberg (*The Lachnocladiaceae and Coniophoraceae of North Europe*: 96 pp., 1985), Boidin & Lanquetin (*Biblthca Mycol.* **114**: 130 pp., 1987), Nakasone & Micales (*Mycol.* **80**: 546, 1988), Stalpers (*Stud. Mycol.* **40**: 185 pp., 1996), Welden (*Revta Biol. trop.* **44** Suppl. 4: 91, 1996), Ginns (*Mycol.* **90**: 19, 1998), Müller *et al.* (*MR* **104**: 1485, 2000), Hibbett & Donoghue (*Syst.*

Biol. **50**: 215, 2001), Wagner (*Mycotaxon* **79**: 235, 2001), Larsson & Larsson (*Mycol.* **95**: 1037, 2003), Larsson (*MR* **111**: 1058, 2007; syn. of *Peniophoraceae*).

Lachnocladiales = Russulales.

Lachnocladium Lév. (1846) nom. cons., Lachnocladiaceae. 5, widespread. See Corner (*Ann. Bot. Mem.* [A monograph of Clavaria and allied genera] **1**: 416, 1950), Stalpers (*Stud. Mycol.* **40**: 85, 1996; key), Larsson & Larsson (*Mycol.* **95**: 1037, 2003; phylogeny).

Lachnodochium Marchal (1895), anamorphic *Pezizomycotina*, Hsp.0eH.10. 1, Europe.

Lachnum Retz. (1795), Hyaloscyphaceae. *c.* 251, widespread. See also *Dasyscyphus*. See Haines & Dumont (*Mycotaxon* **19**: 1, 1984; key 9 long-spored spp.), Raitviir (*Eesti NSV Tead. Akad. Toim.* Biol. seer **36**: 313, 1987; concept), Spooner (*Biblthca Mycol.* **116**, 1987; 37 spp. Australasia), Raitviir & Sacconi (*Mycol. Helv.* **4**: 161, 1991), Haines (*Nova Hedwigia* **54**: 97, 1992; key 13 spp. Guayana), Galán & Raitviir (*Nova Hedwigia* **58**: 453, 1994), Verkley (*Nova Hedwigia* **63**: 215, 1996; ultrastr.), Cantrell & Haines (*MR* **101**: 1081, 1997; trop. spp.), Cantrell & Hanlin (*Mycol.* **89**: 745, 1997; phylogeny), Wu & Haines (*Mycotaxon* **67**: 341, 1998), Zhuang (*Mycotaxon* **69**: 359, 1998; spp. trop. China), Leenurm *et al.* (*Sydowia* **52**: 30, 2000; ultrastr.), Tanaka & Hosoya (*Mycoscience* **42**: 597, 2001; Japan), Zhuang & Hyde (*Mycol.* **93**: 606, 2001; key long-spored spp.), Ye & Zhuang (*Mycosystema* **21**: 340, 2002; phylogeny), Zhuang (*Mycotaxon* **86**: 375, 2003; key Chinese spp.), Zhuang (*Nova Hedwigia* **78**: 425, 2004; on bamboo), Suková (*Czech Mycol.* **57**: 183, 2005; Czech Republic), Wang *et al.* (*Mol. Phylogen. Evol.* **41**: 295, 2006; phylogeny), Wang *et al.* (*Mycol.* **98**: 1065, 2006; phylogeny).

lacinia, a delicate branch of a foliose lichen thallus having an anatomical structure typical of foliose lichens.

laciniate (of an edge, etc.), as if cut into delicate bands (Fig. 23.47).

Laciniocladium Petri (1917), anamorphic *Pezizomycotina*, Hso.0eH.?15. 1, Europe.

Lacrimasporonites R.T. Clarke (1965), Fossil Fungi. 9 (Cretaceous, Paleocene, Tertiary), widespread.

lacrimiform (lacrimoid), like a tear drop.

Lacrymaria Pat. (1887), Psathyrellaceae. 14, widespread. See Hopple & Vilgalys (*Mol. Phylogen. Evol.* **13**: 1, 1999; phylogeny), Cortez & Coelho (*Mycotaxon* **93**: 129, 2005).

Lacrymospora Aptroot (1991), Requienellaceae. 1 (on lichens), Madagascar. Placed in the *Pyrenulaceae* by Erikssson (*in litt.*).

Lactarelis Earle (1909) = Russula fide Singer (*Agaric. mod. Tax.*, 1951).

Lactaria Pers. (1797) [non *Lactaria* Rafin. 1838, *Apocynaceae*] ≡ Lactarius.

Lactariaceae Gäum. (1926) = Russulaceae.

Lactariella J. Schröt. (1889) = Lactarius fide Singer (*Agaric. mod. Tax.*, 1951).

Lactariopsis Henn. (1901) = Lactarius fide Singer (*Agaric. mod. Tax.*, 1951).

Lactarius Pers. (1797), Russulaceae. *c.* 450, widespread (esp. temperate). The milk-caps, *L. deliciosus* (saffron milk cap) and other spp. are edible. The genus is expanded and includes gastroid taxa although requires a split in two genera. See Hesler & Smith (*North American species of Lactarius*, 1979; key North American spp.), Korhonen (*Suomen rouskut*, 1984; key Europ. spp.), Thiers (*Sydowia* **37**: 296, 1984; key spp. W. USA), Verbeken (*Mycotaxon* **66**: 363, 1998; keys African spp.), Hutchison (*Lactarius – Ectomycorrhizal fungi, key genera in profile*, 1999; ecology), Lebel & Trappe (*Mycol.* **92**: 1188, 2000; type studies, gasteroid forms), Miller *et al.* (*Mycol.* **93**: 344, 2001; phylogeny), Péter *et al.* (*MR* **105**: 1231, 2001; molecular phylogeny, ectomycorrhiza), Verbeken & Horak (*Aust. Syst. Bot.* **13**: 649, 2001; Papua New Guinea), Lebel (*Australasian Mycologist* **21**: 4, 2002; Australia sp.), Van Rooij *et al.* (*Nova Hedwigia* **77**: 721, 2003; Benin), Nuytinck *et al.* (*Belg. Jl Bot.* **136**: 145, 2003; European spp.), Nuytinck & Verbeken (*Mycotaxon* **92**: 125, 2005; sect. *Deliciosi* in Europe), Rayner *et al.* (*British Fungus Flora. Agarics and Boleti* **9**: 203 pp., 2005; key Brit. spp.), Nuytinck *et al.* (*Fungal Diversity* **22**: 171, 2006; sect. *Deliciosi* in Asia), Nuytinck *et al.* (*Mycotaxon* **96**: 261, 2006; sect. *Deliciosi* in N & C America), Buyck *et al.* (*MR* **111**: 787, 2007; Madagascar), Le *et al.* (*Fungal Diversity* **24**: 61, 2007; Thailand), Le *et al.* (*Fungal Diversity* **24**: 173, 2007; Thailand), Buyck *et al.* (*Fungal Diversity* **28**: 15, 2008; phylogeny; two monophyletic clades).

lacteous, like milk.

lactescent, becoming like milk.

lactifer, a latex-carrying hypha (Singer, 1960: 33).

lactiferous, having a milk-like juice.

lactiferous hypha, see hypha.

Lactifluus (Pers.) Roussel (1806) ≡ Lactarius.

Lactocollybia Singer (1939), ? Marasmiaceae. 17, widespread (tropical). See Singer (*Fl. Neotrop.* **3**: 55, 1970), Singer *et al.* (*Butlletí. Societat Catalana de Micologia, Barcelona* **13**: 67, 1990; Spain), Reid & Eicker (*Mycotaxon* **66**: 153, 1998; key South African spp.), Contu (*Bollettino dell'Associazione Micologica ed Ecologica Romana* **16**: 9, 2000; Sardegna).

Lactomyces Boulanger (1899) ? = Mucor Fresen. fide Foster (*Chemical activities of fungi*, 1949).

Lactydina Subram. (1978), anamorphic *Pezizomycotina*, Hso.≡ eH.1. 1 (on amoebae), USA. See Subramanian (*Kavaka* **5**: 95, 1977), Rubner (*Stud. Mycol.* **39**, 1996).

lacuna, a hole or hollow.

lacunose, having lacunae.

Lacunospora Cailleux (1969) ? = Apiosordaria fide Hawksworth *et al.* (*Dictionary of the Fungi* edn 8, 1995).

Lacustromyces Longcore (1993), Cladochytriaceae. 1, USA. See Longcore (*CJB* **71**: 415, 1993).

Ladrococcus Locq. (1985), Fossil Fungi. 1, Sahara.

Laestadia Auersw. (1869) [non *Laestadia* Kunth ex Less. 1832, *Compositae*] = Plagiostoma fide Hawksworth *et al.* (*Dictionary of the Fungi* edn 8, 1995).

Laestadiella Höhn. (1918) = Guignardia fide von Arx & Müller (*Beitr. Kryptfl. Schweiz* **11** no. 1, 1954).

Laestadites Mesch. (1892), Fossil Fungi. 1 (Pliocene, Pleistocene), Japan.

Laeticorticium Donk (1956) ≡ Corticium Pers.

Laetifomes Hattori (2001), Polyporaceae. 1, Solomon Is. See Hattori (*Mycoscience* **42**: 26, 2001).

Laetinaevia Nannf. (1932) nom. cons., Helotiales. Anamorph *Eriosporella*. 13, Europe. See Hein (*Willdenowia* Beih. **9**, 1976; key), Nauta & Spooner (*Mycologist* **13**: 65, 1999; Brit. spp.).

Laetiporaceae Jülich (1982) = Polyporaceae.

Laetiporus Murrill (1904), Fomitopsidaceae. 5, widespread. *L. sulphureus* on oak. See Corner (*Beih. Nova Hedwigia* **78**: 181, 1984), Burdsall & Banik (*Harvard Pap. Bot.* **6**: 43, 2001; N. Am.).

Laetisaria Burds. (1979), Corticiaceae. 3, widespread. See Burdsall (*TBMS* **72**: 420, 1979).

laevigate, smooth (Fig. 20.1).

Laeviomeliola Bat. (1960), Meliolaceae. 1, Brazil. See Batista (*Atas Inst. Micol. Univ. Recife* **1**: 224, 1960).

Laeviomyces D. Hawksw. (1981), anamorphic *Pezizomycotina*, Cpd.0eP.19. 2 (on lichens), Europe; N. America. See Hawksworth (*Bull. Br. Mus. nat. Hist.* Bot. **9**: 26, 1981).

Lagarobasidium Jülich (1974), Schizoporaceae. 1, Europe. See Langer (*Biblthca Mycol.* **154**, 1994), Larsson *et al.* (*Mycol.* **98**: 926, 2007).

lageniform, swollen at the base, narrowed at the top; like a Florence flask (Fig. 23.28).

Lageniformia Plunkett (1925) ? = Eutypa fide Petrak (*Sydowia* **5**: 169, 1951).

Lagenomyces Cavalc. & A.A. Silva (1972), anamorphic *Pezizomycotina*, St.0eH.?. 1, Brazil. See Cavalcanti & Silva (*Publicações. Instituto de Micologia da Universidade de Pernambuco* **647**: 28, 1972), Lücking *et al.* (*Lichenologist* **30**: 121, 1998; non-lichenized status).

Lagenula G. Arnaud (1930) [non *Lagenula* Lour. 1790, *Vitaceae*] = Caliciopsis fide Fitzpatrick (*Mycol.* **34**: 464, 1942).

Lagenulopsis Fitzp. (1942), Coryneliaceae. 1 (on *Podocarpus*), widespread (pantropical). See Benny *et al.* (*Bot. Gaz.* **146**: 431, 1985).

lager, see beer, brewing.

Lagerbergia J. Reid (1971) = Ascocalyx fide Müller & Dorworth (*Sydowia* **36**: 193, 1983).

Lagerheima Sacc. (1892), Helotiaceae. 3 or 4, widespread. See Gamundí (*Sydowia* **34**: 82, 1981).

Lagerheimina Kuntze (1891) = Diploschistes fide Hawksworth *et al.* (*Dictionary of the Fungi* edn 8, 1995).

lagynocarpus ascomycetes, Pyrenomycetes (Moreau, 1953).

Lagynodella Petr. (1922) = Cryptosporiopsis fide Wollenweber (*Arb. Biol. Reichsans. Berl.* **22**: 521, 1938).

Lahmia Körb. (1861), Lahmiaceae. 1 (from bark), widespread (north temperate). See Eriksson (*Mycotaxon* **27**: 347, 1986).

Lahmiaceae O.E. Erikss. (1986), Lahmiales. 1 gen. (+ 1 syn.), 1 spp.

Lit.: Eriksson (*Mycotaxon* **27**: 347, 1986), Ostry (*CJB* **64**: 1834, 1986).

Lahmiales O.E. Erikss. (1986). Pezizomycotina. 1 fam., 1 gen., 1 spp. Fam.:

Lahmiaceae

For *Lit.* see under fam.

Lahmiomyces Cif. & Tomas. (1953), Patellariaceae. 1, Italy. See Eriksson (*Mycotaxon* **27**: 347, 1986; posn).

Lalaria R.T. Moore (1990), anamorphic *Taphrina*, Hso.0eH.10. 23, widespread. Yeast phase anamorphs. See Moore *in* Kurtzman & Fell (Eds) (*Yeasts, a taxonomic study* 4th edn: 582, 1998).

Lamb (Ivan Mackenzie; 1911-1990; England, later Argentina, Canada, USA). Assistant for Lichens, British Museum (Natural History), London (1935-1946); Professor of Cryptogams, Tucumán University (1947) [after expedition to Antarctica and the Falklands / Malvinas]; Curator of Cryptogams, Canadian National Museum, Ottawa (1950); Director, Farlow Herbarium, Cambridge, Massachusetts (1954-1974). Noted for his studies on *Stereocaulon*; also for his catalogue of lichen names. Collections are in **BM**, **FH**, **CAN**, **LIL**. *Publs. Index Nominum Lichenum inter Annos 1932 et 1960 Divulgatorum* (1963). *Biogs, obits etc.* Ainsworth (*Brief Biographies of British Mycologists* p. 106, 1996); Stafleu & Cowan (*TL-2* **2**: 735, 1979).

Lambdasporium Matsush. (1971), anamorphic *Pezizomycotina*, Hso.1bH.10. 2, S.E. Asia. See Matsushima (*Bull. natn. Sci. Mus.* Tokyo, B **14**: 467, 1971), Chen *et al.* (*Bot. Bull. Acad. sin.* Taipei **41**: 81, 2000; Taiwan).

Lambertella Höhn. (1918), ? Rutstroemiaceae. Anamorph *Helicodendron*. 63, widespread. See Dumont (*Mem. N. Y. bot. Gdn* **33**, 1971; monogr.), Korf & Zhuang (*Mycotaxon* **24**: 361, 1985; key), Korf & Zhuang (*Mycotaxon* **39**: 477, 1990), Korf & Zhuang (*Mycosystema* **8-9**: 15, 1996; 15 spp. China), Holst-Jensen *et al.* (*Mycol.* **89**: 885, 1997; phylogeny, polyphyly), Zhuang & Zhang (*Taxon* **51**: 769, 2002; China), Wang *et al.* (*Mycol.* **98**: 1065, 2006; phylogeny).

Lambertellinia Korf & Lizoň (1994), Sclerotiniaceae. Anamorph *Idriella*. 3, Japan. See Korf & Lizoň (*Mycotaxon* **50**: 167, 1994).

Lambiella Hertel (1984) = Rimularia fide Hertel (*Mitt. bot. StSamml., München* **23**: 321, 1987).

Lambinonia Sérus. & Diederich (2005), anamorphic *Pezizomycotina*. 1 (lichenicolous), widespread. See Sérusiaux & Diederich (*Lichenologist* **37**: 500, 2005).

Lambottiella (Sacc.) Sacc. (1913) = Lophiostoma fide von Arx & Müller (*Stud. Mycol.* **9**, 1975).

Lambro Racib. (1900), Diaporthales. 1, Indonesia. See Müller & von Arx (*Beitr. Kryptfl. Schweiz* **11** no. 2, 1962), Monod (*Beih. Sydowia* **9**: 1, 1983).

lamella (pl. **lamellae**) (of an agaric), one of the characteristic hymenium-covered vertical plates on the underside of the pileus (Fig. 4B); gill.

lamellate (1) having lamellae; (2) made up of thin plates.

Lamelloporus Ryvarden (1987), Meruliaceae. 1, America (tropical). See Ryvarden (*Mycotaxon* **28**: 529, 1987).

lamellula (pl. **lamellulae**), a small lamella which runs from the edge of the pileus towards the stipe, as in *Russula*.

Lamia Nowak. (1884) = Entomophthora Fresen. fide Remaudière & Keller (*Mycotaxon* **11**: 323, 1980).

lamina (pl. **laminae**), (1) blade; (2) the main part of the thallus in foliose lichens; (3) epithecium + hymenium + subhymenium in an apothecium (Hertel, *Beih. Nova Hedw.* **24**, 1967); (4) (of leaves) the flat surface; **-l**, on the lamina.

Lamproconium (Grove) Grove (1937), anamorphic *Pezizomycotina*, ?.0eP.19. 1, Europe. See Scheuer *et al.* (*Fritschiana* **24**: 36, 2001).

lamprocystidium, see cystidium.

Lamprospora De Not. (1863), Pyronemataceae. *c.* 46 (many bryophilous), widespread (north temperate). See Benkert (*Z. Mykol.* **53**: 195, 1987; 29 spp.), Caillet & Moyne (*Bull. Soc. Hist. nat. Doubs* **84**: 9, 1991; keys), Schumacher (*Sydowia* **45**: 307, 1993; key 13 arctic-alpine spp.), Yao & Spooner (*MR* **99**: 1521, 1995; Brit. spp.), Hansen & Pfister (*Mycol.* **98**: 1029, 2006; phylogeny), Perry *et al.* (*MR* **111**: 549, 2007; phylogeny).

Lampteromyces Singer (1947), Marasmiaceae. 1 (poisonous), Japan; China. See Mo *et al.* (*Mycosystema* **19**: 529, 2000; China), Moncalvo *et al.* (*Syst. Biol.* **49**: 278, 2000; phylogeny).

Lamyella Berl. (1899) ≡ Neolamya.

Lamyella Fr. (1849) = Cytospora fide Clements & Shear (*Gen. Fung.*, 1931).

Lamyxis Raf. (1820) nom. dub., ? Boletales. See Donk (*Persoonia* **1**: 173, 1960).

lanate, like wool; covered with short hair-like processes.

Lanatonectria Samuels & Rossman (1999), Nectriaceae. Anamorph *Actinostilbe*. 4 (twigs and old fungi etc.), widespread (pantropical). See Rossman *et al.* (*Stud. Mycol.* **42**: 248 pp., 1999), Samuels *et al.* (*CBS Diversity Ser.* **4**, 2006; USA), Zhang & Zhuang (*Mycosystema* **25**: 15, 2006; phylogeny).

Lanatosphaera Matzer (1996), Dothideomycetes. 2, Costa Rica. See Matzer (*Mycol. Pap.* **171**: 129, 1996).

Lanatospora Voronin (1986), Microsporidia. 2.

Lanceispora Nakagiri, Okane, Tad. Ito & Katum. (1997), Xylariales. 2, E. Asia; Australia. Perhaps close to *Ceriospora* and *Urosporellopsis*. See Nakagiri *et al.* (*Mycoscience* **38**: 207, 1997), Sarma & Hyde (*Mycoscience* **42**: 97, 2001; Singapore), Jeewon *et al.* (*MR* **107**: 1392, 2003; phylogeny).

Lancisporomyces Santam. (1997), Legeriomycetaceae. 4 (in *Plecoptera*), widespread. See Santamaría (*Mycol.* **89**: 639, 1997), Strongman & White (*CJB* **84**: 1478, 2006; Canada), White *et al.* (*Mycol.* **98**: 333, 2006; USA), Valle *et al.* (*Mycol.* **100**: 149, 2008; Mexico).

Lange (Jakob Emanuel; 1864-1941; Denmark). Teacher, Agricultural Folk High School, Dalum (1888-1918); Principal, Smallholders Agricultural School, Odense (1918-1934). Agaricologist. *Publs.* Studies on the agarics of Denmark I-XII. *Dansk Botanisk Arkiv* (1914-1938) [see M. Lange, *Friesia* **9**: 121, 1969]; *Flora Agaricina Danica* 5 vols (1935-1940). *Biogs, obits etc.* Buchwald (*Botanisk Tidsskrift* **46**: 81, 1942) [portrait]; Pearson (*Mycol.* **39**: 1, 1947) [portrait]; Stafleu & Cowan (*TL-2* **2**: 745, 1979).

Langermannia Rostk. (1838) nom. rej., Agaricaceae. 3, widespread. The basidioma of *L. gigantea* (giant puff-ball) may be 120 cm across and have 10^{13} spores (see record fungi). See Kreisel (*Persoonia* Supplement **14**: 431, 1992), Coetzee & Van Wyk (*Bothalia* **33**: 156, 2003; *Calvatia* sect. *Macrocalvatia*), Coetzee & Van Wyk (*Taxon* **54**: 541, 2005; nomencl.) Nom. rej. vs. *Calvatia* s.l. (q.v.) but used for segregate of it.

Langeron (Maurice Charles Pierre; 1874-1950; France). Doctor of medicine, botanist, parasitologist, and mycologist, Institut de Parasitologie, Paris (1903-1950). Made many important contributions to medical and general mycology. *Publs. Précis de Microscopie* (1913) [edn 7, 1949]; *Précis de Mycologie* (1945) [edn 2, by Vanbreuseghem, 1952]. *Biogs, obits etc.* Vanbreuseghem (*Mycopathologia et Mycologia Applicata* **6**: 58, 1951) [bibliography, portrait]; Stafleu & Cowan (*TL-2* **2**: 750, 1979).

Langeronia Vanbreus. (1950) = Trichophyton fide Ajello (*Sabouraudia* **6**: 153, 1968).

Langloisula Ellis & Everh. (1889) nom. conf., Fungi. See Rogers & Jackson (*Farlowia* **1**: 263, 1943) and, Donk (*Fungus* Wageningen **26**: 3, 1956).

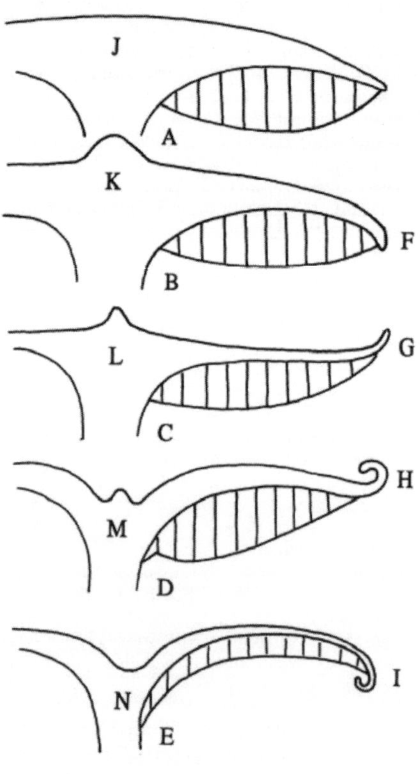

Fig. 19. A-E. Insertion of lamellae (gills). A, free; B, adnexed; C,adnate; D, sinuate; E, decurrent. F-I. Margin of pileus (cap). F, inflexed; G, reflexed; H, revolute; I, involute. J-N. Pileus. J, plano-convex; K, bullate; L, umbonate; M, umbilicate; N, depressed.

languid, feeble; hanging down.

Lanolea Nieuwl. (1916) ≡ Entoloma.

Lanomyces Gäum. (1922) = Cystotheca fide Katumoto (*Rep. Tottori mycol. Inst.* **10**: 437, 1973).

Lanopila Fr. (1848) nom. rej. prop. = Langermannia fide Demoulin & Dring (*Bull. Jard. bot. nat. Belg.* **45**: 361, 1975).

Lanosa Fr. (1825) nom. dub., ? Fungi.

lanose, see lanate.

Lanspora K.D. Hyde & E.B.G. Jones (1986), Halosphaeriaceae. 1 (marine), Seychelles. See Hyde & Jones (*CJB* **64**: 1581, 1986), Hyde & Jones (*Bot. Mar.* **32**: 205, 1989).

lanuginose, see lanate; see also nematogenous.

Lanzia Sacc. (1884), Rutstroemiaceae. Anamorph *Myrioconium*. *c.* 54, widespread. See Sharma & Sharda (*Int. J. Mycol. Lichenol.* **2**: 95, 1985; key 15 spp., India), Spooner (*Biblthca Mycol.* **116**, 1987; 7 spp. Australasia), Baral (*SA* **13**: 113, 1994; synonymy with *Rutstroemia*), Zhuang (*Mycosystema* **8-9**: 15, 1996; 5 spp. China), Holst-Jensen *et al.* (*Mycol.* **89**: 885, 1997; phylogeny), Zhuang (*Mycotaxon* **72**: 325, 1999; spp. trop. China), Simpson & Grgurinovic

(*Australas. Mycol.* **22**: 11, 2003; Australia), Zhuang & Liu (*Mycotaxon* **99**: 123, 2007; phylogeny).

lao-chao, see Fermented food and drinks.

Laocoön J.C. David (1997), anamorphic *Mycosphaerellaceae*, Hso.?.?. 1, Colombia. See David (*Mycol. Pap.* **172**: 116, 1997).

Lappodochium Matsush. (1975), anamorphic *Pezizomycotina*, Hsp.0eH.10. 1, Brazil. See Matsushima (*Icon. microfung. Matsush. lect.*: 91, 1975).

lapsus (Lat.), a slip; **-calami**, a slip of the pen.

Laquearia Fr. (1849), ? Rhytismatales. 1, Europe.

Larger fungi, see Macromycetes.

largest fungi, see longevity, record fungi.

Laricifomes Kotl. & Pouzar (1957) ≡ Fomitopsis fide Ryvarden (*Syn. Fung.* **5**: 171, 1991).

Laricifomitaceae Jülich (1982) = Fomitopsidaceae.

Laricina Velen. (1934), ? Helotiales. 1, former Czechoslovakia.

Laridospora Nawawi (1976) = Monacrosporium fide Nawawi (*TBMS* **66**: 344, 1976), Rubner (*Stud. Mycol.* **39**, 1996).

Larseniella Munk (1942) = Neorehmia fide Barr (*Mycotaxon* **39**: 43, 1990), Samuels & Barr (*CJB* **75**: 2165, 1998).

Larssonia Vidtmann & Sokolova (1994), Microsporidia. 1.

Larssoniella J. Weiser & David (1997), Microsporidia. 2.

Lasallia Mérat (1821), Umbilicariaceae (L). 12, widespread (esp. temperate). See Llano (*A monograph of the lichen family Umbilicariaceae in the Western Hemisphere*, 1950), Llano (*Hvalråd. Skr.* **48**: 112, 1965), Dombrovskaya (*Bot. Zh. SSSR* **63**: 233, 1978), Blum (*Dokl. Akad. Nauk. Ukr. SSR* B **12**: 58, 1986; DNA homology, status), Krog & Swinscow (*Nordic Jl Bot.* **6**: 75, 1986; E. Afr.), Sancho & Balaquer (*An. Jard. bot. Madr.* **46**: 273, 1989; anatomy 4 spp.), Wei & Jiang (*Mycosystema* **2**: 135, 1989; key 8 spp. China), Posner *et al.* (*Z. Naturf.* **46c**: 19, 1991; chemotax.), Wei & Jiang (*The Asian Umbilicariaceae (Ascomycota)* [*Mycosystema, Monogr.* **1**], 1993; keys, descr. 60 spp. Asia), Valladares *et al.* (*CJB* **72**: 415, 1994), Wei (*Proc. First Korea-China Jt Sem. Mycol.*: 5, 1994; molec. syst.), Ivanova *et al.* (*Lichenologist* **31**: 477, 1999; phylogeny), Narui *et al.* (*Bryologist* **102**: 80, 1999; polysaccharides), Cohen (*Lichenologist* **34**: 521, 2002; anthraquinones), Türk & Uhl (*Biblthca Lichenol.* **86**: 465, 2003; Austria), Krzewicka (*Polish Bot. Stud.* **17**, 2004; Poland), Peršoh *et al.* (*Mycol. Progr.* **3**: 103, 2004; asci, phylogeny), Miądlikowska *et al.* (*Mycol.* **98**: 1088, 2006; phylogeny).

Laschia Fr. (1830) nom. rej. = Campanella fide Donk (*Persoonia* **1**: 173, 1960).

Laschia Jungh. (1838) ≡ Junghuhnia.

Lascoderma, see *Lasioderma*.

Lasiella Quél. (1875) = Lasiosphaeria fide Hawksworth *et al.* (*Dictionary of the Fungi* edn 8, 1995).

Lasiobelonis Clem. & Shear (1931) = Lasiobelonium Ellis & Everh. fide Spooner (*Biblthca Mycol.* **116**, 1987).

Lasiobelonium (Sacc.) Sacc. & P. Syd. (1899) ≡ Belonidium fide Korf (*Mycotaxon* **7**: 399, 1978).

Lasiobelonium Ellis & Everh. (1897), Hyaloscyphaceae. *c.* 20, Europe; Australasia. See Korf (*Mycotaxon* **7**: 399, 1978), Raitviir (*Scripta Mycol.* **9**, 1980; key 13 spp.), Spooner (*Biblthca Mycol.* **116**, 1987; 3 spp. Australasia, concept), Baral (*SA* **13**: 113, 1994).

Lasiobertia Sivan. (1978), Apiosporaceae. Anamorph *Melanographium*. 1 (on dead wood), Ghana; Puerto Rico. See Hyde (*Sydowia* **45**: 204, 1993; posn), Wang & Hyde (*Fungal Diversity* **3**: 159, 1999; posn), Huhndorf *et al.* (*MR* **108**: 1384, 2004; phylogeny).

Lasiobolidium Malloch & Cain (1971), Pyronemataceae. 7 (coprophilous), N. Africa; N. America. See Loquin-Linard (*Cryptog. Mycol.* **4**: 283, 1983), van Brummelen *in* Hawksworth (Ed.) (*Ascomycete Systematics. Problems and Perspectives in the Nineties* NATO ASI Series vol. **269 269**: 400, 1994), van Brummelen (*Persoonia* **16**: 425, 1998; posn), Hansen & Pfister (*Mycol.* **98**: 1029, 2006; phylogeny, polyphyly), Hansen *et al.* (*Mycol.* **97**: 1023, 2005; phylogeny), Perry *et al.* (*MR* **111**: 549, 2007; phylogeny).

Lasiobolus Sacc. (1884), Ascodesmidaceae. 11, widespread (temperate). See Bezerra & Kimbrough (*CJB* **53**: 1206, 1975; key), van Brummelen *in* Hawksworth (Ed.) (*Ascomycete Systematics. Problems and Perspectives in the Nineties* NATO ASI Series vol. **269 269**: 400, 1994; posn), Yao (*MR* **100**: 737, 1996; Brit. spp.), Landvik *et al.* (*Mycoscience* **39**: 49, 1998; posn), Prokhorov (*Mikol. Fitopatol.* **32**: 14, 1998; key), Hansen & Pfister (*Mycol.* **98**: 1029, 2006; phylogeny), Perry *et al.* (*MR* **111**: 549, 2007; phylogeny).

Lasiobotrys Kunze (1823), Venturiaceae. Anamorph *Ulocladium*-like. 2 or 2 (on *Lonicera*), widespread (temperate). See Müller (*Adv. Front. Myc. Pl. Path.* **11**, 1981).

Lasiochlaena Pouzar (1990), Fomitopsidaceae. 1, Europe. See Pouzar (*Česká Mykol.* **44**: 92, 1990) = *Ischnoderma* fide, Ryvarden (*Gen. Polyp.*, 1991).

Lasioderma Mont. (1845) ? = Phleogena fide Stalpers (*in litt.*).

Lasiodiplodia Ellis & Everh. (1896), anamorphic *Botryosphaeriaceae*, St.1eP.1. 9, widespread. *L. theobromae* (syn. *Botryodiplodia theobromae*), a common trop. saprobe and wound parasite. See Punithalingam (*IMI Descr. Fungi Bact.* **519**, 1976), Robell & Forester (*Sabouraudia* **14**: 155, 1976; keratomycosis in humans), Punithalingam (*Biblthca Mycol.* **71**, 1980), Yaguchi & Nakamura (*Jl Agr. Sci.* **35**: 282, 1991; TEM of conidial wall), Denman *et al.* (*Stud. Mycol.* **45**: 129, 2000; review), Zhou & Stanosz (*Mycol.* **93**: 516, 2001; phylogeny), Pavlic *et al.* (*Stud. Mycol.* **50**: 313, 2004), Summerbell *et al.* (*Medical Mycology* **42**: 543, 2004; phaeohyphomycosis), Mohali *et al.* (*Forest Pathology* **35**: 385, 2005; endophytic strains), Burgess *et al.* (*Mycol.* **98**: 423, 2006; n. spp. from tropics), Crous *et al.* (*Stud. Mycol.* **55**: 235, 2006; phylogeny, review), Damm *et al.* (*Mycol.* **99**: 664, 2007; S Africa, on *Prunus*), Alves *et al.* (*Fungal Diversity* **28**: 1, 2008; cryptic species), Phillips *et al.* (*Persoonia* **21**: in press, 2008; phylogeny).

Lasiodiplodiella Zambett. (1955), anamorphic *Pezizomycotina*, Hso.1eP.?. 3, widespread (tropical). See Zambettakis (*BSMF* **70**: 229, 1954).

Lasioloma R. Sant. (1952), Ectolechiaceae (L). 8, widespread (tropical). See Lücking & Sérusiaux (*Mycotaxon* **77**: 301, 2001; key).

Lasiolomataceae Hafellner (1984) = Ectolechiaceae.

Lasiomollisia Raitv. & Vesterh. (2006), ? Hyaloscyphaceae. 1, Europe. See Baral *et al.* (*Bulletin Mycologique et Botanique Dauphiné-Savoie* **46** no. 183: 33, 2006), Raitviir (*Fungi Non Delineati* Raro vel

Haud Perspecte et Explorate Descripti aut Definite Picti **31**, 2006).

Lasionectria (Sacc.) Cooke (1884), Bionectriaceae. Anamorph *Acremonium*-like. 6 (dead wood and stem), widespread (temperate). See Rossman *et al.* (*Stud. Mycol.* **42**: 248 pp., 1999).

Lasiophoma Naumov (1916) = Pyrenochaeta fide Sutton (*Mycol. Pap.* **141**, 1977).

Lasiophoma Speg. (1918), anamorphic *Pezizomycotina*, Cpd.0eH.?. 3, widespread. See Sutton (*Mycol. Pap.* **141**, 1977).

Lasiosordaria Chenant. (1919) = Cercophora fide Lundqvist (*Symb. bot. upsal.* **20** no. 1, 1972).

Lasiosordariella Chenant. (1919) = Lasiosphaeria fide Hawksworth *et al.* (*Dictionary of the Fungi* edn 8, 1995).

Lasiosordariopsis Chenant. (1919) = Cercophora fide Lundqvist (*Symb. bot. upsal.* **20** no. 1, 1972).

Lasiosphaera Reichardt (1870) = Langermannia fide Demoulin & Dring (*Bull. Jard. bot. nat. Belg.* **45**: 362, 1975).

Lasiosphaeria Ces. & De Not. (1863), Lasiosphaeriaceae. 60 (from wood), widespread. See Lundqvist (*Symb. bot. upsal.* **20** no. 1, 1972), Hilber & Hilber (*Sydowia* **36**: 105, 1983), Réblová (*Sydowia* **50**: 229, 1998), Candoussau *et al.* (*Mycotaxon* **80**: 201, 2001; France), Taylor *et al.* (*Mycoscience* **42**: 369, 2001; on palms), Hilber & Hilber (*The genus Lasiosphaeria and Allied Taxa*, 2002; review), Huhndorf *et al.* (*Mycol.* **96**: 368, 2004; phylogeny, family limits), Miller & Huhndorf (*MR* **108**: 26, 2004; phylogeny), Miller & Huhndorf (*Mycol.* **96**: 1106, 2004; phylogeny), Zhang *et al.* (*Mycol.* **98**: 1076, 2006; phylogeny).

Lasiosphaeriaceae Nannf. (1932), Sordariales. 36 gen. (+ 30 syn.), 308 spp. The family is paraphyletic.

Lit.: Lundqvist (*Symb. bot. upsal.* **20** no. 1, 1972), Lundqvist (*Symb. bot. upsal.* **20**: 374 pp., 1972), Barr (*Mycotaxon* **39**: 43, 1990) separated the *Tripterosporaceae* from the fam. The previous edition of the *Dictionary* largely followed Barr's circumscription, which has proved to be polyphyletic. The family as circumscribed here is much more tightly defined.

Lit.:, Khan & Krug (*Mycol.* **81**: 862, 1989), Guarro *et al.* (*SA* Reprint of Volumes 1-4 (1982-1985) **10**: 79, 1991), Taylor & White (*Fungal Genetics Newsl.* Suppl. **38**: 26, 1991), Krug & Scott (*CJB* **72**: 1302, 1994), Bell & Mahoney (*Mycol.* **87**: 375, 1995), Bell & Mahoney (*Mycol.* **88**: 163, 1996), Bell & Mahoney (*Mycol.* **89**: 908, 1997), Coppin *et al.* (*Microbiol. Mol. Biol. Rev.* **61**: 411, 1997), Chen *et al.* (*Mycol.* **91**: 84, 1999), Lee & Hanlin (*Mycol.* **91**: 434, 1999), Gams (*Stud. Mycol.* **45**: 192, 2000), Huhndorf *et al.* (*Mycol.* **96**: 368, 2004), Miller & Huhndorf (*Mycol.* **96**: 1106, 2004), Miller & Huhndorf (*MR* **108**: 26, 2004), Cai *et al.* (*Fungal Diversity* **19**: 1, 2005), Cai *et al.* (*MR* **110**: 137, 2006), Cai *et al.* (*MR* **110**: 359, 2006), Lumbsch & Huhndorf (*MR* **111**: 1064, 2007).

Lasiosphaeriella Sivan. (1975), Sordariomycetidae. 2, widespread (pantropical). See Huhndorf & Fernández (*Mycol.* **91**: 544, 1999), Huhndorf *et al.* (*Mycol.* **96**: 368, 2004; phylogeny), Zhang *et al.* (*Mycol.* **98**: 1076, 2006; phylogeny).

Lasiosphaeriopsis D. Hawksw. & Sivan. (1980), Coronophorales. 3 (lichenicolous), widespread. See Eriksson & Santesson (*Mycotaxon* **25**: 569, 1986), Alstrup & Hawksworth (*Meddr Grønland* Biosc. **31**, 1990), Huhndorf *et al.* (*MR* **108**: 1384, 2004), Zhurbenko & Triebel (*Mycol. Progr.* **4**: 317, 2005; on *Pilophorus*).

Lasiosphaeris Clem. (1909), Sordariales. 2, widespread (temperate). See Hawksworth *et al.* (*Dictionary of the Fungi* edn 8, 1995).

Lasiostemma Theiss., Syd. & P. Syd. (1917), Pseudoperisporiaceae. Anamorph *Chaetosticta*. *c.* 15, widespread (tropical). See Farr (*Mycol.* **71**: 243, 1979; sects.), Barr (*Mycotaxon* **64**: 149, 1997).

Lasiostemmella Petr. (1950) = Epipolaeum fide Müller & von Arx (*Beitr. Kryptfl. Schweiz* **11** no. 2, 1962).

Lasiostictella Sherwood (1986), Rhytismatales. 1, France. See Candoussau *et al.* (*Sydowia* **38**: 28, 1986).

Lasiostictis (Sacc. & Berl.) Sacc. (1889) = Naemacyclus fide Di Cosmo *et al.* (*Eur. J. For. Path.* **13**: 206, 1983).

Lasiostroma Griffon & Maubl. (1911) = Phomopsis (Sacc.) Bubák. fide Sutton (*Mycol. Pap.* **141**, 1977).

Lasiothelebolus Kimbr. & Luck-Allen (1974) = Thelebolus fide Lundqvist (*SA* **7**: 77, 1988), van Brummelen (*Persoonia* **16**: 425, 1998).

Lasiothyrium Syd. & P. Syd. (1913), anamorphic *Pezizomycotina*, Cpt.0fP.?. 1, Philippines.

Lasmenia Speg. (1886), anamorphic *Pezizomycotina*, St.0eP.1. 5, S. America; Philippines.

Lasmeniella Petr. & Syd. (1927), anamorphic *Pezizomycotina*, St.0eP.19. 13, widespread (esp. tropical). See Bhat & Hedge (*Acta Bot. Indica* **19**: 90, 1991).

Lasseria Dennis (1960), ? Helotiales. 1, Venezuela. See Dennis (*Kew Bull.* **14**: 434, 1960).

lateral, at the side.

Latericonis G.V. Rao, K.A. Reddy & de Hoog (1984), anamorphic *Pezizomycotina*, Hsp.#eP.10. 1, India. See Rao *et al.* (*Mycotaxon* **19**: 409, 1984).

Lateriramulosa Matsush. (1971), anamorphic *Pezizomycotina*, Hso.1bH.10. 1, Japan; Australasia. See Matsushima (*Microfungi of the Solomon Islands and Papua-New Guinea*: 34, 1971), Matsushima (*Matsush. Mycol. Mem.* **6**, 1989).

Laterispora Uecker, W.A. Ayers & P.B. Adams (1982), anamorphic *Pezizomycotina*, Hso.0fP.1. 1 (on sclerotia), USA. See Uecker *et al.* (*Mycotaxon* **14**: 492, 1982).

lateritiin, see enniatin A.

Laternea Turpin (1822), Phallaceae. 2, America (tropical).

Lateropeltis Shanor (1946) = Kiehlia fide Farr (*Mycol.* **60**: 924, 1968).

Laterotheca Bat. (1963) = Acrogenotheca fide Hughes (*Mycol.* **68**: 693, 1976).

Laterradea Raspail (1824) nom. dub., Agaricales.

latex, a milk-like juice, as in *Lactarius*.

Lathagrium (Ach.) Gray (1821) = Collema F.H. Wigg. fide Degelius (*Symb. bot. upsal.* **13** no. 2, 1954).

Lathraeodiscus Dissing & Sivertsen (1989), Pyronemataceae. 1, Svalbard. See Dissing & Sivertsen (*Mycol.* **80**: 832, 1989).

Latin. Acquaintance with Latin is an asset to any biologist and a working knowledge of the language is particularly useful to systematists because scientific names of all taxa are latinized. Latin descriptions or diagnoses are currently required under the international code governing the naming of fungi (see Nomenclature) to validate the names of fungi at all formal ranks, and many early works on systematic mycology are in Latin, as are descriptions in Saccardo's *Sylloge Fungorum*. Botanical Latin is derived from the classical language with liberal borrowing from

Greek, but with a simplified syntax. Although not generally spoken, differences in stress and pronunciation of latinized names may cause confusion: guidance can be found at the start of Latin dictionaries or in Allen (Ed.) (*Vox Latina*, 1978).

Stearn (*Botanical Latin*, 1966 [edn 4, 1992]) provides an excellent introduction to grammar, syntax, and vocabulary and gives much helpful guidance together with examples of mycological descriptions. This work may be supplemented by Cash (A mycological English-Latin glossary. *Mycol. Mem.* **1**, 1965); Baranov (*Botanical Latin for Plant Taxonomists*, 1968) [reprint 1971]; Borror (*Dictionary of Word Roots and Combining Forms*, 1960) [11th print, 1971]; Clements & Shear (*The Genera of Fungi*, 1931) [useful Latin/English glossary]. Nybakken (*Greek and Latin in Scientific Terminology*, 1960), gives much information on derivation of scientific and medical terms.

Further articles which may be consulted for guidance on the correct formulation of Latin names are Zabinkova (*Taxon* **17**: 19, 1968) [generic stems ending in **-is**]; Nicholson (*Taxon* **43**: 97, 1994) [gender of generic names, particularly those ending in **-ma**]; Manara (*Taxon* **41**: 52, 1992) [geographical epithets]; Manara (*Taxon* **40**: 301, 1991) [gender of generic names]; Nicholson & Brooks (*Taxon* **23**: 163, 1974) [orthography, stems and compounds].

See Classification, Colour, Nomenclature.

latticed, cross-barred; like a network.

lattice-work fungus, basidioma of *Clathrus* spp.

Latzeiia, see *Latzelia.*

Latzelia Zahlbr. (1926) = Epiphloea fide Gyelnik (*Rabenh. Krypt.-Fl.* **9** 2.2, 1940).

Latzinaea Kuntze (1891) ≡ Entoloma.

Laudatea Johow (1884) = Dictyonema C. Agardh ex Kunth fide Parmasto (*Nova Hedwigia* **29**: 99, 1978).

Lauderlindsaya J.C. David & D. Hawksw. (1989) = Normandina fide Aptroot (*Willdenowia* **21**: 263, 1991), Diederich & Sérusiaux (*Lichenologist* **25**: 97, 1993), Aptroot (*Lichenologist* **30**: 501, 1998).

Laurera Rchb. (1841) nom. cons. prop., Trypetheliaceae (L). 20, widespread (tropical). See Letrouit-Galinou (*Revue bryol. lichén.* **26**: 207, 1957), Letrouit-Galinou (*Revue bryol. lichén.* **27**: 66, 1958), Harris (*Acta Amazon.* Supl. **14**: 55, 1984; gen. concept., key 4 spp. Brazil), Upreti & Singh (*Bull. Jard. bot. nat. Belg.* **57**: 367, 1987; key 10 spp. India), Makhija & Patwardham (*Mycotaxon* **31**: 565, 1988; key 17 spp. India), McCarthy & Kantvilas (*Lichenologist* **25**: 51, 1993; Tasmania), McCarthy (*Lichenologist* **27**: 310, 1995; Papua New Guinea), McCarthy (*Flora of Australia* **58** A: 242 pp., 2001), Prado *et al.* (*MR* **110**: 511, 2006; phylogeny), Aptroot *et al.* (*Biblthca Lichenol.* **97**, 2008; Costa Rica).

Laureriella Hepp (1867) = Glypholecia fide Hawksworth *et al.* (*Dictionary of the Fungi* edn 8, 1995).

Laureromyces Cif. & Tomas. (1953) = Laurera fide Hawksworth *et al.* (*Dictionary of the Fungi* edn 8, 1995).

Laurilia Pouzar (1959), Echinodontiaceae. Anamorph *Spiniger*. 2, widespread (northern hemisphere). See Gross (*Mycopath. Mycol. appl.* **24**: 1, 1964), Eriksson & Ryvarden (*Cortic. N. Europ.* **4**: 787, 1976), Chamuris (*Mycol. Mem.* **14**, 1988; key) = Echinodontium (Echinodont.) fide.

Lauriomyces R.F. Castañeda (1990), anamorphic *Pezizomycotina*, Hso.0eH.3. 5, widespread. See Castañeda (*Univ. Waterloo Biol. Ser.* **32**: 21, 1990), Crous & Wingfield (*Sydowia* **46**: 193, 1994; Switzerland), Somrithipol *et al.* (*Nova Hedwigia* **82**: 209, 2006; Thailand), Somrithpol & Jones (*Nova Hedwigia* **84**: 479, 2007; Thailand).

Laurobasidium Jülich (1982), Exobasidiaceae. 1 (on *Laurus*), Europe. See Begerow *et al.* (*Mycol. Progr.* **1**: 187, 2002).

Lauterbachiella Henn. (1898) = Rhagadolobium fide von Arx & Müller (*Stud. Mycol.* **9**, 1975).

Lautisporopsis E.B.G. Jones, Yusoff & S.T. Moss (1998), Halosphaeriaceae. 1 (marine), Atlantic Ocean; Pacific Ocean. See Yusoff *et al.* (*CJB* **72**: 1550, 1994), Jones (*Mycotaxon* **67**: 1, 1998).

Lautitia S. Schatz (1984), ? Phaeosphaeriaceae. 1 (on marine algae), USA. See Schatz (*CJB* **62**: 31, 1984).

Lautospora K.D. Hyde & E.B.G. Jones (1989), Lautosporaceae. 2, Brunei; N. America. See Kohlmeyer *et al.* (*Bot. Mar.* **38**: 165, 1995).

Lautosporaceae Kohlm., Volkm.-Kohlm. & O.E. Erikss. (1995), Pezizomycotina (inc. sed.). 1 gen., 2 spp.

Lit.: Hyde & Jones (*Bot. Mar.* **32**: 479, 1989), Kohlmeyer *et al.* (*Bot. Mar.* **38**: 165, 1995).

Lawalreea Dieder. (1990), anamorphic *Pezizomycotina*, Cpd.0eH.15. 1 (on lichens), Luxembourg. See Diederichs (*Mycotaxon* **37**: 308, 1990).

Laxitextum Lentz (1956), Hericiaceae. 3, widespread. See Lentz (*Sydowia* **14**: 123, 1960), Gross (*Mycopath. Mycol. appl.* **24**: 1, 1964), Chamuris (*Mycol. Mem.* **14**, 1988; key), Ginns & Freeman (*Biblthca Mycol.* **157**, 1994) = Echinodontium (Echinodont.) fide.

Lazarenkoa Zerova (1938), Dothideomycetes. 1 (on *Selaginella*), Siberia.

Lazaroa Gonz. Frag. [not traced] = Phellinus fide Donk (*Verh. K. ned. Akad. Wet.* tweede sect. **62**: 1, 1974).

Lazuardia Rifai (1988), Pyronemataceae. 1, widespread (pantropical). See Rifai (*Mycotaxon* **31**: 239, 1988), Perry *et al.* (*MR* **111**: 549, 2007; phylogeny).

Lazulinospora Burds. & M.J. Larsen (1974) = Amauroderma Murrill fide Kõljalg (*Syn. Fung.* **9**, 1996).

LCP, see PC.

LD, lethal dose; **LD**$_{50}$ (of the concentration of a fungicide, etc.), that which kills 50% of the spores (cells or individuals) of the test organism. Cf. ED.

LE, Komarov Botanical Institute (St Petersburg, Russia); founded 1714; directed by the Academy of Sciences.

leaf curl (1) of peach and almond (*Taphrina deformans*); (2) of cherry (*T. minor*).

Leaia Banker (1906) = Gloiodon fide Banker (*Mycol.* **2**: 10, 1910).

Leandria Rangel (1915), anamorphic *Pezizomycotina*, Hso.#eP.1. 1, S. America.

leather fungi, members of the *Thelephoraceae* s. lat.

Lecanactidaceae Stizenb. (1862) = Graphidaceae.

Lecanactiomyces Cif. & Tomas. (1953) ≡ Lecanactis Körb. fide Hawksworth *et al.* (*Dictionary of the Fungi* edn 8, 1995).

Lecanactis Eschw. (1824) nom. rej. = Phaeographis fide Tehler (*Taxon* **35**: 382, 1986).

Lecanactis Körb. (1855) nom. cons., Roccellaceae (L). *c.* 54, widespread. See Tehler (*Taxon* **35**: 382, 1986; nomencl.), Torrente & Egea (*Biblthca Lichenol.* **32**, 1989), Tehler (*CJB* **68**: 2458, 1990; cladistics),

Tehler (*Willdenowia* **22**: 201, 1992), Egea *et al.* (*Pl. Syst. Evol.* **187**: 103, 1993; *L. grumulosa* group), Tehler & Egea (*Lichenologist* **29**: 397, 1997; phylogeny), Myllys *et al.* (*Bryologist* **101**: 70, 1998; phylogeny), Kantvilas (*Symb. bot. upsal.* **34** no. 1: 183, 2004; Tasmania), Tehler & Irestedt (*Cladistics* **23**: 432, 2007).

Lecanephebe Frey (1929) = Zahlbrucknerella fide Henssen (*Symb. bot. upsal.* **18** no. 1, 1963).

Lecania A. Massal. (1853), Lecanorales (L). *c.* 64, widespread (esp. temperate). Possibly related to *Crocynia*. See Mayrhofer (*Biblthca Lichenol.* **28**, 1988; 19 spp. on rock, Eur.), van den Boom (*Nova Hedwigia* **54**: 229, 1992; key 11 spp. on rock, Netherlands), Ekman & Wedin (*Pl. Biol.* **2**: 350, 2000; phylogeny), Ekman (*MR* **105**: 783, 2001; phylogeny), Boom & Khodosovtsev (*Graphis Scripta* **16**: 1, 2004; Europe, Asia), Lumbsch *et al.* (*Mol. Phylogen. Evol.* **31**: 822, 2004; phylogeny), Boom & Brand (*Lichenologist* **37**: 277, 2005; Europe), Miądlikowska *et al.* (*Mycol.* **98**: 1088, 2006; phylogeny), Naesborg *et al.* (*MR* **111**: 581, 2007; phylogeny).

Lecaniascus Moniez (1887) nom. dub., ? Fungi.

Lecanicillium W. Gams & Zare (2001), anamorphic *Cordycipitaceae*. 15, widespread. See Gams & Zare (*Nova Hedwigia* **72**: 332, 2001), Zare & Gams (*Nova Hedwigia* **73**: 1, 2001), Cortez-Madrigal *et al.* (*BioControl* **48**: 321, 2003; biocontrol of cocoa pests), Kouvelis *et al.* (*Fungal Genetics Biol.* **41**: 930, 2004; mitochondrial genome), Kope & Leal (*Mycotaxon* **94**: 331, 2005; on white pine weevil).

Lecanidiaceae O.E. Erikss. (1981) = Patellariaceae.

Lecanidiales = Patellariales.

Lecanidiella Sherwood (1986), Patellariaceae. 1, USA. See Sherwood (*Sydowia* **38**: 267, 1986), Kutorga & Hawksworth (*SA* **15**: 1, 1997).

Lecanidion Endl. (1830) ≡ Patellaria Fr. fide Hawksworth (*Taxon* **35**: 787, 1986).

Lecanidium A. Massal. (1856) = Pertusaria fide Hawksworth *et al.* (*Dictionary of the Fungi* edn 8, 1995).

Lecaniella Jatta (1889) = Lecania fide Hawksworth *et al.* (*Dictionary of the Fungi* edn 8, 1995).

Lecaniella Vain. (1896) ≡ Gonolecania.

Lecaniocola Brain (1923), anamorphic *Pezizomycotina*, ?.0eH.10. 15 (in scale insects), widespread (tropical).

Lecaniomyces E.A. Thomas (1939) nom. inval. = Lecania fide Hawksworth *et al.* (*Dictionary of the Fungi* edn 8, 1995).

Lecaniopsis (Vain.) Zahlbr. (1926) = Coenogonium Ehrenb. fide Hawksworth *et al.* (*Dictionary of the Fungi* edn 8, 1995), Rivas Platas *et al.* (*Fungal Diversity* **23**: 255, 2006).

Lecanocaulon Nyl. (1860) = Stereocaulon Hoffm. fide Hawksworth *et al.* (*Dictionary of the Fungi* edn 8, 1995).

Lecanocybe Desjardin & E. Horak (1999), Marasmiaceae. 1, Java; Hawaii. See Desjardin & Horak (*Sydowia* **51**: 21, 1999).

Lecanographa Egea & Torrente (1994), Roccellaceae (L). 30, widespread. See Egea & Torrente (*Biblthca Lichenol.* **54**: 116, 1994), Hawksworth (*Taxon* **55**: 528, 2006; nomencl.), Sparrius *et al.* (*Lichenologist* **38**: 27, 2006; SE Asia).

Lecanora Ach. (1809), Lecanoraceae (L). *c.* 552, widespread (esp. temperate). See Poelt (*Mitt. bot. StSamml., München* **2**: 411, 1958), Imshaug & Brodo (*Nova Hedwigia* **12**: 1, 1966; *L. pallida* group), Eigler (*Diss. Bot., Lehre* **4**, 1969; gen. concept), Brodo (*Beih. Nova Hedwigia* **79**: 63, 1984; key 38 spp., N. Am. *subfusca*-group), Leuckert & Poelt (*Nova Hedwigia* **49**: 121, 1989; *L. rupicola*-group), Nimis & Bolognini (*Not. Soc. Lich. Ital.* **6**: 29, 1993; key spp. Italy), Poelt & Grube (*Nova Hedwigia* **57**: 305, 1993; key 22 spp. subgen. *Placodium* in Himalaya), Lumbsch (*J. Hattori bot. Lab.* **77**: 1, 1994; key 46 spp. Australasia), Dickhäuser *et al.* (*Mycotaxon* **56**: 303, 1995; *L. subcarnea* group), Lumbsch *et al.* (*Bryologist* **99**: 269, 1996), Lumbsch *et al.* (*Symb. bot. upsal.* **32** no. 1: 131, 1997; *L. palida*-group), Ryan & Nash (*Nova Hedwigia* **64**: 111, 1997), Ryan & Nash (*Nova Hedwigia* **64**: 393, 1997), Arup & Grube (*Lichenologist* **30**: 415, 1998; DNA subg. *Placodium*), Ryan *in* Glenn *et al.* (Eds) (*Lichenogr. Thomsoniana*: 105, 1998; subgen. *Placodium*), Upreti (*Bryologist* **10**: 256, 1998; Indian spp. *L. subfusca* group), Upreti & Chatterjee (*Feddes Repert.* **109**: 279, 1998; Indian spp. subg. *Placodium*), Guderley (*J. Hattori bot. Lab.* **87**: 131, 1999; *L. subfusca* group), Guderley & Lumbsch (*Lichenologist* **31**: 197, 1999; multispored spp.), Ekman & Wedin (*Pl. Biol.* **2**: 350, 2000; phylogeny), Feige *et al.* (*Biblthca Lichenol.* **75**: 99, 2000; chemistry), Guderley *et al.* (*Bryologist* **103**: 139, 2000; S America), Śliwa & Wetmore (*Bryologist* **103**: 475, 2000; *L. varia* group, N. Am.), LaGreca & Lumbsch (*Bryologist* **104**: 204, 2001; N America), Printzen (*Bryologist* **104**: 382, 2001; Sonoran Desert), Laundon (*Nova Hedwigia* **76**: 83, 2003; *L. varia* group), Lumbsch *et al.* (*Bryologist* **106**: 552, 2003; N America), Grube *et al.* (*MR* **108**: 506, 2004; *L/ rupicola* group), Martínez & Aragón (*Bryologist* **107**: 222, 2004; Spain), Peršoh *et al.* (*Mycol. Progr.* **3**: 103, 2004; asci, phylogeny), Wedin *et al.* (*MR* **109**: 159, 2005; phylogeny), Miądlikowska *et al.* (*Mycol.* **98**: 1088, 2006; phylogeny).

Lecanoraceae Körb. (1855), Lecanorales (L). 23 gen. (+ 26 syn.), 766 spp.

Lit.: Hafellner (*Nova Hedwigia* Beih. **79**: 241, 1984), Gargas *et al.* (*Mol. Biol. Evol.* **12**: 208, 1995), Kärnefelt (*Abstracta Botanica* **21**: 21, 1997), Lumbsch *et al.* (*Symb. bot. upsal.* **32** no. 1: 131, 1997), Ryan & Nash (*Nova Hedwigia* **64**: 393, 1997), Arup & Grube (*Lichenologist* **30**: 415, 1998), Rambold & Hagedorn (*Lichenologist* **30**: 473, 1998), Arup & Grube (*CJB* **78**: 318, 2000), Ekman & Wedin (*Pl. Biol.* **2**: 350, 2000), Grube *et al.* (*MR* **108**: 506, 2004), Miądlikowska *et al.* (*Mycol.* **98**: 1088, 2006; phylogeny), Arup *et al.* (*Mycol.* **99**: 42, 2007; phylogeny), Hofstetter *et al.* (*Mol. Phylogen. Evol.* **44**: 412, 2007; phylogeny).

lecanoralean (1) (of asci), of asci which are essentially bitunicate in structure, generally thick-walled with an especially strongly thickened apex, and in which discharge is by a rostrate eversion of the endoascus; see ascus; (2) (of apothecioid ascomata), see lecanorine.

Lecanorales Nannf. (1932), (±L). Lecanoromycetidae. 26 fam., 269 gen., 5695 spp. Thallus very varied. Ascomata apothecial, with or without a thalline margin, rarely mazaedial. Interascal tissue of paraphyses, usually branched and swollen at the apices, often with a pigmented epithecium, rarely absent. Hymenial gel often prominent. Asci with a single wall layer visible in LM but thick-walled, almost always with a conspicuous thick cap-like apical part, often with an internal beak (ocular chamber) and sometimes also with complex apical structures, ascus walls and/or apical structures often J+; rarely thin-walled and evanescent. Ascospores very varied. Anamorphs pycnidial, poorly known. Mainly lichen-forming (almost all with protococcoid green photobionts), some lichenicolous or saprobes (then esp. on wood in xeric situations).

Probably still the largest order of the *Ascomycota*, but many groups have been removed in recent years. An extended subordinal classification was proposed by Rambold & Triebel (*Bibl. Lich.* **48**, 1992) which recognized 6 suborders, but much of this research has been superseded by sequence-based studies.The following fams are accepted here; all are exclusively lichen-forming unless otherwise indicated. Not all have been studied using molecular methods, and the order may contract further. Fams:

(1) **Aphanopsidaceae**
(2) **Biatorellaceae**
(3) **Brigantiaeaceae**
(4) **Cladoniaceae**
(5) **Crocyniaceae**
(6) **Dactylosporaceae**
(7) **Ectolechiaceae**
(8) **Gypsoplacaceae**
(9) **Haematommataceae**
(10) **Helocarpaceae**
(11) **Lecanoraceae**
(12) **Megalariaceae**
(13) **Miltideaceae**
(14) **Mycoblastaceae**
(15) **Parmeliaceae** (syn. *Hypogymniaceae*, *Usneaceae*)
(16) **Pilocarpaceae** (syn. *Micareaceae*)
(17) **Psoraceae**
(18) **Ramalinaceae** (syn. *Bacidiaceae*, *Megalariaceae*)
(19) **Sphaerophoraceae**
(20) **Stereocaulaceae**
(21) **Tephromelataceae**
(22) **Vezdaeaceae** (±L)

Further references to family treatments are included in the fam. entries and under the type genera.

Lit.: Brodo *et al.* (Eds) (*J. Hattori bot. Lab.* **52**: 303, 1982; papers from symp. 'Evolution in Lichenized Fungi'), Eriksson (1981; nomencl. fams), Hafellner (*Beih. Nova Hedw.* **79**: 241, 1984, fams, asci; *in* Galun (Ed.), *Handbook of lichenology* **3**: 41, 1988; fams; *in* Hawksworth (Ed.), 1994: 315), Henssen & Jahns (1973), Hertel (*Ergebn. ForschUnternehems Nepal Himalaya* **6**: 154, 1977; lecideine gen.), Ozenda & Clauzade (*Les lichens*, 1970; key 2188 spp. France), Poelt (*Bestimmungschlüssel europäischer Flechten*, 1969, & Vězda, *Erganzungsheft* I-II, 1977-81; keys Eur. spp.), Purvis *et al.* (*Lichen flora of Great Britain and Ireland*, 1992), Rambold (*Cryptogamic Botany* **5**: 111, 1995), Rambold & Hagedorn (*Lichenologist* **30**: 473, 1998), Santesson (*Symb. bot. upsal.* **12** (1), 1952; foliicolous taxa), Timdal (*Opera Bot.* **110**, 1991; fam. concepts), Wirth (*Die Flechten Baden-Württembergs*, edn 2, 2 vols, 1995), Zahlbruckner (*Naturl. PflFam.*, edn 2, **8**: 61, 1926; fams and gen.).

Lecanorella Frey (1926) = Koerberiella fide Hafellner (*Beih. Nova Hedwigia* **79**: 241, 1984).

lecanorine (of an apothecium), having an excipulum (q.v.) thallinum (e.g. *Lecanora*, *Parmelia*) (Fig. 20A); lecanoroid.

Lecanoromyces E.A. Thomas ex Cif. & Tomas. (1953) ≡ Lecanora.

Lecanoromycetes O.E. Erikss. & Winka (1997), Pezizomycotina. 12 ord., 77 fam., 630 gen., 14199 spp. The largest and most varied group of lichenized *Ascomycota*. Subcl.:

(1) **Acarosporomycetidae**
(2) **Lecanoromycetidae**
(3) **Ostropomycetidae**

Lit.: Hofstetter *et al.* (*Mol. Phylogenet. Evol.* **44**: 412, 2007), Lumbsch *et al.* (*Mol. Phylogenet. Evol.* **31**: 822, 2004), Lutzoni *et al.* (*Am. J. Bot.* **91**: 1446, 2004), Miądlikowska *et al* (*Mycol.* **98**: 1088, 2006), Wedin *et al.* (*MR* **109**: 159, 2005).

Lecanoromycetidae P.M. Kirk, P.F. Cannon, J.C. David & Stalpers ex Miądl., Lutzoni & Lumbsch (2007), Lecanoromycetes. Thallus very varied. Ascomata almost always apothecial, flat to strongly cup-shaped, with or without a thalline margin, rarely mazaedial. Interascal tissue of paraphyses, usually branched and swollen at the apices, often with a pigmented or J+ epithecium, rarely absent. Asci typically with a single wall layer visible in LM but thick-walled, almost always with a conspicuous thick cap-like apical part, often with complex apical structures, often J+, rarely thin-walled and evanescent. Ascospores very varied. Mainly lichen-forming (with protococcoid green or cyanobacterial photobionts), some lichenicolous or saprobes (then esp. on wood in xeric situations). Ords:

(1) **Lecanorales**
(2) **Peltigerales**
(3) **Teloschistales**

The *Umbilicariales* may belong in this order, along with the *Catillariaceae* and *Rhizocarpaceae*, but relative positions are still uncertain.

Lit.: see under ord. and fam.

Lecanoropsis M. Choisy (1949) = Lecanora fide Hawksworth *et al.* (*Dictionary of the Fungi* edn 8, 1995).

Lecanosticta Syd. (1922), anamorphic *Eruptio*, Cac.≡ eP.19. 3 (on pines), N. America; C. America. See Wolf & Barbour (*Phytopathology* **31**: 61, 1941), Petrak (*Sydowia* **15**: 252, 1961), Evans (*Mycol. Pap.* **153**, 1984), Han *et al.* (*J. Nanjing For. Univ.* **15**: 1, 1991; pathotypes of *L. acicola*), Suto & Ougi (*Mycoscience* **39**: 319, 1998; Japan), Marmolejo (*Mycotaxon* **76**: 393, 2000; Mexico), Verkley & Priest (*Stud. Mycol.* **45**: 123, 2000; review), Crous *et al.* (*Mycol.* **93**: 1081, 2001; phylogeny), Verkley *et al.* (*Mycol.* **96**: 558, 2004; phylogeny).

Lecanostictopsis B. Sutton & Crous (1997), anamorphic *Pezizomycotina*, Cpd.?.?. 4, widespread. See Sutton & Crous (*MR* **101**: 215, 1997).

Leccinellum Bresinsky & Manfr. Binder (2003), Boletaceae. 10, widespread (north temperate). See Bresinsky & Besl (*Regensb. Mykol. Schr.* **11**: 231, 2003).

Leccinum Gray (1821), Boletaceae. *c.* 75, widespread (mainly north temperate). See Lannoy & Estades (*Monographie des Leccinum d'Europe*, 1995), Halling & Mueller (*Mycol.* **95**: 488, 2003; Costa Rica), Den Bakker *et al.* (*Mycol.* **96**: 102, 2004; ITS phylogeny), Den Bakker *et al.* (*New Phytol.* **163**: 201, 2004; ectomycorrhizae, evolution and host specificity), Den Bakker & Noordeloos (*Persoonia* **18**: 511, 2005; rev. Europ. spp.), Fu *et al.* (*Mycotaxon* **96**: 47, 2006; China), Den Bakker *et al.* (*MR* **111**: 663, 2007; phylogeography).

Lecidea Ach. (1803), Lecideaceae (±L). *c.* 427, widespread. *Lecidea* s.str. contains ± 100 species; other species have been redisposed over many segregate genera. See Lowe (*Lloydia* **2**: 225, 1939), Poelt (*Ber. bayer. bot. Ges.* **34**: 82, 1961), Hertel (*Beih. Nova Hedwigia* **24**, 1967), Hertel (*Herzogia* **1**: 25, 1968), Hertel (*Herzogia* **1**: 321, 1969), Thomson *et al.* (*Bryologist* **72**: 137, 1969), Hertel (*Herzogia* **1**: 405, 1970; parasitic spp.), Hertel (*Herzogia* **2**: 231, 1971), Hertel (*Herzogia* **2**: 479, 1973), Hertel (*Herzogia* **3**: 365, 1975), Hertel (*Decheniana* **127**: 37, 1975; key saxic. holarctic spp.), Hertel (*Ergebn. Forsch. Unternehems Nepal Himalaya* **6**: 145, 1977; Asia), Inoue (*J. Sci. Hiroshima Univ.* B(2) **18**: 1, 1982; Japan), Schwab (*Mitt. bot. StSamml., München* **22**: 221, 1986; key 17 spp.), Hertel (*Mitt. bot. StSamml., München* **30**: 297, 1991; arctic spp.), Hertel (*Biblthca Lichenol.* **58**: 137, 1995; key Europe spp.), Casares-Porcel *et al.* (*Lichenologist* **28**: 37, 1996; spp. on gypsum), Stenroos & DePriest (*Am. J. Bot.* **85**: 1548, 1998; DNA), Inoue (*Bull. natn. Sci. Mus.* Tokyo, B **28**: 7, 2002; Hawaii), Leuckert & Hertel (*Biblthca Lichenol.* **86**: 13, 2003; *L. atrobrunnea* complex, Americas), Arup (*Symb. bot. upsal.* **34** no. 1: 39, 2004; Sweden), Buschbom & Mueller (*Mol. Phylogen. Evol.* **32**: 66, 2004; phylogeny), Peršoh *et al.* (*Mycol. Progr.* **3**: 103, 2004; asci, phylogeny), Hafellner (*Fritschiana* **52**: 31, 2006; on *Rhizocarpon*), Hertel (*Central European Lichens*: 19, 2006; Europe), Miądlikowska *et al.* (*Mycol.* **98**: 1088, 2006; phylogeny), Upreti *et al.* (*Mycotaxon* **95**: 323, 2006; India).

Lecideaceae Chevall. (1826), Lecideales (L). 7 gen. (+ 8 syn.), 436 spp.

Lit.: Hertel & Rambold (*Bot. Jb.* **107**: 469, 1985), Hertel (*Biblthca Lichenol.* **25**: 219, 1987), Rambold (*Biblthca Lichenol.* **34**: 345 pp., 1989), Triebel (*Biblthca Lichenol.* **35**: 278 pp., 1989), Pietschmann (*Nova Hedwigia* **51**: 521, 1990), Hertel (*Cryptog. Bot.* **5**: 99, 1995; gen. concepts), Hertel (*Biblthca Lichenol.* **58**: 137, 1995), Hertel (*Symb. bot. upsal.* **32**: 95, 1997), Rambold & Hagedorn (*Lichenologist* **30**: 473, 1998), Kantvilas & McCarthy (*Lichenologist* **31**: 555, 1999), Miądlikowska *et al.* (*Mycol.* **98**: 1088, 2006; phylogeny), Hofstetter *et al.* (*Mol. Phylogen. Evol.* **44**: 412, 2007; phylogeny).

Lecideales Vain. (1934), (L). Lecanoromycetidae. 2 fam., 23 gen., 494 spp. Fams:

(1) **Lecideaceae**

(2) **Porpidiaceae**

For *Lit.* see under fam.

lecideine (of an apothecium), one having no excipulum thallinum (q.v.) and the margin usually consisting only of the excipulum proprium (e.g. *Lecidea*, *Bacidia*) (Fig. 20B).

Lecidella Körb. (1855), Lecanoraceae (L). 79, widespread (esp. temperate). See Hertel & Leuckert (*Willdenowia* **5**: 369, 1969), Knoph (*Biblthca Lichenol.* **36**, 1990; key 15 spp. rock), Leuckert & Knoph (*Lichenologist* **24**: 383, 1992; chloroxanthone spp.), Leuckert & Knoph (*Biblthca Lichenol.* **53**: 161, 1993; chemistry), Inoue (*Bull. natn. Sci. Mus.* Tokyo, B **23**: 127, 1997), Knoph *et al.* (*Mycotaxon* **71**: 163, 1999), Ekman & Wedin (*Pl. Biol.* **2**: 350, 2000; phylogeny), Inoue (*Bull. natn. Sci. Mus.* Tokyo, B **26**: 139, 2000; Japan), Knoph & Leuckert (*Herzogia* **14**: 1, 2000; chemistry), Peršoh *et al.* (*Mycol. Progr.* **3**: 103, 2004; asci, phylogeny), Miądlikowska *et al.* (*Mycol.* **98**: 1088, 2006; phylogeny).

Lecidellomyces E.A. Thomas (1939) nom. inval. = Lecidella fide Hawksworth *et al.* (*Dictionary of the Fungi* edn 8, 1995).

Lecideola A. Massal. (1861) = Lecidella fide Knoph (*Biblthca Lichenol.* **36**, 1990).

Lecideomyces E.A. Thomas ex Cif. & Tomas. (1953) = Adelolecia fide Hawksworth *et al.* (*Dictionary of the Fungi* edn 8, 1995).

Lecideopsella Höhn. (1909), Schizothyriaceae. *c.* 10, widespread (esp. tropical).

Lecideopsis (Almq.) Rehm (1891) = Arthonia fide Hawksworth *et al.* (*Dictionary of the Fungi* edn 8, 1995).

Lecidocollema Vain. (1890) = Homothecium fide Hawksworth *et al.* (*Dictionary of the Fungi* edn 8, 1995).

Lecidoma Gotth. Schneid. & Hertel (1981), Lecanorales (L). 1, Europe. See Pietschmann (*Nova Hedwigia* **51**: 521, 1990; posn), Miądlikowska *et al.* (*Mycol.* **98**: 1088, 2006; phylogeny).

Lecidopyrenopsis Vain. (1907), Lichinaceae (L). 1, Thailand. See Schultz & Büdel (*Lichenologist* **34**: 39, 2002; key).

Lecidora Motyka (1996) nom. inval. = Lecanora fide Hawksworth *et al.* (*Dictionary of the Fungi* edn 8, 1995).

Lecidorina Motyka (1996) nom. inval. = Lecanora fide Hawksworth *et al.* (*Dictionary of the Fungi* edn 8, 1995).

Leciographa A. Massal. (1854) = Opegrapha Ach. See also *Dactylospora*. fide Hafellner (*Beih. Nova Hedwigia* **62**, 1979), Ertz *et al.* (*J. Linn. Soc.* Bot. **144**: 235, 2004).

Leciophysma Th. Fr. (1865), Collemataceae (L). 2, temperate. See Henssen (*Lichenologist* **3**: 29, 1965), Henssen (*Biblthca Lichenol.* **96**: 129, 2007; Subantarctic).

Lecithium Sacc. (1895) ≡ Lecythium.

Lecoglyphis Clem. (1909) = Opegrapha Ach. fide Hawksworth *et al.* (*Dictionary of the Fungi* edn 8, 1995).

Lecophagus M.W. Dick (1990), anamorphic *Pezizales*, Hso.≡ eH.6. 2, N. America; Antarctica. See Dick (*MR* **94**: 347, 1990; key), Morikawa *et al.* (*MR* **97**: 421, 1993; synonymy with *Cephaliophora*), Tanabe *et al.* (*Mycol.* **91**: 830, 1999; ascomycete anamorph & phylogeny), McInnes (*Polar Biol.* **26**: 79, 2003; Antarctica).

Lecotheciaceae Körb. (1855) = Placynthiaceae.

Lecothecium Trevis. (1851) = Placynthium fide Hawksworth *et al.* (*Dictionary of the Fungi* edn 8, 1995).

Lecozonia Trevis. (1857) = Placynthium fide Hafellner (*Beih. Nova Hedwigia* **62**: 94, 1979).

lectotype, see type.

Lectularia Stirt. (1878) = Diploschistes fide Hawksworth *et al.* (*Dictionary of the Fungi* edn 8, 1995).

Lecythea Lév. (1847) = Phragmidium fide Dietel (*Nat. Pflanzenfam.* **6**, 1928) Anamorph name for (II).

lecythiform, like a stoppered bottle; ninepin-shaped (Fig. 23.13).

Lecythispora Chowdhry (1985), anamorphic *Pezizomycotina*. 1. See Chowdhry (*Curr. Sci.* **54**: 469, 1985).

Lecythium Zukal (1893), Sordariomycetes. 1, Austria. Type material is missing and identity is obscure. See Rossman *et al.* (*Stud. Mycol.* **42**: 248 pp., 1999).

Lecythophora Nannf. (1934), Coniochaetaceae. 1 (on

wood or dung), widespread. See Goidanich (*Ent. naz. Cell. e Cart., Roma* **112**, 1938), Weber (*Nova Hedwigia* **74**: 159, 2002; morphology), Weber *et al.* (*Nova Hedwigia* **74**: 187, 2002; phylogeny), García *et al.* (*MR* **110**: 1271, 2006; phylogeny).

Lecythothecium Réblová & Winka (2001), ? Chaetosphaeriaceae. Anamorph *Ellisembia*-like. 1, Europe (central). See Réblová & Winka (*Mycol.* **93**: 478, 2001), Shenoy *et al.* (*MR* **110**: 916, 2006; anamorph).

Leeina Petr. (1923), anamorphic *Pezizomycotina*, St.0eH.?. 1, Philippines.

Legerioides M.M. White (1999), Legeriomycetaceae. 1 (in *Caecidotea*), USA. See White (*Mycol.* **91**: 1022, 1999).

Legeriomyces Pouzar (1972), Legeriomycetaceae. 5 (in *Ephemeroptera*), Canada; Spain; USA. See Lichtwardt (*The Trichomycetes. Fungal associates of arthropods*, 1986; key), Williams & Lichtwardt (*CJB* **71**: 1109, 1993), White (*MR* **110**: 1011, 2006; phylogeny), Strongman (*CJB* **85**: 949, 2007; Canada), Valle (*Mycol.* **99**: 442, 2007; Spain).

Legeriomycetaceae Pouzar (1972), Harpellales. 31 gen. (+ 5 syn.), 158 spp.

Lit.: Moss (*TBMS* **65**: 115, 1975), Moss & Young (*Mycol.* **70**: 944, 1978), Lichtwardt (*The Trichomycetes. Fungal associates of arthropods*: 343 pp., 1986), Lichtwardt (*Fourth International Mycological Congress Abstracts*, 1986), Lichtwardt *et al.* (*TMSJ* **28**: 376, 1987), Longcore (*Mycol.* **81**: 482, 1989), Lichtwardt & Williams (*CJB* **68**: 1045, 1990), Peterson *et al.* (*Mycol.* **83**: 389, 1991), Lichtwardt & Williams (*CJB* **70**: 1196, 1992), Williams & Lichtwardt (*CJB* **71**: 1109, 1993), Benny *in* McLaughlin *et al.* (Eds) (*The Mycota* A Comprehensive Treatise on Fungi as Experimental Systems for Basic and Applied Research **7A**: 147, 2001), Lichtwardt (*Cellular Origin and Life in Extreme Habitats* **4**: 577, 2002), Beard & Adler (*Mycol.* **95**: 317, 2003), Valle & Santamaría (*Mycol.* **96**: 682, 2004).

Legeriosimilis M.C. Williams, Lichtw., M.M. White & Misra (1999), Legeriomycetaceae. 2 (in *Ameletus*), Europe; N. America. See Williams & Lichtwardt (*Mycol.* **91**: 400, 1999), White & Lichtwardt (*Mycol.* **96**: 891, 2004; Norway), White *et al.* (*Mycol.* **98**: 333, 2006; USA).

legitimate (1) (of validly published names or epithets), in accordance with the *Code*, priorable; **illegitimate**, contrary to the *Code*, impriorable. A recognized taxon of uncertain taxonomic position may have more than one legitimate name according to the position or rank given to it; (2) (of mating types), compatible. See Nomenclature.

Leifia Ginns (1998) = Odonticium fide Zmitrovich & Spirin (*Mycena* **6**: 4, 2006).

Leifidium Wedin (1993), Sphaerophoraceae (L). 1, Australia; S. America. See Wedin (*Nord. J. Bot.* **10**: 539, 1991; spore, ascoma ontogeny), Wedin (*Pl. Syst. Evol.* **187**: 213, 1993), Wedin (*Symb. Bot. Upsal.* **31** no. 1, 1995; key, monograph), Tibell (*Symb. bot. upsal.* **32**: 291, 1997; anamorphs), Wedin *et al.* (*Pl. Syst. Evol.* **209**: 75, 1998; DNA).

Leightonia Trevis. (1853) = Endocarpon fide Hawksworth *et al.* (*Dictionary of the Fungi* edn 8, 1995).

Leightonia Trevis. (1861) = Trypethelium fide Hawksworth *et al.* (*Dictionary of the Fungi* edn 8, 1995).

Leightoniella Henssen (1965), Collemataceae (L). 1, Sri Lanka. See Henssen (*Lichenologist* **3**: 39, 1965).

Leightoniomyces D. Hawksw. & B. Sutton (1977), anamorphic *Pezizomycotina*, Hso.0eP.19. 1 (on lichens), Azores; British Isles. See Hawksworth & Sutton (*J. Linn. Soc.* Bot. **75**: 199, 1977).

Leioderma Nyl. (1888), Pannariaceae (L). 9, mostly southern hemisphere. See Galloway & Jørgensen (*Lichenologist* **19**: 345, 1987), Jørgensen & Galloway (*Lichenologist* **21**: 295, 1989; key), Jørgensen & Galloway (*J. Jap. Bot.* **69**: 383, 1994; Japan), Ekman & Jørgensen (*CJB* **80**: 625, 2002; phylogeny), Wiklund & Wedin (*Cladistics* **19**: 419, 2003; phylogeny), Wedin & Wiklund (*Symb. bot. upsal.* **34** no. 1: 469, 2004; phylogeny), Jørgensen (*Lichenologist* **37**: 369, 2005; Malaysia), Jørgensen & Tønsberg (*Bryologist* **108**: 412, 2005; USA).

leiodisc (of an apothecium), having a smooth glazed disc.

Leiogramma Eschw. (1833) ≡ Leiorreuma.

Leiophallus (Fr.) Nees (1858) = Phallus fide Stalpers (*in litt.*).

Leiophloea (Ach.) Gray (1821) = Arthopyrenia fide Riedl (*Sydowia* **15**: 257, 1961), Riedl (*Sydowia* **16**: 263, 1962).

Leiophloea Trevis. (1860) ≡ Leiophloea (Ach.) Gray.

Leiopoda Velen. (1947) ? = Mycena fide Singer (*Agaric. mod. Tax.*, 1951).

Leiorreuma Eschw. (1824), Graphidaceae (L). 1, pantropical. See Archer (*Biblthca Lichenol.* **94**, 2006; Australia), Archer (*Systematics & Biodiversity* **5**: 9, 2007; Solomon Is).

Leiosepium Sacc. (1900) = Sepedonium fide Damon (*Mycol.* **44**: 95, 1952).

Leiosphaerella Höhn. (1919), Xylariales. Anamorph *Beltraniella*. 9, widespread. See Müller & von Arx (*Beitr. Kryptfl. Schweiz* **11** no. 2, 1962), Samuels & Rossman (*Mycotaxon* **28**: 461, 1987), Kang *et al.* (*Mycoscience* **40**: 151, 1999), Jeewon *et al.* (*MR* **107**: 1392, 2003; phylogeny).

leiosporous, having smooth spores.

Leiostigma Kirschst. (1944) = Sordaria fide von Arx & Müller (*Beitr. Kryptfl. Schweiz* **11** no. 1, 1954).

Leiothecium Samson & Mouch. (1975), Eurotiales. 1, Greece. See Samson & Mouchaca (*CJB* **53**: 1364, 1975).

Lejophallus, see *Leiophallus*.

Lejophlea, see *Leiophloea* (Ach.) Gray.

Lejosphaerella, see *Leiosphaerella*.

Lelum Racib. (1900) ? = Kordyana fide Donk (*Reinwardtia* **4**: 117, 1956; affinities).

Lemalis Fr. (1825), ? Helotiales. 3, widespread.

Lemania Bory (1824), Algae. Algae.

Lembidium Körb. (1855) nom. rej. prop. = Anisomeridium fide Hawksworth *et al.* (*Dictionary of the Fungi* edn 8, 1995).

Lembopodia Bat. (1960) = Cirsosia fide von Arx & Müller (*Stud. Mycol.* **9**, 1975).

Lembosia Lév. (1845), Asterinaceae. *c.* 60, widespread (esp. warmer areas). See Mibey & Hawksworth (*Mycol. Pap.* **174**, 1997; Kenya), Sivanesan & Shivas (*Fungal Diversity* **11**: 159, 2002; Australia), Song & Li (*Mycotaxon* **84**: 401, 2002; China).

Lembosiaceae Höhn. (1918) = Asterinaceae.

Lembosidium Speg. (1923) = Lembosia fide von Arx & Müller (*Stud. Mycol.* **9**, 1975).

Lembosiella Sacc. (1891), ? Microthyriaceae. 1, Africa.

Lembosiellina Bat. & H. Maia (1960) = Lembosia fide von Arx & Müller (*Stud. Mycol.* **9**, 1975).

Lembosina Theiss. (1913), ? Asterinaceae. 21, widespread. See Mibey & Hawksworth (*Mycol. Pap.* **174**, 1997; Kenya), Sivanesan & Shivas (*Fungal Diversity* **11**: 159, 2002; Australia).

Lembosiodothis Höhn. (1917) = Echidnodes fide von Arx & Müller (*Stud. Mycol.* **9**, 1975).

Lembosiopeltis Bat. & J.L. Bezerra (1967) = Uleothyrium fide von Arx & Müller (*Stud. Mycol.* **9**, 1975).

Lembosiopsis Theiss. (1918), ? Asterinaceae. 1, N. America. See Farr (*Sydowia* **38**: 65, 1986).

Lembuncula Cif. (1954), anamorphic *Pezizomycotina*, Cpt.0eP.?. 1, Santo Domingo. See Ciferri (*Sydowia* **8**: 259, 1954).

Lemkea Morgan-Jones & R.C. Sinclair (1983), anamorphic *Pezizomycotina*, Hso.0eP.1. 1, S. Africa; N. America. See Morgan-Jones & Sinclair (*Mycol.* **75**: 159, 1983), Chamuris *et al.* (*Mycotaxon* **24**: 319, 1985).

Lemmopsis (Vain.) Zahlbr. (1906), Lichinaceae (L). 3, Europe. See Ellis (*Lichenologist* **13**: 123, 1981; key), Schultz & Büdel (*Lichenologist* **34**: 39, 2002; key).

Lemniscium Wallr. (1827) = Leptogium fide Jørgensen (*Op. Bot.* **45**, 1978).

Lemonniera De Wild. (1894), anamorphic *Helotiales*, Hso.1bH.15. 6 (aquatic), widespread. See Descals *et al.* (*TBMS* **69**: 89, 1977; key), Campbell *et al.* (*MR* **110**: 1025, 2006; phylogeny, polyphyly).

Lempholemma Körb. (1855), Lichinaceae (L). *c.* 33, widespread. See Schiman-Czeika (*Pl. Syst. Evol.* **158**: 283, 1988), Kantvilas & Jørgensen (*Muelleria* **11**: 45, 1998; Australia), Schultz & Büdel (*Lichenologist* **34**: 39, 2002; key), Schultz (*Biblthca Lichenol.* **86**: 155, 2003; Socotra), Schultz (*Lichenologist* **37**: 227, 2005; Yemen).

Lennisia Nieuwl. (1916) ≡ Monochaetia fide Sutton (*Mycol. Pap.* **141**, 1977).

Lenormandia Delise (1841) [non *Lenormandia* Sond. 1845, nom. cons., *Rhodophyceae*] = Normandina fide Aptroot (*Lichenologist* **30**: 501, 1998).

Lentaria Corner (1950), Lentariaceae. 17, widespread. See Sharda (*Biovigyanam* **10**: 131, 1984; Himalayan spp.), Govorova (*Mikol. Fitopatol.* **36**: 24, 2002; Far East of Russia), Dentinger & McLaughlin (*Mycol.* **98**: 746, 2006; phylogeny).

Lentariaceae Jülich (1982), Gomphales. 3 gen., 23 spp.

Lentescospora Linder (1944) nom. dub., Pezizomycotina. See Kohlmeyer & Kohlmeyer (*Marine Mycology*, 1979).

lenthionine, an odorous metabolic product of *Lentinula edodes* (Nishikawa, *Chem. Pharm. Bull., Tokyo* **15**: 756, 1967).

lentic, habitat still water (lakes). Cf. lotic.

lenticular, like a double convex lens in form (Fig. 23.5*c*).

lentiginose (**lentiginous**), having very small spots as though freckled.

Lentinaceae Jülich (1982) = Polyporaceae. *Lentinaceae* is listed as nom. rej. against *Tricholomataceae*.

Lentinaria Pilát (1941) nom. nud. = Lentinellus fide Stalpers (*in litt.*).

Lentinellaceae Locq. (1972) = Auriscalpiaceae.

Lentinellus P. Karst. (1879), Auriscalpiaceae. 15, widespread. See Miller & Stewart (*Mycol.* **63**: 333, 1971; key), Printz (*Svampe* **14**: 59, 1986; key), Stalpers (*Stud. Mycol.* **40**: 88, 1996; key), Miller & Methven (*Mycol.* **92**: 792, 2000; taxonomy), Hughes & Petersen (*Biblthca Mycol.* **198**: 249, 2004; phylogeny), Petersen (*Biblthca Mycol.* **198**: 181, 2004; types), Petersen & Hughes (*Biblthca Mycol.* **198**: 1, 2004; monograph).

Lentinopanus Pilát (1941) = Panus fide Singer (*Agaric. mod. Tax.*, 1951).

Lentinula Earle (1909), Marasmiaceae. 8, widespread (tropical). *L. edodes* is the edible shii-take mushroom. See Pegler (*Kew Bull.* Addit. Ser. **6**, 1977), Pegler (*Sydowia* **36**: 227, 1983; key), Mata & Petersen (*Mycotaxon* **79**: 217, 2001; New World types), Mata & Petersen (*Mycol.* **93**: 1102, 2001; Americas), Zhang *et al.* (*Mycosystema* **24**: 517, 2005; molecular typing).

Lentinus Fr. (1825), Polyporaceae. 40, widespread (esp. subtropical). Wood (e.g. railway sleepers) is attacked by *L. lepideus*. See Pegler (*Kew Bull.* Addit. Ser. **10**: 1, 1983; monogr. s.l.), Hibbett & Vilgalys (*Mycol.* **83**: 425, 1991; s.str.), Hibbett & Vilgalys (*Syst. Bot.* **18**: 409, 1993) Not all species treated by Pegler have been reclassified.

Lentispora Fayod (1889) = Coprinopsis P. Karst. fide Redhead *et al.* (*Mycotaxon* **50**: 203, 2001).

Lentodiellum Murrill (1915) = Lentinus fide Kuyper (*in litt.*).

Lentodiopsis Bubák (1895) = Pleurotus fide Singer (*Agaric. mod. Tax.*, 1951).

Lentodium Morgan (1895) = Lentinus. Abnormal variants of *L. tigrinus*, with which it is sexually intercompatible. fide Rosinsky & Robinson (*Am. J. Bot.* **55**: 242, 1968; arising under the absence of light), Hibbett (*Am. J. Bot.* **81**: 466, 1994; developmental variant with a recessive allele at a single locus).

Lentomita Niessl (1876) = Chaetosphaeria fide Kohlmeyer & Kohlmeyer (*Marine Mycology*, 1979).

Lentomitella Höhn. (1906), Sordariomycetes. 1, widespread (temperate). See Hawksworth *et al.* (*Dictionary of the Fungi* edn 8, 1995), Réblová (*Mycol.* **98**: 68, 2006; phylogeny).

Lentus Lloyd ex Torrend (1920) ≡ Polyporus P. Micheli ex Adans. fide Donk (*Verh. K. ned. Akad. Wet.* tweede sect. **62**: 1, 1974).

Lenzitella Ryvarden (1991) = Lenzitopsis fide Stalpers (*Stud. Mycol.* **35**: 168 pp., 1993).

Lenzites Fr. (1836), Polyporaceae. 6, widespread. *L. sepiaria* causes timber rot. See Gilbertson & Ryvarden (*N. Amer. Polyp.* **1**: 424, 1987).

Lenzitina P. Karst. (1889) ≡ Gloeophyllum fide Donk (*Verh. K. ned. Akad. Wet.* tweede sect. **62**: 1, 1974).

Lenzitites Mesch. (1892), Fossil Fungi. 1 (Miocene), Italy.

Lenzitopsidaceae Jülich (1982) = Thelephoraceae.

Lenzitopsis Malençon & Bertault (1963), Thelephoraceae. 1, Morocco; Spain. See Stalpers (*Stud. Mycol.* **35**: 43, 1993; key).

Leohumicola N.L. Nickers. (2005), anamorphic *Leotiomycetes*. 4, Canada; Chile. See Nickerson (*Stud. Mycol.* **53**: 41, 2005).

Leolophia Klotzsch (1836) nom. inval., Basidiomycota.

Leotia Pers. (1794), Leotiaceae. 4, widespread (temperate). See Tai (*Lloydia* **7**: 146, 1944), Honegger (*Lichenologist* **15**: 57, 1983; TEM), Verkley (*Persoonia* **15**: 405, 1994; asci), Döring & Triebel (*Cryptog. Bryol.-Lichénol.* **19**: 123, 1998; DNA), Landvik *et al.* (*Mycoscience* **39**: 49, 1998), Lizoň *et al.* (*Mycotaxon* **67**: 73, 1998; posn), Gernandt *et al.* (*Mycol.* **93**: 915, 2001; phylogeny), Korf & Lizon (*Czech Mycol.* **52**: 255, 2001; nomencl.), Zhong & Pfister (*Mycol. Progr.* **3**: 237, 2004; phylogeny), Wang *et al.*

(*Mycol.* **98**: 1065, 2006; phylogeny).

Leotiaceae Corda (1842), Leotiales. 7 gen. (+ 11 syn.), 34 spp. The family is much more restricted in extent than in previous editions of this *Dictionary*; see also *Helotiaceae*.

Lit.: Verkley (*Persoonia* **15**: 405, 1994; asci), Gargas & Taylor (*Exp. Mycol.* **19**: 7, 1995), Döring & Triebel (*Cryptog. Bryol.-Lichénol.* **19**: 123, 1998), Gamundí & Romero (*Fl. criptog. Tierra del Fuego* **10**: 130 pp., 1998), Landvik *et al.* (*Mycoscience* **39**: 49, 1998), Lizoň *et al.* (*Mycotaxon* **67**: 73, 1998), Korf (*Mycotaxon* **73**: 493, 1999), Liu *et al.* (*Mol. Biol. Evol.* **16**: 1799, 1999), Zhong & Pfister (*Mycol. Progr.* **3**: 237, 2004), Wang *et al.* (*Mycol.* **98**: 1065, 2006; phylogeny), Wang *et al.* (*Mol. Phylogen. Evol.* **41**: 295, 2006).

Leotiales Korf & Lizoň (2001). Leotiomycetidae. 2 fam., 11 gen., 41 spp. Fams:
(1) **Bulgariaceae**
(2) **Leotiaceae**
For *Lit.* see under fam.

Leotiella Plöttn. (1900) ? = Cudonia fide Clements & Shear (*Gen. Fung.*, 1931), Nannfeldt (*Nova Acta R. Soc. Scient. Upsal.* IV **8**, 1932).

Leotiomycetes O.E. Erikss. & Winka (1997), Pezizomycotina. 5 ord., 19 fam., 641 gen., 5587 spp. The inoperculate discomycetes as traditionally circumscribed, but with the inclusion of the powdery mildews. Ords.:
(1) **Cyttariales**
(2) **Erysiphales**
(3) **Helotiales**
(4) **Leotiales**
(5) **Rhytismatales**
Lit.: Wang *et al.* (*Mycol.* **98**: 1065, 2006).
Still in need of major revision; the *Thelebolales* may belong here also.

Leotiomycetidae, see *Leotiomycetes*.

Lepadolemma Trevis. (1853) = Haematomma fide Hawksworth *et al.* (*Dictionary of the Fungi* edn 8, 1995).

Lepidella E.-J. Gilbert (1925) [non *Lepidella* Tiegh. 1911, *Loranthaceae*] = Amanita Adans. fide Bas (*Persoonia* **5**: 285, 1969).

Lepidocollema Vain. (1890) = Collema F.H. Wigg. fide Swinscow & Krog (*Lichenologist* **18**: 309, 1986), Hawksworth *et al.* (*Dictionary of the Fungi* edn 8, 1995).

Lepidogium Clem. & Shear (1931) ≡ Lepidoleptogium.

Lepidoleptogium A.L. Sm. (1922) = Pannaria fide Hawksworth *et al.* (*Dictionary of the Fungi* edn 8, 1995).

Lepidoma (Ach.) Gray (1821) ≡ Lecidoma.

Lepidoma Link (1809) ≡ Rhizocarpon.

Lepidomyces Jülich (1979), Pterulaceae. 2, Europe. See Jülich (*Persoonia* **10**: 330, 1979).

Lepidonectria Speg. (1910) ? = Ijuhya fide Rossman *et al.* (*Stud. Mycol.* **42**: 248 pp., 1999).

Lepidophyton Trib. (1899) nom. dub., anamorphic *Pezizomycotina*. No spp. Described.

Lepidopterella Shearer & J.L. Crane (1980), Argynnaceae. 1 (aquatic), USA. See Shearer & Crane (*TBMS* **75**: 194, 1980).

Lepidosphaeria Parg.-Leduc (1970), Testudinaceae. 1, Sahara. See Hawksworth (*CJB* **57**: 91, 1979), Kruys *et al.* (*MR* **110**: 527, 2006; phylogeny), Schoch *et al.* (*Mycol.* **98**: 1041, 2006; phylogeny).

Lepidostroma Mägd. & S. Winkl. (1967), Polyporales (L). 1, Europe. See Oberwinkler (*Dtsch. bot. Ges.* N.F. **4**: 139, 1970).

lepidote, covered with small scales.

Lepidotia Boud. (1885), Pezizaceae. 1, USA. See Korf (*Persoonia* **7**: 205, 1973), Norman & Egger (*Mycol.* **91**: 820, 1999; phylogeny), Hansen *et al.* (*Mycol.* **93**: 958, 2001; phylogeny).

Lepidotus Clem. (1902) ≡ Lepiota.

lepiochlorin, an antibacterial antibiotic from *Lepiota* cultivated by the gardening ant (*Cyphomyrmex costatus*); see Hervey & Nair (*Mycol.* **71**: 1064, 1979).

Lepiota (Pers.) Gray (1821), Agaricaceae. Anamorph *Coccobotrys*. *c.* 400, widespread. *L. scobinella* and several pink, red or orange spp. are very poisonous. See Kauffman (*Pap. Mich. Acad. Sci.* **4**: 311, 1924; N. Am. spp.), Beeli (*Bull. Soc. Bot. Belg.* **64**: 206, 1931), Beeli (*Fl. Icon. Champ. Congo* **2**: 29, 1936; Congo spp.), Dennis (*Kew Bull.* **7**: 459, 1952; W. Indian spp.), Aberdeen (*Kew Bull.* **16**: 129, 1962; Australian type studies), Pegler (*Kew Bull.* **27**: 155, 1972; key 34 Sri Lanka spp.), Pegler (*Kew Bull.* Addit. Ser. **6**, 1977; key 30 E. Afr. spp.), Candusso & Lanzoni (*Lepiota s.l.*, 1990; key Europ. spp.), Johnson & Vilgalys (*Mycol.* **90**: 971, 1998; phylogeny), Johnson (*Mycol.* **91**: 443, 1999; phylogeny), Vellinga (*Mycol. Progr.* **2**: 305, 2003; phylogeny), Vellinga (*MR* **108**: 354, 2004; phylogeny).

Lepiotasporites T.C. Huang (1981), Fossil Fungi. 1 (Miocene), Taiwan.

Lepiotella (E.-J. Gilbert) Konrad (1934) = Chamaemyces fide Singer (*Agaric. mod. Tax.* edn 3, 1975).

Lepiotella Rick (1938) ≡ Volvolepiota.

Lepiotophyllum Locq. (1942) nom. inval. [as subgenus?] = Leucocoprinus fide Kirk (*in litt.*).

Lepiotula (Maire) Locq. ex E. Horak (1968) = Lepiota fide Kuyper (*in litt.*).

Lepista (Fr.) W.G. Sm. (1870), Tricholomataceae. *c.* 50, widespread. Blewits. See Bigelow & Smith (*Brittonia* **21**: 144, 1969; as *Clitocybe*), Harmaja (*Karstenia* **14**: 82, 1974), Pegler & Young (*Kew Bull.* **29**: 659, 1974; basidiospore form), Harmaja (*Karstenia* **18**: 49, 1978; gen. division), Chiari (*Bollettino del Circolo Micologico 'Giovanni Carini'* **42**: 38, 2001), Consiglio & Contu (*Rivista di Micologia* **46**: 131, 2003; Italy), Stott *et al.* (*MR* **109**: 205, 2005; molec. analysis of Australian and European isolates).

Lepistella T.J. Baroni & Ovrebo (2007), Tricholomataceae. 1, C. America. See Ovrebo & Baroni (*Fungal Diversity* **27**: 157, 2007).

Lepocolla Eklund (1883) nom. dub., ? Actinobacteria. or anamorphic fungi.

Lepolichen Trevis. (1853) = Coccotrema fide Henssen *in* Brown *et al.* (Eds) (*Lichenology: progress and problems*: 107, 1976), Galloway & Watson-Gandy (*Bryologist* **95**: 227, 1992), Schmitt *et al.* (*Lichenologist* **33**: 315, 2001; phylogeny).

Leporina Velen. (1947) = Coprotus fide van Brummelen *in* Hawksworth (Ed.) (*Ascomycete Systematics. Problems and Perspectives in the Nineties* NATO ASI Series vol. **269 269**: 400, 1994).

Lepra Chevall. (1826) nom. dub., anamorphic *Pezizomycotina*.

Lepra Scop. (1777) nom. rej. = Pertusaria fide Hawksworth *et al.* (*Dictionary of the Fungi* edn 8, 1995).

Lepra Willd. (1787) = Illosporium fide Hawksworth *et al.* (*Dictionary of the Fungi* edn 8, 1995).

Leprantha Dufour ex Körb. (1855) = Arthonia fide

Hawksworth *et al.* (*Dictionary of the Fungi* edn 8, 1995).

Lepraria Ach. (1803) nom. cons., anamorphic *Stereocaulaceae*, ?.?.? (L (sterile)). *c.* 71, widespread. See Laundon (*Lichenologist* **24**: 315, 1992; key 9 Br. spp.), Lohtander (*Ann. bot. fenn.* **31**: 223, 1994; key 11 spp. Finland), Leuckert *et al.* (*Biblthca Lichenol.* **58**: 245, 1995; chemistry), Orange (*British Lichen Society Bulletin* **74**: 1, 1995; UK), Saag & Saag (*Folia cryptog. Estonica* **34**: 55, 1999; Estonia), Kukwa (*Biblthca Lichenol.* **82**: 67, 2001; Poland), Ekman & Tønsberg (*MR* **106**: 1262, 2002; phylogeny), Kukwa (*Ann. bot. fenn.* **39**: 225, 2002), Kukwa *et al.* (*Botanica Lithuanica* **9**: 259, 2003; Russia), Elix & Tønsberg (*Graphis Scripta* **16**: 43, 2004; chemistry), Sipman (*Herzogia* **17**: 23, 2004; tropical spp.), Myllys *et al.* (*Taxon* **54**: 605, 2005; phylogeny), Baruffo *et al.* (*Nova Hedwigia* **83**: 387, 2006; Italy), Kukwa (*Lichenologist* **38**: 293, 2006; Poland), Miądlikowska *et al.* (*Mycol.* **98**: 1088, 2006; phylogeny), Slavíková-Bayerová & Orange (*Lichenologist* **38**: 503, 2006), Flakus & Kukwa (*Lichenologist* **39**: 463, 2007; S America), Nelsen & Gargas (*Nova Hedwigia* **86**: 115, 2008; chemistry, phylogeny).

Leprariaceae Rchb. (1841) = Stereocaulaceae.

Leprieuria Laessøe, J.D. Rogers & Whalley (1989), Xylariaceae. Anamorph *Geniculosporium*. 1, C. & S. America (tropical). See Læssøe *et al.* (*MR* **93**: 152, 1989), San Martín & Rogers (*Mycotaxon* **48**: 179, 1993; Mexico), González *et al.* (*Mycol.* **96**: 675, 2004).

Leprieurina G. Arnaud (1918), anamorphic *Pezizomycotina*, Cpt.0eP.1. 1, S. America; Vietnam. See Reynolds (*Gdns' Bull.* Singapore **51**: 71, 1999).

Leprieurinella Bat. & H. Maia (1961), anamorphic *Pezizomycotina*, Cpt.?.?. 1, Brazil. See Batista & Maia (*Publções Inst. Micol. Recife* **338**: 18, 1961).

Leprocaulon Nyl. (1878), anamorphic *Lecanorales*, ?.?.? (L (sterile)). 10, widespread. See Lamb & Ward (*J. Hattori bot. Lab.* **38**: 499, 1974; key), Marcano *et al.* (*Trop. Bryol.* **13**: 47, 1997; Venezuela), Ekman & Tønsberg (*MR* **106**: 1262, 2002; phylogeny).

Leprocollema Vain. (1890), Lichinaceae (L). 3, widespread. See Magnusson (*Hedwigia* **78**: 219, 1938), Henssen (*SA* **5**: 137, 1986), Schultz & Büdel (*Lichenologist* **34**: 39, 2002; key).

Leproloma Nyl. ex Cromb. (1883), Stereocaulaceae (L (sterile)). 6, widespread. See Laundon (*Lichenologist* **21**: 1, 1989; key), Leuckert & Kümmerling (*Nova Hedwigia* **52**: 17, 1991; chemotax.), Jørgensen *in* Hawksworth (Ed.) (*Ascomycete Systematics. Problems and Perspectives in the Nineties* NATO ASI Series vol. **269 269**: 392, 1994; posn), Leuckert *et al.* (*Biblthca Lichenol.* **58**: 245, 1995; chemistry), Lohtander (*Ann. bot. fenn.* **32**: 49, 1995; Finland), Tønsberg & Jørgensen (*Lichenologist* **29**: 597, 1997; comments on alleged ascoma), Kukwa (*Biblthca Lichenol.* **82**: 67, 2001; Poland), Ekman & Tønsberg (*MR* **106**: 1262, 2002; phylogeny), Kukwa (*Ann. bot. fenn.* **39**: 225, 2002), Myllys *et al.* (*Taxon* **54**: 605, 2005; phylogeny).

Lepronicus Vent. (1799) nom. rej. = Pertusaria fide Hawksworth *et al.* (*Dictionary of the Fungi* edn 8, 1995).

Lepropinacia Vent. (1799) = Buellia fide Hawksworth *et al.* (*Dictionary of the Fungi* edn 8, 1995).

Leproplaca (Nyl.) Nyl. ex Hue (1888) = Caloplaca fide Jørgensen & Tønsberg (*Nordic Jl Bot.* **8**: 293, 1988), Kärnefelt (*Cryptog. Bot.* **1**: 147, 1989), Hafellner & Vězda (*Nova Hedwigia* **55**: 183, 1992).

leprose (of lichens), having the surface or the whole thallus entirely dissolved into soredia (e.g. *Lepraria*) (Fig. 21A).

Leprosis Neck. ex Kremp. (1869) nom. dub., Fungi (L).

Leptascospora Speg. (1918), Meliolaceae. 1, Java.

Leptasteromella Petr. (1968) ? = Leptodothiorella Höhn. fide Sutton (*Mycol. Pap.* **141**, 1977).

Lepteutypa Petr. (1923), Amphisphaeriaceae. Anamorph *Seiridium*. 10, widespread. See Shoemaker & Müller (*CJB* **43**: 1459, 1965), von Arx (*Gen. Fungi Sporul. Cult.* Edn 3, 1981), Nag Raj & Kendrick (*Sydowia* **38**: 178, 1986; key), Kang *et al.* (*Fungal Diversity* **1**: 147, 1998; DNA), Kang *et al.* (*MR* **103**: 53, 1999), Dulymamode *et al.* (*MR* **105**: 247, 2001; key), Jeewon *et al.* (*MR* **107**: 1392, 2003; phylogeny).

Lepteutypella Petr. (1925) = Lepteutypa fide Müller & von Arx *in* Ainsworth *et al.* (Eds) (*The Fungi* **4A**: 87, 1973).

Leptina Bat. & Peres (1960) ? = Discosia fide Sutton (*Mycol. Pap.* **141**, 1977).

Leptinia Juel (1897) = Puccinia fide Dietel (*Nat. Pflanzenfam.* **6**, 1928).

Leptobelonium Höhn. (1924) = Strossmayeria fide Iturriaga (*Mycotaxon* **20**: 169, 1984), Iturriaga & Korf (*Mycotaxon* **36**: 383, 1990).

Leptocapnodium (G. Arnaud) Cif. & Bat. (1963) = Scorias fide Reynolds (*Bull. Torrey bot. Club* **98**: 157, 1971).

Leptochaete Lév. (1846) ≡ Hymenochaete.

Leptochaete Zmitr. & Spirin (2006), Agaricomycetes. 5, widespread.

Leptochidium M. Choisy (1952), Massalongiaceae (L). 2, Europe; N. America. See Choisy (*Bull. mens. Soc. linn. Lyon* **21**: 165, 1952), Burgaz & Martínez (*Nova Hedwigia* **73**: 381, 2001; Iberian peninsula), Jørgensen (*Graphis Scripta* **18**: 19, 2006), Wedin *et al.* (*Lichenologist* **39**: 61, 2007; n. fam.).

Leptochlamys Died. (1921), anamorphic *Pezizomycotina*, St.0fH.10. 1 (on *Musci*), Europe.

Leptocladia Marvanová & Descals (1985) [non *Leptocladia* J. Agardh 1892, *Rhodophyta*] ≡ Stenocladiella.

Leptocorticium Hjortstam & Ryvarden (2002), Corticiaceae. 5, widespread. See Hjortstam & Ryvarden (*Syn. Fung.* **15**: 23, 2002), Nakasone (*Mycol. Progr.* **4**: 251, 2005).

Leptocoryneum Petr. (1925) = Seimatosporium fide Sutton (*Mycol. Pap.* **141**, 1977).

Leptocrea Syd. & P. Syd. (1916) ? = Stigmatula fide von Arx & Müller (*Beitr. Kryptfl. Schweiz* **11** no. 1, 1954).

Leptocucurthis Aptroot (1998), Dacampiaceae (L). 1, Papua New Guinea. See Aptroot & van Iperen (*Nova Hedwigia* **67**: 481, 1998).

leptocystidium, see cystidium.

Leptodendriscum Vain. (1890) = Polychidium fide Henssen (*Symb. bot. upsal.* **18** no. 1, 1963).

leptodermatous (of hyphae), having the outer wall thinner than the lumen; cf. mesodermatous.

Leptodermella Höhn. (1915), anamorphic *Pezizomycotina*, Cac.0eH.1. 1, Europe.

Leptodermopsis Speg. ex Höhn. (1923) nom. dub., anamorphic *Pezizomycotina*. See Falk (*Mykol. Unters.* **1**: 331, 1923), Sutton (*Mycol. Pap.* **141**, 1977).

Leptodiscella Papendorf (1969), anamorphic *Pezizo-*

mycotina, Hsp/Cac.1eH.10. 1, S. Africa; Japan. See Papendorf (*TBMS* **53**: 146, 1969), Udagawa & Toyazaki (*Mycotaxon* **22**: 407, 1985).

Leptodiscus Gerd. (1953) [non *Leptodiscus* Hertwig 1877, *Algae*] ≡ Mycoleptodiscus.

Leptodon Quél. (1886) [non *Leptodon* D. Mohr 1803, nom. cons., *Musci*] ≡ Mycoleptodon.

Leptodontidium de Hoog (1979), anamorphic *Helotiales*, Hso.0eH.10. 10, widespread. See de Hoog (*Taxon* **28**: 347, 1979), Tsuneda *et al.* (*Mycotaxon* **60**: 485, 1996), Tsuneda *et al.* (*CJB* **75**: 1649, 1997; antagonism to *Lentinula*), Grünig *et al.* (*CJB* **80**: 1239, 2002; phylogeny).

Leptodontium de Hoog (1977) [non *Leptodontium* (Müll. Hal.) Hampe ex Lindb. 1864, *Musci*] ≡ Leptodontidium.

Leptodothiora Höhn. (1918) = Dothiora Fr. (1849) fide Froidevaux (*Nova Hedwigia* **23**: 679, 1973).

Leptodothiorella Aa (1973) = Leptodothiorella Höhn. fide Sutton (*Mycol. Pap.* **141**, 1977).

Leptodothiorella Höhn. (1918), anamorphic *Guignardia*, Cpd.0eH.15. 6, widespread (tropical).

Leptodothis Theiss. & Syd. (1914) nom. dub., Fungi. See Müller & von Arx (*Beitr. Kryptfl. Schweiz* **11** no. 2, 1962).

lepto-form, see *Pucciniales*.

Leptogiaceae Körb. (1855) = Collemataceae.

Leptogidiomyces Cif. & Tomas. (1953) ≡ Leptogidium.

Leptogidium Nyl. (1873) = Polychidium fide Henssen (*Symb. bot. upsal.* **18** no. 1, 1963).

Leptogiomyces E.A. Thomas ex Cif. & Tomas. (1953) = Leptogium fide Hawksworth *et al.* (*Dictionary of the Fungi* edn 8, 1995).

Leptogiopsis Müll. Arg. (1882) = Leptogium fide Hawksworth *et al.* (*Dictionary of the Fungi* edn 8, 1995).

Leptogiopsis Nyl. (1884) = Mastodia fide Hawksworth *et al.* (*Dictionary of the Fungi* edn 8, 1995).

Leptogiopsis Trevis. (1880) = Leptogium fide Hawksworth *et al.* (*Dictionary of the Fungi* edn 8, 1995).

Leptogium (Ach.) Gray (1821), Collemataceae (L). *c.* 189, widespread. See Sierk (*Bryologist* **67**: 245, 1964; N. Am.), Jørgensen (*Svensk bot. Tidskr.* **67**: 53, 1973; *Mallotium*), Jørgensen *in* Poelt & Vězda (*Best. europ. Flecht.*, 1977; Eur.), Awasthi (*Geophytology* **8**: 189, 1979; India), Poelt & James (*Lichenologist* **15**: 109, 1983; W. Eur.), Jørgensen (*Lichenologist* **26**: 1, 1994; key small spp. Eur.), Jørgensen (*Symb. bot. upsal.* **32** no. 1: 113, 1997; key hairy spp.), Galloway (*Nova Hedwigia* **69**: 317, 1999; NZ), Jørgensen & Tønsberg (*Bryologist* **102**: 412, 1999; Pacific NW), Guttová (*Lichenologist* **32**: 291, 2000; Europe), Martin *et al.* (*Bryologist* **105**: 358, 2002; Pacific NW), Wiklund & Wedin (*Cladistics* **19**: 419, 2003; phylogeny), Aragón & Otálora (*Nova Hedwigia* **78**: 353, 2004; Iberian peninsula), Miądlikowska & Lutzoni (*Am. J. Bot.* **91**: 449, 2004; phylogeny), Aragón *et al.* (*Nova Hedwigia* **80**: 199, 2005; Iberian peninsula), Miądlikowska *et al.* (*Mycol.* **98**: 1088, 2006; phylogeny), Stenroos *et al.* (*Cladistics* **22**: 230, 2006; phylogeny, photobiont association), Lindström (*Biblthca Lichenol.* **95**: 405, 2007; neotropics), Jørgensen & Kashiwadani (*Lichenologist* **40**: 123, 2008; key, Papua New Guinea).

Leptoglossum (Cooke) Sacc. (1884) = Microglossum Gillet fide Maas Geesteranus (*Persoonia* **3**: 89, 1964).

Leptoglossum P. Karst. (1879) = Arrhenia fide Redhead (*CJB* **62**: 865, 1984).

leptogonidium, a photobiont consisting of small-sized cells (obsol.); cf. gonidium.

Leptographa (Th. Fr.) M. Choisy (1950) = Ptychographa fide Hawksworth *et al.* (*Dictionary of the Fungi* edn 8, 1995).

Leptographa Jatta (1892) = Toninia fide Hawksworth *et al.* (*Dictionary of the Fungi* edn 8, 1995).

Leptographium Lagerb. & Melin (1927), anamorphic *Ophiostoma*, Hso.0eH.21. *c.* 68, widespread (temperate). See Shaw & Hubert (*Mycol.* **44**: 693, 1952), Wingfield (*TBMS* **85**: 81, 1985; gen. relationships), Harrington & Cobb (*Leptographium Root Diseases on Conifers*, 1988), Ayer *et al.* (*J. Nat. Prod.* **52**: 119, 1989; metabolites from *L. wageneri* var. *pseudotsugae*), Zambino & Harrington (*Mycol.* **81**: 122, 1989; isozyme variation), Zambino & Harrington (*Mycol.* **84**: 12, 1992; isozyme characterization), Wingfield *in* Wingfield *et al.* (Eds) (*Ceratocystis* and *Ophiostoma* Taxonomy, Ecology and Pathogenicity: 43, 1993; anamorphs), Strydom *et al.* (*Syst. Appl. Microbiol.* **20**: 295, 1997; ribosomal DNA), Jacobs *et al.* (*CJB* **76**: 1660, 1998), Jacobs *et al.* (*CJB* **76**: 1660, 1998; on conifers), Jacobs *et al.* (*S. Afr. J. Bot.* **65**: 388, 1999; Congo), Jacobs *et al.* (*MR* **104**: 1524, 2000; Russia), Jacobs *et al.* (*Mycoscience* **41**: 595, 2000; Indonesia, N America), Masuya *et al.* (*Mycoscience* **41**: 425, 2000; Japan), Zhou *et al.* (*Mycoscience* **41**: 573, 2000; China), Jacobs & Wingfield (*Leptographium Species: Tree Pathogens, Insect Associates, and Agents of Blue-stain*, 2001), Jacobs *et al.* (*Mycol.* **93**: 380, 2001; Europe), Jacobs *et al.* (*CJB* **79**: 719, 2001; phylogeny), Jacobs *et al.* (*MR* **105**: 490, 2001; on pine), Six *et al.* (*Mycol.* **95**: 781, 2003; associations with bark beetles), Jacobs *et al.* (*MR* **108**: 411, 2004; introduction to N America), Kim *et al.* (*MR* **108**: 699, 2004; on *Pinus*), Masuya *et al.* (*Mycol.* **96**: 548, 2004; on *Prunus*, Japan), Jacobs *et al.* (*MR* **109**: 1149, 2005; phylogeny), Kim *et al.* (*MR* **109**: 275, 2005; Korea), Alamouti *et al.* (*Mycol.* **98**: 149, 2006; Canada), Jacobs *et al.* (*CJB* **84**: 759, 2006; N America).

Leptoguignardia E. Müll. (1955), Botryosphaeriaceae. Anamorph *Dothichiza*-like. 1, France. See Barr (*Prodr. Cl. Loculoasc.*, 1987).

Leptogydiomyces, see *Leptogidiomyces*.

Leptojum Beltr. (1858) = Leptogium fide Hawksworth *et al.* (*Dictionary of the Fungi* edn 8, 1995).

Leptokalpion Brumm. (1977), Thelebolaceae. 2, Asia. See Wang (*Bull. natn. Sci. Mus.* Tokyo, B **7**: 131, 1996), van Brummelen (*Persoonia* **16**: 425, 1998; posn), Hoog *et al.* (*Stud. Mycol.* **51**: 33, 2005).

Leptomassaria Petr. (1914), Xylariaceae. 2, Europe. See Læssøe (*SA* **13**: 43, 1994; ? synonym of *Anthostomella*), Rappaz (*Mycol. Helv.* **7**: 99, 1995).

Leptomelanconium Petr. (1923), anamorphic *Pezizomycotina*, Cac.0-1eP.15. 5, widespread. See Sutton (*The Coelomycetes*, 1980), Hunt (*CJB* **63**: 1157, 1985), Vujanovic & St-Arnaud (*Mycol.* **93**: 212, 2001; on *Abies*).

Leptomeliola Höhn. (1919), Parodiopsidaceae. Anamorph *Ophiotrichum*. 9, widespread (tropical). See Hughes (*Mycol. Pap.* **166**, 1993), Thaung (*Australas. Mycol.* **25**: 5, 2006; Burma).

Leptomyces Mont. (1856) nom. dub. ? = Mycena fide Kuyper (*in litt.*).

Leptomycorhaphis Cif. & Tomas. (1953) nom. rej.

prop. = Leptorhaphis fide Hawksworth *et al.* (*Dictionary of the Fungi* edn 8, 1995).

Leptonema J. Sm. Dalry (1896), Fossil Fungi (mycel.) Fungi. 1 (Carboniferous), British Isles. See Smith (*Trans. Geol. Soc. Glasgow* **10**: 321, 1896).

Leptonia (Fr.) P. Kumm. (1871) = Entoloma fide Kuyper (*in litt.*).

Leptoniella Earle (1909) ≡ Entoloma.

Leptopeltella Höhn. (1917) = Leptopeltis fide Holm & Holm (*Bot. Notiser* **130**: 215, 1977), Shoemaker *et al.* (*Sydowia* **41**: 308, 1989).

Leptopeltidaceae Höhn. ex Trotter (1928), Microthyriales. 6 gen. (+ 8 syn.), 14 spp.

Lit.: von Arx (*Acta Bot. Neerl.* **13**: 182, 1964), Holm & Holm (*Bot. Notiser* **130**: 215, 1977), Eriksson (*Op. Bot.* **60**, 1981), Spooner & Kirk (*MR* **94**: 223, 1990), Cannon (*SA* **15**: 121, 1997).

Leptopeltina Petr. (1947) ≡ Leptopeltinella.

Leptopeltina Speg. (1924) = Stomiopeltis fide von Arx & Müller (*Stud. Mycol.* **9**, 1975).

Leptopeltinella Petr. (1951) ≡ Leptopeltis.

Leptopeltis Höhn. (1917), Leptopeltidaceae. Anamorph *Leptothyrium*-like. 7 or 9, widespread (north temperate). See Holm & Holm (*Bot. Notiser* **130**: 215, 1977; key), Eriksson (*Op. Bot.* **no. 60**: 1, 1981; posn), Cannon (*SA* **15**: 121, 1997).

Leptopeltopsis Petr. (1947) = Leptopeltis fide Holm & Holm (*Bot. Notiser* **130**: 215, 1977).

Leptoperidia Rappaz (1987), Diatrypaceae. 4, widespread. See Rappaz (*Mycol. Helv.* **2**: 285, 1987), Rappaz (*Mycol. Helv.* **3**: 281, 1989), Chacón Zapata (*Brenesia* **62**: 41, 2004; key, Mexico).

Leptopeza G.H. Otth (1871) = Plicaria fide Eckblad (*Nytt Mag. Bot.* **15**: 1, 1968).

Leptopeziza Rostr. (1888) ? = Pyrenopeziza fide Nannfeldt (*Nova Acta R. Soc. Scient. upsal.*, 1932).

Leptophacidium Höhn. (1918) = Guignardia fide von Arx & Müller (*Beitr. Kryptfl. Schweiz* **11** no. 1, 1954).

Leptophoma Höhn. (1915) = Phoma Sacc. fide Sutton (*Mycol. Pap.* **141**, 1977).

Leptophrys Hertwig & Lesser (1885), Cercozoa. q.v.

Leptophyllosticta I.E. Brezhnev (1939), anamorphic *Pezizomycotina*, Cpd.0eH.?. 1, former USSR.

Leptophyma Sacc. (1889) = Helostroma fide Moore (*Mycotaxon* **14**: 13, 1982).

Leptopodia Boud. (1885) = Helvella fide Seaver (*North American Cup Fungi* (Operculates), 1928).

Leptopora Raf. (1808) = Perenniporia fide Merrill (*Index Rafinesq.*, 1949).

Leptoporellus Spirin (2001), Polyporaceae. 5, widespread. See Spirin (*Mycena* **1**: 69, 2001).

Leptoporus Quél. (1886), Polyporaceae. 1, widespread (north temperate). See Ryvarden & Gilbertson (*Europ. Polyp.* **1**: 281, 1993), Yu *et al.* (*Mycosystema* **23**: 596, 2004; China).

Leptopterygium Zahlbr. (1930) = Zahlbrucknerella fide Henssen (*Symb. bot. upsal.* **18** no. 1, 1963).

Leptopuccinia (G. Winter) Rostr. (1902) = Puccinia fide Dietel (*Nat. Pflanzenfam.* **6**, 1928).

lepto-Puccinia, a *Puccinia* of lepto-form; see *Pucciniales*.

Leptorhaphiomyces Cif. & Tomas. (1953) = Leptorhaphis fide Hawksworth *et al.* (*Dictionary of the Fungi* edn 8, 1995).

Leptorhaphis Körb. (1855) nom. cons., ? Naetrocymbaceae (±L). 15, widespread (northern hemisphere). See Aguirre-Hudson (*Bull. Br. Mus. nat. hist.* Bot. **21**: 85, 1991; key), Aguirre-Hudson & Fiol (*Lichenologist* **25**: 207, 1993; Balearic Is).

Leptosacca Syd. (1928), Pezizomycotina. 1, Chile.

Leptosillia Höhn. (1928), Diaporthales. Anamorph *Harpostroma*. 1, Europe. See Hawksworth (*SA* **6**: 136, 1987).

Leptosphaerella (Sacc.) Hara (1913) = Phaeosphaeria fide Shoemaker & Babcock (*CJB* **67**: 1500, 1989), Hawksworth *et al.* (*Dictionary of the Fungi* edn 8, 1995).

Leptosphaerella Speg. (1909), Pezizomycotina. 7, S. America.

Leptosphaeria Ces. & De Not. (1863) nom. cons., Leptosphaeriaceae. Anamorphs *Coniothyrium*, *Phoma*. *c.* 134, widespread. *L. bondari* (areolate spot of citrus in S. Am.), *L. coniothyrium* (rose stem canker, raspberry cane blight), *L. maculans* (anamorph *Phoma lingam*; dry rot of turnips, canker of crucifers). At one time a much more widely circumscribed genus, see segregates esp. *Phaeosphaeria*, many pathogens of grasses. See Müller (*Sydowia* **4**: 185, 1950), Müller (*Sydowia* **5**: 49, 1951; Swiss spp.), Wehmeyer (*Mycol.* **44**: 621, 1952), Holm (*Symb. bot. upsal.* **14** no. 3: 1, 1957), Holm (*Taxon* **24**: 475, 1975; nomencl.), Shoemaker (*CJB* **62**: 2688, 1984; key 64 spp.), Crane & Shearer (*Illinois nat. Hist. Bull.* **34**, 1991; disposition 1689 names), Huhndorf (*Illinois nat. Hist. Bull.* **34**: 479, 1992; spp. on *Rosaceae*), Ahn & Shearer (*CJB* **73**: 573, 1995), Khashnobish *et al.* (*Mycotaxon* **54**: 91, 1995), Morales *et al.* (*MR* **99**: 593, 1995), Khashnobish & Shearer (*MR* **100**: 1341, 1996), Khashnobish & Shearer (*MR* **100**: 1355, 1996; phylogeny), Ahn & Shearer (*CJB* **76**: 258, 1998), Dong *et al.* (*MR* **102**: 151, 1998; DNA), Voigt *et al.* (*J. Phytopath.* **146**: 567, 1998; RAPDs of *L. maculans*), Ahn & Shearer (*Mycol.* **91**: 684, 1999; *L. vagabunda* group), Pongam *et al.* (*Pl. Dis.* **83**: 149, 1999; genetic variation in *L. maculans*), Purwantara *et al.* (*MR* **104**: 772, 2000; genetic diversity in *L. maculans*), Howlett *et al.* (*Fungal Genetics Biol.* **33**: 1, 2001; *L. maculans*, review), Voigt *et al.* (*Microbiol. Res.* **156**: 169, 2001; *L. maculans*, Poland), Câmara *et al.* (*Mycol.* **94**: 630, 2002; phylogeny), Kuusk *et al.* (*J. Phytopath.* **150**: 349, 2002; *L. maculans*, Sweden), Moreno-Rico *et al.* (*Can. J. Pl. Path.* **24**: 69, 2002; *L. maculans*, Mexico), Mendes-Pereira *et al.* (*MR* **107**: 1287, 2003; phylogeny, *L. maculans* complex), Barrins *et al.* (*Australas. Pl. Path.* **33**: 529, 2004; *L. maculans*, Australia), Abo *et al.* (*Phytopathology* **95**: 1391, 2005; genetic diversity), Voigt *et al.* (*Mol. Phylogen. Evol.* **37**: 541, 2005; phylogeny, *L. maculans*), Schoch *et al.* (*Mycol.* **98**: 1041, 2006; phylogeny), Vincenot *et al.* (*Phytopathology* **98**: 321, 2008; Australia).

Leptosphaeriaceae M.E. Barr (1987), Pleosporales. 8 gen. (+ 19 syn.), 302 spp.

Lit.: Holm (*Symb. bot. upsal.* **14** no. 3: 1, 1957), Petersen (*TBMS* **50**: 641, 1967), Shoemaker (*CJB* **62**: 2688, 1984), Barr (*Prodr. Cl. Loculoasc.*: 168 pp., 1987), Shearer *et al.* (*Fourth International Mycological Congress Abstracts*: 156, 1990), Crane & Shearer (*Bull. Ill. St. nat. Hist. Surv.* **34**: 1, 1991), Crane & Shearer (*Illinois nat. Hist. Bull.* **34**, 1991; disposition 1689 names), Huhndorf (*Bull. Ill. nat. Hist. Surv.* **34**: 475, 1992), Yuan & Barr (*Mycotaxon* **52**: 495, 1994), Khashnobish *et al.* (*Mycotaxon* **54**: 91, 1995), Morales *et al.* (*MR* **99**: 593, 1995), Khashnobish &

Shearer (*MR* **100**: 1355, 1996; phylogeny), Mahuku *et al.* (*CJB* **75**: 1485, 1997), Ahn & Shearer (*CJB* **76**: 258, 1998), Dong *et al.* (*MR* **102**: 151, 1998; DNA), Ahn & Shearer (*Mycol.* **91**: 684, 1999), Olivier *et al.* (*Mycol.* **92**: 736, 2000), Purwantara *et al.* (*MR* **104**: 772, 2000), Câmara *et al.* (*Mycol.* **94**: 630, 2002), Boerema *et al.* (*Phoma Identification Manual* Differentiation of Specific and Infra-Specific Taxa in Culture: 470 pp., 2004), Kodsueb *et al.* (*Mycol.* **98**: 571, 2006).

Leptosphaeriopsis Berl. (1892) = Ophiobolus fide Walker (*Mycotaxon* **11**: 1, 1980).

Leptosphaerites Richon (1885), Fossil Fungi. 2 (Tertiary), France; former USSR.

Leptosphaerulina McAlpine (1902), Pleosporales. Anamorph *Pithoascus*. 27, widespread (temperate). See Graham & Luttrell (*Phytopathology* **51**: 680, 1961; key), Crivelli (*Über die heterogene Ascomycetengattung Pleospora Rbh.*, 1983; key), Irwin & Davis (*Aust. J. Bot.* **33**: 233, 1985; Australia), Roux (*TBMS* **86**: 319, 1986; anamorph), Wu & Hanlin (*Mycol.* **84**: 241, 1992; ontogeny), Silva-Hanlin & Hanlin (*MR* **103**: 153, 1999; DNA), Kodsueb *et al.* (*Mycol.* **98**: 571, 2006; phylogeny).

Leptospora Rabenh. (1857), Dothideomycetes. 1, Europe; N. America. See Walker (*Mycotaxon* **11**: 1, 1980).

Leptospora Raf. (1808) nom. dub., Fungi. See Donk (*Persoonia* **1**: 173, 1960).

leptospore (of *Pucciniales*), a teliospore (q.v.) adapted for immediate germination without a dormant period.

Leptosporella Penz. & Sacc. (1897), Sordariomycetes. 2, Java; S. America. See Huhndorf *et al.* (*Mycol.* **96**: 368, 2004; phylogeny).

Leptosporina Chardón (1939), Pezizomycotina. 1, S. America.

Leptosporium (Sacc.) Höhn. (1923) = Fusicolla fide Clements & Shear (*Gen. Fung.*, 1931).

Leptosporium Bonord. (1857) = Vibrissea fide Sánchez & Korf (*Mycol.* **58**: 733, 1966).

Leptosporomyces Jülich (1972), Atheliaceae. 11, widespread (north temperate). See Greslebin & Rajchenberg (*N.Z. Jl Bot.* **41**: 437, 2003; Patagonia).

Leptosporopsis Höhn. (1920) = Leptosphaeria fide Walker (*Mycotaxon* **11**: 1, 1980).

Leptostroma Fr. (1815), anamorphic *Lophodermium*, St.0eH.10. *c.* 200, widespread. See Minter (*CJB* **58**: 906, 1980; spp. on *Pinus*), Sieber-Canavesi *et al.* (*Mycol.* **83**: 89, 1991; ecology endophytic spp.), Guo *et al.* (*MR* **107**: 680, 2003; endophytes), Ortiz-García *et al.* (*Mycol.* **95**: 846, 2003; phylogeny).

Leptostromataceae Sacc. (1884) = Rhytismataceae.

Leptostromella (Sacc.) Sacc. (1884), anamorphic *Pezizomycotina*, St.0e-fH.?. 20, widespread (temperate).

Leptostromites Poinar (2003), Fossil Fungi. 1, Dominican Republic. See Poinar (*MR* **107**: 121, 2003).

Leptoteichion Kleb. (1933) = Phacidiopycnis fide Rupprecht (*Sydowia* **13**: 10, 1959).

Leptothyrella Sacc. (1885), anamorphic *Pezizomycotina*, Cpt.1eH.?. 10, widespread.

Leptothyrina Höhn. (1915), anamorphic *Pezizomycotina*, St.0eH.10. 1, Switzerland.

Leptothyriomyces Kräusel (1929), Fossil Fungi, Asterinaceae. 2 (Miocene), Sumatra.

Leptothyrites Poinar (2003), Fossil Fungi. 1, Dominica. See Poinar (*MR* **107**: 121, 2003).

Leptothyrium Kunze (1823), anamorphic *Pezizomycotina*, Cpt.0eH.15. 2, widespread. See Sutton (*Mycol. Pap.* **141**, 1977).

leptotichous (of tissue), thin-walled.

Leptotrema Mont. & Bosch (1855), Thelotremataceae (L). 1, pantropical. See Hale (*Bull. Br. Mus. nat. hist.* Bot. **8**: 227, 1981), Matsumoto & Deguchi (*Bryologist* **102**: 86, 1999; anamorph), Matsumoto (*J. Hattori bot. Lab.* **88**: 1, 2000; Japan), Kalb (*Mycotaxon* **79**: 319, 2001; Australia), Frisch (*Biblthca Lichenol.* **92**: 3, 2006; monograph).

Leptotrichum Corda (1842) nom. dub., anamorphic *Pezizomycotina*. See Holubová-Jechová (*Sydowia* **45**: 95, 1993).

Leptotrimitus Pouzar (1966) = Incrustoporia fide Donk (*Verh. K. ned. Akad. Wet.* tweede sect. **62**: 1, 1974).

Leptotrochila P. Karst. (1871), ? Dermateaceae. Anamorph *Sporonema*. 16, widespread. See Schüepp (*Phytopath. Z.* **36**: 236, 1959; key), Romaszewska-Salata & Salata (*Scripta Mycologica* Tartu **17**: 102, 1989; Poland), Nauta & Spooner (*Mycologist* **14**: 65, 2000; key British spp.).

Leptotus P. Karst. (1879) = Arrhenia fide Redhead (*CJB* **62**: 865, 1984).

Leptoxyphium Speg. (1918), anamorphic *Aithaloderma*, Hsy.0-1eH-P.15. 2, widespread (tropical). See Hughes (*Mycol.* **68**: 693, 1976), Roquebert & Bury (*CJB* **66**: 2265, 1988; ultrastr. conidiomata), Olejnik & Ingrouille (*MR* **103**: 333, 1999; numerical taxonomy), Reynolds & Faull (*Taxon* **50**: 1183, 2001; nomencl.).

Leptuberia Raf. (1808), Fungi (L). 1, N. America.

Le-Ratia Pat. (1907) [non *Le-Ratia* Broth. & Paris 1909, *Musci*] = Leratiomyces fide Heim (*Revue Mycol.* Paris **33**: 137, 1968).

Leratiomyces Bresinsky & Binder ex Bridge, Spooner, Beever & Park (2008), Strophariaceae. 4, New Caledonia. Basidioma gasteroid. See Bresinsky & Binder (*Z. Mykol.* **64**: 79, 1998; phylogeny).

Lesdainea Harm. (1910) nom. dub., ? Verrucariaceae. 1, France. See Hawksworth *et al.* (*Dictionary of the Fungi* edn 8, 1995), Ertz & Diederich (*Mycol. Progr.* **3**: 229, 2004).

lesion, a wound; a well-marked but limited diseased-area.

Lespiaultinia Zobel (1854) = Tuber fide Trappe (*Mycotaxon* **9**: 247, 1979), Læssøe & Hansen (*MR* **111**: 1075, 2007; phylogeny).

Letendraea Sacc. (1880), Pleosporales. 2, Europe; Africa. See Samuels (*CJB* **51**: 1275, 1973), Rossman (*Mycol. Pap.* **157**, 1987; key), Lumbsch & Lindemuth (*MR* **105**: 901, 2001; phylogeny), Kodsueb *et al.* (*Fungal Diversity* **21**: 105, 2006; phylogeny, morphology).

Letendraeopsis K.F. Rodrigues & Samuels (1994), ? Tubeufiaceae. 1, Brazil. See Rodrigues & Samuels (*Mycol.* **86**: 254, 1994), Crane *et al.* (*CJB* **76**: 602, 1998), Kodsueb *et al.* (*Fungal Diversity* **21**: 105, 2006).

Lethagrium A. Massal. (1853) ≡ Lathagrium.

Letharia (Th. Fr.) Zahlbr. (1892) nom. cons., Parmeliaceae (L). 2, widespread. See Schade (*Ber. bayer. bot. Ges.* **30**: 108, 1954), Schade (*Feddes Repert.* **58**: 179, 1955), Kroken & Taylor (*Mycol.* **93**: 38, 2001; phylogeny), Högberg *et al.* (*Mol. Ecol.* **11**: 1191, 2002; genetic variation), Arnerup *et al.* (*MR* **108**: 311, 2004; population structure), Thell *et al.* (*Symb. bot. upsal.* **34** no. 1: 429, 2004; biogeography).

Lethariella (Motyka) Krog (1976), Parmeliaceae (L). 10, widespread. See Obermayer (*Biblthca Lichenol.* **68**: 45, 1997), Obermayer (*Biblthca Lichenol.* **78**: 321, 2001), Niu *et al.* (*Lichenologist* **39**: 549, 2007; chemistry, China).

Lethariicola Grummann (1969) = Odontotrema fide Lumbsch & Hawksworth (*Biblthca Lichenol.* **38**: 325, 1990; key), Diederich *et al.* (*Lichenologist* **34**: 479, 2002).

Lethariopsis Zahlbr. (1926) = Caloplaca fide Eriksson & Hawksworth (*SA* **10**: 141, 1991).

Letrouitia Hafellner & Bellem. (1982), Letrouitiaceae (L). 15, widespread. See Hafellner (*Nova Hedwigia* **35**: 645, 1983; key), Awasthi & Srivastava (*Proc. Indian Acad. Sci.* Pl. Sci. **99**: 165, 1989; India), Johansson *et al.* (*Mycol. Progr.* **4**: 139, 2005; chemistry), Miądlikowska *et al.* (*Mycol.* **98**: 1088, 2006; phylogeny).

Letrouitiaceae Hafellner & Bellem. (1982), Teloschistales (L). 1 gen., 15 spp.

Lit.: Hafellner & Bellemère (*Nova Hedwigia* **35**: 263, 1982), Awasthi & Srivastava (*Proc. Indian Acad. Sci.* Pl. Sci. **99**: 165, 1989), Kärnefelt (*Cryptog. bot.* **1**: 147, 1989), Kärnefelt (*Lichenologist* **22**: 307, 1990), Kasalicky *et al.* (*CJB* **78**: 1580, 2000), Miądlikowska *et al.* (*Mycol.* **98**: 1088, 2006; phylogeny).

Lettauia D. Hawksw. & R. Sant. (1990), ? Fuscideaceae. 1 (on lichens, esp. *Cladonia*), Europe. See Ihlen & Tønsberg (*Bryologist* **99**: 32, 1996), Etayo (*Biblthca Lichenol.* **84**, 2002; Colombia).

Leucangium Quél. (1883), ? Helvellaceae. 1 (hypogeous), widespread (temperate). Clusters close to *Fischerula* and is perhaps related to *Morchellaceae*. See Li (*Int. J. Pl. Sci.* **158**: 189, 1997; ultrastr.), O'Donnell *et al.* (*Mycol.* **89**: 48, 1997; phylogeny), Hansen & Pfister (*Mycol.* **98**: 1029, 2006; phylogeny), Læssøe & Hansen (*MR* **111**: 1075, 2007; phylogeny).

Leucinocybe, see *Leucoinocybe*.

Leucoagaricus Locq. ex Singer (1948), Agaricaceae. Anamorph *Attamyces*. *c.* 90, widespread. The genus as presently circumscribed is paraphyletic. See Heinemann (*Fl. Illustr. Champ. Afr. Centr.* **2**: 30, 1973; key 18 Afr. spp.), Johnson (*Mycol.* **91**: 443, 1999), Vellinga (*Mycotaxon* **76**: 429, 2000; type studies), Hausknecht & Pidlich-Aigner (*Öst. Z. Pilzk.* **13**: 1, 2004; Austria), Consiglio & Contu (*Bollettino del Circolo Micologico 'Giovanni Carini'* **48**: 16, 2004), Consiglio *et al.* (*Micologia e Vegetazione Mediterranea* **19**: 131, 2004; Italy).

Leucobolbitius (J.E. Lange) Locq. (1952) nom. inval. = Leucocoprinus fide Kuyper (*in litt.*).

Leucobolites Beck (1923) = Gyroporus fide Kuyper (*in litt.*).

Leucocarpia Vězda (1969), Verrucariaceae (L). 1, Europe; Papua New Guinea. See Vězda (*Herzogia* **1**: 188, 1969).

Leucocarpopsis G. Salisb. (1975) = Verrucaria Schrad. fide Purvis *et al.* (*Lichen Flora of Great Britain and Ireland*, 1992).

Leucocintractia M. Piepenbr., Begerow & Oberw. (1999), Anthracoideaceae. 4 (on *Cyperaceae*), widespread. See Pérez *et al.* (*IMI Descr. Fungi Bact.* **153**, 2002), Piepenbring (*Caldasia* **24**: 103, 2002; Colombia), Pérez & Minter (*IMI Descr. Fungi Bact.* **164**, 2005).

Leucoconiella Bat., H. Maia & Peres (1960), Pezizomycotina. 1, Paraguay. See Batista *et al.* (*Brotéria* Sér. Ci. Nat. **29**: 130, 1960).

Leucoconis Theiss. & Syd. (1918), Pezizomycotina. 1, India.

Leucoconius Beck (1923) ≡ Gyroporus.

Leucocoprinaceae Jülich (1982) = Agaricaceae.

Leucocoprinus Pat. (1888), Agaricaceae. *c.* 40, widespread (esp. tropical; in greenhouses in temperate regions). See Johnson (*Mycol.* **91**: 443, 1999; phylogeny), Hausknecht & Pidlich-Aigner (*Öst. Z. Pilzk.* **13**: 1, 2004; Austria), Kumar & Manimohan (*Mycotaxon* **90**: 393, 2004; India).

Leucocortinarius (J.E. Lange) Singer (1945), Tricholomataceae. 1, Europe.

Leucocrea Sacc. & P. Syd. ex Lindau (1900) = Balzania fide Rossman *et al.* (*Stud. Mycol.* **42**: 248 pp., 1999).

Leucocytospora (Höhn.) Höhn. (1927) = Cytospora fide Petrak (*Annls mycol.* **19**: 17, 1921), Adams *et al.* (*Mycol.* **94**: 947, 2002; phylogeny, teleomorph), Adams *et al.* (*Stud. Mycol.* **52**: 146 pp., 2005; phylogeny, morphology).

Leucodecton A. Massal. (1860), Thelotremataceae (L). 1, pantropical. See Thor (*Op. Bot.* **103**, 1990), Frisch (*Biblthca Lichenol.* **92**: 3, 2006; revision), Frisch *et al.* (*Biblthca Lichenol.* **92**: 517, 2006; phylogeny).

Leucodiaporthe M.E. Barr & Lar.N. Vassiljeva (2008), Diaporthaceae. 1, N. America; E. Asia. See Vasilyeva *et al.* (*Mycol.* **99**: 916, 2007).

Leucodochium Syd. & P. Syd. (1917), anamorphic *Pezizomycotina*, Cpd.0eH.?. 1, Philippines.

Leucofomes Kotl. & Pouzar (1957) = Rigidoporus fide Pegler (*The polypores [Bull. BMS Suppl.]*, 1973).

Leucogaster R. Hesse (1882), Albatrellaceae. 20, widespread (north temperate). See Fogel (*CJB* **57**: 1718, 1979), Francis & Bougher (*Australasian Mycologist* **21**: 81, 2002).

Leucogastraceae Moreau ex Fogel (1979) = Albatrellaceae.

Leucogastrales = Boletales. See also under gasteromycetes.

Leucogloea R. Kirschner (2004), anamorphic *Atractiellales*. 1. See Kirschner (*Frontiers in Basidiomycote Mycology*: 177, 2004).

Leucoglossum S. Imai (1942), ? Geoglossaceae. 1, Japan. See Rifai (*Lloydia* **28**: 113, 1965).

Leucogomphidius Kotl. & Pouzar (1972) ≡ Gomphidius.

Leucogramma A. Massal. (1860) = Hemithecium fide Staiger (*Biblthca Lichenol.* **85**, 2002).

Leucogramma G. Mey. (1825) nom. rej. prop. = Graphina fide Staiger (*Biblthca Lichenol.* **85**, 2002).

Leucographa Nyl. (1857) nom. nud., Dothideomycetes (L).

Leucogymnospora Fink (1930), ? Ostropales (L). 1, Puerto Rico.

Leucogyrophana Pouzar (1958), Hygrophoropsidaceae. 13, widespread. See Ginns & Weresub (*Mem. N. Y. bot. Gdn* **28**: 86, 1976; sclerotial spp.), Ginns (*CJB* **56**: 1953, 1978; key), Jarosch & Besl (*Pl. Biol.* **3**: 443, 2001).

Leucogyroporus Snell (1942) = Tylopilus fide Singer (*Farlowia* **2**: 223, 1945).

Leucoinocybe Singer (1943), ? Tricholomataceae. 1, Europe.

Leucolenzites Falck (1909) ≡ Lenzites.

Leucoloma Fuckel (1870) [non *Leucoloma* Brid. 1827) nom. cons., *Musci*] ≡ Octospora.

Leucomyces Earle (1909) = Amanita Pers. fide Singer (*Agaric. mod. Tax.*, 1951).

Leuconectria Rossman, Samuels & Lowen (1993), Nectriaceae. Anamorph *Gliocephalotrichum*. 1 (from leaves and in soil), widespread. See Rehner & Samuels (*CJB* **73** Suppl. 1: S816, 1995; phylogeny), Rossman *et al.* (*Stud. Mycol.* **42**: 248 pp., 1999), Schoch *et al.* (*Stud. Mycol.* **45**: 45, 2000), Rossman *et al.* (*Mycol.* **93**: 100, 2001; phylogeny), Decock *et al.* (*Mycol.* **98**: 488, 2006; French Guiana).

Leuconeurospora Malloch & Cain (1970), ? Pseudeurotiaceae. 2, Europe; Asia. See Udagawa *in* Subramanian (Ed.) (*Taxonomy of fungi* **1**: 225, 1978), von Arx *et al.* (*Beih. Nova Hedwigia* **94**, 1988; posn), Suh & Blackwell (*Mycol.* **91**: 836, 1999; phylogeny), Sogonov *et al.* (*Mycol.* **97**: 695, 2005; phylogeny).

Leucopaxillaceae Jülich (1982) nom. rej. = Tricholomataceae.

Leucopaxillus Boursier (1925), Tricholomataceae. *c.* 15, widespread (temperate; subtropical). See Singer & Smith (*Pap. Mich. Acad. Sci.* **28**: 85, 1943; key 12 spp.), Singer & Smith (*Mycol.* **39**: 725, 1947), Guzmán & Escalona (*Docums Mycol.* **33** no. 132: 37, 2004; *Leucopaxillus gracillimus* variation and distribution).

Leucopenicillifer G.R.W. Arnold (1971), anamorphic *Pezizomycotina*, Hso.0eH.15. 1 (on myxomycete), Russia. See Carmichael *et al.* (*Genera of Hyphomycetes*, 1980; ? = *Paecilomyces*).

Leucopezis Clem. (1909) ? = Humaria Fuckel fide Eckblad (*Nytt Mag. Bot.* **15**: 1, 1968).

Leucophellinus Bondartsev & Singer (1944), Schizoporaceae. 1, Europe. See Parmasto (*Eesti NSV Tead. Akad. Toim.* Biol. seer **32**: 269, 1983).

Leucophlebs, see *Leucophleps*.

Leucophleps Harkn. (1899), Albatrellaceae. 4, widespread (north temperate). See Fogel (*CJB* **57**: 1718, 1979), Montecchi & Sarasini (*Funghi Ipogei d'Europa*: 714 pp., 2000).

Leucopholiota (Romagn.) O.K. Mill., T.J. Volk & Bessette (1996), Tricholomataceae. 1, USA. See Miller *et al.* (*Mycol.* **88**: 138, 1996), Vellinga (*MR* **108**: 354, 2004; posn.).

Leucophomopsis Höhn. (1917) = Phomopsis (Sacc.) Bubák. fide Clements & Shear (*Gen. Fung.*, 1931).

Leucoporus Quél. (1886) ≡ Polyporellus.

Leucopus P. Kumm. (1871) = Cortinarius fide Donk (*Bull. Jard. Bot. Buitenzorg* Sér. 3 **18**: 271, 1949).

Leucorhizon Velen. (1925) = Gastrosporium fide Pilát (*Fl. ČSR* B **1**: 227, 1958).

Leucoscypha Boud. (1885), Pyronemataceae. 10, widespread (temperate). See Svrček (*Česká Mykol.* **28**: 129, 1974), Harmaja (*Karstenia* **17**: 73, 1977), Yao & Spooner (*MR* **99**: 1513, 1995; Brit. spp.), Benkert (*Z. Mykol.* **66**: 181, 2000; typification), Wang & Zhong (*Fungal Science* Taipei **16**: 17, 2001; China), Perry *et al.* (*MR* **111**: 549, 2007; phylogeny).

Leucosphaera Arx, Mukerji & N. Singh (1978) [non *Leucosphaera* Gilg 1897, *Spermatophyta*] ≡ Leucosphaerina.

Leucosphaerina Arx (1987), ? Bionectriaceae. Anamorphs *Acremonium*-like, *Sporothrix*. 1 (coprophilous), India; USA. See Malloch (*Stud. Mycol.* **31**: 107, 1989), Suh & Blackwell (*Mycol.* **91**: 836, 1999; phylogeny), Rossman *et al.* (*Mycol.* **93**: 100, 2001; phylogeny).

Leucosporidiaceae Jülich (1982), Leucosporidiales. 3 gen., 8 spp.

Lit.: Joo (*Korean J. Mycol.* **19**: 258, 1991), Suh & Sugiyama (*J. gen. appl. Microbiol.* Tokyo **39**: 257, 1993), Statzell-Tallman & Fell *in* Kurtzman & Fell (Eds) (*Yeasts, a taxonomic study* 4th edn: 670, 1998), DePriest *et al.* (*CJB* **78**: 1450, 2000), Sampaio *et al.* (*Mycol. Progr.* **2**: 63, 2003).

Leucosporidiales J.P. Samp., M. Weiss & R. Bauer (2003). Microbotryomycetes. 1 fam., 3 gen., 8 spp. With colacosomes, teliospores, and white cultures. Fam.:

Leucosporidiaceae

For *Lit.* see under fam.

Leucosporidiella Samp. (2003), anamorphic *Leucosporidiaceae*. 4, widespread. See Sampaio *et al.* (*Mycol. Progr.* **2**: 63, 2003), Aime *et al.* (*Mycol.* **98**: 896, 2006; phylogeny).

Leucosporidium Fell, Statzell, I.L. Hunter & Phaff (1970), Leucosporidiaceae. 3 (from sea water), Antarctica. See Yamada & Komagata (*J. gen. appl. Microbiol.* Tokyo **33**: 456, 1987), Fell *et al.* (*Int. J. Syst. Evol. Microbiol.* **50**: 1351, 2000; mol. phylogeny), Sampaio *et al.* (*Mycol. Progr.* **2**: 53, 2003), Sampaio (*Frontiers in Basidiomycote Mycology*: 49, 2004).

Leucosporium Corda (1833) nom. dub., anamorphic *Pezizomycotina*. See Holubová-Jechová (*Sydowia* **45**: 97, 1993).

leucosporous, having spores white in the mass.

Leucostoma (Nitschke) Höhn. (1917), Valsaceae. Anamorph *Cytospora*. *c.* 13, widespread (temperate). The genus should probably be placed into synonymy with *Valsa*. See also IMI Descriptions of Fungi and Bacteria nos 1363-1635, 1999. See Kern (*Phytopath. Z.* **40**: 303, 1961), Surve-Iyer *et al.* (*Mycol.* **87**: 471, 1995; isozymes), Wang *et al.* (*Phytopathology* **88**: 376, 1998; DNA), Adams *et al.* (*Mycol.* **94**: 947, 2002; phylogeny, paraphyly), Adams *et al.* (*Stud. Mycol.* **52**: 146 pp., 2005; phylogeny, paraphyly), Zhang *et al.* (*Mycol.* **98**: 1076, 2006; phylogeny).

Leucotelium Tranzschel (1935), Uropyxidaceae. 3 (on *Ranunculaceae* (I); on *Prunus* (II, III) (on *Rosaceae*)), Eurasia (temperate). See Savile (*CJB* **67**: 2983, 1989), Cummins & Hiratsuka (*Illustr. Gen. Rust Fungi* edn 3: 225 pp., 2003; syn. of *Soratаea*), Helfer (*Nova Hedwigia* **81**: 325, 2005; Europ. spp.).

Leucothallia Trevis. (1853) ≡ Sphaerotheca Lév. fide Braun (*SA* **7**: 57, 1988).

Leucothecium Arx & Samson (1973), Gymnoascaceae. 2, Netherlands. See Valldosera *et al.* (*MR* **95**: 243, 1991).

Leucothyridium Speg. (1909) = Cucurbitaria fide Petrak & Sydow (*Annls mycol.* **32**: 1, 1934).

Leucovibrissea (A. Sánchez) Korf (1990), Vibrisseaceae. 1, USA. See Korf (*Mycosystema* **3**: 19, 1990).

Leuliisinea Matsush. (1985), anamorphic *Pezizomycotina*, Hsy.1eH/1eP.19. 1, Taiwan. See Matsushima (*Matsush. Mycol. Mem.* **4**: 11, 1985).

Léveillé (Joseph-Henri; 1796-1870; France). A Parisian medic; Dr med. (1824). Introduced the terms basidium and cystidium; produced an early account of the mycota of large parts of eastern and southeast Europe; early work on powdery mildews. Most collections and papers destroyed during the Franco-Prussian war (1870-1871), but some types in **K**, **L** and [mostly] **PC**. *Publs*. Recherches sur l'hyménium des champignons. *Annales des Sciences Naturelles* Botanique, Série 2 (1837); Observations médicinales et énumerations des plantes recueilliés en Tauride. In

A. Demidoff [ed.], *Voyage dans la Russie Meridionale et la Crimeé, par la Hongrie, la Valachie et la Moldavie* (1842); Organisation et disposition méthodique des espèces qui composent le genre *Erysiphe. Annales des Sciences Naturelles* Botanique, Série 3 (1851); *Iconographie des Champignons de Paulet* (1855). *Biogs, obits etc.* Grummann (1974: 286); Stafleu & Cowan (*TL-2* **2**: 965, 1979).

Leveillea Fr. (1849) [non *Leveillea* Decne. 1839, *Algae*] ? = Phylacia fide Læssøe (*SA* **13**: 43, 1994).

Leveillella Theiss. & Syd. (1915), Asterinaceae. 1, Chile. See Eriksson (*SA* **7**: 78, 1988).

Leveillina Theiss. & Syd. (1915), Dothideomycetes. 2, Africa; S. America.

Leveillinopsis F. Stevens (1924) = Coccostromopsis fide Hyde & Cannon (*Mycol. Pap.* **175**, 1999).

Leveillula G. Arnaud (1921), Erysiphaceae. Anamorph *Oidiopsis*. 23, widespread. See Durrieu & Rostam (*Cryptog. Mycol.* **5**: 279, 1985), Braun (*Beih. Nova Hedwigia* **89**, 1987; key), Gelyuta & Simonian (*Biol. Zh. Armenii* **39**: 20, 1987; subgen. divis.), Palti (*Bot. Rev.* **54**: 423, 1988; monogr.), Cook *et al.* (*MR* **101**: 975, 1997), Takamatsu *et al.* (*Mycoscience* **39**: 441, 1998; phylogeny), Saenz & Taylor (*CJB* **77**: 150, 1999; phylogeny), Khodaparast *et al.* (*MR* **105**: 909, 2001; phylogeny), Takamatsu (*Mycoscience* **45**: 147, 2004; phylogeny), Wang *et al.* (*Mycol.* **98**: 1065, 2006; phylogeny), Khodaparast *et al.* (*MR* **111**: 673, 2007; on monocots), Takamatsu *et al.* (*MR* **112**: 299, 2008; phylogeny).

Leveillulaceae V.P. Gelyuta (1988) = Erysiphaceae.

Levieuxia Fr. (1848) nom. dub., anamorphic *Pezizomycotina*. See Sutton (*Mycol. Pap.* **141**, 1977).

levigate, see laevigate.

Levispora Routien (1957) = Pseudeurotium fide Malloch (*in litt.*).

Lewia M.E. Barr & E.G. Simmons (1986), Pleosporaceae. Anamorph *Alternaria*. 17, widespread. See Simmons (*Mycotaxon* **25**: 287, 1986), Bottalico & Logrieco (*Mycotoxins in Agriculture and Food Safety*: 65, 1998; toxigenic spp.), Dong *et al.* (*MR* **102**: 151, 1998; DNA), Hoog & Horré (*Mycoses* **45**: 259, 2002; from humans), Simmons (*Mycotaxon* **83**: 127, 2002; morphology, anamorphs), Berbee *et al.* (*MR* **107**: 169, 2003; mating types), Kwasna & Kosiak (*MR* **107**: 371, 2003; on *Avena*, key), Kwasna *et al.* (*Mycol.* **98**: 662, 2006; on *Hordeum*), Simmons (*CBS Diversity Ser.* **6**, 2007; revision, nomenclator).

Liaoningnema S.L. Zheng & W. Zhang (1986), Fossil Fungi. 1, China.

Libartania Nag Raj (1979), anamorphic *Helotiales*, St.≡ eH.10. 3, widespread. See Nag Raj (*Coelomycetous Anamorphs with Appendage-bearing Conidia*, 1993), Lee & Crous (*Sydowia* **55**: 115, 2003).

Libellus Lloyd (1913) [non *Libellus* Cleve 1873, *Algae*] = Hymenogloea fide Singer (*Agaric. mod. Tax.*, 1951).

Libertella Desm. (1830), anamorphic *Xylariales*, St.0fH.10. 20, widespread (temperate). Primarily used for anamorphs of *Diatrypaceae*. See Ju & Rogers (*Mycotaxon* **73**: 343, 1999), Mostert *et al.* (*Mycol.* **93**: 146, 2001; on *Vitis*), Acero *et al.* (*Mycol.* **96**: 249, 2004; phylogeny).

Libertiella Speg. & Roum. (1880), anamorphic *Pezizomycotina*, St.0eH.15. 5 (on *Peltigera*), Europe. See Hawksworth (*Bull. Br. Mus. nat. hist.* Bot. **9**: 30, 1981).

Libertina Höhn. (1920) = Phomopsis (Sacc.) Bubák. fide Sutton (*TBMS* **50**: 355, 1967).

liberty cap, basidioma of the hallucinogenic *Psilocybe semilanceata*.

Licentia Pilát (1940) = Lopharia fide Donk (*Taxon* **5**: 69, 1956).

Lichen L. (1753) nom. rej. ≡ Parmelia.

lichen (pl. lichens; pronounced '*lie'ken*') [Lat. *lichen*, from Gk λειχηυ, tree moss] (cen (Welsh), Flechten (Germ.), gil-i-sang (Persian), huidmos and korsmos (Afrikaans), jäkälä (Finnish), kerpés (Lith.), korstmos (Dutch), lav (Danish, Norweg., Swed.), lichène (Ital.), ligen (Afrikaans), líquen and liquen (Portug., Span., S. Am.), lišaj (Croat.), lišejník (Bohem., Czech. Slovak), liszaj and porost (Pol.), mareru (Kikuyu), лишај (Serb.), лишай (Russian), лишей (Bulgar.), ragu (Old English), フイケン tii (Jap.), 582.29 (Universal Decimal Classification)), one of the Lichens (q.v.); (in Medicine), any of various eruptive skin diseases. **- acids**, see Metabolic products; **- alga**, phycobiont (q.v.); photobiont (q.v.); **-biont interaction** see Ahmadjian (*in* Reisser, *Algae and symbiosis*: 675, 1992); **- desert**, the area in a town or around an air pollution source from which all lichens, or at least all foliose and fruticose lichens are absent (obsol.); see Air pollution; **-icolous**, inhabiting lichens; **-iform**, having the form of a lichen; **-iverous**, lichen-eating; **-oglyph**, a centuries-old picture made by Canadian Indians by scraping lichens off the surface of a large vertical rock face; **-oid**, resembling a lichen; **-ologist**, one engaged in the pursuit of lichenology; **-ology**, the scientific study of lichens; **- products**, **- substances**, see Metabolic products.

Lichenagaricus P. Micheli (1729) nom. inval. = Xylaria Hill ex Schrank fide Hawksworth *et al.* (*Dictionary of the Fungi* edn 8, 1995).

lichenen (**lichenin**), an I+ red linear polymer of β-d-glucose with 1,3 and 1,4 linkages in the ratio 3:2; **isolichenan** (isolichenin), an isomer of lichenan, is I+ lilac or lavender. These carbohydrates occur in the walls of the hyphae of many lichen-forming fungi; lichenan tends to occur in higher concentrations and its presence/absence is taxonomically significant (Common, *Mycotaxon* **41**: 67, 1991). See amylomycan, Iodine.

Lichenes (obsol.). Name of a class for all lichen-forming fungi used when these were regarded as quite separate from *Fungi*; *Mycophycophyta*. **- imperfecti** (obsol.), lichen-forming anamorphic fungi; *Deuterolichenes* (q.v.), often used inclusive of lichen-forming fungi in which the sporocarps are unknown and the position is uncertain.

Lichenicolous fungi. Fungi dwelling on or in lichens as parasites (pathogens), commensals (see also parasymbiont) or saprobes; partly gall-inducing. In the ninth edn of this *Dictionary*, it was estimated that about 300 gen. and 1000 spp. of obligately lichenicolous fungi were known; many of the genera being exclusively lichenicolous. Common saprobic moulds are very scarce on lichens with depsides or depsidones, but a wide range of non-obligate species can be recovered by isolation (Petrini *et al.*, *Mycol.* **82**: 441, 1990; Prillinger *et al.*, *Syst. Appl. Microbiol.* **20**: 572, 1997).

Ascomycota, *Agaricomycetes* and anamorphic fungi all include obligately lichenicolous spp. Some lichenicolous fungi can take over the algae from other lichens and form new thalli (Hawksworth, *Bot. J. Linn. Soc.* **96**: 3, 1988).

Many groups are undergoing a critical re-appraisal in what is proving to be one of the most unexplored of all ecological niches occupied by fungi. Some may affect the host chemistry or produce separate novel metabolites (Feige *et al.*, *Cryptog. Bot.* **3**: 101, 1993; Hawksworth *et al.*, *Bibl. Lich.* **53**: 101, 1993). Cell wall degrading enzymes are found to be active in culture depending on lichen metabolites of a possible host lichen (Torzelli & Lawrey, *Mycol.* **87**: 841, 1995). Host-parasite interfaces of some lichenicolous fungi in the *Dacampiaceae* have been investigated (Rios & Grube, *MR* **104**: 1348, 2000), as has genetic variation in *Marchandiomyces corallinus* (Molina *et al.*, *Mycol.* **97**: 454, 2005).

Lit.: **General**: Clauzade *et al.* (*Bull. Soc. linn. Provence, num.-spéc.* **1**, 1989; keys 682 spp.; amendments in *Bull. Ass. Fr. lichén.* **16**(2): 71, 1991), Diederich (*Bibl. Lich.* **61**, 1996; basidiomycetes), Diederich & Christiansen (*Lichenologist* **26**: 47, 1994; basidiomycetes), Grummann (*Bot. Jb.* **80**: 101, 1960; galls), Hafellner (*Herzogia* **6**: 289, 299, **7**: 145, 163, 343, 353, 1982-87, *Nova Hedw.* **48**: 357, 1989), Hawksworth (*Bull. Br. Mus. nat. Hist., Bot.* **6**: 183, 1979; hyphomycetes, *Bull. Br. Mus. nat. Hist., Bot.* **9**: 1, 1981; coelomycetes, *TBMS* **74**: 363, 1980; on *Peltigera*, *J. Hattori bot. Lab.* **52**: 357, 1982; review biol., *J. Hattori bot. Lab.* **52**: 323, 1982; coevolution), Keissler (*Beih. bot. Zbl.* **50**: 380, 1933), Matzer (*Mycol. Pap.* **171**, 1996; on foliicolous lichens), Lawrey & Diedeerich (*Bryologist* **106**: 80, 2003; biodiversity, evolution and interactions), Rambold & Triebel (*Bibl. Lich.* **48**, 1992; inter-lecanoralean associations), Triebel (*Bibl. Lich.* 35, 1989; on lecideoid spp.), Vouaux (*BSMF* **28**: 177, **29**: 33, 399, **30**: 135, 281, 1912-14; keys, descr., monogr.), Zopf (*Hedwigia* **35**: 312, 1896; host index).

Regional: **Europe**, Keissler (*Rabenh. Krypt.-Fl.* **8**, 1930). **Austria**, Hafellner (*Herzogia* **10**: 1, 1994; *Mitt. Naturwiss. Vereins Steiermark* **125**: 73, 1996; *Carinthia II* **187/107**: 457, 1998). **Belgium**, Diederich *et al.* (*Dumortiera* **42**: 17, 1988). **British Isles**, Hawksworth (*Kew Bull.* **30**: 183, 1975, *Notes R. bot. Gdn Edinb.* **36**: 181, 1978, **38**: 165, 1980, **40**: 375, 1982, **43**: 497, 1986, **46**: 391, 1990, *Lichenologist* **15**: 1, 1983; key 218 spp., **26**: 337, 1994). **Denmark**, Alstrup (*Graphis Scripta* **5**: 60, 1993). **France**, Olivier (*Bull. internat. géogr. Bot.* **15-17**, 1905-07), Roux [*et al.*] (*Bull. Mus. Hist. nat. Marseille* **36**: 19, 1976; **37**: 83, 1977, *Bull. Soc. linn. Provence* **36**: 195, 1984, **41**: 117, 1990, **43**: 81, 1992), see also Spain, Portugal (Etayo & Diederich; Navarro-Rosinés & Roux). **Greenland**, Alstrup & Hawksworth (*Meddr Grønl., Biosci.* **31**, 1990; keys 124 spp.), Hansen & Alstrup (*Graphis Scripta* **7**: 33, 1995). **Luxembourg**, Diederich (*Lejennia* **119**: 1, 1986, *Mycotaxon* **37**: 297, 1990). **Norway**, Santesson (*The lichens and lichenicolous fungi of Norway and Sweden*, 1993; checklist), Ihlen (*Graphis Scripta* **7**: 17, 1995; *Lichenologist* **30**: 27, 1998). Papua New Guinea, Aptroot *et al.* (*Bibl. Lich.* **64**: 1997). **Poland**, Faltynowicz (*Polish Bot. Stud.* **6**, 1993; checklist). **Portugal**, see Spain. **Russia**, Zhurbenko & Santeson (*Herzogoa* **12**: 147, 1997). **Sardinia**, Nimis & Poelt (*Studia Geobot.* **7** (Suppl.), 1987). **Spain (Canary Islands)**, Santesson (*Svensk bot. Tidskr.* **54**: 499, 1960), Hafellner & Sancho (*Herzogia* **8**: 363, 1990), Calatayud *et al.* (*Mycotaxon* **55**: 363, 1995), Calatayud & Barreno (*Flechten Follmann*: 397, 1995), Hafellner (*Flechten Follmann*: 427, 1995), Cologne (*Herzogia* **11**: 133, 1995; *Linzer Biol. Beitr.* **27**: 489, 1995; *Fritschiana* **5**: 1, 1995; *Bull. Soc. Nat. Luxemb.* **97**: 93, 1996; *Mycotaxon* **60**: 415, 1996; *Lichenologist* **30**: 103, 1998)), Etato & Diederich (*Flechten Follmann*: 205, 1995), Navarro-Rosines & Roux (*Mycotaxon* **53**: 161, 1995), Etayo & Calatayud (*Ann. Naturhist. Mus. Wien* B **100**: 677, 1998), Matinez & Hafellner (*Mycotaxon* **69**: 271, 1998). **Sweden**, Santesson (*The lichens and lichenicolous fungi of Norway and Sweden*, 1993; checklist), Ihlen (*Graphis Scripta* **7**: 17, 1995; *Lichenologist* **30**: 27, 1998). **former Czechoslovakia**, Vězda (*Česká Myk.* **17**: 149, 1960; **23**: 104, 1969; **24**: 220, 1970). **Middle East**, Wasser & Nero (*Lichen-forming, lichenicolous, and allied fungi of Israel*, 2005). **North America**, Esslinger & Egan (*Bryologist* **98**: 467, 1995), Triebel *et al.* (*Mycotaxon* **42**: 263, 1991; 58 spp.), Zhurbenko *et al.* (*Evansia* **12**: 92, 1995), Alstrup & Cole (*Bryologist* **101**: 221, 1998). **South America**, Etayo (*Bibliotheca lichenologica* **84**: 1, 2002; Colombia). Also often included in fungus and/or lichen catalogues or mycotas. See also Internet. See also Geographical distribution, Lichenicolous lichens, Lichens.

Lichenicolous lichens. Lichens which grow on (or in) other lichens, either as commensals or parasites; 4-biont symbioses (see symbiosis). Each fungal partner in such associations has an independent algal or cyanobacterial partner, whereas no additional photosynthetic partner occurs in the obligately lichenicolous fungi (q.v.). The algae of the lichenicolous lichens occur either inside the host in e.g. *Buellia pulverulenta* on *Physconia distorta* (Hafellner & Poelt, *Phyton* **20**: 129, 1980) or as discrete thalli on its surface as in *Caloplaca epithallina* on 13 host lichens (Poelt, *Bot. Jahrb.* **167**: 457, 1985). Parasitic spp. are known esp. in *Acarospora*, *Diploschistes*, *Rhizocarpon* (Poelt, *Mitt. bot. StSamml., Münch.* **29**: 515, 1990), and *Verrucaria*. See also Poelt & Döbbeler (*Planta* **46**: 467, 1956), Rambold & Triebel (*Bibl. Lich.* **48**, 1992; inter-lecanoralean assocs.).

Lichenobactridium Diederich & Etayo (1995), anamorphic *Pezizomycotina*, Hso.?.?. 1, France. See Etayo & Diederich (*Flechten Follmann* Contributions to Lichenology in Honour of Gerhard Follmann: 212, 1995).

Lichenochora Hafellner (1989) nom. cons. prop., ? Phyllachorales. 25 (lichenicolous), widespread (esp. temperate). See Hafellner (*Nova Hedwigia* **48**: 357, 1992), Hoffmann & Hafellner (*Biblthca Lichenol.* **77**: 1, 2000), Hoffmann *et al.* (*Taxon* **55**: 802, 2006; nomencl.).

Lichenoconium Petr. & Syd. (1927), anamorphic *Pezizomycotina*, Cpd.0eP.19. 20 (on lichens), widespread. See Hawksworth (*Persoonia* **9**: 159, 1977), Hawksworth (*Bull. Br. Mus. nat. hist.* Bot. **9**: 33, 1981), Christiansen (*Graphis Scripta* **5**: 18, 1993; on *Lecanora*), Cole & Hawksworth (*Lichenologist* **36**: 1, 2004; Pacific NW, key).

Lichenodiplis Dyko & D. Hawksw. (1979), anamorphic *Pezizomycotina*, Cpd.1eP.19. 9 (on lichens), Europe; Asia. See Hawksworth (*Bull. Br. Mus. nat. hist.* Bot. **9**: 37, 1981), Berger & Diederich (*Herzogia* **12**: 35, 1996).

Lichenodiplisiella S. Kondratyuk & I. Kudratov (2002), anamorphic *Pezizomycotina*, C?.?.?. 1, Tadzhikistan. See Kondratyuk *et al.* (*Polish Botanical*

Journal **47**: 1, 2002).

Lichenohendersonia Calat. & Etayo (2001), anamorphic *Pezizomycotina*. 3, Spain. See Calatayud & Etayo (*CJB* **79**: 225, 2001).

Lichenoides Hoffm. (1789) ≡ Anaptychia.

Lichenometry. Technique for study of exposure age of rock surfaces based on the size/diameter (proportional to age) of lichen thalli. Pioneered by Beschel in the 1950s (see Raasch (Ed.), *Geology of the arctic* **2**: 1044, 1961) and now used extensively by glaciologists but also applicable to the minimum dating of many stone and other surfaces, and so of value in archaeology. The technique is potentially more precise than radio-carbon dating for exposures less than 500 years old. See Innes (Lichenometry [*Prog. phys. Geogr.* **9**: 187], 1985; *in* Galun, *CRC Handbook of lichenology* **3**: 75, 1988), Luckman (*Can. J. Earth Sci.* **14**: 1804, 1977; Alberta), Matthews (*Norsk geogr. Tidsskr.* **29**: 97, 1975; Jotunheimen), Solomina *et al.* (*Arctic, Antarctic and Alpine Res.* **35**: 129, 2003), Webber & Andrews (*Arctic Alpine Res.* **5**: 293, 1973; review, bibliogr.).

A critical review of the technique is provided by Worsley (*in* Goudie *et al.*, *Geomorphological techniques*: 302, 1981) and a size-frequency approach addressing some criticisms of the technique has been developed by Winchester & Harrison (*Earth surface processes & landforms* **19**: 137, 1994). See Ecology, Growth rates.

Lichenomphalia Redhead, Lutzoni, Moncalvo & Vilgalys (2002), Hygrophoraceae. 8, widespread. See Redhead & Kuyper (*Mycotaxon* **31**: 221, 1978), Redhead & Kuyper (*Arctic Alpine Mycology* **2**: 319, 1987), Lutzoni & Vilgalys (*CJB* **73**: S649, 1995; phylogeny), Lutzoni (*Syst. Biol.* **46**: 373, 1997; phylogeny), Moncalvo *et al.* (*Syst. Biol.* **49**: 278, 2000; phylogeny), Redhead *et al.* (*Mycotaxon* **83**: 38, 2002), Delannoy & Eyssartier (*BSMF* **120**: 403, 2004; *Lichenomphalia* in Svalbard).

Lichenomyces Trevis. (1853) = Plectocarpon fide Hawksworth & Galloway (*Lichenologist* **16**: 85, 1984).

Lichenopeltella Höhn. (1919), Microthyriaceae. *c.* 20 (some on lichens), widespread. See Santesson (*SA* **9**: 15, 1990; nomencl.), Spooner & Kirk (*MR* **94**: 223, 1990; as *Micropeltopsis*), Hariharan *et al.* (*Lichenologist* **28**: 294, 1996; India spp.), Earland-Bennett & Hawksworth (*Lichenologist* **31**: 575, 1999; on *Verrucaria*), Cole & Hawksworth (*Mycotaxon* **83**: 391, 2002; on *Heterodermia*).

Lichenopeziza Zukal (1884), Pezizomycotina (?L). 1, Europe.

Lichenophoma Keissl. (1911) nom. dub., anamorphic *Pezizomycotina*, Cpd.0eH.?. 1 (on lichens), Europe. See Hawksworth (*Bull. Br. Mus. nat. hist.* Bot. **9**: 77, 1981).

Lichenopsis Schwein. (1832) = Stictis fide Hawksworth *et al.* (*Dictionary of the Fungi* edn 8, 1995).

Lichenopuccinia D. Hawksw. & Hafellner (1984), anamorphic *Pezizomycotina*, Hsp.≡ eH.1. 1 (on lichens, esp. *Parmelia*), Austria; British Isles. See Hawksworth & Hafellner (*Nova Hedwigia* Beih. **79**: 373, 1984).

Lichenopyrenis Calat., Sanz & Aptroot (2001), Pleomassariaceae. 1 (lichenicolous), Spain. See Calatayud *et al.* (*MR* **105**: 634, 2001).

Lichenosphaeria Bornet (1873) ? = Didymella fide Henssen (*Symb. bot. upsal.* **18** no. 1, 1963).

Lichenostella Calat. & Etayo (1999), anamorphic *Pezizomycotina*. 1 (lichenicolous), Colombia. See Calatayud & Etayo (*Lichenologist* **31**: 597, 1999).

Lichenosticta Zopf (1898), anamorphic *Pezizomycotina*, Cpd.0eH.15. 1 (on lichens, esp. *Cladonia*), widespread (north temperate). See Hawksworth (*Bull. Br. Mus. nat. hist.* Bot. **9**: 38, 1981).

Lichenostigma Hafellner (1983), Lichenotheliaceae. 2 (on lichens), widespread. See Thor (*Lichenologist* **17**: 269, 1985), Hafellner & Calatayud (*Mycotaxon* **72**: 107, 1999), Calatayud *et al.* (*MR* **106**: 1230, 2002; key), Ihlen (*Lichenologist* **36**: 183, 2004; Scandinavia), Vondrák & Šoun (*Lichenologist* **39**: 211, 2007; review).

Lichenothelia D. Hawksw. (1981), Lichenotheliaceae. 20, widespread. See Henssen (*Biblthca Lichenol.* **25**: 257, 1987).

Lichenotheliaceae Henssen (1986), ? Dothideomycetes (inc. sed.). 2 gen. (+ 1 syn.), 22 spp.

Lit.: Henssen (*Biblthca Lichenol.* **25**: 257, 1987), Navarro-Rosinés & Hafellner (*Mycotaxon* **57**: 211, 1996), Hafellner & Calatayud (*Mycotaxon* **72**: 107, 1999).

Lichenothrix Henssen (1964) = Pyrenothrix fide Eriksson (*Op. Bot.* **60**, 1981).

Lichens. A lichen is a stable self-supporting association of a fungus (mycobiont) and an alga or cyanobacterium (photobiont). More precisely, a lichen is an ecologically obligate, stable mutualism between an exhabitant fungal partner and an inhabitant population of extracellularly located unicellular or filamentous algal or cyanobacterial cells (Hawksworth & Honegger, *in* Williams (Eds) (*Plant galls*: 77, 1994). Many other definitions have been proposed (Hawksworth, *Bot. J. Linn. Soc.* **96**: 3, 1988; review), and Ahmadjian (*The lichen symbiosis*, edn 2, 1993) stresses the formation of a 'thallus or lichenized stroma that may contain unique secondary compounds' as the key feature. The partners may have coevolved (Ahmadjian, *Ann. N.Y. Acad. Sci.* **503**: 307, 1987; Hawksworth, *in* Pirozynski & Hawksworth (Eds), *Coevolution of fungi with plants and animals*: 125, 1988).

Lichens are a biological and not a systematic group, and are unique in that in many (but not all) cases the resulting life form and behaviour differ markedly from that of the isolated components. In most lichens the fungal partners, which do not occur free-living (except in some facultatively lichenized fungi), appear to be responsible for the overall form of the thallus and its fruiting bodies (ascomata), but there is an interplay between both components to produce the final form (see below), and a wide range of biological interactions is involved. These may involve four, five or even more partners (Hawksworth, 1988; review).

Fungal partners: Around 20% of all *Fungi* and 40% of all *Ascomycota* are lichen-forming. Most recent estimates of global diversity suggest that between 17500 and 20000 species are known, with a further 1500 lichenicolous fungi (Feuerer & Hawksworth, *Biodiv. Cons.* **16**: 85, 2007; Galloway, *Biodiv. Conserv.* **1**: 312, 1992, Lawrey & Diederich, *Bryol.* **106**: 80, 2003; Nash & Egan, *Bibl. Lich.* **30**: 11, 1988; Sipman & Aptroot, *MR* **105**: 1433, 2001). Lichens are polyphyletic in origin, as established beyond doubt by molecular methods (Gargas *et al.*, *Science* **269**: 1492, 1995); Hibbett *et al.*, *MR* **111**: 509,

2007). The fungi of most species belong to the *Ascomycota* (q.v.), but only four orders are exclusively lichen-forming. Lichenization also occurs in a few *Basidiomycota* (e.g. *Dictyonema*, *Multiclavula*, *Lichenomphalia*), and also anamorphic fungi (e.g. *Blarneya*, *Cystocoleus*; see Vobis & Hawksworth, *in* Cole & Kendrick (Eds), *The biology of conidial fungi* **1**: 245, 1981; review). However, while many families are exclusively lichen-forming, a mixture of lichen-forming and non-lichen-forming (and also sometimes lichenicolous) species can occur within the same family or even genus (e.g. *Arthonia*, *Arthothelium*, *Mycomicrothelia*, *Toninia*). What is or is not regarded as a 'lichen' is partly a matter of history; for example, *Orbilia*, and *Pezizella* are not usually studied by lichenologists, while some taxa generally regarded as the preserve of lichenologists do not form lichens (e.g. *Arthopyrenia*, *Chaenothecopsis*, *Leptorhaphis*, *Stenocybe*). The existence of genera crossing biological boundaries, and of species which may be viewed as components of primitive lichens (Kohlmeyer & Kohlmeyer, *Marine mycology*, 1979) or facultatively lichen-forming, makes it clear that the lichen method of nutrition (q.v.) is at various stages of evolution in different groups; evolving to be more strongly lichen-forming in some (e.g. some *Leotiomycetes*, *Agaricales*) and less strongly lichen-forming in others (e.g. non-lichen-forming *Arthoniales* and *Lecanorales*). Axenic cultures of the fungal partner can be obtained from ascospores and thallus fragments; Crittenden *et al.* (*New Phytol.* **130**: 267, 1995) had success with 493 spp. (42%) of 1183 attempted. The cultures are generally slow-growing, have little organized structure, and do not produce ascomata (although some form conidiomata).

Algal and cyanobacterial partners: The number of photosynthetic partners involved in lichen formation is relatively small; only 40 genera are represented, 25 algae and 15 cyanobacteria (see Ahmadjian, *Phycologia* **6**: 128, 1967, identification; Ahmadjian, 1993; Hildreth & Ahmadjian, *Lichenologist* **13**: 65, 1981, keys *Trebouxia* spp.; Honegger, *Exper. Phycol.* **1**: 40, 1990; Tschermak-Woess, *in* Galun, *CRC Handbook of lichenology* **1**: 29, 1988). Most belong to genera which are also free-living both in the cyanobacteria (e.g. *Calothrix*, *Gloeocapsa*, *Nostoc*, *Scytonema*, *Stigonema*) and the green algae (*Cephaleuros*, *Coccomyxa*, *Myremecia*, *Trentepohlia*). Only *Trebouxia* (incl. *Pseudotrebouxia*) appears to be primarily lichen-forming, with doubt cast on its free occurrence in nature (Ahmadjian, 1993; Tschermak-Woess, 1988). The same species, but often as different strains, may occur with a very wide range of lichen-forming fungal genera. In addition more than one strain of a single alga (or species of an algal genus) can be involved in the formation of a single lichen thallus. However, the systematics of the algal and cyanobacterial partners of lichens is currently inadequate to make more definite assertions as to specificity. However, in some lichen-forming fungi (e.g. *Lobaria*, *Placopsis*, *Psoroma*, *Stereocaulon*) with an algal partner, cyanobacteria also occur in 'cephalodia' (q.v.) which may, according to the species, be warts on the upper or lower surface or circumscribed areas within the thallus. In some cases cephalodia can be separated and persist as independent lichens (e.g. *Dendriscocaulon*); occasionally, two species with different types of algae previously regarded as distinct have been found joined together as a composite ('chimeroid') thallus showing these 'photomorphs' (q.v.; phycosymbiodemes, photosymbiodeme, phycotypes) are the result of interactions between a single mycobiont and distinct photobionts (Armaleo & Clerc, *Exp. Mycol.* **15**: 1, 1991; James & Henssen, *in* Brown *et al.*, 1976: 27; Stöcker-Wörgötter & Türk, *Crypt. Bot.* **4**: 300, 1994; Tønsberg & Holtan-Hartwig, *Nordic Jl Bot.* **3**: 681, 1983), White & James, *Lichenologist* **20**: 103, 1988).

Nomenclature: Lichens do not have independent scientific names; the fungal and photosynthetic partners each have separate names, and names given to lichens are considered as referring to the fungal partner alone (see Nomenclature). The classification of 'lichens' therefore has to be completely integrated into the system of *Fungi*. The names of lichen-forming fungi have one other special provision in the Code, exemption from the provisions for the naming of pleomorphic fungi. These fungi are covered by the *Index of Fungi* from vol. **5**(4), 1982 onwards. Current nomenclatural practice is consistent with the recognition of lichens as a nutritional rather than a taxonomic group.

Synthesis: Although the bionts can be separated and cultured, resynthesis under laboratory conditions is difficult. Some success has been achieved by lowering the nutrient supply and modifying the moisture requirements, features which suggest that lichens behave as two organisms united in adversity. Synthesis was first achieved experimentally by Stahl in 1877 (repeated by Ahmadjian & Heikkilä, *Lichenologist* **4**: 259, 1970), and has now also been carried out in *Dermatocarpon*, *Usnea* and *Xanthoria* (Ahmadjian, 1993). Novel syntheses have also been attempted; Ahmadjian *et al.* (*Mycol.* **72**: 73, 1980) found that *Cladonia cristatella* could be re-synthesized with 13 different *Trebouxia* isolates, but none of 10 '*Pseudotrebouxia*' isolates or free- living *Pleurastrum*; squamules, pycnidia and short podetia with immature hymenia were obtained and the metabolic products (q.v.) were identical in all cases (Ahmadjian, *The lichen symbiosis*, edn 2, 1993; *in* Cook (Ed.), *Cellular interactions with symbiosis and parasitism*: 3, 1980; *Progress Phycol. Res.* **1**: 179, 1982). Techniques for growing lichens as tissue cultures with intermixed fungi and algae have been developed for about 200 spp. since 1981 (Yamamoto *et al.*, *Bryologist* **96**: 383, 1993).

Structure: The algae or cyanobacteria are either distributed at random, often in a gelatinous matrix, throughout the thallus (**homoiomerous**; unlayered; unstratified), or in a compact layer below the upper or outer cortex (**heteromerous**; layered; stratified). There are a few **filamentous** lichens (*Coenogonium*, *Cystocoleus*, *Racodium*) in which the filamentous form of the alga predominates; in the majority, the fungus forms the outer structure and gives the lichen its shape. Six life-form categories are generally recognized (Fig. 21): shrub- or beard- like (**fruticose**), leaf-like (**foliose**), scale-like (**squamulose**; **placodioid** if rosette-like in surface view with only the tips free), crust-like (**crustose**, **crustaceous**), **filamentous** (see above), and **leprose** (powdery, loose and often powdery aggregations of algal cells entwined with hyphae and with no cortex). Fruticose thalli are the most complex and largest; these may be erect or pendulous, hair- or strap-like and are often richly branched. They have a radial structure arranged around a central cavity (e.g. *Bryoria*) or a tough axial strand (*Usnea*), and are attached to the substratum by small disc-like holdfasts or clusters of rhizoids. Some *Usnea* spp. on trees attain a length of 10 m. In foliose and crustose thalli the structure is essentially dorsiventral and growth most pronounced at the circumference; the former have a well-developed lower cortex and are attached to the substratum by puckering of the underside (*Hypogymnia*), a fine felted tomentum (*Lobaria*), or by coarse bundles of compacted hyphae or rhizines (*Parmelia*). Crustose lichens may be more or less immersed in the substratum and, like squamules, lack a lower cortex.

The surface of crustose species is often characteristically cracked or warted, and may be bounded by a pale or black marginal **prothallus** which contains no photobiont cells. The gelatinous homoiomerous lichens (*Collema*, *Placynthium*) also occur in similar life forms. Any particular life-form is not necessarily confined to a genus, family, or order (e.g. all except the filamentous life-form occur in *Teloschistales*). In a few genera, such as *Cladonia*, *Pilophorus* and *Stereocaulon*, there is a combination of different growth forms: in many species of *Cladonia*, the basal part of the thallus is squamulose, from which arise fruticose, hollow, simple, or branched structures (podetia; q.v.), often also bearing squamules, the apices of which are either pointed (subulate) or bear terminal cups (**scyphi**) on which are borne the ascocarps. In *Stereocaulon* the basal, peltate, or coralloid phyllocladia give rise to simple or branched pseudopodetia (see podetia), also more or less also covered in phyllocladia. The pseudopodetia bear lateral or terminal ascocarps.

A few genera have additional thalline structures, such as aeration pores in the form of **cyphellae** (*Sticta*) and pseudocyphellae (*Punctelia*, *Pseudocyphellaria*); marginal eye-lash-like cilia occur in *Heterodermia* and in some species of *Physcia* and *Parmotrema*. See Beltmann (*Bibl. Lich.* **11**, 1978; Parmeliaceae), Hannemann (*Bibl. Lich.* **1**, 1973; attachment organs), Henssen & Jahns (1973), Jahns (*in* Galun, *CRC Handbook of lichenology* **1**: 95, 1988), Ozenda (*Handb. Pflanzenanatomie* **6** (9), 1963).

Reproduction: Asexual or vegetative reproduction of lichens involving both partners (dual propagules) may be achieved by a wide range of methods (Fig. 22), including: (1) simple fragmentation, (2) development of delimited or widespread areas of cortical breakdown (**soralia**) which contain minute powdery propagules (**soredia**), and (3) the development of numerous small, simple, or branched-coralloid, corticate papillae (**isidia**). These methods involve the dissemination of the united bionts. The position, shape, and sometimes the colour of the soralia and isidia is often diagnostic and is usually accompanied by a marked suppression of ascomatal development (see Phylogeny).

A number of species also develop innate, flask-shaped **pycnidial conidiomata** in which numerous, often characteristically shaped conidia develop. Although conidia of some species germinate in culture their functional role is not fully understood and may, in some species at least, be sexual. Only the mycobiont reproduces sexually, a process culminating in the development of ascomata. The nature of the ascomata depends on the order to which the fungal partner belongs. However, the ascomata are mainly perennial with serial development of the asci. Two main apothecial types are often distinguished: **lecanorine** with a thalline exciple (containing algae and generally the same colour as the main thallus); and **lecideine** only with a true exciple (lacking algae and usually differing in colour from the thallus). In some asexually reproducing species ascomata can be absent or very rare (and then often with few or no well-formed ascospores); see also species pairs.

Establishment: Where dual propagules are the means of dispersal, locating a compatible partner does not present difficulties. However, where several dual propagules of the same or sometimes different lichens start to grow close together they can become intermixed and form interspecific or even intergeneric **mechanical hybrids** (see hybrid); e.g. *Physcia adscendens* + *P. tenella* (Schuster *et al.*, *Lichenologist* **13**: 247, 1985), *P. tenella* + *Xanthoria parietina* (Ott, *Bibl. Lich.* **25**: 81, 1987). The extent to which 'individual' lichen thalli represent the product of a single propagule and belong to one genotype merits investigation.

Where ascospores are the dispersal agent, various strategies for establishment are documened: the ascospore may land on and re-shape an already growing algal or cyanobacterial colony (e.g. *Collema*), invade an established lichen killing the fungal partner and taking over the algae (e.g. *Diploschistes*), land near and out-compete the fungal partner in a dual propagule derived from another lichen, or persist in a loose unstructured association with other algae until a truely compatible algal species arrives (e.g. *Xanthoria*; Ott, *Nordic Jl Bot.* **7**: 219, 1987). See Spore discharge and dispersal.

Interactions between the bionts: In the early stages of synthesis, the partners can be bound by a common sheath (Ahmadjian *et al.*, *Science* **200**: 1062, 1978); similar ultrastructural-level patterns on the surfaces of both partners can enable them to mesh together (Honegger *et al.*, *Lichenologist* **16**: 111, 1984). The actual nature of the cell-to-cell contact varies in different groups; these range from wall-to-wall apposition to a variety of kinds of intraparietal haustoria (see Honegger, *New Phytol.* **103**: 785, 1986; see haustorium. Intracellular haustoria completely penetrating the walls of healthy algal cells are unusual, but the extent of penetration may vary even within a single thallus (Galun *et al.*, *J. Microscopie* **9**: 801, 1970). Dead algal cells occuring in thalli may be used, leading to suggestions that 'controlled parasitism' may more accurately explain the relationship than 'mutualism' (see above, Ahmadjian, 1993). The fungus obtains from the alga growth substances (thiamine, biotin), and products of its photosynthesis, including simple sugars and polyalcohols. Cyanobacterial partners, whether the main photobiont or in cephalodia, fix atmospheric nitrogen which is passed to the fungus. Some green algae involved may be distingushed from non-lichenized relatives by accelerated leakage of sugars across their cell walls in the presence of a compatible fungus.

Whole-thallus physiology: See Kershaw (*Physiological ecology of lichens*, 1985), Vincente *et al.* (Eds) (*Surface physiology of lichens*, 1985), Brown (Ed.) (*Lichen physiology and cell biology*, 1985), Galun (Ed.), 1985), Nutrition, Physiology.

Chemistry: See Chemotaxonomy, chromatography, Metabolic products, Microcrystal tests, Pigments.

Economics: See Air pollution, Antibiotics, Biodeterioration, Bioindication, Dyeing, Edible fungi, Insects and fungi, lichenometry, Medical uses of fungi and lichens, RIEC. See also Llano (*Bot. Rev.* **10** (1), 1944), Richardson (*The vanishing lichens*, 1975; *in* Galun, *CRC Handbook of lichenology* **3**: 93, 1988), Seaward (Ed.) (*Lichen ecology*, 1977).

Lit.: **General**: Ahmadjian (*The lichen symbiosis*, edn 2, 1993), Ahmadjian & Hale (Eds) (*The lichens*, 1974), Bates & Farmer (Eds) (*Bryophytes and lichens in a changing environment*, 1992), Brown *et al.* (Eds) (*Lichenology: progress and problems*, 1978), Dalby *et al.* (Eds) (*Horizons in lichenology* [*Bot. J. Linn. Soc.* **96**], 1988), Galloway (*Tropical Lichens*, 1991), Filho *et al* (*Biologia de liquens*, 2006; in Portuguese), Galun (Ed.) (*Handbook of lichenology*, 3 vols, 1988), Gilbert (*Lichens*, 2000), Hale (*The biology of lichens*, 1983), Hawksworth (*Crypt. Bot.* **4**: 117, 1994; advances 1972-92), Hawksworth & Hill (*The lichen-forming fungi*, 1984), Henssen & Jahns (*Lichenes*, 1973), Lawrey (*Biology of lichenized fungi*, 1984), Lücking (*Foliicolous Lichens*, 1992). Nash (*Lichen Biology*, 2008), Nimis *et al.* (*Monitoring with Lichens. Monitoring Lichens*, 2002), Purvis (*Lichens*, 2000; basic review), Smith (*Lichens*, 1921 [reprint 1975); history), StClair & Seaward (*Biodeterioration of Stone Surfaces*, 2004; lichens as weathering agents), Thor *et al.* (*Contributions to Lichen Taxonomy and Biogeography*, 2004). Zahlbruckner (*Naturl. Pflanzenfam.* **8**: 61, 1926; keys gen.).

Regional: Hawksworth & Ahti (*Lichenologist* **22**: 1, 1990; 1390 refs. by continent and country); the following list is restricted to major modern works, esp. of value in identification. **Africa, East**, Swinscow & Krug (*Macrolichens of East Africa*, 1988). **Antarctica**, Øvstedal & Smith (*Lichens of Antarctica and South Georgia*, 2001), Rédon (*Liquenes Antárticos*, 1985). **Austria**, Türk & Poelt (*Bibliographie de Flechten. Österreich*, 1993). **Australia**: *Flora of Australia* **54** (1992 ongoing), McCarthy (*Catalogue of Australian lichens*, 2003), Rogers (*The genera of Australian lichens (lichenized fungi)*, 1980). **Brazil**, Cáceres (*Corticolous, Crustose and Microfoliose Lichens of Northeast Brazil*, 2007). **British Isles**, Dobson (*Lichens: An Illustrated Guide to the British and Irish Species* edn 5, 2005), Purvis *et al.* (*The lichen flora of Great Britain and Ireland*, 1992; keys 1700 spp.; *Bull. Br. lichen. Soc.* **72** (Suppl.), 1993; checklist). **Canada**, Brodo (*Lichens of the Ottawa region*, edn 2, 1988), Goward *et al.* (*The lichens of British Columbia, Illustrated keys*, 1994); see also N. America. **China**, Wei (*An enumeration of lichens in China*, 1991), Zhao *et al.* (*Prodromus Lichenum Sinicorum*, 1982). **Costa Rica**: Sipman & Umana (*Costa Rica Lichens*, 2002). **Czech Republic**, Vežda & Liska (*Catalogue of Lichens of the Czech Republic*, 1999). **Estonia**, Trass & Randlane (*Eesti suur-samblikud*, 1994). **Europe**, Clauzade & Roux (*Bull. Soc. Bot. Centre-Ouest, n.s.* **7**:1, 1985), Jahns (*Farne, Moose, Flechten Mittel-, Norde- und Westeuropas*, 1980), Poelt (*Bestimmungschfüssel europäischer Flechten*, 1969), Poelt & Vězda (*Bibl. Lich.* **9**: 1, 1977; **16**: 1, 1981). **France**, Ozenda & Clauzade (*Les Lichens. Étude biologique et flore illustrée*, 1970). **Germany**, Wirth (*Die Flechten Baden-Württembergs*, edn 2, 2 vols, 1995; keys 1500 spp., 555 col. pl.). **Greenland**: Hansen & Anderson (*Greenland Lichens*, 1995). **Hungary**, Verseghy (*Magyarország zuzmó flórájának Kézikönyve*, 1994). **Iberian peninsula**, Llimona & Hladun (*Checklist of the Lichens and Lichenicolous Fungi of the Iberian Peninsula and Balearic Islands*, 2001). **India**, Awasthi (*J. Hattori Bot. Lab.* **65**: 207, 1988; macrolichens; *Bibl. Lich.* **40**, 1991; microlichens), Singh & Sinha (*Lichen flora of Nagaland*, 1994; keys, descr.). **Israel**, Wasser & Eviatar (*Lichen-forming, Lichenicolous and Allied Fungi of Israel*, 2005). **Italy**, Nimis (*The lichens of Italy: an annotated catalogue*, 1993), Nimis & Martellos (*Keys to the Lichens of Italy – Terricolous Species*, 2004). **Japan**, Yoshimura (*Lichen flora of Japan in colour*, 1974), Kurokawa (*Checklist of Japanese Lichens*, 2003). **Morocco**, Egea (*Catalogue of Lichenized and Lichenicolous Fungi of Morocco*, 1996). **Netherlands**, Aptroot & van Herk (*Veldgids Korstmossen*, 1994), Brand *et al.* (*Weten. Med. KNNV* **118**, 1988; checklist). **N. America**, Brodo *et al.* (*Lichens of North America*, 2001), Egan (*Bryologist* **90**: 77, 1987), Hale (*How to know the lichens*, edn 2, 1979), Nash *et al.* (*Lichen Flora of the Greater Sonoran Desert Region*, 2 vols, 2002/2004), Thomson (*American Arctic Lichens*. I. *The Macrolichens*, 1984; II. *The Microlichens*, 1997), Vitt *et al.* (*Mosses, lichens and ferns of northwest North America*, 1988). **New Zealand**, Galloway (*Flora of New Zealand lichens* ed 2, 2008). **Norway**, Krog *et al.* (*Lavflora. Norske busk- og bladiav*, 1980); Tønsberg (*Sommerfeltia* **14**: 1, 1992; keys 128 corticolous spp.). **Papua New Guinea**, Aptroot *et al.* (*Lichens and Lichenicolous Fungi from New Guinea*, 1997), Streimann (*Catalogue of the Lichens of Papua New Guinea and Irian Jaya*, 1986). **Poland**, Nowak & Tobolewski (*Porosty Polskie*, 1975), Fałtynowicz (*Lichens, Lichenicolous and Allied Fungi of Poland*, 2003). **Russia**, Abramov (Ed.) (*Opredéltel' lishaínikov SSSR*, 5 vols, 1971-78), Andreev *et al* (*Handbook of the Lichens of Russia* vol. 8, 2003). **Scandinavia**, Santesson *et al* (*Lichen-forming and Lichenicolous Fungi of Fennoscandia*, 2004; checklist), *Nordic Lichen Flora*; vol. **1** (ed. Ahti *et al.*, 1999; calicioid spp.), **2** (ed. Foucard *et al.*, 2002; *Physciaceae*), **3** (ed Ahti *et al.*, 2007; cyanolichens). **Saudi Arabia**, Abu-zinada *et al.* (*Arab Gulf J. Sci. Res., s.p.* **2**: 1, 1986). **S.E. Asia**, Aguirre-Hudson & Wolseley (*J. Hattori Bot. Lab.* **76**: 313, 1994; bibliogr.). **Spain**: Llimona (*História natural des Països Catalans* **5**, *Fongsi líquens*, 1991). **Sweden**, Foucard (*Svensk skorplavs flora*, 1990), Moberg & Holmåsen (*Lavar. Enfälthandbok*, edn 2, 1984), Santesson (*The lichens and lichenicolous fungi of Sweden and Norway*, 1993). **Ukraine**, Oxner (*Flora lishaĭnik[i]v Ukraïni*, **1**, 1956; **2**(1), 1968; **2**(2), 1993), Kondratyuk *et al.* (*Second Checklist of Lichen-forming, Lichenicolous and Allied Fungi of Ukraine*, 1998). **USA**, Hale & Cole (*Lichens of California*, 1988); see also N. America. **Venezuela**, López Figueiras (*Censo de macroliquenes Venezolanos*, 1986).

See also *Ascomycota*, Lichenicolous fungi, Literature, and under individual subjects.

Lichina C. Agardh (1817) nom. cons., Lichinaceae (L). 7, widespread (temperate). See Henssen (*Symb. bot. upsal.* **18** no. 1, 1963), Henssen (*Lichenologist* **4**: 88, 1969), Henssen (*Nova Hedwigia* **15**: 543, 1969), Janson *et al.* (*New Phytol.* **124**: 149, 1993; ultrastr.), Schultz *et al.* (*Pl. Biol.* **3**: 116, 2001; phylogeny),

Schultz & Büdel (*Lichenologist* **34**: 39, 2002; key), Prieto *et al.* (*MR* **112**: 381, 2008; phylogeny, chemistry).

Lichinaceae Nyl. (1854), Lichinales (L). 47 gen. (+ 65 syn.), 307 spp.

Lit.: Henssen & Dobelmann (*Biblthca Lichenol.* **25**: 103, 1987), Henssen & Jørgensen (*Lichenologist* **22**: 137, 1990), Moreno & Egea (*Acta Bot. Malac.* **15**: 19, 1990), Moreno & Egea (*Biología y Taxonomía de la Familia Lichinaceae, con Especial Referencia a las Especies del S.E. Español y Norte de África*: 87 pp., 1991), Moreno & Egea (*Acta Bot. Barcinon.* **41**: 1, 1992; Iberia, N. Afr., keys 14 spp., gen. concepts), Moreno & Egea (*Bull. Soc. linn. Provence* **45**: 291, 1994), Kantvilas & Jørgensen (*Muelleria* **11**: 45, 1998), Jørgensen & Henssen (*Bryologist* **102**: 22, 1999), Schultz *et al.* (*Pl. Biol.* **2**: 482, 2000), Lutzoni *et al.* (*Nature* Lond. **411**: 937, 2001), Schultz *et al.* (*Pl. Biol.* **3**: 116, 2001), Wedin *et al.* (*MR* **109**: 159, 2005), Jørgensen (*Nordic Lichen Flora* **3**: Cyanolichens: 46, 2007), Prieto *et al.* (*MR* **112**: 381, 2008; phylogeny, chemistry).

Lichinales Henssen & Büdel (1986), (L). Lichinomycetes. 3 fam., 53 gen., 350 spp. Thallus crustose, fruticose or foliose, sometimes peltate, often gelatinous. Ascomata eventually apothecial but initially ± perithecial, often formed from pycnidia, opening by a pore, ± sessile or immersed; peridium often not well-defined. Interascal tissue varied; hymenium often J+. Asci thin-walled, without separable wall layers, without well-defined apical structures, usually with a J+ outer gelatinized layer; sometimes evanescent, sometimes polysporous. Ascospores usually hyaline, aseptate. Anamorphs pycnidial. Lichenized with cyanobacteria, widespr. Fams:

(1) **Gloeoheppiaceae**
(2) **Lichinaceae**
(3) **Peltulaceae**

Lit.: Moreno & Egea (*Biología Taxonomía de la Familia Lichinaceae*, 1991), Henssen (*Symb. bot. upsal.* **18**(1), 1963, *Lichenologist* **27**: 261, 1995).

Lichinella Nyl. (1873), Lichinaceae (L). 11, widespread. See Moreno & Egea (*Cryptog. Bryol.-Lichénol.* **13**: 237, 1992; key 7 spp. Spain & N. Afr.), Schultz *et al.* (*Pl. Biol.* **3**: 116, 2001; phylogeny), Schultz & Büdel (*Lichenologist* **34**: 39, 2002; key), Schultz (*Bryologist* **108**: 567, 2005; N America).

Lichingoldia D. Hawksw. & Poelt (1986) nom. rej. = Woessia fide Hawksworth *et al.* (*Dictionary of the Fungi* edn 8, 1995).

Lichiniza Nyl. (1881) = Porocyphus fide Henssen (*Symb. bot. upsal.* **18** no. 1, 1963).

Lichinodium Nyl. (1875), Lichinaceae (L). 2, Europe. See Henssen (*Symb. bot. upsal.* **18** no. 1, 1963), Schultz & Büdel (*Lichenologist* **34**: 39, 2002; key).

Lichinomyces E.A. Thomas ex Cif. & Tomas. (1939) nom. inval. = Lichina fide Hawksworth *et al.* (*Dictionary of the Fungi* edn 8, 1995).

Lichinomycetes Reeb, Lutzoni & Cl. Roux (2004), Pezizomycotina. 1 ord., 3 fam., 53 gen., 350 spp. Ord.:

Lichinales

For *Lit.* see fam.

Lichtheimia Vuill. (1903) = Absidia fide Hesseltine (*Mycol.* **47**: 344, 1955).

Licipenicillium, see *Lysipenicillium*.

Licopolia Sacc., Syd. & P. Syd. (1900), Dothideomycetes. 2, Brazil; Kenya.

Licrostroma P.A. Lemke (1964), Corticiaceae. Anamorph *Michenera*. 1, N. America; Japan. See Lemke (*CJB* **42**: 762, 1964).

Licrostromataceae Jülich (1982) = Corticiaceae.

Lidophia J. Walker & B. Sutton (1974), Dothideomycetes. Anamorph *Dilophospora*. 1, Germany. See Walker (*Mycotaxon* **11**: 38, 1980).

Liebermannia Y.Y. Sokolova, C.E. Lange & J.R. Fuxa (2006), Microsporidia. 2. See Sokolova *et al.* (*J. Invert. Path.* **91**: 168, 2006).

Liesgang phenomenon, see zonation.

life history (**life-cycle**) (in fungi), the stage or series of stages (frequently characterized by different spore states, see States of fungi) between one spore form and the development of the same spore form again. Cf. sex.

ligative hyphae, Pouzar's term for binding hyphae; see hyphal analysis.

Light and fungi. Spore germination, growth, pathogenicity, sporocarp development, sporulation, spore discharge, and movement in fungi may all be influenced by light, and fungi with photoreceptors have been reported. See Alvarez *et al.* (Light converts endosymbiotic fungus to pathogen, influencing seedling survival and host tree recruitment. *Nature precedings*, http://precedings.nature.com/documents/1908/version/1), Carlisle (*Ann. Rev. Pl. Physiol.* **16**: 175, 1965; photobiology of fungi), Kritsky *et al.* (*Molecular Biology* **39**: 514, 2005), Marsh *et al.* (*Pl. Dis. Reptr Suppl.* **261**: 251, 1959; guide to lit.). Cf. coolplate, luminescent fungi, photo-, zonation.

Ligiella J.A. Sáenz (1980), Phallaceae. 1, Costa Rica. See Sáenz (*Mycol.* **72**: 338, 1980).

ligneous (**lignose**), wood-like.

lignicolous, living on or in wood.

Ligniella Naumov (1926) = Discula fide Robak (*Sydowia* **6**: 378, 1952).

Lignincola Höhnk (1955), Halosphaeriaceae. 2 (marine), widespread. See Jones (*CJB* **74** Suppl. 1: S790, 1995; ultrastr.), Spatafora *et al.* (*Am. J. Bot.* **85**: 1569, 1998; phylogenetic analysis), Kong *et al.* (*MR* **104**: 35, 2000; DNA), Pang *et al.* (*Mycol. Progr.* **2**: 29, 2003; polyphyly), Zhang *et al.* (*Mycol.* **98**: 1076, 2006; phylogeny).

lignituber, see papillae.

Lignoscripta B.D. Ryan (2004), Agyriaceae (L). 1, USA. See Nash *et al.* (*Lichen Flora of the Greater Sonoran Desert Region* **2**: 350, 2004).

Lignosus Lloyd ex Torrend (1920), Polyporaceae. 5, widespread. See Ryvarden & Johansen (*Prelim. Polyp. Fl. E. Afr.*: 405, 1980), Douanla-Meli & Langer (*Mycotaxon* **86**: 389, 2003; Cameroon).

ligulate (**liguliform**), flat and narrow; strap-like in form; lorate.

Lilaceophlebia (Parmasto) Spirin & Zmitr. (2004), Meruliaceae. 1, widespread. See Spirin & Zmitrovich (*Nov. sist. Niz. Rast.* **37**: 177, 2004).

Lillicoa Sherwood (1977), Stictidaceae. 1, N. America. See Sherwood (*Mycotaxon* **5**: 57, 1977).

Lilliputeana Sérus. (1989) nom. inval. = Scoliciosporum fide Hawksworth *et al.* (*Dictionary of the Fungi* edn 8, 1995).

Lilliputia Boud. & Pat. (1900) = Roumegueriella fide Malloch & Cain (*CJB* **50**: 61, 1972), Rossman *et al.* (*Stud. Mycol.* **42**: 248 pp., 1999).

Limacella Earle (1909), Amanitaceae. *c.* 20, widespread. See Smith (*Pap. Mich. Acad. Sci.* **30**: 125,

1944; key), Yang & Chou (*Mycotaxon* **83**: 77, 2002; Taiwan), Bhatt *et al.* (*Mycotaxon* **88**: 249, 2003; India).

Limacinia Neger (1895) nom. dub., Fungi. See Hughes (*Mycol.* **68**: 693, 1976) but see, Reynolds (*Mycotaxon* **23**: 153, 1985).

Limaciniaseta D.R. Reynolds (1998), Capnodiaceae. 1, USA. See Reynolds (*Madroño* **45**: 250, 1998).

Limaciniella J.M. Mend. (1925) = Actinocymbe fide von Arx & Müller (*Stud. Mycol.* **9**, 1975).

Limaciniopsis J.M. Mend. (1925) nom. dub., Capnodiales. 1, Hawaii. See von Arx & Müller (*Stud. Mycol.* **9**, 1975).

Limacinula (Sacc. & D. Sacc.) Höhn. (1909) ≡ Limacinula Höhn.

Limacinula Höhn. (1907), Coccodiniaceae. 6, widespread (tropical). See Reynolds (*Mycol.* **63**: 1173, 1973; key), Reynolds (*Mycotaxon* **27**: 377, 1986; phylogeny), Reynolds & Gilbert (*Aust. Syst. Bot.* **18**: 265, 2005; Australia), Reynolds & Gilbert (*Cryptog. Mycol.* **27**: 249, 2006; Panama).

Limacinus (Quél.) Marchand (1896) ? = Hygrophorus fide Kuyper (*in litt.*).

Limacium (Fr.) P. Kumm. (1871) = Hygrophorus fide Singer (*Agaric. mod. Tax.*, 1951).

Limacospora Jørg. Koch & E.B.G. Jones (1995), Halosphaeriaceae. 1 (marine). See Jones *et al.* (*CJB* **73**: 1010, 1995).

Limbalba Nieuwl. (1916) ≡ Wainioa.

limbate (1) edged with another colour; (2) (of a volva), adnate to base of stipe and having a narrow, free, membranous margin (Bas, 1969).

Limboria Ach. (1815) nom. conf., Dothideomycetes.

Limnaiomyces Thaxt. (1900), Laboulbeniaceae. 3, widespread. See Nannfeldt (*Svensk bot. Tidskr.* **43**: 468, 1949).

Limnomyces Lohammar (1953) nom. dub., Fungi. 'phycomycetes'.

Limnoperdaceae G.A. Escobar (1976), Agaricales. 1 gen., 1 spp.

Lit.: Escobar & McCabe (*Mycotaxon* **9**: 48, 1979).

Limnoperdon G.A. Escobar (1976), Limnoperdaceae. 1, USA; Japan; S. Africa; Europe. See Escobar & McCabe (*Mycotaxon* **9**: 48, 1979; gastroid rel. *Cyphellaceae*), Webster *et al.* (*South African Journal of Botany* **59**: 519, 1993; South Africa).

limoniform, lemon-like in form.

Limonomyces Stalpers & Loer. (1982), Corticiaceae. 2, Europe; N. America. See Stalpers & Loerakker (*CJB* **60**: 533, 1982).

Lindau (Gustav; 1866-1923; Germany). Student of Schwendener (q.v.), Berlin (1888); working with Brefeld (q.v.), Münster (1888-1890); Member of staff (1892-1899) then Keeper (1899-1902) then Professor (1902-1923), Royal Botanical Museum and Botanic Garden, Berlin. Organizer of the series *Kryptogamen-Flora der Mark Brandenburg*; with P. Sydow (q.v.) produced a comprehensive bibliography of mycological literature up to 1910. *Publs.* Fungi Imperfecti: Hyphomycetes. *Rabenhorst's Kryptogamen-Flora* 2 vols (1904-1910); Pflanzliche Parasiten. *Sorauer's Handbuch der PflanzenKrankheiten* (1905-1908); Hemiasci. *Kryptogamen-Flora der Mark Brandenburg* (1906); *Hilfsbuch für das Sammeln parasitischer Pilze* (1901); *Hilfsbuch für das Sammeln der Ascomyceten mit Berücksichtigung der Nährpflanzen Deutschlands, Österreich-Ungarns, Belgiens, der Schweiz und der Niederlande* (1903); (with P. Sydow) *Thesaurus Litteraturae Mycologicae* 5 vols (1908-1918) [these volumes are freely available online; see Internet: catalogues & thesauri]. *Biogs, obits etc.* Grummann (1974: 29); Loesener (*Bericht der Deutschen Botanischen Gesellschaft* **41**: (93), 1923); Stafleu & Cowan (*TL-2* **3**: 23, 1981).

Lindauella Rehm (1900), ? Phyllachorales. 1 (from living grass leaves), S. America. Affinities uncertain. See Sherwood (*Mycotaxon* **5**: 1, 1977).

Lindauomyces Koord. (1907) = Arthrobotryum Ces. fide Clements & Shear (*Gen. Fung.*, 1931).

Lindauopsis Zahlbr. (1907) = Caloplaca fide Reidl (*Sydowia* **28**: 166, 1976).

Lindavia Nieuwl. (1916) [non *Lindavia* (F. Schütt) De Toni & Forti 1900, *Algae*] ≡ Scopularia.

Linderia G. Cunn. (1931) [non *Lindera* Thunb. 1783, *Lauraceae*] ≡ Linderiella.

Linderiella G. Cunn. (1942) ≡ Clathrus fide Dring (*Kew Bull.* **35**: 12, 1980).

Linderina Raper & Fennell (1952), Kickxellaceae. 2, widespread. See Benjamin (*Aliso* **4**: 321, 1959), Chang (*TBMS* **50**: 312, 1967), Young (*Ann. Bot.* **33**: 211, 1969), Young (*TBMS* **54**: 15, 1970), Young (*TBMS* **55**: 29, 1970; ultrastr.), Benny & Aldrich (*CJB* **53**: 2325, 1975; ultrastr. spore ontogeny), McKeown *et al.* (*MR* **100**: 821, 1996; ultrastr.), Konova *et al.* (*Microbiology* **74**: 99, 2005; fatty acids).

Linderomyces Singer (1947) = Gloeocantharellus fide Petersen (*The genera Gomphus and Gloeocantharellus in North America*, 1972).

Lindquistia Subram. & Chandrash. (1977), anamorphic *Poronia*, Hsy.0eH.6. 1, widespread. See Subramanian & Chandrashekara (*Boln Soc. argent. Bot.* **17**: 150, 1977), Rogers & Læssøe (*Mycotaxon* **44**: 435, 1992), Rogers *et al.* (*Mycotaxon* **67**: 61, 1998).

Lindquistomyces Aramb., E. Müll. & Gamundí (1982), Xylariales. 1 (from leaves of *Nothofagus*), Argentina. See Arambarri *et al.* (*Sydowia* **35**: 6, 1982).

Lindra I.M. Wilson (1956), Lulworthiaceae. Anamorph *Anguillospora*. 4 (marine), Europe. See Nakagiri (*TMSJ* **25**: 377, 1985), Yusoff *et al.* (*Cryptog. Bot.* **5**: 307, 1995; ultrastr.), Spatafora *et al.* (*Am. J. Bot.* **85**: 1569, 1998; phylogenetic analysis), Kohlmeyer *et al.* (*Mycol.* **92**: 453, 2000; morphology, DNA), Campbell *et al.* (*MR* **109**: 556, 2005; phylogeny), Zhang *et al.* (*Mycol.* **98**: 1076, 2006; phylogeny).

Lindroth, see Liro.

Lindrothia Syd. (1922) = Puccinia fide Arthur (*Manual Rusts US & Canada*, 1934).

Lindsay (William Lauder; 1829-1880; Scotland). Graduated in medicine, University (1852); resident physician, Murray's Royal Institution for the Insane, Perth (1854-1879). A proponent of non-restraint treatments, he studied the practices of psychiatric hospitals in other countries, and mental aspects of animals (*Mind in the Lower Animals* 2 vols, 1879); he regarded natural history as therapeutic and popularized lichenology in particular; a polymath with a rigorous scientific approach he studied dyeing and chemical variation within lichens, and opened up little explored areas of science, including the diversity of lichenicolous fungi and conidiomata in lichens; visited Otago, New Zealand (1861), thereafter publishing profusely not only on lichens and other fungi but also mineralogy and plants from that country (e.g. *Contributions to New Zealand Botany*, 1868) and even colonialism. Collections now in **E**; many new taxa based on specimens loaned from **K** (now in **BM**). *Publs. A Popular History of British Lichens* (1856); Memoirs on the spermagones and pycnides of lichens. *Transactions of the Royal Society of Edinburgh* (1859) [macrolichens]; Observations on new lichenicolous microfungi. *Transactions of the Royal Society of Edinburgh* (1869); *Transactions of the Linnean Society* London (1872) [crustose lichens]. *Biogs, obits etc.* Anon. (*Proceedings of the Royal Society of Edinburgh* **1881-1882**: 736, 1882); Grummann (1974: 376); Hawksworth & Seaward (*Lichenology in the British Isles 1568-1975*, 1977); Stafleu & Cowan (*TL-2* **3**: 63, 1981).

Lindtneria Pilát (1938), Stephanosporaceae. 8, widespread. See Hjortstam (*Mycotaxon* **28**: 19, 1987; key), Larsen (*Mycotaxon* **75**: 131, 2000; new sp.).

Lindtneriaceae Jülich (1982) = Corticiaceae.

line (as a measure of length), 2.1167 mm ($^{1}/_{12}$ inch); **Paris -** (Parisier Linie, P.L.), 2.2558 mm; **Paris inch**, 27.9 mm (fide Mason, *Mycol. Pap.* **3**: 24, 1933); 'p.p.' and 'p.p.p.' (parts per Paris inch) are abbreviations used by Corda.

linear, long and narrow.

Linearistroma Höhn. (1910), Clavicipitaceae. Anamorph *Ephelis*. 1 (on grasses), Brazil; India. See Phelps *et al.* (*Mycotaxon* **48**: 165, 1993).

Lineolata Kohlm. & Volkm.-Kohlm. (1990), Dothideomycetes. 1 (marine), pantropical. See Kohlmeyer & Volkmann-Kohlmeyer (*MR* **94**: 687, 1990).

lineolate, marked with lines.

Lineostroma H.J. Swart (1988), ? Venturiaceae. 1, Australia. See Swart (*TBMS* **91**: 453, 1988).

linguiform, see lingulate.

lingulate, tongue-like in form.

Linhartia Sacc. & P. Syd. (1902) = Psorotheciopsis fide Lücking (*Cryptog. Mycol.* **20**: 193, 1999), Henssen & Lücking (*Ann. bot. fenn.* **39**: 273, 2002).

Link (Johann Heinrich Friedrich; 1767-1851; Germany). Born at Hildesheim, and educated there and in Göttingen [obtaining medical degree there in 1789]; Professor, Rostock (1792-1811); Professor of Botany, Breslau (1811-1815); Professor of Botany and Director of the Botanic Garden and Royal Herbarium, Berlin (1815-1851). He provided early accounts of a range of important fungal genera. *Publs. Observationes in Ordines Plantarum Naturales* **1** (1809) and **2** (1815); Hyphomycetes (1824) and Gymnomycetes (1825) in Linné, *Species Plantarum* edn 4 [by Willdenow] **6**; *Handbuch zur Erkennung der.Gewächse* (1831-1833) [Fungi **3**: 274-486; = Willdenow, *Grundiss der Kräuterkunde.* **3**]. *Biogs, obits etc.* Grummann (1974: 29); Stafleu & Cowan (*TL-2* **3**: 65, 1981).

Linkiella Syd. (1921) = Puccinia fide Arthur (*Manual Rusts US & Canada*, 1934).

Linkomyces Golovin (1958) = Erysiphe fide Braun (*Nova Hedwigia* **34**: 689, 1981).

Linkosia A. Hern. Gut. & B. Sutton (1997), anamorphic *Pezizomycotina*, Hso.?.?. 2, Cuba. See Hernández-Gutiérrez & Sutton (*MR* **101**: 208, 1997), Wu & Zhuang (*Fungal Diversity Res. Ser.* **15**, 2005), Shenoy *et al.* (*MR* **110**: 916, 2006; phylogeny, polyphyly).

Linnaeus (Carl; 1707-1778; Carl von Linné; L.; Sweden). Professor of Medicine and Botany at the University of Uppsala (1741-1778). The father of modern taxonomy. Introduced the binomial system of biological nomenclature, but classified fungi as members of the plant kingdom. In *Species Plantarum* (1753), he drew extensively on work of Dillenius (*Historia Muscorum*, 1742) for treatment of lichen-forming fungi, and of Micheli (1729, q.v.) for other fungi, but recognized only about 170 fungal species. Among generic names used were *Agaricus*, *Clathrus*, *Hydnum*, *Mucor*, *Phallus* and *Tremella*; all lichen-forming fungi were placed in the genus *Lichen* (80 species.) except for some inadvertently placed in *Mucor* and *Tremella*. The enormous influence of this work confirmed a tradition in life sciences to treat fungi as plants. *Species Plantarum* is the nomenclatural starting point for plants and, because of that tradition, fungal names are still governed by the botanical code of nomenclature. This continues to affect development of mycology as a separate discipline. His last pupil was Acharius (q.v.). Collections and personal library are at the Linnean Society of London (**LINN**). *Biogs, obits etc.* Blunt (*The Compleat Naturalist*, 1971); Jørgensen *et al.* (*Botanical Journal of the Linnean Society* **115**: 261, 1994) [typification of 109 names of lichen-forming fungi]; Grummann (1974: 477); Ramsbottom (*Proceedings of the Linnean Society of London* **151**: 280, 1941) [Linnaeus' contribution to mycology]; Stafleu (*Linnaeus and the Linnaeans*, 1971); Stafleu & Cowan (*TL-2* **3**: 71, 1981); Stearn (*in* Linnaeus, *Species plantarum*, 2 vols, 1753) [reprint 1957, biography, reference collections, history of work, etc.]; Uggla (*Linnaeus*, 1957).

Linobolus Syd. & P. Syd. (1917) ? = Tubeufia fide von Arx & Müller (*Stud. Mycol.* **9**, 1975).

Linocarpon Syd. & P. Syd. (1917), Sordariomycetidae. Anamorph *Phialophora*-like. *c.* 42 (mainly on *Palmae*), widespread (tropical). See Hyde (*Sydowia* **44**: 32, 1992; key), Hyde (*J. Linn. Soc.* Bot. **123**: 109, 1997), Dulymamode *et al.* (*MR* **102**: 1331, 1998; Mauritius), Poonyth *et al.* (*Bot. Mar.* **43**: 213, 2000; ultrastr.), Thongkantha *et al.* (*Mycol.* **95**: 360, 2003; key), Yanna & Ho (*MR* **107**: 1305, 2003; ultrastr.), Zhang *et al.* (*Mycol.* **98**: 1076, 2006; phylogeny).

Linochora Höhn. (1910), anamorphic *Phyllachora*, St.0fH.?. 5, widespread (esp. tropical). See Cannon (*Mycol. Pap.* **163**: 302 pp., 1991).

Linochorella Syd. & P. Syd. (1912), anamorphic *Pezizomycotina*, St.≡ eH.?. 1, S. Africa.

Linodochium Höhn. (1909), anamorphic *Pezizomycotina*, Hsp.0fH.10. 2, Europe; N. America. See Dyko & Sutton (*CJB* **57**: 370, 1979), Dulymamode *et al.* (*Mycotaxon* **73**: 313, 1999).

Linopeltis I. Hino & Katum. (1961), Schizothyriaceae. 1, Japan. See Hino & Katumoto (*J. Jap. Bot.* **36**: 99, 1961).

Linopodium Earle (1909) = Mycena fide Singer (*Agaric. mod. Tax.*, 1951).

Linospora Fuckel (1870), Diaporthales. Anamorph *Depazea*. 4, Europe; N. America. See Barr (*Mycol. Mem.* **7**, 1978), Walker (*Mycotaxon* **11**: 1, 1980), Monod (*Beih. Sydowia* **9**: 1, 1983), Zhang & Blackwell (*Mycol.* **93**: 355, 2001; phylogeny).

Linosporoidea R. Keller (1895), Fossil Fungi (? perithecia) ? Fungi. 1 (Miocene), Switzerland.

Linostoma Höhn. (1918) [non *Linostoma* Wall. 1831, *Thymelaeaceae*] ≡ Ophiostoma.

Linostomella Petr. (1925) ? = Endoxyla fide Hawksworth *et al.* (*Dictionary of the Fungi* edn 8, 1995).

Linotexis Syd. & P. Syd. (1917) = Parenglerula fide Deighton (*Mycol. Pap.* **78**, 1960), Müller & von Arx

(*Beitr. Kryptfl. Schweiz* **11** no. 2, 1962).

lipids, esters of higher aliphatic alcohols (e.g. oils, fats, waxes); constituents of fungi (see Weete & Weber, *Lipid biochemistry of fungi and other organisms*, 1980).

lipochroic, see colour.

Lipocystis Cummins (1937), Raveneliaceae. 1 (on *Mimosa* (*Leguminosae*)), West Indies.

Lipomyces Lodder & Kreger-van Rij (1952), Lipomycetaceae. 13 (in soil), N. America; Europe. See Nieuwdorp *et al.* (*Antonie van Leeuwenhoek* **40**: 241, 1974; key), Smith *et al.* (*Int. J. Syst. Bacteriol.* **34**: 80, 1984; ultrastr.), Smith *et al.* (*Antonie van Leeuwenhoek* **68**: 75, 1995), Smith *et al.* (*Antonie van Leeuwenhoek* **68**: 177, 1995; DNA), Gouliamova *et al.* (*Antonie van Leeuwenhoek* **74**: 283, 1998), Smith *et al. in* Kurtzman & Fell (Eds) (*Yeasts, a taxonomic study* 4th edn: 174, 1998), van der Walt *et al.* (*Syst. Appl. Microbiol.* **22**: 229, 1999), Suh *et al.* (*Mycol.* **98**: 1006, 2006; phylogeny), Thanh (*Int. J. Syst. Evol. Microbiol.* **56**: 2009, 2006; Vietnam), Kurtzman *et al.* (*FEMS Yeast Res.* **7**: 1027, 2007; phylogeny), Müller *et al.* (*Can. J. Microbiol.* **53**: 509, 2007; cycloheximide tolerance).

Lipomycetaceae E.K. Novák & Zsolt (1961), Saccharomycetales. 3 gen. (+ 4 syn.), 20 spp.

Lit.: Hoog *et al.* (*Stud. Mycol.* **29**: 131 pp., 1986), van der Walt *et al.* (*Syst. Appl. Microbiol.* **9**: 115, 1987), Cottrell & Kock (*Syst. Appl. Microbiol.* **12**: 291, 1989), Weijman & van der Walt (*Stud. Mycol.* **31**: 193, 1989; cell wall carbohydrates), Kurtzman & Liu (*Curr. Microbiol.* **21**: 387, 1990), Kock *et al.* (*Antonie van Leeuwenhoek* **62**: 251, 1992; phylogeny gen.), Weijman & van der Walt (*Antonie van Leeuwenhoek* **62**: 247, 1992; review), Lomascolo *et al.* (*Can. J. Microbiol.* **40**: 724, 1994), Jansen van Rensburg *et al.* (*Syst. Appl. Microbiol.* **18**: 410, 1995; lipids, DNA), Yamada & Nogawa (*Bull. Fac. Agric. Shizuoka Univ.* **45**: 31, 1995), Gouliamova *et al.* (*Antonie van Leeuwenhoek* **74**: 283, 1998), Kurtzman *in* Kurtzman & Fell (Eds) (*Yeasts, a taxonomic study* 4th edn: 111, 1998), Smith *in* Kurtzman & Fell (Eds) (*Yeasts, a taxonomic study* 4th edn: 248, 1998), Walt *et al.* (*Syst. Appl. Microbiol.* **22**: 229, 1999), Smith *et al.* (*S. Afr. J. Sci.* **96**: 247, 2000), Suh *et al.* (*Mycol.* **98**: 1006, 2006; phylogeny).

Lipospora Arthur (1921) = Tranzschelia fide Arthur (*Manual Rusts US & Canada*, 1934).

lipsanenchyma, primordial tissue of a basidioma, other than the universal veil, covering the hymenium (Reijnders, 1963; Singer, 1962: 29).

lipstick mould, *Sporendonema purpurascens*, an invader of mushroom beds.

liquid nitrogen, see Genetic resource collections.

Lirasporis R. Potonié & Sah (1960), Fossil Fungi ? Fungi. 1 (Miocene), India.

lirella, a long, narrow apothecium as in *Graphis* and *Hysterium*.

Lirellodisca Aptroot (1998), Patellariaceae (L). 1, Papua New Guinea. See Aptroot & Iperen (*Nova Hedwigia* **67**: 485, 1998).

Liro [also 'Lindroth', his step-father's name under which he published up to 1906 when he took the Finnish name Liro] (Johan Ivar; 1872-1943; Finland). First Professor of Plant Biology and Plant Pathology, Helsinki University (1922-1943). A founder of the Vanamo Society (1896). Noted for more than 250 publications on taxonomic mycology and plant pathology, especially on fungi of Finland. *Publs. Die Ustilagineen Finnlands* 2 vols (1924, 1938). *Biogs, obits etc.* Grummann (1974: 620); Jamalainen (*Archivum Societatis Zoologicae Botanicae Fennicae 'Vanamo'* **3**, 1949) [biography, bibliography]; Stafleu & Cowan (*TL-2* **3**: 62, 1981).

Liroa Cif. (1933), Microbotryaceae. 1 (on *Polygonum*), Asia; Indonesia. See Kamat & Viswanathan (*Mycopathologia* **26**: 289, 1965), Vánky (*Illustrated genera of smut fungi*, 1987), Piepenbring (*Fungal Science* **17**: 55, 2002; morphology).

Lirula Darker (1967), Rhytismataceae. 7, widespread. See Darker (*CJB* **45**: 1420, 1967), Gernandt *et al.* (*Mycol.* **93**: 915, 2001; phylogeny), Guo *et al.* (*MR* **107**: 680, 2003; endophytes), Kaneko (*Mycoscience* **44**: 335, 2003; Japan), Ortiz-García *et al.* (*Mycol.* **95**: 846, 2003; phylogeny).

Lisea Sacc. (1877) = Gibberella fide Müller & von Arx (*Beitr. Kryptfl. Schweiz* **11** no. 2, 1962), Rossman *et al.* (*Stud. Mycol.* **42**: 248 pp., 1999).

Lisiella (Cooke & Massee) Sacc. (1891) = Gibberella fide Rossman *et al.* (*Stud. Mycol.* **42**: 248 pp., 1999).

Lister (Arthur, 1830-1908; England). A London wine merchant; amateur mycologist only in retirement (1887 onwards). Working through collections in the British Museum (Natural History), Royal Botanic Gardens, Kew and of de Bary (q.v.) in Strassburg, produced many early papers on, and a definitive monograph of the *Mycetozoa*, jointly with his daughter, G. Lister (q.v.); Fellow of the Royal Society (1898); President of the British Mycological Society (1906). *Publs. A monograph of the Mycetozoa* (1894) [edn 3, 1925]. *Biogs, obits etc.* Grummann (1974: 377); Hawker (*TBMS* **35**: 177, 1952) [portrait]; Lister (*Proceedings of the Royal Society* B **88**: i, 1915); Stafleu & Cowan (*TL-2* **3**: 118, 120, 1981).

Lister (Gulielma, 1860-1949; England). Daughter of A. Lister (q.v.); *de facto* honorary curator of the *Mycetozoa* collections, British Museum (Natural History) London. An eminent English amateur naturalist, noted for her many publications on the *Mycetozoa* and for her artwork, illustrating not only *Mycetozoa* but also various plants; twice President of the British Mycological Society (1912, 1932); the seventy-four manuscript notebooks she produced with her father were bequeathed to the British Mycological Society are are kept in the British Museum (Natural History) London. *Publs. A monograph of the Mycetozoa* edn 3 (1925). *Biogs, obits etc.* Ainsworth (*TBMS* **35**: 188, 1952) [notebooks]; Ainsworth & Balfour-Browne (*Nature* **188** (4748): 362, 1960); Grummann (1974: 377); Hawker (*TBMS* **35**: 177, 1952); Stafleu & Cowan (*TL-2* **3**. 118, 120, 1981); Wakefield (*TBMS* **33**: 165, 1950) [portrait].

Listeromyces Penz. & Sacc. (1901), anamorphic *Pezizomycotina*, Hsp.≡ eP.1. 1, Java; Hawaii. See Goos (*Mycol.* **63**: 213, 1971).

Literature. Mycological literature is vast. In the ninth edn of this *Dictionary*, it was estimated that there are not fewer than 500000 books and papers, with an annual output of *c.* 5000 items distributed through not less than 3500 publications, approximately half in English. The rate of output of paper publications has not diminished, and electronic publication on-line has grown exponentially since the previous edition.

Most literature is ephemeral, but systematic literature is longer-lived and must be readily available (the valid publication of fungal names dates from 1753, see Nomenclature). Much of the literature up to 1930 is listed in compilations by Lindau & Sydow (1908-18) and Ciferri (1957-60), see below [freely available on-line through *Cyberliber* (q.v.)]. Current systematic literature is covered in *Bibliography of Systematic Mycology* (1946-; twice-yearly, indexed to genus). In addition to new work, an increasing amount of archivad systematic literature can be found freely available on-line in various digital and downloadable formats. Internet search engines such as Google, and major international resources such as the CAB ABSTRACTS database are now an essential tool in any mycological studies.

Frequent use has been made here of the general works listed below. These are referred to by author and year or part, e.g. Buller (**5**), Clements & Shear (1931). In addition, literature is given under the names of genera and higher ranks, and special topics (e.g. Geographical distribution, Industrial mycology, Lichens, Macromycetes, Nomenclature, Physiology, Plant pathogenic fungi).

General Textbooks.

Ainsworth, G.C. & Sussman, A.S. (Eds), *The fungi, an advanced treatise*, **1** (The fungal cell), 1965; **2** (The fungal organism), 1966; **3** (The fungal population), 1968; (& Sparrow, F.K.) **4A** and **4B** (A taxonomic review), 1973.

Alexopoulos, C.J., *Introductory mycology*, edn 4 (& Mims, Blackwell), 1996; edn 3 (& Mimms), 1979; edn 2, 1962 [Germ. transl., *Einführung in die Mykologie*, by Farr, 1965]; edn 1, 1952.

Arx, J.A. von, *Pilzkunde*, 1967; edn 3, 1976.

Arx, J.A. von, *The genera of fungi sporulating in pure culture*, 1970; edn 3, 1981 [keys, *c.* 850 gen.].

Burnett, J.H., *Fundamentals of mycology*, edn 2, 1976; edn 1, 1968.

Cannon, P.F. & Kirk, P.M., *Fungal families of the world*, 2007.

Carlile, M.J. & Watkinson, S.C. *The fungi*, 1994.

Deacon, J.W., *Fungal biology*, edn 4, 2005; earlier editions as *Introduction to modern mycology*.

Deshmukh, S.K. & Rai, M.K. (Eds), *Biodiversity of fungi: their role in human life*, 2005.

Dugan, F.M., *The identification of fungi. An illustrated introduction with keys, glossary, and guide to literature*, 2006.

Esser, K., *Kryptogamen*, 1976. [Engl. transl. *Cryptogams*, by M.G. Hackston & J. Webster, 1981].

Esser, K.A. & Lemke, P.A. (Eds), *The Mycota*, **1**, *Growth, differentiaion and sexuality*, 1994; **2**, *Genetics and biotechnology*, 1995; **3**, *Biochemistry and molecular biology*, 1996; **4**, *Environmental and microbial relationships*, 1997; **5**, *Plant relationships*, 1997; **6**, *Animal and human relationships*, 1996; **7A** and **7B**, *Systematics and cell structure*, 2001; **8**, *Biology of the fungal cell*, 2007; **9**, *Fungal associations*, 2000; **10**, *Industrial applications*, 2002; **11**, *Agricultural applications*, 2002; **12**, *Human fungal pathogens*, 2004; **13**, *Fungal genomics*, 2006.

Gams, W., *et al.*, *CBS course of mycology*, 1975; edn 2, 1980.

Ganguly, B.N. & Deshmukh, S.K., [Eds], *Fungi: multifaceted microbes*, 2007.

Govi, G., *Introduzione alla Micologia*, 1986.

Gravesen, S., Frisvad, J.C. & Samson, R.A., *Microfungi*, 1994.

Hawksworth, D.L. (Ed.), *Frontiers in mycology*, 1991.

Herrera, T. & Ulloa, M., *El reino los fungos*, 1990.

Hudson, H.J., *Fungal biology*, 1986.

Ingold, C.T. & Hudson, H.J., *The biology of Fungi*, edn 6, 1993 [a good introduction]; edn 1, 1961.

Kendrick, B., *The fifth kingdom*, edn 3, 2000.

Margulis, L., Corliss, J.O., Melkonian, M. & Chapman, D.J., *Handbook of Protoctista*, 1990.

Moore-Landecker, E., *Fundamentals of the Fungi*, 1972; edn 3, 1990.

Spooner, B.M. & Robers, P.J., *Fungi* New Naturalist Library, 2005.

Webster, J. & Weber, R.S., *Introduction to fungi*, edn 3, 2007; edn 2, 1980; edn 1, 1970.

See also Genetics, Methods, Physiology, Sex.

Bibliographies.

Culberson, W.L., Recent literature on lichens 1-100; Egan, R.S., 101-143; Esslinger, T.L. 144- [*Bryologist* **54-**, 1951-]. Annotated bibliogr., 2-4 times per year.

Grummann, V.J., *Biographisch-bibliographisches Handbuch der Lichenologie*, 1974.

Lindau, G. & Sydow, P., *Thesaurus litteraturae mycologicae et lichenologicae*, 5 vols, 1908-18; *Supplementum*, 1911-1930, 4 vols, by R. Ciferri, 1957-60. Lists books and papers published up to 1930.

Poelt, J., *Systematik der Flechten* [*Fortschr. Bot. Berl.* **17-36**, 1955-74]; by H. Hertel, **38-42**, 1976-80. Annotated bibliogr. by group.

Stafleu, F.A., *Taxonomic literature* [*Regnum veg.* **52**], 1967.

Stafleu, F.A. & Cowan, R.S., *Taxonomic literature*, edn 2 [Regnum veg. **94, 98, 105, 110, 112, 115, 116**), 1976-88; Stafleu & Mennega (*Supplementum*) [*Regnum veg.* **125**, **132**, **130**], 1992-. Exhaustive treatment of major systematic lit. incl. dates publ., collections, types; good mycological and lichenological coverage.

Just's Bot. Jahresb. (63 vols., 1874-1944) included annotated lists of papers on fungi and lichens. Petrak contributed 2185 pp. citing lit. (**55-59**, **63**, 1930-44), and Zahlbruckner prepared comprehensive lichen lists (**13-59**, 1884-1931).

Bibliography of systematic mycology, 1946-; see above

Mycological abstracts, 1967-; see above

Biographies

See under Authors' names.

Catalogues of fungal names. With the exception of names published less than five years before present (for which *Index of Fungi* is the recommended source), virtually all of these names are now freely available on-line through *IndexFungorum* (q.v.), while the catalogues themselves (except recent issues of *Index of Fungi*) are freely available on-line as page images through *Cyberliber* (q.v.).

Index of fungi, 1940-. Catalogue of new fam., gen., spp., vars., ff., comb. novs., nom. novs. publ.; lichens incl. since **4-** (1970-). The *Index* has also published four supplements: *A Supplement to Petrak's lists 1920-1939* (1969), *Lichens 1961-1969* (1972), *Saccardo's omissions* (1985), *Family names* (1989), and *Authors of fungal names* (1992).

Lamb, I.M., *Index nominum lichenum*, 1963. Names publ. 1932-1960.

Petrak, F., Verzeichnis der neuen Arten, Varietäten, Formen, Namen und wichstigsten Synonyme. *Just's bot. Jber.* **48**(3): 184[**1**], **49**(2): 267 [**2**], **56**(2): 291 [**3**], **57**(2): 592 [**4**], **58**(1): 447 [**5**], **60**(1): 449 [**6**], **63**(2): 805 [**7**], 1930- 44; *Index of fungi 1936-1939*,

1950 [**8**]. Usually cited as 'Petrak's lists' nos. **1- 8**; fungal names 1920-40; all reprinted by CABI.
Saccardo, P.A., *Sylloge Fungorum*, 1882-1972, 26 vols [reprints 1944, 1967; microfiche edn, 1972]. Names gen., spp. with Latin descriptions to 1950.
Phycomycetes, vols 7, 9, 11, 14, 16, 17, 21, 24.
Pyrenomycetes, vols 1, 2, 9, 11, 14, 16, 17, 24.
Discomycetes, vols 8, 10, 11, 14, 16, 18, 22, 24.
Uredinales and Ustilaginales, vols 7, 9, 11, 14, 16, 17, 21, 23.
Hymenomycetes, vols 5, 6, 9, 11, 14, 16. 17. 21. 23.
Gasteromycetes, vols 7, 9, 11, 14, 16, 17, 21.
Coelomycetes, vols 3, 10, 11, 14, 16, 18, 22, 25.
Hyphomycetes, vols 4, 10, 11, 14, 16, 18, 22, 25.
Index, vol. 12; Host index, vol. 13; Synonyms, vol. 15; Index to illustrations, vol. 19; Generic index, vol. 18.
[Reed, C.F. & Farr, D.F., *Index*, 1993 (to scientific names).]
Zahlbruckner, A., *Catalogus lichenum universalis*, 10 vols, 1921- 40. Lichen gen. and infraspecific taxa to 1932; incl. full listings of citations of names by later authors.

Other lists of fungal names.
Clements, F.E. & Shear, C.L., *The genera of fungi*, 1931. [generic names; keys; largely of historical interest]
David, J.C., Fungal family names in current use. *Regnum veg.* **126**: 71-91, 1993.
Jong, S.C., *Stedman's ATCC fungus names*, 1993.
Pfeiffer, L.J.G., *Nomenclator botanicus*, 2 vols, 1871-74. [names of genera]
Streinz, W.M., *Nomenclator fungorum*, 1862. [alphabetical listing of genera and species with synonyms]
Steudel, E.G., *Nomenclator botanicus*, **2**, 1824. [especially useful for pre-1821 names].

Catalogues of names of other associated organisms. *Index Kewensis*, 1885-. [seed plants]. See Internet.

Checklists.
Checklists constitute the most basic information about what fungi occur in a particular region, and are essential for all other work with these organisms. Some general checklists are listed here (the list is very incomplete, and some are old). Checklists dealing with individual fungal groups are listed under the appropriate taxon name. Most new fungal checklists are being published digitally. See also Geographical distribution, Internet, Lichens, Macrofungi.

Africa
Alasoadura, S.O., *Occasional publications of the Department of Botany, University of Ibadan* **1**, 1970. [Nigeria]
Doidge, E.M. *Bothalia* **5**, 1950. [Angola, Botswana, Lesotho, Mozambique, South Africa, Swaziland, Zimbabwe]
Piening, L.J. *Bulletin of the Ghana Ministry of Agriculture* **2**, 1961. [Ghana]
Rieuf, P., *Cahiers de la recherche agronomique* **27**, 1969 & **28**, 1970. [Morocco]

Asia
Bilgrami, K.S. *et al.*, *Fungi of India, lists and references*, 1991.
Braun, U., *Schlechtendalia* **3**: 1, 1999. [Mongolia]
Lu, B.-s. *et al.*, *Checklist of Hong Kong fungi*, 2000.
Petch, T. & Bisby, G.R., The fungi of Ceylon. *Peradeniya manual* **6**: 1, 1950. [Sri Lanka]
Qasba, G.N. & Shah, A.M., *Fungi of Jammu, Kashmir & Ladakh*, 1991.
Sesli, E. & Denchev, C.M., *Mycologia Balcanica* **2**: 119, 2005. [Turkey]

Australasia
May, T.W. *et al.*, *Fungi of Australia* **2B**, 2003.

Caribbean
Minter, D.W. *et al.*, *Fungi of the Caribbean, an annotated checklist*, 2001. [insular Caribbean]

Central America
Piepenbring, M., *Checklist of fungi in Panama: preliminary version*, 2006.

Europe
Bontea, V., *Ciuperci parazite şi saprofite din România*, 1985-86. [Rumania]
Cannon *et al.*, *The British Ascomycotina*, 1985.
Jurc, D. *et al.*, *Glive Slovenije / Fungi of Slovenia*, 2005.
Legon, N.W. & Henrici, A., *Checklist of the British and Irish Basidiomycota*, 2005.
Lizoň, P. & Bacigálová, K. Huby / Fungi in Marhold, K. & Hindák, F. (Eds), *Zoznam mižších a vyšších rastlín Slovenska*, 1998 [Slovakia].
Minter, D.W. & Dudka, I.O., *Fungi of Ukraine, a preliminary checklist*, 1996.
Mirek, Z. [Ed.], *Biodiversity of Poland*, 2003-. [vols 6 - lichen-forming fungi, 7 - larger basidiomycetes, 8 - larger ascomycetes, 9 - microfungi, 10 - myxomycetes]
Onofri, S., *Checklist dei funghi Italiani*, 2005. [larger basidiomycetes only]
Zotti, M. & Orsino, F., *Flora Mediterranea* **11**: 115, 2001. [Italy, Liguria]

Middle East
Mouchacca, J., *J. Arid Environment* **60**: 359, 2005.

North America
Connors, J.L., *An annotated index of plant diseases in Canada*, 1967.
Farr, D.F. *et al.*, *Fungi on plants and plant products in the United States*, 1989.
Ginns, J.H., *Compendium of plant disease and decay fungi in Canada, 1960-1980*, 1986.

South America
Da Silva, M. & Minter, D.W., *Mycological Papers* **169**, 1995. [Brazil]
Mujica, R.F. *et al.*, *Flora fungosa Chilena*, edn 2, 1980. [Chile]

Mycotas
Some general mycotas are listed here (the list is very incomplete, and some are old), and they include descriptions of taxa. Mycotas dealing with individual fungal groups are listed under the appropriate taxon name. See also Geographical distribution, Internet, Lichens, Macrofungi.

Asia
Ito, S., [*Mycological flora of Japan*], vols **1-3**, 1936-1964. [in Japanese]
Miura, M., [*Flora of Manchuria and East Mongolia*], vols **1-3**, 1928. [in Japanese]
Prasher, I.B., *Fungi of Bhutan*, 1999.
Sawada, K., [*Formosan Fungi*], vols **1-11**, 1919-1959. [in Japanese]
Teng, S.-c., *Fungi of China*, 1996.
Teterevnikova-Babayan, D.N., *Микофлора Армянской СССР* [*Mycoflora of the Armenian SSR*, 1967-75. [Armenia, in Russian]
Ulyanishchev, B.I., *Микофлора Азербайджана* [*Mycoflora of Azerbaijan*], 1952-62. [in Russian]

Zhuang, W.-y., *Higher fungi of tropical China*, 2001.
various authors, *Флора споровых растений Казахстана* [*Flora of spore plants of Kazakhstan*], 1956-85. [in Russian]

Caribbean

Dennis, R.W.G., *Fungi flora of Venezuela and adjacent countries*, 1970. [Trinidad & Tobago]

Europe

Lind, J., *Danish fungi as represented in the herbarium of E. Rostrup*, 1913.
Möller, F.H., *Fungi of the Faeroes*, 1958.
Migula, W., *Kryptogamen-Flora von Deutschland, Österreich und der Schweiz*, edn 2, 1910-34. [Austria, Germany, Switzerland]
various authors, *Beiträge zur Kryptogamenflora der Schweiz*, 1901-. [Switzerland]
various authors, *Flora Italica cryptogama*, 1905-. [Italy]
various authors, *Kryptogamenflora der Mark Brandenburg*, 1905-. [Germany]
various authors, *Rabenhorst's Kryptogamen-Flora von Deutschland, Deutsch-Österreich und der Schweiz*, edn 2, 1881-1960. [Austria, Germany, Switzerland]
various authors, *Визначник грибів України* [*Guide to the fungi of Ukraine*], 1967-79. [in Ukrainian]

N. America

North American Flora, 1906-.

S. America

Chardón, C.E. & Toro, R.A., *J. Dept. Agric. Porto Rico* **14**: 195, 1930. [Colombia]
Dennis, R.W.G., *Fungi flora of Venezuela and adjacent countries*, 1970. [Guyana, Venezuela]

Methods.

See Methods.

Dictionaries and glossaries

Abiev, S.A. & Byzova, Z.M., *Микологический и фитопатологический словарь справочник* [*Mycological & Phytopathological Dictionary & Resource*], 2003. [in Russian]
Berger, K. (Ed.), *Mykologisches Wörterbuch*, 1980 [Germ., Engl., Fr., Span., Latin, Czech., Polish, Russ. glossary].
Cowan, S.T., *A dictionary of microbial taxonomy*, 1978.
Dörfelt, H., *Bi-Lexicon Mykologue Pilzkunde*, 1988.
Fidalgo, O. & Maria, E.P.K. *Dicionário micologico*, 1967 [*Rickia*, Suppl. 2].
Holliday, P., *A Dictionary of plant pathology*, 1989.
Josserand, M., *La description des champignons supérieurs*, 1952.
Martínez, A.T., Terminos científicos relacionados con los micromicetos. *Revista Ibérica Micol.* **2**: 36, 1985.
Singleton, P. & Sainsbury, D., *Dictionary of microbiology and molecular biology*, edn 2, 1987.
Snell, W.H., *Three thousand mycological terms*, 1936; & Dick, E.A., *A glossary of mycology*, 1957; edn 2, 1971.
Viégas, A.P., *Dicionario Almão-Português de micologia e fitopatologia*, 1958; Viégas, A.P., *Dicionario de fitopatologia e micologia*, 1979.
Ulloa & Hanlin, *Illustrated Dictionary of Mycology*, 2000.
Ulloa & Hanlin, *Nuevo Diccionario Ilustrado de Micología*, 2006 [Spanish edition of preceding work].

See also glossaries in Alexopoulos & Mimms (1979), Clements & Shear (1931), Kendrick (1992), Seaward (Ed.) (*Lichen ecology*, 1977), and Smith (1921), and *Lit.* cited under Nomenclature.

Major works also of historical interest:

Bary, A. de, *Vergleichende Morphologie und Biologie der Pilze* (1884) [Engl. transl. *Comparative morphology and biology of the Fungi, Mycetozoa, and Bacteria*, by Garnsey & Balfour, 1887].
Bessey, E.A., *A text-book of mycology*, 1935; *Morphology and taxonomy of fungi*, 1951 [reprint 1961]. [Many refs. systematic lit.].
Buller, A.H.R., *Researches on fungi*, **1** (1909), **2** (1911), **3** (1924), **4** (1931), **5** (1933), **6** (1934), **7** (1950). See also *TBMS* **57**: 5 (1971).
Chadefaud, M., *Les végé non vasculaires cryptogamie (Traité de botanique systématique*, **3**), 1960 [Fungi pp. 429-902].
Engler, A. & Prantl, K., *Die naturlichen Pflanzenfamilien* **1**, 1*, 1**; edn 2, **2**, **5b**, **6**, **7a**, **8**, 1889-1933. [Dates publ. parts, Stafleu, *Taxon* **21**: 501 (1972).] Descriptions fams. etc., illustr., incl. lichens; multi-authored.
Gäumann, E.A., *Vergleichende Morphologie der Pilze*, 1926. [Engl. transl., revision, *Comparative morphology of fungi*, by C.W. Dodge, 1928].
Gäumann, E.A., *Die Pilze*, 1949 [Engl. transl., *The fungi* by Wynd, 1952]; edn 2, 1965.
Jackson, B.D., *A glossary of botanic terms*, 1900; edn 4, 1928 [reprint 1960].
Kreisel, H., *Grundzuge eines natürlichen Systems der Pilze*, 1969.
Langeron, M., *Précis de mycologie*, 1945; edn 2, by Vanbreuseghem, 1952 [Engl. transl., 2 vol.; *Outline of mycology*, 1965].
Lohwag, H., *Anatomie der Asco- und Basidiomyceten*, 1941 (Linsbauer, Tischler & Pascher, *Handbuch der Pflanzenanatomie*, II, Abt. 3, Teilband c: Eumyceten, Band VI) [reprint 1965].
Martin, G.W., *Outline of the fungi*, 1950. Family keys, this *Dictionary* edn 5: 597-519 (1961).
Moreau, F., *Les champignons*, 2 vols, 1953.
Müller, E. & Loeffler, W., *Mykologie*, 1968; edn 2, 1971 [Engl. transl., *Mycology. An outline for science and medical students*, by B. Kendrick & F. Baerlocher, 1976].
Parker, S.P. (Ed.), *Synopsis and classification of living organisms*, 2 vols, 1982.
Talbot, P.H.B., *Principles of fungal taxonomy*, 1971.
Ubrizsy, G. & Vörös, J., *Mezögazdasági Mykologia*, 1968.
Wolf, F.A. & F.T., *The fungi*, 2 vol., 1947 [reprint, 1970].
See also Clements & Shear (1931) and Saccardo (1882-72) under Catalogues of names.

Mycological periodicals

Acta Mycologica (Warzawa-Łódż, Poland), 1965-.
Acta Mycologica Sinica (Beijing, China), 1982-96; replaced by *Mycosystema*.
African Journal of Mycology and Biotechnology (Cairo, Egypt), 1993-2001.
Annales Mycologici (Berlin, Germany), 1903-44; continued as *Sydowia*.
Atas do Instituto de Micologia. Universidade do Recife (Recife, Brazil), 1960-67.
Australasian Mycologist (Otago, New Zealand), 1996-.
Boletín de la Sociedad Mexicana de Micología (Mexico City, Mexico), 1985-87; continued as *Revista Mexicana de Micología*.

Bulletin Trimestriel de la Société Mycologique de France (Paris, France), 1885- [Indexes **1- 40** (1884-1924), 1934; **41-70** (1925-54), 1972].
Catathelasma (Bratislava, Slovakia), 2001-.
Česká Mykologie (Praha, former Czechoslovakia), 1947-93; continued as *Czech Mycology*.
Coolia (Leiden, Netherlands), 1954-.
Cryptogamie Mycologie (Paris, France), 1980- [in 3 ser., incorporating *Revue bryologique et lichénologique*, 1928-79; *Revue de mycologie*, 1936-79].
Czech Mycology (Praha, Czech Republic), 1994-.
Experimental Mycology (New York, USA), 1977-.
Field Mycology (Amsterdam, Netherlands), 1999-.
Friesia (Copenhagen, Denmark), 1932-80 (1981-, incorporated in *Nordic Journal of Botany*, Copenhagen).
Fungal Diversity (Hong Kong, China), 1998-.
Grevillea (London, UK), 1872-1894.
Hedwigia (Dresden, Germany), 1852-1944.
Herzogia (Vaduz, Liechtenstein), 1968-.
International Journal of Medicinal Mushrooms (Redding, Connecticut, USA), 1999-.
International Journal of Mushroom Sciences (Berkeley, California, USA), 1998-2000; continued as *Micología Aplicada International*.
International Journal of Mycology and Lichenology (Braunschweig, Germany), 1982-1992.
Journal of Mycology (Manhattan, Kansas, USA), 1885-1908.
Kavaka (Chennai, India), 1973-.
Lichenologist (London, UK), 1958-.
Michelia (Patavia, Italy), 1877-1881.
Micología Aplicada International (Berkeley, California, USA), 1998-.
Micologia Italiana (Bologna, Italy), 1972-.
Micología Neotropical Applicada (Puebla, Mexico), 1989-96; continued as *Micología Aplicada International*.
Микология и Фитопатология [*Mycology & Plant Pathology*] (Moscow-St Petersburg, Russia), 1967-.
Mycena (Minsk, Belarus), 2001-.
Mycologia (New York, USA), 1909- [Index, 1-58 (1909-66), 1968].
Mycologia Balcanica (Sofia, Bulgaria), 2004-.
Mycological Papers (Egham, UK), 1925-2000 [*Mycol. Pap.* **100**, Index to Papers 1-99, 1925-65].
Mycological Progress (Tübingen, Germany), 2002-.
Mycological Research (Amsterdam, Netherlands), 1989-.
Mycologist (Amsterdam, Netherlands), 1987-2006.
Mycopathologia (Heidelberg. Germany), 1975-.
Mycopathologia et Mycologia Applicata (den Haag. Netherlands), 1938-1974; continued as *Mycopathologia*.
Mycorrhiza (Heidelberg, Germany), 1992-.
Mycoscience (Tokyo, Japan), 1994-.
Mycosystema (Beijing, China), 1988-.
Mycotaxon (Ithaca, New York, USA), 1974-.
Nova Hedwigia (Lehre, Germany), 1959-.
Persoonia (Leiden, Netherlands), 1959-.
Revista Iberoamericana de Micología (Bilbao, Spain), 1984-.
Revista Mexicana de Micología (Mexico City, Mexico), 1985-.
Revue de Mycologie (Paris, France), 1936-79; continued as *Cryptogamie Mycologie*.
Rheinland-Pfälzisches Pilzjournal (Mittelhof, Germany), 1991-96.
Studies in Mycology (Baarn, Netherlands), 1973-.
Sydowia (Vienna, Austria), 1947-.
Systema Ascomycetum (Egham, UK), 1982-1997.
Transactions of the British Mycological Society (Cambridge, UK) 1898-1988 [Indexes, 1-30 (1898-1946), 1952; 31-40 (1947-57), 1961; 41- 67 (1958-76), 1979; 68-91 (1977-1988), 1991]; continued as *Mycological Research*.
Transactions of the Mycological Society of Japan (Tokyo, Japan), 1956-1993; continued as *Mycoscience*.

Other periodicals

Bryologist (Brooklyn, New York, USA), 1898-. [incl. lichenology]
Новости Систематики Низших Растений [*Novitates Systematicae Plantarum non Vascularum*] (St Petersburg, Russia), 1964-. [incl. fungi]
See also under Bibliographies and Catalogues of names above; and under Medical and veterinary mycology, Plant pathogenic fungi, and Societies and organizations.

litho- (prefix), pertaining to rocks.

Lithocia Gray (1821) = Verrucaria Schrad. fide Hawksworth *et al.* (*Dictionary of the Fungi* edn 8, 1995).

Lithoecis Clem. (1909) = Verrucaria Schrad. fide Hawksworth *et al.* (*Dictionary of the Fungi* edn 8, 1995).

Lithoglypha Brusse (1988), Acarosporaceae (L). 1, Natal. See Brusse (*Bothalia* **18**: 89, 1988).

Lithographa Nyl. (1857), Trapeliaceae (L). 10, widespread. See Hertel & Rambold (*Biblthca Lichenol.* **38**: 145, 1990; key), Lumbsch *et al.* (*MR* **105**: 265, 2001; asci), Coppins & Fryday (*Lichenologist* **38**: 93, 2006; Canada), Fryday & Coppins (*Lichenologist* **39**: 245, 2007; Chile), Lumbsch *et al.* (*MR* **111**: 1133, 2007).

Lithographomyces Cif. & Tomas. (1953) ≡ Lithographa.

Lithogyalideopsis Lücking, Sérus. & Vězda (2005), Gomphillaceae (L). 1, widespread. See Lücking *et al.* (*Lichenologist* **37**: 165, 2005), Lücking *et al.* (*Lichenologist* **38**: 131, 2006).

Lithomyces Viala & Marsais (1930) = Melanospora Corda fide Udagawa & Cain (*CJB* **47**: 1915, 1969).

lithophyte, a plant living on rocks; see saxicolous.

lithophytic, the habit of a lithophyte (q.v.).

Lithopolyporales R.K. Kar, N. Sharma, A. Agarwal & R. Kar (2003), Fossil Fungi, Polyporaceae. 1, India. See Kar *et al.* (*Curr. Sci.* **85**: 37, 2003).

Lithopythium Bornet & Flahault (1891), ? Dothideomycetes (?L). 1, Europe.

Lithosphaeria Beckh. ex Körb. (1863) = Psoroglaena fide Hawksworth *et al.* (*Dictionary of the Fungi* edn 8, 1995).

Lithothelidium M. Choisy (1954) nom. inval. = Polyblastia fide Hawksworth *et al.* (*Dictionary of the Fungi* edn 8, 1995).

Lithothelium Müll. Arg. (1885), Pyrenulaceae (L). 30, widespread (tropical). See Singh (*Curr. Sci.* **55**: 198, 1986), Aptroot (*Biblthca Lichenol.* **44**, 1991), McCarthy (*Lichenologist* **28**: 290, 1996), Harada (*Bryologist* **100**: 204, 1997), McCarthy (*Australasian Lichenology* **49**: 7, 2001; Christmas Island), Aptroot (*Lichenologist* **38**: 541, 2006; China, Thailand, key), Aptroot *et al.* (*Biblthca Lichenol.* **97**, 2008; Costa Rica).

lithotroph, utilizing rocks as nourishment.

Lithouncinula N. Sharma, R.K. Kar, A. Agarwal & R.

Kar (2005), Fossil Fungi. 1 (Maastrichtian).

litmus, a water-soluble mixture of amphoteric dyes obtained from depside-containing lichens, esp. *Roccella montagnei*; the mixture has the property of changing colour at different pH levels, and is widely used as an indicator of acidity. See dyeing.

Litschauerella Oberw. (1966), Hydnodontaceae. 4, widespread. See Eriksson & Ryvarden (*Bot. Notiser* **130**: 461, 1977).

Litschauerellaceae Jülich (1982) = Schizoporaceae.

Litschaueria Petr. (1923) = Helminthosphaeria fide Samuels *et al.* (*Mycol.* **89**: 141, 1997).

Littispora J. Campb., J.L. Anderson & Shearer (2003) = Saagaromyces fide Campbell *et al.* (*Mycol.* **95**: 547, 2003), Pang & Jones (*Nova Hedwigia* **78**: 269, 2004).

littoral, growing on sea or lake shores.

Lituaria Riess (1853) = Helicoma fide Linder (*Ann. Mo. bot. Gdn* **16**: 298, 1929).

lituate, forked, and having the points turned out a little.

liverwort, sometimes used in common names for large foliose lichens, e.g. ash-coloured ground liverwort (*Peltigera canina*).

Livia Velen. (1947), Helotiales. 1, Europe.

Lizonia (Ces. & De Not.) De Not. (1864), Pseudoperisporiaceae. 1, Europe. See Hansford (*Mycol. Pap.* **15**, 1946), Barr (*Mycotaxon* **64**: 149, 1997), Döbbeler (*Nova Hedwigia* **76**: 1, 2003).

Lizoniella Henn. ex Sacc. & D. Sacc. (1905) = Microcyclus fide Barr (*Mycotaxon* **39**: 43, 1990).

Llanoa C.W. Dodge (1968) = Umbilicaria fide Wei & Jiang (*Mycosystema* **2**: 135, 1989).

Llanolichen Tomas. & Cif. (1952) = Placynthium fide Hawksworth *et al.* (*Dictionary of the Fungi* edn 8, 1995).

Llanomyces Cif. & Tomas. (1953) ≡ Llanolichen.

Llimonaea Egea & Torrente (1991), Arthoniales (L). 2, Europe; N. Africa. See Egea & Torrente (*Mycotaxon* **53**: 63, 1995), Grube (*Bryologist* **101**: 377, 1998), Sparrius (*Biblthca Lichenol.* **89**: 141 pp., 2004), Tehler & Irestedt (*Cladistics* **23**: 432, 2007), van den Boom & Brand (*Lichenologist* **39**: 309, 2007; Europe, key).

Llimoniella Hafellner & Nav.-Ros. (1993), Helotiales. 10 (lichenicolous), Europe. See Hafellner & Navarro-Rosinés (*Herzogia* **9**: 769, 1993), Kümmerling *et al.* (*Biblthca Lichenol.* **53**: 147, 1993), Diederich & Etayo (*Lichenologist* **32**: 423, 2000), Kondratyuk *et al.* (*Mycol. Balcanica* **3**: 95, 2006; on *Caloplaca*).

Lloyd (Curtis Gates; 1859-1926; USA). A businessman and amateur mycologist. Noted for his studies on the *Gasteromycetes*, *Xylariaceae* etc. Founded, with his brothers, the Lloyd Library and Museum, Cincinnati (see *Lloydia* **27**: 141, 1964); he poked fun at 'name changers' by marking comments with the icon of a small smiling buddha-like figure and by making certain taxonomic and nomenclatural proposals under the name of a hypothetical 'Professor N.J. McGinty'; such proposals are generally attributed to Lloyd without comment, perhaps wrongly (Donk, *Reinwardtia* **1**: 205, 1951); these robust and interesting views led some, particularly nomenclaturalists, to consider him eccentric. Collections in **BPI** (*Mycol.* **24**: 247, 1932). *Publs. Mycological Notes* 7 vols (1898-1925) [in these privately produced volumes he published results of his mycological investigations; the volumes were indexed, *Bulletin of the Lloyd Library* **32** (mycological series 7): 64 pp., 1932; the new names proposed by Lloyd were indexed by Stevenson & Cash, *Bulletin of the Lloyd Library* **35** (mycological series 8): 209 pp., 1936)]. *Biogs, obits etc.* Fitzpatrick (*Mycol.* **19**: 153, 1927, portrait), Simons (*John Uri Lloyd, His life and works, 1849-1936, with a history of the Lloyd Library*, 1972); Stafleu & Cowan (*TL-2* **3**: 123, 1981).

Lloydella Bres. (1901) = Lopharia fide Donk (*Taxon* **5**: 69, 1956).

Lloydellopsis Pouzar (1959) ≡ Amylostereum.

Lloydia C.H. Chow (1935) [non *Lloydia* Salisb. 1812, *Liliaceae*] ≡ Sinolloydia fide Dring (*Kew Bull.* **35**: 68, 1980).

LN, liquid nitrogen; see Genetic resource collections.

LO-analysis (L, lux [light]; O, obscuritas [darkness]). A method introduced by Erdtman (see *An introduction to pollen analysis*, 1943; edn 2, 1954) for pollen, and used by Payak (*Nature* **184**: 738, 1959, *Mycopath.* **16**: 70, 1962, *Recent advances in palynology*: 27, 1964) for fungus spores, by which different types of surface ornamentation may be distinguished microscopically by the differences in appearance of the spore surface at upper and lower focus. See ornamentation, Fig. 20.

Lobaca Vězda (1983) nom. inval. = Fellhanera fide Vězda (*Folia geobot. phytotax.* **21**: 199, 1986).

Lobaria (Schreb.) Hoffm. (1796), Lobariaceae (L). *c.* 67, widespread. See Hale (*Bryologist* **60**: 35, 1957), Hakulinen (*Ann. bot. fenn.* **1**: 202, 1964), Jordan (*Bryologist* **73**: 669, 1970; internal cephalodia), Yoshimura (*J. Hattori bot. Lab.* **34**: 231, 1971), Jordan (*Bryologist* **76**: 225, 1973; N. Am. spp.), Wei *et al.* (*Acta Mycol. Sin.* Suppl. **1**: 363, 1989; 17 spp. China), Burgaz & Martínez (*Botanica Complutensis* **23**: 59, 1999; Iberian peninsula), Zoller *et al.* (*Mol. Ecol.* **8**: 2049, 1999; genetic diversity), Thomas *et al.* (*Biblthca Lichenol.* **82**: 123, 2001; phylogeny), Hallingbäck (*Svensk bot. Tidskr.* **97**: 26, 2002; conservation), Stenroos *et al.* (*CJB* **81**: 232, 2003; photobionts), Wiklund & Wedin (*Cladistics* **19**: 419, 2003; phylogeny), Miądlikowska & Lutzoni (*Am. J. Bot.* **91**: 449, 2004; phylogeny), Sipman (*Biblthca Lichenol.* **88**: 573, 2004; New Guinea), Walser *et al.* (*Heredity* **93**: 322, 2004; recombination), Miądlikowska *et al.* (*Mycol.* **98**: 1088, 2006; phylogeny).

Lobariaceae Chevall. (1826), Peltigerales (L). 7 gen. (+ 21 syn.), 370 spp.

Lit.: Yoshimura & Hurutani (*Bull. Kochi Gakuen Coll.* **18**: 345, 1987; SEM), Galloway (*Bull. Br. Mus. nat. Hist.* Bot. **17**: 1, 1988), Galloway (*Symbiosis* **11**: 327, 1991), Galloway *et al.* (*Fl. criptog. Tierra del Fuego* **13**, 1995), Galloway *et al.* (*Fl. criptog. Tierra del Fuego* **13**: 78 pp., 1995), Galloway (*Trop. Bryol.* **15**: 117, 1998), Sillet & Goward (*Lichenogr. Thomsoniana* North American Lichenology in Honor of John W. Thomson: 377, 1998), Zoller *et al.* (*Mol. Ecol.* **8**: 2049, 1999), McDonald *et al.* (*Bryologist* **106**: 61, 2003), Stenroos *et al.* (*CJB* **81**: 232, 2003), Walser *et al.* (*Fungal Genetics Biol.* **40**: 72, 2003), Wiklund & Wedin (*Cladistics* **19**: 419, 2003), Miądlikowska & Lutzoni (*Am. J. Bot.* **91**: 449, 2004), Miądlikowska *et al.* (*Mycol.* **98**: 1088, 2006; phylogeny), Hofstetter *et al.* (*Mol. Phylogen. Evol.* **44**: 412, 2007; phylogeny), Jørgensen & Tønsberg (*Nordic Lichen Flora* **3**: Cyanolichens: 77, 2007).

Lobariella Yoshim. (2002), Lobariaceae. 3, widespread. See Nash *et al.* (*Lichen Flora of the Greater Sonoran Desert Region* **1**: 270, 2002), Miądlikowska *et al.* (*Mycol.* **98**: 1088, 2006; phylogeny).

Lobarina Nyl. ex Cromb. (1894), Lobariaceae (L). 1, temperate. See Yoshimura *in* Marcelli & Ahti (Eds) (*Recollecting Edvard August Vainio*: 85, 1998), Stenroos *et al.* (*CJB* **81**: 232, 2003; photobionts).

Lobariomyces E.A. Thomas (1939) nom. inval. ≡ Lobaria.

lobate, lobed.

Lobatopedis P.M. Kirk (1979), anamorphic *Pezizomycotina*, Hso.≡ eP.3. 3, Europe; E. Africa. See Kirk (*TBMS* **73**: 75, 1979), Treigiene & Markovskaja (*Botanica Lithuanica* **9**: 285, 2003; Lithuania).

Lobiolataceae C. Agardh (1821) = Peltigeraceae.

Lobiona H. Kilias & Gotth. Schneid. (1978) = Toninia fide Timdal (*Op. Bot.* **110**, 1991).

Loboa Cif., P.C. Azevedo, Campos & Carneiro (1956) = Paracoccidioides fide Taborda *et al.* (*J. Clin. Microbiol.* **37**: 2031, 1999; synonymy with *Paracoccidioides*).

Lobodirina Follmann (1967) = Roccellina fide Tehler (*Op. Bot.* **70**, 1983).

Lobomyces Borelli (1968) nom. inval. = Lacazia fide Grigoriu *et al.* (*Medical Mycol.*, 1987), Taborda *et al.* (*J. Clin. Microbiol.* **37**: 2031, 1999).

lobomycosis, keloidal blastomycosis caused by *Lacazia loboi*.

Loborhiza A.M. Hanson (1944), Chytridiaceae. 1 (on *Volvox*), USA.

Lobosporangium M. Blackw. & Benny (2004), Mortierellaceae. 1, Mexico; USA. See Malloch (*Mycol.* **59**: 326, 1967), Ranzoni (*Mycol.* **60**: 356, 1968), Benny & Blackwell (*Mycol.* **96**: 144, 2004), White *et al.* (*Mycol.* **98**: 872, 2006; phylogeny).

Lobothallia (Clauzade & Cl. Roux) Hafellner (1991), Megasporaceae (L). 4, widespread (northern hemisphere). See Hafellner (*Acta Bot. Malac.* **16**: 138, 1991), Lumbsch (*Bryologist* **100**: 180, 1997; ontogeny), Schmitt *et al.* (*J. Hattori bot. Lab.* **100**: 753, 2006; phylogeny).

Lobularia Velen. (1934), ? Helotiales. 1, former Czechoslovakia.

lobulate, having small lobes.

Lobulicium K.H. Larss. & Hjortstam (1982), Atheliaceae. 1, Europe. See Larsson & Hjortstam (*Mycotaxon* **14**: 69, 1982).

Locelliderma Tehon (1935) = Hypoderma De Not. fide Hawksworth *et al.* (*Dictionary of the Fungi* edn 8, 1995).

Locellina Gillet (1876) nom. dub. ? = Cortinarius fide Kuyper (*in litt.*).

Lockerbia K.D. Hyde (1994), Sordariomycetes. 1, Australia. See Hyde (*Prodr. fl. neomarch.* **46**: 23, 1994).

locule (loculus), a cavity, esp. one in a stroma.

Loculistroma F. Patt. & Charles (1910), Hypocreales. 1 (on *Phyllostachys*), China. Type material is lost; possibly a synonym of *Aciculisporium*. See Rossman *et al.* (*Stud. Mycol.* **42**: 248 pp., 1999).

Loculoascomycetes Ainsw. (1966) (syn. Ascoloculares, Bitunicatae). This taxon was first proposed by Luttrell (*Mycol.* **47**: 511, 1955), as a subclass for *Ascomycota* with fissitunicately discharging asci, producing ascospores which are generally septate, and borne in unwalled locules (pseudothecia) in ascostromatic ascomata with an ascolocular ontogeny. Orders commonly recognized included the *Myriangiales* (asci globose), *Pleosporales* (asci clavate; pseudoparaphyses present), *Hysteriales* (ascocarp opening by a longitudinal split), *Dothideales* (asci clavate; pseudoparaphyses absent), *Capnodiales* (sooty-moulds; q.v.) and *Microthyriales* (ascomata shield-shaped). This class is not accepted in this edition of the *Dictionary*.

See Barr (*Prodromus to class Loculoascomycetes*, 1987), Lumbsch & Huhndorf (*MR* **111**: 1064, 2007).

Loculohypoxylon M.E. Barr (1976), Dothideomycetes. 1, N. America. See Barr (*Mycotaxon* **3**: 326, 1976), Barr (*Mycotaxon* **82**: 373, 2002; poss. affinities with *Teichospora*).

Loculomycetes. Class introduced by Gorovii (*Dopov. Akad. Nauk. ukr. RSR* B 1977(8): 742) for *Zerovaemyces* (q.v.), an agaric similar to *Coprinus* but in which the gills are replaced by spherical multilocular structures containing spores; each chamber contains one **loculospore**. Considered to be a new type of sexual sporulation in fungi.

loculospore, a spore of *Loculomycetes*, q.v.

Loculotuber Trappe, Parladé & I.F. Alvarez (1993), ? Tuberaceae. 1, Europe (southern); N. Africa. See Alvarez *et al.* (*Mycol.* **84**: 926, 1993), Læssøe & Hansen (*MR* **111**: 1075, 2007; phylogeny).

Lodderomyces Van der Walt (1966), Saccharomycetales. Anamorph *Candida*. 1, USA; S. Africa. Utilizes higher alkanes. See James *et al.* (*Lett. Appl. Microbiol.* **19**: 308, 1994), Kurtzman *in* Kurtzman & Fell (Eds) (*Yeasts, a taxonomic study* 4th edn: 254, 1998), Fitzpatrick *et al.* (*BMC Evolutionary Biology* **6**: 99, 2006; phylogeny), Suh *et al.* (*Mycol.* **98**: 1006, 2006; phylogeny).

Loflammia Vězda (1986), Ectolechiaceae (L). 4 (on leaves), widespread (tropical). See Vězda (*Folia geobot. phytotax.* **21**: 211, 1986), Aptroot *et al.* (*Mycotaxon* **88**: 41, 2003; China, Taiwan).

Loflammiopsis Lücking & Kalb (2000), Ectolechiaceae (L). 1, Brazil. See Lücking & Kalb (*Bot Jahrb. Syst.* **122**: 1, 2000; Brazil).

Logilvia Vězda (1986), Ectolechiaceae (L). 1 (on leaves), widespread (tropical). See Vězda (*Folia geobot. phytotax.* **21**: 211, 1986).

Lohwagia Petr. (1942), ? Phyllachoraceae. 3 (from living leaves and stems), C. & S. America. See von Arx & Müller (*Beitr. Kryptfl. Schweiz* **11** no. 1, 1954).

Lohwagiella Petr. (1970) = Niesslia fide Samuels & Barr (*CJB* **75**: 2165, 1998).

Lojkania Rehm (1905), Fenestellaceae. 10, widespread (temperate). See Barr (*Mycotaxon* **20**: 1, 1984), Yuan & Barr (*Sydowia* **46**: 338, 1995; key 4 spp. China), Lumbsch & Lindemuth (*MR* **105**: 901, 2001; phylogeny), Chen & Hsieh (*Fungal Science* Taipei **18**: 119, 2003; Taiwan).

Loliomyces Maire (1937), anamorphic *Pezizomycotina*. 1, Morocco.

Lolium endophyte. see endophyte.

Lollipopaia Inderbitzin (2001), Diaporthales. 1, Thailand. See Inderbitzin & Berbee (*CJB* **79**: 1100, 2001), Pinruan *et al.* (*Mycol.* **96**: 1163, 2004; phylogeny).

Loma Morrison & Sprague (1981), Microsporidia. 11.

Lomaantha Subram. (1954), anamorphic *Pezizomycotina*, Hso.≡ eP.1. 1, India. See Subramanian (*J. Indian bot. Soc.* **33**: 31, 1954), Wu & Zhuang (*Fungal Diversity Res. Ser.* **15**, 2005).

Lomachashaka Subram. (1956), anamorphic *Pezizomycotina*, Hsp.0eH/1eP.15. 4, W. Africa; India. See Nag Raj (*Mycotaxon* **53**: 311, 1995).

lomasome, a vesicle derived from an intracytoplasmic membrane (Marchant *et al.*, *New Phytol.* **66**: 623, 1967). Cf. plasmalemmasome.

Lomatia (Fr.) P. Karst. (1889) [non *Lomatia* R. Br. 1810, *Proteaceae*] ≡ Lomatina.

Lomatina P. Karst. (1892) = Cytidia fide Cooke (*Beih. Sydowia* **4**: 1, 1961).

Lomentospora Hennebert & B.G. Desai (1974) = Scedosporium fide Hennebert & Desai (*Mycotaxon* **1**: 45, 1974), Lennon *et al.* (*J. Clin. Microbiol.* **32**: 2413, 1994).

Lonchospermella Speg. (1908) nom. dub., anamorphic *Pezizomycotina*. See Sutton (*Mycol. Pap.* **141**, 1977).

Longevity. The life-span of an individual fungus in nature is difficult to ascertain. Mycorrhizal and other soil-dwelling fungi whose mycelia radiate out for huge distances may in general live longer than other other groups; it has been claimed that an 'individual' *Armillaria bulbosa* thallus in Michigan weighs in excess of 10,000 kg, occupies 15 ha, and has been genetically stable for over 1,500 yrs (Smith *et al.*, *Nature* **356**: 428, 1992; see critique by Brasier, *Nature* **356**: 382, 1992). The oldest viable fungi known were isolated from the inner portions of Greenland glacier ice cores, including glacial ice estimated as 140,000 yrs old (Ma *et al*, *Mycol.* **92**: 286, 2000). Some crustose lichens have been claimed to be older than 1,500 yrs, and 3,700 yrs has been presumed for a *Rhizocarpon geographicum* thallus in Alaska (Denton & Karlén, *Arctic Alp. Res.* **5**: 347, 1973). However, the tendency of individuals of some lichen species to coalesce into larger thalli makes such conclusions suspect. Studies on historical surfaces do, however, leave no doubt that some crustose lichens can live several centuries, e.g. a 620 yr *Aspicilia calcarea* thallus in Oxfordshire (Winchester, *Bot. J. Linn. Soc.* **96**: 57, 1988). In contrast, some lichens may exist for a few months or be seasonal (Poelt & Vězda, *Bibl. Lich.* **38**: 377, 1990). Fungal spores, sclerotia and other vegetative structures may retain their viability under natural conditions for periods of minutes or up to 50 years or more. With freeze drying or liquid nitrogen preservation, lyophilization and other culture collection techniques, they may retain viability in the laboratory for a period which, barring accidents and radiation-induced mutations, may be theoretically without limit. See Sussman (*in* Ainsworth & Sussman (Eds), *The Fungi* **3**: 447, 1968; review, refs.), Ellis & Roberson (*Mycol.* **60**: 399, 1968; lyophilized cultures), Hwang *et al.* (*Mycol.* **68**: 377, 1976; at ultra low temp.). See also Fairy rings, Genetic resource collections, Lichenometry, Record fungi.

Longia Syd. (1921) = Haploravenelia fide Dietel (*Nat. Pflanzenfam.* **6**, 1928).

Longia Zeller (1943) ≡ Longula.

longicollous, having long beaks or necks.

longiseptum, see septum.

Longoa Curzi (1927) = Togninia fide Eriksson (*SA* **5**: 139, 1986), Mostert *et al.* (*Mycol.* **95**: 646, 2003).

longuinose, see lanate.

Longula Zeller (1945) = Agaricus L. fide Harding (*Mycol.* **49**: 273, 1957), Vellinga (*MR* **108**: 354, 2004; basidioma gasteroid) See also.

Lopacidia Kalb (1984), Lecideaceae (L). 2, S. America. See Kalb (*Lichenes Neotropici* Fascicle **VIII** (nos 301-350) **8**: 2, 1984).

Lopadiaceae Hafellner (1984) = Ectolechiaceae.

Lopadiomyces Cif. & Tomas. (1953) nom. illegit. ≡ Lopadium.

Lopadiopsidomyces Cif. & Tomas. (1953) ≡ Lopadiopsis.

Lopadiopsis Vain. (1896) = Gyalectidium fide Hawksworth *et al.* (*Dictionary of the Fungi* edn 8, 1995).

Lopadium Körb. (1855) nom. cons., ? Ectolechiaceae (L). *c.* 50, widespread. See Santesson (*Symb. bot. upsal.* **12** no. 1: 1, 1952), Patwardhan & Makhija (*Indian J. Bot.* **4**: 20, 1981; 10 spp.).

Lopadostoma (Nitschke) Traverso (1906), Xylariaceae. Anamorph *Libertella*-like. 6, Europe. See Eriksson (*Svensk bot. Tidskr.* **60**: 315, 1966), Rappaz (*Mycol. Helv.* **7**: 99, 1995), Granmo & Petrini (*Mycol. Helv.* **8**: 43, 1996).

Lopezaria Kalb & Hafellner (1990), Ramalinaceae (L). 1, S. America. See Kalb & Hafellner (*Lichenes Neotropici* Fascicle **XI** (nos 451-475): 2, 1990), Miądlikowska *et al.* (*Mycol.* **98**: 1088, 2006; phylogeny).

Lopharia Kalchbr. & MacOwan (1881), Polyporaceae. 13, widespread. See Hjortstam & Ryvarden (*Syn. Fung.* **4**: 19, 1990; key), Boidin & Gilles (*BSMF* **118**: 91, 2002).

Lophariaceae Boidin, Mugnier & Canales (1998) = Phanerochaetaceae.

Lophiaceae H. Zogg ex Arx & E. Müll. (1975) = Mytilinidiaceae.

Lophidiopsis Berl. (1890) = Lophiostoma fide von Arx & Müller (*Stud. Mycol.* **9**, 1975).

Lophidium P. Karst. (1873) [non *Lophidium* Rich. 1792, *Schizaeaceae*] = Lophium fide Zogg (*Beitr. Kryptfl. Schweiz* **11** no. 3, 1962).

Lophidium Sacc. (1878) ≡ Platystomum.

Lophiella Sacc. (1878), Lophiostomataceae. 1, Europe. See Parbery (*Aust. J. Bot.* **15**: 271, 1967).

Lophionema Sacc. (1883), Lophiostomataceae. 2 or 3, widespread. See Chesters & Bell (*Mycol. Pap.* **120**, 1970).

Lophiosphaera Trevis. (1877) = Lophiostoma fide von Arx & Müller (*Stud. Mycol.* **9**, 1975).

Lophiosphaerella Hara (1948), ? Dothideomycetes. 1, Asia.

Lophiostoma Ces. & De Not. (1863) nom. cons., Lophiostomataceae. Anamorph *Pleurophomopsis*-like. 83, widespread. See Chesters & Bell (*Mycol. Pap.* **120**, 1970; key 23 spp.), Leuchtmann (*Sydowia* **38**: 158, 1986; anamorph), Holm & Holm (*Symb. bot. upsal.* **28** no. 2: 1, 1988; keys 28 spp.), Hyde & Aptroot (*Nova Hedwigia* **66**: 489, 1998; tropical freshwater spp.), Hyde *et al.* (*Nova Hedwigia* **70**: 143, 2000; on palms), Liew *et al.* (*Mol. Phylogen. Evol.* **16**: 392, 2000; phylogeny), Hyde *et al.* (*Fungal Diversity Res. Ser.* **7**: 93, 2002; marine taxa), Liew *et al.* (*Mycol.* **94**: 803, 2002; phylogeny), Tanaka & Harada (*Mycoscience* **44**: 85, 2003; Japan), Schoch *et al.* (*Mycol.* **98**: 1041, 2006; phylogeny), Kodsueb *et al.* (*Botanical Journal of the Linnean Society* **155**: 283, 2007; descr.).

Lophiostomataceae Sacc. (1883), Pleosporales. 15 gen. (+ 13 syn.), 138 spp. Paraphyletic as currently circumscribed.

Lit.: Chesters & Bell (*Mycol. Pap.* **120**, 1970), Holm (*Windahlia* **16**: 49, 1986), Kohlmeyer (*CJB* **64**: 3036, 1986), Holm & Holm (*Symb. bot. upsal.* **28**: 50 pp., 1988), Holm & Holm (*Symb. bot. upsal.* **28** no. 2, 1988; keys 28 spp.), Barr (*N. Amer. Fl.* ser. 2 **13**: 129 pp., 1990), Barr (*Mycotaxon* **45**: 191, 1992), Yuan & Zhao (*Symbiosis* **46**: 112, 1994; China), Berbee (*Mol. Biol. Evol.* **13**: 462, 1996), Checa (*Mycotaxon* **63**: 467, 1997; Iberian spp.), Aptroot (*Nova Hedwigia* **66**: 89, 1998), Hyde *et al.* (*Nova Hedwigia* **70**: 143, 2000), Liew *et al.* (*Mol. Phylogen. Evol.* **16**: 392, 2000), Liew *et al.* (*Mycol.* **94**: 803, 2002), Eriksson & Hawksworth (*Mycol.* **95**: 426, 2003), Schoch *et al.* (*Mycol.* **98**: 1041, 2006; phylogeny), Wang *et al.* (*MR* **111**: 1268, 2007; phylogeny).

Lophiotrema Sacc. (1878), Lophiostomataceae. 12, widespread. See Holm & Holm (*Symb. bot. upsal.* **28** no. 2, 1988), Tanaka & Harada (*Mycoscience* **44**: 115, 2003; Japan), Tang *et al.* (*Persoonia* **18**: 265, 2003; Hong Kong).

Lophiotricha Richon (1885) = Lophiostoma fide von Arx & Müller (*Stud. Mycol.* **9**, 1975).

Lophium Fr. (1823), Mytilinidiaceae. Anamorph *Papulaspora*-like. 4 (on conifers), Europe; America. See Zogg (*Beitr. Kryptfl. Schweiz* **11** no. 3, 1962), Vasil'eva (*Mikol. Fitopatol.* **35**: 15, 2001; Russia), Schoch *et al.* (*Mycol.* **98**: 1041, 2006; phylogeny).

Lophoderma Chevall. (1822) nom. rej. prop. ≡ Hypoderma DC.

Lophodermella Höhn. (1917), Rhytismataceae. 9, widespread. See Millar *in* Peterson (*Recent Research in Conifer Needle Diseases*: 45, 1986), Minter (*Mycopathologia* **121**: 51, 1993).

Lophodermellina Höhn. (1917) = Lophodermium fide Hawksworth *et al.* (*Dictionary of the Fungi* edn 8, 1995), Johnston (*Mycol. Pap.* **176**: 239 pp., 2001).

Lophodermina Höhn. (1917) = Lophodermium fide Hawksworth *et al.* (*Dictionary of the Fungi* edn 8, 1995), Johnston (*Mycol. Pap.* **176**: 239 pp., 2001).

Lophodermium Chevall. (1826) nom. cons., Rhytismataceae. Anamorph *Leptostroma*. 145, widespread. *L. seditiosum* (needle cast of *Pinus sylvestris*. See Darker (*Contr. Arnold Arbor.* **1**: 65, 1932), Tehon (*Ill. biol. Monogr.* **13**, 1935), Darker (*CJB* **45**: 1399, 1967), Morgan-Jones & Hulton (*Mycol.* **71**: 1043, 1979; ontogeny), Minter (*Mycol. Pap.* **147**, 1981; monogr. 16 spp. on pines, key), Cannon & Minter (*Taxon* **30**: 572, 1983; gen. nomencl.), Johnston (*N.Z. Jl Bot.* **27**: 243, 1989; key 21 spp.), Gernandt *et al.* (*Mycol.* **93**: 915, 2001; phylogeny), Johnston (*Aust. Syst. Bot.* **14**: 377, 2001; Australasia), Johnston (*Mycol. Pap.* **176**, 2001; spp. on monocots), Ortiz-García *et al.* (*Mycol.* **95**: 846, 2003; on pines), Hou *et al.* (*CJB* **83**: 37, 2005; on *Juniperus*, China), Lin *et al.* (*Mycosystema* **24**: 1, 2005; China), Wang *et al.* (*Mycol.* **98**: 1065, 2006; phylogeny).

Lophodermopsis Speg. (1910) nom. dub., Pezizomycotina. See Petrak & Sydow (*Annls mycol.* **34**: 29, 1936).

Lophodiscella Tehon (1933) = Colletotrichum fide Sutton (*Mycol. Pap.* **141**, 1977).

Lopholeptosphaeria Sousa da Câmara (1932) nom. dub., Dothideomycetes. No spp. Included.

Lophomerum Ouell. & Magasi (1966), Rhytismataceae. 6, widespread. See Johnston (*Mycotaxon* **33**: 423, 1988; gen. concept).

Lophophacidium Lagerb. (1949), Phacidiaceae. Anamorph *Apostrasseria*. 2, widespread (north temperate). See Corlett & Shoemaker (*CJB* **62**: 1836, 1984).

Lophophyton Matr. & Dassonv. (1899) = Trichophyton fide Carmichael *et al.* (*Genera of Hyphomycetes*, 1980).

Lophothelium Stirt. (1887) = Polycoccum fide Hawksworth (*Notes R. bot. Gdn Edinb.* **36**: 181, 1978).

Lophotrichaceae Seth (1971) = Microascaceae.

lophotrichous (**lophotrichate**), having several flagella at one or both ends.

Lophotrichus R.K. Benj. (1949), Microascaceae. Anamorph *Humicola*-like. 4 (coprophilous), widespread. See von Arx *et al.* (*Beih. Nova Hedwigia* **94**, 1988; key), Lee & Hanlin (*Mycol.* **91**: 434, 1999; DNA).

Loramyces W. Weston (1929), Loramycetaceae. Anamorph *Anguillospora*-like. 2, USA; Europe. See Ingold & Chapman (*TBMS* **35**: 268, 1952), Digby & Goos (*Mycol.* **79**: 821, 1988; posn), Gernandt *et al.* (*Mycol.* **93**: 915, 2001; phylogeny), Wang *et al.* (*Mycol.* **98**: 1065, 2006; phylogeny), Wang *et al.* (*Mol. Phylogen. Evol.* **41**: 295, 2006; phylogeny, links with *Mollisia*).

Loramycetaceae Dennis ex Digby & Goos (1988), Helotiales. 1 gen., 2 spp. Perhaps the most appropriate placement for *Mollisia* and its relatives.

Lit.: Digby & Goos (*Mycol.* **79**: 821, 1987), Gernandt *et al.* (*Mycol.* **93**: 915, 2001), Wang *et al.* (*Mycol.* **98**: 1065, 2006; phylogeny), Wang *et al.* (*Mol. Phylogen. Evol.* **41**: 295, 2006).

Loranitschkia Lar.N. Vassiljeva (1990), Nitschkiaceae. 1 (from wood), Sakhalin. See Vasil'eva (*Mikol. Fitopatol.* **24**: 207, 1990).

Loranthomyces Höhn. (1909) = Actinopeltis fide Spooner & Kirk (*MR* **94**: 223, 1990).

lorate, like a narrow band; strap-like in form; ligulate.

Loratospora Kohlm. & Volkm.-Kohlm. (1993), Planistromellaceae. 1, USA. See Kohlmeyer & Volkmann-Kohlmeyer (*SA* **12**: 7, 1993), Barr (*Mycotaxon* **60**: 433, 1996).

lorchel (**false morel**, **lorel**), the ascoma of *Gyromitra esculenta*, which is poisonous; see gyromitrin.

lorel, see lorchel.

Loreleia Redhead, Moncalvo, Vilgalys & Lutzoni (2002), Agaricomycetes. 3, Europe; N. America. *Hymenochaetales* or *Agaricales* (*Rickenella* clade). See Redhead *et al.* (*Mycotaxon* **82**: 162, 2002).

Loricella Velen. (1934), Helotiales. 2, former Czechoslovakia.

Loten Adans. (1763) nom. dub., ? Pezizomycotina.

lotic, habitat running water (streams). Cf. lentic.

Loweomyces (Kotl. & Pouzar) Jülich (1982), Meruliaceae. 2, widespread. See Jülich (*Persoonia* **11**: 424, 1982).

Loweporus J.E. Wright (1976) = Perenniporia fide Decock (*in litt.*) See, Ryvarden & Eriksson (*Prelim. Polyp. Fl. E. Afr.*: 410, 1980).

lower fungi, *Chytridiomycota*, *Hyphochytriomycota*, *Mycetozoa*, *Oomycota*, *Plasmodiophoromycota* and *Zygomycota*. See also phycomycetes.

Loxophyllum Klotzsch (1831) [non *Loxophyllum* Blume 1826, *Scrophulariaceae*] = Cyclomyces fide Saccardo (*Syll. fung.* **5**: 1, 1887).

Loxospora A. Massal. (1852), Sarrameanaceae (L). *c.* 5, widespread. See Staiger & Kalb (*Biblthca Lichenol.* **59**: 3, 1995), Kantvilas (*Herzogia* **14**: 35, 2000; S hemisphere), Miądlikowska *et al.* (*Mycol.* **98**: 1088, 2006; phylogeny), Nelsen *et al.* (*Lichenologist* **38**: 251, 2006; Costa Rica), Lumbsch *et al.* (*Lichenologist* **39**: 509, 2007; Australia), Lumbsch *et al.* (*Nova Hedwigia* **86**: 105, 2008; phylogeny, chemistry).

Loxosporaceae Kalb & Staiger (1995) = Sarrameanaceae. See Miądlikowska *et al.* (*Mycol.* **98**: 1088, 2006; phylogeny), Lumbsch *et al.* (*Nova Hedwigia* **86**: 105, 2008; phylogeny, chemistry).

Loxosporopsis Henssen (1995), Pertusariaceae (L). 1, USA. See Brodo & Henssen (*Biblthca Lichenol.* **58**: 27, 1995), Schmitt & Lumbsch (*Mol. Phylogen. Evol.* **33**: 43, 2004; chemistry, phylogeny), Lumbsch *et al.* (*Biol. J. Linn. Soc.* **89**: 615, 2006; evolution, chemistry), Schmitt *et al.* (*J. Hattori bot. Lab.* **100**: 753, 2006; phylogeny).

LPS, Instituto de Botánica C. Spegazzini (q.v.; La Plata, Argentina); founded 1935; a part of the Museo de La Plata.

LSD, see lysergic acid.

Luciotrichus R. Galán & Raitv. (1995), Pyronemataceae. 1, Spain. See Galán & Raitviir (*Czech Mycol.* **47**: 271, 1995).

Ludovicia Trevis. (1857) [non *Ludovicia* Coss. 1856, *Leguminosae*] = Baeomyces fide Hawksworth *et al.* (*Dictionary of the Fungi* edn 8, 1995).

Ludwigiella Petr. (1922) = Selenophoma fide Sutton (*Mycol. Pap.* **141**, 1977).

Ludwigomyces Kirschst. (1939), Pezizomycotina. 1, Europe.

Lueckingia Aptroot & Umaña (2006), ? Ramalinaceae (L). 1, Costa Rica. See Aptroot *et al.* (*J. Hattori bot. Lab.* **100**: 617, 2006).

Luellia K.H. Larss. & Hjortstam (1974), Hydnodontaceae. 3, Europe; N. America. See Larsson & Hjortstam (*Svensk bot. Tidskr.* **68**: 59, 1974).

Lulesia Singer (1970), Tricholomataceae. 3, tropical. See Singer (*Fl. Neotrop.* Monogr. **3**: 16, 1970), Lechner *et al.* (*Fungal Diversity* **21**: 131, 2006; Argentina).

Lulwoana Kohlm., Volkm.-Kohlm., J. Campb., Spatafora & Gräfenhan (2005), Lulworthiaceae. Anamorph *Zalerion*. 1, Japan. See Campbell *et al.* (*MR* **109**: 562, 2005).

Lulwoidea Kohlm., Volkm.-Kohlm., J. Campb., Spatafora & Gräfenhan (2005), Lulworthiaceae. 1, Denmark; Japan. See Campbell *et al.* (*MR* **109**: 564, 2005).

Lulworthia G.K. Sutherl. (1916), Lulworthiaceae. 10 (marine), widespread. See Nakagiri (*TMSJ* **25**: 377, 1985), Yusoff *et al.* (*Cryptog. Bot.* **5**: 307, 1995; ultrastr.), Spatafora *et al.* (*Am. J. Bot.* **85**: 1569, 1998; phylogenetic analysis), Kohlmeyer *et al.* (*Mycol.* **92**: 453, 2000; morphology, DNA), Campbell (*Mycol.* **97**: 549, 2005; nomencl.), Campbell *et al.* (*MR* **109**: 556, 2005; phylogeny), Descals (*MR* **109**: 545, 2005; phylogeny), Zhang *et al.* (*Mycol.* **98**: 1076, 2006; phylogeny).

Lulworthiaceae Kohlm., Spatafora & Volkm.-Kohlm. (2000), Lulworthiales. 7 gen. (+ 1 syn.), 27 spp.

Lit.: Nakagiri (*Trans. Mycol. Soc. Japan* **25**: 377, 1984), Kohlmeyer & Volkmann-Kohlmeyer (*Mycol.* **81**: 289, 1989), Yusoff *et al.* (*Cryptog. bot.* **5**: 307, 1995), Spatafora *et al.* (*Am. J. Bot.* **85**: 1569, 1998), Kohlmeyer *et al.* (*Mycol.* **92**: 453, 2000), Campbell *et al.* (*Fungal Diversity Res. Ser.* **7**: 15, 2002), Zuccaro *et al.* (*MR* **107**: 1451, 2003), Inderbitzin *et al.* (*MR* **108**: 737, 2004), Campbell *et al.* (*MR* **109**: 556, 2005), Tang *et al.* (*Ant. v. Leeuwenh.* **91**: 327, 2007).

Lulworthiales Kohlm., Spatafora & Volkm.-Kohlm. (2000). Spathulosporomycetidae. 3 fam., 11 gen., 40 spp. Fams:

(1) **Hispidicarpomycetaceae**
(2) **Lulworthiaceae**
(3) **Spathulosporaceae**

The *Hispidicarpomycetaceae* may also belong here, but molecular data are inadequate at present. *Lit.*: Kohlmeyer *et al.* (*Mycol.* **92**: 453, 2000; morphol., DNA), Kohlmeyer & Kohlmeyer (*Marine Mycology*, 1979), Nakagiri (*TMSJ* **25**: 377, 1984; morphol.), Spatafora *et al.* (*Am. J. Bot.* **85**: 1569, 1998; DNA).

lumen, the central cavity of a cell or other structure.

Luminescent fungi. Some fungi (40 listed by Wassink, 1979), such as species of *Panus* and *Pleurotus*, *Clitocybe illudens* (also mycelium; see Carey, *Mycol.* **66**: 951, 1974), *Armillaria mellea* (mycelium in wood exposed to air), and *Xylaria hypoxylon* (mycelium) and certain bacteria, give out light, sometimes causing the attacked wood or leaves to become luminous. See Berliner (*Mycol.* **53**: 84, 1962), Buller (*Researches* **3, 6**), Calleja & Reynolds (*TBMS* **55**: 149, 1970), Corner (*Mycol.* **42**: 423, 1950; *TBMS* **37**: 256, 1954), Glawe & Solberg (*Mycol.* **81**: 296, 1989; early records), Murrill (*Mycol.* **7**: 131, 1915), Ramsbottom (*Mushrooms and toadstools*, 1953), Wassink (*Meded. Landbouwhogeschool Wageningen* **79** (5), 1979; review, incl. physiol. basis).

lunate, like a new moon; crescentic (Fig. 23.9).

Lundquistia Vánky (2001) = Sporisorium fide Cunnington *et al.* (*Mycol. Balcanica* **2**: 95, 2005).

lungwort, the lichen *Lobaria pulmonaria* from its external resemblance to a human lung; lung lichen; lungs of oak. Formerly used in folk-lore as a cure for pulmonary diseases.

Lunospora Frandsen (1943) = Pseudoseptoria fide Sutton (*Mycol. Pap.* **141**, 1977).

Lunulospora Ingold (1942), anamorphic *Pezizomycotina*, Hso.0fH.11. 1 (aquatic), widespread.

lupinosis, a mycotoxicosis of sheep, caused by *Phoma leptostromiformis*; see van Warmelo & Marasas (*Mycol.* **64**: 316, 1972); teleomorph *Diaporthe* (q.v.).

luteoskyrin, a carcinogenic toxin of *Penicillium islandicum*; a cause of hepatitis (yellow rice disease) in man. See also islandicin; cf. skyrin.

luteous, yellow.

Luttrell (Everett Stanley, 1916-1988; USA). PhD, Duke University (1940); Mycologist and plant pathologist, Georgia Experiment Station (1942-1944, 1949-1966); University of Georgia (1966-1986) where he was later D.W. Brooks Distinguished Professor of Plant Pathology. His *Taxonomy of Pyrenomycetes* (1951) was a major advance in ascomycete systematics, linking the bitunicate ascus to ascolocular ontogeny the importance of which had been stressed by Nannfeldt (q.v.). *Publs.* Loculoascomycetes. *in* Ainsworth, Sparrow & Sussman [eds] *The Fungi* **4A**: 133 (1973) [the first critical keys to all non-lichen-forming loculoascomycete genera]; also detailed accounts of many *Helminthosporium*-like fungi and plant pathogenic ascomycetes. *Biogs, obits etc.* Hanlin & Garrett (*Phytopathology* **78**: 1388, 1988) [portrait]; Hanlin & Mims (*Mycol.* **82**: 9, 1990) [bibliography, portrait]: Glenn (*Inoculum* **45**: 1, 1998) [portrait].

Luttrellia Khokhr. & Gornostaĭ (1978) nom. rej. = Exserohilum fide Kirk (*in litt.*).

Luttrellia Shearer (1978), Halosphaeriaceae. 4 (marine), USA. See Kohlmeyer & Volkmann-Kohlmeyer (*Bot. Mar.* **34**: 1, 1991), Ferrer & Shearer (*Mycol.* **99**: 144, 2007; freshwater spp.).

Lutypha Khurana, K.S. Thind & Berthier (1977), Typhulaceae. 1, India. See Khurana *et al.* (*TBMS* **68**: 480, 1977).

Lutziomyces O.M. Fonseca (1939) ≡ Paracoccidioides.

Luxuriomyces R.F. Castañeda (1988) = Diplococcium fide Sutton (*in litt.*).

Luykenia Trevis. (1860) = Thelenella fide Mayrhofer & Poelt (*Herzogia* **7**: 13, 1985), Hawksworth *et al.* (*Dictionary of the Fungi* edn 8, 1995).

Luzfridiella R.F. Castañeda & W.B. Kendr. (1991), anamorphic *Pezizomycotina*, Hsy.1eP.24/27. 1, Cuba. See Castañeda Ruiz & Kendrick (*Univ. Waterloo Biol. Ser.* **35**: 69, 1991).

Lycogalopsis E. Fisch. (1886), Agaricaceae. 1, widespread (tropical). See Martin (*Lilloa* **4**: 69, 1939).

lycomarasmin, a dipeptide wilt toxin from *Fusarium bulbigenum* f.sp. *lycopersici* (Gäumann *et al.*, *Phytopath. Z.* **16**: 257, 1950, **30**: 87, 1957); **lycomarasmic acid** is a derivative (Gäumann & Naef-Roth, *Phytopath. Z.* **34**: 426, 1959). Cf. fusaric acid.

Lycoperdaceae Chevall. (1826) = Agaricaceae.

Lycoperdales = Agaricales.

Lycoperdastrum Haller ex Kuntze (1891) = Elaphomyces fide Dodge (*Ann. Myc.* **27**: 145, 1929).

Lycoperdastrum P. Micheli (1729) nom. inval. = Scleroderma fide Stalpers (*in litt.*).

Lycoperdellon Torrend (1913) = Ostracoderma fide Hennebert (*Persoonia* **7**: 200, 1973).

Lycoperdites Poinar (2001), Fossil fungi. 1.

Lycoperdodes Haller ex Kuntze (1891) = Pisolithus fide Fischer (*Nat. Pflanzenfam.* **5b**: viii, 1938).

Lycoperdoides, see *Lycoperdodes*.

Lycoperdon Pers. (1801), Agaricaceae. *c.* 50, widespread. The common Puff Balls. See Kreisel (*Khumbu Himal.* [*Ergebn. ForschUnte rnehmens Nepal Himalaya*] **6**: 32, 1969; key spp. pedicellate spores), Ponce de Leon (*Fieldiana, Bot.* **32**: 109, 1970; key (as *Vascellum*)), Smith (*Bull. mens. Soc. linn. Lyon* Num. Spéc. **43**: 407, 1974; N. Am. (as *Vascellum*)), Demoulin (*Mycotaxon* **3**: 275, 1976; key spp. peridial spherocysts), Demoulin (*Revista Biol.* **12**: 65, 1983; key S. Eur.), Homrich & Wright (*CJB* **66**: 1285, 1988; S. Am. (as *Vascellum*)), Kreisel (*Blyttia* **51**: 125, 1993; key (as *Vascellum*)), Kreisel & Calonge (*Mycotaxon* **48**: 13, 1993; as *Bovistella*), Kasuya (*Mycoscience* **45**: 298, 2004; Japan), Baseia (*Mycotaxon* **91**: 81, 2005; Brazil), Blaschke (*Mycologia Bavarica* **7**: 21, 2005; as *Bovistella*), Larsson & Jeppson (*MR* **112**: 4, 2008; phylogeny).

Lycoperdopsis Henn. (1899), Agaricaceae. 1, Asia (tropical). See Pegler & Young (*MR* **98**: 904, 1994).

Lylea Morgan-Jones (1975), anamorphic *Pezizomycotina*, Hso.≡ eP.3. 3, widespread (esp. north temperate). See Holubová-Jechová (*Folia geobot. phytotax.* **13**: 437, 1978), Chang (*Bot. Bull. Acad. sin.* Taipei **40**: 247, 1999).

lymabiont, an organism only found in sewage.

lymaphile, an organism commonly found in sewage (Cooke, *Sydowia, Beih.* **1**, 1957).

lymaphobe, an organism never found in sewage (Cooke, *Sydowia, Beih.* **1**, 1957).

lymaxene, an organism rarely found in sewage (Cooke, *Sydowia, Beih.* **1**, 1957).

Lymphocystidium Weiser (1943), ? Fungi. 1, Europe. or Protozoa. See Weiser (*Zool. Anzeiger* **142**: 200, 1943).

Lyoathelia Hjortstam & Ryvarden (2004), ? Atheliaceae. 1, Canada. See Hjortstam & Ryvarden (*Syn. Fung.* **18**: 10, 2004).

lyocystidium, see cystidium.

Lyomices, see *Lyomyces P. Karst. (1881)*.

Lyomyces P. Karst. (1881) nom. rej. = Hyphoderma Wallr. fide Donk (*Taxon* **5**: 69, 1956).

Lyomyces P. Karst. (1882) = Laeticorticium fide Donk (*Persoonia* **3**: 199, 1964).

Lyonella Syd. (1925), Pezizomycotina. 1, Hawaii.

Lyonomyces T.N. Taylor, Hass & W. Remy (1993), Fossil Fungi. 1 (Devonian), British Isles. See Taylor *et al.* (*Mycol.* **84**: 906 and fig. 19, 1992).

lyophilization, freeze-drying, a technique used to preserve fungal cultures in a state of suspended animation. See Genetic resource collections.

Lyophyllaceae Jülich (1982), Agaricales. 8 gen. (+ 6 syn.), 157 spp.

Lyophyllopsis Sathe & J.T. Daniel (1981), ? Lyophyllaceae. 1, India. See Sathe & Daniel (*Maharashtra Assoc. Cult. Sci.* Monogr. **1**: 87, 1980).

Lyophyllum P. Karst. (1881) nom. cons. prop., Lyophyllaceae. *c.* 40, widespread (north temperate). See Clemençon (*Mycotaxon* **15**: 67, 1982; staining spp.), Consiglio & Contu (*Rivista di Micologia* **45**: 99, 2002; Italy), Hofstetter *et al.* (*MR* **106**: 1043, 2002; phylogeny), Kalamees (*Folia cryptog. Estonica* **40**: 15, 2003; Nordic spp.), Consiglio & Contu (*Micologia e Vegetazione Mediterranea* **20**: 143, 2005; European collybioid spp.).

Lyromma Bat. & H. Maia (1965), anamorphic *Pyrenulales*, St.0fH.? (L). 3, widespread (tropical). See Aptroot *et al.* (*Biblthca Lichenol.* **64**, 1997), Lücking *et al.* (*Lichenologist* **30**: 121, 1998), Lücking & Kalb (*Bot. Jb.* **122**: 1, 2000; Brazil, key).

Lyrommotheca Bat. & T. Herrera (1972) nom. inval. ≡ Lyromma fide Aptroot (*in litt.*), Lücking *et al.* (*Lichenologist* **30**: 121, 1998).

lysergic acid, and derivatives, some of which are hallucinogenic (e.g. lysergic acid diethylamide; LSD) occur in *Claviceps* sclerotia and are the cause of paspalum staggers (q.v.); see also hallucinogenic fungi.

lysigenous, formed by the breaking down of cells. Cf. schizogenous.

Lysipenicillium Bref. (1908) = Roumegueriella fide Malloch & Cain (*CJB* **50**: 61, 1972), Rossman *et al.* (*Stud. Mycol.* **42**: 248 pp., 1999).

lysis (1) dissolution of a cell, e.g. by a lysin, see antigen; (2) (of conidial liberation), secession by the dissolution of the wall of the adjacent cell; cf. fission, fracture.

Lysospora Arthur (1906) = Puccinia fide Dietel (*Nat. Pflanzenfam.* **6**, 1928).

Lysotheca Cif. (1962), anamorphic *Pezizomycotina*, Cpd.0fH.?. 1 (on *Balladynella*), Dominican Republic. See Ciferri (*Atti Ist. bot. Univ. Lab. crittog. Pavia* sér. 5 **19**: 112, 1962).

Lysuraceae Corda (1842) = Phallaceae.

Lysurus Fr. (1823), Phallaceae. 5, widespread (esp. tropical). See Kasuya (*Nippon Kingakukai Kaiho* **45**: 39, 2004; Japan).

M, Botanischen Staatssammlung München (Munich, Germany); founded 1813; funded by the State of Bavaria; see Hertel & Schreiber (*Mitt. Bot. StSamml., Münch.* **28**: 81, 1988; history, collectors index).

MA, see Media.

Maasoglossum K.S. Thind & R. Sharma (1985), ? Geoglossaceae. 1, Himalaya. See Thind & Sharma (*Kavaka* **12**: 37, 1985).

Macabuna Buriticá & J.F. Hennen (1994), anamorphic *Phakopsoraceae*. 11, widespread. Anamorph name for (II). See Cummins & Hiratsuka (*Illustr. Gen. Rust Fungi* edn 3: 225 pp., 2003; syn. of *Calidion*).

macaedium, see mazaedium.

Macalpinia Arthur (1906) = Uromycladium fide Dietel (*Nat. Pflanzenfam.* **6**, 1928).
Macalpinomyces Langdon & Full. (1977), ? Ustilaginaceae. 44 (on *Poaceae*), widespread. See Vánky (*Mycotaxon* **59**: 115, 1996; monogr.), Vánky (*Mycotaxon* **62**: 128, 1997; emend.), Vánky (*Mycotaxon* **69**: 112, 1998; key), Shivas & Vánky (*Mycol. Balcanica* **2**: 101, 2005; Austral.), Vánky (*Mycotaxon* **95**: 1, 2006).
Macbridella Seaver (1909) = Byssosphaeria fide Samuels (*CJB* **51**: 1275, 1973), Barr (*Mycotaxon* **20**, 1984).
Maccagnia Mattir. (1922), ? Hydnangiaceae. 1, Italy.
Macentina Vězda (1973) = Psoroglaena fide Aptroot (*in litt.*), Coppins & Vězda (*Lichenologist* **9**: 47, 1977), Orange (*Lichenologist* **23**: 15, 1991; key), Matzer (*Mycol. Pap.* **171**: 202, 1996), Aragón & Sarrión (*Nova Hedwigia* **77**: 169, 2003; Spain).
Macilvainea Nieuwl. (1916) ≡ Drudeola fide Sutton (*Mycol. Pap.* **141**, 1977).
Mackenziea Yanna & K.D. Hyde (2002), anamorphic *Pezizomycotina*, H?.?.?. 1, Australia. See Yanna & Hyde (*Aust. Syst. Bot.* **15**: 757, 2002).
Mackintoshia Pacioni & Sharp (2000), Cortinariaceae. 1, Zimbabwe. See Pacioni & Sharp (*Mycotaxon* **75**: 225, 2000), Verbeken & Walleyn (*Boll. Gruppo Micol. 'G. Bresadola'* **46**: 87, 2003).
Macmillanina Kuntze (1898) = Disculina fide Sutton (*Mycol. Pap.* **141**, 1977).
Macowania Kalchbr. ex Berk. (1876) [non *Macowania* Oliv. 1870, *Compositae*] ≡ Macowanites.
Macowaniella Doidge (1921), Asterinaceae. 2, S. Africa.
Macowanites Kalchbr. (1882) = Russula.
Macraea Subram. (1953) [non *Macraea* Lindl. 1828, *Geraniaceae*] ≡ Prathigada.
macro- (prefix), long, but commonly used in the sense of mega- (q.v.).
Macroallantina Speer (1987), anamorphic *Delpinoina*, St.0eH.1. 1, France. See Speer (*BSMF* **103**: 14, 1987).
Macrobasis Starbäck (1893) = Leptosphaeria fide von Arx & Müller (*Stud. Mycol.* **9**, 1975).
Macrobiotophthora Reukauf (1912), Ancylistaceae. 2, Germany; Australia. See Reukauf (*Centralbl. Bakt.* Abt. Orig. **63**: 390, 1912), McCulloch (*TBMS* **68**: 173, 1977), Tucker (*Mycotaxon* **13**: 481, 1981; emend., key).
macrocephalic, see septum.
Macrochytrium Minden (1902), ? Chytridiaceae. 1, Europe; N. America.
macroconidium (1) the larger, and generally more diagnostic conidium of a fungus which has microconidia (and sometimes also mesoconidia) in addition; (2) (infrequent), a long large conidium.
Macrocybe Pegler & Lodge (1998), Tricholomataceae. 7, widespread (tropics). See Pegler *et al.* (*Mycol.* **90**: 496, 1998), Stijve (*AMK Mededelingen*: 93, 2004).
macrocyclic, see *Pucciniales* and Table 5.
macrocyst, see cyst.
Macrocystidia Joss. (1934), Marasmiaceae. 5, widespread. See Capellano (*BSMF* **92**: 221, 1971; basidiospore structure), Kühner (*Hyménomycètes agaricoïdes*: 428, 1980), Anderson *et al.* (*Mycologist* **15**: 132, 2001).
Macrocystidiaceae Kühner (1979) = Marasmiaceae.
Macrocystis R. Heim (1931) [non *Macrocystis* C. Agardh 1820, *Algae*] ≡ Macrocystidia.
Macrodendrophoma T. Johnson (1904) nom. inval., anamorphic *Pezizomycotina*. See Sutton (*Mycol. Pap.* **141**, 1977).
Macroderma Höhn. (1917), Rhytismataceae. 2, N. America; Brazil.
Macrodiaporthe Petr. (1920), Melanconidaceae. Anamorph *Melanconium*. 2, Europe; N. America. See Barr (*Mycol. Mem.* **7**, 1978).
Macrodictya A. Massal. (1852) ≡ Lasallia.
Macrodiplis Clem. & Shear (1931) ≡ Macrodiplodiopsis.
Macrodiplodia Sacc. (1884), anamorphic *Pezizomycotina*, C?.1eP.?. 2, Europe. See Sutton (*Mycol. Pap.* **141**, 1977).
Macrodiplodina Petr. (1962) = Ascochyta fide Punithalingam (*Mycol. Pap.* **142**, 1979).
Macrodiplodiopsis Petr. (1922), anamorphic *Pezizomycotina*, St.≡ -#eP.19. 1, Europe; N. America. See Glawe (*Mycol.* **77**: 880, 1985).
macrofungi, see Macromycetes.
macrogonidium, a large gonidium (obsol.); megalogonidium (obsol.).
Macrohilum H.J. Swart (1988), anamorphic *Diaporthales*, Cpd.1eP.19. 1, Australasia. See Swart (*TBMS* **90**: 288, 1988), Crous *et al.* (*Stud. Mycol.* **55**: 53, 2006; phylogeny).
Macrohyporia I. Johans. & Ryvarden (1979), Polyporaceae. 3, widespread (tropical). See Johansen & Ryvarden (*TBMS* **72**: 192, 1979).
Macrolepiota Singer (1948), Agaricaceae. *c.* 30, widespread. *M. procera*, the edible parasol mushroom. See Heinemann (*Bull. Jard. bot. nat. Belg.* **39**: 201, 1969; key 11 Congo spp.), Pegler (*Kew Bull.* Addit. Ser. **6**, 1977; E. Afr. spp.), Candusso & Lanzoni (*Lepiota s.l.*, 1990; key Europ. spp.), Johnson (*Mycol.* **91**: 443, 1998; phylogeny), Johnson & Vilgalys (*Mycol.* **90**: 971, 1998; phylogeny), Vellinga (*Aust. Syst. Bot.* **16**: 361, 2003; Australia), Vellinga (*Coolia* **46**: 177, 2003; The Netherlands), Vellinga *et al.* (*Mycol.* **95**: 442, 2003; phylogeny and taxonomy), Hausknecht & Pidlich-Aigner (*Öst. Z. Pilzk.* **13**: 1, 2004; Austria), Vellinga (*MR* **108**: 354, 2004; phylogeny).
macrolichen, one of the larger lichens of squamulose, foliose, or fruticose habit incl. spp. *Cladonia*, *Parmelia*, *Usnea*, etc.
Macrometrula Donk & Singer (1948), Psathyrellaceae. 1 (in glasshouse), British Isles (introduced).
Macromycetes. Fungi having large (macroscopic) sporocarps; larger fungi. This is an arbitrary division of the fungi which includes most (but far from all) basidiomycetes (mushrooms, puffballs, stinkhorns, toadstools etc.), and a small number of other fungi, mostly ascomycetes (cup fungi, cyttarias, morels, truffles etc.). Although in popular use, it is not always advantageous to apply the term: in conservation, for example, the result is that those working for ascomycete conservation tend to be deprived of their flagship species. Because of the interest shown by amateur mycologists, mycophagists, and others in macrofungi there are for many countries popular or semi-popular texts covering macromycetes. These works, which are appearing for an increasing number of countries, are frequently of a high mycological standard and are useful both to amateur and professional mycologists. They include:

Regional: **Africa**. **General**, Ryvarden *et al.* (*An introduction to the larger fungi of south central Africa*,

1994), Zoberi (*Tropical macrofungi*, 1972). **Benin**, De Kessel *et al.* (*Guide des champignons comestibles du Bénin*, 2002). **Burundi**, Buyck (*Ubwoba: les champignons comestibles de l'ouest de Burundi*, 1994). **Malawi**, Morris (*Common mushrooms of Malawi*, 1987; *Checklist of the macrofungi of Malawi* [*Kirkia* **13**: 323, 1990]). **Morocco**, Malençon & Bertault (*Flore des champignons supérieures du Maroc*, 2 vols, 1975). **Tanzania**, Härkönen *et al.* (*Tanzanian mushrooms: edible, harmful and other fungi*, 2003). **Asia**. **China**, Pfister (*Champignons du Tonkin*, 1985), Yuan & Sun (*中國蕈菌原色圖集* [*Coloured atlas of mushrooms (macrofungi) in China*, in Chinese], 2007). **Japan**, Imazeki *et al.* (*Fungi of Japan* [in Japanese], 1998). **South Korea**, Kim & Kim (*Illustrated Korean mushrooms* [in Korean], 1990), Park & lee (*Korean mushrooms* [in Korean], 1999). **Sri Lanka**, Pegler (*Kew Bull. Addn. Ser.* **12**, 1986). **Australasia**. **Australia**, Aberdeen (*Introduction to the mushrooms. of Queensland*, 1979; *Field companion to Australian fungi*, 1985; *A field guide to Australian fungi*, 2005), Shepherd & Totterdell (*Mushrooms and toadstools of Australia*, 1988), Young (*Common Australian fungi*, 1982), Young *et al.* (*A field guide to the fungi of Australia*, 2005). **New Zealand**, Taylor (*Mushrooms and toadstools in New Zealand*, 1970). **Caribbean**. **Dominican Republic**, Perdomo *et al.* (*Hongos comestibles de la República Dominicana, guía de Campo*, 2007). **Central America**. **Costa Rica**, Halling & Mueller (*Common mushrooms of the Talamanca mountains, Costa Rica*, 2005), Mata (*Macrohongos de Costa Rica* vol. 1, 1999), Mata *et al.* (*Macrohongos de Costa Rica* vol. 2, 2003). **Europe**. **General**, Hansen & Knudsen (*Nordic macromycetes*, 2 vols, 1992-3; Scandinavia only), Gams (*Kleine Kryptogamenflora* Band IIa, Julich, *Nichtblatterpilze, Gallertpilze und Bauchpilze*, Band IIb2, Lange (*Guía de campo de los hongos de Europa*, 1981), Moser, *Phycomyceten und Ascomyceten*, 1963, Bd IIb1, Moser, *Rohrlinge und Blatterpilze*, edn 5, 1983 [English edn, Phillips, 1978]), Pegler *et al.* (*Fungi of Europe*, 1993), Peter (*Das grosse Pilzbuch. Eine Pilzkunde Mitteleuropas*, 1964), *Fungi Europaei*, **1** Cappelli, *Agaricus*, 1984, **2** Alessio, *Boletus*, 1985, suppl. 1991, **3** Riva, *Tricholoma*, 1988, **4** Candusso & Lanzoni, *Lepiota*, 1990, **5** Noordeloos, *Entoloma*, 1992). **British Isles**, Buczacki (*Fungi of Britain and Europe*, 1989), Hofmann (*Field guide to mushrooms and other fungi of Britain and Europe*, 2006), Pegler (*Field guide to mushrooms and toadstools of Britain and Europe*, 1990), Wakefield & Dennis (*Common British fungi*, 1950; edn 2, 1981), Watling (*Identification of larger fungi*, 1973), see also Holden, *Guide to the literature for the identification of British fungi*, edn 4, 1982. **Bulgaria**, Sechanov ([*Fungi of Bulgaria*] edn 2, 1957). **Czech Republic**, Papoušek (*Velký fotoatlas hub* [in Czech], 2004). **Denmark**, Lange (*Illustreret svampeflora*, 1961 [also English, French and German versions]; *Soppflora*, 1991). **France**, Costantin & Dufour (*Nouvelle flore des champignons*, 1891 [and later editions]), Rolland (*Atlas des champignons de France, Suisse et Belgique*, 1910), Heim (*Les champignons d'Europe*, 2 vols, 1957 [edn 2 (1 vol.), 1969]), Maublanc (*Les champignons comestibles et veneneux de France*, edn 6 [by Viennot-Bourgin], 2 vols, 1971), Romagnesi (*Petit atlas des champignons*, 3 vols, 1962-63; *Nouvelle atlas des champignons*, 4 vols, edn 2, 1970), Marchand (*Champignons de Nord et du Midi*, 9 vols, 1972-86), Moreau (*Larousse des champignons*, 1978; *Guide des champignons comestibles et vénéneux*, 1984). **former Czechoslovakia**, Pilát (*Nase Houby*, 1952 [English transl., *Mushrooms*, 1954]; *Houby Ceskoslovenska*, 1969; spp. grouped ecologically). **Germany**, Ricken (*Die Blatterpilze (Agaricaceae) Deutschlands*, 1915), Gerhardt (*Pilze*, 2 vols, 1984-5), Hennig (*Taschenbuch fur Pilzefreunde*, 1964 [Aufl. 2, 1966]), Michael & Hennig (*Handbuch fur Pilzfreunde*, 6 vols, 1964-75 [and later edns by Kreisel]), Kreisel (Ed.) (*Pilzflora der DDR*, 1987). **Greece**, Diamandis (*Τα μανιτάρια της Ελλάδος* [*Fungi of Greece*, in Greek], 1992, edn 2 expected late 2008), Pantidou (*Mushrooms in the forests of Greece*, 1991). **Italy**, Cetto (*Enzyklopadie der Pilze*, 4 vols, 1987-8), Goidanich & Govi (*Funghi e ambiente*, 1982), Pacioni (*I funghi nostrani*, 1980), Venturella (*Funghi di Sicilia*, 1997). **Netherlands**, Arnolds *et al.* (*List of fungi in the Netherlands*, 1989), Arnolds & Veerkamp (*Gids voor de paddestoelen in het meetnet*, 1999), Bas *et al.* (*Flora Agaricina Neerlandica*, 1988-). **Spain**, Calonge (*Setas (Hongos), guía ilustrada*, 1990), Calonge & Reche (*Las setas de la Comunidad de Madrid*, 1991), Llamas Frade & Terrón Alfonso (*Guía de hongos de la península Ibérica*, 2005). **Sweden**, Cortin (*Svampplockarens handbok*, 1942), Nilsson & Persson (*Svampar i naturen*, 2 vols, 1977), Ryman & Holmåasen (*Svampar*, 1984). **Switzerland**, Breitenbach & Kranzlin (*Fungi of Switzerland* **1** Ascomycetes, 1981, **2** Non-gilled fungi, 1986, **3** Boletes and agarics, 1991 [English, French and German edns]), Clémençon (*Les quatre saisons des champignons*, 2 vols, 1980). **Middle East**. **Israel**, Wasser (*Edible and poisonous mushrooms of Israel*, 1995). **N. America**. **General**, Phillips (*Mushrooms of North America*, 1991). **Canada**, Barron (*Mushrooms of Ontario and eastern Canada*, 1999), Groves (*Edible and poisonous mushrooms of Canada*, 1979), Pomerleau (*Mushrooms of eastern Canada and the United States*, 1951; *Flore des champignons de Quebec*, 1980), Schalkwijk-Barendsen (*Mushrooms of western Canada*, 1991). **Mexico**, Blanco *et al.*, in Vega & Bousquets (Eds) (*Hongos Macroscopicos*, 1993). **USA**, Atkinson (*Studies of American fungi: mushrooms, edible, poisonous, etc.*, 1903 [reprint 1961], Christensen (*Common fleshy fungi*, 1946 [edn 3, 1965]), Krieger (*The mushroom handbook*, edn 2, 1936 [reprint 1967]), Lincoff (*Audubon field guide to North American mushrooms*, 1981), Metzler & Metzler (*Texas mushrooms*, 1992), Miller (*Mushrooms of North America*, 1974), Phillips (*Mushrooms of North America*, 1991), Smith (*The mushroom hunter's field guide*, 1958 [edn 3, 1980], *Field guide to western mushrooms*, 1975), States (*Mushrooms and truffles of the south-west*, 1990). **South America**. **Argentina**, Gamundí & Horak (*Hongos de los bosques Andino=Patagónicos*, 1993). **Chile**, Furci & Nascimento (*Fungi austral, guía de campo de los hongos más vistosos de Chile*, 2007). **Colombia**, Franco *et al.* (*Setas de Colombia, Agaricales, Boletales y otros hongos, guía de campo*, 2000).

Coloured plates. The following are important series of coloured reference plates (mostly of agarics but other groups sometimes included): Fries (*Icones selectae Hymenomycetum*, 1867- 84), Cooke (*Illustrations of British fungi* (*Hymenomycetes*), 8 vols,

1881-1891; index, *TBMS* **20**: 33, 1935), Konrad & Maublanc (*Icones selectae fungorum*, 6 vols, 1924-36), *BSMF* **41-** (1925-) includes supplementary coloured plates constituting an Atlas), Bresadola (*Iconographia mycologica*, 28 vols, 1927-60; vol. 2 by Gilbert, + 3 suppl.), Cetto (*Der Grosse Pilzfuhrer*, 3 vols, 1979), Dahncke & Dahncke (*700 Pilze in Farbfotos*, 1980), Lange (*Flora agaricina danica*, 5 vols, 1935-41), Romagnesi (*Nouvelle atlas des champignons*, 4 vols, edn 2, 1970), Reid *et al.* (*Fungorum rariorum icones coloratae*, 1966-), Phillips (*Mushrooms and other fungi of Great Britain and Europe*, 1981), Moser & Julich (*Coloured atlas of Basidiomycetes*, 1986-). See further Bolton, Boudier, Bulliard, Greville and Sowerby. See also Watling & Watling (*A literature guide for identifying mushrooms*, 1980; by fam., gen. and geogr. region), Moser & Jülich (*Farbatlas der Basidiomyceten*, 1985). There is an increasing number of guides becoming available on the internet, although quality is more variable.

macronematous (of conidiophores), morphologically different from vegetative hyphae (cf. micronematous).

Macronemeae, hyphomycetes having conidia unlike the hyphae and the conidiophores. Cf. Micronemeae.

Macronodus G.F. Orr (1977) = Gymnoascus fide von Arx (*Persoonia* **9**: 393, 1977).

Macroön Corda (1833) ? = Helminthosporium fide Saccardo (*Syll. fung.* **4**: 403, 1886). Synonymy is doubtful *fide* Sutton (*in litt.*).

Macrophoma (Sacc.) Berl. & Voglino (1886), Botryosphaeriaceae. See Petrak & Sydow (*Beih. Rep. spec. nov. regn. veg.* **42**, 1927; redispositions into *Diplodia*), Phillips & Lucas (*Sydowia* **49**: 150, 1997), Denman *et al.* (*Stud. Mycol.* **45**: 129, 2000; phylogeny), Slippers *et al.* (*Mycol.* **96**: 83, 2004; *Botryosphaeria dothidea* aggregate), Phillips *et al.* (*Mycopathologia* **159**: 433, 2005; phylogeny).

Macrophomella Died. (1916) = Lasiodiplodia fide Petrak & Sydow (*Feddes Repert.* **42**: 1, 1927).

Macrophomina Petr. (1923), anamorphic *Botryosphaeriaceae*, Cpd.0eH.15. 2, widespread. *M. phaseolina* (syn. *Rhizoctonia bataticola*), an important root-parasite in warmer regions. See Goidànich (*Ann. Sper. agr.* N.S. **1**: 449, 1947), Dhingra & Sinclair (Eds) (*An annotated bibliography of Macrophomina phaseolina, 1905-1976*, 1977; 904 refs.), Punithalingam (*Nova Hedwigia* **38**: 339, 1983; cytology), Jones *et al.* (*CJB* **76**: 694, 1998; minichromosomes), Harrington *et al.* (*Pl. Dis.* **84**: 83, 2000; diagnostics), Mena-Portales *et al.* (*Boln Soc. Micol. Madrid* **25**: 265, 2000; Mexico), Vandemark *et al.* (*Mycol.* **92**: 656, 2000; genetic variation), Almeida *et al.* (*Fitopatol. Brasil* **28**: 279, 2003; Brazil, genetic variation), Jana *et al.* (*Can. J. Microbiol.* **51**: 159, 2005; genetic differentiation), Jana *et al.* (*MR* **109**: 81, 2005; population studies), Crous *et al.* (*Stud. Mycol.* **55**: 235, 2006; phylogeny), Reyes-Franco *et al.* (*J. Phytopath.* **154**: 447, 2006; pathogenicity, genetic variation), Schoch *et al.* (*Mycol.* **98**: 1041, 2006; phylogeny).

Macrophomopsis N.E. Stevens & Baechler (1926) = Lasiodiplodia fide Petrak & Sydow (*Feddes Repert.* **42**: 1, 1927).

Macrophomopsis Petr. (1924) = Fusicoccum fide Pennycook & Samuels (*Mycotaxon* **24**: 445, 1985).

macrophylline (of foliose lichens), having large lobes (obsol.).

Macrophyllosticta Sousa da Câmara (1929) = Phyllosticta fide Sutton (*Mycol. Pap.* **141**, 1977).

macroplasmodium, used for the large plasmodium of *Physarum polycephalum*.

Macroplodia Westend. (1857) nom. rej. = Sphaeropsis Sacc. fide Hawksworth *et al.* (*Dictionary of the Fungi* edn 8, 1995).

Macroplodiella Speg. (1908) = Phoma Sacc. fide Clements & Shear (*Gen. Fung.*, 1931).

Macropodia Fuckel (1870) = Helvella fide Dissing (*Dansk bot. Ark.* **25** no. 1, 1966).

Macropyrenium Hampe ex A. Massal. (1860) = Ocellularia fide Hale (*Bull. Br. Mus. nat. hist.* Bot. **8**: 227, 1981).

Macrorhabdus Tomasz., Logan, Snowden, Kurtzman & Phalen (2003), anamorphic *Saccharomycetales*, H?.?.?. 1 (in bird guts), widespread. See Tomaszewski *et al.* (*Int. J. Syst. Evol. Microbiol.* **53**: 1204, 2003), Suh *et al.* (*Mycol.* **98**: 1006, 2006; phylogeny).

macroscopic, visible without a lens.

Macroscyphus Nees ex Gray (1821) = Helvella fide Dissing (*Dansk bot. Ark.* **25** no. 1, 1966), Rifai (*Reinwardtia* **7**: 376, 1968).

Macroseptoria Petr. (1923) = Trichoseptoria fide Clements & Shear (*Gen. Fung.*, 1931).

Macrospora Fuckel (1870), Pleosporaceae. Anamorph *Nimbya*. 3, Europe. See Simmons (*Sydowia* **41**: 314, 1989; anamorph), Shoemaker & Babcock (*CJB* **70**: 1617, 1992), Barr (*SA* **12**: 27, 1993; posn), Fallah & Shearer (*Mycotaxon* **67**: 85, 1998), Johnson *et al.* (*Mycotaxon* **84**: 413, 2002; N America).

macrospore, a large spore when there are spores of two sizes.

Macrosporiaceae Nann. (1934) = Pleosporaceae.

Macrosporites Renault (1899), Fossil Fungi. 3 (Carboniferous), Germany.

Macrosporium Fr. (1832) nom. rej. prop. = Alternaria fide Hughes (*CJB* **36**: 727, 1958).

Macrostilbum Pat. (1898) nom. dub., anamorphic *Pezizomycotina*, Hsy.0eH.?. 1, Java. See Seifert (*TBMS* **85**: 123, 1985).

Macrothelia M. Choisy (1949) = Amphisphaeria fide Santesson (*Bull. Soc. linn. Lyon* **23**: 103, 1954).

Macrotrichum Grev. (1825), anamorphic *Pezizomycotina*, Hso.0eH/≡ eH.?. 2, British Isles.

Macrotyphula R.H. Petersen (1972), Typhulaceae. 6, widespread. See Berthier (*Bull. mens. Soc. linn. Lyon* Num. Spéc. **46**: 213, 1976), Dentinger & McLaughlin (*Mycol.* **98**: 746, 2006; phylogeny).

Macrovalsaria Petr. (1962), Dothideomycetes. 1, widespread (tropical). See Sivanesan (*TBMS* **65**: 395, 1975), Barr (*Mycotaxon* **49**: 129, 1993; posn).

Macroventuria Aa (1971), Pleosporales. 2, Africa; USA. See Aa (*Persoonia* **6**: 359, 1971), Kodsueb *et al.* (*Mycol.* **98**: 571, 2006; phylogeny).

Macruropyxis Azbukina (1972), Uropyxidaceae. 1 (on *Fraxinus* (*Oleaceae*)), former USSR. See Azbukina (*Komarovskie Ctenija (Moscow & Leningrad)* **19**: 20, 1972).

maculate, spotted; blotched.

Maculatifrondes K.D. Hyde (1996), ? Phyllachorales. Anamorph *Cyclodomus*. 1 (on *Palmae* fronds), Ecuador. See Hyde *et al.* (*MR* **100**: 1509, 1996), Hyde & Cannon (*Mycol. Pap.* **175**: 114, 1999).

Maculatipalma J. Fröhl. & K.D. Hyde (1995), Diaporthales. 1 (on *Palmae*), Australia. See Fröhlich &

Hyde (*MR* **99**: 727, 1995).

maculicole (adj. **-icolous**), an organism living on spots, e.g. leaf spots.

Madura foot, see mycetoma.

maduramycosis, see mycetoma.

Madurella Brumpt (1905), anamorphic *Sordariales*, Hsy.0eP.?. 10 (on humans causing mycetoma, q.v.), widespread (esp. tropical). See Ciferri & Redaelli (*Mycopathologia* **3**: 182, 1941), Ahmed *et al.* (*J. Clin. Microbiol.* **40**: 1031, 2002; Sudan), Ahmed *et al.* (*J. Clin. Microbiol.* **41**: 4537, 2003; clonality), Ellabib *et al.* (*Mycoses* **46**: 321, 2003; Libya), Hoog *et al.* (*Mycoses* **47**: 121, 2004; phylogeny, typification, exclusion of *M. grisea*), Vilela *et al.* (*Mycopathologia* **158**: 415, 2004; Brazil), Sande *et al.* (*J. Clin. Microbiol.* **43**: 4349, 2005; population biology).

Magdalaenaea G. Arnaud (1952), anamorphic *Pezizomycotina*, Hso.1bH.?. 1, Europe; Asia. See Arnaud (*BSMF* **68**: 209, 1952), Sati *et al.* (*Mycotaxon* **81**: 445, 2002; India).

magic mushrooms, typically hallucinogen-containing species of *Psilocybe*, and also *Gymnopilus*, *Panaeolus*, *Conocybe*, and *Amanita muscaria*; hallucinogenic fungi (q.v.).

Magmopsis Nyl. (1875) = Arthopyrenia fide Hawksworth *et al.* (*Dictionary of the Fungi* edn 8, 1995).

Magnaporthaceae P.F. Cannon (1994), Sordariomycetidae (inc. sed.). 13 gen. (+ 14 syn.), 93 spp.

Lit.: Bunting *et al.* (*Phytopathology* **84**: 1097, 1994), Cannon (*SA* **13**: 25, 1994), Augustin *et al.* (*J. Phytopath.* **147**: 109, 1999), Bryan *et al.* (*MR* **103**: 319, 1999), Chen *et al.* (*Mycol.* **91**: 84, 1999), Kumar *et al.* (*Genetics* Bethesda **152**: 971, 1999), Ward & Bateman (*New Phytol.* **141**: 323, 1999), Kusaba *et al.* (*Ann. phytopath. Soc. Japan* **65**: 588, 1999), Ulrich *et al.* (*New Phytol.* **145**: 127, 2000), Winka & Eriksson (*Mycoscience* **41**: 97, 2000), Bussaban *et al.* (*Mycol.* **97**: 1002, 2005), Réblová (*Mycol.* **98**: 68, 2006).

Magnaporthe R.A. Krause & R.K. Webster (1972), Magnaporthaceae. Anamorph *Pyricularia*. 5, widespread (esp. tropical). *M. grisea* (rice blast; anamorph *Pyricularia oryzae*. See Barr (*Mycol.* **69**: 952, 1977), Tsuda & Ueyama (*TMSJ* **19**: 425, 1978; teleomorph formation), Kato (*Ann. phytopath. Soc. Japan* **60**: 266, 1994; *M. grisea* phylogeny), Kato *et al.* (*Ann. phytopath. Soc. Japan* **60**: 175, 1994; *M. grisea* microconidia), Zeigler *et al.* (*Rice blast disease*, 1994), Bunting *et al.* (*Phytopathology* **86**: 398, 1996; ITS), Roumen *et al.* (*Eur. J. Pl. Path.* **103**: 363, 1997; Europ. pops.), George *et al.* (*Phytopathology* **88**: 223, 1998; PCR), Kusaba *et al.* (*Ann. phytopath. Soc. Japan* **64**: 125, 1998; host range), Viji & Gnanamanickam (*Pl. Dis.* **82**: 36, 1998; India), Kumar *et al.* (*Genetics* Bethesda **152**: 971, 1999; population biology, India), Kim *et al.* (*Molecules and Cells* **10**: 127, 2000; microsatellites), Couch & Kohn (*Mycol.* **94**: 683, 2002; differentiation of *M. oryzae*), Farman (*Phytopathology* **92**: 245, 2002; on *Lolium*), Ikeda *et al.* (*Molecular Microbiology* **45**: 1355, 2002; genetics), Gilbert *et al.* (*Applied Mycology and Biotechnology* **4**: [331, 2004; functional genomics), Javan-Nikkhah *et al.* (*Eur. J. Pl. Path.* **110**: 909, 2004; Iran), Rathour *et al.* (*J. Phytopath.* **152**: 304, 2004; India, population structure), Tosa *et al.* (*Phytopathology* **94**: 454, 2004; on *Lolium*), Consolo *et al.* (*Mycopathologia* **160**: 285, 2005; mating types, Argentina), Couch *et al.* (*Genetics* Bethesda **170**: 613, 2005; evolution), Dean (*Nature* Lond. **434** no. 7036: 980, 2005; whole genome), Farman & Kim (*Molecular Plant Pathology* **6**: 287, 2005; telomeres), Piotti *et al.* (*J. Phytopath.* **153**: 80, 2005; Italy), Bussaban *et al.* (*Mycol.* **97**: 1002, 2005; anamorph), Ninh Thuan *et al.* (*Eur. J. Pl. Path.* **114**: 381, 2006; Vietnam), Zhang *et al.* (*Mycol.* **98**: 1076, 2006; phylogeny), Ebbole (*Ann. Rev. Phytopath.* **45**: 437, 2007; model organism).

Magninia M. Choisy (1929) = Lecanora fide Hawksworth *et al.* (*Dictionary of the Fungi* edn 8, 1995).

Magnisphaera J. Campb., J.L. Anderson & Shearer (2003), Halosphaeriaceae. 2, widespread. See Campbell *et al.* (*Mycol.* **95**: 546, 2003), Pang & Jones (*Nova Hedwigia* **78**: 269, 2004).

Magnosporites Rouse (1962), Fossil Fungi ? Fungi. 1 (Paleogene), Canada.

Magnusia Sacc. (1878) [non *Magnusia* Klotzsch 1854, *Begoniaceae*] ≡ Kernia Nieuwl.

Magnusiella Sadeb. (1893) = Taphrina fide Mix (*Kansas Univ. Sci. Bull.* **33**: 1, 1949).

Magnusiomyces Zender (1925), Dipodascaceae. Anamorph *Saprochaete*. 1, widespread. See de Hoog *et al.* (*Stud. Mycol.* **29**, 1986), Hoog & Smith (*Stud. Mycol.* **50**: 489, 2004).

Magnusson (Adolf Hugo; 1885-1964; Sweden). School teacher, Gothenburg (1909-1948). Devoted spare time to lichenology making invaluable contributions to knowledge of the lichens of China, Hawaii (partly with Zahlbruckner, q.v.) and Scandinavia, and to the genera *Acarospora*, *Caloplaca*, *Lecidea* and *Lecanora*. He described about 900 new species in *c.* 150 papers. *Publs*. A monograph of the genus Acarospora. *Kungliga Svenska Vetenskapsakademiens Handlingar* (1929); Acarosporaceae und Thelocarpaceae. *Rabenhorst?s Kryptogamen-Flora von Deutschland, Oesterreich und der Schweiz* (1935). Collection of *c.* 70,000 specimens in Uppsala (**UPS**). *Biogs, obits etc*. Almborn (*Botaniska Notiser* **117**: 428, 1964); Degelius (*Svensk Botanisk Tidskrift* **59**: 393, 1965); Grummann (1974: 478); Stafleu & Cowan (*TL-2* **3**: 247, 1981).

Magnussoniolichen Tomas. & Cif. (1952) = Polyblastia fide Hawksworth *et al.* (*Dictionary of the Fungi* edn 8, 1995).

Magnussoniomyces Cif. & Tomas. (1953) ≡ Magnussoniolichen.

Magoderna Steyaert (1973) = Amauroderma Murrill fide Ryvarden (*Syn. Fung.* **5**: 180, 1991).

Mahabalella B. Sutton & S.D. Patil (1966), anamorphic *Pezizomycotina*, Hso.0eH.15. 3, India; Europe. See Sutton & Patil (*Nova Hedwigia* **11**: 203, 1966), Mel'nik *et al.* (*Mikol. Fitopatol.* **37**: 54, 2003; Austria).

Mahanteshamomyces Hosag. (2004), anamorphic *Asterinaceae*. 1, India. See Hosagoudar (*J. Econ. Taxon. Bot.* **28**, 2004).

Mahevia Lagarde (1917) = Hirsutella fide Kendrick & Carmichae *in* Ainsworth *et al.* (Eds) (*The Fungi* **4A**: 390, 1973).

Mainsia H.S. Jacks. (1931) = Gerwasia fide Cummins & Hiratsuka (*Illustr. Gen. Rust Fungi rev. edit.*, 1983).

Maire (René Charles Joseph Ernest; 1878-1949; France, later Algeria). Professor of Botany, University of Algiers (1911-1949). Noted for his contributions to fungal cytology, the anatomy of *Russula*, and biodiversity studies of the fungi of north Africa, Catalonia and Greece. *Publs*. Fungi Catalaunici. Series altera. Contributions a la flore mycologique de la Catalogne. *Publicaciones. Junta de Ciències Naturals de Barcelona* (1933); (with Werner) Fungi Maroccani. *Mémoires de la Société des Sciences Naturelles du Maroc* (1937); (with J. Politis) Fungi Hellenici. Catalogue raisonné des champignons connus jusqu?ici en Grèce. *Actes de l?Institut Botanique de l?Université d?Athènes* (1940). *Biogs, obits etc.* Feldmann & Guinier (*Bulletin de la Société d'Histoire Naturelle de l'Afrique de Nord* **41**: 65, 1952, bibliography); Grummann (1974: 336); Kühner (*BSMF* **69**: 7, 1953) [portrait]; Stafleu & Cowan (*TL-2* **3**: 257, 1981).

Maireella Syd. ex Maire (1908) = Gibbera fide Müller & von Arx (*Beitr. Kryptfl. Schweiz* **11** no. 2, 1962).

Maireina W.B. Cooke (1961) = Cyphellopsis fide Donk (*Reinwardtia* **1**, 1951).

Maireomyces Feldmann (1941) nom. dub., ? Fungi. See Kohlmeyer & Kohlmeyer (*Marine Mycology*, 1979).

Majewskia Y.B. Lee & K. Sugiy. (1986), Laboulbeniaceae. 1, Japan. See Lee & Sugiyama (*Mycol.* **78**: 289, 1986).

Malacaria Syd. (1930), Tubeufiaceae. 2 (on *Meliola* etc.), pantropical. See Rossman (*Mycol. Pap.* **157**, 1987; key), Johnston (*N.Z. Jl Bot.* **37**: 289, 1999).

malaceoid venation, see veins.

Malacharia Fée (1843) ? = Cerebella fide Hawksworth *et al.* (*Dictionary of the Fungi* edn 8, 1995).

Malacodermis Bubák & Kabát (1912) = Glutinium fide Höhnel (*Sber. Akad. Wiss. Wien* Math.-naturw. Kl., Abt. 1 **125**: 27, 1916).

Malacodermum Marchand (1896) nom. dub., Fungi. See Donk (*Taxon* **6**: 84, 1957).

Malacodon Bataille (1923) nom. dub., Cantharellales. See Donk (*Taxon* **5**: 102, 1956).

malacoid, like mucilage.

Malacosphaeria Syd. (1924) = Eriosphaeria fide Müller & von Arx (*Beitr. Kryptfl. Schweiz* **11** no. 2, 1962).

Malacostroma Höhn. (1920) [non *Malacostroma* Gürich 1906, fossil ? *Algae*] = Phomopsis (Sacc.) Bubák. fide Petrak (*Annls mycol.* **19**: 176, 1921).

Malajczukia Trappe & Castellano (1992), Mesophelliaceae. 8, Australia; New Zealand. See Trappe *et al.* (*Aust. Syst. Bot.* **5**: 618, 1992).

Malassezia Baill. (1889), anamorphic *Malasseziales*. 7 (on humans), widespread. See Anthony *et al.* (*Microb. Ecol. Health Dis.* **7**: 161, 1994; epidemiology), Guillot & Guého (*Antonie van Leeuwenhoek* **67**: 297, 1995; mol. phylogeny), Guillot *et al.* (*Antonie van Leeuwenhoek* **67**: 173, 1995; nomencl.), Guillot *et al.* (*Antonie van Leeuwenhoek* **69**: 337, 1996; taxonomy), Boekhout *et al.* (*Medical Mycol.* **36**: 365, 1998; PFGE, RAPD), Guillot *et al.* (*Medical Mycol.* **36**: 220, 1998; review), Guillot & Bond (*Medical Mycol.* **37**: 295, 1999; review *M. pachydermatis*), Begerow *et al.* (*MR* **104**: 53, 2000; mol. phylogeny), Fell *et al.* (*Int. J. Syst. Evol. Microbiol.* **50**: 1351, 2000; mol. phylogeny), Morishita *et al.* (*Mycopathologia* **161**: 61, 2006; molecular analysis), Xu *et al.* (*Proc. natn Acad. Sci. U.S.A.* **104**: 18730, 2007; phylogeny). See pityriasis versicolor.

Malasseziales R.T. Moore (1980). Ustilaginomycotina. 1 gen., 7 spp. Causing pityriasis versicolor (q.v.) of humans. No families recognized.

Lit. Moore (*Bot. Mar.* **23**: 361, 1980; used for all *Basidioblastomycetes* lacking ballistospores (e.g. *Cryptococcus*, *Rhodotorula*, *Schizoblastosporon*, *Trichosporon*, *Vanrija*), Begerow *et al.* (*MR* **104**: 59, 2000; emend. to those belonging in *Ustilaginomycetes*), Begerow *et al.* (*Mycol.* **98**: 906, 2006; phylogeny).

Malbranchea Sacc. (1882), anamorphic *Myxotrichum*, Hso.0eH.40. 18, widespread. Also anamorphic *Onygenaceae*; teleomorphs *Auxarthron*, *Uncinocarpus*. See Sigler & Carmichael (*Mycotaxon* **4**: 412, 1976; key), Currah (*Mycotaxon* **24**: 1, 1985), Martínez *et al.* (*TBMS* **86**: 490, 1986; conidiogenesis), Sigler *et al.* (*Mycotaxon* **28**: 119, 1987; synanamorph), Okada *et al.* (*Mycoscience* **36**: 385, 1995; ubiquinones, biochemistry), Sugiyama *et al.* (*Mycoscience* **40**: 251, 1999; phylogeny), Sigler *et al.* (*Stud. Mycol.* **47**: 111, 2002; teleomorphs).

Malenconia Bat. & H. Maia (1960) = Coccomyces fide von Arx (*in litt.*).

malformin, a plant-malforming cyclic pentapeptide from *Aspergillus niger* (Takahashi & Curtis, *Plant Physiol.* **36**: 30, 1961).

Malinvernia Rabenh. (1857) = Podospora fide Lundqvist (*Symb. bot. upsal.* **20** no. 1, 1972).

Malleomyces Hallier (1870) nom. dub., anamorphic *Pezizomycotina*. See Rivolta (*Parass. Veg.*, 1884).

Mallochia Arx & Samson (1986), ? Onygenaceae. 1, paleotropical. See von Arx & Samson (*Persoonia* **13**: 185, 1986), Solé *et al.* (*MR* **106**: 388, 2002; phylogeny), Udagawa & Uchiyama (*Stud. Mycol.* **47**: 181, 2002; revision).

Mallotium (Ach.) Gray (1821) = Leptogium fide Hawksworth *et al.* (*Dictionary of the Fungi* edn 8, 1995).

Malmella C.W. Dodge (1933) = Erioderma fide Galloway & Jørgensen (*Lichenologist* **7**: 139, 1975).

Malmeomyces Starbäck (1899), Niessliaceae. 1 (on bamboo), Brazil. See Rossman (*Mycotaxon* **8**: 537, 1979), Rossman *et al.* (*Stud. Mycol.* **42**: 248 pp., 1999).

Malmgrenia Vain. (1939) = Cryptothele fide Jørgensen & Henssen (*Taxon* **39**: 343, 1990).

Malmia M. Choisy (1931) = Rinodina fide Hawksworth *et al.* (*Dictionary of the Fungi* edn 8, 1995).

Malotium Velen. (1934), Helotiales. 9, former Czechoslovakia.

Malthomyces K.D. Hyde & P.F. Cannon (1999), ? Phyllachoraceae. 2 (on *Palmae* fronds), India; Sri Lanka. See Hyde & Cannon (*Mycol. Pap.* **175**: 114, 1999).

malting, see brewing.

maltoryzine, a metabolite of *Aspergillus oryzae* var. *microsporus*; toxic for cattle (Iizuka & Iida, *Nature* **196**: 681, 1962).

Malupa Y. Ono, Buriticá & J.F. Hennen (1992), anamorphic *Pucciniales*. 15, widespread (tropical). Anamorph name for (II). See Ono *et al.* (*MR* **96**: 828, 1992).

Malustela Bat. & J.A. Lima (1960) = Curvularia fide Ellis (*Mycol. Pap.* **106**, 1966).

Malvinia Döbbeler (2003), ? Ostropales. 1, Falkland Islands. See Döbbeler (*Nova Hedwigia* **76**: 19, 2003).

Mamiania Ces. & De Not. (1863), Diaporthales. 3, widespread (north temperate). See Barr (*Mycol. Mem.* **7**, 1978), Monod (*Beih. Sydowia* **9**: 1, 1983).

Mamianiella Höhn. (1917), Diaporthales. Anamorph *Mazzantiella*. 1 (on *Corylus*), Europe; N. America. See Barr (*Mycol. Mem.* **7**, 1978).

Mamillisphaeria K.D. Hyde, S.W. Wong & E.B.G.

Jones (1996), ? Massariaceae. 1, Australia. See Hyde *et al.* (*Nova Hedwigia* **62**: 513, 1996).

Mammaria Ces. ex Rabenh. (1854), anamorphic *Pseudocercophora*, Hso.0eP.10. 1, Europe; N. America. See Hughes (*Beih. Sydowia* **1**: 359, 1957), Park (*TBMS* **60**: 351, 1973), Subramanian & Sekar (*Journal of the Singapore National Academy of Science* **15**: 58, 1986; teleomorph).

Mammariopsis L.J. Hutchison & J. Reid (1988), anamorphic *Pezizomycotina*, Hso.0eH.38; Hso.0-1eH.1/10; Hso.0eH.15. 1, New Zealand. See Hutchison & Reid (*N.Z. Jl Bot.* **26**: 94, 1988).

mammiform, breast-like in form.

Manaustrum Cavalc. & A.A. Silva (1972) = Strigula fide Lücking *et al.* (*Lichenologist* **30**: 121, 1998).

Manginella Bat. & H. Maia (1961), anamorphic *Pezizomycotina*, Cpt.0eP.?. 2, Brazil. See Batista & Maia (*Publções Inst. Micol. Recife* **338**: 23, 1961), Farr (*Mycol.* **78**: 269, 1986).

Manginia Viala & Pacottet (1904) = Sphaceloma fide Jenkins & Bitancourt (*Mycol.* **33**: 338, 1941).

Manginiella, see *Mauginiella*.

Manginula G. Arnaud (1918) nom. dub., anamorphic *Pezizomycotina*. 5 (recent & fossil), N. America; Australia. See Lange (*Aust. J. Bot.* **17**: 565, 1969), Sutton (*Mycol. Pap.* **141**, 1977).

Manginulopsis Bat. & Peres (1963) = Leptothyrium fide von Arx (*K. Ned. Akad. Wet. Amst.* **66**: 172, 1963).

Manglicola Kohlm. & E. Kohlm. (1971), Hypsostromataceae. 2, Guatemala; Guyana. See Huhndorf (*Mycol.* **86**: 266, 1994).

Mangrovispora K.D. Hyde & Nakagiri (1991), ? Phyllachorales. 1 (dead wood, trop. marine), Australia. See Wang & Hyde (*Fungal Diversity* **3**: 159, 1999; placement).

Manikinipollis Krutzsch (1970), Fossil Fungi ? Fungi. 1 (Miocene), former USSR.

Manilaea Syd. & P. Syd. (1914) = Arthonia fide Santesson (*Symb. bot. upsal.* **12** no. 1: 1, 1952).

Manina Adans. (1763) nom. dub., Clavariaceae. See Donk (*Reinwardtia* **2**: 441, 1954).

Manina Banker (1912) = Hericium Pers. fide Donk (*Taxon* **5**: 69, 1956).

manna, see *Sphaerothallia*.

Mannia Trevis. (1857) [non *Mannia* Opiz 1829, *Hepaticae*] = Buellia fide Hawksworth *et al.* (*Dictionary of the Fungi* edn 8, 1995).

mannitol, a polyhydric alcohol, often found as a storage compound in ectotrophic mycorrhizal mantles and lichens.

manocyst (of *Phytophthora*), a projection (receptive papilla) from the oogonium which undergoes fusion with the antheridium.

Manoharachariomyces N.K. Rao, D.K. Agarwal & Kunwar (2005), anamorphic *Pezizomycotina*. 1, India. See Rao *et al.* (*Indian Phytopath.* **58**: 96, 2005).

Manokwaria K.D. Hyde (1993), Xylariales. 1 (on *Palmae*), Australia; Irian Jaya. See Hyde (*Sydowia* **45**: 246, 1993).

mantle, a compact layer of hyphae enclosing short feeder roots of ectomycorrhizal plants, connected to the Hartig (q.v.) net on the inside, and to the extramatrical hyphae (q.v.) on the outside; acts as a nutrient sink.

Manuripia Singer (1960), Marasmiaceae. 1, Bolivia. See Singer (*Sydowia* **14**: 273, 1960).

Manzonia Garov. (1866) = Aspicilia fide Hawksworth *et al.* (*Dictionary of the Fungi* edn 8, 1995).

map lichen, species of *Rhizocarpon* (e.g. *R. geographicum*) which have yellow or yellow-green areolae separated by dark black lines (the prothallus) and forming mosaics. See Lichenometry.

Mapea Boedijn (1957) ≡ Telomapea.

Mapea Pat. (1906) = Uredo fide Berndt (*in litt.*; probably anamorphic *Chaconiaceae*).

Mapletonia B. Sutton (1991), anamorphic *Pezizomycotina*, Cpd.1eH.15. 1, Australia. See Sutton (*Sydowia* **43**: 162, 1991).

Mapping distribution, see Geographical distribution.

Marasmiaceae Roze ex Kühner (1980), Agaricales. 54 gen. (+ 30 syn.), 1590 spp. The family is circumscribed here to include *Marasmiaceae* and *Omphalotaceae* as circumscribed by Matheny *et al.* (2006).

Lit.: Kämmerer *et al.* (*Pl. Syst. Evol.* **150**: 101, 1985), Segedin (*N.Z. Jl Bot.* **31**: 375, 1993), Gordon *et al.* (*MR* **98**: 200, 1994), Miller (*Mycol. helv.* **6**: 91, 1994), Corner (*Beih. Nova Hedwigia* **111**: 1, 1996), Antonín *et al.* (*Mycotaxon* **63**: 359, 1997), Desjardin & Horak (*Biblthca Mycol.* **168**: 152 pp., 1997), Nicholson *et al.* (*Mycol.* **89**: 400, 1997), Hibbett *et al.* (*MR* **102**: 1041, 1998), Hughes & Petersen (*Pl. Syst. Evol.* **211**: 231, 1998), Petersen & Hughes (*Pl. Syst. Evol.* **211**: 217, 1998), Antonín (*Mykol. Listy* **68**: 13, 1999), Desjardin *et al.* (*Sydowia* **52**: 92, 2000), Mata *et al.* (*Mycol.* **93**: 1102, 2001), Kirchmair *et al.* (*Persoonia* **17**: 583, 2002), Moncalvo *et al.* (*Mol. Phylogen. Evol.* **23**: 357, 2002), Saito *et al.* (*Biosc., Biotechn., Biochem.* **66**: 2125, 2002), Takahashi (*Mycoscience* **43**: 343, 2002), Abesha *et al.* (*Mycol.* **95**: 1021, 2003), Antonín (*Mycotaxon* **88**: 53, 2003), Arruda *et al.* (*MR* **107**: 25, 2003), Griffith *et al.* (*N.Z. Jl Bot.* **41**: 423, 2003), Bodensteiner *et al.* (*Mol. Phylogen. Evol.* **33**: 501, 2004), Kirchmair *et al.* (*Mycol.* **96**: 1253, 2004), Mossebo & Antonín (*Czech Mycol.* **56**: 85, 2004), Wilson & Desjardin (*Mycol.* **97**: 667, 2005), Rincones *et al.* (*MR* **110**: 821, 2006).

Marasmiellus Murrill (1915), Marasmiaceae. *c.* 250, widespread. *M. inoderma*, root rot of maize; *M. paspali* var. *americana*, 'borde blanco' of maize. See Sabet *et al.* (*TBMS* **54**: 123, 1970), Singer (*Beih. Nova Hedwigia* **44**: 1, 1973; 134 neotrop. spp., extralimital spp.), Singer (*Beih. Nova Hedwigia* **26**: 847, 1976), Pegler (*Kew Bull.* Addit. Ser. **6**, 1977), Wilson & Desjardin (*Mycol.* **97**: 667, 2005; phylogeny).

Marasmiopsis Henn. (1898) = Phaeomarasmius fide Singer (*Agaric. mod. Tax.*, 1951).

Marasmius Fr. (1836) nom. cons., Marasmiaceae. *c.* 500, widespread (esp. tropical). *M. oreades* (fairy-ring champignon), *M. perniciosus*, (see *Crinipellis*), *M. plicatus* (Sugar-cane root rot). See Singer (*Mycol.* **50**: 103, 1958; key sections), Singer (*Bull. Jard. bot. Brux.* **34**: 317, 1964; key 50 Congo spp.), Singer (*Sydowia* **18**: 106, 1965; keys 154 S. Am. spp.), Gilliam (*Mycotaxon* **4**: 1, 1976; 30 N. Am. spp.), Singer (*Fl. Neotrop.* **17**: 62, 1976; key 233 neotrop. spp.), Pegler (*Kew Bull.* Addit. Ser. **6**, 1977; 43 E. Afr. spp.), Desjardin (*Sydowia* **42**: 17, 1990; culture), Antonín & Noordeloos (*Libri Botanici* **8**, 1993; Europ. spp.), Desjardin *et al.* (*Sydowia* **52**: 92, 2000; Indonesia), Antonín (*Mycotaxon* **85**: 109, 2003; tropical Africa), Antonín (*Mycotaxon* **89**: 399, 2004; sect. *Sicci* in Africa), Mossebo & Antonín (*Czech Mykol.* **56**: 85, 2004; Cameroon), Wilson & Desjardin (*Mycol.* **97**: 667, 2005; phylogeny), Antonín & Buyck (*Fungal Diversity* **23**: 17, 2006; Madagascar, Mascarenes),

Desjardin & Ovrebo (*Fungal Diversity* **21**: 19, 2006; Panamá), Antonín (*Fungus Flora of tropical Africa* **1**, 2007) See horse-hair blight fungi.

Maravalia Arthur (1922), Chaconiaceae. *c.* 35 (on angiosperms), widespread (tropical). See Mains (*Bull. Torrey bot. Club* **66**: 173, 1939), Ono (*Mycol.* **76**: 892, 1984; key), Wingfield *et al.* (*Australas. Pl. Path.* **33**: 327, 2004), Thaung (*Australasian Mycologist* **24**: 29, 2005; Burma).

Marcelleina Brumm., Korf & Rifai (1967), Pezizaceae. 9, widespread (north temperate). See Pfister (*Sydowia* **38**: 235, 1986; N. Am.), Moravec (*Mycotaxon* **30**: 473, 1987; key), Dissing *et al.* (*Taxon* **39**: 130, 1990; nomencl.), Hansen *et al.* (*Mol. Phylogen. Evol.* **36**: 1, 2005; phylogeny), Hansen & Pfister (*Mycol.* **98**: 1029, 2006; phylogeny).

Marceloa Bat. & Peres (1962) = Treubiomyces fide von Arx & Müller (*Stud. Mycol.* **9**, 1975).

marcescent (of basidiomata), withering, drying up *in situ*. Cf. putrescent.

Marchalia Sacc. (1889) nom. conf., Polystomellaceae. See von Arx (*Sydowia* **12**: 400, 1959).

Marchaliella G. Winter ex E. Bommer & M. Rousseau (1891) = Testudina fide Hawksworth (*CJB* **57**: 91, 1979).

Marchandiobasidium Diederich & Schultheis (2003), Corticiaceae. Anamorph *Marchandiomyces*. 1 (on lichens), Europe; N. America. See Diederich *et al.* (*MR* **107**: 524, 2003), Larsson (*MR* **111**: 1040, 2007; phylogeny).

Marchandiomphalina Diederich, Lawrey & Binder (2007), Corticiaceae. 1, Venezuela. See Diederich & Lawrey (*Mycol. Progr.* **6**: 61, 2007).

Marchandiomyces Dieder. & D. Hawksw. (1990), anamorphic *Marchandiobasidium*. 3 (on lichens), widespread. See Sikaroodi *et al.* (*MR* **105**: 453, 2001; phylogenetic posn.), DePriest *et al.* (*MR* **109**: 57, 2005).

Marcosia Syd. & P. Syd. (1916) = Stigmina fide Sutton (*TBMS* **58**: 164, 1972).

Margaretbarromyces Mindell, Stockey, Beard & Currah (2007), Fossil Fungi ? Pleosporales. 1 (Eocene). See Mindell *et al.* (*MR* **111**: 680, 2007).

Margarinomyces Laxa (1930), anamorphic *Pezizomycotina*. 1, widespread. See Schol-Schwarz (*Persoonia* **6**: 63, 1970), Cole & Kendrick (*Mycol.* **65**: 682, 1973), Gams (*Stud. Mycol.* **45**: 187, 2000; links with *Phialophora*), Mostert *et al.* (*Stud. Mycol.* **54**: 115 pp., 2006; key).

Margaritispora Ingold (1942), anamorphic *Helotiales*, Hso.≡ eH.15. 2 (aquatic), Europe. See Marvanová & Descals (*J. Linn. Soc.* Bot. **41**: 1, 1985), Campbell *et al.* (*MR* **110**: 1025, 2006; phylogeny).

marginal frill (of a conidium), the periclinal wall left attached to a spore after secession. Cf. basal frill.

marginal veil (of agarics), an incurving proliferation of the margin of the pileus which protects the developing hymenium. Cf. partial veil.

marginate (1) having a well-marked edge; (2) (of basal bulb of agaric stipe), having a gutter-like rim as in *Leucocortinarius bulbiger*.

margo proprius, see excipulum proprium, proper margin.

margo thallinus, see excipulum thallinum, thalline margin.

Mariaella Šutara (1987) = Suillus Gray fide Kuyper (*in litt.*).

Mariannaea G. Arnaud ex Samson (1974), anamorphic *Nectriaceae*, Hso.0eH.15. 4, widespread. See Samson (*Stud. Mycol.* **6**: 74, 1974), Samuels & Seifert (*Beih. Sydowia* **43**: 264, 1991), Okuda & Yamamoto (*Mycoscience* **41**: 411, 2000; Japan), Luangsa-ard *et al.* (*Mycol.* **96**: 773, 2004; phylogeny), Samuels *et al.* (*CBS Diversity Ser.* **4**, 2006; USA).

Marielliottia Shoemaker (1999), anamorphic *Pleosporaceae*, Hso.?.?. 3, widespread. See Shoemaker (*CJB* **76**: 1559, 1998).

Marine fungi. Fungi are of widespread occurrence in the sea, with over 1500 spp. (excluding coastal lichen-forming fungi) known from this habitat (Hyde *et al.*, *Biodiversity and Conservation* **7**: 1147, 1998). These parasitize marine algae or animals or are saprobes on timber, algae (Kohlmeyer, *Veroeff. Inst. Merresforsch. Bremerhaven* Suppl. **5** 339, 1974, Nakagiri, *Mycologia* **85**: 638, 1993), sea grasses (Cuomo *et al.*, *Prog. Oceanogr.* **21**: 189, 1988), protozoal cysts and corals (Kohlmeyer & Volkmann-Kohlmeyer, *CJB* **68**: 1554, 1990, Raghukumar & Ragukumar, *Mar. Ecol.* **12**: 251, 1991), sea foam (Nakagiri, *IFO Res. Comm.* **14**: 52, 1989) and other substrata; spores of many (esp. *Halosphaeriales*) have special appendages for attachment (Hyde & Jones, *Bot. Mar.* **32**: 205, 1989; 10 types).

Other fungi occur in brackish water (Tubaki & Ito, *Rep. Tottori mycol. Inst.* **10**: 523, 1973), salt marshes (Bayliss Elliott, *Ann. appl. biol.* **17**: 284, 1930; Apinis & Chesters, *TBMS* **47**: 419, 1964), mangrove swamps (Hyde & Jones, *Marine Ecology* **9**: 15, 1988, Tan Leong & Jones, *CJB* **67**: 2686, 1989), in the salt of salt pans (Quinta, *Food Technol.* **22**: 102, 1968) and on sand (Rees, *Bot. Mar.* **23**: 375, 1980, Nakagiri & Tokura, *Trans. mycol. Soc. Japan* **28**: 413, 1987); 27-36 spp. exclusively tropical (Kohlmeyer, *Marine Ecology* **5**: 329, 1984).

A very diverse range of unusual secondary metabolites are produced by marine fungi (San-Martín *et al.*, *J. Chil. Chem. Soc.* **53**: 1377, 2008), including potential herbicides (Motti *et al.*, *Appl. Environ. Microbiol.* **73**: 1921, 2007), and several have shown great potential for use in bioremediation, particularly in decolourizing effluents (Raghukumar, *MR* **104**: 1222, 2000). Marine fungi appear to play an important role in decomposition of organic matter on sandy beaches (Steinke & Lubke, *South African J. Bot.* **69**: 540, 2003) and have the potential to be used as pollution indicators (Taboski *et al.*, *FEMS Microbiol. Ecol.* **53**: 445, 2006).

Lichens form distinctive band-like zones on siliceous rocky shores (Fletcher, *Lichenologist* **5**: 368, 401, 1973, terminology of zones; *in* Brown *et al.*, *Lichenology: progress and problems*: 359, 1976, nutrition; *in* Price *et al.*, *The shore environment* **2**: 789, 1980, compr. review).

Lit.: *General*: Fell *et al.* (*Molecular marine biol. biotech.* **1**: 175, 1992; molecular detection marine microeukaryotes), Hyde & Pointing (Eds) (*Marine Mycology. A Practical Approach*, 2000), Johnson & Sparrow (*Fungi in oceans and estuaries*, 1961 [reprint, 1970]), Jones (Ed.) (*Recent advances in aquatic mycology*, 1976; compr. review), Kohlmeyer (*in* Moss, 1986: 199; checklist marine *Ascomycota*), Kohlmeyer & Kohlmeyer (*Marine Mycology*, 1979; *Icones Fungorum Maris*, 1968), Kohlmeyer & Volkmann-Kohlmeyer (*Bot. Mar.* **34**: 1, 1991), Moss (Ed.) (*The biology of marine fungi*, 1986).

Regional: **Aldabra**, Kohlmeyer & Volkmann-

Kohlmeyer (*Can J. Bot.* **65**: 571, 1987). **Argentina**, Malacalza & Martinez (*Boln Soc. Argent. Bot.* **14**: 57, 1971). **Australia**, Kohlmeyer (*Aust. J. Mar. Freshwater Res.* **42**: 91, 1991), Hyde (*Aust. Syst. Bot.* **3**: 711, 1990). **British Isles**, Fletcher (*Lichenologist* **7**: 1 [siliceous rocky shore lichens], 73 [calcareous and terricolous lichens], 1975; keys Br. spp.). **Brunei**, Hyde (*Bot. J. Linn. Soc.* **98**: 135, 1988). **Denmark**, Koch & Jones (*Svampe* **8**: 49, 1983). **Ecuador, Galapagos**, Kohlmeyer & Volkmann-Kohlmeyer (*Can J. Bot.* **65**: 571, 1987). **India**, Borse (*Trans. mycol. Soc. Japan* **28**: 55, 1987; *Indian Bot. Reptr* **7**: 18, 1988; key 55 spp.). **Malaysia**, Jones & Kuthubutheen (*Sydowia* **41**: 160, 1990). **Pacific**, Volkmann-Kohlmeyer & Kohlmeyer (*Mycol.* **85**: 337, 1993; biogeogr.). **Philippines**, Jones *et al.* (*Asian Marine Biol.* **5**: 103, 1988). **Seychelles**, Hyde & Jones (*Marine Ecol.* **9**: 15, 1988). **South Africa**, Steinke & Jones (*S. Afr. J. Bot.* **59**: 385, 1993). **USA**, Kohlmeyer, (*TMBS* **57**: 473, 1971; New England checklist), Jones (*Bot. J. Linn. Soc.* **91**: 219, 1985); and many other countries.

Marinosphaera K.D. Hyde (1989), Sordariomycetes. 1, Indian Ocean; Pacific Ocean. See Hyde (*CJB* **67**: 3078, 1989), Read *et al.* (*MR* **99**: 1465, 1995; ultrastr.).

Marinospora A.R. Caval. (1966), Halosphaeriaceae. 3 (marine), widespread. See Johnson *et al.* (*Bot. Mar.* **27**: 557, 1984), Jones (*CJB* **74** Suppl. 1: S790, 1995; ultrastr.), Campbell *et al.* (*Fungal Diversity Res. Ser.* **7**: 15, 2002; phylogeny).

Mariona Stempell (1909), Microsporidia. 1.

Marisolaris Jørg. Koch & E.B.G. Jones (1989), ? Dothideomycetes. 1 (marine), Denmark. See Koch & Jones (*CJB* **67**: 1190, 1989).

Mariusia D. Pons & Boureau (1977), Fossil Fungi, Microthyriaceae. 1 (Cretaceous), France.

Maronea A. Massal. (1856), Fuscideaceae (L). 13, widespread. See Magnusson (*Acta horti gotoburg.* **9**: 41, 1934), Singh (*Geophytology* **10**: 34, 1980; India), Brusse (*Bothalia* **19**: 36, 1989; S Africa), Kantvilas (*Biblthca Lichenol.* **78**: 169, 2001; Tasmania), McCarthy & Mallett (*Flora of Australia* **56** A, 2004), Harris (*Opuscula Philolichenum* **3**: 65, 2006; N America), LaGreca (*Lichenologist* **38**: 595, 2006; chemistry), Miądlikowska *et al.* (*Mycol.* **98**: 1088, 2006; phylogeny), Bylin *et al.* (*Biblthca Lichenol.* **96**: 49, 2007; phylogeny).

Maronella M. Steiner (1959), Acarosporaceae (L). 1. See Hawksworth *et al.* (*Dictionary of the Fungi* edn 8, 1995), Hafellner (*Symb. bot. upsal.* **34** no. 1: 87, 2004).

Maroneomyces Cif. & Tomas. (1953) = Maronea.

Maronina Hafellner & R.W. Rogers (1990), Lecanoraceae (L). 2, Australia; S. America. See Hafellner & Rogers (*Biblthca Lichenol.* **38**: 100, 1990), McCarthy & Mallett (*Flora of Australia* **56** A, 2004), Kantvilas & Elix (*Biblthca Lichenol.* **96**: 137, 2007; Australia).

Marssonia J.C. Fisch. (1874) [non *Marssonia* H. Karst. 1860, *Gesneriaceae*] ≡ Marssonina.

Marssoniella Höhn. (1916) [non *Marssoniella* Lemmerm. 1900, *Algae*] = Neomarssoniella fide Hawksworth *et al.* (*Dictionary of the Fungi* edn 8, 1995).

Marssoniella Lemmermann (1900), Microsporidia. 1.

Marssonina Magnus (1906), anamorphic *Diplocarpon, Drepanopeziza*, Cac.1eH.15/19. 78, widespread (esp. temperate). *M. ochroleuca* (chestnut (*Castanea*) leaf spot), *M. rosae* (black spot of rose). See von Arx (*Verh. K. ned. Akad. Wet. Amst.* C **51**: 1, 1957), Sutton *et al.* (*TBMS* **86**: 619, 1986; redisposition), Spiers (*Eur. J. For. Path.* **18**: 140, 1988; key 4 spp. on *Populus*), Spiers (*N.Z. Jl Bot.* **27**: 503, 1989; conidial morphology of poplar spp. on host and in culture), Spiers (*Eur. J. For. Path.* **20**: 154, 1990; conidial morphology variation spp. on *Populus*), Farr (*Mycol.* **85**: 814, 1993; relationship to *Septogloeum*), Wenefrida & Spencer (*Pl. Dis.* **77**: 246, 1993; on *Rosa*), Morgan-Jones & Phelps (*Mycotaxon* **55**: 215, 1995; on *Veratrum*), Spiers (*Eur. J. For. Path.* **28**: 233, 1998; on *Salicaceae*), Han *et al.* (*Theoretical and Applied Genetics* **100**: 614, 2000; genetic variation).

Marssoninites Babajan & Tasl. (1970), Fossil Fungi. 1 (Tertiary), former USSR.

Martella Adans. (1763) = Hericium Pers. fide Donk (*Reinwardtia* **2**: 466, 1954).

Martella Endl. (1836) = Hericium Pers. fide Donk (*Taxon* **5**: 103, 1955).

Martellia Mattir. (1900) = Russula fide Kuyper (*in litt.*).

Martensella Coem. (1863), Kickxellaceae. 1, Europe; N. America. See Jackson & Dearden (*Mycol.* **40**: 168, 1948), Benjamin (*Aliso* **4**: 321, 1959).

Martensiomyces J.A. Mey. (1957), Kickxellaceae. 1, Congo. See Benjamin (*Aliso* **4**: 321, 1959).

Marthamyces Minter (2003), ? Rhytismataceae. 9, widespread. See Minter (*Mycotaxon* **87**: 50, 2003), Johnston (*Aust. Syst. Bot.* **19**: 135, 2006; Australia).

Marthanella States & Fogel (1999), ? Boletales. 1, USA. See States & Fogel (*Mycotaxon* **71**: 424, 1999).

Martin (George Willard; 1886-1971; USA). Professor of Botany, University of Iowa (1929-1955). Noted for studies on the *Tremellales* (q.v.) and *Mycetozoa* (q.v.). He contributed keys to families of fungi for the first five editions of this *Dictionary*. *Publs*. The Tremellales of the north central United States and adjacent Canada. *University of Iowa Studies in Natural History* (1940); (with C.J. Alexopoulos, q.v.) *The Myxomycetes* (1969). *Biogs, obits etc*. Stafleu & Cowan (*TL-2* **3**: 320, 1981); Wells & Lentz (*Mycol.* **65**: 985, 1973, bibliography, portrait).

Martindalia Sacc. & Ellis (1885) = Phleogena fide Barr & Bigelow (*Mycol.* **60**: 456, 1968).

Martinella (Cooke) Sacc. (1892) nom. dub., Fungi. See Sutton (*Mycol. Pap.* **141**, 1977).

Martinellisia V.G. Rao & Varghese (1977), anamorphic *Pezizomycotina*, Hso.1eP.10. 1, India. See Rao & Varghese (*Norw. Jl Bot.* **24**: 279, 1977).

Martinia Whetzel (1942) [non *Martinia* Vaniot 1903, *Compositae*] ≡ Martininia.

Martininia Dumont & Korf (1970), Sclerotiniaceae. Anamorph *Myrioconium*. 1, widespread. See Dumont (*Mycol.* **65**: 175, 1973), Zhuang (*Mycotaxon* **29**: 393, 1987).

Mason (Edmund William; 1890-1975; England). Wounded in the Somme (1916); Mycologist, IMI (1921-1960). His organization of the IMI fungal reference collection set new standards, showing a profound understanding of how mycological information should be handled, amounting almost to prescience of how it should be structured for digitized databases. Noted for his catalytic ideas on use of conidial development to classify hyphomycetes, publishing little himself, he strongly influenced Hughes and M.B. Ellis (q.v.). *Publs*. *Mycological Papers* **1-3** (1925-1941). *Biogs, obits etc*. Booth (*Bulletin of the British*

Mycological Society **9**: 114, 1975); Ellis & Hughes (*TBMS* **66**: 371, 1976) [portrait].

Masonhalea Kärnefelt (1977), Parmeliaceae (L). 1, Arctic. See Kärnefelt *et al.* (*Pl. Syst. Evol.* **183**: 113, 1992), Thell *et al.* (*Mycol. Progr.* **1**: 335, 2002; phylogeny), Miądlikowska *et al.* (*Mycol.* **98**: 1088, 2006; phylogeny).

Masonia G. Sm. (1952) ≡ Masoniella.

Masonia Hansf. (1944) = Dictyonella fide von Arx (*Persoonia* **2**: 421, 1963).

Masoniella G. Sm. (1952) = Scopulariopsis fide Morton & Smith (*Mycol. Pap.* **86**, 1963).

Masoniomyces J.L. Crane & Dumont (1975), anamorphic *Pezizomycotina*, Hso.0eH.10. 1, Mexico; Caribbean. See Crane & Dumont (*CJB* **53**: 847, 1975).

Massalongia Körb. (1855), Massalongiaceae (L). 3, Europe; N. America. See Henssen (*CJB* **41**: 1331, 1963), Jørgensen *in* Hawksworth (Ed.) (*Ascomycete Systematics. Problems and Perspectives in the Nineties* NATO ASI Series vol. **269 269**: 390, 1994; posn), Burgaz & Martínez (*Nova Hedwigia* **73**: 381, 2001; Iberian peninsula), Wiklund & Wedin (*Cladistics* **19**: 419, 2003; phylogeny), Miądlikowska & Lutzoni (*Am. J. Bot.* **91**: 449, 2004; phylogeny), Miądlikowska *et al.* (*Mycol.* **98**: 1088, 2006; phylogeny), Wedin *et al.* (*Lichenologist* **39**: 61, 2007; n. fam.).

Massalongiaceae Wedin, P.M. Jørg. & Wiklund (2007), Peltigerales. 3 gen. (+ 9 syn.), 11 spp.

Lit.: Henssen (*CJB* **41**: 1331, 1963), Miądlikowska & Lutzoni (*Am. J. Bot.* **91**: 449, 2004), Wedin & Wiklund (*Symb. bot. upsal.* **34** no. 1: 469, 2004), Jørgensen (*Nordic Lichen Flora* **3**: Cyanolichens: 87, 2007), Wedin *et al.* (*Lichenologist* **39**: 61, 2007).

Massalongiella Speg. (1880) = Enchnoa fide Petrak & Sydow (*Annls mycol.* **34**: 11, 1936).

Massalongina Bubák (1916), anamorphic *Pezizomycotina*, St.0eH.?. 1, Europe.

Massalongo (Abramo Bartolommeo; 1824-1860; Italy). Professor at the University of Verona. Palaeobotanist and lichenologist, renowned for his contribution to understanding the systematics of lichen-forming fungi with crustose thalli through use of microscopic characters, introducing 114 genera of which many have since been consistently used (e.g. *Arthopyrenia*, *Catillaria*, *Haematomma*, *Pyrenula*, *Solenopsora*) or recently reintroduced (e.g. *Bactrospora*, *Celothelium*, *Psilolechia*). Often involved in controversies with Trevisan (q.v.). Main collections in Verona (**VER**). *Publs. Ricerche sull'Autonomia dei Licheni Crustosi* (1852); *Systema Lichenum Novorum* (1855) [a selection of these works have been reprinted (Lazzarini, *Selezione di Lavori Lichenologici di A.B. Massalongo*, 1991)]. *Biogs, obits etc.* De Toni (*L'Opera Lichenologica di Abramo Massalongo*, 1933); Grummann (1974: 518); Poelt (*in* Lazzarini, 1991: 13); Stafleu & Cowan (*TL-2* **3**: 349, 1981).

Massalongomyces Cif. & Tomas. (1953) ≡ Massalongia.

Massaria De Not. (1844), Massariaceae. Anamorphs *Myxocyclus*, *Neohendersonia*, *Macrodiplodia*. 22, widespread. See Wehmeyer (*Revision of Melanconis*, 1941), Shoemaker & Le Clair (*CJB* **53**: 1568, 1975; type studies), Barr (*Mycotaxon* **9**: 17, 1979; key 4 N. Am. spp.), Barr (*N. Amer. Fl.* ser. 2 **13**: 129 pp., 1990; N America), Liew *et al.* (*Mol. Phylogen. Evol.* **16**: 392, 2000; phylogeny), Treigiene & Rukšeniene (*Botanica Lithuanica* **11**: 55, 2005; Lithuania), Schoch *et al.* (*Mycol.* **98**: 1041, 2006; phylogeny).

Massariaceae Nitschke (1869), Pyrenulales. 6 gen. (+ 3 syn.), *c.* 41 spp.

Lit.: Barr (*Sydowia* **38**: 11, 1985), Eriksson (*SA* **5**: 140, 1986), Barr (*N. Amer. Fl.* ser. 2 **13**: 129 pp., 1990), Hyde (*MR* **96**: 1044, 1992), Hyde *et al.* (*Nova Hedwigia* **62**: 513, 1996), Liew *et al.* (*Mol. Phylogen. Evol.* **16**: 392, 2000).

Massariella Speg. (1880) = Amphisphaeria fide Dzagania (*BSMF* **102**: 199, 1986), Eriksson & Hawksworth (*SA* **7**: 80, 1988).

Massariellops Curzi (1927) = Didymosphaeria fide Aptroot (*Nova Hedwigia* **60**, 1995).

Massarina Sacc. (1883), Massarinaceae. Anamorphs *Acrocalymma*, *Ceratophoma*, *Tetraploa*. 125, widespread. See Bose (*Phytopath. Z.* **41**: 156, 1961; key), Webster (*TBMS* **48**: 449, 1965), Srinivasulu & Sathe (*Sydowia* **26**: 83, 1974; Indian spp.), Leuchtmann (*Sydowia* **37**: 75, 1984), Hyde (*Mycol.* **83**: 839, 1992; key marine spp.), Hyde (*Mycol.* **99**: 291, 1995; list 132 names), Aptroot (*Nova Hedwigia* **66**: 89, 1998; monogr.), Tsui *et al.* (*Mycol.* **91**: 721, 1999; ultrastr.), Au *et al.* (*Bot. Mar.* **44**: 261, 2001; ultrastr.), Hyde *et al.* (*Fungal Diversity Res. Ser.* **7**: 93, 2002; marine taxa), Liew *et al.* (*Mycol.* **94**: 803, 2002; phylogeny), Eriksson & Hawksworth (*Mycol.* **95**: 426, 2003; phylogeny, fam. placement), Tanaka & Harada (*Mycoscience* **44**: 173, 2003; Japan), Döbbeler (*MR* **111**: 1406, 2007), Wang *et al.* (*MR* **111**: 1268, 2007; phylogeny).

Massarinaceae Munk (1956), Pleosporales. 5 gen. (+ 17 syn.), 189 spp. Molecular data have elucidated some relationships within this fam., but more research is needed. It has previously been confused with the *Lophiostomataceae*. See Aptroot (*Nova Hedwigia* **66**: 89, 1998; monogr.), Kodsueb *et al.* (*Botanical Journal of the Linnean Society* **155**: 283, 2007; descr.), Wang *et al.* (*MR* **111**: 1268, 2007; phylogeny).

Massarinula Géneau (1894) = Massarina fide Bose (*Phytopath. Z.* **41**, 1961), Aptroot (*Nova Hedwigia* **66**: 89, 1998).

Massariola Füisting (1868), Dothideomycetes. 3 (on bark), Europe.

Massariopsis Niessl (1876) = Amphisphaeria fide Müller & von Arx *in* Ainsworth *et al.* (Eds) (*The Fungi* **4A**: 87, 1973).

Massariosphaeria (E. Müll.) Crivelli (1983), Pleosporales. 23, widespread. See Crivelli (*Über die heterogene Ascomycetengattung Pleospora Rbh.*, 1983; key 7 spp.), Leuchtmann (*Sydowia* **37**: 75, 1985; key), Holm & Holm (*Symb. bot. upsal.* **28** no. 2: 1, 1988), Huhndorf *et al.* (*Mycotaxon* **37**: 203, 1990), Liew *et al.* (*Mol. Phylogen. Evol.* **16**: 392, 2000; phylogeny), Inderbitzin *et al.* (*Mycol.* **94**: 651, 2002; phylogeny), Tanaka & Harada (*Mycoscience* **45**: 96, 2004; Japan), Wang *et al.* (*MR* **111**: 1268, 2007; phylogeny).

Massariothea Syd. (1939), anamorphic *Pezizomycotina*, St.≡ eP.15. 8, widespread. See Alcorn (*MR* **97**: 429, 1993; key).

Massariovalsa Sacc. (1882), Melanconidaceae. Anamorph *Melanconiopsis*. *c.* 4, widespread. See Wehmeyer (*Revision of Melanconis*, 1941), Petrak (*Sydowia* **19**: 279, 1966), Speer (*BSMF* **102**: 363, 1986).

Massariovalsaceae Hara (1913) = Melanconidaceae.

Massartia De Wild. (1897), ? Zoopagales. 1 (in *Algae*), Java. See Hesseltine (*Mycol.* **47**: 344, 1955).

Massee (George Edward; 1850-1917; England). French Foreign Legion (*c.* 1870s); schoolteacher, Scarborough (*c.* 1880s); freelance worker, Kew (1990s); Cryptogamic Botanist, Royal Botanic Gardens, Kew (1893-1915). First President of the British Mycological Society (1896). *Publs. British Fungus Flora* (1892-1895); *Textbook of Fungi* (1906); *Diseases of Cultivated Plants and Trees* (1910); *British Fungi and Lichens* (1911); (with Crossland) *Fungus Flora of Yorkshire* (1902-1905). *Biogs, obits etc.* Anon. (*Kew Bulletin* **1922**: 335, 1922, bibliography); Grummann (1974: 377); Ramsbottom (*TBMS* **5**: 469, 1917; Ramsbottom (*Journal of Botany* London **55**: 223, 1915); Stafleu & Cowan (*TL-2* **3**: 359, 1981).

Masseea Sacc. (1889), Helotiales. 1, N. America. See Baral (*SA* **13**: 113, 1994).

Masseeëlla Dietel (1895), Pucciniales. *c.* 5 (on dicots, mostly *Euphorbiaceae*), widespread (palaeotropical). See Sathe (*Sydowia* **19**: 187, 1966), Thaung (*Australasian Mycologist* **24**: 29, 2005; Burma) Cf. *Kamatomyces*.

Masseeola Kuntze (1891) ≡ Sparassis.

Masseerina Lloyd (1920) nom. nud., Agaricomycetes. See Donk (*Taxon* **6**: 64, 1957).

Massospora Peck (1879), Entomophthoraceae. 13 (on cicadas), widespread. *M. cicadina* is a parasite of the 17-year cicada in N. Am. See Soper (*Mycotaxon* **1**: 13, 1974; key), Soper *et al.* (*Ann. Ent. Soc. Am.* **69**: 89, 1976; biology), Keller & Petrini (*Sydowia* **57**: 23, 2005; key).

Mastigobasidium Golubev (1999), Leucosporidiaceae. 1, Japan. See Golubev (*Int. J. Syst. Bacteriol.* **49**: 1301, 1999; taxonomy), Fell *et al.* (*Int. J. Syst. Evol. Microbiol.* **50**: 1351, 2000; mol. phylogeny), Aime *et al.* (*Mycol.* **98**: 896, 2006; phylogeny).

Mastigochytrium Lagerh. (1892) ? = Rhizophydium fide Sparrow (*Aquatic Phycomycetes* Edn 2: 1187 pp., 1960).

Mastigocladium Matr. (1911) = Acremonium fide Gams (*Cephalosporium-artige Schimmelpilze*, 1971).

Mastigomyces Imshen. & Kriss (1933), anamorphic *Pezizomycotina*, Hso.?.?. 2, former USSR. See Babjeva & Levin (*Mikrobiologiya* **48**: 541, 1979; neotypification).

Mastigomycètes Imshen. & Kriss (1933), anamorphic *Pezizomycotina*, Hso.?.?. 2, former USSR. See Babjeva & Levin (*Mikrobiologiya* **48**: 541, 1979; neotypification).

Mastigomycotina (obsol.). Used for fungi with motile zoospores now dispersed in the phyla *Chytridiomycota, Hyphochytriomycota, Oomycota, Plasmodiophoromycota*.

Mastigonema Speg. (1926) = Chaetospermum fide Petrak & Sydow (*Annls mycol.* **34**: 11, 1936).

mastigoneme, see flagellum.

Mastigonetron Kleb. (1914) = Harknessia fide Petrak (*Sydowia* **24**: 253, 1971), Sutton (*Mycol. Pap.* **141**, 1977).

mastigopod (of *Mycetozoa*), a swarm cell (obsol.).

Mastigosporella Höhn. (1914), anamorphic *Wuestneiopsis*, Cpd.0eH.19. 2, USA. See Nag Raj (*Coelomycetous Anamorphs with Appendage-bearing Conidia*, 1993).

Mastigosporium Riess (1852), anamorphic *Pezizomycotina*, Hsy.≡ eH.19. 6 (causing leaf spots of grasses), widespread (temperate). See Bollard (*TBMS* **33**: 250, 1950), Hughes (*Mycol. Pap.* **36**, 1951), Austwick (*TBMS* **37**: 161, 1954), Huss *et al.* (*Pflanzenschutzberichte* **49**: 97, 1988; on *Dactylis* spp.), Mayrhofer *et al.* (*Mitt. naturwiss. Ver. Steiermark* **121**: 73, 1991; key), Braun (*Monogr. Cercosporella, Ramularia Allied Genera (Phytopath. Hyphom.)* **1**: 260, 1995; key).

Mastocephalus Battarra ex Kuntze (1891) ≡ Leucocoprinus.

Mastodia Hook. f. & Harv. (1847), Mastodiaceae (L). 6, Antarctica. See Brodo (*Bryologist* **79**: 385, 1976), Gremmen *et al.* (*Lichenologist* **27**: 387, 1995; ecology), Kohlmeyer *et al.* (*Mycol. Progr.* **3**: 51, 2004).

Mastodiaceae Zahlbr. (1907), Pezizomycotina (inc. sed.) (±L). 2 gen. (+ 3 syn.), 8 spp.

Lit.: Kohlmeyer & Volkmann-Kohlmeyer (*Bot. Mar.* **34**: 1, 1991), Sancho & Valladares (*Polar Biol.* **13**: 227, 1993).

Mastoleucomyces Battarra ex Kuntze (1891) nom. nud. = Tricholoma fide Singer (*Agaric. mod. Tax.*, 1951).

Mastomyces Mont. (1848) ≡ Topospora fide Groves (*CJB* **43**: 495, 1965).

Mathurisporites Kalgutkar & Janson. (2000), Fossil Fungi, anamorphic *Ascomycota*. 1, Canada; India. See Kalgutkar & Jansonius (*AASP Contributions Series* **39**: 166, 2000).

matrix (1) the substratum in or on which an organism is living; (2) mucilaginous material in which conidia and some ascospores are produced, influences dissemination, survival, germination etc. See Louis & Cooke (*TBMS* **84**: 661, 1985).

Matruchotia Boulanger (1893) nom. dub., anamorphic *Fungi*. See Donk (*Reinwardtia* **2**: 466, 1954).

Matruchotiella Grigoraki (1924) nom. dub. ≡ Ateleothylax fide Dodge (*Medical Mycology*: 431, 1935).

Matsushimaea Subram. (1978), anamorphic *Pezizomycotina*, Hso.0bP.10. 1, Japan. See Subramanian (*Kavaka* **5**: 96, 1977), Mel'nik & Belomesyatseva (*Mikol. Fitopatol.* **35**: 29, 2001; Russia, Belorus).

Matsushimiella R.F. Castañeda & Heredia (2001), anamorphic *Pezizomycotina*, Hso.?.?. 1, Australia. See Castañeda Ruiz *et al.* (*Cryptog. Mycol.* **22**: 16, 2001).

Matsushimomyces V.G. Rao & Varghese (1979), anamorphic *Pezizomycotina*, Hso.1eH.?10. 1, India. See Rao & Varghese (*Bot. Notiser* **132**: 313, 1979), Castañeda Ruíz *et al.* (*Mycotaxon* **87**: 385, 2003).

Matsusphaeria K.L. Pang & E.B.G. Jones (2003) = Magnisphaera fide Pang *et al.* (*Nova Hedwigia* **77**: 14, 2003), Pang & Jones (*Nova Hedwigia* **78**: 269, 2004).

matsu-take, *Tricholoma matsutake*, an important edible fungus in Japan. Nisikado *et al.* (*Ber. Ohara Inst.* **8**(4), 1941); pine mushroom. 'Matsutake' of N. Am., *T. ponderosum*, *T. murrillianum* (fide Singer, 1961).

Mattickiolichen Tomas. & Cif. (1952) = Buellia fide Hawksworth *et al.* (*Dictionary of the Fungi* edn 8, 1995).

Mattickiomyces Cif. & Tomas. (1953) ≡ Mattickiolichen.

Mattirolella S. Colla (1929), anamorphic *Kathistaceae*, St.0eH.15. 1. See Khan & Kimbrough (*Am. J. Bot.* **61**: 395, 1974), Blackwell *et al.* (*Mycol.* **95**: 987, 2003).

Mattirolia Berl. & Bres. (1889), Thyridiaceae. 1 (from bark), Italy. See Rossman *et al.* (*Stud. Mycol.* **42**: 248 pp., 1999).

Mattirolomyces E. Fisch. (1938), Pezizaceae. 1 (Mycorrhizal under *Cistaceae*), Mediterranean; N. America. See Percudani *et al.* (*Mol. Phylogen. Evol.* **13**: 169, 1999; phylogeny), Díez *et al.* (*Mycol.* **94**: 247, 2002; phylogeny), Healy (*Mycol.* **95**: 765, 2003; USA), Ferdman *et al.* (*MR* **109**: 237, 2005; phylogeny), Kovács *et al.* (*Mycol. Progr.* **6**: 19, 2007; ontogeny, mycorrhiza), Læssøe & Hansen (*MR* **111**: 1075, 2007; phylogeny).

Matula Massee (1888), anamorphic *Aleurocystis*. 2, widespread (tropical). See Martin (*Lloydia* **5**: 158, 1942).

Maublancia G. Arnaud (1918), Microthyriaceae. 6, S. America; S. Africa.

Maublancomyces Herter (1950) = Gyromitra fide Eckblad (*Nytt Mag. Bot.* **15**: 1, 1968).

Mauginiella Cavara (1925), anamorphic *Pezizomycotina*, Hso.0eH.38. 1 (on *Phoenix*), Mediterranean; Middle East. See Sigler & Carmichael (*Mycotaxon* **4**: 349, 1976), von Arx *et al.* (*Sydowia* **34**: 42, 1981; ultrastr.), Abdullah *et al.* (*J. Phytopath.* **153**: 417, 2005; Spain).

Maurinia Niessl (1876) = Anthostomella fide Francis (*Mycol. Pap.* **139**, 1975), Lu & Hyde (*Fungal Diversity Res. Ser.* **4**, 2000).

Mauritiana Poonyth, K.D. Hyde, Aptroot & Peerally (2000), Requienellaceae. 1, Mauritius. See Poonyth *et al.* (*Fungal Diversity* **4**: 101, 2000).

Mauritzia Gyeln. (1935) = Pyrenopsis fide Hawksworth *et al.* (*Dictionary of the Fungi* edn 8, 1995).

Maurodothella G. Arnaud (1918) = Echidnodes fide von Arx & Müller (*Stud. Mycol.* **9**, 1975).

Maurodothina G. Arnaud ex Piroz. & Shoemaker (1970) = Eupelte fide von Arx & Müller (*Stud. Mycol.* **9**, 1975).

Maurodothis Sacc., Syd. & P. Syd. ex Syd. & P. Syd. (1904) = Cycloschizon fide Müller & von Arx (*Beitr. Kryptfl. Schweiz* **11** no. 2, 1962).

Maurya Pat. (1898) nom. dub., Fungi. See Petrak (*Sydowia* **5**: 345, 1951).

Mawsonia C.W. Dodge (1948), ? Lichinaceae (L). 1, Antarctica. See Kärnefelt (*Cryptog. Bot.* **1**: 147, 1989).

Maxillispora Höhn. (1914) = Tetracladium fide Ingold (*TBMS* **25**: 371, 1942).

Mayamontana Castellano, Trappe & Lodge (2007), Stephanosporaceae. 1. See Castellano *et al.* (*Mycotaxon* **100**, 2007).

Mazaediate lichens, Traditionally treated as *Caliciales*, but that order has been shown to be heterogenous using molecular methods.

Lit.: Hutchinson (*Mycol.* **79**: 786, 1987; spp. on polypores), Middelborg & Mattsson (*Sommerfeltia* **5**, 1987; Norway), Nádvornik (*Studia bot. čsl.* **5**: 6, 1942), Pant & Awasthi (*Biovigyanam* **15**: 3, 1989; key 31 spp. India & Nepal), Puntillo (*Webbia* **43**: 145, 1989; key 59 spp. Italy), Santesson (*Ark. Bot.* **30A** (14), 1943; S. Am.), Sato (*Miscnea bryol. lich.* **7**: 39, 1975; gen. concepts), Schmidt (*Mitt. St. Inst. allg. Bot. Hamburg* **13**: 111, 1970), Selva & Tibell (*Bryologist* **102**: 377, 1999, N Am.), Tibell (*Symb. bot. upsal.* **21**(2), 1975; boreal N. Am., *Lichenologist* **13**: 161, 1981; Afr., **14**: 219, 1982; Costa Rica, *Beih. Nova Hedw.* **79**: 597, 1985; system, *Symb. bot. upsal.* **27**(2), 1987; keys 18 gen., 78 spp. Australia, *Bot. J. Linn. Soc.* **116**: 159, 1994; distrib. 162 spp., dispersal patterns, *Flora Neotropica Monograph* **69**, 1996), *Symb. bot. upsal.* **32**: 291, 1997; anams, *Bibl. Lichenol.* **71**: 107, 1998; S Am. spp.; *Nordic Lichen Flora* **1**, 1999), Tobolewski (*Pr. Kom. biol. Poznán* **24** (5), 1966; Poland).

Mazaediothecium Aptroot (1991), Pyrenulaceae (L). 2, Costa Rica; Papua New Guinea. See Eriksson (*Lichenologist* **25**: 307, 1993), Aptroot *et al.* (*Biblthca Lichenol.* **97**, 2008; Costa Rica).

mazaedium, a spore mass formed by an ascoma, as in *Caliciales* and *Onygenaceae*, in which the spores, generally with sterile elements, become free from the asci as a dry, loose powdery mass on the fruiting surface.

Mazosia A. Massal. (1854), ? Roccellaceae (±L). Anamorph *Sporhaplus*. 21 (partly on lichens), widespread (tropical). See Santesson (*Symb. bot. upsal.* **12** no. 1: 1, 1952), Batista *et al.* (*Atas Inst. Micol. Univ. Recife* **5**: 429, 1967), Kalb & Vězda (*Folia geobot. phytotax.* **23**: 199, 1988; key), Eriksson (*SA* **14**: 58, 1995; posn), Lücking & Matzer (*Nova Hedwigia* **63**: 109, 1996), Lücking & Kalb (*Bot. Jb.* **122**: 1, 2000; Brazil), Aptroot *et al.* (*Mycotaxon* **88**: 41, 2003; China, Taiwan), Singh & Pinokiyo (*Lichenologist* **40**: 23, 2008; India).

mazu, see Fermented food and drinks.

Mazzantia Mont. (1855), Diaporthaceae. Anamorph *Mazzantiella*. 6, Europe; America. See Monod (*Beih. Sydowia* **9**: 1, 1983), Castlebury *et al.* (*Mycoscience* **44**: 203, 2003), Zhang *et al.* (*Mycol.* **98**: 1076, 2006; phylogeny).

Mazzantiella Höhn. (1925), anamorphic *Mazzantia*. See Clements & Shear (*Gen. Fung.*, 1931).

McAlpine (Daniel; 1849-1932; Scotland, later Australia). Lecturer, Heriot-Watt College, Edinburgh (1877); lecturer, Melbourne University (1884); Vegetable Pathologist, Department of Agriculture, Victoria (1890). As Vegetable Pathologist, McAlpine occupied the first full-time permanent post in applied mycology in the then British Empire. *Publs. Rusts of Australia, their Structure, Nature and Classification* (1906); *Smuts of Australia, their Structure, Life History, Treatment and Classification* (1910); *Fungous Diseases of Stone-fruit Trees of Australia, and their Treatment* (1902); *Handbook of Fungous Diseases of the Potato in Australia and their Treatment* (1911). *Biogs, obits etc.* Fish (*Annual Review of Phytopathology* **8**: 14, 1970); Stafleu & Cowan (*TL-2* **3**: 207, 1981).

McGinty, see Lloyd.

MCZ, see Media.

MEA, see Media.

Mebarria J. Reid & C. Booth (1989), Melanconidaceae. Anamorph *Harknessia*. 1, USA. See Barr (*Mycotaxon* **41**: 287, 1991).

medallion clamp, a clamp connexion with a space between the main hypha and the hook.

Medastina Dodart [not traced] nom. dub., ? Fungi.

Medeolaria Thaxt. (1922), Medeolariaceae. 1, N. America. See Thaxter (*Proc. Amer. Acad. Arts & Sci.* **57**: 425, 1922).

Medeolariaceae Korf (1982), Medeolariales. 1 gen., 1 spp.

Lit.: Thaxter (*Proc. Amer. Acad. Arts & Sci.* **57**: 425, 1922), Korf (*Mycosystema* **3**: 19, 1990), Pfister & Kimbrough *in* McLaughlin *et al.* (Eds) (*The Mycota* A Comprehensive Treatise on Fungi as Experimental Systems for Basic and Applied Research **7A**: 257, 2001).

Medeolariales Korf (1982). Pezizomycotina. 1 fam., 1 gen., 1 spp. Fam.:

Medeolariaceae

For *Lit.* see under fam.

Media. Media for cultivation of fungi can be prepared from natural, synthetic or semi-synthetic components. Essential requirements for fungal growth include carbon and nitrogen sources, vitamins, minerals, air and water. There is no single standard medium on which all fungi grow. Many can be cultivated by supplying simple sugars however low-nutrient media are often required for obtaining sporulation. Since media can affect colony morphology, pigmentation, sporulation and retention of properties, standardization of formulae is often necessary for the purpose of identification. Defined media formulations are essential in quantitative experiments and investigations that require an element of reproducibility. Many culture media are available commercially, others are prepared in the laboratory. Useful publications include Booth (*in* Booth (Ed.), *Methods in microbiology* **4**: 57, 1971), Jong *et al.* (*ATCC Filamentous fungi*, edn 19, 1996), Atlas (*Handbook of microbiological media*, 1996), Smith *et al.* (Eds) (*The UKNCC biological resource: properties, maintenance and management*, 2001). The following entry provides an introduction to the most commonly used media for cultivation of fungi.

Potato Carrot Agar (PCA) is used for obtaining sporulation in a wide range of fungi. This medium is low in carbohydrates and many fungi are stimulated to sporulate in response to poor nutrient levels. Most Mucorales grow successfully on PCA and Malt Agar (MA). Many anamorphic fungi and ascomycetes also sporulate successfully on Malt Agar (MA), Cornmeal Agar (CMA) and Oatmeal Agar (OA). Hay Infusion Agar is useful as a low nutrient medium for production of conidia. Potato Dextrose Agar (PDA) is rich in carbohydrates stimulating good vegetative growth in many fungi. In the presence of high carbohydrate levels however, sporulation is often reduced. Some basidiomycetes in culture including *Armillaria* and *Ganoderma* have been grown successfully on Malt Extract Agar (MEA) and Yeast Malt Extract (YM). A specialized medium for obtaining sporulation of wood-rotting basidiomycetes in culture was described by Badcock (*TBMS* **25**: 200, 1941). To encourage sporulation in cellulolytic fungi, sterilized pieces of filter paper, wheatstraw or lupin stem can be placed on the surface of weak media such as PCA or Tap Water Agar (TWA). Successful cultivation of chytrids has been obtained using Yeast Peptone Soluble Starch Agar (YPSS). 25% seawater can be added for marine species. The use of Rabbit Dung Agar (RDA) is recommended for obtaining sporulation in coprophilous fungi. A method for culturing pathogenic *Entomophthorales* on coagulated egg-yolk medium was described by Müller-Kögler (*Entomophaga* **3**: 261, 1959). Further culture methods for this group are described by Keller (*Sydowia* **40**: 122, 1987) and Bałazy (*Fungi.T. XXIV. Entomophthorales*, 1993). Lichen-forming fungi produce good vegetative growth on Malt Yeast Extract Agar (MYE).

Many fungi from medical sources require rich media with a high nitrogen content. Dermatophytes grow successfully on media containing peptone and glucose (Sabouraud's Agar). Keratinophylic fungi benefit from the addition of sterilized animal hair or feathers (Al-Doory, 1968). For further details of media used in medical mycology see Kwon-Chung & Bennett (*Medical mycology*, edn 4, 1992), Rippon (*Medical mycology*. The pathogenic fungi and the pathogenic actinomycetes, edn 3, 1988).

Specialized media are required for some genera. For growth of *Aspergillus* and *Penicillium*, Czapek Dox Agar (CZ), Czapek Yeast Autolysate (CYA) and 2% Malt Extract Agar (MEA) are recommended. To distinguish closely related species in these genera, Creatine Sucrose Agar (CREA) is useful. CYA and MEA with 20-40% sucrose are used for cultivation of xerophilic species of *Aspergillus* and those producing *Eurotium* teleomorphs. See Samson & Pitt (Eds) (*Advances in Penicillium and Aspergillus Systematics*, 1985). Sporulation of *Fusarium* can be encouraged by the use of Carnation Leaf Agar (Fischer *et al.*, *Phytopathology* **72**: 151, 1982) or Spezieller Nährstoffarmer Agar (Nirenberg, *Mitt. Biol. Bundesanst. Land. Forstw.* **169**: 1, 1976). Sterilized pieces of filter paper are added to this medium to stimulate sporulation. To study growth rate and colony colour in *Fusarium*, Potato Sucrose Agar (PSA) is recommended. Lima Bean Agar (LBA) and V8 Juice Agar have been used successfully for growing *Pythium* and *Phytophthora*. See Harnish (*Mycol.* 57: 85, 1965), Romero & Gallegly (*Phytopathology* **53**: 899, 1963). *Saprolegniaceae* grow well on Cornmeal Agar and Glucose Yeast Peptone Salts medium (GYPS). Inoculated blocks can be transferred to water baited with sterile hemp seeds to obtain sporangia and oogonia. Most yeasts grow sucessfully on MEA, Glucose Peptone Yeast Extract Agar (GPY) or Yeast Malt Extract Agar (YM). For osmophilic and osmotolerant yeasts, a sugar content of 30-50% is recommended. For further details of media for cultivation of yeasts see Kurtzman & Fell (*The Yeasts, a taxonomic study*, edn 4, 1998), Barnett *et al.* (*The Yeasts: characteristics and identification*, edn 3, 2000).

Bacterial contamination in fungal cultures can be eliminated by the use of media containing antibiotics. Most frequently used is Penicillin G in combination with either streptomycin sulphate or chloramphenicol.

Antiobiotic medium: Penicillin G and Chloramphenicol

Chloramphenicol	150 mg
Penicillin G	150 mg
Distilled Water	10 ml
Ethanol	10 ml

Dissolve 150 mg penicillin G in 10ml distilled water. Filter sterilize using a 0.45 μm pore membrane. Dissolve 150 mg chloramphenicol in 10ml ethanol. To add to media, melt 250 ml agar. When still hot, add 1.25 ml chloramphenicol in ethanol using a sterile pipette. Allow agar to cool to 45°C then add 1.25 ml penicillin G. Swirl thoroughly to mix.

Antibiotic medium: Penicillin G and Streptomycin sulphate

Penicillin G	1.2 g
Streptomycin sulphate	1.0 g
Distilled water	40 ml

Dissolve 1g of streptomycin sulphate powder in sterile water. Add penicillin G and agitate to mix. To add to media, melt 250 ml agar and allow to cool until hand-hot (45°C). Using a sterile pipette, add 2 ml of the antibiotic mixture to the agar and swirl bottle thoroughly to mix.

Carnation Leaf Agar (CLA)

Agar	20 g

Distilled water 1 l
Carnation leaf pieces
Cut carnation leaves into 1cm pieces, allow to dry and sterilize using a few drops of propylene oxide. Add agar to water. Heat until dissolved and autoclave at 121°C for 15 min. Pour plates and add a few sterile leaf pieces when plates are almost set. See Fisher *et al.* (*Phytopatholgy* **72**: 151, 1982).

Corn Meal Agar (CMA)

Maize meal	30 g
Oxoid Agar N° 3	20 g
Tap Water	1 l

Place the maize and water in a saucepan. Heat over a double saucepan until boiling, continue heating for one hour stirring occasionally. Filter the decoction through muslin, add the agar, and boil until it is dissolved. Autoclave at 121°C for 20 min.

Creatine Sucrose Agar (CREA)

Creatine ($1H_2O$)	3.0 g
Sucrose	30.0 g
KCl	0.5 g
$MgSO_4.7H_2O$	0.5 g
$FeSO_4.7H_2O$	0.01 g
$KH_2PO_4.3H_2O$	1.3 g
Bromocresol purple	0.05 g
Agar	15.0 g
Distilled Water	1 l

Dissolve ingedients in water. Autoclave at 121°C for 15 min. Adjust pH to 8.0-8.2 after autoclaving.

Czapek Dox Agar (CZ)

Made with stock Czapek solution

Solution A

Sodium nitrate $NaNO_3$	40.0 g
Potassium chloride KCl	10.0 g
Magnesium sulphate $MgSO_4$ $7H_2O$	10.0 g
Ferrous sulphate $FeSO_4$ $7H_2O$	0.2 g

Dissolve in 1 l distilled water and store in a refrigerator.

Solution B

Di-potassium hydrogen orthophosphate $K_2HPO_4$20 g

Dissolve in 1 l distilled water and store in a refrigerator.

For 1 litre

Stock solution A	50 ml
Stock solution C	50 ml
Distilled water	900 ml
Sucrose (Analar)	30 g
Oxoid Agar N° 3	20 g

Dissolve agar in distilled water then add sucrose and stock solutions just before autoclaving.
To each litre add 1 ml of following stock solutions:
Zinc sulphate $ZnSO_4$ $7H_2O$ Analar 1.0 g in 100 ml distilled water
Cupric sulphate $CuSO_4$ $5H_2O$ Analar 0.5 g in 100 ml distilled water
Autoclave at 121°C for 20 min. See Smith (1954).

Czapek Yeast Autolysate Agar (CYA)

di-Potassium hydrogen phosphate K_2HPO_4	1.0 g
Czapek concentrate	10.0 ml
Oxoid Yeast extract or autolysate	5.0 g
Sucrose (Analar)	30.0 g
Oxoid Agar N° 3	15.0 g
Distilled water	1.0 l

Autoclave at 121°C for 15 min. See Pitt (1973).

Glucose Peptone Yeast Extract Agar (GPY)

Glucose	40 g
Peptone	10 g
Yeast Extract	5 g
Agar	20 g
Distilled Water	1 l

Dissolve ingredients in water. Autoclave at 121°C for 15 min.

Glucose Yeast Peptone Salts (GYPS)

Glucose	5.0 g
Mycological Peptone	0.5 g
Yeast Extract (Oxoid)	0.05 g
KH_2PO_4	0.5 g
$MgSO_4.7H_2O$	0.15 g
Agar	15.0 g
Distilled Water	1 l

Dissolve agar and yeast extract in water. When agar has dissolved, add peptone and glucose. Autoclave at 121°C for 15 min.

Hay Infusion Agar

Hay	50.0 g
Agar	15.0 g
Distilled Water	1 l

Sterilize 50 g hay in 1 litre of water at 121°C for 30 min. Strain through cloth and make up to 1 litre. Adjust pH to 6.2 with K2HPO4. Add 15 g agar to 1 litre of extract. Autoclave at 121°C for 15 min.

Malt Agar (MA)

Toffee barley malt extract	20 g
Oxoid Agar N° 3	20 g
Tap water	1 l

Add malt extract and agar to water in a double saucepan. Heat and stir until dissolved. Adjust pH to 6.5. Autoclave at 121°C for 15 min.

Malt-Czapek Agar (MCZ)

Stock Czapek solution A	50 ml
Stock Czapek solution C	50 ml
Sucrose	30 g
Toffee malt extract	40 g
Oxoid Agar N° 3	20 g
Distilled water	900 ml

Dissolve malt extract and agar in water. Heat over a double saucepan until dissolved. Then add sucrose, when dissolved add stock solutions. Adjust pH to 5.0. Autoclave at 121°C for 20 min.

Malt Extract Agar (MEA) Blakeslee's formulation

Malt extract (powdered Difco or Oxoid)	20 g
Peptone (bacteriological)	1 g
Glucose (Analar)	20 g
Oxoid Agar N° 3	15 g
Distilled water	1 l

Add malt extract, peptone and agar to water. Heat over a double saucepan until dissolved. Add glucose last and stir to dissolve. Sterilize by autoclaving at 121°C for 15 min.

Malt Extract Agar plus sucrose (M20, M40, M60)

Malt extract	20 g
Sucrose (Analar)	200 g
Oxoid Agar N° 3	20 g
Tap water	1 l

For M40 use 400 g sucrose, for M60 use 600g sucrose. Dissolve malt extract and agar in water by heating in a double saucepan. Reduce heat to avoid caramelisation and add sucrose. Stir until dissolved. Autoclave at 121°C for 15 min.

Malt Yeast Extract (MYE)

Malt extract	20 g
Yeast Extract	2 g
Agar	20 g
Distilled Water	1 l

Dissolve ingredients in water. Autoclave at 121°C

for 15 min.

Oat Agar (OA)

Oat Meal ground	30 g
Oxoid Agar N° 3	20 g
Tap water	1 l

Add oat meal to 500 ml of water in a saucepan. Heat for 1 h. To the other 500 ml water add agar and dissolve in a double saucepan. Pass cooked oat meal through a fine strainer and add to agar mixture. Stir thoroughly. Autoclave at 121°C for 20 min.

Potato Carrot Agar (PCA)

Avoid new crop potatoes, which do not make good media. Red Désirée potatoes have been found to be best. Wash, peel and grate vegetables.

Grated potato	20 g
Grated carrot	20 g
Oxoid Agar N° 3	20 g
Tap water	1 l

Boil vegatables for about 1 h in 500 ml tap water, then pass through a fine sieve keeping the liquid. The agar is added to 500 ml of water in a double saucepan. When the agar has dissolved add the strained liquid and stir. Pour through a funnel into bottles. Sterilize at 121°C for 20 min.

Potato Dextrose Agar (PDA)

Avoid new crop potatoes. Red Désirée have been found to be best.

Potatoes	200 g
Oxoid Agar N° 3	20 g
Dextrose	15 g
Tap water	1 l

Scrub potatoes clean and cut into 12 mm cubes (do not peel). Weigh out 200 g and rinse rapidly under a running tap, and drop into 1 l of tap water in a saucepan. Boil until potatoes are soft (about 1 h) then put through blender. Add 20 g of agar, and heat in a double saucepan until dissolved. Then add 15 g of dextrose and stir until dissolved. Make up to 1 l. Pour into bottles, stiring occasionally to ensure that each bottle has a percentage of solid matter. Autoclave at 121°C for 20 min.

Potato Sucrose Agar (PSA)

To make 1 litre of medium.

Potato water	500 ml
Sucrose	20 g
Oxoid Agar N° 3	20 g
Distilled water	500 ml

Heat in double saucepan until agar is dissolved. Autoclave at 121°C for 15 min. 2.3 kg potatoes makes 7 l. Adjust to pH 6.5 with calcium carbonate if necessary.

To make 2 litres of potato water:

Tap water	1125 g
Potato	450 g

Peel and dice potatoes, suspend in double cheesecloth and boil in the tap water until almost cooked.

PSA using powdered potato

Powdered potato	5 g
Sucrose	20 g
Oxoid Agar N° 3	20 g
Distilled water	1 l
Calcium carbonate	5 g

Autoclave at 121°C for 15 min.

Rabbit Dung Agar (RDA)

The rabbit dung must be from wild rabbits, and dried before use.

Oxoid Agar N° 3	15 g
Tap water	1 l

Heat agar to dissolve. Pour into bottles containing the rabbit dung. Autoclave at 126°C for 20 min.

Sabouraud's Agar

Glucose	20 g
Peptone	10 g
Oxoid Agar N° 3	15 g
Water	1 l

Autoclave 114°C for 15 min.

Spezieller Nährstoffarmer Agar (SNA)

Potassium di-hydrogen orthophosphate (KH_2PO_4)	1 g
Potassium nitrate (KNO_3)	1g
Magnesium sulphate ($MgSO_4.7H_2O$)	0.5 g
Potassium chloride (KCl)	0.5 g
Glucose (Analar)	0.2 g
Sucrose (Analar)	0.2 g
Oxoid Agar N° 3	20 g
Distilled Water	1 l

Dissolve all ingredients, except agar, in distilled water. Add agar and heat in a double saucepan until dissolved. Adjust pH to 6-6.5, then dispense into bottles. Autoclave at 121°C for 20 mins.

Tap Water Agar (TWA)

Tap water	1 l
Oxoid Agar N° 3	15 g

Dissolve agar in water. Autoclave at 121°C for 20 min.

V8 Agar (as recommended for *Actinomycetes*)

V8 Vegetable juice	200 ml
Calcium carbonate	4 g
Oxoid Agar N° 3	20 g
Water	800 ml

Adjust to pH 7.3 with KOH Autoclave at 121°C for 20 min. See Galindo & Gallegly (1960).

Yeast Malt Agar (YM)

Yeast Extract	3 g
Malt Extract	3 g
Mycological Peptone	5 g
Glucose	10 g
Agar	20 g
Distilled Water	1 l

Dissolve ingredients in water and adjust pH to 5-6. Add agar and heat until dissolved. Autoclave at 121°C for 15 minutes.

Yeast Peptone Soluble Starch Agar (YPSS)

Yeast Extract	4.0 g
Soluble Starch	15.0 g
KH_2PO_4	1.0 g
$MgSO_4.7H_2O$	0.5 g
Agar	20.0 g
Distilled Water	1 l

Dissolve ingredients in water. Autoclave at 121°C for 15 minutes.

Addition of dried materials to media

Filter papers

Cut filter papers to size (approximately 1 x 0.7 cm). Transfer to glass universal bottles and autoclave at 121°C for 20 min. Using sterilised forceps, transfer pieces of filter paper to agar plates when set.

Wheatstraws

Cut dried wheatstraws into 2 cm lengths and transfer to glass universal bottles, 10 pieces in each. Autoclave at 121°C for 20 min. Using sterilised forceps, add wheatstraws to poured plates before the agar sets.

Medical and veterinary mycology. The scientific

study of fungi that cause diseases in humans and animals, and the characteristics and epidemiology of the diseases they cause. In this *Dictionary*, veterinary mycology is treated as relating to vertebrates only; see also: Insects (and other invertebrates) and fungi. Fungi affect different organs and according to the disorder syndromes, the diseases are named chromoblastomycoses, dermatophytoses, mycetoma, onichoblastomycoses, otomycoses, phaeohyphomycoses, rhinosporidioses, etc. Clinical groupings for fungal infections recognize skin diseases attributed to superficial (e.g. piedra, seborrhoeic dermatitis), cutaneous (e.g. candidiasis of skin, dermatophytosis, ringworm of the scalp) and subcutaneous (e.g. lobomycosis, mucoromycosis, sporotrichomycosis) mycoses; infectious diseases are described as dimorphic (e.g. blastomycosis, histoplasmosis) and opportunistic (e.g. aspergillosis, candidiasis, cryptococcosis) systematic mycoses. The earliest to be described (see Gruby) and best known mycosis is ringworm (tinea), of the skin and hair (caused by gymnoascaceous fungi). The other widespread and systemic mycoses (e.g. coccidioidomycosis, cryptococcosis, histoplasmosis) are frequently fatal.

More than 3400 fungal names have appeared in the medical and veterinary literature (de Hoog *et al.*, *Atlas of clinical fungi*, edn 2, 1126 pp., 2000). Fungi important in medical and veterinary mycology include some members of the following genera: *Absidia*, *Arthrinium*, *Aspergillus*, *Bipolaris*, *Blastomyces*, *Candida*, *Cladophialophora*, *Cladosporium*, *Coccidioides*, *Cryptococcus*, *Exophiala*, *Histoplasma*, *Malassezia*, *Microsporum*, *Mucor*, *Phialophora*, *Penicillium*, *Pneumocystis*, *Pseudallescheria*, *Rhizopus*, *Rhinosporidium*, *Sporothrix*, *Stachybotrys*, *Trichophyton*, *Trichosporon*, *Trichothecium*.

Fungi pathogenic for humans are often polymorphic (see dimorphic) or 'pleomorphic' and this, in addition to the number of species proposed on clinical grounds and without adequate descriptions, made the earlier literature very confused. Most fungi causing serious skin mycoses are extremely variable soil saprobes with an ability to adapt to the environment of human or animal tissues. Primary fungal infections of dimorphic systematic mycoses are mainly pulmonary, following inhalation of conidia. Opportunistic systematic infections are caused by cosmopolitan fungi with a very low level of virulence, and occur when the normal defence mechanism of the host is weakened as, for example, in AIDS patients. Most of the fungal pathogens important in veterinary mycology are the same as those in medical mycology, but there are exceptions. The zoophilic dermatophyte *Microsporum gallinae* causes disease in chickens, but not humans; some of more common infections of the skin and tails of fish are caused by organisms no longer considered to be Fungi, e.g., *Saprolegnia* spp.

Mycoses (with the exception of dermatophytoses) are generally not contagious but originate from fungal populations in both the external and internal environments where many potential pathogens are present as saprobes (e.g. *Candida albicans* in the human mouth, alimentary and genital tracts; *Coccidioides immitis* and *Histoplasma capsulatum* in soil). With the rise in numbers of people who are immunocompromised, more and different opportunistic fungi are being recorded as causing often serious mycoses on such patients; see opportunistic fungi.

Medical antifungal agents belongs to groups of allylamines (butenafine, naftifine, etc.), antimetabolites, azoles (fluconazole, ketoconazole, clotrimazole, itraconazole, oxiconazole, etc.), chitin and glucan synthesis inhibitors, polyenes (different drugs on the basis of nystatin, amphotericin B, etc.) and some systemic (griseofulvin) and topical drugs (Pfaller *et al.*, *J. Clin. Microbiol.* **42**: 3142, 2004; Charlier *et al.*, *J. Antimicrob. Chemoter.* **57**: 384, 2006; Matta *et al.*, *Antimicrob. Agents and Chemother.* **51**: 1573, 2007; etc.). The eukaryotic nature of fungi and their hosts leads to problems with antifungal drug toxicity or cross-reactivity with host molecules during treatment. Therapy for dermatophytoses is by topical application of fungicides, oral administration of griseofulvin, or X-ray epilation. Systemic mycoses, which are frequently difficult to cure, may respond to surgery, X-rays, potassium iodide, or antibiotics (see amphotericin, nystatin). Certain honeys have significant antifungal activity against isolates of *Candida* spp. (Irish *et al.*, *Medical Mycology* **44**: 289, 2006) which cause systemic mycoses and are increasing worldwide (Pfaller & Diekema, *Clin. Microbiol. Rev.* **20**: 133, 2007). A wide range of molecular, immunological and cytological techniques is essential for ongoing research in medical mycology (Irish *et al.*, in Kavanagh (Ed.) *Medical Mycology: Cellular and Molecular Techniques*, 348pp., 2006).

For some important mycoses named after the pathogenic agents see: adiaspiromycosis, aspergillosis, blastomycosis, candidiasis, chromomycosis, *Coccidioides* (coccidioidomycosis), cryptococcosis, dermatophyte (ringworm), epizootic lymphangitis, *Histoplasma* (histoplasmosis), mycosis, mycetoma, *Paracoccidioides* (paracoccidiomycosis), phycomycosis, piedra, rhinosporidiosis, sporotrichosis, tinea; see also Allergy.

The International Society for Human and Animal Mycology (ISHAM) encourages and facilitates the study and practice of all aspects of medical and veterinary mycology; see also Internet, Societies and organizations.

Lit.: *General*: Ainsworth (*Medical mycology. An introduction to its problems*, 1952), Carter & Wise (*Essentials of veterinary bacteriology and mycology*, 2004), Emmons *et al.* (*Henrici's molds, yeasts and actinomycetes*, 1947), Howard (*Pathogenic fungi in humans and animals*, edn 2, 2002), Hungerford *et al.* (*Veterinary mycology*, 1999), Kavanagh (*New Insights in Medical Mycology*, 2008).

Nomenclature: *Nomenclature of fungi pathogenic to man and animals* (*Med. Res. Council Mem.* **23**, edn 4, 1977), ISHAM (*Sabouraudia* **18**: 78, 1980; internat. nomencl., Engl., Fr.), Ellis *et al.* (Eds) (*Descriptions of Medical Fungi*, edn 2, 2007), Grigoriu *et al.* (*Medical mycology*, 1987), Odds *et al.* (Fungal disease nomenclature, *J. med. vet. mycol.* **30**: 1-10, 1992).

Mycology: Brumpt (*Précis de parasitologie*, edn 6, **2**, 1949), Ciferri (*Manuale di micologia medica*, 2 vols, 1958-60), d'Enfert & Hube (*Candida, comparative and functional genomics*, 2008), Dodge (*Medical mycology*, 1935), Ellis *et al* (*Descriptions of medical fungi*, 2007), Evans & Richardson (Eds) (*Medical mycology: A practical approach*, 1989), Fragner (*Parasitische Pilze beim Menschen*, 1958), Gedek (*Hefen als Krankheitserreger bei Tieren*, 1968), Hay (Ed.) (*Tropical fungal infections*, 1989), Howard

(Ed.) (*Fungi pathogenic for humans and animals*, 3 vol., 1983-5), Jacobs & Nall (Eds) (*Antifungal drug therapy*, 1990), Koch (*Leitfadender medizinschen Mykologie*, 1973), Kwon-Chung & Bennett (Eds) *Medical Mycology*, 1992), Larone (*Medically important fungi: a guide to identification*, 1987), Nannizzi (*Repertorio sistematico dei miceti dell'uomo e degli animale*, 1934), Odds (*Candida and Candidosis*, edn 2, 480 pp., 1988), Rippon (*Medical mycology*, 1974 (edn 3, 1988), San Blas & Calderone (*Pathogenic fungi: structural biology and taxonomy*, 2004), San Blas & Calderone (*Pathogenic fungi: host interactions and emerging strategies*, 2004), San Blas & Calderone (*Pathogenic fungi, insights in molecular biology*, 2008), Smith (*Opportunistic mycoses of man and other animals*, 1989), Warnock & Richardson (Eds) (*Fungal infection of the compromised patient*, edn 2, 1990), Wentworth (Ed.) (*Diagnostic procedures for mycotic and parasitic infections*, edn 7, 1988).

Methods: Ajello *et al.* (*Laboratory manual for medical mycology*, edn 2, 1962), Beneke & Rogers (*Medical mycology manual*, edn 3, 1971), Evans (Ed.) (*Serology of fungus infections. A laboratory manual*, 1976), Golvin *et al.* (*Techniques en parasitologie et en mycologie*, 1970), Kavanagh (Ed.) (*Medical Mycology: Cellular and Molecular Techniques*, 2006), Segretain *et al.* (*Diagnostic de laboratoire en mycologie*, edn 3, 1974), Vanbreuseghem (*Guide practique de mycologie médicale et vétérinaire*, 1966).

Humans: Conant *et al.* (*Manual of clinical mycology*, edn 3, 1971), Emmons *et al.* (*Medical mycology* edn 3, 1977), Hildick-Smith *et al.* (*Fungus diseases and their treatment*, 1964), de Hoog *et al.* (*Atlas of clinical fungi*, edn 2, 1126 pp., 2000), Kibbler (*Principles and Practice of Clinical Mycology*, 1996), Land & McCracken (*Handb. Appl. Mycol.: Humans, animals & insects* **2**, 1991; infection in the compromised host), Lewis *et al.* (*An introduction to medical mycology*, edn 4, 1958), 1967), McGinnis (*Laboratory handbook of medical mycology*, 1980), Sarosi & Scott (*Fungal diseases of the lung*, 1986), da Silva Lacaz (*Compendio de micologia medica*.

Other vertebrates: Ainsworth & Austwick (*Fungal diseases of animals*, edn 2, 1973), Chute *et al.* (*A bibliography of avian mycoses* [*Misc. Publ. Me agric. Exp. Stn* **655**], 1962 [edn 3, 1971]), Jungerman & Schwartzman (*Veterinary medical mycology*, 1972), Songer & Post. *Veterinary microbiology. Bacterial and fungal agents of animal diseases*, 2004), Welsh & Hughes (*Fungal diseases of fishes*, 1980).

Bibliography & journals: Ciferri & Redaelli (*Bibliographia mycopathologica 1800-1940*, 2 vols, 1958); journals, *J. de mycologie medicale*, 1991-; *Medical Mycology* (formerly *Sabouraudia* 1961-1986, then *Journal of Medical & Veterinary Mycology*), 1986-1998; *Mikologia Lekarska* [*Medical Mycology*], 1994-; *Mykosen*, 1957-; *Review of medical and veterinary mycology*, 1943- (abstracts; also available in electronic formats 1973-).

History: Ainsworth (*Introduction to the history of medical and veterinary mycology*, 1991), Ainsworth & Stockdale (*RMVM* **19**: 1, 1984; biographical notices).

Websites:
Dr Fungus, www.doctorfungus.org;
Fifth Kingdom, Chapter 23, www.mycolog.com/chapter23.htm;
Fungal Infections Virtual Grand Rounds, http://hstelearning.mit.edu/fi/index.html;
ISHAM, www.isham.org;
Mycology Online, www.mycology.adelaide.edu.au;
http://timm.main.teikyo-u.ac.jp/pfdb/cover/database_eng.html
See also: Voss (*Hautarzt*, **56**: 71-74, 2005).

See also Laboratory safety.

Medical uses of Fungi. At various times different fungi have been put to medical use. *Fomes officinalis* (Agaricum, the female, white, or purging agaric) was a noted 'universal remedy' (see Faull, *Mycol.* **11**: 267, 1919; 'pineapple fungus'). Yeast is a part of some patent medicines and takadiastase has medical uses. *Lycoperdon* spores and capillitium has been used for stopping blood from wounds. *Inonotus obliquus* has been widely used in Siberia for prevention of cancers (Solzhenitsyn, *Cancer ward*, 1968). In the Far East *Cordyceps sinensis* (attached to the larva ('caterpillar') of which it is a parasite), bukuryo (sclerotia of *Pachyma hoelen*) and lingzhi (*Ganoderma lucidum*) are widely used as medicines. Many spp. continue to be used in Chinese traditional medicine (Ying *et al.*, *Icones of medicinal fungi for China*, 1987; 272 spp.). The Bini and Igbo peoples of Nigeria use *Daldinia concentrica* s.l. for curing skin diseases, stomach upsets, ulcers, whooping cough and for prevention of excessive growth of the foetus for easy delivery (Akpaja *et al*, *Int. j. medicinal mushrooms* **7**: 373, 2005). The medieval Doctrine of Signatures, whereby an organism's appearance was considered to indicate the diseased organs it could treat, led to use of a variety of lichens for medicinal purposes (e.g. dog lichen, lungwort); some lichens are still supplied by pharmacies and health shops, either loose or in pastilles, e.g. Iceland moss (*Cetraria islandica*); see Richardson (*in* Galun, *CRC Handbook of lichenology* **3**: 93, 1988; review). For further applications of products from lichens see Fahselt (*Symbiosis* **16**: 117, 1994; review). Few, if any, fungi now, however, appear in lists of officially sanctioned medicines (in Britain, for example, the last fungus to be listed in the British Pharmaceutical Codex was ergot (q.v.), the sclerotia of *Claviceps purpurea*; the Codex has now replaced by the British Pharmacopoeia, in which several preparations from ergot are listed, but not the fungus itself).

Interest in medicinal use of fungi has grown enormously since the ninth edition of this *Dictionary*, with the establishment of the *Int. j. medicinal mushrooms* devoted to the topic. This has been accompanied by intense exploration of chemicals produced by fungi and of pharmaceutical value, for example β-glucans used in cancer therapies (from e.g. *Agaricus subrufescens*), or the secondary metabolite concentricolide, isolated from *Daldinia eschscholzii*, which has potential value for treating HIV-positive patients (Stadler *et al.*, *Phytochemistry* **56**: 787, 2001).

Lit.: Chang & Miles (*Mushrooms: cultivation, nutritional value, medicinal effect, and environmental impact*, edn 2, 2004), Chen & Seviour (*MR* **111**: 635, 2007; review of medicinal importance of fungal β-(1→3), (1→6)-glucans), Didukh *et al.* (*Impact of the family Agaricaceae (Fr.) Cohn on nutrition and medicine*, 2004). See also Antibiotics.

medium, **culture**, a substance or solution for the culture of microorganisms (see Media).

medulla (1) (of lichen thalli), the loose layer of hyphae below the cortex and algal layer; (2) (of sporocarps of macromycetes), the part composed mainly or entirely of longitudinal hyphae.

medullary excipulum (of ascomata), tissue below the generative layer in an apothecium; hypothecium. See excipulum.

Medusamyces G.L. Barron & Szijarto (1990), anamorphic *Pezizomycotina*, Hso.0eH.?. 1 (on rotifers), Canada. See Barron & Szijarto (*Mycol.* **82**: 136, 1990).

Medusina Chevall. (1826) = Hericium Pers. fide Saccardo (*Syll. fung.* **5**: 1, 1887).

Medusomyces Lindau (1913) nom. conf., Fungi. See Lindner (*Ber. dt. bot. Ges.* **31**: 364, 1913).

Medusosphaera Golovin & Gamalizk. (1962) = Erysiphe fide Braun & Takamatsu (*Schlechtendalia* **4**: 3, 2000).

Medusula Eschw. (1824) [non *Medusula* Pers. 1807, *Violaceae*] nom. dub. ? = Sarcographa fide Staiger (*Biblthca Lichenol.* **85**, 2002).

Medusula Tode (1790) nom. dub., anamorphic *Pezizomycotina*. See Holubová-Jechová (*Sydowia* **46**: 244, 1994).

Medusulina Müll. Arg. (1894) = Fissurina fide Staiger (*Biblthca Lichenol.* **85**, 2002), Archer (*Telopea* **10**: 589, 2004; Australia).

mega- (prefix), of great size; large; cf. macro-.

Megacapitula J.L. Chen & Tzean (1993), anamorphic *Pezizomycotina*, Hso.#eP.1. 1, Taiwan. See Chen & Tzean (*MR* **97**: 347, 1993).

Megachytriaceae Sparrow (1943) = Cladochytriaceae.

Megachytrium Sparrow (1931), Cladochytriaceae. 1 (on *Elodea*), N. America.

Megacladosporium Vienn.-Bourg. (1949) = Fusicladium fide Hughes (*Mycol. Pap.* **36**, 1951), Partridge & Morgan-Jones (*Mycotaxon* **85**: 357, 2003), Schubert *et al.* (*Schlechtendalia* **9**, 2003).

Megacollybia Kotl. & Pouzar (1972), Marasmiaceae. 1, Europe. See Kotlaba & Pouzar (*Česká Mykol.* **26**: 220, 1972).

Megalaria Hafellner (1984), Megalariaceae (L). 1, widespread (temperate). See Ekman & Tønsberg (*Bryologist* **99**: 34, 1996), Kantvilas *et al.* (*Lichenologist* **31**: 213, 1999), Ekman (*MR* **105**: 783, 2001; phylogeny).

Megalariaceae Hafellner (1984), Lecanorales (L). 3 gen., 3 spp.

Lit.: Ekman & Tønsberg (*Bryologist* **99**: 34, 1996), Kantvilas *et al.* (*Lichenologist* **31**: 213, 1999), Ekman (*MR* **105**: 783, 2001).

Megaloblastenia Sipman (1983), Megalosporaceae (L). 2, widespread. See Kantvilas (*Lichenologist* **26**: 349, 1994).

Megalocitosporides Wernicke (1892) = Coccidioides fide Dodge (*Medical Mycology*, 1935).

Megalocystidium Jülich (1978), Stereaceae. *c.* 10, widespread. See Boidin *et al.* (*BSMF* **113**: 60, 1997).

Megalodochium Deighton (1960), anamorphic *Pezizomycotina*, Hsp.0eP.1. 2, Africa (tropical). See Deighton (*Mycol. Pap.* **78**: 17, 1960).

megalogonidium (obsol.), a macrogonidium (q.v.).

Megalographa A. Massal. (1860), ? Dothideomycetes. See Staiger (*Biblthca Lichenol.* **85**, 2002).

Megalohypha A. Ferrer & Shearer (2007), Aliquandostipitaceae. 1, Panama; Thailand. See Ferrer *et al.* (*Mycol.* **99**: 456, 2007).

Megalonectria Speg. (1881) = Nectria fide Seifert (*Stud. Mycol.* **27**, 1985), Rossman *et al.* (*Stud. Mycol.* **42**: 248 pp., 1999).

Megalopsora Vain. (1921) = Physcidia fide Kalb & Elix (*Biblthca Lichenol.* **57**: 265, 1995).

Megaloseptoria Naumov (1925), anamorphic *Dothideomycetes*, Cpd.0fH.15. 1, Russia. See Shoemaker (*CJB* **45**: 1297, 1967).

Megalospora A. Massal. (1852) = Mycoblastus fide Hawksworth *et al.* (*Dictionary of the Fungi* edn 8, 1995).

Megalospora Meyen (1843), Megalosporaceae (L). 35, widespread (esp. tropical). See Hafellner & Bellemère (*Nova Hedwigia* **35**: 207, 1982; ultrastr.), Sipman (*Biblthca Lichenol.* **18**, 1983; monogr., key), Kantvilas (*Lichenologist* **26**: 349, 1994; key 16 spp. Australia), Stenroos & DePriest (*Am. J. Bot.* **85**: 1548, 1998; DNA), Miądlikowska *et al.* (*Mycol.* **98**: 1088, 2006; phylogeny), Untari (*Mycotaxon* **97**: 129, 2006; Indonesia), Lücking (*Fungal Diversity* **27**: 103, 2007; key American spp.).

Megalospora Naumov (1927) ≡ Gemmamyces.

Megalosporaceae Vězda ex Hafellner & Bellem. (1982), Teloschistales (L). 3 gen. (+ 5 syn.), 39 spp.

Lit.: Sipman (*Biblthca Lichenol.* **18**, 1983; monogr.), Sipman (*Willdenowia* **15**: 557, 1986), Kantvilas (*Lichenologist* **26**: 349, 1994; key 19 spp. Australia), Stenroos & DePriest (*Am. J. Bot.* **85**: 1548, 1998), Miądlikowska *et al.* (*Mycol.* **98**: 1088, 2006; phylogeny), Hofstetter *et al.* (*Mol. Phylogen. Evol.* **44**: 412, 2007; phylogeny).

Megalosporon, see *Endothrix*.

Megalotremis Aptroot (1991), Trypetheliaceae (L). 10, widespread (tropical). See Aptroot (*Nova Hedwigia* **60**: 325, 1995), Aptroot *et al.* (*Biblthca Lichenol.* **64**, 1997), Aptroot *et al.* (*Biblthca Lichenol.* **97**, 2008; Costa Rica).

Megaloxyphium Cif., Bat. & Nascim. (1956) = Leptoxyphium fide Hughes (*Mycol.* **68**: 693, 1976).

Megaspora (Clauzade & Cl. Roux) Hafellner & V. Wirth (1987), Megasporaceae (L). 2, widespread (north temperate). See Lumbsch *et al.* (*J. Hattori bot. Lab.* **75**: 295, 1994; posn), Stenroos & DePriest (*Am. J. Bot.* **85**: 1548, 1998; DNA), Ivanova & Hafellner (*Biblthca Lichenol.* **82**: 113, 2001; phylogeny), Schmitt *et al.* (*J. Hattori bot. Lab.* **100**: 753, 2006; phylogeny).

Megasporaceae Lumbsch (1994), Pertusariales (L). 3 gen. (+ 10 syn.), 236 spp.

Lit.: Hafellner (*Acta Bot. Malac.* **16**: 133, 1991), Lumbsch *et al.* (*J. Hattori bot. Lab.* **75**: 295, 1994), Stenroos & DePriest (*Am. J. Bot.* **85**: 1548, 1998), Ivanova & Hafellner (*Biblthca Lichenol.* **82**: 113, 2001), Schmitt *et al.* (*J. Hattori bot. Lab.* **100**: 753, 2006), Lumbsch *et al.* (*MR* **111**: 257, 2007; phylogeny).

megaspore, see macrospore.

Megasporoporia Ryvarden & J.E. Wright (1982), Polyporaceae. 8, widespread (pantropical). See Ryvarden & Wright (*Mycotaxon* **16**: 173, 1982), Dai & Wu (*Mycotaxon* **89**: 379, 2004; China).

Megaster Cif., Bat., Nascim. & P.C. Azevedo (1956), anamorphic *Pezizomycotina*, Hso.0bP.1. 2, Brazil. See Ciferri *et al.* (*Publções Inst. Micol. Recife* **48**: 2, 1956).

Megathecium Link (1826) nom. rej. prop. ≡ Ceratostoma Fr.

Megatricholoma G. Kost (1984) = Tricholoma fide

Kuyper (*in litt.*).

Megatrichophyton Neveu-Lem. (1921) = Trichophyton fide Hawksworth *et al.* (*Dictionary of the Fungi* edn 8, 1995).

Mehtamyces Mundk. & Thirum. (1945), Pucciniales. 1 (on *Stereospermum* (*Bignoniaceae*)), Africa; India; Sri Lanka. See Thirumalachar & Mundkur (*Indian Phytopath.* **2**: 193, 1949; syn of *Phragmidiella*), Ramachar & Rao (*Mycol.* **73**: 778, 1981), Cummins & Hiratsuka (*Illustr. Gen. Rust Fungi* edn 3: 225 pp., 2003).

meiocyte, a cell in which meiosis takes place. Cf. gonotokont.

Meionomyces Thaxt. (1931), Laboulbeniaceae. 6, widespread (tropical). See Nannfeldt (*Svensk bot. Tidskr.* **43**: 468, 1949).

meiophase, the part of a life cycle in which a diploid nucleus undergoes reduction.

Meiorganaceae R. Heim ex Jülich (1982) = Coniophoraceae.

Meiorganum R. Heim (1966), Paxillaceae. 2, Malaysia; New Caledonia. See Heim (*Compte rendu hebdomadaire des Sciences de l'Academie des sciences* Paris **261**: 1720, 1965).

meiosporangium, a thick-walled diploid sporangium of certain *Blastocladiales* producing uninucleate, haploid zoospores (**meiospores**, q.v.) (Emerson, 1950); see Dick (*Mycologist* **1**: 166, 1987); cf. mitosporangium.

meiospore (1) a spore from a meiosporangium (q.v.); (2) (of ascomycetes and basidiomycetes), a basidiospore or ascospore which is the product of meiosis (see Kendrick & Watling, *in* Kendrick (Ed.), *The whole fungus* **2**: 473, 1979).

meiotangium, for the sporangium or gametangium in which meiosis occurs. See Corner (*TBMS* **15**: 336, 1931).

Meira Boekhout, Scorzetti, Gerson & Sztejnberg (2003), Exobasidiomycetidae. 3 (from mites), Israel; Japan. See Boekhout *et al.* (*Int. J. Syst. Evol. Microbiol.* **53**: 1655, 2003; phylogeny, links with *Cryptobasidiaceae*), Sampaio (*Frontiers in Basidiomycote Mycology*: 49, 2004), Yasuda (*Mycoscience* **47**: 36, 2006; from Japanese pear), Yasuda *et al.* (*Mycoscience* **47**: 36, 2006).

Meissneria Fée (1837) [non *Meisneria* DC. 1828, *Melastomataceae*] ≡ Laurera.

Meixner test. To detect amatoxins. Express juice from a fresh basidioma onto a piece of newspaper, allow to dry, add one drop of conc. hydrochloric acid (HCl) when a blue colour indicates presence of amatoxins. Adding more than one drop will significantly reduce detection rate when amatoxins are present only at low levels. Always perform the test with controls, using for example a species of *Russula* (not containing amatoxins) and a species known to contain amatoxins (e.g. *Amanita virosa*). See Meixner (*Z. Mykol.* **45**: 137, 1979), Beutler & Vergeer (*Mycol.* **72**: 1142, 1981), Vergeer (*West J. Med.* **138**: 576, 1983).

Melachroia Boud. (1885) = Podophacidium fide Seaver (*North American Cup Fungi* (Inoperculates), 1951).

Melaleuca Pat. (1887) [non *Melaleuca* L. 1767, *Myrtaceae*] ≡ Melanoleuca.

Melampsora Castagne (1843), Melampsoraceae. *c.* 90 (autoecious (*Euphorbia*, *Hypericum*, *Linum*), or heteroecious: (0, I) on conifers (*Abies*, *Larix*, *Pinus*, *Tsuga*) or angiosperms (*Allium*, *Mercurialis*, *Ribes*); (II, III) on *Salicaceae*), widespread (esp. north temperate). *M. lini* on flax (*Linum*). See Newcombe *et al.* (*MR* **104**: 261, 2000; hybrid sp.), Tian *et al.* (*Mycoscience* **45**: 56, 2004; phylog.), Pei *et al.* (*Rust Diseases of Willow and Poplar*: 1, 2005; phylog.), Pei *et al.* (*MR* **109**: 401, 2005; phylog.).

Melampsoraceae Dietel (1897), Pucciniales. 1 gen. (+ 5 syn.), 90 spp.

Lit.: Burdon & Roberts (*Pl. Path.* **44**: 270, 1995), Spiers & Hopcroft (*MR* **100**: 1163, 1996), Brasier (*BioScience* **51**: 123, 2001), Samils *et al.* (*Eur. J. Pl. Path.* **107**: 399, 2001), Cummins & Hiratsuka (*Illustr. Gen. Rust Fungi* edn 3: 225 pp., 2003), Maier *et al.* (*CJB* **81**: 12, 2003), Nakamura *et al.* (*Mycoscience* **44**: 253, 2003), Smith *et al.* (*Mycol.* **96**: 1330, 2004), Tian *et al.* (*Mycoscience* **45**: 56, 2004), Wingfield *et al.* (*Australas. Pl. Path.* **33**: 327, 2004), Pei (*Rust Diseases of Willow and Poplar*: 11, 2005), Pei *et al.* (*MR* **109**: 401, 2005), Pei *et al.* (*Rust Diseases of Willow and Poplar*: 1, 2005), Aime (*Mycoscience* **47**: 112, 2006).

Melampsorella J. Schröt. (1874), Pucciniastraceae. 2 (on *Abies* (0, I); on *Boraginaceae* and *Caryophyllaceae* (II, III)), widespread (north temperate). See Berndt & Oberwinkler (*Mycol.* **89**: 698, 1997).

Melampsoridium Kleb. (1899), Pucciniastraceae. 9 (on *Larix* (0, I; where known) (*Pinaceae*); on *Betulaceae* (II, III) esp.), widespread (esp. north temperate). See Kurkela *et al.* (*Mycol.* **91**: 987, 1999; on birch and alder).

Melampsoropsis (J. Schröt.) Arthur (1906) = Chrysomyxa fide Arthur (*Manual Rusts US & Canada*, 1934).

Melampydiomyces Cif. & Tomas. (1953) ≡ Melampydium.

Melampydium, see *Melampylidium*.

Melampylidium Stirt. ex Müll. Arg. (1894) = Bactrospora fide Egea & Torrente (*Mycotaxon* **53**: 57, 1995).

Melanamphora Lafl. (1976), Diaporthales. Anamorph *Cytosporina*. 2 (from bark), widespread. See Prasil *et al.* (*Česká Mykol.* **28**: 1, 1974; anamorph).

Melanaria Erichsen (1936) = Pertusaria fide Hawksworth *et al.* (*Lichenologist* **12**: 1, 1980), Schmitt *et al.* (*Biblthca Lichenol.* **86**: 147, 2003; phylogeny), Schmitt *et al.* (*Lichenologist* **38**: 411, 2006; phylogeny).

Melanaspicilia Vain. (1909) = Buellia fide Lamb (*Sci. Repts Br. Antarct. Surv.* **61**, 1968).

Melanchlenus Calandron (1953) nom. inval. = Exophiala fide de Hoog (*Stud. Mycol.* **15**, 1977), Hoog *et al.* (*J. Clin. Microbiol.* **41**: 4767, 2003).

Melanconiaceae Corda (1842) = Melanconidaceae.

Melanconiales. (obsol.). Traditionally used for anamorphic fungi with acervular conidiomata. Not accepted by von Höhnel (1923) who treated *Melanconiales* in the *Hyphomycetes*, *Tuberculariales*, or by Sutton (1973, 1980) who discussed relationships of suprageneric taxa in Deuteromycotina. See Anamorphic fungi.

Melanconidaceae G. Winter (1886), Diaporthales. 13 gen. (+ 13 syn.), 104 spp.

Lit.: Wehmeyer (*Revision of Melanconis*, 1941), Barr (*Mycol. Mem.* **7**, 1978), von Arx *in* Kendrick (Ed.) (*The whole fungus* **1**: 201, 1979), Sieber *et al.* (*CJB* **69**: 2170, 1991), Belisario & Onofri (*MR* **99**: 1059, 1995), Orsenigo *et al.* (*Mycotaxon* **67**: 257, 1998), Belisario (*Eur. J. For. Path.* **29**: 317, 1999),

Yanna *et al.* (*Fungal Diversity* **2**: 199, 1999), Castlebury *et al.* (*Mycol.* **94**: 1017, 2002), Rossman *et al.* (*Mycoscience* **48**: 135, 2007; phylogeny).

Melanconidium (Sacc.) Kuntze (1898) ≡ Melanconis.

Melanconiella Sacc. (1882), Diaporthales. See Wehmeyer (*Revision of Melanconis*, 1941), Rossman *et al.* (*Mycoscience* **48**: 135, 2007; review).

Melanconiopsis Ellis & Everh. (1900), anamorphic *Massariovalsa*, Cac.0eP.19. 4, widespread. See Speer (*BSMF* **102**: 363, 1986), Suarez *et al.* (*MR* **104**: 1530, 2000; Argentina).

Melanconis Tul. & C. Tul. (1863), Melanconidaceae. Anamorph *Melanconium*. 28, widespread. *M. juglandis* (die-back of *Juglans*). The genus seems to be polyphyletic. See Wehmeyer (*Revision of Melanconis*, 1941), Castlebury *et al.* (*Mycol.* **94**: 1017, 2002).

Melanconites Göpp. (1852), Fossil Fungi. 1 (Tertiary), Europe.

Melanconium Link (1809), anamorphic *Melanconis*, Cac.0eP.19. *c.* 50, widespread. See Sutton (*Persoonia* **3**: 193, 1964; typification), Sieber *et al.* (*CJB* **69**: 2170, 1991; biochem. charact. *Alnus* isol.), Shamoun & Sieber (*Mycotaxon* **49**: 151, 1993; isozyme, protein patterns *Alnus* isol.), Belisario (*Eur. J. For. Path.* **29**: 317, 1999; on *Juglans*).

Melanelia Essl. (1978), Parmeliaceae (L). 42, widespread (esp. temperate and boreal). See Esslinger (*Taxon* **29**: 692, 1980), Lumbsch *et al.* (*Mycotaxon* **33**: 447, 1988), Kashiwadani *et al.* (*Bull. natn. Sci. Mus.* Tokyo, B **24**: 43, 1998; Japan), Crespo *et al.* (*Lichenologist* **31**: 451, 1999; morphology, chemistry, phylogeny), Guzow-Krzeminska & Wegrzyn (*Lichenologist* **35**: 83, 2003; phylogeny), Chen & Esslinger (*Mycotaxon* **93**: 71, 2005; China), Rico *et al.* (*Lichenologist* **37**: 199, 2005; Iberian peninsula), Blanco *et al.* (*Mol. Phylogen. Evol.* **39**: 52, 2006; phylogeny).

Melanelixia O. Blanco, A. Crespo, Divakar, Essl., D. Hawksw. & Lumbsch (2004), Parmeliaceae (L). 8, widespread. See Blanco *et al.* (*MR*, 2004), Blanco *et al.* (*Mol. Phylogen. Evol.* **39**: 52, 2006; phylogeny), Laundon (*Lichenologist* **38**: 277, 2006; infraspecific taxa), Miądlikowska *et al.* (*Mycol.* **98**: 1088, 2006; phylogeny), Honegger & Zippler (*MR* **111**: 424, 2007; mating systems).

Melaniella R. Bauer, Vánky, Begerow & Oberw. (1999), Melaniellaceae. 2 (on *Selaginella*), India; Jawa; Zimbabwe. See Vánky & Vánky (*Lidia* **5**: 157, 2002; southern Africa).

Melaniellaceae R. Bauer, Vánky, Begerow & Oberw. (1999), Doassansiales. 1 gen., 2 spp.

Lit.: Bauer *et al.* (*CJB* **75**: 1273, 1997), Bauer *et al.* (*Mycol.* **91**: 475, 1999), Vánky (*Mycol. Balcanica* **1**: 175, 2004).

melanin, a black pigment (tyrosine derivative) produced by fungi, animals etc. (Butler *et al.* (*Mycol.* **93**: 1, 2001; pathogenic properties), Wheeler, *TBMS* **81**: 29, 1983; synthesis).

melanized, containing dark brown pigments.

Melanobasidium Maubl. (1906) = Sphaceloma fide Jenkins & Bitancourt (*Mycol.* **33**: 338, 1941).

Melanobasis Clem. & Shear (1931) = Sphaceloma fide Sutton (*Mycol. Pap.* **141**, 1977).

Melanobotrys Rodway (1926) = Xylobotryum fide Clements & Shear (*Gen. Fung.*, 1931).

Melanocarpus Arx (1975), Sordariales. 3 (thermophilic), widespread. See Maheshwari & Kamalam (*J. gen. Microbiol.* **131**: 3017, 1985), Guarro *et al.* (*MR* **100**: 75, 1996).

Melanocephala S. Hughes (1979), anamorphic *Pezizomycotina*, Hso.0eP.1. 5, widespread. See Hughes (*N.Z. Jl Bot.* **17**: 166, 1979), Wu & Zhuang (*Fungal Diversity Res. Ser.* **15**, 2005).

Melanochaeta E. Müll., Harr & Sulmont (1969), Chaetosphaeriaceae. Anamorphs *Sporoschisma*, *Chalara*-like. 4 (from wood etc.), Europe; Sri Lanka. See Réblová (*Czech Mycol.* **50**: 73, 1997), Réblová *et al.* (*Sydowia* **51**: 49, 1999), Sivichai *et al.* (*MR* **104**: 478, 2000), Huhndorf *et al.* (*Mycol.* **96**: 368, 2004; phylogeny), Fernández *et al.* (*Mycol.* **98**: 121, 2006), Zhang *et al.* (*Mycol.* **98**: 1076, 2006; phylogeny).

Melanochlamys Syd. & P. Syd. (1914) = Gilletiella fide von Arx & Müller (*Stud. Mycol.* **9**, 1975).

Melanocryptococcus Della Torre & Cif. (1964) = Cryptococcus Vuill. fide von Arx *et al.* (*Stud. Mycol.* **14**: 1, 1977).

Melanodecton A. Massal. (1860) = Chiodecton fide Thor (*Op. Bot.* **103**, 1990).

Melanodiscus Höhn. (1918), anamorphic *Pezizomycotina*, Hsp.0eP.?. 1, Europe.

Melanodochium Syd. (1938) ? = Sphaceloma fide Sutton (*Mycol. Pap.* **141**, 1977).

Melanodothis R.H. Arnold (1972), Mycosphaerellaceae. Anamorph *Ramularia*-like. 1, Asia; N. America. See Sivanesan (*Bitunicate Ascomycetes and their Anamorphs*, 1984).

Melanogaster Corda (1831) nom. cons., Paxillaceae. 25, widespread. See Zeller & Dodge (*Ann. Mo. bot. Gdn* **23**: 639, 1936), Wang *et al.* (*Mycotaxon* **93**: 315, 2005; Japan).

Melanogastraceae E. Fisch. (1933) = Paxillaceae.

Melanogastrales = Boletales. See also under Hypogeous fungi, gasteromycetes.

Melanogone Wollenw. & Ha. Richt. (1934) = Humicola fide Mason (*Mycol. Pap.* **5**: 113, 1941).

Melanogramma Pers. [not traced] nom. dub., Fungi. See Keissler (*Nyt Mag. naturv.* **66**: 79, 1927).

Melanographa Müll. Arg. (1882), Melaspileaceae (L). See Lendemer (*Bryologist* **106**: 311, 2003).

Melanographium Sacc. (1913), anamorphic *Pezizomycotina*, Hso/Hsy.0eP.10. 9, Asia (tropical). See Ellis (*Dematiaceous Hyphomycetes*, 1971), Srivastava & Morgan-Jones (*Mycotaxon* **57**: 195, 1996; India), Goh & Hyde (*MR* **101**: 1097, 1997; Hong Kong, key), Somrithipol & Jones (*Fungal Diversity* **19**: 137, 2005; Thailand).

Melanohalea O. Blanco, A. Crespo, Divakar, Essl., D. Hawksw. & Lumbsch (2004), Parmeliaceae (L). 20, widespread. See Blanco *et al.* (*MR* **108**: 882, 2004), Divakar & Upreti (*Lichenologist* **37**: 511, 2005), Blanco *et al.* (*Mol. Phylogen. Evol.* **39**: 52, 2006; phylogeny).

Melanolecia Hertel (1981), Lecanorales (L). 8, widespread. See Poelt & Vězda (*Biblthca Lichenol.* **16**, 1981), Buschbom & Mueller (*Mol. Phylogen. Evol.* **32**: 66, 2004; reln with *Farnoldia*).

Melanoleuca Pat. (1897) nom. cons., Tricholomataceae. *c.* 50, widespread. See Bresinsky & Stangl (*Z. Pilzk.* **43**: 145, 1977), Gillman & Miller (*Mycol.* **69**: 927, 1977), Kühner (*Bull. Soc. linn. Lyon* **47**: 12, 1978), Boekhout (*Persoonia* **13**: 397, 1988; Netherlands), Boekhout & Kuyper (*Persoonia* **16**: 253, 1996), Fontenla *et al.* (*Rivista di Micologia* **44**: 27, 2001), Fontenla *et al.* (*Micologia e Vegetazione Mediterranea* **17**: 18, 2002), Fontenla *et al.* (*Rivista *di Micologia* **48**: 113, 2005; spore morphology).

Melanomma Nitschke ex Fuckel (1870), Melanommataceae. Anamorph *Nigrolentilocus*. *c.* 27, widespread. See Chesters (*TBMS* **22**: 116, 1938), Holm (*Symb. bot. upsal.* **14** no. 3: 1, 1957), Réblová (*Czech Mycol.* **50**: 161, 1998), Liew *et al.* (*Mol. Phylogen. Evol.* **16**: 392, 2000; DNA), Inderbitzin & Huang (*Mycoscience* **42**: 187, 2001; China), Kruys *et al.* (*MR* **110**: 527, 2006; phylogeny), Schoch *et al.* (*Mycol.* **98**: 1041, 2006; phylogeny), Wang *et al.* (*MR* **111**: 1268, 2007; phylogeny).

Melanommataceae G. Winter (1885), Pleosporales. 21 gen. (+ 9 syn.), 265 spp. The family is polyphyletic and in need of revision.

Lit.: Boise (*Mycol.* **77**: 230, 1985), Hawksworth & Boise (*Sydowia* **38**: 114, 1985), Barr (*N. Amer. Fl.* ser. 2 **13**: 129 pp., 1990), Huhndorf (*Mycol.* **90**: 527, 1998), Hyde & Fröhlich (*Sydowia* **50**: 81, 1998), Réblová (*Czech Mycol.* **50**: 161, 1998), Hyde & Goh (*Nova Hedwigia* **68**: 251, 1999), Liew *et al.* (*Mol. Phylogen. Evol.* **16**: 392, 2000), Lindemuth *et al.* (*MR* **105**: 1176, 2001), Lumbsch & Lindemuth (*MR* **105**: 901, 2001), Kodsueb *et al.* (*Mycol.* **98**: 571, 2006), Wang *et al.* (*MR* **111**: 1268, 2007; phylogeny).

Melanommatales = Pleosporales.

Melanomphalia M.P. Christ. (1936), Tricholomataceae. 1, Europe. See Christiansen (*Friesia* **1**: 288, 1936), Montag (*Z. Mykol.* **62**: 75, 1996), Aime *et al.* (*Am. J. Bot.* **92**: 74, 2005; most species should be placed in *Crepidotus*), Eyssartier & Boisselet (*BSMF* **120**: 423, 2004).

Melanomyces Syd. & P. Syd. (1917) nom. dub., Fungi. See Petrak (*Sydowia* **1**: 169, 1947).

Melanopelta Kirschst. (1939) = Gnomonia fide Bolay (*Ber. schweiz. bot. Ges.* **81**, 1972).

Melanopeziza Velen. (1939), ? Helotiales. 1, former Czechoslovakia.

Melanophloea P. James & Vězda (1971), Thelocarpaceae (L). 1, British Solomon Islands; Papua New Guinea. See James & Vězda (*Lichenologist* **5**: 89, 1971).

Melanophoma Papendorf & J.W. du Toit (1967), anamorphic *Pezizomycotina*, Cpd.0eH.15. 1, S. Africa. See Papendorf & du Toit (*TBMS* **50**: 503, 1967).

Melanophora Arx (1957) = Sphaceloma fide Jenkins (*Aq. Inst. Biol. S. Paulo* **38**: 83, 1971).

Melanophthalmum Fée (1825) = Strigula fide Hawksworth *et al.* (*Dictionary of the Fungi* edn 8, 1995).

Melanophyllum Velen. (1921), Agaricaceae. 3, widespread. See Velenovský (*České Houby*: 569, 1921), Migliozzi & Zecchin (*Boll. Gruppo Micol. 'G. Bresadola'* **44**: 49, 2001), Vellinga (*MR* **108**: 354, 2004; phylogeny).

Melanoplaca Syd. & P. Syd. (1917) = Dothidasteroma fide Müller & von Arx (*Beitr. Kryptfl. Schweiz* **11** no. 2, 1962).

Melanoporella Murrill (1907) = Nigrofomes fide Lowe (*Tech. Pub. Sta. Univ. Coll. Forestry, Syracuse* **90**, 1966) but used by, Pegler (*The polypores [Bull. BMS Suppl.]*, 1973).

Melanoporia Murrill (1907) = Nigrofomes fide Ryvarden (*Norw. Jl Bot.* **19**: 233, 1972).

Melanoporthe Wehm. (1938) = Diaporthe fide Müller & von Arx (*Beitr. Kryptfl. Schweiz* **11** no. 2, 1962).

Melanops Nitschke ex Fuckel (1870), Botryosphaeriaceae. 1. See Holm (*Taxon* **24**: 475, 1975), Phillips & Pennycook (*Sydowia* **56**: 288, 2004), Crous *et al.* (*Stud. Mycol.* **55**: 235, 2006).

Melanopsamma Niessl (1876), Hypocreales. Anamorphs *Custingophora*-like, *Stachybotrys*. 41 (on old wood etc.), widespread (north temperate). See Barr (*Mycotaxon* **39**: 43, 1990; status, posn), Samuels & Barr (*CJB* **75**: 2165, 1998; key N Am. spp.), Castlebury *et al.* (*MR* **108**: 864, 2004; phylogeny).

Melanopsammella Höhn. (1920), Chaetosphaeriaceae. Anamorphs *Chloridium*, *Gonytrichum*. 1, Europe. See Alcorn (*Mycotaxon* **39**: 361, 1990; phylogeny), Réblová (*Czech Mycol.* **50**: 73, 1997), Réblová *et al.* (*Sydowia* **51**: 49, 1999), Réblová (*Stud. Mycol.* **45**: 149, 2000; review), Fernández & Huhndorf (*Fungal Diversity* **18**: 15, 2005), Fernández *et al.* (*Mycol.* **98**: 121, 2006; phylogeny).

Melanopsammina Höhn. (1919) = Lentomita fide Holm (*Svensk bot. Tidskr.* **62**: 217, 1968).

Melanopsammopsis Stahel (1915) = Microcyclus fide von Arx & Müller (*Stud. Mycol.* **9**, 1975).

Melanopsichiaceae Vánky (2001) = Ustilaginaceae.

Melanopsichium Beck (1894), Ustilaginaceae. 2 (on *Polygonaceae*), widespread. See Zundel (*Mycol.* **35**: 180, 1943), Zundel (*Mycol.* **35**: 654, 1943), Zundel (*Mycol.* **52**: 189, 1961), Vánky (*Illustrated genera of smut fungi*, 1987), Vánky (*Mycol. Balcanica* **2**: 113, 2005).

Melanopus Pat. (1887) = Polyporus P. Micheli ex Adans. fide Singer (*Agaric. mod. Tax.*, 1951).

Melanormia Körb. (1865), ? Helotiales (?L). 1, Germany.

Melanosella Örösi-Pál (1936), anamorphic *Pezizomycotina*, Hso.0eH.?. 1 (yeast-like), widespread. *M. mors-apis* (melanosis of bees).

Melanosorus De Not. (1847) ≡ Rhytisma.

Melanosphaeria Sawada (1922) ? = Sirosphaera fide Petch (*TBMS* **11**: 258, 1926).

Melanosphaerites Grüss (1928), Fossil Fungi. 2 (Devonian), Europe.

Melanospora Corda (1837) nom. cons., Ceratostomataceae. Anamorphs *Harzia*, *Gonytrichum*. 29 (from soil, often fungicolous), widespread. See Doguet (*Botaniste* **39**, 1955), Cannon & Hawksworth (*J. Linn. Soc.* Bot. **84**: 115, 1982; key 12 Br. spp.), Vakili (*MR* **93**: 67, 1989; anamorph), Goh & Hanlin (*Mycol.* **86**: 357, 1994; ontogeny), Goh & Hanlin (*Mycol.* **90**: 655, 1998; TEM), Goh *et al.* (*Fungal Science* Taipei **13**: 1, 1998; anamorph), Jones & Blackwell (*MR* **102**: 661, 1998; DNA), Stchigel *et al.* (*MR* **103**: 1305, 1999), Nitzan *et al.* (*Sydowia* **56**: 281, 2004), Zhang *et al.* (*Mycol.* **98**: 1076, 2006; phylogeny), Schoch *et al.* (*MR* **111**: 154, 2007; phylogeny), Tsui *et al.* (*Mycol.* **99**: 884, 2007; phylogeny, anamorph).

Melanospora Mudd (1861) ≡ Poeltinula.

Melanosporaceae Bessey (1950) = Ceratostomataceae.

Melanosporales N. Zhang & M. Blackw. (2007). Hypocreomycetidae. 1 fam., 12 gen., 63 spp. Fam.:

Ceratostomataceae

For *Lit.* see under fam.

Melanosporites Pampal. (1902), Fossil Fungi. 1 (Miocene), Italy.

Melanosporopsis Naumov (1927) = Melanospora Corda fide Clements & Shear (*Gen. Fung.*, 1931).

melanosporous, black-spored.

Melanostigma Kirschst. (1939) = Herpotrichiella fide Barr (*Rhodora* **78**: 67, 1976).

Melanostroma Corda (1829) = Ceuthospora Grev. fide Nag Raj *in* Sherwood (Ed.) (*Mycotaxon* **5**: 1, 1977).

Melanostromella Petr. (1953) = Antennularia fide Müller & von Arx (*Beitr. Kryptfl. Schweiz* **11** no. 2, 1962).

Melanotaeniaceae Begerow, R. Bauer & Oberw. (1998), Ustilaginales. 3 gen., 14 spp.

Lit.: Ingold (*TBMS* **91**: 712, 1988), Boekhout *et al.* (*Stud. Mycol.* **38**: 175, 1995), Bauer *et al.* (*CJB* **75**: 1273, 1997), Begerow *et al.* (*CJB* **75**: 2045, 1998), Piepenbring *et al.* (*Protoplasma* **204**: 202, 1998), Walker (*MR* **105**: 225, 2001), Vánky (*Mycol. Balcanica* **1**: 175, 2004).

Melanotaenium de Bary (1874), Melanotaeniaceae. 9, widespread. Black spots or swellings on dicotyledonous plants. See Zambettakis & Joly (*BSMF* **88**: 193, 1972; key, numerical taxonomy), Vánky (*Illustrated genera of smut fungi*, 1987), Bauer *et al.* (*CJB* **75**: 1273, 1997), Vánky (*Mycotaxon* **70**: 44, 1999), Vánky (*Mycol. Balcanica* **2**: 169, 2005).

Melanotheca Fée (1837) = Pyrenula Ach. (1809) fide Harris (*Mem. N. Y. bot. Gdn* **49**, 1989).

Melanothecomyces Cif. & Tomas. (1953) = Laurera fide Letrouit-Galinou (*Revue bryol. lichén.* **26**: 207, 1957).

Melanothecopsis C.W. Dodge (1967), Pyrenulaceae (?L). 5, widespread. See Dodge (*Nova Hedwigia* **12**: 308, 1966).

Melanotopelia Lumbsch & Mangold (2008), Thelotremataceae (L). 2, N. America; Australia. See Mangold *et al.* (*Lichenologist* **40**: 39, 2008).

Melanotrema A. Frisch (2006), Thelotremataceae (L). 6, pantropical. See Frisch (*Biblthca Lichenol.* **92**: 382, 2006), Frisch *et al.* (*Biblthca Lichenol.* **92**: 517, 2006; phylogeny).

Melanotrichum Corda (1833) = Trichosporum Vuill. fide Saccardo (*Syll. fung.* **4**: 292, 1886).

Melanotus Pat. (1900) = Psilocybe (Fr.) P. Kumm. fide Kuyper (*in litt.*).

Melanustilospora Denchev (2003), Urocystidaceae. 2 (on *Araceae*), Europe. See Denchev (*Mycotaxon* **87**: 475, 2003).

Melascypha Boud. (1885) = Pseudoplectania fide Seaver (*North American Cup Fungi* (Operculates), 1928).

Melasmia Lév. (1846), anamorphic *Rhytisma*, St.0eH.15. 20, widespread. See Braun (*Arnoldia* **15**: 44, 1998; *M. ulmicola*).

Melaspilea Nyl. (1857), Melaspileaceae (±L). *c.* s.lat. 66, widespread (esp. tropical). See Hafellner *in* Hawksworth (Ed.) (*Ascomycete Systematics. Problems and Perspectives in the Nineties* NATO ASI Series vol. **269 269**: 419, 1994; rels, polyphyly), Vrijmoed *et al.* (*MR* **100**: 291, 1996), Kantvilas & Coppins (*Lichenologist* **29**: 525, 1997), Lendemer (*Bryologist* **106**: 311, 2003; nomencl.).

Melaspileaceae Walt. Watson (1929), Arthoniomycetes (inc. sed.) (±L). 1 gen. (+ 6 syn.), *c.* 66 spp.

Lit.: Hafellner *in* Hawksworth (*Ascomycete Systematics. Problems and Perspectives in the Nineties* NATO ASI Series vol. **269 269**: 419, 1994), Vrijmoed *et al.* (*MR* **100**: 291, 1996).

Melaspileella (P. Karst.) Vain. (1921) ? = Melaspilea fide Hawksworth *et al.* (*Dictionary of the Fungi* edn 8, 1995).

Melaspileomyces Cif. & Tomas. (1953) = Melaspilea fide Hawksworth *et al.* (*Dictionary of the Fungi* edn 8, 1995).

Melastiza Boud. (1885), Pyronemataceae. 14, widespread (north temperate). See Maas Geesteranus (*Persoonia* **4**: 418, 1967; key), Dissing (*Svampe*: 29, 1980; key Danish spp.), Lassuer (*Docums Mycol.* **11** no. 42: 1, 1980; key), Häffner (*Beitr. Kenntn. Pilze Mitteleur.* **2**: 183, 1986; key 9 spp.), Arroyo & Calonge (*Boln Soc. Micol. Madrid* **12**: 23, 1988), Moravec (*Czech Mycol.* **47**: 237, 1994; synonym of *Aleuria*), Spooner & Yao (*Mycotaxon* **53**: 467, 1995), Hansen *et al.* (*Mol. Phylogen. Evol.* **36**: 1, 2005; phylogeny), Hansen & Pfister (*Mycol.* **98**: 1029, 2006; phylogeny), Liu & Zhuang (*Mycosystema* **25**: 546, 2006; phylogeny), Perry *et al.* (*MR* **111**: 549, 2007; phylogeny).

Melastiziella Svrček (1948) = Scutellinia fide Eckblad (*Nytt Mag. Bot.* **15**: 1, 1968).

Melchioria Penz. & Sacc. (1897), Trichosphaeriaceae. 1, Java.

Meliderma Velen. (1920) = Cortinarius fide Singer (*Agaric. mod. Tax.*, 1951).

Melidium Eschw. (1822) = Thamnidium Link fide Benny (*Mycol.* **84**: 834, 1992).

Meliniomyces Hambl. & Sigler (2005), anamorphic *Leotiomycetes*. 3 (mycorrhizal), widespread (north temperate). See Hambleton & Sigler (*Stud. Mycol.* **53**: 16, 2005), Vohník *et al.* (*Czech Mycol.* **59**: 215, 2007; from truffle).

Meliola Fr. (1825), Meliolaceae. *c.* 1297 (from living leaves), widespread (esp. tropical). Each species is commonly given a 'Beeli formula' (q.v.). See Hansford (*Beih. Sydowia* **2**, 1961), Hansford (*Beih. Sydowia* **5**, 1963), Luttrell (*Mycol.* **81**: 192, 1989; ontogeny), Mueller *et al.* (*CJB* **69**: 803, 1991; mucronate hyphopodia), Reynolds (*Mycotaxon* **42**: 99, 1991), Mibey & Hawksworth (*SA* **14**: 25, 1995; capitate hyphopodia), Dianese & Furlanetto (*Progress in Microbial Ecology* Proceedings of the Seventh International Symposium on Microbial Ecology. Santos, São Paulo, Brazil 1995: 207, 1997; Brazil), Hosagoudar *et al.* (*The Meliolineae* A Supplement: 201 pp., 1997; revision), Mibey & Hawksworth (*Mycol. Pap.* **174**: 108 pp., 1997; Kenya), Hu *et al.* (*Flora Fungorum Sinicorum* **11**, 1999; China), Mibey & Cannon (*Cryptog. Mycol.* **20**: 249, 1999; Kenya), Saenz & Taylor (*MR* **103**: 1049, 1999; phylogeny).

Meliolaceae G.W. Martin ex Hansf. (1946), Meliolales. 22 gen. (+ 12 syn.), 1980 spp.

Lit.: Hansford (*Mycol. Pap.* **23**, 1948; W. Afr.), Boedijn (*Persoonia* **1**: 393, 1961; Indonesia), Hansford (*Sydowia* Beih. **2**, 1961; 1814 taxa), Yamamoto (*Spec. Publ. Agric. nat. Taiwan Univ.* **10**: 197, 1961; Taiwan), Hansford (*Sydowia* Beih. **5**, 1963; 1814 figs), Hansford (*Sydowia* **16**: 302, 1963), Stevenson (*Sydowia* **22**: 225, 1969; host index), Goos & Anderson (*Sydowia* **26**: 73, 1974; Hawaii), Hawksworth & Eriksson (*SA* **5**: 142, 1986; status), Mueller *et al.* (*CJB* **69**: 803, 1991; ultrastr. hyphopodia), Mueller *et al.* (*MR* **95**: 1208, 1991), Hosagoudar (*Meliolales of India*: 363 pp., 1996), Hosagoudar (*Meliolales of India*, 1996), Hu *et al.* (*Flora Fungorum Sinicorum* **4**. Meliolales: 270 pp., 1996), Hosagoudar *et al.* (*The Meliolineae* A Supplement, 1997), Hosagoudar *et al.* (*The Meliolineae* A Supplement: 201 pp., 1997), Mibey & Hawksworth (*Mycol. Pap.* **174**, 1997; Kenya), Mibey & Hawksworth (*Mycol. Pap.* **174**: 108 pp., 1997), Saenz & Taylor (*MR* **103**: 1049, 1999).

Meliolales Gäum. ex D. Hawksw. & O.E. Erikss. (1986). Meliolomycetidae. 1 fam., 22 gen., 1980 spp. Fam.:

Meliolaceae
Lit.: Hosagoudar (*Meliolales of India*, 1996), Katumoto & Hosagoudar (*J. Econ. Tax. Bot.* **13**: 615, 1989; lists 169 taxa supplemental to Hansford's monograph), Saenz & Taylor (*MR* **103**: 1049, 1999; phylogeny), Schmiedeknecht (*Wiss. Zeitschr. Friedrich-Schiller Univ. Jena* **38**, 185, 1989; 109 spp. Cuba), and see under *Meliolaceae*.

Meliolaster Doidge (1920) = Amazonia fide Stevens (*Annls mycol.* **25**: 405, 1927).

Meliolaster Höhn. (1918), Asterinaceae. 1, widespread (tropical).

Meliolidium Speg. (1924) = Perisporiopsis Henn. fide Müller & von Arx (*Beitr. Kryptfl. Schweiz* **11** no. 2, 1962).

Meliolina Syd. & P. Syd. (1914), Meliolinaceae. Anamorph *Briania*. 39, widespread. See Reynolds (*Cryptog. Mycol.* **10**: 305, 1989; asci), Hughes (*Mycol. Pap.* **166**, 1993; key), Saenz & Taylor (*MR* **103**: 1049, 1999; phylogeny), Hosagoudar (*Zoos' Print Journal* **17**: 786, 2002; key).

Meliolinaceae S. Hughes (1993), Pezizomycotina (inc. sed.). 2 gen., 40 spp.
Lit.: Reynolds (*Pacific Sci.* **43**: 161, 1989), Reynolds (*Cryptog. Mycol.* **10**: 305, 1989), Hughes (*Mycol. Pap.* **166**: 255 pp., 1993), Hughes (*Mycol. Pap.* **166**, 1993), Johnston (*N.Z. Jl Bot.* **37**: 289, 1999), Saenz & Taylor (*MR* **103**: 1049, 1999; phylogeny).

melioline, one of the *Meliolaceae*, esp. *Meliola* spp.

Meliolinella Hansf. (1946) = Scolionema fide von Arx & Müller (*Stud. Mycol.* **9**, 1975), Hosagoudar (*Persoonia* **18**: 123, 2002).

Meliolinites Selkirk ex Janson. & Hills (1978), Fossil Fungi. 3 (Eocene, Miocene), Australia; USA.

Meliolinopsis Beeli (1920) ≡ Patouillardina G. Arnaud.

Meliolinopsis F. Stevens (1924) = Scolionema fide Hawksworth *et al.* (*Dictionary of the Fungi* edn 8, 1995).

Melioliphila Speg. (1924), Tubeufiaceae. Anamorphs *Chionomyces*, *Eriomycopsis*. 7, widespread (tropical). See Rossman (*Mycol. Pap.* **157**, 1987; key).

Meliolomycetidae, Sordariomycetes. Ord.:
Meliolales
For *Lit.* see fam.

Meliolopsis (Sacc.) Sacc. (1891) nom. dub., Fungi. See Theissen & Sydow (*Annls mycol.* **15**: 465, 1917).

Meliothecium Sacc. (1901) ≡ Myxothecium.

Melittosporiopsis, see *Mellitiosporiopsis*.

Melittosporium, see *Mellitiosporium*.

Mellitiosporiella Höhn. (1919), Rhytismatales. 2, Europe; N. America. See Sherwood (*Mycotaxon* **5**: 1, 1977), Sherwood (*Sydowia* **38**: 267, 1986).

Mellitiosporiopsis Rehm (1900) = Tapellaria fide Hawksworth *et al.* (*Dictionary of the Fungi* edn 8, 1995).

Mellitiosporium Corda (1838), Rhytismatales. 3, Europe; N. America. See Hawksworth & Kinsey (*SA* **14**: 59, 1995; nomencl.).

Meloderma Darker (1967), Rhytismataceae. 4, widespread. Nests within *Lophodermium* in phylogenetic analyses. See Johnston (*Mycotaxon* **33**: 423, 1988; gen. concept), Gernandt *et al.* (*Mycol.* **93**: 915, 2001; phylogeny), Ortiz-García *et al.* (*Mycol.* **95**: 846, 2003; phylogeny).

Melogramma Fr. (1849), Melogrammataceae. Anamorph *Cytosporina*. 3, Europe. See Laflamme (*Sydowia* **28**: 237, 1977), Barr (*Mycol. Mem.* **7**, 1978), Cannon (*SA* **7**: 23, 1988; posn).

Melogrammataceae G. Winter (1886), Diaporthales. 1 gen., 3 spp. Ascomata perithecial, long-necked, clustered within a dark pseudostroma. Asci cylindrical, fairly thick-walled but not fissitunicate, with an apical ring that does not stain in iodine, remaining attached at maturity. Ascospores fusiform to cylindrical, often curved, with several transverse septa, dark brown but with the end cells remaining ± hyaline. Anamorph coelomycetous.

Melomastia Nitschke ex Sacc. (1875), Xylariales. 3 or 4 (bark and wood), widespread. Probably close to *Pleurotrema* as interpreted by Harris & Barr, but the true identity of that genus is uncertain. See Kang *et al.* (*Fungal Diversity* **2**: 135, 1999).

melophase, the part of a life cycle in which a diploid nucleus undergoes reduction.

Melophia Sacc. (1884), anamorphic *Pezizomycotina*, St.0fH.?. 10, widespread. See Sutton (*Mycol. Pap.* **141**, 1977).

Melzericium Hauerslev (1975), Atheliaceae. 3, widespread. See Greslebin & Rajchenberg (*N.Z. Jl Bot.* **41**: 437, 2003; Patagonia).

Melzerodontia Hjortstam & Ryvarden (1980), Corticiaceae. 3, Tanzania. See Hjortstam & Ryvarden (*Mycotaxon* **12**: 177, 1980).

Membranatheca Matsush. (1995) = Amerosporium fide Sutton (*in litt.*).

Membranicium J. Erikss. (1958) = Phanerochaete fide Donk (*Persoonia* **2**: 223, 1962) but see, Hayashi (*Bull. Govt. For. Res. Stn* **88**: 260, 1974).

Membranomyces Jülich (1975), Clavulinaceae. 2, Europe; Canada. See Dentinger & McLaughlin (*Mycol.* **98**: 746, 2006; phylogeny).

membranous (**membranaceous**), like a thin skin or parchment.

Memnoniella Höhn. (1923), anamorphic *Hypocreales*, Hso.0eP.15. 5, widespread. See Verona & Mazzucchetti (*Publ. Ente naz. Cellul. Carta*, 1968), Haugland *et al.* (*Mycol.* **93**: 54, 2001; phylogeny), Photita *et al.* (*Cryptog. Mycol.* **24**: 147, 2003; on *Musa*).

Memnonium Corda (1833) = Trichosporum Vuill. fide Saccardo (*Syll. fung.* **4**: 294, 1886).

memnospore, a spore remaining at its place of origin (Gregory, *in* Madelin (Ed.), *The fungus spore*, 1966). Cf. xenospore.

Mendogia Racib. (1900), Schizothyriaceae. 2, widespread.

Mendoziopeltis Bat. (1959), Micropeltidaceae. 2, C. America. See Batista (*Publções Inst. Micol. Recife* **56**: 434, 1959).

Menegazzia A. Massal. (1854), Parmeliaceae (L). *c.* 78, widespread (esp. south temperate). See Santesson (*Ark. Bot.* **30A** no. 11: 1, 1943), Adler & Calvelo (*Mycotaxon* **59**: 367, 1996), Bjerke & Elvebakk (*Mycotaxon* **78**: 363, 2001; S. Am.), James *et al.* (*Biblthca Lichenol.* **78**: 91, 2001; New Guinea), Bjerke (*Lichenologist* **34**: 503, 2002; neotropics), Aptroot *et al.* (*Bryologist* **106**: 157, 2003; Taiwan), Bjerke (*Lichenologist* **36**: 15, 2004; Japan), Kantvilas & Louwhoff (*Lichenologist* **36**: 103, 2004; Australia), Thell *et al.* (*Mycol. Progr.* **3**: 297, 2004; phylogeny), Bjerke (*Mycotaxon* **91**: 423, 2005; S America), Bjerke & Obermayer (*Nova Hedwigia* **81**: 301, 2005; Tibet), Miądlikowska *et al.* (*Mycol.* **98**: 1088, 2006; phylogeny), Arup *et al.* (*Mycol.* **99**: 42, 2007; phylogeny), Bjerke & Sipman (*J. Linn. Soc.* Bot. **153**: 489, 2007; SE Asia).

Menezesia Torrend (1913), ? Saccharomycetales. 1,

Madeira.

Menidochium R.F. Castañeda & W.B. Kendr. (1990), anamorphic *Pezizomycotina*, Hsp.0eP.15. 1, Cuba. See Castañeda & Kendrick (*Univ. Waterloo Biol. Ser.* **32**: 28, 1990).

Meniscoideisporites Watanabe, H. Nishida & Kobayashi (1999), Fossil Fungi. 1. See Watanabe *et al.* (*Int. J. Pl. Sci.* **160**: 438, 1999).

Menispora Pers. (1822), anamorphic *Chaetosphaeria*, Hso.0eH.15. 13, widespread (esp. temperate). See Hughes (*CJB* **41**: 693, 1963; key), Holubová-Jechová (*Folia geobot. phytotax.* **8**: 317, 1973; key 4 spp.), Constantinescu *et al.* (*MR* **99**: 585, 1995; teleomorph-anamorph connexions), Réblová (*Stud. Mycol.* **45**: 149, 2000; review), Réblová & Winka (*Mycol.* **92**: 939, 2000; phylogeny), Markovskaja (*Botanica Lithuanica* **8**: 63, 2002; Lithuania), Fernández *et al.* (*Mycol.* **98**: 121, 2006; phylogeny), Réblová *et al.* (*MR* **110**: 104, 2006; teleomorph), Zhang *et al.* (*Mycol.* **98**: 1076, 2006; phylogeny).

Menisporella Agnihothr. (1962) = Dictyochaeta fide Sutton (*in litt.*).

Menisporopascus Matsush. (2003), Chaetosphaeriaceae. Anamorph *Menisporopsis*. 1, Japan. See Matsushima (*Matsush. Mycol. Mem.* **10**: 141, 2001).

Menisporopsis S. Hughes (1952), anamorphic *Chaetosphaeriaceae*, Hsy.0-1eH.15. 6, Africa; New Zealand. See Hughes (*Mycol. Pap.* **48**: 59, 1952), Castañeda Ruíz *et al.* (*Mycotaxon* **64**: 335, 1997), Tsui *et al.* (*MR* **103**: 148, 1999), Castañeda Ruíz *et al.* (*Cryptog. Mycol.* **22**: 259, 2001; Venezuela), Matsushima (*Matsush. Mycol. Mem.* **10**, 2001; teleomorph).

Menoidea L. Mangin & Har. (1907), anamorphic *Pezizomycotina*, Hsp.0eH.?. 1, France; British Isles. See Wilson & Waldie (*TBMS* **13**: 151, 1928).

Mensularia Lázaro Ibiza (1916), Hymenochaetaceae. 4, widespread. See Wagner & Fischer (*Mycol.* **94**: 998, 2002).

Merarthonis Clem. (1909) = Arthonia fide Santesson (*Symb. bot. upsal.* **12** no. 1: 1, 1952).

Mercadomyces J. Mena (1988), anamorphic *Pezizomycotina*, Hsy.≡ eP.25. 1, Cuba. See Mena (*Revta Jardín bot. Nac.* Univ. Habana **9**: 52, 1988).

merenchyma, see plectenchyma.

Meria Vuill. (1896), anamorphic *Rhabdocline*, Hso.0eH.?. 2, Europe; N. America. *M. laricis* on *Larix*. See Peace & Holmes (*Oxf. forest. Mem.* **15**, 1933), Drechsler (*Phytopathology* **31**: 773, 1941), Jansson *et al.* (*Antonie van Leeuwenhoek* **50**: 321, 1984; ultrastr., life history), Gernandt *et al.* (*Mycol.* **89**: 735, 1997; teleomorph), Stone & Gernandt (*Mycotaxon* **91**: 115, 2005; phylogeny), Wang *et al.* (*Mycol.* **98**: 1065, 2006; phylogeny).

Merimbla Pitt (1979), anamorphic *Talaromyces*, Hso.0eH.15. 4, widespread. See Pitt (*CJB* **57**: 2394, 1979), Ogawa & Sugiyama (*Integration of Modern Taxonomic Methods for Penicillium and Aspergillus Classification*: 149, 2000; phylogeny), Pitt *et al.* (*Integration of Modern Taxonomic Methods for Penicillium and Aspergillus Classification*: 9, 2000; accepted names), Bills *et al.* (*MR* **105**: 1273, 2001; Mexico).

Meringosphaeria Peyronel (1918) = Acerbiella fide Clements & Shear (*Gen. Fung.*, 1931).

Meripilaceae Jülich (1982), Polyporales. 7 gen. (+ 7 syn.), 57 spp.

Lit.: Larsen & Lombard (*Mycol.* **80**: 612, 1988), Corner (*Beih. Nova Hedwigia* **96**: 218 pp., 1989), Shen *et al.* (*Mycol.* **94**: 472, 2002), Kim *et al.* (*Antonie van Leeuwenhoek* **83**: 81, 2003), Wang *et al.* (*Mycol.* **96**: 1015, 2004), Binder *et al.* (*Systematics and Biodiversity* **3**: 113, 2005).

Meripilus P. Karst. (1882), Meripilaceae. 5, widespread. See Corner (*Beih. Nova Hedwigia* **78**: 193, 1984), Larson & Lombard (*Mycol.* **80**: 612, 1988; key).

Merisma (Fr.) Gillet (1878) ≡ Grifola.

Merisma Pers. (1797) = Thelephora fide Donk (*Reinwardtia* **2**: 435, 1954).

Merismatium Zopf (1898), Verrucariaceae. *c.* 11 (on lichens), Europe. See Triebel (*Biblthca Lichenol.* **35**, 1989).

merismatoid (of a pileus), made up of smaller pilei.

Merismella Syd. (1927), anamorphic *Chaetothyrium*, Cpt.0eH.38. 5, C. America. See Hofmann & Piepenbring (*Fungal Diversity* **22**: 55, 2006; Panama).

Merismodes Earle (1909), Niaceae. 20, widespread. See Singer (*Agaric. mod. Tax.* edn 3, 1975).

merispore, see sporidesm.

Meristacraceae Humber (1989), Entomophthorales. 3 gen. (+ 3 syn.), 7 spp.

Lit.: Tucker (*Mycotaxon* **13**: 481, 1981; key), Humber (*Mycotaxon* **34**: 441, 1989), Saikawa (*CJB* **67**: 2484, 1989), Saikawa & Sakuramata (*Trans. Mycol. Soc. Japan* **33**: 237, 1992), Saikawa *et al.* (*CJB* **75**: 762, 1997).

Meristacrum Drechsler (1940), Meristacraceae. 2, widespread. See Drechsler (*J. Wash. Acad. Sci.* **30**: 250, 1940), Drechsler (*Sydowia* **14**: 98, 1960; spore discharge), Davidson & Barron (*CJB* **51**: 231, 1973), Couch *et al.* (*Proc. natn Acad. Sci. U.S.A.* **76**: 2299, 1979), Tucker (*Mycotaxon* **13**: 481, 1981; key), Prasad & Dayal (*Curr. Sci.* **55**: 321, 1986), Saikawa *et al.* (*CJB* **75**: 762, 1997; ultrastr.), Keller & Petrini (*Sydowia* **57**: 23, 2005; key).

meristem arthrospore, one of the chain of conidia maturing in basipetal succession and originating by apical wall building at the tip of the condiogenous cell; **- blastospore**, a conidium arising either apically or laterally from a conidiogenous cell which elongates through ring wall building at the base (a basauxic conidiophore; Hughes, 1953). See Anamorphic fungi.

meristematic (of conidiophores), see wall-building.

meristogenous (of pycnidia, etc.), formed by growth and division of one hypha; **symphogenous**, formed from a number of hyphae. See Sutton (*in* Ainsworth *et al.* (Eds), *The Fungi* **4A**: 1973).

Meristosporum A. Massal. (1860) = Laurera fide Hawksworth *et al.* (*Dictionary of the Fungi* edn 8, 1995).

Merocinta Pell & E.U. Canning (1993), Microsporidia. 1.

Merodontis Clem. (1909), Helotiales. 1, Java. See Carpenter (*Mem. N. Y. bot. Gdn* **33**, 1981; status).

merogamy, copulation between special sex cells or gametes.

Merolpidiaceae A. Fisch. (1892) = Synchytriaceae.

meront (of *Mycetozoa*), one of the daughter myxamoebae cut off in succession by a parent myxamoeba.

Merophora Clem. (1909) = Umbilicaria fide Hawksworth *et al.* (*Dictionary of the Fungi* edn 8, 1995).

Meroplacis Clem. (1909) = Caloplaca fide Hawksworth *et al.* (*Dictionary of the Fungi* edn 8, 1995).

Merorinis Clem. (1909) = Rinodina fide Zahlbruckner

(*Catalogus Lichenum Universalis* **7**, 1931).

merosporangium (of *Zygomycetes*), a cylindrical outgrowth from the swollen end of a sporangiophore in which a chain-like series of sporangiospores is generally produced. See Benjamin (*Mycol.* **58**: 1, 1966; review, *in* Kendrick (Ed.), *The whole fungus* **2**: 573, 1979).

Merosporium Corda (1831) nom. dub., ? Fungi. See Hughes (*CJB* **36**: 784, 1958), Hennebert (*CJB* **76**: 1596, 1998).

Merostictina Clem. (1909) = Pseudocyphellaria fide Hawksworth *et al.* (*Dictionary of the Fungi* edn 8, 1995).

Merostictis Clem. (1909) = Diplonaevia fide Hein (*Nova Hedwigia* **38**: 669, 1983).

Merrilliopeltis Henn. (1908) = Oxydothis fide Barr (*Mycol.* **68**, 1976), Hyde (*Sydowia* **46**: 265, 1995).

Merugia Rogerson & Samuels (1990), Sordariomycetes. 1 (living petioles of *Palicourea*), Guyana. See Rogerson & Samuels (*Mem. N. Y. bot. Gdn* **64**: 165, 1990).

Meruliaceae P. Karst. (1881), Polyporales. 47 gen. (+ 44 syn.), 420 spp.

Lit.: Reid (*Beih. Nova Hedwigia* **18**: 1090, 1965), Chamuris (*Mycol. Mem.* **14**: 247 pp., 1988; as *Podoscyphaceae*), Corner (*Beih. Nova Hedwigia* **96**: 218 pp., 1989), Tzean & Liou (*Phytopathology* **83**: 1015, 1993; as *Hyphodermataceae*), Nakasone & Sytsma (*Mycol.* **85**: 996, 1993), Nakasone & Burdsall Jr. (*Mycotaxon* **54**: 335, 1995), Chang *et al.* (*Mycol. Monogr.* **10**: 126 pp., 1996; as *Podoscyphaceae*), Legon & Pegler (*Mycologist* **10**: 180, 1996; as *Podoscyphaceae*), Ginns (*Mycol.* **90**: 19, 1998; as *Podoscyphaceae*), Langer (*Folia cryptog. Estonica* **33**: 57, 1998; as *Hyphodermataceae*), Johannesson *et al.* (*MR* **104**: 92, 2000), Parmasto & Hallenberg (*Nordic Jl Bot.* **20**: 105, 2000), Hjortstam (*Mycotaxon* **79**: 181, 2001; as *Hyphodermataceae*), Kim & Jung (*Mycotaxon* **82**: 295, 2002), Nakasone (*Mycotaxon* **81**: 477, 2002), Koker *et al.* (*MR* **107**: 1032, 2003), Mossebo & Ryvarden (*Mycotaxon* **88**: 229, 2003), Nilsson *et al.* (*MR* **107**: 645, 2003; as *Hyphodermataceae*), Dai (*Mycotaxon* **89**: 389, 2004), Douanla-Meli & Langer (*Mycotaxon* **90**: 323, 2004; as *Podoscyphaceae*), Greslebin *et al.* (*Mycol.* **96**: 261, 2004), Hjortstam & Ryvarden (*Syn. Fung.* **18**: 14, 2004), Larsson *et al.* (*MR* **108**: 983, 2004), Binder *et al.* (*Systematics and Biodiversity* **3**: 113, 2005), Larsson (*MR* **111**: 186, 2007; as *Hyphodermataceae*).

Merulicium J. Erikss. & Ryvarden (1976), ? Pterulaceae. 1, Nordic. See Eriksson & Ryvarden (*Cortic. N. Europ.* **4**: 859, 1976).

Merulioporia Bondartsev & Singer (1943) = Perennıporia fide Donk (*Taxon* **5**: 69, 1956), Lowe (*Tech. Pub. Sta. Univ. Coll. Forestry, Syracuse* **90**, 1966) = Merulius (Cortic.) fide.

Meruliopsis Bondartsev (1959), Phanerochaetaceae. 13, widespread. See Bondartsev (*Izv. Akad. Nauk Estonsk. SSR* Ser. Biol. **8**: 274, 1959), Napoli (*Boll. Gruppo Micol. 'G. Bresadola'* **40**: 343, 1997).

Meruliporia Murrill (1942) = Meruliopsis fide Parmasto (*Consp. Syst. Corticiaceae*: 103, 1968).

Merulius Fr. (1821) = Phlebia fide Nakasone & Burdsall (*Mycotaxon* **21**: 241, 1984).

Merulius Haller ex Boehm. (1760) = Cantharellus fide Stalpers (*in litt.*).

Mesniera Sacc. & P. Syd. (1902), Mesnieraceae. 1 (from living leaves), Java. See Eriksson (*Op. Bot.* **60**, 1981), Cannon (*Stud. Mycol.* **31**: 49, 1989), Pearce & Hyde (*Fungal Diversity* **6**: 83, 2001).

Mesnieraceae Arx & E. Müll. (1975), ? Dothideomycetes (inc. sed.). 3 gen., 4 spp.

Lit.: Hyde (*Mycotaxon* **57**: 347, 1996).

Mesobotrys Sacc. (1880) = Gonytrichum fide Hughes (*CJB* **36**: 727, 1958).

mesochroic, see colour.

Mesochytrium B.V. Gromov, Mamkaeva & Pljusch (2000), Chytridiaceae. 1 (on *Chlorococcum*), Russia. See Gromov *et al.* (*Nova Hedwigia* **71**: 159, 2000).

mesodermatous (of hyphae), having the outer wall and lumen of about the same thickness. Cf. leptodermatous.

Mesomycetozoa, Rankless name for the organisms contained in the DRIPs clade (q.v.). See Herr *et al.* (*J. Clin. Microbiol.* **37**: 2750, 1999).

Mesonella Petr. & Syd. (1924) ? = Guignardia fide Hawksworth *et al.* (*Dictionary of the Fungi* edn 8, 1995).

Mesophellia Berk. (1857), Mesophelliaceae. *c.* 15, Australasia. See Beaton & Weste (*TBMS* **79**: 455, 1983; key), Bougher & Lebel (*Aust. Syst. Bot.* **14**: 439, 2001).

Mesophelliaceae Jülich (1982), Hysterangiales. 8 gen. (+ 2 syn.), 33 spp.

Lit.: Beaton & Weste (*TBMS* **82**: 665, 1984), Cribb (*Qd Nat.* **30**: 25, 1990), Dell *et al.* (*New Phytol.* **114**: 449, 1990), Trappe *et al.* (*Aust. Syst. Bot.* **5**: 618, 1992), Trappe *et al.* (*Aust. Syst. Bot.* **9**: 808, 1996), Trappe *et al.* (*Aust. Syst. Bot.* **9**: 773, 1996), Bougher & Lebel (*Aust. Syst. Bot.* **14**: 439, 2001), Trappe & Bougher (*Australas. Mycol.* **21**: 9, 2002).

Mesophelliopsis Bat. & A.F. Vital (1957), Agaricales. 1, Brazil. ? = *Geastrum* (Geastr.) fide Demoulin (*in litt.*).

mesophile, see thermophily.

Mesopsora Dietel (1922) = Melampsora fide Berndt (*in litt.*).

Mesopyrenia M. Choisy (1931) = Arthopyrenia fide Hawksworth *et al.* (*Dictionary of the Fungi* edn 8, 1995).

mesospore (1) a 1-celled teliospore among 2-celled ones; (2) an amphispore (obsol.); (3) the middle layer of a three-layered spore wall.

meta- (prefix), changed in form or position; between; with; after.

Metabasidiomycetidae. Subclass in *Basidiomycota* (Lowy, *Taxon* **17**: 125, 1968), for fungi considered intermediate between *Heterobasidiomycetes* and *Homobasidiomycetes*.

metabasidium, see basidium.

metabiosis, the association of two organisms acting or living one after the other; cf. synergism.

Metabolic products. Fungal metabolites are many and diverse. In addition to those associated with protein synthesis and respiration many additional products ('secondary metabolites') have been isolated and, frequently, chemically defined. Some of these are waste products while others such as antibiotics, pigments and toxins clearly have biological functions. Because of their synthetic abilities fungi are used in industry for the production of alcohol, citric acid and other organic acids, various enzymes, riboflavin, etc. (see Industrial mycology). Molecular genetic and metabolomic approaches are slowly providing a greater insight into the initiation, synthesis and production of secondary metabolites. In *Aspergillus nidulans*, for example, it has been shown how the metabolic model establishes the functional links between genes. David *et al. BMC Genomics* **9**: 163, 2008)

Many fungal products, even when chemically defined, have been given vernacular names derived from the scientific names of the fungi involved, e.g. 'griseofulvin' from *Penicillium griseofulvum*. More than 400 such names were compiled in the sixth edition of this *Dictionary*, but most of these have been omitted since the seventh because of the monograph by Turner (1971) and Turner & Aldridge (1982) although representative antibiotics, hormones, mycotoxins, pigments, and other interesting metabolites are still included. See also Hegnauer (*Chemotaxonomie der Pflanzen* **7**: 277, 1986).

About 400 compounds have been reliably reported from lichens of which about 230 are only known in this biological group (lichen products, lichen substances; Culberson & Elix, *Meth. Pl. Biochem.* **1**: 509, 1989; Huneck, *Beih. Nova Hedw.* **79**: 793, 1984). The lichen substances are mainly derivatives of orcinol and β-orcinol and are weak phenolic acids. The most important groups of these are depsides (e.g. olivetoric acid), depsidones (e.g. physodic acid), and dibenzonfuran derivatives (e.g. usnic acid). Most are colourless but some are brightly coloured: red, yellow, orange or emerald green (e.g. pulvic acid derivatives such as vulpinic acid). These are deposited on the surfaces of hyphae in the medulla and cortex (different substances often occurring in different regions of the thalli; e.g. hymenium, thalline exciple, medulla, cortex) and are produced by the fungal partner; the hypothesis that production was dependent on the presence of an alga (Culberson & Ahmadjian, *Mycol.* **72**: 90, 1980), has not been upheld by later work (Leuckert *et al.*, *Mycol.* **82**: 370, 1990). However, their position and quantitative expression can be affected by the thallus environment (Fahselt, 1994).

Some depsides and depsidones give characteristic colours with 10% caustic potash (K), bleach (C), K followed by C (KC), C followed by KC (CK), iodine (I, q.v.), and *p*-phenylenediamine (P, PD; see Steiner's stable PD solution); diagnostic microcrystal tests (q.v.); and characteristic colours in UV-light. The metabolic products generally characterize particular lichens and are routinely used in identification of antibiotic properties. Their role in the lichen thallus is largely unknown, but may be partly antimicrobial and antifeedent (see Lawrey, *Biology of lichenized fungi*, 1984).

Lit.: Raistrick *et al.* (Studies in the biochemistry of microorganisms [mainly devoted to fungi, 116 parts], *Phil. Trans.* **B220**: 1, 1931; *Biochem. J.* **25-93**, 1931-1964), Asahina & Shibata (*Chemistry of lichen substances*, 1954 [reprint 1971], WHO (*WHO Chron.* **17**: 389, 1963; nomenclature pharmaceutical preparations), Miller (*The Pfizer handbook of microbial metabolites*, 1961), Shibata *et al.* (*List of fungal products*, 1964), Culberson (*Chemical and botanical guide to lichen products*, 1969; *Bryologist* **73**: 177, 1970 [suppl. 1]; *et al.*, Second supplement, 1977), Turner (*Fungal metabolites*, 1971), Turner & Aldridge, (*Fungal metabolites* **II**, 1983), Pidoplichko (*Metabolity pochvennykh mikromitsetov*, 1971; of soil fungi), Weete (*Fungal lipid biochemistry*, 1974), Eugster (*Z. Pilzk.* **39**: 45, 1973; review), Sussman (*Taxon* **23**: 301, 1974; trends in metabolic specialization), Fahselt (*Symbiosis* **16**: 117, 1994), Cole *et al.* (*Handbook of Secondary Fungal Metabolites*, [3 vols], 2003), Keller (*Nature Reviews Microbiology* **12**:937, 2005).

See also antibiotics, Chemotaxonomy, chromatography, dyeing, ergot, Hallucinogenic fungi, hormones, litmus, Mycetism, Mycotoxicoses, Phytotoxic mycotoxins, Pigments, Toxins.

Metabotryon Syd. (1926) = Sphaerellopsis Cooke fide Sutton (*Mycol. Pap.* **141**, 1977).

Metabourdotia L.S. Olive (1957), ? Auriculariales. 1, Tahiti. See Olive (*Am. J. Bot.* **44**: 429, 1957).

Metacapnodiaceae S. Hughes & Corlett (1972), Capnodiales. 6 gen. (+ 2 syn.), 19 spp.

Lit.: Hughes (*N.Z. Jl Bot.* **10**: 225, 1972), Hughes (*Mycol.* **68**: 693, 1976; gen. names, anamorphs), Parbery & Brown (*Microbiology of the Phyllosphere*: 101, 1986), Sugiyama & Amano *in* Sugiyama (Ed.) (*Pleomorphic Fungi: The Diversity and its Taxonomic Implications*: 141, 1987), Reynolds (*CJB* **76**: 2125, 1998; phylogeny).

Metacapnodium Speg. (1918), Metacapnodiaceae. Anamorphs *Capnophialophora*, *Capnobotrys*, *Capnosporium*. *c.* 10, widespread (tropical). See Hughes (*N.Z. Jl Bot.* **10**: 239, 1972), Reynolds (*Mycotaxon* **23**: 153, 1985; synonymy with *Limacinia*), Eriksson & Hawksworth (*SA* **6**: 138, 1987; nomencl.).

metacellulose, a cellulose in certain fungi.

Metachora Syd., P. Syd. & E.J. Butler (1911) = Phyllachora Nitschke ex Fuckel (1870) fide Cannon (*Mycol. Pap.* **163**, 1991).

metachroic, see Colour.

metachromic, giving a red reaction to cresyl blue. See Singer (*The Agaricales (mushrooms) in modern taxonomy*: 77, 1951).

Metacoleroa Petr. (1927) = Venturia Sacc. fide Kruys *et al.* (*MR* **110**: 527, 2006; phylogeny), Crous *et al.* (*Stud. Mycol.* **58**: 185, 2007; phylogeny), Winton *et al.* (*Mycol.* **99**: 240, 2007; phylogeny).

Metacordyceps G.H. Sung, J.M. Sung, Hywel-Jones & Spatafora (2007), Clavicipitaceae. Anamorphs *Metarhizium*, *Nomuraea*, *Paecilomyces*-like, *Pochonia*. 6 (entomogenous), widespread (esp. E. Asia). See Sung *et al.* (*Stud. Mycol.* **57**, 2007).

Metadiplodia Syd. (1937), anamorphic *Pezizomycotina*, Cpd.1eP.19. 38, widespread. See Zambettakis (*BSMF* **70**: 219, 1955), Sutton (*Sydowia* **43**: 264, 1991; redescr. type sp.).

Metadothella Henn. (1904), ? Hypocreales. 1, Peru.

Metadothis (Sacc.) Sacc. (1892) = Dothiora Fr. (1849) fide Saccardo (*Syll. fung.* **12**: 430, 1897).

Metamelanea Henssen (1989), Lichinaceae (L). 2, Europe; N. America. See Henssen (*Lichenologist* **21**: 102, 1989), Schultz & Büdel (*Lichenologist* **34**: 39, 2002; key), Schultz (*Lichenologist* **40**: 81, 2008; UK).

Metameris Theiss. & Syd. (1915), Pleosporales. 3 (on ferns), widespread (northern hemisphere). See Holm & Holm (*Bot. Notiser* **131**: 97, 1978; synonym of *Scirrhia*), Barr (*Mycotaxon* **43**: 371, 1992).

Metanectria Sacc. (1878) = Thelocarpon fide Rossman *et al.* (*Stud. Mycol.* **42**: 248 pp., 1999).

Metapezizella Petr. (1968), Helotiaceae. 1, Mexico. See Petrak (*Sydowia* **20**: 207, 1966).

metaphysis (obsol.) used by Petrak; see paraphysis.

metaplasm, see epiplasm.

Metarhizium Sorokīn (1879), anamorphic *Metacordyceps*, Hsy.0eH.15. 9 (on *Insecta*), widespread. See Tulloch (*TBMS* **66**: 407, 1976; key), Guo *et al.* (*Acta

Mycol. Sin. **5**: 177, 1986), Rombach *et al.* (*Mycotaxon* **27**: 87, 1986; *M. flavo-viride* var. *minus* on leaf- and plant-hoppers on rice), Rombach *et al.* (*TBMS* **88**: 451, 1987; *M. album* on leaf-hoppers and plant-hoppers on rice), Shimizu *et al.* (*J. Invert. Path.* **60**: 185, 1992; electrophoretic karyotype of *M. anisopliae*),), Bridge *et al.* (*J. gen. Microbiol.* **139**: 1163, 1993; morphology, biochem., molecular differentiation), Rakotonirainy *et al.* (*MR* **98**: 225, 1994; rRNA and separation of taxa), Pipe *et al.* (*MR* **99**: 485, 1995; RFLPs in *M. anisopliae*), Mavridou *et al.* (*MR* **102**: 1233, 1998; rRNA analysis), Revankar *et al.* (*J. Clin. Microbiol.* **37**: 195, 1999; in man), Driver *et al.* (*MR* **104**: 134, 2000; revision, phylogeny), Bidochka *et al.* (*Appl. Environm. Microbiol.* **67**: 1335, 2001; ecology), Evans (*Mycology Series* **19**: 517, 2003; biocontrol), Padmavathi *et al.* (*MR* **107**: 572, 2003; telomere fingerprinting), Pantou *et al.* (*Fungal Genetics Biol.* **38**: 159, 2003; genetics), Bidochka & Small (*Insect-Fungal Associations* Ecology and Evolution: 28, 2005; phylogeography), Bidochka *et al.* (*Environmental Microbiology* **7**: 1361, 2005; phylogeny, speciation), Bischoff *et al.* (*Mycol.* **98**: 737, 2006; cryptic speciation), Ghikas *et al.* (*Archs Microbiol.* **185**: 393, 2006; mitochondrial genome), Huang *et al.* (*Mycotaxon* **94**: 181, 2005; phylogeny), Sung *et al.* (*Stud. Mycol.* **57**, 2007).

Metasphaerella Speg. (1924) nom. dub., Fungi. See Petrak & Sydow (*Annls mycol.* **34**: 42, 1936).

Metasphaeria Sacc. (1883) nom. ambig. = Saccothecium fide Hawksworth *et al.* (*Dictionary of the Fungi* edn 8, 1995).

Metasteridium Speg. (1923) nom. dub., ? Meliolaceae. No spp. named. See Eriksson & Hawksworth (*SA* **7**: 59, 1988).

metathallus, assimilative (photobiont-containing) part of a lichen thallus, esp. where there is also a prothallus (q.v.).

Metathyriella Syd. (1927), Schizothyriaceae. 3 or 4, America (tropical). See Reynolds & Gilbert (*Cryptog. Mycol.* **27**: 249, 2006; Panama).

Metazythia Petr. (1950), anamorphic *Pezizomycotina*, Cpd.0eH.15. 1, S. America. See Petrak (*Sydowia* **4**: 373, 1950).

Metazythiopsis M. Morelet (1988), anamorphic *Pezizomycotina*, Cpd.0eH.15. 1, France. See Morelet (*Annales de la Société des Sciences Naturelles et d'Archéologie de Toulon et du Var* **40**: 41, 1988).

Metchnikovella Caullery & Mesnil (1897), Microsporidia. 3.

Methods.

General literature: Booth (Ed.) (*Methods in microbiology* **4**, 1971), Constantinescu (*Metode și tehnici în micologie*, 1974), Fuller (*Lower fungi in the laboratory*, 1978), Hall (*Methods for the Examination of Organismal Diversity in Soils and Sediments*, 1996), Hawksworth (*Mycologist's handbook*, 1974), Hawksworth & Kirsop (Eds) (*Filamentous fungi*, 1985 [source book]), Johanssen (*Plant microtechnique*, 1940), Koch (*Fungi in the laboratory*, 1966), Koneman & Roberts (*Practical laboratory mycology*, edn 3, 1985), Malloch (*Moulds: their isolation, cultivation and identification*, 1981), McLean & Cooke (*Plant science formulae*, 1941), Smith (*An introduction to industrial mycology*, edn 7, 1981), Spector (Ed.) (*Handbook of biological data*, 1956), Stevens (Ed.) (*Mycology guidebook*, 1974), Waller & Lenné (*Plant pathologist's pocketbook*, 2001). See also Lichens, Literature, Medical and veterinary mycology, Plant pathogenic fungi.

Special topics (q.v.):
Abbreviation of authors' names, see Authors' names
Arsenic detection, see *Scopulariopsis*
Auxanogram
Bauch test
Beeli formulae
Biomass determination, see biomass
Block culture, see Culture methods
Cellulolysis adequacy index
Chemotaxonomy
Chromatography
Collection and preservation
Colour nomenclature, see Colour
Continuous culture, see Culture methods
Coolplate
Culture media, see Media
Culture methods
Culture preservation, see Genetic resource collections
Electron microscopy (EM), see Ultrastructure
Electrophoresis
Fixatives
Fungicide testing, see Fungicides
Herbarium beetle control
Herbarium management, see Reference collections
Hydrogen-ion concentration
Iodine
Ionomidotic reaction
Isolation methods
Keys, construction of
Laboratory safety, see Safety, laboratory
Lyophilization, see Genetic resource collections
Media, culture
Meixner test
Melzer's solution, see Stains
Microchemical tests for lichen products, see Metabolic products, microcrystal tests
Microscopy
Mite infestation control, see mites
Molecular biology
Mounting media
Nomenclature
Normal saline solution
Numerical taxonomy
Phytosociology
Preservatives
Protoplasts
Reference collections
RIEC
Scanning electron microscopy (SEM)
Serological Techniques, see Serology
Single-spore isolation, see Isolation methods
Spore
Spore print
Spore trapping, see Air spora
Stains
Statistical methods and design of experiments
Steiner's stable PD solution
Sterilization
Ultrastructure
Zymogram.

Methysterostomella Speg. (1910) ? = Phragmopeltis fide Hawksworth *et al.* (*Dictionary of the Fungi* edn 8, 1995).

metoecious (obsol.) used by de Bary; see heteroecious.

Metraria Cooke & Massee (1891), Agaricaceae. 2, Australia; Nigeria. See Singer (*Agaric. mod. Tax.* 4th ed: 609, 1986; type includes material of *Hebeloma* (*Bolbitiaceae*) and *Amanita* (*Amanitaceae*)).

Metrodia Raithelh. (1971), Agaricaceae. 2, Argentina. See Raithelhuber (*Metrodiana* **2**: xxvii, 1971).

Metschnikowia T. Kamieński (1899), Metschnikowiaceae. Anamorphs *Candida*, *Nectaromyces*. 42 (from sea water and arthropods, esp. associated with nectar), widespread. See Pitt & Miller (*Mycol.* **60**: 682, 1968; anamorphs), Pitt & Miller (*Mycol.* **62**: 462, 1970; parasexual cycle), Batra (*USDA Tech. Bull.* **1469**, 1973; key), Giménez-Jurado (*Syst. Appl. Microbiol.* **15**: 432, 1992), Mendonça-Hagler *et al.* (*Int. J. Syst. Bacteriol.* **43**: 368, 1993; rRNA phylogeny), Giménez-Jurado *et al.* (*Antonie van Leeuwenhoek* **68**: 101, 1995), Lopandic *et al.* (*Syst. Appl. Microbiol.* **19**: 393, 1996), Valente *et al.* (*J. gen. appl. Microbiol.* Tokyo **43**: 179, 1997; ITS), Lachance *et al.* (*Can. J. Microbiol.* **44**: 279, 1998), Miller & Phaff *in* Kurtzman & Fell (Eds) (*Yeasts, a taxonomic study* 4th edn: 256, 1998), Kurtzman & Droby (*Syst. Appl. Microbiol.* **24**: 395, 2001; biocontrol), Lachance *et al.* (*Can. J. Microbiol.* **47**: 103, 2001; from insects), Lachance & Bowles (*FEMS Yeast Res.* **2**: 81, 2002; from beetles), Giménez-Jurado *et al.* (*Int. J. Syst. Evol. Microbiol.* **53**: 1665, 2003; from insects), Marinoni *et al.* (*FEMS Yeast Res.* **3**: 85, 2003; ascospores), Lachance & Bowles (*Stud. Mycol.* **50**: 69, 2004; from beetles), Marinoni & Lachance (*FEMS Yeast Res.* **4**: 587, 2004; speciation), Lachance *et al.* (*Int. J. Syst. Evol. Microbiol.* **55**: 1369, 2005; from beetles, Hawaii), Molnár & Prillinger (*Syst. Appl. Microbiol.* **28**: 717, 2005; phylogeny), Lachance *et al.* (*Int. J. Syst. Evol. Microbiol.* **56**: 1141, 2006; from insects, Africa), Nguyen *et al.* (*MR* **110**: 346, 2006; from insects), Suh *et al.* (*Mycol.* **98**: 1006, 2006; phylogeny), Xue *et al.* (*Int. J. Syst. Evol. Microbiol.* **56**: 2245, 2006; from *Zizyphus*).

Metschnikowiaceae T. Kamieński (1899), Saccharomycetales. 2 gen. (+ 6 syn.), 44 spp.

Lit.: Lachance & Phaff *in* Kurtzman & Fell (Eds) (*Yeasts, a taxonomic study* 4th edn: 148, 1998), Miller & Phaff *in* Kurtzman & Fell (Eds) (*Yeasts, a taxonomic study* 4th edn: 256, 1998), Giménez-Jurado *et al.* (*Int. J. Syst. Evol. Microbiol.* **53**: 1665, 2003), Lachance *et al.* (*FEMS Yeast Res.* **4**: 253, 2003), Lachance *et al.* (*FEMS Yeast Res.* **3**: 97, 2003), Diezmann *et al.* (*J. Clin. Microbiol.* **42**: 5624, 2004), Marinoni & Lachance (*FEMS Yeast Res.* **4**: 587, 2004), Suh *et al.* (*Mycol.* **98**: 1006, 2006; phylogeny).

Metschnikowiella Genkel (1913) = Metschnikowia fide Hawksworth *et al.* (*Dictionary of the Fungi* edn 8, 1995).

metula, a conidiophore branch having phialides, e.g. of *Penicillium* and *Aspergillus* (obsol.).

Metulocladosporiella Crous, Schroers, J.Z. Groenew., U. Braun & K. Schub. (2006), anamorphic *Herpotrichiellaceae*. 3, widespread. See Crous *et al.* (*MR* **110**: 269, 2006).

Metulocyphella Agerer (1983), ? Marasmiaceae. 2, S. America. See Agerer (*Z. Mykol.* **49**: 155, 1983).

Metulodontia Parmasto (1968), Peniophoraceae. 2, widespread. See Jülich (*Persoonia* **8**: 78, 1974).

metuloid, an encrusted cystidium thick-walled at maturity, as in *Peniophora*.

Metuloidea G. Cunn. (1965) = Junghuhnia fide Ryvarden (*Gen. Polyp.*: 184, 1991).

Metus D.J. Galloway & P. James (1987), Cladoniaceae (L). 3, Chile; Australasia. See Stenroos & DePriest (*Am. J. Bot.* **85**: 1548, 1998; DNA), Wedin *et al.* (*Lichenologist* **32**: 171, 2000; phylogeny), Stenroos *et al.* (*Mycol. Progr.* **1**: 267, 2002; phylogeny), Miądlikowska *et al.* (*Mycol.* **98**: 1088, 2006; phylogeny), Messuti *et al.* (*Biblthca Lichenol.* **95**: 471, 2007; Argentina).

Miainomyces Corda (1833) = Sporotrichum fide Saccardo (*Syll. fung.* **4**: 106, 1886).

micaceous (of a pileus surface), covered with bright particles.

Micarea Fr. (1825) nom. cons., Pilocarpaceae (L). *c.* 107, widespread. See Vězda & Wirth (*Folia geobot. phytotax.* **11**: 93, 1976; key 31 spp.), Coppins (*Bull. Br. Mus. nat. hist.* Bot. **11**: 17, 1983; monogr. 45 Eur. spp.), Coppins (*Notes R. bot. Gdn Edinb.* **45**: 161, 1988; Europe), Coppins (*Taxon* **38**: 499, 1989; nomencl.), Coppins & Kantvilas (*Lichenologist* **22**: 277, 1990; Tasmania), Pietschmann (*Nova Hedwigia* **51**: 521, 1990; asci), Coppins (*Biblthca Lichenol.* **58**: 57, 1995; Europe), Coppins (*Lichenologist* **31**: 559, 1999; S Africa), Andersen & Ekman (*MR* **109**: 21, 2005; phylogeny), Miądlikowska *et al.* (*Mycol.* **98**: 1088, 2006; phylogeny).

Micarea Fr. (1825) nom. rej. = Placynthiella Elenkin fide Hawksworth *et al.* (*Dictionary of the Fungi* edn 8, 1995).

Micareaceae Vězda ex Hafellner (1984) = Pilocarpaceae.

Lit.: Coppins & Purvis (*Lichenologist* **19**: 29, 1987), Hawksworth (*CRC Handbook of Lichenology*: 181, 1988), Coppins & Kantvilas (*Lichenologist* **22**: 277, 1990), Pietschmann (*Nova Hedwigia* **51**: 521, 1990), Brodo & Tønsberg (*Acta Bot. Fenn.* **150**: 1, 1994), Triebel *et al.* (*Symb. bot. upsal.* **32**: 323, 1997), Coppins (*Lichenologist* **31**: 559, 1999), Andersen & Ekman (*Lichenologist* **36**: 27, 2004), Andersen & Ekman (*MR* **109**: 21, 2005).

Micheli (Pier Antonio; 1679-1737; Italy). Botanist to Cosmo III (Grand Duke of Tuscany), keeper of the public gardens in Florence, and Professor of Botany, University of Pisa (1706). His additions to the knowledge of fungi, including lichen-forming species, were the greatest made by any one man before Persoon and Fries. He described these discoveries in *Nova Plantarum Genera*, his most important printed work. Some of Micheli's fungal names are still used for common genera (e.g. *Aspergillus*, *Clathrus*, *Mucor* and *Polyporus*). He made a new systematic arrangement with keys to what would now be regarded as genera and species. With the help of the microscope (then coming into use), he was the first to see cystidia on the lamella-edge and between the lamellae of agarics, and he saw the arrangement of spores (which he took to be seeds) in groups of four in the *Agaricaceae*, and in asci of *Pertusaria*. He was the first to make experiments on the culture of moulds by placing spores of *Botrytis*, *Aspergillus* and *Mucor* on freshly-cut bits of melon, quince and pear, noting their growth and development [for an English translation of his observations on culturing moulds, see Buller, *Transactions of the Royal Society of Canada* section 4, series 3, **9**: 1915 (reprinted by Ainsworth, 1976)]. His important discoveries in connexion with flowering plants and bryophytes are, in general, better known. *Publs. Nova Plantarum Genera* (1729) [this book was completed in 1719, but the first part not published till 1729, and the second (the Figures for which are still in existence) was never printed]. *Biogs, obits etc.* Ainsworth (*Introduction to the History of Mycology*, 1976); Hawksworth (introduction, *Nova Plantarum Genera*, 1976 reprint); Stafleu & Cowan (*TL-2* **3**: 446, 1981); Targioni-Tozzetti (*Notizie della Vita e delle Opere di Pier Antonio Micheli*, 1858).

Michenera Berk. & M.A. Curtis (1868), anamorphic *Licrostroma*. 3, pantropical. See Donk (*Taxon* **11**: 89, 1962), Parbery & Rumba (*MR* **95**: 761, 1991).

Micraspis Darker (1963), Helotiales. Anamorph *Periperidium*. 3 (on *Rhytismatales*), N. America; British Isles. See Darker (*CJB* **41**: 1389, 1963).

micro- (prefix), small; one-thousandth (Système International d'Unités); see micron.

microaerophilic, making best growth under lowered oxygen pressure.

Microallomyces R. Emers. & J.A. Robertson (1974), Blastocladiaceae. 1 (from pond sediment), Costa Rica. See Meyer (*Mycol.* **79**: 44, 1987; ultrastr.).

Microanthomyces Grüss (1926) = Candida fide Lodder (*Yeasts, a taxonomic study* 2nd edn, 1970).

Microascaceae Luttr. ex Malloch (1970), Microascales. 20 gen. (+ 18 syn.), 131 spp.

Lit.: Malloch (*Mycol.* **62**: 729, 1970), von Arx *et al.* (*Beih. Nova Hedwigia* **94**, 1988), von Arx *et al.* (*Beih. Nova Hedwigia* **94**: 104 pp., 1988), Barr (*Mycotaxon* **39**: 43, 1990; concept), Dykstra *et al.* (*Mycol.* **81**: 896, 1989), Issakainen *et al.* (*J. Med. Vet. Mycol.* **35**: 389, 1997), Abbott *et al.* (*Mycol.* **90**: 297, 1998), Wedde *et al.* (*Medical Mycology* **36**: 61, 1998), Issakainen *et al.* (*MR* **103**: 1179, 1999), Lee & Hanlin (*Mycol.* **91**: 434, 1999; DNA), Okada *et al.* (*CJB* **76**: 1495, 1998), Hausner *et al.* (*CJB* **78**: 903, 2000), Rainer *et al.* (*J. Clin. Microbiol.* **38**: 3267, 2000), Issakainen *et al.* (*Medical Mycology* **41**: 31, 2003), Rainer & Hoog (*MR* **110**: 151, 2006).

Microascales Luttr. ex Benny & R.K. Benj. (1980). Hypocreomycetidae. 4 fam., 92 gen., 397 spp. Stromata absent. Ascomata solitary, perithecial or cleistothecial, usually black, thin-walled, sometimes with well-developed smooth setae. Interascal tissue absent or rarely of undifferentiated hyphae. Asci ± globose to clavate, very thin-walled, evanescent, 8-spored, sometimes formed in chains. Ascospores hyaline, yellow or reddish brown, aseptate or septate, sometimes curved, sometimes with very inconspicuous germ pores, with or without a sheath. Anamorphs hyphomycetous, prominent. Saprobic from soil or rotting vegetation, sometimes marine, a few opportunistic human and animal pathogens, cosmop. Fams:

(1) **Ceratocystaceae**
(2) **Chadefaudiellaceae**
(3) **Halosphaeriaceae**
(4) **Microascaceae**.

The *Halosphaeriaceae* are often placed separately in their own order, and further molecular studies may confirm this arrangement. *Lit.*: Tang *et al.* (*Ant. v. Leeuwenh.* **91**: 327, 2007).

Microascus Sacc. (1916) ≡ Microdiscus Sacc.

Microascus Zukal (1885), Microascaceae. Anamorphs *Cephalotrichum*, *Scopulariopsis*, *Wardomyces*, *Wardomycopsis*. 19 (coprophilous, on soil etc.), widespread. See Barron *et al.* (*CJB* **39**: 1609, 1961), Corlett (*CJB* **41**: 253, 1963; ontogeny, sexuality), Corlett (*CJB* **44**: 79, 1966; perithecial development), von Arx (*Persoonia* **8**: 191, 1975; key), Nishimura & Miyaji (*Mycopathologia* **90**: 29, 1985; yeast-like state), von Arx *et al.* (*Beih. Nova Hedwigia* **94**, 1988), Hausner *et al.* (*CJB* **71**: 1249, 1993; DNA), Abbott *et al.* (*Mycol.* **90**: 297, 1998; anamorph), Lee & Hanlin (*Mycol.* **91**: 434, 1999; DNA), Abbott & Sigler (*Mycol.* **93**: 1211, 2001; heterothallism), Abbott *et al.* (*Mycol.* **94**: 362, 2002), Issakainen *et al.* (*Medical Mycology* **41**: 31, 2003; anamorph), Mohammedi *et al.* (*European Journal of Clinical Microbiology & Infectious Diseases* **23**: 215, 2004; clinical strain), Zhang *et al.* (*Mycol.* **98**: 1076, 2006; phylogeny).

Microasellaria Tuzet, Manier & Jolivet (1957) nom. dub., Kickxellomycotina. See Lichtwardt (*The Trichomycetes. Fungal associates of arthropods*, 1986).

Microbasidium Bubák & Ranoj. (1914) = Hadrotrichum fide Höhnel (*Sber. Akad. Wiss. Wien* Math.-naturw. Kl., Abt. 1 **125**: 111, 1916).

microbe, a microorganism (q.v.).

microbial (adj.), pertaining to microbes (q.v.).

microbiology, the study of microorganisms (q.v.).

microbiota, all the microorganisms present in the area or habitat specificied, including algae, bacteria and protozoa as well as fungi; see mycobiota.

Microbispora Nonom. & Y. Ohara (1957), Actinobacteria. q.v.

Microblastosporon Cif. (1930), anamorphic *Pezizomycotina*, Hso.0eH.?. 1, Europe.

microbodies, see peroxisome.

Microbotryaceae R.T. Moore (1996), Microbotryales. 5 gen. (+ 2 syn.), 104 spp.

Lit.: Berbee *et al.* (*CJB* **69**: 1795, 1991), Begerow *et al.* (*CJB* **75**: 2045, 1998), Vánky (*Mycotaxon* **67**: 33, 1998), Swann *et al.* (*Mycol.* **91**: 51, 1999), Bucheli *et al.* (*Mol. Ecol.* **10**: 285, 2001), Vánky (*Aust. Syst. Bot.* **14**: 385, 2001), Almaraz *et al.* (*MR* **106**: 541, 2002), Hood *et al.* (*Infect. Genet. Evol.* **2**: 167, 2003), Sampaio *et al.* (*Mycol. Progr.* **2**: 63, 2003), Van Putten *et al.* (*Evolution* Lancaster, Pa. **57**: 766, 2003), Begerow *et al.* (*MR* **108**: 1257, 2004), Giraud (*Heredity* **93**: 559, 2004), Van Putten *et al.* (*J. Evol. Biol.* **18**: 203, 2005), Vánky (*Mycol. Balcanica* **1**: 175, 2004), Lutz *et al.* (*Mycol. Progr.* **4**: 225, 2005), Kemler *et al.* (*BMC Evol. Biol.* **6**: 35, 2006).

Microbotryales R. Bauer & Oberw. (1997). Microbotryomycetes. 2 fam., 9 gen., 114 spp. Fams:

(1) **Microbotryaceae**
(2) **Ustilentylomataceae**

For *Lit.* see under fam.

Microbotryodiplodia Sousa da Câmara (1951) = Microdiplodia Allesch. fide Petrak (*Sydowia* **16**: 353, 1963), Sutton (*Mycol. Pap.* **141**, 1977).

Microbotryomycetes R. Bauer, Begerow, J.P. Samp., M. Weiss & Oberw. (2006), Pucciniomycotina. 4 ord., 4 fam., 25 gen., 208 spp. Host-parasite interactions without deposits of specific fungal vesicles. Ord:

(1) **Heterogastridiales**
(2) **Leucosporidiales**
(3) **Microbotryales**
(4) **Sporidiobolales**

Lit.: Bauer *et al.* (*Mycol. Progr.* **5**: 41, 2006).

Microbotryum Lév. (1847), Microbotryaceae. 87 (on dicots), widespread (temperate). See Deml & Oberwinkler (*Phytopath. Z.* **104**: 345, 1982), Vánky (*Mycotaxon* **67**: 33, 1998; monogr., emend., key), Swann *et al.* (*Mycol.* **91**: 51, 1999; classific.), Almaraz *et al.* (*MR* **106**: 541, 2002; phylogeny), Vánky (*Mycologia Balcanica* **1**: 189, 2004; taxonomy, nomenclature, problems in species delimitation), Lutz *et al.* (*Mycol. Progr.* **4**: 225, 2005; molecular characters), Aime *et al.* (*Mycol.* **98**: 896, 2006; phylogeny).

Microcaliciaceae Tibell (1984), Teloschistales (±L). 1 gen. (+ 2 syn.), 4 spp.

Lit.: Tibell (*Biblthca Lichenol.* **71**: 107 pp., 1998), Tibell (*Nordic Lichen Flora* **1**. Introductory Parts; Calicioid Lichens and Fungi: 20, 1999).

Microcalicium Vain. (1927), Microcaliciaceae. 4 (mainly on lichens), widespread (boreal; temperate). See Tibell (*Bot. Notiser* **131**: 229, 1978; key), Hawksworth (*Bull. Br. Mus. nat. hist.* Bot. **9**: 1, 1981; anamorphs), Tibell (*Symb. bot. upsal.* **27** no. 1: 279 pp., 1987; Australasia), Tibell (*Symb. bot. upsal.* **32** no. 1: 291, 1997; anamorphs), Tibell (*Biblthca Lichenol.* **71**: 107 pp., 1998; S America).

Microcalliomyces Bat. & Cif. (1962) = Microcallis fide von Arx & Müller (*Stud. Mycol.* **9**, 1975).

Microcalliopsis Bat. & Cif. (1962) = Microcallis fide von Arx & Müller (*Stud. Mycol.* **9**, 1975).

Microcallis Syd. (1926), Chaetothyriaceae. 10, widespread (tropical). See Hansford (*Mycol. Pap.* **15**, 1946).

microcephalic, see septum.

Microcephalis Bainier (1882) = Syncephalis fide Benjamin (*Aliso* **4**: 321, 1959).

Microcera Desm. (1848) = Fusarium fide Petch (*TBMS* **7**: 89, 1921), Wollenweber & Reinking (*Die Fusarien*: 6, 1935), Dingley (*Mem. N. Y. bot. Gdn* **49**: 206, 1989).

Microclava F. Stevens (1917), anamorphic *Pezizomycotina*, Hso.1eP.1. 4, pantropical. See Deighton (*TBMS* **52**: 315, 1969; key).

Microcollybia Métrod ex Lennox (1979) = Collybia fide Kuyper (*in litt.*).

Microcollybia Métrod (1952) nom. nud. = Collybia fide Singer (*Agaric. mod. Tax.* edn 3, 1975).

microconidium (1) the smaller conidium of a fungus which also has macroconidia; (2) a spermatium (q.v.).

microcrystal test, method of identification of phenolic metabolites in lichens developed by Asahina (q.v.) involving re-crystalizations on microscope slide from a range of solvents and the formation of salts with diagnostic shapes. The crystals are examined microscopically for identification. Many of these tests are extremely sensitive and can detect some compounds at concentrations close to the resolution possible by thin-layer chromatography (q.v.).

Lit.: Asahina (*J. Jap. Bot.* **12-16**, 1936-40), Hale (1974), Hawksworth (*Bull. Br. lichen Soc.* **28**: 5, 1971), Thomson (*The lichen genus Cladonia in North America*, 1968). See Metabolic products.

microculture, a culture of an organism under the microscope, as in a hanging drop.

Microcyclella Theiss. (1914), Dothideomycetes. 1, Africa.

Microcyclephaeria Bat. (1958), Pezizomycotina. 1, Australia. See Batista (*Revta Biol.* Lisb. **1**: 301, 1958).

microcyclic (of conidiation), germination of spores by the direct function of the conidia without the intervention of mycelial growth (Hanlin, *Mycoscience* **35**: 113, 1994); (of rusts), see *Pucciniales*.

Microcyclus Sacc., Syd. & P. Syd. (1904), ? Mycosphaerellaceae. Anamorphs *Aposphaeria*-like, *Fusicladium*-like, *Pazschkeella*. *c.* 17, widespread (esp. tropical). Almost certainly polyphyletic. *M. ulei* (anamorphs *Aposphaeria ulei*, *Fusicladium macrosporum*); S. Am. leaf blight of *Hevea*). See Cannon *et al.* (*MR* **99**: 353, 1995; key 7 S. Am. spp.), Barr (*Mycotaxon* **60**: 433, 1996).

microcyst, see cyst.

Microcyta Petr. & Syd. (1927) ≡ Ceuthosporella Petr. & Syd. fide Sutton (*Mycol. Pap.* **141**, 1977).

microcytospore (1) an encysted zoospore (planospore); (2) an encysted gametangium (*Blastocladiales*).

Microdiplodia Allesch. (1901), anamorphic *Botryosphaeriaceae*, Cpd.1eP.?. 31, widespread. See Zambettakis (*BSMF* **70**: 219, 1955), Sutton (*Mycol. Pap.* **141**, 1977), Crous *et al.* (*Stud. Mycol.* **55**: 235, 2006; phylogeny).

Microdiplodia Tassi (1902) ? = Microdiplodia Allesch. fide Sutton (*Mycol. Pap.* **141**, 1977).

Microdiplodiites Babajan & Tasl. (1973), Fossil Fungi. 1 (Tertiary), former USSR.

Microdiscula Höhn. (1915), anamorphic *Pezizomycotina*, Cpd.0eH.?. 1, Europe.

Microdiscus Sacc. (1916), Helotiales. 1, N. America.

Microdiscus Steinecke (1916) nom. dub., ? Fungi.

Microdochium Syd. (1924), anamorphic *Monographella*, Hsp.0-≡ eH.21. 10, widespread. *M. nivale* (pink snow mould of turf grasses), *M. panattonianum* (lettuce ring spot). See Sutton *et al.* (*CJB* **50**: 1899, 1972), Sutton & Hodges (*Nova Hedwigia* **27**: 215, 1976), v. Arx (*Sydowia* **34**: 30, 1981), Samuels & Hallett (*TBMS* **81**: 473, 1983; teleomorphs), Litschko & Burpee (*TBMS* **89**: 252, 1987; variation in *M. nivale*), Parman & Price (*Australas. Pl. Path.* **20**: 41, 1991; microsclerotia in *M. panattonianum*), Braun (*Monogr. Cercosporella, Ramularia Allied Genera (Phytopath. Hyphom.)* **1**: 267, 1995; subgen., key), Mahuku *et al.* (*MR* **102**: 559, 1998; on turfgrass, genetics), Glynn *et al.* (*MR* **109**: 872, 2005; speciation), Kwasna & Bateman (*Mycol.* **99**: 765, 2007; UK, key).

Microdothella Syd. & P. Syd. (1914), Dothideales. 2, N. America; Philippines.

Microdothiorella C.A.A. Costa & Sousa da Câmara (1955), anamorphic *Pezizomycotina*, St.0eH.?. 1, Portugal. See Costa & Sousa da Câmara (*Agron. lusit.* **17**: 156, 1955).

Microeccrina Maessen (1955) = Arthromitus fide Manier (*Annls Parasit. hum. comp.* **38**: 1, 1961) See *Arthromitus*.

microendospore, minute cytoplasmic particles behaving like spores in *Ophiostoma ulmi* (Ouellette & Gagnon, *CJB* **38**: 235, 1960).

Microeurotium Ghatak (1936), ? Pezizales. 1 (? coprophilous), British Isles.

Microfilum N. Faye, B.S. Toguebaye & G. Bouix (1991), Microsporidia. 1.

microflora, sometimes used, inappropriately, for all the microorganisms present in a specified site or habitat; see microbiota, mycobiota.

microform, see *Pucciniales* and Table 5.

microfungi, see *Micromycetes*.

Microgemma Ralphs & Matthews (1986), Microsporidia. 4.

Microglaena Körb. (1855) [non *Microglena* Ehrenb. 1831, *Chlorophyta*] = Thelenella fide Eriksson (*Op. Bot.* **60**, 1981), Mayrhofer & Poelt (*Herzogia* **7**: 13, 1985; Europe), Mayrhofer (*Biblthca Lichenol.* **26**: 106 pp., 1987).

Microglaenaceae Servít (1954) = Thelenellaceae.

Microglaenomyces Cif. & Tomas. (1953) = Chroma-

tochlamys fide Mayrhofer & Poelt (*Herzogia* **7**: 13, 1985).
Microglena Lönnr. (1858) ≡ Microglaena.
Microgloeum Petr. (1922), anamorphic *Blumeriella*, Cac.0eH.10. 1, widespread.
Microglossum Gillet (1879), Leotiaceae. 8, widespread (temperate). See Mains (*Mycol.* **47**: 846, 1955; key N Am spp.), Maas Geesteranus (*Persoonia* **3**: 82, 1964), Nitare & Ryman (*Svensk bot. Tidskr.* **78**: 63, 1984; Sweden), Spooner (*Biblthca Mycol.* **116**: 711 pp., 1987; Australasia), Verkley (*Persoonia* **15**: 405, 1994; ultrastr.), Gernandt *et al.* (*Mycol.* **93**: 915, 2001; phylogeny), Wang *et al.* (*Mol. Phylogen. Evol.* **41**: 295, 2006; phylogeny), Wang *et al.* (*Mycol.* **98**: 1065, 2006; phylogeny).
Microglossum Sacc. (1884) = Geoglossum fide Spooner (*in litt.*, 2008).
microgonidia, very small green bodies in lichen hyphae (Minks) (obsol.).
Micrographa Müll. Arg. (1890) ? = Lembosia fide Santesson (*Svensk bot. Tidskr.* **43**: 547, 1949), Müller & von Arx (*Beitr. Kryptfl. Schweiz* **11** no. 2, 1962).
Micrographina Fink (1930) = Mazosia fide Santesson (*Svensk bot. Tidskr.* **43**: 547, 1949).
Micrographomyces Cif. & Tomas. (1953) nom. illegit. ≡ Micrographa.
Microhaplosporella Sousa da Câmara (1949) ? = Aplosporella fide Sutton (*in litt.*).
Microhendersonula Dias & Sousa da Câmara (1952), anamorphic *Pezizomycotina*, St.≡ eP.?. 1, Portugal. See Dias & Sousa da Câmara (*Agron. lusit.* **14**: 118, 1952).
Microhilum H.Y. Yip & A.C. Rath (1989), anamorphic *Cordycipitaceae*, Hso.0eH.10. 1 (on *Oncoptera* larvae), Australia. See Yip & Rath (*J. Invert. Path.* **53**: 361, 1989), Wright & Patel (*MR* **96**: 578, 1992), Sung *et al.* (*Stud. Mycol.* **57**, 2007; phylogeny).
Microides H.Y. Yip & A.C. Rath (1989), anamorphic *Cordycipitaceae*, Hso.0eH.10. 1 (on *Oncoptera* larvae), Australia. See Yip & Rath (*J. Invert. Path.* **53**: 361, 1989), Wright & Patel (*MR* **96**: 578, 1992), Sung *et al.* (*Stud. Mycol.* **57**, 2007; phylogeny).
Microlecia M. Choisy (1949) = Catillaria fide Hafellner (*Beih. Nova Hedwigia* **79**: 241, 1984).
microlichens, lichens in which the whole of their morphological characteristics can be seen only with a magnifier equal or larger than ×10. See Messuti (*Br. Lich. Soc. Bull.* **73**: 49, 1993).
Microlychnus A. Funk (1973) = Gyalideopsis fide Sérusiaux & De Sloover (*Veröff. geobot. Inst., Zürich* **91**: 260, 1986), Lücking *et al.* (*Lichenologist* **37**: 123, 2005).
Micromastia Speg. (1909), Dothideomycetes. 2.
micrometre, one-thousandth of a millimetre (0.001 mm; 1 µm) (Système International d'Unités); formerly often as 'µ'.
Micromium Pers. (1811) ≡ Phlyctis.
Micromma A. Massal. (1860) = Pyrenula Ach. (1814).
Micromonospora Orskov (1923), Actinobacteria. q.v.
Micromphale Gray (1821) nom. rej. = Marasmiellus fide Antonín *et al.* (*Mycotaxon* **63**: 359, 1997; taxonomy) A proposal to reject *Micromphale* against *Marasmiellus* has not been published.
Micromucor (W. Gams) Arx (1984) = Umbelopsis fide von Arx (*Sydowia* **35**: 10, 1982; key), Yip (*TBMS* **86**: 334, 1986), Amano *et al.* (*Mycotaxon* **44**: 257, 1992; chemotaxonomy), Ruiter *et al.* (*MR* **97**: 690, 1993; chemotaxonomy) = Umbelopsis (Mortierell.) fide.
Micromucor Malchevsk. (1939) = Umbelopsis fide Kirk (*in litt.*).
Micromyces P.A. Dang. (1889), Synchytriaceae. 12 (on green algae), widespread. See Sparrow (*Aquatic Phycomycetes* Edn 2: 194, 1960; key), Kadłubowska (*Acta Mycologica* Warszawa **34**: 177, 1999; Poland).
Micromycetes, fungi having small (microscopic) sporocarps; microfungi. Most handbooks for the identification of microfungi are based on systematic groups (e.g. ascomycetes, hyphomycetes, coelomycetes), orders (*Laboulbeniales*, *Saprolegniales*), or genera (*Aspergillus*, *Penicillium*), q.v. Ellis & Ellis (*Microfungi on land plants: an identification handbook*, 1985; and *Microfungi on miscellaneous substrates*, 1988) cover most temperate groups. See also Literature.
Micromycopsidaceae Subram. (1974) = Synchytriaceae.
Micromycopsis Scherff. (1926) ? = Micromyces fide Sparrow (*Aquatic Phycomycetes* Edn 2: 1187 pp., 1960) but see, Subramanian (*Curr. Sci.* **43**: 723, 1974).
Micromyriangium Petr. (1929) = Uleomyces fide von Arx (*Persoonia* **2**: 421, 1963).
micron, see micrometre.
Micronectria Speg. (1885) nom. dub., Sordariomycetes. 3, widespread. See Rossman *et al.* (*Stud. Mycol.* **42**: 248 pp., 1999; ? placement in *Hyponectriaceae*).
Micronectriella Höhn. (1906) = Sphaerulina fide von Arx & Müller (*Stud. Mycol.* **9**, 1975), Rossman *et al.* (*Stud. Mycol.* **42**: 248 pp., 1999).
Micronectriopsis Höhn. (1918) = Linocarpon fide Hyde (*J. Linn. Soc.* Bot. **123**: 109, 1997), Rossman *et al.* (*Stud. Mycol.* **42**: 248 pp., 1999).
Micronegeria, see *Mikronegeria*.
micronematous (**micronemeous**), (1) having hyphae of small diameter; (2) (of conidiophores) similar morphologically to vegetative hyphae (cf. Micronemeae).
Micronemeae, hyphomycetes with conidiophores or conidia like the hyphae (e.g. *Oospora*), or having no hyphae. Cf. Macronemeae.
microorganism, an organism which belongs to a phylum many members of which either cannot be seen with the unaided eye or require microscopic examination and/or growth in pure culture for their identification; a microbe; includes all unicellular prokaryotes and eukaryotes, and also some multicellular eukaryotes, i.e. microscopic algae, bacteria, fungi (including yeasts), protozoa and viruses; sometimes used (incorrectly) only for prokaryotes (bacteria and viruses). See Cowan (*A dictionary of microbial taxonomy*: 162, 1978; inappropriateness of term), Zavarzin (*in* Allsopp *et al.* (Eds), *Microbial diversity and ecosystem function*: 17, 1995; concept).
Micropeltella Syd. & P. Syd. (1913) = Micropeltis fide von Arx & Müller (*Stud. Mycol.* **9**, 1975).
Micropeltidaceae Clem. & Shear (1931), ? Dothideomycetes (inc. sed.). 27 gen. (+ 25 syn.), 186 spp.

Lit.: Batista (*Publções Inst. Micol. Recife* **56**, 1959; 445 spp.), Eboh (*Sydowia* **39**: 37, 1986), Farr (*Mycol.* **79**: 97, 1987), Panwar *et al.* (*Geobios* New Rep. **10**: 41, 1991), Hsieh *et al.* (*MR* **99**: 917, 1995), Hyde (*Sydowia* **49**: 1, 1997), Cannon (*IMI Descr. Fungi Bact.* **141**: [22] pp., 1999).
Micropeltidium Speg. (1919) = Micropeltis fide von

Arx & Müller (*Stud. Mycol.* **9**, 1975).

Micropeltis Mont. (1842), Micropeltidaceae. *c.* 106, widespread (tropical). See Batista (*Publções Inst. Micol. Recife* **56**, 1959), Reynolds & Gilbert (*Aust. Syst. Bot.* **18**: 265, 2005; Australia), Hofmann & Piepenbring (*Fungal Diversity* **22**: 55, 2006; Panama), Reynolds & Gilbert (*Cryptog. Mycol.* **27**: 249, 2006; Panama).

Micropeltopsis Vain. (1921) = Lichenopeltella fide Santesson (*SA* **9**: 15, 1990).

Micropera Lév. (1846) [non *Micropera* Lindl. 1832, *Orchidaceae*] = Foveostroma fide DiCosmo (*CJB* **56**: 1682, 1978).

Microperella Höhn. (1909), anamorphic *Pezizomycotina*, St.1eH.1. 1, Japan.

Micropeziza Fuckel (1870), Helotiales. 5, Europe. See Nannfeldt (*Bot. Notiser* **129**: 323, 1976; key), Nauta & Spooner (*Mycologist* **14**: 65, 2000; UK), Raitviir (*Folia cryptog. Estonica* **40**: 43, 2003; Estonia).

Microphiale (Stizenb.) Zahlbr. (1902) ≡ Dimerella fide Hawksworth *et al.* (*Dictionary of the Fungi* edn 8, 1995), Rivas Platas *et al.* (*Fungal Diversity* **23**: 255, 2006).

Microphiodothis Speg. (1919) = Ophiodothella fide Petrak (*Sydowia* **5**: 352, 1951).

Microphlyctis J. Schröt. (1889) nom. dub., Chytridiales.

Microphoma N.F. Buchw. (1958) = Dothichiza fide Sutton (*Mycol. Pap.* **141**, 1977).

microphylline (of lichen thalli), composed of minute lobes or scales.

Microphyma Speg. (1889) = Phillipsiella fide Müller & von Arx (*Beitr. Kryptfl. Schweiz* **11** no. 2, 1962).

Micropodia Boud. (1885), ? Helotiaceae. 1 or 2, Europe. Some authors place this gen. in the *Hyaloscyphaceae*.

Microporaceae Jülich (1982) = Polyporaceae.

Microporellus Murrill (1905), Polyporaceae. 19, widespread (tropical). See Reid (*Microscopy* **32**: 452, 1975), Decock & Ryvarden (*Czech Mycol.* **54**: 19, 2002).

Microporus P. Beauv. (1805), Polyporaceae. 11, widespread. See Pegler (*The polypores [Bull. BMS Suppl.]*, 1973), Ryvarden & Johansen (*Prelim. Polyp. Fl. E. Afr.*: 429, 1980; key 7 Afr. spp.).

Micropsalliota Höhn. (1914), Agaricaceae. *c.* 40, widespread (esp. tropical). See Heinemann (*BSMF* **106**: 1, 1990; key 73 spp.), Heinmann & Leelavathy (*MR* **95**: 341, 1991; key spp. of Kerala), Vellinga (*MR* **108**: 354, 2004; phylogeny).

Micropuccinia Rostr. (1902) = Puccinia fide Dietel (*Nat. Pflanzenfam.* **6**, 1928) See, Laundon (*Mycol. Pap.* **99**, 1965).

micro-Puccinia, a microform of *Puccinia*.

Micropustulomyces R.W. Barreto (1995), anamorphic *Capnodiales*, Cac.≡ eH.1/4. 1, Brazil. See Barreto *et al.* (*MR* **99**: 779, 1995).

Micropyrenula Vain. (1921) = Microtheliopsis fide Santesson (*Svensk bot. Tidskr.* **43**: 547, 1949).

Micropyxis Seeler (1943) [non *Micropyxis* Duby 1930, *Algae*] ≡ Gelatinopsis.

microsclerid, see cystidium.

microsclerotium, a very small sclerotium, as in *Verticillium dahliae*; pseudosclerotium.

Microscopy.

Light microscopes. The compound microscope has two sets of lenses (objective and eyepiece) which magnify the object at each step. The image has a maximum final magnification of around ×1000. Stereomicroscopes (also known as dissecting microscopes) are lower power and have a long free working distance for manipulation of the specimen. For use in the field, portable instruments are available. For viewing with the compound microscope, a small sample of the subject is mounted on a glass slide with water or a mountant which may contain a stain, and overlaid with a coverslip. Common stains in mycology are cotton blue or fuchsin in lactic acid, or lactic acid alone if no stain is needed. Air bubbles can be eliminated by gentle heating. To keep the slide for examination at a later date, the coverslip must be sealed with a preparatory sealant or nail varnish.

Brightfield: The most common method of viewing subjects in the compound microscope. Light travels from beneath the subject, passing through a condenser and through the subject, illuminating translucent structures.

Differential Interference Contrast (DIC; Nomarski). This system uses special polarizing prisms arranged according to a design by Georges Nomarski; high quality objectives are required in addition to a dedicated condenser and adapter, with a concomitant rise in cost. Using such lenses and prisms results in an almost three dimensional image of the subject, particularly useful for colourless structures such as gelatinous sheaths.

Phase contrast: Special objectives and condenser are required for this technique. The image suffers from optical imperfections, such as a halo, but this effect can be of some value in drawing attention to small objects or details. Low contrast material benefits greatly from this technique, but it can be difficult to measure accurately in phase contrast.

Darkfield: Direct light is prevented from passing through the objective aperture and the image is formed from light scattered by features in the object, the detail appearing bright against a dark background.

Fluorescence: Microscopy in which the image is formed by fluorescence emitted from the specimen when illuminated with ultra-violet radiation. The sample may autofluoresce naturally, but fluorochromes such as mithramycin and calcafluor are typically used to enhance specific structures. Fluorescence microscopy has been important in detecting regions of wall synthesis in fungal hyphae and spores.

Limitations in depth of field with light microscopy can now, to some extent, be overcome when using digital cameras, by photographic the same object several times, each at a slightly different plane of focus, then using proprietory montage software to 'stitch' the resulting images into a single composite image. Among other activities, the software compares the different images and selects in favour of well-defined edges. Where such edges are not present, the result may be unsatisfactory, with 'halo' artefacts.

Electron microscopes. Scanning electron microscopy (SEM; q.v.) provides higher magnification images (to around ×80,000) of the surface morphology of the specimen, and also gives large depth of field. Transmission electron microscopy (TEM) images provide a view similar to that of light microscopy at much higher magnifications (over x100,000) to show ultrastructural details. Specimens are prepared by fixing, dehydrating, staining, embedding in resin and cutting ultrathin sections on an ultramicrotome; see also Ultrastructure.

Lit.: Beckett & Read (*in* Todd (Ed.), *Ultrastructure techniques for microorganisms*, 1986), Bradbury (*An introduction to the optical microscope* (rev. edn), 1989), Ploem & Tanke (*Introduction to fluorescence microscopy*, 1987), Rawlins (*Light microscopy*, 1992), Sanderson (*Biological microtechnique*, 1994), Slayter & Slayter (*Light and electron microscopy*, 1982).

Microscypha Syd. & P. Syd. (1919), Hyaloscyphaceae. 4, Europe. See Huhtinen (*Karstenia* **29**: 45, 1990), Cantrell & Hanlin (*Mycol.* **89**: 745, 1997; DNA), Hosoya & Otani (*Mycoscience* **38**: 171, 1997; Japan), Raitviir (*Scripta Mycol.* **20**, 2004).

Microsebacina P. Roberts (1993), Auriculariales. 2, Europe. See Roberts (*MR* **97**: 473, 1993).

Microsomyces Thaxt. (1931), Laboulbeniaceae. 2, widespread. See Tavares (*Mycol. Mem.* **9**: 627 pp., 1985), Benjamin (*Aliso* **11**: 127, 1986).

Microspatha P. Karst. (1889) = Gyalideopsis fide Sérusiaux & De Sloover (*Veröff. geobot. Inst., Zürich* **91**: 260, 1986), Lücking *et al.* (*Lichenologist* **37**: 123, 2005).

Microsphaera Lév. (1851) ? = Erysiphe fide Braun & Takamatsu (*Schlechtendalia* **4**: 3, 2000), Braun *et al.* (*The Powdery Mildews* A Comprehensive Treatise: 13, 2001; review), Takamatsu (*Mycoscience* **45**: 147, 2004; phylogeny).

Microsphaeropsis Höhn. (1917), anamorphic *Paraphaeosphaeria*, Cpd.0eP.15. 29, widespread. Almost certainly polyphyletic. See Sutton (*Mycol. Pap.* **123**, 1971), Morgan-Jones (*CJB* **52**: 2575, 1974), Morgan-Jones & White (*Mycotaxon* **30**: 177, 1987; ultrastr. conidiogenesis *M. concentricum*), Guarro *et al.* (*Medical Mycol.* **37**: 133, 1999; etiological agent of human skin infection), Câmara *et al.* (*MR* **107**: 516, 2003; teleomorph), Boerema *et al.* (*Phoma Identification Manual* Differentiation of Specific and Infra-Specific Taxa in Culture: 470 pp., 2004), Verkley *et al.* (*Stud. Mycol.* **50**: 323, 2004).

Microsphaeropsis Sousa da Câmara, Oliveira & Luz (1936) nom. dub., anamorphic *Pezizomycotina*.

Microspora Velen. (1934), ? Helotiales. 1, former Czechoslovakia.

microsporangium, secondary sporangium formed from a zoospore cyst, either without any mitoses, or with very few mitoses.

microspore (1) a small spore, where there are spores of two sizes; (2) a spore from a microsporangium (q.v.).

Microsporella Höhn. (1918) ? = Microsphaeropsis Höhn. fide Sutton (*Mycol. Pap.* **141**, 1977).

Microsporidia (**Microspora** V. Sprague 1977), Eukaryota. Earlier classified as *Protozoa* (q.v.), now usually as *Fungi* (q.v.). More than 1300 species and about 170 genera (Bulla & Cheng (Eds), 1976, 1977; Larsson, 1999; Wittner & Weiss (Eds), 1999; Canning & Vávra, 2000; Lom *et al.* (Eds), 2005). Supergeneric classification remains unclear. Eukaryotic, spore-forming, parasitic organisms lacking flagellae and peroxisomes. With polar vesicles (mitosomes) interpreted as remainders of mitochondria (Willams *et al.*, 2002; Vávra, 2005), mitochondrial enzymes (a chaperon protein, HSP70) (Germot *et al.*, 1997; Hirt *et al.*, 1997) and genes related to mitochondrial functions (alpha and beta subunits of PDH) (Fast & Keeling, 2001). With 70S ribosomes, chitin in the spore wall, and atypical Golgi apparatus (Ishihara & Hayashi, 1968; Curgy *et al.*, 1980; Vávra & Larsson, 1999). Life cycle usually comprising vegetative reproduction (called merogony) and sporogony. Nuclei either isolated (monokaryotic) or coupled (diplokaryotic) with closed intranuclear pleuromitosis (Hollande, 1972). In some microsporidia diplokaryotic merogony is followed by a sporogony with isolated nuclei and with sexual processes and meiosis in the shift from diplokaryotic to monokaryotic condition at the beginning of sporogony (Becnel & Andreadis, 1999). Spores of various shapes, with a usually coiled polar filament with layered internal structure, which during hatching is ejected and reorganized to an infection tube through which the infectious cell (the sporoplasm) is injected into the host cell (Weidner 1976, 1982).

Microsporidians are obligate parasites of all groups of animals with a few species parasitizing protozoa. Some species have complex life cycles with 2 or 3 hosts, and producing 2 or 3 different types of spores. They most commonly infect insects, crustaceans and fish (Canning & Lom, 1986; Becnel & Andreadis, 1999). They have been extensively studied as biocontrol agents, and one species (*Nosema locustae*) is commercially available for the control of locusts and grasshoppers (Canning, 1981; Shah & Goettel (Eds), 1999). Of economic importance are among others *Nosema apis* (a cosmopolitan parasite of honey bees) and *Nosema bombycis* (a parasite of silk worms). At least 8 species of microsporidians cause disease in immuno-compromized humans (Kotler & Orenstein, 1999).

Early phylogenies that included microsporidians, when based on ultrastructure and SSU rRNA, had these organisms at the base of the Tree of Life (Cavalier-Smith, 1988; Patterson, 1994; Vossbrinck *et al.*, 1987). Studies of alpha-and beta-tubulin have indicated that microsporidians either are related to organisms higher in the phylogenetic tree (Canning, 1998; Edlind, 1998) or to the *Fungi* (Hirt *et al.*, 1997; Keeling & McFadden, 1998; Keeling *et al.*, 2000; van de Peer *et al.*, 2000). The last alternative is also supported by other genes.

In the official classifications by the Society of Protozologists microsporidians have been ranked in the Subphylum *Cnidosporidia* of the Phylum *Protozoa* (Honigberg *et al.*, 1964), as the Phylum *Microspora* (Levine *et al.*, 1980) or included in (the Kingdom) *Opisthokonta* together with *Fungi*, *Metazoa* and some other organisms (Adl *et al.*, 2005). Microsporidians were for a period included in the Kingdom *Archezoa* (Cavalier-Smith, 1988, 1993) or split between the Infrakingdom *Alveolata* (Subphylum *Manubrispora* – Family *Metchnikovellidae*) and the Kingdom *Fungi* (Phylum *Microsporidia* – all other microsporidians) (Cavalier-Smith, 1998).

The recent classifications treat the supergeneric ranking differently and no classification can accomodate all established taxa (Sprague, 1977; Weiser, 1977; Sprague, 1982; Issi, 1986; Canning 1989; Sprague *et al.*, 1992).

Microsporidiopsis Schereschewsky (1925), Microsporidia. 1. See Schereschewsky (*Ark. Russk. Protistol. Obšč* **3**: 137, 1925).

Microsporidium Balbiani (1884), Microsporidia. 8.

Microsporon (orthographic variant), see *Microsporum*.

Microsporonites K.P. Jain (1968), Fossil Fungi ? Fungi. 1 (Triassic), Argentina. See Pirozynski &

Weresub *in* Kendrick (Ed.) (*The Whole Fungus* **1**: 93, 1979).

Microsporum Gruby (1843), anamorphic *Arthroderma*, Hsy.≡ eH.2. 15 (on humans and other mammals, causing **microsporoses**), widespread. *M. audouinii* (tinea capitis in humans, esp. children), *M. canis* (ringworm in cats, dogs and humans). See Conant (*Arch. Derm. Syph. Chicago* **36**: 781, 1937), Morace *et al.* (*Mycopathologia* **94**: 53, 1986; serotyping by monoclonal antibodies), Vismer *et al.* (*Mycopathologia* **98**: 149, 1987; ultrastr. conidia), Brasch (*Mycoses* **32**: 33, 1989; polymorphic conidia), Tucker & Noble (*J. Med. Vet. Mycol.* **28**: 117, 1990; protein profiles), Demange *et al.* (*J. Med. Vet. Mycol.* **30**: 301, 1992; review), Morganti *et al.* (*European Journal of Epidemiology* **8**: 340, 1992; morphology, biochemistry), Tucker (*Mycoses* **35**: 147, 1992; biotypes), Kawasaki *et al.* (*Mycopathologia* **130**: 11, 1995; phylogeny), Mancianti & Papini (*J. Mycol. Médic.* **7**: 87, 1997; mating types), Kano *et al.* (*Mycoses* **41**: 139, 1998; RAPDs), Simpanya *et al.* (*Mycoses* **41**: 501, 1998; isozymes), Simpanya *et al.* (*Medical Mycology* **36**: 255, 1998; isozymes, New Zealand), Gräser *et al.* (*Medical Mycology* **37**: 105, 1999; phylogeny), Gräser *et al.* (*Medical Mycology* **38**: 143, 2000; phylogeny, morphology), Kano *et al.* (*Veter. Pathol.* **78**: 85, 2001; chitin synthase gene), Liu *et al.* (*Medical Mycology* **39**: 215, 2001; PCR diagnostics), Kaszubiak *et al.* (*Infect. Genet. Evol.* **4**: 179, 2004; population structure, evolution), Untereiner *et al.* (*Mycol.* **96**: 812, 2004; phylogeny), Brilhante *et al.* (*Journal of Applied Microbiology* **99**: 776, 2005; Brazil), Sharma *et al.* (*Antonie van Leeuwenhoek* **89**: 197, 2006; phylogeny).

Microstelium Pat. (1899) = Gomphillus fide Lücking (*Bryologist* **110**: 475, 2007).

Microstella K. Ando & Tubaki (1984), anamorphic *Agaricomycotina*, Hso.1bH.10. 1 (aquatic), Japan. See Ando & Tubaki (*TMSJ* **25**: 34, 1984).

Microsticta Desm. (1849) = Schizothyrium fide Müller & von Arx (*Beitr. Kryptfl. Schweiz* **11** no. 2, 1962).

Microstoma Auersw. (1860) = Valsa Fr. fide Hawksworth *et al.* (*Dictionary of the Fungi* edn 8, 1995).

Microstoma Bernstein (1852), Sarcoscyphaceae. 3 or 8, widespread (esp. temperate). See Boedijn (*Sydowia* **5**: 211, 1951), Otani (*TMSJ* **21**: 149, 1980), Otani (*Rep. Tottori mycol. Inst.* **28**: 251, 1990; Japanese spp.), Landvik *et al.* (*Nordic Jl Bot.* **17**: 403, 1997; DNA), Harrington (*Mycol.* **90**: 235, 1998), Harrington *et al.* (*Mycol.* **91**: 41, 1999; phylogeny), Harada & Kudo (*Mycoscience* **41**: 275, 2000), Wang (*Mycotaxon* **89**: 119, 2004; Taiwan).

Microstroma Niessl (1861), Microstromataceae. 4, widespread. See von Arx (*Gen. Fungi Sporul. Cult.* Edn 3: 104, 1981), Begerow *et al.* (*CJB* **75**: 2045, 1998), Begerow *et al.* (*MR* **105**: 798, 2001).

Microstromataceae Jülich (1982), Microstromatales. 2 gen. (+ 2 syn.), 5 spp.

Lit.: Bauer *et al.* (*CJB* **75**: 1273, 1997), Begerow *et al.* (*CJB* **75**: 2045, 1998), Begerow *et al.* (*MR* **105**: 809, 2001), Begerow *et al.* (*Mycol. Progr.* **1**: 187, 2002), Beer *et al.* (*Stud. Mycol.* **55**: 289, 2006).

Microstromatales R. Bauer & Oberw. (1997). Exobasidiomycetes. 3 fam., 4 gen., 10 spp. Fams:

(1) **Microstromataceae**
(2) **Quambalariaceae**
(3) **Volvocisporiaceae**

For *Lit.* see under fam.

Microtetraspora Thiemann, Pagani & Beretta (1968), Actinobacteria. q.v.

Microthallites Dilcher (1965), Fossil Fungi ? Microthyriaceae. 3 (Eocene, Miocene), India; USA. See Selkirk (*Proc. Linn. Soc. N. S. W.* **100**: 70, 1975), Blanz (*Z. Mykol.* **44**: 91, 1978; annellides in), Hansen (*Grana* **19**: 67, 1980) = Phragmothyrites (Fossil fungi) fide.

Microthecium Corda (1842) = Melanospora Corda. See also *Sphaerodes*. fide von Arx (*Gen. Fungi Sporul. Cult.* Edn 3, 1981).

Microthelia Körb. (1855) nom. rej. = Anisomeridium fide Hawksworth & Sherwood (*Taxon* **30**: 339, 1981), Hawksworth (*Bull. Br. Mus. nat. hist.* Bot. **14**: 1, 1985; redisp. names).

Microtheliomyces Cif. & Tomas. (1953) = Leptorhaphis fide Aguirre-Hudson (*Bull. Br. Mus. nat. hist.* Bot. **21**: 85, 1991).

Microtheliopsidaceae O.E. Erikss. (1981), ? Dothideomycetes (inc. sed.) (L). 1 gen. (+ 5 syn.), 1 spp.

Lit.: Eriksson (*Op. Bot.* **60**, 1981), Lücking (*Mycotaxon* **51**: 69, 1994), Aptroot (*Lichenologist* **30**: 501, 1998), Aptroot (*Lichenologist* **30**: 515, 1998), Lücking *et al.* (*Lichenologist* **30**: 121, 1998).

Microtheliopsidomyces Cif. & Tomas. (1953) ≡ Microtheliopsis.

Microtheliopsis Müll. Arg. (1890), Microtheliopsidaceae (L). 1, C. & S. America; Australasia. See Santesson (*Symb. bot. upsal.* **12** no. 1: 1, 1952), Lücking (*Mycotaxon* **51**: 69, 1994), Aptroot (*Lichenologist* **30**: 515, 1998; posn).

Microthia Gryzenh. & M.J. Wingf. (2006), Cryphonectriaceae. 2, C. America; Azores. See Gryzenhout & Wingfield (*Stud. Mycol.* **55**: 44, 2006).

Microthyriaceae Sacc. (1883), Microthyriales. 54 gen. (+ 40 syn.), 278 spp.

Lit.: Stevens & Ryan (*Ill. biol. monogr.* **17** no. 2: 1, 1939), Doidge (*Bothalia* **4**: 273, 1942; S. Afr.), Swart (*TBMS* **87**: 81, 1986), Kirk & Spooner (*MR* **92**: 335, 1989), Spooner & Kirk (*MR* **94**: 223, 1990), Holm & Holm (*Nordic Jl Bot.* **11**: 675, 1991), Crous & Kendrick (*CJB* **72**: 63, 1994), Hariharan *et al.* (*Lichenologist* **28**: 294, 1996), Hosagoudar *et al.* (*Mycotaxon* **58**: 489, 1996), Matzer (*Mycol. Pap.* **171**: 202 pp., 1996), Ramaley (*Mycotaxon* **70**: 7, 1999).

Microthyriacites Cookson (1947), Fossil Fungi, Microthyriaceae. 8 (Cretaceous, Tertiary), widespread. = Phragmothyrites (Fossil fungi) fide Selkirk (1975).

Microthyriales G. Arnaud (1918), ? Dothideomycetes. 3 fam., 62 gen., 323 spp. Superficial mycelium indistinct and often absent. Ascomata small, strongly flattened, superficial, with a central ostiole or slit(s); peridium brown, the upper wall composed of radiating rows of ± isodiametric cells, the basal layer hyaline, poorly developed or absent. Interascal tissue of pseudoparaphyses, often inconspicuous and/or deliquescing. Asci saccate, fissitunicate, without a clear ocular chamber, not blueing in iodine. Ascospores hyaline or brown, transversely septate, sometimes ciliate, without a sheath. Anamorphs hardly studied. Saprobic or epiphytic on leaves and stems, cosmop. Poorly known, many of the genera are likely to be artificial. The Order as circumscribed here is almost certainly polyphyletic, but the fams. have much shared morphology. Fams.:

(1) **Aulographaceae**
(2) **Leptopeltidaceae**
(3) **Microthyriaceae**

Lit.: Doidge (*Bothalia* **4**: 273, 1942; S. Afr.), Stevens & Ryan (*Ill. biol. monogr.* **17** (2), 1939).

Microthyriella Höhn. (1909) = Schizothyrium fide Müller & von Arx (*Beitr. Kryptfl. Schweiz* **11** no. 2, 1962).

Microthyrina Bat. (1960) = Microthyrium fide von Arx & Müller (*Stud. Mycol.* **9**, 1975).

Microthyriolum Speg. (1917) = Ferrarisia fide Müller & von Arx (*Beitr. Kryptfl. Schweiz* **11** no. 2, 1962).

Microthyris Clem. (1931) ≡ Lichenopeltella.

Microthyrites Pampal. (1902), Fossil Fungi. 1 (Miocene), Italy.

Microthyrium Desm. (1841), Microthyriaceae. Anamorph *Leptothyrium*. 57, widespread. See Ellis (*TBMS* **67**: 382, 1977; key 13 Br. spp.), Doi & Uemura (*Bull. natn. Sci. Mus.* Tokyo, B **11**: 127, 1985), Spooner & Kirk (*MR* **94**: 223, 1990), Holm & Holm (*Nordic Jl Bot.* **11**: 675, 1991), Ramaley (*Mycotaxon* **70**: 7, 1999).

Microtrichella Maessen (1955) = Arthromitus fide Manier & Lichtwardt (*Annls Sci. Nat.* Bot., sér. 12 **9**: 519, 1968).

Microtrichophyton (Castell. & Chalm.) Neveu-Lem. (1921) = Trichophyton fide Hawksworth *et al.* (*Dictionary of the Fungi* edn 8, 1995).

Microtyle Speg. (1919), anamorphic *Pezizomycotina*, Cpd.1eH.?. 1, S. America. See Müller & von Arx (*Beitr. Kryptfl. Schweiz* **11** no. 2: 830, 1962).

Microtypha Speg. (1910) = Arthrinium fide Subramanian (*Hyphomycetes – an Account of Indian Species, except Cercosporae*, 1971).

Microxiphium (Harv. ex Berk. & Desm.) Thüm. (1879), anamorphic *Dennisiella*, Hsy.0eH.?. 2, widespread. See Hughes (*Mycol.* **68**: 693, 1976), Reynolds & Gilbert (*Aust. Syst. Bot.* **18**: 265, 2005; Australia).

Microxyphiella Speg. (1918), anamorphic *Pezizomycotina*, Cpd.1eH.?. 7, widespread (tropical). See Hughes (*Mycol.* **68**: 788, 1976).

Microxyphiomyces Bat., Valle & Peres (1961) = Tricharia Fée fide Lücking *et al.* (*Lichenologist* **30**: 121, 1998).

Microxyphiopsis Bat. (1963), anamorphic *Pezizomycotina*, Cpd.1eH.?. 2, Brazil. See Batista (*Quad. Lab. crittogam., Pavia* **31**: 103, 1963).

Microxyphium, see *Microxiphium*.

Micula Duby (1858) ? = Foveostroma fide Sutton (*The Coelomycetes*, 1980).

Micularia Boedijn (1961), Elsinoaceae. 1, Java. See Boedijn (*Persoonia* **2**: 67, 1961).

Midotiopsis Henn. (1902), Helotiales. 2, America (tropical).

Midotis Fr. (1828) ? = Wynnella fide Nannfeldt (*TBMS* **23**: 239, 1939).

migration pseudoplasmodium (of *Acrasiales*), the migration stage after the massing of the myxamoebae (Raper, *J. agric. Res.* **50**: 135, 1935).

Mikronegeria Dietel (1899), Mikronegeriaceae. 3 (on *Araucaria* (*Araucariaceae*), *Austrocedrus* (*Cupressaceae*) or *Phyllocladus* (*Podocarpaceae*) (0, I); on *Nothofagus* (*Fagacae* or *Fuchsia* (*Onagraceae*) (II, III)), Argentina; Chile; New Zealand. See Butin (*Phytopath. Z.* **64**: 242, 1969), Peterson & Oehrens (*Mycol.* **70**: 321, 1978), Aime (*Mycoscience* **47**: 112, 2006).

Mikronegeriaceae Cummins & Y. Hirats. (1983), Pucciniales. 4 gen. (+ 1 syn.), 13 spp.

Lit.: Sato & Sato (*TBMS* **85**: 223, 1985), Cummins & Hiratsuka (*Illustr. Gen. Rust Fungi* edn 3: 225 pp., 2003), Aime (*Mycoscience* **47**: 112, 2006).

miktohaplont, a haplont made up of cells having genotypically different nuclei (Kniep, 1928); cf. isohaplont.

Miladina Svrček (1972), Pyronemataceae. Anamorph *Actinosporella*. 1, Europe; N. America. See Pfister & Korf (*CJB* **52**: 1643, 1974), Descals & Webster (*TBMS* **70**: 466, 1978), Yao & Spooner (*MR* **99**: 1525, 1995), Hansen & Pfister (*Mycol.* **98**: 1029, 2006; phylogeny).

mildew (1) a plant disease in which the pathogen is seen as a growth on the surface of the host. A **powdery ('true') -** is caused by one of the *Erysiphaceae* (e.g. the American (*Sphaerotheca mors-uvae*) and European - (*Microsphaera grossulariae*) of *Ribes*); a **downy ('false') -** by one of the *Peronosporaceae* (the first may be controlled by sulphur, the second by copper fungicides); a **dark -**, or **black -**, by one of the *Meliolales* or *Capnodiaceae*; (2) the staining, and frequently the breaking up, of cloth and fibres, paint, etc., by fungi and bacteria (cf. mould); (3) a fungus causing (1) or (2).

Milesia F.B. White (1878), anamorphic *Milesina*. 25, widespread. Anamorph name for (II). Teleomorphs also found in *Phakopsora* (Phakops.). See Moss (*Ann. Bot.* **40**: 813, 1926; morphology), Ono *et al.* (*MR* **96**: 828, 1992), Berndt *et al.* (*CJB* **72**: 1084, 1994; ultrastr.).

Milesina Magnus (1909), Pucciniastraceae. Anamorph *Milesia*. 36 (on *Abies* (0, I; where known) (*Pinaceae*); on Pteridophyta (II, III)), widespread (temperate). (22 *Uredo*/*Milesia* species may belong here). See Faull (*Contr. Arnold Arbor.* **2**, 1932), Faull (*J. Arnold Arbor.* **15**: 50, 1934; as '*Milesia*'), Hiratsuka (*Revision of taxonomy of the Pucciniastreae*, 1958), Cummins & Hiratsuka (*Illustr. Gen. Rust Fungi rev. edit.*, 1983), Ono *et al.* (*MR* **96**: 828, 1992; expanded anamorph concept (*Milesia*, II) to include phakopsoraceous species on legumes), Rodríguez & Buriticá (*ASCOLFI Informa* **28**: 12, 2002; Colombia), Wingfield *et al.* (*Australas. Pl. Path.* **33**: 327, 2004; phylog.).

Milesites Ramanujam & Ramachar (1980), Fossil Fungi, Pucciniastraceae. 1. See Ramanujam & Ramachar (*Records of the Geological Survey of India* **113**: 81, 1980).

Military use of fungi. see Bioterrorism and fungi.

milk-cap, basidioma of *Lactarius* spp.

Millardet (Pierre Marie Alexis; 1838-1902; France). Student, University of Heidelberg and University of Freiberg; Professor of Botany, University of Strasbourg (1869); Professor of Botany, University of Nancy (1872); Professor of Botany, University of Bordeaux (1876). Noted for his work protecting vinyards from pests; invented Bordeaux mixture, the world's first commercial fungicide, and used it successfully against grape downy mildew (*Plasmopara viticola*). *Publs*. Sur le traitement du mildew et du rot. *Comptes Rendus Hebdomadaire des Séances de l'Académie des Sciences, Paris* (1885) *Biogs, obits etc.* Ayres (*Mycologist* **18**: 23, 2004).

Millerburtonia Cif. (1951), anamorphic *Plagiostoma*, Cac.0fH.?. 1, Venezuela. See Ciferri (*Mycopath. Mycol. appl.* **6**: 26, 1951).

Milleria Peck (1879) [non *Milleria* L. 1753, *Compositae*] = Testicularia fide Dietel (*Nat. Pflanzenfam.* **6**, 1928).

Milleromyces T.N. Taylor, Hass & W. Remy (1993), Fossil Fungi. 1 (Devonian), British Isles. See Taylor *et al.* (*Mycol.* **84**: 902, 1992).

Milospium D. Hawksw. (1975), anamorphic *Pezizomycotina*, Hso.0eP.1. 1 (on lichens), Europe. See Hawksworth (*Nova Hedwigia* **79**: 373, 1984), Hawksworth (*Taxon* **55**: 528, 2006; nomencl.).

Milowia Massee (1884) nom. dub. ? = Thielaviopsis fide Wakefield & Bisby (*TBMS* **25**: 63, 1941), Nag Raj & Kendrick (*Monograph of Chalara and allied genera*: 43, 1975).

Milowiaceae Nann. (1934) = Ceratocystidaceae.

Miltidea Stirt. (1898), Miltideaceae (L). 1, Australasia. See Hafellner (*Beih. Nova Hedwigia* **79**: 241, 1984), Lumbsch (*J. Hattori bot. Lab.* **83**, 1997), McCarthy & Mallett (*Flora of Australia* **56** A, 2004).

Miltideaceae Hafellner (1984), Lecanorales (L). 1 gen., 1 spp.

Lit.: Lumbsch (*J. Hattori bot. Lab.* **83**: 1, 1997).

Mimema H.S. Jacks. (1931), ? Uropyxidaceae. 2 (on *Cassia*, *Dalbergia* (*Leguminosae*), S. America. See Dianese *et al.* (*MR* **98**: 786, 1994).

Mimeomyces Thaxt. (1912), Laboulbeniaceae. 16, widespread. See Nannfeldt (*Svensk bot. Tidskr.* **43**: 468, 1949).

Mindoa Petr. (1949), anamorphic *Pezizomycotina*, Cpt.1eH.?. 1, S. America.

MINE, Microbial Information Network Europe; see Genetic resource collections.

Miniancora Marvanová & Bärl. (1989), anamorphic *Pezizomycotina*, Hso.1bH.15. 1 (aquatic), Canada. See Marvanová & Bärlocher (*Mycotaxon* **35**: 86, 1989).

minimal medium, the simplest chemically defined medium on which the wild type (prototroph) of a species will grow and which must be supplemented by one or more specific substances for the growth of auxotrophic mutants derived from the wild type.

Minimedusa Weresub & P.M. LeClair (1971), anamorphic *Agaricomycetes*. 1, Cuba. See Weresub & LeClair (*CJB* **49**: 2210, 1971), Diederich & Lawrey (*Mycol. Progr.* **6**: 61, 2007; phylogeny), Lawrey *et al.* (*Mol. Phylog. Evol.* **44**: 778, 2007; phylogeny).

Minimelanolocus R.F. Castañeda & Heredia (2001), anamorphic *Pezizomycotina*, Hso.?.?. 14, widespread. See Castañeda Ruiz *et al.* (*Cryptog. Mycol.* **22**: 7, 2001), Castañeda Ruíz *et al.* (*Mycotaxon* **85**: 231, 2003).

Minimidochium B. Sutton (1970), anamorphic *Pezizomycotina*, Hsp.0eH.15. 3, widespread (tropical). See Sutton (*CJB* **47**: 2095, 1969), Cabello *et al.* (*MR* **102**: 383, 1998; Argentina).

Minksia Müll. Arg. (1882), Roccellaceae (L). 4, widespread (tropical).

Minostroscyta Hjortstam & Ryvarden (2001), Agaricomycetes. 1, Colombia. See Hjortstam & Ryvarden (*Mycotaxon* **79**: 194, 2001).

Mintera Inácio & P.F. Cannon (2003), Parmulariaceae. 1, S. America. See Inácio & Cannon (*MR* **107**: 86, 2003).

Minutoexcipula V. Atienza & D. Hawksw. (1994), anamorphic *Pezizomycotina*, Hsp.1eP.19. 4 (lichenicolous), widespread. See Atienza & Hawksworth (*MR* **98**: 587, 1994), Hafellner (*Herzogia* **10**: 1, 1994), Atienza (*Biblthca Lichenol.* **82**: 141, 2001; Spain).

Minutophoma D. Hawksw. (1981), anamorphic *Pezizomycotina*, Cpd.0eH.15. 1 (on *Chrysothrix*), British Isles. See Hawksworth (*Bull. Br. Mus. nat. Hist.* Bot. **9**: 44, 1981).

Mirandia Toro (1934) nom. dub., Fungi. See Petrak (*Sydowia* **5**: 328, 1951), Müller & von Arx (*Beitr. Kryptfl. Schweiz* **11** no. 2, 1962) See.

Mirandina G. Arnaud ex Matsush. (1975), anamorphic *Helotiales*, Hso.0fH.10. 9, Europe; Papua New Guinea. See Hoog (*Stud. Mycol.* **26**: 1, 1985; synonymy with *Dactylaria*), Goh & Hyde (*MR* **101**: 1265, 1997).

Mirannulata Huhndorf, F.A. Fernández, A.N. Mill. & Lodge (2003), Sordariomycetes. 1, neotropics. See Huhndorf *et al.* (*Sydowia* **55**: 173, 2003).

MIRCEN. Microbial Resource Centre. Since 1975 a network of centres for information on the resources available for diverse aspects of microbiology has been established throughout the world under UNESCO (United Nations Educational, Scientific, and Cultural Organization)/UNEP (United Nations Environment Programme). The mycology MIRCEN is based at CABI Europe, Egham (formerly the site of IMI). See Kirsop & Da Silva (*in* Hawksworth & Kirsop (Eds), *Filamentous fungi*: 173, 1988), *MIRCEN NEWS*.

Mirimyces Nag Raj (1993), anamorphic *Pezizomycotina*, St.0eH.15. 1, Cuba. See Nag Raj (*Coelomycetous Anamorphs with Appendage-bearing Conidia*: 477, 1993).

Miriquidica Hertel & Rambold (1987), Lecanoraceae (L). 23, widespread (esp. arctic-alpine). See Rambold & Schwab (*Nordic Jl Bot.* **10**: 117, 1990; rust-coloured spp.), Timdal (*Bryologist* **96**: 616, 1993), Rambold *et al.* (*Mycotaxon* **58**: 319, 1996; Mexico), Andreev (*Nov. sist. Niz. Rast.* **34**: 82, 2001; Russia), Owe-Larsson & Rambold (*Biblthca Lichenol.* **78**: 335, 2001; sorediate spp.), Andreev (*Biblthca Lichenol.* **88**: 15, 2004; Russia), Arup *et al.* (*Mycol.* **99**: 42, 2007; sister group relations with *Parmeliaceae*).

Mischoblastia A. Massal. (1852) = Rinodina fide Rambold *et al.* (*Pl. Syst. Evol.* **192**: 31, 1994), Marbach (*Biblthca Lichenol.* **74**, 2000).

mischoblastiomorph (of ascospores), ones similar to the polarilocular type but either without a septum or only with an incomplete septum.

Mischolecia M. Choisy (1931) = Rinodina fide Hawksworth *et al.* (*Dictionary of the Fungi* edn 8, 1995).

Misgomyces Thaxt. (1900), Laboulbeniaceae. 4, widespread. See Tavares & Balazuc (*Mycol.* **69**: 1069, 1977), Tavares (*Mycol. Mem.* **9**: 627 pp., 1985), Santamaría (*Mycol.* **87**: 697, 1995).

miso, an oriental food product, used for soups and as a flavouring agent, composed of rice and cereals + soybeans fermented by *Aspergillus oryzae* and *Saccharomyces rouxii* (Hesseltine, *Mycol.* **57**: 168, 1965); see Fermented food and drinks.

Mison Adans. (1763) nom. rej. = Phellinus fide Donk (*Persoonia* **1**: 173, 1960).

Mites. Species of *Tyrophagus* and *Tarsonemus* sometimes infest fungus cultures. They are a serious problem as they transfer spores from culture to culture, causing cross-contamination that may be irredeemable and often introduce bacteria into the culture.

To prevent mite invasion, work surfaces must be kept clean and cultures protected from aerial infestation by storage in cabinets or incubators. To clean work surfaces, wash with a non-fungicidal acaricide (0.2% v/v Actelic was used at IMI). Mites can be detected by scrutiny of cultures at twice-weekly intervals; they appear as white objects, just detectable with the naked eye. Ragged colony margins or growth of contaminant fungi or bacteria forming trails mey denote their presence. If mites are detected, contaminated cultures should be destroyed by autoclaving (121°C for 15 min.).

Where contamination is not severe and the original fungus is sporulating it will normally be possible to recover an uncontaminated culture. If mite-infested cultures cannot be reisolated, they can be stored at –18°C for 1-3 days before being subcultured. This procedure will kill both eggs and adult mites. Fungi which would not survive short-term cold storage may be covered with a layer of mineral oil and subcultured after 24 hours. However, this latter procedure does not kill the eggs.

A cigarette paper fastened on the necks of universal bottles or test-tubes provides an effective barrier against mites. The cigarette papers are sterilized by dry heat at 180°C for 2 h. and aseptically stuck to the rim of the bottle using a copper sulphate/gelatin glue (20 g gelatin; 2 g copper sulphate; 100 mls distilled water) (Snyder & Hansen, 1946).

Standing the cultures on liquid paraffin and treating the wool plugs with kerosene (crude, such as 'tractor vapourizing oil') has proved an effective control. See Area Leão *et al.* (*Mem. Inst. Oswaldo Cruz* **42**: 559, 1946), Smith (*Mycol.* **59**: 600, 1967). Cf. insects, invertebrates and lichens.

mitochondrion, membrane bound intracellular organelle containing enzymes and electron transport chains for oxidative respiration of organic acids and the concomitant production of ATP. Possesses DNA, messenger RNA and small ribosomes and thus capable of protein synthesis.

Mitochytridium P.A. Dang. (1911), ? Endochytriaceae. 2, Europe; N. America. See Couch (*J. Elisha Mitchell scient. Soc.* **51**: 293, 1935).

Mitopeltis Speg. (1921), Micropeltidaceae. 1, S. America. See Petrak & Sydow (*Annls mycol.* **33**: 169, 1935).

Mitoplistophora Codreanu (1966), Microsporidia. 1.

Mitosis in fungi. For reviews see Aist & Morris (*Fungal Genetics and Biol.* **27**: 1, 1999), Fuller (*Internat. Rev. Cytology* **45**: 113, 1976), Heath (*Mycol.* **72**: 229, 1980).

mitosporangium, a thin-walled diploid sporangium of certain *Blastocladiales* producing uninucleate diploid zoospores (**mitospores**) (Emerson, 1950); cf. meiosporangium.

mitospore (1) a spore from a mitosporangium (q.v.); (2) (of ascomycetes and basidiomycetes), any non-basidiosporous or -ascosporous propagule (see Kendrick & Watling, *in* Kendrick (Ed.), *The whole fungus* **2**: 473, 1979). See Anamorphic fungi.

Mitosporic fungi, See Anamorphic fungi.

Mitosporium Clem. & Shear (1931) ≡ Aciculosporium.

Mitrasphaera Dumort. (1822) nom. rej. prop. = Cordyceps fide Hawksworth *et al.* (*Dictionary of the Fungi* edn 8, 1995).

mitrate (**mitriform**), mitre-like in form.

Mitremyces Nees (1816) = Calostoma fide Stalpers (*in litt.*).

mitriform, see mitrate.

Mitrophora Lév. (1846) = Morchella fide Seaver (*North American Cup Fungi* (Operculates), 1928), Eckblad (*Nytt Mag. Bot.* **15**: 1, 1968), Wipf *et al.* (*Can. J. Microbiol.* **45**: 769, 1999).

Mitrorhizopeltis Bat. & Cavalc. (1964) = Tracylla fide Nag Raj (*CJB* **53**: 2435, 1975).

Mitrula Fr. (1821), Helotiales. 8, widespread (temperate). See Verkley (*Persoonia* **15**: 405, 1994; ultrastr.), Wang *et al.* (*Am. J. Bot.* **92**: 1565, 2005; phylogeny, life history), Wang *et al.* (*Mycol.* **98**: 1065, 2006; phylogeny).

Mitrula Pers. (1794) = Heyderia fide Hawksworth *et al.* (*Dictionary of the Fungi* edn 8, 1995).

Mitrulinia Spooner (1987), Sclerotiniaceae. 1, southern hemisphere. See Spooner (*Biblthca Mycol.* **116**, 1987).

Mitruliopsis Peck (1903) = Spathularia fide Durand (*Annls mycol.* **6**: 387, 1908).

Mitteriella Syd. (1933), anamorphic *Schiffnerula*, Hso.≡ eP.10. 3, India; Uganda. See Hughes (*CJB* **61**: 1727, 1984), Hosagoudar (*Zoos' Print Journal* **18**: 1071, 2003).

Miuraea Hara (1948), anamorphic *Mycosphaerella*, Hso.≡ eH.?. 3, Asia. See Hara (*Manual of pests & diseases*: 779, 1948), Deighton (*Mycol. Pap.* **133**: 3, 1973; relationship to *Cercosporella*), Braun (*Monogr. Cercosporella, Ramularia Allied Genera (Phytopath. Hyphom.)* **1**: 218, 1995; key).

Mixia C.L. Kramer (1959), Mixiaceae. 1, Japan; USA. See Kramer (*Stud. Mycol.* **30**: 151, 1987), Nishida *et al.* (*CJB* **73** Suppl. 1: S660, 1995; posn), Sjamsuridzal *et al.* (*J. gen. appl. Microbiol.* Tokyo **48**: 121, 2002; phylogeny).

Mixiaceae C.L. Kramer (1987), Mixiales. 1 gen., 1 spp.
Lit.: Kramer (*Stud. Mycol.* **30**: 151, 1987), Nishida *et al.* (*CJB* **73** Suppl. 1: S660, 1995; posn), Sjamsuridzal *et al.* (*Mycoscience* **40**: 21, 1999), Sjamsuridzal *et al.* (*J. gen. appl. Microbiol.* Tokyo **48**: 121, 2002).

Mixiales R. Bauer, Begerow, J.P. Samp., M. Weiss & Oberw. (2006). Mixiomycetes. 1 fam., 1 gen., 1 spp. Fam.:
Mixiaceae.
For *Lit.* see under fam.

Mixiomycetes R. Bauer, Begerow, J.P. Samp., M. Weiss & Oberw. (2006). Pucciniomycotina. 1 ord., 1 fam., 1 gen., 1 spp. Ord.:
Mixiales
For *Lit.* see fam.

Mixtoconidium Etayo (1995), anamorphic *Pezizomycotina*, Cpd.1eH.19. 1 (on lichens), Spain. See Etayo (*Mycotaxon* **53**: 425, 1995).

Mixtura O.E. Erikss. & J.Z. Yue (1990), Phaeosphaeriaceae. 1, S. America. See Eriksson & Yue (*Mycotaxon* **38**: 201, 1990).

Miyabella S. Ito & Homma (1926) = Synchytrium fide Karling (*Synchytrium*, 1964).

Miyagia Miyabe ex Syd. & P. Syd. (1913), Pucciniaceae. 2 (on *Asteraceae*), Africa; Asia; Europe. See Hiratsuka (*TMSJ* **10**: 89, 1969), Gjærum *et al.* (*Lidia* **5**: 83, 2000; Uganda) Probably belongs to *Puccinia*.

Miyakeomyces Hara (1913), Niessliaceae. 1 (on bamboo leaves), Japan. See Eriksson & Yue (*SA* **8**: 9, 1989), Rossman *et al.* (*Stud. Mycol.* **42**: 248 pp., 1999).

Miyoshia Kawam. (1907) [non *Miyoshia* Makino 1903, *Liliaceae*] ≡ Miyoshiella.

Miyoshiella Kawam. (1929), Chaetosphaeriales. Anamorph *Stanjehughesia*. 1 (from wood), Eurasia (temperate). See Réblová (*Mycotaxon* **71**: 13, 1999),

Shenoy *et al.* (*MR* **110**: 916, 2006; phylogeny).

mizaume, see Fermented food and drinks.

MLC, microbody-liquid globule complex; see *Chytridiales*.

Mniaecia Boud. (1885), Helotiaceae. 3 (on *Hepaticae*), Europe. See Purvis *et al.* (*Lichen Flora of Great Britain and Ireland*, 1992; key), Raspé & De Sloover (*Belg. Jl Bot.* **131**: 251, 1998; Belgium), de Sloover (*Lejeunia* n.s. **166**, 2001).

Mniopetalum Donk & Singer (1962) = Rimbachia fide Kuyper (*in litt.*).

Moana Kohlm. & Volkm.-Kohlm. (1989), Halosphaeriaceae. 1 (marine), Pacific. See Kohlmeyer & Volkmann-Kohlmeyer (*MR* **92**: 410, 1989).

Mobergia H. Mayrhofer & Sheard (1992), Physciaceae (L). 7, C. & N. America. See Mayrhofer *et al.* (*Bryologist* **95**: 438, 1992), Grube & Arup (*Lichenologist* **33**: 63, 2001; phylogeny).

Models of larger fungi. Early models were made by James Sowerby during 1796 and 1815 for his museum. These were later restored by Worthington G. Smith (q.v.) and are now at The Natural History Museum, London (see Tribe, *Bull. BMS* **15**: 161, 1984). The Ware Collection of Glass Models of Plants, made by B. Blaschka in 1929, at the Botanical Museum of Harvard University include a series showing fungal diseases of fruit; the Department of Applied Biology, Cambridge University has a collection of glass models of macro- and microfungi made by W.A.R. Dillon Western (1899-1953) which were exhibited at the 50th Anniversary Meeting of the British Mycological Society (*TBMS* **30**: 21, 1948). The longest more recent series of coloured pottery models (some 200) are those designed by the late Mr and Mrs Lovenzens of Lantz, Nova Scotia, of which there is a complete set in the Nova Scotia Museum, Halifax. Many larger fungi are visually very attractive and various artists who are also interested in mycology produce models, usually from clay, but also from wax, plaster of Paris or various resins, and often using silicone moulds. The 1987 British Mycological Society workshop on modelling larger fungi, led by Dr S. Diamandis, revived interest in production of such models, and the society maintains a collection as a travelling exhibition (Chattaway, Making models of fungi. *Association of British fungus groups journal* **6**: 26, 2002). There are still, however, few if any museums with such models, and perhaps none with models showing microscopic structures of fungi greatly enlarged, although this medium is ideal for educational purposes.

Algorithms to produce images of fungi by computer models have been described by Desbenoit *et al* (Interactive modelling of mushrooms. *Eurographics*, 2004), and animated 3D computerized images of fungi can be purchased from various internet sites. Although apparently mainly for use as components for realistic backgrounds in computer games, models are also being used in airflow studies of spore dispersal (Deering *et al.*, *Mycol.* **93**: 732, 2001).

Modicella Kanouse (1936), Mortierellaceae. 2, N. & S. America. See Gerdemann & Trappe (*Mycol. Mem.* **5**, 1974; key), Benny *et al. in* Sylvia *et al.* (Eds) (*Mycorrhizae in the next decade, practical applications and research priorities*: 311, 1987; syst. posn).

Moelleria Bres. (1896) [non *Moelleria* Scop. 1777, *Algae*] ≡ Moelleriella.

Moelleriella Bres. (1897), Clavicipitaceae. Anamorph *Aschersonia*-like. 40, pantropical. See Rogerson (*Mycol.* **62**: 865, 1970), Sung *et al.* (*Stud. Mycol.* **57**, 2007), Chaverri *et al.* (*Stud. Mycol.* **60**, 2008; phylogeny, neotropics).

Moelleroclavus Henn. (1902), anamorphic *Xylaria*. 1, neotropics; Hawaii. See Clements & Shear (*Gen. Fung.*, 1931), Læssøe (*SA* **13**: 43, 1994; ? synonym of *Xylaria*), Rogers *et al.* (*MR* **101**: 345, 1997).

Moellerodiscus Henn. (1902), Rutstroemiaceae. 1, widespread. See Dumont (*Mycol.* **68**: 233, 1976), Spooner (*Biblthca Mycol.* **116**: 1, 1987), Johnston (*N.Z. Jl Bot.* **40**: 105, 2002; New Zealand).

Moelleropsis Gyeln. (1939), Pannariaceae (L). 2, widespread (north temperate). See Maas (*Proc. Nova Scotia Int. Sci.* **37**: 21, 1987), Jørgensen (*Cryptog. Mycol.* **21**: 49, 2000; Europe), Lumbsch *et al.* (*Lichenologist* **37**: 291, 2005; phylogeny), Wedin *et al.* (*MR* **109**: 159, 2005; phylogeny), Carballal Durán & López de Silanes Vázqez (*Cryptog. Mycol.* **27**: 69, 2006; Iberian peninsula).

Moeszia Bubák (1914) = Cylindrocarpon fide von Arx (*Gen. Fungi Sporul. Cult.*, 1970).

Moesziella Petr. (1927) = Leptopeltis fide Holm & Holm (*Bot. Notiser* **130**: 215, 1977).

Moesziomyces Vánky (1977), Ustilaginaceae. 1 (on *Poaceae*), widespread. *M. bullatus* (syn. *M. penicillariae*), pearl millet seed smut. See Rao & Thacker (*TBMS* **81**: 597, 1983), Vánky (*Nordic Jl Bot.* **6**: 67, 1986), Mordue (*IMI Descr. Fungi Bact.*: no. 1245, 1995), Vánky & Vánky (*Lidia* **5**: 157, 2002; southern Africa), Vánky (*Mycotaxon* **85**: 1, 2003).

Moeszopeltis Petr. (1947) = Leptopeltis fide Holm & Holm (*Bot. Notiser* **130**: 215, 1977).

Mohgaonidium Singhai (1974), Fossil Fungi, anamorphic *Pezizomycotina*. 1 (Eocene), India.

Mohortia Racib. (1909) = Septobasidium fide Couch (*The genus Septobasidium*, 1938).

Molecular biology. Techniques in molecular biology have made a major contribution to understanding of fungal biology and relationships (see Phylogeny; Bruns *et al.*, 1991; Bridge & Hawksworth, 1998; Moncalvo (*in* Xu (Ed.), *Evolutionary Genetics of Fungi*, 1, 2005), also opening new possibilities for their use. Various different molecular techniques are available, and the one(s) most appropriate for a particular task must be selected. Most are DNA based, and the DNA for study first has to be extracted; of the methods available those of Raeder & Broda (*Lett. App. Microbiol.* **1**: 17, 1985) and Zolan & Pukkila (*Molec. Cell. Biol.* **6**: 195, 1986) are now well-tested. The DNA can then be studied by a variety of methods:

DNA-DNA hybridisation involves the separation of the DNA from two fungi to be compared into separate strands and then noting the extent to which they can reassociate; the extent of successful pairing is expressed as % DNA relatedness; formerly used extensively in yeasts and mould-fungi (see Kurtzman, 1985) but now used predominantly in bacterial systematics studies.

DNA probes, labelled fragments of DNA that are used to identify particular regions of DNA in other organisms, including fungi. Typically, a DNA probe will be used to generate RFLPs (see below) from DNA digestions. DNA probes may be constructed for regions specific to individual species and can provide valuable identification tools (e.g. Dobrowolski &

O'Brien, *FEMS Microbiol. Letts* **113**: 43, 1993), currently being harnessed to PCR (q.v.) technology for high throughput studies (e.g. Martin *et al. J. Clin. Microbiol.* **38:**: 3735, 2000).

mol % G + C (guanine + cytosine, **GC**) contents in the DNA are determined by thermal denaturation profiles and expressed as a percentage; widely used in the 1970s (see Kurtzman, 1985) but now largely replaced by more sensitive methods, and now only of historical interest; see base ratio.

RFLP (restriction fragment length polymorphisms), used on both nuclear and mitochondrial DNA involves the use of restriction enzymes (e.g. *Hae*III, *Msp*I) which cut the molecules into fragments at particular sites; the size of the fragments will vary, and these are then separated by electrophoresis (q.v.); electrophoresis of digested total cellular DNA produces a 'smear' of fragments of different sizes and polymorphisms associated with individual genes or gene clusters require hybridization to a labelled DNA probe (q.v.; e.g. Correll *et al.*, *Phytopath.* **83**: 1199, 1993). In some instances, such as PCR products or purified mitochondrial DNA, RFLPs can be detected without probes (e.g. Varga *et al.*, *Can. J. Microbiol.* **40**: 612, 1994). Fungal mitochondrial DNA is very rich in A+T which means that if (good quality) total fungal DNA from a given organism is digested using a G+C cutting enzyme such as Hae III [cuts at –GG'CC-], any bands of >1-2Kb may be presumed mitochondrial in origin [the chromosomal DNA will have been digested to form a background smear]; see Freeman *et al.* (*Exper. Mycol.* **17**: 309, 1993) for rationale.

PCR (polymerase chain reaction), a series of heating and cooling steps that allow for amplification of DNA. DNA is melted and primers attached to the separated single strands; new complementary strands are formed from the primers by the addition of dNTPs in the presence of a thermostable DNA polymerase. Further heating melts this new dimer and the process is repeated through a number of cycles. This method has revolutionised the study of the biology of fungi and other organisms since its inception in the late 1980s. Numerous variations/modifications now exist on/from the basic PCR methodology, allowing investigation of specific genes or loci in a 'targeted' approach (i.e. where the sequence of the gene of interest or its flanking region is known), or alternatively a 'non-targeted' approach that allows population/taxon discrimination on the basis of complex banding patterns. Some of the most significant variations are listed below [see also: DNA fingerprinting; DNA sequencing]. Amplification products may be visualised by use of a suitable stain (normally ethidium bromide – although less toxic alternatives are now available – in the presence of UV light; see Electrophoresis).

reverse transcriptase (RT) PCR. Sometimes confused with real time PCR (see below), this method is used to investigate specific mRNAs by 'converting' the RNA (which is notoriously difficult to handle) 'back' into DNA (called 'cDNA'), using the enzyme reverse transcriptase. It has also been used as an aid to the study of gene expression; e.g. see Reddy & D'Souza (*in* Bridge *et al.* (Eds) *Applications of PCR in Mycology*, 1998: 207; gene expression in lignocellulose degradation in white rot fungi).

real time PCR is a powerful technique that is used in quantification of DNA, gene expression studies, screening environmental samples, for diagnosis of certain disease genes and, increasingly, in food safety & biosecurity analyses. The technique uses double stranded DNA dyes or, for more accurate results, fluorescent reporter probes. See Nolan *et al.* (*Nat. Protoc.* **1**: 1559, 2006; use in mRNA quantification), Nailis *et al.* (*BMC Mol. Biol.* **7**: 25, 2006; use in *Candida albicans* gene expression studies), Kasai *et al.* (*J. Clin. Microbiol.* **44**: 143, 2006; detection of *Candida*).

DNA microarrays are a valuable tool for identifying gene expression and for diagnosing particular diseases such as cancers (see Sambrook & Russell, *Molecular Cloning: A Laboratory Manual*, edn 3, **3**: A10, 2001) and even medically important fungi (e.g. Leinberger *et al.*, *J. Clin. Microbiol.* **43**: 4943, 2005). In the construction of a DNA microarray, relevant DNA fragments (e.g. PCR products) are fixed precisely, by robotic 'printing', to a glass slide (although other supports may be used). Alternatively, arrays of single-stranded oligonucleotides that correspond to known gene sequences may be constructed *in situ*. Either way, the microarrays operate as a form of DNA probe to cDNA targets derived from RNA samples which are mixed with the probe to determine the degree of hybridization. Detection is usually by high resolution laser scanning of fluorophore-labelled probes (see Sambrook & Russell, 2001). This can be viewed semi-quantitatively (see Kazan *et al.*, *Mol. Plant Pathol.* **2**: 177, 2001). [see also *Genomics*].

Lit.: Bennett & Lasure (Eds) (*Gene manipulations in fungi*, 1985), Bennett & Lasure (Eds) (*More gene manipulations in fungi*, 1991), Brambl & Marzluf (Eds) (*The Mycota vol. III: Biochemistry and Molecular Biology* 1996), Bridge & Hawksworth (*Lichenologist* **30**: 307, 1998; use in lichen systematics), Bridge *et al.* (Eds) (*Applications of PCR in Mycology*, 1998), Bridge *et al.* (Eds) (*Molecular Variability of Fungal Pathogens*, 1998), Bruns *et al.* (*Ann. Rev. Ecol. Syst.* **22**: 525, 1991), Burnett (*Fungal Populations and Species*, 2003), Clutterbuck (*in* Gow & Gadd, *The growing fungus*: 255, 1995), Edel (*in* Frisvad *et al.* (Eds), *Chemical Fungal Taxonomy*: 51, 1998), Guthrie & Fink (Eds) (*Guide to yeast genetics and molecular biology*, [*Methods Enzym. Biol.* **194**], 1991), Hawksworth (Ed.) (*Identification and characterization of pest organisms*, 1994), Hibbett (*Trans. mycol. Soc. Japan* **33**: 533, 1992; use in systematics), Kinghorn & Turner (Eds) (*Applied molecular genetics of filamentous fungi*, 1992), Kück (Ed.) (*The Mycota vol. II: Genetics and Biotechnology* 1995), Moore (*Ant. van Leeuw.* **72**: 209, 1997; use in systematics), Oliver (*in* Fox, *Principles of diagnostic techniques in plant pathology*: 153, 1993), Paterson & Bridge (*Biochemical techniques for fungi* [*IMI Techn. Handbk* **1**], 1994), Reynolds & Taylor (Eds) (*The fungal holomorph*, 1993; examples use in systematics), Schots *et al.* (Eds) (*Modern assays for plant pathogenic fungi*, 1994), Rolfs *et al.* (Eds) (*Methods in DNA amplification*, 1994), Talbot (Ed.) (*Molecular and Cellular Biology: A practical approach*, 2001), Towner & Cockayne (*Molecular methods for microbial identification and typing*, 1993), Xu (Ed.), (*Evolutionary Genetics of Fungi*, 1, 2005).

Molgosphaera Dumort. (1822) nom. dub., Pezizomycotina. Used for diverse perithecioid fungi.

Molgosporidium Thor (1930) nom. dub., ? Fungi.

Molinea Doub. & D. Pons (1975), Fossil Fungi, Asterinaceae. 1 (Cretaceous), Colombia.

Moelleriella G. Winter (1886), Elsinoaceae. 13 (on glandular hairs), widespread (tropical).

Mölleropsis (orthographic variant), see *Moelleropsis*.

Molliardiomyces Paden (1984), anamorphic *Sarcoscypha, Phillipsia*, Hso.0eH.10. 9, widespread. See Paden (*CJB* **62**: 211, 1984), Paden (*Mycotaxon* **25**: 165, 1986), Harrington (*Mycotaxon* **38**: 417, 1990; N America), Hansen *et al.* (*Mycol.* **91**: 299, 1999; phylogeny).

Mollicamarops Lar.N. Vassiljeva (2007), Boliniaceae. 1, E. Russia. Questionably distinct from *Camarops*. See Vasiljeva (*Mycotaxon* **99**: 159, 2007).

Mollicarpus Ginns (1984), Polyporaceae. 1, S.E. Asia. See Ryvarden (*Gen. Polyp.*: 186, 1991).

mollicute, see mycoplasma.

Mollisia (Fr.) P. Karst. (1871) nom. cons., Helotiales. Anamorphs *Anguillospora*, *Casaresia*, *Phialophora*-like, *Sirocyphellina*. *c.* 121, widespread. Almost certainly polyphyletic. Molecular data separate this genus and its relatives from the *Dermateaceae*; it may possibly be best accommodated within the *Loramycetaceae*. See Dennis (*Kew Bull.* **1950**: 171, 1950), Thind & Sharma (*Biblthca Mycol.* **91**: 221, 1983; India), Webster *et al.* (*Nova Hedwigia* **57**: 483, 1993; aquatic anam.), Gminder (*Z. Mykol.* **62**: 181, 1996), Gams (*Stud. Mycol.* **45**: 187, 2000; anamorph), Nauta & Spooner (*Mycologist* **14**: 65, 2000; UK), Harrington & McNew (*Mycotaxon* **87**: 141, 2003; phylogeny), Pärtel & Raitviir (*Mycol. Progr.* **4**: 149, 2005; ultrastr.), Gminder (*Czech Mycol.* **58**: 125, 2006), Wang *et al.* (*Mycol.* **98**: 1065, 2006; phylogeny).

Mollisiaster Kirschst. (1939) = Microscypha fide Arendholz (*in litt.*).

Mollisiella (W. Phillips) Massee (1895) ≡ Unguiculariopsis fide Zhuang (*Mycotaxon* **32**: 1, 1988), Huhtinen (*Karstenia* **29**: 45, 1989).

Mollisiella Boud. (1885) = Mollisia fide Korf & Zhuang (*SA* **6**: 139, 1987).

Mollisina Höhn. ex Weese (1926), Hyaloscyphaceae. 11, Europe; Asia. See Arendholz (*Blattbewohnenden Ascomyceten aus der Ordnung der Helotiales*, 1979), Arendholz & Sharma (*Mycotaxon* **20**: 633, 1984), Hosoya & Otani (*Mycoscience* **38**: 187, 1997; Japan), Raitviir (*Scripta Mycologica* Tartu **20**, 2004).

Mollisinopsis Arendh. & R. Sharma (1984), Helotiaceae. 3, India. See Arambarri & Sharma (*Mycotaxon* **20**: 660, 1984).

Mollisiopsis Rehm (1908), Helotiales. 7, Europe. See Dennis (*Persoonia* **2**: 171, 1962), Graddon (*TBMS* **58**: 153, 1972), Svrček (*Česká Mykol.* **41**: 88, 1987), Nauta & Spooner (*Mycologist* **14**: 65, 2000).

Molybdoplaca Nieuwl. (1916) ≡ Steinera.

Monacrosporiella Subram. (1978) = Arthrobotrys fide Scholler *et al.* (*Sydowia* **51**: 89, 1999).

Monacrosporium Oudem. (1885), anamorphic *Orbilia*, Hsy.≡ eP.10. *c.* 68 (on nematodes), widespread. See Subramanian (*J. Indian bot. Soc.* **42**: 292, 1963; key), Cooke (*TBMS* **50**: 517, 1967; key 7 nematode-trapping spp.), Cooke (*TBMS* **53**: 475, 1969), Liu & Zhang (*MR* **98**: 862, 1994; checklist of spp.), Rubner (*Stud. Mycol.* **39**, 1996), Hagedorn & Scholler (*Sydowia* **51**: 27, 1999; phylogeny), Scholler *et al.* (*Sydowia* **51**: 89, 1999; synonym of *Arthrobotrys*), Li *et al.* (*MR* **107**: 888, 2003), Meyer *et al.* (*Mycol.* **97**: 405, 2005; phylogeny, morphology), Kim *et al.* (*Pl. Path. J.* **22**: 174, 2006; Korea), Li *et al.* (*Mycol.* **97**: 1034, 2005; phylogeny).

Monadelphus, see *Monodelphus*.

monads (*Monadineae*), a diverse group of amoeboflagellate *Protozoa* some of which were compiled by Saccardo (*Syll. Fung.* **7**: 453, 1888) based on the monograph by Zopf [Die Pilzthiere oder Schleimpilze. *Encykl. Naturwiss.* **3**(2): 1, 1884]. Also referred to as *Proteomyxa* (Lankester, *Encycl. Brit.*, 9 ed., 1885) or *Hydromyxales* Klein. Those studied belong in different groups of *Protozoa* but many do not appear to have been investigated recently.

Monandromyces R.K. Benj. (1999), Laboulbeniaceae. 11, widespread (esp. tropical). See Benjamin (*Aliso* **18**: 71, 1999).

monandrous (of oospores), formed when only one functioning antheridium is present. Cf. polyandrous.

Monascaceae J. Schröt. (1894), Eurotiomycetidae (inc. sed.). 3 gen. (+ 5 syn.), 16 spp.

Lit.: Bridge & Hawksworth (*Lett. Appl. Microbiol.* **1**: 25, 1985), Hocking (*Handbook of Applied Mycology* Vol. 3. Foods and Feeds: 69, 1991), Landvik & Eriksson (*SA* **12**: 34, 1993; posn.), Cannon *et al.* (*MR* **99**: 659, 1995), Pitt & Hocking (*Fungi and Food Spoilage* Edn 2: 593 pp., 1997), Udagawa & Baba (*Cryptog. Mycol.* **19**: 269, 1998), Chaisrisook (*MR* **106**: 298, 2002), Park & Jong (*Mycoscience* **44**: 25, 2003).

Monascella Guarro & Arx (1986), ? Pyronemataceae. 1, Spain. See Guarro & Arx (*Mycol.* **78**: 869, 1986), Stchigel *et al.* (*MR* **105**: 377, 2001; phylogeny).

Monascostroma Höhn. (1918) = Hendersonia Berk. fide Holm & Holm *in* Laursen *et al.* (Eds) (*Arctic and Alpine Mycology* **2**: 109, 1987), Menendez (*Darwiniana* **31**: 351, 1992; Argentina).

Monascus Tiegh. (1884), Monascaceae. Anamorph *Basipetospora*. 14, widespread. *M. ruber* turns silage light red; *M. purpureus* produces ang-kak and other Asian foods and drinks. See Hawksworth & Pitt (*Aust. J. Bot.* **31**: 51, 1983; key), Bridge & Hawksworth (*Lett. Appl. Microbiol.* **1**: 25, 1985; biochem. tests), Wong & Chien (*Mycol.* **78**: 713, 1986; ultrastr.), Nishikawa *et al.* (*J. gen. appl. Microbiol.* Tokyo **34**: 467, 1988; proteinase), Nishikawa *et al.* (*J. Basic Microbiol.* **29**: 369, 1989; fatty acids), Nishikawa *et al.* (*J. Basic Microbiol.* **33**: 331, 1993; isozymes), Cannon *et al.* (*MR* **99**: 659, 1995; key), Lakrod *et al.* (*MR* **104**: 403, 2000; genetic variation), Chaisrisook (*MR* **106**: 298, 2002; VCGs), Park & Jong (*Mycoscience* **44**: 25, 2003; phylogeny), Park *et al.* (*Bot. Bull. Acad. sin.* Taipei **45**: 325, 2004; phylogeny), Stchigel *et al.* (*Stud. Mycol.* **50**: 299, 2004; key).

monaxial, having one individual stem or axis.

Monera. Kingdom embracing *Bacteria* and *Cyanobacteria* in the five-Kingdom system of Whittaker (1969); see Kingdoms of fungi for current treatment.

Monerolechia Trevis. (1857), Physciaceae (L). 1, widespread. See Hafellner (*Beih. Nova Hedwigia* **62**, 1979), Marbach (*Biblthca Lichenol.* **74**, 2000), Kalb (*Biblthca Lichenol.* **88**: 301, 2004).

Monilia Bonord. (1851) nom. cons., anamorphic *Monilinia*, Hso.0eH.3. *c.* 24, widespread. Formerly widely used by medical mycologists for *Candida*; see candidiasis. See Donk (*Taxon* **12**: 266, 1963; nomencl.), Turian & Bianchi (*Arch. Mikrobiol.* **77**: 262, 1971; conidiation), Anon. (*Taxon* **23**: 419, 1974; nomencl.), Funk (*CJB* **65**: 23, 1987; pleoanamorphism

M. versiformis), Holst-Jensen *et al.* (*Am. J. Bot.* **84**: 686, 1997; phylogeny), Leeuwen *et al.* (*MR* **106**: 444, 2002), Côé *et al.* (*Pl. Dis.* **88**: 1219, 2004; PCR diagnostics), Harada *et al.* (*J. Gen. Pl. Path.* **70**: 297, 2004; Japan), Harada *et al.* (*Mycoscience* **46**: 376, 2005; Japan), Takahashi *et al.* (*Mycoscience* **46**: 106, 2005; Japan).

Monilia Hill ex F.H. Wigg. (1780) nom. rej. prop. = Monilia Bonord. fide Hawksworth *et al.* (*Dictionary of the Fungi* edn 8, 1995).

Monilia Link (1809) nom. rej. = Monilia Bonord. fide Hawksworth *et al.* (*Dictionary of the Fungi* edn 8, 1995).

Moniliaceae Dumort. (1822) = Sclerotiniaceae.

moniliaceous (of mycelium, spores, etc.), hyaline or brightly coloured; mucedinaceous.

Moniliales = blastomycetes + hyphomycetes.

moniliasis, see candidiasis.

Moniliella Stolk & Dakin (1966), anamorphic *Tremellomycetes*, Hso.0eP.3+38. 5, north temperate. See de Hoog (*Stud. Mycol.* **19**, 1979; key), Hoog & Smith *in* Kurtzman & Fell (Eds) (*Yeasts, a taxonomic study* 4th edn: 785, 1998).

moniliform (**monilioid**), having swellings at regular intervals like a string of beads.

Moniliger Letell. [not traced] nom. dub. ? = Penicillium Link fide Hawksworth *et al.* (*Dictionary of the Fungi* edn 8, 1995).

Moniliites Babajan & Tasl. (1973), Fossil Fungi. 1 (Tertiary), former USSR.

Monilinia Honey (1928), Sclerotiniaceae. Anamorph *Monilia*. 28, widespread. See Byrde & Willets (*Brown Rot Fungi of Fruit*, 1977), Harada (*Bull. Fac. Agr. Hirosaki Univ.* **27**: 30, 1977; Japan), Willetts *et al.* (*J. gen. Microbiol.* **103**: 77, 1977; chemotaxonomy), Batra (*Mycotaxon* **8**: 476, 1979; N. Am.), von Arx (*Sydowia* **34**: 13, 1981; anamorphs), Buchwald (*Friesia* **11**: 287, 1987; key 12 spp.), Batra (*World species of Monilinia*, 1991), Holst-Jensen *et al.* (*Am. J. Bot.* **84**: 686, 1997; phylogeny, polyphyly, coevolution), Fulton *et al.* (*Eur. J. Pl. Path.* **105**: 495, 1999; mol. variation), van Leeuwen & van Kesteren (*CJB* **76**: 2042, 1999; morphometrics), Förster & Adaskaveg (*Phytopathology* **90**: 171, 2000; PCR detection), Ioos & Frey (*Eur. J. Pl. Path.* **106**: 373, 2000; infraspecific variation), Gernandt *et al.* (*Mycol.* **93**: 915, 2001; phylogeny), Lane (*Bulletin OEPP* EPPO Bulletin **32**: 489, 2002; synoptic key), Leeuwen *et al.* (*MR* **106**: 444, 2002), Baayen *et al.* (*Bulletin OEPP* EPPO Bulletin **33**: 281, 2003; diagnostic protocol), Côé *et al.* (*Pl. Dis.* **88**: 1219, 2004; PCR diagnostics), Harada *et al.* (*Mycoscience* **46**: 376, 2005; Japan), Takahashi *et al.* (*Mycoscience* **46**: 106, 2005; Japan), Wang *et al.* (*Mycol.* **98**: 1065, 2006; phylogeny).

Moniliophthora H.C. Evans, Stalpers, Samson & Benny (1978), Marasmiaceae. *c.* 10, tropical. *M. perniciosa* (syn *Crinipellis perniciosa*, witches' broom of cacao); *M. roreri* (frosty pod rot of cacao). See Evans (*Phytopath. Pap.* **24**: 1, 1981), Evans *et al.* (*Mycologist* **16**: 148, 2002), Evans *et al.* (*Pl. Path.* **52**: 476, 2003; phylogeny *C. roreri*), Aime & Phillips-Mora (*Mycol.* **97**: 1012, 2005).

Moniliopsis Ruhland (1908) = Rhizoctonia fide Roberts (*Rhizoctonia-Forming Fungi*: 239 pp., 1999).

Monilites Pampal. (1902), Fossil Fungi. 1 (Miocene), Italy.

Monilochaetes Halst. ex Harter (1916), anamorphic *Pezizomycotina*, Hsy.0eP.1. 1, widespread (tropical; esp. USA). *M. infuscans* on *Ipomoea batatas* (sweet potato scurf). See Rong & Gams (*Mycotaxon* **76**: 451, 2000).

Monka Adans. (1763) = Verpa fide Hawksworth *et al.* (*Dictionary of the Fungi* edn 8, 1995).

mono- (prefix), one.

Monoblastia Riddle (1923), Monoblastiaceae (L). 10, widespread (tropical). See Harris (*Some Florida lichens*, 1990), Aptroot (*Bryologist* **94**: 404, 1991), Eriksson (*SA* **11**: 178, 1993), Sérusiaux & Aptroot (*Nova Hedwigia* **67**: 259, 1998; Papua New Guinea), Aptroot *et al.* (*Biblthca Lichenol.* **97**, 2008; Costa Rica).

Monoblastiaceae Walt. Watson (1929), Pyrenulales (L). 6 gen. (+ 8 syn.), 133 spp.

Lit.: Harris (*Some Florida lichens*, 1990), Aptroot (*Bryologist* **94**: 404, 1991), Eriksson & Hawksworth (*SA* **11**: 178, 1993), Sérusiaux & Aptroot (*Nova Hedwigia* **67**: 259, 1998), Aptroot (*Lichenologist* **31**: 641, 1999), Etayo & Lücking (*Lichenologist* **31**: 145, 1999), Hyde & Wong (*MR* **103**: 347, 1999).

monoblastic (of a conidiogenous cell), producing a blastic conidium at one locus.

Monoblastiopsis R.C. Harris & C.A. Morse (2008), Pleosporales. 2, USA. See Harris & Morse (*Opuscula Philolichenum* **5**: 89, 2008).

Monoblepharella Sparrow (1940), Gonapodyaceae. 5, C. & N. America. See Sparrow (*Aquatic Phycomycetes* Edn 2: 721, 1960; key), Steciow & Arambarri (*Nova Hedwigia* **70**: 107, 2000; *Monoblepharella mexicana*), James *et al.* (*Mycol.* **98**: 860, 2006; phylogeny).

Monoblepharidaceae A. Fisch. (1892), Monoblepharidales. 1 gen. (+ 2 syn.), 12 spp.

Lit.: Perrott (*TBMS* **38**: 247, 1955), Noyes Mollicone & Longcore (*Mycol.* **86**: 615, 1994).

Monoblepharidales Sparrow (1942). Monoblepharidomycetes. 4 fam., 5 gen., 26 spp. Thallus hyphal with oogamous reproduction (the only chytrids with non-motile female gametes and smaller, motile, male gametes), or thallus simple and no sexual stage known; zoospores possessing a non-flagellated centriole parallel to the kinetosome, a striated disk extending part way around the kinetosome, microtubules arising from the striated disk and extending into the cytoplasm, rumposome backed by microbody, and numerous lipid globules; aquatic saprobes. Fams:

(1) **Gonapodyaceae**
(2) **Harpochytriaceae**
(3) **Monoblepharidaceae**
(4) **Oedogoniomycetaceae**

Lit.: Sparrow (1960: 713, *in* Ainsworth *et al.* (Eds), *The Fungi* **4B**: 85, 1973), Gauriloff *et al.* (*CJB* **58**: 2098, 1980; ultrastr.), Barr (1990), Mollicone & Longcore (*Mycol.* **86**: 615, 1994, **91**: 727, 1999; zoospore ultrastr.), James *et al.* (*Mycol.* **98**: 860, 2006; molecular phylogeny), Hibbett *et al.* (*MR* **111**: 109, 2007), and see under Familes.

Monoblepharidomycetes J.H. Schaffn. (1909), Chytridiomycota. 1 ord., 4 fam., 5 gen., 26 spp. Ord.: **Monoblepharidales**

Lit.: James *et al.* (*Mycol.* **98**: 860, 2006; molecular phylogeny), and see under Order.

Monoblephariopsis Laib. (1927) = Monoblepharis fide Sparrow (*Aquatic Phycomycetes* Edn 2: 1187 pp., 1960).

Monoblepharis Cornu (1871), Monoblepharidaceae. 12, Europe; N. America. See Perrott (*TBMS* **38**: 247, 1955), Sparrow (*Aquatic Phycomycetes* Edn 2: 727, 1960; key), James *et al.* (*Mycol.* **98**: 860, 2006; phylogeny).

monocarpic (of *Exobasidium* infections), circumscribed and annual (Nannfeldt, 1981: 10), cf. polycarpic, surculicolous.

monocentric (of a chytrid thallus), having one centre of growth and development; see polycentric and cf. reproductocentric.

monocephalic (adj. **monocephalous**), 1-headed.

Monocephalis Bainier (1882) = Syncephalis fide Benjamin (*Aliso* **4**: 321, 1959).

Monoceras Guba (1961) [non *Monoceras* Gothan 1909, fossil *Phanerogamae*] = Seimatosporium fide Sutton (*Mycol. Pap.* **97**, 1964).

Monochaetia (Sacc.) Allesch. (1902), anamorphic *Amphisphaeriaceae*, Cac.≡ eP.19. 28, widespread. See Anon. (*Monograph of Monochaetia and Pestalotia*, 1961), Sutton (*The Coelomycetes*, 1980), Nag Raj (*Coelomycetous Anamorphs with Appendage-bearing Conidia*, 1993), Nag Raj & Mel'nik (*Mycotaxon* **50**: 435, 1994; redispositions), Jeewon *et al.* (*Mol. Phylogen. Evol.* **25**: 378, 2002; phylogeny), Jeewon *et al.* (*MR* **107**: 1392, 2003; phylogeny).

Monochaetiella E. Castell. (1943), anamorphic *Pezizomycotina*, Cac.0eH.15. 1, widespread.

Monochaetiellopsis B. Sutton & DiCosmo (1977), anamorphic *Hypnotheca*, Cac.1eH.?. 2, widespread. See Punithalingam (*Nova Hedwigia* **54**: 255, 1992; nuclei in conidial appendages).

Monochaetina Subram. (1961) = Bleptosporium fide Sutton (*Mycol. Pap.* **88**, 1963).

Monochaetinula Muthumary, Abbas & B. Sutton (1986), anamorphic *Amphisphaeriaceae*, Cpd.≡ eP.19. 6, widespread (tropical). See Bianchinotti (*Mycotaxon* **39**: 455, 1990; Argentina), Nag Raj (*Coelomycetous Anamorphs with Appendage-bearing Conidia*, 1993).

Monochaetiopsis L. Cai, R. Jeewon & K.D. Hyde (2003), anamorphic *Dyrithiopsis*, Cac.?.?. 1, China. See Jeewon *et al.* (*Mycol.* **95**: 913, 2003).

Monochaetites Babajan & Tasl. (1973), Fossil Fungi. 1 (Tertiary), former USSR.

Monochaetopsis Pat. (1931), anamorphic *Pezizomycotina*. 1, N. Africa.

Monochytrium Griggs (1910) ? = Olpidium fide Fitzpatrick (*The lower fungi. Phycomycetes*, 1930).

Monocillium S.B. Saksena (1955), anamorphic *Niesslia, Hyaloseta*, Hso.0eH/1eP.15. *c.* 15, widespread. See Gams (*Cephalosporium-artige Schimmelpilze*, 1971; key 12 spp.), Gams & Turhan (*Mycotaxon* **59**: 343, 1996), Girlanda & Luppi-Mosca (*Mycotaxon* **67**: 265, 1998), Ramaley (*Mycotaxon* **79**: 267, 2001).

monoclinous, having the antheridium on the oogonial stalk. Cf. androgynous.

Monoconidia Roze (1897) ? = Acremonium fide Hawksworth *et al.* (*Dictionary of the Fungi* edn 8, 1995).

Monodelphus Earle (1909) = Omphalotus fide Singer (*Agaric. mod. Tax.*, 1951).

Monodia Breton & Faurel (1970), anamorphic *Pezizomycotina*, Cpd.≡ eH.1. 1 (coprophilous), Chad. See Breton & Faurel (*Revue mycol.* Toulouse **35**: 23, 1970), Nag Raj & DiCosmo (*Univ. Waterloo Biol. Ser.* **20**, 1982).

Monodictyites Barlinge & Paradkar (1982), Fossil Fungi, anamorphic *Pezizomycotina*. 1 (Cretaceous), India.

Monodictys S. Hughes (1958), anamorphic *Dothideomycetes*, Hsp.#eP.1. 27, widespread. Certainly polyphyletic. See Rao & de Hoog (*Stud. Mycol.* **28**: 25, 1986; key), Wedin (*Lichenologist* **25**: 203, 1993; ultrastr.), Rodríguez *et al.* (*Nova Hedwigia* **72**: 201, 2001; Cuba), Zhao & Zhang (*Mycosystema* **23**: 324, 2004; China), Kukwa & Diederich (*Lichenologist* **37**: 217, 2005; lichenicolous spp.), Day *et al.* (*Mycotaxon* **98**: 261, 2006).

Monodidymaria U. Braun (1994), anamorphic *Pezizomycotina*, Hso.≡ eH.1. 5, Canada; India. See Braun (*Monogr. Cercosporella, Ramularia Allied Genera (Phytopath. Hyphom.)* **2**, 1998).

Monodisma Alcorn (1975), anamorphic *Pezizomycotina*, Hso.≡ eH.10. 1, widespread. See Alcorn (*TBMS* **65**: 139, 1975).

monoecism (adj. **monoecious**), having the male and female sex organs on one thallus; cf. dioecism, heterothallism, homothallism.

monogeocentric, see reproductocentric.

Monogrammia F. Stevens (1917) = Titaea fide Damon (*J. Wash. Acad. Sci.* **42**: 365, 1952).

Monographella Petr. (1924), Xylariales. Anamorph *Microdochium*. 6 (pathogenic on cereals), widespread. See Müller (*Revue Mycol.* Paris **41**: 129, 1977; *M. nivalis*), Subramanian & Bhat (*Revue Mycol.* Paris **42**: 293, 1978; ontogeny), Parkinson *et al.* (*TBMS* **76**: 59, 1981; *M. albescens*, rice leaf scald), Samuels & Hallett (*TBMS* **8**: 473, 1983), Müller & Samuels (*Nova Hedwigia* **40**: 113, 1984), Mahuku *et al.* (*MR* **102**: 559, 1998; mol. diversity), Wang & Hyde (*Fungal Diversity* **3**: 159, 1999), Winka & Eriksson (*Mycoscience* **41**: 97, 2000), Jeewon *et al.* (*MR* **107**: 1392, 2003; phylogeny).

Monographos Fuckel (1875) = Scirrhia fide Holm & Holm (*Bot. Notiser* **131**: 97, 1978), Eriksson & Hawksworth (*SA* **11**: 178, 1993).

Monographus Clem. & Shear (1931) ≡ Monographos.

Monoicomyces Thaxt. (1900), Laboulbeniaceae. 48, widespread. See Tavares (*Mycol. Mem.* **9**: 627 pp., 1985), Santamaría (*Mycotaxon* **50**: 89, 1994), Santamaría (*MR* **100**: 1179, 1996), De Kesel (*Sterbeeckia* **25**: 62, 2005; Belgium).

monokaryon, see haplont.

monokaryotic, having genetically identical haploid nuclei; cf. dikaryon.

Monoloculia Hara (1927) = Yoshinagaia fide Sivanesan & Hsieh (*MR* **99**: 1295, 1995).

Monolpidiaceae A. Fisch. (1892) = Olpidiaceae.

monomitic, see hyphal analysis.

monomorphic, having one structure or form; not pleomorphic.

monomycelial (of an isolate), from one spore or hyphal tip.

Monomyces Battarra ex Earle (1909) = Tricholoma fide Singer (*Agaric. mod. Tax.*, 1951).

Mononema Balbiani (1889), ? Kickxellomycotina. 3, Africa; Europe. See Manier *et al.* (*Biol. Gabon.* **8**: 323, 1972), Lichtwardt (*The Trichomycetes. Fungal associates of arthropods*: 343 pp., 1986; posn).

mononematous (of conidiophores), solitary or in tufts or loose fascicles; cf. synnematous.

monophagy (adj. **monophagous**) (of *Chytridiales*), the condition of having the thallus in one host cell; the opposite of polyphagy, in which the thallus branches invade more than one host cell.

monophialidic (of a conidiogenous cell), having one locus through which conidia are produced. Cf. polyphialidic.

monophyllous (of foliose lichens), having a single leaf-like thallus.

monoplanetism (of zoospores in oomycetes), the condition of having one motile phase, with no resting period.

Monoplodia Westend. (1859) = Coniothyrium fide Saccardo & Trotter (*Syll. fung.* **22**, 1913).

monopodial, a type of branching in which a persistant main axis gives off branches, one at a time and frequently in alternate or spiral series.

Monopodium Delacr. (1890) = Acremoniella fide Mason (*Mycol. Pap.* **3**: 29, 1933).

Monoporidicellaesporites Frunzescu & Bacaran (1990), Fossil Fungi. 1. See Frunzescu & Bacaran (*Revue roum. Géol., Géophys. Géogr.* Géol. **34**: table 1 + pl. 1, fig. 16, 1990).

Monoporimonocellasporites Frunzescu & Bacaran (1990), Fossil Fungi. 1. See Frunzescu & Bacaran (*Revue roum. Géol., Géophys. Géogr.* Géol. **34**: table 1 + pl. 1, fig. 7, 1990).

Monoporimonodicellaesporites Frunzescu & Bacaran (1990), Fossil Fungi. 1. See Frunzescu & Bacaran (*Revue roum. Géol., Géophys. Géogr.* Géol. **34**: table 1 + pl. 1, figs 9, 10, 1990).

Monoporimulticellaesporites Frunzescu & Bacaran (1990), Fossil Fungi. 1. See Frunzescu & Bacaran (*Revue roum. Géol., Géophys. Géogr.* Géol. **34**: table 1+ pl. 1, figs 23, 25, 1990).

Monoporisporites Hammen (1954), Fossil Fungi. 15 (Cretaceous, Tertiary), widespread.

Monoporitetracellaesporites Frunzescu & Bacaran (1990), Fossil Fungi. 1. See Frunzescu & Bacaran (*Revue roum. Géol., Géophys. Géogr.* Géol. **34**: table 1 + pl. 1, figs 14, 15, 17, 1990).

Monoporitricellaesporites Frunzescu & Bacaran (1990), Fossil Fungi. 1. See Frunzescu & Bacaran (*Revue roum. Géol., Géophys. Géogr.* Géol. **34**: table 1, 1990).

Monopus Theiss. & Syd. (1915) = Rosenscheldiella fide Hansford (*Mycol. Pap.* **15**, 1946).

Monopycnis Naumov (1916) = Cytospora fide Petrak (*Annls mycol.* **23**: 1, 1925).

monoreproductocentric, see reproductocentric.

Monorhiza Theiss. & Syd. (1915) = Rhagadolobium fide Müller & von Arx (*Beitr. Kryptfl. Schweiz* **11** no. 2, 1962).

Monorhizina Theiss. & Syd. (1915), Microthyriaceae. 1, Sri Lanka.

Monospermella Speg. (1924) = Psorotheciopsis fide Hawksworth *et al.* (*Dictionary of the Fungi* edn 8, 1995).

monospermous, 1-spored.

Monospora Metschn. (1884) [non *Monospora* Hochst. 1841, *Algae*] ≡ Metschnikowia.

Monosporascus Pollack & Uecker (1974), Xylariales. 3 (thermophilic), widespread. See Hawksworth & Ciccarone (*Mycopathologia* **66**: 147, 1978), Martyn *et al.* (*Phytopathology* **82**: 1115, 1992; ascospores, germination), Uematsu *et al.* (*Ann. phytopath. Soc. Japan* **58**: 354, 1992; Japan), Martyn *et al.* (*Phytopathology* **83**: 1347, 1993), Lovic *et al.* (*Phytopathology* **85**: 655, 1995; phylogeny), Waugh *et al.* (*MR* **105**: 745, 2001; SEM), Collado *et al.* (*MR* **106**: 118, 2002; Spain), Sales *et al.* (*Fitopatol. Brasil* **28**: 567, 2003; Brazil).

monospore, used by Corner (*New Phytol.* **46**: 195, 1947) for the one spore maturing on a 2-spored basidium normally bearing 2 dispores (q.v.).

Monosporella Keilin (1920) ≡ Metschnikowia.

Monosporella S. Hughes (1953) ≡ Monotosporella.

monosporic (adj. **monosporous**), 1-spored.

Monosporidium Barclay (1888), Phakopsoraceae. 4 (on *Euphorbiaceae*, *Lauraceae*, *Rubiaceae*), India; China; Japan. Telial aeciospores produce basidia on germination. See Thirumalachar & Kern (*Bull. Torrey bot. Club* **82**: 102, 1955), Buriticá (*Rev. Acad. Colomb. Cienc.* **18** no. 69: 131, 1991), Ono (*Mycoscience* **43**: 421, 2002; nuclear cycle).

Monosporiella Speg. (1918), anamorphic *Pezizomycotina*, Hso.0eH.?. 1, Argentina. See Deighton & Pirozynski (*Mycol. Pap.* **128**, 1972).

Monosporium Bonord. (1851) nom. illegit., anamorphic *Pezizomycotina*. See Hughes (*CJB* **36**: 727, 1958).

Monosporonella Oberw. & Ryvarden (1991) = Oliveonia fide Roberts (*Rhizoctonia-forming fungi*, 1999).

Monostachys Arnaud (1954) = Chloridium fide Gams & Holubová-Jechová (*Stud. Mycol.* **13**, 1976).

Monostichella Höhn. (1916), anamorphic *Drepanopeziza*, Cac.0eH.15. 10, widespread. See von Arx (*Biblthca Mycol.* **24**, 1970).

monostichous, in one line or series.

Monothecium Lib. [non *Monothecium* Hochst. 1842, Acanthaceae] [not traced] nom. nud. et nom. dub. = Mastigosporium fide Saccardo (*Syll. fung.* **4**: 220, 1886).

Monotospora Corda (1837) nom. dub., Pezizomycotina. See Carmichael (*CJB* **40**: 1137, 1962).

Monotospora Sacc. (1880) ≡ Acrogenospora.

Monotosporaceae Vuill. (1912) = Hysteriaceae.

Monotosporella S. Hughes (1958), anamorphic *Pezizomycotina*. 1, widespread. See Ellis (*Mycol. Pap.* **72**, 1959; synonymy with *Brachysporiella*), Rao & de Hoog (*Stud. Mycol.* **28**: 6, 1986), Ranghoo *et al.* (*Mycoscience* **40**: 377, 1999), Tsui *et al.* (*Mycol.* **93**: 389, 2001).

Monotoyella Castell. & Chalm. (1913) nom. dub., anamorphic *Pezizomycotina*. See de Hoog & Hermanides-Nijhof (*Stud. Mycol.* **15**: 186, 1977).

monotretic, see tretic.

Monotretomyces Morgan-Jones, R.C. Sinclair & Eicker (1987) = Corynesporopsis fide Sutton (*in litt.*).

monotrichous (**monotrichiate**) (of bacteria), having one polar flagellum.

Monotrichum Gäum. (1922), anamorphic *Pezizomycotina*, Hso.1eH.?. 1, Celebes; India. See Carmichael *et al.* (*Genera of Hyphomycetes*, 1980; relationships).

Monotropomyces Costantin & Dufour (1921) nom. dub., anamorphic *Pezizomycotina*. See de Hoog & Hermanides-Nijhof (*Stud. Mycol.* **15**: 185, 1977).

monotypic, having only one representative, as a genus having only one species.

monoverticillate (of a penicillus), composed of phialides only.

Montagne (Jean Pierre François Camille; 1784-1866; France). A medic in Napoleon's army; in retirement, an amateur mycologist living in Paris. Noted for his work on fungi of Chile, Cuba, France and other countries, and on plant diseases, particularly as the author of *Botrytis infestans* (= *Phytophthora infestans*) the causal organism of potato late blight. *Publs.* Centuries des plantes cellulaires. *Annales des Sciences Naturelles* Botanique (1837-1860) [reprint 1970; see Stafleu, *Taxon* **19**: 633, 1970]; Botanique. Plantes cellulaires. *In* R. de la Sagra [ed.] *Histoire Physique Politique et Naturelle de l'Ile de Cuba* (1838-1842); Hongos I& II. *In* C. Gay [ed.] *Historia Física y Política de Chile. Botánica* (1853-1854); *Sylloge Generum Specierumque Cryptogamarum* (1856). *Biogs, obits etc.* Stafleu & Cowan (*TL-2* **3**: 557, 1981).

Montagnea Fr. (1836), Agaricaceae. 5, widespread (subtropical dry areas). See Reid & Eicker (*S. Afr. J. Bot.* **57**: 161, 1991), Hopple & Vilgalys (*Mol. Phylogen. Evol.* **13**: 1, 1999; phylogeny), Stasinska & Prajs (*Polish Botanical Journal* **47**: 211, 2002; *Montagnea arenaria* in Poland).

Montagneaceae Singer (1976) = Agaricaceae.

Montagnella Speg. (1881), Dothideomycetes. 9, widespread.

Montagnellina Höhn. (1912) ? = Guignardia fide von Arx & Müller (*Beitr. Kryptfl. Schweiz* **11** no. 1, 1954).

Montagnina Höhn. (1910) ≡ Gibbera fide von Arx & Müller (*Stud. Mycol.* **9**, 1975).

Montagnites Fr. (1838) ≡ Montagnea.

Montagnula Berl. (1896), Montagnulaceae. Anamorph *Aposphaeria*. 24, widespread. The systematic position is doubtful, and it is almost certainly polyphyletic. See Huhndorf (*Brittonia* **44**: 208, 1992), Aptroot (*Nova Hedwigia* **60**: 325, 1995; posn in *Phaeosphaeriaceae*), Ramaley & Barr (*Mycotaxon* **54**: 75, 1995), Hyde *et al.* (*Nova Hedwigia* **69**: 449, 1999; on palms), Barr (*Mycotaxon* **77**: 193, 2001; fam. placement), Liew *et al.* (*Mycol.* **94**: 803, 2002; phylogeny), Schoch *et al.* (*Mycol.* **98**: 1041, 2006; phylogeny).

Montagnulaceae M.E. Barr (2001), Pleosporales. 7 gen. (+ 8 syn.), 100 spp. In need of reappraisal using molecular methods.

Lit.: Barr (*Mycotaxon* **77**: 193, 2001), Liew *et al.* (*Mycol.* **94**: 803, 2002; phylogeny), Wang *et al.* (*MR* **111**: 1268, 2007; phylogeny).

Montemartinia Curzi (1927) = Chaetosphaeria fide Kohlmeyer & Kohlmeyer (*Marine Mycology*, 1979).

Montinia A. Massal. (1855) [non *Montinia* Thunb. 1776, *Saxifragaceae*] ≡ Pyrenocarpon.

Montoyella Castell. & Chalm. (1907) nom. dub., anamorphic *Pezizomycotina*. See de Hoog & Hermanides-Nijhof (*Stud. Mycol.* **15**: 186, 1977).

Moorella P. Rag. Rao & D. Rao (1964), anamorphic *Pezizomycotina*, Hso.0-≡ hP.10. 1, India. See Rao & Rao (*Mycopathologia* **22**: 51, 1964).

Morakotiella Sakay. (2005), Halosphaeriaceae. 1, Italy. See Sakayaroj *et al.* (*Mycol.* **97**: 806, 2005).

Moralesia Urries (1956), anamorphic *Pezizomycotina*, St.1bH.?. 1, Canary Islands. See Urries (*An. Inst. bot. A.J. Cavanilles* **14**: 167, 1956).

Moravecia Benkert, Caillet & Moyne (1987), Pyronemataceae. 2, Europe. See Caillet & Moyne (*Bull. Soc. Hist. nat. Doubs* **84**: 9, 1991), Benkert & Kristiansen (*Z. Mykol.* **65**: 33, 1999; Norway).

Morchella Dill. ex Pers. (1794), Morchellaceae. Anamorph *Costantinella*. *c.* 36, widespread (esp. temperate). The morels, important edible fungi. See Patrignani & Pellegrini (*Atti Accad. naz. Lincei Rc. Sed. solen.* Cl. Sci., Fis., Mat. Nat. **75**: 161, 1983; ultrastr.), Gessner *et al.* (*Mycol.* **79**: 683, 1987; genetics), Yoon *et al.* (*Mycol.* **82**: 227, 1990; population genetics), Amir *et al.* (*MR* **96**: 943, 1992; growth in culture), Rohe (*Biblthca Mycol.* **146**, 1992; genetics), Amir *et al.* (*MR* **97**: 683, 1993; morphology, physiology), Jung *et al.* (*Mycol.* **85**: 677, 1993; ELISA), Bunyard *et al.* (*Exp. Mycol.* **19**: 223, 1995; phylogeny), Bunyard *et al.* (*Mycol.* **86**: 762, 1994; RFLP analysis), Gessner (*CJB* **73** Suppl. 1: S967, 1995; genetics, systematics), Wipf *et al.* (*Appl. Environm. Microbiol.* **62**: 3541, 1996; ITS sequences), Landvik *et al.* (*Nordic Jl Bot.* **17**: 403, 1997; phylogeny), Guzmán & Tapia (*Mycol.* **90**: 705, 1998; Mexico), Parguey-Leduc *et al.* (*Cryptog. Bryol.-Lichénol.* **19**: 277, 1998; asci, ascospores), Harrington *et al.* (*Mycol.* **91**: 41, 1999; phylogeny), Wipf *et al.* (*Can. J. Microbiol.* **45**: 769, 1999; ITS variation), Kellner *et al.* (*Organ. Divers. Evol.* **5**: 101, 2005; phylogeny), Hansen & Pfister (*Mycol.* **98**: 1029, 2006; phylogeny).

Morchellaceae Rchb. (1834), Pezizales. 4 gen. (+ 8 syn.), 49 spp.

Lit.: Jacquetant (*Les Morilles*: 144 pp., 1984), Volk & Leonard (*MR* **94**: 399, 1990), Amir *et al.* (*MR* **96**: 943, 1992), Buscot (*Mycorrhiza* **4**: 223, 1994), Bunyard *et al.* (*Mycol.* **86**: 762, 1994), Bunyard *et al.* (*Exp. Mycol.* **19**: 223, 1995), Landvik *et al.* (*Nordic Jl Bot.* **17**: 403, 1997), O'Donnell *et al.* (*Mycol.* **89**: 48, 1997; phylogeny), Wipf *et al.* (*Cryptog. Mycol.* **18**: 95, 1997), Guzmán & Tapia (*Mycol.* **90**: 705, 1998), Parguey-Leduc *et al.* (*Cryptog. Bryol.-Lichénol.* **19**: 277, 1998), Harrington *et al.* (*Mycol.* **91**: 41, 1999), Wipf *et al.* (*Can. J. Microbiol.* **45**: 769, 1999), Hansen & Pfister (*Mycol.* **98**: 1029, 2006; phylogeny), Læssøe & Hansen (*MR* **111**: 1075, 2007; phylogeny).

Moreaua Liou & H.C. Cheng (1949), Anthracoideaceae. 31 (on *Cyperaceae*), widespread. See Vánky (*Mycotaxon* **74**: 343, 2000), Vánky (*Mycotaxon* **81**: 367, 2002), Vánky & McKenzie (*Fungal Diversity Res. Ser.* **8**: 259 pp., 2002; New Zealand).

morel, the edible ascoma of *Morchella*; **false -**, see lorchel.

Morella Pérez Reyes (1964), ? Olpidiaceae. 1, USA. See Karling (*Bull. Torrey bot. Club* **99**: 223, 1972).

Morellus Eaton (1818) ≡ Phallus.

Morenoella Speg. (1885) = Lembosia fide von Arx & Müller (*Stud. Mycol.* **9**, 1975), Goos (*Mycotaxon* **73**: 455, 1999).

Morenoina Theiss. (1913), ? Asterinaceae. Anamorph *Sirothyriella*. 11, widespread. See Ellis (*TBMS* **74**: 297, 1980; key Brit. spp.), Cannon (*TBMS* **86**: 190, 1986; Pakistan), Hofmann & Piepenbring (*Fungal Diversity* **22**: 55, 2006; Panama).

Morfea (G. Arnaud) Cif. & Bat. (1963) = Polychaeton fide Punithalingam (*Mycol. Pap.* **149**, 1981), Reynolds (*Madroño* **45**: 250, 1998).

Morfea Roze (1867) = Polychaeton fide Hughes (*Mycol.* **68**: 792, 1976).

Morganella Zeller (1948), Agaricaceae. 9, widespread (esp. tropical). See Kreisel & Dring (*Feddes Repert.* **74**: 109, 1967; key 7 spp.), Ponce de Leon (*Fieldiana, Bot.* **34**: 27, 1971), Suárez & Wright (*Mycol.* **88**: 655, 1996; S. Amer. spp.), Krüger & Kreisel (*Mycotaxon* **86**: 169, 2003; *Morganella* subgen. *Apioperdon*), Kreisel & Karasch (*Mycologia Bavarica* **8**: 9, 2005).

moriform, like a mulberry (*Morus*) fruit in form.

Morilla Quél. (1886) ≡ Morchella.

Morinia Berl. & Bres. (1889), anamorphic *Amphisphaeriaceae*, Cac.#eP.?. 2, Europe; Asia. See

Collado *et al.* (*Mycol.* **98**: 616, 2006).

Moriola Norman (1872), Moriolaceae (±L). 4, Europe. See Bachman (*Nyt Mag. naturv.* **64**: 170, 1926), Eriksson (*Op. Bot.* **60**, 1981), Triebel (*Biblthca Lichenol.* **35**, 1989).

Moriolaceae Zahlbr. (1903), ? Dothideomycetes (inc. sed.) (±L). 1 gen., 4 spp.
Lit.: Eriksson (*Op. Bot.* **60**, 1981), Clauzade & Roux (*Bull. Soc. bot. Centre-Ouest* Nouv. sér., num. spec. 7: 893 pp., 1985).

Moriolomyces Cif. & Tomas. (1953), Dothideomycetes. 1, Europe. See Eriksson (*Op. Bot.* **60**, 1981), Eriksson & Hawksworth (*SA* **9**: 27, 1991).

Moriolopis Norman ex Keissl. (1927) nom. inval. = Melanomma fide Keissler (*Rabenh. Krypt.-Fl.* **9** 1(1), 1933).

Morispora Salazar-Yepes, Pardo-Card. & Buriticá (2007), Phragmidiaceae. Anamorph *Gerwasia*. 1, S. America. See Salazar Yepes, M.; Pardo Cardona, V.M.; Buriticá Céspedes, P. (*Caldasia* **29**: 105, 2007).

Moristroma A.I. Romero & Samuels (1991), ? Chaetothyriales. 1, widespread. See Romero & Samuels (*Sydowia* **43**: 228, 1991), Nordén *et al.* (*Mycol. Progr.* **4**: 325, 2005; phylogeny, morphology), Lee *et al.* (*Mycol.* **98**: 598, 2006; phylogeny).

Morobia E. Horak (1979) = Lepiota fide Kuyper (*in litt.*).

Morochkovsky (Semion Filimonovich; 1897-1962; Ukraine). Postgraduate student in mycology and plant pathology (1928-1931) then Research Officer (1931-1938), All Union Scientific Research Institute of the Sugar Beet Industry, Kiev; Senior Research Officer (1938-1939) then Head of the Department of Mycology (1939-1962), M.G. Kholodny Institute of Botany, Kiev. Greatly influenced by Jaczewski (q.v.). Investigated widespread sudden death of horses (1938-1939) [an extremely dangerous occupation when Stalin was purging Ukraine of so-called 'wreckers'], establishing the cause as toxins from fodder contaminated by *Stachybotrys*. Studied factors promoting growth of wild mushrooms to alleviate postwar hunger (1945-1948). Established the series [*Guide to the Identification of Fungi of Ukraine*] [in Ukrainian]. *Publs.* (with Zerova, Lavitska & Smitska) [*Handbook of the Fungi of Ukraine*] (1969) [in Ukrainian, published posthumously]. *Biogs, obits etc.* Smitskaya & Dudka ([*Mikologiya i Fitopatologiya*] **2**: 79, 1968, portrait [in Russian]).

Moronopsis Delacr. (1891) ? = Cheirospora fide Sutton (*Mycol. Pap.* **141**, 1977).

Morosporium Renault & Roche (1898), Fossil Fungi, anamorphic *Pezizomycotina*. 1 (Eocene), France.

morph, form.

morphotype, a group of morphologically differentiated individuals of a species of unknown or of no taxonomic significance.

Morqueria Bat. & H. Maia (1963) = Cirsosia fide von Arx & Müller (*Stud. Mycol.* **9**, 1975).

Morrisiella Saikia & A.K. Sarbhoy (1985), Chaetosphaeriaceae. See Sutton (*in litt.*), Shenoy *et al.* (*MR* **110**: 916, 2006; phylogeny).

Morrisographium M. Morelet (1968), anamorphic *Pezizomycotina*, Hsy.≡ eP.1. 7, widespread. See Illman & White (*CJB* **63**: 423, 1985; key), Murvanishvili & Svanidze (*Soobshch. Akad. Nauk Gruz.* **137**: 573, 1990).

Morthiera Fuckel (1870) = Entomosporium fide Saccardo (*Syll. fung.* **3**: 657, 1884).

Mortierella Coem. (1863), Mortierellaceae. *c.* 85 (mainly in soil), widespread. See Kuhlman (*Mycol.* **64**: 325, 1972), Kuhlman (*Mycol.* **67**: 678, 1975; zygospores), Gams (*Persoonia* **9**: 381, 1977; key spp. in cult.), Benjamin (*Aliso* **9**: 157, 1978), Ansell & Young (*TBMS* **91**: 221, 1988; zygospores), Chen (*Mycosystema* **5**: 23, 1992; Chinese spp.), Degawa & Tokumasu (*Mycoscience* **38**: 387, 1997; zygosporogensis), Kirk (*IMI Descr. Fungi Bact.* **131**, 1997), Degawa & Tokumasu (*Mycol.* **90**: 1040, 1998; n.spp. & zygospores), Degawa & Tokumasu (*MR* **102**: 593, 1998; zygospores), Watanabe *et al.* (*Mycoscience* **39**: 475, 1998), Watanabe *et al.* (*MR* **105**: 506, 2001; key homothallic spp.), Meyer & Gams (*MR* **107**: 339, 2003; delimitation of *Umbellopsis*), Kwasna *et al.* (*MR* **110**: 501, 2006; phylogeny soil isolates), White *et al.* (*Mycol.* **98**: 872, 2006; phylogeny), Idnurm *et al. in* Heitman *et al.* (Eds.) (*Sex in Fungi*: 407, 2007; mating).

Mortierellaceae A. Fisch. (1892), Mortierellales. 6 gen. (+ 5 syn.), 93 spp.
Lit.: O'Donnell *et al.* (*Mycol.* **93**: 286, 2001), Voigt & Wöstemeyer (*Gene* **270**: 113, 2001), Corradi *et al.* (*Fungal Genetics Biol.* **41**: 262, 2004), Tanabe *et al.* (*Mol. Phylogen. Evol.* **30**: 438, 2004), Kwasna *et al.* (*MR* **110**: 501, 2006).

Mortierellales Caval.-Sm. (1998). Mucoromycotina. 1 fam., 6 gen., 93 spp. Fam.:
Mortierellaceae
Lit.: Cavalier-Smith (*Biol. Rev.* **73**: 203, 1998), White *et al.* (*Mycol.* **98**: 860, 2006; molecular phylogeny), Hibbett *et al.* (*MR* **111**: 109, 2007), and see under Family.

Morularia Nann. (1925) ≡ Heterobotrys.

mosaic fungus, a network resembling disorganized dermatophyte mycelium sometimes seen in skin scales cleared in potassium hydroxide. An artifact, fide Weidman, who first described the effect in 1927, or the extracellular deposit of a dermatophyte (fide Dowding, *Arch. Derm. Chicago* **66**: 470, 1952).

Moschomyces Thaxt. (1894) = Compsomyces fide Tavares (*Mycol. Mem.* **9**, 1985).

Moser (Meinhard Michael; 1924-2002; Austria). Researcher, Federal Institute of Forestry Research, Imst, Tirol (1951-1968); Professor of Microbiology, Botanical Institute, Innsbruck (1968-1972); first Director, Institute of Microbiology, Innsbruck (1972-1991). Career began late because of years as a prisoner of war and then Soviet labour camp internee; developed techniques for inoculating trees with ectomycorrhizal fungi; outstanding European agaricologist, specializing in *Cortinarius*. *Publs. Kleine Kryptogamenflora von Mitteleuropa. Basidiomyceten. IIB. Die Röhrlinge, Blätter- und Bauch pilze (Agaricales und Gasteromycetales)* (1955); *Kleine Kryptogamenflora von Mitteleuropa. IIA. Ascomyceten (Schlauchpilze)* (1963). *Biogs, obits etc.* Horak, Peintner & Pöder (*Mycological Progress* **1**: 331, 2002); Horak, Peintner & Pöder (*MR* **107**: 506, 2003) [portrait].

Moserella Pöder & Scheuer (1994), ? Sclerotiniaceae. 1 (on roots), Austria. See Pöder & Scheuer (*MR* **98**: 1334, 1994).

Mosigia Fr. (1845) [non *Mosigia* Spreng. 1826, *Compositae*] = Rimularia fide Hertel & Rambold (*Biblthca Lichenol.* **38**: 145, 1990).

Moss (Stephen Thomas; 1943-2001; England). PhD,

Reading (1972); Research Associate, University of Kansas (1972-1973); Research Fellow (1974-1978) then Lecturer (1978-1981) then Senior Lecturer (1981-1988) then Reader (1988-2001), University of Portsmouth. Expert on Trichomycetes and marine fungi and noted electron microscopist; his teamwork skills made him an outstanding Secretary then President of the British Mycological Society, dying in office. *Publs. Biology of Marine Fungi* (1986). *Biogs, obits etc.* Dick (*MR* **108**: 214, 2004).

Mossopisporites Kalgutkar & Janson. (2000), Fossil Fungi, anamorphic *Ascomycota*. 1, China. See Kalgutkar & Jansonius (*AASP Contributions Series* **39**: 183, 2000).

Mothesia Oddo & Tonolo (1967) = Claviceps fide Hawksworth *et al.* (*Dictionary of the Fungi* edn 8, 1995).

mould (1) a microfungus having a well-marked mycelium or spore mass, esp. an economically important saprobe; **anther -** of clover, *Botrytis anthophila*; **black -**, *Aspergillus niger*; **blue -**, *Penicillium*; of apple, *P. expansum*; of citrus, *P. italicum*; of tobacco, *Peronospora hyoscyami* (syn. *P. tabacina*; Shepherd, *TBMS* **55**: 253, 1970), McKeen (*Blue mold of tobacco*, 1989); **bread -**, *Chrysomilia sitophila*; also used of *Mucorales* on bread; **green -** of citrus, *Penicillium digitatum*; **grey -**, *Botrytis cinerea*; of snowdrop, *B. galanthina*; **pin -**, *Mucor* and other *Zygomycetes*; **plaster -**, brown, *Papulospora byssina*; white, '*Oospora*' (q.v.) *fimicola*; **slime -s**, *Mycetozoa*; **snow -s**, low temp. tolerant pathogens growing on unfrozen soil surface below snow cover, causing diseases of winter cereals, grasses and forage legumes, *Monographella nivalis* (pink snow mould), *Sclerotinia borealis*, *Typhula* spp.; **sooty -s**, *Atichiaceae*, *Capnodiales*, etc.; **tomato leaf -**, *Fulvia fulva*; **water -s**, aquatic *Chytridiomycetes* and *Oomycetes*, esp. *Saprolegniales*; **white -** of sweet pea, *Hyalodendron album*; (2) see mildew (2). Illman (*Mycol.* **62**: 1214, 1970) advocated as American usage 'mould' for a fungus, 'mold' for a shape.

Mouliniea C.P. Robin (1853), Pezizomycotina. 1, Europe.

Mounting media. The appearance of microscopic structures of fungi differs greatly depending on the condition of the material and the mountant used. Ideally and if possible, microscopic examination of fungi should begin with material in a fresh living state and water as the mounting fluid (see Baral, *Mycotaxon* **44**: 333, 1992; vital taxonomy); this will enable structures to be measured without problems of shrinkage, and their natural colour to be observed; it is also useful for observing gelatinous material and non-cellular appendages. Thereafter, if necessary or desired, other mountants including stains may be suitable. Various mounting fluids are available. Stains are often combined with clearing agents such as lactic acid, glycerol and chloral hydrate to produce temporary or permanent mounts. Lactophenol (Amann, 1896) was also formerly used, but is now not recommended because of its carcinogenic properties. See Dring in Booth (Ed.) (*Methods in microbiology* **4**: 96, 1971), Tuite (*Plant pathological methods. Fungi & bacteria*, 1969). Many mountants are adversely affected by exposure to light and should be stored in a dark bottle.

Lactic acid in glycerol

Although less effective than lactophenol for clearing plant tissues, lactic acid has similar properties and can be used neat or combined with glycerol and water for routine examination of many fungi.

Lactic acid	25.0 ml
Glycerol	50.0 ml
Distilled water	25.0 ml

Cotton blue in lactic acid

This is widely used as a standard mountant which rapidly stains the cytoplasm of fungal cells. Permanent mounts can be prepared by sealing with nail varnish.

Cotton blue	0.01 g
Lactic acid (85%)	100.0 ml

Add cotton blue powder to lactic acid. Heat in a glass beaker and stir until dissolved. Allow to cool then filter to remove any sediment. Store in a dark bottle.

Erythrosin B in ammonia

An aqueous mountant which stains fungal cell walls, producing a clear outline of septation in spores and mycelium. It is particularly useful for examining hyaline structures and conidiogenesis in anamorphic fungi. This stain crystallizes within 48 hours and can only be used as a temporary mountant.

Erythrosin B	1.0 g
Ammonia	10.0 ml
Distilled water	90.0 ml

Mix ammonia and distilled water. Add erythrosin B and stir until dissolved.

Lactofuchsin

A red stain produced by combining acid-fuchsin with lactic acid (Carmichael, *Mycologia* **47**: 611, 1955). This has a slightly better refractive index than cotton blue, enabling structures to be seen clearly and is particularly suitable for staining structures prior to photographing.

Acid-fuchsin	0.1 g
Lactic acid	100.0 ml

Polyvinyl alcohol in lactic acid

This mountant has a high refractive index (1.39) resulting in minimal distortion of structures, sets rapidly and requires no sealing, providing an excellent permanent mounting medium for examination of fungi. See Omar *et al.* (*Stain Technology* **53**: 293, 1978).

Polyvinyl alcohol	1.66 g
Lactic acid	10.0 ml
Glycerol	1.0 ml
Distilled water	10.0 ml

Potassium hydroxide (KOH)

Used at 5-10%, potassium hydroxide is useful for softening and clearing fungal tissues. It is most commonly used as a pre-treatment for examining ascomycetes and anamorphic fungi with tough-walled fruit-bodies. After clearing, the KOH can be washed off with water using a dropper or pipette and replaced with a stain. A 5% solution is prepared as follows:

Potassium hydroxide	5.0 g
Distilled water	100.0 ml

Weigh out 5 g potassium hydroxide pellets. Add to distilled water in a glass beaker. Stir until dissolved.

Shear's mounting fluid

A colourless mounting fluid suitable for permanent preparations of myxomycetes, ascomycetes, rusts and smuts and particularly useful for microphotography. See Chupp (*Mycologia* **32**: 269, 1940), Gams *et al.* (*CBS Course of mycology*, edn 3, 1987).

Potassium acetate	3 g

glycerol 60 ml
Ethanol (95%) 90 ml
Distilled water 150 ml

Lugol's iodine

Used as an alternative to Melzer's iodine for testing the amyloid reaction in spores, asci and hymenium of ascomycetes and basidiomycetes. It is particularly useful for examining apical apparatus in asci of lichenised fungi, giving a clearer result than Melzer's iodine.

Iodine 1.0 g
Potassium iodide 2.0 g
Distilled water 100.0 ml

Prepare in a fume cupboard. Dissolve potassium iodide in distilled water in a glass beaker. Add iodine. Cover the beaker and leave to dissolve for 24 hours. Mix thoroughly.

Melzer's iodine

Mainly used for testing the amyloid reaction in sporulating structures of ascomycetes and basidiomycetes. (Melzer, *BSMF* **40**: 78, 1924). Also clears and stains fungal tissues including cell walls and can be used for general examination of fungal structures. Pre-treatment of material with 5-10 % potassium hydroxide is recommended. Permanent mounts can be made by drawing off the reagent with filter paper, replacing with lactic acid or lactophenol and sealing with nail varnish.

Chloral hydrate 22.0 g
Iodine 0.5 g
Potassium iodide 1.5 g
Distilled water 20.0 ml

Prepare in a fume cupboard. Dissolve potassium iodide in distilled water. Add iodine and leave to dissolve for 24 hours. Add chloral hydrate and stir until dissolved. Mix thoroughly.

Water (distilled or tap)

See also stains.

mouse favus, a skin disease in humans (*Trichophyton quinckeanum*).

Moutoniella Penz. & Sacc. (1901), Rhytismataceae. 1, Java. See Sherwood (*Mycotaxon* **5**: 1, 1977).

Mrakia Y. Yamada & Komag. (1987), Cystofilobasidiaceae. 4, Antarctica. See Fell *et al.* (*Int. J. Syst. Evol. Microbiol.* **50**: 1351, 2000; mol. phylogeny), Malosso *et al.* (*Polar Biology* **29**: 552, 2006; maritime Antarctic soils).

Mrazekia L. Léger & E. Hesse (1916), Microsporidia. 6.

MSDN, Microbial Strain Data Network (Cambridge, UK); started 1985; see Krichevsky *et al.* (*in* Hawksworth & Kirsop, 1988: 31).

Mucedinaceae Link (1809) = Moniliaceae. (obsol.).

mucedinaceous, see moniliaceous.

mucedinous, white or pale in colour and mould-like; mucedinoid.

Mucedites C.E. Bertrand & Renault (1896), Fossil Fungi (mycel.) Fungi. 1 (Carboniferous), France.

Mucedo Pers. (1794) ≡ Mucor Fresen.

Muchmoria Sacc. (1906) nom. dub., Fungi. See Hughes (*CJB* **36**: 787, 1958).

Mucidula Pat. (1887) = Oudemansiella fide Singer (*Agaric. mod. Tax.*, 1951).

mucilaginous, sticky when wet; slimy.

Mucilopilus Wolfe (1979) = Fistulinella fide Kuyper (*in litt.*).

Muciporus Juel (1897) = Tulasnella fide Juel (*Ark. Bot.* **14**: 1, 1914), Donk (*Persoonia* **4**: 145, 1966).

Muciturbo P.H.B. Talbot (1989), Pezizaceae. Anamorph *Chromelosporium*. 3 (hypogeous), Australia. Probably close to *Ruhlandiella*. See Warcup & Talbot (*MR* **92**: 95, 1989).

MUCL, Mycothèque de l'Université Catholique de Louvain (Louvain-la-Neuve, Belgium); founded 1894; an institute of the Catholic University of Louvain, with special funding for the genetic resource collection from the Belgian Science Policy Office.

Mucobasispora Mustafa & Abdul-Wahid (1990), anamorphic *Pezizomycotina*, Hso.0eP.35. 1 (from soil), Egypt; Russia. See Mustafa & Abdul-Wahid (*MR* **94**: 131, 1990), Mel'nik *et al.* (*Nov. sist. Niz. Rast.* **28**: 76, 1992).

Mucomassaria Petr. & Cif. (1932), Dothideomycetes. 1, C. America. See Petrak (*Sydowia* **13**: 1, 1956).

Mucophilus Plehn (1920), ? Chytridiales. 1 (in *Cyprinus*), Germany.

Mucor Fresen. (1850) nom. cons., Mucoraceae. *c.* 50, widespread. See Benjamin & Mehrotra (*Aliso* **5**: 235, 1963; azygosporic spp.), Zycha *et al.* (*Mucorales*, 1969), Mehrotra & Mehrotra (*Sydowia* **31**: 94, 1974; azygosporic spp.), Schipper *et al.* (*Persoonia* **8**: 321, 1975; zygospore ornamentation), O'Donnell *et al.* (*CJB* **55**: 2712, 1977; azygospore ultrastr.), Schipper (*Stud. Mycol.* **17**: 48, 1978; key 39 spp.), Stalpers & Schipper (*Persoonia* **11**: 39, 1980; zygospore ornamentation), James & Gauger (*Mycol.* **74**: 744, 1982; genetics), Chen & Zheng (*Acta Mycol. Sin.* **1**: 56, 1986), Hesseltine & Rogers (*Mycol.* **79**: 289, 1987; zygospore formation), Michailides & Spotts (*Mycol.* **80**: 837, 1988; zygospore germination), Ginman & Young (*MR* **93**: 314, 1989; azygospore ultrastr.), Schipper (*Stud. Mycol.* **31**: 151, 1989), Bärschi *et al.* (*MR* **94**: 373, 1991; selective medium), Orlowski (*Microbiol. Rev.* **55**: 234, 1991; dimorphism), Liou *et al.* (*Mycol. Monogr.* **6**, 1993), Weitzman *et al.* (*J. Clin. Microbiol.* **31**: 2523, 1993; mucormycosis), Nagy *et al.* (*Curr. Genet.* **26**: 45, 1994; karyotype), Schipper & Samson (*Mycotaxon* **50**: 475, 1994), Watanabe (*Mycol.* **86**: 691, 1994; key homothallic spp.), Vágvölgyi *et al.* (*Can. J. Microbiol.* **42**: 613, 1996), Zalor *et al.* (*Mycotaxon* **65**: 507, 1996), Kirk (*IMI Descr. Fungi Bact.* **131**, 1997), Michaelides *et al.* (*Mycol.* **89**: 603, 1997; zygosporogenesis), Papp *et al.* (*Antonie van Leeuwenhoek* **72**: 167, 1997; DNA polymorphism), O'Donnell *et al.* (*Mycol.* **93**: 286, 2001; phylogeny), Voigt & Wöstemeyer (*Gene* **270**: 113, 2001; phylogeny), Alves *et al.* (*Revista Brasileira de Botânica* **25**: 147, 2002; Brazil), Tanabe *et al.* (*Mol. Phylogen. Evol.* **30**: 438, 2004; phylogeny), Kwasna *et al.* (*MR* **110**: 501, 2006; phylogeny soil isolates), White *et al.* (*Mycol.* **98**: 872, 2006; phylogeny), Wöstemeyer & Schimek *in* Heitman *et al.* (Eds.) (*Sex in Fungi*: 431, 2007; trisporic acid & mating), Dávila López *et al.* (*Nucleic Acidis Research* **36**: 3001, 2008; spliceosomal RNA gene phylogeny), Nyilasi *et al.* (*Clin. Microbiol. Infect.* **14**: 393, 2008; molecular identification).

Mucor L. (1753) nom. rej. = Calicium fide Hawksworth *et al.* (*Dictionary of the Fungi* edn 8, 1995).

Mucor P. Micheli ex Fr. (1832) nom. rej. prop. = Rhizopus fide Kirk (*Taxon* **35**: 371, 1986).

Mucoraceae Dumort. (1822), Mucorales. 25 gen. (+ 33 syn.), 129 spp.

Lit.: Ellis & Hesseltine (*Sabouraudia* **5**: 59, 1966), Zycha *et al.* (*Mucorales*: 355 pp., 1969), Benny & Benjamin (*Aliso* **8**: 301, 1975), Benny & Benjamin

(*Aliso* **8**: 391, 1976), Schipper (*Stud. Mycol.* **17**: 1, 1978), Schipper (*Persoonia* **14**: 133, 1990), Kimura *et al.* (*J. Med. Vet. Mycol.* **33**: 137, 1995), Munipalli *et al.* (*J. Clin. Microbiol.* **34**: 2601, 1996), Tanabe *et al.* (*Mol. Phylogen. Evol.* **30**: 438, 2004), Iwen *et al.* (*J. Clin. Microbiol.* **43**: 5819, 2005), Kwasna *et al.* (*MR* **110**: 501, 2006), Liou *et al.* (*MR* **111**: 196, 2007).

Mucorales Fr. (1832). Mucoromycotina. 9 fam., 51 gen., 205 spp. Asexual reproduction by multi-spored or few- (to one) spored sporangia (sporangiola), forcibly discharged in *Pilobolus*, sexual reproduction by zygospores; cosmop. saprobes (rarely mycoparasites), or few facultative parasites of plants or animals (incl. humans). Fams:
(1) **Choanephoraceae**
(2) **Cunninghamellaceae**
(3) **Mucoraceae**
(4) **Mycotyphaceae**
(5) **Phycomycetaceae**
(6) **Pilobolaceae**
(7) **Radiomycetaceae**
(8) **Syncephalastraceae**
(9) **Umbelopsidaceae**

Recent molecular evidence (Voigt *et al.*, 1999; O'Donnell *et al.*, 2001) has suggested that the traditional classification is highly artificial and the above can only be considered a temporary, pragmatic solution.

Lit.: Benny (PhD thesis, Claremont-Graduate School, USA, 1973, *CJB* **73**(Suppl. 1): S725, 1995), Benny *et al.* (*Mycotaxon* **22**: 119, 1985), Hesseltine & Ellis (*in* Ainsworth *et al.* (Eds), *The Fungi* **4B**: 187, 1973), Zycha *et al.* (*Mucorales*, 1969), Benjamin (*in* Kendrick (Ed.), *The whole fungus*, 1979), Hesseltine (*Mycol.* **47**: 344, 1955, **57**: 149, 1965; *Mycologist* **5**: 162, 1991; food fermentations), Kirk (*Mycol. Pap.* **152**, 1984), Benny & Benjamin (*Mycol.* **83**: 713, 1992; **85**: 660, 1983), von Arx (*Sydowia* **35**: 10, 1982), Scholer *et al.* (*Fungi pathogenic for humans and animals* **3A**: 9, 1983), Newsham & Gauger (*Exp. Mycol.* **8**: 314, 1984; heterokaryons), Mikawa (*in* Watanabe & Malla (Eds), *Cryptogams of the Himalayas* **1**: 77, 1988), Arnold (*Rev. Jard. Bot. Nac. Univ. Habana* **12**: 121, 1991), Zhou *et al.* (*Mycosystema* **4**: 1, 1991; DNA base composition), Mikawa (*in* Nakaike & Malik (Eds), *Cryptogamic flora of Pakistan*: 119, 1992; **2**: 65, 1993), Orlowski (*in* Wessels & Meinhardt (Eds), *The Mycota* **1**: 143, 1994; dimorphism), Gooday (*in* Wessels & Meinhardt (Eds), *The Mycota* **1**: 401, 1994; sex hormones), Richardson & Shankland (*in* Murray *et al.* (Eds), *Manual of clinical microbiology*, 6th ed, 1995; mucormycosis), Ellis (*in* Ajello & Hay, (eds), Medical Mycology **4**), Collier *et al.* (eds, *Topley & Wilson's Microbiology and Microbial Infections*, ed. 9, p. 247, 1998), Voigt *et al.* (*J. Clin. Microbiol.* **37**: 3957, 1999; phylogeny), O'Donnell *et al.* (*Mycologia* **93**: 286, 2001; phylogeny), White *et al.* (*Mycol.* **98**: 860, 2006; molecular phylogeny), Hibbett *et al.* (*MR* **111**: 109, 2007), Dávila López *et al.* (*Nucleic Acids Research* **36**: 3001, 2008; spliceosomal RNA gene phylogeny), and see under fams.

Mucoralites Patel (1979), Fossil Fungi, Zygomycetes. 1 (Tertiary), India.

Mucoricola Nieuwl. (1916) ≡ Piptocephalis fide Benjamin (*Aliso* **4**: 321, 1959).

Mucorites Mesch. (1898), Fossil Fungi. 1 (Carboniferous), France.

mucormycosis, strictly, a disease of humans or animals caused by one of the *Mucorales* (e.g. *Absidia corymbifera*) but sometimes also applied to infections caused by members of the *Entomophthorales*. Cf. phycomycosis, zygomycosis.

Mucorodium K.M. Zalessky (1915), Fossil Fungi ? Mucorales. 1 (Permian), former USSR.

Mucoromycotina Benny (2007). Zygomycota. 3 ord., 61 gen., 325 spp. Ords:
(1) **Endogonales**
(2) **Mortierellales**
(3) **Mucorales**

Lit: James *et al.* (*Mycol.* **98**: 860, 2006; molecular phylogeny), Hibbett *et al.* (*MR* **111**: 109, 2007), Hoffmann *et al.* (*in* Gherbawy (Ed.) *Current Advances in Molecular Mycology*): in press, 2008).

Mucoromycotina K.M. Zalessky (1915), Fossil Fungi ? Mucorales. 1 (Permian), former USSR.

Mucosetospora M. Morelet (1972), anamorphic *Pezizomycotina*, St.0eH.1. 1, France. See Morelet (*Bulletin de la Société des Sciences naturelles et d'Archéologie de Toulon et du Var* **201**: 4, 1972).

mucronate, pointed; ending in a short, sharp point. (Fig. 23.39).

Mucronella Fr. (1874), Clavariaceae. 8, widespread. See Corner (*Ann. Bot. Mem.* [A monograph of Clavaria and allied genera] **1**, 1950), Stalpers (*Stud. Mycol.* **40**: 93, 1996; key).

Mucronia Fr. (1849) [non *Mucronea* Benth. 1836, *Polygonaceae*] ≡ Mucronella.

Mucronoporus Ellis & Everh. (1889) ≡ Onnia.

Mucrosporium Preuss (1851) = Cladobotryum fide de Hoog (*Persoonia* **10**: 33, 1978).

Muda Adans. (1763) nom. dub., ? Rhodophyta. may alternatively be a member of the Fungi, (L).

Muellerella Hepp (1862), Verrucariaceae. *c.* 16 (on lichens and *Hepaticae*), widespread (esp. north temperate). See Hawksworth (*Bot. Notiser* **132**: 283, 1979), Döbbeler & Triebel (*Bot. Jahrb. Syst.* **107**: 503, 1985; on *Hepaticae*), Matzer (*Nova Hedwigia* **56**: 203, 1993).

Muellerellomyces Cif. & Tomas. (1953) ≡ Muellerella.

Muellerites L. Holm (1968), ? Chaetothyriales. Anamorph *Aureobasidium*-like. 1 (on *Juniperus*), Europe. See Casagrande (*Phytopath. Z.* **66**: 97, 1969), Müller & Magnuson (*Arctic and Alpine Mycology* II: 3, 1987).

Muelleromyces Kamat & Anahosur (1968), Phyllachoraceae. 1 (from leaves), India. See Kang *et al.* (*Fungal Diversity* **2**: 135, 1999).

mu-erh, the edible cultivated *Auricularia* sp., esp. *A. polytricha* (China) and *A. auricula* (Japan) (Chang & Hayes, 1978).

Muhria P.M. Jørg. (1987) = Stereocaulon Hoffm. fide Myllys *et al.* (*Taxon* **54**: 605, 2005), Högnabba (*MR* **110**: 1080, 2006; phylogeny).

Muiaria Thaxt. (1914), anamorphic *Pezizomycotina*, Hso.#eP.1. 4 (on *Insecta*), Africa; Borneo. See Weir & Blackwell (*Insect-Fungal Associations* Ecology and Evolution: 119, 2005).

Muiogone Thaxt. (1914), anamorphic *Pezizomycotina*, Hsp.#eP.1. 1 (on *Insecta*), Africa. See Weir & Blackwell (*Insect-Fungal Associations* Ecology and Evolution: 119, 2005).

Muirella R. Sprague (1959), anamorphic *Pezizomycotina*, Hsp.≡ eH.10. 1, USA. See Braun (*Monogr. Cercosporella, Ramularia Allied Genera (Phytopath.*

Hyphom.) **1**, 1995).

Mukagomyces S. Imai (1940) = Tuber fide Trappe (*Mycotaxon* **9**: 247, 1979), Læssøe & Hansen (*MR* **111**: 1075, 2007; phylogeny).

Mukhakesa Udaiyan & V.S. Hosag. (1992) nom. inval., Xylariales. 1 (wood, aquatic), India. See Kang *et al.* (*Fungal Diversity* **2**: 135, 1999).

Mukorites Grüss (1931), Fossil Fungi. 1.

Müller (Emil; 1920-2008; Switzerland). Graduate in agriculture, Swiss Institute of Technology (1944); Agricultural Advisor, Plantahof Agricultural School, Grisons Canton (1944-1948); PhD, Swiss Institute of Technology (1949); Curator of Fungal Collections (1954-1966) then Lecturer (1966-1970) then Professor (1970-1973) then Regius Professor (1973 onwards), Swiss Institute of Technology (1954 onwards). An outstanding systematist, specializing in the ascomycetes (he was affectionately nicknamed 'Ascus'), collaborating with Ahmad (q.v.) and Arx (q.v.) and producing more than 200 papers; Managing Editor of *Sydowia* for many years. *Publs.* (with Arx) Die Gattungen der amerosporen Pyrenomyceten. *Beiträge zur Kryptogamenflora der Schweiz* (1954); (with Arx) Die Gattungen der didymosporen Pyrenomyceten. *Beiträge zur Kryptogamenflora der Schweiz* (1962); (with Loeffler) *Mycology. An Outline for Science and Medical Students* (1971) [English translation, 1976]; (with Arx) Pyrenomycetes: Meliolales, Coronophorales, Sphaeriales. In Ainsworth, Sparrow & Sussman [eds] *The Fungi. An Advanced Treatise* **4A** (1973). *Biogs, obits etc.* Petrini (*Sydowia* **52**: v, 2000); Petrini & Horak (*Sydowia* **38**: 400, 1985) [portrait].

Müller Argoviensis (Jean; 1828-1896; Switzerland). Student, Zürich (up to 1857); Curator, de Candolle Herbarium (1851-1869); Curator, Delessert Herbarium, Geneva (1869-1896); Director, Geneva Botanic Garden (1870-1874); Professor of Botany, Geneva (1871-1889). Carried out important work on tropical lichens, particularly from Cuba; published over 100 papers on lichen-forming fungi including many new species [reprint *Gesammelte Lichenologische Schriften*, 2 vols, 1967]. Most specimens in **G**, other material in **BM** and **M**. *Publs.* Lichenologische Beiträge. I-XXV. *Flora* (1874-1891). *Biogs, obits etc.* Briquet (*Bulletin de l'Herbier Boissier* **4**: 111, 1896); Grummann (1974: 633); Stafleu & Cowan (*TL-2* **3**: 628, 1981).

multi- (prefix), a great number; many; much.

multiallelic (of an incompatibility system), having more than 2 alleles per locus; cf. biallelic.

Multicellaesporites Elsik (1968), Fossil Fungi. 42 (Tertiary), widespread.

Multicellites Kalgutkar & Janson. (2000), Fossil Fungi, anamorphic *Ascomycota*. 1, widespread. See Kalgutkar & Jansonius (*AASP Contributions Series* **39**: 189, 2000).

Multicladium K.B. Deshp. & K.S. Deshp. (1966), anamorphic *Pezizomycotina*, Hso.0bP.?. 1, India. See Deshpande & Deshpande (*Mycopath. Mycol. appl.* **30**: 185, 1966).

Multiclavula R.H. Petersen (1967), Clavulinaceae. 13 (3 L; some assoc. with *Myxomycetes* and *Musci*), widespread. See Petersen (*Am. midl. Nat.* **77**: 205, 1967; key), Oberwinkler (*Dtsch. bot. Ges.* N.F. **4**: 139, 1970), Petersen & Kan tvilas (*Aust. J. Bot.* **34**: 217, 1986), Poelt & Obermayer (*Herzogia* **8**: 289, 1990; bulbil diaspores).

multifid, having division into a number of parts or lobes.

Multifurca Buyck & V. Hofstetter (2008), Russulaceae. 5, USA; India; Thailand; New Caledonia. See Buyck *et al.* (*Fungal Diversity* **28**: 15, 2008).

multiguttulate, containing many oil-like drops.

Multipatina Sawada (1928), anamorphic *Pezizomycotina*, Hso.?.?. 1, Taiwan. *M. citricola* ('leaf felt' of citrus).

multiperforate, of a septum with many pores connecting compartments. See Cole & Samson (*Patterns of development in conidial fungi*, 1979). (Fig. 22D).

multiseptate, having a number of septa.

multisporous, having a number of spores.

multivesicular bodies, small vesicles limited by a membrane which in *Sclerotinia fructigena* originate from the endoplasmic reticulum and are possibly related to extracellular enzyme secretion (Calonge *et al.*, *J. gen. Microbiol.* **55**: 177, 1969).

Mundkur (Balchendra Bhavanishankar; 1896-1952; India). Student, Presidency College, Madras (*c.* 1920); Assistant Mycologist, Cotton Research Scheme, Dharwar, Bombay (1922-1928); PhD, Iowa State College of Agriculture, Iowa, USA (1929-1931); Assistant Mycologist, Division of Mycology, Indian Agricultural Research Institute, first in Pusa and later in Delhi (1931-1947); Deputy Director of Plant Diseases, Directorate of Plant Protection, Quarantine and Storage (1947); Professor of Botany, Poona University (after 1947). Noted for work on rusts and smuts. *Publs. Fungi and Plant Disease* (1949), (with M.J. Thirumalachar) *Ustilaginales of India* (1952). *Biogs, obits etc.* Joshi *et al.* (*Review of Tropical Plant Pathology* **7**: 91, 1993); Mehta (*Indian Phytopathology* **5**: 1, 1953) [bibliography, portrait]; Stafleu & Cowan (*TL-2* **3**: 660, 1981).

Mundkurella Thirum. (1944), Urocystidaceae. 5 (on *Araliaceae*), S. & E. Asia; N. Amer.; New Zealand. See Savile (*Mycol.* **67**: 273, 1975), Vánky (*MR* **94**: 269, 1990; key), Vánky *et al.* (*N.Z. Jl Bot.* **37**: 329, 1999; key).

Munk pores, Small (*c.* 1 µm) pores, each surrounded by a ring of thickening, between cells of the ascoma wall in the *Nitschkeaceae*.

Munkia Speg. (1886), anamorphic *Hypocreales*, St.0eH.?. 2, S. America. See Marchionatto (*Rev. argent. Agron.* **7**: 172, 1940), Bischoff *et al.* (*Mycol.* **96**: 1088, 2004; placement).

Munkiella Speg. (1885), Polystomellaceae. Anamorph *Lasmenia*. 3, S. America.

Munkiellaceae Luttr. (1973) = Polystomellaceae.

Munkiodothis Theiss. & Syd. (1915) = Rehmiodothis fide Müller & von Arx (*Beitr. Kryptfl. Schweiz* **11** no. 2, 1962).

Munkovalsaria Aptroot (1995), ? Dacampiaceae. Anamorph *Lecythophora*-like. 3, widespread (esp. tropical). See Aptroot (*Nova Hedwigia* **60**: 325, 1995).

Murangium Seaver (1951), Patellariaceae. 1, N. America. See Kutorga & Hawksworth (*SA* **15**: 1, 1997).

Murashkinskija Petr. (1928) = Mytilinidion fide Barr (*SA* **6**: 144, 1986).

Muratella Bainier & Sartory (1913) = Cunninghamella fide Hesseltine (*Mycol.* **47**: 344, 1955).

Murciasporidium Thor (1930), Fungi. 1 (in mites), Spitzbergen.

Muribasidiospora Kamat & Rajendren (1968), Exobasidiaceae. 3, India. See Rajendren (*Mycol.* **61**:

1159, 1969), Rajendren (*Mycopathologia* **41**: 287, 1970; culture) = Exobasidium (Exobasid.) fide, Donk (*Persoonia* **8**: 33, 1974).

Muribasidiosporaceae Kamat & Rajendren (1969) = Exobasidiaceae.

muricate, rough with short, hard outgrowths.

Muricopeltis Viégas (1944), Micropeltidaceae. 1, Brazil. See Petrak (*Sydowia* **5**: 341, 1951).

Muricularia Sacc. (1877) nom. dub., anamorphic *Pezizomycotina*. See Sutton (*Mycol. Pap.* **141**, 1977).

muriculate, delicately muricate.

muriform (of spores), see dictyospore.

muriform cell, a thick-walled, dark, muriform cell (frequently referred to as a sclerotic cell or body), found in tissues affected by chromoblastomycosis (Matsumoto *et al.*, *Mycol.* **76**: 244, 1984).

Murogenella Goos & E.F. Morris (1965) = Coryneum fide Sutton (*TBMS* **86**: 1, 1986).

Muroia I. Hino & Katum. (1958), Lophiostomataceae. 1, Japan. See Hino & Katumoto (*J. Jap. Bot.* **33**: 79, 1958).

Murrill (William Alphonso; 1869-1957; USA). Student, Randolph-Macon College, Ashland, Virginia (1887-1891); Assistant Cryptogamic Botanist, Cornell (up to 1890); Assistant Curator (1904-1909) then Assistant Director (1908-1919) then Curator & Supervisor of Public Instruction (1919-1924), New York Botanic Garden; after a period of ill-health (1924-1926), employed during semi-retirement, University of Florida, Gainesville (1926 onwards). A leading agaricologist; first editor of *Mycol.* (1909-1924); author of more than 500 papers and articles; he also identified and named the causal organism of chestnut blight. His specimens of hymenomycetes (more than 70,000) are in the fungal reference collection of New York Botanic Garden. *Publs.* Agaricaceae. *North American Flora* (1910-1916); Agaricaceae of tropical North America. *Mycol.* (1911-1918 [reprint 1971]). *Biogs, obits etc.* Halling (An annotated index to species and infraspecific taxa of Agaricales & Boletales described by W.A. Murrill. *Memoirs of the New York Botanic Garden* **40**: 1, 1980); Stafleu & Cowan (*TL-2* **3**: 672, 1981); Weber (*Mycol.* **53**: 543, 1961) [bibliography, portrait]; Weber (*Mycologia Index*, 1968); see also the website: Smith-Vikos (*William Alphonso Murrill (1869-1957)* http://sciweb.nybg.org/science2/hcol/intern/murrill1.asp.

Murrilloporus Ryvarden (1985) = Heterobasidion fide Stalpers (*Stud. Mycol.* **40**: 59, 1996).

Musaespora Aptroot & Sipman (1993) = Trypetheliopsis fide Aptroot (*in litt.*), Lücking & Sérusiaux (*Nordic Jl Bot.* **16**: 661, 1996), Aptroot *et al.* (*Biblthca Lichenol.* **64**, 1997), Aptroot *et al.* (*Biblthca Lichenol.* **97**, 2008; Costa Rica).

muscardine fungus (green), *Metarhizium anisopliae*; -- (yellow), *Paecilomyces farinosus*. Pathogens of silkworms and other insects.

muscaridin, and **muscarin(e)**, toxic quaternary ammonium compounds from *Amanita muscaria*; muscarin also from *Inocybe patouillardii*.

muscazone, an insecticidal toxin from *Amanita muscaria*. Cf. tricholomic acid.

Muscia Gizhitsk. (1929), ? Helotiales. 1, former USSR.

Muscicola Velen. (1934), Helotiales. 1, Europe. See Svrček (*Česká Mykol.* **43**: 65, 1989).

muscicolous, growing on *Musci*; see Bryophilous fungi.

muscimol, see pantherine.

Muscodor Worapong, Strobel & W.M. Hess (2001), anamorphic *Xylariaceae*. 3, widespread (tropical). See Worapong *et al.* (*Mycotaxon* **79**: 71, 2001), Daisy *et al.* (*Mycotaxon* **84**: 39, 2002), Worapong *et al.* (*Mycotaxon* **81**: 463, 2002), Sopalun *et al.* (*Mycotaxon* **88**: 239, 2003; Thailand), Ezra *et al.* (*Microbiology* Reading **150**: 4023, 2004).

mushroom (1) an agaric (or a bolete), basidioma, esp. an edible one; (2) any agaric; a macrofungus with a distinctive fruiting body which can be either hypogeous or epigeous, large enough to be seen with the naked eye and to be picked by hand (Chang & Miles, *Mycologist* **6**: 65, 1992). Cf. toadstool.

mushroom bodies (entom.), corpora pedunculata, paired lobes of neurophile in dorsal brain of insects.

Mushroom cultivation. Apart from the button mushroom, *Agaricus bisporus*, several other species are now cultivated, but still only very few at a commercial scale, e.g. enokitake (*Flammulina velutipes*), lawyer's wig or shaggy-mane mushroom (*Coprinus comatus*), lion's mane mushroom (*Hericium erinaceus*), maitake (*Grifola frondosa*), oyster mushrooms (*Pleurotus* spp.), padi straw mushroom (*Volvariella volvacea*), reishi (*Ganoderma lucidum*), shimeji (*Hypsizygus tessulatus*), shii-take (*Lentinula edodes*), snow-fungus (*Tremella fuciformis*); most are cultivated for their culinary, but some for their medicinal or psychotropic value; the market is substantial.

Cultivation of the button mushroom starts from a master culture which is used to inoculate spawn. Well-composted horse manure is used as substratum (or sometimes 'artificial compost'). Gypsum and superphosphate may be added before or during decomposition and the compost has to be turned occasionally. Mushrooms do not need light but for the best results they have to be kept at an even temperature and the beds may be in any place meeting these conditions. Indoor beds, where space is limited, such as those in special houses ('sheds'), are frequently of the 'flat' type and are made up of a 15-23 cm layer of compost in boxes or trays; those outdoors are generally made as ridges about.75 m high. In place of the old method of using natural, 'virgin' spawn to make the 'brick' spawn for spawning the beds, 'pure culture' spawn (made by inoculating a sterile compost with cultures from spores or tissues of a good type of mushroom) is now used for inoculating bricks of horse manure, cow manure, and rich soil, or, better, the beds themselves. Spawning is done when the temperature has become 80°F or less and some 10 days later, when the spawn is 'running', the bed is covered ('cased') with an inch of soil and, if necessary, a layer of straw in addition. Mushrooms are first seen 6-8 weeks after spawning and basidioma production is stimulated by bacteria (Hayes *et al.*, *Ann. appl. Biol.* **64**: 177, 1969). A bed may go on producing for 4 months; the mean weight of mushrooms is about 6 kg for every m^2 of bed, though higher weights are not uncommon. Cultivation of other commercial species can be more complex: *Tremella fuciformis*, for example, which grows on wood, derives its nutrition as a parasite from ascomycetes in the wood, and not from the wood itself.

Lit.: Chang *et al.* (Eds) (*Mushroom biology and mushroom products*, 1993), Ware (*Bull. Min. Agric., Lond.* **34**, edn 4, 1938), Lambert (*Fmr's Bull. US Dep. Agric.* **1875**, 1941; *Bot. Rev.* **4**: 397, 1938),

Maher (Ed.) (*Mushroom Science 13, Science and cultivation of edible fungi*, 1991), Stamets (*Growing gourmet and medicinal mushrooms*, 1993), Stoller (*Pl. Physiol.* **18**: 397, 1943; artificial composts), Atkins (*Mushroom growing today*, edn 5, 1966), Singer & Harris (*Mushrooms and truffles*, 1987), Chang & Hayes (Eds) (*The biology and cultivation of edible mushrooms*, 1978; compr. review).

Mushroom parasites: *Mycogone perniciosa* (white mould, wet bubble; *IMI Descr.* **499**), *Verticillium lamellicola* (dry bubble, brown spot, mole; flock), *V. fungicola*, *Cladobotryum dendroides* (mildew or cobweb diseases; *IMI Descr.* **498**, *Fusarium solani*, *Myceliophthora* (q.v.). See also Hawksworth (*in* Cole & Kendrick (Eds), *The biology of conidial fungi* **1**: 171, 1981), Jeffries & Young (*Interfungal parasitic relationships*, 1994).

Mushroom losses are sometimes caused by the presence of other fungi which inhibit the development of the mushrooms (mushroom weeds). Examples of such fungi ('invaders') are: *Xylaria vaporaria* (= *X. pedunculata*), *Scopulariopsis fimicola* (see *Oospora*) and *Papulospora byssina* (white and brown plaster moulds), *Diehliomyces microsporus* (q.v.), *Clitocybe dealbata*, *Clitopilus augeana*; see also cat's ear. Recent sanitary techniques have eliminated some of these species from commercial production, leading to concern over the conservation status of the mushroom weed itself (Minter, *Descriptions of Fungi and Bacteria*, 1714, 2007).

Mushroom pests: see Ware (*loc. cit.*; *J.S.-E. agric. Coll., Wye* **31**: 15, 1933 *et seq.*).

mushroom stones, mushroom-like effigies perhaps associated with Mayan religious cults, mainly S. Am. (see Lowy, *Mycol.* **63**: 983, 1971).

mushroom sugar, see trehalose.

Musicillium Zare & W. Gams (2007), anamorphic *Plectosphaerellaceae*. 1, pantropical. See Zare *et al.* (*Nova Hedwigia* **85**: 463, 2007).

musiform, banana-shaped (basidiospores in *Exobasidium*, fide Nannfeldt, *Symb. bot. upsal.* **23**: 27, 1981).

must (1) unfermented or fermented grape juice; new wine; (2) = mould.

mutagen, a chemical or physical agent which promotes or increases the mutation rate.

mutant, a strain that differs by an induced or natural mutation of at least one genetic locus.

Mutatoderma (Parmasto) C.E. Gómez (1976), Corticiaceae. 4, widespread. See Gómez (*Boln Soc. argent. Bot.* **17**: 346, 1976).

muticate (adj. **muticous**), having no point; not sharp at the ends.

Mutinus Fr. (1849) nom. cons., Phallaceae. 12, widespread. See Guez & Nagasawa (*Nippon Kingakukai Kaiho* **41**: 75, 2000; Japan).

mutualism, persistent and intimate association between organisms of different size in which the larger organism (the host) utilizes novel or enhanced properties possessed by the smaller partner(s) (symbionts), e.g. lichens, mycorrhizas. See Douglas & Smith (*in* Smith, *Pap. Proc. R. Soc. Tasmania* **123**: 1, 1989), symbiosis.

Muyocopron Speg. (1881), Microthyriaceae. 7, widespread (tropical). See Hyde (*Sydowia* **49**: 1, 1997).

myc- (**mycet-**, **myceto-**, **myco-**) (prefix), pertaining to fungi.

Mycacolium Reinke (1895) nom. inval. = Acolium (Ach.) Gray fide Hawksworth *et al.* (*Dictionary of the Fungi* edn 8, 1995).

mycangium (pl. **mycangia**), a sac or cup-shaped fungal repository of ectodermal origin located in or on an ambrosia beetle (Batra, *Trans. Kansas Acad. Sci.* **66**: 226, 1963).

Mycardothelium, see *Arthothelium*.

Mycarthonia Reinke (1895) ? = Arthonia fide Hawksworth *et al.* (*Dictionary of the Fungi* edn 8, 1995).

Mycarthopyrenia Keissl. (1921) = Arthopyrenia fide Hawksworth *et al.* (*Dictionary of the Fungi* edn 8, 1995).

Mycarthothelium Vain. (1928) = Arthothelium fide Hawksworth *et al.* (*Dictionary of the Fungi* edn 8, 1995).

Mycasterotrema Räsänen (1943) ≡ Asterotrema.

Mycastrum Raf. (1813) = Scleroderma fide de Toni *in* Saccardo (*Syll. fung.* **7**: 134, 1888).

Mycaureola Maire & Chemin (1922), Physalacriaceae. (parasitic on *Rhodophyceae*). See Binder *et al.* (*Am. J. Bot.* **93**: 547, 2006; evolutionary relationships).

Mycelia sterilia Maire & Chemin (1922), Physalacriaceae. (parasitic on *Rhodophyceae*). See Binder *et al.* (*Am. J. Bot.* **93**: 547, 2006; evolutionary relationships).

mycelial cord, a discrete filamentous aggregation of hyphae which, in contrast to a rhizomorph (q.v.), has no apical meristem; syrrotia. Thompson & Rayner (*TBMS* **78**: 193, 1982) prefer not to use 'mycelial strand' for such structures.

mycelial muff, a subterranean hyphal system surrounding a living root (Buscott & Roux, *TBMS* **89**: 249, 1987).

Myceliochytrium A.E. Johanson (1945), Actinobacteria. See Sparrow (*Aquatic Phycomycetes* Edn 2: 1187 pp., 1960), Karling (*Chytriomyc. Iconogr.*, 1977).

Myceliophthora Costantin (1892), anamorphic *Arthroderma, Ctenomyces*, Hso.0eH/1eP.1/3/10/13. 9, widespread (temperate). Also used for anamorphs of *Corynascus*, *Arthroderma*. *M. lutea* (vert-de-gris disease ('mat disease') of mushrooms in culture). See van Oorschot (*Stud. Mycol.* **20**, 1980; key), Currah (*Mycotaxon* **24**: 1, 1985), Guarro & Figueras (*Int. J. Mycol. Lichenol.* **3**: 135, 1986), Stchigel *et al.* (*MR* **104**: 879, 2000).

Mycelites W. Roux (1887), Fossil Fungi ? Fungi. 1 (Devonian – Recent), widespread.

Mycelithe Gasp. (1841), anamorphic *Polyporus*. 1. See Donk (*Verh. K. ned. Akad. Wet.* tweede sect. **62**: 1, 1974).

Mycelium nom. dub., anamorphic *Fungi*. auct. (obsol.). The name *M. radicis* has been used for certain mycorrhizal fungi.

mycelium, a mass of hyphae; the thallus of a fungus; 'spawn'; **mycelioid**, like mycelium. See Jennings & Rayner (Eds) (*The ecology and physiology of the fungal mycelium*, 1984), Gregory (*TBMS* **82**: 1, 1984; review).

Myceloblastanon M. Ota (1924) = Candida fide Diddens & Lodder (*Die anaskosporogenen Hefen* **2**, 1942), von Arx (*Gen. Fungi Sporul. Cult.* Edn 3, 1981; anamorph of *Pichia*).

myceloconidium, see stylospore.

Myceloderma Ducomet (1907), anamorphic *Pezizomycotina*, Hso.0eP/#eP.1. 1 (with pycnothyrial state), Europe.

Mycena (Pers.) Roussel (1806), Mycenaceae. Anamorph *Decapitatus*. *c.* 500, widespread. *M. citricolor* on coffee, etc. is luminous (see Buller, **6**). The genus is largely polyphyletic but a new classification has not yet been proposed. See Kühner (*Le Genre Mycena*, 1938), Smith (*North American species of Mycena [reprint 1971]*, 1947), Métrod (*Les Mycènes de Madagascar*, 1949), Pearson (*Naturalist* Hull: 41, 1955; key Br. spp.), Haluwyn (*Docums Mycol.* **5**: 17, 1972; ecology), Charbonnel (*Docums Mycol.* **7**: 26, 1977; microsc. char.), Pegler (*Kew Bull.* Addit. Ser. **6**, 1977; key 12 E. Afr. spp.), Maas Geesteranus (*Persoonia* **11**: 93, 1980; subdiv.), Maas Geesteranus (*Proc. K. Ned. Akad. Wet.* C **86**: 401, 1983; sect. *Sacchariferae*, *Basipedes*, *Bulbosae*, *Clavulares*, *Exiguae*, *Longisetae*), Maas Geesteranus (*Proc. K. Ned. Akad. Wet.* C **87**: 131, 1984; sect. *Viscipellies*, *Amictae*, *Supinae*, **87**: 413; sect. *Filipedes*), Maas Geesteranus (*Proc. K. Ned. Akad. Wet.* C **88**: 339, 1985; sect. *Mycena*), Maas Geesteranus (*Proc. K. Ned. Akad. Wet.* C **89**: 83, 1986; sect. *Lucentae*, *Carolineses*, *Monticola*), Maas Geesteranus (*Proc. K. Ned. Akad. Wet.* C **89**: 159, 1986; sect. *Polyadelphia*, *Saetulipedes*), Maas Geesteranus (*Proc. K. Ned. Akad. Wet.* C **89**: 279, 1986; sect. *Intermediae*, *Rubromarginatae*), Maas Geesteranus (*Proc. K. Ned. Akad. Wet.* C **91**: 129, 1988; sect. *Fragilipedes*), Maas Geesteranus (*Proc. K. Ned. Akad. Wet.* C **91**: 377, 1988; sect. *Lactipedes*, *Sanguinolentae*, *Galactopoda*, *Crocatae*), Maas Geesteranus (*Proc. K. Ned. Akad. Wet.* C **92**: 89, 1989; sect. *Hygrocyboideae*), Maas Geesteranus (*Proc. K. Ned. Akad. Wet.* C **93**: 163, 1990; sect. *Adonidae*, *Aciculae*, *Oregonenses*), Treu & Agerer (*Mycotaxon* **38**: 279, 1990; culture), Maas Geesteranus (*Proc. K. Ned. Akad. Wet.* C **94**: 81, 1991; sect. *Hiemales*, *Exornatae*), Emmett (*Mycologist* **6**: 72, 114, 164, 1992), Maas Geesteranus (*Mycenas of the Northern Hemisphere*, 1992), Emmett (*Mycologist* **7**: 4, 1993; Br. spp., list & key), Desjardin (*Biblthca Mycol.* **159**: 1, 1995; sect. *Sacchariferae*, worldwide), Maas Geesteranus & Horak (*Biblthca Mycol.* **159**: 143, 1995; Papua New Guinea, New Caledonia ssp.), Desjardin *et al.* (*Fungal Diversity* **11**: 69, 2002; sect. *Longisetae*), Grgurinovic (*Fungal Diversity Res. Ser.* **9**: 329 pp., 2003; Australia), Miersch & Rönsch (*Z. Mykol.* **69**: 123, 2003; sect. *Filipedes*), Robich (*Mycena d'Europa*: 728 pp., 2003; Eur. spp.), Lodge *et al.* (*Mem. N. Y. bot. Gdn* **89**: 131, 2004; sect. *Hygrocyboideae* in Dominican Republic), Miersch *et al.* (*Feddes Repert.* **115**: 43, 2004; key Eur. corticolous species), Robich (*Persoonia* **19**: 1, 2006; key sect. *Fragilipedes* N hemisph.).

Mycenaceae Overeem (1926), Agaricales. 10 gen. (+ 35 syn.), 705 spp.

Lit.: Benny *et al.* (*Mycotaxon* **22**: 119, 1985), Corner (*Gdns' Bull.* Singapore **39**: 103, 1986), Maas Geesteranus (*Verh. K. Akad. Wet.* tweede sect. **1**: 571 pp., 1992), Maas Geesteranus & Horak (*Biblthca Mycol.* **159**: 208, 1995), Jin *et al.* (*Mycotaxon* **79**: 7, 2001), Jin *et al.* (*Mycol.* **93**: 309, 2001), Desjardin *et al.* (*Fungal Diversity* **11**: 69, 2002), Grgurinovic (*Fungal Diversity Res. Ser.* **9**: 329 pp., 2003), Lodge *et al.* (*Mem. N. Y. bot. Gdn* **89**: 131, 2004).

Mycenastraceae Zeller (1948) = Agaricaceae.

Mycenastrum Desv. (1842), Agaricaceae. 1, widespread. See Heim (*Revue Mycol.* Paris **36**: 81, 1971; Patouillard's spp.), Homrich & Wright (*Mycol.* **65**: 779, 1973), Miller *et al.* (*Mycol.* **97**: 530, 2005).

Mycenella (J.E. Lange) Singer (1938), Tricholomataceae. 10, widespread (temperate). See Boekhout (*Persoonia* **12**: 425, 1985; key Eur. spp.), Kühner (*Mycol. helv.* **3**: 331, 1989; alpine spp.), Grgurinovic (*Victorian Naturalist* **110**: 65, 1993; Australia), Robich (*Rivista di Micologia* **40**: 365, 1997; *Mycenella margaritispora*), Komorowska (*Polish Botanical Journal* **50**: 83, 2005; Poland).

Mycenitis, see *Mycetinis*.

Mycenoporella Overeem (1926) c Mycena fide Kuyper (*in litt.*).

Mycenopsis Velen. (1947) ? = Mycena fide Kuyper (*in litt.*).

Mycenula P. Karst. (1889) = Mycena fide Singer (*Agaric. mod. Tax.*, 1951).

Mycepimyce Nieuwl. (1916) = Sphaerellopsis Cooke fide Sutton (*Mycol. Pap.* **141**, 1977).

Mycerema Bat., J.L. Bezerra & Cavalc. (1963), Schizothyriaceae. Anamorph *Plenotrichaius*-like. *c.* 2, widespread (tropical). See Farr (*Mycol.* **79**: 97, 1987).

Myces Paulet (1808) ≡ Fungus Tourn. ex Adans.

mycetal, a fungus or a lichen (obsol.).

Myceteae, See *FUNGI*.

Mycetes (obsol.), (1) *Fungi*. (2) **mycetes**, a general term (obsol.) for minute vegetable organisms or microbes. Hence **mycetology** = mycology; schizomycetes = bacteria; etc.

-mycetes (suffix), indicating the rank of a fungal class (see Classification).

Mycetinis Earle (1909), Marasmiaceae. 8, widespread. See Wilson & Desjardin (*Mycol.* **97**: 667, 2005).

Mycetism (**mycetismus**) (poisoning by larger fungi). This is of common occurrence and has a long history. Some people are allergic (see Allergy) or intolerant to mushrooms and illness may result from eating decayed or mouldy specimens, but serious poisoning is usually the result of eating fungus fruitbodies which contain toxins. About 6 species are both common and deadly poisonous. Symptoms vary from a slight stomach upset to death. It is important to identify the fungus consumed by the patient, as there are several different categories of poisoning. These include:

(1) *Cyclopeptide poisoning* (see amatoxins, phallotoxins) (by *Amanita phalloides*, *A. virosa*, *A. verna*). Symptoms first occur 4-6 h or more after ingestion. Early gasteroenteric symptoms may obscure hepatic and renal damage. This, the most dangerous type, is responsible for most deaths by mushroom poisoning in Eur. and N. Am. but, where early treatment can be made, chances of recovery have been enhanced by the introduction of antiserum therapy and dialysis. Orollanine and Cortinarins (from *Cortinarius* spp.) have an incubation period of 2-20+ days and cause renal failure (see Muchelot & Tebbett, *MR* **94**: 289, 1990).

(2) *Haemolytic poisoning* (*A. rubescens*, *A. vaginata*). Characterized by anaemia resulting from consumption of raw or undercooked mushrooms containing thermolabile haemotoxins.

(3) *Muscarine poisoning* (*A. muscaria*, *A. pantherina*). Early symptoms, within 2 h, include increased perspiration, salivation, dehydration, and nausea.

(4) *Coprine poisoning* (*Coprinus atramentarius*). See antabuse.

(5) *Psychotropic poisoning*. Hallucinations and delirium 2-4 h after ingestion. (a) Ibotenic acid, muscimol group (*A. muscaria*, *A. pantherina*): sleep, torpidity, or coma in extreme cases; (b) Indole group

(psilocin, psilocybin): stimulates psychic perception, see Hallucinogenic fungi.

(6) Gasteroenteric irritants (*Entoloma sinuatum*, *Paxillus involutus*, *Agaricus xanthoderma*, *Boletus satanus*, *Hebeloma crustuliniforme*, many *Tricholoma* spp., *Hypholoma fasciculare*, several *Lactarius* and *Russula* spp.).

The ascomycete *Gyromitra esculenta* contains gyromitrin (q.v.) which causes gasteroenteric discomfort, followed by hepatic and renal attack. This is the only fungus poisoning known to induce fever. The ascoma is edible if used without the cooking water or after drying.

Lit.: Dujarric de la Rivière & Heim (*Les champignons toxiques*, 1938; 600 refs), Heim (*Champignons toxiques et hallucinogènes*, 1963; edn 2, 1978), Arietti & Tomasi (*Funghi velenosi*, 1969), Duffy & Vergeer (*California toxic fungi*, 1977), Lincoff & Mitchell (*Toxic and hallucinogenic mushroom poisoning. A handbook for physicians and mushroom hunters*, 1977), Rumack & Salzman (Eds) (*Mushroom poisoning: diagnosis and treatment*, 1978), Pegler & Watling (*Bull. BMS* **16**: 66, 1982; Br. toxic fungi), Ammirati *et al.* (*Poisonous fungi of the Northern United States and Canada*, 1985), Bresinsky & Besl (*Giftpilze mit einer Entführung in die Pilzbestimmung*, 1985 [Engl. transl. *A colour atlas of poisonous fungi*, 1990]), Oldridge *et al.* (*Wild mushroom and toadstool poisons*, 1989). See also amatoxins, Ergot, Mycotoxicoses, phallotoxins.

mycetismus, see Mycetism.

mycetobionts, fungus-dependent, used of obligate fungus-feeding arthropods.

mycetocyte, see mycetosome.

Mycetodium A. Massal. (1856) ≡ Gomphillus.

mycetology, see *Mycetes*.

mycetoma (maduramycosis, madura foot), a disease, esp. tropical, of humans, mostly adult males, of the foot or other part, usually acquired during agricultural work, resulting in tumefactions and characterized by mycotic granules ('grains') in the infected tissues. Although a clinical entity, many different fungi (**eumycetoma**) and actinomycetes (**actinomycetoma**) are involved. Mycetomas can be roughly classified according to whether the grains are white or yellow (*Aspergillus* spp., *Nocardia madurae*, *Pseudallescheria boydii*, etc.), red (*Streptomyces pelletieri*, *S. somaliensis*) or black (*Madurella mycetomatis*, etc.). See Mahgoub & Murray (*Mycetoma*, 1973).

Mycetophagites Poinar & R. Buckley (2007), Fossil Fungi, Basidiomycota. 1.

mycetophagy, see mycophagy.

mycetophiles, fungus-loving, facultative fungus-feeding arthropods.

mycetophilous, see mycophilic.

mycetosome, a sac-like structure in the gut of Anobiid beetles lined with cells (**mycetocytes**) containing yeast cells.

Mycetosporidium L. Léger & E. Hesse (1905), ? Fungi. 2 (on insects), Europe. See Tate (*Parasitology* **32**: 462, 1940).

mycid, a secondary effect (manifested as eczema, urticaria, etc.) which is an allergic reaction to spores or toxin of a dermatophyte; dermatophytid. Cf. mycosis (2). A mycid may be a trichophytid (caused by *Trichophyton*); microsporid (*Microsporum*); epidermophytid (*Epidermophyton*).

-mycin (suffix), the recommended ending for names coined for antibiotics derived from actinomycetes.

Mycinema C. Agardh (1824) nom. dub. = Corticium Pers. fide Saccardo (*Syll. fung.* 326 **6**: 614, 1888).

myco- (prefix), pertaining to fungi.

Mycoacia Donk (1931), Meruliaceae. 15, widespread. See Ragab (*Mycol.* **43**: 459, 1951), Eriksson & Ryvarden (*Cortic. N. Europ.* **4**: 873, 1976; key Eur. spp.), Hjortstam & Ryvarden (*Syn. Fung.* **18**: 20, 2004).

Mycoaciella J. Erikss. & Ryvarden (1978), Meruliaceae. 5, widespread. See Hjortstam *et al.* (*Kew Bull.* **45**: 303, 1990; key), Hjortstam & Ryvarden (*Syn. Fung.* **18**: 20, 2004).

Mycoalvimia Singer (1981), Tricholomataceae. 1, Brazil. See Singer (*Mycol.* **73**: 504, 1981).

Mycoamaranthus Castellano, Trappe & Malajczuk (1992), Boletaceae. 3, Australasia; Africa; S.E. Asia. See Castellano *et al.* (*Aust. Syst. Bot.* **5**: 613, 1992), Lumyong *et al.* (*Mycol. Progr.* **2**: 323, 2003).

Mycoarachis Malloch & Cain (1970), Bionectriaceae. Anamorph *Acremonium*-like. 1 (coprophilous), widespread. See Valldosera & Guarro (*Nova Hedwigia* **47**: 231, 1988), Rehner & Samuels (*CJB* **73** Suppl. 1: S816, 1995; phylogeny), Ogawa *et al.* (*Mycol.* **89**: 756, 1997; phylogeny), Rossman *et al.* (*Stud. Mycol.* **42**: 248 pp., 1999; posn), Suh & Blackwell (*Mycol.* **91**: 836, 1999; phylogeny).

Mycoarctium K.P. Jain & Cain (1973), ? Pezizales. 2 (coprophilous), Canary Islands; USA. See Korf & Zhuang (*Mycotaxon* **40**: 79, 1991), van Brummelen (*Persoonia* **16**: 425, 1998; posn).

Mycoarthopyrenia Cif. & Tomas. (1953) nom. illegit. ≡ Arthopyrenia.

Mycoarthris Marvanová & P.J. Fisher (2002), anamorphic *Hyaloscyphaceae*. 1, Great Britain. See Marvanová *et al.* (*Nova Hedwigia* **75**: 258, 2002), Baschien *et al.* (*Nova Hedwigia* **83**: 311, 2006; phylogeny).

Mycobacidia Rehm (1890) = Arthrorhaphis fide Hawksworth *et al.* (*Dictionary of the Fungi* edn 8, 1995).

Mycobacillaria Naumov (1915), anamorphic *Pezizomycotina*, Hso.≡ eP.1/10. 1, former USSR.

Mycobanche Pers. (1818) ≡ Mycogone.

Mycobilimbia Rehm (1890), ? Lecanorales (L). *c.* 26, widespread. See Awasthi & Mathur (*Proc. Indian Acad. Sci.* Pl. Sci. **97**: 481, 1987; key 3 spp. India), Hafellner (*Herzogia* **8**: 53, 1989; key 8 spp. Eur.), Pietschmann (*Nova Hedwigia* **51**: 521, 1990; asci), Sarrión *et al.* (*Lichenologist* **35**: 1, 2003; Spain), Buschbom & Mueller (*Mol. Phylogen. Evol.* **32**: 66, 2004; phylogeny), Kantvilas *et al.* (*Lichenologist* **37**: 251, 2005; S Hemisphere), Naesborg *et al.* (*MR* **111**: 581, 2007; phylogeny).

mycobiont, the fungal component of a lichen (Scott, *Nature* **179**: 486, 1957); cf. phycobiont, photobiont.

mycobiota (1) the total fungal inventory of the area under consideration (e.g. all the species present); (2) the fungal mass present (e.g. in a soil sample).

Mycoblastaceae Hafellner (1984), Lecanorales (L). 1 gen. (+ 3 syn.), 10 spp.

Lit.: Śliwa (*Fragm. flor. geobot.* **41**: 491, 1996), Miądlikowska *et al.* (*Mycol.* **98**: 1088, 2006; phylogeny), Arup *et al.* (*Mycol.* **99**: 42, 2007; sister group relations with *Parmeliaceae*), Hofstetter *et al.* (*Mol. Phylogen. Evol.* **44**: 412, 2007; phylogeny).

Mycoblastomyces Cif. & Tomas. (1953) ≡ Mycoblastus.

Mycoblastus Norman (1853) nom. cons., Mycoblastaceae (L). 10, widespread. See Anders (*Hedwigia* **68**: 87, 1928), James (*Lichenologist* **5**: 114, 1971; key 4 Br. spp.), Miądlikowska *et al.* (*Mycol.* **98**: 1088, 2006; phylogeny), Arup *et al.* (*Mycol.* **99**: 42, 2007; sister group relations with *Parmeliaceae*).

Mycobonia Pat. (1894) nom. cons., Gloeophyllaceae. 1, widespread (tropical). See Martin (*Mycol.* **31**: 247, 1939), Corner (*Beih. Nova Hedwigia* **78**: 102, 1984; key).

Mycoboniaceae Jülich (1982) = Gloeophyllaceae.

Mycobystrovia Goujet & Locq. (1979), Fossil Fungi (mycel.) Fungi. 1 (Devonian), Europe.

Mycocalia J.T. Palmer (1961), Agaricaceae. 7, widespread. See Burnett & Boulter (*New Phytol.* **62**: 217, 1963; mating systems), Jeppson (*Agarica* **6**: 228, 1985; key), Malaval (*Docums Mycol.* **32**: 19, 2002), Sarasini (*Gasteromiceti Epigei*: 406 pp., 2005).

Mycocaliciaceae Alf. Schmidt (1970), Mycocaliciales. 6 gen. (+ 4 syn.), 99 spp.

Lit.: Tibell (*Nova Hedwigia* Beih. **79**: 597, 1984), Hutchison (*Mycol.* **79**: 786, 1987), Tibell (*Nordic Jl Bot.* **10**: 221, 1990), Tibell (*Nordic Jl Bot.* **13**: 331, 1993), Tibell (*Ann. bot. fenn.* **33**: 205, 1996), Tibell (*Symb. bot. upsal.* **32**: 291, 1997; anams), Wedin & Tibell (*CJB* **75**: 1236, 1997), Peterson & Rikkinen (*Mycol.* **90**: 1087, 1998), Tibell (*Biblthca Lichenol.* **71**: 107 pp., 1998; S. Am.), Selva & Tibell (*Bryologist* **102**: 377, 1999), Rikkinen & Poinar (*MR* **104**: 7, 2000), Tibell & Wedin (*Mycol.* **92**: 577, 2000), Angeles Vinuesa *et al.* (*MR* **105**: 323, 2001).

Mycocaliciales Tibell & Wedin (2000), (±L). Mycocaliciomycetidae. 2 fam., 8 gen., 108 spp. Thallus immersed, often absent. Ascomata stalked, brown or black. Asci cylindrical, uniformly thin-walled or thick-walled at least at the apex, evanescent or not. Ascospores brown, smooth or ornamented, sometimes in a mazaedial mass. Mainly saprobic on wood or bark, some lichenicolous, fungicolous or ? lichen-forming. Fams.:

(1) **Mycocaliciaceae**
(2) **Sphinctrinaceae**

Lit.: Geiser *et al* (*Mycol.* **98**: 1051, 2006), Tibell & Wedin (*Mycol.* **92**: 577, 2000), Wedin & Tibell (*CJB* **75**: 1236, 1997).

Mycocaliciomycetidae Tibell (2007), Eurotiomycetes. Ord.:

Mycocaliciales
For *Lit.* see fam.

Mycocalicium Vain. (1890), Mycocaliciaceae. Anamorph *Phialophora*-like. 11, widespread (esp. temperate). See Samuels & Buchanan (*N.Z. Jl Bot.* **21**: 163, 1983; anamorph), Tibell (*Beih. Nova Hedwigia* **79**: 597, 1984), Tibell (*Nordic Jl Bot.* **10**: 221, 1990; anamorphs), Vinuesa *et al.* (*MR* **105**: 323, 2001; ITS), Tibell & Vinuesa (*Taxon* **54**: 427, 2005; phylogeny), Geiser *et al.* (*Mycol.* **98**: 1053, 2006; phylogeny), Muñiz & Hladun (*Lichenologist* **39**: 205, 2007; Spain).

Mycocandida Langeron & Talice (1932) = Candida fide Diddens & Lodder (*Die anaskosporogenen Hefen* **2**, 1942).

Mycocarpon S.A. Hutch. (1955), Fossil Fungi ? Ascomycota. 5 (Carboniferous), Europe; USA. See Baxter (*Paleont. Contrib. Univ. Kansas* **77**, 1975), Pirozynski (*Ann. Rev. Phytopath.* **14**: 237, 1976).

mycocecidium, see cecidium.

Mycocentrodochium K. Matsush. & Matsush. (1996), anamorphic *Pezizomycotina*, Hso.?.?. 1, Japan. See Matsushima & Matsushima (*Matsush. Mycol. Mem.* **9**: 34, 1996).

Mycocentrospora Deighton (1972), anamorphic *Pleosporales*, Hso.≡ fH.10. 9, widespread. See Constantinescu (*Revue Mycol.* Paris **42**: 105, 1978; conidial polymorphism), Braun (*Mycotaxon* **48**: 275, 1993; reassessment), Braun (*Monogr. Cercosporella, Ramularia Allied Genera (Phytopath. Hyphom.)* **1**: 224, 1995; key), Marvanová (*Czech Mycol.* **49**: 7, 1996; aquatic spp.), Stewart *et al.* (*MR* **103**: 1491, 1999; phylogeny), Goodwin *et al.* (*Phytopathology* **91**: 648, 2001; ITS seq. data; affinity with *Pleosporales*).

Mycochaetophora Hara & Ogawa (1931), anamorphic *Pezizomycotina*, Hso.0fH.?. 1, Japan.

Mycochlamys S. Marchand & Cabral (1976), anamorphic *Pezizomycotina*, Hso.0eP.10. 1 (from soil), Argentina. See Marchand & Cabral (*Boln Soc. argent. Bot.* **17**: 66, 1976).

Mycociferria Tomas. (1953) ≡ Ciferriolichen.

Mycocitrus Möller (1901), Bionectriaceae. Anamorph *Acremonium*-like. 1 (on living bamboo stems), Asia (tropical). See Rossman *et al.* (*Stud. Mycol.* **42**: 248 pp., 1999).

Mycocladus Beauverie (1900), Syncephalastraceae. 4, widespread (esp. tropical). See Hesseltine & Ellis (*Mycol.* **56**: 569, 1964), Mirza *et al.* (*Mucorales of Pakistan*, 1979), Vánová (*Česká Mykol.* **45**: 25, 1991), Santos *et al.* (*Rev. Iberoam. Micol.* **22**: 174, 2005; preservation, zygospore formation), Hoffman *et al.* (*MR* **111**: 1169, 2007; phylogeny, classification).

Mycoclelandia Trappe & G.W. Beaton (1984), Pezizaceae. 2 (hypogeous), Australia. See Trappe & Beaton (*TBMS* **83**: 536, 1984), Læssøe & Hansen (*MR* **111**: 1075, 2007; phylogeny).

mycoclena (orig. 'micoclena'), term coined by Peyronel (1922) for the 'fungus mantle' of an ectotrophic mycorrhiza having a loose structure; cf. mycoderm.

Mycocoelium Kütz. (1843) nom. dub., ? Fungi.

Mycocoenogonium Cif. & Tomas. (1954) = Coenogonium Ehrenb. fide Hawksworth *et al.* (*Dictionary of the Fungi* edn 8, 1995).

mycocoenosis, the complete assemblage of fungi within a certain plant community and its environment, or in the absence of green plants another defined habitat; see mycosociology, phytosociology.

Mycoconiocybe Reinke (1895) nom. inval. ? = Coniocybe fide Hawksworth *et al.* (*Dictionary of the Fungi* edn 8, 1995).

Mycocoscoma Bref. (1912), Ustilaginales. 1, Europe.

Mycocryptospora J. Reid & C. Booth (1987), Dothideomycetes. 1, Germany. See Reid & Booth (*CJB* **65**: 1333, 1987).

Mycodendron Massee (1891) nom. dub., Agaricomycetes. 'basidiomycetes' ? based on an abnormal polypore (see Donk, *Fungus* **28**: 13, 1958).

mycoderm, coined by Ziegenspeck (1929) for a compact, tissue-like, ectotrophic mycorrhiza; cf. mycolena.

Mycoderma Desm. (1827) nom. illegit. = Pichia fide Hawksworth *et al.* (*Dictionary of the Fungi* edn 8, 1995) but see Wu *et al.* (*FEMS Yeast Res.* **6**: 305, 2006; phylogeny).

Mycoderma Pers. (1822) nom. dub., Pezizomycotina.

mycodextran, an unbranched polysaccharide from *Aspergillus niger*, etc. (Barker *et al.*, *J. chem. Soc.*

1957: 2488, 1957); nigeran; see Bobbitt & Nordin (*Mycol.* **70**: 1201, 1979; as phylogenetic marker).

Mycodidymella C.Z. Wei, Y. Harada & Katum. (1998), Pleosporales. Anamorphs *Blastostroma*, *Mycopappus*-like. 1 (on *Aesculus*), Japan. See Wei *et al.* (*Mycol.* **90**: 334, 1998), Suto & Suyama (*Mycoscience* **46**: 227, 2005).

mycoecology, ecology of fungi; see Ecology.

Mycoëmilia Kurihara, Degawa & Tokum. (2004), Kickxellaceae. 1, Japan. See Kurihara *et al.* (*MR* **108**: 1143, 2004).

Mycoenterolobium Goos (1970), anamorphic *Pezizomycotina*, Hso.#eP.1. 1 (conidia flattened and fan-shaped), widespread (tropical). See Goos (*Mycol.* **62**: 171, 1970).

Mycofalcella Marvanová, Om-Kalth. & J. Webster (1993), anamorphic *Pezizomycotina*, Hso.?.?. 1, British Isles. See Marvanová *et al.* (*Nova Hedwigia* **56**: 402, 1993).

mycoflora, see mycobiota (a more appropriate term as fungi are not plants).

mycofungicides, see Mycopesticides.

Mycogala Rostaf. ex Sacc. (1884) = Orbicula fide Hughes (*Mycol. Pap.* **42**, 1951).

Mycogalopsidaceae Gjurašin (1925) = Pyronemataceae.

Mycogalopsis Gjurašin (1925), Pyronemataceae. 1, Europe.

Mycogelidiaceae W.Y. Zhuang (2007), Basidiomycota (inc. sed.). 1 gen., 1 spp. See Zhuang & He (*Mycosystema* **26**: 339, 2007).

Mycogelidium W.Y. Zhuang (2007), Mycogelidiaceae. 1, China. See Zhuang & He (*Mycosystema* **26**: 339, 2007).

Mycogemma K.M. Zalessky (1915), Fossil Fungi. 1 (Carboniferous), former USSR.

mycogenous, coming from, or living on, fungi.

mycogeography, study of the geographical distribution of fungi (q.v.).

Mycoglaena Höhn. (1909), Dothideomycetes. *c.* 10, Europe; N. America. See Riedl (*Öst. bot. Z.* **119**: 41, 1971; key 6 spp.), Harris (*Mich. Bot.* **12**: 3, 1973), Holm & Holm (*Nordic Jl Bot.* **11**: 675, 1991).

Mycogloea L.S. Olive (1950), Agaricostilbales. Anamorph *Kurtzmanomyces*. 7, widespread. See McNabb (*TBMS* **48**: 187, 1965), Kirschner *et al.* (*Antonie van Leeuwenhoek* **84**: 109, 2003; anam.).

Mycogone Link (1809), anamorphic *Hypomyces*, Hso.1eH.1. 10, widespread. See Mushroom diseases. See Holland *et al.* (*TBMS* **85**: 730, 1985; germination), Põldmaa (*Stud. Mycol.* **45**: 83, 2000; phylogeny).

mycohaemia (**mycohemia**), a condition in which fungi are present in the blood stream.

mycoherbicide (1) a herbicide derived from a fungus; (2) a preparation of fungal spores used in the biocontrol of weeds (see Mycopesticides).

myco-heterotrophic, See Leake (*New Phytologist* **127**: 171, 1994), Domínguez & Sérsic (*Mycologia* **96**: 1143, 2004).

Mycohypallage B. Sutton (1963), anamorphic *Deshpandiella*, St.≡ eP.1. 1, Africa; Sri Lanka. See Sutton (*Mycol. Pap.* **88**: 4, 1963).

mycoin, see patulin.

mycoinsecticide, a fungus used to control insects; see Entomogenous fungi, Mycopesticide.

Mycokidstonia D. Pons & Locq. (1981), Fossil Fungi. 1, France.

Mycokluyveria Cif. & Redaelli (1947) = Pichia fide von Arx *et al.* (*Stud. Mycol.* **14**: 1, 1977), Wu *et al.* (*FEMS Yeast Res.* **6**: 305, 2006; phylogeny).

Mycolachnea Maire (1937) ≡ Humaria Fuckel fide Wu & Kimbrough (*Int. J. Pl. Sci.* **153**: 128, 1992), Eriksson & Hawksworth (*SA* **11**: 179, 1993).

Mycolangloisia G. Arnaud (1918) = Actinopeltis fide von Arx & Müller (*Stud. Mycol.* **9**, 1975).

mycolatry, the worship of fungi (introduced by Wasson, 1980).

Mycolecidia P. Karst. (1888) ? = Dactylospora fide Hawksworth *et al.* (*Dictionary of the Fungi* edn 8, 1995).

Mycolecis Clem. (1909) ? = Dactylospora fide Hawksworth *et al.* (*Dictionary of the Fungi* edn 8, 1995).

Mycoleptodiscus Ostaz. (1968), anamorphic *Omnidemptus*, Hsp.0-1eH.15. 16, widespread. *M. indicus* on vanilla. See Bezerra & Ram (*Fitopatol. Brasil* **12**: 717, 1986), Sutton & Alcorn (*MR* **94**: 564, 1990; key), Alcorn (*Aust. Syst. Bot.* **7**: 591, 1994; incl. review of appressoria), Cannon & Alcorn (*Mycotaxon* **51**: 483, 1994; teleomorph), Ando (*Czech Mycol.* **49**: 1, 1996; Australia), Watanabe *et al.* (*Mycoscience* **38**: 91, 1997; Dominican Republic).

Mycoleptodon Pat. (1897) = Steccherinum fide Donk (*Persoonia* **3**: 199, 1964).

Mycoleptodonoides M.I. Nikol. (1952), Meruliaceae. 4, former USSR. See Nikolajeva (*Bot. Mater. Otd. Sporov. Rast. Bot. Inst. Komarova Akad. Nauk S.S.S.R.* **8**: 117, 1952).

Mycoleptorhaphis Cif. & Tomas. (1953) = Leptorhaphis fide Hawksworth *et al.* (*Dictionary of the Fungi* edn 8, 1995).

Mycolevis A.H. Sm. (1965), Albatrellaceae. 1, N. America. Basidioma gasteroid. See Fogel (*Mycol.* **68**: 1097, 1976), Albee-Scott (*MR* **111**: 653, 2007; phylogenetic placement).

Mycolindtneria Rauschert (1988) = Lindtneria fide Rauschert (*Feddes Repert.* **98**: 660, 1988).

mycoliths, sand grains bound together with mycelium, forming structures 5-6 cm long, esp. by *Melanospora tulasnei*.

Mycological Society of America. Founded in 1932; recognized as the Committee for North America within the International Mycological Association (q.v.); structure comprises individual and corporate members, and an elected executive; organizes annual meetings. Publications: *Inoculum*, *Mycologia*. Website: www.msafungi.org.

mycologist, one engaged in the pursuit of mycology.

mycology, the scientific study of fungi.

mycolysis, the lysis of a fungus.

Mycomalus Möller (1901), Hypocreales. 1, Brazil. See Shao *et al.* (*Taxonomy of Fungi*, 1984), Bischoff *et al.* (*Mycol.* **96**: 1088, 2004).

Mycomater Fr. (1825) nom. dub., anamorphic *Fungi*. Based on mycelium fide Fries (*Summ. veg. Scand.*, 1849).

Mycomedusa R. Heim (1966) = Favolaschia fide Pegler (*Kew Bull.* Addit. Ser. **6**, 1977).

Mycomedusiospora G.C. Carroll & Munk (1964), Lasiosphaeriaceae. 1, Brazil; Costa Rica. See Carroll & Munk (*Mycol.* **56**: 91, 1964).

Mycomelanea Velen. (1947), ? Helotiales. 1, former Czechoslovakia.

Mycomelaspilea Reinke (1895) ? = Melaspilea fide Müller & von Arx (*Beitr. Kryptfl. Schweiz* **11** no. 2, 1962).

Mycomicrothelia Keissl. (1936), Arthopyreniaceae (±L). 44, widespread (esp. tropical). See Hawksworth (*Bull. Br. Mus. nat. hist.* Bot. **14**: 43, 1985; key 25 spp.), Aptroot (*Biblthca Lichenol.* **44**, 1991), Aptroot (*Nova Hedwigia* **60**: 325, 1995), Sérusiaux & Aptroot (*Bryologist* **101**: 144, 1998), Sipman & Aptroot (*Lichenologist* **37**: 307, 2005; gen. limits), Aptroot *et al.* (*Biblthca Lichenol.* **97**, 2008; Costa Rica).

Mycomyces Wyss-Chod. (1927), anamorphic *Pezizomycotina*. 1 (on humans), Europe.

Mycomycophytes, see *Mycophytes*.

mycomyringitis, a fungal inflammation of the eardrum.

mycomysticism, mystical state induced by eating hallucinogenic fungi.

Myconeesia Kirschst. (1936) = Anthostomella See Læssøe (*SA* **13**: 43, 1994; ? synonym of *Xylaria*), Læssøe & Spooner (*Kew Bull.* **49**: 1, 1994), Lu & Hyde (*Fungal Diversity Res. Ser.* **4**, 2000).

myconematicides, see Mycopesticides.

Myconymphaea Kurihara, Degawa & Tokum. (2001), Kickxellaceae. 1, Japan. See Kurihara *et al.* (*MR* **105**: 1398, 2001).

Mycopandora Velen. (1947) = Unguicularia fide Svrček (*Česká Mykol.* **40**: 215, 1986).

Mycopappus Redhead & G.P. White (1985), anamorphic *Redheadia*, Sc.#eH-P.42. 2, north temperate. The genus is clearly polyphyletic, with disparate teleomorph links. See Redhead & White (*CJB* **63**: 1430, 1985), Suto & Kawai (*Mycoscience* **41**: 55, 2000), Suto & Suyama (*Mycoscience* **46**: 227, 2005; teleomorph), Takahashi *et al.* (*Mycoscience* **47**: 388, 2006; Japan).

Mycopara Bat. & J.L. Bezerra (1960), anamorphic *Pezizomycotina*, Cpd.0eH.15. 1 (on *Dimerosporiopsis*), USA. See Batista & Bezerra (*Publções Inst. Micol. Recife* **286**: 17, 1960).

mycoparasitism, the parasitism of one fungus by another (the **mycoparasite**); preferable to hyperparasitism which has been used for the same phenomenon; see Fungicolous fungi.

mycopathology, the study of disease caused by fungi.

Mycopegrapha Vain. (1921) ? = Opegrapha Ach. fide Hawksworth *et al.* (*Dictionary of the Fungi* edn 8, 1995).

Mycopepon Boise (1987), Melanommataceae. 2, widespread (neotropics). See Boise (*SA* **6**: 167, 1987), Boise (*Mycotaxon* **52**: 303, 1994; nomencl.), San Martín González (*Acta Bot. Mexicana* **35**: 9, 1996), Bhattacharya *et al.* (*Mol. Biol. Evol.* **17**: 937, 2000; posn).

Mycopesticides. Are mass-produced, usually commercially formulated and marketed products based on fungi which are pathogens, parasites or antagonists, of arthropod pests (mycoinsecticides), plant parasitic nematodes (myconematicides), weeds (mycoherbicides) or crop pathogens (mycofungicides), applied inundatively, like a chemical pesticide (Hall & Mann (Eds), *Biopesticides: use and delivery*, 1998). Mycopesticides have also been used to further political and military objectives, for example against illegal cultivation of drug plants like coca; such uses are generally secretive, and have raised public concerns (editorial, A very unholy war, *New Scientist* **2203**: 3, 1999; website: www.mycoherbicide.info; see also Biological control, Bioterrorism and fungi, Mycotoxicoses).

The potential of mycoinsecticides was first investigated more than 100 years ago when *Metarhizium anisopliae* was applied against weevil pests in the former USSR (Steinhaus, *Hilgardia* **26**: 107, 1965). A mycopesticide using *Metarhizium anisopliae* has been developed for control of termites (Milner, *in* Lomer & Prior, *Biological control of locusts and grasshoppers* 200, 1992), scarab beetles (Rath, *in* Glare & Jackson, *Use of pathogens in scarab pest management*, 217, 1992) and black vine weevils (Wolfram, *in Proc. V Int. Colloq. Invert. Path. & Microb. Contr.* 2, 1990). *Metarhizium anisopliae* has been field tested against locusts and grasshoppers in Africa (Lomer & Prior, 1992; Driver *et al.*, *MR* **104**: 134, 2000; Cherry *et al.*, *Biocontrol Science & Technology* **9**: 35, 1999) and is currently marketed as Green Muscle^R; see http://www.lubilosa.org/. *Beauveria bassiana* has been extensively employed against a range of crop pests but mainly at the semi-commercial or cottage-industry level, under the trade name Boverin™ in the former USSR (Ferron, *in* Burge, *Microbial control of pests and plant diseases 1970-1980*, 45, 1981), and in a commune-produced system in China (Hussey & Tinsley, *in* Burge, 1981: 785). *Beauveria brongniartii* has been mass-produced and aerially applied against cockchafer pests in Switzerland and, despite problems with fungal stability and formulation (Zimmermann, *in* Glare & Jackson, 1992), is now a registered product (Engerlingspilz^R).

The first commercial mycoinsecticides became available in the 1980s and were based on strains of *Verticillium lecanii* for aphid (Vertalec™) and whitefly (Mycotal™) control in glasshouses (Quinlan, *in* Burge, *Fungi in biological control systems* 19, 1988). *Aschersonia aleyrodis* has also been assessed for control of whitefly pests (Samson & Rombach, *in* Hussey & Scopes, *Biological pest control* 34, 1985). Production and marketing problems are discussed by Samson *et al.* (*Atlas of entomopathogenic fungi*, 1988) and Prior & Moore (*Biocontrol News Inf.* **14**: 31, 1993).

Few myconematicides have been commercialized and, because of problems with quality control, inconsistent performance, potential health hazards and uneconomic application rates, none is currently available on the market (Stirling, *Biological control of plant parasitic nematodes*, 1991). Previous commercial or semi-commercial products have been based on: *Arthrobotrys robusta* (Royal 300^R), *A. superba* (Royal 350^R) and *Verticillium chlamydosporium* (Kerry *et al.*, *Ann. appl. Biol.* **105**: 509, 1984; Kerry, *in* Brown & Kerry, *Principles and practice of nematode control in crops*, 233, 1987).

The use of mycoherbicides is a relatively recent concept but a number of products are already on the market: COLLEGO (*Colletotrichum gloeosporioides* f.sp. *aeschynomenes* for control of northern jointvetch, DeVine (*Phytophthora palmivora*) for control of milkweed vine; CASST (*Alternaria cassiae*) for control of sicklepod; Bio Mal (*C. gloeosporioides* f.sp. *malvae*) against round-leaved mallow; BioChon (*Chondrostereum purpureum*) for stump treatment of woody invasives (DeJong, *Mycologist* **14**: 58, 2000); while several others are in the final stages of commercialization (Charudattan, *in* Burge (Ed.), *Fungi in biological control systems* 86, 1988; TeBeest & Templeton, *Plant Disease* **69**: 6, 1985; Charudattan, *in* TeBeest (Ed.), *Microbial control of weeds*, 24,

1991). Although most are readily culturable, necrotrophic pathogens, marketing of the rust *Puccinia canaliculata*, as a formulated product ('Dr Biosedge') for control of yellow nutsedge has been proposed (Phatak, *Proc. I. Int. Weed Control Conf.* 388, 1992). Considerable research is now being concentrated on fermentation technology and formulation to improve performances of potential mycoherbicides (*Pl. Prot. Quart.* **7** (4): 30 pp., 1992; Auld, *Crop Protection* **12**: 477, 1993); Burge (Ed., *Formulation of microbial biopesticides*, 1998); Green *et al.* (*in* Boland & Kuykendall (Eds), *Plant – microbe interactions and biological control*: 249, 1997).

The first mycofungicide, and indeed one of the first commercially available mycopesticides, arose from the pioneering work of Rishbeth (*Ann. appl. Biol.* **52**: 63, 1963) who investigated the potential of *Phlebia gigantea* for control of butt rot of conifers caused by *Heterobasidion annosum* (Deacon, *Microbial control of plant pests and diseases*, 1983; Campbell, *Biological control of microbial plant pathogens*, 1989). Most work has concentrated on *Trichoderma* and *Gliocladium* as antagonists; reviewed by Lynch & Ebben (*J. appl. Bact. Symp. Suppl.* 1986, 115), Papavizas (*Ann. Rev. Phytopath.* **23**: 23, 1985), Whipps & Lumsden (Eds) (*Biotechnology of fungi for improving plant growth* 1989) and Whipps (*Aspects Appl. Biol.* **24**: 211, 1990). Apart from Binab T, based on *T. viride*, for control of forest diseases, GlioGard™ (now SoilGard™) based on *Trichoderma virens* has been marketed recently for control of *Rhizoctonia solani* and *Pythium ultimum* in horticultural crops (Mink & Walker, *in* Lumsden & Vaughn (Eds), *Pest management: biologically based technologies*, 398, 1993).

Mycophaga F. Stevens (1924) ? = Hyaloderma fide Pirozynski (*Kew Bull.* **31**: 595, 1977).

mycophage (1) a mycophagist; (2) a phage-like antibacterial substance produced by certain actinomycetes.

mycophagist, an eater of fungi.

mycophagy (adj. **mycophagous**), (1) the use of fungi as food; mycetophagy; (2) the lysis of a fungus by a phage.

Mycopharus Petch (1926) = Lysurus fide Dring (*Kew Bull.* **35**: 1, 1980).

mycophilic (1) fond of fungi (or mushrooms); mycetophilous; see Fungicolous fungi; (2) growing on fungi.

mycophobia, fear of mushrooms.

mycophthorous (of a fungus), parasitic on another fungus; see mycoparasitism.

Mycophycias Kohlm. & Volkm.-Kohlm. (1998) = Stigmidium fide Kohlmeyer & Volkmann-Kohlmeyer (*SA* **16**: 2, 1998), Aptroot (*CBS Biodiversity Series* **5**, 2006).

mycophycobiosis, an obligate symbiosis between a systemic (inhabitant) marine fungus and a marine alga in which the alga is the exhabitant and dominates (Kohlmeyer & Kohlmeyer, *Bot. Mar.* **15**: 109, 1972), e.g. *Mycosphaerella ascophylli* on *Ascophyllum nodosum*; cf. Lichens.

Mycophycophila Cribb & J.W. Cribb (1960) = Chadefaudia fide Müller & von Arx *in* Ainsworth *et al.* (Eds) (*The Fungi* **4A**: 87, 1973).

Mycophycophyta. Phylum name used by Margulis (*Symbiosis in cell evolution*, edn 2, 1993) for *all* lichens regardless of the systematic position of either the fungal, algal or cyanobacterial partners; i.e. embracing lichen-forming *Ascomycota* and *Basidiomycota* as well as their photosynthetic partners. ≡ *Lichenes*. See Lichens.

Mycophycophytes, see *Mycophytes*.

Mycophyta, see *Eumycetes*.

Mycophytes (obsol.) = Mycomycophytes (i.e. non-lichenized fungi) + Mycophycophytes (i.e. lichenized fungi) (Marchand, *Énumération méthodique des Mycophytes*, 1896).

Mycoplacographa Reinke (1895) ? = Lithographa fide Hawksworth *et al.* (*Dictionary of the Fungi* edn 8, 1995).

mycoplasm, a symbiotic phase of rust fungus and host protoplasm (Eriksson; now taken as an error); see also mycoplasma.

mycoplasma (1) an intimate relationship between plant-invading fungi or other microorganisms and their host cells (Frank, *Ber. Deutsch. bot. Ges.* **7**: 332, 1889) (obsol.); (2) bacterium-like organisms without a cell wall living inside cells of a host, mollicutes (Krass & Gardner, *Internat. J. Syst. Bact.* **23**: 62, 1973).

Mycoporaceae Zahlbr. (1903), ? Dothideomycetes (inc. sed.). 1 gen. (+ 3 syn.), *c.* 28 spp.

Lit.: Harris (*Mich. Bot.* **12**: 3, 1973), Coppins & James (*Lichenologist* **11**: 27, 1979), Eriksson (*Op. Bot.* **60**, 1981), Lumbsch (*Pl. Biol.* **1**: 321, 1999).

Mycoporellum Müll. Arg. (1884), Dothideomycetes. 7, widespread. See Riedl (*Sydowia* **15**: 257, 1962).

Mycoporis Clem. (1909), Dothideomycetes. 1, Austria.

Mycoporopsis Müll. Arg. (1885), Dothideomycetes (?L). 6, widespread. See Riedl (*Sydowia* **16**: 215, 1963).

Mycoporum Flot. ex Nyl. (1855) nom. cons., Mycoporaceae (±L). *c.* 28, widespread. See Harris (*Mich. Bot.* **12**: 3, 1973), Coppins & James (*Lichenologist* **11**: 27, 1979), Lumbsch (*Pl. Biol.* **1**: 321, 1999; development), Aptroot *et al.* (*Biblthca Lichenol.* **97**, 2008; Costa Rica).

Mycoporum G. Mey. (1825) nom. rej., Pyrenulales.

mycoprotein, fungal protein, e.g. 'Quorn' (q.v.); commercially processed mycelium of non-pathogenic *Fusarium venenatum* (also sometimes and formerly called *Fusarium graminearum* A35) for human consumption (Newark, *Nature* **287**: 6, 1980); see Gray & Staff (*Econ. Bot.* **21**: 341, 1966), Trinci (*MR* **96**: 1, 1992).

Mycopyrenium Hampe ex A. Massal. (1860) = Thelotrema fide Zahlbruckner (*Catalogus Lichenum Universalis* **2**, 1923).

Mycopyrenula Vain. (1921) = Pyrenula Ach. (1809) fide Harris (*Mem. N. Y. bot. Gdn* **49**, 1989).

Mycorhizonium F.E. Weiss (1904), Fossil Fungi (mycorrhizal) Fungi. 1 (Cretaceous), British Isles.

Mycorhynchella Höhn. (1918) = Ceratocystis fide Sutton (*Mycol. Pap.* **141**, 1977).

Mycorhynchidium Malloch & Cain (1971), Pyxidiophoraceae. 1 (coprophilous), Kenya. See Malloch & Cain (*CJB* **49**: 850, 1971).

Mycorhynchus Sacc. & D. Sacc. (1906) = Pyxidiophora fide Lundqvist (*Bot. Notiser* **133**: 121, 1980).

Mycorrhaphiaceae Jülich (1982) = Steccherinaceae.

Mycorrhaphium Maas Geest. (1962), Meruliaceae. 3, USA; Europe; Africa. See Ryvarden (*Mem. N. Y. bot. Gdn* **49**: 344, 1989; key), Mossebo & Ryvarden (*Mycotaxon* **88**: 229, 2003; Africa).

Mycorrhiza (pl. **mycorrhizas**, **mycorrhizae**) (fungus root). A symbiotic, non-pathogenic or feebly or weakly pathogenic association of a fungus and the roots of a plant. Found in most, perhaps 85% of plant species. The resulting dual organism (cf. lichen) was first described in detail by Frank (1895) who observed it on the roots of the major tree species of temperate forests. Frank (1887) then recorded two types of mycorrhiza, which he called: (a) **ectotrophic** (the characteristic mycorrhiza of temperate and boreal forest trees with different basidiomycetes esp. spp. of *Amanita*, *Boletus*, *Cortinarius*, *Russula*, *Suillus*, and some ascomycetes, e.g. *Tuber*), in which the fungus forms a sheath on the surface of the root from which hyphae extend outward into the soil and inwards between the outer cortical cells with which they interface to form a 'Hartig net'; and: (b) **endotrophic** (e.g. of orchid-basidiomycete and ericoid-ascomycete associations) in which the fungal hyphae enter the cortical cells of the root, enveloped by the plasmalemma of host.

The mycorrhiza now known to be most widely distributed both through the plant kingdom (see Trappe, *in* Safir *et al.*, 1987), and geographically, is a type of endo-association formed by *Glomeromycota* and called vesicular-arbuscular (**VA**, or **VAM**). In this, the penetrating hyphae produce finely branched haustorial branches (arbuscules) or coils (pelotons) and vesicles (Schlicht, 1889; Gallaud, 1905; Mosse, 1956). The taxonomic status of VA fungi is reviewed by Morton & Benny (*Mycotaxon* **37**: 471, 1990). VA mycorrhizas are known from the fossil record going back at least 400 million years (Remy *et al.*, *Proc. nat. acad. sci.* **91**: 11841, 1994).

Current classifications of mycorrhizal types avoid use of the suffix 'trophic' and recognize the following categories: **Ectomycorrhiza, Vesicular-arbuscular** (occasionally simply '**arbuscular**'; **AM**), **Ericoid, Orchid, Arbutoid** and **Monotropoid** (Lewis, *Biol. Rev.* **48**: 261, 1973; Read, *CJB* **61**: 985, 1983).

For the fungal partner, the main benefit is usually constant access to a supply of carbohydrates produced by the plant partner's photosynthesis. For the plant partner, the benefit is an enhanced supply of mineral nutrients from the soil, taken up by the hyphae of the fungal partner. These, being more narrow than root hairs, have a larger surface area through which mineral absorption can take place, and can approach more closely to a mineral nutrient source, thereby steepening the diffusion gradient by which the nutrients may move. Mycorrhizal associations are therefore particularly beneficial for plants growing on nutrient poor soils. Fungal partners are known which trap insects and thereby enhance the supply of nitrogen for the plant partner (Klironomos & Hart, *Nature* **410**: 651, 2001).

Early development of some plants, especially those with very small seeds (e.g. orchids) is completely dependent upon establishment of a successful fungal association. This dependence may be retained where, as in *Monotropaceae*, and some members of *Orchidaceae* and *Gentianaceae*, plants lack chlorophyll throughout their lives, and hence continue to require all carbon and most mineral supplies from their fungal symbiont. In green plants the main function of the association differs according to mycorrhizal type. Enhancement of phosphorus (P) supply to the plant is characteristic of VA associations, that of nitrogen (N) of ericoid associations, while both N and P supplies can be supplemented by ecto, ectendo and probably orchid associations. The global distribution of mycorrhizal types largely reflects these functional attributes, selection favouring plants with VA fungi in primarily P limited tropical and temperate grasslands, ectomycorrhizal plants in predominantly N and P limited temperate and boreal forests, and ericoid plants in N-limited tundra (Read, *Experientia* **47**: 376, 1991). In each of these circumstances the associated plants show differing levels of responsiveness depending upon the extent of nutrient limitation and the structure of their root systems. Those with 'fibrous' root systems, e.g. grasses, are less responsive than those with coarsely branched so called 'magnolioid' roots (Baylis, see Sanders *et al.*, 1975).

In the past 50 years, many plantations of northern-hemisphere conifers have been established on poor soils in southern hemisphere countries, and mycorrhizal fungi known to enhance their growth have been deliberately introduced with them. Concern has been voiced that this can lead to carbon depletion, thereby contributing to rather than ameliorating climate change (Chapela *et al*, *Soil biol. biochem.* **33**: 1733, 2001).

Lit.: Agerer (*Colour atlas of ectomycorrhizae*, 1991), Allen (*Ecology of ectomycorrhizae*, 1990), Currah & Zelmer (*Rep. Tottori mycol. Inst.* **30**: 43, 1992; key 15 gen. with orchids), Declerk *et al.* (Eds) (*In vitro culture of mycorrhizas*, 2005), De Roman *et al.* (*MR* **109**: 1063, 2005; revision of ectomycorrhiza descriptions since 1961), Frank (*Ber. dtsch bot. Ges.* **3**: 128, 1885; **5**: 248, 1887), Gallaud (*Rev. Gen. Bot.* **17**: 5, 1905), Gianinnazzi & Schüepp (Eds) (*Impact of arbusclar mycorrhizas on sustainable agriculture and natural ecosystems*, 1994), Harley (*Biology of mycorrhiza*, 1959, edn 2, 1970), Harley & Harley (*A check-list of mycorrhiza in the British flora*, 1987; *New Phytol.* Suppl. **105**), Harley & Smith (*Mycorrhizal Symbiosis*, 1983), Kapulnik & Douds (*Arbuscular mycorrhizas: physiology and function*, 2000), Kelly (*Mycotrophy in plants*, 1950), Khalil *et al.* (*Soil Biol. Biochem.* **26**: 1587, 1994; recovery VAM spores from soil), Marks & Kozlowski (Eds) (*Ectomycorrhizae their physiology and ecology*, 1973), Mejstřík *et al.* (Eds) (*Agric., Ecosyst. Environ.* **28-29**, 1990; Eur. Congr. Mycorrh.), Melin (*Untersuchungen uber die Bedeutung der Baummycorrhiza*, 1925), Morton (*Mycotaxon* **32**: 267, 1988; checklist 126 VA spp.), Mosse (*Ann. Bot. Lond.* **20**: 349, 1956), Norris *et al.* (Eds) (*Techniques for the study of mycorrhiza. Methods in Microbiology* **23** & **24**, 1991), Peterson *et al.* (*Mycorrhizas: anatomy and cell biology*, 2004), Read *et al.* (Eds) (*Mycorrhizas in ecosystems*, 1992), Safir (Ed.) (*Ecophysiology of VA mycorrhizal plants*, 1987), Sanders *et al.* (Eds) (*Endomycorrhiza*, 1975), Schlicht (*Landwirtsch. Jahrb.* **18**: 478, 1889), Sieverding (*VA mycorrhiza management in tropical agroecosystems*, 1991), Trappe (*Bot. Rev.* **28**: 538, 1962; fungus & tree lists, 407 refs.), Varma & Hoch (Eds) (*Mycorrhiza: structure, function, molecular biology and biochemistry*, 1995); see also the journal *Mycorrhiza* (1991-).

Mycosarcoma Bref. (1912) = Ustilago fide Vánky (*in litt.*).

mycosclerid, see cystidium.

mycose, see trehalose.

mycosin, a nitrogenous substance like animal chitin in the cell wall of fungi.

mycosis (pl. **mycoses**), (1) a fungus disease of humans, animals, or, rarely, plants (e.g. tracheomycosis). Mycoses are frequently named after the part attacked (**broncho-**, respiratory tract; **dermato-**, skin; **onycho-**, nails; **oto-**, ear; **pneumo-**, lungs), or the pathogen **blasto-** (q.v.) (*Blastomyces*); **coccidioido-**, coccidioidal granuloma (*Coccidioides immitis*); (2) the first limited infection of a dermatophyte; cf. mycid. Nomenclature, see Odds (*J. med. vet. Mycol.* **30**: 21, 1992). See AIDS, Medical and veterinary mycology.

Mycosisymbrium Carris (1994) = Scolecobasidiella fide Sutton (*in litt.*).

mycosocieties (1) communities of fungi occurring in a special habitat; see Winterhoff (1992), mycosociology; (2) mycological societies (see Societies and organizations).

Mycosociology. the study of fungal communities. Some communities of lichens and plants have a long tradition of being named using rules developed for plant communities (Barkman *et al.*, *Vegetatio* **67**: 145, 1986), and this form of name has also sometimes been used for other fungi, e.g. the association *Clitocybo-Phellodonetum nigrae* (Smarda, *Acta Nat. Acad. Sci. Bohemoslov.* **7**, 1973).

Darimont (*Inst. roy. Sci. nat. Brussels, Mém.* **170**, 1975), introduced a separate system for fungi in which the basic unit was the **sociomycie** (the name for which terminates in '**-ecium**', e.g. *Amanitecium muscariae*) (± = 'association') which are grouped (in ascending order) as 'alliances' ('*-ecion*'; *Boletecion scabri*), 'orders' ('*-ecia*'; *Boleto-Amanitecia*), and 'classes' ('*-ecea*'; *Cortinario-Boletacea*).

The number of works explicitly on this topic remains small, but includes studies in southern Chile (Valenzuela *et al.*, *Mycotaxon* **72**: 217, 1999). Issues relating to mycosociology were reviewed by Hawksworth & Mueller *in* Dighton *et al.*, *The fungal community: its organization and role in the ecosystem*, 2005).

See also Benkert (*Boletus* **2**: 37, 1978; synopsis approach, bibliogr.), Winterhoff (Ed.) (*Fungi in vegetation science* [*Handbook of Vegetation Science* **19**(1)], 1992), mycocoenosis, mycosynusium, Phytosociology.

Mycosphaerangium Verkley (1999), Helotiales. 3 (from dead wood), USA. See Verkley (*Stud. Mycol.* **44**: 180 pp., 1999).

Mycosphaerella Johanson (1884), Mycosphaerellaceae. Anamorphs many, esp. *Cercospora*, *Cercosporella*, *Cercosporidium*, *Passalora*, *Pseudocercospora*, *Ramularia* and *Septoria*. *c.* 647, widespread. A large genus with enormously varied anamorphs. Several large groups have been removed (e.g. *Davidiella*, *Teratosphaeria*, but it may well require further subdivision. See von Arx (*Sydowia* **3**: 28, 1949; anamorph gen.), von Arx (*Proc. K. ned. Akad. Wet.* Ser. C, Biol. Med. Sci. **86**: 15, 1983; anamorph gen.), Evans (*Mycol. Pap.* **153**, 1984; on *Pinus* (now as *Eruptio*)), Sivanesan (*Bitunicate Ascomycetes and their Anamorphs*, 1984), Niyo *et al.* (*Mycol.* **78**: 202, 1986; ultrastructure), Corlett (*Mycotaxon* **31**: 59, 1988; on *Brassicaceae*)), Corlett (*Mycol. Mem.* **18**: 328 pp., 1991; nomenclator), Corlett (*Mycotaxon* **53**: 37, 1995; catalogue names), Crous (*Mycol. Mem.* **21**: 170 pp., 1998; spp. on *Eucalyptus*), Fröhlich & Hyde (*Sydowia* **50**: 171, 1998; spp. on palms), Crous (*MR* **103**: 607, 1999; spp. on *Myrtaceae*), Crous & Corlett (*CJB* **76**: 1523, 1998; spp. on *Platanus*), Silva-Hanlin & Hanlin (*MR* **103**: 153, 1999; DNA), Stewart *et al.* (*MR* **103**: 1491, 1999; DNA), Crous *et al.* (*Stud. Mycol.* **45**: 107, 2000; review), Taylor & Crous (*MR* **104**: 618, 2000; on *Proteaceae*), Verkley & Priest (*Stud. Mycol.* **45**: 123, 2000; *Septoria* anams.), Crous *et al.* (*MR* **105**: 425, 2001; ITS of spp. on *Myrtaceae*), Crous *et al.* (*Mycol.* **93**: 1081, 2001; phylogeny), Goodwin *et al.* (*Phytopathology* **91**: 648, 2001; phylogeny), Milgate *et al.* (*Forest Pathology* **31**: 53, 2001; on *Eucalyptus*), Crous & Mourichon (*Sydowia* **54**: 35, 2002; on banana), Kuijpers & Aptroot (*Nova Hedwigia* **75**: 451, 2002; long-spored spp.), Linde *et al.* (*Petria* **12**: 95, 2002; population structure), Sivanesan & Shivas (*MR* **106**: 355, 2002; Australia), Braun *et al.* (*Mycol. Progr.* **2**: 3, 2003; exclusion of *Davidiella*), Crous & Braun (*CBS Diversity Ser.* **1**: 571 pp., 2003; anamorphs), Crous *et al.* (*Sydowia* **55**: 136, 2003; separation from *Sphaerulina*), Hayden *et al.* (*Pl. Path.* **52**: 703, 2003; on banana), Taylor *et al.* (*MR* **107**: 653, 2003; phylogeny on *Proteaceae*), Zhan *et al.* (*Fungal Genetics Biol.* **38**: 286, 2003; *M. graminicola* genetics), Crous *et al.* (*CBS Diversity Ser.* **2**, 2004; anamorphs), Crous *et al.* (*Stud. Mycol.* **50**: 195, 2004; on *Eucalyptus*), Crous *et al.* (*Stud. Mycol.* **50**: 457, 2004; on *Acacia*), Verkley *et al.* (*MR* **108**: 1271, 2004; typification), Hayden *et al.* (*Phytopathology* **95**: 489, 2005; genetics), Aptroot (*CBS Diversity Ser.* **5**, 2006; monogr.), Crous *et al.* (*Stud. Mycol.* **55**: 99, 2006; on *Eucalyptus*), Hunter *et al.* (*Stud. Mycol.* **55**: 147, 2006; on *Eucalyptus*), Schoch *et al.* (*Mycol.* **98**: 1041, 2006; phylogenetic posn), Crous *et al.* (*Stud. Mycol.* **58**: 1, 2007; phylogeny), Arzanlou *et al.* (*Persoonia* **20**: 19, 2008; on *Musa*).

Mycosphaerellaceae Lindau (1897) nom. cons., Capnodiales. 53 gen. (+ 70 syn.), 6033 spp.

Lit.: Scharen & Sanderson (*Septoria of Cereals* Proceedings of the Workshop held August 2-4, 1983, at Montana State University, Bozeman, Montana: 37, 1985), Corlett (*Mycol. Mem.* **18**: 328 pp., 1991), Johanson *et al.* (*Pl. Path.* **43**: 701, 1994), Cannon *et al.* (*MR* **99**: 353, 1995), David (*Mycol. Pap.* **172**: 157 pp., 1997), Crous (*Mycol. Mem.* **21**: 170 pp., 1998), Caten (*Septoria on Cereals* A Study of Pathosystems: 26, 1999), Crous (*MR* **103**: 607, 1999), Crous & Corlett (*CJB* **76**: 1523, 1998), Cunfer & Ueng (*Ann. Rev. Phytopath.* **37**: 267, 1999), Stewart *et al.* (*MR* **103**: 1491, 1999), Taylor *et al.* (*Ann. Rev. Phytopath.* **37**: 197, 1999), Crous *et al.* (*Stud. Mycol.* **45**: 107, 2000), Verkley & Priest (*Stud. Mycol.* **45**: 123, 2000), Crous *et al.* (*Mycol.* **93**: 1081, 2001), Braun *et al.* (*Mycol. Progr.* **2**: 8, 2003), Crous & Braun (*CBS Diversity Ser.* **1**: 571 pp., 2003), Hunter *et al.* (*MR* **108**: 672, 2004), Verkley *et al.* (*Mycol.* **96**: 558, 2004), Schoch *et al.* (*Mycol.* **98**: 1041, 2006; phylogeny).

Mycosphaerellales P.F. Cannon (2001) = Capnodiales.

Mycosphaerellopsis Höhn. (1918) = Didymella fide Barr (*Mycotaxon* **60**: 433, 1996).

Mycospongia Velen. (1939) nom. dub., ? Agaricales. ? 'gasteromycetes'.

Mycospraguea U. Braun & Rogerson (1993), anamorphic *Pezizomycotina*, Cac.≡ eH.1. 1, USA. See Braun (*Cryptog. bot.* **4**: 108, 1993).

mycostasis (adj. **mycostatic**), inhibition of fungal growth; fungistasis (fungistatic); sporostasis; **mycostatic** in soil, see Dobbs *et al.* (*Nature* **172**: 197, 1953), Parkinson & Waid (Eds) (*The ecology of fungi*, 130, 1960).

mycostatin, trade name for nystatin.

Mycostevensonia Bat. & Cif. (1962) = Treubiomyces fide von Arx & Müller (*Stud. Mycol.* **9**, 1975).

Mycosticta Höhn. (1918), anamorphic *Pezizomycotina*, ?.0eH.?. 1, Europe.

Mycostigma Jülich (1976), ? Atheliaceae. 1, Europe. See Jülich (*Persoonia* **8**: 432, 1976).

mycostratum, see perisporium; Spore wall.

Mycosylva M.C. Tulloch (1973), anamorphic *Pezizomycotina*, Hsy.0eH.3. 3, Europe; Asia (esp. temperate). See Samson & Hintikka (*Karstenia* **14**: 133, 1974).

mycosymbiont, see mycobiont.

mycosymbiosis, symbiosis of two or more fungi (Vainio, 1921; cf. *Gonidiomyces*).

mycosynusium (pl. **-iae**), a part of the total number of fungal communities in a site or region studied; e.g. **macrofungussynusiae** (only macromycetes considered); see Winterhoff (1992), mycosociology.

Mycosyringaceae R. Bauer & Oberw. (1997), Urocystidales. 1 gen., 4 spp.

Lit.: Vánky (*Cryptog. Stud.* **1**: 159 pp., 1987), Vánky (*Mycoscience* **37**: 173, 1996), Bauer & Oberwinkler (*Mycotaxon* **64**: 303, 1997), Bauer *et al.* (*CJB* **75**: 1273, 1997), Piepenbring *et al.* (*Protoplasma* **204**: 170, 1998), Vánky (*MR* **102**: 513, 1998), Vánky (*Mycol. Balcanica* **1**: 175, 2004).

Mycosyrinx Beck (1894), Mycosyringaceae. 4 (on *Vitaceae*), pantropical. See Mordue (*Mycopathologia* **103**: 171, 1988), Vánky (*Mycoscience* **37**: 173, 1996; key).

Mycota, See *FUNGI*.

mycothallus (pl. **-lli**), a mutualistic symbiosis of a fungus with a hepatic (liverwort) or fern gametophyte (Boullard, *Syllogeus* **19**: 1, 1979; *in* Pirozynski & Hawksworth (Eds), *Coevolution of fungi with plants and animals*: 107, 1988).

Mycothamnion Kütz. (1843) nom. dub., ? Fungi.

mycotheca, a distributed set of dried specimens of fungi.

Mycothele Jülich (1976), ? Gloeophyllaceae. 1, New Zealand. See Jülich (*Persoonia* **8**: 452, 1976).

Mycothelocarpon Cif. & Tomas. (1953) nom. illegit. = Thelocarpon.

Mycothyridium E. Müll. (1973) = Thyridium Nitschke fide Hawksworth *et al.* (*Dictionary of the Fungi* edn 8, 1995).

Mycothyridium Petr. (1962), Dothideomycetes. 36, widespread. See Petrak (*Sydowia* **15**: 288, 1972), Eriksson & Hawksworth (*SA* **13**: 191, 1994; nomencl.).

mycotic (esp. of disease), caused by fungi.

Mycotodea Kirschst. (1936), Pezizomycotina. 14 (on *Musci*), widespread. See Hawksworth (*SA* **8**: 71, 1989; posn).

mycotope, a major fungal association of a particular type of woodland (Darimont, 1975); see mycosociology.

Mycotorula H. Will (1916) = Syringospora fide von Arx *et al.* (*Stud. Mycol.* **14**: 1, 1977).

Mycotoruloides Langeron & Talice (1932) = Syringospora fide von Arx *et al.* (*Stud. Mycol.* **14**: 1, 1977).

Mycotoxicoses. Literally fungus poisonings, but in current use limited to poisoning of animals and esp. humans, usually by feed and food products contaminated (and sometimes rendered carcinogenic) by toxin-producing microfungi, also by contact with toxin-containing microfungi in particular environments (e.g. sick-building syndrome). See aflatoxins, citreoviridin, citrinin, fumonisins, islanditoxin, lupinosis, luteoskyrin, lysergic acid, maltoryzine, ochratoxin, patulin, roridins, rubratoxin, satratoxins, slaframine, sporidesmin, sterigmatocystin, tremorgen, trichothecenes, zearalenone. The difficult issue of fungi and fungal toxins as weapons was reviewed by Paterson (*MR* **110**: 1003, 2006). See also Biological control, Bioterrorism and fungi, Mycopesticides.

Lit.: Reviews: Assouline-Dayan *et al.* (*J. Asthma* **39**: 191, 2002), Forgacs & Carll (*Adv. vet. Sci.* **7**: 272, 1962), Wheeler & Luke (*Ann. Rev. Microbiol.* **17**: 223, 1963), Wright (*Ann. Rev. Microbiol.* **22**: 269, 1968). Books: Betina (*Mycetotoxins*, 1989), Castegnaro *et al.* (Eds) (*Mycotoxins, endemic nephropathy and urinary tract tumours*, 1991), Champ *et al.* (Eds) (*Fungi and mycotoxins in stored products*, 1992). Chelkowski (Ed.) (*Fusarium mycotoxins*, 1989), Chelkowski (Ed.) (*Mycotoxin research. Fusarium* **7**(I & II), 1991), Chelkowski (Ed.) (*Cereal grain mycotoxins, fungi and quality in drying and storage*, 1991), Cole (Ed.) (*Modern methods in the analysis structural elucidation of mycotoxins*, 1986), Cole & Cox (Eds) (*Handbook of toxic fungal metabolites*, 1981), Desjardins (*Fusarium mycotoxins: chemistry, genetics and biology*, 2006), Hui *et al.* (Eds, *Foodborne Diseases Handbook* **3**, 2001), Joffe (*Fusarium species; their biology and toxicology*, 1986), Kadis (Ed.) (*Microbial toxins*, **6**, *Fungal toxins*, 1971), Krogh (Ed.) (*Mycotoxins in food*, 1987), Logrieco *et al.* (*Mycotoxins in plant disease*, 2002), Matossian (*Poisons of the past*, 1989), Moreau (*Moisissures toxiques dans l'alimentation*, 1968; edn 2, 1974, 2938 refs), O'Neill *et al.* (Eds) (*Relevance of human cancer of N-Nitroso compounds, tobacco and mycotoxins*, 1991), Sinha & Bhatnagar (Eds) (*Mycotoxins in agriculture and food safety*, 1998), Wogan (Ed.) (*Mycotoxins in foodstuffs*, 1965), Wyllie & Morehouse (Eds) (*Mycotoxic fungi, mycotoxins, mycotoxicoses. An encyclopaedic handbook* **1**, *Mycotoxic fungi and chemistry of mycotoxicoses*, 1977; **2**, *Mycotoxicoses of domestic and laboratory animals, poultry, and aquatic invertebrates and vertebrates*, 1978; **3**, *Mycotoxicoses of man and plants; mycotoxin control and regulatory aspects*, 1978), Smith & Henderson (*Mycotoxins and animal foods*, 1991), Smith & Moss (Eds) (*Mycotoxins*, 1985), Steyn (Ed.) (*The biosynthesis of mycotoxins*, 1980), Upadhyay (*Advances in microbial toxin research and its biotechnological exploitation*, 2002), Xu *et al.* (*Epidemiology of mycotoxin producing fungi*, 2003). See also phytoalexins.

mycotoxicosis, one of the Mycotoxicoses (q.v.).

mycotoxin, see toxin.

Mycotribulus Nag Raj & W.B. Kendr. (1970), anamorphic *Pezizomycotina*, St.0eH.4. 1, pantropical. See Nag Raj & Kendrick (*CJB* **48**: 2219, 1970).

mycotroph, a fungus which obtains its nutrients from another fungus; cf. mycoparasitism and see Fungi on fungi.

mycotrophein, a 'growth factor' from fungi needed by a mycoparasite (Waley & Barnett, *Mycol.* **55**: 209, 1963).

mycotrophic (of plants), having mycorrhiza.

Mycotypha Fenner (1932), Mycotyphaceae. 3, widespread. See Young (*J. gen. Microbiol.* **55**: 243, 1969; sporangiolum ultrastr.), Benny & Benjamin (*Aliso* **8**: 391, 1976), Brain & Young (*Microbios* **25**: 93, 1979; ultrastr.), Benny *et al.* (*Mycotaxon* **22**: 119, 1985;

key), Edelmann & Klomparens (*MR* **99**: 539, 1995; ultrastr), O'Donnell *et al.* (*Mycol.* **93**: 286, 2001; phylogeny), Voigt & Wöstemeyer (*Gene* **270**: 113, 2001; phylogeny).

Mycotyphaceae Benny & R.K. Benj. (1985), Mucorales. 1 gen., 3 spp.
Lit.: Benny *et al.* (*Mycotaxon* **22**: 119, 1985; key), Forst & Prillinger (*Z. Mykol.* **54**: 139, 1988).

Mycousteria M.L. Farr (1986), anamorphic *Pezizomycotina*, Cpt.0eP.?. 2, Brazil. See Farr (*Mycol.* **78**: 280, 1986).

Mycovellosiella Rangel (1917) = Passalora fide Deighton (*Mycol. Pap.* **137**, 1974), Liu & Guo (*Mycosystema* **1**: 241, 1988; 21 spp. from China), Braun (*Monogr. Cercosporella, Ramularia Allied Genera (Phytopath. Hyphom.)* **2**, 1998), Crous (*MR* **103**: 607, 1999; on *Myrtaceae*), Crous *et al.* (*Stud. Mycol.* **45**: 107, 2000; review), Crous *et al.* (*Mycol.* **93**: 1081, 2001; phylogeny), Crous *et al.* (*MR* **105**: 425, 2001; on *Myrtaceae*), Crous & Braun (*CBS Diversity Ser.* **1**: 571 pp., 2003; nomenclator).

mycoviruses, see Viruses in fungi.

Mycowinteria Sherwood (1986), Protothelenellaceae. 1, widespread (temperate). See David (*SA* **6**: 217, 1987), Sherwood-Pike (*Mycotaxon* **28**: 137, 1987), Aptroot & Iperen (*Nova Hedwigia* **67**: 481, 1998; New Guinea).

Mydonosporium Corda (1833) = Cladosporium fide Saccardo (*Syll. fung.* **4**: 354, 1886).

Mydonotrichum Corda (1831) = Helminthosporium fide Saccardo (*Syll. fung.* **4**: 407, 1886).

Myelochroa (Asahina) Elix & Hale (1987), Parmeliaceae (L). 22, widespread (esp. north-east Asia). See Hale (*Smithson. Contr. Bot.* **33**, 1976), Kurokawa & Arakawa (*Bull. Bot. Gdns Toyama* **2**: 23, 1997; Japan), Wang *et al.* (*Mycotaxon* **77**: 25, 2001; China), Blanco *et al.* (*Mol. Phylogen. Evol.* **39**: 52, 2006; phylogeny), Miądlikowska *et al.* (*Mycol.* **98**: 1088, 2006; phylogeny).

Myeloconidiaceae P.M. McCarthy (2001), Ostropales. 2 gen., 5 spp.
Lit.: McCarthy & Elix (*Lichenologist* **28**: 402, 1996), McCarthy (*Flora of Australia* **58** A: 242 pp., 2001), McCarthy *et al.* (*Lichenologist* **33**: 292, 2001).

Myeloconis P.M. McCarthy & Elix (1996), Myeloconidiaceae (L). 4, widespread (tropical). See McCarthy & Elix (*Lichenologist* **28**: 402, 1996).

Myelorrhiza Verdon & Elix (1986), Cladoniaceae (L). 2, Australia. See Stenroos & DePriest (*Am. J. Bot.* **85**: 1548, 1998; DNA), Stenroos *et al.* (*Mycol. Progr.* **1**: 267, 2002; phylogeny).

Myelosperma Syd. & P. Syd. (1915), Myelospermataceae. 4 (on *Palmae*), Brunei; Irian Jaya. See Hyde (*Sydowia* **45**: 241, 1993), Hyde & Wong (*MR* **103**: 347, 1999; fam. rels).

Myelospermataceae K.D. Hyde & S.W. Wong (1999), Xylariales. 1 gen., 4 spp.
Lit.: Hyde (*Sydowia* **45**: 241, 1993), Kang *et al.* (*Fungal Diversity* **1**: 147, 1998), Hyde & Wong (*MR* **103**: 347, 1999).

Myiocopraloa Cif. (1958) = Schizothyrium fide Müller & von Arx (*Beitr. Kryptfl. Schweiz* **11** no. 2, 1962).

Myiocoprella Sacc. (1916) = Rhagadolobium fide Hawksworth *et al.* (*Dictionary of the Fungi* edn 8, 1995).

Myiocopron, see *Muyocopron*.

Myiocoprula Petr. (1955), anamorphic *Pezizomycotina*, Cpt.0eH.?. 1, USA. See Petrak (*Sydowia* **9**: 547, 1955).

Myiophagus Thaxt. ex Sparrow (1939), ? Chytridiales. 2, widespread (north temperate). See Karling (*Am. J. Bot.* **35**: 246, 1948).

Myiophyton Lebert (1857) = Empusa fide Saccardo (*Syll. fung.* **7**: 1, 1888).

Mykoblastus, see *Mycoblastus*.

mykoholz, basidioma of the edible *Kuehneromyces mutabilis* cultivated in Germany (Chang & Hughes, 1978).

Mylitta Fr. (1825) nom. dub., Fungi. Based on bacterial root nodules on *Robinia* fide Mattirolo (*Bull. Soc. bot. ital.* 1924: 13, 1924).

mylitta, a large sclerotium, e.g. that of blackfellows' bread (q.v.).

Mylittopsis Pat. (1895), Basidiomycota. 1, USA; Malaysia. See Rogers & Martin (*Mycol.* **47**: 891, 1955).

myosin, see actin.

Myosporidium E. Baquero, M. Rubio, I.N.S. Moura, N.J. Pieniazek & R. Jordana (2005), Microsporidia. 1. See Baquero *et al.* (*J. Eukary. Microbiol.* **52**: 476, 2005).

Myriadoporus Peck (1884) = Bjerkandera fide Donk (*Persoonia* **1**: 173, 1960).

Myriangiaceae Nyl. (1854), Myriangiales. 4 gen. (+ 8 syn.), 18 spp.
Lit.: Boedijn (*Persoonia* **2**: 63, 1963; Indonesia), von Arx (*Persoonia* **2**: 421, 1963; gen. names), Rao & Pande (*Cryptog. bot.* **3**: 255, 1993), Inácio & Dianese (*MR* **102**: 695, 1998), Winka & Eriksson (*Phylogenetic Relationships Within the Ascomycota Based on 18S rDNA Sequences* Akademisk Avhandling [Thesis (PhD), Department of Ecology and Environmental Science, Umeå University]: [17] pp., 2000), Lumbsch & Lindemuth (*MR* **105**: 901, 2001).

Myriangiales Starbäck (1899). Dothideomycetidae. 3 fam., 18 gen., 157 spp. Stromata crustose or pulvinate, composed of subhyaline or brown thin-walled pseudoparenchymatous tissue, with immersed ascomatal locules or fertile outgrowths of similar tissue containing scattered asci in individual locules, becoming gelatinous at maturity, opening by unordered breakdown of the surface layers. Specialized interascal tissue absent. Asci in a single layer or irregularly disposed, ± globose, sessile, fissitunicate, usually with a poorly defined ocular chamber. Ascospores hyaline to brown, transversely septate or muriform. Anamorphs acervular where known. Fams:
(1) **Cookellaceae**
(2) **Elsinoaceae**
(3) **Myriangiaceae**
Molecular data have confirmed links between the *Elsinoaceae* and *Myriangiaceae*, but relationships of the *Cookellaceae* remain uncertain.
Lit.: Eriksson (*Opera Bot.* **60**, 1981), Schoch *et al* (*Mycol* **98**: 1041, 2006; phylogeny).

Myriangiella Zimm. (1902), ? Schizothyriaceae. *c.* 5, widespread (tropical). See von Arx & Müller (*Stud. Mycol.* **9**, 1975), Hofmann & Piepenbring (*Fungal Diversity* **22**: 55, 2006; Panama).

Myriangina (Henn.) Höhn. (1909) = Uleomyces fide von Arx (*Persoonia* **2**: 421, 1963).

Myrianginella F. Stevens & Weedon (1923) = Uleomyces fide von Arx & Müller (*Stud. Mycol.* **9**, 1975).

Myriangiomyces Bat. (1958) = Saccardia fide von Arx (*Persoonia* **2**: 421, 1963).

Myriangiopsis Henn. (1902), Dothideomycetes. 1, Mexico.

Myriangium Mont. & Berk. (1845), Myriangiaceae. *c.* 8, widespread. See Petch (*TBMS* **10**: 45, 1924; on *Insecta*), Miller (*Mycol.* **30**: 158, 1938), Miller (*Mycol.* **32**: 587, 1940), von Arx (*Persoonia* **2**: 241, 1963), Lumbsch & Lindemuth (*MR* **105**: 901, 2001; phylogeny), Schoch *et al.* (*Mycol.* **98**: 1041, 2006; phylogeny).

Myriapodophila Speg. (1918) ? = Pyxidiophora fide Blackwell (*Mycol.* **86**: 1, 1994).

Myridium Clem. (1909) nom. rej. = Laetinaevia fide Hawksworth *et al.* (*Dictionary of the Fungi* edn 8, 1995).

Myriellina Höhn. (1915), anamorphic *Pezizomycotina*, Cac.≡ eH.15. 2, Europe; Papua New Guinea. See Sankaran & Sutton (*MR* **95**: 1021, 1991).

Myrillium Clem. (1931) nom. dub., ? Onygenales. See Kuehn (*Mycol.* **51**: 665, 1959), Currah (*Mycotaxon* **24**: 1, 1985).

Myrioblastus Trevis. (1857) ≡ Biatorella.

Myrioblepharis Thaxt. (1895) nom. dub., Fungi. Based on a protozoan on *Pythium* or *Phytophthora* fide Waterhouse (*TBMS* **28**: 94, 1945), Koch (*Mycol.* **56**: 436, 1964).

Myriocarpa Fuckel (1870) [non *Myriocarpa* Benth. 1846, *Urticaceae*] ? = Guignardia fide von Arx & Müller (*Beitr. Kryptfl. Schweiz* **11** no. 1, 1954).

Myriocarpium Bonord. (1864) = Leptosphaeria fide Saccardo (*Syll. fung.* **2**: 13, 1883).

Myriocephalum De Not. ex Corda (1842) = Cheirospora fide Hughes (*CJB* **36**: 727, 1958).

Myriococcum Fr. (1823), anamorphic *Agaricomycetes*. 1 or 2, widespread. *M. albomyces* is included in *Melanocarpus*.

Myrioconium Fr. (1823), anamorphic *Agaricomycetes*. 1, widespread. See Stalpers *in* Sugiyama (Ed.) (*Pleomorphic Fungi: The Diversity and its Taxonomic Implications*: 201, 1987).

Myrioconium Syd. (1912), anamorphic *Sclerotinia, Myriosclerotinia*, Cac.0eH.15. 7, Europe; N. America. See Palmer (*Öst. Z. Pilzk.* **4**: 81, 1995; on *Eriophorum*).

Myriodiscus Boedijn (1935), Helotiales. 1, Sumatra. See Liu & Guo (*Acta Mycol. Sin.* Suppl. **1**: 97, 1988).

Myriodontium Samson & Polon. (1978), anamorphic *Pezizomycotina*, Hso.0eH.12. 1, Europe; N. America. See Samson & Polonelli (*Persoonia* **9**: 505, 1978).

Myriogenis Clem. & Shear (1931) ≡ Myriogenospora.

Myriogenospora G.F. Atk. (1894), Clavicipitaceae. 2 (on *Poaceae*, incl. *Saccharum*), America. See Diehl (*USDA agric. Monogr.* **4**, 1950), Luttrell & Bacon (*CJB* **55**: 2090, 1977), Hanlin & Tortolero (*Mycotaxon* **39**: 237, 1990), Phelps & Morgan-Jones (*Mycotaxon* **47**: 41, 1993), White & Glenn (*Am. J. Bot.* **81**: 216, 1994), Glenn *et al.* (*MR* **102**: 483, 1998; DNA), Bacon & White (*Microbial Endophytes*, 2000).

Myriogonium Cain (1948) = Helicogonium fide Kreger-van Rij *in* Ainsworth *et al.* (Eds) (*The Fungi* **4A**: 15, 1973), Carpenter & Krapp (*Mycotaxon* **21**: 487, 1984), Baral (*Nova Hedwigia* **69**: 1, 1999), Suh *et al.* (*Mycol.* **98**: 1006, 2006; phylogeny).

Myriolecis Clem. (1909) = Lecanora fide Hawksworth *et al.* (*Dictionary of the Fungi* edn 8, 1995).

Myrionora R.C. Harris (1988), Lecanoraceae (L). 1, USA. See Harris (*Evansia* **5**: 27, 1988).

Myriophacidium Sherwood (1974), Rhytismataceae. 4, widespread. See Sherwood (*Occ. Pap. Farlow Herb. Crypt. Bot.* **15**, 1980; key), Cannon & Minter (*Mycol. Pap.* **155**, 1986).

Myriophysa Fr. (1849) ? = Atichia fide Hawksworth *et al.* (*Dictionary of the Fungi* edn 8, 1995).

Myriophysella Speg. (1910) = Seuratia fide Hawksworth *et al.* (*Dictionary of the Fungi* edn 8, 1995).

Myriosclerotinia N.F. Buchw. (1947), Sclerotiniaceae. Anamorphs *Myrioconium, Sclerotium*. 10, widespread. See Palmer (*Friesia* **9**: 193, 1969), Schwegler (*Schweiz. Z. Pilzk.* **56**: 49, 1978), Schumacher & Kohn (*CJB* **63**: 1610, 1985; key), Holst-Jensen *et al.* (*Mycol.* **89**: 885, 1997; phylogeny), Saito (*Mycoscience* **39**: 145, 1998), Holst-Jensen *et al.* (*Nordic Jl Bot.* **18**: 705, 1999; phylogeny), Holec *et al.* (*Sydowia* **59**: 57, 2007).

Myriosperma Nägeli (1853) = Sarcogyne Flot. (1851) fide Hawksworth *et al.* (*Dictionary of the Fungi* edn 8, 1995), Jørgensen (*Taxon* **53**: 521, 2004; nomencl.).

Myriospora Nägeli (1853), Acarosporaceae (L). 2. See Hawksworth *et al.* (*Dictionary of the Fungi* edn 8, 1995), Harris & Knudsen (*Opuscula Philolichenum* **3**: 1, 2006).

myriosporous, having many spores. Cf. oligosporous.

Myriostigma G. Arnaud (1925) ≡ Myriostigmella.

Myriostigma Kremp. (1874) nom. rej. = Cryptothecia fide Hawksworth *et al.* (*Dictionary of the Fungi* edn 8, 1995), Lücking *et al.* (*Lichenologist* **38**: 235, 2006).

Myriostigmella G. Arnaud (1952), Dothideomycetes. 1, Brazil. See Müller & von Arx (*Beitr. Kryptfl. Schweiz* **11** no. 2, 1962).

Myriostoma Desv. (1809), Geastraceae. 1, widespread. See Sunhede (*Syn. Fung.* **1**: 534 pp., 1989), Rees *et al.* (*Australasian Mycologist* **24**: 25, 2005; Australia).

Myriotrema Fée (1824), Thelotremataceae (L). 150, widespread (esp. tropical). See Hale (*Mycotaxon* **11**: 130, 1980), Hale (*Bull. Br. Mus. nat. hist.* Bot. **8**: 227, 1981), Matsumoto & Deguchi (*Bryologist* **102**: 86, 1999; anamorphs), Frisch (*Biblthca Lichenol.* **92**: 3, 2006; monogr. African spp.), Frisch *et al.* (*Biblthca Lichenol.* **92**: 517, 2006; phylogeny, polyphyly), Mangold *et al.* (*Nova Hedwigia* **83**: 275, 2006; Australia).

Myriotrichum, see *Myxotrichum*.

Myrmaeciella Lindau (1897), Hypocreales. Anamorph *Patellina*. 1, N. America. See Samuels & Seifert *in* Sugiyama (Ed.) (*Pleomorphic Fungi: The Diversity and its Taxonomic Implications*: 29, 1987), Rossman *et al.* (*Stud. Mycol.* **42**: 248 pp., 1999; allied to *Niessliaceae*).

Myrmaecium Nitschke ex Fuckel (1870), Pezizomycotina. 2, Europe.

Myrmaecium Sacc. (1880) ≡ Myrmaeciella.

Myrmecocystis Harkn. (1899) = Genabea fide Trappe (*TBMS* **65**: 496, 1975), Smith *et al.* (*Mycol.* **98**: 699, 2006), Læssøe & Hansen (*MR* **111**: 1075, 2007; phylogeny).

Myrmecomyces Jouvenaz & Kimbr. (1991), anamorphic *Pezizomycotina*, Hso.0eH.10. 1 (endoparasitic on fire ants), USA; Argentina. See Jouvenaz & Kimbrough (*MR* **95**: 1400, 1991).

myrmecophilous (of fungi), being a covering or food for ants.

Myrmecridium Arzanlou, W. Gams & Crous (2007), anamorphic *Sordariomycetes*. 2, widespread.

Myrophagus, see *Myiophagus*.

Myropyxis Ces. ex Rabenh. (1851), anamorphic *Pezizomycotina*. 1 or 2, Europe.

Myrotheciastrum Abbas & B. Sutton (1988), anamor-

phic *Pezizomycotina*, St.0eP.19. 1, Pakistan; India. See Abbas & Sutton (*TBMS* **91**: 352, 1988).
Myrotheciella Speg. (1910) = Myrothecium fide Tulloch (*Mycol. Pap.* **130**, 1972).
Myrothecium Tode (1790), anamorphic *Hypocreales*, Cpd.0eP.15. 22, widespread. See Tulloch (*Mycol. Pap.* **130**, 1972; key), Nag Raj (*Mycotaxon* **53**: 295, 1995; *M. prestonii* heterogeneous), Schroers *et al.* (*Sydowia* **51**: 114, 1999; teleomorph), Ahrazem *et al.* (*Archs Microbiol.* **173**: 296, 2000; cell wall polysaccharides), Rossman *et al.* (*Mycol.* **93**: 100, 2001; phylogeny), Seifert *et al.* (*Mycotaxon* **87**: 317, 2003; Canada), Watanabe *et al.* (*Mycoscience* **44**: 283, 2003; Japan), Zhang *et al.* (*Mycol.* **98**: 1076, 2006; phylogeny).
Mystrosporiella Munjal & Kulshr. (1969), anamorphic *Pezizomycotina*, Hso.#eP.10. 2, India. See Munjal & Kulshreshtha (*Mycopathologia* **39**: 356, 1969), Sharma *et al.* (*Indian Phytopath.* **59**: 257, 2006).
Mystrosporium Corda (1837) nom. dub., anamorphic *Pezizomycotina*, Hso.#eP.?. 5, widespread (temperate). *M. adustum* (ink disease of iris). See Hughes (*CJB* **36**: 788, 1958).
Mythicomyces Redhead & A.H. Sm. (1986), Psathyrellaceae. 1, widespread (north temperate). See Redhead & Smith (*CJB* **64**: 643, 1986).
Mytilidion Sacc. (1875) ≡ Mytilinidion.
mytiliform, like a mussel shell in form.
Mytilinidiaceae Kirschst. (1924), Pleosporales. 11 gen. (+ 7 syn.), 93 spp.
Lit.: Zogg (*Beitr. Kryptfl. Schweiz* **11** no. 3, 1962; Eur.), Blackwell & Gilbertson (*Mycol.* **77**: 50, 1985), Checa (*Mycotaxon* **63**: 467, 1997), Checa (*Mycotaxon* **62**: 349, 1997; Iberian spp.), Vasilyeva (*Mycoscience* **38**: 341, 1997).
Mytilinidion Duby (1862), Mytilinidiaceae. Anamorph *Septonema*-like. 12, Europe; N. America. See Zogg (*Beitr. Kryptfl. Schweiz* **11** no. 3, 1962), Speer (*BSMF* **102**: 97, 1986), Checa (*Mycotaxon* **63**: 467, 1997; Spain), Vasil'eva (*Mikol. Fitopatol.* **35**: 15, 2001; E Russia).
Mytilodiscus Kropp & S.E. Carp. (1984), Helotiaceae. 1, N. America; Europe. See Kropp & Carpenter (*Mycotaxon* **20**: 365, 1984).
Mytilostoma P. Karst. (1880), Dothideomycetes. 2, Finland. See Mathiassen (*Pyrenomyceter (Ascomyceter) på Salix i Troms*, 1985).
Myxacium (Fr.) P. Kumm. (1871) = Cortinarius fide Kuyper (*in litt.*).
myxamoeba (1) (of *Mycetozoa*), a zoospore after becoming amoeba-like; (2) a myxopod (obsol.).
Myxariaceae Jülich (1982) = Hyaloriaceae.
myxarioid (of basidia), having a stalk-like portion separated by a wall from the globose metabasidial portion, as in *Myxarium* (Donk, *Persoonia* **4**: 232, 1966).
Myxarium Wallr. (1833), Hyaloriaceae. 6, widespread. See Roberts (*Mycotaxon* **69**: 209, 1998), Wells *et al.* (*Frontiers in Basidiomycote Mycology*: 237, 2004).
Myxasterina Höhn. (1909) = Asterina fide Stevens & Ryan (*Ill. biol. monogr.* **17** no. 2, 1939).
Myxobacterales. An order of Bacteria (gliding bacteria) bearing extracellular slime; lyse filamentous and yeast fungi and bacteria in soil, dung and decaying vegetation.
Myxobilimbia Hafellner (2001) = Bilimbia fide Hafellner & Türk (*Stapfia* **76**: 154, 2001), Veldkamp (*Lichenologist* **36**: 191, 2004), Llop (*Lichenologist* **38**: 279, 2006).
Myxocephala G. Weber, Spaaij & Oberw. (1989), ? Sordariomycetes. 1. See Jacobs *et al.* (*CJB* **79**: 110, 2001; phylogeny).
Myxochytridiales, see *Archimycetes*.
Myxocladium Corda (1837) = Cladosporium fide Saccardo (*Syll. fung.* **4**: 364, 1886).
Myxocollybia Singer (1936) = Flammulina fide Singer (*Agaric. mod. Tax.*, 1951).
Myxocybe Fayod (1889) = Hebeloma fide Singer (*Agaric. mod. Tax.*, 1951).
Myxocyclus Riess (1852), anamorphic *Pleomassariaceae*, Hsp/Cac.#eP.1. 1, Europe; N. America. See Tanaka *et al.* (*Mycoscience* **46**: 248, 2005).
Myxocystis Mrázek (1897), Microsporidia. 1.
Myxoderma Fayod ex Kühner (1926) ≡ Limacella.
Myxodictyon A. Massal. (1860) = Brigantiaea fide Hafellner & Bellemère (*Nova Hedwigia* **35**: 237, 1982), Hafellner (*Symb. bot. upsal.* **32** no. 1: 35, 1997).
Myxodictyonomyces Cif. & Tomas. (1953) ≡ Myxodictyon.
Myxodiscus Höhn. (1906) = Leptothyrium fide Höhnel (*Mykol. Unters.* **1**: 301, 1923).
Myxodochium Arnaud (1952) nom. inval., anamorphic *Agaricomycetes*, Hsp.0eH.1. 1 (with clamp connexions), Europe.
Myxofusicoccum Died. (1912) = Pseudodiscula fide Sutton (*Mycol. Pap.* **141**, 1977).
Myxolibertella Höhn. (1903) nom. rej. = Phomopsis (Sacc.) Bubák. fide Sutton (*Mycol. Pap.* **141**, 1977).
myxolichens, synthetically produced loose associations of myxomycete plasmodia and a green alga (*Chlorella*).
Myxomonas Brzez. (1906) nom. dub., Fungi. Based on dead or partly formed cells fide Trzebinski (*Z. PflanKrKh.* **17**: 321, 1908).
Myxomphalia Hora (1960), Tricholomataceae. 4, widespread (north temperate). See Hora (*TBMS* **43**: 453, 1960), Antonín (*Mykol. Listy* **76**: I/1, 2001), Antonín (*Mykol. Listy* **90-91**: 20, 2004; key European species).
Myxomphalos Wallr. (1833) = Agyrium fide Saccardo (*Syll. fung.* **8**: 635, 1889).
Myxomycetes Renault (1895), Fossil Fungi ? Myxogastria. 1 (Carboniferous), France.
myxomyceticolous, growing on *Mycetozoa*. See Fungi on fungi.
Myxomycidium Massee (1901) = Mucronella fide Petersen (*Mycol.* **72**: 301, 1980).
Myxomycites Mesch. (1898) ≡ Myxomycetes (class).
Myxomyriangiaceae Theiss. (1917) = Elsinoaceae.
Myxomyriangium Theiss. (1913) = Saccardinula fide von Arx (*Persoonia* **2**: 421, 1963).
Myxonema Corda (1837) [non *Myxonema* Fr. 1825, *Algae*] ≡ Myxonyphe.
Myxonyphe Rchb. (1841) nom. dub., Fungi.
Myxoparaphysella Caball. (1941), anamorphic *Pezizomycotina*, Cac.0eH.?. 1, Spain.
Myxophacidiaceae Petr. (1922) = Ascodichaenaceae.
Myxophacidiella Höhn. (1917) = Pseudophacidium fide von Arx & Müller (*Beitr. Kryptfl. Schweiz* **11** no. 1, 1954).
Myxophacidium Höhn. (1917) = Pseudophacidium fide von Arx & Müller (*Beitr. Kryptfl. Schweiz* **11** no. 1, 1954), DiCosmo *et al.* (*Mycotaxon* **21**: 68, 1984).
Myxopholis Locq. (1979) = Cortinarius fide Kuyper (*in litt.*).

Myxophora Döbbeler & Poelt (1978), Dothideomycetes. 1 (lichenicolous), widespread. See Döbbeler (*Biblthca Lichenol.* **58**: 73, 1995), Hoffmann & Hafellner (*Biblthca Lichenol.* **77**: 1, 2000).

Myxopuntia Mont. (1846) = Leptogium fide Hawksworth *et al.* (*Dictionary of the Fungi* edn 8, 1995).

Myxormia Berk. & Broome (1850) = Myrothecium fide Tulloch (*Mycol. Pap.* **130**, 1972).

myxospore, a myxomycete spore (obsol.).

Myxosporella Sacc. (1881), anamorphic *Pezizomycotina*, Cac.0eH.23. 1, Italy. See Sutton (*Mycol. Pap.* **141**, 1977).

Myxosporidiella Negru (1960), anamorphic *Pezizomycotina*, Ccu.0eH.?. 1, Roumania. See Negru (*Stud. Cercet. Biol. Acad. romana* **11**: 11, 1960).

Myxosporina Höhn. (1927) = Rhodesia fide Sutton (*Mycol. Pap.* **141**, 1977).

Myxosporium Link (1825) nom. conf., anamorphic *Pezizomycotina*. See Höhnel (*Z. Garungs.* **5**: 191, 1915), Weindlmayr (*Sydowia* **17**: 74, 1964), Weindlmayr (*Sydowia* **18**: 26, 1965), Weindlmayr (*Sydowia* **19**: 193, 1966).

myxosporium, see perisporium; Spore wall.

Myxostomella Syd. (1931) = Campoa fide Müller & von Arx (*Beitr. Kryptfl. Schweiz* **11** no. 2, 1962).

Myxostomellina Syd. (1931), anamorphic *Pezizomycotina*, St.0eH.?. 1, Philippines.

Myxotheca Ferd. & Winge (1910) = Cryptothecia fide Hawksworth *et al.* (*Dictionary of the Fungi* edn 8, 1995), Lücking *et al.* (*Lichenologist* **38**: 235, 2006).

Myxotheciella Petr. (1959) = Scolionema fide Müller & von Arx (*Beitr. Kryptfl. Schweiz* **11** no. 2, 1962).

Myxothecium Kunze (1823) = Meliola fide Hansford (*Beih. Sydowia* **2**, 1961).

Myxothyriaceae Höhn. (1909) = Asterinaceae.

Myxothyriopsis Bat. & A.F. Vital (1956), anamorphic *Pezizomycotina*, Cpd.0eH.?. 1, Brazil. See Batista & Vital (*An. Soc. Biol. Pernambuco* **14**: 90, 1956).

Myxothyrium Bubák & Kabát (1915), anamorphic *Pezizomycotina*, St.0eH.15. 1, Europe.

Myxotrichaceae Locq. ex Currah (1985), Leotiomycetidae (inc. sed.). 5 gen. (+ 7 syn.), 60 spp.

Lit.: Currah (*Mycotaxon* **24**: 1, 1985), Rosing (*Mycol.* **77**: 920, 1985), Currah (*SA* **7**: 1, 1988; key gen.), Dalpé (*New Phytol.* **113**: 523, 1989; mycorrhizas), Caretta & Piontelli (*Mycopathologia* **140**: 77, 1997), Hambleton *et al.* (*Mycol.* **90**: 854, 1998; anams), Currah *et al.* (*Karstenia* **39**: 65, 1999), McLean *et al.* (*New Phytol.* **144**: 351, 1999), Monreal *et al.* (*CJB* **77**: 1580, 1999), Sugiyama *et al.* (*Mycoscience* **40**: 251, 1999; phylogeny), Chambers *et al.* (*MR* **104**: 168, 2000), Gibas *et al.* (*Stud. Mycol.* **47**: 131, 2002), Tsuneda & Currah (*Mycol.* **96**: 627, 2004), Wang *et al.* (*Mycol.* **98**: 1065, 2006; phylogeny).

Myxotrichella (Sacc.) Sacc. (1899) = Myxotrichum fide Kendrick & Carmichae *in* Ainsworth *et al.* (Eds) (*The Fungi* **4A**: 390, 1973).

Myxotrichellaceae Nann. (1934) = Myxotrichaceae.

Myxotrichum Kunze (1823), Myxotrichaceae. Anamorph *Oidiodendron*. 12, widespread. See Orr *et al.* (*CJB* **41**: 1463, 1963; key), Hughes (*CJB* **46**: 939, 1968; synonymy), Currah (*Mycotaxon* **24**: 1, 1985), Dalpé (*MSA News* **38**: 22, 1987; anamorph), Udagawa & Uchiyama (*Mycoscience* **40**: 291, 1999), Rice & Currah (*MR* **106**: 1463, 2002), Tsuneda & Currah (*Mycol.* **96**: 627, 2004; ontogeny), Rice & Currah (*Stud. Mycol.* **53**: 83, 2005; anamorphs), Wang *et al.* (*Mycol.* **98**: 1065, 2006; phylogeny).

Myxozyma Van der Walt, Weijman & Arx (1981), anamorphic *Lipomycetaceae*, Hso.0eH.3/4. 5, N. America; S. Africa. See Yamada (*J. gen. appl. Microbiol.* Tokyo **32**: 259, 1986; coenzyme Q system), van der Walt *et al.* (*Syst. Appl. Microbiol.* **9**: 121, 1987; key 4 spp.), Cottrell & Kock (*Syst. Appl. Microbiol.* **12**: 291, 1989; history, delimitation, phenotypic characters), Kurtzman & Liu (*Curr. Microbiol.* **21**: 387, 1990), Spaaij *et al.* (*Syst. Appl. Microbiol.* **13**: 182, 1990), Jansen van Rensburg *et al.* (*Syst. Appl. Microbiol.* **18**: 410, 1995), Kurtzman *in* Kurtzman & Fell (Eds) (*Yeasts, a taxonomic study* 4th edn: 592, 1998), Suh *et al.* (*Mycol.* **98**: 1006, 2006; phylogeny), Kurtzman *et al.* (*FEMS Yeast Res.* **7**: 1027, 2007; phylogeny).

Myxyphe (orthographic variant), see *Myxonyphe*.

Myzeloblastanon, see *Myceloblastanon*.

nacreous, like mother-of-pearl.

Nadelspora Olson, Tiekotter & Reno (1994), Microsporidia. 1.

Nadsonia Syd. (1912), Saccharomycetales. 2, widespread (north temperate). See Golubev *et al.* (*Antonie van Leeuwenhoek* **55**: 369, 1989; key), Yamada *et al.* (*J. gen. appl. Microbiol.* Tokyo **38**: 585, 1992; phylogeny), Miller & Phaff *in* Kurtzman & Fell (Eds) (*Yeasts, a taxonomic study* 4th edn: 268, 1998), Schweigkofler *et al.* (*Organ. Divers. Evol.* **2**: 1, 2002; phylogeny), Suh *et al.* (*Mycol.* **98**: 1006, 2006; phylogeny).

Nadsoniella Issatsch. (1914), anamorphic *Herpotrichiellaceae*, Hso.0eH.6. 1 (from sea water), former USSR. See Haase *et al.* (*Stud. Mycol.* **43**: 80, 1999; phylogeny).

Nadsoniomyces Kudrjanzev (1932) = Trichosporon fide Carmo-Sousa *in* Lodder (Ed.) (*Yeasts, a taxonomic study* 2nd edn, 1970).

Nadvornikia Tibell (1984), Thelotremataceae (L). 3, widespread. See Harris (*Some Florida lichens*, 1990; posn), Lumbsch *et al.* (*Symb. bot. upsal.* **34** no. 1: 9, 2004; phylogeny).

Naegelia Rabenh. (1844) nom. dub., Fungi.

Naegiella (Rehm) Clem. (1909) = Eupropolella fide Défago (*Sydowia* **21**: 1, 1967).

Naemacyclus Fuckel (1873), Leotiomycetes. 6, widespread. See Cannon & Minter (*Mycol. Pap.* **155**: 123 pp., 1986), Gernandt *et al.* (*Mycol.* **93**: 915, 2001; phylogeny), Johnston (*Aust. Syst. Bot.* **19**: 135, 2006), Wang *et al.* (*Mycol.* **98**: 1065, 2006; phylogeny).

Naemaspora Pers. (1796) nom. conf., anamorphic *Pezizomycotina*. See Sutton (*Mycol. Pap.* **141**, 1977).

Naemaspora Sacc. (1880) = Roscoepoundia fide Sutton (*Mycol. Pap.* **141**, 1977).

Naemaspora Willd. (1787) ≡ Bombardia.

Naematelia Fr. (1822) nom. dub., Basidiomycota. 1. Based on parasitized *Stereum* fide Bandoni (*Am. midl. Nat.* **66**: 319, 1961).

Naematoloma P. Karst. (1879) = Psilocybe (Fr.) P. Kumm. fide Kuyper (*in litt.*).

Naemosphaera (Sacc.) Sacc. (1892) = Naemosphaera P. Karst. fide Sutton (*in litt.*).

Naemosphaera P. Karst. (1888), anamorphic *Pezizomycotina*, Cpd.0eP.?. 1, Finland.

Naemosphaerella Höhn. (1923), anamorphic *Pezizomycotina*. 2, Brazil; Japan. See Petrak (*Sydowia* **6**: 302, 1952).

Naemospora Roth ex Kuntze (1898) nom. dub., anamorphic *Pezizomycotina*.

Naemostroma Höhn. (1919) = Septoriella fide Sutton *in* Ainsworth *et al.* (Eds) (*The Fungi* **4A**, 1973).

Naetrocymbaceae Höhn. ex R.C. Harris (1995), Pleosporales (±L). 5 gen. (+ 16 syn.), 79 spp.

Lit.: Aguirre & Hawksworth (*Biblthca Lichenol.* **25**: 249, 1987), Aguirre-Hudson & Fiol (*Lichenologist* **25**: 207, 1993), Harris (*More Florida Lichens*, 1995), Harris (*More Florida Lichens*: 192 pp., 1995), Hyde & Wong (*MR* **103**: 347, 1999).

Naetrocymbe Bat. & Cif. (1963) = Limacinula Höhn. fide Eriksson (*Op. Bot.* **60**, 1981).

Naetrocymbe Körb. ex Körb. (1865), Naetrocymbaceae (L). 12. See Eriksson (*Opera Bot.* **60**: 111, 1981; concept), Harris (*More Florida Lichens*, 1995).

Naevala B. Hein (1976), Helotiales. 6, Europe. See Hein (*Willdenowia* Beih. **9**, 1976), Nauta & Spooner (*Mycologist* **13**: 65, 1999; Brit. spp.).

Naevia Fr. (1825) = Arthonia fide Hawksworth *et al.* (*Dictionary of the Fungi* edn 8, 1995).

Naevia Fr. (1849) ≡ Naevala.

Naeviella (Rehm) Clem. (1909), Helotiales. 3, Europe. See Nannfeldt (*Sydowia* **35**: 162, 1982), Scheuer (*Biblthca Mycol.* **123**: 274, 1988), Suková (*Czech Mycol.* **56**: 63, 2004).

Naeviopsis B. Hein (1976), Helotiales. 15, Europe. See Suková *et al.* (*Czech Mycol.* **55**: 223, 2003).

Naganishia Goto (1963) = Cryptococcus Vuill. fide Phaff & Fell *in* Lodder (Ed.) (*Yeasts, a taxonomic study* 2nd edn: 1088, 1970).

Nagrajia R.F. Castañeda & W.B. Kendr. (1991), anamorphic *Pezizomycotina*, Cpd.0eH.15. 1, Cuba. See Castañeda Ruiz & Kendrick (*Univ. Waterloo Biol. Ser.* **35**: 76, 1991).

Nagrajomyces Melnik (1984), anamorphic *Pezizomycotina*, Cpd.#eP.19. 1, Russia. See Melnik (*Mikol. Fitopatol.* **18**: 9, 1984).

Naiadella Marvanová & Bandoni (1987), anamorphic *Classicula*, Hso.0eH.1. 1 (aquatic), Canada; former Czechoslovakia. See Marvanová & Bandoni (*Mycol.* **79**: 579, 1987).

Nailisporites T.C. Huang (1981), Fossil Fungi. 1 (Miocene), Taiwan.

Naïs Kohlm. (1962), Halosphaeriaceae. 2 (marine and freshwater), Europe; N. America. See Crane & Shearer (*TBMS* **86**: 509, 1986), Hyde *et al.* (*Mycoscience* **40**: 165, 1999), Kong *et al.* (*MR* **104**: 35, 2000; DNA), Pang *et al.* (*Mycol. Progr.* **2**: 29, 2003).

Nakaiomyces Kobayasi (1939) nom. dub., Fungi. Based on a parasitized *Tremella* fide Olive (*Bull. Torrey bot. Cl.* **85**: 99, 1958).

Nakaseomyces Kurtzman (2003), Saccharomycetaceae. 2, widespread. See Kurtzman (*FEMS Yeast Res.* **4**: 240, 2003), Suh *et al.* (*Mycol.* **98**: 1006, 2006; phylogeny).

Nakataea Hara (1939) = Pyricularia fide Sivanesan (*in litt.*), Bussaban *et al.* (*Mycol.* **97**: 1002, 2005; phylogeny, morphology).

Nakatopsis Whitton, McKenzie & K.D. Hyde (2001), anamorphic *Pezizomycotina*. 2, Malaysia. See Whitton *et al.* (*Fungal Diversity* **8**: 165, 2001).

Nakazawaea Y. Yamada, K. Maeda & Mikata (1994), Saccharomycetales. 1, USA. See Kurtzman *in* Kurtzman & Fell (Eds) (*Yeasts, a taxonomic study* 4th edn: 273, 1998), Suh *et al.* (*Mycol.* **98**: 1006, 2006; phylogeny).

Nalanthamala Subram. (1956), anamorphic *Rubrinectria*, Hsp.0eH.15. 2, pantropical. See Subramanian (*J. Indian bot. Soc.* **35**: 478, 1956), Schroers *et al.* (*Mycol.* **97**: 375, 2005; phylogeny, morphology).

Namakwa Hale (1988), Parmeliaceae (L). 2, S. Africa. See Thell *et al.* (*Mycol. Progr.* **3**: 297, 2004; phylogeny), Thell *et al.* (*J. Hattori bot. Lab.* **100**: [797], 2006).

nameko, basidioma of the edible *Pholiota nameko* cultivated in Japan (Chang & Hayes, 1978).

names (1) of fungi, see Common names, Nomenclature; (2) Escallon (*Precis de myconymie*, 1989; meaning of scientific names).

Nannenga-Bremekamp (Neeltje Elizabeth; 1916-1996; Netherlands). An amateur and authority on *Mycetozoa*, noted for her acute observation; contributed to understanding of certain difficult genera (*Cribraria*, *Diderma*, *Didymium*, *Licea*, *Reticularia* and the *Stemonitidaceae* in general); her illustrative work was exemplary, and her book became a worldwide standard on the subject. *Publs. De Nederlandse Myxomyceten* (1974); *A Guide to Temperate Myxomycetes* (1991). *Biogs, obits etc.* Ing (*Mycologist* **10**: 191, 1996) [portrait]; Rammeloo (*Coolia* **40**: 1, 1997).

Nannfeldt (John Axel Frithiof; 1904-1985; Sweden). Assistant Professor (*c.* 1932-1939) then Professor and Director of the Botanical Garden and Herbarium (1939-1970), University of Uppsala. He produced critical accounts of numerous groups of discomycete fungi, and later some pyrenomycetes (especially *Sordariales* and *Xylariales*) and smuts. He was noted for his innovative thinking, and his dedication to mycology: he continued to dictate the text of his final paper, from a hospital oxygen tent. *Publs.* Studien über die Morphologie und Systematik der nichlichenisierten inoperculaten Discomyceten. *Nova Acta Regiae Societatis Scientiarum Upsaliensis* (1932) [this work established two main types of ascomycete ontogeny: ascohymenial and ascolocular (see also Luttrell)]; (with E. Melin) *Researches into the Blueing of Ground Wood-pulp* (1934) [a pioneering investigation into blue-stain fungi]. *Biogs, obits etc.* Grummann (1974: 495); Holm (*Mycol.* **78**: 692, 1986).

Nannfeldtia Petr. (1947), ? Leptopeltidaceae. 1, Austria. See Scheuer (*Biblthca Mycol.* **123**, 1988).

Nannfeldtiella Eckblad (1968) = Pseudombrophila fide Harmaja (*Ann. bot. fenn.* **16**: 159, 1979), Nannfeldt (*Sydowia* **35**: 166, 1982), van Brummelen (*Libri Botanici* **14**, 1995), Hansen *et al.* (*Mycol.* **97**: 1023, 2005), Yao & Spooner (*Fungal Diversity* **22**: 267, 2006).

Nannfeldtiomyces Vánky (1981), Doassansiaceae. 2 (on *Sparganium*), Europe; N. America. See Vánky (*Sydowia* **34**: 171, 1981).

Nannizzia Stockdale (1961) = Arthroderma fide Weitzman *et al.* (*Mycotaxon* **25**: 505, 1986), Kawasaki *et al.* (*Mycopathologia* **118**: 95, 1992), Hirai *et al.* (*Antonie van Leeuwenhoek* **83**: 11, 2003).

Nannizziopsis Currah (1985), Onygenaceae. Anamorph *Chrysosporium*. 3, widespread. See Guarro *et al.* (*Mycotaxon* **42**: 193, 1991; key), Udagawa & Uchiyama (*Mycoscience* **40**: 291, 1999), Sugiyama & Mikawa (*Mycoscience* **42**: 413, 2001), Solé *et al.* (*MR* **106**: 388, 2002; phylogeny), Thomas *et al.* (*Medical Mycology* **40**: 143, 2002; from crocodile).

Nanomyces Thaxt. (1931), Laboulbeniaceae. 3, Pacific. See Nannfeldt (*Svensk bot. Tidskr.* **43**: 468, 1949).

Nanoschema B. Sutton (1980), anamorphic *Pezizomycotina*, St.1eH.15. 1, Solomon Islands. See Sutton

(*The Coelomycetes*: 589, 1980).

Nanoscypha Denison (1972), Sarcoscyphaceae. Anamorph *Molliardiomyces*. c. 10, widespread. See Das & Pant (*Indian Phytopath.* **37**: 294, 1984; Indian spp.), Pfister (*Mem. N. Y. bot. Gdn* **49**: 339, 1989), Harrington (*Mycol.* **90**: 235, 1998; DNA), Harrington *et al.* (*Mycol.* **91**: 41, 1999; phylogeny), Hansen & Pfister (*Mycol.* **98**: 1029, 2006; phylogeny), Perry *et al.* (*MR* **111**: 549, 2007; phylogeny).

Nanostictis M.S. Christ. (1954), Stictidaceae. 2 (on lichens, esp. *Peltigera*), Denmark; N. America. See Sherwood (*Sydowia* **38**: 267, 1986).

Nanstelocephala Oberw. & R.H. Petersen (1990), Cortinariaceae. 1, USA. See Oberwinkler & Petersen (*Mycol.* **82**: 783, 1990).

Naohidea Oberw. (1990), Naohideales. 1, widespread (north temperate). See Piątek (*Polish Botanical Journal* **47**: 49, 2002; rediscov. in Poland).

Naohideales R. Bauer, Begerow, J.P. Samp., M. Weiss & Oberw. (2006). Cystobasidiomycetes. 1 gen., 1 spp. No familes recognized.

Naohidemyces S. Sato, Katsuya & Y. Hirats. (1993), Pucciniastraceae. 3 (on *Tsuga* (0, I); on *Ericaceae* (II, III)), north temperate. See Sato *et al.* (*TMSJ* **34**: 48, 1993).

Naothyrsium Bat. (1960), anamorphic *Pezizomycotina*, Cpt.1eP.1. 1, Brazil. See Batista (*Publções Inst. Micol. Recife* **250**: 5, 1960).

Napamichum J.I.R. Larsson (1990), Microsporidia. 3.

Napicladium Thüm. (1875) = Fusicladium fide Hughes (*CJB* **36**: 727, 1958), Schubert *et al.* (*Schlechtendalia* **9**, 2003).

napiform, turnip-like in form (Fig. 23.26).

Napomyces Setch. ex Clem. & Shear (1931) ≡ Daleomyces fide Korf (*Mycol.* **48**: 711, 1956).

Naranus Ts. Watan. (1995), anamorphic *Pezizomycotina*. 1, Japan. See Watanabe (*MR* **99**: 806, 1995).

Narasimhania Thirum. & Pavgi (1952), Doassansiaceae. 1 (on *Alisma*), India; Mali. See Vánky (*Sydowia* **34**: 167, 1981).

Narasimhella Thirum. & P.N. Mathur (1966) = Gymnoascus fide von Arx (*Persoonia* **13**: 173, 1986), Solé *et al.* (*Stud. Mycol.* **47**: 141, 2002; phylogeny).

Nascimentoa Cif. & Bat. (1956) = Spiropes fide Ellis (*Mycol. Pap.* **114**, 1968).

nassace (**nasse**), the finger-like protrusion of the inner part of a bitunicate ascus into the inner tunica; internal apical beak; see ascus.

nasse, see nassace.

Natalia Fr. (1847) [non *Natalia* Hochst. 1841, *Melianthaceae*] ≡ Levieuxia.

Natantispora J. Campb., J.L. Anderson & Shearer (2003), Halosphaeriaceae. 2, USA. See Campbell *et al.* (*Mycol.* **95**: 543, 2003), Sakayaroj *et al.* (*Mycol.* **97**: 804, 2005; key).

Natarajania Pratibha & Bhat (2006), anamorphic *Pezizomycotina*. 1 (from decaying leaves), India.

native bread, see blackfellows' bread.

Nattrassia B. Sutton & Dyko (1989) = Fusicoccum fide Sutton & Dyko (*MR* **93**: 483, 1989), Roeijmans *et al.* (*J. Med. Vet. Mycol.* **35**: 181, 1997; molecular taxonomy), Sigler *et al.* (*J. Clin. Microbiol.* **35**: 433, 1997; in man), Farr *et al.* (*Mycol.* **97**: 730, 2005; synonymy).

Naucoria (Fr.) P. Kumm. (1871), Strophariaceae. 30 (usually associated with *Alnus*), widespread. Because of nomenclatural complications the name *Alnicola* is often used for this genus. See Orton (*TBMS* **43**: 308, 1960), Reid (*TBMS* **82**: 191, 1984; key 17 Br. spp.), Moreau (*Fungal Diversity* **20**: 121, 2005; as *Alnicola*), Moreau *et al.* (*Mol. Phyl. Evol.* **38**: 794, 2006; phylogeny, as *Alnicola*).

Naufragella Kohlm. & Volkm.-Kohlm. (1998), Halosphaeriaceae. 2 (marine), Europe; N. America. See Kohlmeyer & Volkmann-Kohlmeyer (*SA* **16**: 9, 1998).

Naumov (Nikolai Aleksandrovich; 1888-1959; Russia). Student, Department of Natural Sciences, St Petersburg University (1906-1910); Assistant, Bureau of Mycology and Phytopathology (1910-1919); organizer and Head of Phytopathology Station, Tsarskoye Selo (1919-1931); Head, Phytopathology Department (1930-1935) then Director (1935-1954), All Union Institute of Plant Protection; Lenin Prize (1954). Directed development of plant pathology (particularly diseases of brassicas, cereals and potato), and research on the mechanism of fungicides in the difficult conditions of the Soviet Union under Stalin and World War II; produced textbooks highly influential in the Soviet Union and China. *Publs*. [*Tables of the Identification of Mucoraceae*] (1915) [in Russian]; [*Methods of Microscopic Investigations in Phytopathology*] (1932) [in Russian]; [*Methods of Mycological and Phytopathological Investigations*] (1937) [in Russian]; [*Mycoflora of Leningrad Oblast'*] vols 1 & 2 (1954, 1964) [in Russian, published posthumously]. *Biogs, obits etc*. Khokhryakov ([*Mikologiya i Fitopatologiya*] **2**: 167, 1968) [portrait, in Russian]; Khokhryakov, Novotelnova & Potlaichuk ([*Mikologiya i Fitopatologiya*] **3**: 197, 1969) [in Russian].

Naumovela Kravtzev (1955), Pezizomycotina. 2 (on wood), former USSR. See Kravtzev (*Trudy Inst. Bot., Alma-Ata* **2**: 153, 1955).

Naumovia Dobrozr. (1928) = Rosenscheldia fide von Arx & Müller (*Stud. Mycol.* **9**, 1975).

Naumovia Kurtzman (2003) ≡ Naumovozyma.

Naumoviella Novot. (1950) = Mortierella fide Hesseltine (*Mycol.* **47**: 344, 1955).

Naumovozyma Kurtzman (2008), Saccharomycetaceae. 2. See Kurtzman (*FEMS Yeast Res.* **4**: 240, 2003; as *Naumovia*), Suh *et al.* (*Mycol.* **98**: 1006, 2006; phylogeny; as *Naumovia*).

Nautosphaeria E.B.G. Jones (1964), Halosphaeriaceae. 1 (marine), British Isles. See Kohlmeyer & Volkmann-Kohlmeyer (*Bot. Mar.* **34**: 1, 1991).

Navaneethospora V.G. Rao (1994), anamorphic *Pezizomycotina*, Hso.?.?. 1, India. See Rao (*Advances in Mycology and Aerobiology* (Dr S.T. Tilak Commemoration Volume). *Current Trends in Life Sciences* **20**: 134, 1994).

navel, see umbilicus.

Navicella Fabre (1879), Massariaceae. 5, widespread. See Eriksson (*Op. Bot.* **60**, 1981), Holm & Holm (*Symb. bot. upsal.* **28** no. 2, 1988), Barr (*N. Amer. Fl.* ser. 2 **13**: 129 pp., 1990; posn), Aptroot & van Iperen (*Nova Hedwigia* **67**: 481, 1998), Elshafie *et al.* (*Sydowia* **57**: 19, 2005; Costa Rica).

navicular (**naviculate**), boat-like in form; cymbiform (Fig. 23.32).

Navisporus Ryvarden (1980), Polyporaceae. 6, widespread (pantropical). See Ryvarden & Johansen (*Prelim. Polyp. Fl. E. Afr.*: 429, 1980), Decock & Herrera Figueroa (*Cryptog. Mycol.* **21**: 153, 2000; Cuba), Ryvarden & Iturriaga (*Mycol.* **95**: 1066, 2003; Venezuela).

Nawawia Marvanová (1980), anamorphic *Pezizomy-*

cotina, Hso.0eH.15. 2 (aquatic), widespread. See Kuthubutheen *et al.* (*CJB* **70**: 96, 1992), Hyde *et al.* (*MR* **100**: 810, 1996; S Africa), Mel'nik & Hyde (*Mikol. Fitopatol.* **40**: 411, 2006; E Russia).

Necator Massee (1898), anamorphic *Erythricium*. 1, S.E. Asia. See Brooks & Sharples (*Ann. appl. Biol.* **2**: 58, 1915).

Necium Arthur (1907) = Melampsora fide Davis (*Transactions of the Wisconsin Academy of Sciences, Arts and Letters* **18**: 78, 1915).

Necol technique, see Mounting media.

necral layer, a layer of horny dead fungal hyphae with indistinct lumina in or near the cortex of lichens; **epinecral layer** if above the algal layer, **hyponecral layer** if below.

Necraphidium Cif. (1951), anamorphic *Pezizomycotina*, Hso.0-1eH.?. 1 (on larvae of *Callipterus*), Italy. See Ciferri (*Mycopath. Mycol. appl.* **6**: 25, 1951).

necrophagous, saprobic.

necrophyte, an organism living on dead material (Münch); cf. perthophyte, saprophyte.

Necrosis Paulet (1793) = Ustilago fide Dietel (*Nat. Pflanzenfam.* **6**, 1928).

necrosis, death of plant cells, esp. when resulting in the tissue becoming dark in colour; commonly a symptom of fungus infection.

necrotroph, a parasite that derives its energy from dead cells of the host (Thrower, *Phytopath. Z.* **56**: 258, 1966). Cf. biotroph.

nectar, the sticky, sometimes sweet, secretion (esp. of pycnia of rusts) in which spores or spermatia may be freed, and which has an attraction for insects.

Nectaromyces Syd. & P. Syd. (1919) = Metschnikowia fide von Arx *et al.* (*Stud. Mycol.* **14**: 1, 1977), Giménez-Jurado (*Syst. Appl. Microbiol.* **15**: 432, 1992; teleomorph).

Nectaromycetaceae Cif. & Redaelli (1929) = Metschnikowiaceae.

Nectria (Fr.) Fr. (1849) nom. cons., Nectriaceae. Anamorphs *Tubercularia*, *Acremonium*-like. 82 (from decaying wood, bark etc.), widespread. *N. cinnabarina* (coral spot of woody plants. See also Fungicolous fungi. See Lohman & Watson (*Lloydia* **6**: 77, 1943; spp. on hardwoods), Booth (*Mycol. Pap.* **73**, 1959; UK, keys), Passeur (*Sydowia* **27**: 7, 1975; morphology, substratum), Perrin (*BSMF* **92**: 335, 1976; key Eur. spp.), Doyle (*Mycol.* **70**: 355, 1978; metabolites), Rossman (*Mycol. Pap.* **150**, 1983; phragmosporous spp.), Dargan *et al.* (*Nova Hedwigia* **42**: 109, 1986; Himalayas), Samuels (*Brittonia* **40**: 306, 1988), Samuels (*Mem. N. Y. bot. Gdn* **48**: 1, 1988; fungicolous spp.), Rossman (*Mem. N. Y. bot. Gdn* **49**: 253, 1989; key 28 spp.; *N. cinnabarina* agg.), Rossman (*The Fungal Holomorph: Mitotic, Meiotic and Pleomorphic Speciation in Fungal Systematics*: 149, 1993; review), Samuels & Brayford (*Sydowia* **46**: 75, 1994; spp. with striate ascospores), Rossman (*Mycol.* **88**: 1, 1996; phylogeny, review), Rossman *et al.* (*Stud. Mycol.* **42**: 248 pp., 1999; generic revision), Zhang *et al.* (*Mycol.* **98**: 1076, 2006; phylogeny).

Nectriaceae Tul. & C. Tul. (1865), Hypocreales. 57 gen. (+ 64 syn.), 646 spp.

Lit.: Brayford & Samuels (*Mycol.* **85**: 612, 1993), Samuels & Brayford (*Sydowia* **45**: 55, 1993), Rehner & Samuels (*CJB* **73**: S816, 1995), Nirenberg & O'Donnell (*Mycol.* **90**: 434, 1998), O'Donnell *et al.* (*Mycol.* **90**: 465, 1998), Victor *et al.* (*MR* **102**: 273, 1998), Aoki & O'Donnell (*Mycol.* **91**: 597, 1999), Arie *et al.* (*Mycoscience* **40**: 311, 1999), Leslie (*Pl. Path. J.* **15**: 259, 1999), Möller *et al.* (*J. Phytopath.* **147**: 497, 1999), Rossman *et al.* (*Stud. Mycol.* **42**: 248 pp., 1999), Baayen *et al.* (*Phytopathology* **90**: 891, 2000), Lieckfeldt & Seifert (*Stud. Mycol.* **45**: 35, 2000), O'Donnell et al. (*Proc. natn Acad. Sci. U.S.A.* **97**: 7905, 2000), Rossman (*Stud. Mycol.* **45**: 27, 2000), Schoch *et al.* (*Stud. Mycol.* **45**: 45, 2000), Crous (*Taxonomy and Pathology of Cylindrocladium (Calonectria) and Allied Genera*: 278 pp., 2002), Brayford *et al.* (*Mycol.* **96**: 572, 2004), Crous *et al.* (*Stud. Mycol.* **55**: 213, 2006).

Nectricladiella Crous & C.L. Schoch (2000), Nectriaceae. Anamorph *Cylindrocladiella*. 2, widespread (tropical). See Schoch *et al.* (*Stud. Mycol.* **45**: 45, 2000), Crous (*Taxonomy and Pathology of Cylindrocladium (Calonectria) and Allied Genera*: 278 pp., 2002; revision).

Nectriella Nitschke ex Fuckel (1870), Bionectriaceae. Anamorphs *Acremonium*-like, *Kutilakesa*, *Dendrodochium*-like. 35 (on dead plant tissues or fungicolous), widespread (temperate). See Lowen (*Mycotaxon* **39**: 461, 1990; concept), Rossman *et al.* (*Stud. Mycol.* **42**: 248 pp., 1999), Ramaley (*Mycotaxon* **90**: 181, 2004; on *Agavaceae*).

Nectriella Sacc. (1877) ≡ Pseudonectria fide Rossman *et al.* (*Stud. Mycol.* **42**: 248 pp., 1999).

NectrioidaceaeSacc. (1884).

Lit.: Petch (*TBMS* **26**: 53, 1943; genera), Sutton *in* Ainsworth *et al.* (Eds) (*The Fungi* **4A**, 1973; discussion family names).

Nectriopsis Maire (1911), Bionectriaceae. Anamorphs *Acremonium*-like, *Rhopalocladium*. 58 (on fungi and myxomycetes), widespread. See Samuels (*Mem. N. Y. bot. Gdn* **48**, 1988), Rossman *et al.* (*Stud. Mycol.* **42**: 248 pp., 1999), Rossman *et al.* (*Mycol.* **93**: 100, 2001; phylogeny), Samuels *et al.* (*CBS Diversity Ser.* **4**, 2006; USA).

needle cast (of conifers), loss of leaves caused by spp. of *Hypoderma*, *Lophodermium*, *Rhabdocline*, or other *Rhytismatales*.

Neelakesa Udaiyan & V.S. Hosag. (1992), Sordariomycetes. 1 (submerged wood), India. See Udaiyan & Hosagoudar (*J. Econ. Taxon. Bot.* **15**: 649, 1992).

Nees (von Esenbeck, Christian Gottfried Daniel; 1776-1858; Germany). Medical degree, Giesen (1800); Professor of Botany, Erlangen (1818-1819); Professor of Botany, Bonn (1819-1830); Professor of Botany, Breslau (1831-1851) [dismissed for alleged 'moral turpitude', but more probably because of involvement in a movement for emancipation of the under-privileged]. Produced influential early work on fungal taxonomy. *Publs. Das System der Pilze und Schwämme* (1817) [366 colour figures on 44 plates]; *Horae Physicae Berolinensis* (1820); also various papers jointly with his brother (T.F.L. Nees, q.v.). *Biogs, obits etc.* Grummann (1974: 34); Stafleu & Cowan (*TL-2* **3**: 705, 1981).

Nees (von Esenbeck, Theodor Friedrich Ludwig; 1787-1837; Germany, later Switzerland, Netherlands). Pharmaceutical Assistant, Basle (1811-1816), then Hanau (1816-1817); Inspector, Botanic Garden, Leiden (1817-1819); Garden Inspector and Lecturer (1819-1822) then Professor of Pharmacy (1822-1827) then Regius Professor of Pharmacy (1827-1833) then Co-Director of Botanic Garden (1833 onwards), University of Bonn. Produced influential early work on fungal taxonomy. *Publs*. (with Henry) *Das System der Pilze* (1837, with coloured plates) [Abt. 2 by Bail (1858)]; also various papers jointly his brother (C.G.D. Nees, q.v.). *Biogs, obits etc.* Grummann (1974: 34); Stafleu & Cowan (*TL-2* **3**: 712, 1981).

Neesiella Kirschst. (1935) [non *Neesiella* Schiffn. 1893, *Hepaticae*] ≡ Myconeesia.

Negeriella Henn. (1897), anamorphic *Pezizomycotina*, Hsy.≡ eP.?. 2, S. America. See Morris (*Mycopathologia* **33**: 183, 1967).

Neilreichina Kuntze (1891) ≡ Polytrichia.

Nelliemelba J.I.R. Larsson (1983), Microsporidia. 1.

Nemaclada J. Sm. (1896), Fossil Fungi (mycel.) Fungi. 1 (Carboniferous), British Isles.

Nemacola A. Massal. (1855) nom. dub., ? Dothideomycetes (L). Possibly an alga.

Nemadiplodia Zambett. (1955) = Lasiodiplodia fide Petrak (*Sydowia* **16**: 353, 1963).

Nemania Gray (1821), Xylariaceae. Anamorph *Geniculosporium*. 44, widespread. See Pouzar (*Česká Mykol.* **39**: 129, 1985), Pouzar (*Česká Mykol.* **39**: 15, 1985), Petrini & Rogers (*Mycotaxon* **26**: 40, 1986; key, as *Hypoxylon serpens* group), Læssøe (*SA* **13**: 113, 1994), Granmo *et al.* (*Sommerfeltia* **27**, 1999), Sánchez-Ballesteros *et al.* (*Mycol.* **92**: 964, 2000; phylogeny), Stadler *et al.* (*Mycotaxon* **77**: 379, 2001; chemistry), Ju & Rogers (*Nova Hedwigia* **74**: 75, 2002; revision), Ju *et al.* (*Mycol.* **97**: 562, 2005), Ju *et al.* (*Mycol.* **99**: 612, 2007; phylogeny), Tang *et al.* (*MR* **111**: 392, 2007; Thailand).

Nemaplana J. Sm. (1896), Fossil Fungi (mycel.) ? Fungi. 1 (Carboniferous), British Isles.

Nemaria Navas (1909) = Roccella fide Hawksworth *et al.* (*Dictionary of the Fungi* edn 8, 1995).

Nemariaceae Navas (1910) = Roccellaceae.

Nematidia Stirt. (1879) = Eremothecella fide Sérusiaux (*SA* **11**: 39, 1992).

Nematococcomyces C.L. Hou, M. Piepenbr. & Oberw. (2004), Rhytismataceae. 1, China. See Hou *et al.* (*Mycol.* **96**: 1381, 2004).

Nematococcus Kütz. (1833) nom. dub., Fungi.

Nematocolla Link (1833) = Myxosporium fide Link (*Handbuck zur Erkennung der Nutzbarsten und am Häufigsten Vorkommenden Gewächse* **3**, 1833).

Nematoctonus Drechsler (1941), anamorphic *Hohenbuehelia*, Hso.0-1eP.10+1. 16 (with clamp connexions, on nematodes), widespread. See Giuma *et al.* (*TBMS* **60**: 49, 1973; nematoxins), Thorn & Barron (*Mycotaxon* **25**: 321, 1986; key 14 spp.), Rubner (*Stud. Mycol.* **39**, 1996).

nematodes (fungi pathogenic to), see Mycopesticides, Predacious fungi, *Zoopagaceae*.

nematogenous, of conidiogenous cells arising at all levels from single hyphae; lanose. See Gams (*Cephalosporium-artige Schimmelpilze*, 1971).

Nematogonum Desm. (1834), anamorphic *Pezizomycotina*, Hso.0eH.3. 2 (contact mycoparasites), widespread (temperate). See Gams (*Revue Mycol.* Paris **39**: 273, 1975), Walker & Minter (*TBMS* **77**: 299, 1981).

Nematographium Goid. (1935), anamorphic *Pezizomycotina*, Hsy.0eH.?. 5, Europe.

Nematoloma, see *Naematoloma*.

Nematomyces Faurel & Schotter (1966) = Gliomastix fide von Arx (*Gen. Fungi Sporul. Cult.*, 1970).

Nematonostoc Nyl. ex Elenkin (1934), Algae. Algae.

nematophagous, nematode-feeding; see Predacious fungi. Nematophagous fungi are also reported as fossils (Jansson & Poinar, *TBMS* **87**: 471, 1986).

Nematophagus Mekht. (1975) = Arthrobotrys fide Scholler *et al.* (*Sydowia* **51**: 89, 1999).

Nematora Fée (1825) = Strigula fide Santesson (*Symb. bot. upsal.* **12** no. 1: 1, 1952).

Nematospora Peglion (1897) = Eremothecium fide Batra (*USDA Tech. Bull.* **1469**, 1973), Kurtzman (*J. Industr. Microbiol.* **14**: 523, 1995), de Hoog *et al. in* Kurtzman & Fell (Eds) (*Yeasts, a taxonomic study* 4th edn: 201, 1998).

Nematospora Tassi (1904) ≡ Giulia.

Nematosporaceae E.K. Novák & Zsolt (1961) = Eremotheciaceae.

Nematostigma Syd. & P. Syd. (1913), Pseudoperisporiaceae. 1, S. Africa. See Hansford (*Mycol. Pap.* **15**, 1946), Petrak (*Sydowia* **3**: 251, 1949).

Nematostoma Syd. & P. Syd. (1914), Pseudoperisporiaceae. Anamorph *Chaetosticta*. *c.* 12, widespread (tropical). See Hansford (*Mycol. Pap.* **15**, 1946), Petrak (*Sydowia* **3**: 251, 1949), Rossman (*Mycol. Pap.* **157**, 1987), Barr (*Mycotaxon* **64**: 149, 1997).

Nematothecium Syd. & P. Syd. (1912), ? Pseudoperisporiaceae. Anamorph *Atractilina*. *c.* 2, widespread (tropical). See Rossman (*Mycol. Pap.* **157**, 1987).

nematotoxin, a metabolic product toxic to nematodes, e.g. **nematoctonins** from *Nematoctonus* spp. (Kennedy & Taplin, *TBMS* **70**: 140, 1978); see also Mycopesticides.

Nemecomyces Pilát (1933) = Pholiota fide Singer (*Agaric. mod. Tax.* edn 3, 1975).

nemin, a principle from a nematode-free culture filtrate of *Neoplectana glaseri*, which caused *Arthrobotrys conoides* to differentiate traps (Pramer & Stoll, *Science, N.Y.* **129**: 966, 1959); see Predacious fungi.

nemoral, living in woods or groves.

Nemostroma, see *Naemostroma*.

Nemozythiella Höhn. (1925), anamorphic *Pezizomycotina*, St.1eH.?. 1, Europe.

neo- (prefix), new.

Neoalpakesa Punith. (1981), anamorphic *Pezizomycotina*, Cpd.0eH.19. 1, British Isles. See Punithalingam (*Nova Hedwigia* **34**: 137, 1981).

Neoarachnotheca Ulfig, Cano & Guarro (1997), Onygenaceae. 2, Australasia; S. Africa. See Cano *et al.* (*Antonie van Leeuwenhoek* **72**: 149, 1997).

Neoarbuscula B. Sutton (1983), anamorphic *Pezizomycotina*, Hso.#eP.42. 1, widespread (tropical). See Subramanian & Vittal (*CJB* **51**: 977, 1973).

Neoarcangelia Berl. (1900) = Pleurostoma fide Shear (*Mycol.* **29**: 355, 1937).

Neoballadyna Boedijn (1961) = Dysrhynchis fide Müller & von Arx (*Beitr. Kryptfl. Schweiz* **11** no. 2, 1962).

Neobarclaya Sacc. (1899), anamorphic *Pezizomycotina*, Cac.1eP.19. 1, USA. See Nag Raj (*Coelomycetous Anamorphs with Appendage-bearing Conidia*, 1993).

Neobarya Lowen (1986), ? Clavicipitaceae. Anamorph *Acremonium*-like. 3 (parasitic on agarics), widespread. See Seifert (*SA* **5**: 310, 1986; status), Rossman *et al.* (*Stud. Mycol.* **42**: 248 pp., 1999; possible synonym of *Berkelella*), Candoussau *et al.* (*Sydowia* **59**: 179, 2008; monogr.).

Neobroomella Petr. (1947), Amphisphaeriaceae. Anamorph *Pestalotiopsis*. 1, Syria. See Eriksson (*SA* **5**: 145, 1986), Kang *et al.* (*MR* **103**: 53, 1999).

Neobulgaria Petr. (1921), ? Helotiaceae. Anamorph

Myrioconium-like. 8, Europe. See Verkley (*Persoonia* **15**: 3, 1992; ultrastr.), Landvik *et al.* (*Mycoscience* **37**: 237, 1996; phylogeny), Wang *et al.* (*Mycol.* **98**: 1065, 2006; phylogeny).

Neocallimastigaceae I.B. Heath (1983), Neocallimastigales. 6 gen. (+ 1 syn.), 20 spp.

Lit.: Heath *et al.* (*CJB* **61**: 295, 1983), Heath *et al.* (*Fourth International Mycological Congress Abstracts*, 1989), Coleman *in* Levett (Ed.) (*Anaerobic microbiology. A practical approach*, 1991), Dore & Stahl (*CJB* **69**: 1964, 1991; RNA, phylogeny), Li & Heath (*Fifth International Colloquium on Invertebrate Pathology and Microbial Control* [Adelaide, Australia, 20-24 August 1990] Proceedings and Abstracts, 1992), Li *et al.* (*CJB* **71**: 393, 1993), Ho & Barr (*Mycol.* **87**: 655, 1995).

Neocallimastigales J.L. Li, I.B. Heath & L. Packer (1993). Neocallimastigomycetes. 1 fam., 6 gen., 20 spp. Thallus mono- or polycentric; zoospores mono- or polyflagellate, without mitochondria; kinetosomes with a skirt, spur, cylinder and microtubule root, lacking props; anaerobic saprobes occurring in the guts of herbivores. Fam.:

Neocallimastigaceae

Lit.: Heath (*CJB* **67**: 2815, 1989; emend.), Heath *et al.* (*CJB* **71**: 393, 1993), Li & Heath (*CJB* **70**: 1738, 1992; RNA, cladistics), Trinci *et al.* (*MR* **98**: 129, 1994, review), Ho & Barr (*Mycol.* **87**: 655, 1995; taxonomy), James *et al.* (*Mycol.* **98**: 860, 2006; molecular phylogeny), Hibbett *et al.* (*MR* **111**: 109, 2007), and see under Familes.

Neocallimastigomycetes M.J. Powell (2007), Neocallimastigomycota. 1 ord., 1 fam., 6 gen., 20 spp. Ord.:

Neocallimastigales

Lit.: James *et al.* (*Mycol.* **98**: 860, 2006; molecular phylogeny), and see under Order.

Neocallimastigomycota M.J. Powell (2007), Fungi. 1 class., 1 ord., 1 fam., 6 gen., 20 spp. Class:

Neocallimastigales

Lit.: James *et al.* (*Mycol.* **98**: 860, 2006; molecular phylogeny), and see under Orders.

Neocallimastix Vavra & Joyon ex I.B. Heath (1983), Neocallimastigaceae. 5 (in rumen of sheep and cattle), widespread. See Orpin & Munn (*TBMS* **86**: 178, 1986; emend.), Wubah *et al.* (*CJB* **69**: 835, 1991; morphology), Wubah *et al.* (*Mycol.* **83**: 40, 1991; resistant body), Chen *et al.* (*FEMS Microbiol. Lett.* **221**: 227, 2003; genetics).

Neocapnodium W. Yamam. (1955) = Phragmocapnias fide Reynolds (*Mycotaxon* **8**: 917, 1979).

Neocarpenteles Udagawa & Uchiy. (2002), Trichocomaceae. 1. See Udagawa & Uchiyama (*Mycoscience* **43**: 4, 2002).

Neocatapyrenium H. Harada (1993), Verrucariaceae (L). 4, widespread. See Harada (*Nat. Hist. Res.* **2**: 129, 1993), Breuss (*Bryologist* **108**: 537, 2005; N America), Gueidan *et al.* (*MR* **111**: 1145, 2007; phylogeny).

Neochaetospora B. Sutton & Sankaran (1991), anamorphic *Pezizomycotina*, Cpd.≡ eH.1. 1 (coprophilous), Sahara. See Sutton & Sankaran (*MR* **95**: 768, 1991).

Neoclaviceps J.F. White, Bills, S.C. Alderman & Spatafora (2001), Clavicipitaceae. 1, Costa Rica. See White *et al.* (*Mycol.* **93**: 90, 2001).

Neoclitocybe Singer (1962), Tricholomataceae. 10, widespread (esp. tropical). See Singer (*Sydowia* **15**: 55, 1961), Liew *et al.* (*Mol. Phylogen. Evol.* **16**: 392, 2000; phylogeny).

Neococcomyces Y.R. Lin, C.T. Xiang & Z.Z. Li (1999), Rhytismataceae. 1, China. See Lin *et al.* (*Mycosystema* **18**: 357, 1999), Gao & Hou (*Nova Hedwigia* **82**: 123, 2006).

Neocoleroa Petr. (1934) = Wentiomyces fide von Arx & Müller (*Stud. Mycol.* **9**, 1975).

Neocordyceps Kobayasi (1984), ? Cordycipitaceae. 1 (entomogenous), Japan. See Fukatsu (*Nippon Kingakukai Kaiho* **40**: 34, 1999), Fukatsu & Nikoh (*Mycology Series* **19**: 311, 2003), Sung *et al.* (*Stud. Mycol.* **57**, 2007).

Neocosmospora E.F. Sm. (1899), Nectriaceae. Anamorph *Acremonium*-like. 13, widespread (esp. tropical). See also *Haematonectria*. See Cannon & Hawksworth (*TBMS* **82**: 673, 1984; key), Udagawa *et al.* (*Sydowia* **41**: 349, 1989; key 7 spp.), Rossman *et al.* (*Stud. Mycol.* **42**: 248 pp., 1999), O'Donnell (*Mycol.* **92**: 919, 2000; phylogeny), Cornely *et al.* (*Emerging Infectious Diseases* **7**: 149, 2001; clinical strain).

Neocryptospora Petr. (1959), Pezizomycotina. 1, Brazil. See Petrak (*Sydowia* **13**: 51, 1959).

Neocudoniella S. Imai (1941), ? Helotiaceae. 2, Japan; Canada. See Kohn *et al.* (*Mycol.* **78**: 934, 1986).

Neodasyscypha Spooner (1987) nom. inval. = Neodasyscypha Suková & Spooner.

Neodasyscypha Suková & Spooner (2005), Hyaloscyphaceae. 2, widespread (temperate). See Spooner (*Biblthca Mycol.* **116**: 711 pp., 1987), Eriksson & Hawksworth (*SA* **7**: 83, 1988; status), Suková (*Czech Mycol.* **57**: 139, 2005).

Neodeightonia C. Booth (1970), Botryosphaeriaceae. 1. See Punithalingam (*Mycol. Pap.* **119**, 1969), von Arx & Müller (*Stud. Mycol.* **9**, 1975), Phillips *et al.* (*Persoonia* **21**: in press, 2008; phylogeny).

Neodimerium Petr. (1950) = Epipolaeum fide Müller & von Arx (*Beitr. Kryptfl. Schweiz* **11** no. 2, 1962).

Neodiplodina Petr. (1954), anamorphic *Pezizomycotina*, St.1eH.?. 1, Australia. See Petrak (*Sydowia* **8**: 36, 1954).

Neoerysiphe U. Braun (1999), Erysiphaceae. Anamorph *Oidium* subgen. *Striatoidium*. 5, widespread. See Braun (*Beih. Nova Hedwigia* **89**, 1987; key (as *Erysiphe* sect. *Galeopsidis*)), Braun (*Schlechtendalia* **3**: 50, 1999; taxonomy), Bolay (*Cryptogamica Helvetica* **20**, 2005; Switzerland), Voytyuk *et al.* (*Mycotaxon* **97**: 247, 2006; Israel), Wang *et al.* (*Mycol.* **98**: 1065, 2006; phylogeny).

Neofabraea H.S. Jacks. (1913), Dermateaceae. Anamorphs *Cryptosporiopsis*, *Phlyctema*. 4, widespread. Molecular data do not support separation from *Pezicula*. See Verkley (*Stud. Mycol.* **44**, 1999), Abeln *et al.* (*Mycol.* **92**: 685, 2000; phylogeny), Cunnington (*Australas. Pl. Path.* **33**: 453, 2004; Australia), Gariépy *et al.* (*Can. J. Pl. Path.* **27**: 118, 2005; Pacific NW), Wang *et al.* (*Mycol.* **98**: 1065, 2006; phylogeny).

Neoflageoletia J. Reid & C. Booth (1966), Hyponectriaceae. 1 (on bamboo stem), Philippines. See Reid & Booth (*CJB* **44**: 445, 1966).

Neofracchiaea Teng (1938), Scortechiniaceae. 1, China; USA. See Huhndorf *et al.* (*MR* **108**: 1384, 2004; phylogeny).

Neofuckelia Zeller & Goodd. (1935), anamorphic *Pezizomycotina*, St.0eH.?. 1, N. America.

Neofuscelia Essl. (1978) = Xanthoparmelia fide Elix (*Bryologist* **96**: 359, 1993; close to *Xanthoparmelia*) but see also Elix (*Flora of Australia* **55**: 68, 1994),

Krug & Sang (*Flechten Follmann* Contributions to Lichenology in Honour of Gerhard Follmann: 263, 1995; Namibia), Rico & Manrique (*Lazaroa* **16**: 9, 1995; Spain), Crespo & Cubero (*Lichenologist* **30**: 369, 1998; phylogeny), Elix (*Mycotaxon* **71**: 431, 1999; S hemisphere), Elix (*Australasian Lichenology* **51**: 7, 2002; chemistry), Elix (*Australasian Lichenology* **53**: 14, 2003; Australia), Giordani *et al.* (*Lichenologist* **35**: 377, 2003; Italy), Blanco *et al.* (*Taxon* **53**: 959, 2004; phylogeny, synonymy with *Xanthoparmelia*), Thell *et al.* (*Mycol. Progr.* **3**: 297, 2004; phylogeny), Blanco *et al.* (*Mol. Phylogen. Evol.* **39**: 52, 2006; phylogeny).

Neofusicoccum Crous, Slippers & A.J.L. Phillips (2006), anamorphic *Botryosphaeriaceae*, H?.?.?. 13, widespread. See Morgan-Jones & White (*Mycotaxon* **30**: 117, 1987), Darvas (*Phytophylactica* **23**: 295, 1991; S Africa), Jacobs & Rehner (*Mycol.* **90**: 601, 1998; morphology, ITS), Smith & Stanosz (*Mycol.* **93**: 505, 2001), Phillips *et al.* (*Sydowia* **54**: 59, 2002), Denman *et al.* (*Mycol.* **95**: 294, 2003; on *Proteaceae*), Slippers *et al.* (*Stud. Mycol.* **50**: 343, 2004; on *Eucalyptus*), van Niekerk *et al.* (*Mycol.* **96**: 781, 2004; on *Vitis*), Barber *et al.* (*MR* **109**: 1347, 2005; synanamorph), Slippers *et al.* (*Mycol.* **97**: 99, 2005; on *Mangifera*), Crous *et al.* (*Stud. Mycol.* **55**: 247, 2006), Slippers *et al.* (*Pl. Path.* **56**: 128, 2007; on *Rosaceae*).

Neogeotrichum O. Magalh. (1932) = Trichosporon fide Lodder & Kreger-van Rij (*Yeasts, a taxonomic study*, 1952).

Neogibbera Petr. (1947) = Acantharia fide Müller & von Arx *in* Ainsworth *et al.* (Eds) (*The Fungi* **4A**: 87, 1973).

Neogodronia Schläpf.-Bernh. (1969) = Encoeliopsis fide Korf *in* Ainsworth *et al.* (Eds) (*The Fungi* **4A**: 249, 1973).

Neogymnomyces G.F. Orr (1970), Onygenaceae. 1, USA. See Orr (*CJB* **48**: 1065, 1970), Currah (*Mycotaxon* **24**: 1, 1985), Currah (*SA* **7**: 1, 1988; key).

Neogyromitra S. Imai (1932) = Gyromitra fide Eckblad (*Nytt Mag. Bot.* **15**: 1, 1968).

Neohaplomyces R.K. Benj. (1955), Laboulbeniaceae. 3, widespread. See Benjamin (*Aliso* **3**: 189, 1955), Santamaría & Rossi (*Pl. Biosystems* **133**: 163, 1999; Mediterranean).

Neohendersonia Petr. (1921), anamorphic *Pezizomycotina*, St.≡ eP.1/19. 3, widespread. See Sutton (*The Coelomycetes*, 1980).

Neohenningsia Koord. (1907) = Hydropisphaera fide Rossman *et al.* (*Stud. Mycol.* **42**: 248 pp., 1999).

Neoheppia Zahlbr. (1909), Peltulaceae (L). 2, Brazil; Zaïre. See Büdel (*Mycotaxon* **54**: 137, 1995).

Neoheteroceras Nag Raj (1993), anamorphic *Pezizomycotina*, Cpd/Cac.≡ eP.19. 1, widespread (temperate). See Nag Raj (*Coelomycetous Anamorphs with Appendage-bearing Conidia*: 539, 1993), Hüseyin *et al.* (*Mycotaxon* **94**: 241, 2005), Yonezawa & Tanaka (*Mycoscience* **49**: 152, 2008; Japan).

Neohoehnelia Theiss. & Syd. (1918) = Dysrhynchis fide Müller & von Arx (*Beitr. Kryptfl. Schweiz* **11** no. 2, 1962).

Neohygrocybe Herink (1958) = Hygrocybe fide Singer (*Agaric. mod. Tax.* edn 2, 1962).

Neohygrophorus Singer ex Singer (1962), Tricholomataceae. 1, N. America. See Redhead *et al.* (*Mycotaxon* **76**: 321, 2000), Matheny *et al.* (*Mycol.* **98**: 982, 2006; phylogeny).

Neohypodiscus (Lloyd) J.D. Rogers, Y.M. Ju & Laessøe (1994), ? Boliniaceae. 3, widespread. See Rogers *et al.* (*Mycol.* **86**: 684, 1994).

Neojohnstonia B. Sutton (1983), anamorphic *Pezizomycotina*, Hso.1eP.1/10. 1, pantropical. See Sutton (*TBMS* **81**: 407, 1983), Gusmão & Grandi (*Mycotaxon* **80**: 97, 2001; Brazil).

Neokeissleria Petr. (1920) = Melanconis fide Müller & von Arx *in* Ainsworth *et al.* (Eds) (*The Fungi* **4A**: 87, 1973), Barr (*Mycol. Mem.* **7**, 1978).

Neokellermania Punith. (1981) = Pseudorobillarda M. Morelet fide Nag Raj (*Coelomycetous Anamorphs with Appendage-bearing Conidia*, 1993).

Neokneiffia Sacc. (1898) = Hyphoderma Wallr. fide Donk (*Persoonia* **3**: 199, 1964).

Neolamya Theiss. & Syd. (1918), Pezizomycotina. 1 (on lichens, esp. *Peltigera*), Europe. See Kümmerling & Alstrup (*Graphis Scripta* **3**: 120, 1992; Scandinavia), Zhurbenko & Davydov (*Folia cryptog. Estonica* **37**: 109, 2000; Siberia).

Neolecta Speg. (1881), Neolectaceae. 3, north temperate; Brazil. See Redhead (*CJB* **55**: 301, 1977; ecology), Landvik *et al.* (*SA* **11**: 107, 1993; phylogeny), Landvik (*MR* **100**: 199, 1996), Sjamsuridzal *et al.* (*Mycoscience* **38**: 267, 1997), Liu *et al.* (*Mol. Biol. Evol.* **16**: 1799, 1999; phylogeny), Landvik *et al.* (*Mycol.* **93**: 1151, 2001; phylogeny), Landvik *et al.* (*MR* **107**: 1021, 2003; morphology, ultrastr.), Sugiyama *et al.* (*Mycol.* **98**: 996, 2006; phylogeny).

Neolectaceae Redhead (1977), Neolectales. 1 gen. (+ 2 syn.), 3 spp.

Lit.: Redhead (*CJB* **55**: 301, 1977), Landvik *et al.* (*SA* **11**: 107, 1993), Landvik (*MR* **100**: 199, 1996), Sugiyama (*Mycoscience* **39**: 487, 1998), Liu *et al.* (*Mol. Biol. Evol.* **16**: 1799, 1999), Berbee & Taylor *in* McLaughlin *et al.* (Eds) (*The Mycota* A Comprehensive Treatise on Fungi as Experimental Systems for Basic and Applied Research **7A**: 229, 2001), Landvik *et al.* (*Mycol.* **93**: 1151, 2001), Sugiyama *et al.* (*Mycol.* **98**: 996, 2006; phylogeny), Sugiyama *et al.* (*Mycol.* **98**: 996, 2006).

Neolectales Landvik, O.E. Erikss., Gargas & P. Gust. (1993). Neolectomycetidae. 1 fam., 1 gen., 3 spp. Fam.:

Neolectaceae

For *Lit.* see under fam.

Neolectomycetes O.E. Erikss. & Winka (1997), Taphrinomycotina. 1 ord., 1 fam., 1 gen., 3 spp. Contains only one genus, *Neolecta*, which has some outward similarities to members of the *Pezizomycetes*, but clusters with the *Taphrinomycetes* when molecular data are analysed. Ord.:

Neolectales

For *Lit.* see fam.

Neolectomycetidae, see *Neolectomycetes*.

Neolentinus Redhead & Ginns (1985), Polyporaceae. 11, widespread. See Redhead & Ginns (*TMSJ* **26**: 357, 1985), Lechner & Wright (*Mycotaxon* **82**: 281, 2002; N. Am.).

Neolentiporus Rajchenb. (1995), Fomitopsidaceae. 1, S. America. See Rajchenberg (*Nordic Jl Bot.* **15**: 105, 1995).

Neolichina Gyeln. (1940) = Lichina fide Hawksworth *et al.* (*Dictionary of the Fungi* edn 8, 1995).

Neoligniella Naumov (1951), anamorphic *Pezizomycotina*, St.0eH.?. 1, former USSR. See Naumov (*Bot. Mater. Otd. Sporov. Rast. Bot. Inst. Komarova Akad. Nauk S.S.S.R.* **7**: 116, 1951).

Neolinocarpon K.D. Hyde (1992) = Linocarpon fide Yanna & Ho (*MR* **107**: 1305, 2003; ultrastr.).

Neolysurus O.K. Mill., Ovrebo & Burk (1991), Phallaceae. 1, Costa Rica. See Miller *et al.* (*MR* **95**: 1230, 1991).

Neomarssoniella U. Braun (1991), anamorphic *Pezizomycotina*, Cac.1eH.?. 2, Europe; N. America. See Braun (*Nova Hedwigia* **53**: 304, 1991).

Neomelanconium Petr. (1940), anamorphic *Pezizomycotina*, St.0eP.19. 2, Europe; Africa.

Neomichelia Penz. & Sacc. (1901) = Pithomyces fide Ellis (*Mycol. Pap.* **76**, 1960).

Neomunkia Petr. (1947), anamorphic *Hypocreales*, St.0eH.?. 1, S. America. See Bischoff *et al.* (*Mycol.* **96**: 1088, 2004; phylogeny).

Neonaumovia Schwarzman (1959) = Lophophacidium fide Korf *in* Ainsworth *et al.* (Eds) (*The Fungi* **4A**: 249, 1973).

Neonectria Wollenw. (1917), Nectriaceae. Anamorph *Cylindrocarpon*. 33, widespread. *N. galligena* (canker and eye rot of apple and pear; Flack & Swinburne, *TBMS* **68**: 186, 1977; host range). See Brayford & Samuels (*Mycol.* **85**: 612, 1993; *Cylindrocarpon* anamorphs), Samuels & Brayford (*Sydowia* **45**: 55, 1993; phragmosporous spp.), Rossman *et al.* (*Stud. Mycol.* **42**: 248 pp., 1999), Schoch *et al.* (*Stud. Mycol.* **45**: 45, 2000), Mantiri *et al.* (*CJB* **79**: 334, 2001; phylogeny), Langrell (*MR* **106**: 280, 2002; molecular detection), Seifert *et al.* (*Phytopathology* **93**: 1533, 2003; on ginseng), Brayford *et al.* (*Mycol.* **96**: 572, 2004; spp. without microconidia), Halleen *et al.* (*Stud. Mycol.* **50**: 431, 2004; on *Vitis*), Castlebury *et al.* (*CJB* **84**: 1417, 2006; phylogeny on *Fagus*), Hirooka & Kobayashi (*Mycoscience* **48**: 53, 2007; Japan).

Neonorrlinia Syd. (1923) = Cercidospora fide Hafellner & Obermayer (*Cryptog. Bryol.-Lichénol.* **16**: 177, 1995).

Neonosemoides N. Faye & B.S. Togubaye (1996), Microsporidia. 1.

Neonothopanus R.H. Petersen & Krisai (1999), Marasmiaceae. 1, S. America. See Petersen & Krisai-Greilhuber (*Persoonia* **17**: 210, 1999).

Neoovularia U. Braun (1992), anamorphic *Pezizomycotina*, Hsp.0eP.10. 3, Europe; Japan. See Braun (*Monogr. Cercosporella, Ramularia Allied Genera (Phytopath. Hyphom.)* **2**: 342, 1998; key).

Neoparodia Petr. & Cif. (1932), Parodiopsidaceae. Anamorph *Chuppia*-like or *Sarcinella*-like. 1, West Indies.

Neopatella Sacc. (1908) = Selenophoma fide Petrak (*Sydowia* **5**: 354, 1951).

Neopaxillus Singer (1948), Serpulaceae. 5, C. & S. America. See Singer *et al.* (*Beih. Nova Hedwigia* **98**: 17, 1990).

Neopeckia Sacc. (1883), Pleosporales. 1, N. America. See Bose (*Phytopath. Z.* **41**, 1961; syn. with *Herpotrichia*), Barr (*Mycotaxon* **20**: 1, 1984), von Arx & Müller (*Sydowia* **37**: 6, 1984; syn. with *Herpotrichia*), Chen & Hsieh (*Sydowia* **56**: 24, 2004).

Neopeltella Petr. (1950), ? Schizothyriaceae. 1, S. America. See Petrak (*Sydowia* **4**: 329, 1950).

Neopeltis Petr. (1949) = Microcallis fide von Arx & Müller (*Stud. Mycol.* **9**, 1975).

Neopeltis Syd. (1937), anamorphic *Pezizomycotina*, St.0eH.?. 1, C. America.

Neoperezia I.V. Issi & Voronin (1979), Microsporidia. 1.

Neopericonia Kamal, A.N. Rai & Morgan-Jones (1983), anamorphic *Pezizomycotina*, Hso.#eP.10. 1, India. See Kamal *et al.* (*Mycotaxon* **18**: 15, 1983).

Neopetromyces Frisvad & Samson (2000), Trichocomaceae. Anamorph *Aspergillus*. 1, Philippines. See Frisvad & Samson (*Stud. Mycol.* **45**: 201, 2000).

Neophacidium Petr. (1950), ? Rhytismatales. 2, Pakistan; S. America. See Petrak (*Sydowia* **4**: 333, 1950).

Neophaeosphaeria M.P.S. Câmara, M.E. Palm & A.W. Ramaley (2003), Pleosporales. Anamorph *Coniothyrium*-like. 5 (on *Agavaceae*), widespread. See Câmara *et al.* (*MR* **107**: 519, 2003).

Neophoma Petr. & Syd. (1927), anamorphic *Pezizomycotina*, St.0eH.?. 2, Europe; N. America.

Neophyllis F. Wilson (1891), Sphaerophoraceae (L). 2, Australasia. See Ahti (*Regnum veg.* **128**: 58, 1993), Döring *et al.* (*Aust. J. Bot.* **47**: 783, 1999), Wedin & Döring (*MR* **103**: 1131, 1999; phylogeny), Döring & Wedin (*Pl. Biol.* **2**: 361, 2000; phylogeny, morphology), Wedin *et al.* (*Lichenologist* **32**: 171, 2000; phylogeny), Miądlikowska *et al.* (*Mycol.* **98**: 1088, 2006; phylogeny).

Neophyscia M. Choisy (1959) nom. nud. = Physcia fide Hawksworth *et al.* (*Dictionary of the Fungi* edn 8, 1995).

Neopiptostoma Kuntze (1898) ≡ Piptostoma.

Neoplaconema B. Sutton (1977), anamorphic *Pezizomycotina*, St.0eH.15. 2, Germany; Australia. See Yuan & Mohammed (*MR* **101**: 1531, 1997; n.sp. Australia).

Neoplacosphaeria Petr. (1921) = Sphaeriothyrium fide Petrak (*Annls mycol.* **22**: 1, 1924).

Neopsoromopsis Gyeln. (1940), ? Parmeliaceae (L). 1, Argentina.

Neopycnodothis Tak. Kobay. (1965) = Cytoplea fide Sutton (*Mycol. Pap.* **141**, 1977).

Neoramularia U. Braun (1991), anamorphic *Pezizomycotina*, St.0-1eH.3. 9, widespread. See Braun (*Monogr. Cercosporella, Ramularia Allied Genera (Phytopath. Hyphom.)* **2**: 351, 1998; key), Braun *et al.* (*Australas. Pl. Path.* **34**: 509, 2005; Australia).

Neoravenelia Long (1903) = Ravenelia fide Arthur (*Manual Rusts US & Canada*, 1934).

Neorehmia Höhn. (1902), Trichosphaeriaceae. Anamorph *Tritirachium*-like. 2, Austria. See Müller & Samuels (*Sydowia* **35**: 143, 1982), Barr (*Mycotaxon* **39**: 43, 1990), Samuels & Barr (*CJB* **75**: 2165, 1998), Rossman *et al.* (*Stud. Mycol.* **42**: 248 pp., 1999).

Neosaccardia Mattir. (1921) ? = Scleroderma fide Guzmán (*Ciencia* **25**: 195, 1967).

Neosartorya Malloch & Cain (1973), Trichocomaceae. Anamorph *Aspergillus*. 12, widespread. See Takada *et al.* (*TMSJ* **27**: 415, 1986; mating), Guarro & Figueras (*Boln Soc. Micol. Madrid* **11**: 217, 1987; ontogeny), Peterson (*MR* **96**: 547, 1992; 6 spp. DNA homology), Udagawa *et al.* (*Mycoscience* **37**: 217, 1996; toxigenic spp.), Geiser *et al.* (*Mycol.* **90**: 831, 1998; DNA), Samson (*Contributions to Microbiology* **2**: 5, 1999; review), Varga *et al.* (*Antonie van Leeuwenhoek* **77**: 235, 2000; phylogeny), Takada *et al.* (*Mycoscience* **42**: 361, 2001; Africa), Klich (*Identification of Common Aspergillus Species*: 116 pp., 2002), Horie *et al.* (*Mycoscience* **44**: 397, 2003; Brazil), Balajee *et al.* (*J. Clin. Microbiol.* **43**: 5996, 2005; comparison with *Aspergillus fumigatus*), Dyer & Paoletti (*Medical Mycology* **43** Suppl. 1: S7, 2005; genetics), Nierman (*Nature* Lond. **438** no. 7071: 1151, 2005; genome sequence), Rydholm *et al.* (*Eu-*

karyotic Cell **5**: 650, 2006; population structure), Samson *et al.* (*Medical Mycology* **44** Suppl. 1: S133, 2006; species concepts), Wortman *et al.* (*Medical Mycology* **44** Suppl. 1: S3, 2006; phylogeny), Samson *et al.* (*Stud. Mycol.* **59**: 147, 2007; phylogeny, chemistry, morphology), Samson *et al.* (*Stud. Mycol.* **59**: 147, 2007; section *Fumigati*), Hong *et al.* (*Antonie van Leeuwenhoek* **93**: 87, 2008).

Neoscytalidium Crous & Slippers (2006), anamorphic *Botryosphaeriaceae*, H?.?.?. 1, tropical. See Crous & Slippers (*Stud. Mycol.* **55**: 244, 2006).

Neosecotium Singer & A.H. Sm. (1960), Agaricaceae. 2, N. America; Africa. Basidioma gasteroid. See Singer & Smith (*Madroño* **15**: 154, 1960).

neosexual, see protosexual.

Neoskofitzia Schulzer (1880) nom. dub., Sordariomycetes. 5, widespread. See Rossman (*Stud. Mycol.* **42**, 1999).

Neosolorina (Gyeln.) Räsänen (1943) = Solorina fide Hawksworth *et al.* (*Dictionary of the Fungi* edn 8, 1995).

Neospegazzinia Petr. & Syd. (1936), anamorphic *Pezizomycotina*, Cpd.0fH.?. 1, S. America.

Neosphaeropsis Petr. (1921) ? = Sphaeropsis Sacc. fide Hawksworth *et al.* (*Dictionary of the Fungi* edn 8, 1995).

Neosporidesmium Mercado & J. Mena (1988), anamorphic *Pezizomycotina*, Hsy.≡ eP.1. 1, Cuba. See Mercado & Mena (*Acta Bot. Cubana* **59**: 2, 1988).

Neostomella Syd. (1927), Asterinaceae. 4, C. America. See Hansford (*Mycol. Pap.* **15**, 1946).

Neotapesia E. Müll. & Hütter (1963), Helotiales. 3, Europe. See Korf *in* Ainsworth *et al.* (Eds) (*The Fungi* **4A**: 249, 1973), Nannfeldt (*TBMS* **67**: 283, 1976).

Neotestudina Segretain & Destombes (1961), Testudinaceae. 3, widespread. *N. rosatii* (mycetoma of foot). See Hawksworth & Booth (*Mycol. Pap.* **135**, 1974), Hawksworth (*CJB* **57**: 91, 1979), Kruys *et al.* (*MR* **110**: 527, 2006; phylogeny).

Neothyridaria Petr. (1934), Pezizomycotina. 1, Europe.

neotony, a process in which normal development of cells, except those involved in reproduction, is arrested; results in a sexually mature organism with juvenile features; see Kreisel (*in* Hawksworth (Ed.), *Frontiers in mycology*: 69, 1991; in fungal phylogeny); cf. pedogenesis.

Neotremella Lowy (1979), Tremellaceae. 1, Mexico. See Lowy (*Boletín de la Sociedad Mexicana de Micología* **13**: 224, 1979).

Neotrichophyton Castell. & Chalm. (1919) = Trichophyton fide Hawksworth *et al.* (*Dictionary of the Fungi* edn 8, 1995).

Neotrotteria Sacc. (1921), Coronophorales. See Nannfeldt (*Svensk bot. Tidskr.* **69**: 49, 1975), Nannfeldt (*Svensk bot. Tidskr.* **69**: 289, 1975), Subramanian & Sekar (*Kavaka* **18**: 19, 1993).

Neottiella (Cooke) Sacc. (1889), Pyronemataceae. *c.* 13, Europe. See Caillet & Moyne (*BSMF* **103**: 179, 1987; synonym of *Octospora*), Caillet & Moyne (*Bull. Soc. Hist. nat. Doubs* **84**: 9, 1991; keys), Kristiansen (*SA* **14**: 61, 1995), Moravec (*Czech Mycol.* **49**: 149, 1997; sect. *Neottiellae*), Benkert (*Z. Mykol.* **64**: 153, 1998; Germany), Hansen & Pfister (*Mycol.* **98**: 1029, 2006; phylogeny), Perry *et al.* (*MR* **111**: 549, 2007; phylogeny).

Neottiopezis Clem. (1903) ≡ Neottiella fide Clements (*Gen. Fung.*: 90, 1909).

Neottiospora Desm. (1843), anamorphic *Pezizomycotina*, Cpd.0eH/1eP.15. 2, widespread.

Neottiosporella Höhn. ex Falck (1923) nom. dub., anamorphic *Pezizomycotina*. No spp. included.

Neottiosporina Subram. (1961), anamorphic *Pezizomycotina*, Cpd.≡ eH.1. 10, widespread. See Sutton & Alcorn (*Aust. J. Bot.* **22**: 517, 1974; key), Sutton & Wu (*MR* **99**: 831, 1995; on *Sorghum*).

Neottiosporis Clem. & Shear (1931) ≡ Neottiosporella.

neotype, see type.

Neotyphodium Glenn, C.W. Bacon & Hanlin (1996), anamorphic *Epichloë*. 4 (endosymbiotic in grasses), widespread (esp. temperate). See Glenn *et al.* (*Mycol.* **88**: 369, 1996; phylogeny), Glenn & Bacon (*Neotyphodium/Grass Interactions Proceedings of the Third International Symposium on Acremonium/Grass Interactions, held May 28-31, 1997, in Athens, Georgia*: 53, 1997; chemistry), Kuldau *et al.* (*Mycol.* **89**: 431, 1997; phylogeny), Leuchtmann (*Neotyphodium/Grass Interactions Proceedings of the Third International Symposium on Acremonium/Grass Interactions, held May 28-31, 1997, in Athens, Georgia*: 93, 1997; ecology), White (*Neotyphodium/Grass Interactions Proceedings of the Third International Symposium on Acremonium/Grass Interactions, held May 28-31, 1997, in Athens, Georgia*: 27, 1997), Wilkinson & Schardl (*Neotyphodium/Grass Interactions Proceedings of the Third International Symposium on Acremonium/Grass Interactions, held May 28-31, 1997, in Athens, Georgia*: 13, 1997; mutualism), Leuchtmann & Schardl (*MR* **102**: 1169, 1998; genetics, phylogeny), Miles *et al.* (*Appl. Environm. Microbiol.* **64**: 601, 1998; Australasia), Naffaa *et al.* (*Ann. appl. Biol.* **132**: 211, 1998; Europe), White & Reddy (*Mycol.* **90**: 226, 1998; phylogeny), Kuldau *et al.* (*Mycol.* **91**: 776, 1999; hybridisation), Schardl & Leuchtmann (*Mycol.* **91**: 95, 1999; N America), Tredway *et al.* (*MR* **103**: 1593, 1999; phylogeny), Lane *et al.* (*Microbial Endophytes*: 341, 2000; coevolution, chemistry), Moon *et al.* (*Mycol.* **92**: 1103, 2000; evolution), Schardl (*Fungal Genetics Biol.* **33**: 69, 2001; on *Festuca*), Moon *et al.* (*Mycol.* **94**: 694, 2002; S hemisphere), Brem & Leuchtmann (*Evolution* Lancaster, Pa. **57**: 37, 2003; on *Bromus*), Moon *et al.* (*Mol. Ecol.* **13**: 1455, 2004; hybridisation), Sullivan & Faeth (*Mol. Ecol.* **13**: 649, 2004; gene flow), Yokoyama *et al.* (*FEMS Microbiol. Lett.* **264**: 182, 2006; genetics), Moon *et al.* (*Mycol.* **99**: 895, 2007; on *Stipeae* and *Meliceae*).

Neotyphula Wakef. (1934), Basidiomycota. 1, Guyana. See Martin (*Lloydia* **11**: 121, 1948).

Neournula Paden & Tylutki (1969), Chorioactidaceae. 2, USA; Europe; N. Africa. See Paden (*Mycol.* **64**: 457, 1972; nomencl.), Hansen & Pfister (*Mycol.* **98**: 1029, 2006; phylogeny), Pfister *et al.* (*MR* **112**: 513, 2008).

Neoventuria Syd. & P. Syd. (1919), Dothideomycetes. 1, S. America.

Neovossia Körn. (1879), Tilletiaceae. 1 (on *Poaceae*), widespread. See Vánky (*Illustrated genera of smut fungi*, 1987), Castlebury *et al.* (*Mycol.* **97**: 888, 2005; phylog.) cf. *Tilletia*.

Neoxenophila Apinis & B.M. Clark (1974) = Aphanoascus fide Cano & Guarro (*MR* **94**, 1990), Cano *et al.* (*Stud. Mycol.* **47**: 153, 2002).

Neozimmermannia Koord. (1907) = Glomerella fide von Arx & Müller (*Beitr. Kryptfl. Schweiz* **11** no. 1,

1954).

Neozygitaceae Ben Ze'ev, R.G. Kenneth & Uziel (1987), Entomophthorales. 3 gen., 21 spp.
Lit.: Ben-Ze'ev *et al.* (*Mycotaxon* **28**: 313, 1987), Humber (*Mycotaxon* **34**: 441, 1989), Keller (*Sydowia* **43**: 39, 1991), Delalibera *et al.* (*Can. J. Microbiol.* **50**: 579, 2004), Delalibera *et al.* (*Mycol.* **96**: 1002, 2004), Keller & Petrini (*Sydowia* **57**: 23, 2005).

Neozygites Witlaczil (1885), Neozygitaceae. 18 (on *Arthropoda*), widespread. See Keller & Wuest (*Entomophaga* **28**: 123, 1983), Butt & Heath (*Eur. J. Cell. Biol.* **46**: 499, 1988; cell division), Butt & Humber (*Protoplasma* **151**: 115, 1989; mitosis), Keller (*Sydowia* **43**: 39, 1991; key Swiss spp.), Dick *et al.* (*Mycol.* **84**: 729, 1992), Steenberg *et al.* (*J. Invert. Path.* **68**: 97, 1996), Keller (*Sydowia* **49**: 118, 1997), Keller & Steenberg (*Sydowia* **49**: 21, 1997), Delalibera *et al.* (*Mycol.* **96**: 1002, 2004; *Neozygites tanajoae* n.sp. pathogenic on cassava green mite), Keller & Petrini (*Sydowia* **57**: 23, 2005; key), White *et al.* (*Mycol.* **98**: 872, 2006; phylogeny).

Neozythia Petr. (1958), anamorphic *Helotiales*, St.0eH.38. 1, Iran. See Petrak (*Sydowia* **11**: 351, 1957).

Nephlyctis Arthur (1907) = Prospodium fide Dietel (*Nat. Pflanzenfam.* **6**, 1928).

Nephrochytrium Karling (1938), Endochytriaceae. 4, N. America.

Nephroma Ach. (1809), Nephromataceae (L). 36, widespread. See Wetmore (*Publs Mich. St. Univ. Mus., ser. biol.* **1**: 369, 1960; N. Am.), White & James (*Lichenologist* **20**: 103, 1988; key 15 S. temp. spp.), Eriksson & Strans (*SA* **14**: 33, 1995; molec. phylogeny), Goffinet & Bayer (*Fungal Genetics Biol.* **21**: 228, 1997; DNA), Burgaz & Martínez (*Cryptog. Mycol.* **20**: 225, 1999; Iberian spp.), Lohtander *et al.* (*MR* **106**: 777, 2002; phylogeny), Wiklund & Wedin (*Cladistics* **19**: 419, 2003; phylogeny), Miądlikowska & Lutzoni (*Am. J. Bot.* **91**: 449, 2004; phylogeny), Louwhoff (*Muelleria* **22**: 3, 2005; Australia), Miądlikowska *et al.* (*Mycol.* **98**: 1088, 2006; phylogeny), Piercey-Normore *et al.* (*Lichenologist* **38**: 441, 2006; Pacific NW).

Nephromataceae Wetmore ex J.C. David & D. Hawksw. (1991), Peltigerales (L). 1 gen. (+ 6 syn.), 36 spp.
Lit.: James & White (*Lichenologist* **19**: 215, 1987), White & James (*Lichenologist* **20**: 103, 1988), Galloway (*Symbiosis* **11**: 327, 1991), Eriksson & Strand (*SA* **14**: 33, 1995), Eriksson & Strans (*SA* **14**: 33, 1995; molec. phylogeny), Goffinet & Bayer (*Fungal Genetics Biol.* **21**: 228, 1997; DNA), Goffinet & Goward (*Lichenogr. Thomsoniana* North American Lichenology in Honor of John W. Thomson: 41, 1998), Burgaz & Martínez (*Cryptog. Mycol.* **20**: 225, 1999), Wiklund & Wedin (*Cladistics* **19**: 419, 2003), Buschbom & Mueller (*Mol. Phylogen. Evol.* **32**: 66, 2004), Miądlikowska & Lutzoni (*Am. J. Bot.* **91**: 449, 2004), Miądlikowska *et al.* (*Mycol.* **98**: 1088, 2006; phylogeny), Vitikainen (*Nordic Lichen Flora* **3**: Cyanolichens: 91, 2007).

Nephromatomyces E.A. Thomas (1939) nom. inval. = Nephroma fide Hawksworth *et al.* (*Dictionary of the Fungi* edn 8, 1995).

Nephromiomyces E.A. Thomas ex Cif. & Tomas. (1953) = Nephroma fide Hawksworth *et al.* (*Dictionary of the Fungi* edn 8, 1995).

Nephromium Nyl. (1860) = Nephroma fide Hawksworth *et al.* (*Dictionary of the Fungi* edn 8, 1995).

Nephromopsis Müll. Arg. (1891), Parmeliaceae (L). 26, Asia. See Kärnefelt *et al.* (*Pl. Syst. Evol.* **183**: 113, 1992), Randlane & Saag (*Mycotaxon* **44**: 485, 1992), Randlane (*Cryptog. Bryol.-Lichénol.* **16**: 35, 1995), Randlane & Saag (*Cryptog. Bryol.-Lichénol.* **19**: 175, 1998), Chen & Gao (*Mycotaxon* **77**: 491, 2001), Thell *et al.* (*Mycol. Progr.* **4**: 303, 2005; revision).

Nephromyces Giard (1888), ? Cladochytriaceae. 3 (in ascidians), Europe. See Saffo & Nelson (*CJB* **61**: 3230, 1983), Beakes *in* Coombs *et al.* (Eds) (*Evolutionary Relationships Among Protozoa*: 351, 1998), Cavalier-Smith *in* Coombs *et al.* (Eds) (*Evolutionary Relationships Among Protozoa*: 375, 1998).

Nephromyces Sideris (1927) ≡ Rhizidiocystis.

Nephrospora Loubière (1923) = Microascus Zukal fide von Arx (*Persoonia* **8**: 191, 1975).

Nepotatus Lloyd (1925) = Scleroderma fide Stevenson & Cash (*Bull. Lloyd Libr. Mus.* **35**: 130, 1936).

Neptunella K.L. Pang & E.B.G. Jones (2003), Halosphaeriaceae. 1. See Pang *et al.* (*Mycol. Progr.* **2**: 35, 2003).

Nereiospora E.B.G. Jones, R.G. Johnson & S.T. Moss (1983), Halosphaeriaceae. Anamorph *Monodictys*-like. 2 (marine), widespread (northern hemisphere). See Mouzouras & Jones (*CJB* **63**: 2444, 1985), Jones & Moss (*SA* **6**: 179, 1987), Hyde & Jones (*Bot. Mar.* **32**: 205, 1989; ultrastr.), Anderson *et al.* (*Mycol.* **93**: 897, 2001; phylogeny), Chatmala *et al.* (*Fungal Diversity Res. Ser.* **7**: 59, 2002; anamorph), Campbell *et al.* (*Mycol.* **95**: 530, 2003; phylogeny).

nervicolous, living on veins of leaves or stems.

Nervostroma Narumi & Y. Harada (2006), Sclerotiniaceae. Anamorph *Cristulariella*. 1, widespread (temperate). See Narumi-Saito *et al.* (*Mycoscience* **47**: 351, 2007).

Nesolechia A. Massal. (1856), Parmeliaceae. 1 (lichenicolous), widespread. See Triebel *et al.* (*Bryologist* **98**: 71, 1995), Peršoh & Rambold (*Mycol. Progr.* **1**: 43, 2002; phylogeny), Doré *et al.* (*Lichenologist* **38**: 425, 2006; biometrics), Crespo *et al.* (*Mol. Phylogen. Evol.* **44**: 812, 2007; phylogeny).

Nesolechiaceae Arnold (1858) = Parmeliaceae.

Nesophloea Fr. (1849) nom. dub., Fungi.

Neta Shearer & J.L. Crane (1971), anamorphic *Pezizomycotina*, Hso.1-≡ eH.10. 6 (terrestrial and aquatic), widespread (north temperate). See de Hoog (*Stud. Mycol.* **26**: 44, 1985; key), Castañeda Ruiz & Heredia (*Mycotaxon* **76**: 131, 2000; Mexico).

Neural networks (Artificial Neural Networks, ANNs), allow for the identification or classification of input patterns (characters). In practice this approach generally requires a considerable amount of data in order to set up the network. Final placement is arrived at after the passage of the input values through a series of nodes arranged in layers. The method has not yet been used widely with fungal data, but has potential for more use, particularly in classification and data mining; examples include Acuña *et al.* (*Biotechnology techniques* **12**: 515, 1998; estimation of fungal biomass), Boddy & Morris (*Binary* **5**: 17, 1991), Bridge *et al.* (*in* Hawksworth (Ed.), *Identification and characterization of pest organisms*: 153, 1994), Morgan *et al* (*MR* **102**: 975, 1998), Morris *et al.* (*MR* **96**: 967, 1992), Nie *et al.* (*J. Phytopath.* **155**: 364, 2007; identification of *Fusarium*). For methodology see Boddy *et al.* (*Binary* **2**: 179, 1990), Specht (*Neural networks 3: 109, 1990), Zaruda (*Artificial neural systems*, 1992). Successful use requires a good understanding of underlying theory. See Numerical taxonomy.

Neuroecium Kunze (1823) nom. dub., Fungi. Not a fungus fide Juel (*Dansk. bot. Ark.* **5**: 1, 1928).

Neuronectria Munk (1957) ≡ Hydropisphaera fide Rossman *et al.* (*Stud. Mycol.* **42**: 248 pp., 1999).

Neurophyllum, see *Neurophyllum*.

Neuropogon Nees & Flot. (1835) = Usnea fide Lamb (*Index Nom. Lich.*, 1963), Articus (*Taxon* **53**: 925, 2004), Wirtz *et al.* (*Taxon* **55**: 367, 2006; phylogeny, synonymy with *Usnea*).

Neurospora Shear & B.O. Dodge (1927), Sordariaceae. Anamorph *Chrysonilia*. 51, widespread. See mould, bread, Genetics. See Frederick *et al.* (*Mycol.* **61**: 1083, 1970; key), Perkins & Turner (*Exp. Mycol.* **12**: 91, 1988; natural populations), Raju (*Mycol.* **80**: 825, 1988; multipored ascus apices), Perkins (*Genetics* Bethesda **130**: 687, 1992; history as a model organism), Raju (*MR* **96**: 241, 1992; genetic control of sexual cycle), Beatty *et al.* (*MR* **98**: 1309, 1994; genetics), Skupski*et al.* (*Fungal Genetics Biol.* **21**: 153, 1997; phylogeny), Pöggeler (*Curr. Genet.* **36**: 222, 1999; mating type genes), Turner *et al.* (*Fungal Genetics Biol.* **32**: 67, 2001; natural populations, review), Dettman *et al.* (*Evolution* Lancaster, Pa. **57**: 2703, 2003; phylogeny), Dettman *et al.* (*Evolution* Lancaster, Pa. **57**: 2721, 2003; species concepts), Galagan (*Nature* Lond. **422** no. 6934: 859, 2003; genome sequence), Braun *et al.* (*Applied Mycology and Biotechnology* **4**: 295, 2004; genetics review), Dettman & Taylor (*Genetics* Bethesda **168**: 1231, 2004; microsatellites), García *et al.* (*MR* **108**: 1119, 2004; wide concept including *Gelasinospora*), Miller & Huhndorf (*Mol. Phylogen. Evol.* **35**: 60, 2005; phylogeny, wall structure), Dettman *et al.* (*Mycol.* **98**: 436, 2006; *N. discreta* agg.), Jacobson *et al.* (*Mycol.* **98**: 550, 2006; biogeography), Zhang *et al.* (*Mycol.* **98**: 1076, 2006; phylogeny), Wik *et al.* (*BMC Evolutionary Biology* **8**: 109, 2008; mating type genes).

neurotoxin, a toxin which affects the nervous system.

Nevodovsky (Gavril Stepanovich; 1874-1952; Ukraine, later Georgia, Kazakhstan, Russia). Schoolteacher (1897-1903); student, Novo-Aleksandiysky Institute of Agriculture and Forestry, Pulavy, now Poland (1903-1911); Head, Phytopathological Laboratory, Tbilisi Botanic Garden (1911-1913); mycologist, Smelianskaya Mycological-Entomological Station, later the Ukrainian Scientific Research Institute of Sugar Beet (1913-1930); mycologist, Alma Ata Station of Plant Protection (1930-1940); Lecturer, Tomsk Forest Technical School, and mycologist, Tomsk Zonal Flax Experiment Station (1940-1946). A pioneer mycologist in the Caucasus and Central Asia, particularly noted for his compilation of exsiccati. Collections are in the Kazakh State University, Alma Ata (**AA**) [about 17,000 specimens] and the V.I. Komarov Botanical Institute, St Petersburg (**LE**). *Publs.* [Fungal diseases of cultivated and wild useful plants of the Caucasus in 1912. Second year]. *Trudy Tiflisskogo Botanicheskogo Sada* (1914) [in Russian] [*Flora of the Sporing Plants of Kazakhastan. Rust Fungi*] (1956) [published posthumously, in Russian].

Nevrophyllum Pat. (1886) [non *Neurophyllum* Torrey & A. Gray 1840, *Umbelliferae*] ≡ Gomphus Pers.

Newinia Thaung (1973), ? Phakopsoraceae. 3 (on *Bignoniaceae*), Asia; Africa. See Cummins & Hiratsuka (*Illustr. Gen. Rust Fungi* edn 3: 225 pp., 2003; placed in *Uropyxidaceae*) Cf. *Phragmidiella*.

Nia R.T. Moore & Meyers (1961), Niaceae. 3 (marine), widespread (northern hemisphere). See Doguet (*BSMF* **85**: 93, 1969; cultures, development), Doguet (*BSMF* **84**: 343, 1969), Kohlmeyer & Kohlmeyer (*Marine Mycology*, 1979), Leightley & Eaton (*TBMS* **73**: 35, 1979), Jones & Jones (*MR* **97**: 1, 1993), Rosello *et al.* (*MR* **97**: 68, 1993; key), Binder *et al.* (*Mycol.* **93**: 679, 2001; phylogeny).

Niaceae Jülich (1982), Agaricales. 6 gen. (+3 syn.), 56 spp.

Lit.: Kohlmeyer & Kohlmeyer (*Marine Mycology*: 690 pp., 1979), Jones & Jones (*MR* **97**: 1, 1993), Barata *et al.* (*MR* **101**: 687, 1997), Binder *et al.* (*Mycol.* **93**: 679, 2001), Bodensteiner *et al.* (*Mol. Phylogen. Evol.* **33**: 501, 2004), Sivichai & Jones (*Sydowia* **56**: 132, 2004).

Nicholsoniella Kuntze (1891) ≡ Libertiella.

nidose (**nidorose**), having an unpleasant smell.

Nidula V.S. White (1902), Agaricaceae. 4, widespread.

nidulant, lying free in a cavity.

Nidularia Bull. (1791) nom. rej. = Cyathus fide Stalpers (*in litt.*).

Nidularia Fr. (1817) nom. cons., Agaricaceae. 3, widespread. See Cejp & Palmer (*Česká Mykol.* **17**: 125, 1963; key), Baseia & Milanez (*Revista Brasileira de Botânica* **24**: 479, 2001; Brazil), Sarasini (*Gasteromiceti Epigei*: 406 pp., 2005).

Nidulariaceae Dumort. (1822) = Agaricaceae.

Nidulariales = Agaricales.

Nidulariopsis Greis (1935), Geastraceae. 2, Europe; N. America. See Zeller (*Mycol.* **40**: 639, 1948), Miller & Miller (*Gasteromycetes. Morphological and Development Features with Keys to the Orders, Families, and Genera*: 157 pp., 1988).

Nidulispora Nawawi & Kuthub. (1990), anamorphic *Pezizomycotina*, Hso.1bH.1. 1 (aquatic), Malaysia. See Nawawi & Kuthubutheen (*Mycotaxon* **36**: 329, 1990), Ho *et al.* (*Mycol.* **92**: 582, 2000).

Niebla Rundel & Bowler (1978), Ramalinaceae (L). 1, California; Mexico. Perhaps synonymous with *Ramalina*. See Bowler *et al.* (*Phytologia* **77**: 23, 1994), Marsh & Nash (*Phytologia* **76**: 458, 1994; Mexico), Riefner *et al.* (*Mycotaxon* **54**: 397, 1995; USA), Nash *et al.* (*Lichen Flora of the Greater Sonoran Desert Region* **2**, 2004), Miądlikowska *et al.* (*Mycol.* **98**: 1088, 2006; phylogeny).

NIEC, see RIEC.

Nielsenia Syd. (1921) = Uromyces fide Dietel (*Nat. Pflanzenfam.* **6**, 1928).

Niessl (von Meyendorf, Gustav; 1839-1919; Italy, later Austria, Moravia). Born in Verona; Geometrist, Vienna (1857-1865); Professor of Geodesy and Astronomy, University of Brno (1865-1906); retired to Vienna (1906 onwards). A major contributor to early understanding of the mycotas of central Europe. Most collections are in **M**. *Publs.* Vorarbeiten zu einer Kryptogamenflora von Mähren und Oest Schlesiens. II. Pilze und Myxomyceten. *Verhandlungen des Naturforschender Vereins in Brunn* (1865); Über Myxomyceten. *Verhandlungen des Naturforschender Vereins in Brunn* (1868); Notizen über neue und kritische Pyrenomyceten. *Verhandlungen des Naturforschender Vereins in Brunn* (1875). *Biogs, obits etc.* Grummann (1974: 436); Nožička (*Česká Mykologie* **18**: 185, 1964) [portrait].

Niesslella Höhn. (1919) = Micropeziza fide Nannfeldt

(*Bot. Notiser* **129**: 323, 1976).

Niesslella Speg. (1880) ? = Sporormiella fide Hawksworth *et al.* (*Dictionary of the Fungi* edn 8, 1995).

Niesslia Auersw. (1869), Niessliaceae. Anamorph *Monocillium*. 23, widespread. See Gams (*Cephalosporium-artige Schimmelpilze*, 1971), Samuels & Barr (*CJB* **75**: 2165, 1998), Tretiach (*Nova Hedwigia* **75**: 357, 2002; on lichens), Castlebury *et al.* (*MR* **108**: 864, 2004; phylogeny), Zhang *et al.* (*Mycol.* **98**: 1076, 2006; phylogeny).

Niessliaceae Kirschst. (1939), Hypocreales. 12 gen. (+ 4 syn.), 51 spp.

Lit.: Barr (*Mycotaxon* **39**: 43, 1990), Scheuer (*MR* **97**: 543, 1993), Crous *et al.* (*Mycol.* **88**: 789, 1996), Gams & Turhan (*Mycotaxon* **59**: 343, 1996), Samuels & Barr (*CJB* **75**: 2165, 1997), Rossman *et al.* (*Stud. Mycol.* **42**: 248 pp., 1999), Castlebury *et al.* (*MR* **108**: 864, 2004).

nietsuki, a product of failure in the drying of *Lentinula edodes* basidiomata (Kawai & Kawai, *Rep. Tottori mycol. Inst.* **1**: 29, 1961).

nigeran, see mycodextran.

Nigredo (Pers.) Roussel (1806) nom. rej. prop. ≡ Uredo.

Nigrococcus E.K. Novák & Zsolt (1961) [non *Nigrococcus* Castell. & Chalm. 1919, *Bacteria*] = Phaeococcus fide Hawksworth *et al.* (*Dictionary of the Fungi* edn 8, 1995).

Nigrocornus Ryley & Langdon (2003), Clavicipitaceae. 1, widespread (old world). See Ryley (*Mycology Series* **19**: 266, 2003), Ryley (*Mycotaxon* **95**: 97, 2006).

Nigrocupula Sawada (1944) = Graphiola fide Mordue (*in litt.*).

Nigrodiplodia Kravtzev (1955) nom. dub., anamorphic *Pezizomycotina*. See Sutton (*Mycol. Pap.* **141**, 1977).

Nigrofomes Murrill (1904), Polyporaceae. 1, widespread (pantropical). See Corner (*Beih. Nova Hedwigia* **86**: 124, 1987), Roberts & Ryvarden (*Kew Bull.* **61**: 55, 2006; Cameroon).

Nigrofomitaceae Jülich (1982) = Polyporaceae.

Nigrohydnum Ryvarden (1987), Polyporales. 1, S. America. See Ryvarden (*Mycotaxon* **28**: 531, 1987).

Nigrolentilocus R.F. Castañeda & Heredia (2001), anamorphic *Melanomma*, Hso.?.?. 5, widespread. See Castañeda Ruiz *et al.* (*Cryptog. Mycol.* **22**: 13, 2001).

Nigromacula Etayo (2002), anamorphic *Pezizomycotina*, H?.?.?. 2, Colombia; British Isles. See Etayo (*Biblthca Lichenol.* **84**: 87, 2002).

Nigromammilla K.D. Hyde & J. Fröhl. (2003), Sordariomycetes. 1, Hong Kong. See Hyde & Fröhlich (*Cryptog. Mycol.* **24**: 17, 2003).

Nigropogon Coker & Couch (1928) = Entoloma fide Kuyper (*in litt.*).

Nigroporus Murrill (1905), Polyporaceae. 3, widespread. See Pegler (*The polypores [Bull. BMS Suppl.]*, 1973), Ryvarden & Iturriaga (*Mycol.* **95**: 1066, 2003; Venezuela).

Nigropuncta D. Hawksw. (1981), anamorphic *Pezizomycotina*, Cpd.0eP.23. 1 (on lichens), Austria. See Hawksworth (*Bull. Br. Mus. nat. Hist.* Bot. **9**: 46, 1981).

Nigrosabulum Malloch & Cain (1970), Hypocreales. Anamorph *Acremonium*-like. 1 (coprophilous), widespread. See Suh & Blackwell (*Mycol.* **91**: 836, 1999; phylogeny), Rossman *et al.* (*Mycol.* **93**: 100, 2001; phylogeny).

Nigrosphaeria N.L. Gardner (1905) ? = Melanospora Corda fide von Arx & Müller (*Beitr. Kryptfl. Schweiz* **11** no. 1, 1954).

Nigrospora Zimm. (1902), anamorphic *Khuskia*, Hso.0eP.1. 6, widespread. *N. oryzae* on maize (*Zea*) and other hosts. See Standen (*Iowa St. Coll. J. Sci.* **17**: 263, 1943), Minter (*Proc. Indian Acad. Sci.* Pl. Sci. **94**: 281, 1985).

Nimbomollisia Nannf. (1983) = Niptera fide Baral (*SA* **13**: 113, 1994).

Nimbospora J. Koch (1982), Halosphaeriaceae. 3 (marine), pantropical. See Hyde & Jones (*CJB* **63**: 611, 1985), Read *et al.* (*Phil. Trans. Roy. Soc.* London ser. B **339**: 483, 1993; ultrastr.), Jones (*CJB* **74** Suppl. 1: S790, 1995; ultrastr.), Spatafora *et al.* (*Am. J. Bot.* **85**: 1569, 1998; phylogenetic analysis), Campbell *et al.* (*Mycol.* **95**: 530, 2003; phylogeny), Zhang *et al.* (*Mycol.* **98**: 1076, 2006; phylogeny).

nimbospore, a spore having a gelatinous, apparently many-layered wall, e.g. *Histoplasma capsulatum* (Nielsen & Evans, *J. Bact.* **68**: 261, 1954).

Nimbya E.G. Simmons (1989), anamorphic *Macrospora*, Hso.≡ eP.26. 12, widespread. See Simmons (*Sydowia* **41**: 316, 1989), Harada *et al.* (*Ann. phytopath. Soc. Japan* **58**: 766, 1992; on *Eleocharis*), Johnson *et al.* (*Mycotaxon* **84**: 413, 2002; on *Scirpus*), Pryor & Bigelow (*Mycol.* **95**: 1141, 2003; phylogeny), Zhao & Zhang (*Fungal Diversity* **19**: 201, 2005; China).

Nimisia Kärnefelt & A. Thell (1993) = Himantormia fide Kärnefelt & Thell (*Lichenologist* **25**: 370, 1993), Thell *et al.* (*Mycol. Progr.* **1**: 335, 2002; phylogeny), Fryday (*Lichenologist* **37**: 313, 2005; nomencl.), Thell *et al.* (*Biblthca Lichenol.* **95**: 531, 2007; phylogeny).

Nimisiostella Calat., Barreno & O.E. Erikss. (1997), Lecanorales (L). 1, Spain. See Calatayud *et al.* (*SA* **15**: 112, 1997).

Niopsora A. Massal. (1861) = Caloplaca fide Hawksworth *et al.* (*Dictionary of the Fungi* edn 8, 1995).

Niorma A. Massal. (1861) = Teloschistes fide Zahlbruckner (*Catalogus Lichenum Universalis* **7**, 1931).

Niospora Kremp. (1869) ≡ Niopsora.

Nipholepis Syd. (1935), Arthoniales. 1, S. America.

Nipicola K.D. Hyde (1992), Sordariomycetes. 2, Brunei; Hong Kong. See Eriksson & Hawksworth (*SA* **13**: 198, 1994; posn), Hyde (*Sydowia* **46**: 257, 1995), Hyde & Taylor (*Nova Hedwigia* **63**: 417, 1996).

Niptera Fr. (1849), Helotiales. Anamorph *Cystodendron*. 5, Europe. See Dennis (*Kew Bull.* **26**: 439, 1972), Nannfeldt (*Sydowia* **38**: 194, 1986), Nauta & Spooner (*Mycologist* **14**: 65, 2000; UK).

Nipterella Starbäck ex Dennis (1962), ? Helotiaceae. 3, Europe; N. America. See Starbäck ex Dennis (*Persoonia* **2**: 189, 1962).

nitid (adj. **nitidous**), smooth and clear; lustrous.

nitrophilous, having a preference for habitats rich in nitrogen; chionophilous; **nitrophobous**, having a preference for habitats poor in nitrogen.

Nitschkia G.H. Otth ex P. Karst. (1873), Nitschkiaceae. *c.* 22 (from wood), widespread. See Nannfeldt (*Svensk bot. Tidskr.* **65**: 49, 1971), Subramanian & Sekar (*Kavaka* **18**: 19, 1993), Bianchinotti (*Mycol.* **96**: 911, 2004; Argentina), Huhndorf *et al.* (*MR* **108**: 1384, 2004; phylogeny), Schoch *et al.* (*MR* **111**: 154, 2007; phylogeny).

Nitschkiaceae Nannf. (1932), Coronophorales. 16 gen. (+ 27 syn.), 59 spp.

Lit.: Sivanesan (*TBMS* **62**: 40, 1974), Nannfeldt (*Svensk bot. Tidskr.* **69**: 49, 1975), Nannfeldt (*Svensk bot. Tidskr.* **69**: 289, 1975), Corlett & Krug (*CJB* **62**: 2561, 1984), Jensen (*Mycol.* **77**: 688, 1985), Eriksson & Santesson (*Mycotaxon* **25**: 569, 1986), Subramanian & Sekar (*Kavaka* **18**: 19, 1990), Hyde (*Nova Hedwigia* **61**: 141, 1995), Linde & Botha (*S. Afr. J. Bot.* **63**: 66, 1997), Petersen (*Bot. Mar.* **40**: 71, 1997), Navarro-Rosinés *et al.* (*Bull. Soc. linn. Provence* **50**: 233, 1999), Bianchinotti (*Mycol.* **96**: 911, 2004), Huhndorf *et al.* (*Mycol.* **96**: 368, 2004), Huhndorf *et al.* (*MR* **108**: 1384, 2004).

Nitschkiopsis Nannf. & R. Sant. (1975) = Niesslia fide Samuels & Barr (*CJB* **75**: 2165, 1998).

Nivatogastrium Singer & A.H. Sm. (1959), Strophariaceae. 4, N. America. Basidioma gasteroid. See Singer & Smith (*Brittonia* **11**: 224, 1959).

Niveostoma Svrček (1988) = Solenopezia fide Raitviir *et al.* (*Sydowia* **43**: 219, 1991).

nm, nanometer; one billionth of a metre.

noble rot, a condition in which the mould *Botrytis* grows on overripe grapes. A rich, sweet wine is made in small quantities from the affected grapes (Sauternes, Trockenbeerenlauslese, Botrytis-wine).

Nocardia Trevis. (1889), Actinobacteria. q.v.

Nocardioides Prauser (1976), Actinobacteria. q.v.

Nocardiopsis (Brocq-Rouss.) J. Mey. (1976), Actinobacteria. q.v.

Nochascypha Agerer (1983), Marasmiaceae. 6, S. America. See Agerer (*Mitt. bot. StSamml.* Münch. **19**: 262, 1983), Bodensteiner & Agerer (*Mycol. Progr.* **2**: 297, 2003).

node cell, see hyphopodium.

Nodobryoria Common & Brodo (1995), Parmeliaceae (L). 3, N. America; Greenland. See Common & Brodo (*Bryologist* **98**: 198, 1995).

nodose-septum, see clamp-connexion.

Nodotia Hjortstam (1987) = Hypochnicium fide Hjortstam & Larsson (*Windahlia* **21**, 1994).

nodular bodies (of dermatophytes), rounded bodies made up of massed hyphae.

Nodularia Peck (1872) [non *Nodularia* Link ex Lyngb. 1819, *Algae*] = Aleurodiscus fide Rogers & Jackson (*Farlowia* **1**: 263, 1943).

Nodulisporium Preuss (1849), anamorphic *Xylaria, Annulohypoxylon, Biscogniauxia, Daldinia, Pulveria, Thamnomyces*, Hso.0eP.10. *c.* 200, widespread. Most anamorphic taxa are unnamed. See Deighton (*TBMS* **85**: 391, 1985), Glawe & Rogers (*CJB* **64**: 1493, 1986), Polishook *et al.* (*Mycol.* **93**: 1125, 2001; biogeography, chemistry).

nodulose (of spores), having broad-based, blunt, wart-like excrescences.

Nodulosphaeria Rabenh. (1858) nom. cons., Phaeosphaeriaceae. *c.* 48, widespread. See Holm (*Svensk bot. Tidskr.* **55**: 63, 1961), Crivelli (*Über die heterogene Ascomycetengattung Pleospora Rbh.*, 1983), Barr & Holm (*Taxon* **33**: 109, 1984; nomencl.), Shoemaker (*CJB* **62**: 2730, 1985; key 23 spp.), Shoemaker & Babcock (*CJB* **65**: 1921, 1987; key 5-septate spp.), Ahn & Shearer (*CJB* **76**: 258, 1998; on *Ranunculaceae*).

Nodulospora Marvanová & Bärl. (2000), anamorphic *Basidiomycota*. 1 (freshwater), Canada. See Marvanová & Bärlocher (*Mycotaxon* **75**: 416, 2000).

nodum (pl. **noda**) (in phytosociology), particular well-defined plant communities. See also Phytosociology.

Nohea Kohlm. & Volkm.-Kohlm. (1991), Halosphaeriaceae. 1 (marine), Hawaii. See Kohlmeyer & Volkmann-Kohlmeyer (*SA* **10**: 121, 1991), Spatafora *et al.* (*Am. J. Bot.* **85**: 1569, 1998; phylogeny), Campbell *et al.* (*Mycol.* **95**: 530, 2003; phylogeny), Zhang *et al.* (*Mycol.* **98**: 1076, 2006; phylogeny).

Nolanea (Fr.) P. Kumm. (1871) = Entoloma fide Kuyper (*in litt.*).

Nolleria C.B. Beard, J.F. Butler & J.J. Becnel (1990), Microsporidia. 1.

nomen (Latin), name; **- ambiguum**, one having different senses; **- anamorphosis**, see States of fungi; **- confusum**, one of a taxonomic group based on two or more different elements; **- conservandum**, one authorized for used by a decision of an International Botanical Congress (see Nomenclature); **- - propositum**, one put up for conservation; **- dubium**, one of uncertain sense; **- holomorphosis**, see States of fungi; **- illegitimum (nom. illegit.)**, a validly published name contravening particular Articles in the *Code* (see Nomenclature); **- invalidum (nom. inval.)**, one not validly published (see Nomenclature); **- monstrositatis**, one based on an abnormality; **- novum**, a new name; a replacement [nom. nov. should replace the author's name only at the first publication]; **- nudum**, one for a taxon having no description; **- provisorium**, one proposed provisionally; **- rejiciendum**, one rejected by a Botanical Congress. A generic name may be a nomen ambiguum (etc.), but a binomial under such a name may be without ambiguity. **- species** (of bacteria), a type species.

Nomenclature. The allocation of scientific names to units which a systematist considers merit formal recognition. The nomenclature of fungi (including chromistans, lichen-forming fungi, slime-moulds and yeasts) is governed by the *International code of botanical nomenclature* (latest edn, McNeill *et al.* (Eds) [*Regnum veg.* **138**], 2006), as adopted by each International Botanical Congress. The tradition of treating fungi, chromistans and slime-moulds under the 'Botanical' Code, although none have been classified in the Kingdom *Plantae* for many years has been criticized by some and supported by others, but undoubtedly exerts a strong influence on mycology. Any proposals to change the Code are published in *Taxon*, debated, and voted on at the Nomenclature Section of such a Congress. As rules can change in different editions, the latest should always be consulted. The Congress appoints the Committee for Fungi (CF) which advises on action to be taken on proposals concerned with fungi. The Code aims to provide a stable method of naming taxonomic groups, avoiding and rejecting use of names which may cause error or ambiguity or throw science into confusion.

The Code comprises six **Principles**, 62 **Articles** (which are mandatory), **Recommendations** (non-mandatory but good practice) and various Appendices. The Code is designed to allow any taxon to have as many correct names as there may be opinions as to its classification. When the taxonomic decisions have been taken, the Code provides the rules to determine the name that should be applied; each taxon in a given position and rank can have only one nomenclaturally correct name.

In determining the correct name for a taxon, five steps must be followed in what is in effect a nomenclatural filter.

Effective publication (A): To be effectively published names must be in printed matter (journals, books) distributed to the public or at least botanical institutions through sale, exchange, or gift. See Nicolson (*Taxon* **29**: 485, 1980) for categories and special provisions.

Valid publication (B): To be validly published, names of newly described taxa must also simultaneously fulfil the following requirements: (1) have a correct form; (2) have a description or diagnosis (in Latin (q.v.) after 1 Jan. 1935); (3) be accepted by the author and comply with any relevant special provisions elsewhere in the Code; (4) have a clear indication of rank (after 1 Jan. 1953); and (5) indicate the type (after 1 Jan. 1958) and its place of conservation (after 1 Jan. 1990). Replacement names and combinations have to give full bibliographic details of the place of publication of and cite the replaced name or basionym (after 1 Jan. 1953). Names which are not validly published (**nom. inval.**) need not be considered further. Many are knowingly included in this *Dictionary* as an alert so that future workers can avoid them and also as they can be encountered in the literature. The inclusion of a name in the *Dictionary* does not therefore automatically mean that the name is validly published and available for use.

Typification (C): The linking of each name to a nomenclatural type is the keystone of stability in the application of names. All ranks from fam. downwards are ultimately based on a single collection; e.g. *Erysiphaceae* on *Erysiphe* on *E. polygoni* on a single collection. A holotype is required (see type) but where none exists, the order of priority is isotype - lectotype - neotype. Should the type not be critically identifiable, an interpretative type (epitype) can be designated. An epitype may also be used to change the status of names of pleomorphic fungi that formerly applied to the anamorph only to be usable for all morphs (i.e. the holomorph). See 'type' for definitions of kinds of types. Type specimens must be dried specimens, specimens in a liquid preservative, dried cultures preserved in a fungal dried reference collection, cultures preserved in a metabolically inactive state (e.g. freeze-dried, in liquid nitrogen) or microscopic preparations; actively growing cultures are not permitted. If a specimen cannot be preserved an illustration or description can suffice. If a type collection is mixed, one part must be selected as lectotype.

Legitimacy (D): Validly published names not in accordance with certain provisions of the Code are **illegitimate (nom. illegit.)** and to be rejected, i.e. (1) **superfluous**, i.e. including the type of a name that should have been used; or (2) **homonyms**, i.e. spelled like a previously validly published name.

Priority (E): Priority of publication determines the correct name for a taxon. The **correct name** (i.e. name in accordance with the Code) of a species is the combination of the earliest available legitimate epithet in the same rank with the correct generic name. This may change if the generic placement or rank is altered; i.e. a species can have more than one correct name according to different taxonomies. The principle of priority can only be set aside through **conservation**, or where the rules on pleomorphic fungi apply (see below). Names for a taxon other than the correct one are **synonyms**; **heterotypic synonyms** (also known as facultative or taxonomic synonyms) are names based on different nomenclatural types, and **homotypic synonyms** (obligate or nomenclatural synonyms) are names based on the same type. Where a name change is considered desirable as a result of to former misinterpretation or competition between unfamiliar (but priorable) and familiar names, conservation proposals must be published in *Taxon*, receive a positive vote from the Special Committee for Fungi and be ratified at the next nomenclatural Congress.

In the code, the starting point for nomenclature is Linnaeus' *Species plantorum* (1 May 1753), with later exceptions for some groups. Prior to 1981, different groups of fungi had dates later than 1753: 31 Dec. 1801 (Persoon, *Synopsis methodica fungorum*) for *Gasteromycetes* (s.l.), *Pucciniales* and *Ustilaginales*; and 1 Jan. 1821 (Fries, *Systema mycologicum* **1** (1)) for the remaining fungi (other than slime-moulds and lichen-forming species). Reasons for the change in starting point dates are discussed in Demoulin *et al.* (*Taxon* **30**: 52, 1981). Names used in the previous starting point books of Fries and Persoon are **sanctioned** and not affected by, and take priority over, homonymous and synonymous names published earlier (listed by Gams, *Mycotaxon* **19**: 219, 1984). Names given to lichens are ruled as applying only to their fungal components, i.e. to the lichen-forming fungus. Algae in lichens have separate names, and the composite 'lichen' strictly has no name. Many names ending in *-myces* introduced for lichen mycobionts by Thomas (*Beitr. Kryptog.-Fl. Schweiz.* **9** (1), 1939) and Ciferri & Tomaselli (*Atti Ist. Bot. Univ. Lab. crittog. Pavia* ser. V **10**, 1953) are thus superfluous (see Lücking & Hawksworth, *Taxon* **56**: 1274, 2007).

Pleomorphic fungi:

The Code permits the different states of fungi (q.v.), other than lichen-forming fungi, with pleomorphic life-cycles to be given separate names; it does not apply to fungi in which pleomorphism is not confirmed (Reynolds & Taylor, *Taxon* **41**: 98, 1992). Whether a name can be considered as that of an anamorph or teleomorph depends on its original description and nomenclatural type, *not* the genus in which it was placed; if a teleomorph is present, the name automatically refers to that morph even if the anamorph is also evident (e.g. *Penicillium brefeldianum* B.O. Dodge (1933) included both states in its type and description and is thus a teleomorph name that has correctly been combined as *Eupenicillium brefeldianum* (B.O. Dodge) Stolk & D.B. Scott (1967); the name of the anamorph is *P. dodgei* Pitt (1980), the type and description of which are only of the anamorph). The correct name of a holomorph is that of its teleomorph.

Name changing:

In order to avoid name changes for nomenclatural reasons through a strict application of the Code, special provisions have been made. Names of families, genera and species can be **conserved** (**nomina conservanda, nom. cons.**) against names threatening their retention, and names in any rank can be **rejected** (**nomina rejicienda, nom. rej.**) if their use would cause disadvantageous nomenclatural change. If a name has been widely used for a taxon in a sense conflicting with its type, it is not to be used unless and until a proposal to deal with it has been submitted and the name rejected. If a particular work would cause instability if it were used as a source of valid names, it can be proposed for inclusion in a list of **suppressed works**. Proposals for conservation, rejection, or suppression have to be published in *Taxon* and voted on by appropriate Committees. Appendices to the Code list names and works in these categories.

Lists of Names in Current Use (NCUs) were developed through a Committee appointed by the 1994 Congress; those published include those for names of fungal families (David, *Regnum Veg.* **126**: 71, 1993; NCU-1), genera (Greuter *et al.*, *Regnum Veg.* **129**, 1993, NCU-3), and species in *Cladoniaceae* (Ahti, *Regnum Veg.* **128**: 58, 1983) and *Trichocomaceae* (Pitt & Samson, *Regnum Veg.* **128**: 13, 1993; subject of a special Nomenclatural Section Resolution). The 1993 Congress, noting improvements being made in the systems of nomenclature to promote stability, urged taxonomists 'to avoid displacing well established names for purely nomenclatural reasons' (see Greuter & Nicolson, *Taxon* **42**: 925, 1993; Hawksworth, *SA* **12**: 1, 1993).

Authorities:

The authority of a name, usually abbreviated (see Authors' names) is cited after a name for precision and is intended to be a much abbreviated bibliographic reference. If a species is move from one genus to another, or has its rank changed, the original author, i.e. of the basionym (q.v.), is given in brackets and the one making the change cited outside the brackets (e.g. *Fusarium poae* (Peck) Wollenw., based on the earlier *Sporotrichum poae* Peck). Where a name not validly published by one author is taken up and validated by a second, '**ex**' is used to link the names of the two authors. In the case of names sanctioned by Persoon or Fries (see above), a colon (**:**) can be used where it is considered approporiate to indicate the special status of that name or epithet (see Korf, *Mycol.* **74**: 250, 1982; *Mycotaxon* **14**: 476, 1982); 'ex', 'per' or '[ ]' had previously been used for devalidated names taken up again after the former later starting dates; '**in**' was sometimes formerly used where one author contributed an account of a taxon to the work of another but this was ruled to be part of the bibliographic reference by the 1994 Code and does not form a part of author citations. The Code now emphasizes that the authority is not part of the name itself and there is no reason to use it in works other than formal systematic treatments, unless homonyms are involved.

Registration:

The time and effort needed to maintain formal nomenclators of fungal names (i.e. the *Index of Fungi*, see under *Literature*, and its web-based counterpart *Index Fungorum*, see Internet) are considerable, and scientific names can currently be effectively published in almost any non-ephemeral publicly available printed product. Some of these may only become apparent to the taxonomic community years after publication, potentially causing great confusion and nomenclatural instability. There have been several initiatives over the past ten years or so, developing partly from the Names in Current Use scheme, to make registration of fungal (and plant) names mandatory so that priority of publication is determined by the date of receipt of the information or publication by a Registration Authority. A similar scheme is already in use for the Bacteriological *Code*. The proposals received a very mixed response from the nomenclatural community, with concerns expressed about the potential for nomenclatural censorship and over the possible continuing use of non-registered names in contravention of the Code. References in the Tokyo *Code* (1994) to registration as a future requirement were summarily removed by the St Louis Congress (2000). Some pilot schemes have nevertheless taken place. Under defined conditions, unique IndexFungorum-compatible identifier numbers for new scientific fungal names can now be obtained via the internet (at the time of writing, the only 'registration' website in operation is **Mycobank**, www.mycobank.org). Although the numbers themselves currently have no formal nomenclatural status, the names 'registered' in this manner are then passed automatically to *Index Fungorum*. Such a system ('registration' websites issuing numbers on behalf of a single internationally recognized nomenclator) may suit the needs of mycology, but ultimately innovations in Internet searching may make all of these initiatives superfluous, especially as the average age continues to increase of those who have in-depth knowledge of nomenclature, and the ability to write Latin decriptions declines.

Other Codes:

Separate Codes exist for zoology, bacteriology, cultivated plants, and viruses. These are independent of the Botanical Code, but there has been extensive discussion as to how harmonization between the Codes can be improved (Hawksworth *et al.*, 1994). See also ambiregnal organisms. A draft International Code of Bionomenclature (the *BioCode*) was planned to apply to names in all groups introduced after (provisionally) 1 Jan. 2000 (see Hawksworth, *Taxon* **44**: 447, 1995) but its proponents have to date been unable to gain sufficient support to see it into reality. A further challenge to the *status quo* has been development of a draft *PhyloCode* (http://www.ohiou.edu/phylocode/) which proposes a formal nomenclature system for nodes and branches of phylogenies. There is effectively an almost infinite number of ranks in a phylogenetic classification if these are measured by evolutionary age, and it is often difficult to map these onto a traditional system. Molecular tools will increasingly dominate research into systematics, diagnostics etc., and it is important to embrace new technologies while building on rather than rejecting current knowledge.

Lit.: Davis & Heywood (*Principles of angiosperm taxonomy*, 1963), McNeill *et al.* (*Regnum Veg.* **146**, 1994; Vienna Code), Hawksworth (*Mycologist's handbook*, 1974 [incl. relevant parts of 1972 Code with mycological examples, glossary]; (Ed.), *Improving the stability of names; needs and options*, [*Regnum Veg.* **123**], 1991; *Bot. J. Linn. Soc.* **109**: 543, 1992; *A draft glossary of terms used in bionomenclature*, [IUBS Monogr. **8**], 1994 [1,175 entries]), Hawksworth *et al.* (*Towards a harmonized bionomenclature for life on Earth*, [*Biol. Internat., Sp. Issue* **30**, 1994), Jeffrey (*Biological nomenclature*, edn 3, 1989; general survey all Codes), Weresub (*Sydowia, Beih.* **8**: 416, 1979; history, problems).

Nomuraea Maubl. (1903), anamorphic *Clavicipitaceae*, Hso.0eH.15. 2, widespread. See Samson (*Stud. Mycol.* **6**, 1974), Tzean *et al.* (*Atlas of Entomopathogenic Fungi from Taiwan*, 1997), Boucias *et al.* (*Biological Control* **19**: 124, 2000; genetic variation), Han *et al.* (*Lett. Appl. Microbiol.* **34**: 376, 2002; phylogeny), Suwannakut *et al.* (*J. Invert. Path.* **90**: 169, 2005; on noctuids), Sung *et al.* (*Stud. Mycol.* **57**, 2007; phylogeny).

nonfissitunicate (of asci), ones in which discharge does not involve a separation of the wall layers; see ascus.
non-target organisms, organisms found with or near those being treated with chemical or biological control agents.
Norlevinea Vávra (1984), Microsporidia. 1.
Normal saline solution, sodium chloride (NaCl), 8.5 g, water, 1000.0 ml.
Normandina Nyl. (1855), Verrucariaceae (L). 7, widespread. See also *Lauderlindsaya*. See Henssen *in* Brown *et al.* (Eds) (*Lichenology: progress and problems*: 107, 1976), Tschermak-Woess (*Nova Hedwigia* **35**: 63, 1981; photobiont), Aptroot (*Willdenowia* **21**: 263, 1991), Eriksson & Hawksworth (*SA* **11**: 67, 1992), Aptroot *et al.* (*Biblthca Lichenol.* **97**, 2008; Costa Rica).
Normandinomyces Cif. & Tomas. (1953) = Normandina fide Aptroot (*Lichenologist* **30**: 501, 1998).
Normkultur (of cultures or states), one in which all the forms characteristic of a fungus are present and of good development; **Ankultur**, one of poor development; **Jungkultur**, one when young; **Hochkultur**, one when mature; **Altkultur**, one over-mature; **Abkultur**, a degenerate culture. These names were first used by Appel & Wollenweber (*Ark. biol. BundAnst. Land-u. Forstw.* **8**(1): 209, 1910) for *Fusarium*.
Norrlinia Theiss. & Syd. (1918), Verrucariales. 1 (on lichens), Europe. See Hawksworth (*TBMS* **74**: 363, 1980), Santesson (*Nordic Jl Bot.* **9**: 97, 1989), Eriksson & Hawksworth (*SA* **8**: 109, 1990), Lumbsch *et al.* (*Mol. Phylogen. Evol.* **31**: 822, 2004; phylogeny), Geiser *et al.* (*Mycol.* **98**: 1053, 2006; phylogeny).
Norrlinia Vain. (1921) ≡ Neonorrlinia.
Nosema Nägeli (1857), Microsporidia. 81.
Nosemoides Vinckier (1975), Microsporidia. 3. See Vinckier (*J. Protozool.* **22**: 170, 1975).
Nosophloea Fr. (1849), anamorphic *Pezizomycotina*. 3, Europe.
Nostoclavus Paulet (1791) nom. dub., Agaricomycetes. 'basidiomycetes', inc. sed.
Nostocotheca Starbäck (1899) = Molleriella fide von Arx (*Persoonia* **2**: 421, 1963).
Notarisiella (Sacc.) Clem. & Shear (1931) ≡ Pseudonectria fide Rossman *et al.* (*Stud. Mycol.* **42**: 248 pp., 1999).
notate (of surfaces), marked by straight or curved lines.
Nothadelphia Degawa & W. Gams (2004), Zygomycota. 1, Japan. See Degawa & Gams (*Stud. Mycol.* **50**: 569, 2004).
Nothocastoreum G.W. Beaton (1984), Mesophelliaceae. 1, Australia. See Beaton (*TBMS* **82**: 666, 1984).
Nothoclavulina Singer (1970), anamorphic *Arthrosporella*, Hsy.0eH.38. 1, Argentina. See Singer (*Fl. Neotrop.* Monogr. **3**: 18, 1970).
Nothocorticium Gresl. & Rajchenb. (1999), Corticiaceae. 1, Argentina. See Greslebin & Rajchenberg (*Mycotaxon* **70**: 372, 1999).
Nothodiscus Sacc. (1917) nom. dub., Fungi. See Petrak (*Annls mycol.* **38**: 253, 1940).
Nothojafnea Rifai (1968), Pyronemataceae. 2, Australia; S. America. See Gamundí *et al.* (*Mycologist* **13**: 84, 1999; col. pls).
Notholepiota E. Horak (1971), Agaricaceae. 1, New Zealand. Basidioma gasteroid. See Horak (*N.Z. Jl Bot.* **9**: 463, 1971), Vellinga (*MR* **108**: 354, 2004; posn.).
Nothomitra Maas Geest. (1964), Helotiales. 3, widespread. Position uncertain, formerly placed in the *Geoglossaceae*. See Zhuang & Wang (*Mycotaxon* **63**: 307, 1997), Wang *et al.* (*Mycol.* **94**: 641, 2002).
Nothopanus Singer (1944), Marasmiaceae. 2, widespread (tropical). See Moncalvo *et al.* (*Syst. Biol.* **49**: 278, 2000; phylogeny), Mossebo *et al.* (*Mycotaxon* **76**: 267, 2000; Cameroon).
Nothopatella Sacc. (1895) = Botryodiplodia fide Petrak & Sydow (*Beih. Rep. spec. nov. regn. veg.* **42**, 1927).
Nothophacidium J. Reid & Cain (1962), Helotiales. 1, N. America. *N. phyllophilum* (snow blight of conifers). See Smerlis (*CJB* **44**: 563, 1966).
Nothopodospora Mirza (1963) nom. inval. = Arnium fide Lundqvist (*Symb. bot. upsal.* **20** no. 1, 1972).
Nothoporpidia Hertel (1984) = Lecidea fide Hertel (*Mitt. bot. StSamml., München* **23**: 321, 1987).
Nothoravenelia Dietel (1910), Phakopsoraceae. 2 (on *Burseraceae*, *Euphorbiaceae*), Japan; Russian far east; Africa. See Thirumalachar (*Sydowia* **5**: 23, 1951).
Nothorhytisma Minter, P.F. Cannon, A.I. Romero & Peredo (1998), Rhytismataceae. 1, S. America (temperate). See Minter *et al.* (*SA* **16**: 27, 1998).
Nothospora Peyronel (1913), anamorphic *Pezizomycotina*, Hso.0eH.?. 1, Italy.
Nothostrasseria Nag Raj (1983), anamorphic *Teratosphaeria*, Cpd.1eP.15. 1, Australia. See Nag Raj (*CJB* **61**: 23, 1983), Crous *et al.* (*Stud. Mycol.* **58**: 1, 2007; teleomorph).
Nothostroma Clem. (1909) = Tomasellia fide Hawksworth *et al.* (*Dictionary of the Fungi* edn 8, 1995).
Notocladonia S. Hammer (2003), Cladoniaceae (L). 2, Australasia. See Hammer (*Bryologist* **106**: 162, 2003).
Notolecidea Hertel (1984), Lecanorales (L). 1, Antarctica. See Hertel (*Nova Hedwigia* Beih. **79**: 440, 1984), Buschbom & Mueller (*Mol. Phylogen. Evol.* **32**: 66, 2004; phylogeny).
Notothyrites Cookson (1947), Fossil Fungi, Microthyriaceae. 12 (Cretaceous, Tertiary), widespread. See Eriksson (*Ann. bot. fenn.* **15**: 122, 1978).
NRRL, Agricultural Research Service Culture Collection (Peoria, Ill, USA); founded 1941; a part of the United States Department of Agriculture's (USDA) Agricultural Research Service; see Kurtzman (*Enzyme Microbiol. Tech.* **8**: 328, 1986).
nubilated, cloudy and semi-opaque as viewed by transmitted light.
nuclear cap (of *Blastocladiaceae*), a body at one side of the nucleus of a zoospore or gamete.
Nuclear division. See Beccard & Pfeffer (*Protoplasma***174**: 62, 1993; status in arbuscular mycorrhizal fungi), Gladfelter (*Current opinion in microbiology* **9**: 547, 2006; asynchronous mitosis in multinucleated fungal hyphae), Heath (Ed.) (*Nuclear division in fungi*, 1978; review), Plamann (*J. Genetics* **75**: 351, 1996; review); **- status**, of fungal cells (fide Jinks & Simchen, *Nature* **210**: 778, 1966) may be considered (1) numerically: mono-, di-, multi-karyon (= uni-, bi-, multinucleate) and (2) by gene content: homokaryon or heterokaryon according as to whether the nuclei are genetically identical or not. See Chromosome numbers.
Nucleophaga P.A. Dang. (1895), ? Olpidiaceae. 2 (on *Amoeba*), France. See Karling (*Bull. Torrey bot. Club* **99**: 223, 1972).

Nucleospora R.P. Hedrick, J.M. Groff & D.V. Baxa (1991), Microsporidia. 2.

Nudispora J.I.R. Larsson (1990), Microsporidia. 1. See Larsson (*J. Protozool.* **37**: 310, 1990).

Numbers of fungi. Numbers of fungi. The development of reliable estimates of both the numbers of accepted species of fungi and the actual number on Earth is fraught with uncertainties. The difficulties involved in the enumeration of fungi as individuals, kinds (i.e. as genera and species) or even names are considerable (Ainsworth, 1968). The early edns of this *Dictionary* estimated the number of known 'good' species to be 38,000, rising to 50,000 in the fifth edn. For the seventh edn of this *Dictionary*, the numbers of genera and species accepted were derived by upwards summation of the totals from the generic entries; this gave a total of 64,200 species. For the eighth edn the number was 72,065, and for the ninth edn the number was 80060. The editors of that edition commented that this figure was within the limits of error expected in an exercise of this type. Since the start of 2001, new species of fungi have been catalogued in the *Index of Fungi* at the rate of c. 1200 each year. Assuming that mycologists still inadvertently redescribe already known species at the rate of about 2.5:1 (Hawksworth, 1991) this means that an additional c. 2,200 'good' species would be expected to have been added in the period 2001-2007; i.e. a

TABLE 4. The Numbers of Fungi.

	gen.	spp.	gen.	spp.
FUNGI				
Ascomycota			5674	64056
Arthoniomycetes	78	1608		
Dothideomycetes	1302	19010		
Eurotiomycetes	281	3401		
Laboulbeniomycetes	151	2072		
Lecanoromycetes	630	14199		
Neolectomycetes	1	3		
Orbiliomycetes	12	288		
Pezizomycetes	200	1684		
Pneumocystidomycetes	1	5		
Saccharomycetes	95	915		
Schizosaccharomycetes	2	5		
Sordariomycetes	1119	10564		
Taphrinomycetes	8	140		
incertae sedis	1794	10162		
Basidiomycota			1586	31503
Agaricomycetes	1147	20951		
Agaricostilbomycetes	10	47		
Atractiellomycetes	10	34		
Classiculomycetes	2	2		
Cryptomycocolacomycetes	2	2		
Cystobasidiomycetes	7	14		
Dacrymycetes	9	101		
Entorrhizomycetes	2	15		
Exobasidiomycetes	53	597		
Microbotryomycetes	25	208		
Mixiomycetes	1	1		
Pucciniomycetes	190	8016		
Tremellomycetes	50	377		
Ustilaginomycetes	62	1113		
incertae sedis	16	25		
Chytridiomycota			105	706
Monoblepharidomycetes	5	26		
Chytridiomycetes	98	678		
incertae sedis	2	2		
Zygomycota			168	1065
Entomophthoromycotina	23	277		
Kickxellomycotina	59	264		
Mucoromycotina	61	325		
Zoopagomycotina	22	190		
incertae sedis	3	9		
TOTAL			**75337**	**97330**

total of 82,060 species. The figure of just over 97,000 in Table 4 obtained by upwards addition in entries in this edn of the *Dictionary* is substantially larger. Reasons for this include possible double counting (because of the on-going process of incorporating the conidial fungi in a single holomorphic system), and an increase in the number of species accepted in some groups (perhaps partly as a result of molecular studies); in addition, it is now evident that the earlier estimates were probably very conservative. These figures show a clear upward trend over more than half a century in the number of species discovered, described and accepted by the scientific community.

It is also important to note that amongst the over 338,000 names at species rank proposed for fungi, many have never been reassessed since their first description nor have they been adopted by modem workers. Names in many accepted genera, especially some of the larger, have not been critically revised; in this edn of the *Dictionary* we have not added in all species names proposed and catalogued in the *Index of Fungi* in the absence of the opinions of specialists. In the absence of a world checklist of accepted fungi (a topic starting to be addressed through the IUBS/IUMS SPECIES 2000 project), the possibility that as many as 100,000 'good' species are already described seems quite believable, and as many as 150,000 (Rossman, 1994) 'good' species already described cannot be excluded.

Estimating the total number of fungi on Earth is even more problematic. Bisby & Ainsworth (1943) estimated that there are at least 100,000 species of fungi. Martin (*Proc. Iowa Acad. Sci.* **58**: 175, 1951) considered this estimate 'excessively conservative' and suggested that the number of species of fungi is at least as great as the number of 'good' species of phanerogams, then believed to be not less than 250,000. Based on extrapolations from three independent data sets (ratios of the numbers of fungi in all habitats to plants in the British Isles, numbers restricted to particular hosts, and a community studied in depth) Hawksworth (1991) suggested that a 'conservative' working figure of c. 1.5 million be adopted. Although Hammond (1992) recommended 1 million for general use, this 1.5 million estimate has been supported by subsequent data (Hawksworth, 1993; Cannon & Hawksworth, 1995). Schmit & Mueller (*Biodiversity conservation* **16**: 99, 2007) suggested that 712,000 species of fungi worldwide was a very conservative lower limit of global fungal diversity. Currently Hawksworth's estimate of 1.5 million species remains the most widely accepted (Heywood, *Global biodiversity assessment*, 1995; Rossman, 1994).

The gap between described and estimated species of fungi is immense, and new species are regularly found in all countries of the world. Of the 16,013 new species recorded in the *Index of Fungi* in 1981-90, 51% were from countries outside the tropics; the individual countries providing most species were India and the USA (c. 10% each) (Hawksworth, 1993). In the tropics, around 15-25% of the fungi collected in short studies can be expected to be undescribed, the percentage rising to 60-85% in more prolonged intensive investigations (depending on the groups and habitats). Examination of more recent issues of the *Index of Fungi* suggests this pattern continues. Pirozynski (*Mycol. Pap.* 129, 1972) reported a ratio of micro-fungi *alone* to plants of at least 3:1 (and possibly 5:1) in Tanzania, and detailed (but still incomplete) site inventories in temperate regions yield ratios of 3-4:1 when all groups are considered (Hawksworth, 1991).

Lit.: Ainsworth (*in* Ainsworth & Sussman, *The Fungi* **3**: 505, 1968), Bisby (*Am. J. Bot.* **20**: 246, 1933), Bisby & Ainsworth (*TBMS* **26**: 16, 1943), Cannon & Hawksworth (*Adv. Pl. Path.* **11**: 277, 1995), Hammond (in Groombridge, *Global biodiversity*: 17, 1992), Hawksworth (*MR* **95**: 641, 1991; *in* Isaac *et al.*, *Aspects of tropical mycology*: 265, 1994), Kirk (*MR* 104: 516, 2000), Pascoe (*in* Short, *History of systematic mycology in Australia*: 259, 1990), Rossman (*in* Hawksworth (Ed.), *Identification and characterization of pest organisms*: 35, 1994).

Numerical taxonomy (phenetics, taxometrics). Now largely replaced by phylogenetic analysis (q.v.). The entry from the previous edn of this *Dictionary*, retained for historical interest, follows. The derivation of computer-based assessments of resemblance, for both classification and identification, has not been as widely used for fungi as for bacteria. Examples of this approach applied to fungal classification are the papers by Bridge *et al.* (*J. Gen. Microbiol.* **135**: 2941, 1989; *Penicillium*), Hõls & Raitviir (*Scripta mycol.* **6**, 1974; morphometrics), Ibrahim & Threlfall (*Proc. R. Soc.* **B165**: 362, 1966; graminicolous *Helminthosporium*), Joly (*BSMF* **85**: 213, 1969; *Alternaria*), Kendrick & Proctor (*CJB* **42**:65, 1964; anamorphic fungi), Kendrick & Weresub (*Syst. Zool.* **15**: 307, 1966; ordinal level in basidiomycetes), Kiefer (*Mycol.* **71**: 343, 1979; methodology of Machol & Singer, *Nova Hedw.* **21**: 353, 1971), Parker-Rhodes & Jackson (*in* Cole (Ed.), 1969: 181; ecology of basidiomycetes). For examples of identification schemes see Barnett *et al.* (*Yeasts: characteristics and identification*, edn 2, 1990), Bridge *et al.* (*Mycol. Pap.* **165**, 1992), Pankhurst (*Nature* **227**: 1269, 1970; *Biological identification*, 1978; key generation).

For methods see Bridge & Sackin (*Mycopath.* **115**: 105, 1991), Carmichael & Sneath (*Syst. Zool.* **18**:402, 1969), Cole (Ed.) (*Numerical taxonomy*, 1969), Cutbill (Ed.) (*Data processing in biology and geology*, 1971), Felsenstein (*Numerical taxonomy*, 1983), Sneath & Sokal (*Numerical taxonomy*, edn 2, 1973).

Nummospora E. Müll. & Shoemaker (1964), anamorphic *Pezizomycotina*, Cpd.#eP.?. 1, Switzerland. See Müller & Shoemaker (*Nova Hedwigia* **7**: 1, 1964), Wu *et al.* (*MR* **102**: 179, 1998).

Nummularia Tul. & C. Tul. (1863) [non *Nummularia* Hill 1756, *Primulaceae*] ≡ Biscogniauxia fide Jong & Benjamin (*Mycol.* **63**: 862, 1971; N. Am. spp.), Ju *et al.* (*Mycotaxon* **66**: 1, 1998).

Nummulariella Eckblad & Granmo (1978) = Biscogniauxia fide Pouzar (*Česká Mykol.* **33**: 129, 1979), Læssøe (*SA* **13**: 43, 1994; ? synonym of *Xylaria*).

Nummularioidea (Cooke & Massee) Lloyd (1924) = Camillea fide Læssøe *et al.* (*MR* **93**: 121, 1989), Læssøe (*SA* **13**: 43, 1994; ? synonym of *Xylaria*).

Nummularoidea, see *Nummularioidea*.

Numulariola House (1925) ≡ Biscogniauxia.

nurse cells (in *Scleroderma*), hyphae supplying food material to spores which have come away from the basidia.

TABLE 4 (cont.). The Numbers of protozoan and chromistan analogues.

	gen.	spp.	gen.	spp.
PROTOZOA				
Percolozoa (*Acrasida*)			6	14
Amoebozoa			82	1019
Dictyostelia	4	93		
Myxogastria	62	888		
Protostelia	16	38		
Cercozoa (*Plasmodiophorida*)			15	50
Choanozoa (*Amoebidiales, Eccrinales*)			19	78
Incertae sedis			3	4
TOTAL			**125**	**1165**
CHROMISTA				
Hyphochytriomycota	6	24		
Labyrinthista	12	56		
Oomycota	106	956		
Incertae sedis	2	3		
TOTAL			**1264**	**1036**

Nusia Subram. (1995), anamorphic *Pezizomycotina*, Hso.≡ eP.24. 2, Singapore. See Subramanian (*Cryptog. Mycol.* **14**: 109, 1993), Subramanian (*Kavaka* **20/21**: 57, 1992/1993).

nutant, nodding.

nutrilite, any organic compound necessary in small amounts for the nutrition of an organism (Williams, *Biol. Rev.* **16**: 49, 1941).

nutriocyte (of *Ascosphaera*), the inflated part of the ascogonium which eventually develops into a spore cyst (Spiltoir & Olive, *Mycol.* **47**: 240, 1955).

Nutrition. Fungi are able to degrade and subsequently metabolize many widely different materials. Some parasites (obligate parasites, e.g. *Pucciniales*, *Erysiphales*, *Peronosporaceae*) have such special needs that full development takes place only on the right host; but growth of other parasites, like that of most fungi, will take place on a synthetic medium (see Methods). The growth of fungi is dependent on carbon (C), hydrogen (H), oxygen (O), nitrogen (N), potassium (K), phosphorus (P), magnesium (Mg), and sulphur (S), together with very small amounts of iron (Fe), zinc (Zn), copper (Cu), and/or possibly other minor (or trace) elements. Calcium (Ca) is probably a necessary element but it is not always possible to demonstrate that this is so. In addition, complex 'growth substances' are sometimes needed.

Fungi are heterotrophic in needing their carbon in a complex (organic) form. In general, aliphatic carbon compounds (esp. carbohydrates) are more readily used by fungi than aromatic ones. Nevertheless, basidiomycete fungi are the key organisms in breaking down the ubiquitous aromatic polymer lignin. There are some yeasts which can grow on the one carbon methanol. Some fungi are dependent on organic nitrogen (esp. amino acids and proteins), others can use ammonium or nitrate nitrogen. Fungi are able to adapt and regulate their metabolism according to the nutrients available. The complex bios (q.v.) was the first '**growth substance**' to be noted for fungi, yeast making better growth with it. Thiamin [aneurin, vitamin B_1] is one of the substances necessary for the growth of some fungi; for certain species the complete molecule is needed but others are able to synthesize thiamin if given one, or both, of its components (thiazole and pyrimidine). Fungi are often able to tolerate high concentrations of toxic metals and high concentrations of salt and sugar (which can be lethal to many other organisms) and the species which are able to do this cause spoilage of food.

Lichen-forming fungi obtain the carbohydrates they require in the form of sugars and sugar-alcohols (polyols) produced by the algal partner; the nature of the mobile carbohydrate depends on the kind of alga present (e.g. glucose with *Nostoc*, ribitol with *Myrmecia* and *Trebouxia*, erythritol with *Trentepohlia*); cyanobacteria fix atmospheric nitrogen. The carbohydrates are commonly stored as mannitol by the fungal component. The mineral requirements of lichen fungi are met by ions dissolved in rain and from the deposition of dust; in some cases diffusion from the substrate can occur but is usually very limited in extent.

See Jennings (*The physiology of fungal nutrition*, 1995), Lichens, Physiology.

NY, New York Botanical Garden (Bronx, New York, USA); founded 1891; a private institution.

Nyctalidaceae Jülich (1982) nom. rej. = Lyophyllaceae.

Nyctalina Arnaud (1952), anamorphic *Agaricomycetes*, Hso.0eH.?. 1 (with clamp connexions), Europe. See Arnaud (*BSMF* **68**: 189, 1952).

Nyctalis Fr. (1825) = Asterophora fide Stalpers (*in litt.*).

Nyctalospora E.F. Morris (1972), anamorphic *Pezizomycotina*, Hsy.≡ eP.1. 1, Costa Rica. See Morris (*Mycol.* **64**: 890, 1972).

Nycteromyces Thaxt. (1917), Laboulbeniaceae. 1, S. America.

Nyctomyces Hartig (1833), Fossil Fungi (mycel.) Fungi. 4 (Tertiary), Europe; N. Africa.

Nylander (William; 1822-1899; Finland). Practising medic (1852-1858; Professor of Botany, Helsinki University (1858-1863); living in Paris (1863 onwards). Visited England to examine Hooker's lichen collection [now in the British Museum (Natural History), London **BM**] (1857); published *c.* 330 works [of which the first few were on ants]; introduced the use of chemical reagents (C, I and K) into lichenology (see Metabolic products); described over 5,000 new species of lichen-forming fungi from all parts of the world. Main collections in Helsinki (**H**), but material in most major European institutions with mycological collections. *Publs. Synopsis Methodica Lichenum Omnium hucusque Cognitorum, Praemissa Introductione Lingua Gallica* **1** (1), **1** (2) and **2** (1858, 1860, 1863); Lichenes Scandinaviae sive prodromus lichenographiae Scandinaviae. *Notiser ur Sällskapets pro Fauna et Flora Fennica Förhandlingar* (1861); *Lichenes Japoniae. Accedunt Observationibus Lichenes Insulae Labuan* (1890). *Biogs, obits etc.* Ahti [ed.] (*Collected Papers of William Nylander* 6 vols, 1967-1990) [biography in **1**: viii, 1990]; Grummann (1974: 611); Hue (*Bulletin de la Société Botanique de France* **46**: 153, 1899) [bibliography, portrait]; Stafleu & Cowan (*TL-2* **3**: 788, 1981).

Nylanderaria Kuntze (1891) = Letharia fide Hawksworth *et al.* (*Dictionary of the Fungi* edn 8, 1995).

Nylanderiella Hue (1914) = Siphula Fr. (1831) fide Hawksworth *et al.* (*Dictionary of the Fungi* edn 8, 1995).

Nylanderopsis Gyeln. (1935) = Heppia fide Henssen (*Acta Bot. Fenn.* **150**: 57, 1994).

Nymanomyces Henn. (1899), Rhytismataceae. 1, Java.

Nypaella K.D. Hyde & B. Sutton (1992), anamorphic *Pezizomycotina*, Cpd.0eH.15. 1, Brunei. See Hyde & Sutton (*MR* **96**: 210, 1992).

Nyssopsora Arthur (1906), Raveneliaceae. 9 (on dicots, esp. *Araliales*), widespread. See Lütjeharms (*Blumea*, 1937), Lohsomboon *et al.* (*MR* **94**: 907, 1990; key).

Nyssopsorella Syd. (1921) = Triphragmiopsis fide Dietel (*Nat. Pflanzenfam.* **6**, 1928).

nystatin (mycostatin), an antibiotic from the actinomycete *Streptomyces noursei*; antifungal, widely used against *Candida albicans* infections of man. See Brown & Hazen (*Trans. N.Y. Acad. Sci.* ser. 2 **19**: 447, 1957), Baldwin (*The fungus fighters*, 1981).

Nyungwea Sérus., Eb. Fischer & Killmann (2006), ? Arthoniales (L). 1, E. Africa. See Sérusiaux *et al.* (*Lichenologist* **38**: 115, 2006).

O, Botanical Garden and Museum, University of Oslo (Norway); founded 1812.

OA, see Media.

oak-moss (oakmoss, oak moss), *Evernia prunastri* (mousse de chêne); extracts of which are used in perfumes to reduce the rate of evaporation of other ingredients. Cf. tree hair.

ob- (prefix), inversely or oppositely.

obclavate, inversely clavate (widest at the base) (Fig. 23.17).

Obconicum Velen. (1939), ? Helotiales. 2, former Czechoslovakia.

Obelidium Nowak. (1877), Chytridiaceae. 3, widespread (north temperate).

Obeliospora Nawawi & Kuthub. (1990), anamorphic *Pezizomycotina*, Hso.0bH.15. 1 (on submerged decaying wood), Malaysia. See Nawawi & Kuthubutheen (*Mycotaxon* **37**: 395, 1990), Kuthubutheen & Nawawi (*MR* **98**: 677, 1994), Wu & McKenzie (*Fungal Diversity* **12**: 223, 2003).

Oberwinkleria Vánky & R. Bauer (1995), Tilletiaceae. 1 (in seeds of *Poaceae*), Venezuela. See Vánky & Bauer (*Mycotaxon* **53**: 363, 1995).

oblate, with the equatorial diameter greater than the polar diameter; **- spheroidal**, shaped like the earth.

obligate (1) necessary; essential; (2) (of a parasite), living as a parasite in nature, sometimes of one that has not been cultured on laboratory media, cf. facultative; see synonym.

oblique septum, see septum.

oblong (of spores), twice as long as wide and having somewhat truncate ends (Fig. 23.4*b*, *c*); **- ellipsoid** (of spores), rounded-oblong; having long sides parallel and ends almost hemispherical.

Oblongichytrium R. Yokoy. & D. Honda (2007), Thraustochytriaceae. 3 (marine), widespread. See Yokoyama, R.; Honda, D. (*Mycoscience* **48**: 199, 2007).

Obolarina Pouzar (1986), Xylariaceae. Anamorph *Rhinocladiella*-like. 1, Europe. See Hawksworth (*SA* **13**: 198, 1985), Candoussau & Rogers (*Mycotaxon* **39**: 345, 1990), Læssøe & Spooner (*Kew Bull.* **49**: 1, 1994; as *Rosellinia*), Eriksson (*SA* **14**: 61, 1995; posn), Nordén & Sunhede (*Svensk bot. Tidskr.* **94**: 331, 2001; ecology).

obovate, inversely ovate.

obovoid, inversely ovoid (Fig. 23.12).

obpyriform, the reverse of pyriform (Fig. 23.15).

Obryzaceae Körb. (1855), Sordariomycetes (inc. sed.). 1 gen., 3 spp.

Lit.: Hoffmann & Hafellner (*Biblthca Lichenol.* **77**: 181 pp., 2000).

Obryzum Wallr. (1825), Obryzaceae. 3 (lichenicolous, on *Leptogium*), Europe. See Eriksson (*Op. Bot.* **60**, 1981), Hoffmann & Hafellner (*Biblthca Lichenol.* **77**: 1, 2000).

Obscurodiscus Raitv. (2002), Helotiales. 1, Europe. See Raitviir (*Mycotaxon* **81**: 49, 2002).

obsolete (1) (of organs or parts), rudimentary or absent; (2) (of terms), no longer in use.

Obstipipilus B. Sutton (1968), anamorphic *Pezizomycotina*, Cac.1eP.19. 1, India. See Sutton (*CJB* **46**: 187, 1968).

Obstipispora R.C. Sinclair & Morgan-Jones (1979), anamorphic *Pezizomycotina*, Hso.0fH.1. 1 (aquatic), USA. See Sinclair & Morgan-Jones (*Mycotaxon* **8**: 152, 1979).

obsubulate, very narrow; pointed at the base and a little wider at the tip.

Obtectodiscus E. Müll., Petrini & Samuels (1980), Helotiales. 2, Switzerland; S. America. See Samuels & Rogerson (*Acta Amazon.* Supl. **14**: 81, 1984), Scheuer (*Biblthca Mycol.* **123**: 274, 1988).

obtuse (1) rounded or blunt (Fig. 23.4*b*); (2) greater than a right angle.

occluded, closed; often used of the lumina of hyphae or pseudoparenchymatous cells.

Occultifur Oberw. (1990), Cystobasidiaceae. 4, widespread (north temperate). See Fell *et al.* (*Int. J. Syst. Evol. Microbiol.* **50**: 1351, 2000; mol. phylogeny).

Occultitheca J.D. Rogers & Y.-M. Ju (2003), Xylariaceae. 1, Costa Rica. See Rogers & Ju (*Sydowia* **55**: 359, 2003).

Oceanites Kohlm. (1977), ? Halosphaeriaceae. 1, Atlantic. See Kohlmeyer (*Revue Mycol.* Paris **41**: 193,

1977).

Ocellaria (Tul. & C. Tul.) P. Karst. (1871) = Pezicula Tul. & C. Tul. fide Verkley (*Stud. Mycol.* **44**: 180 pp., 1999), Abeln (*Mycol.* **92**: 685, 2000; phylogeny).

Ocellariella Petr. (1947) ? = Naevia Fr. (1849) fide Korf *in* Ainsworth *et al.* (Eds) (*The Fungi* **4A**: 249, 1973).

ocellate, having rounded marks, like eyes.

Ocellis Clem. (1909) = Ocellularia fide Zahlbruckner (*Catalogus Lichenum Universalis* **2**, 1923).

Ocellularia G. Mey. (1825) nom. cons., Thelotremataceae (L). 249, widespread (esp. tropical). See Hale (*Mycotaxon* **11**: 130, 1980; limits), Nagarkar *et al.* (*Biovigyanam* **14**: 24, 1988; key 32 spp. India), Matsumoto & Deguchi (*Bryologist* **102**: 86, 1999; anamorphs), Sipman (*Biblthca Lichenol.* **86**: 177, 2003; Singapore), Frisch (*Biblthca Lichenol.* **92**: 3, 2006; Africa), Frisch & Kalb (*Biblthca Lichenol.* **92**: 371, 2006; anatomy), Frisch *et al.* (*Biblthca Lichenol.* **92**: 517, 2006; phylogeny), Mangold *et al.* (*Biblthca Lichenol.* **96**: 193, 2007; Australia).

ocellus, an eyespot functioning as a lens and concentrating light rays on a sensitive spot.

Ochotrichobolus Kimbr. & Korf (1983), Pezizales. 1, USA. See Brummelin (*Persoonia* **16**: 425, 1998; possible placement in *Pyronemataceae*).

Ochraceospora Fiore (1930) nom. dub., Hypocreales. Possibly an earlier name for *Haematonectria* (*Nectriaceae*), but type material is lost. See Rossman *et al.* (*Stud. Mycol.* **42**: 248 pp., 1999).

ochratoxin (**A**, **B**), toxins of *Aspergillus ochraceus*, *Penicillium viridicatum*, etc.; the cause of nephrotoxicosis in sheep, cattle, and pigs; also carcinogenic and has been found in coffee.

Ochrocladosporium Crous & U. Braun (2007), anamorphic *Pleosporales*. 2, Sweden; Germany. See Crous *et al.* (*Stud. Mycol.* **58**: 33, 2007).

Ochroconis de Hoog & Arx (1974), anamorphic *Pezizomycotina*, Hso.1eP.11. 8, widespread. See de Hoog (*Stud. Mycol.* **26**: 51, 1985), Fukushiro *et al.* (*J. Med. Vet. Mycol.* **24**: 175, 1986; subcutaneous abscesses), Sides *et al.* (*J. Med. Vet. Mycol.* **29**: 317, 1991; phaeohyphomycosis), Padhye *et al.* (*J. Med. Vet. Mycol.* **32**: 141, 1994; in cat), Schaumann & Priebe (*CJB* **72**: 1629, 1994; in salmon), Horré *et al.* (*Stud. Mycol.* **43**: 194, 1999; DNA, physiology).

Ochroglossum S. Imai (1955) = Microglossum Gillet fide Maas Geesteranus (*Persoonia* **3**: 89, 1964).

Ochrolechia A. Massal. (1852), Ochrolechiaceae (L). 58, widespread (esp. temperate). See Verseghy (*Beih. Nova Hedwigia* **1**, 1962; monogr., keys), Poelt (*Ergebn. Forsch.-Unternehmen Nepal Himal.* **1**: 251, 1966; Himalaya), Brodo (*CJB* **66**: 1264, 1988; N. Am.), Awasthi & Tewari (*Kavaka* **15**: 23, 1989; key 12 spp. Indian subcont.), Schmitz *et al.* (*Acta Bot. Fenn.* **150**: 153, 1994; gen. concept), Archer (*Biblthca Lichenol.* **69**, 1997; Australia), Boqueras *et al.* (*Cryptog. Mycol.* **20**: 303, 1999; Iberian spp.), Messuti & Lumbsch (*Biblthca Lichenol.* **75**: 33, 2000; S America), Schmitt & Lumbsch (*Mol. Phylogen. Evol.* **33**: 43, 2004; phylogeny, chemistry), Miądlikowska *et al.* (*Mycol.* **98**: 1088, 2006; phylogeny), Schmitt *et al.* (*J. Hattori bot. Lab.* **100**: 753, 2006; phylogeny).

Ochrolechiaceae R.C. Harris ex Lumbsch & I. Schmitt (2006), Pertusariales. 2 gen. (+ 2 syn.), 61 spp.

Lit.: Brodo (*CJB* **69**: 733, 1991), Lumbsch *et al.* (*Biblthca Lichenol.* **57**: 355, 1995), Schmitt *et al.* (*Biblthca Lichenol.* **86**: 147, 2003), Schmitt *et al.* (*J. Hattori bot. Lab.* **100**: 753, 2006).

Ochromitra Velen. (1934) = Pseudorhizina fide Petrak (*Sydowia* **1**: 61, 1947).

Ochronectria Rossman & Samuels (1999), Bionectriaceae. 1, widespread (tropical). See Rossman *et al.* (*Stud. Mycol.* **42**: 53, 1999), Rossman *et al.* (*Mycol.* **93**: 100, 2001; phylogeny).

Ochroporus J. Schröt. (1888) = Phellinus fide Donk (*Persoonia* **1**: 173, 1960).

Ochropsora Dietel (1895), ? Uropyxidaceae. 4 (on dicots), widespread (north temperate). See Soong (*Flora* **133**: 345, 1939; morphology), Cummins & Hiratsuka (*Illustr. Gen. Rust Fungi* edn 3: 225 pp., 2003; under *Chaconiaceae*), Ono (*Mycoscience* **47**: 145, 2006; life cycle, basidium morph.).

Ochrosphaera Sawada (1959), Pezizomycotina. 1, Taiwan. See Sawada (*Special Publication College of Agriculture, National Taiwan University* **8**: 50, 1959).

Ochrosporellus (Bondartseva & S. Herrera) Bondartseva & S. Herrera (1992) = Phellinus fide Ryvarden (*Syn. Fung.* **5**: 312, 1991).

ochrosporous, having yellow or yellow-brown spores.

Ocostaspora E.B.G. Jones, R.G. Johnson & S.T. Moss (1983), Halosphaeriaceae. 1 (marine), widespread. See Jones *et al.* (*Bot. Mar.* **26**: 353, 1983), Pang *et al.* (*CJB* **82**: 485, 2004).

Ocotomyces H.C. Evans & Minter (1985), ? Rhytismatales. Anamorph *Uyucamyces*. 1 (on *Pinus*), Honduras. See Evans & Minter (*TBMS* **84**: 57, 1985).

Octaviania Vittad. (1831), Boletaceae. 15, widespread.

Octavianiaceae Locq. ex Pegler & T.W.K. Young (1979) = Boletaceae.

Octavianina, see *Octaviania*.

octo- (in combinations), 8.

Octojuga Fayod (1889) = Crepidotus fide Kuyper (*in litt.*).

Octomyces Mello & L.G. Fern. (1918) ? = Saccharomyces fide Batra *in* Subramanian (Ed.) (*Taxonomy of fungi* **1**: 187, 1978).

octophore, see haerangium.

Octopodotus Kohlm. & Volkm.-Kohlm. (2003), anamorphic *Pezizomycotina*. 1, USA. See Kohlmeyer & Volkmann-Kohlmeyer (*Mycol.* **95**: 117, 2003).

octopolar (of incompatability systems), having 3 loci, as in *Psathyrella coprobia* (Jurand & Kemp, *Genetical Res., Cambr.* **22**: 125, 1973); cf. tetrapolar.

Octospora Hedw. (1789), Pyronemataceae. *c.* 84 (often bryophilous), widespread. See Dennis & Itzerott (*Kew Bull.* **28**: 5, 1973; key), Khare & Tewari (*Mycol.* **67**: 972, 1975), Dennis & Itzerott (*Kew Bull.* **31**: 497, 1977; W. Eur.), Döbbeler (*Nova Hedwigia* **31**: 817, 1980), Döbbeler & Itzerott (*Nova Hedwigia* **34**: 127, 1981; biology), Caillet & Moyne (*BSMF* **103**: 277, 1987; key 32 spp.), Caillet & Moyne (*Bull. Soc. Hist. nat. Doubs* **84**: 9, 1991; keys), Wang (*NMNS Spec. Publ.*, 1992; key, N. American spp.), Yao & Spooner (*MR* **100**: 175, 1996; key Brit. spp.), Moravec (*Czech Mycol.* **49**: 149, 1997; sect. *Neottiellae*), Benkert (*Z. Pilzk.* **7**: 39, 1998), Jakobson *et al.* (*Karstenia* **38**: 1, 1998), Khare (*Nova Hedwigia* **77**: 445, 2003), Benkert & Brouwer (*Persoonia* **18**: 381, 2004; Netherlands), Hansen & Pfister (*Mycol.* **98**: 1029, 2006; phylogeny), Perry *et al.* (*MR* **111**: 549, 2007; phylogeny).

octospore, one spore of an 8-spored ascus.

Octosporea Flu (1911), Microsporidia. 3.

Octosporella Döbbeler (1980), Pyronemataceae. 7 (on mosses and liverworts), Europe; Venezuela. See Döbbeler (*Mitt. bot. StSamml., München* **16**: 471, 1980), Yao & Spooner (*Kew Bull.* **51**: 193, 1996), Yao *et al.* (*Nova Hedwigia* **82**: 483, 2006; key).

Octosporomyces Kudrjanzev (1960) = Schizosaccharomyces fide Sipiczki *in* Nasim *et al.* (Eds) (*Molecular biology of the fission yeast*, 1989), Kurtzman & Robnett (*Yeast* Chichester **7**: 61, 1991).

Octosporonites Locq. & Sal.-Cheb. (1980), Fossil Fungi. 1, Cameroon.

octosporous, producing spores in 8s.

Octotetraspora I.V. Issi, Kadyrova, Pushkar, Khodzhaeva & S.V. Krylova (1990), Microsporidia. 1.

ocular chamber, see ascus.

Oculimacula Crous & W. Gams (2003), Helotiales. Anamorph *Helgardia*. 2 (causing eyespot disease of cereals), widespread. See Crous *et al.* (*Eur. J. Pl. Path.* **109**: 845, 2003), Walsh *et al.* (*J. Phytopath.* **153**: 715, 2005; molecular diagnostics).

Odontia Fr. (1835) = Steccherinum fide Donk (*Persoonia* **3**: 199, 1964) but used by, Furukawa (*Bull. Govt For. Exp. Stn* **261**, 1974).

OdontiaGray (1821) nom. dub. fide Stalpers (*in litt.*).

Odontia Pers. (1794) nom. rej. = Caldesiella fide Stalpers (*in litt.*).

Odonticium Parmasto (1968), Agaricomycetes. 6, widespread. *Hymenochaetales* or *Agaricales* (*Rickenella* clade). See Parmasto (*Consp. System. Corticiac.*: 126, 1968), Zmitrovich (*Mikol. Fitopatol.* **35**: 9, 2001).

Odontina Pat. (1887) = Steccherinum fide Donk (*Taxon* **5**: 69, 1956).

Odontiochaete Rick (1940) nom. dub., Cantharellales. See Donk (*Taxon* **5**: 107, 1956).

Odontiopsis Hjortstam & Ryvarden (1980), Schizoporaceae. 2, widespread (tropical). See Hjortstam & Ryvarden (*Mycotaxon* **12**: 180, 1980).

Odontium Raf. (1817) ≡ Caldesiella fide Donk (*Taxon* **5**: 107, 1956).

Odontodictyospora Mercado (1984), anamorphic *Pezizomycotina*, Hso.#eH.11. 1, Cuba. See Mercado (*Acta Bot. Cubana* **22**: 1, 1984).

odontoid, tooth-like; dentate.

Odontoschizon Syd. & P. Syd. (1914) = Patinella fide Clements & Shear (*Gen. Fung.*, 1931).

Odontotrema Nyl. (1858), Odontotremataceae. *c.* 18, widespread (temperate). See Sherwood-Pike (*Mycotaxon* **28**: 137, 1987), Diederich *et al.* (*Lichenologist* **34**: 479, 2002; lichenicolous spp.), Wedin *et al.* (*Lichenologist* **37**: 67, 2005; phylogeny).

Odontotremataceae D. Hawksw. & Sherwood (1982), Ostropales. 15 gen. (+ 7 syn.), 78 spp.

Lit.: Sherwood-Pike (*Mycotaxon* **28**: 137, 1987; cellulases), Triebel (*Biblthca Lichenol.* **35**: 278 pp., 1989), Lumbsch & Hawksworth (*Biblthca Lichenol.* **38**: 325, 1990), Döbbeler (*Nova Hedwigia* **62**: 61, 1996), Holien & Triebel (*Lichenologist* **28**: 307, 1996), Wedin *et al.* (*Lichenologist* **37**: 67, 2005).

Odontotremella Rehm (1912) ≡ Odontura.

Odontura Clem. (1909), Odontotremataceae. 1, Europe. See Sherwood *et al.* (*TBMS* **75**: 479, 1980), Sherwood-Pike (*Mycotaxon* **28**: 137, 1987).

odour, see Smell.

Oedemium Link (1824), anamorphic *Chaetosphaerella*, Hso.1-≡ eP.9. 2, widespread. See Hughes & Hennebert (*CJB* **41**: 773, 1963), Réblová (*Mycotaxon* **70**: 387, 1999; teleomorph).

Oedemocarpus Trevis. (1857) = Mycoblastus fide Hafellner (*Beih. Nova Hedwigia* **79**: 241, 1984).

Oedipus Bataille (1908) = Boletus Fr. fide Singer (*Farlowia* **2**: 223, 1945).

oedocephaloid, having a swelling at the end or tip, as conidiophores of *Oedocephalum* and *Cunninghamella*.

Oedocephalum Preuss (1851), anamorphic *Pezizales*, Hso.0eH.6. 9, widespread. Used for anamorphs of various genera in the *Pezizaceae* and *Pyronemataceae*. See Stalpers (*Proc. K. Ned. Akad. Wet.* C **77**: 383, 1974; key), Watanabe (*Mycol.* **83**: 524, 1991; Japan), Norman & Egger (*Mycol.* **91**: 820, 1999; phylogeny).

Oedogoniomyces Tak. Kobay. & M. Ôkubo (1954), Oedogoniomycetaceae. 1, widespread (north tropical). See Emerson & Whisler (*Arch. Mikrobiol.* **61**: 195, 1968), James *et al.* (*Mycol.* **98**: 860, 2006; phylogeny).

Oedogoniomycetaceae D.J.S. Barr (1990), Monoblepharidales. 1 gen., 1 spp.

Lit.: Emerson & Whisler (*Archs Microbiol.* **61**: 195, 1968), Barr *in* McLaughlin *et al.* (Eds) (*The Mycota* A Comprehensive Treatise on Fungi as Experimental Systems for Basic and Applied Research **7A**: 93, 2001).

Oedomyces Sacc. ex Trab. (1894) = Physoderma fide Karling (*Lloydia* **13**: 29, 1950) = Urophlyctis (Physodermat.) fide, Ciferri (*Atti Ist. bot. Univ. Lab. crittog. Pavia* sér. 5 **20**: 246, 1963).

Oedothea Syd. (1930), anamorphic *Pezizomycotina*, Hsp.1eP.19. 1, S. America.

oenology, the study of wines; see wine making.

Oerskovia Prauser, M.P. Lechev. & H. Lechev. (1970), Actinobacteria. q.v.

Oesophagomyces Manier & Ormières (1980), Fungi. 1, France. See Manier & Ormières (*Annls Sci. Nat.* Zool., sér. 13 **2**: 154, 1980).

Oevstedalia Ertz & Diederich (2004), Verrucariaceae (L). 1, Antarctic. See Ertz & Diederich (*Mycol. Progr.* **3**: 232, 2004).

Ogataea Y. Yamada, K. Maeda & Mikata (1994), Saccharomycetales. Anamorph *Candida*. 1, widespread. See Mikata & Yamada (*Res. Commun.* Inst. Ferm., Osaka **17**: 99, 1995), Kurtzman *in* Kurtzman & Fell (Eds) (*Yeasts, a taxonomic study* 4th edn: 273, 1998), Morais *et al.* (*FEMS Yeast Res.* **5**: 81, 2004; phylogeny), Suh *et al.* (*Mycol.* **98**: 1006, 2006; phylogeny), Limtong *et al.* (*Int. J. Syst. Evol. Microbiol.* **58**: 302, 2008; phylogeny).

Ohleria Fuckel (1868), Melanommataceae. Anamorph *Monodictys*. 4 (on wood), widespread. See Samuels (*N.Z. Jl Bot.* **18**: 515, 1980).

Ohleriella Earle (1902), ? Delitschiaceae. 1, N. America. See Ahmed & Cain (*CJB* **50**: 419, 1972; synonym of *Sporormiella*), Barr (*SA* **6**: 142, 1987), Barr (*N. Amer. Fl.* ser. 2 **13**: 129 pp., 1990), Barr (*Mycotaxon* **76**: 105, 2000).

Oichitonium Durieu & Mont. [not traced] nom. dub., ? Fungi.

-oid (suffix), like; having the form of. Most of the many mycological terms ending in this suffix (e.g. achlyoid, as in *Achlya*; daedaleoid, as *Daedalea*) have not been compiled in this *Dictionary*.

Oideum Ehrenb. (1818) ≡ Oidium Link (1824).

Oidiaceae Link (1826) = Erysiphaceae.

Oidiodendron Robak (1932), anamorphic *Byssoascus, Myxotrichum*, Hsy.0eP.38. 28, widespread (north temperate). See Barron (*CJB* **40**: 589, 1962; key), Morrall (*CJB* **46**: 204, 1968), Currah *et al.* (*CJB* **71**: 1481, 1993; conidiogenesis), Hambleton *et al.* (*Mycol.* **90**: 854, 1998; phylogeny), Lacourt *et al.* (*New Phytol.* **149**: 565, 2001; phylogeny), Rice & Currah (*Mycotaxon* **79**: 383, 2001; physiology, morphology), Rice & Currah (*MR* **106**: 1463, 2002; ecology), Calduch *et al.* (*Stud. Mycol.* **50**: 159, 2004; Spain), Rice & Currah (*Stud. Mycol.* **53**: 83, 2005; review).

oidiomycin, an antigen prepared from *Candida albicans*, esp. for skin testing.

oidiophore, a structure producing oidia.

Oidiopsis Scalia (1902), anamorphic *Leveillula*, Hso.0eH.1. 7, widespread. See Braun (*Beih. Nova Hedwigia* **89**: 700 pp., 1987; monogr.), Saenz & Taylor (*CJB* **77**: 150, 1999; phylogeny), Liberato *et al.* (*Australas. Pl. Path.* **34**: 409, 2005; on *Euphorbia*, Australasia).

oidiospore, see oidium.

Oidites Mesch. (1892), Fossil Fungi. 2 (Oligocene), Baltic.

oidium (pl. **oidia**), (1) spermatia formed on hyphal branches, esp. in heterothallic hymenomycetes; (2) flat-ended conidia formed by the breaking up (usually centripetally) of a hypha into cells, as in *Geotrichum candidum*; arthrospore; (3) a mildew.

Oidium Link (1809) nom. rej. = Oidium Link (1824) fide Hawksworth *et al.* (*Dictionary of the Fungi* edn 8, 1995).

Oidium Link (1824) nom. cons., anamorphic *Erysiphe*, Hso.0eH.23. *c.* 212, widespread. See Carmichael *et al.* (*Genera of Hyphomycetes*, 1980), Braun (*Beih. Nova Hedwigia* **89**: 700 pp., 1987; monogr.), Gorter (*Phytophylactica* **20**: 113, 1988; S Africa), Ialongo (*Mycotaxon* **47**: 193, 1993; statistical characterization), Cook *et al.* (*MR* **101**: 975, 1997; SEM, host range), Braun (*Schlechtendalia* **3**: 48, 1999; gen. concept), Saenz & Taylor (*CJB* **77**: 150, 1999; phylogeny), Braun *et al.* (*The Powdery Mildews* A Comprehensive Treatise: 13, 2001; review), Cunnington *et al.* (*Australas. Pl. Path.* **32**: 421, 2003; molecular identification), Cunnington *et al.* (*Australas. Pl. Path.* **33**: 281, 2004; on *Fabaceae*, Australia), Bolay (*Cryptogamica Helvetica* **20**, 2005), Limkaisang *et al.* (*Mycoscience* **46**: 220, 2005; on *Hevea*), Cook *et al.* (*MR* **110**: 672, 2006; on Catalpa).

Oidium Sacc. (1880) nom. rej. prop. = Oidium Link (1824) fide Hawksworth *et al.* (*Dictionary of the Fungi* edn 8, 1995).

oidization, dikaryotization by the fusion of an oidium with a haploid hypha.

Oidospora Will. (1878), Fossil Fungi ? Fungi. 1 (Carboniferous).

Ojibwaya B. Sutton (1973), anamorphic *Pezizomycotina*, Ccu.0eP.23. 1, widespread. See Sutton (*Mycol. Pap.* **132**: 82, 1973), Mel'nik & Belomesyatseva (*Mikol. Fitopatol.* **35**: 40, 2001; Belorus).

Okeanomyces K.L. Pang & E.B.G. Jones (2004), Halosphaeriaceae. 1 (marine), pantropical. See Pang *et al.* (*J. Linn. Soc.* Bot. **146**: 228, 2004).

old man's beard, see beard moss.

Oleina Tiegh. (1887), ? Saccharomycetales. 2, Europe.

Oleinis Clem. (1931) ≡ Oleina fide Batra *in* Subramanian (Ed.) (*Taxonomy of fungi* **1**: 187, 1978).

oleocystidium, see cystidium.

oleoso-locular (of spores), having cells like drops of oil.

Oletheriostrigula Huhndorf & R.C. Harris (1996), Strigulaceae. 1, widespread (esp. mediterranean). See Huhndorf & Harris (*Brittonia* **48**: 551, 1996), Liew *et al.* (*Mycol.* **94**: 803, 2002; phylogeny).

Oligoporus Bref. (1888), Polyporaceae. Anamorph *Ptychogaster*. 10, Europe. See Niemelä & Kinnunen (*Karstenia* **45**: 76, 2005).

Oligosporidium Codreanu-Balcescu, Codreanu & Traciuc (1981), Microsporidia. 1.

oligosporous, having few spores. Cf. myriosporous.

Oligostroma Syd. & P. Syd. (1914) = Mycosphaerella fide von Arx & Müller (*Stud. Mycol.* **9**, 1975), Aptroot (*CBS Diversity Ser.* **5**, 2006).

oligotrophic, poor in nutrients; cf. eutrophic.

Olivea Arthur (1917), Chaconiaceae. 8 (on dicots, esp. *Verbenaceae*), widespread (tropical). See Ono & Hennen (*TMSJ* **24**: 369, 1983).

Oliveonia Donk (1958), ? Auriculariales. Anamorph *Oliveorhiza*. 5, widespread. See Roberts (*Rhizoctonia-forming fungi*, 1999), Grosse-Brauckmann (*Z. Mykol.* **68**: 135, 2002; Germany), Kotiranta & Saarenoksa (*Ann. bot. fenn.* **42**: 237, 2005; Finland).

Oliveoniaceae P. Roberts (1998) ? = Auriculariales.

Oliveorhiza P. Roberts (1998), anamorphic *Oliveonia*. 1, Europe. See Roberts (*Rhizoctonia-forming fungi*, 1999).

Olla Velen. (1934), Hyaloscyphaceae. *c.* 12, Europe. See Baral (*SA* **13**: 113, 1994; concept), Baral (*Nova Hedwigia* **69**: 1, 1999).

Ollula Lév. (1863) = Tubercularia Tode fide Sutton (*Mycol. Pap.* **141**, 1977).

Olpidiaceae J. Schröt. (1889), Chytridiomycetes (inc. sed.). 5 gen. (+ 7 syn.), 45 spp.

Lit.: Sampson (*TBMS* **23**: 199, 1939), Sparrow (*Aquatic Phycomycetes* Edn 2: 1187 pp., 1960), Sahtiyanci (*Archs Microbiol.* **41**: 187, 1962), Karling (*Chytriomyc. Iconogr.*: 414 pp., 1977), Lange & Insunza (*TBMS* **69**: 377, 1977), Barr (*CJB* **58**: 2380, 1980).

Olpidiaster Pascher (1917) ≡ Asterocystis.

Olpidiella Lagerh. (1888) = Olpidium fide Minden (*Krypt.-Fl. Brandenburg augr. Gebiete*, 1911).

Olpidium (A. Braun) J. Schröt. (1886), Olpidiaceae. *c.* 40 (in algae, aquatic fungi, rotifers etc.), widespread. *O. brassicae* (lettuce big vein virus vector); *O. uredinis* on rust spores. See Sampson (*TBMS* **23**: 199, 1939), Litvinov (*Trudy Bot. Inst. Akad. nauk URSS* ser. 2 **12**: 188, 1959), Sparrow (*Aquatic Phycomycetes* Edn 2: 128, 1960; key), Sahtiyanci (*Arch. Mikrobiol.* **41**: 187, 1962), Karling (*Chytriomyc. Iconogr.*, 1977), Lange & Insunza (*TBMS* **69**: 377, 1977), Jiang & Hiruki (*Journal of Microbiological Methods* **26**: 87, 1996; PCR on *Olpidium radicale*), Glockling (*MR* **102**: 206, 1998; on rotifer), Koganezawa *et al.* (*Japanese Journal of Phytopathology* **70**: 307, 2004; on crucifers), James *et al.* (*Mycol.* **98**: 860, 2006; molecular phylogeny suggests relationship with *Basidiobolus*).

Olpitrichum G.F. Atk. (1894), anamorphic *Pezizomycotina*, Hso.0eH.10. 3, widespread. See Holubová-Jechová (*Folia geobot. phytotax.* **9**: 425, 1974), Chen *et al.* (*J. Phytopath.* **153**: 124, 2005; China).

Omalycus Raf. (1814) nom. rej. ? = Calvatia fide Stalpers (*in litt.*).

Ombrophila Fr. (1849), Helotiaceae. *c.* 11, widespread. See Verkley (*Persoonia* **15**: 3, 1992; ultrastr.), Gamundí & Romero (*Fl. criptog. Tierra del Fuego* **10**: 130 pp., 1998), Verkley (*Nova Hedwigia* **77**: 271, 2003; ultrastr.), Wang *et al.* (*Mycol.* **98**:

1065, 2006; phylogeny).

Ombrophila Quél. (1892) nom. conf., Fungi. See Donk (*Persoonia* **4**: 219, 1966).

Omega B. Sutton & Minter (1988), anamorphic *Pezizomycotina*, Ccu.0fH.19. 1, Greece. See Sutton & Minter (*TBMS* **91**: 715, 1988).

Ommatomyces Kohlm., Volkm.-Kohlm. & O.E. Erikss. (1995), Clypeosphaeriaceae. 3 (marine, on *Juncus*), N. America. See Kohlmeyer *et al.* (*Mycol.* **87**: 532, 1995), Wang *et al.* (*Fungal Diversity* **4**: 125, 2000).

Ommatospora Bat. & Cavalc. (1964) = Microclava fide Deighton (*TBMS* **52**: 315, 1969).

Ommatosporella Bat., J.L. Bezerra & Poroca (1967), anamorphic *Pezizomycotina*, Hso.1eP.1. 1, Brazil. See Batista *et al.* (*Atas Inst. Micol. Univ. Pernambuco* **5**: 424, 1967).

Omnidemptus P.F. Cannon & Alcorn (1994), Magnaporthaceae. Anamorph *Mycoleptodiscus*. 1, Australia. See Cannon & Alcorn (*Mycotaxon* **51**: 483, 1994).

omnivorous (of parasites), attacking a number of different hosts.

Omoriza Paulet (1812) nom. conf., Fungi. Used for a range of *Ascomycota* and *Basidiomycota*. See Donk (*Taxon* **6**: 85, 1957).

Omorrhiza, see *Omoriza*.

Omphalaria A. Massal. (1855) nom. rej. = Anema fide Hawksworth *et al.* (*Dictionary of the Fungi* edn 8, 1995).

Omphalaria R. Girard & Dunal ex Nyl. (1855) = Thyrea fide Hawksworth *et al.* (*Dictionary of the Fungi* edn 8, 1995).

Omphalariaceae Körb. (1855) = Lichinaceae.

Omphalia (Fr.) Staude (1857) ≡ Omphalina.

Omphalia (Pers.) Gray (1821) [non *Omphalea* L. 1759) nom. cons., *Euphorbiaceae*] = Pseudoclitocybe fide Donk (*Beih. Nova Hedwigia* **5**: 203, 1963), Redhead & Weresub (*Mycol.* **70**: 556, 1978).

Omphaliaster Lamoure (1971), Tricholomataceae. 7, widespread (north temperate). See Lamoure (*Svensk bot. Tidskr.* **65**: 281, 1971).

Omphalina Quél. (1886) nom. cons., Tricholomataceae. *c.* (s.lat.) 50, widespread (esp. temperate). The genus is polyphyletic. See Bigelow (*Mycol.* **62**: 1, 1970; key (s.lat.)), Lutzoni & Vilgalys (*CJB* **73**: S649, 1995; phylogeny), Lutzoni (*Syst. Biol.* **46**: 373, 1997; phylogeny), Moncalvo *et al.* (*Syst. Biol.* **49**: 278, 2000; phylogeny), Redhead *et al.* (*Mycotaxon* **83**: 19, 2002; phylogeny (s.str.)).

Omphaliopsis (Noordel.) P.D. Orton (1991) = Entoloma fide Kuyper (*in litt.*).

Omphalius Roussel (1806) = Pseudoclitocybe fide Kuyper (*in litt.*) A proposal to conserve *Pseudoclitocybe* over *Omphalius* has not been published.

Omphalocystis Balbiani (1889), Fungi. 1 (on *Cryptopus*), France.

Omphalodiella Henssen (1991), Parmeliaceae (L). 1, Argentina. See Henssen (*Lichenologist* **23**: 334, 1991).

Omphalodina M. Choisy (1929) = Rhizoplaca fide Hawksworth *et al.* (*Dictionary of the Fungi* edn 8, 1995).

omphalodisc (1) an orbicular conical shaped disk; (2) (of *Umbilicaria* [*Omphalodiscus*]), an apothecium with a central knob of sterile hyphae.

Omphalodiscus Schol. (1934) = Umbilicaria fide Hawksworth *et al.* (*Dictionary of the Fungi* edn 8, 1995).

Omphalodium Meyen & Flot. (1843), Parmeliaceae (L). *c.* 4, N. & S. America. See Henssen (*Lichenologist* **24**: 27, 1992), Thell *et al.* (*Mycol. Progr.* **1**: 335, 2002; phylogeny).

Omphalodium Rabenh. (1845) = Umbilicaria fide Hawksworth *et al.* (*Dictionary of the Fungi* edn 8, 1995).

Omphalomyces Battarra ex Earle (1909) = Russula fide Singer (*Agaric. mod. Tax.*, 1951).

Omphalophallus Kalchbr. (1883) = Phallus fide Stalpers (*in litt.*).

Omphalopsis Earle (1909) [non *Omphalopsis* Grev. 1863, *Algae*] ≡ Xeromphalina.

Omphalora T.H. Nash & Hafellner (1990), Parmeliaceae (L). 1, N. America. See Henssen (*Lichenologist* **24**: 27, 1992).

Omphalosia Neck. ex Kremp. (1869) = Umbilicaria fide Hawksworth *et al.* (*Dictionary of the Fungi* edn 8, 1995).

Omphalospora Theiss. & Syd. (1915), Dothideaceae. Anamorph *Podoplaconema*. 4, Europe. See Obrist (*Phytopath. Z.* **35**: 383, 1959).

Omphalotaceae Bresinsky (1985) = Marasmiaceae.

Omphalotus Fayod (1889), Marasmiaceae. 5, widespread. *O. olearius* is luminescent when fresh, and poisonous (Maretic, *Toxicon* **13**: 379, 1975). See Bigelow *et al.* (*Mycotaxon* **3**: 363, 1976; discussion), Kirchmair *et al.* (*Persoonia* **17**: 583, 2002; chemotaxonomy, morphology), Kirchmair *et al.* (*Mycol.* **96**: 1253, 2004; phylogeny).

Onakawananus Radforth (1958), Fossil Fungi (mycel.) Fungi. 1 (Cretaceous), Canada.

Onchopus P. Karst. (1879) = Coprinus fide Redhead *et al.* (*Mycotaxon* **50**: 203, 2001).

Oncidium Nees (1823) [non *Oncidium* Sw. 1800, *Orchidaceae*] ≡ Myxotrichum.

Oncobasidium P.H.B. Talbot & Keane (1971) = Thanatephorus. *O. theobromae*, vascular-streak dieback. fide Keane & Prior (*Phytopath. Pap.* **33**, 1991), Roberts (*Rhizoctonia-forming fungi*, 1999).

Oncobyrsa C. Agardh (1827) nom. dub., ? Fungi.

Oncocladium Wallr. (1833), anamorphic *Gymnoascus*, Hso.0eH.38. 1 (from soil), Europe; Canada. See Hughes (*CJB* **46**: 941, 1968), Sigler *et al.* (*Mycotaxon* **28**: 119, 1987; relationship to *Malbranchea flava*).

oncom merah (**oncom hitah**), Javanese fermented soya bean products in which the principal fungi are *Neurospora sitophila* and *Rhizopus oligosporus*, respectively (Hedger, *Bull. BMS* **12**: 53, 1978); see Fermented food and drinks.

Oncomyces Klotzsch (1843) = Auricularia fide Saccardo (*Syll. fung.* **6**: 762, 1888).

Oncopodiella G. Arnaud ex Rifai (1965), anamorphic *Pezizomycotina*, Hso.#eP.10. 7, Europe. See Rifai (*Persoonia* **3**: 409, 1965), Révay (*Mycotaxon* **56**: 479, 1995; Hungary), Zhao & Zhang (*Nova Hedwigia* **81**: 421, 2005; China).

Oncopodium Sacc. (1904), anamorphic *Pezizomycotina*, Hsy.#eP.1. 7, Europe; N. America. See Hudson (*TBMS* **44**: 406, 1961), Sutton (*Mycol.* **70**: 793, 1978), Castañeda Ruíz *et al.* (*Mycotaxon* **61**: 319, 1997), Zhao & Zhang (*Mycosystema* **22**: 351, 2003; China).

Oncopus, see *Onchopus*.

Oncospora Kalchbr. (1880), anamorphic *Pezizomycotina*, Ccu.1eP.1. 7, widespread. See Chevassut

(*BSMF* **106**: 107, 1990; genus rev.).

Oncosporella P. Karst. (1887), anamorphic *Pezizomycotina*, Ccu.≡ eH.10. 1, Finland.

Oncosporomyces Bat. (1965) nom. dub., anamorphic *Pezizomycotina*, St.0fH.? (L). 1, Brazil. See Lücking *et al.* (*Lichenologist* **30**: 121, 1998).

Oncostroma Bat. & Marasas (1966), anamorphic *Pezizomycotina*, Cpt.0eH.?. 1, S. Africa. See Batista & Marasas (*Bothalia* **9**: 209, 1966).

Ondiniella E.B.G. Jones, R.G. Johnson & S.T. Moss (1984), Halosphaeriaceae. 1 (marine), widespread. See Jones *et al.* (*Bot. Mar.* **27**: 136, 1984).

Onnia P. Karst. (1889), Hymenochaetaceae. 5, widespread. See Wagner & Fischer (*Mycol.* **94**: 998, 2002).

ontjom, an Indonesian fermented food prepared from peanut press cake, surface inoculated with *Neurospora sitophila*.

ontomycosis, see mycosis.

Ontostheca Bat. (1963) = Eudimeriolum fide von Arx (*in litt.*).

Ontotelium Syd. (1921) = Uromyces fide Dietel (*Nat. Pflanzenfam.* **6**, 1928).

Onychocola Sigler (1990), anamorphic *Arachnomyces*, Hso.0-1eH.38. 1 (from humans), widespread. See Sigler (*J. Med. Vet. Mycol.* **28**: 409, 1990), Sigler *et al.* (*J. Med. Vet. Mycol.* **32**: 275, 1994; teleomorph), Midgley & Moore (*Revta Iberoamer. Micol.* **15**: 113, 1998; review), Gibas *et al.* (*Stud. Mycol.* **47**: 131, 2002; phylogeny), Gibas *et al.* (*Stud. Mycol.* **50**: 525, 2004; mating type).

onychomycosis, see mycosis.

Onychophora W. Gams, P.J. Fisher & J. Webster (1984), anamorphic *Pezizomycotina*, Hso.0eH.15. 1 (coprophilous), British Isles. See Gams *et al.* (*TBMS* **82**: 174, 1984), Hoog *et al.* (*Stud. Mycol.* **43**: 107, 1999; phylogeny).

Onygena Pers. (1800), Onygenaceae. 5 (on feathers, bones, etc.), Europe; N. America. See Rammeloo (*Dumort.* **6**: 1, 1977), Currah (*Mycotaxon* **24**: 1, 1985), Sugiyama *et al.* (*Stud. Mycol.* **47**: 5, 2002; phylogeny).

Onygenaceae Berk. (1857), Onygenales. 23 gen. (+ 16 syn.), 134 spp.

Lit.: Currah (*Mycotaxon* **24**: 1, 1985), Currah (*SA* **7**: 1, 1988), Cano & Guarro (*MR* **94**: 355, 1990), Cano *et al.* (*MR* **100**: 343, 1996), Guého *et al.* (*Mycoses* **40**: 69, 1997), Hoog *et al.* (*Medical Mycology* **36** Suppl. 1: 52, 1998), Peterson & Sigler (*J. Clin. Microbiol.* **36**: 2918, 1998), Sigler *et al.* (*CJB* **76**: 1624, 1998), Sugiyama *et al.* (*Mycoscience* **40**: 251, 1999), Bialek *et al.* (*J. Clin. Microbiol.* **38**: 3190, 2000), Sugiyama *et al.* (*Stud. Mycol.* **47**: 5, 2002), Untereiner *et al.* (*Stud. Mycol.* **47**: 25, 2002), Untereiner *et al.* (*Mycol.* **96**: 812, 2004), Stchigel & Guarro (*MR* **111**: 1100, 2007).

Onygenales Cif. ex Benny & Kimbr. (1980). Eurotiomycetidae. 4 fam., 52 gen., 271 spp. Stromata absent. Ascomata formed from coiled initials, cleistothecial, sometimes agrregated, rarely stipitate, pale; peridium composed of loosely woven usually thick-walled hyphae, sometimes with complex appendages. Interascal tissue absent. Asci ? formed from croziers, ± globose, small, evanescent, 8-spored. Ascospores small, often brightly coloured, usually oblate, often ornamented, esp. with equatorial ridges. Anamorphs prominent, hyphomycetous, arthric. Keratinophilic or cellulolytic, some parasitic on humans and other animals (see ringworm, dermatophytes), also in soil, cosmop.

The *Myxotrichaceae* is now excluded from the *Onygenales*. Fams:

(1) **Ajellomycetaceae**
(2) **Arthrodermataceae**
(3) **Gymnoascaceae**
(4) **Onygenaceae**

Lit.: Currah (*Mycotaxon* **24**: 1, 1985; *SA* **7**: 1, 1988; key gen.; *in* Hawksworth, 1994: 281), Geiser *et al.* (*Mycol* **98**: 1051, 2006), Geiser & LoBuglio (*in* McLaughlin *et al.* (Eds), *The Mycota* **7A**: 201, 2001), Landvik *et al.* (*Mycoscience* **37**: 237, 1997; phylogeny), Malloch & Cain (*CJB* **50**: 61, 1972), Orr (*Mycotaxon* **5**: 283, 1977; gen. septal swellings, *et al.*, **5**: 466, 1977; gen. discoid-oblate spores), Sigler & Carmichael (*Mycotaxon* **4**: 349, 1976; anamorphs), Sugiyama *et al.* (*Mycoscience* **40**: 251, 1999; phylogeny), Takizawa *et al.* (*Mycoscience* **35**: 327, 1994; ubiquinones).

Onygenopsis Henn. [not traced], Sordariomycetes. 1, widespread (tropical). See Petch (*Ann. R. bot. Gdns Peradeniya* **5**: 265, 1912), Eriksson & Hawksworth (*SA* **5**: 147, 1986; posn).

Oochytrium Renault (1895), Fossil Fungi, Chytridiomycetes. 1 (Carboniferous), France.

oocyst, the product of aposporous spore development in *Oomycetes* (Dick, *New Phytol.* **71**: 1151, 1972).

oogamy, heterogamy in which the gametes are a non-motile egg and a small, motile sperm.

Oogaster Corda (1854) = Tuber fide Fischer (*Nat. Pflanzenfam.* **5b**: viii, 1938), Læssøe & Hansen (*MR* **111**: 1075, 2007; phylogeny).

oogenesis, the development of the oogonium after being fertilized.

oogoniols, *Achlya* hormones which induce oogonial formation. Cf. antheridiol.

oogonium (**oogone**), uninucleate or coenocytic cell producing female gametes (oospheres).

Oolithinia M. Choisy & Werner (1932) = Protoblastenia fide Hawksworth *et al.* (*Dictionary of the Fungi* edn 8, 1995).

Oomyces Berk. & Broome (1851), Acrospermaceae. 1, Europe. See Petch (*J. Bot., Lond.* **75**: 217, 1937), Eriksson (*Op. Bot.* **60**, 1981; posn).

ooplasm (of *Peronosporales*), the protoplasm, at the centre of the oogonium, which becomes the oosphere; cf. periplasm and gonoplasm.

ooplast (of *Saprolegniaceae*), a large membrane bound inclusion of oospores formed by the fusion of dense body vesicles. See Howard & Moore (*Bot. Gaz.* **131**: 311, 1970).

oosphere (of *Oomycetes*), the female gamete; the 'egg' of the oogonium; **compound** -, one having many functional nuclei.

Oospora Wallr. (1833) ≡ Oidium Link (1824) fide Donk (*Taxon* **12**: 270, 1963), Sigler & Carmichael (*Mycotaxon* **4**: 349, 1976).

oospore (of *Oomycetes*), the resting spore from a fertilized oosphere; a like structure produced by parthenogenesis.

Oosporidium Stautz (1931), anamorphic *Saccharomycetes*, Hso.0eH.10. 1, Europe; N. America. See Smith *in* Kurtzman & Fell (Eds) (*Yeasts, a taxonomic study* 4th edn: 598, 1998).

Oosporoidea Sumst. (1913) = Geotrichum fide Carmichael (*Mycol.* **49**: 820, 1957).

Oostroma Bonord. (1864) = Pseudovalsa fide Læssøe

(*SA* **13**: 43, 1994).

Oothecium Speg. (1919) = Asterostomella fide Farr (*Biblthca Mycol.* **35**, 1973).

Oothyrium Syd. (1939), anamorphic *Pezizomycotina*, Cpt.0eH.?. 1, Africa.

Oovorus Entz (1930) nom. dub., Fungi. 'phycomycetes'.

Opadorhiza T.F. Andersen & R.T. Moore (1996), anamorphic *Sebacina*. 1, N. America. See Moore *in* Sneh *et al.* (Eds) (*Rhizoctonia Species* Taxonomy, Molecular Biology, Ecology, Pathology and Disease Control: 25, 1996).

Opasterinella Speg. (1917) = Asterinella fide von Arx & Müller (*Stud. Mycol.* **9**, 1975).

Opeasterina Speg. (1919) = Asterina fide Müller & von Arx (*Beitr. Kryptfl. Schweiz* **11** no. 2, 1962).

Opegrapha Ach. (1809) nom. cons., Roccellaceae (±L). *c.* 361 (some on lichens), widespread. See Santesson (*Symb. bot. upsal.* **12** no. 1: 1, 1952; foliicolous spp.), Sérusiaux (*Lichenologist* **17**: 1, 1985; spp. with goniocysts), Torrente-Paños (*Cryptog. Mycol.* **8**: 159, 1987; asci), Torrente & Egea (*Biblthca Lichenol.* **32**, 1989), Tehler (*CJB* **68**: 2458, 1990; cladistics), Hawksworth (*J. Linn. Soc.* Bot. **109**: 543, 1992; gen. nomencl.), Hafellner (*Herzogia* **10**: 1, 1994; key 19 spp. on lichens), Letrouit-Galinou *et al.* (*Bull. Soc. linn. Provence* **45**: 389, 1994; ultrastr.), Bartók & Crisan (*Studia Universitatis Babeş-Bolyai* Ser. Biol. **49**: 33, 2004; Romania), Ertz *et al.* (*J. Linn. Soc.* Bot. **144**: 235, 2004; spp. on *Pertusaria* and *Ochrolechia*), Ertz & Diederich (*Lichenologist* **39**: 143, 2007; muriform-spored spp.).

Opegrapha Humb. (1793) nom. rej. = Graphis fide Hawksworth *et al.* (*Dictionary of the Fungi* edn 8, 1995).

Opegraphaceae Stizenb. (1862) = Roccellaceae.

Opegraphales M. Choisy ex D. Hawksw. & O.E. Erikss. (1986) = Arthoniales.

Opegraphella Müll. Arg. (1890) = Opegrapha Ach. fide Santesson (*Symb. bot. upsal.* **12** no. 1: 1, 1952).

Opegraphellomyces Cif. & Tomas. (1953) = Leptopeltis.

Opegraphites Debey & Ettingsh. (1859), Fossil Fungi, Ascomycota (L). 1 (Cretaceous), Belgium.

Opegraphoidea Fink (1933) = Opegrapha Ach. fide Hawksworth *et al.* (*Dictionary of the Fungi* edn 8, 1995).

Opegraphomyces E.A. Thomas ex Cif. & Tomas. (1953) = Opegrapha Ach. fide Hawksworth *et al.* (*Dictionary of the Fungi* edn 8, 1995).

Opercularia Stirt. (1878) [non *Opercularia* Gaertn. 1788, *Rubiaceae*] = Phyllobathelium fide Santesson (*Symb. bot. upsal.* **12** no. 1: 1, 1952).

operculate (of an ascus or sporangium), opening by an apical lid to discharge the spores, as in the ascus of the *Pezizales*; see ascus; cf. inoperculate.

Operculella Khesw. (1941) = Phacidiopycnis fide Sutton (*Mycol. Pap.* **141**, 1977).

operculum, a cover or lid.

Opethyrium Speg. (1919) nom. dub., Fungi. No spp. listed.

ophiobolin (cochliobolin), an antibiotic from *Cochliobolus miyabeanus* and *C. heterostrophus*; antibacterial, antifungal, anti-*Trichomonas vaginalis* (Ishibashi, *J. agric. Chem. Soc. Japan* **35**: 257, 1961); phytotoxic to rice (Orsenigo, *Phytopath. Z.* **29**: 189, 1957). See Tsuda *et al.* (*Tetrahedron Lett.* **35**: 3369, 1967; nomencl.).

Ophiobolus Riess (1854), Leptosphaeriaceae. Anamorphs *Coniothyrium*, *Phoma*, *Rhabdospora*. *c.* 101, widespread. *O. herpotrichus* (on cereals), *O. heterostrophus*. See Berthet & Korf (*Nat. can.* **96**: 247, 1969), Shoemaker (*CJB* **54**: 2365, 1976; key 31 Can. spp.), Walker (*Mycotaxon* **11**: 1, 1980; type, disposition 60 names), Scheuer (*Mycotaxon* **47**: 67, 1993; nomencl.), Kruys *et al.* (*MR* **110**: 527, 2006; phylogeny).

Ophiocapnocoma Bat. & Cif. (1963) = Metacapnodium fide Reynolds (*Mycotaxon* **23**: 153, 1985; = synonymy with *Limacinia*), Eriksson & Hawksworth (*SA* **6**: 124, 1987).

Ophiocapnodium Speg. (1918) ? = Euantennaria fide von Arx & Müller (*Stud. Mycol.* **9**, 1975).

Ophiocarpella Theiss. & Syd. (1915) = Sphaerulina fide Barr (*Contr. Univ. Mich. Herb.* **9**, 1972).

Ophioceras Sacc. (1883), Magnaporthaceae. 15 (wood etc., many spp. aquatic), widespread. See Conway & Barr (*Mycotaxon* **5**: 376, 1977), Chen *et al.* (*Mycol.* **91**: 84, 1999; phylogeny), Shearer *et al.* (*Mycol.* **91**: 145, 1999), Tsui *et al.* (*Mycoscience* **42**: 321, 2001; tropical spp.).

Ophiochaeta (Sacc.) Sacc. (1895) ≡ Acanthophiobolus.

Ophiociliomyces Bat. & I.H. Lima (1955), Pseudoperisporiaceae. 4 (from living leaves), widespread (tropical). See Barreto & Evans (*MR* **98**: 1107, 1994).

Ophiocladium Cavara (1893) = Ramularia Unger fide Sutton & Waller (*TBMS* **90**: 55, 1988).

Ophiocordyceps Petch (1931), Ophiocordycipitaceae. Anamorphs *Syngliocladium*, *Hymenostilbe*, *Paraisaria*. *c.* 140 (on insects), widespread. See Seifert & Boulay (*Mycol.* **96**: 929, 2004), Chaverri *et al.* (*Mycol.* **97**: 433, 2005), Sung *et al.* (*Stud. Mycol.* **57**, 2007; phylogeny).

Ophiocordycipitaceae G.H. Sung, J.M. Sung, Hywel-Jones & Spatafora (2007), Hypocreales. 10 gen. (+ 7 syn.), 260 spp. See Sung *et al.* (*Stud. Mycol.* **57**: 1, 2007; phylogeny, monogr.), Sung *et al.* (*Mol. Phylogenet. Evol.* **44**: 1204, 2007; phylogeny).

Ophiodeira Kohlm. & Volkm.-Kohlm. (1988), Halosphaeriaceae. 1 (marine), Caribbean. See Kohlmeyer & Volkmann-Kohlmeyer (*CJB* **66**: 2062, 1988), Spatafora *et al.* (*Am. J. Bot.* **85**: 1569, 1998; phylogeny), Campbell *et al.* (*Mycol.* **95**: 530, 2003).

Ophiodendron Arnaud (1952) nom. dub., anamorphic *Pezizomycotina*. See Hennebert (*TBMS* **51**: 13, 1968).

Ophiodictyon Sacc. & P. Syd. (1902) = Trichothelium fide Hawksworth *et al.* (*Dictionary of the Fungi* edn 8, 1995).

Ophiodothella (Henn.) Höhn. (1910), Phyllachoraceae. Anamorph *Acerviclypeatus*. 29, widespread (tropical). See Boyd (*Mycol.* **26**: 456, 1934), Hanlin (*Mycotaxon* **39**: 1, 1990), Hanlin *et al.* (*Mycotaxon* **44**: 103, 1992; key 26 spp.), Pearce & Hyde (*MR* **97**: 1272, 1993; spp. from Australia), Hyde & Cannon (*Mycol. Pap.* **175**: 114 pp., 1999; on palms), Hanlin *et al.* (*Mycoscience* **43**: 321, 2002; Venezuela), Hanlin (*Mycol.* **95**: 506, 2003; conidiomatal development).

Ophiodothis Sacc. (1883) ? = Balansia fide Diehl (*USDA agric. Monogr.* **4**, 1950), Ryley (*Mycotaxon* **95**: 97, 2006).

Ophiogene Petr. (1931) = Nematothecium fide von Arx & Müller (*Stud. Mycol.* **9**, 1975).

Ophiogloea Clem. (1903) = Vibrissea fide Sánchez &

Korf (*Mycol.* **58**: 733, 1966).

Ophiognomonia (Sacc.) Sacc. (1899), Diaporthales. 3, Europe; N. America. See Barr (*Mycol. Mem.* **7**, 1978), Monod (*Beih. Sydowia* **9**: 1, 1983), Wilson *et al.* (*Mycol.* **89**: 537, 1997).

Ophioirenina Sawada & W. Yamam. (1959), Meliolaceae. 1 (from leaves), India; Taiwan. See Patil & Mahamulkar (*Indian Phytopath.* **52**: 245, 1999).

Ophiomassaria Jacz. (1894), Pezizomycotina. 1 (on *Alnus*), Europe. See von Arx & Müller (*Stud. Mycol.* **9**, 1975).

Ophiomeliola Starbäck (1899), ? Parodiopsidaceae. 1 or 2, widespread (tropical). Material of the type species is effete fide Cannon (*in litt.*, 2001); possibly a synonym of *Ophioparodia* but probably based on discordant elements.

Ophionectria Sacc. (1878), Nectriaceae. Anamorph *Antipodium*. 2, widespread (tropical). See Rossman (*Mycol.* **69**: 355, 1977; key redisposed spp.), Subramanian & Bhat (*Kavaka* **6**: 55, 1979; anamorph, ontogeny), Rossman (*Mycol. Pap.* **150**, 1983), Rossman *et al.* (*Stud. Mycol.* **42**: 159, 1999), Castlebury *et al.* (*MR* **108**: 864, 2004; phylogeny).

Ophioparma Norman (1852), Ophioparmaceae (L). 10, widespread (northern hemisphere). See Rogers & Hafellner (*Lichenologist* **20**: 167, 1988), Kalb & Staiger (*Biblthca Lichenol.* **58**: 191, 1995; America), Skult (*Ann. bot. fenn.* **34**: 291, 1997), May *in* Glenn *et al.* (Eds) (*Lichenogr. Thomsoniana*: 77, 1998; N. Am.), Martínez & Aragón (*Bryologist* **106**: 528, 2003; Spain), Miądlikowska *et al.* (*Mycol.* **98**: 1088, 2006; phylogeny), Lumbsch *et al.* (*Nova Hedwigia* **86**: 105, 2008; phylogeny, chemistry).

Ophioparmaceae R.W. Rogers & Hafellner (1988), Umbilicariales (L). 3 gen. (+ 1 syn.), 25 spp.

Lit.: Rogers & Hafellner (*Lichenologist* **20**: 167, 1988), Kalb & Staiger (*Biblthca Lichenol.* **58**: 191, 1995), Printzen & Rambold (*Herzogia* **12**: 23, 1996), Skult (*Ann. bot. fenn.* **34**: 291, 1997), Wedin *et al.* (*MR* **109**: 159, 2005), Miądlikowska *et al.* (*Mycol.* **98**: 1088, 2006; phylogeny), Lumbsch *et al.* (*Nova Hedwigia* **86**: 105, 2008; phylogeny, chemistry).

Ophioparodia Petr. & Cif. (1932), Parodiopsidaceae. Anamorph *Septoideum*. 1, West Indies.

Ophiopeltis J.V. Almeida & Sousa da Câmara (1903) ? = Micropeltis fide Clements & Shear (*Gen. Fung.*, 1931).

Ophiopodium Arnaud (1954) nom. inval. = Grallomyces fide Deighton & Pirozynski (*Mycol. Pap.* **105**, 1966).

Ophiorosellinia J.D. Rogers, A. Hidalgo, F.A. Fern. & Huhndorf (2004), Xylariaceae. 1, Costa Rica. See Rogers *et al.* (*Mycol.* **96**: 172, 2004).

Ophiosira Petr. (1955), anamorphic *Pezizomycotina*, St.0fH.?. 1, Austria. See Petrak (*Sydowia* **9**: 510, 1955).

Ophiosphaerella Speg. (1909), Phaeosphaeriaceae. Anamorph *Scolecosporiella*. *c*. 9, widespread. See Walker (*Mycotaxon* **11**: 1, 1980), Dong *et al.* (*MR* **102**: 151, 1998), Wetzel *et al.* (*Pl. Dis.* **83**: 1160, 1999; distrib.), Wetzel *et al.* (*MR* **103**: 981, 1999; N. Am.), Câmara *et al.* (*Mycol.* **92**: 317, 2000; phylogeny), Hsiang *et al.* (*CJB* **81**: 307, 2003; mating types), Iriarte *et al.* (*Pl. Dis.* **88**: 1341, 2004; genetic diversity), Kaminski *et al.* (*Pl. Dis.* **89**: 980, 2005; molecular detection), Kaminski *et al.* (*Pl. Dis.* **90**: 146, 2006; genetic diversity), Schoch *et al.* (*Mycol.* **98**: 1041, 2006; phylogeny).

Ophiosphaeria Kirschst. (1906) = Acanthophiobolus fide von Arx & Müller (*Stud. Mycol.* **9**, 1975).

Ophiosporella Petr. (1947) = Phloeosporina fide Sutton (*Mycol. Pap.* **141**, 1977).

Ophiostoma Syd. & P. Syd. (1919), Ophiostomataceae. Anamorphs *Graphilbum*, *Leptographium*, *Pesotum*, *Sporothrix*, *Hyalorhinocladiella*, *Phialocephala*. 188 (in wood and bark, usually associated with beetles), widespread. *O. novo-ulmi* (Dutch elm disease; Brasier *Mycopath.* **115**: 151, 1991). See van Wyk & Wingfield (*Mycol.* **83**: 698, 1991; ascospores), Berbee & Taylor (*Exp. Mycol.* **16**: 87, 1992; 18S rRNA), Hausner *et al.* (*Mycol.* **84**: 870, 1992; posn), Wolfaardt *et al.* (*S. Afr. J. Bot.* **58**: 277, 1992; synoptic key, database), Grylls & Seifert *in* Wingfield *et al.* (Eds) (*Ceratocystis* and *Ophiostoma* Taxonomy, Ecology and Pathogenicity: 261, 1993; key), Hausner *et al.* (*CJB* **71**: 52, 1993; gen. concept), Seifert *et al.* *in* Wingfield *et al.* (Eds) (*Ceratocystis* and *Ophiostoma* Taxonomy, Ecology and Pathogenicity: 269, 1993), Wingfield *et al.* (*Ceratocystis* and *Ophiostoma* Taxonomy, Ecology and Pathogenicity, 1993; many papers), Spatafora & Blackwell (*MR* **91**: 1, 1994; posn), Benade *et al.* (*CJB* **74**: 891, 1996), Benade *et al.* (*Mycotaxon* **68**: 251, 1998; anam.), Jacobs *et al.* (*MR* **102**: 289, 1998), Hintz (*Gene* **237**: 215, 1999; chitin synthase gene), Wingfield *et al.* (*MR* **103**: 1616, 1999; phylogeny), Hausner *et al.* (*CJB* **78**: 903, 2000; phylogeny), Okada *et al.* (*Stud. Mycol.* **45**: 169, 2000; anams), Brasier & Kirk (*MR* **105**: 547, 2001; subspecies of *O. novo-ulmi*), Harrington *et al.* (*Mycol.* **93**: 111, 2001; phylogeny, *O. piceae* complex), Kim & Breuil (*MR* **105**: 331, 2001; sibling species), Marais & Wingfield (*MR* **105**: 240, 2001; spp. on *Protea*), Beer *et al.* (*MR* **107**: 469, 2003; *O. piceae* complex), Beer *et al.* (*Mycol.* **95**: 434, 2003; phylogeny of *O. stenoceras* complex), Lim *et al.* (*FEMS Microbiol. Lett.* **237**: 89, 2004; *O. clavigerum* agg.), Zhou *et al.* (*Mycol.* **96**: 1306, 2004; typification, phylogeny), Roets *et al.* (*Stud. Mycol.* **55**: 199, 2006; on *Proteaceae*), Spatafora *et al.* (*Mycol.* **98**: 1018, 2006; phylogeny), Zhang *et al.* (*Mycol.* **98**: 1076, 2006; phylogeny), Zipfel *et al.* (*Stud. Mycol.* **55**: 75, 2006; generic limit).

Ophiostomataceae Nannf. (1932), Ophiostomatales. 12 gen. (+ 4 syn.), 341 spp.

Lit.: Harrington (*Leptographium Root Diseases on Conifers*: 1, 1988), Malloch & Blackwell (*CJB* **68**: 1712, 1990), Brasier (*Mycopathologia* **115**: 151, 1991), Bates *et al.* (*MR* **97**: 449, 1993), Grylls & Seifert *in* Wingfield *et al.* (Eds) (*Ceratocystis* and *Ophiostoma* Taxonomy, Ecology and Pathogenicity: 261, 1993), Hausner *et al.* (*CJB* **71**: 1249, 1993), Hausner *et al.* (*CJB* **71**: 52, 1993), Samuels (*Ceratocystis* and *Ophiostoma* Taxonomy, Ecology and Pathogenicity: 15, 1993), Seifert & Okada (*Ceratocystis* and *Ophiostoma* Taxonomy, Ecology and Pathogenicity: 27, 1993), Wingfield *et al.* (*Ceratocystis* and *Ophiostoma* Taxonomy, Ecology and Pathogenicity, 1993), Wyk *et al.* (*Ceratocystis* and *Ophiostoma* Taxonomy, Ecology and Pathogenicity: 133, 1993), Benade *et al.* (*MR* **101**: 1108, 1997), Blackwell & Jones (*Biodiv. Cons.* **6**: 689, 1997), Okada *et al.* (*Stud. Mycol.* **45**: 169, 2000), Gorton *et al.* (*MR* **108**: 759, 2004), Zipfel *et al.* (*Stud. Mycol.* **55**: 75, 2006).

Ophiostomatales Benny & Kimbr. (1980). Sordariomycetidae. 1 fam., 12 gen., 341 spp. Stromata absent.

Ascomata perithecial, rarely cleistothecial, hyaline or black, thin-walled, membranous, usually long-necked, with ostiolar setae. Interascal tissue absent. Asci small, evanescent, formed in chains. Ascospores usually small, hyaline, mostly aseptate, often with eccentric wall thickening or sheaths. Anamorphs hyphomycetous, very varied. Necrotrophs of a wide range of plants, many economically important; some arthropod-associated, a few coprophilous; cosmop. Fam.:

Ophiostomataceae

Ceratocystis (*Ceratocystidaceae*) has almost indistinguishable teleomorphs, but quite different asexual forms, and is now considered to be allied to the *Microascales*.

Lit.: Wingfield *et al.* (*Ceratocystis and Ophiostoma*, 1993; *in* Hawksworth, *Ascomycete Systematics*: 333, 1994).

Ophiostomella Petr. (1925) ? = Scopinella fide Cannon (*in litt.*).

Ophiotexis Theiss. (1916) = Schweinitziella fide von Arx (*Acta Bot. Neerl.* **7**: 503, 1958).

Ophiotrichia Berl. (1893) = Acanthophiobolus fide Walker (*Mycotaxon* **11**: 1, 1980).

Ophiotrichum Kunze (1849) nom. dub., anamorphic *Pezizomycotina*, Hso.≡ eP.?. 2, Europe; West Indies. See Nannfeldt (*in litt.*).

Ophiovalsa Petr. (1966) = Cryptosporella fide Glawe & Jensen (*Mycotaxon* **25**: 645, 1986), Reid & Booth (*CJB* **65**: 1320, 1987), Castlebury *et al.* (*Mycol.* **94**: 1017, 2002; phylogeny), Mejía *et al.* (*MR* **112**: 23, 2008; phylogeny).

Ophisthomastigomycota,see *Chytridiomycota*.

Ophryomyces L. Léger & E. Hesse (1909) nom. dub., ? Fungi.

Ophthalmidium Eschw. (1824) = Porina Müll. Arg. fide Hawksworth *et al.* (*Dictionary of the Fungi* edn 8, 1995).

Opisteria (Ach.) Vain. (1909) ≡ Nephroma.

opisthokont, having one or more flagella at the posterior end.

Opisthokonta, Rankless term introduced by Cavalier-Smith & Chao (*Proc. R. Soc. Lond.* B **261**: 1, 1995) to include Animalia, Choanozoa and Fungi. The group is characterized by the presence of a single posterior cilium (or flagellum) together with non-discoid mitochondrial cristae. Based on current molecular phylogenies it would seem that the DRIPs clade and the Microsporidia should belong here also. This name was also used by Copeland (1956), at the rank of Phylum, for the Chytridiomycetes.

Oplophora Syd. (1921) = Nyssopsora fide Dietel (*Nat. Pflanzenfam.* **6**, 1928).

Oplotheciaceae Bat. & Cif. (1963) = Trichosphaeriaceae.

Oplotheciopsis Bat. & Cif. (1963) = Neorehmia fide Samuels & Barr (*CJB* **75**: 2165, 1998).

Oplothecium P. Syd. (1923) = Neorehmia fide Samuels & Barr (*CJB* **75**: 2165, 1998).

opportunistic (of fungi), normally saprobic and frequently common but on occasion able to cause disease in or grow on a host (esp. of humans or other animals), rendered susceptible by some predisposing factor(s).

opsis-form, see *Pucciniales*.

Opuntiella L. Léger & M. Gauthier (1932) nom. dub., Harpellales.

Oramasia Urries (1956) = Vermiculariopsiella fide Nawawi *et al.* (*Mycotaxon* **37**: 173, 1990).

Oraniella Speg. (1909), Pleosporales. 1, S. America. See Bose (*Phytopath. Z.* **41**, 1961), Aptroot (*Nova Hedwigia* **66**: 89, 1998).

Orbicula Cooke (1871), Pyronemataceae. 1, widespread (north temperate). See Hughes (*Mycol. Pap.* **42**, 1951), Campbell *et al.* (*Mycologist* **5**: 113, 1991), Hansen & Pfister (*Mycol.* **98**: 1029, 2006; phylogeny), Hansen *et al.* (*Mycol.* **97**: 1023, 2005; phylogeny), Perry *et al.* (*MR* **111**: 549, 2007; phylogeny).

Orbilia Fr. (1849), Orbiliaceae (±L). Anamorphs *Arthrobotrys*, *Dactylella*, *Dicranidion*, *Dwayaangam*, *Helicoön*, *Monacrosporium*, *Trinacrium*. *c.* 58 (on wood etc., sometimes nematode-trapping), widespread. See Benny *et al.* (*CJB* **56**: 2006, 1978; biol.), Spooner (*Biblthca Mycol.* **116**, 1987; 7 spp. Australasia), Baral (*SA* **13**: 113, 1994; concept), Pfister (*Mycol.* **86**: 451, 1994; anamorph), Liou & Tzean (*Mycol.* **89**: 876, 1997; DNA), Pfister (*Mycol.* **89**: 1, 1997; review), Kohlmeyer *et al.* (*Mycol.* **90**: 303, 1998), Webster *et al.* (*MR* **102**: 99, 1998; anamorph), Hagedorn & Scholler (*Sydowia* **51**: 27, 1999; DNA), Spatafora *et al.* (*Mycol.* **98**: 1018, 2006; phylogeny), Wang *et al.* (*Mol. Phylogen. Evol.* **41**: 295, 2006; phylogeny).

Orbiliaceae Nannf. (1932), Orbiliales (±L). 12 gen. (+ 12 syn.), 288 spp.

Lit.: Spooner (*Biblthca Mycol.* **116**, 1987; key gen.), Spooner (*Biblthca Mycol.* **116**: 711 pp., 1987), Pfister (*Mycol.* **86**: 451, 1994; anamorphs), Liou & Tzean (*Mycol.* **89**: 876, 1997), Pfister (*Mycol.* **89**: 1, 1997; review), Sugiyama (*Mycoscience* **39**: 487, 1998), Webster *et al.* (*MR* **102**: 99, 1998), Hagedorn & Scholler (*Sydowia* **51**: 27, 1999), Harrington *et al.* (*Mycol.* **91**: 41, 1999), Tehler *et al.* (*Mycol.* **92**: 459, 2000), Eriksson *et al.* (*Myconet* **9**: 1, 2003), Yang *et al.* (*Proc. natn Acad. Sci. U.S.A.* **104**: 8379, 2007; evolution).

Orbiliales Baral, O.E. Erikss., G. Marson & E. Weber (2003). Orbiliomycetes. 1 fam., 12 gen., 288 spp. Fam.:

Orbiliaceae

For *Lit.* see under fam.

Orbiliaster Dennis (1954) = Orbilia fide Baral (*SA* **13**, 1994).

Orbiliella Kirschst. (1938) = Orbilia fide Korf *in* Ainsworth *et al.* (Eds) (*The Fungi* **4A**: 249, 1973).

Orbiliomycetes O.E. Erikss. & Baral (2003), Pezizomycotina. 1 ord., 1 fam., 12 gen., 288 spp. Ord.:

Orbiliales

For *Lit.* see ord. and fam.

Orbiliomycetidae, see *Orbiliomycetes*.

Orbiliopsis (Sacc.) Syd. & P. Syd. (1924), ? Helotiales. 1, New Zealand.

Orbiliopsis Höhn. (1926) ? = Parorbiliopsis fide Spooner & Dennis (*Sydowia* **38**: 294, 1986).

orbilla (pl. **orbillae**), an apothecium (obsol.), (Sprengel, *Intr. study cryptog.*, 1807).

Orbiliopsis Velen. (1934), ? Helotiales. 1, former Czechoslovakia.

Orbimyces Linder (1944), anamorphic *Pezizomycotina*, Hso.0bP.10. 1 (marine), USA. See Kohlmeyer & Volkmann-Kohlmeyer (*Bot. Mar.* **34**: 1, 1991).

Orcadia G.K. Sutherl. (1915), Pezizomycotina. 1 (on marine *Algae*), British Isles; Norway. Possibly allied to the *Pezizales*, but the ecology is quite different. See Kohlmeyer & Kohlmeyer (*Marine Mycology*: 454, 1979), Rossman *et al.* (*Stud. Mycol.* **42**: 248 pp.,

1999).

Orcella Kuntze ex Earle (1909) ≡ Clitopilus.

Orceolina Hertel (1970), Trapeliaceae (L). 1, subantarctic islands. See Hertel (*Vortrag. Gesamtgeb. Bot.* ser. 2 **4**: 182, 1970), Lumbsch (*J. Hattori bot. Lab.* **83**: 1, 1997), Poulsen *et al.* (*Lichenologist* **33**: 323, 2001; phylogeny, morphology), Schmitt *et al.* (*Mycol.* **95**: 827, 2003; phylogeny), Miądlikowska *et al.* (*Mycol.* **98**: 1088, 2006; phylogeny), Lumbsch *et al.* (*MR* **111**: 1133, 2007).

Orcheomyces Burgeff (1909) nom. dub., Basidiomycota. Used for sterile mycelium associated with orchids; most of these belong in *Ceratobasidium* or *Thanatephorus*.

Orchesellaria Manier ex Manier & Lichtw. (1968), Asellariaceae. 4 (in *Collembola*), widespread. See Moss (*TBMS* **65**: 115, 1975; ultrastr.), Manier (*Revue Mycol.* Paris **43**: 341, 1979), Lichtwardt (*The Trichomycetes. Fungal associates of arthropods*, 1986; key), Valle (*Fungal Diversity* **21**: 167, 2006; Spain), White *et al.* (*Mycol.* **98**: 872, 2006; phylogeny).

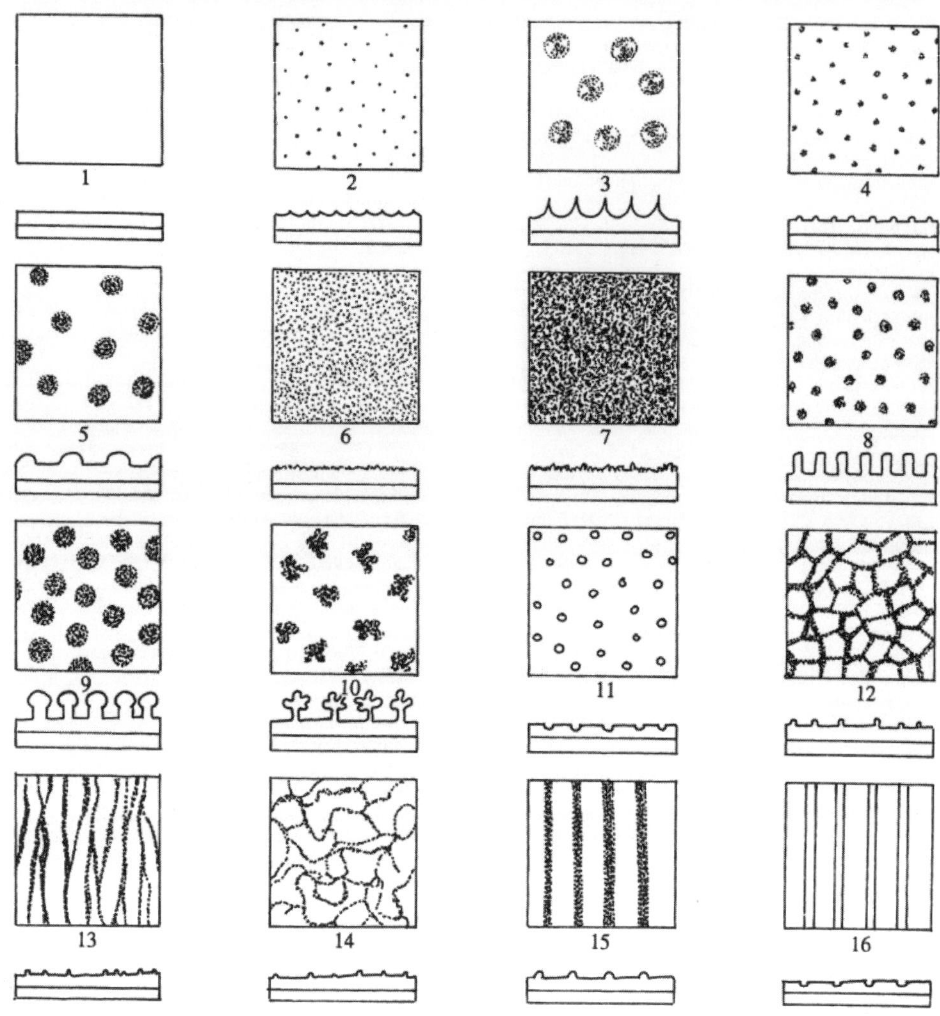

Fig. 20. Terminology for ornamentation (upper in surface view, lower in side view). 1, smooth (laevigate); 2, spinose (echinate); 3, aculeate; 4, granulate; 5, punctate (tuberculate); 6, verruculose; 7, verrucose; 8, baculate; 9, capitate; 10, irregularly projecting (see Sato & Sato, *Trans. Mycol. Soc. Jap.* **23**: 51, 1982, for additional terms, esp. for aeciospores); 11, foveate; 12, reticulate; 13, striate; 14, rugose; 15, cicatricose; 16, canaliculate.

orchil (**orchill**), see Dyeing.

orchinol (and **hircinol**), dihydrophenanthrenes produced by orchids as a response to infection by mycorrhizal fungi (Gäumann, *Phytopath. Z.* **49**: 212, 1964).

orculiform, see polarilocular.

Ordonia Racib. (1909), Septobasidiaceae. 2, N. Americe; Indonesia. See Oberwinkler (*Op. Bot.* **100**: 185, 1989).

Ordospora J.I.R. Larsson, D. Ebert & J. Vavra (1997), Microsporidia. 1. See Larsson *et al.* (*Eur. J. Protist.* **33**: 432, 1997).

Ordovicimyces M.K. Elias (1966), Fossil Fungi ? Fungi. 2 (Ordovician, ? Recent), USA.

Ordus K. Ando & Tubaki (1983), anamorphic *Pezizomycotina*, Hso.1bH.1. 1, Japan. See Ando & Tubaki (*TMSJ* **24**: 274, 1983), Ando *et al.* (*Mycoscience* **38**: 95, 1997), Watanabe *et al.* (*Mycoscience* **40**: 383, 1999).

orellanine, see Mycetism (1).

Oreophylla Cif. (1954) = Passalora fide Ciferri (*Sydowia* **8**: 253, 1954), Braun *et al.* (*N.Z. Jl Bot.* **37**: 297, 1999), Crous & Braun (*CBS Diversity Ser.* **1**: 571 pp., 2003).

Organizations, see Societies and organizations.

Ormathodium Syd. (1928) nom. dub., anamorphic *Pezizomycotina*. See Deighton (*Mycol. Pap.* **137**: 4, 1974).

Ormieresia Vivarès, Bouix & Manier (1977), Microsporidia. 1.

Ormomyces I.I. Tav. (1985), Laboulbeniaceae. 1, Africa; Indonesia. See Tavares (*Mycol. Mem.* **9**: 266, 1985).

ornamented (of organs, esp. spores), having the surface marked or sculptured with lines, wrinkles, warts, striations, ridges, reticulations, fibrils, scales etc.; not smooth. See Fig. 20 for terminology; also basidiospore, LO-analysis.

Ornasporonites Ramanujam & Rao (1979), Fossil Fungi. 1 (Miocene), India.

Ornatinephroma Gyeln. (1934) = Nephroma fide Hawksworth *et al.* (*Dictionary of the Fungi* edn 8, 1995).

Ornatispora K.D. Hyde, Goh, Joanne E. Taylor, J. Fröhl. (1999), Sordariomycetes. Anamorph *Didymostilbe*-like. 5, S.E. Asia. See Hyde *et al.* (*MR* **103**: 1432, 1999).

Ornatisporites M.G. Parsons & G. Norris (1999), Fossil Fungi. 1. See Parsons & Norris (*Palaeontographica* Abt. B **250**: 117, 1999).

Ornatopyrenis Aptroot (1991) = Mycomicrothelia fide Aptroot *et al.* (*Biblthca Lichenol.* **64**, 1997), Sipman & Aptroot (*Lichenologist* **37**: 307, 2005).

Ornithascus Velen. (1934) = Saccobolus fide van Brummelen (*Persoonia* Suppl. **1**: 1, 1967).

ornithocoprophilous, preferring habitats rich in bird droppings.

Oropogon Th. Fr. (1861), Parmeliaceae (L). 30, widespread. See Esslinger (*Syst. Bot. Monogr.* **28**, 1989; key), Chen (*Acta Mycol. Sin.* **15**: 173, 1996; China).

Orphanocoela Nag Raj (1989), anamorphic *Pezizomycotina*, Cpd.#eH-P.1. 3 (on *Poaceae*), widespread. See Nag Raj (*CJB* **67**: 3176, 1989).

Orphanomyces Savile (1974), ? Anthracoideaceae. 3 (on leaves of *Cyperaceae*), Europe; N. Americe; Asia. See Vánky (*Europ. Smut Fungi*: 570 pp., 1994; key), Vánky & McKenzie (*Fungal Diversity Res. Ser.* **8**: 259 pp., 2002; New Zealand).

Orphella L. Léger & M. Gauthier (1932), Legeriomycetaceae. 7 (in *Plecoptera*), Europe; N. America. See Lichtwardt (*The Trichomycetes. Fungal associates of arthropods*, 1986), Williams & Lichtwardt (*Mycol.* **79**: 473, 1987), Lichtwardt *et al.* (*Mycol.* **83**: 214, 1991), Santamaría & Girbal (*MR* **102**: 174, 1998; Spain), Lichtwardt *et al.* (*Mycol.* **93**: 764, 2001; Canada), Strongman & White (*CJB* **84**: 1478, 2006; Canada), Valle & Santamaría (*Mycol.* **97**: 1335, 2005; zygospores), White *et al.* (*Mycol.* **98**: 872, 2006; some species with possible affinities to *Kickxellales*).

Orphniospora Körb. (1874), ? Fuscideaceae (L). 3, widespread. See Hertel & Rambold (*Mitt. bot. StSamml., München* **27**: 111, 1988; posn).

Orphniosporaceae Bellem. & Hafellner (1984) ? = Fuscideaceae.

Orphniosporomyces Cif. & Tomas. (1953) ≡ Orphniospora.

Orphnodactylus Malloch & A. Mallik (1998), Phyllachoraceae. Anamorph *Hysterodiscula*. 2 (from living stems), widespread (circumboreal). See Malloch & Malik (*CJB* **76**: 1265, 1998).

Orpinomyces D.J.S. Barr, H. Kudo, Jakober & K.J. Cheng (1989), Neocallimastigaceae. 2 (anaerobic rumen fungi), widespread. See Ho & Barr (*Mycol.* **87**: 656, 1995; emend.), Ho *et al.* (*Fungal Diversity* **4**: 37, 2000).

Orromyces B. Sur & G.R. Ghosh (1987), Gymnoascaceae. Anamorph *Chrysosporium*-like. 1, India. See Sur & Ghosh (*Kavaka* **13**: 63, 1985).

orthidium (obsol.), conidiomata of *Pyrenotrichum*, sometimes incorrectly listed as if a generic name. See Hawksworth (*Bull. Br. Mus. nat. Hist., Bot.* **9**: 59, 1981).

Orthobellus A.A. Silva & Cavalc. (1973), ? Schizothyriaceae. 2, S. America. See Silva & Cavalcanti (*Publções Inst. Micol. Recife* **691**: 4, 1973).

Orthochaeta Sawada (1943) = Erysiphe fide Zheng (*Mycotaxon* **22**: 209, 1985).

orthographic variant, a variant spelling; under the Code the original spelling must be retained but typographic or orthographic errors should be corrected; orthographic variants should not be listed as synonyms, but when two or more generic names are so similar as to be confused they are treated as orthographic variants or, when based on different types, as homonyms. See Nomenclature.

Orthomyces Steinkr., Humber & J.B. Oliv. (1998), Entomophthoraceae. 1, USA. See Steinkraus *et al.* (*J. Invert. Path.* **72**: 5, 1998), Keller & Petrini (*Sydowia* **57**: 23, 2005; key).

Orthomycotina Steinkr., Humber & J.B. Oliv. (1998), Entomophthoraceae. 1, USA. See Steinkraus *et al.* (*J. Invert. Path.* **72**: 5, 1998), Keller & Petrini (*Sydowia* **57**: 23, 2005; key).

Orthoscypha Syd. (1927) = Pocillum fide Petrak (*Sydowia* **5**: 328, 1951).

Orthosomella E.U. Canning, Wigley & R.J. Barker (1983), Microsporidia. 2.

Orthothelohania Codreanu & Codreanu-Balcescu (1974), Microsporidia. 1.

orthotrophy, Corda's (1842) term for the condition in which a basidiospore primordium develops at the apex of the apophysis in contrast to **heterotrophy** when the primordium develops laterally.

Oscarbrefeldia Holterm. (1898), Saccharomycetes. 1, Java.

oscule (obsol.) a pore of a rust spore (Tulasne).

-osis (suffix), condition of; state caused by.

osmophily (adj. **-ilic**, **-ilous**), making growth under conditions of high osmotic pressure, as in *Xeromyces* (q.v.) and some yeasts on conc. sugar solutions. See Moustafa (*Can. J. Microbiol.* **21**: 1573, 1975; osmophilous spp. of Kuwait).

Osmoporus Singer (1944) = Gloeophyllum fide Donk (*Verh. K. ned. Akad. Wet.* tweede sect. **62**: 1, 1974).

osmotolerant, capable of growing in high osmotic pressure, e.g. some yeasts and filamentous fungi on concentrated sugar solutions.

osmotrophic, exhibiting absorptive nutrition as do fungi.

Osoriomyces Terada (1981), Laboulbeniaceae. 1, Taiwan. See Terada (*Mycotaxon* **13**: 412, 1981), Santamaría (*Mycol.* **96**: 761, 2004).

Ospriosporium Corda (1831) nom. dub., ? Fungi.

Ossicaulis Redhead & Ginns (1985), Lyophyllaceae. 2, widespread (north temperate). See Redhead & Ginns (*TMSJ* **26**: 362, 1985), Contu (*Bollettino dell'Associazione Micologica ed Ecologica Romana* **15**: 3, 1999), Clémençon (*Mycol. Progr.* **4**: 167, 2005; rhizomorph anatomy *Ossicaulis lignatilis*).

Osteina Donk (1966) = Postia fide Ryvarden (*Syn. Fung.* **5**, 1991).

Osteomorpha G. Arnaud ex Watling & W.B. Kendr. (1979), anamorphic *Sistotrema*. 1, Europe. See Markovskaja *et al.* (*Botanica Lithuanica* **8**: 179, 2002; Lithuania).

ostiole (sing. **ostiolum**), the schizogenous, paraphysis-lined cavity, ending in a pore, in the papilla or neck of a perithecium (Miller, *Mycol.* **20**: 196, 1928); any pore by which spores are freed from an ascigerous or pycnidial fruit-body. See von Arx (*K. ned. Akad. Wet.*, C, **76**: 289, 1973; taxonomic importance in *Ascomycota*).

Ostracoblabe Bornet & Flahault (1891), ? Fungi. 1, Europe. *O. implexa* (hinge disease of European oysters). See Alderman *in* Jones (Ed) (*Recent Advances in Aquatic Mycology*: 223, 1973; morphology, disease), Eckblad & Kirstiansen (*MR* **94**: 706, 1990) Excluded from *Chromista* by Dick (*in press*.

Ostracoderma Fr. (1825), anamorphic *Peziza*, St.0eH.6. 3, widespread. See Hennebert (*Persoonia* **7**: 183, 1973), Martinez & Godeas (*Mycotaxon* **65**: 453, 1997).

Ostracodermataceae Malençon (1964) = Pezizaceae.

Ostracodermidium Mukerji (1973), anamorphic *Pezizomycotina*, Hso.0eP.6. 1, India. See Mukerji (*TMSJ* **14**: 175, 1973).

Ostracodermis Locq. (1982), Fossil Fungi. 1, Libya.

Ostreichnion Duby (1862), Mytilinidiaceae. 3, widespread. See Barr (*Mycotaxon* **3**: 81, 1975; key), Checa & Blanco (*Mycotaxon* **94**: 225, 2005; Spain).

Ostreion Sacc. (1883) ≡ Ostreichnion.

Ostreionella Seaver (1926) = Actidium fide von Arx & Müller (*Stud. Mycol.* **9**, 1975).

Ostreola Darker (1963), Mytilinidiaceae. 4, Canada; India. See Tilak & Kale (*Indian Phytopath.* **21**: 289, 1968; key).

Ostropa Fr. (1825), Stictidaceae. 1, widespread (north temperate). See Sherwood (*Mycotaxon* **5**: 1, 1977).

Ostropaceae Rehm (1888) = Stictidaceae.

ostropalean (of asci), ones essentially unitunicate in structure and with a thickened apex penetrated by a narrow pore; see ascus.

Ostropales Nannf. (1932), (±L). Ostropomycetidae. 11 fam., 131 gen., 2753 spp. Stromatic development usually weak, reduced to intramatrical hyphae, or absent, or with usually crustose thalli. Ascomata ± apothecial, often deeply immersed and appearing perithecial, with rather varied peridial structure. Interascal tissue of ± simple paraphyses, occ. gelatinized. Asci usually narrow, cylindrical, with a well-developed capitate apical thickening often pierced by a J- pore. Ascospores varied, often filiform. Anamorphs pycnidial, varied, sometimes absent. Usually saprobic on stems, bark or wood, some lichenized with green algae or lichenicolous, cosmop. A polymorphic order in both biology and morphology, and molecular studies suggest that further subdivision may be justified. Fams:

(1) **Asterothyriaceae**
(2) **Coeonogoniaceae**
(3) **Gomphillaceae**
(4) **Graphidaceae**
(5) **Gyalectaceae**
(6) **Myeloconidiaceae**
(7) **Odontotremataceae**
(8) **Phlyctidaceae**
(9) **Porinaceae**
(10) **Stictidaceae**
(11) **Thelotremataceae**.

Lit.: Grube *et al.* (*MR* **108**: 1111, 2004), Lücking *et al.* (*Mycol.* **96**: 283, 2004), Lumbsch *et al.* (*MR* **111**: 257, 2007; phylogeny), Sherwood (*Mycotaxon* **5**: 1, 1977; monogr., keys), Winka *et al.* (*Lichenologist* **30**: 455, 1998; DNA; lichenized taxa).

Ostropella (Sacc.) Höhn. (1918), Melanommataceae. 5, widespread. See Huhndorf (*Mycol.* **85**: 490, 1993), Vasil'eva (*Mycoscience* **38**: 341, 1997), Huhndorf (*Mycol.* **90**: 527, 1998).

Ostropomycetidae Reeb, Lutzoni & Cl. Roux (2004), Lecanoromycetes. The second major subdivision of the *Lecanoromycetes*, including a number of non-lichenized taxa. Ords:

(1) **Agyriales**
(2) **Baeomycetales**
(3) **Ostropales**
(4) **Pertusariales**

Lit.: Grube *et al.* (*MR* **108**: 1111, 2004), Hofstetter *et al.* (*Mol. Phylogenet. Evol.* **44**: 412, 2007), Lumbsch *et al.* (*Mol. Phylogenet. Evol.* **31**: 822, 2004), Lumbsch *et al.* (*MR* **111**: 257, 2007; phylogeny), Lutzoni *et al.* (*Am. J. Bot.* **91**: 1446, 2004), Miądlikowska *et al.* (*Mycol.* **98**: 1088, 2006), Wedin *et al.* (*MR* **109**: 159, 2005).

Oswaldia Rangel (1921) = Apiosphaeria fide Müller & von Arx *in* Ainsworth *et al.* (Eds) (*The Fungi* **4A**: 87, 1973).

Oswaldina Rangel (1921), anamorphic *Apiosphaeria*, St.0fH.15. 1, Brazil. See Dianese *et al.* (*Sydowia* **46**: 233, 1994; reinstatement gen.).

Oswaldoa Bat. & I.H. Lima (1959) ? = Myriangiella fide Hawksworth *et al.* (*Dictionary of the Fungi* edn 8, 1995).

Otagoa Lloyd (1922) nom. nud., ? Fungi.

Otidea (Pers.) Bonord. (1851), Pyronemataceae. *c.* 23, widespread (north temperate). See Kanouse (*Mycol.* **41**: 660, 1949), Harmaja (*Karstenia* **14**: 138, 1974; gen. concept), Cao *et al.* (*Mycol.* **82**: 734, 1990; key Chinese spp.), Landvik *et al.* (*Nordic Jl Bot.* **17**: 403, 1997; DNA), Harrington *et al.* (*Mycol.* **91**: 41, 1999; phylogeny), Hansen & Pfister (*Mycol.* **98**: 1029, 2006; phylogeny), Liu & Zhuang (*Fungal Diversity* **23**: 181, 2006), Perry *et al.* (*MR* **111**: 549, 2007; phylogeny).

Otideaceae Eckblad (1968) = Pyronemataceae.

Lit.: Landvik *et al.* (*Nordic Jl Bot.* **17**: 403, 1997).

Otidella Sacc. (1889) ≡ Caloscypha.
Otideopsis B. Liu & J.Z. Cao (1987), Pyronemataceae. 2, China; Europe. See Moravec (*Mycol. Helv.* **3**: 135, 1988), Liu & Zhuang (*Fungal Diversity* **23**: 181, 2006), Liu & Zhuang (*Mycosystema* **25**: 546, 2006; phylogeny).
Otomyces Hallier (1869) nom. dub., anamorphic *Pezizomycotina*.
Otomyces Wreden (1874) ? = Aspergillus fide Hawksworth *et al.* (*Dictionary of the Fungi* edn 8, 1995).
otomycosis, see mycosis.
Otthia Nitschke ex Fuckel (1870), Botryosphaeriaceae. Anamorph *Diplodia*-like. *c.* 11, widespread. See Scheinpflug (*Ber. schweiz. bot. Ges.* **68**: 325, 1958), Van Niekerk *et al.* (*Mycol.* **96**: 781, 2004; phylogeny), Phillips *et al.* (*Mycol.* **97**: 513, 2005).
Otthiella (Sacc.) Sacc. & D. Sacc. (1905) = Otthia fide Barr (*Rept. Kevo Subarct. Res. Stn* **11**: 12, 1974).
Otwaya G.W. Beaton (1978), Hyaloscyphaceae. 1 (on *Nothofagus*), Australia. See Beaton (*TBMS* **71**: 219, 1978).
Oudemans (Cornelius Anton Jan Abraham; 1825-1906; Netherlands). Medical degree, Leiden (1847); Lecturer, Rotterdam Medical School (1848-1859); Professor of Medicine & Botany, Amsterdam Athenaeum (1859-1877); Professor of Botany & Pharmacognosy, Amsterdam University (1877-1896); General Secretary, Royal Netherlands Academy of Sciences (1879-1895). Produced a great and still valuable catalogue of European fungi arranged by associated organisms. *Publs. Catalogue Raisonné des Champignons des Pays-Bas* (1905); *Enumeratio Systematica Fungorum* 5 vols (1919-1925) [published posthumously]. *Biogs, obits etc.* Moll (*Bericht der Deutschen Botanischen Gesellschaft* **26a**: 12, 1908).
Oudemansia Speg. (1880) [non *Oudemansia* Miq. 1854, *Byttneriaceae*] ≡ Oudemansiella.
Oudemansiella Speg. (1881), Physalacriaceae. *c.* 15, widespread (tropical; temperate). See Pegler & Young (*TBMS* **87**: 583, 1987; key), Baroni & Ortiz (*Mycotaxon* **82**: 269, 2002; Puerto Rico), Mizuta (*Nippon Kingakukai Kaiho* **46**: 25, 2005; Japan).
Outhovia Nieuwl. (1916) ≡ Scopularia.
Ovadendron Sigler & J.W. Carmich. (1976), anamorphic *Pezizomycotina*, Hso.0eH.38. 1 (from humans), Europe. See Sigler & Carmichael (*Mycotaxon* **4**: 391, 1976).
oval, widely elliptical (Fig. 23.1).
ovariicolous, living in ovaries.
ovate (of a surface [or sometimes a solid]), **ovoid** (of a solid) (Fig. 23.11), like a hen's egg with the narrower end at the top (Fig. 23.1).
Ovavesicula T.G. Andreadis & J.L. Hanula (1987), Microsporidia. 1. See Andreadis & J.L. Hanula (*J. Protozool.* **34**: 21, 1987), Vossbrinck & Andreadis (*J. Invert. Path.* **96**: 270, 2007; phylogeny).
Overeemia Arnaud (1954) = Brooksia fide Deighton & Pirozynski (*Mycol. Pap.* **105**, 1966).
Oviascoma Y.J. Yao & Spooner (1996), Pyronemataceae. 1, British Isles. See Yao & Spooner (*MR* **100**: 102, 1996).
Ovinus (Lloyd) Torrend (1920) ≡ Albatrellus.
Ovipleistophora M. Pekkarinen, J. Lom & F. Nilsen (2002), Microsporidia. 1. See Pekkarinen *et al.* (*Dis. Aq. Organ.* **48**: 133, 2002).
Ovosphaerella Laib. (1922) = Mycosphaerella fide von Arx & Müller (*Stud. Mycol.* **9**, 1975).
Ovularia Sacc. (1880) = Ramularia Unger fide Hughes (*TBMS* **32**: 34, 1949).
Ovulariella Bubák & Kabát (1912) nom. nud., anamorphic *Pezizomycotina*.
Ovulariopsis Pat. & Har. (1900), anamorphic *Phyllactinia*, Hsy.0eH.1. 9, widespread. See Braun (*Beih. Nova Hedwigia* **89**: 700 pp., 1987).
Ovularites A.C. Whitford (1916), Fossil Fungi. 1 (Cretaceous), USA.
Ovulinia F.A. Weiss (1940), Sclerotiniaceae. Anamorph *Ovulitis*. 2, Europe; N. America. See Holst-Jensen *et al.* (*Mycol.* **89**: 885, 1997; phylogeny), Holst-Jensen *et al.* (*Nordic Jl Bot.* **18**: 705, 1999; phylogeny).
Ovulitis N.F. Buchw. (1970), anamorphic *Ovulinia*, Hso.0eH.10. 2, USA; British Isles. See Buchwald (*Friesia* **9**: 328, 1970).
Oxneria S.Y. Kondr. & Kärnefelt (2003) = Xanthomendoza fide Kondratyuk & Kärnefelt (*Ukrainian Jour. Bot.* **60**: 427, 2003), Gaya *et al.* (*MR* **112**: 528, 2008).
Oxodeora K.D. Hyde & P.F. Cannon (1999), Phyllachoraceae. 1 (living palm leaves), Caribbean. See Hyde & Cannon (*Mycol. Pap.* **175**: 114, 1999).
oxydated (of crustose lichens), having thalli tinged rust-red by iron oxides; oxydized.
Oxydontia L.W. Mill. (1933) ≡ Sarcodontia fide Donk (*Taxon* **5**: 69, 1956).
Oxydothis Penz. & Sacc. (1898), Xylariales. Anamorph *Selenosporella*. 41 (saprobic on palms etc.), widespread (esp. tropical). See Hyde (*Sydowia* **45**: 226, 1993; Australia), Hyde (*Sydowia* **45**: 204, 1993), Hyde (*Sydowia* **46**: 265, 1994; revision), Wong & Hyde (*Fungal Diversity* **2**: 181, 1999; ultrastr.), Shenoy *et al.* (*Nova Hedwigia* **80**: 511, 2005; Hong Kong), Hidayat *et al.* (*Fungal Diversity* **23**: 159, 2006; phylogeny).
Oxyflavus Ryvarden (1973) = Leucophellinus fide Ryvarden (*Syn. Fung.* **5**, 1991).
Oxyporus (Bourdot & Galzin) Donk (1933), Agaricomycetes. 10, widespread. *Hymenochaetales* or *Agaricales* (*Oxyporus* clade). See Gilbertson & Ryvarden (*N. Amer. Polyp.* Vol. **2** *Megasporoporia - Wrightoporia*: 437, 1987), Ryvarden (*Syn. Fung.* **5**: 363 pp., 1991), Martín & Montón (*Boletín de Sanidad Vegetal, Plagas* **30**: 93, 2004; molecular identification *Oxyporus latemarginatus*).
Oxysporium Lév. (1863) = Helminthosporium fide Saccardo (*Syll. fung.* **10**: 613, 1892).
Oxystoma Eschw. (1824) = Graphis fide Hawksworth *et al.* (*Dictionary of the Fungi* edn 8, 1995).
oyster mushroom (**- cap fungus**), basidiomata of the edible *Pleurotus ostreatus*.
Ozocladium Mont. (1851) = Polistroma fide Hawksworth *et al.* (*Dictionary of the Fungi* edn 8, 1995).
Ozonium Link (1809) nom. dub., anamorphic *Agaricomycotina*. Used for anamorphs of *Coprinus* and similar genera. See Hughes (*CJB* **36**: 727, 1958).
P, see Metabolic products, Steiner's Stable PD solution.
Paathramaya Subram. (1956), anamorphic *Pezizomycotina*, Hsp.0eP.10. 3, India. See Bhat (*TBMS* **85**: 101, 1985; key).
Pachnocybaceae Oberw. & R. Bauer (1989), Pachnocybales. 1 gen., 1 spp.
Lit.: Oberwinkler & Bandoni (*CJB* **60**: 1726, 1982), Kropp & Corden (*Mycol.* **78**: 334, 1986), Kleven & McLaughlin (*CJB* **67**: 1336, 1989), Oberwinkler & Bauer (*Sydowia* **41**: 224, 1989), McLaughlin *et al.* (*Mycol. Soc. Amer. Newsl.* **41**: 28, 1990),

Berres *et al.* (*Mycol.* **87**: 821, 1995), Begerow *et al.* (*CJB* **75**: 2045, 1998).

Pachnocybales R. Bauer, Begerow, J.P. Samp., M. Weiss & Oberw. (2006). Pucciniomycetes. 1 fam., 1 gen., 1 spp. Fam.:
Pachnocybaceae
For *Lit.* see under fam.

Pachnocybe Berk. (1836), Pachnocybaceae. 1, Europe. See Begerow *et al.* (*CJB* **75**: 2045, 1998; phylogeny), Bauer *et al.* (*Mycol. Progr.* **5**: 41, 2006).

Pachnodium H.P. Upadhyay & W.B. Kendr. (1975) = Pesotum fide Upadhyay & Kendrick (*Mycol.* **67**: 802, 1975), Okada *et al.* (*CJB* **76**: 1495, 1998).

Pachnolepia A. Massal. (1855) = Arthonia fide Hawksworth *et al.* (*Dictionary of the Fungi* edn 8, 1995).

Pachyascaceae Poelt ex P.M. Kirk, P.F. Cannon & J.C. David (2001), Lecanoromycetes (inc. sed.) (L). 1 gen., 1 spp.
Lit.: Poelt (*Sydowia* **38**: 241, 1985).

Pachyascus Poelt & Hertel (1968), Pachyascaceae (L). 1 (in leaf axils of *Andreaea*), Europe. See Poelt & Hertel (*Ber. dt. bot. Ges.* **81**: 210, 1968), Poelt (*Sydowia* **38**: 241, 1985).

Pachybasidiella Bubák & Syd. (1915) = Aureobasidium fide de Hoog & Hermanides-Nijhof (*Stud. Mycol.* **15**, 1977).

Pachybasium Sacc. (1885) = Trichoderma Pers. (1794) fide Hughes (*CJB* **36**: 727, 1958), Bissett (*CJB* **69**: 2373, 1991).

Pachycladina Marvanová (1987), anamorphic *Pezizomycotina*, Hso.1bH.10. 1 (from foam), Europe; N. America. See Marvanová (*TBMS* **87**: 617, 1986), Marvanová & Bärlocher (*MR* **102**: 750, 1998; Canada).

Pachycudonia S. Imai (1950), ? Helotiales. 3, Japan. See Imai (*Bot. Mag.* Tokyo **63**: 235, 1950).

Pachyderma Schulzer (1876) [non *Pachyderma* Blume 1826, *Oleaceae*] = Mycenastrum fide Stalpers (*in litt.*).

pachydermatous (1) thick-skinned; (2) (of hyphae), having the outer wall thicker than the lumen.

Pachydisca Boud. (1885), ? Helotiaceae. 1, Europe. See Dumont (*Mycol.* **67**: 162, 1975).

Pachydiscula Höhn. (1915) = Cryptosporiopsis fide Petrak (*Annls mycol.* **21**: 182, 1923).

Pachyella Boud. (1907), Pezizaceae. *c.* 19, widespread. See Pfister (*CJB* **51**: 2009, 1973; key), Pfister & Candoussau (*Mycotaxon* **13**: 457, 1981), Häffner (*Rheinl.-Pfälz. Pilzj.* **2**: 118, 1992), Pfister (*Mycotaxon* **54**: 393, 1995), Norman & Egger (*Mycol.* **91**: 820, 1999; phylogeny), Hansen *et al.* (*Mycol.* **93**: 958, 2001; phylogeny), Harrington & Pfister (*Harvard Pap. Bot.* **6**: 113, 2001; in culture), Hansen *et al.* (*Mol. Phylogen. Evol.* **36**: 1, 2005; phylogeny).

Pachykytospora Kotl. & Pouzar (1963), Polyporaceae. 3, widespread. See Kotlaba & Pouzar (*Česká Mykol.* **33**: 129, 1979; key), Piątek (*Ann. bot. fenn.* **42**: 23, 2005; *Pachykytospora nanospora*).

Pachykytosporaceae Jülich (1982) = Polyporaceae.

Pachylepyrium Singer (1958), Strophariaceae. 7, widespread (north temperate). See Singer (*Sydowia* **11**: 321, 1957), Moser (*Hoppea* **61**: 267, 2000).

Pachyma Fr. (1822), anamorphic *Macrohyporia, Lentinus, Polyporus*. 1, N. America. See Weber (*Mycol.* **21**: 113, 1929; sclerotia), Ginns & Lowe (*CJB* **61**: 1672, 1984).

Pachyospora A. Massal. (1852), Hymeneliaceae (L). 1, widespread (arctic-alpine). See Hertel (*Willdenowia* **6**: 225, 1971).

Pachypatella Theiss. & Syd. (1915), Parmulariaceae. 2, Asia.

Pachyphiale Lönnr. (1858), Gyalectaceae (L). 7, widespread. See Vězda (*Sborn. Vysoké Školy Zeměd. Lesn. Brno* C **1**: 22, 1958), Vězda (*Ergebn. Forsch. Unternehmens Nepal Himalaya* **6**: 127, 1974; key 5 spp.), Alvarez Andrés (*Nova Hedwigia* **77**: 139, 2003; Iberian peninsula), Kauff & Büdel (*Bryologist* **108**: 272, 2005; ontogeny).

Pachyphlodes Zobel (1854) = Pachyphloeus fide Fischer (*Nat. Pflanzenfam.* **5b**: viii, 1938).

Pachyphloeus Tul. & C. Tul. (1844), Pezizaceae. 11 (hypogeous), widespread. See Gilkey (*N. Amer. Fl.* **2**, 1954; key N. Am. spp.), Lange (*Dansk bot. Ark.* **16**: 32, 1956; key Eur. spp.), Norman & Egger (*Mycol.* **91**: 820, 1999; phylogeny), Percudani *et al.* (*Mol. Phylogen. Evol.* **13**: 169, 1999; phylogeny), Hansen *et al.* (*Mycol.* **93**: 958, 2001; phylogeny), Calonge *et al.* (*Boll. Gruppo Micol. 'G. Bresadola'* **45**: 51, 2002; Spain), Colgan & Trappe (*Mycotaxon* **90**: 281, 2004; USA), Frank *et al.* (*Mycotaxon* **98**: 253, 2006; USA), Tedersoo *et al.* (*New Phytol.* **170**: 581, 2006; mycorrhizas), Læssøe & Hansen (*MR* **111**: 1075, 2007; phylogeny).

Pachyphysis R.C. Harris & Ladd (2007), Porpidiaceae (L). 1, USA. See Harris & Ladd (*Opuscula Philolichenum* **4**: 57, 2007).

pachypleurous, thick-walled.

Pachyrhytisma Höhn. (1917) = Rhytisma fide Nannfeldt (*Nova Acta R. Soc. Scient. upsal.*, 1932).

Pachysacca Syd. (1930), Dothideaceae. 1, Australia. See Swart (*TBMS* **79**: 261, 1982).

Pachysolen Boidin & Adzet (1958), Saccharomycetales. 1, Europe. See Suh *et al.* (*Mycol.* **98**: 1006, 2006; phylogeny).

Pachyspora Kirschst. (1906) = Delitschia fide von Arx & Müller (*Stud. Mycol.* **9**, 1975).

Pachysporaria (Malme) M. Choisy (1949) ? = Rinodina fide Marbach (*Biblthca Lichenol.* **74**, 2000).

Pachysterigma Johan-Olsen ex Bref. (1888) = Tulasnella fide Rogers (*Annls mycol.* **31**: 183, 1933).

Pachythyrium G. Arnaud ex Spooner & P.M. Kirk (1990), Microthyriaceae. 1, Europe. See Spooner & Kirk (*MR* **94**: 223, 1990).

Pachytichospora Van der Walt (1978) = Kazachstania fide Augustyn *et al.* (*Syst. Appl. Microbiol.* **13**: 44, 1990; fatty acid analysis), Kurtzman & Robnett (*FEMS Yeast Res.* **3**: 417, 2003).

Pachytrichum Syd. (1925) = Periconia fide Linder (*Mycol.* **29**: 659, 1937).

Pachytrype Berl. ex M.E. Barr, J.D. Rogers & Y.M. Ju (1993), ? Calosphaeriaceae. Anamorph *Cytospora*-like. 2 (bark and wood), widespread. See Barr *et al.* (*Mycotaxon* **48**: 529, 1993), Fernández *et al.* (*Mycol.* **96**: 175, 2004; Costa Rica), Mostert *et al.* (*Stud. Mycol.* **54**: 115 pp., 2006).

Pacispora Oehl & E. Sieverd. (2004), Pacisporaceae. 3, widespread. See Oehl & Sieverding (*J. Appl. Bot.* Angew. Bot. **78**: 74, 2004), Redecker & Raab (*Mycol.* **98**: 885, 2006; phylogeny), Walker *et al.* (*MR* **111**: 253, 2007; nomencl.).

Pacisporaceae C. Walker, Błaszk., A. Schüssler & Schwarzott (2004), Diversisporales. 1 gen. (+ 1 syn.), 3 spp.
Lit.: Azcon-Aguilar & Barea (*Mycorrhiza* **6**: 457, 1996), van der Heijden *et al.* (*Nature* Lond. **396**: 69, 1998), Schüßler *et al.* (*MR* **105**: 1413, 2001), Oehl &

Sieverding (*J. Appl. Bot.* Angew. Bot. **78**: 74, 2004), Walker & Schüssler (*MR* **108**: 982, 2004), Walker *et al.* (*MR* **108**: 707, 2004), Redecker (*Glomeromycota* Arbuscular mycorrhizal fungi and their relative(s), *in* The Tree of Life Web Project, http://tolweb.org: [unpaginated], 2005).

Pactilia Fr. (1837), anamorphic *Pezizomycotina*, Hsp.?.?. 2 or 3, Europe; S. America.

PAD, Istituto di Botanica e Fisiologia Vegetale, Padova (Italy); founded 1837; a state institution; see Montemartini Corte (*in* Minelli (Ed.), *The botanical garden of Padua 1545-1995*: 271, 1995; mycol. collns esp. Saccardo).

paddy (paddi) straw mushroom, see straw mushrooms.

Padixonia Subram. (1972), anamorphic *Xylaria*, Hsy.0eH.9. 1, Ghana. The connection with termiticolous *Xylaria* needs confirmation. See Subramanian (*Curr. Sci.* **41**: 282, 1972), Dixon (*J. Linn. Soc.* Bot. **91**: 203, 1985).

Paecilomyces Bainier (1907), anamorphic *Byssochlamys*, Hso.0eH.15. *c.* 76, widespread. See also *Isaria*. See Samson (*Stud. Mycol.* **6**, 1974; key), Castro *et al.* (*J. Med. Vet. Mycol.* **28**: 15, 1990; review of *Paecilomyces* in human infections), Mountfort & Rhodes (*Appl. Environm. Microbiol.* **57**: 1963, 1991; anaerobic growth of marine *P. lilacinus*), Yaguchi *et al.* (*Mycoscience* **37**: 55, 1996; ubiquinones), Cantone & Vandenberg (*MR* **102**: 209, 1998; genetic diversity), Domenech *et al.* (*Microbiology* Reading **145**: 2789, 1999; cell wall chemistry), Azevedo *et al.* (*Scientia Agricola* **57**: 729, 2000; genetic diversity), Pitt *et al.* (*Integration of Modern Taxonomic Methods for Penicillium and Aspergillus Classification*: 9, 2000; accepted names), Oborník *et al.* (*Can. J. Microbiol.* **47**: 813, 2001; polyphyly), Haugland *et al.* (*Syst. Appl. Microbiol.* **27**: 198, 2004; quantitative PCR), Luangsa-ard *et al.* (*Mycol.* **96**: 773, 2004; polyphyly, generic limits), Luangsa-ard *et al.* (*Kasetsart Journal* Natural Sciences **38**: 94, 2004; Thailand), Luangsa-ard *et al.* (*MR* **109**: 581, 2005; removal of entomopathogenic spp.).

Paepalopsis J.G. Kühn (1882), Urocystidaceae. 3 (on *Primulaceae*), Europe; N. Americe; Asia. See Kirk (*in litt.*).

Pagidospora Drechsler (1960), anamorphic *Agaricomycetes*, Hsy.0eH.1/41. 1 (with clamp connexions, on amoebae), USA. See Drechsler (*Sydowia* **14**: 246, 1960).

Paidania Racib. (1909) = Erikssonia fide von Arx & Müller (*Beitr. Kryptfl. Schweiz* **11** no. 1, 1954).

Paipalopsis, see *Paepalopsis*.

Palaeachlya F.R.S. Duncan (1876), Fossil Fungi, Oomycota. 2 (Silurian – Recent), widespread.

Palaeancistrus R.L. Dennis (1970), Fossil Fungi, Agaricomycetes. 1 (Carboniferous), USA.

Palaeoagaracites Poinar & Buckley (2007), Fossil Fungi, Agaricomycetes. 1.

Palaeoamphisphaerella Ramanujam & Srisailam (1980), Fossil Fungi, Ascomycota. 2 (Miocene), India.

Palaeoasterina S. Mitra, Bera & Manju Banerjee (2002), Fossil Fungi. 1, India. See Mitra *et al.* (*Phytomorphology* **52**: 286, 2002).

Palaeocephala Singer (1962), Agaricales. 1, Sierra Leone. Belongs either to the *Marasmiaceae* or the *Physalacriaceae*. See Singer (*Sydowia* **15**: 60, 1961), Antonín (*Fungus Flora of tropical Africa* **1**, 2007).

Palaeocercospora S. Mitra & Manju Banerjee (2000), Fossil Fungi, Pezizomycotina. 1, India. See Mitra & Banerjee (*Journal of Mycopathological Research* **38**: 8, 2000).

Palaeocirrenalia Ramanujam & Srisailam (1980), Fossil Fungi, anamorphic *Pezizomycotina*. 2 (Miocene), India.

Palaeoclavaria Poinar & A.E. Br. (2003), Fossil Fungi. 1, Myanmar. See Poinar & Brown (*MR* **107**: 765, 2003).

Palaeocolletotrichum S. Mitra & Manju Banerjee (2000), Fossil Fungi, Pezizomycotina. 1, India. See Mitra & Banerjee (*Journal of Mycopathological Research* **38**: 10, 2000).

Palaeocybe Dörfelt & Striebich (2000), Fossil Fungi, Psathyrellaceae. 1, Germany. See Dörfelt & Striebich (*Z. Mykol.* **66**: 31, 2000).

Palaeocytosphaera R.B. Singh & G.V. Patil (1980), Fossil Fungi, anamorphic *Pezizomycotina*. 1 (Cretaceous), India.

Palaeodikaryomyces Dörfelt (1998), Fossil Fungi. 1 (Triassic), Germany. See Dörfelt & Schäfer (*Z. Mykol.* **64**: 145, 1998).

Palaeodiplodites Watanabe, H. Nishida & Kobayashi (1999), Fossil Fungi. 1. See Watanabe *et al.* (*Int. J. Pl. Sci.* **160**: 440, 1999).

Palaeofibula J.M. Osborn, T.N. Taylor & J.F. White (1989), Fossil Fungi. 1 (middle Triassic), Antarctica.

Palaeoglomus D. Redecker, Kodner & L.E. Graham (2002), Fossil Fungi, Glomerales. 1, USA. See Redecker *et al.* (*Mycotaxon* **84**: 35, 2002).

Palaeoleptosphaeria Barlinge & Paradkar (1982), Fossil Fungi, Ascomycota. 1 (Cretaceous), India.

Palaeomycelites Bystrov (1956), Fossil Fungi (mycel.) Fungi. 1 (Devonian), former USSR.

Palaeomyces Renault (1896), Fossil Fungi (mycorrhizal endomycobiont) Fungi. 3 (Carboniferous), France; British Isles. See Butler (*TBMS* **22**: 274, 1939).

Palaeomycites Mesch. (1902) ≡ Palaeomyces.

Palaeopede D.E. Ether. (1899), Fossil Fungi ? Fungi. 1 (Devonian), Australia.

Palaeopericonia C.G. Ibáñez & Zamuner (1996), Fossil Fungi. 1, Argentina. See Ibáñez & Zamuner (*Mycotaxon* **59**: 138, 1996).

Palaeoperone D.E. Ether. (1891), Fossil Fungi ? Fungi. 1 (Permo-Carboniferous), Australia.

Palaeophoma Singhai (1975), Fossil Fungi, anamorphic *Pezizomycotina*. 1 (Eocene), India.

Palaeophthora Singhai (1978), Fossil Fungi, Oomycota. 1 (Eocene), India. See Singhai (*Palaeobotanist* **25**: 481, 1978).

Palaeosclerotium G.W. Rothwell (1972), Fossil Fungi ? Ascomycota. 1 (Carboniferous), USA. See Dennis (*Science* N.Y. **192**: 66, 1976), Singer (*Mycol.* **69**: 850, 1977), Pirozynski & Weresub *in* Kendrick (Ed.) (*The Whole Fungus* **1**: 687, 1979).

Palaeosordaria Sahni & H.S. Rao (1943), Fossil Fungi, Xylariales. 1 (Tertiary), India.

Palambages Whetzel (1961), Fossil Fungi (? sclerotia) Fungi. 3 (Cretaceous, Tertiary), Canada; Europe; Malaysia.

Palavasciaceae Manier ex Manier & Lichtw. (1968), Eccrinales. 1 gen., 3 spp.

Palawania Syd. & P. Syd. (1914), Microthyriaceae. 1, Asia.

Palawaniella Doidge (1921), Parmulariaceae. 5, S. America; S. Africa. See Inácio *et al.* (*Mycol. Progr.*

4: 133, 2005).

Palawaniopsis Bat., Cif. & Nascim. (1959), anamorphic *Pezizomycotina*, Cpt.0eH.?. 1, S. Africa. See Batista *et al.* (*Mycopath. Mycol. appl.* **11**: 75, 1959).

Paleoarcyria M. Jacq.-Fél., C.N. Mill. & Locq. (1983), Fossil Fungi. 1, France.

Paleobasidiospora Locq. (1983), Fossil Fungi. 1, France.

Paleoblastocladia W. Remy, T.N. Taylor & Hass (1994), Fossil Fungi. 1 (lower Devonian). See Remy *et al.* (*Am. J. Bot.* **81**: 690, 1994).

Paleocatenaria Locq. (1983), Fossil Fungi. 1, British Isles.

Paleoguttulina Locq. & Mišík (1983), Fossil Fungi. 1, former Czechoslovakia.

Paleomastigomycetes, Form class name proposed (Taylor *et al. Mycol.* **84**: 901, 1992) for fossil chytrid-like fungi.

paleomycology, the study of fossil fungi (q.v.).

Paleopyrenomycites Taylor, Hass, Kerp, M. Krings & Hanlin (2004), Fossil Fungi. 1, British Isles. See Taylor *et al.* (*Mycol.* **96**: 1404, 2004).

Paleoserenomyces Currah, Stockey & B.A. LePage (1998), Fossil Fungi. 1. See Currah *et al.* (*Mycol.* **90**: 668, 1998).

Paleoslimacomyces Kalgutkar & Sigler (1995), Fossil Fungi, anamorphic *Pezizomycotina.* 1 (Palaeocene – Eocene), Canada. See Kalgutkar & Sigler (*MR* **99**: 521, 1995).

Palifer Stalpers & P.K. Buchanan (1991), Schizoporaceae. 4, widespread. See Hjortstam & Ryvarden (*Syn. Fung.* **22**: 9, 2007).

Paliphora Sivan. & B. Sutton (1985), anamorphic *Pezizomycotina*, Hso.0-1eH.27. 3, Australia; Malaysia. See Sivanesan & Sutton (*TBMS* **85**: 249, 1985), Kuthubutheen (*TBMS* **89**: 270, 1987), Alcorn (*Mycotaxon* **59**: 145, 1996).

palisade fungi, see *Agaricomycotina.*

palisade plectenchyma, plectenchyma in the cortex of a lichen thallus composed of hyphae arranged perpendicular to the surface.

palisade-cells (of lichens), the end cells of the hyphae of a fastigiate cortex.

palisoderm (of basidiomata), an outer layer composed of several strata of cells or hyphal tips; see derm.

pallid, light-coloured; pale.

pallisadoplectenchyma, see plectenchyma.

palmate, having lobes radiating from a common centre but not extending to the point of insertion. Cf. chiroid, digitate.

Palmella Lyngb. (1819), Algae. Algae.

Palmellathyrites Locq., D. Pons & Sal.-Cheb. (1981), Fossil Fungi. 1, France.

Palmicola K.D. Hyde (1993), Xylariales. 1, Australia; Ecuador. Affinities uncertain. See Hyde (*Sydowia* **45**: 15, 1993), Goh & Hyde (*MR* **100**: 714, 1996).

Palmomyces K.D. Hyde, J. Fröhl. & Joanne E. Taylor (1998) nom. illegit., ? Phyllachorales. 1 (on living *Palmae* leaf), Australia. See Hyde *et al.* (*Sydowia* **50**: 21, 1998), Wang & Hyde (*Fungal Diversity* **3**: 159, 1999; posn).

Palmomyces Maire (1926) ≡ Andreaeana.

Palomyces Höhnk (1955) = Halosphaeria fide Kohlmeyer (*CJB* **50**: 1951, 1972).

paludal, living in wet places (marshes).

Palynomorphites L.R. Moore (1963), Fossil Fungi. 1 (Carboniferous), British Isles.

Pampolysporium Magnus (1900) = Guignardia fide Eriksson (*SA* **5**: 147, 1986).

Panaeolina Maire (1933), Agaricales. 2, widespread. Belongs either to the *Inocybaceae* or the *Psathyrellaceae.* See Gerhardt (*Biblthca Botanica* **147**: 149 pp., 1996), He (*Mycosystema* **23**: 150, 2004; China).

Panaeolopsis Singer (1969), Agaricaceae. 4, widespread. See Singer (*Nova Hedwigia* Beihefte **29**: 367, 1969).

Panaeolus (Fr.) Quél. (1872) nom. cons., Agaricales. *c.* 15 (coprophilous), widespread. Belongs either to the *Inocybaceae* or the *Psathyrellaceae.* See Hora (*Naturalist* Hull: 77, 1957), Ola'h (*Revue Mycol.* Mém. Hors Sér. **10**, 1970), Pegler (*Kew Bull.* Addit. Ser. **6**, 1977; E. Afr. spp.), Gerhardt (*Taxonomische Revision der Gattungen Panaeolus und Panaeolina*, 1996), Bon & Courtecuisse (*Docums Mycol.* **32**: 75, 2003; key).

Panama disease, A disease of banana caused by *Fusarium oxysporum* f.sp. *cubense.* See Stover (*Phytopath. Pap.* **4**, 1962), Stover & Buddenhagen (*Fruits* **41**: 175-191, 1986), Pegg & Langdon (*in* Persley & De Langhe (Eds), *Banana and plantain breeding strategies*: 119, 1987), Ploetz (Ed.) (*Fusarium wilt of banana*, 1990, *Int. Jl of Pest Manag.* **40**: 326, 1994).

Panchanania Subram. & N.G. Nair (1966), anamorphic *Pezizomycotina*, Hsy.1eP.10. 1, India. See Bhat (*TBMS* **85**: 101, 1985; synonymy with *Paathramaya*).

Pandanicola K.D. Hyde (1994), Xylariales. 2, pantropical. See Hyde (*Sydowia* **46**: 35, 1994), Lu & Hyde (*Mycoscience* **41**: 83, 2000).

Pandora Humber (1989), Entomophthoraceae. 32, widespread. See Humber (*Mycotaxon* **34**: 441, 1989), Miller & Keil (*Mycotaxon* **38**: 227, 1990), Keller & Petrini (*Sydowia* **57**: 23, 2005; key), Keller (*Sydowia* **59**: 75, 2007; Switzerland, n.sp.).

Panellaceae Jülich (1982) = Mycenaceae.

Panellus P. Karst. (1879), Mycenaceae. *c.* 55, widespread. See Miller (*Mich. Bot.* **9**: 17, 1970), Burdsall & Miller (*Beih. Nova Hedwigia* **51**, 1975), Corner (*Gdns' Bull.* Singapore **39**: 103, 1986; Malaysian spp.), O'Kane *et al.* (*Mycol.* **82**: 595, 1990; bioluminescence), Hibbett *et al.* (*Nature* **407**: 506, 2000; relationship with *Clavariaceae* suggested), Jin *et al.* (*Mycotaxon* **79**: 7, 2001; phylogeny), Jin *et al.* (*Mycol.* **93**: 309, 2001; biogeography *Panellus stypticus*).

Pannaria Delise ex Bory (1828), Pannariaceae (L). 51, widespread (esp. tropical). See Jørgensen (*Op. Bot.* **45**, 1978; Eur. spp.), Jørgensen (*J. Hattori bot. Lab.* **76**: 197, 1994), Jørgensen (*N.Z. Jl Bot.* **37**: 257, 1999; NZ spp.), Jørgensen (*Nova Hedwigia* **71**: 405, 2000; subg. *Chrysopannaria*), Jørgensen (*Bryologist* **103**: 670, 2000; N America), Jørgensen (*Biblthca Lichenol.* **78**: 109, 2001; Australia), Jørgensen & Kashiwadani (*J. Jap. Bot.* **76**: 1, 2001; Japan), Ekman & Jørgensen (*CJB* **80**: 625, 2002; phylogeny), Jørgensen (*Lichenologist* **35**: 11, 2003; Africa), Jørgensen (*Nova Hedwigia* **76**: 245, 2003; isidiate taxa), Wiklund & Wedin (*Cladistics* **19**: 419, 2003; phylogeny), Jørgensen (*Biblthca Lichenol.* **88**: 229, 2004; S hemisphere), Jørgensen & Arvidsson (*Symb. bot. upsal.* **34** no. 1: 113, 2004; Ecuador), Jørgensen & Sipman (*Nova Hedwigia* **78**: 311, 2004; S America), Miądlikowska & Lutzoni (*Am. J. Bot.* **91**: 449, 2004; phylogeny), Jørgensen & Sipman (*J. Hattori bot. Lab.* **100**: 695, 2006; New Guinea), Miądlikowska *et al.* (*Mycol.* **98**: 1088, 2006; phylogeny).

Pannariaceae Tuck. (1872), Peltigerales (L). 18 gen.

(+ 8 syn.), 328 spp.

Lit.: Galloway & James (*Lichenologist* **17**: 173, 1985), Galloway & Jørgensen (*Lichenologist* **19**: 345, 1987), Jørgensen & Galloway (*Lichenologist* **21**: 295, 1989), Lumbsch & Kothe (*Mycotaxon* **43**: 277, 1992), Øvstedal & Smith (*Cryptog. Bryol.-Lichénol.* **14**: 337, 1993), Jørgensen (*J. Jap. Bot.* **69**: 383, 1994), Henssen (*Symb. bot. upsal.* **32**: 75, 1997), Jørgensen (*Lichenologist* **30**: 533, 1998), Jørgensen (*N.Z. Jl Bot.* **37**: 257, 1999), Jørgensen & Wedin (*Lichenologist* **31**: 341, 1999), Makhija & Adawadkar (*Mycotaxon* **71**: 323, 1999), Jørgensen (*Bryologist* **103**: 104, 2000), Jørgensen (*Bryologist* **103**: 670, 2000; N. Am.), Ekman & Jørgensen (*CJB* **80**: 625, 2002), Wiklund & Wedin (*Cladistics* **19**: 419, 2003), Wedin & Wiklund (*Symb. bot. upsal.* **34** no. 1: 469, 2004), Miądlikowska *et al.* (*Mycol.* **98**: 1088, 2006; phylogeny), Hofstetter *et al.* (*Mol. Phylogen. Evol.* **44**: 412, 2007; phylogeny), Jørgensen (*Nordic Lichen Flora* **3**: Cyanolichens: 96, 2007).

Pannariella (Vain.) Gyeln. (1935) = Heppia fide Hawksworth *et al.* (*Dictionary of the Fungi* edn 8, 1995).

Pannoparmelia (Müll. Arg.) Darb. (1912), Parmeliaceae (L). 5, widespread. See Sato (*Miscnea bryol. lichen., Nichinan* **4**: 45, 1966), Yoshimura (*Biblthca Lichenol.* **25**: 185, 1987), Yoshimura & Elix (*J. Hattori bot. Lab.* **74**: 287, 1993), Galloway & Sammy (*Flora of Australia* **55**: 86, 1994), Blanco *et al.* (*Mol. Phylogen. Evol.* **39**: 52, 2006; phylogeny).

pannose (**panniform**), having the appearance of felt or woollen cloth.

Pannucia P. Karst. (1879) = Psathyrella fide Singer (*Agaric. mod. Tax.*, 1951).

Pannularia Nyl. (1886) = Placynthium fide Jørgensen (*Op. Bot.* **45**, 1986).

Panomycetes Nyl. (1886) = Placynthium fide Jørgensen (*Op. Bot.* **45**, 1986).

Panorbis J. Campb., J.L. Anderson & Shearer (2003), Halosphaeriaceae. 1. See Campbell *et al.* (*Mycol.* **95**: 544, 2003).

pantherine, a metabolite of *Amanita pantherina*, etc., toxic to humans and flies (*Musca* sp.); muscimol; amanita factor B.

pantonemic (of e.g. *Oomycota*, *Hyphochytriomycota*), presumed to have evolved from algae. Pantonemic 'fungi' exhibit the diaminopimelate pathway for lysine synthesis; the dominant sterols are cholesterols (not ergosterol); and the storage product is mycolaminarin (not glycogen).

Pantonemomycota. Phylum proposed by Olive (*The mycetozoans*, 1975) for the fungi placed in *Oomycota* and *Hyphochytriomycota* in this *Dictionary*.

Pantospora Cif. (1938) = Pseudocercospora fide Deighton (*Mycol. Pap.* **140**, 1976), Crous & Braun (*CBS Diversity Ser.* **1**: 571 pp., 2003).

Panus Fr. (1838) nom. cons., Polyporaceae. 25 (on wood), widespread. See Corner (*Beih. Nova Hedwigia* **69**, 1981), Pegler (*Kew Bull.* Addit. Ser. **10**, 1983; world monogr., as *Lentinus* subgen. *Panus*), Hibbett & Vilgalys (*Mycol.* **83**: 425, 1991; phylogeny), Hibbett & Vilgalys (*Syst. Bot.* **18**: 409, 1993; phylogeny) See also Luminescent fungi.

Paoayensis Cabanela, Jeewon & K.D. Hyde (2007), Sordariomycetes. 1, Philippines. See Cabanela *et al.* (*Cryptog. Mycol.* **28**: 301, 2007).

paper (fungi on), see Biodeterioration.

papilionaceous, variegated; mottled; marked with different colours as the lamellae of certain *Panaeolus* spp.

Papilionospora V.G. Rao & B. Sutton (1976), anamorphic *Pezizomycotina*, Hsy.0eP.10. 1, Myanmar; Italy. See Rao & Sutton (*Kavaka* **3**: 23, 1975).

Papilionovela Aptroot (1997), Ostropales. 1 (from bark), Papua New Guinea. See Aptroot (*MR* **101**: 266, 1997).

Papiliotrema J.P. Samp., M. Weiss & R. Bauer (2002), Tremellaceae. 1, Portugal. See Sampaio *et al.* (*Mycol.* **94**: 875, 2002), Sampaio (*Frontiers in Basidiomycote Mycology*: 49, 2004).

papilla, a small rounded process.

papillae (of host tissue), localized wall thickenings on the inner surface of plant cell walls at sites penetrated by fungi; callosity, lignituber, callus, 'sheath', are whole or part synonyms (Aist, *Ann. Rev. Phytopath.* **14**: 145, 1976).

Papillaria J. Kickx f. (1835) nom. rej. prop. ≡ Pycnothelia.

papillate, having a papilla (Fig. 23.40).

Pappimyces B. Sutton & Hodges (1975), anamorphic *Pezizomycotina*, Hso.0fH.6. 1, Brazil. See Sutton & Hodges (*Nova Hedwigia* **26**: 527, 1975).

Papulare Tomas. & Cif. (1952) ≡ Phragmothele.

Papularia Fr. (1825) [non *Papularia* Forssk. 1775, *Aizoaceae*] = Arthrinium fide Ellis (*Mycol. Pap.* **103**, 1965).

Papulariomyces Cif. & Tomas. (1953) ≡ Papulare.

Papulaspora Preuss (1851), anamorphic *Chaetomium, Melanospora*, Hso.#eP.1. *c.* 24, widespread. *P. byssina* is the brown plaster mould of mushroom beds; wood pulp is damaged by other spp.; *P. stoveri* is a mycoparasite of *Rhizoctonia*. The hyphae bear 'papulospores' (q.v.), small rounded propagules comprising one or more enlarged central cells surrounded by a sheath of smaller cells. Forms having 'bulbils' (q.v.) are excluded as *Burgoa* and *Minimedusa* by Weresub & LeClair (*CJB* **49**: 2203, 1971; gen. key). See Tubaki *et al.* (*TMSJ* **32**: 31, 1991).

Papulasporites Shete & A.R. Kulk. (1978), Fossil Fungi. 1 (Tertiary), India.

Papulosa Kohlm. & Volkm.-Kohlm. (1993), Papulosaceae. 1 (*Juncus* culms), USA. See Kohlmeyer & Volkmann-Kohlmeyer (*SA* **11**: 95, 1993), Winka & Eriksson (*Mycoscience* **41**: 97, 2000; DNA), Zhang *et al.* (*Mycol.* **98**: 1076, 2006; phylogeny).

Papulosaceae Winka & O.E. Erikss. (2000), Sordariomycetidae (inc. sed.). 1 gen., 1 spp.

Lit.: Winka & Eriksson (*Mycoscience* **41**: 97, 2000; DNA).

papulose, covered with pimples or pustules.

Papulospora, see *Papulaspora*.

papulospore, asexual spore in e.g. *Papulaspora sepedonioides* (Weresub & LeClair, *CJB* **49**: 2203, 1971).

Papulosporonites Schmied. & A.J. Schwab (1964), Fossil Fungi. 1 (Eocene), Germany.

Papyrodiscaceae Jülich (1982) = Corticiaceae.

Papyrodiscus D.A. Reid (1979), ? Corticiaceae. 1, Papua New Guinea. See Reid (*Beih. Sydowia* **8**: 333, 1979).

Paraaoria R.K. Verma & Kamal (1987), anamorphic *Pezizomycotina*, St.0eP.19. 1, Nepal. See Verma & Kamal (*TBMS* **87**: 645, 1986).

Paraarthrocladium Matsush. (1993), anamorphic *Pezizomycotina*, Hso.≡ eH.1/10. 1, Peru. See Matsushima (*Matsush. Mycol. Mem.* **7**: 58, 1993).

Parabotryon Syd. (1926) ? = Eudarluca fide Müller &

von Arx (*Beitr. Kryptfl. Schweiz* **11** no. 2, 1962).

Paracainiella Lar.N. Vassiljeva (1983), Amphisphaeriaceae. 1 (on *Dryas*), Siberia. See Kang *et al.* (*MR* **103**: 53, 1999).

paracapillitium (of *Lycoperdales*), a capillitium composed of thin-walled, hyaline, septate hyphae in contrast to a true capillitium of thick-walled, brown, aseptate hyphae (Kreisel, 1962).

Paracapnodium Speg. (1909) = Scorias fide Hughes (*Mycol.* **68**: 693, 1976).

Paracarpidium Müll. Arg. (1883) = Endocarpon fide Zahlbruckner (*Catalogus Lichenum Universalis* **1**, 1921).

paracephalodium (pl. **-ia**), a hyphal mat covering cyanobacteria arising from a squamulose lichen thallus which has a green alga as the photobiont (Poelt & Mayrhofer, *Pl. Syst. Evol.* **158**: 265, 1988).

Paraceratocladium R.F. Castañeda (1987), anamorphic *Pezizomycotina*, Hso.1eH.15. 2, pantropical. See Castañeda (*Fungi Cubenses* **2**: 8, 1987), Goh & Hyde (*Nova Hedwigia* **71**: 95, 2000; Malaysia), Whitton *et al.* (*Fungal Diversity* **7**: 175, 2001; Seychelles), Calduch *et al.* (*Mycol.* **94**: 1071, 2002; Brazil).

Paracercospora Deighton (1979) = Pseudocercospora. Widely used for the causal organisms of Sigatoka disease of banana. fide Deighton (*Mycol. Pap.* **144**: 47, 1979), Crous *et al.* (*Mycol.* **93**: 1081, 2001; phylogeny), Crous & Braun (*CBS Diversity Ser.* **1**: 571 pp., 2003).

Paracesatiella Petr. (1929) = Schweinitziella fide von Arx (*Acta Bot. Neerl.* **7**: 503, 1958).

Parachinomyces Thaung (1979), anamorphic *Pezizomycotina*, Hso.≡ eH.10. 1 (on *Acroconidiellina*), Myanmar. See Thaung (*TBMS* **72**: 333, 1979).

Parachionomyces Thaung (1979), anamorphic *Pezizomycotina*, Hso.≡ eH.10. 1 (on *Acroconidiellina*), Myanmar.

Parachnopeziza Korf (1978), Hyaloscyphaceae. 8, widespread. See Arendholz & Sharma (*Mycotaxon* **20**: 633, 1984), Korf & Zhuang (*Mycotaxon* **22**: 483., 1985), Gamundí & Giaiotti (*Sydowia* **46**: 12, 1994), Zhuang (*Mycotaxon* **69**: 359, 1998; 2 spp. China).

Paracoccidioidaceae Redaelli & Cif. (1937) = Ajellomycetaceae.

Paracoccidioides F.P. Almeida (1930), anamorphic *Ajellomycetaceae*, Hso.0eH.3. 1 (on humans), S. America. *P. brasiliensis* (paracoccidioidomycosis; q.v.). See Restrepo *et al.* (*RMVM* **8**: 97, 1973; 183 refs), Takeo *et al.* (*MR* **94**: 1118, 1990; cytoplasmic and plasma membrane ultrastructure), San-Blas & San-Blas (*Jap. J. Med. Mycol.* **32**: 75, 1991; morphogenesis), Franco *et al.* (Eds) (*Paracoccidioidomycosis*, 1994), Le Clerc *et al.* (*J. Med. Vet. Mycol.* **32**: 331, 1994; syst. posn), Vidal *et al.* (*Revista do Instituto de Medicina Tropical de São Paulo* **37**: 43, 1995; from armadillo), San-Blas *et al.* (*Microbiology* Reading **143**: 197, 1997; biometrics), Sandhu *et al.* (*J. Clin. Microbiol.* **35**: 1894, 1997; molecular detection), Calcagno *et al.* (*J. Clin. Microbiol.* **36**: 1733, 1998; biogeography), Cano *et al.* (*J. Clin. Microbiol.* **36**: 742, 1998; karyotypes), Hamilton (*Medical Mycology* **36**: 351, 1998; serological diagnostics), Wanke & Londero (*Topley & Wilson's Microbiology and Microbial Infections* Edn 9. Vol. **4** Medical Mycology: 395, 1998; review), Montoya *et al.* (*Medical Mycology* **37**: 219, 1999; karyotypes), Sano *et al.* (*Mycopathologia* **143**: 165, 1998; comparison between human and armadillo strains), Bialek *et al.* (*J. Clin. Microbiol.* **38**: 3190, 2000; phylogeny), Niño-Vega *et al.* (*Medical Mycology* **38**: 437, 2000; geographical isolation), Hahn *et al.* (*Revta Iberoamer. Micol.* **19**: 49, 2002; RAPD characterization), San-Blas *et al.* (*Medical Mycology* **40**: 225, 2002; molecular approaches), Feitosa *et al.* (*Fungal Genetics Biol.* **39**: 60, 2003; chromosomes), Hebeler-Barbosa *et al.* (*J. Clin. Microbiol.* **41**: 5733, 2003; phylogeny), Nascimento *et al.* (*J. Clin. Microbiol.* **42**: 5007, 2004; microsatellites), Borba *et al.* (*Antonie van Leeuwenhoek* **88**: 257, 2005; polyphasic characterization), Oliveira *et al.* (*MR* **109**: 707, 2005; cDNA), San-Blas *et al.* (*J. Clin. Microbiol.* **43**: 4255, 2005; molecular detection), Geiser *et al.* (*Mycol.* **98**: 1053, 2006; phylogeny), Matute *et al.* (*J. Clin. Microbiol.* **44**: 2153, 2006; microsatellites, phylogenetic species), Matute *et al.* (*Mol. Biol. Evol.* **23**: 65, 2006; cryptic speciation), Soares *et al.* (*Mycopathologia* **165**: 179, 2008; review).

paracoccidiomycosis, see *Paracoccidioides*.

Paraconiothyrium Verkley (2004), anamorphic *Montagnulaceae*. Anamorph *Paraphaeosphaeria*. 6, widespread. See Verkley *et al.* (*Stud. Mycol.* **50**: 327, 2004), Damm *et al.* (*Persoonia* **20**: 9, 2008; on woody plants).

Paracoreomyces R.A. Poiss. (1929) = Coreomyces fide Thaxter (*Memoirs of the American Academy of Arts and Sciences* **16**: 1, 1931).

Paracostantinella Subram. & Sudha (1989), anamorphic *Pezizomycotina*, Hso.0eH.10. 1, India. See Subramanian & Sudha (*Kavaka* **15**: 12, 1987).

Paracryptophiale Kuthub. & Nawawi (1994), anamorphic *Pezizomycotina*, Hso.#eH-P.15. 2, Malaysia; China. See Kuthubutheen & Nawawi (*MR* **98**: 125, 1994), Wu & Sutton (*Fungal Diversity* **14**: 265, 2003).

Paracudonia Petr. (1927) = Roesleria fide Nannfeldt (*Nova Acta R. Soc. Scient. upsal.*, 1932).

Paracytospora Petr. (1925), anamorphic *Pezizomycotina*, St.0eH.?. 1, N. America.

Paradactylaria Subram. & Sudha (1989), anamorphic *Pezizomycotina*, Hsy.1eH.10. 1, India. See Subramanian & Sudha (*Kavaka* **15**: 16, 1987).

Paradactylella Matsush. (1993), anamorphic *Pezizomycotina*, Hso.≡ eH.2. 1, Peru. See Matsushima (*Matsush. Mycol. Mem.* **7**: 59, 1993).

Paradendryphiopsis M.B. Ellis (1976), anamorphic *Pezizomycotina*, Hso.≡ eP.25. 3, Europe; S. Africa. See Révay (*Stud. Bot. Hung.* **18**: 65, 1985).

Paradictyoarthrinium Matsush. (1996), anamorphic *Pezizomycotina*, Hso.?.?. 1, Transvaal. See Matsushima (*Matsush. Mycol. Mem.* **9**: 18, 1996).

Paradidymella Petr. (1927) = Discostroma fide Dennis (*British Ascomycetes*, 1968), Hawksworth & Sivanesan (*TBMS* **67**: 39, 1976).

Paradidymobotryum C.J.K. Wang & B. Sutton (1984), anamorphic *Pezizomycotina*, Hsy.1eP.4. 1, USA. See Wang & Sutton (*Mycol.* **76**: 570, 1984).

Paradiplodia Speg. ex Trotter (1931), anamorphic *Pezizomycotina*, St.1eP.?. 7, widespread. See Zambettakis (*BSMF* **70**: 219, 1955; as *Paradiplodiella*), Petrak (*Sydowia* **16**, 1963).

Paradiplodiella Zambett. (1955) = Botryodiplodia fide Sutton (*Mycol. Pap.* **141**, 1977).

Paradischloridium Bhat & B. Sutton (1985), anamorphic *Pezizomycotina*, Hso.≡ eP.15. 1, Ethiopia; New Zealand. See Bhat & Sutton (*TBMS* **84**: 723, 1985).

Paradiscina Benedix (1969) = Gyromitra fide Benedix

(*Kulturpflanz.* **19**: 163, 1972), Harmaja (*Karstenia* **15**: 33, 1976).
Paradiscula Petr. (1941), anamorphic *Pezizomycotina*, Cpd.0eH.15. 1, Europe. See Morgan-Jones (*Mycotaxon* **2**: 167, 1975).
Paradoxa Mattir. (1935), Tuberaceae. 1 (hypogeous), Italy. See Læssøe & Hansen (*MR* **111**: 1075, 2007; phylogeny).
Paradoxomyces Matzer (1996), Arthoniaceae (L). 1 (on lichens), Indonesia. See Matzer (*Mycol. Pap.* **171**: 93, 1996), Grube (*Bryologist* **101**: 377, 1998).
Paraeccilia Largent (1994) = Entoloma fide Kuyper (*in litt.*).
Paraepicoccum Matsush. (1993), anamorphic *Pezizomycotina*, Hsp.#eP.2. 1, Peru. See Matsushima (*Matsush. Mycol. Mem.* **7**: 59, 1993).
Paraeutypa Subram. & Ananthap. (1988) = Leptoperidia fide Rappaz (*Mycol. Helv.* **3**: 281, 1989).
Parafulvia Kamal, A.N. Rai & Morgan-Jones (1983), anamorphic *Pezizomycotina*, Hso.≡ eP.10. 1, India. See Kamal *et al.* (*Mycotaxon* **18**: 67, 1983).
Paragaeumannomyces Matsush. (2003), Sordariomycetes. 1, Japan. See Matsushima (*Matsush. Mycol. Mem.* **10**: 156, 2001).
Paraglomeraceae J.B. Morton & D. Redecker (2001), Paraglomerales. 1 gen., 3 spp.
Lit.: Azcon-Aguilar & Barea (*Mycorrhiza* **6**: 457, 1996), van der Heijden *et al.* (*Nature* Lond. **396**: 69, 1998), Morton & Redecker (*Mycol.* **93**: 183, 2001), Schüßler *et al.* (*MR* **105**: 1413, 2001), Redecker (*Glomeromycota* Arbuscular mycorrhizal fungi and their relative(s). Version 01 July 2005. http://tolweb.org/Glomeromycota/28715/2005.07.01 in The Tree of Life Web Project, http://tolweb.org: [unpaginated], 2005).
Paraglomerales C. Walker & A. Schüssler (2001). Glomeromycetes. 1 fam., 1 gen., 3 spp. Fam.: **Paraglomeraceae**
For *Lit.* see under fam.
Paraglomus J.B. Morton & D. Redecker (2001), Paraglomeraceae. 3, widespread. See Morton & Redecker (*Mycol.* **93**: 188, 2001), Redecker & Raab (*Mycol.* **98**: 885, 2006; phylogeny), Walker *et al.* (*MR* **111**: 253, 2007; nomencl.).
Paragranatisporites Kalgutkar & Janson. (2000), Fossil Fungi. 1.
Paragranatisporites Zhong Y. Zhang (1980), Fossil Fungi. 1. See Zhang (*Acta Palaeont. Sin.* **19**: 298, 1980).
paragynous (of *Pythiaceae*), having the antheridium at the side of the oogonium.
Paragyrodon (Singer) Singer (1942), Paxillaceae. 1, N. America. See Bruns *et al.* (*Mol. Ecol.* **7**: 257, 1998; phylogeny).
Parahaplotrichum W.A. Baker & Partridge (2001), anamorphic *Botryobasidium*. 1, USA. See Partridge *et al.* (*Mycotaxon* **77**: 360, 2001).
Paraharknessia Matsush. (2003), anamorphic *Pezizomycotina*, Hso.?.?. 1, Japan. See Matsushima (*Matsush. Mycol. Mem.* **10**: 30, 2001).
Parahelminthosporium Subram. & Bhat (1989) = Polytretophora fide Sutton (*in litt.*).
Parahendersonia A.W. Ramaley (1995), anamorphic *Chaetoplea*, Cpd.?.?. 2, USA. See Ramaley (*Aliso* **14**: 152, 1995).
Parahyalotiopsis Nag Raj (1976), anamorphic *Pezizomycotina*, Cpd.≡ eP.19. 1, Myanmar. See Nag Raj (*CJB* **54**: 1370, 1976).
Parahydraeomyces Speg. (1915) = Hydraeomyces fide Siemaszko & Siemaszko (*Polsk. Pismo Entom.* **12**: 125, 1933).
Paraisaria Samson & B.L. Brady (1983), anamorphic *Ophiocordyceps*, Hsy.0eH.15. 1 (on larvae of *Hepialus*), British Isles; Germany. See Samson & Brady (*TBMS* **81**: 285, 1983).
Paralaestadia Sacc. ex Vain. (1921) nom. rej. prop. ? = Lichenochora fide Hoffmann & Hafellner (*Biblthca Lichenol.* **77**: 1, 2000).
Paralepista Raithelh. (1981) = Lepista fide Kuyper (*in litt.*).
Paraleptonia (Romagn. ex Noordel.) P.D. Orton (1991) = Entoloma fide Kuyper (*in litt.*).
Paralethariicola Calatayud, Etayo & Diederich (2001), Odontotremataceae. 1, Spain. See Calatayud *et al.* (*Lichenologist* **33**: 478, 2001), Diederich *et al.* (*Lichenologist* **34**: 479, 2002).
Paraliomyces Kohlm. (1959), Pleosporales. 1 (marine), India. See Read *et al.* (*CJB* **70**: 2223, 1993; ultrastr.), Tam *et al.* (*Bot. Mar.* **46**: 487, 2003; phylogeny).
Paramassariothea Subram. & Muthumary (1979), anamorphic *Pezizomycotina*, St.≡ eH.19. 1, India. See Subramanian & Muthumary (*CJB* **57**: 2427, 1979).
Paramazzantia Petr. (1927) = Mazzantia fide Barr (*Mycol. Mem.* **7**: 232 pp., 1978).
Paramenisporopsis Matsush. (2003), anamorphic *Pezizomycotina*, Hso.?.?. 1, Japan. See Matsushima (*Matsush. Mycol. Mem.* **10**: 26, 2001).
Paramicrothallites K.P. Jain & R.C. Gupta (1970), Fossil Fungi, Microthyriaceae. 2 (Miocene), India.
Paramitra Benedix (1962) = Peziza Fr. fide Nannfeldt (*Ann. bot. fenn.* **3**: 309, 1966).
Paramoeciella Zebrowski (1936), Fossil Fungi ? Chytridiomycetes. 1 (Cambrian to ? Recent), Australia.
paramorph, a neutral term proposed by Huxley (*The new systematics*: 37, 1940) for any form differing from the mean of the group; advocated for use in fossil fungi lacking characters essential for their proper placement, but believed to have affinities with particular non-fossil groups by Reynolds (*Mycotaxon* **23**: 141, 1985).
paramorphogen, a compound (e.g. validomycin A, L-sorbose) able to induce a reversible morphological change in fungi (Tatum *et al.*, *Science* **109**: 509, 1949, Jejewolo *et al.*, *TBMS* **91**: 653, 661, 1988).
Paramphisphaeria F.A. Fern., J.D. Rogers, Y.M. Ju, Huhndorf & L. Umaña (2004), Xylariaceae. 1, Costa Rica. See Fernández *et al.* (*Mycol.* **96**: 175, 2004).
Paramyces Oehm (1937) nom. inval. = Ptychogaster fide Donk (*Taxon* **11**: 94, 1962).
Paranectria Sacc. (1878), Bionectriaceae. 3 (lichenicolous), Europe. See Hawksworth & Pirozynski (*CJB* **55**: 2555, 1977; nomencl.), Hawksworth (*Notes R. bot. Gdn Edinb.* **40**: 375, 1982), Rossman *et al.* (*Stud. Mycol.* **42**: 248 pp., 1999).
Paranectriella (Henn. ex Sacc. & D. Sacc.) Höhn. (1910), Tubeufiaceae. Anamorphs *Araneomyces*, *Titaea*. 7, widespread (tropical). See Hawksworth & Pirozynski (*CJB* **55**: 2555, 1977; nomencl.), Rossman (*Mycol. Pap.* **157**, 1987; key).
Paraniesslia K.M. Tsui, K.D. Hyde & Hodgkiss (2001), Niessliaceae. 1, Hong Kong. See Tsui *et al.* (*Mycol.* **93**: 1002, 2001).
Paranosema Y.Y. Sokolova, V.V. Dolgikh, E.V. Morzhina, E.S. Nassonova, I.V. Issi, R.S. Terry, J.E. Ironside, J.E. Smith & C.R. Vossbrinck (2003), Microsporidia. 2. See Sokolova *et al.* (*J. Invert. Path.*

84: 159, 2003).

Paranthostomella Speg. (1910) ? = Anthostomella fide Francis (*Mycol. Pap.* **139**, 1975), Læssøe (*SA* **13**: 43, 1994), Lu & Hyde (*Fungal Diversity Res. Ser.* **4**, 2000).

Paraparmelia Elix & J. Johnst. (1986) = Xanthoparmelia fide Elix & Johnston (*Mycotaxon* **32**: 399, 1988; 15 spp. S. Hemisph.), Elix (*Bryologist* **96**: 359, 1993), Elix (*Biblthca Lichenol.* **80**, 2001; revision), Elix (*Mycotaxon* **87**: 395, 2003), Blanco *et al.* (*Taxon* **53**: 959, 2004).

Parapaxillus Singer (1942) nom. inval.; nom. prov. = Paxillus fide Kuyper (*in litt.*).

Parapeltella Speg. (1919) = Micropeltis fide von Arx & Müller (*Stud. Mycol.* **9**, 1975).

Parapericonia M.B. Ellis (1976), anamorphic *Pezizomycotina*, Hsp.0eP.1/10. 2, Africa; Australia. See Alcorn & Kirk (*TBMS* **85**: 561, 1985; amendment gen.).

Parapericoniella U. Braun, Heuchert & K. Schub. (2006), anamorphic *Pezizomycotina*, H?.?.?. 1, W. Africa. See Braun *et al.* (*Schlechtendalia* **13**: 59, 2006).

Paraphaeoisaria de Hoog & Morgan-Jones (1978), anamorphic *Pezizomycotina*, Hso.0eH.10. 1, USA. See de Hoog & Morgan-Jones (*Mycotaxon* **7**: 133, 1978).

Paraphaeosphaeria O.E. Erikss. (1967), Montagnulaceae. Anamorph *Paraconiothyrium*. 23, Europe; N. America. See Crivelli (*Über die heterogene Ascomycetengattung Pleospora Rbh.*, 1983; key), Shoemaker & Babcock (*CJB* **63**: 1284, 1985; key), Ramaley (*Mycotaxon* **61**: 347, 1997), Câmara *et al.* (*MR* **105**: 41, 2001; morphology, DNA), Câmara *et al.* (*MR* **107**: 516, 2003; revision, phylogeny), Verkley *et al.* (*Stud. Mycol.* **50**: 323, 2004; anamorph).

Paraphelaria Corner (1966), Pucciniomycotina. 2, Java; south Pacific. See Corner (*Persoonia* **4**: 346, 1966).

Paraphelariaceae Jülich (1982) = Auriculariaceae.

Paraphialocephala Budathoki (1997), anamorphic *Pezizomycotina*, Hso.?.?. 1, Nepal. See Budathoki (*Himalayan Microbial Biodiversity* Part **1** *Recent Researches in Ecology, Environment and Pollution* **10**: 122, 1997).

Paraphoma Morgan-Jones & J.F. White (1983) = Phoma Sacc. fide van der Aa *et al.* (*Stud. Mycol.* **32**: 1, 1990).

paraphysis (pl. **paraphyses**), a sterile upward growing, basally attached hyphal element in a hymenium, esp. in ascomycetes where they are generally filiform, unbranched or branched, and the free ends frequently make an epithecium over the asci, see hamathecium; in basidiomycetes, see hyphidium; **apical paraphyses**, the downward growing hyphae with free tips in the centrum of hypocrealean fungi (Luttrell, *TBMS* **48**: 135, 1965), periphysoids; **asco-**, a multicellular diploid storage hypha originating from the base of the ascus in *Erysiphaceae* (Speer, *Sydowia* **27**: 1, 1976); **paraphysate**, having paraphyses.

paraphysoid (1) (of ascomycetes), see hamathecium; (2) (of basidiomycetes), a sterile accessory hymenial structure; see basidiole, cystidiole, hyphidium; **-network**, branched and anastomosing paraphysoids surrounding asci in some ascolocular ascomycetes.

Paraphysorma A. Massal. (1852) nom. rej. = Staurothele fide Hawksworth *et al.* (*Dictionary of the Fungi* edn 8, 1995).

Paraphysotheca Bat. (1961) = Schizothyrium fide von Arx & Müller (*Stud. Mycol.* **9**, 1975).

Paraphysothele Zschacke (1934) = Thelidium fide Hawksworth *et al.* (*Dictionary of the Fungi* edn 8, 1995).

Parapithomyces Thaung (1976), anamorphic *Pezizomycotina*, Hso.#eP.10. 2, Myanmar; Australia. See Alcorn (*Aust. Syst. Bot.* **5**: 711, 1992).

Paraplacidiopsis Servít (1953) = Placidiopsis fide Hawksworth *et al.* (*Dictionary of the Fungi* edn 8, 1995).

Paraplacodium Motyka (1996) nom. inval. = Aspicilia fide Hawksworth *et al.* (*Dictionary of the Fungi* edn 8, 1995).

paraplectenchyma, plectenchyma (q.v.) composed of cells which have isodiametric lumina and unthickened walls (see Yoshimura & Shimada, *Bull. Kochi Gakuen Jun. Coll.* **11**: 13, 1980).

Parapleistophora I.V. Issi, Kadyrova, Pushkar, Khodzhaeva & S.V. Krylova (1990), Microsporidia. 1.

Parapleurotheciopsis P.M. Kirk (1982), anamorphic *Pezizomycotina*, Hso.≡ eH-P.3. 3, British Isles; Japan; Cuba. See Castañeda Ruíz *et al.* (*Mycotaxon* **59**: 203, 1996).

Parapolyporites Tanai (1987), Fossil Fungi. 1, Japan.

Paraporpidia Rambold & Pietschm. (1989), Porpidiaceae (L). 3, Australasia; China; Indonesia. See Rambold & Pietschmann (*Biblthca Lichenol.* **34**: 243, 1989).

Parapterulicium Corner (1952), Pterulaceae. 3, Brazil. See Corner (*Ann. Bot., Lond.* n.s. **16**: 285, 1952).

Parapyrenis Aptroot (1991), Requienellaceae. 8, widespread (tropical). See Eriksson (*in litt.*), Aptroot (*Nova Hedwigia* **60**: 325, 1995; key).

Parapyricularia M.B. Ellis (1972), anamorphic *Pezizomycotina*, Hso.≡ eP.10. 2, Asia; Brazil. See Ellis (*Mycol. Pap.* **131**: 12, 1972), Silva *et al.* (*Mycotaxon* **92**: 43, 2005).

Pararobillarda Matsush. (1996), anamorphic *Pezizomycotina*, Cpd.?.?. 1, Cape Province. See Matsushima (*Matsush. Mycol. Mem.* **9**: 18, 1996).

Parasaccharomyces Beurm. & Gougerot (1909) nom. rej. = Candida fide Hawksworth *et al.* (*Dictionary of the Fungi* edn 8, 1995).

Parasarcopodium Melnik, S.J. Lee & Crous (2004), anamorphic *Pezizomycotina*. 1, S. Africa. See Mel'nik *et al.* (*Mycol. Progr.* **3**: 22, 2004).

Parascedosporium Gilgado, Gené, Cano & Guarro (2007), anamorphic *Microascaceae*. 1. See Gilgado *et al.* (*Int. J. Syst. Evol. Microbiol.* **57**: 2171, 2007).

Parasclerophoma Petr. (1924) ? = Dothichiza fide Clements & Shear (*Gen. Fung.*, 1931).

Parascorias J.M. Mend. (1930) = Rizalia fide Pirozynski (*Kew Bull.* **31**: 595, 1977).

Parascutellinia Svrček (1975), Pyronemataceae. 3, Europe. See Benkert (*Gleditschia* **13**: 147, 1985), Donadini (*Boll. Gruppo Micol. 'G. Bresadola'* **29**: 273, 1986), Schumacher (*Mycotaxon* **33**: 149, 1988), Hansen & Pfister (*Mycol.* **98**: 1029, 2006; phylogeny), Perry *et al.* (*MR* **111**: 549, 2007; phylogeny).

parasexual cycle. A mechanism discovered by Pontecorvo (q.v.) & Roper in 1952 in filamentous fungi by which re-combination of hereditary properties is based not on sexual reproduction (meiosis) but on the mitotic cycle. The essential features of the process are: (1) the production of diploid nuclei in a heterokaryotic haploid mycelium; (2) the multiplication of the diploid nuclei along with haploid nuclei in a heterokaryotic mycelium; (3) the sorting out of a diploid homokaryon; (4) segregation and recombination by crossing-over at mitosis; (5) haploidization of the diploid nuclei.

The results are similar to those achieved by meiosis but instead of a regular sequence of events in time as in the meiotic cycle, in the parasexual cycle the various processes may all be occurring at one time in one mycelium, at rates which have been estimated. The parasexual cycle may (as in *Aspergillus nidulans*) or may not (*A. niger*) be accompanied by a sexual cycle. Although the details of the mechanism of the parasexual cycle are unknown, Pontecorvo and his school obtained data from the parasexual cycle for mapping the 8 chromosomes of *Aspergillus nidulans*, results which showed good agreement with those for the same fungus obtained by the analysis of the sexual cycle. See Pontecorvo *et al.* (*Advances in genetics* **5**: 141, 1953), Pontecorvo & Käfer (*Advances in Genetics* **9**: 71, 1958). For other aspects see Pontecorvo (*Ann. Rev. Microbiol.* **10**: 393, 1956), Pontecorvo *et al.* (*J. gen. Microbiol.* **8**: 198, 1953; *Aspergillus niger*, **11**: 94, 1954; *Penicillium chrysogenum*), Buxton (*J. gen. Microbiol.* **15**: 133, 1956; *Fusarium oxysporum*). See protosexual.

Parasiphula Kantvilas & Grube (2006), Coccotremataceae (L). 1, Australasia. See Grube & Kantvilas (*Lichenologist* **38**: 241, 2006).

parasite, an organism living on or in, and obtaining its nutrients from, its host, another living organism; a biotroph or necrotroph. Cf. parasymbiont, pathogen, symbiosis.

Parasitella Bainier (1903), Mucoraceae. 1 (facultative parasite of *Mucorales*), widespread (north temperate). See Schipper (*Stud. Mycol.* **17**: 65, 1978), Kellner *et al.* (*Mycologist* **5**: 120, 1990; parasitism, genetic transfer), Burmester & Wöstemeyer (*Curr. Genet.* **26**: 456, 1994; genetics), Voigt & Wöstemeyer (*Gene* **270**: 113, 2001; phylogeny), Schultze *et al.* (*Gene* **348**: 33, 2005; sexuality & parasitism), Wöstemeyer & Schimek *in* Heitman *et al.* (Eds.) (*Sex in Fungi*: 431, 2007; trisporic acid & mating).

parasol mushroom, basidioma of the edible *Macrolepiota procera* (syn. *Lepiota procera*).

Parasola Redhead, Vilgalys & Hopple (2001), Psathyrellaceae. 20, widespread. See Jørgensen & Stalpers (*Taxon* **50**: 909, 2001; nomencl.), Redhead *et al.* (*Taxon* **50**: 203, 2001; nomencl.), Keirle *et al.* (*Fungal Diversity* **15**: 33, 2004; Hawaii), Gams (*Taxon* **54**: 520, 2005; nomencl.).

parasoredium, soredium-like structure, originally used for a structured plectenchyma in the upper thallus layer of the *Umbilicaria hirsuta* aggr. (see Codogno *et al.*, *Pl. Syst. Evol.* **165**: 55, 1989).

Parasphaeria Syd. (1924) nom. dub. ? = Massarina fide von Arx & Müller (*Stud. Mycol.* **9**, 1975), Aptroot (*Nova Hedwigia* **66**: 89, 1998).

Parasphaeropsis Petr. (1953), anamorphic *Pezizomycotina*, Cpd.0fH.?. 1, Hawaii. See Petrak (*Sydowia* **7**: 364, 1953).

Paraspora Grove (1884), anamorphic *Pezizomycotina*, Hsp.≡ eH.?. 2, Europe.

Parastempellia Khodzhaeva (1988), Microsporidia. 1.

Parastenella J.C. David (1991), anamorphic *Pezizomycotina*, Hso.≡ eP.10. 1, USA. See Morgan-Jones (*Mycotaxon* **66**: 421, 1998), Schubert & Braun (*Mycol. Progr.* **4**: 101, 2005).

Parastereopsis Corner (1976), Cantharellaceae. 1, Borneo. See Corner (*Nova Hedwigia* **27**: 331, 1976).

Parasteridiella H. Maia (1960) = Asteridiella fide Hughes (*in litt.*).

Parasteridium Speg. (1923) nom. dub., ? Meliolaceae. No species included. See Eriksson & Hawksworth (*SA* **7**: 59, 1988).

Parasterina Theiss. & Syd. (1917) = Asterina fide Doidge (*Bothalia* **4**: 273, 1942).

Parasterinella Speg. (1924), Asterinaceae. 2, S. America. See Arambarri (*Sydowia* **38**: 1, 1986).

Parasterinopsis Bat. (1960), Asterinaceae. 3, widespread (tropical). See Batista (*Atas Inst. Micol. Univ. Recife* **1**: 327, 1960).

Parastigmatea Doidge (1921), ? Polystomellaceae. 3, widespread (tropical).

Parastigmatellina Bat. & C.A.A. Costa (1959), anamorphic *Pezizomycotina*, Cpt.0eH.?. 1, Philippines; Australia. See Batista & Costa (*Mycopath. Mycol. appl.* **11**: 61, 1959), Reynolds & Gilbert (*Aust. Syst. Bot.* **18**: 265, 2005).

parasymbiont (obsol.), a fungus or lichen living on a lichen thallus but not causing any obvious damage (Zopf, *Ber. dtsch. bot. Ges.* **15**: 90, 1897); a commensalistic or gall-forming lichenicolous fungus (Hawksworth, *Bot. J. Linn. Soc.* **96**: 3, 1988). See Fungicolous fungi, Lichenicolous fungi.

Parasympodiella Ponnappa (1975), anamorphic *Pezizomycotina*, Hso.≡ eH.39. 6, widespread. See Tokumasu (*TMSJ* **28**: 19, 1987), Castañeda Ruíz *et al.* (*Mycotaxon* **65**: 93, 1997).

Parataeniellaceae Manier & Lichtw. (1968), Eccrinales. 2 gen., 7 spp.

Paratalaromyces Matsush. (2003), Trichocomaceae. Anamorph *Penicillium*. 1, Taiwan. See Matsushima (*Matsush. Mycol. Mem.* **10**: 111, 2001).

Paratetraploa M.K.M. Wong & K.D. Hyde (2002), anamorphic *Pezizomycotina*, H?.?.?. 1, Hong Kong. See Wong *et al.* (*Cryptog. Mycol.* **23**: 196, 2002).

Parathalle Clem. (1909) = Arthrorhaphis fide Poelt & Hafellner (*Phyton* Horn **17**: 213, 1976).

parathecium (1) (of apothecia), the outside hyphal layer, esp. if darker in colour (obsol.); (2) ectal excipulum (q.v.).

Paratheliaceae Zahlbr. (1903) = Pyrenulaceae.

Parathelium Nyl. (1862) = Pyrenula Ach. (1809) fide Upreti & Singh (*J. Econ. Taxon. Bot.* **10**: 236, 1987; India), Harris (*Mem. N. Y. bot. Gdn* **49**, 1989).

Parathelohania Codreanu (1966), Microsporidia. 9.

Paratomenticola M.B. Ellis (1976), anamorphic *Pezizomycotina*, Hso.≡ eP.10. 1, USA. See Ellis (*More Dematiaceous Hyphomycetes*: 175, 1976).

Paratorulopsis E.K. Novák & Zsolt (1961) = Candida fide Lodder (*Yeasts, a taxonomic study* 2nd edn, 1970).

Paratrichaegum Faurel & Schotter (1966) = Epicoccum fide Kendrick & Carmichae *in* Ainsworth *et al.* (Eds) (*The Fungi* **4A**: 390, 1973).

Paratrichaptum Corner (1987), Schizoporaceae. 1, Sumatra. See Corner (*Beih. Nova Hedwigia* **86**: 136, 1987), Hattori (*Mycoscience* **42**: 19, 2001).

Paratricharia Lücking (1997), Gomphillaceae (L). 1, Costa Rica. See Lücking (*Biblthca Lichenol.* **65**: 77, 1997).

Paratrichoconis Deighton & Piroz. (1972), anamorphic *Pezizomycotina*, Hso.≡ eH.11. 4, widespread (esp. tropical).

Paratrichophaea Trigaux (1985), Pyronemataceae. 4,

Europe; N. America. See Pfister (*Mycol.* **80**: 515, 1988; key 3 spp.), Perry *et al.* (*MR* **111**: 549, 2007; phylogeny).

paratype, see type.

Paraulocladium R.F. Castañeda (1986), anamorphic *Pezizomycotina*, Hso.#eP.26. 1, Cuba. See Castañeda (*Fungi Cubenses* **1**: 12, 1986).

Parauncinula S. Takam. & U. Braun (2005), Erysiphaceae. 2, Japan. See Takamatsu *et al.* (*Mycoscience* **46**: 14, 2005), Wang *et al.* (*Mycol.* **98**: 1065, 2006; phylogeny).

Paravalsa Ananthap. (1990), ? Valsaceae. 1, India. See Ananthapadmanaban (*MR* **94**: 275, 1990).

paraxonemal body (**PAR**), proteinaceous structure restricted to particular region of the flagellum in chromophytes and euglenoids; also manifest as flagellar spines in male gametes of oogamous phaeophytes.

Parberya C.A. Pearce & K.D. Hyde (2001), Phyllachoraceae. 2, Australia. See Pearce & Hyde (*Fungal Diversity* **6**: 83, 2001).

Parencoelia Petr. (1950), Helotiaceae. 4 (on *Phyllachoraceae*), C. America; West Indies. See Zhuang (*Mycotaxon* **32**: 85, 1988), Zhuang (*Mycotaxon* **38**: 117, 1990).

Parendomyces Queyrat & Laroche (1909) nom. rej. = Candida fide Hawksworth *et al.* (*Dictionary of the Fungi* edn 8, 1995).

Parenglerula Höhn. (1910), Englerulaceae. 5, widespread (tropical).

parenthesome, a curved double membrane (which may be perforate, imperforate, or vesiculate; see Moore, *Bot. Marina* **23**: 362, 1980) on each side of a dolipore septum (Moore & McAlear, *Am. J. Bot.* **49**: 86, 1962); see Fig. 13 'dolipore'; septal pore cap.

Parepichloë J.F. White & P.V. Reddy (1998), Clavicipitaceae. 7 (Epibiotic on grasses), pantropical. See White & Reddy (*Mycol.* **90**: 226, 1998), White *et al.* (*Stud. Mycol.* **45**: 95, 2000).

Pareutypella Y.M. Ju & J.D. Rogers (1996), Xylariales. 2, Taiwan. See Ju & Rogers (*Mycol.* **88**: 6, 1996).

parietal, fixed to the wall, e.g. of asci in a perithecium.

parietin (**physcion**), a bright yellow-orange to red anthraquinone pigment (in *Teloschistaceae*) giving a crimson-purple reaction with K (q.v.). Also known in some species of *Aspergillus*, *Penicillium*, *Polygonum* (Angiosp.), *Rheum* (Angiosp.) and *Ventilago* (Angiosp.). See Metabolic products, Pigments.

pariobasidium, see basidium.

Paris inch (**Paris line**), see line.

Parkerella A. Funk (1976) = Lahmia fide Eriksson (*Mycotaxon* **27**: 347, 1986).

Parksia E.K. Cash (1945) = Chloroscypha fide Korf *in* Ainsworth *et al.* (Eds) (*The Fungi* **4A**: 249, 1973).

Parmastomyces Kotl. & Pouzar (1964), Fomitopsidaceae. 7, widespread. See Kotlaba & Pouzar (*Reprium nov. Spec. Regni. veg.* **69**: 138, 1964).

Parmathyrites K.P. Jain & R.C. Gupta (1970), Fossil Fungi, Microthyriaceae. 2 (Tertiary), India.

Parmelaria D.D. Awasthi (1987), Parmeliaceae (L). 2, Himalaya. See Blanco *et al.* (*Mycol.* **97**: 150, 2005; phylogeny), Blanco *et al.* (*Mol. Phylogen. Evol.* **39**: 52, 2006; phylogeny).

Parmelia Ach. (1803) nom. cons., Parmeliaceae (L). 1322 s.lat.; s.str. 95, widespread (esp. temperate). This genus has been segregated into many smaller genera, not all of which are of value. See also *Lit.* under segregate names, esp. *Bulbothrix*, *Flavoparmelia*, *Hypotrachyna*, *Melanelia*, *Parmelina*, *Parmotrema*, *Pseudoparmelia*, *Punctelia*, *Relicina* and *Xanthoparmelia*. See Krog (*J. Hattori bot. Lab.* **52**: 303, 1982; conidia, gen. concepts), Galloway & Elix (*N.Z. Jl Bot.* **21**: 397, 1983), Hale (*Smithson. Contr. bot.* **66**, 1987; monogr. s. str., key), Eriksson & Hawksworth (*SA* **8**: 72, 1989; typification), Adler (*Mycotaxon* **38**: 331, 1990; key to segregates), Elix (*Bryologist* **96**: 359, 1993), Eriksson & Hawksworth (*SA* **13**: 146, 1994; concept), Hawksworth *in* Hawksworth (Ed.) (*Ascomycete Systematics. Problems and Perspectives in the Nineties* NATO ASI Series vol. **269**: 383, 1994; gen. concepts), Kurokawa (*J. Jap. Bot.* **69**: 121, 1994; Japan spp.), Kurokawa (*J. Jap. Bot.* **69**: 204, 1994; Japan spp.), Kurokawa (*J. Jap. Bot.* **69**: 61, 1994; Japan spp.), Crespo *et al.* (*Lichenologist* **29**: 275, 1997; DNA), Crespo & Cubero (*Lichenologist* **30**: 369, 1998; segregate genera), Crespo *et al.* (*Pl. Syst. Evol.* **217**: 177, 1999; genetics, recolonization), Hale & DePriest (*Bryologist* **102**: 462, 1999; checklist parmelioid species), Wedin *et al.* (*MR* **103**: 1152, 1999; phylogeny), Smith & Hyde (*Fungal Diversity* **7**: 89, 2001; mtDNA phylogeny), Adler & Calvelo (*Mitteilungen aus dem Institut für Allgemeine Botanik Hamburg* **30-32**: 9, 2002; S America), Crespo *et al.* (*MR* **106**: 788, 2002; phylogeny, *P. saxatilis*), Thell *et al.* (*Mycol. Progr.* **1**: 335, 2002; phylogeny), Molina *et al.* (*Lichenologist* **36**: 37, 2004; phylogeny), Thell *et al.* (*Mycol. Progr.* **3**: 297, 2004; phylogeny), Thell *et al.* (*Symb. bot. upsal.* **34** no. 1: 429, 2004; Scandinavia), Blanco *et al.* (*Mol. Phylogen. Evol.* **39**: 52, 2006; phylogeny), Miądlikowska *et al.* (*Mycol.* **98**: 1088, 2006; phylogeny), Honegger & Zippler (*MR* **111**: 424, 2007; mating systems).

Parmeliaceae Zenker (1827), Lecanorales (L). 88 gen. (+ 38 syn.), 2489 spp.

Lit.: Beltman (*Biblthca Lichenol.* **11**, 1978; thallus str.), Awasthi (*J. Hattori bot. Lab.* **65**: 207, 1989), Adler (*Mycotaxon* **38**: 331, 1990; key gen. *Parmelia* s.l.), Grgurinovic (Ed.) (*Flora of Australia* **32**: 41, 1990), Kärnefelt & Thell (*Pl. Syst. Evol.* **180**: 181, 1992), Elix (*Bryologist* **96**: 359, 1993; syn. key 64 gen.), Smith (*Bryologist* **96**: 326, 1993; key 115 spp. Hawaii), Orchard (Ed.) (*Flora of Australia* **55**, 1994; keys 31 gen.), Thell *et al.* (*Cryptog. Bot.* **5**: 120, 1995; ascus types), Elix (*Biblthca Lichenol.* **62**: 150 pp., 1996), Calvelo *in* Marcelli & Seaward (Eds) (*Lichenology in Latin America. History, Current Knowledge and Applications* [Proceedings of GLAL-3, Terceiro Encontro do Grupo Latino-Americano de Liquenólogos, São Paulo, Brazil, 24-28 September, 1997]: 117, 1998; keys to S Am. spp.), Crespo & Cubero (*Lichenologist* **30**: 369, 1998; DNA), Ingold (*Mycologist* **12**: 161, 1998), Kärnefelt (*Cryptog. Bryol.-Lichénol.* **19**: 93, 1998; rels with *Teloschistales*), Kärnefelt *et al.* (*Nova Hedwigia* **67**: 71, 1998; usneoid spp.), Lumbsch (*Lichenologist* **30**: 357, 1998), Mattsson & Wedin (*Lichenologist* **30**: 463, 1998), Saag (*Dissertationes Biologicae Universitatis Tartuensis* **34**, 1998; cetrarioid genera), Crespo *et al.* (*Lichenologist* **31**: 451, 1999; comparison of data types), Crespo *et al.* (*Pl. Syst. Evol.* **217**: 177, 1999), Grube *et al.* (*Curr. Genet.* **35**: 536, 1999), Mattsson & Wedin (*Lichenologist* **31**: 431, 1999), Stevens (*Biblthca Lichenol.* **72**: 128, 1999; usneoid taxa, Australia), Thell (*Lichenologist* **31**: 441, 1999), Wedin *et al.* (*MR* **103**: 1152, 1999; phylogeny), Esslinger

(*Bryologist* **103**: 568, 2000; S Afr. brown-spored spp.), Crespo *et al.* (*Taxon* **50**: 807, 2001), Kurokawa & Lai (*Mycotaxon* **77**: 225, 2001; Taiwan), Thell *et al.* (*Mycol. Progr.* **3**: 297, 2004), Blanco *et al.* (*Mycol.* **97**: 150, 2005), Blanco *et al.* (*Mol. Phylogen. Evol.* **39**: 52, 2006), Divakar *et al.* (*Mol. Phylogen. Evol.* **40**: 448, 2006), Miądlikowska *et al.* (*Mycol.* **98**: 1088, 2006; phylogeny), Arup *et al.* (*Mycol.* **99**: 42, 2007; sister group relations with *Parmeliaceae*), Crespo *et al.* (*Mol. Phylogen. Evol.* **44**: 812, 2007), Hofstetter *et al.* (*Mol. Phylogen. Evol.* **44**: 412, 2007; phylogeny).

Parmeliella Müll. Arg. (1862), Pannariaceae (L). *c.* 92, widespread (esp. tropical). See Jørgensen (*Op. bot. Soc. bot. Lund* **44**, 1978; Eur.), Jørgensen (*Lichenologist* **30**: 533, 1998), Makhija & Adawadkar (*Mycotaxon* **71**: 323, 1999; Andamans), Jørgensen (*Lichenologist* **32**: 139, 2000; C. & S. Am.), Jørgensen & Kashiwadani (*J. Jap. Bot.* **76**: 288, 2001; Japan), Ekman & Jørgensen (*CJB* **80**: 625, 2002; phylogeny), Jørgensen (*Bryologist* **106**: 121, 2003; subtropics), Wedin & Wiklund (*Symb. Bot. Upsal.* **34** no. 1, 2004), Miądlikowska *et al.* (*Mycol.* **98**: 1088, 2006; phylogeny).

Parmelina Hale (1974), Parmeliaceae (L). 22, widespread (esp. Europe; Asia). See Hale (*Smithson. Contr. bot.* **33**, 1976), Elix & Johnston (*Brunonia* **9**: 155, 1986; Australasia), Elix (*Flora of Australia* **55**: 124, 1994), Thell *et al.* (*Symb. bot. upsal.* **34** no. 1: 429, 2004; Scandinavia), Blanco *et al.* (*Mol. Phylogen. Evol.* **39**: 52, 2006; phylogeny), Argüello *et al.* (*Biol. J. Linn. Soc.* **91**: 455, 2007; cryptic species), Honegger & Zippler (*MR* **111**: 424, 2007; mating systems).

Parmelinella Elix & Hale (1987), Parmeliaceae (L). 4, widespread (esp. tropical). See Elix (*SA* **7**: 110, 1988; nomencl.), Elix (*Flora of Australia* **55**: 130, 1994), Blanco *et al.* (*Mol. Phylogen. Evol.* **39**: 52, 2006; phylogeny), Divakar *et al.* (*Mol. Phylogen. Evol.* **40**: 448, 2006; phylogeny, reln with *Bulbothrix*).

Parmelinopsis Elix & Hale (1987) = Hypotrachyna fide Elix & Hale (*Mycotaxon* **29**: 242, 1987), Adler & Elix (*Mycotaxon* **43**: 283, 1992; Argentina), Elix (*Flora of Australia* **55**: 131, 1994), Crespo *et al.* (*Lichenologist* **31**: 451, 1999; phylogeny, chemistry), Thell *et al.* (*Mycol. Progr.* **3**: 297, 2004; phylogeny), Masson (*Cryptog. Mycol.* **26**: 205, 2005; France), Blanco *et al.* (*Mol. Phylogen. Evol.* **39**: 52, 2006; phylogeny), Divakar *et al.* (*Mol. Phylogen. Evol.* **40**: 448, 2006; phylogeny, synonymy).

Parmeliomyces E.A. Thomas ex Cif. & Tomas. (1953) = Parmelina fide Hawksworth *et al.* (*Dictionary of the Fungi* edn 8, 1995).

Parmeliopsis (Nyl. ex Stizenb.) Nyl. (1866) nom. cons., Parmeliaceae (L). 8, widespread (north temperate). See Meyer (*Mycol.* **74**: 592, 1982), Hinds (*Mycotaxon* **72**: 271, 1999; N. Am.), Crespo *et al.* (*Taxon* **50**: 807, 2001; mtDNA phylogeny), Tehler & Källersjö (*Lichenologist* **33**: 403, 2001), Thell *et al.* (*Symb. bot. upsal.* **34** no. 1: 429, 2004; Scandinavia), Blanco *et al.* (*Mol. Phylogen. Evol.* **39**: 52, 2006; phylogeny).

Parmentaria Fée (1824) = Pyrenula Ach. (1809) fide Upreti & Singh (*Candollea* **43**: 109, 1988; key 11 spp. India), Harris (*Mem. N. Y. bot. Gdn* **49**, 1989).

Parmentariomyces Cif. & Tomas. (1953) ≡ Parmentaria.

Parmentieria Trevis. (1860) ≡ Parmentaria.

Parmocarpus Trevis. (1861) = Xanthoria fide Hawksworth *et al.* (*Dictionary of the Fungi* edn 8, 1995).

Parmophora M. Choisy (1950) = Umbilicaria fide Hawksworth *et al.* (*Dictionary of the Fungi* edn 8, 1995).

Parmosticta Nyl. (1875) ≡ Crocodia.

Parmostictina Nyl. (1875) nom. rej. = Pseudocyphellaria fide Hawksworth *et al.* (*Dictionary of the Fungi* edn 8, 1995).

Parmotrema A. Massal. (1860), Parmeliaceae (L). *c.* 348, widespread (esp. tropical). See Hale (*Contr. US natn Herb.* **36**: 193, 1965), Krog & Swinscow (*Bull. Br. Mus. nat. hist.* Bot. **9**: 143, 1981), Krog & Swinscow (*Lichenologist* **15**: 127, 1983), Louwhoff & Elix (*Biblthca Lichenol.* **73**, 1999; New Guinea), Louwhoff & Crisp (*Bryologist* **103**: 541, 2000), Kurokawa & Lai (*Mycotaxon* **77**: 225, 2001; Taiwan), Louwhoff (*Biblthca Lichenol.* **78**: 223, 2001; biogeography), Eliasaro & Donha (*Revista Brasileira de Botânica* **26**: 239, 2003; Brazil), Blanco *et al.* (*Mycol.* **97**: 150, 2005; phylogeny), Divakar *et al.* (*Lichenologist* **37**: 55, 2005; phylogeny), Egan *et al.* (*Bryologist* **108**: 402, 2005; Americas), Blanco *et al.* (*Mol. Phylogen. Evol.* **39**: 52, 2006; phylogeny), Miądlikowska *et al.* (*Mycol.* **98**: 1088, 2006; phylogeny), Ohmura *et al.* (*Bryologist* **109**: 43, 2006; genetics).

Parmotremopsis Elix & Hale (1987), Parmeliaceae (L). 2, C. America; West Indies. See Elix & Hale (*Mycotaxon* **29**: 243, 1987), Louwhoff & Crisp (*Bryologist* **103**: 541, 2000; phylogeny, ? synonymy with *Parmotrema*).

Parmularia Lév. (1846), Parmulariaceae. 2, S. America. See Batista & Vital (*Atas Inst. Micol. Univ. Recife* **1**: 159, 1960), Inácio & Cannon (*MR* **107**: 82, 2003).

Parmularia Nilson (1907) = Lecanora fide Hawksworth *et al.* (*Dictionary of the Fungi* edn 8, 1995), Ryan & Nash (*Nova Hedwigia* **64**: 111, 1997).

Parmulariaceae E. Müll. & Arx ex M.E. Barr (1979), ? Dothideomycetes (inc. sed.). 34 gen. (+ 24 syn.), 119 spp.

Lit.: Stevens & Ryan (*Ill. biol. monogr.* **17** no. 2: 1, 1939), Doidge (*Bothalia* **4**: 273, 1942; S. Afr.), von Arx & Müller (*Stud. Mycol.* **9**, 1975; connexion), Butin & Marmolejo (*Revta Mex. Micol.* **4**: 9, 1988), Vasil'eva (*Mikol. Fitopatol.* **22**: 388, 1988), Zhuang (*Mycotaxon* **32**: 85, 1988), Sivanesan & Sinha (*MR* **92**: 246, 1989), Inácio (*A monograph of the* Parmulariaceae [Thesis (PhD), University of London]: 417 pp., 2003), Inácio & Cannon (*MR* **107**: 82, 2003).

Parmulariella Henn. (1904), Dothideomycetes. 1, S. America.

Parmulariopsella Sivan. (1970), Parmulariaceae. 1, Sierra Leone. See Sivanesan (*TBMS* **55**: 509, 1970).

Parmulariopsis Petr. (1954), Parmulariaceae. 1, Borneo. See Petrak (*Sydowia* **8**: 186, 1954).

parmuliform, shield-shaped with the margins slightly upturned.

Parmulina Theiss. & Syd. (1914), Parmulariaceae. 5, Asia; America.

Parodiella Speg. (1880), Parodiellaceae. Anamorph *Ascochytopsis*. 1 (on *Leguminosae*), widespread. See Sydow & Petrak (*Annls mycol.* **29**: 190, 1931), von Arx & Müller (*Stud. Mycol.* **9**, 1975).

Parodiellaceae Theiss. & Syd. ex M.E. Barr (1987), Pleosporales. 2 gen. (+ 3 syn.), 5 spp.

Lit.: Cannon (*IMI Descr. Fungi Bact.* **141**: [22] pp.,

1999).

Parodiellina Henn. ex G. Arnaud (1918), Parodiopsidaceae. Anamorph *Exosporinella*. 1, widespread (tropical).

Parodiellina Viégas (1944) = Botryostroma fide von Arx & Müller (*Stud. Mycol.* **9**, 1975).

Parodiellinaceae G. Arnaud (1918) = Parodiopsidaceae.

Parodiellinopsis Hansf. (1946) ? = Chevalieropsis fide von Arx & Müller (*Stud. Mycol.* **9**, 1975).

Parodiodia Bat. (1960) = Apiosporina Höhn. fide Müller & von Arx (*Beitr. Kryptfl. Schweiz* **11** no. 2, 1962).

Parodiopsidaceae Toro (1952), ? Dothideomycetes (inc. sed.). 20 gen. (+ 32 syn.), 94 spp.

Lit.: Sivanesan (*Mycol. Pap.* **146**: 38 pp., 1981), von Arx & Müller (*Sydowia* **37**: 6, 1984), Farr (*Sydowia* **38**: 65, 1985), Hughes (*Mycol. Pap.* **166**: 255 pp., 1993), Rodríguez Hernández (*Revta Jardín bot. Nac.* Univ. Habana **21**: 127, 2000).

Parodiopsis Maubl. (1915) = Perisporiopsis Henn. fide Müller & von Arx (*Beitr. Kryptfl. Schweiz* **11** no. 2, 1962).

Paropodia Cif. & Bat. (1956) = Triposporiopsis fide Hughes (*Mycol.* **68**: 693, 1976).

Paropsis Speg. (1918) = Treubiomyces fide Hawksworth *et al.* (*Dictionary of the Fungi* edn 8, 1995).

Parorbiliopsis Spooner & Dennis (1986), Helotiaceae. *c.* 5, Europe. See Svrček (*Česká Mykol.* **45**: 134, 1992), Svrček (*Česká Mykol.* **46**: 149, 1993).

parsimonious (parsimony), see Cladistics.

part spore, one of the 1-celled spores resulting from the breaking up of a 2 or more celled ascospore.

parthenogamy, the state of an oospore formed with a diploid nucleus resulting from restitution at telophase I or telophase II of meiosis (note that the parthenogametic state cannot be presumed from the absence of an antheridium).

parthenogenesis, the apomitic development of haploid cells (Gäumann).

parthenomixis, see parthenogamy.

Parthenope Velen. (1934), ? Helotiales. 1, former Czechoslovakia.

parthenospore, an oospore (aboospore) or zygospore (azygospore) produced by parthenogenesis.

partial veil (or **inner veil**) (of agarics), a layer of tissue, developed from the stipe, which joins the stipe to the pileus edge during hymenium development, and which later may become an annulus or cortina; = velum (Persoon).

partridge wood (1) wood attacked by a pocket rot, e.g. one caused by *Stereum frustulatum*; [(2) wood of *Caesalpinia*].

Paruephaedria Zukal (1891) = Dactylospora fide Döbbeler & Triebel (*Bot. Jahrb. Syst.* **107**: 503, 1985).

Parvacoccum R.S. Hunt & A. Funk (1988), Rhytismataceae. 1, Canada. See Hunt & Funk (*Mycotaxon* **33**: 51, 1988).

Parvobasidium Jülich (1975), Cystostereaceae. 3, widespread. See Jülich (*Persoonia* **8**: 302, 1975).

Parvodontia Hjortstam & Ryvarden (2004), ? Cystostereaceae. 1, Brazil. See Hjortstam & Ryvarden (*Syn. Fung.* **18**: 28, 2004).

Parvomyces Santam. (1995), Laboulbeniaceae. 1, Spain. See Santamaría (*MR* **99**: 1071, 1995).

Parvosympodium A.W. Ramaley (2002), anamorphic *Pezizomycotina*, H?.?.?. 1, USA. Synanamorph *Stomiopeltopsis*. See Ramaley (*Mycotaxon* **83**: 329, 2002).

Parvulago R. Bauer, M. Lutz, Piątek, Vánky & Oberw. (2007), Ustilaginaceae. 1 (on *Eleocharis* (*Cyperaceae*)), Europe (southern). See Bauer *et al.* (*MR* **111**: 1199, 2007).

Paryphydria, see *Paruephaedria*.

Pasania fungus, the edible *Cortinellus shiitake*. See shiitake.

Paschelkiella Sherwood (1987), Odontotremataceae. 1, Europe; N. America. See Sherwood-Pike (*Mycotaxon* **28**: 137, 1987).

Pascherinema De Toni (1936) nom. dub., anamorphic *Pezizomycotina*.

Pasithoe M.J. Decne. (1840) ≡ Paulia Fée.

paspalitrem, a tremorgenic mycotoxin from *Phomopsis* sp. (Bills *et al.*, *MR* **96**: 977, 1992).

Paspalomyces Linder (1933), anamorphic *Pezizomycotina*, Hso.1eP.10. 1, USA.

paspalum staggers, see ergot.

passage, the experimental infection of a host with a parasite which is subsequently re-isolated; [a method used to increase the virulence of the parasite].

Passalora Fr. (1849), anamorphic *Mycosphaerella*, Hso.1-≡ eP.10. 581, Europe; N. America. *P. graminis* (leaf spot of cerials and grasses). See Deighton (*Mycol. Pap.* **112**, 1967; key), Medeiros & Dianese (*Mycotaxon* **51**: 509, 1994; key), Braun (*Schlechtendalia* **2**: 1, 1999), Braun (*Cryptog. Mycol.* **20**: 155, 1999; gen. concepts), Stewart *et al.* (*MR* **103**: 1491, 1999), Crous *et al.* (*Stud. Mycol.* **45**: 107, 2000; review), Crous *et al.* (*Mycol.* **93**: 1081, 2001; phylogeny), Goodwin *et al.* (*Phytopathology* **91**: 648, 2001; phylogeny), Braun *et al.* (*Mycol. Progr.* **2**: 3, 2003; phylogeny), Crous & Braun (*CBS Diversity Ser.* **1**: 571 pp., 2003; nomenclator), Crous *et al.* (*Mycosphaerella Leaf Spot Diseases of Bananas: Present Status and Outlook* Proceedings of the 2nd International Workshop on *Mycosphaerella* Leaf Spot Diseases Held in San José, Costa Rica, 20-23 May 2002: 43, 2003; polyphasic systematics), Crous *et al.* (*Stud. Mycol.* **50**: 457, 2004; on *Acacia*), Crous *et al.* (*Stud. Mycol.* **50**: 195, 2004; on *Eucalyptus*), Hunter *et al.* (*Stud. Mycol.* **55**: 147, 2006; on *Eucalyptus*).

Passeriniella Berl. (1891) nom. dub., Dothideomycetes. 2, Europe. See Kohlmeyer & Volkmann-Kohlmeyer (*Bot. Mar.* **34**: 1, 1991), Khashnobish & Shearer (*MR* **100**: 1341, 1996).

Passerinula Sacc. (1875), Pleosporales. 1 or 2 (on stromata of other fungi), Europe. See Rossman *et al.* (*Stud. Mycol.* **42**: 248 pp., 1999).

Pasteur (Louis; 1822-1895; France). Born in Dole, Jura; student, École Normale Supérieure, Paris; Professor of Physics, Dijon Lycée (1848); Professor of Chemistry, Stassbourg University (1849); Dean, Faculty of Sciences, University of Lille (1854); Director of Scientific Studies, École Normale Supérieure, Paris (1855). The work of this noted scientist, who was by training a chemist, has strongly affected biological thought and work. Among Pasteur's early investigations was a study of the isomeric forms of tartaric acid from which he discovered that only one of the two forms (the *d*-) in a racemic (*dl*-) mixture was used by '*Penicillium glaucum*'. This gave him an interest in fermenting processes and from 1854 onwards he made detailed studies on wine and beer making. One important outcome of this work was demonstration by experiment of the general principle 'omne vivum e vivo' (every living thing from a living thing). In 1865-1869, as a result of an investigation suggested by the chemist Dumas, he made clear the nature of the 'pébrine' and 'flacherie' diseases of silkworms and methods for their control. Pasteur's most important work was a development from Jenner's vaccination against small-pox (1796). His studies on chicken cholera (*Pasteurella aviseptica*), anthrax (*Bacillus anthracis*), and rabies in dogs and man made it clear that active immunization of man and animals against pathogenic organisms and viruses could be a generally applicable method. In 1888 the Institut Pasteur opened in Paris for development of his discoveries. *Publs*. Note relative au *Penicillium glaucum* et à la dissymétrie moléculaire des produit organiques naturelles. *Comptes Rendus Hebdomadaire des Séances de l'Académie des Sciences, Paris* (1860); *Études sur le Vin* (1866); *Études sur le Vinaigre* (1868); *Études sur la Maladie des Vers à Soie* (1870); *Études sur la Bière* (1876). *Biogs, obits etc*. Debré & Forster (*Louis Pasteur*, 1998); Dubos (*Louis Pasteur,*., 1951); Tiner (*Louis Pasteur, Founder of Modern Medicine*, 1990); Vallery-Radot (*Vie de Pasteur*, 1900) [English translation Devonshire, 1911].

Pasteur effect, Increased respiration and decreased fermentation in the presence of oxygen and vice versa in the absence of oxygen; **- filter**, an unglazed porcelain tube for sterilization by filtration; **- pipette**, a short length of glass tubing with one end drawn out into a sealed capillary and the other plugged with cotton wool and sterilized, the tip of the capillary is broken before use; **pasteurization**, freeing a liquid or other material from pathogenic microorganisms by heat.

Patella F.H. Wigg. (1780) nom. rej. ≡ Scutellinia fide Korf (*Taxon* **35**: 378, 1986).

Patellaria Fr. (1822), Patellariaceae. *c*. 10, widespread (esp. temperate). See Butler (*Mycol.* **331**: 612, 1939), Tilak & Srinivasulu (*Beih. Nova Hedwigia* **47**: 459, 1974; Indian spp.), Bellemère *et al.* (*Cryptog. Mycol.* **7**: 47, 1986; asci), Hawksworth (*Taxon* **35**: 787, 1986; nomencl.), Kutorga & Hawksworth (*SA* **15**: 1, 1997).

Patellaria Hoffm. (1789) = Diploschistes fide Santesson *in* Eriksson (*Op. Bot.* **60**, 1981).

Patellaria Pers. (1794) = Caloplaca fide Riedl (*Taxon* **27**: 302, 1978).

Patellariaceae Corda (1838), Patellariales. 15 gen. (+ 7 syn.), 38 spp.

Lit.: Bellemère (*Annls Sci. Nat.* Bot., sér. 12 **12**: 429, 1971; ultrastr.), Kamat *et al. in* Subramanian (Ed.) (*Taxonomy of fungi*, 1978; ascocarp structure), Bellemère *et al.* (*Cryptog. Mycol.* **7**: 113, 1986), Kutorga & Hawksworth (*SA* **15**: 1, 1997), Magnes *et al.* (*Mycotaxon* **68**: 321, 1998), Silva-Hanlin & Hanlin (*MR* **103**: 153, 1999; DNA), Liew *et al.* (*Mol. Phylogen. Evol.* **16**: 392, 2000).

Patellariales D. Hawksw. & O.E. Erikss. (1986), (±L). Dothideomycetes. 1 fam., 15 gen., 38 spp. Thallus crustose or immersed within the host tissues, or absent. Stromata absent or weakly developed, composed of intramatrical hyphae. Ascomata erumpent, eventually apothecial, sometimes strongly cupulate, sometimes elongated, the margin well-developed, frequently inrolled when dry. Interascal tissue of narrow anastomosing pseudoparaphyses, initially attached at apex and base but eventually becoming free at the apex. Asci cylindrical, fissitunicate, J-, often with a well-developed ocular chamber. Ascospores hyaline or brown, transversely septate or muriform, usually without a sheath. Anamorphs coelomycetous where known. Saprobes on bark or wood or lichenized with green algae, a few lichenicolous, cosmop. Fam.:

Patellariaceae (syn. *Lecanidiaceae*)

Lit.: Bellemère (*Ann. Sci. nat., Bot.* sér. 12 **12**: 429, 1971; asci), Honegger (*Ber. dtsch. bot. Ges.* **91**: 579, 1978, *Lichenologist* **12**: 157, 1980; asci), Kamat (*in* Subramanian (Ed.), *Taxonomy of fungi* **1**: 292, 1978; ontogeny), Pirozynski & Reid (*CJB* **44**: 655, 1966).

Patellariopsis Dennis (1964), Helotiales. 5, Europe; Australia. See Dennis (*Kew Bull.* **19**: 114, 1964), Nauta & Spooner (*Mycologist* **14**: 65, 2000; UK).

Patellea (Fr.) Sacc. (1884) = Xylogramma fide Nannfeldt (*Nova Acta R. Soc. Scient. upsal.*, 1932).

patelliform, like a round plate having a well-marked edge.

Patellina Grove ex Petch (1943) nom. dub., anamorphic *Pezizomycotina*. See Sutton (*Mycol. Pap.* **141**, 1977).

Patellina Speg. (1880) = Catinula fide Höhnel (*Sber. Akad. Wiss. Wien* Math.-naturw. Kl., Abt. 1 **124**: 49, 1915).

Patellonectria Speg. (1919) = Aspidothelium fide Santesson (*Symb. bot. upsal.* **12** no. 1: 1, 1952).

patent, stretching out; spreading.

Patent protection. The principle under which an 'inventor' of a new product or process by publicly disclosing details of his/her 'invention' is granted for a limited period a legally enforceable right to exclude others from exploiting the invention. Procedures, requirements and rights vary considerably in different countries. In general, it is possible to patent discoveries (and under certain circumstances organisms, including fungi) which are novel, show evidence of an inventive step, and are of utility (industrial applicability). The procedure for taking out a patent is complex and professional advice should be sought. It is recommended that the patent disclosure procedure should include the depositing in a genetic resource collection of a culture any fungus or other organism which forms part of the 'invention'. The Budapest Treaty allows for a single deposit in a recognized International Depository Authority (IDA), removing the need for deposits to be made in all countries where patent protection is sought. For fungi, recognized depositories include the American Type Culture Collection (www.atcc.org), Centralbureau voor Schimmelcultures (www.cbs.knaw.nl) and IMI (q.v.).

The Convention on Biological Diversity (1992) also has implications for the patenting of genetic material from nature (Fritze, 1994).

Lit.: *Budapest treaty on the international recognition of the deposit of microorganisms for the purpose of patent procedures* (World Intellectual Property Organization, 1981), Bousfield (*in* Hawksworth & Kirsop (Eds), *Filamentous fungi* : 115, 1988), Crespi (*Patents: a basic guide to patenting in biotechnology* [*Cambridge studies in biotechnology* **6**], 1988), Fritze (*in* Kirsop & Hawksworth (Eds), *The biodiversity of micro-organisms and the role of microbial resource centres*: 37, 1994). See Bioprospecting.

Patescospora Abdel-Wahab & El-Shar. (2002), Aliquandostipitaceae. 1, Egypt. See Pang *et al.* (*MR* **106**: 1033, 2002), Campbell *et al.* (*CJB* **85**: 873, 2007;

phylogeny).

pathogen, a parasite able to cause disease in a particular host or range of hosts (*TBMS* **33**: 155, 1950); **-ic**, disease-causing or able to be so; **-icity**, the condition of being pathogenic.

pathotoxin, see toxin.

pathotype, see pathovar.

pathovar (pv.) (of bacteria), an infrasubspecific subdivision characterized by a pathogenic reaction in one or more hosts; a pathotype (not recommended) See Young *et al.* (*ROPP* **70**: 213, 1991).

Patila Adans. (1763) ≡ Oncomyces.

Patinella Sacc. (1875), Helotiales. 3, widespread. See Spooner (*Biblthca Mycol.* **116**, 1987).

Patinellaria H. Karst. (1885), Helotiaceae. 1, Europe; Mexico. See Dennis (*Mycol. Pap.* **62**, 1956), Medel & Chaćon (*Rev. Mex. Micol.* **4**: 251, 1988).

Patouillard (Narcisse Théophile; 1854-1926; France). A pharmacist, Paris (1881-1922); Preparateur de Cryptogamie, École Supérieure de Pharmacie, Paris (1893-1900); Assistant de la Chaire de Cryptogamie du Muséum d'Histoire Naturelle, Paris (1922 onwards). Noted for his work on anatomy and biology of larger fungi, and for contributions to the mycotas of north Africa, the Caribbean, France, North America, Oceania, South America and Spain. Collections mostly in **FH**, but some types in **PC**. *Publs. Les Hyménomycètes d'Europe* (1887); *Essai Taxonomique sur les Familles et les Genres des Hyménomycètes* (1900) [reprint, 1963]; also more than 200 papers [reprint, 3 vols, 1978]. *Biogs, obits etc.* Heim (*Annales de Cryptogamie Exotique* **1**: 25, 1928); Heim (*Revue de Mycologie* **36** (2): i, 1972) [history, portrait]; Grummann (1974: 292); Mangin (*BSMF* **43**: 8, 1927) [bibliography, portrait]; Pfister (Annotated index to fungi described by N. Patouillard. *Contributions of Reed Herbarium* **25**, 1977); Stafleu & Cowan (*TL-2* **3**: 94, 1981).

Patouillardea Roum. (1885) = Dendrodochium fide Höhnel *in* Falk (*Mykol. Unters.* **1**: 360, 1923).

Patouillardiella Speg. (1889), anamorphic *Pezizomycotina*, Hsp.1eH.?. 1, Brazil. See Petrak & Sydow (*Annls mycol.* **34**: 11, 1936).

Patouillardina Bres. (1906), Auriculariales. 1, widespread (tropical). See Martin (*Mycol.* **31**: 507, 1939), Donk (*Taxon* **7**: 238, 1958).

Patouillardina G. Arnaud (1917) = Meliolaster Höhn. fide Hawksworth *et al.* (*Dictionary of the Fungi* edn 8, 1995).

Patouillardinaceae Jülich (1982) Auriculariales.

Patoullardiella, see *Patouillardiella*.

Patriciomyces D. Hawksw. (2001), anamorphic *Pezizomycotina*. 1, USA. See Cole & Hawksworth (*Mycotaxon* **77**: 327, 2001).

patulin (clavicin, clavatin, claviformin, expansin, mycoin, penicidin), a mycotoxin from *Aspergillus clavatus*, *Penicillium patulum*, *P. claviforme*, *P. expansum*, etc.; also antibacterial and antifungal (Raistrick *et al.*, *Lancet* **2**: 625, 1943); toxic to plants and animals (carcinogenic to mice) and the cause of neurotoxicosis in cattle; can also occur in apple and pear juice (from *P. expansum*).

Pauahia F. Stevens (1925), Meliolaceae. 1 (from leaves), Hawaii. See Hughes (*Mycol.* **87**: 702, 1995).

Paucithecium Lloyd (1923), Xylariaceae. 1, Brazil. See Læssøe (*SA* **13**: 43, 1994; ? synonym of *Xylaria*).

Paulia Fée (1836), Lichinaceae (L). 10, widespread (tropical). See Henssen (*Lichenologist* **18**: 201, 1986; key), Henssen & Tretiach (*Nova Hedwigia* **60**: 297, 1995), Schultz *et al.* (*Bryologist* **102**: 61, 1999; Socotra), Schultz & Büdel (*Lichenologist* **34**: 39, 2002; key), Schultz & Büdel (*Lichenologist* **35**: 151, 2003).

Paulia Lloyd (1916) ≡ Xenosoma.

Paullicorticium J. Erikss. (1958), Hydnaceae. 5, N. America; Europe. See Liberta (*Brittonia* **13**: 219, 1962; key).

Paurocotylis Berk. ex Hook. f. (1855), Pyronemataceae. 1 (hypogeous), Europe; New Zealand. See Dennis (*Kew Bull.* **30**: 345, 1974), Pegler *et al.* (*British truffles*, 1993), Crous *et al.* (*S. Afr. J. Bot.* **62**: 89, 1996), Landvik *et al.* (*Nordic Jl Bot.* **17**: 403, 1997; DNA), O'Donnell *et al.* (*Mycol.* **89**: 48, 1997; phylogeny), Liu & Zhuang (*Mycosystema* **25**: 546, 2006; phylogeny), Læssøe & Hansen (*MR* **111**: 1075, 2007; phylogeny), Perry *et al.* (*MR* **111**: 549, 2007; phylogeny).

Paxillaceae Lotsy (1907), Boletales. 9 gen. (+ 15 syn.), 78 spp.

Lit.: Beaton *et al.* (*Kew Bull.* **40**: 573, 1985), Liu *et al.* (*Acta Mycol. Sin.* **8**: 210, 1989), Montecchi & Lazzari (*Micol. Ital.* **18**: 33, 1989), Besl *et al.* (*Z. Mykol.* **62**: 195, 1996), Calonge & Siquier (*Boln Soc. Micol. Madrid* **23**: 91, 1998), Bresinsky *et al.* (*Pl. Biol.* **1**: 327, 1999), Hahn & Agerer (*Nova Hedwigia* **69**: 241, 1999), Jarosch & Bresinsky (*Pl. Biol.* **1**: 701, 1999), Kretzer & Bruns (*Mol. Phylogen. Evol.* **13**: 483, 1999), Hönig *et al.* (*Mycorrhiza* **9**: 315, 2000), Claridge *et al.* (*Aust. Syst. Bot.* **14**: 273, 2001), Grubisha *et al.* (*Mycol.* **93**: 82, 2001), Binder & Bresinsky (*Mycol.* **94**: 85, 2002), Bruns & Shefferson (*CJB* **82**: 1122, 2004), Becerra *et al.* (*Mycorrhiza* **15**: 7, 2005), Nouhra *et al.* (*Mycol.* **97**: 598, 2005).

Paxillogaster E. Horak (1966), Boletaceae. 1, S America. Basidioma gasteroid. See Horak & Moser (*Nova Hedwigia* **10**: 332, 1966).

Paxillopsis E.-J. Gilbert (1931) nom. nud. = Paxillus fide Singer (*Agaric. mod. Tax.*, 1951).

Paxillopsis J.E. Lange (1940) ≡ Clitopilus.

Paxillus Fr. (1836), Paxillaceae. 15 (ectomycorrhizal), widespread. See Singer (*Nova Hedwigia* **7**: 93, 1964; S. Am. spp.), McNabb (*N.Z. Jl Bot.* **7**: 349, 1969; NZ spp.), Corner (*Nova Hedwigia* **20**: 793, 1970; Malaysia), Watling (*British Fungus Flora* **1**, 1970; key Brit. spp.), Horak (*Sydowia* **32**: 154, 1980; Australasia), Fries (*Mycotaxon* **24**: 403, 1985; taxonomy), Singer *et al.* (*Beih. Nova Hedwigia* **98**: 12, 1990; key C. Am. spp.), Bruns *et al.* (*Mol. Ecol.* **7**: 257, 1998; phylogeny) The saprotrophic species are referred to *Tapinella*, Wallander & Söderström (*Paxillus – Ectomycorrhizal fungi, key genera in profile*, 1999; ecology).

Paxina Kuntze (1891) ≡ Acetabula.

Payosphaeria W.F. Leong (1990), Hypocreales. 1 (wood, marine), Malaysia. See Leong *et al.* (*Bot. Mar.* **33**: 511, 1990).

Pazschkea Rehm (1898) = Tapellaria fide Hawksworth *et al.* (*Dictionary of the Fungi* edn 8, 1995).

Pazschkeella Syd. & P. Syd. (1901), anamorphic *Microcyclus*, St.1eH.?. 2, Brazil; Philippines.

PC, Laboratoire de Cryptogamie, Museum National d'Histoire Naturelle (Paris, France); founded 1904; supported by the Ministry for Education; genetic resource collection **LCP**; see Anon. (*La Chaire de Cryptogamie du Museum National d'Histoire Naturelle*, 1954).

PCA, see Media.

PCP, see *Pneumocystis*.

PCR, see Molecular biology.
PD, see Metabolic products; Steiner's Stable PD solution.
PDA, see Media.
peach leaf curl, leaf hypertrophy caused by *Taphrina deformans*.
peat mould, *Chromelosporium fulvum*.
Peccania A. Massal. ex Arnold (1858) nom. cons., Lichinaceae (L). 8, Europe; N. Africa. See Moreno & Egea (*Acta Bot. Barcinon.* **41**: 1, 1992), Schultz & Büdel (*Mycotaxon* **84**: 21, 2002), Schultz & Büdel (*Lichenologist* **35**: 151, 2003; phylogeny).
Peccaniomyces Cif. & Tomas. (1953) ≡ Peccania.
Peccaniopsis M. Choisy (1949) = Peccania fide Hawksworth *et al.* (*Dictionary of the Fungi* edn 8, 1995).
Peck (Charles Horton; 1833-1917; USA). Student, Union College, Schenectady (1855-1859); teacher, Sandlake and Albany (1859-1867); Botanist to the State Cabinet of Natural History [later the New York State Museum], Albany, New York (1867-1913). He proposed more than 2,700 fungal names [for list 1867-1908, see Peck, *Bulletin of the New York State Museum* **131**: 59, 1909; 1905-1915, see Gilbertson, *Mycol.* **54**: 460, 1962]. *Publs.* Report of the Botanist. *Annual Report on the New York State Museum of Natural History* (1872-1913) [reprint, 6 vols, 1980]. *Biogs, obits etc.* Atkinson (*Botanical Gazette* **65**: 103, 1918); Barr *et al.* (An annotated catalogue of the pyrenomycetes described by Peck. *Bulletin of the New York State Museum* **459**, 1986]); Burnham (*Mycol.* **11**: 33, 1919); Grummann (1974: 194); Jenkins (*Mycotaxon* **7**: 23, 1978, *Amanita* types); Lloyd (*Mycological Notes* **4**: 509, 1912) [portrait]; Petersen (*Mycol.* **68**: 304, 1976) [cantharelloid species]; Stafleu & Cowan (*TL-2* **4**: 135, 1983).
Peckia Clinton (1878) [non *Peckia* Vell. 1825, *Myrsinaceae*] ≡ Drudeola.
Peckiella (Sacc.) Sacc. (1891) = Hypomyces fide Rogerson & Samuels (*Mycol.* **81**: 413, 1, 1989), Rossman *et al.* (*Stud. Mycol.* **42**: 248 pp., 1999).
Peckifungus Kuntze (1891) ≡ Appendiculina.
Peckiomyces Sacc. & Trotter (1913) nom. nud., Fungi. See Welch (*Mycol.* **18**: 82, 1926).
pecky cypress, decay of *Taxodium distichum* by *Stereum taxodii* (Davidson *et al.*, *Mycol.* **52**: 260, 1961).
pectic enzymes, components of the macerating enzymes of a number of fungal parasites; two types distinguished by Wood (*Physiological plant pathology*, 1967: 138; q.v. for details) are **pectinesterases** (syn. pectinmethylesterases), specific enzymes which saponify methyl ester groups of pectinic acid; and **polygalacturonases**, which break polygalacturonide chains by hydrolysis at the glycosidic linkages.
pectinate, like the teeth of a comb.
pectinate hyphae, comb-like hyphae of e.g. *Microsporum audouinii*.
Pectinotrichum Varsavsky & G.F. Orr (1971), Onygenaceae. Anamorph *Chrysosporium*. 1, Venezuela. See Sugiyama *et al.* (*Mycoscience* **40**: 251, 1999; DNA), Sugiyama & Mikawa (*Mycoscience* **42**: 413, 2001; phylogeny).
Pediascus Chardón & Toro (1934) = Vestergrenia Rehm fide von Arx & Müller (*Stud. Mycol.* **9**, 1975).
pedicel, a small stalk.
pedicellate, having a pedicel.
Pedilospora Höhn. (1902) = Dicranidion fide Hughes (*CJB* **31**: 607, 1953).
pedogamy, pseudomixis between mature and immature cells, e.g. copulation between a yeast mother cell and its bud.
pedogenesis (1) reproduction in young or immature organisms; (2) soil formation (see Weathering).
Pedumispora K.D. Hyde & E.B.G. Jones (1992), Diaporthales. 1 (on *Rhizophora*), Seychelles. See Hyde & Jones (*MR* **96**: 78, 1992).
Peethambara Subram. & Bhat (1978), Hypocreales. Anamorph *Didymostilbe*. 1 (from decaying wood), widespread (pantropical). See Rossman *et al.* (*Stud. Mycol.* **42**: 248 pp., 1999), Rossman *et al.* (*Mycol.* **93**: 100, 2001; phylogeny), Castlebury *et al.* (*MR* **108**: 864, 2004; phylogeny), Zhang *et al.* (*Mycol.* **98**: 1076, 2006; phylogeny).
Peethasthabeeja P. Rag. Rao (1981), anamorphic *Pezizomycotina*, Hso.0eP.10. 1, India. See Rao (*Indian J. Bot.* **4**: 132, 1981).
peg, see hyphal peg.
Peglerochaete Sarwal & Locq. (1983), Tricholomataceae. 1, Sikkim. See Sarwal & Locquin (*Compt. Rend. Congr. Natl. Soc. Savantes* Sec. Sci., 1983: 193, 1983).
Pegleromyces Singer (1981), Tricholomataceae. 1, Brazil. See Singer (*Mycol.* **73**: 500, 1981).
Peglionia Goid. (1937) = Gyrothrix fide Hughes & Pirozynski (*N.Z. Jl Bot.* **9**: 39, 1971).
Pegmatheca E.I. Hazard & Oldacre (1975), Microsporidia. 2.
peitschengeissel (obsol.) unornamented flagellum.
Pekilo, protein from *Paecilomyces* sp. used in animal feeds.
Pelicothallos Dilcher (1965), Fossil Fungi, Microthyriaceae. 1 (Eocene), USA. See Pirozynski (*Ann. Rev. Phytopath.* **14**: 237, 1976), Sherwood-Pike (*Lichenologist* **17**: 114, 1985).
pellet, a three dimensional colony in a liquid culture (particularly a shaken culture).
pellicle (1) outermost living layer lying below any non-living secreted material, containing plasma membrane plus underlying epiplasm or other membranes and may show ridges, folds or distinct crests; (2) a growth on the surface of a liquid culture; (3) (of agaric basidiomata), a detachable, skin-like cuticle of the pileus.
pellicular veil, a very thin partial veil of a sporophore not having a stipe (Singer, 1962). Cf. cortina.
Pellicularia Cooke (1876) = Ceratobasidium fide Roberts (*Rhizoctonia-forming fungi*, 1999).
pelliculose, like a thin crust, as is the hymenial layer in *Thelephoraceae*.
Pellidiscus Donk (1959), Inocybaceae. 3, Europe. See Donk (*Persoonia* **1**: 89, 1959), Hjortstam (*Windahlia* **15**: 59, 1985).
Pellionella (Sacc.) Petch (1924) nom. inval., anamorphic *Pezizomycotina*. Cited in *ING* (1979) but Petch (*Ann. R. bot. Gdn Peradeniya* **8**: 170, 1924) did not make the transfer.
Pellionella Sacc. (1902), anamorphic *Pezizomycotina*, St.1eP.1. 1, Java.
pellis, the cellular cortical layers, not belonging to the veils, of a basidioma (Bas, *Persoonia* **5**: 327, 1969); cuticle. Bas distinguished different layers of the pellis as **supra-**, **medio-**, and **sub-**, and different topographies of the pellis as **pilei-**, etc.
Pelloporus Quél. (1886) ≡ Coltricia.
Pellucida Dulym., Sivan., P.F. Cannon & Peerally (2001), Hyponectriaceae. 1, Mauritius. See Dulyma-

mode *et al.* (*MR* **105**: 247, 2001).

pellucid-striate (of a pileus), having a somewhat transparent top so that the gills are seen through it as rays.

Pelodiscus Clem. (1901) nom. dub., ? Fungi. See Seaver (*North American Cup Fungi* (Operculates), 1928; synonym of *Patella*), Eckblad (*Nytt Mag. Bot.* **15**: 1, 1968).

Peloronectria Möller (1901) nom. dub., ? Hypocreales. 1 or 2, S. America. Type material is lost. See Rossman *et al.* (*Stud. Mycol.* **42**: 248 pp., 1999).

Peloronectriella Yoshim. Doi (1968) = Nectriopsis fide Rossman *et al.* (*Stud. Mycol.* **42**: 248 pp., 1999).

peloton, see mycorrhiza.

Peltaster Syd. & P. Syd. (1917), anamorphic *Dothideomycetes*, Cpt.0eH.?. 6, Philippines; S. America. See Johnson *et al.* (*Mycol.* **88**: 114, 1996), Williamson & Sutton (*Pl. Dis.* **84**: 714, 2000; pathology), Williamson *et al.* (*Mycol.* **96**: 885, 2004), Batzer *et al.* (*Mycol.* **97**: 1268, 2005; phylogeny).

Peltasterales (obsol.). Anamorphic fungi with superficial shield-like conidiomata; anamorphs of *Microthyriaceae*. Batista & Ciferri (*Mycopath.* **11**: 6, 1959; 6 fam., keys, host index), Batista & Ciferri (*Atti Ist. bot. Univ. Pavia* ser. 5 **16**: 86, 1959).

Peltasterella Bat. & H. Maia (1959), anamorphic *Pezizomycotina*, Cpt.0eH.?. 6, widespread (tropical). See Batista & Maia (*Mycopath. Mycol. appl.* **11**: 37, 1959).

Peltasterinostroma Punith. (1975), anamorphic *Pezizomycotina*, Cpt.0eH.15. 1, British Isles. See Punithalingam (*TBMS* **64**: 541, 1975).

Peltasteropsis Bat. & H. Maia (1959), anamorphic *Pezizomycotina*, Cpt.0eH.?. 7, widespread (tropical). See Batista & Maia (*Mycopath. Mycol. appl.* **11**: 29, 1959).

peltate, in the form of a round plate with a stalk from the centre of the underside (Fig. 23.29).

Peltella Syd. & P. Syd. (1917) = Muyocopron fide von Arx & Müller (*Stud. Mycol.* **9**, 1975).

Peltidea Ach. (1803) = Peltigera fide Hawksworth *et al.* (*Dictionary of the Fungi* edn 8, 1995).

Peltideaceae Körb. (1846) = Peltigeraceae.

Peltideomyces E.A. Thomas (1939) nom. inval. = Peltigera fide Hawksworth *et al.* (*Dictionary of the Fungi* edn 8, 1995).

Peltidium Kalchbr. (1862) [non *Peltidium* Zollik. 1820, *Compositae*] ≡ Pachyella.

Peltigera Willd. (1787) nom. cons., Peltigeraceae (L). *c.* 91, widespread. See Thomson (*Am. midl. Nat.* **44**: 1, 1950), Kurokawa *et al.* (*Bull. natn. Sci. Mus. Tokyo*, B **9**: 101, 1966), Chen (*Acta Mycol. Sin.* **5**: 18, 1986; key 28 spp., China), Holtan-Hartwig (*Sommerfeltia* **15**, 1993; key 17 spp. Norway), Vitikainen (*Acta Bot. Fenn.* **152**: 1, 1994; key 29 spp. Eur.), Eriksson & Strand (*SA* **14**: 33, 1995; DNA), Goward *et al.* (*CJB* **73**: 91, 1995; key 28 N. Am. spp.), Martínez & Burgaz (*Ann. bot. fenn.* **33**: 223, 1996; anatomy), Goffinet & Bayer (*Fungal Genetics Biol.* **21**: 228, 1997; DNA, species pairs), Martínez & Burgaz (*Clementeana* **3**: 29, 1997; key), Martínez Moreno (*Ruizia* **15**, 1999; Iberian spp.), Miądlikowska & Lutzoni (*Int. J. Pl. Sci.* **161**: 925, 2000; phylogeny), Vitikainen (*Mitteilungen aus dem Institut für Allgemeine Botanik Hamburg* **30-32**: 297, 2002; S America, Antarctica), Goffinet *et al.* (*Bryologist* **106**: 349, 2003; phylogeny, morphospecies), Martínez *et al.* (*Lichenologist* **35**: 301, 2003; biogeography), Miądlikowska *et al.* (*Mycol.* **95**: 1181, 2003; phylogeny), Miądlikowska *et al.* (*Mycol.* **98**: 1088, 2006; phylogeny).

Peltigeraceae Dumort. (1822), Peltigerales (L). 2 gen. (+ 15 syn.), 101 spp.

Lit.: Hawksworth (*J. Linn. Soc.* Bot. **96**: 3, 1988), Galloway (*Symbiosis* **11**: 327, 1991), Czeczuga & Upreti (*Bull. bot. Surv. India* **32**: 80, 1990), Holtan-Hartwig (*Sommerfeltia* **15**: 1, 1993), Smith & Øvstedal (*Lichenologist* **26**: 209, 1994), Vitikainen (*Acta Bot. Fenn.* **152**: 1, 1994), Eriksson & Strand (*SA* **14**: 33, 1995), Goward *et al.* (*CJB* **73**: 91, 1995), Goffinet & Bayer (*Fungal Genetics Biol.* **21**: 228, 1997), Galloway (*Cryptog. Bryol.-Lichénol.* **19**: 137, 1998), Miądlikowska & Lutzoni (*Int. J. Pl. Sci.* **161**: 925, 2000), Miądlikowska *et al.* (*Mycol.* **95**: 1181, 2003), Wiklund & Wedin (*Cladistics* **19**: 419, 2003), Miądlikowska & Lutzoni (*Am. J. Bot.* **91**: 449, 2004), Miądlikowska *et al.* (*Mycol.* **98**: 1088, 2006; phylogeny), Hofstetter *et al.* (*Mol. Phylogen. Evol.* **44**: 412, 2007; phylogeny), Vitikainen (*Nordic Lichen Flora* **3**: Cyanolichens: 113, 2007).

Peltigerales Walt. Watson (1929), (L). Lecanoromycetidae. 8 fam., 50 gen., 1208 spp. Thallus usually foliose, occ. crustose, fruticose or squamulose, corticate on the upper or both surfaces, the lower surface sometimes characteristically areolate. Ascomata apothecial, initially immersed, usually with a covering layer which splits to expose the hymenium at a late stage. Interascal tissue of simple paraphyses. Asci usually with a thickened apex (usually not capitate), a J+ apical ring, a well-developed ocular chamber and a J+ gelatinized outer coat. Ascospores hyaline or brown, often elongated, often multiseptate. Anamorphs pycnidial. Lichenized with green algae or cyanobacteria, if the former then often with cyanobacteria in cephalodia, widespr. esp. temp. Fams:

(1) **Coccocarpiaceae**
(2) **Collemataceae**
(3) **Lobariaceae** (syn. *Stictaceae*)
(4) **Massalongiaceae**
(5) **Nephromataceae**
(6) **Pannariaceae**
(7) **Peltigeraceae** (syn. *Solorinaceae*)
(8) **Placynthiaceae** (syn. *Lecotheciaceae*)

Two suborders have been recognized based on molecular data, the *Peltigerineae* (containing *Lobariaceae*, *Massalongiaceae*, *Nephromataceae* and *Peltigeraceae*) and *Collematineae* which contains the other 4 fams.

Lit.: Galloway (*Symbiosis* **11**: 327, 1991; chemical evol.), Hawksworth (*J. Hattori bot. Lab.* **52**: 323, 1982; evolutionary significance), Hofstetter *et al.* (*Mol. Phylogenet. Evol.* **44**: 412, 2007), Honegger (*Lichenologist* **10**: 47, 1978; asci), Letrouit-Galinou & Lallement (*Lichenologist* **5**: 59, 1971; asci, ontogeny), Matsubara *et al.* (*Bryologist* **102**: 196, 1999; chemistry), Miądlikowska & Lutzoni (*Am. J. Bot.* **91**: 449, 2004), Miądlikowska *et al.* (*Mycol.* **98**: 1088, 2006), Wiklund & Wedin (*Cladistics* **19**: 419, 2003; phylogeny).

Peltigeromyces Möller (1901), Helotiales. 3, Africa. See Le Gal (*Bull. Jard. Bot. État* **29**: 73, 1959).

Peltigeromyces E.A. Thomas ex Cif. & Tomas. (1953) ≡ Peltigera.

Peltigeropsis V. Marcano & A. Morales (1994) nom. inval., Peltigerales (L). 1, Europe. See Marcano *et al.* (*Libro de Resúmenes, Sesiones Técnicas, VI Congreso Latinoamericano de Botánica* Mar del Plata,

Argentina, 2 al 8 de Octubre de 1994: 192, 1994), Eriksson & Hawksworth (*SA* **15**: 139, 1997).

Peltiphylla M. Choisy (1950) = Psora Hoffm. (1796) fide Hawksworth *et al.* (*Dictionary of the Fungi* edn 8, 1995).

Peltistroma Henn. (1904), anamorphic *Pezizomycotina*, Cpt.0eP.?. 1, Brazil.

Peltistromella Höhn. (1907), anamorphic *Pezizomycotina*, Cpt.≡ eP.?. 2, Brazil; Sri Lanka.

Peltophora Clem. (1909) ≡ Peltigera.

Peltophoromyces Clem. (1909) ≡ Peltigeromyces A. Møller.

Peltopsis Bat. (1960) = Muyocopron fide von Arx & Müller (*Stud. Mycol.* **9**, 1975).

Peltosoma Syd. (1925), anamorphic *Pezizomycotina*, Cpt.≡ eP.?1. 1, Philippines. See Sutton & Pascoe (*MR* **92**: 210, 1989; re-evaluation).

Peltosphaeria Berl. (1888) = Julella fide Mayrhofer (*Biblthca Lichenol.* **26**, 1987).

Peltostromellina Bat. & A.F. Vital (1959), anamorphic *Pezizomycotina*, Cpd.0eH.?. 1, Sri Lanka. See Batista & Vital (*Revista Soc. Brazil agron.* **13**: 79, 1959).

Peltostromopsis Bat. & A.F. Vital (1959), anamorphic *Pezizomycotina*, Cpt.1eP.?. 1, Sri Lanka. See Batista & Vital (*Revista Soc. Brazil agron.* **13**: 77, 1959).

Peltula Nyl. (1853), Peltulaceae (L). 31, widespread (esp. temperate). See Wetmore (*Ann. Mo. Bot. Gdn* **57**: 158, 1970; N. Am.), Swinscow & Krog (*Norw. Jl Bot.* **26**: 213, 1979; key 10 E. Afr. spp.), Büdel & Henssen (*Int. J. Mycol. Lichenol.* **2**: 235, 1986), Filson (*Muelleria* **6**: 495, 1988; 10 Australian spp.), Egea (*Biblthca Lichenol.* **31**, 1989; W. Eur. & N. Afr.), Büdel & Elix (*Biblthca Lichenol.* **67**: 3, 1997), Schultz *et al.* (*Pl. Biol.* **3**: 116, 2001), Büdel & Schultz (*Biblthca Lichenol.* **86**: 225, 2003; drought resistance), Schultz & Büdel (*Lichenologist* **35**: 151, 2003), Miądlikowska *et al.* (*Mycol.* **98**: 1088, 2006; phylogeny).

Peltulaceae Büdel (1986), Lichinales (L). 3 gen., 35 spp.

Lit.: Büdel (*Biblthca Lichenol.* **25**: 209, 1987; ontogeny), Filson (*Muelleria* **6**: 495, 1988), Büdel (*Biblthca Lichenol.* **38**: 47, 1990), Upreti & Büdel (*J. Hattori bot. Lab.* **68**: 279, 1990), Büdel (*Mycotaxon* **54**: 137, 1995), Reeb *et al.* (*Mol. Phylogen. Evol.* **32**: 1036, 2004), Jørgensen (*Nordic Lichen Flora* **3**: Cyanolichens: 132, 2007).

Peltularia R. Sant. (1944), Coccocarpiaceae (L). 2, S. America; Australia. See Jørgensen & Galloway (*Lichenologist* **16**: 189, 1984), Henssen & Jørgensen (*Biblthca Lichenol.* **78**: 85, 2001; S America).

Pemphidium Mont. (1840), Xylariales. 6 (monocots), S. America (tropical). See Petrak (*Sydowia* **7**: 354, 1953), Hyde (*Sydowia* **45**: 204, 1993; posn), Hyde (*Sydowia* **48**: 122, 1996).

Penardia Cash (1904), Cercozoa. q.v.

penatin (corylophillin, notatin, penicillin B), an antibiotic from *Penicillium notatum* and *P. chrysogenum*; antibacterial (Kocholary, *J. Bact.* **44**: 143, 1942).

pendant (1) (of a fruticose lichen), one hanging downwards; (2) (in an ascus) see ascus.

Pendulispora M.B. Ellis (1961), anamorphic *Tubeufia*, Hso.#eP.10. 1, Venezuela. See Ellis (*Mycol. Pap.* **82**: 41, 1961).

penicidin, see patulin.

Penicillaria Chevall. (1826) [non *Penicillaria* Willd. 1809, *Gramineae*] = Pterula fide Corner (*Ann. Bot. Mem.* [A monograph of Clavaria and allied genera] **1**, 1950).

penicillate, like a little brush.

penicillic acid, a tetronic acid from *Penicillium puberulum*, *P. cyclopium*, etc. and antibacterial (Birkinshaw *et al.*, *Biochem. J.* **30**: 394, 1936).

Penicillifer Emden (1968), anamorphic *Viridispora*, Hso.≡ eP.32/33. 7, widespread. See Samuels (*Mycol.* **81**: 347, 1989; key), Polishook *et al.* (*Mycol.* **83**: 797, 1991).

penicillin, one of a group of antibiotic substances (q.v.) produced by *Penicillium chrysogenum* (syn. *P. notatum*), active against Gram+ bacteria and of low toxicity to humans. Discovered by Fleming (*Br. J. exp. Path.* **10**: 226, 1929). See Hare (*The birth of penicillin*, 1970), *Penicillin 1929-43* [*Br. med. Bull.* **2**, 1944], Fleming (Ed.) (*Penicillin; its practical application*, 1946), Waksman (*Microbial antagonisms and anti-bacterial substances*, edn 2, 1947), Knox (*Nature* **192**: 492, 1961; survey of new penicillins).

penicillin B, **B**, see penatin; **F**, see flavicin; **dihydro-F**, see gigantic acid; **N**, see cephalosporin N. See Knox (1961) under penicillin.

penicillinase, a bacterial enzyme which inhibits the action of penicillin.

Penicilliopsis Solms (1887), Trichocomaceae. Anamorphs *Stilbodendron*, *Sarophorum*, *Pseudocordyceps*. 3, widespread (tropical). See Udagawa & Takada (*Bull. natn. Sci. Mus.* Tokyo, B **14**: 501, 1971), Samson & Seifert *in* Samson & Pitt (Eds) (*Advances in Penicillium and Aspergillus systematics* **102**: 397, 1985), Kuraishi *et al.* (*MR* **95**: 705, 1991; ubiquinones), Pitt *et al.* (*Integration of Modern Taxonomic Methods for Penicillium and Aspergillus Classification*: 9, 2000; nomenclator), Hsieh & Ju (*Mycol.* **94**: 539, 2002).

Penicillites Mesch. (1892), Fossil Fungi. 1 (Oligocene), Europe.

Penicillium Fr. (1832) ? = Botrytis fide Hawksworth *et al.* (*Taxon* **25**: 665, 1976).

Penicillium Link (1809), anamorphic *Eupenicillium*, *Talaromyces*, Hso.0eP.15. 304, widespread. Common moulds, many first named on cultural, biochemical, or pathogenic characters. See Fig. 30. *P. expansum* (blue mould of apples), *P. digitatum* and *P. italicum* (green and blue moulds of citrus fruits), *P. gladioli* (storage rot of *Gladiolus*). See Pitt (*CJB* **52**: 2231, 1974; sclerotial spp.), Pitt (*The genus Penicillium and its teleomorphic states Eupenicillium and Talaromyces*, 1979; revision), Frisvad & Filtenborg (*Appl. Environm. Microbiol.* **46**: 1301, 1983; secondary metabolites and classification), Frisvad (*Toxigenic Fungi: their toxins and health hazard*: 98, 1984; chemistry and taxonomy), Pitt (*Toxigenic Fungi: their toxins and health hazard*: 107, 1984; physiology and taxonomy), Bridge (*J. gen. Microbiol.* **131**: 1887, 1985; characterization methods), Hawksworth (*Advances in Penicillium and Aspergillus Systematics* **102**: 3, 1985; typification), Peberdy (*Biology of Industrial Microorganisms*: 407, 1985; biology), Pitt (*Laboratory Guide to common Penicillium species*, 1985), Pitt & Hocking (*Fungi and Food Spoilage*, 1985), Anné (*Advances in Penicillium and Aspergillus Systematics* **102**: 337, 1985; protoplast hybridization), Bridge *et al.* (*Advances in Penicillium and Aspergillus Systematics* **102**: 281, 1985; polyphasic systematics), Bridge *et al.* (*TBMS* **87**: 389, 1986; intra-strain variation), Deng *et al.* (*J. Med. Vet. Mycol.* **24**: 383, 1986; clinical sp.), Fassatiová

(*Moulds and Filamentous Fungi in Technical Microbiology* Progress in industrial microbiology **22**, 1986), Fassatiová (*Advances in Penicillium and Aspergillus Systematics* **102**: 105, 1985; media and morphology), Frisvad (*Advances in Penicillium and Aspergillus Systematics* **102**: 311, 1985; chemistry), Ho & Smith (*J. gen. Microbiol.* **132**: 3479, 1986; effect of CO_2 on morphology), Minter *et al.* (*Advances in Penicillium and Aspergillus Systematics* **102**: 71, 1985; terminology), Pitt (*Advances in Penicillium and Aspergillus Systematics* **102**: 93, 1985; diagnostic methods), Pitt *et al.* (*Food Microbiology* **3**: 363, 1986; white cheese moulds), Polonelli *et al.* (*Advances in Penicillium and Aspergillus Systematics* **102**: 267, 1985; serological diagnosis), Ramírez (*Advances in Penicillium and Aspergillus Systematics* **102**: 445, 1985; recombination), Ramírez (*Cryptog. Mycol.* **7**: 181, 1986; morphological species concept), Seifert & Samson (*Advances in Penicillium and Aspergillus Systematics* **102**: 143, 1985; synnematous spp.), Stolk & Samson (*Advances in Penicillium and Aspergillus Systematics* **102**: 163, 1985; classification), Wicklow (*Advances in Penicillium and Aspergillus Systematics* **102**: 255, 1985; ecology), Williams & Pitt (*Advances in Penicillium and Aspergillus Systematics* **102**: 129, 1985; key), Bridge *et al.* (*J. gen. Microbiol.* **133**: 995, 1987; phenotype, DNA content), Cruickshank & Pitt (*Microbiological Sciences* **4**: 14, 1987; isoenzymes), Cruickshank & Pitt (*Mycol.* **79**: 614, 1987; enzyme electrophoresis), El-Banna *et al.* (*Syst. Appl. Microbiol.* **10**: 42, 1987; mycotoxin production), Frisvad *et al.* (*CJB* **65**: 765, 1987; from rodent seed caches), Peberdy (*Penicillium and Acremonium* [Biotechnology Handbooks] **1**, 1987), Pitt (*Appl. Environm. Microbiol.* **53**: 266, 1987; ochratoxin-producing spp.), Polonelli *et al.* (*Appl. Environm. Microbiol.* **53**: 872, 1987; serological diagnosis, cheese-associated spp.), Drouhet *et al.* (*Bulletin de la Société Française de Mycologie Médicale* **17**: 77, 1988; clinical sp.), Petruccioli *et al.* (*Mycol.* **80**: 726, 1988; extracellular enzymes), Bridge & Hudson (*J. Linn. Soc.* Bot. **99**: 11, 1989; DNA content), Bridge *et al.* (*J. gen. Microbiol.* **135**: 2967, 1989; polyphasic systematics), Bridge *et al.* (*J. gen. Microbiol.* **135**: 2941, 1989; polyphasic systematics), Fuhrmann *et al.* (*Can. J. Microbiol.* **35**: 1043, 1989; immunology), Head *et al.* (*Exp. Mycol.* **13**: 203, 1989; ultrastr.), Kozakiewicz (*J. Linn. Soc.* Bot. **99**: 273, 1989; conidial ornamentation), Mills *et al.* (*Can. J. Pl. Path.* **11**: 357, 1989; mycotoxins from wheat), Paterson *et al.* (*J. gen. Microbiol.* **135**: 2979, 1989; polyphasic systematics), Pitt (*Journal of Applied Bacteriology* Symposium Supplement Series no. 18 **67**: 37S, 1989; review), Ström *et al.* (*Journal of Applied Bacteriology* **66**: 461, 1989; chemosystematics), Berny & Hennebert (*Modern Concepts in Penicillium and Aspergillus Classification* [NATO ASI Series A: Life Sciences] **185**: 49, 1990; *P. expansum*), Bridge (*Modern Concepts in Penicillium and Aspergillus Classification* [NATO ASI Series A: Life Sciences] **185**: 283, 1990; physiology), Fassatiová & Kubatová (*Modern Concepts in Penicillium and Aspergillus Classification* [NATO ASI Series A: Life Sciences] **185**: 149, 1990; section *Divaricatum*), Filtenborg & Frisvad (*Modern Concepts in Penicillium and Aspergillus Classification* [NATO ASI Series A: Life Sciences] **185**: 27, 1990; identification of mixed cultures), Frisvad & Filtenborg (*Mycol.* **81**: 837, 1989; chemotaxonomy), Frisvad & Filtenborg (*Modern Concepts in Penicillium and Aspergillus Classification* [NATO ASI Series A: Life Sciences] **185**: 373, 1990; synoptic key), Frisvad & Filtenborg (*Modern Concepts in Penicillium and Aspergillus Classification* [NATO ASI Series A: Life Sciences] **185**: 159, 1990; sect. *Furcatum*), Hawksworth (*Modern Concepts in Penicillium and Aspergillus Classification* [NATO ASI Series A: Life Sciences] **185**: 75, 1990; nomencl.), Paterson & Kemmelmeier (*Journal of Chromatography* **511**: 195, 1990; chemistry), Pitt (*PENNAME* A Computer Key to *Penicillium* Species, 1990; computer key), Pitt & Samson (*Stud. Mycol.* **32**: 77, 1990; review), Pitt *et al.* (*Syst. Appl. Microbiol.* **13**: 304, 1990; monoverticillate spp.), Reenen-Hoekstra *et al.* (*Modern Concepts in Penicillium and Aspergillus Classification* [NATO ASI Series A: Life Sciences] **185**: 173, 1990; *P. funiculosum* complex), Samson & Pitt (*Modern Concepts in Penicillium and Aspergillus Classification* [NATO ASI Series A: Life Sciences], 1990), Stolk *et al.* (*Modern Concepts in Penicillium and Aspergillus Classification* [NATO ASI Series A: Life Sciences] **185**: 121, 1990; terverticillate spp.), Andersen (*Antonie van Leeuwenhoek* **60**: 115, 1991; *P. brevicompactum*, chemistry), Kuraishi *et al.* (*MR* **95**: 705, 1991; ubiquinones), Pitt (*Jap. J. Med. Mycol.* **32** Suppl.: 31, 1991; chemotaxonomy), Pitt *et al.* (*Phytopathology* **81**: 1108, 1991; on apples), Schubert & Kriesel (*Persoonia* **14**: 341, 1991; ubiquinones), Bridge *et al.* (*Mycol. Pap.* **165**, 1992; computer key), Drouhet (*J. Mycol. Médic.* **3**: 195, 1993; penicilliosis), LoBuglio *et al.* (*Mycol.* **85**: 592, 1993; phylogeny, sexuality), Pitt (*Journal of Applied Bacteriology* **75**: 559, 1993; media, species differentiation), Pitt (*The Fungal Holomorph: Mitotic, Meiotic and Pleomorphic Speciation in Fungal Systematics*: 107, 1993; speciation), Pitt & Samson (*Regnum veg.* **128**: 13, 1993; list of accepted names), Christensen *et al.* (*MR* **98**: 635, 1994; colour standards), Holmes *et al.* (*Phytopathology* **84**: 719, 1994; on citrus), Lobuglio *et al.* (*MR* **98**: 250, 1994; phylogeny, synnematous spp.), Lund & Frisvad (*MR* **98**: 481, 1994; *P. aurantiogriseum*, chemistry), Okuda (*Mycol.* **86**: 259, 1994; morphology in culture), Tzean *et al.* (*Mycol. Monogr.* **9**, 1994; Taiwan), Berbee *et al.* (*Mycol.* **87**: 210, 1995; phylogeny), Larsen & Frisvad (*MR* **99**: 1167, 1995; volatile metabolites), LoBuglio & Taylor (*J. Clin. Microbiol.* **33**: 85, 1995; *P. marneffei* diagnostics), Mills *et al.* (*Mycopathologia* **130**: 23, 1995; toxigenic spp., Canada), Pitt (*CJB* **73** Suppl. 1: S768, 1995; morphology), Boysen *et al.* (*Microbiology* Reading **142**: 541, 1996; *P. roqueforti* agg.), Banke *et al.* (*MR* **101**: 617, 1997; isoenzymes), Mekha *et al.* (*Mycoscience* **38**: 97, 1997; RAPDs), Smedsgaard & Frisvad (*Biochemical Systematics and Ecology* **25**: 51, 1997; mass spectroscopy), Cao *et al.* (*J. Clin. Microbiol.* **36**: 3028, 1998; serological diagnosis), Lopes da Silva *et al.* (*FEMS Microbiol. Lett.* **164**: 303, 1998; fatty acid profiles), Vanittanakom *et al.* (*Medical Mycology* **36**: 169, 1998; PCR diagnosis), Ahmad & Malloch (*Mycol.* **91**: 1031, 1999; physiology), Christensen *et al.* (*MR* **103**: 527, 1999; *P. miczynskii* group), Skouboe *et al.* (*MR* **103**: 873, 1999; phylogeny), Banke & Rosendahl (*Integration of Modern Taxonomic Methods for Penicillium and Aspergillus Classification*: 199, 2000; genetic markers), Boysen *et al.* (*Appl. Environm. Microbiol.* **66**:

1523, 2000; *P. roqueforti* group), Dörge *et al.* (*J. Microbiol. Meth.* **41**: 121, 2000; image analysis), Frisvad *et al.* (*Integration of Modern Taxonomic Methods for Penicillium and Aspergillus Classification*: 113, 2000; conidiophore ornamentation), Peterson (*Integration of Modern Taxonomic Methods for Penicillium and Aspergillus Classification*: 163, 2000; phylogeny), Pitt & Samson (*Integration of Modern Taxonomic Methods for Penicillium and Aspergillus Classification*: 51, 2000; typification), Pitt *et al.* (*Integration of Modern Taxonomic Methods for Penicillium and Aspergillus Classification*: 9, 2000; accepted names), Scott *et al.* (*Integration of Modern Taxonomic Methods for Penicillium and Aspergillus Classification*: 225, 2000; DNA fingerprinting), Seifert & Giuseppin (*Integration of Modern Taxonomic Methods for Penicillium and Aspergillus Classification*: 259, 2000; cycloheximide tolerance), Seifert & Louis-Seize (*Integration of Modern Taxonomic Methods for Penicillium and Aspergillus Classification*: 189, 2000; phylogeny), Seifert *et al.* (*Integration of Modern Taxonomic Methods for Penicillium and Aspergillus Classification*: 239, 2000; physiology), Tuthill *et al.* (*Mycol.* **93**: 298, 2001; *P. simplicissimum*), Vossler (*Clinical Microbiology Newsletter* **23**: 25, 2001; clinical diagnostics), Castella *et al.* (*Syst. Appl. Microbiol.* **25**: 74, 2002; ochratoxigenic spp.), Yuen *et al.* (*Archs Microbiol.* **179**: 339, 2003; *P. marneffei* genome), Fisher *et al.* (*J. Clin. Microbiol.* **42**: 5065, 2004; *P. marneffei* population structure), Frisvad & Samson (*Stud. Mycol.* **49**: 1, 2004; polyphasic systematics), Frisvad *et al.* (*Stud. Mycol.* **49**: 201, 2004; mycotoxins), Haugland *et al.* (*Syst. Appl. Microbiol.* **27**: 198, 2004; quantitative PCR), La Guerche *et al.* (*Curr. Microbiol.* **48**: 405, 2004; from grapes), Peterson (*MR* **108**: 434, 2004; phylogeny), Samson & Frisvad (*Stud. Mycol.* **49**, 2004; revision of subgen. *Penicillium*), Samson *et al.* (*Introduction to Food- and Airborne Fungi* edn 7, 2004; identification of common spp.), Samson *et al.* (*Stud. Mycol.* **49**: 175, 2004; phylogeny), Smedsgaard *et al.* (*Stud. Mycol.* **49**: 243, 2004; mass spectroscopy), Frisvad *et al.* (*Journal of Applied Microbiology* **98**: 684, 2005; ochratoxigenic sp.), Peterson *et al.* (*Mycol.* **97**: 659, 2005; on coffee), Dupont *et al.* (*Rev. Iberoam. Micol.* **23**: 145, 2006; RFLP diagnostics), Geiser *et al.* (*Mycol.* **98**: 1053, 2006; phylogeny), Peterson (*Rev. Iberoam. Micol.* **23**: 134, 2006; phylogeny), Vega *et al.* (*Mycol.* **98**: 31, 2006; endophytic spp., ochratoxins), Seifert *et al.* (*Proc. natn Acad. Sci. U.S.A.* **104**: 3901, 2007; phylogenetic barcodes), Horn & Peterson (*Mycol.* **100**: 12, 2008; on *Aspergillus* heads).

penicillus, the brush-like conidiogenous apparatus of *Penicillium* and related genera composed of a stipe bearing a tuft of conidiogenous cells and other elements formerly on the rami and metulae (Fig. 30).

Penidiella Crous & U. Braun (2007), anamorphic *Teratosphaeriaceae*. 7, widespread. See Crous *et al.* (*Stud. Mycol.* **58**: 1, 2007), Mafia *et al.* (*Mycol. Progr.* **7**: 49, 2007).

Peniophora Cooke (1879), Peniophoraceae. 62, widespread. See Eriksson (*Symb. bot. upsal.* **10** no. 5, 1950), Slysh (*Tech. Pub. Sta. Univ. Coll. Forestry, Syracuse* **83**, 1960; N.Y. State), Boidin (*Bull. Soc. linn. Lyon* **34**: 161, 1965), Boidin (*Bull. Soc. linn. Lyon* **34**: 213, 1965; key), Hayashi (*Bull. Govt For. Exp. Stn* **260**, 1974; Japan), Leger (*BSMF* **92**: 377, 1976; electrophoresis) Many authors use *Peniophora* s.s. for *Peniophora* sect. *Coloratae*; see, Leger & Poncet (*BSMF* **92**: 229, 1976; numerical tax.), Eriksson *et al.* (*Cortic. N. Europ.* **5**: 916, 1978; key 21 Eur. spp.) ? ≡ Corticium Pers. (Cortic.), Wu (*Mycotaxon* **85**: 187, 2003; Taiwan).

Peniophoraceae Lotsy (1907), Russulales. 7 gen. (+ 6 syn.), 88 spp.

Lit.: Donk (*Reinwardtia* **1**, 1951; gen. names), Reid (*Beih. Nova Hedwigia* **18**: 484 pp., 1965; stipitate steroid spp.), Liberta (*Mycol.* **60**: 827, 1968; *Peniophora*), Ginns (*Mycol.* **60**: 1211, 1969; *Merulius*), Rattan (*Biblthca Mycol.* **60**, 1977; NW Himalayas), Jülich & Stalpers (*Verh. Kon. Ned. Akad. Wetensch., Afd. Natuurk.* sect. 2 **74**, 1980; temp. N. Hemisph.), Boidin *et al.* (*BSMF* **107**: 91, 1991; trop. spp.), Wu & Chen (*Bull. natn. Mus. Nat. Sci.* Taiwan **4**: 101, 1993), Hallenberg *et al.* (*MR* **100**: 179, 1996), Ginns (*Mycol.* **90**: 19, 1998), Yurchenko (*Mikol. Fitopatol.* **34**: 37, 2000), Larsson & Larsson (*Mycol.* **95**: 1037, 2003), Hjortstam & Ryvarden (*Syn. Fung.* **18**: 20, 2004), Larsson *et al.* (*MR* **108**: 983, 2004), Binder *et al.* (*Systematics and Biodiversity* **3**: 113, 2005), Küffer & Senn-Irlet (*Mycol. Progr.* **4**: 77, 2005).

Peniophorella P. Karst. (1889) nom. dub., Agaricomycetes. 35, widespread. *Hymenochaetales* or *Agaricales* (*Rickenella* clade). See Larsson (*MR* **111**, 2007).

Peniophorina Höhn. (1917) nom. dub., Fungi. See Donk (*Reinwardtia* **1**: 216, 1951).

penitrem A, a mycotoxin produced by some *Penicillium* spp., including *P. cyclopium*, which affects the nervous system, causing tremors; see neurotoxin, tremorgen.

Pennella Manier (1968), Legeriomycetaceae. 7 (in *Diptera*), widespread. See Lichtwardt (*The Trichomycetes. Fungal associates of arthropods*, 1986; key), Williams & Lichtwardt (*CJB* **68**: 1045, 1990), Sato (*Mycoscience* **43**: 33, 2002; ultrastr. *Pennella angustispora*), Nelder *et al.* (*Fungal Diversity* **22**: 121, 2006; ecology and taxonomy), White (*MR* **110**: 1011, 2006; phylogeny).

Penomyces Giard ex Sacc. & Trotter (1913) nom. dub., anamorphic *Pezizomycotina*, Hso.?.?. 2, Europe. See Petch (*TBMS* **19**: 190, 1935).

Pentagenella Darb. (1897), Roccellaceae (L). 3, Chile. See Follmann *et al.* (*J. Hattori bot. Lab.* **85**: 245, 1998), Tehler & Irestedt (*Cladistics* **23**: 432, 2007).

Pentaposporium Bat. (1957) ? = Tripospermum fide Hughes (*Mycol.* **68**: 810, 1976).

Pentasporium, see *Pentaposporium*.

Penzigia Sacc. (1888) = Xylaria Hill ex Schrank fide Læssøe (*SA* **8**: 25, 1989), Læssøe (*SA* **13**: 43, 1994).

Penzigina Kuntze (1891) ≡ Eriosphaeria.

Penzigomyces Subram. (1992), anamorphic *Pezizomycotina*, Hso.≡ eP.19. 13, widespread. See Subramanian (*Proc. Indian natn Sci. Acad.* Part B. Biol. Sci. **58**: 186, 1992), Mercado-Sierra *et al.* (*Mycol.* **96**: 424, 2004; Spain).

Peplopus (Quél.) Quél. ex Moug. & Ferry (1887) ≡ Suillus Gray.

per (in author citations) (obsol.), see Nomenclature.

percurrent (1) extending throughout the entire length, as of the columella of a gasteromycete basidioma; (2) growing through in the direction of the long axis, as of a conidial germ tube emerging through the hilum or of a proliferation growing through the tip of the conidiogenous cell (Luttrell, 1963).

Perelegamyces R.F. Castañeda & W.B. Kendr. (1990), anamorphic *Pezizomycotina*, Hso.0eP.11. 1, Cuba. See Castañeda & Kendrick (*Univ. Waterloo Biol. Ser.* **32**: 34, 1990).

perennial, living for a number of years.

Perenniporia Murrill (1942) nom. cons. prop., Polyporaceae. *c.* 60, widespread. See Donk (*Persoonia* **1**: 266, 1960; nomencl. of *Poria*), Lowe (*Mycol.* **55**: 453, 1963; key 162 trop. spp., as *Poria*), Wright (*Mycol.* **56**: 694, 1964; dextrinoid reaction; as *Poria*), Lowe (*Tech. Pub. Sta. Univ. Coll. Forestry, Syracuse* **90**, 1966; syns, key 133 N. Am. spp., as *Poria*), Ryvarden (*Polyp. N. Eur.* **2**: 305, 1978; key Eur. spp.), Ryvarden & Johansen (*Prelim. Polyp. Fl. E. Afr.*: 463, 1980; key 10 Afr. spp.), Decock & Mossebo (*Systematics and Geography of Plants* **71**: 607, 2002; African taxa), Decock & Mossebo (*Systematics and Geography of Plants* **73**: 161, 2003; African taxa), Decock & Ryvarden (*MR* **107**: 94, 2003; key Afr. spp.), Decock & Stalpers (*Taxon* **55**: 227, 759, 2006; nomencl.).

Perenniporiaceae Jülich (1982) = Polyporaceae.

Perenniporiella Decock & Ryvarden (2003), Polyporaceae. 3, S. America. See Decock & Ryvarden (*MR* **107**: 94, 2003).

Peresia H. Maia (1960) = Colletotrichum fide Sutton (*Mycol. Pap.* **141**, 1977).

Peresiopsis Bat. (1960) = Yamamotoa fide von Arx & Müller (*Stud. Mycol.* **9**, 1975).

Perezia L. Léger & Duboscq (1909), Microsporidia. 9.

perfect state, see States of fungi.

Perforaria Müll. Arg. (1891) = Coccotrema fide Hawksworth *et al.* (*Dictionary of the Fungi* edn 8, 1995).

Perforariomyces Cif. & Tomas. (1953) ≡ Perforaria.

perforation lysis, the process by which degradation of resistant fungus propagules in soil is initiated (Old & Wong, *Soil Biol. Biochem.* **8**: 285, 1976; review).

perfume, see oak-moss, smell.

pergameneous (**pergamenous**, **pergamentaceous**), like paper.

Periamphispora J.C. Krug (1989), Lasiosphaeriaceae. 1, Spain. See Krug (*Mycol.* **81**: 475, 1989), Huhndorf *et al.* (*Mycol.* **96**: 368, 2004).

Periascomycetes. Class introduced for *Protascus*, *Protomyces*, *Ashbya*, and other 'primitive' ascomycetes (Moreau, 1953).

Periaster Theiss. & Syd. (1917) = Erikssonia fide Petrak (*Annls mycol.* **29**: 390, 1931).

Peribotryon Fr. (1832) nom. rej. = Chrysothrix fide Hawksworth *et al.* (*Dictionary of the Fungi* edn 8, 1995).

Pericauda.

Perichlamys Clem. & Shear (1931) ≡ Kuntzeomyces.

Pericladium Pass. (1875), ? Ustilaginaceae. 3 (forming galls on *Grewia*), Africa; Asia; Australia. See Mordue (*Mycopathologia* **103**: 173, 1988), Vánky (*Mycotaxon* **95**: 1, 2006).

periclinal, curved in the direction of, or parallel to, the surface or the circumference; cf. anticlinal; **-thickening**, zone of increased material surrounding the protoplasmic channel at the apex of a 'phialide' (see Sutton, *The Coelomycetes*, 1980).

Pericoccis Clem. (1931) = Leptogium fide Rossman *et al.* (*Stud. Mycol.* **42**: 248 pp., 1999).

Pericoelium Bonord. (1851) = Ustilago fide Saccardo (*Syll. fung.* **7**: 1, 1888).

Periconia Tode (1791), anamorphic *Pleosporales*, Hso.0eP.10. 36, widespread. See Mason & Ellis (*Mycol. Pap.* **56**, 1953; Brit. spp., key), Bunning & Griffiths (*TBMS* **78**: 147, 1984; ultrastr.), Carmarán & Novas (*Fungal Diversity* **14**: 67, 2003; spp. described by Spegazzini).

Periconiella Sacc. (1885), anamorphic *Mycosphaerellaceae*, Hso.0eP/≡ eP.10. 50 (some fungicolous), widespread. See Ellis (*Mycol. Pap.* **111**, 1967; key), McKenzie (*Mycotaxon* **39**: 229, 1990), Priest (*MR* **95**: 924, 1991; spp. on *Proteaceae*), Braun (*Feddes Repert.* **115**: 50, 2004; on ferns), Kirschner & Piepenbring (*Mycol. Progr.* **7**: 21, 2008; Panama, generic limits).

Pericystales = Ascosphaerales.

Pericystis Betts (1912) [non *Pericystis* J. Agardh 1847, Algae] ≡ Ascosphaera.

Peridermium (Link) J.C. Schmidt & Kunze (1817) nom. cons., anamorphic *Cronartium, Chrysomyxa, Pucciniastrum*. *c.* 20 (on gymnosperms (I)), widespread. Anamorph name for (I). See Arthur & Kern (*Bull. Torrey bot. Club* **33**: 403, 1906), Peterson (*Bull. Torrey bot. Club* **94**: 511, 1967; pine stem rust), Laundon (*Taxon* **25**: 186, 1976; taxonomy) Cf. *Endocronartium*.

peridermium, an aecium as in the form genus *Peridermium*.

peridiole (**peridiolum**) (esp. of *Nidulariaceae*), a division of the gleba having a separate wall, frequently acting as a unit for distribution.

Peridiomyces H. Karst. (1843) nom. dub., anamorphic *Pezizomycotina*.

Peridiopsora Kamat & Sathe (1969), anamorphic *Milesina*. 1 (on *Adelocaryum* (*Boraginaceae*)), India. Anamorph name for (II).

Peridiospora C.G. Wu & S.J. Lin (1997), Endogonaceae. 2, Brazil; Taiwan. See Wu & Lin (*Mycotaxon* **64**: 180, 1997), Goto & Maia (*Mycotaxon* **96**: 327, 2006; Brazil).

Peridiothelia D. Hawksw. (1985), Pleomassariaceae. 3, Europe. See Barr (*Mycotaxon* **43**: 371, 1992; posn).

Peridipes Buriticá & J.F. Hennen (1994) nom. inval., Pucciniales. 2 (on *Arachis*, *Hymenaea*), S. America. Anamorph name for (II). See Cummins & Hiratsuka (*Illustr. Gen. Rust Fungi* edn 3: 225 pp., 2003; as syn. of *Milesia*).

peridium, the wall or limiting membrane of a sporangium or other fruit-body, an excipule; **peridial cells** (esp. of aecia), the cells of the peridium.

Peridoxylon Shear (1923) = Camarops fide Nannfeldt (*Svensk bot. Tidskr.* **66**: 335, 1972), Dargan & Thind (*J. Indian bot. Soc.* **63**: 177, 1984; India).

Peridoxylum Clem. & Shear (1931) ≡ Peridoxylon.

perifulcrium, the wall of a pycnidium in a lichen thallus (obsol.).

Perigrapha Hafellner (1996), Roccellaceae (L). 1, Europe; Australasia. See Hafellner (*Nova Hedwigia* **63**: 174, 1996), Grube (*Bryologist* **101**: 377, 1998), Ertz *et al.* (*Biblthca Lichenol.* **91**: 155 pp., 2005).

Periline Syd. (1939) = Antennularia fide Müller & von Arx (*Beitr. Kryptfl. Schweiz* **11** no. 2, 1962).

Perinidium Cromb. (1870) ≡ Pyrenidium.

Periola Fr. (1822), anamorphic *Pezizomycotina*, Hsp.?.?. 1 or 2, Europe; America.

Periolopsis Maire (1913) = Sarcopodium fide Sutton (*TBMS* **76**: 97, 1981).

Periperidium Darker (1963), anamorphic *Micraspis*, St.≡ eH.15. 1, Canada. See Darker (*CJB* **41**: 1392, 1963).

Peripherostoma Gray (1821) nom. rej. ≡ Daldinia.
periphysis (pl. **periphyses**), an upward pointing hypha inside, or near, the ostiole of a perithecium, pycnidium, or pycnium; see hamathecium.
periphysoid, see hamathecium.
periphyton, the 'assemblage of organisms growing upon free surfaces of submerged objects in water and covering them with a slimy coat.' Young (1945) fide Cooke (*Bot. Rev.* **22**: 616, 1956).
periplasm (of *Peronosporales*), the outer, non-functional protoplasm of an oogonium or antheridium; cf. gonoplasm, ooplasm.
Perirhiza Karling (1946) = Catenophlyctis fide Karling (*Chytriomyc. Iconogr.*, 1977).
Perischizon Syd. & P. Syd. (1914), Parmulariaceae. 1, S. Africa.
Perisperma, see *Pyrisperma*.
Perisphaeria Roussel (1806) = Hypoxylon Bull. fide Læssøe (*SA* **13**: 43, 1994).
perispore (perisporium), sheath outside the true spore wall. See Harmaja (*Karstenia* **14**: 123, 1974; cyanophilic in *Pezizales*); ascospore, spore wall.
Perisporiacites Félix (1894), Fossil Fungi, Ascomycota. 2 (Cretaceous, Eocene), former USSR. See Sahni & Rao (*Proc. Indian natn Sci. Acad.* Part B. Biol. Sci. **13**: 45, 1943).
perisporial sac, a perispore forming a loose envelope around the spore as in *Coprinus* sp.
Perisporiales = Pleosporales.
Perisporiella Henn. (1902) = Hypocrella fide Clements & Shear (*Gen. Fung.*, 1931).
Perisporina Henn. (1904) = Perisporiopsis Henn. fide Müller & von Arx (*Beitr. Kryptfl. Schweiz* **11** no. 2, 1962).
Perisporiopsella Bat., J.L. Bezerra, Castr. & Matta (1964) = Pilgeriella fide von Arx & Müller (*Stud. Mycol.* **9**, 1975).
Perisporiopsidaceae E. Müll. & Arx (1962) = Parodiopsidaceae.
Perisporiopsis F. Stevens (1917) ≡ Stevensea.
Perisporiopsis Henn. (1904), Parodiopsidaceae. Anamorph *Septoidium*. *c.* 16, widespread (tropical).
Perisporites Pampal. (1902), Fossil Fungi. 2 (Miocene), Italy.
Perisporium Fr. (1821) nom. dub., Pezizomycotina. Sensu Corda (1842) = Preussia.
perisporium, see perispore.
Peristemma Syd. (1921) = Miyagia fide Cummins & Hiratsuka (*Illustr. Gen. Rust Fungi rev. edit.*, 1983).
peristome, an edging round an opening, esp. of basidiomata of certain gasteromycetes.
Peristomialis (W. Phillips) Boud. (1907) = Ijuhya fide Rossman *et al.* (*Stud. Mycol.* **42**: 248 pp., 1999).
Peristomium Lechmere (1912) = Microascus Zukal fide von Arx (*Persoonia* **8**: 191, 1975).
perithecium (pl. **perithecia**), a subglobose or flask-like ostiolate ascoma; sometimes limited to ascohymenial types formed from the development of an ascogonium (not of stromatic origin), but now widely used as a general term regardless of the ontogenetic type. See Cherepanova (*Vestn. Leningrad Univ. Biol.* **3**: 39, 1986; types and evol. pathways).
Peritrichospora Linder (1944) = Corollospora fide Kohlmeyer (*Ber. dt. bot. Ges.* **75**: 125, 1962).
peritrichous (peritrichiate), having hairs or flagella all over the surface.
Perizomatium Syd. (1927) = Phaeofabraea fide Pfister (*Occ. Pap. Farlow Herb. Crypt. Bot.* **11**, 1977).
Perizomella Syd. (1927), anamorphic *Pezizomycotina*, St.0eP.1. 1, C. America.
Pernicivesicula E.K.C. Bylén & J.I.R. Larsson (1994), Microsporidia. 1. See Bylén & Larsson (*Eur. J. Protist.* **30**: 139, 1994).
Perona Pers. (1825) = Hemimycena fide Kuyper (*in litt.*) *Perona* is a superfluous name change for *Helotium* Tode, nom. utique rej.
peronate, sheathed; having a boot or covering, esp. of the lower part of a stipe covered by a volva or veil.
Peroneutypa Berl. (1902), Diatrypaceae. 1, widespread. See Rappaz (*Mycol. Helv.* **2**: 285, 1987), Carmarán *et al.* (*Fungal Diversity* **23**: 67, 2006; morphology).
Peroneutypella Berl. (1902) ≡ Scoptria.
Peronosporites W.G. Sm. (1877), Fossil Fungi. 3 (Carboniferous, Miocene), Italy; British Isles.
Peronosporoides E.W. Berry (1916), Fossil Fungi, Peronosporales. 1 (Oligocene), USA. See Berry (*Mycol.* **8**: 76, 1916).
Peronosporoides J. Sm. Dalry (1896), Fossil Fungi (mycel.) Fungi. 1 (Carboniferous), British Isles. See Smith (*Trans. Geol. Soc. Glasgow* **10**: 321, 1896).
Peroschaeta Bat. & A.F. Vital (1957), Dothideomycetes. 1, Philippines. See Batista & Vital (*An. Soc. Biol. Pernambuco* **15**: 500, 1957).
peroxisome, one of the subcellular organelles which have indispensable functions in the metabolism of *n*-alkalenes, fatty acids, methanol, and several nitrogen-containing compounds in eukaryotic microorganisms. See Tanaka & Ueda (*MR* **97**: 1025, 1993).
Perrotia Boud. (1901), Hyaloscyphaceae. *c.* 19, widespread. See Dennis (*Persoonia* **2**: 182, 1962), Spooner (*Biblthca Mycol.* **116**, 1987; 5 spp. Australasia), Cantrell & Hanlin (*Mycol.* **89**: 745, 1997; DNA), Wang (*Mycotaxon* **72**: 461, 1999), Zhuang & Hyde (*Mycol.* **93**: 606, 2001; Hong Kong), Zhuang & Yu (*Nova Hedwigia* **73**: 261, 2001; China), Verkley (*Mycotaxon* **89**: 205, 2004; New Zealand).
Perrotiella Naumov (1916) = Hydropisphaera fide Rossman *et al.* (*Stud. Mycol.* **42**: 248 pp., 1999).
perrumpent, breaking through. Cf. erumpent.
Persiciospora P.F. Cannon & D. Hawksw. (1982), Ceratostomataceae. 4, widespread. See Horie *et al.* (*Mycotaxon* **25**: 229, 1986), Krug (*Mycol.* **80**: 414, 1988; key 3 spp.).
persistent (1) (of interascal tissues) still evident at maturity; (2) (of spores), non-deciduous; (3) (of teliospore pedicels), remaining firmly attached to the spore after liberation.
Persoon (Christiaan Hendrik; 1761-1836; South Africa, later Germany, France). Born in Cape Province; at school, Lingen-on-Ems, Germany (1775); student in Halle (1783), then Leiden (1786) and Göttingen and Erlangen (1799), moving to Paris (*c.* 1803), living there for the rest of his life. With E.M. Fries (q.v.) the founder of Mycology. There had been little development in the taxonomy of fungi from its low level at the time of Linnaeus till the work of Persoon began to appear in 1793. His early works, including *Observationes Mycologicae* (1795-1799) and the *Synopsis Methodica Fungorum* (1801) formed the framework on which Fries and all later systematists based their classifications. Names of fungi belonging in the *Pucciniales*, *Ustilaginales* and 'gasteromycetes' which appear in *Synopsis Methodica Fungorum* have sanctioned status (see Nomenclature). Persoon's working conditions in Paris were hard, and his later work was

overshadowed by the simultaneous publication by E.M. Fries (q.v.) of the *Systema Mycologicum*. Persoon's collection, a very important one because it has nomenclatural types of a great number of common species, was taken over by the Dutch government before his death and is now in Leiden (**L**). *Publs. Observationes Mycologicae* (1795-1799); *Synopsis Methodica Fungorum* (1801) [reprint 1952; index by G.H. Lünemann, 1808]; *Mycologia Europaea* (1822-1828). *Biogs, obits etc.* Ainsworth (*Nature* **193**: 22, 1962); Grummann (1974: 704); Lloyd (*Mycological Notes* **1**: 158, 1903); Lloyd (*Mycological Notes* **7**: 1301, 1924); Petersen (*Kew Bulletin* **31**: 695, 1977); Stafleu & Cowan (*TL-2* **4**: 178, 1983).

Persooniana Britzelm. (1897) = Tyromyces fide Donk (*Verh. K. ned. Akad. Wet.* tweede sect. **62**: 1, 1974).

Persooniella Syd. (1922) = Puccinia fide Dietel (*Nat. Pflanzenfam.* **6**, 1928).

Perspicinora Riedl (1990) = Koerberiella fide Coppins (*SA* **10**: 48, 1991).

perthophyte (**perthotroph**), a necrophyte on dead tissues of living hosts (Münch); cf. saprophyte.

Pertusaria DC. (1805) nom. cons., Pertusariaceae (L). 525, widespread. See Brodo (*CJB* **66**: 1264, 1988), Takeshita *et al.* (*Pl. Syst. Evol.* **165**: 49, 1989; phycobionts), Kantvilas (*Lichenologist* **22**: 289, 1990; Tasmania), Archer (*Telopea* **4**: 165, 1991; 28 spp., Australia), Archer (*Mycotaxon* **41**: 223, 1991; key 63 Australian spp.), Niebel-Lohmann & Feuerer (*Mitteilungen aus dem Institut für Allgemeine Botanik Hamburg* **24**: 199, 1992; Germany), Archer (*Biblthca Lichenol.* **53**: 1, 1993; subgen.), Archer & Elix (*N.Z. Jl Bot.* **31**: 111, 1993; Australasia), Awasthi & Srivastava (*Bryologist* **96**: 210, 1993; India), Archer *et al.* (*Mycotaxon* **56**: 387, 1995; Papua New Guinea), Elix *et al.* (*Flechten Follmann* Contributions to Lichenology in Honour of Gerhard Follmann: 15, 1995; chemistry), Elix *et al.* (*Mycotaxon* **53**: 273, 1995), Archer (*Biblthca Lichenol.* **69**, 1997; monogr. Australian spp.), Elix *et al.* (*Mycotaxon* **64**: 17, 1997; Australasia), Archer & Messuti (*Nova Hedwigia* **67**: 403, 1998; Argentina), Ladd & Wilhelm (*Lichenogr. Thomsoniana* North American Lichenology in Honor of John W. Thomson: 89, 1998; USA), Lumbsch *et al.* (*Bryologist* **102**: 215, 1999; N America), Messuti & Archer (*Bryologist* **102**: 208, 1999; Falkland Is), Lumbsch & Schmitt (*Lichenologist* **33**: 161, 2001; phylogeny), Lumbsch & Schmitt (*Lichenology* **1**: 37, 2002; phylogeny), Boqueras & Llimona (*Mycotaxon* **88**: 471, 2003; Iberian peninsula), Schmitt & Lumbsch (*Mol. Phylogen. Evol.* **33**: 43, 2004; phylogeny, chemistry), Zhao *et al.* (*Bryologist* **107**: 531, 2004; key, China), Lumbsch *et al.* (*Biol. J. Linn. Soc.* **89**: 615, 2006; micromorphology, chemistry), Miądlikowska *et al.* (*Mycol.* **98**: 1088, 2006; phylogeny), Schmitt *et al.* (*J. Hattori bot. Lab.* **100**: 753, 2006; phylogeny, fam. placement), Messuti *et al.* (*Lichenologist* **39**: 227, 2007; Zimbabwe).

Pertusariaceae Körb. ex Körb. (1855), Pertusariales (L). 2 gen. (+ 24 syn.), 526 spp.

Lit.: Letrouit-Galinou (*Revue bryol. lichén.* **34**: 413, 1966; ontogeny), Dibben (*Milwaukee Public Mus., Publs Biol. Geol.* **5**, 1980; N. Am.), Dibben (*J. Hattori bot. Lab.* **52**: 343, 1982; evol. trends), Honegger (*Lichenologist* **14**: 205, 1982; asci), Oshio (*J. Sci. Hiroshima Univ.* B(2): 76, 1991), Lumbsch *et al.* (*Biblthca Lichenol.* **57**: 355, 1995), Archer (*Biblthca Lichenol.* **69**: 249 pp., 1997), Lumbsch *et al.* (*Bryologist* **102**: 215, 1999), Lumbsch & Schmitt (*Lichenologist* **33**: 161, 2001), Schmitt *et al.* (*Biblthca Lichenol.* **86**: 147, 2003), Lumbsch *et al.* (*Mol. Phylogen. Evol.* **31**: 822, 2004), Schmitt & Lumbsch (*Mol. Phylogen. Evol.* **33**: 43, 2004), Wedin *et al.* (*MR* **109**: 159, 2005), Miądlikowska *et al.* (*Mycol.* **98**: 1088, 2006; phylogeny), Schmitt *et al.* (*J. Hattori bot. Lab.* **100**: 753, 2006), Hofstetter *et al.* (*Mol. Phylogen. Evol.* **44**: 412, 2007; phylogeny), Lumbsch *et al.* (*MR* **111**: 257, 2007; phylogeny).

Pertusariales M. Choisy ex D. Hawksw. & O.E. Erikss. (1986), (L). Ostropomycetidae. 5 fam., 15 gen., 901 spp. Thallus crustose, rarely lobate or minutely foliose. Ascomata apothecial, often deeply cupulate, usually initially immersed, opening widely or with a poroid aperture and appearing perithecial, usually with a well-developed thalline margin. Interascal tissue of basal paraphyses, sometimes also with apical paraphyses in the poroid taxa. Asci short, widely cylindrical, with a thick multilayered usually J+ wall, the apex often more strongly thickened, releasing spores through a ± vertical apical split, often less than 8-spored. Ascospores very large, hyaline, aseptate, with a very thick multilayered wall. Anamorphs pycnidial. Lichenized with green algae, cosmop. Fams:

(1) **Coccotremataceae**
(2) **Icmadophilaceae**
(3) **Megasporaceae**
(4) **Ochrolechiaceae**
(5) **Pertusariaceae**

Many spp. reproduce mainly by soredia or isidia, forming 'species-pairs' with their meiotic counterparts. *Lit.*: Lumbsch *et al.* (*MR* **111**: 257, 2007; phylogeny).

Pertusariomyces E.A. Thomas ex Cif. & Tomas. (1953) ≡ Pertusaria.

pervious (of lichenized scyphi), open or perforate basally.

Pesavis Elsik & Janson. (1974), Fossil Fungi, anamorphic *Pezizomycotina*. 2 (Paleogene), widespread. See Smith & Crane (*J. Linn. Soc.* Bot. **79**: 243, 1979).

Pesotum J.L. Crane & Schokn. (1973), anamorphic *Ophiostoma*, Hsy.0eH.10. 16, widespread. See Wingfield *et al.* (*MR* **95**: 1328, 1991; conidial ontogeny), Okada *et al.* (*CJB* **76**: 1495, 1998; phylogeny), Hausner *et al.* (*CJB* **78**: 903, 2000; phylogeny), Geldenhuis *et al.* (*Fungal Diversity* **15**: 137, 2004), Yamaoka *et al.* (*Mycoscience* **45**: 277, 2004; Japan).

pest control, in reference collections of specimens, see Hall (*Taxon* **37**: 885, 1988).

Pestalopezia Seaver (1942), Helotiaceae. 3, N. America.

Pestalosphaeria M.E. Barr (1975), Amphisphaeriaceae. Anamorph *Pestalotiopsis*. 11, widespread. See Nag Raj (*Mycotaxon* **22**: 52, 1985; key), Zhu *et al.* (*Acta Agric. Univ. Zhejiang* **16**: 163, 1990; key 7 spp.), Kang *et al.* (*Fungal Diversity* **1**: 147, 1998; DNA), Kang *et al.* (*MR* **103**: 53, 1999), Kobayashi *et al.* (*Mycoscience* **42**: 211, 2001), Jeewon *et al.* (*MR* **107**: 1392, 2003; phylogeny).

Pestalotia De Not. (1841), anamorphic *Broomella*, Ccu.≡ eP.19. 1, Italy. See Steyaert (*Bull. Jard. bot. Brux.* **19**: 285, 1949; *s.str.* monotypic), Guba (*Monograph of Monochaetia and Pestalotia*, 1961; 222 spp. accepted, key), Sutton (*CJB* **47**: 2083, 1969), Jeewon *et al.* (*MR* **107**: 1392, 2003).

Pestalotiopsis Steyaert (1949), anamorphic *Pestalosphaeria*, Cac.≡ eP.19. 162, widespread. *P. theae* (grey blight of tea; *CMI Descr.* **318**, 1971). See Steyaert (*TBMS* **36**: 181, 1953), Sutton (*Mycol. Pap.* **80**, 1961; conidium development), Zhu *et al.* (*Acta Agr. Univ. Zhejiangensis* **16**: 173, 1990; soluble protein patterns), Zhu *et al.* (*Mycotaxon* **40**: 129, 1991; teleomorphs in China), Zhu *et al.* (*Acta Mycol. Sin.* **10**: 273, 1991; China), Suto & Kobayashi (*TMSJ* **34**: 323, 1993; Japan), Nag Raj & Mel'nik (*Mycotaxon* **50**: 435, 1994), Keller (*Atlas des Basidiomycetes* Vus aux Microscopes Électroniques, 1997; taxol production), Morgan *et al.* (*MR* **102**: 975, 1998; neural networks), Watanabe *et al.* (*Mycoscience* **39**: 71, 1998; conidiomatal ontogeny), Hopkins & McQuilken (*Eur. J. Pl. Path.* **106**: 77, 2000; UK), Murugan & Muthumary (*Mycotaxon* **79**: 455, 2001; ultrastr.), Jeewon *et al.* (*Mol. Phylogen. Evol.* **25**: 378, 2002; phylogeny), Jeewon *et al.* (*MR* **107**: 1392, 2003; phylogeny), Jeewon *et al.* (*Mol. Phylogen. Evol.* **27**: 372, 2003; phylogeny, morphology), Ebenezer *et al.* (*Journal of Mycology and Plant Pathology* **34**: 794, 2004; fatty acid profiles), Jeewon *et al.* (*Fungal Diversity* **17**: 39, 2004; phylogeny, host specificity), Sergeeva *et al.* (*Australas. Pl. Path.* **34**: 255, 2005; on *Vitis*), Keith *et al.* (*Pl. Dis.* **90**: 16, 2006; on guava), Lee *et al.* (*Stud. Mycol.* **55**: 175, 2006; on *Restionaceae*), Hu *et al.* (*Fungal Diversity* **24**: 1, 2007; phylogeny).

Pestalozzia, see *Pestalotia*.

Pestalozziella Sacc. & Ellis ex Sacc. (1882), anamorphic *Pezizomycotina*, Cpd.0eH.10. 4, widespread (temperate). See Nag Raj & Kendrick (*CJB* **50**: 607, 1972).

Pestalozzina (Sacc.) Sacc. (1895) ≡ Zetiasplozna fide Nag Raj (*Coelomycetous Anamorphs with Appendage-bearing Conidia*, 1993).

Pestalozzina P. Karst. & Roum. (1890) ? = Bartalinia fide Nag Raj (*Coelomycetous Anamorphs with Appendage-bearing Conidia*, 1993).

Pestalozzites E.W. Berry (1917), Fossil Fungi. 2 (Oligocene, Miocene), USA.

Petaloides Lloyd ex Torrend (1920) = Polyporus P. Micheli ex Adans. fide Donk (*Persoonia* **1**: 173, 1960).

Petalosporus G.R. Ghosh, G.F. Orr & Kuehn (1963) = Arachniotus fide von Arx (*Persoonia* **9**: 393, 1977).

Petasodes Clem. (1909) = Phomopsis (Sacc.) Bubák. fide Sutton (*Mycol. Pap.* **141**, 1977).

Petasospora Boidin & Abadie (1955) = Pichia fide Yamada *et al.* (*Biosc., Biotechn., Biochem.* **60**: 818, 1996), Kurtzman *in* Kurtzman & Fell (Eds) (*Yeasts, a taxonomic study* 4th edn: 273, 1998).

Petch (Thomas; 1870-1948; England, later Sri Lanka). Schoolmaster, Hull and King's Lynn (1890s); mycologist, Royal Botanic Gardens, Peradeniya (1905-1924); founder and first Director, Tea Research Institute (1925-1928). A leading student of tropical mycology (many works published in *Annals of the Royal Botanic Gardens, Peradeniya*, 1906-1926), entomogenous fungi (q.v.) and plant pathology; married a daughter of Charles Plowright and on retirement lived in Plowright's old house at North Wootton, near King's Lynn. His specimens are in the fungal reference collections of the Royal Botanic Gardens, Kew (**K**) and Peradeniya; slides in the British Museum (Natural History), London (**BM**). *Publs. Diseases and Pests of the Rubber Tree* (1921);*Diseases and Pests of the Tea Bush* (1923); British *Hypocreales*. *TBMS* **21**, **25**, **27** (1938-1945). *Biogs, obits etc.* Ainsworth (*TBMS* **67**: 179, 1976) [portrait]; Stafleu & Cowan (*TL-2* **4**: 168, 1983).

Petchiomyces E. Fisch. & Mattir. (1938), Pyronemataceae. 2, N. America; Sri Lanka. See Eckblad (*Nytt Mag. Bot.* **15**: 1, 1968), Læssøe & Hansen (*MR* **111**: 1075, 2007; phylogeny).

Petelotia Pat. (1924) = Acanthonitschkea fide Nannfeldt (*Svensk bot. Tidskr.* **66**: 49, 1975).

Peterjamesia D. Hawksw. (2006), Roccellaceae. 2, widespread. See Hawksworth (*Lichenologist* **38**: 187, 2006).

Petersonia Cummins & Y. Hirats. (2003) nom. inval., anamorphic *Mikronegeria*. 1 (on *Araucaria*), Chile. Anamorph name for (I).

Petractis Fr. (1845), Ostropales (L). 5, widespread. See Vězda (*Preslia* **37**: 127, 1965), Kauff & Lutzoni (*Mol. Phylogen. Evol.* **25**: 138, 2002; phylogeny), Kauff & Büdel (*Bryologist* **108**: 272, 2005; ontogeny), Miądlikowska *et al.* (*Mycol.* **98**: 1088, 2006; phylogeny).

Petrak (Franz; 1886-1973; Austria). Born in Mährisch-Weisskirchen [now Hranice, Czech Republic]; student, Vienna (1906-1913); school teacher (1913-1916); Austrian army (1916-1918); private scientist, living by selling exsiccati (1918-1938); contract employee, Natural History Museum, Vienna (1938-1951). Eminent and prolific taxonomist (especially ascomycetes and anamorphic fungi), bibliographer and indexer (see Literature); collector, distributing thousands of exsiccati specimens; contributor to the mycotas of many parts of the world; founded *Sydowia* (1947) to replace *Annales Mycologici*. *Publs. Petrak's Lists* (1930-1950) [these volumes are freely available on-line; see Internet: catalogues & thesauri]; see also full bibliography in Samuels. *Biogs, obits etc.* Arx (*Persoonia* **9**: 95, 1976); Grummann (1974: 938); Lohwag (*Beihefte zur Sydowia* **1**: IV, 1957) [portrait]; Riedl (*Sydowia* **26**: xxix, 1974); Samuels (*An Annotated Index to the Mycological Writings of Franz Petrak* **1-5** A-O, 1981-1986) [bibliography and biography in vol. 1]; Stafleu & Cowan *TL-2* **4**: 204, 1983).

Petrakia Syd. & P. Syd. (1913), anamorphic *Pezizomycotina*, Hsp.#eP.1. 3, Europe; Asia. See van der Aa (*Acta Bot. Neerl.* **17l**: 221, 1968), Carmichael *et al.* (*Genera of Hyphomycetes*, 1980; synonymy with *Echinosporium*), Wong *et al.* (*Cryptog. Mycol.* **23**: 195, 2002).

Petrakiella Syd. (1924), ? Sordariomycetes. 1 (from bark), Brazil. Affinities uncertain, no recent research is available.

Petrakina Cif. (1932), ? Asterinaceae. 2, C. America. See Farr & Palm (*Mycotaxon* **24**: 275, 1985).

Petrakiopeltis Bat., A.F. Vital & Cif. (1957), ? Microthyriaceae. 1, Guyana. See Batista *et al.* (*Publções Inst. Micol. Recife* **90**: 10, 1957).

Petrakiopsis Subram. & K.R.C. Reddy (1968), anamorphic *Pezizomycotina*, Hsp.1bH.1. 1, India. See Subramanian & Reddy (*Sydowia* **20**: 340, 1966).

Petrakomyces Subram. & K. Ramakr. (1953) = Ciliochora fide Nag Raj (*Coelomycetous Anamorphs with Appendage-bearing Conidia*, 1993).

Petriella Curzi (1930), Microascaceae. Anamorphs *Graphium*, *Scopulariopsis*, *Scedosporium*. 5, Europe; N. America. See Barron *et al.* (*CJB* **39**: 837, 1961; key), Corlett (*CJB* **44**: 79, 1966; ontogeny), Valmaseda *et al.* (*CJB* **65**: 1802, 1987; concept), von Arx *et al.* (*Beih. Nova Hedwigia* **94**, 1988), Issakainen *et al.* (*J. Med. Vet. Mycol.* **35**: 389, 1997; DNA), Issakainen *et al.* (*MR* **103**: 1179, 1999), Lee & Hanlin (*Mycol.* **91**: 434, 1999; DNA), Okada *et al.* (*CJB* **76**: 1495, 1999), Issakainen *et al.* (*Medical Mycology* **41**: 31, 2003; phylogeny), Kwasna *et al.* (*Acta Mycologica* Warszawa **40**: 267, 2005; morphology, phylogeny), Rainer & Hoog (*MR* **110**: 151, 2006; phylogeny), Zhang *et al.* (*Mycol.* **98**: 1076, 2006; phylogeny).

Petriellidium Malloch (1970) = Pseudallescheria fide McGinnis *et al.* (*Mycotaxon* **14**: 94, 1982).

Petriellopsis Gilgado, Cano, Guarro & Gené (2007), Microascaceae. 1, Argentina. See Gilgado *et al.* (*Int. J. Syst. Evol. Microbiol.* **57**: 2171, 2007).

Petromyces Malloch & Cain (1973), Trichocomaceae. Anamorph *Aspergillus*. 2, widespread. See Kuraishi *et al.* (*NATO ASI Series A: Life Sciences* **185**: 407, 1990; ubiquinones), Udagawa *et al.* (*Mycotaxon* **52**: 207, 1994), Samson (*Contributions to Microbiology* **2**: 5, 1999; review), Frisvad & Samson (*Stud. Mycol.* **45**: 201, 2000), Varga *et al.* (*Antonie van Leeuwenhoek* **77**: 83, 2000), McAlpin & Wicklow (*Can. J. Microbiol.* **51**: 765, 2005; physiology, ontogeny), McAlpin & Wicklow (*Can. J. Microbiol.* **51**: 1039, 2005; genetic diversity).

Petrona Adans. (1763) ? = Schizophyllum fide Donk (*Beih. Nova Hedwigia* **5**: 220, 1962).

petrophilous, see saxicolous.

Petrosphaeria Stopes & H. Fujii (1910), Fossil Fungi (mycel.) Fungi. 1 (Cretaceous), Japan.

Peyerimhoffiella Maire (1916), Laboulbeniaceae. 1, N. Africa. See Nannfeldt (*Svensk bot. Tidskr.* **43**: 468, 1949).

Peylia Opiz (1857) = Botryosporium Corda fide Saccardo & Trotter (*Syll. fung.* **22**: 1252, 1913).

Peyritschiella Thaxt. (1890), Laboulbeniaceae. 47, widespread. See De Kesel (*Belg. Jl Bot.* **131**: 176, 1998; Belgium).

Peyritschiellaceae Thaxt. (1908) = Laboulbeniaceae.

Peyronelia Cif. & Gonz. Frag. (1927), anamorphic *Glyphium*, Hso.#eP.4. 4, America; Europe.

Peyronelina G. Arnaud (1952) nom. inval., anamorphic *Pezizomycotina*.

Peyronelina P.J. Fisher, J. Webster & D.F. Kane (1976), anamorphic *Pezizomycotina*, Hso.#bH.1. 1 (aquatic), British Isles. See Fisher *et al.* (*TBMS* **67**: 351, 1976), Nakagiri & Ito (*Res. Commun.* Inst. Ferm., Osaka **18**: 57, 1995; conidial ontogeny).

Peyronellaea Goid. (1952) = Phoma Sacc. fide Boerema *et al.* (*Persoonia* **4**: 47, 1965), Boerema (*Persoonia* **15**: 197, 1993).

Peyronellula Malan (1952) = Emericellopsis fide Stolk (*TBMS* **38**: 419, 1955).

Pezicula Paulet (1791) nom. rej. = Craterellus fide Cannon & Hawksworth (*Taxon* **32**: 477, 1983).

Pezicula Tul. & C. Tul. (1865) nom. cons., Dermateaceae. Anamorph *Cryptosporiopsis*. *c.* 37, widespread (esp. temperate). See Groves (*Can. J. Res.* C **17**: 125,, 1939), Dennis (*Kew Bull.* **29**: 158, 1974; key Brit. spp.), Verkley (*Stud. Mycol.* **44**: 180 pp., 1999; monogr., key 26 spp., phylogeny), Abeln (*Mycol.* **92**: 685, 2000; phylogeny), Vrålstad *et al.* (*New Phytol.* **155**: 131, 2002; phylogeny), Pärtel & Raitviir (*Mycol. Progr.* **4**: 149, 2005; ulttrastr.), Wang *et al.* (*Mycol.* **98**: 1065, 2006; phylogeny).

Peziotrichum (Sacc.) J. Lindau (1900), anamorphic *Podonectria*, Hsy.0eP.?. 1 (entomogenous), Australia to Sri Lanka. See Petch (*TBMS* **12**: 44, 1927), Downing (*Mycol.* **45**: 938, 1953).

Peziza Dill. ex Fr. (1822), Pezizaceae. Anamorphs *Chromelosporium*, *Oedocephalum*. 104, widespread. Apparently highly polyphyletic with two main clades broadly corresponding to species with *Chromelosporium* and *Oedocephalum* anamorphs. See Donadini (*Bull. Soc. linn. Provence* **31**: 9, 1978), Donadini (*Docums Mycol.* **9** no. 36: 1, 1979; key), Donadini (*Docums Mycol.* **11** no. 41: 25, 1980), Donadini (*Docums Mycol.* **14** no. 56: 39, 1980), Donadini (*BSMF* **96**: 239, 1980), Donadini (*BSMF* **96**: 247, 1980; key *nivalis* group), Donadini (*Docums Mycol.* **11** no. 53: 57, 1984), Donadini (*Bull. Soc. linn. Provence* **36**: 153, 1985; spores 74 spp.), Moravec (*Agarica* **6**: 56, 1985), Dyby & Kimbrough (*Bot. Gaz.* **148**: 283, 1987; ascospore ontogeny), Moravec & Spooner (*TBMS* **90**: 43, 1988; brown-spored spp.), Turnau (*TBMS* **91**: 338, 1988; asci), Pfister (*Mycotaxon* **41**: 505, 1991; tropical spp.), Pfister (*Mycotaxon* **43**: 171, 1992; on snow banks), Pant (*Journal of Mycopathological Research* **31**: 21, 1993; India), Pant & Tewari (*Journal of Mycopathological Research* **33**: 37, 1995; India, key), Yao *et al.* (*SA* **14**: 17, 1995; nomencl.), Wang (*Bull. natn. Mus. Nat. Sci.* Taiwan **8**: 57, 1996; Taiwan), Landvik *et al.* (*Nordic Jl Bot.* **17**: 403, 1997; phylogeny), Norman & Egger (*Mycol.* **88**: 986, 1996; phylogeny), Norman & Egger (*Mycol.* **91**: 820, 1999; phylogeny), Hansen *et al.* (*Mycol.* **93**: 958, 2001; phylogeny), Hansen *et al.* (*MR* **106**: 879, 2002; phylogeny), Hansen *et al.* (*Mol. Phylogen. Evol.* **36**: 1, 2005; phylogeny), Hansen & Pfister (*Mycol.* **98**: 1029, 2006; phylogeny), Læssøe & Hansen (*MR* **111**: 1075, 2007; phylogeny).

Peziza Fuckel (1870) = Aleuria Fuckel fide Hawksworth *et al.* (*Dictionary of the Fungi* edn 8, 1995).

Peziza L. (1753) = Cyathus fide Dennis (*Kew Bull.* **37**: 643, 1983).

Pezizaceae Dumort. (1829), Pezizales. 31 gen. (+ 40 syn.), 230 spp.

Lit.: Curry & Kimbrough (*Mycol.* **75**: 781, 1983; septal ultrastr.), Moravec (*Mycotaxon* **30**: 473, 1987), Kimbrough (*Mem. N. Y. bot. Gdn* **49**: 326, 1989; delimitation), Kimbrough *et al.* (*Bot. Gaz.* **152**: 408, 1991), Maia *et al.* (*Mycotaxon* **57**: 371, 1996), Landvik *et al.* (*Nordic Jl Bot.* **17**: 403, 1997; DNA), Norman & Egger (*Mycol.* **88**: 986, 1996), Harrington *et al.* (*Mycol.* **91**: 41, 1999; phylogeny), Norman & Egger (*Mycol.* **91**: 820, 1999; phylogeny), Doveri *et al.* (*Docums Mycol.* **30**: 3, 2000), Bougher & Lebel (*Aust. Syst. Bot.* **14**: 439, 2001), Hansen *et al.* (*Mycol.* **93**: 958, 2001), Hansen *et al.* (*MR* **106**: 879, 2002), Tehler *et al.* (*MR* **107**: 901, 2003), Hansen *et al.* (*Mol. Phylogen. Evol.* **36**: 1, 2005), Hansen & Pfister (*Mycol.* **98**: 1029, 2006; phylogeny), Læssøe & Hansen (*MR* **111**: 1075, 2007; phylogeny).

Pezizales J. Schröt. (1894). Pezizomycetidae. 16 fam., 199 gen., 1683 spp. Operculate discomycetes; cup-fungi. Stroma absent. Ascomata apothecial or cleistothecial, rarely absent, often large, discoid, cupulate or ± globose, sometimes stalked, often brightly coloured; excipulum usually thick-walled, fleshy or membranous, composed of thin-walled pseudoparenchymatous cells. Interascal tissue of simple or moniliform paraphyses, often pigmented and swollen at the apices, absent in some cleistothecial taxa. Asci elongated, persistent, thin-walled, usually without

obvious apical thickening, opening by a circular pore (operculum) or vertical split, the wall sometimes blueing in iodine, the asci cylindrical to ± globose and usually indehiscent in cleistothecial taxa. Ascospores usually ellipsoidal, aseptate, hyaline to strongly pigmented, often ornamented, usually without a sheath. Anamorphs hyphomycetous where known, usually with sympodial proliferation. Saprobes on soil and very rotten wood, some coprophilous, mycorrhizal, some hypogeous (and then mycorrhizal); esp. temp. Fams:
(1) **Ascobolaceae**
(2) **Ascodesmidaceae**
(3) **Caloscyphaceae**
(4) **Carbomycetaceae**
(5) **Chorioactidaceae**
(6) **Discinaceae**
(7) **Glaziellaceae**
(8) **Helvellaceae**
(9) **Karstenellaceae**
(10) **Morchellaceae**
(11) **Pezizaceae**
(12) **Pyronemataceae** (syn. *Aleuriaceae*, *Humariaceae*, *Otideaceae*)
(13) **Rhizinaceae**
(14) **Sarcoscyphaceae**
(15) **Sarcosomataceae**
(16) **Tuberaceae**

Lit.: van Brummelen (*Persoonia* **10**: 113, 1978; *in* Reynolds, 1981; ascus ultrastr., *in* Hawksworth (Ed.), 1994: 303), Cabello (*Boln. Soc. Argent. Bot.* **25**: 394, 1988; *Sarcoscyphineae*), Eckblad (*Nytt. Mag. Bot.* **15**: 1, 1968; nomencl. gen., fam., etc.), Hansen & Pfister (*Mycol.* **98**: 1029, 2006; phylogeny), Harmaja (*Karstenia* **14**: 109, 123, 1974; tetranucleate spored, cyanophilic perispored spp.), Harrington *et al.* (*Mycol.* **91**: 41, 1999; phylogeny), Hennebert & Bellemère (1979; anamorphs), Korf (*Mycol.* **64**: 937, 1972; synoptic key, gen. names, *in* Ainsworth *et al.* (Eds), *The Fungi* **4A**: 249, 1973; gen. keys, lit.), Landvik *et al.* (*Nordic J. Bot.* **17**: 403, 1997; DNA), Norman & Egger (*Mycol.* **91**: 820, 1999; phylogeny), Pegler *et al.* (*British truffles*, 1993), Percudani *et al.* (*Mol. Phylog. Evol.* **13**: 169, 1999; DNA), Pfister (*Mycol.* **65**: 326, 1973; psilopezioid gen.), Pfister & Kimbrough (in McLaughlin *et al.* (Eds), *The Mycota* **7A**: 257, 2001), Schumacher & Jenssen, *Arctic Alp. Fungi* **4**, 1992; montane spp.); Svrček (*Česká Myk.* **30**: 129, 135, 1976; *Sborník narod. Muz. Praza* **32B**: 115, 1976; Velenovský spp.), Thind & Kaushal (*in* Subramanian (Ed.), *Taxonomy of fungi* **1**: 283, 1978; tissue types), Trappe (*Mycotaxon* **9**: 297, 1979; hypogeous taxa); Weber (*Bibl. Mycol.* **140**, 1992; reprod. system).

Regional: **Australasia**: Rifai (*Verh. K. ned. Akad. wet.* ser. 2 **57**(3), 1968). **Caribbean**: Pfister (*J. Agric. Univ. Puerto Rico* **58**: 358, 1974). **Denmark**: Petersen (*Op. Bot.* **77**, 1985; ecology). **N. America**: Larsen & Denison (*Mycotaxon* **7**: 68, 1978; checklist). **Scandinavia**: Dissing (in Hanse & Knudsen (eds), *Nordic Macromycetes* **1**, 2000). **Tierra del Fuego**: Gamundí (*Fl. cript. Tierra del Fuego* **10** (3), 1975). **Ukraine**: Smits'ka (*Petsitsovi gribi Ukraini*, 1975).

See also under *Ascomycota*, Discomycetes, Macrofungi, truffle.

Pezizasporites T.C. Huang (1981), Fossil Fungi. 1 (Miocene), Taiwan.

Pezizella Fuckel (1870) = Calycina fide Hawksworth *et al.* (*Dictionary of the Fungi* edn 8, 1995).

Pezizella P. Karst. (1872) ≡ Pezizula.

Pezizellaceae Velen. (1934) = Hyaloscyphaceae.

Pezizellaster Höhn. (1917) = Lachnum fide Dennis (*Mycol. Pap.* **32**, 1949).

Pezizites Göpp. & Berendt (1845), Fossil Fungi. 3 (Tertiary), Europe.

Pezizomycetes O.E. Erikss. & Winka (1997), Pezizomycotina. 1 ord., 16 fam., 200 gen., 1684 spp. Ord.: **Pezizales**

For *Lit.* see ord. and fam.

Pezizomycetidae Locq. (1974), see *Pezizomycetes*.

Pezizomycotina O.E. Erikss. & Winka (1997), see *Pezizomycetes*.

Pezizula P. Karst. (1871) = Thelebolus fide Kimbrough & Korf (*Am. J. Bot.* **54**: 9, 1967).

Pezolepis Syd. (1925), Helotiales. 2, widespread (tropical).

Pezoloma Clem. (1909), Leotiaceae. 13, widespread (temperate). See Korf (*Phytologia* **21**: 201, 1971), Gamundí & Romero (*Fl. criptog. Tierra del Fuego* **10**, 1998), Lizoň *et al.* (*Mycotaxon* **67**: 73, 1998; posn), Korf (*Mycotaxon* **73**: 493, 1999; subgen. *Phaeopezoloma*).

Pezomela Syd. (1928), ? Helotiales. 1, Chile.

PFGE, see pulsed field gel-electrophoresis.

Pfistera Korf & W.Y. Zhuang (1991) = Peziza Fr. fide Brummelen (*Cryptog. Bryol.-Lichénol.* **19**: 257, 1998).

pH, see Hydrogen-ion concentration.

Phacellium Bonord. (1860), anamorphic *Mycosphaerella*, Hsy.≡ eP.10. 22, widespread. See Braun (*Nova Hedwigia* **50**: 499, 1990), Braun (*Monogr. Cercosporella, Ramularia Allied Genera (Phytopath. Hyphom.)* **2**: 316, 1998; key).

Phacellula Syd. (1927), Exobasidiales. 1, C. America; Greece. See Seifert & Bandoni (*Sydowia* **53**: 156, 2001).

Phacidiaceae Fr. (1849), Helotiales. 7 gen. (+ 7 syn.), 148 spp.

Lit.: Di Cosmo *et al.* (*Mycotaxon* **21**: 1, 1984), DiCosmo *et al.* (*Mycotaxon* **21**: 1, 1984), Kramer (*Stud. Mycol.* **30**: 151, 1987), Roll-Hansen (*Eur. J. For. Path.* **17**: 311, 1987), Roll-Hansen (*Eur. J. For. Path.* **19**: 237, 1989), Gernandt *et al.* (*Mycol.* **93**: 915, 2001).

Phacidiales = Helotiales.

Phacidiella P. Karst. (1884), anamorphic *Pyrenopeziza*, St.0eH.40. 5, widespread.

Phacidiella Potebnia (1912) ≡ Potebniamyces.

Phacidina Höhn. (1917), ? Leptopeltidaceae. 1, Europe. See Nannfeldt (*Nova Acta R. Soc. Scient. upsal.*, 1932).

Phacidiopsis Geyl. (1887), Fossil Fungi. 1 (Tertiary), Indonesia. = Phacidites fide Meschinelli (1892).

Phacidiopsis Hazsl. (1873) = Triblidium fide Saccardo (*Syll. fung.* **8**: 804, 1889).

Phacidiopycnis Potebnia (1912), anamorphic *Potebniamyces*, St.0eH.15. 4, widespread. *P. pseudotsugae* (syn. *Phomopsis pseudotsugae*), canker of *Pseudotsuga* and other conifers. See Rupprecht (*Sydowia* **13**: 10, 1959), Sutton (*The Coelomycetes*, 1980).

Phacidiostroma Höhn. (1917) = Phacidium fide Sherwood (*in litt.*).

Phacidiostromella Höhn. (1917) nom. dub., ? Dothideomycetes.

Phacidites Mesch. (1892), Fossil Fungi. 19 (Tertiary), Europe.

Phacidium Fr. (1815) nom. cons., Phacidiaceae. Anamorphs *Apostrasseria*, *Ceuthospora*. 27, widespread (temperate). *P. infestans* (snow blight of conifers). See DiCosmo *et al.* (*Mycotaxon* **21**: 1, 1984; 26 spp.), Roll-Hansen (*Eur. J. For. Path.* **17**: 311, 1987), Roll-Hansen (*Eur. J. For. Path.* **19**: 237, 1989; review).

Phacobolus Fr. (1849) = Stictis fide Eriksson & Hawksworth (*SA* **5**: 150, 1986).

Phacodothis, see *Placodothis*.

Phacopeltis Petch (1919) = Vizella fide von Arx & Müller (*Stud. Mycol.* **9**, 1975).

Phacopsis Tul. (1852), Parmeliaceae. 21 (on lichens), widespread. See Hafellner (*Herzogia* **7**: 343, 1987), Triebel *et al.* (*Bryologist* **98**: 71, 1995; key), Peršoh & Rambold (*Mycol. Progr.* **1**: 43, 2002; phylogeny).

Phacorhiza Pers. (1822) = Typhula fide Corner (*Ann. Bot. Mem.* [A monograph of Clavaria and allied genera] **1**, 1950).

Phacostroma Petr. (1955), anamorphic *Pezizomycotina*, Cac.0eH.15. 1, former Czechoslovakia. See Petrak (*Sydowia* **9**: 527, 1955).

Phacostromella Petr. (1955), anamorphic *Pezizomycotina*, St.0eH.15. 1, Germany. See Petrak (*Sydowia* **9**: 480, 1955).

Phacothecium Trevis. (1857) = Arthonia fide Hawksworth *et al.* (*Dictionary of the Fungi* edn 8, 1995).

Phacotiella Vain. (1927) = Sphinctrina fide Tibell (*Beih. Nova Hedwigia* **79**: 597, 1984).

Phacotium (Ach.) Trevis. (1821) ≡ Phacotrum.

Phacotrum Gray (1821) = Calicium fide Tibell (*Beih. Nova Hedwigia* **79**: 597, 1984).

Phaeangella (Sacc.) Massee (1895) ? = Phibalis fide Korf & Kohn (*Mem. N. Y. bot. Gdn* **28**: 109, 1976).

Phaeangellina Dennis (1955), Helotiaceae. 1, Europe. See Dennis (*Kew Bull.* **1955**: 360, 1955).

Phaeangium (Sacc.) Sacc. (1902) [non *Phaeangium* Pat. 1894] = Pezicula Tul. & C. Tul. fide Verkley (*Stud. Mycol.* **44**: 180 pp., 1999).

Phaeangium Pat. (1894), Pyronemataceae. 1, N. Africa; Arabian peninsula. See Alsheikh & Trappe (*CJB* **61**: 1923, 1983), Læssøe & Hansen (*MR* **111**: 1075, 2007; phylogeny).

Phaeaspis Clem. & Shear (1931) = Vizella fide von Arx & Müller (*Stud. Mycol.* **9**, 1975).

Phaeaspis Kirschst. (1939) = Phomatospora fide Francis (*Mycol. Pap.* **139**, 1975).

Phaedropezia Le Gal (1953) = Acervus fide Pfister (*Occ. Pap. Farlow Herb. Crypt. Bot.* **8**, 1975).

Phaeidium Clem. (1931) ≡ Laaseomyces.

phaeo- (prefix), dark-coloured or swarthy, esp. of spores; cf. Anamorphic fungi.

Phaeoacremonium W. Gams, Crous & M.J. Wingf. (1996), anamorphic *Togninia*, Hso.?.?. 14, widespread. See Crous *et al.* (*Mycol.* **88**: 789, 1996).

Phaeoannellomyces McGinnis & Schell (1985), anamorphic *Chaetothyriales*, Hso.0eH.19. 2 (from humans), widespread. See McGinnis & Schell (*Sabouraudia* **23**: 182, 1985).

Phaeoantenariella Cavalc. (1969) nom. dub., anamorphic *Pezizomycotina*. Based on mycelium fide Hawksworth (*Bull. Br. Mus. nat. Hist., Bot.* **9**: 78, 1981).

Phaeoaphelaria Corner (1953), Aphelariaceae. 1, Australia. See Corner (*Ann. Bot. Mem.* [A monograph of Clavaria and allied genera, addenda] **17**: 357, 1953).

Phaeoapiospora (Sacc. & P. Syd.) Theiss. & Syd. (1915) = Anisomyces Theiss. & Syd. fide Petrak (*Sydowia* **1**: 35, 1947).

Phaeobarlaea Henn. (1903) nom. nud. = Plicaria fide Eckblad (*Nytt Mag. Bot.* **15**: 1, 1968).

Phaeoblastophora Partr. & Morgan-Jones (2002), anamorphic *Pezizomycotina*, H?.?.?. 2. See Partridge & Morgan-Jones (*Mycotaxon* **83**: 339, 2002), Crous *et al.* (*MR* **110**: 264, 2006; key).

Phaeobotryon Theiss. & Syd. (1915), Botryosphaeriaceae. 4. See Phillips *et al.* (*Persoonia* **21**: in press, 2008).

Phaeobotryosphaeria Speg. (1908), Botryosphaeriaceae. 1. See Phillips *et al.* (*Persoonia* **21**: in press, 2008).

Phaeobotrys M. Calduch, Gené & Guarro (2002), anamorphic *Pezizomycotina*, H?.?.?. 1, Africa; Peru. See Calduch *et al.* (*Mycol.* **94**: 127, 2002).

Phaeobulgaria Seaver (1932) ≡ Bulgaria fide Korf (*Mycol.* **49**: 102, 1957).

Phaeocalicium Alb. Schmidt (1970), Mycocaliciaceae. *c.* 10, widespread (temperate). See Tibell (*Beih. Nova Hedwigia* **79**: 597, 1984), Titov (*Bot. Zh. SSSR* **71**: 384, 1986; key 4 spp.), Hutchison (*Mycol.* **79**: 786, 1987; N America), Tibell (*Symb. bot. upsal.* **27** no. 1: 279 pp., 1987; Australasia), Tibell (*Ann. bot. fenn.* **33**: 205, 1996; N Europe), Tibell (*Biblthca Lichenol.* **71**: 107 pp., 1998; S America), Titov *et al.* (*Botanicheskiĭ Zhurnal* **87**: 60, 2002), Tibell & Thor (*J. Hattori bot. Lab.* **94**: 205, 2003; Japan), Berglund *et al.* (*Symb. bot. upsal.* **34** no. 1: 49, 2004; Scandinavia).

Phaeocandelabrum R.F. Castañeda, Heredia, Saikawa (2007), anamorphic *Pezizomycotina*. 3, widespread.

Phaeocapnias Cif. & Bat. (1963) = Euantennaria fide Hughes (*Mycol.* **68**: 693, 1976).

Phaeocapnodinula Clem. & Shear (1931) ≡ Phaeocapnodinula Speg.

Phaeocapnodinula Speg. (1924) = Phaeostigme fide Hawksworth *et al.* (*Dictionary of the Fungi* edn 8, 1995).

Phaeocarpus Pat. (1887) [non *Phaeocarpus* Mart. & Zucc. 1824, *Sapindaceae*] ≡ Chromocyphella.

Phaeochaetia Bat. & Cif. (1962) ≡ Aithaloderma.

Phaeochora Höhn. (1909), Phaeochoraceae. 5 (on *Palmae*), widespread (tropical to subtropical). See Müller (*Sydowia* **86**: 1, 1965), Carrai-Giovanni d'Aghiano (*Sperimentozione Applicata* **4**: 4, 1992; col. pls), Cannon (*SA* **11**: 181, 1993), Hyde *et al.* (*SA* **15**: 117, 1997), Hyde & Cannon (*Mycol. Pap.* **175**: 114, 1999; gen. delim.).

Phaeochoraceae K.D. Hyde, P.F. Cannon & M.E. Barr (1997), ? Phyllachorales. 4 gen., 19 spp.

Lit.: Barr *et al.* (*Mycol.* **81**: 47, 1989), Carrai & D'Agliano (*Inftore fitopatol.* **42**: 17, 1992), Hyde *et al.* (*SA* **15**: 117, 1997), Hyde *et al.* (*SA* **15**: 117, 1997), Hyde & Cannon (*Mycol. Pap.* **175**: 114 pp., 1999; spp. on palms).

Phaeochorella Theiss. & Syd. (1915), Phyllachoraceae. 6 (from living leaves), S. Africa; Philippines. See Hyde (*TMSJ* **32**: 265, 1991).

Phaeochoropsis K.D. Hyde & P.F. Cannon (1999), Phaeochoraceae. 4 (on *Palmae*), N. & S. America. See Hyde & Cannon (*Mycol. Pap.* **175**, 1999).

Phaeociboria Höhn. (1918) = Lambertella fide Dumont (*Mem. N. Y. bot. Gdn* **33**, 1971).

Phaeociliospora Bat. & Peres (1967) = Ciliochorella fide Sutton (*Mycol. Pap.* **141**, 1977).

Phaeoclavulina Brinkmann (1897) = Ramaria Fr. ex Bonord. fide Corner (*Ann. Bot. Mem.* [A monograph of Clavaria and allied genera] **1**, 1950).

Phaeococcomyces de Hoog (1979), anamorphic *Herpotrichiellaceae*, Hso.0eH/1eP.3. 4, widespread. See de Hoog (*Taxon* **28**: 348, 1979), McGinnis *et al.* (*Sabouraudia* **23**: 179, 1985), Sugiyama *et al.* (*J. gen. appl. Microbiol.* Tokyo **33**: 197, 1987; ubiqiuinones), Butler *et al.* (*Can. J. Microbiol.* **35**: 728, 1989), Hoog *et al.* (*Antonie van Leeuwenhoek* **68**: 43, 1995; physiology), Rogers *et al.* (*Stud. Mycol.* **43**: 122, 1999; phylogeny), Untereiner & Naveau (*Mycol.* **91**: 67, 1999; phylogeny).

Phaeococcomycetaceae McGinnis & Schell (1985) = Herpotrichiellaceae.

Phaeococcus de Hoog (1977) [non *Phaeococcus* Borzí 1892, *Algae*] ≡ Phaeococcomyces.

Phaeocollybia R. Heim (1931) nom. cons., Cortinariaceae. *c.* 50, widespread (esp. temperate). See Singer (*Fl. Neotrop.* **4**: 3, 1970; monogr. neotrop. spp.), Smith & Trappe (*Mycol.* **64**: 1141, 1972; key N. Am. spp.), Horak (*Sydowia* **29**: 28, 1978), Singer (*Mycol. Helv.* **2**: 247, 1987; key Costa Rica spp.), Norvell (*MR* **102**: 615, 1998; ecology), Norvell (*Mycotaxon* **90**: 241, 2004; western North America spp.).

Phaeoconis Clem. (1909) ≡ Nigrospora.

Phaeocoriolellus Kotl. & Pouzar (1957) = Gloeophyllum fide David (*BSMF* **84**: 119, 1968).

Phaeocrella Réblová, L. Mostert, W. Gams & Crous (2004), anamorphic *Togniniella*. 1, New Zealand. See Réblová *et al.* (*Stud. Mycol.* **50**: 545, 2004), Mostert *et al.* (*Stud. Mycol.* **54**: 115 pp., 2006; phylogeny).

Phaeocreopsis Sacc. & P. Syd. ex Lindau (1900) = Valsaria fide Ju *et al.* (*Mycotaxon* **58**: 419, 1996).

Phaeocryptopus Naumov (1915), Dothideales. Anamorph *Rhizosphaera*. 5 or 6 (on conifers), Europe; N. America. *P. gaeumannii* (Swiss needle cast of *Pseudotsuga taxifolia* in Eur.), now not considered congeneric. See Butin (*Phytopath. Z.* **68**: 269, 1970), Stone & Carroll (*Sydowia* **38**: 317, 1985; ontogeny *P. gaeumannii*), Hambleton *et al.* (*Mycol.* **95**: 959, 2003; phylogeny), Winton *et al.* (*Mycol.* **99**: 240, 2007; phylogeny).

Phaeocyphella Pat. (1900) ≡ Chromocyphella.

Phaeocyphella Speg. (1909) = Chromocyphella fide Singer (*Agaric. mod. Tax.* edn 3, 1975).

Phaeocyphellopsis W.B. Cooke (1961) = Merismodes fide Reid (*Persoonia* **3**: 97, 1964).

Phaeocyrtidula Vain. (1921), ? Dothideomycetes. 2, Europe.

Phaeocyrtis Vain. (1921) = Merismatium fide Triebel (*Biblthca Lichenol.* **35**, 1989).

Phaeocytosporella G.L. Stout (1930) = Phaeocytostroma fide Petrak (*Annls mycol.* **39**: 252, 1941).

Phaeocytostroma Petr. (1921), anamorphic *Pezizomycotina*, St.0eP.15. 4, widespread. See Sutton (*The Coelomycetes*, 1980), Levič & Petrovič (*Mycopathologia* **140**: 149, 1997; α-conidia & β-conidia).

Phaeodactylella Udaiyan (1992), anamorphic *Pezizomycotina*, Hso.≡ eP.10. 1 (from water cooling tower), India. See Udajyan (*J. Econ. Taxon. Bot.* **15**: 629, 1991).

Phaeodactylium Agnihothr. (1968), anamorphic *Pezizomycotina*, Hso.0eH/≡ eH.10. 2, paleotropical. See Agnihothrudu (*Proc. Indian Acad. Sci.* series B **68**: 206, 1968), Calduch *et al.* (*Mycol.* **94**: 127, 2002).

Phaeodaedalea M. Fidalgo (1962) = Daedalea fide Reid (*Mem. N. Y. bot. Gdn* **28**: 183, 1976).

Phaeodepas D.A. Reid (1961), Marasmiaceae. 2, Venezuela. See Reid (*Kew Bull.* **15**: 273, 1961).

Phaeoderris (Sacc.) Höhn. (1907) = Leptosphaeria fide von Arx & Müller (*Stud. Mycol.* **9**, 1975).

Phaeodiaporthe Petr. (1920) = Prosthecium fide Barr (*Mycol. Mem.* **7**, 1978).

Phaeodictyon M. Choisy (1929) = Anthracothecium fide Hawksworth *et al.* (*Dictionary of the Fungi* edn 8, 1995).

Phaeodimeriella Speg. (1908), Pseudoperisporiaceae. Anamorph *Cicinnobella*. *c.* 13 (on *Asterinaceae*, *Meliolaceae*), widespread (tropical). See Hansford (*Mycol. Pap.* **15**, 1946), Barr (*Mycotaxon* **64**: 149, 1997), Ahn & Crane (*CJB* **82**: 1625, 2004).

Phaeodimeriella Theiss. (1912) = Phaeodimeriella Speg. fide Hawksworth *et al.* (*Dictionary of the Fungi* edn 8, 1995).

Phaeodimeris Clem. & Shear (1931) ≡ Phaeodimeriella Speg.

Phaeodiscula Cub. (1891), anamorphic *Pezizomycotina*. 1, Europe. See Petch (*Ann. Bot.* Lond. **22**: 389, 1908).

Phaeodiscus L.R. Batra (1968) = Lambertella fide Dumont (*Mem. N. Y. bot. Gdn* **33**, 1971).

Phaeodochium M.L. Farr (1968) ? = Hymenopsis fide Sutton (*in litt.*).

Phaeodomus Höhn. (1909), anamorphic *Pezizomycotina*, St.0eP.19. 3, Cuba; S. America.

Phaeodon J. Schröt. (1888) = Hydnellum P. Karst. (1879) fide Donk (*Taxon* **5**: 69, 1956).

Phaeodothiopsis Theiss. & Syd. (1914) nom. dub., Pezizomycotina. 3, widespread (tropical). See Müller & von Arx (*Beitr. Kryptfl. Schweiz* **11** no. 2, 1962).

Phaeodothiora Petr. (1948) = Saccothecium fide Eriksson & Hawksworth (*in litt.*), von Arx & Müller (*Stud. Mycol.* **9**, 1975).

Phaeodothis Syd. & P. Syd. (1904), Montagnulaceae. 5, widespread. See Aptroot (*Nova Hedwigia* **60**: 325, 1995; key), Liew *et al.* (*Mol. Phylogen. Evol.* **16**: 392, 2000; phylogeny), Câmara *et al.* (*MR* **107**: 516, 2003; phylogeny), Schoch *et al.* (*Mycol.* **98**: 1041, 2006; phylogeny).

Phaeofabraea Rehm (1909), Helotiaceae. 2, S. America; Taiwan. See Pfister (*Occ. Pap. Farlow Herb. Crypt. Bot.* **11**, 1977; key).

Phaeogalera Kühner (1973) = Galerina fide Singer (*Nova Hedwigia* **26**: 436, 1975).

Phaeoglabrotricha W.B. Cooke (1961) = Pellidiscus fide Reid (*Persoonia* **3**: 97, 1964).

Phaeoglaena Clem. (1909) nom. dub., ? Dothideomycetes (?L). No spp. included. See Mayrhofer & Poelt (*Herzogia* **7**: 13, 1985).

Phaeoglossum Petch (1922), Geoglossaceae. 1, Sri Lanka. No recent research is available.

Phaeographidomyces Cif. & Tomas. (1953) = Phaeographis fide Hawksworth *et al.* (*Dictionary of the Fungi* edn 8, 1995).

Phaeographina Müll. Arg. (1882) nom. illegit., Graphidaceae. *c.* 117. The correct name for most of the species appears to be *Thecographa* A. Massal. See Wirth & Hale (*Contr. US natn Herb.* **36**: 63, 1963; C. Am.), Nakanishi (*J. Sci. Hiroshima Univ.* B(2) **11**: 51, 1966; Japan), Awasthi & Singh (*Kavaka* **1**: 87, 1974; India), Archer & Elix (*Mycotaxon* **72**: 91, 1999; chemistry), Archer (*Telopea* **8**: 461, 2000; Australia), Archer (*Telopea* **9**: 329, 2001; Australia), Archer (*Mycotaxon* **83**: 361, 2002; Guadalcanal), Nakanishi *et al.* (*Bull. natn. Sci. Mus.* Tokyo, B **28**: 107, 2002; Vanuatu), Staiger (*Biblthca Lichenol.* **85**, 2002; monogr.), Archer (*Systematics & Biodiversity* **5**: 9, 2007; Solomon Is), Lücking *et al.* (*Taxon* **56**:

1296, 2007; nomencl.; ≡ Pliariona).

Phaeographinomyces Cif. & Tomas. (1953) = Phaeographina fide Hawksworth *et al.* (*Dictionary of the Fungi* edn 8, 1995).

Phaeographis Müll. Arg. (1882) nom. cons. prop., Graphidaceae (L). *c.* 226, widespread (esp. tropical). See Wirth & Hale (*Contr. US natn Herb.* **36**: 63, 1963; C. Am.), Nakanishi (*J. Sci. Hiroshima Univ.* B(2) **11**: 51, 1966; Japan), Singh & Awasthi (*Bull. Bot. Surv. India* **21**: 97, 1981; key 28 Indian spp.), Ernst & Hauck (*Herzogia* **10**: 39, 1994; Germany), Archer & Elix (*Mycotaxon* **72**: 91, 1999; chemistry), Archer (*Telopea* **8**: 461, 2000; Australia), Archer (*Telopea* **9**: 663, 2001; Australia), Staiger (*Biblthca Lichenol.* **85**, 2002; revision), López de Silanes & Álvarez (*Nova Hedwigia* **77**: 147, 2003; Iberian peninsula), Archer (*Biblthca Lichenol.* **94**, 2006; Australia), Staiger *et al.* (*MR* **110**: 765, 2006; phylogeny), Archer (*Systematics & Biodiversity* **5**: 9, 2007; Solomon Is), Lücking *et al.* (*Taxon* **56**: 1296, 2007; nomencl.).

Phaeographopsis Sipman (1997), Graphidaceae (L). 3, pantropical. See Aptroot *et al.* (*Biblthca Lichenol.* **64**: 129, 1997), Kalb (*Biblthca Lichenol.* **88**: 301, 2004), Flakus & Wilk (*J. Hattori bot. Lab.* **99**: 307, 2006).

Phaeogyroporus Singer (1944) = Phlebopus fide Kuyper (*in litt.*).

Phaeoharziella Loubière (1924) ? = Arthrinium fide Sutton (*in litt.*).

Phaeohelotium Kanouse (1935), Helotiaceae. *c.* 16, widespread (north temperate). See Dennis (*Kew Bull.* **25**: 335, 1971; key Brit. spp.), Baral & Krieglsteiner (*Beih. Sydowia* **6**, 1985), Huhtinen (*Karstenia* **29**: 45, 1990), Svrček (*Česká Mykol.* **46**: 149, 1993), Gamundí & Messuti (*MR* **110**: 493, 2006; key S hemisphere spp.).

Phaeohendersonia Höhn. (1918) nom. dub., Pezizomycotina. ≡ 'Hendersonia *sensu* Sacc.' (Ascomycetes, inc. sed.) fide Sutton (*in litt.*).

Phaeohiratsukaea Udagawa & Iwatsu (1990), anamorphic *Pezizomycotina*, Hso.0eP.15. 1 (from a stained closet), Japan. See Udagawa & Iwatsu (*Rep. Tottori Mycol. Inst.* **28**: 151, 1990).

Phaeohydnochaete Lloyd (1916), Hymenochaetaceae. 1, N. America.

Phaeohygrocybe Henn. (1901) = Russula fide Pegler (*in litt.*).

Phaeohymenula Petr. (1954), anamorphic *Pezizomycotina*, Hsp.0eH.?. 1, Australia. See Petrak (*Sydowia* **8**: 77, 1954).

phaeohyphomycosis, a mycotic infection in humans or other animals caused by a dematiaceous fungus (Ajello *et al.*, *Mycol.* **66**: 490, 1974); cf. hyalohyphomycosis.

Phaeoidiomyces Dorn.-Silva & Dianese (2004), anamorphic *Pezizomycotina*. 1, Brazil. See Dornelo-Silva & Dianese (*Mycol.* **96**: 881, 2004).

Phaeoisaria Höhn. (1909), anamorphic *Pezizomycotina*, Hsy.0eH.10. 15, widespread. Anamorph links with *Scoptria* require confirmation. See de Hoog & Papendorf (*Persoonia* **8**: 407, 1976; key), Müller & Samuels (*Sydowia* **35**: 143, 1982; teleomorph), Mercado Sierra *et al.* (*Mycotaxon* **63**: 369, 1997; Cuba), Siboe *et al.* (*Mycotaxon* **73**: 283, 1999; Kenya), Guarro *et al.* (*J. Clin. Microbiol.* **38**: 2434, 2000; keratomycosis), Castaneda Ruiz *et al.* (*Cryptog. Mycol.* **23**: 9, 2002; Cuba).

Phaeoisariopsis Ferraris (1909) = Pseudocercospora fide Ellis (*MDH*, 1976), Deighton (*MR* **94**: 1096, 1990; redisposition spp. from *Cercospora*), Walker & White (*MR* **95**: 1005, 1991; on *Wikstroemia*), Busogoro *et al.* (*Eur. J. For. Path.* **105**: 559, 1999; mol. biol.), Crous *et al.* (*Mycol.* **93**: 1081, 2001; phylogeny), Mahuku *et al.* (*Pl. Path.* **51**: 594, 2002; genetic diversity, C America), Crous & Braun (*CBS Diversity Ser.* **1**: 571 pp., 2003; reassignment of names), Sartorato (*J. Phytopath.* **152**: 385, 2004; genetic diversity, Brazil), Wagara *et al.* (*J. Phytopath.* **152**: 235, 2004; genetic diversity, Kenya), Crous *et al.* (*Stud. Mycol.* **55**: 173, 2006; phylogeny, synonymy with *Pseudocercospora*).

Phaeolabrella Speg. (1912), anamorphic *Pezizomycotina*, Cac.0eH.1. 1, S. America. See Petrak & Sydow (*Annls mycol.* **33**: 157, 1935).

Phaeolaceae Jülich (1982) = Polyporaceae.

Phaeolepiota Maire ex Konrad & Maubl. (1928), Agaricaceae. 1, widespread (north temperate). Probably a separate family is justified. See Stijve & Andrey (*Australasian Mycologist* **21**: 24, 2002).

Phaeolimacium Henn. (1899) ≡ Oudemansiella.

Phaeolopsis Murrill (1905) = Phylloporia fide Ryvarden (*Syn. Fung.* **5**, 1991).

Phaeolus (Pat.) Pat. (1900), Fomitopsidaceae. 2, widespread. See Pegler (*The polypores [Bull. BMS Suppl.]*, 1973), Simpson & May (*Australas. Pl. Path.* **31**: 99, 2002; Australia).

Phaeomacropus Henn. (1899) = Helvella fide Eckblad (*Nytt Mag. Bot.* **15**: 1, 1968).

Phaeomarasmius Scherff. (1897), Inocybaceae. *c.* 20, widespread. See Singer (*Schweiz. Z. Pilzk.* **34**: 44, 1956; monogr.), Singer (*Schweiz. Z. Pilzk.* **34**: 53, 1956; monogr.), Horak (*Sydowia* **32**: 167, 1979; key 8 Papua New Guinea spp.).

Phaeomarsonia (orthographic variant), see *Phaeomarssonia* Speg.

Phaeomarssonia Bubák (1915) = Didymosporina fide Sutton (*Mycol. Pap.* **141**, 1977).

Phaeomarssonia Speg. (1908) nom. dub., anamorphic *Pezizomycotina*. See Petrak & Sydow (*Annls mycol.* **34**: 32, 1936).

Phaeomassaria Speg. (1880) ? = Massaria fide Hawksworth *et al.* (*Dictionary of the Fungi* edn 8, 1995).

Phaeomeris Clem. (1909) = Spheconisca fide Hawksworth *et al.* (*Dictionary of the Fungi* edn 8, 1995).

Phaeomoniella Crous & W. Gams (2000), anamorphic *Herpotrichiellaceae*. 1, widespread. See Crous & Gams (*Phytopath. Mediterr.* **39**: 113, 2000), Gams (*Stud. Mycol.* **45**: 187, 2000; morphology), Groenewald *et al.* (*S. Afr. J. Sci.* **96**: 43, 2000; PCR diagnostics), Groenewald *et al.* (*MR* **105**: 651, 2001; phylogeny), Whiting *et al.* (*Mycotaxon* **92**: 351, 2005; culture, phylogeny), Lee *et al.* (*Mycol.* **98**: 598, 2006; pine endophytes), Mostert *et al.* (*Australas. Pl. Path.* **35**: 453, 2006; genetic diversity), Mostert *et al.* (*Stud. Mycol.* **54**: 115 pp., 2006).

Phaeomonilia R.F. Castañeda, Heredia & R.M. Arias (2007), anamorphic *Pezizomycotina*. 2, Cuba; Mexico. See Castañeda Ruiz *et al.* (*Mycotaxon* **100**: 327, 2007).

Phaeomonostichella Keissl. ex Petr. (1941), anamorphic *Pezizomycotina*, Cpd/Cac.0eP.?. 1, China. See Sutton (*Mycol. Pap.* **141**, 1977).

Phaeomycena R. Heim ex Singer & Digilio (1966), Tricholomataceae. 5, Africa; Asia. See Singer & Digilio (*Lilloa* **25**: 175, 1966).

Phaeomyces E. Horak (2005), Inocybaceae. 2, Europe.

See Horak (*Röhrlinge und Blätterpilze in Europa*, 2005).

phaeomycosis, see phaeohyphomycoses.

Phaeonaevia L. Holm & K. Holm (1977), Helotiales. 1 (on *Rubus*), widespread (northern Europe). See Holm & Holm (*Kew Bull.* **31**: 576, 1977).

Phaeonectria (Sacc.) Sacc. & Trotter (1913) = Calostilbe fide Rossman *et al.* (*Stud. Mycol.* **42**: 248 pp., 1999).

Phaeonectriella R.A. Eaton & E.B.G. Jones (1971), Halosphaeriaceae. 1 (submerged wood), widespread. See Lowen (*SA* **5**: 150, 1986; posn), Chen *et al.* (*Mycol.* **91**: 84, 1999; phylogeny), Hyde *et al.* (*Mycoscience* **40**: 165, 1999), Rossman *et al.* (*Stud. Mycol.* **42**: 248 pp., 1999), Campbell *et al.* (*Mycol.* **95**: 530, 2003; phylogeny).

Phaeonema Kütz. (1843) nom. dub., ? Fungi.

Phaeonematoloma (Singer) Bon (1994) = Pholiota fide Kuyper (*in litt.*).

Phaeopeltis Clem. (1909) ≡ Phaeosaccardinula.

Phaeopeltis Petch (1919) ≡ Phaeaspis Clem. & Shear.

Phaeopeltium Clem. & Shear (1931) ≡ Phaeopeltosphaeria.

Phaeopeltosphaeria Berl. & Peglion (1892), Dothideomycetes. 2, widespread.

Phaeopezia (Vido) Sacc. (1884) = Peziza Fr. fide Eckblad (*Nytt Mag. Bot.* **15**: 1, 1968).

Phaeophacidium Henn. & Lindau (1897), Rhytismatales. 1, America.

Phaeophlebia W.B. Cooke (1956) = Punctularia fide Talbot (*Bothalia* **7**: 131, 1958).

Phaeophleospora Rangel (1916), anamorphic *Mycosphaerellaceae*. 18. See also *Kirramyces* for species on eucalypts. See Sutton (*Mycol. Pap.* **141**, 1977), Crous *et al.* (*S. Afr. J. Bot.* **63**: 111, 1997), Crous (*Mycol. Mem.* **21**: 170 pp., 1998), Taylor & Crous (*Fungal Diversity* **3**: 153, 1999; on *Proteaceae*), Crous *et al.* (*Stud. Mycol.* **45**: 107, 2000; review), Crous *et al.* (*Mycol.* **93**: 1081, 2001; phylogeny), Taylor *et al.* (*S. Afr. J. Bot.* **67**: 39, 2001; on *Proteaceae*), Hunter *et al.* (*MR* **108**: 672, 2004; S Africa), Hunter *et al.* (*Stud. Mycol.* **55**: 147, 2006; phylogeny), Andjic *et al.* (*MR* **111**: 1184, 2007; phylogeny).

Phaeophloeosporella Crous & B. Sutton (1997), anamorphic *Pezizomycotina*, Cpd.?.?. 1, S. Africa. See Crous & Sutton (*S. Afr. J. Bot.* **63**: 281, 1997).

Phaeopholiota Locq. & Sarwal (1983), Agaricaceae. 1, Sikkim. See Locquin & Sarwal (*Compt. Rend. Congr. Natl. Soc. Savantes* Sec. Sci., 1983: 193, 1983).

Phaeophomatospora Speg. (1909) = Anthostomella fide Petrak & Sydow (*Annls mycol.* **23**: 209, 1925), Francis (*Mycol. Pap.* **139**, 1975), Lu & Hyde (*Fungal Diversity Res. Ser.* **4**, 2000).

Phaeophomopsis Höhn. (1917), anamorphic *Pezizomycotina*, St.0eP.?. 1, France. See Sutton (*Mycol. Pap.* **141**, 1977).

Phaeophragmeriella Hansf. (1944) = Leptomeliola fide Hughes (*Mycol. Pap.* **166**, 1993).

Phaeophragmocauma F. Stevens (1931) = Dermatodothis fide Müller (*Sydowia* **28**: 149, 1976).

Phaeophycopsis Bat. & Peres (1967) = Seuratia fide Meeker (*CJB* **53**: 2462, 1975).

Phaeophyscia Moberg (1977), Physciaceae (L). 28, widespread. See Esslinger (*Mycotaxon* **7**: 283, 1978; N. Am.), Hafellner *et al.* (*Herzogia* **5**: 39, 1979), Hale (*Lichenologist* **15**: 157, 1983), Moberg (*Nordic Jl Bot.* **3**: 509, 1983; E. Africa), Kashiwadani (*Bull. natn. Sci. Mus.* Tokyo, B **10**: 127, 1984; Japan), Moberg (*Op. Bot.* **121**: 281, 1993; S. Am.), Moberg (*Nordic Jl Bot.* **15**: 319, 1995; Far East), Lohtander *et al.* (*Mycol.* **92**: 728, 2000; Scandinavia, phylogeny), Dahlkild *et al.* (*Bryologist* **104**: 527, 2001; photobionts), Grube & Arup (*Lichenologist* **33**: 63, 2001; phylogeny), Helms *et al.* (*Mycol.* **95**: 1078, 2003; phylogeny), Simon *et al.* (*J. Mol. Evol.* **60**: 434, 2005; introns), Miądlikowska *et al.* (*Mycol.* **98**: 1088, 2006; phylogeny).

Phaeopolynema Speg. (1912) = Hymenopsis fide Sutton (*Mycol. Pap.* **141**, 1977).

Phaeopolystomella Bat. & H. Maia (1960) = Microdothella fide von Arx & Müller (*Stud. Mycol.* **9**, 1975).

Phaeoporotheleum (W.B. Cooke) W.B. Cooke (1961), Cyphellaceae. 2, Cuba; Argentina. See Cooke (*Beih. Sydowia* **4**: 129, 1961).

Phaeoporus Bataille (1908) = Tylopilus fide Kuyper (*in litt.*).

Phaeoporus J. Schröt. (1888) ? = Inonotus fide Donk (*Verh. K. ned. Akad. Wet.* tweede sect. **62**: 1, 1974).

Phaeopterula (Henn.) Sacc. & D. Sacc. (1905) = Pterula fide Corner (*Ann. Bot. Mem.* [A monograph of Clavaria and allied genera] **1**, 1950).

Phaeopyxis Rambold & Triebel (1990), Helotiales. 4 (mainly on lichens), widespread (north temperate). See Rambold & Triebel (*Notes R. bot. Gdn Edinb.* **46**: 375, 1990).

Phaeoradulum Pat. (1900), Boletales. 1, West Indies.

Phaeoramularia Munt.-Cvetk. (1960) = Passalora fide Ellis (*More Dematiaceous Hyphomycetes*, 1976), Deighton (*Mycol. Pap.* **144**: 26, 1979), Braun (*Monogr. Cercosporella, Ramularia Allied Genera (Phytopath. Hyphom.)* **2**: 382, 1998; key), Braun (*Schlechtendalia* **2**: 1, 1998), Crous & Braun (*CBS Diversity Ser.* **1**: 571 pp., 2003).

Phaeorhytisma Henn. (1899) ? = Criella fide Müller (*Sydowia* **12**: 160, 1959; non-sporulating).

Phaeorobillarda Bat. & J.L. Bezerra (1961) = Robillarda Sacc. fide Nag Raj (*Coelomycetous Anamorphs with Appendage-bearing Conidia*, 1993).

Phaeorrhiza H. Mayrhofer & Poelt (1979), Physciaceae (L). 2, widespread. See Mayrhofer & Poelt (*Nova Hedwigia* **30**: 783, 1979), Elvebakk & Moberg (*Lichenologist* **34**: 311, 2002; Chile), Helms *et al.* (*Mycol.* **95**: 1078, 2003; phylogeny), Golubkova (*Nov. sist. Niz. Rast.* **37**: 200, 2004; Russia).

Phaeosaccardinula Henn. (1905), Chaetothyriaceae. 14, widespread (tropical). See Eriksson & Yue (*Mycotaxon* **22**: 269, 1985).

Phaeoschiffnerula Theiss. (1914) = Schiffnerula. fide Petrak [not traced]. fide Petrak (*Annls mycol.* **25**: 193, 1927).

Phaeoschizophyllum W.B. Cooke (1962) = Schizophyllum fide Donk (*Persoonia* **3**: 199, 1964).

Phaeosclera Sigler, Tsuneda & J.W. Carmich. (1981), anamorphic *Dothideomycetes*, Hso.#eP.1. 1, Canada. See Sigler *et al.* (*Mycotaxon* **12**: 461, 1981), Hambleton *et al.* (*Mycol.* **95**: 959, 2003; phylogeny).

Phaeosclerotinia Hori (1916), Sclerotiniaceae. Anamorph *Monilia*. 1, Japan.

Phaeoscopulariopsis M. Ota (1928) nom. provis. = Scopulariopsis fide Morton & Smith (*Mycol. Pap.* **86**, 1963).

Phaeoscutella Henn. (1904) nom. dub., Fungi.

Phaeoscypha Spooner (1984), Hyaloscyphaceae. Anamorph *Chalara*-like. 1, British Isles. See Spooner

(*Kew Bull.* **38**: 574, 1984).

Phaeoseptoria Speg. (1908), anamorphic *Phaeosphaeria*, Cpd.≡ fP.?. 19, widespread. See Petrak (*Annls mycol.* **39**: 292, 1941), Punithalingam (*Nova Hedwigia* **32**: 585, 1980; key), Knipscheer *et al.* (*S. Afr. For. J.* **154**: 56, 1990; *P. eucalypti* in S. Afr.), Walker *et al.* (*MR* **96**: 911, 1992; redisp. of *P. eucalypti*, redescr. of *P. papayae*), Punithalingam & Spooner (*MR* **101**: 292, 1997; on lichens), Câmara *et al.* (*Mycol.* **94**: 630, 2002; phylogeny).

Phaeosiphonia Kütz. (1849) nom. dub., ? Fungi.

Phaeosolenia Speg. (1902), ? Inocybaceae. 8, S. America. See Cooke (*Sydowia* Beih. **4**, 1961), Siepe & Kasparek (*Z. Mykol.* **68**: 153, 2002; *Phaeosolenia densa* from Germany).

Phaeosperma (Sacc.) Traverso (1906) = Valsaria fide Shear (*Mycol.* **30**: 589, 1938).

Phaeosperma Nitschke ex Fuckel (1870), ? Dothideomycetes. 1 (on *Alnus*), Switzerland.

Phaeosperma Nitschke ex G.H. Otth (1869) [non *Phaiosperma* Raf. 1883] = Camarops fide Nannfeldt (*Svensk bot. Tidskr.* **66**: 335, 1972).

Phaeosphaera Bat. & Cif. (1963), anamorphic *Pezizomycotina*, Cpd.0eP.?. 2, Philippines; USA. See Batista & Ciferri (*Quaderno Ist. Bot. Univ. Pavia* **31**: 142, 1963).

Phaeosphaerella P. Karst. (1888) = Venturia Sacc. fide von Arx & Müller (*Stud. Mycol.* **9**, 1975).

Phaeosphaeria I. Miyake (1909), Phaeosphaeriaceae. Anamorphs *Phaeoseptoria*, *Stagonospora*. *c.* 81 (esp. on *Poaceae*), widespread. See Holm (*Symb. bot. upsal.* **14** no. 3: 1, 1957), Eriksson (*Ark. Bot.* ser. 2 **6**: 339, 1967), Koponen & Mäkelä (*Ann. bot. fenn.* **12**: 141, 1975; Finland), Otani (*Bull. natn. Sci. Mus.* Tokyo, B **2**: 87, 1976; Japan), Leuchtmann (*Sydowia* **37**: 75, 1984; key 45 spp.), Leuchtmann *in* Laursen *et al.* (Eds) (*Arctic and Alpine Mycology* **2**: 153, 1987; key arctic alpine 16 spp.), Shoemaker & Babcock (*CJB* **67**: 1500, 1989; monogr., keys), Khashnobish & Shearer (*MR* **100**: 1355, 1996; phylogeny), Halama *in* Lucas*et al.* (Eds) (*Septoria on Cereals*: 70, 1999), Silva-Hanlin & Hanlin (*MR* **103**: 153, 1999), Czembor & Arseniuk (*MR* **104**: 919, 2000; recombination), Câmara *et al.* (*Mycol.* **94**: 630, 2002; phylogeny), Fukuhara (*Mycoscience* **43**: 375, 2002; on *Phragmites*), Eriksson & Hawksworth (*Mycol.* **95**: 426, 2003; phylogeny), Ueng *et al.* (*Curr. Genet.* **43**: 121, 2003; mating type genes), Ueng *et al.* (*Plant Pathology Bulletin* Taichung **12**: 255, 2003; on cereals), Solomon *et al.* (*Eur. J. Pl. Path.* **110**: 763, 2004; mating types), Stoykov (*Mycol. Balcanica* **1**: 125, 2004; Bulgaria), Tanaka (*Mycoscience* **45**: 377, 2004; Japan), Verkley *et al.* (*Mycol.* **96**: 558, 2004), Malkus *et al.* (*FEMS Microbiol. Lett.* **249**: 49, 2005; on cereals), Schoch *et al.* (*Mycol.* **98**: 1041, 2006; phylogeny), Adhikari *et al.* (*Phytopathology* **98**: 101, 2008; population biology, USA).

Phaeosphaeriaceae M.E. Barr (1979), Pleosporales. 19 gen. (+ 28 syn.), 394 spp.

Lit.: Leuchtmann (*Sydowia* **37**: 75, 1984), Shoemaker (*CJB* **62**: 2730, 1984), Shoemaker & Babcock (*CJB* **63**: 1284, 1985), Boise (*Mem. N. Y. bot. Gdn* **49**: 308, 1989), Shoemaker & Babcock (*CJB* **67**: 1500, 1989), Barr (*Mycotaxon* **43**: 371, 1992; keys 14 N. Am. gen.), Khashnobish & Shearer (*MR* **100**: 1355, 1996), Ramaley (*Mycotaxon* **61**: 351, 1997), Caten (*Septoria on Cereals* A Study of Pathosystems: 26, 1999), Cunfer & Ueng (*Ann. Rev. Phytopath.* **37**: 267, 1999), Silva-Hanlin & Hanlin (*MR* **103**: 153, 1999), Liew *et al.* (*Mol. Phylogen. Evol.* **16**: 392, 2000), Câmara *et al.* (*MR* **105**: 41, 2001), Câmara *et al.* (*Mycol.* **94**: 630, 2002), Câmara *et al.* (*MR* **107**: 516, 2003), Kodsueb *et al.* (*Mycol.* **98**: 571, 2006), Schoch *et al.* (*Mycol.* **98**: 1041, 2006; phylogeny).

Phaeosphaeriopsis M.P.S. Câmara, M.E. Palm & A.W. Ramaley (2003), Pleosporales. Anamorph *Phaeostagonospora*. 6, widespread. See Câmara *et al.* (*MR* **107**: 519, 2003).

Phaeospora Hepp ex Stein (1879), ? Verrucariaceae. 15 (on lichens), widespread.

Phaeosporella Keissl. (1922) ≡ Phaeosphaerella.

Phaeosporis Clem. (1909), ? Sordariales. 2, France. See Hawksworth (*SA* **6**: 145, 1987), Krug *et al.* (*Mycol.* **86**: 581, 1994).

Phaeosporobolus D. Hawksw. & Hafellner (1986), anamorphic *Pezizomycotina*, St.0eP.3. 1 (on lichens), Europe; USA. See Hawksworth & Hafellner (*Nova Hedwigia* **43**: 525, 1986).

Phaeostagonospora A.W. Ramaley (1997), anamorphic *Phaeosphaeriopsis*, Cpd.?.?. 1, USA. See Ramaley (*Mycotaxon* **61**: 351, 1997).

Phaeostagonosporopsis Woron. (1925) = Stenocarpella fide Sutton (*Mycol. Pap.* **141**, 1977).

Phaeostalagmus W. Gams (1976), anamorphic *Chaetosphaeria*, Hso.0eH.15. 7, widespread. See Sutton & Mel'nik (*MR* **96**: 908, 1992), Réblová (*Stud. Mycol.* **50**: 171, 2004), Fernández & Huhndorf (*Fungal Diversity* **18**: 15, 2005; teleomorph).

Phaeosticta Trevis. (1869) nom. rej. = Pseudocyphellaria fide Hawksworth *et al.* (*Dictionary of the Fungi* edn 8, 1995).

Phaeostigme Syd. & P. Syd. (1917), Pseudoperisporiaceae. *c.* 20, widespread (tropical). See Hughes (*Mycol. Pap.* **166**, 1993).

Phaeostilbella Höhn. (1919) = Myrothecium fide Tulloch (*Mycol. Pap.* **130**, 1972).

Phaeostoma Arx & E. Müll. (1954) [non *Phaeostoma* Spach 1835, *Onagraceae*] ≡ Arxiomyces.

Phaeostomiopeltis Bat. & Cavalc. (1963) = Haplopeltheca fide von Arx & Müller (*Stud. Mycol.* **9**, 1975).

Phaeotellus Kühner & Lamoure (1972) = Omphalina fide Kuyper (*in litt.*).

Phaeotheca Sigler, Tsuneda & J.W. Carmich. (1981), anamorphic *Capnodiales*, Hsp.0eP.endoconidia. 2, widespread. See DesRochers & Ouellette (*CJB* **72**: 808, 1994; sp. inhibiting *Ophiostoma ulmi*), Hoog *et al.* (*Antonie van Leeuwenhoek* **71**: 289, 1997; from humidifier), Hoog *et al.* (*Stud. Mycol.* **43**: 31, 1999; phylogeny), Zalar *et al.* (*Stud. Mycol.* **43**: 49, 1999; revision), Tsuneda *et al.* (*Mycol.* **96**: 1136, 2004; endoconidiogenesis).

Phaeothecium, see *Phacothecium*.

Phaeothecoidea Crous (2007), anamorphic *Mycosphaerellaceae*. 1, Australia. See Crous *et al.* (*Stud. Mycol.* **58**: 1, 2007), Crous *et al.* (*Fungal Diversity* **26**: 1, 2007).

Phaeothrombis Clem. (1909) = Thrombium fide Hawksworth *et al.* (*Dictionary of the Fungi* edn 8, 1995).

Phaeothyriolum Syd. (1938), Microthyriaceae. 1 (on *Eucalyptus*), Australia. See Swart (*TBMS* **87**: 81, 1986).

Phaeothyrium Petr. (1947), anamorphic *Pezizomycotina*, Cpt.0eP.?. 1, China.

Phaeotomasellia Katum. (1981), Dothideomycetes. 1,

Uganda. See Katumoto (*J. Jap. Bot.* **56**: 389, 1981).

Phaeotrabutia Orejuela (1941) = Phyllachora Nitschke ex Fuckel (1870) fide von Arx & Müller (*Beitr. Kryptfl. Schweiz* **11** no. 1, 1954).

Phaeotrabutiella Theiss. & Syd. (1915) = Phyllachora Nitschke ex Fuckel (1870) fide Hyde & Cannon (*Mycol. Pap.* **175**, 1999).

Phaeotrametaceae O.F. Popoff ex Piątek (2005) = Polyporaceae.

Phaeotrametes Lloyd ex J.E. Wright (1966), Polyporaceae. 1, southern hemisphere. See Wright (*Mycol.* **58**: 529, 1966), Piątek & Cabala (*Mycotaxon* **91**: 173, 2005; Poland).

Phaeotrema Müll. Arg. (1887), ? Dothideomycetes. 14. See Salisbury (*Nova Hedw.* **29**: 405, 1978), Matsumoto & Deguchi (*Bryologist* **102**: 86, 1999; anamorphs), Matsumoto (*J. Hattori bot. Lab.* **88**: 1, 2000).

Phaeotremella Rea (1912) = Tremella Pers. fide Donk (*Taxon* **7**: 238, 1958).

Phaeotrichaceae Cain (1956), Pleosporales. 3 gen., 7 spp.

Lit.: Cain (*CJB* **34**: 675, 1956), Ebersohn & Eicker (*S. Afr. J. Bot.* **58**: 145, 1992), Khan & Krug (*Proceedings of the Thirteenth Plenary Meeting of AETFAT* Zomba, Malawi, 2-11 April 1991, Vol. 1: 755, 1994), Lumbsch & Lindemuth (*MR* **105**: 901, 2001), Kruys *et al.* (*MR* **110**: 527, 2006).

Phaeotrichoconis Subram. (1956), anamorphic *Pezizomycotina*, Hso.≡ eP.26. 2, widespread. See Subramanian (*Proc. Indian Acad. Sci.* series B **44**: 2, 1956).

Phaeotrichosphaeria Sivan. (1983), ? Xylariales. Anamorph *Endophragmiella*. 4 (from wood etc.), India; British Isles; New Zealand. Perhaps related to *Iodosphaeria*. See Barr (*Mycotaxon* **39**: 43, 1990; posn), Réblová (*Mycotaxon* **71**: 13, 1999; posn), Romero & Carmarán (*Persoonia* **18**: 253, 2003).

Phaeotrichum Cain & M.E. Barr (1956), Phaeotrichaceae. 2, widespread. See Cain (*CJB* **34**: 675, 1956), Barr (*Mycotaxon* **76**: 105, 2000), Kruys *et al.* (*MR* **110**: 527, 2006; phylogeny), Schoch *et al.* (*Mycol.* **98**: 1041, 2006; phylogeny).

Phaeotrombis, see *Phaeothrombis*.

Phaeotrype Sacc. (1920) = Diatrype fide Petrak (*Annls mycol.* **23**: 46, 1925).

Phaeoxyphiella Bat. & Cif. (1963), anamorphic *Capnodium*, Cpd.≡ eP.?. 7, widespread. See Hughes (*Mycol.* **68**: 693, 1976).

Phaeoxyphium Bat. & Cif. (1963) nom. dub., anamorphic *Pezizomycotina*. See Sutton (*Mycol. Pap.* **141**, 1977).

Phaeoxyphium Bat. & J.L. Bezerra (1960) ≡ Phaeoxyphium Bat. & Cif.

Phaffia M.W. Mill., Yoney. & Soneda (1976), anamorphic *Xanthophyllomyces*, Hso.0eH.3/19. 1, Japan; USA. See Nagy *et al.* (*FEMS Microbiol. Lett.* **123**: 315, 1994; PGFE), Adrio *et al.* (*Curr. Genet.* **27**: 447, 1995; PFGE), Johnson & Schroeder (*Stud. Mycol.* **38**: 81, 1995; production astaxanthin), Varga *et al.* (*Int. J. Syst. Bacteriol.* **45**: 173, 1995; isozymes, RAPD, RFLP), Kucsera *et al.* (*Antonie van Leeuwenhoek* **73**: 163, 1998; life cycle), Fell *et al.* (*Int. J. Syst. Evol. Microbiol.* **50**: 1351, 2000; mol. phylogeny), Fell *et al.* (*The Mycota, A Comprehensive Treatise on Fungi as Experimental Systems for Basic and Applied Research* **7** B: 3, 2001).

Phaffomyces Y. Yamada (1997), Phaffomycetaceae. 3 (on cacti), widespread. See Yamada *et al.* (*Biosc., Biotechn., Biochem.* **63**: 827, 1999), Suh *et al.* (*Mycol.* **98**: 1006, 2006; phylogeny).

Phaffomycetaceae Y. Yamada, H. Kawas., Nagats., Mikata & T. Seki (1999), Saccharomycetales. 3 gen., 7 spp.

Lit.: Yamada *et al.* (*Bull. Fac. Agric. Shizuoka Univ.* **47**: 23, 1997), Yamada *et al.* (*Biosc., Biotechn., Biochem.* **63**: 827, 1999).

Phagodinium Kristiansen (1993) nom. dub., ? Fungi. 1. See Kristiansen (*Arch. Protistenk.* **143**: 213, 1993).

phagosome, a membrane surrounding an endosymbiont to form a distinctive structure, as *Nostoc*-containing vesicles of *Geosiphon*.

phagotrophic, feeding by ingestion, engulfing food.

Phakopsora Dietel (1895), Phakopsoraceae. *c.* 110 (on dicots esp. (› 30 families)), widespread (tropical). *P. pachyrhizi* (soybean rust), *P. euvitis* (vine (*Vitis*) rust); *P. gossypii* (cotton rust). See Thirumalachar & Kern (*Mycol.* **41**: 283, 1949), Cummins & Hiratsuka (*Illustr. Gen. Rust Fungi rev. edit.*, 1983), Ono *et al.* (*MR* **96**: 825, 1992), Ono (*Mycol.* **92**: 154, 2000), Frederick *et al.* (*Phytopathology* **92**: 217, 2002; PCR soybean rust detection), Berndt (*Mycol. Progr.* **4**: 339, 2005; Tanzania), Aime (*Mycoscience* **47**: 112, 2006; family relationships).

Phakopsoraceae Cummins & Hirats. f. (1983), Pucciniales. 18 gen. (+ 8 syn.), 205 spp.

Lit.: Gjærum *et al.* (*Lidia* **5**: 13, 2000), Ono (*Mycol.* **92**: 154, 2000), Frederick *et al.* (*Phytopathology* **92**: 217, 2002), Cummins & Hiratsuka (*Illustr. Gen. Rust Fungi* edn 3: 225 pp., 2003), Hüseyın & Selçuk (*Pakist. J. Bot.* **36**: 203, 2004), Tessmann *et al.* (*Fitopatol. Brasil* **29**: 338, 2004), Wingfield *et al.* (*Australas. Pl. Path.* **33**: 327, 2004), Aime (*Mycoscience* **47**: 112, 2006).

Phalacrichomyces R.K. Benj. (1992), Laboulbeniaceae. 2, Venezuela. See Benjamin (*Aliso* **13**: 428, 1992), Santamaria (*MR* **99**: 1071, 1995).

phalacrogenous, of conidiogenous cells arising at the same level from single hyphae to form a turf-like layer; velvety. See Gams (*Cephalosporium-artige Schimmelpilze*, 1971).

Phalangispora Nawawi & J. Webster (1982), anamorphic *Pezizomycotina*, Hsp.1bH.10. 2, pantropical. See Kuthubutheen (*TBMS* **89**: 414, 1987), Keshava Prasad & Bhat (*Mycotaxon* **83**: 405, 2002).

Phallaceae Corda (1842), Phallales. 21 gen. (+ 46 syn.), 77 spp.

Lit.: Boedijn (*Bull. Jard. bot. Buitenz. ser. 3* **12**: 71, 1932; Indonesia), Calonge (*Boln Soc. Micol. Castell.* **10**: 59, 1985), Beaton & Malajczuk (*TBMS* **86**: 478, 1986), Boa (*Mycologist* **2**: 107, 1988), Fan *et al.* (*Beih. Nova Hedwigia* **108**: 72 pp., 1994), Marren (*British Wildlife* **6**: 366, 1995), Kreisel (*Czech Mycol.* **48**: 273, 1996), Stijve (*Australas. Mycol. Newsl.* **16**: 11, 1997), Pine *et al.* (*Mycol.* **91**: 944, 1999), Bougher & Lebel (*Aust. Syst. Bot.* **14**: 439, 2001), Humpert *et al.* (*Mycol.* **93**: 465, 2001), Baseia *et al.* (*Mycotaxon* **85**: 77, 2003).

Phallales E. Fisch. (1898). Phallomycetidae. 2 fam., 26 gen., 88 spp. There may be some justification is recognizing *Gomphales* and *Phallales* as previously circumscribed at sub-ordinal level. Fams:

(1) **Claustulaceae**
(2) **Phallaceae**

For *Lit.* see fam.; see also under gasteromycetes.

Phallobata G. Cunn. (1926), Trappeaceae. 1, Australia.

Phalloboletus Adans. (1763) ≡ Morchella.

Phallogaster Morgan (1893), Phallogastraceae. 1, widespread (north temperate; ? introd.).

Phallogastraceae Locq. (1974) nom. inval., Hysterangiales. 2 gen. (+ 1 syn.), 14 spp.

phalloid, one of the *Phallales*.

Phalloidastrum Battarra (1755) nom. inval. = Phallus fide Stalpers (*in litt.*).

phallolysin, a protein of *Amanita phalloides*, has cytolytic effects *in vitro*, and is toxic to animals (Odenthal *et al.*, *Naunyn-Schmiederberg's Arch. Pharmacol.* **290l**: 133, 1975).

Phallomyces Bat. & Valle (1961) = Echinoplaca fide Lücking *et al.* (*Lichenologist* **30**: 121, 1998).

Phallomycetidae K. Hosaka, Castellano & Spatafora (2007), Agaricomycotina. Ords:

(1) **Geastrales**
(2) **Gomphales**
(3) **Hysterangiales**
(4) **Phallales**

For *Lit.* see ord. and fam.

phallotoxins, cyclic heptapeptides (phallcidin, phalloidin, phallicin, phallicidin, phallin B) toxic to humans from *Amanita phalloides* etc., see Wieland (*Science* **159**: 950, 1968; *Peptides of Amanita mushrooms*, 1986). Cf. amatoxins.

Phallus Junius ex L. (1753), Phallaceae. 18, widespread. See Baseia *et al.* (*Mycotaxon* **85**: 77, 2003; Brazil).

Phalodictyum Clem. (1909) = Rhizocarpon fide Hawksworth *et al.* (*Dictionary of the Fungi* edn 8, 1995).

Phalomia Nieuwl. (1916) ≡ Omphalia (Fr.) Staude.

Phalostauris Clem. (1909) ≡ Willeya.

Phalothrix Clem. (1909) = Unguicularia fide Nannfeldt (*Nova Acta R. Soc. Scient. upsal.*, 1932).

Phaneroascus Baudyš (1919) = Cookella fide von Arx (*Persoonia* **2**: 421, 1963).

Phanerochaetaceae Jülich (1982), Polyporales. 19 gen. (+ 9 syn.), 249 spp.

Phanerochaete P. Karst. (1889), Phanerochaetaceae. Anamorph *Necator*. *c.* 65, widespread. See Donk (*Persoonia* **2**: 223, 1962), Eriksson *et al.* (*Taxon* **27**: 299, 1978; nomencl.), Eriksson *et al.* (*Cortic. N. Europ.* **5**: 987, 1978; key 12 Eur. spp.), Burdsall (*Mycol. Mem.* **10**: 165, 1985; key 46 spp.), Wu (*Acta Bot. Fenn.* **142**: 37, 1990; key Taiwan spp.), Martinez *et al.* (*Nature Biotechnology* **22**: 695, 2004; *Phanerochaete chrysosporium* genome), Martínez & Nakasone (*Sydowia* **57**: 94, 2005; Uruguay).

Phanerococculus Cif. (1954), anamorphic *Koordersiella*, Cpd.≡ eH.?. 1, Santo Domingo. See Ciferri (*Sydowia* **8**: 265, 1954).

Phanerococcus Theiss. & Syd. (1918) ? = Koordersiella fide Hansford (*Mycol. Pap.* **15**, 1946).

Phanerocorynella Höhn. (1923) ≡ Exosporiella.

Phanerocoryneum Höhn. (1923) = Clasterosporium fide Sutton (*Mycol. Pap.* **141**, 1977).

Phaneromyces Speg. & Hara (1889), Phaneromycetaceae. 1, S. America. See Gamundí & Spinedi (*Sydowia* **38**: 106, 1986), Kutorga & Hawksworth (*SA* **15**: 1, 1997).

Phaneromycetaceae Gamundí & Spinedi (1985), Pezizomycotina (inc. sed.). 1 gen., 1 spp.

Lit.: Gamundí & Spinedi (*Sydowia* **38**: 106, 1985), Kutorga & Hawksworth (*SA* **15**: 1, 1997).

phaneroplasmodium, see plasmodium.

Phanosticta Clem. (1909) = Pseudocyphellaria fide Hawksworth *et al.* (*Dictionary of the Fungi* edn 8, 1995).

Phanotylium Clem. (1909) = Tremotylium fide Hawksworth *et al.* (*Dictionary of the Fungi* edn 8, 1995).

Pharcidia Körb. (1865) = Stigmidium fide Santesson (*Svensk bot. Tidskr.* **54**: 499, 1960), Roux & Triebel (*Bull. Soc. linn. Provence* **45**: 451, 1994).

Pharcidiella (Sacc.) Clem. & Shear (1931) = Phaeospora fide Hawksworth *et al.* (*Dictionary of the Fungi* edn 8, 1995).

Pharcidiopsis Sacc. (1905) = Stigmidium fide Keissler (*Rabenh. Krypt.-Fl.* **8**, 1930).

pharmaceutical prospecting, see Bioprospecting, Biotechnology, Screening.

Pharus Petch (1919) [non *Pharus* P. Browne 1756, *Gramineae*] ≡ Mycopharus.

Phascolomyces Boedijn ex Benny & R.K. Benj. (1976), Syncephalastraceae. 1, widespread. See Benny & Benjamin (*Aliso* **8**: 391, 1976), Jeffries & Young (*CJB* **56**: 747, 1978; sporangiolum ultrastr.), Jeffries & Young (*CJB* **56**: 2449, 1978; response to mycoparasitism), Balasubramanian & Manocha (*CJB* **64**: 2441, 1986; biochemistry), Benny *et al.* *in* McLaughlin *et al.* (Eds) (*The Mycota* A Comprehensive Treatise on Fungi as Experimental Systems for Basic and Applied Research **7A**: 113, 2001), Voigt & Wöstemeyer (*Gene* **270**: 113, 2001; phylogeny).

phaseolin, a phytoalexin (q.v.) from bean (*Phaseolus vulgaris*).

Phasya Syd. (1934) = Venturia Sacc. fide Müller & von Arx (*Beitr. Kryptfl. Schweiz* **11** no. 2, 1962).

Phaulomyces Thaxt. (1931), Laboulbeniaceae. 14, widespread. See De Kesel (*Mycotaxon* **50**: 191, 1994), Majewski (*Acta Mycologica* Warszawa **34**: 7, 1999; Poland), Santamaría & Rossi (*Pl. Biosystems* **133**: 163, 1999; Mediterranean).

Phellinaceae Jülich (1982) = Hymenochaetaceae.

Phellinidium (Kotl.) Fiasson & Niemelä (1984), Hymenochaetaceae. 3, Europe. See Fiasson & Niemelä (*Karstenia* **24**: 25, 1984), Wagner & Fischer (*MR* **105**: 773, 2001), Wagner & Fischer (*Mycol.* **94**: 998, 2002; phylogeny).

Phellinites Singer & S. Archang. (1958), Fossil Fungi, Basidiomycota. 1 (Jurassic), Argentina.

Phellinus Quél. (1886), Hymenochaetaceae. *c.* 180, widespread. s. str. *c.* 30 species. See Ryvarden & Johansen (*Prelim. Polyp. Fl. E. Afr.*: 129, 1980; key 62 Afr. spp.), Wright & Blumenfeld (*Mycotaxon* **21**: 413, 1984; key 26 spp. Argentina), Larsen & Coldo-Poulle (*Syn. Fung.* **3**, 1990; world key), Wagner & Fischer (*MR* **105**: 773, 2001), Wagner & Fischer (*Mycol.* **94**: 998, 2002; phylogeny), Ryvarden (*Syn. Fung.* **19**: 229 pp., 2004).

Phellodon P. Karst. (1881), Bankeraceae. 16, Europe; N. America. See Stalpers (*Stud. Mycol.* **35**: 168 pp., 1993; key), Arnolds (*Coolia* **46** 3, Suppl.: 96 pp., 2003; Netherlands and Belgium), Parfitt *et al.* (*MR* **111**: 761, 2007; molecel. phylog. British spp.).

Phellomyces A.B. Frank (1898) = Colletotrichum fide Husz (*Z. Pflanzenkr. Pflanzensch.* **44**: 186, 1934).

Phellomycetes Renault (1896), Fossil Fungi (mycel.) Fungi. 1 (Carboniferous), France.

Phellomycites Mesch. (1896) ≡ Phellomycetes.

phellophagy, ability to attack cork cells (Speer, *Mycotaxon* **21**: 235, 1984).

Phellopilus Niemelä, T. Wagner & M. Fisch. (2001), Hymenochaetaceae. 1, widespread. See Niemelä *et al.* (*Ann. bot. fenn.* **38**: 53, 2001).

Phellorinia Berk. (1843), Phelloriniaceae. 1, widespread (subtropical dry areas). See Malençon (*Annals Cryptog. Exot.* **8**: 5, 1935), Long (*Lloydia* **9**: 132, 1946), Kreisel (*Česká Mykol.* **15**: 195, 1961), Martin *et al.* (*Cryptogamie, Mycologie* **21**: 3, 2000; phylogeny), Sarasini (*Rivista di Micologia* **46**: 7, 2003; mediterranean sp.), Fan & Liu (*Mycosystema* **23**: 306, 2004; China), Sarasini (*Gasteromiceti Epigei*: 406 pp., 2005).

Phelloriniaceae Ulbr. (1951), Agaricales. 2 gen. (+ 5 syn.), 2 spp.
Lit.: Wright *et al.* (*Cryptog. Mycol.* **14**: 77, 1993), Coetzee *et al.* (*Bothalia* **27**: 117, 1997), Moreno *et al.* (*Mycotaxon* **64**: 393, 1997), Calonge (*Fl. Mycol. Iberica* **3**: 271 pp., 1998), Martín *et al.* (*Cryptog. Mycol.* **21**: 3, 2000), Dios *et al.* (*Mycotaxon* **84**: 265, 2002).

Phellostroma Syd. & P. Syd. (1914), Pezizomycotina. 1, Philippines.

Phelonites Fresen. (1861), Fossil Fungi. 1 (Miocene), Germany.

Phelonitis Chevall. (1826) nom. dub., ? Dothideomycetes.

Phenacopodium Debey (1849) = Melanospora Corda fide Mussat (*Syll. fung.* **15**: 279, 1901).

Pherima Raf. (1819) ≡ Phorima.

pheromone, a substance secreted to the outside by an individual and received by a second individual of the same species, in which it induces a specific reaction, e.g. a definite behaviour or developmental process (Karlson & Luscher, *Nature* **183**: 55, 1959).

Phiala Raf. (1815) nom. dub., Fungi. No spp. included.

Phialastrum Sunhede (1989), Geastraceae. 1, Africa (tropical). See Sunhede (*Syn. Fung.* **1**: 66, 1989).

Phialea (Fr.) Gillet (1879) nom. dub., Pezizomycotina. See Dumont & Korf (*Taxon* **26**: 598, 1977).

Phialemonium W. Gams & McGinnis (1983), anamorphic *Cephalothecaceae*, Hso.0eH.15. 3, widespread. See King *et al.* (*J. Clin. Microbiol.* **31**: 1804, 1993; re-evaluation), Guarro *et al.* (*J. Clin. Microbiol.* **37**: 2493, 1999; clinical cases), Gavin *et al.* (*J. Clin. Microbiol.* **40**: 2207, 2002; endocarditis), Mostert *et al.* (*Stud. Mycol.* **54**: 115 pp., 2006; key), Weinberger *et al.* (*Medical Mycology* **44**: 253, 2006; endophthalmitis), Yaguchi *et al.* (*Mycotaxon* **96**: 309, 2006; phylogeny).

Phialetea Bat. & Nascim. (1960) = Grallomyces fide Deighton & Pirozynski (*Mycol. Pap.* **105**, 1966).

Phialicorona Subram. (1993) nom. inval. = Kionochaeta fide Sutton (*in litt.*).

phialide (after Vuillemin), a cell which develops one or more (the **polyphialide** of Hughes, *Mycol. Pap.* **45**, 1951) open ended conidiogenous loci from which a basipetal succession of conidia, **phialospores**, develops without an increase in length of the phialide itself (Hughes, *loc. cit.*); cf. annellophore; sterigma. In some fungi, e.g. *Acremonium*, the phialide may be the conidiophore; more frequently the phialide is either an end cell of a conidiophore or attached to a conidiophore (or **phialophore**). See Roquebert (*Rev. Myc.* **40**: 417, 1976; review), terminus phialospore, and Minter *et al.* (*TBMS* **81**: 109, 1983; history).

phialidic (of conidiogenesis, obsol.), the sort of conidiogenesis in which each conidium (**phialoconidium**, phialidic conidium, phialospore) originates by the laying down of new wall material not from existing walls or layers of the wall of the conidiogenous cell (**phialide**). A *basipetal* succession of conidia is formed from a *fixed* conidiogenous locus (cf. tretic). **mono-**, **poly-**, (of phialides), producing conidia through a single opening or a sympodial irregular or synchronous succession of openings, respectively, in the conidiogenous cell wall.

Phialina Höhn. (1926), Hyaloscyphaceae. 13, widespread (north temperate). See Huhtinen (*Karstenia* **29**: 545, 1990; key), Baral (*Z. Mykol.* **59**: 3, 1993; ? synonymy with *Calycellina*), Huhtinen & Scheuer (*Öst. Z. Pilzk.* **4**: 1, 1995), Raitviir & Schneller (*Sydowia* **55**: 306, 2003), Raitviir (*Scripta Mycol.* **20**, 2004).

Phialisphaera Dumort. (1822) nom. dub., Pezizomycotina.

Phialoarthrobotryum Matsush. (1975), anamorphic *Pezizomycotina*, Hsy.≡ eP.15. 1, Japan. See Matsushima (*Icon. microfung. Matsush. lect.*: 111, 1975), Okada & Tubaki (*Sydowia* **39**: 148, 1986).

Phialoascus Redhead & Malloch (1977), ? Endomycetaceae. 1, Canada. Has been linked to *Cephaloascus*, but similarities are probably superficial. See Malloch & Hoog *in* Kurtzman & Fell (Eds) (*Yeasts, a taxonomic study* 4th edn: 197, 1998), Suh *et al.* (*Mycol.* **98**: 1006, 2006; phylogeny).

Phialocephala W.B. Kendr. (1961), anamorphic *Vibrisseaceae*, Hso.0eP.15. 20, widespread. See Wingfield *et al.* (*TBMS* **89**: 509, 1987; classification), Currah & Tsuneda (*TMSJ* **34**: 345, 1993; sporulation *P. fortinii* in culture), Onofri *et al.* (*MR* **98**: 745, 1994; key), Kowalski & Kehr (*CJB* **73**: 26, 1995; n.spp.), Kirschner & Oberwinkler (*Sydowia* **50**: 205, 1998; n.sp.), Vujanovic *et al.* (*Mycol.* **92**: 571, 2000; on orchids), Jacobs *et al.* (*CJB* **79**: 110, 2001), Grünig *et al.* (*CJB* **80**: 1239, 2002; phylogeny), Vrålstad *et al.* (*New Phytol.* **155**: 131, 2002; phylogeny), Grünig *et al.* (*MR* **107**: 1332, 2003; population biology), Jacobs *et al.* (*Mycol.* **95**: 637, 2003; phylogeny), Grünig *et al.* (*Fungal Genetics Biol.* **41**: 676, 2004; cryptic species), Piercey *et al.* (*MR* **108**: 955, 2004; genetic diversity, Canada), Grünig *et al.* (*Fungal Genetics Biol.* **43**: 410, 2006; population genetics, Sweden), Wang *et al.* (*Mycol.* **98**: 1065, 2006; phylogeny), Grünig *et al.* (*Mycol.* **100**: 47, 2008; cryptic speciation).

Phialocladus Kreisel (1972) nom. inval. ≡ Escovopsis.

Phialoconidiophora M. Moore & F.P. Almeida (1937) = Fonsecaea fide de Hoog (*Stud. Mycol.* **15**, 1977), Hoog *et al.* (*Medical Mycology* **42**: 405, 2004).

Phialocorona Subram. (1995) = Kionochaeta fide Sutton (*in litt.*).

Phialocybe P. Karst. (1879) ? = Crepidotus fide Singer (*Agaric. mod. Tax.*, 1951).

Phialogangliospora Udaiyan & V.S. Hosag. (1992), anamorphic *Pezizomycotina*, Hso.0eH.40+15. 1 (from cooling tower), India. See Udaiyan & Hosagoudar (*J. Econ. Taxon. Bot.* **15**: 654, 1991).

Phialogeniculata Matsush. (1971), anamorphic *Chaetosphaeriaceae*, Hso.≡ eH.15. 1, Guadaloupe. See Kuthubutheen & Nawawi (*MR* **95**: 1220, 1991; synonymy with *Dictyochaeta*).

Phialographium H.P. Upadhyay & W.B. Kendr. (1974), anamorphic *Ophiostoma*. See Wingfield *et al.* (*MR* **95**: 1328, 1991), Harrington *et al.* (*Mycol.* **93**: 111, 2001; morphology, phylogeny), Zhou *et al.* (*Mycol.* **96**: 1306, 2004).

Phialomyces P.C. Misra & P.H.B. Talbot (1964), anamorphic *Pezizomycotina*, Hso.0eH/1eP.32/33. 2, widespread. See Misra & Talbot (*CJB* **42**: 1287,

1964), Castañeda Ruiz & Gams (*Mycotaxon* **42**: 239, 1991), Delgado Rodriguez & Decock (*Mycol.* **95**: 896, 2003).

Phialophaeoisaria Matsush. (1995), anamorphic *Pezizomycotina*, Hsy.0eH.16. 1, Japan. See Matsushima (*Matsush. Mycol. Mem.* **8**: 29, 1995).

Phialophora Medlar (1915), anamorphic *Herpotrichiellaceae*, Hso.0eH/1eP.15. 2, widespread. *P. verrucosa* (a cause of chromoblastomycosis (q.v.) in humans). The genus in its traditional circumscription is grossly polyphyletic; see segregates including especially *Cadophora*, *Harpophora*, *Lecythophora* and *Phaeoacremonium*. See Cain (*CJB* **30**: 338, 1952), Schol-Schwarz (*Persoonia* **6**: 63, 1970; key), Gams & Holubová-Jechová (*Stud. Mycol.* **13**, 1976), Yamamoto *et al.* (*Ann. phytopath. Soc. Japan* **56**: 584, 1990; isozyme polymorphism in *P. gregata*), Kobayashi *et al.* (*Ann. phytopath. Soc. Japan* **57**: 225, 1991; f.spp. in *P. gregata*), Yamamoto *et al.* (*TMSJ* **34**: 465, 1993; RFLPs in *P. gregata*), Yan *et al.* (*Mycol.* **87**: 72, 1995; rDNA supports morphol. spp. separation), de Hoog *et al.* (*Stud. Mycol.* **43**: 107, 1999; *P. verrucosa* complex), Untereiner & Naveau (*Mycol.* **91**: 67, 1999), Gams (*Stud. Mycol.* **45**: 187, 2000; review, polyphyly), Hoog *et al.* (*Mycoses* **43**: 409, 2000; in humans).

phialophore, see phialide.

Phialophorophoma Linder (1944), anamorphic *Pezizomycotina*, Cpd.0eH.15. 1 (marine), Europe; USA. See Hyde (*Beih. Sydowia* **43**: 31, 1991).

Phialophoropsis L.R. Batra (1968) ? = Ambrosiella fide Sutton & Brady (*TBMS* **72**: 337, 1979), Gebhardt *et al.* (*MR* **109**: 687, 2005).

Phialopsis Körb. (1855) = Gyalecta Ach. fide Hawksworth *et al.* (*Dictionary of the Fungi* edn 8, 1995).

Phialoscypha Raitv. (1977) = Phialina fide Huhtinen (*Karstenia* **29**: 545, 1990).

Phialoselanospora Udaiyan (1992), anamorphic *Pezizomycotina*, Hso.0eH.15. 1 (from water cooling tower), India. See Udajyan (*J. Econ. Taxon. Bot.* **15**: 627, 1991).

Phialospora Raf. (1832) ≡ Cucurbitaria.

phialospore, see phialide.

Phialosporostilbe Mercado & J. Mena (1985), anamorphic *Pezizomycotina*, Hsy.0eH.15. 1, pantropical. See Mercado & Mena (*Revta Jardín bot. Nac.* Univ. Habana **6**: 57, 1985), Shirouzu & Harada (*Mycoscience* **45**: 390, 2004).

Phialostele Deighton (1969), anamorphic *Pezizomycotina*, Hsy/Hsp.0eH.15. 1, Africa. See Deighton (*Mycol. Pap.* **117**: 11, 1969).

Phialotubus R.Y. Roy & Leelav. (1966), anamorphic *Pezizomycotina*, Hso.0eH.32/33. 1, India. See Roy & Leelavathy (*TBMS* **49**: 495, 1966).

Phibalis Wallr. (1833) nom. rej. = Encoelia fide Korf & Kohn (*Mem. N. Y. bot. Gdn* **28**: 109, 1976).

Philately. More than 1400 postage stamps, of over 100 countries, illustrating 575 spp. fungi issued up to 1996 are catalogued by McKenzie (*Collect fungi on stamps*, 1997). Macromycetes predominate but some medically important fungi, lichens and mycorrhizas are covered. See also Ing (*Bull. BMS* **10**: 32, 1976), Moss & Dunkley (*Bull. BMS* **15**: 61, 1981), Moss (*Mycologist* **6**: 68, 1992, **7**: 28, 1993), Coetzee (*Mycologist* **7**: 29, 1993).

Phillippiregis Cif. & Tomas. (1953) ≡ Polyblastidea.

Phillipsia Berk. (1881) nom. cons., Sarcoscyphaceae. Anamorph *Molliardiomyces*. *c.* 17, widespread (subtropical; tropical). See Denison (*Mycol.* **61**: 289, 1969; key C. Am. spp.), Paden (*CJB* **55**: 2685, 1977), Paden (*Mycotaxon* **25**: 165, 1986; anamorph), Romero & Gamundí (*Darwiniana* **27**: 43, 1986; key 4 spp. Argentina, SEM), Li & Kimbrough (*CJB* **74**: 10, 1996; ontogeny), Hansen *et al.* (*Mycol.* **91**: 299, 1999; phylogeny, SEM), Weinstein *et al.* (*Mycol.* **94**: 673, 2002; phylogeny), Prasad & Pant (*Journal of Mycopathological Research* **41**: 99, 2003), Zhuang (*Mycotaxon* **86**: 291, 2003; China), Perry *et al.* (*MR* **111**: 549, 2007; phylogeny).

Phillipsiella Cooke (1878), Phillipsiellaceae. 7, America. See Rossman *et al.* (*Sydowia* **46**: 66, 1994).

Phillipsiellaceae Höhn. (1909), ? Dothideomycetes (inc. sed.). 1 gen. (+ 2 syn.), 7 spp.

Lit.: Rossman *et al.* (*Sydowia* **46**: 66, 1994) suggest a possible relationship with the *Schizothyriaceae*.

Lit.:, Kohlmeyer *et al.* (*CJB* **76**: 470, 1998).

Philliscidiopsis, see *Phylliscidiopsis*.

Philobryon Döbbeler (1988), Dothideomycetes. 1 (on *Bryophyta*), Papua New Guinea. See Döbbeler (*Pl. Syst. Evol.* **158**: 335, 1988).

Philocopra Speg. (1880) = Podospora fide Lundqvist (*Symb. bot. upsal.* **20** no. 1, 1972).

Philonectria Hara (1914), Dothideomycetes. 3, Japan; Uganda. See Eriksson & Hawksworth (*SA* **6**: 146, 1987).

Philophora Wallr. (1833) = Rhizopus fide Hesseltine (*Mycol.* **47**: 344, 1955).

Phlebia Fr. (1821), Meruliaceae. *c.* 50, widespread. See Cooke (*Mycol.* **48**: 386, 1956), Corner (*Gdns' Bull.* Singapore **25**: 355, 1971; Malaysian spp.), Ginns (*CJB* **54**: 100, 1976; disposition spp.), Nakasone & Burdsall (*Mycotaxon* **21**: 241, 1984; synonymy with *Merulius*), Wu (*Acta Bot. Fenn.* **142**: 25, 1990; key Taiwan spp.), Spirin & Zmitrovich (*Novosti Sistematiki Nizshikh Nov. sist. Niz. Rast.* **37**: 166, 2004).

Phlebiaceae Jülich (1982) = Meruliaceae.

Phlebiella P. Karst. (1890), Polyporales. *c.* 20, widespread. See Stalpers (*in litt.*), Hjortstam *et al.* (*Cortic. N. Europ.* **8**, 1988; key Eur. spp.) = *Xenasmatella* fide.

Phlebiopsis Jülich (1978), Phanerochaetaceae. 11, widespread. See Burdsall (*Mycol. Mem.* **10**, 1985), Vainio & Hantula (*Mycol.* **92**: 436, 2000; genetics *Phlebiopsis gigantea*), Hjortstam *et al.* (*Syn. Fung.* **20**: 42, 2005; Venezuela) = *Phanerochaete* fide.

Phlebogaster Fogel (1980), Claustulaceae. 2, Canary Islands; Asia. See Fogel (*Contr. Univ. Mich. Herb.* **14**: 79, 1980).

Phlebomarasmius R. Heim (1967) ? = Xeromphalina fide Singer (*Agaric. mod. Tax.* edn 3, 1975).

Phlebomycena R. Heim (1966) = Mycena fide Singer (*Agaric. mod. Tax.*, 1951).

Phlebonema R. Heim (1929), ? Agaricaceae. 1, Madagascar. See Heim (*Compte rendu hebdomadaire des Sciences de l'Academie des sciences* Paris **188**: 1568, 1929).

Phlebophora Lév. (1841) nom. dub. = Tricholoma fide Boedijn (*Sydowia* **5**: 211, 1951).

Phlebophyllum R. Heim (1969), ? Agaricales. 1, Gabon. See Heim (*Revue Mycol.* Paris **33**: 38, 1968).

Phlebopus (R. Heim) Singer (1936), Boletinellaceae. 12 (saprotrophs, possibly ectomycorrhizal with exotic trees), widespread (pantropical). Saprotrophs, under exceptional conditions possibly ectomycorrhizal with exotic trees. See Groves (*Mycol.* **54**: 319, 1962), Groves (*Fl. Illustr. Champ. Afr. centr.* **7**: 128, 1980),

Groves (*Mycotaxon* **15**: 384, 1982), Thoen & Ducousso (*Agric., Ecosyst. Envir.* **28**: 519, 1989; possibly ectomycorrhizal), Bruns *et al.* (*Mol. Ecol.* **7**: 257, 1998; phylogeny), Deschamps & Moreno (*Mycotaxon* **72**: 205, 1999; Argentina).

Phleboscyphus Clem. (1903) = Helvella fide Eckblad (*Nytt Mag. Bot.* **15**: 1, 1968).

Phlebriella, see *Phlebiella*.

Phlegmacium (Fr.) Wünsche (1877) = Cortinarius fide Kauffman (*N. Amer. Fl.* **10**, 1932) but used by, Moser (*Die Gattung Phlegmacium*, 1960).

Phlegmatium Fr. (1819) nom. dub., Agaricomycetes. 'basidiomycetes', inc. sed. See Horníček (*Mykol. Sborn.* **4**: 121, 1984).

Phlegmophiale Zahlbr. (1926) = Arthonia fide Santesson (*Symb. bot. upsal.* **12** no. 1: 1, 1952).

Phlegographa, see *Flegographa*.

Phleogena Link (1833), Phleogenaceae. 1, widespread (north temperate). See Donk (*Persoonia* **4**: 160, 1966), McNabb *in* Ainsworth *et al.* (Eds) (*The Fungi* **4B**: 303, 1973), Holec (*Mykologické Listy* **84-85**: 33, 2003; Czech Republic), Aime *et al.* (*Mycol.* **98**: 896, 2006; phylogeny).

Phleogenaceae Gäum. (1926), Atractiellales. 6 gen. (+ 6 syn.), 30 spp.

Lit.: Oberwinkler & Bandoni (*CJB* **60**: 1726, 1982), Oberwinkler & Bandoni (*Mycol.* **74**: 634, 1982), Oberwinkler & Bauer (*Sydowia* **41**: 224, 1989), Cook (*Mycologist* **8**: 107, 1994), Hladun (*Clementeana* **4**: 48, 1999), Kirschner *et al.* (*Mycol.* **91**: 542, 1999), Bandoni & Inderbitzin (*Czech Mycol.* **53**: 265, 2002), Bauer *et al.* (*Mycol.* **95**: 756, 2003), Bauer *et al.* (*Mycol. Progr.* **5**: 41, 2006).

Phloeochora Höhn. (1917) = Phloeospora fide Sutton (*Mycol. Pap.* **141**, 1977).

Phloeoconis Fr. (1849), anamorphic *Pezizomycotina*, Hso.?.?. 3, widespread (temperate).

Phloeopannaria Zahlbr. (1941) = Psoroma fide Hawksworth *et al.* (*Dictionary of the Fungi* edn 8, 1995).

Phloeopeccania J. Steiner (1902), Lichinaceae (L). 1, Arabia; N. America. See Moreno & Egea (*Acta Bot. Barcinon.* **41**: 1, 1992), Schultz & Büdel (*Lichenologist* **34**: 39, 2002; key), Schultz & Mies (*Nova Hedwigia* **77**: 73, 2003; Socotra), Schultz & Büdel (*Bryologist* **108**: 520, 2005; Mexico).

Phloeopeccaniomyces Cif. & Tomas. (1953) ≡ Phloeopeccania.

Phloeoscoria Wallr. (1825) ≡ Polymorphum.

Phloeospora Wallr. (1833), anamorphic *Mycosphaerella*, Cac.≡ eH.19. *c.* 160, widespread. See Sutton (*The Coelomycetes*, 1980), Evans *et al.* (*MR* **97**: 59, 1993; biocontrol), Fatehi & Punithalingam (*Kew Bull.* **54**: 571, 1999; on *Lens*), Verkley & Priest (*Stud. Mycol.* **45**: 123, 2000; review), Verkley *et al.* (*Mycol.* **96**: 558, 2004; phylogeny), Feau *et al.* (*Mol. Phylogen. Evol.* **40**: 808, 2006; phylogeny).

Phloeosporella Höhn. (1924), anamorphic *Blumeriella*, Cac.≡ eH.10. 13, widespread. See Sutton (*The Coelomycetes*, 1980), Williamson & Bernard (*CJB* **66**: 2048, 1988; teleomorph), Sutton *et al.* (*MR* **100**: 979, 1996; N America), Yip (*Australas. Pl. Path.* **26**: 26, 1997; on *Eucalyptus*), Taylor & Crous (*MR* **104**: 618, 2000; on *Proteaceae*).

Phloeosporina Höhn. (1924), anamorphic *Pezizomycotina*, Cac.≡ eH.?. 2, Europe (eastern). See Sameva (*Mycol. Balcanica* **1**: 55, 2004; Bulgaria).

Phloepeccania Henssen (1990), Lichinaceae (L). 1, Australasia. See Henssen (*Lichenes Cyanophili et Fungi Saxicolae Exsiccati* Fascicle **2** (nos 1-25): [10], 1990).

Phlogicylindrium Crous, Summerb. & Summerell (2006), anamorphic *Amphisphaeriaceae*. 2, Australia. See Summerell *et al.* (*Fungal Diversity* **23**: 340, 2006).

Phlogiotis Quél. (1886) = Guepinia Fr. fide Donk (*Persoonia* **4**: 185, 1966).

Phlyctaena, see *Phlyctema*.

Phlyctaeniella Petr. (1922), anamorphic *Pezizomycotina*, St.0fH.15. 1 or 3, Europe; Australia.

Phlyctella Kremp. (1876) = Phlyctis fide Galloway & Guzmán Grimaldi (*Lichenologist* **20**: 393, 1988).

Phlyctellomyces Cif. & Tomas. (1953) ≡ Phlyctella.

Phlyctema Desm. (1847), anamorphic *Neofabraea*, St.0eH.15. *c.* 30, widespread. See Sutton (*The Coelomycetes*, 1980), Spiers & Hopcroft (*N.Z. Jl Bot.* **28**: 67, 1990; TEM *P. vagabunda*), Verkley (*Stud. Mycol.* **44**: 180 pp., 1999; review), Abeln *et al.* (*Mycol.* **92**: 685, 2000; phylogeny), Jong *et al.* (*MR* **105**: 658, 2001; phylogeny), Gariépy *et al.* (*MR* **107**: 528, 2003; phylogeny).

Phlyctibasidium Jülich (1974), Agaricomycetidae. 1, Europe. See Vězda (*Folia geobot. phytotax.* **21**: 208, 215, 1986).

Phlyctidaceae Poelt ex J.C. David & D. Hawksw. (1991), Ostropales (L). 2 gen. (+ 9 syn.), 13 spp.

Lit.: Brusse (*Bothalia* **17**: 182, 1987), Galloway & Guzmán Grimaldi (*Lichenologist* **20**: 393, 1988), Wedin *et al.* (*MR* **109**: 159, 2005), Miądlikowska *et al.* (*Mycol.* **98**: 1088, 2006; phylogeny).

Phlyctidia Müll. Arg. (1895) = Phlyctis fide Galloway & Guzmán Grimaldi (*Lichenologist* **20**: 393, 1988).

Phlyctidiaceae Sparrow (1942) = Chytridiaceae.

Phlyctidium (A. Braun) Rabenh. (1868) [non *Phlyctidium* Wallr. 1833] = Rhizophydium fide Longcore (*in litt.*).

Phlyctidium Müll. Arg. (1888) [non *Phlyctidium* Wallr. 1833] = Calenia fide Hawksworth *et al.* (*Dictionary of the Fungi* edn 8, 1995).

Phlyctidium Wallr. (1833) = Spilocaea fide Sutton (*Mycol. Pap.* **141**, 1977).

Phlyctidomyces E.A. Thomas ex Cif. & Tomas. (1953) ≡ Phlyctis.

Phlyctis (Wallr.) Flot. (1850) nom. cons., Phlyctidaceae (L). *c.* 12, widespread. See Erichsen (*Hedwigia* **70**: 216, 1930), Galloway & Guzmán Grimaldi (*Lichenologist* **20**: 393, 1988), Wedin *et al.* (*MR* **109**: 159, 2005; phylogeny).

Phlyctochytrium J. Schröt. (1892), Chytridiaceae. *c.* 30, widespread (temperate). See Sparrow (*Aquatic Phycomycetes* Edn 2: 324, 1960; key), Barr (*CJB* **62**: 1171, 1984), Letcher & Powell (*Nova Hedwigia* **80**: 135, 2005).

Phlyctomia A. Massal. (1860) = Phlyctis fide Hawksworth *et al.* (*Dictionary of the Fungi* edn 8, 1995).

Phlyctorhiza A.M. Hanson (1946), Chytridiaceae. 1, America.

Phlyctospora Corda (1841) = Scleroderma fide Stalpers (*in litt.*).

Phoebus R.C. Harris & Ladd (2007), Roccellaceae (L). 1, USA. See Harris & Ladd (*Opuscula Philolichenum* **4**: 57, 2007).

phoenicoid fungi, fungi growing amongst the ashes of former fires (Carpenter & Trappe, *Mycotaxon* **23**: 203, 1985); see Pyrophilous fungi.

Phoenicostroma Syd. (1925) = Coccostromopsis fide Hawksworth *et al.* (*Dictionary of the Fungi* edn 8,

1995), Hyde & Cannon (*Mycol. Pap.* **175**, 1999).

Pholidotopsis Earle (1909) = Galerina fide Singer (*Agaric. mod. Tax.*, 1951).

Pholiota (Fr.) P. Kumm. (1871) nom. cons., Strophariaceae. *c.* 150 (lignicolous causing heartwood rot), widespread (esp. temperate). See Smith & Hesler (*The North American species of Pholiota*, 1968; 205 taxa, keys), Jacobsson (*Mycotaxon* **36**: 95, 1989; culture), Klán *et al.* (*Mycotaxon* **36**: 249, 1989; culture), Jacobsson (*Windahlia* **19**: 1, 1990; key Eur. spp.), Holec (*Libri Botanici* **20**: 220 pp., 2001), Bon & Roux (*Docums Mycol.* **33**: 3, 2003; key), Chang *et al.* (*Australasian Mycologist* **24**: 53, 2006; Tasmania) See nameko.

Pholiotella Speg. (1889) nom. rej. = Conocybe fide Kuyper (*in litt.*).

Pholiotina Fayod (1889), Bolbitiaceae. *ca*, cosmopolitan. See Singer (*Agaric. mod. Tax.* edn 2, 1962; 16 spp.).

Phoma Fr. (1821) nom. rej. = Plagiostoma fide Hawksworth *et al.* (*Dictionary of the Fungi* edn 8, 1995).

Phoma Sacc. (1880) nom. cons., anamorphic *Didymella*, Cpd.0-1eH.15. *c.* 140, widespread. See Boerema (*Persoonia* **3**: 9, 1963; typification), Dorenbosch (*Persoonia* **6**: 1, 1970; soil spp.), Boerema & Dorenbosch (*Stud. Mycol.* **3**, 1973; fruit rotting spp.), Boerema & Bollen (*Persoonia* **8**: 111, 1975; sep. from *Ascochyta*), Boerema (*TBMS* **67**: 289, 1976; spp. studied by Dennis), Sutton (*The Coelomycetes*, 1980; key 27 spp.), Hawksworth (*Bull. Br. Mus. nat. hist.* Bot. **9**: 49, 1981; lichenicolous spp.), Morgan-Jones *et al.* (*Mycotaxon* **16**: 403, 1983; studies on *Phoma*), Rajak & Rai (*J. Econ. Taxon. Bot.* **7**: 588, 1986; key to spp. in culture), Monte & Garcia-Acha (*TBMS* **91**: 133, 1988; germination of *P. betae*), Monte & Garcia-Acha (*TBMS* **90**: 659, 1988; ultrastr. conidiogenesis *P. betae*), Pons (*Fitopat. Venez.* **3**: 34, 1990; spp. on *Saccharum*), Upadhyay *et al.* (*CJB* **68**: 2059, 1990; *P. cyperi* phytotoxin production), van der Aa *et al.* (*Stud. Mycol.* **32**: 1, 1990; sect. classn), Monte *et al.* (*Mycopathologia* **115**: 89, 1991; integrated systematics), de Gruyter & Nordeloos (*Persoonia* **15**: 71, 1992; sect. *Phoma*), Schäfer & Wöstemeyer (*J. Phytopath.* **136**: 124, 1992; aggressive and non-aggressive strains of *P. lingam* separated by PCR), Boerema (*Persoonia* **15**: 197, 1993; sect. *Peyronellaea*), de Gruyter *et al.* (*Persoonia* **15**: 369, 1993; sect. *Phoma*, taxa with small conidia), Nordeloos *et al.* (*MR* **97**: 1343, 1993; dendritic crystals and taxonomy), Boerema *et al.* (*Persoonia* **15**: 431, 1994; sect. *Plenodomus*), Boerema (*Mycotaxon* **64**: 321, 1997; infrageneric names), de Gruyter *et al.* (*Persoonia* **16**: 471, 1998), Voigt *et al.* (*J. Phytopath.* **146**: 567, 1998; population study), Boerema & Gruyter (*Persoonia* **17**: 273, 1999), Boerema *et al.* (*Persoonia* **17**: 281, 1999), Miric *et al.* (*Aust. J. agric. Res.* **50**: 325, 1999; on *Helianthus*), Aa *et al.* (*Persoonia* **17**: 435, 2000), Pedras & Biesenthal (*Can. J. Microbiol.* **46**: 685, 2000; chemistry), Voigt *et al.* (*Microbiol. Res.* **156**: 169, 2001; cryptic species), Abeln *et al.* (*MR* **106**: 419, 2002; population genetics), Gruyter (*Persoonia* **18**: 85, 2002), Gruyter & Boerema (*Persoonia* **17**: 541, 2002), Gruyter *et al.* (*Persoonia* **18**: 1, 2002), Hollingsworth *et al.* (*Mycotaxon* **81**: 331, 2002; cultural plasticity), Larsen *et al.* (*Pl. Dis.* **86**: 928, 2002; molecular detection), Boerema (*Persoonia* **18**: 153, 2003), Mendes-Pereira *et al.* (*MR* **107**: 1287, 2003; phylogeny), Arenal *et al.* (*Mycotaxon* **89**: 465, 2004; *Epicoccum* synanamorph), Barrins *et al.* (*Australas. Pl. Path.* **33**: 529, 2004; genetic diversity, Australia), Boerema *et al.* (*Phoma Identification Manual* Differentiation of Specific and Infra-Specific Taxa in Culture: 470 pp., 2004; monograph), Hawksworth & Cole (*Lichenologist* **36**: 7, 2004; lichenicolous spp.), Pethybridge *et al.* (*Australas. Pl. Path.* **33**: 173, 2004; on *Chrysanthemum*), Balmas *et al.* (*Eur. J. Pl. Path.* **111**: 235, 2005; *P. tracheiphila*), Torres *et al.* (*Mycotaxon* **93**: 333, 2005; separation from *Plenodomus*), Voigt *et al.* (*Mol. Phylogen. Evol.* **37**: 541, 2005; phylogeny), Zhou *et al.* (*Mycol.* **97**: 612, 2005; biocontrol), Schoch *et al.* (*Mycol.* **98**: 1041, 2006; phylogeny).

Phomachora Petr. & Syd. (1925), anamorphic *Pezizomycotina*, St.0eH.?. 2 or 3, America; Australia.

Phomachorella Petr. (1947), anamorphic *Pezizomycotina*, St.0eH.?. 1, S. Africa. See Swart (*TBMS* **48**: 463, 1965).

Phomatosphaeropsis Ribaldi (1953) = Sphaeropsis Sacc. fide Sutton (*The Coelomycetes*, 1980).

Phomatospora Sacc. (1875), Xylariales. Anamorph *Sporothrix*-like. 37 (on dead stems etc., sometimes aquatic), widespread. See Rappaz (*Mycotaxon* **45**: 323, 1992; anamorph), Barr (*Mycotaxon* **51**: 191, 1994), Fallah & Shearer (*Mycol.* **90**: 323, 1998), Jeewon *et al.* (*MR* **107**: 1392, 2003; phylogeny).

Phomatosporella Tak. Kobay. & K. Sasaki (1982), anamorphic *Phomatospora*, St.0eH.?. 1, Japan. See Kobayashi & Sasaki (*TMSJ* **23**: 254, 1982).

Phomatosporopsis Petr. (1925) = Phomatospora fide von Arx & Müller (*Beitr. Kryptfl. Schweiz* **11** no. 1, 1954).

phomin, a cytostatic antibiotic from *Phoma* sp. (S 298) (Rothweiler & Tamm, *Experientia* **22**: 750, 1966); cytochalasin B (q.v.).

Phomites Fritel (1910), Fossil Fungi. 2 (Cretaceous, Paleocene), France; India.

Phomopsella Höhn. (1920) = Phomopsis (Sacc.) Bubák. fide Petrak (*Annls mycol.* **23**: 1, 1925).

phomopsin, see lupinosis.

Phomopsina Petr. (1922) = Phoma Sacc. fide Clements & Shear (*Gen. Fung.*, 1931).

Phomopsioides M.E.A. Costa & Sousa da Câmara (1954) = Phomopsis (Sacc.) Bubák. fide Sutton (*Mycol. Pap.* **141**, 1977).

Phomopsis (Sacc.) Bubák (1905) nom. cons., anamorphic *Diaporthe*, St.0eH.15. *c.* 234, widespread. See Uecker (*Mycol. Mem.* **13**, 1988; list sp. names), Tomaz *et al.* (*Publ. Lab. Patol. Veg. 'Ver. de Alm.' Lisboa*: 54, 1989; *P. mali* on almonds), Brayford (*MR* **94**: 691 and 745, 1990; variation and vegetative incompatibility in spp. from *Ulmus*), Wechtl (*Linzer biol. Beitr.* **22**: 161, 1990; spp. on *Compositae* and *Umbelliferae*), Shivas *et al.* (*MR* **95**: 320, 1991; variation in *P. leptostromiformis* from lupin), Uecker & Johnson (*Mycol.* **83**: 192, 1991; spp. on *Asparagus*), Rehner & Uecker (*CJB* **72**: 1666, 1994; DNA and species concepts), Muntañola-Cvetković *et al.* (*J. Phytopath.* **144**: 285, 1996; growth patterns and incompatibility), Zhou *et al.* (*Mycosystema* **17**: 199, 1998; ontogeny), Farr *et al.* (*Mycol.* **91**: 1008, 1999; on *Prunus*), Kanematsu *et al.* (*Ann. phytopath. Soc. Japan* **65**: 264, 1999; Japan), Phillips (*Mycol.* **91**: 1001, 1999; on *Vitis*), Sutton *et al.* (*J. Clin. Microbiol.* **37**: 807, 1999; from human), Kajitani & Kanematsu (*Mycoscience* **41**: 111, 2000; on *Vitis*, Japan), Kanematsu *et al.* (*J. Gen. Pl. Path.* **66**: 191, 2000;

phylogeny, on fruit trees), Scheper *et al.* (*MR* **104**: 226, 2000; teleomorph, on *Vitis*), Mostert *et al.* (*Sydowia* **53**: 227, 2001; on *Protea*), Mostert *et al.* (*Mycol.* **93**: 146, 2001; on *Vitis*, S Africa), Roy *et al.* (*Mycopathologia* **150**: 15, 2001; on *Glycine*), Farr *et al.* (*Mycol.* **94**: 494, 2002; on *Vaccinium*, USA), Says-Lesage *et al.* (*Phytopathology* **92**: 308, 2002; on *Helianthus*, France), Castlebury *et al.* (*Mycoscience* **44**: 203, 2003; phylogeny, teleomorph), Phillips (*Sydowia* **55**: 274, 2003; on *Foeniculum*), Rekab *et al.* (*MR* **108**: 393, 2004; on *Helianthus*), Adams *et al.* (*Stud. Mycol.* **52**: 146 pp., 2005), Schilder *et al.* (*Pl. Dis.* **89**: 755, 2005; on *Vitis*, N America), Murali *et al.* (*Can. J. Microbiol.* **52**: 673, 2006; endophytes).

Phomopsis Sacc. & Roum. (1884) nom. rej. prop., Pezizomycotina.

Phomyces Clem. (1931), anamorphic *Pezizomycotina*, Cpd.0eH.?. 1 (on *Meliola*), Brazil.

Phorcys Niessl (1876) = Amphisphaeria fide Hawksworth *et al.* (*Dictionary of the Fungi* edn 8, 1995).

Phorima Raf. (1830) nom. dub. = Hexagonia Fr. fide Cooke (*Spec. Publ. Div. Myc. Dis. Surv. US Department of Agriculture* **3**, 1953).

phorophyte, the 'host' tree of an epiphyte.

photo- (prefix), pertaining to light.

photobiont, a photosynthetic symbiont in a lichen which may be a eukaryotic alga (phycobiont), see Gärtner (*in* Reisser (Ed.), *Algae and symbiosis*: 325, 1992; review systematics), or a cyanobacterium (bactobiont, cyanobiont) (Ahmadjian, *Internat. Lich. Newsl.* **15** (2): 19, 1982), Büdel (*in* Reisser (Ed.), *Algae and symbiosis*: 301, 1992; review systematics).

photomorph, an organism whose form is determined by the nature of its photosynthesis (Laundon, *Taxon* **44**: 387, 1995); see also Lichens, phycotype, phototype, photosymbiodeme, phycosymbiodeme.

photophilous, having a preference for well-illuminated habitats; cf. heliophilous, anheliophilous.

photophobous, having a preference for shaded habitats.

photosporogenic, requiring light for sporogenesis.

photosymbiodeme, a replacement term for phycosymbiodeme to allow for one biont being a cyanobacterium (Stocker-Wörgötter & Türk, *Crypt. Bot.* **4**: 300, 1994); a photomorph. See Lichens.

phototaxis, movement (e.g. of zoospores) influenced by light.

phototropism, see tropism.

Phragmaspidium Bat. (1960), ? Microthyriaceae. 3, widespread. See Batista (*Publções Inst. Micol. Recife* **260**: 109, 1960).

Phragmeriella Hansf. (1946), Pseudoperisporiaceae. 1 (on *Meliolaceae*), Africa.

Phragmidiaceae Corda (1837), Pucciniales. 14 gen. (+ 13 syn.), 164 spp.

Lit.: Eriksson (*Mycotaxon* **15**: 249, 1982), Wahyuno *et al.* (*Mycoscience* **42**: 519, 2001), Ono (*Mycoscience* **43**: 37, 2002), Wahyuno *et al.* (*Mycoscience* **43**: 159, 2002), Cummins & Hiratsuka (*Illustr. Gen. Rust Fungi* edn 3: 225 pp., 2003), Maier *et al.* (*CJB* **81**: 12, 2003), Wingfield *et al.* (*Australas. Pl. Path.* **33**: 327, 2004), Ritz *et al.* (*MR* **109**: 603, 2005), Aime (*Mycoscience* **47**: 112, 2006), Gomez *et al.* (*MR* **110**: 423, 2006).

Phragmidiella Henn. (1905), Phakopsoraceae. *c.* 8 (on *Bignoniaceae*, *Meliaceae*), Africa; India; S. America. See Ramachar & Rao (*Mycol.* **73**: 778, 1981).

Phragmidiites Babajan & Tasl. (1970), Fossil Fungi. 1 (Tertiary), former USSR.

Phragmidiopsis (G. Winter) Mussat (1901) = Phragmidium fide Berndt (*in litt.*).

Phragmidium Link (1816), Phragmidiaceae. Anamorph *Physonema*. *c.* 110 (on *Rosa*, *Rubus*, *Potentilla* (*Rosaceae*) autoecious), widespread (esp. temperate). *P. violaceum* (blackberry (*Rubus*) rust), *P. rubi-idaei* (raspberry rust) and *P. mucronatum* (rose (*Rubus*) rust). See Dietel (*Hedwigia* **44**: 112, 1905), Dietel (*Hedwigia* **44**: 330, 1905), Cummins (*Mycol.* **23**: 433, 1931), Hiratsuka *et al.* (*Rep. Tottori Mycol. Inst.* **18**: 53, 1980), Wei (*Mycosystema* **1**: 179, 1988), Wahyuno *et al.* (*Mycoscience* **42**: 519, 2001; morph. urediniosp. & telios.), Helfer (*Nova Hedwigia* **81**: 325, 2005; Europ. spp.), Ritz *et al.* (*MR* **109**: 603, 2005; evolut.).

Phragmitensis K.M. Wong, Poon & K.D. Hyde (1998), Hyponectriaceae. 1 (on dead stems of *Phragmites*), Hong Kong. See Wong *et al.* (*Bot. Mar.* **41**: 379, 1998), Wong *et al.* (*Fungal Diversity* **2**: 175, 1999).

Phragmiticola Sherwood (1987), ? Helotiales. 1, Europe. See Magnes (*Biblthca Mycol.* **165**, 1997).

Phragmobasidiomycetes R.T. Moore (1980), see *Tremellomycetes*; now placed within the *Agaricomycotina*.

Phragmobasidiomycetidae Gäum. (1949), see *Tremellomycetes*; now placed within the *Agaricomycotina*.

phragmobasidium, see basidium.

Phragmocalosphaeria Petr. (1923), Calosphaeriaceae. 1, Czech Republic. See Wehmeyer (*Revision of Melanconis*, 1941).

Phragmocapnias Theiss. & Syd. (1918), Capnodiaceae. Anamorph *Conidiocarpus*. *c.* 2, widespread (tropical). See Reynolds (*Mycotaxon* **8**: 917, 1979), Reynolds (*Mycotaxon* **27**: 377, 1986), Reynolds & Gilbert (*Aust. Syst. Bot.* **18**: 265, 2005; Australia), Reynolds & Gilbert (*Cryptog. Mycol.* **27**: 249, 2006; Panamá).

Phragmocarpella Theiss. & Syd. (1915) = Phyllachora Nitschke ex Fuckel (1870) fide Petrak (*Annls mycol.* **29**: 349, 1931).

Phragmocauma Theiss. & Syd. (1915) = Phyllachora Nitschke ex Fuckel (1870) fide Cannon (*SA* **7**: 111, 1988).

Phragmocephala E.W. Mason & S. Hughes (1951), anamorphic *Pezizomycotina*, Hso/Hsy.≡ eP.1. 9, widespread. See Hughes (*N.Z. Jl Bot.* **17**: 163, 1979), Holubová-Jechová (*Folia geobot. phytotax.* **21**: 173, 1986; Czechoslovakia).

Phragmodiaporthe Wehm. (1941), Melanconidaceae. 2 or 4, N. America. See Barr (*Mycol. Mem.* **7**, 1978).

Phragmodimerium Petr. & Cif. (1932) = Philonectria fide von Arx & Müller (*Stud. Mycol.* **9**, 1975).

Phragmodiscus Hansf. (1947), Lasiosphaeriaceae. 1 or 2 (on bamboo), Africa (tropical). See Eriksson & Yue (*SA* **8**: 17, 1989), Huhndorf *et al.* (*Mycol.* **96**: 368, 2004).

Phragmodochium Höhn. (1924), anamorphic *Pezizomycotina*, Hsp.≡ eH.?. 1, Java.

Phragmodothella Theiss. & Syd. (1915) = Discostroma fide Hawksworth *et al.* (*Dictionary of the Fungi* edn 8, 1995).

Phragmodothidea Dearn. & Barthol. (1926) = Scirrhia fide Hawksworth *et al.* (*Dictionary of the Fungi* edn 8, 1995).

Phragmodothis Theiss. & Syd. (1914) = Dothidea fide Barr (*Mycotaxon* **43**: 371, 1992).

Phragmogibbera Samuels & Rogerson (1990), Venturiaceae. Anamorph *Stigmina*. 1, Venezuela. See Samuels & Rogerson (*Mem. N. Y. bot. Gdn* **64**: 165, 1990).

Phragmogloeum Petr. (1954), anamorphic *Pezizomycotina*, Cpd.≡ eP.?. 1, Australia. See Petrak (*Sydowia* **8**: 158, 1954).

Phragmographium E.F. Morris (1966) ≡ Morrisographium.

Phragmographum Henn. (1905) = Opegrapha Ach. fide Hawksworth *et al.* (*Dictionary of the Fungi* edn 8, 1995).

Phragmonaevia Rehm (1896) nom. dub., Helotiales. See Sherwood (*Mycotaxon* **5**: 1, 1977), Hawksworth & Santesson (*Biblthca Lichenol.* **38**: 121, 1990; lichenicolous taxa excluded).

Phragmopeltheca L. Xavier (1976) = Porina Müll. Arg. fide Lücking *et al.* (*Lichenologist* **30**: 121, 1998).

Phragmopelthecaceae L. Xavier (1976) = Porinaceae.

Phragmopeltis Henn. (1904), anamorphic *Pezizomycotina*, Cpt.1eP.?. 5, America (tropical).

Phragmoporthe Petr. (1934), Gnomoniaceae. 1, Europe; N. America. See Barr (*Mycol. Mem.* **7**, 1978), Monod (*Beih. Sydowia* **9**: 1, 1983).

Phragmopsora Magnus (1875) = Pucciniastrum fide Sydow & Sydow (*Monographia Uredinearum seu Specierum Omnium ad hunc usque Diem Descriptio et Adumbratio Systematica* **3**, 1915).

Phragmopyxine Clem. (1909) = Pyxine fide Hawksworth *et al.* (*Dictionary of the Fungi* edn 8, 1995).

Phragmopyxis Dietel (1897), Uropyxidaceae. 4 (on *Leguminosae*, esp. *Faboideae*), America (tropical); Africa.

Phragmoscutella Woron. & Abramov (1926), Dothideomycetes. 1, former USSR.

Phragmospathula Subram. & N.G. Nair (1966), anamorphic *Pezizomycotina*, Hsp.≡ eP.19. 2, India; Cuba. See Subramanian & Nair (*Antonie van Leeuwenhoek Ned. Tijdschr. Hyg.* **32**: 384, 1966), Mercado Sierra *et al.* (*Mycol.* **89**: 304, 1997; revision).

Phragmospathulella J. Mena & Mercado (1986), anamorphic *Pezizomycotina*, Hso.≡ eP.28. 1, Cuba. See Mena & Mercado (*Revta Jardín bot. Nac.* Univ. Habana **7**: 32, 1986).

Phragmosperma Theiss. & Syd. (1917), Dothideomycetes. See von Arx & Müller (*Stud. Mycol.* **9**, 1975), Aptroot (*Nova Hedwigia* **66**: 89, 1998).

phragmospore, differs from an amerospore (q.v.) and didymospore (q.v.) in having 2 to many transverse septa. See Anamorphic fungi.

Phragmosporonema Moesz (1924) ≡ Diplosporonema.

Phragmostachys Costantin (1888) = Sterigmatobotrys fide Bisby (*TBMS* **26**: 138, 1943).

Phragmostele Clem. (1909) ≡ Pucciniostele.

Phragmostilbe Subram. (1959) = Arthrosporium fide Wang (*Mycol.* **64**: 1175, 1972).

Phragmotaenium R. Bauer, Begerow, A. Nagler & Oberw. (2001), Tilletiariaceae. 1 (on *Poaceae*), India. See Bauer *et al.* (*MR* **105**: 423, 2001).

Phragmotelium Syd. (1921), Pucciniales. *c.* 10 (on *Rubus* (*Rosaceae*)), Asia; Australia. See Thirumalachar (*Proc. Indian natn Sci. Acad.* Part B. Biol. Sci. **15**: 186, 1942), Cummins & Hiratsuka (*Illustr. Gen. Rust Fungi rev. edit.*, 1983; syn. of *Phragmidium*).

Phragmothele Clem. (1909) = Thelidium fide Hawksworth *et al.* (*Dictionary of the Fungi* edn 8, 1995).

Phragmothyriella Höhn. (1912) ≡ Myriangiella.

Phragmothyriella Speg. (1919) nom. dub., Fungi. See Petrak (*Sydowia* **5**: 169, 1951).

Phragmothyrites W.N. Edwards (1922), Fossil Fungi. 13 (Tertiary), widespread. See Selkirk (*Proc. Linn. Soc. N. S. W.* **100**: 70, 1975).

Phragmothyrium Höhn. (1912) = Microthyrium fide von Arx & Müller (*Stud. Mycol.* **9**, 1975).

Phragmotrichum Kunze (1823), anamorphic *Pezizomycotina*, Cac/St.≡ -#eP.23. 4, Europe. See Sutton & Pirozynski (*TBMS* **48**: 349, 1965), Shabunin (*Mikol. Fitopatol.* **31**: 52, 1997; Russia).

Phragmoxenidiaceae Oberw. & R. Bauer (1990), Tremellales. 2 gen., 7 spp.

Lit.: Oberwinkler *et al.* (*Syst. Appl. Microbiol.* **13**: 186, 1990), Roberts (*Windahlia* **22**: 15, 1995).

Phragmoxenidium Oberw. (1990), Phragmoxenidiaceae. 1, Europe. See Oberwinkler (*Syst. Appl. Microbiol.* **13**: 187, 1990).

Phragmoxyphium Bat. & Cif. (1963) = Ciferrioxyphium fide Hughes (*Mycol.* **68**: 693, 1976).

Phrototecha, see *Prototheca*.

Phruensis Pinruan (2004), Diaporthales. 1, Thailand. See Pinruan *et al.* (*Mycol.* **96**: 1165, 2004).

Phthora d'Hérelle (1909), Pezizomycotina. 1 (on *Coffea*), Guatemala.

Phurmomyces Thaxt. (1931), Ceratomycetaceae. 1, Asia. See Nannfeldt (*Svensk bot. Tidskr.* **43**: 468, 1949).

phyco- (prefix), pertaining to algae.

Phycoascus Möller (1901) ? = Pyronema fide Eckblad (*Nytt Mag. Bot.* **15**: 1, 1968).

phycobiont, the algal partner in a lichen (Scott, *Nature* **179**: 486, 1957), photobiont (q.v.); cf. mycobiont.

Phycodiscis Clem. (1909) = Knightiella fide Hawksworth *et al.* (*Dictionary of the Fungi* edn 8, 1995).

phycolichenes (obsol.), lichens in which the vegetative thallus morphology is determined by the photobiont and which are of uncertain systematic position as the sporocarps are unknown (e.g. *Cystocoleus*, *Racodium*).

Phycomater Fr. (1825) nom. dub., ? Fungi.

Phycomelaina Kohlm. (1968), ? Phyllachorales. 1 (on *Laminaria*), Europe; N. America. Affinities are uncertain. See Schatz (*Mycol.* **75**: 762, 1983).

Phycomyces Kunze (1823), Phycomycetaceae. 3, widespread. See Benjamin & Hesseltine (*Mycol.* **51**: 751, 1959; key), Carlile (*J. gen. Microbiol.* **28**: 161, 1962; sporangiophore phototropism), Cerdá-Olmedo & Lipson (*Phycomyces*, 1987; review, lit.), Ootaki *et al.* (*Mycoscience* **37**: 427, 1996; mating response), Yamazaki & Ootaki (*Mycoscience* **37**: 269, 1996; zygospore formation), Yamazaki & Ootaki (*MR* **100**: 984, 1996; progametangia), James *et al.* (*Nature* **443**: 818, 2006; phylogeny), Wöstemeyer & Schimek *in* Heitman *et al.* (Eds.) (*Sex in Fungi*: 431, 2007; trisporic acid & mating), Dávila López *et al.* (*Nucleic Acids Research* **36**: 3001, 2008; spliceosomal RNA gene phylogeny).

Phycomycetaceae Arx (1982), Mucorales. 2 gen., 6 spp.

Lit.: Benjamin & Hesseltine (*Mycol.* **51**: 751, 1959), Yamazaki & Ootaki (*Mycoscience* **37**: 269, 1996).

Phycomycetes J. Schröt. (1892) (obsol.). Class formerly used for fungi now treated in *Chromista* (q.v.) and some *Fungi* (*Chytridiomycota* and *Zygomycota*);

best used only as a trivial term, 'phycomycetes'; an approx. syn. of 'lower fungi'.

Phycomycites D.E. Ellis (1915), Fossil Fungi. 3 (Jurassic), Germany; British Isles.

phycomycosis, a general term for a disease of humans or animals caused by a phycomycete. Cf. mucormycosis, zygomycosis.

Phycomycotera. Superdivision used by Moore (1971) for all fungi with non-septate hyphae (i.e. *Oomycota* and *Zygomycota* in the *Dictionary*, this edition); cf. *Septomycotera*.

phycophilous, growing with or on algae.

Phycopsis L. Mangin & Pat. (1912) [non *Phycopsis* (Fisch.-Oost.) Rothpletz 1896, *Algae*] = Seuratia fide Meeker (*CJB* **53**: 2462, 1975).

Phycorella Döbbeler (1980), Dothideomycetes. 1 (biotrophic parasite of *Scytonema*), Australia. See Döbbeler (*Sydowia* **33**: 33, 1980).

Phycosiphon A. Massal. (1859) ≡ Brachycarphium.

phycosymbiodeme, joined lichen thalli with a single mycobiont but different photobionts (Renner & Galloway, *Mycotaxon* **16**: 197, 1982); photomorph, phycotype.

phycosymbiont, phycobiont.

phycotrophic (of fungi), obtaining nutrients from algae (Dobbs, *Lichenologist* **4**: 323, 1970).

phycotype, see type.

Phylacia Lév. (1845), Xylariaceae. Anamorph *Geniculosporium*. 7, America (tropical). See Dennis (*Kew Bull.*: 320, 1957), Pérez-Silva (*Boln. Soc. mex. Micol.* **6**: 9, 1972), Silveira & Rogers (*Acta Amazon.* Supl. **15**: 7, 1985; key 4 spp. Brazil), Rodrigues & Samuels (*Mem. N. Y. bot. Gdn* **49**: 290, 1989), Hladki (*Lilloa* **41**: 9, 2004; Argentina), Medel *et al.* (*Mycotaxon* **97**: 279, 2006; Mexico), Stadler & Fournier (*Revta Iberoamer. Micol.* **23**: 160, 2006; chemistry), Bitzer *et al.* (*MR* **112**: 251, 2008; phylogeny, chemistry).

Phylaciaceae Speer (1980) = Xylariaceae.

Phylacteria (Pers.) Pat. (1887) ≡ Scyphopilus.

Phylacteriaceae Imazeki (1953) = Thelephoraceae.

Phyllachora Nitschke ex Fuckel (1867) nom. rej. = Scirrhia fide Holm (*Taxon* **24**: 475, 1973).

Phyllachora Nitschke ex Fuckel (1870) nom. cons., Phyllachoraceae. Anamorph *Linochora*. *c*. 944 (from living leaves), widespread. See Parbery & Langdon (*Aust. J. Bot.* **12**: 265, 1964; species concepts), Parbery (*Aust. J. Bot.* **15**: 271, 1967; key), Parbery (*Aust. J. Bot.* **19**: 207, 1971; spp. on *Gramineae*), Kamat *et al.* (*Univ. Agric. Sci. Hebbal Monogr.* **4**, 1978; key 88 Indian spp.), Parbery *in* Subramanian (Ed.) (*Taxonomy of fungi* **1**: 263, 1978; fungi on), Cannon (*Mycol. Pap.* **163**, 1991; spp. on *Leguminosae*), Cannon (*MR* **100**: 1409, 1996; spp. on *Rosaceae*), Cannon & Evans (*MR* **103**: 577, 1999; spp. on *Erythroxylaceae*), Hyde & Cannon (*Mycol. Pap.* **175**: 114, 1999; spp. on palms), Pearce *et al.* (*Fungal Diversity* **3**: 123, 1999; spp. on *Asclepiadaceae*), Winka & Eriksson (*Mycoscience* **41**: 97, 2000; DNA), Pearce *et al.* (*Aust. Syst. Bot.* **14**: 283, 2001; on *Proteaceae*), Bentes *et al.* (*Fungal Diversity* **12**: 1, 2003; on *Xanthium*), Johnston & Cannon (*N.Z. Jl Bot.* **42**: 921, 2004; New Zealand), Pearce & Hyde (*Fungal Diversity Res. Ser.* **17**: 308 pp., 2006; Australia).

Phyllachoraceae Theiss. & Syd. (1915), Phyllachorales. 51 gen. (+ 54 syn.), 1173 spp.

Lit.: Cannon (*Mycol. Pap.* **163**: 302 pp., 1991), Hanlin *et al.* (*Mycotaxon* **44**: 103, 1992), Cannon *in* McKey & Sprent (Eds) (*The Nitrogen Factor. Advances in Legume Systematics* **5**: 179, 1994; coevolution), Cannon (*MR* **100**: 1409, 1996; spp. on *Rosaceae*), Cannon (*Biodiversity of Tropical Microfungi*: 255, 1997; diversity), Malloch & Mallik (*CJB* **76**: 1265, 1998), Silva-Hanlin & Hanlin (*Mycoscience* **39**: 97, 1998), Silva-Hanlin & Hanlin (*Mycoscience* **39**: 97, 1998; review), Cannon & Evans (*MR* **103**: 577, 1999), Hyde & Cannon (*Mycol. Pap.* **175**: 114, 1999; spp. on palms), Hyde & Cannon (*Mycol. Pap.* **175**: 114 pp., 1999), Pearce *et al.* (*Fungal Diversity* **3**: 123, 1999), Pearce *et al.* (*Aust. Syst. Bot.* **14**: 283, 2001; spp. on *Proteaceae*), Pearce & Hyde (*Fungal Diversity Res. Ser.* **17**: 308 pp., 2006).

Phyllachorales M.E. Barr (1983). Sordariomycetes. 2 fam., 63 gen., 1226 spp. Stromata absent to well-developed, immersed in plant tissue, often clypeate, usually black. Ascomata perithecial, usually thin-walled, the ostioles periphysate. Peridium usually composed of thin-walled compressed hyaline or brown tissue, sometimes irregular in form. Interascal tissue of simple rather wide thin-walled paraphyses, sometimes deliquescing. Asci ± cylindrical, thin-walled, not fissitunicate, persistent, usually with an inconspicuous J- apical ring. Ascospores usually hyaline, aseptate, occ. ornamented. Anamorphs usually coelomycetous, spermatial or disseminative. Appressoria usually formed as adhesion/penetration structures. Mostly biotrophic, some necrotrophic and a few saprobic, on leaves, stems, and roots, widespr. esp. trop. Fams:

(1) **Phaeochoraceae**,

(2) **Phyllachoraceae**

Lit.: Cannon (*SA* **7**: 23, 1987; posn), Eriksson & Hawksworth (*SA* **11**: 181, 1993), Winka & Eriksson (*Mycoscience* **41**: 97, 2000; DNA).

Phyllachorella Syd. (1914), ? Dothideales. 1, India. See von Arx & Müller (*Beitr. Kryptfl. Schweiz* **11** no. 1, 1954; syn. with *Vestergrenia*), Kar & Maity (*Mycol.* **63**: 1024, 1971).

Phyllactinia Lév. (1851), Erysiphaceae. Anamorph *Ovulariopsis*. 40, widespread. *P. guttata* (mildew of hazel (*Corylus*), birch (*Betula*), and other trees). See Braun (*Beih. Nova Hedwigia* **89**, 1987; key), Hirata & Takamatsu (*Mycoscience* **37**: 283, 1996; phylogeny), Takamatsu *et al.* (*Mycoscience* **39**: 441, 1998; phylogeny), Saenz & Taylor (*CJB* **77**: 150, 1999; phylogeny), Mori *et al.* (*Mycol.* **92**: 74, 2000; phylogeny), Shin & Lee (*Mycotaxon* **83**: 301, 2002; morphology), Takamatsu (*Mycoscience* **45**: 147, 2004; phylogeny), Wang *et al.* (*Mycol.* **98**: 1065, 2006; phylogeny), Takamatsu *et al.* (*MR* **112**: 299, 2008; phylogeny).

Phyllerites Mesch. (1892), Fossil Fungi ? Fungi. 16 (Tertiary), Europe.

Phyllerium Fr. (1832) nom. dub., Fungi. Based on leaf outgrowths fide Fries (*Syst. mycol.* **3**: 523, 1832).

Phylleutypa Petr. (1934), Phyllachoraceae. 2 (from living stems), Africa; north temperate. See Malloch & Malik (*CJB* **76**: 1265, 1998).

phyllidium, lichen propagule formed by abstriction of a leaf-like or scale-like portion of the thallus. See Poelt (*Flora, Jena* **169**: 23, 1980) (Fig. 22F).

Phyllis F. Wilson (1889) [non *Phyllis* L. 1753, *Rubiaceae*] ≡ Neophyllis.

Phylliscaceae Th. Fr. (1860) = Lichinaceae.

Phylliscidiopsis Sambo (1937), Lichinaceae (L). 1, Ethiopia. See Moreno & Egea (*Acta Bot. Barcinon.* **41**: 1, 1992).

Phylliscidium Forssell (1885), Lichinaceae (L). 1, Brazil. See Moreno & Egea (*Acta Bot. Barcinon.* **41**: 1, 1992), Schultz & Büdel (*Lichenologist* **34**: 39, 2002; key).

Phylliscíella Henssen (1984), Lichinaceae (L). 3, southern hemisphere. See Moreno & Egea (*Acta Bot. Barcinon.* **41**: 1, 1992), Schultz & Büdel (*Lichenologist* **34**: 39, 2002; key).

Phylliscum Nyl. (1855), Lichinaceae (L). 3, widespread. See Schultz & Büdel (*Lichenologist* **34**: 39, 2002; key), Wedin *et al.* (*MR* **109**: 159, 2005; phylogeny).

Phyllobaeis Kalb & Gierl (1993), Baeomycetaceae (L). 5, widespread (tropical). See Gierl & Kalb (*Herzogia* **9**: 593, 1993; concept), Stenroos & DePriest (*Am. J. Bot.* **85**: 1548, 1998), Peršoh *et al.* (*Mycol. Progr.* **3**: 103, 2004; asci), Miądlikowska *et al.* (*Mycol.* **98**: 1088, 2006; phylogeny).

Phyllobatheliaceae Bitter & F. Schill. (1927), Pezizomycotina (inc. sed.) (L). 2 gen. (+ 2 syn.), 9 spp.
Lit.: Harris (*More Florida Lichens*, 1995)
Lit.:, Aptroot *et al.* (*Biblthca Lichenol.* **64**, 1997), Aptroot *et al.* (*Biblthca Lichenol.* **64**: 220 pp., 1997), Lücking (*Trop. Bryol.* **13**: 87, 1997), Lücking (*Trop. Bryol.* **15**: 45, 1998).

Phyllobathelium (Müll. Arg.) Müll. Arg. (1890), Phyllobatheliaceae (L). 8, C. & S. America. See Santesson (*Symb. bot. upsal.* **12** no. 1: 1, 1952).

Phylloblastia Vain. (1921), ? Strigulaceae (L). 3, pantropical. See Santesson (*Symb. bot. upsal.* **12** no. 1: 1, 1952), Aptroot *in* Galloway (Ed.) (*Systematics, conservation and ecology of tropical lichens*: 253, 1991), McCarthy (*Flora of Australia* **58** A: 242 pp., 2001), Herrera-Campos *et al.* (*Lichenologist* **36**: 309, 2004; Mexico), Sérusiaux *et al.* (*Lichenologist* **39**: 103, 2007; Europe, Madeira).

Phylloboletellus Singer (1952), Boletaceae. 1, C and S America. See Kuyper (*in litt.*), Singer & Digilio (*Lilloa* **25**: 438, 1952) ? = Boletellus (Bolet.) fide.

Phyllobolites Singer (1942), Boletaceae. 1, S. America (tropical). See Singer (*Annls mycol.* **40**: 59, 1942).

Phyllobrassia Vain. (1921) = Chroodiscus fide Hawksworth *et al.* (*Dictionary of the Fungi* edn 8, 1995).

Phyllocarbon Lloyd (1921) = Polyozellus fide Donk (*Beih. Nova Hedwigia* **5**: 228, 1962).

Phyllocardium Korshikov (1927), Algae.

Phyllocarpos Poir. (1813) = Cladonia fide Hawksworth *et al.* (*Dictionary of the Fungi* edn 8, 1995).

Phyllocaulon (Tuck.) Vain. (1909) = Stereocaulon Hoffm. fide Zahlbruckner (*Catalogus Lichenum Universalis* **4**, 1927).

Phyllocelis Syd. (1925), Pezizomycotina. 1 or 2, C. America.

Phyllocharis Fée (1825) = Strigula fide Hawksworth *et al.* (*Dictionary of the Fungi* edn 8, 1995).

phyllocladia, the granular, verrucose, coralloid, squamuliform, digitate, peltate, or foliaceous parts of the thallus of *Stereocaulon* which contain the photobiont.

Phyllocratera Sérus. & Aptroot (1997), Phyllobatheliaceae (L). 1, Papua New Guinea. See Aptroot *et al.* (*Biblthca Lichenol.* **64**, 1997).

Phyllocrea Höhn. (1918), Phyllachoraceae. 1 (from living leaves), S. America. See Müller & von Arx (*Beitr. Kryptfl. Schweiz* **11** no. 2, 1962).

Phyllodontia P. Karst. (1883) = Cerrena fide Donk (*Verh. K. ned. Akad. Wet.* tweede sect. **62**: 1, 1974).

Phylloedia Fr. (1849) ≡ Phylloedium.

Phylloedium Fr. (1825), anamorphic *Pezizomycotina*, St.0eP.?. 1, Europe. See Sutton (*Mycol. Pap.* **141**, 1977).

Phyllogaster Pegler (1969), Agaricaceae. 1, Ghana. Basidioma gasteroid. See Pegler & Young (*Proc. K. Ned. Akad. Wet.* Ser. C, Biol. Med. Sci. **72**: 222, 1969).

Phyllogloea Lowy (1961), Phragmoxenidiaceae. 6, widespread (tropical). See Roberts (*Mycotaxon* **87**: 187, 2003; Dominican Republic).

Phyllographa (Müll. Arg.) Räsänen (1943) nom. inval. = Opegrapha Ach. fide Hawksworth *et al.* (*Dictionary of the Fungi* edn 8, 1995).

Phyllohendersonia Tassi (1902), anamorphic *Pezizomycotina*, Cpd.≡ eP.?. 28, widespread. See Sutton (*Mycol. Pap.* **141**, 1977).

Phyllomyces Lloyd (1921) = Cordierites fide Zhuang (*Mycotaxon* **31**: 261, 1988).

Phyllonochaeta Gonz. Frag. & Cif. (1927) nom. dub., Fungi. See Petrak (*Annls mycol.* **32**: 317, 1934).

Phyllopeltula Kalb (2001), Peltulaceae (L). 2, Kenya; Venezuela. See Kalb (*Biblthca Lichenol.* **78**: 158, 2001).

Phyllopezis Petr. (1949), ? Helotiales. 1, S. America.

Phyllophiale R. Sant. (1952) = Porina Müll. Arg. fide Lücking *et al.* (*Lichenologist* **30**: 121, 1998), Lücking & Cáceres (*Lichenologist* **31**: 349, 1999).

Phyllophthalmaria (Müll. Arg.) Zahlbr. (1905) ? = Leptosphaeria fide Santesson (*Symb. bot. upsal.* **12** no. 1: 1, 1952).

phylloplane, the leaf surface; Last & Deighton (*TBMS* **48**: 83, 1965), the non-parasitic biotas of the leaf surface.

Phylloporia Murrill (1904), Hymenochaetaceae. 7, widespread (pantropical). See Ryvarden & Johansen (*Prelim. Polyp. Fl. E. Afr.*: 230, 1980; key), Wagner & Fischer (*MR* **105**: 773, 2001), Wagner & Ryvarden (*Mycol. Progr.* **1**: 105, 2002; phylogeny and taxonomy).

Phylloporina (Müll. Arg.) Müll. Arg. (1890) = Porina Müll. Arg. fide Hawksworth *et al.* (*Dictionary of the Fungi* edn 8, 1995).

Phylloporina C.W. Dodge (1948), Pezizomycotina (L). 1, Kerguelen.

Phylloporis Clem. (1909), Strigulaceae (L). 1, widespread. See Vězda (*Folia geobot. phytotax.* **19**: 177, 1984), Hawksworth (*SA* **5**: 151, 1986), Lücking & Lücking (*Herzogia* **11**: 143, 1995; Costa Rica), Malcolm & Galloway (*New Zealand Lichens* Checklist, Key, and Glossary, 1997).

Phylloporthe Syd. (1925), Diaporthales. 1, America (tropical). See Barr (*Mycol. Mem.* **7**, 1978), Cannon (*Fungal Diversity* **7**: 17, 2001).

Phylloporus Quél. (1888), Boletaceae. *c.* 50, cosmopolitan (mostly tropical). See Kuyper (*in litt.*), Corner (*Nova Hedwigia* **20**: 793, 1970; Malaysia), Singer & Gómez (*Brenesia* **22**: 163, 1984; Costa Rica) ? = Boletus (Bolet.) fide.

Phyllops Raf. (1817) nom. dub., Fungi.

Phyllopsora Müll. Arg. (1894), Ramalinaceae (L). 81, widespread (esp. tropical). See Swinscow & Krog (*Lichenologist* **13**: 203, 1981), Brako (*Mycotaxon* **35**: 1, 1989), Timdal & Krog (*Mycotaxon* **77**: 57, 2001; Africa), Upreti *et al.* (*Biblthca Lichenol.* **86**: 185, 2003; India), Elix (*Australasian Lichenology* **59**: 23, 2006; Australia).

Phyllopsoraceae Zahlbr. (1905) = Ramalinaceae.

Phyllopta (Fr.) Fr. (1825) nom. dub., ? Trichosporonaceae. See Donk (*Taxon* **7**: 239, 1958), Donk (*Persoo-

nia **4**: 306, 1966).

Phyllopyrenia C.W. Dodge (1948) nom. dub. ? = Coccotrema.

Phyllopyreniaceae Zahlbr. (1903) = Coccotremataceae.

Phyllosphaera Dumort. (1822) nom. rej. prop. ≡ Phyllosticta.

phyllosphere, the zone immediately surrounding a leaf; frequently used in the sense of phylloplane (q.v.). See Preece & Dickinson (Eds) (*Ecology of leaf surface micro-organisms*, 1970), Dickinson & Preece (Eds) (*Microbiology of aerial plant surfaces*, 1976), Blakeman (Ed.) (*Microbial ecology of the phylloplane*, 1981), Fokkema & van der Heuvel (Eds) (*Microbiology of the phyllosphere*, 1986). Cf. rhizoplane; spermoplane.

Phyllosticta Pers. (1818) nom. cons., anamorphic *Guignardia*, Cpd.0eH.1/19. 92, widespread. See van der Aa (*Stud. Mycol.* **5**, 1973), Punithalingam (*Mycol. Pap.* **136**, 1974), Punithalingam & Woodhams (*Nova Hedwigia* **36**: 151, 1982; conidial appendages), Yip (*MR* **93**: 489, 1989; 5 n.spp. from Australia), Petrini *et al.* (*Sydowia* **43**: 148, 1991; key spp. on conifers), Leuchtmann *et al.* (*MR* **96**: 287, 1992; isozyme polymorphism in endophytic spp.), Muthumary *et al.* (*Mycotaxon* **47**: 147, 1993; ontogeny), Okane *et al.* (*CJB* **79**: 101, 2001; on *Ericaceae*), Zhou & Stanosz (*Mycol.* **93**: 516, 2001; phylogeny), Aa & Vanev (*A Revision of the Species Described in Phyllosticta*, 2002; nomenclator), Baayen *et al.* (*Phytopathology* **92**: 464, 2002; *G. citricarpa*), Pandey *et al.* (*MR* **107**: 439, 2003; endophytes), Crous *et al.* (*Stud. Mycol.* **55**: 235, 2006; phylogeny).

Phyllostictaceae Fr. (1849) = Botryosphaeriaceae.

Phyllostictella Tassi (1901) ? = Microsphaeropsis Höhn. fide Sutton (*Mycol. Pap.* **141**, 1977).

Phyllostictina Syd. & P. Syd. (1916) = Phyllosticta fide van der Aa (*Stud. Mycol.* **5**, 1973).

Phyllostictites Babajan & Tasl. (1970), Fossil Fungi. 1 (Tertiary), former USSR.

Phyllothelium Trevis. (1861) = Trypethelium fide Hawksworth *et al.* (*Dictionary of the Fungi* edn 8, 1995).

Phyllotopsis E.-J. Gilbert & Donk ex Singer (1936), Agaricales. 5, widespread (temperate). Belongs to *Marasmiaceae* or *Tricholomataceae*. See Johannesen (*Blekksoppen* **26**: 23, 1998).

Phyllotremella Lloyd (1920) = Resupinatus fide Kuyper (*in litt.*).

Phyllotus P. Karst. (1879) = Resupinatus fide Kuyper (*in litt.*) Lectotypification has been contested; under another lectotypification it becomes the correct name for *Pleurocybella*.

Phylogenetic analysis. Systematics methods that aim to reconstruct the genealogical descent of organisms by means of objective and repeatable analysis, and from this pattern to propose a falsifiable hypothesis of natural classification or phylogeny. The theory is largely based on concepts introduced by the German entomologist Willi Hennig (1913-1976) which revolutionized systematic research through emphasis on evolutionary processes rather than shared characters. Hennig himself referred to his methodology as phylogenetic systematics, but for a time proponents of his methods preferred the term **cladistics**, derived from the Greek *κλάδος*, branch. Since around 2000 that term has fallen into disuse, and almost all researchers refer to their discipline as phylogenetics, or phylogenetic analysis.

Phylogenetic methodologies are based upon three fundamental assumptions: that taxa are united into natural groups on the basis of shared derived characters (**synapomorphies**); that all groups recognized must descend from a single ancestor (i.e. they are monophyletic) and that the most **parsimonious** pattern (that is the one requiring the least number of steps to resolve the relationships of the taxa) is the one most likely to be correct. Groups defined using shared primitive characters (**symplesiomorphies**) are likely to be paraphyletic and thus not include all the descendants of an ancestor. Groups delimited that do not share a common ancestor are polyphyletic and often result from convergent (non-homologous) characters. The product of phylogenetic analysis is a branching diagram (**phylogenetic tree** or **cladogram**, frequently referred to informally as a **tree**) which shows the pattern of relationships between the organisms based on the characters used.

Phylogenetic analysis methods were initially developed using morphology data sets and manual calculation, but now molecular data are used almost exclusively and powerful computers are required. Genes or gene fragments are amplified from extracted DNA using specific primers, and sequenced using equipment that is usually highly automated and optimized for high throughput. The resulting sequences must first be **aligned**, taking into account insertions and deletions and variations in overall sequence length. This is done using a computer package such as ClustalW (Chenna *et al.*, *Nucleic Acids Res.* **31**: 3497, 2003; Larkin *et al.*, *Bioinformatics* **23**: 2947, 2007; http://www.clustal.org), either down-loaded or via a web-based service. Many users prefer manually to adjust the alignment that is generated, to compensate for ambiguities in the sequence traces etc. The aligned sequences are then trimmed to a common length and output (typically in **NEXUS** format) so they can be read by phylogenetics packages. These use complex statistical algorithms to generate phylogenetic trees, often with numerous alternative settings to allow for differences in sequence characteristics, presumed evolutionary models, use of molecular clocks etc.

Currently, the three most widely used algorithms for phylogenetic analysis are:

Parsimony analysis (using the PAUP package; Swofford, *PAUP*: phylogenetic analyses using parsimony (*and other methods)*, 2002).

Maximum likelihood estimation (e.g. Stamatakis *et al.*, *Bioinformatics* **21**: 456, 2005).

Bayesian analysis (using the MrBayes package; Huelsenbeck & Ronquist, *Bioinformatics* **17**: 754, 2001; Ronquist *et al.*, http://www.mrbayes.csit.fsu.edu/).

In some cases the output is restricted to a text-based depiction of the resulting trees, which can be visualized using e.g. **Treeview** (http://taxonomy.zoology.gla.ac.uk/rod/treeview/). There are several freely available integrated software packages that make phylogenetic analysis somewhat more user-friendly, including **MESQUITE** (version 2.5: http://mesquiteproject.org/mesquite/mesquite/) and **MEGA** (Tamura *et al.*, *Mol. Biol. Evol.* **24**: 1596, 2007; version 4: http://www.megasoftware.net/).

Correct interpretation of phylogenetic trees is critically important: all the tree generation methods are based on statistical assumptions that may not be justifiable in all cases, and they merely represent the most probable evolutionary pathway by which the organisms included have come into being. Trees must be **rooted** with the inclusion and specification of an **outgroup** – a taxon or group of taxa that are known to be outside of but related to the group in question. Phylogenetic analyses involve repeated analyses of the data, sometimes with many thousands of replicates, each of which produces a subtly different branching pattern or branch lengths. A **consensus** tree represents the most likely evolutionary pathway in statistical terms, and the degree of likelihood that this is the correct **topology** (or branching pattern) can also be measured statistically. This may be achieved using the **bootstrap** resampling technique (Felsenstein, *Evolution* **39**: 783, 1985; Nei & Kumar,

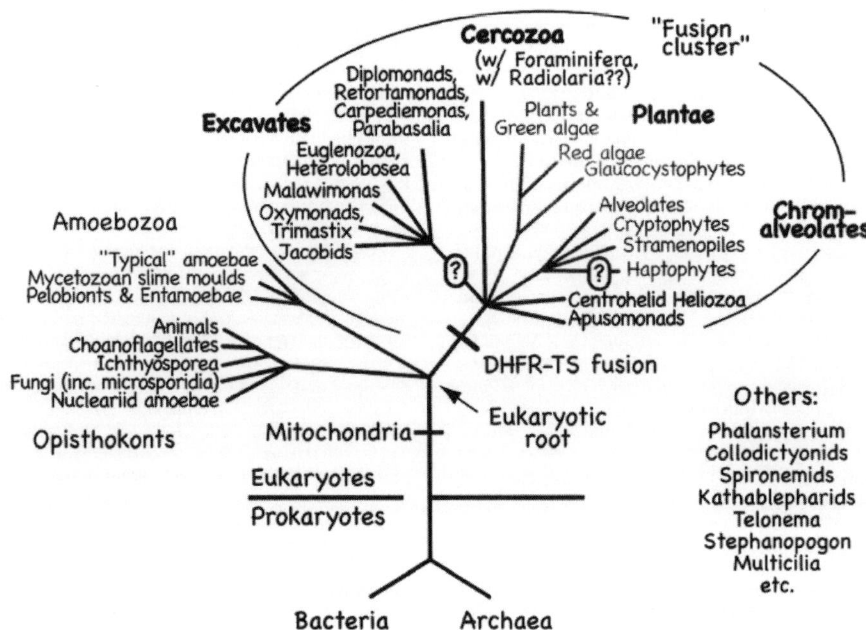

Fig. 31. Phylogenetic tree of eukaryotes based Simpson & Rogers, Eukaryotic evolution: getting to the root of the problem (*Current Biology* **12**: R691-R693, 2002).

Molecular evolution and phylogenetics, 2000) by comparing trees with those generated after random replacement of nucleotides within the sequences. The proportion of individual branching events with identical structure provides a measure of robustness of the analysis; a bootstrap value of less than about 70% is considered by most to be an unreliable indicator of relationships. Bootstrapping may be used in conjunction with several phylogenetic analysis methods, including parsimony and maximum likelihood analysis. Bayesian analysis is assessed using posterior probabilities (PP; the posterior distribution of trees, see Ronquist *et al.*, 2005); here strong support for a topology can be inferred from a PP figure of more than 0.95. Results from any of these techniques are only as reliable as the original data, and problems with alignment in particular can lead to serious anomalies.

Many different genes and gene fragments have been explored as potential indicators of evolutionary pathways for fungi. They show varying degrees of conservation, allowing their use for analysis of relationships at different levels of the taxonomic hierarchy. The most widely used remain ribosomal genes within nuclei or mitochondria; these are present in multiple copies which makes amplification easier (although there are instances of sequence heterogeneity within organisms). Of these, the 18S (SSU) fragment is relatively conserved and has been widely used to infer relationships at family level and above, and the 25S (LSU) fragment is more variable and is useful at genus or species level. The two ITS (internal transcribed spacer) regions of ribosomal DNA which are more variable again, are most widely used to investigate species relationships, although there are many cases where the level of variation is insufficient to distinguish reliably between taxa commonly accepted at species level. The search for the most effective gene for use in species-level diagnosis in fungi continues (**genetic barcoding**; see e.g. Seifert *et al.* (*Proc. Nat. Acad. Sci* **104**: 3901, 2007; barcoding of fungi), Kress & Erickson (*PNAS* **105**: 2761, 2008; introduction to barcoding), http://barcoding.si.edu/ and http://www.barcodinglife.org/). Other widely used genes in phylogenetic research include those coding for α-tubulin, transcription elongation factor-1α (EF-1α), RNA polymerase B (RPB), glyceraldehyde 3-phosphate dehydrogenase (GAPDH) and mitochondrial ATP synthase F0 subunit 6 (ATP6). Primers for amplification of many of these genes may be found at the AFTOL website (http://aftol.org/).

Lit.: Forey *et al.* (*Cladistics: a practical course in systematics*, 1992); Linder (*S. Afr. J. Bot.* **54**: 208, 1988, review for botanists); Scotland *et al.* (Eds) (*Models in phylogeny reconstruction*, 1994); Page & Holmes (*Molecular evolution: a phylogenetic approach*, 1998); Nei & Kumar, *Molecular evolution and phylogenetics*, 2000); Felsenstein (*Inferring phylogenies*, 2004), See also websites associated with the software packages detailed above, and numerous articles posted on GENBANK (http://www.ncbi.nlm.nih.gov/) and EMBL (http://www.embl.de/).

See also Classification, Coevolution, Kingdoms of fungi, Phylogeny.

Phylogeny. Mycologists' views on phylogeny, the evolution of a group of organisms through time, have been based traditionally on comparative morphology, cytology, hyphal wall chemistry, ultrastructure, and to a lesser degree fossils (see Fossil fungi). The advent of cladistic and molecular approaches has led to major changes in understanding the relationships of fungal phyla.

The three kingdoms including fungi as traditionally circumscribed (see Kingdoms of fungi) are shown by molecular studies of 16S-like ribosomal RNA sequences to be phylogenetically remote (Fig. 31).

Current data mostly support the initial indications that *Animalia* and *Fungi* are sister groups and most probably arose from a choanozoan-like protozoan ancestor (see Adl *et al*, *J. Eukaryotic Microbiol.* **52**: 399, 2005; Baldauf *et al.*, *Science* **290**: 972, 2000; Cavalier-Smith, in Rayner, Brasier & Moore (Eds), *Evolutionary biology of the fungi*: 339, 1987, in McLaughlin *et al.* (Eds) *The Mycota* **7A**: 3, 2001); Cracraft & Donaghue (Eds), *Assembling the Tree of Life*, 2004; Steenkamp *et al.*, *Mol. Biol. Evol.* **23**: 93, 2006; van der Peer *et al.*, *J. Mol. Evol.* **51**: 565, 2000; Wainright *et al.*, *Science* **260**: 340, 1993). These two kingdoms have been informally grouped with the members of the DRIPs clade [DRIPs is an acronym of the original members of the clade: *Dermocystidium*, rosette agent, *Ichthyophonus*, and *Psorospermium*], microsporidians and the choanoflagellates in the Opisthokonta (q.v.) (see Baldauf *et al.*, *Science* **290**: 972, 2000; Philip *et al.*, *Mol. Biol. Evol.* **22**: 1175, 2005), though the most authors now regard the *Microsporidia* (q.v.) as members of the Kingdom *Fungi* (see e.g. Gill & Fast, *Gene* **375**: 103, 2006). No molecular data have been found to support the hypothesis for a red algal (*Rhodophyta*) ancestry of the *Ascomycota* and *Basidiomycota* which was strongly argued for by Demoulin (see *Bot. Rev.* **40**: 315, 1974, *BioSystems* **18**: 347, 1985).

'Fungi' in the other two kingdoms seem to have arisen polyphyletically. Within the Protozoa (q.v.) the fungus-like groups are all slime moulds but these various groups are not closely related and even such taxa as the acrasids are now thought to be quite different in origin. Molecular data are now showing where these groups belong. The Acrasids are now included among the amoeboid flagellate protozoa (*Heterolobosea*); the true slime moulds, protostelids and dictyostelids cluster together with another amoeboid protozoan group, *Rhizopoda*, which is informally known as the Ramicristates and the plasmodiophorids cluster together with the amoeboid, photosynthetic chlorarachnids and the cercomonads (a diverse group of soil flagellates). Other taxa such as *Fonticula* and the copromyxids are now thought to be independent lineages (Patterson, 1999).

Major new research elucidating relationships within the true fungi has been published in the last five years; see especially Hibbett *et al.* (*MR* **111**: 509, 2007), James *et al.* (*Nature* **443**: 818, 2006), Liu *et al.*, *BMC Evol. Biol.* **6**: 74, 2006), Lutzoni *et al.* (*Am. J. Bot.* **91**: 1446, 2004) and papers in the *Deep Hypha* issue of *Mycologia* (*Mycol.* **98** (6), 2006, publ. 2007). Within the Kingdom *Fungi*, analyses of molecular data strongly support the monophyly of the *Ascomycota* and *Basidiomycota* (recognized as Subkingdom *Dikarya*), and also led to acceptance of five further phyla – the *Blastocladiomycota*, *Chytridiomycota*, *Glomeromycota*, *Microsporidia* and *Neocallimastigomycota*. The *Zygomycota* as accepted in the previous edn of this *Dictionary* is no longer considered to be a viable taxonomic unit, with its components including the *Mucorales*, *Endogonales*, '*Trichomycetes*' and others shown to be only distantly related. The new system promises a robust framework within which to develop classifications of lower-order groups.

Insights into evolution of the phylum *Ascomycota* can be found in the apparently basal position of *Neolecta*, associated in a grouping with *Pneumocystis*, *Schizosaccharomyces* and *Taphrinales*, sometimes referred to as the archiascomycetes. These morphologically disparate taxa are here placed within subphylum *Taphrinomycotina*, alongside the ascomycetous yeast subphylum *Saccharomycotina* and the *Pezizomycotina* (syn. *Ascomycotina*) that contains the vast bulk of filamentous taxa. Major changes since the last edn of this *Dictionary* include remodelling of the major lichenized groups, with the most speciose taxon now named the *Lecanoromycetes*, which includes the subclasses *Lecanoromycetidae* and *Ostropomycetidae*. *Aspergillus*, *Penicillium* and similar fungi are confirmed as related to morphologically disparate groups including the black yeasts (*Chaetothyriales*) and some pyrenocarpous lichenized taxa (*Verrucariales*). Studies have underlined time and again that many of the morphological structures used in traditional classifications of the *Ascomycota* have evolved on multiple occasions, presumably in response to common evolutionary pressures.

Even more dramatic changes have resulted in the new classification of the *Basidiomycota*. Three familiar groups are recognized as subphyla, the *Pucciniomycotina* (rusts and relatives), *Ustilaginomycotina* (smuts etc.) and *Agaricomycotina* (agarics, polypores, 'gasteromycetes', jelly fungi etc.). Within the *Agaricomycotina* many of the traditional groups based on anatomical structure have been shown to be artificial. One of the most ancient lineages of the *Agaricomycetes* appears to be the *Cantharellales*, which includes *Rhizoctonia* and *Tulasnella* as well as the chanterelles and their relatives. The agarics and boletes cluster together within the *Agaricomycetidae*, earthstars, phalloid taxa and many basidiomycetous truffles form a distinct lineage (the *Phallomycetidae*) and the relatives of *Russula* include *Hericium* and

Stereum.

The *Chytridiomycota* constitute the main phylum of the *Fungi* that possesses motile zoospores. Their systematic arrangement has also undergone profound change as a result of molecular phylogenetic research, with the *Blastocladiomycota* now separated at phylum level (James *et al*., *Mycol*. **98**: 860, 2006). The morphologically reduced genus *Rozella* appears to represent the most basal clade of all the true *Fungi*, though the position of the *Microsporidia* (strange protozoan-like fungi that lack mitochondria) remains to be stabilized. Major groups within the *Chytridiomycota* include the *Neocallimastigales* (rumen fungi) as well as the chytrids and *Spizellomycetales*. Many of the groups need further study, and further nomenclatural changes are likely.

In most, but not all analyses the *Glomerales* are basal to the ascomycete and basidiomycete lineage, and the group has now been elevated to phylum level as the *Glomeromycota*. This is particularly significant given their suggested role as mycorrhizal fungi associated with early land plants and the recent report of fossil glomeralean fungi from mid-Ordovician deposits (c. 460 Myr ago) (see Redecker *et al*., *Science* **289**: 1920, 2000; Simon *et al*., *Nature* **363**: 67, 1993).

The *Zygomycota* are no longer recognized as a phylogenetically coherent entity (see Sugiyama, *Mycoscience* **39**: 487, 1998; James *et al*., *Can. J. Bot.* **78**: 336, 2000, White *et al*., *Mycol*. **98**: 872, 2006). Further research is necessary, but there are strong indications that even the zygomycetes as traditionally recognized are paraphyletic with the *Glomeromycota*, *Ascomycota* and *Basidiomycota*. Two of the groups traditionally treated as trichomycetes are now understood to have protist affinities, and the *Entomophthorales* in their usual circumscription are now shown to be polyphyletic. Formerly included among the trichomycetes, *Amoebidium* and its relatives have been suggested to be a member of the DRIPs clade based on molecular evidence (see Ustinova *et al*., *Protist* **151**: 253, 2000; White *et al*., *Mycol*. **98**: 872, 2006), confirming the observations of Lichtwardt (*Trichomycetes*: 279, 1986), who thought that *Amoebidium* had arisen independently from a protozoan ancestor and was unrelated to any other *Trichomycetes*.

The predominantly autotrophic *Chromista* contain two separate heterotrophic lineages that include 'fungal' taxa. Both arose from an ancestral heterokont flagellate with chloroplasts. One lineage comprises the oomycetes and hyphochytrids with its sister taxon the marine flagellate, *Developayella*. The other lineage leads to the labyrinthulids and thraustochytrids (the slime nets), largely characterized by the presence of bothrosomes. Molecular data tend to support the morphological groups recognized in the oomycetes, now reflected in two subclasses: *Peronosporomycetidae* and *Saprolegniomycetidae*, although the position of *Rhipidium*, treated as a separate subclass by Dick (*Can. J. Bot*. **73**: S712, 1995) is uncertain since it appears basal in the oomycete phylogeny (see Peterson & Rosendahl, *MR* **104**: 1295, 2000).

The poor representation of fungi in the fossil record has stimulated much research in recent years using molecular methods to estimate the dates of major evolutionary radiations within the group. Statistical methods have been refined in recent years to take into account different rates of nucleotide substitution in different lineages (Peterson *et al*., *Proc. Nat. Acad. Sci.* USA **101**: 6536, 2004). Berbee & Taylor (*CJB* **71**: 1114, 1993) calculated that the three main fungal phyla diverged from the *Chytridiomycota* approximately 550 Myr ago. The *Ascomycota*/*Basidiomycota* split most probably occurred at about 400 Myr ago after plants invaded the land and the dicot/monocot divergence occurred (Taylor & Berbee, *Mycol*. **98**: 838, 2006), but possibly as early as 1500 Myr ago if the fungal fossil *Paleopyrenomycetes devonicus* belongs to the *Sordariomycetes* (Taylor *et al*., *Mycol*. **97**: 269, 2006; Padovan *et al*., *J. Mol. Evol*. **60**: 726, 2005) and 1800Myr ago based on calibrations derived from assumed divergence of mammals and birds 300Myr ago. Many ascomycetous yeasts and moulds (e.g. *Eurotiales*) evolved after the origin of angiosperms, probably in the last 200 Myr. See also **Fossil fungi**.

Fungi are associated with some of the earliest remains of land plants, and may well have played a crucial role in enabling life out of water. Church (*J. Bot., Lond*, **59**: 7, 1921) considered that 'lichens' might be transmigrants, the earliest colonizers of land (see also Corner, *The life of plants*, 1964). Land plants may, however, have their origins in a green alga-oomycete symbiosis (Pirozynski & Malloch, *BioSystems* **6**: 153, 1975), and it has even been argued that plants are of biphyletic origin with some parts of the fungal genome (Jørgensen, *BioSystems* **31**: 193, 1993) or even 'inside-out' lichens as they contain organelles questionably of fungal origin (Atstatt, *Ecology* **69**: 17, 1988).

Revolutions in the development of automated gene sequencing have provided enormously more data with which to test evolutionary hypotheses compared with even five years ago. The first complete human genome to be sequenced, containing over 3 billion base pairs, was presented only in 2003, but by June 2008 genome sequencing projects were under way or completed for over 70 fungal species (see http://www.ncbi.nlm.nih.gov/Genomes/). The challenge is now to move from this *embarras de richesses* in terms of data to its effective analysis. It is now common practice to analyse phylogenies using multiple gene sequences (as with the studies presented in *Mycol*. **98** (6), 2006). One recent paper has mined data from 34 complete genome sequences to produce a phylogeny of the Fungi based on 29 universally conserved protein-encoding gene sequences (Cornell *et al*., *Genome Res*. **17**: 1809, 2007).

Phylogenies may also focus on the other end of the taxonomic spectrum, at or below the species level. Molecular methods have revolutionized study of population biology, biogeography and microevolutionary processes. New methods of DNA extraction mean that unculturable taxa can be included in phylogenetic analyses, and dried material from fungal reference collections etc. can be used more extensively. Many morphospecies are now shown to be aggregates of taxa with distinct metabolic, ecological or pathological properties, though the level of the taxonomic hierarchy at which these should be recognized is debatable. A further major benefit of the molecular phylogenetic revolution is the ability to integrate anamorphic taxa effectively within the mainstream fungal classifications, and researchers are increasingly blurring the nomenclatural distinction between sexual and asexual morphs.

See Classification, Coevolution, Kingdoms of fungi, Phylogenetic analysis.

phylum, a taxonomic rank between kingdom and class; a division; see Classification.

Phymatium Chevall. (1826) = Elaphomyces fide Fries (*Syst. mycol.* **3**: 57, 1829).

Phymatodiscus Speg. (1919) = Myriangium fide von Arx (*Persoonia* **2**: 421, 1963).

Phymatomyces Kobayasi (1937) = Barssia fide Trappe (*Mycotaxon* **9**: 247, 1979).

Phymatopsis Tul. ex Trevis. (1857) = Abrothallus fide Keissler (*Rabenh. Krypt.-Fl.* **8**, 1930).

Phymatosphaeria Pass. (1875) = Myriangium fide von Arx (*Persoonia* **2**: 421, 1963).

Phymatosphaeriaceae Speg. (1888) = Myriangiaceae.

Phymatostroma Corda (1837) nom. dub. = Leucosporium fide Holubová-Jechová (*Sydowia* **45**: 97, 1993).

Phymatotrichopsis Hennebert (1973), anamorphic *Rhizinaceae*, Hso.0eH.6. 1, N. America. *P. omnivorum*, root rot of cotton and other plants. See Mouton (*Rev. Myc. suppl. colon.* 2 **18**: 69, 1953), Baniecki & Bloss (*Mycol.* **61**: 1054, 1969), Hennebert (*Persoonia* **7**: 199, 1973), Marek (*Phytopath.* **95**: S65, 2005; posn. in *Pezizales*).

Phymatotrichum Bonord. (1851) = Botrytis fide Hennebert (*Persoonia* **7**: 183, 1973).

Physalacria Peck (1882), Physalacriaceae. 30, widespread (tropical; southern hemisphere). See Berthier (*Biblthca Mycol.* **98**, 1985; world monogr.), Horak & Desjardin (*Aust. J. Bot.* **7**: 153, 1994; taxonomy), Antonín & Mossebo (*Mycotaxon* **83**: 419, 2002; Africa), Tanaka *et al.* (*Mycoscience* **45**: 143, 2004; Australasia).

Physalacriaceae Corner (1970), Agaricales. 11 gen. (+ 12 syn.), 169 spp. *Physalacriaceae* is listed as nom. rej. against *Tricholomataceae*.

Lit.: Berthier (*Biblthca Mycol.* **98**: 128 pp., 1985), Horak & Desjardin (*Aust. Syst. Bot.* **7**: 162, 1994), Antonín & Mossebo (*Mycotaxon* **83**: 419, 2002), Moncalvo *et al.* (*Mol. Phylogen. Evol.* **23**: 357, 2002), Poinar & Brown (*MR* **107**: 765, 2003), Tanaka *et al.* (*Mycoscience* **45**: 143, 2004), Wilson & Desjardin (*Mycol.* **97**: 667, 2005), Binder *et al.* (*Am. J. Bot.* **93**: 547, 2006), Dentinger & McLaughlin (*Mycol.* **98**: 746, 2006).

Physalidiella Rulamort (1990), anamorphic *Pezizomycotina*, Hso.0bP.1. 2, Italy; Cuba. See Rulamort (*Bull. Soc. bot. Centre-Ouest* Nouv. sér. **21**: 512, 1990).

Physalidiopsis R.F. Castañeda & W.B. Kendr. (1990), anamorphic *Pezizomycotina*, Hso.0bP.1. 1, Cuba. See Castañeda & Kendrick (*Univ. Waterloo Biol. Ser.* **32**: 36, 1990).

Physalidium Mosca (1965) [non *Physalidium* Fenzl 1866, *Cruciferae*] ≡ Physalidiella.

Physalospora Niessl (1876), Hyponectriaceae. 36 (on dead leaves etc.), widespread. The genus has been confused with *Botryosphaeria* (Botryosph.), and is probably polyphyletic as currently circumscribed. See Barr (*Mycol.* **68**: 611, 1976; nomencl.), Kohlmeyer *et al.* (*Bot. Mar.* **38**: 175, 1995), Hoffmann & Hafellner (*Biblthca Lichenol.* **77**: 1, 2000; excl. spp. on lichens), Sivanesan & Shivas (*Fungal Diversity* **9**: 169, 2002).

Physalosporella Speg. (1910) nom. dub., Pezizomycotina. See Petrak & Sydow (*Annls mycol.* **32**: 28, 1934).

Physalosporina Woron. (1911) = Stigmatula fide Cannon (*Mycol. Pap.* **163**, 1991).

Physalosporopsis Bat. & H. Maia (1955), Dothideomycetes. 1, Brazil. See Müller & von Arx (*Beitr. Kryptfl. Schweiz* **11** no. 2, 1962).

Physcia (Schreb.) Michx. (1803), Physciaceae (L). *c.* 73, widespread. See Awasthi (*J. Indian bot. Soc.* **39**: 1, 1960; India), Frey (*Ber. schweiz. bot. Ges.* **73**: 389, 1963), Thomson (*Beih. Nova Hedwigia* **7**, 1963; N. Am.), Moberg (*Symb. bot. upsal.* **22** no. 1, 1977), Moberg (*Nordic Jl Bot.* **6**: 843, 1986; key 21 spp. E. Afr.), Moberg & Hansen (*Meddr Grønland* Biosc. **22**, 1986; keys), Moberg (*Herzogia* **8**: 249, 1989; Eur.), Moberg (*Nordic Jl Bot.* **10**: 319, 1990; key 34 spp. C. & S. Am.), Moberg (*Symb. bot. upsal.* **32**: 163, 1997; Sonoran Desert), Stenroos & DePriest (*Am. J. Bot.* **85**: 1548, 1998; DNA), Myllys *et al.* (*Mycol.* **93**: 335, 2001; species pair concept), Elvebakk & Moberg (*Lichenologist* **34**: 311, 2002; Chile), Wedin *et al.* (*Taxon* **51**: 655, 2002; phylogeny), Helms *et al.* (*Mycol.* **95**: 1078, 2003; phylogeny), Moberg (*Symb. bot. upsal.* **34** no. 1: 257, 2004; S Africa), Simon *et al.* (*J. Mol. Evol.* **60**: 434, 2005; rDNA introns), Miądlikowska *et al.* (*Mycol.* **98**: 1088, 2006; phylogeny), Türk & Obermayer (*Central European Lichens*: 119, 2006; Austria), Honegger & Zippler (*MR* **111**: 424, 2007; mating systems).

Physciaceae Zahlbr. (1898) nom. cons., Teloschistales (L). 17 gen. (+ 27 syn.), 512 spp.

Lit.: Hafellner *et al.* (*Herzogia* **5**: 39, 1979; key gen.), Aptroot (*Fl. Guianas, E,* **1**, 1987; 48 spp. Guianas), Aptroot (*Cryptog. Bryol.-Lichénol.* **9**: 141, 1988), Moberg & Hansen (*Meddr Grønland* Biosc. **22**: 389, 1988; keys 17 foliose spp. Greenland), Aptroot & Berendsen (*Proc. K. Ned. Akad. Wet.* Ser. C, Biol. Med. Sci. **92**: 409, 1989; pruina), Rambold *et al.* (*Pl. Syst. Evol.* **192**: 31, 1994; ascus types), Giralt & Llimona (*Mycotaxon* **62**: 175, 1997), Nimis & Tretiach (*Bryologist* **100**: 217, 1997), Nordin (*Symb. bot. upsal.* **32**: 195, 1997; ultrastr.), Esslinger & Bratt (*Lichenogr. Thomsoniana* North American Lichenology in Honor of John W. Thomson: 25, 1998), Stenroos & DePriest (*Am. J. Bot.* **85**: 1548, 1998; DNA), Chen & Wang (*Mycotaxon* **73**: 335, 1999), Moberg & Nash (*Bryologist* **102**: 1, 1999), Esslinger (*Bull. Calif. Lichen Soc.* **7**: 1, 2000), Giralt *et al.* (*Lichenologist* **32**: 105, 2000), Lohtander *et al.* (*Mycol.* **92**: 728, 2000; phylogeny, Scandinavia), Mayrhofer *et al.* (*Muelleria* **12**: 169, 1999), Wedin *et al.* (*CJB* **78**: 246, 2000; rels with *Caliciaceae*), Dahlkild *et al.* (*Bryologist* **104**: 527, 2001), Grube & Arup (*Lichenologist* **33**: 63, 2001), Nordin & Mattsson (*Lichenologist* **33**: 3, 2001; phylogeny), Scheidegger *et al.* (*Lichenologist* **33**: 25, 2001; evolutionary trends), Anon. *in* Anon. (Ed.) (*Nordic Lichen Flora* **2**: *Physciaceae*, 2002), Wedin *et al.* (*Taxon* **51**: 655, 2002), Helms *et al.* (*Mycol.* **95**: 1078, 2003), Tibell (*MR* **107**: 1403, 2003), Crespo *et al.* (*Taxon* **53**: 405, 2004), Cubero *et al.* (*MR* **108**: 498, 2004), Lumbsch *et al.* (*Mol. Phylogen. Evol.* **31**: 822, 2004), Obermayer *et al.* (*Symb. bot. upsal.* **34** no. 1: 327, 2004), Miądlikowska *et al.* (*Mycol.* **98**: 1088, 2006; phylogeny), Hofstetter *et al.* (*Mol. Phylogen. Evol.* **44**: 412, 2007; phylogeny).

Physcidia Tuck. (1862), Ramalinaceae (L). 7, widespread. See Kalb & Elix (*Biblthca Lichenol.* **57**: 265, 1995; key).

Physciella Essl. (1986) ? = Phaeophyscia fide Hawksworth (*SA* **5**: 151, 1986).

Physciomyces E.A. Thomas ex Cif. & Tomas. (1953) = Physcia fide Hawksworth *et al.* (*Dictionary of the Fungi* edn 8, 1995).

physcion, see parietin.

Physciopsis M. Choisy (1950) = Hyperphyscia fide Hafellner *et al.* (*Herzogia* **5**: 39, 1979).

Physconia Poelt (1965) nom. cons., Physciaceae (L). *c.* 26, widespread. See Moberg (*Symb. bot. upsal.* **22** no. 1, 1977), Moberg (*Nordic Jl Bot.* **7**: 719, 1987; key 3 spp., E. Afr.), Esslinger (*Mycotaxon* **51**: 91, 1994; N America), Nordin (*Symb. bot. upsal.* **32** no. 1: 195, 1997; ultrastr.), Esslinger (*Bull. Calif. Lichen Soc.* **7**: 1, 2000; California), Lohtander *et al.* (*Mycol.* **92**: 728, 2000; Scandinavia), Nordin & Mattsson (*Lichenologist* **33**: 3, 2001; phylogeny), Chen & Hu (*Mycotaxon* **86**: 185, 2003; China), Helms *et al.* (*Mycol.* **95**: 1078, 2003; phylogeny), Cubero *et al.* (*MR* **108**: 498, 2004; phylogeny, revision), Miądlikowska *et al.* (*Mycol.* **98**: 1088, 2006; phylogeny), Türk & Obermayer (*Central European Lichens*: 119, 2006; Austria), Divakar *et al.* (*MR* **111**: 1311, 2007; phylogeny, anatomy).

physiologic form (Stakman & Levine), see physiologic race.

physiologic race, one of a group of forms alike in morphology but unlike in certain cultural, physiological, biochemical, pathological, or other characters. The use of this term in place of biologic form, etc. was a recommendation of the International Botanical Congress, 1935. The term 'race' has been used in different senses by plant pathologists, the classical approach of using tests on differential hosts now being replaced by smaller flexible groups of host genotypes permitting characterization by virulence. See Caten (*in* Wolfe & Caten, *Populations of plant pathogens: their dynamics and genetics*: 21, 1987; review), Dennis (*Proc. Linn. Soc. Lond.* **163**: 47, 1952; taxonomic treatment), Johnson (*Biol. Rev.* **28**: 105, 1953; variation), Classification, special form, Species.

Physiology. Fungal physiology is the study of their mechanical, physical and biochemical functions, their adaptations to stresses and mechanisms for their development. Good introductory texts are: Berry (*Physiology of industrial fungi*, 1988) and Garraway & Evans (*Fungal nutrition and physiology*, 1984). Griffin (*Fungal physiology*, 2nd ed, 1996) summarizes fungal physiology from the viewpoint of the physical and chemical factors and processes behind the functioning of fungi, integrating discoveries in molecular genetics with conventional whole organism and biochemical approaches. At the more encyclopaedic level, Smith & Berry (Eds) (*The filamentous fungi*, 4 vols, 1975-83) and volumes of Esser & Lemke (Eds) (*The Mycota*, 1994-, e.g. Esser *et al.* (Eds), *The Mycota* **8**: 1, 2001) are recommended, though now some are now dated. Specialist treatments include: Ayers & Boddy (*Water, fungi and plants*, 1986), Bennett & Ciegler (*Secondary metabolism and differentiation in fungi*, 1983), Boddy *et al.* (*Nitrogen, phosphorus and sulphur utilisation by fungi*, 1989), Burnett & Trinci (*Fungal walls and hyphal growth*, 1989), Calvo *et al.*, *Microbiol. Mol. Biol. Rev.* **66**: 447, 2002; relations between secondary metabolism and fungal development), Cooke & Whipps (*Ecophysiology of fungi*, 1983), Cox (*In*: *In* Robson *et al.* (Eds) *Exploitation of Fungi*, 2007), Elliott (*Reproduction in fungi: genetical and physiological aspects*, 1994), Frankland *et al.* (*Decomposer basidiomycetes: their biology and ecology*, 1982), Jennings (*Stress tolerance of fungi*, 1993; *The physiology of fungal nutrition*, 1995), Jennings & Rayner (*The ecology and physiology of the fungal mycelium*, 1984), Kershaw (*Physiological ecology of lichens*, 1985), Lewis (*MR* **95**: 897, 1991; sugars), Moore *et al.* (*Developmental biology of higher fungi*, 1985), Smith (*Fungal differentiation*, 1983); Szaniszlo (*Fungal dimorphism*, 1985), Vicente *et al.* (*Surface physiology of lichens*, 1985), Weete (*Lipid biochemistry of fungi and other organisms*, 1981), Winkelmann & Winge (*Metal ions in fungi*, 1994), Yu & Keller (*Annual Review of Phytopathology* **43**: 437, 2005), Zhdanova & Vasilevskaya (*Melanin biosynthesizing fungi in extreme environments*, 1988). Special studies and reviews devoted to physiology of different ecological groups of fungi include: **anaerobic fungi**, study of physiology of fungal development in anaerobic conditions (Lübbehüsen *et al.*, *MR* **107**: 223, 2003); **marine fungi**, especially, their growth, enzyme production and secondary metabolites (Jensen & Fenical, *In*: Hyde, K.D. (Ed.) *Fungi in marine environments. Fungal diversity research series* **7**: 293, 2002; Kis-Papo *et al.*, *Microbial ecology* **45**: 83, 2003; Molitoris & Schaumann, *In*: *The biology of marine fungi*, 35, 1986); **mycorrhizal fungi**, the physiological interaction within the root (regulation of host defences, bidirectional transport of nutrients, modification of gene expression, carbon metabolism) and effects of the mycorrhiza upon the whole plant (Bago *et al.*, *Plant physiology* **124**: 949, 2000; Bago *et al.*, *Plant Physiology* **128**: 108, 2002; Kapulnik & Douds *Arbuscular mycorrhizas: physiology and function*, 2000; Le Quéré *et al.*, *Mol. plant microbe interactions MPMI* **18**: 659, 2005; Olsson *et al.*, *Appl. Envir. Microbiol.* **71**: 2592, 2005; Trepanier *et al.*, *Appl. Envir. Microbiol.* **71**: 5341, 2005), contribution of arbuscular mycorrhiza in maintaining antioxidant and carbon metabolism in plants under water stress (Goicoechea *et al.*, *J. Plant Physiol.* **162**: 27, 2005), biofertilizing, bioprotecting and bioregulating role of arbuscular mycorrhizal fungi in plant growth and health promotion in agricultural systems (Gianinazzi *et al.*, *Mycorrhizal technology in agriculture: from genes to bioproducts*, 2002); **soil fungi**, their nutritional requirments and place in ecosystems (Cullings *et al.*, *Appl. Environ. Microbiol.* **71**: 1996, 2005); **thermophilic and thermotolerant fungi**, especially their ezymes (Bhabhra *et al.*, *Infect. Immun.* **72**: 4731, 2004; Maheshwari *et al.*, *Microbiology and Molecular Biology Reviews* **64**: 461, 2000; Sandgren *et al.*, *Protein Sci.* **12**: 2782, 2003).

Bakers' yeast is by far the most studied fungus, physiologically and biochemically (see *Saccharomyces*). Because yeasts in general are easy to handle experimentally in comparison with other fungi, there is an extensive literature on their physiology and biochemistry, the best entries into which are Berry *et al.* (*Yeast biotechnology*, 1987), Rose & Harrison (*The yeasts* **3-4**, 1989-91) and van der Heijden (Ed.) (*Mycorrhizal Ecology*, 2002). Adaptation mechanisms of yeast cells to environmental stress, specific adaptations of their growth and metabolism to nutrient depletion, osmotic and heat shock, salt and oxidative stress are discussed by Hohmann & Mager (*Yeast stress responses*, 2003), Guerra *et al.* (*Microbiology* **151**: 805, 2005). For results of molecular biology studies using *Saccharomyces cerevisiae*, and some other yeast species as model eukaryotes to study their growth mechanisms, division, elongation, pseudohyphal growth, mating and hormonal attraction see Madhani, *From a to α: yeast as a model for cellular differentiation*, 2007). Summarized data on the physiology, cell biology and genetics of the dimorphic fungus *Yarrowia lipolytica* are given by Barth & Gaillardin (*FEMS Microbiology Reviews* **19**: 219, 1997).

Advances in Microbial Physiology, *Ann. Rev. Phytopath.* and *Microbiol. Mol. Biol. Rev.* contain significant review articles. Because of the power of molecular biology, many of key physiological findings about fungi, albeit for a restricted number of genera (particularly *Aspergillus*, *Neurospora*, *Saccharomyces*), are in such journals as *J. biol. Chem.* and *Mol. cell. Biol.*

General reviews of lichen physiology are given by Ahmadjian (*The lichen symbiosis*, edn 2, 1993), Ahmadjian & Hale (Eds) (*The lichens*, 1974), Brown *et al.* (Eds) (*Lichenology: progress and problems*, 1976), Brown (Ed.) (*Lichen physiology and cell biology*, 1985), Hale (*The biology of lichens*, 1983), Kappen (Ed.) (New species and novel aspects in ecology and physiology of lichens. In honour of O.L. Lange. *In*: Bibliotheca Lichenologica **67**: 1, 1997), Nash (*Lichen Biology*, 1996), Quispel (*Encycl. pl. physiol.* **11**: 577, 1959), Seaward (Ed.) (*Lichen ecology*, 1977; ecophysiology), Smith (*Biol. Rev.* **37**: 537, 1962). All these publications have extensive bibliographies.

See also Air pollution, Anaerobic fungi, Antibiotics, Diurnal cycles, Lichens, Light and fungi, Metabolic products, Nutrition, Pigments, Spore discharge, Symbiosis, etc.

Physisporinus P. Karst. (1889), Meripilaceae. 2, Europe; Uganda. See Gilbertson & Ryvarden (*N. Amer. Polyp.* **2**: 628, 1987), Ipulet & Ryvarden (*Syn. Fung.* **20**: 87, 2005; Uganda).

Physisporus Chevall. (1826) nom. rej. prop. ≡ Perenniporia.

Physma A. Massal. (1854), Collemataceae (L). *c.* 12, widespread. See Degelius (*Svensk bot. Tidskr.* **49**: 136, 1955), Verdon & Elix (*Acta Bot. Fenn.* **150**: 209, 1994; key 4 Australasian spp.), Jørgensen & Aptroot (*Lichenologist* **34**: 441, 2002; Taiwan).

Physmaceae Walt. Watson (1929) = Collemataceae.

Physmatomyces E.A. Thomas ex Cif. & Tomas. (1953) = Physma.

Physmatomyces Rehm (1900), ? Helotiales. 1, Brazil.

Physocladia Sparrow (1932), Cladochytriaceae. 1, N. America.

Physocystidium Singer (1962), Tricholomataceae. 1, Trinidad. See Singer (*Persoonia* **2**: 410, 1962).

Physoderma Wallr. (1833), Physodermataceae. 50, widespread. *P. maydis* (maize brown spot), *P. alfalfae* (on *Medicago*). See Karling (*Lloydia* **13**: 29, 1950), Sparrow (*TMSJ* **3**: 16, 1962), Sparrow (*TBMS* **60**: 340, 1973; type), Lange & Olson (*Protoplasma* **102**: 323, 1980), Hadar (*C.R. Acad. Sci. Paris* Sér. III **294**: 329, 1982).

Physodermataceae Sparrow (1952), Blastocladiales. 1 gen. (+ 2 syn.), 50 spp.

Lit.: Hadar (*Compt. Rend. Hebd. Seances Acad. Sci.* Ser. 3 **294**: 329, 1982; in *Spizellomycetales* but this doubtful as TEM unclear), Barr *in* McLaughlin *et al.* (Eds) (*The Mycota* A Comprehensive Treatise on Fungi as Experimental Systems for Basic and Applied Research **7A**: 93, 2001).

Physodontia Ryvarden & H. Solheim (1977), Hymenochaetales. 1, Europe. See Ryvarden & Solheim (*Mycotaxon* **6**: 375, 1977), Larsson (*MR* **111**: 1056, 2007).

Physokermincola Brain (1923) nom. dub., anamorphic *Saccharomycetes*.

Physomitra Boud. (1885) ≡ Gyromitra.

Physomyces Harz (1890) = Monascus fide Hawksworth *et al.* (*Dictionary of the Fungi* edn 8, 1995).

Physonema Lév. (1847), anamorphic *Phragmidium*. 1, widespread (temperate). Anamorph name for (I). See Laundon (*Mycol. Pap.* **102**, 1965).

Physopella Arthur (1906) = Phakopsora. Anamorph name for (II). fide Cummins & Hiratsuka (*Illustr. Gen. Rust Fungi* edn 3: 225 pp., 2003).

Physorhizophidium Scherff. (1926), Chytridiaceae. 1 (on diatoms), Europe; N. America.

Physospora Fr. (1835) nom. dub. = Rhinotrichum fide Sumstine (*Mycol.* **3**: 45, 1911), Donk (*Taxon* **11**: 95, 1962).

Physosporella Höhn. (1919), Hyponectriaceae. 1, Germany. See Barr (*Mycol.* **68**: 611, 1976), Cannon (*MR* **100**: 1409, 1996; as *Plectosphaera*).

Physosporellaceae Höhn. ex M.E. Barr (1976) = Hyponectriaceae.

phytoalexin, a metabolite produced by a plant in response to infection by a fungus or other pathogen (or by an abiotic factor) inhibitory to the invading pathogen. Reviews: Cruickshank, *Ann. Rev. Phytopath.* **1**: 351, 1963; Kuć, *Ann. Rev. Phytopath.* **10**: 207, 1972; Ingram, *Bot. Rev.* **38**: 343, 1972. Cf. antibiotics.

Phytoalexins include: capsidiol, glyceollin, ipomearone, kievitone, phaseollin, pisatin, rishitin, wyerone (q.v.).

phytoalternarin (**A**, **B**, **C**), host-specific toxins produced by *Alternaria kikuchiana* (black spot of Japanese pears; *Pyrus serotina*) (Park *et al.*, *Physiol. Pl. Pathol.* **9**: 167, 1976).

Phytoconis Bory (1797) nom. rej. prop. = Lichenomphalia fide Kuyper (*in litt.*) The name has been used for the anamorph of *Lichenomphalia*. See, Gams (*Öst. bot. Z.* **109**: 376, 1962), Poelt & Jülich (*Herzogia* **1**: 331, 1969), Oberwinkler (*Ber. Deut. Bot. Ges.* **4**: 139, 1970), Redhead & Kuyper (*Arctic Alpine Mycology* **2**: 319, 1987), Redhead & Kuyper (*Mycotaxon* **31**: 221, 1988), Jørgensen & Ryman (*Taxon* **38**: 305, 1989).

Phytocordyceps C.H. Su & H.H. Wang (1986) = Cordyceps fide Su & Wang (*Mycotaxon* **26**: 337, 1986), Sung *et al.* (*Stud. Mycol.* **57**, 2007).

phytolysine, a plant tissue macerating enzyme from *Plowrightia ribesia* (Neaf-Roth *et al.*, *Phytopath. Z.* **40**: 283, 1961).

Phytomyxinae. An order for *Plasmodiophora*, *Phytomyxa*, *Sorosphaera* and *Tetramyxa*.

phytopathology, the branch of science for plant disease; see Plant pathogenic fungi.

Phytosociology. see Mycosociology.

Phytotoxic mycotoxins (toxins injurious to plants; see also phytotoxin (2) under Toxins). May be host specific (e.g. helminthosporoside, phytoalternarins, victorin, q.v.) or non-host specific (e.g. alternaric acid, baccatin, chaetoglobosin, culmomarasmin, diaporthin, fusaric acid, fusicoccin, helminthosporal, macrocyclic trichothecenes, ophiobolin (cochliobolin), skyrin, tentoxin, zearalenone, q.v.).

Fungi producing phytotoxic mycotoxins have potential as mycoherbicides for control of weeds. Phytotoxic mycotoxins biologically active against *Trifolium repens* and *Pueraria montana* have been found among metabolites of a *Sclerotinia* sp. (Yeon-Kyu Hong, *Plant Pathology Journal* **20**: 52, 2004). Macrocyclic trichothecenes produced by *Myrothecium verrucaria* exhibit phytotoxicity to both weeds and crops (beans, corn, tobacco, wheat) causing a range of damage (Cutler & Jarvis, *Environmental and Experimental Botany* **25**: 115, 1985; Abbas *et al.*, *Phytochemistry* **59**: 309, 2002). These mycotoxins including epiroridin, epiisororidin, roridins, trichoverrins and verrucarins are effective against *Lemna pausicostata* and *Pueraria montana* (Abbas *et al.*, *Phytochemistry* **58**: 269, 2001). *Fusarium* toxins such as enniatin and fumonisin, as well as fusaric and dehydrofusaric acids, reportedly have herbicidal properties (Hershenhorn *et al.*, *Plant Science* **86**: 155, 1992; Zonno & Vurro, *Phytoparasitica* **30**: 519, 2002). Seed germination of *Orobanche ramosa* is inhibited by fusaric acids, and *Striga hermonthica*, a serious pest of African crops, is totally suppressed by other fusaric metabolites (Idris *et al.*, *Phytopathologia Mediterranea* **42**: 65, 2003). Mycotoxins produced by *Alternaria alternata*, *Fusarium* species, *Myrothecium verrucaria* and *Phomopsis foeniculi* showed phytotoxic effects, i.e. chlorosis, necrosis and other symptoms, particularly on seedlings (Lamprecht *et al.*, *Phytopathology* **84**: 383, 1994; Corsaro *et al.*, *Carbohydr. Res.* **308**: 349, 1998; Abbas *et al.*, *Phytochemistry* **59**: 309, 2002). Phytotoxic mycotoxins may inhibit translation in eukaryotic ribosomes (Masuda *et al.*, *J. Experimental Botany* **58**: 1617, 2007). Trichoverrins, atranone and some *Fusarium* toxins are also weakly cytotoxic for mammals (Ueno & Ueno, *Toxicology, Biochemistry and Pathology of Mycotoxins*, 107, 1978; Abbas *et al.*, *Phytochemistry* **59**: 309, 2002).

Reviews: Pringle & Scheffer (*Ann. Rev. Phytopath.* **2**: 133, 1964), Wright (*Ann. Rev. Microbiol.* **22**: 269, 1968), Wood *et al.* (Eds) (*Phytotoxins in plant disease*, 1972), Strobel (*Ann. Rev. Pl. Physiol.* **25**: 541, 1974), Durbin (Ed.) (*Toxins in plant diseases*, 1981), Uraguche & Yamazaki [eds] (*Toxicology, Biochemistry and Pathology of Mycotoxins*, 1978).

Lit.: Abbas *et al.* (*Phytochemistry* **58**: 269, 2001; *Phytochemistry* **59**: 309, 2002), Corsaro *et al.* (*Carbohydr. Res.* **308**: 349, 1998), Cutler & Jarvis (*Environmental and Experimental Botany* **25**: 115, 1985), Hershenhorn *et al.* (*Plant Science* **86**: 155, 1992), Idris *et al.* (*Phytopathologia Mediterranea* **42**: 65, 2003), Lamprecht *et al.* *Phytopathology* **84**: 383, 1994), Masuda *et al.* (*J. Experimental Botany* **58**: 1617, 2007), Ueno & Ueno (Toxicology and biochemistry of mycotoxins. In Uraguche & Yamazaki [eds] *Toxicology, Biochemistry and Pathology of Mycotoxins*, 107, 1978), Yeon-Kyu Hong (*Plant Pathology Journal* **20**: 52, 2004), Zonno & Vurro (*Phytoparasitica* **30**: 519, 2002).

phytotoxin, see toxin.

Picardella I.I. Tav. (1985), Laboulbeniaceae. 2, Europe. See Tavares (*Mycol. Mem.* **9**: 276, 1985), Santamaría (*Revta Ibér. Micol.* **6**: 13, 1989).

Piccolia A. Massal. (1856), Lecanorales (L). 1, pantropical. See Hafellner (*Biblthca Lichenol.* **58**: 107, 1995), Hawksworth *et al.* (*Dictionary of the Fungi* edn 8, 1995), Hafellner (*Symb. bot. upsal.* **34** no. 1: 87, 2004).

Piceomphale Svrček (1957), Helotiales. Basal to a clade including *Rustroemiaceae* and *Sclerotiniaceae*. See Holst-Jensen *et al.* (*Mycol.* **89**: 885, 1997; DNA), Gernandt *et al.* (*Mycol.* **93**: 915, 2001; phylogeny), Wang *et al.* (*Mycol.* **98**: 1065, 2006; phylogeny).

Pichia E.C. Hansen (1904), Pichiaceae. 112, widespread. Highly polyphyletic, soon to be subdivided. See Poncet (*Mycopathologia* **57**: 99, 1975; numerical tax.), Miller *et al.* (*Syst. Appl. Microbiol.* **12**: 191, 1989; OFAGE), Kurtzman (*Mycol.* **84**: 72, 1992; DNA-relatedness), Yamada *et al.* (*Biosc., Biotechn., Biochem.* **58**: 1245, 1994; gen. concept), Kurtzman *in* Kurtzman & Fell (Eds) (*Yeasts, a taxonomic study* 4th edn: 273, 1998), Suzuki & Nakase (*J. gen. appl. Microbiol.* Tokyo **45**: 239, 1999), Kurtzman (*Int. J. Syst. Evol. Microbiol.* **50**: 395, 2000), Mikata & Ueda-Nishimura (*Antonie van Leeuwenhoek* **77**: 159, 2000), Masih *et al.* (*FEMS Microbiol. Lett.* **202**: 109, 2001; biocontrol), Ueda-Nishimura & Mikata (*Antonie van Leeuwenhoek* **79**: 371, 2001), Schweigkofler *et al.* (*Organ. Divers. Evol.* **2**: 1, 2002; phylogeny), Suh & Blackwell (*FEMS Yeast Res.* **5**: 87, 2004; beetle associates), Sutar *et al.* (*Journal of Medical Microbiology* **53**: 119, 2004; strain typing), Friel *et al.* (*Journal of Applied Microbiology* **98**: 783, 2005; genetic diversity), Suh *et al.* (*Mycol.* **98**: 1006, 2006; phylogeny), Villa-Carvajal *et al.* (*Antonie van Leeuwenhoek* **90**: 171, 2006; molecular identification), Wu *et al.* (*FEMS Yeast Res.* **6**: 305, 2006; genetic diversity), Rivera *et al.* (*Biol. J. Linn. Soc.* **93**: 475, 2008; experimental evolution).

Pichiaceae Zender (1925), Saccharomycetales. 6 gen. (+ 18 syn.), 133 spp.

Lit.: Sugiyama *et al.* (*Mycol.* **98**: 996, 2006), Suh *et al.* (*Mycol.* **98**: 1006, 2006; phylogeny).

Picoa Vittad. (1831), Pezizales. 4 (hypogeous), widespread. See Montecchi (*Micologia e Vegetazione Mediterranea* **13**: 131, 1999), Moreno *et al.* (*MR* **104**: 378, 2000), Popolizio (*Riv. Micol.* **49**: 155, 2006; Italy), Læssøe & Hansen (*MR* **111**: 1075, 2007; phylogeny).

Picromyces Battarra ex Earle (1909) ≡ Hebeloma.

Pidoplitchkoviella Kiril. (1975), Xylariales. 1 (from soil), British Islesraine. See von Arx *et al.* (*Beih. Nova Hedwigia* **94**: 104, 1988), Suh & Blackwell (*Mycol.* **91**: 836, 1999; phylogeny).

piedra, black, see *Piedraia*; **white -**, see *Trichosporon*.

Piedraia Fonseca & Leão (1928), Piedraiaceae. Anamorph *Trichosporon*. 2 (on hair), widespread (tropical). *P. hortae* (**black piedra** of humans). See Boedijn (*Mycopathologia* **11**: 354, 1959), van Uden *et al.* (*Revta bras. Biol.* **3**: 271, 1963), Takashio & Vanbreuseghem (*Mycol.* **63**: 612, 1971), de Hoog & Guého *in* Ajello & Hay (Eds) (*Topley & Wilson's Microbiology and Microbial Infections* edn 9 **4**: 191, 1998), Schoch *et al.* (*Mycol.* **98**: 1041, 2006; phylogeny).

Piedraiaceae Viégas ex Cif., Bat. & S. Camposa (1956), Capnodiales. 1 gen. (+ 4 syn.), 2 spp.

Lit.: Hoog & Guého (*Topley & Wilson's Microbiology and Microbial Infections* Edn 9. Vol. 4. Medical Mycology: 191, 1998), Kane & Summerbell (*Manual of Clinical Microbiology*: 1275, 1999), Lumbsch & Lindemuth (*MR* **105**: 901, 2001), Schoch *et al.* (*Mycol.* **98**: 1041, 2006; phylogeny).

Piemycus Raf. (1813) ≡ Piesmycus.

Piersonia Harkn. (1899) = Choiromyces fide Trappe

(*TBMS* **65**: 496, 1975).

Piesmycus Raf. (1808) ? = Bovista fide Stalpers (*in litt.*).

Pietria fungaia, see stone-fungus.

Piggotia Berk. & Broome (1851), anamorphic *Venturiaceae*, Cac.0eP.19. 3, widespread (temperate). See Sutton (*The Coelomycetes*, 1980), Braun (*Arnoldia* **15**: 44, 1998), Johnston (*N.Z. Jl Bot.* **37**: 703, 1999).

Pigments. The text of this entry is unrevised and remains the same as for the ninth edition. Characteristic pigments are produced by a wide variety of fungi. Species of *Drechslera* s.l. give hydroxyanthraquinones (e.g. helminthosporin ('maroon' in colour), catenarin (red), cynodontin ('bronze'), tritisporin (red brown)); the similar compound, erythroglaucin (red) is produced by forms of *Aspergillus glaucus* which gives in addition, auroglaucin (orange) and flavoglaucin (yellow). Among other pigments, investigations have been made on aurofusarin (orange yellow) and rubrofusarin from *Fusarium culmorum*; aurantin (yellow) and oosporin (purple-brown with ferric chloride ($FeCl_3$) from *Chrysonilia sitophila* (teleomorph *Neurospora sitophila*); boletol (blue) from *Boletus luridus* and other species; citromycetin, chrysogenin, citrinin, fulvic acid, and other yellow pigments from *Penicillium*. Many mycotoxins are pigmented, e.g. naphthoquinones from *Penicillium* and *Aspergillus*. 'Yellow rice' refers to the colour of rice infected by pigment and toxin producing species of *Penicillium*. Important fungal sex hormones such as trisporic acid from *Blakeslea trispora* are carotenoid pigments (Turner & Aldridge, *Fungal metabolites* II, 1983). Pigments from fungi are involved in biodeterioration of commodities, such as paint. The Chinese red rice (ang-kak, q.v.) is used for colouring food. See Wolf (*J. Elisha Mitch. Sci. Soc.* **89**: 184, 1973; synthesis, 261 refs).

Many pigments occur in lichens and may be red (e.g. rhodophyscin), yellow (e.g. usnic acid (q.v.), stictaurin), orange (e.g. parietin, q.v.), bright emerald green (e.g. vulpinic acid), or brown (e.g. *Parmelia*-brown). The chemical structures of pigments in thalli are mainly known whilst those of apothecial tissues (e.g. epithecium colours) are less understood. Lichen pigments are mainly pulvic acid derivatives, usnic acids, and anthraquinones.

See see Carotene, Dyeing, Metabolic products, Wood attacking fungi.

Pila Speg. (1923) [non *Pila* C.E. Bertrand & Renault 1892, fossil *Algae*] ≡ Gastropila.

Pilacre Fr. (1829) nom. dub., Agaricomycetes. See Donk (*Persoonia* **4**: 325, 1966).

Pilacrella J. Schröt. (1887) = Atractiella fide Oberwinkler & Bauer (*Sydowia* **41**: 224, 1989).

Pilaira Tiegh. (1875), Mucoraceae. 4 (esp. coprophilous), widespread (north temperate). See Zycha *et al.* (*Mucorales*, 1969; key), Mil'ko (*Mikol. Fitopatol.* **4**: 262, 1970), Fletcher (*J. Biol. Educ.* **5**: 229, 1971), Fletcher (*TBMS* **61**: 553, 1973; phototropism), Wood & Cooke (*TBMS* **86**: 672, 1986; parasitism), Wood & Cooke (*TBMS* **88**: 247, 1987; nutrition), Voigt & Wöstemeyer (*Gene* **270**: 113, 2001; phylogeny).

Pilát (Albert; 1903-1974; Czech). Student of Velenovský (q.v.), Charles University (1926); Scientific Assistant and Research Officer (1930-1948) then Head, Botanical Department (1948-1965) then Head, Mycological Department (1965-1974), National Museum, Prague. An authority on basidiomycetes, particularly wood-inhabiting and mycorrhizal species, particularly of Europe and temperate Asia; rediscovered the lost collections of Corda (q.v.). *Publs. Atlas des Champignons de l'Europe* (1934-1948) [including *Pleurotus* (1935); Polyporaceae (1936-1942); *Agaricus* (1951)]; *Houby Československa ve Svém Životním Prostředí* (1969). *Biogs, obits etc.* Kotlaba & Pouzar (*TBMS* **65**: 163, 1975) [portrait]; Singer (*Mycol.* **67**: 445, 1975) [portrait]; Stafleu & Cowan (*TL-2* **4**: 265, 1983).

Pilatia Velen. (1934) = Urceolella fide Raschle (*Sydowia* **29**: 170, 1977).

Pilatoporus Kotl. & Pouzar (1990) = Fomitopsis fide Ryvarden (*Syn. Fung.* **5**: 201, 1991).

pileate, having a pileus.

pileipellis, see pellis.

pileocystidium, see cystidium.

Pileodon P. Roberts & Hjortstam (1998), Gloeophyllales. 2, Brunei; Philippines. See Hjortstam *et al.* (*Kew Bull.* **53**: 817, 1998), Nakasone (*Sydowia* **56**: 258, 2004).

Pileolaria Castagne (1842), Pileolariaceae. *c.* 15 (on *Anacardiaceae*), widespread. See Hüseyın & Selçuk (*Pakist. J. Bot.* **36**: 203, 2004; Turkey).

Pileolariaceae Cummins & Y. Hirats. (1983), Pucciniales. 4 gen. (+ 3 syn.), 34 spp.

Lit.: Gardner & Hodges (*Mycol.* **77**: 575, 1985), Hiratsuka (*Mycotaxon* **31**: 517, 1988), Dianese *et al.* (*Fitopatol. Brasil* **18**: 342, 1993), Chen *et al.* (*Mycoscience* **37**: 91, 1996), Walker (*Australasian Mycologist* **20**: 3, 2001), Cummins & Hiratsuka (*Illustr. Gen. Rust Fungi* edn 3: 225 pp., 2003), Wingfield *et al.* (*Australas. Pl. Path.* **33**: 327, 2004), Aime (*Mycoscience* **47**: 112, 2006).

pileolus, a small pileus.

pileus, the hymenium-supporting part of the basidioma of non-resupinate *Agaricomycetes*; cap (Fig. 4A).

Pilgeriella Henn. (1900), Parodiopsidaceae. Anamorph *Septoidium*. 2, S. America.

Pilidiella Petr. & Syd. (1927), anamorphic *Schizoparme*. 12, widespread. See Sutton (*The Coelomycetes*, 1980), Castlebury *et al.* (*Mycol.* **94**: 1017, 2002; phylogeny), Van Niekerk *et al.* (*MR* **108**: 283, 2004; phylogeny), Rossman *et al.* (*Mycoscience* **48**: 135, 2007; review).

Pilidium Kunze (1823), anamorphic *Discohainesia*, St.0eH.15. 2, widespread. See Sutton (*The Coelomycetes*, 1980), Palm (*Mycol.* **83**: 787, 1992; synanamorph *Hainesia*), Rossman *et al.* (*Mycol. Progr.* **3**: 275, 2004; phylogeny), Wang *et al.* (*Mycol.* **98**: 1065, 2006; phylogeny).

Piligena Schumach. (1821) = Onygena fide Saccardo (*Syll. fung.* **20**: 230, 1911).

Pilimelia W.D. Kane (1966), Actinobacteria. q.v.

Piline Theiss. (1917) = Perisporiopsis Henn. fide Müller & von Arx (*Beitr. Kryptfl. Schweiz* **11** no. 2, 1962).

Pillulasclerotes Stach & Pickh. (1957), Fossil Fungi. 1 (Carboniferous), Germany.

Pilobolaceae Corda (1842), Mucorales. 2 gen. (+ 4 syn.), 6 spp.

Lit.: Kirk & Benny (*TBMS* **75**: 123, 1980), Hu *et al.* (*Mycosystema* **2**: 111, 1989), Cacialli *et al.* (*Funghi e Ambiente* **78–79**: 13, 1998), Voigt & Wöstemeyer (*Gene* **270**: 113, 2001).

Pilobolus Tode (1784), Pilobolaceae. 5 (coprophilous), widespread. See Grove *in* Buller (*Researches on Fungi* **6**: 190, 1934), Zycha *et al.* (*Mucorales*, 1969),

Nand & Mehrotra (*Sydowia* **30**: 283, 1978; spp.), Bourret & Keierleber (*Arch. Mikrobiol.* **126**: 43, 1980; spore germination), Foos & Royer (*CJB* **65**: 1063, 1987; growth temp), Hu *et al.* (*Mycosystema* **2**: 111, 1989; revision), Eysker (*Res. Vet. Sci.* **50**: 29, 1991; dispersal of cattle lungworm), Horie *et al.* (*Mycoscience* **39**: 463, 1998; gravitropic response), Voigt & Wöstemeyer (*Gene* **270**: 113, 2001; phylogeny), White *et al.* (*Mycol.* **98**: 872, 2006; phylogeny).

Pilocarpaceae Zahlbr. (1905), Lecanorales (L). 11 gen. (+ 11 syn.), 259 spp.

Lit.: Awasthi & Mathur (*Proc. Indian Acad. Sci.* Pl. Sci. **97**: 481, 1987), Kalb & Vězda (*Nova Hedwigia* **51**: 435, 1990), Lücking (*Nova Hedwigia* **52**: 267, 1991), Lücking *et al.* (*Bot. Acta* **107**: 393, 1994), Cáceres (*Mycotaxon* **71**: 383, 1999), Thor *et al.* (*Symb. bot. upsal.* **32**: 72 pp., 2000), Andersen & Ekman (*Lichenologist* **36**: 27, 2004), Andersen & Ekman (*MR* **109**: 21, 2005).

Pilocarpon Vain. (1890) = Byssoloma fide Hawksworth *et al.* (*Dictionary of the Fungi* edn 8, 1995).

Pilocintractia Vánky (2004), ? Anthracoideaceae. 2 (in seeds of *Fimbristylis* (*Cyperaceae*)), S. Asia; Australia; C. & S. America. See Vánky (*Mycol. Balcanica* **1**: 169, 2004), Vánky (*Mycotaxon* **95**: 1, 2006; key).

Pilocratera Henn. (1891) ≡ Cookeina.

Piloderma Jülich (1969), Atheliaceae. 6, widespread. See Jülich (*Willdenowia* Beih. **7**, 1972).

Pilodermataceae Jülich (1982) = Atheliaceae.

Pilonema Nyl. (1860) = Polychidium fide Hawksworth *et al.* (*Dictionary of the Fungi* edn 8, 1995).

Pilopeza Fr. (1849) ≡ Psilopezia.

Pilophora Wallr. (1833) [non *Pilophora* Jacq. 1802, *Palmae*] = Rhizopus fide Hesseltine (*Mycol.* **47**: 344, 1955) See *Crinofera*.

Pilophoraceae Stizenb. (1862) = Cladoniaceae.

Pilophoron (Tuck.) Tuck. (1858) = Pilophorus fide Hawksworth *et al.* (*Dictionary of the Fungi* edn 8, 1995).

Pilophorum Nyl. (1857) ≡ Pilophorus.

Pilophorus Th. Fr. (1857), Cladoniaceae (L). 11, widespread (esp. temperate). See Jahns (*Lichenologist* **4**: 199, 1970; Eur.), Hawksworth *et al.* (*Taxon* **21**: 327, 1972; nomencl.), Jahns (*Mycotaxon* **13**: 289, 1981; monogr., key), Hertel & Rambold (*Pl. Syst. Evol.* **158**: 289, 1989; gen. concept), Stenroos & DePriest (*Am. J. Bot.* **85**: 1548, 1998; phylogeny), Wedin & Döring (*MR* **103**: 1131, 1999; DNA), Wedin *et al.* (*Lichenologist* **32**: 171, 2000; phylogeny), Stenroos *et al.* (*Mycol. Progr.* **1**: 267, 2002; phylogeny), Harada & Yoshimura (*Lichenology* **3**: 11, 2004; nomencl.), Myllys *et al.* (*Taxon* **54**: 605, 2005; phylogeny), Miądlikowska *et al.* (*Mycol.* **98**: 1088, 2006; phylogeny).

Piloporia Niemelä (1982), Polyporaceae. 2, widespread. See Niemelä (*Karstenia* **22**: 13, 1982).

Pilosace (Fr.) Quél. (1873) nom. dub., Agaricales. See Singer (*Agaric. mod. Tax.*, 1951).

pilose, covered with hairs.

Pilosporella E.I. Hazard & Oldacre (1975), Microsporidia. 2.

Pilula Harker, Sarjeant & Caldwell (1990), Fossil Fungi. 1.

Pilula Massee (1910) = Dimerina fide Clements & Shear (*Gen. Fung.*, 1931).

Pilulina Arnaud (1954), anamorphic *Pezizomycotina*, Hso.0eP.?. 1, France.

Pimina Grove (1888) = Zygosporium fide Mason (*Mycol. Pap.* **5**, 1941).

Piminella Arnaud (1954) = Gonytrichum fide Gams & Holubová-Jechová (*Stud. Mycol.* **13**, 1976).

Pinacisca A. Massal. (1854) = Hymenelia fide Hawksworth *et al.* (*Dictionary of the Fungi* edn 8, 1995).

Pinatubo J.B. Manandhar & Mew (1996), anamorphic *Pezizomycotina*, Hso.?.?. 1, Philippines. See Manandhar & Mew (*Mycotaxon* **60**: 203, 1996).

Pindara Velen. (1934) = Helvella. Clusters within *Helvella*, but that genus is polyphyletic. fide Svrček (*Česká Mykol.* **1**: 45, 1947), Kristiansen (*Agarica* **5**: 105, 1984), Landvik *et al.* (*Mycol.* **91**: 278, 1999).

pine moss, species of *Alectoria* and *Bryoria*.

pine mushroom, see matsu-take.

Pinnaticoemansia Kurihara & Degawa (2006), Kickxellaceae. 1, Japan. See Kurihawa & Degawa (*Mycoscience* **47**: 205, 2006).

Pinoyella Castell. & Chalm. (1919) = Trichophyton fide Hawksworth *et al.* (*Dictionary of the Fungi* edn 8, 1995).

Pinuzza Gray (1821) = Suillus Gray fide Singer (*Farlowia* **2**: 223, 1945).

Pionnotes Fr. (1849) = Fusarium fide Wollenweber & Reinking (*Die Fusarien*: 6, 1935).

pionnotes (of *Fusarium*), a spore mass having a fat-like or grease-like appearance; **pseudo-**, see sporodochium.

Pionospora Th. Fr. (1859) = Pertusaria fide Hawksworth *et al.* (*Dictionary of the Fungi* edn 8, 1995), Boqueras & Llimona (*Mycotaxon* **88**: 471, 2003).

Piperivora Siboe, P.M. Kirk & P.F. Cannon (1999), anamorphic *Pezizomycotina*, Hso.?.?. 1, Kenya. See Siboe *et al.* (*Mycotaxon* **73**: 298, 1999).

Piptarthron Mont. ex Höhn. (1918), anamorphic *Planistroma*, Cpd.0eH/≡ eH.1. 7 (on *Agavaceae*), widespread. See Sutton (*The Coelomycetes*, 1980), Sutton (*TBMS* **81**: 407, 1983), Ramaley (*Mycotaxon* **42**: 63, 1991; teleomorph), Ramaley (*Mycotaxon* **55**: 255, 1995; key 6 spp.), Ramaley (*Mycotaxon* **66**: 509, 1998).

Piptocephalidaceae J. Schröt. (1886), Zoopagales. 3 gen. (+ 9 syn.), 70 spp.

Lit.: Benjamin (*Aliso* **4**: 321, 1959), Benjamin (*Aliso* **5**: 273, 1963), Benjamin (*Mycol.* **48**: 1, 1966), Patil & Patil (*Indian Phytopath.* **47**: 217, 1994), Tanabe *et al.* (*Mol. Phylogen. Evol.* **30**: 438, 2004).

Piptocephalis de Bary (1865), Piptocephalidaceae. 24 (mycoparasites esp. of *Mucorales*), widespread. See Dobbs & English (*TBMS* **37**: 375, 1954; *P. xenophila* on non-mucoralean hosts), Leadbetter & Mercer (*TBMS* **39**: 17, 1956), Leadbetter & Mercer (*TBMS* **40**: 109, 1957; zygospores), Benjamin (*Aliso* **4**: 321, 1959), Berry (*Mycol.* **51**: 824, 1961; parasitism), Zycha *et al.* (*Mucorales*, 1969; key), Kuzuha (*J. Jap. Bot.* **51**: 123, 1976), Kirk (*TBMS* **70**: 335, 1978), Jeffries (*Mycol.* **71**: 209, 1979), Manocha (*CJB* **63**: 772, 1985; mycoparasitism), Manocha & McCullough (*CJB* **63**: 967, 1985; mycoparasitism), Gräfenhan (*A Taxonomic Revision of the Genus Piptocephalis (Fungi)*, 1998), Ho (*Taiwania* **49**: 188, 2004; Taiwan), Tanabe *et al.* (*J. gen. appl. Microbiol.* Tokyo **51**: 267, 2005; phylogeny), White *et al.* (*Mycol.* **98**: 872, 2006; phylogeny).

Piptoporaceae Jülich (1982) = Fomitopsidaceae.

Piptoporus P. Karst. (1881), Fomitopsidaceae. 4, widespread. *P. betulinus*, razor strop fungus, on birch. See Corner (*Beih. Nova Hedwigia* **78**: 140, 1984).

Piptostoma Berk. & Broome (1875) nom. dub., Fungi. See Petch (*Ann. R. bot. Gdns Peradeniya* **6**: 341, 1917) ? lapsus for *Piptostomum*.

Piptostomum Lév. (1845) nom. dub., anamorphic *Pezizomycotina*. See Höhnel (*Sber. Akad. Wiss. Wien* Math.-naturw. Kl., Abt. 1 **119**: 617, 1910).

Piptostroma Fr. (1849) ≡ Piptostomum.

Pirella Bainier (1882), Mucoraceae. 2, widespread. See Benny (*A taxonomic revision of the Thamnidiaceae (Mucorales)*, 1973), Benny & Schipper (*Mycol.* **84**: 52, 1992), Voigt & Wöstemeyer (*Gene* **270**: 113, 2001; phylogeny), Liu (*Mycosystema* **23**: 301, 2004; China).

Pirex Hjortstam & Ryvarden (1985), Meruliaceae. 1, N. America. See Hjortstam & Ryvarden (*Mycotaxon* **24**: 287, 1985).

Piricauda Bubák (1914), anamorphic *Pezizomycotina*, Hso.#eP.24. 4, neotropics. See Hughes (*CJB* **38**: 921, 1960), Mercado Sierra *et al.* (*Mycotaxon* **63**: 155, 1997; Mexico), Mercado Sierra *et al.* (*MR* **109**: 723, 2005; monogr.).

Piricaudilium Hol.-Jech. (1988), anamorphic *Pezizomycotina*, Hso.#eP.24. 1, Cuba. See Holubová-Jechová (*Česká Mykol.* **42**: 200, 1988), Heredia Abarca *et al.* (*Mycotaxon* **64**: 203, 1997).

Piricaudiopsis J. Mena & Mercado (1987), anamorphic *Pezizomycotina*, Hso.#eP.24. 1, Cuba. See Mena & Mercado (*Acta Bot. Cubana* **51**: 1, 1987).

Piricularia, see *Pyricularia*.

piricularin, a phytotoxin from *Pyricularia oryzae* (Togashi *et al.*, *Ann. phytopath. Soc. Japan* **25**: 142, 1960).

piriform, see pyriform.

Piriformospora Sav. Verma, Aj. Varma, Rexer, G. Kost & P. Franken (1998), Sebacinales. 1, India. See Verma *et al.* (*Mycol.* **90**: 897, 1998), Begerow *et al.* (*MR* **108**: 1257, 2004; posn.).

Piringa Speg. (1910) [non *Piringa* Juss. 1820, *Rubiaceae*] = Camarosporium fide Petrak & Sydow (*Annls mycol.* **33**: 157, 1935).

Pirispora Faurel & Schotter (1966), anamorphic *Pezizomycotina*, Cpd.0eH.?. 1, Algeria. See Faurel & Schotter (*Revue Mycol.* Paris **30**: 341, 1965).

Piriurella Cookson & Eisenack (1979), Fossil Fungi. 2 (Cretaceous – Tertiary), Australia; Canada. ? = Alternaria (Hyphom.) fide Smith & Chaloner (*N. Jb. Geol. Paläont. Mh.* **11**: 701, 1979).

Pirobasidium Höhn. (1902) = Coryne fide Groves & Wilson (*Taxon* **16**: 35, 1967).

Pirogaster Henn. (1901) = Scleroderma fide Demoulin (*in litt.*).

Piromonas E. Liebet. (1910), Chytridiomycetes. 3, Europe. See Orpin (*J. gen. Microbiol.* **99**: 107, 1977; life history).

Piromyces J.J. Gold, I.B. Heath & Bauchop (1988), Neocallimastigaceae. 8 (anaerobic rumen fungi), widespread. See Gaillard-Martinie *et al.* (*Curr. Microbiol.* **24**: 159, 1992), Ho *et al.* (*Fungal Diversity* **4**: 37, 2000), Wubah (*Biodiversity of Fungi* Inventory and Monitoring Methods: 501, 2004).

Pirostoma (Fr.) Fuckel (1870) nom. dub., anamorphic *Pezizomycotina*. See Sutton (*Mycol. Pap.* **141**, 1977).

Pirostoma (Fr.) Sacc. (1896) nom. dub., anamorphic *Pezizomycotina*. See Sutton (*Mycol. Pap.* **141**, 1977).

Pirostomella Sacc. (1914), anamorphic *Pezizomycotina*, Hsy.0eP.?. 2, Philippines.

Pirottaea Sacc. (1878), Helotiales. 26, widespread (esp. north temperate). See Nannfeldt (*Symb. bot. upsal.* **25** no. 1: 1, 1985; key 23 spp.), Gamundí (*Fl. criptog. Tierra del Fuego* **10**, 1987), Johnston (*N.Z. Jl Bot.* **36**: 645, 1998; NZ spp.), Nauta & Spooner (*Mycologist* **14**: 65, 2000; key British spp.).

Pirozynskia Subram. (1972), anamorphic *Maurodothina*, Hso.1eP.24. 2, Canada; USA. See Subramanian (*Curr. Sci.* **41**: 711, 1972).

Pirozynskiella S.J. Hughes (2007), Pezizomycotina. 1 (on sooty moulds), pantropical. See Hughes (*Mycol.* **99**: 628, 2007).

pisatin, a phytoalexin (q.v.) from pea (*Pisum sativum*).

Pisocarpiaceae Corda (1838) = Pisolithaceae.

Pisocarpium Link (1808) ≡ Pisolithus.

Pisolithaceae Ulbr. (1928) = Sclerodermataceae.

Pisolithus Alb. & Schwein. (1805), Sclerodermataceae. 5 (ectomycorrhizal), widespread. See Marx (*Can. J. Microbiol.* **23**: 217, 1977; ecology), Martin *et al.* (*New Phytol.* **139**: 341, 1998; taxonomy), Chambers & Cairney (*Pisolithus – Ectomycorrhizal fungi, key genera in profile*, 1999; ecology), Martin *et al.* (*New Phytol.* **153**: 345, 2002; phylogeography), Kanchanaprayudh *et al.* (*Mycoscience* **44**: 287, 2003; phylogeny), Thomas *et al.* (*Mycotaxon* **87**: 405, 2003; hypogeous species from Australia).

Pisomyxa Corda (1837), ? Eurotiales. 2, S. America; Madeira.

Pistillaria Fr. (1821), Typhulaceae. *c.* 50, widespread (north temperate). See Berthier (*Bull. mens. Soc. linn. Lyon* Num. Spéc. **46**, 1976), Siepe (*Z. Mykol.* **65**: 187, 1999) = Typhula (Typhul.) fide.

Pistillina Quél. (1880), Typhulaceae. 4, widespread (north temperate). See Corner (*Ann. Bot. Mem.* [A monograph of Clavaria and allied genera] **1**, 1950).

Pithoascaceae Benny & Kimbr. (1980) = Microascaceae.

Pithoascina Valmaseda, A.T. Martínez & Barrasa (1987) = Eremomyces fide Malloch & Sigler (*CJB* **66**: 1929, 1988).

Pithoascus Arx (1973), Microascaceae. Anamorphs *Arthrographis*-like, *Scopulariopsis*. 6, widespread. See von Arx (*Persoonia* **7**: 367, 1973; key), Malloch (*SA* **6**: 147, 1987; status), Valmaseda *et al.* (*CJB* **65**: 1802, 1987; concept), von Arx *et al.* (*Beih. Nova Hedwigia* **94**, 1988), Abbott *et al.* (*Mycol.* **94**: 362, 2002; synonymy with *Microascus*), Issakainen *et al.* (*Medical Mycology* **41**: 31, 2003; phylogeny).

Pithomyces Berk. & Broome (1875), anamorphic *Leptosphaerulina*, Hso.≡ -#P.1. 20, widespread. See facial eczema. See Ellis (*Mycol. Pap.* **76**, 1960; key), Rao & de Hoog (*Stud. Mycol.* **28**: 1, 1986; key), Roux (*TBMS* **86**: 319, 1986; teleomorph), Morgan-Jones (*Mycotaxon* **30**: 29, 1987; synonymy), Brewer *et al.* (*Proc. N.S. Inst. Sci.* **38**: 73, 1989; production of sporidesmin and sporidesmolides), Zhang & Zhang (*Mycotaxon* **85**: 241, 2003; China).

Pithosira Petr. (1949), anamorphic *Pezizomycotina*, Hsp.0eP.?. 1, S. America. See von Arx (*Persoonia* **11**, 1981; redescr.).

Pithospermum Mont. ex Berk. & Broome (1856) = Sporoschisma fide Mussat (*Syll. fung.* **15**: 294, 1901).

Pithya Fuckel (1870), Sarcoscyphaceae. Anamorph *Molliardiomyces*. 5, widespread (north temperate). See Nannfeldt (*Svensk bot. Tidskr.* **43**: 468, 1949), Denison (*Mycol.* **64**: 616, 1972), Meléndez-Howell *et al.* (*Anales Inst. Biol. Univ. Nac. México* **60**: 9, 1990), Harrington *et al.* (*Mycol.* **91**: 41, 1999; phylogeny).

Pithyella Boud. (1885), Hyaloscyphaceae. 8, widespread. See Korf & Zhuang (*Mycotaxon* **29**: 1, 1987), Galán *et al.* (*MR* **98**: 1137, 1994), Döbbeler (*Feddes Repert.* **115**: 5, 2004; on bryophytes).

Pittierodothis Chardón (1939) = Protoscypha fide Kutorga & Hawksworth (*SA* **15**: 1, 1997).

Pittostroma Kowalski & T.N. Sieber (1992), anamorphic *Pezizomycotina*. 1, Poland. See Kowalski & Sieber (*MR* **96**: 685, 1992).

Pitya, see *Pithya*.

pityriasis versicolor ('tinea versicolor'), a superficial skin disease of humans (*Malassezia furfur*).

Pityrosporum Sabour. (1904) = Malassezia fide Keddie (*XIII Int. Congr. Derm. (Munich)*: 867, 1968) See pityriasis versicolor.

Placella Syd. (1938), anamorphic *Pezizomycotina*, Cpt.0eH.?. 1, Australia.

Placentaria Auersw. ex Rabenh. (1851) ? = Periola fide Hawksworth *et al.* (*Dictionary of the Fungi* edn 8, 1995).

Placidiopsis Beltr. (1858), Verrucariaceae (L). 20, widespread. See Thomson (*Bryologist* **90**: 27, 1987; key 2 N. Am. spp.), Breuss (*Nova Hedwigia* **58**: 229, 1994; key 2 N. Afr. spp.), Breuss (*Öst. Z. Pilzk.* **5**: 65, 1996; revision), Gueidan *et al.* (*MR* **111**: 1145, 2007; phylogeny).

Placidium A. Massal. (1855), Verrucariaceae (L). 30, widespread. See Brusse (*Annln naturh. Mus. Wien* Ser. B, Bot. Zool. **98**: 35, 1996; key), Gueidan *et al.* (*MR* **111**: 1145, 2007; phylogeny).

Placoasterella Sacc. ex Theiss. & Syd. (1915), Asterinaceae. 3, Africa; Australia. See Tyson & Griffiths (*TBMS* **66**: 249, 1976; morphology, ultrastr.).

Placoasterina Toro (1930), ? Asterinaceae. 1, S. America.

Placocarpus Trevis. (1860), Verrucariaceae (L). 1, Europe. See Breuss (*Pl. Syst. Evol.* **148**: 313, 1985), Gueidan *et al.* (*MR* **111**: 1145, 2007; phylogeny).

Placocrea Syd. (1939), ? Mycosphaerellaceae. 1, Ecuador. See Petrak (*Sydowia* **6**: 336, 1952).

Placoderma (Ricken) Ulbr. (1928) ≡ Piptoporus fide Donk (*Verh. K. ned. Akad. Wet.* tweede sect. **62**: 1, 1974).

Placodes Quél. (1886) ≡ Fomes.

Placodiella (Zahlbr.) Szatala (1941) = Solenopsora fide Hafellner (*Beih. Nova Hedwigia* **79**: 241, 1984).

placodioid (of a crustose lichen thallus), disc-shaped with plicate lobes at the circumference (Fig. 21C).

placodiomorph, a 2-celled spore with a thickened septum which may or may not have a pore. Cf. polarilocular.

Placodiomyces E.A. Thomas ex Cif. & Tomas. (1953) = Lecanora fide Hawksworth *et al.* (*Dictionary of the Fungi* edn 8, 1995).

Placodion Adans. (1763) nom. rej. = Peltigera fide Hawksworth *et al.* (*Dictionary of the Fungi* edn 8, 1995).

Placodiplodia Bubák (1916), anamorphic *Pezizomycotina*, St.1eP.15. 2, Philippines. See Buchanan (*Mycol. Pap.* **156**, 1987).

Placodium (Ach.) DC. (1805) = Caloplaca fide Hawksworth *et al.* (*Dictionary of the Fungi* edn 8, 1995).

placodium, see stroma, thyriothecium.

Placodium Weber ex F.H. Wigg. (1780) = Xanthoria fide Hawksworth *et al.* (*Dictionary of the Fungi* edn 8, 1995).

Placodium Hill (1773) nom. dub., Fungi (L). See Laundon (*in litt.*).

Placodothis Syd. (1928), ? Pezizomycotina. 1, West Indies.

Placographa Th. Fr. (1860) = Lithographa fide Hawksworth *et al.* (*Dictionary of the Fungi* edn 8, 1995).

Placoheppia (Zahlbr.) Oxner (1956) nom. inval. = Heppia fide Hawksworth *et al.* (*Dictionary of the Fungi* edn 8, 1995).

placoid, see placodioid.

Placolecania (J. Steiner) Zahlbr. (1906) = Solenopsora fide Hawksworth *et al.* (*Dictionary of the Fungi* edn 8, 1995).

Placolecanora Räsänen (1940) ≡ Protoparmeliopsis.

Placolecidaceae Hafellner (1984) = Catillariaceae.

Placolecis Trevis. (1857), Catillariaceae (L). 1, Europe. See Schneider (*Biblthca Lichenol.* **13**, 1980), Makryi (*Botanicheskiĭ Zhurnal* **88**: 123, 2003; Russia).

Placomaronea Räsänen (1944), Candelariaceae (L). 2 or 3, S. America. See Osorio (*Bryologist* **77**: 463, 1974), Poelt (*Phyton* Horn **16**: 189, 1974; key), Westberg *et al.* (*MR* **111**: 1277, 2007; phylogeny).

Placomelan Cif. (1962), anamorphic *Dothidasteroma*, Cpt.0eH.?. 1, Dominican Republic. See Ciferri (*Atti Ist. bot. Univ. Lab. crittog. Pavia* sér. 5 **19**: 124, 1962).

placomycetoid, pileus with diam : stipe ratio ‹1. Cf. campestroid.

Placonema (Sacc.) Petr. (1921), anamorphic *Pezizomycotina*, St.0-1eH.19. 3, S. America. See Nag Raj (*Coelomycetous Anamorphs with Appendage-bearing Conidia*, 1993).

Placonemina Petr. (1921), anamorphic *Pezizomycotina*, St.1eH.15. 1, former Czechoslovakia.

Placoparmelia Henssen (1992), Parmeliaceae (L). 1, Argentina. See Henssen (*Lichenologist* **24**: 134, 1992), Calvelo (*Lichenology in Latin America. History, Current Knowledge and Applications* [Proceedings of GLAL-3, Terceiro Encontro do Grupo Latino-Americano de Liquenólogos, São Paulo, Brazil, 24-28 September, 1997]: 117, 1998; key).

Placopezizia Höhn. (1916) = Leptotrochila fide Schüepp (*Phytopath. Z.* **36**: 236, 1959).

Placophomopsis Grove (1921) = Phomopsis (Sacc.) Bubák. fide Sutton (*Mycol. Pap.* **141**, 1977).

Placopsis (Nyl.) Linds. (1867), Trapeliaceae (L). 56, widespread. See Lamb (*Lilloa* **13**: 151, 1947), Brodo (*Biblthca Lichenol.* **57**: 59, 1995; key 3 N. Am. spp.), Moberg & Carlin (*Symb. bot. upsal.* **31** no. 3: 319, 1996; Norway), Moberg & Carlin (*Lichenologist* **31**: 647, 1999; Africa), Galloway (*Biblthca Lichenol.* **78**: 49, 2001; New Zealand), Lumbsch *et al.* (*MR* **105**: 16, 2001; phylogeny), Makarova & Shapiro (*Nov. sist. Niz. Rast.* **35**: 167, 2001; Russia), Galloway (*Mitteilungen aus dem Institut für Allgemeine Botanik Hamburg* **30-32**: 79, 2002; S America, key), Schmitt *et al.* (*Mycol.* **95**: 827, 2003; phylogeny), Galloway (*Biblthca Lichenol.* **88**: 147, 2004; New Zealand), Miądlikowska *et al.* (*Mycol.* **98**: 1088, 2006; phylogeny), Galloway & Arvidsson (*Biblthca Lichenol.* **96**: 87, 2007; Ecuador), Lumbsch *et al.* (*MR* **111**: 1133, 2007).

Placopyrenium Breuss (1987), Verrucariaceae (L). 14, Europe; Asia. See Ménard & Roux (*Mycotaxon* **53**: 129, 1995), Gueidan *et al.* (*MR* **111**: 1145, 2007; phylogeny).

Placosoma Syd. (1924), Asterinaceae. 1, New Zealand. See Hughes (*N.Z. Jl Bot.* **16**: 311, 1978).

Placosphaerella Pat. (1897) nom. conf., anamorphic *Pezizomycotina*. See Petrak (*Sydowia* **5**: 195, 1951).

Placosphaeria (De Not.) Sacc. (1880), anamorphic

Dothideales, St.0eH.?. 50, widespread.

Placosphaerina Maire (1917), anamorphic *Pezizomycotina*, St.≡ eH.?. 1, Algeria.

Placostroma Theiss. & Syd. (1914) = Phyllachora Nitschke ex Fuckel (1870) fide Cannon (*Mycol. Pap.* **163**, 1991).

Placostromella Petr. (1947), Dothideomycetes. 2, China. See Inácio *et al.* (*Mycol. Progr.* **4**: 133, 2005).

Placothallia Trevis. (1853) = Rinodina fide Hawksworth *et al.* (*Dictionary of the Fungi* edn 8, 1995).

Placothea Syd. (1931), anamorphic *Pezizomycotina*, St.0eH.?. 1, Philippines.

Placotheliomyces Cif. & Tomas. (1953) ≡ Placothelium.

Placothelium Müll. Arg. (1893), Verrucariaceae (L). 1, USA.

Placothyrium Bubák (1916), anamorphic *Pezizomycotina*, St.0fH.10. 1, Europe.

Plactogene Theiss. (1917) = Dimerina fide von Arx & Müller (*Stud. Mycol.* **9**, 1975).

Placuntium Ehrenb. (1818) = Rhytisma fide Minter (*in litt.*), Eckblad & Torkelsen (*Agarica* **7**: 60, 1986).

Placynthiaceae Å.E. Dahl (1950) nom. cons., Peltigerales (L). 5 gen. (+ 15 syn.), 32 spp.

Lit.: Henssen (*Symb. bot. upsal.* **32**: 75, 1997), Henssen & Tønsberg (*Bryologist* **103**: 108, 2000), Wiklund & Wedin (*Cladistics* **19**: 419, 2003), Miądlikowska & Lutzoni (*Am. J. Bot.* **91**: 449, 2004), Wedin & Wiklund (*Symb. bot. upsal.* **34** no. 1: 469, 2004), Miądlikowska *et al.* (*Mycol.* **98**: 1088, 2006; phylogeny), Jørgensen (*Nordic Lichen Flora* **3**: Cyanolichens: 134, 2007).

Placynthiella Elenkin (1909), Trapeliaceae (L). 7, widespread. See Coppins & James (*Lichenologist* **16**: 241, 1984; key), Coppins *et al.* (*Lichenologist* **19**: 93, 1987; nomencl.), Lumbsch (*J. Hattori bot. Lab.* **83**: 1, 1997), Schmitt *et al.* (*Mycol.* **95**: 827, 2003; phylogeny), Lendemer (*Opuscula Philolichenum* **1**: 75, 2004; N America), Miądlikowska *et al.* (*Mycol.* **98**: 1088, 2006; phylogeny), Lumbsch *et al.* (*MR* **111**: 1133, 2007).

Placynthiella Gyeln. (1939) = Placynthiella Elenkin fide Hawksworth *et al.* (*Dictionary of the Fungi* edn 8, 1995).

Placynthiomyces Cif. & Tomas. (1953) ≡ Placynthium.

Placynthiopsis Zahlbr. (1932), Placynthiaceae (L). 1, central Africa.

Placynthium (Ach.) Gray (1821), Placynthiaceae (L). 25, widespread (esp. temperate). See Gyelnik (*Rabenh. Krypt.-Fl.* **9** 2.2, 1940; Eur.), Henssen (*CJB* **41**: 1687, 1963; N. Am.), Weber (*Muelleria* **3**: 250, 1970; Australia), Burgaz *et al.* (*Cryptog. Mycol.* **23**: 367, 2002; Iberian peninsula), Miądlikowska & Lutzoni (*Am. J. Bot.* **91**: 449, 2004; phylogeny), Miądlikowska *et al.* (*Mycol.* **98**: 1088, 2006; phylogeny).

plage (1) a smooth, paler-coloured, or colourless spot on a surface; (2) (of basidiospores), esp. a smooth spot above the hilar appendage.

Plagiocarpa R.C. Harris (1973) = Lithothelium fide Aptroot (*Biblthca Lichenol.* **44**, 1991).

Plagiographis C. Knight & Mitt. (1860) ? = Opegrapha Ach. fide Redinger (*Feddes Repert.* **43**: 49, 1938).

Plagiolagynion Schrantz (1962) = Oxydothis fide Barr (*Mycol.* **68**, 1976), Hyde (*Sydowia* **46**: 265, 1995).

Plagionema Subram. & K. Ramakr. (1953) = Ciliochorella fide Chona & Munjal (*Indian Phytopath.* **8**: 74, 1955).

Plagiophiale Petr. (1955), Diaporthales. 2, Europe; N. America. See Barr (*Mycol. Mem.* **7**, 1978), Barr (*Mycotaxon* **41**: 287, 1991).

Plagiorhabdus Shear (1907) = Strasseria fide Höhnel (*Mitt. bot. Inst. tech. Hochsch. Wien* ser. 2 **2**: 26, 1925).

Plagiosphaera Petr. (1941), Sordariomycetes. 5, widespread. See Walker (*Mycotaxon* **11**: 94, 1980), Barr (*Mycotaxon* **39**: 43, 1990; posn).

Plagiostigme Syd. (1925), Diaporthales. 1, C. America. See Petrak (*Sydowia* **18**: 382, 1965).

Plagiostigmella Petr. (1949), anamorphic *Pezizomycotina*, Cpd.0eH.?. 1, S. America.

Plagiostoma Fuckel (1870), Gnomoniaceae. Anamorphs *Asteroma*, *Discula*. 22, Europe; N. America. See Bolay (*Ber. schweiz. bot. Ges.* **81**: 398, 1972), Barr (*Mycol. Mem.* **7**, 1978; key 13 spp.), Monod (*Beih. Sydowia* **9**: 1, 1983), Barr (*Mycotaxon* **41**: 287, 1991; N. Am. spp.), Fröhlich & Hyde (*MR* **99**: 727, 1995), Okada & Katumoto (*Mycoscience* **41**: 615, 2000; Japan), Castlebury *et al.* (*Mycol.* **94**: 1017, 2002; phylogeny), Sogonov *et al.* (*Sydowia* **57**: 102, 2005; phylogeny), Zhang *et al.* (*Mycol.* **98**: 1076, 2006; phylogeny).

Plagiostomella Höhn. (1917) = Plagiostoma fide Barr (*Mycol. Mem.* **7**: 232 pp., 1978).

Plagiostromella Höhn. (1917), Dothideomycetes. 4, India; Japan. See Bose & Müller (*Sydowia* **31**: 1, 1979; key).

Plagiothecium Schrantz (1961) [non *Plagiothecium* Bruch & Schimp. 1851, *Musci*] ≡ Plagiolagynion.

Plagiothelium Stirt. (1881) = Anthracothecium fide Aptroot (*Australasian Lichenology* **60**: 35, 2007).

Plagiotrema Müll. Arg. (1885) = Pseudopyrenula fide Harris (*Lichenographia Thomsoniana*: 133, 1998).

Plagiotus Kalchbr. ex Roum. (1879) nom. inval. = Anthracophyllum fide Kuyper (*in litt.*).

Planctomyces M. Gimesi (1924) nom. dub., ? Fungi. See Wawrik (*Sydowia* **6**: 443, 1952), Skerman *et al.* (*Approved list of bacterial names*, 1980), Schlesner & Stackebrandt (*Syst. Appl. Microbiol.* **8**: 174, 1986; assigned to new order *Planctomycetales*) Also considered a bacterium by.

plane, flat.

Planetella Savile (1951), Anthracoideaceae. 1 (on *Cyperaceae*), Canada. Resembles *Anthracoidea*. See Savile (*CJB* **29**: 326, 1951), Vánky (*Cryptog. Stud.* **1**: 159 pp., 1987), Piepenbring (*Frontiers in Basidiomycote Mycology*: 117, 2004).

planetism (of oomycetes), the condition of having motile stages.

Planistroma A.W. Ramaley (1991), ? Planistromellaceae. Anamorph *Piptarthron*. 2 (on *Yucca*), N. America. See Ramaley (*Mycotaxon* **42**: 63, 1991), Eriksson & Hawksworth (*SA* **12**: 38, 1993; posn), Ramaley (*Mycotaxon* **66**: 509, 1998).

Planistromella A.W. Ramaley (1993), Planistromellaceae. Anamorph *Kellermania*. 7, USA. See Eriksson & Hawksworth (*SA* **12**: 38, 1993; posn), Ramaley (*Mycotaxon* **47**: 259, 1993), Ramaley (*Mycotaxon* **66**: 509, 1998), Sivanesan & Shivas (*Fungal Diversity* **9**: 175, 2002; on *Opuntia*; key).

Planistromellaceae M.E. Barr (1996), ? Dothideales. 5 gen. (+ 1 syn.), 30 spp.

Lit.: Ramaley (*Mycotaxon* **55**: 255, 1995), Barr (*Mycotaxon* **60**: 433, 1996), Ramaley (*Mycotaxon* **66**: 509, 1998).

Planobispora Thiemann & Beretta (1968), Actinobacteria. q.v.

planoconvex, a flat zygote.

planocyte (**planont**), a motile cell.

planogamete, a motile gamete; zoogamete.

Planomonospora Thiemann, Pagani & Beretta (1967), Actinobacteria. q.v.

planont, see planocyte.

Planoroccellaceae Elenkin (1929) = Roccellaceae.

planospore, a zoospore.

planozygote (1) a motile zygote; (2) a flat zygote (of a pileus or spore), convex but somewhat flat.

Plant pathogenic fungi. A large proportion of fungal species live in association with plants, as saprobes on plant surfaces or dead plant tissue, as mutualists in mycorrhizal associations or as endophytes or parasites of living plants. Some, e.g. *Pucciniales*, *Erysiphaceae*, are obligate parasites of plants. Many parasitic fungi are highly phytopathogenic and induce severe and frequently economically important diseases. They are the most important agents of plant disease; more than 60 per cent of the literature on plant diseases is devoted to fungal infections. Plant pathogens are represented in all the major groups of fungi and some groups, e.g. *Ustilaginales* (smuts), *Pucciniales* (rusts), *Peronosporaceae* (downy mildews), *Erysiphaceae* (powdery mildews), are known by the common names of the diseases for which they are responsible. Other groups important as plant pathogens are basidiomycetes (frequently wood-attacking), and a wide range of ascomycetes, represented particularly on living tissues by their anamorphs. (The teleomorph of a pathogen frequently develops only on dead plant tissue in the terminal phase of the disease.)

The symptoms of fungus diseases in plants are diverse. They may be roughly classified as **necrosis** - death of tissues (cf. anthracnose, blight, canker, damping-off, scab, shot hole); **wilting** - loss of turgor due to the mechanical plugging of the vessels by hyphae or to the action of a toxin secreted by the pathogen; **hyperplasia** - overgrowth (cf. galls, witches' brooms); and **hypoplasia** - dwarfing (and chlorosis).

Transmission of plant pathogenic fungi may be by air (either dry or moist, as in droplet transmission), water, soil, seeds and other propagating material, plant debris, and insects and other animal agents. Humans have spread many important diseases within one country, from one country to another, and from one part of the world to another by transporting infected material in the course of trade.

Control of parasitic disease is (fide Whetzel) based on exclusion, eradication, protection, and immunization. Quarantine procedures, inspection at ports, and legislation governing the movement of plants and plant products, the growth of susceptible plants under conditions unfavourable for the development of the pathogen, and the use of disease-free seed or planting material are based on exclusion. Eliminating ('roguing') diseased plants, the cutting-out of diseased parts, heat treatment of seed material, and the use of eradicant fungicides on infected plants (i.e. against the pathogen) are examples of methods based on eradication. Among protective measures are the dusting or spraying of healthy plants with fungicides. The use of immune or resistant varieties is an example of 'immunization', but active immunization, which is so important in the control of many diseases of animals and man, is of little value against plant diseases. The three main strategies for plant disease control are: phytosanitation (quarantine, removal of inoculum sources, crop rotation, clean seed etc.); protection (use of fungicides, avoidance of predisposing cultural conditions) and resistance (use of disease resistance cultivars). The new sciences of genetic engineering and genomics are making a huge impact on development of resistant cultivars, e.g. Kamoun & Smart (*Plant disease* **89**: 692, 2005; review of genomics and late blight of potato and tomato), Punja (2004).

Lit.: The literature on plant disease is large, but well covered by review journals and databases (see below). See also the CABI *Crop protection compendium* CD. The standard general phytopathological texts all deal with fungal diseases, as do most monographs on diseases of particular crop plants. Here a selection has been made of general works and review journals. For additional references see Moreau (*Rev. mycol., Paris* **27**: 41, 1962), *RAM* **46**: 1 and 113, 1967 (select bibliogr.), Waller & Lenné (*Plant pathologist's pocketbook*, 2001). See also Fungicides, gene-for-gene, Genetics, Geographical distribution, Mycotoxicoses, physiologic race, Seed-borne fungi, Special forms.

General: Agrios (*Plant pathology*, 1969; edn 3, 1985), Alford (*Pest and disease management handbook*, 2000), APS (*Compendia of plant diseases*, series), Brown & Brotzman (*Phytopathogenic fungi: a scanning electron stereoscopic survey*, 1979), Bruell (*Soilborne plant pathogens*, 1987), Buczacki & Harris (*Collins photoguide to pests, diseases and disorders of garden plants*, 1998), Butler & Jones (*Plant pathology*, 1949), Gaumann (*Principles of plant infection*, 1950) [trans. of *Pflanzliche Infektionslehre*, 1946; edn 2, 1951], Heald (*Manual of plant diseases*, 1926, edn 2, 1933), Holton *et al.* (Eds) (*Plant pathology, problems and progress, 1908-1958*, 1959), Horsfall & Cowling (Eds) (*Plant disease, an advanced treatise* 5 vols, 1977-80), Horsfall & Dimond (Eds) (*Plant pathology, an advanced treatise*, 3 vols, 1959- 60), Horst (*Westcott's plant disease handbook*, edn 6, 2001), Kronstad (Ed.) (*Fungal pathology*, 2000), Schumann (*Plant diseases: their biology and social impact*, 1991), Schumann & D'Arcy (*Essential plant pathology*, 2006; textbook with CD), Smith *et al.* (*European handbook of plant diseases*, 1988), Stakman & Harrar (*Principles of plant pathology*, 1957), Strange (*Introduction to plant pathology*, 480 pp., 2003), Tarr (*Principles of plant pathology*, 1971), Vanderplank (*Plant diseases: epidemics and control*, 1963), Vidhyasekaran (*Concise encyclopaedia of plant pathology*, 619 pp., 2004), Walker (*Plant pathology*, 1950; edn 3, 1969), Wood (*Physiological plant pathology*, 1967).

Grasses: Murray *et al.* (*A colour handbook of diseases of small grain cereal crops*, 1998), Smiley *et al.* (Eds) (*Compendium of turfgrass diseases*, edn 3, 2005.

Tropical: Holliday (*Fungus diseases of tropical crops*, 1980), Ploetz (*Diseases of tropical fruit crops*, 2003), Ploetz (*Plant disease* **91**: 644, 2007; review), Roger (*Phytopathologie des pays chauds*, 3 vol. 1951-54), Thurston (*Tropical plant diseases*, 1984), Weber (*Bacterial and fungal diseases of plants in the tropics*, 1973).

Forest pathology: Boyce (*Forest pathology*, edn 3, 1961), Browne (*Pests and diseases of forest plantation trees. An annotated list of the principle species occurring in the British Commonwealth*, 1968), *Dis-*

eases of widely planted forest trees (1964; lists diseases and their distribution), *Forestry abstracts* 1939- [also as TREE-CD], Anse & Lewis (Eds) (*Compendium of conifer diseases*, 1997), Hepting (*Diseases of forest and shade trees in the United States* [*Agric. Handb. Forest Serv. US* **386**], 1971), Jones & Benson (*Diseases of woody ornamentals and trees in nurseries*, 2001), Lundquist & Hamelin (Eds) (*Forest pathology: from genes to landscape*, 2005), Peace (*Pathology of trees and shrubs with special reference to Britain*, 1961), Sinclair & Lyon (*Diseases of trees and shrubs* edn 2, 2005). See also Wood-attacking fungi.

Resistance: Hang (*Plant pathogenesis and resistance; biochemistry and physiology of plant-microbe interactions*, 2001), Punja (Ed.) (*Fungal disease resistance in plants: biochemistry, molecular biology and genetic engineering*, 2004), Talbot (Ed.) (*Plant pathogen interactions*, 2004).

Legislation and quarantine: Ebbels (*Principles of plant health and quarantine*, 2003), Ebbels & King (Eds) (*Plant health*, 1979), CABI/EPPO (*Quarantine pests from Europe*, 1992), McGee (*Plant pathogens and the worldwide movement of seeds*, 1997).

Methods: Babcock *et al.* (*Plant disease* **91**: 476; review of genetic resources for plant pathology), Fox (*Principles of diagnostic techniques in plant pathology*, 1993), Ingram & Helgeson (Eds) (*Tissue culture methods for plant pathologists*, 1980), Riker & Riker (*Introduction to research on plant diseases*, 1936), Schots *et al.* (Eds) (*Modern assays for plant pathogenic fungi: identification, detection and quantification*, 1994), Waller & Lenné (*Plant pathologist's pocketbook*, 2001).

Identification: Arx (*Plant pathogenic fungi*, 1987), Brandenburger (*Parasitische Pilze au Gefasspłanzen in Europa*, 1985), Brooks (*Plant diseases*, edn 2, 1953), Farr *et al.* (*Fungi on plants and plant products in the United States*, 1989), Hodge (*The fungal spore and disease initiation in plants and animals*, 1991), Holliday (*A dictionary of plant pathology*, *IMI Descriptions of fungi and bacteria* (1964-), Kreisel (*Die phytopathogen-Grosspilze Deutschlands*, 1961), Lindquist (*Revta Fac. Agron. Univ. nac. La Plata* **43**: 1, 1967; key), *Names of British plant diseases and their control* [*Phytopath. Pap.* **28**, 1984], 1989), Rossman *et al.* (*A literature guide for the identification of plant pathogenic fungi*, 1987), Stevens (*The fungi which cause plant disease*, 1913; *Plant disease fungi*, 1925 [reprint, 1966]), Viennot-Bourgin (*Les champignons parasites des plantes cultivées*, 1949).

History: Ainsworth (*Introduction to the history of plant pathology*, 1981), Large (*The advance of the fungi*, 1940), Parris (*A chronology of plant pathology*, 1968), Schumann (*Plant diseases: their biology and social impact*, 1991), Whetzel (*An outline of the history of phytopathology*, 1918), *Phytopathological classics* (1926-).

Serial publications and review journals: *Advances in plant pathology*, 1982-95; *Annual review of phytopathology*, 1963-; *Eur. j. forest path.*, 1971-1999, 2000- as *Forest path.*; *Eur. j. plant path.*, 1895-; *Plant disease reporter*, 1917-1979, 1980- as *Plant disease*; *Plant pathology*, 1952-; *Review of applied mycology*, 1922-69 (abstracts; *Index* **1-40** (1922-61), 1968),1970- as *Review of plant pathology* [also available in a variety of electronic formats, incl. PEST-CD; see Literature].

plaque, a clear area in a bacterial colony caused by localized viral lysis; also applied to similar areas in fungal cultures (e.g. Riegman & Wessels, *TBMS* **75**: 325, 1980).

Plasia Sherwood (1981), anamorphic *Xylogramma*, St.≡ eH.15/16. 1, Europe. See Sherwood (*TBMS* **77**: 197, 1981).

plasma membrane, see plasmalemma.

plasmalemma, outer membrane composed of phospholipids and proteins which surrounds a cell and regulates exchange of materials between the cell and its environment; the limiting (boundary) membrane of the cytoplasm; cell, cytoplasmic, or plasma membrane.

plasmalemmasome, an intracytoplasmic vesicle (formed by invagination of the plasmalemma) filled with tubular diverticula. Cf. lomasome.

plasmamembranic vesicle, sporangial vesicle, the membrane of which is the same membrane bounding the protoplasm within the sporangium, c.f. homohylic vesicle, precipitative vesicle.

plasmatoogosis (of *Pythiaceae*), a bud-like outgrowth (like a prosporangium in form) in the host tissue (Sideris, *Mycol.* **23**: 255, 1931).

plasmodesma (pl. **-ta**), an isthmus-like strand of protoplasm connecting adjacent cells.

plasmodia, see plasmodium.

plasmodic granules, very small, dark-coloured particles on the surface of the peridium, and frequently of the spores, in the *Cribrariaceae*; dictydine granules.

plasmodiocarp (of *Mycetozoa*), a sessile and vein-like sporangium, like part of the larger veins of a plasmodium.

plasmodium (pl. **plasmodia**) (of *Myxogastria*), a multinucleate, motile mass of protoplasm bounded by a plasma membrane but lacking a wall, generally reticulate, characteristic of the growth phase; Alexopoulos (*Mycol.* **52**: 17, 1961) distinguishes: **protoplasmodium**, an undifferentiated microscopic plasmodium which gives rise to a single sporangium, as in *Echinosteliales*; **aphanoplasmodium**, a plasmodium composed of a network of undifferentiated strands of non-granular protoplasm, as in *Stemonitis*; **phaneroplasmodium**, a plasmodium composed of a well-differentiated advancing fan and thick strands of granular protoplasm exhibiting ecto- and endoplasmic regions, as in *Physarales*; **aggregate plasmodium**, **pseudoplasmodium** (of *Acrasiales*), a structure formed by the massing of separate myxamoebae or cells before reproduction (a grex); **filoplasmodium**, the net-like pseudoplasmodium of the *Labyrinthulales*.

plasmogamospore, aeciospore (Laundon, *TBMS* **58**: 345, 1972).

plasmogamy, fusion of two cells or plasmodial cytoplasms without karyogamy (nuclear fusion) or a precursor to karyogamy.

Plasmophagus De Wild. (1895), Chytridiales. 3 (in algae), Europe. See Dick (*Straminipilous Fungi*: 670 pp., 2001).

Platisma Hoffm. (1790) = Cetraria fide Santesson & Culberson (*Bryologist* **69**: 100, 1966).

Platisma P. Browne ex Adans. (1763) = Ramalina fide Santesson & Culberson (*Bryologist* **69**: 100, 1966).

Platismatia W.L. Culb. & C.F. Culb. (1968), Parmeliaceae (L). 10, widespread. See Culberson & Culberson (*Contr. US natn Herb.* **34**: 449, 1968; monogr.),

Thell *et al.* (*Cryptog. Bryol.-Lichénol.* **19**: 307, 1998; phylogeny), Wedin *et al.* (*MR* **103**: 1152, 1999; phylogeny), Miądlikowska *et al.* (*Mycol.* **98**: 1088, 2006; phylogeny).

Platisphaera Dumort. (1822) nom. rej. ≡ Lophiostoma fide Holm (*Taxon* **24**: 475, 1975).

Platistoma, see *Platystoma*.

Platistomum, see *Platystomum*.

Platycarpa Couch (1949), Eocronartiaceae. 2 (on *Pteridophyta*), America. See Couch (*Mycol.* **41**: 427, 1949), Oberwinkler & Bandoni (*TBMS* **83**: 369, 1984).

Platycarpium P. Karst. (1905) = Fusamen fide von Arx (*Verh. K. ned. Akad. Wet. Amst.* C **51**: 1, 1957).

Platychora Petr. (1925), Venturiaceae. Anamorph *Piggotia*. 2, widespread (north temperate). See Sivanesan (*Bitunicate Ascomycetes and their Anamorphs*, 1984), Winton *et al.* (*Mycol.* **99**: 240, 2007; phylogeny).

Platygloea J. Schröt. (1887), Platygloeaceae. *c.* 2, widespread. See Bandoni (*Mycol.* **48**: 821, 1956; key), Dueñas (*Nova Hedwigia* **72**: 441, 2001; Iberian).

Platygloeaceae Racib. (1909), Platygloeales. 4 gen. (+ 5 syn.), 6 spp.

Lit.: Berres *et al.* (*Mycol.* **87**: 821, 1995), Bandoni (*Mycoscience* **39**: 31, 1998), Oberwinkler *et al.* (*Kew Bull.* **54**: 763, 1999), Chen & Oberwinkler (*Mycotaxon* **76**: 279, 2000), Dueñas (*Nova Hedwigia* **72**: 441, 2001), Bauer *et al.* (*Mycol.* **96**: 960, 2004), Lutz *et al.* (*Mycol.* **96**: 614, 2004), Lutz *et al.* (*Mycol.* **96**: 1316, 2004), Lutz *et al.* (*MR* **108**: 227, 2004), Nakamura *et al.* (*MR* **108**: 641, 2004), Wingfield *et al.* (*Australas. Pl. Path.* **33**: 327, 2004).

Platygloeales R.T. Moore (1990). Pucciniomycetes. 2 fam., 11 gen., 19 spp. Fams:
(1) **Eocronartiaceae**
(2) **Platygloeaceae**
For *Lit.* see under fam.

platygonidia, photobionts occurring in stellately or orbicular spreading colonies (e.g. *Cephaleuros*) (obsol.).

Platygramma G. Mey. (1825) nom. rej. prop. = Phaeographis fide Lücking *et al.* (*Taxon* **56**: 1296, 2007; nomencl.).

Platygramma Leight. (1854) = Enterographa fide Hawksworth *et al.* (*Dictionary of the Fungi* edn 8, 1995).

Platygramme Fée (1874), Graphidaceae (L). 9, pantropical. See Staiger (*Biblthca Lichenol.* **85**, 2002), Nakanishi *et al.* (*Bull. natn. Sci. Mus.* Tokyo, B **29**: 83, 2003; Japan), Archer (*Telopea* **11**: 59, 2005; Australia), Archer (*Systematics & Biodiversity* **5**: 9, 2007; Solomon Is).

Platygrapha Berk. & Broome (1870) [non *Platygrapha* Nyl. 1855] ? = Ocellularia fide Sherwood (*Mycotaxon* **3**: 233, 1976).

Platygrapha Nyl. (1855) = Schismatomma fide Tehler (*Op. Bot.* **118**: 1, 1993).

Platygraphis Hook. f. (1867) ≡ Platygrapha Nyl.

Platygraphomyces Cif. & Tomas. (1953) ≡ Platygrapha Nyl.

Platygraphopsis Müll. Arg. (1887), Ostropales (L). 1, America (tropical); Papua New Guinea. See Awasthi (*Biblthca Lichenol.* **40**, 1991; Indian subcontinent).

Platypeltella Petr. (1929), Microthyriaceae. 2, C. America. See Farr (*Mycotaxon* **15**: 448, 1982).

platyphyllous, broadly lobed.

Platyphyllum Vent. (1799) nom. rej. ≡ Cetraria.

Platysma Hill (1773) = Ramalina fide Santesson & Culberson (*Bryologist* **69**: 100, 1966).

platysmoid (1) (of lichen thalli), loosely attached foliose thalli with ascending lobes, as in *Platismatia* (obsol.); (2) (of tissues), a scleroplectenchyma in which the hyphae are brown (Yoshimura & Shimada, *Bull. Kochi Gakuen Jun. Coll.* **11**: 13, 1980).

Platysphaera Trevis. (1877) ≡ Platisphaera.

Platysphaeria, see *Platisphaera*.

Platyspora Wehm. (1961) = Comoclathris fide Shoemaker & Babcock (*CJB* **70**: 1617, 1992), Barr (*SA* **12**: 27, 1993).

Platysporoides (Wehm.) Shoemaker & C.E. Babc. (1992), Pleosporaceae. 12, widespread. See Shoemaker & Babcock (*CJB* **70**: 1617, 1992), Barr (*SA* **11**: 27, 1993).

Platysticta Cooke (1889) = Stictis fide Sherwood (*Mycotaxon* **6**: 215, 1977).

Platystoma (Fr.) Bonord. (1851) = Lophiostoma fide Mussat (*Syll. fung.* **15**: 295, 1901).

Platystomaceae J. Schröt. (1894) = Lophiostomataceae.

Platystomum Trevis. (1877) = Lophiostoma fide Chesters & Bell (*Mycol. Pap.* **120**, 1970), Eriksson (*SA* **10**: 144, 1991), Abdel-Wahab & Jones (*Mycoscience* **41**: 379, 2000).

Platythecium Staiger (2002), Graphidaceae. 17, widespread. See Staiger (*Biblthca Lichenol.* **85**: 385, 2002), Nakanishi *et al.* (*Bull. natn. Sci. Mus.* Tokyo, B **29**: 83, 2003; Japan), Adawadkar & Makhija (*Mycotaxon* **92**: 387, 2005; India), Archer (*Biblthca Lichenol.* **94**, 2006; Australia), Archer (*Systematics & Biodiversity* **5**: 9, 2007; Solomon Is).

Pleamphisphaeria Höhn. (1919) ≡ Titanella fide Harris (*Mem. N. Y. bot. Gdn* **49**, 1989).

Plearthonis Clem. (1909) = Arthonia fide Hawksworth *et al.* (*Dictionary of the Fungi* edn 8, 1995).

Plecopteromyces Lichtw., Ferrington & López-Lastra (1999), Legeriomycetaceae. 3 (in *Plecoptera*), Australia; S. America. See Lichtwardt *et al.* (*Mycol.* **91**: 1060, 1999), Ferrington *et al.* (*Mycol.* **97**: 254, 2005; Tasmania), White (*MR* **110**: 1011, 2006; phylogeny).

Plecostoma Desv. (1809) ≡ Geastrum.

Plecotrichum Corda (1831) nom. dub., anamorphic *Pezizomycotina*. See Hughes (*CJB* **36**: 798, 1958).

Plectania Fuckel (1870), Sarcosomataceae. Anamorph *Conoplea*. 15, widespread (north temperate). See Korf (*Mycol.* **49**: 107, 1957), Paden (*Fl. Neotrop.* **37**: 1, 1983; 4 sect.), Li & Kimbrough (*CJB* **73**: 1761, 1995; ultrastr.), Zhuang & Wang (*Mycotaxon* **67**: 355, 1998; spp. China), Harrington *et al.* (*Mycol.* **91**: 41, 1999; phylogeny), Medel & Chacón (*Acta Bot. Mexicana* **50**: 11, 2000; Mexico), Calonge & Mata (*Mycotaxon* **81**: 237, 2002; Costa Rica), Perry *et al.* (*MR* **111**: 549, 2007; phylogeny).

Plectascales. *Ascomycota* with ascoma in which the asci were irregularly distributed; sometimes extended and including *Erysiphales*, *Eurotiales* (q.v.) and *Myriangiaceae*. See also *Plectomycetes*.

plectenchyma, a thick tissue formed by hyphae becoming twisted and fixed together; synchyma (Vuillemin); it is **prosenchyma (proso-)** when the hyphal elements are seen to be hyphae; **pseudoparenchyma (para-)**, when they are not; Vuillemin (1912) distinguished as **merenchyma** tissue derived by cell division in several planes, and Degelius (1954) used **euthyplectenchyma** for hyphal tissue with no cellular structure. Yoshimura & Shimada (*Bull. Kochi Gakuen Jun. Coll.* **11**: 13, 1980) key nine categories of plectenchyma, **chalaroplectenchyma** (hyphal walls not united, lumina wide), **pallisadoplectenchyma** (hyphae parallel, not coherent, with intercellular spaces), **platysmoid** (a scleroplectenchyma with brown hyphal walls), **scleroplectenchyma** (cell-walls thickened, lumina narrow), **scleroprosoplectenchyma** (hyphae parallel, cohering, walls thick, lumina narrow), and **serioplectenchyma** (hyphae parallel, cohering, walls not thick, lumina wide).

Plectobasidiales Gäum. (1926). See *Melanogastrales*, *Sclerodermatales* and *Tulostomatales*.

Plectocarpon Fée (1825), Roccellaceae. 10 (on lichens, esp. *Peltigerales*), widespread. See Hawksworth & Galloway (*Lichenologist* **16**: 85, 1984), Diederich & Etayo (*Nordic Jl Bot.* **14**: 589, 1994; key 7 spp.), Ertz *et al.* (*Bryologist* **106**: 465, 2003), Ertz *et al.* (*Biblthca Lichenol.* **91**: 155 pp., 2005; revision).

Plectodiscella Woron. (1914) = Elsinoë fide Jenkins (*J. agric. Res.* **44**: 689, 1932).

Plectodiscellaceae Woron. (1914) = Elsinoaceae.

Plectolitus Kohlm. (1960) = Amylocarpus fide Kohlmeyer & Kohlmeyer (*Syn. Pl. Marine Fungi*, 1971), Crumlish & Curran (*Mycologist* **8**: 83, 1994).

Plectomyces Thaxt. (1931), Ceratomycetaceae. 1, Philippines. See Nannfeldt (*Svensk bot. Tidskr.* **43**: 468, 1949).

Plectomycetes. Class of *Ascomycota* with ± globose non-ostiolate ascomata; formerly frequently used for *Elaphomycetales*, *Erysiphales*, *Eurotiales*, *Meliolales*, *Microascales* and *Onygenales* but now recognized as heterogeneous, the non-ostiolate habit having repeatedly evolved. Reintroduced by Berbee & Taylor (1992) in a restricted sense (see *Ascomycota*).

Lit.: Benny & Kimbrough (*Mycotaxon* **12**: 1, 1980; keys 90 gen., 6 orders, 12 fams, excluded gen. names), Malloch (*in* Kendrick (Ed.), *The whole fungus* **1**: 153, 1979; *in* Reynolds (Ed.), 1981: 73, placement of *all* fams in orders also characterized by ostiolate or apothecioid taxa), Stchigel & Guarro (*MR* **111**: 1100, 2007; review of cleistothecial taxa).

Plectomyriangium C. Moreau & M. Moreau (1950) = Lecideopsella fide Eriksson & Hawksworth (*SA* **9**: 15, 1991).

Plectonaemella Höhn. (1915), anamorphic *Pezizomycotina*, Cpd.0eH.?. 1, Europe.

plectonematogenous, of conidiogenous cells arising from rope-like strands of interwoven hyphae (not from single hyphae), the strands intertwined and not synnematous; funiculose. See Gams (*Cephalosporium-artige Schimmelpilze*, 1971).

Plectopeltis Syd. (1927), anamorphic *Pezizomycotina*, Cpt.0eH.?. 1, C. America; Australia. See Reynolds & Gilbert (*Aust. Syst. Bot.* **18**: 282, 2005).

Plectophoma Höhn. (1907) = Asteromella fide Sutton (*Mycol. Pap.* **141**, 1977).

Plectophomella Moesz (1922), anamorphic *Pezizomycotina*, St.0eH.15. 3, Europe. See Guarro *et al.* (*J. Clin. Microbiol.* **44**: 4279, 2006; from human).

Plectophomopsis Petr. (1922), anamorphic *Pezizomycotina*, St.0eH.?. 1, Europe.

Plectopsora A. Massal. (1860) = Lempholemma fide Hawksworth *et al.* (*Dictionary of the Fungi* edn 8, 1995).

Plectopycnis Bat. & A.F. Vital (1959), anamorphic *Dothideomycetes*. 1, neotropics. See von Arx (*in litt.*), Farr (*Mycol.* **78**: 269, 1986).

Plectosclerotes Stach & Pickh. (1957), Fossil Fungi. 2 (Carboniferous), Germany.

Plectosira Petr. (1929), anamorphic *Pezizomycotina*, Cpd/St.0eH.?. 1, Europe.

Plectosphaera Theiss. (1917) = Phyllachora Nitschke ex Fuckel (1870) fide Cannon (*Mycol. Pap.* **163**, 1991), Summerell *et al.* (*Fungal Diversity* **23**: 323, 2006; culture of *P. eucalypti*, link to *Amphisphaeriaceae*).

Plectosphaeraceae Theiss. (1916) = Phyllachoraceae.

Plectosphaerella Kirschst. (1938) ≡ Plectosphaerina.

Plectosphaerella Kleb. (1929), Plectosphaerellaceae. Anamorph *Plectosporium*. 2, widespread. See Gams & Gerlagh (*Persoonia* **5**: 177, 1968), Uecker (*Mycol.* **85**: 470, 1993; ontogeny), Palm *et al.* (*Mycol.* **87**: 397, 1995), Pitt & Gams (*Nova Hedwigia* **81**: 311, 2005), Sung *et al.* (*Stud. Mycol.* **57**, 2007), Zare *et al.* (*Nova Hedwigia* **85**: 463, 2007).

Plectosphaerellaceae W. Gams, Summerb. & Zare (2007), Hypocreomycetidae (inc. sed.). 5 gen. (+ 1 syn.), 56 spp.

Lit.: Zare *et al.* (*Nova Hedwigia* **85**: 463, 2007).

Plectosphaerina Kirschst. (1938) = Omphalospora fide von Arx & Müller (*Stud. Mycol.* **9**, 1975).

Plectosporium M.E. Palm, W. Gams & Nirenberg (1995), anamorphic *Plectosphaerella*, Hso.1eH.15/16/19. 1, widespread. See Palm *et al.* (*Mycol.* **87**: 398, 1995), Gams *et al.* (*Taxon* **53**: 193, 2004; nomencl.), Pitt & Gams (*Nova Hedwigia* **81**: 311, 2005; on *Alisma*), Sung *et al.* (*Stud. Mycol.* **57**, 2007), Zare *et al.* (*Nova Hedwigia* **85**: 463, 2007).

Plectronidiopsis Nag Raj (1979), anamorphic *Pezizomycotina*, Cac.1eH.19. 1, Chile. See Nag Raj (*CJB* **57**: 1397, 1979).

Plectronidium Nag Raj (1977), anamorphic *Pezizomycotina*, Cac.0eH.19. 4, widespread. See Sutton & Pascoe (*TBMS* **87**: 249, 1986), Alcorn & Sutton (*Aust. Syst. Bot.* **12**: 169, 1999; Australia).

Plectrothrix Shear (1902), anamorphic *Pezizomycotina*, Hso.0eH.?. 1, N. America.

Pleiobolus, see *Plejobolus*.

Pleiochaeta (Sacc.) S. Hughes (1951), anamorphic *Pezizomycotina*, Hso.≡ eH-P.10. 4, widespread. *P. setosa* (lupin leaf spot). See Castañeda Ruíz *et al.* (*Mycotaxon* **57**: 457, 1996), Yang & Sweetingham (*Australas. Pl. Path.* **31**: 273, 2002; on legumes).

Pleiopatella Rehm (1908), Helotiales. 1, N. America.

Pleiopyrenis, see *Pleopyrenis*.

pleiosporous, having many spores.

Pleiostictis Rehm (1882) = Mellitiosporium fide Sherwood (*Mycotaxon* **5**: 1, 1977).

Pleiostomella Syd. & P. Syd. (1917) = Mendogia fide von Arx & Müller (*Stud. Mycol.* **9**, 1975).

Pleiostomellina Bat., J.L. Bezerra & H. Maia (1964), Dothideomycetes. 1, Brazil. See Batista *et al.* (*Portugaliae Acta Biologica* Série B **7**: 373, 1964).

pleioxeny, the condition of plurivorous parasitism.

Pleistophora Gurley (1893), Microsporidia. 34.

Pleistosporidium Codreanu-Balcescu & Codreanu (1982), Microsporidia. 1.

Plejobolus (E. Bommer, M. Rousseau & Sacc.) O.E. Erikss. (1967) = Robergea fide Sherwood-Pike (*in litt.*).

Plenocatenulis Bat. & Cif. (1959), anamorphic *Pezizomycotina*, Cpt.0eH.?. 2, USA. See Batista & Ciferri (*Mycopath. Mycol. appl.* **11**: 63, 1959).

Plenodomus Preuss (1851), anamorphic *Leptosphaeriaceae*. 1, widespread. See Boerema & Kesteren (*Per-

soonia **3**: 17, 1963), Boerema *et al.* (*TBMS* **77**: 61, 1981), Câmara *et al.* (*Mycol.* **94**: 630, 2002; phylogeny), Boerema *et al.* (*Phoma Identification Manual* Differentiation of Specific and Infra-Specific Taxa in Culture: 470 pp., 2004; monograph), Torres *et al.* (*Mycotaxon* **93**: 333, 2005; phylogeny, separation from *Phoma*).

Plenophysa Syd. & P. Syd. (1920), anamorphic *Pezizomycotina*, Cpt.0eH.?. 1 or 2, Philippines; Sri Lanka.

Plenotrichaius Bat. & Valle (1961), anamorphic *Pezizomycotina*, Cpt.0eH.?. 1, Brazil. See Batista & Valle (*Publções Inst. Micol. Recife* **337**: 9, 1961), Farr (*Mycol.* **78**: 269, 1986).

Plenotrichella Bat. & A.F. Vital (1959) = Myxothyriopsis fide von Arx (*in litt.*), Reynolds & Gilbert (*Aust. Syst. Bot.* **18**: 265, 2005; as '*Plenotrichiella*').

Plenotrichopsis Bat. (1961), anamorphic *Pezizomycotina*, Cpt.0eH.38. 1, Brazil. See Batista (*Publções Inst. Micol. Recife* **338**: 33, 1961).

Plenotrichum Syd. (1927), anamorphic *Pezizomycotina*, Cpt.0eH.39. 2, C. America; China.

Plenozythia Syd. & P. Syd. (1916), anamorphic *Pezizomycotina*, Cpd.0eH.?. 1 or 2, Asia.

pleoanamorphy, having more than one anamorph. See Hennebert (*in* Sugiyama (Ed.), *Pleomorphic fungi: the diversity and its taxonomic implications*, 1987), States of fungi.

Pleochaeta Sacc. & Speg. (1881), Erysiphaceae. Anamorph *Streptopodium*. 6, widespread. See Kimbrough (*Mycol.* **55**: 608, 1963; development), Kimbrough & Korf (*Mycol.* **55**: 619, 1963), Braun (*Beih. Nova Hedwigia* **89**, 1987; key), Mori *et al.* (*Mycol.* **92**: 74, 2000; phylogeny), Takamatsu (*Mycoscience* **45**: 147, 2004; phylogeny), Kiss *et al.* (*MR* **110**: 1301, 2006; on *Celtis*), Wang *et al.* (*Mycol.* **98**: 1065, 2006; phylogeny), Takamatsu *et al.* (*MR* **112**: 299, 2008; phylogeny).

Pleochroma Clem. (1909) ≡ Candelariella.

Pleococcum Desm. & Mont. (1849) nom. dub., anamorphic *Pezizomycotina*. See Höhnel (*Sber. Akad. Wiss. Wien* Math.-naturw. Kl., Abt. 1 **119**: 617, 1910).

Pleoconis Clem. (1909) = Peccania fide Hawksworth *et al.* (*Dictionary of the Fungi* edn 8, 1995).

Pleocouturea G. Arnaud (1911), anamorphic *Pezizomycotina*. 1, France. See Sutton (*Mycol. Pap.* **141**, 1977).

Pleocryptospora J. Reid & C. Booth (1969), Diaporthales. 1, Brazil. See Reid & Booth (*CJB* **47**: 1055, 1969).

Pleocyta Petr. & Syd. (1927) = Phaeocytostroma fide Sutton (*Mycol. Pap.* **97**, 1964).

Pleodothis Clem. (1909) = Sydowia fide von Arx & Müller (*Stud. Mycol.* **9**, 1975).

Pleogibberella Sacc. ex Berl. & Voglino (1886), Nectriaceae. 2 (on *Palmae* fruits), India; Brazil. See Rossman *et al.* (*Stud. Mycol.* **42**: 248 pp., 1999).

Pleoglonis Clem. (1909) ≡ Hariotia.

Pleolecis Clem. (1909) ≡ Steinia.

Pleolpidium A. Fisch. (1892) = Rozella fide Sparrow (*Aquatic Phycomycetes* Edn 2: 1187 pp., 1960).

Pleomassaria Speg. (1880), Pleomassariaceae. Anamorph *Prosthemium*. 4, widespread (temperate). See Barr (*Mycotaxon* **49**: 129, 1993; synonymy with *Splanchnonema*), Hantula *et al.* (*MR* **102**: 1509, 1998; anamorph), Paavolainen *et al.* (*Mycol.* **92**: 253, 2000; biological species), Lumbsch & Lindemuth (*MR* **105**: 901, 2001; phylogeny), Tanaka *et al.* (*Mycoscience* **46**: 248, 2005; Japan), Kruys *et al.* (*MR* **110**: 527, 2006; phylogeny).

Pleomassariaceae M.E. Barr (1979), Pleosporales. 12 gen. (+ 2 syn.), 86 spp.

Lit.: Barr (*Mycotaxon* **15**: 349, 1982; keys 24 spp.), Barr (*Mycotaxon* **49**: 129, 1993; cladistic analysis, key gen.), Liew *et al.* (*Mol. Phylogen. Evol.* **16**: 392, 2000), Lumbsch & Lindemuth (*MR* **105**: 901, 2001), Shoemaker *et al.* (*Can. J. Pl. Path.* **25**: 384, 2003).

Pleomeliola (Sacc.) Sacc. (1899), ? Meliolaceae. 1 or 2, Europe; N. America.

Pleomelogramma Speg. (1909), ? Herpotrichiellaceae. 1, S. America. See Petrak & Sydow (*Annls mycol.* **32**: 21, 1934).

Pleomeris Syd. (1921) = Puccinia fide Dietel (*Nat. Pflanzenfam.* **6**, 1928).

Pleomerium Speg. (1918), Meliolaceae. 1 (from leaves), Brazil.

pleomorphic (1) of fungi having more than one independent form or spore-stage in the life cycle, especially of holomorphs comprising a teleomorph and one or more anamorphs; polymorphic; see Savile (*Mycol.* **61**: 1161, 1970), Sugiyama (Ed.) (*Pleomorphic fungi: their diversity and its taxonomic implications*, 1987), States of fungi; (2) (of dermatophytes), changes due to 'degeneration' in culture.

pleomorphism, the condition of being pleomorphic.

Pleonectria Sacc. (1876) = Nectria fide Rossman *et al.* (*Stud. Mycol.* **42**: 248 pp., 1999).

pleont, any one of the two or more states of a pleomorphic fungus (Delphino, 1887).

Pleophalis Clem. (1909) = Spheconisca fide Hawksworth *et al.* (*Dictionary of the Fungi* edn 8, 1995).

Pleophragmia Fuckel (1870), Sporormiaceae. 1, Europe. See von Arx & Müller (*Stud. Mycol.* **9**, 1975), Réblová & Svrcek (*Czech Mycol.* **49**: 193, 1997; Czech Republic).

Pleopsidium Körb. (1855), Acarosporaceae (L). 3, widespread. See Hafellner (*Nova Hedwigia* **56**: 281, 1993), Hafellner (*Cryptog. bot.* **5**: 99, 1995), Obermayer (*Ann. bot. fenn.* **33**: 231, 1996; Tibet), Lutzoni *et al.* (*Am. J. Bot.* **91**: 1446, 2004; phylogeny), Reeb *et al.* (*Mol. Phylogen. Evol.* **32**: 1036, 2004; phylogeny), Crewe *et al.* (*MR* **110**: 521, 2006; phylogeny), Miądlikowska *et al.* (*Mycol.* **98**: 1088, 2006; phylogeny), Reeb *et al.* (*J. Mol. Evol.* **64**: 285, 2007; evolution), Knudsen *et al.* (*Opuscula Philolichenum* **5**: 1, 2008; S America).

Pleopus Paulet (1808) ? = Gyromitra fide Hawksworth *et al.* (*Dictionary of the Fungi* edn 8, 1995).

Pleopyrenis Clem. (1909) = Pyrenopsis fide Hawksworth *et al.* (*Dictionary of the Fungi* edn 8, 1995).

Pleoravenelia Long (1903) = Ravenelia fide Dietel (*Nat. Pflanzenfam.* **6**, 1928).

Pleorinis Clem. (1909) = Rinodina fide Hawksworth *et al.* (*Dictionary of the Fungi* edn 8, 1995).

Pleoscutula Vouaux (1913), ? Leotiomycetes. 2 (lichenicolous), neotropics. See Hafellner (*Herzogia* **6**: 289, 1983), Diederich & Etayo (*Lichenologist* **32**: 423, 2000; key), Etayo (*Biblthca Lichenol.* **84**, 2002; Colombia).

Pleoseptum A.W. Ramaley & M.E. Barr (1995), ? Leptosphaeriaceae. 1 (on *Yucca*), N. America. See Ramaley & Barr (*Mycotaxon* **54**: 75, 1995).

Pleosphaerellula Naumov & Czerepan. (1952), Dothideomycetes. 1 (on *Cornus*), former USSR. See Naumov & Czerepanova (*Bot. Mater. Otd. Sporov.*

Rast. Bot. Inst. Komarova Akad. Nauk S.S.S.R. **8**: 149, 1952).

Pleosphaeria Henssen (1964) nom. illegit. = Pyrenothrix fide Henssen (*Ber. dt. bot. Ges.* **77**: 320, 1964), Herrera-Campos *et al.* (*Mycol.* **97**: 356, 2005).

Pleosphaeria Speg. (1881), Pezizomycotina. 15, widespread. See Petrak (*Annls mycol.* **38**: 197, 1940).

Pleosphaeropsis Died. (1916) = Aplosporella fide Petrak & Sydow (*Beih. Rep. spec. nov. regn. veg.* **42**, 1927).

Pleosphaeropsis Vain. (1921) ≡ Norrlinia Vain.

Pleosphaerulina Pass. (1891) = Saccothecium fide Hawksworth *et al.* (*Dictionary of the Fungi* edn 8, 1995).

Pleospilis Clem. (1909) = Spirographa fide Santesson (*The lichens and lichenicolous fungi of Sweden and Norway*, 1993).

Pleospora Rabenh. ex Ces. & De Not. (1863) nom. cons., Pleosporaceae. Anamorph *Stemphylium*. *c.* 63, widespread. *P. herbarum* (anamorph *Stemphylium botryosum*), plurivorous, causes leaf spots of legumes, lettuce and storage fruit spot of apple. See Müller (*Sydowia* **5**: 248, 1951), Wehmeyer (*Monogr. Pleospora segreg.*, 1961; keys), Crivelli (*Über die heterogene Ascomycetengattung Pleospora Rbh.*, 1983; gen. concept, keys), Simmons (*Mycotaxon* **25**: 287, 1986), Simmons (*Sydowia* **38**: 284, 1986), Holm & Holm (*Sydowia* **45**: 167, 1993; 9 spp. Svalbard), Henssen (*SA* **13**: 202, 1995; ontogeny), Dong *et al.* (*MR* **102**: 151, 1998; phylogeny), Silva-Hanlin & Hanlin (*MR* **103**: 153, 1999; DNA), Simmons (*Harvard Pap. Bot.* **6**: 199, 2001), Câmara *et al.* (*Mycol.* **94**: 660, 2002; phylogeny), Kodsueb *et al.* (*Mycol.* **98**: 571, 2006; phylogeny), Schoch *et al.* (*Mycol.* **98**: 1041, 2006; phylogeny), Wang *et al.* (*MR* **111**: 1268, 2007; phylogeny).

Pleosporaceae Nitschke (1869), Pleosporales. 36 gen. (+ 37 syn.), 769 spp.

Lit.: Holm (*Symb. bot. upsal.* **14** no. 3: 1, 1957), Wehmeyer (*Monogr. Pleospora segreg.*, 1961; keys), Corlett (*Nova Hedwigia* **24**: 347, 1975; ontogeny), Shoemaker & Babcock (*CJB* **65**: 373, 1987), Sivanesan (*Mycol. Pap.* **158**: 261 pp., 1987), Crous *et al.* (*MR* **99**: 1098, 1995), Borchardt *et al.* (*Phytopathology* **88**: 322, 1998), Checa (*Mycotaxon* **68**: 205, 1998; Iberian spp.), Dong *et al.* (*MR* **102**: 151, 1998; phylogeny), Shimizu *et al.* (*J. gen. appl. Microbiol.* Tokyo **44**: 251, 1998), Turgeon & Berbee (*Molecular Genetics of Host-Specific Toxins in Plant Disease*: 153, 1998), Berbee *et al.* (*Mycol.* **91**: 964, 1999), Shoemaker (*CJB* **76**: 1559, 1998), Silva-Hanlin & Hanlin (*MR* **103**: 153, 1999), Yun *et al.* (*Proc. natn Acad. Sci. U.S.A.* **96**: 5592, 1999), Farr *et al.* (*Mycol.* **92**: 145, 2000), Olivier *et al.* (*Mycol.* **92**: 736, 2000), Pryor & Gilbertson (*MR* **104**: 1312, 2000), Câmara *et al.* (*Mycol.* **94**: 660, 2002), Serdani *et al.* (*MR* **106**: 561, 2002), Berbee *et al.* (*MR* **107**: 169, 2003), Pryor & Bigelow (*Mycol.* **95**: 1141, 2003), Peever *et al.* (*Mycol.* **96**: 119, 2004), Kodsueb *et al.* (*Mycol.* **98**: 571, 2006), Schoch *et al.* (*Mycol.* **98**: 1041, 2006; phylogeny), Wang *et al.* (*MR* **111**: 1268, 2007; phylogeny).

Pleosporales Luttr. ex M.E. Barr (1987). Pleosporomycetidae. 23 fam., 332 gen., 4764 spp. Thallus or stroma absent or rarely poorly developed. Ascomata perithecial or rarely cleistothecial, sometimes clypeate, variable in form but usually ± globose, thick-walled, immersed or erumpent, black, normally opening by a well-developed lysigenous ostiole, sometimes hairy or setose; peridium usually thick, composed of pseudoparenchymatous cells. Interascal tissue of cellular or trabeculate pseudoparaphyses, often immersed in J- gel. Asci usually ± cylindrical, fissitunicate, with a well-developed ocular chamber, rarely also with a poorly defined ring, not blueing in iodine. Ascospores hyaline to brown, septate, thin or thick-walled, sometimes muriform, often with a gelatinous sheath. Anamorphs hyphomycetous or coelomycetous, varied and often prominent. Fams.:

(1) **Arthopyreniaceae**
(2) **Corynesporascaceae**
(3) **Cucurbitariaceae**
(4) **Dacampiaceae**
(5) **Delitschiaceae**
(6) **Diademaceae**
(7) **Didymosphaeriaceae**
(8) **Fenestellaceae**
(9) **Leptosphaeriaceae**
(10) **Lophiostomataceae**
(11) **Massarinaceae**
(12) **Melanommataceae**
(13) **Montagnulaceae**
(14) **Mytilinidiaceae**
(15) **Naetrocymbaceae**
(16) **Parodiellaceae**
(17) **Phaeosphaeriaceae**
(18) **Phaeotrichaceae**
(19) **Pleomassariaceae**
(20) **Pleosporaceae**
(21) **Sporormiaceae**
(22) **Testudinaceae**
(23) **Venturiaceae**

Several further fams may belong here, including *Zopfiaceae*, but more extensive molecular data are needed.

Lit.: Checa (*Mycotaxon* **68**: 205, 1998; Iberian spp.), Kruys et al. (*MR* **110**: 527, 2006; coprophilous taxa), Liew *et al.* (*Mol. Phylog. Evol.* **16**: 392, 2000; interascal tissue, phylogeny), Schoch *et al* (*Mycol.* **98**: 1041, 2006; phylogeny), Silva-Hanlin & Hanlin (*MR* **103**(2): 153, 1999), Wang *et al.* (*MR* **111**: 1268, 2007; phylogeny), Winka *et al.* (*Mycol.* **90**: 822, 1998; phylogeny).

Pleosporites Y. Suzuki (1910), Fossil Fungi. 2 (Cretaceous), India; Japan.

Pleosporomycetidae C.L. Schoch, Spatafora, Crous & Shoemaker (2007), Dothideomycetes. Ord.:
Pleosporales

For *Lit.* see fam.

Pleosporonites R.T. Lange & P.H. Sm. (1971), Fossil Fungi. 1 (Eocene), Australia.

Pleosporopsis Oerst. (1866) ? = Coniochaeta fide Petrini (*Sydowia* **44**: 169, 1993), Læssøe (*SA* **13**: 43, 1994).

Pleostigma Kirschst. (1939) = Teichosporella fide Petrak (*Annls mycol.* **38**: 297, 1940).

Pleostomella, see *Pleiostomella*.

Pleothelis M. Choisy (1949) = Muellerella fide Hawksworth *et al.* (*Dictionary of the Fungi* edn 8, 1995).

Pleotrichiella Sivan. (1984), Dothideomycetes. 1, Australia. See Sivanesan (*TBMS* **83**: 531, 1984).

Pleovalsa Kirschst. (1936) = Fenestella fide Petrak (*Annls mycol.* **38**: 189, 1940).

plerotic, of an oospore which completely fills the oogonial cavity; but also of an oospore which occupies ›65% of the oogonium volume (Dick *et al.*, 1992).

Plesiospora Drechsler (1971), anamorphic *Pezizomycotina*, Hso.0eH.1. 1, USA; Japan; British Isles. See Drechsler (*Sydowia* **24**: 174, 1970), Glockling & Holbrook (*Mycologist* **17**: 150, 2003).

pleuracrogenous, formed at the end and on the sides.

Pleurage Fr. (1849) ≡ Schizothecium.

Pleurella E. Horak (1971), Tricholomataceae. 1, New Zealand. See Horak (*N.Z. Jl Bot.* **9**: 477, 1971).

pleuro- (in combination), side, at the side.

Pleuroascus Massee & E.S. Salmon (1901), Pseudeurotiaceae. 10, widespread. Familial placement uncertain, probably close to *Connersia*. See Lodha *in* Subramanian (Ed.) (*Taxonomy of fungi* **1**: 241, 1978), Barrasa & Moreno (*An. Jard. bot. Madr.* **41**: 31, 1984), Suh & Blackwell (*Mycol.* **91**: 836, 1999; phylogeny), Sogonov *et al.* (*Mycol.* **97**: 695, 2005; phylogeny).

Pleurobasidium Arnaud (1951) nom. inval., Thelephoraceae. 1, Europe.

pleurobasidium, see basidium.

Pleurobotrya Berk. [not traced] ? = Botrytis fide Hawksworth *et al.* (*Dictionary of the Fungi* edn 8, 1995).

Pleurocatena G. Arnaud ex Aramb. (1981) nom. inval. ≡ Pleurocatena G. Arnaud ex Aramb., Gamundí, W. Gams & G.R.W. Arnold fide Arambarri (*Darwiniana* **23**: 329, 1981).

Pleurocatena G. Arnaud ex Aramb., Gamundí, W. Gams & G.R.W. Arnold (2007), anamorphic *Pyxidiophora*, Hso.0eP.32/33. 1 (fungicolous), Argentina; Europe. See Kwasna & Bateman (*MR* **102**: 1487, 1998; UK), Gams & Arnold (*Nova Hedwigia* **84**: 381, 2007).

Pleuroceras Riess (1854), Gnomonia. Anamorphs *Asteroma*, *Cylindrosporella*, *Septogloeum*. 12, widespread (north temperate). See Barr (*Mycol. Mem.* **7**, 1978), Monod (*Beih. Sydowia* **9**: 1, 1983), Butin & Wulf (*Sydowia* **40**: 38, 1988; anamorph), Barr (*Mycotaxon* **41**: 287, 1991; N. Am. spp.), Zhang & Blackwell (*Mycol.* **93**: 355, 2001; phylogeny).

Pleurocolla Petr. (1924), anamorphic *Nectriaceae*, Hsp.0eH.?. 1, Europe. See Kirschner (*Frontiers in Basidiomycote Mycology*: 165, 2004; anam.; type stud.).

Pleurocollybia Singer (1947), Tricholomataceae. 5, America; Asia. See Singer (*Mycol.* **39**: 80, 1947), Pegler (*Kew Bull.* **43**: 53, 1988), Baroni (*Mycotaxon* **103**: 353, 2008; key).

Pleurocybe Müll. Arg. (1884) = Bunodophoron fide Wedin (*Pl. Syst. Evol.* **187**: 213, 1993).

Pleurocybella Singer (1947), Marasmiaceae. 1, widespread (north temperate). See Singer (*Mycol.* **39**: 80, 1947), Matsumoto *et al.* (*Mycoscience* **46**: 370, 2005; ITS sequences).

Pleurocybomyces Cif. & Tomas. (1953) = Bunodophoron.

pleurocystidium, see cystidium.

Pleurocystis Bonord. (1851) ≡ Helicostylum.

Pleurocytospora Petr. (1923), anamorphic *Thyridium*, St.0eH.15. 2, Europe; China. See Leuchtmann & Müller (*Bot. Helv.* **96**: 283, 1986), Sun *et al.* (*Mycosystema* **22**: 12, 2003).

Pleurodesmospora Samson, W. Gams & H.C. Evans (1979), anamorphic *Pezizomycotina*, Hso.0eH.16. 1 (on *Insecta*), Sri Lanka; China. See Li & Han (*Acta Mycol. Sin.* **10**: 166, 1991; China).

Pleurodiscula Höhn. (1926), anamorphic *Pezizomycotina*, St.0eH.?. 1, Europe.

Pleurodiscus Petr. (1931), anamorphic *Pezizomycotina*, Cpd.?.?. 1, Austria.

Pleurodomus Petr. (1934), anamorphic *Pezizomycotina*, St.0eH.?. 1, Siberia.

Pleurodon Quél. ex P. Karst. (1881) ≡ Auriscalpium.

Pleuroflammula Singer (1946), Inocybaceae. 10, America; Asia. See Horak (*Persoonia* **9**: 439, 1978; key 10 spp.), Bon & Roux (*Docums Mycol.* **33**: 3, 2003; key).

Pleurogala Redhead & Norvell (1993) = Lactarius fide Miller *et al.* (*Mycol.* **98**: 960, 2006).

pleurogenous, formed on the side.

Pleurographium Goid. (1935) = Nodulisporium fide Sutton (*in litt.*), de Hoog & von Arx (*Kavaka* **1**: 55, 1973).

Pleuromycenula Singer (1974) = Rimbachia fide Redhead (*CJB* **62**: 865, 1984).

Pleuronaema Höhn. (1917) = Cytospora fide Gvritishvili (*Mikol. Fitopatol.* **3**: 207, 1969).

Pleuropedium Marvanová & S.H. Iqbal (1973), anamorphic *Pezizomycotina*, Hso.1bH.?. 3 (aquatic), Europe; N. America. See Marvanová & Bärlocher (*MR* **102**: 750, 1998; Canada).

Pleurophoma Höhn. (1914), anamorphic *Pezizomycotina*, St.0eH.15. 8, widespread. See Boerema *et al.* (*Phoma Identification Manual* Differentiation of Specific and Infra-Specific Taxa in Culture: 470 pp., 2004).

Pleurophomella Höhn. (1914) = Sirodothis fide Sutton & Funk (*CJB* **53**: 521, 1975).

Pleurophomopsis Petr. (1924), anamorphic *Astrosphaeriella*, Cpd.0eH.15. 7, widespread. See Padhye *et al.* (*J. Clin. Microbiol.* **35**: 2136, 1997; medical), Padhye *et al.* (*Medical Mycology* **42**: 129, 2004; from human), Tanaka & Harada (*Mycoscience* **46**: 114, 2005; teleomorph).

Pleurophragmium Costantin (1888), anamorphic *Pezizomycotina*, Hso.0eH/≡ eH.10. 21, widespread. See Hughes (*CJB* **36**: 796, 1958), Ellis (*Mycol. Pap.* **114**: 42, 1968), de Hoog (*Stud. Mycol.* **26**: 1, 1985; synonymy with *Dactylaria*), Paulus *et al.* (*Fungal Diversity* **14**: 143, 2003).

Pleuroplaconema Petr. (1923), anamorphic *Pezizomycotina*, St.0eH.15. 2, Europe; Asia. See Sun & Guo (*Mycosystema* **25**: [543], 2006; endophyte).

Pleuroplacosphaeria Syd. (1928), anamorphic *Pezizomycotina*, St.0eH.?. 1, Chile.

Pleuropus (Pers.) Roussel (1806) nom. rej. = Panus fide Kuyper (*in litt.*).

Pleuropyxis Corda (1837) nom. conf., anamorphic *Pezizomycotina*. See Hughes (*CJB* **36**: 798, 1958).

Pleurosordaria Fernier (1954) = Arnium fide Lundqvist (*Symb. bot. upsal.* **20** no. 1, 1972).

pleurosporous, having spores on the sides, e.g. a basidium of the *Pucciniales*.

Pleurosticta Petr. (1931), Parmeliaceae (L). 2, Europe. See Esslinger (*Taxon* **29**: 692, 1980; synonymy with *Melanelia*), Lumbsch *et al.* (*Mycotaxon* **33**: 447, 1988), Mattsson & Wedin (*Lichenologist* **30**: 463, 1998; phylogeny, morphology), Wedin *et al.* (*MR* **103**: 1152, 1999; phylogeny), Thell *et al.* (*Symb. bot. upsal.* **34** no. 1: 429, 2004; Scandinavia), Blanco *et al.* (*Mol. Phylogen. Evol.* **39**: 52, 2006; phylogeny), Miądlikowska *et al.* (*Mycol.* **98**: 1088, 2006; phylogeny).

Pleurostoma Tul. & C. Tul. (1863), Pleurostomataceae. Anamorph *Pleurostomophora*. 4 (on wood and bark), Europe; N. America. See Barr (*Mycol.* **77**: 549,

1985), Barr *et al.* (*Mycotaxon* **48**: 529, 1993), Réblová *et al.* (*Stud. Mycol.* **50**: 533, 2004), Vijaykrishna *et al.* (*Stud. Mycol.* **50**: 387, 2004; anamorph), Mostert *et al.* (*Stud. Mycol.* **54**: 115 pp., 2006; phylogeny).

Pleurostomataceae Réblová, L. Mostert, W. Gams & Crous (2004), Calosphaeriales. 2 gen. (+ 1 syn.), 7 spp.
Lit.: Barr (*Mycol.* **77**: 549, 1985), Romero & Minter (*TBMS* **90**: 457, 1988), Romero & Samuels (*Beih. Sydowia* **43**: 228, 1991), Mostert *et al.* (*Mycol.* **95**: 646, 2003), Réblová *et al.* (*Stud. Mycol.* **50**: 533, 2004), Vijaykrishna *et al.* (*Stud. Mycol.* **50**: 387, 2004).

Pleurostomophora Vijaykr., L. Mostert, Jeewon, W. Gams, K.D. Hyde & Crous (2004), anamorphic *Pleurostoma*. 3, widespread. See Vijaykrishna *et al.* (*Stud. Mycol.* **50**: 390, 2004), Mostert *et al.* (*Stud. Mycol.* **54**: 115 pp., 2006).

Pleurostromella Petr. (1922), anamorphic *Gibberidea*, St.0eH.?. 15, Europe.

Pleurotaceae Kühner (1980), Agaricales. 6 gen. (+ 10 syn.), 94 spp.
Lit.: Vilgalys (*Fungi and Lichens in the Baltic Region* Abstracts, Twelth International Conference on Mycology and Lichenology: 133, 1993), Corner (*Gdns' Bull.* Singapore **46**: 49, 1994), Corner (*Gdns' Bull.* Singapore **46**: 1, 1994), Hibbett & Thorn (*Mycol.* **86**: 696, 1994), Liou & Tzean (*Mycol.* **89**: 876, 1997), Albertó *et al.* (*Mycol.* **90**: 142, 1998), Zervakis (*Mycol.* **90**: 1063, 1998), Gonzalez & Labarère (*Microbiology (Reading)* **146**: 209, 2000), Thorn *et al.* (*Mycol.* **92**: 241, 2000; phylogeny), Fazio & Albertó (*Mycotaxon* **77**: 117, 2001), Hibbett & Binder (*Proc. R. Soc. Lond.* B. Biol. Sci. **269**: 1963, 2002), Moncalvo *et al.* (*Mol. Phylogen. Evol.* **23**: 357, 2002), Capelari & Fungaro (*MR* **107**: 1050, 2003), Lechner *et al.* (*Mycol.* **96**: 845, 2004), Zervakis *et al.* (*Microbiology (Reading)* **150**: 715, 2004), De Gioia *et al.* (*MR* **109**: 71, 2005).

Pleurotellus Fayod (1889) = Crepidotus fide Aime *et al.* (*Am. J. Bot.* **92**: 74, 2005).

Pleurotheciopsis B. Sutton (1973), anamorphic *Pezizomycotina*, Hso.≡ eH.3. 6, widespread. See Castañeda Ruíz & Iturriaga (*Mycotaxon* **70**: 63, 1999), Castañeda Ruiz *et al.* (*Mycotaxon* **77**: 1, 2001).

Pleurothecium Höhn. (1919), anamorphic *Carpoligna*, Hso.≡ eH-P.10. 4, widespread. See Goos (*Mycol.* **61**: 1048, 1970), Hyde & Goh (*MR* **102**: 739, 1998), Fernández *et al.* (*Mycol.* **91**: 251, 1999; teleomorph).

Pleurotheliopsis Zahlbr. (1922) = Pyrenula Ach. (1809) fide Singh & Upreti (*Geophytology* **16**: 261, 1986; India), Harris (*Mem. N. Y. bot. Gdn* **49**, 1989).

Pleurothelium Müll. Arg. (1877) = Pyrenula Ach. (1814) fide Aptroot (*in litt.*).

Pleurothelium Müll. Arg. (1885) = Pyrenula Ach. (1809) fide Hawksworth *et al.* (*Dictionary of the Fungi* edn 8, 1995).

Pleurothyriella Petr. & Syd. (1925), anamorphic *Pezizomycotina*, Cac.0eH.?. 1, Europe.

Pleurothyrium Bubák (1916), anamorphic *Pezizomycotina*, St.0fH.10. 1, Europe.

Pleurotopsis (Henn.) Earle (1909) = Resupinatus. Belongs to *Mycenaceae* or *Marasmiaceae*. See Hughes *et al.* (*Mycol.* **90**: 595, 1998; biogeography).

Pleurotrema Müll. Arg. (1885) = Lithothelium fide Aptroot (*Biblthca Lichenol.* **44**, 1991), Barr (*Mycotaxon* **51**: 191, 1994).

Pleurotremataceae Walt. Watson (1929) = Pyrenulaceae.
Lit.: Barr (*Mycotaxon* **51**: 191, 1994; circumscribed with 5 genera with non-fissitunicate asci (*Daruvedia*, *Melomastia*, *Phomatospora*, *Pleurotrema* and *Saccardoella*)).

Pleurotrematomyces Cif. & Tomas. (1957) ≡ Pleurotrema.

Pleurotus (Fr.) P. Kumm. (1871) nom. cons., Pleurotaceae. Anamorph *Antromycopsis*. 20 (esp. on wood), widespread. *P. ostreatus*, the Oyster Cap, and several other species are edible and commercially cultivated. See Hilber (*Biblthca Mycol.* **87**: 1, 1982; monogr.), Barron & Thorn (*CJB* **64**: 774, 1987; ecology), Thorn *et al.* (*TMSJ* **34**: 449, 1993; ecology), Vilgays & Sun (*Proc. natn Acad. Sci. U.S.A.* **91**: 4599, 1994; speciation), Zervakis & Balis (*MR* **100**: 717, 1996; key Europ. spp.), Zervakis (*Mycol.* **90**: 1063, 1998; taxonomy), Albertó *et al.* (*Persoonia* **18**: 55, 2002), Li & Yao (*Mycosystema* **23**: 345, 2004; phylogeny), Zervakis *et al.* (*Microbiology* Reading **150**: 715, 2004; phylogeny, biogeography and speciation), Stajic *et al.* (*Mycotaxon* **93**: 247, 2005; genetics).

Pleurovularia R. Kirschner & U. Braun (2002), anamorphic *Pezizomycotina*, H?.?.?. 1, E. Asia. See Kirschner *et al.* (*Mycoscience* **43**: 16, 2002).

Pliariona A. Massal. (1860) = Thecaria fide Hawksworth *et al.* (*Dictionary of the Fungi* edn 8, 1995).

Plicaria Fuckel (1870), Pezizaceae. Anamorph *Chromelosporium*. *c.* 10, widespread (temperate). Clusters within *Peziza*, but that genus is itself polyphyletic. See Hirsch (*Agarica* **6**: 241, 1985; key 4 Eur. spp.), Egger (*Mycotaxon* **29**: 183, 1986), Norman & Egger (*Mycol.* **88**: 986, 1996), Norman & Egger (*Mycol.* **91**: 820, 1999; phylogeny), Hansen *et al.* (*Mycol.* **93**: 958, 2001; phylogeny), Spooner (*Czech Mycol.* **52**: 259, 2001), Hansen *et al.* (*Mol. Phylogen. Evol.* **36**: 1, 2005; phylogeny).

Plicariaceae Velen. (1934) = Pezizaceae.

Plicariella (Sacc.) Rehm (1894) = Plicaria fide Spooner (*Czech Mycol.* **52**: 259, 2001).

plicate, folded into pleats; **plica**, a pleat.

Plicatura Peck (1872), Agaricales. 1, widespread (north temperate). Belongs to *Amylocorticiaceae* or *Tricholomataceae*.

Plicaturaceae Jülich (1982) = Atheliaceae.

Plicaturella Murrill (1910) = Boletinellus fide Kuyper (*in litt.*).

Plicaturopsis D.A. Reid (1964), Agaricales. 2, Europe; New Zealand. Belongs to *Amylocorticiaceae* or *Tricholomataceae*.

Plistophora, see *Pleistophora*.

Plocaria Nees (1820) nom. rej. prop., Algae.

Plochmopeltidella J.M. Mend. (1925) = Chaetothyrina fide von Arx & Müller (*Stud. Mycol.* **9**, 1975).

Plochmopeltinites Cookson (1947), Fossil Fungi. 2 (Oligocene, Miocene), India.

Plochmopeltis Theiss. (1914), Schizothyriaceae. 5, pantropical. See Gómez Acosta & Clavel Calzado (*Rev. Jardín Bot. Nac. Univ. la Habana* **17-18**: 137, 1998; Cuba), Reynolds & Gilbert (*Aust. Syst. Bot.* **18**: 265, 2005; Australia).

Plochmothea Syd. (1939) = Xenostomella fide von Arx & Müller (*Stud. Mycol.* **9**, 1975).

Ploettnera Henn. (1899), Helotiales. 5, Europe. See Hein (*Willdenowia* Beih. **9**, 1976; key), Nauta & Spooner (*Mycologist* **13**: 65, 1999; Brit. spp.).

Ploettnerula Kirschst. (1924) = Pirottaea fide Nannfeldt (*Nova Acta R. Soc. Scient. upsal.*, 1932).

Ploioderma Darker (1967), Rhytismataceae. Anamorph *Cryocaligula*. 5, India; N. America. See Hou & Liu (*Acta Mycol. Sin.* **12**: 99, 1993), Lin & Hou (*Acta Mycol. Sin.* **14**: 175, 1995), Hansen & Lewis (*Compendium of Conifer Diseases*, 1997), Jewell (*Forest Pathology* **31**: 33, 2001; histopathology).

Plokamidomyces Bat., C.A.A. Costa & Cif. (1957), anamorphic *Trichopeltheca*, Hso.0eH.15. 1, New Zealand. See Batista *et al.* (*Publções Inst. Micol. Recife* **90**: 15, 1957).

Plowrightia Sacc. (1883), Dothioraceae. *c.* 7, widespread. *P. ribesia* (black pustule of gooseberry and currants (*Ribes*)). See Wakefield (*TBMS* **24**: 286, 1940), Barr (*Harvard Pap. Bot.* **6**: 25, 2001), Winton *et al.* (*Mycol.* **99**: 240, 2007; phylogeny).

Plowrightiella (Sacc.) Trotter (1926) = Sydowia fide von Arx & Müller (*Stud. Mycol.* **9**, 1975).

Pluesia Nieuwl. (1916) ≡ Maurya.

plug (in an ascus), see ascus.

Plunkettomyces G.F. Orr (1977) = Arachniotus fide von Arx (*Persoonia* **9**: 393, 1977).

Pluricellaesporites Hammen (1954), Fossil Fungi. 30 (Cretaceous, Tertiary), widespread.

Pluricellulites Hammen (1954), Fossil Fungi ? Fungi. 1.

plurilocular (1) (of ascospores), many-celled; (2) (of stromata), having several locules.

Pluriporus F. Stevens & R.W. Ryan (1925) = Dothidella fide Müller & von Arx (*Beitr. Kryptfl. Schweiz* **11** no. 2, 1962).

Plurisperma Sivan. (1970), ? Verrucariales. 1, Pakistan. See Matzer (*Nova Hedwigia* **56**: 203, 1993).

plurivorous, attacking a number of hosts or substrates; not specialized.

Pluteaceae Kotl. & Pouzar (1972), Agaricales. 4 gen. (+ 8 syn.), 364 spp.

Lit.: Miller & Horak (*Mycol.* **84**: 64, 1992), Banerjee & Sundberg (*Mycotaxon* **53**: 189, 1995), Weiss *et al.* (*CJB* **76**: 1170, 1998), Drehmel *et al.* (*Mycol.* **91**: 610, 1999), Hibbett *et al.* (*Nature* **407**: 506, 2000; as syn. of *Amanitaceae*), Moncalvo *et al.* (*Syst. Biol.* **49**: 278, 2000), Grgurinovic (*Aust. Syst. Bot.* **14**: 395, 2001), Bougher & Lebel (*Aust. Syst. Bot.* **15**: 514, 2002), Moncalvo *et al.* (*Mol. Phylogen. Evol.* **23**: 357, 2002), Seok *et al.* (*Mycobiology* **30**: 183, 2002), Simmons *et al.* (*Persoonia* **17**: 563, 2002), Bhatt & Miller (*Mem. N. Y. bot. Gdn* **89**: 33, 2004), Oda *et al.* (*MR* **108**: 885, 2004), Zhang *et al.* (*Fungal Diversity* **17**: 219, 2004).

Pluteolus (Fr.) Gillet (1876) = Bolbitius fide Singer (*Agaric. mod. Tax.*, 1951).

Pluteopsis Fayod (1889) = Psathyrella fide Singer (*Agaric. mod. Tax.*, 1951).

Pluteus Fr. (1836), Pluteaceae. *c.* 300 (esp. on wood), widespread. See Singer (*TBMS* **39**: 222, 1956; key), Smith & Stuntz (*Lloydia* **21**: 115, 1959; key S. Am. spp.), Singer (*Sydowia* **15**: 114, 1962; suppl. S. Am. spp.), Homola (*Mycol.* **64**: 1211, 1972; sect. *Celluloderma* N. Am., keys), Horak (*Fl. Illustr. Champ. Afr. Centr.* **6**: 107, 1976; key 10 C. Afr. spp.), Pegler (*Kew Bull.* Addit. Ser. **6**, 1977; key 6 E. Afr. spp.), Vellinga & Schreurs (*Persoonia* **12**: 337, 1985; key W. Eur. spp.), Orton (*British fungus flora* **4**: 4, 1986; key Brit. spp.), Contu (*Mycol. helv.* **11**: 137, 2001; Sardinia), Pradeep *et al.* (*Mycotaxon* **83**: 59, 2002; India), Wartchow *et al.* (*Mycotaxon* **96**: 241, 2006; Brazil).

Pneumatospora B. Sutton, Kuthub. & Muid (1984) = Minimedusa fide Diederich & Lawrey (*Mycol. Progr.* **6**: 61, 2007).

Pneumocystidaceae O.E. Erikss. (1994), Pneumocystidales. 1 gen., 5 spp.

Lit.: Edman *et al.* (*Nature* Lond. **334**: 519, 1988), Eriksson (*SA* **13**: 165, 1994), Lu *et al.* (*J. Clin. Microbiol.* **32**: 2904, 1994), Eriksson (*SA* **13**: 165, 1995), Sugiyama *et al.* *in* Colwell *et al.* (Eds) (*Microbial Diversity in Time and Space*: 41, 1996), Sjamsuridzal *et al.* (*Mycoscience* **38**: 267, 1997; phylogeny), Dei-Cas *et al.* (*J. Mycol. Médic.* **8**: 1, 1998), Mazars & Dei-Cas (*FEMS Immunol. Med. Microbiol.* **22**: 75, 1998), Stringer (*FEMS Immunol. Med. Microbiol.* **22**: 15, 1998), Wakefield *et al.* (*Medical Mycology* **36**: 183, 1998), Chin *et al.* (*J. Eukary. Microbiol.* **46**: 95S, 1999), Frenkel (*J. Eukary. Microbiol.* **46**: 89S, 1999), Palmer *et al.* (*Appl. Environm. Microbiol.* **66**: 4954, 2000), Demanche *et al.* (*J. Clin. Microbiol.* **39**: 2126, 2001), Norris *et al.* (*Clin. Diagn. Lab. Immunol.* **10**: 1037, 2003), Cushion *et al.* (*Mycol.* **96**: 429, 2004), Guillot *et al.* (*Mol. Phylogen. Evol.* **31**: 988, 2004), Keely *et al.* (*Microbiology (Reading)* **150**: 1153, 2004), Robberts *et al.* (*J. Clin. Microbiol.* **42**: 1505, 2004), Sugiyama *et al.* (*Mycol.* **98**: 996, 2006; phylogeny).

Pneumocystidales O.E. Erikss. (1994). Pneumocystidomycetidae. 1 fam., 1 gen., 5 spp. Fam.:

Pneumocystidaceae

For *Lit.* see under fam.

Pneumocystidomycetes O.E. Erikss. & Winka (1997). Taphrinomycotina. 1 ord., 1 fam., 1 gen., 5 spp. Clusters within the *Taphrinomycotina* and contains a single genus of obligate mammalian parasites. Ord.:

Pneumocystidales

For *Lit.* see under fam.

Pneumocystidomycetidae, see *Pneumocystidomycetes*.

Pneumocystis P. Delanoë & Delanoë (1912), Pneumocystidaceae. 5, widespread. *P. carinii* causes pneumonia (PCP) in immunocompromized (esp. AIDS) patients. See Hong *et al.* (*J. Clin. Microbiol.* **28**: 1785, 1990; karyotypes), Kovacs *et al.* (*Exp. Parasitol.* **71**: 60, 1990; characterization of dihydrofolate reductase), Mackenzie (*Rev. Iberoam. Micol.* **7**: 3, 1990; aetiology), Tamburrini *et al.* (*J. Clin. Microbiol.* **31**: 2788, 1993; rapid diagnosis by PCR), Wakefield *et al.* (*Mol. Microbiol.* **8**: 426, 1993; basidiomycete affinity), Eriksson (*SA* **13**: 165, 1995), Stringer (*Clin. Microbiol. Rev.* **9**: 489, 1996; review), Dei-Cas *et al.* (*J. Mycol. Médic.* **8**: 1, 1998; host specificity), Dei-Cas *et al.* (*FEMS Immunol. Med. Microbiol.* **22**: 185, 1998; ultrastr.), Lee *et al.* (*J. Clin. Microbiol.* **36**: 734, 1998; ITS), Stringer (*FEMS Immunol. Med. Microbiol.* **22**: 15, 1998; genome), Wakefield (*FEMS Immunol. Med. Microbiol.* **22**: 5, 1998; genetic diversity), Wakefield *et al.* (*Medical Mycology* **36** Suppl. 1: 183, 1998; host specificity), Rabodonirina *et al.* (*J. Clin. Microbiol.* **37**: 127, 1999; nested PCR detection), Baldauf *et al.* (*Science* N.Y. **290** no. 5493: 972, 2000; phylogeny), Denis *et al.* (*Medical Mycology* **38**: 289, 2000; genetic diversity), Durand-Joly *et al.* (*Medical Mycology* **38**, 2000; from monkey), Palmer *et al.* (*Appl. Environm. Microbiol.* **66**: 4954, 2000; population structure in rats), Demanche *et al.* (*J. Clin. Microbiol.* **39**: 2126, 2001; phylogeny, coevolution), Kaiser *et al.* (*J. Microbiol. Meth.* **45**: 113, 2001; PCR analysis), Palladino *et al.* (*Diagn. Microbiol. Infect. Dis.* **39**: 233, 2001; molecular detection), Smulian (*Fungal Genetics Biol.* **34**: 145, 2001; genetic diversity), Guillot *et al.* (*J. Eukary. Microbiol.* Suppl.: 113S, 2001; phylogeny), Stringer *et al.* (*J. Eukary. Microbiol.* Suppl.: 184S, 2001; nomencl., species concepts), Hugot *et al.* (*Syst. Biol.* **52**: 735, 2003; phylogeny), Keely *et al.* (*J. Eukary. Microbiol.* **50** Suppl.: 624, 2003; evolution), Cushion (*J. Eukary. Microbiol.* **51**: 30, 2004; comparative genomics), Cushion *et al.* (*Mycol.* **96**: 429, 2004; from rat), Guillot *et al.* (*Mol. Phylogen. Evol.* **31**: 988, 2004; from monkey), Keely *et al.* (*Microbiology* Reading **150**: 1153, 2004; from mouse), James (*Nature* Lond. **443** no. 7113: 818, 2006; phylogeny), Slaven *et al.* (*J. Eukary. Microbiol.* **53** Suppl. 1: S89, 2006; whole genome), Sugiyama *et al.* (*Mycol.* **98**: 996, 2006; phylogeny).

pneumomycosis, see mycosis.

Pochonia Bat. & O.M. Fonseca (1965), Clavicipitaceae. See Barron & Onions (*CJB* **44**, 1966; synonym of *Diherterospora*), Gams & Zare (*Nova Hedwigia* **72**: 334, 2001), Zare *et al.* (*Nova Hedwigia* **73**: 51, 2001; revision), Morton *et al.* (*MR* **107**: 198, 2003; genetic diversity), Stadler *et al.* (*Mycol. Progr.* **2**: 95, 2003; chemistry), Sung *et al.* (*Stud. Mycol.* **57**, 2007).

Pocillaria P. Browne (1756) ≡ Lentinus.

Pocillopycnis Dyko & B. Sutton (1979), anamorphic *Pezizomycotina*, St.0fH.10. 1, Sweden. See DiCosmo (*Mycotaxon* **10**: 288, 1980).

Pocillum De Not. (1863), Helotiaceae. 2 or 3, Europe; America. See Petrak (*Sydowia* **5**: 345, 1951).

pocket plums, plums swollen then 'mummified' by *Taphrina pruni*.

pocket rot, localized rotting of trunks of trees or roots by wood-destroying fungi.

Pocosphaeria (Sacc.) Berl. (1892) = Nodulosphaeria fide Hawksworth *et al.* (*Dictionary of the Fungi* edn 8, 1995).

Pocsia Vězda (1975), ? Verrucariales (L). 3, widespread (tropical). See Hafellner & Kalb (*Biblthca Lichenol.* **57**: 161, 1995; posn), McCarthy (*Lichenologist* **31**: 141, 1999; Australia), Herrera-Campos & Lücking (*Lichenologist* **34**: 211, 2002; Mexico).

Poculinia Spooner (1987), Sclerotiniaceae. 1 (on *Nothofagus*), Tasmania. See Spooner (*Biblthca Mycol.* **116**, 1987).

Poculopsis Kirschst. (1935), Helotiaceae. 1, Europe.

Poculum Velen. (1934), Rutstroemiaceae. *c.* 20, widespread. See Holst-Jensen *et al.* (*Mycol.* **89**: 885, 1997; phylogeny).

Podabrella Singer (1945) = Termitomyces fide Kuyper (*in litt.*).

Podaleuris Clem. (1909) = Peziza Fr. fide Eckblad (*Nytt Mag. Bot.* **15**: 1, 1968).

Podaxaceae Corda (1842) = Agaricaceae.

Podaxales = Agaricales.

Podaxis Desv. (1809), Agaricaceae. 10, widespread (subtropical dry areas). See Morse (*Mycol.* **25**: 1, 1933), Martinez (*Boln Soc. argent. Bot.* **14**: 73, 1971; Argentina), McKnight & Stransky (*Mycol.* **72**: 195, 1980), McKnight (*Mycol.* **77**: 24, 1985; key small-spored spp.), Hopple & Vilgalys (*Mol. Phylogen. Evol.* **13**: 1, 1999; phylogeny), Dios *et al.* (*Mycotaxon* **80**: 453, 2001; Argentina), Keirle *et al.* (*Fungal Diversity* **15**: 33, 2004; Hawaiian Islands), Vellinga (*MR* **108**: 354, 2004; phylogeny).

Podaxon, see *Podaxis*.

podetium, lichenized stem-like portion (stipe, or discopodium) bearing the hymenial discs and sometimes conidiomata in a fruticose apothecium (Ahti, *Lichenologist* **14**: 105, 1982), esp. as in *Cladonia*. Cf. pseudopodetium.

Podisoma Link (1809) = Gymnosporangium fide Saccardo (*Syll. fung.* **7**: 737, 1888).

Podobactridium Penz. & Sacc. ex Petch (1916) = Bactridium fide Hughes (*N.Z. Jl Bot.* **4**: 522, 1966).

Podobelonium (Sacc.) Sacc. & D. Sacc. (1906) = Crocicreas fide Nannfeldt (*Nova Acta R. Soc. Scient. upsal.*, 1932).

Podocapsa Tiegh. (1887), Saccharomycotina. 1, Europe. Position uncertain.

Podocapsium Clem. (1909) ≡ Podocapsa.

Podochytrium Pfitzer (1870), Chytridiaceae. 7 (on diatoms or saprobes), widespread (north temperate). See Canter (*J. Linn. Soc.* Bot. **63**: 47, 1970).

Podoconis Boedijn (1933) = Sporidesmium fide Ellis (*Mycol. Pap.* **70**, 1958).

Podocratera Norman (1861) nom. rej. ≡ Tholurna.

Podocrea (Sacc.) Lindau (1897) = Podostroma fide Rossman *et al.* (*Stud. Mycol.* **42**: 248 pp., 1999), Chamberlain *et al.* (*Karstenia* **44**: 1, 2004).

Podocrella Seaver (1928), Clavicipitaceae. Anamorph *Harposporium*. 4 (on rotten wood), widespread (esp. tropical). See Rossman *et al.* (*Stud. Mycol.* **42**: 248 pp., 1999), Chaverri *et al.* (*Mycol.* **97**: 433, 2005), Chaverri *et al.* (*Mycol.* **97**: 1225, 2005).

Podocystis Fr. (1849) nom. rej. prop. = Melampsora fide Saccardo (*Syll. Fung.* **18**: 812, 1906).

Pododimeria E. Müll. (1959), ? Pseudoperisporiaceae. 4, Europe; N. America. See Luttrell & Barr (*Am. J. Bot.* **65**: 251, 1978; key).

Podofomes Pouzar (1966), Polyporaceae. 3. See David *et al.* (*BSMF* **99**: 361, 1983).

Podohydnangium G.W. Beaton, Pegler & T.W.K. Young (1984) = Laccaria fide Kropp & Mueller (*Laccaria – Ectomycorrhizal fungi, key genera in profile*, 1999; phylogeny, congeneric with *Laccaria*).

Podonectria Petch (1921), Tubeufiaceae. Anamorphs *Peziotrichum*, *Tetracrium*. 11 (on scale insects), widespread (esp. tropical). See Pirozynski (*Kew Bull.* **31**: 595, 1977), Rossman (*Mycotaxon* **7**: 163, 1978; key), Rossman (*Mycol. Pap.* **157**, 1987), Kodsueb *et al.* (*Fungal Diversity* **21**: 105, 2006; review).

Podonectrioides Kobayasi & Shimizu (1983) nom. inval. = Podonectria fide Hawksworth *et al.* (*Dictionary of the Fungi* edn 8, 1995).

Podophacidium Niessl (1868), Helotiales. 2, Europe; N. America. See Seaver (*Mycol.* **31**: 350, 1939), Raitviir & Järv (*Proc. Est. Acad. Sci. Biol. Ecol.* **46**: 94, 1997), Pärtel & Raitviir (*Mycol. Progr.* **4**: 149, 2005; asci).

Podoplaconema Petr. (1921), anamorphic *Omphalospora*, St.0eH.15. 1, Europe.

Podoporia P. Karst. (1892) = Oligoporus fide Ryvarden (*Syn. Fung.* **5**, 1991).

Podoscypha Pat. (1900), Meruliaceae. *c.* 35, widespread (esp. tropical). See Reid (*Beih. Nova Hedwigia* **18**: 150, 1965; key), Douanla-Meli & Langer (*Mycotaxon* **90**: 323, 2004).

Podoscyphaceae D.A. Reid (1965), Polyporales.

Podoserpula D.A. Reid (1963), Amylocorticiaceae. 1, widespread. See Reid (*Kew Bull.* **16**: 437, 1963).

Podosordaria Ellis & Holw. (1897), Xylariaceae. Anamorph *Geniculosporium*-like. 23 (esp. coprophilous), widespread. See Krug & Cain (*CJB* **52**: 589, 1974; key), Rogers *et al.* (*Mycol.* **84**: 166, 1992),

Krug & Jeng (*CJB* **73**: 65, 1995; key 17 spp.), Rogers *et al.* (*Mycotaxon* **67**: 61, 1998).

Podosphaera Kunze (1823), Erysiphaceae. Anamorph *Oidium* subgen. *Fibroidium*. 73, widespread. *P. leucotricha* (mildew of apple and other pome fruits), *P. clandestina* (hawthorn (*Crataegus*) mildew), *P. macularis* (hop mildew), *P. mors-uvae* (American mildew of currant and gooseberry (*Ribes*)), *P. pannosa* (rose mildew). See Braun (*Beih. Nova Hedwigia* **89**, 1987; key), Hirata *et al.* (*CJB* **78**: 1521, 2000; ITS, coevolution), Takamatsu (*MR* **104**: 1303, 2000), Takamatsu (*Mycoscience* **45**: 147, 2004; phylogeny), Cunnington *et al.* (*MR* **109**: 357, 2005; on *Prunus*), Leus *et al.* (*J. Phytopath.* **154**: 23, 2006; on *Rosaceae*), Wang *et al.* (*Mycol.* **98**: 1065, 2006; phylogeny), Voytyuk *et al.* (*Nova Hedwigia* **85**: 277, 2007; Israel).

Podospora Ces. (1856) nom. cons., Lasiosphaeriaceae. 94 (coprophilous), widespread. See Mirza & Cain (*CJB* **47**: 1999, 1969; key), Lundqvist (*Symb. bot. upsal.* **20** no. 1, 1972), Krug & Khan (*CJB* **67**: 1174, 1989; culture, records), Schmidt (*Biblthca Mycol.* **127**, 1989; *P. anserina*), Guarro *et al.* (*SA* **10**: 79, 1991), Bell & Mahoney (*Mycol.* **87**: 375, 1995), Bell & Mahoney (*Mycol.* **88**: 163, 1996; development), Bell & Mahoney (*Mycol.* **89**: 908, 1997), Wang (*Mycotaxon* **76**: 373, 2000; Taiwan), Lorenzo & Havrylenko (*Mycol.* **93**: 1221, 2001; Argentina), Huhndorf *et al.* (*Mycol.* **96**: 368, 2004; phylogeny), Miller & Huhndorf (*Mol. Phylogen. Evol.* **35**: 60, 2005; anatomy, phylogeny), Fitzpatrick *et al.* (*BMC Evolutionary Biology* **6**: 99, 2006; whole genome phylogeny), Zhang *et al.* (*Mycol.* **98**: 1076, 2006; phylogeny).

Podosporiella Ellis & Everh. (1894), anamorphic *Pezizomycotina*, Hsy.≡ eP.?. 1, N. America; Australia.

Podosporium Bonord. (1851) = Aplosporella fide Sutton (*Mycol. Pap.* **141**, 1977).

Podosporium Lév. (1847) = Melampsora fide Dietel (*Nat. Pflanzenfam.* **6**, 1928).

Podosporium Sacc. & Schulzer (1884) = Aplosporella fide Saccardo (*Syll. fung.* **3**: 1, 1884).

Podosporium Schwein. (1832), anamorphic *Pezizomycotina*, Hsy.≡ eP.24. 15, widespread (esp. tropical). See Ellis (*Dematiaceous Hyphomycetes*: 291, 1971), Ellis (*More Dematiaceous Hyphomycetes*: 303, 1976), Rong (*Phytophylactica* **24**: 103, 1992), Chen & Tzean (*MR* **97**: 637, 1993; Taiwan).

Podostictina Clem. (1909) = Pseudocyphellaria fide Galloway (*Bull. Br. Mus. nat. hist.* Bot. **17**, 1988).

Podostroma P. Karst. (1892), Hypocreaceae. Anamorph *Trichoderma*. 11 (from decaying wood), widespread. See Boedijn (*Annls mycol.* **36**: 314, 1938), Rossman *et al.* (*Stud. Mycol.* **42**: 248 pp., 1999), Wang & Liu (*Mycosystema* **21**: 156, 2002; China), Chamberlain *et al.* (*Karstenia* **44**: 1, 2004).

Podostrombium Kunze ex Rchb. (1828) nom. inval., Clavariaceae. See Donk (*Taxon* **6**: 110, 1957).

Podotara Malcolm & Vězda (1996), Lecanoromycetes (L). 1, New Zealand. See Malcolm & Vězda (*Folia geobot. phytotax.* **31**: 263, 1996).

Podoxyphiomyces Bat., Valle & Peres (1961) nom. dub., anamorphic *Pezizomycotina*, St.0eH.? (L). 1, Brazil. See Lücking *et al.* (*Lichenologist* **30**: 121, 1998).

Podoxyphium Speg. (1918) = Conidiocarpus fide Hughes (*Mycol.* **68**: 693, 1976).

Poecilosporium, see *Poikilosporium*.

Poecylomyces, see *Paecilomyces*.

Poelt (Josef; 1924-1995; Germany). Bavarian; doctorate (1950) then member of academic staff (1950-1965), University of Munich; Professor of Systematic Botany, Free University of Berlin (1965-1972); Professor and Director of the Botanical Garden, University of Graz (1972-1991), where he remained active after retirement. His wide-ranging interests included lichen-forming fungi, rusts and smuts, and oomycetes, as evident in over 320 publications, including pioneering keys to European lichens; he travelled extensively, making important contributions to Himalayan lichens (1966-1977) and Mediterranean (especially Greek) mycotas; renowned for a determination to tackle 'difficult' groups, for insights into lichen biology, phylogeny and structure, and for work on lichenicolous and bryophilous lichens; an inspiring teacher, his PhD students included P. Döbbeler, J. Hafellner, B. Hein, H. Hertel, W. Jülich, H. Mayrhofer and F. Oberwinkler; editor/co-editor of journals including *Herzogia* and *Nova Hedwigia*; also worked with bryophytes. *Publs. Bestimmungschlussel Europäischer Flechten* (1969); (A. Vězda) *Erganzungsheft I-II* (1977, 1981); (with Mayrhofer) Die saxicolen Arten der Flechtengattung *Rinodina* in Europa. *Bibliotheca Lichen* (1979); (with Hinteregger) Beiträge zur kenntnis der flechtenflora des Himalaya. VII die gattungen *Caloplaca*, *Fulgensia* und *Ioplaca* (mit englischem bestimmungsschlüssel). *Bibliotheca Lichenologica* (1993). *Biogs, obits etc.* De Priest (*Inoculum* **46**(3): 3, 1995); Grummann (1974: 36); Hertel (*International Lichenological Newsletter* **26**: 25, 1993); Hertel & Oberwinkler (*Bericht der Bayerischen Botanischen Gesellschaft* **66-67**: 327, 1996) [bibliography, obituary].

Poeltia Petr. (1972) [non *Poeltia* Grolle 1966, *Hepaticae*] ≡ Poeltiella.

Poeltiaria Hertel (1984), Porpidiaceae (L). 3, S. America; Australasia. See Hertel (*Nova Hedwigia* Beih. **79**: 430, 1984).

Poeltidea Hertel & Hafellner (1984), Porpidiaceae (L). 1, Kerguelen; Australasia. See Hertel & Hafellner (*Nova Hedwigia* Beih. **79**: 462, 1984).

Poeltiella Petr. (1974) = Hyalocrea fide Rossman (*Mycol. Pap.* **157**, 1987).

Poeltinula Hafellner (1984), Rhizocarpaceae (L). 2, Europe. See Hawksworth (*SA* **10**: 36, 1991; nomencl.).

Poetschia Körb. (1861), Patellariaceae. 4, Europe; S. America. See Hafellner (*Beih. Nova Hedwigia* **62**, 1979), Kutorga & Hawksworth (*SA* **15**: 1, 1997).

Pogonomyces Murrill (1904) = Hexagonia Fr. fide Pegler (*The polypores [Bull. BMS Suppl.]*, 1973).

Pogonospora Petr. (1957) ? = Endoxylina fide Müller & von Arx (*Beitr. Kryptfl. Schweiz* **11** no. 2, 1962).

Poikiloderma Füisting (1868) = Amphisphaeria fide Eriksson & Hawskworth (*SA* **7**: 80, 1988).

Poikilosperma Bat. & J.L. Bezerra (1961), anamorphic *Pezizomycotina*, Cpt.0eH.?. 1, Brazil. See Batista & Bezerra (*Publções Inst. Micol. Recife* **340**: 22, 1961).

Poikilosporium Dietel (1897) = Thecaphora fide Dietel (*Nat. Pflanzenfam.* **6**, 1928).

Poisonous fungi. Diverse fungi produce toxins (q.v.) which affect humans, animals and plants. For larger fungi poisonous for humans, see **Mycetisms**; microfungi which produce toxins affecting humans and higher animals, see Mycotoxicoses, plants, Phytotoxic mycotoxins. See also Bresinsky & Besl (*A col-*

our atlas of poisonous fungi: a handbook for pharmacists, doctors and biologists, 1989). See also Antibiotics, Hallucinogenic fungi.

Poisonous lichens, see Edible fungi.

Poitrasia P.M. Kirk (1984), Choanephoraceae. 1, widespread (tropical). See Kirk (*TBMS* **68**: 429, 1977; zygospore ultrastr.), Kirk (*Mycol. Pap.* **152**: 61 pp., 1984; monogr.), Voigt & Wöstemeyer (*Gene* **270**: 113, 2001; phylogeny).

polar (of bacteria, spores, etc.), at the ends or poles.

Polar and alpine mycology. Mycotas characteristic for cooler regions have been recognized at least since the description by Greville in 1821 of the mountain top species *Amanita nivalis*. Fungi from all major taxonomic groups are represented in these specialist mycotas: distribution of *Lachnellula pini* seems to be governed by the mean winter temperature (Kurkela & Norokorpi, *Eur. J. Forest Pathol.* **9**: 65, 1979); *Phacidium infestans* and various other ascomycetes exploit snow-cover to develop on their hosts. Cold adaptations in arctic and antarctic fungi are discussed by Robinson (*New Phytologist* **151**: 341, 2001). Specialist polar marine fungi exist (Pang *et al.*, *Mycol.* **100**: 291, 2008). Fungal spores have been recorded during flights over the north pole (Polunin & Kelly, *Nature* **170**: 314, 1952). Cryptoendolithic fungi have been extensively investigated in antarctic dry deserts by Friedmann (q.v.) and others, partly as a paradigm for extra-terrestrial life. Hypogeous fungi play a significant role in plant colonization of alpine areas formerly covered by retreating glaciers (Cazares & Trappe, *Mycol.* **86**: 507, 1994). Some myxomycetes (known as nivicolous myxomycetes) are specialist colonizers of areas uncovered by snow-melt (see Singer *et al.*, *Cryptogamie Mycologie* **22**: 79, 2001). Many of these organisms are likely to be highly susceptible to climate change. See also Laursen *et al.* (Eds) (*Arctic and alpine mycology* **1**, 1985; **2**, 1987), Gulden *et al.* (*Arctic and alpine fungi* **1**, 1985; **2**, 1988; **3**, 1990; **4**, 1992), Longton (*Biology of polar bryophytes and lichens*, 1988), Trappe (*Mycol.* **80**: 1, 1988; alpine fungi). See also Climate change, Lichenometry, Space exploration and fungi.

polar-diblastic, see polarilocular.

polaribilocular, polarilocular (q.v.) with two cells.

polarilocular (of ascospores), bicellular and the two cells separated by a central perforated septum; orculiform, polaribilocular, polar-diblastic. See Sheard (*Lichenologist* **3**: 328, 1967). (Fig. 22B).

Polaroscyphus Huhtinen (1987), Hyaloscyphaceae. 1, Svalbard. See Huhtinen *in* Laursen *et al.* (Eds) (*Arctic and Alpine Mycology* **2**: 123, 1987), Raitviir (*Scripta Mycologica* Tartu **20**, 2004).

poleophilous, town loving; sometimes used of lichens which thrive in urban areas (e.g. *Lecanora conizaeoides* in Eur.).

Polhysterium Speg. (1912) ? = Hysterographium fide Clements & Shear (*Gen. Fung.*, 1931).

Polioma Arthur (1907), Pucciniaceae. 4 (on *Geraniaceae*, *Lamiaceae*), N. & S. America. See Baxter & Cummins (*Bull. Torrey bot. Club* **78**: 51, 1951).

Poliomella Syd. (1922) = Puccinia fide Laundon (*Mycol. Pap.* **102**, 1965).

Poliomopsis A.W. Ramaley (1987), ? Uropyxidaceae. 1 (on *Thermopsis* (*Leguminosae*)), USA. See Ramaley (*Mycotaxon* **28**: 361, 1987), Cummins & Hiratsuka (*Illustr. Gen. Rust Fungi* edn 3: 225 pp., 2003; excluded from *Pucciniales*).

Poliotelium Syd. (1922) = Uromyces fide Berndt (*in litt.*) See, Mains (*Bull. Torrey bot. Club* **66**: 173, 1939).

Polistophthora Lebert (1858) = Cordyceps fide Tulasne & Tulasne (*Select. fung. carpol.* **3**: 4, 1865).

Polistroma Clemente (1807), Gyalectaceae (L). 1, America (tropical). See Frisch (*Biblthca Lichenol.* **92**: 1, 2006).

Pollaccia E. Bald. & Cif. (1937) = Fusicladium fide Funk (*Can. J. Pl. Path.* **11**: 353, 1989; *P. borealis*), Wu & Sutton (*MR* **99**: 983, 1995; relationships of *P. mandshurica*, key), Schubert *et al.* (*Schlechtendalia* **9**, 2003), Beck *et al.* (*Mycol. Progr.* **4**: 111, 2005; phylogeny).

Poloniodiscus Svrček & Kubička (1967) = Ameghiniella fide Zhuang (*Mycotaxon* **31**: 261, 1988), Gamundí (*MR* **95**: 1131, 1991).

poly- (prefix), a great number; many.

Polyactidaceae Corda (1838) = Sclerotiniaceae.

Polyactis Link (1809) = Botrytis fide Hennebert (*Persoonia* **7**: 183, 1973).

Polyadosporites Hammen (1954), Fossil Fungi. 6 (Cretaceous, Paleocene), Colombia.

Polyancora Voglmayr & Yule (2006), Xylariales. 1 (aquatic), Malaysia. See Voglmayr & Yule (*MR* **110**: 1247, 2006).

Polyandromyces Thaxt. (1920), Laboulbeniaceae. 1, widespread. See Nannfeldt (*Svensk bot. Tidskr.* **43**: 468, 1949).

polyandrous (of oospores), formed when more than one functioning antheridium is present; cf. monandrous.

Polyangium Link (1809) nom. dub., Predibacteria.

Polyascomyces Thaxt. (1900), Laboulbeniaceae. 1, British Isles. See Tavares (*Mycol.* **65**: 929, 1973), Tavares (*Mycol. Mem.* **9**: 627 pp., 1985).

polyascous, having many asci; esp. having the asci in one hymenium, not separated by sterile bands.

Polyblastia A. Massal. (1852) nom. cons., Verrucariaceae (L). *c.* 85, widespread. See Servít (*Československé lišejníky čeledi Verrucariaceae*, 1954), Swinscow (*Lichenologist* **5**: 92, 1971; Brit. spp.), McCarthy (*Muelleria* **8**: 269, 1995; Australia), Ceynowa-Gieldon (*Acta Mycologica* Warszawa **33**: 299, 1998; Poland), Roux *et al.* (*Mycotaxon* **84**: 1, 2002), Geiser *et al.* (*Mycol.* **98**: 1053, 2006; phylogeny), Gueidan *et al.* (*MR* **111**: 1145, 2007; phylogeny).

polyblastic (of conidiogenous cell), producing blastic conidia at several conidiogenous loci, either synchronously or irregularly.

Polyblastidea (Zschacke) Tomas. & Cif. (1952) = Polyblastia fide Hawksworth *et al.* (*Dictionary of the Fungi* edn 8, 1995).

Polyblastiomyces Cif. & Tomas. (1953) = Polyblastia fide Hawksworth *et al.* (*Dictionary of the Fungi* edn 8, 1995).

Polyblastiomyces E.A. Thomas (1939) nom. inval. = Staurothele fide Hawksworth *et al.* (*Dictionary of the Fungi* edn 8, 1995).

Polyblastiopsis Zahlbr. (1903) = Julella fide Mayrhofer (*Biblthca Lichenol.* **26**, 1987), Aptroot & Boom (*Mycotaxon* **56**: 1, 1995).

Polybulbophiale Goh & K.D. Hyde (1998), anamorphic *Pezizomycotina*, Hso.?.?. 1 (on palms), Brunei. See Goh & Hyde (*Mycotaxon* **69**: 145, 1998).

Polycarpella Theiss. & Syd. (1918) = Muellerella fide Hawksworth *et al.* (*Dictionary of the Fungi* edn 8, 1995).

polycarpic (of *Exobasidium* infections), systemic (or circumscribed) and perennial (Nannfeldt, 1981: 15); cf. monocarpic.

Polycarpum Stempel [not traced] nom. dub., Fungi. Protozoa or fungi.

Polycauliona Hue (1908) = Caloplaca fide Hawksworth *et al.* (*Dictionary of the Fungi* edn 8, 1995), Kärnefelt (*Cryptog. Bryol.-Lichénol.* **19**: 93, 1998), Arup & Mayrhofer (*Lichenologist* **32**: 359, 2000).

Polycellaesporonites A. Chandra, R.K. Saxena & Setty (1984), Fossil Fungi. 1, Arabian Sea.

Polycellaria H.D. Pflug (1965), Fossil Fungi. 1 (Tertiary), USA.

polycentric, having a number of centres of growth and development and more than one reproductive organ, as in the *Cladochytriaceae* (Karling, *Mycol.* **36**: 528, 1934); see monocentric; cf. reproductocentric.

Polycephalomyces Kobayasi (1941), anamorphic *Berkelella*, Hsy.0eH.15. 6, widespread. See Seifert (*Stud. Mycol.* **27**: 168, 1985; key), Bischoff *et al.* (*Mycotaxon* **86**: 433, 2003; phylogeny), Stensrud *et al.* (*MR* **109**: 41, 2005; phylogeny).

polycephalous, many-headed.

Polycephalum Kalchbr. & Cooke (1880) nom. dub., Predibacteria. See Seifert (*TBMS* **85**: 123, 1985).

Polychaetella Speg. (1918), anamorphic *Capnodium*, Hso.?.?. 3, Europe; N. America. See Sutton (*Mycol. Pap.* **141**, 1977).

Polychaeton (Pers.) Lév. (1846) nom. dub., ? Capnodiaceae. 8. See Hughes (*Mycol.* **68**: 693, 1976), Sutton (*Mycol. Pap.* **141**, 1977; nomencl.), Sharma & Singh (*Acta Bot. Indica* **24**: 117, 1997; citrus sooty mould), Manoharachary *et al.* (*Journal of Mycology and Plant Pathology* **33**: 212, 2003; India).

Polychidium (Ach.) Gray (1821), Massalongiaceae (L). 6, widespread. See Henssen (*Symb. bot. upsal.* **18** no. 1, 1963), Burgaz & Martínez (*Nova Hedwigia* **73**: 381, 2001; Iberian peninsula), Miądlikowska *et al.* (*Mycol.* **98**: 1088, 2006; phylogeny), Xiao *et al.* (*Lichenology* **5**: 53, 2006; China), Wedin *et al.* (*Lichenologist* **39**: 61, 2007; n. fam.).

polychotomous, having an apex dividing simultaneously into more than two branches (Corner, 1950).

Polychytrium Ajello (1942), Cladochytriaceae. 1, USA; Brazil.

Polycladium Ingold (1959), anamorphic *Pezizomycotina*, Hso.1bH.1. 1 (aquatic), British Isles. See Ingold (*TBMS* **42**: 114, 1959).

Polyclypeolina Bat. & I.H. Lima (1959), Aulographaceae. 1, Uganda. See Batista & Lima (*Publções Inst. Micol. Recife* **56**: 457, 1959).

Polyclypeolum Theiss. (1914) = Schizothyrium fide Müller & von Arx (*Beitr. Kryptfl. Schweiz* **11** no. 2, 1962).

Polycoccum Saut. ex Körb. (1865), Dacampiaceae. Anamorph *Cyclothyrium*. *c.* 49 (on lichens), widespread. See Hawksworth & Diederich (*TBMS* **90**: 293, 1988; key 23 spp.), van der Aa (*Stud. Mycol.* **31**: 15, 1989; anamorph), Kocourková & Berger (*Czech Mycol.* **51**: 171, 1999), Atienza *et al.* (*Lichenologist* **35**: 125, 2003; Spain, key).

Polycornum Malcolm & Vězda (1995), Porinaceae (L). 1, New Zealand. See McCarthy & Malcolm (*Lichenologist* **29**: 1, 1997), McCarthy (*Biblthca Lichenol.* **87**, 2003; nomenclator).

Polycyclina Theiss. & Syd. (1915), Parmulariaceae. 1, S. America.

Polycyclinopsis Bat., A.F. Vital & I.H. Lima (1958), Microthyriaceae. 1, Brazil. See Batista *et al.* (*Revta Biol. Lisb.* **1**: 284, 1958).

Polycyclus Höhn. (1909), Parmulariaceae. 2, S. America. See Petrak (*Sydowia* **4**: 533, 1950).

Polycystis Lév. (1846) nom. rej. = Urocystis fide Vánky (*in litt.*).

Polycytella C.K. Campb. (1987) = Scedosporium fide Campbell (*J. Med. Vet. Mycol.* **25**: 302, 1987), Borman *et al.* (*Medical Mycology* **44**: 33, 2006).

polydactyloid venation, see veins.

Polydesmia Boud. (1885), Hyaloscyphaceae. Anamorph *Brefeldochium*. 7, widespread. See Korf (*Mycotaxon* **7**: 457, 1978), Raitviir & Galán (*Mycotaxon* **53**: 447, 1995; key), Huhtinen & Santesson (*Lichenologist* **29**: 205, 1997), Zhuang (*Mycotaxon* **72**: 325, 1999), Zhuang (*Mycol.* **92**: 593, 1999), Verkley (*Nova Hedwigia* **80**: 503, 2005; anamorph).

Polydesmus Mont. (1845), anamorphic *Pezizomycotina*. 7. See Hernández-Gutiérrez & Sutton (*MR* **101**: 201, 1997), Shoemaker & Hambleton (*CJB*: in press, 2001), Shoemaker & Hambleton (*CJB* **79**: 592, 2001).

Polydiscidium Wakef. (1934), Helotiaceae. 1, Guyana; Africa. See Dennis (*Bull. Jard. bot. Brux.* **31**: 154, 1961).

Polydiscina Syd. (1930), ? Helotiales. 1, Bolivia.

Polydispyrenia E.U. Canning & E.I. Hazard (1982), Microsporidia. 1.

polyenergid, coenocytic.

Polyetron Bat. & Peres (1963), anamorphic *Pezizomycotina*, Cpt.0-1eH-P.?. 1, Brazil. See Batista & Peres (*Publções Inst. Micol. Recife* **394**: 6, 1963).

Polygaster Fr. (1823) nom. dub., ? Agaricales. ? 'gasteromycetes'.

Polyhyphaethyrites R. Srivast. & R.K. Kar (2004), Fossil Fungi, Ascomycota. 1, India. See Srivastava & Kar (*Curr. Sci.* **87**: 867, 2004).

Polylagenochromatia Sousa da Câmara (1929) ? = Polystigmina fide Hawksworth *et al.* (*Dictionary of the Fungi* edn 8, 1995).

Polylobatispora Matsush. (1996), anamorphic *Pezizomycotina*, Hso.?.?. 2, Malaysia. See Matsushima (*Matsush. Mycol. Mem.* **9**: 21, 1996).

Polymarasmius Murrill (1915) = Marasmius fide Singer (*Agaric. mod. Tax.*, 1951).

Polymeridium (Müll. Arg.) R.C. Harris (1980), Trypetheliaceae (L). 19, widespread (esp. tropical). See Harris (*Bolm Mus. paraense 'Emílio Goeldi'* sér. bot. **7**: 619, 1993; key), Aptroot *et al.* (*Biblthca Lichenol.* **97**, 2008; Costa Rica).

polymorphic, having different forms; pleomorphic.

polymorphic species, species with a series of intergrading morphological features; e.g. resulting from inbreeding or automictic sexual reproduction.

Polymorphomyces Coupin (1914) = Geotrichum fide Kendrick & Carmichae *in* Ainsworth *et al.* (Eds) (*The Fungi* **4A**: 390, 1973).

Polymorphum Chevall. (1822), anamorphic *Ascodichaena*, St.0eH/1eP.1. 1, widespread (temperate). See Hawksworth & Punithalingam (*TBMS* **60**: 501, 1973), Hawksworth (*Taxon* **32**: 212, 1983; nomencl.).

Polymyces Battarra ex Earle (1909) ≡ Armillaria.

Polynema Lév. (1846), anamorphic *Clavicipitaceae*, Ccu.1eH/1eP.15. 8, widespread. See Nag Raj (*Coelomycetous Anamorphs with Appendage-bearing Conidia*, 1993), Vujanovic *et al.* (*Mycol.* **91**: 136, 1999; n.sp. from Canada).

Polyopeus A.S. Horne (1920) = Phoma Sacc. fide Boerema & Dorenbosch (*Persoonia* **6**: 49, 1970), Gruyter *et al.* (*Persoonia* **18**: 1, 2002).
Polyorus, see *Polyozus*.
polyoxin antibiotics, inhibit chitin synthesis, e.g. polyoxin D inhibits synthetase formation (Endo *et al.*, *J. Bact.* **104**: 189, 1970).
Polyozellus Murrill (1910), Thelephoraceae. 1, USA; Japan. See Imazeki (*Mycol.* **45**: 555, 1953), Bigelow (*Mycol.* **70**: 707, 1978), Yang (*Zhongguo Shiyongjun* **11**: 1 [unnumbered page], 1992; edibility).
Polyozosia A. Massal. (1855) = Lecania fide Hafellner (*Beih. Nova Hedwigia* **79**: 214, 1984).
Polyozus P. Karst. (1881) = Tremellodendropsis fide Donk (*Persoonia* **4**: 184, 1966).
Polypaecilum G. Sm. (1961), anamorphic *Thermoascus*, Hso.0eH.19. 3, widespread. See Piontelli *et al.* (*Bol. Micol.* **4**: 155, 1989; hyalohyphomycosis caused by *P. insolitum*).
Polypedia Bat. & Peres (1959), Micropeltidaceae. 1, Brazil. See Batista & Peres (*Atti Ist. bot. Univ. Lab. crittog. Pavia* sér. 5 **16**: 118, 1959).
Polypera Pers. (1818) ≡ Pisolithus.
polyphagous, see polyphagy.
Polyphagus Nowak. (1877), Chytridiaceae. *c.* 9, widespread. See Bartch (*Mycol.* **37**: 553, 1945).
polyphagy (adj. **polyphagous**), see monophagy.
polyphialidic (of a conidiogenous cell), having more than one conidiogenous locus at which conidia are produced. Cf. monophialidic.
Polyphlyctis Karling (1968), Chytridiaceae. 1 or 2, widespread. See Karling (*Sydowia* **20**: 86, 1966).
polyphyllous (of foliose lichen thalli), having many connected leaf-like lobes.
Polypilus P. Karst. (1881) ≡ Grifola.
polyplanetism, sequence of two or more motile flagellate phases with interspersed mobile aplanosporic phases in the zoosporic part of the life history; the aplanosporic phase may be naked or as a walled cyst; motile phases may be monomorphic or dimorphic.
Polyplocium Berk. (1843) = Gyrophragmium fide Zeller (*Mycol.* **35**: 409, 1943).
polyploidy (in fungi), see Chromosome numbers.
Polyporaceae Fr. ex Corda (1839), Polyporales. 92 gen. (+ 79 syn.), 636 spp.

Lit.: Pilát (*Atlas Champ. Eur.* III *Polyporaceae* **1**, 1936; Europe), Pilát (*Atlas Champ. Eur.* III *Polyporaceae* **2**, 1942; Europe), Donk (*Reinwardtia* **1**, 1951; gen. lists), Bondartsev (*[Polyporaceae of European Russia and the Caucasus]*, 1953; former USSR), Corner (*Ann. Bot. Mem.* [A monograph of Clavaria and allied genera, addenda] **17**: 152, 1953; construction of polypores), Overholts (*Polyporaceae of the United States, Alaska, & Canada*, 1953; N. America), Nobles (*CJB* **40**: 987, 1958; cultural characters), Cooke (*Lloydia* **22**: 163, 1959), Lowe & Gilbertson (*J. Elisha Mitchell scient. Soc.* **77**: 43, 1961; keys gen., 293 spp. S.E. USA), Teixeira (*Biol. Rev.* **37**: 51, 1962; taxon. rev.), Jahn (*Westfäl. Pilzbr.* **4**, 1963; C. Eur.), Lowe (*Mycol.* **55**: 1, 1963; rev.), Cunningham (*Polyporaceae of New Zealand* [*N.Z. Dep. sci. industr. Res. Bull.* **164**], 1965; Trinidad & Tobago), Fidalgo & Fidalgo (*Mycol.* **58**: 862, 1967; former USSR), Boquiren (*Mycol.* **63**: 937, 1971; as *Epitheliaceae*), Domanski (*Fungi, Polyporaceae* 2 vols, 1972-73; [Engl. transl.]), Ryvarden (*Norw. Jl Bot.* **19**: 229, 1972; E. Africa checklist), Pegler (*The polypores [Bull. BMS Suppl.]*, 1973; key 123 gen., Brit. spp.), Donk (*Checklist of European polypores* [*Verh. K. ned. Akad. Wet.* **62**], 1974; Europe), Nuss (*Biblthca Mycol.* **45**, 1975; ecology, sporulation), Ryvarden (*The Polyporaceae of North Europe* **1**, 1976; Europe), Martin & Gilbertson (*Mycotaxon* **7**: 337, 1978; key 99 wood-rotting spp. N. America), Ryvarden (*The Polyporaceae of North Europe* **2**, 1978; New Zealand), Ryvarden (*TBMS* **73**: 9, 1979; as *Grammotheleaceae*), Nuss (*Hoppea* **39**: 127, 1980; structure, taxonomy), Ryvarden & Johansen (*A preliminary polypore flora of East Africa*, 1980; Europe), Corner (*Ad Polyporaceae I* [Beih. Nova Hedwigia] **75**, 1983), Corner (*Ad Polyporaceae II* [Beih. Nova Hedwigia] **75**, 1983; N. America), Corner (*Ad Polyporaceae III* [Beih. Nova Hedwigia] **78**: 92, 1984), Gilbertson & Ryvarden (*N. Amer. Polypores* 2 vol., 1986-7; E. Africa), De (*Int. J. Mycol. Lichenol.* **4**: 59, 1989; as *Epitheliaceae*), Pouzar (*Česká Mykol.* **44**: 92, 1990; as *Haplopilaceae*), Hibbett & Donoghue (*CJB* **73**: S853, 1995; as *Haplopilaceae*), Vampola & Pouzar (*Czech Mycol.* **48**: 315, 1996; as *Haplopilaceae*), Quanten (*Op. bot. Belg.* **11**: 352 pp., 1997; as *Grammotheleaceae*), Ginns (*Mycol.* **90**: 19, 1998; as *Epitheliaceae*), Chang & Chou (*MR* **103**: 674, 1999; as *Haplopilaceae*), Ko & Jung (*Antonie van Leeuwenhoek* **75**: 191, 1999), Parmasto & Hallenberg (*Karstenia* **40**: 129, 2000), Rajchenberg (*Karstenia* **40**: 143, 2000; as *Haplopilaceae*), Carranza & Ruiz-Boyer (*Harvard Pap. Bot.* **6**: 57, 2001), Hibbett & Donoghue (*Syst. Biol.* **50**: 215, 2001), Kim & Jung (*Mycobiology* **29**: 73, 2001), Ko *et al.* (*Mycol.* **93**: 270, 2001; as *Haplopilaceae*), Loguercio-Leite & Gonçalves (*Mycotaxon* **79**: 285, 2001; as *Haplopilaceae*), Binder & Hibbett (*Mol. Phylogen. Evol.* **22**: 76, 2002; as *Haplopilaceae*), Dai *et al.* (*Ann. bot. fenn.* **39**: 169, 2002), Hong *et al.* (*Mycol.* **94**: 823, 2002), Ko & Jung (*Mol. Phylogen. Evol.* **23**: 112, 2002), Ko & Jung (*Mycotaxon* **82**: 315, 2002), Moncalvo *et al.* (*Mol. Phylogen. Evol.* **23**: 357, 2002), Decock & Ryvarden (*MR* **107**: 94, 2003), Greslebin & Rajchenberg (*N.Z. Jl Bot.* **41**: 437, 2003; as *Epitheliaceae*), Kotlaba & Pouzar (*Czech Mycol.* **55**: 7, 2003; as *Grammotheleaceae*), Silveira *et al.* (*MR* **107**: 597, 2003), Suhara *et al.* (*Mycotaxon* **86**: 335, 2003; as *Haplopilaceae*), Valderrama *et al.* (*Antonie van Leeuwenhoek* **84**: 289, 2003), Krüger & Gargas (*Feddes Repert.* **115**: 530, 2004), Piątek *et al.* (*Acta Mycologica* Warszawa **39**: 25, 2004; as *Haplopilaceae*), Tomšovský & Homolka (*Nova Hedwigia* **79**: 425, 2004), Binder *et al.* (*Systematics and Biodiversity* **3**: 113, 2005).
Polyporales Gäum. (1926), Agaricomycetes. 13 fam., 216 gen., 1801 spp. Fams:
(1) **Cystostereaceae**
(2) **Fomitopsidaceae**
(3) **Ganodermataceae**
(4) **Grammotheleaceae.**
(5) **Limnoperdaceae**
(6) **Meripilaceae**
(7) **Meruliaceae**
(8) **Phanerochaetaceae**
(9) **Polyporaceae**
(10) **Sparassidaceae**
(11) **Steccherinaceae**
(12) **Tubulicrinaceae**
(13) **Xenasmataceae**

Lit.: Ryvarden (*Genera of polypores* [*Synopsis Fungorum* **5**], 1991), Teixeira (*Genera of Polypo-

raceae, an objective approach [*Bol. Chácara Bot. Itu* **1**], 1994), and under fams.

Polyporasclerotes Stach & Pickh. (1957), Fossil Fungi. 1 (Carboniferous), Germany.

polypore, one of the *Polyporaceae*.

Polyporellus P. Karst. (1879) = Polyporus P. Micheli ex Adans. fide Donk (*Verh. K. ned. Akad. Wet.* tweede sect. **62**: 1, 1974).

Polyporisporites Hammen (1954), Fossil Fungi. 1 (Cretaceous – Paleogene), Colombia.

Polyporites Daugherty (1941), Fossil Fungi. 1 (Triassic), USA.

Polyporites Lindl. & Hutton (1833), Fossil Fungi. 12 (Carboniferous to Quaternary), Europe; USA.

Polyporoletus Snell (1936), Albatrellaceae. 1, N. & S. America. See Gilbertson & Ryvarden (*N. Amer. Polyp.* **2**: 636, 1987).

Polyporus (Pers.) Gray (1821) = Laetiporus fide Donk (*Persoonia* **1**: 263, 1960).

Polyporus P. Micheli ex Adans. (1763), Polyporaceae. 26 (mostly on wood), widespread. The sclerotioid anamorph (*Mylitta australis*) of *P. mylittae* is 'blackfellows' bread' of Australia. See Stahl (*Biblthca Mycol.* **50**, 1970; genetics sporocarp formation *P. ciliatus*), Pouzar (*Česká Mykol.* **26**: 82, 1972), Ryvarden (*Polyp. N. Eur.* **2**: 378, 1978; key 8 Eur. spp.), Nuss (*Hoppea* **39**: 127, 1980), Corner (*Beih. Nova Hedwigia* **78**: 12, 1984; key subgen. classif.), Núñez & Ryvarden (*Syn. Fung.* **10**: 33, 1995; world monogr.), Ko & Jung (*Mycotaxon* **82**: 315, 2002; phylogenetic evaluation of *Polyporus s. str.*), Silveira & Wright (*MR* **106**: 1323, 2002; southern S. Am. spp, mating tests), Silveira *et al.* (*MR* **107**: 597, 2003; southern S. Am. spp, isoenzyme analysis), Krüger & Gargas (*Feddes Repert.* **115**: 530, 2004; phylogeny).

Polyporus Pers. (1821) ≡ Laetiporus.

Polypyrenula D. Hawksw. (1985), Requienellaceae. 1, West Indies. See Aptroot (*Biblthca Lichenol.* **44**, 1991; posn in *Pyrenulaceae*).

Polyrhina Sorokīn (1876) ≡ Harposporium fide Karling (*Mycol.* **30**: 512, 1938).

Polyrhizium Giard (1889) ? = Cladosporium fide Petch (*TBMS* **19**: 190, 1935).

Polyrhizon Theiss. & Syd. (1914), ? Venturiaceae. 2, India; S. Africa.

Polysaccopsis Henn. (1898) = Urocystis fide Vánky (*Illustrated genera of smut fungi*, 1987).

Polysaccum F. Desp. & DC. (1807) = Pisolithus fide Stalpers (*in litt.*).

Polyschema H.P. Upadhyay (1966), anamorphic *Pezizomycotina*, Hso.≡ eP.24/27. 17, widespread. See Ellis (*More Dematiaceous Hyphomycetes*, 1976), Castañeda Ruíz *et al.* (*Mycotaxon* **57**: 451, 1996), Castañeda Ruíz *et al.* (*Cryptog. Mycol.* **21**: 215, 2000; Cuba).

Polyschistes J. Steiner (1898) = Diploschistes fide Hawksworth *et al.* (*Dictionary of the Fungi* edn 8, 1995).

Polyscytalina Arnaud (1954), anamorphic *Pezizomycotina*, Hso.1eP.3. 1, France. *P. pustulans* (syn. *Oospora pustulans*), skin spot of potato tubers. See Ciferri & Caretta (*Mycopathologia* **16**: 304, 1962), Holubová-Jechová (*Fol. Geobot. Phyt. Praha* **13**: 433, 1976; synonym of *Hormiactella*).

Polyscytalum Riess (1853), anamorphic *Pezizomycotina*, Hso.0-1eH.3. 12, widespread. See Sutton & Hodges (*Nova Hedwigia* **28**: 487, 1976), Crous *et al.* (*MR* **110**: 264, 2006; key), Crous *et al.* (*Stud. Mycol.* **58**: 185, 2007; phylogeny).

polysidia, specialized dual vegetative propagules in certain *Pyxine* spp. formed in depressions at the tips of coral-like structures (**polysidiangia**) which recall isidia but are not themselves propagules (Kalb, *Bibl. Lich.* **24**, 1987).

Polyspora Laff. (1921) [non *Polyspora* Sweet ex Don 1831, *Theaceae*] = Aureobasidium fide Hermanides-Nijhof (*Stud. Mycol.* **15**: 141, 1977).

Polysporella Woron. (1916), Fungi. See Voronikhin (*Izvestiya Kavkazskago Muzeya* **10**: 7, 1916).

Polysporidiella Petr. (1960), Dothideomycetes. 1, Iran. See Petrak (*Sydowia* **14**: 355, 1960).

Polysporidium Syd. & P. Syd. (1908) = Guignardia fide Eriksson (*SA* **5**: 197, 1986).

Polysporina Vězda (1978) nom. cons., Acarosporaceae (L). 9, widespread. See Kantvilas (*Lichenologist* **30**: 551, 1998), Lutzoni *et al.* (*Am. J. Bot.* **91**: 1446, 2004; phylogeny), Reeb *et al.* (*Mol. Phylogen. Evol.* **32**: 1036, 2004; phylogeny), Knudsen (*Opuscula Philolichenum* **2**: 17, 2005; USA), Crewe *et al.* (*MR* **110**: 521, 2006; phylogeny), Kantvilas & Seppelt (*Lichenologist* **38**: 109, 2006; Antarctica), Miądlikowska *et al.* (*Mycol.* **98**: 1088, 2006; phylogeny).

Polysporinopsis Vězda (2002) = Acarospora fide Vězda (*Lichenes Rariores Exsiccati* **48**: 2, 2002), Crewe *et al.* (*MR* **110**: 521, 2006).

polysporous, many-spored.

Polystema Raf. (1815) nom. dub., Fungi. No spp. included.

Polystictaceae Rea (1922) = Hymenochaetaceae.

Polystictoides Lázaro Ibiza (1916) ≡ Inonotus.

Polystictus Fr. (1851) ≡ Coltricia.

Polystigma DC. (1815), Phyllachoraceae. Anamorph *Rhodosticta*. 24 (living leaves of *Rosaceae*), widespread (temperate). See Hyde & Cannon (*Aust. Syst. Bot.* **5**: 415, 1992), Cannon (*MR* **100**: 1409, 1996; monogr.), Cannon & Evans (*MR* **103**: 577, 1999).

Polystigmataceae Höhn. ex Nannf. (1932) = Phyllachoraceae.

Polystigmatales = Phyllachorales.

Polystigmella Jacz. & Natalina (1931) = Polystigma fide Cannon (*in litt.*), Cannon (*MR* **100**: 1409, 1996; monogr.).

Polystigmina Sacc. (1884), anamorphic *Polystigma*, St.0fH.10. 2 (on *Rosaceae*), Europe. See Cannon (*MR* **100**: 1409, 1996; monogr. teleom.).

Polystigmites A. Massal. (1857), Fossil Fungi. 1 (Miocene), Italy.

Polystoma Gray (1821) ≡ Myriostoma.

Polystomella Speg. (1888) = Dothidella fide Müller & von Arx (*Beitr. Kryptfl. Schweiz* **11** no. 2, 1962), von Arx & Müller (*Stud. Mycol.* **9**, 1975).

Polystomellaceae Theiss. & P. Syd. (1915), ? Dothideomycetes (inc. sed.). 3 gen. (+ 5 syn.), 9 spp.
Lit.: Swart (*TBMS* **89**: 483, 1987).

Polystomellina Bat. & A.F. Vital (1958), ? Microthyriaceae. 1, Brazil. See Batista & Vital (*Revta Biol.* Lisb. **1**: 280, 1958).

Polystomellomyces Bat. (1959), anamorphic *Pezizomycotina*, Cpd.0eH.?. 1, Greece. See Batista (*An. Soc. Biol. Pernambuco* **16**: 148, 1959).

Polystomellopsis F. Stevens (1923), Dothideomycetes. 1, S. America. See Ciferri & Batista (*Ann. Soc. Biol. Pernambuco* **14**: 79, 1956).

Polystratorictus Matsush. (1993), anamorphic *Pezizomycotina*, Hso.≡ eP.15. 2, Peru. See Matsushima (*Matsush. Mycol. Mem.* **7**: 62, 1993).

Polystroma Fée (1824) ≡ Polistroma.

Polysynnema Constant. & Seifert (1988), anamorphic *Pezizomycotina*, Hsy.0eH.10. 1, Hawaii; Grenada. See Constantinescu & Seifert (*TBMS* **90**: 332, 1988).

Polythecium Bonord. (1861) = Fusicoccum fide Saccardo (*Syll. fung.* **3**: 1, 1884).

Polythelis Arthur (1906) = Tranzschelia fide Arthur (*Manual Rusts US & Canada*, 1934).

Polythelis Clem. (1909) ≡ Polypyrenula.

Polythrinciella Bat. & H. Maia (1960), anamorphic *Pezizomycotina*, Hsp.1eP.?. 1, Brazil. See Batista & Maia (*Publções Inst. Micol. Recife* **283**: 20, 1960).

Polythrinciopsis J. Walker (1966), anamorphic *Pezizomycotina*, Hso.1eH.10. 1, Australia.

Polythrincium Kunze (1817), anamorphic *Mycosphaerella*, Hso.1eP.10. 1 (on *Trifolium*), widespread (temperate).

Polythyrium Syd. (1929) = Neostomella fide Müller & von Arx (*Beitr. Kryptfl. Schweiz* **11** no. 2, 1962).

Polytolypa J.A. Scott & Malloch (1993), ? Ajellomycetaceae. 1 (coprophilous), Canada. See Scott *et al.* (*Mycol.* **85**: 503, 1993), Untereiner *et al.* (*Stud. Mycol.* **47**: 25, 2002; phylogeny), Untereiner *et al.* (*Mycol.* **96**: 812, 2004).

polytomous, dividing into many branches, usually at one node or point.

polytretic, see tretic.

Polytretophora Mercado (1983), anamorphic *Pezizomycotina*, Hso.1eP.27. 2, Cuba; S.E. Asia. See Kuthubutheen & Nawawi (*MR* **95**: 623, 1991), Whitton *et al.* (*Mycoscience* **42**: 555, 2001).

Polytrichia Sacc. (1882) = Pyrenophora fide von Arx & Müller (*Stud. Mycol.* **9**, 1975), Sivanesan (*Mycol. Pap.* **158**: 261 pp., 1987).

Polytrichiella M.E. Barr (1972) = Capronia fide Müller *et al.* (*TBMS* **88**: 63, 1987).

polyxeny, see pleioxeny.

Pomatomyces Oerst. (1864) = Thekopsora fide Sydow & Sydow (*Monogr. Ured.*, 1915) Anamorph name for (I).

Pompholyx Corda (1834) = Scleroderma fide Guzmán (*Darwiniana* **16**: 270, 1970).

Pontecorvo (Guido; 1907-1999; Italy). Undergraduate (1924-1928) then postgraduate student, Pisa University; Agricultural Advisory Service, Tuscany (-1938); left Italy in 1938 anticipating anti-Semitism; PhD, Edinburgh University (*c.* 1938-1940); Academic staff (1941-1952), then Reader (1952-1955), then Professor of Genetics (1955-1968), Glasgow University; Honorary Director, UK Medical Research Council Unit of Cell Genetics (1966-1968); Royal Society (1955). Noted for his work on genetics, at first with the insect *Drosophila* then, as a pioneer in mycology, with *Emericella nidulans* (anamorph *Aspergillus nidulans*), and finally with human genetics. His review of *Aspergillus* genetics became a handbook for the science, and he proposed the term 'parasexual cycle'. *Publs.* (with Roper, Hemmons, Macdonald & Bufton) The genetics of *Aspergillus nidulans*. *Advances in Genetics* **5**: 141, 1953). *Biogs, obits etc.* Roper (*MR* **104** (10): 1276, 2000).

Pontogeneia Kohlm. (1975), Pezizomycotina. 7 (marine), widespread. See Kohlmeyer & Kohlmeyer (*Marine Mycology*, 1979), Kohlmeyer & Volkmann-Kohlmeyer (*Bot. Mar.* **34**: 1, 1991).

Pontomyxa Topsent (1892), Hydromyxales. q.v.

Pontoporeia Kohlm. (1963) = Zopfia fide Malloch & Cain (*CJB* **50**: 61, 1972).

Population biology. The study of patterns of distribution and variation in space and time within fungal species and the interpretation of these patterns in terms of genetic, developmental and environmental influences on phenotype and modes of proliferation. Some means of distinguishing between individual population components is necessary, and this can cause difficulties in view of the indeterminacy (indefinite growth potential) and capacity of anastomosis of fungal mycelia. However, it is usually possible to determine whether fungal samples have the same or different genetic origins (i.e. arise from the same or different 'genets') on the basis of somatic and/or sexual incompatibility, phenotypic differences (including isoenzyme polymorphism) or DNA polymorphisms. The latter are sometimes regarded as most definitive in that they do not depend on gene expression. However, they can be relatively expensive to detect, and identify differences and similarities at varied levels of resolution which may or may not be correlated with 'biological' entities or groups.

Often a variety of approaches is needed before deciding on which, or which combination, of the above approaches resolves population components at the level most appropriate to the biological question being addressed. Once population components have been located, their distribution can be quantified by mapping their extent and varied content within regional boundaries. The resulting information can help provide meaningful answers to fundamentally important questions concerning where the organisms (and their offspring) are and what they do there, how they arrived, whether they will persist and how they are likely to change in character and/or distribution. It is also vital for assessment of the levels at which population variation should be recognized for taxonomic purposes.

Distributional patterns in fungal populations can be interpreted in terms of resource relationships, modes of development and modes of reproduction. Resource relationships, the ways in which fungi interact with those non-living or living materials (substrata, hosts) which provide them with a source of organic nutrients, are determined both by selective influences and by the mode of fungal arrival (propagules or migratory mycelium) at colonization sites. Genets capable of arriving only as propagules have 'resource-unit-restricted' regional boundaries; those with migratory mycelium are 'non-unit-restricted' and can become very large. At the individual level, the heterogeneity of microenvironmental conditions encountered by a fungal genet may be reflected in the range of developmental modes or alternative phenotypes that it produces, allowing it to vary functional role with circumstances. At the population level, whether reproduction is clonal (non-recombinatorial) or diversifying (recombinatorial) may partly reflect habitat heterogeneity. The abilities of fungi to vary both their developmental pattern and degree of commitment to clonal and recombinatorial modes of reproduction has important implications for understanding gene flow and genetic and epigenetic diversity within and betwen populations. These abilities are also a crucial consideration when attempting to predict or interpret the evolutionary responses of fungi to environmental change.

In the fungi, work on mathematical modelling of populations has concentrated on attempts to predict outbreaks of plant diseases. See Bearchell *et al.* (*Proc. nat. acad. sci.* **102**: 5438, 2005), Gilligan & Kleczkowski (*Philosophical trans.: biol. sci.* **352**: 591, 1997).

Lit.: Brasier (*Adv. Pl. Pathol.* **5**: 53, 1986, *Nature* **332**: 438, 1988), Carroll & Wicklow (Eds) (*The fungal community*, edn 2, 1992), Jacobson *et al.* (*PNAS* **90**: 9159, 1993), McDonald *et al.* (*Ann. Rev. Phytopathol.* **27**: 77, 1989), Rayner (*Mycol.* **83**: 48, 1991, *Ann. Rev. Phytopathol.* **29**: 305, 1993), Rayner & Todd (*Adv. Bot. Res.* **7**: 333, 1979), Smith *et al.* (*Nature* **356**: 428, 1992).

pore (1) a small opening, as in tretic (q.v.) conidiogenesis; (2) in **- fungi** (*Polyporaceae* and *Boletaceae*), the mouth of a tube.

Poria Adans. (1763) ≡ Polyporus P. Micheli ex Adans. fide Decock & Stalpers (*Taxon* **55**: 759, 2006; syn. of *Albatrellus*).

Poria P. Browne (1756) ≡ Polyporus P. Micheli ex Adans.

Poria Pers. (1794) = Perenniporia fide Decock & Stalpers (*Taxon* **55**: 227, 759, 2006).

Poriaceae Locq. (1957) = Polyporaceae.

Poriales = Polyporales.

poricidal (of asci), see ascus.

poricin, an antitumour antibiotic from *Poria corticola* (Ruelius *et al.*, *Arch. Biochem. Biophys.* **125**: 126, 1968).

poriform, pore-like.

Porina Ach. (1809) nom. cons., Porinaceae. *c.* 336, widespread (esp. tropical). See Meylan (*Bulletin de la Société Vaudoise des Sciences Naturelles* **57** no. 228: 359, 1932; Ecuador), Singh (*Revue Bryol. Lichenol.* **37**: 973, 1971; India), Janex-Favre (*Cryptog. Bryol.-Lichénol.* **2**: 253, 1981; ontogeny), McCarthy (*Biblthca Lichenol.* **52**, 1993; S. Hemisph. saxicolous spp., S hemisphere), McCarthy (*Nova Hedwigia* **59**: 509, 1994; corticolous spp., Australia), McCarthy (*Nova Hedwigia* **58**: 391, 1994; corticolous spp., Australia), Makhija *et al.* (*J. Econ. Taxon. Bot.* **18**: 521, 1994; India), Malcolm *et al.* (*Mycotaxon* **55**: 353, 1995; Australasia), Lücking (*Bot. Acta* **109**: 248, 1996; *P. rufula* aggr.), McCarthy (*Taxon* **45**: 533, 1996; nomencl.), McCarthy (*Lichenologist* **29**: 229, 1997; Australia), McCarthy & Malcolm (*Lichenologist* **29**: 1, 1997; generic limits), Lücking & Vezda (*Willdenowia* **28**: 181, 1998; *P. epiphylla* aggr.), McCarthy (*Lichenologist* **31**: 239, 1999; Thailand), McCarthy (*Lichenologist* **32**: 1, 2000; key saxicolous spp.), McCarthy (*Biblthca Lichenol.* **87**, 2003; checklist), Grube *et al.* (*MR* **108**: 1111, 2004; phylogeny), Lücking (*Biblthca Lichenol.* **88**: 409, 2004; key foliicolous spp.), Baloch & Grube (*MR* **110**: 125, 2006; phylogeny), Sérusiaux *et al.* (*Lichenologist* **39**: 15, 2007; Macaronesia, key).

Porina Müll. Arg. (1883) = Pertusaria.

Porinaceae Rchb. (1828), Ostropales. 4 gen. (+ 6 syn.), 378 spp.

Lit.: McCarthy (*Nova Hedwigia* **59**: 509, 1994), Hafellner & Kalb (*Biblthca Lichenol.* **57**: 161, 1995), Lücking (*Bot. Acta* **109**: 248, 1996), McCarthy & Elix (*Lichenologist* **28**: 402, 1996), Aptroot *et al.* (*Biblthca Lichenol.* **64**: 220 pp., 1997), Lücking & Ferraro (*Lichenologist* **29**: 217, 1997), McCarthy & Malcolm (*Lichenologist* **29**: 1, 1997), Lücking (*Trop. Bryol.* **15**: 45, 1998), Lücking & Vezda (*Willdenowia* **28**: 181, 1998), Lücking & Cáceres (*Lichenologist* **31**: 349, 1999), McCarthy & Kantvilas (*Lichenologist* **32**: 247, 2000; Tasmania, biogeogr.), McCarthy (*Flora of Australia* **58** A: 242 pp., 2001), Grube *et al.* (*MR* **108**: 1111, 2004), Lücking (*Biblthca Lichenol.* **88**: 409, 2004), Lücking & Cáceres (*MR* **108**: 571, 2004), Baloch & Grube (*MR* **110**: 125, 2006).

Porinella R. Sant. (2004) = Caprettia fide Vězda (*Acta Musei Richnoviensis* Sect. Natur. **11**: 62, 2004), Galloway (*Flora of New Zealand. Lichens* edn 2 **1**, 2007).

Porinopsis Malme (1928) = Aspidothelium fide Hawksworth *et al.* (*Dictionary of the Fungi* edn 8, 1995).

Porinula (Nyl. ex Hue) Flagey (1896) = Porina Müll. Arg. fide Galloway (*Flora of New Zealand. Lichens* edn 2 **1**, 2007).

Porinula Vězda (1975) nom. illeg. = Caprettia fide Harris *in* Hafellner & Kalb (*Biblthca Lichenol.* **57**: 161, 1995), Sérusiaux & Lücking (*Biblthca Lichenol.* **86**: 161, 2003), Galloway (*Flora of New Zealand. Lichens* edn 2 **1**, 2007).

Poriodontia Parmasto (1982), Schizoporaceae. 1, former USSR. See Parmasto (*Mycotaxon* **14**: 103, 1982).

Poroauricula McGinty (1917) = Favolaschia fide Donk (*Persoonia* **1**: 173, 1960).

Porobeltraniella Gusmão (2004), anamorphic *Pezizomycotina*. 2. See Gusmão (*Mycol.* **96**: 151, 2004).

Porocladium Descals (1976), anamorphic *Pezizomycotina*, Hso.1bH.24. 1 (aquatic), British Isles. See Descals (*TBMS* **67**: 211, 1976).

poroconidium, see tretic.

Poroconiochaeta Udagawa & Furuya (1979) = Coniochaeta fide Udagawa & Furuya (*TMSJ* **20**: 5, 1979), Læssøe (*SA* **13**: 43, 1994; posn), García *et al.* (*Mycol.* **95**: 525, 2003; Russia), García *et al.* (*MR* **110**: 1271, 2006; phylogeny), Zhang *et al.* (*Mycol.* **98**: 1076, 2006; phylogeny).

Porocyphaceae Körb. (1855) = Lichinaceae.

Porocyphus Körb. (1855), Lichinaceae (L). 8, widespread. See Henssen (*Symb. bot. upsal.* **18** no. 1, 1963), Schultz & Büdel (*Lichenologist* **34**: 39, 2002; key).

Porodaedalea Murrill (1905), Hymenochaetaceae. 10, widespread. See Fiasson & Niemalä (*Karstenia* **24**: 14, 1984).

Porodiscella Viégas (1944) ? = Xylaria Hill ex Schrank fide Læssøe (*SA* **13**: 43, 1994).

Porodisculaceae Jülich (1982) = Polyporaceae.

Porodisculus Murrill (1907), Fistulinaceae. 1, USA. See Ginns (*CJB* **75**: 220, 1997; systematics, cultural characters *Porodisculus pendulus*), Bodensteiner *et al.* (*Mol. Phylogen. Evol.* **33**: 501, 2004).

Porodiscus Lloyd (1919), Pezizomycotina. 1, Brazil.

Porodiscus Murrill (1903) [non *Porodiscus* Grev. 1863, *Algae*] ≡ Porodisculus.

Porodothion Fr. (1825) ≡ Porothelium.

Porogramme (Pat.) Pat. (1900), Polyporaceae. 3, widespread (tropical). See Lowe (*Pap. Mich. Acad. Sci.* **49**: 33, 1964; key), Ryvarden (*TBMS* **73**: 9, 1979), Ryvarden & Johansen (*Prelim. Polyp. Fl. E. Afr.*: 34, 1980; key).

poroid, of a form resembling *Polyporus*; specifically, with a pore-like hymenium.

Poroidea Göttinger ex G. Winter (1884) = Craterocolla fide Donk (*Persoonia* **4**: 165, 1966).

Poroisariopsis M. Morelet (1971), anamorphic *Pezizomycotina*, Hso.≡ eP.24. 2, N. America. See Morelet (*Bulletin de la Société des Sciences naturelles et* *d'Archéologie de Toulon et du Var* **195**: 7, 1971).

Porolaschia Pat. (1900) = Favolaschia fide Singer (*Farlowia* **2**: 223, 1945).

Poroleprieuria M.C. González, Hanlin, Ulloa & E. Aguirre (2004), Xylariaceae. 1, Mexico. See González *et al.* (*Mycol.* **96**: 676, 2004).

Poromarasmius Binder et al. (2006) nom. nud., Marasmiaceae. 1. See Binder *et al.* (*Amer. J. Bot.* **93**: 547, 2006).

Poromycena Overeem (1926) = Mycena fide Singer (*Agaric. mod. Tax.*, 1951).

Poronea Raf. (1815) nom. dub., Fungi. No spp. included.

Poronia Willd. (1787), Xylariaceae. Anamorph *Lindquistia*. 4 (mostly coprophilous), widespread. See Jong & Rogers (*Mycol.* **61**: 853, 1969), Paden (*CJB* **56**: 1774, 1976; culture, anamorph), Lohmeyer & Benkert (*Z. Mykol.* **54**: 93, 1988), Rogers & Læssøe (*Mycotaxon* **44**: 435, 1992; as *Podosordaria*), San Martín & Rogers (*Mycotaxon* **48**: 179, 1993; Mexico), Rogers *et al.* (*Mycotaxon* **67**: 61, 1998; delimitation), Winka & Eriksson (*Mycoscience* **41**: 97, 2000; DNA), Ju & Rogers (*MR* **105**: 1134, 2001).

Poronidulus Murrill (1904), Polyporaceae. 1, N. America. See Gilbertson & Ryvarden (*N. Amer. Polyp.* **2**: 773, 1987; relationship with *Trametes*).

Poroniopsis Speg. (1922) = Hypocreodendron fide Lindquist & Wright (*Darwiniana* **11**: 598, 1959).

Poropeltis Henn. (1904), anamorphic *Pezizomycotina*, Cpt.0eP.?. 1, S. America.

Porophilomyces U. Braun (2000), anamorphic *Pezizomycotina*. 1. See Braun (*Schlechtendalia* **5**: 42, 2000).

Porophora G. Mey. (1825) ≡ Ascidium Fée fide Hafellner & Kalb (*Biblthca Lichenol.* **57**: 161, 1995).

Porophora Zenker ex Göbelez (1827) = Trypethelium fide Hawksworth *et al.* (*Dictionary of the Fungi* edn 8, 1995).

Porophoromyces Thaxt. (1926), Laboulbeniaceae. 1, Africa. See Nannfeldt (*Svensk bot. Tidskr.* **43**: 468, 1949).

Poroptyche Beck (1888) ? = Perenniporia fide Donk (*Verh. K. ned. Akad. Wet.* tweede sect. **62**: 1, 1974).

Porosphaera Dumort. (1822) nom. dub., Pezizomycotina.

Porosphaerella E. Müll. & Samuels (1982), ? Coniochaetales. Anamorphs *Cordana*, *Pseudobotrytis*. 1 (on rotten wood), widespread. See Müller & Samuels (*Sydowia* **35**: 150, 1982), Réblová *et al.* (*Sydowia* **51**: 49, 1999), Réblová & Winka (*Mycol.* **92**: 939, 2000; phylogeny), Fernández & Huhndorf (*Fungal Diversity* **17**: 11, 2004), Réblová & Seifert (*MR* **111**: 287, 2007; phylogeny).

Porosphaerellopsis Samuels & E. Müll. (1982), Hypocreomycetidae. Anamorph *Sporoschismopsis*. 1, Brazil. See Müller & Samuels (*Sydowia* **35**: 150, 1982), Réblová *et al.* (*Sydowia* **51**: 49, 1999), Huhndorf *et al.* (*Mycol.* **96**: 368, 2004; phylogeny).

Porosphaeria Samuels & E. Müll. (1979) ≡ Porosphaerellopsis.

porospore, see tretic.

Porostereum Pilát (1936), Phanerochaetaceae. 15, widespread. See Hjortstam & Ryvarden (*Syn. Fung.* **4**: 25, 1990; key), Ryvarden (*Syn. Fung.* **18**: 76, 2004).

Porostigme Syd. & P. Syd. (1917) = Phaeostigme fide Müller & von Arx (*Beitr. Kryptfl. Schweiz* **11** no. 2, 1962).

Porosubramaniania Hol.-Jech. (1985), anamorphic *Pezizomycotina*, Hso.1eP.27. 2, former Czechoslovakia; India. See Holubová-Jechová (*Proc. Indian Acad. Sci.* Pl. Sci. **94**: 253, 1985).

Porotenus Viégas (1960), Uropyxidaceae. 4 (on *Memora* (*Bignoniaceae*)), C. America; Brazil. See Viégas (*Bragantia* **19**: xcviii, 1960).

Porotheleaceae Murrill (1916) = Meripilaceae.

Porotheleum Fr. (1818) = Hydnopolyporus fide Kirk (*in litt.*).

Porothelium Eschw. (1824), ? Pyrenulales (L). 4, widespread (tropical).

Porphyrellus E.-J. Gilbert (1931) = Tylopilus fide Singer (*Agaric. mod. Tax.* edn 2, 1962) but see Smith & Thiers (*The Boletes of Michigan*, 1971), Wolfe & Petersen (*Mycotaxon* **7**: 152, 1978; nomencl. suprasp. taxa).

Porphyriospora A. Massal. (1852) = Polyblastia fide Hawksworth *et al.* (*Dictionary of the Fungi* edn 8, 1995).

Porphyrosoma Pat. (1927), ? Hypocreales. 1 (on *Amphisphaeria*), Madagascar. See Rossman (*Stud. Mycol.* **42**, 1999; type material lost).

Porpidia Körb. (1855), Porpidiaceae (L). 27, widespread (temperate). See Hertel (*Beih. Nova Hedwigia* **79**: 399, 1984; Antarct.), Hertel & Knoph (*Mitt. bot. StSamml., München* **20**: 467, 1984), Gowan (*Bryologist* **92**: 25, 1989; key 21 spp. N. Am.), Gowan & Ahti (*Ann. bot. fenn.* **30**: 53, 1993; key 15 spp. Scand.), Buschbom & Mueller (*Mol. Phylogen. Evol.* **32**: 66, 2004; phylogeny), Fryday (*Lichenologist* **37**: 1, 2005; Europe), Buschbom & Mueller (*Mol. Biol. Evol.* **23**: 574, 2006; evolution), Miądlikowska *et al.* (*Mycol.* **98**: 1088, 2006; phylogeny).

Porpidiaceae Hertel & Hafellner (1984), Lecideales (L). 16 gen. (+ 6 syn.), 58 spp.

Lit.: Brodo & Hertel (*Herzogia* **7**: 493, 1987), Gowan (*Syst. Bot.* **14**: 77, 1989; chemistry), Gowan (*Bryologist* **92**: 25, 1989), Hafellner (*Herzogia* **8**: 53, 1989), Rambold (*Biblthca Lichenol.* **34**: 345 pp., 1989), Hertel (*Mitt. bot. StSamml.* Münch. **28**: 211, 1990), Pietschmann (*Nova Hedwigia* **51**: 521, 1990), Brodo (*Bryologist* **98**: 609, 1995), Rambold & Hagedorn (*Lichenologist* **30**: 473, 1998), Reeb *et al.* (*Mol. Phylogen. Evol.* **32**: 1036, 2004), Miądlikowska *et al.* (*Mycol.* **98**: 1088, 2006; phylogeny), Hofstetter *et al.* (*Mol. Phylogen. Evol.* **44**: 412, 2007; phylogeny).

Porpoloma Singer (1952), Tricholomataceae. *c.* 12 (esp. under *Nothofagus*), S. America. See Singer (*Sydowia* **6**: 198, 1952), Desjardin & Hemmes (*Harvard Pap. Bot.* **6**: 85, 2001; Hawaiian Islands), Laskibar *et al.* (*Docums Mycol.* **30** no. 120: 47, 2001).

Porpomyces Jülich (1982), Hydnodontaceae. 5, widespread. See Spirin & Zmitrovich (*Karstenia* **43**: 80, 2003).

Porrectotheca Matsush. (1996), anamorphic *Pezizomycotina*, Cpd.?.?. 1, Japan. See Matsushima (*Matsush. Mycol. Mem.* **9**: 21, 1996).

Portalia V. González, Vánky & Platas (2007), ? Anthracoideaceae. 1 (in flowers of *Scirpus*, *Cyperaceae*), Europe; Asia. See González *et al.* (*Fungal Diversity* **27**: 45, 2007).

Portalites Hemer & Nygreen (1967), Fossil Fungi. 1 (Carboniferous), Libya.

porter, ale (q.v.) produced by fermenting malt that has been charred by heat.

Porterula Speg. (1920) = Asteromella fide Clements & Shear (*Gen. Fung.*, 1931).

Posadasia Cantón (1898) = Coccidioides fide Dodge (*Medical Mycology*, 1935).

posterior (1) at or in the direction of the back; (2) (of a lamella), the end at or near the stipe.

Postia Fr. (1874), Fomitopsidaceae. 30, widespread. See Yao *et al.* (*FEMS Microbiol. Lett.* **242**: 109, 2005; *Postia caesia* complex).

Potamomyces K.D. Hyde (1995), Dothideomycetes. 1, Australia. See Hyde (*Nova Hedwigia* **61**: 132, 1995).

Potebnia (Andrei Aleksandrovich; 1870-1919; Ukraine). Student, Kharkiv University (1890-1894); military service (1894-1896); Assistant to the Phylloxera Committee, Department of Agriculture, Bessarabia [now Moldova] (1896-1897); student in Berlin, Bern and Paris (1897-1898); Botanist, Nikita Botanic Garden, Yalta (1898-1903); Lecturer, Kharkiv University (1903-1912) [with interruptions when drafted for military service in the Russian-Japan war (1904-1905) and for a sabbatical in Hamburg Botanical Institute (1907)]; Head, Department of Plant Pathology, Kharkiv Regional Agricultural Experimental Station (1913). Carried out early explorations of Ukraine's fungal diversity; developed and spread in the Russian speaking world ideas about linking conidial and ascomycetous fungi. *Publs.* [*Gribnye Parazity Vysshikh Rasteniy Khar?kovskoy i Smezhnykh Guberniyi. I. Bakterii, Amebovidnye Organizmy i Nizshye Griby*] (1915) [in Russian]; *Gribnye Parazity Vysshikh Rasteniy. II. Golosumchatye, Muchnistorosianye i Discomitsety* (1916) [in Russian]. *Biogs, obits etc.* Tatarenko ([*Mikologiya i Fitopatologiya*] **4**: 493, 1970) [in Russian].

Potebniamyces Smerlis (1962), Bulgariaceae. Anamorph *Phacidiopycnis*. 1, Europe; N. America. *P. discolor* (apple bark canker). See Hahn (*Mycol.* **49**: 226, 1957), Xiao *et al.* (*Mycol.* **97**: 464, 2005; USA), Wang *et al.* (*Mycol.* **98**: 1065, 2006; phylogeny, reln with *Phacidium*).

Potoromyces Müll. bis ex Hollós (1902) = Mesophellia fide Stalpers (*in litt.*).

Potridiscus Döbbeler & Triebel (2000), Leotiaceae. 1, Australia. See Döbbeler & Triebel (*Hoppea* **61**: 72, 2000).

Potriphila Döbbeler (1996), Odontotremataceae. 3, Austria. See Döbbeler (*Haussknechtia* Beih. **9**: 79, 1999).

Pouzarella Mazzer (1976) ≡ Pouzaromyces.

Pouzaromyces Pilát (1953) = Entoloma fide Kuyper (*in litt.*).

Pouzaroporia Vampola (1992), Meruliaceae. 1, N. America. See Vampola (*Česká Mykol.* **46**: 59, 1992).

Powellia Bat. & Peres (1964), anamorphic *Pezizomycotina*. 1, Jamaica. See Batista & Peres (*Publções Inst. Micol. Recife* **398**: 4, 1964).

Powellomyces Longcore, D.J.S. Barr & Désauln. (1995), Spizellomycetaceae. 2, Canada. See Longcore *et al.* (*CJB* **73**: 1389, 1995).

ppb, parts per billion; a measure of concentration.

ppm, parts per million; a measure of concentration.

Prachtflorella Matr. (1903) ≡ Gonatobotrys.

praemorse (of the stipe base), as if broken off; truncate.

Pragmoparopsis Höhn. (1917) = Colpoma fide Holm & Holm (*Symb. bot. upsal.* **21** no. 3, 1977).

Pragmopora A. Massal. (1855), Helotiaceae. Anamorphs *Pragmopycnis*, *Hormonema*-like. 6, Europe; N. America. See Groves (*CJB* **45**: 169, 1967; key), de Hoog & McGinnis (*Stud. Mycol.* **30**: 187, 1987; anamorph).

Pragmopycnis B. Sutton & A. Funk (1975), anamorphic *Pragmopora*, St.0eH.16. 1, Canada. See Sutton & Funk (*CJB* **53**: 522, 1975).

Prasiola Menegh. (1838), Algae. Algae.

Prataprajella Hosag. (1992), Meliolaceae. 2 (from leaves), India; Taiwan. See Hosagoudar (*Nova Hedwigia* **55**: 223, 1992), Song *et al.* (*Jour. Shandong Agric. Univ.* **30**: 49, 1999).

Pratella (Pers.) Gray (1821) ≡ Agaricus L.

Prathigada Subram. (1956), anamorphic *Pezizomycotina*, Hso.≡ eP.10. 7, Asia. See Subramanian (*Journal of Madras Univ.* B **26**: 366, 1956), Furlanetto & Dianese (*MR* **103**: 1203, 1999; Brazil), Crous & Braun (*CBS Diversity Ser.* **1**: 571 pp., 2003).

Prathoda Subram. (1956) = Alternaria fide Cejp & Deighton (*Mycol. Pap.* **117**, 1969).

precipitative vesicle, sporangial vesicle, the membrane of which is formed after extrusion of naked sporangial protoplasm, whether partially cleaved into planonts or not, possibly as a precipitation reaction between periprotoplasmic colloids and the environment; amorphous or fibrillar, c.f. homohylic vesicle, plasmamembranic vesicle.

precipitin, see antigen.

Predacious fungi. The ability to trap and then parasitize or kill amoebae, nematodes, rotifers and other small terrestrial or aquatic animals has evolved several times in various unrelated groups of fungi, with >200 + spp. involved. The most important are *Zoopagales* (q.v.); others are hyphomycetes (*Arthrobotrys*, *Harposporium*, *Monacrosporium*, *Rotiferophthora*, *Verticillium*), anamorphic basidiomycetes (*Nematoctonus*) or members of the *Saprolegniales* (*Sommerstorffia*). Study of these fungi is particularly associated with the mycologist Drechsler who produced a long series of beautifully illustrated accounts of them, mostly in *Mycologia*. Various types of trap are produced, including adhesive hyphal nets or pegs to which the prey become stuck, and special hyphal rings in which individual cells suddenly expand in response to touch, thereby constricting their prey. Comandon & de Fonbrune (see Lloyd & Madison, *Discovery* **7**: 303, 1946) studied the trapping mechanism of *Arthrobotrys*, etc. by cinemicrography.

Lit.: Barron (*The nematode-destroying fungi*, 1977), Barron (*CJB* **63**: 211, 1985, *CJB* **69**: 494, 1991; rotifer endoparasites), Carris & Glawe (*Fungi colonizing cysts of Heterodera glycines*, 1990), Castner (*CJB* **46**: 764, 1966; key to 14 nematode-destroying spp. with constricting rings), Cooke & Godfrey (*TBMS* **47**: 61, 1964; key to nematode-destroying fungi), Dayal (*Predaceous fungi*, 2000), Dix & Webster (*Fungal ecology*, 1995), Dolfus (*Parasites. des helminthes*, 1946), Duddington (*Bot. Rev.* **21**: 377, 1955, (& Wyborn) *Bot. Rev.* **38**: 545, 1972, *Biol. Rev.* **31**: 152, 1956; *TBMS* **38**: 97, 1955; technique, *Friendly fungi*, 1956), Dürschner (*Mitt. biol. Bundes. Land-Forstw.* 217, 1983; nematode endoparasites), Gams (*Neth. J. Plant Pathol.* **94**: 123, 1988; nematophagous species of *Verticillium*), Gray (*Biol. Rev.* **62**: 245, 1987; review), Ilyaletdinov *et al.* (*Griby Gifomitsety Regulyatory Chislennosti Paraziticheskikh Nematod*, 1990, in Russian), Oorschot (*Stud. Mycol.* **26**: 61, 1985), Poinar & Jansson (Ed.) (*Diseases of nematodes* **2**, 1988; 150+ fungi known), Soprunov (*Predacious hyphomycetes and their application to the control of pathogenic nematodes*, 1966 [transl. 1958, Russ. edn]), Verona & Lepidi (*Agric. ital.* A, **71**: 204, 1971; identif., review), Newell *et al.* (*Bull. mar. Sci.* **17**: 177, 1977; keys to marine nematode-destroying spp.).

Predaldinia P. Briot, Lar.-Coll. & Locq. (1983), Fossil Fungi. 1, Australia.

PREM, National Collection of Fungi of the Republic of South Africa (Pretoria, South Africa); founded 1908; a government institute; genetic resource collection **PPRI**; see Baxter (*S. Afr. J. Sci.* **82**: 348, 1986).

Premyxomyces Locq. (1979), Fossil Fungi. 1, France.

Preservation. To keep alive, retaining quality and condition, preventing deterioration (*ex situ* conservation). See Genetic resource collections, Reference collections.

Preservatives. *General preservative for reference specimens*: 5% formaldehyde (40%) in water; or 25 ml formaldehyde (40%) and 150 ml alcohol (95%) in 1000 ml water.

Preuss (Carl Gottlieb Traugott; 1795-1855; Germany). Apotheker and Sanitätsrath of Hoyerswerda (1834-1855); died of apoplexy. Between 1843 and 1855 described 344 new species (including approximately 40 new genera) from central Germany. *Publs*. Die Pilze Deutschlands. In J. Sturm, *Deutschlands Flora* Abth. 3 (1848-1862) [part published posthumously]; Uebersicht untersuchter Pilze, besonders aus der Umgegend von Hoyerswerda. *Linnaea* (1851-1855) [a series of papers in vols 24-26]. *Biogs, obits etc.* Jülich (*Willdenowia* **7**: 261, 1974) [list of collections]; Meyer (*Willdenowia* **1**: 573, 1956); Stafleu & Cowan (*TL-2* **4**: 396, 1983).

Preussia Fuckel (1867), Sporormiaceae. 51 (coprophilous, soil etc.), widespread. See Cain (*CJB* **39**: 1633, 1961; key), Valldosera & Guarro (*Boln Soc. Micol. Madrid* **14**: 81, 1990; key 20 spp. inc. *Sporormiella*), Guarro *et al.* (*Nova Hedwigia* **64**: 177, 1997; Iraq), Arenal *et al.* (*Fungal Diversity* **20**: 1, 2005), Kruys *et al.* (*MR* **110**: 527, 2006; phylogeny), Schoch *et al.* (*Mycol.* **98**: 1041, 2006; phylogeny).

Preussiaster Kuntze (1891) ≡ Cordana fide Ellis (*Dematiaceous Hyphomycetes*, 1971).

Preussiella Lodha (1978) [non *Preussiella* Gilg 1897, *Spermatophyta*] = Pycnidiophora fide Hawksworth *et al.* (*Dictionary of the Fungi* edn 8, 1995).

Prévost (Isaac-Bénédict; 1755-1819; France). Professor of Philosophy, Faculté de Théologie Protestante, Montauban, Departement du Lot (1810 onwards). An amateur scientist; through his experiments on wheat bunt (*Tilletia caries*), he was the first to demonstrate pathogenicity of a fungus. *Publs*. *Mémoire sur la Cause Immèdiate de la Carie ou Charbon des Blés* (1807) [English translation *Phytopathological Classics* **6**, 1939, biography]. *Biogs, obits etc.* Keitt (*Phytopathology* **46**: 1, 1956); Stafleu & Cowan (*TL-2* **4**: 398, 1983).

Priapus Raf. (1808) ? = Lycoperdon Pers. fide Donk (*Taxon* **5**: 109, 1956).

Prillieuxia Sacc. & P. Syd. (1899) = Tomentella Pat. fide Donk (*Persoonia* **3**: 199, 1964).

Prillieuxina G. Arnaud (1918), Asterinaceae. Anamorph *Leprieurina*. 44, widespread (esp. tropical). See Doidge (*Bothalia* **4**: 273, 1942).

primary, first; first-formed; **- homothallism**, see homothallism; **- mycelium** (of basidiomycetes), the haploid mycelium from a basidiospore; **- septum**, see septum; **- squamules**, the first formed squamules of *Cladonia* from which the podetia arise; **- universal veil**, see protoblem; **- uredo** (uredium, uredinium), see *Pucciniales*.

primordial, first in order of appearance; pertaining to the earliest stages of development; **- covering** or **cuticle** = blematogen; **- hypha**, intensely coloured hyphae of the epicutis in *Russula* (Melzer, 1934); **- shaft**, the monaxial basidioma initial, as in *Clavariaceae* (Corner, 1950); **- tissue**, undifferentiated tissue of a basidioma initial; cf. lipsanenchyma; **- veil**, protoblem; **primordium**, the earliest stage of development of an organ.

primospore, a spore very like a cell of the thallus (MacMillan).

Pringsheimia Schulzer (1866), Dothideales. 1, widespread. See Holm (*in litt.*), Yurlova *et al.* (*Stud. Mycol.* **43**: 63, 1999; phylogeny).

Pringsheimiella Couch (1939) [non *Pringsheimiella* Höhn. 1920, *Algae*] ≡ Dictyomorpha.

priorable, see legitimate.

Prismaria Preuss (1851), anamorphic *Pezizomycotina*, Hso.1bH.?. 2, Europe.

Pritzeliella Henn. (1903) = Penicillium Link fide Seifert & Samson *in* Samson & Pitt (Eds) (*Advances in Penicillium and Aspergillus systematics* **102**: 143, 1985).

Proabsidia Vuill. (1903) = Absidia fide Hesseltine (*Mycol.* **47**: 344, 1955).

Proactinomyces (K. Lehm. & Haag) H.L. Jensen (1934) = Nocardia fide Hawksworth *et al.* (*Dictionary of the Fungi* edn 8, 1995).

probasidium, see basidium.

Probilimbia Vain. (1899) ≡ Mycobilimbia.

Proboscispora Punith. (1984), anamorphic *Pezizomycotina*, Cpd.1eH.10. 1, Malaysia. See Punithalingam (*Nova Hedwigia* **39**: 63, 1984).

Proboscispora S.W. Wong & K.D. Hyde (1999) nom. illegit. ≡ Pseudoproboscispora.

Procandida E.K. Novák & Zsolt (1961) = Syringospora fide von Arx *et al.* (*Stud. Mycol.* **14**: 1, 1977).

Proceropycnis M. Villarreal, Arenal, V. Rubio, Begerow, R. Bauer, R. Kirschner & Oberw. (2006), Phleogenaceae. 1, Spain; E. Asia. See Villarreal *et al.* (*Mycol.* **98**: 641, 2006), Hausner *et al.* (*Mycotaxon* **103**: 279, 2008).

pro-diploidization hypha, a hypha which may be diploidized, cf. flexuous hypha.

progametangium (of *Mucorales*), a hyphal branch forming a gametangium and suspensor cell (Fig. 27C).

progamones, a group of sex hormones of zygomycetes (Reschke & Plempel, *Z. Pflanzenphysiol.* **67**: 343, 1972).

prohybrid, a mycelium having additional nuclei from hyphal fusions (after Dodge, *Mycol.* **28**: 407, 1936).

Prokaryota. See *Eukaryota*.

prokaryote (adj. **-otic**), an organism lacking membrane-limited nuclei and not exhibiting mitosis, e.g. bacteria; cf. eukaryote.

prolate (of a spore, sporocarp, etc.), elongated in the direction of the poles; **- spheroidal**, only slightly so. See subglobose.

proliferation, successive development of new parts, esp. of new sporangia within the old wall in *Oomycota*, or of new wall material in conidiogenous cells.

proliferin, an anti-tubercle bacillus antibiotic from *Aspergillus proliferans* (Gupta & Viswanathan, *Antibiot. & Chemotherapy* **5**: 496, 1955).

Proliferobasidium J.L. Cunn. (1976), Brachybasidiaceae. 1 (on *Heliconia*), West Indies. See Begerow *et al.* (*Mycol. Progr.* **1**: 187, 2002).

Proliferodiscus J.H. Haines & Dumont (1983), Hyaloscyphaceae. 7, widespread. See Spooner (*Biblthca Mycol.* **116**, 1987; 4 spp. Australasia), Cantrell & Hanlin (*Mycol.* **89**: 745, 1997; DNA).

Prolisea Clem. (1931) = Cercidospora fide Müller & von Arx (*Beitr. Kryptfl. Schweiz* **11** no. 2, 1962).

Prolixandromyces R.K. Benj. (1970), Laboulbeniaceae. 6, C. America; Europe. See Benjamin (*Aliso* **7**: 174, 1970), Tavares (*Mycol. Mem.* **9**: 627 pp., 1985), Santamaria (*Mycotaxon* **32**: 433, 1988).

Promicromonospora Krassiln., Kalakout. & Kirillova (1961), Actinobacteria. q.v.

promitosis, the special type of nuclear division during the growth stage in *Plasmodiophoraceae*; cruciform division.

promycelium, Tulasne's term for the germ tube of the teliospore (*Pucciniales*) or ustilospore (*Ustilaginales*) from which **promycelial spores** (Plowright) (sporidia) are produced. The teliospore has been interpreted as a probasidium, the - as a metabasidium (after septation a phragmobasidium) and the ballistisporous promycelial spores basidiospores, as have the non-ballistosporic smut sporidia (see Donk, *K. ned. Akad. Wet.* C **76**: 109, 1973).

Promycetes. See *Pucciniales* and *Ustilaginales* (Clements & Shear, 1931).

Pronectria Clem. (1931), Bionectriaceae. Anamorph *Acremonium*-like. 19 (lichenicolous and algicolous), widespread. See Lowen (*Mycotaxon* **39**: 461, 1990), Rossman *et al.* (*Stud. Mycol.* **42**: 248 pp., 1999).

proper exciple (**margin**), see excipulum (proprium).

prophialide, see metula; primary sterigma.

Prophytroma Sorokīn (1877), anamorphic *Pezizomycotina*, Hsy.0eP.?. 1, former USSR.

Propolidium Sacc. (1884), ? Rhytismatales. 2, Africa; N. America. See Sherwood (*Mycotaxon* **5**: 1, 1977).

Propolina Sacc. (1884), ? Dothideales. 1, Europe.

Propoliopsis Rehm (1914), ? Stictidaceae. 1, Asia; N. America. See Sherwood (*Mycotaxon* **5**: 1, 1977).

Propolis Fr. (1849), ? Rhytismataceae. *c.* 14, widespread. See Sherwood (*Mycotaxon* **5**: 320, 1977; key), Johnston (*N.Z. Jl Bot.* **24**: 84, 1986; key 4 NZ spp.), Spooner (*Kew Bull.* **45**: 451, 1990), Minter (*Mycotaxon* **87**: 43, 2003).

Propolomyces Sherwood (1977) ≡ Propolis.

Proprioscypha Spooner (1987), Hyaloscyphaceae. 1 (on *Eucalyptus*), Australia. See Spooner (*Biblthca Mycol.* **116**, 1987).

Propythium M.K. Elias ex Janson. & Hills (1979), Fossil Fungi, Oomycota. 1 (Carboniferous), USA.

Prosaccharomyces E.K. Novák & Zsolt (1961) = Saccharomycopsis Schiønning fide von Arx *et al.* (*Stud. Mycol.* **14**: 1, 1977).

prosenchyma, see plectenchyma.

Prosopidicola Crous & C.L. Lennox (2004), anamorphic *Cryphonectriaceae*. 1, Mexico; USA. See Lennox *et al.* (*Stud. Mycol.* **50**: '187' [191], 2004).

prosoplectenchyma, see plectenchyma.

prosorus (of *Chytridiales*), a cell giving a group of sporangia (the sorus).

Prospodium Arthur (1907), Uropyxidaceae. *c.* 85 (on *Bignoniaceae*, some on *Verbenaceae*), America (tropical & subtropical); India. See Cummins (*Annls mycol.* **35**: 15, 1937; morphology), Cummins (*Lloydia* **3**: 1, 1940; key), Hernández & Hennen (*Mycol.* **95**: 728, 2003).

prosporangium (in *Oomycota*), a sporangium-like body which puts out a vesicle (sporangium) in which zoospores may undergo development and from which they are freed; presporangium.

Prosporobolomyces E.K. Novák & Zsolt (1961) = Sporobolomyces fide Lodder (*Yeasts, a taxonomic study* 2nd edn, 1970).

Prosthecium Fresen. (1852), Diaporthales. Anamorph *Stilbospora*. 12 (from bark), Europe; N. America. See Merezhko (*Ukr. Bot. Zh.* **45**: 57, 1988), Jaklitsch & Voglmayr (*Stud. Mycol.* **50**: 229, 2004; phylogeny).

Prosthemiella Sacc. (1881), anamorphic *Pezizomycotina*, Cac.1bH.?. 3, widespread.

Prosthemium Kunze (1817), anamorphic *Pleomassaria, Splanchnonema*, Cac.0bP.1. 6, widespread (temperate). See Hantula *et al.* (*MR* **102**: 1509, 1998; teleomorph), Paavolainen *et al.* (*Mycol.* **92**: 253, 2000; cryptic spp.), Tanaka *et al.* (*Mycoscience* **46**: 248, 2005; Japan).

Prostratus Sivan., W.H. Hsieh & Chi Y. Chen (1993), Diaporthales. 1 (from leaves), Taiwan. See Sivanesan *et al.* (*MR* **97**: 1179, 1993).

Protascus Wolk (1913) ≡ Wolkia.

Protasia Racib. (1900) nom. nud., ? Fungi.

Protendomycopsis Windisch (1965) = Trichosporon fide Carmo-Sousa *in* Lodder (Ed.) (*Yeasts, a taxonomic study* 2nd edn, 1970).

proteoglycan, an antitumour metabolite from *Coriolus pubescens* active against sarcoma-180.

Proteomyces Moses & Vianna (1913) = Trichosporon fide Diddens & Lodder (*Die anaskosporogenen Hefen* **2**, 1942).

Proteophiala Cif. (1958), anamorphic *Melanospora*, Hso.0eH.15. 3, widespread. See Ciferri (*Sydowia* **11**: 289, 1957).

proteophilous fungi, fungi associated with ammonia-rich (e.g. urea affected) soils (Sagara, *Trans. mycol. Soc. Japan* **14**: 41, 1973).

proterospore, a spore formed at the start of the sporulation period in *Ganoderma* and able to germinate easily without passing through the gut of a fly larva (Nuss, *Pl. Syst. Evol.* **141**: 53, 1982).

prothallus, see hypothallus.

prothecium, a primitive or rudimentary perithecium, as in the *Gymnoascaceae*.

Protista. See *Protoctista*.

proto- (prefix), primitive; primordial.

Protoabsidia Naumov (1935) = Absidia fide Hesseltine (*Mycol.* **47**: 344, 1955).

protoaecium, a haploid structure which, after diploidization, becomes a fruiting structure.

Protoascon L.R. Batra, Segal & R.W. Baxter (1964), Fossil Fungi ? Mucorales. 1 (Carboniferous), USA. See Baxter (*Paleont. Contrib. Univ. Kansas* **77**, 1975), Pirozynski (*Ann. Rev. Phytopath.* **14**: 237, 1976).

Protobagliettoa Servít (1955) = Verrucaria Schrad. fide Hawksworth *et al.* (*Dictionary of the Fungi* edn 8, 1995).

protobasidium, see basidium.

Protoblastenia (Zahlbr.) J. Steiner (1911), Psoraceae (L). 14, widespread. See Pant (*J. Bombay nat. Hist. Soc.* **85**: 658, 1988; India), Kainz & Rambold (*Biblthca Lichenol.* **88**: 267, 2004; phylogeny, Europe), Miądlikowska *et al.* (*Mycol.* **98**: 1088, 2006; phylogeny).

protoblem, a loose flocculent mycelial layer covering the universal veil, as in *Amanita*; primordial veil.

Protocalicium Woron. (1927) = Chaenothecopsis fide Titov (*Folia cryptog. Estonica* **32**: 127, 1998).
Protocolletotrichum R.K. Kar, Neeta Sharma & U.K. Verma (2004), Fossil Fungi. 1.
protoconidium, see hemispore.
Protocoronis Clem. & Shear (1931) = Aureobasidium fide Sutton (*in litt.*).
Protocoronospora G.F. Atk. & Edgerton (1907) = Aureobasidium fide Hermanides-Nijhof (*Stud. Mycol.* **15**: 141, 1977).
Protocrea Petch (1937), Hypocreaceae. Anamorphs *Acremonium*-like, *Verticillium*. 4 (on decaying wood and old basidiomycetes), widespread (temperate). See Rossman *et al.* (*Stud. Mycol.* **42**: 248 pp., 1999), Põldmaa (*Stud. Mycol.* **45**: 83, 2000).
Protocreopsis Yoshim. Doi (1977), Bionectriaceae. Anamorph *Acremonium*-like. 9 (from decaying plant material), widespread (tropical). See Doi (*Bull. natn. Sci. Mus.* Tokyo, B **4**: 113, 1978), Rossman *et al.* (*Stud. Mycol.* **42**: 248 pp., 1999).
Protoctista (Protista), kingdom of eukaryotic microorganisms exclusive of the Kingdoms *Animalia*, *Fungi*, and *Plantae*, but including *Myxomycota* and *Oomycota* and related groups (Margulis *et al.* (Eds), *Handbook of Protoctista*, 1991). Molecular data has shown the 'kingdom' to be exceptionally heterogenous and recent authors accept *Chromista* and *Protozoa* instead. See *Chromista*, Kingdoms of fungi, *Protozoa*.
Protocucurbitaria Naumov (1951), Pleosporales. 1, former USSR. See Naumov (*Bot. Mater. Otd. Sporov. Rast. Bot. Inst. Komarova Akad. Nauk S.S.S.R.* **7**: 110, 1951).
Protodaedalea Imazeki (1955) = Elmerina.
Protodontia Höhn. (1907) = Stypella fide Roberts (*Mycotaxon* **69**: 209, 1998).
Protoerysiphe N. Sharma, R.K. Kar, A. Agarwal & R. Kar (2005), Fossil Fungi, Erysiphales. See Sharma *et al.* (*Micropaleontology* **51**: 73, 2005).
Protogaster Thaxt. (1934), Protogastraceae. 1 (on roots of *Viola*), USA.
Protogastraceae Zeller (1934), ? Boletales. 1 gen., 1 spp.
Lit.: Zeller (*Ann. Mo. bot. Gdn* **21**: 231, 1934), Zeller & Walker (*Mycol.* **27**: 573, 1935), Reijnders (*MR* **104**: 900, 2000).
Protogautieria A.H. Sm. (1965), Gomphaceae. 2, N. America. See Smith (*Mycopath. Mycol. appl.* **26**: 393, 1965).
Protogenea Kobayasi (1964) = Hydnocystis fide Trappe (*TBMS* **65**: 496, 1975), Læssøe & Hansen (*MR* **111**: 1075, 2007; phylogeny).
Protoglossum Massee (1891), Cortinariaceae. 8, widespread. See May (*Muelleria* **8**: 287, 1995), Vidal (*Revista Catalana de Micologia* **24**: 287, 2002).
protogonidium, the first of a series of gonidia (obsol.).
Protograndinia Rick (1933) = Patouillardina Bres. fide Rogers (*Mycol.* **31**: 513, 1939).
Protohydnum Möller (1895), Auriculariales. 3, widespread. See Martin (*Lloydia* **4**: 262, 1941), Roberts & Spooner (*Kew Bull.* **53**: 631, 1998; Brunei Darussalam).
protohymenial, having a primitive hymenium (Maire).
protologue, everything associated with a name on its first publication, i.e. diagnosis, description, references, synonymy, geographical data, citation of specimens, discussion, comments, illustrations (see Stearn, *in* Linnaeus, *Species plantarum* [Reprint] **1**: 126, 1957). See also Nomenclature.
Protomarasmius Overeem (1927) ? = Marasmius fide Kuyper (*in litt.*).
Protomerulius Möller (1895), Auriculariales. *c.* 4, widespread. See Ryvarden (*Synopsis Fungorum* **5**, 1991).
Protomicarea Hafellner (2001), Psoraceae (L). 1, Europe. See Hafellner & Türk (*Stapfia* **76**: 156, 2001).
Protomycena Hibbett, D. Grimaldi & Donoghue (1997) nom. inval., Fossil Fungi. 1, Dominican Republic. See Hibbett *et al.* (*Am. J. Bot.* **84**: 984, 1997).
Protomyces Unger (1832), Protomycetaceae. *c.* 10 (biotrophic on plants), widespread. See Reddy & Kramer (*Mycotaxon* **3**: 1, 1975; key), Sjamsuridzal *et al.* (*Mycoscience* **38**: 267, 1997; phylogeny), Nishida *et al.* (*J. Mol. Evol.* **46**: 442, 1998; introns), Sugiyama (*Mycoscience* **39**: 487, 1998; phylogeny), Schweigkofler *et al.* (*Organ. Divers. Evol.* **2**: 1, 2002; phylogeny), Lopandic *et al.* (*Mycol. Progr.* **4**: 205, 2005; phylogeny), Sugiyama *et al.* (*Mycol.* **98**: 996, 2006; phylogeny).
Protomycetaceae Gray (1821), Taphrinales. 6 gen., 22 spp.
Lit.: Reddy & Kramer (*Mycotaxon* **3**: 1, 1975), Blanz & Unseld (*Stud. Mycol.* **30**: 247, 1987), Nishida & Sugiyama (*Mol. Biol. Evol.* **10**: 431, 1993), Döbbeler (*Nova Hedwigia* **60**: 171, 1995), Sugiyama *et al.* (*Microbial Diversity in Time and Space* Proceedings of the International Symposium, October 24-26, 1994, Tokyo, Japan: 41, 1996), Sjamsuridzal *et al.* (*Mycoscience* **38**: 267, 1997; phylogeny), Ahearn *et al. in* Kurtzman & Fell (Eds) (*Yeasts, a taxonomic study* 4th edn: 600, 1998), Kurtzman *in* Kurtzman & Fell (Eds) (*Yeasts, a taxonomic study* 4th edn: 353, 1998), Kurtzman & Blanz *in* Kurtzman & Fell (Eds) (*Yeasts, a taxonomic study* 4th edn: 69, 1998), Sugiyama (*Mycoscience* **39**: 487, 1998; phylogeny).
Protomycetales Luttr. ex D. Hawksw. & O.E. Erikss. (1986) = Taphrinales.
Protomycites Mesch. (1892), Fossil Fungi. 1 (Carboniferous), British Isles.
Protomycocladus Schipper & Samson (1994), Syncephalastraceae. 1, Pakistan. See Schipper & Samson (*Mycotaxon* **50**: 487, 1994), Voigt & Wöstemeyer (*Gene* **270**: 113, 2001; phylogeny).
Protomycopsis Magnus (1905), Protomycetaceae. 5 (from living plants), widespread. See Reddy & Kramer (*Mycotaxon* **3**: 1, 1975; key), Haware & Pavgi (*Caryologia* **30**: 313, 1977; development), Piepenbring & Bauer (*Mycol.* **89**: 924, 1997; excl. spp.).
protonym (in nomenclature), a name effectively but not validly published after the starting point for the group (Donk, *Persoonia* **1**: 175, 1960).
Protopannaria (Gyeln.) P.M. Jørg. & S. Ekman (2000), Pannariaceae (L). 5, widespread. See Jørgensen (*Bryologist* **103**: 699, 2000), Jørgensen (*Cryptog. Mycol.* **22**: 67, 2001; Antarctic), Ekman & Jørgensen (*CJB* **80**: 625, 2002; phylogeny), Wiklund & Wedin (*Cladistics* **19**: 419, 2003; phylogeny), Wedin & Wiklund (*Symb. bot. upsal.* **34** no. 1: 469, 2004; phylogeny), Miądlikowska *et al.* (*Mycol.* **98**: 1088, 2006; phylogeny).
Protoparmelia M. Choisy (1929), Parmeliaceae (L). 20, widespread. See Sancho & Crespo (*Actas Simp. Nac. bot. Crypt.* **6**: 441, 1987; 3 spp. Spain), Miyawaki (*Hikobia* **11**: 29, 1991), Poelt & Leuckert

(*Nova Hedwigia* **52**: 39, 1991; key 6 spp., mostly on lichens), Hafellner *et al. in* Hawksworth (Ed.) (*Ascomycete Systematics. Problems and Perspectives in the Nineties* NATO ASI Series vol. **269 269**: 379, 1994; posn), Crespo *et al.* (*Taxon* **50**: 807, 2001; phylogeny), Grube *et al.* (*MR* **108**: 506, 2004), Brodo & Aptroot (*CJB* **83**: 1075, 2005; corticolous spp., N America), Barbero *et al.* (*Mycotaxon* **97**: 299, 2006; Spain), Arup *et al.* (*Mycol.* **99**: 42, 2007; sister group relations with *Parmeliaceae*).

Protoparmeliopsis M. Choisy (1929) = Lecanora fide Hawksworth *et al.* (*Dictionary of the Fungi* edn 8, 1995), Grube *et al.* (*MR* **108**: 506, 2004).

Protopeltis Syd. (1927) = Myriangiella fide von Arx & Müller (*Stud. Mycol.* **9**, 1975).

protoperithecium, a young but walled perithecium before ascus formation (Ellis, *Mycol.* **51**: 416, 1960).

Protophallaceae Zeller (1939) = Phallaceae.

Protophallus Murrill (1910) = Protubera fide Furtado & Dring (*TBMS* **50**: 500, 1967).

protophyte, see antithetic.

Protopistillaria Rick (1933) = Eocronartium fide Martin (*Lloydia* **5**: 158, 1942).

Protoplacodium Motyka (1995) nom. inval. = Lobothallia fide Aptroot (*in litt.*).

protoplasmodium, see plasmodium.

protoplast, traditionally the totality of the living cell constituents, whether walled or not, but now frequently used for the cell protoplasm after experimental removal of the cell wall, a usage to which Brenner *et al.* (*Nature* **181**: 1713, 1958) proposed to restrict the term. Fungal protoplasts are proving to have value in the study of cell organelles, biochemistry, genetics; they also have potential applications in biotechnology, esp. since protoplasts from different strains or even species can be fused (first achieved in 1975) into an aggregate protoplast and produce a heterokaryon with changed properties. See Villanueva (*in* Ainsworth & Sussman (Eds), *The Fungi* **2**: 3, 1966), Perberdy (*Sci. Progr.* **60**: 73, 1972; review methods), Villanueva *et al.* (Eds) (*Yeast, mould and protoplasts*, 1973), Perberdy *et al.* (Eds) (*Microbial and plant protoplasts*, 1976), Peberdy & Ferenczy (Eds) (*Fungal protoplasts. Applications in biochemistry and genetics*, 1985), Peberdy (*Microbiol. Sci.* **4**: 108, 1987; review). **sphaeroplast** a - enclosed by a modified or fragmentary cell wall. [**gymnoplast** was used by Küster (1935) for plant cells without a cell wall and Frey-Wyssling (*Nature* **216**: 516, 1967) prefered **semi-gymnoplast** to sphaeroplast.].

Protoradulum Rick (1933), ? Auriculariales. 1, Brazil. See Rick (*Egatea* **18**: 348, 1933).

Protoroccella L.M. Sánchez-Pinto & M. Schulz (2001) = Protoroccella Follmann fide Grube (*Bryologist* **101**: 377, 1998).

Protoroccella Follmann (2001), Roccellaceae (L). 2, Chile. See Follmann (*J. Hattori bot. Lab.* **90**: 261, 2001).

Protoschistes M. Choisy (1928) = Diploschistes fide Hawksworth *et al.* (*Dictionary of the Fungi* edn 8, 1995).

Protoscypha Syd. (1925), Protoscyphaceae. 1, widespread (tropical). See Petrak (*Annls mycol.* **32**: 317, 1934), Kutorga & Hawksworth (*SA* **15**: 1, 1997).

Protoscyphaceae Kutorga & D. Hawksw. (1997), ? Dothideomycetes (inc. sed.). 1 gen. (+ 1 syn.), 1 spp.

Lit.: Kutorga & Hawksworth (*SA* **15**: 1, 1997), Magnes (*Biblthca Mycol.* **165**: 177 pp., 1997).

protosexual (of yeasts or other organisms), having diploid or dikaryotic cells which produce haploid or unisexual cells in the absence of fructifications or sexual spores; in contrast to parasexual (q.v.), which is redefined to cover organisms having both protosexual and sexual cycles, and **neo-sexual**, for organisms having a sexual but not a protosexual cycle (Wickerham, *Mycol.* **56**: 254, 1964, cf. **57**: 134, 1965, **58**: 943, 1967).

protospore (1) the multinucleated mass of cytoplasm cut out by primary cleavage planes, to be followed by further cleavage to form the uninucleate spores of *Phycomyces* and other *Mucorales* (Harper), and the sporangiospores of *Coccidioides*; (2) (of *Synchytriaceae*), a 1-nucleate portion of protoplasm which becomes the sporangium.

Protostegia Cooke (1880), anamorphic *Pezizomycotina*, Cpd.0fH.1. 1, S. Africa. See Dyko *et al.* (*Mycol.* **71**: 918, 1979).

Protostegiomyces Bat. & A.F. Vital (1955), anamorphic *Pezizomycotina*, Ccu.0fH.?. 1 (on *Lembosia*), Brazil. See Batista & Vital (*An. Soc. Biol. Pernambuco* **13**: 94, 1955).

protosterigma, see basidium.

Protostroma Bat. (1957), anamorphic *Pezizomycotina*, St.0eH.?. 1, USA. See Batista (*Revta Biol.* Lisb. **1**: 109, 1957).

Prototaxaceae Hueber (2001), Fungi (inc. sed.). See Fossil fungi.

Prototaxites Dawson (1859), Fossil Fungi. 2 (Silurian-Devonian). See Boyce *et al.* (*Geology* **35**: 399, 2007; carbon isotopic range), Kibby (*Field Mycology* **9**: 48, 2008).

Prototheca W. Krüger (1894) nom. dub., Algae. ? Achloric algae. See Arnold & Ahearn (*Mycol.* **64**: 265, 1972; key), Smith (*Mycopath.* **71**: 95, 1980), Capriotti (*Arch. Microbiol.* **42**: 409, 1962; in fish). Associated with disease in humans (**protothecosis**), bovine mastitis, etc.; see Sudman (*Am. J. clin. Path.* **61**: 10, 1974). *P. richardsii* has been transferred to a new genus, *Aneurofeca*, in the *Icthyosporea* (DRIPs clade), see Baker *et al.* [*Microbiology* **145**: 1777, 1999].

protothecium, an incompletely differentiated ascoma containing neither asci nor ascospores (Shoemaker, *CJB* **34**: 641, 1955).

Protothelenella Räsänen (1943), Protothelenellaceae (±L). 10 (on lichens and *Musci*), Europe; N. America. See Mayrhofer (*Herzogia* **7**: 313, 1987; key), Fryday (*Bryologist* **107**: 231, 2004; USA), Schmitt *et al.* (*Mycol.* **97**: 362, 2005; phylogeny).

Protothelenellaceae Vězda, H. Mayrhofer & Poelt (1985), Ostropomycetidae (inc. sed.) (±L). 2 gen. (+ 2 syn.), 11 spp.

Lit.: Sherwood-Pike & Boise (*Brittonia* **38**: 35, 1986), David (*SA* **6**: 217, 1987), Mayrhofer (*Herzogia* **7**: 313, 1987), Schmitt *et al.* (*Mycol.* **97**: 362, 2005), Lumbsch *et al.* (*MR* **111**: 257, 2007; phylogeny).

Protothyrium G. Arnaud (1917), Parmulariaceae. 4, widespread (tropical).

Prototrema M. Choisy (1928) ? = Thelotrema fide Hawksworth *et al.* (*Dictionary of the Fungi* edn 8, 1995).

Prototremella Pat. (1888) = Tulasnella fide Rogers (*Annls mycol.* **31**: 183, 1933).

prototroph, see wild type.

prototunicate (of asci), basically unitunicate, but with the wall lysing at or before maturity and lacking differentiated apical structures (e.g. *Saccharomycetales*); such asci may develop in an hymenium or be distributed randomly in the interior of the ascoma. See ascus.

Prototylium M. Choisy (1929) ? = Pseudopyrenula fide Hawksworth *et al.* (*Dictionary of the Fungi* edn 8, 1995).

Protounguicularia Raitv. & R. Galán (1986), Hyaloscyphaceae. 5, Europe. See Huhtinen (*Mycotaxon* **30**: 9, 1987), Baral (*SA* **13**: 113, 1994; ? synonymy with *Olla*).

protouredinium, see protoaecium.

Protousnea (Motyka) Krog (1976), Parmeliaceae (L). 7, S. America. See Kärnefelt *et al.* (*Nova Hedwigia* **67**: 71, 1998), Calvelo *et al.* (*Mycotaxon* **85**: 277, 2003; Argentina), Calvelo *et al.* (*Bryologist* **108**: 1, 2005; revision), Czeczuga *et al.* (*Feddes Repert.* **116**: 195, 2005; carotenoids).

Protoventuria Berl. & Sacc. (1887), Venturiaceae. Anamorph *Fusicladium*-like. 1, widespread. See Müller & von Arx (*Beitr. Kryptfl. Schweiz* **11** no. 2, 1962), Barr (*Prodr. Cl. Loculoasc.*, 1987), Richiteanu & Bontea (*Rev. Roum. Biol.* biol. vég. **32**: 15, 1987; nomencl., on *Ericaceae*), Barr (*Sydowia* **41**: 25, 1989; key 23 N. Am. spp.), Pascoe & Sutton (*Aust. Syst. Bot.* **3**: 281, 1990; Australia), Carris & Poole (*Mycol.* **85**: 93, 1993; USA), Winton *et al.* (*Mycol.* **99**: 240, 2007; phylogeny).

Protubera Möller (1895), Phallogastraceae. 13, widespread (tropical; subtropical). See Furtado & Dring (*TBMS* **50**: 500, 1967), Castellano & Beever (*N.Z. Jl Bot.* **32**: 305, 1994; NZ spp.).

protuberate (of conidia), having short projections.

Protuberella S. Imai & A. Kawam. (1958), Phallaceae. 1, Japan. See Malloch (*Mycotaxon* **34**: 133, 1989).

pruinose, having a frost-like or flour-like surface covering of **pruina**. Often caused by calcium oxalate hydrates on lichen thalli; see Wadsten & Moberg (*Lichenologist* **17**: 239, 1985).

Prunulus Gray (1821) = Mycena fide Donk (*Beih. Nova Hedwigia* **5**, 1962).

PSA, see Media.

Psalidosperma Syd. & P. Syd. (1914) = Ypsilonia fide Clements & Shear (*Gen. Fung.*, 1931).

Psaliota, see *Psalliota*.

Psalliota (Fr.) P. Kumm. (1871) ≡ Agaricus L.

Psalliotina Velen. (1939) = Psathyrella fide Pilát (*Agaricalium Europaeorum Clavis Dichotomica*, 1951).

Psammina Sacc. & M. Rousseau ex E. Bommer & M. Rousseau (1891), anamorphic *Pezizomycotina*, Cac.0bP.1 (±L). 8 (mostly lichenicolous or algicolous), Europe. See Earland-Bennett & Hawksworth (*Lichenologist* **31**: 579, 1999), Earland-Bennett & Hawksworth (*Lichenologist* **37**: 191, 2005; key).

Psammocoparius Delile ex De Seynes (1863) nom. inval. = Psathyrella fide Kuyper (*in litt.*).

Psammomyces Lebedeva (1932) = Galeropsis fide Zeller (*Mycol.* **35**: 409, 1943).

Psammospora Fayod (1893) nom. rej. ≡ Melaleuca.

Psathyra (Fr.) P. Kumm. (1871) [non *Psathyra* Spreng. 1818, *Rubiaceae*] = Psathyrella fide Singer (*Agaric. mod. Tax.*, 1951).

Psathyrella (Fr.) Quél. (1872), Psathyrellaceae. *c.* 400, widespread. See Smith (*Mem. N. Y. bot. Gdn* **24**, 1972; 414 N. Am. spp.), Pegler (*Kew Bull.* Addit. Ser. **6**, 1977; E. Afr.), Kits van Waveren (*Persoonia* Suppl. **2**: 1, 1985; monogr. Dutch, French, Br. spp.), Enderle & Hübner (*Beitr. Kenntn. Pilze Mitteleur.* **14**: 53, 2005), Vašutová (*Czech Mycol.* **58**: 1, 2006; Czech Republic, Slovakia), Padamsee *et al.* (*Mol. Phylog. Evol.* **46**: 415, 2008; phylogeny), Vašutová *et al.* (*MR* **112**: in press, 2008; phylogeny).

Psathyrellaceae Vilgalys, Moncalvo & Redhead (2001), Agaricales. 12 gen. (+ 14 syn.), 746 spp. Generic delimitation in the *Psathyrellaceae* is in need of revision considering the extensive polyphyly in *Psathyrella* and the redisposition of coprinoid species to it.

Lit.: Smith (*Mem. N. Y. bot. Gdn* **24**: 633 pp., 1972), Bogart (*Mycotaxon* **10**: 155, 1979), Kits van Waveren (*Persoonia* Suppl. **2**: 300 pp., 1985), Enderle & Bender (*Z. Mykol.* **56**: 19, 1990), Moncalvo *et al.* (*Syst. Biol.* **49**: 278, 2000), Redhead *et al.* (*Taxon* **50**: 203, 2001; redisposition of coprinoid species), Adamsee *et al.* (*Mol. Phyl. Evol.* **46**: 415, 2008; phylogeny), Vašutová *et al.* (*MR* **112**: in press, 2008).

Psathyrodon Maas Geest. (1977) = Beenakia fide Parmasto & Ryvarden (*Windahlia* **18**: 39, 1990).

Psathyromyces Bat. & Peres (1964) = Tricharia Fée fide Lücking *et al.* (*Lichenologist* **30**: 121, 1998).

Psathyrophlyctis Brusse (1987), Phlyctidaceae (L). 1, S. Africa. See Brusse (*Bothalia* **17**: 182, 1987).

Psecadia Fr. (1849) ≡ Cytospora fide Sutton (*Mycol. Pap.* **141**, 1977).

Pselaphidomyces Speg. (1917), Laboulbeniaceae. 1, N. America. See Nannfeldt (*Svensk bot. Tidskr.* **43**: 468, 1949).

Pselliophora P. Karst. (1879) nom. rej. = Coprinus fide Redhead *et al.* (*Mycotaxon* **50**: 203, 2001).

Pseudacoliomyces Cif. & Tomas. (1953) ≡ Pseudacolium.

Pseudacolium Stizenb. ex Clem. (1909) = Cyphelium Ach. fide Hawksworth *et al.* (*Dictionary of the Fungi* edn 8, 1995), Tibell (*MR* **107**: 1403, 2003).

Pseudaegerita J.L. Crane & Schokn. (1981), anamorphic *Hyaloscypha*, Hsp.0eP.3. 10, widespread. See Crane & Schoknecht (*Mycol.* **73**: 78, 1981), Abdullah *et al.* (*MR* **109**: 590, 2005).

Pseudaleuria Lusk (1987), Pyronemataceae. 1, USA. See Lusk (*Mycotaxon* **30**: 417, 1987), Moravec (*Acta Musei Moraviae* Scientiae Biologicae **88**: 37, 2003), Perry *et al.* (*MR* **111**: 549, 2007; phylogeny).

Pseudallescheria Negr. & I. Fisch. (1944), Microascaceae. Anamorphs *Graphium*, *Scedosporium*. 8, widespread. *P. boydii*complex on humans (white grain mycetoma). See Campbell & Smith (*Mycopathologia* **78**: 145, 1982; conidiogenesis), McGinnis *et al.* (*Mycotaxon* **14**: 94, 1982), von Arx *et al.* (*Beih. Nova Hedwigia* **94**, 1988), Issakainen *et al.* (*J. Med. Vet. Mycol.* **35**: 389, 1997; DNA), Wedde *et al.* (*Medical Mycology* **36**: 61, 1998; PCR-based identification), Issakainen *et al.* (*MR* **103**: 1179, 1999; DNA), Okada *et al.* (*CJB* **76**: 1495, 1999; DNA), Paul & Masih (*FEMS Microbiol. Lett.* **189**: 61, 2000; molecular variation), Issakainen *et al.* (*Medical Mycology* **41**: 31, 2003; phylogeny), Riddell *et al.* (*Mycoses* **47**: 442, 2004; disseminated infection), Gilgado *et al.* (*J. Clin. Microbiol.* **43**: 4930, 2005; phylogeny), Guarro *et al.* (*Medical Mycology* **44**: 295, 2006; review), Rainer & Hoog (*MR* **110**: 151, 2006; phylogeny), Zeng *et al.* (*Medical Mycology* **45**: 547, 2007; infraspecific variation), Rainer *et al.* (*Antonie van Leeuwenhoek* **93**: 315, 2008; selective isolation).

Pseudapiospora Petr. (1928) = Pseudomassaria fide Müller & von Arx (*Beitr. Kryptfl. Schweiz* **11** no. 2,

1962).

Pseudarctomia Gyeln. (1939), ? Lichinaceae (L). 1, France. See Moreno & Egea (*Acta Bot. Barcinon.* **41**: 1, 1992).

Pseudarthopyrenia Keissl. (1935) = Pyrenocollema fide Harris (*Lichenogr. Thomsoniana*: 133, 1998).

Pseudascozonus Brumm. (1985), Thelebolaceae. 1 (coprophilous), France. See Brummelen (*Proc. Indian Acad. Sci.* Pl. Sci. **94**: 363, 1985), Brummelin (*Persoonia* **16**: 425, 1998; ultrastr.), Hoog *et al.* (*Stud. Mycol.* **51**: 33, 2005).

Pseudasterodon Rick (1959) nom. dub., Agaricomycetes. See Donk (*Persoonia* **3**: 199, 1964).

Pseudephebe M. Choisy (1930), Parmeliaceae (L). 2, widespread (montane-arctic; bipolar). See Brodo & Hawksworth (*Op. bot. Soc. bot. Lund* **42**, 1977), Kantvilas (*Flora of Australia* **55**: 162, 1994), Crespo *et al.* (*Taxon* **50**: 807, 2001; phylogeny), Thell *et al.* (*Mycol. Progr.* **1**: 335, 2002; phylogeny), Thell *et al.* (*Mycol. Progr.* **3**: 297, 2004; phylogeny).

Pseuderiospora Keissl. (1924) = Eriosporella fide Sutton (*Mycol. Pap.* **141**, 1977).

Pseuderiospora Petr. (1959) ≡ Suttoniella.

Pseudeurotiaceae Malloch & Cain (1970), Leotiomycetes (inc. sed.). 6 gen. (+ 1 syn.), 26 spp.

Lit.: Booth (*Mycol. Pap.* **83**, 1961), Malloch & Cain (*CJB* **48**: 1815, 1970; genera), Lodha *in* Subramanian (Ed.) (*Taxonomy of fungi* **1**: 241, 1978), von Arx *et al.* (*Beih. Nova Hedwigia* **94**: 104 pp., 1988), Suh & Blackwell (*Mycol.* **91**: 836, 1999; phylogeny), Sogonov *et al.* (*Mycol.* **97**: 695, 2005), Wang *et al.* (*Mycol.* **98**: 1065, 2006; phylogeny), Stchigel & Guarro (*MR* **111**: 1100, 2007).

Pseudeurotium J.F.H. Beyma (1937), Pseudeurotiaceae. Anamorph *Teberdinia*. 5 (on rotten wood etc.), widespread. See Booth (*Mycol. Pap.* **83**, 1961; key), Suh & Blackwell (*Mycol.* **91**: 836, 1999; phylogeny), Sogonov *et al.* (*Mycol.* **97**: 695, 2005; anamorph), Wang *et al.* (*Mycol.* **98**: 1065, 2006; phylogeny).

Pseudevernia Zopf (1903), Parmeliaceae (L). 5 or 6, widespread. See Hale (*Bryologist* **71**: 1, 1968), Legaz *et al.* (*Cryptog. Bryol.-Lichénol.* **6**: 343, 1985; anatomy), López Redondo & Manrique Reol (*An. Jard. bot. Madr.* **46**: 295, 1989; chemical races, Spain), Strobl *et al.* (*Biblthca Lichenol.* **53**: 251, 1993; protein composition), Czeczuga & Christensen (*Feddes Repert.* **105**: 473, 1994; carotenoids), Kärnefelt *et al.* (*Nova Hedwigia* **67**: 71, 1998; anatomy), Crespo *et al.* (*Lichenologist* **31**: 451, 1999; anatomy, chemistry, phylogeny), Crespo *et al.* (*Taxon* **50**: 807, 2001; phylogeny), Wei & Abbas (*Mycosystema* **22**: 26, 2003; China), Miądlikowska *et al.* (*Mycol.* **98**: 1088, 2006; phylogeny), Honegger & Zippler (*MR* **111**: 424, 2007; mating systems).

Pseudhaplosporella Speg. (1920) = Botryodiplodia fide Petrak & Sydow (*Feddes Repert.* **42**: 1, 1927).

Pseudhydnotrya E. Fisch. (1897) = Geopora fide Fischer (*Nat. Pflanzenfam.* **5b**: viii, 1938).

pseudo- (prefix), false; spurious.

Pseudoabsidia Bainier (1903) = Absidia fide Hesseltine (*Mycol.* **47**: 344, 1955).

Pseudoacrodictys W.A. Baker & Morgan-Jones (2003), anamorphic *Pezizomycotina*. 8, widespread. See Baker & Morgan-Jones (*Mycotaxon* **85**: 373, 2003), Somrithipol & Jones (*Sydowia* **55**: 365, 2003; Thailand).

pseudoaethium (of *Mycetozoa*), a group of separate sporangia looking like an aethalium.

Pseudoamauroascus Cano, M. Solé & Guarro (2002), Onygenaceae. 1, Australia. See Cano *et al.* (*Stud. Mycol.* **47**: 175, 2002).

pseudoamyloid, see dextrinoid.

pseudoangiocarpous (of a basidioma), hymenial surface at first exposed but later covered by an incurving pileus margin and/or excrescences from the stipe (Singer, 1975: 26).

Pseudoanguillospora S.H. Iqbal (1974), anamorphic *Pezizomycotina*, Hso.0fH.1/10. 2, Europe. See Iqbal (*Biologia* Lahore **20**: 11, 1974), Marvanová (*Czech Mycol.* **49**: 7, 1996).

Pseudoarachniotus Kuehn (1957) = Gymnascella fide von Arx (*Gen. Fungi Sporul. Cult.* Edn 3, 1981), Currah (*Mycotaxon* **24**: 1, 1985).

Pseudoaristastoma Suj. Singh (1979), anamorphic *Pezizomycotina*, Cpd.≡ eH.?. 1, India. See Singh (*Sydowia* **31**: 238, 1978).

Pseudoarmillariella (Singer) Singer (1956), Agaricales. 2, N. America; C. America. Belongs to *Hygrophoraceae* or *Tricholomataceae*. See Singer (*Mycol.* **48**: 725, 1956), Redhead *et al.* (*Mycotaxon* **83**: 19, 2002).

Pseudoarthonia Marchand (1896) = Arthonia fide Hawksworth *et al.* (*Dictionary of the Fungi* edn 8, 1995).

Pseudoasperisporium U. Braun (2000), anamorphic *Pezizomycotina*. 2. See Braun (*Schlechtendalia* **5**: 72, 2000), Schubert & Braun (*Fungal Diversity* **20**: 187, 2005).

Pseudoauricularia Kobayasi (1982), Agaricaceae. 1, Papua New Guinea. See Kobayasi (*TMSJ* **22**: 421, 1981).

Pseudobaeomyces M. Satô (1940), Icmadophilaceae (L). 1, Asia. See Stenroos & DePriest (*Am. J. Bot.* **85**: 1548, 1998; DNA), Stenroos *et al.* (*Mycol. Progr.* **1**: 267, 2002).

Pseudobaeospora Singer (1942), Tricholomataceae. *c.* 20, widespread. See Singer (*Mycol.* **60**: 13, 1963), Horak (*Revue Mycol.* Paris **29**: 72, 1964; key), Bas (*Persoonia* **18**: 115, 2002; Europe), Bas (*Persoonia* **18**: 163, 2003).

Pseudobalsamia E. Fisch. (1907) = Balsamia fide Trappe (*TBMS* **65**: 496, 1975), Læssøe & Hansen (*MR* **111**: 1075, 2007; phylogeny).

Pseudobasidiospora Dyko & B. Sutton (1978), anamorphic *Pezizomycotina*, Cpd.0eH.1. 1 (aquatic), Australia; USA. See Dyko & Sutton (*Nova Hedwigia* **29**: 168, 1977).

Pseudobasidium Tengwall (1924) = Arthrinium fide Ellis (*Mycol. Pap.* **103**, 1965).

Pseudobeltrania Henn. (1902), anamorphic *Pezizomycotina*, Hso.0eP.10. 7, widespread (esp. tropical). See Sutton (*TBMS* **55**: 506, 1970), Zucconi (*MR* **95**: 1017, 1991).

Pseudoboletus Šutara (1991), Boletaceae. 2 (in parasitic association with *Scleroderma* and *Pisolithus*), north temperate. See Šutara (*Česká Mykol.* **45**: 1, 1991), Šutara (*Czech Mycol.* **57**: 1, 2005; anatomical characters).

Pseudobotrytis Krzemien. & Badura (1954), anamorphic *Porosphaerella*, Hso.1eP.10. 1, widespread. See Mel'nik (*Mikol. Fitopatol.* **32**: 32, 1998), Fernández & Huhndorf (*Fungal Diversity* **17**: 11, 2004; teleomorph), Yamaguchi *et al.* (*Mycoscience* **45**: 9, 2004; chemistry).

Pseudobuellia B. de Lesd. (1907) = Rinodina fide

Hawksworth *et al.* (*Dictionary of the Fungi* edn 8, 1995).

Pseudocalicium Marchand (1896) = Chaenothecopsis fide Hawksworth *et al.* (*Dictionary of the Fungi* edn 8, 1995).

Pseudocalopadia Lücking (1999), Ectolechiaceae (L). 1, Costa Rica. See Lücking (*Phyton* Horn **39**: 142, 1999).

Pseudocamptoum Gonz. Frag. & Cif. (1925) = Melanographium fide Ellis (*Mycol. Pap.* **93**, 1963).

Pseudocanalisporium R.F. Castañeda & W.B. Kendr. (1991), anamorphic *Pezizomycotina*, Hsp.#eP.1. 1, Cuba. See Castañeda Ruiz & Kendrick (*Univ. Waterloo Biol. Ser.* **35**: 89, 1991).

pseudocapillitium (of *Mycetozoa*), a sterile structure in the fruit-body which has had no connexion with the sporogenous protoplasm.

Pseudocenangium A. Knapp (1924) = Ascocoryne fide Dennis (*Mycol. Pap.* **62**, 1956).

Pseudocenangium P. Karst. (1886), anamorphic *Pezizomycotina*, Ccu.0fH.19. 2, Europe; N. America. See Dyko & Sutton (*CJB* **57**: 370, 1979), Mel'nik *et al.* (*Mikol. Fitopatol.* **40**: 510, 2006).

Pseudocercophora Subram. & Sekar (1986), Lasiosphaeriaceae. Anamorph *Mammaria*. 1, India. See Subramanian & Sekar (*Journal of the Singapore National Academy of Science* **15**: 58, 1986).

Pseudocercospora Speg. (1910) nom. cons. prop., anamorphic *Mycosphaerella*, Hso.0fP.10. *c.* 1124, widespread (esp. tropical). See Deighton (*Mycol. Pap.* **140**, 1976), Deighton (*TBMS* **88**: 365, 1987), Guo & Liu (*Mycosystema* **2**: 225, 1989; China), Anon. (*Progr. Rep. Asian Veg. Res. Develop. Centre, 1990*: 150, 1991; in vitro data on *P. fuligena*), Guo & Hsieh (*The genus Pseudocercospora in China*, 1995), Braun (*Monogr. Cercosporella, Ramularia Allied Genera (Phytopath. Hyphom.)* **2**: 397, 1998; relat. to *Cercostigmina* and *Uwebraunia*, key to sections), Stewart *et al.* (*MR* **103**: 1491, 1999; phylogeny), Crous *et al.* (*Stud. Mycol.* **45**: 107, 2000; review), Gonzalez & Pons (*MR* **104**: 1507, 2000; ontogeny), Crous *et al.* (*Mycol.* **93**: 1081, 2001; phylogeny), Crous *et al.* (*MR* **105**: 425, 2001; on *Myrtaceae*), Beilharz *et al.* (*Mycotaxon* **82**: 397, 2002; on *Kennedia*), Braun & Dick (*N.Z. Jl For. Sci.* **32**: 221, 2002; on *Eucalyptus*), Crous & Mourichon (*Sydowia* **54**: 35, 2002; on *Musa*), Crous & Braun (*CBS Diversity Ser.* **1**: 571 pp., 2003; nomenclator), Taylor *et al.* (*MR* **107**: 653, 2003; on *Proteaceae*), Crous *et al.* (*Stud. Mycol.* **50**: 457, 2004; on *Acacia*), Ávila *et al.* (*MR* **109**: 881, 2005; on *Olea*), Braun & Crous (*Taxon* **55**: 803, 2006; nomencl.), Hunter *et al.* (*Stud. Mycol.* **55**: 147, 2006; on *Eucalyptus*).

Pseudocercosporella Deighton (1973), anamorphic *Mycosphaerella*, Hso.0fH.10. 119, widespread. See Inman *et al.* (*MR* **95**: 1334, 1991; *Mycosphaerella* teleomorph of *P. capsellae*), Braun (*Monogr. Cercosporella, Ramularia Allied Genera (Phytopath. Hyphom.)* **1**: 129, 1995; monogr., keys), Braun (*Monogr. Cercosporella, Ramularia Allied Genera (Phytopath. Hyphom.)* **2**: 405, 1998), Braun (*Schlechtendalia* **1**: 41, 1998), Braun (*Monogr. Cercosporella, Ramularia Allied Genera (Phytopath. Hyphom.)* **2**, 1998), Braun *et al.* (*Mycol. Progr.* **2**: 197, 2003), Crous & Braun (*CBS Diversity Ser.* **1**: 571 pp., 2003; nomenclator), Hunter *et al.* (*Stud. Mycol.* **55**: 147, 2006; on *Eucalyptus*).

Pseudocercosporidium Deighton (1973), anamorphic *Mycosphaerellaceae*, Hso.0fP.10. 1, Venezuela. See Deighton (*Mycol. Pap.* **133**: 55, 1973), Crous *et al.* (*Mycol.* **93**: 1081, 2001), Crous & Braun (*CBS Diversity Ser.* **1**: 571 pp., 2003).

Pseudochaete T. Wagner & M. Fisch. (2002), Hymenochaetaceae. 1, widespread. See Wagner & Fischer (*Mycol. Progr.* **1**: 100, 2002).

Pseudochaetosphaeronema Punith. (1979), anamorphic *Pleosporales*, Cpd.0eH.15. 1 (from humans), Venezuela. See Punithalingam (*Nova Hedwigia* **31**: 126, 1979), Hoog *et al.* (*Mycoses* **47**: 121, 2004; phylogeny).

Pseudochuppia Kamal, A.N. Rai & Morgan-Jones (1984), anamorphic *Pezizomycotina*, Hso.#eP.1/10. 1, India. See Kamal *et al.* (*Mycol.* **76**: 163, 1984).

Pseudociboria Kanouse (1944), Sclerotiniaceae. 1, widespread. See Dumont (*Mycol.* **66**: 706, 1974), Schumacher (*Mycotaxon* **38**: 233, 1990).

Pseudocladosporium U. Braun (1998) = Fusicladium fide Braun (*Monogr. Cercosporella, Ramularia Allied Genera (Phytopath. Hyphom.)* **2**: 392, 1998), Braun *et al.* (*Mycol. Progr.* **2**: 3, 2003; phylogeny), Schubert *et al.* (*Schlechtendalia* **9**, 2003), Beck *et al.* (*Mycol. Progr.* **4**: 111, 2005; phylogeny), Crous *et al.* (*Stud. Mycol.* **58**: 185, 2007; phylogeny).

Pseudoclathrosphaerina Voglmayr (1997), anamorphic *Pezizomycotina*, Hso.?.?. 1, USA. See Voglmayr & Krisai-Greilhuber (*Mycol.* **89**: 943, 1997).

Pseudoclathrus B. Liu & Y.S. Bau (1980), Phallaceae. 1, China. See Liu & Bau (*Mycotaxon* **10**: 293, 1980).

pseudocleistothecium, see discrete body.

Pseudoclitocybe (Singer) Singer (1956), Tricholomataceae. 10, widespread (north temperate; S. America). See Singer (*Mycol.* **48**: 725, 1956), Watling & Turnbull (*British Fungus Flora*, 1998; UK).

Pseudococcidioides O.M. Fonseca (1928) = Coccidioides fide Dodge (*Medical Mycology*, 1935).

Pseudocochliobolus Tsuda, Ueyama & Nishih. (1978) = Cochliobolus fide Alcorn (*Mycotaxon* **16**: 353, 1983), Sivanesan (*Mycol. Pap.* **158**: 261 pp., 1987).

Pseudocoelomomyces E.A. Nam & Dubitskiĭ (1977) ≡ Tabanomyces.

Pseudocollema Kanouse & A.H. Sm. (1940) = Byssonectria fide Sivertsen (*SA* **9**: 23, 1990), Pfister (*Mycol.* **85**: 952, 1993).

pseudocolumella (of *Physaraceae*), lime-knots massed like a columella at the centre of the sporangium.

Pseudocolus Lloyd (1907), Phallaceae. 2, widespread (tropical; subtropical introduced). See Dring (*Kew Bull.* **35**: 1, 1980).

Pseudoconiocybe Marchand (1896) = Coniocybe fide Hawksworth *et al.* (*Dictionary of the Fungi* edn 8, 1995).

Pseudoconium Petr. (1969), anamorphic *Pezizomycotina*, Cpd.0eP.?. 1, Europe. See Petrak (*Sydowia* **22**: 385, 1968).

Pseudoconocybe Hongo (1967) = Conocybe fide Watling (*Kew Bull.* **31**: 593, 1977).

Pseudocoprinus Kühner (1928) = Coprinellus fide Redhead *et al.* (*Mycotaxon* **50**: 203, 2001).

Pseudocordyceps Hauman (1936), anamorphic *Penicilliopsis*, Hsy.0eH.?. 1, Taiwan; Zaire. See Hsieh & Ju (*Mycol.* **94**: 539, 2002; teleomorph).

Pseudocornicularia Gyeln. (1933) = Coelocaulon fide Kärnefelt (*Op. Bot.* **86**, 1986).

pseudocortex (of lichen thalli), a false cortex, used for the outer layer of the pseudopodetia in *Pycnothelia papillaria*.

Pseudocraterellus Corner (1958) = Craterellus fide Peterson (*Česká Mykol.* **29**: 199, 1975).

Pseudocryptosporella J. Reid & C. Booth (1969), Diaporthales. 1, Venezuela. See Reid & Booth (*CJB* **47**: 1058, 1969).

pseudocyphella (pl. **-e**), an opening in the cortex of lichens where the medulla is exposed to the open air but lacking specialized cells surrounding the cavity; they provide valuable taxonomic characters in e.g. *Alectoria, Bryoria, Pseudocyphellaria*.

Pseudocyphellaria Vain. (1890) nom. cons., Lobariaceae (L). *c.* 170, widespread (esp. south temperate). See Coppins & James (*Lichenologist* **11**: 139, 1979; key Eur. spp.), Galloway & James (*Lichenologist* **12**: 291, 1980; NZ), Galloway *et al.* (*Lichenologist* **15**: 135, 1983; NZ), Kondratyuk & Galloway (*Biblthca Lichenol.* **57**: 327, 1985; lichenicolous spp.), Galloway (*Lichenologist* **18**: 105, 1986; S. Am. non-glabrous spp.), Galloway & James (*Nova Hedwigia* **42**: 423, 1986; Delise's spp.), Galloway (*Bull. Br. Mus. nat. hist.* Bot. **17**: 1, 1988; key 48 spp.), Galloway & Laundon (*Taxon* **37**: 48, 1988; gen. nomencl.), Galloway & James (*Biblthca Lichenol.* **46**, 1992; 53 S. Am. spp.), Czeczuga *et al.* (*J. Hattori bot. Lab.* **87**: 277, 1999; carotenoids), Kantvilas & Elix (*Muelleria* **12**: 217, 1999; Tasmania), Thomas *et al.* (*Biblthca Lichenol.* **82**: 123, 2001; phylogeny, New Zealand), Summerfield *et al.* (*New Phytol.* **155**: 121, 2002; photobionts), Wiklund & Wedin (*Cladistics* **19**: 419, 2003; phylogeny), Miądlikowska & Lutzoni (*Am. J. Bot.* **91**: 449, 2004; phylogeny), Wedin & Wiklund (*Symb. bot. upsal.* **34** no. 1: 469, 2004; phylogeny), Miądlikowska *et al.* (*Mycol.* **98**: 1088, 2006; phylogeny), Summerfield & Eaton-Rye (*New Phytol.* **170**: 597, 2006; N and S hemisphere spp. synonymous).

pseudocystidium (1) (in agarics), see cystidium; (2) (of *Entomophthora*), the organ penetrating the insect cuticle allowing conidiophores to emerge.

Pseudocytoplacosphaeria Punith. & Spooner (2002), anamorphic *Pezizomycotina*. 1, British Isles. See Punithalingam & Spooner (*Kew Bull.* **57**: 534, 2002).

Pseudocytospora Petr. (1923), anamorphic *Pezizomycotina*, St.0eH.?. 1, Europe.

Pseudodasyscypha Velen. (1939) = Merismodes fide Kuyper (*in litt.*).

Pseudodeconica Overeem (1927) nom. nud. = Agrocybe fide Singer (*Agaric. mod. Tax.*, 1951).

Pseudodelitschia J.N. Kapoor, Bahl & S.P. Lal (1976) = Neotestudina fide Hawksworth (*CJB* **57**: 91, 1979).

Pseudodermatosorus Vánky (1999), Doassansiaceae. 2 (on *Alismataceae*), Venezuela. See Vánky (*Mycotaxon* **71**: 213, 1999).

Pseudodescolea Raithelh. (1980) = Descolea fide Bougher (*MR* **94**: 287, 1990).

Pseudodiaporthe Speg. (1909) = Massarina fide Bose (*Phytopath. Z.* **41**, 1961), Aptroot (*Nova Hedwigia* **66**: 89, 1998).

pseudodiblastic (of ascospores), having oil-drops at the poles so that they superficially resemble polarilocular spores (q.v.).

Pseudodichomera Höhn. (1918), anamorphic *Pezizomycotina*, St.#eP.?. 3, Europe. See Arnold & Russell (*Mycol.* **52**: 509, 1960).

Pseudodictya Tehon & G.L. Stout (1929) nom. dub., anamorphic *Pezizomycotina*. See Sutton (*Mycol. Pap.* **141**, 1977).

Pseudodictyosporium Matsush. (1971), anamorphic *Massarinaceae*, Hso.#eP.24. 2, Papua New Guinea. See Matsushima (*Bull. natn. Sci. Mus.* Tokyo, B **14**: 473, 1971), Tsui *et al.* (*Fungal Diversity* **21**: 157, 2006; phylogeny), Cai *et al.* (*Persoonia* **20**: 53, 2008; phylogeny).

Pseudodidymaria U. Braun (1993), anamorphic *Pezizomycotina*, Hsp.1eP.10/12. 2, USA; Japan. See Braun (*Monogr. Cercosporella, Ramularia Allied Genera (Phytopath. Hyphom.)* **2**, 1998), Ono & Kobayashi (*Mycoscience* **46**: 352, 2005).

Pseudodidymella C.Z. Wei, Y. Harada & Katum. (1997), Pleosporales. Anamorph *Pycnopleiospora*. 1 (on *Fagus*), Japan. See Wei *et al.* (*Mycol.* **89**: 494, 1997).

Pseudodidymium R. Michel, Walochnik & Aspöck (2003), Didymiaceae. 1. See Michel *et al.* (*Acta Protozool.* **42**: 331, 2003).

Pseudodimerium Petr. (1924) = Phaeostigme fide Hansford (*Mycol. Pap.* **15**, 1946).

Pseudodiplodia (P. Karst.) Sacc. (1884), anamorphic *Pezizomycotina*, Cpd.1eP.15. *c.* 45, widespread. See Buchanan (*Mycol. Pap.* **156**, 1987), Dianese & Furlanetto (*Fitopatol. Brasil* **18** (Supl.): 344, 1993; Brazil).

Pseudodiplodia Speg. (1920) [non *Pseudodiplodia* (P. Karst) Sacc. 1884] = Botryodiplodia fide Petrak & Sydow (*Feddes Repert.* **42**: 1, 1927).

Pseudodiplodiella Bender (1932) = Botryodiplodia fide Sutton (*Mycol. Pap.* **141**, 1977).

Pseudodiscinella Dennis (1956) ≡ Ciliatula.

Pseudodiscosia Hösterm. & Laubert (1921) = Heteropatella fide Sutton (*Mycol. Pap.* **141**, 1977), Leuchtmann (*Mycotaxon* **28**: 261, 1987).

Pseudodiscula Laubert (1911), anamorphic *Pezizomycotina*, St.0eH.?. 2, Europe. See Sutton (*Mycol. Pap.* **141**, 1977).

Pseudodiscus Arx & E. Müll. (1959), Saccardiaceae. 1, Europe. See von Arx & Müller (*Sydowia* **13**: 64, 1959).

Pseudodoassansia (Setch.) Vánky (1981), Doassansiaceae. 2 (on *Hydrocleys, Sagittaria*), N. & S. America. See Vánky (*Mycotaxon* **78**: 265, 2001).

Pseudoecteinomyces W. Rossi (1977), Euceratomycetaceae. 1, Africa; Asia. See Rossi (*Mycol.* **69**: 1075, 1977), Tavares (*Mycol. Mem.* **9**: 627 pp., 1985).

Pseudoendococcus Marchand (1896) = Endococcus fide Hawksworth *et al.* (*Dictionary of the Fungi* edn 8, 1995).

Pseudoepicoccum M.B. Ellis (1971), anamorphic *Pezizomycotina*, Hsp.0eP.10. 1, widespread (tropical). See Ellis (*Dematiaceous Hyphomycetes*: 270, 1971).

pseudoepithecium, an amorphous or granular layer overlying paraphyses in an apothecium and in which their tips are immersed, but not forming a separate tissue.

Pseudofarinaceus Battarra ex Kuntze (1891) ≡ Amanitopsis.

Pseudofarinaceus Earle (1909) = Volvariella fide Kuyper (*in litt.*).

Pseudofavolus Pat. (1900), Polyporaceae. 4, widespread (pantropical). See Núñez & Ryvarden (*Syn. Fung.* **10**: 68, 1995).

pseudofissitunicate (of asci), see ascus.

Pseudofistulina O. Fidalgo & M. Fidalgo (1963), Fistulinaceae. 1, Brazil. See Gilbertson & Ryvarden (*N. Amer. Polyp.*: 261, 1987; syn. of *Fistulina*).

Pseudofomes Lázaro Ibiza (1916) = Phellinus fide

Donk (*Verh. K. ned. Akad. Wet.* tweede sect. **62**: 1, 1974).

Pseudofumago Briosi & Farneti (1906) nom. dub., anamorphic *Pezizomycotina*. See de Hoog & Hermanides-Nijhof (*Stud. Mycol.* **15**: 186, 1977).

Pseudofungi. A subdivision in the *Chromista*; including the fungi treated as belonging to the oomycetes in a broad sense, and as *Oomycota*, *Hyphochytriomycota*, *Labyrinthulales* and *Thraustochytriales* in this edition of the *Dictionary*; see Cavalier-Smith (*Progr. Phycol. Res.* **4**: 309, 1986), Pseudomycotina.

Pseudofusarium Matsush. (1971) = Fusarium fide Booth & Sutton (*in litt.*).

Pseudofuscophialis Sivan. & H.S. Chang (1995), anamorphic *Pezizomycotina*, Hso.≡ eP.15. 1, Taiwan. See Sivanesan & Chang (*MR* **99**: 711, 1995).

Pseudofusicoccum Mohali, Slippers & M.J. Wingf. (2006), anamorphic *Botryosphaeriaceae*. 1, Venezuela. See Mohali *et al.* (*Stud. Mycol.* **55**: 249, 2006).

Pseudofusidium Deighton (1969) = Acremonium fide Gams (*Cephalosporium-artige Schimmelpilze*, 1971).

Pseudogaster Höhn. (1907), anamorphic *Pezizomycotina*, Hsy.0eP.?. 1 (on bark), Brazil. See Seifert (*TBMS* **85**: 123, 1985).

Pseudogenea Buchholz (1901) = Genabea fide Trappe (*TBMS* **65**: 496, 1975), Læssøe & Hansen (*MR* **111**: 1075, 2007; phylogeny).

Pseudogibellula Samson & H.C. Evans (1973), anamorphic *Cordycipitaceae*, Hsy.0eH.32/33. 1, Ghana. See Samson & Evans (*Acta Bot. Neerl.* **22**: 524, 1973), Evans (*Trichomycetes and Other Fungal Groups* Robert W. Lichwardt Commemoration Volume: 119, 2001; review).

Pseudogliomastix W. Gams (1985), anamorphic *Wallrothiella*, Hso.0eP.15. 1, Italy. See Gams (*Proc. Indian Acad. Sci.* Pl. Sci. **94**: 279, 1985), Gams (*Stud. Mycol.* **45**: 187, 2000), Mostert *et al.* (*Stud. Mycol.* **54**: 115 pp., 2006; key).

Pseudogliophragma Phadke & V.G. Rao (1980), anamorphic *Pezizomycotina*, Hsy.0eH.10. 1, India. See Phadke & Rao (*Norw. Jl Bot.* **27**: 127, 1980).

Pseudogloeosporium Jacz. (1917) = Kabatia fide Sutton (*Mycol. Pap.* **141**, 1977).

Pseudogomphus R. Heim (1970), Gomphaceae. 1, Gabon. See Heim (*Revue Mycol.* Paris **34**: 344, 1969).

Pseudographiella E.F. Morris (1966), anamorphic *Pezizomycotina*, Hsy.1eH.?. 3, widespread. See Morris (*Mycopath. Mycol. appl.* **28**: 97, 1966), Illman *et al.* (*Mycol.* **77**: 662, 1985).

Pseudographis Nyl. (1855) nom. cons., Triblidiaceae. 3, widespread (temperate). See Magnes (*Biblthca Mycol.* **165**, 1997).

Pseudographium Höhn. (1915), anamorphic *Pezizomycotina*, Hsy.?.?. 1, USA. See Sutton (*Mycol. Pap.* **141**, 1977).

Pseudographium Jacz. (1898) ≡ Sphaerographium fide Sutton (*Mycol. Pap.* **141**, 1977).

Pseudoguignardia Gutner (1927) = Physalospora fide Müller & von Arx (*Beitr. Kryptfl. Schweiz* **11** no. 2, 1962).

Pseudogyalecta Vězda (1975) nom. rej. = Badimia fide Lücking & Vězda (*Taxon* **44**: 227, 1995).

Pseudogymnoascus Raillo (1929), Pseudeurotiaceae. Anamorph *Geomyces*. 7, widespread. See Samson (*Acta Bot. Neerl.* **21**: 517, 1972), Orr (*Mycotaxon* **8**: 165, 1979), Ito & Yokoyama (*IFO Res. Comm.* **13**: 83, 1987), Sogonov *et al.* (*Mycol.* **97**: 695, 2005; phylogeny), Jiang & Yao (*Mycotaxon* **94**: 55, 2005; phylogeny, ontogeny), Rice & Currah (*Mycol.* **98**: 307, 2006; phylogeny), Wang *et al.* (*Mycol.* **98**: 1065, 2006; phylogeny).

Pseudogymnopilus Raithelh. (1974), ? Strophariaceae. 1, S. America. See Raithelhuber (*Hong. Argentin.* **1**: 148, 1974).

Pseudogyrodon Heinem. & Rammeloo (1983) = Gyrodon fide Kuyper (*in litt.*).

Pseudohalonectria Minoura & T. Muroi (1978), Magnaporthaceae. 9 (on wood, freshwater), widespread. See Shearer (*CJB* **67**: 194, 1989; key, posn), Chen *et al.* (*Mycol.* **91**: 84, 1999; phylogeny), Promputtha *et al.* (*Cryptog. Mycol.* **25**: 43, 2004; Thailand), Shenoy *et al.* (*Cryptog. Mycol.* **26**: 123, 2005).

Pseudohansenula E.K. Novák & Zsolt (1961) nom. dub., Fungi. See Lodder (*Yeasts, a taxonomic study* 2nd edn: 236, 1970).

Pseudohansenula Mogi (1939) nom. dub., Fungi.

Pseudohansfordia G.R.W. Arnold (1970), anamorphic *Pezizomycotina*, Hso.1-≡ eH.10. 15 (on fungi), widespread. See de Hoog (*Persoonia* **10**: 57, 1978), Eicker *et al.* (*Bot. Bull. Acad. Sin.* **31**: 205, 1990; disease of *Auricularia mesenterica*).

Pseudohansfordia S.M. Reddy & Bilgrami (1975), anamorphic *Pezizomycotina*, Hso.0eH.?. 2, India. See de Hoog (*Persoonia* **10**: 58, 1978).

Pseudohaplis Clem. & Shear (1931) = Botryodiplodia fide Sutton (*Mycol. Pap.* **141**, 1977).

Pseudohaplosporella, see *Pseudhaplosporella*.

Pseudoharpella Ferrington, M.M. White & Lichtw. (2003), Harpellales. 1, USA. See Ferrington *et al.* (*Aquatic Insects* **25**: 86, 2003), White (*MR* **110**: 1011, 2006; phylogeny).

Pseudohelicomyces Garnica & E. Valenz. (2000), anamorphic *Pezizomycotina*. 2, Chile. See Valenzuela & Garnica (*MR* **104**: 739, 2000).

Pseudohelotium Fuckel (1870), Helotiaceae. *c.* 50, Europe. See Arendholz (*Mycotaxon* **36**: 283, 1989; nomencl.).

Pseudohendersonia Crous & M.E. Palm (1999), anamorphic *Pezizomycotina*, Cpd.?.?. 1, S. Africa. See Crous & Palm (*MR* **103**: 1302, 1999), Crous *et al.* (*CBS Diversity Ser.* **2**, 2004).

Pseudohepatica P.M. Jørg. (1993), Pezizomycotina (L). 1 (ascomata unknown), Venezuela. See Jørgensen (*Bryologist* **96**: 435, 1993), Marcano *et al.* (*Trop. Bryol.* **18**: 203, 2000).

Pseudoheppia Zahlbr. (1903), ? Lichinaceae (L). 1, Europe.

pseudoheterothallism, see heterothallism.

Pseudohiatula (Singer) Singer (1938), Tricholomataceae. *c.* 5, widespread (tropical). See Singer (*Persoonia* **2**: 407, 1962).

Pseudohiatulaceae Grgur. (2000) = Tricholomataceae.

Pseudohydnotrya E. Fisch. (1897) = Geopora fide Fischer (*Nat. Pflanzenfam.* **5b**: viii, 1938), Læssøe & Hansen (*MR* **111**: 1075, 2007; phylogeny).

Pseudohydnum P. Karst. (1868), Auriculariales. 1, widespread. See Donk (*Persoonia* **3**: 199, 1964), Hjortstam *et al.* (*Kew Bull.* **45**: 319, 1990), Roberts & Spooner (*Kew Bull.* **53**: 631, 1998; Brunei Darussalam).

Pseudohydnum Rick (1904) = Hydnodon fide Donk (*Taxon* **5**: 69, 1956).

Pseudohygrocybe (Bon) Kovalenko (1988) = Hygrocybe fide Kuyper (*in litt.*).

Pseudohygrophorus Velen. (1939) nom. dub.,

Tricholomataceae. 1, Europe.

Pseudohypocrea Yoshim. Doi (1972), Hypocreaceae. Anamorph *Acremonium*-like. 1 (on decaying wood and bark), N. America. See Rossman *et al.* (*Stud. Mycol.* **42**: 248 pp., 1999).

pseudoidia, separated hyphal cells able to be germinated (Bensaude).

Pseudoidium Y.S. Paul & J.N. Kapoor (1986) = Oidium Link (1824) fide Sutton (*in litt.*), Takamatsu (*Mycoscience* **45**: 147, 2004).

Pseudoinonotus T. Wagner & M. Fisch. (2001), Hymenochaetaceae. 4, widespread. See Wagner & Fischer (*MR* **105**: 773, 2001).

pseudoisidium (pl. **-ia**), (1) an outgrowth from the surface of a lichen thallus resembling an isidium (e.g. *Gyalideopsis*; see Vězda, *Folia geobot. phytotax.* **14**: 48, 1979); (2) isidium without photosynthetic cells in *Pseudocyphellaria* (see Galloway, *Bull. Br. Mus. nat. Hist., Bot.* **17**: 1, 1988).

Pseudokarschia Velen. (1934) = Dactylospora fide Hafellner (*Beih. Nova Hedwigia* **62**, 1979).

Pseudolachnea Ranoj. (1910), anamorphic *Pezizomycotina*, Cac/Ccu.1-≡ eH.15. 5, widespread. See Sutton (*The Coelomycetes*, 1980).

Pseudolachnea Velen. (1934) [non *Pseudolachnea* Ranoj. 1910] ? = Hyalopeziza fide Eckblad (*Nytt Mag. Bot.* **15**: 174, 1978), Huhtinen (*Öst. Z. Pilzk.* **10**: 1, 2001; redispositions).

Pseudolachnella Teng (1936) = Pseudolachnea Ranoj. fide Sutton (*Mycol. Pap.* **141**, 1977).

Pseudolachnum Velen. (1934), ? Helotiales. 1, former Czechoslovakia.

Pseudolagarobasidium J.C. Jang & T. Chen (1985), Phanerochaetaceae. 2, widespread. See Stalpers (*Folia Cryptog. Estonica* **33**: 133, 1998).

Pseudolasiobolus Agerer (1983), Tricholomataceae. 1, tropical. See Agerer (*Mitt. bot. StSamml.* Münch. **19**: 279, 1983).

Pseudolecanactis Zahlbr. (1907), ? Roccellaceae (L). 1, Samoa.

Pseudolecidea Clauzade & Cl. Roux (1984) ≡ Claurouxia.

Pseudolecidea Marchand (1896) = Abrothallus fide Hawksworth *et al.* (*Dictionary of the Fungi* edn 8, 1995).

Pseudolembosia Theiss. (1913), Parmulariaceae. 2, Australia; C. America. See Petrak (*Sydowia* **8**: 297, 1954), Inácio & Cannon (*MR* **107**: 82, 2003), Inácio *et al.* (*Mycol. Progr.* **4**: 133, 2005).

Pseudoleptogium Jatta (1900) = Polychidium fide Hawksworth *et al.* (*Dictionary of the Fungi* edn 8, 1995).

Pseudoleptogium Müll. Arg. (1885) = Leptogium fide Degelius (*Svensk bot. Tidskr.* **37**: 65, 1943).

Pseudolignincola Chatmala & E.B.G. Jones (2006), Halosphaeriaceae. Anamorph *Humicola*-like. 1 (marine), Thailand. See Chatmala & Jones (*Nova Hedwigia* **83**: 225, 2006).

Pseudolizonia Pirotta (1889) = Lizonia fide von Arx & Müller (*Stud. Mycol.* **9**, 1975).

Pseudoloma J.L. Matthews, A.M.V. Brown, K. Larison, J.K. Bishop-Stewart, P. Rogers & M.L. Kent (2001), Microsporidia. 1. See Matthews *et al.* (*J. Eukary. Microbiol.* **48**: 227, 2001).

Pseudolycoperdon Velen. (1947) = Bovista fide Kreisel (*Beih. Nova Hedwigia* **25**: 63, 1967).

Pseudolyophyllum (Singer) Raithelh. (1979) = Clitocybe fide Kuyper (*in litt.*).

Pseudomassaria Jacz. (1894), Hyponectriaceae. Anamorph *Beltraniella*. 20 (from dead plant tissues), widespread. See Barr (*Mycol.* **56**: 841, 1964; keys), Barr (*Mycol.* **68**: 611, 1976; nomencl.), Hyde *et al.* (*Sydowia* **50**: 21, 1998), Wang & Hyde (*Fungal Diversity* **3**: 159, 1999).

Pseudomassariella Petr. (1955) = Leiosphaerella fide Müller & von Arx (*Beitr. Kryptfl. Schweiz* **11** no. 2, 1962).

Pseudombrophila Boud. (1885), Pyronemataceae. 28 (esp. coprophilous), widespread. See van Brummelen (*Libri Botanici* **14**, 1995; monogr., key), Hansen & Pfister (*Mycol.* **98**: 1029, 2006; phylogeny), Hansen *et al.* (*Mycol.* **97**: 1023, 2005; phylogeny), Yao & Spooner (*Fungal Diversity* **22**: 267, 2006), Perry *et al.* (*MR* **111**: 549, 2007; phylogeny).

Pseudomelasmia Henn. (1902) = Phyllachora Nitschke ex Fuckel (1870) fide Höhnel (*Sber. Akad. Wiss. Wien* Math.-naturw. Kl., Abt. 1 **119**: 54, 1910).

Pseudomeliola Speg. (1889), ? Parodiellaceae. 4, America (tropical). Poorly known and heterogenous. See von Arx (*Acta Bot. Neerl.* **7**: 503, 1958), Cannon (*Mycol. Pap.* **163**: 302 pp., 1991).

Pseudomeria G.L. Barron (1980), anamorphic *Clavicipitaceae*, Hso.0eH.10. 1 (on nematodes), Canada. See Barron (*CJB* **58**: 443, 1980).

Pseudomerulius Jülich (1979), Tapinellaceae. 3, widespread. See Jülich (*Persoonia* **10**: 350, 1979).

Pseudomicrocera Petch (1921) = Fusarium fide Wollenweber & Reinking (*Die Fusarien*: 7, 1935).

Pseudomicrodochium B. Sutton (1975), anamorphic *Hypocreales*, Hso.1-≡ eH.15. 9, widespread. See Sutton *et al.* (*Mycopathologia* **114**: 159, 1991; from humans), Castañeda Ruíz *et al.* (*Mycotaxon* **68**: 23, 1998), Decock *et al.* (*Antonie van Leeuwenhoek* **84**: 209, 2003; phylogeny).

Pseudomitrula Gamundí (1980), Helotiaceae. 1, S. America. See Gamundí (*Fl. criptog. Tierra del Fuego* **10**, 1987).

pseudomixis (**-gamy**), the type of fertilization in which the copulating elements are not special sexual cells.

Pseudomonilia A. Geiger (1910) nom. rej. = Candida fide Hawksworth *et al.* (*Dictionary of the Fungi* edn 8, 1995).

Pseudomorfea Punith. (1981), Dothideomycetes. Anamorph *Chaetasbolisia*. 1, India; Argentina. See Venedikian (*Boln Soc. argent. Bot.* **25**: 495, 1988).

pseudomorph, a stroma made up of plant parts kept together by plectenchyma.

pseudomycelium (of *Candida*, etc.), loosely united, catenulate groups of cells (see Zobl, *RAM* **23**: 177, 1944).

Pseudomycena Cejp (1929) = Mycena fide Smith (*North American Species of Mycena*, 1947).

Pseudomycoderma H. Will (1916), anamorphic *Saccharomycetales*, Hso.0eH.?. 1, Europe.

Pseudomycoporon Marchand (1896) ? = Mycoporum Flot. ex Nyl. fide Hawksworth *et al.* (*Dictionary of the Fungi* edn 8, 1995).

pseudomycorrhiza, see mycorrhiza.

Pseudomycotina, see mycorrhiza.

Pseudonaevia Dennis & Spooner (1993), Helotiales. 1, British Isles. See Dennis & Spooner (*Persoonia* **15**: 177, 1993).

Pseudonectria Seaver (1909), Nectriaceae. Anamorph *Volutella*. 5 (on leaves and twigs), Europe; America. See Rossman *et al.* (*Mycol.* **85**: 685, 1993; key), Rossman *et al.* (*Stud. Mycol.* **42**: 248 pp., 1999),

Zhang *et al.* (*Mycol.* **98**: 1076, 2006; phylogeny).

Pseudonectriella Petr. (1959) = Catabotrys fide Rossman *et al.* (*Stud. Mycol.* **42**: 248 pp., 1999).

Pseudoneottiospora Faurel & Schotter (1965), anamorphic *Pezizomycotina*, Cpd.1bH.19. 2 (coprophilous), Algeria; Italy. See Faurel & Schotter (*Revue Mycol.* Paris **29**: 278, 1964).

Pseudoneurospora Dania García, Stchigel & Guarro (2004), Sordariaceae. 1, Thailand. See García *et al.* (*MR* **108**: 1139, 2004).

Pseudoniptera Velen. (1947) = Hymenoscyphus fide Gminder (*Czech Mycol.* **58**: 125, 2006).

Pseudonitschkia Coppins & S.Y. Kondr. (1995), ? Dacampiaceae. 1 (on *Parmotrema*), widespread. See Coppins & Kondratyuk (*Edinb. J. Bot.* **52**: 229, 1995).

Pseudonocardia Henssen (1957), Actinobacteria. q.v.

Pseudonosema E.U. Canninga, D. Refardtb, C.R. Vossbrinckc, B. Okamurad & A. Curry (2002), Microsporidia. 1. See Canninga *et al.* (*Eur. J. Protist.* **38**: 247, 2002).

Pseudoolla Velen. (1934) = Olla fide Baral (*SA* **13**: 113, 1994).

Pseudoomphalina (Singer) Singer (1956), Tricholomataceae. *c.* 5, widespread (north temperate). See Singer (*Mycol.* **48**: 725, 1956), Watling & Turnbull (*British Fungus Flora*, 1998; UK), Contu (*Micologia e Vegetazione Mediterranea* **18**: 61, 2003; Sardinia).

pseudooperculate (of asci), ones which are essentially unitunicate in structure and with a thickened apical cap which splits completely away at discharge (e.g. *Odontotrema*); see ascus.

Pseudopannaria (B. de Lesd.) Zahlbr. (1924), ? Lecideaceae (L). 1, France.

Pseudopapulaspora N.D. Sharma (1977), anamorphic *Pezizomycotina*, Hso.#eP.1. 1, India. See Sharma (*J. Indian bot. Soc.* **56**: 100, 1977).

pseudoparaphyses (of ascomycetes), see hamathecium; (of basidiomycetes), see hyphidium.

pseudoparenchyma, see plectenchyma.

Pseudoparmelia Lynge (1914), Parmeliaceae (L). 27, pantropical. See Hale (*Smithson. Contr. bot.* **31**, 1976), Hale (*Mycotaxon* **25**: 603, 1986), Elix & Nash (*Bryologist* **100**: 482, 1998).

Pseudoparodia Theiss. & Syd. (1917), Venturiaceae. 1 (on *Vaccinium*), S. America. See Petrak (*Sydowia* **1**: 169, 1947).

Pseudoparodiella F. Stevens (1927), Venturiaceae. Anamorph *Spilodochium*. 1, Costa Rica. See Sivanesan (*TBMS* **86**: 187, 1986).

Pseudopatella Sacc. (1884) = Cystotricha fide Höhnel (*Sber. Akad. Wiss. Wien* Math.-naturw. Kl., Abt. 1 **119**: 617, 1910).

Pseudopatella Speg. (1891) = Botryodiplodia fide Petrak & Sydow (*Feddes Repert.* **42**: 1, 1927).

Pseudopatellina Höhn. (1908), anamorphic *Pezizomycotina*, St.0eH.?. 1, Europe.

Pseudopaulia M. Schultz (2002), Lichinaceae. 1, Socotra. See Schultz (*Mycotaxon* **82**: 446, 2002), Schultz & Büdel (*Lichenologist* **35**: 151, 2003; phylogeny).

Pseudopeltis L. Holm & K. Holm (1978), ? Helotiales. 1 (on *Dryopteris*), Sweden. See Holm & Holm (*Bot. Notiser* **131**: 97, 1978).

Pseudopeltistroma Katum. (1975), anamorphic *Pezizomycotina*, Cpt.0eP.?. 1, Japan. See Katumoto (*Bulletin of the Faculty of Agriculture, Yamaguchi University* **26**: 99, 1975).

Pseudopeltula Henssen (1995), Gloeoheppiaceae (L). 3, Mexico; Caribbean. See Henssen (*Lichenologist* **27**: 261, 1995), Schultz (*Bryologist* **110**: 286, 2007; USA).

pseudoperidium, a false peridium; covering membrane of the aecium in the *Pucciniales*.

Pseudoperis Clem. & Shear (1931) ≡ Pseudoperisporium.

Pseudoperisporiaceae Toro (1926), ? Dothideomycetes (inc. sed.). 25 gen. (+ 39 syn.), 228 spp.

Lit.: Farr (*Mycol.* **55**: 226, 1963; on *Pinaceae*), Farr (*Mycol.* **58**: 221, 1966; on *Gramineae*), Farr (*Mycol.* **71**: 243, 1979; on *Asteraceae*), Farr (*Mycol.* **76**: 793, 1984; on *Rubiaceae*), Farr (*Sydowia* **38**: 65, 1985), Döbbeler (*Arctic and Alpine Mycology* II: 87, 1987), Sivanesan (*SA* **6**: 201, 1987), Barr (*Mycotaxon* **64**: 149, 1997), Döbbeler (*Biodiv. Cons.* **6**: 721, 1997), Valinsky *et al.* (*Appl. Environm. Microbiol.* **68**: 5999, 2002), Lumbsch *et al.* (*Mol. Phylogen. Evol.* **34**: 512, 2005).

Pseudoperisporium Toro (1926) = Lasiostemma fide Farr (*Mycol.* **71**: 250, 1979).

Pseudoperitheca Elenkin (1922), Pezizomycotina (L). 1, former USSR.

pseudoperithecium (of *Laboulbeniales*), a perithecium-like structure in which the asci and spores become free.

Pseudopestalotia Elenkin & Ohl (1912) = Truncatella fide Steyaert (*Bull. Jard. Bot. État* **19**: 285, 1949), Sutton (*Mycol. Pap.* **141**, 1977).

Pseudopetrakia M.B. Ellis (1971), anamorphic *Pezizomycotina*, Hsy.#eP.10. 1, India. See Ellis (*Mycol. Pap.* **125**: 3, 1971), Sutton & Alcorn (*Proc. R. Soc. Qd.* **95**: 41, 1984).

Pseudopezicula Korf (1986), Helotiaceae. Anamorph *Phialophora*-like. 2, Europe; N. America. See Korf *et al.* (*Mycotaxon* **26**: 457, 1986), Pearson *et al.* (*Pl. Dis.* **72**: 796, 1988; disease), Reiss & Zinkernagel (*Z. PflKrankh. PflPath. PflSchutz* **101**: 212, 1994; heterothallism), Verkley (*Stud. Mycol.* **44**: 180 pp., 1999).

Pseudopeziza Fuckel (1870), ? Dermateaceae. Anamorph *Gloeosporidiella*. 3, widespread. *P. trifolii* (clover leaf spot). See Schüepp (*Phytopath. Z.* **36**: 224, 1959; key), Thite & Nagaraja (*Geophytology* **20**: 17, 1990; ontogeny), Reiss & Zinkernagel (*Z. PflKrankh. PflPath. PflSchutz* **101**: 212, 1994; heterothallism), Nauta & Spooner (*Mycologist* **14**: 65, 2000; key British spp.).

Pseudopezizites Fiore (1932), Fossil Fungi. 1 (Eocene), Italy.

Pseudophacidiaceae Rehm (1887) ? = Ascodichaenaceae.

Pseudophacidium P. Karst. (1885), ? Ascodichaenaceae. 5, widespread. See Minter *in* Capretti *et al.* (Eds) (*Shoot and Foliage Diseases in Forest Trees*: 65, 1995), Yuan *et al.* (*Australas. Pl. Path.* **29**: 215, 2000).

Pseudophaeolus Ryvarden (1975) = Laetiporus fide Westhuizen (*Bothalia* **11**: 143, 1973), Ofosu-Asiedu (*TBMS* **65**: 285, 1975), Ryvarden (*Gen. Polyp.*: 135, 1991) See.

Pseudophaeoramularia U. Braun (1997) = Pseudocercospora fide Braun & Melnik (*Trudy Botanicheskogo Instituta im. V.L. Komarova* **20**: 18, 1997), Crous & Braun (*CBS Diversity Ser.* **1**: 571 pp., 2003).

Pseudophaeotrichum Aue, E. Müll. & C. Stoll (1969) = Neotestudina fide Hawksworth (*CJB* **57**: 91, 1979).

pseudophialide, a cell bearing a sporangiolum in the

Kickxellaceae.

Pseudophloeosporella U. Braun (1993), anamorphic *Pezizomycotina*, Cac.≡ eH.10. 1, Japan. See Braun (*Cryptog. bot.* **4**: 110, 1993).

Pseudophoma Höhn. (1916) = Chaetosphaeronema fide Petrak (*Annls mycol.* **42**: 58, 1944).

Pseudophomopsis Höhn. (1926) = Phomopsis (Sacc.) Bubák. fide Clements & Shear (*Gen. Fung.*, 1931).

Pseudophragmotrichum W.P. Wu, B. Sutton & Gange (1998), anamorphic *Pezizomycotina*, Cpd.?.?. 1, Cuba. See Wu *et al.* (*MR* **102**: 179, 1998).

Pseudophyllachora Speg. (1919) nom. dub., Fungi. See Petrak (*Sydowia* **5**: 350, 1951).

Pseudophysalospora Höhn. (1918) = Physalospora fide Barr (*Mycol.* **68**, 1976).

Pseudophyscia Müll. Arg. (1894) ≡ Heterodermia.

Pseudophysciaceae Tomas. (1949) = Physciaceae.

pseudophyse, see cystidium.

pseudophysis, see hyphidium.

Pseudopileum Canter (1963), Chytridiaceae. 1 (on *Mallomonas*), British Isles. See Canter (*TBMS* **46**: 309, 1963).

pseudopionnotes, see sporodochium.

Pseudopiptoporus Ryvarden (1980), Polyporaceae. 2, E. Africa; India. See Ryvarden & Johansen (*Prelim. Polyp. Fl. E. Afr.*: 523, 1980), Decock & Ryvarden (*Nova Hedwigia* **77**: 199, 2003).

Pseudopithyella Seaver (1928), Sarcoscyphaceae. 2, widespread. See Dissing & Raitviir (*Eesti NSV Tead. Akad. Toim.* Biol. seer **23**: 104, 1974), Donadini *et al.* (*Cryptog. Mycol.* **10**: 283, 1989; asci), Harrington *et al.* (*Mycol.* **91**: 41, 1999; phylogeny), Hansen & Pfister (*Mycol.* **98**: 1029, 2006; phylogeny).

Pseudoplacodium Motyka (1996) nom. inval. = Protoparmelia fide Hawksworth *et al.* (*Dictionary of the Fungi* edn 8, 1995).

pseudoplasmodium, see plasmodium.

Pseudoplea Höhn. (1918) = Leptosphaerulina fide Barr (*Contr. Univ. Mich. Herb.* **9**, 1972).

Pseudoplectania Fuckel (1870), Sarcosomataceae. 3 or 4, Europe; N. America. See Donadini (*Mycol. Helv.* **2**: 217, 1987; 4 spp.), Li & Kimbrough (*CJB* **73**: 1761, 1995; ultrastr.), Landvik *et al.* (*Nordic Jl Bot.* **17**: 403, 1997; DNA), Harrington *et al.* (*Mycol.* **91**: 41, 1999; phylogeny), Hansen & Pfister (*Mycol.* **98**: 1029, 2006; phylogeny), Perry *et al.* (*MR* **111**: 549, 2007; phylogeny).

Pseudopleistophora Sprague (1977), Microsporidia. 1.

Pseudopleospora Petr. (1920), Dothideomycetes. 3, Europe. See Crivelli (*Über die heterogene Ascomycetengattung Pleospora Rbh.*, 1983; key).

pseudopodetium, a lichenized podetium-like structure of vegetative origin, ascogonia arising on this not on the pre-formed granular or squamulose thallus initials (e.g. *Cladia*, *Stereocaulon*).

pseudopodium (of *Mycetozoa*), a protoplasmic process from a myxamoeba or plasmodium.

Pseudopolyporus Hollick (1910), Fossil Fungi ? Basidiomycota. 1 (Carboniferous), USA.

Pseudopolystigmina Murashk. (1928), anamorphic *Pezizomycotina*, St.0eH.?. 1, Russia.

Pseudoproboscispora Punith. (1999), Annulatascaceae. 1 (on submerged wood), widespread. See Wong & Hyde (*MR* **103**: 81, 1999), Campbell *et al.* (*Mycol.* **95**: 41, 2003; phylogeny).

Pseudoprotomyces Gibelli (1874) = Phloeoconis fide Saccardo (*Syll. fung.* **14**: 1197, 1899).

Pseudopuccinia Höhn. (1925) = Stigmina fide Laundon (*Mycol. Pap.* **99**: 14, 1965).

pseudopycnidium (obsol.), a pycnidium-like structure of hyphal tissue, as in certain anamorphic fungi.

Pseudopyrenula Müll. Arg. (1883), Trypetheliaceae (±L). 10, widespread (tropical). See Riedl (*Sydowia* **16**: 215, 1963), Aptroot (*Nova Hedwigia* **66**: 89, 1998), Aptroot (*Trop. Bryol.* **14**: 25, 1998), Harris (*Lichenogr. Thomsoniana*: 133, 1998), Prado *et al.* (*MR* **110**: 511, 2006; phylogeny), Aptroot *et al.* (*Biblthca Lichenol.* **97**, 2008; Costa Rica).

Pseudopythium Sideris (1930) nom. nud., Fungi. See Merlich (*Mycol.* **24**: 453, 1932).

Pseudoramonia Kantvilas & Vězda (2000), Thelotremataceae. 2, Australia; Venezuela. See Kantvilas & Vězda (*Lichenologist* **32**: 343, 2000).

Pseudoramularia Matsush. (1983), anamorphic *Pezizomycotina*, Hso.≡ eH.3. 2, Uganda; Pacific islands. See Matsushima (*Matsush. Mycol. Mem.* **3**: 13, 1983).

Pseudorbilia Ying Zhang, Z.F. Yu, Baral & K.Q. Zhang (2007), Orbiliaceae. 1, China. See Zhang *et al.* (*Fungal Diversity* **26**: 305, 2007).

pseudorhiza, a rooting base, as in *Collybia radicata* (Buller, **4**).

Pseudorhizina Jacz. (1913), Discinaceae. 2, N. America. See Harmaja (*Karstenia* **13**: 48, 1973), Abbott & Currah (*Mycotaxon* **62**: 1, 1997; N. Am.), O'Donnell *et al.* (*Mycol.* **89**: 48, 1997; phylogeny), Holec & Beran (*Czech Mycol.* **59**: 51, 2007; Europe).

Pseudorhizinaceae Harmaja (1974) = Discinaceae.

Pseudorhizopogon Kobayasi (1983), anamorphic *Pezizomycotina*, Cpd.0eH.?23. 1, Japan. See Kobayasi (*J. Jap. Bot.* **58**: 175, 1983).

Pseudorhynchia Höhn. (1909) = Trichosphaeria fide Samuels & Barr (*CJB* **75**: 2165, 1998), Dulymamode *et al.* (*MR* **105**: 247, 2001).

Pseudorhytisma Juel (1895), Rhytismataceae. 1, widespread. See Schüepp (*Phytopath. Z.* **36**: 262, 1959).

Pseudorobillarda M. Morelet (1968), anamorphic *Pezizomycotina*, Cpd.≡ eH.15. 11, widespread. See Nag Raj (*Coelomycetous Anamorphs with Appendage-bearing Conidia*, 1993), Bianchinotti (*MR* **101**: 1233, 1997), Vujanovic & St-Arnaud (*Mycol.* **95**: 955, 2003; key), Plaingam *et al.* (*Nova Hedwigia* **80**: 335, 2005; Thailand).

Pseudorobillarda Nag Raj, Morgan-Jones & W.B. Kendr. (1972) = Pseudorobillarda M. Morelet fide Nag Raj *et al.* (*CJB* **51**: 688, 1973).

Pseudosaccharomyces Briosi & Farneti (1906), anamorphic *Pezizomycotina*, HSo.0eH.?. *c.* 20, Italy.

Pseudosaccharomyces Klöcker (1912) [non *Pseudosaccharomyces* Briosi & Farneti 1906] ≡ Kloeckera fide Cadez *et al.* (*FEMS Yeast Res.* **1**: 279, 2002).

Pseudosagedia (Müll. Arg.) M. Choisy (1949), Porinaceae (L). 1, widespread. See Hafellner & Kalb (*Biblthca Lichenol.* **57**: 161, 1995; status), McCarthy & Malcolm (*Lichenologist* **29**: 1, 1997), Harris (*Opuscula Philolichenum* **2**: 15, 2005; N America).

Pseudosarcophoma Urries (1952) ? = Selenophoma fide Sutton (*Mycol. Pap.* **141**, 1977).

Pseudoschizothyra Punith. (1980), anamorphic *Pezizomycotina*, Cpt.1eH.15. 1, Myanmar. See Punithalingam (*Nova Hedwigia* **31**: 890, 1979).

Pseudosclerophoma Petr. (1923) = Phoma Sacc. fide Sutton (*Mycol. Pap.* **141**, 1977).

pseudosclerotium, a compacted mass of intermixed substratum (soil, stones, etc.) held together by mycelium, as in *Polyporus tuberaster* (see stone-fungus;

also zone lines).

Pseudoscypha J. Reid & Piroz. (1966), ? Hysteriaceae. 1 (on *Abies*), Canada. See Reid & Pirozynski (*CJB* **44**: 351, 1966).

Pseudoseptoria Speg. (1910), anamorphic *Pezizomycotina*, Cpd.0eH.19. 7 (on *Poaceae*), widespread (temperate). See Sutton (*The Coelomycetes*, 1980).

pseudoseptum (1) (obsol.) a protoplasmic or vacuolar membrane looking like a septum (= distoseptum, distoseptate) as in *Corynespora*; (2) (of *Blastocladiales*), a septum having pores.

pseudosetae (false setae), the upturned free-ends of context hyphae in the hymenium of *Duportella*.

Pseudosigmoidea K. Ando & N. Nakam. (2000), anamorphic *Pezizomycotina*. 1, USA. See Ando & Nakamura (*J. gen. appl. Microbiol.* Tokyo **46**: 55, 2000).

Pseudosolidum Lloyd (1923), Ascoporiaceae. Anamorph *Plectophomella*-like. 1, S. America. See Samuels & Romero (*Bol. Mus. Paraense Emílio Goeldi* ser. bot. **7**: 263, 1991; as *Ascoporia*), Kutorga & Hawksworth (*SA* **15**: 1, 1997; as *Ascoporia*), Rossman *et al.* (*Stud. Mycol.* **42**: 248 pp., 1999).

Pseudosphaerella Höhn. (1911) = Microcyclus fide von Arx & Müller (*Stud. Mycol.* **9**, 1975).

Pseudosphaeria Höhn. (1907) = Wettsteinina fide Müller & von Arx (*Beitr. Kryptfl. Schweiz* **11** no. 2, 1962), Shoemaker & Babcock (*CJB* **65**: 373, 1987).

Pseudosphaeriales = Pleosporales.

Pseudosphaerialites Venkatach. & R.K. Kar (1969), Fossil Fungi ? Ascomycota. 1 (Tertiary), India.

Pseudosphaerophorus M. Satô (1968) = Bunodophoron fide Wedin (*Pl. Syst. Evol.* **187**: 213, 1993).

Pseudospiropes M.B. Ellis (1971), anamorphic *Strossmayeria*, Hso.≡ eP.10. 26, widespread. See Ellis (*Dematiaceous Hyphomycetes*, 1971), Ellis (*More Dematiaceous Hyphomycetes*, 1976), Hyde & Goh (*MR* **102**: 739, 1998).

Pseudospora Schiffn. (1931), Algae.

pseudospore (1) (of *Acrasiales*), an encysted myxamoeba; (2) (of *Ustilaginales*), a basidiospore (obsol.); (3) a chlamydospore, as in *Rhizoctonia rubi*.

Pseudostegia Bubák (1906), anamorphic *Pezizomycotina*, Cac.0eH.15. 1, N. America.

pseudostem (of gasteromycete basidiomata), spongy tissue in which hyphae are not orientated parallel to the stipe axis (Dring, 1973).

Pseudostemphylium (Wiltshire) Subram. (1961) = Ulocladium fide Simmons (*Mycol.* **59**: 80, 1967).

Pseudostictis Lambotte (1887) = Cryptodiscus fide Sherwood (*Mycotaxon* **6**: 215, 1977).

Pseudostilbella Munt.-Cvetk. & Gómez-Bolea (1995), anamorphic *Pezizomycotina*. 1, Morocco. See Muntañola-Cvetkovic & Gómez-Bolea (*CJB* **73**: 591, 1995).

Pseudostracoderma A.E. Martinez & Godeas (1997), anamorphic *Pezizomycotina*, Hso.?.?. 1. See Martinez & Godeas (*Mycotaxon* **65**: 455, 1997).

pseudostroma, a stroma formed of thalline tissue and remnants of host tissue (see Eriksson, *Opera bot.* **60**: 14, 1981), an aggregation of perithecial ascomata into a pustule some partly of bark cells altered by the fungus (see Johnson, *Ann. Mo bot. Gdn* **27**: 31, 1940), a coelomycetous conidioma of fungal and host tissue (see Sutton, *The Coelomycetes*, 1980), Fig. 10S). cf. substroma.

Pseudostypella McNabb (1969), Auriculariaceae. 1, New Zealand. See McNabb (*N.Z. Jl Bot.* **7**: 259, 1969).

Pseudotaeniolina J.L. Crane & Schokn. (1986), anamorphic *Capnodiales*, Hso.0-1eP.3. 1, Iran. See Crane & Schoknecht (*Mycol.* **78**: 88, 1986), De Leo *et al.* (*Antonie van Leeuwenhoek* **83**: 351, 2003), Kurzai *et al.* (*Mycoses* **46**: 141, 2003), Selbmann *et al.* (*Stud. Mycol.* **51**: 1, 2005; phylogeny).

Pseudotapesia Velen. (1939), ? Helotiales. 1, former Czechoslovakia.

pseudothecium (1) an ascostromatic ascoma having asci in numerous unwalled locules, as in loculoascomycetes; cf. euthecium; (2) a protoperithecium.

Pseudothelephora Lloyd (1919) = Thelephora fide Donk (*Taxon* **6**: 17, 1957).

Pseudothiella Petr. (1928), Phyllachoraceae. Anamorph *Pseudothiopsella*. 1 (from living leaves), Brazil. See Hyde (*TMSJ* **32**: 265, 1991).

Pseudothiopsella Petr. (1928), anamorphic *Pseudothiella*, St.0eH.?. 1, Brazil.

Pseudothis Theiss. & Syd. (1914), Diaporthales. 1, America (tropical). See Cannon (*Fungal Diversity* **7**: 17, 2001).

Pseudothyridaria Petr. (1925) = Valsaria fide Clements & Shear (*Gen. Fung.*, 1931), Ju *et al.* (*Mycotaxon* **58**: 419, 1996).

Pseudothyrium Höhn. (1927), anamorphic *Pezizomycotina*, St.0eH.15. 1, Europe.

Pseudotis (Boud.) Boud. (1907) = Otidea fide Eckblad (*Nytt Mag. Bot.* **15**: 1, 1968).

Pseudotomentella Svrček (1958), Thelephoraceae. 12, widespread. See Larsen (*Mycol.* **66**: 167, 1974; key), Stalpers (*Stud. Mycol.* **35**: 49, 1993; key), Kõljalg (*Syn. Fung.* **9**: 41, 1996; key eurasian spp.), Martini & Hentic (*BSMF* **119**: 19, 2003; France).

Pseudotorula Subram. (1958), anamorphic *Pezizomycotina*, Hso.≡ fP.3. 1, India. See Subramanian (*J. Indian bot. Soc.* **37**: 58, 1958).

Pseudotracya Vánky (1999), Doassansiaceae. 1 (on *Ottelia* (*Hydrocharatiaceae*)), Australia. See Vánky (*Mycotaxon* **71**: 216, 1999).

Pseudotracylla B. Sutton & Hodges (1976), anamorphic *Pezizomycotina*, Cpt.0eH.15. 2, Brazil; USA. See Carris (*Mycol.* **84**: 534, 1992), Carris (*Mycotaxon* **50**: 93, 1994).

Pseudotrametes Bondartsev & Singer (1944) = Trametes fide Donk (*Verh. K. ned. Akad. Wet.* tweede sect. **62**: 1, 1974).

Pseudotremellodendron D.A. Reid (1957) = Tremellodendropsis fide Corner (*TBMS* **49**: 205, 1966).

Pseudotrichia Kirschst. (1939), Pleosporales. 6, widespread (north temperate). See Barr (*Mycotaxon* **20**: 1, 1984), Huhndorf (*Mycol.* **86**: 134, 1994; key 4 spp. neotropics), Liew *et al.* (*Mol. Phylogen. Evol.* **16**: 392, 2000; phylogeny), Câmara *et al.* (*MR* **107**: 516, 2003; phylogeny).

Pseudotrichoconis W.A. Baker & Morgan-Jones (2001), anamorphic *Pezizomycotina*. 1, Europe. See Baker *et al.* (*Mycotaxon* **79**: 367, 2001).

Pseudotrochila Höhn. (1917), ? Rhytismatales. 1, Java.

Pseudotryblidium Rehm (1890), Helotiales. 1, Europe.

Pseudotrype Henn. (1900) = Eutypella fide Höhnel (*Sber. Akad. Wiss. Wien* Math.-naturw. Kl., Abt. 1 **119**: 926, 1910).

Pseudotthia Henn. (1900) = Gibbera fide von Arx & Müller (*Stud. Mycol.* **9**, 1975).

Pseudotulasnella Lowy (1964), Tulasnellaceae. 1, Guatemala. See Lowy (*Mycol.* **56**: 696, 1964).

Pseudotulostoma O.K. Mill. & T. Henkel (2001),

Elaphomycetaceae. 2, Japan; Guyana. See Miller *et al.* (*MR* **105**: 1269, 2001), Asai *et al.* (*Bull. natn. Sci. Mus.* Tokyo, B **30**: 1, 2004), Geiser *et al.* (*Mycol.* **98**: 1053, 2006; phylogeny), Henkel *et al.* (*Mycorrhiza* **16**: 241, 2006; mycorrhiza).

Pseudotyphula Corner (1953), Marasmiaceae. 1, N. America. See Berthier (*Biblthca Mycol.* **98**: 89, 1985).

Pseudovalsa Ces. & De Not. (1863), Pseudovalsaceae. Anamorph *Coryneum*. 4, widespread (north temperate). See Wehmeyer (*Revision of Melanconis*, 1941), Barr (*Mycol. Mem.* **7**, 1978).

Pseudovalsaceae M.E. Barr (1978), Diaporthales. 3 gen. (+ 2 syn.), 26 spp. Much reduced in diversity compared with its traditional circumscription. See Wehmeyer (*Revision of Melanconis*, 1941), Barr (*Mycol. Mem.* **7**, 1978), Castlebury *et al.* (*Mycol.* **94**: 1017, 2002), Rossman *et al.* (*Mycoscience* **48**: 135, 2007; phylogeny).

Pseudovalsaria Spooner (1986), Boliniaceae. 4, Europe; China. See Spooner (*TBMS* **86**: 401, 1986), Barr (*Mycotaxon* **46**: 45, 1993), Barr (*Mycotaxon* **51**: 191, 1994; posn), Andersson *et al.* (*SA* **14**: 1, 1995; posn), Kang *et al.* (*Mycoscience* **40**: 151, 1999; posn).

Pseudovalsella Höhn. (1918), Pseudovalsaceae. Anamorph *Hendersonula*. 1 (from bark), widespread. See Barr (*Mycol. Mem.* **7**, 1978).

Pseudovirgaria H.D. Shin, U. Braun, Arzanlou & Crous (2007), ? Capnodiales. 1 (on rusts), Korea. See Arzanlou *et al.* (*Stud. Mycol.* **58**: 57, 2007).

Pseudovularia Speg. (1910) = Ramularia Unger fide Deighton (*TBMS* **59**: 419, 1972).

Pseudoxenasma K.H. Larss. & Hjortstam (1976), Russulaceae. 1, Europe. See Larsson & Hjortstam (*Mycotaxon* **4**: 307, 1976), Larsson & Larsson (*Mycol.* **95**: 1037, 2003; phylogeny).

Pseudoxylaria Boedijn (1959) = Xylaria Hill ex Schrank fide Læssøe (*SA* **13**: 43, 1994).

Pseudoyuconia Lar.N. Vassiljeva (1983), Pleosporaceae. 1, Germany. See Ahn & Shearer (*CJB* **73**: 573, 1995; as *Barrella*).

Pseudozyma Bandoni (1985), anamorphic *Ustilaginaceae*. 9, Canada; China. See Hajlaoui *et al.* (*Phytopathology* **82**: 583, 1992; biocontrol, as *Stephanoascus*), Boekhout (*J. gen. appl. Microbiol.* Tokyo **41**: 359, 1995; taxonomy), Hoegh *et al.* (*CJB* **73**: 869, 1995; lipases), Dik *et al.* (*Eur. J. Pl. Path.* **104**: 413, 1998; biocontrol), Kurtzman & Fell (*Yeasts, a taxonomic study* 4th edn, 1998), Fell *et al. in* McLaughlin *et al.* (Eds) (*The Mycota* A Comprehensive Treatise on Fungi as Experimental Systems for Basic and Applied Research **7B**: 3, 2001; China), Wang *et al.* (*Int. J. Syst. Evol. Microbiol.* **56**: 289, 2006; China).

Pseudozythia Höhn. (1902), anamorphic *Pezizomycotina*, Cpd.0eH.?. 1, Europe.

Psiammopomopiospora Locq. & Sal.-Cheb. (1980), Fossil Fungi. 3, Cameroon.

Psiamspora Locq. & Sal.-Cheb. (1980), Fossil Fungi. 1, Cameroon.

Psidimobipiospora Locq. & Sal.-Cheb. (1980), Fossil Fungi. 4, Cameroon.

Psilachnum Höhn. (1926), Hyaloscyphaceae. 27, widespread. See Dennis (*Persoonia* **2**: 171, 1962), Raitviir (*Scripta Mycol.* **1**: 1, 1970; key), Sharma (*Nova Hedwigia* **46**: 369, 1988; 4 spp. on *Pteridophyta*), Galán & Raitviir (*Mycotaxon* **72**: 163, 1999), Raitviir (*Scripta Mycol.* **20**, 2004).

Psilainaperturites Y.K. Mathur (1966), Fossil Fungi. 1, India.

Psiloboletinus Singer (1945), Suillaceae. 1, Asia (temperate). See Smith (*Mycol.* **58**: 332, 1966).

Psilobotrys Sacc. (1879) = Chloridium fide Hughes (*CJB* **36**: 727, 1958).

psilocin, a hallucinatory indole derivative from *Psilocybe mexicana* (Hoffman *et al.*, *Experientia* **14**: 11, 397, 1958). See also psilocybin, hallucinogenic fungi.

Psilocistella Svrček (1977), Hyaloscyphaceae. 10, Europe. See Huhtinen (*Karstenia* **29**: 45, 1990), Raitviir (*Scripta Mycol.* **20**, 2004).

Psilocybe (Fr.) P. Kumm. (1871), Strophariaceae. *c.* 300, widespread. Several species of subgen. *Psilocybe* hallucinogenic, producing psilocybin, psilocin and baeocystin (Stamets, *Psilocybin mushrooms of the world*, 1996; Guzmán, *International Journal of Medicinal Mushrooms* **7**: 410, 2005); the edible *P. rugoso-annulata* ('strophaire') cultivated in Eur. In the present edition the genus is very widely circumscribed (Smith, *Taxon* **28**: 19, 1979; Kühner, *Les Hyménomycètes agaricoïdes*: 249, 1980; Noordeloos, *Persoonia* **16**: 127, 1995), whereas Singer (*Agaricales mod. Taxon.*, 4th ed., 1986) recognized a number of smaller segregate genera (*Hypholoma*, *Melanotus*, *Psilocybe*, *Stropharia*). Recognition of these segregate genera is unsatisfactory, a number of species having been transferred from one genus to another and moved back again, depending on the significance attached on various characters. Recognition of one broadly defined genus is not entirely satisfactory, and the older genera are still maintained at subgeneric or sectional level. Moncalvo *et al.* (*Mol. Phylog. Evol.* **23**: 357, 2002) distributed the taxon over at least 8 clades, and 6 of these were also recognised by Bridge *et al.* (*Mycotaxon* **103**: 109, 2008). The taxon in a broad sense also includes the gasteroid genera *Leratiomyces* and *Weraroa*. See Guzmán (*Beih. Nova Hedwigia* **74**, 1983, monograph Psilocybe, *Bibl. Mycol.* **159**: 91, 1995; suppl. to monogr., *International Journal of Medicinal Mushrooms* **7**: 305, 2005), Noordeloos (*Öst. Z. PIzk.* **10**: 115, 2001; sect. *Psilocybe*), Boekhout *et al.* (*MR* **106**: 1251, 2002; taxonomy sect. *Psilocybe*). See also Hallucinogenic Fungi.

Psilocybe Fayod (1889) = Panaeolina fide Singer & Smith (*Mycol.* **38**: 287, 1946).

psilocybin, a hallucinatory indole derivative from *Psilocybe mexicana*. See Bazanté (*Rev. Mycol.* **36**: 25, 1971; action). See also psilocin, hallucinogenic fungi.

Psilodiporites C.P. Varma & Rawat (1963), Fossil Fungi. 7, Cameroon; India.

Psiloglonium Höhn. (1918) = Glonium fide von Arx & Müller (*Stud. Mycol.* **9**, 1975).

Psilolechia A. Massal. (1860), Lecanorales (L). 4, widespread. See Coppins & Purvis (*Lichenologist* **19**: 29, 1987; key), Yoshimura & Harada (*Lichenology* **3**: 41, 2004; Japan), Andersen & Ekman (*MR* **109**: 21, 2005; phylogeny).

Psilonia Fr. (1825) ? = Volutella fide Hawksworth *et al.* (*Dictionary of the Fungi* edn 8, 1995).

Psiloniaceae Corda (1837) = Nectriaceae.

Psiloniella Costantin (1888) = Catenularia fide Mason (*Mycol. Pap.* **5**: 120, 1941).

Psiloparmelia Hale (1989), Parmeliaceae. 12, S. America; S. Africa. See Elix & Nash (*Bryologist* **95**: 377, 1992; key), Lumbsch *et al.* (*Mycotaxon* **45**: 489, 1992; anatomy), Thell *et al.* (*J. Hattori bot. Lab.* **100**:

797, 2006; phylogeny), Crespo *et al.* (*Mol. Phylogen. Evol.* **44**: 812, 2007; phylogeny).

Psilopezia Berk. (1847), Pezizales. 5, widespread. See Pfister (*Am. J. Bot.* **60**: 355, 1973; key), Zhuang (*Mycotaxon* **61**: 3, 1997), Hansen & Pfister (*Mycol.* **98**: 1029, 2006; phylogeny), Perry *et al.* (*MR* **111**: 549, 2007; phylogeny).

Psilophana Syd. (1939), ? Helotiales. 1, Ecuador.

Psilosphaeria Cooke (1879), Pezizomycotina. 1. The genus needs typification.

Psilospora Rabenh. (1856) ≡ Polymorphum fide Hawksworth & Punithalingam (*TBMS* **60**: 501, 1973).

Psilosporina Died. (1913) ≡ Polymorphum fide Hawksworth & Punithalingam (*TBMS* **60**: 501, 1973).

Psilothecium Clem. (1903), ? Helotiales. 1, N. America.

Psilothecium Fuckel (1866) nom. dub., Pezizomycotina. '*Fusarium*-like'. See Sutton (*Mycol. Pap.* **141**, 1977), Crous & Braun (*CBS Diversity Ser.* **1**: 571 pp., 2003).

Psora Hoffm. (1789) nom. rej. = Physcia fide Hawksworth & Sherwood (*Taxon* **30**: 338, 1981).

Psora Hoffm. (1796) nom. cons., Psoraceae (L). 31, widespread. See Schneider (*Biblthca Lichenol.* **13**, 1980; key), Timdal (*Nordic Jl Bot.* **4**: 525, 1984; gen. concept), Timdal (*Bryologist* **89**: 253, 1987; key 18 spp. N. Am.), Kainz & Rambold (*Biblthca Lichenol.* **88**: 267, 2004; phylogeny), Miądlikowska *et al.* (*Mycol.* **98**: 1088, 2006; phylogeny).

Psora Link (1833) = Toninia fide Hawksworth *et al.* (*Dictionary of the Fungi* edn 8, 1995).

Psoraceae Zahlbr. (1898), Lecanorales (L). 6 gen. (+ 5 syn.), 51 spp.

Lit.: Brusse (*Lichenologist* **17**: 267, 1985), Timdal (*Bryologist* **89**: 253, 1986), Hertel & Rambold (*Pl. Syst. Evol.* **158**: 289, 1988), Pant (*J. Bombay nat. Hist. Soc.* **85**: 658, 1988), Lumbsch & Kothe (*Nova Hedwigia* **57**: 19, 1993), Timdal (*Cryptog. Bryol.-Lichénol.* **15**: 171, 1994), Buschbom & Mueller (*Mol. Phylogen. Evol.* **32**: 66, 2004), Peršoh *et al.* (*Mycol. Progr.* **3**: 103, 2004), Miądlikowska *et al.* (*Mycol.* **98**: 1088, 2006; phylogeny), Hofstetter *et al.* (*Mol. Phylogen. Evol.* **44**: 412, 2007; phylogeny).

Psorella Müll. Arg. (1894) ? = Bacidia fide Brako (*Mycotaxon* **35**: 1, 1989).

Psorinia Gotth. Schneid. (1980), Lecanoraceae (L). 2, Europe; former USSR. See Schneider (*Biblthca Lichenol.* **13**: 128, 1979).

Psoroglaena Müll. Arg. (1891), Verrucariaceae (L). 9, widespread. See Eriksson (*SA* **11**: 11, 1992), Eriksson & Hawksworth (*SA* **14**: 65, 1995), Henssen (*Biblthca Lichenol.* **57**: 161, 1995; key), Grube (*Nova Hedwigia* **68**: 241, 1999; asci), Harada (*Lichenology* **2**: 5, 2003; Japan), Aptroot *et al.* (*Biblthca Lichenol.* **97**, 2008; Costa Rica).

Psoroglaenomyces Cif. & Tomas. (1953) ≡ Psoroglaena.

Psorographis Clem. (1909) = Acanthothecis fide Hawksworth *et al.* (*Dictionary of the Fungi* edn 8, 1995).

Psoroma Ach. ex Michx. (1803), Pannariaceae (L). *c.* 58, widespread (esp. south temperate). See Jørgensen (*Op. bot. Soc. bot. Lund* **45**, 1978), Quilhot *et al.* (*J. Nat. Prod.* **52**: 191, 1989; chemistry), Jørgensen & Wedin (*Lichenologist* **31**: 341, 1999; cephalodiate spp.), Jørgensen (*Bryologist* **103**: 670, 2000; N America), Ekman & Jørgensen (*CJB* **80**: 625, 2002; phylogeny), Wiklund & Wedin (*Cladistics* **19**: 419, 2003; phylogeny), Jørgensen (*Biblthca Lichenol.* **88**: 229, 2004; S hemisphere), Miądlikowska & Lutzoni (*Am. J. Bot.* **91**: 449, 2004; phylogeny), Carballal Durán & López de Silanes Vázqez (*Cryptog. Mycol.* **27**: 69, 2006; Iberian peninsula), Miądlikowska *et al.* (*Mycol.* **98**: 1088, 2006; phylogeny).

Psoromaria Nyl. ex Hue (1891) = Psoromidium fide Galloway & James (*Lichenologist* **17**: 173, 1985).

Psoromatomyces Cif. & Tomas. (1953) ≡ Psoroma.

Psoromella Gyeln. (1940), Parmeliaceae (L). 1, Argentina.

Psoromidium Stirt. (1877), Pannariaceae (L). 2, S. America; Australasia. See Galloway & James (*Lichenologist* **17**: 173, 1985; key), Ekman & Jørgensen (*CJB* **80**: 625, 2002; phylogeny), Jørgensen (*Biblthca Lichenol.* **88**: 229, 2004).

Psoromopsis Nyl. (1863) ? = Phyllopsora fide Hawksworth *et al.* (*Dictionary of the Fungi* edn 8, 1995).

Psoropsis Nyl. ex Zwackh (1883) nom. inval. = Porocyphus fide Hawksworth *et al.* (*Dictionary of the Fungi* edn 8, 1995).

Psorotheciella Sacc. & P. Syd. (1902) = Asterothyrium Müll. Arg. fide Hawksworth *et al.* (*Dictionary of the Fungi* edn 8, 1995).

Psorotheciopsis Rehm (1900), Asterothyriaceae (L). 5, pantropical. See Santesson (*Symb. bot. upsal.* **12** no. 1: 1, 1952), Lücking (*Willdenowia* **29**: 299, 1999; Ecuador), Henssen & Lücking (*Ann. bot. fenn.* **39**: 273, 2002; morphology, anatomy, ontogeny).

Psorothecium A. Massal. (1860) nom. dub., Fungi. See Sipman (*Willdenowia* **15**: 557, 1986).

Psorotichia A. Massal. (1855), Lichinaceae (L). *c.* 52, widespread. See Moreno & Egea (*Bull. Soc. linn. Provence* **45**: 291, 1994), Schultz & Büdel (*Lichenologist* **34**: 39, 2002; key), Schultz & Büdel (*Lichenologist* **35**: 151, 2003; phylogeny), Schultz (*Bryologist* **110**: 286, 2007; USA).

Psorotichiella Werner (1955), ? Lecanorales (L). 1, Lebanon. See Werner (*Bull. Soc. bot. Fr.* **102**: 350, 1955).

Psorotichiomyces Cif. & Tomas. (1953) ≡ Psorotichia.

Psorula Gotth. Schneid. (1980), ? Psoraceae (L). 3, widespread. See Pietschmann (*Nova Hedwigia* **51**: 521, 1990; posn).

Psorulaceae Hafellner (1984) = Psoraceae.

psychrophile, see thermophily.

psychrotolerant, growing at temperatures below 10°C (opt. below 20°C).

Psyllidomyces Buchner (1912), ? Saccharomycetales. 1 (in insects), Europe.

psylocybin, see psilocybin.

Ptechetelium Oberw. & Bandoni (1984), Eocronartiaceae. 1, Ecuador. See Oberwinkler & Bandoni (*TBMS* **83**: 645, 1984).

pterate, having wings; alate.

Pteridiosperma J.C. Krug & Jeng (1979), Ceratostomataceae. 1, Japan. See Krug & Jeng (*Mycotaxon* **10**: 41, 1979).

Pteridiospora Penz. & Sacc. (1897), Dothideomycetes. 4, widespread. See Filer (*Mycol.* **61**: 167, 1969).

Pteridomyces Jülich (1979), Atheliaceae. 7. See Jülich (*Persoonia* **10**: 331, 1979).

Pteroconium Sacc. ex Grove (1914), anamorphic *Apiospora*, Hsp.0eP.37. 3, widespread.

Pterodinia Chevall. (1837) = Botrytis fide Hawksworth *et al.* (*Dictionary of the Fungi* edn 8, 1995).

Pteromaktron Whisler (1963), Legeriomycetaceae. 1

(in *Ephemeroptera*), USA. See Lichtwardt (*The Trichomycetes. Fungal associates of arthropods*, 1986).

Pteromyces E. Bommer, M. Rousseau & Sacc. (1906), ? Helotiales. 1, Europe.

Pteromycula P.F. Cannon (1997), Dothideomycetes. 1, Great Britain. See Cannon (*SA* **15**: 121, 1997).

Pterophyllus Lév. (1844) nom. rej. = Pleurotus fide Stalpers (*in litt.*).

Pteropus R.W. Ham (2005), Fossil Fungi, Pezizomycotina. 1 (Maastrichtian), Belgium. See van der Ham & Dortangs (*Review of Palaeobotany and Palynology* **136**: 60, 2005).

Pterospora Métrod (1949) [non *Pterospora* Nutt. 1818, *Pyrolaceae*] ≡ Tetrapyrgos.

Pterosporidium W.H. Ho & K.D. Hyde (1996), Phyllachoraceae. 2 (living mangrove leaves), widespread. See Ho & Hyde (*CJB* **74**: 1826, 1996).

Pterula Fr. (1832), Pterulaceae. *ca* 50, widespread (esp. tropical). See Corner (*Ann. Bot. Mem.* **1**: 394, 1950; *Deflexula*), Pine *et al.* (*Mycol.* **91**: 944, 1999; phylogeny), Roberts (*Kew Bull.* **54**: 517, 1999; Cameroon), Munkacsi *et al.* (*Proc. R. Soc. Lond.* B **271**: 177, 2004; phylogeny, co-evolution with fungus-growing ants).

Pterulaceae Corner (1970), Agaricales. 12 gen. (+ 6 syn.), 99 spp.

Lit.: Petersen (*Bull. N.Z. Dept. Sci. Industr. Res., Pl. Dis. Div.* **263**: 143, 1988), Perez-Moreno & Villarreal (*Micol. Neotrop. Aplic.* **2**: 123, 1989), Cripps & Caesar (*Mycotaxon* **69**: 153, 1998), Pine *et al.* (*Mycol.* **91**: 944, 1999), Roberts (*Kew Bull.* **54**: 517, 1999), Roberts & Spooner (*Kew Bull.* **55**: 843, 2000), Munkacsi *et al.* (*Proc. R. Soc. Lond.* B. Biol. Sci. **271**: 1777, 2004).

Pterulicium Corner (1950), Pterulaceae. 1, S.E. Asia. See Corner (*Ann. Bot. Mem.* [A monograph of Clavaria and allied genera] **1**: 689, 1950).

Pterulopsis Wakef. & Hansf. (1943), anamorphic *Pezizomycotina*, Hsy.0fH.?. 1, Uganda.

Pterygellaceae Jülich (1982) = Cantharellaceae.

Pterygellus Corner (1966), Cantharellaceae. 5, Asia (tropical). See Corner (*Monogr. Cantharelloid Fungi*: 166, 1966).

Pterygiopsidomyces Cif. & Tomas. (1953) ≡ Pterygiopsis.

Pterygiopsis Vain. (1890), Lichinaceae (L). 12, widespread (arid regions). See Henssen (*Symb. bot. upsal.* **18** no. 1, 1963), Jørgensen (*Lichenologist* **22**: 213, 1990), Moreno & Egea (*Bull. Soc. linn. Provence* **45**: 291, 1994), Schultz & Büdel (*Lichenologist* **34**: 39, 2002; key), Schultz & Büdel (*Lichenologist* **35**: 151, 2003; phylogeny), Schultz (*Biblthca Lichenol.* **88**: 555, 2004; Yemen), Schultz (*Bryologist* **109**: 68, 2006; USA, Mexico).

Pterygium Nyl. (1854) = Placynthium fide Hawksworth *et al.* (*Dictionary of the Fungi* edn 8, 1995).

Pterygosporopsis P.M. Kirk (1983), anamorphic *Pezizomycotina*, Hso.0eP.4. 1, British Isles. See Kirk (*Mycotaxon* **18**: 285, 1983).

Ptychella Roze & Boud. (1879), Bolbitiaceae. 1, Europe. See Singer (*Agaric. mod. Tax.* 4th ed: 851, 1986; possible teratological variant of *Agrocybe*).

Ptychogaster Corda (1838), anamorphic *Oligoporus*. 6, widespread. See Donk (*Verh. K. ned. Akad. Wet.* tweede sect. **62**: 1, 1974), Sigler & Carmichael (*Mycotaxon* **4**: 394, 1976), Stalpers (*Karstenia* **40**: 167, 2000).

Ptychographa Nyl. (1874), Agyriaceae (L). 2, Europe. See Redinger (*Rabenh. Krypt.-Fl.* **9** 2.1: 217, 1938), McCune (*Bryologist* **100**: 239, 1997; N America), Schmitt *et al.* (*Mycol.* **95**: 827, 2003; phylogeny), Baloch & Grube (*MR* **110**: 125, 2006; phylogeny), Lumbsch *et al.* (*MR* **111**: 1133, 2007).

Ptychographomyces Cif. & Tomas. (1953) ≡ Ptychographa.

Ptychopeltis Syd. (1927) = Calothyriopsis fide Müller & von Arx (*Beitr. Kryptfl. Schweiz* **11** no. 2, 1962), von Arx & Müller (*Stud. Mycol.* **9**, 1975).

Ptychoverpa Boud. (1907) = Verpa fide Eckblad (*Nytt Mag. Bot.* **15**: 1, 1968).

ptyophagous (of endotrophic mycorrhiza), the young hyphae rupturing and extruding plasmal masses (**ptyosomes**) which are digested by the host cells (Burgeff, 1924); **tolypophagous**, the penetrating hyphae killed and digested by the host (Burgeff, 1924); **thamnisophagous**, forming haustorial arbuscles which are finally digested by the host (Burgeff, 1938). Cf. halmophagous.

pubescent, having soft hairs.

Pubigera Baral, Gminder & Svrček (1995) nom. inval., Hyaloscyphaceae. 1, Europe. See Baral *et al.* (*Docums Mycol.* **25** nos 98-100: 47, 1995).

Puccinella Fuckel (1860) [non *Puccinella* Parl. 1848) nom. cons., *Gramineae*] = Uromyces fide Dietel (*Nat. Pflanzenfam.* **6**, 1928).

Puccinia Pers. (1794), Pucciniaceae. *c.* 4000 (on angiosperms, 1 on ferns), widespread. Heteroecious or autoecious, macro- or micro-cyclic; teliospores 2-celled, though in some species (e.g. *P. heterospora* on *Malvaceae*) most of the spores are 1-celled (mesospores), or spores several-celled by transverse septa ('*Rostrupia*'). There are important pathogens of cereals (*P. hordei*) (barley, brown or leaf rust; I on *Ornithogalum*), *P. coronata* (oat, crown rust; I on *Rhamnus*), *P. striiformis* (syn. *P. glumarum*) (yellow or stripe rust), *P. graminis* (black or stem rust; I on *Berberis*), *P. recondita* (syn. *P. dispersa*) (rye, brown, or leaf rust; I on *Anchusa*) and *P. perplexans* f.sp. *triticina* (syn. *P. triticina*) (wheat, brown, or leaf rust; I on *Isopyrum*); *P. polysora* (maize rust), *P. kuehnii*, *P. melanocephata* (sugarcane), groundnut (*P. arachidis*), sunflower (*P. helianthi*), and a great number of other crop plants. See Lehmann *et al.* (*Der Schwarzrost*, 1937; *P. graminis*), Chester (*Cereal rusts* , 1946; *P. recondita*), Cammack (*TBMS* **41**: 89, 1958; *P. polysora*), Cummins (*Rust fungi of cereals, grasses and bamboos*, 1971), Johnson *et al.* (*TBMS* **58**: 475, 1972; *P. striiformis*, physiologic races), Urban (*Česká Mykol.* **28**: 80, 1974; *P. recondita*), Cummins (*Rust fungi on legumes and composites in North America*, 1978), Bushnell & Roelfs (Eds) (*The cereal rusts* **1** Origin, specificity, structure and physiology, 1984), Bushnell & Roelfs (Eds) (*The cereal rusts* **2** Diseases, distribution, epidemiology, control, 1985), Mordue (*TBMS* **84**: 758, 1985; *P. kuehnii*, *P. melanocephala*, grasses), Urban & Marková (*Acta Univ. Carol. Biol.* **37**: 93, 1994; *P. coronata*), Buriticá & Pardo-Cardona (*Revista de la Academia Colombiana de Ciencias Exactas, Fisicas y Naturales* **20**: 183, 1996; Colombia), Swann *et al.* (*The Mycota, A Comprehensive Treatise on Fungi as Experimental Systems for Basic and Applied Research* **7** B: 37, 2001), Virtudazo *et al.* (*Mycoscience* **42**: 167, 2001; sugarcane), Berndt (*Frontiers in Basidiomycote Mycology*: 185, 2004; Costa Rican),

Wingfield *et al.* (*Australasian Plant Pathology* **33**: 327-335, 2004; phylogeny), Helfer (*Nova Hedwigia* **81**: 325, 2005; on *Rosaceae* in Europe), Hennen *et al.* (*Catalogue of the Species of Plant Rust Fungi (Uredinales) of Brazil*: 490 pp., 2005), Aime (*Mycoscience* **47**: 112, 2006; phylogeny) Probably not separable from *Uromyces*.

Pucciniaceae Chevall. (1826), Pucciniales. 20 gen. (+ 50 syn.), *c.* 4938 spp.

Lit.: Lee & Kakishima (*Mycoscience* **40**: 109, 1999), Virtudazo *et al.* (*J. Gen. Pl. Path.* **67**: 28, 2001), Ono (*Mycoscience* **43**: 421, 2002), Cummins & Hiratsuka (*Illustr. Gen. Rust Fungi* edn 3: 225 pp., 2003), Maier *et al.* (*CJB* **81**: 12, 2003), Weber *et al.* (*MR* **107**: 15, 2003), Anikster *et al.* (*Phytopathology* **94**: 569, 2004), Araya *et al.* (*Pl. Dis.* **88**: 830, 2004), Chung *et al.* (*Mycoscience* **45**: 1, 2004), Chung *et al.* (*Mycoscience* **45**: 233, 2004), Kosman *et al.* (*Phytopathology* **94**: 632, 2004), Wingfield *et al.* (*Australas. Pl. Path.* **33**: 327, 2004), Anikster *et al.* (*Mycol.* **97**: 474, 2005), Wood & Crous (*MR* **109**: 387, 2005), Aime (*Mycoscience* **47**: 112, 2006), Maier *et al.* (*MR* **111**: 176, 2007), van der Merwe *et al.* (*MR* **111**: 163, 2007).

Pucciniales Clem. & Shear (1931). Pucciniomycetes. 14 fam., 166 gen., 7798 spp. The rust fungi or rusts. Mycelium (without clamp connexions) generally intercellular (frequently with haustoria), limited to parts of leaves or other aerial organs of the host ('local' infection), sometimes perennial, if systemic overwintering in roots or other parts; cosmop. on seed plants and ferns, frequently causing major disease; obligate parasites but axenic culture reported for *Gymnosporangium juniperi-virginianae* (Cutter, *Mycol.* **51**: 248, 1959), *Puccinia graminis* f.sp. *tritici* (Williams *et al.*, *Phytopath.* **56**: 1418, 1966; **57**: 326, 1967; *TBMS* **57**: 129, 137, 1971), *Melampsora lini* (Turel, *CJB* **47**: 821, 1969), *Uromyces dianthi* (Jones, *TBMS* **58**: 29, 1972) and others, are now known to be possible from all states of the life cycle (Narisawa *et al.*, *Trans. Mycol. Soc. Japan* **33**: 35, 1992).

Rusts have up to five spore states (frequently numbered **0-IV**; these roman numerals can be ambiguous unless restricted to a morphological system. See Holm, *Notes R. Bot. Gard. Edin.* **44**: 433, 1987) (Fig. 25A). Traditionally the spore terminology was based on morphology. Arthur coined contractions for the original terms (e.g. telium for teleutosorus) and later linked them to the nuclear events in the life cycle, as did Cummins (1959). Other authors have used either the long or short spellings with interchanged definitions and much confusion resulted. In attempts to unite the merits of both schemes Laundon (*TBMS* **50**: 189, 1964; **58**: 344, 1972) proposed a basically morphological terminology which incorporated nuclear events by adding qualifiers where desirable and Holm (1987) devised a compromise in which 'short' terms were linked to nuclear events and 'long' terms to morphology. Essentially, Hiratsuka (*Mycol.* **65**: 432, 1973, *Rep. Tottori mycol. Inst.* **12**: 99, 1975), who, following Cummins, relates the spore states to the nuclear cycle, is followed here. Hennen & Hennen (*Biologico* São Paulo **62**: 113, 2000) provide a review of the terminology applied to the sori and life cycles of rust fungi from 1729 to 2000.

0. Spermatia (sing. -ium; pycniospores), monokaryotic gametes produced in **Spermogonia** (sing. -ium; pycnia; morphological types of spermogonia, see Hiratsuka & Hiratsuka, *Rept. Tottori Mycol. Inst.* **18**: 257, 1980) which are variable in form and position, contain a palisade of sporogenous cells which produce spores in nectar exuded through the ostiole and may have periphyses and flexuous hyphae (q.v.). Savile (*Mycotaxon* **33**: 387, 1988) emphasized that spermogonia are 'hermaphroditic' structures.

I. Aeciospores (aecidiospores, plasmogamospores), produced in **aecia** (sing. -ium; aecidiosori; morphological types of aecia see Sato & Sato, *TBMS* **85**: 223, 1985), are unicellular, non-repeating vegetative spores, usually resulting from dikaryotization (and thus usually associated with pycnia), which germinate to give dikaryotic mycelium. Aeciospores (aecial aeciospore (I^{I}), Laundon) are typically catenulate, thin-walled, and verrucose but sometimes they resemble typical urediniospores when they are designated **uredinioid aeciospores** by Cummins (= aecial urediniospores (II^{I}), Laundon; primary uredospores, Winter).

II. Urediniospores (uredospores, urediospores (for orthography see Savile, *Mycol.* **60**: 459, 1968), summer spores, red rust spores), repeating vegetative spores (which give urediniospores again or teliospores), usually on dikaryotic mycelium, in **uredinia** (uredosori, uredia; morphological types of uredinia, see Sathe, *Kavaka* **5**: 59, 1977; Hiratsuka & Sato, *in* Scott & Chakravorty (Eds), 1982). Typical urediniospores are unicellular, pedicellate, deciduous, with the pigmented echinulate wall showing two or more germ pores. Rarely they resemble typical aeciospores when they are designated **aecidioid urediniospores** by Cummins (= uredinial aeciospores (I^{II}), Laundon). **Amphispores** (II^{II};X) or resting urediniospores are produced by some rusts. These spores generally have thicker and darker walls than normal urediniospores.

III. Teliospores (teleutospores, teleutosporodesma, winter spores, black rust spores), produced in **telia** (sing. -ium; teleutosori; ontogeny and morphology, see Hiratsuka, *Mycotaxon* **31**: 517, 1988), are basidia-producing spores. Telia and teliospores, which characterize the teleomorph of rust fungi, show wide morphological variation but typically teliospores are resting spores, 2- or more celled, sessile or pedicellate but not deciduous, and the thick wall is variously ornamented. Rarely they resemble typical aeciospores when they are designated **aecidioid teliospores** by Cummins (= telial aeciospores (I^{III}), Laundon). Teliospores that germinate immediately, especially in species of genera that usually show dormancy, may be termed leptospores.

IV. Basidiospores (sporidia) are haploid, unicellular, thin-walled, short-lived spores produced on 2-4-celled **basidia** (sing. -ium; promycelium, metabasidium) after meiosis and liberated from sterigmata by abjection (Buller, **3**).

Hughes (*CJB* **48**: 2147, 1970) studied the development (ontogeny) of rust spores and concluded that in his hyphomycete spore terminology O spores are phialospores; I, meristem arthrospores; II, sympodioconidia (or, less common, meristem arthrospores); III, terminal chlamydospore-like cells, sympodioconidia, or meristem arthrospores. Savile (*in* Kendrick (Ed.), *The whole fungus* **2**: 547, 1979) has considered the evolution of anamorphs in rusts.

Rust life cycles (Petersen, *Bot. Rev.* **40**: 453, 1974) vary according to which stages are present or absent. Special terms designate the different life cycles as

TABLE 5. Nomenclature of rust life-cycles.

Example	Schröter (1894-7)	Arthur (1925)	Laundon (1974)	Durrieu (1979)	Cummins & Hiratsuka (2003)
Puccinia graminis	0 I II III IV eu-form	0 I II III IV macrocyclic	0 I II III IV macrocyclic	macrocyclic	0 I-II III IV heteromacrocyclic
P. helianthi					0 I II III IV automacrocyclic
P. punctiformis	0 II III IV brachy-form	0 I II III IV macrocyclic	0 II^{I} II^{II} III IV brachycyclic	brachycyclic	
Gymnosporangium sp. (most spp.)	0 I III IV opsis-form	0 I III IV macrocyclic	0 I III IV demicyclic	opsicyclic	0 I-III IV heterodemicyclic
Gymnoconia peckiana					0 I III IV autodemicyclic
Coleosporium sp.[a]	0 I II III IV eu-form	0 I II III IV macrocyclic	0 I^{I} II^{II} III IV demicyclic	opsicyclic (pseudopsicyclic)	
P. lagenophorae[b]	(0) I III IV opsis-form	(0 I) II III IV macrocyclic	(0 I^{I}) II^{II} III IV demicyclic	opsicyclic (pseudopsicyclic)	
P. heterospora[c]	(0) III IV micro-form	(0) III IV macrocyclic	(0) III IV demicyclic	microcyclic	0 III IV microcyclic
Endophyllum sp.	0 I IV endo-form(s)	0 III IV microcyclic	0 I^{III} IV microcyclic	endocyclic	0 III IV endocyclic
P. chryanthemi	II III IV hemi-form				

[a] The uredinial aecia of *Coleosporium* spp. have been traditionally known as uredosori because of their association with telia in a heteroecious life-cycle.
[b] The uredinial aecia of such rusts as this have not generally been recognized on account of their being morphologically indistinguishable from ordinary aecia. The pycnia and true aecia are rarely found.
[c] Rusts like *P. malvacearum* with spores which germinate immediately were referred to as leptoforms by Schröter.)

shown in Table 8 but have been applied in slightly different ways (see Durrieu, *BSMF* **95**: 379, 1979). Hiratsuka *et al.* (*Rust flora of Japan*, 1992) revert to terminology similar to Schröter.

A rust fungus may be **autoecious** (Fig. 25B) with its life cycle on one host (or group of closely related collateral hosts) or **heteroecious** (Fig. 25A) with O and I on one sort of host and II and III (or I^{II}, III or III only) on another sort (i.e. it has alternate hosts generally living in the same plant association).

The life-cycle of a rust is generally constant, though there may be no development of O, II, or sometimes I, because of weather or other conditions. A species with I, II, III, but not O, is sometimes given the name **cata-species.** If there is no knowledge of III, the form-genus (e.g. *Aecidium*, *Uredo*) is, however, still included in the Uredinales.

Physiologic specialization of *Uredinales* has had much attention. *Puccinia graminis* s.l. has 7 races (formae speciales) (*tritici*, *avenae*, etc.), and there are about 250 physiologic races of *P. graminis* f.sp. *tritici*, and so on. Races are determined by the use of differential hosts.

Nuclear cycle: A rust may be heterothallic or homothallic. In a heterothallic macrocyclic species a basidium has two + and two – basidiospores (see Sex). A + (or –) spore, after infection of the right host, gives a haploid mycelium, pycnia with + (–) spermatia, and protoaecia. If taken (frequently by an insect) to a flexuous hypha of a – (+) pycnium, a + (–) spermatium may put out a 'peg' to make a connexion, its nucleus goes into (spermatizes) the hypha, and by division gives nuclei for the diploidization of cells down to the protoaecia. The cells of a protoaecium undergo conjugate division and an aecium with aeciospores is produced. An aeciospore and its mycelium, and urediniospores and their mycelia, are dikaryotic. There is nuclear fusion in the teliospore, meiosis in the basidium.

In homothallic species, where pycnia are not necessary and are frequently not present, the dikaryophase has its start from two cell nuclei at some point or points in the life-cycle. Nuclear fusion and reduction are as in heterothallic species. Following Dietel (1928) two families, *Melampsoraceae* and *Pucciniaceae*, have frequently been recognized. Concepts have been refined by subsequent workers (see Hennen & Buriticá, *Rept. Tottori Mycol. Inst.* **18**: 43, 1980). In this edition of the *Dictionary* fourteen familes are recognized as they are in current use by Cummins & Hiratsuka (1983) and in the major compilation Hiratsuka *et al.* (*Rust Flora of Japan*, 1992). Fams:

(1) **Chaconiaceae**
(2) **Coleosporiaceae**
(3) **Cronartiaceae**

(4) **Melampsoraceae**
(5) **Mikronegeriaceae**
(6) **Phakopsoraceae**
(7) **Phragmidiaceae**
(8) **Pileolariaceae**
(9) **Pucciniaceae**
(10) **Pucciniastraceae**
(11) **Pucciniosiraceae**
(12) **Raveneliaceae**
(13) **Uncolaceae**
(14) **Uropyxidaceae**

Placement of some genera is controversial and still others cannot be accommodated in any of the accepted families due to absence of key structures (especially pycnia, regarded as conservative and therefore valuable at the higher levels of classification) or otherwise incomplete information on their characters. It is acknowledged that family circumscription requires further investigation.

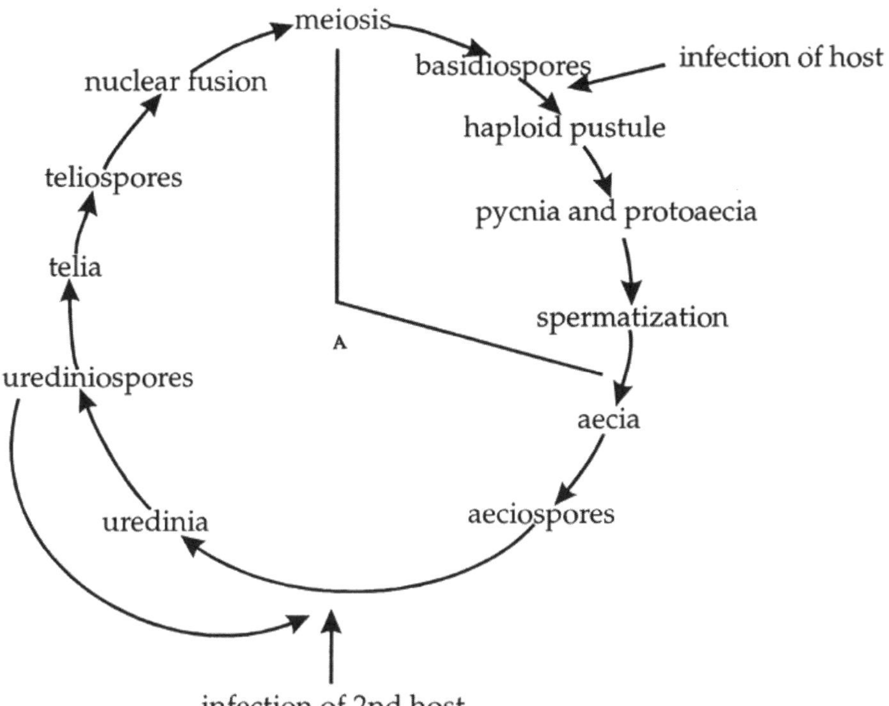

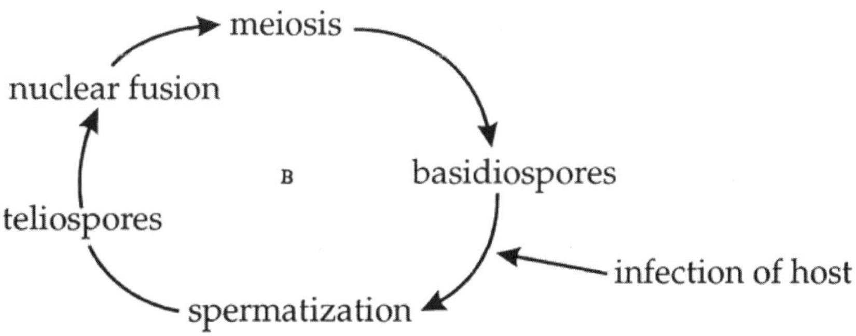

Fig. 25. Life cycles of A, a macrocyclic heteroecious rust; B, a microcyclic autoecious rust

Control: Sulphur is used against a number of rusts. Destruction of an alternate host (e.g. *Berberis* for *Puccinia graminis*) is sometimes of value. The development of resistant varieties of plants is ever in view.

Lit. (*General*): Sydow (*Monographia Uredinearum*, 4 vols, 1902-24), Dietel (*Naturl. PflFam.* **6**, 1928), Arthur (*Plant rusts*, 1929), Guyot (*Les Uredinées*, 1939-57), Thirumalachar & Mundkur (*Indian Phytopath.* **2**: 65, **3**: 4, 203, 1949-50; gen.), Hiratsuka (*Revision of taxonomy of the Pucciniastreae*, 1958), Cummins (*Illustrated genera of rust fungi*, 1959; keys, bibliogr. [edn 2, Cummins & Hiratsuka, 1983]; *The rust fungi of cereals, grasses and bamboos*, 1971, keys; *Rust fungi on legumes and composites in North America*, 1978), Staples & Wynn (*Bot. Rev.* **31**: 537, 1965; urediniospore physiology), Preece & Hick (*Introductory scanning electron microscope atlas of rust fungi*, 1990), Ziller (*Tree Rusts of Western Canada*, 1974), Dupias (*BSMF* **87**: 129, 1971; biogeogr.), Hart (*Cladistics* **4**: 339, 1988; coevolution), Laundon (*Mycol. Pap.* **89**, **91**, **102**, 1963-65; *Acanthaceae-Amaryllidaceae*; 99 gen. names), Leppik (*Mycol.* **45**: 46, phylogeny conifer rusts; **48**: 637, *Gymnosporangium*; **51**: 512, grass rusts; **53**: 378, stem rusts; 1953-61; *Annls bot. fenn.* **9**: 135, 1972; evolutionary specialization), Littlefield (*Biology of rust fungi*, 1981), Littlefield & Heath (*Ultrastructure of rust fungi*, 1979), Petersen (*Bot. Rev.* **40**: 453, 1975; life cycle), Savile (*Evol. Biol.* **9**: 137, 1976; *Rept. Tottori Mycol. Inst.* **28**: 15, 1990; evolution with hosts, *Nova Hedw.* **57**: 269, 1993; cladistics), Scott & Chakravorty (Eds) (*The rust fungi*, 1982), Bushnell & Roelfs (Eds) (*The cereal rusts* **1** *Origin, specificity, structure and physiology*, 1984; **2** *Diseases, distribution, epidemiology, control*, 1985), Zhang, Dickinson & Pryor (*Ann. Rev. Phytopathol.* **32**: 115, 1994; double-stranded RNAs), Vogler & Bruns (*in* Reynolds & Taylor (Eds), *The fungal holomorph*: 273, 1993; molecular approaches), Swertz (*Stud. Mycol.* **36**, 1994; morphology of urediniospore germlings), Fell *et al.* (*Int. J. Syst. Evol. Microbiol.* **50**: 1351, 2000; mol. phylogeny basidiomycetous yeast), and (esp.) under Fams.

Serials: *Uredineana*, 1939 on (Paris); *Cereal rusts bulletin*, 1973 on (Wageningen).

Regional: **Australia,** McAlpine (*Rusts of Australia*, 1906). **Canada,** see USA. **Central Europe,** Gäumann (*Beitr. Kryptog.-fl. Schweiz.* **12**, 1959). **Austria,** Poelt (*Catalogus Florae Austriae* **3(1)** *Uredinales*, 1985). **former Czechoslovakia,** Markova & Urban (*Novit. Bot. Univ. Carol.* **3**: 25, 1987). **Finland,** Liro (*Uredineae Fennicae*, 1908), Makinen (*Annls bot. fenn.* **1**: 214, 1964). **Germany,** Braun (*Feddes Repert.* **93**: 213, 1982, Brandenburger (*Regensb. Mykol. Schriften* **3**, 1994). **British Isles,** Wilson & Henderson (*British rust fungi*, 1966), Henderson & Bennell (*Notes R. bot. Gdn Edinb.* **37**: 475, 1979; **38**: 184, 1980); see also Grove, Plowright. **Iceland,** Jørstad (*Skr. norske Vidensk Akad.* I *Mat. Nat.* n.s. **2**, 1952). **Indonesia,** Boedijn (*Nova Hedw.* **1**: 463, 1960). **Italy,** Ciferri & Camara (*Ist. Bot. Univ. Pavia Quaderno* **23**, 1962; list). **Jamaica,** Dale (*Mycol. Pap.* **60,** 1955). **Japan,** Hiratsuka *et al.* (*The rust flora of Japan*, 1992). **Madagascar,** Bouriquet & Bassino (*Prodr. Flor. mycol. Madagascar* **5**, 1966). **Majorca and Minorca,** Jørstad (*Skr. norske VidenskAkad.*, I *Mat. Nat.* n.s. **2**, 1962). **Morocco,** Guyot & Malençon (*Uredinées du Maroc*, I, 1957). **Mexico,** León-Gallegos & Cummins (*Uredinales (Royas) de Mexico* **1+2**, 1981). **Nepal,** Durrieu (*Mycol.* **79**: 90, 1987). **New Caledonia,** Huguenin (*BSMF* **82**: 248, **83**: 941, 1966-68). **New Zealand,** Cunningham (*Rust fungi of New Zealand*, 1931), Dingley (*N.Z. Jl Bot.* **15**: 29, 1977). **Nigeria,** Eboh (*Mycol.* **70**: 1077; **73**: 445; **76**: 179, 1978-84). **Norway,** Jørstad (*Nytt. Mag. Bot.* **8**: 103, list; **9**: 61, distribution; **11**: 27, life cycles, etc., 109, distribution, 1960-64), Gjaerum (*Nordeus Rustsopper*, 1974). **Malawi,** Bisby & Wiehe (*Mycol. Pap.* **51**, 1953). **Pakistan,** Ahmad (*Biologia* **2**: 26, 1956). **Poland,** Majewski (*Flora Polska* **9,11** *Uredinales* **1**, 1977, **2**, 1979). **Romania,** Savulescu (*Monografia Uredinalelor.*, 2 vols, 1953). **Scandinavia,** Hylander, Jørstad & Nannfeldt (*Op. bot. Soc. bot. Lund* **1**, 1953). **Spain,** Gonzales-Fragoso (*Flora Iberica. Uredinales*, 2 vols, 1924-25). **South Africa,** Doidge (*Bothalia* **2**, **3**, **4**, 1927-48). **USA,** Arthur (*Manual of rusts in the United States and Canada*, 1934 [suppl. by Cummins, 1962]), Cummins & Stevenson (*Pl. Dis. Reptr, Suppl.* **240,** 1956; list), Gardner & Hodges (*Pacific Science* **43**: 41, 1989; Hawaii). **Cuba,** Urban (*Rept. Tottori Mycol. Inst.* **28**: 37, 1990). **Argentina,** Lindqvist (*Royas de la Republica Argentina*, 1981). **Brazil,** Hennen, Hennen & Figueiredo (*Arq. Inst. Biol., Saõ Paulo* **49**(Suppl. 1), 1982). **former USSR,** Tranzschel (*Conspectus Uredinalium U.R.S.S.*, 1939), Kuprevich & Tranzschel ([*Crypt. Pl. USSR* **4**], 1957 [Engl. transl. 1970]. **Ukraine,** Gutsevich ([*Survey of the rust fungi of the Crimea*], 1952 [Russ.]). **Kazakhstan,** Nevodovsky (*Flora sporovykh rastenii Kazakhstana* [I. Uredinales], 1956), Kuprevich & Ulyanishchev ([*Keys to the rust fungi of USSR*] **1**, 1975, **2**, 1978). **Lithuania,** Minkevičius & Ignatavičiute (*Lietuvos Grybai* **5**, 1991). **Tadzhikistan,** Korbonskaya ([*Rust fungi of Tadzhikistan*], 1954). *Melampsoraceae*).

Pucciniasporonites Ramanujam & Ramachar (1980), Fossil Fungi, Pucciniaceae. 1. See Ramanujam & Ramachar (*Records of the Geological Survey of India* **113**: 82, 1980).

Pucciniastraceae Gäum. ex Leppik (1972), Pucciniales. 11 gen. (+ 2 syn.), 158 spp.

Lit.: Yamaoka & Katsuya (*Trans. Mycol. Soc. Japan* **28**: 155, 1987), Berndt *et al.* (*CJB* **72**: 1084, 1994), Berndt & Oberwinkler (*Mycol.* **89**: 698, 1997), Kurkela *et al.* (*Mycol.* **91**: 987, 1999), Motokura *et al.* (*Res. Bull. Pl. Protect. Serv.* Japan **35**: 103, 1999), Cummins & Hiratsuka (*Illustr. Gen. Rust Fungi* edn 3: 225 pp., 2003), Maier *et al.* (*CJB* **81**: 12, 2003), Aime (*Mycoscience* **47**: 112, 2006), Liang *et al.* (*Mycoscience* **47**: 137, 2006).

Pucciniastrum G.H. Otth (1861), Pucciniastraceae. *c.* 34 (on *Abies*, *Picea*, *Tsuga* (0, I) (*Pinaceae*); on dicots, orchids? (II, III)), widespread (esp. north temperate). See Hiratsuka (*Revision of taxonomy of the Pucciniastreae*, 1958), Cummins & Hiratsuka (*Illustr. Gen. Rust Fungi rev. edit.*, 1983), Liang *et al.* (*Mycoscience* **47**: 137, 2006; phylog.) Cf. *Thekopsora*, *Calyptospora*, which are sometimes included.

Puccinidia Mayr (1890) = Rostrupia fide Saccardo (*Syll. fung.* **9**: 316, 1891) = Puccinia fide, Cummins & Hiratsuka (*Illustr. Gen. Rust Fungi rev. edit.*, 1983).

Pucciniola L. Marchand (1829) nom. rej. = Uromyces fide Berndt (*in litt.*).

Pucciniomycetes R. Bauer, Begerow, J.P. Samp., M.

Weiss (2006), Pucciniomycotina. 5 ord., 21 fam., 190 gen., 8016 spp. Ords:
(1) **Helicobasidiales**
(2) **Pachnocybales**
(3) **Platygloeales**
(4) **Pucciniales**
(5) **Septobasidiales**
For *Lit.* see ord. and fam.

Pucciniomycotina R. Bauer, Begerow, J.P. Samp, M. Weiss & Oberw. (2006), Basidiomycota. 8 class., 18 ord., 36 fam. Class.:
(1) **Agaricostilbomycetes**
(2) **Atractiellomycetes**
(3) **Classiculomycetes**
(4) **Cryptomycocolacomycetes**
(5) **Cystobasidiomycetes**
(6) **Microbotryomycetes**
(7) **Mixiomycetes**
(8) **Pucciniomycetes**
For *Lit.* see ord. and fam.

Pucciniopsis Speg. (1888) nom. ambig., anamorphic *Pezizomycotina*. See Sutton (*TBMS* **60**: 525, 1973).

Pucciniosira Lagerh. (1892), Pucciniosiraceae. *c.* 15 (on angiosperms (0, III)), America; Africa; Asia. May include *Gambleola*. See Buriticá (*Rev. Acad. Colomb. Cienc.* **19**: 131, 1991), Gjærum *et al.* (*Lidia* **5**: 18, 2000; Uganda).

Pucciniosiraceae Cummins & Y. Hirats. (1983), Pucciniales. 10 gen. (+ 7 syn.), 57 spp.
Lit.: Gjærum *et al.* (*Lidia* **5**: 18, 2000), Hernández (*Mycotaxon* **76**: 329, 2000), Cummins & Hiratsuka (*Illustr. Gen. Rust Fungi* edn 3: 225 pp., 2003), Wingfield *et al.* (*Australas. Pl. Path.* **33**: 327, 2004), Aime (*Mycoscience* **47**: 112, 2006).

Pucciniospora Speg. (1886), anamorphic *Pezizomycotina*, Cpd.1eH.?. 1, S. America.

Pucciniostele Tranzschel & Kom. (1899), Phakopsoraceae. *c.* 4 (on *Astilbe* (*Saxifragaceae*)), Asia. Cf. *Scalarispora*. See Cummins & Thirumalachar (*Mycol.* **45**: 572, 1953), Koursanov *et al.* (*Bull. Soc. Nat. Moscou. Cinenc.* **19** (no. 69): 131, 1991).

Puccinites Ettingsh. (1853), Fossil Fungi. 3 (Cretaceous, Tertiary), Europe; USA.

Puciola De Bert. (1976) = Dicyma fide von Arx (*Gen. Fungi Sporul. Cult.* Edn 3, 1981).

puff-ball, basidioma of the *Lycoperdales*.

puffing, a phenomenon in which thousands of asci in an apothecial ascoma discharge their ascospores simultaneously, producing a visible cloud.

Puiggariella Speg. (1881) = Strigula fide Santesson (*Symb. bot. upsal.* **12** no. 1: 1, 1952).

Puiggarina Speg. (1919) = Phyllachora Nitschke ex Fuckel (1870) fide Cannon (*Mycol. Pap.* **163**, 1991).

Pulcherricium Parmasto (1968) = Terana fide Parmasto (*Consp. System. Corticiac.*: 132, 1968).

Pulchromyces Hennebert (1973), anamorphic *Pezizomycotina*, Hso.0eH.7. 1 (coprophilous), Ghana. See Pfister *et al.* (*Mycotaxon* **1**: 137, 1974).

Pulicispora Vedmed, S.V. Krylova & I.V. Issi (1991), Microsporidia. 1.

Pulina Adans. (1763) nom. rej. ≡ Lepraria.

Pullospora Faurel & Schotter (1965), anamorphic *Pezizomycotina*, Cpd.0eH.15. 2 (coprophilous), Algeria; USA. See Nag Raj (*Coelomycetous Anamorphs with Appendage-bearing Conidia*, 1993).

Pullularia Berkhout (1923) = Aureobasidium fide Cannon (*Mycopathologia* **111**: 75, 1990), Hawksworth *et al.* (*Dictionary of the Fungi* edn 8, 1995).

pullulation, budding, as in yeasts.

Pulmonaria Hoffm. (1789) [non *Pulmonaria* L. 1753, *Boraginaceae*] = Lobaria fide Hawksworth *et al.* (*Dictionary of the Fungi* edn 8, 1995).

Pulmosphaeria Joanne E. Taylor, K.D. Hyde & E.B.G. Jones (1996), Xylariales. 1 (on *Palmae*), Australia. Perhaps related to *Linocarpon*. See Taylor *et al.* (*Sydowia* **48**: 255, 1996).

Pulparia P. Karst. (1866) nom. rej. = Pulvinula. See also *Marcelleina*. fide Dissing (*Mycotaxon* **32**: 365, 1988), Dissing *et al.* (*Taxon* **39**: 130, 1990; nomencl.), Hawksworth *et al.* (*Dictionary of the Fungi* edn 8, 1995).

pulque, a Mexican alcoholic drink made by yeast fermentation of the juice of *Agave* spp.; *Lactobacillus* and *Leuconostoc* spp. add acidity and viscosity; when distilled yields the spirit **tequila**.

pulsed field gel-electrophoresis, see Electrophoresis.

pulveraceo-delitescent, covered with a layer of powdery granules.

Pulveraria Ach. (1803) nom. rej. prop. = Chrysothrix fide Hawksworth *et al.* (*Dictionary of the Fungi* edn 8, 1995).

Pulverariaceae Schltdl. (1824) = Chrysothricaceae.

Pulveria Malloch & Rogerson (1977) = Pyrenomyxa fide Læssøe (*SA* **13**: 43, 1994; nomencl.).

Pulveroboletus Murrill (1909), Boletaceae. 25, cosmopolitan. See Singer (*Am. midl. Nat.* **37**: 1, 1947), Gómez (*Revta Biol. trop.* **44** Suppl. 4: 59, 1996; Costa Rica), Zang *et al.* (*Mycotaxon* **80**: 481, 2001; China).

Pulverolepiota Bon (1993) = Cystolepiota fide Kuyper (*in litt.*).

pulverulent, powdered; as if powdered over.

Pulvinaria Bonord. (1851), Pezizomycotina. 1, Europe.

Pulvinaria Rodway (1918) ≡ Waydora.

Pulvinaria Velen. (1934) ? = Pachyella fide Pfister (*CJB* **51**: 2009, 1973).

pulvinate, cushion-like in form.

Pulvinella A.W. Ramaley (2001), anamorphic *Pezizomycotina*. 1, USA. See Ramaley (*Mycotaxon* **79**: 52, 2001).

Pulvinodecton Henssen & G. Thor (1998), Roccellaceae (L). 2, tropical. See Henssen & Thor (*Nordic Jl Bot.* **18**: 95, 1998; ontogeny), Thor (*J. Jap. Bot.* **77**: 47, 2002; Japan, Taiwan).

Pulvinotrichum Gamundí, Aramb. & Giaiotti (1981), anamorphic *Pezizomycotina*, Hso.0-1eH.15. 2, Argentina; Australia. See Summerbell *et al.* (*CJB* **67**: 577, 1989; synonymy with *Cylindrodendrum*), Sutton (*Sydowia* **41**: 330, 1989), Crous *et al.* (*Mycotaxon* **50**: 441, 1994), Wu *et al.* (*Mycoscience* **38**: 11, 1997).

Pulvinula Boud. (1885) nom. cons., Pezizales. 27, widespread. See Pfister (*Occ. Pap. Farlow Herb. Crypt. Bot.* **9**, 1976; key), O'Donnell & Hooper (*CJB* **56**: 101, 1978; ontogeny), Kaushal (*Kavaka* **9**: 21, 1981), Korf & Zhuang (*Mycotaxon* **20**: 607, 1984), Dissing (*Mycotaxon* **32**: 365, 1988; nomencl.), Korf & Zhuang (*Mycotaxon* **40**: 79, 1991), Liu (*Acta Mycol. Sin.* **10**: 185, 1991), Yao & Spooner (*MR* **100**: 193, 1996), Yao & Spooner (*MR* **100**: 883, 1996; Brit. spp.), Landvik *et al.* (*Nordic Jl Bot.* **17**: 403, 1997; DNA), Hansen & Pfister (*Mycol.* **98**: 1029, 2006; phylogeny), Liu & Zhuang (*Mycosystema* **25**: 546, 2006; phylogeny), Perry *et al.* (*MR* **111**: 549, 2007; phylogeny).

Pumilus Viala & Marsais (1934), Pezizomycotina. 1 (on *Vitis*), Europe.

punctate, marked with very small spots (Fig. 20.5); **puncta**, small spots.

Punctelia Krog (1982), Parmeliaceae (L). 42, widespread. See Galloway & Elix (*N.Z. Jl Bot.* **22**: 441, 1984; Australasia, key), Modenesi (*Nova Hedwigia* **45**: 423, 1987; histochemistry), Wilhelm & Ladd (*Mycotaxon* **44**: 495, 1992; key 13 N. Am. spp.), Longán *et al.* (*Mycotaxon* **74**: 367, 2000), van Herk & Aptroot (*Lichenologist* **32**: 233, 2000; Eur.), Aptroot (*Bryologist* **106**: 317, 2003; sorediate spp., N America), Crespo *et al.* (*Lichenologist* **36**: 299, 2004; Iberian peninsula), Thell *et al.* (*Folia cryptog. Estonica* **41**: 115, 2005; phylogeny), Blanco *et al.* (*Mol. Phylogen. Evol.* **39**: 52, 2006; phylogeny), Miądlikowska *et al.* (*Mycol.* **98**: 1088, 2006; phylogeny).

punctiform (of rust sori, bacterial colonies, etc.), very small, but seen without a lens.

Punctillina Toro (1934), anamorphic *Pezizomycotina*, Cpt.0eH.?. 1, S. America.

Punctillum Petr. & Syd. (1924), Dothideomycetes. 2 (on *Musci*), widespread. See Döbbeler (*Beih. Nova Hedwigia* **79**: 203, 1984).

Punctodiporites C.P. Varma & Rawat (1963), Fossil Fungi, anamorphic *Ascomycota*. 1. See Varma & Rawat (*Grana Palynologica* **4**: 136, 1963).

Punctonora Aptroot (1997), Lecanoraceae (L). 1, Papua New Guinea. See Aptroot *et al.* (*Biblthca Lichenol.* **64**: 151, 1997).

Punctularia Pat. (1895), Corticiaceae. 2, widespread. See Niemelä (*Sienilehti* **55**: 67, 2003; *P. strigosozonata* in Finland and Estonia).

Punctulariaceae Donk (1964) = Corticiaceae.

punk, see touchwood or amadou; **punky**, soft and tough.

Pureke P.R. Johnst. (1991), Rhytismataceae. 1, New Zealand. See Johnston (*N.Z. Jl Bot.* **29**: 395, 1991).

Pustularia Bonord. (1851) nom. dub., Pezizomycotina. Used for a wide range of taxa.

Pustularia Fuckel (1870) ≡ Pustulina.

pustule, a blister-like, frequently erumpent, spot or spore-mass.

Pustulina Eckblad (1968) = Tarzetta fide Rogers *et al.* (*Mycol.* **63**: 1084, 1971).

Pustulipora P.F. Cannon (1982), Ceratostomataceae. 1, British Isles. See Cannon (*Mycotaxon* **15**: 523, 1982), Ranghoo & Hyde (*MR* **103**: 938, 1999).

Putagraivam Subram. & Bhat (1978) = Didymostilbe Henn. fide Hawksworth *et al.* (*Dictionary of the Fungi* edn 8, 1995).

putrescent (of basidiomata), decaying, rotting. Cf. marcescent.

Puttemansia Henn. (1902), Tubeufiaceae. Anamorphs *Guelichia*, *Tetracrium*. 7 (on fungi), widespread (tropical). See Rossman (*Mycol. Pap.* **157**, 1987; key), Kodsueb *et al.* (*Fungal Diversity* **21**: 105, 2006).

Puttemansiella Henn. (1908) nom. dub., Fungi.

pv., see pathovar.

Pycnidiales (obsol.), see *Sphaeropsidales* (Anam. fungi).

Pycnidiella Höhn. (1915), anamorphic *Sarea*, St.0eH.15. 1 (on resin), widespread. See Hawksworth & Sherwood (*CJB* **59**: 357, 1981).

Pycnidioarxiella Punith. & N.D. Sharma (1980), anamorphic *Pezizomycotina*, Cpd.1eP.2. 1, India. See Punithalingam & Sharma (*Nova Hedwigia* **31**: 893, 1979).

Pycnidiochaeta Sousa da Câmara (1950) = Dinemasporium fide Sutton (*Mycol. Pap.* **141**, 1977).

Pycnidiopeltis Bat. & C.A.A. Costa (1959), anamorphic *Pezizomycotina*, Cpt.0eH.?. 1, USA; Brazil. See Batista & Costa (*Mycopath. Mycol. appl.* **11**: 83, 1959).

Pycnidiophora Clum (1956), Sporormiaceae. Anamorph *Phoma*-like. 2 (from soil etc.), widespread. See Cain (*CJB* **39**: 1633, 1961; as *Preussia*), Berbee (*Mol. Biol. Evol.* **13**: 462, 1996; DNA, as *Westerdykella*), Barr (*Mycotaxon* **76**: 105, 2000), Kruys *et al.* (*MR* **110**: 527, 2006; phylogeny).

pycnidiospore, a conidium in or from a pycnidium (obsol.).

Pycnidiostroma F. Stevens (1927) = Phomachora fide Petrak (*Annls mycol.* **27**: 324, 1929).

pycnidium (pl. **-ia**), a frequently ± flask-shaped conidioma of fungal tissue with a circular or longitudinal ostiole, the inner surface of which is lined entirely or partially by conidiogenous cells; pycnidial conidioma (Fig. 10A-F).

pycniospore (of *Pucciniales*), a spore from a pycnium; spermatium; sometimes used in error for pycnidiospore.

Pycnis Bref. (1881), anamorphic *Pezizomycotina*, Cpd.0eH.?. 1, Europe.

pycnium (in *Pucciniales*), the pycnidium-like haploid fruit-body, or spermogonium. See Hiratsuka & Cummins (*Mycol.* **55**: 487, 1963), Savile (*Mycol.* **63**: 1089, 1971).

pycnoascocarp, an ascoma arising from a pycnidial conidioma.

Pycnocalyx Naumov (1916) = Bothrodiscus fide Sutton (*Mycol. Pap.* **141**, 1977).

Pycnocarpon Theiss. (1913), Dothideomycetes. 4, India; Philippines.

Pycnochytrium (de Bary) J. Schröt. (1892) = Synchytrium fide Fitzpatrick (*The lower fungi. Phycomycetes*, 1930).

Pycnociliospora Bat. (1962) = Strigula fide Lücking *et al.* (*Lichenologist* **30**: 121, 1998).

pycnoconidium, see pycnidiospore (obsol.).

Pycnodactylus Bat., A.A. Silva & Cavalc. (1967), anamorphic *Pezizomycotina*, Cpd.#eP.?. 1, Brazil. See Batista *et al.* (*Atas Inst. Micol. Univ. Pernambuco* **5**: 140, 1967).

Pycnodallia Kohlm. & Volkm.-Kohlm. (2001), anamorphic *Pezizomycotina*. 1, USA. See Kohlmeyer & Volkmann-Kohlmeyer (*MR* **105**: 500, 2001).

Pycnoderma Syd. & P. Syd. (1914), Cookellaceae. 2, widespread (tropical). See Petrak (*Sydowia* **1**: 108, 1947).

Pycnodermella Petr. (1947) = Saccardinula fide von Arx (*Persoonia* **2**: 421, 1963).

Pycnodermellina Bat. & H. Maia (1957) ? = Echinoplaca fide von Arx & Müller (*Stud. Mycol.* **9**, 1975).

Pycnodermina Petr. (1954) ? = Pycnoderma fide von Arx & Müller (*Stud. Mycol.* **9**, 1975).

Pycnodon Underw. (1898) ≡ Neokneiffia.

Pycnodothis F. Stevens (1924) nom. dub., Fungi. See Petrak (*Sydowia* **5**: 169, 1951).

Pycnofusarium Punith. (1973) = Fusarium fide Sutton (*TBMS* **86**: 1, 1986).

pycnogonidium, see pycnidiospore (obsol.), pycniospore, or stylospore (obsol.).

Pycnographa Müll. Arg. (1890) = Parmularia Lév. fide Santesson (*Svensk bot. Tidskr.* **43**: 547, 1949).

Pycnoharknessia Matsush. (1996), anamorphic *Pezizomycotina*, Cpd.?.?. 1, Pakistan. See Matsushima

(*Matsush. Mycol. Mem.* **9**: 23, 1996).

Pycnomma Syd. (1924), anamorphic *Pezizomycotina*, St.0eH.?. 1, Canary Islands.

Pycnomoreletia Rulamort (1990), anamorphic *Pezizomycotina*, St.≡ eP.1. 2, Africa; Pakistan. See Rulamort (*Bull. Soc. bot. Centre-Ouest* Nouv. sér. **21**: 512, 1990).

Pycnopeltis Syd. & P. Syd. (1916) = Saccardinula fide von Arx & Müller (*Stud. Mycol.* **9**, 1975).

Pycnopeziza W.L. White & Whetzel (1938), Sclerotiniaceae. Anamorph *Acarosporium*. 4, Europe; N. America. See Whetzel & White (*Mycol.* **32**: 616, 1940), Holst-Jensen *et al.* (*Mycol.* **89**: 885, 1997; phylogeny), Holst-Jensen *et al.* (*Mycol.* **96**: 135, 2004).

Pycnopleiospora C.Z. Wei, Y. Harada & Katum. (1997), anamorphic *Pezizomycotina*, Hso.?.?. 1, Japan. See Wei *et al.* (*Mycol.* **89**: 496, 1997).

Pycnopodium Corda (1842) ≡ Pilobolus fide Hesseltine (*Mycol.* **47**: 344, 1955).

Pycnoporellus Murrill (1905), Fomitopsidaceae. 2, widespread. See Ryvarden (*Polyp. N. Eur.* **2**, 1978), Piątek (*Polish Botanical Journal* **48**: 131, 2003; Poland).

Pycnoporus P. Karst. (1881), Polyporaceae. 4, widespread. See Nobles & Frew (*CJB* **40**: 987, 1962), Roberts & Ryvarden (*Kew Bull.* **61**: 55, 2006; Cameroon).

Pycnopsammina Diederich & Etayo (1995) = Psammina fide Etayo & Diederich (*Flechten Follmann* Contributions to Lichenology in Honour of Gerhard Follmann: 215, 1995), Earland-Bennett & Hawksworth (*Lichenologist* **37**: 191, 2005).

Pycnora Hafellner (2001), Lecanoromycetes. 4, widespread. See Hafellner & Türk (*Stapfia* **76**: 157, 2001), Kubiak *et al.* (*Botanica Lithuanica* **9**: 371, 2003; Poland), Wedin *et al.* (*MR* **109**: 159, 2005; phylogeny), Crewe *et al.* (*MR* **110**: 521, 2006; phylogeny).

Pycnorostrum Golovin (1950) nom. dub., anamorphic *Pezizomycotina*. See Sutton (*Mycol. Pap.* **141**, 1977).

pycnosclerotium, a more or less hard-walled structure resembling a pycnidial conidioma but having no spores.

Pycnoseynesia Kuntze (1898), anamorphic *Pezizomycotina*, Cpt.0eH.?. 2, widespread (tropical).

pycnosis (of *Microthyriaceae*), the process by which a part of the stroma is arched up and becomes thick while an ascigerous hymenium is formed under it.

pycnospore, formerly occasionally used for pycniospore or pycnidiospore (obsol.).

Pycnosporium Siegel (1909) nom. dub., anamorphic *Pezizomycotina*.

Pycnostemma Syd. (1927) = Poropeltis fide Clements & Shear (*Gen. Fung.*, 1931).

Pycnostroma Clem. (1909) = Munkia fide Höhnel (*Sber. Akad. Wiss. Wien* Math.-naturw. Kl., Abt. 1 **120**: 379, 1911).

pycnostroma, see stroma.

Pycnostysanus Lindau (1904) = Sorocybe. See also *Seifertia*. fide Kaneko *et al.* (*Ann. phytopath. Soc. Japan* **54**: 323, 1988), Partridge & Morgan-Jones (*Mycotaxon* **83**: 335, 2002).

pycnothecium (of *Microthyriaceae*), an ascoma formed by pycnosis.

Pycnothele Sommerf. (1826) ≡ Dufourea.

Pycnothelia (Ach.) Dufour (1821), Cladoniaceae (L). 2, widespread. See Laundon (*Lichenologist* **18**: 169, 1986; nomencl.), Ahti (*Regnum veg.* **128**: 58, 1993), Stenroos & DePriest (*Am. J. Bot.* **85**: 1548, 1998; DNA), Hammer (*Am. J. Bot.* **87**: 33, 2000; ontogeny), Miądlikowska *et al.* (*Mycol.* **98**: 1088, 2006; phylogeny), Zhou *et al.* (*J. Hattori bot. Lab.* **100**: 871, 2006; phylogeny).

Pycnotheliomyces Cif. & Tomas. (1953) ≡ Pycnothelia.

Pycnothera N.D. Sharma & G.P. Agarwal (1974), anamorphic *Pezizomycotina*, Cpt.0eH.38. 2, India. See Punithalingam (*Nova Hedwigia* **31**: 95, 1979).

Pycnothyriella Bat. (1952), anamorphic *Pezizomycotina*, Cpt.0eH.?. 1, Brazil. See Batista (*Bol. Secr. Agric. (Pernambuco)* **19**: 110, 1952).

Pycnothyrium Died. (1913), anamorphic *Pezizomycotina*, Cpt.0eH.?. 6, Europe; Philippines. See Holm & Holm (*Bot. Notiser* **130**: 115, 1977; sensu v. Arx (1964) = *Leptopeltis*), Sutton (*Mycol. Pap.* **141**, 1977).

pycnothyrium, a superficial flattened shield-shaped conidioma with radiate upper and sometimes lower walls; pycnothyrial conidioma. See Hughes (*Mycol. Pap.* **50**: 7, 1953). Characteristic of *Microthyriaceae* (q.v.) etc. (Fig. 10G-M).

Pycnovellomyces R.F. Castañeda (1987), anamorphic *Agaricomycetes*, Cpd.0eH.41. 1 (with clamp connexions), Cuba. See Nag Raj *et al.* (*CJB* **67**: 3386, 1989; redescription).

Pygmaea Stackh. (1809) nom. rej. = Lichina fide Hawksworth *et al.* (*Dictionary of the Fungi* edn 8, 1995).

Pyonema Nieuwl. (1916) ≡ Myxonyphe.

Pyramidospora Sv. Nilsson (1962), anamorphic *Pezizomycotina*, Hso.1bH.1. 2 (aquatic), widespread. See Nilsson (*Svensk bot. Tidskr.* **56**: 358, 1962), Kendrick (*CJB* **81**: 75, 2003; morphogenesis).

Pyrenastromyces Cif. & Tomas. (1953) ? = Pyrenula Ach. (1809) fide Hawksworth *et al.* (*Dictionary of the Fungi* edn 8, 1995).

Pyrenastrum Eschw. (1824) = Pyrenula Ach. (1814) fide Harris (*Mem. N. Y. bot. Gdn* **49**, 1989).

Pyrenidiaceae Zahlbr. (1898) = Dacampiaceae.

Pyrenidiomyces Cif. & Tomas. (1953) ≡ Pyrenidium.

Pyrenidium Nyl. (1865), Dacampiaceae. 3 (on lichens), widespread. See Hawksworth (*TBMS* **80**: 547, 1983).

Pyreniella Theiss. (1916) = Botryosphaeria fide von Arx & Müller (*Beitr. Kryptfl. Schweiz* **11** no. 1, 1954).

Pyrenillium, see *Pyrenyllium*.

Pyreniococcus Wheldon & A. Wilson (1915) ? = Phaeospora fide Hawksworth *et al.* (*Dictionary of the Fungi* edn 8, 1995).

Pyreniopsis Kuntze (1898) ≡ Trichoderma Pers. (1794).

Pyrenisperma, see *Pyrisperma*.

Pyrenium Tode (1790) = Trichoderma Pers. (1794) fide Hawksworth *et al.* (*Dictionary of the Fungi* edn 8, 1995).

pyrenium, a pyrenomycete ascoma (obsol.).

Pyrenobotrys Theiss. & Syd. (1914), Venturiaceae. 3, America; Scandinavia. See Eriksson (*Svensk bot. Tidskr.* **68**: 223, 1974), Barr (*Sydowia* **41**: 25, 1989; key 3 N. Am. spp.).

pyrenocarp, a perithecium (s.l., q.v.); used colloquially as a term for any fungus with a perithecium-like ascoma.

Pyrenocarpon Trevis. (1855), Lichinaceae (L). 1,

Europe. See Jørgensen & Henssen (*Taxon* **39**: 343, 1990), Jørgensen (*SA* **10**: 51, 1991), Schultz & Büdel (*Lichenologist* **34**: 39, 2002; key).

pyrenocarpous, see pyrenocarp.

Pyrenochaeta De Not. (1849), anamorphic *Herpotrichia, Leptosphaeria*, Cpd.0eH.15. 16, widespread. See Last *et al.* (*Ann. appl. Biol.* **57**: 95, 1966), Last *et al.* (*Ann. appl. Biol.* **62**: 55, 1968), Last *et al.* (*Ann. appl. Biol.* **64**: 449, 1969; *P. lycopersici*, tomato brown rot), Schneider (*Mitt. biol. BundAnst. Ld- u. Forstw.* **189**, 1979), Schneider *in* Subramanian (Ed.) (*Taxonomy of fungi* **2**: 513, 1984; disposition of spp.), Datnoff *et al.* (*TBMS* **87**: 297, 1986; sclerotial state), Ferreira *et al.* (*J. Phytopath.* **133**: 289, 1991; variability in *P. terrestris* using isozyme polymorphism etc.), Sieber (*MR* **99**: 274, 1995; *P. ligniputridi* in *Abies* butt rot), Grondona *et al.* (*MR* **101**: 1405, 1997), Padhye & McGinnis *in* Murray *et al.* (Eds) (*Manual of Clinical Microbiology*: 1318, 1999), Infantino *et al.* (*MR* **107**: 707, 2003; Italy), Boerema *et al.* (*Phoma Identification Manual* Differentiation of Specific and Infra-Specific Taxa in Culture: 470 pp., 2004), Infantino & Pucci (*Eur. J. Pl. Path.* **112**: 337, 2005; molecular detection), Schoch *et al.* (*Mycol.* **98**: 1041, 2006; phylogeny).

Pyrenochaetella P. Karst. ex Höhn. (1917) nom. dub., anamorphic *Pezizomycotina*. See Sutton (*Mycol. Pap.* **141**, 1977).

Pyrenochaetina Syd. & P. Syd. (1916) = Parodiella fide Müller & von Arx (*Beitr. Kryptfl. Schweiz* **11** no. 2, 1962).

Pyrenochium Link (1833), ? Dothideomycetes. 1, Europe.

Pyrenocollema Reinke (1895), Xanthopyreniaceae (L). 1, Europe; N. America. Most species are transferred to *Collemopsidium*. See Puymaly (*Botaniste* **36**: 331, 1952), Tucker & Harris (*Bryologist* **83**: 1, 1980), Harris (*More Florida Lichens*, 1995), Aptroot & van den Boom (*Cryptog. Bryol.-Lichénol.* **19**: 193, 1998), Harada (*Bryologist* **102**: 50, 1999), McCarthy & Kantvilas (*Herzogia* **14**: 39, 2000; Tasmania), Kohlmeyer *et al.* (*Mycol. Progr.* **3**: 51, 2004), Mohr *et al.* (*MR* **108**: 515, 2004).

Pyrenocyclus Petr. (1955), Dothideomycetes. 1, Hawaii. See Petrak (*Sydowia* **9**: 515, 1955).

Pyrenodermium Bonord. (1851) = Hypoxylon Bull. fide Læssøe (*SA* **13**: 43, 1994).

Pyrenodesmia A. Massal. (1853) nom. rej. = Caloplaca fide Hawksworth *et al.* (*Dictionary of the Fungi* edn 8, 1995).

Pyrenodiscus Petr. (1927) = Diplonaevia fide Hein (*Nova Hedwigia* **38**: 669, 1983).

Pyrenodium Fée (1837) = Asterothelium fide Zahlbruckner (*Catalogus Lichenum Universalis* **1**, 1921).

Pyrenodochium Bonord. (1851) = Diatrype fide Hawksworth *et al.* (*Dictionary of the Fungi* edn 8, 1995).

Pyrenogaster Malençon & Riousset (1977) = Schenella fide Estrada-Torres *et al.* (*Mycol.* **97**: 139, 2005).

Pyrenographa Aptroot (1991), Requienellaceae. 1, Australia; S.E. Asia. Placed in the *Pyrenulaceae* by Eriksson (*in litt.*). See Alias *et al.* (*MR* **100**: 580, 1996).

pyrenomycete, one of the *Pyrenomycetes*.

Pyrenomycetes. Class of *Ascomycota*; used in various senses, but mostly for fungi with perithecioid ascomata which are also ascohymenial in ontogeny and have unitunicate asci, often with an apical annulus (e.g. *Diaporthales, Hypocreales, Sordariales, Xylariales*). The *Dothideales, Erysiphales, Meliolales* and *Laboulbeniales* also included by some authors.

The class is not generally accepted but was reintroduced by Berbee & Taylor (1992) in a restricted sense (see *Ascomycota*). However, the term 'pyrenomycetes' still has value as a colloquial term for all ascomycetes with flask-shaped ascomata. See Hanlin (*in* Parker, 1982, **1**: 225), Rogers (*in* Hawksworth, 1994: 321).

Pyrenomyxa Morgan (1895), Xylariaceae. 3, N. America; Asia. See Læssøe (*SA* **13**: 43, 1994; nomencl.), Stadler *et al.* (*Mycol.* **97**: 1129, 2005; key, chemistry), Bitzer *et al.* (*MR* **112**: 251, 2008; phylogeny, chemistry).

Pyrenopeziza Fuckel (1870), Helotiales. Anamorph *Cylindrosporium*. *c.* 59, widespread. See Nannfeldt (*Nova Acta R. Soc. Scient. upsal.*, 1932; 32 spp.), Gremmen (*Fungus* Wageningen **28**: 37, 1958; key sects.), Hütter (*Phytopath. Z.* **33**: 1, 1958; key), Majer *et al.* (*Pl. Path.* **47**: 22, 1998; genetic variation), Foster *et al.* (*Physiol. Mol. Pl. Pathol.* **55**: 111, 1999; PCR), Paavolainen *et al.* (*MR* **104**: 611, 2000), Paavolainen *et al.* (*Mycol.* **93**: 258, 2001; population genetics), Foster *et al.* (*Eur. J. Pl. Path.* **108**: 379, 2002; molecular diagnosis), Vrålstad *et al.* (*New Phytol.* **155**: 131, 2002; phylogeny), Pärtel & Raitviir (*Mycol. Progr.* **4**: 149, 2005; ultrastr.).

Pyrenopezizopsis Höhn. (1917) = Cenangiopsis Rehm fide Dennis (*Persoonia* **2**: 171, 1962).

Pyrenophora Fr. (1849), Pleosporaceae. Anamorph *Drechslera*. 8, widespread. Plant pathogens include: *P. chaetomioides*, anamorph *Drechslera avenacea* (oat leaf spot and seedling blight); *P. graminea*, anamorph *D. graminea* (barley leaf stripe); *P. teres*, anamorph *D. teres* (net blotch of barley). See Ammon (*Phytopath. Z.* **47**: 269, 1963; keys), Shoemaker (*CJB* **39**: 901, 1963; status), Sutton (*Taxon* **21**: 319, 1972; nomencl.), Crivelli (*Über die heterogene Ascomycetengattung Pleospora Rbh.*, 1983; key), Sivanesan (*Mycol. Pap.* **158**: 261 pp., 1987; anamorph), Scott (*Mycopathologia* **116**: 29, 1991; on *Hordeum*), Medd (*Review of Plant Pathology* **71**: 891, 1992; *P. semeniperda*), Peever & Milgroom (*CJB* **72**: 915, 1994; population structure), Stevens *et al.* (*Diagnosis and Identification of Plant Pathogens. Proceedings of the 4th International Symposium of the European Foundation for Plant Pathology*: 461, 1997; phylogeny), Dong *et al.* (*MR* **102**: 151, 1998; phylogeny), Pecchia *et al.* (*FEMS Microbiol. Lett.* **166**: 21, 1998; IGS analysis), Liew *et al.* (*Mol. Phylogen. Evol.* **16**: 392, 2000; phylogeny), Wu *et al.* (*Can. J. Pl. Path.* **25**: 82, 2003; genetic variation), Friesen *et al.* (*Phytopathology* **95**: 1144, 2005; population genetics), Serenius *et al.* (*MR* **109**: 809, 2005; genetic variation, mating types), Kodsueb *et al.* (*Mycol.* **98**: 571, 2006; phylogeny), Schoch *et al.* (*Mycol.* **98**: 1041, 2006; phylogeny).

Pyrenophoraceae M.E. Barr (1979) = Pleosporaceae.

Pyrenophoromyces Cif. & Tomas. (1953) ≡ Pyrenophorum.

Pyrenophoropsis C. Ramesh (1988), Fungi. 1, India. See Eriksson & Hawksworth (*SA* **8**: 75, 1989).

Pyrenophorum Tomas. & Cif. (1952) = Thelidium fide Hawksworth *et al.* (*Dictionary of the Fungi* edn 8, 1995).

Pyrenopolyporus Lloyd (1917) = Daldinia fide Miller (*Monogr. World spec. Hypoxylon*, 1961), Læssøe (*SA*

13: 43, 1994; ? synonym of *Xylaria*).

Pyrenopsidaceae Th. Fr. (1860) = Lichinaceae.

Pyrenopsidium (Nyl.) Forssell (1885) = Cryptothele fide Henssen & Büdel (*Beih. Nova Hedwigia* **79**: 381, 1984), Moreno & Egea (*Biología y taxonomía de la familia Lichinaceae*: 87 pp., 1991).

Pyrenopsis (Nyl.) Nyl. (1858) nom. cons., Lichinaceae (L). *c.* 41, widespread. See Jørgensen & Henssen (*Taxon* **39**: 343, 1990; nomencl.), Schultz & Büdel (*Lichenologist* **34**: 39, 2002; key).

Pyrenostigme Syd. (1926), Dothideomycetes. 2, C. America. See Hansford (*Mycol. Pap.* **15**, 1946).

Pyrenotea Fr. (1821) nom. rej. = Lecanactis Körb. fide Hawksworth & David (*Taxon* **38**: 493, 1989).

Pyrenothamnia Tuck. (1883) = Endocarpon fide Santesson (*Svensk bot. Tidskr.* **43**: 547, 1949).

Pyrenothamniaceae Zahlbr. (1898) = Verrucariaceae.

Pyrenothamniomyces Cif. & Tomas. (1953) ≡ Pyrenothamnia.

Pyrenothea, see *Pyrenotea*.

Pyrenotheca Pat. (1886) = Myriangium fide von Arx & Müller (*Stud. Mycol.* **9**, 1975).

Pyrenothricaceae Zahlbr. (1926), Pezizomycotina (inc. sed.) (L). 2 gen. (+ 1 syn.), 2 spp.

Lit.: Eriksson (*Systematic Botany*: 560, 1981)

Lit.:, Eriksson (*Op. Bot.* **60**, 1981), Rogers (*Flora of Australia* **54**: 65, 1992), Herrera-Campos *et al.* (*Mycol.* **97**: 356, 2005).

Pyrenothrix Riddle (1917), Pyrenothricaceae (L). 1, N. America; New Zealand. See Eriksson (*Op. Bot.* **60**, 1981), Tschermak-Woess *et al.* (*Pl. Syst. Evol.* **143**: 293, 1983; ultrastr.), Herrera-Campos *et al.* (*Mycol.* **97**: 356, 2005; Mexico).

Pyrenotrichum Mont. (1843), anamorphic *Badimia, Calopadia, Lasioloma, Sporopodium, Tapellaria*, St.0f-bH.15 (L). 2, widespread (esp. tropical). See Hawksworth (*Bull. Br. Mus. nat. hist.* Bot. **9**: 59, 1981), Sérusiaux (*Lichenologist* **18**: 1, 1986; conidioma campylidia), Kalb & Vězda (*Folia geobot. phytotax.* **22**: 286, 1987; nomencl.), Lücking *et al.* (*Lichenologist* **30**: 121, 1998), Lücking & Santesson (*Bryologist* **105**: 57, 2002).

Pyrenotrochila Höhn. (1917) = Trochila fide Dennis (*British Ascomycetes*, 1968).

Pyrenowilmsia R.C. Harris & Aptroot (1991), Pyrenulaceae (L). 1, S. Africa. See Aptroot (*Biblthca Lichenol.* **44**: 75, 1991).

Pyrenula Ach. (1809) nom. rej. = Thelidium fide Hawksworth *et al.* (*Dictionary of the Fungi* edn 8, 1995).

Pyrenula Ach. (1814) nom. cons., Pyrenulaceae (L). *c.* 200, widespread (esp. tropical). See Singh & Upreti (*Geophytology* **17**: 75, 1987; key 21 spp. Andaman Isl.), Upreti & Singh (*Geophytology* **18**: 67, 1988; key 14 spp. Sri Lanka), Harris (*Mem. N. Y. bot. Gdn* **49**: 74, 1989; gen. concept, key 47 N. Am. spp.), Upreti (*Bull. Soc. bot. Fr.* Let. bot. **3**: 241, 1991; key 12 spp. *P. brunnea* group), Upreti (*Feddes Repert.* **103**: 279, 1992; 24 spp. India), Aptroot *et al.* (*Biblthca Lichenol.* **64**, 1997; key Papua New Guinea spp.), Upreti (*Nova Hedwigia* **66**: 557, 1998; key Indian spp.), Etayo & Aptroot (*Lichenologist* **35**: 233, 2003; Panama), Grube *et al.* (*MR* **108**: 1111, 2004; phylogeny), Geiser *et al.* (*Mycol.* **98**: 1053, 2006; phylogeny), Aptroot *et al.* (*Biblthca Lichenol.* **97**, 2008; Costa Rica).

Pyrenulaceae Rabenh. (1870), Pyrenulales (±L). 10 gen. (+ 12 syn.), 284 spp.

Lit.: Hawksworth (*Nova Hedwigia* **43**: 1, 1986), Harris (*Mem. N. Y. bot. Gdn* **49**: 74, 1989), Aptroot (*Biblthca Lichenol.* **44**, 1991), Aptroot (*Biblthca Lichenol.* **44**: 120, 1991), Eriksson & Hawksworth (*SA* **11**: 71, 1992; concept), Tibell (*Fl. Neotrop.* Monogr. **69**: 78 pp., 1996), Upreti (*Nova Hedwigia* **66**: 557, 1998), Harada (*Lichenologist* **31**: 567, 1999), Hyde & Wong (*MR* **103**: 347, 1999), Bhattacharya *et al.* (*Mol. Biol. Evol.* **17**: 1971, 2000), Lumbsch *et al.* (*Symb. bot. upsal.* **34** no. 1: 9, 2004), Prado *et al.* (*MR* **110**: 511, 2006).

Pyrenulales Fink ex D. Hawksw. & O.E. Erikss. (1986), (±L). Chaetothyriomycetidae. 6 fam., 42 gen., 538 spp. Thallus (where present) crustose, often immersed, often inconspicuous; ascomata perithecial, immersed or erumpent, ± globose or flattened, often clypeate, sometimes aggregated into stromata, the peridium usually thick-walled, usually composed of small-celled pseudoparenchymatous tissue, sometimes crystalline; interascal tissue initially of thin-walled anastomosing pseudoparaphyses, subsequently ± unbranched paraphyses often developing from the base of the ascoma, occasionally only of true paraphyses; asci cylindrical, persistent, usually multilayered and fissitunicate, sometimes with a conspicuous apical ring, not blueing in iodine; ascospores hyaline or brown, septate, sometimes muriform, often with strongly thickened septa and angular lumina, sometimes with a sheath. Anamorphs pycnidial where known. Saprobes on bark, or lichenized with green algae (esp. *Trentepohliaceae*), mainly trop.

Fams:

(1) **Celotheliaceae**

(2) **Massariaceae**

(3) **Monoblastiaceae** (syn. *Acrocordiaceae*)

(4) **Pyrenulaceae** (syn. *Pleurotremataceae*)

(5) **Requienellaceae**

(6) **Xanthopyreniaceae**

Lit.: Aptroot (*Biblthca Lichenol.* **44**, 1991), Del Prado *et al.* (*MR* **110**: 511, 2006; phylogeny), Lutzoni (*Am. J. Bot.* **91**: 114, 2004).

Pyrenulella Fink (1935) ≡ Pharcidiella.

Pyrenulomyces E.A. Thomas ex Cif. & Tomas. (1953) = Pyrenula Ach. (1814) fide Hawksworth *et al.* (*Dictionary of the Fungi* edn 8, 1995).

Pyrenyllium Clem. (1909), anamorphic *Pezizomycotina* (?L). 1. See Aguirre-Hudson (*Bull. Br. Mus. nat. hist.* Bot. **21**: 85, 1991).

Pyrgidiomyces Cif. & Tomas. (1953) ≡ Pyrgidium.

Pyrgidium Nyl. (1867), Sphinctrinaceae (?L). 1, widespread. See Tibell (*Lichenologist* **14**: 219, 1982), Tibell (*Symb. bot. upsal.* **27** no. 1, 1987), Tibell (*Symb. bot. upsal.* **32**: 291, 1997; anamorphs), Tibell & Wedin (*Mycol.* **92**: 577, 2000).

Pyrgillocarpon Nádv. (1942) = Pyrgillus fide Aptroot (*Biblthca Lichenol.* **44**, 1991).

Pyrgillomyces Cif. & Tomas. (1953) ≡ Pyrgillus.

Pyrgillus Nyl. (1858), Pyrenulaceae (L). 4, widespread. See Tibell (*Beih. Nova Hedwigia* **79**: 597, 1984), Lumbsch *et al.* (*Symb. bot. upsal.* **34** no. 1: 9, 2004; phylogeny), Geiser *et al.* (*Mycol.* **98**: 1053, 2006; phylogeny), Aptroot *et al.* (*Biblthca Lichenol.* **97**, 2008; Costa Rica).

Pyrgostroma Petr. (1951), anamorphic *Pezizomycotina*, St.≡ eP.?. 1, USA. See Petrak (*Sydowia* **5**: 484, 1951).

Pyricularia Sacc. (1880), anamorphic *Magnaporthe*,

Hso.≡ eP.10. 14, widespread. *P. oryzae* (sometimes as *P. grisea*; conservation to be sought), rice blast. See Ou (Ed.) (*The rice blast disease*, 1965), Vales *et al.* (*L'Agron. Tropic.* **41**: 242, 1986; electrophoresis in *P. oryzae* identification), Leung & Williams (*CJB* **65**: 112, 1987; nuclear division and chromosomes), Rossman *et al.* (*Mycol.* **82**: 509, 1990; nomencl. rice blast pathogen), Zhu *et al.* (*Mycosystema* **5**: 89, 1992; DNA fingerprinting and pathotype identification), Zeigler *et al.* (*Rice blast disease*, 1994), George *et al.* (*Phytopathology* **88**: 223, 1998), Kusaba *et al.* (*Ann. phytopath. Soc. Japan* **64**: 125, 1998), Babujee & Gnanamanickam (*Curr. Sci.* **78**: 248, 2000; population analysis), Kato *et al.* (*J. Gen. Pl. Path.* **66**: 30, 2000; genetics, pathogenicity), Kusaba *et al.* (*Ann. phytopath. Soc. Japan* **65**: 588, 1999; genetic diversity), Couch & Kohn (*Mycol.* **94**: 683, 2002; separation of *P. oryzae* and *P. grisea*), Farman & Kim (*Molecular Plant Pathology* **6**: 287, 2005; genetics), Hirata *et al.* (*MR* **111**: 799, 2007; speciation).

Pyriculariopsis M.B. Ellis (1971), anamorphic *Pezizomycotina*, Hso.≡ eP.10. 5, widespread (tropical). See Lai & Gao (*Acta Mycol. Sin.* **10**: 79, 1991; synonymy of *Gonatopyricularia*), Bussaban *et al.* (*Mycol.* **97**: 1002, 2005; phylogeny).

pyriform, pear-like in form. Cf. obpyriform (Fig. 23.14).

Pyriomyces Bat. & H. Maia (1965), anamorphic *Byssoloma*, Cpd.0eH.? (L). 1, Brazil. See Lücking *et al.* (*Lichenologist* **30**: 121, 1998).

Pyripnomyces Cavalc. (1972), anamorphic *Pezizomycotina*, Cpt.1bH.? (L). 1, Brazil. See Cavalcanti (*Publicações. Instituto de Micologia da Universidade de Pernambuco* **647**: 20, 1972), Lücking *et al.* (*Lichenologist* **30**: 121, 1998).

Pyrisperma Raf. (1808) nom. dub., Agaricomycetes. 'gasteromycetes'. = Hymenogaster (Hymenogastr.) fide Soehner, 1962; or 'discomycetes', hypogeous).

Pyrispora, see *Pyrisperma*.

Pyrobolus Kuntze (1891) ≡ Eurotium.

Pyrochroa Eschw. (1824) = Phaeographis fide Hawksworth *et al.* (*Dictionary of the Fungi* edn 8, 1995).

Pyroctonum Prunet (1897) = Cladochytrium fide Karling (*Chytriomyc. Iconogr.*, 1977).

Pyrofomes Kotl. & Pouzar (1964), Polyporaceae. 6, widespread. See Decock & Ryvarden (*MR* **103**: 1138, 1970), Ryvarden & Johansen (*Prelim. Polyp. Fl. E. Afr.*: 528, 1980; key 3 Afr. spp.), Piątek & Cabala (*Mycotaxon* **91**: 173, 2005; Poland).

Pyrographa Fée ex A. Massal. (1860) nom. rej. prop. = Phaeographis fide Lücking *et al.* (*Taxon* **56**: 1296, 2007; nomencl.).

Pyronema Carus (1835), Pyronemataceae. 2, widespread. See Moore & Korf (*Bull. Torrey bot. Club* **90**: 33, 1963; key), Hansen & Pfister (*Mycol.* **98**: 1029, 2006; phylogeny), Perry *et al.* (*MR* **111**: 549, 2007; phylogeny).

Pyronemataceae Corda (1842), Pezizales. 80 gen. (+ 55 syn.), 662 spp.

Lit.: Trappe (*Mycotaxon* **9**: 297, 1979) separated the *Geneaceae* from cleistothecial *Pyronemataceae*, based primarily on a hollow rather than solid or chambered ascoma. Molecular data do not as yet provide a convincing picture of evolution within the *Pyronemataceae*, but it is likely that further subdivision will occur.

Lit.:, Kimbrough & Curry (*Mycol.* **78**: 735, 1986; septal ultrastr.), Kimbrough & Curry (*Mycol.* **78**: 407, 1986), Zhuang & Korf (*Mycotaxon* **26**: 361, 1986), Benkert (*Z. Mykol.* **53**: 195, 1987), Moore (*Mycotaxon* **29**: 91, 1987), Hohmeyer (*Mitt. Arb. Pilzk. Niederrhein* **6**: 32, 1988; synoptic key *Aleuriae*), Kimbrough (*Mem. N. Y. bot. Gdn* **49**: 323, 1989) Many previous authors have separated the *Humariaceae* from this fam., based largely on the presence of carotenoid pigments.

Lit.:, Kimbrough (*Mem. N. Y. bot. Gdn* **49**: 326, 1989; limits), Moravec (*Mycotaxon* **36**: 169, 1989), Schumacher (*Op. bot.* **101**: 107 pp., 1990), Wu & Kimbrough (*Bot. Gaz.* **152**: 421, 1991), Wu & Kimbrough (*Int. J. Pl. Sci.* **153**: 128, 1992; ascosporogenesis) and under *Pezizales*, Pegler *et al.* (*British truffles*, 1993; hypogeous taxa), Landvik *et al.* (*Nordic Jl Bot.* **17**: 403, 1997; DNA), Harrington *et al.* (*Mycol.* **91**: 41, 1999), Norman & Egger (*Mycol.* **91**: 820, 1999; phylogeny), Hansen *et al.* (*Mol. Phylogen. Evol.* **36**: 1, 2005), Hansen & Pfister (*Mycol.* **98**: 1029, 2006; phylogeny), Hansen *et al.* (*Mycol.* **97**: 1023, 2005), Liu & Zhuang (*Fungal Diversity* **23**: 181, 2006), Læssøe & Hansen (*MR* **111**: 1075, 2007; phylogeny), Perry *et al.* (*MR* **111**: 549, 2007; phylogeny).

Pyronemella (Vido) Sacc. (1889) = Lasiobolus fide Pfister (*Mycol.* **76**: 843, 1984).

pyrophilous, growing on burnt ground, steam sterilized soil etc.; carbonicolous; **- fungi**, fireplace fungi, phoenicoid fungi (see Ramsbottom, *Mushrooms and toadstools*: 231, 1953; list of some larger fungi, Webster *et al.*, *TBMS* **47**: 445, 1964; discomycetes); see also *Pyronema*.

Pyropolyporus Murrill (1903) ≡ Phellinus.

Pyropyxis Egger (1984), Pyronemataceae. Anamorph *Dichobotrys*. 1, N. America. See Egger (*CJB* **62**: 705, 1984).

Pyrotheca E. Hesse (1935), Microsporidia. 3.

pyroxylophilous, living on burnt wood.

Pyrrhoderma Imazeki (1966), Hymenochaetaceae. 2, Japan. See Ryvarden (*Syn. Fung.* **5**, 1991), Wagner & Fischer (*Mycol.* **94**: 998, 2002; phylogeny).

Pyrrhoglossum Singer (1944), Cortinariaceae. 11, widespread (tropical). See Horak (*Op. bot.* **100**: 115, 1989), Horak & Desjardin (*Mem. N. Y. bot. Gdn* **89**: 61, 2004).

Pyrrhospora Körb. (1855), Lecanoraceae (L). 28, widespread. See Hafellner (*Herzogia* **9**: 725, 1993; key), Ekman & Wedin (*Pl. Biol.* **2**: 350, 2000; phylogeny), Spribille & Hauck (*Bryologist* **106**: 560, 2003; N America), Elix (*Australasian Lichenology* **55**: 26, 2004; Australia), Elix & Kantvilas (*Australasian Lichenology* **57**: 6, 2005; Australia), Miądlikowska *et al.* (*Mycol.* **98**: 1088, 2006; phylogeny).

pythiosis, a cosmopoliitan disease of horses, cattle, and dogs caused by *Pythium insidiosum* (De Cock *et al.*, *J. clin. Microbiol.* **25**: 344, 1987). See also Mendoza *et al.* (*J. med. vet. Mycol.* **26**: 5, 1988).

Pythites Pampal. (1902), Fossil Fungi. 1 (Eocene), Italy.

pyxidate, provided with a lid, pertaining to, of having the character of a box; box-like.

Pyxidiophora Bref. & Tavel (1891), Pyxidiophoraceae. Anamorphs *Chalara*-like, *Thaxteriola*. 17 (mainly coprophilous), widespread. See Lundqvist (*Bot. Notiser* **133**: 121, 1980), Barrasa & Moreno (*Cryptog. Mycol.* **4**: 251, 1983; sects.), Blackwell *et al.* (*Science* **232**: 993, 1986; anamorphs), Blackwell & Malloch (*CJB* **67**: 2552, 1989; life-histories), Blackwell *et al.*

(*MR* **92**: 397, 1989; spp. on mites), Blackwell (*Mycol.* **86**: 1, 1994; life-cycle, molec. relationships), Weir & Blackwell (*MR* **105**: 1182, 2001; phylogeny), Kirschner (*Mycol. Progr.* **2**: 209, 2003; Europe).

Pyxidiophoraceae G.R.W. Arnold (1971), Pyxidiophorales. 5 gen. (+ 13 syn.), 22 spp.
Lit.: Lundqvist (*Bot. Notiser* **133**: 121, 1980), Blackwell *et al.* (*Science* N.Y. **232**: 993, 1986), Webster & Hawksworth (*TBMS* **87**: 77, 1986), Blackwell & Malloch (*Mem. N. Y. bot. Gdn* **49**: 23, 1989), Blackwell & Malloch (*CJB* **67**: 2552, 1989), Blackwell *et al. in* Wingfield *et al.* (Eds) (*Ceratocystis* and *Ophiostoma* Taxonomy, Ecology and Pathogenicity: 105, 1993; rels), Blackwell & Jones (*Biodiv. Cons.* **6**: 689, 1997; biology), Henk *et al.* (*Mycol.* **95**: 561, 2003).

Pyxidiophorales P.F. Cannon (2001). Laboulbeniomycetidae. 1 fam., 5 gen., 22 spp.
The order is confirmed as a member of *Laboulbeniomycetidae* by molecular data (Blackwell, *Mycol.* **86**: 1, 1994). Fam.:
Pyxidiophoraceae
Lit.: see under Fam. and *Pyxidiophora*.

Pyxidium Hill (1771) = Cladonia fide Hawksworth *et al.* (*Dictionary of the Fungi* edn 8, 1995).

Pyxine Fr. (1825), Caliciaceae (L). 51, widespread (tropical). See Imshaug (*Trans. Am. microsc. Soc.* **76**: 246, 1957), Swinscow & Krog (*Norw. Jl Bot.* **22**: 43, 1975; E. Afr.), Kashiwadani (*J. Jap. Bot.* **52**: 137, 1977; Japan), Kashiwadani (*Bull. natn. Sci. Mus.* Tokyo, B **3**: 63, 1977; Papua New Guinea), Awasthi (*Phytomorphology* **30**: 359, 1982; key 21 spp. India), Rogers (*Aust. J. Bot.* **34**: 131, 1986; key 15 spp. Australia), Rogers (*Brunonia* **9**: 229, 1986; Australia), Kalb (*Biblthca Lichenol.* **24**, 1987; key 24 spp. Brazil), Sammy (*Nuytsia* **6**: 279, 1988; key 4 spp. W. Australia), Kalb (*Herzogia* **10**: 61, 1994; key 21 spp. Australia), Nordin & Mattsson (*Lichenologist* **33**: 3, 2001; phylogeny), Scheidegger *et al.* (*Lichenologist* **33**: 25, 2001; evolution), Helms *et al.* (*Mycol.* **95**: 1078, 2003; phylogeny), Hu & Chen (*Mycotaxon* **86**: 445, 2003; China), Moberg (*Symb. bot. upsal.* **34** no. 1: 257, 2004; S Africa), Miądlikowska *et al.* (*Mycol.* **98**: 1088, 2006; phylogeny).

Pyxineaceae Stizenb. (1862) = Caliciaceae.

Q, the ratio of length to breadth of elongate spores of agarics; spores ellipsoidal or ovoid when Q = 2, ellipsoidal-oblong, fusoid, cylindrical, etc. when Q = › 2 (Singer, 1962: 68). Cf. E, sporograph.

Quadracaea Lunghini, Pinzari & Zucconi (1996), anamorphic *Pezizomycotina*, Hso.?.?. 1, Italy. See Lunghini *et al.* (*Mycotaxon* **60**: 103, 1996), Wu & Zhuang (*Fungal Diversity Res. Ser.* **15**, 2005).

quadrangular, see rhomboidal.

Quadricladium Nawawi & Kuthub. (1989), anamorphic *Pezizomycotina*, Hso.1bH.10. 1 (aquatic), Malaysia. See Nawawi & Kuthubutheen (*Mycotaxon* **34**: 489, 1989).

Quadrispora Bougher & Castellano (1993), Cortinariaceae. 3, Australia. See Bougher & Castellano (*Mycol.* **85**: 285, 1993).

Quadrisporomyces Sekunova (1960) = Schizosaccharomyces fide Batra *in* Subramanian (Ed.) (*Taxonomy of fungi* **1**: 187, 1978).

Quambalaria J.A. Simpson (2000), Quambalariaceae. 4, widespread. See Simpson (*Australasian Mycologist* **19**: 60, 2000), Beer *et al.* (*Stud. Mycol.* **55**: 289, 2006; mol. phylog.).

Quambalariaceae Z.W. Beer, Begerow & R. Bauer (2006), Microstromatales. 1 gen., 4 spp.
Lit.: Simpson (*Australas. Mycol.* **19**: 57, 2000), Begerow *et al.* (*MR* **105**: 809, 2001), Beer *et al.* (*Stud. Mycol.* **55**: 289, 2006).

Quasiconcha M.E. Barr & M. Blackw. (1981), Mytilinidiaceae. Anamorph *Chalara*-like. 1, USA. See Blackwell & Gilbertson (*Mycol.* **77**: 50, 1985; anamorph), Barr (*N. Amer. Fl.* ser. 2 **13**: 129 pp., 1990).

Quasidiscus B. Sutton (1991), anamorphic *Pezizomycotina*, Ccu.0eH.15. 1, Australia. See Sutton (*Sydowia* **43**: 276, 1991).

Quasiphloeospora B. Sutton, Crous & Shamoun (1996), anamorphic *Mycosphaerellaceae*. 1, USA. See Sutton *et al.* (*MR* **100**: 979, 1996), Crous & Braun (*CBS Diversity Ser.* **1**: 571 pp., 2003).

Quaternaria Tul. & C. Tul. (1863), Diatrypaceae. 2, widespread (temperate). See Rappaz (*Mycol. Helv.* **2**: 285, 1987).

Quatsinoporites S.Y. Sm., Currah & Stockey (2004), Fossil Fungi. 1, Canada. See Smith *et al.* (*Mycol.* **96**: 181, 2004).

Queenslandia Bat. & H. Maia (1959), anamorphic *Pezizomycotina*, Cpt.0eH.?. 4, widespread. See Batista & Maia (*Mycopath. Mycol. appl.* **11**: 75, 1959).

Queirozia Viégas & Cardoso (1944) = Pleochaeta fide Braun (*Mycotaxon* **15**: 150, 1982), Liberato *et al.* (*MR* **110**: 567, 2006; anamorph).

Quélet (Lucien; 1832-1899; France). Student, Montbéliard; medical student then laboratory technician, Strasbourg (*c.* 1854); practising doctor, Montbéliard (*c.* 1854 onwards. An amateur contributing significantly to the mycota of France; first President of the Société Mycologique de France (1885); decorated for services as a volunteer doctor during the Franco-Prussian war (1870-1871); described as a 'rough diamond' particularly after the premature death of his son. *Publs. Les Champignons du Jura et des Voges* (1872-1875) [in which a number of Fries' sub-genera of the *Agaricaceae* were first used as genera; reprint 1964 + 1972 suppl., 1872-1902; index to reprint, *BSMF* **88**: cxi, 1973]; (with Cooke, q.v.) *Clavis Synoptica Hymenomycetum Europaeorum* (1878); *Encheiridion Fungorum in Europa Media et Praesertim in Gallia Vigentium* (1886); *Flore Mycologique de la France.* (1888) [reprint 1962]. *Biogs, obits etc.* Boudier (*BSMF* **15**: 321, 1899); Gilbert (*BSMF* **65**: 5, 1949) [bibliography]; Grummann (1974: 294); Mangin & Chomette (*Essai d'une Table de Concordance des Principales Espèces Mycologiques avec la Flore de France et des Pays Limitrophes de Lucien Quélet* (1906) [reprint 1963]); Stafleu & Cowan (*TL-2* **4**: 453, 1983).

Queletia Fr. (1872), Agaricaceae. 2, widespread. See Wright (*Cryptog. Bot.* **1**: 26, 1989).

quellkörper, a mucilaginous mass of thick-walled cells within the ascoma of *Nitschkeaceae*; believed to induce rupture of the ascoma.

Quercella Velen. (1921) nom. rej. = Phaeocollybia fide Kuyper (*in litt.*).

Questieria G. Arnaud (1918) = Schiffnerula fide von Arx & Müller (*Stud. Mycol.* **9**, 1975).

Questieriella G. Arnaud ex S. Hughes (1983), anamorphic *Schiffnerula*, Hso.≡ eP.1. 10, widespread. See Hughes (*CJB* **61**: 1729, 1983), Hughes (*Mycol.* **82**: 657, 1990), Hosagoudar (*Zoos' Print Journal* **18**: 1071, 2003).

Quezelia Faurel & Schotter (1965), anamorphic *Pezizomycotina*, Cpt.1eH.?. 1 (coprophilous), Sahara. See Faurel & Schotter (*Revue Mycol.* Paris **30**: 161, 1964).

Quilonia K.P. Jain & R.C. Gupta (1970), Fossil Fungi, anamorphic, *Pezizomycotina*. 1 (Miocene), India.

quinine fungus, *Fomes officinalis*.

Quintaria Kohlm. & Volkm.-Kohlm. (1991), Lophiostomataceae. 1 (marine), Belize; Australasia. See Kohlmeyer & Volkmann-Kohlmeyer (*Aust. J. mar. Freshwat. Res.* **42**: 91, 1991), Hyde & Goh (*Nova Hedwigia* **68**: 251, 1999).

Quorn, The trade name under which mycoprotein (q.v.) manufactured from *Fusarium graminearum* by Marlow Foods (a subsidiary of Rank Hovis McDougall and ICI) is marketed.

Rabdosporium Chevall. (1826) = Cheirospora fide Hughes (*CJB* **36**: 800, 1958).

Rabenhorst (Gottlob Ludwig; 1806-1881; Germany). Student, Belzig and Berlin (1822-1830); a pharmacist in Lukau (1830-1840); doctorate, Jena (1841); private scientists, Dresden (1840-1875) and Meissen (1875-1881). An outstanding collector, publishing large series of exsiccati; first editor of *Hedwigia* (1852-1878). *Publs. Klotzschii Herbarium Vivum Mycologicum* (1842-1855) [nearly 2,000 fungal exsiccati]; *Deutschlands Kryptogamen-Flora* (1844-1853); 2 Aufl. *Kryptogamen-Flora von Deutschland, Oesterreich und der Schweiz* 2 Aufl. (1884-1932) [a vast work, initiated by him and continuing long after his death]; *Lichenes Europaei Exsiccati* (1855-1879) [nearly 1,000 exsiccati]; *Fungi Europaei Exsiccati* (1859 onwards) [see Kohlmeyer, *Beihefte zur Nova Hedwigia* **4**, 1963; Stevenson, *Taxon* **16**: 112, 1967, numbering of specimens]. *Biogs, obits etc.* Grummann (1974: 38); Richter (*Hedwigia* **20**: 113, 1881); Stafleu & Cowan (*TL-2* **4**: 460, 1983).

Rabenhorstia Fr. (1849), anamorphic *Hercospora*, St.0eH.1. 10, widespread (esp. Europe).

Rabenhorstiites Teterevn. & Tasl. (1977), Fossil Fungi. 1 (Tertiary), former USSR.

Rabenhorstinidium R.B. Singh & G.V. Patil (1980), Fossil Fungi, anamorphic *Pezizomycotina*. 1 (Cretaceous), India.

race, see physiologic race and Classification.

Racemella Ces. (1861) = Cordyceps fide Tulasne & Tulasne (*Select. fung. carpol.* **3**: 4, 1865).

Racemosporium Moreau & Moreau (1941) = Arthrinium fide Ellis (*Mycol. Pap.* **103**, 1965).

Rachicladosporium Crous, U. Braun & C.F. Hill (2007), anamorphic *Capnodiales*. 1, New Zealand. See Crous *et al.* (*Stud. Mycol.* **58**: 33, 2007).

Rachidicola K.D. Hyde & J. Fröhl. (1995), Dothideomycetes. 1, Hong Kong. See Hyde (*Sydowia* **47**: 217, 1995).

rachis, a geniculate or zig-zag holoblastic extension of a conidiogenous cell (as in *Tritirachium*) resulting from sympodial conidiogenous cell development; **rachiform** (of conidiogenous cells), having a rachis; cf. raduliform.

Rachisia Lindner (1913) = Fusarium fide Hawksworth *et al.* (*Dictionary of the Fungi* edn 8, 1995).

Raciborski (Marjan; 1863-1917; Poland). Student, München (1894); Indonesia (1896-1900); Professor of Botany, Agricultural Academy, Dublany (1900-1909) then Professor and Director, Plant Physiological Institute (1909-1912), L'viv; Professor of Botany, Cracow University (1912-1917). Pioneer mycologist in Indonesia; pioneer of nature conservation in Poland; also wrote papers on fungi of Poland and Switzerland. *Publs. Parasitische Algen und Pilze Java?s* (1900). *Biogs, obits etc.* Goebel (*Bericht der Deutschen Botanischen Gesellschaft* **35**: (97), 1918; bibliography); Grummann (1974: 437); Kornasia (*Marian Raciborski*, 1986); Stafleu & Cowan (*TL-2* **4**: 529, 1983).

Raciborskiella Höhn. (1909) = Strigula fide Harris (*in litt.*), Roux & Sérusiaux (*Bull. Soc. linn. Provence* **46**: 91, 1995; ascospores), Roux *et al.* (*Biblthca Lichenol.* **90**: 96 pp., 2004).

Raciborskiella Speg. (1919) ≡ Trichopeltella.

Raciborskiomyces Siemaszko (1925) = Epipolaeum fide Müller & von Arx (*Beitr. Kryptfl. Schweiz* **11** no. 2, 1962).

racket (**racquette**) **cell** (of dermatophytes), a hyphal cell having a swelling at one end; cf. hypha.

Racoblenna A. Massal. (1852) = Placynthium fide Hawksworth *et al.* (*Dictionary of the Fungi* edn 8, 1995).

Racoblennaceae Arnold (1858) = Placynthiaceae.

Racodium Fr. (1829) nom. cons., anamorphic *Capnodiales*, Hso.0eP.? (±L). 1, widespread. See Hawksworth & Riedl (*Taxon* **26**: 208, 1977; nomencl.), Kantvilas (*Muelleria* **16**: 65, 2002; Australia), Sakamoto & Miyamaoto (*Forest Pathology* **35**: 1, 2005; Japan), Tribe *et al.* (*Mycologist* **20**: 171, 2006), Muggia *et al.* (*MR* **112**: 50, 2008; phylogeny).

Racodium Pers. (1794) nom. rej. = Rhinocladiella Nannf. See cellar fungus. fide Hawksworth & Riedl (*Taxon* **26**: 208, 1977), Tribe *et al.* (*Mycologist* **20**: 171, 2006).

Racoplaca Fée (1825) = Strigula fide Hawksworth *et al.* (*Dictionary of the Fungi* edn 8, 1995).

Racospermyces J. Walker (2001), ? Raveneliaceae. 6 (on *Acacia* (*Leguminosae*)), Australia; E. Asia; Hawaii. See Walker (*Australasian Mycologist* **20**: 3, 2001; monogr., key), Scholler & Aime (*Mycoscience* **47**: 159, 2006; syn. of *Endoraecium*).

Racovitziella Döbbeler & Poelt (1978), Dothideomycetes. 1 (on *Musci*), Austria. See Döbbeler & Poelt (*Mitt. bot. StSamml.* Münch. **14**: 206, 1978).

racquette cell, see racket cell.

Radaisiella Bainier (1910) = Botryosporium Corda fide Mason (*Mycol. Pap.* **2**: 27, 1928).

Raddetes P. Karst. (1887) nom. rej. = Conocybe fide Kuyper (*in litt.*).

radial (of lichen thalli), radially symmetrical in transverse section (e.g. *Alectoria*, *Bryoria*, *Coelocaulon*, *Usnea*).

radiate, spreading from a centre.

Radiation and fungi. Fungi like all living organisms are influenced by ionizing and non-ionizing radiation.

Ionizing radiation affects fungal DNA both directly and indirectly and causes a wide range of damage. Harmful effects, including chromosomal breaks and elevated mutation frequencies were identified during satellite and space studies with different fungi (Dublin & Volz, *Origins of Life and Evolution of Biospheres* **4**: 223, 1973; Volz *et al.*, *Mycopath. mycol. Appl.* **54**: 221, 1974). Spore germination and subsequent hyphal growth of soil microfungi in the presence of pure gamma or mixed beta and gamma radiation is influenced by the origin of the isolate. Fungi isolated from background radiation sources show inhibition or no response in germination when experimentally irradiated, but isolates from sites with elevated mixed radiation showed stimulation in spore germination and enhanced hyphal growth. Prior long-term exposure to radiation influences fungal decomposition of radionuclide-bearing resources in the environment (Tugay *et al.*, *Mycologia* **98**: 521, 2006).

Ecosystems throughout the world have been contaminated with radionuclides as a result of above ground nuclear testing, nuclear reactor accidents and nuclear power generation. The reactor disaster at the Chernobyl Nuclear Power Station in April 1986, in particular, led to huge amounts of radioactive material being distributed over a large area of Europe, stimulating studies of radiation effects on fungi. Radioisotopes characteristic of nuclear fission, such as ^{137}Cs and ^{90}Sr, were released into the environment, becoming more concentrated as they move up the food chain and often becoming human health hazards. Following Chernobyl, lichens (mainly *Cladonia rangiferina*) of northern Scandinavia accumulated so much radioactivity that reindeer feeding on them were considered dangerous for human consumption. Basidiomycete fruit bodies are known to accumulate radioactive nuclides (Haselwandter, *Health Phys.* **4**: 713, 1978; Bakken & Olsen, *Can. J. Microbiol.* **36**: 704, 1990; Grodzinskaya *et al.*, *Sydowia* **10**: 88, 1995). Concentrations of radioactive ^{134}Cs, ^{137}Cs in mushrooms vary in different regions with, e.g. a maximum of 2860 Bq kg- (dry wt.) for France (Kirchner & Daillant, *Science of the Total Environment* **222**: 63, 1998). Radionuclides accumulate in mushroom fruit bodies under experimental conditions in the following order: ^{83}Rb ^{65}Zn ^{54}Mn ^{22}Na ^{75}Se and ^{85}Sr ^{60}Co ^{88}Y, ^{102}Rh, ^{139}Ce ^{143}Ce, ^{144}Pm ^{146}Pm, ^{153}Gd and ^{173}Lu ^{175}Hf (Tadaaki Ban-Nai *et al.*, *J. Radiation Res.* **38**: 213, 1997).

Reports of high concentrations of fallout radiocaesium in basidiomycete fruit bodies after Chernobyl and speculation that fungi could be long-term ^{137}Cs accumulators led to special studies which indicated that uptake of Cs by common fungi of upland grassland ranged from 44 to 235 nmol Cs g^{-1} d.w. h^{-1} with more than 40% of the Cs taken up from environment being bound within the hyphae. Immobilization of radiocaesium fallout by the soil fungal component is discussed by Dighton *et al.* (*MR* **95**: 1052, 1991). Study during 1996-1998 of ^{137}Cs concentration in mushrooms from one of the regions of Ukraine contaminated by radiation revealed higher accumulations in fruit bodies of *Paxillus involutus*, *Sarcodon imbricatus*, *Xerocomus subtomentosus* and *Cantharellus cibarius* than in mycelium from soil, with ratios ranging from 0.1 to 66 and a mean of 9.9. Fungal mycelium accumulated from 0.1 to 50% ^{137}Cs compared with sampled soil (Vinichuk & Johanson *J. environ. Radioactivity* **64**: 27, 2002). Another study revealed higher levels of ^{90}Sr transfer to fruit bodies of some mushrooms from soils of sampling sites than $^{239+240}$Pu and ^{241}Am transfer. The Available Transfer Factor (ATF) index, based on the the ratio between fruit-body content and the amount of radionuclide available for transfer in soil, has been proposed as a measure of the actual process of transfer. *Hebeloma cylindrosporum*and *Lycoperdon perlatum* were found to have $^{239+240}$Pu and ^{241}Am ATF values greater than or similar to those of ^{90}Sr, and such fungi have been suggested as suitable as bioindicators for radioactive nuclides in a forest ecosystems (Baeza *et al.*, *Radiochimica Acta* **94**: 75, 2006). *Ganoderma lucidum* has also been proposed as an indicator for ^{137}Cs environmental monitoring because of its high levels of accumulation (Tran Van & Le Duy, *J. radioanalytical nucl. chem.* **155**: 451, 1991).

Biosorption of radioactive elements is exhibited by different fungi. *Rhizopus arrhizus* takes up to 180 mg U^{6}/g of uranium from aqueous solutions. Uranium adsorption in cell-wall chitin occurs simultaneously and rapidly; precipitation of uranyl hydroxide within the chitin microcrystalline cell-wall occurs at a lower rate (Tsezos & Volesky, *Biotechnol. Bioeng. Symp.* **24**: 385, 1982). There is experimental evidence for fungal transformations of uranium solids and production of secondary mycogenic uranium minerals. Saprotrophic, ericoid and ectomycorrhizal fungi dissolve uranium oxides and accumulate uranium within the mycelium to over 80 mg (g dry weight)(-1) biomass (Fomina *et al.*, *Environ. Microbiol.* **9**: 1696, 2007). Absorption and adsorption of radionuclides from 'hot particles' originating from the Chernobyl accident and from atomic bomb testing sites has also been observed. Studied species and strains of soil fungi overgrew and dissolved the 'hot particles' after prolonged contact. Accumulation was generally more intensive for ^{152}Eu than for ^{137}Cs by a factor of about 2.6–134. The main factors influencing Cs and Eu accumulation in soil fungi seem to be fungal species, fungal strain, and size and composition of the 'hot particle' (Zhdanova *et al.*, *J. environ. radioactivity* **67**: 119, 2003).

Microfungi have been found to be highly resilient to exposure to ionizing radiation. Free-living and mycorrhizal fungi can colonize thermodynamically unstable depleted uranium surfaces and transform metallic depleted uranium into uranyl phosphate minerals (Fomina *et al.*, *Curr. Biol.* **18**: R375, 2008). Radioresistance of some fungal species has been linked to the presence of melanin, which has been shown to have emerging properties of acting as an energy transporter for metabolism (Dighton *et al.*, *FEMS Microbiology Letters* **281**: 109, 2008; review). Sustained exposure of microfungi to radiation appears to have resulted in formerly unknown adaptive features, such as directed growth of fungi toward sources of ionizing radiation. Soil microfungi isolated from around the Chernobyl Nuclear Power Station are able to grow into and decompose 'hot particles' in the carbon based radioactive graphite from the reactor. Long-term investigations of about 2000 strains of 200 species of 98 genera revealed that both beta and gamma radiation promote directional growth of hyphae towards the source of radiation (Zhdanova *et al.*, *MR* **108**: 1089, 2004). Recent research raises the possibility that some fungi use the pigment melanin to extract energy from ionizing radiation, such as gamma radiation, for 'radiotrophic' growth. There is speculation that these properties of energy capture may have potential use in genetic engineering for development of biofuels, bioremediation of sites with radioactive contamination, and similar applications (Dadachova *et al.*, *PLoS ONE* **2** (5): e457, 2007).

Non-ionizing radiation effects are electromagnetic and do not carry enough energy per quantum to ionize atoms and molecules of fungi. Among them are near ultraviolet (UV), infrared and visible light radiation, etc. Some lethal effects of UV radiation on fungus spores are known (Zahl *et al.*, *J. gen. physiol.* **22**:

689, 1939; Dimond & Duggar, *Proc. nat. acad. sci. USA* **27**: 459, 1941; Ghar *et al.*, *Biocontrol Sci. Technol.* **16**: 451, 2006). UV radiation trigger sporogenesis in many fungi, and action spectra constructed for fungal sporogenesis typically have peaks at 280-290 nm (Leach & Trione, *Photochem. Photobiol.* **5**: 621, 1966; Leach, *In*: Booth (Ed.) *Methods in microbiology* **4**: 609, 1971) or near 360, 440, and 480 nm (Curtis, *Plant Physiol.* **49**: 235, 1972). Shortwave UV-B radiation (280-315 nm) influences distribution of fungi on leaf surfaces and its increase directly affects abundance of specific phylloplane fungi. Yeasts and yeast-like fungi such as *Aureobasidium pullulans* and *Sporobolomyces roseus* are reduced by supplemental levels of UV radiation, but *Alternaria* spp. and *Microdochium nivale* are less affected, while the most tolerant are *Cladosporium* spp. and *Epicoccum nigrum* (Newsham *et al.*, *New phytologist* **136**: 287, 1997). Sometimes UV radiation influence is unpredictable (Duguay & Klironomos *Appl. soil ecol.* **14**: 157, 2000). Solar UV radiation has been observed to reduce fungal growth on Antarctic soil surfaces. Exposure to solar radiation of >287 nm reduced hyphal extension rates of *Geomyces pannorum*, *Phoma herbarum*, *Pythium* sp., *Verticillium* sp., and *Mortierella parvispora*, each isolated from Antarctic terrestrial habitats. Short-wave solar UV radiation of between 287 and 305 nm inhibited growth of hyphae on and below the surface of an agar medium in a day, but radiation of ≥345 nm only reduced growth of surface hyphae (Hughes *et al.*, *Appl. environ. microbiol.* **69**: 1488, 2003).

Radiatispora Matsush. (1995), anamorphic *Pezizomycotina*, Hso.1bH.1. 1, China. See Matsushima (*Matsush. Mycol. Mem.* **8**: 32, 1995).

radicating (of stipes), like a root; rooting (Fig. 4H).

Radiciseta Sawada & Katsuki (1959), anamorphic *Pezizomycotina*, Hso.≡ eH.?. 1, Taiwan. See Sawada & Katsuki (*Publ. Coll. Agric. Nat. Taiwan Univ.* **8**: 205, 1959).

Radiigera Zeller (1944), Geastraceae. 4, widespread. See Kers (*Bot. Notiser* **129**: 173, 1976), Domínguez & Castellano (*Mycol.* **88**: 863, 1996), Calonge (*Boletín de la Sociedad Micológica de Madrid* **22**: 105, 1997).

Radiogaster Lloyd (1924) nom. inval. ? = Hymenogaster fide Stalpers (*in litt.*).

Radiomyces Embree (1959), Radiomycetaceae. 3, USA. See Benjamin (*Aliso* **4**: 523, 1960), Ellis & Hesseltine (*Mycol.* **66**: 85, 1974), Kitz *et al.* (*Mycol.* **74**: 110, 1982), Benny & Khan (*Scanning Microsc.* **2**: 1199, 1988; ultrastr.), Benny & Samson (*Mem. N. Y. bot. Gdn* **49**: 11, 1989; ultrastr.), Benny & Benjamn (*Mycol.* **83**: 713, 1991; key), Voigt & Wöstemeyer (*Gene* **270**: 113, 2001; phylogeny), White *et al.* (*Mycol.* **98**: 872, 2006; phylogeny).

Radiomycetaceae Hesselt. & J.J. Ellis (1974), Mucorales. 3 gen. (+ 1 syn.), 5 spp.

Lit.: Kitz *et al.* (*Sabouraudia* **18**: 115, 1980), Benny & Khan (*Scanning Microscopy* **2**: 1199, 1988; ultrastr., sporangiolar appendage chemistry), Benny & Samson (*Mem. N. Y. bot. Gdn* **49**: 11, 1989; ultrastr.), Tanphaichitr *et al.* (*Mycoses* **33**: 303, 1990), Benny & Benjamin (*Mycol.* **83**: 713, 1991; key), Benny & Benjamin (*Mycol.* **83**: 713, 1991), Meis *et al.* (*J. Clin. Microbiol.* **32**: 3078, 1994), Ribes *et al.* (*Clin. Microbiol. Rev.* **13**: 236, 2000), Chakrabarti *et al.* (*J. Clin. Microbiol.* **41**: 783, 2003).

Radiomycopsis Pidopl. & Milko (1971) = Radiomyces fide Benny & Benjamin (*Mycol.* **83**: 713, 1991).

Radotinea Velen. (1934), ? Helotiales. 1, former Czechoslovakia.

radula spore (radulaspore, radulospore), one of the slimy spores borne over the surface of ascospores as in *Nectria coryli* while still in the ascus; **dry - -** = sympodulospore (Mason, 1933, 1937).

Radulaceae Gäum. (1926) = Valsaceae.

Radulidium Arzanlou, W. Gams & Crous (2007), anamorphic *Pezizomycotina*. 1. See Arzanlou *et al.* (*Stud. Mycol.* **58**: 57, 2007).

raduliform (of conidiogenous cells), the elongating conidiogenous axis resulting from holoblastic sympodial development, clavate or somewhat inflated rather than zig-zag; cf. rachiform.

Raduliporus Spirin & Zmitr. (2006), Polyporaceae. 1.

Radulochaete Rick (1940) nom. dub., Cantharellales. 'basidiomycetes' fide Donk (*Persoonia* **3**: 199, 1964).

Radulodon Ryvarden (1972), Meruliaceae. 9, widespread. See Ryvarden (*Česká Mykol.* **30**: 38, 1976; key), Hjortstam *et al.* (*Kew Bull.* **45**: 303, 1990; key).

Radulomyces M.P. Christ. (1960), Pterulaceae. 10, widespread. See Christiansen (*Dansk bot. Ark.* **19**: 230, 1960), Gilbertson & Nakasone (*Mycol.* **95**: 467, 2003; key Hawaiian spp.).

Radulum Fr. (1825) = Xenotypa fide Donk (*Taxon* **5**: 109, 1956).

Raesaeneniolichen Tomas. & Cif. (1952) = Polyblastia fide Hawksworth *et al.* (*Dictionary of the Fungi* edn 8, 1995).

Raesaeneniomyces Cif. & Tomas. (1953) ≡ Raesaeneniolichen.

Raffaelea Arx & Hennebert (1965), anamorphic *Ophiostoma*, Hso.0eH.10. 12, widespread. See Scott & du Toit (*TBMS* **55**: 181, 1970), Kubono & Ito (*Mycoscience* **43**: 255, 2002; Japan), Gebhardt & Oberwinkler (*Antonie van Leeuwenhoek* **88**: 61, 2005; ontogeny).

ragi, a starter for arrack, etc. composed of small balls of rice flour containing *Mucor*, *Rhizopus*, yeasts and bacteria (Hesseltine, *Mycol.* **57**: 163, 1965).

Ragnhildiana Solheim (1931) = Passalora fide Deighton (*Mycol. Pap.* **137**, 1974), Crous & Braun (*CBS Diversity Ser.* **1**: 571 pp., 2003).

Raizadenia S.L. Srivast. (1981), anamorphic *Pezizomycotina*, Hsy.≡ eP.1. 1, India. See Srivastava (*Indian Phytopath.* **34**: 335, 1981).

Rajapa Singer (1945) = Termitomyces fide Singer (*Agaric. mod. Tax.*, 1951).

Ramakrishnanella Kamat & Ullasa ex Ullasa (1970), anamorphic *Pezizomycotina*, Cpd.0eP.?. 1, India. See Kamat & Ullasa ex Ullasa (*Bull. Torrey bot. Club* **97**: 255, 1970).

Ramakrishnania Ramachar & Bhagyan. (1979), Pucciniaceae. 1 (on *Ixora* (*Rubiaceae*)), India. See Cummins & Hiratsuka (*Illustr. Gen. Rust Fungi* edn 3: 225 pp., 2003; syn. of *Gymnoconia*) Probably a *Puccinia* with atypical sporogeneous cells.

Ramalea Nyl. (1866), Cladoniaceae (L). 4, widespread. See Stenroos *et al.* (*Mycol. Progr.* **1**: 267, 2002; phylogeny), Printzen & Kantvilas (*Biblthca Lichenol.* **88**: 541, 2004; phylogeny).

Ramaleaceae Elenkin (1929) = Cladoniaceae.

Ramalia Bat. (1957), anamorphic *Dothideomycetes*. See Sutton & Pascoe (*Aust. Syst. Bot.* **1**: 79, 1988; synonymy with *Fusicladium*), Braun & Mouchacca

(*MR* **104**: 1009, 2000).

Ramalina Ach. (1809) nom. cons., Ramalinaceae (L). *c.* 246, widespread. See Howe (*Bryologist* **16**: 65, 1913), Howe (*Bryologist* **17**: 1, 1914), Krog & Swinscow (*Norw. Jl Bot.* **23**: 153, 1976; E. Afr.), Krog & James (*Norw. Jl Bot.* **24**: 15, 1977; Fennoscandia, UK), Krog & Østhagen (*Norw. Jl Bot.* **27**: 255, 1980; Canary Is.), Stevens (*Bull. Br. Mus. nat. hist.* Bot. **16**: 107, 1987; key 20 spp. Australia), Kashiwadani & Kalb (*Lichenologist* **25**: 1, 1993; key 22 spp. Brazil), Blanchon *et al.* (*J. Hattori bot. Lab.* **79**: 43, 1996; NZ), Groner & LaGreca (*Lichenologist* **29**: 441, 1997; DNA), LaGreca (*Bryologist* **102**: 602, 1999; DNA), Rosato & Scutari (*Mycotaxon* **74**: 141, 2000; Argentina), Alvarez *et al.* (*Cryptog. Mycol.* **22**: 271, 2001; chemistry, Spain), Kashiwadani & Nash (*Mycotaxon* **83**: 385, 2002; Mexico), Kashiwadani *et al.* (*J. Jap. Bot.* **77**: 351, 2002; Hawaii), Cordeiro *et al.* (*Phytochem.* **63**: 967, 2003; chemistry), Bannister *et al.* (*N.Z. Jl Bot.* **42**: 121, 2004; New Zealand), Joneson *et al.* (*Bryologist* **107**: 98, 2004; Kuril Is), Sparrius (*Biblthca Lichenol.* **89**, 2004; monogr.), Stocker-Wörgötter *et al.* (*Bryologist* **107**: 152, 2004; chemistry), Kashiwadani *et al.* (*Bull. natn. Sci. Mus.* Tokyo, B **32**: 161, 2006; China), Miądlikowska *et al.* (*Mycol.* **98**: 1088, 2006; phylogeny), Aptroot & Bungartz (*Lichenologist* **39**: 519, 2007; Galápagos Is), Honegger & Zippler (*MR* **111**: 424, 2007; mating systems), Kashiwadani & Moon (*Biblthca Lichenol.* **96**: 145, 2007; Indonesia), Werth & Sork (*Am. J. Bot.* **95**: 568, 2008; genetic structure).

Ramalinaceae C. Agardh (1821), Lecanorales (L). 36 gen. (+ 27 syn.), 679 spp. Treated in a broad sense including *Bacidiaceae* and relatives.

Lit.: Stevens (*Bull. Br. Mus. nat. Hist.* Bot. **16**: 107, 1987), Culberson *et al.* (*Bryologist* **93**: 167, 1990), Rambold & Triebel (*Notes R. bot. Gdn Edinb.* **46**: 375, 1990), Culberson *et al.* (*Am. J. Bot.* **80**: 1472, 1993), Kashiwadani & Kalb (*Lichenologist* **25**: 1, 1993), Blanchon *et al.* (*J. Hattori bot. Lab.* **79**: 43, 1996), Spjut (*Sida* Bot. Misc. **14**: 208 pp., 1996), Groner & LaGreca (*Lichenologist* **29**: 441, 1997), Lumbsch (*Lichenologist* **30**: 357, 1998), LaGreca (*Bryologist* **102**: 602, 1999), LaGreca & Lumbsch (*Lichenologist* **33**: 172, 2001), Cordeiro *et al.* (*MR* **108**: 489, 2004), Stocker-Wörgötter *et al.* (*Bryologist* **107**: 152, 2004), Miądlikowska *et al.* (*Mycol.* **98**: 1088, 2006; phylogeny), Hofstetter *et al.* (*Mol. Phylogen. Evol.* **44**: 412, 2007; phylogeny), Næsborg *et al.* (*MR* **111**: 581, 2007; phylogeny).

Ramalinomyces E.A. Thomas ex Cif. & Tomas. (1953) ≡ Ramalina.

Ramalinopsis (Zahlbr.) Follmann & Huneck (1969), Ramalinaceae (L). 1, Hawaii. See Riedl (*Sydowia* **28**: 134, 1977).

Ramalinora Lumbsch, Rambold & Elix (1995), Lecanoraceae (L). 1, Australia. See LaGreca & Lumbsch (*Lichenologist* **33**: 172, 2001).

Ramalodium Nyl. (1879), Collemataceae (L). 5, Japan; New Caledonia; Tasmania. See Henssen (*Lichenologist* **3**: 29, 1965), Henssen (*Nova Hedwigia* **68**: 117, 1999).

Ramaraomyces N.K. Rao, Manohar. & Goos (1989), anamorphic *Pezizomycotina*, Hsy.≡ eH.1. 1, India. See Rao *et al.* (*Mycol.* **81**: 790, 1989).

Ramaria Fr. ex Bonord. (1851) nom. cons., Gomphaceae. *c.* 220, widespread. See Corner (*Ann. Bot. Mem.* [A monograph of Clavaria and allied genera] **1**: 124, 1950), Corner (*Ann. Bot. Mem.* [A monograph of Clavaria and allied genera] **1**: 542, 1950), Mann & Stuntz (*Ramaria of western Washington*, 1973; keys), Petersen (*Biblthca Mycol.* **43**, 1975; subgen. *Lentoramaria*), Petersen (*Biblthca Mycol.* **79**, 1981; subgen. *Echinoramaria*), Humpert *et al.* (*Mycol.* **93**: 465, 2001; molecular phylogenetics), Christan & Hahn (*Z. Mykol.* **71**: 7, 2005; systematics), Nouhra *et al.* (*Mycorrhiza* **15**: 55, 2005; mycorrhizal spp.).

Ramaria Holmsk. (1790) nom. rej. = Clavulinopsis fide Donk (*Persoonia* **1**: 25, 1959).

Ramariaceae Corner (1970) = Gomphaceae.

Ramaricium J. Erikss. (1954), Gomphaceae. 3, widespread. See Ginns (*Bot. Notiser* **132**: 93, 1979), Villegas *et al.* (*Fungal Diversity* **18**: 157, 2005; sporal characters).

ramarioid, with a form similar to that of the basidioma of *Ramaria*.

Ramariopsis (Donk) Corner (1950), Clavariaceae. 44, widespread. See Petersen (*Mycol.* **61**: 557, 1967; key N. Am. spp.), Pegler & Young (*TBMS* **84**: 207, 1985; basidiospore structure), García-Sandoval *et al.* (*Mycotaxon* **82**: 323, 2002; Mexico), García-Sandoval *et al.* (*Mycotaxon* **94**: 265, 2005; phylogeny).

Ramasricellites Kalgutkar & Janson. (2000), Fossil Fungi, anamorphic *Ascomycota*. 1, Tibet. See Kalgutkar & Jansonius (*AASP Contributions Series* **39**: 259, 2000).

Ramboldia Kantvilas & Elix (1994), Lecanoraceae (L). 9, Australia; Papua New Guinea. See Kantvilas *et al.* (*Lichenologist* **31**: 213, 1999), Kantvilas & Elix (*Lichenologist* **39**: 135, 2007; key), Kalb *et al.* (*Nova Hedwigia* **86**: 23, 2008; phylogeny).

Ramgea Brumm. (1992), Thelebolaceae. 1, Netherlands. See Brummelen (*Persoonia* Suppl. **14**: 577, 1992), Brummelin (*Persoonia* **16**: 425, 1998), Hoog *et al.* (*Stud. Mycol.* **51**: 33, 2005).

Ramicandelaber Y. Ogawa, S. Hayashi, Degawa & Yaguchi (2001), Kickxellaceae. 2, Japan. See Ogawa *et al.* (*Mycoscience* **42**: 193, 2001), Kurihara *et al.* (*MR* **108**: 1143, 2004; n.sp. Japan), Ogawa *et al.* (*Nippon Kingakukai Kaiho* **46**: 13, 2005; phylogeny).

Ramicephala Voglmayr & Delg.-Rodr. (2003), anamorphic *Pezizomycotina*. 1 (aeroaquatic), Cuba. See Voglmayr & Delgado-Rodríguez (*MR* **107**: 237, 2003).

Ramichloridium Stahel ex de Hoog (1977), anamorphic *Mycosphaerellaceae*, Hso.0eP.10. 27, widespread. The genus is polyphyletic. See de Hoog *et al.* (*TBMS* **81**: 485, 1983; relationships with *Stenella* and *Veronaea*), Campbell & Al-Hedaithy (*J. Med. Vet. Mycol.* **31**: 325, 1993; *R. mackenziei*, phaeohyphomycosis of the brain), Samuels & Candoussau (*Nova Hedwigia* **62**: 47, 1996; teleomorph), Untereiner (*Mycol.* **89**: 120, 1997; teleomorph), Sutton *et al.* (*J. Clin. Microbiol.* **36**: 708, 1998; phaeohyphomycosis), Haase *et al.* (*Stud. Mycol.* **43**: 80, 1999; phylogeny), Geiser *et al.* (*Mycol.* **98**: 1053, 2006; phylogeny), Arzanlou *et al.* (*Stud. Mycol.* **58**: 57, 2007; phylogeny, polyphyly).

Ramicola Velen. (1929) = Simocybe fide Kuyper (*in litt.*).

ramicolous, living on branches.

Ramoconidiifera B. Sutton, Carmarán & A.I. Romero (1996), anamorphic *Pezizomycotina*, Hso.?.?. 1, Argentina. See Sutton *et al.* (*MR* **100**: 1338, 1996).

ramoconidium, an apical branch of a conidiophore which secedes and functions as a conidium, as in

Cladosporium and *Subramaniomyces*.

Ramonia Stizenb. (1862), ? Gyalectaceae (L). 17, widespread. See Vězda (*Folia geobot. phytotax.* **2**: 311, 1967), Coppins (*Lichenologist* **19**: 409, 1987; key 4 Br. spp.), Coppins *et al.* (*Graphis Scripta* **6**: 89, 1994; Sweden), Lücking *et al.* (*Mycol.* **96**: 283, 2004; phylogeny).

Ramophialophora M. Calduch, Stchigel, Gené & Guarro (2004), anamorphic *Sordariales*. 1, Spain. See Calduch *et al.* (*Stud. Mycol.* **50**: 84, 2004).

Ramosiella Syd. & P. Syd. (1917) = Anhellia fide von Arx (*Persoonia* **2**: 421, 1963).

Ramosphaerella Laib. (1921) = Mycosphaerella fide Hawksworth *et al.* (*Dictionary of the Fungi* edn 8, 1995).

Ramsaysphaera H.D. Pflug (1978), Fossil Fungi, anamorphic *Pezizomycotina*. 1 (*c.* 3400 million yr old Swartkoppie chert), S. Africa.

Ramsbottom (John; 1885-1974; England). Taxonomist (1910-1930) then Keeper of Botany (1930 onwards), British Museum (Natural History). Wrote extensively on many aspects of mycology including taxonomy, nomenclature, and history (q.v.) and did much to popularize the study of fungi. *Publs. Edible Fungi* (1944); *Poisonous Fungi* (1944); *Mushrooms and Toadstools; a Study of the Activities of Fungi* (1953). *Biogs, obits etc.* Gregory (*TBMS* **65**: 1, 1975) [portrait]; Grummann (1974: 411); Stafleu & Cowan (*TL-2* **4**: 574, 1983).

Ramsbottomia W.D. Buckley (1923), Pyronemataceae. 3, widespread (boreal; temperate). See Benkert & Schumacher (*Agarica* **6**: 28, 1985), Caillet & Moyne (*Bull. Soc. Hist. nat. Doubs* **84**: 9, 1991; keys), Kullman & van Brummelen (*Persoonia* **15**: 93, 1992), Yao & Spooner (*MR* **99**: 1521, 1995; Brit. spp.), Perry *et al.* (*MR* **111**: 549, 2007; phylogeny).

Ramularia Sacc. (1880) = Ramularia Unger fide Hawksworth *et al.* (*Dictionary of the Fungi* edn 8, 1995).

Ramularia Unger (1833) nom. cons., anamorphic *Mycosphaerella*, Hso.0eH/≡ eH.3/10. 138 (causing leaf spots), widespread (esp. temperate). *R. vallisumbrosae* (*Narcissus* white mould). See Hughes (*TBMS* **32**: 50, 1949), Vimba (*The flora of the genus Ramularia Sacc. in the Latvian SSR*, 1970), Braun (*Int. J. Mycol. Lichenol.* **3**: 271, 1988), Deighton (*TBMS* **90**: 330, 1988; spp. on *Salix*), Sutton & Waller (*TBMS* **90**: 55, 1988; spp. on *Gramineae*), Braun (*Nova Hedwigia* **47**: 335, 1989), Braun (*Nova Hedwigia* **53**: 291, 1991; related genera), Braun & Sutton (*Taxon* **112**: 656, 1991; conservation), Braun (*Monogr. Cercosporella, Ramularia Allied Genera (Phytopath. Hyphom.)* **1**, 1995), Braun (*Monogr. Cercosporella, Ramularia Allied Genera (Phytopath. Hyphom.)* **2**: 11, 1998; monogr., subgen. keys), Crous *et al.* (*Mycol.* **93**: 1081, 2001; phylogeny), Goodwin *et al.* (*Phytopathology* **91**: 648, 2001; phylogeny), Braun *et al.* (*Australas. Pl. Path.* **34**: 509, 2005; checklist, Australia), Havis *et al.* (*FEMS Microbiol. Lett.* **256**: 217, 2006; on *Hordeum*).

Ramulariaceae Nann. (1934) = Mycosphaerellaceae.

Ramulariites Babajan & Tasl. (1973), Fossil Fungi. 1 (Tertiary), former USSR.

Ramulariopsis Speg. (1910), anamorphic *Mycosphaerella*, Hso.1-≡ eH.3. 4, S. America. See Deighton (*TBMS* **59**: 185, 1972), Braun (*Monogr. Cercosporella, Ramularia Allied Genera (Phytopath. Hyphom.)* **2**: 313, 1998; key).

Ramulariospora Bubák (1914) = Phacidiella P. Karst. fide Sutton (*Mycol. Pap.* **141**, 1977).

Ramularisphaerella Kleb. (1918) = Mycosphaerella fide von Arx & Müller (*Stud. Mycol.* **9**, 1975).

Ramularites Pia (1927), Fossil Fungi. 1 (Oligocene), Europe.

Ramulaspera Lindr. (1902), anamorphic *Pezizomycotina*, Hso.0eH.?. 2, Europe. See Griffiths (*TBMS* **40**: 232, 1957).

Ramulina Velen. (1947) [non *Ramulina* Thurm. 1863, fossil ? *Algae*] = Pseudombrophila fide Svrček (*Acta Mus. Nat. Prag.* **32B**: 115, 1976).

Ramulispora Miura (1920), anamorphic *Mycosphaerellaceae*, Cpd.≡ eH.10. 18, widespread (temperate). See also *Oculimacula* and *Helgardia*. See Rawla (*TBMS* **60**: 285, 1973), von Arx (*Proc. K. ned. Akad. Wet.* Ser. C, Biol. Med. Sci. **86**: 36, 1983; syn. with *Pseudocercosporella*), Wallwork & Spooner (*TBMS* **91**: 703, 1988; *Tapesia* teleomorph of *P. herpotrichoides*), Creighton (*Pl. Path.* **38**: 484, 1989; identification of W-type and R-type isolates of *P. herpotrichoides*), Nicholson *et al.* (*Pl. Path.* **40**: 584, 1991; DNA markers for classification of *P. herpotrichoides*), Fitt *in* Singh *et al.* (Eds) (*Plant Diseases of International importance: diseases of Cereals and Pulses*, 1992), Priestley *et al.* (*Pl. Path.* **41**: 591, 1992; comparison of isoenzyme and DNA markers for *P. herpotrichoides*), Thomas *et al.* (*J. Gen. Microbiol.* **138**: 2305, 1992; identification of R- and W-types of *P. herpotrichoides* by DNA probes), Frei & Wenzel (*J. Phytopath.* **139**: 229, 1993; genomic DNA gene probes), Hocart *et al.* (*MR* **97**: 967, 1993; parasexuality), Nicholson & Rezanoor (*MR* **98**: 13, 1994; identification of pathotypes of *P. herpotrichoides* by random amplified polymorphic DNA), Braun (*Monogr. Cercosporella, Ramularia Allied Genera (Phytopath. Hyphom.)* **1**, 1995), Robbertse *et al.* (*S. Afr. J. Bot.* **61**: 43, 1995; spp. causing eyespot of wheat), Stewart *et al.* (*MR* **103**: 1491, 1999), Kwasna *et al.* (*MR* **104**: 765, 2000; UK), Crous *et al.* (*Mycol.* **93**: 1081, 2001; phylogeny), Goodwin *et al.* (*Phytopathology* **91**: 648, 2001; phylogeny), Crous *et al.* (*Eur. J. Pl. Path.* **109**: 841, 2003; phylogeny, separation of eyespot disease-causing spp.).

ramus (1) (pl. **-i**), a branch (Lat.); (2) (of a penicillus), a cell bearing a verticil of 'metulae' and phialides.

ramycin, see fusidic acid.

rangiferoid, branched like a reindeer's horn.

Ranojevicia Bubák (1910), anamorphic *Pezizomycotina*, Hsp.0eH.?. 1, Europe.

Rapacea E. Horak (1999) = Cortinarius fide Peintner *et al.* (*Mycotaxon* **83**: 447, 2002).

RAPD, see Molecular biology.

Raperia Subram. & Rajendran (1976) = Aspergillus fide Samson (*Stud. Mycol.* **18**: 1, 1979), von Arx (*Mycotaxon* **26**: 119, 1986; teleom. and affinities with *Onygenaceae*), Pitt *et al.* (*Integration of Modern Taxonomic Methods for Penicillium and Aspergillus Classification*: 9, 2000; synonym).

Raphanozon P. Kumm. (1871) nom. rej. ≡ Telamonia.

raphe (of *Chaetomella*), the longitudinal dehiscence mechanism. See Di Cosmo & Cole (*CJB* **58**: 1129, 1980; ultrastr. development).

raphides, needle-shaped crystals; as hyphae inside some lichen thalli.

Raphiospora A. Massal. (1853) nom. rej. = Arthrorhaphis fide Jørgensen & Santesson (*Taxon* **42**: 881,

1993).

Rasutoria M.E. Barr (1987), Euantennariaceae. 4, widespread (tropical). Perhaps better placed in *Mycosphaerellaceae*, but molecular data are inconclusive. See Barr (*Mycotaxon* **29**: 501, 1987), Winton *et al.* (*Mycol.* **99**: 240, 2007; phylogeny).

Ratnagiriathyrites R.K. Saxena & N.K. Misra (1990), Fossil Fungi, Ascomycota. 1. See Saxena & Misra (*Palaeobotanist* **38**: 268, 1989).

Ravenelia Berk. (1853), Raveneliaceae. *c.* 200 (on *Leguminosae* esp. *Mimosoideae*, 1 on *Zygophyllaceae*), widespread (tropical to subtropical). *R. esculenta* is edible. See Narasimhan & Thirumalachar (*Phytopath. Z.* **41**: 97, 1961; *R. esculenta*) See, Hiremath & Pavgi (*Mycopathologia* **59**: 95, 1976; morphology), Cummins (*Rust fungi on legumes and composites in North America*, 1978; key 61 spp.), El-Gazzar (*Symb. Bot. Upsal.* **44**: 182, 1979), Mordue (*TBMS* **90**: 473, 1988; SEM of III), Hennen & Cummins (*Rep. Tottori Mycol. Inst.* **28**: 1, 1990), Berndt (*MR* **101**: 23, 1997; haustorial types), Rezende & Dianese (*Fitopatol. Brasil* **26**: 627, 2001; Brazilian cerrado), Hernández & Hennen (*MR* **106**: 954, 2002; Argentina), Wood (*South African Journal of Botany* **72**: 534, 2006; southern African).

Raveneliaceae Leppik (1972), Pucciniales. 26 gen. (+ 14 syn.), *c.* 323 spp.
Lit.: Savile (*CJB* **67**: 2983, 1989), Reenen & Merwe (*Phytophylactica* **24**: 127, 1992), Berndt (*MR* **101**: 23, 1997), Hennen *et al.* (*Mycol.* **90**: 1079, 1998), Gjærum *et al.* (*Lidia* **5**: 91, 2000), Hennen (*Mycol.* **92**: 315, 2000), Hernández & Hennen (*MR* **106**: 954, 2002), Ono (*Mycoscience* **43**: 421, 2002), Cummins & Hiratsuka (*Illustr. Gen. Rust Fungi* edn 3: 225 pp., 2003), Aime (*Mycoscience* **47**: 112, 2006).

Ravenelites Ramanujam & Ramachar (1980), Fossil Fungi, Raveneliaceae. 1. See Ramanujam & Ramachar (*Records of the Geological Survey of India* **113**: 83, 1980).

Ravenelula Speg. (1881), ? Lecanorales. 1 or 2, America.

ray fungi, members of the *Actinomycetes* (Bacteria), q.v.

razor-strop fungus, basidioma of *Piptoporus betulinus*.

RDA, see Media.

Readeriella Syd. (1908), anamorphic *Teratosphaeriaceae*, Cpd.0eP.19. 1, widespread. See Macauley & Thrower (*TBMS* **48**: 105, 1965), Crous *et al.* (*Stud. Mycol.* **50**: 195, 2004; phylogeny), Decock (*Cryptog. Mycol.* **26**: 143, 2005; French Guiana), Crous *et al.* (*Stud. Mycol.* **55**: 99, 2006; phylogeny), Hunter *et al.* (*Stud. Mycol.* **55**: 147, 2006; phylogeny), Crous *et al.* (*Stud. Mycol.* **58**: 1, 2007; phylogeny).

Rebentischia P. Karst. (1869), Tubeufiaceae. 3, Europe; N. America. See Barr (*Mycotaxon* **12**: 137, 1980; key), Crane *et al.* (*CJB* **76**: 602, 1998), Kodsueb *et al.* (*Fungal Diversity* **21**: 105, 2006).

receptacle, an axis having one or more organs, as the stem in *Phallales*; any hymenium-supporting structure.

receptive body, a small branched or unbranched process from the stroma (as in *Stromatinia gladioli*) able to be 'spermatized' by microconidia (see Drayton, *Mycol.* **26**: 46, 1934); **receptive hypha** = flexuous hypha, trichogyne, and possibly other like structures.

Rechingeria Servít (1931) ? = Lichinella fide Hawksworth *et al.* (*Dictionary of the Fungi* edn 8, 1995).

Rechingeriella Petr. (1940), ? Zopfiaceae. 2, Europe; Asia. See Hawksworth (*CJB* **57**: 91, 1979).

Recifea Bat. & Cif. (1962) = Microcallis fide von Arx (*in litt.*).

recognition (of mutualistic symbionts), the process by which two compatible potential symbionts initiate the relationship.

Reconditella Matzer & Hafellner (1990), Sordariales. 1 (on lichens, esp. *Physconia*), Europe. See Matzer & Hafellner (*Biblthca Lichenol.* **37**: 46, 1990).

Record fungi. *Heaviest fruit body*, *Laetiporus sulphureus*, 45.4 kg, New Forest, Hants, UK (found 1990); *heaviest mycelium*, *Armillaria bulbosus*, 100+ tonnes, Michigan, USA, reported April 1992 (see Longevity); *highest spore count*, 161,037 fungus spores per m^3, Cardiff 1971; *largest area occupied*, *Armillaria ostoyae* occupying 600 ha in Washington State, USA; *largest edible*, *Langermannia gigantea*, circumference 2.64 m, 22 kg, from Canada, 1987; *largest fruit body*, *Rigidoporus ulmarius*, 163 cm long × 140 cm wide × 50 cm tall, circumference 4.8 m, in the grounds of the former IMI (q.v.), Kew, Surrey, UK (measurements in Feb. 1995 when still growing; unweighed); *most poisonous*, *Amanita phalloides*, 5-7 mg amanitoxins lethal; *oldest spores* (see Longevity); *oldest thallus*, *Rhizocarpon geographicum* (see Longevity); *tallest* (see Fossil fungi). See Glenday (Ed.) (*Guinness world records*, 2008), also acidophily, alkilophily, Longevity, spore discharge, thermophily, water activity.

Recticharella D. Scheer (1944) = Asellaria fide Scheer (*Arch. Protistenk.* **114**: 343, 1972).

Rectipilus Agerer (1973), Marasmiaceae. 9, widespread. See Wei & Dai (*Mycosystema* **23**: 437, 2004; China).

Rectispora J.I.R. Larsson (1990), Microsporidia. 1. See Larsson (*Eur. J. Protist.* **26**: 55, 1990).

recurved (of a pileus), convexo-expanded (q.v.).

Red List, see Conservation.

red rice, see ang-kak; **- bread mould**, see *Chrysonilia*; **- rust**, (1) urediniospore state of rusts, especially of cereals; (2) (of tea), the alga *Cephaleuros*; **- truffle**, *Melanogaster variegatus*.

Redaellia Cif. (1930) = Aspergillus fide Alecrim (*Ann. Fac. Med. Univ. Recife* **18**: 81, 1958).

Redbia Deighton & Piroz. (1972), anamorphic *Pezizomycotina*, Hso.≡ eH.10. 2 (on *Puccinia*), widespread (tropical). See Deighton & Pirozynski (*Mycol. Pap.* **128**: 83, 1972).

Reddellomyces Trappe, Castellano & Malajczuk (1992), Tuberaceae. 4, Australasia; Mediterranean. See Hansen & Pfister (*Mycol.* **98**: 1029, 2006; phylogeny), Læssøe & Hansen (*MR* **111**: 1075, 2007; phylogeny).

Redheadia Y. Suto & Suyama (2005), Sclerotiniaceae. Anamorph *Mycopappus*. 1, Japan. See Suto & Suyama (*Mycoscience* **46**: 228, 2005).

Redingeria A. Frisch (2006), Thelotremataceae (L). 5, pantropical. See Frisch (*Biblthca Lichenol.* **92**: 402, 2006).

Redonia C.W. Dodge (1973), ? Physciaceae (L). 1, S. America. See Dodge (*Lich. Fl. Antarct. Cont.*: 353, 1973).

Reduviasporonites L.R. Wilson (1962), Fossil Fungi. 1 (Permian), USA.

Reference collections. For fungi, reference collections are essential to secure the application of scientific names (see Nomenclature), to serve as vouchers supporting the occurrence of species or material used in research, and as the raw materials for new systematic and other investigations. As for animals and plants, fungal reference collections are also, no less than art galleries and great libraries, an irreplaceable part of the cultural heritage of each country, although this is rarely recognized. Reference collections of fungi are of two main types: (1) Genetic resource collections (q.v.; culture collections) hold material in a living or metabolically inactive state from which they can be resuscitated; and (2) Reference collections of non-living material, including dried specimens attached to plant, rock or other substrata, dried cultures, microscope slides, colour transparencies, drawings and paintings, and ultrastructural mounts. Collections of dried fungal material have often been referred to inappropriately as herbaria (q.v.), but as fungi are not plants, and not always kept with plants, the more general term is preferable, particularly where living and dried material are stored by the same institution.

An immense amount of information related to knowledge of the distribution and ecology of fungi is stored in the world's mycological collections; this is increasingly being captured in computerized databases so that it can be made more accessible, and a substantial but still small proportion of that information is already available on-line (see Internet).

Most institutional collections have unique acronyms (generally retained unchanged despite name changes of the institutions), a selection of which are included in this edn of the *Dictionary* (e.g. **BPI**, **DAOM**, **IMI**, **K**, **L**, **LE**, **UPS**); collections of living cultures are given acronyms by the World Data Center for Microorganisms (see Genetic resource collections) - where possible the same acronym is used under both systems, and compatibility is maintained with acronyms used for plant reference collections. Acronyms are commonly associated with collection accession numbers in mycological publications to make it clear what material was employed.

Lit.: **General** (see also Internet: Fungal reference collections): Hall & Minter (*International mycological directory*, edn 3, 1994), Hawksworth (*Mycologist's handbook*, 1974), Laundon (*Lichenologist* **11**: 1, 1974).

Regional: **Australia**, Walker (*Mycological herbaria and culture collections in Australia*, 1980); **British Isles**, Hawksworth & Seaward (*Lichenology in the British Isles 1568-1975*, 1977), Kent & Allen (*British and Irish herbaria*, 1984); **Germany**, Scholz (*Boletus* **12**(2): 33, 1989; lichens); **Israel**, Ferber (*Israel national collections of natural history*, 1985); **Italy**, Ciferri (*Taxon* **1**: 126, 1952), Tretiach & Passadore (*Notiz. Soc. Lich. Ital.* **3** (*Suppl.*), 1990); **New Zealand**, Wright (*N.Z. Jl Bot.* **22**: 323, 1984); **former USSR**, Parmasto (*Taxon* **34**: 359, 1985); **Other useful works**: Bartz *et al.* (*Museums of the world*, edn 4, 1992), Bridson & Forman (*The herbarium handbook*, edn 2, 1992), Fosberg & Sachet (*Manual for tropical herbaria* [*Regnum. veg.* **39**], 1965), Holmgren *et al.* (*Index herbariorum. Part* 1. *Herbaria of the world*, edn 8 [*Regnum veg.* **120**], 1990), Pinniger (*Biodet. Abstr.* **5**: 125, 1991; insect control in), Stafleu & Cowan (*TL-2*, 7 vols. [+ 3 suppl.], 1976-).

See also Collection and preservation, Genetic resource collections, herbarium beetle.

reflexed (of an edge), turned up or back (Fig. 19G).

Refractohilum D. Hawksw. (1977), anamorphic *Pezizomycotina*, Hso.0eH/≡ eH.19. 6 (on lichens), widespread (temperate). See Hawksworth (*J. Linn. Soc.* Bot. **75**: 204, 1977), Castañeda Ruíz *et al.* (*Mycotaxon* **68**: 23, 1998).

Regiocrella P. Chaverri & K.T. Hodge (2006), Clavicipitaceae. 2, Cameroon; China. See Chaverri & Hodge (*Mycol.* **97**: 1232, 2005).

Rehm (Heinrich; 1828-1916; Germany). Student, Erlangen, München, Heidelberg; Dr med. (1852); medical studies, Prague then Vienna (1853-1854); physician (1854-1898). In his time, the foremost expert on ascomycetes, particularly discomycetes, particularly from central Europe, but also from North America, the Philippines and South America. Most specimens in the fungal reference collection in Stockholm (**S**). *Publs*. Ascomyceten: Hysteriaceen und Discomyceten. *Rabenhorst?s Kryptogamen-Flora von Deutschland, Oesterreich und der Schweiz* (1896). *Biogs, obits etc.* Anon. (*Österreichische Botanische Zeitschrift* **66**: 212, 1916); Grummann (1974: 39); Stafleu & Cowan (*TL-2* **4**: 652, 1983).

Rehmia Kremp. (1861) = Rhizocarpon fide Hawksworth *et al.* (*Dictionary of the Fungi* edn 8, 1995).

Rehmiella G. Winter (1883) = Gnomonia fide Barr (*Mycol. Mem.* **7**: 232 pp., 1978).

Rehmiellopsis Bubák & Kabát (1910) = Delphinella fide Barr (*Contr. Univ. Mich. Herb.* **9**, 1972).

Rehmiodothis Theiss. & Syd. (1914), Phyllachoraceae. 7 (on living leaves of *Melastomataceae*), Asia. Possibly belongs rather to the Dothideomycetes. See Katumoto (*TMSJ* **32**: 37, 1991), Pearce & Hyde (*Fungal Diversity Res. Ser.* **17**: 308 pp., 2006; Australia).

Rehmiomycella E. Müll. (1962), Dothideomycetes. 1, Brazil. See Müller (*Beitr. Kryptfl. Schweiz* **11** no. 2: 601, 1962).

Rehmiomyces (Sacc. & P. Syd.) Syd. (1904) ≡ Rehmiomycella.

Rehmiomyces Henn. (1904) = Dictyonia fide Clements & Shear (*Gen. Fung.*, 1931).

Reichlingia Diederich & Scheid. (1996), anamorphic *Pezizomycotina*, Hso.?.?. 1 (lichenicolous), Germany. See Diederich & Scheidegger (*Bull. Soc. Nat. luxemb.* **97**: 4, 1996).

Reimnitzia Kalb (2001), Thelotremataceae (L). 1, widespread. See Kalb (*Mycotaxon* **79**: 325, 2001), Frisch (*Biblthca Lichenol.* **92**: 3, 2006).

reindeer lichen (-moss), mainly *Cladonia stellaris* and *C. rangiferina*, species grazed by reindeer and caribou. Partially digested thalli from stomach contents of freshly-killed reindeer are highly nutritious, rich in vitamins A and B, and considered a delicacy by some native peoples. Affected by radiation pollution, particularly from Chernobyl, leading to food-chain concerns (Skuterud *et al.*, *J. Env. Radioactivity* **83**: 231, 2005).

Reinkella Darb. (1897) = Roccella fide Choisy (*Bull. Soc. bot. Fr.* **104**: 333, 1957), Tehler (*CJB* **68**: 2458, 1990; cladistics), Follmann (*Biblthca Lichenol.* **67**: 14, 1997), Tehler *et al.* (*Symb. bot. upsal.* **32** no. 1: 255, 1997; on *Roccella*), Myllys *et al.* (*Bryologist* **101**: 70, 1998; phylogeny).

Reinkellomyces Cif. & Tomas. (1953) ≡ Reinkella.

Reisneria Velen. (1922) = Gloeophyllum fide Donk (*Verh. K. ned. Akad. Wet.* tweede sect. **62**: 1, 1974).

Relhanum Gray (1821) ≡ Verpa.

Relicina (Hale & Kurok.) Hale (1974), Parmeliaceae (L). 48, widespread. See Hale (*Smithson. Contr. bot.* **26**, 1975; monogr.), Elix (*Biblthca Lichenol.* **62**, 1996; revision), Elix (*Mycotaxon* **69**: 129, 1998; key), Noicharoen *et al.* (*Mycotaxon* **85**: 325, 2003; Thailand), Blanco *et al.* (*Mol. Phylogen. Evol.* **39**: 52, 2006; phylogeny), Divakar *et al.* (*Mol. Phylogen. Evol.* **40**: 448, 2006; phylogeny).

Relicinopsis Elix & Verdon (1986), Parmeliaceae (L). 5, S. Africa; Australasia. See Elix & Verdon (*Mycotaxon* **27**: 281, 1986), Divakar & Upreti (*Parmelioid Lichens in India* (A Revisionary Study), 2005).

religion (use of mushrooms in), see entheogen, ethnomycology, hallucinogenic fungi, soma.

Remersonia Samson & Seifert (1997), anamorphic *Pezizomycotina*, Hso.?.?. 1, Europe. See Seifert *et al.* (*CJB* **75**: 1160, 1997).

Remispora Linder (1944), Halosphaeriaceae. 6 (marine), widespread. See Johnson *et al.* (*Bot. Mar.* **27**: 557, 1984), Manimohan *et al.* (*MR* **97**: 1190, 1993).

Remleria Raitv. (2004), Hyaloscyphaceae. 1, widespread. See Raitviir (*Scripta Mycologica* Tartu **20**: 109, 2004).

remote (of lamellae), proximal end free and at some distance from the stipe.

Renatobasidium Hauerslev (1993), Auriculariales. 1, Europe. See Hauerslev (*Mycotaxon* **49**: 228, 1993).

reniform, kidney-like in form; fabiform (Fig. 23.7).

Reniforma Pore & Sorenson (1990), anamorphic *Microbotryales*. 1 (basidiomycetous yeast), USA. See Fell *et al.* (*Int. J. Syst. Evol. Microbiol.* **50**: 1351, 2000; mol. phylogeny), Scorzetti *et al.* (*FEMS Yeast Research* **2**: 495, 2002; phylogeny), Aime *et al.* (*Mycol.* **98**: 896, 2006; phylogeny).

Renispora Sigler & J.W. Carmich. (1979), Onygenaceae. Anamorph *Chrysosporium*. 1, USA. See Sugiyama *et al.* (*Mycoscience* **40**: 251, 1999; DNA), Sugiyama & Mikawa (*Mycoscience* **42**: 413, 2001; phylogeny).

repand (of a pileus), having a waved edge which is turned back.

repeating spore, a spore which gives rise to the same type of mycelium as that on which it developed.

repetition (spore germination by), producing a new spore like the first.

Repetobasidiaceae Jülich (1982) = Hydnaceae.

Repetobasidiellum J. Erikss. & Hjortstam (1981), Hydnaceae. 1, widespread (northern Europe). See Pegler & Roberts (*Mycologist* **14**: 52, 2000; UK).

Repetobasidiopsis Dhingra & Avneet P. Singh (2006), Polyporales. 1, India. See Dhingra & Singh (*Mycotaxon* **97**: 116, 2006).

Repetobasidium J. Erikss. (1958), Agaricomycetes. 11, widespread. *Hymenochaetales* or *Agaricales* (*Rickenella* clade). See Eriksson (*Symb. bot. upsal.* **16** no. 1: 67, 1958).

Repetophragma Subram. (1992), anamorphic *Dothideomycetes*, Hso.≡ eP.19. 11, widespread. See Subramanian (*Proc. Indian natn Sci. Acad.* Part B. Biol. Sci. **58**: 185, 1992), Shenoy *et al.* (*MR* **110**: 916, 2006; phylogeny).

reproductocentric (of *Chytridiales*), having development of one (**mono-**, **monogenocentric**) or more reproductive structures at the centre of gravity of the thallus; genocentric. See Karling (*Am. J. Bot.* **19**: 54, 1932). Cf. mono- and polycentric.

Requienella Fabre (1883), Requienellaceae (±L). 3, widespread (north temperate). See Boise (*Mycol.* **78**: 37, 1986), Aptroot (*Biblthca Lichenol.* **44**, 1991; monogr.).

Requienellaceae Boise (1986), Pyrenulales. 7 gen. (+ 2 syn.), 21 spp.

Lit.: Boise (*Mycol.* **78**: 37, 1986), Harris (*Mem. N. Y. bot. Gdn* **49**: 74, 1989), Aptroot (*Biblthca Lichenol.* **44**, 1991), Alias *et al.* (*MR* **100**: 580, 1996), Poonyth *et al.* (*Fungal Diversity* **4**: 101, 2000).

Resendea Bat. (1961), ? Microthyriaceae. 2, S. America. See Batista (*Brotéria* Sér. Ci. Nat. **30**: 87, 1961).

Resinicium Parmasto (1968), Agaricomycetes. 5, widespread. *Hymenochaetales* or *Agaricales* (*Rickenella* clade). See Parmasto (*Consp. System. Corticiac.*: 97, 1968).

Resinomycena Redhead & Singer (1981), Mycenaceae. 8, N. America. See Redhead & Singer (*Mycotaxon* **13**: 151, 1981), Antonín & Noordeloos (*A monograph of the genera Hemimycena,* , 2004).

Resiomeria J.I.R. Larsson (1986), Microsporidia. 1. See Larsson (*Protistologica* **22**: 379, 1986).

resistance, the power of an organism to overcome, completely or in some degree, the effect of a pathogen or other damaging factor. **acquired -**, a non-inherited resistance response in a normally susceptible host following a predisposing treatment. See immunity, axeny.

resistant sporangium, see meiosporangium.

resting sporangium, see meiosporangium.

resting spore (1) a spore germinating after a resting period (frequently after overwintering), as does an oospore or a teliospore; a 'winter spore'; (2) an encysted zygote formed after the fusion of gametes, meiosis occuring on germination to produce planospores which encyst to produce gametes (*Blastocladiales*).

Restiosporium Vánky (2000), Websdaneaceae. 21 (in seeds of *Restionaceae*), Australia. See Vánky (*Mycotaxon* **74**: 343, 2000), Vánky & McKenzie (*Fungal Diversity Res. Ser.* **8**: 259 pp., 2002; New Zealand), Vánky & Shivas (*Fungal Diversity* **14**, 2003; Australia), Vánky (*Mycol. Balcanica* **3**: 19, 2006; key).

Resupinataceae Jülich (1982) nom. rej. = Tricholomataceae.

resupinate (of basidiomata), flat on the substrate with the hymenium on the outer side.

Resupinatus Nees ex Gray (1821) nom. rej., Tricholomataceae. *c.* 20, widespread. See Thorn *et al.* (*Mycol.* **92**: 241, 2000; phylogeny), Thorn *et al.* (*Mycol.* **97**: 1140, 2005; Puerto Rico).

retention (of surfaces), the ability of a surface to hold a fungicide or other crop protectant. Cf. adherance.

Retiarius D.L. Olivier (1978), anamorphic *Pezizomycotina*, Hso.1bH.10. 2 (on pollen), S. Africa. See Olivier (*TBMS* **71**: 193, 1978).

Retiboletus Manfr. Binder & Bresinsky (2002), Boletaceae. 5, north temperate. See Binder & Bresinsky (*Feddes Repert.* **113**: 36, 2002).

Reticellites D.L.E. Glass, D.D. Br. & Elsik (1987), Fossil Fungi. 1 (upper Eocene), USA. See Ethridge Glass *et al.* (*Pollen Spores* **28**: 414, 1986).

Reticularia Baumg. (1790) = Lobaria fide Hawksworth *et al.* (*Dictionary of the Fungi* edn 8, 1995).

reticulate, like a net; netted (Fig. 20.12).

Reticulatisporonites Elsik (1968), Fossil Fungi. 1 (Paleocene), USA.

Reticulocephalis Benny, R.K. Benj. & P.M. Kirk (1992), Sigmoideomycetaceae. 2, USA; British Isles.

See Benny *et al.* (*Mycol.* **84**: 635, 1992; key).

Reticulosphaeria Sivan. & Bahekar (1982), ? Xylariales. 1 (from wood), India. See Sivanesan & Bahekar (*TBMS* **78**: 547, 1982), Kang *et al.* (*MR* **103**: 1621, 1999).

Reticulosporidium Thor (1930) nom. dub., ? Fungi.

Retidiporites C.P. Varma & Rawat (1963), Fossil Fungi ? Fungi. 1 (Eocene), India.

Retigerus Raddi (1829) = Phallus fide Stalpers (*in litt.*).

Retihelicosporonites Ramanujam & Rao (1979), Fossil Fungi, anamorphic *Pezizomycotina*. 1 (Miocene), India. ? Fossil *Hiospira*.

Retinocyclus Fuckel (1871) ≡ Tromera A. Massal. ex Körb.

Retocybe Velen. (1947) = Delicatula fide Kuyper (*in litt.*).

retraction septum, see septum (adventitious).

Retroa P.F. Cannon (1991), Phyllachoraceae. 2 (on living legume leaves), S. America. See Cannon (*Mycol. Pap.* **163**, 1991).

Retroconis de Hoog & Bat. Vegte (1989), anamorphic *Pezizomycotina*, Hso.0eP.3. 1, old world. See de Hoog & Batenburg-van der Vegte (*Stud. Mycol.* **31**: 99, 1989).

retroculture, a reisolate of a pathogen from a host into which it had been experimentally introduced.

retroneme, a bipartite tubular hair.

retrorse, backward.

Retrostium Nakagiri & Tad. Ito (1997), ? Spathulosporaceae. 1 (on marine red algae), Japan. See Nakagiri & Ito (*Mycol.* **89**: 484, 1997).

revalidated, see devalidated.

revolute, having the edge rolled back or up (Fig. 19H).

Reyesiella Sacc. (1917) ≡ Anthomycetella.

Reymanella Marcink. (1979), Fossil Fungi. 1 (Jurassic), Poland.

RFLP, see Molecular biology.

Rhabdium P.A. Dang. (1903) [non *Rhabdium* Wallr. 1833, *Algae*] = Harpochytrium fide Atkinson (*Ann. mycol.* **1**: 479, 1903).

Rhabdoclema Syd. (1939), anamorphic *Pezizomycotina*, Cpd.0eH.?. 2, S. America.

Rhabdocline Syd. (1922), Hemiphacidiaceae. Anamorphs *Meria*, *Rhabdogloeum*. 3, widespread (north temperate). *R. pseudotsugae* (needle cast of *Pseudotsuga menziesii*). See Parker & Reid (*CJB* **47**: 1533, 1969; infraspecific taxa), Sherwood-Pike *et al.* (*CJB* **64**: 1849, 1986), Stone (*CJB* **65**: 2614, 1987; devel.), Gernandt *et al.* (*Mycol.* **89**: 735, 1997; DNA), Gernandt *et al.* (*Mycol.* **93**: 915, 2001; phylogeny), Stone & Gernandt (*Mycotaxon* **91**: 115, 2005; revision), Wang *et al.* (*Mycol.* **98**: 1065, 2006; phylogeny).

Rhabdodiscus Vain. (1921) = Ocellularia fide Hale (*Bull. Br. Mus. nat. hist.* Bot. **8**: 227, 1981).

Rhabdogloeopsis Petr. (1925), anamorphic *Pezizomycotina*, Cac.0eH.10. 1, USA.

Rhabdogloeum Syd. (1922), anamorphic *Rhabdocline*, Cac.0eH.15. 3, N. America. See Stone & Gernandt (*Mycotaxon* **91**: 115, 2005).

Rhabdomyces Balbiani (1889), anamorphic *Pezizomycotina*, Hso.?.?. 1, France.

Rhabdopsora Müll. Arg. (1888), Verrucariaceae (L). 1, Brazil. See Hawksworth (*SA* **9**: 24, 1991).

Rhabdopsoromyces Cif. & Tomas. (1953) ≡ Rhabdopsora.

Rhabdospora (Durieu & Mont. ex Sacc.) Sacc. (1884) nom. cons., anamorphic *Mycosphaerellaceae*, Cpd.≡ fH.?. 151, widespread. See Verkley & Priest (*Stud. Mycol.* **45**: 123, 2000; status).

Rhabdosporium, see *Rabdosporium*.

Rhabdostroma Syd. & P. Syd. (1916) = Apiospora fide Barr (*Mycol.* **68**, 1976).

Rhabdostromella Höhn. (1915), anamorphic *Pezizomycotina*, St.0eH.?. 1, Europe.

Rhabdostromellina Höhn. (1917) = Coleophoma fide Sutton (*Mycol. Pap.* **141**, 1977).

Rhabdostromina Died. (1921) = Coleophoma fide Petrak (*Annls mycol.* **27**: 324, 1929).

Rhabdothyrella Höhn. (1917) = Leptothyrium fide Clements & Shear (*Gen. Fung.*, 1931).

Rhabdothyrium Höhn. (1915) = Leptothyrium fide Clements & Shear (*Gen. Fung.*, 1931).

Rhachomyces Thaxt. (1895), Laboulbeniaceae. *c.* 71, widespread. See Balazuc (*Annls Soc. ent. Fr.* N.S. **6**: 677, 1970), Tavares (*Mycol. Mem.* **9**: 627 pp., 1985), Santamaría (*Nova Hedwigia* **68**: 351, 1999; Spain), Hughes *et al.* (*Mycol.* **96**: 1355, 2004; New Zealand), Rossi (*Nova Hedwigia* **83**: 129, 2006; on cave beetles), Santamaría & Faille (*Nova Hedwigia* **85**: 159, 2007; Pyrenees).

Rhacodiella Peyronel (1919) = Myrioconium Syd. fide Kendrick & Carmichae *in* Ainsworth *et al.* (Eds) (*The Fungi* **4A**: 390, 1973).

Rhacodiopsis Donk (1975) ≡ Racodium Fr.

Rhacodium, see *Racodium Fr.*

Rhacophyllus Berk. & Broome (1871), anamorphic *Coprinus*. 1, Asia; Africa. See Petch (*TBMS* **11**: 238, 1926), Redhead *et al.* (*Taxon* **49**: 789, 2000; anamorphic of *Coprinopsis clastophyllus*).

Rhadinomyces Thaxt. (1893) ? = Corethromyces fide Tavares (*Mycol. Mem.* **9**: 627 pp., 1985).

rhagadiose, having deep cracks.

Rhagadolobium Henn. & Lindau (1896), Parmulariaceae. 10 (on *Pteridophyta*), widespread (esp. tropical). See Dingley (*N.Z. Jl Bot.* **10**: 74, 1972; NZ), Swart (*TBMS* **91**: 581, 1988; on *Dicksonia*).

Rhagadostoma Körb. (1865), Nitschkiaceae. 6 (lichenicolous), Europe. See Hawksworth (*TBMS* **74**: 363, 1980), Navarro-Rosinés & Hladun (*Bull. Soc. linn. Provence* **45**: 431, 1994), Navarro-Rosinés *et al.* (*Bull. Soc. linn. Provence* **50**: 233, 1999).

Rhagadostomella Etayo (2002), Nitschkiaceae. 1 (lichenicolous), Colombia. See Etayo (*Biblthca Lichenol.* **84**: 109, 2002).

Rhagidiasporidium Thor (1930) nom. dub., ? Fungi.

Rhamphoria Niessl (1876), ? Annulatascaceae. Anamorph *Phaeoisaria*-like. 9 (from wood), widespread (esp. temperate). See Sivanesan (*TBMS* **67**: 469, 1976), Réblová & Winka (*Mycol.* **93**: 478, 2001; DNA), Campbell & Shearer (*Mycol.* **96**: 822, 2004; phylogeny), Réblová (*Mycol.* **98**: 68, 2006).

Rhamphosphaeria Kirschst. (1936), Pezizomycotina. 1, Europe.

Rhamphospora D.D. Cunn. (1888), Rhamphosporaceae. 1 (on leaf of *Nymphaeaceae*), widespread. See Reid (*TBMS* **52**: 25, 1969), Vánky (*Illustrated genera of smut fungi*, 1987).

Rhamphosporaceae R. Bauer & Oberw. (1997), Doassansiales. 1 gen., 1 spp.

Lit.: Vánky (*Cryptog. Stud.* **1**: 159 pp., 1987), Bauer *et al.* (*CJB* **75**: 1273, 1997), Begerow *et al.* (*CJB* **75**: 2045, 1998), Begerow *et al.* (*MR* **106**: 1392, 2002), Begerow *et al.* (*Mycol. Progr.* **1**: 187, 2002).

Rhaphidicyrtis Vain. (1921), Pyrenulales (L). 1, widespread (temperate). See Aguirre-Hudson (*Bull. Br. Mus. nat. hist.* Bot. **21**: 85, 1991).

Rhaphidophora Ces. & De Not. (1863) [non *Rhaphidophora* Hassk. 1842, *Araceae*] = Gaeumannomyces fide Walker (*Mycotaxon* **11**: 1, 1980).

Rhaphidospora Fr. (1849) [non *Rhaphidospora* Nees 1832, *Acanthaceae*] ≡ Rhaphidophora.

Rheophila Cépède & F. Picard (1907) = Peyritschiella fide Cépède & Picard (*Bull. Sci. Fr. Belg.* **42**: 251, 1908).

Rheumatopeltis F. Stevens (1927) = Trabutia fide von Arx & Müller (*Stud. Mycol.* **9**, 1975).

Rhexoacrodictys W.A. Baker & Morgan-Jones (2002), anamorphic *Pezizomycotina*, H?.?.?. 4, widespread. See Baker *et al.* (*Mycotaxon* **82**: 98, 2002), Baker & Morgan-Jones (*Mycotaxon* **85**: 371, 2003).

Rhexoampullifera P.M. Kirk (1982), anamorphic *Pezizomycotina*, Hso.≡ eP.40. 1, British Isles; Brazil. See Kirk (*TBMS* **78**: 299, 1982), Castañeda *et al.* (*Mycol.* **93**: 168, 2001).

Rhexocercosporidium U. Braun (1994), anamorphic *Helotiales*, Hso.≡ eP.11. 1, Europe; Canada. See Braun (*Monogr. Cercosporella, Ramularia Allied Genera (Phytopath. Hyphom.)* **1**, 1995), Reeleder *et al.* (*Phytopathology* **96**: 1243, 2006; on ginseng), Reeleder (*Mycol.* **99**: 91, 2007; on ginseng).

Rhexodenticula W.A. Baker & Morgan-Jones (2001), anamorphic *Pezizomycotina*. 2, Cuba; S. Africa. See Baker *et al.* (*Mycotaxon* **79**: 363, 2001), Mel'nik *et al.* (*Mycol. Progr.* **3**: 19, 2004; S Africa).

Rhexographium M. Morelet (1995) = Graphium fide Morelet (*Annales de la Société des Sciences Naturelles et d'Archéologie de Toulon et du Var* **47**: 90, 1995), Jacobs *et al.* (*Mycol.* **95**: 714, 2003).

rhexolytic, secession of conidia involving the circumscissile splitting of the periclinal wall of the cell below the basal conidial septum (Cole & Samson, *Patterns of development in conidial fungi*, 1979) rather than the septum itself; fracture. Cf. schizolytic.

Rhexophiale Th. Fr. (1860) = Sagiolechia fide Hawksworth *et al.* (*Dictionary of the Fungi* edn 8, 1995), Henssen (*Biblthca Lichenol.* **58**: 123, 1995).

Rhexoprolifer Matsush. (1996), anamorphic *Pezizomycotina*, Hso.?.?. 1, Transvaal. See Matsushima (*Matsush. Mycol. Mem.* **9**: 24, 1996).

Rhexosporium Udagawa & Furuya (1977), Sordariales. 1, Japan. See Udagawa & Furuya (*TMSJ* **18**: 302, 1977).

Rhexothecium Samson & Mouch. (1975), Eremomycetaceae. Anamorph *Trichosporiella*-like. 1, Egypt; Kenya. See Malloch & Sigler (*CJB* **66**: 1929, 1988).

Rhinocephalum Kamyschko (1961) = Arthrinium fide Kendrick & Carmichae *in* Ainsworth *et al.* (Eds) (*The Fungi* **4A**: 390, 1973).

Rhinocladiella Kamyschko (1960) [non *Rhinocladiella* Nannf. 1934] = Chrysosporium fide Kendrick & Carmichae *in* Ainsworth *et al.* (Eds) (*The Fungi* **4A**: 390, 1973).

Rhinocladiella Nannf. (1934), anamorphic *Capronia*, Hso.0eH.10. 10, widespread. See cellar fungus. See de Hoog (*Stud. Mycol.* **15**, 1977), Onofri & Castagnola (*Mycotaxon* **18**: 337, 1983; gen. synonymy), Iwatsu *et al.* (*Mycotaxon* **28**: 199, 1987; conidiogenesis), Untereiner (*Mycol.* **89**: 120, 1997; teleomorph), Haase *et al.* (*Stud. Mycol.* **43**: 80, 1999; phylogeny), Untereiner & Naveau (*Mycol.* **91**: 67, 1999; phylogeny), Hoog *et al.* (*J. Clin. Microbiol.* **41**: 4767, 2003), Tribe *et al.* (*Mycologist* **20**: 171, 2006).

Rhinocladiopsis Kamyschko (1961) = Chrysosporium fide Carmichael (*in litt.*).

Rhinocladium Sacc. & Marchal (1885), anamorphic *Pezizomycotina*, Hso.0eP.1. 9, widespread. See Mercado Sierra & Castañeda Ruiz (*Acta Bot. Cubana* **50**: 1, 1987; Cuba), Furlanetto & Dianese (*MR* **100**: 244, 1996; Brazil), Mercado Sierra *et al.* (*Mycotaxon* **63**: 155, 1997; Mexico).

Rhinodinomyces E.A. Thomas ex Cif. & Tomas. (1953) = Rinodina fide Hawksworth *et al.* (*Dictionary of the Fungi* edn 8, 1995).

rhinosporidiosis, polypoid growths in the nose and other organs of humans, horse, etc., caused by *Rhinosporidium seeberi* [formerly classified among the fungi, now considered to be a protozoan].

Rhinotrichella G. Arnaud ex de Hoog (1977), anamorphic *Pezizomycotina*, Hso.0eP.10. 2, Japan. See Matsushima & Matsushima (*Matsush. Mycol. Mem.* **8**: 45, 1995), Crous *et al.* (*S. Afr. J. Bot.* **62**: 89, 1996).

Rhinotrichum Corda (1837) nom. dub., anamorphic *Pezizomycotina*. See Carmichael *et al.* (*Genera of Hyphomycetes*, 1980).

Rhipidiomyces Thaxt. (1926), Laboulbeniaceae. 1, S. America. See Nannfeldt (*Svensk bot. Tidskr.* **43**: 468, 1949).

Rhipidium Wallr. (1833) nom. rej. ≡ Schizophyllum.

Rhipidocarpon (Theiss.) Theiss. & Syd. (1915), Parmulariaceae. 1, Java; Philippines.

Rhipidocephalum Trail (1888), anamorphic *Pezizomycotina*, Hso.0eH.?. 1, British Isles.

Rhipidonema Mattir. (1881) = Dictyonema C. Agardh ex Kunth fide Parmasto (*Nova Hedwigia* **29**: 99, 1978).

Rhipidonematomyces Cif. & Tomas. (1954) ≡ Rhipidonema.

Rhizidiaceae J. Schröt. (1886) = Chytridiaceae.

Rhizidiocystis Sideris (1929), ? Chytridiales. 1, Hawaii.

Rhizidiopsis Sparrow (1933) = Podochytrium fide Sparrow (*Aquatic Phycomycetes* Edn 2: 1187 pp., 1960).

Rhizidium A. Braun (1856), Chytridiaceae. *c.* 17, widespread. See Sparrow (*Aquatic Phycomycetes* Edn 2: 1187 pp., 1960; key).

Rhizina Fr. (1815), Rhizinaceae. 1 or 2, widespread (north temperate). See Abbott & Currah (*Mycotaxon* **62**: 1, 1997; N. Am.), O'Donnell *et al.* (*Mycol.* **89**: 48, 1997; phylogeny), Vasiliauskas & Stenlid (*Mycol.* **93**: 447, 2001), Hansen & Pfister (*Mycol.* **98**: 1029, 2006; phylogeny).

rhizina (pl. **-ae**), a root-like hair or thread; the attachment organs of many foliose lichens (e.g. *Parmelia*); rhizine; they may be divided into several types, for details see Gyelnik (*Bot. Közl.* **24**: 122, 1927), Hale & Kurokawa (*Contr. U.S. natn Herb.* **36**: 122, 1964), Hannemann (*Bibl. Lich.* **1**, 1973). Cf. hyphal net.

Rhizinaceae Bonord. (1851), Pezizales. 2 gen., 3 spp.

Lit.: Weber & Bresinsky (*Persoonia* Suppl. **14**: 553, 1992), Abbott & Currah (*Mycotaxon* **62**: 1, 1997), O'Donnell *et al.* (*Mycol.* **89**: 48, 1997; phylogeny), Harrington *et al.* (*Mycol.* **91**: 41, 1999; phylogeny), Lygis *et al.* (*Mycol.* **97**: 788, 2005), Hansen & Pfister (*Mycol.* **98**: 1029, 2006; phylogeny), Perry *et al.* (*MR* **111**: 549, 2007; phylogeny).

rhizinose-strand ('Rhizinenstränge'), a rhizine-like organ of attachment in squamulose (placodioid) lichens (e.g. *Toninia*), which is tough and much branched. At least 3 types occur. See Poelt & Baumgartner (*Öst. bot. Z.* **111**: 1, 1964), Hannemann (*Bibl. Lich.* **1**, 1973). Cf. hyphal net.

Rhizoblepharia Rifai (1968), Pyronemataceae. 2, Australia; Jamaica. See Erb (*Phytologia* **24**: 11, 1972).

Rhizocalyx Petr. (1928), Helotiaceae. Anamorph *Rhizothyrium*. 1 (on *Abies*), Siberia.

Rhizocarpaceae M. Choisy ex Hafellner (1984), Rhizocarpales (±L). 4 gen. (+ 14 syn.), 228 spp.

Lit.: Timdal & Holtan-Hartwig (*Graphis Scripta* **2**: 41, 1988), Poelt (*Mitt. bot. StSamml.* Münch. **29**: 515, 1990), Feuerer (*Biblthca Lichenol.* **39**: 218 pp., 1991), Rambold *et al.* (*Cryptog. Bryol.-Lichénol.* **19**: 247, 1998), Stenroos & DePriest (*Am. J. Bot.* **85**: 1548, 1998), Ihlen & Ekman (*Biol. J. Linn. Soc.* **77**: 535, 2002), Buschbom & Mueller (*Mol. Phylogen. Evol.* **32**: 66, 2004), Grube *et al.* (*MR* **108**: 1111, 2004), Ihlen (*MR* **108**: 533, 2004), Wedin *et al.* (*MR* **109**: 159, 2005), Miądlikowska *et al.* (*Mycol.* **98**: 1088, 2006; phylogeny).

Rhizocarpales Miądl. et al. (2007). Lecanoromycetidae. 2 fam., 12 gen., 489 spp. Fams:

(1) **Catillariaceae**
(2) **Rhizocarpaceae**

For *Lit.* see under fam.

Rhizocarpon Ramond ex DC. (1805), Rhizocarpaceae (±L). *c.* 224, widespread (esp. temperate and boreal). See also map lichen. See Runemark (*Op. bot. Soc. bot. Lund* **2**, 1956; yellow spp. Eur.), Thomson (*Nova Hedwigia* **14**: 421, 1967; arctic), Feurer (*Ber. bayer. bot. Ges.* **49**: 59, 1978; C. Eur.), Hertel & Leuckert (*Herzogia* **5**: 25, 1979; lichenicolous spp.), Honegger (*Lichenologist* **12**: 157, 1980; ultrastr., posn), Geyer *et al.* (*Pl. Syst. Evol.* **145**: 41, 1984; key *R. superficale* group), Innes (*Boreas* **14**: 83, 1985; lichenometry nomencl.), Timdal & Holtan-Hartwig (*Graphis Scripta* **2**: 41, 1988; key Scandinavian spp.), Poelt (*Mitt. bot. StSamml., München* **29**: 515, 1990; key 28 parasitic spp.), Rambold *et al.* (*Cryptog. Bryol.-Lichénol.* **19**: 247, 1998; asci), Stenroos & DePriest (*Am. J. Bot.* **85**: 1548, 1998; DNA), Fryday (*Lichenologist* **32**: 207, 2000; UK), Fryday (*Lichenologist* **34**: 451, 2002; UK), Ihlen (*MR* **108**: 533, 2004; non-yellow spp.), Miądlikowska *et al.* (*Mycol.* **98**: 1088, 2006; phylogeny).

Rhizocarponomyces Cif. & Tomas. (1953) = Rhizocarpon fide Hawksworth *et al.* (*Dictionary of the Fungi* edn 8, 1995).

Rhizochaete Gresl., Nakasone & Rajchenb. (2004), Phanerochaetaceae. 6, widespread. See Greslebin *et al.* (*Mycol.* **96**: 261, 2004).

Rhizocladosporium Crous & U. Braun (2007), anamorphic *Helotiales*. 1, Japan. See Crous *et al.* (*Stud. Mycol.* **58**: 33, 2007).

Rhizoclosmatium H.E. Petersen (1903), Chytridiaceae. 4, widespread.

Rhizoctonia DC. (1805) nom. cons., anamorphic *Thanatephorus*. *c.* 1, widespread. *R. solani* s.l. is a serious pathogen of a wide variety of crops. In a wide sense *Rhizoctonia* has also been used for mycelial anamorphs belonging to *Athelia* (*Fibulorhizoctonia*), *Helicobasidium* (*Thanatophytum*), *Sebacina* (*Opadorhiza*), *Tulasnella* (*Epulorhiza*), *Waitea* (*Chrysorhiza*), and even to ascomycetous species (*Ascorhizoctonia*). See Parmeter (Ed.) (*Rhizoctonia solani. Biology and pathogenicity*, 1970), Carling *et al.* (*Phytopathology* **77**: 1609, 1987; new anastomosis group AG-9), Moore (*Mycotaxon* **29**: 91, 1987; segregated genera), Ogoshi (*Ann. Rev. Phytopath.* **25**: 125, 1987; ecology and pathology), Liu *et al.* (*Phytopathology* **79**: 1205, 1989; isozyme phylogeny), Mordue *et al.* (*MR* **92**: 78, 1989; integrated approach to taxonomy), Yang *et al.* (*MR* **93**: 429, 1989; sclerotial morphogenesis), Andersen (*Mycotaxon* **37**: 25, 1990; hyphal morphology), Cruickshank (*MR* **94**: 938, 1990; pectic zymograms), Jabaji-Hare *et al.* (*Can. J. Pl. Path.* **12**: 393, 1990; anastomosis group relationships using cloned DNA probes), Liu *et al.* (*Can. J. Pl. Path.* **12**: 376, 1990; *R. solani* group 2 relationships by isozyme analysis), Burpee & Martin (*Pl. Dis.* **76**: 112, 1991; spp. associated with turfgrasses), Cubeta *et al.* (*Phytopathology* **81**: 1395, 1991; anastomosis groups characterized by RA analysis of amplified rRNA gene), Kellens & Peumans (*MR* **95**: 1235, 1991; biochemistry and serology of lectins), Sneh *et al.* (*Identification of Rhizoctonia species*, 1991), Wako *et al.* (*J. Gen. Microbiol.* **137**: 2817, 1991; unique DNA plasmid pRS64 associated with chromosomal DNA), Johnk & Jones (*Phytopathology* **83**: 278, 1993; analysis of fatty acids), Andersen & Stalpers (*Mycotaxon* **51**: 437, 1994; check list of 119 names), Vilgalys & Cubeta (*Ann. Rev. Phytopath.* **32**: 135, 1994; molecular systematics), Sneh *et al.* (Eds) (*Rhizoctonia species: taxonomy, molecular biology, ecology, pathology and disease control*, 1996), Roberts (*Rhizoctonia-forming fungi*, 1999), Weerasena *et al.* (*MR* **108**: 649, 2004; DNA probe and a PCR based diagnostic assay), Matsumoto (*Mycoscience* **46**: 319, 2005; fatty acids and anastomosis relationships), González *et al.* (*Mol. Phylogen. Evol.* **40**: 459, 2006; ribosomal DNA and β-tubulin sequences).

Rhizodiscina Hafellner (1979), Patellariaceae. 1 (on wood), widespread (temperate). See Kutorga & Hawksworth (*SA* **15**: 1, 1997).

Rhizogaster Reinsch (1875) nom. dub., ? Fungi.

Rhizogene Syd. & P. Syd. (1921), Venturiaceae. 1 (on *Symphoricarpos*), N. America. See Hafellner (*Nova Hedwigia* **27**: 903, 1976).

Rhizohypha Chodat & Sigr. (1911), anamorphic *Pezizomycotina*, Sc.-.-. 1 (mycorrhizal), Europe.

rhizoid, a root-like structure consisting of anucleate filaments; branched, extension of a chytrid thallus acting as a feeding organ (Karling, *Am. J. Bot.* **19**: 44, 1932), cf. haustorium, rhizina, and holdfast; **-al**, having, or made up of, rhizoids.

Rhizolecia Hertel (1984), ? Lecideaceae (L). 1, New Zealand. See Hertel (*Nova Hedwigia* Beih. **79**: 427, 1984).

Rhizomarasmius R.H. Petersen (2000), Physalacriaceae. 2. See Petersen (*Mycotaxon* **75**: 333, 2000).

rhizomorph, a root-like aggregation of hyphae having a well-defined apical meristem (cf. mycelial cord) and frequently differentiated into a rind of small dark-coloured cells surrounding a central core of elongated colourless cells. See Snider (*Mycol.* **51**: 693, 1961; *Armillaria mellea*), Jacques-Felix (*BSMF* **83**: 1, 1967; *Marasmius*, **84**: 161, 1969; agarics).

Rhizomorpha Ach. (1809) nom. dub., Fungi (?L).

Rhizomorpha Roth (1791) nom. dub. ? = Armillaria fide Donk (*Taxon* **11**: 97, 1962; rhizomorphs of *Armillaria mellea*).

Rhizomorphaceae Chevall. (1826) = Marasmiaceae.

Rhizomorphites Göpp. (1848), Fossil Fungi. 3 (Carboniferous, Jurassic), Europe; USA.

Rhizomucor Lucet & Costantin (1900), Mucoraceae. 6, widespread. See Schipper (*Stud. Mycol.* **17**: 53, 1978), Zheng & Chen (*Mycosystema* **4**: 45, 1991; key), Deploey (*Mycol.* **84**: 77, 1992; spore germination), Rajak *et al.* (*Mycopathologia* **118**: 109, 1992; keratinolysis), Zheng & Chen (*Mycosystema* **6**: 1, 1993; key), Zheng & Hu (*Mycotaxon* **56**: 455, 1995; endophyte), Zheng & Jiang (*Mycotaxon* **56**: 455, 1995), Roux *et al.* (*Proc. Microsc. Soc. S. Afr.* **26**: 42, 1996; ultrastr.), Vastag *et al.* (*J. Clin. Microbiol.* **36**: 2155, 1998; identification), Vágvölgyi *et al.* (*MR* **103**: 318, 1999; status of *R. tauricus*), Voigt & Wöstemeyer (*Gene* **270**: 113, 2001; phylogeny), Iwen *et al.* (*J. Clin. Microbiol.* **43**: 5819, 2005; molecular identification Molecular identification), Nyilasi *et al.* (*Clin. Microbiol. Infect.* **14**: 393, 2008; molecular identification), Richardson & Lass-Flörl (*Clin. Microbiol. Infect.* **14** Suppl. 4: 5, 2008; mucormycosis, review).

rhizomycelium, a rhizoidal system which resembles mycelium, e.g. thallus of the *Cladochytriaceae* (Karling, *Am. J. Bot.* **19**: 53, 1932).

Rhizomyces Thaxt. (1896), Laboulbeniaceae. 10, Africa. See Nannfeldt (*Svensk bot. Tidskr.* **43**: 468, 1949).

Rhizonema Thwaites (1849) = Dictyonema C. Agardh ex Kunth fide Parmasto (*Nova Hedwigia* **29**: 99, 1978).

Rhizophagites Rosend. (1943), Fossil Fungi. 3 (Cretaceous, Pleistocene), N. America. See Butler (*TBMS* **22**: 274, 1939) = Glomus (Glom.) fide, Yao *et al.* (*Kew Bull.* **50**: 349, 1995).

Rhizophagus P.A. Dang. (1896) ? = Glomus fide Gerdemann & Trappe (*Mycol. Mem.* **5**, 1974).

Rhizophidites Daugherty (1941), Fossil Fungi, Chytridiomycetes. 1 (Triassic), USA.

Rhizophila K.D. Hyde & E.B.G. Jones (1989), Pezizomycotina. 1 (marine), Seychelles. See Hyde & Jones (*Mycotaxon* **34**: 527, 1989), Kohlmeyer & Volkmann-Kohlmeyer (*Bot. Mar.* **34**: 1, 1991).

Rhizophlyctidaceae H.E. Petersen (1909) = Spizellomycetaceae.

Rhizophlyctis A. Fisch. (1892), ? Spizellomycetaceae. *c.* 10, widespread. See Sparrow (*Aquatic Phycomycetes* Edn 2: 438, 1960; key), Barr & Désaulniers (*CJB* **64**: 561, 1986), Blackwell & Powell (*Mycotaxon* **70**: 213, 1999).

Rhizophoma Petr. & Syd. (1927) = Rhizosphaera fide Kobayasi (*Bull. Govt For. Exp. Stn* **204**: 91, 1969).

Rhizophydiaceae Letcher (2006), Rhizophydiales. 1 gen. (+ 6 syn.), 100 spp.

Rhizophydiales Letcher (2006), Chytridiomycetes. 3 fam., 5 gen., 104 spp. Fams:
(1) **Kappamycetaceae**
(2) **Rhizophydiaceae**
(3) **Terramycetaceae**

Lit.: Letcher *et al.* (*MR* **110**: 898, 2006), James *et al.* (*Mycol.* **98**: 860, 2006; molecular phylogeny), Hibbett *et al.* (*MR* **111**: 109, 2007), and see under Familes.

Rhizophydium Schenk (1858), Rhizophydiaceae. *c.* 100, widespread. See Sparrow (*Aquatic Phycomycetes* Edn 2: 231, 1960; key), Chen & Chien (*Taiwania* **41**: 105, 1996; zoospore ultrastr.), Letcher *et al.* (*Mycol.* **96**: 162, 2004; North America, Australia).

Rhizophyton Zopf (1887) = Rhizophydium fide Minden (*Krypt.-Fl. Brandenburg augr. Gebiete*, 1911).

Rhizoplaca Zopf (1905), Lecanoraceae (L). 11, widespread. See Leuckert *et al.* (*Nova Hedwigia* **28**: 71, 1977), Weber (*Mycotaxon* **8**: 559, 1979), Wei (*Acta Mycol. Sin.* **3**: 207, 1984; key 4 spp. China), McCune (*Bryologist* **90**: 6, 1987; N. Am.), Roux *et al.* (*CJB* **71**: 1660, 1993; posn), Arup & Grube (*CJB* **78**: 318, 2000; polyphyly), Wei & Wei (*Mycosystema* **24**: 24, 2005; *R. chrysolecua* group), Zhou & Wei (*Mycosystema* **25**: 376, 2006; polyphyly, reln with *Rhizoplacopsidaceae*), Zhou *et al.* (*Mycol.* **98**: 57, 2006; genetic variation).

Rhizoplacopsidaceae J.C. Wei & Q.M. Zhou (2006), Umbilicariales. 1 gen., 1 spp. See Wei & Zhou (*Mycosystema* **25**: 381, 2006).

Rhizoplacopsis J.C. Wei & Q.M. Zhou (2006), Rhizoplacopsidaceae. 1, China. See Wei & Zhou (*Mycosystema* **25**: 381, 2006).

rhizoplane, the surface of a root. See Lynch (Ed.) (*The rhizosphere*, 1990).

rhizoplast, see blepharoplast.

Rhizopodella (Cooke) Boud. (1885) ≡ Plectania.

rhizopodium, a branched process (pseudopodium) from a plasmodium. Cf. filopodium.

Rhizopodomyces Thaxt. (1931), Laboulbeniaceae. 7, America. See Benjamin (*Aliso* **9**: 379, 1979; key), Tavares (*Mycol. Mem.* **9**: 627 pp., 1985).

Rhizopodopsis Boedijn (1959), Mucoraceae. 1, Java. See Boedijn (*Sydowia* **12**: 220, 1958).

Rhizopogon Fr. (1817), Rhizopogonaceae. *c.* 150 (ectomycorrhizal with *Pinaceae*, one with *Adenostoma* (*Rosaceae*)), widespread (north temperate; introduced with pines in southern hemisphere. See Smith & Zeller (*Mem. N. Y. bot. Gdn* **14**, 1966; key N. Am. spp.), Hosford & Trappe (*TMSJ* **29**: 63, 1988; Japan), Martín (*The genus Rhizopogon in Europe*, 1996), Johannesson & Martín (*Mycotaxon* **71**: 267, 1999; phylogeny), Molina *et al.* (*Rhizopogon – Ectomycorrhizal fungi, key genera in profile*, 1999; ecology), Grubisha *et al.* (*Mycol.* **93**: 82, 2001; phylogeny), Martín (*Mycotaxon* **78**: 191, 2001), Grubisha *et al.* (*Mycol.* **94**: 607, 2002), Grubisha *et al.* (*Mycotaxon* **93**: 345, 2005).

Rhizopogonaceae Gäum. & C.W. Dodge (1928), Boletales. 3 gen. (+ 3 syn.), 152 spp.

Lit.: Bruns *et al.* (*Nature* **339**: 140, 1989; relat. with *Suillus*), Martín (*Edic. Espec. Soc. Catalana Micol.* **5**: 173 pp., 1996), Martín *et al.* (*MR* **102**: 855, 1998), Johannesson & Martín (*Mycotaxon* **71**: 267, 1999), Molina *et al.* (*Ectomycorrhizal Fungi*: 129, 1999), Grubisha *et al.* (*Mycol.* **93**: 82, 2001), Jarosch (*Biblthca Mycol.* **191**: 158 pp., 2001), Martín (*Karstenia* **41**: 23, 2001), Grubisha *et al.* (*Mycol.* **94**: 607, 2002), Martín & Raidl (*Mycotaxon* **84**: 221, 2002), Kretzer *et al.* (*Mycol.* **95**: 480, 2003), Kretzer *et al.* (*New Phytol.* **161**: 313, 2004).

Rhizopogoniella Soehner (1953) = Hymenogaster fide Gross *et al.* (*Beih. Sydowia* **2**: 210, 1980).

Rhizopus Ehrenb. (1821) nom. cons., Mucoraceae. *c.* 9, widespread. *R. stolonifer* (syn. *R. nigricans*) is a common saprobe and facultative parasite of mature fruits and vegetables. See Dabinett & Wellman (*CJB* **51**: 2053, 1973; numerical taxonomy), Gauger (*J. Gen. Microbiol.* **101**: 211, 1977; genetics), Seviour *et al.* (*CJB* **61**: 2374, 1983), Schipper (*Stud. Mycol.* **25**: 1, 1984), Schipper & Stalpers (*Stud. Mycol.* **25**: 20, 1984), Yuan & Jong (*Mycotaxon* **20**: 397, 1984), Ellis (*Mycol.* **77**: 243, 1985), Schipper *et al.* (*J. Gen. Microbiol.* **131**: 2359, 1985; hybridization), Sevior *et al.* (*TBMS* **84**: 701, 1985), Ellis (*Mycol.* **78**: 508,

1986; DNA), Ho (*Trans. Mycol. Soc. Rep. China* **3**: 73, 1988), Huang & Yu (*Mycosystema* **1**: 61, 1988; electrophoresis), Polonelli *et al.* (*Antonie van Leeuwenhoek* **54**: 5, 1988; antigens), Chen & Chen (*Taiwania* **35**: 191, 1990), Liou *et al.* (*Mycol. Monogr.* **3**, 1990), Liou *et al.* (*TMSJ* **32**: 535, 1991; SEM of sporangiospores), El Ghaouth *et al.* (*MR* **96**: 769, 1992; post harvest), Frye & Reinhardt (*Mycopathologia* **124**: 139, 1993), Jong & McManus (*Mycotaxon* **47**: 161, 1993; computer coding), Schipper & Samson (*Mycotaxon* **50**: 475, 1994), Ho (*Fung. Sci.* **10**: 29, 1995), Takagi & Horiuchi (*Food Biotechnology, Microorganisms*: 535, 1995), Schipper *et al.* (*J. Med. Vet. Mycol.* **34**: 199, 1996; mucormycosis & azygosporogenesis), Weitzman *et al.* (*Mycotaxon* **59**: 217, 1996), Winkler *et al.* (*J. Clin. Microbiol.* **34**: 2585, 1996; mucormycosis), Ho (*Bot. Bull. Acad. Sin.* Taipei **39**: 269, 1998; ultrastr.), Zheng & Chen (*Mycotaxon* **69**: 181, 1998), Vágvölgyi *et al.* (*Antonie van Leeuwenhoek* **86**: 181, 2004; Genetic variability in *Rhizopus stolonifer*), Partida-Martinez & Hertweck (*Nature* **437**: 884, 2005; endosymbiotic bacterium (*Berkholderia*) produces rhizoxin), Abe *et al.* (*Biosc., Biotechn., Biochem.* **70**: 2387, 2006; phylogeny), Machouart *et al.* (*J. Clin. Microbiol.* **44**: 805, 2006; PCR-restriction fragment length polymorphism), Liu *et al.* (*Sydowia* **59**: 235, 2007; phylogeny), Zheng *et al.* (*Sydowia* **59**: 273, 2007; monogr.), Nyilasi *et al.* (*Clin. Microbiol. Infect.* **14**: 393, 2008; molecular identification), Richardson & Lass-Flörl (*Clin. Microbiol. Infect.* **14** Suppl. 4: 5, 2008; mucormycosis, review).

Rhizopycnis D.F. Farr (1998), anamorphic *Dothideomycetes*, Cpd.?.?. 1, USA; Europe. See Farr *et al.* (*Mycol.* **90**: 291, 1998), Armengol *et al.* (*Pl. Path.* **52**: 68, 2003; Spain), Ghignone *et al.* (*Eur. J. Pl. Path.* **109**: 861, 2003; molecular diagnostics).

Rhizoscyphus W.Y. Zhuang & Korf (2004), Helotiaceae. 2, widespread. See Zhang & Zhuang (*Nova Hedwigia* **78**: 481, 2004).

Rhizosiphon Scherff. (1926), Chytridiales. 3, Europe.

Rhizosphaera L. Mangin & Har. (1907), anamorphic *Phaeocryptopus*, Cpd.0eH.15. 6 (on conifers), widespread. *R. kalkhoffii* (on pine needles). See Petrak (*Annls mycol.* **36**: 9, 1938), Sutton (*The Coelomycetes*, 1980), Juzwik (*Pl. Dis.* **77**: 630, 1993; *R. kalkhoffii* on *Picea*), Butin & Kehr (*MR* **104**: 1012, 2000), Tsuneda *et al.* (*Mycol.* **96**: 1128, 2004; phylogeny).

Rhizosphaerella Höhn. (1917) ? = Phoma Sacc. fide Sutton (*Mycol. Pap.* **141**, 1977).

Rhizosphaerina B. Sutton (1986), anamorphic *Pezizomycotina*, St.0eP.15. 1, Australia; S.E. Asia. See Sutton (*Sydowia* **38**: 332, 1985).

rhizosphere, the region immediately surrounding a root and influenced by its presence, particularly from proteins and sugars released by the plant; the range of fungi, bacteria and other microorganisms is frequently richer than that of the soil away from a root. See Katznelson *et al.* (*Bot. Rev.* **14**: 543, 1948).

Rhizosporium Rabenh. (1844) ? = Phloeoconis fide Hawksworth *et al.* (*Dictionary of the Fungi* edn 8, 1995).

Rhizostilbella Wolk (1914), anamorphic *Ascobolus*, Hsy.?0eH.?. 1, Java. See Seifert (*Stud. Mycol.* **27**: 1, 1985; used for anamorph, *Sphaerostilbe repens*, of *Nectria mauritiicola*).

Rhizostroma Fr. (1819) nom. dub., anamorphic *Pezizomycotina*. See Donk (*Taxon* **11**: 98, 1962).

Rhizotexis Theiss. & Syd. (1917), Englerulaceae. 1, Natal.

Rhizothyriaceae Tehon ex Bat. & Cif. (1959) = Helotiaceae.

Rhizothyrium Naumov (1915), anamorphic *Rhizocalyx*, Cpt.≡ eH.?. 2, Russia; Chile.

Rhodactina Pegler & T.W.K. Young (1989), Boletaceae. 1, India. See Yang *et al.* (*Mycotaxon* **96**: 133, 2006).

Rhodesia Grove (1937), anamorphic *Helotiales*, Cac.0eH.15. 1, Germany.

Rhodesiopsis B. Sutton & R. Campb. (1979), anamorphic *Pezizomycotina*, Cac.0eH.19. 1, British Isles; Australia. See Sutton & Campbell (*Nova Hedwigia* **30**: 289, 1978).

Rhodoarrhenia Singer (1964), Bolbitiaceae. 8, widespread (tropical). See Singer (*Sydowia* **17**: 142, 1964; key).

Rhodobolites Beck (1923) = Tylopilus fide Singer (*Farlowia* **2**: 223, 1945).

Rhodocarpon Lönnr. (1858) ≡ Endocarpon.

Rhodocephalus Corda (1837) ? = Aspergillus fide Thom (*Manual of the Aspergilli*, 1945).

Rhodochytrium Lagerh. (1893), ? Chlorophyta. 1, widespread (north temperate). Possibly a member of the Algae.

Rhodococcus Zopf (1891), Actinobacteria. q.v.

Rhodocollybia Singer (1939), Marasmiaceae. *c.* 20, widespread. See Antonín & Noordeloos (*Libri Botanici* **17**, 1997; Eur. spp.), Antonín *et al.* (*Mycotaxon* **63**: 359, 1997; taxonomy), Mata *et al.* (*Mycol. Progr.* **3**: 337, 2004; neotropical montane forests spp.), Wilson & Desjardin (*Mycol.* **97**: 667, 2005; phylogeny).

Rhodocybe Maire (1926) = Entoloma.

Rhodocybella T.J. Baroni & R.H. Petersen (1987) = Entoloma fide Kuyper (*in litt.*).

Rhodocyphella W.B. Cooke (1961) = Stigmatolemma fide Donk (*Beih. Nova Hedwigia* **5**, 1962).

Rhodofomes Kotl. & Pouzar (1990) = Fomitopsis fide Ryvarden (*Syn. Fung.* **5**: 218, 1991).

Rhodogaster E. Horak (1964) = Entoloma. Basidioma gasteroid. fide Kuyper (*in litt.*), Horak (*Sydowia* **17**: 190, 1964), Horak & Moreno (*Sydowia* **50**: 187, 1998; *Rhodogaster calongei* sp. nov. from northern Spain).

Rhodomyces Wettst. (1885) = Phaffia fide Kreger-van Rij (Ed.) (*Yeasts, a taxonomic study* 3rd edn, 1984), Boekhout *et al.* (*Taxon* **48**: 147, 1999; nomencl.) = Sporidiobolus (Sporidiobol.) fide.

Rhodonia Niemelä (2005), Polyporaceae. 1, Europe. See Niemelä (*Karstenia* **45**: 79, 2005).

Rhodopaxillus Maire (1913) = Lepista fide Singer (*Agaric. mod. Tax.*, 1951).

Rhodopeziza Hohmeyer & J. Moravec (1995), Pezizaceae. 1, Tierra del Fuego. See Moravec (*Czech Mycol.* **47**: 260, 1995).

Rhodophana Kühner (1971) = Rhodocybe fide Kuyper (*in litt.*).

Rhodophyllaceae Singer (1951) = Entolomataceae.

Rhodophyllus Quél. (1886) = Entoloma fide Hesler (*Entoloma in southeastern North America*, 1967; key 200 spp.), Largent & Benedict (*Madroño* **21**: 32, 1971; gen. defin.), Romagnesi (*Bull. Soc. linn. Lyon* **43**: 325, 1974), Singer (*Agaric. mod. Tax.* edn 3, 1975), Kühner (*BSMF* **93**: 445, 1978; alpine spp.), Romagnesi & Gilles (*Beih. Nova Hedwigia* **59**, 1979;

key 200 spp. Ivory Coast etc.).

Rhodoporus Quél. ex Bataille (1908) ≡ Tylopilus.

Rhodoscypha Dissing & Sivertsen (1983), Pyronemataceae. 1, widespread. See Lohmeyer (*Z. Mykol.* **50**: 147, 1984), Hohmeyer (*Mitt. Arbeitsgem. Pilzk. Niederrhein* **6**: 11, 1988), Perry *et al.* (*MR* **111**: 549, 2007; phylogeny).

Rhodoseptoria Naumov (1913) = Polystigmina fide Clements & Shear (*Gen. Fung.*, 1931).

Rhodosporidium Banno (1967), Sporidiobolales. Anamorph *Rhodotorula*. 6 (often from sea water), widespread. See Fell *et al.* (*Can. J. Microbiol.* **19**: 643, 1973; key), Swann & Taylor (*Stud. Mycol.* **38**: 147, 1995; mol. phylogeny), Fell *et al.* (*Int. J. Syst. Evol. Microbiol.* **50**: 1351, 2000; mol. phylogeny), Sampaio *et al.* (*Int. J. Syst. Evol. Microbiol.* **51**: 687, 2001; polyphasic taxonomy), Aime *et al.* (*Mycol.* **98**: 896, 2006; phylogeny).

rhodosporous, having light-red spores.

Rhodosporus J. Schröt. (1889) ≡ Pluteus.

Rhodosticta Woron. (1911), anamorphic *Polystigma*, St.0eH.15. 3, Asia; USA.

Rhodotarzetta Dissing & Sivertsen (1983), Pyronemataceae. 1 (on burnt ground), Europe. See Hohmeyer (*Mitt. Arbeitsgem. Pilzk. Niederrhein* **6**: 11, 1988), Yao & Spooner (*MR* **106**: 1243, 2002), Hansen & Pfister (*Mycol.* **98**: 1029, 2006; phylogeny), Perry *et al.* (*MR* **111**: 549, 2007; phylogeny).

Rhodothallus Bat. & Cif. (1959), anamorphic *Pezizomycotina*, Cpt.-.-. 1, Brazil. See Batista & Ciferri (*Mycopath. Mycol. appl.* **11**: 96, 1959).

Rhodothrix Vain. (1921) = Nectria. Material has not been examined recently. fide Santesson (*Svensk bot. Tidskr.* **43**: 547, 1949).

Rhodotorula F.C. Harrison (1927), anamorphic *Rhodosporidium, Erythrobasidium*. *c.* 40, widespread. See Fell *et al.* (*Stud. Mycol.* **38**: 129, 1995; mol. phylogeny, 2000; mol. phylogeny), Suh *et al.* (*J. gen. appl. Microbiol.* Tokyo **42**: 1, 1996; mol. phylogeny), Golubev & Churkina (*Mikrobiologiya* **66**: 254, 1997; mycocons), Sampaio *et al.* (*Can. J. Microbiol.* **45**: 491, 1999; utilization aromatic compounds), Gadanho *et al.* (*CJB* **47**: 213, 2001; polyphasic taxonomy), Nagahama *et al.* (*Int. J. Syst. Evol. Microbiol.* **53**: 897, 2003; n.sp. from deep-sea floor).

Rhodotus Maire (1926), Physalacriaceae. 1, Europe; N. America. See Pegler & Young (*Kew Bull.* **30**: 19, 1975; spore ornament.), Kühner (*Hyménomycètes agaricoïdes*, 1980), Legon & Pegler (*Mycologist* **5**: 147, 1991).

Rhodoveronaea Arzanlou, W. Gams & Crous (2007), anamorphic *Annulatascaceae*. 1, Germany. See Arzanlou *et al.* (*Stud. Mycol.* **58**: 57, 2007).

Rhodozyma Phaff, M.W. Mill., Yoney. & Soneda (1972), anamorphic *Xanthophyllomyces*. 1, Alaska; Japan. See Kurtzman & Fell (*Yeasts, a taxonomic study* 4th edn, 1998).

Rhombiella Liro (1939), anamorphic *Thecaphora*. 1.

rhomboidal, resembling an equilateral not right-angled parallelogram (a rhomboid); quadrangular (Fig. 23.22).

Rhombostilbella Zimm. (1902), anamorphic *Pezizomycotina*, Hsy.0eH.10. 2 (on *Capnodiaceae*), Java; Brazil. See Pirozynski (*Mycol. Pap.* **90**, 1963; key), Kendrick (*CJB* **81**: 75, 2003).

Rhopalidium Mont. (1856) = Alternaria fide Höhnel (*Sber. Akad. Wiss. Wien* Math.-naturw. Kl., Abt. 1 **119**: 670, 1910).

Rhopalocladium Schroers, Samuels & W. Gams (1999), anamorphic *Nectriopsis*, Hso.?.?. 1 (on myxomycete), USA. See Schroers *et al.* (*Mycol.* **91**: 375, 1999), Schroers (*Stud. Mycol.* **46**: 1, 2001).

Rhopaloconidium Petr. (1952) = Miuraea fide Braun (*Monogr. Cercosporella, Ramularia Allied Genera (Phytopath. Hyphom.)* **1**, 1995).

Rhopalocystis Grove (1911) = Aspergillus fide Raper & Fennell (*The genus Aspergillus*, 1965).

Rhopalogaster J.R. Johnst. (1902), Rhizopogonaceae. 1, USA.

Rhopalomyces Corda (1839), Helicocephalidaceae. 7, widespread. See Costantin (*Bull. Soc. bot. Fr.* **33**: 489, 1886), Thaxter (*Bot. Gaz.* **16**: 14, 1891), Berlese (*BSMF* **10**: 94, 1892), Marchal (*Revue Mycol.* Paris **15**: 7, 1893), Drechsler (*Bull. Torrey bot. Club* **82**: 473, 1955), Ellis (*Mycol.* **55**: 183, 1963), Barron (*CJB* **51**: 2505, 1973; biology), Stalpers (*Proc. K. ned. Akad. Wet.* Ser. C, Biol. Med. Sci. **77**: 383, 1974), Benjamin *in* Kendrick (Ed.) (*The Whole Fungus* **2**: 573, 1979), Barron (*Mycol.* **72**: 427, 1980; biology), Cano *et al.* (*Nova Hedwigia* **49**: 427, 1989), James *et al.* (*Nature* **443**: 818, 2006; phylogeny), White *et al.* (*Mycol.* **98**: 872, 2006; phylogeny).

Rhopalomyces Harder & Sörgel (1938) [non *Rhopalomyces* Corda 1839] = Blastocladiella fide Couch & Whiffen (*Am. J. Bot.* **29**: 582, 1942).

Rhopalophlyctis Karling (1945), Chytridiaceae. 1, Brazil; N. America.

Rhopalopsis Cooke (1883) ≡ Kretzschmaria.

Rhopalospora, see *Ropalospora*.

Rhopalostroma D. Hawksw. (1977), Xylariaceae. Anamorph *Nodulisporium*-like. 11, Africa; Asia. See Hawksworth & Whalley (*TBMS* **84**: 560, 1985), Whalley *et al.* (*Botanical Journal of Scotland* **50**: 185, 1998; Thailand), Stadler *et al.* (*MR* **108**: 239, 2004; chemistry), Triebel *et al.* (*Nova Hedwigia* **80**: 25, 2005; phylogeny).

Rhopographella (Henn.) Sacc. & Trotter (1913), Pezizomycotina. 1, Brazil.

Rhopographina Theiss. & Syd. (1915) = Ophiodothella fide von Arx & Müller (*Stud. Mycol.* **9**, 1975).

Rhopographus Nitschke ex Fuckel (1870), ? Pleosporales. 6 (on *Pteridophyta*), widespread. See Obrist (*Phytopath. Z.* **35**: 367, 1959), Barr (*Mycotaxon* **43**: 371, 1992; posn).

Rhymbocarpus Zopf (1896), Odontotremataceae. 9 (on lichens, esp. *Rhizocarpon*), Europe. See Coppins *et al.* (*SA* **10**: 51, 1991), Diederich & Etayo (*Lichenologist* **32**: 423, 2000).

Rhymovis Pers. ex Rabenh. (1844) ≡ Paxillus.

Rhynchodiplodia Briosi & Farneti (1906), anamorphic *Pezizomycotina*, Cpd.1eP.?. 1, Italy.

Rhynchogastrema B. Metzler & Oberw. (1989), Rhynchogastremataceae. 1, Europe. See Metzler & Oberwinkler (*Syst. Appl. Microbiol.* **12**: 281, 1989).

Rhynchogastremataceae Oberw. & B. Metzler (1989), Tremellales. 1 gen., 1 spp.

Lit.: Metzler *et al.* (*Syst. Appl. Microbiol.* **12**: 280, 1989), Sampaio *et al.* (*Mycol.* **94**: 874, 2002).

Rhynchomelas Clem. (1909) = Melanospora Corda fide von Arx & Müller (*Beitr. Kryptfl. Schweiz* **11** no. 1, 1954).

Rhynchomeliola Speg. (1884), ? Chaetothyriomycetidae. 3, S. America; Australasia. See Müller & von Arx (*Beitr. Kryptfl. Schweiz* **11** no. 2, 1962), Lee *et al.* (*Mycol.* **95**: 902, 2003).

Rhynchomyces Sacc. & Marchal ex Marchal (1885) ≡

Mycorhynchus.

Rhynchomyces Willk. (1866), anamorphic *Pezizomycotina*, Hso.≡ eP.?. 1, Europe.

Rhynchonectria Höhn. (1902) ? = Pyxidiophora fide Hawksworth & Webster (*TBMS* **68**: 329, 1977), Lundqvist (*Bot. Notiser* **133**: 121, 1980), Malloch & Blackwell (*CJB* **68**: 1712, 1990), Rossman *et al.* (*Stud. Mycol.* **42**: 248 pp., 1999; type material no longer extant).

Rhynchophoma P. Karst. (1884) = Ceratostomella fide Petrak (*Sydowia* **7**: 298, 1953), Verkley (*Nova Hedwigia* **75**: 433, 2002; revision).

Rhynchophoromyces Thaxt. (1908), Ceratomycetaceae. 8, widespread. See Majewski & Sugiyama (*TMSJ* **27**: 425, 1986; Borneo), Santamaría (*Nova Hedwigia* **68**: 351, 1999; Iberian peninsula).

Rhynchophorus Hollós (1926) = Ceratopycnis fide Clements & Shear (*Gen. Fung.*, 1931).

Rhynchoseptoria Unamuno (1940), anamorphic *Pezizomycotina*, Cpd.≡ eH.?. 1, Morocco.

Rhynchosphaeria (Sacc.) Berl. (1891), Pezizomycotina. 7, widespread.

Rhynchosporina Arx (1957), anamorphic *Pezizomycotina*, Hso.0eH.15. 2, N. America. See von Arx (*Verh. K. ned. Akad. Wet. Amst.* C **51**: 19, 1957), Wu *et al.* (*Mycoscience* **38**: 11, 1997).

Rhynchosporium Heinsen ex A.B. Frank (1897), anamorphic *Helotiales*, Hso.1eH.19. 3, widespread (temperate). *R. secalis* (barley and rye leaf blotch). See Owen (*TBMS* **41**: 99, 1958), Owen (*TBMS* **46**: 604, 1963), Williams & Owen (*TBMS* **60**: 223, 1973; physiol. specialization on barley), Ryan *et al.* (*Rhynchosporium secalis A keyword index to the literature*, 1987), Goodwin *et al.* (*Phytopathology* **80**: 1330, 1990; nomencl. *R. secalis* pathotypes), Beer (*Zentralbl. Mikrobiol.* **146**: 339, 1991; review *R. secalis*), Goodwin *et al.* (*MR* **97**: 49, 1993; isozyme variation), Braun (*Monogr. Cercosporella, Ramularia Allied Genera (Phytopath. Hyphom.)* **1**: 254, 1995; key), McDonald *et al.* (*Phytopathology* **89**: 639, 1999; genetic diversity, Australia), Robbertse & Crous (*S. Afr. J. Sci.* **96**: 391, 2000; S Africa), Newton *et al.* (*Z. PflKrankh. PflPath. PflSchutz* **108**: 446, 2001; UK), Goodwin (*MR* **106**: 645, 2002; posn), Crous *et al.* (*Eur. J. Pl. Path.* **109**: 841, 2003; phylogeny), Korff *et al.* (*J. Phytopath.* **152**: 106, 2004; genetic diversity).

rhynchosporous, having beaked spores.

Rhynchostoma P. Karst. (1870), ? Chaetothyriomycetidae. Anamorph *Arthropycnis*. 11, widespread. See Müller & von Arx (*Beitr. Kryptfl. Schweiz* **11** no. 2, 1962), Constantinescu & Tibell (*Nova Hedwigia* **55**: 169, 1992; anamorph), Lee *et al.* (*Mycol.* **95**: 902, 2003; phylogeny).

Rhynchostomopsis Petr. & Syd. (1923) = Amphisphaeria fide Aptroot (*Stud. Mycol.* **37**, 1995).

Rhynchostrigula Bat., J.L. Bezerra & Cavalc. (1966) = Strigula fide Lücking *et al.* (*Lichenologist* **30**: 170, 1998).

Rhynchotheca Kleb. (1933) nom. dub., Fungi. See Sutton (*Mycol. Pap.* **141**, 1977).

Rhyncomeliola, see *Rhynchomeliola*.

Rhyparobius, see *Ryparobius*.

Rhytidenglerula Höhn. (1918), Englerulaceae. Anamorph *Capnodiastrum*. *c.* 10, widespread. See Castlebury *et al.* (*Mycotaxon* **54**: 461, 1995).

Rhytidhysterium Sacc. (1883) ≡ Rhytidhysteron.

Rhytidhysteron Speg. (1881), Patellariaceae. Anamorphs *Diplodia*-like, *Aposphaeria*-like. 4, widespread (esp. tropical). See Samuels & Müller (*Sydowia* **32**: 277, 1979), Kutorga & Hawksworth (*SA* **15**: 1, 1997), Silva-Hanlin & Hanlin (*MR* **103**: 153, 1999; DNA), Liew *et al.* (*Mol. Phylogen. Evol.* **16**: 392, 2000; phylogeny).

Rhytidiella Zalasky (1968), ? Cucurbitariaceae. Anamorph *Phaeoseptoria*. 3, Canada; Sweden. See Aguirre-Hudson (*Bull. Br. Mus. nat. hist.* Bot. **21**: 85, 1991), Johnston (*N.Z. Jl Bot.* **45**: 151, 2007).

Rhytidocaulon Nyl. ex Elenkin (1916) nom. rej. prop. = Letharia fide Hawksworth *et al.* (*Dictionary of the Fungi* edn 8, 1995).

Rhytidopeziza Speg. (1885) = Rhytidhysteron fide von Arx & Müller (*Stud. Mycol.* **9**, 1975).

Rhytidospora Jeng & Cain (1977), Ceratostomataceae. 3, widespread. See Krug & Jeng (*Mycotaxon* **10**: 41, 1979), Guarro (*Mycol.* **75**: 927, 1983), Valldosera *et al.* (*MR* **95**: 243, 1991; SEM).

Rhytisma Fr. (1818), Rhytismataceae. Anamorph *Melasmia*. 18, widespread. *R. acerinum* and *R. punctatum* (tar spot of *Acer*), *R. salicinum* (of *Salix*). See also air pollution. See Duravetz & Morgan-Jones (*CJB* **49**: 1267, 1971; ontogeny), Rath (*Rivista Micol.* **35**: 43, 1992; key 6 Italian spp.), Hudler & Jensen-Tracy (*Mycotaxon* **68**: 405, 1998), Landvik *et al.* (*Mycoscience* **39**: 49, 1998; DNA), Gernandt *et al.* (*Mycol.* **93**: 915, 2001; phylogeny), Vasil'eva (*Mikol. Fitopatol.* **36**: 17, 2002; E Russia), Hou & Piepenbring (*Mycopathologia* **159**: 299, 2005; China), Wang *et al.* (*Mycol.* **98**: 1065, 2006; phylogeny).

Rhytismataceae Chevall. (1826) nom. cons., Rhytismatales. 55 gen. (+ 30 syn.), 728 spp.

Lit.: Minter (*Mycol. Pap.* **147**: 54 pp., 1981), Cannon & Minter (*Mycol. Pap.* **155**, 1986; Indian subcont., key 14 gen.), Cannon & Minter (*Mycol. Pap.* **155**: 123 pp., 1986), Minter *in* Peterson (Ed.) (*Recent research on conifer needle diseases*: 71, 1986; illustr. 14 gen.), Johnston (*N.Z. Jl Bot.* **29**: 395, 405, 1991; NZ spp.), Johnston (*Mycotaxon* **52**: 221, 1994; ascospore sheaths), Johnston (*Aust. Syst. Bot.* **13**: 199, 2000), Stone *et al.* (*Microbial Endophytes*: 3, 2000), Gernandt *et al.* (*Mycol.* **93**: 915, 2001), Johnston (*Mycol. Pap.* **176**: 239 pp., 2001), Johnston (*Aust. Syst. Bot.* **14**: 377, 2001), Deckert & Melville (*MR* **105**: 991, 2002), Ortiz-García *et al.* (*Mycol.* **95**: 846, 2003), Ganley *et al.* (*Proc. natn Acad. Sci. U.S.A.* **101**: 10107, 2004), Sokolski *et al.* (*Mycol.* **96**: 1261, 2004), Wang *et al.* (*Mycol.* **98**: 1065, 2006; phylogeny).

Rhytismatales M.E. Barr ex Minter (1986). Leotiomycetes. 3 fam., 83 gen., 795 spp. Ascomata apothecial, immersed, sometimes erumpent, opening by longitudinal or radial splits, often within black clypeate pseudostromatic tissue. Interascal tissue of simple paraphyses, often anastomosing near the base, often with mucous coating, often swollen at the apices. Asci cylindrical, thin-walled, usually not differentiated at the apex, rarely blueing in iodine, often releasing spores through an irregular split. Ascospores usually hyaline and aseptate, often elongated, often with a mucous sheath. Anamorphs coelomycetous. Saprobes and necrotrophic parasites, sometimes endophytic, on leaves, also on bark and wood, mainly temp. Fams:

(1) **Ascodichaenaceae**
(2) **Cudoniaceae**

(3) **Rhytismataceae** (syn. *Phacidiaceae* auct.)
Lit.: Cannon & Minter (*Mycol. Pap.* **155**: 1986; Indian subcont.), Darker (*Contr. Arnold Arb.* **1**, 1932; on conifers, *CJB* **45**: 1399, 1967; gen. revis.), Hunt & Ziller (*Mycotaxon* **6**: 481, 1978; host-gen. keys on conifers), Livsey & Minter (*CJB* **72**: 549, 1994; circumscription, fams), Minter (*in* Capretti *et al.* (Eds), *Shoot and foliage diseases in forest trees*: 65, 1995; Eur. spp. on conifers, illustr., key, 20 gen.), Minter & Cannon (*TBMS* **83**: 65, 1984; ascospore discharge), Pirozynski & Weresub (*in* Kendrick (Ed.), *The whole fungus* **1**: 93, 1979; biogeogr.), Sherwood (*Occ. Pap. Farlow Herb.* **15**, 1980; key 13 gen.), Tehon (*Ill. Biol. Monogr.* **13** (4), 1935; *Lophodermium* s.l.).

Rhytismella P. Karst. (1884) ≡ Cliostomum.

Rhytismites Mesch. (1892), Fossil Fungi. 27 (Cretaceous, Tertiary), Europe.

Rhytismopsis Geyl. (1887), Fossil Fungi. 1 (Eocene), Indonesia. = Rhytismites (Fossil fungi) fide Meschinelli (1892).

Ribaldia Cif. (1954) = Asteroma fide Sutton (*Mycol. Pap.* **141**, 1977).

Ricasolia A. Massal. (1855) = Solenopsora fide Aptroot (*in litt.*).

Ricasolia De Not. (1846) = Lobaria fide Hawksworth *et al.* (*Dictionary of the Fungi* edn 8, 1995).

Ricasoliomyces E.A. Thomas ex Cif. & Tomas. (1953) ≡ Ricasolia De Not.

Riccia L. (1753), Hepaticae. Hepaticae.

Riccoa Cavara (1903) = Heydenia fide Höhnel (*Sber. Akad. Wiss. Wien* Math.-naturw. Kl., Abt. 1 **124**: 56, 1915).

Richonia Boud. (1885), Zopfiaceae. 1, France. See Hawksworth (*CJB* **57**: 91, 1979).

Richoniella Costantin & L.M. Dufour (1916) = Entoloma. Basidioma gasteroid. fide Kuyper (*in litt.*), Dring & Pegler (*Kew Bull.* **32**: 563, 1978; key), Sarasini (*Rivista di Micologia* **37**: 19, 1994).

Richoniellaceae Jülich (1982) = Entolomataceae.

Rick (Johann; 1869-1946; Austria, later Brazil). School teacher, Feldkirch (1894-1898); theological student, Valkenburg, Netherlands (1899-1902) [priest]; teacher (1903-1915) then social worker (1915-1929) then Professor of Theology (1929-1942), São Leopoldo, Rio Grande do Sul; in São Salvador, Rio Grande do Sul (1942-1946). With Theissen (q.v.) a pioneer explorer of the mycota of Brazil. Collections are widely distributed (**B**, **BPI**, **CUP**, **FH**, **IAC**, **IACM**, **K**, **MICH**, **PACA**, **R**, **RB**, **S**, **SFPA** and **SI**). *Publs.* His publications are distributed in *Annales Mycologici*, *Broteria* and *Iheringia*. *Biogs, obits etc.* Lloyd (*Mycological Notes by C.G. Lloyd* **72**: 1286, 1924); Stafleu & Cowan (*TL-2* **4**: 780, 1983); Torrend, in Lloyd (*Mycological Notes by C.G. Lloyd* **53**: 749, 1918).

Rickella Locq. (1952) ≡ Volvolepiota.

Rickenella Raithelh. (1973), Agaricomycetes. 5, widespread. *Hymenochaetales* or *Agaricales* (*Rickenella* clade). See Kost (*La Famiglia delle Tricholomataceae*, 1986; perforate parenthosome), Moncalvo *et al.* (*Syst. Biol.* **49**: 278, 2000; possible posn. in russuloid clade), Redhead *et al.* (*Mycotaxon* **82**: 151, 2002; posn. in hymenochaetoid clade), Redhead *et al.* (*Mycotaxon* **83**: 19, 2002; posn. in hymenochaetoid clade), Antonín & Noordeloos (*A monograph of the genera Hemimycena*, , 2004).

Rickia Cavara (1899), Laboulbeniaceae. 132, widespread (esp. tropical). See Tavares (*Mycol. Mem.* **9**: 627 pp., 1985), Weir (*MR* **102**: 327, 1998; Sulawesi).

Rickiella Syd. (1904) = Phillipsia fide Korf (*Aust. J. Bot.* Suppl. Ser. **10**: 77, 1983), Pfister (*Mycotaxon* **29**: 329, 1987).

Riclaretia Peyronel (1915), anamorphic *Pezizomycotina*, Hsp.0eH.15. 1, Europe.

Ricnophora Pers. (1825) = Phlebia fide Donk (*Taxon* **6**: 113, 1957).

Riddlea C.W. Dodge (1953) = Laurera fide Harris (*Acta Amazon.* Supl. **14**: 55, 1984).

RIEC (Revised Index of Ecological Continuity). The percentage occurrence of up to a maximum of 20 out of a total list of 30 selected old-forest indicator lichens that require a continuity of mature trees to persist in a site. This value has proved useful in identifying woodland of particular antiquity and long environmental continuity in the UK. See Rose (*in* Brown *et al.* (Eds), *Lichenology: progress and problems*: 279, 1976). An *NIEC* (New Index of Ecological Continuity) for use in wider areas of W. Europe uses 'main' (counted to 70) plus 'bonus' species to give a *T* value (Rose, *in* Bates & Farmer (Eds), *Bryophytes and lichens in a changing environment*: 211, 1992).

Riedera Fr. (1849) nom. dub., ? Helotiales. 1, Russia.

Riessia Fresen. (1852), anamorphic *Agaricomycetes*, Hsy.1bH.1. 4, widespread. See Goos (*Mycol.* **59**: 718, 1967; dikaryotic basidiomycete).

Riessiella Jülich (1985), anamorphic *Agaricomycetes*, Hso.≡ eP.1. 2 (with clamp connexions), S.E. Asia. See Jülich (*Int. J. Mycol. Lichenol.* **2**: 127, 1985).

Rigidoporaceae Jülich (1982) = Meripilaceae.

Rigidoporopsis I. Johans. & Ryvarden (1979) = Amylosporus fide Stalpers (*Stud. Mycol.* **40**: 129, 1996).

Rigidoporus Murrill (1905), Meripilaceae. *c.* 40, widespread. See Ryvarden & Johansen (*Prelim. Polyp. Fl. E. Afr.*: 537, 1980; key 6 Afr. spp.), Corner (*Beih. Nova Hedwigia* **86**: 152, 1987; key Malaysia spp.), Gilbertson & Ryvarden (*N. Amer. Polyp.* **2**: 693, 1987; key temp. spp.), Corner (*Mycologist* **9**: 127, 1995; gen. concept), Legon (*Mycologist* **17**: 156, 2003), Ryvarden & Iturriaga (*Mycol.* **95**: 1066, 2003; Venezuela) See also record fungi.

Rikatlia P.F. Cannon (1993), ? Phyllachoraceae. 1 (from living leaves), E. Africa. See Cannon (*SA* **11**: 83, 1993).

Rileya A. Funk (1979), anamorphic *Pezizomycotina*, Cpd/St.≡ eH.19. 1, Canada. See Funk (*CJB* **57**: 7, 1979).

Rimaconus Huhndorf, F.A. Fernández, J.E. Taylor & K.D. Hyde (2001), Sordariomycetes. 1. Perhaps belonging to the *Pleurotremataceae* sensu Barr (q.v.). See Huhndorf *et al.* (*Mycol.* **93**: 1073, 2001), Huhndorf *et al.* (*Fungal Diversity* **20**: 59, 2005).

Rimbachia Pat. (1891), Tricholomataceae. *c.* 10, widespread (tropical). See Singer (*Boln Soc. argent. Bot.* **10**: 210, 1963; key), Redhead (*CJB* **62**: 865, 1984; key), Redhead *et al.* (*Mycotaxon* **83**: 19, 2002; posn. in hymenochaetoid clade), Senn-Irlet & Moreau (*Czech Mycol.* **54**: 145, 2003; alpine spp.).

Rimelia Hale & A. Fletcher (1990) = Parmotrema fide Hale & Fletcher (*Bryologist* **93**: 23, 1990), Czeczuga & Kashiwadani (*Bull. natn. Sci. Mus.* Tokyo, B **19**: 113, 1993; chemistry), Eliasaro & Adler (*Mycotaxon* **66**: 127, 1998; Brazil), Louwhoff (*Biblthca Lichenol.* **78**: 223, 2001; biogeography), Thell *et al.* (*Mycol. Progr.* **3**: 297, 2004; phylogeny), Blanco *et al.* (*Mycol.* **97**: 150, 2005; phylogeny), Carbonero *et al.* (*Phytochem.* **66**: 929, 2005; chemistry), Divakar *et* *al.* (*Lichenologist* **37**: 55, 2005; phylogeny), Blanco *et al.* (*Mol. Phylogen. Evol.* **39**: 52, 2006; phylogeny).

Rimeliella Kurok. (1991) = Canomaculina fide Elix (*Mycotaxon* **65**: 475, 1997).

Rimella Raf. (1819) nom. dub., ? Agaricomycetes. ? 'gasteromycetes'.

rimose (1) cracked; (2) (of a pileus surface), cracked; originally, cracked in all directions (the recommended usage); frequently, cracked by radial fissures as in *Inocybe*. Cf. rimulose.

Rimula Velen. (1934), ? Helotiales. 1, former Czechoslovakia.

Rimularia Nyl. (1868), Trapeliaceae (L). 13, widespread (temperate). See Muhr & Tønsberg (*Nordic Jl Bot.* **8**: 649, 1989), Hertel & Rambold (*Biblthca Lichenol.* **38**: 145, 1990; key), Rambold & Printzen (*Mycotaxon* **44**: 453, 1992; N America), Lumbsch (*J. Hattori bot. Lab.* **83**: 1, 1997), Coppins & Kantvilas (*Biblthca Lichenol.* **78**: 35, 2001), Timdal (*Bryologist* **105**: 219, 2002), Coppins & Fryday (*Lichenologist* **38**: 93, 2006), Lumbsch *et al.* (*MR* **111**: 1133, 2007).

Rimulariaceae Hafellner (1984) = Trapeliaceae.
Lit.: Lumbsch *et al.* (*MR* **105**: 265, 2001), Lumbsch *et al.* (*MR* **111**: 1133, 2007).

rimulose, having small cracks. Cf. rimose.

rind, sometimes used for the firm outer layer of a rhizomorph or other organ; cortex (q.v.).

ring (1) see annulus (Fig. 4C); (2) (of liquid cultures, esp. of bacteria), growth at the surface, sticking to the glass; **- wall building**, see wall building.

Ringueletium J.J. Garcia (1990), Microsporidia. 1.

ringworm, see tinea.

Rinia Penz. & Sacc. (1901) = Erikssonia fide Petrak (*Annls mycol.* **29**: 390, 1931).

Rinodina (Ach.) Gray (1821), Physciaceae (L). *c.* 265, widespread. See Sheard (*Lichenologist* **3**: 328, 1967; Brit. spp.), Lamb (*Br. Antarct. Surv. Sci. Rep.* **61**, 1968; Antarctica), Mayrhofer & Poelt (*Biblthca Lichenol.* **12**, 1979; Eur. saxic. spp.), Poelt & Mayrhofer (*Beih. Sydowia* **8**: 312, 1979; spore types), Hecklau *et al.* (*Herzogia* **5**: 489, 1981; chemistry), Mayrhofer (*J. Hattori bot. Lab.* **55**: 327, 1984; Eur., key 95 spp.), Mayrhofer (*Beih. Nova Hedwigia* **79**: 571, 1984; key 20 spp. Austral.), Giralt & Matzer (*Lichenologist* **26**: 319, 1994; S. Eur.), Giralt *et al.* (*Mycotaxon* **50**: 47, 1994; pannarin-cont. spp.), Giralt *et al.* (*Lichenologist* **27**: 3, 1995; S. Eur.), Sheard (*Herzogia* **11**: 115, 1995; disjunct distr.), Matzer & Mayrhofer (*Bothalia* **26**: 11, 1996; S. Afr.), Giralt & Llimona (*Mycotaxon* **62**: 175, 1997; Iberian spp.), Giralt *et al.* (*Mycotaxon* **61**: 103, 1997; Benelux), Sarv (*Folia Cryptog. Estonica* **31**: 30, 1997; Estonia), Matzer *et al.* (*N.Z. Jl Bot.* **36**: 175, 1998), Grube & Arup (*Lichenologist* **33**: 63, 2001; phylogeny), Nordin & Mattsson (*Lichenologist* **33**: 3, 2001; polyphyly), Sheard & Mayrhofer (*Bryologist* **105**: 645, 2002; N America), Wedin *et al.* (*Taxon* **51**: 655, 2002; phylogeny), Helms *et al.* (*Mycol.* **95**: 1078, 2003; phylogeny, polyphyly), Simon *et al.* (*J. Mol. Evol.* **60**: 434, 2005; introns), Kaschik (*Biblthca Lichenol.* **93**, 2006; S hemisphere), Lendemer & Sheard (*Bryologist* **109**: 562, 2006), Miądlikowska *et al.* (*Mycol.* **98**: 1088, 2006; phylogeny), Mayrhofer & Sheard (*Biblthca Lichenol.* **96**: 229, 2007; *R. archaea* group).

Rinodinella H. Mayrhofer & Poelt (1978), Physciaceae (L). 2, Europe. See Giralt & Llimona (*Mycotaxon* **62**: 175, 1997; Iberian spp.), Grube & Arup (*Lichenologist* **33**: 63, 2001; morphology, phylogeny), Helms *et al.* (*Mycol.* **95**: 1078, 2003; phylogeny), Kaschik (*Biblthca Lichenol.* **93**, 2006).

Rinomia Nieuwl. (1916) ≡ Morinia.

Riopa D.A. Reid (1969) = Perenniporia fide Donk (*Verh. K. ned. Akad. Wet.* tweede sect. **62**: 1, 1974).

Ripartitella Singer (1947), Agaricaceae. 1, America (tropical). Probably a separate family is justified. See Singer (*Mycol.* **39**: 85, 1947), Halling & Franco-Molano (*Mycol.* **88**: 666, 1996; Costa Rica), Hofstetter (*MR* **106**: 1043, 2002; posn.).

Ripartites P. Karst. (1879), Tricholomataceae. 5, widespread. See Huijsman (*Persoonia* **1**: 335, 1960), Pegler & Young (*Kew Bull.* **29**: 659, 1974; spores), Urbonas (*Fungi and Lichens in the Baltic Region* Abstracts, Twelth International Conference on Mycology and Lichenology: 130, 1993; posn.).

Ripexicium Hjortstam (1995), Corticiaceae. 1, Solomon Islands. See Hjortstam (*Mycotaxon* **54**: 191, 1995).

rishitin (**rishitinol**), terpenoid phytoalexins (q.v.) from potato (*Solanum tuberosum*).

Rivilata Kohlm., Volkm.-Kohlm. & O.E. Erikss. (1998), Saccardiaceae. 1, USA. See Kohlmeyer *et al.* (*CJB* **76**: 470, 1998).

Rivulicola K.D. Hyde (1997), ? Lasiosphaeriaceae. 2, Australia. See Hyde *et al.* (*Mycol.* **92**: 1019, 2000), Ranghoo *et al.* (*Mycol.* **92**: 1019, 2000).

rivulose, marked with lines like little rivers.

Rizalia Syd. & P. Syd. (1914), Sordariomycetes. 2 (on *Meliolaceae*), widespread (tropical). See Pirozynski (*Kew Bull.* **31**: 595, 1977).

Rizaliopsis Bat., Castr., J.L. Bezerra & Matta (1964) = Rizalia fide Pirozynski (*Kew Bull.* **31**: 595, 1977).

RNA, see Molecular biology.

Roannaisia T.N. Taylor, Galtier & Axsmith (1994), Fossil Fungi. 1 (Carboniferous), France. See Taylor *et al.* (*Review of Palaeobotany and Palynology* **83**: 256, 1994).

Robakia Petr. (1952), anamorphic *Pezizomycotina*, St.0eH.?. 1, Norway. See Petrak (*Sydowia* **6**: 372, 1952).

Robergea Desm. (1847), Stictidaceae. 8, widespread. See Sherwood (*Mycotaxon* **5**: 1, 1977).

Robertomyces Starbäck (1905) = Bagnisiella fide Nannfeldt (*Nova Acta R. Soc. Scient. upsal.*, 1932).

Robillarda Castagne (1845) nom. rej. = Pestalotiopsis fide Nag Raj & Morgan-Jones (*Taxon* **21**: 535, 1972), Nag Raj (*Coelomycetous Anamorphs with Appendage-bearing Conidia*, 1993).

Robillarda Sacc. (1882) nom. cons., anamorphic *Pezizomycotina*, Cpd.1eP.10. 5, widespread. See Nag Raj (*Coelomycetous Anamorphs with Appendage-bearing Conidia*, 1993).

Robillardiella S. Takim. (1943) nom. dub., ? Dothideales.

Robincola Velen. (1947), ? Helotiales. 1, former Czechoslovakia.

Robledia Chardón (1929) = Botryostroma fide Müller & von Arx (*Beitr. Kryptfl. Schweiz* **11** no. 2, 1962).

Roburnia Velen. (1947), ? Helotiales. 1, former Czechoslovakia.

Roccella DC. (1805) nom. cons., Roccellaceae (L). 26, widespread. See Tavares (*Revta Fac. Cienc. Lisb.* sect. C **6**: 125, 1958; dyeing), Tehler (*CJB* **68**: 2458, 1990; cladistics), Follmann *et al.* (*J. Hattori bot. Lab.* **75**: 345, 1994), Myllys *et al.* (*Mol. Phylogen. Evol.*

12: 295, 1999), Myllys *et al.* (*Curr. Genet.* **36**: 79, 1999; DNA), Follmann & Werner (*Biblthca Lichenol.* **75**: 1, 2000; evolution), Follmann (*J. Hattori bot. Lab.* **90**: 251, 2001; key S American spp.), Follmann (*Mitteilungen aus dem Institut für Allgemeine Botanik Hamburg* **30-32**: 61, 2002; biogeography), Tehler *et al.* (*Symb. bot. upsal.* **34** no. 1: 405, 2004; phylogeny, Europe, Macaronesia, Mediterranean), Lumbsch *et al.* (*Mol. Phylogen. Evol.* **34**: 512, 2005; phylogeny), Tehler (*Biblthca Lichenol.* **95**: 517, 2007), Tehler & Irestedt (*Cladistics* **23**: 432, 2007).

Roccellaceae Chevall. (1826), Arthoniales (±L). 46 gen. (+ 62 syn.), 779 spp.

Lit.: Darbishire (*Bibl. bot.* **45**, 1898), Follmann (*Nova Hedwigia* **31**: 285, 1979), Torrente & Egea (*Biblthca Lichenol.* **32**, 1989; Mediterranean, N. Afr.), Torrente & Egea (*Biblthca Lichenol.* **32**: 282 pp., 1989), Tehler (*CJB* **68**: 2458, 1990), Thor (*Op. bot.* **103**: 92 pp., 1991), Egea *et al.* (*Pl. Syst. Evol.* **187**: 103, 1993), Feige *et al.* (*Cryptog. bot.* **3**: 101, 1993), Follman *et al.* (*Herzogia* **9**: 653, 1993; chemistry), Lücking & Matzer (*Nova Hedwigia* **63**: 109, 1996), Tehler & Egea (*Lichenologist* **29**: 397, 1997), Tehler *et al.* (*Symb. bot. upsal.* **32**: 255, 1997), Grube (*Bryologist* **101**: 377, 1998), Henssen & Thor (*Nordic Jl Bot.* **18**: 95, 1998), Lohtander *et al.* (*Bryologist* **101**: 404, 1998), Lohtander *et al.* (*Lichenologist* **30**: 341, 1998), Myllys *et al.* (*Bryologist* **101**: 70, 1998), Follmann & Peine (*J. Hattori bot. Lab.* **87**: 259, 1999), Myllys *et al.* (*Lichenologist* **31**: 461, 1999; DNA), Sparrius (*Biblthca Lichenol.* **89**: 141 pp., 2004), Tehler *et al.* (*Symb. bot. upsal.* **34** no. 1: 405, 2004), Ertz *et al.* (*Biblthca Lichenol.* **91**: 155 pp., 2005).

Roccellaria Darb. (1897) nom. rej. prop. = Roccellina See Tehler (*CJB* **68**: 2458, 1990; cladistics), Grube (*Bryologist* **101**: 377, 1998), Follmann (*Mitteilungen aus dem Institut für Allgemeine Botanik Hamburg* **30-32**: 61, 2002; biogeography), Tehler (*Taxon* **56**: 254, 2007; nomencl.).

Roccellina Darb. (1898) nom. cons. prop., Roccellaceae (L). 32, widespread. See Tehler (*Op. Bot.* **70**, 1983; key), Tehler (*CJB* **68**: 2458, 1990; cladistics), Tehler (*Acta Bot. Fenn.* **150**: 185, 1994; phylogeny), Lohtander *et al.* (*Lichenologist* **30**: 341, 1998; dispersal), Myllys *et al.* (*Lichenologist* **31**: 461, 1999; phylogeny), Follmann (*J. Hattori bot. Lab.* **90**: 251, 2001; key S American spp.), Follmann (*Mitteilungen aus dem Institut für Allgemeine Botanik Hamburg* **30-32**: 61, 2002; biogeography), Tehler (*Taxon* **56**: 254, 2007; nomencl.), Tehler & Irestedt (*Cladistics* **23**: 432, 2007).

Roccellinastraceae Hafellner (1984) = Pilocarpaceae.

Roccellinastrum Follmann (1968), ? Pilocarpaceae (L). 3, Chile; Tasmania. See Kantvilas (*Lichenologist* **22**: 79, 1990), Andersen & Ekman (*MR* **109**: 21, 2005; phylogeny).

Roccellodea Darb. (1932) = Roccella fide Weber (*Mycotaxon* **27**: 451, 1986), Follmann (*Akt. lichenol. Mitt.* Essen **5**: 20, 1994), Follmann (*J. Hattori Bot. Lab.* **85**: 257, 1998), Follmann *et al.* (*J. Hattori bot. Lab.* **85**: 245, 1998).

Roccellographa J. Steiner (1902), Roccellaceae (L). 1, Socotra. See Tehler (*CJB* **68**: 2458, 1990; cladistics), Myllys *et al.* (*Lichenologist* **31**: 461, 1999; phylogeny).

Roccellographomyces Cif. & Tomas. (1953) ≡ Roccellographa.

Roccellomyces E.A. Thomas ex Cif. & Tomas. (1953) ≡ Roccella fide Hawksworth *et al.* (*Dictionary of the Fungi* edn 8, 1995).

Roccellopsis Elenkin (1929) = Roccella fide Eriksson & Hawksworth (*SA* **6**: 150, 1987).

Rocellinastrum (orthographic variant), see *Roccellinastrum*.

rock hair, pendent brown, grey to black species of *Bryoria* which resemble human hair.

rock tripe, edible lichens of the genus *Umbilicaria*; *U. esculenta*, 'Iwatake', is still eaten in Japan; for details of its use see Sato (*Nova Hedw.* **16**: 505, 1969), Mattick (*Nova Hedw.* **16**: 511, 1969).

rodlet, structural unit of conidial and some hyphal walls composed of particles *c.* 50 Å diam. arranged in linear series (Hess *et al.*, *Mycol.* **60**: 290, 1968).

Rodríguez Hernández (Miguel; 1949-2003; Cuba). Lecturer, Universidad Central de Las Villas (1973-1974); Lecturer, Universidad de la Habana and mycologist, then Head of the Mycological Laboratory, then Scientific Director, Jardín Botánico Nacional, Havana (1974-2003). An expert on tropical epiphyllous ascomycetes. As first and second President of the Asociación Latino-Americana de Micología (1990-1993, 1993-1996), and as organizing committee chairman for the first and second Congreso Latino-Americano de Micología (Havana, 1993, 1996), he played a pivotal role in developing an infrastructure for Latin American mycology. His later collaboration in digitizing fungal records from the Caribbean resulted in production of one of the largest regional checklists ever produced in mycology. *Publs.* (with Minter & Mena Portales) *Fungi of the Caribbean, an Annotated Checklist* (2001). *Biogs, obits etc.* Guzmán (*Inoculum* **55**: 43, 2004).

Rodwaya Syd. & P. Syd. (1901) = Gyrodon fide Singer (*Farlowia* **2**: 223, 1945).

Rodwayella Spooner (1986), Hyaloscyphaceae. 4, widespread (temperate). See Spooner in Dennis (*Fungi Hebrid.*: 383, 1986), Spooner (*Biblthca Mycol.* **116**, 1987).

Roesleria Thüm. & Pass. (1877), Roesleriaceae. 1 (on *Vitis* roots), Europe; N. America. See Arnaud (*Annls Épiphyt.* **16**: 235, 1930), Redhead (*CJB* **62**: 2514, 1985), Nieder (*Pflanzensch.* **3**: 24, 1987), Yao & Spooner (*Kew Bull.* **54**: 683, 1999).

Roesleriaceae Y.J. Yao & Spooner (1999), Pezizomycotina (inc. sed.). 2 gen. (+ 1 syn.), 3 spp.

Lit.: Redhead (*CJB* **62**: 2514, 1984), Véghelyi (*Acta phytopath. entom. Hung.* **24**: 293, 1989), Yao & Spooner (*Kew Bull.* **54**: 683, 1999).

Roeslerina Redhead (1985), Roesleriaceae. 2, Europe; N. America. See Redhead (*CJB* **62**: 2514, 1985), Yao & Spooner (*Kew Bull.* **54**: 683, 1999).

Roestelia Rebent. (1804), anamorphic *Gymnosporangium*. 14, widespread. Anamorph name for (I). See Kern (*Revised taxonomic account of Gymnosporangium*, 1973), Lee & Kakishima (*Mycoscience* **40**: 109, 1999; ultrastr.).

roestelioid (of an aecium), long and tube-like, as in *Gymnosporangium*.

Rogellia Döbbeler (1999), Odontotremataceae. 1, Tierra del Fuego. See Döbbeler (*Haussknechtia* Beih. **9**: 79, 1999), Döbbeler (*MR* **111**: 1406, 2007).

Rogergoosiella A. Hern. Gut. & J. Mena (1996), anamorphic *Pezizomycotina*, Hso.?.?. 1, Cuba. See Hernández Gutiérrez & Mena Portales (*MR* **100**:

1483, 1996).

Rogersella Liberta & A.J. Navas (1978), Schizoporaceae. 1, widespread. See Langer (*Biblthca Mycol.* **154**, 1994; sub *Hyphodontia*).

Rogersia Shearer & J.L. Crane (1976) = Filosporella fide Crane & Shearer (*Mycotaxon* **6**: 27, 1977).

Rogersiomyces J.L. Crane & Schokn. (1978), Sporidiobolaceae R.T. Moore. 1, USA; Taiwan. See Crane & Schoknecht (*Am. J. Bot.* **65**: 903, 1978), Kirschner & Chen (*Sydowia* **55**: 86, 2003; Taiwan).

Rogerson (Clark Thomas; 1918-2001; USA). Assistant in Plant Pathology, Utah State University (1938-1941); Technical Sergeant, US Army Evacuation Hospital, Solomon & Philippine Islands (1942-1945); Teaching Assistant with Fitzpatrick (q.v.), Cornell University (1946-1950); Assistant Professor then Associate Professor, Kansas State University (1950-1958); Curator of Cryptogamic Botany, New York Botanic Garden (1958 onwards). Noted for his studies of the *Hypocreales* where he was a pioneer in 'whole fungus' studies, assiduously connecting anamorphs with their teleomorphs; also contributed greatly to knowledge of the mycota of Utah. *Publs.* The hypocrealean fungi. *Mycol.* (1970); (with Samuels & Doi) *Hypocreales. Memoirs of the New York Botanic Garden* (1990); (with Rossman, Samuels & Lowen) Genera of *Bionectriaceae*, *Hypocreaceae*, and *Nectriaceae* (*Hypocreales*, *Ascomycetes*). *Studies in Mycology* (1999). *Biogs, obits etc.* Samuels & Fogel (*Mycol.* **95** (4): 773-779, 2003) [portrait].

Rogersonia Samuels & Lodge (1996), Hypocreaceae. 1, Puerto Rico. See Samuels & Lodge (*Sydowia* **48**: 250, 1996), Rossman *et al.* (*Stud. Mycol.* **42**: 248 pp., 1999).

rohr, extra-cellular infection apparatus of plasmodiophorids. See also schlauch and stachel.

Roigiella R.F. Castañeda (1984), anamorphic *Pezizomycotina*, Hsy.1eH.10. 1, Cuba. See Castañeda (*Revta Jardín bot. Nac.* Univ. Habana **5**: 62, 1984).

Rolfidium Moberg (1986), Ramalinaceae (L). 1, Sri Lanka. See Moberg (*Lichenologist* **18**: 305, 1986), Timdal (*Op. bot.* **110**, 1991), Ekman (*Op. bot.* **127**: 148 pp., 1996).

Rollandina Pat. (1905) = Arachniotus fide Roy *et al. in* Subramanian (Ed.) (*Taxonomy of fungi* **1**: 215, 1978), Ghosh *et al.* (*Mycotaxon* **10**: 21, 1979; SEM), Sugiyama & Mikawa (*Mycoscience* **42**: 413, 2001; phylogeny).

Rolueckia Papong, Thammath. & Boonpr. (2008), Gomphillaceae (L). 3. See Papong *et al.* (*Nova Hedwigia* **86**: 201, 2008).

Romanoa Thirum. (1954), ? Hypocreales. 1 (in soil), Italy. See Thirumalachar (*Rendiconti Ist. Sup. Sanitä* (Rome) **17**: 1326, 1954).

Romellia Berl. (1900) = Togninia fide Barr (*Mycol.* **77**: 549, 1985), Mostert *et al.* (*Stud. Mycol.* **54**: 115 pp., 2006), Réblová & Mostert (*MR* **111**: 299, 2007; phylogeny).

Romellia Murrill (1904) ≡ Phaeolus.

Romellina Petr. (1955), Pezizomycotina. 1 (on insects), Java. See Petrak (*Sydowia* **9**: 597, 1955).

Romjularia Timdal (2007), Porpidiaceae (L). 1, Europe; USA. See Timdal (*Lichen Flora of the Greater Sonoran Desert Region* **3**: 287, 2007).

Ronnigeria Petr. (1947), Leptopeltidaceae. 1, Europe. See Holm & Holm (*Bot. Notiser* **130**: 115, 1977).

root nodules, of legumes are caused by nitrogen-fixing bacteria of the genus *Rhizobium*; those of spp. of *Alnus*, *Elaeagnus*, *Hippophaë* and *Myrica* by members of the *Plasmodiophorales* fide Hawker & Fraymouth (*J. gen. Microbiol.* **5**: 369, 1951).

Root rots of cereals. Various fungi are implicated, including *Cochliobolus sativus*, *Fusarium graminearum*, *Gaeumannomyces graminis*, *Monographella nivalis*, *Pythium* and *Rhizoctonia* spp. Successive cropping of the same cereal may exacerbate losses which, in the case of take-all disease cause by *G. graminis*, have been reported as up to 60% of the crop. Biological control using various fungi (e.g. *Idriella bolleyi* and *Phialophora graminicola*) may be possible. See: Deacon (*Plant pathology* **22**: 149, 1973), Duczek (*Can. J. Plant Path.* **19**: 402, 1997), Garrett (*Root disease fungi*, 1944, *Biology of root-infecting fungi*, 1956, *Pathogenic root-infecting fungi*, 1970), Simmonds (*Bot. Rev.* **7**: 308, 1941; **19**: 131, 1953). See also take-all.

Ropalospora A. Massal. (1860), Ropalosporaceae (L). 9, widespread (arctic-alpine). See Hertel (*Mitt. bot. StSamml., München* **17**: 537, 1981), Purvis *et al.* (*Lichen Flora of Great Britain and Ireland*, 1992; synonymy with *Fuscidea*), Ekman (*Bryologist* **96**: 582, 1993), Kantvilas (*Biblthca Lichenol.* **78**: 169, 2001; Tasmania), Bylin *et al.* (*Biblthca Lichenol.* **96**: 49, 2007; phylogeny).

Ropalosporaceae Hafellner (1984), Lecanoromycetidae (inc. sed.). 1 gen., 9 spp.

Lit.: Hafellner (*Beih. Nova Hedwigia* **79**: 241, 1984), Bylin *et al.* (*Biblthca Lichenol.* **96**: 49, 2007; phylogeny).

Ropalosporia, see *Ropalospora*.

Roridella E. Horak (2005) ≡ Roridomyces.

roridins, terpinoid toxins of *Myrothecium roridum* and *M. verrucaria*; the cause of dendrochiotoxicosis (ill-thrift) in sheep, pigs, and humans.

Roridomyces Rexer (1994), Mycenaceae. 7, widespread (temperate). See Rexer (*Die Gattung Mycena s.l., Studien zu Ihrer Anatomie, Morphologie und Systematik*: [322] pp., 1994), Horak (*Röhrlinge und Blätterpilze in Europa, Fussend auf Moser, 5 Aufl. (1983): Kleine Kryptogamenflora Band 2, Teil b2. Gustav Fischer Verlag* Bestimmungsschlüssel für *Polyporales* (*p.p.*), *Boletales*, *Agaricales*, *Russulales*, 2005; as *Roridella*).

roridous, covered with drops of liquid like dew.

Rosaria N. Carter (1922), Fungi. 1.

Rosasporina Beneš (1956), Fossil Fungi. 1 (Carboniferous), former Czechoslovakia.

Roscoepoundia Kuntze (1898), anamorphic *Pezizomycotina*, St.0eH.?. 1, Europe.

Rosellinia De Not. (1844), Xylariaceae. Anamorphs *Dematophora*, *Geniculosporium*. 130, widespread. Root rots caused by *R. aquila* (mulberry, *Morus*), *R. arcuata* (tea), *R. bunodes* (tropical crops), *R. necatrix* (with *Dematophora* state) (apple, vine (*Vitis*), etc.), *R. pepo* (cacao), *R. quercina* (oak, *Quercus*). See Dargan & Thind (*Mycol.* **71**: 1010, 1979; key Indian spp.), Francis (*Sydowia* **38**: 75, 1985), Petrini *et al.* (*Sydowia* **41**: 257, 1989), Ju & Rogers (*Mycol.* **82**: 342, 1990), Matzer & Hafellner (*Biblthca Lichenol.* **37**, 1990; lichenicolous taxa, now excluded), Ofong *et al.* (*MR* **95**: 189, 1991; biology, pathogenicity), Sousa & Whalley (*Beih. Sydowia* **43**: 281, 1991), Bermúdez & Carranza-Morse (*Revta Biol. trop.* **40**: 43, 1992; anamorph), Petrini (*Sydowia* **44**: 169, 1992; temperate spp.), Læssøe & Spooner

(*Kew Bull.* **49**: 1, 1993), Pande & Rao (*Czech Mycol.* **48**: 177, 1995; India), San Martín González & Rogers (*Mycotaxon* **53**: 115, 1995; Mexico), Stadler *et al.* (*MR* **105**: 1191, 2001; chemistry), Schena *et al.* (*Eur. J. Pl. Path.* **108**: 355, 2002; molecular diagnostics), Petrini (*N.Z. Jl Bot.* **41**: 71, 2003; New Zealand), Schena & Ippolito (*Journal of Plant Pathology* **85**: 15, 2003; rtPCR), Petrini & Petrini (*MR* **109**: 569, 2005; morphology), Bahl *et al.* (*Mycol.* **97**: 1102, 2005; phylogeny), Ju *et al.* (*Mycol.* **99**: 612, 2007; phylogeny).

Roselliniella Vain. (1921), Sordariomycetidae. 12 (lichenicolous), widespread. See Matzer & Hafellner (*Biblthca Lichenol.* **37**, 1990), Hoffmann & Hafellner (*Biblthca Lichenol.* **77**: 1, 2000).

Roselliniomyces Matzer & Hafellner (1990), Sordariomycetidae. 1 (on lichens, *Trichothelium*), Costa Rica. See Matzer (*Cryptog. Mycol.* **14**: 11, 1993), Matzer (*Mycol. Pap.* **171**: 202 pp., 1996).

Roselliniopsis Matzer & Hafellner (1990), Sordariomycetidae. 5 (on lichens), Africa; north temperate. See Matzer & Hafellner (*Biblthca Lichenol.* **37**, 1990), Matzer (*Cryptog. Mycol.* **14**: 11, 1993).

Rosellinites Mesch. (1892), Fossil Fungi. 2 (Tertiary), Europe.

Rosellinites R. Potonié (1893), Fossil Fungi. 1 (Oligocene), Germany.

Rosellinula R. Sant. (1986), Dothideomycetes. 3 (on lichens), widespread. See Santesson (*SA* **5**: 311, 1986).

Rosenscheldia Speg. (1885), Dothideomycetes. 3 or 4, widespread. See Holm (*Svensk bot. Tidskr.* **62**: 217, 1968).

Rosenscheldiella Theiss. & Syd. (1915), Venturiaceae. *c.* 13, widespread (tropical). See Hansford (*Mycol. Pap.* **15**, 1946), Swart (*TBMS* **58**: 417, 1972), Sivanesan & Shivas (*Fungal Diversity* **11**: 151, 2002).

Rosenschoeldia L. Holm (1968) ≡ Rosenscheldia.

Roseodiscus Baral (2006), Helotiaceae. 3, widespread. See Baral & Krieglsteiner (*Acta Mycologica* Warszawa **41**: 11, 2006).

Roseofavolus T. Hatt. (2003), Polyporaceae. 1, Singapore. See Stalpers (*in litt.*), Hattori (*Mycoscience* **44**: 457, 2003) = *Grifola* fide.

Roseograndinia Hjortstam & Ryvarden (2005), Phanerochaetaceae. 1, widespread. See Hjortstam & Ryvarden (*Syn. Fung.* **20**: 40, 2005).

Rossiomyces R.K. Benj. (2001), Laboulbeniaceae. 1, Poland. See Benjamin (*Aliso* **19**: 132, 2000).

Rossmania Lar.N. Vassiljeva (2001) = Tunstallia fide Vasilyeva (*Mycoscience* **42**: 401, 2001).

Rostafiński (Józef Thomasz; 1850-1928; Poland). Born in Warsaw; student (under E.A. Strasburger), Jena; student under de Bary (q.v.), Halle; Lecturer (1876-1881) then Professor of Botany (1881-1924), Cracow University. Noted for his work on *Mycetozoa*; also worked with Algae. *Publs. Śluzowce (Mycetozoa) Monografia* (1874) [re-issued with supplement, 1875]. *Biogs, obits etc.* Kulczynski (*Acta Societatis Botanicorum Poloniae* **6**: 391, 1929); Stafleu & Cowan (*TL-2* **4**: 908, 1983).

Rostafinskia Speg. (1880) nom. dub., ? Dothideomycetes.

Rostania Trevis. (1880) = Collema F.H. Wigg. fide Hawksworth *et al.* (*Dictionary of the Fungi* edn 8, 1995).

Rostkovites P. Karst. (1881) = Suillus Gray fide Singer (*Farlowia* **2**: 223, 1945).

rostrate (1) beaked; (2) (of asci), see ascus; bent tip of macroconidia of *Microsporum canis* and other anamorphic fungi.

Rostraureum Gryzenh. & M.J. Wingf. (2005), Cryphonectriaceae. 2, widespread. See Gryzenhout *et al.* (*MR* **109**: 1039, 2005), Gryzenhout *et al.* (*Mycol.* **98**: 239, 2006; phylogeny).

Rostrella Fabre (1879) = Ceratocystis fide de Hoog (*Stud. Mycol.* **7**, 1974).

Rostrella Zimm. (1900) = Ceratocystis fide Bakshi (*Mycol. Pap.* **35**, 1951).

Rostrocoronophora Munk (1953) = Gnomonia fide Bolay (*Ber. schweiz. bot. Ges.* **81**, 1972).

Rostronitschkia Fitzp. (1919), Diatrypaceae. 1, Puerto Rico. See Petrak (*Sydowia* **5**: 169, 1951), Rappaz (*Mycol. Helv.* **2**: 285, 1987).

Rostrosphaeria Tehon & E.Y. Daniels (1927) = Botryosphaeria fide von Arx & Müller (*Beitr. Kryptfl. Schweiz* **11** no. 1, 1954).

Rostrospora Subram. & K. Ramakr. (1952) = Colletotrichum fide Nag Raj (*CJB* **51**: 2463, 1973).

rostrum, any beak-like process.

Rostrup (Emil; 1831-1907; Denmark). School teacher at Skaarup, southern Funen (1858-1883); lecturer (1883-1902) then Professor and Chief Consulting Pathologist (1902-1907), Royal Veterinary and Agricultural College, Copenhagen. Made significant contribution to knowledge of the fungi of Denmark; also a pioneer plant pathologist, particularly in forest pathology. Collections in Botanical Museum, Copenhagen (**C**) [catalogued by Lind, *Danish Fungi, as Represented in the Herbarium of E. Rostrup*, 1913]. *Publs. Plantepatologi. Haandbog i Læren om Plantesygdomme for Landbrugere Havebrugere og Skovbrugere* (1902). *Biogs, obits etc.* Grummann (1974: 679); Lind (*Danish Fungi, as Represented in the Herbarium of E. Rostrup*, pp. 1-9, 1913); Ravn (*Bericht der Deutschen Botanischen Gesellschaft* **26A**: (47), 1909); Rosenvinge (*Botanisk Tidsskrift* **28**: 185, 1908) [portrait]; Shear (*Phytopathology* **12**: 1, 1922) [portrait]; Stafleu & Cowan (*TL-2* **4**: 912, 1983).

Rostrupia Lagerh. (1889) = Puccinia fide Berndt (*in litt.*).

Rostrupiella Jørg. Koch, K.L. Pang & E.B.G. Jones (2007), Lulworthiaceae. 1, Denmark; USA. See Koch *et al.* (*Bot. Mar.* **50**: 294, 2007).

rostrupioid (of *Pucciniales*), having teliospores as in *Rostrupia*.

rosulate, in a rosette.

Rosulomyces S. Marchand & Cabral (1976), anamorphic *Pezizomycotina*, Hso.0eH.39. 1, Argentina. See de Hoog (*Stud. Mycol.* **15**, 1977).

Rota Bat., Cif. & Nascim. (1959), anamorphic *Pezizomycotina*, Cpt.1eH.?. 1, Brazil. See Batista *et al.* (*Mycopath. Mycol. appl.* **11**: 58, 1959).

Rotaea Ces. ex Schltdl. (1851), anamorphic *Pezizomycotina*, Hsy.≡ eH.?. 1, Europe.

Rotiferophthora G.L. Barron (1991), anamorphic *Cordycipitaceae*, Hso.0eH.15/16. 22 (on rotifers), widespread. See Glockling & Dick (*MR* **98**: 833, 1994), Glockling (*MR* **102**: 1142, 1998; Japan), Zare *et al.* (*Nova Hedwigia* **73**: 51, 2001).

rots (types of), see Wood-attacking fungi.

Rotula (Müll. Arg.) Müll. Arg. (1890) [non *Rotula* Lour. 1790, *Boraginaceae*] ≡ Mazosia.

Rotula Raf. (1815) nom. dub., Agaricales. See Merrill

(*Index Rafinesq*., 1949).

Rotularia (Vain.) Zahlbr. (1923) [non *Rotularia* Sternb. 1825, fossil *Phanerogamae*] = Mazosia fide Hawksworth *et al.* (*Dictionary of the Fungi* edn 8, 1995).

Roumegueria (Sacc.) Henn. (1908), Dothideomycetes. *c.* 2, widespread (tropical).

Roumegueriella Speg. (1880), Bionectriaceae. Anamorph *Gliocladium*-like. 3 (coprophilous and in mushroom beds), widespread (temperate). See Malloch & Cain (*CJB* **50**: 61, 1972), Udagawa *et al.* (*Mycoscience* **35**: 409, 1994), Rehner & Samuels (*CJB* **73** Suppl. 1: S816, 1995; phylogeny), Rossman *et al.* (*Stud. Mycol.* **42**: 248 pp., 1999), Rossman *et al.* (*Mycol.* **93**: 100, 2001; rDNA phylogeny), Zhang *et al.* (*Mycol.* **98**: 1076, 2006; phylogeny).

Roumeguerites P. Karst. (1879) = Hebeloma fide Singer (*Agaric. mod. Tax.*, 1951).

Roussoella Sacc. (1888), Didymosphaeriaceae. Anamorph *Cytoplea*. 20 (mostly on *Palmae* and bamboos), widespread (tropical). See Aptroot (*Nova Hedwigia* **60**: 325, 1995; key), Hyde *et al.* (*MR* **100**: 1522, 1996), Ju *et al.* (*Mycotaxon* **58**: 419, 1996), Hyde (*MR* **101**: 609, 1997), Kang *et al.* (*Fungal Diversity* **1**: 147, 1998; DNA), Zhou *et al.* (*Cryptog. Mycol.* **24**: 191, 2003; China).

Roussoellopsis I. Hino & Katum. (1965), Dothideomycetes. 3 (on bamboo), Japan. See Hino & Katumoto (*J. Jap. Bot.* **40**: 86, 1965).

Royella R.S. Dwivedi (1960) nom. nud. = Dichotomomyces fide Scott (*TBMS* **55**: 313, 1970).

Royoporus A.B. De (1996), Polyporaceae. 3, widespread. See De (*Mycotaxon* **60**: 143, 1996).

Royoungia Castellano, Trappe & Malajczuk (1992), Boletaceae. 1, Australia. Basidioma gasteroid. See Castellano *et al.* (*Aust. Syst. Bot.* **5**: 613, 1992).

Rozella Cornu (1872), Fungi. 22, widespread. Possible the earliest diverging lineage of the *Fungi*. See Sparrow (*Aquatic Phycomycetes* Edn 2: 167, 1960; key), Held (*Bot. Rev.* **47**: 451, 1981), James *et al.* (*Mycol.* **98**: 860, 2006; phylogeny).

Rozia Cornu (1872) [non *Rozea* Besch. 1872, *Musci*] ≡ Rozella.

Rozites P. Karst. (1879) = Cortinarius fide Peintner *et al.* (*Mycotaxon* **83**: 447, 2002).

r-selection, adaptation to the rapid colonization and exploitation of newly opened or uncolonized habitats; in fungi generally involving large numbers of usually small and asexually produced and short-lived propagules, e.g. conidia, soredia; cf. K-selection. See Andrews, *in* Carroll & Wicklow (1992: 119), Armstrong (*Ecology* **57**: 953, 1976), Boyce (*Ann. Rev. Ecol.Syst.* **15**: 427, 1984), population biology, secondary species.

Rubelia Nieuwl. (1916) ≡ Sphaerosporula.

Rubeolarius Raith. (1981) = Clitocybe fide Kirk (*in litt.*).

Rubetella Tuzet, Rioux & Manier ex Manier (1964) = Smittium fide Lichtwardt (*Am. J. Bot.* **51**: 836, 1964).

Rubigo (Pers.) Roussel (1806) = Uredo fide Berndt (*in litt.*).

Rubikia H.C. Evans & Minter (1985), anamorphic *Pezizomycotina*, St.#eP.1. 1, pantropical. See Evans & Minter (*TBMS* **84**: 57, 1985), Dulymamode *et al.* (*MR* **102**: 1242, 1998; Mauritius).

Rubinoboletus Pilát & Dermek (1969) = Chalciporus fide Kuyper (*in litt.*).

rubratoxin B, a toxic metabolite of *Penicillium rubrum* P-13 (Hayes & Wilson, *Appl. Microbiol.* **16**: 1163, 1968) causing hepatitis in cattle and pigs.

Rubrinectria Rossman & Samuels (1999), Nectriaceae. Anamorph *Nalanthamala*. 1 (dead wood etc.), widespread (pantropical). See Rossman *et al.* (*Stud. Mycol.* **42**: 248 pp., 1999), Schroers *et al.* (*Mycol.* **97**: 375, 2005; phylogeny, anamorph).

Rubromadurella Talice (1935) = Madurella fide Ciferri & Redaelli (*Mycopathologia* **3**: 182, 1941).

Rubroporus Log.-Leite, Ryvarden & Groposo (2002), Polyporaceae. 1, Brazil. See Loguercio-Leite *et al.* (*Mycotaxon* **83**: 224, 2002).

Rubrotricha Lücking, Sérus. & Vězda (2005), Gomphillaceae (L). 1, Ecuador. See Lücking *et al.* (*Lichenologist* **37**: 165, 2005).

ruderal (1) living in waste places; (2) (of fungi) having a high growth rate, rapidly germinating spores, and a short life expectancy due to exhaustion of the available nutrients; cf. zymogenous (see autochthonous), sugar fungus.

Rudetum Lloyd (1919) = Septobasidium fide Stevenson & Cash (*Bull. Lloyd Libr. Mus.* **35**: 47, 1936).

Ruggieria Cif. & Montemart. (1958), anamorphic *Pezizomycotina*, St.0fH.?. 1, Italy. *R. glaucescens* on *Citrus* (melanose).

Rugosaria Raf. (1833) nom. inval. ≡ Gemmularia.

rugose, wrinkled (Fig. 20.14). Cf. rugulose.

Rugosomyces Raithelh. (1979) = Calocybe fide Kuyper (*in litt.*).

Rugosospora Heinem. (1973), Agaricaceae. 2, widespread (tropical). See Heinemann (*Bull. Jard. Bot. Nat. Belg.* **43**: 12, 1973).

rugulose, delicately wrinkled. Cf. rugose.

Ruhlandiella Henn. (1903), Pezizaceae. 1, widespread (probably native to Australia). See also *Muciturbo*. See Dissing & Korf (*Mycotaxon* **12**: 290, 1980), Warcup & Talbot (*MR* **92**: 95, 1989), Galán & Moreno (*Mycotaxon* **68**: 265, 1998; Europe), Hansen *et al.* (*Mycol.* **93**: 958, 2001; phylogeny), Hansen *et al.* (*Mol. Phylogen. Evol.* **36**: 1, 2005; phylogeny), Læssøe & Hansen (*MR* **111**: 1075, 2007; phylogeny).

rum, see spirits.

rumen fungi, see anaerobic fungi.

Ruminomyces Y.W. Ho (1990) = Anaeromyces fide Letcher (*in litt.*).

Rumpomycetes. Class within the *Chytridiomycota* distinguished by the presence of a rumposome (q.v.) and including *Chytridiales* and *Monoblepharidales* but excluding *Spizellomycetales* (Cavalier-Smith, *in* Rayner *et al.* (Eds), *Evolutionary biology of the fungi*: 339, 1987). See James *et al.* (2000).

rumposome, an organelle in zoospores of certain *Chytridiomycota* located close to the cell wall; tooth-like in section and honey-comb like in surface view; see *Rumpomycetes*.

rupestral (**rupestrine**), living on walls or rocks; cf. saxicolous.

Rupinia Speg. & Roum. (1879) [non *Rupinia* L. f. 1782, *Hepaticae*] = Heydenia fide Saccardo (*Syll. fung.* **4**: 625, 1886).

Rusavskia S.Y. Kondr. & Kärnefelt (2003) = Xanthoria fide Kondratyuk & Kärnefelt (*Ukrainian Jour. Bot.* **60**: 427, 2003), Gaya *et al.* (*MR* **112**: 528, 2008).

Ruspoliella Sambo (1937) = Solorina fide Hawksworth *et al.* (*Dictionary of the Fungi* edn 8, 1995).

Russula Pers. (1796), Russulaceae. *c.* 750, widespread. Gastroid forms are polyphyletic and have formerly been morphologically recognised as a separate genus.

See Crawshay (*The spore ornamentation of the Russulas*, 1930), Pearson (*Naturalist* Hull: 85, 1948; key Brit. spp.), Schaeffer (*Russula Monographie*, 1951), Hesler (*Mem. Torrey bot. Club* **21**, 1960), Blum (*Les Russules*, 1962; France), Hesler (*Mycol.* **53**: 605, 1962; N. Am. types), Smith (*Mycol.* **55**: 435, 1963; key), Watson (*TBMS* **49**: 11, 1966; pigments), Romagnesi (*Les Russules d'Europe et Afrique du Nord*, 1967), Rayner (*Bull. BMS* **2**: 76, 1968; keys Brit. spp.), Rayner (*Bull. BMS* **3**: 59, 1969; keys Brit. spp.), Rayner (*Bull. BMS* **3**: 89, 1969; keys Brit. spp.), Rayner (*Bull. BMS* **4**: 19, 1970; descr. Brit. spp.), Burge (*Mycol.* **71**: 977, 1979; spore structure), Pegler & Young (*TBMS* **72**: 353, 1979), Rayner (*Keys to British species of Russula* edn 3, 1985), Sarnari (*Monografia illustrata del genere Russula in Europa*, 1998), Miller *et al.* (*Mycol.* **93**: 344, 2001; phylogeny), Lebel (*N.Z. Jl Bot.* **40**: 489, 2002; New Zealand gasteroid spp.), Miller & Buyck (*MR* **106**: 259, 2002; phylogeny European spp.), Lebel (*Aust. Syst. Bot.* **16**: 401, 2003; australasian spp.), Moënne-Loccoz & Reumaux (*Les Russules Émétiques, Prolégomènes à Une Monographie des Emeticinae d'Europe et d'Amérique du Nord*: 264 pp., 2003), Miller (*Mycotaxon* **89**: 283, 2004; southeastern USA), Buyck (*Cryptog. Mycol.* **26**: 85, 2005; African spp.), Buyck *et al.* (*Boll. Gruppo Micol. 'G. Bresadola'* **46**: 57, 2003; Costa Rica).

Russulaceae Lotsy (1907), Russulales. 5 gen. (+ 18 syn.), 1243 spp.

Lit.: Singer & Smith (*Mem. Torrey bot. Club* **21**, 1960), Smith (*Mycol.* **54**: 626, 1962), Pegler & Young (*TBMS* **72**: 353, 1973), Beaton *et al.* (*Kew Bull.* **33**: 669, 1984; Austr. spp.), Mueller & Gardes (*MR* **95**: 592, 1991), Mueller (*Fieldiana, Bot.* **30**: 158 pp., 1992), Bougher *et al.* (*MR* **97**: 613, 1993; generic delimitation), Mueller & Ammirati (*Am. J. Bot.* **80**: 322, 1993), Henrion *et al.* (*Mol. Ecol.* **3**: 571, 1994), Mueller *et al.* (*Mycol.* **85**: 890, 1993), Buyck (*Russulales News* **3**: 3, 1995), Sweeney *et al.* (*MR* **100**: 1515, 1996), Mueller (*Revta Biol. trop.* **44**: 131, 1996), Mueller (*Mycotaxon* **61**: 205, 1997), Buyck & Horak (*Mycol.* **91**: 532, 1999), Gherbi *et al.* (*Mol. Ecol.* **8**: 2003, 1999), Kropp & Mueller (*Laccaria – Ectomycorrhizal fungi, key genera in profile*, 1999; generic classification), Martín *et al.* (*MR* **103**: 203, 1999), Martin *et al.* (*Microbiology (Reading)* **145**: 1605, 1999), Hibbett *et al.* (*Nature* **407**: 506, 2000), Moncalvo *et al.* (*Syst. Biol.* **49**: 278, 2000), Bougher & Lebel (*Aust. Syst. Bot.* **14**: 439, 2001), Fiore-Donno & Martin (*New Phytol.* **152**: 533, 2001), Miller *et al.* (*Mycol.* **93**: 344, 2001; phylogeny), Binder & Bresinsky (*Mycol.* **94**: 85, 2002), Binder & Hibbett (*Mol. Phylogen. Evol.* **22**: 76, 2002), Eberhardt (*Mycol. Progr.* **1**: 201, 2002), Miller & Buyck (*MR* **106**: 259, 2002), Moncalvo *et al.* (*Mol. Phylogen. Evol.* **23**: 357, 2002), Desjardin (*Mycol.* **95**: 148, 2003), Larsson & Larsson (*Mycol.* **95**: 1037, 2003), Lebel (*Aust. Syst. Bot.* **16**: 401, 2003), Miller & Henkel (*Mem. N. Y. bot. Gdn* **89**: 297, 2004), Shimono *et al.* (*Mycoscience* **45**: 303, 2004), Wang *et al.* (*Nova Hedwigia* **79**: 511, 2004).

Russulales Kreisel ex P.M. Kirk, P.F. Cannon & J.C. David (2001). Agaricomycetes. 12 fam., 80 gen., 1767 spp. Ectomycorrhizal, saprobic or pathogenic on trees, basidiomata epigeous or partly hypogeous, cosmopolitan.

The order *Russulales* has the same circumscription as the russuloid clade in Hibbett & Thorn (*The Mycota* **7B**, 2001). It contains lamellate, poroid, hydnoid and gastroid representatives. The classification is based on Miller *et al.* (*Mycol.* **98**: 960, 2006). Fams:

(1) **Albatrellaceae**
(2) **Amylostereaceae**
(3) **Auriscalpiaceae**
(4) **Bondarzewiaceae**
(5) **Echinodontiaceae**
(6) **Hericiaceae**
(7) **Hybogasteraceae**
(8) **Lachnocladiaceae**
(9) **Peniophoraceae**
(10) **Russulaceae**
(11) **Stephanosporaceae**
(12) **Stereaceae**

Lit.: Heim (*TBMS* **30**: 161, 1948; phylogeny and classification), Malençon (*Recueil trav. crypt. L. Mangin*: 377, 1931), Pegler & Young (*TBMS* **72**: 353, 1979; classification, spore structure), Singer *et al.* (*Beih. Nova Hedw.* **77**, 1983), Singer & Smith (*Mem. Torrey bot. Cl.* **21**: 1, 1960; gasteroid genera), Reijnders (*Persoonia* **9**: 65, 1976; development), Beaton *et al.* (*Kew Bull.* **39**: 669, 1984; Austral. gasteroid spp.), Bourdot & Galzin (1927), Donk (1964: 255), Hjortstam & Ryvarden (*Corticiaceae of North Europe*, 8 vols., 1973-88), Jülich (*Willdenowia Beih.* **7**, 1972; *Athelieae*), Parmasto (*Conspectus Systematis Corticiacearum*, 1968; 11 subfam.), Hjortstam (*Windahlia* **17**: 55, 1987; gen. & spp. checklist), Maekawa (*Rep. Tottori Mycol. Inst.* **31**: 1, 1993; Japanese spp.), and under fams.

Russulina J. Schröt. (1889) = Russula fide Singer & Smith (*A monograph on the genus Galerina Earle*, 1964).

Russuliopsis J. Schröt. (1889) ≡ Laccaria.

rust (1) a disease caused by one of the *Pucciniales*; (2) one of the *Pucciniales*; (3) a disease with 'rusty' symptoms; **black (stem) -** of cereals, *Puccinia graminis*; **blister -** of *Pinus* and *Ribes*, *Cronartium ribicola*; **brown (leaf) -** of barley, *P. hordei*; of rye and wheat, *P. recondita*; **crown -** of oats, *P. coronata*; **red -**, (1) urediniospore state of cereal rusts, esp. *Puccinia graminis*; (2) (of tea) the alga *Cephaleuros*; **white -** (1) (esp *Cruciferae*) = white blister; (2) (of *Chrysanthemum*) = *P. horiana*; **yellow (stripe) -** of cereals, *P. striiformis*.

Ruthea Opat. (1836) ≡ Paxillus.

Ruthiaceae Bat., J.A. Lima & M.A. Tatlasse{?} (1962) = Strigulaceae.

Rutola J.L. Crane & Schokn. (1978), anamorphic *Pezizomycotina*, Hso.≡ eP.1. 1, widespread. See Crane & Schoknecht (*CJB* **55**: 3015, 1977).

Rutstroemia P. Karst. (1871) nom. cons., Rutstroemiaceae. Anamorphs *Myrioconium*, *Phialophora*-like. *c.* 75, widespread. See Holm (*TBMS* **67**: 333, 1976), Korf & Dumont (*Mycotaxon* **5**: 517, 1977; status), Holm (*Mycotaxon* **7**: 139, 1978; status), Kohn & Schumacher (*Mycotaxon* **18**: 531, 1983), Kohn & Schumacher (*Taxon* **33**: 507, 1984; nomencl.), Baral (*SA* **13**: 113, 1994; concept), Holst-Jensen *et al.* (*Mycol.* **89**: 885, 1997; phylogeny), Wang *et al.* (*Mycol.* **98**: 1065, 2006; phylogeny), Wang *et al.* (*Mol. Phylogen. Evol.* **41**: 295, 2006; phylogeny).

Rutstroemiaceae Holst-Jensen, L.M. Kohn & T. Schumach. (1997), Helotiales. 7 gen. (+ 8 syn.), 223 spp.

Lit.: Korf & Zhuang (*Mycotaxon* **24**: 361, 1985),

Baral (*Z. Mykol.* **53**: 119, 1987), Spooner (*Biblthca Mycol.* **116**: 711 pp., 1987), Kohn & Grenville (*CJB* **67**: 371, 1989; anatomy), Zhuang (*Mycosystema* **8-9**: 15, 1995), Holst-Jensen *et al.* (*Mycol.* **89**: 885, 1997; phylogeny), Schumacher & Holst-Jensen (*Mycoscience* **38**: 55, 1997), Zhuang & Wang (*Mycotaxon* **64**: 449, 1997), Wang *et al.* (*Mycol.* **98**: 1065, 2006; phylogeny), Wang *et al.* (*Mol. Phylogen. Evol.* **41**: 295, 2006).

Ruzenia O. Hilber ex A.N. Mill. & Huhndorf (2004), Lasiosphaeriaceae. Anamorph *Selenosporella*-like. 2. See Miller & Huhndorf (*MR* **108**: 26, 2004; separation from *Lasiosphaeria*).

Ryparobius Boud. (1869) = Thelebolus fide Kimbrough & Korf (*Am. J. Bot.* **54**: 9, 1967).

Ryssospora Fayod (1889) = Pholiota fide Kuyper (*in litt.*).

Ryvardenia Rajchenb. (1994), Polyporaceae. 2, widespread. See Rajchenberg (*Nordic Jl Bot.* **14**: 436, 1994).

SA, see Media.

Saagaromyces K.L. Pang & E.B.G. Jones (2003), Halosphaeriaceae. 3, widespread. See Pang *et al.* (*Mycol. Progr.* **2**: 35, 2003).

Sabalicola K.D. Hyde (1995), ? Xylariaceae. 1, USA. See Hyde (*Nova Hedwigia* **60**: 596, 1995).

Sablicola E.B.G. Jones, K.L. Pang & Vrijmoed (2004), Halosphaeriaceae. 1, China. See Pang *et al.* (*CJB* **82**: 486, 2004).

Sabouraud (Raymond; 1864-1938; France). Voluntary military service, Lille (1885); student, Hôpital Cochin, Paris (1887); intern, Hôpital St Louis, Paris (1889-1890); medic, Hôpital St Antoine, Paris (1891-1894); Head of Laboratory (1894-1897) then Head of the Ecole Lailler (1897-1928), Hôpital St Louis, Paris. A dermatologist of Paris who revived and amplified the work of Gruby (q.v.); noted for his monumental researches on the dermatophytes; introduced procedures which shortened the treatment period for *tinea capitis* from 2 years to 3 months removing the need for segregated schools for infected children; his name is familiar to mycologists through 'Sabouraud agar'; also an amateur sculptor. *Publs. Les Teignes* (1910). *Biogs, obits etc.* Grigoraki (*Mycopathologia et Mycologia Applicata* **2**: 171, 1939) [bibliography, portrait]; Pautrier (*Annales de Dermatologie*, Série 7, **9**: 275, 1938) [portrait]; Pignot (*Mycopathologia et Mycologia Applicata* **7**: 348, 1956) [portrait].

Sabouraudiella Boedijn (1953) = Trichophyton fide Hawksworth *et al.* (*Dictionary of the Fungi* edn 8, 1995).

Sabouraudites M. Ota & Langeron (1923) ≡ Microsporum.

Saccardaea Cavara (1894) nom. dub., anamorphic *Pezizomycotina*. See Tulloch (*Mycol. Pap.* **130**: 36, 1972).

Saccardia Cooke (1878), Saccardiaceae. 3, widespread (subtropical). See von Arx (*Persoonia* **2**: 421, 1963).

Saccardiaceae Höhn. (1909), ? Dothideomycetes (inc. sed.). 11 gen. (+ 7 syn.), 35 spp.

Lit.: Barr (*Mycotaxon* **29**: 501, 1987), Hsieh *et al.* (*MR* **101**: 897, 1997), Inácio & Dianese (*MR* **102**: 695, 1998).

Saccardinula Speg. (1885), Elsinoaceae. 5, widespread (tropical). See Barr (*Mem. N. Y. bot. Gdn* **62**, 1990).

Saccardo (Pier Andrea; 1845-1920; Italy). Doctorate, University of Padua (1867); Professor of Natural History (1869) then Professor of Botany and Prefect of the Botanic Gardens (1879), University of Padua. Established *Michelia* (1877-1881), a short-lived journal important for the number of new fungal taxa described in it; distributed exsiccati, prepared an early mycological colour chart (see Colour) and made a special study of pyrenomycetes and anamorphic fungi; his outstanding contribution was the *Sylloge Fungorum Omnium hucusque Cognitorum*, a huge compendium on which all systematic mycology rests. By Saccardo's time the amount of printed work on systematic mycology was great; much was hard to obtain, and it was in no order; for these reasons, Saccardo took up the work of listing with Latin descriptions, all the genera and species of fungi of which there was then knowledge; to do this it was frequently necessary to make new arrangements of old groups, and 'emendations' of numbers of genera and species; he made special use of his system of 'spore groups' (see Anamorphic fungi) and his classification has had to be taken into account by all later systematic mycologists; most of the genera and species of non-lichen-forming fungi described up to 1920 are listed [these volumes are freely available on-line; see Internet: catalogues & thesauri]. His specimens are in the fungal reference collection at Padua (**PAD**) [see Gola, *L'erbario di P.A. Saccardo, Catalogo*, 1930]. *Publs. Fungi Veneti Novi vel Critici* (1873-1878) [a series of individual papers published in *Atti della Società Veneto-Trentina di Scienze Naturali*, *Nuovo Giornale Botanico Italiano* and *Michelia*]; *Fungi Italici Autographice Delineati* (1877-1886) [1,500 small coloured figs]; (with P. Sydow (q.v.), A.N. Berlese (q.v.), G.B. de Toni, Mussat, D. Saccardo [his sone], Traverso & others) *Sylloge Fungorum Omnium hucusque Cognitorum* vols 1-23 (1882-1925) [vols 24-26 were issued after his death, between 1926 and 1972]. *Biogs, obits etc.* Grummann (1974: 521); Paganelli, *in* Minelli [ed.] (*The Botanical Garden of Padua 1545-1995*: 118, 1995); Stafleu & Cowan (*TL-2* **4**: 1023, 1983); Zalin & Lazzari (*Carteggio Bresadola-Saccardo*, 1987) [correspondence].

Saccardoa Trevis. (1869) nom. rej. = Pseudocyphellaria fide Hawksworth *et al.* (*Dictionary of the Fungi* edn 8, 1995).

Saccardoella Speg. (1879), Sordariomycetes. 2 or 19, widespread. See Hawksworth & Eriksson (*SA* **10**: 52, 1991), Hyde (*Mycol.* **84**: 803, 1992), Barr (*Mycotaxon* **51**: 191, 1994; posn), Fallah & Shearer (*Mycol.* **93**: 566, 2001), Cai *et al.* (*Mycotaxon* **84**: 255, 2002; Philippines).

Saccardomyces Henn. (1904) = Schweinitziella fide Pirozynski (*Kew Bull.* **31**: 595, 1977).

saccate, like a sac or bag.

Saccharata Denman & Crous (2004), Botryosphaeriaceae. Anamorphs *Fusicoccum*-like, *Diplodia*-like. 1, S. Africa. See Crous *et al.* (*CBS Diversity Ser.* **2**: 104, 2004).

Saccharicola D. Hawksw. & O.E. Erikss. (2003), Massarinaceae. 2, Africa; S.E. Asia. See Eriksson & Hawksworth (*Mycol.* **95**: 431, 2003).

Saccharomyces Meyen ex E.C. Hansen (1838), Saccharomycetaceae. 26, widespread. *S. cerevisiae* (with 25 races), 'brewer's yeast', is used in beer- and bread-making, and other fermentations. See von Arx *et al.* (*Stud. Mycol.* **14**: 1, 1977), Martini & Kurtzman (*Int. J. Syst. Bacteriol.* **35**: 508, 1985; DNA homology), Stewart & Russell *in* Demain & Solomon (Eds) (*Biology of industrial micro-organisms*: 511, 1985; review biology), Kock *et al.* (*Appl. Microbiol. Biotechn.* **23**: 499, 1986; fatty acids), Martini & Martini (*Antonie van Leeuwenhoek* **53**: 77, 1987; DNA reassociation), Herskowitz (*Microbiol. Rev.* **52**: 536, 1988; *S. cerevisiae* life-cycle, 166 refs.), Martini & Martini *in* Cantarelli & Lanzarini (Eds) (*Biotechnology applications in beverage production*: 1, 1989; nomencl. domesticated spp.), Tuite & Oliver (Eds) (*Saccharomyces* [*Biotech. Handb.*] **4**, 1991), Barnett (*Yeasts* **8**: 1, 1992; review taxonomy), Vaughan-Martini (*Syst. Appl. Microbiol.* **16**: 113, 1993; key), Dujon *et al.* (*Nature* **369**: 371, 1994; complete DNA sequence chromosome XI, and refs others), Ando *et al.* (*Biosc., Biotechn., Biochem.* **60**: 1070, 1996; rDNA), Hampsey (*Yeast* Chichester **13**: 1099, 1997; phenotypes), James *et al.* (*Int. J. Syst. Bacteriol.* **47**: 453, 1997; phylogeny), Naumov *et al.* (*Syst. Appl. Microbiol.* **20**: 595, 1997; biological species), Oda *et al.* (*Yeast* Chichester **13**: 1243, 1997), Rinaldi *et al.* (*Mol. Biol. Evol.* **14**: 200, 1997; mitochondria), Cherry *et al.* (*Nucl. Acids Res.* **26**: 73, 1998; genome database), Cohn *et al.* (*Curr. Genet.* **33**: 83, 1998; telomere sequences), Gouliamova & Hennebert (*Mycotaxon* **66**: 337, 1998; phylogeny), McCullough *et al.* (*J. Clin. Microbiol.* **36**: 1035, 1998; ITS), Montrocher *et al.* (*Int. J. Syst. Bacteriol.* **48**: 295, 1998; phylogeny), Vaughan-Martini & Martini *in* Kurtzman & Fell (Eds) (*Yeasts, a taxonomic study* 4th edn: 358, 1998), Dlauchy *et al.* (*Syst. Appl. Microbiol.* **22**: 445, 1999; restriction analysis), Duarte *et al.* (*Int. J. Syst. Bacteriol.* **49**: 1907, 1999; enzymes), Mäntynen *et al.* (*Syst. Appl. Microbiol.* **22**: 87, 1999; sour dough yeasts), Oda *et al.* (*Curr. Microbiol.* **38**: 61, 1999; rDNA sequence), Petersen *et al.* (*Int. J. Syst. Bacteriol.* **49**: 1925, 1999; karyotypes), Casaregola *et al.* (*Int. J. Syst. Evol. Microbiol.* **51**: 1607, 2001; genome analysis), Hansen & Jakobsen (*Int. J. Food Microbiol.* **69**: 59, 2001; in cheese), Hennequin *et al.* (*J. Clin. Microbiol.* **39**: 551, 2001; microsatellite strain typing), Oda & Fujisawa (*Biosc., Biotechn., Biochem.* **65**: 164, 2001; genetic variation), Pérez *et al.* (*FEMS Microbiol. Lett.* **205**: 375, 2001; molecular techniques), Pérez *et al.* (*Lett. Appl. Microbiol.* **33**: 461, 2001; microsatellites), Lopes *et al.* (*FEMS Yeast Res.* **1**: 323, 2002; hybridisation), Mitterdorfer *et al.* (*Proteomics* **2**: 1532, 2002; proteomics), Sebastiani *et al.* (*Research in Microbiology* **153**: 53, 2002; hybridisation), Cliften *et al.* (*Science* N.Y. **300** no. 5629: 71, 2003; functional genomics), Kurtzman (*FEMS Yeast Res.* **4**: 233, 2003; phylogeny), Kurtzman & Robnett (*FEMS Yeast Res.* **3**: 417, 2003; phylogeny), Špírek *et al.* (*FEMS Yeast Res.* **3**: 363, 2003; chromosomal evolution), Christie *et al.* (*Nucl. Acids Res.* **32** Database issue: D311, 2004; genome database), Kellis *et al.* (*Nature* Lond. **428** no. 6983: 617, 2004; evolution), Llanos *et al.* (*Syst. Appl. Microbiol.* **27**: 427, 2004; clinical strains), Fay & Benavides (*PLoS Genetics* **1**: e5 [66, 2005; population biology), Legras *et al.* (*Int. J. Food Microbiol.* **102**: 73, 2005; microsatellites), Aa *et al.* (*FEMS Yeast Res.* **6**: 702, 2006; population structure), Fares *et al.* (*Mol. Biol. Evol.* **23**: 245, 2006; phylogenetic methodology), Suh *et al.* (*Mycol.* **98**: 1006, 2006; phylogeny).

Saccharomycetaceae G. Winter (1881), Saccharomycetales. 14 gen. (+ 14 syn.), 121 spp.

Lit.: von Arx (*Antonie van Leeuwenhoek* **38**: 289, 1972), Kreger-van Rij *in* Ainsworth *et al.* (Eds) (*The Fungi* **4A**, 1973), Redhead & Malloch (*CJB* **55**: 1701, 1977), von Arx *et al.* (*Stud. Mycol.* **14**: 1, 1977; gen. names), Batra *in* Subramanian (Ed.) (*Taxonomy of fungi* **1**: 187, 1978), Gams *et al.* (*CBS Course of Mycology* edn 2: 59 pp., 1980), Kreger-van Rij (Ed.) (*Yeasts, a taxonomic study* 3rd edn, 1984), Viljoen *et al.* (*Antonie van Leeuwenhoek* **52**: 45, 1986; fatty acid composition) and also under Yeasts (q.v.), Ando *et al.* (*Biosc., Biotechn., Biochem.* **60**: 1063, 1996), Ando *et al.* (*Biosc., Biotechn., Biochem.* **60**: 1070, 1996), Hudson *et al.* (*Genome Res.* **7**: 1169, 1997), James *et al.* (*Int. J. Syst. Bacteriol.* **47**: 453, 1997), James *et al.* (*Int. J. Syst. Bacteriol.* **48**: 591, 1998), Kurtzman *in* Kurtzman & Fell (Eds) (*Yeasts, a taxonomic study* 4th edn: 111, 1998; key), Kurtzman & Blanz *in* Kurtzman & Fell (Eds) (*Yeasts, a taxonomic study* 4th edn: 69, 1998; rDNA), Kurtzman & Fell (*Yeasts, a taxonomic study* 4th edn: 3, 1998), Belloch *et al.* (*Int. J. Syst. Evol. Microbiol.* **50**: 405, 2000), Fischer *et al.* (*Nature* Lond. **405**: 451, 2000), Mortimer (*Genome Res.* **10**: 403, 2000), Cliften *et al.* (*Science* N.Y. **300** no. 5629: 71, 2003), Kurtzman & Robnett (*FEMS Yeast Res.* **3**: 417, 2003), Dujon *et al.* (*Nature* Lond. **430**: 35, 2004), Edwards-Ingram *et al.* (*Genome Res.* **14**: 1043, 2004), Suh *et al.* (*Mycol.* **98**: 1006, 2006; phylogeny), Wu *et al.* (*FEMS Yeast Research* **8**: 641, 2008; phylogeny, hybridization).

Saccharomycetales Kudryavtsev (1960). Saccharomycetidae. 13 fam., 88 gen., 906 spp. Mycelium absent or poorly developed, where present usually with septa which have a series of minute pores rather than a single simple pore; vegetative cells proliferating by budding (blastically) or by fission (thallically), the walls usually lacking chitin except around bud scars, sometimes with J+ gel; ascomata absent; asci formed singly or in chains, sometimes not strongly differentiated morphologically from vegetative cells, usually at least eventually evanescent; ascospores varied in shape, sometimes with equatorial or asymmetric thickenings. The teleomorphic ascomycetous yeasts, cosmop. in a very wide range of habitats. Fams:

(1) **Ascoideaceae**
(2) **Cephaloascaceae**
(3) **Dipodascaceae**
(4) **Endomycetaceae**
(5) **Eremotheciaceae**
(6) **Lipomycetaceae**
(7) **Metschnikowiaceae**
(8) **Phaffomycetaceae**
(9) **Pichiaceae**
(10) **Saccharomycetaceae**
(11) **Saccharomycodaceae**
(12) **Saccharomycopsidaceae**
(13) **Trichomonascaceae**

Lit.: Eriksson *et al.* (*SA* **11**: 119, 1993), Kurtzman (*in The Yeasts, a taxonomic Study*: 111, 1998; key), Kurtzman & Blanz (*in* Kurtzman & Fell (Eds) *The Yeasts, a taxonomic Study*: 69, 1998; rDNA), Viljoen & Kock (*Syst. Appl. Microbiol.* **14**: 178, 1991; pyrolysis g.l.c. identification of yeasts).

Saccharomycetes Grüss (1928), Fossil Fungi ? Fungi. 915 (Devonian), Spitsbergen.

Saccharomycetes G. Winter (1880), Saccharomycotina. 1 ord., 13 fam., 95 gen., 915 spp. Subcl.: **Saccharomycetidae**

Almost all of the non-filamentous (yeast-forming) *Ascomycota* are placed here.

Lit.: Barnett *et al.* (*Yeasts: Characteristics and Identification* edn 3, 2000), Kurtzman & Fell (Eds, *The Yeasts. A Taxonomic Study* edn 4, 1998), Kurtzman & Sugiyama (*in* McLaughlin*et al.* (Eds), *The Mycota* **7A**: 179, 2001), Suh *et al.* (*Mycol.* **98**: 1006, 2006; phylogeny).

Saccharomycetidae Tehler (1988), see *Saccharomycetes*.

Saccharomycodaceae Kudrjanzev (1960), Saccharomycetales. 3 gen. (+ 8 syn.), 18 spp.

Lit.: Simmons & Ahearn (*Mycol.* **77**: 660, 1985), Tredoux *et al.* (*Syst. Appl. Microbiol.* **9**: 299, 1987), Golubev *et al.* (*Antonie van Leeuwenhoek* **55**: 369, 1989), Yamada *et al.* (*J. gen. appl. Microbiol.* Tokyo **38**: 585, 1992), Boekhout *et al.* (*Int. J. Syst. Bacteriol.* **44**: 781, 1994), Mikata & Nakase (*Microbiol. Culture Coll.* **13**: 97, 1997), Miller & Phaff *in* Kurtzman & Fell (Eds) (*Yeasts, a taxonomic study* 4th edn: 268, 1998), Miller & Phaff *in* Kurtzman & Fell (Eds) (*Yeasts, a taxonomic study* 4th edn: 372, 1998), Phaff & Miller *in* Kurtzman & Fell (Eds) (*Yeasts, a taxonomic study* 4th edn: 409, 1998), Smith *in* Kurtzman & Fell (Eds) (*Yeasts, a taxonomic study* 4th edn: 214, 1998), Esteve-Zarzoso *et al.* (*Int. J. Syst. Bacteriol.* **49**: 329, 1999), Cadez *et al.* (*FEMS Yeast Res.* **1**: 279, 2002), Cadez *et al.* (*Int. J. Syst. Evol. Microbiol.* **53**: 1671, 2003), Kurtzman & Robnett (*FEMS Yeast Res.* **3**: 417, 2003), Suh *et al.* (*Mycol.* **98**: 1006, 2006; phylogeny).

Saccharomycodes E.C. Hansen (1904), Saccharomycodaceae. 2, widespread (north temperate). See Yamada *et al.* (*J. gen. appl. Microbiol.* Tokyo **38**: 585, 1992; phylogeny), Miller & Phaff *in* Kurtzman & Fell (Eds) (*Yeasts, a taxonomic study* 4th edn: 372, 1998), Suh *et al.* (*Mycol.* **98**: 1006, 2006; phylogeny).

Saccharomycopsidaceae Arx & Van der Walt (1987), Saccharomycetales. 1 gen. (+ 5 syn.), 13 spp.

Lit.: Ota & Morishita (*Microbios* **73**: 149, 1993), Kurtzman & Robnett (*CJB* **73**: S824, 1995), Yamada *et al.* (*Biosc., Biotechn., Biochem.* **60**: 1303, 1996), Kurtzman & Smith *in* Kurtzman & Fell (Eds) (*Yeasts, a taxonomic study* 4th edn: 374, 1998), Esteve-Zarzoso *et al.* (*Int. J. Syst. Bacteriol.* **49**: 329, 1999), Kurtzman (*Mycotaxon* **71**: 241, 1999), Suh *et al.* (*Mycol.* **98**: 1006, 2006; phylogeny).

Saccharomycopsis Guilliermond. (1912) ≡ Cyniclomyces.

Saccharomycopsis Schiønning (1903), Saccharomycopsidaceae. 13, widespread (temperate). See Yamada *et al.* (*Biosc., Biotechn., Biochem.* **60**: 1303, 1996; phylogeny), Kurtzman & Smith *in* Kurtzman & Fell (Eds) (*Yeasts, a taxonomic study* 4th edn: 374, 1998), Kurtzman (*Mycotaxon* **71**: 241, 1999), Suh *et al.* (*Mycol.* **98**: 1006, 2006; phylogeny).

Saccharomycotina Schiønning (1903), Saccharomycopsidaceae. 13, widespread (temperate). See Yamada *et al.* (*Biosc., Biotechn., Biochem.* **60**: 1303, 1996; phylogeny), Kurtzman & Smith *in* Kurtzman & Fell (Eds) (*Yeasts, a taxonomic study* 4th edn: 374, 1998), Kurtzman (*Mycotaxon* **71**: 241, 1999), Suh *et al.* (*Mycol.* **98**: 1006, 2006; phylogeny).

Saccharopolyspora J. Lacey & Goodfellow (1975), Actinobacteria. q.v.

Saccisporonites Kalgutkar & Janson. (2000), Fossil Fungi, anamorphic *Ascomycota*. 1, USA. See Kalgutkar & Jansonius (*AASP Contributions Series* **39**: 266, 2000).

Saccoblastia Möller (1895), Saccoblastiaceae. *c.* 3. See Jülich (*Persoonia* **9**: 39, 1976).

Saccoblastiaceae Jülich (1982), Atractiellales. 2 gen., 4 spp.

Lit.: Bauer *et al.* (*Mycol. Progr.* **5**: 41, 2006).

Saccobolus Boud. (1869), Ascobolaceae. 27, widespread. See van Brummelen (*Persoonia* Suppl. **1**: 1, 1967; key), van Brummelen (*Persoonia* **8**: 421, 1976), Kaushal & Virdi (*Willdenowia* **16**: 269, 1986; Himalayan spp.), Dissing (*Op. Bot.* **100**: 43, 1989), Landvik *et al.* (*Nordic Jl Bot.* **17**: 403, 1997; DNA), Ramos *et al.* (*Mycotaxon* **74**: 447, 2000; isozymes), Giménez *et al.* (*Cryptog. Mycol.* **26**: 3, 2005; Argentina), Hansen & Pfister (*Mycol.* **98**: 1029, 2006; phylogeny).

Saccomorpha Elenkin (1912) = Placynthiella Elenkin fide Hawksworth *et al.* (*Dictionary of the Fungi* edn 8, 1995).

Saccomyces Serbinow (1907), ? Chytridiaceae. 2, Europe.

Saccopodium Sorokīn (1877), ? Cladochytriaceae. 1, former USSR.

Saccotheciaceae Bonord. (1864) = Dothioraceae.

Saccothecium Fr. (1836), Dothioraceae. Anamorph *Aureobasidium*-like. *c.* 9, widespread. See Froidevaux (*Nova Hedwigia* **23**: 679, 1973; key), Holm (*Taxon* **24**: 475, 1975; status).

Sachsia Bay (1894), ? Saccharomycetales. 1, Europe.

Sachsia Lindner (1895) nom. dub., Pezizomycotina.

Sacidium Nees (1823) = Pilobolus fide von Höhnel (*Sber. Akad. Wiss. Wien* Math.-naturw. Kl., Abt. 1 **119**: 617, 1910).

Sacidium Sacc. (1880) = Pilobolus fide Höhnel (*Sber. Akad. Wiss. Wien* Math.-naturw. Kl., Abt. 1 **119**: 617, 1910) See, Sutton (*Mycol. Pap.* **141**, 1977).

Sackea Rostk. (1844) = Bovista fide Stalpers (*in litt.*).

Sadasivanella Agnihothr. (1964) = Neottiosporina fide Sutton & Alcorn (*Aust. J. Bot.* **22**: 517, 1974).

Sadasivania Subram. (1957), anamorphic *Pezizomycotina*, Hso.0eP.3. 3, widespread. See Sundberg & Wicklow (*Mycol.* **65**: 925, 1973), Markovskaya & Treigiene (*Mikol. Fitopatol.* **38**: 52, 2004).

saddle fungus, ascomata of *Helvella* spp.

saddle-back fungus (**dryad's saddle**), basidioma of *Polyporus squamosus*.

Saeenkia Kudrjanzev (1960) = Saccharomycodes fide Lodder (*Yeasts, a taxonomic study* 2nd edn, 1970).

Safety, Laboratory. Health & Safety legislation for places of work appropriate to the country in which the laboratory is operating should always be consulted. In addition to normal good laboratory practice (for example wearing suitable protective clothing, and prohibition of eating and smoking), further precautions should be taken to avoid laboratory hazards specific to mycology (e.g. see the European Community Directive 2000/54/CE on protection of workers from risks related to exposure to biological agents at work, website: http://europa.eu/eur-lex/pri/en/oj/dat/2000/l_262/l_26220001017en00210045.pdf; the UK Advisory Committee on Dangerous Pathogens, website: www.hse.gov.uk/aboutus/meetings/acdp).

It must be remembered that many fungi pathogenic for humans are 'opportunistic', widespread saprobes, and their full potential to effect human health is rarely known (see Smith, *Opportunistic mycoses of man and other animals*, 1989). Fungi and other organisms can enter the body through the mouth, lungs, broken or unbroken skin, and the conjunctiva. However, in a laboratory the route of infection may not be the same as when the disease is acquired elsewhere. Potential hazards may be greater in the laboratory than in nature if fungi are being grown in substantial numbers, transferred from one container to another, and manipulated, increasing the risk of infection. The main routes of infection are accidential inoculation, accidential ingestion, splashing into the face and eyes, spillage and direct contact. Good laboratory practice to keep cultures pure normally prevents their escape to cause infection, but note that the conditions provided for the growth of a particular fungus may also be suitable for the growth of a potentially hazardous contaminant.

Special precautions should be taken when harvesting bulk cultures or large quantities of spores. Fungi may also cause allergic reactions or mycotoxicoses (poisoning) and therefore contact with them and the materials or equipment they have been in contact with must be avoided.

Great care must be taken in handling the pathogens of coccidioidomycosis, histoplasmosis, and other systemic mycoses for there are many instances of laboratory infections by these fungi resulting in illness or even death (see Hanel & Kruse, *Misc. Publ. Fort Detrich, Dept. Army* **27**, 1967; laboratory acquired mycoses).

In the European Community, the European Directive 2000/54/CE allocates fungi (and other organisms) to four hazard groups. Similar categories exist in other countries. See also *Biosafety in microbiological and biomedical laboratories* edn 4, 1999; downloadable in PDF format from www.cdc.gov/od/ohs/biosfty/bmbl4/bmbl4toc.htm; *Categorization of pathogens according to hazard and categories of containment*, edn 2, 1990):

Group 1: A biological agent that is most unlikely to cause human disease.

Group 2: A biological agent that may cause human disease and which might be a hazard to laboratory workers but is unlikely to spread in the community. Laboratory exposure rarely produces infection and effective prophylaxis or treatment is available.

Group 3: A biological agent that may cause severe human disease and present a serious hazard to laboratory workers. It may present a risk of spread in the community but there is usually effective prophylaxis or treatment.

Group 4: A biological agent that causes severe human disease and is a serious hazard to laboratory workers. It may present a high risk of spread in the community and there is usually no effective prophylaxis or treatment.

Most fungi belong to category 1. The European Union directive (see above) lists the following species in category 2: *Aspergillus fumigatus*, *Candida* spp., *Cryptococcus neoformans*, *Emmonsia parva*, *Epidermophyton floccosum*, *Fonsecaea* spp., *Madurella* spp., *Microsporum* spp., *Neotestudina rosatii*, *Penicillium marneffii*, *Scedosporium apiospermum*, *Sporothrix schenkii* and *Trichophyton* spp. Only 5 fungi are included in category 3: *Blastomyces dermatitidis*, *Cladophialophora bantiana*, *Histoplasma capsulatum*, *Coccidioides immitis* and *Paracoccidioides brasiliensis* [*Note*: teleomorph names only listed where the fungi are pleomorphic; see generic entries for those of the anamorphs]. No fungi are placed in category 4. Other classifications of hazardous organisms are available (e.g. US Public Health Service, *Classification of etiologic agents on the basis of hazard*, 1974 [also as an appendix to *Biosafety in microbiological and biomedical laboratories*, see above]; World Health Organization, *Laboratory biosafety manual*, edn 3, 2004; as a downloadable PDF file from www.who.int/csr/delibepidemics/WHO_CDS_CSR_LYO_2004_11/en/).

Increasing containment levels are prescribed by Health and Safety regulations for these hazard categories in most countries, and information on those which apply should be sought from appropriate national agencies before starting work with listed organisms.

Lit.: Collins (*Biologist* **23**: 83, 1976; biological hazards), Fuscaldo *et al.* (*Laboratory safety. Theory and practice*, 1980; general, biological, medical), Collins (*Laboratory-acquired infections*, 1983), Miller (*Laboratory safety: principles and practices*, 1986), Simpson & Simpson (*The COSHH regulations: a practical guide*, 1991), Stricoff & Walters (*Handbook of laboratory health and safety*, edn 2, 1995).

See also Allergy, Human and veterinary mycology, Mycetism, Mycotoxins.

saffron milk-cap, basidioma of the edible *Lactarius deliciosus*.

Sagedia A. Massal. (1852) nom. rej. prop. = Porina Müll. Arg. fide Hawksworth *et al.* (*Dictionary of the Fungi* edn 8, 1995).

Sagedia Ach. (1809) nom. rej. = Aspicilia fide Laundon & Hawksworth (*Taxon* **37**: 478, 1988).

Sagediomyces Cif. & Tomas. (1953) = Strigula fide Harris (*in litt.*).

Sagediopsis (Sacc.) Vain. (1921), Adelococcaceae. 7 (on lichens), widespread (esp. temperate). See Alstrup & Hawksworth (*Meddr Grønland* Biosc. **31**, 1990; nomencl.), Hafellner (*Herzogia* **9**: 749, 1993), Hoffmann & Hafellner (*Biblthca Lichenol.* **77**: 1, 2000).

Sagema Poelt & Grube (1993), Lecanoraceae (L). 1, Nepal. See Grube & Poelt (*Graphis Scripta* **5**: 69, 1993).

sagenetosome, see sagenogen.

Sagenidiopsis R.W. Rogers & Hafellner (1987), Arthoniaceae (L). 2, Australia; S. America. See Egea *et al.* (*Lichenologist* **27**: 351, 1995; posn), Grube (*Bryologist* **101**: 377, 1998; phylogeny).

Sagenidium Stirt. (1877), Roccellaceae (L). 3, S. America; Australia. See Henssen *et al.* (*Lichenologist* **11**: 263, 1979), Grube (*Bryologist* **101**: 377, 1998; phylogeny), Kantvilas (*Symb. bot. upsal.* **34** no. 1: 183, 2004; Tasmania).

sagenogen (sagenetosome), see bothrosome.

sagenogenetosome, see bothrosome.

sagenogens, Olive's (1975) modification of Porter's (1972) 'sagenogenetosomes', organelles from which the ectoplasmic nets of *Labyrinthulales* are produced.

Sagenoma Stolk & G.F. Orr (1974), Trichocomaceae. Anamorph *Sagenomella*. 2, Australia; Japan. See Ueda & Udagawa (*Mycotaxon* **20**: 499, 1984), Malloch *in* Samson & Pitt (Eds) (*Advances in Penicillium and Aspergillus systematics* **102**: 365, 1985).

Sagenomella W. Gams (1978), anamorphic *Sagenoma*, Hso.0eH/1eP.3. 11, widespread. See Gams (*Persoonia* **10**: 100, 1978), Gené *et al.* (*J. Clin. Microbiol.* **41**: 1722, 2003; infection of dog).

Sageria A. Funk (1975), Helotiaceae. Anamorph *Asconidium*. 1 (on *Tsuga*), Canada. See Funk (*CJB* **53**:

1196, 1975), Verkley (*Stud. Mycol.* **44**: 180 pp., 1999; ultrastr.).

Sagiolechia A. Massal. (1854), ? Gomphillaceae (L). 6, Europe. See Vězda (*Folia geobot. phytotax.* **2**: 383, 1968), Henssen (*Biblthca Lichenol.* **58**: 123, 1995).

Sagittospora Lubinsky (1955), Chytridiales. 1 (on *Eudiplodinium* Protozoa in goat rumen), Pakistan. See Lubinsky (*Can. J. Microbiol.* **1**: 683, 1955).

Sagrahamala Subram. (1972), anamorphic *Pezizomycotina*, Hso.0eH/1eP.15. 10, widespread. See Subramanian (*Curr. Sci.* **41**: 48, 1972).

Saitoa Rajendran & Muthappa (1980) = Neosartorya fide Samson *in* Samson & Pitt (Eds) (*Advances in Penicillium and Aspergillus systematics* **102**: 365, 1985; coenzyme-Q system).

Saitoella Goto, Sugiy., Hamam. & Komag. (1987), anamorphic *Protomycetaceae*, Hso.0eH.3/10. 1 (from soil), Himalaya. See Goto *et al.* (*J. gen. appl. Microbiol. Tokyo* **33**: 76, 1987), Nishida & Sugiyama (*Mol. Biol. Evol.* **10**: 431, 1993), Sjamsuridzal *et al.* (*Mycoscience* **38**: 267, 1997), Ahearn *et al. in* Kurtzman & Fell (Eds) (*Yeasts, a taxonomic study* 4th edn: 600, 1998), Sugiyama *et al.* (*Mycol.* **98**: 996, 2006; phylogeny).

Saitomyces Ricker (1906) ≡ Actinocephalum.

Sakaguchia Y. Yamada, K. Maeda & Mikata (1994), Cystobasidiomycetes. 1 (from sea water), Antarctica. See Fell *et al.* (*Int. J. Syst. Evol. Microbiol.* **50**: 1351, 2000; mol. phylogeny), Scorzetti *et al.* (*FEMS Yeast Research* **2**: 495, 2002).

Sakireeta Subram. & K. Ramakr. (1957) = Tiarosporella fide Sutton (*Mycol. Pap.* **141**, 1977).

Saksenaea S.B. Saksena (1953), Radiomycetaceae. 1, widespread. See Ellis & Hesseltine (*Mycol.* **66**: 85, 1974), Ajello *et al.* (*Mycol.* **68**: 53, 1976; on humans), Ellis & Ajello (*Mycol.* **74**: 144, 1982), Chien *et al.* (*TMSJ* **33**: 443, 1992), Mathews *et al.* (*J. Mycol. Médic.* **3**: 95, 1993; causing zygomycosis, lit. review), Pillai & Ahmed (*Curr. Sci.* **65**: 291, 1993), Voigt & Wöstemeyer (*Gene* **270**: 113, 2001; phylogeny), Prabhu & Patel (*Clin. Microbiol. Infect.* **10** Suppl. 1: 31, 2004; zygomycosis, review).

Saksenaeaceae Hesselt. & J.J. Ellis (1974) = Radiomycetaceae.
Lit.: Kirk (*in litt.*).

Saliastrum Kujala (1946), anamorphic *Pezizomycotina*, Hso.0bP.1. 1, Europe; Canada. See Redhead & Perrin (*CJB* **50**: 2083, 1972), Sutton (*Mycol. Pap.* **141**, 1977).

Salmonia S. Blumer & E. Müll. (1964) = Brasiliomyces fide Braun (*Nova Hedwigia* **34**: 704, 1981).

Salmonomyces Chidd. (1959) = Erysiphe fide Zheng (*Mycotaxon* **22**: 209, 1985).

Salsuginea K.D. Hyde (1991), Dothideomycetes. 1 (marine), Thailand. See Eriksson & Hawksworth (*SA* **11**: 73, 1992; posn).

saltation (of fungi), mutation; dissociation; see Variation in fungi.

Samarospora Rostr. (1892), ? Eurotiales. 1, Europe.

Samarosporella Linder (1944) nom. dub., Fungi. See Kohlmeyer (*CJB* **50**: 1951, 1972).

Samboa Tomas. & Cif. (1952) nom. dub. ? = Buellia fide Hawksworth *et al.* (*Dictionary of the Fungi* edn 8, 1995).

Samboamyces Cif. & Tomas. (1953) ≡ Samboa.

Sambucina Velen. (1947), ? Helotiales. 1, former Czechoslovakia.

sambucinin, see enniatin.

Sampaioa Gonz. Frag. (1923) = Mycoglaena fide Checa (*Mycotaxon* **63**: 467, 1997).

Samuelsia Chaverri & K.T. Hodge (2008), Clavicipitaceae. 5, neotropics. See Chaverri *et al.* (*Stud. Mycol.* **60**, 2008; phylogeny).

Samukuta Subram. & K. Ramakr. (1957) = Neottiospora fide Nag Raj (*CJB* **51**: 2463, 1973).

Sand dune fungi. Sand dunes have a characteristic mycota of both microfungi (Moreau & Moreau, *Rev. Mycol.* **6**: 49, 1941, Dickinson & Kent, *TBMS* **58**: 269, 1972) and macrofungi (Andersson, *Op. bot. Soc. bot. Lund* **2**, 1950, Bon, *BSMF* **86**: 79, 1970, Picardy; **88**: 15, 1972; *Lepiota*, Courtecuisse, *Docums Mycol.* fasc. 57-58, 66, 1984-1986, Høiland, *Blyttia* **33**: 127, 1975, **35**: 139, 1977, **36**: 69, 1978, Jaiswal & Rodrigues (*Curr. Sci.* **80**: 827, 2001; mycorrhizal fungi in coastal dunes of Goa, India), Singer (*Mycopathologia* **34**: 129, 1968; on south Atlantic coast from Uruguay to Bahía Blanca, Argentina), Watling & Rotheroe, *Proc. R. Soc. Edinb.* **96**: 111, 1989) which, like that of flowering plants, exhibits a zonation from the shore sand to fixed dune (Wallace, *Trans. Devon. Ass. Advmt Sci.* **86**: 201, 1954; Cordoba *et al.* (*Mycoscience* **42**: 379, 2001; diversity gradients of mycorrhizal fungi in coastal dunes of Brazil)). Macrofungi play a role in facilitating dune colonization by flowering plants and in sustaining coastal dune systems (Logan *et al.*, *Aust. J. Pl. Physiol.* **16**: 141, 1989; mycorrhizal fungi enhancing Australian coastal dune ecosystems), both by the action of fungal hyphae in accretion of sand particles and as mycorrhizal symbionts (see Rotheroe, *in* Pegler *et al.* (Eds), *Fungi of Europe*, 1993). Characteristic lichen communities frequently develop on stable sand dunes and are usually dominated by species of *Cladonia*; succession in the lichen communities is correlated with the stability of the substratum (see James *et al.*, *in* Seaward (Ed.), *Lichen ecology*: 400, 1977). See also Ecology, Marine fungi.

Sandersoniomyces R.K. Benj. (1968), Laboulbeniaceae. 1, USA. See Benjamin (*Aliso* **10**: 345, 1983), Tavares (*Mycol. Mem.* **9**: 627 pp., 1985).

sandimmum, see cyclosporin A.

Sanjuanomyces R.F. Castañeda & W.B. Kendr. (1991), anamorphic *Pezizomycotina*, Hso.≡ eP.24. 1, Cuba. See Castañeda Ruiz & Kendrick (*Univ. Waterloo Biol. Ser.* **35**: 93, 1991).

Santapauella Mundk. & Thirum. (1945) = Phragmidiella fide Thirumalachar & Mundkur (*Indian Phytopath.* **2**: 193, 1949).

Santapauinda Subram. (1994), anamorphic *Pezizomycotina*, Hso.#eP.1. 1, India. See Subramanian (*Kavaka* **20/21**: 58, 1992/1993).

Santessonia Hale & Vobis (1978), Caliciaceae (L). 6, Africa; S. America. See Sérusiaux & Wessels (*Mycotaxon* **19**: 479, 1984; key), Stenroos & DePriest (*Am. J. Bot.* **85**: 1548, 1998; DNA), Nordin & Mattsson (*Lichenologist* **33**: 3, 2001; phylogeny), Helms *et al.* (*Mycol.* **95**: 1078, 2003; phylogeny), Follmann (*J. Hattori bot. Lab.* **100**: 651, 2006; revision, S America).

Santessoniella Henssen (1997), ? Pannariaceae (L). 11, widespread (temperate). See Henssen (*Symb. bot. upsal.* **32**: 75, 1997), Henssen (*Lichenologist* **32**: 57, 2000; S America), Henssen & Kantvilas (*Lichenologist* **32**: 149, 2000; Tasmania), Jørgensen (*Biblthca Lichenol.* **88**: 229, 2004), Spribille *et al.* (*Biblthca Lichenol.* **96**: 287, 2007; N America).

Santessoniolichen Tomas. & Cif. (1952) = Naetrocymbe Körb. fide Harris (*More Dematiaceous Hyphomycetes*, 1995).

Santessoniomyces Cif. & Tomas. (1953) ≡ Santessoniolichen.

Santiella Tassi (1900) = Passeriniella fide Sutton & Sellar (*CJB* **44**: 1505, 1966).

Sappinia P.A. Dang. (1896) nom. dub., Protozoa. Excluded from *Mycetozoa* by Olive (1975: 6).

saprobe (**saprogen**, **saprotroph**), an organism using dead organic material as food, and commonly causing its decay (saprobe is the preferred term for fungi); **saprobic**, **saprogen-ic** (**-ous**), **saprophilous**, **saprotrophic** (adj.). See Hudson (*Fungal saprophytism*, edn 2, 1980), saprophyte.

Saprochaetaceae Coker & Shanor ex D.T.S. Wagner & Dawes (1970) = Dipodascaceae.

Saprochaete Coker & Shanor (1939), anamorphic *Magnusiomyces*, Hso.0fH.10. 14, widespread. See Wagner & Dawes (*J. Elisha Mitchell scient. Soc.* **55**: 163, 1939), Hoog & Smith (*Stud. Mycol.* **50**: 489, 2004; revision, phylogeny).

Saprogaster Fogel & States (2001), Phallales. 1, USA. See Fogel & States (*Mycotaxon* **80**: 317, 2001).

saprogen, a saprobe (q.v.).

Saprophragma K.B. Deshp. & K.S. Deshp. (1966), anamorphic *Pezizomycotina*, Hso.≡ eH.?. 1, India. See Deshpande & Deshpande (*Mycopathologia* **30**: 200, 1966).

saprophyte, a plant feeding by external digestion of dead organic matter; commonly misapplied to fungi where saprobe (q.v.) is the preferred term.

Saprotaphrina Verona & Rambelli (1962) nom. nud., anamorphic *Pezizomycotina*. Used for the yeast phase of *Taphrina*.

saprotroph (1) a saprobe (q.v.); (2) a necrophyte on dead material which is not part of a living host (Münch), cf. perthophyte.

Sapucchaka K. Ramakr. (1956), ? Microthyriaceae. 1, India. See von Arx & Müller (*Stud. Mycol.* **9**, 1975).

Sarawakus Lloyd (1924), Hypocreaceae. Anamorphs *Gliocladium*-like, *Trichoderma*-like, *Verticillium*-like. 11 (from wood and bark), widespread (tropical). See Samuels & Rossman (*Mycol.* **84**: 26, 1992; key), Rossman *et al.* (*Stud. Mycol.* **42**: 248 pp., 1999).

Sarbhoy (Ashok Kumar; 1939-2006; India). BSc (1956), MSc (1959), PhD (1963), Assistant Professor of Botany (1962-1965), Allahabad University; Commonwealth Mycological Institute (1965-1967); Mycologist (1969-1972), Senior Mycologist (1972-1986), Principal Scientist (1986-1997), Professor (1997-1999), Emeritus Scientist (2000-2003), Indian Agricultural Research Institute, New Delhi. Responsible for development of the *Indian Type Culture Collection*. *Principal mycological publications*. (with L.V. Gangawane & D.K. Agarwal) *Compendium of Soil Borne Plant Pathogens* (1987); (with S.P. Raychaudhuri & A. Johnston) *History of Plant Pathology of South East Asia* (1996); *Textbook of Mycology* (2006). *Biogs, obits etc.* Smith (*World Federation for Culture Collections Newsletter* January 2007) [portrait].

Sarbhoyomyces Saikia (1981), anamorphic *Pezizomycotina*, Hsy.0eH.?. 1, India. See Saikia (*Curr. Sci.* **50**: 949, 1981).

Sarcanthia Raf. (1817) nom. dub., Fungi. No spp. included.

Sarcinella Sacc. (1880), anamorphic *Schiffnerula*, Hso.#eP.1. 27, widespread. See Hughes (*CJB* **61**: 1727, 1984), Hughes (*Mycol.* **82**: 657, 1990), Hosagoudar (*Zoos' Print Journal* **18**: 1071, 2003).

Sarcinellaceae Nann. (1934) = Englerulaceae.

sarciniform, bundle-like, as the dictyospore of *Stemphylium botryosum*.

Sarcinodochium Höhn. (1905), anamorphic *Pezizomycotina*, Hsp.#eH.?. 1, Europe.

Sarcinomyces Lindner (1898), anamorphic *Eurotiomycetes*, Hso.#eH.21. 2, widespread. See Jong & King (*Mycotaxon* **3**: 397, 1976), Hermanides-Nijhof (*Stud. Mycol.* **15**: 173, 1977), Wollenzien *et al.* (*Antonie van Leeuwenhoek* **71**: 281, 1997; Mediterranean), Uijthoff *et al.* (*Medical Mycology* **36**: 143, 1998; phylogeny), Sterflinger *et al.* (*Stud. Mycol.* **43**: 5, 1999; phylogeny, ecology), Sterflinger & Prillinger (*Antonie van Leeuwenhoek* **80**: 275, 2001; Austria), Bogomolova & Minter (*Mycotaxon* **86**: 195, 2003).

Sarcinomyces Oho (1926) ≡ Sarcinosporon fide Jong & King (*Mycotaxon* **3**: 397, 1976).

Sarcinosporon D.S. King & S.C. Jong (1975), anamorphic *Pezizomycotina*, Hso.0eH.10. 1 (from humans), widespread. See King & Jong (*Mycotaxon* **3**: 92, 1975).

Sarcinulella B. Sutton & Alcorn (1983) = Anisomeridium fide Purvis *et al.* (*Lichen Flora of Great Britain and Ireland*, 1992).

sarcodimitic, see Hyphal analysis.

Sarcodon Quél. ex P. Karst. (1881), Bankeraceae. 36, widespread. See Maas Geesteranus & Nannfeldt (*Svensk bot. Tidskr.* **63**: 401, 1969), Singer (*Beih. Nova Hedwigia* **77**, 1983), Stalpers (*Stud. Mycol.* **35**: 168 pp., 1993; key), Schmidt-Stohn (*Boletus* **24**: 48, 2000; *Sarcodon imbriatus* and *S. squamosus*).

Sarcodontaceae Bondartsev & Singer ex Singer (1983) = Bankeraceae.

Sarcodontia Schulzer (1866), Meruliaceae. 2, widespread (boreal). See Stalpers (*Folia Cryptog. Estonica* **33**: 133, 1998).

Sarcographa Fée (1824), Graphidaceae (L). *c.* 74, widespread (tropical). See Pant (*Geophytology* **20**: 48, 1991; 3 spp., India), Nakanishi *et al.* (*Bull. natn. Sci. Mus.* Tokyo, B **28**: 107, 2002; Vanuatu), Staiger (*Biblthca Lichenol.* **85**, 2002; monogr.), Archer (*Telopea* **10**: 589, 2004; Australia), Archer (*Biblthca Lichenol.* **94**, 2006; Australia), Staiger *et al.* (*MR* **110**: 765, 2006; phylogeny), Archer (*Systematics & Biodiversity* **5**: 9, 2007; Solomon Is).

Sarcographina Müll. Arg. (1887), Graphidaceae (L). 2, widespread. See Pant (*Geophytology* **20**: 48, 1991; India), Archer (*Telopea* **10**: 589, 2004; Australia), Archer (*Biblthca Lichenol.* **94**, 2006; Australia), Staiger *et al.* (*MR* **110**: 765, 2006; phylogeny), Lücking *et al.* (*Biblthca Lichenol.* **95**: 429, 2007; Costa Rica).

Sarcographomyces Cif. & Tomas. (1953) = Sarcographa fide Hawksworth *et al.* (*Dictionary of the Fungi* edn 8, 1995).

Sarcogyne Flot. (1850) nom. rej. = Polysporina fide Hawksworth *et al.* (*Dictionary of the Fungi* edn 8, 1995).

Sarcogyne Flot. (1851) nom. cons., Acarosporaceae (L). 30, widespread. See Jørgensen & Santesson (*Taxon* **42**: 881, 1993; nomencl.), Seppelt *et al.* (*Lichenologist* **30**: 249, 1998), Lutzoni *et al.* (*Am. J. Bot.* **91**: 1446, 2004; phylogeny), Reeb *et al.* (*Mol. Phylogen. Evol.* **32**: 1036, 2004; phylogeny), Crewe *et al.* (*MR* **110**: 521, 2006; phylogeny), Miądlikowska *et al.* (*Mycol.* **98**: 1088, 2006; phylogeny, polyphyly).

Sarcoleotia S. Ito & S. Imai (1934), ? Geoglossaceae. 1, widespread. See Schumacher & Sivertsen *in* Laursen *et al.* (Eds) (*Arctic and Alpine Mycology* **2**: 163, 1987), Jumpponen *et al.* (*CJB* **75**: 2228, 1998), Wang *et al.* (*Mycol.* **98**: 1065, 2006; phylogeny).

Sarcoloma Locq. (1979) = Hebeloma fide Kuyper (*in litt.*).

Sarcomelas Raf. (1817) nom. dub., Fungi.

Sarcomyces Massee (1891), Helotiales. 1, America.

Sarcomyxa P. Karst. (1891), Marasmiaceae. 1, widespread.

Sarconemus Raf. (1815) nom. dub., ? Fungi.

Sarconiptera Raitv. (2003), Dermateaceae. 1, Greenland. See Raitviir (*Mycotaxon* **87**: 363, 2003).

Sarcopaxillus Zmitr., V. Malysheva & E. Malysheva (2004), Paxillaceae. 1, Europe (northern). See Zmitrovich *et al.* (*Folia Cryptogamica Petropolitana* **1**: 52, 2004).

Sarcophoma Höhn. (1916), anamorphic *Discosphaerina*, St.0eH.15. 1, widespread. See van der Aa (*Persoonia* **8**: 283, 1975).

Sarcopodium Ehrenb. (1818), anamorphic *Pezizomycotina*, Hsp.0-1eH.15. 11, widespread. See Ellis (*More Dematiaceous Hyphomycetes*, 1976), Sutton (*TBMS* **76**: 97, 1981), Watanabe (*Mycol.* **85**: 520, 1993; Japan), Guarro *et al.* (*J. Clin. Microbiol.* **40**: 3071, 2002; from human).

Sarcoporia P. Karst. (1894), Fomitopsidaceae. 7, widespread. See Ko *et al.* (*Mycol.* **93**: 270, 2001; misapplied), Niemelä & Kinnunen (*Karstenia* **45**: 77, 2005; correct name for *Parmastomyces*) see *Erastia*.

Sarcopyrenia Nyl. (1858), ? Verrucariaceae (±L). 3 (on lichens), widespread (northern hemisphere). See Navarino-Rosinés & Hladún (*Candollea* **45**: 469, 1990), Aguirre-Hudson (*Bull. Br. Mus. nat. hist.* Bot. **21**: 85, 1991).

Sarcopyreniaceae Nav.-Ros. & Cl. Roux (1998) = Verrucariaceae.

Sarcopyreniomyces Cif. & Tomas. (1953) ≡ Sarcopyrenia fide Hawksworth *et al.* (*Dictionary of the Fungi* edn 8, 1995).

Sarcopyreniopsis Cif. & Tomas. (1953) ≡ Sarcopyrenia.

Sarcorhopalum Rabenh. (1851) = Taphrina fide Mix (*Kansas Univ. Sci. Bull.* **33**: 1, 1949).

Sarcosagium A. Massal. (1856), Thelocarpaceae (L). 1, widespread (temperate). See Poelt & Vězda (*Best. europ. Flecht.*, 1977), Reeb *et al.* (*Mol. Phylogen. Evol.* **32**: 1036, 2004; phylogeny).

Sarcoscypha (Fr.) Boud. (1885), Sarcoscyphaceae. Anamorph *Molliardiomyces*. *c.* 28, widespread (north temperate). See Denison (*Mycol.* **64**: 609, 1972), Baral (*Z. Mykol.* **50**: 117, 1984; key 5 spp.), Harrington (*Mycotaxon* **38**: 417, 1990; 3 spp. N. Am.), Benkert (*Gleditschia* **19**: 173, 1991), Zhuang (*Mycosystema* **5**: 65, 1993; SEM), Harrington (*Harvard Pap. Bot.* **10**: 53, 1997; 5 spp.), Harrington & Potter (*Mycol.* **89**: 258, 1997; ITS), Harrington (*Mycol.* **90**: 235, 1998; DNA), Harrington *et al.* (*Mycol.* **91**: 41, 1999; phylogeny), Zhuang (*Mycotaxon* **76**: 1, 2000; distrib.), Weinstein *et al.* (*Mycol.* **94**: 673, 2002; phylogeny), Hansen *et al.* (*Mol. Phylogen. Evol.* **36**: 1, 2005; phylogeny), Hansen & Pfister (*Mycol.* **98**: 1029, 2006; phylogeny), Perry *et al.* (*MR* **111**: 549, 2007; phylogeny).

Sarcoscyphaceae Le Gal ex Eckblad (1968), Pezizales. 13 gen. (+ 8 syn.), 102 spp.

Lit.: Paden (*Mycotaxon* **25**: 165, 1986), Cabello (*Boln Soc. argent. Bot.* **25**: 395, 1988; numerical taxonomy), Pfister (*Mem. N. Y. bot. Gdn* **49**: 339, 1989), Bellemère *et al.* (*Cryptog. Mycol.* **11**: 203, 1990; ultrastr.), Benkert (*Gleditschia* **19**: 173, 1991; keys German spp.), Butterfill & Spooner (*Mycologist* **9**: 20, 1995), Li & Kimbrough (*Int. J. Pl. Sci.* **156**: 841, 1995; septa), Li & Kimbrough (*CJB* **74**: 10, 1996), Wang & Zhuang (*Mycosystema* **8-9**: 39, 1995), Durrieu *et al.* (*Acta Bot. Yunn.* **19**: 128, 1997), Harrington & Potter (*Mycol.* **89**: 258, 1997), Landvik *et al.* (*Nordic Jl Bot.* **17**: 403, 1997; DNA), Harrington (*Mycol.* **90**: 235, 1998), Hansen *et al.* (*Mycol.* **91**: 299, 1999), Harrington *et al.* (*Mycol.* **91**: 41, 1999; phylogeny), Weinstein *et al.* (*Mycol.* **94**: 673, 2002), Hansen & Pfister (*Mycol.* **98**: 1029, 2006; phylogeny), Iturriaga & Pfister (*Mycotaxon* **95**: 137, 2006).

Sarcosoma Casp. (1891), Sarcosomataceae. Anamorph *Verticicladium*. 5, widespread (north temperate and tropical). See Korf (*Mycol.* **49**: 102, 1957), Landvik *et al.* (*Nordic Jl Bot.* **17**: 403, 1997; DNA), Harrington *et al.* (*Mycol.* **91**: 41, 1999; phylogeny), Köpcke *et al.* (*Phytochem.* **60**: 709, 2002; chemistry), Cittadini & Lunghini (*Micologia e Vegetazione Mediterranea* **20**: 155, 2005; conservation), Hansen & Pfister (*Mycol.* **98**: 1029, 2006; phylogeny).

Sarcosomataceae Kobayasi (1937), Pezizales. 10 gen. (+ 8 syn.), 57 spp.

Lit.: Otani (*TMSJ* **21**: 149, 1980; key Jap. spp.), Paden (*Fl. Neotrop.* Monogr. **37**, 1983; key), Donadini (*Mycol. helv.* **2**: 217, 1987), Bellemère *et al.* (*Cryptog. Mycol.* **11**: 203, 1990), Benkert (*Gleditschia* **19**: 173, 1991; keys E. German spp.), Cao *et al.* (*Mycol.* **84**: 261, 1992), Li & Kimbrough (*Int. J. Pl. Sci.* **156**: 841, 1995; septa), Li & Kimbrough (*CJB* **73**: 1761, 1995), Li & Kimbrough (*CJB* **74**: 1651, 1996), Landvik *et al.* (*Nordic Jl Bot.* **17**: 403, 1997; DNA), Zhuang & Wang (*Mycotaxon* **67**: 355, 1998), Harrington *et al.* (*Mycol.* **91**: 41, 1999; phylogeny), Köpcke *et al.* (*Phytochem.* **60**: 709, 2002), Hansen & Pfister (*Mycol.* **98**: 1029, 2006; phylogeny).

Sarcosphaera Auersw. (1869), Pezizaceae. 1 (hypogeous), N. America; Europe. See Trappe (*Mycotaxon* **2**: 109, 1975), Brandrud *et al.* (*Blyttia* **44**: 113, 1986), Landvik *et al.* (*Nordic Jl Bot.* **17**: 403, 1997; DNA), Hansen *et al.* (*Mycol.* **93**: 958, 2001; phylogeny), Bohlin (*MR* **108**: 3, 2004; phylogeny), Hansen & Pfister (*Mycol.* **98**: 1029, 2006; phylogeny), Tedersoo *et al.* (*New Phytol.* **170**: 581, 2006; ectomycorrhiza).

Sarcostroma Cooke (1871), anamorphic *Griphosphaerioma*. 1, widespread. See Sutton (*Mycol. Pap.* **138**, 1975), Nag Raj (*Coelomycetous Anamorphs with Appendage-bearing Conidia*, 1993), Jeewon *et al.* (*Mol. Phylogen. Evol.* **25**: 378, 2002; phylogeny), Ono & Kobayashi (*Mycoscience* **44**: 109, 2003; Japan, teleomorph), Lee *et al.* (*Stud. Mycol.* **55**: 175, 2006; S Africa).

Sarcostromella Boedijn (1959) = Camarops fide Nannfeldt (*Svensk bot. Tidskr.* **66**: 335, 1972).

Sarcostromellaceae Boedijn (1959) = Boliniaceae.

sarcotrimitic, see Hyphal analysis.

Sarcotrochila Höhn. (1917), Hemiphacidiaceae. Anamorph *Rhabdogloeopsis*. 4 (on conifers), N. America; Europe. See Korf (*Mycol.* **54**: 12, 1962), Dennis (*British Ascomycetes, addenda and corrigenda*: 22, 1981), Gernandt *et al.* (*Mycol.* **93**: 915, 2001; phylogeny), Stone & Gernandt (*Mycotaxon* **91**: 115, 2005).

Sarcoxylon Cooke (1883), Xylariaceae. 2, Asia; Africa. See Rogers (*Mycol.* **73**: 28, 1981), Læssøe (*SA* **8**: 25, 1989).

Sarcoxylum Clem. & Shear (1931) nom. nud. ≡ Sarcoxylon.

Sarea Fr. (1825), ? Leotiomycetes. Anamorphs *Epithyrium, Pycnidiella*. 2 (on resin), widespread (temperate). See Hawksworth & Sherwood (*CJB* **59**: 357, 1981), Reeb *et al.* (*Mol. Phylogen. Evol.* **32**: 1036, 2004; phylogeny), Lumbsch *et al.* (*MR* **111**: 1133, 2007).

Sarocladium W. Gams & D. Hawksw. (1976), anamorphic *Hypocreales*, Hso.0eH.15. 3 (on *Oryza*), Asia; USA. See Boa & Brady (*TBMS* **89**: 161, 1987), Chen *et al.* (*Acta Mycol. Sin.* Suppl. **1**: 318, 1987; purple sheath disease of rice), Bridge *et al.* (*Pl. Path.* **38**: 239, 1989; *Bambusa* blight), Joe & Manibhushanrao (*Z. PflKrankh. PflPath. PflSchutz* **102**: 291, 1995; chemistry), Bills *et al.* (*MR* **108**: 1291, 2004; chemistry, phylogeny), Ayyadurai *et al.* (*Curr. Microbiol.* **50**: 319, 2005; genetic variation).

Sarophorum Syd. & P. Syd. (1916), anamorphic *Penicilliopsis*. See Malloch & Cain (*CJB* **50**: 2613, 1972), Pitt & Samson (*Regnum veg.* **128**: 13, 1993), Ogawa & Sugiyama (*Integration of Modern Taxonomic Methods for Penicillium and Aspergillus Classification*: 149, 2000; phylogeny), Hsieh & Ju (*Mycol.* **94**: 539, 2002; teleomorph).

Sarrameana Vězda & P. James (1973), ? Fuscideaceae (L). 3, widespread. See Vězda & Kantvilas (*Lichenologist* **20**: 179, 1988), Aptroot & Sipman (*Willdenowia* **20**: 221, 1991), Kantvilas & Vězda (*Nordic Jl Bot.* **16**: 325, 1996).

Sarrameanaceae Hafellner (1984), Ostropomycetidae (inc. sed.). 1 gen., 5 spp. See Lumbsch *et al.* (*Lichenologist* **39**: 509, 2007; Australia), Lumbsch *et al.* (*Nova Hedwigia* **86**: 105, 2008).

Sartorya Vuill. (1927) nom. dub., Pezizomycotina. See also *Neosartorya*. See Malloch & Cain (*CJB* **50**: 61, 1972).

Sartvellia Berk. (1857) ≡ Dasyspora.

Satchmopsis B. Sutton & Hodges (1975), anamorphic *Pezizomycotina*, Ccu.0eH.15. 3, widespread. See Sutton & Pascoe (*TBMS* **88**: 169, 1987), Crous *et al.* (*Stud. Mycol.* **55**: 53, 2006).

Sathropeltis Bat. & Matta (1959) = Myriangiella fide von Arx & Müller (*Stud. Mycol.* **9**, 1975).

satratoxins, toxins of *Stachybotrys atra* (syn. *S. alternans*); the cause of stachybotryotoxicosis in farm animals and humans.

saturnine (of ascospores), having a flat edge round the middle (as in some *Hansenula* spp.).

Saturnispora Z.W. Liu & Kurtzman (1991), Pichiaceae. 4, widespread. See Kurtzman *in* Kurtzman & Fell (Eds) (*Yeasts, a taxonomic study* 4th edn: 387, 1998), Morais *et al.* (*Int. J. Syst. Evol. Microbiol.* **55**: 1725, 2005), Kurtzman (*FEMS Yeast Res.* **6**: 288, 2006), Suh *et al.* (*Mycol.* **98**: 1006, 2006).

Saturnomyces Cain (1956) = Emericellopsis fide Gilman (*Manual of soil fungi*, 1956).

Satwalekera D. Rao, V.G. Rao & P.Rag. Rao (1970) = Torula fide Kendrick & Carmichae *in* Ainsworth *et al.* (Eds) (*The Fungi* **4A**: 390, 1973).

Satyrus Bosc (1811) = Phallus fide Stalpers (*in litt.*).

Sauvageautia Har. (1892) = Urosporella fide Hawksworth *et al.* (*Dictionary of the Fungi* edn 8, 1995).

Savile (Douglas Barton Osborne; 1909-2000; Ireland). BA & MSc, Macdonald College, Quebec, Canada (1928-1934); PhD, University of Michigan (1935-1939); Royal Canadian Airforce (1939-1943); Division of Botany and Plant Pathology, Ottawa, 1943-1974; Fellow of the Royal Society of Canada; DSc, McGill University (1978); MSA Distinguished Mycologist (1988). Pioneer exploring the mycota of northern Canada, making major contributions on rust biodiversity and co-evolution of fungi and plants. *Publs.* An extensive series of papers on rusts and co-evolution, mainly in *CJB*, *Canadian Field Naturalist* and *Mycol.*, with many contributions to *Fungi Canadenses*. *Biogs, obits etc.* Parmelee (*Mycol.* **93**: 807, 2001).

Savoryella E.B.G. Jones & R.A. Eaton (1969), ? Sordariomycetidae. 11, Europe; Asia. See Jones & Hyde (*Bot. Mar.* **35**: 83, 1992; posn), Read *et al.* (*CJB* **71**: 273, 1993; ultrastr.), Hyde (*Mycoscience* **35**: 59, 1994), Ho *et al.* (*MR* **101**: 803, 1997), Chang *et al.* (*MR* **102**: 709, 1998; Taiwan).

Săvulescu (Traian; 1889-1963; Romania). Professor of Systematic Botany and Phytopathology, Agricultural College, Herastrau-Bucarest (1922). Mycologist and plant pathologist, noted for his studies on rusts and smuts. *Publs. Monographia Uredinalelor din R.P.R.* 2 vols (1953); *Ustilaginalele din R.P.R.* 2 vols (1957). *Biogs, obits etc.* Urban (Česká Mykologie 17: 163, 1963); Bontea (*Review of Tropical Plant Pathology* **7**: 35, 1993); Sandu-Ville (*Sydowia* **18**: 1, 1965) [bibliography, portrait]; Stafleu & Cowan (*TL-2* **5**: 91, 1985).

Savulescua Petr. (1959), Diaporthales. 1, Puerto Rico. See Petrak (*Acad. Republ. Pop. Romine*: 591, 1959).

Savulescuella Cif. (1959), anamorphic *Doassansiaceae*. 3.

Sawada (Kaneyoshi; 1888-1950; Japan). Pioneer of Asian mycology. Described over 2500 species of fungi from Taiwan. *Publs.* [Descriptive Catalogue of Formosan Fungi] [*Special Bulletin of the Agricultural Experiment Station of the Government of Formosa*], 10 vols (1919-1944) [in Japanese]. *Biogs, obits etc.* Stafleu & Cowan (*TL-2* **5**: 92, 1985).

Sawadaea Miyabe (1914), Erysiphaceae. Anamorph *Oidium* subgen. *Octagoidium*. 9, widespread. *S. bicornis* (syn. *Uncinula aceris*; powdery mildew of sycamore, *Acer*). See Zheng & Chen (*Acta Microbiol. Sin.* **20**: 35, 1980), Braun (*Beih. Nova Hedwigia* **89**, 1987), Mori *et al.* (*Mycoscience* **41**: 437, 2000; phylogeny), Cunnington *et al.* (*Australas. Mycol.* **23**: 37, 2004; Australia), Hirose *et al.* (*MR* **109**: 912, 2005; phylogeny), Wang *et al.* (*Mycol.* **98**: 1065, 2006; phylogeny).

Sawadaia, see *Sawadaea*.

saxicolous, growing on rocks. A few fungi are able to live on saxicolous substrata (Kobluk & Kahle, *Bull. Can. Pet. Geol.* **25**: 208, 1977; bibliogr.); *Lichenothelia* (q.v.) is primarily saxicolous. Rocks are one of the main substrata for lichens and the lit. on the latter is vast, see refs. cited under Ecology. Lichens can also change the mineral composition of rocks through the action of oxalic acid, etching being visible by SEM (see Syers & Iskandar, *in* Ahmadjian & Hale (Eds), *The Lichens*: 225, 1974, Jones *et al.*, *Lichenologist* **12**: 277, 1980). See also Biodeterioration.

scab, as delimited by Jenkins (*Phytopath.* **23**: 389, 1933), a plant disease having hyperplastic scab-like lesions; cf. anthracnose. - of apple (*Venturia in-*

aequalis), cherry (*V. cerasi*), pear (*V. pirina*); cereals (*Gibberella zeae*, frequently as *G. saubinettii*, and *Fusarium* spp.), citrus (*Elsinoë fawcettii*), peach (*Fusicladium carpophilum*); **powdery -** of potato (*Spongospora subterranea*).

Scabradiporites Y.K. Mathur (1966), Fossil Fungi. 1 (Paleocene), India. See Mathur (*Quarterly Journal of the Geological, Mining & Metallurgical Society of India* **38**: 43, 1966).

scabrid, rough with delicate and irregular projections.

Scabropezia Dissing & Pfister (1981), Pezizaceae. 3. See Hirsch (*Agarica* **6**: 241, 1985; key Eur. spp.), Norman & Egger (*Mycol.* **91**: 820, 1999; phylogeny), Hansen *et al.* (*Mycol.* **93**: 958, 2001; phylogeny), Spooner (*Czech Mycol.* **52**: 259, 2001; as *Plicariella*), Hansen *et al.* (*Mol. Phylogen. Evol.* **36**: 1, 2005; phylogeny), Læssøe & Hansen (*MR* **111**: 1075, 2007; phylogeny).

scabrous, rough.

Scalaria Lázaro Ibiza (1916) = Phellinus fide Wagner & Fischer (*MR* **105**: 777, 2001).

Scalarispora Buriticá & J.F. Hennen (1994), Phakopsoraceae. 1 (on *Ampelopsis* (*Vitaceae*)), Taiwan. See Buriticá & Hennen (*Revta Acad. colomb. cienc. exact. fis. nat.* **19**: 61, 1994) Cf. *Pucciniostele*.

Scalenomyces I.I. Tav. (1985), Laboulbeniaceae. 1, Sardinia. See Tavares (*Mycol. Mem.* **9**: 313, 1985).

Scalidium Hellb. (1867) = Scoliciosporum fide Poelt & Vězda (*Biblthca Lichenol.* **16**, 1981), Santesson (*Lichens of Norway and Sweden*, 1984).

Scanning electron microscopy (SEM). The scanning electron microscope images a sample surface by scanning it with a high-energy beam of electrons in a raster scan pattern; the electrons interact with the sample surface to produce signals with information about its topography. The resulting images are at much higher magnification and resolution than possible with light microscopy, and with much greater depth of field. Samples are coated prior to examination with a thin layer of gold, palladium or aluminium, or with more recent technology may be examined frozen and even uncoated. When linked to a computer, the analogue signal can be converted to digital form, allowing specimens to be measured, counted and mapped. The technique has been very widely used in mycology to resolve fungal structures, particularly appressoria, asci, basidia, conidiogenous cells, hyphae, fruitbodies and spores.

Lit.: Heywood (Ed.) (*Scanning electron microscopy: systematic and evolutionary applications*, 1971), Claugher (Ed.) (*Scanning electron microscopy in taxonomy and functional morphology*, 1990). See Microscopy, Ultrastructure.

Scaphidiomyces Thaxt. (1912), Laboulbeniaceae. 4, S. America; West Africa. See Nannfeldt (*Svensk bot. Tidskr.* **43**: 468, 1949).

Scaphidium Clem. (1901), anamorphic *Pezizomycotina*, St.1eH.19. 1, USA.

Scaphis Eschw. (1824) = Graphis fide Staiger (*Biblthca Lichenol.* **85**, 2002).

Scaphophoeum Ehrenb. ex Wallr. (1833) ≡ Schizophyllum.

scar (1) (of yeasts), **bud -**, on parent cell; **birth -**, on daughter cell; (2) (of anamorphic fungi), at conidiogenous locus and conidial base/apex, left after secession of conidium.

scariose, thin; paper-like.

scarlet (elf) cup, the ascoma of *Sarcoscypha coccinea*.

Scedosporium Sacc. ex Castell. & Chalm. (1919), anamorphic *Pseudallescheria*, Hso.0eH.19. 4 (on humans, other animals and from hay), widespread (subtropical). See Kendrick & Carmichae *in* Ainsworth *et al.* (Eds) (*The Fungi* **4A**: 390, 1973), Salkin *et al.* (*J. Clin. Microbiol.* **26**: 498, 1988; *S. inflatum*, an emerging pathogen), Dykstra *et al.* (*Mycol.* **81**: 896, 1989; TEM of conidiogenesis), Issakainen *et al.* (*J. Med. Vet. Mycol.* **35**: 389, 1997; SSU rDNA), Wedde *et al.* (*Medical Mycol.* **36**: 61, 1998; PCR-based identification), Issakainen *et al.* (*MR* **103**: 1179, 1999; relationship with *Petriella*), Okada *et al.* (*CJB* **76**: 1495, 1998; phylogeny), Rainer *et al.* (*J. Clin. Microbiol.* **38**: 3267, 2000; molecular variability), Zouhair *et al.* (*Journal of Medical Microbiology* **50**: 925, 2001; strain typing), Solé *et al.* (*Medical Mycology* **41**: 293, 2003; strain typing), Gilgado *et al.* (*J. Clin. Microbiol.* **43**: 4930, 2005; phylogeny), Guarro *et al.* (*Medical Mycology* **44**: 295, 2006; review), Rainer & Hoog (*MR* **110**: 151, 2006; phylogeny).

Scelobelonium (Sacc.) Höhn. (1905) = Crocicreas fide Carpenter (*Mem. N. Y. bot. Gdn* **33**: 1, 1981).

Scelophoromyces Thaxt. (1912), Laboulbeniaceae. 1, America. See Nannfeldt (*Svensk bot. Tidskr.* **43**: 468, 1949).

Scenidium (Klotzsch) Kuntze (1898) = Hexagonia Fr. fide Jülich (*Persoonia* **12**: 107, 1984).

Scenomyces F. Stevens (1927), anamorphic *Pezizomycotina*, Cpt.-.-. 1, Panama.

Scepastocarpus Santam. (2004), Laboulbeniaceae. 1, Spain. See Santamaría (*Mycol.* **96**: 764, 2004).

Sceptrifera Deighton (1965), anamorphic *Pezizomycotina*, Hsy.1eP.10/12. 1, Philippines. See Deighton (*Mycol. Pap.* **101**: 37, 1965).

Sceptromyces Corda (1831) = Aspergillus fide Engelke (*Hedwigia* **41**: 219, 1902).

Schachtia Schulzer (1866) [non *Schachtia* H. Karst. 1859, *Rubiaceae*] = Fenestella fide Hawksworth *et al.* (*Dictionary of the Fungi* edn 8, 1995).

Schadonia Körb. (1859), Ramalinaceae (L). 3, widespread (north temperate). See Ekman (*Op. bot.* **127**: 148 pp., 1996), Upreti & Nayaka (*Mycotaxon* **97**: 275, 2006; India).

Schadoniaceae Hafellner (1984) = Ramalinaceae.

Schaereria Körb. (1855) nom. cons., Schaereriaceae (L). 11, widespread (esp. temperate). See Schneider (*Biblthca Lichenol.* **13**, 1980), Eriksson & Hawksworth (*SA* **12**: 42, 1993; posn), Lunke *et al.* (*Bryologist* **99**: 53, 1996), Kantvilas (*Lichenologist* **31**: 231, 1999; Tasmania), Fryday & Common (*Bryologist* **104**: 109, 2001; Falkland Is), Lumbsch *et al.* (*Mol. Phylogen. Evol.* **31**: 822, 2004; phylogeny), Schmull & Spribille (*Lichenologist* **37**: 527, 2005), Wedin *et al.* (*MR* **109**: 159, 2005; phylogeny).

Schaereriaceae M. Choisy ex Hafellner (1984), Ostropomycetidae (inc. sed.). 1 gen. (+ 1 syn.), 11 spp.

Lit.: Lumbsch *et al.* (*MR* **111**: 257, 2007; phylogeny), Lumbsch *et al.* (*MR* **111**: 1133, 2007).

Scharifia Petr. (1955), Pezizomycotina. 1, Iran. See Petrak (*Sydowia* **9**: 448, 1955).

Schasmaria (Ach.) Gray (1821) = Cladonia fide Hawksworth *et al.* (*Dictionary of the Fungi* edn 8, 1995).

scheda (pl. **-ae**), **schedula** (pl. **-ae**), (Latin), label(s), esp. printed labels relating to sets of dried specimens (exsiccati; q.v.).

Scheleobrachea S. Hughes (1958) = Pithomyces fide

Ellis (*Dematiaceous Hyphomycetes*, 1971).

Schenckiella Henn. (1893), Saccardiaceae. 1, S. America.

Schenella T. Macbr. (1911), Geastraceae. 4, widespread. See Estrada-Torres *et al.* (*Mycol.* **97**: 139, 2005).

Scherffelia Sparrow (1933) [non *Scherffelia* Pascher 1911, *Algae*] ≡ Scherffeliomyces.

Scherffeliomyces Sparrow (1934), Chytridiaceae. 1, Europe; N. America.

Scherffeliomycopsis Geitler (1962), Chytridiaceae. 1 (on *Coleochaete*), Austria; Poland. See Skirgiełło & Szymańska (*Feddes Repert.* **113**: 128, 2002; Poland).

Schiffnerula Höhn. (1909), Englerulaceae. Anamorphs *Questieriella*, *Sarcinella*. *c.* 36, widespread (esp. warmer areas). See Hansford (*Mycol. Pap.* **15**, 1946), Hughes (*CJB* **62**: 2213, 1984), Hughes (*Mycol.* **82**: 657, 1990), Hosagoudar (*Zoos' Print Journal* **18**: 1071, 2003).

Schinzia Nägeli (1842) [non *Dennstätt*, 1818] = Entorrhiza fide Vánky (*Illustrated genera of smut fungi*, 1987).

Schinzinia Fayod (1889), Agaricaceae. 1, E. Africa.

Schismatomma Flot. & Körb. ex A. Massal. (1852), Roccellaceae (±L). 11, widespread. See Torrente & Egea (*Biblthca Lichenol.* **32**, 1989), Tehler (*CJB* **68**: 2458, 1990; cladistics), Tehler (*Op. Bot.* **118**: 1, 1993; key), Tehler (*Mycotaxon* **51**: 31, 1994), Grube (*Bryologist* **101**: 377, 1998; phylogeny), Myllys *et al.* (*Lichenologist* **31**: 461, 1999; phylogeny), Follmann (*Mitteilungen aus dem Institut für Allgemeine Botanik Hamburg* **30-32**: 61, 2002; biogeography), Lumbsch *et al.* (*Mol. Phylogen. Evol.* **34**: 512, 2005; phylogeny), Ertz & Diederich (*Bryologist* **109**: 415, 2006; lichenicolous habit, Mexico), Tehler & Irestedt (*Cladistics* **23**: 432, 2007).

Schismatommatomyces Cif. & Tomas. (1953) = Schismatomma.

Schistodes Theiss. (1918) = Perisporiopsis Henn. fide Müller & von Arx (*Beitr. Kryptfl. Schweiz* **11** no. 2, 1962).

Schistophoron Stirt. (1876), ? Graphidaceae (L). 2, widespread. See Tibell (*Lichenologist* **14**: 219, 1982), Tibell (*Beih. Nova Hedwigia* **79**: 597, 1984), Aptroot & Sipman (*Biblthca Lichenol.* **96**: 21, 2007; key).

Schistoplaca Brusse (1987) = Lecanora fide Lumbsch & Feige (*Mycotaxon* **52**: 429, 1994), Ekman (*Op. bot.* **127**: 148 pp., 1996).

Schistostoma Stirt. (1878) = Graphis fide Salisbury (*Lichenologist* **6**: 126, 1974).

Schizacrospermum Henn. (1899) nom. dub., Fungi. See Walker (*Mycotaxon* **11**: 1, 1980).

schizidium, a propagule formed by upper layers of a lichen thallus splitting off as scale-like segments from the main lobes (e.g. the lobule-like structures in *Fulgensia bracteata* subsp. *deformis*). See Poelt (*Flora, Jena* **159**: 23, 1980). Fig. 22G.

Schizmaturus (Corda) Kalchbr. (1880) = Lysurus fide Dring (*Kew Bull.* **35**: 1, 1980).

schizobiont, bacteria once considered to be additional symbionts of lichens.

Schizoblastosporion Cif. (1930), anamorphic *Saccharomycetales*, Hso.0eH.10. 2, widespread. See Smith *in* Kurtzman & Fell (Eds) (*Yeasts, a taxonomic study* 4th edn: 602, 1998), Schweigkofler *et al.* (*Organ. Divers. Evol.* **2**: 1, 2002), Suh *et al.* (*Mycol.* **98**: 1006, 2006; phylogeny).

Schizoblastosporon Hara (1936) ≡ Schizoblastosporion.

Schizocapnodium Fairm. (1921), Nitschkiaceae. 1 (from wood), USA; India. See Nannfeldt (*Svensk bot. Tidskr.* **65**: 49, 1971; syn. of *Nitschkia*), Subramanian & Sekar (*Kavaka* **18**: 19, 1993), Huhndorf *et al.* (*MR* **108**: 1384, 2004).

Schizocephalum Preuss (1852) = Haplographium fide Saccardo (*Syll. fung.* **4**: 306, 1886).

Schizochora Syd. & P. Syd. (1913), Phyllachoraceae. 3 (from living leaves), widespread. See von Arx & Müller (*Beitr. Kryptfl. Schweiz* **11** no. 1, 1954).

Schizochorella Höhn. (1918) = Hypoderma De Not. fide Minter (*in litt.*).

Schizoderma Ehrenb. (1818) = Leptostroma fide Sutton (*Mycol. Pap.* **141**, 1977).

Schizoderma Kunze (1825), anamorphic *Pezizomycotina*, Cac.?.?. 1, Germany.

Schizodiplodia Zambett. (1955) = Didymosporiella fide Sutton (*Mycol. Pap.* **141**, 1977).

Schizodiscus Brusse (1988), Porpidiaceae (L). 1, Natal. See Brusse (*Bothalia* **18**: 94, 1988).

schizogenous, formed by cracking or splitting, cf. lysigenous.

schizogony, the process of division of a schizont.

Schizographa Nyl. (1857) nom. dub., Fungi.

Schizolaboulbenia Middelh. (1957) = Laboulbenia fide Benjamin *in* Thaxter (*Mem. Am. Acad. Arts Sci. 1896-1931* **12-16**, 1971).

schizolytic, secession of conidia involving a splitting of the delimiting septum so that one half of the cross-wall becomes the base of the seceding conidium and the other half remains at the apex of the conidiogenous cell (Cole & Samson, *Patterns of development in conidial fungi*, 1979). Cf. rhexolytic.

Schizoma Nyl. ex Cromb. (1894) = Thyrea fide Henssen (*Symb. bot. upsal.* **18** no. 1, 1963).

Schizomeromyces Thaxt. (1931) = Clematomyces fide Tavares (*Mycol. Mem.* **9**, 1985).

Schizonella J. Schröt. (1877), Anthracoideaceae. 4 (on *Cyperaceae*), widespread (esp. Europe; N. America). See Vánky (*Mycotaxon* **69**: 102, 1998; key).

Schizonia Pers. (1828) ≡ Schizophyllum.

schizont, a vegetative thallus, having no wall, which undergoes simple or multiple division.

Schizontopeltis Bat. & H. Maia (1962) = Schizothyrium fide von Arx & Müller (*Stud. Mycol.* **9**, 1975).

Schizoparmaceae Rossman (2007), Diaporthales. 3 gen. (+ 5 syn.), 32 spp.

Lit.: Rossman *et al.* (*Mycoscience* **48**: 135, 2007; phylogeny).

Schizoparme Shear (1923), Schizoparmaceae. Anamorph *Pilidiella*. 7 (from leaves), widespread. See Samuels *et al.* (*Mycotaxon* **46**: 459, 1993), Van Niekerk *et al.* (*MR* **108**: 283, 2004; phylogeny, anamorph), Rossman *et al.* (*Mycoscience* **48**: 135, 2007; review).

Schizopelte Th. Fr. (1875), Roccellaceae (L). 1, N. America. See Tehler (*CJB* **68**: 2458, 1990; cladistics), Grube (*Bryologist* **101**: 377, 1998), Myllys *et al.* (*Bryologist* **101**: 70, 1998), Myllys *et al.* (*Lichenologist* **31**: 461, 1999; phylogeny).

Schizopeltis Bat. & I.H. Lima (1959) = Schizothyrium fide Müller & von Arx (*Beitr. Kryptfl. Schweiz* **11** no. 2, 1962).

Schizopeltomyces Cif. & Tomas. (1953) ≡ Schizopelte.

Schizophoma Kleb. (1933) ? = Sclerophoma fide Sutton (*Mycol. Pap.* **141**, 1977).

Schizophyllaceae Quél. (1888), Agaricales. 2 gen. (+

11 syn.), 7 spp.

Lit.: Cooke (*Mycol.* **53**: 575, 1961; *Schizophyllum*), Donk (*Persoonia* **3**: 199, 1964), Nuss (*Hoppea* **39**: 127, 1980), Stalpers (*Persoonia* **13**: 495, 1988), Sigler *et al.* (*J. Med. Vet. Mycol.* **35**: 365, 1997), Ginns (*Mycol.* **90**: 19, 1998), James *et al.* (*Evolution* Lancaster, Pa. **53**: 1665, 1999), Buzina *et al.* (*J. Clin. Microbiol.* **39**: 2391, 2001), James & Vilgalys (*Mol. Ecol.* **10**: 471, 2001), Kano *et al.* (*J. Clin. Microbiol.* **40**: 3535, 2002), Moncalvo *et al.* (*Mol. Phylogen. Evol.* **23**: 357, 2002), Bodensteiner *et al.* (*Mol. Phylogen. Evol.* **33**: 501, 2004), Larsson *et al.* (*MR* **108**: 983, 2004), Wei & Dai (*Mycosystema* **23**: 437, 2004).

Schizophyllales = Agaricales.

Schizophyllum Fr. (1815), Schizophyllaceae. 6 (wood rot; pathogenic for humans), widespread. See Watling & Sweeney (*Sabouraudia* **12**: 214, 1974; cultural chars.), Buzina *et al.* (*J. Clin. Microbiol.* **39**: 2391, 2001; molecular identification), Baron *et al.* (*J. Clin. Microbiol.* **44**: 3042, 2006; diagnosis).

Schizophyllus, see *Schizophyllum*.

Schizopora Velen. (1922) nom. rej., Schizoporaceae. 4, widespread. See Ryvarden & Johansen (*Prelim. Polyp. Fl. E. Afr.*: 315, 1980), Langer (*Biblthca Mycol.* **154**, 1994; sub *Hyphodontia*), Paulus *et al.* (*MR* **104**: 1155, 2000), Lim & Jung (*Mycobiology* **29**: 194, 2001; Korean spp.).

Schizoporaceae Jülich (1982), Hymenochaetales. 14 gen. (+ 6 syn.), 109 spp.

Lit.: Hallenberg (*Mycotaxon* **27**: 361, 1986), Nakasone (*Mycol. Mem.* **15**: 412 pp., 1990), Langer (*Biblthca Mycol.* **154**: 298 pp., 1994), Ginns (*Mycol.* **90**: 19, 1998), Langer (*Folia cryptog. Estonica* **33**: 57, 1998), Paulus *et al.* (*MR* **104**: 1155, 2000), Hjortstam (*Windahlia* **24**: 1, 1998), Hjortstam & Ryvarden (*Syn. Fung.* **15**: 8, 2002), Greslebin & Rajchenberg (*N.Z. Jl Bot.* **41**: 437, 2003), Larsson *et al.* (*MR* **108**: 983, 2004), Martín & Montón (*Bol. Sanid. Veg.* Plagas **30**: 93, 2004), Binder *et al.* (*Systematics and Biodiversity* **3**: 113, 2005), Küffer & Senn-Irlet (*Mycol. Progr.* **4**: 77, 2005).

Schizosaccharis Clem. & Shear (1931) ≡ Schizosaccharomyces.

Schizosaccharomyces Lindner (1893), Schizosaccharomycetaceae. 1 or 4, widespread. See Bridge & May (*J. gen. Microbiol.* **130**: 1921, 1984; numerical taxonomy), Kreger-van Rij (Ed.) (*Yeasts, a taxonomic study* 3rd edn, 1984; key), Mikata & Banno (*IFO Res. Comm.* **13**: 45, 1987; ascospores), Barnett *et al.* (*Yeasts: Characteristics and Identification* 2nd edn, 1990), Cox *in* Wheals *et al.* (Eds) (*Yeast* Chichester **6**: 7, 1995; genetic analysis), Munz & Kohli *in* Wheals *et al.* (Eds) (*Yeast* Chichester **6**: 583, 1995; chromosome maps), Sipiczki (*Antonie van Leeuwenhoek* **68**: 119, 1995; phylogeny), Sprague *in* Wheals *et al.* (Eds) (*Yeast* Chichester **6**: 411, 1995; mating types), Vaughan-Martini & Martini *in* Kurtzman & Fell (Eds) (*Yeasts, a taxonomic study* 4th edn: 391, 1998), Wood (*Nature* Lond. **415** no. 6874: 871, 2002; genome), Lopandic *et al.* (*Mycol. Progr.* **4**: 205, 2005; phylogeny), Suh *et al.* (*Mycol.* **98**: 1006, 2006; phylogeny).

Schizosaccharomycetaceae Beij. ex Klöcker (1905), Schizosaccharomycetales. 2 gen. (+ 3 syn.), 5 spp.

Lit.: Eriksson *et al.* (*SA* **11**: 119, 1993), Egel *in* McLaughlin *et al.* (Eds) (*The Mycota* A Comprehensive Treatise on Fungi as Experimental Systems for Basic and Applied Research **1**: 251, 1994), Munz & Kohli (*The Yeasts* Vol. 6. Yeast Genetics. Edn 2: 583, 1995), Sipiczki (*Antonie van Leeuwenhoek* **68**: 119, 1995), Jeffery *et al.* (*Antonie van Leeuwenhoek* **72**: 327, 1997), Kurtzman & Blanz *in* Kurtzman & Fell (Eds) (*Yeasts, a taxonomic study* 4th edn: 69, 1998), Sugiyama (*Mycoscience* **39**: 487, 1998), Vaughan-Martini & Martini *in* Kurtzman & Fell (Eds) (*Yeasts, a taxonomic study* 4th edn: 391, 1998), Esteve-Zarzoso *et al.* (*Int. J. Syst. Bacteriol.* **49**: 329, 1999), Mikata & Yamada (*Res. Commun.* Inst. Ferm., Osaka **19**: 41, 1997), Wood (*Nature* Lond. **415** no. 6874: 871, 2002), Sugiyama *et al.* (*Mycol.* **98**: 996, 2006; phylogeny).

Schizosaccharomycetales O.E. Erikss. (1994). Schizosaccharomycetidae. 1 fam., 2 gen., 5 spp. Fam.:

Schizosaccharomycetaceae

The name has been used by Prillinger *et al.* (*Z. Mykol.* **56**: 219, 1990) to include *Protomycetales* and *Taphrinales* which they considered intermediate between *Ascomycota* and *Basidiomycota*.

Lit.: Eriksson *et al.* (*SA* **11**: 119, 1993), Jeffery *et al.* (*Antonie van Leeuwenhoek* **72**: 327, 1997; lipids), Kurtzman & Blanz (*The Yeasts, a taxonomic Study*: 69, 1998), Sjamsuridzal *et al.* (*Mycoscience* **38**: 267, 1997), Sugiyama (*Mycoscience* **39**: 487, 1998).

Schizosaccharomycetes O.E. Erikss. & Winka (1997), Taphrinomycotina. 1 ord., 1 fam., 2 gen., 5 spp. The fission yeasts. Ord.:

Schizosaccharomycetales

For *Lit.* see ord. and fam.

Schizosaccharomycetidae, see *Schizosaccharomycetes*.

Schizospora Dietel (1895) [non *Schizospora* Reinsch 1875, *Algae*] = Pucciniosira fide Dietel (*Nat. Pflanzenfam.* **6**, 1928).

Schizosporaceae Dietel (1897) = Pucciniosiraceae.

Schizostege Theiss. (1917) [non *Schizostege* W.F. Hillebr. 1888, *Pteridophyta*] = Saccothecium fide Hawksworth *et al.* (*Dictionary of the Fungi* edn 8, 1995).

Schizostoma Ces. & De Not. ex Sacc. (1878) [non *Schizostoma* Ehrenb. ex Lév. 1846] = Xenolophium fide Holm & Yue (*Acta Mycol. Sin.* Suppl. **1**: 82, 1986), Huhndorf (*Mycol.* **85**: 490, 1993), Barr & Mathiassen (*Mycotaxon* **69**: 159, 1998).

Schizostoma Ehrenb. ex Lév. (1846), Agaricaceae. 1, widespread (subtropical dry areas). See Long & Stouffer (*Mycol.* **35**: 21, 1943), Lunghini (*Bollettino dell'Associazione Micologica ed Ecologica Romana* **17**: 22, 2001), Sarasini (*Rivista di Micologia* **46**: 7, 2003; mediterranean).

Schizothecium Corda (1838) nom. rej., Lasiosphaeriaceae. 1 (coprophilous), widespread. See Lundqvist (*Symb. bot. upsal.* **20** no. 1, 1972), Hawksworth *et al.* (*Dictionary of the Fungi* edn 8, 1995), Huhndorf *et al.* (*Mycol.* **96**: 368, 2004; phylogeny), Cai *et al.* (*Fungal Diversity* **19**: 1, 2005; phylogeny), Miller & Huhndorf (*Mol. Phylogen. Evol.* **35**: 60, 2005; phylogeny, wall structure), Hu *et al.* (*Cryptog. Mycol.* **27**: 89, 2006; China).

Schizothyra Bat. & C.A.A. Costa (1957), anamorphic *Pezizomycotina*, Cpt.0eH.?. 1, Brazil. See Batista & Costa (*An. Soc. Biol. Pernambuco* **15**: 409, 1957).

Schizothyrella Thüm. (1880), anamorphic *Pezizomycotina*. 5, Europe; USA.

Schizothyrellina Petr. (1922) nom. nud. = Phacidiella P. Karst. fide Sutton (*Mycol. Pap.* **141**, 1977).

Schizothyriaceae Höhn. ex Trotter, Sacc., D. Sacc. & Traverso (1928), Capnodiales. 16 gen. (+ 28 syn.), 52 spp. More molecular data are needed, but placement of the family seems clear.

Lit.: von Arx & Müller (*Stud. Mycol.* **9**, 1975), Eriksson (*Op. Bot.* **60**, 1981), Farr (*Mycol.* **79**: 97, 1987), Williamson & Sutton (*Pl. Dis.* **84**: 714, 2000), Batzer *et al.* (*Mycol.* **100**: 232, 2008; phylogeny).

Schizothyrina Bat. & I.H. Lima (1959) = Schizothyrium fide Müller & von Arx (*Beitr. Kryptfl. Schweiz* **11** no. 2, 1962).

Schizothyrioma Höhn. (1917), Helotiales. 2, Europe. See Holm (*Svensk bot. Tidskr.* **65**: 208, 1971), Nauta & Spooner (*Mycologist* **14**: 65, 2000; UK).

Schizothyrium Desm. (1849), Schizothyriaceae. Anamorph *Zygophiala*. *c*. 13, widespread (esp. tropical). *S. pomi* (fly speck of apple and pear). See von Arx & Müller (*Stud. Mycol.* **9**, 1975), Katumoto (*TMSJ* **27**: 1, 1986), Petrini *et al.* (*Mycol. helv.* **3**: 263, 1989), Kohlmeyer *et al.* (*CJB* **76**: 467, 1998), Batzer *et al.* (*Mycol.* **97**: 1268, 2005; phylogeny), Batzer *et al.* (*Mycol.* **100**: 232, 2008; phylogeny).

Schizothyropsis Bat. & A.F. Vital (1960), anamorphic *Pezizomycotina*, Cpt.0eH.?. 1, Paraguay. See Batista & Vital (*Atas Inst. Micol. Univ. Recife* **1**: 355, 1960).

Schizotorula Krassiln. (1954) nom. inval., Fungi. A lapsus for *Schizotorulopsis*.

Schizotorulopsis Cif. (1930) nom. dub., Bacteria. ? Based on bacteria.

Schizotrichella E.F. Morris (1956) = Colletotrichum fide Sutton (*Mycol. Pap.* **141**, 1977).

Schizotrichum McAlpine (1903), anamorphic *Pezizomycotina*, Hso.0fH.10. 2, Australia; Bermuda.

schizotype, a syntype taken to be the type of name by the exclusion of other syntypes; an implicit lectotype, but not acceptable as a formal typification; see type.

Schizoxylon Pers. (1810), Stictidaceae. 33, widespread (esp. temperate). See Sherwood (*Mycotaxon* **6**: 215, 1977; key), Johnston (*Mycotaxon* **24**: 349, 1985; anamorph), Wedin *et al.* (*Lichenologist* **37**: 67, 2005; phylogeny).

Schizoxylum Pers. (1822) ≡ Schizoxylon.

schlauch, open-ended extension of the rohr, orientated toward the cytoplasm of encysted zoospore of plasmodiophorids. See also rohr and stachel.

Schmitzomia Fr. (1849) = Stictis fide Sherwood (*Mycotaxon* **6**: 215, 1977).

Schnablia Sacc. & P. Syd. (1899), ? Helotiales. 1, Europe.

Schneepia Speg. (1885) = Parmularia Lév. fide Müller & von Arx (*Beitr. Kryptfl. Schweiz* **11** no. 2, 1962), Inácio & Cannon (*MR* **107**: 82, 2003).

Schoenbornia Bubák (1906) = Hymenopsis fide Sutton (*Mycol. Pap.* **141**, 1977).

Schoenleinium Johan-Olsen (1897) = Trichophyton fide Kendrick & Carmichae *in* Ainsworth *et al.* (Eds) (*The Fungi* **4A**: 390, 1973).

Schrakia Hafellner (1979), ? Patellariaceae. 1, N. & S. America. See Kutorga & Hawksworth (*SA* **15**: 1, 1997).

Schroedera D.J. Morris & A. Adams (2002), Microsporidia. 2. See Morris & Adams (*Acta Protozool.* **41**: 395, 2002).

Schroeterella Syd. (1922) [non *Schroeterella* Herzog 1916, *Musci*] = Puccinia fide Dietel (*Nat. Pflanzenfam.* **6**, 1928).

Schroeteria G. Winter (1881), anamorphic *Pezizomycotina*. 6 (in seeds of *Veronica*), Asia; Europe. smut-like. See Vánky (*Sydowia* **34**: 157, 1981; monogr.), Nagler *et al.* (*Mycol.* **81**: 884, 1989; ascomycete affinities), Vánky (*Mycol. Balcanica* **1**: 175, 2004).

Schroeteriaster Magnus (1896), Pucciniales. 1 (on *Ranunculus* (I) (*Ranunculaceae*); on *Rumex* (II, III) (*Polygonaceae*)), Europe. See Mains (*Annls mycol.* **32**: 256, 1934) = Uromyces (Puccin.) fide, Cummins & Hiratsuka (*Illustr. Gen. Rust Fungi rev. edit.*, 1983).

Schulzeria Bres. & Schulzer (1886) nom. dub. ? = Leucoagaricus fide Kuyper (*in litt.*).

Schvartsman (Sofiya Ruvinovna; 1912-1975; Ukraine, later Russia, Kazakhstan). Laboratory assistant, Adamy Aul, Adygea Autonomous Region (1930); student, Moscow State University (1931-1940); Senior Lecturer in Botany, Kazakh State University, Alma Ata (1940-1961); founder (1941) then Head (1943-1961) of the Department of Lower Plants, Botanical Institute of the Kazakh SSR Academy of Sciences. Major contributions to the mycota of Kazakhstan, as initiator, organizer and editor-in-chief of the [*Flora of Cryptogamic Plants of Kazakhstan*] [in Russian]. *Publs.* (with various collaborators) [*Flora of Cryptogamic Plants of Kazakhstan*] 8 vols (1960-1973) [in Russian]. *Biogs, obits etc.* Vasyagina & Byzova ([*Mikologiya i Fitopatologiya*] **12**: 176, 1978 [in Russian]).

Schwanniomyces Klöcker (1909), Saccharomycetales. 6. See Kurtzman & Robnett (*Yeast* Chichester **7**: 61, 1991), Yamada *et al.* (*J. gen. appl. Microbiol.* Tokyo **37**: 523, 1991), Nakase *et al.* *in* Kurtzman & Fell (Eds) (*Yeasts, a taxonomic study* 4th edn: 157, 1998).

Schwarzmannia Pisareva (1968), anamorphic *Pezizomycotina*, Cac.1eP.19. 1, Kazakhstan. See Pisareva (*Botanicheskie Materialy* Gerbariya Instituta Botaniki, Akademiya Nauk Kazakhskoĭ SSR **5**: 72, 1968).

Schweinitz (Lewis David von; 1780-1834; USA). Born, Bethlehem, Pennsylvania; educated in Saxony, Germany (1798-1808); church administrator, Salem, North Carolina (1812 onwards). The first important mycologist of his country, Schweinitz learned mycology in Saxony with Albertini, continuing this interest on return to the USA in 1812; a careful observer who had one of the best compound microscopes of his day, he sent parts of his specimens to co-workers in Europe and America. *Publs.* (with Albertini) *Conspectus Fungorum in Lusatiae* (1805); *Synopsis Fungorum Carolinae Superioris* (1822); *Synopsis Fungorum in America Boreali* (1832) [an account of all the fungi (about 4,000) then known in North America]. *Biogs, obits etc.* Rogers (*Mycol.* **69**: 223, 1977); Shear (*Plant World* **5**: 45, 1902) [portrait]; Shear & Stevens (*Mycol.* **9**: 191, 1917); Shear & Stevens (*Memoirs of the Torrey Botanical Club* **16**: 119, 1921, Schweinitz-Torrey correspondence); Stafleu & Cowan (*TL-2* **5**: 437, 1985).

Schweinitzia Grev. (1823) [non *Schweinitzia* Elliott ex Nuttall 1818, *Pyrolaceae*] = Podaxis fide Kuyper (*in litt.*).

Schweinitzia Massee (1895) ? = Velutaria fide Nannfeldt (*Nova Acta R. Soc. Scient. upsal.*, 1932).

Schweinitziella Speg. (1888), Sordariomycetes. 3 (on fungi), widespread (tropical). See Pirozynski (*Kew Bull.* **31**: 595, 1977).

Schwendener (Simon; 1829-1919; Switzerland). Student, Geneva then Zürich (1856-1857) [under C. Nägeli]; Assistant (1857-1860), then Lecturer (1860-

1867) Munich; Professor of Botany, Basle (1867-1877); Professor of Botany, Tübingen (1877-1878); Professor of Botany and Director of Botanical Institute, Berlin (1878-1909). Elucidated true nature of the components of lichen symbiosis. *Publs. Verhandlungen der Schweizerischen Naturforschenden Gesellschaft in Rheinfelden* Aarau **51**: 88, 1867; *Die Algentypen der Flechtengonidien*, 1869. *Biogs, obits etc.* Ainsworth (*Introduction to the History of Mycology*, 1976); Vines (*Proceedings of the Linnean Society* 1919-20: 47, 1921); Stafleu & Cowan (*TL-2* **5**: 443, 1985).

Scindalma Hill ex Kuntze (1898) = Phellinus fide Pegler (*The polypores [Bull. BMS Suppl.]*, 1973).

Sciniatosporium Kalchbr. ex Morgan-Jones (1971) ≡ Seimatosporium fide Sutton (*TBMS* **58**: 164, 1972).

Sciodothis Clem. (1909) = Tomasellia fide Harris (*in litt.*).

Scipionospora E.K.C. Bylén & J.I.R. Larsson (1995), Microsporidia. 1.

Scirrhia Nitschke ex Fuckel (1870), ? Capnodiales. Anamorph *Lecanosticta*. 9, widespread. *S. acicola* (brown needle blight of pine). See Siggers (*Tech. Bull. U.S. Dept. Agric.* **870**, 1944), Obrist (*Phytopath. Z.* **35**: 370, 1959), Barr (*Contr. Univ. Mich. Herb.* **9**: 523, 1972), von Arx & Müller (*Stud. Mycol.* **9**, 1975; ? = *Metameris* p.p.), Butin (*Sydowia* **38**: 20, 1986; anam.), Kohlmeyer *et al.* (*CJB* **74**: 1830, 1996).

Scirrhiachora Theiss. & Syd. (1915), Mycosphaerellaceae. 1, Europe. See von Arx & Müller (*Stud. Mycol.* **9**, 1975), Aptroot (*CBS Diversity Ser.* **5**, 2006).

Scirrhiella Speg. (1885) = Apiospora fide Barr (*Mycol.* **68**, 1976).

Scirrhiopsis Henn. (1905) nom. dub., Fungi.

Scirrhodothis Theiss. & Syd. (1915) = Metameris fide Barr (*Mycotaxon* **43**: 371, 1992).

Scirrhophoma Petr. (1941), anamorphic *Pezizomycotina*, St.0eH.?. 1, Europe.

Scirrhophragma Theiss. & Syd. (1915) = Metameris fide Barr (*Mycotaxon* **43**: 371, 1992).

scissile (of the flesh of a pileus), separating into horizontal layers.

Scitovszkya Schulzer (1866) = Mucor Fresen. fide Hesseltine (*Mycol.* **47**: 344, 1955).

Sclerangium Lév. (1848) = Scleroderma fide Stalpers (*in litt.*).

sclerobasidium, see basidium.

sclerocarps, sclerotium-like modified ascomata permanently lacking a sexual capacity and acting as sclerotia, as in *Varicosporina ramulosa* (Kohlmeyer & Charles, *CJB* **59**: 1787, 1981).

Sclerochaeta Höhn. (1917) = Chaetopyrena Pass. fide Höhnel (*Hedwigia* **60**: 129, 1918).

Sclerochaetella Höhn. (1917) = Diploplenodomus fide Petrak (*Annls mycol.* **42**: 58, 1944).

Sclerocleista Subram. (1972), Trichocomaceae. Anamorph *Aspergillus*. 2. See von Arx (*Gen. Fungi Sporul. Cult.* Edn 3, 1981), Pitt & Samson (*Regnum veg.* **128**: 13, 1993), Bainbridge (*Integration of Modern Taxonomic Methods for Penicillium and Aspergillus Classification*: 373, 2000; phylogeny), Tamura *et al.* (*Integration of Modern Taxonomic Methods for Penicillium and Aspergillus Classification*: 357, 2000; phylogeny), Klich (*Identification of Common Aspergillus Species*: 116 pp., 2002).

Sclerococcum Fr. (1825), anamorphic *Pezizomycotina*, Hsp.#eP.1/10. 15 (on lichens), Europe. See Hawksworth (*TBMS* **65**: 219, 1975), Hawksworth (*Bull. Br. Mus. nat. hist.* Bot. **6**: 181, 1979), Diederich & Scholz (*Biblthca Lichenol.* **57**: 113, 1995), Etayo & Calatayud (*Annln naturh. Mus. Wien* Ser. B, Bot. Zool. **100**: 677, 1998; Spain).

Scleroconidioma Tsuneda, Currah & Thormann (2000), anamorphic *Dothideomycetes*. 1, Canada. See Tsuneda *et al.* (*CJB* **78**: 1295, 2000), Tsuneda *et al.* (*Mycol.* **93**: 1164, 2001; ontogeny), Hambleton *et al.* (*Mycol.* **95**: 959, 2003; phylogeny), Tsuneda *et al.* (*Mycol.* **96**: 1128, 2004; phylogeny).

Scleroconium Syd. (1935), anamorphic *Pezizomycotina*, Hsp.0eH.6. 1, S. America.

Sclerocrana Samuels & L.M. Kohn (1987), Sclerotiniaceae. 1, New Zealand. See Samuels & Kohn (*Sydowia* **39**: 202, 1987).

Sclerocystis Berk. & Broome (1873), Glomeraceae. 1, widespread (tropical). See Almeida & Schenk (*Mycol.* **82**: 703, 1990; revision), Wu (*Mycotaxon* **47**: 25, 1993), Wu (*Mycotaxon* **49**: 327, 1993; revision), Wu & Chen (*TMSJ* **34**: 283, 1993), Wang *et al.* (*Acta Mycol. Sin.* **15**: 161, 1996), Yao *et al.* (*Genera of Endogonales*: 229 pp., 1996), Redecker *et al.* (*Mycol.* **92**: 282, 2000; phylogeny, *S. coremioides* transferred to *Glomus*), Degawa (*Memoirs of the National Science Museum* Tokyo **37**: 119, 2001; Japan).

Sclerodepsis Cooke (1890) = Trametes fide Ryvarden (*Gen. Polyp.*: 222, 1991).

Scleroderma Pers. (1801), Sclerodermataceae. *c.* 30, widespread. See Guzmán (*Darwiniana* **16**: 233, 1970; key), Rifai (*TMSJ* **28**: 97, 1987; Malesia, key), Sarasini (*Boll. Gruppo Micol. 'G. Bresadola'* **34**: 119, 1991; Italy), Richter (*Mycotaxon* **45**: 461, 1992; cultures), Sims *et al.* (*Mycotaxon* **56**: 403, 1995; key), Jeffries (*Scleroderma – Ectomycorrhizal fungi, key genera in profile*, 1999; ecology), Poumarat (*Bulletin Semestriel de la Fédération des Associations Mycologiques Méditerranéennes* **15**: 40, 1999; Europ., key), Watling (*Mycoscience* **47**: 18, 2006).

Sclerodermataceae Corda (1842), Boletales. 6 gen. (+ 23 syn.), 39 spp.

Lit.: Dick (*TBMS* **57**: 417, 1971), Gill & Watling (*Pl. Syst. Evol.* **154**: 225, 1986), Richter (*Mycotaxon* **45**: 461, 1992), Orlovich & Ashford (*Protoplasma* **178**: 66, 1994), Pegler *et al.* (*British Puffballs, Earthstars and Stinkhorns* An Account of the British Gasteroid Fungi: 255 pp., 1995), Sims *et al.* (*Mycotaxon* **56**: 403, 1995), Bruns *et al.* (*Mol. Ecol.* **5**: 257, 1998), Junghans *et al.* (*Mycorrhiza* **7**: 243, 1998), Sims *et al.* (*MR* **103**: 449, 1999), Guzmán & Ovrebo (*Mycol.* **92**: 174, 2000), Díez *et al.* (*New Phytol.* **149**: 577, 2001), Binder & Bresinsky (*Mycol.* **94**: 85, 2002), Cairney (*New Phytol.* **153**: 199, 2002), Kasuya *et al.* (*Mycoscience* **43**: 475, 2002), Martin *et al.* (*New Phytol.* **153**: 345, 2002), Hitchcock *et al.* (*MR* **107**: 699, 2003), Kanchanaprayudh *et al.* (*Mycoscience* **44**: 287, 2003), Moyersoen *et al.* (*New Phytol.* **160**: 569, 2003), Watling & Sims (*Mem. N. Y. bot. Gdn* **89**: 93, 2004), Watling (*Mycoscience* **47**: 18, 2006).

Sclerodermatales = Boletales.

Sclerodermatopsis Torrend (1923) ? = Xylaria Hill ex Schrank fide Læssøe (*SA* **13**: 43, 1994).

Scleroderris (Fr.) Bonord. (1851) = Godronia fide Groves (*CJB* **43**: 1195, 1965).

Sclerodiscus Pat. (1890), anamorphic *Pezizomycotina*, St.0eP.?. 1, Asia.

Sclerodon P. Karst. (1889) ≡ Gloiodon.

Sclerodothiorella Died. (1912), anamorphic *Pezizomycotina*, St.0eH.?. 1, Europe.
Sclerodothis Höhn. (1918) = Leptosphaeria fide von Arx & Müller (*Stud. Mycol.* **9**, 1975).
Sclerogaster R. Hesse (1891), Sclerogastraceae. 10, Europe; America. See Dodge & Zeller (*Ann. Mo. bot. Gdn* **23**: 565, 1937), Fogel (*Mycol.* **82**: 655, 1990; N. Am. key).
Sclerogastraceae Locq. ex P.M. Kirk (2008), Geastrales. 1 gen., 10 spp.
Scleroglossum Hara (1948) = Scleromitrula fide Korf *in* Ainsworth *et al.* (Eds) (*The Fungi* **4A**: 249, 1973).
Scleroglossum Pers. (1820) = Acrospermum fide Hawksworth *et al.* (*Dictionary of the Fungi* edn 8, 1995).
Sclerogone Warcup (1990), Endogonaceae. 1, Australia. See Warcup *in* Sanders *et al.* (Eds) (*Endomycorrhizas*: 53, 1975; culture), Warcup (*MR* **94**: 173, 1990), Yao *et al.* (*Kew Bull.* **50**: 349, 1995), Yao *et al.* (*Genera of Endogonales*: 229 pp., 1996).
Sclerographa (Vain.) Zahlbr. (1923) = Opegrapha Ach. fide Upreti & Singh (*Beitr. Biol. Pfl.* **62**: 233, 1987).
Sclerographiopsis Deighton (1973), anamorphic *Pezizomycotina*, Hsy.≡ eP.3. 1, Sierra Leone. See Sutton & Pascoe (*Aust. J. Bot.* **35**: 183, 1987; cf. *Tandonella*).
Sclerographium Berk. (1854), anamorphic *Pezizomycotina*, Hsy.#eP.10. 4, Asia; Africa. See Deighton (*Mycol. Pap.* **78**, 1960).
Scleroma Fr. (1838) = Pleurotus fide Kuyper (*in litt.*) *Pleurotus* is not yet conserved against *Scleroma*.
Scleromeris Syd. (1926), anamorphic *Pezizomycotina*, St.0eP.?. 1, C. America.
Scleromitra Corda (1829) = Pistillaria Fr. fide Corner (*Ann. Bot. Mem.* [A monograph of Clavaria and allied genera] **1**, 1950).
Scleromitrula S. Imai (1941), Rutstroemiaceae. Anamorph *Myrioconium*. 6, widespread. See Kohn & Nagasawa (*TMSJ* **25**: 127, 1984), Spooner (*Biblthca Mycol.* **116**, 1987), Holst-Jensen *et al.* (*Mycol.* **89**: 885, 1997; phylogeny, as *Verpatinia*), Schumacher & Holst-Jensen (*Mycoscience* **38**: 55, 1997; phylogeny, syn. *Verpatinia*), Wang *et al.* (*Am. J. Bot.* **92**: 1565, 2005; phylogeny), Wang *et al.* (*Mol. Phylogen. Evol.* **41**: 295, 2006; phylogeny), Wang *et al.* (*Mycol.* **98**: 1065, 2006; phylogeny).
Scleromium Linds. (1869) nom. inval. ≡ Pseudographis.
Scleroparodia Petr. (1934) = Ascochytopsis fide Sutton (*Mycol. Pap.* **141**, 1977).
Scleropezicula Verkley (1999), Helotiales. Anamorph *Cryptosympodula*. 1, Canada; Europe. See Verkley (*Stud. Mycol.* **44**: 132, 1999), Verkley *et al.* (*MR* **107**: 689, 2003; phylogeny).
Sclerophoma Höhn. (1909), anamorphic *Sydowia, Xenomeris*, St.0eH.15. 30, widespread. See Hermanides-Nijhof (*Stud. Mycol.* **15**: 166, 1977; as *Dothichiza*), Müller & Hallaksela (*MR* **102**: 1190, 1998).
Sclerophomella Höhn. (1917) = Phoma Sacc. fide Sutton (*Mycol. Pap.* **141**, 1977).
Sclerophomina Höhn. (1917) = Phoma Sacc. fide Sutton (*Mycol. Pap.* **141**, 1977).
Sclerophora Chevall. (1826), Coniocybaceae (L). 6, widespread (temperate). See Tibell (*Beih. Nova Hedwigia* **79**: 597, 1984), Yao & Spooner (*Kew Bull.* **54**: 683, 1999).
Sclerophoraceae Körb. (1846) = Coniocybaceae.
Sclerophytomyces, see *Peterjamesia*.
Sclerophyton Eschw. (1824), Roccellaceae (L). 27, widespread. See Egea & Torrente (*Bryologist* **98**: 207, 1995; N. Am.), Makhija & Adawadkar (*Lichenologist* **34**: 347, 2002; India), Archer (*Mycotaxon* **87**: 85, 2003; Australasia), Sparrius (*Biblthca Lichenol.* **89**, 2004; monogr.), Sparrius *et al.* (*Lichenologist* **37**: 285, 2005; soredia), Hawksworth (*Lichenologist* **38**: 187, 2006; nomencl.).
Sclerophytonomyces Cif. & Tomas. (1953) ≡ Sclerophyton.
Sclerophytonomyces Cif. & Tomas. ex Sparrius (2004) nom. inval. = Peterjamesia fide Sparrius (*Biblthca Lichenol.* **89**: 84, 2004), Hawksworth (*Lichenologist* **38**: 187, 2006).
Scleroplea (Sacc.) Oudem. (1900) = Pyrenophora fide Anon. (*Sydowia* **5**: 248, 1951).
scleroplectenchyma, plectenchyma (q.v.) composed of very thick-walled conglutinate cells. See Yoshimura & Shimada (*Bull. Kochi Gakuen Jun. Coll.* **11**: 13, 1990); stereome.
Scleropleella Höhn. (1918) = Leptosphaerulina fide Barr (*Contr. Univ. Mich. Herb.* **9**, 1972).
scleroprosoplectenchyma, see plectenchyma.
Scleropycnis Syd. & P. Syd. (1911), anamorphic *Pezizomycotina*, St.0eH.15. 2, Europe. See Sutton & Livsey (*TBMS* **88**: 271, 1987; gen. redescription).
Scleropycnium Heald & C.E. Lewis (1912) ? = Phomopsis (Sacc.) Bubák. fide Sutton (*Mycol. Pap.* **141**, 1977).
Scleropyrenium H. Harada (1993), Verrucariaceae (L). 2, Japan. See Harada (*Nat. Hist. Res.* **2**: 131, 1993), Harada (*Lichenology* **1**: 76, 2002).
Sclerosphaeropsis Bubák (1914) nom. dub., anamorphic *Pezizomycotina*. See Sutton (*Mycol. Pap.* **141**, 1977).
Sclerosporis Stach & A. Chandra (1956), Fossil Fungi. 1 (Carboniferous, Tertiary), Europe. See Stach & Pickhardt (*Palaeontologische Zeitschrift* **31**: 140, 1957).
Sclerostagonospora Höhn. (1917), anamorphic *Pleosporales*. See Petrak (*Annls mycol.* **23**: 4, 1925), Sutton (*The Coelomycetes*, 1980), Crous & Palm (*MR* **103**: 1299, 1999).
Sclerostilbum Povah (1932) = Tilachlidiopsis fide Stalpers (*CJB* **69**: 6, 1990).
Sclerotelium Syd. (1921) = Puccinia fide Dietel (*Nat. Pflanzenfam.* **6**, 1928).
Sclerotheca Bubák & Vleugel (1908) [non *Sclerotheca* DC. 1839, *Campanulaceae*] = Camarosporium fide Sutton (*The Coelomycetes*, 1980).
sclerothionine, a plant growth promoting metabolic product of *Sclerotinia libertiana* (Matsu & Satomura, *Agr. biol. Chem.* **32**: 611, 1968).
Sclerothyrium Höhn. (1918) = Microsphaeropsis Höhn. fide Sutton (*Mycol. Pap.* **141**, 1977).
Sclerotiaceae Dumort. (1822) = Typhulaceae.
sclerotic cell, see muriform cell.
Sclerotiella A.K. Sarbhoy & A. Sarbhoy (1975), anamorphic *Pezizomycotina*, Sc.-.-. 1, India. See Sarbhoy & Sarbhoy (*Maharashtra Vidnyan Mandir, Patrika* **9**: 91, 1974).
Sclerotinia Fuckel (1870) nom. cons., Sclerotiniaceae. Anamorphs *Myrioconium, Sclerotium*. 14, widespread. *S. minor* and *S. sclerotiorum* (plurivorous), *S. trifoliorum* (clover rot). See Kohn (*Mycotaxon* **9**: 365, 1979; monogr., key, 259 epithets), Spooner

(*Biblthca Mycol.* **116**, 1987), Kohli *et al.* (*Mol. Ecol.* **4**: 69, 1995; clonality), Kohli & Kohn (*Mol. Ecol.* **5**: 773, 1996; mitochondrial haplotypes), Cubeta *et al.* (*Phytopathology* **87**: 1000, 1997; clonality), Holst-Jensen *et al.* (*Mycol.* **89**: 885, 1997; phylogeny), Willetts (*MR* **101**: 939, 1997; evolution, review), Holst-Jensen *et al.* (*Nordic Jl Bot.* **18**: 705, 1999; phylogeny), Gernandt *et al.* (*Mycol.* **93**: 915, 2001; phylogeny), Freeman *et al.* (*Eur. J. Pl. Path.* **108**: 877, 2002; molecular diagnostics), Meinhardt *et al.* (*Fitopatol. Brasil* **27**: 211, 2002; genetic diversity, Brazil), Kim & Cho (*Pl. Path. J.* **19**: 69, 2003; on *Cruciferae*), Viji *et al.* (*Pl. Dis.* **88**: 1269, 2004; on turfgrass, N America), Ekins *et al.* (*MR* **110**: 1193, 2006; homothallism), Wang *et al.* (*Mycol.* **98**: 1065, 2006; phylogeny).

Sclerotiniaceae Whetzel (1945), Helotiales. 47 gen. (+ 28 syn.), 284 spp.

Lit.: Whetzel (*Mycol.* **37**: 648, 1945; fams.), Buchwald (*Kgl. Veter. Landboh. Aarsskr.*, 1949), Dumont & Korf (*Mycol.* **63**: 157, 1971; gen. nomencl.), Schwegler (*Schweiz. Z. Pilzk.* **56**: 49, 1978; genetics), Kohn (*Mycotaxon* **9**: 365, 1979), Schumacher & Kohn (*CJB* **63**: 1610, 1985), Spooner (*Biblthca Mycol.* **116**, 1987; gen. concepts), Kohn (*TBMS* **91**: 639, 1988; protein electroph.), Kohn & Grenville (*CJB* **67**: 371, 1989; anatomy, histochem.), Kohn & Grenville (*CJB* **67**: 394, 1989; ultrastr.), Batra (*Mycol. Mem.* **16**: 246 pp., 1991), Novak & Kohn (*Appl. Environm. Microbiol.* **57**: 525, 1991; protein electrophor.), Verkley (*MR* **97**: 179, 1993; ultrastr.), Zhuang (*Acta Mycol. Sin.* **13**: 13, 1994; China), Anderson & Kohn (*Ann. Rev. Phytopath.* **33**: 369, 1995), Cubeta *et al.* (*Phytopathology* **87**: 1000, 1997), Holst-Jensen *et al.* (*Am. J. Bot.* **84**: 686, 1997), Holst-Jensen *et al.* (*Mycol.* **89**: 885, 1997; phylogeny), Schumacher & Holst-Jensen (*Mycoscience* **38**: 55, 1997), Willetts (*MR* **101**: 939, 1997; evolution, review), Anderson & Kohn (*Trends in Ecology & Evolution* **13**: 444, 1998), Fulton *et al.* (*Eur. J. Pl. Path.* **105**: 495, 1999), Giraud *et al.* (*Phytopathology* **89**: 967, 1999), Holst-Jensen *et al.* (*Nordic Jl Bot.* **18**: 705, 1998), Holst-Jensen *et al.* (*Mol. Biol. Evol.* **16**: 114, 1999; evolution), Taylor *et al.* (*Ann. Rev. Phytopath.* **37**: 197, 1999), Förster & Adaskaveg (*Phytopathology* **90**: 171, 2000), Côté *et al.* (*Mycol.* **96**: 240, 2004), Wang *et al.* (*Mycol.* **98**: 1065, 2006; phylogeny), Wang *et al.* (*Mol. Phylogen. Evol.* **41**: 295, 2006).

Sclerotiomyces Woron. (1926) ? = Sclerotium fide Stalpers (*in litt.*).

Sclerotiopsis Speg. (1882) = Pilidium fide Petrak (*Sydowia* **5**: 328, 1951), Palm (*Mycol.* **83**: 787, 1991).

Sclerotites A. Massal. (1857), Fossil Fungi. 16 (Tertiary), Europe; USA.

Sclerotites E. Jeffrey & Chrysler (1906), Fossil Fungi. 6 (Tertiary), Europe; USA.

Sclerotium Tode (1790), anamorphic *Athelia, Myriosclerotinia, Sclerotinia, Typhula*. 100, widespread. *S. cepivorum* (onion (*Allium*) white rot), *S. oryzae* (on rice), *S. rolfsii* (plurivorous; teleomorph *Athelia rolfsii*), *S. tuliparum* (grey bulb rot of tulip), *S. delphinii*, *S. microsclerotium*. A number of the earlier spp. are anamorphs of *Basidiomycota*, *Ascomycota*, etc. See Aycock (*Tech. Bull. N. Carol. agric. Exp. Stn* **174**, 1966; *S. rolfsii*), Backhouse & Stewart (*TBMS* **89**: 561, 1987; histochemistry), Hanlin & Tortolero (*CJB* **67**: 1852, 1989; morphology *S. coffeicola*), Punja & Damiani (*Mycol.* **88**: 694, 1996), Cartwright *et al.* (*Mycol.* **89**: 163, 1997; mycoparasite of *S. oryzae*), Almeida *et al.* (*J. Phytopath.* **149**: 493, 2001; genotypic diversity *S. rolfsii*), Okabe & Matsumoto (*MR* **107**: 164, 2003; phylogeny).

sclerotium (1) a firm, frequently rounded, mass of hyphae with or without the addition of host tissue or soil, normally having no spores in or on it (cf. bulbil, stroma); see Willetts (*Biol. Rev.* **46**: 387, 1971; survival, **47**: 515, 1972; morphogenesis), Willetts & Bullock (*MR* **96**: 801, 1992; developm. biology). A sclerotium may give rise to a fruit body, a stroma (as in ergot), or mycelium. See blackfellows' bread, stone fungi, tuckahoe, Anamorphic fungi. (2) (of *Mycetozoa*), the firm, resting condition of a plasmodium.

Sclerozythia Petch (1937), anamorphic *Pezizomycotina*, Cpd.0eH.?. 1, British Isles.

Sclerozythia Petr. (1955) [non *Sclerozythia* Petch 1937] ≡ Neozythia.

scobiculate, in fine grains, like sawdust.

scocho, see Fermented food and drinks.

Scodellina Gray (1821) ≡ Peziza Fr.

Scolecactis Clem. (1909) = Bactrospora fide Egea & Torrente (*Cryptog. Bryol.-Lichénol.* **14**: 329, 1993).

Scoleciasis Roum. & Fautrey (1889) nom. dub. = Leptosphaeria fide Sutton (*in litt.*), Pennycook & McKenzie (*Mycotaxon* **82**: 145, 2002).

Scoleciocarpus Berk. (1843) = Arachnion fide Demoulin (*Nova Hedwigia* **21**: 646, 1972).

scolecite, see Woronin's hypha.

Scolecites (orthographic variant), see *Skolekites*.

Scolecobasidiella M.B. Ellis (1971), anamorphic *Pezizomycotina*, Hso.1eP.10. 1, Somalia. See Ellis (*Mycol. Pap.* **125**: 12, 1971).

Scolecobasidium E.V. Abbott (1927), anamorphic *Pezizomycotina*, Hso.1eP.11. 15, widespread. See Barron & Busch (*CJB* **40**: 77, 1962; key), Graniti (*G. bot. ital.* n.s. **69**: 364, 1963; gen. concept), de Hoog & von Arx (*Kavaka* **1**: 55, 1973; monotypic), de Hoog (*Stud. Mycol.* **26**: 51, 1985; key), Horré *et al.* (*Stud. Mycol.* **43**: 194, 1999), Liu & Zhang (*Mycosystema* **25**: 386, 2006).

Scolecobasis Clem. & Shear (1931) ≡ Scolecobasidium.

Scolecobonaria Bat. (1962), Dothideomycetes. 2, Taiwan; USA. See Batista (*Beih. Sydowia* **3**: 97, 1962).

Scolecoccoidea F. Stevens (1927) = Coccodiella fide Müller & von Arx *in* Ainsworth *et al.* (Eds) (*The Fungi* **4A**: 87, 1973), Pirozynski (*Mycol.* **65**: 164, 1973).

Scolecodochium K. Matsush. & Matsush. (1996), anamorphic *Pezizomycotina*, Hso.?.?. 1, Malaysia. See Matsushima & Matsushima (*Matsush. Mycol. Mem.* **9**: 36, 1996).

Scolecodothis Theiss. & Syd. (1914) = Ophiodothella fide Hawksworth *et al.* (*Dictionary of the Fungi* edn 8, 1995).

Scolecodothopsis F. Stevens (1924) = Diatractium fide Cannon (*MR* **92**: 327, 1989).

Scoleconectria Seaver (1909) = Nectria fide Rossman *et al.* (*Stud. Mycol.* **42**: 248 pp., 1999).

Scolecopeltella Speg. (1923) = Micropeltis fide von Arx & Müller (*Stud. Mycol.* **9**, 1975).

Scolecopeltidella J.M. Mend. (1925) nom. dub., Pezizomycotina. See Petrak (*Sydowia* **5**: 190, 1951), von Arx & Müller (*Stud. Mycol.* **9**, 1975).

Scolecopeltidium F. Stevens & Manter (1925), ? Microthyriaceae. *c*. 80, widespread (tropical). See Reynolds & Gilbert (*Aust. Syst. Bot.* **18**: 265, 2005; Australia).

Scolecopeltis Speg. (1889) = Micropeltis fide von Arx & Müller (*Stud. Mycol.* **9**, 1975).

Scolecopeltium Clem. & Shear (1931) ≡ Scolecopeltidium.

Scolecopeltopsis Höhn. (1909) = Micropeltis fide von Arx & Müller (*Stud. Mycol.* **9**, 1975).

scolecospore, a spore resembling an amerospore (q.v.) but with or without septa and a length/width ratio › 15:1. See Anamorphic fungi.

Scolecosporiella Höhn. (1923) nom. nud., Fungi. 1. See Sutton (*CJB* **46**: 189, 1968; key).

Scolecosporiella Petr. (1921), anamorphic *Phaeosphaeria*, Cpd.≡ -#eP.1. 6, widespread. See Sutton (*CJB* **46**: 189, 1968; key), Sutton (*The Coelomycetes*, 1980), Nag Raj (*Coelomycetous Anamorphs with Appendage-bearing Conidia*, 1993).

Scolecosporites R.T. Lange & P.H. Sm. (1971), Fossil Fungi. 1 (Paleocene), Australia.

Scolecosporium, see *Scolicosporium*.

Scolecostigmina U. Braun (1999), anamorphic *Capnodiales*, Hso.?.?. 19, widespread. See Braun *et al.* (*N.Z. Jl Bot.* **37**: 323, 1999).

Scolecostroma Bat. & Peres (1960) = Lophium fide von Arx & Müller (*Stud. Mycol.* **9**, 1975).

Scolecotheca Søchting & B. Sutton (1997), anamorphic *Pezizomycotina*, Cpd.?.?. 1, Europe (northern). See Søchting & Sutton (*MR* **101**: 1366, 1997).

Scolecotrichum, see *Scolicotrichum*.

Scolecoxyphium Cif. & Bat. (1956), anamorphic *Capnodium*, Hsp.0eH.?. 4, widespread (tropical). See Hughes (*Mycol.* **68**: 693, 1976).

Scolecozythia Curzi (1927), anamorphic *Pezizomycotina*, St.0fH.?. 1, Italy.

Scoliciosporomyces Cif. & Tomas. (1953) = Scoliciosporum fide Hawksworth *et al.* (*Dictionary of the Fungi* edn 8, 1995).

Scoliciosporum A. Massal. (1852), Lecanorales (L). 13, widespread (temperate). See Vězda (*Folia geobot. phytotax.* **13**: 397, 1978; key), Lumbsch *et al.* (*Mol. Phylogen. Evol.* **31**: 822, 2004; phylogeny), Andersen & Ekman (*MR* **109**: 21, 2005; phylogeny), Miądlikowska *et al.* (*Mycol.* **98**: 1088, 2006; phylogeny).

Scolicosporium Lib. ex Roum. (1880), anamorphic *Asteromassaria*, Cac.≡ eP.1. 4, Europe. See Sutton (*Mycol. Pap.* **138**, 1975), Spooner & Kirk (*TBMS* **78**: 247, 1982; relationship with *Excipularia*), Constantinescu (*Mycotaxon* **42**: 467, 1991).

Scolicotrichum Kunze (1817) nom. conf., anamorphic *Pezizomycotina*. See also *Passalora*. See Höhnel (*Zbl. Bakt.* Abt. II **60**: 1, 1932), Hughes (*CJB* **36**: 727, 1958).

Scoliocarpon Nyl. (1858), Pezizomycotina. 1, N. America.

Scolionema Theiss. & Syd. (1918), Parodiopsidaceae. 1, S.E. Asia.

Scoliotidium Bat. & Cavalc. (1963), anamorphic *Pezizomycotina*, Cpt.≡ eH.?. 1, Brazil. See Batista & Cavalcanti (*Publções Inst. Micol. Recife* **395**: 11, 1963).

Scopaphoma Dearn. & House (1925), anamorphic *Pezizomycotina*, Cpd.1eH.1. 1, USA.

Scopella Mains (1939) = Maravalia fide Ono (*Mycol.* **76**: 892, 1984).

Scopellopsis T.S. Ramakr. & K. Ramakr. (1947) = Maravalia fide Thirumalachar & Mundkur (*Indian Phytopath.* **2**: 193, 1949).

Scopinella Lév. (1849), Sordariales. 7, widespread (temperate). See Cannon & Hawksworth (*J. Linn. Soc.* Bot. **84**: 115, 1982; key), Tsuneda (*Rep. Tottori mycol. Inst.* **22**: 82, 1984; ontogeny), Tsuneda *et al.* (*TMSJ* **26**: 221, 1985; SEM, posn), Stchigel *et al.* (*Mycol.* **98**: 815, 2006).

Scoptria Nitschke (1867) nom. rej. = Eutypella. Anamorph links with *Phaeoisaria* require confirmation. fide Rappaz (*Mycol. Helv.* **2**: 285, 1987).

Scopula Arnaud (1952) = Gloiosphaera fide Wang (*Mycol.* **63**: 890, 1971).

Scopularia Preuss (1851) nom. dub., Pezizomycotina. See Kendrick (*CJB* **42**: 1119, 1964).

Scopulariella Gjaerum (1971), anamorphic *Pezizomycotina*, Hsy.1eH.4. 1, Denmark. See Gjaerum (*Friesia* **9**: 416, 1971).

Scopulariopsis Bainier (1907), anamorphic *Microascus*, Hso.0eH.19. 22 (saprobic and pathogenic to animals), widespread. *S. fimicola*, white plaster mould of mushroom beds; *S. brevicaulis* has been used for the detection of arsenic and was proposed as an agent of cot death (q.v.). See Raper & Thom (*Manual of the Penicillia*, 1949), Morton & Smith (*Mycol. Pap.* **86**, 1963; descr. & illustr. 18 spp.), Abbott *et al.* (*Mycol.* **90**: 297, 1998; teleomorph), Abbott & Sigler (*Mycol.* **93**: 1211, 2001; heterothallism), Abbott *et al.* (*Mycol.* **94**: 362, 2002), Issakainen *et al.* (*Medical Mycology* **41**: 31, 2003; phylogeny).

Scopulina Lév. (1846) [non *Scopulina* Dumort. 1882, *Hepaticae*] ≡ Scopinella.

Scopulodontia Hjortstam (1998), Russulales. 2. See Hjortstam *et al.* (*Kew Bull.* **53**: 820, 1998), Nakasone (*Cryptog. Mycol.* **24**: 131, 2003).

Scopuloides (Massee) Höhn. & Litsch. (1908), Meruliaceae. 5, widespread. See Boidin *et al.* (*Cryptog. Mycol.* **14**: 195, 1993; key Eur. spp.).

Scoriadopsis J.M. Mend. (1930), ? Capnodiaceae. 1, S. America.

Scorias Fr. (1825), Capnodiaceae. Anamorphs *Conidiocarpus*, *Scolecoxyphium*. 3, widespread (tropical). See Reynolds (*Mycotaxon* **8**: 417, 1979), Reynolds (*Mycotaxon* **27**: 377, 1986), Reynolds (*CJB* **76**: 2125, 1998; phylogeny), Schoch *et al.* (*Mycol.* **98**: 1041, 2006; phylogeny).

scorpioid, a branching system in which the laterals are curved so that they all appear to arise from one side of the main stem, as in *Cladonia arbuscula*.

Scorpiosporium S.H. Iqbal (1974), anamorphic *Pezizomycotina*, Hso.0bP.1. 4, British Isles. See Iqbal (*Biologia* Lahore **20**: 17, 1974).

Scortechinia Sacc. (1885), Scortechiniaceae. 1 (from wood etc.), widespread. See Nannfeldt (*Svensk bot. Tidskr.* **69**: 49, 1975), Nannfeldt (*Svensk bot. Tidskr.* **69**: 289, 1975), Subramanian & Sekar (*Kavaka* **18**: 19, 1993), Huhndorf *et al.* (*MR* **108**: 1384, 2004; phylogeny), Schoch *et al.* (*MR* **111**: 154, 2007; phylogeny).

Scortechinia Sacc. (1891) = Scortechinia Sacc. (1885). Typified differently from *Scortechinia* Sacc. (1885), but the two types are conspecific. fide Nannfeldt (*Svensk bot. Tidskr.* **69**: 49, 289, 1975).

Scortechiniaceae Huhndorf, A.N. Mill. & F.A. Fernández (2004), Coronophorales. 3 gen., 3 spp. See Huhndorf *et al.* (*MR* **108**: 1384, 2004; phylogeny), Schoch *et al.* (*MR* **111**: 154, 2007; phylogeny).

Scortechiniella Arx & E. Müll. (1954) = Nitschkia fide Nannfeldt (*Svensk bot. Tidskr.* **66**: 49, 1975).

Scortechiniellopsis Sivan. (1974) = Nitschkia fide Nannfeldt (*Svensk bot. Tidskr.* **66**: 49, 1975).

Scorteus Earle (1909) = Marasmius fide Pennington (*North Amer. Flora* **9**: 250, 1915).

Scothelius Bat., J.L. Bezerra & Cavalc. (1965), anamorphic *Pezizomycotina*, Cpt.0eH.?. 1, Brazil. See Batista *et al.* (*Atas Inst. Micol. Univ. Pernambuco* **2**: 302, 1965).

Scotiosphaeria Sivan. (1977), Pezizomycotina. 1, British Isles. See Sivanesan (*TBMS* **69**: 119, 1977).

Scotoderma Jülich (1974), Stereaceae. 1, British Isles. See Jülich (*Proc. K. Ned. Akad. Wet.* Ser. C, Biol. Med. Sci. **77**: 149, 1974).

Scotomyces Jülich (1978), Ceratobasidiaceae. 1, widespread. See Roberts (*Rhizoctonia-forming fungi*, 1999), Wojewoda (*Acta Mycologica* Warszawa **38**: 3, 2003; Poland).

screening, routine testing of organisms or chemical substances for a particular property (e.g. for antibiotic production, fungicidal effect, etc.). See Dreyfuss (*Sydowia* **39**: 22, 1987; pharmaceutical prospecting), Bioprospecting.

scrobiculate (1) roughened; resembling sawdust; (2) (of lichens), coarsely pitted; foveolate.

scrupose, rough with very small hard points.

Sculptolumina Marbach (2000), Caliciaceae (L). 2, pantropical. See Marbach (*Biblthca Lichenol.* **74**: 296, 2000).

scutate, like a round plate or shield.

Scutellaria Baumg. (1790) nom. dub., Lecanorales.

Scutelliformis Salazar-Yepes, Pardo-Card. & Buriticá (2007), Phragmidiaceae. Anamorph *Gerwasia*. 1, S. America. See Salazar Yepes, M.; Pardo Cardona, V.M.; Buriticá Céspedes, P. (*Caldasia* **29**: 105, 2007).

Scutellinia (Cooke) Lambotte (1887) nom. cons., Pyronemataceae. 66, widespread (esp. north temperate). See Kaushal *et al.* (*Biblthca Mycol.* **91**: 583, 1983; key 14 spp. India), Schumacher (*Mycotaxon* **33**: 149, 1988; 144 excl. names), Schumacher (*Op. Bot.* **101**: 101, 1990; monogr.), Wu & Kimbrough (*Bot. Gaz.* **152**: 421, 1992; ascospores), van Brummelen (*Persoonia* **15**: 129, 1993; ultrastr.), Landvik *et al.* (*Nordic Jl Bot.* **17**: 403, 1997; DNA), Paal *et al.* (*Persoonia* **16**: 491, 1998; Netherlands), Wang (*Bull. natn. Mus. Nat. Sci.* Taiwan **11**: 119, 1998; Taiwan), Matocec (*Mycotaxon* **76**: 481, 2000; Croatia), Hansen & Pfister (*Mycol.* **98**: 1029, 2006; phylogeny), Perry *et al.* (*MR* **111**: 549, 2007; phylogeny).

Scutelloidea Tim (1971) = Rosenscheldiella fide von Arx & Müller (*Stud. Mycol.* **9**, 1975).

Scutellospora C. Walker & F.E. Sanders (1986), Gigasporaceae. 32, widespread. See Koske & Walker (*Mycol.* **77**: 702, 1985), Koske & Walker (*Mycotaxon* **27**: 219, 1986), Walker & Saunders (*Mycotaxon* **27**: 169, 1986), Morton & Koske (*Mycol.* **80**: 520, 1988), Koske & Halvorson (*Mycol.* **81**: 927, 1989), Spain *et al.* (*Mycotaxon* **35**: 219, 1989), Błaszkowski (*Mycol.* **83**: 537, 1991), Walker *et al.* (*Cryptog. Mycol.* **14**: 279, 1993), Franke & Morton (*CJB* **72**: 122, 1994; revision, phylogeny), Koske & Gemma (*Mycol.* **87**: 678, 1995), Morton (*Mycol.* **87**: 127, 1995; development, phylogeny), Thingstrup *et al.* (*MR* **99**: 1225, 1995; detection), Yao *et al.* (*Genera of Endogonales*: 229 pp., 1996), Walker *et al.* (*Ann. Bot.* **82**: 721, 1998), Bever & Morton (*Am. J. Bot.* **86**: 1209, 1999; genetic variation), Redecker *et al.* (*Fungal Genetics Biol.* **28**: 238, 1999; phylogeny), Stürmer & Morton (*MR* **103**: 949, 1999), Souza *et al.* (*MR* **109**: 697, 2005; morphological, ontogenetic and molecular characterization of *Scutellospora reticulata*), Dotzler *et al.* (*Mycol. Progr.* **5**: 178, 2006; germination shields in *Scutellospora* from the 400 million-year-old Rhynie chert), Redecker & Raab (*Mycol.* **98**: 885, 2006; phylogeny), Silva *et al.* (*Mycotaxon* **94**: 293, 2005; key).

Scutellosporites Dotzler, M. Krings, T.N. Taylor & Agerer (2006), Fossil Fungi. 1.

Scutellum Speg. (1881) nom. dub., Fungi. See Müller & von Arx (*Beitr. Kryptfl. Schweiz* **11** no. 2, 1962).

scutellum, see thyriothecium.

Scutiger Paulet (1793) nom. inval., Albatrellaceae. 1. See Donk (*Verh. K. ned. Akad. Wet.* tweede sect. **62**: 1, 1974).

Scutigeraceae Bondartsev & Singer (1941) = Albatrellaceae.

Scutisporium Preuss (1851) = Stemphylium fide Saccardo (*Syll. fung.* **4**: 519, 1886).

Scutisporus K. Ando & Tubaki (1985), anamorphic *Pezizomycotina*, Hso.#eH.1/10. 1, Japan. See Ando & Tubaki (*TMSJ* **26**: 153, 1985).

Scutobelonium Graddon (1984), Helotiales. 1, British Isles. See Graddon (*TBMS* **83**: 377, 1984), Nauta & Spooner (*Mycologist* **14**: 65, 2000).

Scutomollisia Nannf. (1976), Helotiales. *c.* 14, widespread (northern & central Europe). See Graddon (*TBMS* **74**: 265, 1980), Hein & Scheuer (*Sydowia* **38**: 125, 1986), Nauta & Spooner (*Mycologist* **14**: 65, 2000).

Scutomyces J.L. Bezerra & Cavalc. (1972) = Microtheliopsis fide Lücking *et al.* (*Lichenologist* **30**: 121, 1998).

Scutopeltis Bat. & H. Maia (1957), anamorphic *Pezizomycotina*, Cpt.0eH.?. 1, Ghana. See Batista & Maia (*Revta Biol.* Lisb. **1**: 123, 1957).

Scutopycnis Bat. (1957), anamorphic *Pezizomycotina*, Cpt.0eH.?. 1, USA. See Batista (*An. Soc. Biol. Pernambuco* **15**: 422, 1957).

Scutoscypha Graddon (1980) = Calycellina fide Baral & Krieglsteiner (*Beih. Sydowia* **6**, 1985), Raitviir (*Scripta Mycol.* **20**, 2004).

Scutula Tul. (1852) nom. cons., Ramalinaceae (±L). 13 (in lichens), widespread. See Hawksworth (*TBMS* **74**: 363, 1980), Rambold & Triebel (*Biblthca Lichenol.* **48**, 1992), Jørgensen & Santesson (*Taxon* **42**: 881, 1993; nomencl.), Triebel *et al.* (*Symb. bot. upsal.* **32**: 323, 1997), Andersen & Ekman (*MR* **109**: 21, 2005; phylogeny), Wedin *et al.* (*Lichenologist* **39**: 329, 2007; N Hemisphere, key).

Scutularia P. Karst. (1885) = Xylogramma fide Nannfeldt (*Nova Acta R. Soc. Scient. upsal.*, 1932).

Scutulopsis Velen. (1934), ? Helotiales. 1, former Czechoslovakia.

scutulum (pl. **-a**), cup-like crust or mat of hyphae produced in the follicles of the scalp or the body in infections by *Trichophyton schoenleinii*.

Scypharia (Quél.) Quél. (1886) [non *Scypharia* Miers 1860, *Rhamnaceae*] = Sarcoscypha fide Denison (*Mycol.* **64**: 616, 1972).

Scyphiphorus Vent. (1799) ≡ Pyxidium Hill.

scyphoid, cup-like.

Scyphophora Gray (1821) ≡ Pyxidium Hill.

Scyphophorus Ach. ex Michx. (1803) ≡ Pyxidium Hill.

Scyphopilus P. Karst. (1881) = Thelephora fide Killermann *in* Engler & Prantl (*Naturl. Pflanzenfam.* **2**, 1928).

Scyphorus Raf. (1814) ≡ Pyxidium Hill.

Scyphorus Raf. (1815) ≡ Scyphophora.

Scyphospora L.A. Kantsch. (1928), anamorphic *Apiospora*, Cac.0eP.1. 1, widespread. See Sivanesan (*TBMS* **81**: 313, 1983; teleomorph).

Scyphostroma Starbäck (1899) = Subicularium fide Eriksson (*SA* **8**: 77, 1989).

scyphus, a cup-like apex of a lichenized podetium, as in *Cladonia fimbriata*.

Scytalidium Pesante (1957), anamorphic *Helotiales*, Hso.0-1eP.38. 18, widespread. A polyphyletic genus; see also *Neofusicoccum*. See Sigler & Carmichael (*Mycotaxon* **4**: 395, 1976; key 6 spp.), Sigler & Wang (*Mycol.* **82**: 399, 1990), Straatsma & Samson (*MR* **97**: 321, 1993; taxonomy of *S. thermophilum*), Marriott *et al.* (*J. Clin. Microbiol.* **35**: 2949, 1997; from human), Roeijmans *et al.* (*J. Med. Vet. Mycol.* **35**: 181, 1997; DNA), Currah *et al.* (*Karstenia* **39**: 65, 1999; from roots), Lacaz *et al.* (*Revista do Instituto de Medicina Tropical de São Paulo* **41**: 319, 1999; synanamorphs, onychomycosis), Lyons *et al.* (*MR* **104**: 1431, 2000; molecular variation), Machouart-Dubach *et al.* (*FEMS Microbiol. Lett.* **208**: 187, 2001; phylogeny), Dunn *et al.* (*J. Clin. Microbiol.* **41**: 5817, 2003; from human), Machouart *et al.* (*FEMS Microbiol. Lett.* **238**: 455, 2004; phylogeny), Hambleton & Sigler (*Stud. Mycol.* **53**: 1, 2005; phylogeny), Crous *et al.* (*Stud. Mycol.* **55**: 99, 2006; phylogeny).

Scytenium Gray (1821) = Leptogium fide Hawksworth *et al.* (*Dictionary of the Fungi* edn 8, 1995).

Scytinopogon Singer (1945), Clavariaceae. 4, Europe. See Zhang & Yang (*Mycosystema* **22**: 663, 2003), García-Sandoval *et al.* (*Mycotaxon* **89**: 185, 2004; systematics), García-Sandoval *et al.* (*Mycotaxon* **94**: 265, 2005).

Scytinopogonaceae Jülich (1982) = Clavariaceae.

Scytinostroma Donk (1956), Lachnocladiaceae. 32, widespread. See Boidin & Lanquetin (*Biblthca Mycol.* **114**, 1987; key), Stalpers (*Stud. Mycol.* **40**: 103, 1996; key), Larsson & Larsson (*Mycol.* **95**: 1037, 2003; phylogeny).

Scytinostromella Parmasto (1968), Russulales. 5, widespread. See Freeman & Petersen (*Mycol.* **71**: 85, 1979; key), Ginns & Freeman (*Biblthca Mycol.* **157**, 1994; N. Am. spp.), Spirin (*Novosti Sistematiki Nizshikh Nov. sist. Niz. Rast.* **36**: 66, 2002; Russia).

Scytinotopsis Singer (1936) nom. nud. = Resupinatus fide Singer (*Agaric. mod. Tax.*, 1951).

Scytinotus P. Karst. (1879) = Panellus fide Singer (*Agaric. mod. Tax.*, 1951).

Scytonema C. Agardh ex Bornet & Flahault (1887), Algae. Algae.

Scytopezis Clem. (1903) nom. dub., Fungi. See Eckblad (*Nytt Mag. Bot.* **15**: 1, 1968).

SEA, see Media.

Searchomyces B.S. Mehrotra & M.D. Mehrotra (1963) = Amblyosporium fide Pirozynski (*CJB* **47**: 325, 1969).

Seaver (Fred Jay; 1877-1970; USA). Student, Morningside College, Sioux City, Iowa (1902); PhD, Iowa State University (1912); Director of Laboratories (1908-1911) then Curator (1911-1943) then Head Curator (1943-1948), New York Botanic Garden. Associate Editor (1909-1924) then Editor (1925-1932) then Editor-in-Chief (1935-1947) of *Mycol.*; author of 125 papers, mainly on discomycete systematics; surveyed mycotas of the West Indies with Chardón (q.v.) and Whetzel (q.v.). *Publs.* Hypocreales. *North American Flora* (1910); *North American Cup Fungi (Operculates)* (1928) [suppl. edn, 1942]; *North American Cup Fungi (Inoperculates)* (1951). *Biogs, obits etc.* Rogerson (*Mycol.* **65**: 721, 1973) [portrait]; Stafleu & Cowan (*TL-2* **5**: 463, 1985).

Seaverinia Whetzel (1945), Sclerotiniaceae. Anamorph *Verrucobotrys*. 1, N. America.

Sebacina Tul. & C. Tul. (1871), Sebacinaceae. Anamorph *Opadorhiza*. 9, widespread. See Glen *et al.* (*Mycorrhiza* **12**: 243, 2002; mycorhizal with *Eucalyptus*), Selosse *et al.* (*Mol. Ecol.* **11**: 1831, 2002; ectomycorrhizae), Roberts (*Sydowia* **55**: 348, 2003; N. America), Weiss *et al.* (*MR* **108**: 1003, 2004; mycorrhizae).

Sebacinaceae K. Wells & Oberw. (1982), Sebacinales. 8 gen. (+ 5 syn.), 29 spp.

Lit.: Stalpers & Andersen *in* Sneh *et al.* (Eds) (*Rhizoctonia Species* Taxonomy, Molecular Biology, Ecology, Pathology and Disease Control: 58, 1996), Weiss & Oberwinkler (*MR* **105**: 403, 2001), Selosse *et al.* (*New Phytol.* **155**: 183, 2002), Roberts (*Sydowia* **55**: 348, 2003), McCormick *et al.* (*New Phytol.* **163**: 425, 2004), Walker & Parrent (*Mem. N. Y. bot. Gdn* **89**: 291, 2004), Weiss *et al.* (*MR* **108**: 1003, 2004).

Sebacinales M. Weiss, Selosse, Rexer, A. Urb. & Oberw. (2004). Agaricomycetes. 1 fam., 9 gen., 30 spp. Fam.:

Sebacinaceae

For *Lit.* see under fam.

Sebacinella Hauerslev (1977) = Oliveonia fide Roberts (*Rhizoctonia-forming fungi*, 1999).

seceding (1) (of lamellae), at first joined to the stipe (adnate), then free; separating from the stipe; (2) (of conidia), at first attached to the conidiogenous cell, then separating by schizolysis or rhexolysis; secession.

Secoliga Norman (1853) = Gyalecta Ach. fide Hafellner (*Beih. Nova Hedwigia* **79**: 241, 1984).

Secoligella (Müll. Arg.) Vain. (1896) = Aspidothelium fide Hawksworth *et al.* (*Dictionary of the Fungi* edn 8, 1995).

secondary metabolite, a compound, generally produced in a phase subsequent to growth, which is not an essential intermediary of the central metabolism, often with an unusual chemical structure and found as a mixture of closely related chemicals (see *Secondary metabolites: their structure and function* [*CIBA Foundation Symp.* **171**], 1992); e.g. antibiotics, Metabolic products, mycotoxins. See Stadler & Keller (Eds) (Fungal secondary metabolite research. *MR* **112**, 2008). See also extrolite, metabolic products.

secondary mycelium (of *Agaricomycotina*), the dikaryotic mycelium resulting from plasmogamy in the primary mycelium (q.v.); mycelium developed from the base of a basidioma (Corner).

secondary spores (of basidiomycetes), spores other than basidiospores.

Secotiaceae Tul. (1845) = Agaricaceae.

secotioid (of *Agaricomycotina*), the margin of the pileus does not break free from the stipe (or if it does the pileus never fully expands), lamellae convoluted and anastomosed, basidiospores not ballistosporic; phylogeny of a **- syndrome**, see Thiers (*Mycol.* **76**: 1,

1984).

Secotium Kunze (1840), Agaricaceae. *c.* 10, widespread (esp. warm and arid areas). See Singer & Smith (*Madroño* **15**: 152, 1960).

sectoring, 'mutation' or selection in plate cultures resulting in one or more sectors of the culture having a changed form of growth.

secund, having parts directed to one side only; cf. scorpioid.

Sedecula Zeller (1941), ? Agaricales. 1, USA. See Jülich (*Biblthca Mycol.* **85**, 1992).

Seed-borne fungi. Are frequently important in connexion with the transmission of plant diseases, esp. into new areas. Examples include are: *Ascochyta pisi*, *Colletotrichum lindemuthianum*, *Microdochium panattonianum*, *Gloeotinea temulenta*, *Phoma betae*, *P. lingam*, *Septoria apiicola*, *Sphaerella linorum*, *Uromyces betae* and *Ustilago nuda* in the seed (controlled by the use of hot water sterilization or sometimes by seed disinfectants); *Polyspora lini*, *Puccinia antirrhini*, *Tilletia tritici*, *Urocystis agropyri*, *Ustilago avenae*, *U. hordei*, and other smuts as spores on the surface of the seed (controlled by the use of seed disinfectants); *Claviceps purpurea*, *Sclerotinia trifoliorum* and *Sclerotium rolfsii* as sclerotia with the seed.

Lit.: Agarwal & Sinclair (Eds) (*Principles of seed pathology*, 2 vols, 1987-9), Agarwal *et al.* (*Phytopath. Pap.* **30**, 1989; testing of rice), Doyer (*Manual of seed-borne diseases*, 1938), Dykstra (*Bot. Rev.* **27**: 445, 1961; production resistant seed), 1991), Jeffs (Ed.) (*Seed treatment*, 1986), Malone & Muskett (*Proc. Internat. Seed Testing Assn* **29** (2), 1964; descr. 77 spp.), Martin (Ed.) (*Seed treatment: progress and prospects. British Crop Protection Council Monograph* **57**, 1994), McGee (*Maize diseases: a reference source for seed technologists*, 2nd edn, 1990), McGee (*Soybean diseases: a reference source for seed technologists*, 1992), McGee (*Plant pathogens and the worldwide movement of seeds*, 1997), IRRI (*Rice seed health*, 1988), Neergaard (*Seed pathology*, 2 vols, 1977; monogr.), Orton (*Bull. Virg. agric. Exp. Stn* **245**, 1931; bibliogr.), Singh & Mathur (*Histopathology of seed-borne infections*, 2004), Singh *et al.* (*An illustrated manual on identification of some seed-borne aspergilli, fusaria, penicillia and their mycotoxins*, Watanabe (*Pictorial atlas of soil and seed fungi: morphologies of cultured fungi*, 2002).

Segestrella Fr. (1831) ≡ Segestria.

Segestria Fr. (1825) = Porina Müll. Arg. fide Hawksworth *et al.* (*Dictionary of the Fungi* edn 8, 1995), McCarthy & Malcolm (*Lichenologist* **29**: 1, 1997).

segment (of a dictyospore), a part of the spore cut off by an A-trans-septum (Eriksson, *Opera bot.* **60**, 1981); see septum.

segregate (in taxonomy), a taxon based on part of an earlier taxon.

Seifertia Partr. & Morgan-Jones (2002), anamorphic *Pezizomycotina*, H?.?.?. 1, widespread. *S. azaleae* (*Rhododendron* leaf and bud scorch). See Kaneko *et al.* (*Ann. phytopath. Soc. Japan* **54**: 323, 1988; as *Pycnostysanus*), Partridge & Morgan-Jones (*Mycotaxon*, 2002).

Seimatoantlerium Strobel, E. Ford, J. Yi Li, J. Sears, Sidhu & W.M. Hess (1999) = Pestalotiopsis fide Bashyal *et al.* (*Mycotaxon* **72**: 33, 1999), Strobel *et al.* (*Syst. Appl. Microbiol.* **22**: 432, 1999).

Seimatosporiella Abbas, B. Sutton & Ghaffar (1998) nom. inval., anamorphic *Pezizomycotina*, Hso.?.?. 1, Pakistan. See Abbas *et al.* (*Pakist. J. Bot.* **30**: 271, 1998).

Seimatosporiopsis B. Sutton, Ghaffer & Abbas (1972), anamorphic *Pezizomycotina*, Cpd.≡ eP.19. 1, Pakistan. See Sutton *et al.* (*TBMS* **59**: 295, 1972), Abbas *et al.* (*Pakist. J. Bot.* **34**: 449, 2002).

Seimatosporium Corda (1833), anamorphic *Discostroma*, Cac/Ccu.≡ eP.19. 47, widespread. See Sutton (*The Coelomycetes*, 1980), Kang *et al.* (*MR* **103**: 53, 1999), Jeewon *et al.* (*Mol. Phylogen. Evol.* **25**: 378, 2002; phylogeny), Jeewon *et al.* (*MR* **107**: 1392, 2003; phylogeny), Hatakeyama & Harada (*Mycoscience* **45**: 106, 2004), Zimowska (*Phytopathologia Polonica* **34**: 41, 2004).

seiospore, a dry dispersal spore.

Seiridiella P. Karst. (1890) ? = Phragmotrichum fide Sutton & Pirozynski (*TBMS* **48**: 363, 1965).

Seiridina Höhn. (1930) = Seimatosporium fide Sutton (*Mycol. Pap.* **97**, 1964).

Seiridium Nees (1816), anamorphic *Lepteutypa, Blogiascospora*, Cac.≡ eP.19. 30, widespread. See Nag Raj & Kendrick (*Sydowia* **38**: 179, 1986), Chou (*Eur. J. For. Path.* **19**: 435, 1989; spp. on *Cupressus*), Nag Raj (*Coelomycetous Anamorphs with Appendage-bearing Conidia*, 1993), Viljoen *et al.* (*Exp. Mycol.* **17**: 323, 1993; spp. on cypress compared by sequence data), Nag Raj & Mel'nik (*Mycotaxon* **50**: 435, 1994), Barnes *et al.* (*Pl. Dis.* **85**: 317, 2001; on *Cupressus*), Jeewon *et al.* (*Mol. Phylogen. Evol.* **25**: 378, 2002; phylogeny), Jeewon *et al.* (*MR* **107**: 1392, 2003; phylogeny).

Seirophora Poelt (1983), Teloschistaceae (L). 11, north temperate. See Poelt (*Flora* Jena **174**: 440, 1983), Frödén & Lassen (*Lichenologist* **36**: 289, 2004; generic limits), Frödén & Litterski (*Graphis Scripta* **17**: 22, 2005).

Seismosarca Cooke (1889) = Auricularia fide Donk (*Persoonia* **4**: 158, 1966).

Selenaspora R. Heim & Le Gal (1936), Sarcosomataceae. 1, Europe; N. America. See Weber (*Mycol.* **87**: 90, 1, 1995; posn).

Selenodriella R.F. Castañeda & W.B. Kendr. (1990), anamorphic *Pezizomycotina*, Hso.0eH.?12. 5, Cuba; N. America. See Crous *et al.* (*S. Afr. J. Bot.* **62**: 89, 1996), Castañeda Ruiz *et al.* (*Mycotaxon* **85**: 211, 2003).

Selenophoma Maire (1907), anamorphic *Discosphaerina*, Cpd.0eH/1eP.16. 12, widespread (temperate). See Sutton (*The Coelomycetes*, 1980), Yurlova *et al.* (*Stud. Mycol.* **43**: 63, 1999).

Selenophomites Babajan & Tasl. (1973), Fossil Fungi. 1 (Tertiary), former USSR.

Selenophomopsis Petr. (1924) = Selenophoma fide von Arx (*Verh. K. ned. Akad. Wet. Amst.* C **51**: 1, 1957).

Selenosira Petr. (1957), anamorphic *Pezizomycotina*, Cpd.0eH.?. 1, Pakistan. See Petrak (*Sydowia* **10**: 278, 1957).

Selenosporella G. Arnaud ex MacGarvie (1969), anamorphic *Sordariomycetes*, Hso.0eH.?16. 5, widespread. Often synanamorphs of dematiaceous anamorphic fungi. See Sutton & Hodges (*Nova Hedwigia* **28**: 487, 1977), Sutton & Hodges (*Nova Hedwigia* **29**: 602, 1978), Hughes (*N.Z. Jl Bot.* **17**: 139, 1979), Castañeda-Ruíz & Guarro (*CJB* **76**: 1584, 1998), Wang & Sutton (*CJB* **76**: 1608, 1998), Miller & Huhndorf (*MR* **108**: 26, 2004).

Selenosporium Corda (1837) = Fusarium fide Holubová-Jechová *et al.* (*Sydowia* **46**: 247, 1994; revision).

Selenosporopsis R.F. Castañeda & W.B. Kendr. (1991), anamorphic *Pezizomycotina*, Hso.0fH.16. 1, Cuba. See Castañeda Ruiz & Kendrick (*Univ. Waterloo Biol. Ser.* **35**: 100, 1991).

Selenotila Lagerh. (1892), anamorphic *Saccharomycetes*, Hso.0eH.?. 2, America; Europe. See Yarrow (*Antonie van Leeuwenhoek* **35**: 418, 1969; *S. nivalis* in snow), Yarrow (*Stud. Mycol.* **14**: 29, 1977; nom. dub.).

Selenozyma Yarrow (1977), anamorphic *Saccharomycetes*, Hso.0eH.4. 2 (from humans and bovines), Netherlands. See Yarrow (*Stud. Mycol.* **14**: 29, 1977).

Selinia P. Karst. (1876), Bionectriaceae. Anamorph *Tubercularia*-like. 2 (coprophilous), widespread. See Udagawa (*TMSJ* **21**: 283, 1980), Khan & Krug (*Mycol.* **81**: 653, 1989), Ranalli & Mercuri (*Mycotaxon* **53**: 109, 1995; anamorph), Rossman *et al.* (*Stud. Mycol.* **42**: 248 pp., 1999), Rossman *et al.* (*Mycol.* **93**: 100, 2001; phylogeny).

Seliniana Kuntze (1891) ≡ Selinia fide Rossman *et al.* (*Stud. Mycol.* **42**: 248 pp., 1999).

Seliniella Arx & E. Müll. (1955) = Ascobolus fide Müller & von Arx (*Beitr. Kryptfl. Schweiz* **11** no. 2, 1962).

SEM, scanning electron microscopy (q.v.).

Semenovaia Voronin & I.V. Issi (1986), Microsporidia. 1.

semi- (prefix), half. Cf. hemi-.

Semidelitschia Cain & Luck-Allen (1969), Delitschiaceae. 2 (coprophilous), widespread. See Cain & Luck-Allen (*Mycol.* **61**: 580, 1969), Bell & Mahoney (*Muelleria* **15**: 3, 2001; Australia).

Semifissispora H.J. Swart (1982), Dothideomycetes. 3 (on *Eucalyptus*), Australia. See Swart (*TBMS* **78**: 259, 1982).

semifissitunicate, see ascus.

Semigyalecta Vain. (1921), Gyalectaceae (L). 1, S.E. Asia. See Santesson (*Symb. bot. upsal.* **12** no. 1: 1, 1952), Aptroot *et al.* (*Mycotaxon* **88**: 41, 2003).

semigymnoplast, see protoplast.

Semilecanora Motyka (1996) nom. inval. = Lecanora fide Hawksworth *et al.* (*Dictionary of the Fungi* edn 8, 1995).

semimacronematous, only slightly different from vegetative hyphae; frequently ascending, rarely erect. See macronematous.

Semiomphalina Redhead (1984), ? Hygrophoraceae (L). 1, Papua New Guinea. See Redhead (*CJB* **62**: 886, 1984), Redhead *et al.* (*Mycotaxon* **83**: 19, 2002; likely sister taxon to *Lichenomphalia*).

Semipseudocercospora J.M. Yen (1983), anamorphic *Mycosphaerellaceae*. 2. See Crous & Braun (*CBS Diversity Ser.* **1**: 571 pp., 2003).

Semisphaeria K. Holm & L. Holm (1991), Dothideomycetes. 1, Norway; Sweden. See Holm & Holm (*Nordic Jl Bot.* **11**: 686, 1991).

Senegalosporites S. Jardiné & Magloire (1965), Fossil Fungi. 2 (Cretaceous, Miocene), Senegal; Taiwan.

senescence (of fungi), the degeneration which makes indefinite propagation of certain fungi in culture impossible. See Holliday (*Nature* **221**: 1224, 1969; *Neurospora*, *Podospora*).

Senoma A.V. Simakova, T.F. Pankova,Y.S. Tokarev & I.V. Issi (2005), Microsporidia. 1. See Simakova *et al.* (*Protistology* **4**: 143, 2005).

sensitive, reacting with severe symptoms to the attack of a given pathogen (*TBMS* **33**: 155, 1950); **sensitivity**, the tendency of an organism attacked by a disease to give more or less strong symptoms; sensibility (Wilbrink).

sensu lato (s.l.), in a broad sense; **- stricto** (s.str.), in a narrow sense.

separating, see seceding; **- cell**, a cell between a conidium and a conidiogenous cell, involved in rhexolytic secession, q.v.

Sepedoniaceae Fr. (1832) = Hypocreaceae.

Sepedonium Link (1809), anamorphic *Hypomyces*, Hso.0eH.1. 14, widespread (temperate). See Rudakov (*Mik. Grib. Biol. Prakt. Znach.*, 1981; key), Sahr *et al.* (*Mycol.* **91**: 935, 1999), Ng *et al.* (*Clinical Microbiology Newsletter* **25**: 20, 2003; from human; 108889; host-parasite reln).

septal pore apparatus, see dolipore septum (Fig. 13) (- - **swelling**), parenthesome (- - **cap**); **- plug**, an occlusion of a septal pore.

Septaria Fr. (1819) nom. rej. ≡ Septoria Sacc.

Septata A. Cali, D.P. Kotler & J.M. Orenstein (1993), Microsporidia. 1. See Cali *et al.* (*J. Eukary. Microbiol.* **40**: 111, 1993).

septate, having septa.

Septatium Velen. (1934) = Hymenoscyphus fide Dennis (*British Ascomycetes*, 1978).

Septobasidiaceae Racib. (1909), Septobasidiales. 7 gen. (+ 6 syn.), 179 spp.

Lit.: Bandoni (*Stud. Mycol.* **38**: 13, 1995), Begerow *et al.* (*CJB* **75**: 2045, 1998), Gómez & Kisimova-Horovitz (*Mycotaxon* **80**: 255, 2001), Gómez & Henk (*Lankesteriana* **4**: 75, 2004), Wingfield *et al.* (*Australas. Pl. Path.* **33**: 327, 2004), Henk (*Mycol.* **97**: 908, 2005).

Septobasidiales Couch ex Donk (1964). Pucciniomycetes. 1 fam., 7 gen., 179 spp. Basidioma not gelatinous; complex relationship with insect hosts. Fam.:
Septobasidiaceae

Lit.: Donk (1951-63, VIII, *Taxon* **7**: 164, 193, 236, 1958, **12**: 165, 1963).

Septobasidium Pat. (1892) nom. cons., Septobasidiaceae. Anamorph *Johncouchia*. 170 (symbiotic with scale insects), widespread (esp. warmer areas). See Couch (*The genus Septobasidium*, 1938), Dykstra (*CJB* **52**: 971, 1974; ultrastr.), Gómez & Henk (*Lankesteriana* **4**: 75, 2004; nomencl.).

Septocarpus Zopf (1888) = Podochytrium fide Fitzpatrick (*The lower fungi. Phycomycetes*, 1930).

Septochora Höhn. (1917) = Helhonia fide Sutton (*SA* **6**: 151, 1987).

Septochytrium Berdan (1939), Cladochytriaceae. 4, N. America.

Septocladia Coker & F.A. Grant (1922) = Allomyces fide Coker (*The Saprolegniaceae with Notes on Other Water Molds*, 1923).

Septocolla Bonord. (1851) = Dacrymyces fide Saccardo (*Syll. fung.* **6**: 798, 1888).

Septocylindriaceae Nann. (1934) = Mycosphaerellaceae.

Septocylindrium Bonord. ex Sacc. (1880) = Ramularia Unger fide Braun (*Monogr. Cercosporella, Ramularia Allied Genera (Phytopath. Hyphom.)* **2**: 11, 1998).

Septocyta Petr. (1927), anamorphic *Pezizomycotina*, St.0fH.10. 3, widespread. See Priest (*Fungi of Australia: Septoria* **0**, 2006).

Septocytella Syd. (1929), anamorphic *Pezizomycotina*,

St.0fH.?. 1, China.

Septodochium Matsush. (1971), anamorphic *Pezizomycotina*, Hsp.≡ eH.15. 1, Solomon Islands. See Matsushima (*Microfungi of the Solomon Islands and Papua-New Guinea*: 54, 1971).

Septodothideopsis Henn. (1904) nom. dub., anamorphic *Pezizomycotina*. See Höhnel (*Sber. Akad. Wiss. Wien* Math.-naturw. Kl., Abt. 1 **119**: 877, 1910).

Septofusidium W. Gams (1971), anamorphic *Nectriaceae*, Hso.1eH.?33. 5, widespread. See Gams (*Cephalosporium-artige Schimmelpilze*: 147, 1971), Luangsa-ard *et al.* (*Mycol.* **96**: 773, 2004; phylogeny).

Septogloeum Sacc. (1880), anamorphic *Pezizomycotina*, Cac.≡ eH.15. 2, widespread. See Sutton & Pollack (*Mycopathologia* **52**: 331, 1974), Sutton & Webster (*TBMS* **83**: 59, 1984).

Septoidium G. Arnaud (1921), anamorphic *Alina*, Hso.≡ eP.19. 7, widespread (tropical).

Septolpidium Sparrow (1933), Chytridiales. 1 (in diatoms), British Isles.

Septomazzantia Theiss. & Syd. (1915) = Diaporthe fide Petrak (*Sydowia* **8**: 294, 1954).

Septomycetes. Class within the *Orthomycotina* (q.v.) including *Brachybasidiales*, *Exobasidiales*, *Septobasidiales* and *Tilletiales* (Cavalier-Smith, *in* Rayner *et al.* (Eds), *Evolutionary biology of the fungi*, 1987).

Septomycotera. Class within the *Orthomycotina* (q.v.) including *Brachybasidiales*, *Exobasidiales*, *Septobasidiales* and *Tilletiales* (Cavalier-Smith, *in* Rayner *et al.* (Eds), *Evolutionary biology of the fungi*, 1987).

Septomyrothecium Matsush. (1971), anamorphic *Hypocreales*, Hsp.0eP.15. 1, Papua New Guinea. See Matsushima (*Bull. natn. Sci. Mus.* Tokyo, B **14**: 469, 1971).

Septomyxa Sacc. (1884) = Diplodina fide Sutton (*Mycol. Pap.* **141**, 1977).

Septomyxella (Höhn.) Höhn. (1923), anamorphic *Pezizomycotina*, Cac.1eH.?. 1, Italy.

Septonema Corda (1837), anamorphic *Dothideomycetes*, Hso.≡ eP.3 (±L). 13, widespread. See Hughes (*Naturalist* Hull: 173, 1951), Hughes (*Naturalist* Hull: 7, 1952), Holubová-Jechová (*Folia geobot. phytotax.* **13**: 422, 1978), Grondona *et al.* (*MR* **101**: 1489, 1997; ecology).

Septopatella Petr. (1925), anamorphic *Pezizomycotina*, Ccu.0fH.10. 1, N. America; Europe. See Dyko & Sutton (*CJB* **57**: 370, 1979).

Septopezizella Svrček (1987), Helotiaceae. 1, Europe. See Svrček (*Česká Mykol.* **41**: 88, 1987).

Septoplaca Petr. (1964) = Piptarthron fide Sutton (*The Coelomycetes*, 1980).

Septorella Allesch. (1897) = Fusarium fide Höhnel (*Sber. Akad. Wiss. Wien* Math.-naturw. Kl., Abt. 1 **121**: 339, 1912).

Septorella Berk. [not traced] ? nom. nud., anamorphic *Pezizomycotina*. See Sutton (*Mycol. Pap.* **141**, 1977).

Septoria Fr. (1819) nom. rej. prop., anamorphic *Pezizomycotina*.

Septoria Sacc. (1884) nom. cons., anamorphic *Mycosphaerella*, Cpd.0fH.10. *c.* 1072, widespread. *S. apiicola* (celery (*Apium*) leaf spot), *S. azaleae* (azalea (*Rhododendron*) leaf scorch), *S. gladioli* (hard rot and leaf spot of *Gladiolus*), *S. lycopersici* (tomato leaf spot), *S. nodorum* (wheat glume blotch), *S. tritici* (wheat leaf spot). See MacMillan & Plunkett (*J. agric. Res.* **64**: 547, 1942; spore structure and germination), Jörstad (*Skr. Utg. Norske Vidensk.-Akad. Oslo* N.S. **22**: 1, 1965), Jörstad (*Skr. Utg. Norske Vidensk.-Akad. Oslo* N.S. **24**: 1, 1967; Norwegian spp. on dicots and *Gramineae*), Sutton (*The Coelomycetes*, 1980; refs), Constantinescu (*TBMS* **83**: 383, 1984; spp. on *Betulaceae*), Scharen & Sanderson (*Septoria of Cereals*: 37, 1985), Andrianova (*Mikol. Fitopatol.* **20**: 259, 1986; taxonomic criteria), Andrianova (*Mikol. Fitopatol.* **21**: 393, 1987; subgen. classif.), Sutton & Pascoe (*TBMS* **89**: 521, 1987; spp. on *Acacia*), Teterevnikova-Babayan (*Griby roda Septoria*, 1987; spp. former USSR), McDonald & Martinez (*Phytopathology* **79**: 1186, 1989), Sutton & Pascoe (*Stud. Mycol.* **31**: 177, 1989; spp. from Australia), Cooley *et al.* (*MR* **94**: 145, 1990; cotransformation in *S. nodorum*), McDonald & Martinez (*Phytopathology* **80**: 1368, 1990; RLFP's and *S. tritici*), Farr (*Mycol.* **83**: 611, 1991; spp. on *Cornus*), Newton & Caten (*Pl. Path.* **40**: 546, 1991; strains of *S. nodorum* adapted to wheat or barley), Andrianova (*Mikol. Fitopatol.* **26**: 425, 1992; spp. described by Hollós), Farr (*Sydowia* **44**: 13, 1992; spp. on *Fabaceae* tribe *Genisteae*), Verkley (*Mycol.* **90**: 189, 1998; ultrastr.), Caten *in* Lucas *et al.* (Eds) (*Septoria on Cereals* A Study of Pathosystems: 26, 1999; genetics), Carlier *et al.* (*Phytopathology* **90**: 884, 2000; on *Musa*), Cunfer (*Can. J. Pl. Path.* **22**: 332, 2000; on *Gramineae*), Verkley & Priest (*Stud. Mycol.* **45**: 123, 2000; review), Goodwin *et al.* (*Phytopathology* **91**: 648, 2001; phylogeny), Schnieder *et al.* (*Eur. J. Pl. Path.* **107**: 285, 2001; genetic diversity), Linde *et al.* (*Petria* **12**: 95, 2002; population structure), Goodwin *et al.* (*Molecular and General Genetics* **269**: 1, 2003; mating types), Banke *et al.* (*Fungal Genetics Biol.* **41**: 226, 2004; molecular variation), Verkley & Starink-Willemse (*Mycol. Progr.* **3**: 315, 2004; phylogeny, on *Asteraceae*), Verkley *et al.* (*Mycol.* **96**: 558, 2004; phylogeny), Feau *et al.* (*J. Mol. Evol.* **64**: 489, 2007; phylogeny).

Septoriaceae W.B. Cooke (1983) = Mycosphaerellaceae.

Septoriella Oudem. (1889), anamorphic *Dothideomycetes*, St.≡ eP.1. 16, widespread. See Nag Raj (*Coelomycetous Anamorphs with Appendage-bearing Conidia*, 1993).

Septoriomyces Cavalc. & A.A. Silva (1972) = Phyllobathelium fide Lücking *et al.* (*Lichenologist* **30**: 121, 1998).

Septoriopsis F. Stevens & Dalbey (1918) ≡ Cercoseptoria.

Septoriopsis Gonz. Frag. & M.J. Paúl (1915) ? = Septoria Sacc. fide Hawksworth *et al.* (*Dictionary of the Fungi* edn 8, 1995).

Septoriopsis Höhn. (1920) ≡ Groveolopsis.

Septorisphaerella Kleb. (1918) = Mycosphaerella fide von Arx & Müller (*Stud. Mycol.* **9**, 1975).

Septosperma Whiffen ex R.L. Seym. (1971), Chytridiaceae. 5 (on chytrids), widespread (temperate). See Blackwell & Powell (*Mycotaxon* **42**: 43, 1991).

Septosphaerella Laib. (1922) = Mycosphaerella fide Hawksworth *et al.* (*Dictionary of the Fungi* edn 8, 1995).

Septosporium Corda (1831), anamorphic *Pezizomycotina*, Hso.#eP.1. 5, Europe; N. America.

Septothyrella Höhn. (1911), anamorphic *Pezizomycotina*, Cpt.≡ eH.?. 5, Africa.

Septotinia Whetzel ex J.W. Groves & M.E. Elliott (1961), Sclerotiniaceae. Anamorph *Septotis*. 2, widespread (temperate). See Groves & Elliott (*Mycol.* **29**:

134, 1961).

Septotis N.F. Buchw. ex Arx (1970), anamorphic *Septotinia*, Cac.≡ eH.39. 2, N. America; Europe. See Sutton (*Mycol.* **72**: 208, 1980).

Septotrapelia Aptroot & Chaves (2007), Pilocarpaceae (L). 1, Costa Rica. See Aptroot *et al.* (*J. Hattori bot. Lab.* **100**: 617, 2006).

Septotrichum Corda (1840) nom. dub., Plantae. Based on a leaf outgrowth fide Bonorden (1851).

Septotrullula Höhn. (1902), anamorphic *Pezizomycotina*, Hsp.≡ eH.38. 1, widespread. See Sutton & Pirozynski (*TBMS* **48**: 355, 1965).

septum (pl. **septa**), a cell wall or partition; Talbot (*Taxon* **17**: 622, 1968) distinguishes **primary -**, a septum formed in direct association with nuclear division (by constriction, mitosis, or meiosis) separating the daughter cells and having a pore (see Markham, *MR* **98**: 1089, 1994; review) which may be modified as a dolipore (Fig. 13) (q.v.; in basidiomycetes) or be associated with Woronin bodies (in ascomycetes; see Kimbrough, *in* Hawksworth, 1994: 127, ultrastr.), and **adventitious -** (retraction or retaining septum, 'cloison de retrait'), a septum formed in the absence of, or independently of, nuclear division, esp. in association with the movement of cytoplasm from one part of the fungus to another; primary septa are characteristic of higher fungi, adventitious septa of lower fungi where nuclear division is by constriction; **angular -** (Eriksson, *Ark. Bot.* II **6**: 339, 1967), oblique septum; **longiseptum**, a longitudinal septum within a spore (Reynolds, *Mycol.* **63**: 1173, 1971); **oblique -**, a septum within a segment of a spore arising at an oblique angle to that delimiting the segment (Eriksson, *Opera bot.* **60**, 1981); **trans-**, a transverse septum within a spore (Reynolds, 1971), which may be an **A trans-septum** (forming a segment) or a **B trans-septum** (laid down in a segment after division by a longiseptum; never in macrocephalic spores) (Eriksson, 1981). Ascospores in which the septation proceeds from the primary septum towards the poles, so that immature spores have longer end cells are termed **macrocephalic** (Fig. 22A); those in which it proceeds by each trans-septum dividing a segment into two of ± equal lengths are **microcephalic** (Fig. 22B) (Eriksson, 1967, 1981), Curry & Kimbrough (*Mycol.* **75**: 781, 1983; *Pezizaceae*). See also Anamorphic fungi, euseptum, distoseptum, multiperforate septum, and Fig. 22.

Sepultaria (Cooke) Boud. (1885) = Geopora fide Burdsall (*Mycol.* **60**: 504, 1968), Yao & Spooner (*MR* **100**: 72, 1996), Læssøe & Hansen (*MR* **111**: 1075, 2007; phylogeny).

Sepultariella Kutorga (2000), Pyronemataceae. 1, widespread. See Kutorga (*Lietuvos Grybai* **3**: 188, 2000).

sequestrate, fungal fruit-bodies which have evolved from having exposed hymenia and forcibly discharged spores to a closed or even hypogeous habit in which spores are retained in the fruit-body until it decays or is eaten by an animal vector. Many sequestrate taxa can be clearly recognized as being derived from specific spore-discharging ancestors, e.g. *Rhizopogon* from *Suillus*.

Serda Adans. (1763) nom. rej. = Gloeophyllum fide Donk (*Verh. K. ned. Akad. Wet.* tweede sect. **62**: 1, 1974).

Serendipita P. Roberts (1993), Auriculariales. 8, widespread. See Roberts (*MR* **97**: 474, 1993), Trichies (*BSMF* **118**: 351, 2002; France n.sp.).

Serenomyces Petr. (1952), Phaeochoraceae. 5 (on *Palmae*), widespread (tropical). See Barr *et al.* (*Mycol.* **81**: 47, 1989; key), Hyde *et al.* (*SA* **15**: 117, 1997), Hyde & Cannon (*Mycol. Pap.* **175**: 114, 1999).

Sericeocybe Rob Henry (1993) nom. inval. = Cortinarius fide Kuyper (*in litt.*).

Sericeomyces Heinem. (1978) = Leucoagaricus fide Kuyper (*in litt.*).

sericeous, like silk.

Seriella Fr. (1849) nom. dub., Pezizomycotina.

serioplectenchyma, see plectenchyma.

Serology. Serological methods, which depend on the ability of a fungus to act as an antigen (q.v.), have two main mycological applications: (1) for the identification of fungi or investigations into the relationships between different fungi; (2) for the diagnosis of infections of humans and other vertebrates by fungal pathogens.

For identification, a preparation of the unknown fungus is tested with antiserum prepared against a named fungus when a reaction indicates that the two fungi have antigens in common and are therefore conspecific or closely related. An antiserum prepared from one fungus may, however, not infrequently also give a reaction ('cross-reaction') with other fungi indicating the possession of some antigens in common. This may be taken to be evidence of phylogenetic relatedness. *Phymatotrichum omnivorum* antiserum, for example, gives a reaction with certain gasteromycetes. Because of such cross-reactions and the difficulty of obtaining purified fungal antigens serological identification of fungi is in a less developed state than that for bacteria and viruses.

Serological diagnosis of infection (or past infection) of humans or animals by fungal pathogens is a medical mycological technique of proven utility for a number of important mycoses. In N. Am., for example, millions of humans and domestic and farm animals are marked for life by antibodies induced by subclinical or mild attacks of coccidioidomycosis and histoplasmosis as revealed by testing the reaction of the skin or blood serum to coccidioidin and histoplasmin (q.v.), antigens prepared from the pathogens (See also blastomycin, oidiomycin, spherulin, sporotrichin.).

The antigen-antibody (antiserum) reaction may be tested in several ways. The most useful techniques are based on agglutination or precipitation (see antigen). In the first the antigenic particles are usually whole fungal cells but an alternative is to adsorb the antigen onto particles of latex or charcoal. In the second, the reaction is demonstrated by the formation of a precipitate when the antiserum (antibody) is added to a tube containing the corresponding antigen in solution or when antibody and antigen are allowed to diffuse towards one another in a clear agar gel. See also anaphylaxis, complement-fixation, ELISA.

Lit.: Chester (*Quart. Rev. Biol.* **12**: 19, 165, 294, 1937; plants), Seeliger (*Mykologische Serodiagnostik*, 1958), Mathews (*Plant virus serology*, 1956; methods), Proctor (*Progress in microbiological techniques*: 213, 1967), Kaufman (*Manual of clinical mycology*, 1975), Evans (Ed.) (*Serology of fungal infections and farmer's lung*, 1976).

serous (of latex), like serum; watery; opalescent.

Serpentisclerotes Beneš (1959), Fossil Fungi. 1 (Carboniferous), former Czechoslovakia.

Serpula (Pers.) Gray (1821), Serpulaceae. 2, widespread. *S. lacrymans* (syn. *Merulius lacrymans*), the dry rot or 'house' fungus. See Cooke (*Mycol.* **49**: 197, 1957; key), Harmsen (*Friesia* **6**: 233, 1960; taxonomy, culture) Wood-attacking fungi, Anon. (*Leafl. For. Prod. Res. Lab.*: 6, 1964), White *et al.* (*MR* **105**: 447, 2001; intraspecific variation), Zmitrovich (*Novosti Sistematiki Nizshikh Nov. sist. Niz. Rast.* **35**: 70, 2001), Zmitrovich & Spirin (*Mikol. Fitopatol.* **36**: 11, 2002), Legon (*Mycologist* **17**: 156, 2003; *Serpula himantioides*), Huckfeldt & Schmidt (*Mycologist* **20**: 42, 2006; key european strand-forming house-rot fungi).

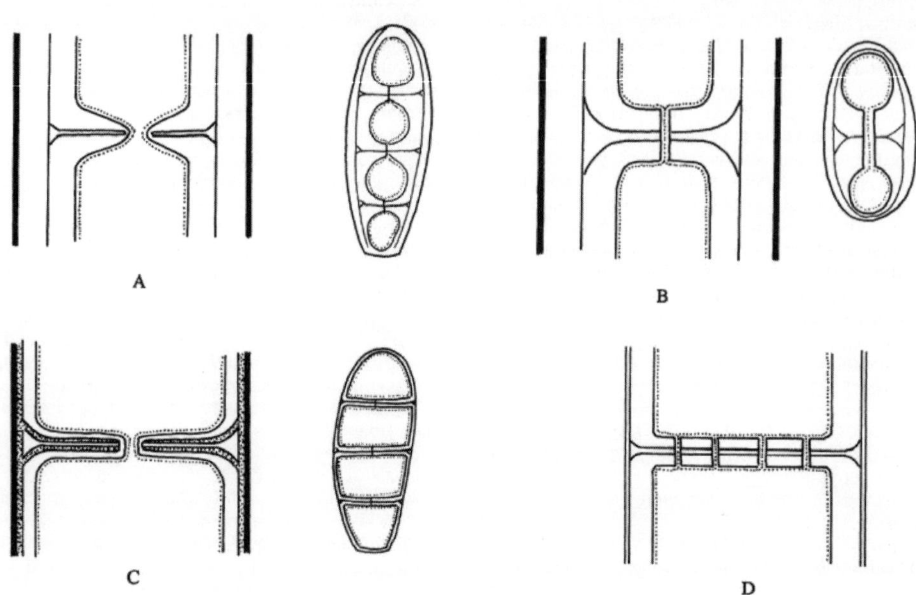

Fig. 34. Types of septum. A, distoseptum (pseudoseptum); B, distoseptum (polarilocular); C, euseptum; D, multiperforate septum. See also dolipore septum.

Serpulaceae Jarosch & Bresinsky (2001), Boletales. 4 gen. (+ 4 syn.), 20 spp.

Serpulomyces (Zmitr.) Zmitr. (2002), Amylocorticiaceae. 1, Europe. See Zmitrovich & Spirin (*Mikol. Fitopatol.* **36**: 20, 2002).

serrate, edged with teeth, like a saw (Fig. 23.46).

serrulate, delicately toothed.

Servazziella J. Reid & C. Booth (1987), Xylariales. 1 (on *Araucaria*), Australasia. See Eriksson & Hawksworth (*SA* **6**: 250, 1987; posn).

Servitia M.S. Christ. & Alstrup (2001), Verrucariaceae (L). 1, Greenland. See Alstrup & Hansen (*Graphis Scripta* **12**: 41, 2001).

Sesia Adans. (1763) nom. rej. ≡ Gloeophyllum fide Murrill (*Bull. Torrey bot. Club* **31**: 415, 1904).

Sesquicillium W. Gams (1968) = Clonostachys fide Veenbas-Rijks (*Acta Bot. Neerl.* **19**: 323, 1970), Samuels (*Mem. N. Y. bot. Gdn* **49**: 266, 1989; teleomorph), Rossman *et al.* (*Mycol.* **93**: 100, 2001; phylogeny), Schroers (*Stud. Mycol.* **46**: 1, 2001; synonymy).

sessile, having no stem.

Sessiliospora D. Hawksw. (1979), anamorphic *Pezizomycotina*, Hso.≡ eP.24. 1 (on lichens), Malaysia. See Hawksworth (*Bull. Br. Mus. nat. Hist.* Bot. **6**: 250, 1979).

seta (pl. **-ae**) (Lat., a bristle), (1) a stiff hair, generally thick-walled and dark in colour; in hyphomycetes, see Dev Rao (*in* Subramanian (Ed.), *Taxonomy of fungi*: 397, 1984); (2) (in hymenomycetes), a sterile hyphal end, thick-walled, darkening in KOH sol., found frequently projecting from the hymenium in xanthochroic basidiomata (Fig. 18A). Lentz (1954) distinguished 'seta', 'embedded seta', and 'stellate seta' (asteroseta) (Fig. 18C). Smith (1966) treated the last as a cystidium; see cystidium (3).

setaceous (Lat., setaceus), bristle-like; cf. setose.

Setaria Ach. ex Michx. (1803) nom. rej. prop. ≡ Bryoria.

Setchellia Magnus (1896) = Doassansia fide Dietel (*Nat. Pflanzenfam.* **6**, 1928).

Setchelliogaster Pouzar (1958), Agaricales. 5, widespread (warm dry areas). Belongs to the *Bolbitiaceae* or *Cortinariaceae*. See Singer & Smith (*Madroño* **15**: 73, 1959), Lago *et al.* (*Mycotaxon* **78**: 37, 2001), Martín & Moreno (*Mycotaxon* **78**: 257, 2001), Francis & Bougher (*Australasian Mycologist* **23**: 1, 2004; Australia).

Setella Syd. & P. Syd. (1916) = Dysrhynchis fide von Arx & Müller (*Stud. Mycol.* **9**, 1975).

Seticyphella Agerer (1983), Cyphellaceae. 3, Europe. See Agerer (*Mitt. bot. StSamml.* Münch. **19**: 282, 1983).

Setiferotheca Matsush. (1995), Chaetomiaceae. 1

(from soil), Japan; Venezuela. See Matsushima (*Mycol. Mem.* **8**, 1995), Silva & Hanlin (*Mycoscience* **37**: 261, 1997; as *Chaetomidium*).

setiform, see setaceous.

Setigeroclavula R.H. Petersen (1988), Clavariaceae. 1, New Zealand. See Petersen (*Bull. N.Z. Dept. Sci. Industr. Res., Pl. Dis. Div.* **263**: 143, 1988).

Setocampanula Sivan. & W.H. Hsieh (1989), Trichosphaeriaceae. 1 (from bark), Taiwan. See Sivanesan & Hsieh (*MR* **93**: 83, 1989).

Setodochium Bat. & Cif. (1957) ? = Wiesneriomyces fide Maniotis & Strain (*Mycol.* **60**: 203, 1968).

Setoerysiphe Y. Nomura (1984) = Erysiphe fide Braun (*Monogr. Cercosporella, Ramularia Allied Genera (Phytopath. Hyphom.)* **1**, 1995), Braun *et al.* (*The Powdery Mildews* A Comprehensive Treatise: 13, 2001).

Setogyroporus Heinem. & Rammeloo (1982), Boletaceae. 1, tropical Africa. See Heinemann & Rammeloo (*Bulletin du Jardin Botanique National de Belgique* **52**: 481, 1982).

Setolibertella Punith. & Spooner (1999), anamorphic *Pezizomycotina*. 1, Azores. See Punithalingam & Spooner (*Kew Bull.* **54**: 562, 1999).

Setomelanoma, see *Setomelanomma*.

Setomelanomma M. Morelet (1980), Pleosporales. 1, France; USA. See Morelet (*Bulletin de la Société des Sciences naturelles et d'Archéologie de Toulon et du Var* **227**: 15, 1980), Rossman *et al.* (*CJB* **80**: 1209, 2002; phylogeny), Kruys *et al.* (*MR* **110**: 527, 2006; phylogeny).

Setomyces Bat. & Peres (1961) nom. inval. = Tricharia Fée fide Lücking *et al.* (*Lichenologist* **30**: 121, 1998).

Setopeltis Bat. & A.F. Vital (1959), Micropeltidaceae. 1, pantropical. See von Arx & Müller (*Stud. Mycol.* **9**, 1975), Reynolds & Gilbert (*Aust. Syst. Bot.* **18**: 265, 2005), Reynolds & Gilbert (*Cryptog. Mycol.* **27**: 249, 2006).

Setophiale Matsush. (1995), anamorphic *Pezizomycotina*, Hso.0eH.15. 1, Peru. See Matsushima (*Matsush. Mycol. Mem.* **8**: 35, 1995).

Setoscypha Velen. (1934) = Phialina fide Huhtinen (*Karstenia* **29**: 545, 1990).

setose (Lat. setosus; bristly), covered with bristles; cf. setaceous, setulose.

Setosphaeria K.J. Leonard & Suggs (1974), Pleosporaceae. Anamorph *Exserohilum*. 8 (on *Poaceae*), widespread. See Borchardt *et al.* (*Eur. J. Pl. Path.* **104**: 611, 1998), Borchardt *et al.* (*J. Phytopath.* **146**: 451, 1998; population biology), Silva-Hanlin & Hanlin (*MR* **103**: 153, 1999; DNA), Olivier *et al.* (*Mycol.* **92**: 736, 2000; phylogeny), Zhang & Berbee (*Mycol.* **93**: 1048, 2001; phylogeny), Kodsueb *et al.* (*Mycol.* **98**: 571, 2006; phylogeny).

Setosporella Mustafa & Abdul-Wahid (1989), anamorphic *Pezizomycotina*, Hso.0eP.?36. 1 (from soil), Egypt. See Mustafa & Abdul-Wahid (*MR* **93**: 227, 1989).

Setosynnema D.E. Shaw & B. Sutton (1985), anamorphic *Pezizomycotina*, Hsy.0fH.10. 1 (aero-aquatic), Australia; Papua New Guinea. See Shaw & Sutton (*J. Linn. Soc.* Bot. **91**: 33, 1985).

setula (Lat., a little bristle), (1) a delicate hair-like appendage arising from a conidium, as in *Dinemasporium*; (2) (in hymenomycetes), a thick-walled, pigmented, terminal element of a tramal cystidium.

setule (in hymenomycetes), a thin-walled, rarely pigmented, usually lageniform cystidium on the pileus or stipe.

Setulipes Antonín (1987), Marasmiaceae. *c.* 25, widespread (north temperate). See Antonín *et al.* (*Mycotaxon* **63**: 359, 1997; taxonomy), Antonín (*Mycotaxon* **88**: 53, 2003; tropical African spp.), Wilson & Desjardin (*Mycol.* **97**: 667, 2005; phylogeny; in *Gymnopus* clade), Antonín (*Monograph of Marasmius, Gloiocephala, Plalaeocephala and Setulipes in Tropical Africa*, 2007), Antonín & Buyck (*MR* **111**: 919, 2007; tropical African spp.; key).

setulose (Lat., setulosus), covered with fine bristles or hairs; cf. setaceous.

Seuratia Pat. (1904), Seuratiaceae (?L). Anamorph *Atichia*. 5, widespread. See Meeker (*CJB* **53**: 2462, 1975), Meeker (*CJB* **53**: 2485, 1975; key, morphology), Reynolds & Gilbert (*Aust. Syst. Bot.* **18**: 265, 2005; Australia).

Seuratiaceae Vuill. ex M.E. Barr (1987), Pezizomycotina (inc. sed.). 3 gen. (+ 10 syn.), 12 spp.

Lit.: Meeker (*CJB* **53**: 2464, 1975), Eriksson (*Mycotaxon* **15**: 249, 1982), Rodríguez Hernández & Camino Vilaró (*Revta Jardín bot. Nac.* Univ. Habana **6**: 61, 1985), Sutton (*Sydowia* **38**: 324, 1985), Barr (*Mycotaxon* **29**: 501, 1987), Sivanesan & Hsieh (*MR* **99**: 1295, 1995).

Seuratiopsis Woron. (1934), ? Seuratiaceae. 1, Crimea.

sewage fungus, *Leptomitus lacteus*, found in polluted water, sometimes blocking sewage filters.

Sex. In some fungi reproduction is by asexual methods only, in others it is parthenogenic, but in most there is a sexual process. Probably one third of all fungi have more than one type of reproduction, frequently in two well-marked phases (the 'teleomorph' and 'anamorph') which may have separate names (see Nomenclature).

As in the *Algae* and *Bryophyta*, the sexual or assimilative phase is generally haploid (i.e. a gametophytic generation), though in basidiomycetes (see under) the cells are commonly dikaryotic. In most Fungi (but not Mycetozoa) nuclear fusion is not till shortly before meiosis and the development of 'sexual' spores. For a life-cycle summary of an hypothetical typical filamentous fungus see Fig. 36, and for *Pucciniales* see Fig. 25.

Fungi having sex organs are commonly monoecious, infrequently dioecious. If sex organs are not present or (as in basidiomycetes and *Mucorales*) are of like morphology the fungus is (1) homothallic (q.v.), when one haploid mycelium is able to give a sporocarp or (2) heterothallic (q.v.), when there are two or more haploid mycelial types and diploidization must be effected for sporocarp development to take place. Heterothallic mycelia may be different (sexually dimorphic) or, more commonly, alike in form so that, as the 'sexes' may be determined only by chemical methods, they are generally named + and -.

In bipolar species sporocarp development is dependent on two factors, and in tetrapolar species on four. When segregation in a tetrapolar species takes place at the second division of meiosis one basidium may give all four types of spore, but when segregation is at the first division only two types (A_1B_1 and A_2B_2 or A_1B_2 and A_2B_1) are produced by one basidium. Normal basidiocarps are generally produced only when the secondary mycelium resulting from the fusion of primary mycelia is of the genotype $A_1B_1A_2B_2$ (Buller, 1941). These various genetical devices favour or ensure outbreeding or inbreeding. As Esser has pointed out incompatibility is of 2 main types: (1) **homogenic incompatibility** (e.g. bipolar and tetrapolar systems) in which mating is prevented between strains having the same factor(s) so that inbreeding is prevented and outbreeding encouraged, and (2) **heterogenic incompatibility** in which mating is prevented between strains having different factors so that outbreeding is prevented and inbreeding encouraged.

Two genes determine mating type and, in *Coprinus cinereus*, these have been extensively studied, with more than 12,000 mating types already identified.

Lit.: Burnett (*Fungal populations and species*, 2003: 130), Casselton & Olesnicky (*Microbiol. mol. biol. rev.* **62**: 55, 1998; molecular genetics of mating recognition in basidiomycetes), Esser & Raper (Eds) (*Incompatibility in fungi*, 1965), Heitman *et al.* (*Sex in fungi: molecular determination and evolutionary implications*, 2007), Raper (*in* Wenrich (Ed.), *Sex in micro-organisms*, 1954, *Mycol.* **51**: 107, 1959, **55**: 79, 1963, *Genetics of sexuality in higher fungi*, 1966), Stabens (*in* Gow & Gadd (Eds), *The growing ungus*: 383, 1995). See also Genetics, hormones, parasexual cycle.

Seychellomyces Matsush. (1981), anamorphic *Pezizomycotina*, Hso.≡ eP.23. 1, Seychelles. See Matsushima (*Matsush. Mycol. Mem.* **2**: 14, 1981).

Seynesia Sacc. (1883), Xylariales. 2, widespread (tropical). See Barr (*Mycotaxon* **39**: 43, 1990; posn), Læssøe (*SA* **13**: 43, 1994), Hyde (*Sydowia* **47**: 199, 1995), Kang *et al.* (*Fungal Diversity* **2**: 135, 1999), Guo & Hyde (*Nova Hedwigia* **72**: 461, 2001; Hong Kong), Zhang *et al.* (*Mycol.* **98**: 1076, 2006; phylogeny).

Seynesiella G. Arnaud (1918), Microthyriaceae. 4, widespread. See Sivanesan & Shivas (*Fungal Diversity* **11**: 151, 2002), Belomesyatseva (*Mycena* **4**, 2004).

Seynesiola Speg. (1919) = Arnaudiella fide Stevens & Ryan (*Ill. biol. monogr.* **17** no. 2, 1939), Müller & von Arx (*Beitr. Kryptfl. Schweiz* **11** no. 2, 1962).

Seynesiopeltis F. Stevens & R.W. Ryan (1925), Microthyriaceae. 1, Hawaii.

Seynesiopsis Henn. (1904), anamorphic *Pezizomycotina*, Cpt.1eP.?. 1, S. America.

Seynesiospora Bat. (1960) = Cyclotheca fide von Arx & Müller (*Stud. Mycol.* **9**, 1975).

shaggy ink cap (or **mane**), basidioma of the edible *Coprinus comatus*.

Shanorella R.K. Benj. (1956), Onygenaceae. Anamorph *Chrysosporium*. 1, USA; Europe. See Benjamin (*Aliso* **3**: 319, 1956), Sugiyama & Mikawa (*Mycoscience* **42**: 413, 2001; phylogeny), Solé *et al.* (*MR* **106**: 388, 2002; phylogeny), Untereiner *et al.* (*Stud. Mycol.* **47**: 25, 2002; phylogeny).

Shanoria Subram. & K. Ramakr. (1956) = Strigula fide Sutton (*Kew Bull.* **31**: 461, 1977), Eriksson & Hawksworth (*SA* **11**: 56, 1992).

Shanoriella Bat. & Cif. (1962) = Limacinula Höhn. fide Reynolds (*Bull. Torrey bot. Club* **98**: 157, 1971).

shape, configuration or form, total effect produced by the outline of the structure (Fig. 23).

Shawiella Hansf. (1957), anamorphic *Pezizomycotina*, Ccu.≡ eP.1. 1, Papua New Guinea. See Hansford (*Proc. Linn. Soc. N. S. W.* Ser. 2 **82**: 226, 1957).

shaybah, Saudi Arabian name for *Parmelia austrosinensis* used in cooking.

Shear (Cornelius Lott; 1865-1956; USA). Plant pathologist and mycologist (1901-1923) then Head of the Division of Mycology and Disease Survey (1923-1935), USDA Bureau of Plant Industry, Beltsville, Maryland. Second President of the Mycological Society of America. Noted for his studies in diseases of cranberry and other fruits, pyrenomycetes and many publications on general mycology and nomenclature. *Publs.* (with Clements) *Genera of Fungi* (1931). *Biogs, obits etc.* Petrak (*Sydowia* **11**: 1, 1957) [bibliography, portrait]; Stafleu & Cowan (*TL-2* **5**: 585, 1985); Stevenson (*Mycol.* **49**: 283, 1957) [bibliography, portrait]; Stevenson (*Taxon* **6**: 7, 1957) [portrait]; Stevenson *Phytopathology* **47**: 321, 1957).

Shearia Petr. (1924), anamorphic *Pleomassaria*, St.#eP.19. 2, widespread. See Barr (*Mycotaxon* **15**: 349, 1982; teleomorph), Tubaki *et al.* (*TMSJ* **24**: 121, 1983; cultural characters), Tanaka *et al.* (*Mycoscience* **46**: 248, 2005; Japan).

Sheariella Petr. (1952), anamorphic *Pezizomycotina*, St.0eP.1. 1, Hawaii. See Petrak (*Sydowia* **6**: 302, 1952).

Shecutia Nieuwl. (1916) ≡ Libertiella.

shield-lichens, formerly applied to lichens having large apothecia (obsol.).

shii-take, *Lentinula edodes*. For cultivation the wood is inoculated with spawn. Basidiomata are produced after 2 years, then there are two crops a year for a number of years. Alternatively, basidiomata are obtained 12 weeks after inoculation of a mixture of hardwood sawdust, nutrients and water (Jones & Liu, *Mycologist* **4**: 121, 1990). See Singer & Harris (*Mushrooms and truffles*, 1987), Maher (Ed.) (*Mushroom Science* **13**, *Science and cultivation of edible fungi*, 1991), Stamets (*Growing gourmet and medicinal mushrooms*, 1993), Chang & Hayes (*Biology and cultivation of edible mushrooms*, 1978), Przybylowicz & Donoghue (*Shiitake growers handbook*, 1988), Tan & Moore (*MR* **96**: 1077, 1992; *in vitro* cultivation).

Shimizuomyces Kobayasi (1981), Clavicipitaceae. 2, Japan. See Kobayasi (*Bull. natn. Sci. Mus.* Tokyo, B **7**: 1, 1981), Fukatsu & Nikoh (*Mycology Series* **19**: 311, 2003), Sung *et al.* (*Stud. Mycol.* **57**, 2007).

Shiraia Henn. (1900), Pleosporales. 1, China; Japan. See Amano (*Bull. natn. Sci. Mus.* Tokyo, B **6**: 55, 1980), Amano (*TMSJ* **24**: 35, 1983), Tsuda *et al.* (*TMSJ* **30**: 493, 1989; synanamorph), Cheng *et al.* (*Journal of Basic Microbiology* **44**: 339, 2004; phylogeny), Kruys *et al.* (*MR* **110**: 527, 2006; phylogeny).

Shiraiella Hara (1914) = Mycocitrus fide Rossman *et al.* (*Stud. Mycol.* **42**: 248 pp., 1999).

Shitaker Lloyd (1924) nom. nud. ≡ Lentinula.

Shivomyces Hosag. (2004), anamorphic *Asterinaceae*. 1, India. See Hosagoudar (*J. Econ. Taxon. Bot.* **28**, 2004).

shoe-string fungus, the honey agaric, *Armillaria mellea*.

Shomea Bat. & J.L. Bezerra (1964) ? = Geastrumia fide Sutton (*in litt.*).

Shortensis Dilcher (1965), Fossil Fungi, Micropeltidaceae. 1 (Eocene), USA. = Manginula fide Lange (*Austr. J. Bot.* **17**: 565, 1969), = Vizella (Dothid.) fide Selkirk (1972).

shot-hole, a leaf spot disease characterized by holes made by the dead parts dropping out; of *Antirrhinum* (*Heteropatella antirrhini*); of peach (*Stigmina carpo-*

phila); of cherry (*Blumeriella jaapii*).

shoyu (soy sauce), an oriental sauce of soybeans and wheat fermented by *Aspergillus*, yeasts, and bacteria (Hesseltine, *Mycol.* **57**: 174, 1965). See also ketjap.

Shropshiria F. Stevens (1927) = Munkia fide Marchionatto (*Rev. Argent.-agron.* **7**: 172, 1940).

Shrungabeeja V.G. Rao & K.A. Reddy (1981), anamorphic *Pezizomycotina*, Hso.#eP.19. 1, India. See Rao & Reddy (*Indian J. Bot.* **4**: 109, 1981).

Shuklania J.N. Dwivedi (1959), Fossil Fungi, Pucciniales. 1 (Tertiary), India.

Siamia V. Robert, C. Decock & R.F. Castañeda (2000), anamorphic *Pezizomycotina*. 1, Thailand. See Robert *et al.* (*MR* **104**: 893, 2000). **Sibirina** G.R.W. Arnold (1970), anamorphic *Hypomyces*, Hso.1eH.15. 7 (mycoparasitic), widespread. See Gams *et al.* (*CJB* **76**: 1570, 1998), Põldmaa (*Mycol.* **95**: 921, 2003), Watanabe *et al.* (*Mycoscience* **44**: 411, 2003; Japan).

sibling species, accepted synonym for cryptic species; species which are genetically distinct and not interbreeding, but are not morphologically distinct.

sicyospore, a thick-walled storage cell (obsol.).

siderophore, a metabolic product of a fungus (or other organism) which binds iron and facilitates its transport from the environment into the microbial cell. See Winkelmann (*in* Hawksworth (Ed.), *Frontiers in mycology*: 49, 1991; review roles).

SIDS, see cot death.

Siegertia Körb. (1860) = Rhizocarpon fide Hawksworth *et al.* (*Dictionary of the Fungi* edn 8, 1995).

Siemaszkoa I.I. Tav. & T. Majewski (1976), Laboulbeniaceae. 7, Europe; S. America. See Majewski (*Polish Bot. Stud.* **2**: 219, 1991; key).

sigatoka, A major disease of banana (*Musa*) caused by *Mycosphaerella musicola* (anamorph *Cercospora musae*). See *IMI Descriptions* **414**, 1974; *IMI Map* **7**; Meredith (*Phytopath. Pap.* **11**, 1970; monogr.).

Sigmatomyces Sacc. & P. Syd. (1913), anamorphic *Pezizomycotina*, Hsp.0eH.?. 1, Philippines.

Sigmogloea Bandoni & J.C. Krug (2000), Tremellales. 1, Canada. See Bandoni & Krug (*Mycoscience* **41**: 376, 2000).

sigmoid, curved like the letter 'S' (Fig. 23.6).

Sigmoidea J.L. Crane (1968), anamorphic *Corollospora*, Hso.≡ eH.15. 2, widespread. See Crane (*Am. J. Bot.* **55**: 998, 1968), Ando & Nakamura (*J. gen. appl. Microbiol.* Tokyo **46**: 51, 2000), Marvanová & Hywel-Jones (*Cryptog. Mycol.* **21**: 13, 2000; Thailand).

Sigmoideomyces Thaxt. (1891), Sigmoideomycetaceae. 1, N. America; Europe. See Benny *et al.* (*Mycol.* **84**: 635, 1992; key).

Sigmoideomycetaceae Benny, R.K. Benj. & P.M. Kirk (1992), Zoopagales. 3 gen., 6 spp.

Lit.: Benny *et al.* (*Mycol.* **84**: 615, 1992; key), Tanabe *et al.* (*Mol. Phylogen. Evol.* **16**: 253, 2000).

Sigridea Tehler (1993), Roccellaceae (L). 5, widespread. See Tehler (*Nova Hedwigia* **57**: 417, 1993; key), Myllys *et al.* (*Bryologist* **101**: 70, 1998; phylogeny), Follmann (*J. Hattori bot. Lab.* **90**: 251, 2001; key S American spp.), Ertz *et al.* (*Biblthca Lichenol.* **91**: 155 pp., 2005).

sikyotic cell (the 'Schröpfkopf-Zelle' of Burgeff, 1924), the terminal cell of a *Parasitella simplex* hypha by which the parasite anchors itself to its host (*Absidia glauca*). After several days this organ differentiates into a **sikospore** (sikyotic spore). See Kellner *et al.* (*Mycologist* **5**: 120, 1991).

Sillia P. Karst. (1873), Sydowiellaceae. 1 or 3, Europe; America. See Barr (*Mycol. Mem.* **7**, 1978), Rossman *et al.* (*Mycoscience* **48**: 135, 2007; review).

silver ear, the basidioma of the edible *Tremella fuciformis*. See *Tremella*.

silver leaf, of plum and other fruit trees (*Chondrostereum purpureum*.

simblospore, Langeron's (1945) term for zoospore, q.v.

Simblum Klotzsch ex Hook. (1831) = Lysurus fide Dring (*Kew Bull.* **35**: 1, 1980).

Simocybe P. Karst. (1879), Inocybaceae. 25, widespread. See Singer (*Sydowia* **15**: 71, 1962), Singer (*Beih. Nova Hedwigia* **44**: 485, 1973; key 18 neotrop. spp.), Pegler & Young (*Kew Bull.* **30**: 225, 1975; 4 Brit. spp.), Horak (*Sydowia* **32**: 123, 1979; key 4 Papua New Guinea spp.), Horak (*N.Z. Jl Bot.* **18**: 187, 1980; key 6 NZ spp.), Reid (*TBMS* **82**: 221, 1984; key Brit. spp.), Aime *et al.* (*Am. J. Bot.* **92**: 74, 2005; phylogeny and taxonomy).

Simoninus Roum. (1879) ≡ Chaenocarpus Spreng.

Simonyella J. Steiner (1902), ? Roccellaceae (L). 1, W. Asia. See Tehler (*CJB* **68**: 2458, 1990; cladistics), Feige *et al.* (*Flora* **187**: 159, 1992; anatomy), Myllys *et al.* (*Lichenologist* **31**: 461, 1999; phylogeny).

simple, unbranched; having no divisions.

Simplicillium W. Gams & Zare (2001), anamorphic *Torrubiella*. 4, widespread. See Zare & Gams (*Nova Hedwigia* **73**: 38, 2001), Sung *et al.* (*Stud. Mycol.* **57**, 2007).

Simuliomyces Lichtw. (1972), Legeriomycetaceae. 1 (in *Diptera*), widespread. See Lichtwardt (*The Trichomycetes. Fungal associates of arthropods*, 1986; key), Nelder *et al.* (*Fungal Diversity* **22**: 121, 2006; ecology and taxonomy).

Simuliospora Khodzhaeva, S.V. Krylova & I.V. Issi (1990), Microsporidia. 2.

Singer (Rolf; 1906-1994; Germany, later Austria, Spain, Russia, USA, Argentina, Chile). Born at Schliersee in south Germany, studying chemistry at the University of Munich; PhD student of Wettstein, Vienna (1928-1931); subsequently lived and worked in Barcelona (1933-1935), Leningrad [St Petersburg] (1935-1940) [Dr Biol. Sci.], Harvard (1941-1947), Tucumán (1948-1960), Buenos Aires (1960-1969), Santiago (1967-1968), and Chicago (1968-1994) [associate of the Field Museum and University of Illinois]. Made major contributions to the overall understanding of *Agaricales* and to neotropical agaricology, describing some 2,450 new fungi in about 400 publications in 9 languages; also made important contributions on ethnomycology, mycorrhizas, nomenclature, and truffles. His collections are distributed among at least 42 institutions. *Publs. The Agaricales in Modern Taxonomy*, 1951 [edn 4, 1986, the most recent, treats 230 genera]. *Biogs, obits etc.* Mueller (*Mycol.* **87**: 144, 1995); M. Singer (*Mycologists and Other Taxa*, 1984).

Singera Bat. & J.L. Bezerra (1960) = Vermiculariopsiella fide Sutton (*Mycol. Pap.* **141**, 1977).

Singerella Harmaja (1974) ≡ Singerocybe.

Singeriella Petr. (1959) = Blasdalea fide von Arx (*in litt.*).

Singerina Sathe & S.D. Deshp. (1981), Agaricaceae. 1, India. See Sathe & Deshpande (*Maharashtra Assoc. Cult. Sci.* Monogr. **1**: 35, 1980).

Singerocybe Harmaja (1988) = Clitocybe fide Kuyper (*in litt.*).

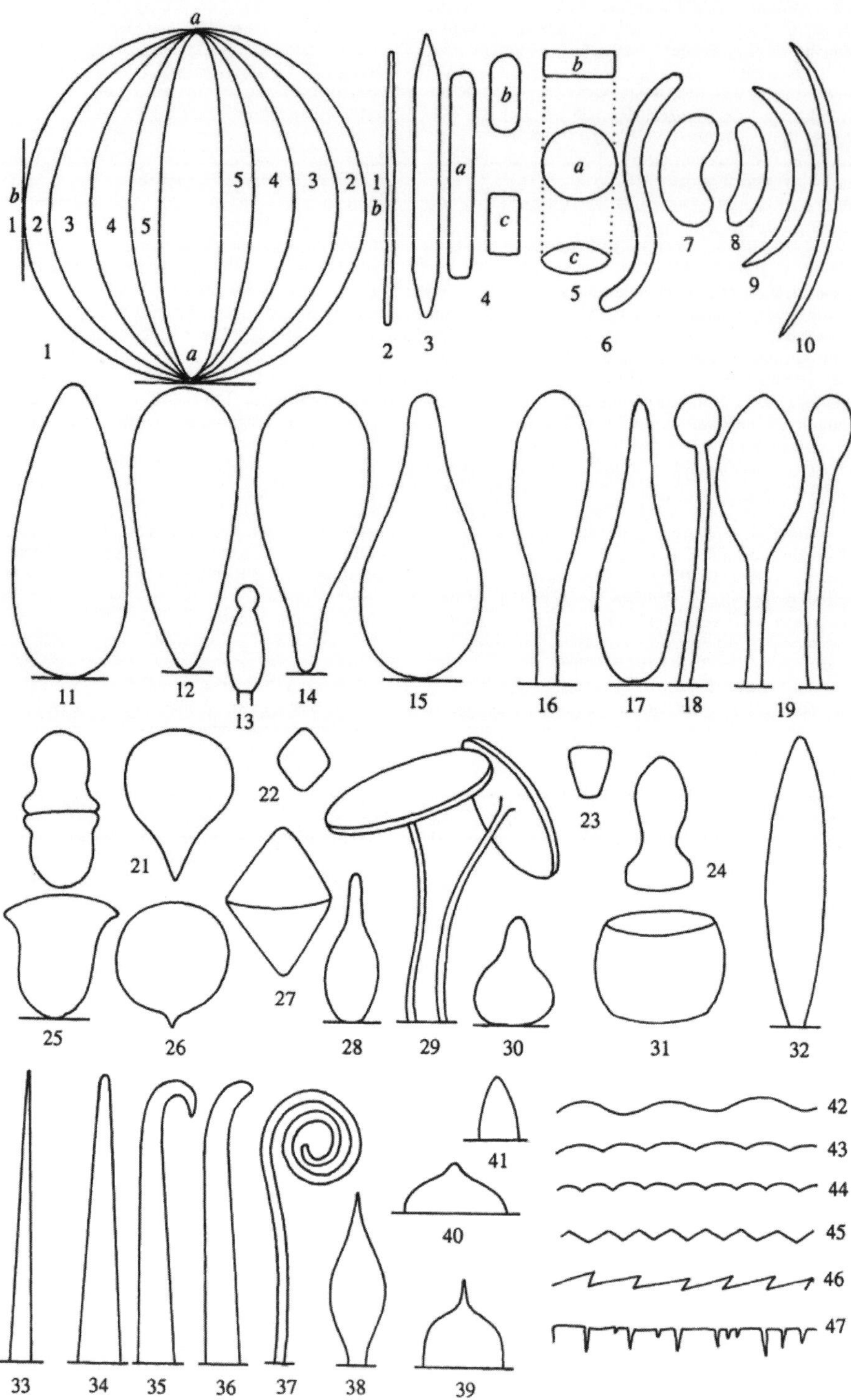
a
b
1 2 3 4 5
5 4 3 2 1
b
a
1
2 3
a
b
c
4
b
a
c
5
6
7
8
9
10
11
12
13
14
15
16
17
18
19
21
22
23
24
25
26
27
28
29
30
31
32
33
34
35
36
37
38
39
40
41
42
43
44
45
46
47

Fig. 23. Shapes. 1, shapes based on the sphere and ellipsoids, adapted from Payak (*Mycopath.* **16**: 72, 1962). Terms in parentheses are used by Erdtman (*An introduction to pollen analysis*, 1954). Ratios are those of Bas (*Persoonia* **5**: 321, 1969).

	axis aa	axis bb	length:breadth
1 by 1	globose (spherical)	(spherical)	1.0-1.05
1 to 2	subglobose (prolate spheroidal)	(oblate spheroidal)	1.05-1.15
2 to 3	broadly ellipsoidal (subprolate)	(suboblate)	1.15-1.3
3 to 4	ellipsoidal (prolate)	(oblate)	1.3-1.6; elongate
4 by 4	oval (perprolate)	(peroblate)	
5 by 5	fusiform		

2, filiform; 3, acerose; 4, cylindrical, restricted by Bas (1969) to cylinders with a length:breadth ratio of 2.0-3.0; a, bacilliform (1:b >3.0), b, c, oblong, b, apices rounded (obtuse), c, apices truncate; 5, a, discoid or lenticular in surface view, b, discoid in side view, c, lenticular in side view; 6, sigmoid; 7, reniform (fabiform); 8, allantoid; 9, lunate (crescentic); 10, falcate; 11, ovoid; 12, obovoid; 13, lecythiform; 14, pyriform; 15, obpyriform; 16, clavate; 17, obclavate; 18, capitate; 19, spathulate; 20, bicampanulate; 21, turbinate; 22, quadrangular (rhomboidal); 23, cuneiform; 24, dolabriform; 25, campanulate; 26, napiform; 27, biconic; 28, lageniform; 29, peltate; 30, ampulliform; 31, doliiform; 32, cymbiform (navicular); 33, acicular; 34, subulate; 35, hamate (uncinate); 36, corniform; 37, circinate; 38, ventricose. Apices. 39, mucronate; 40, papillate; 41, acute. Edges. 42, sinuate; 43, crenate; 44, crenulate; 45, dentate; 46, serrate; 47, laciniate. See also Systematics Association Committee for Descriptive Terminology (*Taxon* **9**: 245, 1960; list of works, **11**: 145, 1962; terminology of simple symmetrical plane shapes, chart).

Singeromyces M.M. Moser (1966), Boletaceae. 1, Argentina. Basidioma gasteroid. See Moser (*Nova Hedwigia* **10**: 331, 1965).

Sinoboletus M. Zang (1992), Boletaceae. 10, China. See Zang (*Mycotaxon* **45**: 223, 1992), Zang *et al.* (*Mycosystema* **25**: 366, 2006).

Sinodidymella J.Z. Yue & O.E. Erikss. (1985), Dacampiaceae. 2, China; N. America. See Yue & Eriksson (*Mycotaxon* **24**: 293, 1985), Barr (*Mycotaxon* **46**: 387, 1993), Barr (*Mycotaxon* **82**: 373, 2002).

Sinolloydia C.H. Chow (1936) = Lysurus fide Dring (*Kew Bull.* **35**: 1, 1980).

Sinosphaeria J.Z. Yue & O.E. Erikss. (1987) = Thyridium Nitschke fide Eriksson & Yue (*SA* **8**, 1989), Taylor *et al.* (*Sydowia* **49**: 94, 1997).

Sinotermitomyces M. Zang (1981) = Termitomyces fide Wei *et al.* (*Fungal Diversity* **21**: 225, 2006).

sinuate (1) (of lamellae), notched at the proximal end at junction with stipe (Fig. 19D) (cf. adnate); emarginate; (2) (of an edge), undulating (Fig. 23.42).

Siphomycetes. See *Phycomycetes*.

siphon (obsol.), an aseptate hypha (Vuillemin, 1912).

Siphonaria H.E. Petersen (1903), Chytridiaceae. 3, Europe; America. See Karling (*Am. J. Bot.* **32**: 580, 1945), Dogma (*Nova Hedwigia* **25**: 107, 1974).

Siphonia Fr. (1820) [non *Siphonia* Rich. ex Schreb. 1791, *Euphorbiaceae*] ≡ Dufourea.

Siphula Fr. (1824) nom. rej. ≡ Dufourea.

Siphula Fr. (1831) nom. cons., Icmadophilaceae (L). *c.* 33, widespread. See Santesson (*Svensk Natur.*: 176, 1964), Bendz *et al.* (*Acta Chem. Scand.* **19**: 1250, 1965), Kantvilas (*Herzogia* **12**: 7, 1996), Kantvilas (*Herzogia* **13**: 119, 1998), Stenroos & DePriest (*Am. J. Bot.* **85**: 1548, 1998), Platt & Spatafora (*Lichenologist* **31**: 409, 1999; DNA), Kantvilas (*Biblthca Lichenol.* **82**: 37, 2001), Ekman & Tønsberg (*MR* **106**: 1262, 2002; phylogeny), Kantvilas & Elix (*Herzogia* **15**: 1, 2002; S America), Kantvilas (*Austrobaileya* **6**: 949, 2004; Australia), Grube & Kantvilas (*Lichenologist* **38**: 241, 2006; phylogeny), Miądlikowska *et al.* (*Mycol.* **98**: 1088, 2006; phylogeny).

Siphulaceae Rchb. (1837) = Icmadophilaceae.

Siphulastrum Müll. Arg. (1889), Pannariaceae (L). 4, widespread (temperate). See Jørgensen (*Lichenologist* **30**: 533, 1998).

Siphulella Kantvilas, Elix & P. James (1992), Icmadophilaceae (L). 1, Tasmania. See Stenroos & DePriest (*Am. J. Bot.* **85**: 1548, 1998; DNA).

Siphulina (Hue) C.W. Dodge (1965) = Pannaria fide Dodge (*Trans. Am. microsc. Soc.* **84**: 510, 1965), Jørgensen (*Cryptog. Mycol.* **26**: 37, 2005).

Sipmania Egea & Torrente (1994), Roccellaceae (L). 1, Fiji. See Egea & Torrente (*Biblthca Lichenol.* **54**: 165, 1994), Grube (*Bryologist* **101**: 377, 1998).

sirenin, a hormone secreted by the female gamete of *Allomyces* which attracts male gametes.

Sirentyloma Henn. (1895) ? = Phyllachora Nitschke ex Fuckel (1870) fide Hawksworth *et al.* (*Dictionary of the Fungi* edn 8, 1995).

Sirexcipula Bubák (1907), anamorphic *Pezizomycotina*, Cpd.0eH/1eP.15. 2, Europe; N. America.

Sirexcipulina Petr. (1923) = Topospora fide Sutton (*Mycol. Pap.* **141**, 1977).

Sirobasidiaceae Lindau (1897), Tremellales. 2 gen., 11 spp.

Lit.: Benkert (*Mykol. MittBl.* **34**: 79, 1991), Ingold (*MR* **99**: 1187, 1995), Bandoni & Boekhout *in* Kurtzman & Fell (Eds) (*Yeasts, a taxonomic study* 4th edn: 705, 1998), Bandoni (*CJB* **76**: 1540, 1998), Fell *et al.* (*Int. J. Syst. Evol. Microbiol.* **50**: 1351, 2000), Sampaio *et al.* (*Mycol.* **94**: 874, 2002).

Sirobasidium Lagerh. & Pat. (1892), Sirobasidiaceae. 8, widespread (esp. tropical). See Boedijn (*Bull. Jard. bot. Buitenz.* ser. 3 **13**: 266, 1934), Lowy (*Mycol.* **48**: 324, 1956; key), Kobayashi (*TMSJ* **4**: 29, 1962), Moore (*Antonie van Leeuwenhoek* **45**: 113, 1979; septal ultrastr.), Fell *et al.* (*Int. J. Syst. Evol. Microbiol.* **50**: 1351, 2000; mol. phylogeny), Weiss *et al.* (*Frontiers in Basidiomycote Mycology*: 7, 2004).

Sirococcus Preuss (1855), anamorphic *Diaporthales*, St.1eH.15. 3, widespread (temperate). See Sutton (*The Coelomycetes*, 1980), Furnier *et al.* (*CJB* **77**:

783, 1999), Smith *et al.* (*Forest Pathology* **33**: 141, 2003; host-related variation in conifers), Kirisits *et al.* (*Forest Pathology* **37**: 40, 2007; Bhutan), Konrad *et al.* (*Forest Pathology* **37**: 22, 2007; genetic variation), Mejía *et al.* (*MR* **112**: 23, 2008; phylogeny), Rossman *et al.* (*Forest Pathology* **38**: 47, 2008; phylogeny).

Sirocrocis Kütz. (1843) nom. dub., ? Fungi.

Sirocyphis Clem. (1909), anamorphic *Pezizomycotina*, Cpd.0eH.?. 1, USA.

Sirodesmites Pia (1927), Fossil Fungi. 1 (Oligocene), Europe.

Sirodesmium De Not. (1849) = Coniosporium fide Hughes (*CJB* **36**: 805, 1958), Hughes & Crane (*Kavaka* **32**: 27, 2004).

Sirodiplospora Naumov (1915) = Topospora fide Groves (*CJB* **43**: 1195, 1965).

Sirodochiella Höhn. (1925), anamorphic *Pezizomycotina*, Cpd.0eH.?. 1, Europe.

Sirodomus Petr. (1947) = Sirexcipula fide Sutton (*Mycol. Pap.* **141**, 1977).

Sirodothis Clem. (1909), anamorphic *Tympanis*, St.0eH.15. 4, widespread. See Sutton & Funk (*CJB* **53**: 521, 1975).

Sirogloea Petr. (1923), anamorphic *Pezizomycotina*, St.0eH.?. 1, Europe.

Sироligniella Naumov (1926), anamorphic *Pezizomycotina*, St.0eH.3. 1, former USSR.

Siropatella Höhn. (1903) nom. dub., anamorphic *Pezizomycotina*. See Sutton (*Mycol. Pap.* **141**, 1977).

Siropeltis Arx & R. Garnier (1960) = Sphaerobolus fide Stalpers (*in litt.*).

Sirophoma Höhn. (1917), anamorphic *Pezizomycotina*, Cpd.0eH.?. 2 or 3, Europe.

Siroplacodium Petr. (1940), anamorphic *Pezizomycotina*, St.0eH.15. 5, Europe; Asia.

Siroplaconema Petr. (1922) ? = Ceuthospora Grev. fide Hawksworth *et al.* (*Dictionary of the Fungi* edn 8, 1995).

Siropleura Petr. (1934), anamorphic *Pezizomycotina*, St.0eH.?. 1, Europe.

Siroscyphella Höhn. (1910) = Pseudocenangium P. Karst. fide Dyko & Sutton (*CJB* **57**: 370, 1979).

Siroscyphellina Petr. (1923), anamorphic *Pezizomycotina*, St.0eH.?. 2, Austria; N. America. See Wehmeyer (*Sydowia* **6**: 433, 1952).

Sirosiphon Kütz. (1843), Algae.

Sirosperma Syd. & P. Syd. (1916), anamorphic *Pezizomycotina*, Cpd.0eH.23. 2, Papua New Guinea; Florida.

Sirosphaera Syd. & P. Syd. (1913), anamorphic *Pezizomycotina*, Cpd/St.0eH.23. 1, Philippines.

Sirosporium Bubák & Serebrian. (1912), anamorphic *Mycosphaerellaceae*, Hso.≡ eP.10. 25, widespread. See Ellis (*Dematiaceous Hyphomycetes*, 1971), Ellis (*More Dematiaceous Hyphomycetes*, 1976), Braun (*Cryptog. Mycol.* **20**: 155, 1999; gen. concepts), Crous & Braun (*CBS Diversity Ser.* **1**: 571 pp., 2003).

Sirosporonaemella Naumov (1951), anamorphic *Pezizomycotina*, Cpd.0eH.15. 1, Russia. See Naumov (*Bot. Mater. Otd. Sporov. Rast. Bot. Inst. Komarova Akad. Nauk S.S.S.R.* **7**: 114, 1951).

Sirostromella Höhn. (1916) = Pseudodiscula fide Sutton (*Mycol. Pap.* **141**, 1977).

Sirothecium P. Karst. (1887), anamorphic *Pezizomycotina*, St.0bP.1. 2, Finland; Azores. See Sutton (*Proc. Indian Acad. Sci.* Pl. Sci. **94**: 229, 1985).

Sirothyriella Höhn. (1910), anamorphic *Pezizomycotina*, Cpt.0eH.?. 1, Europe.

Sirothyrium Syd. & P. Syd. (1916), anamorphic *Pezizomycotina*, Cpt.0eH.?. 1, India.

Sirotrema Bandoni (1986), Tremellaceae. 3, widespread (north temperate). See Reid & Minter (*TBMS* **72**: 345, 1979; mycoparasitic on *Lophodermium*).

Sirozythia Höhn. (1904), anamorphic *Pezizomycotina*, Cpd.0eH.?. 2, Europe.

Sirozythiella Höhn. (1909), anamorphic *Pezizomycotina*, St.1-≡ eH.?. 1, Europe.

Sisostrema, see *Sistotrema Fr.*

Sistotrema Fr. (1821), Hydnaceae. Anamorphs *Burgoa*, *Ingoldiella*, *Osteomorpha*. 46, widespread. See Hallenberg (*Mycotaxon* **21**: 389, 1984), Boidin & Gilles (*BSMF* **110**: 185, 1994; key), Greslebin (*MR* **105**: 1392, 2001; key 10 spp Argentina).

Sistotrema Pers. (1794) ≡ Cerrena.

Sistotrema Raf. (1820) ≡ Sistotrema Fr.

Sistotremastrum J. Erikss. (1958), Hydnodontaceae. 3, widespread. See Eriksson (*Symb. bot. upsal.* **16** no. 1: 62, 1958).

Sistotremataceae Jülich (1982) = Hydnaceae.

Sistotremella Hjortstam (1984), Hydnodontaceae. 4, widespread. See Eriksson *et al.* (*Cortic. N. Europ.* **7**: 1379, 1984).

Sitochora H.B.P. Upadhyay (1964), anamorphic *Pezizomycotina*, Hso.1eH.?. 1 (on *Phyllachora*), Brazil. See Upadhyay (*Publções Inst. Micol. Recife* **414**: 4, 1964).

Sivanesania W.H. Hsieh & Chi Y. Chen (1996), Botryosphaeriaceae. 1, Taiwan. See Hsieh & Chen (*MR* **98**: 44, 1994), Hsieh & Chen (*MR* **100**: 1106, 1996).

Sivanesaniella Gawande & Agarwal (2004), Venturiaceae. 1, India. See Gawande & Agarwal (*Indian Phytopath.* **57**: 231, 2004).

skeletal hyphae, see hyphal analysis.

Skeletocutis Kotl. & Pouzar (1958), Polyporaceae. *c.* 30, widespread. See David (*Naturaliste can.* **109**: 235, 1982), Gilbertson & Ryvarden (*N. Amer. Polyp.* **2**: 711, 1987), Pieri & Rivoire (*Bulletin Semestriel de la Fédération des Associations Mycologiques Méditerranéennes* **26**: 3, 2004; Europe), Spirin (*Karstenia* **45**: 103, 2005), Roberts & Ryvarden (*Kew Bull.* **61**: 55, 2006; Cameroon).

skeletocystidium, see cystidium (1).

Skeletohydnum Jülich (1979), Polyporaceae. 1, New Zealand. See Jülich (*Persoonia* **10**: 331, 1979).

Skelophoromyces Thaxt. (1931) ≡ Scelophoromyces.

Skepperia Berk. (1857), Thelephoraceae. 5, widespread (esp. tropical).

Skepperiella Pilát (1927), Marasmiaceae. 4, widespread. See Pegler (*Kew Bull.* **28**: 257, 1973).

Skierka Racib. (1900), ? Pileolariaceae. 10 (on *Geraniales*, *Sapindales*, *Rhamnales*), widespread (tropical). See Mains (*Mycol.* **31**: 175, 1939).

skiophilous, shade loving. Cf. photophobous.

Skirgiellia A. Batko (1978) = Rozella fide Longcore (*in litt.*).

Skolekites Norman (1852) nom. rej. = Toninia fide Hafellner (*Beih. Nova Hedwigia* **79**: 241, 1984).

Skoteinospora Bat. (1962) = Euantennaria fide von Arx & Müller (*Stud. Mycol.* **9**, 1975).

Skottsbergiella Petr. (1927), Diaporthales. 1, Juan Fernández. See Petrak (*Sydowia* **24**: 264, 1971).

Skvortzovia Bononi & Hjortstam (1987), ? Agaricomycetes. 1, S. America. *Hymenochaetales* or *Agaricales* (*Rickenella* clade). See Bononi & Hjortstam

(*Mycotaxon* **28**: 12, 1987).

Skyathea Spooner & Dennis (1986), Helotiaceae. 1 (on *Hedera*), British Isles. See Spooner & Dennis (*Sydowia* **38**: 294, 1986).

skyrin, an orange-yellow wilt toxin of *Cryphonectria parasitica* (Gäumann *et al.*, *Phytopath. Z.* **36**: 116, 1959). Cf. luteoskyrin.

Skyttea Sherwood, D. Hawksw. & Coppins (1981), Odontotremataceae. 31 (on lichens), widespread. See Coppins *et al.* (*SA* **10**: 51, 1991; gen. concept), Diederich & Etayo (*Lichenologist* **32**: 423, 2000), Iturriaga & Hawksworth (*Mycol.* **96**: 925, 2004; N America).

Skyttella D. Hawksw. & R. Sant. (1988), Helotiales. 1 (on lichens, *Peltigera*), Europe; N. America. See Hawksworth & Santesson (*Graphis Scripta* **2**: 33, 1988), Diederich & Etayo (*Lichenologist* **32**: 423, 2000).

slaframine, a toxin of *Rhizoctonia leguminicola*; the cause of the slobber syndrome in livestock.

Slide Making. The following mounting procedures are those most frequently used for microscopic examination of fungi. For details of methods see Dring in Booth (Ed.) (Methods in microbiology 4: 95, 1971) Waller & Lenné (*Plant pathologist's pocketbook*, 2001), Gams *et al.* (*CBS Course of mycology*, edn 3, 1987).

Squash mount

The most widely used preparation for microscopic examination of fungi. A small portion of fungal material is removed from a culture or specimen using a mounted needle or scalpel blade and transferred to a drop of a suitable mountant on a microscope slide (see mounting fluid and stains). If necessary the material can be teased apart using a second needle and a coverslip applied. To release contents from fruit-bodies, light pressure on the coverslip is required. Air bubbles are released by gentle heating of the slide over a spirit lamp or bunsen flame. If necessary slides can be made permanent by sealing the cover slip with two layers of nail varnish. To apply this successfully, excess mounting fluid must first be removed from around the cover slip.

An improved technique involving use of a double cover-glass for preparing permanent microscope mounts was described by Kohlmeyer & Kohlmeyer (*Mycologist* **10**: 107, 1996).

Agar block mount

Used to observe undisturbed structures in anamorphic fungi (e.g. Penicillium, Aspergillus and Fusarium) grown on low-nutrient, translucent media. See Samson *et al.* (*Introduction to Food-borne fungi*, edn 5, 1996). From a sporulating area of the colony an agar block 0.5-1cm in diameter is removed with a sterilized scalpel blade and placed fungus side up on a microscope slide. A drop of water is placed on the block and a cover slip applied. The slide can then be examined under a light microscope.

Cellophane tape mount

These are used to prepare rapid mounts of fungi from leaf surfaces and agar cultures. See Butler & Mann (*Phytopathology* **49**: 231, 1959). A short length of clear adhesive single-sided tape is cut and the sticky side pressed gently on to the fungal material. The tape with fungus attached is then lowered, fungus side down, on to a drop of mounting fluid on a microscope slide and is pressed down firmly. If required a second drop of mountant can be applied to the top surface of the tape and a cover slip applied.

Slide culture

A culture grown directly on the surface of a slide and cover slip. Mainly used to examine conidial development in anamorphic fungi but also useful for observing appressoria in the genus *Colletotrichum*. A square block of nutrient agar is placed on a sterile microscope slide supported on a bent glass rod in a Petri dish which serves as a damp chamber. The four sides of the block are inoculated with the fungus and a sterile cover slip is lowered on to the block. Sterile water is then added to the base of the dish. When sporulation has occurred, the cover slip with attached mycelium is removed and carefully lowered, fungus side down on to a drop of mountant on a slide. The agar block is removed from the slide, leaving mycelium attached to the glass. A drop of stain is applied to this and a clean cover slip added thus two mounts can be prepared from each slide culture. Air bubbles can be expelled by heating the slide gently on a hot plate overnight. Mounts are made permanent by sealing with nail varnish if required. See Riddell (*Mycol.* **42**: 265, 1950).

Film culture

Using sterilized tweezers, add molten agar to one side of a coverslip so that it is covered by a thin film. When this has cooled, place it agar downwards on an inoculated block of nutrient agar as for slide cultures. After the fungus has grown into the agar film, the coverslip may be removed and mounted on a separate microscope slide with or without mounting fluid.

Hand-cut sections

For examining the structure and arrangement of fruit-bodies in plant tissues, vertical sections can be cut by hand. See Fox (*Principles of diagnostics in plant pathology*, 1993). Material to be cut is supported in elder pith, cork or polystyrene and thin sections cut with a razor blade (it is easier to cut thinner sections using a double-edge razor blades, but the additional upward-pointing blade makes this more hazardous). Sections are transferred with a mounted needle to a drop of mountant on a microscope slide and a cover slip applied.

Microtome-cut sections

Use of a freezing microtome provides the most satisfactory method for obtaining thin vertical sections of fungal fruit-bodies from plant tissues or cultures. See Waller & Lenné (*Plant pathologist's pocketbook*, 516 pp., 2001). Material is soaked for several hours in gum arabic solution. Dry specimens are best soaked overnight. The microtome stage is cooled to around -20°C with a supply of CO_2 or freon gas and a few drops of glue placed on the stage and allowed to freeze. This provides a platform on which to mount the material. The specimen is placed on the frozen glue block with the aid of a mounted needle and aligned carefully. A layer of glue is then allowed to freeze around the specimen. Required section thickness is set and sections are cut by rocking the blade across the specimen. Using a fine paint brush, sections are transferred to water contained in a watch glass. A fine mounted needle is used to transfer sections to a drop of mountant on a slide. A cover slip is then applied and the mount made permanent if required by sealing with nail varnish.

Slimacomyces Minter (1986), anamorphic *Pezizomycotina*, Hso.0-≡ hP.1. 2, British Isles. See Minter (*Bull. Br. mycol. Soc.* **20**: 23, 1986).

slime, a wet, generally sticky, substance; mucus; **- flux**, a thick liquid from the stems or branches of trees made up of, or having a connexion with, fungi and bacteria (Ogilvie, *TBMS* **9**: 167, 1924; Stautz, *Phytopath. Z.* **3**: 163, 1931); **- moulds**, the *Acrasiomycota*, *Dictyosteliomycota*, and *Mycetozoa*; **- spore**, a spore that becomes separated with slime from the cell producing it (Mason, 1937), cf. dry spore. See Gloiosporae.

slug (in slime moulds), the aggregated pseudoplasmodium of *Dictyosteliomycota*.

Smardaea Svrček (1969), Pyronemataceae. 7, widespread (north temperate). See Dissing (*Sydowia* **38**: 35, 1986), Cao *et al.* (*Acta Mycol. Sin.* **9**: 282, 1990), Hansen *et al.* (*Mycol.* **93**: 958, 2001; phylogeny), Hansen *et al.* (*Mol. Phylogen. Evol.* **36**: 1, 2005; phylogeny), Perry *et al.* (*MR* **111**: 549, 2007; phylogeny).

Smarodsia Raitv. & Vimba (2006), Pyronemataceae. 1, Latvia. See Vimba & Raitviir (*Folia cryptog. Estonica* **42**: 91, 2006).

Smell. See Chiron & Michelot (*Cryptogamie Mycologie* **26**: 299, 2005; detailed review). It seems likely that many fungi produce aromatic volatile chemicals, but these may be in quantities insufficient for detection by the human nose. Those which can be detected are often characteristic and sometimes of great economic value. Given that importance, investigation of fungal smell has been surprisingly limited. The odours of truffles and other underground fungi attract various animals including insects, rodents and larger mammals which are participants in a mutualistic arrangement for spore dispersal (see Animal mycophagists). In the case of truffles, the smell is particularly prized in gastronomy, and there has been extensive research into the volatile chemicals responsible (e.g. Pacioni *et al.*, *MR* **94**: 201, 1990); animals can be trained to find truffles using their scent (or even using potatoes injected with the main pure chemical ingredient). Components of some perfumes are obtained from actinomycetes, lichens, and truffles. The spicy ester, methyl cinnamate, produced by *Tricholoma matsutake*, while attractive to humans, is a powerful slug repellent (Wood & Lefevre, *Biochemical systematics and ecology* **35**: 634, 2007). Badcock (*TBMS* **23**: 188, 1939) describes the smells of wood-attacking fungi in culture. The distinctive 'coconut' smell of some *Trichoderma* colonies in pure culture is familiar to many laboratory mycologists. The smell of 'stinking' smut (bunt) of wheat (*Tilletia laevis*) is that of trimethylamine, $(CH_3)_3N$ (Hanna *et al.*, *J. biol. Chem.* **97**: 351, 1932; *Phytopath.* **22**: 978, 1932). A smell of mould is often associated with natural decaying materials, such as forest leaf litter, and with decaying building materials and other artefacts. A sweet smell has been reported in association with the spruce endophyte, *Tryblidiopsis pinastri*, although its function remains unknown (Livsey & Minter, *Can. J. Bot.* **72**: 549, 1994). Some fungi, e.g. ergots and rusts, produce sweet smells which may attract insects for dispersal of the spores (Steinebrunner *et al.*, *J. Chemical Ecol.* **34**: 1573, 2008); this may be associated with floral mimicry (Roy, *Nature* **362**: 56, 1993). Other fungi, e.g. *Phallus impudicus*, produce strong odours usually repulsive to humans, but attractive to insects associated with spore dispersal. The odours of basidiomycete species are often of diagnostic value, although in field mycology evaluation can be rather subjective. Gilbert (*Méthode de mycologie descriptive*, 1934; *BSMF* **48**: 241, 1932; **50**: 25, 1935) made a classification of the smells of natural basidiomata (see also Imler, *BSMF* **68**: 400, 1952; Josserand, 1952: 63; Locquin, *Petite flore des champignons de France* **1**: 97, 1956). See also Taste.

Smeringomyces Thaxt. (1908), Laboulbeniaceae. 4, widespread. See Weir & Rossi (*CJB* **75**: 791, 1997; New Zealand).

Smith (Alexander Hanchett; 1904-1986; USA). University of Michigan (1934-1972). Agaricologist; monographed *Mycena* (q.v.); see also Macromycetes, USA. *Biogs, obits etc.* Rogers (*Mushroom* **5**: 17, 1987); Thiers (*Mycol.* **79**: 811, 1987) [bibliography, portrait]; Stafleu & Cowan (*TL-2* **5**: 662, 1985).

Smith (Annie Lorrain; 1854-1937; Scotland, later England). Originally a governess; an 'unofficial worker', British Museum (Natural History), London. Her first publications were on seaweeds; she was associated with early work on seed testing; prepared exhibits on microfungi for the British Museum (Natural History) London; President of the British Mycological Society (1907); she completed J.M. Crombie's *A monograph of lichens found in Great Britain* (2 vols, 1894, 1911) and is best known for her work on lichen-forming fungi. *Publs. A Monograph of the British Lichens* (vol. 1, 1918; vol. 2, 1926); *A Handbook of British Lichens* (1921) [for a time the standard English introduction to the subject; reprint 1975]. *Biogs, obits etc.* Gepp & Rendle (*Journal of Botany* London **75**: 328, 1937); Grummann (1974: 240); Stafleu & Cowan (*TL-2* **5**: 663, 1985).

Smith (Erwin Frink; 1854-1927; USA). Plant pathologist and bacteriologist, USDA (1886-1927). First interested in peach yellows, later specializing in bacterial pathogens of plants. *Publs. Bacteria in Relation to Plant Disease* 3 vols (1911-1914); *Introduction to Bacterial Diseases of Plants* (1920). *Biogs, obits etc.* Rogers (*Erwin Frink Smith. A story of North American Plant Pathology*, 1952); Stafleu & Cowan (*TL-2* **5**: 669, 1985); Verma (*Review of Tropical Plant Pathology* **7**: 25, 1993).

Smith (Worthington George; 1835-1917; England). Architect, wood engraver and book illustrator; artist for the *Gardeners' Chronicle* (1869-1910) and *Floral Magazine* (1870-1876). An amateur mycologist and antiquarian who regularly attended meetings of the Woolhope Club (a precursor of the British Mycological Society), making humorous reports in *Gardeners' Chronicle* and a series of cartoons in other periodicals. A founder member and early President of the British Mycological Society, who also restored the models of fungi in the British Museum made by J. Sowerby (q.v.). *Biogs, obits etc.* Ainsworth (*Mycologist* **4**(1): 32, 1990) [portrait]; Stafleu & Cowan (*TL-2* **5**: 706, 1985).

Smithiogaster J.E. Wright (1975), Agaricaceae. 1, Argentina. Basidioma gasteroid. See Wright (*Nova Hedwigia* Beihefte **51**: 360, 1975), Salusso (*Docums Mycol.* **23** no. 91: 13, 1993).

Smithiomyces Singer (1944), Agaricaceae. 2, America (tropical); Europe (introduced). See Singer (*Mycol.* **36**: 366, 1944).

Smithiozyma Van der Walt, Kock & Y. Yamada (1995) = Lipomyces fide Gouliamova *et al.* (*Antonie van Leeuwenhoek* **74**: 283, 1998), Smith *in* Kurtzman & Fell (Eds) (*Yeasts, a taxonomic study* 4th edn: 248,

1998), Kurtzman *et al.* (*FEMS Yeast Res.* **7**: 1027, 2007; phylogeny).

Smittium R.A. Poiss. (1937), Legeriomycetaceae. 75 (in *Diptera*), widespread. See Williams (*Mycol.* **75**: 242, 1983; trichospore germination and zygospores), Lichtwardt (*The Trichomycetes. Fungal associates of arthropods*, 1986; biology, key), Williams & Lichtwardt (*Mycol.* **79**: 832, 1987), Horn (*Mycol.* **81**: 742, 1989; ultrastr.), Horn (*MR* **93**: 303, 1989; spore extrusion), Sato *et al.* (*TMSJ* **30**: 51, 1989; ultrastr.), Horn (*Exp. Mycol.* **14**: 113, 1990; physiology & spore extrusion), Lichtwardt & Williams (*CJB* **68**: 1057, 1990), Williams & Lichtwardt (*CJB* **68**: 1045, 1990), Lichtwardt & Williams (*CJB* **70**: 1193, 1992), Lichtwardt & Williams (*Mycol.* **84**: 392, 1992), Lichtwardt & Williams (*Mycol.* **84**: 384, 1992), Valle & Santamaría (*Mycol.* **96**: 682, 2004; Iberian peninsula), White & Lichtwardt (*Mycol.* **96**: 891, 2004; Norway), Strongman & Xu (*Mycol.* **98**: 479, 2006; China), White (*MR* **110**: 1011, 2006; phylogeny), Strongman (*CJB* **85**: 949, 2007; Canada).

smut (1) a disease caused by one of the *Ustilaginales*, esp. a member of the *Ustilaginaceae* (cf. bunt); (2) a smut fungus. **- spore**, a chlamydospore of a **-** fungus; ustilospore; ustospore; **covered -**, a smut in which the mature spore mass keeps for a time within a covering of host (or fungal) tissue, frequently till after the sorus becomes free from the host: of barley, oats *Ustilago segetium* (syn. *U. hordei*, *U. kolleri*); sorghum, *Sporisorium sorghi*; wheat (**bunt, stinking -**), *Tilletia caries* and *T. laevis* (syn. *T. foetida*); (dwarf bunt), *T. controversa*. **'fig -'**, *Aspergillus niger*; **flag -** of wheat, see stripe smut. **loose -**, a smut in which the spores are as an uncovered mass of powder becoming free from the host plant by wind and rain; of barley and wheat, *Ustilago segetum* var. *tritici* (syn. *U. nuda*); oats, *U. segetum* var. *avenae*; sorghum, *S. cruenta*. **stripe -**, of grasses, *U. striiformis*; rye, *Urocystis occulta*; wheat, *U. agropyri*.

sociation, society, see Phytosociology.

Societies and organizations. In very general terms, mycological societies function at three levels: international, national and local. The *International Mycological Association* (IMA) is global in scope. It constitutes the Section for General Mycology of the International Union of Biological Societies (IUBS), the primary biological component of the International Council of Scientific Unions (ICSU). The IMA also organizes the periodic International Mycological Congresses (q.v.), and has Regional Committees which are mostly organized through a series of affiliated continental-level mycological societies: the African Mycological Association, Asociación Latinoamericana de Micología, Australasian Mycological Society, European Mycological Association, International Mycological Association Committee for Asia and Mycological Society of America. Although the geographical area each covers is still in places ill-defined, these organize periodic continental-level congresses. Within each continent there are various societies functioning at a regional or national level. Some of these societies in practice are international in outlook and may themselves be affiliated to the *IMA*. Many countries, particularly in Africa, Asia, the Caribbean and Oceania, do not have any national mycological society, and some countries have no mycological societies at all, although in others there may be large numbers of local societies.

In many parts of the world there are parallel structures and societies for lichen-forming fungi (the International Association of Lichenologists is associated with the IUBS and participates in International Mycological Congresses), edible and medicinal fungi (the International Society for Mushroom Science arranges independent international congresses), fungi of medical and veterinary significance (the International Society for Human and Animal Mycology arranges independent international congresses), and for fungal plant pathogens (the International Society for Plant Pathology is associated with the IUBS); there are also certain mycological societies which cannot be categorized geographically because they are defined by their specialist interests, such as the International Mycorrhiza Society.

The picture is complicated by the fact that some botanical and microbiological societies also have mycological sections, and some executive functions relating purely to mycology are administered by those bodies. Issues related to the formal system of naming fungi are considered through the Permanent Committee on Fungi which reports through the General Committee on Botanical Nomenclature to International Botanical Congresses (see Nomenclature); the International Association for Plant Taxonomy also covers fungi. Within microbiology, mycological interests are represented by the Mycology Division of the International Union of Microbiological Societies. In addition there are two IUBS/IUMS inter-Union commissions: the International Commission on the Taxonomy of Fungi, and the World Federation for Culture Collections.

Since the previous edn of this *Dictionary*, the number of mycological societies has grown substantially. The *International Mycological Directory*, which has listed these societies since 1971, is now available on-line (see Internet), and should be consulted for up-to-date information.

See also BioNET-INTERNATIONAL, MIRCEN.

sociomycie, see mycosociology.

soft rot, a decomposition of plant parts (fruits, roots, stems, etc.) by fungi or bacteria resulting in the tissues becoming soft.

Soil fungi. The soil normally has large numbers of fungi (mostly anamorphic fungi, *Mucorales*, *Ascomycota*, *Chytridiomycetes* and *Oomycetes*) accompanied by even larger numbers of bacteria (and actinomycetes), although with the exception of rhizosphere soil, the fungal biomass almost everywhere exceeds that of bacteria. Soil fungi are important for keeping the soil fertile, though some may cause disease, particularly of plants. Some components of the soil mycota are cosmopolitan. These often opportunistic species grow when and where suitable nutrition can be found. They are the fungi most frequently isolated from soil using conventional techniques, and have been termed ruderals. Although cosmopolitan, environment, geographical location, climate and the soil physico-chemical properties have an effect on their proportions within the soil: for example *Penicillium* is more frequent than *Aspergillus* in temperate areas while in warmer areas the opposite is true. The mycelium of basidiomycetes is most frequent under trees and contributes to a second main category of soil fungi, the mycorrhizal species. Usually more than half of the fungal biomass is of basidiomycetes, although these are missed by most isolation techniques.

Peat bogs have large numbers of fungi (Dooley & Dickinson, *Irish J. agric. Res.* **10**: 195, 1971). See also Sand dune fungi.

The chief work of soil fungi relates to decomposition and recycling organic residues. Fungi are important in the early stages of the decomposition of plant material, immobilizing large amounts of nitrogen (N) as protein. Cellulose and lignin are the most important plant cell wall components that are mainly decomposed in litter and soil by fungi. For the latter mainly basidiomycetes but also some ascomycetes are responsible. Lichenized fungi with cyanobacterial partners take a part in the nitrogen cycle, as these fix nitrogen from the air.

There are complex biological relations between the different soil organisms themselves (see Antagonism) and between them and plants (see Mycorrhiza, rhizosphere) on which they may be parasites. Reinking and Mann (fide Garrett, 1939) grouped soil fungi as, (1) having a wide distribution (the 'soil inhabitants') and (2) the 'soil invaders', having a limited distribution, e.g. most parasitic root-infecting fungi.

A common method for the investigation of the soil biota is that of plating a series of dilutions of the soil (for problems in quantification see James and Sutherland, *Can. J. Res.* C **17**: 73, 97, **18**: 347, 435, 1939-40). Other techniques used are 'direct observation' (Conn, *Soil Sci.* **26**: 257, 1928), the Cholodny slide (Conn, *Tech. Bull. N.Y. Sta. agric. Exp. Stn* **204**, 1932), and Warcup's soil plate (*Nature* **166**: 117, 1950). Washing roots (Harley & Waid, *TBMS* **38**: 105, 1955) or soil particles (Parkinson & Williams; Gams & Domsch, *Arch. Microbiol.* **58**: 134, 1967; Bååth, *CJB* **66**: 1566, 1988), Bills & Polishook (*Mycol.* **86**: 187, 1994) shifts the balance in favour of slow-growing and poorly sporulating fungi that may be active as mycelium in the soil.

Lit.: Allsopp *et al.* (Eds) (*Microbial diversity and ecosystem function.* pp. 321-336, 2000), Arora *et al.* (*Handbook of Applied Mycology* **1**, 1991), Baker & Snyder (Eds) (*Ecology of soil-borne plant pathogens*, 1965), Barron (*The genera of hyphomycetes from soil*, 1968), Boland & Kuykendall (Eds) (*Plant-Microbe Interactions and Biological Control.* 1998), Bruehl (*Soil-borne plant pathogens*, 1987), Brussaard & Ferrera-Cerrato (*Soil ecology in sustainable agricultural systems.* 1997), Burges (*Micro-organisms in the soil*, 1958; gen. ecol. survey), Carroll & Wicklow (Eds) (*The fungal community*, edn 2, 1992), Cooke & Rayner (*Ecology of saprotrophic fungi*, 1984), Davet & Rouxel (*Detection and isolation of soil fungi*, 2000), Dehne *et al.* (Eds) (*Diagnosis and identification of plant pathogens.* pp. 503-505, 1997), Dix & Webster (*The ecology of fungi*, 1995), Domsch & Gams (*Pilze aus Agrarböden*, 1970 [Engl. transl; Hudson, *Fungi in agricultural soils*, 1972]), Domsch *et al.* (*Compendium of soil fungi*, 2 vols, 1980, [supplemented reprint, 1993]; keys, descr., monogr. 400 spp., 6593 refs), Durbin (*Bot. Rev.* **27**: 522, 1961; isolation techniques), Frankland (*Trans. mycol. Soc. Japan* **31**: 89, 1990), Gams (*in* Winterhoff (Ed.), *Fungi in Vegetation Science.* pp. 182-223, 1992), Garrett (*Soil fungi and soil fertility*, edn 2, 1981), Gilman (*A manual of soil fungi*, edn 2, 1957; systematics), Gray & Williams (*Soil micro-organisms*, 1971), Griffin (*Ecology of soil fungi*, 1972), Hall (Ed.) (*Managing soilborne plant pathogens.* pp. 250-278, 1996), Hillocks & Waller (Eds) (*Soilborne Diseases of Tropical Crops.* pp. 3-16, 1997), Hornby (Ed.) (*Biological control of soil-borne plant pathogens*, 1990), Jensen *et al.* (Eds) (*Microbial communities in soil*, 1986), Johnson & Curl (*Methods for studying soil microflora-plant disease relationships*, edn 2, 1972), Litvinov (*Key to microscopic soil fungi*, 1967 [Russ.]), Lockwood (*Biol. Rev.* **52**: 1, 1977; fungistasis), Müller *et al.* (Eds) (*Measuring and monitoring biological diversity: Standard methods for Fungi.* 2000), Parkinson & Waid (Eds) (*The ecology of soil fungi*, 1960), Pfenning (*in* Hyde (Ed.), *Biodiversity of tropical microfungi.* pp. 341-365, 1997), Rayner & Boddy (*Fungal decomposition of wood, its biology and ecology*, 1988), Rossen *et al.* (Eds) (*Fungal identification techniques.* 1996), Schippers & Gams (Eds) (*Soil-borne plant pathogens*, 1979), Seifert & Gams (*in* McLaughlin *et al.* (Eds), *The Mycota*, **7A**. pp. 307-347, 2000), Singleton *et al.* (*Methods for research on soilborne phytopathogenic fungi*, 1992), Sneh *et al.* (Eds) (*Rhizoctonia species.* pp. 423-432, 1996), Tjamos *et al.* (Eds) (*Biocontrol of plant diseases.* pp. 75-78, 1992), Wainwright (*TBMS* **90**: 159, 1988), Waksman (*Principles of soil microbiology*, edn 2, 1931; *Soil microbiology*, 1952), Watanabe (*Pictorial atlas of soil and seed fungi: morphologies of cultured fungi*, 2002). See dermatophytes, Isolation methods, Medical and Veterinary Mycology, Mycorrhiza, Root rots.

soiling, the disfiguring of paint, fruit, or other materials by pigmented moulds.

Solanella Vaňha (1910), ? Helotiales. 1, Europe.

Soleella Darker (1967), Rhytismataceae. 4, N. America; China. See Darker (*CJB* **45**: 1427, 1967), Hou *et al.* (*CJB* **83**: 37, 2005), Hou *et al.* (*Nova Hedwigia* **83**: 511, 2006).

soleiform, shaped like the sole of a shoe, i.e. elongate-ellipsoid, with one cell much larger and broader than the other.

Solenarium Spreng. (1827) = Glonium fide Fries (*Syst. mycol.* **2**: 594, 1823).

Solenia Hill ex Kuntze (1898) ≡ Pinuzza.

Solenia Pers. (1794) [non *Solena* Lour. 1790, *Cucurbitaceae*] ≡ Henningsomyces Kuntze.

Solenodonta Castagne (1845) = Puccinia fide Saccardo (*Syll. fung.* **18**: 824, 1906).

Solenographa A. Massal. (1860) = Diorygma fide Hawksworth *et al.* (*Dictionary of the Fungi* edn 8, 1995), Kalb *et al.* (*Symb. bot. upsal.* **34** no. 1: 133, 2004).

Solenopezia Sacc. (1889), Hyaloscyphaceae. 7, widespread. See Raitviir *et al.* (*Sydowia* **43**: 219, 1991), Raitviir *et al.* (*Docums Mycol.* **25**: 359, 1995), Cantrell & Hanlin (*Mycol.* **89**: 745, 1997; DNA).

Solenoplea Starbäck (1901) = Camarops fide Nannfeldt (*Svensk bot. Tidskr.* **66**: 335, 1972).

Solenopsora A. Massal. (1855), Lecanorales (L). 19, widespread. See Verdon & Rambold (*Mycotaxon* **69**: 399, 1998), Kantvilas (*Lichenologist* **36**: 113, 2004; Tasmania), Guttová *et al.* (*Central European Lichens*: 85, 2006; morphology, chemistry), Naesborg *et al.* (*MR* **111**: 581, 2007; phylogeny).

Solheimia E.F. Morris (1967), anamorphic *Pezizomycotina*, Hsy.0eP.?. 1, C. America. See Morris (*Mycopathologia* **33**: 181, 1967), Bills *et al.* (*Sydowia* **46**: 1, 1994).

Solicorynespora R.F. Castañeda & W.B. Kendr. (1990), anamorphic *Pezizomycotina*, Hso.≡ eP.24. 8, widespread. See Castañeda Ruiz & Kendrick (*Univ.

Waterloo Biol. Ser. **33**: 38, 1990), Castañeda Ruíz (*Mycotaxon* **59**: 449, 1996; Cuba).

Soloacrospora W.B. Kendr. & R.F. Castañeda (1991), anamorphic *Pezizomycotina*, Hso.1eH.12. 1, Cuba. See Castañeda Ruiz & Kendrick (*Univ. Waterloo Biol. Ser.* **35**: 102, 1991), Castañeda Ruíz *et al.* (*Nova Hedwigia* **64**: 473, 1997).

solopathogenic (of a smut such as *Ustilago zeae*), a pathogenic monosporidial line (Christensen).

Solorina Ach. (1808), Peltigeraceae (L). 10, widespread. See Hue (*Mém. Soc. natn Sci. nat. math. Cherbourg* **38**: 1, 1911), Thomson & Thomson (*Bryologist* **87**: 151, 1984; SEM), Jahns *et al.* (*Biblthca Lichenol.* **57**: 161, 1995; ontogeny), Galloway (*Cryptog. Bryol.-Lichénol.* **19**: 137, 1998; NZ), Martínez & Burgaz (*Ann. bot. fenn.* **35**: 137, 1998; Europe), Miądlikowska & Lutzoni (*Am. J. Bot.* **91**: 449, 2004; phylogeny), Wang *et al.* (*Lichenology* **4**: 1, 2005; photomorph), Miądlikowska *et al.* (*Mycol.* **98**: 1088, 2006; phylogeny).

Solorinaceae Bayrh. (1851) = Peltigeraceae.

Solorinaria (Vain.) Gyeln. (1935), Lichinaceae (L). 2, Europe.

Solorinella Anzi (1860) nom. rej. = Gyalidea fide Thor (*Nordic Jl Bot.* **4**: 823, 1985), Vězda & Poelt (*Phyton Horn* **30**: 47, 1990), Farkas & Lökös (*Acta Bot. Fenn.* **150**: 21, 1994; distrib.), Henssen & Lücking (*Ann. bot. fenn.* **39**: 273, 2002; ontogeny, morphology), Aptroot & Lücking (*Biblthca Lichenol.* **86**: 53, 2003; morphology).

Solorinellaceae Vězda & Poelt (1990) = Asterothyriaceae. See Miądlikowska *et al.* (*Mycol.* **98**: 1088, 2006; phylogeny).

Solorinellomyces Cif. & Tomas. (1953) ≡ Solorinella.

Solorinina Nyl. (1884) = Solorina fide Hawksworth *et al.* (*Dictionary of the Fungi* edn 8, 1995).

Solorinomyces E.A. Thomas ex Cif. & Tomas. (1953) = Solorina.

Solosympodiella Matsush. (1971), anamorphic *Pezizomycotina*, Hso.0-1eH.10. 7, Papua New Guinea. See de Hoog (*Stud. Mycol.* **26**: 54, 1985; key).

Soloterminospora Matsush. (1996), anamorphic *Pezizomycotina*, Hso.?.?. 1, Cape Province. See Matsushima (*Matsush. Mycol. Mem.* **9**: 25, 1996).

Solutoparies Whiffen ex W.H. Blackw. & M.J. Powell (1998), Chytridiaceae. 1, USA. See Blackwell & Powell (*Mycotaxon* **67**: 464, 1998).

soma (1) body, excluding reproductive parts or phase; (2) in Aryan religion, *Amanita muscaria* (Allegro, *The sacred mushroom and the cross*, 1970; Wasson, *Soma: the divine mushroom of immortality*, 1968, *The wondrous mushroom*, 1980; Whittier, *The brewing of soma*, 1872); **-tic**, pertaining to the soma. See also ethnomycology, hallucinogenic fungi.

Somatexis Toro (1934) = Aphanostigme fide von Arx & Müller (*Stud. Mycol.* **9**, 1975).

somatogamy, fusion of somatic (vegetative) cells or hyphae involving plasmogamy but not karyogamy.

Somion Adans. (1763) nom. rej. = Spongipellis fide Donk (*Verh. K. ned. Akad. Wet.* tweede sect. **62**: 1, 1974).

Sommerfeltia Flörke ex Sommerf. (1827) nom. rej. prop. = Solorina fide Zahlbruckner (*Catalogus Lichenum Universalis* **3**, 1925).

Sonderhenia H.J. Swart & J. Walker (1988), anamorphic *Mycosphaerella*, Cpd.≡ eP.19. 1, widespread. See Swart & Walker (*TBMS* **90**: 640, 1988), Crous (*Mycol. Mem.* **21**: 170 pp., 1998), Verkley & Priest (*Stud. Mycol.* **45**: 123, 2000), Crous *et al.* (*MR* **105**: 425, 2001; phylogeny), Hunter *et al.* (*MR* **108**: 672, 2004; S Africa), Hunter *et al.* (*Stud. Mycol.* **55**: 147, 2006; phylogeny).

sooty moulds, used for fungi forming dense dark mats of hyphae on living leaves in the tropics, esp. *Capnodiales* (fams: *Asterinaceae*, *Antennulariaceae*, *Capnodiaceae*, *Chaetothyriaceae*, *Euantennariaceae*, *Meliolinaceae*, *Metacapnodiaceae*), *Chaetothyriales* (fam.: *Chaetothyriaceae*), *Meliolales* (fam.: *Meliolaceae*), and their anamorphs. See Hughes (*Mycol.* **68**: 693, 1976; genera and interrelationships). Also *Lit.* under *Dothideales*.

Sopagraha Subram. & Sudha (1979) = Arachnophora fide Sutton (*in litt.*).

Sophronia Pers. (1827) [non *Sophronia* Licht. ex Roem. & Schult. 1817, *Iridaceae*] = Phallus fide Stalpers (*in litt.*).

Soppittiella Massee (1892) = Cristella fide Donk (*Persoonia* **4**: 329, 1966).

soralium (pl. **-ia**), decorticate portions of a lichen thallus where **soredia** are located. Usually formed from medullary tissues thrusting upwards through the cortical layers and so sometimes with the chemistry of the medulla rather than of the cortex. Soralia may be **diffuse** (the upper surface of the lichen becoming a continuous soredial mass), or **delimited** (i.e. confined to well-defined areas), and can be classified according to where they originate. They can arise on tubercles (**tuberculate -**) or as fissures (**fissural -**) in some genera (Hawksworth, *Lichenologist* **5**: 181, 1972). Soralia can arise at the tips of isidia, and in some taxa the soralia can contain a mixture of soredia and isidia-like structures.

Lit.: Du Rietz (*Svensk bot. Tidskr.* **18**: 371, 1924; soralium types).

See Spore discharge and dispersal, Spore germination.

Sorataea Syd. (1930), Uropyxidaceae. *c.* 6 (on *Leguminosae*), S. America; Asia; W. Africa. See Eboh & Cummins (*Mycol.* **72**: 203, 1980), Cummins & Hiratsuka (*Illustr. Gen. Rust Fungi rev. edit.*, 1983).

Sorauer (Paul Carl Moritz; 1839-1916; Germany). Student, Berlin (1862); Doctorate, Rostock (1867); Assistant, Agricultural Museum, Berlin (1867-1868); Assistant, Agricultural Experiment Station, Dahme, Brandenburg (1868-1872); Head, Experimental Station for Plant Physiology, Imperial Cider Institute, Proskau (1872-1893); retired to Berlin. First editor of *Zeitschrift für Pflanzenkrankheiten* (1891-1916). *Publs. Handbuch der Pflanzenkrankheiten* (1874) [2 Aufl., 1886; 3 Aufl. (with Lindau & Rehm), 1905-1908]; *Atlas der Pflanzenkrankheiten* (1887 1893). *Biogs, obits etc.* Stafleu & Cowan (*TL-2* **5**: 749, 1985); Wittmack (*Bericht der Deutschen Botanischen Gesellschaft* **34**: (50), 1917).

Sordaria Ces. & De Not. (1863) nom. cons., Sordariaceae. 12 (esp. coprophilous), widespread. See Lundqvist (*Symb. bot. upsal.* **20** no. 1, 1972), Read & Beckett (*CJB* **63**: 281, 1985; ascoma ontogeny), Barrasa *et al.* (*Persoonia* **13**: 83, 1986; narrow-spored spp.), Guarro & von Arx (*Persoonia* **13**: 369, 1987; key 14 spp., list spp. names), Beatty *et al.* (*MR* **98**: 1309, 1994; genetics), Pöggeler (*Curr. Genet.* **36**: 222, 1999; mating type genes), Huhndorf *et al.* (*Mycol.* **96**: 368, 2004; phylogeny), Miller & Huhndorf (*Mol. Phylogen. Evol.* **35**: 60, 2005; phylogeny), Cai *et al.* (*MR* **110**: 137, 2006; phylogeny), Rottenberg *et* *al.* (*Mycol. Balcanica* **3**: 1, 2006; genetic diversity), Zhang *et al.* (*Mycol.* **98**: 1076, 2006; phylogeny).

Sordariaceae G. Winter (1885), Sordariales. 8 gen. (+ 7 syn.), 94 spp.

Lit.: Lundqvist (*Symb. bot. upsal.* **20** no. 1, 1972), Read & Beckett (*CJB* **63**: 281, 1985), Perkins & Raju (*Exp. Mycol.* **10**: 323, 1986), Perkins & Turner (*Exp. Mycol.* **12**: 91, 1988), Glass *et al.* (*Exp. Mycol.* **14**: 274, 1990), Perkins (*Genetics* Bethesda **130**: 687, 1992), Beatty *et al.* (*MR* **98**: 1309, 1994; genetics), Krug *et al.* (*Mycol.* **86**: 250, 1994), Skupski *et al.* (*Fungal Genetics Biol.* **21**: 153, 1997), Pöggeler (*Curr. Genet.* **36**: 222, 1999), Dettman *et al.* (*Fungal Genetics Biol.* **34**: 49, 2001), Dettman & Taylor (*Genetics* Bethesda **168**: 1231, 2004), García *et al.* (*MR* **108**: 1119, 2004), Huhndorf *et al.* (*Mycol.* **96**: 368, 2004), Jacobson *et al.* (*Mycol.* **96**: 66, 2004), Cai *et al.* (*MR* **110**: 137, 2006), Dettman *et al.* (*Mycol.* **98**: 436, 2006).

Sordariales Chadef. ex D. Hawksw. & O.E. Erikss. (1986). Sordariomycetidae. 5 fam., 97 gen., 854 spp. Stromata absent or rarely present as subicular tissue. Ascomata perithecial or cleistothecial, thin- or thick-walled, often hairy, membranous to carbonaceous, olivaceous to black. Interascal tissue composed of wide thin-walled evanescent paraphyses, or lacking altogether. Asci cylindrical or clavate, persistent or evanescent, not fissitunicate, when persistent usually with a small J- apical ring. Ascospores ± always with at least one dark cell, with germ pore, often with a gelatinous sheath or appendages. Anamorphs usually absent or spermatial. Saprobes on rotting wood and soil, coprophilous, some fungicolous, many cellulolytic. Fams:

(1) **Cephalothecaceae**
(2) **Chaetomiaceae** (syn. *Achaetomiaceae*)
(3) **Lasiosphaeriaceae** (syn. *Tripterosporaceae*)
(4) **Sordariaceae** (syn. *Neurosporaceae*)

Lit.: von Arx & Müller (1954), von Arx (1981, *Sydowia* **34**: 13, 1982; fams.), von Arx *et al.* (*Beih. Nova Hedw.* **94**, 1988), Huhndorf *et al.* (*Mycol.* **96**: 368, 2004; phylogeny), Lumbsch & Huhndorf (*MR* **111**: 1064, 2007), Lundqvist (*Symb. bot. upsal.* **20**(1), 1972), Moreau (*Les genres Sordaria et Pleurage*, 1953), Müller & von Arx (1973).

Sordariella J.N. Kapoor, S.P. Lal & Bahl (1975) = Herpotrichia fide von Arx & Müller (*Sydowia* **37**: 6, 1984).

Sordariomycetes O.E. Erikss. & Winka (1997), Pezizomycotina. 15 ord., 64 fam., 1119 gen., 10564 spp. Contains most non-lichenized perithecial ascomycetes with non-fissitunicate asci. Subclass:

(1) **Hypocreomycetidae**
(2) **Sordariomycetidae**
(3) **Xylariomycetidae**

Lit.: Tang *et al.* (*Ant. v. Leeuwenh.* **91**: 327, 2007), Zhang *et al.* (*Mycol.* **98**: 1076, 2006).

Sordariomycetidae O.E. Erikss. & Winka (1997), Sordariomycetes. Ords:

(1) **Boliniales**
(2) **Calosphaeriales**
(3) **Chaetosphaeriales**
(4) **Coniochaetales**
(5) **Diaporthales**
(6) **Ophiostomatales**
(7) **Sordariales**

Lit.: von Arx & Müller (*Beitr. Kryptogfl. Schweiz* **11**(1), 1954), Barr (*Mycotaxon* **39**: 43, 1990), Müller & von Arx (*Beitr. Kryptogfl. Schweiz* **11**(2), 1962), Müller & von Arx (*in* Ainsworth *et al.* (Eds), *The Fungi* **4A**, 1973), Samuels & Blackwell (*in* McLaughlin *et al.* (Eds), *The Mycota* **7A**: 221, 2001).

soredium (pl. **-ia**), a non-corticate combination of phycobiont cells and fungal hyphae having the appearance of a powdery granule, and capable of reproducing a lichen vegetatively (Fig. 22J-L). For their liberation and dispersal see Bailey (*J. Linn. Soc., Bot.* **59**: 479, 1966, *Revue bryol. lichén.* **36**: 314, 1969, *in* Brown *et al.* (Eds), *Lichenology: progress & problems*: 215, 1976). Cf. soralium.

Soredospora Corda (1837) = Fumago fide Hughes (*CJB* **36**: 727, 1958).

Sörensen coefficient (**K**). A numerical estimate of the affinity between two biotas; $K = {}^{200xc}/_{a+b}$, where a = species in one region, b = species in second region, and c = number of species in common. See Poore (*J. Ecol.* **43**: 606, 1955).

Soreymatosporium Sousa da Câmara (1930) = Stemphylium fide Wiltshire (*TBMS* **21**: 211, 1938).

Sorica Giesenh. (1904) = Caliciopsis fide Fitzpatrick (*Mycol.* **42**: 464, 1942).

sorocarp (of *Acrasiales*), a stalked fruiting structure.

Sorochytriaceae Dewel (1985), Blastocladiales. 1 gen., 1 spp.

Lit.: Dewel *et al.* (*CJB* **63**: 1525, 1985), Dewel & Dewel (*CJB* **68**: 1968, 1990).

Sorochytrium Dewel (1985), Sorochytriaceae. 1 (on *Milnesium* (tardigrade)), USA. See Dewel & Dewel (*CJB* **68**: 1968, 1990; ultrastr.).

Sorocybe Fr. (1849), anamorphic *Herpotrichiellaceae*, Hso.0eP.3. 1 (on resin), widespread. See Hughes (*CJB* **36**: 727, 1958), Partridge & Morgan-Jones (*Mycotaxon* **83**: 335, 2002), Seifert *et al.* (*Stud. Mycol.* **58**: 235, 2007; phylogeny).

sorocyst (of *Acrasiales*), a sorocarp lacking a stalk.

Sorokina Sacc. (1892), Helotiales. *c.* 6, widespread (pantropical). See Dennis (*Kew Bull.* **13**: 321, 1958), Spooner *et al.* (*Kew Bull.* **53**: 237, 1998).

Sorokinella J. Fröhl. & K.D. Hyde (2000), Helotiales. 2, Australia; Hong Kong. See Fröhlich & Hyde (*Fungal Diversity Res. Ser.* **3**: 122, 2000).

sorophore (of *Acrasiales*), a sorocarp stalk.

Sorosporella Sorokīn (1888), anamorphic *Hypocreales*, Hso.0eH.?. 1 (on insects), widespread (north temperate). See Speare (*J. agric. Res.* **18**: 399, 1920), Pendland & Boucias (*Mycopathologia* **99**: 25, 1987; SEM and TEM), Evans & Shah (*MR* **106**: 737, 2002; ecology, morphology).

Sorosporium F. Rudolphi (1829) nom. rej. = Thecaphora fide Vánky (*in litt.*) cf. *Sporisorium*.

Sorosporonites X. Mu (1977), Fossil Fungi. 1, China.

Sorothelia Körb. (1865) ? = Endococcus fide Hawksworth *et al.* (*Dictionary of the Fungi* edn 8, 1995).

sorus, a fruiting structure in certain fungi, esp. the spore mass in *Pucciniales* and *Ustilaginales*; a group of fruit-bodies, as in *Synchytriaceae*.

Sousa da Câmara (Manuel; 1871-1955; Portugal). Member of staff (from 1901), Lecturer (from 1905), then Senior Lecturer and Researcher (working on fungal phytopathology), then Professor of Parasitology & Phytopathology (1914-1941), also Head of Laboratory of Plant Pathology, Vice-Director (1912-1918) and Director (1918-1942), Institututo Superior de Agronomia e Veterinária, Lisbon. An outstanding taxonomist of fungi significant in plant pathology, particularly coelomycetes; he was the author or co-author of 491 taxa; through publication of his series of reports and a comprehensive catalogue of fungi (1902-1955) he made a significant contribution to the knowledge of the mycota of Portugal. *Publs. Mycetes aliquot Lusitaniae* I-XII (1936-1951); *Fungi Lusitaniae* I-XIII (1952-1955); *Catalogus Systematicus Fungorum Omnium Lusitaniae*, 2 vols. (1956-1958). *Biogs, obits etc.* Natividade (*Sep. Bol. Soc. Port. Ciências Naturais*, 2a Serie, **5**: I, 1955); Oliveira (*Agronomia Lusitana* **17**: I, 1955) [portrait]; Stafleu & Cowan (*TL-2* **5**: 758, 1985).

Sowerby (James; 1757-1822; England). Born, London; trained as an artist, became a teacher of drawing and painting. Famous illustrator of fungi and plants; also prepared models of larger edible and poisonous fungi, now in the British Museum (Natural History), London; advised British government regarding fungal decay of wooden warships during the Napoleonic wars; his children and their descendants included a dozen naturalists and illustrators. *Publs. Coloured Figures of English Fungi or Mushrooms* 3 vols (1795-1803) [suppl. 1809-1815; for dates of publication see Ramsbottom *TBMS* **18**: 167, 1933]. *Biogs, obits etc.* Ainsworth (*Mycologist* **2**: 125, 1988) [portrait]; Anon. (*Oxford Dictionary of National Biography* concise edn **3**: 2802, 1993); Grummann (1974: 381); Stafleu & Cowan (*TL-2* **5**: 759, 1985).

Sowerbyella Nannf. (1938), Pyronemataceae. 16, Europe; China. See Jeppson (*Göteborgs Svampklubbs Årsskr*: 9, 1980; key 3 spp.), Moravec (*Mycol. Helv.* **1**: 427, 1985; key 9 spp.), Moravec (*Česká Mykol.* **42**: 193, 1988; key 12 spp.), Klofac & Voglmayr (*Öst. Z. Pilzk.* **12**: 141, 2003; Austria), Hansen & Pfister (*Mycol.* **98**: 1029, 2006; phylogeny), Yao & Spooner (*Fungal Diversity* **22**: 267, 2006; UK), Perry *et al.* (*MR* **111**: 549, 2007; phylogeny).

soy sauce, see shoyu.

Spadicesporium V.N. Boriss. & Dvoïnos (1982) ? = Cladosporium fide Sutton (*in litt.*).

Spadicoides S. Hughes (1958), anamorphic *Tengiomyces*, Hso.≡ eP.27. 42, widespread. See Wang (*Mem. N. Y. bot. Gdn* **28**: 218, 1976), Sinclair *et al.* (*TBMS* **85**: 736, 1985), Kuthubutheen & Nawawi (*MR* **95**: 163, 1991), Goh & Hyde (*CJB* **76**: 1698, 1998), Réblová (*Mycotaxon* **70**: 387, 1999; teleomorph), Ho *et al.* (*Mycol.* **94**: 302, 2002; SE Asia), Cai *et al.* (*Sydowia* **56**: 222, 2004; China).

Spadonia Fr. (1829) nom. dub., ? Fungi.

Spalovia Nieuwl. (1916) ≡ Spermotrichum.

spalted (of wood), with dark zone lines (q.v.) due to the interactions of different fungal colonies, and with paler, darker or otherwise differently coloured areas marked off by those zone lines; as caused, e.g. by *Diatrype disciformis*; generally brittle, short-grained, and easily breaking through decay; if harvested before the decay is too advanced, such wood may be decorative and highly prized for some non-structural woodwork, such as wooden bowls or other artefacts.

Sparassidaceae Herter (1910), Polyporales. 2 gen. (+ 1 syn.), 6 spp.

Lit.: Donk (*Persoonia* **3**: 199, 1964), Burdsall & Miller (*Mycotaxon* **31**: 199, 1988), Burdsall & Miller (*Mycotaxon* **31**: 591, 1988), Desjardin *et al.* (*Mycol.* **96**: 1010, 2004), Wang *et al.* (*Mycol.* **96**: 1015, 2004), Dai *et al.* (*Mycol.* **98**: 584, 2006).

Sparassiella Schwarzman (1964), Sparassidaceae. 1, former USSR. See Schwarzman (*Flora Sporovykh Rasteniĭ Kazakhstana* **4**: 159, 1964).

Sparassis Fr. (1819), Sparassidaceae. 5, widespread (north temperate). *Sparassis crispa* is edible. See Burdsall & Miller (*Mycotaxon* **31**: 199, 1988), Desjardin *et al.* (*Mycol.* **96**: 1010, 2004; Thailand), Wang *et al.* (*Mycol.* **96**: 1015, 2004), Blanco-Dios *et al.* (*MR* **110**: 1227, 2006; Spain).

sparassoid, composed of interlaced flabelliform branches forming ball-like structures recalling *Sparassis* basidiomata. Known also in *Pezizales* (Korf, *Rept Tottori mycol. Inst.* **10**: 389, 1973).

sparassol, orsellinic acid monomethyl ether from *Sparassis ramosa*; antifungal.

Sparrow (Frederick Kroeber; 1903-1977; USA). PhD, Harvard (1929); Instructor, Dartmouth College (1929-1932); Research Fellow, Woods Hole Oceanographic Institution (1934-1936); Assistant Professor (1936-1949) then Professor of Botany (1949-1973) and Director of the Biological Station (1968-1973), University of Michigan, Ann Arbor. President of the second International Mycological Congress, Tampa, Florida (1977). *Publs. Aquatic Phycomycetes* (1943) [edn 2, 1960]; (with Ainsworth & Sussman) *The Fungi, an Advanced Treatise* 4 vols (1965-1973). *Biogs, obits etc.* Grummann (1974: 297); Paterson (*Mycol.* **70**: 213, 1978) [bibliography, portrait]; Stafleu & Cowan (*TL-2* **5**: 775, 1985).

Sparrowia Willoughby (1963), Chytridiaceae. 2 (on ? *Pythium* and chytrids), British Isles; USA. See Willoughby (*Nova Hedwigia* **5**: 336, 1963).

Sparsitubaceae Jülich (1982) = Polyporaceae.

Sparsitubus L.W. Hsu & J.D. Zhao (1980), Polyporaceae. 1, China. See Hsu & Zhao (*Acta Microbiol. Sin.* **20**: 236, 1980).

Spartiella Tuzet & Manier ex Manier (1968), Legeriomycetaceae. 2 (in *Ephemeroptera*), Europe; C. America. See Lichtwardt (*The Trichomycetes. Fungal associates of arthropods*, 1986), Lichtwardt (*Revta Biol. trop.* **45**: 1349, 1997; Costa Rica).

Spathaspora Nguyen, S.O. Suh & M. Blackw. (2006), Saccharomycetales. Anamorph *Candida*. 1 (associated with bark-boring beetles), USA. See Nguyen *et al.* (*MR* **110**: 1237, 2006).

Spathularia Pers. (1797), Cudoniaceae. 12, widespread (north temperate). See Maas Geesteranus (*Proc. K. ned. Akad. Wet.* Ser. C, Biol. Med. Sci. **75**: 243, 1972), Benkert (*Gleditschia* **10**: 141, 1983), Stellmacher (*Abh. Naturhist. Ges. Nüremberg* **40**: 78, 1985), Ohenoja (*Op. Bot.* **100**: 193, 1989; nomencl.), Landvik *et al.* (*Mycoscience* **37**: 237, 1997; phylogeny), Döring & Triebel (*Cryptog. Bryol.-Lichénol.* **19**: 123, 1998; phylogeny), Gernandt *et al.* (*Mycol.* **93**: 915, 2001; phylogeny), Wang *et al.* (*Mycol.* **94**: 641, 2002), Wang *et al.* (*Mycol.* **98**: 1065, 2006; phylogeny), Wang *et al.* (*Mol. Phylogen. Evol.* **41**: 295, 2006; phylogeny).

Spathulariopsis Maas Geest. (1972) = Spathularia fide Otani (*Rep. Crypt. Study Nepal*: 75, 1982), Wang *et al.* (*Mycol.* **94**: 641, 2002).

spathulate, like a spoon in form (Fig. 23.19).

Spathulea Fr. (1825) ≡ Spathularia.

Spathulina Pat. (1900) nom. dub., Agaricomycetes.

Spathulospora A.R. Caval. & T.W. Johnson (1965), Spathulosporaceae. 5 (on *Ballia* spp.), widespread. See Kohlmeyer & Kohlmeyer (*Marine Mycology*, 1979; key), Walker *et al.* (*TBMS* **73**: 193, 1979; host-parasite interface), Inderbitzin *et al.* (*MR* **108**: 737, 2004; phylogeny), Campbell *et al.* (*MR* **109**: 556, 2005; phylogeny), Zhang *et al.* (*Mycol.* **98**: 1076,

2006; phylogeny).

Spathulosporaceae Kohlm. (1973), Lulworthiales. 2 gen., 6 spp.

Lit.: Kohlmeyer (*Mycol.* **65**: 614, 1973), Kohlmeyer & Kohlmeyer (*Mycol.* **67**: 629, 1975), Kohlmeyer & Volkmann-Kohlmeyer (*Bot. Mar.* **34**: 1, 1991), Nakagiri & Ito (*Mycol.* **89**: 484, 1997), Inderbitzin *et al.* (*MR* **108**: 737, 2004).

Spathulosporales Kohlm. (1973) = Lulworthiales.

Spathulosporomycetidae, Locq. (1984), see *Sordariomycetes*.

Spatulosporonites Z.C. Song (1999), Fossil Fungi. 1. See Song *et al.* (*Fossil Spores and Pollen of China* **1**: 825, 1999).

spawn (1) (n.), mycelium, esp. that used for starting mushroom cultures (q.v.); (2) (v.), to put inoculum (spawn) into a mushroom bed or other substrate.

special form (**forma specialis**, **f. sp.**; **formae speciales**, **ff. spp.**, pl.), an informal rank in a Classification (q.v.) not regulated by the Code (see Nomenclature) and used for parasitic fungi characterized from a physiological standpoint (e.g. by the ability to cause disease in particular hosts) but scarcely or not at all from a morphological standpoint; e.g. *Fusarium oxysporum* f.sp. *cubense* (Panama disease of banana) and f.sp. *elaeidis* (wilt of oil palm); first used by Eriksson (*Ber. dtsch bot. Ges.* **12**: 292, 1894) in rusts, and now extensively used in some genera (over 120 ff. spp. in *F. oxysporum*; Armstrong & Armstrong *in* Nelson *et al.* (Eds), *Fusarium: diseases, biology, taxonomy*: 391, 1981), although not always in a parallel manner; best viewed as a temporary category while further research as to taxonomic status is in progress (Hawksworth, *in* Hawksworth (Ed.), *The identification and characterization of pest organisms*: 93, 1994); comparable to 'pathovar' as used for plant pathogenic bacteria (Young *et al.*, *ROPP* **70**: 211, 1991). See Plant pathology.

species (1) (colloquially), one sort of organism; (2) (scientifically), the lowest principal rank in the nomenclatural hierarchy (see Classification), consisting of two elements (a binomial): a generic name and a species epithet.

There is much debate as to how species should be defined, and a variety of concepts have been proposed: (1) **morpho -** (**morphological -**, **phenetic -**), the traditional approach recognizing units that could be delimited on the basis of morphological characters, and ideally by discontinuities in several such; (2) **biological -** (**cryptic-**, **sibling -**), actually or potentially interbreeding populations reproductively isolated from other such groups, whether or not they are distinguishable morphologically; (3) **phylogenetic -** (**evolutionary -**), based on measureable differences in biochemical, molecular or any other characters assessed by cladistic analysis (see Cladistics) and especially well-suited for groups in which no sexual reproduction is known (e.g. **clonal -**, anamorphic fungi); (4) **ecological -**, based on adaptation to particular niches rather than reproductive isolation (e.g. to particular hosts); (5) **polythetic -**, based on a combination of characters, not all of which are necessarily present in each individual. The **aggregate -** (**aggr.**), rarely used in mycology, has been used for groups of closely related morphospecies only distinguishable with difficulty.

The morphospecies concept has predominated in mycology, although in the past some mycologists have frequently taken the host of a parasitic fungus into account (in some cases giving different specific names to like forms on unlike hosts). Population studies (see Population biology) and molecular data are, however, increasingly showing that many morphospecies comprise several biological and(or) phylogenetic species; it is likely that species characterized other than by morphology will increasingly be recognized, especially where these cause different plant diseases (see also special forms).

In lichen-forming fungi, **- pairs** (**Artenpaare**) are also recognized. Otherwise identical species one of which is fertile (the **primary -**), and the other reproduces vegetatively (the **secondary -**). Primary and secondary species generally have identical secondary metabolites, but the secondary species tend to have much wider geographical ranges. Molecular data are starting to chellenge this interpretation of speciation. See Culberson & Culberson (*Science* **180**: 196, 1973), Poelt (*Vortr. bot. Ges.*, n.f. **4**: 187, 1970, *Bot. Notiser* **125**: 77, 1972), Mattsson & Lumbsch (*Taxon* **38**: 238, 1989), Tehler (*Taxon* **31**: 708, 1982), Myllys *et al.* (*Mycol.* **93**: 335, 2001; molecular characterization).

Lit.: Andersson (*Taxon* **39**: 375, 1990; ecological spp.), Brasier (*in* Rayner *et al.* (Eds), *Evolutionary biology of the fungi*: 231, 1987; dynamics fungal speciation), Burges (*in* Lousley (Ed.), *Species studies in the British flora*: 65, 1955), Ciferri (*Ann. Myc.* **30**: 122, 1932), Claridge & Boddy (*in* Hawksworth (Ed.), *Identification and characterization of pest organisms*: 261, 1994; recognition systems), Burnett (*TBMS* **81**: 1, 1983; speciation in fungi), Ereshefsky (Ed.) (*The units of evolution: essays on the nature of species*, 1992; collected key papers), Hawksworth (*Mycologist's handbook*, 1974), Heywood (*Taxon* **27**: 26, 1963; aggregate spp.), Mason (*TBMS* **24**: 115, 1940), Perkins (*in* Bennett & Lasure (Eds), *More gene manipulations in fungi*: 3, 1991), Poelt (*in* Hawksworth (Ed.), *Ascomycete systematics*: 273, 1994; concepts in lichens), Regenmortel (*Intervirol.* **31**: 241, 1990; polythetic spp.), Rieppel (*in* Scotland *et al.* (Eds), *Models in phylogeny reconstruction*: 31, 1994; phylogenetic spp.).

Speerschneidera Trevis. (1861), Ramalinaceae (L). 1, N. America. See Hafellner & Egan (*Lichenologist* **13**: 11, 1981), Ekman (*Op. bot.* **127**: 148 pp., 1996; N America), Lumbsch *et al.* (*Mol. Phylogen. Evol.* **31**: 822, 2004; phylogeny), Lumbsch *et al.* (*Mol. Phylogen. Evol.* **34**: 512, 2005; phylogeny).

Spegazzini (Carlo Luigi; 1858-1926; Italy, later Argentina). Acting professor (*c.* 1880-1884) then Assistant Professor of Botany (1884-1887), Universidad de Buenos Aires; Director of the Provincial University, La Plata (1887-1912); Director, Ministry of Agriculture (1912 onwards). Father of South American mycology, with pioneering works on the fungi of Argentina, Brazil, Chile, Paraguay and Uruguay; after Saccardo (q.v.) responsible for more nomenclatural novelties than any other mycologist; shipwrecked, Tierra del Fuego (1882); the Spegazzini Glacier in southern Argentina, and the Asociación Spegazzini (the Argentine society for non-medical mycology) are named in his honour. Collections in the Instituto de Botánica C. Spegazzini, La Plata (**LPS**). *Publs* [see also Farr, An annotated list of Spegazzini's fungus taxa. *Bibliotheca Mycologica* **35**, 2 vols, 1978]. Fungi Argentini. *Anales de la Sociedad Científica*

Argentina (1880-1882); Fungi Guaranitici. *Anales de la Sociedad Científica Argentina* (1883-1889); Fungi Patagonici. *Boletín de la Academia Nacional de Ciencias* Córdoba (1887); Fungi Fuegiani. *Boletín de la Academia Nacional de Ciencias* Córdoba (1887). *Biogs, obits etc.* Grummann (1974: 522); Molfino (*Anales de la Sociedad Científica Argentina* **108**: 7, 1929); Murrill (*Mycol.* **16**: 200, 1924); Stafleu & Cowan (*TL-2* **5**: 776, 1985); see also *Spegazzini - Dibujos de Hongos, una Biblioteca Digitalizada / Spegazzini's Drawings of Fungi, a Digitized Library* www.cybertruffle.org.uk/spegazzini.

Spegazzinia Sacc. (1879), anamorphic *Pezizomycotina*, Hso/Hsp.0bP.27. 7, widespread. See Hughes (*Mycol. Pap.* **50**, 1953), Roquebert (*Revue Mycol.* Paris **42**: 309, 1978; conidiogenesis), Chen & Tzean (*Fungal Science* Taipei **15**: 81, 2000; Taiwan).

Spegazziniella Bat. & I.H. Lima (1959) ? = Myriangiella fide Luttrell *in* Ainsworth *et al.* (Eds) (*The Fungi* **4A**: 135, 1973).

Spegazzinites Félix (1894), Fossil Fungi. 1 (Neogene, Pleistocene), widespread.

Spegazzinula Sacc. (1883) ≡ Dubitatio fide Rossman *et al.* (*Stud. Mycol.* **42**: 248 pp., 1999).

Speira Corda (1837) = Dictyosporium fide Guéguen (*BSMF* **21**: 98, 1905).

Speiropsis Tubaki (1958), anamorphic *Pezizomycotina*, Hso.1bH.10. 7, widespread. See Petersen (*Mycol.* **55**: 26, 1963), Kuthubutheen & Nawawi (*TBMS* **89**: 584, 1987; key), Keshava Prasad & Bhat (*Mycotaxon* **82**: 127, 2002; India).

Spelaeomyces Fresen. (1863) ? = Daedalea fide Donk (*Verh. K. ned. Akad. Wet.* tweede sect. **62**: 1, 1974).

sperm, a male sex cell, typically motile; **-atiophore**, a spermatia-producing or -supporting structure; **-atium** (pl. **-a**), a 'sex' (+ or -) cell, e.g. a pycniospore; a microconidium in discomycetes and pyrenomycetes; a non-motile gamete, as in *Laboulbeniales*; **-atization**, the placing of spermatia on structures (receptive hyphae, etc.) for diploidization; **-odochidium**, a fruitbody having spermodochia in a lysigenous cavity in the suscept tissue (Whetzel, *Mycol.* **35**: 337, 1943); **-odochium**, a spermogonium having no wall (Whetzel, *Mycol.* **29**: 135, 1937); **-ogonium** (**-agone**, **-agonium**), a walled structure in which spermatia are produced, as in ascomycetes; a pycnium of a rust; a lichen pycnidium (obsol.).

Spermatodermia Wallr. (1833) ≡ Spermodermia.

Spermatodium Fée ex Trevis. (1860) ≡ Endophis fide Aguirre-Hudson (*Bull. Br. Mus. nat. hist.* Bot. **21**: 85, 1991), Hafellner & Kalb (*Biblthca Lichenol.* **57**: 161, 1995).

Spermatoloncha Speg. (1908), anamorphic *Pezizomycotina*, Hso.1bH.10. 1, S. America; Africa.

Spermochaetella Cif. (1954), anamorphic *Pezizomycotina*, Cpd.1eP.?. 1, S. America. See Ciferri (*Sydowia* **8**: 245, 1954).

Spermodermia Tode (1790) ? = Hypoxylon Bull. fide Höhnel (*Mykol. Unters.* **1**: 362, 1923).

Spermoedia Fr. (1822) = Claviceps fide Hawksworth *et al.* (*Dictionary of the Fungi* edn 8, 1995).

Spermophthora S.F. Ashby & W. Nowell (1926) = Eremothecium fide de Hoog *et al. in* Kurtzman & Fell (Eds) (*Yeasts, a taxonomic study* 4th edn: 201, 1998).

Spermophthoraceae Guilliermond. (1928) = Metschnikowiaceae.

Spermophthorales = Saccharomycetales.

Spermophyllosticta Kamilov (1972) nom. nud., anamorphic *Mycosphaerellaceae*.

spermoplane, the surface of a seed; **spermosphere**, the microhabitat around a seed in soil (Verona, *Ann. Inst. Pasteur* **105**: 75, 1963). Cf. phylloplane; rhizoplane.

Spermospora R. Sprague (1948), anamorphic *Pezizomycotina*, Hso.≡ eH.1/10. 10, widespread. See Deighton (*TBMS* **51**: 41, 1968), MacGarvie & O'Rourke (*Iran. Jl agric. Res.* **8**: 151, 1969), Braun (*Monogr. Cercosporella, Ramularia Allied Genera (Phytopath. Hyphom.)* **1**: 236, 1995; key).

Spermosporella Deighton (1969), anamorphic *Pezizomycotina*, Hso.1-≡ eH.10. 3 (on *Meliola*), Sierra Leone; Uganda. See Deighton (*Mycol. Pap.* **118**: 21, 1969), Matsushima & Matsushima (*Matsush. Mycol. Mem.* **8**: 45, 1995), Johnston (*N.Z. Jl Bot.* **37**: 289, 1999; contrast with *Annellospermosporella*).

Spermosporina U. Braun (1993) nom. rej. = Plectosporium fide Braun (*Monogr. Cercosporella, Ramularia Allied Genera (Phytopath. Hyphom.)* **1**: 245, 1995; key), Gams *et al.* (*Taxon* **53**: 193, 2004; nomencl.).

Spermotrichum Kuntze (1898), anamorphic *Pezizomycotina*, Ccu.0fH.?. 1, Paraguay.

Sphacelia Lév. (1827), anamorphic *Claviceps*, Cpd.0eH.?. 7, widespread. See Loveless (*TBMS* **47**: 205, 1964; identification), Frederickson *et al.* (*MR* **93**: 497, 1989; conidiation), White (*Neotyphodium/Grass Interactions Proceedings of the Third International Symposium on Acremonium/Grass Interactions, held May 28-31, 1997, in Athens, Georgia*: 27, 1997), Hodge (*Mycology Series* **19**: 75, 2003), Pažoutová *et al.* (*MR* **108**: 126, 2004), Walker (*Australas. Pl. Path.* **33**: 211, 2004).

Sphaceliopsis Speg. (1910) = Ephelis fide Diehl (*USDA agric. Monogr.* **4**, 1950).

sphacelium, the structure forming conidia in *Claviceps* from which the sclerotium develops (Fredrickson *et al.*, *MR* **95**: 1101, 1991).

Sphaceloma de Bary (1874), anamorphic *Elsinöe*, Cac.0eH.16. 52, widespread. Causing anthracnose and scab diseases. *S. rosarum* (rose leaf spot). See Jenkins & Bitancourt (*Mycol.* **33**: 338, 1941), Sivanesan & Gómez (*TBMS* **85**: 370, 1985), Gonzalez & Pons (*Ernstia*: 1, 1986; *S. manihoticola*, disease), Spiers & Hopcroft (*N.Z. Jl Bot.* **30**: 353, 1992; ontogeny), Song & Koh (*Korean Jl Pl. Path.* **14**: 303, 1998; on *Citrus*), Alvarez & Molina (*Pl. Dis.* **84**: 423, 2000; on *Manihot*), Swart *et al.* (*Mycol.* **93**: 366, 2001; on *Proteaceae*), Alvarez *et al.* (*Pl. Dis.* **87**: 1322, 2003; on *Manihot*), Schoch *et al.* (*Mycol.* **98**: 1041, 2006; phylogeny).

Sphacelotheca de Bary (1884), Microbotryaceae. 7 (on *Polygonaceae*), widespread. See Langdon & Fullerton (*Mycotaxon* **6**: 421, 1978; gen. delimitation), Vánky & Oberwinkler (*Beih. Nova Hedwigia* **107**, 1994) cf. *Sporisorium*.

Sphaerangium Seaver (1951) [non *Sphaerangium* Schimp. 1860, *Musci*] = Pezicula Tul. & C. Tul. fide Verkley (*Stud. Mycol.* **44**: 180 pp., 1999).

Sphaerella (Fr.) Rabenh. (1856) [non *Sphaerella* Sommerf. 1824, *Algae*] ≡ Mycosphaerella fide Corlett (*Mycol. Mem.* **18**, 1991; list of names), Aptroot (*CBS Diversity Ser.* **5**, 2006; nomenclator).

Sphaerella Ces. & De Not. (1863) = Mycosphaerella fide Hawksworth *et al.* (*Dictionary of the Fungi* edn 8, 1995).

Sphaerellaceae Nitschke (1869) = Mycosphaerellaceae.

Sphaerellopsis Cooke (1883), anamorphic *Eudarluca*, St.1-≡ eH.15/19. 3 (on *Uredinales*), widespread. See Mel'nik (*Nov. Sist. niz. Rast.* **13**: 91, 1976), Sutton (*Mycol. Pap.* **141**, 1977), Sutton (*The Coelomycetes*, 1980), Nag Raj (*Coelomycetous Anamorphs with Appendage-bearing Conidia*, 1993), Yuan *et al.* (*MR* **102**: 866, 1998; on *Phragmidium violaceum*), Driessen *et al.* (*Australas. Pl. Path.* **33**: 463, 2004; Australia), Liesebach & Zaspel (*Forest Pathology* **34**: 293, 2004; genetic diversity on willow rusts), Liesebach & Zaspel (*Rust Diseases of Willow and Poplar*: 231, 2005; Europe), Nischwitz *et al.* (*MR* **109**: 421, 2005; phylogeny, host specialization), Bayon *et al.* (*MR* **110**: 1200, 2006; genetic diversity).

Sphaerellopsis Kleb. (1918) = Venturia Sacc. fide Holm (*Taxon* **24**: 275, 1975), von Arx & Müller (*Stud. Mycol.* **9**, 1975).

Sphaerellothecium Zopf (1897), Mycosphaerellaceae. 4 (on lichens), widespread. See Roux & Triebel (*Bull. Soc. linn. Provence* **45**: 451, 1994; descr. 4 spp.).

Sphaeria Haller (1768) nom. rej. = Hypoxylon Bull. Formerly used for most fungi with either perithecia or pycnidia. fide Wakefield (*TBMS* **24**: 286, 1940), Donk (*Regnum veg.* **34**: 16, 1964).

Sphaeriaceae Fr. (1825) = Xylariaceae. s. str. (obsol.).

Sphaerialea Sousa da Câmara (1926) ? = Sphaerulina fide Hawksworth *et al.* (*Dictionary of the Fungi* edn 8, 1995).

Sphaeriales S. str., see *Xylariales*; s. lat., see *Sordariomycetes*.

Sphaerialites Venkatach. & R.K. Kar (1969), Fossil Fungi ? Fungi. 1 (Tertiary), India.

Sphaericeps Welw. & Curr. (1868) = Battarrea fide Cunningham (*Gast. Austr. N.Z.*: 131, 1944).

Sphaeridiobolus (Boud.) Boud. (1885) = Ascobolus fide van Brummelen (*Persoonia* Suppl. **1**: 1, 1967).

Sphaeridium Fresen. (1852), anamorphic *Pezizomycotina*, Hsy.0eH.?. 5, widespread.

SphaerioidaceaeSacc. (1884). (syn. *Phomaceae*, *Sphaeropsidaceae*). See *Sphaeropsidales*.

Sphaeriopsis Geyl. (1887), Fossil Fungi, Pyronemataceae. 1 (Eocene), Indonesia. = Sphaerites (Fossil fungi) fide Meschinelli (1892).

Sphaeriostromella Bubák (1916), anamorphic *Pezizomycotina*, St.0fH.?. 1, Europe.

Sphaeriothyrium Bubák (1916), anamorphic *Pezizomycotina*, St.0eH.15. 2, Europe.

Sphaerites Unger (1850), Fossil Fungi. 110 (Cretaceous, Tertiary), Europe; N. America.

Sphaerobasidioscypha Agerer (1983), Cyphellaceae. 2, New Zealand; Venezuela. See Agerer (*Mitt. bot. StSamml.* Münch. **19**: 294, 1983).

Sphaerobasidium Oberw. (1965), Hydnodontaceae. 4, widespread. See Oberwinkler (*Sydowia* **19**: 57, 1965).

Sphaerobolaceae J. Schröt. (1889) = Geastraceae.

Sphaerobolales = Geastrales.

Sphaerobolus Tode (1790), Geastraceae. 2 (on wood or coprophilous), widespread. See Ingold (*TBMS* **58**: 179, 1972; review), Flegler (*Mycol.* **76**: 944, 1984; cultural char.), Geml *et al.* (*Mycol.* **97**: 680, 2005; systematics).

Sphaerocarpus Ehrh. (1793), Lecanorales (L). 1, Europe.

Sphaerocephalum Weber ex F.H. Wigg. (1780) ? = Calicium fide Hawksworth *et al.* (*Dictionary of the Fungi* edn 8, 1995).

Sphaerocephalus Battarra ex Earle (1909) [non *Sphaerocephalus* Lag. ex DC. 1812, *Compositae*] = Tricholoma fide Singer (*Agaric. mod. Tax.*, 1951).

Sphaerochaetia Bat. & Cif. (1962) = Chaetothyrium fide von Arx & Müller (*Stud. Mycol.* **9**, 1975).

Sphaerocista Preuss (1852) = Topospora fide Groves (*CJB* **43**: 1195, 1965).

Sphaerocladia Stüben (1939) = Blastocladiella fide Couch & Whiffen (*Am. J. Bot.* **29**: 582, 1942).

Sphaerocolla P. Karst. (1892), anamorphic *Pezizomycotina*, Hsp.0eH.3. 2, Europe; S. America. See Donk (*Taxon* **11**: 99, 1962).

Sphaerocordyceps Kobayasi (1981), ? Clavicipitaceae. 2, widespread. See Kobayasi (*Bull. natn. Sci. Mus.* Tokyo, B **7**: 2, 1981).

Sphaerocreas Sacc. & Ellis (1882) = Glomus fide Gerdemann & Trappe (*Mycol. Mem.* **5**, 1974).

Sphaerocybe Magrou & Marneffe (1946), anamorphic *Pezizomycotina*, Hsy.?.?. 1, Europe.

sphaerocysts, globose cells in tissues of fungi, e.g. *Russula* and *Lactarius*.

Sphaeroderma Fuckel (1875) = Melanospora Corda fide Cannon & Hawksworth (*J. Linn. Soc.* Bot. **84**: 115, 1982).

Sphaerodermatella Seaver (1909) ? = Coniochaeta fide Hawksworth *et al.* (*Dictionary of the Fungi* edn 8, 1995).

Sphaerodermella Höhn. (1907) = Coniochaeta fide von Arx & Müller (*Beitr. Kryptfl. Schweiz* **11** no. 1, 1954).

Sphaerodes Clem. (1909), Ceratostomataceae. 12, widespread. See Cannon & Hawksworth (*J. Linn. Soc.* Bot. **84**: 115, 1982; key), García *et al.* (*Stud. Mycol.* **50**: 63, 2004; Spain), Schoch *et al.* (*MR* **111**: 154, 2007; phylogeny).

Sphaerodothella C.A. Pearce & K.D. Hyde (2001), Phyllachoraceae. 1, Australia. See Pearce & Hyde (*Fungal Diversity* **6**: 83, 2001), Pearce & Hyde (*Fungal Diversity Res. Ser.* **17**: 308 pp., 2006).

Sphaerodothis (Sacc. & P. Syd.) Shear (1909), Phyllachoraceae. 21 (on palm leaves), S.E. Asia. Now restricted to palm-inhabiting species. See Joly (*Bull. Res. Counc. Israel* D **10**: 187, 1961; key), Cannon (*Stud. Mycol.* **31**: 49, 1989; on *Gramineae*), Hyde & Cannon (*Mycol. Pap.* **175**: 114, 1999).

Sphaerognomonia Potebnia ex Höhn. (1917) = Apiosporopsis fide Reid & Dowsett (*CJB* **68**: 2398, 1990).

Sphaerognomoniella Naumov & Kusnezowa (1952), ? Diaporthales. 1, former USSR. See Naumov & Kusnezowa (*Bot. Mater. Otd. Sporov. Rast. Bot. Inst. Komarova Akad. Nauk S.S.S.R.* **8**: 153, 1952).

Sphaerographium Sacc. (1884), anamorphic *Helotiales*, St.0fH.15. 11, widespread (north temperate). See Verkley (*Mycol.* **93**: 205, 2001), Verkley (*Nova Hedwigia* **75**: 433, 2002; revision), Rossman *et al.* (*Mycol. Progr.* **3**: 275, 2004; phylogeny).

Sphaerolina Fuckel (1860) = Schizoxylon fide Nannfeldt (*Nova Acta R. Soc. Scient. upsal.*, 1932), Sherwood (*Mycotaxon* **6**: 215, 1977).

Sphaeromma H.B.P. Upadhyay (1964), anamorphic *Pezizomycotina*, Cpd.≡ eH.?. 1 (lichenicolous), Brazil. See Upadhyay (*Publções Inst. Micol. Recife* **402**: 4, 1964), Matzer (*Mycol. Pap.* **171**: 202 pp., 1996).

Sphaeromonas E. Liebet. (1910), Chytridiomycetes. 1, Europe. See Orpin (*J. gen. Microbiol.* **94**: 270, 1976).

Sphaeromphale A. Massal. (1854) = Staurothele fide Hawksworth *et al.* (*Dictionary of the Fungi* edn 8,

1995).

Sphaeromphale Rchb. (1828) ≡ Segestria.

Sphaeromyces G. Arnaud (1954) nom. inval., anamorphic *Pezizomycotina*.

Sphaeromyces Mont. (1845) ? = Aspergillus.

Sphaeromycetella G. Arnaud (1954) = Chloridium fide Gams & Holubová-Jechová (*Stud. Mycol.* **13**, 1976).

Sphaeromyxa Spreng. (1827) ≡ Sphaeronaema.

Sphaeronaema Fr. (1815), anamorphic *Pezizomycotina*, St.0eH.?. 50, widespread. See Jaczewski (*Nouv. Mem. Soc. nat. Moscou* **15**: 277, 1898), Petch (*TBMS* **25**: 167, 1941; monotypic), Sutton (*Mycol. Pap.* **141**, 1977).

Sphaeronaemella P. Karst. (1884), Microascales. Anamorph *Gabarnaudia*. 1 or 5 (on fungi and dung), widespread (north temperate). See Cannon & Hawksworth (*J. Linn. Soc.* Bot. **84**: 115, 1982), Hausner *et al.* (*CJB* **71**: 52, 1993; posn), Seifert *et al. in* Wingfield *et al.* (Eds) (*Ceratocystis* and *Ophiostoma* Taxonomy, Ecology and Pathogenicity: 269, 1993), Rossman *et al.* (*Stud. Mycol.* **42**: 248 pp., 1999), Hausner & Reid (*CJB* **82**: 752, 2004; phylogeny).

Sphaeronaemina Höhn. (1917) ≡ Sphaeronaema.

Sphaeronemopsis Speg. (1910) = Melanospora Corda fide von Arx & Müller (*Beitr. Kryptfl. Schweiz* **11** no. 1, 1954).

Sphaeropezia Sacc. (1884) = Odontotrema fide Sherwood-Pike (*Mycotaxon* **28**: 137, 1987).

Sphaeropeziella P. Karst. (1885) ? = Xylogramma fide Hawksworth *et al.* (*Dictionary of the Fungi* edn 8, 1995).

Sphaerophoma Petr. (1924), anamorphic *Pezizomycotina*, St.0eH.?. 1, N. America.

Sphaerophoraceae Fr. (1831), Lecanorales (L). 5 gen. (+ 10 syn.), 32 spp.

Lit.: Henssen *et al.* (*Bot. Acta* **105**: 457, 1992), Wedin (*Pl. Syst. Evol.* **187**: 213, 1993; cladistics), Wedin (*Acta univ. upsal.* **77**, 1994; keys 23 S. Hemisph. spp.), Gargas *et al.* (*Science* N.Y. **268**: 1492, 1995), Wedin (*Symb. bot. upsal.* **31**: 102 pp., 1995; S. hemisphere), Chen (*Acta Mycol. Sin.* **15**: 105, 1996), Tibell (*Symb. bot. upsal.* **32** no. 1: 291, 1997), Stenroos & DePriest (*Am. J. Bot.* **85**: 1548, 1998), Wedin *et al.* (*Pl. Syst. Evol.* **209**: 75, 1998), Wedin *et al.* (*Pl. Syst. Evol.* **209**: 75, 1998), Döring *et al.* (*Aust. J. Bot.* **47**: 783, 1999), Sarrión *et al.* (*Mycotaxon* **71**: 169, 1999), Wedin & Döring (*MR* **103**: 1131, 1999; phylogeny), Döring & Wedin (*Pl. Biol.* **2**: 361, 2000), Wedin *et al.* (*Lichenologist* **32**: 171, 2000; phylogeny), Högnabba & Wedin (*Cladistics* **19**: 224, 2003), Miądlikowska *et al.* (*Mycol.* **98**: 1088, 2006; phylogeny), Hofstetter *et al.* (*Mol. Phylogen. Evol.* **44**: 412, 2007; phylogeny).

Sphaerophoron, see *Sphaerophorus*.

Sphaerophoronomyces Cif. & Tomas. (1953) ≡ Sphaerophorus.

Sphaerophoropsidaceae Elenkin (1929) = Cladoniaceae.

Sphaerophoropsis Vain. (1890), Cladoniaceae (L). 2, S. America; New Zealand. See Ahti (*Lichenology in Latin America. History, Current Knowledge and Applications* [Proceedings of GLAL-3, Terceiro Encontro do Grupo Latino-Americano de Liquenólogos, São Paulo, Brazil, 24-28 September, 1997]: 109, 1998).

Sphaerophorum Schrad. (1794) nom. rej. prop. = Sphaerophorus fide Hawksworth *et al.* (*Dictionary of the Fungi* edn 8, 1995).

Sphaerophorus Pers. (1794) nom. cons., Sphaerophoraceae (L). 8, widespread. See Wedin (*Nordic Jl Bot.* **10**: 539, 1990; ontogeny), Wedin (*Lichenologist* **24**: 119, 1992; S. hemisph.), Wedin (*Pl. Syst. Evol.* **187**: 213, 1993; cladistics), Wedin (*Symb. Bot. Upsal.* **31** no. 1, 1995; key, monograph), Tibell (*Symb. bot. upsal.* **32**: 291, 1997; anamorphs), Wedin & Döring (*MR* **103**: 1131, 1999; phylogeny), Döring & Wedin (*Pl. Biol.* **2**: 361, 2000; ontogeny, anatomy), Wedin *et al.* (*Lichenologist* **32**: 171, 2000; phylogeny), Högnabba & Wedin (*Cladistics* **19**: 224, 2003; phylogeny), Miądlikowska *et al.* (*Mycol.* **98**: 1088, 2006; phylogeny).

Sphaerophragmiaceae Cummins & Y. Hirats. (1983) = Raveneliaceae.

Sphaerophragmium Magnus (1891), Raveneliaceae. 18 (on *Annonaceae*, most on *Leguminosae*), widespread (tropical). See Hiremath & Pavgi (*Norw. Jl Bot.* **21**: 17, 1974; morphology), Lohsomboon *et al.* (*MR* **98**: 907, 1994; key), Piepenbring (*Mycol. Progr.* **4**: 161, 2005; Panama).

sphaeroplast, see protoplast.

Sphaeropleum Link (1833) nom. dub., ? Fungi.

Sphaeroporalites Hemer & Nygreen (1967), Fossil Fungi ? Fungi. 1, Libya.

Sphaeropsidales (obsol.). Traditionally used for anamorphic fungi with pycnidial conidiomata (*Phomales*, *Phyllostictales*, etc.). Not accepted by Sutton (1973, 1980) or by Nag Raj (1981).

Sphaeropsis Flot. (1847) ≡ Thelocarpon.

Sphaeropsis Lév. (1842) nom. rej., anamorphic *Pezizomycotina*. See Hawksworth & Sherwood (*Taxon* **30**: 338, 1981).

Sphaeropsis Sacc. (1880) nom. cons., anamorphic *Botryosphaeriaceae*, Cpd.1eP.1. 3 (assoc. with canker or die-back), widespread. *S. sapinea* on conifers. See Phillips *et al.* (*Persoonia* **21**: in press, 2008).

Sphaeropus Paulet (1793) = Onygena fide Saccardo (*Syll. fung.* **20**: 230, 1911).

Sphaeropyxis Bonord. (1864) = Coniochaeta fide Petrini (*Sydowia* **44**: 169, 1993).

Sphaerosoma Klotzsch (1839), Pyronemataceae. 3 (hypogeous), N. America; Europe. See Gamundí (*Sydowia* **28**: 339, 1976; gen. concept), Læssøe & Hansen (*MR* **111**: 1075, 2007; phylogeny).

Sphaerosomataceae Bail (1858) = Pyronemataceae.

Sphaerosperma Preuss (1853) = Diatrypella fide Saccardo (*Syll. fung.* **1**: 203, 1882).

Sphaerospora (Sacc.) Sacc. (1889) [non *Sphaerospora* Klatt 1864, *Iridaceae*] ≡ Sphaerosporula.

Sphaerosporella (Svrček) Svrček & Kubička (1961), Pyronemataceae. 2, widespread. See Häffner (*Beitr. Kenntn. Pilze Mitteleur.* **3**: 413, 1987), Yao & Spooner (*Kew Bull.* **51**: 385, 1996), Landvik *et al.* (*Nordic Jl Bot.* **17**: 403, 1997; DNA), Wang (*Mycotaxon* **80**: 197, 2001; Taiwan), Hansen & Pfister (*Mycol.* **98**: 1029, 2006; phylogeny), Perry *et al.* (*MR* **111**: 549, 2007; phylogeny).

Sphaerosporiites Babajan & Tasl. (1977), Fossil Fungi. 1 (Tertiary), former USSR.

Sphaerosporium Schwein. (1832), anamorphic *Pezizomycotina*, Hso.0eH.3. 2, widespread. See Damon & Downing (*Mycol.* **46**: 214, 1954), Partridge & Morgan-Jones (*Mycotaxon* **84**: 69, 2002).

Sphaerosporula Kuntze (1898) = Scutellinia fide Denison (*Mycol.* **51**: 605, 1959).

Sphaerostilbe Tul. & C. Tul. (1861) = Nectria fide

Rogerson (*Mycol.* **62**: 865, 1970), Rossman *et al.* (*Stud. Mycol.* **42**: 248 pp., 1999).

Sphaerostilbella (Henn.) Sacc. & D. Sacc. (1905), Hypocreaceae. Anamorph *Gliocladium*. 4 (decaying fungi and wood), widespread. See Seifert (*Stud. Mycol.* **27**, 1985), Rossman *et al.* (*Stud. Mycol.* **42**: 248 pp., 1999), Põldmaa (*Stud. Mycol.* **45**: 83, 2000), Rossman *et al.* (*Mycol.* **93**: 100, 2001; phylogeny), Castlebury *et al.* (*MR* **108**: 864, 2004; phylogeny), Zhang *et al.* (*Mycol.* **98**: 1076, 2006; phylogeny).

Sphaerostylidium (A. Braun) Sorokīn (1882) = Rhizophydium fide Fischer (*Rabenh. Krypt.-Fl.* **4** 2.1: 25, 1892).

Sphaerothallia Nees ex Eversm. (1831) nom. rej. = Aspicilia. Includes the 'manna lichens'; *S. esculenta* may be one of the types of manna in The Bible, and can be used in bread production. fide Follmann & Crespo (*An. Inst. bot. A.J. Cavanilles* **31**: 325, 1974), Rogers *in* Seaward (Ed.) (*Lichen ecology*: 211, 1977; thalli unattached and blown in the wind), Crespo & Barreno (*Acta Bot. Malac.* **4**: 55, 1978), Hafellner (*Acta Bot. Malac.* **16**: 133, 1991).

Sphaerotheca Desv. (1817) nom. rej. = Aecidium fide Berndt (*in litt.*).

Sphaerotheca Lév. (1851) nom. cons. = Podosphaera fide Saenz & Taylor (*CJB* **77**: 150, 1999; phylogeny), Braun & Takamatsu (*Schlechtendalia* **4**: 3, 2000), Braun *et al.* (*Schlechtendalia* **7**: 45, 2001), Takamatsu (*Mycoscience* **45**: 147, 2004; phylogeny).

Sphaerothyrium Wallr. (1833) ≡ Eustegia.

Sphaerotrachys Fayod (1889) = Cortinarius fide Kauffman (*N. Amer. Fl.* **10**, 1932).

Sphaerozone Zobel (1854), Pezizaceae. 4 (hypogeous), widespread. See Beaton & Weste (*TBMS* **71**: 164, 1978), Dissing (*Mycotaxon* **30**: 2, 1980), Hansen *et al.* (*Mycol.* **93**: 958, 2001; phylogeny), Læssøe & Hansen (*MR* **111**: 1075, 2007; phylogeny).

Sphaerozosma Corda (1842) = Sphaerosoma fide Hawksworth *et al.* (*Dictionary of the Fungi* edn 8, 1995).

Sphaerula Pat. (1883) = Pistillaria Fr. fide Corner (*Ann. Bot. Mem.* [A monograph of Clavaria and allied genera] **1**, 1950).

Sphaerulina Sacc. (1878), ? Mycosphaerellaceae. 41, widespread. *S. rehmiana* (rose leaf scorch), *S. taxi* (yew (*Taxus*) leaf spot), *S. rubi* (raspberry (*Rubus*) leaf spot). See Boerema (*Neth. Jl Pl. Path.* **69**: 76, 1963), Evans *et al.* (*MR* **97**: 59, 1993), Crous *et al.* (*Sydowia* **55**: 136, 2003; polyphyly).

Sphaerulomyces Marvanová (1977), anamorphic *Pezizomycotina*, Hso.0eH.3. 1 (aquatic), Slovakia. See Marvanová (*TBMS* **68**: 485, 1977).

Sphagnicola Velen. (1934) ≡ Ciliatula.

Sphagnomphalia Redhead, Moncalvo, Vilgalys & Lutzoni (2002) ≡ Gyroflexus.

Sphaleromyces Thaxt. (1894), Laboulbeniaceae. 3, widespread. See Nannfeldt (*Svensk bot. Tidskr.* **43**: 468, 1949).

Spheconisca (Norman) Norman (1876), ? Verrucariaceae (L). 20, Europe. See Bachmann (*Nyt Mag. naturv.* **64**: 170, 1926).

Spheconiscomyces Cif. & Tomas. (1953) = Spheconisca fide Hawksworth *et al.* (*Dictionary of the Fungi* edn 8, 1995).

Sphenospora Dietel (1892), Raveneliaceae. 8 (mostly on monocots), America (tropical); Africa. See Linder (*Mycol.* **36**: 464, 1944), Olive (*Mycol.* **39**: 409, 1947; cytology), Pereira *et al.* (*Brazilian Journal of Microbiology* **33**: 155, 2002; on orchid).

spheridium, see capitulum.

Spherites Dijkstra (1949), Fossil Fungi. 1 (Devonian), Netherlands.

spheroplast, see protoplast.

Spheropsis Raf. (1815) nom. dub.; nom. nov. for '*Sphaeria* Pers.', ? Pezizomycotina. No spp. included.

Spherospora J.J. Garcia (1991), Microsporidia. 1.

spherule (1) a sporangium-like structure in *Coccidioides* (Baker & Mrak, *Am. J. trop. Med.* **21**: 589, 1941); (2) a multinucleate cell of a resting myxomycete plasmodium.

spherulin, spherule phase coccidioidin (q.v.) (Scalarone *et al.*, *Sabouraudia* **11**: 222, 1973).

Sphinctrina Fr. (1825), Sphinctrinaceae (±L). 8 (on lichens), widespread. See Löfgren & Tibell (*Lichenologist* **11**: 109, 1979; key 5 Eur. spp.), Tibell (*Beih. Nova Hedwigia* **79**: 597, 1984), Wedin & Tibell (*CJB* **75**: 1236, 1997; phylogeny), Tibell (*Nordic Lichen Flora* **1**. Introductory Parts; Calicioid Lichens and Fungi: 20, 1999; Nordic spp.), Tibell & Vinuesa (*Taxon* **54**: 427, 2005; phylogeny), Geiser *et al.* (*Mycol.* **98**: 1053, 2006; phylogeny).

Sphinctrinaceae M. Choisy (1950), Mycocaliciales (±L). 2 gen. (+ 2 syn.), 9 spp.

Lit.: Tibell (*Symb. bot. upsal.* **27** no. 1: 279 pp., 1987), Otto & Krebs (*Boletus* **17**: 97, 1993), Wedin & Tibell (*CJB* **75**: 1236, 1997; phylogeny), Sarrión *et al.* (*Mycotaxon* **71**: 169, 1999), Tibell & Wedin (*Mycol.* **92**: 577, 2000).

Sphinctrinella Nádv. (1942) = Mycocalicium fide Tibell (*Beih. Nova Hedwigia* **79**: 597, 1984), Tibell *et al.* (*Mycotaxon* **87**: 3, 2003).

Sphinctrinopsis Woron. (1927), Caliciaceae. 1 (on *Pertusaria*), former USSR. See Tibell (*Beih. Nova Hedwigia* **79**: 597, 1984).

Sphinctrosporium Kunze (1828) = Cladotrichum fide Saccardo (*Syll. fung.* **4**: 374, 1886).

Sphondylocephalum Stalpers (1974), anamorphic *Pezizomycotina*, Hso.0eH.6. 1, N. America. See Stalpers (*Proc. K. Ned. Akad. Wet.* Ser. C, Biol. Med. Sci. **77**: 400, 1974).

Sphyridiomyces E.A. Thomas ex Cif. & Tomas. (1953) = Baeomyces fide Hawksworth *et al.* (*Dictionary of the Fungi* edn 8, 1995).

Sphyridium Flot. (1843) = Baeomyces fide Hawksworth *et al.* (*Dictionary of the Fungi* edn 8, 1995).

Spicaria Harting (1846) nom. conf., anamorphic *Pezizomycotina*. See Brown & Smith (*TBMS* **40**: 22, 1957).

Spicariopsis R. Heim (1939) ? = Paecilomyces fide Samson (*Stud. Mycol.* **6**, 1974).

Spicellum Nicot & Roquebert (1976), anamorphic *Hypocreales*, Hso.0eH.10. 1, France. See Nicot & Roquebert (*Revue Mycol.* Paris **39**: 272, 1975), Seifert *et al.* (*Mycol.* **89**: 250, 1997).

Spicularia Chevall. (1826) ≡ Exidia.

Spicularia Pers. (1822), anamorphic *Pezizomycotina*, Hso.0eP.?. 1 or 2, Europe; America.

spicule (**spiculum**) (Tulasne; obol.), see sterigma.

Spiculogloea P. Roberts (1996), Spiculogloeales. 4, Europe. See Roberts (*Mycotaxon* **60**: 112, 1996).

Spiculogloeales R. Bauer, Begerow, J.P. Samp., M. Weiss & Oberw. (2006). Agaricostilbomycetes. 1 gen., 4 spp. No familes recognized.

spiculospore, a spore formed at the tip of a pointed structure often elongate and so resembling a spike, as in *Hirsutella* and *Akanthomyces* (Subramanian, *Curr. Sci.* **31**: 410, 1962).

Spiculostilbella E.F. Morris (1963) = Phaeoisaria fide de Hoog & Papendorf (*Persoonia* **8**: 408, 1976).

spiculum (pl. **spicula**), see sterigma.

Spilobolus Link (1833) = Amphisphaerella fide Gams (*Taxon* **43**: 265, 1994).

Spilocaea Fr. (1819) nom. rej. = Fusicladium fide Ellis (*DH*, 1971), Ellis (*MDH*, 1976), Braun *et al.* (*Taxon* **51**: 557, 2002; nomencl.), González-Lamothe *et al.* (*FEMS Microbiol. Lett.* **210**: 149, 2002; phylogeny), Schubert *et al.* (*Schlechtendalia* **9**, 2003), Beck *et al.* (*Mycol. Progr.* **4**: 111, 2005; phylogeny).

Spilodium A. Massal. (1856) = Arthonia fide Hawksworth (*TBMS* **74**: 363, 1980).

spilodium, introduced by Lettau (*Beih. Feddes Rep.* **69**: 62, 1932) for the minute round blackish structures on the thallus of *Dirina massiliensis* caused by *Milospium graphideorum*.

Spilodochium Syd. (1927), anamorphic *Pseudoparodiella*, Hsp.0-1eP.?. 3, C. America; Australia. See Sivanesan (*TBMS* **86**: 187, 1986; teleomorph).

Spiloma Ach. (1803) = Xylographa fide Hawksworth *et al.* (*Dictionary of the Fungi* edn 8, 1995).

Spilomataceae Chevall. (1826) = Trapeliaceae.
Lit.: Lumbsch *et al.* (*MR* **111**: 1133, 2007).

Spilomela (Sacc.) Keissl. (1920) ≡ Pleospilis.

Spilomium Nyl. (1858) ≡ Sclerococcum fide Hawksworth (*Taxon* **55**: 528, 2006; nomencl.).

Spilomyces Petr. & Syd. (1927) = Neottiospora fide Sutton (*Mycol. Pap.* **141**, 1977).

Spilonema Bornet (1856), Coccocarpiaceae (L). 4, widespread. See Henssen (*Symb. bot. upsal.* **18** no. 1, 1963).

Spilonematopsis Å.E. Dahl (1950) = Ephebe fide Henssen (*Symb. bot. upsal.* **18** no. 1, 1963).

Spilonematopsis Vain. (1909) nom. nud. = Spiloma fide Henssen (*Symb. bot. upsal.* **18** no. 1, 1963).

Spilonemella Henssen & Tonsberg (2000), Coccocarpiaceae (L). 2, Japan; USA. See Henssen & Tønsberg (*Bryologist* **103**: 108, 2000).

Spilonemopsis Vain. (1909) nom. nud. = Spilonema fide Henssen (*Symb. bot. upsal.* **18** no. 1, 1963).

Spilopezis Clem. (1909) = Pseudopeziza fide Nannfeldt (*Nova Acta R. Soc. Scient. upsal.*, 1932).

Spilopodia Boud. (1885), Helotiales. Anamorphs *Discogloeum*, *Melanodiscus*. 4, Europe. See Müller (*Sydowia* **41**: 219, 1989; excl. sp.).

Spilopodiella E. Müll. (1989), Helotiales. 1, Switzerland. See Müller (*Sydowia* **41**: 219, 1989).

Spilosphaeria Rabenh. (1857) = Septoria Sacc. fide Saccardo (*Syll. fung.* **4**: 474, 1884).

Spilosphaerites A. Massal. (1857), Fossil Fungi. 2 (Miocene), Italy.

Spilosticta Syd. (1923) = Venturia Sacc. fide von Arx & Müller (*Stud. Mycol.* **9**, 1975).

Spinalia Vuill. (1904), ? Dimargaritales. 1, Europe. See Benjamin (*Aliso* **4**: 321, 1959), Benjamin *in* Kendrick (Ed.) (*The Whole Fungus* **2**: 573, 1979), Wrzocek & Gajowniczek (*Acta Mycologica* Warszawa **33**: 269, 1998; Poland).

spindle, see fuseau.

spindle pole body, microtubule organizing center, functionally homologous to the animal cell centrosome.

spindle-organ, see turbinate cell.

spine, a narrow sharply pointed process; **spinule**, a small spine (in lichens see Fig. 22C); **spiny**, having spines; **spinose** (dim. **spinulose**), delicately spiny. Fig. 20.2.

Spinellus Tiegh. (1875), Phycomycetaceae. 3 (parasitic on agarics, esp. *Mycena*), widespread (north temperate). See Zycha *et al.* (*Mucorales*, 1969; key), Overton (*Notes on the genus Spinellus with comparisons to other Zygomycetes found in fungicolous associations with Basidiomycetes*, 1997), Voigt & Wöstemeyer (*Gene* **270**: 113, 2001; phylogeny).

Spiniger Stalpers (1974), anamorphic *Heterobasidion, Dichostereum, Laurilia*. 2, widespread. See Carmichael *et al.* (*Genera of Hyphomycetes*, 1980; refs).

Spinomyces Bat. & Peres (1961) nom. inval. ? = Echinoplaca fide Lücking *et al.* (*Lichenologist* **30**: 121, 1998).

Spinosporonites R.K. Saxena & S. Khare (1991), Fossil Fungi. 1, India. See Saxena & Khare (*Geophytology* **21**: 40, 1991).

Spinulosphaeria Sivan. (1974), Chaetosphaerellaceae. 1 (from wood), widespread. See Nannfeldt (*Svensk bot. Tidskr.* **69**: 289, 1975), Barr (*Mycotaxon* **39**: 43, 1990; posn), Subramanian & Sekar (*Kavaka* **18**: 19, 1993; placement in *Nitschkiaceae*), Huhndorf *et al.* (*Mycol.* **96**: 368, 2004; phylogeny).

Spinulospora Deighton (1973), anamorphic *Pezizomycotina*, Hsp.0eH.1. 1 (on *Puccinia*), W. Africa. See Deighton (*TBMS* **61**: 195, 1973).

spiral hypha, a hypha ending in a spiral or helical coil, as in *Trichophyton* (Davidson & Gregory, *TBMS* **21**: 98, 1937).

Spiralia Grigoraki (1925) [non *Spiralia* Toula 1900, fossil ? *Algae*] = Trichophyton fide Hawksworth *et al.* (*Dictionary of the Fungi* edn 8, 1995).

Spiralotrichum H.S. Yates (1918) nom. dub., anamorphic *Pezizomycotina*. See Mulder (*TBMS* **65**: 518, 1975).

Spiralum J.L. Mulder (1975), anamorphic *Pezizomycotina*, Hso.0-≡ hP.10. 2, Asia. See Mulder (*TBMS* **65**: 518, 1975).

Spirechina Arthur (1907) = Kuehneola fide Jackson (*Mycol.* **23**: 96, 1931).

spirits, high alcohol-content drinks obtained by distilling grape wine (**brandy**), malted barley (**whisky**), fermented rye or maize worts (**gin**), or fermented molasses (**rum**). See also pulque.

Spirodactylon R.K. Benj. (1959), Kickxellaceae. 1, USA. See Benjamin (*Aliso* **4**: 408, 1959).

Spirodecospora B.S. Lu, K.D. Hyde & W.H. Ho (1998), Xylariaceae. 2, Hong Kong; Russia. See Lu *et al.* (*Fungal Diversity* **1**: 170, 1998), Mel'nik & Hyde (*Fungal Diversity* **12**: 151, 2003).

Spiroglugea L. Léger & E. Hesse (1924), Microsporidia. 1.

Spirogramma Ferd. & Winge (1909) = Xylaria Hill ex Schrank fide Cannon (*SA* **6**: 171, 1987).

Spirographa Zahlbr. (1903), Odontotremataceae. 3 (on lichens), Europe. See Santesson (*The lichens and lichenicolous fungi of Sweden and Norway*, 1993), Holien & Triebel (*Lichenologist* **28**: 307, 1996).

Spirographomyces Cif. & Tomas. (1953) ≡ Spirographa.

Spirogyromyces Tzean & G.L. Barron (1981), ? Kickxellomycotina. 1 (in gut of *Rhabditis*), Canada. See Tzean & Barron (*CJB* **59**: 1861, 1981), Barron (*Biodiversity of Fungi* Inventory and Monitoring Methods: 435, 2004).

Spiroidium Saito (1949) nom. dub., Fungi. q.v.

Spiromastix Kuehn & G.F. Orr (1962), ? Ajellomycetaceae. 5, widespread. See Currah & Locquin-Linard (*CJB* **66**: 1135, 1988; key 2 spp.), Sugiyama *et al.* (*Mycoscience* **40**: 251, 1999; DNA), Udagawa & Uchiyama (*Mycoscience* **40**: 291, 1999), Sugiyama & Mikawa (*Mycoscience* **42**: 413, 2001; phylogeny), Untereiner *et al.* (*Stud. Mycol.* **47**: 25, 2002; phylogeny), Untereiner *et al.* (*Mycol.* **96**: 812, 2004), Geiser *et al.* (*Mycol.* **98**: 1053, 2006; phylogeny).

Spiromyces R.K. Benj. (1963), Kickxellaceae. 2, USA. See Benjamin (*Aliso* **5**: 273, 1963), Mikawa (*TMSJ* **16**: 146, 1975), O'Donnell *et al.* (*Mycol.* **90**: 624, 1998), James *et al.* (*Nature* **443**: 818, 2006; phylogeny), White *et al.* (*Mycol.* **98**: 872, 2006; phylogeny).

Spiropes Cif. (1955), anamorphic *Pezizomycotina*, Hso.≡ eP.10. 30 (on *Meliolaceae*), widespread (esp. tropical). See Ellis (*Mycol. Pap.* **114**, 1968; key), Seifert & Hughes (*N.Z. Jl Bot.* **38**: 489, 2000).

Spirosphaera Beverw. (1953), anamorphic *Helotiales*, Hso.0-≡ hH.1. 8 (aquatic), Europe; Japan. See Hennebert (*TBMS* **51**: 13, 1968), Abdullah *et al.* (*Nova Hedwigia* **43**: 507, 1986; 2 spp. from Japan), Voglmayr (*CJB* **75**: 1772, 1997), Marvanová & Bärlocher (*Mycotaxon* **68**: 33, 1998), Voglmayr (*Stud. Mycol.* **50**: 221, 2004), Wang *et al.* (*Mycol.* **98**: 1065, 2006; phylogeny).

Spirospora L. Mangin & Vincens (1920) = Acrospeira fide Wiltshire (*TBMS* **21**: 233, 1938).

Spirospora Scherff. (1926), Microsporidia. 1.

Spirotremesporites Dueñas (1979), Fossil Fungi. 3 (Eocene – Pleistocene), Colombia; USA.

Spirotrichum Saito ex J.F.H. Beyma (1940) = Tritirachium fide Langeron (*Revue Mycol.* Paris **14**: 133, 1949), Paulitz & Menge (*Mycol.* **76**: 99, 1984).

Spitzenkörper. A labile vesicle-rich and actin-rich structure present in growing tips of most, but not all, fungi, whose behaviour correlates with tip growth rate and direction (Grove & Bracker, *J. Bact.* **104**: 989, 1970, López-Franco & Bracker, *Protoplasma* **195**: 90, 1996, Bourett & Howard, *Protoplasma* **163**: 199, 1991). See also Virag & Harris (*MR* **110**: 4, 2006; review of molecular research).

Spizellomyces D.J.S. Barr (1980), Spizellomycetaceae. 8, widespread. See Barr (*CJB* **58**: 2384, 1980).

Spizellomycetaceae D.J.S. Barr (1980), Spizellomycetales. 6 gen. (+ 1 syn.), 25 spp.

Lit.: Longcore *et al.* (*CJB* **73**: 1389, 1995), Chen *et al.* (*Bot. Bull. Acad. sin.* Taipei **41**: 73, 2000), Barr *in* McLaughlin *et al.* (Eds) (*The Mycota* A Comprehensive Treatise on Fungi as Experimental Systems for Basic and Applied Research **7A**: 93, 2001).

Spizellomycetales D.J.S. Barr (1980). Chytridiomycetes. 2 fam., 7 gen., 27 spp. Monocentric; development endogenous or exogenous; saprobic or parasitic, predominantly from soil on organic substrata including other fungi, cellulose, keratin, pollen, plant debris; cosmop.

A segregate from *Chytridiales* (q.v.) which they resemble except: zoospores of most species have more than one lipid globule, they may move in an amoeboid fashion, rhizoid tips are blunt. The order was defined using ultrastructural characters of the zoospore (Barr, *CJB* **58**: 2380, 1980). Fams:

(1) **Caulochytriaceae**
(2) **Olpidiaceae**
(3) **Spizellomycetaceae**

Lit.: Barr (*CJB* **62**: 1171, 1984; key spp. in culture), Barr & Désaulniers (*CJB* **64**: 561, 1986), James *et al.* (*Mycol.* **98**: 860, 2006; molecular phylogeny), Hibbett *et al.* (*MR* **111**: 109, 2007), and see under Familes.

Spizomycetes. Class within the *Chytridiomycota* (Cavalier-Smith, *in* Rayner *et al.* (Eds), *Evolutionary biology of the fungi*: 339, 1987) including only *Spizellomycetales*.

Splanchnomyces Corda (1831) = Rhizopogon fide Kuyper (*in litt.*).

Splanchnomycetaceae Corda (1842) = Rhizopogonaceae.

Splanchnonema Corda (1829), Pleomassariaceae. Anamorph *Myxocyclus*. 35, Europe; N. America. See Barr (*Mycotaxon* **15**: 349, 1982), Barr (*Mycotaxon* **42**: 129, 1993; key 27 N. Am. spp.), Tanaka *et al.* (*Mycoscience* **46**: 248, 2005; Japan).

Splanchospora L.N. Vasil'eva (1998), Pyrenulales. 1, Europe. See Vasil'eva (*Nizshie Rasteniya, Griby i Mokhoobraznye Dalnego Vostoka Rossii* Griby. Tom **4**. Pirenomitsety i Lokuloaskomitsety **4**: 237, 1998).

splash (splashing) cup, an open cup-like structure (as in *Cladonia*, *Cyathus*, and the liverwort *Marchantia*), from which the reproductive bodies are discharged by falling drops of water. See Brodie (*CJB* **29**: 593, 1951); bird's nest fungi.

Spogotteria Dyko & B. Sutton (1979) = Plectronidium fide Sutton & Pascoe (*TBMS* **87**: 249, 1986).

Spolverinia A. Massal. (1856) = Phyllactinia. Also used for lichenicolous fungi. fide Junell (*Svensk bot. Tidskr.* **58**: 55, 1964), Hoffmann & Hafellner (*Biblthca Lichenol.* **77**: 181 pp., 2000).

Spondylocladiella Linder (1934), anamorphic *Pezizomycotina*, Hso.≡ eP.24. 1 (on *Corticium*), Europe; N. America.

Spondylocladiopsis M.B. Ellis (1963), anamorphic *Pezizomycotina*, Hso.≡ eH.10. 1, Europe. See Ellis (*Mycol. Pap.* **87**: 15, 1963), Treigiene & Markovskaja (*Botanica Lithuanica* **9**: 285, 2003).

Spondylocladium Mart. ex Sacc. (1880) = Stachylidium fide Hughes (*CJB* **36**: 747, 1958).

Spongiasclerotes Stach & Pickh. (1957), Fossil Fungi. 1 (Carboniferous), Germany. See Stach & Pickhardt (*Palaeontologische Zeitschrift* **31**: 140, 1957).

Spongioides Lázaro Ibiza (1916) nom. dub., Polyporales.

spongiostratum, used for the hypothallus of *Anzia* and *Pannoparmelia* (Hannemann, *Bibl. Lich.* **1**, 1973). See Henssen & Dobelmann (*Bibl. Lich.* **25**: 103, 1987).

Spongiosus Lloyd ex Torrend (1920) ≡ Phaeolus.

Spongipellis Pat. (1887), Polyporaceae. 8, widespread. See Gilbertson & Ryvarden (*N. Amer. Polyp.* **2**: 723, 1987), Ipulet & Ryvarden (*Syn. Fung.* **20**: 87, 2005; Uganda).

Spongiporus Murrill (1905) = Postia fide Jülich (*Persoonia* **11**: 423, 1982).

Sponheimeria Kirschst. (1941) = Lasiobelonium Ellis & Everh. fide Baral (*SA* **13**: 113, 1994).

Spooneromyces T. Schumach. & J. Moravec (1989), Pyronemataceae. 3, Europe. See Schumacher & Moravec (*Nordic Jl Bot.* **9**: 425, 1989), Perry *et al.* (*MR* **111**: 549, 2007; phylogeny).

spora, the spore content of a particular place or ecological niche; of air, see Air spora.

sporabola, the curve made by a basidiospore after discharge from its sterigma (Buller, *Researches* **1**; Ingold, 1971: 111).

Sporacestra A. Massal. (1860) ≡ Bacidia fide Coppins (*Bull. Br. Mus. nat. hist.* Bot. **11**: 203, 1983).

Sporadospora Reinsch (1875) nom. dub., Fungi. Based on appressoria. See Racovitza (*Mém. Mus. natn Hist. nat. Paris* Bot. **10**: 223, 1959; synonym of *Pseudonectria*), Lowen (*SA* **9**: 26, 1991).

sporangial vesicle, vesicle produced at the mouth of the sporangium during planont maturation and discharge. See homohylic vesicle, plasmamembranic vesicle, precipitative vesicle.

sporangiocyst (of *Chytridiales*), a resting sporangium (A. Fischer); cf. cystosorus.

sporangiolum (sporangiole), (1) (of *Mucorales*), a small sporangium with or without a columella, generally having a small number of spores; (2) a degenerating arbuscule (Janse, 1897) (obsol.).

sporangiophore, thallus element (usually morphologically differentiated) subtending one or more sporangia.

sporangiosorus, group of spherical sporangia fused together and formed from a single plasmodium; also one or more lobed sporangia formed from a single plasmodium.

sporangiospore, walled spore produced in a sporangium; **primary infestation -** (of *Eccrinales*), 1–4-nucleate, thick-walled spore which serves to transmit an infestation from one host individual to another after passage through the gut; **secondary infestation -** (of *Eccrinales*), multinucleate, thin-walled spore which germinates in the same gut as where they were produced.

sporangium (sporange), an organ enclosing endogenously generated spore(s), the walls of the spore(s) not being derived from the supporting or containing structure.

Sporastatia A. Massal. (1854), Catillariaceae (L). 4, widespread. See Grube & Poelt (*Fragm. Flor. geobot.* Suppl. **2**: 113, 1993), Buschbom & Mueller (*Mol. Phylogen. Evol.* **32**: 66, 2004; phylogeny), Miądlikowska *et al.* (*Mycol.* **98**: 1088, 2006; phylogeny).

spore, a general term for a reproductive structure in fungi, bacteria, and cryptogamic plants. In fungi, a differentiated morphological form which may be: (a) specialized for dissemination; (b) produced in response to, and resistant to, adverse conditions; and/or (c) produced during or as a result of a sexual or asexual reproductive process. Commonly 1-celled, but in fungi frequently a multicelled structure (e.g. phragmospore, spore ball, etc.) which is in effect a group of 1-celled spores because every cell may produce one or more germ tubes. Thick- or thin-walled, pigmented or not, motile or non-motile.

More attention has been given to the spore than to any other fungal structure. Spore morphology (e.g. flagellation of zoospores) and development (e.g. of ascospores and basidiospores; sexual and asexual spores) provide basic taxonomic criteria and biologically spores may be differentiated into groups disseminated by wind, water, insects and other animals, etc., and those which allow a fungus to survive conditions unfavourable for growth (e.g. resting spores), although one type of spore may serve several functions (Sutton, *TBMS* **86**: 1, 1986). Vuillemin (1912) defined spores morphologically (e.g. motile and non-motile) and biologically and borrowed or coined terms for his different spore types. Most of these terms have never been in current use but the naming of spores continues (see Spore terminology).

Lit.: Chapela (*Sydowia* **43**: 1, 1991; measurement with Coulter counter), De Toledo (*Mycotaxon* **52**: 259, 1994; descriptions of), Madelin (Ed.) (*The fungus spore*, 1966), Weber & Hess (Eds) (*The fungal spore. Form and function*, 1976), Turian & Hohl (Eds) (*The fungal spore: morphogenetic controls*, 1981), Cole & Hoch (*The fungal spore and disease initiation in plants and animals*, 1991), Beakes (*in* Gow & Gadd (Eds), *The growing fungus* 339, 1995; sporulation), and the following entries.

spore ball, a unit of dispersal comprised of a more or less firmly aggregated group of spores (e.g. *Sorosporium*, *Tolyposporium*) or spores and sterile cells (e.g. *Urocystis*).

Spore charge. Airborne spores carry either a positive or negative electrostatic charge, fide Buller (**1**: 192, 1909). This charge may play an important role in dispersal and, in the case of filiform spores, may result in a change in the spore's shape: if the ends are attracted to each other, the spore will have the aerodynamics of a quoit rather than a javelin. See Allitt (*TBMS* **72**: 147, 1979), Webster *et al.* (*TBMS* **91**: 193, 1988), Gregory (*Nature* **180**: 330, 1957; *Ganoderma applanatum*, etc.), Swinbank *et al.* (*Ann. Bot., Lond.* **28**: 239, 1964; *Serpula*).

Spore discharge and dispersal. A wide range of mechanisms for spore discharge and dispersal can be found in the fungi. Most involve air, animals, water or some combination of these. To be discharged, spores must escape from, or overcome the adhesive forces attaching them to the cell(s) in or on which they developed, and in many cases also the fruitbody in or on which those cells were produced.

Air dispersal. Most spores dispersed in this way do not travel far, but a tiny proportion may travel for thousands of km. This usually implies a very low rate of successful colonization by the dispersed spores. Spores must cross the laminar boundary layer of static air that surrounds all surfaces and enter the turbulent layer beyond. Many mechanisms have developed in fungi to enable their spores readily to become airborne.

Dry passive methods of air dispersal are independent of water availability and so can occur in the absence of rain and at low humidities. Some (*Agaricales*, *Myxomycetes*) have tall sporophores which lift spores well into or through the laminar boundary layer. Release and dispersal without mechanical disturbance can occur at low wind speeds in myxomycetes with raised fruiting bodies, e.g., *Dictydium*, and with fungi producing cup-shaped fruitbodies (e.g. *Cyttaria*, *Lachnellula*, *Peziza*, many lichen-forming species) in which double eddies may occur as a result of wind blowing across the surface. However, removal of conidia from conidiophores by wind requires speeds of at least 0.4-2.0 m/s, which are seldom achieved in crops except in gusts. Passive shedding as a result of gravity or in convection currents remains of uncertain importance.

Other passive methods of air dispersal involve mechanical disturbance such as rainsplash. Although water is required to activate them, rain tap and puff & bellows mechanisms are often used to disperse dry spores. In the first, rain striking a stem causes vibration which loosens spores while the cushion of air which precedes the splash disturbs the laminar boundary layer and blows spores into the air. Bellows mechanisms are seen in *Lycoperdon* and similar fungi which have a thin, flexible, unwettable peridium covering the spore mass which is depressed momentarily by raindrops or drips from foliage, expelling a puff of air carrying millions of spores from the aperture. Similarly, open cup-shaped structures ('splash cups', q.v.) produced by bird's nest fungi (*Nidulariales*), e.g., *Cyathus* spp., *Crucibulum vulgare*, and podetia lined with soredia produced by *Cladonia* use the energy of raindrops to project their contents up to 1 m.

Mist pickup has been described for *Cladosporium*. Rain and drip splash are important for fungi with spores embedded in slimy masses, e.g., *Colletotrichum lindemuthianum*, and for other non-slimy species whose spores are easily released by rain, e.g. *Venturia inaequalis*, or which also have active dispersal mechanisms, e.g. *Phytophthora infestans*.

Bubble scavenging concentrates particles, e.g. spores of aquatic fungi, from water suspension and these may then be projected into the air when surface bubbles burst. Droplets from breaking waves and waterfalls fulfil a similar function.

Active mechanisms activated by changing humidity and drying are seen in the violent twisting movements of some *Oomycota* and in water rupture mechanisms. Release through hygroscopic movement is characteristic of sporangiophores of *Phytophthora infestans* and *Peronospora hyoscyami* (syn. *P. tabacina*) and may also serve to loosen spores of some hyphomycetes, e.g. *Botrytis cinerea*, *Cladosporium*, *Drechslera*, before subsequent deflation. Water rupture occurs when tension is sufficient for the cohesion between water molecules or their adhesion to cell walls to fail resulting in the sudden formation of a gas bubble. A spore-bearing cell which had been collapsing as water was withdrawn by evaporation then suddenly returns to its former size as the gas bubble forms, flicking off the spores. The catapult effect may be enhanced by differential thickening of the conidiophore, as seen in *Deightoniella torulosa* and *Zygosporium oscheoides*.

Squirt gun mechanisms are characteristic of many *Ascomycota*, in which the spores are ejected explosively from turgid cells to a distance of 1-2 cm for small ascospores and up to 60 cm for larger ascospores, e.g. *Ascobolus immersus*. This type of discharge may result in the spore becoming electrostatically charged (see Spore charge). Asci may discharge singly from perithecia or many simultaneously from apothecia, e.g., *Sclerotinia sclerotiorum* producing a visible puff. About 0.2 mm rainfall is necessary for discharge of ascospores from perithecia of *Venturia inaequalis* in apple leaves or of *Mycosphaerella pinodes* in pea straw but at least 1.2 mm for ascospores from stromata of *Eutypa armeniacae*. Ascospores are usually dispersed as single spores but in *Pyrenopeziza brassicae* groups, usually of four spores, may be dispersed together.

Squirting mechanisms are also found in *Pilobolus* spp., *Basidiobolus ranarum*, *Entomophthora muscae*, *Pyricularia oryzae* and *Nigrospora sphaerica*. The sudden rounding of turgid cells can cause spores of *Conidiobolus coronata* to travel up to 4 cm and also occurs in *Epicoccum nigrum*, *Arthrinium cuspidatum*, *Sclerospora philippinensis* and the anamorph of *Xylaria furcata*.

Discharge of ballistospores and basidiospores is activated by water. The formation of a drop of liquid at the hilum end of the spore is characteristic; this is termed Buller's drop (q.v.) and has a crucial role in spore discharge. The spore is ejected 0.01-0.02 cm, sufficient for it to be able to fall freely between the gills or through pores into the free air below.

Water dispersal. Aquatic phycomycetes (*Chytridiomycota*, *Chromista*) may be dispersed in water as self-motile zoospores. Some water-dispersed spores (e.g. conidia of *Rubikia*) have many empty (vacated) cells and the air trapped within these cells may enable them to function as float chambers.

Animal dispersal. The success rate of colonization is often higher with animal-dispersed spores than with air- or water-dispersed spores. Arthropods are probably the dominant animal group involved in fungal spore dispersal. Examples include insects carrying the anamorphs of *Claviceps* spp. to new host plants, flies dispersing spores of *Phallus*, and laboulbeniomycetes being transmitted from one insect to another by touch. Other animal groups are also, however, important. Truffles and other hypogeous fungi are usually dispersed when pigs, rodents or other mammals drop fruitbody fragments while feeding. *Pilobolus* and other coprophilous fungi may be spread by animals and even require ingestion for their spores to germinate. However, they may also have an airborne dispersal phase. Humans may aid discharge and dispersal of spores through disturbance caused by agricultural practices such as mechanical harvesting and threshing.

Dispersal of spores depends on their size, shape, roughness and other characteristics and, for air dispersed spores, on wind speed and turbulence. Spore dispersal mechanisms determine the periodicity of spore release (see Air spora). Spore size and surface roughness affect the rate at which spores fall through the air and their ability to impact on stems and other obstacles. Fall rate or terminal velocity is a function of the square of the spore radius while the efficiency with which spores impact onto obstacles increases with their size. Many spores in crops and similarly dense vegetation either fall to the ground or impact onto stems and fail to escape into the air. Most that do escape are deposited within 100 m and few travel long distances. Spore clouds are spread both horizontally and vertically by eddies as they travel downwind from their source and concentrations decrease following theories of diffusion. Nevertheless, clouds of spores released in successive days and nights have been identified up to 600 km downwind of the British coast (Hirst & Hurst, *Symp. Soc. gen. Microbiol.* **17**: 307, 1967). Dispersal is ended by impaction of particles onto obstacles and by washout in rain.

It should be remembered that successful dispersal ends in deposition and results in colonization of new substratum. Many spores therefore have adaptations enabling them to adhere to suitable surfaces (e.g. mucous sheaths, radiating arms).

Lit.: Buller (**1-6**), Edmonds (*Aerobiology, the ecological systems approach*, 1979), Gregory (*TBMS* **21**: 26, 1945, *Microbiology of the atmosphere*, edn 2, 1973), Gregory & Monteith (*Airborne microbes*, 1967), Ingold (*Fungal spores, their liberation and dispersal*), 1971), Malloch & Blackwell (*in* Carroll & Wicklow, *The fungal community*, edn 2 : 147, 1992).

spore dormancy, see Spore germination.

Spore germination. Germination and dormancy in fungi have been reviewed by Sussman (*in* Ainsworth & Sussman (Eds), *The Fungi* **2**: 733, 1966), Sussman

& Douthit (*Ann. Rev. Pl. Physiol.* **24**: 311, 1973), and Gottlieb (*The germination of fungus spores*, 1978). A simple descriptive kinetics model of filamentous fungi spore germination is given by Bosch *et al.* (*Process biochemistry* **30**: 599, 1995).

Germination usually involves production of one or more germ tubes accompained by synthesis of new wall material. Protoplasm pressure and enzymic degradation of the spore wall results in emergence of the germ-tube through the spore wall. Fungal spore germination is influenced mainly by temperature, humidity, daylight and various types of radiation (Schenck *et al.*, *Mycol.* **67**: 1189, 1975; Jack & Tansey, *Mycol.* **69**: 109, 1977). Spores of basidiomycetes and some of conidial fungi are distinctly stimulated by mucin that is chemically similar to the natural spore matrix (Mahuku & Goodwin, *European Jurnal of Plant Pathology* **104**: 849, 1998). Oxygen is required for swelling of spores and the process is energy dependent. Thus aeration is necessary for formation of germ tubes during cultivation in solid state fermentation (Augur *et al.*, *Micología Aplicada Internacional* **19**: 7, 2007). Various llight wavelengths may affect the capacity of fungal spores (e.g. *Puccinia graminis*) to germinate (Calpouzos & Chang, *Plant Physiology* **47**: 729, 1971; Lucas *et al.*, *Plant Physiology* **56**: 847, 1975). Darkness increases germination of *Alternaria alternata* spores (Hatzipapas *et al.*, *MR* **106**: 1349, 2002). Radiation normally inhibits spore germination, but fungal strains with long-term exposure may show enhanced germination rates (Tugay *et al.*, *Mycol.* **98**: 521, 2006). See also Radiation and fungi.

Molecular and biochemical events during early stages of spore germination involve cAMP signalling (Fillinger *et al.*, *Mol. Microbiol.* **44**: 1001, 2002), net synthesis of lipids, with large production and later net degradation of free fatty acids (Mumford & Pappelis, *Mycopatologia* **64**: 63, 1978; Beilby & Kidby, *Journal of Lipid Research* **21**: 739, 1980). Germination can be highly dependent on the ratio of non-nutritive (not metabolizable), non-toxic carbohydrates such as L-sorbose and cellulose. Time of the spore germination is poorly correlated with water activity (Pham *et al.*, *J. Agric. Food Chem.* **47**: 4976, 1999).

Inhibition of fungal spore germination has been studied for biocontrol of plant diseases. Some plant extracts (e.g. of *Cichorium intybus*, *Eryngium creticum*, *Melia azedarach* and *Salvia fruticosa*) can suppress spore germination in species of *Alternaria*, *Botrytis*, *Cladosporium*, *Fusarium* and *Penicillium* (Abou-Jawdah, *et al.* (*J. Agric. Food Chem.* **50**: 3208, 2002; Chu *et al.*, *Plant Disease* **90**: 858, 2006). Similarly, pollen and stigma extracts of *Sorghum* can suppress spore germination of *Claviceps africana* (Prom, *Asian Journal of Plant Pathology* **1**: 12, 2007). Mandibular gland secretions of leaf-cutting ants have been shown to inhibit spore germination of conidial ascomycetes and zygomycetes which occur in their nests (Marsaro *et al.*, *Neotrop. Entomol.* **30**: 403, 2001; Rodrigues *et al.*, *Braz. J. Microbiol.* **39**: 64, 2008).

In lichenized fungi the ascospores germinate readily in water films at moderate temperatures. Percentage germination is lowest in areas of high air pollution (Kofler *et al.*, *Mém. Soc. bot. Fr.* 1968: 219, 1969). In *Xanthoria parietina* ascospores remain dormant until contact with the photobiont is made (Werner, *Mém. Soc. Sci. nat. phys. Maroc* **27**: 7, 1931). See Lichens.

Lit.: Abou-Jawdah *et al.* (*J. Agric. Food Chem.* **50**: 3208, 2002), Augur *et al.* (*Micología Aplicada Internacional* **19**: 7, 2007), Beilby & Kidby (*Journal of Lipid Research* **21**: 739, 1980), Bosch *et al.* (*Process biochemistry* **30**: 599, 1995), Calpouzos & Chang (*Plant Physiology* **47**: 729, 1971), Chu *et al.* (*Plant Disease* **90**: 858, 2006), Fillinger *et al.* (*Mol. Microbiol* **44**: 1001, 2002), Hatzipapas *et al.* (*Mycological Research* **106**: 1349, 2002), Jack & Tansey (*Mycol.* **69**: 109, 1977), Lucas *et al.* (*Plant Physiology* **56**: 847, 1975), Mahuku & Goodwin (*Eur. J. Plant Path.* **104**: 849, 1998), Marsaro *et al.* (*Neotrop. Entomol.* **30**: 403, 2001), Merrill (*Ann. Rev. Phytopath.* **8**: 281, 1970), Mumford & Pappelis (*Mycopathologia* **64**: 63, 1978), Pham *et al.* (*J. Agric. Food Chem.* **47**: 4976, 1999), Prom (*Asian J. Plant Path* **1**: 12, 2007), Rodrigues *et al.* (*Braz. J. Microbiol.* **39**: 64, 2008), Schenck *et al.* *Mycol.* **67**: 1189, 1975), Tugay *et al.* (*Mycol.* **98**: 521, 2006). See also *Lit.* under Spore discharge and dispersal.

spore groups, for Saccardo's spore groups see Anamorphic fungi. Also see Conidial nomenclature, Sphaeriales.

spore horn, see cirrus.

spore longevity. see Longevity.

spore print, the deposit of basidioma obtained by allowing spores from a basidiocarp to fall onto a sheet of paper (white or coloured) placed below the lamellae or pores. See Singer (1975: 5).

spore specific gravity, Buller (**1**: 153, 1909) determined the Specific Gravity (SG) of various agaric spores as 1.02-1.43.

spore terminology. see aboo**spore**, acro-, adia-, aecio-, aleurio-, alpha-, amero-, amphi-, annello- (under annellidic), aplano-, arthro-, asco-, azygo-, ballisto-, basidio-, beta- -, blasto-, botryo-aleurio-, botryoblasto-, chlamydo-, clostero-, conidiole, conidium, cyst, deuteroconidium, dia-, dictyochlamydo-, dictyoporo-, dictyo-, didymo-, diploconidium, dispersal -, di-, dry -, ecto-, endoconidium, endo-, fragmentation -, fuseau, ganglio-, gasteroconidium, gastero-, gemma, gonosphere, haploconidium, helico-, hemi-asco-, hemi-, isthmo-, loculo- (see Loculomycetes), macroconidium, macro-, meio-, memno-, meri-, meristem arthro-, meristem blasto-, meso-, microconidium, micro-endo-, micro-, mischoblastiomorph, mito-, mono-, myceloconidium, myxo-, nimbo-, oidium, oo-, papulo-, part -, partheno-, peri-, phialoconidium (under phialidic), phragmo-, placodiomorph, plasmagamo-, polarilocular-, poroconidium, poro-, primo-, promycelial -, protero-, protoconidium, proto-, pseudoidium, pseudo-, pycnidio-, pycnio-, pycno-, radula -, ramoconidium, repeating -, resting -, scoleco-, secondary -, seio-, sicyo-, simblo-, slime -, smut -, spiculo-, sporangio-, sporidesm, sporidiole, sporidium, stalagmo-, statismo-, stauro-, stylo-, summer -, sympodioconidium, synchrono-, teleutosporodesm, telio-, terminus -, tetra-, texo-, thallo-, theca-, tretoconidium (under tretic), tricho-, ustilo-, usto-, winter -, xeno-, zoo-, zygo-.

Spore wall. Conventional or electron microscopy shows the spore wall to be layered. The terminology of these layers by different authors is somewhat confused (see the comparison by Payak, 1964: 33). Five layers (the spore wall proper; **eusporium**) which have been distinguished are, from within outwards:

(1) **endosporium** (endospore, corium), which is usually thin and is the last to develop during sporogenesis; (2) **episporium**, the thick, fundamental layer which determines the shape of the spore; (3) **exosporium** (exospore, epitunica, trachytectum, tunica), a layer derived from (2) but chemically distinct and frequently responsible for the ornamentation; (4) **perisporium** (= mucostratum, myxosporium), a layer, frequently fugacious, enveloping the whole spore and limited by (5), the hardly visible **ectosporium** (sporothecium). On this disappearance of (4) and (5) (3) is the outer spore layer. (1)-(3) are thus the spore wall proper; (4) and (5) of extrasporal origin. See Heim (*Rev. Mycol.* **27**: 199, 1962), Payak (*in* Nair (Ed.), *Recent advances in palynology*, 1964), Pyatt (*TBMS* **52**: 167, 1969; lichens), Pegler & Young (*TBMS* **72**: 356, 1979; basidiospores). See Fig. 5; cf. Cell wall chemistry.

Sporendocladia G. Arnaud ex Nag Raj & W.B. Kendr. (1975), anamorphic *Ceratocystidaceae*, Hso.0eH.22. 7, widespread. See Wingfield *et al.* (*TBMS* **89**: 609, 1987; separation from *Phialocephala*), Crous & Wingfield (*Sydowia* **46**: 193, 1994; Switzerland), Onofri *et al.* (*MR* **98**: 745, 1994; key), Jacobs *et al.* (*Mycol.* **95**: 637, 2003; phylogeny).

Sporendonema Desm. (1827), anamorphic *Pezizomycotina*, Hso.0eP.40. 3, widespread. See Sigler & Carmichael (*Mycotaxon* **4**: 376, 1976).

Sporhaplus H.B.P. Upadhyay (1964), anamorphic *Pezizomycotina*, Cpd.0eH.?. 1 (lichenicolous), Brazil. See Hawksworth (*Bull. Br. Mus. nat. Hist.* Bot. **9**: 1, 1981), Matzer (*Mycol. Pap.* **171**: 202 pp., 1996).

Sporhelminthium Speg. (1918) = Clasterosporium fide Hughes (*CJB* **36**: 727, 1958).

Sporichthya M.P. Lechev., H. Lechev. & Holbert (1968), Actinobacteria. q.v.

sporidesm (**sporodesm**), a compound spore or spore-ball, the components of which are merispores. See teleutosporodesm.

Sporidesmiella P.M. Kirk (1982), anamorphic *Melanommataceae*, Hso.1-≡ eP.19. 18, widespread. See Zhang *et al.* (*Mycotaxon* **18**: 243, 1983; conidiogenesis), Kuthubutheen & Nawawi (*MR* **97**: 1305, 1993; review), Castañeda Ruíz *et al.* (*MR* **102**: 548, 1998), Wu & Zhuang (*Fungal Diversity Res. Ser.* **15**, 2005), Shenoy *et al.* (*MR* **110**: 916, 2006; phylogeny).

sporidesmin (and **sporodesmolides**), toxins (oligopeptides) of *Pithomyces chartarum* (teleomorph *Leptosphaerulina chartarum*); the cause of facial eczema in sheep and cattle, esp. in NZ.

Sporidesmina Subram. & Bhat (1989), Xylariales. See Sutton (*in litt.*), Shenoy *et al.* (*MR* **110**: 916, 2006; phylogeny).

Sporidesmiopsis Subram. & Bhat (1989) = Brachysporiella fide Sutton (*in litt.*), Wu & Zhuang (*Fungal Diversity Res. Ser.* **15**, 2005; China).

Sporidesmium Link (1809), anamorphic *Pleosporales*, Hso.≡ -#eP.19. 113, widespread. See Ellis (*Mycol. Pap.* **70**, 1958; key), Hughes (*CJB* **36**: 807, 1958), Moore (*Mycol.* **50**: 681, 1958), Bullock *et al.* (*CJB* **67**: 313, 1989; morphology, histochem. and germination of *S. sclerotivorum*), Subramanian (*Proc. Indian natn Sci. Acad.* Part B. Biol. Sci. **58**: 179, 1992; segregation), Réblová (*Mycotaxon* **71**: 13, 1999), Wu & Zhuang (*Fungal Diversity Res. Ser.* **15**, 2005; China), Shenoy *et al.* (*MR* **110**: 916, 2006; phylogeny, polyphyly).

Sporidiaceae R.T. Moore (1980) = Sporidiobolaceae R.T. Moore.

Sporidiales R.T. Moore (1980), see *Sporidiobolales*.

Sporidiobolaceae Kobayasi (1961) nom. inval. = Sporidiobolaceae R.T. Moore.

Sporidiobolaceae R.T. Moore (1980), Sporidiobolales. 3 gen., 10 spp.

Lit.: Boekhout & Nakase *in* Kurtzman & Fell (Eds) (*Yeasts, a taxonomic study* 4th edn: 828, 1998), Fell & Statzell-Tallman *in* Kurtzman & Fell (Eds) (*Yeasts, a taxonomic study* 4th edn: 678, 1998), Kurtzman & Fell (*Yeasts, a taxonomic study* 4th edn, 1998; taxonomy), Statzell-Tallman & Fell *in* Kurtzman & Fell (Eds) (*Yeasts, a taxonomic study* 4th edn: 693, 1998), Fell *et al.* (*Int. J. Syst. Evol. Microbiol.* **50**: 1351, 2000; mol. phylogeny), Hamamoto & Nakase (*Int. J. Syst. Evol. Microbiol.* **50**: 1373, 2000), Biswas *et al.* (*Int. J. Syst. Evol. Microbiol.* **51**: 1191, 2001), Nagahama *et al.* (*Antonie van Leeuwenhoek* **80**: 101, 2001), Bai *et al.* (*Int. J. Syst. Evol. Microbiol.* **52**: 2309, 2002), Fell *et al.* (*FEMS Yeast Res.* **1**: 265, 2002), Scorzetti *et al.* (*FEMS Yeast Res.* **2**: 495, 2002), Libkind *et al.* (*Antonie van Leeuwenhoek* **84**: 313, 2003), Sampaio *et al.* (*Mycol. Progr.* **2**: 63, 2003), Wang & Bai (*FEMS Yeast Res.* **4**: 579, 2004).

Sporidiobolales Doweld (2001). Microbotryomycetes. 1 fam., 7 gen., 83 spp. Fam.:

Sporidiobolaceae

Lit.: Moore (1980), Boekhout *et al.* (*Can. J. Microb.* **39**: 276, 1993), Swann & Taylor (*Stud. Mycol.* **38**: 147, 1995; mol. phylogeny, taxonomy). See also *Lit.* for Yeast.

Sporidiobolus Nyland (1950), Sporidiobolaceae R.T. Moore. Anamorphs *Rhodomyces*, *Sporobolomyces*. 7, widespread. See Bandoni *et al.* (*CJB* **49**: 683, 1971).

sporidiolae, spore-like bodies produced inside **sporidiomata**, perithecium-like structures, in *Kathistes* (Malloch & Blackwell, *CJB* **68**: 1712, 1990).

sporidiole, a small spore.

Sporidiomycetes. Class used by Moore (1980) for *Sporidiales*, q.v.

sporidium (pl. **sporidia**), (1) a basidiospore of the *Pucciniales* and *Ustilaginales* or, in the latter, any spore other than an ustilospore; (2) ascospore (obsol.).

Sporisorium Ehrenb. ex Link (1825), Ustilaginaceae. *c.* 320 (on *Poaceae*), widespread. *S. sorghi* and *S. cruentum* (covered and loose smuts of sorghum); *S. reilianum* (head smut of maize (*Zea*) and sorghum). Many graminicolous smuts originally described as *Sorosporium*, *Sphacelotheca* or *Ustilago* belong here. See Langdon & Fullerton (*Mycotaxon* **6**: 421, 1978), Vánky (*Europ. Smut Fungi*: 570 pp., 1994), Vánky & McKenzie (*Fungal Diversity Res. Ser.* **8**: 259 pp., 2002; New Zealand), Vánky (*Mycotaxon* **2**: 169, 2005; Europe).

Sporoacania A. Massal. (1855) ? = Lecidea fide Hawksworth *et al.* (*Dictionary of the Fungi* edn 8, 1995).

Sporoblastia Trevis. (1851) = Cliostomum fide Hawksworth *et al.* (*Dictionary of the Fungi* edn 8, 1995).

Sporobolomyces Kluyver & C.B. Niel (1924), anamorphic *Sporidiobolus*. *c.* 25, widespread. See Fell *et al.* (*Stud. Mycol.* **38**: 129, 1995; mol. phylogeny, 2000; mol. phylogeny), Nakase *et al.* (*Stud. Mycol.* **38**: 163, 1995; mol. phylogeny), Swann & Taylor (*Stud. Mycol.* **38**: 147, 1995; mol. phylogeny), Takashima & Nakase (*Mycoscience* **41**: 357, 2000; Thailand), Bai *et al.* (*International Journal of Systematic and Evolutionary Microbiology* **52**: 2309, 2002), Sampaio (*Frontiers in Basidiomycote Mycology*: 49, 2004), Libkind *et al.* (*International Journal of Systematic and Evolutionary Microbiology* **55**: 503, 2005; Argentina), Nagahama (*Journal of General and Applied Microbiology, Tokyo* **52**: 37, 2006), Lumbsch *et al.* (*MR* **111**: 1133, 2007).

Sporobolomycetales (mirror or shadow yeasts), anamorphic yeasts with basidiomycetous affinities characteristically producing ballistospores following multiplication by budding (e.g. *Bullera*, *Itersonilia*, *Sporobolomyces*, *Tilletiopsis*). Accepted in *Basidioblastomycetes* by Moore (1980). See Yeast for *Lit.*, also Martin (*Univ. Iowa Stud. nat. Hist.* **19**(3), 1952).

Sporocadaceae Corda (1842) = Amphisphaeriaceae.

Sporocadus Corda (1839), anamorphic *Amphisphaeriaceae*. 1, widespread. See Shoemaker & Müller (*CJB* **42**: 403, 1964), Brockmann (*Sydowia* **28**: 275, 1976), Morelet (*Ann. Soc. Sci. nat. Arch. Toulon & Var* **37**: 233, 1985), Sergeeva *et al.* (*Australas. Pl. Path.* **34**: 255, 2005).

sporocarp, a general term for spore-bearing organs; fruit-body (q.v.). Used esp. of *Acrasiomycota*, *Mycetozoa* and *Endogonaceae*.

Sporocarpon Will. (1878), Fossil Fungi ? Ascomycota. 5 (Carboniferous), Europe; USA. See Hutchinson (*Ann. Bot.* **19**: 428, 1955), Baxter (*Paleont. Contrib. Univ. Kansas* **77**, 1975), Pirozynski (*Ann. Rev. Phytopath.* **14**: 237, 1976).

Sporocephalium Chevall. (1826) = Acladium fide Hughes (*CJB* **36**: 727, 1958).

Sporocephalum Arnaud (1952) ? = Oedocephalum fide Kendrick & Carmichae *in* Ainsworth *et al.* (Eds) (*The Fungi* **4A**: 390, 1973).

Sporochytriaceae A. Fisch. (1892) = Rhizophydiaceae.

Sporocladium Chevall. (1826) ≡ Cladosporium fide Hughes (*CJB* **36**: 727, 1958).

sporocladium, a special sporogenous branch in the *Kickxellaceae*.

Sporoclema Tiesenh. (1912) nom. dub., Fungi. 'phycomycetes'. See Kendrick & Carmichael *in* Ainsworth *et al.* (Eds) (*The Fungi* **4A**: 323, 1973).

Sporoctomorpha J.V. Almeida & Sousa da Câmara (1903), Pezizomycotina. 1, Europe.

Sporocybe Fr. (1825) = Periconia fide Mason & Ellis (*Mycol. Pap.* **56**, 1953), Carmarán & Novas (*Fungal Diversity* **14**: 67, 2003).

Sporocybomyces H. Maia (1967) = Echinoplaca fide Lücking *et al.* (*Lichenologist* **30**: 121, 1998).

sporocyst, a cyst producing asexual spores.

Sporocystis Morgan (1902), anamorphic *Pezizomycotina*, Hsp.#eH.?. 2, N. America; Sri Lanka.

Sporoderma Mont. (1856) = Trichoderma Pers. (1794) fide Höhnel (*Sber. Akad. Wiss. Wien* Math.-naturw. Kl., Abt. 1 **119**: 671, 1910).

Sporodesmium, see *Sporidesmium*.

Sporodictyon A. Massal. (1852), Verrucariaceae (L). 3, Europe.

Sporodinia Link (1824) ≡ Syzygites.

Sporodiniella Boedijn (1959), Mucoraceae. 1 (on eggs, larvae & imago of *Membracidae* (*Diptera*, *Homoptera*, *Hymenoptera* & *Lepidoptera*; apparently parasitic), pantropical. See Evans & Samson (*CJB* **55**: 2981, 1977), Gbaja & Young (*Microbios* **42**: 263, 1985; ultrastr.), Chien & Hwang (*Mycoscience* **38**: 343, 1997), Voigt & Wöstemeyer (*Gene* **270**: 113, 2001; phylogeny), Ali-Shtayeh *et al.* (*Mycopathologia* **156**: 235, 2002; Palestine).

Sporodiniopsis Höhn. (1903) nom. dub., anamorphic *Pezizomycotina*. See Hawksworth (*Mycol. Pap.* **126**, 1971).

sporodochium, conidioma, typical of the *Tuberculariaceae* (q.v.) in which the spore mass is supported by a superficial cushion-like (pulvinate) mass of short conidiophores and pseudoparenchyma; **pionnote - (pseudopionnotes**, Sherbakoff, 1915) (of *Fusarium*), minute sporodochia near the surface of the substrate having no stroma, the spores forming a continuous slimy layer. Cf. acervulus, pionnotes.

Sporodum Corda (1837) = Periconia fide Mason & Ellis (*Mycol. Pap.* **56**, 1953).

sporogenesis, spore development.

sporogenous, producing, having or supporting spores; cf. conidiogenous.

Sporoglena Sacc. (1894), anamorphic *Pezizomycotina*, Hsy.0eP.?. 1, Papua New Guinea.

sporograph, the straight-line graph obtained by plotting the ratio (E) of the length (D) to width (d) against the length of the basidiospores of a species of agaric (Corner, *New Phytol.* **46**: 196, 1947). Cf. Q.

sporoma (pl. **-omata**), a multicellular structure specially developed to produce spores.

Sporomega Corda (1842), Rhytismataceae. 1, Europe; N. America. See Eriksson (*Symb. bot. upsal.* **19** no. 4, 1970).

Sporonema Desm. (1847), anamorphic *Godronia*, St.0eH.15. 15, widespread (north temperate). See Limber (*Mycol.* **47**: 389, 1955), Sutton (*The Coelomycetes*, 1980).

Sporonites R. Potonié (1931), Fossil Fungi ? Fungi. 1.

sporont, a thallus on which spores will be produced.

Sporopachydermia Rodr. Mir. (1978), Saccharomycetales. Anamorph *Candida*. 3, widespread. See Yamada *et al.* (*J. gen. appl. Microbiol.* Tokyo **38**: 179, 1992), Lachance & Phaff *in* Kurtzman & Fell (Eds) (*Yeasts, a taxonomic study* 4th edn: 395, 1998), Lachance *et al.* (*Int. J. Syst. Evol. Microbiol.* **51**: 237, 2001; phylogeny, species concepts), Suh *et al.* (*Mycol.* **98**: 1006, 2006; phylogeny).

Sporophaga Harkn. (1899) nom. dub., ? Fungi.

Sporophagomyces K. Põldmaa & Samuels (1999), Hypocreaceae. 3 (fungicolous), widespread. See Põldmaa *et al.* (*CJB* **77**: 1765, 1999), Põldmaa & Samuels (*Sydowia* **56**: 79, 2004; Thailand).

sporophagous, feeding on spores, as in certain thrips species on fungus spores (Ananthakrishnan *et al.*, *Proc. Ind. Acad. Sci.* Anim. Ser. **92**: 95, 1983), where the dimensions of the mouth parts are related to the sizes of the spores eaten (Ananthakrishnan & Dhileepan, *Proc. Ind. Acad. Sci.* Anim. Ser. **93**: 243, 1984); species lists (Ananthakrishnan *et al.*, 1984).

Sporophiala P. Rag. Rao (1970), anamorphic *Pezizomycotina*, Hso.≡ eP.3. 3, India; USA. See Rao (*Mycopathologia* **41**: 317, 1970).

Sporophleum Nees ex Link (1824) = Arthrinium fide Hughes (*CJB* **36**: 727, 1958).

Sporophlyctidium Sparrow (1933), Chytridiaceae. 2, USA; North Africa.

Sporophlyctis Serbinow (1907), ? Chytridiaceae. 2 (on *Draparnaldia* and *Chaetophora*), widespread.

Sporophora Luteraan (1952), anamorphic *Pezizomycotina*, Hso.0eH.3. 1, France. See Luteraan (*Rev. Biol. gen. theor. appl.* **2**: 2, 1952).

sporophore (1) a spore-producing or -supporting structure, esp. a conidiophore; (2) (of macrofungi), ascoma, basidioma (see Fig. 4); cf. hymenophore; a basidium (sensu Berkeley).

Sporophormis Malloch & Cain (1973) ≡ Warcupiella.

Sporophysa (Sacc.) Vain. (1921) = Verrucaria Schrad. fide Keissler (*Rabenh. Krypt.-Fl.* **8**: 446, 1930).

Sporophyta, see *Cryptogamia*.

sporophyte, the diploid or asexual phase in the life-cycle of a plant; diplont; diplophase. Cf. gametophyte.

sporoplasm, the spore-producing protoplasm within the epiplasm in a sporangium or ascus (Guilliermond).

Sporopodiopsis Sérus. (1997), Ectolechiaceae (L). 2, S.E. Asia. See Sérusiaux (*Abstracta Botanica* **21**: 145, 1997).

Sporopodium Mont. (1851), Ectolechiaceae (L). 24, widespread (tropical). See Santesson (*Symb. bot. upsal.* **12** no. 1: 1, 1952), Hafellner (*Beih. Nova Hedwigia* **79**: 241, 1984), Elix *et al.* (*Biblthca Lichenol.* **58**: 81, 1995; chemistry), Lücking & Lumbsch (*Mycotaxon* **78**: 23, 2001; chemistry), Sérusiaux *et al.* (*Mycotaxon* **103**: 255, 2008; key, chemistry).

Sporormia De Not. (1845), Sporormiaceae. 3 (coprophilous or on wood), widespread. See Ahmed & Cain (*CJB* **50**: 419, 1972), Dissing (*Persoonia* Suppl. **14**: 389, 1992), Berbee (*Mol. Biol. Evol.* **13**: 462, 1996; DNA), Kruys *et al.* (*MR* **110**: 527, 2006; phylogeny).

Sporormiaceae Munk (1957), Pleosporales. 10 gen. (+ 7 syn.), 143 spp.

Lit.: von Arx & van der Aa (*TBMS* **89**: 117, 1987; key 6 gen.), Dissing (*Persoonia* Suppl. **14**: 389, 1992), Berbee (*Mol. Biol. Evol.* **13**: 462, 1996; DNA), Hyde & Steinke (*Mycoscience* **37**: 99, 1996), Guarro *et al.* (*MR* **101**: 305, 1997), Silva-Hanlin & Hanlin (*MR* **103**: 153, 1999), Barr (*Mycotaxon* **76**: 105, 2000; keys, *Delitschiaceae*), Liew *et al.* (*Mol. Phylogen. Evol.* **16**: 392, 2000), Arenal *et al.* (*Mycotaxon* **89**: 137, 2004), Kruys *et al.* (*MR* **110**: 527, 2006), Schoch *et al.* (*Mycol.* **98**: 1041, 2006; phylogeny), Wang *et al.* (*MR* **111**: 1268, 2007; phylogeny).

Sporormiella Ellis & Everh. (1892), Sporormiaceae. *c.* 80 (mainly coprophilous), widespread. Used as indicators of megafauna in Quaternary lake deposits. See also *Preussia*. See Ahmed & Cain (*CJB* **50**: 419, 1972), Davis (*Quatern. Res.* **28**: 290, 1987), Barr (*N. Amer. Fl.* ser. 2 **13**: 129 pp., 1990), Eriksson & Hawksworth (*SA* **10**: 144, 1991), Berbee (*Mol. Biol. Evol.* **13**: 462, 1996; DNA), Lumbsch & Lindemuth (*MR* **105**: 901, 2001; phylogeny), Burney *et al.* (*Proc. natn Acad. Sci. U.S.A.* **100**: 10800, 2003; Holocene extinctions in Madagascar), Prokhorov & Armenskaya (*Mikol. Fitopatol.* **37**: 27, 2003; Russia), Arenal *et al.* (*Mycotaxon* **89**: 137, 2004; biometrics), Schoch *et al.* (*Mycol.* **98**: 1041, 2006; phylogeny).

Spororminula Arx & Aa (1987), Sporormiaceae. 1 (coprophilous), Canary Islands. See von Arx & van der Aa (*TBMS* **89**: 117, 1987), Barr (*Mycotaxon* **76**: 105, 2000).

Sporormiopsis Breton & Faurel (1964) = Sporormiella fide von Arx (*Persoonia* **7**: 367, 1973), Barr (*Mycotaxon* **76**: 105, 2000).

Sporoschisma Berk. & Broome (1847), anamorphic *Melanochaeta*, Hsy.≡ eP.22. 10, widespread. See Nag Raj & Kendrick (*Monogr. Chalara Allied Genera*, 1975), Goh *et al.* (*MR* **101**: 1295, 1997; key), Ho *et al.* (*CJB* **76**: 1614, 1998), Sivichai *et al.* (*MR* **104**: 478, 2000; Thailand).

Sporoschismataceae Nann. (1934) = Chaetosphaeriaceae.

Sporoschismatites Babajan & Tasl. (1977), Fossil Fungi. 1 (Tertiary), former USSR.

Sporoschismopsis Hol.-Jech. & Hennebert (1972), anamorphic *Porosphaerellopsis*, Hso.≡ eP.22. 5, Europe; N. America. See Goh *et al.* (*MR* **101**: 1295, 1997; key).

Sporoschizon Riedl (1960), ? Naetrocymbaceae (L). 1, Austria. See Harris (*More Florida Lichens*, 1995; ? synonym of *Naetrocymbe*).

Sporostachys Sacc. (1919) = Melanographium fide Saccardo (*Syll. fung.* **25**: 936, 1931), Somrithipol & Jones (*Fungal Diversity* **19**: 137, 2005).

sporostasis (adj. **sporostatic**), inhibition of spore germination; cf. mycostasis; sporostatic products of fungi, see Robinson *et al.* (*TBMS* **51**: 113, 1968).

Sporostigma Grube (2001), Arthoniaceae (L). 1, USA. See Grube (*Lichenologist* **33**: 388, 2001).

sporothallus, a thallus producing spores; cf. gametothallus.

Sporotheca Corda (1829) ? = Melanconis fide Mussat (*Syll. fung.* **15**: 398, 1901).

sporothecium (1) the tip of a basidium bearing basidiospores when the basidiospores are sessile (Clémençon, *Z. Pilzk.* **36**: 113, 1970); (2) see spore wall.

Sporothrichum, see *Sporotrichum*.

Sporothrix Hektoen & C.F. Perkins (1901), anamorphic *Ophiostoma*. *c.* 40, widespread. *S. schenckii*, from humans. See Thibaut (*Annls Parasit. hum. comp.* **46**: 93, 1971; *S. schenckii*, see sporotrichosis), De Hoog (*Stud. Mycol.* **7**, 1974), Kreisel & Schauer (*J. Basic Mycol.* **25**: 653, 1985; use of C sources), Findlay & Vismer (*Mycopathologia* **96**: 115, 1986; SEM of morphology), Smith & Batenburg-van der Vegte (*J. gen. appl. Microbiol.* Tokyo **32**: 549, 1986; ultrastr. septa), Takeda *et al.* (*Mycopathologia* **116**: 9, 1991; phylogeny, molecular epidemiology), de Hoog *in* Wingfield *et al.* (Eds) (*Ceratocystis* and *Ophiostoma* Taxonomy, Ecology and Pathogenicity: 53, 1993; anamorphs of *Ophiostoma*), Benade *et al.* (*MR* **101**: 1108, 1997; conidiogenesis), Okada *et al.* (*CJB* **76**: 1495, 1998; phylogeny), Ishizaki *et al.* (*Medical Mycology* **38**: 433, 2000; DNA, Africa, Australia), Lin *et al.* (*Mycopathologia* **148**: 69, 1999; DNA, China), Mesa-Arango *et al.* (*J. Clin. Microbiol.* **40**: 3004, 2002; from human, genetic diversity), de Beer *et al.* (*Mycol.* **95**: 434, 2003; clinical spp., phylogeny), Halmschlager & Kowalski (*Mycol. Progr.* **2**: 259, 2003; on *Quercus* roots), Hu *et al.* (*J. Clin. Microbiol.* **41**: 1414, 2003; molecular detection), Marimon *et al.* (*J. Clin. Microbiol.* **44**: 3251, 2006; phylogeny), Marimon *et al.* (*J. Clin. Microbiol.* **45**: 3198, 2007; new clinical spp.).

Sporotrichella P. Karst. (1887) = Fusarium fide Hughes (*CJB* **36**: 727, 1958).

Sporotrichites Göpp. & Berendt (1945), Fossil Fungi. 4 (Oligocene, Miocene), Baltic.

Sporotrichopsis Guég. (1911) nom. inval. = Sporothrix.

Sporotrichopsis Stalpers (2000), anamorphic *Abortiporus*. 1, widespread.

sporotrichosis, a lymphatic disease in humans and animals caused by *Sporothrix schenckii*. See de Beurmann & Gougerot (*Les Sporotrichoses*, 1912), Norden (*Acta Path. Microbiol. Scand., Suppl.* **84**,

1951).

Sporotrichum Link (1809), anamorphic *Laetiporus, Phanerochaete, Pycnoporellus*, Hso.0eH.10. 3, widespread. See von Arx (*Persoonia* **6**: 179, 1971), Donk (*Verh. K. ned. Akad. Wet.* tweede sect. **62**: 1, 1974), Stalpers (*Stud. Mycol.* **24**: 15, 1984; list 339 epithets) = Poria (Polyp.) fide.

spot anthracnose, see anthracnose.

Spraguea Weissenberg (1976), Microsporidia. 1.

Spragueola Massee (1896) = Neolecta fide Korf (*Phytologia* **21**: 201, 1971).

Spumatoria Massee & E.S. Salmon (1901), ? Ophiostomataceae. 1, Europe. See Malloch & Blackwell (*CJB* **68**: 1712, 1990).

Spumula Mains (1935), Raveneliaceae. 6 (on *Leguminosae*, mainly *Mimosoideae*), America (tropical); Africa; Indonesia; Philippines. See Berndt (*Pl. Syst. Evol.* **200**: 79, 1996).

spunk, see touchwood, amadou.

Squamacidia Brako (1989), Ramalinaceae (L). 1, widespread (tropical). See Brako (*Mycotaxon* **35**: 6, 1989), Okamoto & Iwatsuki (*J. Jap. Bot.* **66**: 292, 1991; Japan), Ekman (*Op. bot.* **127**: 148 pp., 1996; N America).

Squamanita Imbach (1946), Tricholomataceae. 10, widespread. See Bas (*Persoonia* **3**: 331, 1965; key), Harmaja (*Karstenia* **27**: 71, 1987), Læssøe & Rosendahl (*MR* **98**: 88, 1994; mycoparasitism), Redhead *et al.* (*CJB* **72**: 1812, 1994), Vizzini & Girlanda (*Allionia* **35**, 1997; mycoparasitism), Mondiet *et al.* (*MR* **111**: 599, 2007; mycoparasitism).

Squamaphlegma Locq. (1979) = Cortinarius fide Kuyper (*in litt.*).

Squamaria Hoffm. (1789) [non *Squamaria* Ludw. 1757, *Orobanchaceae*] = Cetraria fide Hawksworth *et al.* (*Dictionary of the Fungi* edn 8, 1995).

Squamarina Poelt (1958), Stereocaulaceae (L). 28, widespread. See Poelt & Krüger (*Feddes Repert.* **81**: 187, 1970), Stenroos & DePriest (*Am. J. Bot.* **85**: 1548, 1998; DNA), Peršoh *et al.* (*Mycol. Progr.* **3**: 103, 2004; asci, photobionts), Miądlikowska *et al.* (*Mycol.* **98**: 1088, 2006; phylogeny).

Squamarinaceae Hafellner (1984) = Stereocaulaceae.

Squamariomyces E.A. Thomas (1939) nom. inval. = Lecanora fide Hawksworth *et al.* (*Dictionary of the Fungi* edn 8, 1995).

Squamella S. Hammer (2001), Cladoniaceae (L). 1, Australia. See Hammer (*Bryologist* **104**: 561, 2001).

Squammaria DC. (1805) = Squamarina fide Hawksworth *et al.* (*Dictionary of the Fungi* edn 8, 1995).

squamose, having scales.

Squamotubera Henn. (1903) ? = Sarcoxylon fide Rogers (*Mycol.* **73**: 28, 1981).

squamule, a small scale.

squamulose (1) having small scales; (2) growth form of a lichen thallus (Fig. 21C).

squarrose, rough with scales.

St Anthony's Fire, see ergot.

St George's mushroom, Basidioma of the edible *Calocybe gambosum*.

stachel, bullet-like structure in the rohr with its pointed end orientated towards the appressorium and the host cell wall in plasmodiophorids. See also rohr and schlauch.

Stachybotryella Ellis & Barthol. (1902), anamorphic *Pezizomycotina*, Hso.0eP.?. 2, N. America.

Stachybotryna Tubaki & T. Yokoy. (1971), anamorphic *Pezizomycotina*, Hso.0eH.15. 3, widespread. See Tubaki & Yokoyama (*TMSJ* **12**: 18, 1971), Calduch *et al.* (*Mycol.* **94**: 355, 2002; Spain, key).

stachybotryotoxin, a mycotoxin produced by *Stachybotrys* growing on hay; implicated in serious poisoning of horses (stachyobotrytoxicosis).

Stachybotrys Corda (1837), anamorphic *Hypocreales*, Hsy.0eP.15. *c.* 44, widespread. Toxin-producing spp. are frequently encountered in environmental surveys of damp buildings. See Barron (*CJB* **39**: 1566, 1961), Verona & Mazzucchetti (*Microfunghi della cellulosa e della carta I generi Stachybotrys e Memnoniella*, 1968), Jong & Davis (*Mycotaxon* **3**: 409, 1976; spp. in cult.), McKenzie (*Mycotaxon* **41**: 179, 1991; spp. from *Freycinetia*), Haugland *et al.* (*Mycol.* **93**: 54, 2001; phylogeny), Andersen *et al.* (*Mycol.* **94**: 392, 2002; from water-damaged buildings), Andersen *et al.* (*Mycol.* **95**: 1227, 2003; chemotypes), Koster *et al.* (*CJB* **81**: 633, 2003; phylogeny), Taylor *et al.* (*Can. J. Pl. Path.* **25**: 49, 2003; molecular detection), Castlebury *et al.* (*MR* **108**: 864, 2004; phylogeny), Pinruan *et al.* (*Fungal Diversity* **17**: 145, 2004; key), Zhang *et al.* (*Mycol.* **98**: 1076, 2006; phylogeny).

Stachycoremium Seifert (1986), anamorphic *Pezizomycotina*, Hsy.1eH.16. 1, USA; E Asia. See Okada (*Microbiol. Culture Coll.* **12**: 17, 1996; conidiogenesis).

Stachylidium Link (1809), anamorphic *Pezizomycotina*, Hso.0eP.15. 2, widespread. See Hughes (*TBMS* **34**: 551, 1951).

Stachylina L. Léger & M. Gauthier (1932), Harpellaceae. 29 (in *Diptera*), widespread. See Lichtwardt (*Mycol.* **64**: 167, 1972), Moss (*Mycol.* **66**: 173, 1974; cytology), Moss *in* Fuller & Lovelock (Eds) (*Microbial ultrastructure*: 279, 1976; ultrastr.), Lichtwardt (*The Trichomycetes. Fungal associates of arthropods*, 1986; key), Lichtwardt *et al.* (*TMSJ* **28**: 376, 1987), Lichtwardt & Williams (*Mycol.* **80**: 400, 1988; zygospores), Lichtwardt & Williams (*CJB* **68**: 1057, 1990), Williams & Lichtwardt (*CJB* **68**: 1045, 1990), White & Lichtwardt (*Mycol.* **96**: 891, 2004; Norway), White (*MR* **110**: 1011, 2006; phylogeny), White *et al.* (*Mycol.* **98**: 333, 2006; USA), Valle (*Mycol.* **99**: 442, 2007; Spain).

Stachylinoides Lichtw. & López-Lastra (1999), Harpellaceae. 1 (in *Diptera*), S. America. See Lichtwardt *et al.* (*Mycol.* **91**: 1069, 1999).

Stachyomphalina H.E. Bigelow (1979) = Gamundia fide Kuyper (*in litt.*).

stage, a phase of the life cycle (q.v.). See States of fungi.

Staginospora Trivedi & C.L. Verma (1971), Fossil Fungi, anamorphic *Pezizomycotina*. 1 (Tertiary), India.

Stagnicola Redhead & A.H. Sm. (1986), Strophariaceae. 1, widespread (north temperate). See Redhead & Smith (*CJB* **64**: 645, 1986).

Stagonopatella Petr. (1927), anamorphic *Pezizomycotina*, St.0fH.?. 1, Europe.

Stagonopsis Sacc. (1884), anamorphic *Pezizomycotina*, Cpd.≡ eH.?. 4, widespread (temperate). See Petrak (*Sydowia* **3**: 139, 1949).

Stagonospora (Sacc.) Sacc. (1884) nom. cons., anamorphic *Phaeosphaeria*, Cpd.≡ eH.1. *c.* 209, widespread. *S. nodorum* and *S. avenae*, important cereal pathogens. The genus is poylphyletic as currently circumscribed, and often confused with *Septoria* (q.v.). See Castellani & Germano (*Ann. Fac. Sci. Agric. Univ. Torino* **10**: 1, 1977; 77 graminicolous spp.),

Philipson (*New Phytol.* **113**: 127, 1989; ultrastr.), Crane & Shearer (*Bull. Ill. St. nat. Hist. Surv.* **34**: 1, 1991; nomenclator), McDonald *et al.* (*Phytopathology* **84**: 250, 1994; genetic variation), Ueng & Chen (*Phytopathology* **84**: 800, 1994; RFLPs), Beck & Ligon (*Phytopathology* **85**: 319, 1995; PCR detection), Ueng *et al.* (*Phytopathology* **85**: 44, 1995), Czembor & Arseniuk (*Journal of Applied Genetics* **37**: 239, 1996; RAPDs), Cunfer (*Pl. Dis.* **81**: 427, 1997), Newton *et al.* (*Mycol.* **90**: 215, 1998; mating types, VCGs), Ueng *et al.* (*MR* **102**: 607, 1998), Caten (*Septoria on Cereals* A Study of Pathosystems: 26, 1999; molecular genetics), Cunfer & Ueng (*Ann. Rev. Phytopath.* **37**: 267, 1999), Eyal (*Septoria on Cereals* A Study of Pathosystems: 1, 1999; pathology), Scharen (*Septoria and Stagonospora Diseases of Cereals* A Compilation of Global Research. Proceedings of the Fifth International *Septoria* Workshop, September 20-24, 1999, CIMMYT, Mexico: 19, 1999), Cunfer (*Can. J. Pl. Path.* **22**: 332, 2000), Câmara *et al.* (*Mycol.* **94**: 630, 2002; phylogeny), Ueng *et al.* (*Curr. Genet.* **43**: 121, 2003; mating type genes), Douaiher *et al.* (*Sydowia* **56**: 39, 2004; ontogeny), Solomon *et al.* (*Eur. J. Pl. Path.* **110**: 763, 2004; mating types, Australia), Malkus *et al.* (*FEMS Microbiol. Lett.* **249**: 49, 2005; phylogeny), Schoch *et al.* (*Mycol.* **98**: 1041, 2006; phylogeny), Sommerhalder *et al.* (*Phytopathology* **96**: 234, 2006; genetics).

Stagonosporella Tassi (1902) = Stagonospora fide Saccardo (*Syll. fung.* **18**, 1906).

Stagonosporina Tassi (1902), anamorphic *Pezizomycotina*, Cpd.≡ eH.?. 1, widespread.

Stagonosporites Babajan & Tasl. (1970), Fossil Fungi. 1 (Tertiary), former USSR.

Stagonosporopsis Died. (1912) = Ascochyta fide Petrak (*Annls mycol.* **23**: 5, 1925), Boerema *et al.* (*Phoma Identification Manual* Differentiation of Specific and Infra-Specific Taxa in Culture: 470 pp., 2004).

Stagonostroma Died. (1914) = Fusarium fide Sutton (*Mycol. Pap.* **141**, 1977).

Stagonostromella Petr. & Syd. (1927), anamorphic *Pezizomycotina*, St.≡ eH.?. 1, Brazil.

Staheliella Emden (1974), anamorphic *Pezizomycotina*, Hso.0eP.39. 1, Surinam. See Emden (*Acta Bot. Neerl.* **23**: 251, 1974).

Staheliomyces E. Fisch. (1921), Phallaceae. 1, C. & S. America.

Staibia Bat. & Peres (1966), Leptopeltidaceae. 1, Brazil. See Batista & Peres (*Atas Inst. Micol. Univ. Pernambuco* **3**: 142, 1966).

Stains. See also mounting media. Stains are used to highlight fungal structures for viewing with the compound light microscope. Cotton blue (also known as aniline blue) is the most popular stain for the microscopic examination of most groups of fungi and is frequently used at 0.1% with lactic acid or (formerly, but now not recommended) lactophenol to produce permanent mounts. Acid fuchsin, a red stain, is combined with lactic acid to produce lactofuchsin. This gives excellent clarity and is also suitable for permanent preparations for most groups of fungi (Carmichael, *Mycol.* **47**: 611, 1955). Erythrosin B in aqueous ammonia gives good differentiation between cell walls and cytoplasm (ideal for elucidating conidial ontogeny) but crystallizes rapidly and is therefore only suitable in temporary mounts. Useful red stains for fungi also include congo red (Gurr, *Encyclopaedia of microscopic stains*, 1960), orseillin BB (Alcorn & Yeager, *Stain Technology* **12**: 158, 1937) and rosazurin (Locquin & Langeron, Handbook of microscopy, 1983). Diazonium blue B is used for staining yeasts (Kurtzman & Fell, The Yeasts, a taxonomic study, edn 4, 1998) but has also been used as a stain for filamentous fungi (Hopsu-Havu *et al.*, *Mykosen* **10**: 23, 1967), Summerbell (*Mycol.* **77**: 587, 1985). Use of trypan blue as a stain for fungi was described by Boedijn (*Stain Technology* **31**: 115, 1965). Cresyl blue is used for staining spore-walls of basidiomycetes (Locquin, *BSMF* **68**: 170, 1952). For use of toluidine blue O in polychromatic staining of fungal structures, see Ghemawat (*Physiol. Plant Path.* **11**: 251, 1977). Safranin O is frequently used as a counter-stain but can also be used as a rapid nuclear stain for fungi (Bandoni, *Mycol.* **71**: 873, 1979). Black stains include chlorazole-black (Armitage, *TBMS* **27**: 131, 1944) and nigrosin, which can be used in combination with picric acid (picro-nigrosin) to stain fungal structures (Fleming & Smith, *TBMS* **27**: 13, 1944). Nigrosin is also useful as a negative stain for examining mucilaginous material on the surface of spores and hyphae.

Staining methods for differentiation of fungi in plant tissues include the use of Bruzzese and Hasan's technique (Brusseze & Hasan, *Plant Path.* **32**: 335, 1983), periodic acid-Schiff's technique (Dring, *New Phytol.* **54**: 277, 1955) and thionin and orange G (Stoughton, *Ann. Appl. Biol.* **17**: 162, 1930). For further details of staining methods for fungi in plant tissues see Tuite (*Plant pathological methods. Fungi and bacteria*, 1969) and Fox (*Principles of diagnostic techniques in plant pathology*, 1993).

To differentiate fungi in animal tissues, good results are obtained with the periodic acid-Schiff stain (Kligman & Mescon, *J. Bact.* **60**: 415, 1950), Gridley's technique (Gridley, *Am. J. Clin. Path.* **23**: 303, 1953), Grocott's modification of the Gomori methenamine silver stain (Grocott, *Am. J. Clin. Path.* **25**: 975, 1955) and the standard haematoxylin-eosin stain. Brown and Brenn's modification of Gram's stain is used for examination of *Actinomyces* and *Nocardia* (Brown & Brenn, *Bull. Johns Hopkins Hosp.* **48**: 69, 1931). *Cryptococcus* is differentiated using Mayer's mucicarmine stain see Mallory (*Pathological techniques*, 1942). For further details of staining methods in medical fungi see Kwon-Chung & Bennett (*Medical mycology*, edn 4, 1992) and Rippon (*Medical mycology: the pathogenic fungi and the pathogenic actinomycetes*, 1982).

For nuclear staining of fungi, the Giemsa stain is frequently used (Hrushovetz, *CJB* **34**: 321, 1956), Duncan & Galbraith (*Stain Technology* **48**: 107, 1973). Techniques using carmine as a nuclear stain include the Feulgen-carmine technique (McIntosh, *Stain Technology* **29**: 29, 1954) and the propiono-carmine squash technique (Lu, *CJB* **40**: 843, 1962). A rapid nuclear stain for studying ascocarp development in *Lophodermium* and *Rhytisma* was described by Hulton & Morgan-Jones (*Mycol.* **66**: 881, 1974).

For testing the amyloid reaction in basidiomycetes and both lichenized and non-lichenized ascomycetes, Melzer's and Lugol's iodine are used (see mounting fluids). For discussion see Kohn & Korf (*Mycotaxon* **3**: 165, 1975 and Baral (*Mycotaxon* **29**: 399, 1987).

Stakman (Elvin Charles; 1885-1979; USA). Student,

University of Minnesota (1902-1906); school teacher (1906-1909); Assistant (1909-1913) then Section Head (1913-1940) then Head (1940-1953), Department of Vegetable [later Plant] Pathology, University of Minnesota; also for many years Director of the Federal Rust Laboratory, St Paul, Minnesota. Stakman is noted for his work on cereal rusts and their physiologic races (stemming from his doctoral thesis) and, in particular, for leading the campaign to eradicate barberry, the alternate host of black-stem rust, from major wheat-growing areas of the USA. A cultured and versatile man, author of more than 300 papers and co-author of two books. *Publs.* A study in cereal rusts. Physiological races. *Minnesota Agricultural Experiment Station Bulletin* **138** (1914); (with J.G. Harrar) *Principles of Plant Pathology* (1957). *Biogs, obits etc.* Anon. *Aurora Sporalis. E.C. Stakman Day issue 17 May 1979*, 1979) [Department of Plant Pathology, University of Minnesota]; Christensen (*Phytopathology* **69**: 195, 1979) [portrait]; Christensen (*E.C. Stakman, Statesman of Science*, 1984) [bibliography, biography, portrait]; Stafleu & Cowan (*TL-2* **5**: 830, 1985); Wilcoxson & Kommedahl (*Review of Tropical Plant Pathology* **7**: 223, 1993).

Stakmania Kamat & Sathe (1968) = Phakopsora fide Cummins & Hiratsuka (*Illustr. Gen. Rust Fungi rev. edit.*, 1983).

Stalactocolumella S. Imai (1950) nom. nud. ≡ Circulocolumella.

Stalagmites Theiss. & Syd. (1914), Nectriaceae. 1 (branches), Brazil. See Rossman *et al.* (*Stud. Mycol.* **42**: 248 pp., 1999).

Stalagmochaetia Cif. & Bat. (1963), anamorphic *Pezizomycotina*, Cpd.0eH.?. 1, USA. See Ciferri & Batista (*Quad. Lab. crittogam., Pavia* **31**: 187, 1963).

stalagmoid (of spores, **stalagmospores**), like a long tear or drop.

staling substances, substances produced by an organism which slow up or stop its growth (isoantagonism) (see Brown, *Ann. Bot., Lond.* **37**: 106, 1923); **inhibitory substances** are similar substances which retard or inhibit the growth of *other* organisms (e.g. penicillin) (hetero-antagonism; cf. antibiotic substances). See Porter & Carter (*Bot. Rev.* **4**: 165, 1938).

Stalpersia Parmasto (2001), Auriscalpiaceae. 1, Primorye. See Parmasto (*Folia cryptog. Estonica* **38**: 51, 2001).

Stamnaria Fuckel (1870), Helotiaceae. Anamorph *Titaeospora*. 3, Europe; N. America. See Seaver (*Mycol.* **28**: 186, 1936), Künkele *et al.* (*Mycol. bavarica* **7**: 3, 2005; Germany).

stane crottle, a crottle (q.v.) growing on stone.

Stanglomyces Raithelh. (1986), Tricholomataceae. 1, S. America. See Raithelhuber (*Metrodiana* **14**: 37, 1985).

Stanhughesia Constant. (1989), anamorphic *Ceramothyrium*, Hso.1bH.1. 3, Sweden. See Constantinescu (*Stud. Mycol.* **31**: 71, 1989).

Staninwardia B. Sutton (1971), anamorphic *Chaetothyriales*, Cac.1eP.38. 1, S.E. Asia; Australia; Indian Ocean. See Sutton (*TBMS* **57**: 540, 1971), Summerell *et al.* (*Fungal Diversity* **23**: 323, 2006).

Stanjehughesia Subram. (1992), anamorphic *Miyoshiella, Umbrinosphaeria*, Hso.≡ eP.1. 5, widespread. See Subramanian (*Proc. Indian natn Sci. Acad.* Part B. Biol. Sci. **58**: 184, 1992), Wu & Zhuang (*Fungal Diversity Res. Ser.* **15**, 2005; China), Shenoy *et al.* (*MR* **110**: 916, 2006; phylogeny).

Stanjemonium W. Gams, O'Donnell, Schroers & M. Chr. (1999), anamorphic *Bionectriaceae*, Hso.?.?. 4, France; Iraq; USA. See Gams *et al.* (*CJB* **76**: 1579, 1998), Rossman *et al.* (*Mycol.* **93**: 100, 2001; phylogeny).

Staphlosporonites Sheffy & Dilcher (1971), Fossil Fungi. 5 (Eocene, Tertiary), China; USA.

Staphylotrichum J. Mey. & Nicot (1957), anamorphic *Pezizomycotina*, Hso.0eH.1. 1, widespread. See Maciejowska & Williams (*Mycol.* **55**: 221, 1963), Udagawa (*Tropical Mycology*: 149, 1997; Chile).

Starbaeckia Rehm ex Starbäck (1890), Helotiales. 1, Europe. See Nannfeldt (*Nova Acta R. Soc. Scient. upsal.*, 1932).

Starbaeckiella (Sacc. & P. Syd.) Syd. & P. Syd. (1919) = Pyrenula Ach. (1809) fide Harris (*Mem. N. Y. bot. Gdn* **49**, 1989).

Starkeyomyces Agnihothr. (1956) = Myrothecium fide Tulloch (*Mycol. Pap.* **130**, 1972).

Starmera Y. Yamada, Higashi, S. Ando & Mikata (1997), ? Phaffomycetaceae. 3, widespread. See Yamada *et al.* (*Biosc., Biotechn., Biochem.* **63**: 827, 1999), Starmer *et al.* (*FEMS Yeast Res.* **3**: 441, 2003), Suh *et al.* (*Mycol.* **98**: 1006, 2006; phylogeny).

Starmerella C.A. Rosa & Lachance (1998), Saccharomycetales. Anamorph *Candida*. 1, Canada. See Rosa & Lachance (*Int. J. Syst. Bacteriol.* **48**: 1413, 1998), Teixeira *et al.* (*Int. J. Syst. Evol. Microbiol.* **53**: 339, 2003), Suh *et al.* (*Mycol.* **98**: 1006, 2006; phylogeny).

starters, the pure cultures or mixtures of fungi and bacteria used for starting fermenting processes. Pure culture starters are used in beer making, etc., and frequently in butter and cheese making. Examples of the more complex type are ragi (q.v.) used for Javanese arak (an alcoholic drink from rice starch), Chinese rice (Mingen or Men) (*Rhizopus oryzae*) the starter for rice wine, the Japanese Koji (preparations of *Aspergillus* used in soy and other fermenting processes), and kephir grains, a mixture of yeasts and bacteria which is the starter for kephir. See Fermented foods and beverages.

States of fungi. Since publication of the Tulasnes' *Selecta carpologia fungorum* (1861-1865), it has been accepted that many fungi are pleomorphic, that is, one fungus may produce several sorts of spores which may or may not be coincident in time and may or may not be produced after a nuclear fusion followed by meiosis, a sequence that may be interpreted as sexual. The state characterized by sexual spores (ascospores, basidiospores, etc.) has traditionally been designated the **'perfect' state** (or stage), that characterized by asexual spores (conidia) or the absence of spores the **'imperfect' state** (or stage). Under the code governing scientific names of fungi it is permissible to treat both 'perfect' and 'imperfect' states as species designated by latinized binomials, but when a 'perfect' and 'imperfect' species have been established to be different states of the same fungus the binomial applied to the 'perfect' state also covers that of the 'imperfect' and takes precedence. To increase precision in terminology for states of pleomorphic fungi, Hennebert & Weresub (*Mycotaxon* **6**: 207, 1977) introduced new nouns and adjectives: **holomorph**, for the whole fungus in all its morphs, phases, stages or states; **teleomorph**, for the sexual ('perfect') state (e.g. that characterized by ascomata or basidiomata); and **anamorph**, for the asexual ('imperfect') state (e.g. that characterized only by presence or absence of conidia). **Synanamorph** is applied to any one of two or more anamorphs which have the same teleomorph (Gams, *Mycotaxon* **15**: 459, 1982). A holomorph includes a teleomorph and frequently one, or rarely more anamorphs. The term **ana-holomorph** has been used for an 'imperfect' fungus (anamorphic fungus) which appears to lack a 'perfect' state (teleomorph). A name applied under the Code to both perfect and imperfect states of a fungus is a **nomen holomorphosis**; that to the imperfect state a **nomen anamorphosis**. It has also been suggested that **anamorph-genus** and **anamorph-species** should replace the terms 'form genus' and 'form species'.

The morph terminology (which has been adopted for pleomorphic fungi in this *Dictionary*) is discussed in detail in Kendrick (Ed.) (*The whole fungus*, 2 vols, 1979) which also includes a reprint of Hennebert & Weresub's paper. A more recent discussion of the issues is presented in Seifert *et al.* (*Stud. Mycol.* **45**, 2000). See Nomenclature.

statismospore, a spore not forcibly discharged. Cf. ballistospore.

Statistical methods and design of experiments, see Johnston & Booth (Eds) (*Plant pathologist's pocketbook*: 353 edn 2, 1983).

statolon, an antiviral substance (which induces interferon formation) from *Penicillium stoloniferum*; the active principle of which is considered to be RNA of viral origin (Banks *et al.*, *Nature* **218**: 542, 1968).

Stauriella Sivichai & E.B.G. Jones (2004), anamorphic *Pezizomycotina*. 1, Thailand. See Sivichai & Jones (*Sydowia* **56**: 132, 2004).

Staurochaeta Sacc. (1875) ? = Staurophoma fide Sutton (*Mycol. Pap.* **141**, 1977).

Staurolemma Körb. (1867), Collemataceae (L). 3, widespread. See Jørgensen & Henssen (*Graphis Scripta* **5**: 12, 1993), Henssen (*Nova Hedwigia* **68**: 117, 1999; Australia), Jørgensen & Henssen (*Bryologist* **102**: 22, 1999), Jørgensen (*Bryologist* **107**: 392, 2004; N America).

Stauronema (Sacc.) Syd., P. Syd. & E.J. Butler (1916), anamorphic *Pezizomycotina*, Ccu.0eH.19. 6, widespread. See Nag Raj (*Coelomycetous Anamorphs with Appendage-bearing Conidia*, 1993), Abbas *et al.* (*Pakist. J. Bot.* **34**: 117, 2002).

Stauronematopsis Abbas, B. Sutton & Ghaffar (2002), anamorphic *Pezizomycotina*, C?.?.?. 1, USA. See Abbas *et al.* (*Pakist. J. Bot.* **34**: 118, 2002).

Staurophallus Mont. (1845) nom. dub., Phallaceae.

Staurophoma Höhn. (1907), anamorphic *Pezizomycotina*, Cpd.0eH.15. 1, S. America; Hong Kong. See Morgan-Jones *et al.* (*Univ. Waterloo Biol. Ser.* **4**, 1972), Yanna *et al.* (*Sydowia* **50**: 139, 1998).

Staurosphaeria Rabenh. (1858) = Karstenula fide Saccardo (*Syll. fung.* **3**, 1883).

staurospore (stauroconidium), a non-septate or septate spore with more than one axis; axes not curved through more than 180° (cf. helicospore); protuberances present and ›$^1/_4$ spore body length (cf. amerospore). See Anamorphic fungi.

Staurothele Norman (1852) nom. cons., Verrucariales (L). *c.* 72, widespread. See Swinscow (*Lichenologist* **2**: 152, 1963; Brit. spp.), Thomson (*Bryologist* **94**: 351, 1991; key 17 spp., N. Am.), Harada (*Nat. Hist. Res.* **2**: 39, 1992; Japan), McCarthy (*Muelleria* **8**: 275, 1995; Australia), Geiser *et al.* (*Mycol.* **98**: 1053, 2006; phylogeny), Harada & Wang (*Lichenology* **5**: 13, 2006; China), Gueidan *et al.* (*MR* **111**: 1145, 2007; phylogeny), Lumbsch *et al.* (*MR* **111**: 257, 2007; phylogeny).

Stearophora L. Mangin & Viala (1905), Pezizomycotina. 1, N. Africa.

Steccherícium D.A. Reid (1963), Bondarzewiaceae. 4, widespread (tropics). See Reid (*Kew Bull.* **17**: 270, 1963), Hattori (*Mycoscience* **44**: 453, 2003; type studies).

Steccherinaceae Parmasto (1968) = Meruliaceae.

Steccherinum Gray (1821), Meruliaceae. 33, widespread. See Maas Geesteranus (*Persoonia* **7**: 443, 1974), Saliba & David (*Cryptog. Mycol.* **9**: 93, 1988; key Eur. spp.), Legon & Roberts (*Czech Mycol.* **54**: 7, 2002; England), Yuan & Dai (*Mycotaxon* **93**: 173, 2005; China).

Steganopycnis Syd. & P. Syd. (1916) = Seynesia fide Petrak (*Annls mycol.* **25**: 258, 1927), Læssøe (*SA* **13**: 43, 1994).

Steganosporium, see *Stegonsporium*.

Stegasphaeria Syd. & P. Syd. (1916), Mesnieraceae. 2 (from living leaves), Philippines; Africa. See Müller & von Arx (*Beitr. Kryptfl. Schweiz* **11** no. 2, 1962).

Stegasphaeriaceae Syd. & P. Syd. (1916) = Mesnieraceae.

Stegastroma Syd. & P. Syd. (1916) = Anisomyces Theiss. & Syd. fide Müller & von Arx *in* Ainsworth *et al.* (Eds) (*The Fungi* **4A**: 87, 1973).

Stegia Fr. (1818) [non *Stegia* DC. 1805, *Malvaceae*] ≡ Stegilla.

Stegiacantha Maas Geest. (1966), ? Meruliaceae. 1, Madagascar. See Maas Geesteranus (*Proc. K. ned. Akad. Wet.* C **69**: 317, 1966).

Stegilla Rchb. (1828) nom. dub., Fungi. See Sutton & Pirozynski (*TBMS* **46**: 517, 1964).

Stegites Mesch. (1892), Fossil Fungi. 2 (Tertiary), Europe.

Stegobolus Mont. (1845), Thelotremataceae (L). 1, pantropical. See Hale (*Bull. Br. Mus. nat. hist.* Bot. **8**: 227, 1981), Frisch & Kalb (*Biblthca Lichenol.* **92**: 371, 2006; revision).

Stegocintractia M. Piepenbr., Begerow & Oberw. (1999), Anthracoideaceae. 5 (on *Juncaceae*). See Piepenbring (*Bot. Jb.* **24**: 241, 2003).

Stegolerium Strobel, W.M. Hess & E. Ford (2001), anamorphic *Pezizomycotina*. 1, Venezuela. See Strobel *et al.* (*Mycotaxon* **78**: 356, 2001).

Stegonsporiopsis Van Warmelo & B. Sutton (1981), anamorphic *Pezizomycotina*, St.#eP.15. 1, N. America. See van Warmelo & Sutton (*Mycol. Pap.* **145**: 17, 1981).

Stegonsporium Corda (1827), anamorphic *Splanchnonema*, Cac.#eP.19. 3, widespread. See van Warmelo & Sutton (*Mycol. Pap.* **145**, 1981).

Stegopeziza Höhn. (1917) = Hysterostegiella fide Hein (*Nova Hedwigia* **38**: 669, 1983).

Stegopezizella Syd. (1924) = Sarcotrochila fide Korf (*Mycol.* **54**: 12, 1962).

Stegophora Syd. & P. Syd. (1916), Sydowiellaceae. Anamorph *Cylindrosporella*. 1 (on *Ulmus*), N. America. See Petrak (*Annls mycol.* **38**: 267, 1940), McGranahan & Smalley (*Phytopathology* **74**: 1300, 1984; culture), European and Mediterranean Plant Protection Organization (*Bulletin OEPP* EPPO Bulletin **35**: 416, 2005), Rossman *et al.* (*Mycoscience*

48: 135, 2007; review).

Stegophorella Petr. (1947), Pezizomycotina. 1, Ecuador.

Stegothyrium Höhn. (1918), ? Microthyriaceae. 1, Europe.

Steinera Zahlbr. (1906), Coccocarpiaceae (L). 4, Australasia. See Henssen & James (*Bull. Br. Mus. nat. hist.* Bot. **10**: 227, 1982), Wiklund & Wedin (*Cladistics* **19**: 419, 2003; phylogeny), Lumbsch *et al.* (*Mol. Phylogen. Evol.* **31**: 822, 2004; phylogeny), Miądlikowska & Lutzoni (*Am. J. Bot.* **91**: 449, 2004; phylogeny).

Steiner's stable PD solution, *p*-phenylenediamine, 1 g; sodium sulphite (Na_2SO_3), 10 g liquid detergent, *c.* 10 drops; water, 100 ml. Stable for at least 6 months; see Scott (*Lichenologist* **1**: 88, 1958). See also Metabolic products.

Steinhausia Sprague, Ormières & Manier (1972), Microsporidia. 2.

Steinia Körb. (1873), Aphanopsidaceae (L). 2, Europe; Australia. See Faurel & Schotter (*Bull. Soc. Hist. nat. Afr. Noire* **25**: 126, 1954), Hafellner (*Cryptog. Bot.* **5**: 99, 1995), Printzen & Rambold (*Lichenologist* **27**: 91, 1995), Kantvilas & McCarthy (*Lichenologist* **31**: 555, 1999).

Steirochaete A. Braun & Casp. (1853) = Colletotrichum fide Southworth (*J. Mycol.* **6**: 115, 1890).

Stelechotrichum Ritgen (1831) nom. inval. ≡ Cephalotrichum Link.

steliogen, the structure which gives rise to the sporocarp stalk in protostelids.

Stella Massee (1889) [non *Stella* Medik. 1787, *Leguminosae*] = Scleroderma fide Demoulin (*in litt.*).

Stellasclerotes Beneš (1959), Fossil Fungi. 1 (Carboniferous), former Czechoslovakia.

stellate, like a star in form; **- seta**, a compound seta having several radiating arms; asterophysis.

Stellatospora Tad. Ito & Nakagiri (1994), Chaetomiaceae. 1, Japan. See Ito & Nakagiri (*Mycoscience* **35**: 413, 1994).

Stellatostroma, see *Asterostroma*.

Stellifera Léman (1824) nom. dub., anamorphic *Agaricaceae*.

Stellifraga Alstrup & Olech (1993), Pezizomycotina. 1 (on lichens, esp. *Cladonia*), Spitsbergen. See Alstrup & Olech (*Polish Polar Research* **14**: 39, 1993).

Stelligera R. Heim ex Doty (1948) = Lachnocladium fide Corner (*Ann. Bot. Mem.* [A monograph of Clavaria and allied genera] **1**, 1950).

Stellomyces Morgan-Jones, R.C. Sinclair & Eicker (1987), anamorphic *Pezizomycotina*, Hso.0eH.6. 1, S. Africa; India. See Morgan-Jones *et al.* (*Mycotaxon* **28**: 447, 1987), Keshavaprasad & Bhat (*Mycotaxon* **84**: 61, 2002).

Stellopeltis Bat. & A.F. Vital (1959), anamorphic *Pezizomycotina*, Cpt.0eH.?. 2, Brazil. See Batista & Vital (*Mycopath. Mycol. appl.* **11**: 88, 1959).

Stellosetifera Matsush. (1996), Pseudoperisporiaceae. Anamorph *Penicillifer*. 1, Malaysia. Possibly related to *Wentiomyces*. See Matsushima (*Mycol. Mem.* **9**, 1996).

Stellospora Alcorn & B. Sutton (1984), anamorphic *Pezizomycotina*, Hso.0eH.15. 1 (on *Appendiculella*), Australia; Philippines. See Alcorn & Sutton (*Mycotaxon* **20**: 45, 1984).

Stellothyriella Bat. & Cif. (1959), anamorphic *Pezizomycotina*, Cpt.1eH.?. 1, USA. See Batista & Ciferri (*Mycopath. Mycol. appl.* **11**: 62, 1959).

Stemastrum Raf. (1808) nom. dub., Agaricales. 'gasteromycetes'.

Stemmaria Preuss (1851) nom. dub. ? = Mycosylva fide Seifert (*Sydowia* **45**: 103, 1993).

Stemmatomyces Thaxt. (1931), Laboulbeniaceae. 3, widespread. See Nannfeldt (*Svensk bot. Tidskr.* **43**: 468, 1949).

Stempellia L. Léger & E. Hesse (1910), Microsporidia. 4.

Stemphyliites Babajan & Tasl. (1973), Fossil Fungi. 1 (Neogene), former USSR.

Stemphyliomma Sacc. & Traverso (1911) = Pithomyces fide Kirk (*TBMS* **80**: 449, 1983).

Stemphyliopsis A.L. Sm. (1901) = Stemphylium. An albino mutant. fide Petch (*TBMS* **23**: 146, 1939), Barron & Onions (*CJB* **44**: 861, 1966).

Stemphyliopsis Speg. (1910) [non *Stemphyliopsis* A.L. Sm. 1910] ≡ Stemphyliomma.

Stemphylium Wallr. (1833), anamorphic *Pleospora*, Hso.#eP.19. 43, widespread. See Wiltshire (*TBMS* **21**: 211, 1938), Simmons (*Mycol.* **61**: 1, 1969), Simmons (*Sydowia* **38**: 284, 1986), Simmons (*Mem. N. Y. bot. Gdn* **49**: 305, 1989), Basallote-Ureba *et al.* (*Pl. Path.* **48**: 139, 1999; on *Allium*), Cho & Yu (*Pl. Path. J.* **16**: 328, 2000; on *Trifolium*, Korea), Pryor & Gilbertson (*MR* **104**: 1312, 2000; phylogeny), Simmons (*Harvard Pap. Bot.* **6**: 199, 2001; teleomorph), Câmara *et al.* (*Mycol.* **94**: 660, 2002; phylogeny), Mehta *et al.* (*Curr. Microbiol.* **44**: 323, 2002; DNA), Zhang *et al.* (*Mycotaxon* **85**: 247, 2003; China), Inderbitzin *et al.* (*Proc. natn Acad. Sci. U.S.A.* **102**: 11390, 2005; mating systems), Kodsueb *et al.* (*Mycol.* **98**: 571, 2006; phylogeny), Schoch *et al.* (*Mycol.* **98**: 1041, 2006; phylogeny).

Stenella Syd. (1930), anamorphic *Mycosphaerella*, Hso.≡ eP.10. 155, widespread. See Mulder (*TBMS* **65**: 514, 1975), de Hoog *et al.* (*TBMS* **81**: 485, 1983; relationship to *Ramichloridium*), Braun (*Cryptog. Mycol.* **20**: 155, 1999), Crous *et al.* (*Mycol.* **93**: 1081, 2001; phylogeny), Braun *et al.* (*Mycol. Progr.* **2**: 197, 2003), Crous & Braun (*CBS Diversity Ser.* **1**: 571 pp., 2003), Pretorius *et al.* (*Sydowia* **55**: 286, 2003; on *Citrus*), Crous *et al.* (*Stud. Mycol.* **50**: 457, 2004; on *Acacia*), Crous *et al.* (*Stud. Mycol.* **55**: 99, 2006; on *Eucalyptus*), Crous *et al.* (*Stud. Mycol.* **58**: 33, 2007; posn).

Stenellopsis B. Huguenin (1966), anamorphic *Mycosphaerellaceae*, Hsp.≡ eP.10. 5, New Caledonia. See Huguenin (*BSMF* **81**: 695, 1965), Crous *et al.* (*Mycol.* **93**: 1081, 2001), Crous & Braun (*CBS Diversity Ser.* **1**: 571 pp., 2003).

Stenellopsis Morgan-Jones (1980) [non *Stenellopsis* B. Huguenin 1966] ≡ Parastenella.

Stenhammara A. Massal. ex Zahlbr. (1924) = Lecidea fide Hertel (*Beih. Nova Hedwigia* **24**, 1967).

Stenhammara Flot. ex Körb. (1855) ≡ Stenhammarella.

Stenhammarella Hertel (1967), Porpidiaceae (L). 1, Europe; central Asia. See Hertel (*Nova Hedwigia* Beih. **24**: 124, 1967), Buschbom & Mueller (*Mol. Phylogen. Evol.* **32**: 66, 2004; phylogeny).

Stenocarpella Syd. & P. Syd. (1917), anamorphic *Diaporthales*, Cpd.1-≡ eP.15. 2, widespread (tropical). See Sutton (*The Coelomycetes*, 1980), Dorrance *et al.* (*Pl. Dis.* **83**: 675, 1999; culture, isozymes), Xia & Achar (*J. Phytopath.* **149**: 35, 2001; molecular detection), Crous *et al.* (*Stud. Mycol.* **55**: 235, 2006; phylogeny).

Stenocephalopsis Chamuris & C.J.K. Wang (1998), anamorphic *Pezizomycotina*, Hso.?.?. 1, widespread. See Chamuris & Wang (*Mycol.* **90**: 464, 1998).

Stenocephalum Chamuris & C.J.K. Wang (1990), anamorphic *Pezizomycotina*, Hso.0eH/1eP.10. 1, widespread. See Chamuris & Wang (*Mycol.* **82**: 530, 1990), Chamuris & Wang (*Mycol.* **90**: 464, 1998).

Stenocladiella Marvanová & Descals (1987), anamorphic *Pezizomycotina*, Hso.1bH.10. 2 (aquatic), Europe. See Marvanová & Descals (*TBMS* **89**: 507, 1987).

Stenocybe Nyl. ex Körb. (1855), Mycocaliciaceae. *c.* 14, widespread. See Tschermak-Woess (*Lichenologist* **10**: 69, 1978; biology), Tibell (*Beih. Nova Hedwigia* **79**: 597, 1984), Peterson & Rikkinen (*Mycol.* **90**: 1087, 1998), Tibell & Wedin (*Mycol.* **92**: 577, 2000), Titov *et al.* (*Botanicheskiĭ Zhurnal* **87**: 60, 2002), Döbbeler & Feuerer (*Biblthca Lichenol.* **88**: 91, 2004; on liverwort), Tibell & Vinuesa (*Taxon* **54**: 427, 2005; phylogeny), Geiser *et al.* (*Mycol.* **98**: 1053, 2006; phylogeny).

Stenocybella Vain. (1927) = Calicium fide Tibell (*Beih. Nova Hedwigia* **79**: 597, 1984).

Stenographa Mudd (1861) = Thalloloma fide Hawksworth *et al.* (*Dictionary of the Fungi* edn 8, 1995).

Stenospora Deighton (1969), anamorphic *Pezizomycotina*, Hso.0fH.10. 1 (on *Puccinia*), Sierra Leone. See Deighton (*Mycol. Pap.* **118**: 22, 1969).

Stephanoascus M.T. Sm., Van der Walt & Johannsen (1976) = Trichomonascus fide Traquair *et al.* (*CJB* **66**: 926, 1988), Giménez-Jurado *et al.* (*Syst. Appl. Microbiol.* **17**: 237, 1994), Smith & Hoog *in* Kurtzman & Fell (Eds) (*Yeasts, a taxonomic study* 4th edn: 400, 1998), Ueda-Nishimura & Mikata (*Int. J. Syst. Evol. Microbiol.* **52**: 463, 2002), Kurtzman & Robnett (*FEMS Yeast Res.* **7**: 141, 2007).

Stephanocyclos Hertel (1983), Porpidiaceae (L). 1, Antarctica. See Hertel (*Lecideaceae exsiccatae* Fascicle **V**: no. 96, 1983), Buschbom & Mueller (*Mol. Phylogen. Evol.* **32**: 66, 2004; phylogeny).

stephanocyst, a structure, typically bicellular (basal cell cup-like, terminal cell globose), found in certain basidiomycetes. See Burdsall (*Mycol.* **61**: 915, 1969).

Stephanoma Wallr. (1833), anamorphic *Hypomyces*, Hso.0bP.1. 3, widespread. See Butler & McCain (*Mycol.* **60**: 955, 1968), Põldmaa (*Stud. Mycol.* **45**: 83, 2000), Samuels *et al.* (*CBS Diversity Ser.* **4**, 2006; USA).

Stephanomyces Speg. (1917) = Cucujomyces fide Thaxter (*Memoirs of the American Academy of Arts and Sciences* **16**: 1, 1931).

Stephanonectria Schroers & Samuels (1999), Bionectriaceae. 1, British Isles. See Schroers *et al.* (*Sydowia* **51**: 116, 1999), Castlebury *et al.* (*MR* **108**: 864, 2004; phylogeny).

Stephanophallus MacOwan [not traced] = Anthurus fide Saccardo (*Syll. fung.* **7**: 23, 1888).

Stephanophoron Nádv. (1942) = Nadvornikia fide Tibell (*Beih. Nova Hedwigia* **79**: 597, 1984).

Stephanophorus Flot. (1843) = Leptogium fide Hawksworth *et al.* (*Dictionary of the Fungi* edn 8, 1995).

Stephanopus M.M. Moser & E. Horak (1975), Cortinariaceae. 5, S. America. See Moser & Horak (*Beih. Nova Hedwigia* **52**: 608, 1975; key).

Stephanospora Pat. (1914), Stephanosporaceae. 4, Eurasia; New Zealand. See Oberwinkler & Horak (*Pl. Syst. Evol.* **131**: 157, 1979), Martín *et al.* (*Mycotaxon* **90**: 133, 2004; relationship with *Lindtneria trachyspora*), Larsson (*MR* **111**: 1040, 2007), Lawrey *et al.* (*Mol. Phylog. Evol.* **44**: 778, 2007).

Stephanosporaceae Oberw. & E. Horak (1979), Agaricales. 5 gen. (+ 3 syn.), 21 spp.

Lit.: Oberwinkler & Horak (*Pl. Syst. Evol.* **131**: 157, 1979), Pegler & Young (*TBMS* **72**: 353, 1979), Beaton *et al.* (*Kew Bull.* **40**: 573, 1985), Lebel & Castellano (*Mycol.* **94**: 327, 2002), Martín *et al.* (*Mycotaxon* **90**: 133, 2004).

Stephanosporium Dal Vesco (1961), anamorphic *Pezizomycotina*, Hso.0eP.39. 1, widespread. See Dal Vesco (*Allionia* **7**: 182, 1961).

Stephanotheca Syd. & P. Syd. (1914), Elsinoaceae. 2, Philippines.

Stephanothecaceae Petr. (1931) = Elsinoaceae.

Stephembruneria R.F. Castañeda (1988), anamorphic *Pezizomycotina*, Hso.≡ eP.15. 1, Cuba. See Castañeda (*Fungi Cubenses* **3**: 14, 1988).

Stephensia Tul. (1845), Pyronemataceae. 7 (hypogeous), widespread. See de Vries (*Coolia* **28**: 96, 1985), Fontana & Giovannetti (*Mycotaxon* **29**: 37, 1987; anamorph), Trappe *et al.* (*Mycotaxon* **64**: 431, 1997), Læssøe & Hansen (*MR* **111**: 1075, 2007; phylogeny), Perry *et al.* (*MR* **111**: 549, 2007; phylogeny).

Stephosia Bat. & H. Maia (1967) = Porina Müll. Arg. fide Lücking *et al.* (*Lichenologist* **30**: 121, 1998).

Sterbeeckia Dumort. (1822) ≡ Craterellus.

Stercum, see *Stereum*.

Stereaceae Pilát (1930), Russulales. 22 gen. (+ 3 syn.), 125 spp.

Lit.: Donk (*Persoonia* **3**: 199, 1964), Reid (*Beih. Nova Hedwigia* **18**: 484 pp., 1965; stipitate steroid spp.), Jahn (*Westfäl. Pilzbr.* **8**: 69, 1971), Rattan (*The resupinate Aphyllophorales of the North West Himalayas* [*Bibl. Mycol.* **60**], 1977), Jülich & Stalpers (*Verh. Kon. Ned. Akad. Wetensch.* Afd. Natuurk. sect. 2 **74**, 1980), Ginns (*Mycol.* **90**: 19, 1998), Tabata *et al.* (*Mycoscience* **41**: 585, 2000), Larsson & Hallenberg (*Mycol.* **93**: 907, 2001), Miller *et al.* (*Mycol.* **93**: 344, 2001), Wu *et al.* (*Mycol.* **93**: 720, 2001), Binder & Hibbett (*Mol. Phylogen. Evol.* **22**: 76, 2002), Slippers *et al.* (*Mol. Ecol.* **11**: 1845, 2002), Larsson & Larsson (*Mycol.* **95**: 1037, 2003), Slippers *et al.* (*South African Journal of Science* **99**: 70, 2003), Larsson *et al.* (*MR* **108**: 983, 2004), Binder *et al.* (*Systematics and Biodiversity* **3**: 113, 2005), Küffer & Senn-Irlet (*Mycol. Progr.* **4**: 77, 2005).

Stereales = Russulales.

Sterellum P. Karst. (1889) = Peniophora fide Weresub & Gibson (*CJB* **38**: 833, 1960).

Stereocaulaceae Chevall. (1826), Lecanorales (L). 6 gen. (+ 15 syn.), 247 spp.

Lit.: Jørgensen & Jahns (*Notes R. bot. Gdn Edinb.* **44**: 581, 1987), Smith & Øvstedal (*Polar Biol.* **11**: 91, 1991), Jahns *et al.* (*Biblthca Lichenol.* **58**: 181, 1995), Fryday & Coppins (*Lichenologist* **28**: 513, 1996), Kivistö (*Sauteria* **9**: 25, 1998), Sipman (*Cryptog. Bryol.-Lichénol.* **19**: 229, 1998), Stenroos & DePriest (*Am. J. Bot.* **85**: 1548, 1998; phylogeny), Stenroos & DePriest (*Curr. Genet.* **33**: 124, 1998), Wedin & Döring (*MR* **103**: 1131, 1999; DNA), Wedin *et al.* (*Lichenologist* **32**: 171, 2000; phylogeny), Ekman & Tønsberg (*MR* **106**: 1262, 2002), Printzen & Kantvilas (*Biblthca Lichenol.* **88**: 541, 2004), Myllys *et al.* (*Taxon* **54**: 605, 2005), Högnabba (*MR* **110**: 1080, 2006), Miądlikowska *et al.* (*Mycol.* **98**: 1088, 2006; phylogeny), Hofstetter *et al.* (*Mol. Phylogen. Evol.*

44: 412, 2007; phylogeny).

Stereocauliscum Nyl. (1865) ? = Micarea Fr. (1825) [nom. cons.] fide Hawksworth *et al.* (*Dictionary of the Fungi* edn 8, 1995).

Stereocaulomyces E.A. Thomas ex Cif. & Tomas. (1953) ≡ Stereocaulon Hoffm.

Stereocaulon (Schreb.) Schrad. (1794) nom. rej. = Pertusaria fide Hawksworth *et al.* (*Dictionary of the Fungi* edn 8, 1995).

Stereocaulon Hoffm. (1796) nom. cons., Stereocaulaceae (L). 137, widespread. See Lamb (*J. Hattori bot. Lab.* **43**: 191, 1978; keys), Lamb (*J. Hattori bot. Lab.* **44**: 209, 1978; keys), Smith & Øvstedal (*Polar Biol.* **11**: 91, 1991; 6 Antarctic spp.), Jahns *et al.* (*Biblthca Lichenol.* **58**: 181, 1995; ontogeny), Sipman (*Cryptog. Bryol.-Lichénol.* **19**: 229, 1998; New Guinea), Stenroos & DePriest (*Am. J. Bot.* **85**: 1548, 1998; phylogeny), Wedin & Döring (*MR* **103**: 1131, 1999; DNA), Wedin *et al.* (*Lichenologist* **32**: 171, 2000; phylogeny), Stenroos *et al.* (*Mycol. Progr.* **1**: 267, 2002; phylogeny), Myllys *et al.* (*Taxon* **54**: 605, 2005; phylogeny), Högnabba (*MR* **110**: 1080, 2006; phylogeny), Miądlikowska *et al.* (*Mycol.* **98**: 1088, 2006; phylogeny).

Stereocaulum Clem. (1909) = Pertusaria fide Hawksworth *et al.* (*Dictionary of the Fungi* edn 8, 1995).

Stereochlamydomyces Cif. & Tomas. (1953) ≡ Stereochlamys.

Stereochlamys Müll. Arg. (1885) = Trichothelium fide Hawksworth *et al.* (*Dictionary of the Fungi* edn 8, 1995).

Stereocladium Nyl. (1875) = Stereocaulon Hoffm. fide Hawksworth *et al.* (*Dictionary of the Fungi* edn 8, 1995).

Stereoclamydomyces, see *Stereochlamydomyces*.

Stereocrea Syd. & P. Syd. (1917), Clavicipitaceae. 1 (on grasses), widespread (paleotropics). See Rossman *et al.* (*Stud. Mycol.* **42**: 248 pp., 1999), Bischoff & White (*Mycology Series* **19**: 125, 2003).

Stereofomes Rick (1928), Lachnocladiaceae. 5, S. America; Japan. See Donk (*Taxon* **6**: 114, 1957; a nom. dub.), Boidin & Lanquetin (*Biblthca Mycol.* **114**, 1987).

Stereogloeocystidium Rick (1940) = Podoscypha fide Boidin (*Revue Mycol.* Paris **24**: 197, 1959).

Stereolachnea Höhn. (1917) = Scutellinia fide Denison (*Mycol.* **51**: 605, 1959).

stereome (of lichens), a scleroplectenchyma which forms the main supporting tissue of the thallus, as in *Alectoria*, *Bryoria*, and *Cladonia*.

Stereonema Kütz. (1836) = Lecidea fide Hawksworth et al. (*Dictionary of the Fungi* edn 8, 1995).

Stereopeltis Franzoni & De Not. (1861) = Sarcogyne Flot. (1851) fide Hawksworth *et al.* (*Dictionary of the Fungi* edn 8, 1995).

Stereophyllum P. Karst. (1889) [non *Stereophyllum* Mitt. 1859, *Musci*] ≡ Cyphellostereum.

Stereopodium Earle (1909) ≡ Mycena.

Stereopsis D.A. Reid (1965), Meruliaceae. 13, widespread (esp. tropical). See Boidin *et al.* (*BSMF* **66**: 445, 1998; sub *Phanerochaetaceae*), Douanla-Meli & Langer (*Mycotaxon* **90**: 323, 2004).

Stereosorus Sawada (1943) = Burrillia fide Ling (*Mycol.* **41**: 252, 1949).

Stereosphaeria Kirschst. (1939), Clypeosphaeriaceae. 1 (bark etc.), widespread (temperate). See Barr (*Sydowia* **41**, 1989), Hyde *et al.* (*Sydowia* **50**: 21, 1998), Kang *et al.* (*Mycoscience* **40**: 151, 1999).

Stereostratum Magnus (1899), Pucciniaceae. 2 (on *Bambusoideae* (*Poaceae*)), China; Japan. See Thirumalachar (*Mycol.* **52**: 690, 1961).

Stereum Hill ex Pers. (1794), Stereaceae. 27, widespread. *S. purpureum* is included in *Chondrostereum*. See Talbot (*Bothalia* **6**: 303, 1954; S. Afr.), Boidin (*Rev. Mycol.* **23**: 318, 1958), Boidin (*Rev. Mycol.* **24**: 197, 1959), Boidin (*Bull. Soc. linn. Lyon* **28**: 205, 1959), Boidin (*Bull. Jard. bot. Brux.* **30**: 51, 1960; Congo), Boidin (*Bull. Jard. bot. Brux.* **30**: 283, 1960; Congo), Welden (*Mycol.* **63**: 796, 1971; key genus s.s.), Chamuris (*Mycotaxon* **22**: 105, 1985; key N. Am. spp.), Eicker & Louw (*South African Journal of Botany* **64**: 30, 1998; S. Afr.).

Stericium Raf. (1819) nom. dub., Fungi. No spp. included.

sterigma (pl. **-ata**), (1) (of a basidium, q.v.), an extension of the metabasidium composed of a basal filamentous or inflated part (the **proto-**; epibasidium) and an apical spore-bearing projection (the **spiculum**) (Fig. 8F); (2) (of *Aspergillus*, etc.) [a usage not recommended], phialide (**secondary -**); prophialide (**primary -**); metula; (3) (of lichens), a spermatiophore (Nylander).

Sterigmatobotrys Oudem. (1886), anamorphic *Pezizomycotina*, Hso.≡ eH.10. 3, widespread (esp. north temperate). See Ellis (*Dematiaceous Hyphomycetes*, 1971), Jong & Davies (*Norw. Jl Bot.* **18**: 177, 1971), Sutton (*Mycol. Pap.* **132**, 1973), Chang. (*MR* **95**: 1142, 1991).

sterigmatocystin, a carcinogenic hepatotoxin (xanthone derivative) from *Aspergillus versicolor* (Van der Walt & Purchase, *Brit. J. exp. Path.* **51**: 183, 1970); precursor of aflatoxin B1 (Singh & Hsich, *Appl. environ. Microbiol.* **31**: 743, 1976).

Sterigmatocystis C.E. Cramer (1859) = Aspergillus fide Raper & Fennell (*The genus Aspergillus*, 1965).

Sterigmatomyces Fell (1966), anamorphic *Agaricostilbaceae*, Hso.0eH.10. 2, Atlantic Ocean; Indian Ocean. See Lodder (*Yeasts, a taxonomic study* 2nd edn: 1229, 1970), Yamada *et al.* (*J. gen. appl. Microbiol.* Tokyo **32**: 157, 1986; enzyme systems), van der Walt *et al.* (*Antonie van Leeuwenhoek* **53**: 137, 1987; key to 3 spp.), Yamada *et al.* (*Agric. Biol. Chem.* **53**: 2993, 1989; phylogeny), Guého *et al.* (*Int. J. Syst. Bacteriol.* **40**: 60, 1990; partial rRNA sequences), Kurtzman (*Int. J. Syst. Bacteriol.* **40**: 56, 1990; DNA relatedness), Fell *et al.* (*Int. J. Syst. Evol. Microbiol.* **50**: 1351, 2000; mol. phylogeny).

Sterigmatosporidium G. Kraep. & U. Schulze (1983), Cuniculitremaceae. 1, Germany. See Fell *et al.* (*Int. J. Syst. Evol. Microbiol.* **50**: 1351, 2000; mol. phylogeny).

sterile (1) not producing spores or a sporocarp; (2) free from living microorganisms; sterilized.

Sterilization. Making free from living microorganisms, may be done by chemical (see Fungicides) or physical methods. Heat is the most widely used physical agent. Dry heat may be used for glass and other materials; death of even the most resistant bacterial spores takes place within an hour in a hot air oven at 160°C. A flame may frequently be used for the surface sterilization of instruments. Moist heat, especially if the pressure is increased, is even better. Most media may be made sterile by autoclaving for 15 min at 121°C, but when such severe heating is not possible 'discontinuous steaming' (steaming at atmospheric pressure for 20 min every day for 3 days) is frequently used. In addition, steam is of use for the 'sterilization' (partial (or incomplete) sterilization) of soil for controlling soil-living pathogens, and hot water (about 50°C) has an important application for the control of diseases such as loose smut of wheat (*Ustilago tritici*) in which the pathogen is inside the seed. Though cold will not let growth take place, even great cold may not be lethal to micro-organisms.

Among other physical agents are ultra-violet light (to which the sterilizing effect to sunlight is due) but certain fluorescent dyes, such as eosin, are able to make bacteria sensitive to light of longer wave lengths (Blum, *Physiol. Rev.* **12**: 23, 1932), electricity (the effect of low frequency currents being possibly that of the heat or the nascent oxygen (O) or chlorine (Cl) produced), X-rays, radium emanation, and supersonic waves. Desiccation (drying) is frequently lethal, specially to the vegetative phase, but some fungal and bacterial spores are very resistant. It is sometimes possible to make a liquid sterile by filtering out the micro-organisms with a filter of unglazed porcelain. See Rahn (*Bact. Rev.* **9**: 1, 1945; physical methods), Russell *et al.* (*Principles and practice of disinfection, preservation and sterilization*, 688 pp., 2004), Sykes (*Disinfection and sterilization. Theory and practice*, 1958, *in* Norris & Ribbons (Eds), *Methods in microbiology* **1**: 77, 1969).

Sterrebekia Link (1816) [non *Sterbeckia* Schreb. 1789, *Papilionaceae*] = Scleroderma fide Stalpers (*in litt.*).

Stevensea Trotter (1926) = Diplotheca Starbäck fide von Arx & Müller (*Stud. Mycol.* **9**, 1975).

Stevensiella Trotter (1928) ≡ Diatractium.

Stevensomyces E.F. Morris & Finley (1965), anamorphic *Pezizomycotina*, Hsy.0eH.?. 1, Panama. See Morris & Finley (*Mycol.* **57**: 483, 1965).

Stevensonula Petr. (1952), anamorphic *Pezizomycotina*, Ccu.1eP.1. 1, USA; New Caledonia. See Sutton (*Nova Hedwigia* **26**: 1, 1975).

Stevensula Speg. (1924) = Leptomeliola fide Hughes (*Mycol. Pap.* **166**, 1993).

Steyaertia Bat. & H. Maia (1960) = Asterolibertia fide Müller & von Arx (*Beitr. Kryptfl. Schweiz* **11** no. 2, 1962).

Sthughesia M.E. Barr (1987), Capnodiales. See Eriksson & Hawksworth (*SA* **7**: 91, 1988), Winton *et al.* (*Mycol.* **99**: 240, 2007; phylogeny).

stichobasidium, see basidium.

Stichoclavaria Ulbr. (1928) ≡ Holocoryne fide Pine *et al.* (*Mycol.* **91**: 994, 1999; considered the correct name for *Multiclavula*).

Stichoclavariaceae Ulbr. (1928) = Clavariaceae.

Stichodothis Petr. (1927) = Auerswaldiella fide von Arx & Müller (*Beitr. Kryptfl. Schweiz* **11** no. 1, 1954).

Stichomyces Thaxt. (1901), Laboulbeniaceae. 7, widespread. See Santamaria i del Campo (*L'ordre Laboulbenials (Fungi, Ascomycotina) a la Península Ibèrica i Illes Ballears [Thesis]*: 669 pp., 1990).

Stichophoma Kleb. (1933) ? = Sclerophoma fide Sutton (*Mycol. Pap.* **141**, 1977).

Stichopsora Dietel (1899) = Coleosporium fide Sydow & Sydow (*Monographia Uredinearum seu Specierum Omnium ad hunc usque Diem Descriptio et Adumbratio Systematica* **3**, 1915).

Stichoramaria Ulbr. (1928) ≡ Clavulina.

Stichospora Petr. (1927), anamorphic *Pezizomycotina*, St.0eH/1eP.?. 1, Europe.

Stichus D.E. Ether. (1904), Fossil Fungi (mycel.) ? Fungi. 1 (Cretaceous), Australia.

Sticta (Schreb.) Ach. (1803), Lobariaceae (L). *c.* 114, widespread (esp. tropical). See Joshi & Awasthi (*Biol. Mem.* **7**: 165, 1982; key 13 spp., India), Galloway (*Lichenologist* **26**: 223, 1994; key 12 spp. S. Am.), Galloway (*Lichenologist* **29**: 105, 1997; NZ), Galloway (*Trop. Bryol.* **15**: 117, 1998; Australia), Piovano *et al.* (*Biochemical Systematics and Ecology* **28**: 589, 2000; chemistry), Thomas *et al.* (*Biblthca Lichenol.* **82**: 123, 2001; phylogeny), Tønsberg & Goward (*Bryologist* **104**: 12, 2001; N America), McDonald *et al.* (*Bryologist* **106**: 61, 2003; USA), Wiklund & Wedin (*Cladistics* **19**: 419, 2003; phylogeny), Miądlikowska & Lutzoni (*Am. J. Bot.* **91**: 449, 2004; phylogeny), Takahashi *et al.* (*Lichenology* **2**: 174, 2004; China), Miądlikowska *et al.* (*Mycol.* **98**: 1088, 2006; phylogeny).

Stictaceae Stizenb. (1862) = Lobariaceae.

Stictidaceae Fr. (1849), Ostropales. 20 gen. (+ 22 syn.), 156 spp.

Lit.: Sherwood (*Mycotaxon* **5**: 1, 1977; monogr., keys), Johnston (*Mycotaxon* **24**: 349, 1985), Winka *et al.* (*Lichenologist* **30**: 455, 1998; DNA), Lücking *et al.* (*Mycol.* **96**: 283, 2004), Lumbsch *et al.* (*Mol. Phylogen. Evol.* **31**: 822, 2004), Miądlikowska *et al.* (*Mycol.* **98**: 1088, 2006; phylogeny), Wedin *et al.* (*MR* **110**: 773, 2006), Hofstetter *et al.* (*Mol. Phylogen. Evol.* **44**: 412, 2007; phylogeny).

Stictina Nyl. (1860) nom. rej. = Pseudocyphellaria fide Galloway (*Bull. Br. Mus. nat. hist.* Bot. **17**, 1988).

Stictis Pers. (1800), Stictidaceae. Anamorph *Stictospora*. *c.* 68, widespread. See Sherwood (*Mycotaxon* **5**: 1, 1977; key), Sherwood (*Mycotaxon* **6**: 215, 1977), Winka *et al.* (*Lichenologist* **30**: 455, 1998; DNA), Lumbsch *et al.* (*Organ. Divers. Evol.* **1**: 99, 2001; phylogeny), Kauff & Lutzoni (*Mol. Phylogen. Evol.* **25**: 138, 2002; phylogeny), Wedin *et al.* (*New Phytol.* **164**: 459, 2004; ecology, phylogeny), Wedin *et al.* (*Lichenologist* **37**: 67, 2005; phylogeny, lichen status), Miądlikowska *et al.* (*Mycol.* **98**: 1088, 2006; phylogeny), Wedin *et al.* (*MR* **110**: 773, 2006; Scandinavia).

Stictochorella Höhn. (1917) = Asteromella fide Clements & Shear (*Gen. Fung.*, 1931).

Stictochorellina Petr. (1922) = Asteromella fide Sutton (*Mycol. Pap.* **141**, 1977).

Stictoclypeolum Rehm (1904) = Asterothyrium Müll. Arg. fide Hawksworth *et al.* (*Dictionary of the Fungi* edn 8, 1995).

Stictographa Mudd (1861) = Melaspilea fide Hawksworth *et al.* (*Dictionary of the Fungi* edn 8, 1995).

Stictomyces E.A. Thomas ex Cif. & Tomas. (1953) ≡ Lobaria.

Stictopatella Höhn. (1918), anamorphic *Pezizomycotina*, Ccu.0eH.?. 1, Europe.

Stictophacidium Rehm (1888), ? Stictidaceae. Anamorphs *Ebollia*, *Coleophoma*-like. 1, Europe. See Sherwood (*Mycotaxon* **5**: 1, 1977).

Stictosepta Petr. (1964), anamorphic *Pezizomycotina*, St.0fH.10. 1, former Czechoslovakia. See Petrak (*Sydowia* **17**: 230, 1964), Treigiene (*Botanica Lithuanica* **12**: 131, 2006).

Stictosphaeria Tul. & C. Tul. (1863) = Diatrype fide Hawksworth *et al.* (*Dictionary of the Fungi* edn 8, 1995).

Stictospora Cif. (1957), anamorphic *Stictis*, St.0eH.?. 1, Europe. See Ciferri (*Atti Ist. bot. Univ. Lab. crittog. Pavia* sér. 5 **14**: 276, 1957).

Stictostroma Höhn. (1917) ? = Placuntium fide Sherwood (*Mycotaxon* **5**: 1, 1977).
Stigeosporium C. West (1916) ? = Glomus fide Gerdemann & Trappe (*Mycol. Mem.* **5**, 1974).
Stigmagora Trevis. (1853) = Ocellularia fide Hawksworth *et al.* (*Dictionary of the Fungi* edn 8, 1995).
Stigmastoma Bat. & H. Maia (1960) = Pycnothyrium fide von Arx & Müller (*Stud. Mycol.* **9**, 1975).
Stigmatea Fr. (1849) ≡ Stigmea Fr.
Stigmateacites S.L. Zheng & W. Zhang (1986), Fossil Fungi. 2, China.
Stigmatella Berk. & M.A. Curtis (1857) ? = Chondromyces fide Hawksworth *et al.* (*Dictionary of the Fungi* edn 8, 1995).
Stigmatella Mudd (1861) = Sclerophyton fide Hawksworth *et al.* (*Dictionary of the Fungi* edn 8, 1995).
Stigmatellina Bat. & H. Maia (1960), anamorphic *Pezizomycotina*, Cpt.0eH.?. 1, Belgium. See Batista & Maia (*Nova Hedwigia* **2**: 474, 1960).
Stigmateopsis Bat. (1960) = Placonema fide Sutton (*Kew Bull.* **31**: 461, 1977), Nag Raj (*Coelomycetous Anamorphs with Appendage-bearing Conidia*, 1993).
Stigmatidium G. Mey. (1825) = Enterographa fide Hawksworth *et al.* (*Dictionary of the Fungi* edn 8, 1995).
Stigmatisphaera Dumort. (1822) nom. dub., Pezizomycotina. Used for diverse perithecioid fungi.
Stigmatochroma Marbach (2000), Caliciaceae (L). 7, widespread. See Marbach (*Biblthca Lichenol.* **74**: 304, 2000).
stigmatocyst, see hyphopodium.
Stigmatodothis Syd. & P. Syd. (1914), Micropeltidaceae. 1, Philippines; Australia. See Reynolds & Gilbert (*Aust. Syst. Bot.* **18**: 265, 2005; Australia).
Stigmatolemma Kalchbr. (1882) = Resupinatus fide Thorn *et al.* (*Mycol.* **97**: 1140, 2005).
Stigmatomassaria Munk (1953) = Splanchnonema fide Barr (*Mycotaxon* **15**: 349, 1982), Barr (*Mycotaxon* **46**, 1993).
Stigmatomma Körb. (1855) = Staurothele fide Hawksworth *et al.* (*Dictionary of the Fungi* edn 8, 1995).
Stigmatomyces H. Karst. (1869), Laboulbeniaceae. 135 (on *Diptera*), widespread. See Santamaría & Rossi (*An. Jard. bot. Madr.* **51**: 33, 1993; Iberian spp.), Hedström (*MR* **98**: 403, 1994; neotropics), Weir & Rossi (*MR* **99**: 841, 1995; key Brit. spp.), Weir & Blackwell (*MR* **105**: 1182, 2001; phylogeny), Hughes *et al.* (*Mycol.* **96**: 834, 2004; New Zealand, New Caledonia), Rossi *et al.* (*MR* **109**: 271, 2005; in amber), Rossi & Weir (*Mycol.* **99**: 139, 2007).
stigmatomycosis (of cotton (*Gossypium*) bolls, *Phaseolus*, and other plants), damage caused by insect-inoculated fungi such as *Nematospora gossypii* and *N. coryli*. See Leach (*Insect transmission of plant diseases*, 1940), Frazer (*Ann. appl. Biol.* **31**: 271, 1944).
Stigmatopeltis Doidge (1927) = Vizella fide von Arx & Müller (*Stud. Mycol.* **9**, 1975).
Stigmatophragmia Tehon & G.L. Stout (1929), Micropeltidaceae. 1, N. America.
stigmatopodium (stigmopodium), see hyphopodium.
Stigmatopsis Traverso (1906) = Cryptosphaeria Ces. & De Not. fide Barr (*Mycol. Mem.* **7**, 1978).
Stigmatoscolia Bat. & Peres (1960) = Lophodermium fide von Arx (*in litt.*).
Stigmatula (Sacc.) Syd. & P. Syd. (1901), Phyllachoraceae. Anamorph *Rhodosticta*. 10 (on *Leguminosae*), widespread (temperate). See Cannon (*Mycol. Pap.* **163**, 1991), Cannon *in* McKey, D. & Sprent, J. (Eds) (*The Nitrogen Factor. Advances in Legume Systematics* **5**: 179, 1994; coevolution), Cannon (*Mycopathologia* **135**: 37, 1996).
Stigme Syd. & P. Syd. (1917) = Dimerina fide Müller & von Arx (*Beitr. Kryptfl. Schweiz* **11** no. 2, 1962).
Stigmea Bonord. (1864) = Dothiora Fr. (1849) fide Hawksworth *et al.* (*Dictionary of the Fungi* edn 8, 1995).
Stigmea Fr. (1836), Dothideomycetes. 20, widespread.
Stigmella Lév. (1842), anamorphic *Pezizomycotina*, Cpd.#eP.1. 27, widespread. See Hughes (*Mycol. Pap.* **49**, 1952).
Stigmidium Trevis. (1860), Mycosphaerellaceae. *c.* 75 (on lichens), widespread. See David & Hawksworth (*SA* **5**: 158, 1986; posn), Roux & Triebel (*Bull. Soc. linn. Provence* **45**: 451, 1994; key 17 spp.), Calatayud & Triebel (*Nova Hedwigia* **69**: 439, 1999; Spain), Calatayud & Triebel (*Biblthca Lichenol.* **78**: 27, 2001), Calatayud & Triebel (*Lichenologist* **35**: 103, 2003), Roux & Triebel (*Mycotaxon* **91**: 133, 2005; anatomy), Crous *et al.* (*Stud. Mycol.* **58**: 1, 2007).
Stigmina Sacc. (1880) = Pseudocercospora. Polyphyletic, with species linked to diverse *Dothideomycetes*. See Philipson (*New Phytol.* **113**: 127, 1989; ultrastr.), Sutton & Pascoe (*MR* **92**: 210, 1989; taxonomic re-evaluation), Sutton (*Arnoldia* **14**: 33, 1997), Crous *et al.* (*Mycol.* **93**: 1081, 2001; phylogeny), Crous & Braun (*CBS Diversity Ser.* **1**: 571 pp., 2003), Braun & Crous (*Taxon* **55**: 803, 2006; nomencl.).
Stigmochora Theiss. & Syd. (1914), Phyllachoraceae. 2 or 9 (living legume leaves), S. America. African species are not congeneric. See Cannon (*Mycol. Pap.* **163**, 1991), Cannon (*Mycopathologia* **120**: 61, 1992).
Stigmopeltella Syd. (1927) = Stigmopeltis fide Clements & Shear (*Gen. Fung.*, 1931).
Stigmopeltis Syd. (1927), anamorphic *Pezizomycotina*, Cpt.0fH.?. 2, C. America.
Stigmopeltopsis Peres (1961) = Myxothyriopsis fide von Arx (*in litt.*).
Stigmopsis Bubák (1914) = Cheiromyces fide Moore (*Mycol.* **50**: 681, 1959).
Stigonema C. Agardh ex Bornet & Flahault (1887), Algae. Algae.
stilbaceous (obsol.), having synnemata; synnematous (q.v.).
Stilbechrysomyxa M.M. Chen (1984), Coleosporiaceae. 3 (on *Rhododendron* (II, III) (*Ericaceae*)), Asia. See Chen (*Forest Fungi Phytogeography, Forest Fungi Phytogeography of China, North America, and Siberia and International Quarantine of Tree Pathogens*: 495 pp., 2003) Probably belonging to *Chrysomyxa*.
Stilbella Lindau (1900) nom. cons., anamorphic *Hypocreales*, Hsy.0eH.15. *c.* 44 (from soil, coprophilous, on *Insecta*), widespread. See Morris (*Western Ill. Univ. Ser. biol. Sci.* **3**, 1963), Benjamin (*Taxon* **17**: 521, 1968), Sutton (*Mycol. Pap.* **132**: 18, 1973), Seifert (*Stud. Mycol.* **27**: 1, 1985; key), Seifert & Samuels (*Mycol.* **89**: 512, 1997), Bischoff *et al.* (*Mycotaxon* **86**: 433, 2003).
Stilbellula Boedijn (1951), anamorphic *Pezizomycotina*, Hsy.0eH.?. 1, Java. See Boedijn (*Sydowia* **5**: 227, 1951), Seifert (*Mem. N. Y. bot. Gdn* **59**: 109, 1990).
Stilbites Pia (1927), Fossil Fungi. 1 (Eocene), Baltic.

Stilbochalara Ferd. & Winge (1910) = Chalara fide Nag Raj & Kendrick (*Monogr. Chalara Allied Genera*, 1975).

Stilbocrea Pat. (1900), Bionectriaceae. Anamorphs *Stilbella*-like, *Acremonium*-like. 6 (dead plant tissues and fungal stromata), widespread. See Seifert (*Stud. Mycol.* **27**, 1985), Rossman *et al.* (*Stud. Mycol.* **42**: 248 pp., 1999), Castlebury *et al.* (*MR* **108**: 864, 2004; phylogeny).

Stilbodendron Syd. & P. Syd. (1916), anamorphic *Penicilliopsis*, Hsy.0eP.38. 1, Africa. See Hsieh & Ju (*Mycol.* **94**: 539, 2002).

Stilbodendrum Bonord. (1851) = Syzygites fide Kirk (*in litt.*).

Stilbohypoxylon Henn. (1902), Xylariaceae. 10, widespread (esp. subtropical). See Læssøe (*SA* **13**: 43, 1994; ? synonym of *Xylaria*), Rogers & Ju (*MR* **101**: 135, 1997), Hladki & Romero (*Sydowia* **55**: 65, 2003), Petrini (*Sydowia* **56**: 51, 2004; revision), Ju *et al.* (*Mycol.* **99**: 612, 2007; phylogeny).

stilboid, a sterile, basidioma-like structure (as in *Mycena citricolor* and other agarics) which functions as a propagule (Singer, 1962: 25); gemma (Buller). Cf. carpophoroid.

Stilbomyces Ellis & Everh. (1895) nom. dub., anamorphic *Pezizomycotina*. Based on hyphophores of a lichen. See Seifert (*Sydowia* **45**: 103, 1993).

Stilbonectria P. Karst. (1889) = Nectria fide Rossman (*Mycol. Pap.* **150**, 1983), Rossman *et al.* (*Stud. Mycol.* **42**: 248 pp., 1999).

Stilbopeziza Speg. (1908), ? Helotiales (?L). 1, S. America.

Stilbophoma Petr. (1942), anamorphic *Pezizomycotina*, St.0eH.15. 2, India; Africa. See Sutton (*The Coelomycetes*, 1980).

Stilbospora Pers. (1794), anamorphic *Pezizomycotina*, Cpd/Cac.≡ eP.19. 16, widespread.

Stilbothamnium Henn. (1896), anamorphic *Trichocomaceae*. See Thom & Raper (*Manual of the Aspergilli*, 1945), Roquebert & Nicot (*Advances in Penicillium and Aspergillus Systematics* **102**: 221, 1985), Wicklow *et al.* (*Mycotaxon* **34**: 249, 1989), Dupont *et al.* (*Modern Concepts in Penicillium and Aspergillus Classification* [NATO ASI Series A: Life Sciences] **185**: 335, 1990), Hsieh & Ju (*Mycol.* **94**: 539, 2002).

Stilbotulasnella Oberw. & Bandoni (1982), Cantharellales. 1, Hawaii. See Bandoni & Oberwinkler (*CJB* **60**: 1875, 1982).

Stilbum Tode (1790), Chionosphaeraceae. *c.* 10, widespread (temperate). See Donk (*Taxon* **7**: 236, 1958; typification), McLaughlin *et al.* (*Am. J. Bot.* **91**: 808, 2004; phylogeny; cf. *Bensingtonia*).

stilbum, the erect synnema (q.v.) of *Stilbella* with its head of slime spores.

stink horns, basidiomata of certain *Phallales*.

Stioclettia Dennis (1975), Diaporthales. 1 (on *Luzula*), British Isles. See Dennis (*Kew Bull.* **30**: 362, 1975).

stipe, a stalk (Fig. 4D, F-H).

Stipella L. Léger & M. Gauthier (1932), Legeriomycetaceae. 2 (in *Diptera*), Europe. See Moss (*TBMS* **54**: 1, 1970), Lichtwardt (*The Trichomycetes. Fungal associates of arthropods*, 1986), Nelder *et al.* (*Fungal Diversity* **22**: 121, 2006; ecology and taxonomy), Valle (*Mycol.* **99**: 442, 2007; Spain).

Stipinella, see *Stypinella*.

stipitate, stalked.

Stipitochaete Ryvarden (1985) = Hymenochaete fide Léger (*Biblthca Mycol.* **171**: 1, 1998).

Stipitocyphella G. Kost (1998), Marasmiaceae. 1, Kenya. See Kost (*MR* **102**: 505, 1998).

Stipiza Raf. (1815) nom. nud., Fungi. See Merrill (*Index Rafinesq.*, 1949).

Stiptophyllum Ryvarden (1973), Polyporaceae. 1, Brazil. See Ryvarden (*Norw. Jl Bot.* **20**: 4, 1973).

Stirtonia A.L. Sm. (1926), Arthoniaceae (L). *c.* 16, widespread (tropical). See Santesson (*Symb. bot. upsal.* **12** no. 1: 1, 1952), Makhija & Patwardhan (*Mycotaxon* **67**: 287, 1998).

Stirtoniella D.J. Galloway, Hafellner & Elix (2005), Ramalinaceae (L). 1, Australasia. See Galloway *et al.* (*Lichenologist* **37**: 263, 2005).

Stirtoniopsis Groenh. (1938) nom. inval., Arthoniales (L). 2, Java; Morocco.

stock (of basidiomycetes), a dikaryotic mycelium (fide Raper, 1966). Cf. strain.

stolon, a 'runner', as in *Rhizopus*.

Stomatisora J.M. Yen (1971) = Chrysocelis fide Cummins & Hiratsuka (*Illustr. Gen. Rust Fungi rev. edit.*, 1983).

Stomatogene Theiss. (1917), Parodiopsidaceae. 2, N. America. See Farr (*Mem. N. Y. bot. Gdn* **49**: 70, 1989).

Stomatogenella Petr. (1955), Pezizomycotina. 1, Australia. See Petrak (*Sydowia* **9**: 507, 1955).

stomatopodium (stomopodium), a hyphal branch (an appressorium; cf. hyphopodium) or 'plug' above or in a stoma.

Stomiopeltella Theiss. (1914) = Stomiopeltis fide von Arx & Müller (*Stud. Mycol.* **9**, 1975), Reynolds & Gilbert (*Cryptog. Mycol.* **27**: 249, 2006).

Stomiopeltina Bat. (1963) = Metathyriella fide von Arx & Müller (*Stud. Mycol.* **9**, 1975).

Stomiopeltis Theiss. (1914), Micropeltidaceae. Anamorph *Sirothyriella*. 25, widespread. See Batista (*Publções Inst. Micol. Recife* **56**, 1959), Ellis (*TBMS* **68**: 157, 1977; key Brit. spp.), Reynolds & Gilbert (*Aust. Syst. Bot.* **18**: 265, 2005; Australia), Batzer *et al.* (*Mycol.* **97**: 1268, 2005; phylogeny).

Stomiopeltites Alvin & M.D. Muir (1970), Fossil Fungi. 1 (Cretaceous), British Isles.

Stomiopeltopsis Bat. & Cavalc. (1963), Micropeltidaceae. 1, Brazil. See Batista & Cavalcanti (*Publções Inst. Micol. Recife* **392**: 30, 1963), Ramaley (*Mycotaxon* **83**: 327, 2002).

Stomiotheca Bat. (1959), Micropeltidaceae. 1, Brazil. See Batista (*Publções Inst. Micol. Recife* **56**: 453, 1959).

stone rag (stone raw), *Parmelia saxatilis*.

stone-fungus, the hard pseudosclerotium of *Polyporus tuberaster*; Pietraia fungaia. On being watered, an edible basidioma is produced. The Canadian tuckahoe is the same species (Vanterpool & Macrae, *CJB* **29**: 147, 1951).

stopper, the *Neurospora* phenotype characterized by irregular cycles of cessation and renewal of growth.

Straggaria Reinsch (1888) nom. dub., ? Fungi.

strain (1) a group of clonally related individuals or cells. See Yoder *et al.* (*Phytopath.* **76**: 383, 1986); (2) a homokaryotic mycelium (fide Raper, 1966), cf. stock.

Straminella M. Choisy (1929) = Lecanora fide Brodo & Elix (*Biblthca Lichenol.* **53**: 19, 1993).

straminipile (straminopile), colloquial noun for an organism bearing tripartite tubular hairs (Patterson, 1989).

straminipilous (straminopilous), bearing tripartite

tubular hairs; applicable to flagella and/or cells, whether uniflagellate, multiflagellate or non-flagellate (the auxiliary cyst of *Saprolegnia* which bears a tuft of tripartite tubular hairs would be a straminipilous cyst), cf. flimmergeissel; **- fungi**, see *Chromista*, *Straminipila*.

strand plectenchyma, plectenchyma in strands forming supporting tissues in a lichen thallus.

strangle fungus, *Epichloë typhina*; see choke.

Strangospora Körb. (1860), Lecanorales (L). *c.* 10, widespread (north temperate). See Harris *et al.* (*Evansia* **5**: 26, 1988; posn), Hafellner (*Cryptog. bot.* **5**: 99, 1995), Reeb *et al.* (*Mol. Phylogen. Evol.* **32**: 1036, 2004; phylogeny), Miądlikowska *et al.* (*Mycol.* **98**: 1088, 2006; phylogeny).

Strangulidium Pouzar (1967) = Oligoporus fide Donk (*Verh. K. ned. Akad. Wet.* tweede sect. **62**: 1, 1974).

Strasseria Bres. & Sacc. (1902), anamorphic *Pezizomycotina*, St.0eH.15. 1, widespread. *S. geniculata* (black rot of apple). See Dennis (*Gdnrs' Chron.* **114**: 221, 1943), Parmelee & Cauchon (*CJB* **57**: 1660, 1979), Sutton (*The Coelomycetes*, 1980).

Strasseriopsis B. Sutton & Tak. Kobay. (1970), anamorphic *Pezizomycotina*, St.1eH.15. 1, Japan. See Sutton & Kobayashi (*Mycol.* **61**: 1068, 1969).

Stratiphoromyces Goh & K.D. Hyde (1998), anamorphic *Pezizomycotina*, Hso.?.?. 2, Brunei; Andaman Islands. See Goh & Hyde (*MR* **102**: 1149, 1998).

Stratisporella Hafellner (1979), ? Patellariaceae. 1 (on lichens, *Tremotylium*), Angola. See Kutorga & Hawksworth (*SA* **15**: 1, 1997).

stratose thallus, a lichen thallus having the tissue in horizontal layers.

Strattonia Cif. (1954), Lasiosphaeriaceae. 9 (coprophilous), widespread. See Lundqvist (*Symb. bot. upsal.* **20** no. 1, 1972; key), Barr (*Mycotaxon* **39**: 43, 1990; posn), Abdullah *et al.* (*Nova Hedwigia* **69**: 211, 1999), Huhndorf *et al.* (*Mycol.* **96**: 368, 2004; phylogeny), Miller & Huhndorf (*Mol. Phylogen. Evol.* **35**: 60, 2005; phylogeny), Cai *et al.* (*MR* **110**: 359, 2006; phylogeny), Zhang *et al.* (*Mycol.* **98**: 1076, 2006; phylogeny).

straw mushrooms (paddy straw or Chinese mushroom), the edible *Volvariella volvacea* and *V. diplasia*. These agarics are widely used in the tropics. In Myanmar, where *V. diplasia* is cultured, wet rice ('paddy') straw is made into a bed about 1 × 1 × 5 m which is inoculated with 'pure culture' spawn (cf. mushroom culture), and kept wet. Mushrooms are first seen after 2-3 weeks; 4 kg or so being obtained from one bed. An air temperature of at least 21°C is necessary. See Thet & Seth (*Indian Fmg* **1**: 332, 1940), Chang (*Econ. Bot.* **31**: 374, 1977), Sukara *et al.* (*Bull. BMS* **19**: 129, 1985), Edible fungi.

Streblema Chevall. (1826) nom. dub., anamorphic *Xylariaceae*. 1, France. Composed only of stroma zone lines.

Streblocaulium Chevall. (1837) = Conoplea fide Kendrick & Carmichae *in* Ainsworth *et al.* (Eds) (*The Fungi* **4A**: 390, 1973).

Streblomyces Thaxt. (1920) = Nycteromyces fide Tavares (*Mycol. Mem.* **9**, 1985).

Streimannia G. Thor (1991), Roccellaceae (L). 1, Australia. See Thor (*Op. bot.* **103**: 84, 1991), Grube (*Bryologist* **101**: 377, 1998).

Strelitziana Arzanlou & Crous (2006), anamorphic *Chaetothyriales*. 1, S. Africa. See Arzanlou & Crous (*Fungal Planet* **no. 8**: [1], 2006).

Streptobotrys Hennebert (1973), anamorphic *Streptotinia*, Hso.0eH.7. 3, N. America. See Hennebert (*Persoonia* **7**: 191, 1973), Hong *et al.* (*Pl. Path. J.* **20**: 192, 2004).

Streptomyces Waksman & Henrici (1943), Actinobacteria. q.v.

streptomycin, a broad spectrum aminoglycoside antibiotic produced by *Streptomyces griseus*. Active against *Mycobacterium tuberculosis*, staphylococci, some gram-negative bacteria, and inhibiting vegetative growth of some fungi.

Streptopodium R.Y. Zheng & G.Q. Chen (1978), anamorphic *Pleochaeta*, Hso.0eH.1. 5, widespread. See Zheng & Chen (*Acta Microbiol. Sin.* **18**: 183, 1978), Saenz (*McIlvainea* **13**: 33, 1998; phylogeny), Liberato *et al.* (*MR* **108**: 1185, 2004; on *Carica*), Kiss *et al.* (*MR* **110**: 1301, 2006).

Streptosporangium Couch (1955), Actinobacteria. q.v.

Streptotheca Vuill. (1887) = Ascozonus fide Bergman & Shanor (*Mycol.* **49**: 879, 1957), Eckblad (*Nytt Mag. Bot.* **15**: 1, 1968).

Streptothrix Corda (1839) = Conoplea fide Hughes (*CJB* **36**: 727, 1958).

Streptotinia Whetzel (1945), Sclerotiniaceae. Anamorph *Streptobotrys*. 2, USA.

Streptotrichites Mesch. (1892), Fossil Fungi. 1 (Oligocene), Baltic.

Streptoverticillium E. Bald. (1958), Actinobacteria. q.v.

Striadiporites C.P. Varma & Rawat (1963), Fossil Fungi. 3 (Tertiary), India; USA.

Striadyadosporites Dueñas (1979), Fossil Fungi. 2 (Pleistocene), Colombia.

Striainaperturites Y.K. Mathur (1966), Fossil Fungi. 1.

Striatasclerotes Stach & Pickh. (1957), Fossil Fungi. 1 (Carboniferous), Germany. See Stach & Pickhardt (*Palaeontologische Zeitschrift* **31**: 140, 1957).

striate, marked with delicate lines, grooves or ridges (Fig. 20.13).

Striatodecospora D.Q. Zhou, K.D. Hyde & B.S. Lu (2000), Xylariaceae. 1, Hong Kong. See Zhou *et al.* (*Mycotaxon* **76**: 141, 2000).

Striatosphaeria Samuels & E. Müll. (1979), Chaetosphaeriaceae. Anamorph *Dictyochaeta*. 1 (from dead wood), C. & S. America. See Samuels & Müller (*Sydowia* **31**: 126, 1978), Réblová *et al.* (*Sydowia* **51**: 49, 1999), Réblová (*Stud. Mycol.* **45**: 149, 2000; review), Réblová & Winka (*Mycol.* **92**: 939, 2000; phylogeny), Huhndorf *et al.* (*Mycol.* **96**: 368, 2004; phylogeny), Fernández *et al.* (*Mycol.* **98**: 121, 2006; phylogeny).

Striatospora I.V. Issi & Voronin (1986), Microsporidia. 1.

Strickeria Körb. (1865), Pezizomycotina. *c.* 9, widespread. See Wakefield (*TBMS* **24**: 282, 1940), Eriksson (*SA* **10**: 144, 1991).

Stridiporosporites Ke & Shi (1978), Fossil Fungi. 6 (Oligocene, Tertiary), China.

Striglia Adans. (1763) = Daedalea fide Donk (*Persoonia* **1**: 284, 1960).

Strigopodia Bat. (1957), Euantennariaceae. Anamorphs *Antennatula*, *Capnophialophora*, *Hyphosoma*, *Racodium*. 3, widespread (temperate-boreal). See Hughes (*CJB* **46**: 1009, 1968), Reynolds (*Mycotaxon* **27**: 377, 1986).

strigose, rough with sharp-pointed hairs; hispid.

Strigula Fr. (1823), Strigulaceae (L). Anamorph *Dis-*

cosiella. *c*. 118 (many foliicolous), widespread (esp. tropical). See Santesson (*Symb. bot. upsal.* **12** no. 1: 1, 1952; foliicolous spp.), Margot (*Lichenologist* **9**: 51, 1977; host relationship), Nag Raj (*CJB* **59**: 2519, 1981; asci, pycnidia), Hawksworth (*Taxon* **35**: 787, 1986; nomencl.), McCarthy *et al.* (*Lichenologist* **28**: 239, 1996), McCarthy (*Lichenologist* **29**: 513, 1997), Sérusiaux (*Bryologist* **101**: 147, 1998), Aptroot & Lücking (*MR* **105**: 510, 2001; on ferns), Roux *et al.* (*Biblthca Lichenol.* **90**: 96 pp., 2004), Schmitt *et al.* (*Mycol.* **97**: 362, 2005; phylogeny), Geiser *et al.* (*Mycol.* **98**: 1053, 2006; phylogeny), Tretiach & Rinino (*Nova Hedwigia* **83**: 451, 2006), Aptroot *et al.* (*Biblthca Lichenol.* **97**, 2008; Costa Rica).

Strigulaceae Zahlbr. (1898), Chaetothyriomycetidae (inc. sed.) (L). 4 gen. (+ 22 syn.), 123 spp.

Lit.: Lücking (*Nova Hedwigia* **52**: 267, 1991), Lücking & Lücking (*Herzogia* **11**: 143, 1995), Huhndorf & Harris (*Brittonia* **48**: 551, 1996), McCarthy *et al.* (*Lichenologist* **28**: 239, 1996), Lücking *et al.* (*Lichenologist* **30**: 121, 1998), Sérusiaux (*Bryologist* **101**: 147, 1998), Roux *et al.* (*Biblthca Lichenol.* **90**: 96 pp., 2004), Schmitt *et al.* (*Mycol.* **97**: 362, 2005).

Strigulomyces Cif. & Tomas. (1953) = Strigula fide Hawksworth *et al.* (*Dictionary of the Fungi* edn 8, 1995).

Strilia Gray (1821) = Coltricia fide Parmasto (*in litt.*) See, Donk (*Verh. K. ned. Akad. Wet.* tweede sect. **62**: 49, 1974).

Striodiplodia Zambett. (1955) nom. inval. = Lasiodiplodia fide Sutton (*Mycol. Pap.* **141**, 1977), Verkley & Aa (*Mycotaxon* **65**: 113, 1997).

Strionemadiplodia Zambett. (1955) nom. inval., anamorphic *Pezizomycotina*, Cpd.1eP.?. 3, widespread. See Zambettakis (*BSMF* **70**: 233, 1954), Verkley & Aa (*Mycotaxon* **65**: 113, 1997).

Striosphaeropsis Verkley & Aa (1997), anamorphic *Pezizomycotina*, Cpd.?.?. 1, Papua New Guinea. See Verkley & Aa (*Mycotaxon* **65**: 115, 1997).

strobiliform, like a fir-cone in form.

Strobilofungus Lloyd (1915) = Boletellus fide Singer (*Farlowia* **2**: 223, 1945).

Strobilomyces Berk. (1851), Boletaceae. *c*. 20, cosmopolitan. See Ying & Ma (*Acta Mycol. Sin.* **4**: 95, 1985; China), Lakhanpal & Sharma (*Kavaka* **16**: 27, 1988; NW Himalaya), Watling *et al.* (*British Fungus Flora. Agarics and Boleti* Rev. & Enl. Edn **1**: 173 pp., 2005; Brit. sp.).

Strobilomycetaceae E.-J. Gilbert (1931) = Boletaceae.

Strobiloscypha N.S. Weber & Denison (1995), Pezizales. 1, USA. Perhaps a member of the *Sarcosomataceae*, but molecular data do not clearly support this. See Weber & Denison (*Mycotaxon* **54**: 129, 1995), Harrington *et al.* (*Mycol.* **91**: 41, 1999; phylogeny), Hansen & Pfister (*Mycol.* **98**: 1029, 2006; phylogeny).

Strobilurus Singer (1962), Physalacriaceae. 10, widespread (temperate). See Redhead (*CJB* **58**: 68, 1980).

stroma (pl. **stromata**), a mass or matrix of vegetative hyphae, with or without tissue of the host or substrate, sometimes sclerotium-like in form, in or on which spores or fruit bodies bearing spores are produced. Many ascomycetes (esp. *Xylariales*) and anamorphic fungi have stromata; a few *Pucciniales* and other fungi. **ecto-** (**epi-**, Fuisting), a -, normally conidial, formed in the periderm and frequently breaking through the bark; **endo-** (**ento-**) (**hypho-**, Fuisting), a perithecial - formed under the **ecto-**; **eu-**, one of fungal tissue only (Fig. 10P-R); **pseudo-** see under pseudostroma. Ruhland's name for the ostiolar disc is **placodium**, forms (as in *Diatrype*) from the **endo-** being **ento-placodial**, those (at least in part) from the **ecto-** being **ecto-placodial**. A species having ecto- and endo- is **diplostromatic**; one with only one **haplostromatic**. See Miller (*Mycol.* **20**: 188, 1928). Cf. sclerotium.

Stromaster Höhn. (1930), Phyllachoraceae. 1, S. America. Affinities are uncertain. See Petrak (*Sydowia* **5**: 354, 1951).

Stromatella Henssen (1989), Lichinaceae (L). 1, Bermuda. See Henssen (*Lichenologist* **21**: 111, 1989), Schultz & Büdel (*Lichenologist* **34**: 39, 2002; key).

Stromateria Corda (1837) = Tubercularia Tode fide Saccardo (*Syll. fung.* **4**: 646, 1886).

Stromatinia (Boud.) Boud. (1907), Sclerotiniaceae. Anamorph *Sclerotium*. 7, Europe; N. America. See Schumacher (*Agarica* **5**: 111, 1984), Holst-Jensen *et al.* (*Mycol.* **89**: 885, 1997; phylogeny).

Stromatocrea W.B. Cooke (1952), anamorphic *Hypocreopsis*, Hsp.0eH.15. 1, USA. See Petrak (*Sydowia* **6**: 336, 1952), Cauchon & Quellette (*Mycol.* **56**: 453, 1964).

Stromatocyphella W.B. Cooke (1961), Marasmiaceae. 3, N. America. See Cooke (*Beih. Sydowia* **4**: 104, 1961).

Stromatographium Höhn. (1907), anamorphic *Fluviostroma*, Hsy.0eP.15. 1, widespread (tropical). The genus is polyphyletic; some species are referable to the *Diaporthales*. See Seifert (*CJB* **65**: 2196, 1987), Okada *et al.* (*CJB* **76**: 1495, 1998; 18S rDNA), Verkley (*Mycol.* **93**: 205, 2001), Decock *et al.* (*Antonie van Leeuwenhoek* **88**: 231, 2005; phylogeny).

stromatolites (lichen), laminar calcretes formed abiotically in rock and sometimes wrongly interpreted as fossils (Klappa, *Sediment. Petrol.* **49**: 387, 1979).

Stromatoneurospora S.C. Jong & E.E. Davis (1973), Xylariaceae. 2, widespread (tropical). See Dennis (*Revista Biol.* **1**: 175, 1958), Jong & Davis (*Mycol.* **65**: 458, 1973), Rogers *et al.* (*Mycol.* **84**: 166, 1992).

Stromatopogon Zahlbr. (1897), anamorphic *Pezizomycotina*, Cpd.#eH.?15. 3 (on lichens, *Usnea*), Australia. See Diederich (*Lichenologist* **24**: 371, 1992; redescr.), Diederich & Sérusiaux (*Biblthca Lichenol.* **86**: 103, 2003), Etayo & Breuss (*Öst. Z. Pilzk.* **13**: 277, 2004).

Stromatopycnis A.F. Vital (1956), anamorphic *Pezizomycotina*, Cpd.0eH.?. 1, Brazil. See Vital (*Publções Inst. Micol. Recife* **15**: 5, 1956).

Stromatoscypha Donk (1951) ≡ Porotheleum.

Stromatoscyphaceae Jülich (1982) nom. illegit. = Schizophyllaceae.

Stromatosphaeria Grev. (1824) nom. rej. ≡ Daldinia.

Stromatostilbella Samuels & E. Müll. (1980) ≡ Stromatographium fide Seifert (*CJB* **65**: 2196, 1987).

Stromatostysanus Höhn. (1919), anamorphic *Pezizomycotina*, Hsy.0eH.?. 1, Europe. See Braun (*Sydowia* **45**: 81, 1993).

Stromatothecia D.E. Shaw & D. Hawksw. (1971), Odontotremataceae. 1 (on *Nothofagus*), Papua New Guinea. See Shaw & Hawksworth (*Proc. Papua New Guinea Sci. Soc.* **22**: 24, 1971), Sherwood-Pike (*Mycotaxon* **28**: 137, 1987).

Stromatothelium Trevis. (1861), ? Pyrenulales (L). 4, widespread (tropical).

Stromne Clem. (1909) ≡ Engleromyces.

Strongwellsea A. Batko & Weiser (1965), Entomophthoraceae. 2, USA. See Batko & Weiser (*J. Invert. Path.* **7**: 455, 1965), Weiser & Batko (*Folia Parasit.* **13**: 144, 1966), Humber (*Mycol.* **68**: 1042, 1976; emend., key), Remaudière & Keller (*Mycotaxon* **11**: 323, 1980; syn. of *Erynia*), Humber (*Mycotaxon* **15**: 167, 1982), Eilenberg *et al.* (*Entomophaga* **37**: 65, 1992; isolation), Eilenberg & Michelsen (*J. Invert. Path.* **73**: 189, 1999; host range), Keller & Petrini (*Sydowia* **57**: 23, 2005; key), Keller (*Sydowia* **59**: 75, 2007; Switzerland, n.sp.).

Strongyleuma Vain. (1927) = Chaenothecopsis fide Tibell (*Beih. Nova Hedwigia* **79**: 597, 1984).

Strongylium (Ach.) Gray (1821) = Calicium fide Tibell (*Beih. Nova Hedwigia* **79**: 597, 1984).

Strongylopsis Vain. (1927) = Microcalicium fide Tibell (*Bot. Notiser* **131**: 229, 1978).

Strongylothallus Bat. & Cif. (1959), anamorphic *Pezizomycotina*, Cpt.-.-. 1, Brazil. See Batista & Ciferri (*Mycopath. Mycol. appl.* **11**: 97, 1959).

Stropharia (Fr.) Quél. (1872) = Psilocybe (Fr.) P. Kumm. fide Kuyper (*in litt.*).

Strophariaceae Singer & A.H. Sm. (1946), Agaricales. 18 gen. (+ 46 syn.), 1316 spp.

Lit.: Fogel (*Mycol.* **77**: 72, 1985), Pegler & Young (*Notes R. bot. Gdn Edinb.* **44**: 437, 1987), Watling & Gregory (*British Fungus Flora* **5**, 1987), Jacobsson (*Windahlia* **19**: 1, 1990), Bougher & Castellano (*Mycol.* **85**: 285, 1993), Johnston & Buchanan (*N.Z. Jl Bot.* **33**: 379, 1995), Holec (*Mykol. Listy* **57**: 1, 1996), Bresinsky & Binder (*Z. Mykol.* **64**: 79, 1998), Kytövuori (*Karstenia* **39**: 11, 1999), Reijnders (*MR* **104**: 900, 2000), Bougher & Lebel (*Aust. Syst. Bot.* **14**: 439, 2001), Peintner *et al.* (*Am. J. Bot.* **88**: 2168, 2001), Binder & Bresinsky (*Mycol.* **94**: 85, 2002), Boekhout *et al.* (*MR* **106**: 1251, 2002), Moncalvo *et al.* (*Mol. Phylogen. Evol.* **23**: 357, 2002), Bon & Roux (*Docums Mycol.* **33**: 3, 2003), Clémençon & Roffler (*Mycol. Progr.* **2**: 235, 2003), Matsumoto *et al.* (*Mycoscience* **44**: 197, 2003), Francis & Bougher (*Australasian Mycologist* **23**: 1, 2004), Guzmán & Kasuya (*Mycoscience* **45**: 295, 2004), Peintner *et al.* (*Mycol.* **96**: 1042, 2004), Walther *et al.* (*MR* **109**: 525, 2005), Matheny *et al.* (*Mycol.* **98**: 982, 2006; phylogeny), Moreau *et al.* (*Mol. Phylogen. Evol.* **38**: 794, 2006).

Stropholoma (Singer) Balletto (1989) nom. inval. = Psilocybe (Fr.) P. Kumm. fide Kuyper (*in litt.*).

Strossmayeria Schulzer (1881), Helotiaceae. Anamorph *Pseudospiropes*. 16, widespread. See Iturriaga & Korf (*Mycotaxon* **36**: 119, 1990; monogr.), Hosoya (*Memoirs of the National Science Museum* Tokyo **34**: 241, 2000; Japan).

Strumella Fr. (1849), anamorphic *Urnula*, Hsp.0eP.?. 8, widespread. *S. coryneoidea*, canker of oak (*Quercus*) and sometimes of other trees. See Wolf (*Mycol.* **50**: 837, 1959).

Strumellopsis Höhn. (1909), anamorphic *Pezizomycotina*, St.0eP.?. 2, Java.

Stuartella Fabre (1879), Dothideomycetes. Anamorph *Bactrodesmium*. 3, Europe; N. America. See LaFlamme & Müller (*Sydowia* **29**: 278, 1977), Funk & Shoemaker (*CJB* **61**: 2277, 1983).

stuffed (of a stipe), having the inside of a different structure from that of the outer layer (Fig. 4F).

stupose, of tissue formed from hyphae which are not gelatinized.

Stygiomyces Coppins & S.Y. Kondr. (1995), anamorphic *Pezizomycotina*, Cpd.0fH.15. 1 (on *Pseudocyphellaria*), Tasmania. See Coppins & Kondratyuk (*Edinb. J. Bot.* **52**: 229, 1995).

Stylaspergillus B. Sutton, Alcorn & P.J. Fisher (1982), anamorphic *Pezizomycotina*, Hso.0fH.15. 1, widespread. Synanamorph *Parasympodiella*. See Tokumasu (*TMSJ* **28**: 19, 1987).

Stylina Syd. (1921), Graphiolaceae. 1 (on *Livistona*), China.

Stylobates Fr. (1837) nom. dub., Agaricales.

Stylodothis Arx & E. Müll. (1975), Dothideaceae. 2, widespread. See von Arx & Müller (*Stud. Mycol.* **9**: 11, 1975), Lumbsch & Lindemuth (*MR* **105**: 901, 2001; phylogeny), Lumbsch *et al.* (*Mol. Phylogen. Evol.* **34**: 512, 2005; phylogeny), Prado *et al.* (*MR* **110**: 511, 2006; phylogeny), Schoch *et al.* (*Mycol.* **98**: 1041, 2006; phylogeny).

Styloletendraea Weese (1924) nom. nud. = Nectria. Anamorph name. fide Hawksworth *et al.* (*Dictionary of the Fungi* edn 8, 1995).

Stylonectria Höhn. (1915) = Cosmospora fide Rossman *et al.* (*Stud. Mycol.* **42**: 248 pp., 1999).

Stylonectriella Höhn. (1915) ? = Nectriella Nitschke ex Fuckel fide Sutton (*Mycol. Pap.* **141**, 1977).

Stylopage Drechsler (1935), Zoopagaceae. 17, widespread. See Drechsler (*Mycol.* **27**: 197, 1935), Drechsler (*Mycol.* **27**: 206, 1935), Drechsler (*Mycol.* **28**: 241, 1936), Drechsler (*Mycol.* **30**: 137, 1938), Drechsler (*Mycol.* **31**: 388, 1939), Drechsler (*Mycol.* **37**: 1, 1945), Drechsler (*Mycol.* **40**: 85, 1948), Duddington (*Ann. Bot.* Lond. **17**: 127, 1953), Duddington (*Mycol.* **47**: 245, 1955), Peach & Juniper (*TBMS* **38**: 431, 1955), Dyal (*Sydowia* **27**: 293, 1976; keys parasites of nematodes and amoebae, bibliogr.), Wood (*TBMS* **80**: 368, 1983), Saikawa (*Mycol.* **78**: 309, 1986; ultrastr.), Blackwell & Malloch (*Mycol.* **83**: 360, 1991; life history), Mo & Bi (*Mycosystema* **20**: 129, 2001; China).

stylospore (1) a spore on a pedicel or hypha, esp. a urediniospore (obsol.); (2) an elongated pycnidiospore (obsol.); (3) the sporangiolum (the 'Stielgemmen' of Linnemann, 1941) of *Mortierella*.

Stypella Möller (1895), Auriculariales. 4, widespread. See Martin (*Stud. nat. Hist. Univ. Iowa* **16**: 143, 1934), Donk (*Persoonia* **4**: 241, 1966), Roberts (*Mycotaxon* **69**: 209, 1998; key), Weiss & Oberwinkler (*MR* **105**: 403, 2001; phylogeny).

Stypinella J. Schröt. (1888) ≡ Helicobasidium.

Stysanopsis Ferraris (1909) = Cephalotrichum Link fide Hughes (*CJB* **36**: 727, 1958).

Stysanus Corda (1837) = Cephalotrichum Link fide Hughes (*CJB* **36**: 727, 1958).

suaveolent, having a sweet smell.

sub- (prefix), under; below; frequently in the sense of approximating to the condition qualified, slightly, somewhat.

Subbaromyces Hesselt. (1953), ? Pleosporales. Anamorph *Hyalobelemnospora*. 2 (on filter beds), N. America; India. See Cole *et al.* (*CJB* **52**: 2453, 1974; development, ultrastr.), Malloch & Blackwell (*CJB* **68**: 1712, 1990), Blackwell *et al.* (*Mycol.* **95**: 987, 2003; phylogeny).

subcentric, see centric.

subcutis, see cutis.

subglobose, not quite spherical (Fig. 23.1).

subhymenium, generative tissue below the hymenium; sometimes used as equivalent to medullary exciple or hypothecium.

Subhysteropycnis Wedin & Hafellner (1998), anamorphic *Arthonia*, Cpd.?.? (L). 1, Chile. See Wedin & Hafellner (*Lichenologist* **30**: 69, 1998).

Subicularium M.L. Farr & Goos (1989), anamorphic *Pezizomycotina*, Hsp.#eP.1. 1, Venezuela. See Eriksson (*SA* **8**: 77, 1989).

Subiculicola Speg. (1924) = Melioliphila fide Rossman (*Mycol. Pap.* **157**, 1987).

subiculum (**subicule**), a net-, wool-, or crust-like growth of mycelium under fruit-bodies.

Submersisphaeria K.D. Hyde (1996), Annulatascaceae. 5 (submerged wood), Australia. See Hyde (*Nova Hedwigia* **62**: 171, 1996), Wong *et al.* (*SA* **16**: 17, 1998), Campbell *et al.* (*Mycol.* **95**: 41, 2003; phylogeny), Pinnoi *et al.* (*Sydowia* **56**: 72, 2004; key).

suboperculate, see ascus.

Subramanella H.C. Srivast. (1962) = Phomopsis (Sacc.) Bubák. fide Sutton (*Mycol. Pap.* **141**, 1977).

Subramania D. Rao & P. Rag. Rao (1964), anamorphic *Pezizomycotina*, Hso.0eP.3. 1, India. See Rao & Rao (*Trans. Am. microsc. Soc.* **83**: 403, 1964).

Subramanianospora Narayanan, J.K. Sharma & Minter (2003), anamorphic *Pezizomycotina*. 1, India. See Narayanan *et al.* (*Indian Phytopath.* **56**: 160, 2003).

Subramaniomyces Varghese & V.G. Rao (1980), anamorphic *Pezizomycotina*, Hso.0eP.3. 3, India. See Varghese & Rao (*Kavaka* **7**: 83, 1979), Crous *et al.* (*MR* **110**: 264, 2006).

Subramaniula Arx (1985), Chaetomiaceae. 2, S. Africa; India. See von Arx (*Proc. Indian Acad. Sci.* Pl. Sci. **94**: 341, 1985), Cannon (*TBMS* **86**: 56, 1986; key), von Arx *et al.* (*Beih. Nova Hedwigia* **94**, 1988).

substrate (**substratum**), although these two terms are frequently treated as synonyms by mycologists both have useful special senses: (1) **substrate**: (in enzymology) is applied to the substance on which an enzyme acts and in microbiology to the substances (e.g. culture medium constituents) utilized by a microorganism for growth in distinction to the material; (2) **substratum** (pl. **substrata**): (in ecology), the material on which an organism is growing or to which it is attached.

substroma, pseudostroma (q.v.) in which the vegetative hyphae of the host predominate (Johnston, *Ann. Mo. bot. Gdn* **27**: 31, 1940).

Subulariella Höhn. (1915) nom. dub., anamorphic *Pezizomycotina*. See Sutton (*Mycol. Pap.* **141**, 1977).

subulate, slender and tapering to a point; awl-shaped (Fig. 23.34).

Subulicium Hjortstam & Ryvarden (1979), Hymenochaetales. 3, widespread (north temperate). See Hjortstam & Ryvarden (*Mycotaxon* **9**: 511, 1979).

Subulicystidiaceae Jülich (1982) = Hyphodermataceae.

Subulicystidium Parmasto (1968), Hydnodontaceae. Anamorph *Aegeritina*. 7, widespread. See Jülich (*Persoonia* **8**: 187, 1975), Duhem & Michel (*Cryptog. Mycol.* **22**: 163, 2001).

Subulispora Tubaki (1971), anamorphic *Pezizomycotina*, Hso.0fH.10. 12, widespread (north temperate). See Sutton (*TBMS* **71**: 167, 1978), de Hoog (*Stud. Mycol.* **26**: 54, 1985; key), Castañeda Ruíz *et al.* (*Mycotaxon* **67**: 9, 1998), Marvanová & Laichmanová (*Fungal Diversity* **26**: 241, 2007; key).

subuniversal veil, see protoblem.

Sucinaria Syd. (1925) = Coccodiella fide Müller & von Arx *in* Ainsworth *et al.* (Eds) (*The Fungi* **4A**: 87, 1973).

Sufa Adans. (1763) = Lycoperdon Pers. fide Mussat (*Syll. fung.* **15**: 408, 1901).

sufu (Chinese cheese), an oriental food composed of *Actinomucor* or *Mucor* fermented soybeans (Hesseltine, *Mycol.* **57**: 164, 1965, *Mycologist* **5**: 162, 1991); see Fermented food and drinks.

sugar fungus, a fungus attacking decaying substances and only able to utilize simple sugars, amino acids, and other relatively simple organic compounds (Thom & Morrow, *J. Bact.* **33**: 77, 1937).

Sugiyamaella Kurtzman & Robnett (2007), Trichomonascaceae. 1. See Kurtzman (*FEMS Yeast Res.* **7**: 1046, 2007; from forests), Kurtzman & Robnett (*FEMS Yeast Res.* **7**: 141, 2007).

Sugiyamaemyces I.I. Tav. & Balazuc (1989), Laboulbeniaceae. 1, Borneo. See Tavares & Balazuc (*Mycotaxon* **34**: 566, 1989).

Suillaceae Besl & Bresinsky (1997), Boletales. 3 gen. (+ 15 syn.), 54 spp.

Lit.: Samson & Fortin (*Mycol.* **80**: 382, 1988), Baura *et al.* (*Mycol.* **84**: 592, 1992), Bruns *et al.* (*Mol. Ecol.* **7**: 257, 1998), Dahlberg & Finlay (*Ectomycorrhizal Fungi*: 33, 1999), Noordeloos (*Coolia* **43**: 1, 2000), Noordeloos (*Coolia* **43**: 75, 2000), Manian *et al.* (*FEMS Microbiol. Lett.* **204**: 117, 2001), Binder & Hibbett (*Mol. Phylogen. Evol.* **22**: 76, 2002), Moncalvo *et al.* (*Mol. Phylogen. Evol.* **23**: 357, 2002), Ding & Wen (*Nova Hedwigia* **76**: 459, 2003), Muller *et al.* (*Mol. Ecol.* **16**: 165, 2006).

Suillellus Murrill (1909) = Boletus Fr. fide Singer (*Farlowia* **2**: 223, 1945).

Suillosporium Pouzar (1958), Botryobasidiaceae. 2, Europe; N. America. See Eriksson *et al.* (*Cortic. N. Europ.* **7**, 1984), Langer & Langer (*Frontiers in Basidiomycote Mycology*: 303, 2004; Ecuador).

Suillus (Haller) Kuntze (1898) = Boletus Fr. fide Stalpers (*in litt.*).

Suillus Gray (1821), Suillaceae. 50 (associated with conifers), widespread (north temperate; introduced in southern hemisphere). See Pantidou & Groves (*CJB* **44**: 1371, 1966; cultural studies), Thiers (*Beih. Nova Hedwigia* **51**, 1975; key N. Am. spp.), Bruns & Palmer (*J. Mol. Evol.* **28**: 349, 1989; evolution), Baura *et al.* (*Mycol.* **84**: 592, 1992; gastroid form), Engel (*Schmier- und Filzröhrlinge s.l. in Europe*, 1996; key Eur. spp.), Kretzer *et al.* (*Mycol.* **88**: 776, 1996; taxonomy, phylogeny), Kretzer & Bruns (*Mycol.* **89**: 586, 1997; gastroid form), Dahlberg & Finlay (*Suillus – Ectomycorrhizal fungi, key genera in profile*, 1999; ecology), Wu *et al.* (*Mol. Phylog. Evol.* **17**: 37, 2000; phylogeny, biogeography), Consiglio (*Rivista di Micologia* **44**: 99, 2001), Estades & Lannoy (*Bulletin Mycologique et Botanique Dauphiné-Savoie* **44**: 3, 2004; European spp.), Šutara (*Czech Mycol.* **57**: 1, 2005; Central European genera), Watling *et al.* (*British Fungus Flora. Agarics and Boleti* Rev. & Enl. Edn **1**: 173 pp., 2005; British spp.).

Suillus P. Karst. (1882) ≡ Gyroporus.

Suillus P. Micheli ex Adans. (1763) = Boletus Fr. fide Donk (*Persoonia* **8**: 279, 1975).

Sulcaria Bystrek (1971), Parmeliaceae (L). 3, Asia; USA. See Brodo & Hawksworth (*Op. bot. Soc. bot. Lund* **42**, 1977), Awasthi & Awasthi (*Candollea* **40**: 305, 1985; India, Nepal), Brodo (*Mycotaxon* **27**: 113, 1986; USA), Peterson *et al.* (*Bryologist* **101**: 112, 1998; USA), Obermayer & Elix (*Biblthca Lichenol.*

86: 33, 2003; Tibet, chemistry).
sulcate, grooved.
Sulcatisclerotes Beneš (1959), Fossil Fungi. 2 (Carboniferous), former Czechoslovakia.
Sulcatistroma A.W. Ramaley (2005), Calosphaeriales. 1, USA. See Ramaley (*Mycotaxon* **93**: 140, 2005).
Sulcispora, see *Sulcospora*.
Sulcospora Shoemaker & C.E. Babc. (1989), Phaeosphaeriaceae. 1, Europe. See Shoemaker & Babcock (*CJB* **67**: 1500, 1989).
Sulcopyrenula H. Harada (1999), Pyrenulaceae. 4, widespread. See Harada (*Lichenologist* **31**: 567, 1999).
Sulcospora Kohlm. & Volkm.-Kohlm. (1993) ≡ Aropsiclus.
sulcus, a furrow or groove.
sulphur polypore (**sulphur shelf-mushroom**), basidioma of the edible *Laetiporus sulphureus*.
Sulphurina Pilát (1953) = Lindtneria fide Hjortstam (*in litt.*).
summer spore, a spore germinating without resting, frequently living only a short time; cf. resting spore.
Sungaiicola Fryar & K.D. Hyde (2004), Sordariomycetes. 1, Brunei. See Fryar & Hyde (*Cryptog. Mycol.* **25**: 250, 2004).
super- (prefix), above; used in combination with the names of taxonomic categories to give additional ranks in (e.g. superfamily); only recently permitted under the botanical Code.
superficial, on the surface of the substratum.
superior (of an annulus), near the top of the stipe.
supine (of fructifications), closely applied to the substratum.
supra- (prefix), above; **-generic**, all taxonomic ranks above that of genus; **-specific**, all taxonomic ranks above that of species; also used at other ranks of the hierarchy; cf. infra-.
suprahilar plage (of basidiospores, esp. of *Lactarius* and *Russula*), the area above the hilar appendage on which the eusporial ornamentation is lacking or reduced (Kühner, 1926).
surculicolous (of *Exobasidium* infections), monocarpic (q.v.) and systemic in annual shoots (Nannfeldt, 1981).
Surculiseries Okane, Nakagiri & Tad. Ito (2001), anamorphic *Pezizomycotina*. 1, Japan. See Okane *et al.* (*Mycoscience* **42**: 116, 2001).
surfactant, an agent which reduces the surface tension of a liquid, e.g. detergent.
suscept, a living organism which is **susceptible** to (able to be attacked by; non-immune to) a given disease, pathogen, or toxin.
suspensor, a hypha supporting a gamete, gametangium, or esp. a zygospore (Fig. 27C).
Sutravarana Subram. & Chandrash. (1977), anamorphic *Pezizomycotina*, Hsy.0eH.10. 1, India. See Subramanian & Chandrashekara (*CJB* **55**: 251, 1977).
Suttonia S. Ahmad (1961) [non *Suttonia* A. Rich. 1832, *Myrsinaceae*] ≡ Suttoniella.
Suttoniella S. Ahmad (1961), anamorphic *Pezizomycotina*, St.1bH.16. 3, Asia; Australia. See Sutton (*TBMS* **59**: 285, 1972), Punithalingam (*MR* **107**: 917, 2003; cytology), Hoyo & Gómez-Bolea (*Mycotaxon* **89**: 39, 2004).
Suttonina H.C. Evans (1984), anamorphic *Pezizomycotina*, Cac.1-≡ eP.38. 1, Guatemala. See Evans (*Mycol. Pap.* **153**: 92, 1984).
Svrcekia Kubička (1960) = Boudiera Cooke fide Dissing (*SA* **6**: 153, 1987).
Svrcekomyces J. Moravec (1976) = Pseudombrophila fide van Brummelen (*Libri Botanici* **14**: 1, 1995).
Swampomyces Kohlm. & Volkm.-Kohlm. (1987), Hypocreomycetidae. 2, pantropical. See Hyde & Nakagiri (*Sydowia* **44**: 122, 1992), Read *et al.* (*MR* **99**: 1465, 1995; ultrastr.), Abdel-Wahab *et al.* (*Fungal Diversity* **8**: 35, 2001; Egypt), Sakayaroj *et al.* (*Bot. Mar.* **48**: 395, 2005; phylogeny), Schoch *et al.* (*MR* **111**: 154, 2007; phylogeny).
swarm-cell (of *Mycetozoa* and some *Chytridiales*), a motile cell acting, before or after division, as an isogamete.
swarm-spore (**swarmer**), see zoospore.
Syamithabeeja Subram. & Natarajan (1976), anamorphic *Pezizomycotina*, Hso.0eH.18. 1, India. See Subramanian & Natarajan (*Mycol.* **67**: 1213, 1975).
Sychnoblastia Vain. (1921) = Thelopsis fide Hawksworth *et al.* (*Dictionary of the Fungi* edn 8, 1995).
Sychnogonia Körb. (1855) nom. rej. = Muellerella fide Hawksworth *et al.* (*Dictionary of the Fungi* edn 8, 1995).
Sychnogonia Trevis. (1860) [non *Sychnogonia* Körb. 1855] = Thelopsis fide Hawksworth *et al.* (*Dictionary of the Fungi* edn 8, 1995).
sycosis, a fungus disease of the hair follicles; esp. of the face; ringworm of the beard.
Sydow (Hans; 1879-1946; Germany). Son of P. Sydow (q.v.); employed at the Dresden Bank, Berlin (1904-1937); collected in Costa Rica (1924-1925), Venezuela (1927-1928) and Ecuador (1937-1938). An amateur; founder and Editor *Annales Mycologici* (1903-1944); in addition to works produced with his father, contributed greatly to development of mycotas particularly in Central and South America. *Publs.* (with P. Sydow) *Mycotheca Germanica* (1903-1942) [exsiccati, nos 1-550 issued jointly with P. Sydow]; *Monographia Uredinearum* (1902-1924). *Biogs, obits etc.* Grummann (1974: 49); Petrak (*Sydowia* **2**: 1, 1948 [portrait]); Stafleu & Cowan (*TL-2* **6**: 129, 1986).
Sydow (Paul; 1851-1925; Germany). Schoolmaster, Berlin; father of H. Sydow (q.v.). Major contributions to information handling in mycology through indexes and catalogues; with his son, the author of many works describing new species of ascomycetes, rusts and smuts. *Publs. Die Flechten Deutschlands* (1887); Index vols (**12-13**) to Saccardo's *Sylloge*; (with Lindau) *Thesaurus Literaturae Mycologicae* (1908-1918); *Uredineen* (1888-1924) [exsiccati]; *Ustilagineen* (1894-1915) [exsiccati]; (with H. Sydow, q.v.) *Monographia Uredinearum* (1902-1924). *Biogs, obits etc.* Stafleu & Cowan (*TL-2* **6**: 132, 1986).
Sydowia Bres. (1895), Dothioraceae. Anamorphs *Dothichiza*, *Sclerophoma*, *Hormonema*-like. 10, Europe; N. America. See Barr (*Contr. Univ. Mich. Herb.* **9**: 523, 1972), Froidevaux (*Nova Hedwigia* **23**: 679, 1973), Funk & Finck (*CJB* **66**: 212, 1988), Yurlova *et al.* (*Stud. Mycol.* **43**: 63, 1999; DNA), Barr (*Harvard Pap. Bot.* **6**: 25, 2001), Goodwin *et al.* (*Phytopathology* **91**: 648, 2001; phylogeny), Tsuneda *et al.* (*Mycol.* **96**: 1128, 2004; phylogeny), Schoch *et al.* (*Mycol.* **98**: 1041, 2006; phylogeny).
Sydowiella Petr. (1923), Sydowiellaceae. 3 (from bark), Europe; N. America. See Barr (*Mycotaxon* **41**: 287, 1991), Rossman *et al.* (*Mycoscience* **48**: 135,

2007; review).

Sydowiellaceae Lar.N. Vassiljeva (1987), Diaporthales. 7 gen., 17 spp. Further research is needed on the circumscription of this fam. fide Vasilyeva (*Mycoscience* **42**: 401, 2001), Rossman *et al.* (*Mycoscience* **48**: 135, 2007; phylogeny).

Sydowiellina Bat. & I.H. Lima (1959) = Myriangiella fide von Arx & Müller (*Stud. Mycol.* **9**, 1975).

Sydowina Petr. (1923) = Lojkania fide Barr (*Mycotaxon* **20**, 1984).

Sydowinula Petr. (1923) = Acanthonitschkea fide Nannfeldt (*Svensk bot. Tidskr.* **69**: 49, 1975).

Sylviacollaea Cif. (1963), anamorphic *Pezizomycotina*, St.0eH.15. 1 (on termites), Dominican Republic. See Ciferri (*Atti Ist. bot. Univ. Lab. crittog. Pavia* sér. 5 **20**: 246, 1963).

sym-, see **syn-**.

Symbiosis. Associations between unlike organisms, generally ones persisting for long periods (relative to the generation time of the interacting organisms); apparently first used by the lichenologist A.B. Frank in 1877 (often credited to de Bary, 1879) who later coined the word mycorrhiza (q.v.). At times equated with mutualism (q.v.), but correctly also covering parasitic (harmful) and commensalistic (unharmful) associations (Ahmadjian & Paracer, 1987).

Symbioses involving fungi are frequent, abundant and widespread. Mutualistic symbioses include endophytes, lichens, mycorrhizas and rumen fungi, as well as cultivation of fungi by ambrosial insects, ants, humans and termites. Parasitic symbioses include fungal diseases of animals, other fungi and plants, but also parasitism of fungi by plants such as orchids, and consumption of fungi by various types of animal.

In mutualistic lichen symbioses, there has been debate as to which partner might be regarded as host (e.g. de Bary regarded the algae in lichens as the host, and Douglas, 1994, the fungus as host to the algae); to circumvent this controversy, Law & Lewis (1983) used the neutral terms **exhabitant** (the organism forming the outer tissues) and **inhabitant** (the enclosed organism).

In lichens, Poelt (*Abstr. IMC2*, 1977) recognized: **two-membered -**, one alga + one fungus, **three-membered -**, one alga + two fungi (lichenicolous fungi; q.v.), or two algae + one fungus (cephalodium; q.v.); and **four-membered -**, two algae + two fungi (lichenicolous lichens); Hawksworth (*Bot. J. Linn. Soc.* **96**: 3, 1988) used **'-biont'** (q.v.) instead and discussed further fungal/algal interactions (see also Rambold & Triebel, *Bibl. Lich.* **48**, 1992).

Lit.: Ahmadjian & Paracer (*Symbiosis, an introduction to biological associations*, 1986), Carroll (*in* Carroll & Wicklow, *The fungal community*, edn 2 : 327, 1992; fungal mutualisms), Cook *et al.* (Eds) (*Symp. Soc. exp. Biol.* **29**, 1975), Cooke (*The biology of symbiotic fungi*, 1977), Cook *et al.* (Eds) (*Cellular interactions in symbiosis and parasitism*, 1980), Douglas (*Symbiotic interactions*, 1994), Law & Lewis (*Biol. J. Linn. Soc.* **20**: 249, 1983), Lewis (*Biol. Rev.* **48**: 261, 1973; *in* Rayner *et al.* (Eds), *Evolutionary biology of the fungi*: 161, 1987), Margulis & Fester (Eds) (*Symbiosis as a source of evolutionary innovation*, 1991), Sapp (*Evolution by association: a history of symbiosis*, 1984). See Antagonism, Coevolution, Endophyte, Lichens, Mycorrhiza.

Symbiotaphrina Kühlw. & Jurzitza ex W. Gams & Arx (1980), anamorphic *Pezizomycotina*, Hso.0eH.?. 2 (beetle mycetomas), Germany. See Jones *et al.* (*MR* **103**: 542, 1999; syst. posn), Schweigkofler *et al.* (*Organ. Divers. Evol.* **2**: 1, 2002; phylogeny), Lopandic *et al.* (*Mycol. Progr.* **4**: 205, 2005; phylogeny).

sympatric, occurring in the same geographical region. Cf. allopatric.

Symperidium Klotzsch (1843) = Aecidium fide Saccardo (*Syll. fung.* **18**: 829, 1906).

Symphaeophyma Speg. (1912), Parmulariaceae. 1, S. America. See Inácio & Cannon (*MR* **107**: 82, 2003).

Symphaster Theiss. & Syd. (1915), Asterinaceae. 2, S. America; S. Africa. See Hosagoudar *et al.* (*Journal of Mycopathological Research* **39**: 61, 2001).

symphogenous, see meristogenous.

Symphyosira Preuss (1853), anamorphic *Symphyosirinia*, Hsy.≡ eH.39. 4, Europe.

Symphyosirinia E.A. Ellis (1956), Helotiaceae. Anamorph *Symphyosira*. 4, Europe; N. America. See Baral (*SA* **13**: 113, 1994; posn), Baral (*Z. Mykol.* **60**: 211, 1994; key).

Symphysos Bat. & Cavalc. (1967), anamorphic *Pezizomycotina*, Cpt.0eH.?. 1, Brazil. See Batista & Cavalcanti (*Atas Inst. Micol. Univ. Pernambuco* **5**: 202, 1967).

symplastic, entering living cells; cf. apoplastic.

Symplectromyces Thaxt. (1908), Laboulbeniaceae. 1, widespread. See Benjamin (*Aliso* **10**: 345, 1983), Tavares (*Mycol. Mem.* **9**: 627 pp., 1985).

symplesiomorphy (-ies), see Cladistics.

Symplocia A. Massal. (1854) nom. rej. ≡ Crocynia.

Sympodia (R. Heim) W.B. Cooke (1952) nom. inval. = Marasmius fide Kirk (*in litt.*).

sympodial (of conidiogenous cells), characterized by continued growth, after the main axis has produced a terminal spore, by the development of a succession of apices each of which originates below and to one side of the previous apex.

Sympodiella W.B. Kendr. (1958), anamorphic *Pezizomycotina*, Hso.0eH.39. 3, Europe; Cuba. See Kendrick (*TBMS* **41**: 519, 1958).

Sympodina Subram. & Lodha (1964) = Veronaea fide von Arx (*Gen. Fungi Sporul. Cult.*, 1970).

Sympodiocladium Descals (1982), anamorphic *Pezizomycotina*, Hso.0eH.1. 1 (aquatic), British Isles. See Descals (*TBMS* **78**: 427, 1982).

Sympodioclathra Voglmayr (1997), anamorphic *Pezizomycotina*, Hso.?.?. 1, USA. See Voglmayr & Krisai-Greilhuber (*Mycol.* **89**: 945, 1997).

sympodioconidium (**sympodulospore**), a spore produced on a sympodula.

Sympodiomyces Fell & Statzell (1971) = Blastobotrys fide Statzell-Tallman & Fell *in* Kurtzman & Fell (Eds) (*Yeasts, a taxonomic study* 4th edn: 603, 1998), Carreiro *et al.* (*Int. J. Syst. Evol. Microbiol.* **54**: 1891, 2004), Kurtzman & Robnett (*FEMS Yeast Res.* **7**: 141, 2007).

Sympodiomycopsis Sugiy., Tokuoka & Komag. (1991), anamorphic *Microstromataceae*. 1 (from nectar), Japan. See Begerow *et al.* (*MR* **104**: 53, 2000; mol. phylogeny).

Sympodiophora G.R.W. Arnold (1970) = Pseudohansfordia G.R.W. Arnold fide de Hoog (*Persoonia* **10**: 67, 1978), Arnold (*Abstracts, XI Congress of European Mycologists* Kew, England, 7-11 September 1992: 3, 1992).

Sympodioplanus R.C. Sinclair & Boshoff (1997), anamorphic *Pezizomycotina*, Hso.?.?. 1, S. Africa; S.

America. See Sinclair *et al.* (*Mycotaxon* **64**: 366, 1997).

Sympodomyces R.K. Benj. (1973), Laboulbeniaceae. 1, Papua New Guinea. See Benjamin (*Aliso* **8**: 1, 1973), Tavares (*Mycol. Mem.* **9**: 627 pp., 1985).

sympodula, a sympodial conidiogenous cell.

Sympoventuria Crous & Seifert (2007), ? Venturiaceae. Anamorph *Sympodiella*-like. 1, S. Africa. See Crous *et al.* (*Fungal Diversity* **25**: 19, 2007), Crous *et al.* (*Stud. Mycol.* **58**: 185, 2007; phylogeny).

syn- (sym-) (in compounds), growing together; adhesion; aggregation.

Synalissa Fr. (1825), Lichinaceae (L). 5, Europe; N. America. See Moreno & Egea (*Acta Bot. Barcinon.* **41**: 1, 1992), Schultz & Büdel (*Mycotaxon* **84**: 21, 2002), Schultz & Büdel (*Lichenologist* **34**: 39, 2002; key).

Synalissina Nyl. (1885) = Lempholemma fide Zahlbruckner (*Nat. Pflanzenfam.* **8**: 167, 1926).

Synallisopsis Nyl. ex Stizenb. (1882) nom. inval. = Pyrenopsis fide Hawksworth *et al.* (*Dictionary of the Fungi* edn 8, 1995).

synanamorph, see States of fungi.

Synandromyces Thaxt. (1912), Laboulbeniaceae. 10, widespread. See Nannfeldt (*Svensk bot. Tidskr.* **43**: 468, 1949).

Synaphia Nees & T. Nees ex Rchb. (1841) nom. dub., Fungi.

synapomorphy (-ies), see Cladistics.

Synaptomyces Thaxt. (1912), Ceratomycetaceae. 1, S. America.

synaptonemal complex, proteinaceous, longitudinally aligned structure which usually unites homologous chromosomes during the prophase of meiosis.

Synaptospora Cain (1957), ? Coniochaetaceae. 4, Europe. See Barr (*Mycotaxon* **39**: 43, 1990; posn), Matzer (*Cryptog. Mycol.* **14**: 11, 1993), Huhndorf *et al.* (*Sydowia* **51**: 176, 1999), Réblová (*Sydowia* **54**: 248, 2002), García *et al.* (*MR* **110**: 1271, 2006).

Synarthonia Müll. Arg. (1891), Arthoniales (L). 1, C. America.

Synarthoniomyces Cif. & Tomas. (1953) ≡ Synarthonia.

Synascomycetes. Class (Gäumann, *Die Pilze*, 1949) for *Ascomycota* with spores formed in a spore sac ('synascus', q.v.), interpreted as a group of fused asci; included fungi now in *Ascosphaerales* and *Protomycetales* (here included in *Taphrinales*).

synascus, the gametangium of *Ascosphaera* (Varitchak, 1933).

Syncarpella Theiss. & Syd. (1915), Cucurbitariaceae. Anamorph *Syntholus*. 7, widespread. See Barr & Boise (*Mem. N. Y. bot. Gdn* **49**: 298, 1989), Ramaley & Barr (*Mycotaxon* **65**: 499, 1997).

Syncephalastraceae Naumov ex R.K. Benj. (1959), Mucorales. 8 gen. (+ 1 syn.), 22 spp.

Lit.: Benjamin (*Aliso* **4**: 321, 1959), Zheng *et al.* (*Mycosystema* **1**: 35, 1988).

Syncephalastrum J. Schröt. (1886), Syncephalastraceae. 2, widespread. See Benjamin (*Aliso* **4**: 321, 1959), Misra (*Mycotaxon* **3**: 51, 1975), Hobot & Gull (*Protoplasma* **107**: 339, 1981; germination), Schipper & Stalpers (*Persoonia* **12**: 81, 1983), Chen & Huang (*Mycosystema* **1**: 53, 1988; electrophoresis), Zheng *et al.* (*Mycosystema* **1**: 39, 1988; key), Benny *et al. in* McLaughlin *et al.* (Eds) (*The Mycota* A Comprehensive Treatise on Fungi as Experimental Systems for Basic and Applied Research **7A**: 113, 2001), O'Donnell *et al.* (*Mycol.* **93**: 286, 2001; phylogeny), Voigt & Wöstemeyer (*Gene* **270**: 113, 2001; phylogeny), Nyilasi *et al.* (*Clin. Microbiol. Infect.* **14**: 393, 2008; molecular identification).

Syncephalidium Badura (1963) = Syncephalis fide Benjamin (*Mycol.* **48**: 1, 1966).

Syncephalis Tiegh. & G. Le Monn. (1873), Piptocephalidaceae. *c.* 45 (mycoparasites of *Mucorales*), widespread. See Benjamin (*Aliso* **4**: 321, 1959), Zycha *et al.* (*Mucorales*, 1969; key), Kuzuha (*TMSJ* **14**: 237, 1973), Hunter & Butler (*Mycol.* **67**: 863, 1975), Baker *et al.* (*Mycol.* **69**: 1008, 1977; host range & culture), Kuzuha (*J. Jap. Bot.* **55**: 343, 1980; zygospore formation), Bawcutt (*TBMS* **80**: 219, 1983), Benjamin (*Aliso* **11**: 1, 1985), Can *et al.* (*Nova Hedwigia* **49**: 427, 1989), Ginman & Young (*Mycologist* **5**: 19, 1991; ultrastr.), Gruhn & Petzold (*Can. J. Microbiol.* **37**: 355, 1991), Patil & Patil (*Indian Phytopath.* **43**: 217, 1994), Ho (*Taiwania* **48**: 53, 2003; Taiwan), Tanabe *et al.* (*J. gen. appl. Microbiol.* Tokyo **51**: 267, 2005; phylogeny), White *et al.* (*Mycol.* **98**: 872, 2006; phylogeny), Ho & Benny (*Botanical Studies* **48**: 319, 2007; n.sp., Taiwan), Ho & Benny (*Botanical Studies* **49**: 45, 2008; n.sp., Taiwan).

Syncephalopsis Boedijn (1959) = Syncephalis fide Benjamin (*Mycol.* **48**: 1, 1966).

Syncesia Taylor (1836), Roccellaceae (L). 17, widespread (esp. S. America). See Tehler (*Cryptog. Bot.* **3**: 139, 1993), Tehler (*Fl. Neotrop.* **75**, 1997), Myllys *et al.* (*Bryologist* **101**: 70, 1998; morphology, phylogeny), Myllys *et al.* (*Lichenologist* **31**: 461, 1999; phylogeny), Tehler & Irestedt (*Cladistics* **23**: 432, 2007).

Synchaetomella Decock & Seifert (2005), anamorphic *Diaporthales*. 1, Singapore. See Decock *et al.* (*Antonie van Leeuwenhoek* **88**: 234, 2005).

synchronized culture, a culture manipulated so that division of all the component cells is simultaneous; see Williamson & Scopes (*Symp. Soc. gen. Microbiol.* **11**: 217, 1961; *Saccharomyces cerevisiae*).

Synchronoblastia Uecker & F.L. Caruso (1988), anamorphic *Pezizomycotina*, Cpd.0eH.6. 1, USA. See Uecker & Caruso (*Mycol.* **80**: 345, 1988).

synchronospore, a spore produced simultaneously with other neighbouring spores.

Synchytriaceae J. Schröt. (1892), Chytridiales. 5 gen. (+ 5 syn.), 136 spp.

Lit.: Percival (*Zentbl. Bakt. ParasitKde* Abt. II **25**: 440, 1910), Karling (*Synchytrium*: 470 pp., 1964), Barr (*Geobios* New Rep., 1980; emmend.), Raghavendra & Pavgi (*Indian Phytopath.* **46**: 36, 1993), Hampson *et al.* (*Mycol.* **86**: 733, 1994), Barr *in* McLaughlin *et al.* (Eds) (*The Mycota* A Comprehensive Treatise on Fungi as Experimental Systems for Basic and Applied Research **7A**: 93, 2001).

Synchytrium de Bary & Woronin (1863), Synchytriaceae. *c.* 120 (plant parasites), widespread. *S. aureum* reported from 198 host spp. (123 gen., 34 fam.); Karling experimentally infected 1465 spp. in more than 918 gen. of 176 fam. with *S. macrosporum*, *S. endobioticum* (potato wart disease). See Karling (*Synchytrium*, 1964; keys), Karling (*Adv. Frontier Pl. Sci.* **29**: 1, 1972), Raghavendra & Pavgi (*Indian Phytopath.* **46**: 36, 1993; on *Cucurbitaceae*), Hampson *et al.* (*Mycol.* **86**: 733, 1994; Ultrastr. *S. endobioticum*), Niepold & Stachewicz (*Zeitschrift für Pflanzenkrankheiten und Pflanzenschutz* **111**: 313, 2004; PCR-detection of *S. endobioticum*).

Syncladium Rabenh. (1859), anamorphic *Pezizomycotina*, Cpd.?.?. 1, widespread. See Hughes (*Mycol.* **68**: 693, 1976).
Syncleistostroma Subram. (1972) nom. conf. = Petromyces fide Malloch & Cain (*CJB* **51**: 1647, 1973).
Syncoelium Wallr. (1833), ? Algae. See Hughes (*CJB* **36**: 747, 1958).
Syncollesia Nees (1823) ? = Fumago fide Hawksworth *et al.* (*Dictionary of the Fungi* edn 8, 1995).
Syncomista Nieuwl. (1916) ≡ Toninia.
syncytium, see coenocyte.
Syndiplodia Peyronel (1915) = Microdiplodia Allesch. fide Sutton (*Mycol. Pap.* **141**, 1977).
syndrome, a complex of symptoms, especially that which constitutes the picture of a disease.
Synechoblastus Trevis. (1853) = Collema F.H. Wigg. fide Hawksworth *et al.* (*Dictionary of the Fungi* edn 8, 1995).
synecology, ecology of communities.
synergism, two organisms or environmental factors acting simultaneously to effect a change greater than either could alone; e.g. the increase of fungicidal value in certain mixtures of fungicides, fungicides and non-toxic materials, or of air pollutants. Cf. metabiosis.
syngamy, fertilization; the fusion of male and female cells to form a zygote.
Syngenosorus Trevis. (1860) = Tomasellia fide Hawksworth *et al.* (*Dictionary of the Fungi* edn 8, 1995).
Syngliocladium Petch (1932), anamorphic *Ophiocordyceps*, Hsy.0eH.?. 5, widespread (north temperate). See Petch (*TBMS* **25**: 262, 1942), Pendland & Boucias (*Mycopathologia* **99**: 25, 1987), Hodge *et al.* (*Mycol.* **90**: 743, 1998), Evans & Shah (*MR* **106**: 737, 2002).
Synglonium Penz. & Sacc. (1897) ? = Nymanomyces fide Höhnel (*Annls mycol.* **16**: 154, 1918).
synisonym, one of two or more names having the same basionym (Donk, *Persoonia* **1**: 175, 1960).
synkaryon, a diploid zygote nucleus.
synkaryotic (of a nucleus), having *2n* chromosomes.
synnema (pl. **synnemata**), a conidioma composed of a more or less compacted group of erect and sometimes fused conidiophores bearing conidia at the apex only or on both apex and sides. Seifert (*Stud. Mycol.* **27**: 1, 1985) distinguished 3 types: **determinate -**, with a terminal, non-elongated conidiogenous zone, growth ceasing after sporulation has begun, e.g. *Stilbella*; **indeterminate -**, with an elongated fertile zone, sometimes covering the whole conidioma, growth continuing after sporulation, e.g. *Doratomyces*; **compound -**, branched in which determinate or indeterminate branches are formed on a branched or unbranched axis, e.g. *Tilachlidiopsis*. Anatomical stipe types found in the three groups include **parallel -**, of primarily parallel hyphae; **intricate -**, of primarily or entirely textura intricata; **basistromatic -**, with well-defined basal stromata; **amphistromatic -**, well-defined basal stroma, stipe of parallel hyphae and an apical dome of textura angularis to globulosa with conidiogenous cells; **cupulate -**, conidiogenous zone concave. See Fig. 38.
synnema coremium, sometimes used for synnemata with looser fascicles as in 'coremioid' spp. of *Penicillium* and *Aspergillus* (obsol.).
Synnemadiplodia Zambett. (1955) = Botryodiplodia fide Petrak (*Sydowia* **16**: 353, 1963).
Synnemapestaloides T. Handa & Y. Harada (2004), anamorphic *Amphisphaeriaceae*. 1, Japan. See Handa *et al.* (*Mycoscience* **45**: 138, 2004).
Synnemaseimatoides K. Matsush. & Matsush. (1996), anamorphic *Pezizomycotina*, Hso.?.?. 1, Japan. See Matsushima & Matsushima (*Matsush. Mycol. Mem.* **9**: 38, 1996).
Synnematium Speare (1920) = Hirsutella fide Evans & Samson (*TBMS* **79**: 431, 1982).
Synnematomyces Kobayasi (1981), anamorphic *Pezizomycotina*, Hsy.0eH.?. 1, Japan. See Kobayasi (*J. Jap. Bot.* **56**: 287, 1981).
Synnematomycetes, anamorphic fungi having synnemata (Höhnel, 1923); *Coremiales* (Potebnia, 1909), (obsol.).
synnematous (**synnematogenous**), having synnemata. See Seifert (*Stud. Mycol.* **27**: 1, 1985, *Stilbella* and allies, *Mem. N.Y. bot. Gdn* **59**: 109, 1990, keys).
Synnemellisia N.K. Rao, Manohar. & Goos (1989), anamorphic *Pezizomycotina*, Hsy.0eH.15. 1, India. See Rao *et al.* (*Mycol.* **80**: 896, 1988).
Synnmukerjiomyces Aneja & R. Kumar (1999), anamorphic *Pezizomycotina*, H?.?.?. 1, India. See Aneja & Kumar (*Advances in Microbial Biotechnology*: 1, 1999).
Synomyces Arthur (1924) = Coleosporium fide Dietel (*Nat. Pflanzenfam.* **6**, 1928).
synonym, another name for a species or group, esp. a later or illegitimate name not currently employed for the taxon. If two or more names are based on the same type they are **homotypic** (**nomenclatural**, 'obligate', ≡) synonyms, if on different types they may be **heterotypic** (**taxonomic**, 'facultative', =) synonyms. See Nomenclature, and cf. basionym, homonym, orthographic synisonym, variant.
Synostomella Syd. (1927) = Cyclotheca fide Müller & von Arx (*Beitr. Kryptfl. Schweiz* **11** no. 2, 1962).
Synostomina Petr. (1949), anamorphic *Pezizomycotina*, St.0eH.?. 1, S. America.
Synpeltis Syd. & P. Syd. (1917) = Cyclotheca fide Müller & von Arx (*Beitr. Kryptfl. Schweiz* **11** no. 2, 1962).
Synpenicillium Costantin (1888) = Doratomyces fide Morton & Smith (*Mycol. Pap.* **86**, 1963).
Synphragmidium F. Strauss (1853) = Speira fide Saccardo (*Syll. fung.* **4**: 514, 1886).
Synsphaeria Bonord. (1851), Pezizomycotina. 1, Europe.
Synsphaeridium G. Playford (1981), Fossil Fungi. 1, Western Australia.
Synsporium Preuss (1849) = Stachybotrys fide Hughes (*CJB* **36**: 727, 1958).
Synsterigmatocystis Costantin (1888) nom. inval. = Gibellula fide Hawksworth *et al.* (*Dictionary of the Fungi* edn 8, 1995).
Syntexis Theiss. (1916) ? = Mollisia fide Petrak (*Annls mycol.* **26**: 399, 1928).
Synthetospora Morgan (1892) = Stephanoma fide Mattirolo (*Annals R. Accad. Agric. Torino* **79**: 190, 1936).
Syntholus A.W. Ramaley & M.E. Barr (1997), anamorphic *Syncarpella*, Cpd.?.?. 1, USA. See Ramaley & Barr (*Mycotaxon* **65**: 501, 1997).
syntype, see type.
synzoospore, multinucleate zoospore with many sets of flagella.
Syphosphaera Dumort. (1822), Pezizomycotina. 1, Europe.
Syrigosis Neck. ex Kremp. (1869) = Sphaerophorus

fide Hawksworth *et al.* (*Dictionary of the Fungi* edn 8, 1995).

Syringospora Quinq. (1868) nom. rej. = Candida fide van der Walt (*Mycopathologia* **40**: 231, 1970), von Arx *et al.* (*Stud. Mycol.* **14**: 1, 1977).

Syrropeltis Bat., J.L. Bezerra & Matta (1964), ? Dothideales. 1 (on *Xylopia*), Brazil. See Batista *et al.* (*Portugaliae Acta Biologica* Série B **7**: 376, 1964).

syrrotium (pl. **-ia**), term coined by Falk (1912) for the mycelial cord of *Merulius*. See Thompson (*in* Jennings & Rayner (Eds), *The ecology and physiology of the fungal mycelium*: 185, 1984).

Syspastospora P.F. Cannon & D. Hawksw. (1982), Ceratostomataceae. 2 (saprobic and/or fungicolous), widespread. See Horie *et al.* (*Mycotaxon* **25**: 229, 1986), García *et al.* (*Mycol.* **94**: 862, 2002).

Systematics. also **Biosystematics** (q.v.). The study of the relationships and classification of organisms and the processes by which they have evolved and are maintained (includes the subdisciplines of nomenclature and taxonomy), see Hawksworth (Ed.) (*Prospects in systematics*, 1988), Minelli (*Biological systematics: the state of the art*, 1993), Ross (*Biological systematics*, 1974), Stevens (*The development of biological systematics*, 1995). The word also has other uses outside biology. See also Classification, Nomenclature, Phylogenetic analysis.

systemic (1) (of a parasite), spreading throughout the host; (2) (of a fungicide), absorbed, esp. by the roots, and translocated to other parts of the plant.

Systenostrema E.I. Hazard & Oldacre (1975), Microsporidia. 3.

Systremma Theiss. & Syd. (1915), Dothideaceae. 1, widespread. See von Arx & Müller (*Stud. Mycol.* **9**, 1975), Gams (*Taxon* **54**: 520, 2005; nomencl.).

Systremmopsis Petr. (1923), anamorphic *Pezizomycotina*, St.0eH.?. 1, Europe. See Hoog & McGinnis (*Stud. Mycol.* **30**: 187, 1987).

Syzygites Ehrenb. (1818), Mucoraceae. 1 (on decaying *Boletaceae* (esp. *Boletus*), widespread (north temperate). See Hesseltine (*Lloydia* **20**: 228, 1957), Ekpo & Young (*Microbios* **10**: 63, 1979; ultrastr.), Kaplan & Goos (*Mycol.* **74**: 684, 1982; zygospore formation), Kovacs (*Syzygites megalocarpus (Mucoraceae, Mucorales, Zygomycetes) and its host-parasite association in the contiguous United States*, 1995), Weete *et al.* (*J. Am. Oil Chem. Soc.* **75**: 1367, 1998; fatty acids), Voigt & Wöstemeyer (*Gene* **270**: 113, 2001; phylogeny).

Syzygospora G.W. Martin (1937), Carcinomycetaceae. 16, widespread. If the type species is anamorphic, then the remaining species belong in *Carcinomyces* or *Christiansenia*. See Donk (*Taxon* **11**: 101, 1962; ? nom. anam.), Ginns (*Mycol.* **78**: 619, 1986; key), Diederich (*Biblthca Mycol.* **61**: 29., 1996), Hauerslev (*Mycotaxon* **72**: 465, 1999), Harmaja (*Memoranda Societatis pro Fauna et Flora Fennica* **79**: 73, 2003; Finland), Kotiranta & Miettinen (*Acta Mycologica* Warszawa **41**: 21, 2006; Finland n.sp.).

Syzygosporaceae Jülich (1982) = Carcinomycetaceae.

Szczawinskia A. Funk (1984), anamorphic *Pilocarpaceae*, St.0fH.15 (L). 1, N. America; Papua New Guinea. See Aptroot *et al.* (*Biblthca Lichenol.* **64**, 1997), Holien & Tønsberg (*Lichenologist* **34**: 369, 2002), Andersen & Ekman (*MR* **109**: 21, 2005).

T, see RIEC.

T-2 toxin, a mycotoxin of the trichothecene group (q.v.) produced by some *Fusarium* spp., causing alimentary toxic aleukia.

Tabanispora H.I. Bykova, Y.Y. Sokolova & I.V. Issi (1987), Microsporidia. 2.

Tabanomyces Couch, R.V. Andrejeva, Laird & Nolan (1979) = Meristacrum fide Tucker (*Mycotaxon* **13**: 481, 1981).

Tachaphantium Bref. (1888) = Platygloea fide Bandoni (*Mycol.* **48**: 821, 1956).

Taeniola Bonord. (1851) nom. illegit. = Hormiscium fide Saccardo (*Syll. fung.* **4**: 263, 1886), Hughes (*CJB* **36**: 727, 1958).

Taeniolella S. Hughes (1958), anamorphic *Glyphium*, Hso.≡ eP.3/4. *c.* 48, widespread. See Sutton (*TBMS* **54**: 255, 1970; teleomorphs), Ellis (*More Dematiaceous Hyphomycetes*, 1976), Hawksworth (*Bull. Br. Mus. nat. hist.* Bot. **6**: 183, 1979; on lichens), Zhurbenko (*Folia Cryptog. Estonica* **32**: 153, 1998), Gulis & Marvanová (*Mycotaxon* **72**: 237, 1999), Diederich & Zhurbenko (*Graphis Scripta* **12**: 37, 2001; nomencl.), Jones & Eaton (*Mycoscience* **43**: 201, 2002).

Taeniolina M.B. Ellis (1976), anamorphic *Pezizomycotina*, Hso.≡ eP.3. 1, Europe; Cuba. See Ellis (*More Dematiaceous Hyphomycetes*: 61, 1976).

Taeniophora P. Karst. (1886) = Phragmotrichum fide Sutton & Pirozynski (*TBMS* **48**: 349, 1965).

Taeniospora Marvanová (1977), anamorphic *Fibulomyces*, Hso.1bH.1. 2 (with clamp connexions), Czech Republic. See Marvanová & Stalpers (*TBMS* **89**: 489, 1987; key, teleomorphs).

Tainosphaeria F.A. Fernández & Huhndorf (2005), Chaetosphaeriaceae. 1, Puerto Rico. See Fernández & Huhndorf (*Fungal Diversity* **18**: 44, 2005), Fernández *et al.* (*Mycol.* **98**: 121, 2006; phylogeny).

Taiwanascaceae Sivan. & H.S. Chang (1997) = Niessliaceae.

Taiwanascus Sivan. & H.S. Chang (1997), Niessliaceae. 1 (from dead wood), Taiwan. See Samuels & Barr (*CJB* **75**: 2165, 1997), Sivanesan & Chang (*MR* **101**: 176, 1997).

Taiwanoporia T.T. Chang & W.N. Chou (2003), Agaricomycetidae. 1, Taiwan. See Chang & Chou (*Mycol.* **95**: 1215, 2003).

take-all, a cereal disease (*Gaeumannomyces graminis*).

Talaromyces C.R. Benj. (1955), Trichocomaceae. Anamorphs *Penicillium*, *Geosmithia*-like, *Merimbla*. 42, widespread. See Pitt (*The genus Penicillium and its teleomorphic states Eupenicillium and Talaromyces*, 1979; keys), Frisvad *et al.* (*Antonie van Leeuwenhoek* **57**: 179, 1990; chemotaxonomy), Prieto *et al.* (*MR* **99**: 69, 1995; polysaccharides), Yaguchi *et al.* (*Mycoscience* **37**: 55, 1996; ubiquinones), Ogawa *et al.* (*Mycol.* **89**: 756, 1997; DNA), Ogawa & Sugiyama (*Integration of Modern Taxonomic Methods for Penicillium and Aspergillus Classification*: 149, 2000; phylogeny), Pitt *et al.* (*Integration of Modern Taxonomic Methods for Penicillium and Aspergillus Classification*: 9, 2000; accepted names), Heredia *et al.* (*Mycol.* **93**: 528, 2001; Mexico), Yaguchi *et al.* (*Cryptog. Mycol.* **26**: 133, 2005), Geiser *et al.* (*Mycol.* **98**: 1053, 2006; phylogeny).

Talbotiomyces Vánky, R. Bauer & Begerow (2007), Entorrhizaceae. 1 (galls on roots of dicots), S. Africa. See Vánky *et al.* (*Mycol. Balcanica* **4**: 11, 2007).

Talekpea Lunghini & Rambelli (1979), anamorphic *Pezizomycotina*, Hso.1eP.19. 1, Ivory Coast. See Lunghini & Rambelli (*Micol. Ital.* **8**: 23, 1979).

Talpapellis Alstrup & M. Cole (1998), anamorphic

Pezizomycotina, Hso.?.?. 1, Canada. See Alstrup & Cole (*Bryologist* **10**: 227, 1998).

Tamnidium, see *Thamnidium Link*.

Tamsiniella S.W. Wong, K.D. Hyde, W.H. Ho & S.J. Stanley (1998), Dothideomycetes. 1, Australia. See Wong *et al.* (*CJB* **76**: 334, 1998).

Tandonea, see *Tandonia*.

Tandonella S.S. Prasad & R.A.B. Verma (1970) = Passalora fide Sutton & Pascoe (*Aust. J. Bot.* **35**: 183, 1987; on *Olearia*), Crous *et al.* (*Mycol.* **93**: 1081, 2001; phylogeny), Crous & Braun (*CBS Diversity Ser.* **1**: 571 pp., 2003).

Tandonia M.D. Mehrotra (1991), anamorphic *Pezizomycotina*, St.0eH.15. 1, India. See Mehrotra (*MR* **95**: 1074, 1991).

Tanglella Höhn. (1918), ? Helotiales. 1, Europe.

Tania Egea, Torrente & Sipman (1995), Arthoniaceae (L). 1, Malaysia. See Egea *et al.* (*Lichenologist* **27**: 352, 1995), Grube (*Bryologist* **101**: 377, 1998), Harada & Yamamoto (*Lichenology* **5**: 89, 2006).

tao-cho (**tao-si**), see hamanatto.

tapé, an Indonesian fermented food prepared by the action of *Rhizopus oryzae* and *Endomyces chodatii* on rice (Swan Djien Ko, *Appl. Microbiol.* **23**: 976, 1972); **- ketala** (peuyeum), a Javanese food obtained by the fermentation of cassava tubers with *Mucor javanicus* (Hedger, *Bull. BMS* **12**: 54, 1978). See also Fermented food and drinks.

Tapeinosporium Bonord. (1853) = Septocylindrium fide Saccardo (*Syll. fung.* **4**: 226, 1886).

Tapellaria Müll. Arg. (1890), ? Ectolechiaceae (L). 12, widespread (tropical). See Santesson (*Symb. bot. upsal.* **12** no. 1: 1, 1952), Lücking (*Phyton* Horn **39**: 131, 1999; Costa Rica), Lücking (*Willdenowia* **29**: 299, 1999; Ecuador).

Tapellariopsis Lücking (1999), Ectolechiaceae (L). 1, Costa Rica. See Lücking (*Phyton* Horn **104**: 148, 1992).

Tapesia (Pers.) Fuckel (1870) nom. rej., Helotiales. *c.* 25, widespread. Species associated with cereal disease are now separated as *Oculimacula*. Often synonymized with *Mollisia*. See Baral & Krieglsteiner (*Beih. Z. Mykol.* **6**, 1985; synonymy with *Mollisia*), Gamundí (*Fl. criptog. Tierra del Fuego* **10**: 126 pp., 1986; Argentina), Gminder (*Z. Mykol.* **62**: 181, 1996), Harrington & McNew (*Mycotaxon* **87**: 141, 2003), Gminder (*Czech Mycol.* **58**: 125, 2006).

Tapesina Lambotte (1887), Hyaloscyphaceae. Anamorph *Chalara*-like. 1, Europe. See Svrček (*Česká Mykol.* **41**: 193, 1987), Baral (*Z. Mykol.* **68**: 117, 2002).

Taphria Fr. (1821) = Taphrina fide Fries (*Syst. mycol.* **3**: 520, 1832).

Taphridium Lagerh. & Juel (1902), Protomycetaceae. 2, widespread. See Reddy (*Mycotaxon* **3**: 1, 1975; key).

Taphrina Fr. (1815), Taphrinaceae. Anamorph *Lalaria*. 95 (biotrophic on plants), widespread. *T. populina* (poplar leaf blister), *T. betulina* (witches' broom of birch, *T. bullata* (pear leaf blister), *T. cerasi* (witches' broom of cherry), *T. caerulascens* (oak leaf curl), *T. deformans* (peach leaf curl), *T. insititiae* (witches' broom of plum), *T. minor* (cherry leaf curl), *T. pruni* (pocket plums). See Mix (*Kansas Univ. Sci. Bull.* **33**: 1, 1949), Kramer (*Mycol.* **52**: 295, 1960; dev. and nuclear behaviour), Snider & Kramer (*Mycol.* **66**: 743, 754, 1974; 754, 1974; numerical taxonomy), Bacigálová (*Czech Mycol.* **50**: 107, 1997; spp. on *Betula*), Sjamsuridzal *et al.* (*Mycoscience* **38**: 267, 1997; phylogeny), Sugiyama (*Mycoscience* **39**: 487, 1998; phylogeny), Schweigkofler *et al.* (*Organ. Divers. Evol.* **2**: 1, 2002; phylogeny), Bacigálová *et al.* (*Mycol. Progr.* **2**: 179, 2003; on *Alnus*), Karatygin (*Mikol. Fitopatol.* **37**: 26, 2003; key), Rodrigues & Fonseca (*Int. J. Syst. Evol. Microbiol.* **53**: 607, 2003; phylogeny), Inácio *et al.* (*FEMS Yeast Res.* **4**: 541, 2004; anamorphs, Portugal), Tavares *et al.* (*Eur. J. Pl. Path.* **110**: 973, 2004; molecular detection), Sugiyama *et al.* (*Mycol.* **98**: 996, 2006; phylogeny), Hansen *et al.* (*MR* **111**: 592, 2007; on *Nothofagus*).

Taphrinaceae Gäum. (1928) nom. cons., Taphrinales. 2 gen. (+ 7 syn.), 118 spp.

Lit.: Kramer (*Stud. Mycol.* **30**: 151, 1987), Nishida & Sugiyama (*Mol. Biol. Evol.* **10**: 431, 1993), Bacigálová (*Czech Mycol.* **50**: 107, 1997), Sjamsuridzal *et al.* (*Mycoscience* **38**: 267, 1997), Moore *in* Kurtzman & Fell (Eds) (*Yeasts, a taxonomic study* 4th edn: 582, 1998), Sugiyama (*Mycoscience* **39**: 487, 1998), Rodrigues & Fonseca (*Int. J. Syst. Evol. Microbiol.* **53**: 607, 2003), Inácio *et al.* (*FEMS Yeast Res.* **4**: 541, 2004), Sugiyama *et al.* (*Mycol.* **98**: 996, 2006; phylogeny).

Taphrinales Gäum. & C.W. Dodge (1928). Taphrinomycetidae. 2 fam., 8 gen., 140 spp. Mycelium subcuticular or subepidermal, composed of dikaryotic ascogenous cells, sometimes forming thick-walled smooth or ornamented resting spores, vegetative tissue ± lacking; ascomata absent; interascal tissue absent; asci formed either directly from ascogenous cells or with a separating stalk cell, forming internally or in a palisade on the surface of the host tissue, ± cylindrical, the end often truncate, ± persistent, usually 8-spored, the ascospores discharged simultaneously; ascospores hyaline, aseptate, globose or ellipsoidal, sometimes discharged in a single mass. Anamorph yeast-like, monokaryotic, formed from budding ascospores. Biotrophic on plants, usually causing hyperplasia (galls, witches' brooms) or lesions. Fams:

(1) **Protomycetaceae**
(2) **Taphrinaceae**

Cell walls are two-layered and conidiogenesis (ascospore budding) is basidiomycete-like. Has similarities with the red yeasts, *Exobasidiales* and *Ustilaginales*, but are now accepted as an ancestral lineage of the *Ascomycota*.

Lit.: Kramer (*in* Ainsworth *et al.* (Eds), *The Fungi* **4A**: 33, 1973; keys gen., *Stud. Mycol.* **30**: 151, 1987), von Arx *et al.* (1982; relationships, ultrastr., wall chemistry), Sugiyama (*Mycoscience* **39**: 487, 1998; phylogeny), Sugiyama *et al.* (*Mycol.* **98**: 996, 2006; phylogeny), Sjamsuridzal *et al.* (*Mycoscience* **38**: 267, 1997; phylogeny), Swann *et al.* (*Mycol.* **91**: 51, 1999; contrast with *Basidiomycota*).

Taphrinomycetes O.E. Erikss. & Winka (1997). Taphrinomycotina. 1 ord., 2 fam., 8 gen., 140 spp. Subcl.:

Taphrinomycetidae

For *Lit.* see ord. and fam.

Taphrinomycetidae, see *Taphrinomycetes*.

Taphrinomycotina O.E. Erikss. & Winka (1997), Taphrinomycetes. Ord.:

Taphrinales

For *Lit.* see fam.

Taphrophila Scheuer (1988), Tubeufiaceae. Anamorph *Mirandina*. 2, Europe. See Scheuer (*MR* **95**: 811, 1991; key), Crane *et al.* (*CJB* **76**: 602, 1998), Réblová & Barr (*Sydowia* **52**: 286, 2000), Réblová & Barr (*Sydowia* **52**: 258, 2000), Kodsueb *et al.* (*Fungal Diversity* **21**: 105, 2006).

Tapinella E.-J. Gilbert (1931), Tapinellaceae. 3. See Watling *et al.* (*British Fungus Flora. Agarics and Boleti* Rev. & Enl. Edn **1**: 173 pp., 2005; UK).

Tapinellaceae C. Hahn (1999), Agaricales. 2 gen. (+ 1 syn.), 4 spp.
Lit.: Hahn (*Sendtnera* **6**: 115, 1999).

Tapinia (Fr.) P. Karst. (1879) [non *Tapinia* Steud. 1841, *Anacardiaceae*] ≡ Tapinella.

tapuy, a rice wine indigenous to the Philippines; similar to saki.

Tarbertia Dennis (1974), Arthoniales. 1, British Isles. See Dennis (*Kew Bull.* **29**: 176, 1974).

Tardivesicula J.I.R. Larsson & E.K.C. Bylén (1992), Microsporidia. 1. See Larsson & Bylén (*Eur. J. Protist.* **28**: 25, 1992).

Tarichium Cohn (1875), Entomophthoraceae. *c.* 30, widespread. See MacLeod & Müller-Kögler (*Mycol.* **62**: 33, 1970; key), Humber (*Mycotaxon* **34**: 441, 1989), Keller (*Sydowia* **58**: 38, 2006; Switzerland).

Tarsodisporus Bat. & A.A. Silva (1965), anamorphic *Pezizomycotina*, Cpt.0eP.?. 1, Brazil. See Batista & Silva (*Atas Inst. Micol. Univ. Pernambuco* **2**: 249, 1965).

tartareous, having a thick rough crumbling surface.

Tartufa (Gray) Kuntze (1891) ≡ Choiromyces.

Tarzetta (Cooke) Lambotte (1888), Pyronemataceae. 9, widespread (north temperate). See Dumont & Korf (*Mycol.* **63**: 165, 1971; nomencl.), Rogers *et al.* (*Mycol.* **63**: 1084, 1971), Harmaja (*Karstenia* **14**: 138, 1974; gen. concept), Lazzari (*Micol. ital.* **2**: 20, 1984; nomencl.), Senn-Irlet (*Beitr. Kenntn. Pilze Mitteleur.* **5**: 191, 1989; key 6 spp.), Landvik *et al.* (*Nordic Jl Bot.* **17**: 403, 1997; DNA), Medardi (*Boll. Gruppo Micol. 'G. Bresadola'* **42**: 7, 1999; Italy), Wu & Kimbrough (*Int. J. Pl. Sci.* **162**: 1075, 2001; ontogeny), Yao & Spooner (*MR* **106**: 1243, 2002; UK (as *Tazzetta*)), Hansen & Pfister (*Mycol.* **98**: 1029, 2006; phylogeny), Liu & Zhuang (*Mycosystema* **25**: 546, 2006; phylogeny), Perry *et al.* (*MR* **111**: 549, 2007; phylogeny).

Tasmidella Kantvilas, Hafellner & Elix (1999), Megalariaceae (L). 1, Tasmania. See Kantvilas *et al.* (*Lichenologist* **31**: 213, 1999).

Tassia Syd. & P. Syd. (1919), anamorphic *Pezizomycotina*, Cpt.0eH.?. 2, Brazil; Europe.

Taste. Five basic tastes are now generally recognized: bitter, salty, sour, sweet and savoury. Of these, savoury (also known by the Japanese word umami) is most important in mycology. This is the taste contributed by glutamates and characteristic of fermented foods, many of which are produced using fungi (see Fermented food and drinks). Tastes in fungi combine with smells (q.v.) to produce flavours which are significant in edible fungi, and as an aid to identification. Fresh basidiomata of some agarics have characteristic flavours (e.g. acrid, or apricot or garlic, peppery, etc.), often described as tastes, which are sometimes used to aid identification; see Gilbert (*Méthode de mycologie descriptive*, 1934), Josserand (1952: 68), Locquin (*Petite flore des champignons de France*, 1956: 100). Tasting unknown fungi should be practised with caution.

Tatraea Svrček (1993), Helotiaceae. 1, Europe. See Baral *et al.* (*Öst. Z. Pilzk.* **8**: 71, 1999).

Tauromyces Cavalc. & A.A. Silva (1972) = Gyalectidium fide Lücking *et al.* (*Lichenologist* **30**: 121, 1998).

Tausonia Babeva (1998), anamorphic *Cystofilobasidiaceae*. 1, Tadzhikistan. See Bab'eva (*Mikrobiologiya* **67**: 231, 1998).

Tavaresiella T. Majewski (1980), ? Laboulbeniaceae. 4, widespread. See Benjamin (*Aliso* **13**: 559, 1993).

Tawdiella K.B. Deshp. & K.S. Deshp. (1966), anamorphic *Pezizomycotina*, Hso.#eP.?. 1, India. See Deshpande & Deshpande (*Mycopathologia* **28**: 207, 1966).

tawny grisette, basidioma of the edible *Amanita fulva*.

taxis (frequently a suffix), a movement of a plasmodium or zoospore as a reaction to a one-sided stimulus; + (positive) when the movement is in the direction of the stimulus, – (negative) when away from the stimulus. The following tactic movements of zoospores have been described; (1) **chemotaxis**, in response to root exudates (*Pythium* spp.) and amino acids (*Allomyces* spp.); (2) **gravitaxis** (*Phytophthora palmivora*); (3) **electrotaxis** (*Phytophthora palmivora*; Morris *et al.*, *Plant Cell & Environment* **15**: 645, 1992).

taxol, an antitumor diterpenoid used in treatment of some cancers originally obtained from bark of Pacific yew (*Taxus brevifolia*), but also produced by the endophytic fungus *Taxomyces andreanae* (Stierle *et al.*, *Science* **260**: 214, 1993).

Taxomyces Strobel, A. Stierle, D. Stierle & W.M. Hess (1993), anamorphic *Pezizomycotina*, Hso.1bH.1. 1 (endophytic), USA. See Strobel *et al.* (*Mycotaxon* **47**: 73, 1993).

taxon (pl. **taxa**, **taxons**), a taxonomic group of any rank (Code, Art. 1). See Lam (*Taxon* **6**: 213, 1957; history and usage). See Classification.

taxonomy, the science of classification, in biology the arrangment of organisms into a classification; **idio-**, of organisms; **syn-**, of communities. See Classification, Nomenclature, systematics.

Tazzetta, see *Tarzetta*.

TDP, thermal death point; generally used for a 10 min. application of heat.

tea fungus, a symbiotic association of yeasts (*Saccharomycodes ludwigii*) and bacteria (esp. *Acetobacter xylinum*). See Stadelmann (*Sydowia* **11**: 380, 1958; bibliogr., *Zbl. Bakt.* Abt. I **180**: 401, 1961), Kappel & Anken (*The Mycologist* **7**: 12, 1993; analysis); cf. teekwass, tibi, ginger beer plant.

Teberdinia Sogonov, W. Gams, Summerbell & Schroers (2005), anamorphic *Pseudeurotium*, H?.?.?. 1, Karacheyevo-Cherkessiya. See Sogonov *et al.* (*Mycol.* **97**: 698, 2005).

Tectacervulus A.W. Ramaley (1992), anamorphic *Pezizomycotina*, Cac.0eH.15. 1, USA. See Ramaley (*Mycotaxon* **43**: 438, 1992).

Tectella Earle (1909), ? Mycenaceae. 3, widespread (north temperate). See Cavet (*Bull. Trimestr. Féd. Mycol. Dauphiné-Savoie* **34**: 32, 1994), Jin *et al.* (*Mycotaxon* **79**: 7, 2001; phylogeny).

Tectimyces L.G. Valle & Santam. (2002), Legeriomycetaceae. 2 (in *Habroleptoides*), Spain. See Valle & Santamaria (*MR* **106**: 842, 2002).

teekwass, a Russian drink obtained by fermenting tea with a symbiotic mixture of *Acetobacter xylinum* and *Schizosaccharomyces pombe*. Cf. ginger beer plant, tea fungus, tibi.

Tegillum Mains (1940) = Olivea fide Ono & Hennen (*TMSJ* **24**: 369, 1978).

Tegoa Bat. & Fonseca (1961) nom. inval.; nom. dub., anamorphic *Pezizomycotina*. See Lücking *et al.* (*Lichenologist* **30**: 121, 1998).

Teichospora Fuckel (1870), ? Dacampiaceae. *c.* 23 (on wood), widespread. See Yuan & Barr (*Mycotaxon* **54**: 111, 1995), Barr (*Mycotaxon* **82**: 373, 2002).

Teichosporaceae M.E. Barr (2002) ? = Dacampiaceae. Described to contain non-lichenicolous taxa of the *Dacampiaceae*, but is almost certainly polyphyletic in its original circumscription. Molecular data are lacking.

Lit.: Barr (*Mycotaxon* **82**: 373, 2002).

Teichosporella (Sacc.) Sacc. (1895), Dothideomycetes. 1 or 2, widespread (temperate). See Kutorga & Hawksworth (*SA* **15**: 1, 1997).

Teichosporina (G. Arnaud) Cif. & Bat. (1962) = Limacinula Höhn. fide von Arx & Müller (*Stud. Mycol.* **9**, 1975).

Telamonia (Fr.) Wünsche (1877) nom. cons. = Cortinarius fide Kauffman (*N. Amer. Fl.* **10**, 1932).

teleblem (teleoblema), see universal veil.

Telebolus Lindau (1905) = Thelebolus fide Hawksworth *et al.* (*Dictionary of the Fungi* edn 8, 1995).

Telemeniella Bat. (1955) = Phyllachora Nitschke ex Fuckel (1870) fide Hyde & Cannon (*Mycol. Pap.* **175**, 1999).

teleomorph, see States of fungi.

teleutosorus, see telium.

Teleutospora Arthur & Bisby (1921) = Uromyces fide Arthur (*Manual Rusts US & Canada*, 1934).

teleutospore, see teliospore.

Teleutosporites Mesch. (1902), Fossil Fungi ? Pucciniales. 4 (Carboniferous, Tertiary), France; Malaysia. But see Pirozynski (*Biol. Mem.* **1**: 104, 1976).

teleutosporodesm, Donk's (*Proc. Kon. Nederl. Akad. Wetensch.* C **75**: 385, 1972) term for teliospore (obsol.).

Telimena Racib. (1900), Phyllachoraceae. 11 (from living leaves), widespread. See Müller (*Sydowia* **27**: 74, 1975), Sivanesan (*TBMS* **88**: 473, 1987).

Telimenella Petr. (1940), Phyllachoraceae. 2 (from living leaves), widespread (temperate). See Barr (*Mycol.* **69**: 952, 1977), Sivanesan (*TBMS* **88**: 473, 1987).

Telimeniella, see *Telemeniella*.

Telimenochora Sivan. (1987), Phyllachoraceae. 1 (from living leaves), Mexico; Puerto Rico. See Sivanesan (*TBMS* **88**: 473, 1987).

Telimenopsis Petr. (1950) = Telimena fide Sivanesan (*TBMS* **88**: 473, 1987).

Telioclipeum Viégas (1962), Pezizomycotina. 1 (on *Aspidiosperma*), Brazil. See Viégas (*Bragantia* **21**: 260, 1962).

Teliomycetes, see *Pucciniomycetes*.

teliospore (teleutospore; teleutosporidesm, Donk (1973)), the spore (commonly a winter or resting spore) of the *Pucciniales* (or *Ustilaginales*) from which the basidium is produced (Fig. 8H).

Teliosporeae. A subclass (Bessey, 1935) for *Pucciniomycetes* and *Ustilaginomycetes*.

telium, a sorus producing teliospores.

Telligia Hendr. (1948), anamorphic *Pezizomycotina*, Hso.#eP.?. 1, Congo.

Teloconia Syd. (1921) = Phragmidium fide Dietel (*Nat. Pflanzenfam.* **6**, 1928).

Telogalla Nik. Hoffm. & Hafellner (2000), Verrucariaceae. 1 (lichenicolous), Europe. See Hoffmann & Hafellner (*Biblthca Lichenol.* **77**: 1, 2000).

Telomapea G.F. Laundon (1967) = Maravalia fide Cummins & Hiratsuka (*Illustr. Gen. Rust Fungi rev. edit.*, 1983).

Telomyxa L. Léger & E. Hesse (1910), Microsporidia. 1.

Teloschistaceae Zahlbr. (1898), Teloschistales (L). 12 gen. (+ 41 syn.), 644 spp.

Lit.: Filson (*Muelleria* **2**: 65, 1969; Australia), Santesson (*Vortr. bot. Ges.* n.f. **4**: 5, 1970; chemotaxonomy), Honegger (*Lichenologist* **10**: 47, 1978; asci), Poelt & Hafellner (*Mitt. bot. StSamml., München* **16**: 503, 1980; key gen.), Bellemère & Letrouit-Galinou (*Cryptog. Bryol.-Lichénol.* **3**: 95, 1982; asci), Awasthi (*Proc. Indian Acad. Sci.* Pl. Sci. **96**: 227, 1986), Bellemère *et al.* (*Cryptog. Bryol.-Lichénol.* **7**: 189, 1986), Kärnefelt (*Monogr. Syst. Bot.* Miss. Bot. Gdn **25**: 439, 1988), Almborn (*Nordic Jl Bot.* **8**: 521, 1989), Kärnefelt (*Cryptog. bot.* **1**: 147, 1989), Kärnefelt (*Lichenologist* **22**: 307, 1990), Kärnefelt *in* Galloway (Ed.) (*Tropical lichens*: 105, 1991), Poelt & Hinteregger (*Biblthca Lichenol.* **50**, 1993; Himalayas), Poelt & Hinteregger (*Biblthca Lichenol.* **50**: 247 pp., 1993), Arup (*Bryologist* **98**: 129, 1995), Castello (*Cryptog. Bryol.-Lichénol.* **16**: 79, 1995), Kondratyuk & Kärnefelt (*Lichenologist* **29**: 425, 1997), Lindblom (*J. Hattori bot. Lab.* **83**: 75, 1997), Franc & Kärnefelt (*Graphis Scripta* **9**: 49, 1998), Kärnefelt (*Cryptog. Bryol.-Lichénol.* **19**: 93, 1998), Lumbsch (*Chemical Fungal Taxonomy*: 345, 1998), Stenroos & DePriest (*Am. J. Bot.* **85**: 1548, 1998; phylogeny), Westberg & Kärnefelt (*Lichenologist* **30**: 515, 1998), Wetmore & Kärnefelt (*Bryologist* **10**: 230, 1998), Gaya *et al.* (*Am. J. Bot.* **90**: 1095, 2003), Søchting & Lutzoni (*MR* **107**: 1266, 2003), Honegger *et al.* (*MR* **108**: 480, 2004), Lumbsch *et al.* (*Mol. Phylogen. Evol.* **31**: 822, 2004), Miądlikowska *et al.* (*Mycol.* **98**: 1088, 2006; phylogeny), Hofstetter *et al.* (*Mol. Phylogen. Evol.* **44**: 412, 2007; phylogeny), Gaya *et al.* (*MR* **112**: 528, 2008; phylogeny).

Teloschistales D. Hawksw. & O.E. Erikss. (1986), (L). Lecanoromycetidae. 5 fam., 66 gen., 1954 spp. Thallus varied, usually foliose or fruticose, often brightly coloured. Ascomata apothecial, usually strongly concave, usually orange or reddish, usually with a well-developed thalline margin, often with anthraquinone pigments (K+ crimson). Interascal tissue of usually unbranched paraphyses, the apices often capitate. Asci cylindrical, persistent, the apex strongly thickened, without separable wall layers, with a conspicuous outer J+ layer, sometimes also with an internal J+ structure, releasing ascospores through an apical split. Ascospores mostly hyaline, often with very thick septa, reduced lumina, and polarilocular. Anamorphs pycnidial. Lichenized with green algae, cosmop., esp. on nutrient-rich or basic substrata.

Pycnidia with numerous small locules lined by ± doliiform conidiogenous cells support the separation from *Lecanorales* as well as characters of the ascomata and chemistry. Fams:

(1) **Caliciaceae** (syn. *Buelliaceae*, *Pyxinaceae*)
(2) **Letrouitiaceae**
(3) **Megalosporaceae**
(4) **Physciaceae**
(5) **Teloschistaceae** (syn. *Caloplacaceae*)

Two suborders have been suggested, the *Teloschistineae* which corresponds to the traditionally circumscribed order, and the *Physciineae* which contains the *Caliciaceae* and *Physciaceae*. *Lit.*: Kärnefelt

(*Crypt. Bot.* **1**: 147, 1989; gen. names, phylogeny), Miądlikowska *et al* (*Mycol.* **98**: 1088, 2006).

Teloschistes Norman (1853), Teloschistaceae (L). *c.* 33, widespread. See Thomson (*Am. midl. Nat.* **41**: 706, 1949; N. Am.), Filson (*Muelleria* **2**: 65, 1969; Australia), Almborn (*Nordic Jl Bot.* **8**: 521, 1989; key 9 spp., C. & S. Afr.), Stenroos & DePriest (*Am. J. Bot.* **85**: 1548, 1998; phylogeny), Martín *et al.* (*Fungal Genetics Biol.* **40**: 252, 2003; introns), Søchting & Lutzoni (*MR* **107**: 1266, 2003; phylogeny), Frödén & Lassen (*Lichenologist* **36**: 289, 2004; generic limits), Frödén & Kärnefelt (*Biblthca Lichenol.* **95**: 183, 2007; Africa), Gaya *et al.* (*MR* **112**: 528, 2008; phylogeny).

Teloschistomyces Cif. & E.A. Thomas (1953) = Teloschistes.

Telospora Arthur (1906) = Uromyces fide Arthur (*Manual Rusts US & Canada*, 1934).

TEM, transmission electron microscope. See ultrastructure.

Temerariomyces B. Sutton (1993), anamorphic *Pezizomycotina*, Hso.0eP.?3. 1, Malawi. See Sutton (*Mycol. Pap.* **167**: 62, 1993).

Temnospora A. Massal. (1860) = Chrysothrix fide Laundon (*Lichenologist* **13**: 101, 1981).

tempeh (tempé), an oriental food composed of *Rhizopus oligosporus* fermented soybeans (Hesseltine, *Mycol.* **57**: 154, 1965).

tempeh-bongrek, an oriental food composed of *Rhizopus* and manioc prepared in Malaysia.

tenacle, see haerangium.

Teng (Shu-chun; 1902-1970; China). Student, Cornell University, New York, USA (1923-1928); Lecturer, Lingnan University (1928-1933); Lecturer, Jinling University and Nanking University (1930s); surveying to establish forest stations around the upper Yellow River to control erosion (early 1940s); member of staff, Research Institute of Biology, Academia Sinica (late 1940s). Member, Academia Sinica (1948); establishing Agricultural Colleges, northeast China (1950-1955); Member, Chinese Academy of Sciences (1955); Director, Mycology Section, Microbiology Institute (*c.* 1955 onwards); a victim of the 'cultural revolution', denounced as a 'counter-revolutionary academic authority', imprisoned and tortured (1966-1970). Produced the first major treatment of the fungi of China available in English; he studied Chinese native lore on fungi but his research, which pre-dated ethnomycology (q.v.) elsewhere, including the manuscript of a book of 400,000 words and 600 colour paintings was confiscated during the cultural revolution and, having not re-appeared when he was 'rehabilitated' (1978), must be presumed destroyed. *Publs. Chung-kuo Ti Chen-chun* [*Fungi of China*] (1963) [Chinese translation; original English version, Mycotaxon Ltd, 1996]. *Biogs, obits etc.* Zhuang Deng (Foreword, in *Fungi of China* ix-xiii, 1996) [biography written by Teng's daughter].

Tengiomyces Réblová (1999), Helminthosphaeriaceae. Anamorph *Spadicoides*. 1 (from wood), India; China. See Réblová (*Mycotaxon* **70**: 387, 1999), Réblová (*Sydowia* **51**: 223, 1999).

Tenorea Tornab. (1848) [non *Tenorea* Rafin. 1814, *Rutaceae*, etc.] = Teloschistes fide Hawksworth *et al.* (*Dictionary of the Fungi* edn 8, 1995).

tentoxin, a chlorosis-inducing cyclic tetrapeptide toxin from *Alternaria alternata* (Saad *et al.*, *Phytopath.* **60**: 415, 1970).

Tenuicutites C.E. Bertrand (1898), Fossil Fungi ? Chytridiomycetes. 1 (Carboniferous), France.

teonanácatl, see hallucinogenic fungi.

Tephrocybe Donk (1962), Lyophyllaceae. *c.* 40, widespread (esp. temperate). See Orton (*Bull. BMS* **18**: 114, 1984; key Brit. spp.), Grilli (*Docums Mycol.* **24** no. 95: 27, 1994; Italy), Basso & Candusso (*Bollettino dell'Associazione Micologica ed Ecologica Romana* **16**: 45, 2000; pileipellis struct. *Tephrocybe rancida*).

Tephromela M. Choisy (1929), Tephromelataceae (L). *c.* 20, widespread. See Hertel & Rambold (*Bot. Jahrb. Syst.* **107**: 469, 1985), Poelt & Grube (*Nova Hedwigia* **57**: 1, 1993; key 5 spp. Himalaya), Rambold (*Sendtnera* **1**: 281, 1993), Haugan & Timdal (*Graphis Scripta* **6**: 17, 1994), Ekman & Wedin (*Pl. Biol.* **2**: 350, 2000; phylogeny), Elix & Kalb (*Australasian Lichenology* **58**: 27, 2006; Australia), Miądlikowska *et al.* (*Mycol.* **98**: 1088, 2006; phylogeny), Arup *et al.* (*Mycol.* **99**: 42, 2007; sister group relations with *Parmeliaceae*).

Tephromelataceae Hafellner (1984), Lecanorales. 3 gen., 26 spp. See Miądlikowska *et al.* (*Mycol.* **98**: 1088, 2006; phylogeny), Arup *et al.* (*Mycol.* **99**: 42, 2007; sister group relations with *Parmeliaceae*).

Tephrophana Earle (1909) = Marasmius fide Singer (*Agaric. mod. Tax.*, 1951) but used by, Métrod (*BSMF* **75**: 184, 1959).

Tephrosticta (Sacc., Syd. & P. Syd.) Syd. & P. Syd. (1913) = Phaeosaccardinula fide von Arx & Müller (*Stud. Mycol.* **9**, 1975).

tequila, see pulque.

Teracosphaeria Réblová & Seifert (2007), Sordariomycetidae. Anamorph *Phialophora*-like. 1, New Zealand. See Réblová & Seifert (*MR* **111**: 287, 2007).

Terana Adans. (1763), Phanerochaetaceae. 1.

teratogenic, causing abnormalities in fetus growth.

Teratomyces Thaxt. (1893), Laboulbeniaceae. 9, widespread. See Benjamin (*Aliso* **10**: 345, 1983), Tavares (*Mycol. Mem.* **9**: 627 pp., 1985).

Teratonema Syd. & P. Syd. (1917) = Nitschkia fide Nannfeldt (*Svensk bot. Tidskr.* **66**: 49, 1975).

Teratoschaeta Bat. & O.M. Fonseca (1967), ? Dothideomycetes. Anamorph ? *Ampullifera*. 1 (on lichens), Brazil. See Hawksworth (*Bull. Br. Mus. nat. hist.* Bot. **6**: 183, 1979; anamorph), Matzer (*Mycol. Pap.* **171**: 202 pp., 1996).

Teratosperma Syd. & P. Syd. (1909), anamorphic *Pezizomycotina*, Hso.≡ eP.1. 11, widespread. See Ellis (*Mycol. Pap.* **69**, 1957; key), Hughes (*N.Z. Jl Bot.* **17**: 139, 1979), Wu & Zhuang (*Fungal Diversity Res. Ser.* **15**, 2005; China).

Teratosphaeria Syd. & P. Syd. (1912), Teratosphaeriaceae. Anamorph *Trimmatostroma*-like. 37, southern hemisphere. See Müller & Oehrens (*Sydowia* **35**: 138, 1982), Taylor & Crous (*MR* **104**: 618, 2000), Taylor *et al.* (*MR* **107**: 653, 2003), Crous *et al.* (*Stud. Mycol.* **58**: 1, 2007; phylogeny).

Teratosphaeriaceae Crous & U. Braun (2007), Capnodiales. 10 gen., 67 spp.

Lit.: Crous *et al.* (*Stud. Mycol.* **58**: 1, 2007; phylogeny).

teratum (pl. **-ta**), an abnormal modification; for terata in lichens see Grummann (*Feddes Repert., Beih.* **122**, 1941). See also gall.

terebrate, having scattered perforations.

terebrator (of lichens), a trichogyne (Lindau).

Terenodon Maas Geest. (1971), Gomphaceae. 1, Japan. See Maas Geesteranus (*Verh. K. ned. Akad. Wet.* tweede sect. **60** no. 3: 45, 1971).

terete, cylindrical; frequently circular in section but narrowing to one end.

Teretispora E.G. Simmons (2007), Pleosporaceae. 1, Europe. See Simmons (*Alternaria: an Identification Manual*, 2007).

terfas, see *Terfezia*.

Terfezia (Tul. & C. Tul.) Tul. & C. Tul. (1851), Pezizaceae. 12 (hypogeous, esp. under *Helianthemum*, *Cistus*, etc.), widespread (arid regions). The ascocarps (terfas, kamés) are edible. Nests within *Pezizaceae* in phylogenies, but more research is needed. See Janex-Favre & Parguey-Leduc (*Cryptog. Mycol.* **6**: 87, 1985; asci, ascospores), Weete *et al.* (*Can. J. Microbiol.* **31**: 1127, 1985; chemistry), Bokhary (*Arab Gulf Journal of Scientific Research* B (Agricultural and Biological Sciences) **5**: 245, 1987; Saudi Arabia), Zhang (*Micologia e Vegetazione Mediterranea* **7**: 39, 1992; China), Percudani *et al.* (*Mol. Phylogen. Evol.* **13**: 169, 1999; phylogeny), Kovács *et al.* (*Folia Microbiol.* Praha **46**: 423, 2001; molecular variation), Díez *et al.* (*Mycol.* **94**: 247, 2002; phylogeny), Aviram *et al.* (*Antonie van Leeuwenhoek* **85**: 169, 2004; multiple ITS seqences), Hansen & Pfister (*Mycol.* **98**: 1029, 2006; phylogeny).

Terfeziaceae E. Fisch. (1897) = Pezizaceae.

Lit.: Trappe (*Mycotaxon* **9**: 247, 1979), Abdullah *et al.* (*Int. J. Mycol. Lichenol.* **4**: 9, 1989), Trappe (*Mem. N. Y. bot. Gdn* **49**: 336, 1989), Zhang (*SA* **11**: 31, 1992), Alvarez *et al.* (*Mycol.* **84**: 926, 1992), Khabar *et al.* (*Cryptog. Mycol.* **15**: 187, 1994), Taylor *et al.* (*MR* **99**: 874, 1995), Gutierrez *et al.* (*Cah. Opt. Méditerr.* **20**: 139, 1996), O'Donnell *et al.* (*Mycol.* **89**: 48, 1997; phylogeny), O'Donnell *et al.* (*Mycol.* **89**: 48, 1997), Martin *et al.* (*Mycorrhiza Manual* Springer Lab Manual: 463, 1998), Norman & Eggers (*Mycol.* **91**: 820, 1999; phylogeny), Percudani *et al.* (*Mol. Phylogen. Evol.* **13**: 169, 1999; phylogeny), Kovács *et al.* (*Folia Microbiol.* Praha **46**: 423, 2001), Díez *et al.* (*Mycol.* **94**: 247, 2002), Ferdman *et al.* (*MR* **109**: 237, 2005).

Terfeziopsis Harkn. (1899) = Tuber fide Fischer (*Nat. Pflanzenfam.* **5b**: viii, 1938), Læssøe & Hansen (*MR* **111**: 1075, 2007; phylogeny).

terminus (**phialo**) **spore**, a phialospore of a 1-spored phialide, i.e. one terminating the growth of the phialide (Mason, 1933).

Termitaria Thaxt. (1920), anamorphic *Kathistaceae*, St.0eH.15. 2 or 3, America; Sardinia. See Khan & Kimbrough (*Am. J. Bot.* **61**: 395, 1974), Blackwell *et al.* (*Mycol.* **95**: 987, 2003; phylogeny), Ensaf *et al.* (*Cryptog. Mycol.* **27**: 219, 2006).

Termitariaceae Cif. (1963) = Kathistaceae.

Termitariopsis M. Blackw., Samson & Kimbr. (1980), anamorphic *Kathistaceae*, Hsp.≡ eH.?. 1 (on termites), USA. See Blackwell *et al.* (*Mycotaxon* **12**: 98, 1980), Blackwell *et al.* (*Mycol.* **95**: 987, 2003).

Termite fungi. Species of *Termitomyces* coexist in a mutualistic relationship with termite ants. Traditionally this has been described in terms of the termite cultivating fungi in their nests as food (Heim, *Termites et champignons*, 1977), but it may be more accurate to say that the fungus modifies cellulosic materials gathered by the ants making them more digestible for the ants; this has been described as an extracorporeal digestive system to which the ants have 'outsourced' cellulose digestion; the alternative view that the fungus is employing termites to provide it with a suitable environment and eliminate competitors has also been put forward (Turner, *Natural History* **111**: 62, 2002). In Africa, fruitbodies of *Termitomyces* are prized as food (see Edible fungi). Termites are also attacked by fungi, see Blackwell & Kimbrough (*Mycol.* **70**: 1279, 1979; key gen. termite-infesting fungi). See also: Ochiel *et al.* (*Mycologist* **11**: 7, 1997; ecology, pathogens), Suh *et al.* (*Mycol.* **90**: 611, 1998; pathogens, taxonomy), Thomas (*TBMS* **84**: 519, 1985; isolation medium); Van der Westhuizen & Eicker (*MR* **94**: 923, 1991; key S. Afr. spp.), Zobel & Grace (*Mycol.* **82**: 289, 1990; on *Reticulitermes*).

Termiticola E. Horak (1979), Agaricaceae. 1 (on termite nests), Papua New Guinea. See Horak (*Beih. Sydowia* **8**: 207, 1979), Vellinga (*MR* **108**: 354, 2004; phylogeny; clades with *Leucocoprinus*).

Termitomyces R. Heim (1942), Lyophyllaceae. Anamorph *Termitosphaera*. *c.* 30 (in nests of *Macrotermitinae*), Africa; S.E. Asia. See Heim (*Termites et champignons*, 1977), Pegler (*Kew Bull.* Addit. Ser. **6**, 1977; key), Jing & Ma (*Acta Mycol. Sin.* **4**: 103, 1985; key Chinese spp.), Aanen *et al.* (*PNAS* **99**: 1487, 2002; coevolution with termites), Rouland-Lefevre *et al.* (*Mol. Phylog. Evol.* **22**: 423, 2002; phylogeny), Frøslev *et al.* (*MR* **107**: 1277, 2003; phylogeny), Mossebo *et al.* (*Bull. Soc. Mycol. Fr.* **118**: 195, 2003; Cameroon), Wei *et al.* (*MR* **108**: 108, 2004; key), Tang *et al.* (*Mycotaxon* **94**: 93, 2005; revision Indian spp.), Tang *et al.* (*Mycotaxon* **95**: 285, 2006; revision Chinese spp.), Wei *et al.* (*Fungal Diversity* **21**: 225, 2006).

Termitosphaera Cif. (1935), anamorphic *Termitomyces*. 1, widespread (tropical).

Terramyces Letcher (2006), Terramycetaceae. 1. See Letcher *et al.* (*MR* **110**: 898, 2006).

Terramycetaceae Letcher (2006), Rhizophydiales. 2 gen., 2 spp.

terrestrial, growing on land as opposed to in water. Cf. terricolous.

terricolous, growing on the ground. Cf. terrestrial.

Terriera B. Erikss. (1970), Rhytismataceae. 16, Europe. See Eriksson (*Symb. bot. upsal.* **19**: 58, 1970), Johnston (*Mycotaxon* **87**: 1, 2003), Ortiz-García *et al.* (*Mycol.* **95**: 846, 2003; phylogeny), Wang *et al.* (*Mycol.* **98**: 1065, 2006; phylogeny).

Terrostella Long (1945) ≡ Geasteroides.

terverticillate (of a penicillus), having branching at three levels, i.e. having rami bearing metulae and phialides.

tessellate, marked with a mosaic design; chequered.

Testicularia Klotzsch (1832), Anthracoideaceae. 3 (on *Cyperaceae*), Africa; N. & S. America; West Indies. See Vánky & Piątek (*Mycol. Balcanica* **3**: 163, 2006; monogr., key).

Testudina Bizz. (1885), Testudinaceae. 1, Europe. See Hawksworth (*CJB* **57**: 91, 1979).

Testudinaceae Arx (1971), Pleosporales. 5 gen. (+ 3 syn.), 7 spp.

Lit.: Hawksworth (*CJB* **57**: 91, 1979), Hawksworth (*SA* **6**: 153, 1987), Sivanesan (*Mycopathologia* **114**: 59, 1991), LoBuglio *et al.* (*Mol. Phylogen. Evol.* **6**: 287, 1996), Padhye & McGinnis (*Manual of Clinical Microbiology*: 1318, 1999), Kruys *et al.* (*MR* **110**: 527, 2006), Schoch *et al.* (*Mycol.* **98**: 1041, 2006; phylogeny).

Testudomyces Cano, M. Solé & Guarro (2002), Gymnoascaceae. 1, Spain. See Solé *et al.* (*Stud. Mycol.* **47**: 150, 2002).

Tetena Raf. (1806) nom. dub., Fungi.

Teterevnikova-Babayan (Darya Nikolayevna; 1904-1988; Russia, later Armenia). Student, Leningrad Agricultural Institute (1921-1926); Researcher, Phytopathology Station, Tsarskoye Selo (1926-1929); Plant Pathologist (1929-1934) then Head of Department of Plant Pathology (1934-1938), Station of Plant Protection, People's Commissariat of Agriculture, Armenian SSR; Chair of Plant Protection, Armenian Agricultural Institute (1938-48); Head, Department of Plant Protection, Institute of Wine-Making and Viticulture, Armenian Academy of Sciences (1942-1946); Professor of Botany, Yerevan State University (1948-1964); Corresponding Member, Armenian Academy of Sciences (1960). Although starting as a plant pathologist, she was noted for early work on the mycota of Armenia; initiated and organized the multi-volume series *Mycoflora of Armenia*, writing two of the volumes, and editing four. *Publs.* [*Rust fungi of cultural and wild plants of Armenian SSR*] (1959) [in Russian]; [Rust Fungi]. *Mikoflora Armyansko? SSR* (1977) [in Russian]; [Sphaeropsidales]. *Mikoflora Armyansko? SSR* (1983) [in Russian]; [*Fungi of the Genus Septoria in the USSR*] (1987) [in Russian]. *Biogs, obits etc.* Anon. ([*Mikologiya i Fitopatologiya*] **10**: 75, 1976 [in Russian]); Gorlenko, Melik-Khachatryan, Osipyan & Tomilin ([*Mikologiya i Fitopatologiya*] **23**: 186, 1989) [portrait, in Russian]).

tetra (prefix), four; **-cytes**, the spores resulting from meiosis; **-d**, a group of four; **-polar** (of incompatibility systems), having 2 loci; bifactorial; cf. bipolar; **-spore**, see dispore; **-tomic**, 4-times furcate at one node.

Tetrabrachium Nawawi & Kuthub. (1987), anamorphic *Pezizomycotina*, Hso.1bH.1. 1 (aquatic), Malaysia. See Nawawi & Kuthubutheen (*Mycotaxon* **29**: 291, 1987).

Tetrabrunneospora Dyko (1978), anamorphic *Pezizomycotina*, Hso.0bP.1. 1 (aquatic), USA. See Dyko (*TBMS* **70**: 414, 1978), Marvanová & Bärlocher (*Czech Mycol.* **53**: 1, 2001).

Tetrachaetum Ingold (1942), anamorphic *Pezizomycotina*, Hso.1bH.1. 1 (aquatic), widespread. See Laitung *et al.* (*Mol. Ecol.* **13**: 1679, 2004; genetic diversity).

Tetrachia Berk. & M.A. Curtis (1882) nom. nud. = Spegazzinia fide Saccardo (*Syll. fung.* **12**: 775, 1897).

Tetrachia Sacc. (1921) = Spegazzinia fide Boedijn (*in litt.*).

tetrachotomous, with four branches arising from the same point.

Tetrachytrium Sorokīn (1874), ? Chytridiomycetes. 1, Europe.

Tetracium, see *Tetracrium*.

Tetracladium De Wild. (1893), anamorphic *Helotiales*, Hso.1bH.1. 7 (aquatic), widespread. See Petersen (*Mycol.* **54**: 140, 1962; key), Roldán *et al.* (*MR* **93**: 452, 1989; culture characters, key), Nikolcheva & Bärlocher (*Czech Mycol.* **53**: 285, 2002; phylogeny), Baschien *et al.* (*Nova Hedwigia* **83**: 311, 2006; phylogeny).

Tetracoccosporium Szabó (1905), anamorphic *Pezizomycotina*, Hso.#eP.1. 2, widespread.

Tetracolium Kunze ex Link (1824) nom. dub. = Torula fide Petch (*TBMS* **24**: 56, 1940), Hughes (*CJB* **36**, 1958).

Tetracrium Henn. (1902), anamorphic *Puttemansia*, Hsp.1bH.?. 3 (on insects), widespread (tropical). See Martin (*Pacific Sci.* **2**: 71, 1948).

Tetracytum Vanderw. (1945) = Cylindrocladium fide Kendrick & Carmichae *in* Ainsworth *et al.* (Eds) (*The Fungi* **4A**: 390, 1973).

Tetradia T. Johnson (1904) nom. dub., anamorphic *Pezizomycotina*. See Sutton (*Mycol. Pap.* **141**, 1977).

Tetragoniomyces Oberw. & Bandoni (1981), Tetragoniomycetaceae. 1, N. America; Europe. See Oberwinkler & Bandoni (*CJB* **59**: 1034, 1981), Put *et al.* (*Sterbeeckia* **19**: 23, 2000).

Tetragoniomycetaceae Oberw. & Bandoni (1981), Tremellales. 1 gen., 1 spp.

Lit.: Oberwinkler & Bandoni (*CJB* **59**: 1034, 1981), Bandoni (*Stud. Mycol.* **30**: 87, 1987), Oberwinkler (*Stud. Mycol.* **30**: 61, 1987), Clémençon (*Mycol. helv.* **4**: 53, 1990), Put *et al.* (*Sterbeeckia* **19**: 23, 2000).

Tetramelaena (Trevis.) C.W. Dodge (1971) = Hyperphyscia fide Hafellner *et al.* (*Herzogia* **5**: 39, 1979).

Tetramelas Norman (1853), Caliciaceae (L). 1, widespread. See Hawksworth *et al.* (*Lichenologist* **12**: 85, 1980), Marbach (*Biblthca Lichenol.* **74**, 2000), Nordin (*Lichenologist* **36**: 355, 2004), Nordin & Tibell (*Lichenologist* **37**: 491, 2005).

Tetrameronycha Speg. ex W. Rossi & M. Blackw. (1990), anamorphic *Pezizomycotina*, Hso.0eH.1. 1 (on *Forficulidae*), W. Africa. See Rossi & Blackwell (*Mycol.* **82**: 138, 1990).

Tetramicra R.A. Matthews & A.B.F. Matthews (1980), Microsporidia. 1.

Tetranacriella Kohlm. & Volkm.-Kohlm. (2001), anamorphic *Pezizomycotina*, C?.?.?. 1, USA. See Kohlmeyer & Volkmann-Kohlmeyer (*Bot. Mar.* **44**: 152, 2001).

Tetranacrium H.J. Huds. & B. Sutton (1964), anamorphic *Pezizomycotina*, St.1bH.1. 1, West Indies; India. See Hudson & Sutton (*TBMS* **47**: 202, 1964), Punithalingam (*MR* **107**: 917, 2003; cytology).

Tetrandromyces Thaxt. (1912), Laboulbeniaceae. 6, widespread. See Rossi & Santamaria (*Mycol.* **92**: 786, 2000).

Tetrapisispora Ued.-Nishim. & Mikata (1999), Saccharomycetaceae. 6, widespread. See Ueda-Nishimura & Mikata (*Int. J. Syst. Bacteriol.* **49**: 1915, 1999), Kurtzman *et al.* (*Stud. Mycol.* **50**: 397, 2004), Sumpradit *et al.* (*Int. J. Syst. Evol. Microbiol.* **55**: 1735, 2005), Suh *et al.* (*Mycol.* **98**: 1006, 2006; phylogeny).

Tetraploa Berk. & Broome (1850), anamorphic *Massarina*, Hso.#eP.1. 9, widespread. See Ellis (*TBMS* **32**: 246, 1949), Rifai *et al.* (*Reinwardtia* **10**: 419, 1988; Javan spp.), Scheuer (*MR* **95**: 126, 1991; teleomorph), Hatakeyama *et al.* (*Mycoscience* **46**: 196, 2005; Japan).

Tetraposporium S. Hughes (1951), anamorphic *Pezizomycotina*, Hso.0bP.1. 2, Africa; USA. See Hughes (*Mycol. Pap.* **46**: 25, 1951).

Tetrapyrgos E. Horak (1987), Marasmiaceae. 16, widespread (tropical). See Horak (*Sydowia* **39**: 101, 1986), Bulakh (*Mikol. Fitopatol.* **37**: 23, 2003; Russian Far East), Wilson & Desjardin (*Mycol.* **97**: 667, 2005; phylogeny).

Tettigomyces Thaxt. (1915), Ceratomycetaceae. 16, widespread. See Ye & Shen (*Acta Mycol. Sin.* **11**: 285, 1992; key 10 spp. China).

Tettigorhyza G. Bertol. (1875) ? = Cordyceps fide Hawksworth *et al.* (*Dictionary of the Fungi* edn 8, 1995).

texospore, ascospore coated with a layer of cells of paraphysal origin, as in *Texosporium* (Tibell & Hofsten, *Mycol.* **60**: 557, 1968).

Texosporium Nádv. ex Tibell & Hofsten (1968), Caliciaceae (L). 1, USA. See Tibell & Hofsten (*Mycol.* **60**: 557, 1968), Wedin *et al.* (*Taxon* **51**: 655, 2002), Tibell (*J. Hattori bot. Lab.* **100**: 809, 2006).

Textotheca Matsush. (1996), anamorphic *Pezizomycotina*, Cpd.?.?. 1, Cape Province. See Matsushima (*Matsush. Mycol. Mem.* **9**: 28, 1996).

textura, see tissue types.

Thaelaephora, see *Thelephora*.

Thailandia Vardhan. (1959) = Candida fide Orr & Kuehn (*Mycol.* **63**: 191, 1971).

Thailandiomyces Pinruan, Sakay., K.D. Hyde & E.B.G. Jones (2008), Diaporthales. Anamorph *Craspedodidymum*. 1, Thailand. See Pinruan *et al.* (*Fungal Diversity* **29**: 89, 2008).

thalamium, asci + hamathecium (obsol.).

thalassiomycetes, fungi living in marine environments; see Marine fungi.

Thalassoascus Ollivier (1926), Dothideomycetes. 3 (on *Algae*), widespread. See Kohlmeyer (*Mycol.* **73**: 833, 1981).

Thalassochytrium Nyvall, Pedersén & Longcore (1999), Chytridiomycota. 1 (on marine red alga), China. See Nyvall *et al.* (*Journal of Phycology* **35**: 176, 1999).

Thalassogena Kohlm. & Volkm.-Kohlm. (1987), Halosphaeriaceae. 1 (marine), Belize. See Kohlmeyer & Volkmann-Kohlmeyer (*SA* **6**: 223, 1987), Kohlmeyer & Volkmann-Kohlmeyer (*Bot. Mar.* **34**: 1, 1991).

Thalespora Chatmala & E.B.G. Jones (2006), Halosphaeriaceae. 1, Thailand. See Chatmala & Jones (*Nova Hedwigia* **83**: 228, 2006).

thallic (of conidiogenesis), one of the two basic sorts (cf. blastic) in which any enlargement of the recognizable conidial initial occurs *after* the initial has been delimited by one or more septa. The conidium is differentiated from a *whole* cell. **entero-**, thallic conidiogenesis in which the outer wall of the sporogenous cell is not involved in the formation of the spore wall (as for sporangiospores).

thalline exciple (margin) (margin), see excipulum thallinum.

Thallinocarpon Å.E. Dahl (1950) ? = Lichinella fide Henssen (*Ber. dt. bot. Ges.* **92**: 483, 1980).

Thallisphaera Dumort. (1822) nom. dub., Pezizomycotina. Used for diverse perithecioid fungi.

Thallochaete Theiss. (1913) = Aphanopeltis fide Hawksworth *et al.* (*Dictionary of the Fungi* edn 8, 1995).

thalloconidium, a propagule produced and seceded directly from the lower cortex and(or) rhizines of certain *Umbilicaria* spp.; similar structures arise from the prothallus in *Protoparmelia*, *Rhizoplaca* and *Sporastatia* (Poelt & Obermayer, *Herzogia* **8**: 273, 1990); thalloconidia are dark brown, smooth to rugged, with 2-3 wall layers, and consist of one to 2500 cells. See Hestmark (*Nord. Jl Bot.* **9**: 547, 1990; ultrastr., occurrence); see also thallyles.

thallodic, of, pertaining to, or belonging to a thallus (Weresub & LeClair, *CJB* **49**: 2203, 1971).

Thalloedematomyces E.A. Thomas ex Cif. & Tomas. (1953) ≡ Thalloidima.

Thalloidima A. Massal. (1852) nom. rej. = Toninia fide Hawksworth *et al.* (*Dictionary of the Fungi* edn 8, 1995).

Thalloidimatomyces, see *Thalloedematomyces*.

Thalloloma Trevis. (1853), Graphidaceae (L). 1, widespread. See Hawksworth *et al.* (*Dictionary of the Fungi* edn 8, 1995), Staiger (*Biblthca Lichenol.* **85**, 2002), Kalb *et al.* (*Symb. bot. upsal.* **34** no. 1: 133, 2004), Archer (*Biblthca Lichenol.* **94**, 2006; Australia), Archer (*Systematics & Biodiversity* **5**: 9, 2007; Solomon Is).

Thallomicrosporon Benedek (1964) = Microsporum fide Ajello (*Sabouraudia* **6**: 153, 1968).

Thallomyces H.J. Swart (1975), Parmulariaceae. 1, Australia. See Swart (*TBMS* **65**: 84, 1975).

Thallophyta, see *Fungi*; **thallophyte**, one of the *Thallophyta*.

Thallospora L.S. Olive (1948), anamorphic *Pezizomycotina*, Hso.1bH.1. 1, USA.

thallospore (1) an asexual spore having neither conidiophore nor conidiogenous cell, or one which is not separated from the hypha or conidiogenous cell producing it; i.e. an arthrospore, blastospore, or chlamydospore (and aleuriospore) (after Vuillemin; see Mason, 1933); (2) a thalloconidium (q.v.).

thallus, the vegetative body of a thallophyte; for thallus types in lichens see Lichens; **heteromerous-**, a layered thallus; **homoiomerous-**, an unlayered thallus.

thallyles, minute thallus-like propagules produced on the underside of certain *Umbilicaria* thalli (Krog & Swinscow, *Nordic Jl Bot.* **6**: 75, 1986); see also thalloconidia.

Thamnidiaceae Fitzp. (1930) = Mucoraceae.
Lit.: Kirk (*in litt.*).

Thamnidium Link (1809), Mucoraceae. 1, widespread. See Benny (*Mycol.* **84**: 834, 1992), Roux & Botha (*Proc. Elect. Microsc. Soc. S. Afr.* **24**: 64, 1994; ultrastr.), Voigt & Wöstemeyer (*Gene* **270**: 113, 2001; phylogeny), Ho (*Fungal Science* Taipei **17**: 87, 2002; Taiwan).

Thamnidium Tuck. ex Schwend. (1860) = Lichina fide Hawksworth *et al.* (*Dictionary of the Fungi* edn 8, 1995).

thamniscophagous, see ptyophagous.

Thamnium Vent. (1799) nom. rej. = Roccella fide Ahti (*Taxon* **33**: 330, 1984).

Thamnocephalis Blakeslee (1905), Sigmoideomycetaceae. 3, N. America; India. See Mehrotra (*Zbl. Bakt.* Abt. II **117**: 425, 1964), Benny *et al.* (*Mycol.* **84**: 635, 1992; key), Tanabe *et al.* (*Mol. Phylogen. Evol.* **16**: 253, 2000; phylogeny).

Thamnochrolechia Aptroot & Sipman (1991), Agyriaceae (L). 1, Papua New Guinea. See Lumbsch *et al.* (*Biblthca Lichenol.* **57**: 355, 1995), Schmitt *et al.* (*Biblthca Lichenol.* **86**: 147, 2003; phylogeny).

Thamnogalla D. Hawksw. (1980), Pezizomycotina. 1 (on *Thamnolia*), widespread (northern hemisphere). See Hafellner & Sancho (*Herzogia* **8**: 363, 1990; posn), Hoffmann & Hafellner (*Biblthca Lichenol.* **77**: 1, 2000).

Thamnolecania (Vain.) Gyeln. (1933), Lecanorales (L). 5, widespread (sub-Antarctic). Perhaps allied to *Crocyniaceae*. See Ekman (*MR* **105**: 783, 2001; phylogeny).

Thamnolia Ach. ex Schaer. (1850) nom. cons., Icmadophilaceae (L). 4, widespread (montane). Sterile. See Ihlen (*Graphis Scripta* **7**: 17, 1995; Norway), Kärnefelt & Thell (*Biblthca Lichenol.* **58**: 213, 1995;

genetic variation), Ekman & Tønsberg (*MR* **106**: 1262, 2002; phylogeny), Santesson (*Symb. bot. upsal.* **34**: 393, 2004), Miądlikowska *et al.* (*Mycol.* **98**: 1088, 2006; phylogeny).

Thamnoliaceae Zahlbr. (1898) = Icmadophilaceae.

Thamnoma, see *Thamnonoma*.

Thamnomyces Ehrenb. (1820), Xylariaceae. Anamorph *Nodulisporium*. 4 or 5, widespread (esp. tropical). See Dennis (*Kew Bull.* **12**: 297, 1957), Dennis (*Bull. Jard. bot. Brux.* **31**: 150, 1961), Samuels & Müller (*Sydowia* **33**: 274, 1980; relationships), San Martín & Rogers (*Mycotaxon* **53**: 115, 1995; Mexico), Stadler *et al.* (*MR* **108**: 239, 2004; chemistry), Stadler & Fournier (*Revta Iberoamer. Micol.* **23**: 160, 2006).

Thamnonoma (Tuck.) Gyeln. (1933) = Caloplaca fide Hawksworth *et al.* (*Dictionary of the Fungi* edn 8, 1995).

Thamnostylum Arx & H.P. Upadhyay (1970), Syncephalastraceae. 4, widespread (esp. warmer areas). See Benny & Benjamin (*Aliso* **8**: 301, 1975; key), Botha *et al.* (*S. Afr. J. Bot.* **63**: 104, 1997), Voigt & Wöstemeyer (*Gene* **270**: 113, 2001; phylogeny), Ho (*Fungal Science* Taipei **17**: 87, 2002; Taiwan).

Thanatephorus Donk (1956), Ceratobasidiaceae. Anamorph *Rhizoctonia*. 12, widespread. *T. cucumeris* (stat. mycel. *Rhizoctonia solani*) is now used to replace *Corticium solani*. See Talbot (*Persoonia* **3**: 371, 1965), Currah & Zelmer (*Rep. Tottori mycol. Inst.* **40**: 43, 1992; 4 spp. with orchids), Roberts (*Rhizoctonia-forming fungi*, 1999), Elbakali *et al.* (*Phytopath. Mediterr.* **42**: 167, 2003; on potato), Toda *et al.* (*J. Gen. Pl. Path.* **70**: 270, 2004; *Rhizoctonia solani* ITS), Ciampi *et al.* (*Eur. J. Pl. Path.* **113**: 183, 2005; *Rhizoctonia solani* intraspecific evolution).

Thanatophytum Nees (1816), anamorphic *Helicobasidium*. 1, widespread. *T. crocorum* (a pathogen of *Crocus*; mort de saffran). See Lutz *et al.* (*MR* **108**: 227, 2004).

Thaptospora B. Sutton & Pascoe (1987), anamorphic *Pezizomycotina*, Ccu.0eH.15. 1, Australia. See Sutton & Pascoe (*TBMS* **88**: 174, 1987).

Tharoopama Subram. (1956), anamorphic *Pezizomycotina*, Hsy.0eP.10. 2, India; Panama. See Subramanian (*J. Indian bot. Soc.* **35**: 84, 1956), Sureshkumar *et al.* (*Journal of Mycology and Plant Pathology* **36**: 8, 2006).

Thaumasiomyces Thaxt. (1931), Ceratomycetaceae. 3, W. Africa; Borneo. See Zhou *et al.* (*MR* **105**: 919, 2001; DNA).

Thaxter (Roland; 1858-1932; USA). Assistant Professor (1891) then Professor of Cryptogamic Botany (1901-1919), Harvard University; Honorary Curator of the Farlow Herbarium (1919 onwards). His great contribution was the detailed and meticulous study of the *Laboulbeniales* (q.v.), previously almost completely overlooked. *Publs*. Contribution towards a monograph of the Laboulbeniaceae I-V. *Memoirs of the American Academy of Arts and Sciences* (1896-1931). *Biogs, obits etc*. Grummann (1974: 197); Pfister (*Mycotaxon* **20**: 225, 1984; index to non-*Laboulbeniales* names); Stafleu & Cowan (*TL-2* **6**: 231, 1986); Weston (*Mycol.* **25**: 69, 1933) [bibliography, portrait]; Weston (*Phytopathology* **23**: 565, 1933).

Thaxteria Giard (1892) = Laboulbenia fide Thaxter (*Proc. Amer. Acad. Arts & Sci.* **30**: 471, 1895).

Thaxteria Sacc. (1891), Nitschkiaceae. 2, widespread. See Booth & Müller (*TBMS* **58**: 73, 1972), Nannfeldt (*Svensk bot. Tidskr.* **69**: 204, 1975; posn), Huhndorf *et al.* (*Mycol.* **96**: 368, 2004), Huhndorf *et al.* (*MR* **108**: 1384, 2004).

Thaxteriella Petr. (1924), Tubeufiaceae. 11, widespread. See Rossman (*Mycol. Pap.* **157**, 1987), Crane *et al.* (*CJB* **76**: 602, 1998), Kodsueb *et al.* (*Fungal Diversity* **21**: 105, 2006), Spatafora *et al.* (*Mycol.* **98**: 1018, 2006; phylogeny), Tsui *et al.* (*Mycol.* **98**: 94, 2006).

Thaxteriellopsis Sivan., Panwar & S.J. Kaur (1977) = Chaetosphaerulina fide Crane *et al.* (*CJB* **76**: 602, 1998).

Thaxterina Sivan., R.C. Rajak & R.C. Gupta (1988), Tubeufiaceae. 1, India. See Sivanesan *et al.* (*TBMS* **90**: 662, 1988), Crane *et al.* (*CJB* **76**: 602, 1998), Kodsueb *et al.* (*Fungal Diversity* **21**: 105, 2006).

Thaxteriola Speg. (1918), anamorphic *Pyxidiophora*. 1, widespread. See Blackwell *et al.* (*Mycol.* **78**: 605, 1986), Blackwell *et al.* (*Science* N.Y. **232**: 993, 1986), Simpson & Stone (*Mycotaxon* **30**: 1, 1987), Blackwell & Malloch (*Mem. N. Y. bot. Gdn* **49**: 23, 1989), Blackwell *et al.* (*MR* **92**: 397, 1989), Hawksworth *et al.* (*Dictionary of the Fungi* edn 8, 1995).

Thaxterogaster Singer (1951) = Cortinarius. Formerly recognized as an independent sequestrate genus. fide Peintner *et al.* (*Mycotaxon* **81**: 177, 2002).

Thaxterogastraceae Singer (1962) = Cortinariaceae.

Thaxterosporium Ben Ze'ev & R.G. Kenneth (1987), Neozygitaceae. 1, widespread. See Ben-Ze'ev *et al.* (*Mycotaxon* **28**: 323, 1987) = Neozygites (Entomophthor.) fide, Keller (*Sydowia* **43**: 39, 1991).

theca, see ascus (obsol.).

Thecamycetes, see *Ascomycota* (Marchand, 1896).

Thecaphora Fingerh. (1836) nom. cons., Glomosporiaceae. Anamorphs *Thecaphorella*, *Rhombiella*. 57 (on *Leguminosae*, *Compositae*, *Convolvulaceae*, etc.), widespread. *T. solani*, potato (*Solanum*) smut; *T. frezzii*, peanut (*Arachis*) smut. Frequently has an anamorph. See Zambettakis & Joly (*BSMF* **91**: 71, 1975; numerical taxonomy), Durán (*CJB* **60**: 1512, 1982), Vánky (*Illustrated genera of smut fungi*, 1987), Vánky (*Cryptog. Stud.* **1**: 159 pp., 1987), Mordue (*Mycopathologia* **103**: 177, 1988), Vánky (*TBMS* **32**: 148, 1991), Vánky (*Taxon* **47**: 153, 1998), Vánky (*Fungal Diversity* **6**: 131, 2001; as *Glomosporium*), Piątek (*Mycotaxon* **92**: 33, 2005; as *Kochmania*), Vánky *et al.* (*Mycol. Progr.*: [in press], 2008).

Thecaphorella H. Scholz & I. Scholz (1988), anamorphic *Thecaphora*. 1. See Vánky (*Europ. Smut Fungi*: 228, 1994).

Thecaria Fée (1824), Graphidaceae (L). 1, pantropical. See Hawksworth *et al.* (*Dictionary of the Fungi* edn 8, 1995), Staiger (*Biblthca Lichenol.* **85**, 2002), Archer (*Biblthca Lichenol.* **94**, 2006), Archer (*Systematics & Biodiversity* **5**: 9, 2007; Solomon Is).

thecaspore, see ascospore (obsol.).

Theciopeltis F. Stevens & Manter (1925) = Micropeltis fide Clements & Shear (*Gen. Fung.*, 1931).

thecium, the part of an apothecium containing the asci between the epithecium and hypothecium; sometimes used for the whole sporocarp or as equivalent to hymenium.

Theclospora Harkn. (1884) = Emericella fide Peek & Solheim (*Mycol.* **50**: 844, 1958).

Thecographa A. Massal. (1860), Graphidaceae. 1,

paleotropical. See Lücking *et al.* (*Taxon* **56**: 1296, 2007; nomencl.).

Thecopsora, see *Thekopsora*.

Thecostroma Clem. (1909) = Bloxamia fide Clements & Shear (*Gen. Fung.*, 1931).

Thecotheus Boud. (1869), ? Ascobolaceae. 17, widespread (esp. temperate). See Kimbrough (*Mycol.* **61**: 107, 1969; key), Krug & Khan (*Mycol.* **79**: 200, 1987; key 10 spp.), Aas (*A World-Monograph of the Genus Thecotheus*, 1992), Kimbrough *in* Hawksworth (Ed.) (*Ascomycete Systematics. Problems and Perspectives in the Nineties* NATO ASI Series vol. **269 269**: 398, 1994; posn), Prokhorov (*Mikol. Fitopatol.* **31**: 27, 1997), Landvik *et al.* (*Mycoscience* **39**: 49, 1998; phylogeny), Yao & Spooner (*Kew Bull.* **55**: 451, 2000; UK), Nagao *et al.* (*Mycol.* **95**: 688, 2003; Australia), Hansen & Pfister (*Mycol.* **98**: 1029, 2006; phylogeny).

Thedgonia B. Sutton (1973), anamorphic *Helotiales*, Hsp.1-≡ eH.39. 3, widespread (north temperate). See Yoshikawa & Yokoyama (*TMSJ* **33**: 177, 1992), Braun (*Monogr. Cercosporella, Ramularia Allied Genera (Phytopath. Hyphom.)* **1**: 211, 1995; key), Crous *et al.* (*Mycol.* **93**: 1081, 2001; phylogeny), Crous & Braun (*CBS Diversity Ser.* **1**: 571 pp., 2003), Crous *et al.* (*Stud. Mycol.* **58**: 33, 2007; phylogeny).

Theissen (Ferdinand; 1877-1919; Germany, later Brazil). Stationed in São Leopoldo, Brazil (1902-1908); studied theology and became priest, Valkenburg and Innsbruck (1909-1912); schoolmaster, Feldkirch (1914-1919); died as a result of a climbing accident in the Voralberg Alps while collecting. With Rick (q.v.) he made significant contributions to tropical mycology, particularly for ascomycetes of Brazil. *Publs. Die Gattung Asterina* (1913); (with H. Sydow, q.v.) Die Dothideales. *Annales Mycologici* (1915). *Biogs, obits etc.* Sydow (*Annales Mycologici* **17**: 134, 1919, bibliography); Stafleu & Cowan (*TL-2* **6**: 239, 1986).

Theissenia Maubl. (1914), Xylariaceae. Anamorph *Nodulisporium*. 4, pantropical. See Dennis (*Bull. Jard. bot. Brux.* **34**: 231, 1961), Ju *et al.* (*Mycol.* **95**: 109, 2003; monogr.), Ju *et al.* (*Mycol.* **99**: 612, 2007; phylogeny).

Theissenula Syd. & P. Syd. (1914) nom. dub., Fungi. See Hansford (*Mycol. Pap.* **15**, 1946).

Thekopsora Magnus (1875), Pucciniastraceae. Anamorph *Pomatomyces*. *c.* 11 (on *Picea*, *Tsuga* (0, I) (*Pinaceae*; on dicots (II, III)), widespread (north temperate). See Hiratsuka (*Revision of taxonomy of the Pucciniastreae*, 1958), Cummins & Hiratsuka (*Illustr. Gen. Rust Fungi rev. edit.*, 1983; as *Pucciniastrum*), Cummins & Hiratsuka (*Illustr. Gen. Rust Fungi* edn 3: 225 pp., 2003; accepted).

Thelactis Mart. (1821) = Mucor Fresen. fide Fries (*Syst. mycol.* **3**, 1832).

Thelebolaceae Eckblad (1968), Thelebolales. 9 gen. (+ 9 syn.), 46 spp.

Lit.: Kimbrough (*CJB* **44**: 685, 1966), Kimbrough & Korf (*Am. J. Bot.* **54**: 9, 1967), Moravec (*Česká Mykol.* **25**: 150, 1971), Czymmek & Klomparens (*CJB* **70**: 1669, 1992), Momol *et al.* (*SA* **14**: 91, 1996), Brummelin (*Persoonia* **16**: 425, 1998), Brummelin & Kristiansen (*Persoonia* **17**: 119, 1998), Landvik *et al.* (*Mycoscience* **39**: 49, 1998; phylogeny), Prokhorov (*Mikol. Fitopatol.* **32**: 40, 1998), van Brummelen (*Persoonia* **16**: 425, 1998; ultrastr.) and under *Thelebolus*, Wang (*Bull. natn. Mus. Nat. Sci.* Taiwan **12**: 49, 1999), Stchigel *et al.* (*MR* **105**: 377, 2001).

Thelebolales P.F. Cannon (2001), ? Leotiomycetes. 1 fam., 9 gen., 46 spp. Stromata absent. Ascomata minute, ± globose or pulvinate, at least initially cleistothecial, the excipulum hyaline, poorly developed, ± glabrous. Interascal tissue poorly developed, composed of simple paraphyses. Asci ± ellipsoidal, often multispored, ± persistent, opening with a rather irregular vertical split. Ascospores usually small, hyaline, smooth or with ornamentation formed as an elaboration of an initially homogeneous secondary wall layer. Anamorphs unknown. Saprobic, usually coprophilous, widespr. Fam.:

Thelebolaceae

For *Lit.* see under fam.

Thelebolus Tode (1790), Thelebolaceae. 10 (psychrophilic, coprophilous), widespread (north temperate). See Cooke & Barr (*Mycol.* **56**: 763, 1964; asci and *Erysiphaceae*), Wicklow & Malloch (*Mycol.* **63**: 118, 1971), Kimbrough (*Mycol.* **73**: 1, 1981; ascus structure), Czymmek & Klomparens (*CJB* **70**: 1669, 1992; ascospores), Momol & Kimbrough (*SA* **13**: 1, 1994; posn), Momol *et al.* (*SA* **14**: 91, 1996; posn), Landvik *et al.* (*Mycoscience* **39**: 49, 1998), Gernandt *et al.* (*Mycol.* **93**: 915, 2001; phylogeny), Stchigel *et al.* (*MR* **105**: 377, 2001; phylogeny), Hoog *et al.* (*Stud. Mycol.* **51**: 33, 2005; phylogeny, ecology), Wang *et al.* (*Mycol.* **98**: 1065, 2006; phylogeny).

Thelenella Nyl. (1855), Thelenellaceae (L). 33, widespread. See Mayrhofer (*Biblthca Lichenol.* **26**, 1987; key), Mayrhofer & McCarthy (*Muelleria* **7**: 333, 1991; key 12 saxicolous spp.), Harris (*More Florida Lichens*, 1995), Aptroot (*Fungal Diversity* **2**: 43, 1999), Fryday & Coppins (*Lichenologist* **36**: 89, 2004), Lücking *et al.* (*Lichenologist* **39**: 187, 2007).

Thelenellaceae O.E. Erikss. ex H. Mayrhofer (1987), Pezizomycotina (inc. sed.) (±L). 2 gen. (+ 10 syn.), 51 spp.

Lit.: Mayrhofer & Poelt (*Herzogia* **7**: 13, 1985), Barr (*Sydowia* **38**: 11, 1985), Mayrhofer (*Biblthca Lichenol.* **26**: 106 pp., 1987), Mayrhofer (*Biblthca Lichenol.* **26**, 1987), Kalb (*Flechten Follmann* Contributions to Lichenology in Honour of Gerhard Follmann: 249, 1995), Harada (*J. Nat. Hist. Mus. Inst.* Chiba **5**: 91, 1999), Hyde & Wong (*MR* **103**: 347, 1999), Schmitt *et al.* (*Mycol.* **97**: 362, 2005).

Thelenidia Nyl. (1886), ? Dothideomycetes (L). 1, Greenland; Switzerland. See Topham & Swinscow (*Lichenologist* **4**: 294, 1970).

Thelenidiomyces Cif. & Tomas. (1953) ≡ Thelenidia.

Thelephora Ehrh. ex Willd. (1787), Thelephoraceae. *c.* 50, widespread. See Corner (*Beih. Nova Hedwigia* **27**, 1968; monogr.), Stalpers (*Stud. Mycol.* **35**: 168 pp., 1993; key), Zecchin (*Rivista di Micologia* **48**: 243, 2005).

Thelephoraceae Chevall. (1826), Thelephorales. 12 gen. (+ 20 syn.), 171 spp.

Lit.: Svrček (*Sydowia* **14**: 170, 1960; subfam. *Tomentelloideae*, keys), Cunningham (*The Thelephoraceae of Australia and New Zealand* [*Bull. DSIR N.Z.* **145**], 1963; Cuba), Reid (*Beih. Nova Hedwigia* **18**: 1090, 1965; stipitate steroid spp.), Corner (*Beih. Nova Hedwigia* **27**, 1968), Jülich & Stalpers (*Verh. Kon. Ned. Akad. Wetensch.* Afd. Natuurk. sect. 2 **74**, 1980; N. Hemisph.), Stalpers (*Stud. Mycol.* **35**: 168 pp., 1993), Kõljalg (*Syn. Fung.* **9**: 213 pp., 1996).

Thelephorales Corner ex Oberw. (1976). Agaricomycetes. 2 fam., 18 gen., 269 spp. Fams:
(1) **Bankeraceae**
(2) **Thelephoraceae**
For *Lit.* see under fam.

Thelephorella P. Karst. (1889) nom. dub., Thelephorales. See Donk (*Taxon* **6**: 17, 1957).

Theleporus Fr. (1847), Grammotheleaceae. 3, widespread (tropical). See Ryvarden (*TBMS* **73**: 9, 1979), Roberts & Ryvarden (*Kew Bull.* **61**: 55, 2006; Cameroon).

Thelidea Hue (1902) = Knightiella fide Rambold *et al.* (*Biblthca Lichenol.* **53**: 217, 1993).

Thelidiaceae Walt. Watson (1929) = Verrucariaceae.

Thelidiella Fink (1933), Pezizomycotina. 1 (on lichens), N. America.

Thelidiola C.W. Dodge (1968) = Muellerella fide Castello & Nimis (*Biblthca Lichenol.* **57**: 71, 1995).

Thelidiomyces Cif. & Tomas. (1953) ≡ Thelidium.

Thelidiopsis Vain. (1921), Verrucariaceae (L). 1, Europe; Asia.

Thelidium A. Massal. (1855), Verrucariaceae (L). *c.* 111, widespread. See Servít (*Československé lišejníky čeledi Verrucariaceae*, 1954), Harada (*Hikobia* **12**: 133, 1996; Japan), Harada (*Hikobia* **12**: 289, 1998), Grube (*Nova Hedwigia* **68**: 241, 1999; asci), Harada & Wang (*Lichenology* **3**: 47, 2004), Harada & Wang (*Lichenology* **5**: 23, 2006), Gueidan *et al.* (*MR* **111**: 1145, 2007; phylogeny).

Thelignya A. Massal. (1855), Lichinaceae (L). 1, Europe. See Henssen (*Symb. bot. upsal.* **18** no. 1, 1963), Jørgensen & Henssen (*Taxon* **39**: 343, 1990), Schultz & Büdel (*Lichenologist* **34**: 39, 2002; key).

Thelis Clem. (1931) ≡ Hanseniaspora.

Thelocarpaceae Zukal (1893), Pezizomycotina (inc. sed.). 3 gen. (+ 12 syn.), 25 spp. See Salisbury (*Lichenologist* **3**: 175, 1966), Reeb *et al.* (*Mol. Phylogen. Evol.* **32**: 1036, 2004; phylogeny).

Thelocarpella Nav.-Ros. & Cl. Roux (1999), Acarosporaceae. 1, France. See Navarro-Rosinés *et al.* (*CJB* **77**: 835, 1999), Lutzoni *et al.* (*Am. J. Bot.* **91**: 1446, 2004; phylogeny), Reeb *et al.* (*Mol. Phylogen. Evol.* **32**: 1036, 2004; phylogeny), Wedin *et al.* (*MR* **109**: 159, 2005; phylogeny).

Thelocarpon Nyl. (1853), Thelocarpaceae (±L). 23, Europe; N. America. See Salisbury (*Lichenologist* **3**: 175, 1966), Poelt & Hafellner (*Phyton* Horn **17**: 67, 1975), Kocourková-Horáková (*Czech Mycol.* **50**: 271, 1998), Rossman *et al.* (*Stud. Mycol.* **42**: 248 pp., 1999), Aptroot & Sparrius (*Lichenologist* **32**: 513, 2000), Reeb *et al.* (*Mol. Phylogen. Evol.* **32**: 1036, 2004; phylogeny).

Thelocarponomyces Cif. & Tomas. (1953) ≡ Thelocarpon.

Thelocarpum Clem. (1909) ≡ Thelocarpon.

Thelochroa A. Massal. (1855) = Pyrenocarpon fide Jorgensen & Henssen (*Taxon* **39**: 343, 1990).

Thelococcum Nyl. ex Hue (1888) = Thelocarpon fide Hawksworth *et al.* (*Dictionary of the Fungi* edn 8, 1995).

Thelographis Nyl. (1857) nom. nud. = Graphis fide Hawksworth *et al.* (*Dictionary of the Fungi* edn 8, 1995).

Thelohania Henneguy (1892), Microsporidia. 23.

Thelomma A. Massal. (1860), Caliciaceae (L). 7, widespread. See Tibell (*Bot. Notiser* **129**: 221, 1976), Tibell (*Beih. Nova Hedwigia* **79**: 597, 1984), Wedin & Tibell (*CJB* **75**: 1236, 1997), Ekman & Wedin (*Pl. Biol.* **2**: 350, 2000; phylogeny), Andersen & Ekman (*Lichenologist* **36**: 27, 2004; phylogeny).

Thelomphale Flot. (1863) = Thelocarpon fide Hawksworth *et al.* (*Dictionary of the Fungi* edn 8, 1995).

Thelophora Clem. (1902) ≡ Thelephora.

Thelopsidomyces Cif. & Tomas. (1953) ≡ Thelopsis.

Thelopsis Nyl. (1855) nom. cons., Ostropales (L). 9, Europe; N. America. See Vězda (*Folia geobot. phytotax.* **4**: 363, 1968), Sherwood (*Mycotaxon* **5**: 1, 1977), McCarthy (*Muelleria* **7**: 313, 1991; Australia), Renobales *et al.* (*Lichenologist* **28**: 105, 1996; Spain), Breuss & Schultz (*Lichenologist* **39**: 35, 2007; Socotra, key).

Theloschisma Trevis. (1860) = Phaeographis fide Staiger (*Biblthca Lichenol.* **85**, 2002).

Theloschistes Th. Fr. (1861) ≡ Teloschistes.

Theloschistomyces Cif. & Tomas. (1953) ≡ Theloschistes.

Thelospora, see *Theclospora*.

Thelotrema Ach. (1803), Thelotremataceae (L). *c.* 126, widespread (esp. tropical). See Salisbury (*Lichenologist* **5**: 262, 1972; *lepadinum*-group), Hale (*Smithson. Contr. bot.* **16**, 1974; Dominican Republic), Salisbury (*Lichenologist* **7**: 59, 1975; gen. concept), Hale (*Smithson. Contr. bot.* **38**, 1978; Panama), Hale (*Smithson. Contr. bot.* **40**, 1980; limits), Hale (*Bull. Br. Mus. nat. hist.* Bot. **8**: 227, 1981; Sri Lanka), Purvis *et al.* (*Biblthca Lichenol.* **58**: 335, 1995; Eur.), Matsumoto & Deguchi (*Bryologist* **102**: 86, 1999; anamorph), Matsumoto (*J. Hattori bot. Lab.* **88**: 1, 2000; Japan), Galloway (*Australasian Lichenology* **49**: 16, 2001), Homchantara & Coppins (*Lichenologist* **34**: 113, 2002; SE Asia), Lücking *et al.* (*Mycol.* **96**: 283, 2004; phylogeny), Frisch (*Biblthca Lichenol.* **92**: 3, 2006; Africa), Frisch *et al.* (*Biblthca Lichenol.* **92**: 517, 2006; phylogeny), Miądlikowska *et al.* (*Mycol.* **98**: 1088, 2006; phylogeny), Mangold *et al.* (*Biblthca Lichenol.* **95**: 459, 2007; Australia), Mangold *et al.* (*Lichenologist* **40**: 39, 2008; phylogeny).

Thelotremataceae Stizenb. (1862), Ostropales (L). 21 gen. (+ 37 syn.), 660 spp.
Lit.: Hale (*Mycotaxon* **11**: 130, 1980), Hale (*Bull. Br. Mus. nat. Hist.* Bot. **8**: 227, 1981; gen. concepts), Lumbsch (*Nova Hedwigia* **56**: 227, 1993), Lumbsch & Tehler (*Bryologist* **101**: 398, 1998), Matsumoto & Deguchi (*Bryologist* **102**: 86, 1999; anams), Kantvilas & Vězda (*Lichenologist* **32**: 325, 2000; Tasmania), Kantvilas & Vězda (*Lichenologist* **32**: 343, 2000), Matsumoto (*J. Hattori bot. Lab.* **88**: 1, 2000), Kauff & Lutzoni (*Mol. Phylogen. Evol.* **25**: 138, 2002), Martín *et al.* (*Lichenologist* **35**: 27, 2003), Kalb *et al.* (*Symb. bot. upsal.* **34** no. 1: 133, 2004), Lumbsch *et al.* (*Mol. Phylogen. Evol.* **31**: 822, 2004), Lumbsch *et al.* (*Symb. bot. upsal.* **34** no. 1: 9, 2004), Frisch *et al.* (*Biblthca Lichenol.* **92**: 556 pp., 2006), Miądlikowska *et al.* (*Mycol.* **98**: 1088, 2006; phylogeny), Staiger *et al.* (*MR* **110**: 765, 2006), Rivas Platas *et al.* (*Biodiversity & Conservation* **17**: in press, 2008; indicator of undisturbed tropical forests).

Thelotrematomyces E.A. Thomas ex Cif. & Tomas. (1953) ≡ Thelotrema.

Themisia Velen. (1939), ? Helotiales. 1, former Czechoslovakia.

Thermoactinomyces Tsikl. (1899), Bacteria. q.v.

Thermoascaceae Apinis (1967), Eurotiales. 2 gen., 8 spp. See Hambleton *et al.* (*Stud. Mycol.* **53**: 29, 2005; phylogeny).

Thermoascus Miehe (1907), Thermoascaceae. Anamorph *Polypaecilum*. 5, widespread. See also *Coonemeria*. See Apinis (*TBMS* **50**: 573, 1967), Ellis (*TBMS* **76**: 457, 1981; ultrastr.), Ellis (*TBMS* **76**: 467, 1981; ultrastr.), Malloch *in* Samson & Pitt (Eds) (*Advances in Penicillium and Aspergillus systematics* **102**: 365, 1985), Landvik *et al.* (*Mycoscience* **37**: 237, 1997; DNA), Mouchacca (*Cryptog. Mycol.* **18**: 19, 1997; review), Luangsa-ard *et al.* (*Mycol.* **96**: 773, 2004; phylogeny), Hambleton *et al.* (*Stud. Mycol.* **53**: 29, 2005; phylogeny).

thermodury, withstanding high temperature, esp. when in a dormant state, e.g. as spores. Cf. thermophily.

Thermoidium Miehe (1907) = Malbranchea fide Sigler & Carmichael (*Mycotaxon* **4**: 441, 1976).

Thermomonospora Henssen (1957), Actinobacteria. q.v.

Thermomucor Subraham., B.S. Mehrotra & Thirum. (1977), Mucoraceae. 1, India. See Subrahamanyam *et al.* (*Georgia J. Sci.* **35**: 1, 1977), Schipper (*Antonie van Leeuwenhoek* **45**: 275, 1979), Voigt & Wöstemeyer (*Gene* **270**: 113, 2001; phylogeny).

Thermomyces Tsikl. (1899), anamorphic *Eurotiales*, Hso.0eP.1. 4 (from soil), widespread. See Pugh *et al.* (*TBMS* **47**: 115, 1964), Jensen *et al.* (*MR* **97**: 665, 1993; growth kinetics), Mouchacca (*Cryptog. Mycol.* **18**: 19, 1997; review), Hambleton *et al.* (*Stud. Mycol.* **53**: 29, 2005; phylogeny).

thermophily, making active growth at high temperature. Cf. thermodury. Fungi may be classified as **thermophiles** (adj. **-ilic**), growth at 20-50+°C (opt. 40-50+°C). See Cooney & Emerson (*Thermophilic fungi*, 1964; descriptions), Emerson (*in* Ainsworth & Sussman (Eds), *The Fungi* **3**: 105, 1968), Crisan (*Mycol.* **65**: 1170, 1973; concepts), Bilaĭ & Zakharchenko (*Opredelitel' Termofill'nykh Gribov*, 1987; keys, illustr. 38 spp.); **thermotolerant fungi**, e.g. *Aspergillus fumigatus*, *Absidia ramosa*, max. *c.* 50°C, min. well below 20°C; **mesophiles** (adj. **-ilic**), growth 10-40°C (opt. 20-35°C); **psychrophiles** (adj. **-ilic**), growth below 10°C (opt. below 20°C).

Thermophymatospora Udagawa, Awao & Abdullah (1986), anamorphic *Ganoderma*. 1 (with clamp connexions, from soil), Iraq. See Udagawa *et al.* (*Mycotaxon* **27**: 100, 1986).

Thermutis Fr. (1825), Lichinaceae (L). 1, Europe. See Henssen (*Symb. bot. upsal.* **18** no. 1, 1963), Schultz & Büdel (*Lichenologist* **34**: 39, 2002; key).

Thermutomyces Cif. & Tomas. (1953) ≡ Thermutis.

Thermutopsis Henssen (1990), Lichinaceae (L). 1, Antigua. See Henssen (*Lichenologist* **22**: 254, 1990), Schultz & Büdel (*Lichenologist* **34**: 39, 2002; key).

Therrya Sacc. (1882), Rhytismataceae. 7, widespread. See Reid & Cain (*CJB* **39**: 1117, 1961), Yuan & Mohammed (*Mycotaxon* **64**: 173, 1997), Johnston (*Aust. Syst. Bot.* **14**: 377, 2001).

Thielavia Zopf (1876), Chaetomiaceae. 31, widespread. See Malloch & Cain (*Mycol.* **65**: 1055, 1973), von Arx (*Stud. Mycol.* **8**, 1975), von Arx *et al.* (*Biblthca Mycol.* **94**, 1988), Mouchacca (*Cryptog. Mycol.* **18**: 19, 1997; thermophilic spp.), Lee & Hanlin (*Mycol.* **91**: 434, 1999; DNA), Stchigel *et al.* (*MR* **106**: 975, 2002; phylogeny), Stchigel *et al.* (*Mycol.* **95**: 1218, 2003; Antarctica), Sun *et al.* (*Mycosystema* **24**: 318, 2005; China), Cai *et al.* (*MR* **110**: 137, 2006; phylogeny).

Thielaviella Arx & T. Mahmood (1968) = Boothiella fide Hawksworth *et al.* (*Dictionary of the Fungi* edn 8, 1995).

Thielaviopsis Went (1893), anamorphic *Ceratocystis*, Hso.0eP.38. 12, widespread. *T. basicola* (root rot of tobacco and other plants). See Johnson (*J. agric. Res.* **7**: 289, 1916; hosts), Rawlings (*Ann. Mo. bot. Gdn* **27**: 561, 1940; culture), Punja & Sun (*CJB* **77**: 1801, 1999), Paulin & Harrington (*Stud. Mycol.* **45**: 209, 2000; phylogeny), Paulin-Mahady *et al.* (*Mycol.* **94**: 62, 2002; phylogeny), Geldenhuis *et al.* (*MR* **110**: 306, 2006; phylogeography).

Thind (Kartar Singh; 1917-1991; India). PhD, University of Wisconsin, Madison, USA (1948); Senior Lecturer (1949-1957) then Reader (1957-1962) then Professor (1962-1967) then Senior Professor (1967-1977) in Botany (Mycology and Plant Pathology) and Head of the Botany Department (1976-1977) then Professor (1977-1980) then Emeritus Professor (1980 onwards), Panjab University, Chandigarh. Explored the mycota of the northwest Himalaya over many years, leading a team of younger mycologists, organizing broad coverage of all taxonomic groups and ensuring a continuous output of papers reporting discoveries. *Publs. Clavariaceae of India* (1961); *Myxomycetes of India* (1977). *Biogs, obits etc.* Subramanian (*Current Science* **63**: 151, 1992).

Thindia Korf & Waraitch (1971), Sarcoscyphaceae. 1 (on *Cupressus*), India. See Cabello (*Boln Soc. argent. Bot.* **25**: 395, 1988; numerical taxonomy).

Thindiomyces Arendh. & R. Sharma (1983), Helotiaceae. 1, Bhutan. See Arambarri & Sharma (*Mycotaxon* **17**: 486, 1983).

Thirumalacharia Rathaiah (1981), anamorphic *Pezizomycotina*, Hso.1eP.10. 1, India. See Rathaiah (*Mycol.* **72**: 1210, 1980).

Thirumalachariella Sathe (1975) = Phragmidiella fide Cummins & Hiratsuka (*Illustr. Gen. Rust Fungi rev. edit.*, 1983).

Tholomyces Matsush. (2003), anamorphic *Pezizomycotina*, Cpd.?.?. 1, Honshu. See Matsushima (*Matsush. Mycol. Mem.* **10**: 94, 2001).

Tholurna Norman (1861) nom. cons., Caliciaceae (L). 1, Europe; N. America. See Tibell (*MR* **107**: 1403, 2003; phylogeny), Miądlikowska *et al.* (*Mycol.* **98**: 1088, 2006; phylogeny).

Tholurnaceae Elenkin (1929) = Caliciaceae.

Tholurnomyces Cif. & Tomas. (1953) ≡ Tholurna.

tholus, see ascus.

Thom (Charles; 1872-1956; USA). PhD, University of Missouri (1899); Mycologist at the Storrs (Connecticut) Experiment Station (1904-1913); Mycologist in Charge, Microbiological Laboratory (1913-1927) then Head, Division of Soil Microbiology (1927-1934), Bureau of Chemistry, then Head, Bureau of Plant Industry (1934-1942), US Department of Agriculture, Washington. Noted for his studies on fungi in dairy products and the soil, particularly *Penicillium* and *Aspergillus* (q.v.); his work influenced the establishment of standards for food handling and processing in the USA; he played an important role in development of penicillin during World War II. *Publs. The Penicillia* (1930); (with M.B. Church) *The Aspergilli* (1926); (with K.B. Raper) *A Manual of the Aspergilli* (1945); (with K.B. Raper) *A Manual of the Penicillia* (1949). *Biogs, obits etc.* Raper (*Mycol.* **49**: 134, 1957) [bibliography, portrait]; Stafleu & Cowan (*TL-2* **6**: 268, 1986).

Thomiella C.W. Dodge (1935) = Gonatobotryum fide Kendrick & Carmichae *in* Ainsworth *et al.* (Eds)

(*The Fungi* **4A**: 390, 1973).

Thoracella Oudem. (1900), anamorphic *Pezizomycotina*, St.0eH.10. 1, Europe.

Thozetella Kuntze (1891), anamorphic *Chaetosphaeriaceae*, Hsp.0eH.15. 12, widespread. See Pirozynski & Hodges (*CJB* **51**: 157, 1973), Sutton & Cole (*TBMS* **81**: 97, 1983), Waipara *et al.* (*N.Z. Jl Bot.* **34**: 517, 1996), Markovskaja & Treigiene (*Botanica Lithuanica* **7**: 93, 2001; Lithuania), Allegrucci *et al.* (*Mycotaxon* **90**: 275, 2004; Argentina), Paulus *et al.* (*Mycol.* **96**: 1074, 2004; Australia, phylogeny, key).

Thozetellopsis Agnihothr. (1958) = Thozetella fide Pirozynski & Hodges (*CJB* **51**: 157, 1973).

Thozetia Berk. & F. Muell. (1881) [non *Thozetia* F. Muell. ex Benth. 1868, *Asclepiadaceae*] ≡ Thozetella.

Thrauste Theiss. (1916), Englerulaceae. 1 or 2, Java; Philippines.

thread blight (1) a disease caused by species of *Corticium* and *Marasmius* having mycelium running as well-marked threads over the leaves and stems of tropical plants; (2) a fungus causing - -; see Petch (*Ann. R. bot. Gdns Peradeniya* **9**: 1-43, 1924). Cf. horse-hair blight fungi.

Thrinacospora Petr. (1948), anamorphic *Pezizomycotina*, St.≡ eH.1. 1, Ecuador.

Thripomyces Speg. (1915), Ceratomycetaceae. 1, Italy. See Nannfeldt (*Svensk bot. Tidskr.* **43**: 468, 1949).

Thrombiaceae Poelt & Vězda ex J.C. David & D. Hawksw. (1991), Pezizomycotina (inc. sed.) (L). 1 gen. (+ 4 syn.), 1 spp.

Lit.: Schmitt *et al.* (*Mycol.* **97**: 362, 2005).

Thrombium Wallr. (1831), Thrombiaceae (L). 1, Europe; N. America. See Swinscow (*Lichenologist* **2**: 276, 1964), Schmitt *et al.* (*Mycol.* **97**: 362, 2005; phylogeny).

Thrombocytozoons Tchacarof (1963) = Candida fide Desser & Barta (*Can. J. Microbiol.* **34**: 1096, 1988).

thrush, a throat and genital disease of humans caused by *Candida albicans*.

thryptogen (**thryptophyte**), an organism increasing the sensitivity of a suscept to outside factors, e.g. to cold (Langer, 1936).

Thryptospora Petr. (1947), Dothideomycetes. 1, Syria.

Thuchomyces Hallbauer & Jahns (1977), Fossil Fungi ? Fungi (L). 1 (Precambrian), S. Africa. Formed abiotically fide Cloud (*Palaeobiology* **2**: 351, 1976). See also Klappa (*Sediment. Petrol.* **49**: 387, 1979).

Thuemenella Penz. & Sacc. (1898), Xylariaceae. Anamorph *Nodulisporium*. 1, widespread. See Corlett (*Mycol.* **77**: 272, 1985), Samuels & Rossman (*Mycol.* **84**: 26, 1992), Rossman *et al.* (*Stud. Mycol.* **42**: 248 pp., 1999).

Thuemenia Rehm (1878) = Botryosphaeria fide von Arx & Müller (*Stud. Mycol.* **9**, 1975).

Thuemenidium Kuntze (1891) = Geoglossum fide Eckblad (*Nytt Mag. Bot.* **10**: 137, 1963), Gamundí (*Kew Bull.* **31**: 731, 1976), Benkert (*Delitschia* **10**: 141, 1983; key), Nitare (*Windahlia* **14**: 37, 1984), Spooner (*Biblthca Mycol.* **116**: 1, 1987).

Thujacorticium Ginns (1988), Cyphellaceae. 1, Canada. See Ginns (*Mycol.* **80**: 69, 1988).

Thwaitesiella Massee (1892) = Lopharia fide Donk (*Taxon* **6**: 17, 1957).

Thyrea A. Massal. (1856), Lichinaceae (L). *c.* 23, widespread. See Henssen (*Ber. dt. bot. Ges.* **81**: 176, 1968), Yoshimura (*J. Jap. Bot.* **43**: 354, 1968), Yoshimura (*J. Jap. Bot.* **43**: 500, 1968), Asahina (*J. Jap. Bot.* **45**: 65, 1970), Moreno & Egea (*Lichenologist* **24**: 215, 1992), Moreno & Egea (*Acta Bot. Barcinon.* **41**: 1, 1992; Eur. N. Afr.), Schultz *et al.* (*Lichenologist* **33**: 211, 2001; S America), Schultz *et al.* (*Pl. Biol.* **3**: 116, 2001; phylogeny), Schultz & Büdel (*Lichenologist* **34**: 39, 2002; key), Schultz & Büdel (*Lichenologist* **35**: 151, 2003).

Thyriascus Schulzer (1877) nom. dub., Fungi.

Thyridaria Sacc. (1875), Dothideomycetes. Anamorph *Cyclothyrium*. 23, widespread. See Wehmeyer (*Lloydia* **4**: 241, 1941), Barr (*Mycotaxon* **88**: 271, 2003; placement in *Didymosphaeriaceae*).

Thyridella (Sacc.) Sacc. (1895), Pezizomycotina. 3, widespread.

Thyridiaceae J.Z. Yue & O.E. Erikss. (1987), Sordariomycetidae (inc. sed.). 5 gen. (+ 5 syn.), 22 spp.

Lit.: Leuchtmann & Müller (*Bot. Helv.* **96**: 283, 1986), Eriksson & Yue (*SA* **8**: 9, 1989), Samuels & Rogerson (*Stud. Mycol.* **31**: 145, 1989), Eriksson & Yue (*SA* **10**: 57, 1990), Taylor *et al.* (*Sydowia* **49**: 94, 1997), Miller & Huhndorf (*Mol. Phylogen. Evol.* **35**: 60, 2005).

Thyridium Fuckel (1870) [non *Thyridium* Mitt. 1868, *Musci*] ≡ Mycothyridium E. Müll.

Thyridium Nitschke (1867), Thyridiaceae. Anamorph *Pleurocytospora*. 17, widespread. See Esfandiari & Petrak (*Sydowia* **4**: 11, 1950), Leuchtmann & Müller (*Bot. Helv.* **96**: 283, 1986), Eriksson & Yue (*SA* **8**: 9, 1989).

Thyrinula Petr. & Syd. (1924), anamorphic *Aulographina*, Cpt.0fH.15. 1, S. Africa; Australia. See Swart (*TBMS* **90**: 286, 1988).

Thyriochaetum Frolov (1968) ? = Amerosporium fide Sutton & Sarbhoy (*TBMS* **66**: 297, 1976).

Thyriodictyella Cif. (1962), Micropeltidaceae. 1, Dominican Republic. See Ciferri (*Atti Ist. bot. Univ. Lab. crittog. Pavia* sér. 5 **19**: 129, 1962).

Thyriopsis Theiss. & Syd. (1915), Asterinaceae. 1, widespread. See Câmara & Dianese (*Fitopatol. Brasil* **18** (Supl.): 346, 1993).

Thyriostroma Died. (1913) = Leptostroma fide Höhnel (*Sber. Akad. Wiss. Wien* Math.-naturw. Kl., Abt. 1 **124**: 49, 1915).

Thyriostromella Bat. & C.A.A. Costa (1959), anamorphic *Pezizomycotina*, Cpt.1eH.?. 1, Malaysia. See Batista & Costa (*Mycopath. Mycol. appl.* **11**: 15, 1959).

thyriothecium, an inverted flattened ascoma, having the wall ('scutellum', 'placodium') more or less radial in structure, and lacking a basal plate, e.g. *Microthyrium*; cf. catathecium.

Thyrococcum (Sacc.) Sacc. (1913) nom. dub., anamorphic *Pezizomycotina*. See von Höhnel (*Sber. Akad. Wiss. Wien* Math.-naturw. Kl., Abt. 1 **124**: 49, 1915).

Thyrodochium Werderm. (1924) = Stemphylium fide Wiltshire (*TBMS* **21**: 211, 1938).

Thyronectria Sacc. (1875) nom. dub., ? Sordariomycetes. 1, Italy. See Rossman (*Mem. N. Y. bot. Gdn* **49**: 253, 1989).

Thyronectroidea Seaver (1909), Thyridiaceae. 1 (from bark), USA. See Rossman *et al.* (*Stud. Mycol.* **42**: 248 pp., 1999).

Thyrosoma Syd. (1921) = Cyclotheca fide Müller & von Arx (*Beitr. Kryptfl. Schweiz* **11** no. 2, 1962).

Thyrospora Kirschst. (1938), Dothideales. 1, Germany.

Thyrospora Tehon & E.Y. Daniels (1925) = Stemphylium fide Wiltshire (*TBMS* **21**: 211, 1938), Simmons (*Mycotaxon* **88**: 163, 2003).

Thyrostroma Höhn. (1911), anamorphic *Dothidotthia*, Hsp.≡ -#eP.19. 19, widespread. See Sutton & Pascoe (*MR* **92**: 210, 1989; relationship with *Stigmina*), Yuan & Old (*MR* **94**: 573, 1990; *T. eucalypti* on *Eucalyptus*), Sutton (*Arnoldia* **14**: 33, 1997), Ramaley (*Mycotaxon* **94**: 127, 2005; teleomorph).

Thyrostromella Höhn. (1919), anamorphic *Pezizomycotina*, Hsp.#eP.10. 2, Australia; Europe.

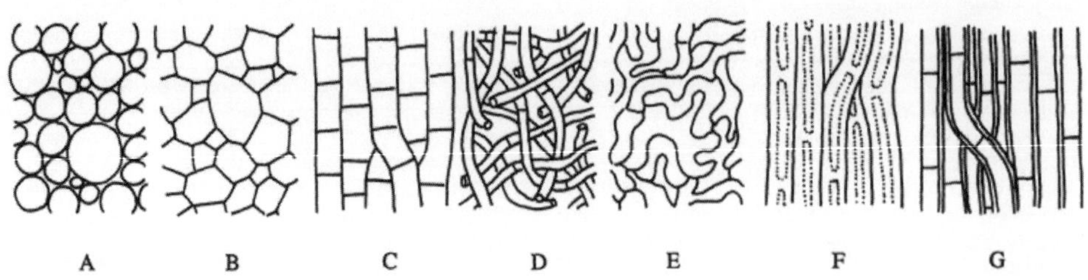

Fig. 24. Hyphal tissue (textura) types (Korf, 1958). A, textura globulosa; B, textura angularis; C, textura prismatica; D, textura intricata; E, textura epidermoidea; F, textura oblita; G, textura porrecta.

Thyrostromella Syd. (1924) [non *Thyrostromella* Höhn. 1919] = Thyrostroma fide Ellis (*Dematiaceous Hyphomycetes*, 1971), Sutton & Pascoe (*MR* **92**: 212, 1989).

Thyrotheca Kirschst. (1944), Pezizomycotina. 1, N. America.

Thyrsidiella Höhn. ex Höhn. (1909), anamorphic *Pezizomycotina*. 2, Europe.

Thyrsidina Höhn. (1905), anamorphic *Pezizomycotina*, Cac.#eH.1. 1, Europe.

Thyrsidium Mont. (1849) ≡ Myriocephalum.

thyrsus (pl. **thyrsi**), (1) a type of inflorescence (Bot.); (2) the densely branched apices of some lichens, e.g. *Cladonia stellaris*.

thysanoblastic (of conidiogenesis), when 'the whole of the upper surface of the conidiogenous cell takes part in the process of conidium formation and secession, and both schizolysis and rhexolysis occur alternately in successively seceding conidia' (Roux & van Warmelo, *MR* **92**: 225, 1989).

Thysanophora W.B. Kendr. (1961), anamorphic *Trichocomaceae*, Hso.0eP.15. 9, Europe; N. America. See Stolk & Hennebert (*Persoonia* **5**: 189, 1968), Iwamoto *et al.* (*Mycoscience* **43**: 169, 2002; phylogeny), Peterson & Sigler (*MR* **106**: 1109, 2002), Iwamoto *et al.* (*Mycol.* **97**: 1238, 2005; evolution).

Thysanophoron Stirt. (1883) = Sphaerophorus fide Wedin (*Pl. Syst. Evol.* **187**: 213, 1993).

Thysanopyxis Rabenh. ex Bonord. (1864) = Volutella fide Saccardo (*Syll. fung.* **4**: 684, 1886).

Thysanorea Arzanlou, W. Gams & Crous (2007), anamorphic *Herpotrichiellaceae*. 1, Papua New Guinea. See Arzanlou *et al.* (*Stud. Mycol.* **58**: 57, 2007).

Thysanothecium Mont. & Berk. (1846), Cladoniaceae (L). 2, Australasia; Asia. See Galloway (*Nova Hedwigia* **28**: 499, 1977), Stenroos (*Ann. bot. fenn.* **25**: 207, 1988; Melanesia), Ahti (*Regnum veg.* **128**: 58, 1993), Wei *et al.* (*Mycosystema* **7**: 23, 1994; China), Hammer (*Biblthca Lichenol.* **78**: 75, 2001; ontogeny).

Tiarospora Sacc. & Marchal (1885), anamorphic *Phaeosphaeriaceae*, Cpd.1eH.1. 3, Europe.

Tiarosporella Höhn. (1924), anamorphic *Darkera*, Cpd.0eH.1. 10, Europe; N. America. See Sutton (*The Coelomycetes*, 1980), Roux *et al.* (*MR* **94**: 254, 1990; conidiomatal and conidial ontogeny), Hyde (*Aust. Syst. Bot.* **6**: 169, 1993), Nag Raj (*Coelomycetous Anamorphs with Appendage-bearing Conidia*, 1993), Kohlmeyer & Volkmann-Kohlmeyer (*Mycotaxon* **59**: 79, 1996), Karadžic (*Eur. J. For. Path.* **28**: 145, 1998), Müller & Hantula (*MR* **102**: 1163, 1998; diversity).

Tiarosporellivora Punith. (1981), anamorphic *Pezizomycotina*, Cpd.0eH.15. 1 (in *Tiarosporella*), Germany. See Punithalingam (*Nova Hedwigia* **35**: 26, 1981).

Tibellia Vězda & Hafellner (1992), Ramalinaceae (L). 1, Australia. See Hafellner & Vezda (*Nova Hedwigia* **55**: 186, 1992), Ekman (*Op. bot.* **127**: 148 pp., 1996).

tibi, Swiss drink derived from a 15% sucrose sol. (+ other ingredients) by fermentation with 'Tibi grains', a symbiotic association of *Betabacterium vermiforme* and *Saccharomyces intermedius*. Cf. gingerbeer plant, tea fungus, teekwass.

Tichodea Körb. (1848) = Collema F.H. Wigg. fide Hawksworth *et al.* (*Dictionary of the Fungi* edn 8, 1995).

Tichospora A. Massal. ex Horw. (1912) nom. inval. ? = Psorotichia fide Hawksworth *et al.* (*Dictionary of the Fungi* edn 8, 1995).

Tichothecium Flot. (1850) = Verrucaria Schrad. fide Hawksworth (*Bot. Notiser* **132**: 283, 1979).

tichus, peripheral layer of cells of perithecial walls forming a dark protective layer as in *Pleospora herbarum* (Groenhart, *Persoonia* **4**: 11, 1965).

Ticogloea G. Weber, Spaaij & W. Gams (1994), anamorphic *Pezizomycotina*, Hso.0eP.10. 1, widespread. See Weber *et al.* (*MR* **98**: 663, 1994), Halmschlager & Kowalski (*Öst. Z. Pilzk.* **9**: 67, 2000).

Ticomyces Toro (1952), Meliolaceae. 1 (from leaves), C. America. See Toro (*J. Agric. Univ. Puerto Rico* **36**: 48, 1952).

Ticonectria Döbbeler (1998), ? Nectriaceae. 1 (liverworts), Costa Rica. See Döbbeler (*Nova Hedwigia* **66**: 325, 1998).

Tieghemella Berl. & De Toni (1888) = Absidia fide Hesseltine (*Mycol.* **47**: 344, 1955).

Tieghemiomyces R.K. Benj. (1959), Dimargaritaceae. 2 (mycoparasites of *Mucorales*), India; Poland; USA. See Benjamin (*Aliso* **5**: 11, 1961; key), Misra & Gupta (*Mycotaxon* **7**: 204, 1978; India), Beblowska (*Acta Mycologica* Warszawa **27**: 271, 1991; Poland).

tiger's milk, *Polyporus sacer*; used as a medicine in Malaysia.

Tigria Trevis. (1853) = Erysiphe fide Braun (*SA* **7**: 57, 1988).

Tilachlidiopsis Keissl. (1924), anamorphic *Dendrocollybia*, Hsy.0eH.41. 1 (with clamp connexions), widespread (north temperate). See Mains (*Bull. Torrey bot. Club* **78**: 122, 1951), Stalpers *et al.* (*CJB* **69**: 6, 1991; gen. revision).

Tilachlidium Preuss (1851), anamorphic *Hypocreales*, Hsp.0eH.15. 1, widespread. See Gams (*Cephalosporium-artige Schimmelpilze*, 1971).

Tilakidium Vaidya, C.D. Naik & Rathod (1986), ? Hypocreaceae. 1, India. Affinities uncertain, the description is inadequate for proper characterization.

Tilakiella Srinivas. (1973), Dothideomycetes. 1, India. See Srinivasalu (*Indian Phytopath.* **25**: 292, 1972).

Tilakiopsis V.G. Rao (1994), anamorphic *Pezizomycotina*, Hso.?.?. 1, India. See Rao (*Advances in Mycology and Aerobiology* (Dr S.T. Tilak Commemoration Volume). *Current Trends in Life Sciences* **20**: 132, 1994).

Tilakomyces Sathe & Vaidya (1979) = Eutypella fide Rappaz (*Mycol. Helv.* **3**:: 281, 1989).

Tilletia Tul. & C. Tul. (1847), Tilletiaceae. Anamorph *Tilletiella*. *c.* 175 (on *Poaceae*), widespread. Basidiospores produced at the apex of an aseptate basidium; few to many in number, when 1-nucleate, often conjugating with one another, giving rise to ballistosporic 'secondary conidia', but in some spp. basidiospores are multinucleate and do not conjugate. Species delimited on morphological characters do not always fit with those based on molecular data, which are also not unanimous. *T. caries* (syn. *T. tritici*) and *T. laevis* (syn. *T. foetida*) (wheat bunt); *T. controversa* (dwarf bunt of wheat, barley, rye and several grasses); *T. horrida* (syn. *Neovossia horrida*) (rice smut); *T. indica* (syn. *N. indica*) (kernal bunt of wheat). See Buller & Vanterpool Buller (*Researches on Fungi* **5**, 1933), Holton & Heald (*Bunt or stinking smut of wheat*, 1941), Durán & Fischer (*The genus Tilletia*, 1961; keys, species delimited on morphological characters), Kollmorgen *et al.* (*TBMS* **75**: 461, 1980), Durán (*Ustilaginales of Mexico*: 331 pp., 1987; basidiospores), Ingold (*TBMS* **88**: 75, 1987), Russell & Mills (*Phytopath.* **84**: 576, 1994), Shi *et al.* (*Phytopath.* **86**: 311, 1996), Pimentel *et al.* (*Mycol.* **90**: 1017, 1998), Pimentel *et al.* (*Phytopath.* **90**: 376, 1998), Pimentel *et al.* (*Mycol.* **92**: 411, 2000).

Tilletiaceae J. Schröt. (1887), Tilletiales. 6 gen. (+ 1 syn.), 186 spp.

Lit.: Kollmagen *et al.* (*TBMS* **75**: 461, 1980), Durán (*Ustilaginales of Mexico. Taxonomy, Symptomatology, Spore Germination and Basidial Cytology*: 331 pp., 1987), Ingold (*TBMS* **88**: 75, 1987), Buller & Vanterpool *in* Buller (*Mycol.* **82**: 276, 1990), Begerow *et al.* (*CJB* **75**: 2045, 1998), Palm (*Mycol.* **91**: 1, 1999), McDonald *et al.* (*Pl. Dis.* **84**: 1121, 2000), Pimentel *et al.* (*Phytopathology* **90**: 376, 2000), Pimentel *et al.* (*Mycol.* **92**: 411, 2000), Levy *et al.* (*Phytopathology* **91**: 935, 2001), Josefsen & Christiansen (*MR* **106**: 1287, 2002), Vánky & McKenzie (*Fungal Diversity Res. Ser.* **8**: 259 pp., 2002), Chesmore *et al.* (*Bulletin OEPP* EPPO Bulletin **33**: 495, 2003), Anon. (*Bulletin OEPP* EPPO Bulletin **34**: 219, 2004), Cunnington & Shivas (*Australas. Mycol.* **22**: 53, 2004), Vánky (*Mycol. Balcanica* **1**: 175, 2004), Castlebury *et al.* (*Mycol.* **97**: 888, 2005), Tan & Murray (*MR* **110**: 203, 2006).

Tilletiales Kreisel ex R. Bauer & Oberw. (1997). Exobasidiomycetes. 1 fam., 6 gen., 186 spp. Fam.:
Tilletiaceae

For *Lit.* see under fam.

Tilletiaria Bandoni & B.N. Johri (1972), Tilletiariaceae. 1 (saprobic), Canada. See Begerow *et al.* (*MR* **104**: 53, 2000; mol. phylogeny).

Tilletiariaceae R.T. Moore (1980), Georgefischeriales. 3 gen., 7 spp.

Lit.: Vánky (*Cryptog. Stud.* **1**: 159 pp., 1987), Ingold (*MR* **96**: 987, 1992), Takashima & Nakase (*J. gen. appl. Microbiol.* Tokyo **42**: 421, 1996), Bauer *et al.* (*CJB* **75**: 1273, 1997), Begerow *et al.* (*CJB* **75**: 2045, 1998), Boekhout *in* Kurtzman & Fell (Eds) (*Yeasts, a taxonomic study* 4th edn: 703, 1998), Begerow *et al.* (*MR* **104**: 53, 2000; mol. phylogeny), Bauer *et al.* (*MR* **105**: 423, 2001).

Tilletiella Zambett. (1970) nom. inval. = Tilletia.

Tilletiopsis Derx (1948), anamorphic *Entyloma, Melanotaenium*. 7, widespread. See Boekhout (*Stud. Mycol.* **33**: 1, 1991; species revision).

Tilotus Kalchbr. (1881) nom. dub. ? = Phyllotopsis fide Singer (*Agaric. mod. Tax.* edn 3, 1975).

Timdalia Hafellner (2001), Lecanoromycetes (L). 1. See Hafellner & Türk (*Stapfia* **76**: 158, 2001), Reeb *et al.* (*Mol. Phylogen. Evol.* **32**: 1036, 2004; phylogeny), Crewe *et al.* (*MR* **110**: 521, 2006; phylogeny).

Timgrovea Bougher & Castellano (1993), Agaricomycetes. 5, Australasia; China. See Bougher & Castellano (*Mycol.* **85**: 288, 1993), Francis & Bougher (*Australasian Mycologist* **21**: 81, 2002).

Tinctoporellus Ryvarden (1979), Polyporaceae. 2, widespread (pantropical). See Ryvarden (*TBMS* **73**: 18, 1979), Ryvarden & Iturriaga (*Mycol.* **95**: 1066, 2003; Venezuela).

Tinctoporia Murrill (1907) = Porogramme fide Ryvarden (*TBMS* **73**: 1, 1979).

tinder fungus, *Fomes fomentarius*; **false - -**, *Phellinus igniarius*. Cf. amadou.

tinea, ringworm or other skin diseases in humans or animals caused by various parasitic fungi (esp. dermatophytes). **- barbae** (- sycosis), beard ringworm; **- capitis** (- tonsurans), head ringworm; **- corporus** (**- circinata**), body ringworm; **- cruris**, groin ringworm; **- favosa** = favus (q.v.); **- imbricata** (Tokelau) (*Trichophyton concentricum*); **- nigra**, pigmented cutaceous infection caused by a dematiaceous fungus; **- nodosa** = piedra (q.v.); **- pedis**, 'athlete's foot', foot ringworm; **- unguium**, ringworm of the nails; **- versicolor** = pityriasis versicolor (q.v.).

Tingiopsidium Werner (1939) = Koerberia fide Henssen (*Symb. bot. upsal.* **18** no. 1, 1963).

tinophyses (Groenhart, *Persoonia* **4**: 11, 1965), paraphysoids; see hamathecium.

Tirispora E.B.G. Jones & Vrijmoed (1994), Halosphaeriaceae. 2 (marine), Hong Kong; India. See Jones *et al.* (*CJB* **72**: 1373, 1994), Sarma & Hyde (*Australas. Mycol.* **19**: 52, 2000; India), Pang *et al.* (*Nova Hedwigia* **77**: 1, 2003; phylogeny).

Tirisporella E.B.G. Jones, K.D. Hyde & Alias (1996), Dothideomycetes. Anamorph *Phialophora*-like. 1, widespread. See Jones *et al.* (*CJB* **74**: 1489, 1996).

Tirmania Chatin (1892), Pezizaceae. 3 (hypogeous, mycorrhizal), widespread. See Trappe (*TBMS* **57**: 185, 1971), Malençon (*Persoonia* **7**: 261, 1973), Alsheikh & Trappe (*TBMS* **81**: 83, 1983), Moreno *et al.* (*MR* **104**: 378, 2000), Díez *et al.* (*Mycol.* **94**: 247, 2002; phylogeny), Hansen *et al.* (*Mol. Phylogen. Evol.* **36**: 1, 2005; phylogeny), Læssøe & Hansen (*MR* **111**: 1075, 2007; phylogeny).

tissue types, Korf distinguished the types of hyphal tissues in discomycetes as different **textura**'s and this is now applied to all ascomycetes and coelomycetes. Tissue (textura) types (from Korf, *Sci. Rep. Yokohama nat. Univ.* II **7**: 13, 1958; which is derived from Starbäck, 1895). See also Dargan (*Nova Hedw.* **44**: 489, 1987; *Xylariaceae*), plectenchyma. See Fig. 24.

Titaea Sacc. (1876), anamorphic *Paranectriella*, Hso.1bH.10. 9 (on fungi), widespread. See Sutton (*TBMS* **83**: 399, 1984; key), Pelaez *et al.* (*Mycotaxon* **70**: 55, 1999; Spain), Kirschner & Piepenbring (*Fungal Diversity* **21**: 93, 2006).

Titaeella G. Arnaud ex K. Ando & Tubaki (1985), anamorphic *Agaricomycetes*, Hso.1bH.1. 1 (with clamp connexions, aquatic), Europe; Japan. See Ando & Tubaki (*TMSJ* **26**: 151, 1985).

Titaeopsis B. Sutton & Deighton (1984), anamorphic *Pezizomycotina*, Hso.1bH.10. 1 (on leaf ascomycetes), Uganda. See Sutton & Deighton (*TBMS* **83**: 409, 1984).

Titaeospora Bubák (1916), anamorphic *Pezizomycotina*, Cac.1eH.15. 2, widespread (temperate).

Titaeosporina Luijk (1920) = Asteroma fide Sutton (*Mycol. Pap.* **141**, 1977).

Titanella Syd. & P. Syd. (1919) = Pyrenula Ach. (1809) fide Barr (*Mycotaxon* **9**: 17, 1979), Harris (*Mem. N. Y. bot. Gdn* **49**, 1989), Checa & Blanco (*Mycotaxon* **91**: 353, 2005).

Titania Berl. (1891) [non *Titania* Endl. 1833, *Orchidaceae*] ≡ Fremineavia.

Tjibodasia Holterm. (1898) = Platygloea fide Martin (*Lloydia* **11**: 119, 1948).

toadstool, a basidioma, esp. an inedible one, of an agaric or a bolete; there is no unequivocal definition and the term is best avoided; in 1959 there was a controversial court case as to whether soup made from *Boletus edulis* should be referred to as 'toadstool soup' (Small, *Pl. Sci. Bull.* **21**(3): 34, 1975). Cf. mushroom.

Tode (Heinrich Julius; 1733-1797; Germany). President of the Synod of Wittenburg. An early mycologist who gave names to such common ascomycete genera as *Acrospermum*, *Hysterium*, *Myrothecium*, *Stilbum*, *Sphaerobolus*, and *Volutella* and gave accounts of 54 species. of *Sphaeria*. *Publs. Fungi Mecklenburgenses Selecti* (1790-1791). *Biogs, obits etc.* Grummann (1974: 50); Stafleu & Cowan (*TL-2* **6**: 382, 1986).

Tofispora G. Langer (1994) = Thanatephorus fide Roberts (*Rhizoctonia-forming fungi*, 1999).

tofu, see sufu.

Togaria W.G. Sm. (1908) = Agrocybe fide Singer (*Agaric. mod. Tax.* edn 3, 1975).

Togninia Berl. (1900), Togniniaceae. Anamorph *Phaeoacremonium*. 11, widespread. See Barr *et al.* (*Mycotaxon* **48**: 529, 1993), Mostert *et al.* (*Mycol.* **95**: 646, 2003; anamorph), Pascoe *et al.* (*Phytopath. Mediterr.* **43**: 51, 2004; Australia), Réblová *et al.* (*Stud. Mycol.* **50**: 533, 2004), Rooney-Latham *et al.* (*Pl. Dis.* **89**: 177, 2005; USA), Mostert *et al.* (*Stud. Mycol.* **54**: 115 pp., 2006; revision), Mostert *et al.* (*Phytopath. Mediterr.* **45** Suppl.: S12, 2006), Réblová & Mostert (*MR* **111**: 299, 2007; phylogeny).

Togniniaceae Réblová, L. Mostert, W. Gams & Crous (2004), ? Diaporthales. 2 gen. (+ 2 syn.), 25 spp.

Lit.: Barr (*Mycol.* **77**: 549, 1985), Mostert *et al.* (*Mycol.* **95**: 646, 2003), Pascoe *et al.* (*Phytopath. Mediterr.* **43**: 51, 2004), Réblová *et al.* (*Stud. Mycol.* **50**: 533, 2004), Vijaykrishna *et al.* (*Stud. Mycol.* **50**: 387, 2004), Fischer *et al.* (*Mycotaxon* **92**: 85, 2005), Mostert *et al.* (*Stud. Mycol.* **54**: 115 pp., 2006), Rossman *et al.* (*Mycoscience* **48**: 135, 2007; phylogeny).

Togniniella Réblová, L. Mostert, W. Gams & Crous (2004), Calosphaeriaceae. Anamorph *Phaeocrella*. 1, New Zealand. See Réblová *et al.* (*Stud. Mycol.* **50**: 543, 2004), Mostert *et al.* (*Stud. Mycol.* **54**: 115 pp., 2006).

Tolediella Viégas (1943) = Phyllachora Nitschke ex Fuckel (1870) fide Petrak (*Sydowia* **5**: 340, 1951).

tolerant (of an organism), giving little reaction to infection by a pathogen or to the effect of other factors (e.g. tolerant of heat, of a virus).

Tolypocladium W. Gams (1971), anamorphic *Elaphocordyceps*, Hso.0eH.15. 11, widespread. See also cyclosporin. See Riba *et al.* (*J. Invert. Path.* **48**: 362, 1986; isozyme analysis), von Arx (*Mycotaxon* **25**: 153, 1986), Holubová-Jechová (*Mykol. Listy* **31**: 8, 1988; key), von Arx (*Persoonia* **13**: 467, 1988), Aarnio & Agathos (*Appl. Microbiol. Biotechn.* **33**: 435, 1990; pigmented variants in *T. inflatum*), Rakotonirainy (*J. Invert. Path.* **57**: 17, 1991; rRNA sequence comparison with *Beauveria*), Stimberg *et al.* (*Appl. Microbiol. Biotechn.* **37**: 485, 1992; electrophoretic karyotyping), Dreyfuss & Gams (*Taxon* **43**: 660, 1994; nomencl.), Doggett & Porter (*Mycol.* **88**: 720, 1996; teleomorph), Fang *et al.* (*Mycosystema* **17**: 40, 1998), Stensrud *et al.* (*MR* **109**: 41, 2005; phylogeny), Sung *et al.* (*Stud. Mycol.* **57**, 2007).

Tolypoderma Thirum. & M.J. O'Brien (1978) nom. inval. = Moesziomyces fide Vánky (*Europ. Smut Fungi*: 163, 1994).

Tolypomyria Preuss (1852) = Trichoderma Pers. (1794) fide Hughes (*Friesia* **9**: 64, 1969).

tolypophagous, see ptyophagous.

Tolyposporella G.F. Atk. (1897), Tilletiariaceae. 5 (on *Poaceae*), N. & S. America; Africa; Asia. See Thirumalachar *et al.* (*Mycol.* **59**: 389, 1967).

Tolyposporidium Thirum. & Neerg. (1978) = Moesziomyces fide Vánky (*Carpathian Ustilaginales*, 1985).

Tolyposporium Woronin ex J. Schröt. (1887), Anthracoideaceae. 2 (on *Cyperaceae*, *Juncaceae*), Europe; New Zealand; N. America. See Vánky (*Bot. Notiser* **130**: 131, 1977), Vánky (*Mycotaxon* **74**: 344, 2000).

Tomasellia A. Massal. (1856), ? Naetrocymbaceae (±L). *c.* 50, widespread (esp. tropical). See Keissler (*Rabenh. Krypt.-Fl.* **9**, 1937), Harris (*More Florida Lichens*, 1995; delimitation).

Tomaselliella Cif. (1952) = Arthonia fide Hawksworth *et al.* (*Dictionary of the Fungi* edn 8, 1995).

Tomaselliellomyces Cif. (1953) ≡ Tomaselliella.

Tomaselliomyces Cif. & Tomas. (1953) ≡ Tomasellia.

Tomelin (Boris Anatol'evich; 1928-2008; Russia). Graduate, Kharkiv State University, Ukraine (1953); student, then laborant, then mycologist, then Head of the Laboratory of Applied Mycology, then Head of the Department of Ecology of Fungi, then Professor of Mycology, V.I. Komarov Botanical Institute (1954-2005), St Petersburg. Carried out pioneering mycological exploration of boreal and oriental Russia (the Amur valley, Komi, the Taimyr peninsula, the northern Urals, Yakutia and the Russian Far East) described in a series of papers in various Russian language journals; known for work on *Mycosphaerella*; also studied diseases of cotton in Kazakhstan (1960s). *Publs*. [*Identification Book of the Genus Mycosphaerella Johans.*] (1979) [in Russian]. *Biogs, obits etc.* Karatygin, Dudka, Andrianova & Lebedeva ([*Mikologiya i Fitopatologiya*] **32**: 89, 1998) [portrait, in Russian].

Tomentella Johan-Olsen (1888) [non *Tomentella* Pat. 1887, nom. cons.] = Tomentella Pat. fide Donk (*Taxon* **6**: 119, 1957).

Tomentella P. Karst. (1889) [non *Tomentella* Pat. 1887, nom. cons.] = Trechispora fide Stalpers (*in litt.*; or = *Phlebiella*).

Tomentella Pers. ex Pat. (1887) nom. cons., Thelephoraceae. 80 (ectomycorrhizal), widespread. See Wakefield (*TBMS* **53**: 161, 1969; key 40 Br. spp.), Larsen (*Mycol. Mem.* **4**, 1974; key 72 spp.), Stalpers (*Stud. Mycol.* **35**: 168 pp., 1993; key), Kõljalg (*Syn. Fung.* **9**: 213 pp., 1996; key), Jakucs *et al.* (*Mycorrhiza* **Mycorrhiza**: 247, 2005).

Tomentellaceae Warm. (1890) = Thelephoraceae.

Tomentellago Hjortstam & Ryvarden (1988) = Amaurodon fide Kõljalg (*Syn. Fung.* **9**, 1996).

Tomentellastrum Svrček (1958) = Tomentella Pat. fide Donk (*Taxon* **12**: 113, 1963).

Tomentelleopsis Orlova (1959) ? = Chromelosporium fide Sutton (*in litt.*).

Tomentellina Höhn. & Litsch. (1906) = Tomentella Pat. fide Kõljalg (*Syn. Fung.* **9**, 1996).

Tomentellopsis Hjortstam (1970), Thelephoraceae. 7, widespread. See Hjortstam (*Svensk bot. Tidskr.* **68**: 51, 1974), Kõljalg (*Syn. Fung.* **9**, 1996; key), Kõljalg *et al.* (*Mycol. Progr.* **1**: 81, 2002; phylogeny).

Tomenticola Deighton (1969), anamorphic *Pezizomycotina*, Hso.≡ eP.10. 1, Africa (tropical). See Deighton (*Mycol. Pap.* **117**: 20, 1969).

Tomentifolium Murrill (1903) ≡ Phyllotopsis.

Tomentoporus Ryvarden (1973) = Microporus fide Reid (*Microscopy* **32**: 448, 1975).

tomentose, having a covering of soft, matted hairs (a **tomentum**; downy.

Tomeoa I. Hino (1954), Dothideomycetes. 1, Japan. See Hino (*Bulletin of the Faculty of Agriculture, Yamaguchi University* **5**: 241, 1954).

Tomophagus Murrill (1905) = Ganoderma fide Furtado (*Mycol.* **57**: 599, 1965).

Tompetchia Subram. (1985), anamorphic *Pezizomycotina*, Hso.#eP.1. 1 (on scale insects), widespread. See Subramanian (*Kavaka* **12**: 66, 1984).

Tonduzia F. Stevens (1927) [non *Tonduzia* Pittier 1908, *Apocynaceae*] ≡ Dontuzia.

Toninia A. Massal. (1852) nom. cons., Catillariaceae (±L). 82 (a few on lichens), widespread. Perhaps better subsumed within the *Ramalinaceae*. See Timdal (*Op. Bot.* **110**, 1991; key), Lumbsch *et al.* (*Mol. Phylogen. Evol.* **31**: 822, 2004; phylogeny), Miądlikowska *et al.* (*Mycol.* **98**: 1088, 2006; phylogeny).

Toniniopsis Frey (1926), Ramalinaceae (L). 1, Europe.

tonophily (adj. **-ilic**, **-ilous**), the ability to grow under conditions of high osmotic pressure.

Topelia P.M. Jørg. & Vězda (1984), ? Stictidaceae (L). 6, Europe; N. America. See Jørgensen & Vězda (*Nova Hedwigia* Beih. **79**: 502, 1984), Tretiach & Vezda (*Lichenologist* **24**: 107, 1992; Italy).

Topeliopsis Kantvilas & Vězda (2000), Thelotremataceae (L). 8, widespread. See Kantvilas & Vězda (*Lichenologist* **32**: 347, 2000), Kalb (*Mycotaxon* **79**: 319, 2001; Australia), Frisch (*Biblthca Lichenol.* **92**: 3, 2006; Africa), Frisch & Kalb (*Lichenologist* **38**: 37, 2006), Frisch *et al.* (*Biblthca Lichenol.* **92**: 517, 2006; phylogeny), Mangold *et al.* (*Lichenologist* **40**: 39, 2008; phylogeny).

Tophora Fr. (1825) ≡ Byssus.

Topospora Fr. (1836), anamorphic *Godronia*, St.≡ eH.15. 6, widespread. See Verkley (*Nova Hedwigia* **75**: 433, 2002; phylogeny).

Tormentella H.D. Pflug (1966), Fossil Fungi. 1, USA.

Tornabea Østh. (1980), Physciaceae (L). 2, widespread. See Kurokawa (*J. Jap. Bot.* **37**: 289, 1962), Nimis & Tretiach (*Bryologist* **100**: 217, 1997; revision), Nordin & Mattsson (*Lichenologist* **33**: 3, 2001; phylogeny), Scheidegger *et al.* (*Lichenologist* **33**: 25, 2001; evolution), Helms *et al.* (*Mycol.* **95**: 1078, 2003; phylogeny), Miądlikowska *et al.* (*Mycol.* **98**: 1088, 2006; phylogeny).

Tornabenia A. Massal. (1853) [non *Tornabenea* Parl. 1850, *Umbelliferae*] = Anaptychia fide Hawksworth *et al.* (*Dictionary of the Fungi* edn 8, 1995).

Tornabenia Trevis. (1853) [non *Tornabenea* Parl. 1850, *Umbelliferae*] ≡ Tornabea.

Tornabeniopsis Follmann (1980) nom. inval. ≡ Tornabea.

Toroa Syd. (1926), Pseudoperisporiaceae. 2, widespread (tropical). See Hansford (*Mycol. Pap.* **15**, 1946).

Torpedospora Meyers (1957), Hypocreomycetidae. 2 (marine), widespread. See Meyers (*Mycol.* **49**: 496, 1957), Sakayaroj *et al.* (*Bot. Mar.* **48**: 395, 2005; revision, phylogeny), Schoch *et al.* (*MR* **111**: 154, 2007; phylogeny).

Torrendia Bres. (1902) = Amanita Pers. fide Kuyper (*in litt.*); basidioma gasteroid.

Torrendiaceae Jülich (1982) = Amanitaceae.

Torrendiella Boud. & Torrend (1911), Sclerotiniaceae. 12, widespread. See Spooner (*Biblthca Mycol.* **116**, 1987), Galán *et al.* (*Mycotaxon* **48**: 229, 1993; posn), Zhuang (*Mycotaxon* **72**: 325, 1999), Johnston & Gamundí (*N.Z. Jl Bot.* **38**: 493, 2000; 6 spp. on *Nothofagus*), Malaval (*Bulletin Semestriel de la Fédération des Associations Mycologiques Méditerranéennes* **28**: 41, 2005; France).

Torrentispora K.D. Hyde, W.H. Ho, E.B.G. Jones, K.M. Tsui & S.W. Wong (2000), Annulatascaceae. 3 (from wood, aquatic), Hong Kong. See Hyde *et al.* (*MR* **104**: 1399, 2000), Lee *et al.* (*Fungal Diversity* **16**: 87, 2004; ultrastr.).

Torrubia Lév. (1863) ≡ Cordyceps.

Torrubiella Boud. (1885), Cordycipitaceae. Anamorphs *Akanthomyces*, *Gibellula*, *Lecanicillium*, *Simplicillium*. 70 (on *Arachnida* and *Coccida*), widespread (tropical). See Petch (*TBMS* **27**: 81, 1944), Samson *et al.* (*Atlas of Entomopathogenic Fungi*, 1988), Hywel-Jones (*MR* **99**: 330, 1995; Thailand), Hywel-Jones (*MR* **101**: 143, 1997; Thailand), Hywel-Jones *et al.* (*MR* **101**: 1242, 1997; anamorphs), Tzean *et al.* (*MR* **102**: 1350, 1998; Taiwan), Zare & Gams (*Nova Hedwigia* **73**: 1, 2001; anamorph), Bischoff & White (*Stud. Mycol.* **50**: 89, 2004), Sung *et al.* (*Stud. Mycol.* **57**, 2007).

Torsellia Fr. (1849) = Cytospora fide Défago (*Phytopath. Z.* **14**: 103, 1944).
torsive, spirally twisted.
Torula Pers. (1794), anamorphic *Pezizomycotina*, Hso.0eP.3. 7, widespread. *T. ligniperda* causes stain in hardwoods. See Rao & de Hoog (*Persoonia* **8**: 201, 1975; key), Crane & Schoknecht (*Mycol.* **78**: 86, 1986), Casares-Porcel *et al.* (*Lichenologist* **28**: 37, 1996; spp. on gypsum), Crane (*Mycotaxon* **80**: 109, 2001; nomenclator), Hughes & Crane (*Mycol.* **98**: 141, 2006).
Torulaspora Lindner (1904), Saccharomycetaceae. 3, widespread. See van der Walt & Johansen (*CSIR Res. Rept.* **325**, 1975), Kreger-van Rij (Ed.) (*Yeasts, a taxonomic study* 3rd edn, 1984), Yamada *et al.* (*J. gen. appl. Microbiol.* Tokyo **37**: 503, 1991; molec. relations), Oda & Tonomura (*Lett. Appl. Microbiol.* **21**: 190, 1995; karyotypes), James *et al.* (*Int. J. Syst. Bacteriol.* **46**: 189, 1996; ITS), Smole Možina *et al.* (*Lett. Appl. Microbiol.* **24**: 311, 1997; PCR), Kurtzman *in* Kurtzman & Fell (Eds) (*Yeasts, a taxonomic study* 4th edn: 404, 1998), Suh *et al.* (*Mycol.* **98**: 1006, 2006; phylogeny).
Torulella Gyeln. (1939) nom. dub., anamorphic *Pezizomycotina*. See Hawksworth (*Bull. Br. Mus. nat. hist.* Bot. **6**: 181, 1979).
Torulina Sacc. & D. Sacc. (1906) = Gliomastix fide Dickinson (*Mycol. Pap.* **115**, 1968).
Torulites Grüss (1927), Fossil Fungi. 1 (Tertiary), Europe.
Torulites Pia (1927), Fossil Fungi. 1 (Oligocene), Europe.
Toruloidea Sumst. (1913) = Oospora fide Clements & Shear (*Gen. Fung.*, 1931).
Torulomyces Delitsch (1943), anamorphic *Trichocomaceae*, Hso.0eH.15. 6, Europe; N. America. See Barron (*Mycol.* **59**: 716, 1967), Ando *et al.* (*Mycoscience* **39**: 313, 1998; key).
Torulopsidosira Geitler (1955), ? Algae. See Batra *in* Subramanian (Ed.) (*Taxonomy of fungi* **1**: 187, 1978; an achlorotic alga).
Torulopsiella Bender (1932), anamorphic *Pezizomycotina*, Hso.0eP.38. 2, S. America. See Hughes (*N.Z. Jl Bot.* **10**: 232, 1972).
Torulopsis Berl. (1894) = Candida fide van der Walt & Johannsen (*Antonie van Leeuwenhoek* **40**: 281, 1974), Yarrow & Meyer (*Int. J. Syst. Bacteriol.* **28**: 611, 1978), Odds *et al.* (*J. gen. Microbiol.* **136**: 761, 1990; synonymy with *Candida*).
Torulopsis Oudem. (1903) = Gliomastix fide Sutton (*in litt.*).
Torulopsis Speg. (1918) ≡ Torulopsiella.
torulose (**torulous**), cylindrical but having swellings at intervals; moniliform.
torulosis, see cryptococcosis.
Tothia Bat. (1960), ? Microthyriaceae. 1, Hungary. See Batista (*Annls hist.-nat. Mus. natn. hung.* **52**: 105, 1960).
Tothiella Vánky (1999) = Thecaphora fide Vánky (*Mycotaxon* **89**: 55, 2004).
totipotent, bisexual.
touchwood (1) wood rotted by fungi (esp. *Polyporus squamosus*); (2) *Fomes fomentarius* or *Phellinus igniarius* basidiomata or the tinder ('amadou', q.v.) made from them.
Tovariella Syd. (1930), Helotiales. 1, S. America.
toxic, of, caused by, or acting as, poison; **-ity**, the power of acting as a poison.
Toxicocladosporium Crous & U. Braun (2007), anamorphic *Capnodiales*. 1, Surinam. See Crous *et al.* (*Stud. Mycol.* **58**: 33, 2007).
toxigenic, toxin producing.
toxin, a non-enzymic metabolite of one organism which is injurious to another (cf. antibiotic); **myco-**, a toxin produced by a fungus, esp. one affecting humans or animals (see Mycetisms, Mycotoxicoses, Phytotoxic mycotoxins); **patho-** (Wheeler & Luke, *Ann. Rev. Phytopath.* **17**: 223, 1963), see vivotoxin; **phyto-**, (1) a toxin produced by a plant (cf. phytoalexins); (2) (frequently, but better avoided), a toxin injurious to plants (see Phytotoxic mycotoxins); **vivo-**, a toxin 'produced in the infected host by the pathogen and/or its host, which functions in the production of disease, but is not itself the initial inciting agent' (Dimond & Waggoner, *Phytopath.* **43**: 229, 1953); pathotoxin.
toxiphilous, favouring a polluted habitat (e.g. *Lecanora conizaeoides* in area of high sulphur dioxide pollution), cf. poleophilous; **toxiphobous**, not tolerating such a habitat, e.g. *Usnea* spp.); **toxitolerant**, tolerant of toxins. See Air pollution.
Toxoglugea L. Léger & E. Hesse (1924), Microsporidia. 3.
Toxonema Léger & Hesse (1922), Microsporidia. 1.
Toxospora Kudo (1925), Microsporidia. 1.
Toxospora Voronin (1993), Microsporidia. 1.
Toxosporiella B. Sutton (1986), anamorphic *Pezizomycotina*, Cpd.≡ eP.1. 1, Australia. See Sutton (*TBMS* **86**: 16, 1986).
Toxosporiopsis B. Sutton & Sellar (1966), anamorphic *Pezizomycotina*, St.≡ eP.1. 3, widespread. See Sutton & Dyko (*MR* **93**: 476, 1989), Wu (*Acta Mycol. Sin.* **12**: 205, 1993; n.sp.).
Toxosporium Vuill. (1896), anamorphic *Pezizomycotina*, Hsp.≡ eP.1. 2, widespread. See Sutton (*Mycol. Pap.* **138**, 1975).
Toxotrichum G.F. Orr & Kuehn (1964) = Myxotrichum fide Apinis (*Mycol. Pap.* **96**, 1964).
trabecula (pl. **-ae**; adj. **-ate**), (1) a lamella primordium; (2) (of *Gymnoglossum* and other gasteromycetes), plates of undifferentiated primordial tissue in the developing gleba forming a branch of a dendroid columella; (3) (of pseudoparaphyses), paraphysoids, tinophyses, see hamathecium.
Trabecularia Bonord. (1857) = Merulius Fr. fide Donk (*Reinwardtia* **1**: 199, 1951).
Trabutia Sacc. & Roum. (1881), ? Phyllachoraceae. Anamorph *Baeumleria*. 1, widespread. See Barr (*Mycol.* **79**: 188, 1987; key 3 N. Am. spp.), von Arx (*SA* **6**: 213, 1987), Cannon (*Mycopathologia* **135**: 37, 1996).
Trabutiella F. Stevens (1920) ≡ Diatractium.
Trabutiella Theiss. & Syd. (1914) = Phyllachora Nitschke ex Fuckel (1870) fide von Arx & Müller (*Beitr. Kryptfl. Schweiz* **11** no. 1, 1954).
trace elements, see nutrition of fungi.
tracheomycosis, see hadromycosis.
Trachipleistophora W.S. Hollister, E.U. Canning, E. Weidner, A.S. Field, J. Kench & D.J. Marriott (1996), Microsporidia. 2.
Trachyderma (Imazeki) Imazeki (1952) = Ganoderma fide Donk (*Persoonia* **1**: 173, 1960).
Trachyderma Norman (1853) = Pannaria fide Jørgensen (*Op. Bot.* **45**, 1978).
Trachylia Fr. (1817) = Arthonia fide Hawksworth *et al.* (*Dictionary of the Fungi* edn 8, 1995).

Trachylia Tuck. (1848) = Cyphelium Ach. fide Tibell (*Symb. bot. upsal.* **27** no. 1: 279 pp., 1987), Hawksworth *et al.* (*Dictionary of the Fungi* edn 8, 1995).

Trachypus Bataille (1908) [non *Trachypus* Reinw. & Hornsch. 1826, *Musci*] ≡ Leccinum.

Trachyspora Fuckel (1861), Phragmidiaceae. 6 (on *Alchemilla* (*Rosaceae*)), widespread (esp. north temperate; Africa; Java). See Gäumann (*Boissiera* **7**: 105, 1943), Gjaerum & Cummins (*Mycotaxon* **15**: 420, 1982), Helfer (*Nova Hedwigia* **81**: 325, 2005; Europ. spp.).

Trachysporella Syd. (1921) = Trachyspora fide Dietel (*Nat. Pflanzenfam.* **6**, 1928).

trachytectum, see exosporium; Spore wall.

Trachythyriolum Speg. (1919) nom. conf., anamorphic *Pezizomycotina*. See Petrak & Sydow (*Annls mycol.* **33**: 157, 1935).

Trachyxylaria Möller (1901) = Xylobotryum fide Clements & Shear (*Gen. Fung.*, 1931).

Tracya Syd. & P. Syd. (1901), Doassansiaceae. Anamorph *Tracyella*. 2 (on aquatic plants), widespread (northern hemisphere). See Spooner & Legon (*Mycologist* **20**: 90, 2006; British sp.).

Tracyella Zambett. (1970) nom. inval. = Tracya fide Vánky (*in litt.*).

Tracylla (Sacc.) Tassi (1904), anamorphic *Pezizomycotina*, Cpt.0eH.15/16. 2, widespread. See Petrak (*Sydowia* **1**: 202, 1947), Sutton (*Sydowia* **43**: 264, 1991), Nag Raj (*Coelomycetous Anamorphs with Appendage-bearing Conidia*, 1993).

Trailia G.K. Sutherl. (1915), ? Halosphaeriaceae. 1 (on marine *Ascophyllum*), British Isles. See Christensen (*Mycol.* **74**: 226, 1982; synoptic keys *A. ochraceus* and *A. nidulans* groups).

Trailia Syd. (1922) [non *Trailia* G.K. Sutherl. 1915] = Puccinia fide Dietel (*Nat. Pflanzenfam.* **6**, 1928).

trama, the layer of hyphae in the central part of a lamella of an agaric, a spine of *Hydnaceae*, or the dissepiment between pores in a polypore. Cf. context.

Trametaceae Boidin, Mugnier & Canales (1998) = Polyporaceae.

Trametella Pinto-Lopes (1952) = Coriolopsis fide Ryvarden (*Syn. Fung.* **5**, 1991).

Trametes Fr. (1836), Polyporaceae. *c.* 50 (on wood), widespread. Living trees attacked by *T. pini* and others. See Haddow (*TBMS* **22**: 182, 1938; *T. pini*), Ryvarden (*Polyp. N. Eur.* **2**: 421, 1978; key 8 Eur. spp.), Ryvarden & Johansen (*Prelim. Polyp. Fl. E. Afr.*: 555, 1980; key 20 Afr. spp. s.lat.), Tomšovský & Homolka (*Nova Hedwigia* **79**: 425, 2004; Mating tests in *Trametes versicolor*), Tomšovský *et al.* (*Nova Hedwigia* **82**: 269, 2006; Molecular phylogeny European spp.), Zhang *et al.* (*Mycosystema* **25**: 23, 2006; molecular taxonomy based on ITS sequences).

Trametites Mesch. (1892), Fossil Fungi. 3 (Cretaceous, Pliocene), Europe.

Transeptaesporites V.S. Ediger (1981), Fossil Fungi. 5 (upper Oligocene), Turkey; USA.

transformations, in fungi; see Biotechnology, Genetic engineering.

trans-septum, see septum.

Tranzschel (Woldemar Andrejevich; 1868-1942; Russia). Assistant, Forestry Institute, St Petersburg; Assystent for Systematic Botany, University of Warsaw (1898); Curator, Department of Cryptogams, St Petersburg Botanical Museum (1900); Professor, Institute for Applied Zoology and Plant Pathology, Leningrad. Noted for his work on rusts of Russia and adjacent countries [formerly the Soviet Union]. *Publs. Conspectus Uredinalium URSS* (1939). *Biogs, obits etc.* Azbukina ([*Mikologiya i Fitopatologiya*] **27**: 91, 1993) [in Russian]; Stafleu & Cowan (*TL-2* **6**: 433, 1986).

Tranzschelia Arthur (1906), Uropyxidaceae. 12 (heteroecious on *Ranunculaceae* (0, I); on *Prunus* (II, III) (*Rosaceae*), or autoecious on *Ranunculaceae*), widespread (esp. north temperate). *T. discolor*, *T. pruni-spinosae* (esp. on cultivated *Prunus*). See Blumer (*Phytopath. Z.* **38**: 355, 1960), Bennell & Henderson (*TBMS* **71**: 271, 1978; spore devel.), López-Franco & Hennen (*System. Bot.* **15**: 560, 1990; American spp.), Helfer (*Nova Hedwigia* **81**: 325, 2005; Europ. spp.).

Tranzscheliella Lavrov (1936), ? Ustilaginaceae. 16 (on *Poaceae* (esp. *Stipa*)), widespread. See Gutner (*Golovnevye griby*, 1941; ? = *Ustilago*), Vánky (*Mycotaxon* **89**: 55, 2004).

Tranzschel's Law. This states in essence that the telia of microcyclic rust species adopt the habit of the parent macrocyclic species and occur on the aecial host plants of the latter. There are no proven exceptions, the only one firmly suspected (*Chrysomyxa arctostaphyli*) having been disproved (Peterson, *Science, N.Y.* **134**: 468, 1961). The significance of haustorial type in relation to this law has been investigated (Quilliam & Shattock, *Plant pathology* **52**: 104, 2003).

Trapelia M. Choisy (1929) nom. cons., Trapeliaceae (L). 12, widespread (temperate). See Hertel (*Herzogia* **1**: 111, 1968), Hertel (*Vortr. bot. Ges.* n.f. **4**: 171, 1970), Honegger (*Lichenologist* **14**: 205, 1982; asci), Coppins & James (*Lichenologist* **16**: 241, 1984), Lumbsch (*J. Hattori bot. Lab.* **83**: 1, 1997), Lumbsch *et al.* (*MR* **105**: 16, 2001; phylogeny), Miądlikowska *et al.* (*Mycol.* **98**: 1088, 2006; phylogeny), Lumbsch *et al.* (*MR* **111**: 1133, 2007).

Trapeliaceae M. Choisy ex Hertel (1970), Baeomycetales. 9 gen. (+ 13 syn.), 103 spp. Contains most lichenized species formerly placed in the *Agyriaceae*.

Lit.: Lumbsch *et al.* (*MR* **105**: 265, 2001), Lumbsch *et al.* (*MR* **111**: 1133, 2007).

Trapelina Motyka (1996) nom. inval. ≡ Trapelia.

Trapeliopsis Hertel & Gotth. Schneid. (1980), Agyriaceae (L). 17, widespread (esp. temperate). See Coppins & James (*Lichenologist* **16**: 241, 1984), Coppins (*SA* **7**: 93, 1988), Lumbsch (*J. Hattori bot. Lab.* **83**: 1, 1997), Lumbsch *et al.* (*MR* **105**: 16, 2001; phylogeny), Miądlikowska *et al.* (*Mycol.* **98**: 1088, 2006; phylogeny).

Traponora Aptroot (1997), Lecanoraceae (L). 1, Papua New Guinea. See Aptroot *et al.* (*Biblthca Lichenol.* **64**: 199, 1997).

Trappea Castellano (1990), Trappeaceae. 3, China; Europe; N. America. See Xu & Luo (*Mycosystema* **22**: 191, 2003; China).

Trappeaceae P.M. Kirk (2008), Hysterangiales. 2 gen., 4 spp.

Traquairia Carruth. (1873), Fossil Fungi, Ascomycota. 2, USA. See Smith (*Am. J. Bot.* **70**: 387, 1983).

Traversoa Sacc. & Syd. (1913) = Lasiodiplodia fide Sutton (*Mycol. Pap.* **141**, 1977).

Trechinothus E.C. Martini & Trichies (2004), Agaricomycetes. 1, Europe. See Martini & Trichiès (*Mycotaxon* **90**: 262, 2004).

Trechispora P. Karst. (1890), Hydnodontaceae. 46, widespread. See Liberta (*CJB* **51**: 1871, 1973), Larsson (*The Genus Trechispora (Corticiaceae, Basidio-*

mycetes) Dissertation. Universität Göteborg, 1992).

Trechisporales K.H. Larss. (2007). Agaricomycetes. 1 fam., 15 gen., 105 spp. Fam.:

Hydnodontaceae

For *Lit.* see under fam.

tree hair (1) lichens, esp. fruticose spp. (*Bryoria*, *Usnea*), etc. growing on tree trunks (obsol.); (2) the lichen *Pseudevernia furfuracea* (mousse d'arbre); a source of perfume. Cf. oak-moss.

trehalose, a reserve disaccharide (α-D-glucopyanosyl-α-D-glucopyanoside) of fungi (esp. yeasts) and lichens which is hydrolyzed by the enzyme **trehalase**; mycose.

Treleasia Speg. (1896) = Pyxidiophora fide Petch (*Annls mycol.* **34**: 74, 1936), Sutton (*Mycol. Pap.* **141**, 1977), Rossman *et al.* (*Stud. Mycol.* **42**: 248 pp., 1999).

Treleasiella Speg. (1896) nom. dub., anamorphic *Pezizomycotina*. See Sutton (*Mycol. Pap.* **141**, 1977).

trellis rust, on pear (*Gymnosporangium fuscum*); on juniper (*G. sabinae*).

Tremateia Kohlm., Volkm.-Kohlm. & O.E. Erikss. (1995), ? Pleosporaceae. Anamorph *Phoma*-like. 1 (on *Juncus*), USA. See Kohlmeyer *et al.* (*Bot. Mar.* **38**: 165, 1995).

Trematomyces Schrantz (1961) = Requienella fide Boise (*Mycol.* **78**, 1986).

Trematophlyctis Pat. (1918), ? Chytridiales. 1 (on *Leptodesmia* (legume)), Madagascar.

Trematophoma Petr. (1924), anamorphic *Pezizomycotina*, Cpd.0eH.19. 1, Austria.

Trematophora Eisenack (1965), Fossil Fungi ? Fungi. 1, Sweden.

Trematosphaerella Kirschst. (1906) ? = Phaeosphaeria fide Holm (*Symb. bot. upsal.* **14** no. 3: 1, 1957), Hara (*Monogr. Rice Diseases*, 1959), Shoemaker & Babcock (*CJB* **67**: 1500, 1989).

Trematosphaeria Fuckel (1870), ? Pleomassariaceae. Anamorph *Zalerion*-like. 15, widespread. See Boise (*Mycol.* **77**: 230, 1985; key), Fisher & Webster (*Nova Hedwigia* **54**: 77, 1992; anamorph), Lumbsch & Lindemuth (*MR* **105**: 901, 2001; phylogeny), McKeown *et al.* (*MR* **105**: 615, 2001; ultrastr.), Tanaka *et al.* (*Fungal Diversity* **19**: 145, 2005; revision, Japan), Schoch *et al.* (*Mycol.* **98**: 1041, 2006; phylogeny, polyphyly).

Trematosphaeriopsis Elenkin (1901), Pleosporales. 1 (lichenicolous). See Theissen & Sydow (*Annls mycol.* **13**: 603, 1915), Hafellner (*Mycotaxon* **80**: 381, 2001).

Trematosphaeris Clem. & Shear (1931) ≡ Trematosphaeriopsis.

Trematosphaerites Grüss (1924), Fossil Fungi ? Fungi. 1 (Devonian), Spitzbergen.

Trematosphaerites Mesch. (1892), Fossil Fungi. 1 (Tertiary), Europe. See Bužek & Holy (*Sbor. geol. ved.* **4**: 108, 1964).

Trematostoma (Sacc.) Shear (1942) = Exarmidium fide Barr & Boise (*Mycotaxon* **23**, 1985), Aptroot (*Nova Hedwigia* **66**: 89, 1998).

Trematovalsa Jacobesco (1906), ? Diaporthales. 1, France.

trembling fungi, the *Tremellales*.

Tremella Dill. ex L. (1753) nom. inval., Cyanophyta.

Tremella Pers. (1794) nom. cons., Tremellaceae. *c.* 90 (parasites on other fungi), widespread. The edible *T. fuciformis* ('Silver ear') is cultivated in China. See Donk (*Persoonia* **4**: 179, 1966; regional lists), Bandoni & Bisalputra (*CJB* **49**: 27, 1971; ultrastr. of haplonts), Barnett *et al.* (*Yeasts: Characteristics and Identification* 2nd edn, 1990; yeast states 10 spp.), Diederich & Christiansen (*Lichenologist* **26**: 47, 1994; on lichens), Roberts (*Mycologist* **15**: 146, 2001; British spp.), Roberts (*Kew Bulletin* **58**: 763, 2003; Cameroon), Diederich (*Opuscula Philolichenum* **4**: 11, 2007; on lichens).

Tremellacantha Jülich (1980) = Protohydnum fide Hjortstam *et al.* (*Kew Bull.* **45**: 319, 1990).

Tremellaceae Fr. (1821), Tremellales. 17 gen. (+ 14 syn.), 238 spp.

Lit.: Bandoni (*Stud. Mycol.* **30**: 87, 1987), Barnett *et al.* (*Yeasts: Characteristics and Identification* 2nd edn, 1990; yeast states), Zugmaier *et al.* (*Mycol.* **86**: 49, 1994), Bandoni & Boekhout *in* Kurtzman & Fell (Eds) (*Yeasts, a taxonomic study* 4th edn: 705, 1998), Banno & Yamada *in* Kurtzman & Fell (Eds) (*Yeasts, a taxonomic study* 4th edn: 768, 1998), Boekhout & Nakase *in* Kurtzman & Fell (Eds) (*Yeasts, a taxonomic study* 4th edn: 731, 1998), Chen (*Biblthca Mycol.* **174**: 225 pp., 1998), Bandoni & Ginns (*CJB* **76**: 1544, 1998), Chen *et al.* (*Mycoscience* **40**: 137, 1999), Chen *et al.* (*Mycotaxon* **77**: 215, 2001), Weiss & Oberwinkler (*MR* **105**: 403, 2001), Sampaio *et al.* (*Mycol.* **94**: 874, 2002), Scorzetti *et al.* (*FEMS Yeast Res.* **2**: 495, 2002), Yan *et al.* (*Mycosystema* **21**: 47, 2002).

Tremellales Fr. (1821). Tremellomycetes. 9 fam., 38 gen., 341 spp. Mycoparasites. Fams:

(1) **Carcinomycetaceae**
(2) **Cuniculitremaceae**
(3) **Hyaloriaceae**
(4) **Phragmoxenidiaceae**
(5) **Rhynchogastremataceae**
(6) **Sirobasidiaceae**
(7) **Tetragoniomycetaceae**
(8) **Tremellaceae**
(9) **Trichosporonaceae**

Lit.: Martin (1952), Donk (1951-1963; VIII, 1966: 218), Lowy (*Fl. Neotrop.* **6**, 1971; *Nova Hedw.* **19**: 407, 1971; key neotrop. spp.), Raitviir (*in* Parmasto, *Zhivaya priroda Dal'nego Vostoka*: 84, 1971; E. former USSR), McNabb (*in* Ainsworth *et al.* (Eds), *The Fungi* **4B**, 1973), Bandoni (*Trans. Mycol. Soc. Japan* **25**: 489, 1984; review, classific.; *Stud. Mycol.* **30**: 87, 1987; review), Wells (*Mycol.* **86**: 18, 1994; review, classif.), Wells & Bandoni (*Mycota* **VIIb**: 85, 2000; taxonomy), and under fams.

Tremellastrum Clem. (1909) ≡ Crepidotus.

Tremellidium Petr. (1927), anamorphic *Pezizomycotina*, St.0eH.?. 1, Europe.

Tremellina Bandoni (1986), anamorphic *Tremellales*. 1, USA. See Bandoni (*Windahlia* **16**: 53, 1986).

Tremellochaete Raitv. (1964) = Exidia fide Donk (*Persoonia* **4**: 166, 1966).

Tremellodendron G.F. Atk. (1902), Sebacinaceae. *c.* 10, N. America; S. America; Europe. See Bodman (*Am. midl. Nat.* **27**: 203, 1942), Walker & Parrent (*Memoirs of the New York Botanical Garden* **89**: 291, 2004; mycorrhizal).

Tremellodendropsidaceae Jülich (1982) = Auriculariales.

Tremellodendropsis (Corner) D.A. Crawford (1954), Auriculariales. 7, widespread. See Petersen (*Mycotaxon* **29**: 45, 1987), Wendland (*Boletus* **18**: 102, 1994), Weiss & Oberwinkler (*MR* **105**: 403, 2001).

Tremellodiscus Lloyd (1925) nom. nud. ≡ Ruhlandiella fide Dissing & Korf (*Mycotaxon* **12**: 290, 1980).

Tremellodon (Pers.) Fr. (1874) ≡ Pseudohydnum P. Karst.

Tremellogaster E. Fisch. (1924), Diplocystidiaceae. 1, Surinam; Guyana. See Linder (*Mycol.* **22**: 265, 1930).

Tremellogastraceae Zeller (1948) = Diplocystidiaceae.

tremelloid (1) like jelly or wet gelatin; gelatinous; (2) *Tremella*-like.

Tremellomycetes Hibbett, Matheny & Manfr. Binder (2007), Agaricomycotina. 3 ord., 11 fam., 50 gen., 377 spp. Ords:
(1) **Cystofilobasidiales**
(2) **Filobasidiales**
(3) **Tremellales**
Lit.: Hibbett *et al.* (*MR* **111**: 509, 2007).

Tremellomycetidae, see *Tremellomycetes*.

Tremellopsis Pat. (1903) = Crepidotus fide Singer (*Lilloa* **13**: 59, 1947).

Tremelloscypha D.A. Reid (1979), Sebacinaceae. 2, Australia; Jamaica; USA. See Reid (*Beih. Sydowia* **8**: 332, 1979), Weiss & Oberwinkler (*Mycol.* **74**: 325, 1982), Weiss & Oberwinkler (*MR* **105**: 403, 2001; posn, phylogeny).

Tremellostereum Ryvarden (1986), Sebacinaceae. 1, Caribbean. See Ryvarden (*Mycotaxon* **27**: 321, 1986).

Tremiscus (Pers.) Lév. (1846) = Guepinia Fr. fide Legon *et al.* (*Checklist of the British & Irish Basidiomycota*, 2005).

Tremolecia M. Choisy (1953), Lecanorales (L). 5, widespread. See Hertel (*Ergebn. Forsch Unternehmens Nepal Himalaya* **6**: 150, 1977), Makarova (*Nov. sist. Niz. Rast.* **27**: 107, 1990; Russia), Buschbom & Mueller (*Mol. Phylogen. Evol.* **32**: 66, 2004; phylogeny), Türk & Uhl (*Biblthca Lichenol.* **88**: 691, 2004; Austria), Wedin *et al.* (*MR* **109**: 159, 2005; phylogeny).

tremorgen, a mycotoxin inducing a neurotoxicosis (tremor) in humans and higher animals, e.g. fumitremorgin, verruculotoxin.

Tremotyliomyces Cif. & Tomas. (1953) = Thelotrema.

Tremotylium Nyl. (1865), Thelotremataceae (L). 6, widespread. See Redinger (*Ark. Bot.* **28A** no. 8: 1, 1936).

Trenomyces Chatton & F. Picard (1908), Laboulbeniaceae. 11, widespread. See Nannfeldt (*Svensk bot. Tidskr.* **43**: 468, 1949).

Trentepohlia Mart. (1817), Algae. Algae.

Tretendophragmia Subram. (1994) = Diplococcium fide Sutton (*in litt.*).

tretic (of conidiogenesis), the sort of conidiogenesis in which each conidium (**tretoconidium**, tretic conidium, poroconidium, porospore) is delimited by an extension of the inner wall of the conidiogenous cell. Tretoconidia are solitary or in acropetal chains (cf. phialidic). **mono-**, **poly-**, (of conidiogenous cells), producing tretoconidia by the extrusion of the inner wall through one or several channels, respectively.

Tretocephala Subram. (1995), anamorphic *Pezizomycotina*, Hso.0eP.24. 1, Singapore. See Subramanian (*Cryptog. Mycol.* **13**: 65, 1992), Subramanian (*Kavaka* **20/21**: 57, 1992/1993).

Tretophragmia Subram. & Natarajan (1974), anamorphic *Pezizomycotina*, Hsy.≡ eP.24. 2, India. See Subramanian & Natarajan (*Proc. Natl. Inst. Sci. India* B, Biol. Sci. **39**: 550, 1973), Holubová-Jechová & Mercado Sierra (*Česká Mykol.* **40**: 142, 1986).

Tretopileus B.O. Dodge (1946), anamorphic *Corticiaceae*, Hsy.#eP.42. 3, USA; Africa. See Deighton (*Mycol. Pap.* **78**, 1960), Okada *et al.* (*Asian Network on Microbial Researches. Proceedings of International Conference on Asian Network on Microbial Researches* NATO ASI Series vol. 269: 537, 1998), Okada *et al.* (*Mycoscience* **39**: 21, 1998; phylogeny), Lawrey *et al.* (*Amer. J. Bot.* **95**: 816, 2008).

Tretospeira Piroz. (1972), anamorphic *Pezizomycotina*, Hso.#eP.24. 1, Uganda. See Pirozynski (*Mycol. Pap.* **129**: 58, 1972).

Tretospora M.B. Ellis (1976), anamorphic *Balladynopsis*, Hso.≡ eP.24. 7, widespread. See Khan *et al.* (*Mycotaxon* **49**: 477, 1993; key), Chaudhary *et al.* (*Mycotaxon* **57**: 201, 1996), Hosagoudar *et al.* (*Zoos' Print Journal* **18**: 967, 2003).

Tretovularia Deighton (1984), anamorphic *Pezizomycotina*, Hso.0eH.15. 1, widespread (northern Europe). See Braun (*Monogr. Cercosporella, Ramularia Allied Genera (Phytopath. Hyphom.)* **2**, 1998).

Treubiomyces Höhn. (1909), Chaetothyriaceae. 2, widespread. See Pohlad & Reynolds (*Mycol.* **66**: 521, 1974), Pohlad (*CJB* **67**: 40, 1989), Rossman *et al.* (*Stud. Mycol.* **42**: 248 pp., 1999).

Trevisan (Vittore, Earl of San Leon; 1818-1897; Italy). Born of a wealthy family in Padova; Professor of Natural History and Popular Physics in Padova (1851-1853); from 1853 onwards worked from his estate in Mason, then moved to Monza and Milan where he died. Trevisan devoted much of his time to advancing study of a wide range of cryptogams, including plant pathogens and lichen-forming fungi; he was especially concerned in using new microscopic features, introducing 75 generic names for lichen-forming taxa and making numerous critical and nomenclatural remarks on the works of Körber (q.v.) and especially Massalongo (q.v.). His collections were transferred to the University of Genoa, but all destroyed in World War II. *Biogs, obits etc.* Lazzarin [ed.] (*L' Opera Lichenologica di Vittore Trevisan*, 1994) [reprint collected works, including biography]; Grummann (1974: 524); Stafleu & Cowan (*TL-2* **6**: 480, 1986).

tri- (in combination), three, triple.

Triactella Syd. (1921) = Hapalophragmium fide Cummins & Hiratsuka (*Illustr. Gen. Rust Fungi rev. edit.*, 1983) See, Lohsomboon *et al.* (*MR* **96**: 461, 1992), Hennen *et al.* (*Mycol.* **92**: 312, 2000).

Triacutus G.L. Barron & Tzean (1981), anamorphic *Pezizomycotina*, Hso.1bH.1. 1 (on bdelloid rotifer), Canada. See Barron & Tzean (*CJB* **59**: 1207, 1981).

Triadelphia Shearer & J.L. Crane (1971), anamorphic *Pezizomycotina*, Hso.≡ eP.1. 13 (aquatic and terrestrial), widespread. See Constantinescu & Samson (*Mycotaxon* **15**: 472, 1982; key), von Arx (*TBMS* **85**: 566, 1985; related genera), Révay (*Acta Bot. Hung.* **33**: 67, 1987), Tzean & Chen (*Mycol.* **81**: 626, 1989; synoptic key).

Triainomyces W. Rossi & A. Weir (1998), Laboulbeniaceae. 1 (on pill-millipedes), New Zealand. See Rossi & Weir (*Mycol.* **90**: 282, 1998).

Triandromyces Thaxt. (1931) = Tetrandromyces fide Tavares (*Mycol. Mem.* **9**, 1985).

Triangularia Boedijn (1934), Lasiosphaeriaceae. 6, widespread. See Cain & Farrow (*CJB* **34**: 689, 1956; key), von Arx & Hennebert (*BSMF* **84**: 423, 1969), Guarro & Cano (*TBMS* **91**: 587, 1988; key 5 spp.), Guarro *et al.* (*SA* Reprint of Volumes 1-4 (1982-1985) **10**: 79, 1991), Huhndorf *et al.* (*Mycol.* **96**: 368,

2004; phylogeny), Miller & Huhndorf (*Mol. Phylogen. Evol.* **35**: 60, 2005; phylogeny), Cai *et al.* (*MR* **110**: 359, 2006; phylogeny).

Triblidiaceae Rehm (1888), Triblidiales. 3 gen. (+ 5 syn.), 11 spp.
Lit.: Eriksson (*SA* **11**: 1, 1992), Magnes (*Biblthca Mycol.* **165**: 177 pp., 1997).

Triblidiales O.E. Erikss. (1992). Pezizomycotina. 1 fam., 3 gen., 11 spp. Fam.:
Triblidiaceae
Lit.: Eriksson (*SA* **11**: 1, 1992), Magnes (*Bibl. Mycol.* **165**, 1997).

Triblidium Rebent. (1804), Triblidiaceae. 1 or 6, Europe. See Eriksson (*Op. Bot.* **60**, 1981), Magnes (*Biblthca Mycol.* **165**, 1997).

Tribolites W.H. Bradley ex Janson. & Hills (1976), Fossil Fungi, anamorphic *Pezizomycotina*. 1 (Eocene), USA.

Tribolospora D.A. Reid (1966), anamorphic *Pezizomycotina*, St.0eH.15. 1, Australia. See Reid (*Aust. J. Bot.* **14**: 31, 1966).

Tribulatia Joanne E. Taylor, K.D. Hyde, E.B.G. Jones (2003), Sordariomycetes. 1, Queensland. See Taylor & Hyde (*Fungal Diversity Res. Ser.* **12**: 189, 2003).

Tricella Long (1912) = Phragmopyxis fide Dietel (*Nat. Pflanzenfam.* **6**, 1928).

Tricellaesporonites Sheffy & Dilcher (1971), Fossil Fungi. 2 (Eocene), USA.

Tricellula Beverw. (1954), anamorphic *Pezizomycotina*, Hsy.≡ eH.10. 4, Europe; USA. See Haskins (*Can. J. Microbiol.* **4**: 279, 1958), Petersen (*Bull. Torrey bot. Club* **89**: 287, 1962).

Tricellulortus Matsush. (1995) = Minimedusa fide Diederich & Lawrey (*Mycol. Progr.* **6**: 61, 2007).

Triceromyces T. Majewski (1980), Laboulbeniaceae. 5, widespread. See Tavares (*Mycol. Mem.* **9**: 627 pp., 1985), Benjamin (*Aliso* **11**: 245, 1986; key 6 spp.), Benjamin (*Aliso* **17**: 1, 1998).

Trichaegum Corda (1837), anamorphic *Pezizomycotina*, Hso.#eP.?. 3, Europe; N. America.

Trichaleurina Rehm (1903) = Scutellinia fide Eckblad (*Nytt Mag. Bot.* **15**: 1, 1968).

Trichaleuris Clem. (1909) = Scutellinia fide Eckblad (*Nytt Mag. Bot.* **15**: 1, 1968).

Trichamelia Bat. (1960), Asterinaceae. 1, Brazil. See Batista (*Publções Inst. Micol. Recife* **295**: 9, 1960), Hosagoudar *et al.* (*Journal of Mycopathological Research* **39**: 61, 2001).

Trichangium Kirschst. (1935), Helotiales. 1, Europe.

Trichaptum Murrill (1904), Polyporaceae. 20, widespread (pantropical). See Macrae (*CJB* **45**: 1371, 1966), Corner (*Beih. Nova Hedwigia* **86**: 197, 1987; key), Hattori (*Mycoscience* **46**: 303, 2005; type studies Corner species).

Tricharia Boud. (1885) ≡ Tricharina.

Tricharia Fée (1825), Gomphillaceae (L). 69, widespread (tropical). See Sérusiaux (*Mycol.* **76**: 108, 1984; key), Vězda & Poelt (*Folia geobot. phytotax.* **22**: 179, 1987), Lücking (*Biblthca Lichenol.* **65**: 1, 1997), Ferraro (*Fungal Diversity* **15**: 153, 2004; hypophores), Lücking *et al.* (*Mycol.* **96**: 283, 2004; phylogeny), Lücking *et al.* (*Lichenologist* **37**: 123, 2005; phylogeny).

Tricharina Eckblad (1968), Pyronemataceae. Anamorph *Ascorhizoctonia*. 13, widespread (temperate). See Yang & Korf (*Mycotaxon* **24**: 467, 1985; key), Yang & Kristiansen (*Mycotaxon* **35**: 313, 1989; anamorph), Egger (*CJB* **74**: 773, 1996; DNA), Barrera & Romero (*Mycotaxon* **77**: 31, 2001; Argentina), Hansen & Pfister (*Mycol.* **98**: 1029, 2006; phylogeny), Liu & Zhuang (*Mycosystema* **25**: 546, 2006; phylogeny), Tedersoo *et al.* (*New Phytol.* **170**: 581, 2006; mycorrhiza), Perry *et al.* (*MR* **111**: 549, 2007; phylogeny).

Trichaster Czern. (1845) = Geastrum fide Staněk *in* Pilát (Ed.) (*Fl. ČSR* **B, 1**: 480, 1958).

Trichasterina G. Arnaud (1918), Asterinaceae. 9, widespread (tropical). See Song *et al.* (*Mycosystema* **21**: 309, 2002; China).

trichidium, see sterigma.

Trichobacidia Vain. (1921) nom. dub., Fungi. See Santesson (*Symb. bot. upsal.* **12** no. 1: 1, 1952).

Trichobasis Lév. (1849) = Uredo fide Saccardo (*Syll. fung.* **7**: 838, 1887) See, Laundon (*Mycotaxon* **3**: 133, 1975).

Trichobelonium (Sacc.) Rehm (1896) = Belonopsis fide Aebi (*Nova Hedwigia* **23**: 49, 1972), Nannfeldt (*Sydowia* **38**: 194, 1985).

Trichobolbus Bat. (1964), anamorphic *Pezizomycotina*, Cpd.0eP.?. 1, Brazil. See Batista (*Riv. Patol. veg., Pavia* sér. 3 **4**: 559, 1964).

Trichobolus (Sacc.) Kimbr. & Cain (1967), Pezizales. 4, Europe; N. America. Perhaps close to *Lasiobolus*. See Krug (*CJB* **51**: 1497, 1973), Samuelson & Kimbrough (*Mycol.* **70**: 1191, 1978), Olsen (*Blyttia* **45**: 117, 1987; Norway), Brummelin (*Persoonia* **16**: 425, 1998), Prokhorov (*Mikol. Fitopatol.* **32**: 40, 1998; Russia).

Trichobotrys Penz. & Sacc. (1901), anamorphic *Pezizomycotina*, Hso.0eP.3. 3, Asia (tropical). See Morgan-Jones *et al.* (*Mycotaxon* **30**: 345, 1987), D'Souza & Bhat (*Mycotaxon* **80**: 105, 2001).

Trichocarpus P. Karst. (1889) [non *Trichocarpus* Schreb. 1789, *Tiliaceae*] = Amylostereum fide Donk (*Persoonia* **3**: 199, 1964).

Trichocephalum Costantin (1888) = Periconia fide Mason & Ellis (*Mycol. Pap.* **56**, 1953).

Trichoceridium R.A. Poiss. (1932) ? = Smittium fide Lichtwardt (*The Trichomycetes. Fungal associates of arthropods*, 1986).

Trichochora Theiss. & Syd. (1915) nom. dub., Fungi. See Petrak (*Annls mycol.* **27**: 324, 1929).

Trichocicinnus (Sacc.) Höhn. (1926) = Chaetosticta fide Sutton (*Mycol. Pap.* **141**, 1977).

Trichocintractia M. Piepenbr. (1995), Anthracoideaceae. 1 (on *Rhynchospora* (*Cyperaceae*)), widespread (tropical). See Piepenbring (*CJB* **73**: 1089, 1995), Piepenbring (*Revta Biol. trop.* **49**: 411, 2001; Panama), Pérez & Minter (*IMI Descr. Fungi Bact.* **164**, 2005).

Trichocladia (de Bary) Neger (1901) = Erysiphe fide Braun & Takamatsu (*Schlechtendalia* **4**: 3, 2000).

Trichocladia Stirt. (1882) [non *Trichocladia* Harv. 1836, *Algae*] = Heterodea fide Hawksworth *et al.* (*Dictionary of the Fungi* edn 8, 1995).

Trichocladium Harz (1871), anamorphic *Chaetomiaceae*, Hso.≡ eP.1. 24, widespread. The genus is highly polyphyletic. See Hughes (*TBMS* **35**: 152, 1952), Hughes (*N.Z. Jl Bot.* **7**: 153, 1969), Pidoplichko & Kirilenko (*Mikol. Fitopatol.* **6**: 510, 1972; key), Kohlmeyer & Volkmann-Kohlmeyer (*Mycotaxon* **53**: 349, 1995; key 7 aquatic spp.), Goh & Hyde (*Fungal Diversity* **2**: 101, 1999; synopsis), Chang (*Fungal Science* Taipei **16**: 35, 2001; teleomorph), Jones *et al.* (*Fungal Diversity* **7**: 49, 2001), Chatmala *et al.* (*Fungal Diversity Res. Ser.* **7**: 59,

2002), Mantle *et al.* (*MR* **110**: 1371, 2006; phylogeny).

Trichocollonema Höhn. (1902) = Zignoëlla fide von Höhnel (*Annls mycol.* **15**, 1917).

Trichocoma Jungh. (1838), Trichocomaceae. Anamorph *Penicillium*. 1, widespread (subtropical). See Martin (*Mycol.* **29**: 620, 1937), Kuraishi *et al.* (*MR* **95**: 705, 1991; ubiquinones), Ogawa & Sugiyama (*Integration of Modern Taxonomic Methods for Penicillium and Aspergillus Classification*: 149, 2000; phylogeny), Geiser *et al.* (*Mycol.* **98**: 1053, 2006; phylogeny).

Trichocomaceae E. Fisch. (1897), Eurotiales. 39 gen. (+ 45 syn.), 881 spp.

Lit.: Raper & Fennell (*The genus Aspergillus*, 1965), Malloch & Cain (*CJB* **50**: 2613, 1972), Malloch & Cain (*CJB* **51**: 1647, 1973; gen. names), Weijman (*Antonie van Leeuwenhoek* **42**: 315, 1976), Pitt (*The genus Penicillium and its teleomorphic states Eupenicillium and Talaromyces*, 1979), Malloch *in* Arai (Ed.) (*Filamentous micro-organisms*: 37, 1985; key gen.), Malloch *in* Samson & Pitt (Eds) (*Advances in Penicillium and Aspergillus Systematics* **102**: 365, 1985), Malloch (*Advances in Penicillium and Aspergillus Systematics* **102**: 365, 1985), Pitt & Samson (*Regnum veg.* **128**: 13, 1993; names in use), Pitt (*CJB* **73** Suppl. 1: S768, 1995), Horie *et al.* (*Mycoscience* **37**: 323, 1996), Landvik *et al.* (*Mycoscience* **37**: 237, 1996), Ogawa *et al.* (*Mycol.* **89**: 756, 1997), Frisvad *et al.* (*Chemical Fungal Taxonomy*: 289, 1998), Geiser *et al.* (*Mycol.* **90**: 831, 1998), Domenech *et al.* (*Microbiology* Reading **145**: 2789, 1999), Skouboe *et al.* (*MR* **103**: 873, 1999), Frisvad & Samson (*Stud. Mycol.* **45**: 201, 2000), Frisvad *et al.* (*Integration of Modern Taxonomic Methods for Penicillium and Aspergillus Classification*: 265, 2000), Kuraishi *et al.* (*Antonie van Leeuwenhoek* **77**: 179, 2000), Ogawa & Sugiyama (*Integration of Modern Taxonomic Methods for Penicillium and Aspergillus Classification*: 149, 2000), Peterson (*Integration of Modern Taxonomic Methods for Penicillium and Aspergillus Classification*: 163, 2000), Peterson (*Integration of Modern Taxonomic Methods for Penicillium and Aspergillus Classification*: 323, 2000), Pitt *et al.* (*Integration of Modern Taxonomic Methods for Penicillium and Aspergillus Classification*: 9, 2000), Tamura *et al.* (*Integration of Modern Taxonomic Methods for Penicillium and Aspergillus Classification*: 357, 2000), Klich (*Identification of Common Aspergillus Species*: 116 pp., 2002), Klich *et al.* (*Mycol.* **95**: 1252, 2003), Samson *et al.* (*Stud. Mycol.* **49**: 175, 2004), Scott *et al.* (*Mycol.* **96**: 1095, 2004), Hong *et al.* (*Mycol.* **97**: 1316, 2005), Peterson (*Mycol.* **100**: 205, 2008; 4-locus phylogeny).

Trichoconiella B.L. Jain (1976) = Alternaria fide Sutton (*in litt.*).

Trichoconis Clem. (1909), anamorphic *Pezizomycotina*, Hso.≡ eH.1. 15 (on fungi), widespread. See Deighton & Pirozynski (*Mycol. Pap.* **128**, 1972; keys), Baker *et al.* (*Mycotaxon* **79**: 361, 2001).

Trichoconium Corda (1837) = Melanconium fide Saccardo (*Syll. fung.* **3**: 1, 1884).

Trichocrea Marchal (1891) nom. dub., anamorphic *Pezizomycotina*. See Minter & Caine (*TBMS* **74**: 434, 1980).

Trichoctosporea J.I.R. Larsson (1994), Microsporidia. 1. See Larsson (*Arch. Protistenk.* **144**: 147, 1994).

trichocyst, a subpellicular organelle of many ciliates and dinoflagellates; sometimes an offensive weapon able to disable prey, sometimes an anchoring device.

Trichodelitschia Munk (1953), Phaeotrichaceae. 4, widespread. See Lundqvist (*Svensk bot. Tidskr.* **58**: 267, 1964), Barr (*Prodr. Cl. Loculoasc.*, 1987; posn), Eriksson (*SA* **7**: 93, 1988), Ebersohn & Eicker (*S. Afr. J. Bot.* **58**: 145, 1992), Kruys *et al.* (*MR* **110**: 527, 2006; phylogeny).

trichoderm (of basidiomata), an outer layer composed of hair-like elements projecting from the surface (Furtado, *Mycol.* **57**: 599, 1965). See derm.

Trichoderma Pers. (1794), anamorphic *Hypocrea*, Hso.0eH.15. 102 (esp. in soil), widespread. See Rifai (*Mycol. Pap.* **116**, 1969; key), Doi *et al.* (*Bull. natn. Sci. Mus.* Tokyo, B **10**: 73, 1984), Barak *et al.* (*Can. J. Microbiol.* **31**: 810, 1985; antigens), Eveleigh (*Biology of Industrial Microorganisms*: 487, 1985; review), Min *et al.* (*Korean J. Mycol.* **13**: 221, 1985; chromosomes), Doi & Doi (*Bull. natn. Sci. Mus.* Tokyo, B **12**: 1, 1986; nomenclator), Doi *et al.* (*Bull. natn. Sci. Mus.* Tokyo, B **13**: 1, 1987), Elad *et al.* (*TBMS* **88**: 259, 1987; ultrastr.), Doi *et al.* (*Bull. natn. Sci. Mus.* Tokyo, B **15**: 27, 1989; conidium ornamentation), Meyer & Plaskowitz (*Mycol.* **81**: 312, 1989; SEM), Stasz *et al.* (*Mycol.* **81**: 391, 1989; cladistics), Toyama & Toyama (*Agric. Biol. Chem.* **54**: 2331, 1990; genetics), Meyer (*Phytopathology* **81**: 1240, 1991), Meyer (*Appl. Environm. Microbiol.* **57**: 2269, 1991; DNA), Nevalainen *et al.* (*Molecular Industrial Mycology: Systems and Applications for Filamentous Fungi* Mycology Series **8**: 129, 1991; gene expression), Samuels *et al.* (*Petria* **1**: 121, 1991; teleomorph), Bissett (*CJB* **69**: 2357, 1991), Bissett (*CJB* **69**: 2418, 1991), Bissett (*CJB* **70**: 639, 1992), Bissett (*CJB* **69**: 2373, 1991), Zimand *et al.* (*Phytoparasitica* **21**: 166, 1993; RAPDs), García *et al.* (*Revta Iberoamer. Micol.* **11**: S48, 1994; numerical taxonomy), Mills & Muthumeenakshi (*Ecology of Plant Pathogens*: 135, 1994; from mushroom compost), Samuels *et al.* (*Mycol.* **86**: 421, 1994), Tschen & Li (*Mycoscience* **35**: 257, 1994; protoplast generation), Zimand *et al.* (*MR* **98**: 531, 1994; RAPDs), Leuchtmann *et al.* (*Mycol.* **88**: 384, 1996; isozymes), Samuels (*MR* **100**: 923, 1996; review), Samuels & Lodge (*Mycol.* **88**: 302, 1996), Gómez *et al.* (*Molecular & General Genetics* **256**: 127, 1997; genetic diversity), Grondona *et al.* (*Appl. Environm. Microbiol.* **63**: 3189, 1997; physiology, biochemistry), Kuhls *et al.* (*Mycol.* **89**: 442, 1997; phylogeny), Turner *et al.* (*MR* **101**: 449, 1997), Castle *et al.* (*Appl. Environm. Microbiol.* **64**: 133, 1998; N America, on mushrooms), Gams & Bissett (*Trichoderma and Gliocladium* Vol. **1**. Basic Biology, Taxonomy and Genetics: 3, 1998; morphology), Gams & Meyer (*Mycol.* **90**: 904, 1998), Harman *et al.* (*Trichoderma and Gliocladium* Vol. **1**. Basic Biology, Taxonomy and Genetics: 243, 1998; genetics), Kindermann *et al.* (*Fungal Genetics Biol.* **24**: 298, 1998; phylogeny), Lieckfeldt *et al.* (*Trichoderma and Gliocladium* Vol. **1**. Basic Biology, Taxonomy and Genetics: 35, 1998; phylogeny), Muthumeenakshi *et al.* (*MR* **102**: 385, 1998; on mushrooms), Ospina-Giraldo *et al.* (*Mycol.* **90**: 76, 1998; on mushrooms), Samuels *et al.* (*Stud. Mycol.* **41**, 1998), Schickler *et al.* (*MR* **102**: 373, 1998; chitinases), Kuhls *et al.* (*Medical Mycology* **37**: 25, 1999; from humans), Lieckfeldt *et al.* (*Appl. Environm. Microbiol.* **65**: 2418, 1999), Lieckfeldt *et al.* (*CJB* **76**: 1507, 1998), Lübeck *et al.* (*MR* **103**: 289,

1999), Ospina-Giralda *et al.* (*Phytopathology* **89**: 308, 1999; phylogeny), Samuels *et al.* (*Sydowia* **51**: 71, 1999), Hermosa *et al.* (*Appl. Environm. Microbiol.* **66**: 1890, 2000; biocontrol strains), Lieckfeldt *et al.* (*Microbiol. Res.* **155**: 7, 2000; endochitinases), Lieckfeldt *et al.* (*Mycol.* **92**: 374, 2000), Lübeck *et al.* (*FEMS Microbiol. Lett.* **185**: 129, 2000; from buildings), Samuels *et al.* (*MR* **104**: 760, 2000), Chaverri *et al.* (*Mycol.* **93**: 1113, 2001), Kullnig *et al.* (*MR* **105**: 770, 2001; biocontrol), Lieckfeldt *et al.* (*MR* **105**: 313, 2001; phylogeny), Thrane *et al.* (*FEMS Microbiol. Lett.* **203**: 249, 2001; HPLC), Chaverri & Samuels (*Mycol. Progr.* **1**: 283, 2002), Kullnig-Gradinger *et al.* (*MR* **106**: 757, 2002; phylogeny), Lee & Hseu (*Can. J. Microbiol.* **48**: 831, 2002), Samuels *et al.* (*Mycol.* **94**: 146, 2002; from mushrooms), Chaverri & Samuels (*Stud. Mycol.* **48**, 2003; revision), Chaverri *et al.* (*Mol. Phylogen. Evol.* **27**: 302, 2003; phylogeny), Chaverri *et al.* (*Mycol.* **95**: 1100, 2003), Dodd *et al.* (*Mycol.* **95**: 27, 2003), Grinyer *et al.* (*Curr. Genet.* **45**: 163, 2004; proteomics), Hermosa *et al.* (*MR* **108**: 897, 2004; genetic diversity), Lu *et al.* (*Mycol.* **96**: 310, 2004), Rey *et al.* (*Applied Mycology and Biotechnology* **4**: [225], 2004; genetics), Druzhinina *et al.* (*Fungal Genetics Biol.* **42**: 813, 2005; barcoding), Kopchinskiy *et al.* (*MR* **109**: 658, 2005; molecular identification), Druzhinina *et al.* (*Mycoscience* **47**: 55, 2006; review), Jaklitsch *et al.* (*Stud. Mycol.* **56**: 135, 2006), Overton *et al.* (*Stud. Mycol.* **56**: 1, 2006), Overton *et al.* (*Stud. Mycol.* **56**: 39, 2006), Samuels (*Phytopathology* **96**: 195, 2006; review), Samuels *et al.* (*Stud. Mycol.* **56**: 67, 2006), Samuels *et al.* (*MR* **110**: 381, 2006; from cacao).

Trichoderma Pers. (1801) nom. conf., Fungi.

Trichodermia Hoffm. (1795) = Trichothecium fide Saccardo (*Syll. fung.* **4**: 178, 1886).

Trichodesmium Chevall. (1826) nom. rej. prop. ≡ Graphiola.

Trichodiscula Vouaux (1910), anamorphic *Pezizomycotina*. 1 (on cardboard), France.

Trichodiscus Kirschst. (1924) [non *Trichodiscus* Welsford 1912, *Algae*] ≡ Dennisiodiscus.

Trichodochium Syd. (1927), anamorphic *Pezizomycotina*, Hsp.1eP.19. 2, C. America; India. See Ellis (*Mycol. Pap.* **111**, 1967).

Trichodothella Petr. (1946), Venturiaceae. 1, Europe.

Trichodothis Theiss. & Syd. (1914), Venturiaceae. 3, N. America; Africa. See Barr (*Sydowia* **41**: 25, 1989).

Trichoduboscqia L. Léger (1926), Microsporidia. 1.

Trichodytes Kleb. (1898) nom. dub., anamorphic *Pezizomycotina*. See Sutton (*Mycol. Pap.* **141**, 1977).

Trichofusarium Bubák (1906) = Fusarium fide Booth (*The genus Fusarium*, 1971).

Trichoglossum Boud. (1885), Geoglossaceae. 20, widespread (esp. temperate). See Mains (*Mycol.* **46**: 61, 1954; key N. Am. spp.), Rifai (*Lloydia* **28**: 113, 1965; key Javanese spp.), Benkert (*Mykol. MittBl.* **20**: 47, 1977; Germany), Spooner (*Biblthca Mycol.* **116**, 1987; key Australasian spp.), Verkley (*Persoonia* **15**: 405, 1994; asci), Zhuang & Wang (*Mycotaxon* **63**: 307, 1997), Zhuang (*Fl. Fung. Sinicorum* **8**, 1998; 15 spp. China), Wang (*Bull. natn. Mus. Nat. Sci.* Taiwan **13**: 147, 2001; Taiwan), Wang *et al.* (*Mycol.* **94**: 641, 2002; phylogeny), Wang *et al.* (*Mycol.* **98**: 1065, 2006; phylogeny), Wang *et al.* (*Mol. Phylogen. Evol.* **41**: 295, 2006; phylogeny).

trichogyne, the receptive hypha of the female organ, esp. in certain ascomycetes.

Trichohelotium Killerm. (1935), Helotiales. 2, Europe.

Trichohleria Sacc. (1908) ? = Trichosphaerella fide von Arx & Müller (*Stud. Mycol.* **9**, 1975), Samuels & Barr (*CJB* **75**: 2165, 1998).

Tricholechia A. Massal. (1853) ≡ Byssoloma.

Tricholeconium Corda (1837) = Sarcopodium fide Saccardo (*Syll. fung.* **4**: 312, 1886).

Tricholoma (Fr.) Staude (1857) nom. cons., Tricholomataceae. *c.* 200, widespread (esp. north temperate). See Horak (*Sydowia* **17**: 153, 1964; key 11 S. Am. spp.), Bon (*Docums Mycol.* **3**: 1, 1971), Gulden (*Musseronflora Slekton Tricholoma sensu lato*, 1972), Haluwyn (*Docums Mycol.* **4**: 43, 1972; ecology), Bon (*Docums Mycol.* **6**: 165, 1974-76), Bon (*Docums Mycol.* **4**: 55, 1974-76), Bon (*Docums Mycol.* **5**: 111, 1974-76), Bon (*Encycl. Mycol.* **36**, 1984), Riva (*Tricholoma*, 1988) See matsutake, Riva (*Fungi Europaei* **3** A: 626, 2003), Matsushita *et al.* (*Mycoscience* **46**: 90, 2005; genetic relationship of *Tricholoma matsutake* and *T. nauseosum*), Jacobsson *et al.* (*Mycotaxon* **95**: 195, 2006; new sp. Fennoscandia).

Tricholomataceae R. Heim ex Pouzar (1983) nom. cons., Agaricales. 78 gen. (+ 57 syn.), 1020 spp.

Lit.: Mankel *et al.* (*Microbiol. Res.* **153**: 377, 1999), Noordeloos (*Coolia* **42**: 163, 1999), Hwang & Kim (*Curr. Microbiol.* **40**: 250, 2000), Kalamees (*Folia cryptog. Estonica* **38**: 13, 2001), Suh *et al.* (*Mycobiology* **29**: 179, 2001), Moncalvo *et al.* (*Mol. Phylogen. Evol.* **23**: 357, 2002), Lian *et al.* (*Mycorrhiza* **13**: 27, 2003), Chapela & Garbelotto (*Mycol.* **96**: 730, 2004), Matsushita *et al.* (*Mycoscience* **46**: 90, 2005), Murata & Babasaki (*Mycorrhiza* **15**: 381, 2005).

Tricholomella Zerova ex Kalamees (1992) = Calocybe fide Kuyper (*in litt.*).

tricholomic acid, an insecticidal amino-acid derivative produced by *Tricholoma muscarium* (Takemoto, *Jap. J. Pharm. Chem.* **33**: 252, 1961). Cf. muscazone.

Tricholomopsis Singer (1939), Agaricales. *c.* 30, widespread. Belongs to *Tricholomataceae* or *Marasmiaceae*. See Smith (*Brittonia* **12**: 41, 1960), Liu (*Acta Mycol. Sin.* **13**: 181, 1994; China).

Tricholosporum Guzmán (1975), Tricholomataceae. 5, widespread. See Guzmán (*Boletín de la Sociedad Mexicana de Micología* **9**: 61, 1975), Contu & Mua (*Rivista di Micologia* **43**: 249, 2000; European spp.), Guzmán *et al.* (*Docums Mycol.* **33** no. 131: 23, 2004).

Trichomaris Hibbits, G.C. Hughes & Sparks (1981), Halosphaeriaceae. 1 (on tanner crab), Alaska. See Kohlmeyer & Volkmann-Kohlmeyer (*Bot. Mar.* **34**: 1, 1991), Sakayaroj *et al.* (*Mycol.* **97**: 804, 2005; phylogeny).

Trichomatomyces Dorn.-Silva & Dianese (2004), anamorphic *Pezizomycotina*. 1, Brazil. See Dornelo-Silva & Dianese (*Mycol.* **96**: 879, 2004).

Trichomerium Speg. (1918), Capnodiaceae. Anamorph *Tripospermum*. 23, widespread. See Hughes (*Mycol.* **68**, 1976; synonymy with *Phragmocapnias*), Reynolds (*Mycotaxon* **14**: 189, 1982), Reynolds (*Mycotaxon* **27**: 377, 1986).

Trichometasphaeria Munk (1953), Lophiostomataceae. Anamorph *Ascochyta*-like. 7, N. America. See Barr (*Mycotaxon* **45**: 191, 1992).

Trichomonascaceae Locq. (1972), Saccharomycetales. 5 gen., 21 spp.

Lit.: Kurtzman & Robnett (*FEMS Yeast Res.* **7**: 141, 2007).

Trichomonascus H.S. Jacks. (1948), Trichomonascaceae. Anamorph *Blastobotrys*. 1 (on *Corticium*), Canada. See Kurtzman & Blanz *in* Kurtzman & Fell (Eds) (*Yeasts, a taxonomic study* 4th edn: 69, 1998; rDNA), Malloch & de Hoog *in* Kurtzman & Fell (Eds) (*Yeasts, a taxonomic study* 4th edn: 197, 1998), Kurtzman & Robnett (*FEMS Yeast Res.* **7**: 141, 2007).

Trichomyces Malmsten (1848) = Trichophyton fide Dodge (*Medical Mycology*, 1935).

Trichomycetes Alexop. (1962), Zygomycota. A polyphyletic taxon as traditionally circumscibed at this rank. Manier & Lichtwardt (1968) recognized 4 Ords but two, the *Amoebidales* and *Eccrinales*, are now place in the *Protozoa* and the other two, the *Asellariales* and *Harpellales* and referred to the *Kickxellomycotina*.

Lit.: Lichtwardt (*The Trichomycetes. Fungal associates of arthropods*, 1986, *in* Ainsworth *et al.* (Eds), *The Fungi* **4B**: 651, 1973, *in* Jones (Ed.), *Recent advances in aquatic mycology*, 1976; keys, *Mycol.* **65**: 1, 1973; phylogeny, *in* Parker (Ed.), 1982, **1**: 195), Cavalier-Smith (*in* Coombs *et al.*, *Evolutionary relationships among Protozoa*, 1998), Moss (*TBMS* **65**: 115, 1975; phylogeny), Moss & Young (*Mycol.* **70**: 944, 1978; phylogeny), Benjamin (*in* Kendrick (Ed.), *The whole fungus* **2**: 573, 1979; phylogeny), Porter & Smiley (*Exp. Mycol.* **3**: 188, 1979; phylogeny), Lichtwardt *et al.* (*Trans. mycol. Soc. Japan* **28**: 359, 1987; Japan).

Reviews: Duboscq *et al.* (*Arch. Zool. Exp. Gén.* **86**: 29, 1948), Manier (*Ann. Sci. nat., Bot.* sér. 12, **10**: 565, 1969), Moss (*in* Batra (Ed.), *Insect-fungus symbiosis*: 175, 1979), Lichtwardt & Williams (*CJB* **66**: 1259, 1988; spp. diversity & distrib., **68**: 1057, 1990; Australia), Williams & Lichtwardt (*CJB* **68**: 1045, 1990; NZ).

trichomycin, an antiobiotic from *Streptomyces hachijoensis* (Hosoya *et al.*, 1955); antifungal (esp. against *Candida albicans*) and anti-*Trichomonas*.

Trichonectria Kirschst. (1907), Bionectriaceae. 15 (rotten wood, often associated with lichen thalli or liverworts), widespread. See Döbbeler (*Mitt. bot. StSamml., München* **14**: 1, 1978; on *Musci*), Hawksworth (*Notes R. bot. Gdn Edinb.* **36**: 181, 1978), Hawksworth (*Lichenologist* **12**: 100, 1980; nomencl.), Rossman *et al.* (*Stud. Mycol.* **42**: 248 pp., 1999).

Trichonosema E.U. Canninga, D. Refardtb, C.R. Vossbrinckc, B. Okamurad & A. Curry (2002), Mi crosporidia. 2. See Canninga *et al.* (*Eur. J. Protist.* **38**: 247, 2002).

Trichopeltella Höhn. (1910), ? Microthyriaceae. 1, Java.

Trichopeltheca Bat., C.A.A. Costa & Cif. (1958), Euantennariaceae. Anamorphs *Plokamidomyces*, *Trichothallus*. 2, Asia; Pacific. See Hughes (*N.Z. Jl Bot.* **10**: 230, 1972).

Trichopeltidaceae Theiss. (1914) = Microthyriaceae.

Trichopeltina Theiss. (1914), ? Microthyriaceae. 2, S. America.

Trichopeltinaceae Bat., C.A.A. Costa & Cif. (1958) = Microthyriaceae.

Trichopeltinites Cookson (1947), Fossil Fungi. 4 (Cretaceous, Tertiary), widespread.

Trichopeltis Speg. (1889) = Trichothyrium fide Spooner & Kirk (*MR* **94**: 223, 1990).

Trichopeltium Clem. (1909) ≡ Trichopeltulum fide Sutton (*Mycol. Pap.* **141**, 1977).

Trichopeltopsis Höhn. (1909) = Trichothyrium fide Spooner & Kirk (*MR* **94**: 223, 1990).

Trichopeltospora Bat. & Cif. (1958), ? Microthyriaceae. 1, Brazil. See Batista & Ciferri (*Publções Inst. Micol. Recife* **90**: 17, 1958).

Trichopeltula Theiss. (1914) = Trichothyrium fide Clements & Shear (*Gen. Fung.*, 1931).

Trichopeltulum Speg. (1889), anamorphic *Pezizomycotina*, Cpt.0eH.?. 1, Brazil. See Petrak & Sydow (*Annls mycol.* **33**: 173, 1935), Reynolds & Gilbert (*Aust. Syst. Bot.* **18**: 265, 2005).

Trichopeltum Bat., Cif. & C.A.A. Costa (1957), ? Microthyriaceae. 1, Hawaii. See Batista *et al.* (*Publções Inst. Micol. Recife* **90**: 20, 1958).

Trichopeziza Fuckel (1870), Hyaloscyphaceae. 26, widespread (esp. temperate). See Raitviir (*Eesti NSV Tead. Akad. Toim.* Biol. seer **36**: 313, 1987), Verkley (*Nova Hedwigia* **63**: 215, 1996; ultrastr.), Cantrell & Hanlin (*Mycol.* **89**: 745, 1997; DNA).

Trichopezizella Dennis ex Raitv. (1969) = Lachnum fide Haines (*Mycol.* **66**: 216, 1974; key), Spooner (*Biblthca Mycol.* **116**, 1987; status).

Trichophaea Boud. (1885), Pyronemataceae. Anamorph *Dichobotrys*. *c.* 25, widespread (north temperate). See Kanouse (*Mycol.* **50**: 128, 1958; key), Svrček (*Česká Mykol.* **28**: 129, 1974), Svrček (*Česká Mykol.* **31**: 69, 1977), Waraitch (*TBMS* **68**: 303., 1977), Pant (*Norw. Jl Bot.* **27**: 31, 1980), Thind & Kaushal (*Kavaka* **7**: 47, 1980), Korf & Zhuang (*Mycotaxon* **22**: 483, 1985), Zhuang & Korf (*Mycotaxon* **35**: 297, 1989), Häffner & Christian (*Z. Mykol.* **57**: 161, 1991), Korf & Zhuang (*Mycotaxon* **40**: 413, 1991), Egger (*CJB* **74**: 773, 1996; DNA), Wu & Kimbrough (*Int. J. Pl. Sci.* **157**: 595, 1996; ultrastr.), Landvik *et al.* (*Nordic Jl Bot.* **17**: 403, 1997; DNA), Jamoni (*Funghi e Ambiente* **76**: 13, 1998), Liu & Zhuang (*Mycosystema* **25**: 546, 2006), Perry *et al.* (*MR* **111**: 549, 2007; phylogeny).

Trichophaeopsis Korf & Erb (1972), Pyronemataceae. 3, widespread. See Häffner & Krieglsteiner (*Z. Mykol.* **57**: 167, 1991), Landvik *et al.* (*Nordic Jl Bot.* **17**: 403, 1997; DNA), Hansen & Pfister (*Mycol.* **98**: 1029, 2006; phylogeny), Perry *et al.* (*MR* **111**: 549, 2007; phylogeny).

Trichophila Oudem. (1889) ? = Piedraia fide Sutton (*Mycol. Pap.* **141**, 1977).

Trichophyma Rehm (1904), ? Arthoniales. 2 (on lichens), tropical. See Santesson (*Svensk bot. Tidskr.* **43**: 547, 1949), Matzer (*Mycol. Pap.* **171**, 1996).

Trichophysalospora Lebedeva (1933) = Physalospora fide von Arx & Müller (*Beitr. Kryptfl. Schweiz* **11** no. 1, 1954).

trichophytin, an antigen prepared from dermatophytes, esp. for use in skin testing. Commercial trichophytin is usually a mixture of antigens of several spp. of *Trichophyton* and *Microsporum*.

Trichophyton Malmsten (1848), anamorphic *Arthroderma*, Hso.≡ eP.2. 22 (on humans and animals, causing trichophytoses, and in soil and river sediments), widespread. *T. mentagrophytes* (syns. *T. asteroides*, *T. gypseum*); *T. schoenleinii* (human favus); *T. concentricum* (tinea imbricata); *T. verrucosum* (cattle ringworm). See also tinea. See Dodge (*Medical Mycology*, 1935), Ajello (*Sabouraudia* **6**: 153, 1968), Ajello & Padhye

(*Mykosen* **30**: 258, 1987; stimulation of macroconidia in culture), Padhye *et al.* (*J. Med. Vet. Mycol.* **25**: 195, 1987; mating behaviour), Mochizuki *et al.* (*J. Med. Vet. Mycol.* **28**: 191, 1990; taxonomy of *T. interdigitale*), Symoens *et al.* (*Mycoses* **32**: 652, 1990; isoelectric focusing to differentiate *T. mentagrophytes* and *T. interdigitale*), Devliotou-Panagiotidou *et al.* (*Mycoses* **35**: 375, 1992; *T. rubrum* in Greece). See also tinea), Nishio *et al.* (*Mycopathologia* **117**: 127, 1992; phylogeny by mitochondrial DNA), Kano *et al.* (*Curr. Microbiol.* **37**: 236, 1998; chitin synthase), Khosravi & Abedian (*Journal of the Faculty of Veterinary Medicine, University of Tehran* **53**: 11, 1998; protein electrophoresis), Makimura *et al.* (*J. Clin. Microbiol.* **36**: 2629, 1998; phylogeny), Gräser *et al.* (*Medical Mycology* **37**: 315, 1999), Gräser *et al.* (*Medical Mycology* **37**: 105, 1999; phylogeny), Kane & Summerbell (*Manual of Clinical Microbiology*: 1275, 1999; review), Kano *et al.* (*Mycoses* **42**: 71, 1999; molecular identification), Makimura *et al.* (*J. Clin. Microbiol.* **37**: 807, 1999; phylogeny), Gräser *et al.* (*J. Clin. Microbiol.* **38**: 3329, 2000; phylogeny), Kano *et al.* (*Medical Mycology* **38**: 47, 2000; chitin synthase), Kano *et al.* (*Mycopathologia* **146**: 111, 1999; phylogeny), Kim *et al.* (*Mycoses* **44**: 157, 2001; RAPDs), Liu *et al.* (*Journal of Medical Microbiology* **51**: 117, 2002; molecular identification), Summerbell *et al.* (*Stud. Mycol.* **47**: 75, 2002; biological species concept), Untereiner *et al.* (*Stud. Mycol.* **47**: 25, 2002; phylogeny), Gaedigk *et al.* (*J. Clin. Microbiol.* **41**: 5478, 2003; genetic diversity), Kaszubiak *et al.* (*Infect. Genet. Evol.* **4**: 179, 2004; population biology), Ohst *et al.* (*J. Clin. Microbiol.* **42**: 4444, 2004; microsatellites), Brilhante *et al.* (*Journal of Medical Microbiology* **55**: 1583, 2006; from dog).

Trichophytonaceae Vuill. ex E.K. Novák & Galgoczy (1969) = Arthrodermataceae.

Trichopilus (Romagn.) P.D. Orton (1991) = Entoloma fide Kuyper (*in litt.*).

Trichoplacia A. Massal. (1853) nom. dub., ? Dothideomycetes. See Zahlbruckner (*Nat. Pflanzenfam.* **8**: 61, 1926).

Trichopsora Lagerh. (1892), Pucciniosiraceae. 1 (on *Tournefortia* (0, III) (*Boraginaceae*)), Ecuador.

Trichoramalina Rundel & Bowler (1974) ? = Ramalina fide Hawksworth *et al.* (*Dictionary of the Fungi* edn 8, 1995).

Trichoscypha (Cooke) Sacc. (1889) ≡ Cookeina.

Trichoscypha Boud. (1885) [non *Trichoscypha* Hook. f. 1862, *Anacardiaceae*] ≡ Trichoscyphella.

Trichoscyphella Nannf. (1932) = Lachnellula fide Dennis (*Persoonia* **2**: 171, 1962).

Trichoseptoria Cavara (1892), anamorphic *Pezizomycotina*, Cpd.0fH.?. 2, Europe; N. America. *T. fructigena* (soft rot of apples).

Trichosia Bat. & R. Garnier (1960) = Trichasterina fide von Arx & Müller (*Stud. Mycol.* **9**, 1975).

Trichoskytale Corda (1842) ≡ Trichocoma.

Trichosperma Speg. (1888) [non *Trichospermum* Blume 1825, *Tiliaceae*] ≡ Spermotrichum.

Trichospermella Speg. (1912), Pezizomycotina. 1, S. America. See Petrak & Sydow (*Annls mycol.* **33**: 175, 1935).

Trichosphaera Dumort. (1822) nom. dub., Pezizomycotina. Used for diverse fungi with flask-like sporocarps in stromata.

Trichosphaerella E. Bommer, M. Rousseau & Sacc. (1891), Niessliaceae. Anamorph *Acremonium*-like. 2, Europe; Africa. See Samuels & Barr (*CJB* **75**: 2165, 1998), Rossman *et al.* (*Stud. Mycol.* **42**: 248 pp., 1999).

Trichosphaeria Fuckel (1870), Trichosphaeriaceae. 27 (from wood and bark), widespread. See Réblová *et al.* (*Sydowia* **51**: 49, 1999), Réblová (*Mycol.* **98**: 68, 2006; phylogeny).

Trichosphaeriaceae G. Winter (1885), Trichosphaeriales. 11 gen. (+ 11 syn.), 65 spp.

Lit.: Seifert (*CJB* **65**: 2196, 1987), Réblová (*Mycotaxon* **70**: 387, 1999), Réblová (*Mycotaxon* **70**: 421, 1999), Réblová (*Mycotaxon* **71**: 45, 1999), Réblová *et al.* (*Sydowia* **51**: 49, 1999), Réblová *et al.* (*Sydowia* **51**: 49, 1999), Gams (*Stud. Mycol.* **45**: 192, 2000), Campbell & Shearer (*Mycol.* **96**: 822, 2004), Réblová & Seifert (*Mycol.* **96**: 343, 2004).

Trichosphaeriales M.E. Barr (1983). Sordariomycetes. 1 fam., 14 gen., 73 spp. Stromata absent, or reduced to a hyphal subiculum. Ascomata perithecial, superficial, often aggregated, ± globose, black, often thick-walled, usually setose, the ostiole papillate, periphysate. Interascal tissue of narrow persistent thin-walled true paraphyses. Asci cylindrical, persistent, thin-walled, not fissitunicate, often with a small J- apical ring. Ascospores variously shaped, hyaline to versicolored, usually septate and rarely muriform, sometimes with polar pores, sometimes fragmenting at the septa, without germ pores, sometimes with a sheath. Anamorphs varied, hyphomycetous, with simple pigmented conidiophores, often with tretic conidiogenesis and often with complex conidia. Saprobic esp. on wood and bark, occ. on other fungi, cosmop. Fam.:

Trichosphaeriaceae

For *Lit.* see under fam.

Trichosphaeropsis Bat. & Nascim. (1960), Pezizomycotina. 1, Brazil. See Batista & Nascimento (*Atas Inst. Micol. Univ. Recife* **1**: 299, 1960).

trichospore, a caducous, dehiscent, monosporous sporangium with basal appendages characteristic of the *Harpellales* (Moss & Lichtwardt, *CJB* **54**: 2346, 1976).

Trichosporiella Kamyschko (1960), anamorphic *Laetinaevia*, Hso.0eH.1/10. 4, widespread. See van Oorschot (*Stud. Mycol.* **20**, 1980), de Hoog *et al.* (*Antonie van Leeuwenhoek* **51**: 79, 1985).

Trichosporites Félix (1894), Fossil Fungi. 1 (Cretaceous), Sweden.

Trichosporodochium Dorn.-Silva & Dianese (2004), anamorphic *Pezizomycotina*. 1, Brazil. See Dornelo-Silva & Dianese (*Mycol.* **96**: 881, 2004).

Trichosporon Behrend (1890), anamorphic *Trichosporonaceae*. *c.* 20, widespread. *T. beigelii* (**white piedra** of man). See Herbrecht *et al.* (*J. Mycol. Médic.* **3**: 129, 1993; medical), de Hoog *et al.* (*Mycotaxon* **63**: 345, 1997; nomenclature), Müller *et al.* (*Microbiology* **144**: 1721, 1998; ultrastr.), Sugita *et al.* (*Mycoscience* **39**: 7, 1998; mol. phylogeny), Sugita *et al.* (*J. Clin. Microbiol.* **36**: 1458, 1998; PCR detection), Barnett *et al.* (*Yeasts: Characteristics and Identification* 3rd edn, 2000), Fell *et al.* (*Int. J. Syst. Evol. Microbiol.* **50**: 1351, 2000; mol. phylogeny), Zhao *et al.* (*Mycosystema* **21**: 533, 2002; 26S rDNA), Middelhoven (*Mycoses* **46**: 7, 2003; clinically relevant species), Biswas *et al.* (*Journal of Clinical Microbiology* **43**: 5171, 2005; phylogeny).

Trichosporonaceae Nann. (1934), Tremellales. 2 gen.

(+ 8 syn.), 21 spp.

Trichosporonales Boekhout & Fell (2001) = Tremellales.

Trichosporonoides Haskins & J.F.T. Spencer (1967), anamorphic *Tremellomycetes*. 6, widespread. See de Hoog (*Stud. Mycol.* **15**: 20, 1979; gen. revision), Ramirez (*Mycopathologia* **108**: 25, 1989; key), Inglis *et al.* (*Mycol.* **84**: 555, 1992), de Hoog & Smith *in* Kurtzman & Fell (Eds) (*Yeasts, a taxonomic study* 4th edn: 873, 1998).

Trichosporum Fr. (1825) nom. conf., nom. illeg., Ascomycota. See Hughes (*CJB* **36**: 727, 1958; nomencl.).

Trichosporum Vuill. (1901) [non *Trichosporum* D. Don 1822, *Gesneriaceae*] ? = Piedraia fide Hawksworth *et al.* (*Dictionary of the Fungi* edn 8, 1995).

Trichosterigma Petch (1923) [non *Trichosterigma* Klotzsch & Garcke 1859, *Euphorbiaceae*] = Hirsutella fide Petch (*TBMS* **9**: 93, 1923).

Trichostroma Corda (1829) nom. dub., anamorphic *Pezizomycotina*. See Hughes (*CJB* **36**: 747, 1958).

Trichostroma Link (1826) nom. dub., anamorphic *Pezizomycotina*. No spp. Included.

Trichothallaceae Bat. & Cif. (1959) = Euantennariaceae.

Trichothallus F. Stevens (1925), anamorphic *Trichopeltheca*, Cpt.-.-. 1, Hawaii. See Reynolds & Gilbert (*Aust. Syst. Bot.* **18**: 282, 2005).

Trichotheca P. Karst. (1887) ? = Bloxamia fide Sutton (*Mycol. Pap.* **141**, 1977).

trichothecenes, toxins (scirpenes) of *Fusarium tricinctum*, *F. sporotrichioides*, *F. poae*, *Trichothecium*, etc.; the cause of alimentary toxic aleukia in farm animals and humans. See Beasley (*Trichothecene mycotoxicosis*, 2 vols, 1991).

trichothecin, an antifungal metabolic product of *Trichothecium roseum* (Freeman & Morrison, *Biochem. J.* **44**: 1, 1949).

Trichotheciopsis J.M. Yen (1979) = Trichothecium fide Sutton (*in litt.*).

Trichothecium Link (1809), anamorphic *Hypocreales*, Hso.1eH.34. 5, widespread. *T. roseum* (pink rot of apples). See Park (*TBMS* **39**: 239, 1956; conidia), Rifai & Cooke (*TBMS* **49**: 147, 1966; status), Sesan (*Probl. Prot. Plant.* **13**: 381, 1985; bibliogr.), Seifert *et al.* (*Mycol.* **89**: 250, 1997; phylogeny).

Trichotheliaceae Bitter & F. Schill. (1927) = Porinaceae.

Trichotheliales Hafellner & Kalb (1995) = Ostropales.

Trichotheliomyces Cif. & Tomas. (1953) ≡ Trichothelium.

Trichothelium Müll. Arg. (1885), Porinaceae (L). 40, widespread (tropical). See Santesson (*Symb. bot. upsal.* **12** no. 1: 1, 1952), Vězda (*Nova Hedwigia* **58**: 123, 1995; key 4 spp.), McCarthy & Malcolm (*Lichenologist* **29**: 1, 1997), Lücking (*Nova Hedwigia* **66**: 375, 1998), McCarthy (*Lichenologist* **33**: 393, 2001; Christmas Island), McCarthy (*Biblthca Lichenol.* **87**, 2003), Grube *et al.* (*MR* **108**: 1111, 2004; phylogeny), Lücking (*Biblthca Lichenol.* **88**: 409, 2004; key), Lücking & Cáceres (*MR* **108**: 571, 2004; corticolous spp.), Baloch & Grube (*MR* **110**: 125, 2006; phylogeny).

Trichothyriaceae Theiss. (1914) = Microthyriaceae.

Trichothyriella Theiss. (1914), Microthyriaceae. 1, S.E. Asia. See Spooner & Kirk (*MR* **94**: 223, 1990).

Trichothyrina (Petr.) Petr. (1950) = Lichenopeltella fide Santesson (*SA* **9**: 15, 1991).

Trichothyrinula Petr. (1950), Microthyriaceae. 1, Ecuador. See Petrak (*Sydowia* **4**: 171, 1950).

Trichothyriomyces Bat. & H. Maia (1955), Microthyriaceae. 1, Brazil. See Spooner & Kirk (*MR* **94**: 223, 1990).

Trichothyriopsis Theiss. (1914), Microthyriaceae. 3 or 4, Indonesia; Brazil. See Spooner & Kirk (*MR* **94**: 223, 1990).

Trichothyrites Rosend. (1943), Fossil Fungi. 2 (Eocene, Pleistocene), British Isles; USA. See Smith (*The mushroom hunter's field guide* edn 3, 1980).

Trichothyrium Speg. (1889), Microthyriaceae. Anamorphs *Hansfordiella*, *Isthmospora*. 12, widespread (subtropical). See Hughes (*Mycol. Pap.* **50**, 1953), Spooner & Kirk (*MR* **94**: 223, 1990).

trichotomous, with three branches arising from the same point.

Trichotrema Clem. (1909) ? = Pleurotrema fide Hawksworth *et al.* (*Dictionary of the Fungi* edn 8, 1995).

Trichotuzetia J. Vavra, J.I.R. Larsson & Baker (1997), Microsporidia. 1.

Trichozygospora Lichtw. (1972), Legeriomycetaceae. 1 (in *Diptera*), USA; Europe. See Moss & Lichtwardt (*CJB* **54**: 2346, 1976), Moss & Lichtwardt (*CJB* **55**: 3099, 1977; ultrastr.), Lichtwardt (*The Trichomycetes. Fungal associates of arthropods*, 1986).

Trichurus Clem. (1896), anamorphic *Microascaceae*, Hsy.0eP.19. 5, widespread (temperate). See Issakainen *et al.* (*Medical Mycology* **41**: 31, 2003; phylogeny).

Tricladiella K. Ando & Tubaki (1984), anamorphic *Pezizomycotina*, Hso.1bH.1. 1 (aquatic), Japan. See Ando & Tubaki (*TMSJ* **25**: 41, 1984).

Tricladiomyces Nawawi (1985), anamorphic *Agaricomycetes*. 1 (aquatic, dolipore septa), Malaysia. See Nawawi (*J. Linn. Soc.* Bot. **91**: 58, 1985), Nawawi & Kuthubutheen (*TBMS* **90**: 670, 1988).

Tricladiopsis Descals (1982), anamorphic *Pezizomycotina*, Hso.1bH.10. 2 (aquatic), British Isles. See Descals (*TBMS* **78**: 417, 1982).

Tricladiospora Nawawi & Kuthub. (1988), anamorphic *Pezizomycotina*, Hso.1bH.19. 3, Malaysia. See Nawawi & Kuthubutheen (*TBMS* **90**: 484, 1988).

Tricladium Ingold (1942), anamorphic *Hymenoscyphus, Cudoniella*, Hso.1bH.1/10. 10 (aquatic), widespread. See Petersen (*Mycol.* **54**: 135, 1962; key), Ando & Kawamoto (*TMSJ* **26**: 471, 1985), Marvanová & Bandoni (*TBMS* **85**: 747, 1985), Webster *et al.* (*Nova Hedwigia* **60**: 493, 1995; teleomorph), Baschien *et al.* (*Nova Hedwigia* **83**: 311, 2006; phylogeny).

Triclinum Fée (1825), Pannariaceae (L). 1, widespread (tropical). See Jørgensen (*Op. Bot.* **45**, 1978).

Tricornia Pell & E.U. Canning (1992), Microsporidia. 1.

Tricornispora Bonar (1967) = Tridentaria fide Kendrick (*in litt.*).

Tridens Massee (1901), ? Rhytismataceae. 1, N. America.

Tridentaria Preuss (1852), anamorphic *Pezizomycotina*, Hso.1bH.1. 2, widespread (temperate). See Drechsler (*J. Wash. Acad. Sci.* **27**: 391, 1937), van der Aa & Oorschot (*Persoonia* **12**: 415, 1985; redisposition spp.).

Trifurcospora K. Ando & Tubaki (1988), anamorphic *Pezizomycotina*, Hso.1bH.10. 1 (aquatic), Japan; USA. See Ando & Tubaki (*TMSJ* **28**: 471, 1987).

Triglyphium Fresen. (1852), anamorphic *Pezizomycotina*, Hso.1bH.?. 2, widespread. See Marvanová & Bärlocher (*Czech Mycol.* **53**: 1, 2001; Canada).

Trigonia J.F.H. Beyma (1933) [non *Trigonia* Aubl. 1775, *Trigoniaceae*] ≡ Triangularia.

Trigonipes Velen. (1939) = Clitocybe fide Singer (*Agaric. mod. Tax.* edn 3, 1975).

Trigonopsis Schachner (1929), anamorphic *Saccharomycetales*, Hso.0eH.?. 1, Europe. See Sentheshanmuganathan & Nickerson (*J. gen. Microbiol.* **27**: 437, 1962; nutrition, form), Yarrow *in* Kurtzman & Fell (Eds) (*Yeasts, a taxonomic study* 4th edn: 605, 1998), Kurtzman & Robnett (*FEMS Yeast Res.* **7**: 141, 2007).

Trigonosporium Tassi (1900), anamorphic *Pezizomycotina*, Cpd.0eP.?. 2, Australasia. See Sutton (*Mycol. Pap.* **123**, 1971).

Trihyphaecites Peppers (1970), Fossil Fungi. 1 (Carboniferous), USA.

Trihyphites Kalgutkar & Janson. (2000), Fossil Fungi, anamorphic *Ascomycota*. 1, China. See Kalgutkar & Jansonius (*AASP Contributions Series* **39**: 305, 2000).

trimerous, in threes.

trimitic, see hyphal analysis.

Trimitiella Dhingra (2006), Agaricomycetes. 1. See Dhingra (*Mycotaxon* **97**: 126, 2006).

Trimmatostroma Corda (1837), anamorphic *Helotiales*, Hsp.#eP.36. 18, widespread. See Ellis (*More Dematiaceous Hyphomycetes*, 1976), Zalar *et al.* (*Stud. Mycol.* **43**: 57, 1999), Hawksworth & Cole (*Fungal Diversity* **11**: 87, 2002), Schoch *et al.* (*Mycol.* **98**: 1041, 2006; phylogeny), Crous *et al.* (*Stud. Mycol.* **58**: 1, 2007), Winton *et al.* (*Mycol.* **99**: 240, 2007; phylogeny).

Trimmatothele Norman ex Zahlbr. (1903) = Verrucaria Schrad. fide Ertz & Diederich (*Mycol. Progr.* **3**: 229, 2004).

Trimmatothelopsis Zschacke (1934) = Acarospora fide Ertz & Diederich (*Mycol. Progr.* **3**: 229, 2004).

Trimorphomyces Bandoni & Oberw. (1983), Tremellaceae. 1, N. America. See Kang *et al.* (*Nucl. Acids Res.* **20**: 5229, 1992; 5S rRNA sequence), Bandoni & Boekhout (*The Yeasts, A Taxonomic Study*: 705, 1998), Scorzetti (*FEMS Yeast Research* **2**: 495, 2002; phylogeny).

Trinacriaceae Nann. (1934) = Orbiliaceae.

Trinacrium Riess (1852), anamorphic *Orbilia*, Hso.1bH.1. 13, Europe; America. See Tzean & Chen (*MR* **93**: 391, 1989; comparison of spp.), Chen & Tzean (*Fungal Science* Taipei **14**: 111, 1999; Taiwan), Soosamma *et al.* (*Mycol.* **93**: 1200, 2001; India).

Triophthalmidium (Müll. Arg.) Gyeln. (1933) = Caloplaca fide Hawksworth *et al.* (*Dictionary of the Fungi* edn 8, 1995).

Triparticalcar D.J.S. Barr (1980), Spizellomycetaceae. 1, Arctic. See Barr (*CJB* **58**: 2386, 1980).

tripartite tubular hair, filamentous appendage on the flagellum composed of a tapered solid base, a hollow cylindrical shaft and one or more terminal filaments. See straminipile.

Tripedotrichum G.F. Orr & Kuehn (1964) = Gymnoascus fide Benny & Kimbrough (*Mycotaxon* **12**: 1, 1980).

Triphragmiopsis Naumov (1914), ? Raveneliaceae. 3 (on *Berberidaceae*, *Ranunculaceae*, *Larix* (*Pinaceae*)), Europe; Asia. See Monoson (*Mycopathologia* **52**: 115, 1974), Lohsomboon *et al.* (*TMSJ* **31**: 335, 1990).

Triphragmium Link (1825), ? Raveneliaceae. 3 (on *Filipendula* (*Rosaceae*)), widespread (north temperate). See Cummins & Hiratsuka (*Illustr. Gen. Rust Fungi rev. edit.*, 1983; under *Sphaerophragmiaceae*), Lohsomboon *et al.* (*TMSJ* **31**: 215, 1990), Helfer (*Nova Hedwigia* **81**: 325, 2005; Europ. spp.) Probably *Phragmidiaceae*.

Triplicaria P. Karst. (1889), anamorphic *Hypoxylon*, Hsp.0eH.10. 1, Finland. See Petrak (*Sydowia* **7**: 299, 1953).

Triplosporium (Thaxt.) A. Batko (1964) = Entomophthora Fresen. fide Krejzova (*Česká Mykol.* **30**: 207, 1976) = Neozygites (Entomophthor.) fide, Remaudière & Keller (*Mycotaxon* **11**: 323, 1980).

Tripoconidium Subram. (1978), anamorphic *Pezizomycotina*, Hso.1bH.1. 1 (on nematodes), USA. See Subramanian (*Kavaka* **5**: 95, 1978).

Tripocorynelia Kuntze (1898) ≡ Tripospora.

Triporicellaesporites Ke & Shi (1978), Fossil Fungi. 3 (Paleocene, Tertiary), China.

Triporisporites Hammen (1954), Fossil Fungi. 1 (Cretaceous), Colombia.

Triporisporonites Sheffy & Dilcher (1971), Fossil Fungi. 1 (Eocene), USA.

Tripospermum Speg. (1918), anamorphic *Trichomerium*, Hso.0bP.1. 10, widespread. See Hughes (*Mycol. Pap.* **46**, 1951), Ando (*MR* **98**: 879, 1994; behaviour in culture).

Tripospora Sacc. (1886), Coryneliaceae. 4 (on *Podocarpus*), S. Africa; S. America. See Benny *et al.* (*Bot. Gaz.* **146**: 431, 1985; key), Catania & Romero (*MR* **105**: 1020, 2001; key).

Triposporiaceae Nann. (1934) = Asterinaceae.

Triposporina Höhn. (1912), anamorphic *Pezizomycotina*, Hso.1bH.19. 2, Java; N. America. See Deighton & Pirozynski (*Mycol. Pap.* **128**: 96, 1972).

Triposporiopsidaceae S. Hughes (1976) = Capnodiaceae.

Triposporiopsis W. Yamam. (1955) = Trichomerium fide Reynolds (*Mycotaxon* **14**: 189, 1982).

Triposporium Corda (1837), anamorphic *Batistinula*, Hso.0bP.19. 13, widespread. See Hughes (*Mycol. Pap.* **46**, 1951), Kuthubutheen & Nawawi (*MR* **95**: 158, 1991).

Triposporonites Sheffy & Dilcher (1971), Fossil Fungi. 1 (Eocene), USA.

Tripterospora Cain (1956) = Zopfiella fide Guarro *et al.* (*SA* **10**: 79, 1991), Cai *et al.* (*MR* **110**: 359, 2006).

Tripterosporaceae Cain (1956) = Lasiosphaeriaceae.

Tripterosporella Subram. & Lodha (1968), Lasiosphaeriaceae. 1 (coprophilous), India; Africa. See Subramanian & Lodha (*Curr. Sci.* **37**: 246, 1968).

triquetrous, three-edged, three-cornered.

Triramulispora Matsush. (1975), anamorphic *Pezizomycotina*, Hso.1bH.1. 3, Japan. See Matsushima (*Icon. microfung. Matsush. lect.*: 158, 1975).

Triscelophorus Ingold (1944), anamorphic *Pezizomycotina*, Hso.1bH.1. 4 (aquatic), widespread. See Petersen (*Mycol.* **54**: 162, 1962; key).

Triscelosporium Nawawi & Kuthub. (1987), anamorphic *Pezizomycotina*, Hso.0bP.2. 1 (aero-aquatic), Malaysia. See Nawawi & Kuthubutheen (*Mycotaxon* **29**: 285, 1987).

trisporic acid C, a hydroxy-keto acid from mated cultures of *Blakeslea trispora* able to induce carotenogenesis in separate strains (Caglioti *et al.*, *Chimica Industria* **46**: 961, 1964).

tristichous, in three rows.

Trisulcosporium H.J. Huds. & B. Sutton (1964), anamorphic *Pezizomycotina*, Hso.1bH.1. 1 (aquatic), British Isles. See Hudson & Sutton (*TBMS* **47**: 200, 1964).

Tritirachium Limber (1940), anamorphic *Pezizomycotina*, Hso.0eH.10. 3 (some on humans), widespread. See MacLeod (*CJB* **32**: 818, 1954), de Hoog (*Stud. Mycol.* **1**, 1972; 2 spp. accepted).

trivial name (1) the zoological term for specific epithet; (2) a common name for a chemical, see Fungal metabolites.

Trochila Fr. (1849), Helotiales. Anamorph *Cryptocline*. *c.* 15, widespread (temperate). See Greenhalgh & Jones (*TBMS* **47**: 311, 1964), Nauta & Spooner (*Mycologist* **14**: 65, 2000; key British spp.).

Trochodium Syd. & P. Syd. (1920) = Uromyces fide Cummins & Hiratsuka (*Illustr. Gen. Rust Fungi rev. edit.*, 1983).

Trochoideomyces Thaxt. (1931), Laboulbeniaceae. 1, Indonesia. See Nannfeldt (*Svensk bot. Tidskr.* **43**: 468, 1949).

Trochophora R.T. Moore (1955), anamorphic *Pezizomycotina*, Hso.0-≡ hP.10. 1, India; S.E. Asia. See Goos (*Mycol.* **78**: 744, 1986).

Trogia Fr. (1836), Marasmiaceae. *c.* 20, widespread (tropical). See Corner (*Monogr. Cantharelloid Fungi*, 1966), Corner (*Gdns' Bull.* Singapore, Suppl. 2, 1991; Asian species, very broad generic concept), Wilson & Desjardin (*Mycol.* **97**: 667, 2005; polyphyletic, sister group to marasmioid clade).

Troglobiomyces Pacioni (1980) = Hirsutella fide Samson *et al.* (*Persoonia* **12**: 123, 1984).

troglobiotic, living in caves.

Troglomyces S. Colla (1932), Laboulbeniaceae. 1, Europe. See Rossi & Balazuc (*Revue Mycol.* Paris **41**: 525, 1977), Tavares (*Mycol. Mem.* **9**: 627 pp., 1985).

troll, a goblin or dwarf in Scandinavian mythology, said to carry off naughty children; dolls were traditionally made of *Alectoria*, *Bryoria* and *Usnea* spp. and given to children to remind them to behave.

Trolliomyces Ulbr. (1938) ≡ Teloconia.

Trombetta Adans. (1763) nom. rej. ≡ Craterellus.

Tromera A. Massal. ex Körb. (1865) = Sarea fide Hawksworth & Sherwood (*CJB* **59**: 357, 1981).

Tromera, see *Tromera A. Massal. ex Körb.*

Tromeropsis Sherwood (1981), Pezizomycotina. 1, Finland. See Sherwood (*CJB* **59**: 370, 1981).

troop, a group of sporocarps (esp. basidiomata), generally from one mycelium.

trophocyst (of *Pilobolus*), a hyphal swelling from which a sporangiophore is produced.

trophogonium (**trophogone**) (of ascomycetes), an antheridium of which the only use is supplying food (Dangeard).

Tropism (frequently as a suffix). A turning or growth in response to an environmental stimulus. The response is dependent on the direction of the stimulus and can be + (positive) when towards or – (negative) when away from the stimulus. Among the more important tropisms seen in fungi are:

(1) **autotropism**, an avoidance (–) response between neighbouring hyphae which in part is responsible for the spacing of hyphae at the colony margin (Trinci, *in* Jennings & Rayner (Eds), *The ecology and physiology of the fungal mycelium*: 23, 1984).

(2) **chemotropism**, a reaction to a chemical, e.g. oxygen (Robinson, *New Phytologist* **72**: 1349, 1973), a hormone (Banbury, *J. Exper. Bot.* **6**: 235, 1975) or (in the case of water moulds but not other fungi), nutrients (Musgrove *et al.*, *J. Gen. Microbiol.* **101**: 65, 1977) or toxic metals (Fomina *et al.*, *MR* **107**: 861, 2003). *Microdochium bolleyi* exhibited tropism toward oxygen released from roots or air-filled tubes in an anoxic environment (Damm *et al.*, *FEMS Microbiology Ecology* **45**: 293, 2003); *Trichophyton rubrum* reacts positively to oxygen (Behzadi & Behzadi, *Turkiye Klinikleri J. Med. Sci.* **26**: 607, 2006); the dry rot fungus, *Serpula lacrymans*, however has linear tropism away from the air supply, and applying a controlled air flow results in mycelium shrivelling, discolouring and becoming susceptible to attack by moulds (Low *et al.*, *Holzforschung* **53**: 129, 1999). Species of *Aphanomyces*, *Phytophthora*, *Pythium* and the nematode-trapping fungus *Arthrobotrys oligospora* are characterized by positive chemotropism to root diffusates due to receptor-based recognition systems (Deacon, *New Phytologist* **133**: 135, 1996; Bordallo *et al.*, *New Phytologist* **154**: 491, 2002). Chemotropism observed in arbuscular mycorrhizal fungi may represent an important mechanism functional to host root location, appressorium formation and symbiosis establishment (Sbrana & Giovannetti, *Mycorrhiza* **15**: 539, 2005). Toxic metals stimulate negative tropism. Hyphae of *Geotrichum candidum*, *Gliocladium roseum*, *Humicola grisea* and *Trichoderma viride* curl and grow away from Cu and Cd (Fomina *et al.*, *FEMS Microbiol. Lett.* **193**: 207, 2000).

(3) **galvanotropism**, a reaction to an electrical field (McGillivray & Gow, *J. Gen. Microbiol.* **132**: 2515, 1986).

(4) **geotropism** (**gravitropism**), a reaction to gravity, e.g. sporangiophores of *Mucorales* and stipes, gills and tubes of fruit bodies of *Basidiomycota* (Moore, *New Phytol.* **117**: 3, 1991). A mathematical model for gravitropic reactions of fruit bodies has been made for *Coprinus cinereus*, with potential as a predictive tool for further analysis of mushroom gravitropism (Meskauskas*et al.*, *New Phytologist* **143**: 387, 1999).

(5) **phototropism**, a reaction of conidiophores, sporangiophores, asci, stipes etc. to light (Bergman *et al.*, *Bact. Rev.* **33**: 99, 1969; Carlile, *J. gen. Microbiol.* **28**: 161, 1962; Shropshire, *Physiological Reviews* **43**: 38, 1963). Classical examples are fungi of the genera *Phycomyces* and *Pilobolus* which have spore-bearing structures showing a phototropic response, bending towards a light source to facilitate dispersal. *Trichophyton rubrum* exhibits negative tropism to ultraviolet light (Behzadi & Behzadi, *Turkiye Klinikleri J. Med. Sci.* **26**: 607, 2006).

(6) **thigmotropism**, a reaction of germ tubes and hyphae to plant and other sufaces (Dickson, *Phytopath. Z.* **66**: 38, 1969; Kwon & Hoch, *Exp. Mycol.* **15**: 116, 1991). Almost all rust fungi show topographical sensing of stomata location by uredospore germ tubes and respond by producing an appressorium. Rust fungi respond to different surface microheights, reflecting adaptations to different host plants: the elevation of stomatal guard cells may provide the signal for production of an appressorium (Allen *et al.*, *Phytopathology* **81**: 323, 1991). Surface topography provides

signals for growth orientation in *Uromyces appendiculatus* (Hoch *et al.*, *Science* **235**: 1659, 1987). Tests with this fungus on silicon wafers demonstrated that germ tubes produce appressoria in the presence of a ridge/groove of about 0.5 µm high, and heights above 1 µm elicit no fungal response (Kwon & Hoch, 1991). Germ tubes of *Puccinia graminis* grow perpendicular to the rows of wheat leaf cells or even to an inert replica of a wheat leaf (Read *et al.*, *Perspectives in Plant Cell Recognition* pp. 137-172, 1992). Germ tube thigmotropism, induction and formation of appressoria following topographic signals have been observed for *Cochliobolus sativus* on artificial replicas of barley leaf surfaces (Clay *et al.*, *Protoplasma* **178**: 1615, 1994).

(6) **ecotropism**, a reaction to a particular ecological niche (close to **chemotropism** and **host specificity**, see **gene for gene**, **phytoalexin**, **axeny**). Dermatophytes (such as *Microsporum* spp. or *Trichophyton* spp.) exhibit tropism in the presence of cutaneous adnexa such as skin, nails or hair (Martino *et al.*, *Proceedings of the 8th Congress of the World Rabbit Science Association* Mexico, 576, 2004); *Paracoccidioides brasiliensis* exhibits tropism to lymphoid tissues (Brummer *et al.*, *Clin. Microbiol. Rev.* **6**: 89, 1993); germinating spores of *Verticillium biguttatum* show tropism towards the host-fungus hyphae of *Rhizoctonia solani*, leading to invasion and production of haustoria within the living host hypha (Boogert & Deacon, *Eur. J. Plant Path.* **100**: 137, 1994).

(7) **radiotropism**, the reaction of germ tubes and hyphae to a source of ionizing radiation (Zhdanova *et al.*, *MR* **108**: 1089, 2004). Free-living and mycorrhizal fungi can colonize depleted uranium and transform it into uranyl phosphate (Fomina *et al.*, *Curr. Biol.* **18**: 375, 2007). Some microfungi can grow into and decompose radioactive graphite and direct their growth toward sources of radioactivity. This response was studied for 2000 strains of 200 species of 98 genera of soil microfungi, most of which expressed positive tropism towards sources of beta and gamma radiation (See also Radiation and fungi).

Lit.: Carlile (*Ann. Rev. Plant Phys.* **16**: 175, 1965); Deacon (*Modern Mycology*, edn 3, 303 pp., 1997); Moore (*Int. J. Medicinal Mushrooms* **7**: 79, 2005).

Troposporella P. Karst. (1892), anamorphic *Pezizomycotina*, Hsp.0-≡ hP.1. 2, widespread (north temperate). See Sutton (*Mycol. Pap.* **132**, 1973).

Troposporium Harkn. (1884), anamorphic *Pezizomycotina*, Hsp.0-≡ hH.1. 1, N. America. See Peek & Solheim (*Mycol.* **50**: 847, 1959).

Troposporopsis S.R. Whitton, McKenzie & K.D. Hyde (1999), anamorphic *Pezizomycotina*. 2, widespread. See Whitton *et al.* (*Fungal Diversity* **3**: 174, 1999).

Trotteria Sacc. (1919) = Actinopeltis fide Sutton (*Mycol. Pap.* **141**, 1977).

Trotterula Speg. (1921) = Chaetothyrium fide Petrak & Sydow (*Annls mycol.* **32**: 6, 1934), Hughes (*Mycol.* **68**: 693, 1976).

TRTC, Cryptogamic Herbarium, Royal Ontario Museum (Toronto, Canada); founded 1887; transferred from the University of Toronto to the Museum in 1992.

Truffle. An ascoma, generally subterranean, of *Tuber* or other *Pezizales* or *Elaphomycetales* (see below), or a basidioma of *Hymenogastrales*; over 180 truffle-forming spp. are known. **Burgundy** -, *Tuber uncinatum*; **desert** -, *Mattirolomyces*, *Terfezia* and *Tirmania*; **false** -, *Hymenogaster*; **hart's** -, *Elaphomyces*; **Périgord (French)** -, *T. melanosporum*; **red** -, *Melanogaster variegatus*; **summer-**, *T. aestivum* (the best British sp.); **white-**, *Choiromyces meandriformis*; **white Piedmont** -, *T. magnatum*; **white winter** -, *T. hiemalbum*; **winter** -, *T. brumale*; **yellow** -, *Mattirolomyces*, *Terfezia* and *Tirmania*.

Growth of many truffles is best in well-drained, calcareous soils, and there is, probably always, an association between truffles and the roots of certain trees, esp. oaks (*Quercus petraea*, *Q. robur*, and the evergreen *Q. ilex* and *Q. coccifera*, among others) or, for desert-truffles, *Cistus* or *Helianthemum*. The position of the truffles is sometimes given by cracks in the soil, by the 'scorched' look of the plants over the truffles, by the look of the 'truffle trees', or by noting the habitat of the truffle fly, *Anistoma cinnamomea*; but they are generally looked for with the help of trained dogs or pigs. The ascomata themselves can be regarded as a complex ecosystem in that they frequently host mycelium of other fungi (Pacioni *et al.*, *MR* **111**: 1450, 2007).

For over one hundred years truffle culture has been undertaken in France where 'truffières' have been started by planting oak trees in good places, by inoculating the soil with soil from under truffle trees or with truffle tissue, and (the best method) by planting young trees taken from soil in which the truffle fungus is present. Truffles are first produced under such trees after 7 to 15 years; then generally for 20 to 30 or more years. In the south of France and Italy the crop is taken from December to March. Over the last about twenty years truffle culture technology has improved and some successful truffle plantations have been established well outside the natural range of these fungi, for example in Australia and New Zealand. There is also, however, concern about competition from exotic truffles, e.g. *Tuber indicum* from China. In addition to their use as food, truffles have been used in liqueur making, for scenting tobacco, and in certain perfumes; the widespread belief that some may be aphrodisiacs, together with their culinary value, high cost, strange ecology, and association in the public eye with chocolate products has resulted in them enjoying a mystique and a positive image almost unique among fungi, making them ideal as charismatic species for the conservation movement.

See Malençon (*Rev. Myc.* n.s. **3**(Mém. hors sér.), **1**, 1938, *Persoonia* **7**: 261, 1973), Kaltenbach (*Int. Rev. Agric.* **26**: 267T, 1935), Singer (*Mushrooms and truffles*, 1961), Chang & Hayes (Eds) (*Biology and cultivation of edible mushrooms*, 1978), Torini (*Jl Tartufo e la sua coltivazione*, 1984; cultivation), Bokhary (*Arab Gulf J. scient. Res., agric. biol. sci.* **B5**: 245, 1987; desert truffles).

The ascomycete truffles were formerly placed in a special order, the *Tuberales*, characterized by fleshy to leathery, ± globose ascomata with the hymenium lining a single or complex series of locules. They are now believed to be the result of the convergent evolution of various discomycete lines and are mainly referred to the *Pezizales* by Trappe (*Mycotaxon* **9**: 247, 1979) except for the isolated *Elaphomycetales* (for *Elaphomyces* only).

Lit.: Fischer (*Nat. PflFam.* **5b**, viii, 1938), Gilkey

(*Ore. St. Monogr. Stud. Bot.* **1**, 1939, *Mycol.* **46**: 783, 1954; keys, *N. Am. Flora, ser.* 2, **1**: 1, 1954; N. Am.), Lange (*Dansk Bot. Arkiv.* **16**, 1956; Denmark), Hawker (*Phil. Trans.* **B237**: 453, 1954), Dennis (1968: 71; Br. Isl.), Trappe (*TBMS* **57**: 87, 1971; *Terfeziaceae, Carbomycetaceae, Mycotaxon* **2**: 109, 1975; gen. names, **9**: 297, 1979; re-classification), Ławrynowicz (*Fl. Polska, Grzyby* **18**, 1988; keys 75 spp. Poland, SEM). Trappe *et al.* (*Austr. Syst. Bot.* **5**: 597, 613, 617, 631, 693, 1992; Australian spp.), Montecchi & Lazzari (*Funghi ipogei*, 1993; col. pls, keys, bibliogr.), Pegler *et al.* (*British truffles*, 1993; keys, descriptions, illustrations, UK spp.).

Truittella Karling (1949), ? Endochytriaceae. 1, USA.

Trullula Ces. (1852), anamorphic *Pezizomycotina*. 15, widespread (north temperate). See Sutton (*Mycol. Pap.* **141**, 1977; nomencl.).

truncate, ending abruptly, as though with the end cut off horizontally (Fig. 23.4*c*).

Truncatella Steyaert (1949), anamorphic *Broomella*, Cac.≡ eP.19. 10, widespread. See Watanabe *et al.* (*Nippon Kingakukai Kaiho* **39**: 21, 1998; characters in cult.), Jeewon *et al.* (*Mol. Phylogen. Evol.* **25**: 378, 2002; phylogeny), Jeewon *et al.* (*MR* **107**: 1392, 2003; phylogeny).

Truncicola Velen. (1934) = Hyaloscypha fide Huhtinen (*Karstenia* **29**: 545, 1990).

Truncocolumella Zeller (1939), Suillaceae. 3, N. America; Africa (tropical). See Smith & Singer (*Brittonia* **11**: 205, 1959).

Truncocolumellaceae Agerer (1999) = Suillaceae.

Truncospora Pilát ex Pilát (1953) = Perenniporia fide Ryvarden (*Syn. Fung.* **5**, 1991).

Tryblidaria (Sacc.) Rehm (1903), ? Patellariaceae. 4 or 9, widespread (subtropical). See Kutorga & Hawksworth (*SA* **15**: 1, 1997), Magnes (*Biblthca Mycol.* **165**, 1997).

Tryblidiaceae, see *Triblidiaceae*.

Tryblidiella Sacc. (1883) = Rhytidhysteron fide von Arx & Müller (*Stud. Mycol.* **9**, 1975).

Tryblidiopsis P. Karst. (1871), Rhytismataceae. Anamorph *Tryblidiopycnis*. 1, widespread (north temperate). See Gams (*Taxon* **41**: 99, 1992; nomencl.), Livsey & Minter (*CJB* **72**: 549, 1994), Magnes (*Arnoldia* **10**: 17, 1995), Magnes (*Biblthca Mycol.* **165**: 177 pp., 1997), Gernandt *et al.* (*Mycol.* **93**: 915, 2001; phylogeny).

Tryblidiopycnis Höhn. (1918), anamorphic *Tryblidiopsis*, St.0fH.10. 1, widespread.

Tryblidis Clem. (1909) ≡ Tryblidiopsis.

Tryblidium Wallr. (1833) ≡ Triblidium.

Tryblis Clem. (1931), Odontotremataceae. 2, Europe. See Magnes (*Biblthca Mycol.* **165**, 1997).

trypacidin, an antitrypanosome antibiotic from *Aspergillus fumigatus* (Balan *et al.*, *J. Antibiotics* A **16**: 157, 1963).

Trypetheliaceae Zenker (1827), Trypetheliales (L). 13 gen. (+ 18 syn.), 192 spp.

Lit.: Harris (*Acta Amazon.* Supl. **14**: 55, 1984; fam. concept, key 9 gen.), Makhija & Patwardhan (*Mycotaxon* **31**: 565, 1988), Makhija & Patwardhan (*Biovigyanam* **15**: 61, 1989; India), Harris (*Bolm Mus. paraense 'Emilio Goeldi'* sér. bot. **7**: 619, 1991), Makhija & Patwardhan (*J. Hattori bot. Lab.* **73**: 183, 1993), McCarthy (*Lichenologist* **27**: 310, 1995), Harris (*Lichenogr. Thomsoniana* North American Lichenology in Honor of John W. Thomson: 133, 1998), Del Prado *et al.* (*MR* **110**: 511, 2006).

Trypetheliales Lücking, Aptroot & Sipman (2008). Dothideomycetes. 1 fam., 13 gen., 192 spp. Fam.: **Trypetheliaceae**

For *Lit.* see under fam.

Trypetheliomyces Cif. & Tomas. (1953) = Trypethelium fide Hawksworth *et al.* (*Dictionary of the Fungi* edn 8, 1995).

Trypetheliopsis Asahina (1937), Monoblastiaceae (L). 6, widespread (tropical). fide Lücking & Sérusiaux (*Nordic Jl Bot.* **16**: 661, 1996), Aptroot *et al.* (*Biblthca Lichenol.* **64**, 1997), Aptroot *et al.* (*Biblthca Lichenol.* **97**, 2008; Costa Rica).

Trypethelium Spreng. (1804) nom. cons., Trypetheliaceae (L). *c.* 50, widespread (esp. tropical). See Lambright & Tucker (*Bryologist* **83**: 170, 1980; ultrastr., biology), Harris (*Acta Amazon.* Supl. **14**: 55, 1984; key 11 spp. Amazonia), Makhija & Patwardhan (*J. Hattori bot. Lab.* **73**: 183, 1993; key 30 spp. India), Prado *et al.* (*MR* **110**: 511, 2006; phylogeny), Schoch *et al.* (*Mycol.* **98**: 1041, 2006; phylogeny), Aptroot *et al.* (*Biblthca Lichenol.* **97**, 2008; Costa Rica).

Tryssglobulus B. Sutton & Pascoe (1987), anamorphic *Pezizomycotina*, Hso.0eP.16. 1, Australia. See Sutton & Pascoe (*TBMS* **88**: 44, 1987).

Tsuchiyaea Y. Yamada, H. Kawas., Itoh, I. Banno & Nakase (1988), anamorphic *Tremellaceae*, Hso.0eH.1. 1, S. Africa. See Yamada *et al.* (*Agric. Biol. Chem.* **53**: 2993, 1989; phylogeny), Yamada & Banno (*The Yeasts, A Taxonomic Study*: 878, 1998).

Tubaki (Keisuke; 1924-2005; Japan). Student, Tokyo University of Agriculture (1948); DSc, Hiroshima University (1959); Institute for Fermentation, Osaka (1961-1976), becoming Deputy Director (1974); Professor, Institute of Biological Sciences, University of Tsukuba (1976-1988). An expert on hyphomycetes, particularly their systematics and role in nature. He, like Barron, contributed to the classification system for anamorphic fungi initiated by Hughes (see Conidiogenesis), which became known as the Hughes-Tubaki-Barron system. Author of over 200 papers and an outstanding ambassador of Japanese mycology. President, Mycological Society of Japan (1985-1987), Chairman of the IMA Committee for Asia (q.v.) (1983-1990), and General Secretary of the 3rd International Mycological Congress (q.v.). *Publs.* Studies on Japanese hyphomycetes. *J. Hattori Bot. Lab.* **20**: 142, 1958. *Biogs, obits etc.* Bandoni (*Mycoscience* **46**: 381, 2005).

Tubakia B. Sutton (1973), anamorphic *Dicarpella*, Cpt.0eH/1eH.15. 5, widespread (north temperate). See Limber & Cash (*Mycol.* **37**: 129, 1945), Yokoyama & Tubaki (*Res. Commun.* Inst. Ferm., Osaka **5**: 43, 1971), Glawe & Crane (*Mycotaxon* **29**: 101, 1987; *T. dryina* revised), Belisario (*Inftore fitopatol.* **40**: 54, 1990; disease caused by *T. dryina*), Belisario (*Mycotaxon* **41**: 147, 1991; teleomorph), Munkvold & Neely (*CJB* **69**: 1865, 1991; development of *T. dryina* in host tissue).

Tubaria (W.G. Sm.) Gillet (1876), Inocybaceae. *c.* 20, widespread (temperate). See Romagnesi (*Revue Mycol.* Paris **5**: 29, 1940), Romagnesi (*Revue Mycol.* Paris **8**: 26, 1943), Volders (*Sterbeeckia* **21-22**: 3, 2002; Belgium spp.), Horak & Moreau (*BSMF* **120**: 215, 2004; France spp.).

Tubariella E. Horak & Hauskn. (2002), Bolbitiaceae. 1, Papua New Guinea. See Horak & Hausknecht (*Öst. Z. Pilzk.* **11**: 213, 2002).

Tubariopsis R. Heim (1931), Bolbitiaceae. 1, Madagascar. See Heim (*Inocybe*: 62, 1931).

Tuber P. Micheli ex F.H. Wigg. (1780), Tuberaceae. 86 (hypogeous), widespread. See truffles. See Gross (*Z. Pilzk.* **41**: 143, 1975; key 13 spp.), Parguey-Leduc & Janex-Favre (*Revue Mycol.* Paris **41**: 1, 1977; ontogeny), Janex-Favre & Parguey-Leduc (*Cryptog. Mycol.* **4**: 353, 1984; spore ultrastr.), Mostecchi & Lazzari (*Boll. Gruppo Micol. 'G. Bresadola'* **27**: 196, 1984; col. pls), Parguey-Leduc *et al.* (*C.R. Acad. Sci. Paris* Sér. III **301**: 143, 1985; ontogeny), Parguey-Leduc *et al.* (*Cryptog. Mycol.* **8**: 173, 1987; ascoma), Parguey-Leduc *et al.* (*CJB* **65**: 1491, 1987; ascospores), Parguey-Leduc *et al.* (*BSMF* **105**: 227, 1989; ascoma ontogeny), Gross (*Docums Mycol.* **21**: 1, 1991; key 30 spp., Europe), Parguey-Leduc *et al.* (*Cryptog. Mycol.* **12**: 165, 1991; ontogeny), Pegler *et al.* (*British truffles*, 1993; key 11 Br. spp.), Trappe *et al.* (*Mycotaxon* **60**: 365, 1996), O'Donnell *et al.* (*Mycol.* **89**: 48, 1997; phylogeny), Amicucci *et al.* (*Mol. Ecol.* **7**: 273, 1998; ITS primers), Mello *et al.* (*New Phytol.* **141**: 511, 1999; primers), Neuner-Plattner *et al.* (*MR* **103**: 403, 1999; immunology), Percudani *et al.* (*Mol. Phylogen. Evol.* **13**: 169, 1999; phylogeny), Roux *et al.* (*FEMS Microbiol. Lett.* **180**: 147, 1999; ITS), Amicucci *et al.* (*FEMS Microbiol. Lett.* **189**: 265, 2000; molecular identification), Giomaro *et al.* (*Mycorrhiza* **10**: 107, 2000; mycorrhizas), Frizzi *et al.* (*MR* **105**: 365, 2001; isozymes), Mello *et al.* (*Microbiol. Res.* **155**: 279, 2001), Murat *et al.* (*New Phytol.* **164**: 401, 2004; evolution), Paolocci *et al.* (*FEMS Microbiol. Lett.* **235**: 109, 2004), Patil *et al.* (*Journal of Mycology and Plant Pathology* **34**: 778, 2004; phylogeography), Urban *et al.* (*MR* **108**: 749, 2004; anamorph), Mello *et al.* (*Environmental Microbiology* **7**: 55, 2005; genetic diversity), Zhang *et al.* (*FEMS Microbiol. Lett.* **245**: 85, 2005; phylogeny), Baciarelli-Falini *et al.* (*Mycorrhiza* **16**: 475, 2006; novel morphotypes), Bertini *et al.* (*Microbiol. Res.* **161**: 59, 2006; molecular detection), Chen *et al.* (*Mycotaxon* **94**: 1, 2005; key), Hansen & Pfister (*Mycol.* **98**: 1029, 2006; phylogeny), Paolocci *et al.* (*Appl. Environm. Microbiol.* **72**: 2390, 2006; life cycle), Suz *et al.* (*FEMS Microbiol. Lett.* **254**: 251, 2005; molecular detection), Tedersoo *et al.* (*New Phytol.* **170**: 581, 2006; mycorrhizas), Wang *et al.* (*MR* **110**: 1034, 2006; phylogeny), Chen & Liu (*Mycol.* **99**: 475, 2007; China), Læssøe & Hansen (*MR* **111**: 1075, 2007; phylogeny).

Tuberaceae Dumort. (1822), Pezizales. 7 gen. (+ 13 syn.), 111 spp.

Lit.: Beaton & Malajczuk (*TBMS* **86**: 503, 1986), Trappe *et al.* (*Aust. Syst. Bot.* **5**: 597, 1992), Pegler *et al.* (*British truffles*, 1993), Comandini & Pacioni (*Mycotaxon* **63**: 77, 1997), Landvik *et al.* (*Nordic Jl Bot.* **17**: 403, 1997), O'Donnell *et al.* (*Mycol.* **89**: 48, 1997; phylogeny), Amicucci *et al.* (*Mol. Ecol.* **7**: 273, 1998), Bertini *et al.* (*New Phytol.* **139**: 565, 1998), Harrington *et al.* (*Mycol.* **91**: 41, 1999), Mello *et al.* (*New Phytol.* **141**: 511, 1999), Percudani *et al.* (*Mol. Phylogen. Evol.* **13**: 169, 1999; phylogeny), Roux *et al.* (*FEMS Microbiol. Lett.* **180**: 147, 1999), Amicucci *et al.* (*FEMS Microbiol. Lett.* **189**: 265, 2000), Mello *et al.* (*Mycol.* **92**: 326, 2000), Zambonelli *et al.* (*Mycotaxon* **74**: 57, 2000), Iotti *et al.* (*New Phytol.* **155**: 499, 2002), Mello *et al.* (*Envir. Microbiol.* **4**: 584, 2002), Mabru *et al.* (*Int. J. Food Microbiol.* **94**: 33, 2004), Paolocci *et al.* (*FEMS Microbiol. Lett.* **235**: 109, 2004), Urban *et al.* (*MR* **108**: 749, 2004), Halász *et al.* (*Mycol. Progr.* **4**: 281, 2005), Hansen & Pfister (*Mycol.* **98**: 1029, 2006; phylogeny), Læssøe & Hansen (*MR* **111**: 1075, 2007; phylogeny).

Tuberales G. Winter (1884) = Pezizales. See also *Elaphomycetaceae* and truffles.

Tuberaster Boccone (1697) ≡ Polyporus P. Micheli ex Adans.

Tubercolarites Barsanti (1903), Fossil Fungi, anamorphic *Pezizomycotina*. 1 (Carboniferous), Italy.

Tubercularia Weber ex F.H. Wigg. (1780) nom. rej. prop. = Dibaeis fide Gierl & Kalb (*Herzogia* **9**: 593, 1993).

Tubercularia Tode (1790) nom. cons., anamorphic *Nectria*, Hsp.0eH.15. 26, widespread. See Seifert (*Stud. Mycol.* **27**: 95, 1985), Wang *et al.* (*FEMS Microbiol. Lett.* **193**: 249, 2000; taxol production), Rossman *et al.* (*Mycol.* **93**: 100, 2001; phylogeny).

Tuberculariaceae Fr. (1825) = Nectriaceae. (obsol.). Anamorphic fungi (hyphomycetes) with conidiomata.

Tuberculariella Höhn. (1915) = Cryptosporiopsis fide Petrak (*Sydowia* **19**: 227, 1966).

Tuberculariopsis Höhn. (1909), anamorphic *Pezizomycotina*, Hsp.0eH.10. 1, Java.

Tubercularis Clem. & Shear (1931) ≡ Tuberculariopsis.

Tubercularites Arcang. (1903), Fossil Fungi. 1 (Carboniferous), Italy.

tuberculate solarium, see solarium.

tubercule (**tubercle**), a small wart-like process; **tuberculate**, having tubercles, syn. of punctate (Fig. 20.5);.

Tuberculina Tode ex Sacc. (1880), anamorphic *Helicobasidium*, Hsp.0eH.15. *c.* 10 (on rusts), widespread. See Bauer *et al.* (*Mycol.* **96**: 960, 2004), Lutz *et al.* (*Mycol.* **96**: 1316, 2004), Lutz *et al.* (*MR* **108**: 227, 2004).

Tuberculinia Velen. (1922) nom. nud., Agaricomycetidae. See Stalpers (*in litt.*).

Tuberculis Clem. & Shear (1931) ≡ Tuberculariella.

Tuberculispora Deighton & Piroz. (1972), anamorphic *Pezizomycotina*, Hso.1bH.15. 1 (on *Irenopsis*), Jamaica. See Deighton & Pirozynski (*Mycol. Pap.* **128**: 92, 1972).

Tuberculopsis Sacc. (1914) nom. nud., Fungi. See Saccardo (*Annls mycol.* **12**: 303, 1914).

Tuberculostoma Sollm. (1864) = Robergea fide Sherwood (*Mycotaxon* **6**: 215, 1977).

Tuberium Raf. (1815) ≡ Tuber.

Tuberosurculus Paulet (1791) nom. dub., Pezizomycotina. Used for diverse taxa, incl. *Clavicipitaceae* and *Xylariaceae*.

Tubeufia Penz. & Sacc. (1898), Tubeufiaceae. Anamorphs *Helicoma*, *Helicosporium*, *Monodictys*-like. 29, widespread (esp. tropical). See Rossman (*Mycol. Pap.* **157**, 1987; key *Tubeufia* s.l.), Crane *et al.* (*CJB* **76**: 602, 1998), Kodsueb *et al.* (*Mycol.* **96**: 667, 2004; Hong Kong), Kodsueb *et al.* (*Fungal Diversity* **21**: 105, 2006; phylogeny), Schoch *et al.* (*Mycol.* **98**: 1041, 2006; phylogeny), Tsui *et al.* (*Mycol.* **98**: 94, 2006; phylogeny), Tsui *et al.* (*Mycol.* **99**: 884, 2007; phylogeny, anamorph).

Tubeufiaceae M.E. Barr (1979), Dothideomycetes (inc. sed.). 32 gen. (+ 17 syn.), 241 spp.

Lit.: Barr (*Mycotaxon* **12**: 137, 1980), Goos (*Mycol.* **78**: 744, 1986), Goos (*Mycol.* **79**: 1, 1987), Rossman (*Mycol. Pap.* **157**, 1987; key gen.), Rossman (*Mycol. Pap.* **157**: 71 pp., 1987), Goos

(*Mycol.* **81**: 356, 1989), Scheuer (*MR* **95**: 811, 1991), Rodrigues & Samuels (*Mycol.* **86**: 254, 1994), Untereiner *et al.* (*MR* **99**: 897, 1995), Crane *et al.* (*CJB* **76**: 602, 1998; saprobic gen.), Inderbitzin *et al.* (*Am. J. Bot.* **88**: 54, 2001), Lumbsch & Lindemuth (*MR* **105**: 901, 2001), Kodsueb *et al.* (*Mycol.* **96**: 667, 2004), Kodsueb *et al.* (*Fungal Diversity* **21**: 105, 2006), Schoch *et al.* (*Mycol.* **98**: 1041, 2006; phylogeny), Tsui *et al.* (*Mycol.* **98**: 94, 2006).

Tubipeda Falck (1923) ≡ Leptopodia.

Tubiporus P. Karst. (1881) ≡ Boletus Fr.

Tubolachnum Velen. (1934), ? Helotiales. 2, former Czechoslovakia.

Tubosaeta E. Horak (1967), Boletaceae. 5, Africa (tropical), Asia. See Horak (*Ber. schweiz. bot. Ges.* **77**: 367, 1967), Zang (*Mycosystema* **20**: 8, 2001; China).

Tubulicium Oberw. (1965), Hydnodontaceae. 7, widespread. See Oberwinkler (*Sydowia* **19**: 53, 1965), Hjortstam (*Windahlia* **24**: 1, 1998), Dämon & Hausknecht (*Öst. Z. Pilzk.* **12**: 129, 2003; on tree ferns in Europe).

Tubulicrinaceae Jülich (1982) Schizoporaceae.

Tubulicrinis Donk (1956), Tubulicrinaceae. 31, widespread. See Weresub (*CJB* **39**: 1456, 1961), Hayashi (*Bull. Govt For. Exp. Stn* **260**, 1974), Hjortstam (*Windahlia* **24**: 1, 1998).

Tubulicrinopsis Hjortstam & Kotir. (2007), Agaricomycetes. 4, Europe. See Kotiranta, H.; Hjortstam, K.; Miettinen, O.; Kulju, M. (*Ann. bot. fenn.* **44**: 128, 2007).

Tubulinosema C. Franzen, S. Fischer, J. Schroeder, J. Schölmerich & S. Schneuwly (2005), Microsporidia. 3. See Franzen *et al.* (*J. Eukary. Microbiol.* **52**: 141, 2005).

Tubulixenasma Parmasto (1965) ≡ Tubulicium.

Tuburcinia Fr. (1832) [non *Tuburcinia* Woronin 1882] nom. rej. ≡ Urocystis.

Tuburcinia Woronin (1882) = Urocystis fide Vánky (*Europ. Smut Fungi*: 280, 1994).

Tuburciniella Zambett. (1970) = Urocystis fide Vánky (*in litt.*) Anamorph name only.

Tucahus Raf. (1830) ≡ Gemmularia.

tuckahoe (or Indian bread), the sclerotia (*Pachyma cocos*) of *Poria cocos* (Weber, *Mycol.* **21**: 113, 1929), USA. **Canadian -**, see stone-fungus.

Tuckerman (Edward; 1817-1886; USA). Student, Union College, Schenectady, New York (1834-1837), then Harvard University (1837-1840); visited Europe (1841-1842) [greatly influenced by E.M. Fries (q.v.)]; Curator of collections, Union College, Schenectady (1843-1844); Professor of Oriental History (1855) and Professor of Botany (1858-1886), Harvard University. Father of American lichenology. *Publs. Genera Lichenum* (1872); *Synopsis of the North American Lichens* 2 vols (1882, 1888). *Biogs, obits etc.* Culberson (*Collected Lichenological Papers of Edward Tuckerman* 2 vols, 1964); Grummann (1974: 197); Reid (*Mycotaxon* **26**: 3, 1986); Stafleu & Cowan (*TL-2* **6**: 523, 1986).

Tuckermanella Essl. (2003), Parmeliaceae (L). 6, Mexico; USA. See Esslinger (*Mycotaxon* **85**: 135, 2003), Thell *et al.* (*Mycol. Progr.* **4**: 303, 2005).

Tuckermanopsis Gyeln. (1933), Parmeliaceae (L). *c.* 28, widespread. See Kärnefelt *et al.* (*Pl. Syst. Evol.* **183**: 113, 1992), Kärnefelt *et al.* (*Bryologist* **96**: 394, 1993), Thell (*Folia Cryptog. Estonica* **32**: 113, 1998), Kärnefelt & Thell (*Biblthca Lichenol.* **78**: 193, 2001), Thell *et al.* (*Mycol. Progr.* **1**: 335, 2002; phylogeny), Thell *et al.* (*Symb. bot. upsal.* **34** no. 1: 429, 2004; Scandinavia), Miądlikowska *et al.* (*Mycol.* **98**: 1088, 2006; phylogeny).

Tuckneraria Randlane & A. Thell (1994), Parmeliaceae (L). 5, Asia. See Ando (*TMSJ* **33**: 223, 1992; phylogeny), Randlane *et al.* (*Acta Bot. Fenn.* **150**: 144, 1994), Thell *et al.* (*J. Hattori bot. Lab.* **78**: 237, 1995).

Tulasne (Charles, 1816-1884; France). Student of medicine, Paris (1840); practising doctor (1840-1865); retired to Hyères (1865-1884). Collaborated with his brother, L.R. Tulasne (q.v.), particularly by illustrating their joint major works; these illustrations set a standard rarely matched by later mycologists. *Publs.* (with L.R. Tulasne) *Selecta Fungorum Carpologia* 3 vols (1861-1865) [English translation Grove, 1931; biographical notes in vol. 1]. *Biogs, obits etc.* Stafleu & Cowan (*TL-2* **6**: 529, 1986).

Tulasne (Louis René, 1815-1885; France). Student of law; public notary (1835-1839); private scientists, Paris (1839-1865); Research Associate, Muséum d'Histoire Naturelle, Paris (1842-1865); retired to Hyères (1865-1885). The 'reconstructor of mycology'; in more than 50 papers he made additions to the knowledge of ergots, lichen-forming fungi, pyrenomycetes, rusts, smuts and subterranean fungi; some work with brother C. Tulasne (q.v.) who prepared the illustrations (their joint works explicitly promoted their respect for a divine creator); an important new concept introduced in the *Selecta Fungorum Carpologia*, a work which has had a great effect on mycology, was that of pleomorphism in fungi. *Publs. Fungi Hypogaei. Histoire et Monographie des Champignons Hypogés* (1851); Mémoire pour servir à l'histoire organographique et physiologique des lichens. *Annales des Sciences Naturelles* Botanique, Série 3 (1852); (with C. Tulasne) *Selecta Fungorum Carpologia* 3 vols (1861-1865) [English translation Grove, 1931; biographical notes in vol. 1]. *Biogs, obits etc.* Stafleu & Cowan (*TL-2* **6**: 530, 1986).

Tulasneinia Zobel ex Corda (1854) ≡ Terfezia.

Tulasnella J. Schröt. (1888), Tulasnellaceae. Anamorph *Epulorhiza*. *c.* 50, widespread. See Jülich & Jülich (*Persoonia* **9**: 49, 1976), Currah & Zelmer (*Rep. Tottori mycol. Inst.* **30**: 43, 1992; 6 spp. with orchids), Roberts (*MR* **98**: 1431, 1994; key Eur. spp.).

Tulasnellaceae Juel (1897), Cantharellales. 3 gen. (+ 4 syn.), 54 spp.

Lit.: Roberts (*MR* **98**: 1235, 1994), Stalpers & Andersen *in* Sneh *et al.* (Eds) (*Rhizoctonia Species* Taxonomy, Molecular Biology, Ecology, Pathology and Disease Control: 58, 1996), Ginns (*Mycol.* **90**: 19, 1998), Greslebin & Rajchenberg (*MR* **105**: 1149, 2001), Gleason & McGee (*Australasian Mycologist* **21**: 12, 2002), Kottke *et al.* (*MR* **107**: 957, 2003), Ma *et al.* (*MR* **107**: 1041, 2003), Kristiansen *et al.* (*Mol. Phylogen. Evol.* **33**: 251, 2004), Larsson *et al.* (*MR* **108**: 983, 2004), McCormick *et al.* (*New Phytol.* **163**: 425, 2004).

Tulasnellales Gäum. (1926) = Cantharellales.

Tulasnia Lesp. [not traced] = Terfezia fide Saccardo (*Syll. fung.* **8**: 902, 1889).

Tulasnodea Fr. (1849) ≡ Tulostoma.

Tulostoma Pers. (1794), Agaricaceae. *c.* 80, widespread (esp. dry regions). See Long (*Mycol.* **36**: 320, 1944; orthogr.), Wright (*Pap. Mich. Acad. Sci.* **40**:

79, 1955; sp. characters), Maas Geesteranus (*Gorteria* **5**: 189, 1971; Netherlands), Wright (*The genus Tulostoma (Gasteromycetes)* [*Bibl. Mycol.* **113**], 1987), Calonge & Wright (*Boln Soc. Micol. Madrid* **13**: 119, 1989; Spain), Moncalvo *et al.* (*Mol. Phylogen. Evol.* **23**: 357, 2002), Sarasini (*Rivista di Micologia* **46**: 7, 2003; mediterranian), Esqueda *et al.* (*Mycotaxon* **90**: 409, 2004), Masuya & Asai (*Bull. natn. Sci. Mus.* Tokyo, B **30**: 9, 2004), Fan *et al.* (*Mycosystema* **24**: 340, 2005; China).

Tulostomataceae E. Fisch. (1900) = Agaricaceae.

Tulostomatales = Agaricales.

tumid, swollen; inflated.

Tumidapexus D.A. Crawford (1954), Aphelariaceae. 1, New Zealand. See Crawford (*Trans. & Proc. Roy. Soc. New Zealand* **82**: 626, 1954).

Tumularia Descals & Marvanová (1987), anamorphic *Pezizomycotina*, Hso.1-≡ eH.21. 2 (aquatic), Europe. See Marvanová & Descals (*TBMS* **89**: 506, 1987), Bussaban *et al.* (*Mycol.* **97**: 1002, 2005; phylogeny).

Tunbridge ware, Ornaments made from coloured wood, usually using marquetry techniques; the green wood used is that of deciduous trees attacked by *Chlorociboria aeruginascens*.

tunic, see exospore.

tunica, a coat, esp. a thin white membrane round the peridiole in most species of the *Nidulariaceae*. See also spore wall (2) and basidiospore.

Tunicago B. Sutton & Pollack (1977), anamorphic *Pezizomycotina*, Cpd.1eP.15. 2, USA; Australia. See Sutton & Pollack (*CJB* **55**: 326, 1977), Alcorn & Sutton (*Aust. Syst. Bot.* **12**: 169, 1999; Australia).

Tunicatispora K.D. Hyde (1990), Halosphaeriaceae. 1 (marine), Australia. See Hyde (*Aust. Syst. Bot.* **3**: 711, 1990), McKeown *et al.* (*MR* **100**: 1247, 1996; ultrastr.), Pang *et al.* (*Nova Hedwigia* **77**: 1, 2003).

Tunicopsora Suj. Singh & P.C. Pandey (1971) = Kweilingia fide Cummins & Hiratsuka (*Illustr. Gen. Rust Fungi rev. edit.*, 1983).

Tunstallia Agnihothr. (1961), ? Sydowiellaceae. 1, India. *T. aculeata* (thorny stem blight of tea, *Camellia*). fide Vasilyeva (*Mycoscience* **42**: 401, 2001).

Tupia L. Marchand (1830) nom. rej. ≡ Icmadophila fide Gierl & Kalb (*Herzogia* **9**: 593, 1993).

turbid, not clear; cloudy.

turbinate, like a top in form (Fig. 23.21). **- organ** or **cell** (of *Cladochytriaceae*), a swelling on the vegetative thallus (see Karling, *Am. J. Bot.* **18**: 528, 1931); spindle-organ.

Turbinellus Earle (1909) = Gomphus Pers. fide Donk (*Beih. Nova Hedwigia* **6**: 291, 1962).

Tureenia J.G. Hall (1915) = Arthrinium fide Höhnel (*Mitt. bot. Inst. tech. Hochsch. Wien* ser. 2 **2**: 9, 1925).

turgid, tightly swollen.

Turgidosculum Kohlm. & E. Kohlm. (1972), Mastodiaceae. 2 (on algae), widespread. See Kohlmeyer & Kohlmeyer (*Marine Mycology*, 1979), Schatz (*Mycol.* **72**: 110, 1980), Kohlmeyer & Volkmann-Kohlmeyer (*Bot. Mar.* **46**: 285, 2003).

turgor pressure, of mycelium, see Adebayo *et al.* (*TBMS* **57**: 145, 1971).

Turturconchata J.L. Chen, T.L. Huang & Tzean (1999), anamorphic *Pezizomycotina*, Hso.?.?. 2, Taiwan. See Chen *et al.* (*MR* **103**: 830, 1999).

Tuzetia Maurand, Fize, Fenwick & Michel (1971), Microsporidia. 4.

TWA, see Media.

twist, a disease of cereals and grasses (*Dilophospora alopecuri*).

Tylochytrium Karling (1939) ≡ Phlyctidium (A. Braun) Rabenh.

Tylodon Banker (1902) ≡ Radulum fide Donk (*Taxon* **12**: 155, 1963).

Tylomyces Cortini (1921), anamorphic *Pezizomycotina*. 1, Italy.

Tylophoma Kleb. (1933) = Dothichiza fide Petrak (*Sydowia* **10**: 201, 1957).

Tylophorella Vain. (1890), Arthoniales (L). 1, tropical. See Egea & Tibell (*Nordic Jl Bot.* **13**: 207, 1993), Aptroot & Tibell (*Mycotaxon* **65**: 339, 1997), Grube (*Bryologist* **101**: 377, 1998; phylogeny).

Tylophorellomyces Cif. & Tomas. (1953) ≡ Tylophorella.

Tylophoron Nyl. ex Stizenb. (1862), Lecanorales (L). 3, widespread (tropical). See Tibell (*Lichenologist* **14**: 219, 1982), Tibell (*Beih. Nova Hedwigia* **79**: 597, 1984), Tibell (*Symb. bot. upsal.* **27** no. 1, 1987), Tibell (*MR* **95**: 290, 1991; anamorph), Tibell (*MR* **107**: 1403, 2003), Aptroot *et al.* (*Biblthca Lichenol.* **97**, 2008; Costa Rica).

Tylophoropsis Sambo (1938) nom. dub., ? Caliciaceae. 1, Africa.

Tylopilus P. Karst. (1881), Boletaceae. *c.* 75, widespread. See Smith (*Mycol.* **60**: 954, 1968), Wolfe (*Biblthca Mycol.* **69**, 1979; N. Am.), Henkel (*Mycol.* **91**: 655, 1999; Guyana), Chen *et al.* (*Taiwania* **49**: 109, 2004; Taiwan), Fu *et al.* (*Mycotaxon* **96**: 41, 2006; China).

Tylosperma Donk (1957) [non *Tylosperma* Botsch. 1952, *Rosaceae*] ≡ Tylospora.

Tylospora Donk (1960), Atheliaceae. 2, widespread. See Donk (*Taxon* **9**: 220, 1960).

Tylosporaceae Jülich (1982) = Atheliaceae.

Tylostoma, see *Tulostoma*.

Tylothallia P. James & H. Kilias (1981), ? Ramalinaceae (L). 1, Europe. See James & Kilias (*Herzogia* **5**: 409, 1981), Ekman (*MR* **105**: 783, 2001).

Tympanella E. Horak (1971), Bolbitiaceae. 1, New Zealand. Basidioma gasteroid. See Horak (*N.Z. Jl Bot.* **9**: 485, 1971), Soop (*Cortinarioid Fungi of New Zealand, An Iconography and Key*: 80 pp., 2003; key).

Tympanicysta Malme (1980), Fossil Fungi. 2 (late Permian – early Triassic), widespread.

Tympanis Tode (1790), Helotiaceae. Anamorph *Sirodothis*. 29, widespread (temperate). See Groves (*CJB* **30**: 571, 1952; monogr.), Ouellette & Pirozynski (*CJB* **52**: 1889, 1974; key), Yao & Spooner (*Kew Bull.* **51**: 187, 1996; Brit. spp.).

Tympanopsis Starbäck (1894) = Nitschkia fide Nannfeldt (*Svensk bot. Tidskr.* **66**: 49, 1975), Huhndorf *et al.* (*MR* **108**: 1384, 2004).

Tympanosporium W. Gams (1974), anamorphic *Pezizomycotina*, Hso.0eH.19. 1 (on *Tubercularia*), Netherlands. See Gams (*Antonie van Leeuwenhoek Ned. Tijdschr. Hyg.* **40**: 478, 1974).

Type (in nomenclature). The element on which the descriptive matter fulfilling the conditions of valid publication of a scientific name is based, or is considered to have been based, and which fixes the application of the name; e.g. a family name on a genus, a generic name on a species (a **- species**; see also nomen species), a specific name generally a **- specimen**, which may be a slide, sometimes on a **- culture** (incorrectly for fungi if still metabolically active), a Figure, or a description. Numerous terms with the suffix '-type' have been used in nomenclature, both formally and informally (*see* Hawksworth, *A draft glossary of terms used in bionomenclature*, [IUBS Monogr. **9**], 1994), and only a selection of those most used by mycologists are included here: **epi-** a specimen or illustration used to serve as an interpretive type where the existing type material is inadequate for the precise application of the name; **ex-type**, out of the type, used especially for living cultures where the holotype is a dried culture or one preserved in a metabolically inactive state; **holo-**, the single element on which the describing author based a name; **iso-**, a duplicate or part of the type collection (other than the holotype) [(in immunology), part of the imunoglobin molecule from the mouse used in the characterization of sera]; [**histo-**, a reaction between different types of cells]; **lecto-**, an element selected in a later work from the original material where no holotype was designated; **mono-**, the only species included in a genus when first described; **neo-**, specimen or other material designated as nomenclatural type when all the original material is missing; **para-**, any specimen other than the holotype on which the first account of a species or other group is based; **patho-**, see pathovar; **phyco-**, each of the morphologically distinct structures derived by symbiosis between a single mycobiont and different photobionts (Swinscow, *Lichenologist* **9**: 89, 1977; see Lichens); **syn-** one of several elements cited by an author when originally proposing a name but where no holotype was selected ; **topo-**, a later collection from the original locality; **typo-**, the specimen used to prepare an illustration where the latter is the type. See Nomenclature, wild type.

Type culture collections, see Genetic resource collections.

Typhella L. Léger & M. Gauthier (1935) nom. nud. ? = Smittium fide Manier & Lichtwardt (*Annls Sci. Nat.* Bot., sér. 12 **9**: 519, 1968).

Typhoderma Gray (1821) nom. dub., ? Fungi.

Typhodium Link [not traced] = Epichloë fide Tulasne & Tulasne (*Select. fung. carpol.* **3**: 23, 1865).

Typhula (Pers.) Fr. (1818), Typhulaceae. 68, widespread (temperate). *T. incarnata* injures cereals under snow. See Berthier (*Bull. mens. Soc. linn. Lyon* Num. Spéc. **45**, 1976), Siepe (*Beitr. Kenntn. Pilze Mitteleur.* **13**: 47, 2000).

Typhulaceae Jülich (1982), Agaricales. 6 gen. (+ 11 syn.), 229 spp.

Lit.: Brodie (*Lejeunia* n.s. **112**: 1, 1984), Metzler (*CJB* **66**: 1321, 1988), Villarreal & Pérez-Moreno (*Micol. Neotrop. Aplic.* **4**: 119, 1991), Matsumoto *et al.* (*Eur. J. Pl. Path.* **102**: 431, 1996), Hsiang *et al.* (*Pl. Dis.* **83**: 788, 1999), Pine *et al.* (*Mycol.* **91**: 944, 1999), Hsiang & Wu (*MR* **104**: 16, 2000), Larsson *et al.* (*MR* **108**: 983, 2004), Vergara *et al.* (*MR* **108**: 1283, 2004).

Typhulochaeta S. Ito & Hara (1915), Erysiphaceae. 4, Asia; N. America. See Solheim *et al.* (*J. Elisha Mitchell scient. Soc.* **84**: 236, 1968), Braun (*Beih. Nova Hedwigia* **98**, 1987; key), Mori *et al.* (*Mycol.* **92**: 74, 2000; phylogeny), Takamatsu (*Mycoscience* **45**: 147, 2004; phylogeny).

typonym, a name having the same type as another name which is neither its basionym nor synisonym (Donk, *Persoonia* **1**: 175, 1960).

Tyrannosorus Unter. & Malloch (1995), Dothideomycetes. Anamorph *Helicodendron*. 1 (on wood), Canada; Pakistan. See Untereiner *et al.* (*MR* **99**: 897, 1995), Schoch *et al.* (*Mycol.* **98**: 1041, 2006; phylogeny).

Tyridiomyces W.A. Wheeler (1907), ? Saccharomycetales. 1 (from ants). See Schultz *et al.* (*Insect-Fungal Associations* Ecology and Evolution: 149, 2005).

Tyrodon P. Karst. (1881) ≡ Hydnum.

Tyromyces P. Karst. (1881), Polyporaceae. *c.* 30, widespread. See David (*Bull. Soc. linn. Lyon* **49**: 596, 1980; key 16 Afr. spp.), Gilbertson & Ryvarden (*N. Amer. Polyp.* **2**: 781, 1987; key N. Am. spp.), Spirin (*Mycena* **1**: 64, 2001), Hattori (*Mycoscience* **44**: 453, 2003; type studies Corner spp.), Ipulet & Ryvarden (*Syn. Fung.* **20**: 79, 2005; tropical Afr.).

UAMH, University of Alberta Microfungus Collection and Herbarium (Edmonton, Alberta, Canada); founded 1960; see Sigler (*J. Ind. Microbiol.* **13**: 191, 1994).

Uberispora Piroz. & Hodges (1973), anamorphic *Pezizomycotina*, Hso.0bP.1. 3, widespread. See Castañeda Ruíz *et al.* (*Nova Hedwigia* **64**: 473, 1997).

ubiquinones, a class of terpenoid lipids involved in electron transport and of potential use in fungal systematics due to variation in the isoprenoid side chain in some taxa. See Kuraishi *et al.* (*MR* **95**: 705, 1991).

Ubrizsya Negru (1965), anamorphic *Pezizomycotina*, St.0eH.?. 1, Rumania. See Negru (*Acta Bot. Hung.* **11**: 217, 1965).

Ucographa A. Massal. (1860) = Pragmopora fide Nannfeldt (*Nova Acta R. Soc. Scient. upsal.*, 1932).

Udeniomyces Nakase & Takem. (1992), anamorphic *Cystofilobasidiaceae*. 4, Japan; Europe. See Nakase *et al.* (*Stud. Mycol.* **38**: 163, 1995; mol. phylogeny), Suh *et al.* (*Microbiology* **141**: 901, 1995; mol. phylogeny), Fell *et al.* (*Int. J. Syst. Evol. Microbiol.* **50**: 1351, 2000; mol. phylogeny), Niwata *et al.* (*International Journal of Systematic and Evolutionary Microbiology* **52**: 1887, 2002; n.sp. Hungary).

Ugola Adans. (1763), anamorphic *Asterophora*. 3. See Stalpers (*Stud. Mycol.* **24**: 80, 1984; anamorph name), Redhead & Seifert (*Taxon* **50**: 243, 2001).

ulcerose, ulcer-like.

Ulea J. Schröt. (1892) nom. inval. = Uleiella.

Uleiella J. Schröt. (1894), Uleiellaceae. 2 (on *Araucaria*), S. America. See Thirumalachar (*Bull. Torrey bot. Club* **76**: 339, 1949), Thirumalachar (*Indian Phytopath.* **3**: 4, 1950), Vánky (*MR* **102**: 513, 1998), Vánky (*Fungal Diversity* **6**: 131, 2001).

Uleiellaceae Vánky (2001), Ustilaginales. 1 gen. (+ 1 syn.), 2 spp.

Lit.: Butin & Peredo (*Biblthca Mycol.* **101**: 100 pp., 1986), Vánky (*MR* **102**: 513, 1998), Vánky (*Fungal Diversity* **6**: 131, 2001).

Uleodothella Syd. & P. Syd. (1921) = Tomasellia fide Müller & von Arx (*Beitr. Kryptfl. Schweiz* **11** no. 2, 1962).

Uleodothis Theiss. & Syd. (1915), Venturiaceae. 4, S. America.

Uleomyces Henn. (1895), Cookellaceae. 10 (on rusts and ascomycetes), widespread. See Eriksson & Yue (*Mycotaxon* **38**: 201, 1990).

Uleomycina Petr. (1954) = Elsinoë fide von Arx (*Persoonia* **2**: 421, 1963).

Uleopeltis Henn. (1904) = Mendogia fide Clements & Shear (*Gen. Fung.*, 1931).

Uleoporthe Petr. (1941), Diaporthales. 1 (from leaves), America (tropical). See Barr (*Mycol. Mem.* **7**, 1978), Cannon (*Fungal Diversity* **7**: 17, 2001).

Uleothyrium Petr. (1929), Asterinaceae. Anamorph *Septothyrella*. 1, Brazil.

Uljanishchev (Valery Ivanovich; 1898-1996; Russia, later Azerbaijan). Laboratory assistant, Institute of Plant Protection, Leningrad (1924-1927); Lecturer, Azerbaijan Agricultural Institute (1927-1931); founder then head, Station of Plant Protection, Azerbaijan SSR (1927-1938); Head of Department of Lower Plants, Institute of Botany, Azerbaijan SSR Academy of Sciences (1938-1993); Order of Lenin and Lenin Prize (1964). Pioneer work on epidemiology and control of smut fungi, particularly on cereals, in the Caucasus; also on the mycota of Azerbaijan. *Publs.* [*Mycoflora of Azerbaijan*] vols 1-4 (1952-1967) [in Russian]; [*Guide to the Identification of Smut Fungi of the USSR*] (1968) [in Russian]; (with collaborators) [*Guide for Identification of Fungi of Transcaucasia* vol. 1 (1985) [in Russian]. *Biogs, obits etc.* Anon. ([*Mikologiya i Fitopatologiya*] **13**: 439, 1976, portrait [in Russian]); Guseinov ([*Mikologiya i Fitopatologiya*] **31**: 92, 1997, portrait [in Russian]).

Ulocladium Preuss (1851), anamorphic *Pleosporaceae*, Hso.#eP.26. 15, widespread. See Simmons (*Mycol.* **59**: 77, 1967; key), Bottalico & Logrieco *in* Sinha & Bhatnagar (Eds) (*Mycotoxins in Agriculture and Food Safety*: 65, 1998; mycotoxins), Simmons (*CJB* **76**: 1533, 1998), Pryor & Gilbertson (*MR* **104**: 1312, 2000; phylogeny), Câmara *et al.* (*Mycol.* **94**: 660, 2002; phylogeny), Hoog & Horré (*Mycoses* **45**: 259, 2002; from humans), Pryor & Bigelow (*Mycol.* **95**: 1141, 2003; phylogeny), Hong *et al.* (*Fungal Genetics Biol.* **42**: 119, 2005; allergens, phylogeny), Zhang & Zhang (*Mycosystema* **25**: 516, 2006; China), Xue & Zhang (*Sydowia* **59**: 161, 2007).

Ulocodium A. Massal. (1855) = Catillaria fide Zahlbruckner (*Catalogus Lichenum Universalis* **4**, 1926).

Ulocolla Bref. (1888) = Exidia fide Donk (*Persoonia* **4**: 166, 1966).

Ulocoryphus Michaelides, L. Hunter & W.B. Kendr. (1982), anamorphic *Pezizomycotina*, Hsy.0fH.19. 1, New Zealand. See Michaelides *et al.* (*Mycotaxon* **14**: 61, 1982).

Uloploca Kleb. (1933) nom. dub., anamorphic *Pezizomycotina*. See Sutton (*Mycol. Pap.* **141**, 1977).

Uloporus Quél. (1886) = Gyrodon fide Patouillard (*Essai taxonomique sur les familles et les genres des Hyménomycètes*, 1900).

Uloseia Bat. (1963) ? = Chaetothyrium fide Hughes (*Mycol.* **68**: 693, 1976).

Ulospora D. Hawksw., Malloch & Sivan. (1979), Testudinaceae. 1, India. See Hawksworth *et al.* (*CJB* **57**: 96, 1979), Hoog *et al.* (*Mycoses* **47**: 121, 2004), Kruys *et al.* (*MR* **110**: 527, 2006; phylogeny), Schoch *et al.* (*Mycol.* **98**: 1041, 2006; phylogeny).

Ultrastructure. Studies with the electron microscope have shown fungi to have a typical eukaryotic structure but the nuclei of some show unusual features. Transmission electron microscopy is usually used for studying fungal ultrastructure, but the scanning electron microscope may also sometimes be applicable (Tsuchiya *et al.*, *Mycol.* **92**: 208, 2004). See Hawker (*Rev. Biol.* **40**: 52, 1965), Bracker (*Ann. Rev. Phytopath.* **5**: 343, 1967; reviews), Beckett *et al.* (*An atlas of fungal ultrastructure*, 1975), Cole & Samson (*Patterns of development in conidial fungi*, 1979), Hess (*Shokubutsu Byogai Kenkyi* **8**: 71, 1973; of germination), Littlefield & Heath (*Ultrastructure of rust fungi*, 1979), Mimms (*Mycol.* **83**: 1, 1991; plant pathogens). For some organelles of which the structure has been elucidated by electron microscopy see: apical granule, concentric bodies, dictyosome, dolipore septum, flagellum, parenthesome, Spitzenkörper. Lichen ultrastructure, see Jacobs & Ahmadjian (*J. Phycol.* **5**: 227, 1969; review), Hale (*in* Brown *et al.*, *Lichenology: progress and problems*: 1, 1976; scanning electron microscope).

See also ascus, Microscopy, Scanning electron microscopy.

Uluguria Vězda (2004), Gomphillaceae (L). 1, Tanzania. See Vězda (*Czech Mycol.* **56**: 150, 2004).

Ulva L. (1753), Algae. Sensu Agardh p.p. = Phycomyces (Mucor.) fide Fries (1832).

Ulvella (Nyl.) Trevis. (1880) nom. dub., Pezizomycotina. See Santesson (*Symb. bot. upsal.* **12** no. 1: 1, 1952).

Umbellidion B. Sutton & Hodges (1975), anamorphic *Pezizomycotina*, Hso.0eH.10. 1, Brazil. See Sutton & Hodges (*Nova Hedwigia* **26**: 529, 1975).

Umbellula E.F. Morris (1955) = Pseudobotrytis fide Subramanian (*Proc. Indian natn Sci. Acad.* Part B. Biol. Sci. **43**: 277, 1956).

Umbelopsidaceae W. Gams & W. Mey. (2003), Mucorales. 1 gen. (+ 2 syn.), 13 spp.

Lit.: O'Donnell *et al.* (*Mycol.* **93**: 286, 2001), Meyer & Gams (*MR* **107**: 339, 2003), Sugiyama *et al.* (*Mycoscience* **44**: 217, 2003), Mahoney *et al.* (*MR* **108**: 107, 2004), Tanabe *et al.* (*Mol. Phylogen. Evol.* **30**: 438, 2004), Kwasna *et al.* (*MR* **110**: 501, 2006).

Umbelopsis Amos & H.L. Barnett (1966), Umbelopsidaceae. 13 (in soil), widespread. See von Arx (*Sydowia* **35**: 10, 1982), Yip (*TBMS* **86**: 334, 1986), Yip (*TBMS* **87**: 243, 1986), Kendrick *et al.* (*Mycotaxon* **51**: 15, 1994), Voigt & Wöstemeyer (*Gene* **270**: 113, 2001; phylogeny), Meyer & Gams (*MR* **107**: 339, 2003; delimitation), Sugiyama *et al.* (*Mycoscience* **44**: 217, 2003; n.sp.), Tanabe *et al.* (*J. gen. appl. Microbiol.* Tokyo **51**: 267, 2005; phylogeny), James *et al.* (*Nature* **443**: 818, 2006; phylogeny), Kwasna *et al.* (*MR* **110**: 501, 2006; phylogeny soil isolates).

Umbilicaria Hoffm. (1789) nom. cons., Umbilicariaceae (L). *c.* 97, widespread (esp. temperate and arctic). See Frey (*Ber. schweiz. bot. Ges.* **59**: 427, 1949), Llano (*A monograph of the lichen family Umbilicariaceae in the Western Hemisphere*, 1950), Hakulinen (*Ann. bot. Soc. Zool.-Bot. Fenn. Vanamo* **32** no. 6, 1962), Llano (*Hvalråd. Skr.* **48**: 112, 1965), Henssen (*Vortr. bot. Ges.* n.f. **4**: 103, 1970; ontogeny), Crespo & Sancho (*An. Inst. bot. A.J. Cavanilles* **35**: 79, 1978; Spain, adaptation, convergence), Krog & Swinscow (*Nordic Jl Bot.* **6**: 75, 1986; E. Africa), Filson (*Muelleria* **6**: 335, 1987; key 5 spp. Antarctica), Wei & Jiang (*Mycosystema* **2**: 135, 1989; key 29 spp. China, 1993; key 50 spp. Asia), Hestmark (*Lichenologist* **23**: 361, 1991; anamorph-teleomorph relationships), Hestmark (*Lichenologist* **23**: 343, 1991; anamorph-teleomorph relationships), Hestmark (*MR* **96**: 1033, 1992; conidiogenesis), Sipman & Topham (*Nova Hedwigia* **54**: 63, 1992; key 6 spp., Colombia), Wei (*Mycosystema* **5**: 1, 1993), Ivanova *et al.* (*Lichenologist* **31**: 477, 1999; phylogeny), Ramstad & Hestmark (*Mycol.* **93**: 453, 2001; population structure), Peterson (*Bull. Calif. Lichen Soc.* **10**: 10, 2003; USA), Krzewicka (*Polish Bot. Stud.* **17**, 2004;

Poland), Krzewicka & Smykla (*Polar Biol.* **28**: 15, 2004; Antarctica), Lumbsch *et al.* (*Mol. Phylogen. Evol.* **31**: 822, 2004; phylogeny), Ott *et al.* (*Lichenologist* **36**: 227, 2004; Antarctica), Galloway & Sancho (*Australasian Lichenology* **56**: 16, 2005; New Zealand), Miądlikowska *et al.* (*Mycol.* **98**: 1088, 2006; phylogeny), Hestmark *et al.* (*Mycol.* **99**: 207, 2007; colonization).

Umbilicariaceae Chevall. (1826), Umbilicariales (L). 2 gen. (+ 14 syn.), 109 spp.

Lit.: Wei & Jiang (*Mycosystema* **2**: 135, 1989; cluster analysis), Hestmark (*Nordic Jl Bot.* **9**: 547, 1990), Hestmark (*Somerfeltia* Suppl. **3**, 1991; sexual strategies), Ascaso *et al.* (*Cryptog. Bryol.-Lichénol.* **13**: 335, 1992), Hageman & Fahselt (*Lichenologist* **24**: 91, 1992), Hestmark (*MR* **96**: 1033, 1992), Wei & Jiang (*The Asian Umbilicariaceae (Ascomycota)* [*Mycosystema, Monogr.* **1**], 1993; keys, descr. 60 spp. Asia), Fahselt & Hageman (*Symbiosis* **16**: 95, 1994), Codogno (*Nova Hedwigia* **60**: 479, 1995; distr.), Fahselt *et al.* (*Bryologist* **98**: 118, 1995), Narui *et al.* (*Bryologist* **99**: 199, 1996), Hestmark (*Biblthca Lichenol.* **68**: 195, 1997), Ivanova *et al.* (*Lichenologist* **31**: 477, 1999; phylogeny), Narui *et al.* (*Bryologist* **102**: 80, 1999; chemistry), Sancho (*Clementeana* **4**: 42, 1999; key), Romeike *et al.* (*Mol. Biol. Evol.* **19**: 1209, 2002), Ott *et al.* (*Lichenologist* **36**: 227, 2004), Reeb *et al.* (*Mol. Phylogen. Evol.* **32**: 1036, 2004), Wedin *et al.* (*MR* **109**: 159, 2005), Miądlikowska *et al.* (*Mycol.* **98**: 1088, 2006; phylogeny), Hofstetter *et al.* (*Mol. Phylogen. Evol.* **44**: 412, 2007; phylogeny), Lumbsch *et al.* (*MR* **111**: 257, 2007; phylogeny).

Umbilicariales J.C. Wei & Q.M. Zhou (2007). Lecanoromycetes. 5 fam., 13 gen., 191 spp. Fams:
(1) **Elixiaceae**
(2) **Fuscideaceae**
(3) **Ophioparmaceae**
(4) **Rhizoplacopsidaceae**
(5) **Umbilicariaceae**
For *Lit.* see under fam.

Umbilicariomyces Cif. & Tomas. (1953) ≡ Lasallia.

umbilicate, having a small hollow; esp. of a pileus having a hollow on the top above the stipe (Fig. 19M).

umbilicus (1) the central hold-fast occurring in some foliose lichens (e.g. *Umbilicaria*), navel, umbo; (2) the pore in the perispore of an ascospore (Eriksson, *Opera bot.* **60**, 1981).

umbo, a central swelling like the boss at the centre of a shield; esp. one on top of a pileus above the stipe; see also umbilicus; **-nate**, having an umbo (Fig. 19L).

Umbrinosphaeria Réblová (1999), Chaetosphaeriaceae. Anamorph *Sporidesmium*-like. 1 (from wood), widespread (temperate). See Réblová (*Mycotaxon* **71**: 13, 1999), Shenoy *et al.* (*MR* **110**: 916, 2006; phylogeny).

Unamunoa Urries (1942) = Rechingeriella fide Petrak (*Sydowia* **23**: 265, 1970).

Uncigera Sacc. (1885) = Cylindrotrichum fide Réblová & Gams (*Czech Mycol.* **51**: 1, 1999).

uncinate (**uncate**), (1) hooked (Fig. 23.35); (2) of gill insertion, near sinuate (q.v.).

Uncinia Velen. (1934) [non *Uncinia* Pers. 1807, *Cyperaceae*] ≡ Hamatocanthoscypha.

Unciniella K. Holm & L. Holm (1977) ≡ Hamatocanthoscypha.

Uncinocarpus Sigler & G.F. Orr (1976), ? Onygenaceae. Anamorph *Malbranchea*. 3, USA. The human pathogen *Coccidioides immitis* may be closely related. See Currah (*Mycotaxon* **24**: 1, 1985), Pan *et al.* (*Microbiol.* Reading **140**: 1481, 1994), Sigler *et al.* (*CJB* **76**: 1624, 1999), Koufopanou *et al.* (*Mol. Biol. Evol.* **18**: 1246, 2001), Guarro & Cano (*Stud. Mycol.* **47**: 1, 2002; phylogeny), Sugiyama *et al.* (*Stud. Mycol.* **47**: 5, 2002; phylogeny).

Uncinula Lév. (1851) = Erysiphe fide Braun & Takamatsu (*Schlechtendalia* **4**: 3, 2000), Mori *et al.* (*Mycoscience* **41**: 437, 2000; phylogeny), Takamatsu (*Mycoscience* **45**: 147, 2004; phylogeny), Takamatsu *et al.* (*Mycoscience* **46**: 9, 2005).

Uncinulella Hara (1936) = Erysiphe fide Braun & Takamatsu (*Schlechtendalia* **4**: 3, 2000).

Uncinuliella R.Y. Zheng & G.Q. Chen (1979) = Erysiphe fide Braun & Takamatsu (*Schlechtendalia* **4**: 3, 2000), Takamatsu (*Mycoscience* **45**: 147, 2004; phylogeny).

Uncinulites Pampal. (1902), Fossil Fungi ? Fungi. 1 (Miocene), Italy. See Salmon (*J. Bot., Lond.* **41**: 127, 1903).

Uncinulopsis Sawada (1916) = Pleochaeta fide Kimbrough & Korf (*Mycol.* **55**: 619, 1963).

Uncispora R.C. Sinclair & Morgan-Jones (1979), anamorphic *Pezizomycotina*, Hso.≡ eP.1. 1, USA. See Sinclair & Morgan-Jones (*Mycotaxon* **8**: 140, 1979).

Uncobasidium Hjortstam & Ryvarden (1978), Meruliaceae. 1, Norway. See Hjortstam & Ryvarden (*Mycotaxon* **7**: 407, 1978).

Uncol Buriticá & Rodríguez (2000), Uncolaceae. 1 (on *Pteridophyta*), S. America. See Buriticá & Rodríguez (*Revta Acad. colomb. cienc. exact. fís. nat.* **24**: 112, 2000), Cummins & Hiratsuka (*Illustr. Gen. Rust Fungi* edn 3: 225 pp., 2003; excluded from *Pucciniales*).

Uncolaceae Buriticá (2000), Pucciniales. 2 gen., 3 spp.

Lit.: Rodríguez & Buriticá (*ASCOLFI Informa* **28**: 12, 2002), Cummins & Hiratsuka (*Illustr. Gen. Rust Fungi* edn 3: 225 pp., 2003).

under cortex, lower cortex in foliose lichens.

Underwoodia Peck (1890), Pezizales. 3 (from soil), widespread. See Harmaja (*Karstenia* **14**: 102, 1974), O'Donnell *et al.* (*Mycol.* **89**: 48, 1997; phylogeny), Landvik *et al.* (*Mycol.* **91**: 278, 1999), Hansen & Pfister (*Mycol.* **98**: 1029, 2006; phylogeny).

Underwoodina Kuntze (1891) ≡ Aschersonia Mont.

undulate, wavy.

Unger (Franz; 1800-1870; Austria). Medic, Stockerau (1828-1830); forensic physician, Kitzbühel; Professor of Botany & Zoology, Graz (1835-1849); Professor of Plant Anatomy & Physiology, Vienna (1849-1866). Contributed to the development of ideas in plant pathology, although still regarding fungi as the result rather than the cause of disease. *Publs. Die Exantheme der Pflanzen* (1833). *Biogs, obits etc.* Anon. (*Journal of Botany* London **8**: 192, 1870); Stafleu & Cowan (*TL-2* **6**: 594, 1986).

Unguicularia Höhn. (1905), Hyaloscyphaceae. 7, Europe. See Raschle (*Sydowia* **29**: 170, 1977; key), Korf & Kohn (*Mycotaxon* **10**: 503, 1980), Inman *et al.* (*Pl. Path.* **41**: 646, 1992; on *Brassica*), Baral (*SA* **13**: 113, 1994; posn), Raitviir (*Scripta Mycologica* Tartu **20**, 2004).

Unguiculariella K.S. Thind & R. Sharma (1990), Hyaloscyphaceae. 1, Bhutan. See Thind & Sharma (*Proc. Indian Acad. Sci.* Pl. Sci. **100**: 279, 1990),

Raitviir (*Scripta Mycol.* **20**, 2004).

Unguiculariopsis Rehm (1909), Helotiaceae. Anamorph *Deltosperma*. 20 (on lichens and other fungi), widespread. See Spooner (*Biblthca Mycol.* **116**, 1987), Zhuang (*Mycotaxon* **32**: 1, 1988; key), Alstrup & Hawksworth (*Meddr Grønland* Biosc. **31**: 3, 1990), Kondrateva *et al.* (*Acta Bot. Fenn.* **150**: 93, 1994), Kondrateva & Galloway (*Biblthca Lichenol.* **58**: 235, 1995), Etayo & Diederich (*Bull. Soc. Nat. Luxemb.* **97**: 93, 1996), Zhuang (*MR* **104**: 507, 2000; 2 spp. China).

Unguiculella Höhn. (1906), Hyaloscyphaceae. *c.* 17, Europe. See Dennis (*Mycol. Pap.* **32**, 1949), Raitviir (*Mikol. Fitopatol.* **21**: 200, 1987), Huhtinen (*Karstenia* **29**: 45, 1990), Raitviir (*Scripta Mycol.* **20**, 2004), Huhtinen & Spooner (*Mycologist* **19**: 59, 2005; UK).

Ungularia Lázaro Ibiza (1916) ≡ Piptoporus.

ungulate, shaped like a horse's hoof.

Ungulina Pat. (1900) ≡ Fomes.

unialgal (of cultures of lichen photobionts), ones in which a single algal species is present but which may also contain bacteria, fungi, or other organisms.

Unicellomycetales, yeasts (Kudriavtsev, 1954).

Unikaryon E.U. Canning, P.F. Lai & J.K. Lie (1974), Microsporidia. 8. See Canning *et al.* (*J. Protozool.* **21**: 19, 1974).

union, see Phytosociology.

unipolar, at one end only (esp. of a bacterial cell).

uniseriate, in one row.

Uniseta Ciccar. (1948), anamorphic *Cryptodiaporthe*, St.1eP.19. 1, USA.

Unisetosphaeria Pinnoi, E.B.G. Jones, McKenzie & K.D. Hyde (2003), Trichosphaeriaceae. 1, Thailand. See Pinnoi *et al.* (*Mycoscience* **44**: 377, 2003).

universal veil (of agarics and gasteromycetes), a layer of tissue covering the basidioma while development takes place; teleblem; blematogen; cf. volva; **primary - -** = protoblem.

unorientated, not arranged in any particular direction.

unstratified (of lichen thalli), not layered; homoiomerous (q.v.).

Uperhiza Bosc (1811) = Melanogaster fide Stalpers (*in litt.*).

UPS, Botanical Museum, University of Uppsala (Uppsala, Sweden); founded 1785; genetic resource collection **UPSC**.

Uraecium Arthur (1933), anamorphic *Pucciniales*. 12, widespread (esp. tropical). Anamorph name for (I). See Berndt (*Nova Hedwigia* **75**: 415, 2002) Established for uredium-like aecia.

Urceola Quél. (1886), Helotiales. 1, Europe.

Urceolaria Ach. (1803) [non *Urceolaria* Molina ex Brandis 1786, *Gesneriaceae*] nom. rej. = Diploschistes fide Hawksworth *et al.* (*Dictionary of the Fungi* edn 8, 1995).

Urceolaria Bonord. (1851), Pezizales. 100, Europe.

Urceolaria Hook. (1821) [non *Urceolaria* Molina ex Brandis 1786, *Gesneriaceae*] = Aspicilia fide Hawksworth *et al.* (*Dictionary of the Fungi* edn 8, 1995).

Urceolariaceae Chevall. (1826) = Thelotremataceae.

Urceolariomyces Cif. & Tomas. (1953) ≡ Diploschistes fide Hawksworth *et al.* (*Dictionary of the Fungi* edn 8, 1995).

urceolate, pitcher-like in form.

Urceolella Boud. (1885), Hyaloscyphaceae. 23, widespread. See Raschle (*Sydowia* **29**: 170, 1977; key), Korf & Kohn (*Mycotaxon* **10**: 503, 1980), Huhtinen (*Karstenia* **27**: 8, 1988), Raitviir & Galán (*Sydowia* **45**: 34, 1993), Huhtinen & Raitviir (*Mycotaxon* **62**: 453, 1997), Raitviir (*Scripta Mycologica* Tartu **20**, 2004).

Urceolina Tuck. (1875) [non *Urceolina* Rchb. 1828, nom. cons., *Amaryllidaceae*] ≡ Orceolina.

Urceolus Velen. (1939) [non *Urceolus* Mereschk. 1879, *Algae*] = Resupinatus fide Stalpers (*in litt.*).

Uredendo Buriticá & J.F. Hennen (1994), anamorphic *Phakopsoraceae*. 1 (on *Tripsacum* (*Poaceae*)), America (tropical). Anamorph name for (II).

Uredinales G. Winter (1880) = Pucciniales.

Uredinaria Chevall. (1826) nom. conf., anamorphic *Pezizomycotina*. See Sutton (*Mycol. Pap.* **141**, 1977).

Uredinella Couch (1937), Septobasidiaceae. 2 (on *Insecta*), widespread. See Couch (*Mycol.* **33**: 405, 1941).

Urediniomycetes D. Hawksw., B. Sutton & Ainsw. (1983) = Pucciniomycetes.

urediniospore (**uredinospore**, **urediospore**), see *Pucciniomycetes*.

Uredinites Velen. (1889), Fossil Fungi ? Pucciniales. 1 (Cretaceous), former Czechoslovakia.

uredinium (**uredium**, **uredosorus**), see *Urediniales*.

Uredinophila Rossman (1987), Tubeufiaceae. 2 (on rusts), widespread. See Rossman (*Mycol. Pap.* **157**, 1987), Kodsueb *et al.* (*Fungal Diversity* **21**: 105, 2006; phylogeny).

Uredinopsis Magnus (1893), Pucciniastraceae. *c.* 25 (on *Abies* (0, I; where known) (*Pinaceae*); *Pteridophyta* (II, III)), widespread (esp. north temperate). See Faull (*J. Arnold Arbor.* **19**: 402, 1938), Faull (*Contr. Arnold Arbor.* **11**: 1, 1938), Hiratsuka (*Revision of taxonomy of the Pucciniastreae*, 1958).

Uredinula Speg. (1880) = Tuberculina fide Spegazzini (*Anal. Soc. cient. argent.* **10**: 64, 1880).

Uredites Babajan & Tasl. (1970), Fossil Fungi. 1 (Tertiary), former USSR.

Uredo Pers. (1801), anamorphic *Pucciniales*. *c.* 500, widespread (esp. tropical). Anamorph name for (II).

uredoconidium (of *Cumminsiella*), see Kuhnholtz-Lordat (*RAM* **26**: 469, 1947).

Uredopeltis Henn. (1908), Phakopsoraceae. 5 (on *Bignoniaceae*, *Burseraceae*, *Euphorbiaceae*), Africa, Australia, India; America (tropical). See Laundon (*TBMS* **46**: 503, 1963), Walker & Shivas (*Australas. Pl. Path.* **33**: 41, 2004; nomencl. sp. on *Grewia*).

Uredostilbe Buriticá & J.F. Hennen (1994), anamorphic *Phakopsoraceae*. 1 (on *Annona* (*Annonaceae*)), Honduras. Anamorph name for (II).

Urnobasidium Parmasto (1968) = Sistotrema Fr. fide Eriksson *et al.* (*Cortic. N. Europ.* **7**, 1984).

Urnula Fr. (1849), Sarcosomataceae. Anamorph *Strumella*. 6, Europe; N. America. See Dissing (*Mycol.* **73**: 272, 1981; key), Bellemère *et al.* (*Cryptog. Mycol.* **11**: 203, 1990; ultrastr.), Landvik *et al.* (*Nordic Jl Bot.* **17**: 403, 1997; DNA), Brunelli (*Bres. Nuova Ser.* **40**: 199, 1998), Harrington *et al.* (*Mycol.* **91**: 41, 1999; phylogeny), Köpcke *et al.* (*Phytochem.* **60**: 709, 2002; chemistry), Hansen & Pfister (*Mycol.* **98**: 1029, 2006; phylogeny).

Urnularia P. Karst. (1866) nom. rej. prop. = Chaetosphaeria fide Kohlmeyer & Kohlmeyer (*Marine Mycology*, 1979).

Urobasidium Giesenh. (1893) = Zygosporium fide Mason (*Mycol. Pap.* **5**: 134, 1941).

Uroconis Clem. (1909) ≡ Urohendersonia.

Urocystidaceae Begerow, R. Bauer & Oberw. (1998),

Urocystidales. 7 gen. (+ 8 syn.), 185 spp.
Lit.: Vánky (*MR* **94**: 269, 1990), Bauer *et al.* (*CJB* **75**: 1273, 1997), Begerow *et al.* (*CJB* **75**: 2045, 1998), Piepenbring *et al.* (*Protoplasma* **204**: 202, 1998), Ershad (*Rostaniha* **1**: 151, 2000), Vánky (*Aust. Syst. Bot.* **14**: 385, 2001), Denchev (*Mycotaxon* **87**: 475, 2003), Vánky (*Mycol. Balcanica* **1**: 175, 2004).

Urocystidales R. Bauer & Oberw. (1997). Ustilaginomycetidae. 4 fam., 1 gen., 4 spp. Fams:
(1) **Doassansiopsidaceae**
(2) **Glomosporiaceae**
(3) **Melanotaeniaceae**
(4) **Urocystidaceae**
For *Lit.* see under fam.

Urocystis Rabenh. ex Fuckel (1870) nom. cons., Urocystidaceae. Anamorph *Paepalopsis*. *c.* 170, widespread. Spore balls comprising spores completely or incompletely covered by sterile cells. Pathogens of anemone (*U. anemones*), cabbage (*Brassica*); *U. brassicae*, onion (*Allium*; *U. magica*, syn. *U. cepulae*), autumn crocus (*Colchicum*; *U. colchici*), gladioli (*Gladiolus*; *U. gladiolicola*), rye (*Secale*; *U. occulta*), *Ranunculus* (*U. ranunculi*), *Trientalis* (*U. trientalis*), violet (*Viola*; *U. violae*), wheat (*Triticum*; *U. agropyri* (syn. *U. tritici*), flag smut). See Liro (*Ann. Univ. Åbo.* A **1** no. 1, 1922; monograph, as *Tuburcinia*), Vánky (*Europ. Smut Fungi*: 570 pp., 1994), Piepenbring (*Caldasia* **24**: 103, 2002; Colombia, key), Vánky & McKenzie (*Fungal Diversity Res. Ser.* **8**: 259 pp., 2002; New Zealand), Wang & Piepenbring (*Mycol. Progr.* **1**: 399, 2002; China), Wolczanska & Rozwalka (*Polish Botanical Journal* **50**: 93, 2005; Poland).

Urocystites Babajan & Tasl. (1970), Fossil Fungi. 1 (Tertiary), former USSR.

Urohendersonia Speg. (1902), anamorphic *Pezizomycotina*, Cpd.≡ eP.19. 5, S. America; India. See Nag Raj & Kendrick (*CJB* **49**: 1853, 1971; key), Roux & Warmelo (*MR* **92**: 223, 1989; ontogeny), Masilamani & Muthumary (*Acta Bot. Indica* **23**: 157, 1995; ultrastr.).

Urohendersoniella Petr. (1955), anamorphic *Pezizomycotina*, Cpd.#eP.19. 1, Australia. See Nag Raj (*CJB* **67**: 3169, 1989; revision).

Uromyces (Link) Unger (1832) nom. cons., Pucciniaceae. *c.* 800 (on angiosperms, esp. species-rich on *Leguminosae*: *Faboideae*), widespread. *U. appendiculatus* (*Phaseolus*), *U. fabae* (*Vicia*), *U. trifolii* (*Trifolium*), *U. striatus* (*Medicago*). See Tranzschel (*Annls mycol.* **8**: 1, 1910), Guyot (*Les Urédinées. Genre Uromyces*, 1938), Guyot (*Les Urédinées. Genre Uromyces*, 1951), Guyot (*Les Urédinées. Genre Uromyces*, 1957), Hiratsuka (*Rep. Tottori Mycol. Inst.* **10**, 1973), El-Gazzar *in* Polhill & Raven (Eds) (*Advances in Legume Systematics* **1(ii)**: 979, 1981) Probably not separable from *Puccinia*.

Uromycetites Braun (1840), Fossil Fungi ? Pucciniales. 1 (Triassic), Germany.

Uromycladium McAlpine (1905), Pileolariaceae. 7 (on *Acacia*, *Albizia* (*Leguminosae*)), Australasia (and spread with cultivated hosts). See Burges (*Proc. Linn. Soc. N. S. W.* **59**: 212, 1934; morphology).

Uromycodes Clem. (1909) ≡ Schroeteriaster.

Uromycopsis Arthur (1906) = Uromyces fide Arthur (*N. Amer. Fl.* **7**: 440, 1921).

Urophiala Vuill. (1910) = Zygosporium fide Mason (*Mycol. Pap.* **5**: 134, 1941).

Urophlyctis J. Schröt. (1886) = Physoderma fide Karling (*Chytriomyc. Iconogr.*, 1977).

Urophlyctites Magnus (1903), Fossil Fungi. 1 (Carboniferous), Europe. See Weiss (*New Phytol.* **3**: 68, 1904).

Urophora Sommerf. ex Arnold (1899) = Bacidia fide Hawksworth *et al.* (*Dictionary of the Fungi* edn 8, 1995).

Uropolystigma Maubl. (1920), Phyllachorales. 1 (in living leaves), Brazil.

Uropyxidaceae Cummins & Y. Hirats. (1983), Pucciniales. 15 gen. (+ 9 syn.), 143 spp.
Lit.: Eriksson (*Systematic Botany*: 560, 1981), Ferreira & Hennen (*Mycol.* **78**: 795, 1986), Ono (*CJB* **72**: 1178, 1994), Barreto *et al.* (*MR* **99**: 779, 1995), Hennen & Sotão (*Sida* **17**: 173, 1996), Cummins & Hiratsuka (*Illustr. Gen. Rust Fungi* edn 3: 225 pp., 2003), Maier *et al.* (*CJB* **81**: 12, 2003), Wingfield *et al.* (*Australas. Pl. Path.* **33**: 327, 2004), Aime (*Mycoscience* **47**: 112, 2006).

Uropyxis J. Schröt. (1875), Uropyxidaceae. *c.* 15 (mostly on *Faboideae* (autoecious) (*Leguminosae*)), America; Africa; Taiwan. See Baxter (*Mycol.* **51**: 210, 1959; key).

Urospora Fabre (1879) [non *Urospora* Aresch. 1866, nom. cons., *Algae*] ≡ Urosporella.

Urospora Fayod (1889) = Panellus fide Singer (*Agaric. mod. Tax.* edn 3, 1975).

Urosporella G.F. Atk. (1897), Xylariales. 1 or 2 (from stems), Europe; N. America. Possibly close to *Ceriospora*. See Barr (*Mycol.* **58**: 690, 1966), Kang *et al.* (*Mycoscience* **40**: 151, 1999).

Urosporellina E. Horak (1968) = Panellus fide Pegler (*in litt.*).

Urosporellopsis W.H. Hsieh, Chi Y. Chen & Sivan. (1994), Xylariales. 1 (from wood), Taiwan. See Cannon (*in litt.*), Hsieh *et al.* (*MR* **98**: 101, 1994).

Urosporium Fingerh. (1836) nom. dub., anamorphic *Pezizomycotina*.

Ursicollum Gryzenh. & M.J. Wingf. (2006), anamorphic *Cryphonectriaceae*. 1, USA. See Gryzenhout & Wingfield (*Stud. Mycol.* **55**: 44, 2006).

Urupe Viégas (1944), Meliolaceae. 1 (from leaves), Brazil.

US, US National Herbarium (Washington, DC, USA); founded 1868; a part of the Smithsonian Institution.

Uslaria Nieuwl. (1916) ≡ Gautieria.

Usnea Dill. ex Adans. (1763), Parmeliaceae (L). *c.* 338, widespread. See Asahina (*Lichens of Japan* **3**, 1956), Swinscow & Krog (*Norw. Jl Bot.* **25**: 221, 1978), Swinscow & Krog (*Lichenologist* **11**: 207, 1979; E. Afr., keys), Walker (*Bull. Br. Mus. nat. hist.* Bot. **13**: 1, 1985; *Neuropogon*, key 15 spp.), Awasthi (*J. Hattori Bot. Lab.* **61**: 333, 1986; key 54 spp. India), Tavares (*Mycotaxon* **30**: 39, 1987; *U. strigosa*, N. Am., key), Clerc (*Lichenologist* **29**: 209, 1997), Clerc & Herrera-Campos (*Bryologist* **100**: 281, 1997; saxicolous N. Am. spp.), Clerc (*Lichenologist* **30**: 321, 1998; species concepts), Halonen *et al.* (*Bryologist* **101**: 36, 1998; British Columbia), Herrera-Campos *et al.* (*Bryologist* **10**: 303, 1998; Mexico), Tavares & Sanders *in* Glenn *et al.* (Eds) (*Lichenogr. Thomsoniana*: 171, 1998; USA), Heibel *et al.* (*Am. J. Bot.* **86**: 753, 1999; genetic diversity), Stevens (*Biblthca Lichenol.* **72**: 128, 1999; Australia), Halonen (*Bryologist* **103**: 38, 2000; Pacific NW), Ohmura & Kashiwadani (*J. Jap. Bot.* **75**: 164, 2000; E Asia), Herrera-Campos *et al.* (*Bryologist* **104**: 235, 2001;

Mexico), Ohmura (*J. Hattori bot. Lab.* **90**: 1, 2001; Japan, Taiwan), Articus *et al.* (*MR* **106**: 412, 2002; phylogeny), Ohmura (*J. Hattori bot. Lab.* **92**: 231, 2002; phylogeny), Tavares (*Constancea* **83**, 2002; nomencl.), Articus (*Taxon* **53**: 925, 2004; phylogeny), Mallavadhani *et al.* (*Biochemical Systematics and Ecology* **32**: 95, 2003; chemistry), Ohmura & Kanda (*Lichenologist* **36**: 217, 2004), Clerc (*Lichenologist* **38**: 191, 2006; Azores), Miądlikowska *et al.* (*Mycol.* **98**: 1088, 2006; phylogeny), Wirtz *et al.* (*Taxon* **55**: 367, 2006; phylogeny), Torra & Randlane (*Lichenologist* **39**: 415, 2007; Baltic region), Wirtz *et al.* (*MR* **112**: 472, 2008; phylogeny, biogeography).

Usneaceae Fée ex Zenker (1827) = Parmeliaceae.

Usneomyces E.A. Thomas ex Cif. & Tomas. (1953) ≡ Usnea fide Hawksworth *et al.* (*Dictionary of the Fungi* edn 8, 1995).

usnic acid, yellow dibenzofuran derivative occurring in lichens (e.g. some spp. of *Cladonia*, *Usnea*) which has some antibiotic properties (anti-Gram + bacterial; antifungal). Usually occurs in lichen cortices. See antibiotics, Lichen products, Pigments.

Ussurithyrites Krassilov (1967), Fossil Fungi, Ascomycota. 1 (Cretaceous), former USSR. See Krasilov (*Rannemelovaya Flora Yuzhnogo Primor'ya* i Ee Znachenie Dlya Stratigrafii: 90, 1967).

Ustacystis Zundel (1945), Urocystidaceae. 2 (on *Rosaceae*), N. America; Europe. See Bauer *et al.* (*Mycol.* **87**: 18, 1995; ultrastr.).

Ustalia Fr. (1825) ≡ Pyrochroa.

Ustanciosporium Vánky (1999), ? Anthracoideaceae. 22 (on *Cyperaceae*), widespread. See Piepenbring (*Nova Hedwigia* **70**: 289, 2000; monogr.), Pérez & Minter (*IMI Descr. Fungi Bact.* **164**, 2005).

Usteria Bat. & H. Maia (1962) [non *Usteria* Willd. 1790, *Loganiaceae*] ≡ Mycousteria.

ustic acid (1) a hydroxyquinol from *Aspergillus ustus* (*Biochem. J.* **48**: 53, 1951); (2) an antimycobacterial product from *Ustilago maydis* (Haskins et al., *Can. J. Microbiol.* **1**: 749, 1955).

ustilagic acids, metabolic products from *Ustilago maydis*; antifungal and antibacterial (Haskins & Thorn, *CJB* **29**: 585, 1951).

Ustilagidium Herzberg (1895) = Ustilago fide Dietel (*Nat. Pflanzenfam.* **6**, 1928).

Ustilaginaceae Tul. & C. Tul. (1847), Ustilaginales. 17 gen. (+ 12 syn.), 607 spp.

Lit.: Spooner (*TBMS* **85**: 540, 1985), Vánky & Oberwinkler (*Nova Hedwigia* Beih. **107**: 96 pp., 1994), Bauer *et al.* (*CJB* **75**: 1273, 1997), Begerow *et al.* (*CJB* **75**: 2045, 1998), Vánky (*Aust. Syst. Bot.* **14**: 385, 2001), Piepenbring *et al.* (*Mycol. Progr.* **1**: 71, 2002), Menzies *et al.* (*Phytopathology* **93**: 167, 2003), Stoll *et al.* (*CJB* **81**: 976, 2003), Austin *et al.* (*Functional & Integrative Genomics* **4**: 207, 2004), Begerow *et al.* (*MR* **108**: 1257, 2004), Jackson (*Evolution* Lancaster, Pa. **58**: 1909, 2004), Vánky (*Mycol. Balcanica* **1**: 175, 2004), Vánky (*Mycol. Balcanica* **1**: 175, 2004), Stoll *et al.* (*MR* **109**: 342, 2005).

Ustilaginales G. Winter (1880). Ustilaginomycetidae. 8 fam., 49 gen., 851 spp. Fams:
(1) **Anthracoideaceae**
(2) **Cintractiaceae**
(3) **Clintramraceae**
(4) **Geminaginaceae**
(5) **Melanopsichiaceae**
(6) **Uleiellaceae**
(7) **Ustilaginaceae**
(8) **Websdaneaceae**

The *Ustilaginales*, which have long been regarded as a distinct group, have a mycelial parasitic phase and are yeast-like when cultured *in vitro*. Their taxonomic status has always been uncertain. Brefeld (*Unters. Gesammtgeb. Mykol.* **12**, 1895) introduced *Hemibasidiomycetes* (*Hemibasidii*) for them (the rusts being assigned to the *Protobasidiomycetes*) and although *Hemibasidiomycetes* was subsequently used in wider senses to include the rusts, Donk (see *Basidiomycota*) again restricted it to the smuts. Von Arx (1967) classified the smuts and the associated saprobic yeasts together with other yeasts in a class *Endomycetes* and in this he was followed by Kreisel (1969) who restricted the *Ustilaginales* to the *Ustilaginaceae*. Moore (*Ant. v. Leeuwenh.* **38**: 567, 1972) on the basis of ultrastructural studies and other considerations treated the group as two classes, *Ustomycetes* and *Sporidiomycetes*, of a division (*Ustomycota*) intermediate between ascomycetes and the basidiomycetes with which smuts have been traditionally associated. The current concept of *Ustomycetes* (q.v.) is broad and includes 7 orders, with the smuts divided into 2 families. Characters of the promycelium in these families are sufficiently distinct to indicate possible ordinal rank, but elevation requires additional support from molecular and ultrastructural data at present available for few species. In many species ustilospore germination (metabasidium development) has never been observed (or is mycelial) and some investigators (e.g. Vánky, 1994) therefore prefer not to emphasize the family division.

The smut fungi occur typically as host-specific endophytes which, but for 2 spp. of *Melanotaenium* on *Selaginella*, are parasitic on flowering plants, esp. *Gramineae* and *Cyperaceae*. Sori are commonly limited to the ovary, anthers, inflorescence, or leaves (*Entyloma*) and stem of the host, though the root is attacked by *Entorrhiza*. The mycelium of delicate hyaline hyphae made up of 1-, 2-, or multinucleate cells which may be throughout the plant or only at the points of infection; it is commonly annual but sometimes, as in *Ustilago segetum* var. *avenae* on *Arrhenatherum*, perennial. The hyphae are generally intercellular (in *U. maydis*) and frequently with haustoria and sometimes clamp connexions; dolipore septa lacking (Moore, *Ant. v. Leeuwenh.* **38**: 567, 1972); conidia may be formed on the surface of the host (esp. in *Entyloma*) although in most genera only the smut spores (chlamydospores, brand spores, resting spores, pseudospores, teliospores, ustospores, ustilospores) and basidiospores. Ustilospores (the preferred term, introduced by Donk, 1972), when mature, are generally exposed as a dark powder (sometimes as in *Anthracoidea* and *Cintractia* they are compacted); less frequently they are enclosed within host tissue or are unpigmented. The spores may be in ones (*Ustilago*, *Tilletia*) or twos (*Mycosyrinx*, *Schizonella*), or in balls made up of fertile spores only (*Sorosporium*, *Tolyposporium*, *Thecaphora*), a sterile layer covering fertile spores (*Urocystis*, *Doassansia*), or fertile spores covering sterile tissue (*Testicularia*, *Doassansiopsis*). Every mature spore has one diploid nucleus (Sampson, *TBMS* **23**: 1, 1939), and limited by a thin endospore and a thicker smooth or ornamented exospore (Zogg & Schwinn, *TBMS* **57**: 403, 1971; Mordue, *TBMS* **87**: 407, 1986). At germination meiosis takes place and a promycelium (basidium, hemi-, or metabasidium, germ tube) having 4 or more basidiospores (sporidia (q.v.), sporidiola, 'conidia', promycelial spores) is produced.

Smuts are facultative saprobes after ustilospore development when their growth on culture media is mycelial or yeast-like (and composed of budding cells variously known as sporidia or sprout cells). Completion of the life cycle in culture is infrequent. There is segregation for sex in the promycelium and conjugation between two basidiospores, a basidiospore and a promycelial cell, two cells of one promycelium, or infrequently, two promycelia (Durán, *Ustilaginales of Mexico*, 1987; Ingold, *Nova Hedw.* **55**: 153, 1992). Many physiologic races are on record. Hybridization of smuts takes place and methods for its experimental study have been perfected. See Whitehouse (*TBMS* **34**: 340, 1951; heterothallism), Carris & Gray (*Mycol.* **86**: 157, 1994), Bakkuren & Kronstad (*Plant Cell* **5**: 123, 1993).

Species have frequently been based on physiologic races and a number of these become conspecific if morphological characters are used for the differentiation of species.

The three chief types of infection are: (1) seedling infection from ustilospores on the seed; (2) seedling infection by mycelium in the seed as a result of ustilospore germination on the stigma at flowering time; (3) infection by wind-borne sporidia from promycelia among decaying plant material (as for *Ustilago maydis*, *Entyloma*). Dusting seed with a fungicide is of use against (1), hot water seed treatment against (2), and spraying and dusting susceptible plants is a possible control for (3). Resistant varieties are used whenever possible.

Lit.: General: Bauer *et al.* (*CJB* **75**: 1273, 1997; ultrastr., classific.), Bauer *et al.* (*in* McLaughlin *et al.* (Eds), *Mycota* **7**(2): 57, 2000; *Ustilaginomycetes*), Begerow *et al.* (*CJB* **75**: 2045, 1997; LS rDNA analyses, classific.), Dietel (*Nat. Pflanzenfam.* **2** Aufl. 6, 1928), Fischer (*The smut fungi: a guide to the literature with bibliography*, 1951; 3,300 ref.), Zundel (*The Ustilaginales of the World*, 1953; descript., syn.), Fischer & Holton (*Biology and control of smut fungi*, 1957; general account), Fischer & Shaw (*Phytopath.* **43**: 181, 1953; speciation), Savulescu (*Sydowia* Beih. **1**: 64, 1957; gen. key), Holm (*Svensk. bot. Tidskr.* **55**: 585, 1961; phylogeny), Savile (*Rep. Tottori mycol. Inst.* **28**: 15, 1990; coevolution), Thirumalachar (*Indian Phytopath.* **19**: 3, 1966; gen. rev.), Vánky (*Illustrated genera of smut fungi*, 1987; 2nd ed., in press; *Trans. Mycol. Soc. Japan* **32**: 281, 1991; spore morphology in the taxonomy; *MR* **102**: 513, 1998; spore-ball-forming smut fungi; *Mycotaxon* **70**: 35, 1999; *Fungal Diversity* **6**: 131, 2001; classific.), Zambettakis (*Rev. Mycol.* Paris **34**: 399, 1970, gen. key; *Bot. Rev.* **6**: 389, **19**: 187, **31**: 114, 1940-65; *Ann. Rev. Phytopath.* **6**: 213, 1968, specialization, genetics, variation).

Regional: **Africa**, Zambettakis (*BSMF* **86**: 305, 1971; c. 300 spp., key gen., host index [also as reprint, *Les Ustilaginales des plantes d'Afrique*]; suppl., *BSMF* **95**: 393, 1979). **Argentina**, Hirschhorn (*Las Ustilaginales de la flora Argentina*, 1985). **Azerbaydzhan**, Uljanishchev ([*Fungal flora of Azerbaydzhan*. **I**, Smut fungi], 1952; Russ.). **Australia**, McAlpine (*The smuts of Australia*, 1910). **Baltic Region**, Ignatavičiute ([*Smuts of the Baltic Region*], 1975). **Brazil**, Viégas (*Bragantia* **4**: 739, 1944). **Carpathians**, Vánky (*Carpathian Ustilaginales*, 1985). **China**, Lee Ling (*Mycol. Pap.* **11**, 1945; *Mycol.* **41**: 252, 1949), Wang ([*Ustilaginales of China*], 1963; Chin.), Guo (Ustilaginaceae, *in Flora Fungorum Sinicorum* **12**, 2001). **Colombia**, Molina-Valero (*Caldesia* **13**: 49, 1980), **Costa Rica**, Piepenbring (*Nova Hedwigia*, Beiheft **113**, 1996). **Cuba**, Piepenbring & Hernández (*Revista Jard. Bot. Nac.* **19**: 121, 1998). **Former Czechoslovakia**, Bubák (*Die Pilze Böhmens* **2**, 1916). **Denmark**, Rostrup (*Ustilagineae Daniae*, 1890). **Europe**, Vánky (*European smut fungi*, 1994). **Finland**, Liro (*Die Ustilagineen Finnlands*, 2 vols, 1924-38). **France**, Viennot-Bourgin (*Encyclopédie Mycologique* **25**, 2 vols, 1956). **Germany**, Scholz & Scholz (*Englera* **8**, 1988; *Verh. Bot. Vereins Berlin Brandenburg* **133**: 343, 2000). **Great Britain**, Ainsworth & Sampson (*The British smut fungi*, 1950), Mordue & Ainsworth (*Mycol. Pap.* **154**, 1984); see also Plowright. **Hungary**, Moesz (*A Kárpát-Medence Üszöggombái. Les Ustilaginales du bassin du Carpathes*, 1950; Hung., Fr. summ.), Vánky, Gönczöl & Tóth (*Acta Bot. Acad. Sci. Hungar.* **28**: 255, 1982). **India**, Mundkur & Thirumalachar (*Ustilaginales of India*, 1952). **Italy**, Ciferri (*Flora Italica* **7** (17), 1938). **Japan**, Ito (*Mycological flora of Japan*. **2**. Basidiomycetes. **1**. Ustilaginales, 1936), Kakishima (*Mem. Inst. Agr. For. Univ. Tsukuba* **1**, 1982). **Kazakhstan**, Schvarzman ([*Flora of the sporing plants of Kazakhstan*. **2**. Smut fungi], 1960; Russ.). **Mexico**, Durán (*Ustilaginales of Mexico*, **1987**). **New Zealand**, Cunningham (*Trans. N.Z. Inst.* **55**: 397, 1924; suppl. *Trans. N.Z. Inst.* **57**, **59**, **61**), Vánky & McKenzie (*Smut fungi of New Zealand*, in press). **N. America**, Clinton (*Proc. Boston Soc. Nat. Hist.* **31**: 329, 1904; *North American Flora* **7**, 1906), Fischer (*Manual of North American smut fungi*, 1953). **Norway**, Jørstad (*Nytt Mag. Bot.* **10**: 85, 1963; excl. *Anthracoidea*). **Pakistan**, Ahmad (*Mycol. Pap.* **64**, 1956). **Poland**, Kochman (*Planta Polonica* **4**, 1936), Kochman & Majewski (*Flora Polska, Mycota* **5**, 1973). **Romania**, Săvulescu (*Ustilaginalele*, 2 vols, 1957). **S. Africa**, Zundel (*Bothalia* **3**: 283, 1938). **Sweden**, Lindeberg (*Symb. bot. upsal.* **16** (2), 1959; excl. *Anthracoidea*). **Switzerland**, Schellenber (*Die Brandpilze der Schweiz*, 1911), Zogg (*Cryptog. Helv.* **16**: 277, 1985]). **former USSR**, Gutner (*Golovnevye griby*, 1941), Ul'yanischev ([*Key to the smut fungi of the USSR*, 1968; Russ.]), Ramazanova *et al.* (*Flora gribov Uzbekistana*.**4**. Golonevye griby, 1987), Karatygin & Azbukina (*Definitorium fungorum URSS*, Ustilaginales, Fasc. **1**, 1989), Azbukina & Karatygin Azbukina (*Definitorium fungorum Rossiae*, Ustilaginales, Fasc. **2**, 1995). **former Yugoslavia**, Lindtner (*Bull. Mus. Hist. nat. Pays Serbe* ser. B **304**: 1, 1950).

Ustilaginoidea Bref. (1895), anamorphic *Hypocreales*, Hso.0eP.1. 6 (on *Poaceae*), widespread (subtropical). See Wang & Bai (*Mycosystema* **16**: 257, 1997), Bischoff *et al.* (*Mycol.* **96**: 1088, 2004; phylogeny), Zhou *et al.* (*Acta phytopath. sin.* **34**: 442, 2004; population structure), Yokoyama *et al.* (*FEMS Microbiol. Lett.* **264**: 182, 2006; mating types).

Ustilaginoidella Essed (1911) = Fusarium fide Brandes (*Phytopathology* **9**: 373, 1919).

Ustilaginomycetes R. Bauer, Oberw. & Vánky (1997), Basidiomycota. 3 ord., 12 fam., 62 gen., 1113 spp. smut fungi. Ords:

(1) **Urocystidiales**
(2) **Ustilaginales**
Usually plant parasites, with very few exceptions on angiosperms. The soma of the smut fungi consists of saprobic haploid cells and parasitic dikaryotic hyphae. Host-parasite interactions with deposits of specific fungal vesicles. Basidiospores in sori. The hyphae are septate, branched, usually intercellular, in some groups also intracellular, in some taxa emitting haustoria into the host cells. In many cases hyphae are perennial, and usually giving little if any evidence of their presence before spore formation sets in. At spore formation karyogamy takes place and diploid spores are produced. The smut spores or teliospores (ustilospores, ustospores, chlamydospores, probasidia) are the organs of dispersion and of resistance. They are formed in the sori which consist of host tissues, of the spore mass, and in several genera also of modified fungal cells or tissues. Sori may be produced in the roots, stems, leaves, inflorescence, flowers, anthers, ovaries, etc. The spores may be single, in pairs, or aggregated in more or less persistent spore balls. The spore balls may be composed entirely of fertile spores, or of a combination of fertile spores, sterile cells, and/or hyphae. At germination of the diploid spore a basidium (promycelium, ustidium) is formed and meiosis takes place. The basidia give rise to haploid basidiospores (sporidia), or more rarely to haploid hyphae. These, or sometimes compatible basidial cells fuse in pairs to restore the parasitic dikaryophase. Frequently, the haploid basidiospores are capable of prolonged reproduction on non-living substrata as yeast-like cells or also as hyphal anamorphs developed from basidiospores, capable to produce forcibly discharged ballistoconidia. The limits toward pure saprobic fungi is not as sharp as was thought earlier and there is even a pure saprobic smut fungus, *Tilletiaria*, known only from culture. The infection may be localized, restricted to a certain organ or part of it, or generalized (systemic). Several species within a few genera may possess anamorphs (conidial states). Ultrastructural and molecular biological investigations showed that there are groups of microfungi closely related with the smut fungi which do not have teliospores (*Microstromatales*, *Malesseziales*, *Exobasidiales*). On the other hand, the *Microbotryales*, with teliospores and life cycle as all other smut fungi, are more closely related with the rust fungi than with other smut fungi. There are even human pathogens belonging to the *Ustilaginomycetes* (*Malessezia*). The number of 'true smut fungus species' (possessing teliospores) is about 1450, distributed in 77 genera. The taxonomic status of the smuts has often changed. Brefeld (*Unters. Gesammtgeb. Mykol.*: 12, 1895) introduced *Hemibasidiomycetes* (*Hemibasidii*) for the smuts (the rusts being assigned to the *Protobasidiomycetes*) and although *Hemibasidiomycetes* was subsequently used in wider senses to include the rusts, Donk (see *Basidiomycota*) again restricted this taxon to the smuts. Von Arx (1967) classified the smuts and the associated saprobic yeasts together with other yeasts in a class *Endomycetes* and in this he was followed by Kreisel (1969) who restricted the *Ustilaginales* to the *Ustilaginaceae*. Moore (*Ant. v. Leeuwenhoek* **38**: 567, 1972) on the basis of ultrastructural studies and other considerations treated the group as two classes, *Ustomycetes* and *Sporidiomycetes*, of a new phylum (*Ustomycota*) intermediate between ascomycetes and the basidiomycetes with which smuts have been traditionally associated. 5S rRNA analyses (Blanz & Gottschalk, 1984), supported also by biochemical analyses (Prillinger *et al.*, 1991) demonstrated an isolated position of some smut fungi, classified today into the *Microbotryales* within the *Pucciniomycetes*. Bauer *et al.* (1997) elaborated a new classification, based on ultrastructural characters. This was slightly modified and perfected by Begerow *et al.* (1997/1998), Piepenbring *et al.* (1999), Begerow *et al.* (2000), Vánky (2001), Bauer *et al.* (2001), based on molecular and classical morphological studies.

Ustilaginomycetidae Jülich (1981), see *Ustilaginomycetes*.

Ustilaginomycotina Doweld (2001), see *Ustilaginomycetes*.

Ustilaginula Clem. (1909) ≡ Ustilagopsis.

Ustilagites Babajan & Tasl. (1970), Fossil Fungi. 1 (Tertiary), former USSR.

Ustilago (Pers.) Roussel (1806), Ustilaginaceae. *c.* 200 (on *Poaceae*), widespread. Important diseases of cereals, the causal organisms biologically distinct but morphologically close and variously treated as species, varieties or even physiologic races: *U. segetum* (covered smut of barley (*Hordeum*; *U. hordei*) and oats (*Avena*; *U. kolleri*, *U. levis*)); *U. segetum* var. *avenae* (loose smut of oats; *U. avenae*); *U. segetum* var. *tritici* (loose smut of wheat; *U. nuda*, and barley); *U. scitaminea* (sugarcane); *U. maydis* (syn. *U. zeae*, maize (*Zea*)); pathogens of various grasses: *U. bromivora* (esp. on *Bromus*), *U. hypodytes*, *U. striiformis* (stripe smut); foxtail millet (*Setaria italica*), *U. crameri*; bermuda grass (*Cynodon dactylon*), *U. cynodontis*; mannagrass (*Glyceria*), *U. filiformis*; reed (*Phragmites*), *U. grandis*; pampagrass (*Cortaderia*), *U. quitensis* (*U. cortaderiae*), crabgrass (*Digitaria*), *U. syntherismae*; barnyard grass (*Echinochloa crusgalli*), *U. trichophora*. *U. esculenta* on *Zizania* and *U. maydis* are edible. See Osner (*Bull. Cornell agric. exp. Stn*: 381, 1916; *U. striiformis*), Ingold (*MR* **93**: 405, 1989; Ustilospore germination), Trione (*MR* **94**: 489, 1990; *U. scitaminea*), Mordue (*Mycopath.* **16**: 227, 1991; *U. esculenta*).

Ustilagopsis Speg. (1880) = Sphacelia fide Langdon (*TBMS* **36**: 74, 1953).

Ustilentyloma Savile (1964), Ustilentylomataceae. 4 (on *Poaceae*), Europe; N. America; Greenland. See Vánky (*Mycotaxon* **78**: 319, 2001; key), Aime *et al.* (*Mycol.* **98**: 896, 2006; phylogeny).

Ustilentylomataceae R. Bauer & Oberw. (1997), Microbotryales. 3 gen., 9 spp.

Lit.: Denchev (*Mycotaxon* **55**: 243, 1995), Piepenbring *et al.* (*Pl. Syst. Evol.* **199**: 62, 1996), Bauer *et al.* (*CJB* **75**: 1273, 1997), Begerow *et al.* (*CJB* **75**: 2045, 1998), Piepenbring (*Revta Biol. trop.* **49**: 411, 2001), Vánky (*Aust. Syst. Bot.* **14**: 385, 2001), Vánky (*Mycol. Balcanica* **1**: 175, 2004), Kemler *et al.* (*BMC Evol. Biol.* **6**: 35, 2006).

ustilospore, Donk's (*K. ned. Akad. Wet.* C **76**: 111, 1973) name for a smut spore; **ustospore** (Moore, *Ant. v. Leeuwenhoek* **38**: 579, 1972).

Ustomycetes, see *Ustilaginomycetes*.

Ustomycota R.T. Moore (1972). Phylum for *Septomycetes* rather like *Saccharomycetales* and *Pucciniales* but not satisfactorily either (Moore, *Ant. v. Leeuwenhoek* **38**: 567, 1972); assimilative phase often yeast-

like. See *Ustilaginomycetes*.
ustospore, see ustilospore.
Ustulina Tul. & C. Tul. (1863) = Kretzschmaria fide Martin (*J S. Afr. Bot.* **36**: 73, 1970), Læssøe (*SA* **13**: 43, 1994), Rogers & Ju (*Mycotaxon* **68**: 345, 1998).
Ustulinites Kirschst. (1925), Fossil Fungi. 1, Germany.
Utharomyces Boedijn (1959), Pilobolaceae. 1 (esp. coprophilous), widespread (subtropical). See Kirk & Benny (*TBMS* **75**: 123, 1980), Voigt & Wöstemeyer (*Gene* **270**: 113, 2001; phylogeny), Delgado Ávila *et al.* (*Revista Científica* **15**: 159, 2005; Venezuela).
Uthatobasidium Donk (1956) = Thanatephorus fide Roberts (*Rhizoctonia-forming fungi*, 1999).
Utraria Quél. (1873) ≡ Lycoperdon Pers. fide Demoulin (*Persoonia* **7**: 152, 1973).
Utriascus Réblová (2003), Sordariales. 1, Czech Republic. Possibly related to the *Coryneliaceae*. See Réblová (*Mycol.* **95**: 128, 2003).
utricle, the bladder-like covering of certain fungi, e.g. *Dendrogaster*.
utriform, bag-like.
Uvarispora Goos & Piroz. (1975), anamorphic *Pezizomycotina*, Hso.1bH.1. 1, Panama. See Goos & Pirozynski (*CJB* **53**: 2930, 1975).
Uvasporina Beneš (1956), Fossil Fungi. 1 (Carboniferous), former Czechoslovakia.
Uwebraunia Crous & M.J. Wingf. (1996) = Dissoconium fide Crous *et al.* (*Sydowia* **51**: 155, 1999; cf. *Dissoconium*), Crous *et al.* (*MR* **105**: 425, 2001; phylogeny), Crous *et al.* (*Mycol.* **93**: 1081, 2001; phylogeny), Crous *et al.* (*Stud. Mycol.* **50**: 195, 2004; phylogeny), Crous (*in litt.*, 2008).
Uyucamyces H.C. Evans & Minter (1985), anamorphic *Ocotomyces*, Ccu.0eH.1. 1, Honduras. See Evans & Minter (*TBMS* **84**: 68, 1985).
V8A, see Media.
VA, VAM (of mycorrhizas), vesicular-arbuscular; see Mycorrhiza.
vagant (of lichens), see vagrant.
Vaginaria (Forq.) Sacc. (1887) [non *Vaginaria* Rich. ex Pers. 1805, *Cyperaceae*] = Amanitopsis fide Kuyper (*in litt.*).
Vaginarius Roussel (1806) nom. rej. ≡ Amanitopsis.
Vaginata Nees ex Gray (1821) nom. rej. = Amanita Pers. fide Kuyper (*in litt.*).
Vaginatispora K.D. Hyde (1995), Lophiostomataceae. 1, Australia. See Hyde (*Nova Hedwigia* **61**: 234, 1995), Liew *et al.* (*Mycol.* **94**: 803, 2002; phylogeny).
Vagnia D. Hawksw. & Miądl. (1997), anamorphic *Pezizomycotina*, Cpd.?.?. 1, Poland. See Hawksworth & Miądlikowska (*Lichenologist* **29**: 45, 1997).
vagrant (of lichens), unattached; erratic; vagant. See Rosentreter (*Bryologist* **96**: 333, 1993; N. Am.).
Vainio [**'Wainio'**] (Edvard August; 1853-1929; Finland). PhD (188), then academic staff (1880-1922), Helsinki University; Turku (1922 onwards). Acquired an uncontested position as the 'Grand Old Man of lichenology' (Lynge). Collections of over 33,000 specimens now in Turku (**TUR**). *Publs.* Monographia Cladoniarum universalis, I-III. *Acta Societatis pro Fauna et Flora Fennica* **4** (1887), **10** (1894), **14** (1897) [reprint 1978]; Lichenes Insularum Philippinarum, I-IV. *Philippine Journal of Science* Section C **4** (1909), **8** (1913), *Annales Academiae Scientiarum Fennicae* Section A **15** (1921), **19** (1923); Lichens du Brésil, I-II. *Acta Societatis pro Fauna et Flora Fennica* **7** (1890); Lichenographia Fennica I-IV. *Acta Societatis pro Fauna et Flora Fennica* **49** (1921), **53** (1922), **57** (1927), **57** (1934); also produced papers on lichens of central Africa, east Asia, Siberia, etc. *Biogs, obits etc.* Alava (*Publications from the Herbarium. University of Turku* **1**, 1986) [journey to Brazil]; Alava (*Publications from the Herbarium. University of Turku* **2**, 1988) [types in **TUR**]; Grummann (1974: 612); Linkola (*Acta Societatis pro Fauna et Flora Fennica* **57** (3), 1934) [bibliography, portrait]; Magnusson (*Annales de Cryptogamie Exotique* **3**: 5, 1930) [bibliography, portrait]; Stafleu & Cowan (*TL-2* **6**: 636, 1986).
Vainiocora, see *Wainiocora*.
Vainiona Werner (1943) ≡ Neonorrlinia.
Vainionia Räsänen (1943) nom. inval. = Calicium fide Hawksworth *et al.* (*Dictionary of the Fungi* edn 8, 1995).
Vainionora Kalb (1991), Lecanoraceae (L). 5, S. America. See Lumbsch *et al.* (*Bryologist* **99**: 269, 1996), Kalb (*Biblthca Lichenol.* **88**: 301, 2004).
Vairimorpha Pilley (1976), Microsporidia. 11.
Vakrabeeja Subram. (1957) ≡ Nakataea fide Ellis (*Dematiaceous Hyphomycetes*, 1971).
Valdensia Peyronel (1923), anamorphic *Valdensinia*, Hso.0fP.1. 1, Italy. See Redhead & Perrin (*CJB* **50**: 409, 1972), Redhead & Perrin (*CJB* **50**: 2083, 1972).
Valdensinia Peyronel (1953), Sclerotiniaceae. Anamorph *Valdensia*. 1, Europe; N. America. See Norvell & Redhead (*Can. J. For. Res.* **24**: 1981, 1994), Holst-Jensen *et al.* (*Mycol.* **89**: 885, 1997; phylogeny).
Valentinia Velen. (1939) = Xeromphalina fide Kuyper (*in litt.*).
Valetoniella Höhn. (1909), Niessliaceae. 3 (dead wood or fungicolous), widespread (esp. tropics). See Samuels & Barr (*CJB* **75**: 2165, 1998).
Valetoniellopsis Samuels & M.E. Barr (1998), Niessliaceae. Anamorph *Acremonium*-like. 1 (dead *Palmae*), N. America. See Samuels & Barr (*CJB* **75**: 2165, 1998), Zhang & Blackwell (*MR* **106**: 148, 2002; phylogeny).
valid (of names), published in accord with the Code Arts 29-45; such names may be illegitimate or legitimate; **pre-** (of names or authors), published before 1753, the starting point for the nomenclature of fungi under the Code; devalidated; cf. Nomenclature.
Valsa Adans. (1763) nom. rej. = Diatrype fide Cannon & Hawksworth (*Taxon* **32**: 478, 1983), Læssøe (*SA* **13**: 43, 1994).
Valsa Fr. (1849) nom. cons., Valsaceae. Anamorph *Cytospora*. 70 (from bark), widespread. See Défago (*Phytopath. Z.* **14**: 103, 1944), Barr (*Mycol. Mem.* **7**, 1978), Spielman (*CJB* **63**: 1355, 1985; N. Am.), Dargan & Sharma (*Kavaka* **19**: 27, 1, 1994; Indian spp.), Farr *et al.* (*Sydowia* **53**: 185, 2001; phylogeny), Adams *et al.* (*Australas. Pl. Path.* **35**: 521, 2006; Australia), Kalkanci *et al.* (*Medical Mycology* **44**: 531, 2006; from human), Zhang *et al.* (*Mycol.* **98**: 1076, 2006; phylogeny).
Valsaceae Tul. & C. Tul. (1861), Diaporthales. 16 gen. (+ 23 syn.), 217 spp.

Lit.: Barr (*Iranian Journal of Plant Pathology*, 1978), Monod (*Beih. Sydowia* **9**: 1, 1983), Spielman (*CJB* **63**: 1355, 1985), Gille (*Arch. phytopath. Pflanz.* **26**: 237, 1990), Vasil'eva (*Pyrenomycetes of the Russian Far East* **1**, 1993; keys), Vasil'eva (*Pyrenomycetes of the Russian Far East* **2**, 1994; keys), Wang *et al.* (*Phytopathology* **88**: 376, 1998), Adams *et al.*

(*Mycol.* **94**: 947, 2002), Castlebury *et al.* (*Mycol.* **94**: 1017, 2002), Adams *et al.* (*Stud. Mycol.* **52**: 146 pp., 2005), Rossman *et al.* (*Mycoscience* **48**: 135, 2007; phylogeny).

Valsales = Diaporthales.

Valsaria Ces. & De Not. (1863), Diaporthales. 41, widespread. See Petrak (*Sydowia* **15**: 299, 1962), Glawe (*Mycol.* **77**: 62, 1985; anamorphs), Ju *et al.* (*Mycotaxon* **58**: 419, 1996), Waldner (*Rheinl.-Pfälz. Pilzj.* **5-6**: 119, 1995).

Valsarioxylon Höhn. (1929) nom. nud., Pezizomycotina.

Valsella Fuckel (1870), Valsaceae. Anamorph *Cytospora*. 2 (on *Salix*), Europe. See Barr (*Mycol. Mem.* **7**, 1978), Zhang *et al.* (*Mycol.* **98**: 1076, 2006; phylogeny).

Valseutypella Höhn. (1919), Valsaceae. Anamorph *Cytospora*. 2, Europe; N. America. See Hubbes (*Phytopath. Z.* **39**: 389, 1960), Checa *et al.* (*Mycotaxon* **25**: 323, 1988), Checa & Martínez (*Mycotaxon* **36**: 43, 1989; anamorph).

valsoid, having groups of perithecia with their beaks pointing inward (convergent), or even parallel to the surface, as in *Valsa*. Cf. eutypoid.

Valsonectria Speg. (1881), Hypocreales. 6 (living and dead plant tissues), widespread. See Huhndorf (*Mycol.* **84**: 642, 1992), Seifert & Samuels (*Mycol.* **89**: 512, 1997), Huhndorf *et al.* (*Mycol.* **96**: 368, 2004; phylogeny).

VAM fungi, vesicular-arbuscular mycorrhizal fungi; VA fungi; see Mycorrhiza.

Vamsapriya Gawas & Bhat (2006), anamorphic *Pezizomycotina*. 1, Karnataka. See Gawas & Bhat (*Mycotaxon* **94**: 150, 2006).

van Tieghem cell, a ring of glass or other material, fixed to a glass slide, over which is placed a coverglass with a 'hanging drop' of the microorganism under investigation. See Duggar (*Fungous diseases of plants*, 1909).

Vanakripa Bhat, W.B. Kendr. & Nag Raj (1993), anamorphic *Pezizomycotina*, Hsp.0-1eP.1. 2, India. See Bhat & Kendrick (*Mycotaxon* **49**: 76, 1993), Pinnoi *et al.* (*Nova Hedwigia* **77**: 213, 2003), Tsui *et al.* (*Mycol.* **95**: 124, 2003), Castañeda Ruíz *et al.* (*Mycotaxon* **91**: 339, 2005).

Vanbeverwijkia Agnihothr. (1961), anamorphic *Pezizomycotina*, Hsp.0-≡ hH.15. 1, Assam; USA. See Agnihothrudu (*TBMS* **44**: 51, 1961), Shearer & Crane (*Mycol.* **63**: 249, 1971).

Vanbreuseghemia Balab. (1965) ≡ Keratinomyces.

Vandasia Velen. (1922) nom. dub., ? Phallales.

Vanderbylia D.A. Reid (1973), Polyporaceae. 5, widespread. See Sikombwa & Piearce (*Bull. BMS* **19**: 124, 1985; medicinal use *V. ungulata*), Corner (*Beih. Nova Hedwigia* **86**: 241, 1987; key), Decock & Ryvarden (*Mycol.* **91**: 386, 1999), Decock (*Mycol.* **93**: 774, 2001; S.E. Asia), Decock & Masuka (*Systematics and Geography of Plants* **73**: 161, 2003; Africa).

Vanderwaltia E.K. Novák & Zsolt (1961) = Kloeckeraspora fide Batra *in* Subramanian (Ed.) (*Taxonomy of fungi* **1**: 187, 1978), von Arx (*Gen. Fungi Sporul. Cult.* Edn 3, 1981).

Vanderwaltozyma Kurtzman (2003), Saccharomycetaceae. 2, Europe; S. Africa. See Kurtzman (*FEMS Yeast Res.* **4**: 242, 2003), Lopandic *et al.* (*Mycol. Progr.* **4**: 205, 2005), Suh *et al.* (*Mycol.* **98**: 1006, 2006; phylogeny).

Vanderystiella Henn. (1908), anamorphic *Pezizomycotina*, Cac.0eP.15. 1, Africa.

Vanhallia L. Marchand (1828) = Chaetomium fide Mussat (*Syll. fung.* **15**: 449, 1901).

Vanibandha Manohar., N.K. Rao, Kunwar & D.K. Agarwal (2006), anamorphic *Pezizomycotina*. 1, India. See Manoharachary *et al.* (*Indian Phytopath.* **59**: 212, 2006).

Vankya D. Ershad (2000), Urocystidaceae. 2 (on *Liliaceae*), widespread. See Bacigálová *et al.* (*Polish Botanical Journal* **50**: 145, 2005).

Vanrija R.T. Moore (1980) = Apiotrichum fide Moore (*MR* **95**: 639, 1991).

Vanrijia R.T. Moore (1980) = Cryptococcus Vuill.

Vanromburghia Holterm. (1898) ? = Marasmius fide Singer (*Agaric. mod. Tax.* edn 2: 808, 1962).

Vanterpoolia A. Funk (1982), anamorphic *Pezizomycotina*, Hsp.1bH.3. 1, Canada. See Funk (*CJB* **60**: 973, 1982).

Vantieghemia Kuntze (1891) ≡ Syncephalis.

Vanudenia Bat. & H. Maia (1963) = Schizothyrium fide von Arx & Müller (*Stud. Mycol.* **9**, 1975).

Vararia P. Karst. (1898), Lachnocladiaceae. 54, widespread. See Boidin & Lanquetin (*BSMF* **91**: 457, 1975), Boidin & Lanquetin (*BSMF* **92**: 247, 1976), Boidin & Lanquetin (*Mycotaxon* **6**: 277, 1977), Pascoe *et al.* (*TBMS* **82**: 723, 1984; white rot of *Rubus*), Stalpers (*Stud. Mycol.* **40**: 103, 1996; key), Larsson & Larsson (*Mycol.* **95**: 1037, 2003; phylogeny).

Vargamyces Tóth (1980), anamorphic *Pezizomycotina*, Hso.≡ eP.1/10. 1 (aquatic). See Gönczöl *et al.* (*Mycotaxon* **39**: 301, 1990; growth and development in culture).

Variation in fungi, The text of this entry is unrevised and remains the same as for the ninth edition. Variation in fungi may be the result of environment and not heritable (modifications, Bauer) or may be heritable and (1) determined by the coming together or separating of heritable factors (as in combinations) or (2) not so determined (mutations). Modifications ('temporary' or 'reversible' variations, such as changes in mycelial growth and pathogenicity) frequently take place in culture, sometimes in association with non-reversible (discontinuous) variations of a mutation (saltation, dissociation) type. The composition of the medium may have an effect on frequency of saltation; *Fusarium*, for example, saltating less readily on a poor medium, *Aspergillus* more readily in a mannitol-nitrite solution.

The polymorphic variation (Snyder & Hansen) in a species in nature may be so great that the forms are put in three, four or more genera. The existence of the 'dual phenomenon' (Hansen) in anamorphic fungi, has the effect of giving two kinds of culture.

Lit.: Brierley (*Proc. Int. Congr. Pl. Sci.* **2**, 1929), Snyder & Hansen (*Proc. 6th Pacific Sci. Congr.* **4**, 1940), Hansen (*Mycol.* **30**: 242, 1938), Day (*Ann. Rev. Microbiol.* **14**: 1, 1960; plant pathogenic fungi), Sugiyama (Ed.) (*Pleomorphic fungi*, 1987).

Varicellaria Nyl. (1858), Ochrolechiaceae (L). 3, widespread. See Erichsen (*Rabenh. Krypt.-Fl.* **9** 2.2: 687, 1935; Eur.), Oshio (*J. Sci. Hiroshima Univ.* B(2) **12**: 81, 1968; Japan), Schmitt *et al.* (*Biblthca Lichenol.* **86**: 147, 2003; phylogeny), Schmitt & Lumbsch (*Mol. Phylogen. Evol.* **33**: 43, 2004; chemistry, phylogeny), Lumbsch *et al.* (*Biol. J. Linn. Soc.* **89**: 615, 2006), Schmitt *et al.* (*J. Hattori bot. Lab.* **100**: 753, 2006; phylogeny).

Varicellariomyces Cif. & Tomas. (1953) = Varicellaria fide Hawksworth *et al.* (*Dictionary of the Fungi* edn 8, 1995).

Varicosporina Meyers & Kohlm. (1965), anamorphic *Corollospora*, Hso.1bH.10. 2 (marine), widespread (tropical to subtropical). See Nakagiri (*TBMS* **90**: 265, 1988; conidia form and function), Spatafora *et al.* (*Am. J. Bot.* **85**: 1569, 1998; phylogeny), Zhang *et al.* (*Mycol.* **98**: 1076, 2006; phylogeny).

Varicosporium W. Kegel (1906), anamorphic *Hymenoscyphus*, Hso.1bH.1. 5, widespread. See Tubaki (*TBMS* **49**: 345, 1966; teleomorph), Nawawi (*TBMS* **63**: 27, 1974), Baschien *et al.* (*Nova Hedwigia* **83**: 311, 2006).

variecolin, an anti-tubercle bacillus antibiotic from *Aspergillus variecolor* (teleomorph, *Emericella variecolor*) (Gupta & Viswanathan, *Antibiot. & Chemother.* **5**: 496, 1955).

variety, see Classification.

Variocladium Descals & Marvanová (1999), anamorphic *Pezizomycotina*, Hso.?.?. 1, British Isles. See Descals *et al.* (*CJB* **76**: 1658, 1998).

Variolaria Bull. [not traced] nom. dub., ? Fungi.

Variolaria Gray (1821) nom. dub., Fungi.

Variolaria Pers. (1794) nom. rej. = Pertusaria.

variolarioid, having powdery or granular tubercules.

Varisulcosporites Rouse & Mustard (1997), Fossil Fungi. 1, British Columbia. See Rouse & Mustard (*Palynology* **21**: 208, 1997).

Varmasporites Kalgutkar & Janson. (2000), Fossil Fungi, anamorphic *Ascomycota*. 1, India. See Kalgutkar & Jansonius (*AASP Contributions Series* **39**: 309, 2000).

Vascellum F. Smarda (1958) = Lycoperdon Pers. See Ponce de Leon (*Fieldiana, Bot.* **32**: 109, 1970; key), Smith (*Bull. Soc. linn. Lyon* num. spéc. **43**: 407, 1974; N America), Homrich & Wright (*CJB* **66**: 1285, 1988; S America), Kreisel (*Blyttia* **51**: 125, 1993; key), Larsson & Jeppson (*MR* **112**: 4, 2008; synonym of *Lycoperdon*).

Vasculomyces S.F. Ashby (1913), anamorphic *Pezizomycotina*, Hso.-.-. 1, West Indies. See Ciferri (*Sydowia* **8**: 258, 1954).

Vasudevella Chona, Munjal & Bajaj (1957), anamorphic *Pezizomycotina*, Cpd.1eH.19. 1, India. See Nag Raj (*CJB* **51**: 1337, 1973).

Vavraia J. Weiser (1977), Microsporidia. 6.

vector, an organism that carries and transmits a pathogen to the host which it attacks, such as an insect carrying fungal mycelium or spores (e.g. *Scolytus* spp. beetles transmitting *Ophiostoma novo-ulmi* to *Ulmus* trees). A **fungus -** carries a virus infection with it when moving from its existing host to a new host; some plant viruses are transmitted to new host plants by association with root-infecting fungi; the vector phase is the motile zoospores in *Olpidium* sp. transmitting *Tombusvirus* and *Necrovirus* spp. and the lettuce big vein and tobacco stunt viruses; *Polymyxa graminis*, *P. betae* and *Spongospora subterranea* transmit viruses in the genus *Furovirus*, and *P. graminis* transmit viruses in the genus *Bymovirus* (see Viruses); **plasmid -**, a plasmid constructed to include a particular gene sequence, and inserted into another fungus where its properties are expressed (see Genetic engineering).

vegetable caterpillar, a mummified lepidopteran larva from which arise teleomorph stromata of *Cordyceps* spp.

vegetative, see assimilative.

vegetative compatibility, the ability of vegetative hyphae to anastomose and form a stable heterokaryon. This ability is restricted genetically by the vegetative incompatibility system such that hyphae differing at one or more loci, termed 'vegetative compatibility' (vc) or 'heterokaryon compatibility (het) loci, are unable to form a stable heterokaryon. Mycelia sharing the same vc loci belong to the same vegetative compatibility group (vcg). Vcg's have been used to determine genetic structures of fungal populations. In most fungi, the vegetative compatibility system is independant of the mating system, which controls sexual compatibility (see sex).

Lit.: Anagnostakis (*Exp. Mycol.* **1**: 306, 1977), Beadle & Coonradlt (*Genetics* **29**: 291, 1944), Brayford (*MR* **94**: 745, 1990), Carlisle (*in* Rayner *et al.* (Eds), *Evolutionary biology of the fungi*, 1987), Caten (*J. Gen. Microbiol.* **72**: 221, 1972), Coates *et al.* (*TBMS* **76**: 41, 1981), Leslie (*Ann. Rev. Phytopath.* **31**: 127, 1993), Mylyk (*Genetics* **83**: 275, 1976), Puhalla (*CJB* **63**: 179, 1985), Todd & Rayner (*Sci. Progress* Oxford **66**: 331, 1980).

veil, see annulus (**apical -**, **hymenial -**), cortina, **marginal -**, **partial -** (**inner -**), **pellicular -**, protoblem (**primordial -**), **universal -**.

veins (of lichens), strands of tissue on the lower surface of foliose lichens, esp. *Peltigera* where they may replace a lower cortex. Gyelnik (*Bot. Közlemén.* **24**: 122, 1927) distinguished 2 types: **caninoid -** where the strands are separated to the tips of the lobes; **polydactyloid -** where the strands are confluent towards the tips of the lobes. Maas Geesteranus also recognized **malaceoid -**, where the undersurface has a few whitish interstices faintly indicating venation.

velar, pertaining to a veil.

Velenovský (Josef; 1858-1949; Czech). Extraordinary Professor of Phytopathology (1892) then Professor of Phytopathology & Plant Systematics (1898-1927) and Director of Botanic Garden of Charles University (1900-1927), Charles University, Prague. Expert on discomycetes. *Publs. České Houby.* 5 parts (1920-1922); *Monographia Discomycetum Bohemiae* 2 vols (1934). Work. *Biogs, obits etc.* Lloyd (*Mycological Notes by C.G. Lloyd* **75**: 1349, 1925); Pilát (*Česká Mykologie* **3**: 65, 1949, portrait) [in Czech]; Kotlaba & Skalický (*Preslia* **30**: 327, 1938).

Veligaster Guzmán (1970) = Scleroderma fide Demoulin & Dring (*Bull. Jard. bot. nat. Belg.* **45**: 343, 1975).

Vellosiella Rangel (1915) [non *Vellosiella* Baill. 1887, *Scrophulariaceae*] ≡ Mycovellosiella.

Velolentinus Overeem (1927) nom. nud. = Panus fide Pegler (*Kew Bull.* Addit. Ser. **10**, 1983).

Velomycena Pilát (1953) = Galerina fide Singer (*Agaric. mod. Tax.* edn 3, 1975).

Veloporphyrellus L.D. Gómez & Singer (1984), Boletaceae. 1, C America. See Gómez & Singer (*Brenesia* **22**: 293, 1984), Watling & Turnbull (*Edinb. J. Bot.* **49**: 343, 1993; South and East Central Africa).

Veloporus Quél. (1888) = Boletus Fr. fide Killermann *in* Engler & Prantl (*Naturl. Pflanzenfam.* **2**, 1928).

velum, see veil.

Velutaria Fuckel (1870), Hyaloscyphaceae. 1, Europe. See Korf (*Mycol.* **45**: 298, 1953).

Velutarina Korf ex Korf (1971), Helotiaceae. 2, widespread. See Korf ex Korf (*Phytologia* **21**: 201, 1971), Zhuang *et al.* (*Mycosystema* **19**: 478, 2000; phylogeny).

Veluticeps (Cooke) Pat. (1894), Gloeophyllaceae. 9, widespread. *V. berkeleyi* (wood rot of *Pinus*). See Gilbertson *et al.* (*Mycol.* **60**: 29, 1968), Hjortstam & Tellería (*Mycotaxon* **37**: 53, 1990), Nakasone (*Mycol.* **82**: 622, 1990; key), Nakasone (*Sydowia* **56**: 258, 2004).

velutinate (**velutinous**), thickly covered with delicate hairs; like velvet; see phalacrogenous.

Velutipila D. Hawksw. (1987), anamorphic *Pezizomycotina*, Hso.0fH.10. 1 (on *Algae*), Austria. See Hawksworth (*Notes R. bot. Gdn Edinb.* **44**: 555, 1987).

Venenarius Earle (1909) ≡ Amanita Pers.

venose, having veins.

ventral, front, or lower surface; the surface facing the axis, cf. dorsal; frequently used for the lower surface of foliose lichens.

ventricose, swelling out in the middle or at one side; inflated (Fig. 23.38).

Ventrographium H.P. Upadhyay, Cavalc. & A.A. Silva (1986), anamorphic *Pezizomycotina*, Hsy.0fH.15. 1, Brazil. See Upadhyay *et al.* (*Mycol.* **78**: 494, 1986).

Venturia De Not. (1844) nom. rej. ≡ Protoventuria.

Venturia Sacc. (1882) nom. cons., Venturiaceae. Anamorph *Fusicladium*. 57, widespread. *V. cerasi* (cherry scab), *V. chlorospora* (willow (*Salix*) scab and canker), *V. inaequalis* (apple scab), *V. pirina* (pear scab) (see *Fusicladium*). See Sivanesan (*Biblthca Mycol.* **59**, 1977; key), Morelet (*Cryptog. Mycol.* **6**: 101, 1985; key 10 spp. on *Populus*), Barr (*Sydowia* **41**: 25, 1989; key 39 N. Am. spp.), Schnabel *et al.* (*Phytopathology* **89**: 100, 1999; DNA), Tenzer & Gessler (*Eur. J. Pl. Path.* **105**: 545, 1999; mol. var.), Ishii & Yanase (*MR* **104**: 755, 2000; spp. on pears), Kasanen *et al.* (*MR* **105**: 338, 2001; on *Populus*), Stehmann *et al.* (*Phytopathology* **91**: 633, 2001; genetics), Ahlholm *et al.* (*Evolution* Lancaster, Pa. **56**: 1566, 2002; genetic diversity), Newcombe (*MR* **107**: 108, 2003; on *Populus*), Beck *et al.* (*Mycol. Progr.* **4**: 111, 2005; phylogeny), Spatafora *et al.* (*Mycol.* **98**: 1018, 2006; phylogeny), Crous *et al.* (*Stud. Mycol.* **58**: 185, 2007; phylogeny), Winton *et al.* (*Mycol.* **99**: 240, 2007; phylogeny).

Venturiaceae E. Müll. & Arx ex M.E. Barr (1979), Pleosporales. 36 gen. (+ 48 syn.), 306 spp.

Lit.: Fischer (*Rabenh. Krypt.-Fl.* **4** 2.1: 25, 1892), Barr (*CJB* **46**: 799, 1968; N. Am.), Sivanesan (*Bitunicate Ascomycetes and their Anamorphs*, 1984), Sivanesan (*Bitunicate Ascomycetes and their Anamorphs*: 700 pp., 1984), Barr (*Prodr. Cl. Loculoasc.*, 1987), Samuels *et al.* (*Brittonia* **40**: 392, 1988), Barr (*Sydowia* **41**: 25, 1989; keys 12 gen. N. Am.), Hughes & Arnold (*Mem. N. Y. bot. Gdn* **49**: 198, 1989), Pascoe & Sutton (*Aust. Syst. Bot.* **3**: 281, 1990), Carris & Poole (*Mycol.* **85**: 93, 1993), Hughes & Seifert (*Sydowia* **50**: 192, 1998), Schnabel *et al.* (*Phytopathology* **89**: 100, 1999), Tenzer & Gessler (*Eur. J. Pl. Path.* **105**: 545, 1999), Untereiner & Naveau (*Mycol.* **91**: 67, 1999), Ishii & Yanase (*MR* **104**: 755, 2000), Braun *et al.* (*Mycol. Progr.* **2**: 8, 2003), Newcombe (*MR* **107**: 108, 2003), Crous *et al.* (*Stud. Mycol.* **58**: 185, 2007; phylogeny).

Venturiella Speg. (1909) [non *Venturiella* Müll. Hal. 1875, *Musci*] ≡ Neoventuria.

Venturiocistella Raitv. (1978), Hyaloscyphaceae. 7, widespread. See Baral (*Z. Mykol.* **59**: 3, 1993; key), Hosoya & Harada (*Mycoscience* **40**: 401, 1999), Raitviir (*Scripta Mycologica* Tartu **20**, 2004; key).

Venturiola.

Venularia Pers. (1822) nom. nud., ? Fungi.

Venustisporium R.F. Castañeda & Iturr. (1999), anamorphic *Pezizomycotina*, Hso.?.?. 1, Venezuela. See Castañeda-Ruíz & Iturriaga (*Mycotaxon* **72**: 455, 1999).

Venustocephala Matsush. (1995), anamorphic *Pezizomycotina*, Hso.0eH.9. 1, Ecuador. See Matsushima (*Matsush. Mycol. Mem.* **8**: 41, 1995).

Venustosynnema R.F. Castañeda & W.B. Kendr. (1990), anamorphic *Pezizomycotina*, Hsy.0eH.15. 1, Cuba. See Castañeda & Kendrick (*Univ. Waterloo Biol. Ser.* **32**: 45, 1990).

Veracruzomyces Mercado, Guarro, Heredia & J. Mena (2002), anamorphic *Pezizomycotina*. 1, Mexico. See Mercado-Sierra *et al.* (*Nova Hedwigia* **75**: 534, 2002).

Veralucia D.R. Reynolds & P.H. Dunn (1982) = Cephaleuros fide Reynolds & Dunn (*Mycol.* **76**: 719, 1984).

Veramyces Matsush. (1993), anamorphic *Pezizomycotina*, Hso.≡ eP.3/10. 1, Peru. See Matsushima (*Matsush. Mycol. Mem.* **7**: 71, 1993).

Veramyces Subram. (1993) [non *Veramyces* Matsush. 1993, anamorphic Ascomycota] nom. inval. = Veramycina fide Hawksworth *et al.* (*Dictionary of the Fungi* edn 8, 1995).

Veramycina Subram. (1995), anamorphic *Chaetosphaerella*, Hso.0eH.15. 1, Himalaya. See Subramanian (*Kavaka* **20/21**: 58, 1992/1993), Réblová (*Mycotaxon* **70**: 387, 1999).

Verdipulvinus A.W. Ramaley (1999), anamorphic *Pezizomycotina*, Hso.?.?. 1, USA. See Ramaley (*Mycol.* **91**: 132, 1999).

Verlandea Bat. & Cif. (1957) = Stomiopeltis fide von Arx & Müller (*Stud. Mycol.* **9**, 1975).

Verlotia Fabre (1879) = Heptameria fide Lucas & Sutton (*TBMS* **57**: 283, 1971).

Vermicularia Tode (1790) [non *Vermicularia* Moench 1802, *Verbenaceae*] = Colletotrichum fide Duke (*TBMS* **13**: 156, 1928), Sutton *in* Bailey & Jeger (Eds) (*Colletotrichum: biology, pathology and control*, 1992).

Vermiculariella Oudem. (1898) nom. conf., Fungi. See Petrak (*Annls mycol.* **42**: 58, 1944).

Vermiculariopsiella Bender (1932), anamorphic *Echinosphaeria*, Hsp.0eH.15. 10, widespread. See Nawawi *et al.* (*Mycotaxon* **37**: 173, 1990; key), Mel'nik (*Mikol. Fitopatol.* **33**: 169, 1999), Puja *et al.* (*Cryptog. Mycol.* **27**: 11, 2006; teleomorph).

Vermiculariopsis Höhn. (1918) ≡ Vermiculariopsiella.

Vermiculariopsis Torrend (1912) nom. conf., anamorphic *Pezizomycotina*. See Sutton (*Mycol. Pap.* **141**, 1977).

vermiform, worm-like.

Vermilacinia Spjut & Hale (1995) ? = Ramalina fide Aptroot (*in litt.*).

Vermispora Deighton & Piroz. (1972), anamorphic *Pezizomycotina*, Hso.≡ eH.10. 4 (on *Irenopsis*), widespread. See Deighton & Pirozynski (*Mycol. Pap.* **128**: 87, 1972), Burghouts & Gams (*Mem. N. Y. bot. Gdn* **49**: 57, 1989), Rajashekhar *et al.* (*Mycol.* **83**: 230, 1991).

Vermisporium H.J. Swart & M.A. Will. (1983), anamorphic *Pezizomycotina*, Cac.0fH.19. 14, widespread. See Nag Raj (*Coelomycetous Anamorphs with Appendage-bearing Conidia*, 1993).

Veronaea Cif. & Montemart. (1957), anamorphic *Pezizomycotina*, Hso.1eP.10. 24 (parasitic on plants, also from humans), widespread. See Ellis (*More Dematiaceous Hyphomycetes*, 1976; key), de Hoog *et al.* (*TBMS* **81**: 485, 1983; relationship to *Ramichloridium*), Wang *et al.* (*Acta Mycol. Sin.* **10**: 159, 1991; *V. botryosa* from humans), Matsushita *et al.* (*J. Clin. Microbiol.* **41**: 2219, 2003; China), Sutton *et al.* (*J. Clin. Microbiol.* **42**: 2843, 2004; USA).

Veronaella Subram. & K.R.C. Reddy (1975), anamorphic *Pezizomycotina*, Cpt.1bH.10. 1, India. See Subramanian & Reddy (*Kavaka* **2**: 27, 1974).

Veronaeopsis Arzanlou & Crous (2007), anamorphic *Venturiaceae*. 1, S. Africa. See Arzanlou *et al.* (*Stud. Mycol.* **58**: 57, 2007), Crous *et al.* (*Stud. Mycol.* **58**: 185, 2007; phylogeny).

Veronaia Benedek (1961), Eurotiales. 3, widespread. See Benedek (*Mycopathologia* **14**: 115, 1961).

Veronidia Negru (1964), anamorphic *Pezizomycotina*, Cpd.0eH.?. 1, Rumania. See Negru (*Mycopath. Mycol. appl.* **23**: 241, 1964).

Verpa Sw. (1815), Morchellaceae. *c.* 5, widespread (north temperate). See O'Donnell *et al.* (*Mycol.* **89**: 48, 1997; phylogeny), Wipf *et al.* (*Can. J. Microbiol.* **45**: 769, 1999; ITS diversity), Hansen *et al.* (*Mycol.* **93**: 958, 2001; phylogeny).

Verpatinia Whetzel & Drayton (1945) = Scleromitrula fide Schumacher & Holst-Jensen (*Mycoscience* **38**: 55, 1997).

verruca (pl. **verrucae**), a wart-like swelling.

Verrucaria Weber ex F.H. Wigg. (1780) = Lecanora fide Hawksworth *et al.* (*Dictionary of the Fungi* edn 8, 1995).

Verrucaria Schrad. (1794) nom. cons., Verrucariaceae (L). *c.* 387 (some on lichens), widespread (esp. north temperate). See Swinscow (*Lichenologist* **4**: 34, 1968; freshwater spp.), Fletcher (*Lichenologist* **7**: 1, 1975; coastal spp.), Zehetleitner (*Nova Hedwigia* **29**: 683, 1978; spp. on lichen spp.), Navarro-Rosinés & Roux (*Bull. Soc. linn. Provence* **39**: 129, 1988; key 8 spp. on lichens), Hawksworth (*Lichenologist* **21**: 23, 1989; Brit., aquatic), McCarthy (*Lichenologist* **27**: 105, 1995; aquatic, Australia), Breuss (*Linzer biol. Beitr.* **30**: 831, 1998; key corticolous & lichenicolous spp.), Bermúdez (*Clementeana* **4**: 52, 1999; key Galician spp.), Hoffmann & Hafellner (*Biblthca Lichenol.* **77**: 1, 2000), McCarthy (*Lichenologist* **34**: 207, 2002; Australia), Aragón & Sarrión (*Nova Hedwigia* **77**: 169, 2003; Spain), Halda (*Acta Musei Richnoviensis* Sect. Natur. **10**, 2003; calcicolous spp.), Orange (*Lichenologist* **36**: 173, 2004; GB, Ireland), Sparrius (*Biblthca Lichenol.* **89**: 141 pp., 2004), Schmitt *et al.* (*Mycol.* **97**: 362, 2005; phylogeny), Geiser *et al.* (*Mycol.* **98**: 1053, 2006; phylogeny), Gueidan *et al.* (*MR* **111**: 1145, 2007; phylogeny).

Verrucaria Scop. (1777) nom. rej. = Dibaeis fide Gierl & Kalb (*Herzogia* **9**: 593, 1993).

Verrucariaceae Zenker (1827), Verrucariales (±L). 46 gen. (+ 98 syn.), 931 spp.

Lit.: Thomson (*Bryologist* **90**: 27, 1987), Aptroot (*Willdenowia* Beih. **21**: 263, 1991), Thomson (*Bryologist* **94**: 351, 1991), Ménard & Roux (*Mycotaxon* **53**: 129, 1995), Breuss (*Annln naturh. Mus. Wien* Ser. B, Bot. Zool. **98**: 35, 1996), Breuss (*Öst. Z. Pilzk.* **5**: 65, 1996), Harada (*Hikobia* **12**: 133, 1996), Breuss (*Annln naturh. Mus. Wien* Ser. B, Bot. Zool. **100**: 671, 1998), Grube (*Nova Hedwigia* **68**: 241, 1999), Hyde & Wong (*MR* **103**: 347, 1999), Heiðmarsson (*MR* **107**: 459, 2003), Andjic *et al.* (*MR* **111**: 1184, 2007; phylogeny).

Verrucariales Mattick ex D. Hawksw. & O.E. Erikss. (1986), (±L). Chaetothyriomycetidae. 2 fam., 54 gen., *c.* 1032 spp. Thallus usually crustose, rarely squamulose or umbilicate, rarely absent. Ascomata perithecial, ± globose, sometimes strongly aggregated, ± superficial or immersed, usually black, thin-walled, sometimes clypeate. Interascal tissue of gelatinized pseudoparaphysis-like hyphae, often evanescent and sometimes altogether absent, the ostiole periphysate. Asci clavate, thin-walled, sometimes with two layers visible near the apex but these not separable, sometimes blueing in iodine, persistent or evanescent. Ascospores usually ellipsoidal, aseptate to muriform, hyaline or brown. Anamorphs pycnidial. Lichenized with green algae, a few lichenicolous or saprobic, cosmop. Fams:

(1) **Adelococcaceae**
(2) **Verrucariaceae**

The *Dermatocarpaceae* are sometimes treated as distinct (e.g. Wanger, *CJB* **65**: 2441, 1987). Gen. concepts still largely based on ascospore colour and septation and in need of a modern revision taking into account ascomatal structure etc.

Lit.: Doppelbauer (*Planta* **53**: 246, 1959; ontogeny), Geiser *et al.* (*Mycol* **98**: 1051, 2006), Grube (*Nova Hedwigia* **68**: 241, 1999; asci), Janex-Favre (1970), McCarthy (*Muelleria* **7**: 317, 1991; Australia), Servít (*Československé lišejníky čelidi Verrucariaceae*, 1954), Swinscow (*Lichenologist* **1-5**, 1960-71; many papers, UK).

Verrucariella S. Ahmad (1967), anamorphic *Pezizomycotina*, Cpd.1eP.?. 1, Pakistan. See Ahmad (*Biologia* Lahore **13**: 41, 1967).

Verrucarina (Hue) Zahlbr. (1931) = Coccotrema fide Hawksworth *et al.* (*Dictionary of the Fungi* edn 8, 1995).

verrucarioid (of asci), see ascus.

Verrucariomyces E.A. Thomas ex Cif. & Tomas. (1953) = Verrucaria Schrad. fide Hawksworth *et al.* (*Dictionary of the Fungi* edn 8, 1995).

Verrucarites Göpp. (1845), Fossil Fungi, Ascomycota (?L). 2 (Tertiary), Europe. See Watelet (*Descr. Pl. Foss.*, 1866).

Verrucaster Tobler (1912) nom. dub., anamorphic *Pezizomycotina*. See Hawksworth (*Bull. Br. Mus. nat. hist.* Bot. **9**: 88, 1981).

Verrucispora D.E. Shaw & Alcorn (1967) [non *Verrucospora* E. Horak, 3 Aug. 1967] ≡ Verrucisporota.

Verrucisporota D.E. Shaw & Alcorn (1993), anamorphic *Mycosphaerellaceae*, Hso.≡ eP.10. 2, Australia; Papua New Guinea. See Shaw & Alcorn (*Aust. Syst. Bot.* **6**: 273, 1993), Beilharz & Pascoe (*Mycotaxon* **82**: 357, 2002), Crous & Braun (*CBS Diversity Ser.* **1**: 571 pp., 2003).

Verrucobotrys Hennebert (1973), anamorphic *Seaverinia*, Hso.0eP.7. 1, N. America. See Hennebert (*Persoonia* **7**: 193, 1973).

Verrucocladosporium K. Schub., Aptroot & Crous (2007), anamorphic *Capnodiales*. 1 (lichenicolous), British Isles. See Crous *et al.* (*Stud. Mycol.* **58**: 33, 2007).

Verrucophragmia Crous, M.J. Wingf. & W.B. Kendr. (1994), anamorphic *Pezizomycotina*, Hso.≡ eP.10. 1, S. Africa. See Crous *et al.* (*Mycol.* **86**: 448, 1994).

verrucose, having small rounded processes or 'warts'

(Fig. 20.7).

Verrucospora E. Horak (1967), Agaricaceae. 2, Africa (tropical). See Horak (*Ber. schweiz. bot. Ges.* **77**: 363, 1967), Pegler (*Kew Bull.* Addit. Ser. **6**: 384, 1977).

Verrucosporaceae Jülich (1982) = Agaricaceae.

Verrucula J. Steiner (1896), Verrucariaceae (L). 20 (lichenicolous). See Gueidan *et al.* (*MR* **111**: 1145, 2007; phylogeny), Navarro-Rosinés *et al.* (*Bull. Soc. linn. Provence* **38**: 133, 2007; key).

Verruculina Kohlm. & Volkm.-Kohlm. (1990), Testudinaceae. 1 (marine), Liberia. See Kohlmeyer & Volkmann-Kohlmeyer (*MR* **94**: 685, 1990), Schoch *et al.* (*Mycol.* **98**: 1041, 2006; phylogeny).

Verruculopsis Gueidan, Nav.-Ros. & Cl. Roux (2007), Verrucariaceae (±L). 3 (sometimes lichenicolous), Europe. See Navarro-Rosinés *et al.* (*Bull. Soc. linn. Provence* **38**: 133, 2007).

verruculose, delicately verrucose (Fig. 20.6).

verruculotoxin, a tremorgenic toxin from *Penicillium verruculosum* (Cole *et al.*, *Toxicol. appl. Pharmacol.* **31**: 465, 1975).

Versicolorisporium Sat. Hatak., Kaz. Tanaka & Y. Harada (2008), anamorphic *Pezizomycotina*. 1, Japan. See Hatakeyama *et al.* (*Mycoscience* **49**: 211, 2008).

versiform, of different forms; changing form with age.

Versiomyces Whalley & Watling (1989) = Daldinia fide Læssøe (*SA* **13**: 43, 1994).

Versipellis Quél. (1886) nom. rej. ≡ Boletus Fr.

vertex (1) the top of an organ; (2) pileus (obsol.).

Vertexicola K.D. Hyde, V.M. Ranghoo & S.W. Wong (2000), Annulatascaceae. 1, Philippines; Hong Kong. See Hyde *et al.* (*Mycol.* **92**: 1019, 2000).

Verticicladiella S. Hughes (1953) = Leptographium fide Kendrick (*CJB* **40**: 771, 1962), Wingfield (*TBMS* **85**: 81, 1985), Harrington (*Leptographium Root Diseases on Conifers*: 1, 1988).

Verticicladium Preuss (1851), anamorphic *Desmazierella*, Hso.0eP.10. 1, Europe. See Hughes (*Mycol. Pap.* **43**, 1951).

Verticicladus Matsush. (1993), anamorphic *Pezizomycotina*, Hso.≡ eP.1. 1, Peru. See Matsushima (*Matsush. Mycol. Mem.* **7**: 71, 1993).

verticillate, having parts in rings (**verticils**); whorled.

Verticilliaceae Nann. (1934) = Plectosphaerellaceae.

Verticilliastrum Dasz. (1912) ? = Trichoderma Pers. (1794) fide Sutton (*in litt.*).

Verticilliodochium Bubák (1914) = Clonostachys fide Schroers *et al.* (*Mycol.* **91**: 365, 1999).

Verticilliopsis Costantin (1892) nom. dub., anamorphic *Pezizomycotina*. See Gams (*Cephalosporium-artige Schimmelpilze*: 18, 1971).

Verticillis Clem. & Shear (1931) ≡ Verticilliodochium.

Verticillium Nees (1816), anamorphic *Plectosphaerellaceae*, Hso.0eH.15. 51, widespread. *V. alboatrum* and *V. dahliae* cause wilt disease (hadromycosis) in many plants. See also *Lecanicillium*. See Rudolph (*Hilgardia* **5**: 197, 1931; host list), Isaac (*TBMS* **32**: 137, 1949), Rudolph (*Plant Dis. Reptr* Suppl. **244**, 1957), Rudolph (*Plant Dis. Reptr* Suppl. **255**, 1959; deciduous fruit trees), Isaac (*Ann. Rev. Phytopath.* **5**: 201, 1967; speciation), Isaac (*IMI Descr. Fungi Bact.* **255-256**, 1970), Anon. (*IMI Descr. Fungi Bact.* **498**, 1976), Gams & van Zaayen (*Neth. Jl Pl. Path.* **88**: 57, 1982; fungicolous spp., keys sects., 9 spp.), Evans & Samson (*Fundamental and Applied Aspects of Invertebrate Pathology*: 186, 1986; spp. taxonomy on invertebrate hosts), Gams (*Neth. Jl Pl. Path.* **94**: 123, 1988; key to 9 nematophagous spp.), Heale *in* Ingrams & Williams (Eds) (*Advances in Plant Pathology* **6**: 291, 1988; vascular wilt spp.), Carder (*MR* **92**: 297, 1989; cellulase isoenzyme patterns of 5 spp.), Chen & Fu (*Acta Mycol. Sin.* **8**: 123, 1989), Carder & Barbara (*MR* **95**: 935, 1991; RFLPs in 6 spp.), Jun *et al.* (*J. Gen. Microbiol.* **137**: 1437, 1991; integrated taxonomic approaches), Typas *et al.* (*FEMS Microbiol. Lett.* **95**: 157, 1992; RFLP analysis), Williams *et al.* (*Mycopathologia* **119**: 101, 1992; biochemical and physiological aids ident. sect. *Nigrescentia*), Okoli *et al.* (*Pl. Path.* **43**: 33, 1994; RFLPs and host adaptation), Morton *et al.* (*MR* **99**: 257, 1995; rRNA intergenic regions in *V. alboatrum* and *V. dahliae*), Rowe (*Phytoparasitica* **23**: 31, 1995; relationships between spp. and subspp. groups), Elena & Paplomatas (*Pl. Path.* **47**: 635, 1998; VCGs), Bidochka *et al.* (*Microbiology* Reading **145**: 955, 1999; phylogeny), Koralev & Katan (*MR* **103**: 65, 1999; VCGs), Zhu *et al.* (*Mycosystema* **18**: 366, 1999; Australia), Pramateftaki *et al.* (*Fungal Genetics Biol.* **29**: 19, 2000; DNA), Zare *et al.* (*Nova Hedwigia* **71**: 465, 2000; phylogeny), Gams & Zare (*Nova Hedwigia* **72**: 329, 2001), Korolev *et al.* (*Eur. J. Pl. Path.* **107**: 443, 2001; genetic diversity), Sung *et al.* (*Nova Hedwigia* **72**: 311, 2001; phylogeny), Mahuku & Platt (*Molecular Plant Pathology* **3**: 71, 2002; genetic variation), Collins *et al.* (*Phytopathology* **93**: 364, 2003; genetic diversity), Zare (*Rostaniha* **4**: 37, 2003; review), Zare *et al.* (*MR* **108**: 576, 2004), Collins *et al.* (*Pl. Path.* **54**: 549, 2005; genetic diversity), Gams *et al.* (*Taxon* **54**: 179, 2005; nomencl.), Pantou *et al.* (*Curr. Genet.* **50**: 125, 2006; mitochondrial genome), Zhang *et al.* (*Mycol.* **98**: 1076, 2006; phylogeny), Zare *et al.* (*Nova Hedwigia* **85**: 463, 2007).

Verticimonosporium Matsush. (1971), anamorphic *Pezizomycotina*, Hso.0eH.15. 2, E. & S.E. Asia. See Matsushima (*Microfungi of the Solomon Islands and Papua-New Guinea*: 68, 1971), Wang *et al.* (*Mycotaxon* **92**: 197, 2005; China).

Vertixore V.A.M. Mill. & Bonar (1941) = Aithaloderma fide von Arx & Müller (*Stud. Mycol.* **9**, 1975).

Vesicladiella Crous & M.J. Wingf. (1994), anamorphic *Bionectriaceae*, Hsp.0eH.15. 1, Australia. See Crous *et al.* (*Mycotaxon* **50**: 454, 1994).

vesicle (1) a bladder-like sac; (2) (of *Aspergillus*), the swollen apex of the conidiophore; (3) (of *Pythium*), the evanescent extra-sporangial structure in which zoospores are differentiated; **homohylic -**, sporangial vesicle, the wall of which is continuous with, and of the same material as the wall layer, or one of the wall layers of the sporangium; **vesiculose**, made from or full of vesicles.

vesicular bodies, thin-walled vesicles in the subhymenium of certain hymenomycetes (mostly *Thelephoraceae*) (Overholts); vesicular-arbuscular type of mycorrhiza, see Mycorrhiza.

Vesicularia I. Schmidt (1974) [non *Vesicularia* P. Micheli ex Targ. Tozz. 1826, *Algae*] = Basramyces fide Piaggio (*Cryptog. Mycol.* **10**: 173, 1989), Hawksworth *et al.* (*Dictionary of the Fungi* edn 8, 1995).

Vesiculomyces E. Hagstr. (1977) = Gloiothele fide Ginns & Freeman (*Biblthca Mycol.* **157**, 1994).

Vestergrenia (Sacc. & Syd.) Died. (1913) = Phomopsis (Sacc.) Bubák. fide Sutton (*Mycol. Pap.* **141**, 1977).

Vestergrenia Rehm (1901), ? Dothideaceae. 10, widespread. See von Arx & Müller (*Beitr. Kryptfl. Schweiz* **11** no. 1, 1954), Pawar & Kapoor (*Indian Phytopath.* **40**: 435, 1988; Indian spp.).

Vestergrenopsis Gyeln. (1940), Placynthiaceae (L). 2, widespread. See Henssen (*CJB* **41**: 1359, 1963), Jørgensen (*J. Jap. Bot.* **76**: 50, 2001; Japan).

Vestigium Piroz. & Shoemaker (1972), anamorphic *Pezizomycotina*, Cac.1bH.1. 1, USA. See Pirozynski & Shoemaker (*CJB* **50**: 1163, 1972).

Veterinary mycology, see Medical mycology.

Vezdaea Tscherm.-Woess & Poelt (1976), Vezdaeaceae (±L). 11, Europe. See Poelt & Döbbeler (*Bot. Jb.* **96**: 328, 1975), Giralt *et al.* (*Herzogia* **9**: 715, 1993; key 10 spp.), Scheidegger (*Cryptog. Bot.* **5**: 163, 1995; reproductive strategy).

Vezdaeaceae Poelt & Vězda ex J.C. David & D. Hawksw. (1991), Lecanorales (±L). 1 gen., 11 spp.
Lit.: Coppins (*Lichenologist* **19**: 167, 1987), Hawksworth (*J. Linn. Soc.* Bot. **96**: 3, 1988), Giralt *et al.* (*Herzogia* **9**: 715, 1993), Scheidegger (*Cryptog. bot.* **5**: 163, 1995).

viable, living; able to make growth.

Vialaea Sacc. (1896), Vialaeaceae. 2 (twigs etc.), widespread. See Redlin (*Sydowia* **41**: 296, 1989), Cannon (*MR* **99**: 367, 1995).

Vialaeaceae P.F. Cannon (1995), Sordariomycetes (inc. sed.). 1 gen. (+ 1 syn.), 2 spp.
Lit.: Redlin (*Sydowia* **41**: 296, 1989), Cannon (*MR* **99**: 367, 1995).

Vialina Curzi (1935) = Phoma Sacc. fide Sutton (*Mycol. Pap.* **141**, 1977).

Vibrissea Fr. (1822), Vibrisseaceae. Anamorph *Anavirga*. *c.* 30 (aquatic or semiaquatic), widespread. See Sanchéz & Korf (*Mycol.* **58**: 733, 1966; sects.), Sanchéz (*J. Agric. Univ. Puerto Rico* **51**: 79, 1967), Beaton & Weste (*TBMS* **69**: 323, 1977; Australia), Sherwood (*Mycotaxon* **5**: 1, 1977; posn), Korf (*Mycosystema* **3**: 19, 1991; concept), Iturriaga (*Mycotaxon* **54**: 1, 1995), Iturriaga (*Mycotaxon* **61**: 215, 1997), Wang *et al.* (*Mycol.* **98**: 1065, 2006; phylogeny, polyphyly), Wang *et al.* (*Mol. Phylogen. Evol.* **41**: 295, 2006; phylogeny).

Vibrisseaceae Korf (1990), Helotiales. 6 gen. (+ 5 syn.), 59 spp.
Lit.: Baral (*Z. Mykol.* **53**: 119, 1987), Hamad & Webster (*Sydowia* **40**: 60, 1987), Kohn (*Mem. N. Y. bot. Gdn* **49**: 112, 1989), Korf (*Mycosystema* **3**: 19, 1991; concept), Iturriaga (*Mycotaxon* **61**: 215, 1997), Wang *et al.* (*Am. J. Bot.* **92**: 1565, 2005), Wang *et al.* (*Mycol.* **98**: 1065, 2006; phylogeny), Wang *et al.* (*Mol. Phylogen. Evol.* **41**: 295, 2006).

victorin, a toxin from *Drechslera victoriae* which induces symptoms of leaf blight of oats.

victotoxinine, a constituent of victorin (Pringle & Brown, *Nature* **181**: 1205, 1958, *Phytopath.* **50**: 324, 1960).

Viégas (Ahmés Pinto; 1906-1986; Brazil). Born in Piracicaba, São Paulo; graduated in agronomy, ESALQ-Piracicaba (1932); PhD, Cornell University, New York, USA (1938). Made important contributions to the study of fungi, particularly microfungi, from the Atlantic forests of Brazil, incl. 12 gen. nov.; also linguist. *Publs. Indice dos Fungos da América do Sul* (1961); *Dicionário de Fitopatologia e Micologia* (1979); also see papers in *Bragantia* **3-19** (1943-1960). *Biogs, obits etc.* Anon. (*Summa Phytopathologica* **13**: iv, 1987); Grummann (1974: 772).

Viegasella Inácio & P.F. Cannon (2003), Parmulariaceae. 1, Brazil. See Inácio & Cannon (*MR* **107**: 82, 2003).

Viegasia Bat. (1951), Asterinaceae. 4, America (tropical); Africa. See Batista (*Bol. Secr. Agric. (Pernambuco)* **18**: 32, 1951).

Viennotidea, see *Viennotidia*.

Viennotidia Negru & Verona ex Rogerson (1970) = Sphaeronaemella fide Hawksworth & Cannon (*J. Linn. Soc.* Bot. **84**: 115, 1982; key), Hutchinson & Reid (*N.Z. Jl Bot.* **26**: 63, 1988), Dissing *in* Hawksworth (Ed.) (*Ascomycete Systematics. Problems and Perspectives in the Nineties* NATO ASI Series vol. **269 269**: 375, 1994; status), Hausner & Reid (*CJB* **82**: 752, 2004; phylogeny).

Viennotiella Negru (1964) ? = Heteropatella fide Sutton (*Mycol. Pap.* **141**, 1977).

villi (sing. **villus**), long soft hairs.

villose (**villous**), covered with villi, which are not matted; cf. tomentose.

Vinculum R.Y. Roy, R.S. Dwivedi & P.K. Khanna (1965) = Barnettella fide Rao & Rao (*Indian Phytopath.* **26**: 233, 1973).

vinescent, turning wine-red.

violet root rot, of a number of plants (*Helicobasidium purpureum*).

virescent, turning green.

Virgaria Nees (1816), anamorphic *Ascovirgaria*, Hso.0eP.10. 1, widespread. See Rogers & Ju (*CJB* **80**: 478, 2002; teleomorph).

Virgariella S. Hughes (1953), anamorphic *Pezizomycotina*, Hso.0eP.10. 11, widespread. See Sutton (*Sydowia* **44**: 321, 1993; related genera), Delgado-Rodríguez & Mena-Portales (*Cryptog. Mycol.* **24**: 153, 2003; Cuba).

Virgasporium Cooke (1875) = Cercospora Fresen. fide Saccardo (*Syll. fung.* **4**: 435, 1886).

virgate, banded; streaked.

Virgatospora Finley (1967), anamorphic *Pezizomycotina*, Hsy.≡ eP.?. 2, Panama. See Finley (*Mycol.* **59**: 538, 1967).

Virgella Darker (1967), Rhytismataceae. 1, N. America. See Darker (*CJB* **45**: 1419, 1967).

Viridiannula Etayo (2002), Pezizomycotina. 1.

viridin, an antibiotic from *Gliocladium virens*; antifungal (Brian *et al.*, *Ann. appl. Biol.* **33**: 190, 1946).

Viridispora Samuels & Rossman (1999), Nectriaceae. Anamorph *Penicillifer*. 4 (bark, wood etc.), widespread (tropical). See Rehner & Samuels (*CJB* **73** Suppl. 1: S816, 1995; phylogeny), Rossman *et al.* (*Stud. Mycol.* **42**: 248 pp., 1999), Rossman *et al.* (*Mycol.* **93**: 100, 2001; phylogeny), Samuels *et al.* (*CBS Diversity Ser.* **4**, 2006; USA).

virose (1) poisonous; (2) having a strong and unpleasant smell.

virulence, the degree or measure of pathogenicity; **virulent**, strongly pathogenic.

Viruses in fungi (mycoviruses). The text of this entry is unrevised and remains the same as for the ninth edition. Many fungi are infected by viruses which typically have double-stranded RNA genomes but 'mycoviruses' with DNA or single-stranded RNA genomes are also known. Those viruses so far classified are in the families *Barnaviridae* (gen. *Barnavirus*) and *Totiviridae* (gen. *Totivirus*) or the genus *Rhizidiovirus*. Unclassified viruses are known which infect fungi in the genera *Agaricus*, *Allomyces*, *Aspergillus*, *Colletotrichum*, *Gaeumannomyces*,

Helminthosporium, *Lentinus* and *Periconia*.

Most viruses of fungi induce only inapparent effects. However some notable effects are hypovirulence of chestnut blight fungus (see *Hypovirus*) and die-back of cultivated mushrooms.

Some fungi act as vectors in the transmission of plant viruses from plant to plant (see vectors).

The naming of viruses has undergone extensive revision in recent years, rendering much of the early literature difficult to interpret. The main viruses infecting fungi (see Murphy *et al.*, 1995) are:

Barnavirus (fam. *Barnaviridae*): bacilliform virions containing a single molecule of positive-sense single-stranded RNA, *c.* 4.4 kb in size; infecting *Agaricus* spp.

Chrysovirus (fam. *Partitiviridae*): isometric virions containing 1 of 3 molecules of linear 2-stranded RNA about 3 kbp in size; infecting *Penicillium* and probably *Helminthosporium*.

Hypovirus (fam. *Hypoviridae*): genomes of linear 2-stranded RNA of 10 to 13 kbp in size, no virions are formed by infected cells contain lipid vesicles which contain the genome RNA; infecting *Cryphonectria parasitica* and causing hypovirulence.

Partitivirus (fam. *Partitiviridae*): isometric particles and genomes which comprise 2 molecules of linear 2-stranded RNA, 1.4 to 2.2 kpb in size; infecting *Agaricus*, *Aspergillus*, *Gaeumannomyces*, *Penicillium*, *Rhizoctonia* and probably *Diplocarpon* and *Phialophora*; the infections are latent.

Rhizidiovirus (fam. uncertain): isometric virions which contain a single molecule of *c.* 25 kbp linear 2-stranded DNA; infecting *Rhizidiomyces*.

Totivirus (fam. *Totiviridae*): isometric virions which contain a single molecule of linear 2-stranded RNA, 4.6 to 6.7 kbp in size; infecting *Helminthosporium*, *Saccharomyces*, *Ustilago* and probably *Aspergillus*, *Gaeumannomyces* and *Mycogone*; the infections are often latent.

Lit.: Adams (*Ann. appl. biol.* **118**: 479, 1991; transmission plant viruses by fungi), Buck (Ed.) (*Fungal virology*, 1986), Cooper & Asher (Eds) (*Viruses with fungal vectors*, 1985), Hillings & Stone (*Ann. Rev. Phytopath.* **9**: 93, 1971), Koltin & Leibowitz (Eds) (*Viruses of fungi and simple procaryotes*, 1988), Lemke (*in* Molitoris *et al.* (Eds), *Fungal viruses*: 2, 1979; coevolution), Murphy *et al.* (*Virus taxonomy - The classification and nomenclature of viruses* [*Archives of Virology, Suppl.* **10**], 1995; taxonomy).

Vischia C.W. Dodge (1971) = Coccocarpia fide Arvidsson (*Op. Bot.* **67**, 1983).

viscid, slimy, sticky, glutinous, lubricous, mucilaginous, viscous. Cf. gelatinous, ixo-.

Viscipellis (Fr.) Quél. (1886) ≡ Suillus Gray.

Viscomacula R. Sprague (1951), anamorphic *Pezizomycotina*, Hso.0eH.?. 1, USA. See Sprague (*Mycol.* **42**: 758, 1950).

Visculus Earle (1909) = Pholiota fide Singer (*Agaric. mod. Tax.*, 1951).

Vismaya V.V. Sarma & K.D. Hyde (2001), Diaporthales. 1, Hong Kong. See Sarma & Hyde (*Nova Hedwigia* **73**: 247, 2001).

Visvesvaria N.J. Pieniazek, T.J. Kurtti, A.J. da Silva, S.B. Slemenda, H. Moura, I.S. Moura, F.J. Bornay-Llinares, D.A. Schwartz & R.A. Wirtz (1997), Microsporidia. 2.

Vitalia Cif. & Bat. (1962) = Dennisiella fide Hughes (*Mycol.* **68**: 693, 1976).

viteline, yellow like egg yolk.

Vitreostroma P.F. Cannon (1991), Phyllachoraceae. 1 (living legume leaves), montane paleotropics. See Müller (*Trans. Bot. Soc. Edinb., Suppl.*: 69, 1986; as *Diachora*), Cannon (*Mycol. Pap.* **163**, 1991).

Vittadinion Zobel (1854) = Tuber fide Fischer (*Nat. Pflanzenfam.* **5b**: viii, 1938).

Vittadinula (Sacc.) Clem. & Shear (1931) ≡ Sphaerodes.

Vittaforma Silveira & E.U. Canning (1994), Microsporidia. 1.

Vittalia Gawas & Bhat (2007), Pezizomycotina. 1, India. See Gawas & Bhatt (*Mycotaxon* **100**: 295, 2007).

vittate, having longitudinal lines, bands, or ridges.

Vittatispora P. Chaudhary, J. Campb., D. Hawksw. & K.N. Sastry (2006), Ceratostomataceae. 1, India. See Chaudhary *et al.* (*Mycol.* **98**: 461, 2006).

Vivantia J.D. Rogers, Y.M. Ju & Cand. (1996), Xylariaceae. Anamorph *Nodulisporium*-like. 1 (from wood), Guadeloupe. See Rogers *et al.* (*MR* **100**: 669, 1996).

Vivianella (Sacc.) Sacc. (1898) = Lophiostoma fide Hawksworth *et al.* (*Dictionary of the Fungi* edn 8, 1995).

vivotoxin, see toxin.

Vizella Sacc. (1883), Vizellaceae. Anamorph *Manginula*. 10, widespread (tropical). See Petrak (*Sydowia* **8**: 294, 1954), Batista & Ciferri (*Beih. Sydowia* **1**: 325, 1957), Swart (*TBMS* **57**: 455, 1971), Selkirk (*Proc. Linn. Soc. N. S. W.* **97**: 141, 1972; fossil taxa), Johnston (*N.Z. Jl Bot.* **38**: 629, 2000; New Zealand), Cunnington (*Mycotaxon* **93**: 135, 2005; Australia).

Vizellaceae H.J. Swart (1971), ? Dothideomycetes (inc. sed.). 3 gen. (+ 10 syn.), 12 spp.

Lit.: Swart (*TBMS* **57**: 455, 1971), Sivanesan & Sutton (*TBMS* **85**: 239, 1985), Farr (*Mycol.* **79**: 97, 1987), Gadgil (*N.Z. Jl For. Sci.* **25**: 107, 1995), Taylor & Crous (*IMI Descr. Fungi Bact.* **135**: [26] pp., 1999), Cannon (*IMI Descr. Fungi Bact.* **141**: [22] pp., 1999).

Vizellopsis Bat., J.L. Bezerra & T.T. Barros (1969), Dothideomycetes. 1, New Caledonia. See Batista *et al.* (*Publções Inst. Micol. Recife* **637**: 5, 1969).

Vladracula P.F. Cannon, Minter & Kamal (1986), Rhytismataceae. 1, Asia. See Cannon & Minter (*Mycol. Pap.* **155**, 1986), Gupta & Fotedar (*Journal of Mycology and Plant Pathology* **34**: 158, 2004).

Vleugelia J. Reid & C. Booth (1969), ? Diaporthales. 1, Sweden. See Reid & Booth (*CJB* **47**: 1057, 1969).

Voeltzkowiella Henn. (1908) ? = Bulgaria fide Korf (*Mycol.* **49**: 102, 1957).

Voglinoana Kuntze (1891) nom. dub., anamorphic *Pezizomycotina*. See Lindau (*Rabenh. Krypt.-Fl.* **1**: 673, 1907; sub *Cystophora*), Hughes (*CJB* **36**: 824, 1958).

Volkartia Maire (1909), Protomycetaceae. 1, Europe. See Reddy & Kramer (*Mycotaxon* **3**: 1, 1975).

Volucrispora Haskins (1958), anamorphic *Pezizomycotina*, Hso.1bH.10. 3, widespread. See Petersen (*Bull. Torrey bot. Club* **89**: 287, 1962), Hawksworth (*TBMS* **67**: 51, 1976), Roldan (*Mycotaxon* **42**: 297, 1991).

Volutella Fr. (1832) nom. cons., anamorphic *Pseudonectria*, Hsp.0eH.15. 20, widespread. See Samuels *et al.* (*CBS Diversity Ser.* **4**, 2006).

Volutellaceae Nann. (1934) = Nectriaceae.

Volutellaria Sacc. (1886), anamorphic *Pezizomycotina*, Hsp.0eH.?. 1, N. America.
Volutellis Clem. & Shear (1931), anamorphic *Pezizomycotina*, Hso.0eH.?. 1, Congo.
Volutellopsis Speg. (1910), anamorphic *Pezizomycotina*, Ccu.≡ eH.?. 1, S. America.
Volutellopsis Torrend (1913) [non *Volutellopsis* Speg. 1910] ≡ Volutellis.
Volutellospora Thirum. & P.N. Mathur (1965) = Chaetomella fide Petrak (*Sydowia* **18**: 373, 1965).
volutin, a reserve material of fungi, esp. yeasts, seen as electron-dense granules; metachromatic polymetaphosphate material. See Nagel (*Bot. Rev.* **14**: 174, 1948).
Volutina Penz. & Sacc. (1901) ? = Volutella.
Volva Adans. (1763) = Volvariella fide Donk (*Beih. Nova Hedwigia* **5**: 297, 1962) The name *Volva* has not yet been proposed for rejection against *Volvariella*.
volva (of agarics and gasteromycetes), the cup-like lower part of the universal veil round the base of the mature stipe or receptacle (Fig. 4E); sometimes = universal veil, which is the preferred usage, fide Bas (*Persoonia* **5**: 304, 1969), who should be consulted for details of volva types and terminology.
Volvaria (Fr.) P. Kumm. (1871) ≡ Volvariopsis.
Volvaria DC. (1805) = Gyalecta Ach. fide Hawksworth *et al.* (*Dictionary of the Fungi* edn 8, 1995).
Volvariella Speg. (1899), Pluteaceae. *c.* 50, widespread. *V. volvacea* and *V. diplasia*, the straw (or 'paddy straw') mushrooms (q.v.) are edible. See Shaffer (*Mycol.* **49**: 545, 1957; N. Am.), Orton (*Bull. mens. Soc. linn. Lyon* Num. Spéc. **43**: 313, 1974; Europ.), Heinemann (*Fl. Illustr. Champ. Afr. Centr.* **4**: 75, 1975; C. Afr.), Pegler (*Kew Bull.* Addit. Ser. **6**, 1977; E. Afr.), Heinemann (*Fl. Illustr. Champ. Afr. Centr.* **6**: 119, 1978), Orton (*British Fungus Flora* **4**: 61, 1986), Desjardin & Hemmes (*Harvard Pap. Bot.* **6**: 85, 2001; Hawaii), Seok *et al.* (*Mycobiology* **30**: 183, 2002; Korea).
Volvariopsis Murrill (1911) = Volvariella fide Singer (*Agaric. mod. Tax.*, 1951).
Volvarius Roussel (1806) = Volvariella. The name Volvarius has not yet been proposed for rejection against Volvariella. fide Donk (*Beih. Nova Hedwigia* **5**, 1962).
Volvella E.-J. Gilbert & Beeli (1941) = Amanita Pers. fide Pegler (*in litt.*).
Volvigerum (E. Horak & M.M. Moser) R. Heim (1966) = Cortinarius. Basidioma gasteroid. fide Peintner *et al.* (*Mycotaxon* **81**: 177, 2002; basidioma gasteroid).
Volvoamanita (Beck) E. Horak (1968) = Amanita Pers. fide Kuyper (*in litt.*).
Volvoboletus Henn. (1898) = Amanita Pers. fide Kuyper (*in litt.*) See, Ulbrich (*Ber. dt. bot. Ges.* **57**: 389, 1939; status).
Volvocisporiaceae Begerow, R. Bauer & Oberw. (2001), Microstromatales. 1 gen., 1 spp.
Lit.: Begerow *et al.* (*MR* **105**: 809, 2001), Beer *et al.* (*Stud. Mycol.* **55**: 289, 2006).
Volvocisporium Begerow, R. Bauer & Oberw. (2001), Volvocisporiaceae. 1 (on leaves of *Triumfetta rhomboidea*), India. See Begerow *et al.* (*MR* **105**: 809, 2001).
Volvolepiota Singer (1959) = Macrolepiota fide Vellinga & Yang (*Mycotaxon* **85**: 183, 2003).
Volvopolyporus Lloyd ex Sacc. & Trotter (1912) = Coltricia fide Pilát (*Atlas Champ. Eur.* III *Polyporaceae* **2**, 1942).
Volvycium Raf. (1808) nom. dub., Agaricomycetes. ? 'gasteromycetes'.
Vonarxella Bat., J.L. Bezerra & Peres (1965), Saccardiaceae. 1 or 2, Brazil. See Batista *et al.* (*Riv. Patol. veg., Pavia* Sér. 4 **1**: 61, 1965).
Vonarxia Bat. (1960), anamorphic *Pezizomycotina*, Cpt.0fH.?. 1, Brazil. See Batista (*Publções Inst. Micol. Recife* **283**: 5, 1960), Aa & von Arx (*Persoonia* **13**: 127, 1986).
Vorarlbergia Grummann (1969) = Epigloea fide Döbbeler (*Beih. Nova Hedwigia* **79**: 203, 1984).
Vossia Thüm. (1879) [non *Vossia* Wall. & Griff. 1836, *Gramineae*] ≡ Neovossia.
Vouauxiella Petr. & Syd. (1927), anamorphic *Pezizomycotina*, Cpd.0-1eP.38. 5 (on lichens), Europe; S. America. See Sutton (*The Coelomycetes*, 1980), Hawksworth (*Bull. Br. Mus. nat. hist.* Bot. **9**: 64, 1981).
Vouauxiomyces Dyko & D. Hawksw. (1979), anamorphic *Abrothallus*, Cpd.0eP.19. 5 (on *Lecanorales*), widespread. See Hawksworth (*Bull. Br. Mus. nat. hist.* Bot. **9**: 67, 1981), Kondratyuk (*Muelleria* **9**: 93, 1996; Australasia), Suija (*Ann. bot. fenn.* **43**: 193, 2006).
Vrikshopama D. Rao & P. Rag. Rao (1964) ? = Dematophora fide Sutton (*in litt.*).
Vuillemin (Paul; 1861-1932; France). Student, University of Nancy (*c.* 1884-1889); Dr Sci., Sorbonne (1892); Professor in the Faculty of Medicine, Nancy (1895-1932). Noted for his innovative proposal to classify anamorphic fungi by their conidial development (he introduced the terms phialide and aleuriospore); these ideas strongly influenced Mason (q.v.), Hughes and many other mycologists (see Conidial nomenclature). *Publs. Les bases actuelles de la systématique en mycologie* (1907); Matériaux pour une classification rationelle des Fungi Imperfecti. *Comptes Rendu Hebdomadaire des Séances de l'Académie des Sciences* (1910); *Beauveria*, nouveau genre des Verticillariées. *Bulletin de la Société Botanique de France* (1912); *Les Champignons; Essai de Classification* (1912); also various papers on diseases of man, animals and plants. *Biogs, obits etc.* Grummann (1974: 356); Joyeux (*Mycopathologia et Mycologia Applicata* **3**: 64, 1941) [bibliography]; Potron (*BSMF* **67**: 42, 1951) [portrait]; Stafleu & Cowan (*TL-2* **6**: 801, 1986).
Vuilleminia Maire (1902), Corticiaceae. 10, widespread. See Boidin *et al.* (*BSMF* **110**: 91, 1994; key).
Vulpicida Mattsson & M.J. Lai (1993), Parmeliaceae (L). 6, widespread (Arctic to northern temperate). See Mattsson (*Op. Bot.* **119**, 1993; key), Thell & Miao (*Ann. bot. fenn.* **35**: 275, 1998; phylogeny), Wedin *et al.* (*MR* **103**: 1152, 1999; phylogeny), Thell *et al.* (*Mycol. Progr.* **1**: 335, 2002; phylogeny), Thell *et al.* (*Mycol. Progr.* **3**: 297, 2004; phylogeny), Randlane & Saag (*Folia cryptog. Estonica* **41**: 89, 2005; biogeography), Miądlikowska *et al.* (*Mycol.* **98**: 1088, 2006; phylogeny), Binder & Ellis (*Lichenologist* **40**: 63, 2008; conservation).
W, Naturhistorisches Museum, Wien (Vienna, Austria); founded 1748; formerly known as the Hofnaturalienkabinett und Naturhistorisches Hofmuseum; a state institution.
Wadeana Coppins & P. James (1978), Catillariaceae (L). 2, Europe. See Hertel & Rambold (*Biblthca

Lichenol. **38**: 145, 1990; posn), Lumbsch (*J. Hattori bot. Lab.* **83**: 1, 1997).

Wageria F. Stevens & Dalbey (1918) = Balladyna fide Sivanesan (*Mycol. Pap.* **146**, 1981).

Waihonghopes Yanna & K.D. Hyde (2002), anamorphic *Pezizomycotina*, H?.?.?. 1, Queensland. See Yanna & Hyde (*Aust. Syst. Bot.* **15**: 763, 2002).

Wainio, see Vainio.

Wainioa Nieuwl. (1916) ≡ Pilocarpon.

Wainiocora Tomas. (1950) = Dictyonema C. Agardh ex Kunth fide Parmasto (*Nova Hedwigia* **29**: 99, 1978).

Waitea Warcup & P.H.B. Talbot (1962), Corticiaceae. Anamorph *Chrysorhiza*. 2, Australia; Switzerland. See Roberts (*Rhizoctonia-forming fungi*, 1999), Toda *et al.* (*Pl. Dis.* **89**: 536, 2005; bentgrass disease), Larsson (*MR* **111**: 1040, 2007; phylogeny).

Wakefield (Elsie Maud; 1886-1972; England). Student, Somerville College, Oxford (ca 1904-1908, making her one of the earlier women graduates of this university); Assistant Cryptogamic Botanist (1910-1915) working with Massee (q.v.), then Cryptogamic Botanist (*c.* 1915-1951), Royal Botanic Gardens, Kew [the first graduate mycologist at Kew and a pioneer woman in the UK scientific civil service]. With Cotton she monographed *Clavaria* but her main interest was resupinate basidiomycetes. She was a skilful painter of watercolours and some of her publications included her artwork. She was Secretary of the British Mycological Society for 17 years and President in 1928, and was regarded as influential. *Publs.* Edible and poisonous fungi [6th edn] *Bulletin of the Ministry of Agriculture and Fisheries* (1945); *Observer's Book of Common Fungi* (1954); (with R.W.G. Dennis, q.v.) *Common British fungi* (1950). *Biogs, obits etc.* Anon. (*TBMS* **49**: 355, 1966) [portrait, 80th birthday]; Blackwell (*TBMS* **60**: 167, 1973) [bibliography, portrait]; Stafleu & Cowan (*TL-2* **7**: 24).

Wakefieldia Corner & Hawker (1953), Boletaceae. 2, Europe, Asia. See Corner & Hawker (*TBMS* **36**: 130, 1953), Gori (*Boll. Gruppo Micol. 'G. Bresadola'* **40**: 239, 1997).

Wakefieldia G. Arnaud (1954) = Anaphysmene fide Sutton (*TBMS* **59**: 285, 1972).

Wakefieldiomyces Kobayasi (1981) = Podocrella fide Rossman *et al.* (*Stud. Mycol.* **42**: 248 pp., 1999), Chaverri *et al.* (*Mycol.* **97**: 433, 2005).

Waksmania H. Lechev. & M.P. Lechev. (1957), Actinobacteria. q.v.

Waldemaria Bat., H. Maia & Cavalc. (1960) [non *Waldemaria* Klotzsch 1862, *Ericaceae*] = Arachniotus fide Orr & Kuehn (*Mycol.* **63**: 191, 1971), Currah (*Mycotaxon* **24**: 1, 1985).

Walkeromyces Thaung (1976) = Mycovellosiella fide Deighton (*Mycol. Pap.* **144**, 1979).

wall building, descriptive of hyphal growth in which cell wall material is produced by certain ultrastructural secretory bodies in the cytoplasm. Three types of wall building may be distinguished (Fig. 26): **apical -** in which the bodies are concentrated at the hyphal tip, producing new wall by distal growth, forming a cylindrical hypha in which the youngest wall material is at the tip; **ring -**, in which the bodies are concentrated adjacent to the cell wall at some point below the tip, in the shape of an imaginary ring, producing new wall by proximal growth, forming a cylindrical hypha in which the youngest wall material is always at the base; **diffuse -**, in which the bodies occur throughout the cytoplasm at a low concentration, producing lateral growth (i.e. swelling of the cylindrical hypha) by alteration of pre-existing wall.

The terms assume special significance in conidial development, where they have been used to clarify the concepts of thallic and blastic. Apical wall building occurs in *Geniculosporium*, *Cladosporium* and *Scopulariopsis*, and 'phialides' where conidia are produced in gummy masses (e.g. *Trichoderma*) or false chains (e.g. *Mariannaea*). Ring wall building occurs in 'phialides' with conidia in true chains (e.g. *Penicillium*, *Chalara*), in so-called meristem arthrospores (e.g. *Wallemia*) and in conidiogenous cells of basauxic fungi (e.g. *Arthrinium*). Diffuse wall building occurs simultaneously with or shortly after apical or ring wall building in most of the preceding examples, but its occurrence is much delayed or even absent in thallic development (e.g. *Geotrichum*). Wall building is a preferable term to meristem which implies growth by cell division rather than within a single cell. See Minter *et al.* (*TBMS* **79**: 75, 1982, **80**: 39, 1983, **81**: 109, 1983). Also see Anamorphic fungi.

Wallemia Johan-Olsen (1887), anamorphic *Wallemiaceae*, Hso.0eH.?38. 2, widespread. *W. sebi* is osmophilic. See Cole & Samson (*Patterns of development in conidial fungi*: 103, 1979; conidiogenesis), Moore (*Antonie van Leeuwenhoek* **52**: 183, 1986; interpreted as a teleomycete), Moore *in* Sneh *et al.* (Eds) (*Rhizoctonia species* Taxonomy, Molecular Biology, Ecology, Pathology and Disease Control: 13, 1996; as fam. *Wallemiaceae*), Zalar *et al.* (*Antonie van Leeuwenhoek* **87**: 311, 2005; taxonomy and phylogeny).

Wallemiaceae R.T. Moore (1996), Wallemiales. 1 gen. (+ 2 syn.), 2 spp.

Lit.: Hashmi & Morgan-Jones (*CJB* **51**: 1669, 1973), Moore (*Antonie van Leeuwenhoek* **52**: 183, 1986), Moore *in* Sneh *et al.* (Eds) (*Rhizoctonia Species* Taxonomy, Molecular Biology, Ecology, Pathology and Disease Control: 25, 1996), Zalar *et al.* (*Antonie van Leeuwenhoek* **87**: 311, 2005).

Wallemiales Zalar, de Hoog & Schroers (2005). Wallemiomycetes. 1 fam., 1 gen., 2 spp. Fam.:

Wallemiaceae

For *Lit.* see under fam.

Wallemiomycetes Zalar, de Hoog & Schroers (2005), Basidiomycota. 1 ord., 1 fam., 1 gen., 2 spp. Ord.:

Wallemiales

For *Lit.* see fam.

Wallrothiella Sacc. (1882), ? Coniochaetales. Anamorph *Gliomastix*. 1, widespread. See Eriksson & Hawksworth (*SA* **10**: 59, 1991), Samuels & Barr (*CJB* **75**: 2165, 1998), Réblová & Winka (*Mycol.* **92**: 939, 2000), Réblová & Winka (*Mycol.* **93**: 478, 2001), Réblová & Seifert (*Mycol.* **96**: 343, 2004), García *et al.* (*MR* **110**: 1271, 2006).

Waltiozyma H.B. Mull. & Kock (1986) = Lipomyces fide Kurtzman & Liu (*Curr. Microbiol.* **21**: 287, 1991).

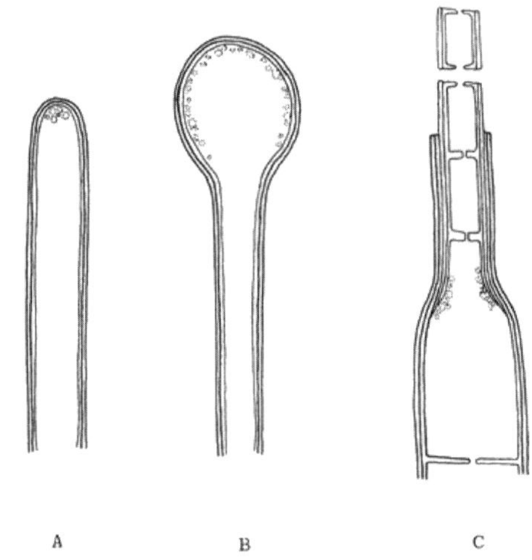

Fig. 26. Wall building in relation to conidiogenesis. A, apical; B, diffuse; C, ring. After Minter *et al.* (*TBMS* **80**: 39, 1983).

Waltomyces Y. Yamada & Nakase (1985) = Lipomyces fide Gouliamova *et al.* (*Antonie van Leeuwenhoek* **74**: 283, 1998), Smith *in* Kurtzman & Fell (Eds) (*Yeasts, a taxonomic study* 4th edn: 248, 1998), Kurtzman *et al.* (*FEMS Yeast Res.* **7**: 1027, 2007; phylogeny).

Waltonia Saho (1970), Helotiales. Anamorph *Chondropodium*-like. 1, Japan. See Saho (*TMSJ* **11**: 5, 1970).

Walzia Sorokīn (1871) ? = Penicillium Link fide Costantin (*Les mucedinées simples*: 201, 1888).

wandering lichens, lichens with an epigeic habit (e.g. *Parmelia afrorevoluta*; see Paulson & Hastings, *Knowledge* **37**: 319, 1914). Cf. manna.

Wangiella McGinnis (1977) = Exophiala fide Matsumoto *et al.* (*J. Med. Vet. Mycol.* **28**: 437, 1990), Hoog *et al.* (*Antonie van Leeuwenhoek* **65**: 143, 1994; life cycle), Hawksworth *et al.* (*Dictionary of the Fungi* edn 8, 1995), Hoog *et al.* (*J. Med. Vet. Mycol.* **33**: 355, 1995; synanamorph), Spatafora *et al.* (*J. Clin. Microbiol.* **33**: 1322, 1995; DNA), Haase *et al.* (*Stud. Mycol.* **43**: 80, 1999; phylogeny), McKemy *et al.* (*Mycol.* **91**: 200, 1999; emend. genus), Méndez *et al.* (*Revta Iberoamer. Micol.* **16**: 114, 1999; Argentina), Rogers *et al.* (*Stud. Mycol.* **43**: 122, 1999; phylogeny).

Warcupia Paden & J.V. Cameron (1972), Pyronemataceae. 1, Canada. See Paden & Cameron (*CJB* **50**: 999, 1972), Perry *et al.* (*MR* **111**: 549, 2007; phylogeny).

Warcupiella Subram. (1972), Trichocomaceae. Anamorph *Aspergillus*. 1, Brunei. See von Arx (*Mycotaxon* **26**: 119, 1986; synonym of *Hamigera*), Chang *et al.* (*J. gen. appl. Microbiol.* Tokyo **37**: 289, 1991; DNA), Tamura *et al.* (*Integration of Modern Taxonomic Methods for Penicillium and Aspergillus Classification*: 357, 2000; phylogeny).

Ward (Harry Marshall; 1854-1906; England). BA, Cambridge (1879); Government Cryptogamist, Ceylon (1880); Fellow, then Demonstrator, then Assistant Lecturer, University of Manchester (1981-1885); Chair of Botany, Royal Indian Engineering College (1885-1895); Chair of Botany, Cambridge (1895-1906); DSc (1892). Published classical studies on coffee rust which set new standards for tropical plant pathology; introduced ideas of de Bary (q.v.) to Britain; FRS (1888). *Publs.* Researches on the life history of *Hemileia vastatrix*, the fungus of 'coffee leaf disease'. *J. Linn. Soc. London* **19**: 299 (1882); *Timber and Some of its Diseases* (1894). *Biogs, obits etc.* Ayres (*Harry Marshall Ward and the Fungal Thread of Death*, 2005).

Wardia J.F. Hennen & M.M. Hennen (2003) nom. inval. = Hemileia fide Ritschel (*Biblthca Mycol.* **200**: 132 pp., 2005) Anamorph name for (II).

Wardina G. Arnaud (1918) = Asterolibertia fide Müller & von Arx (*Beitr. Kryptfl. Schweiz* **11** no. 2, 1962).

Wardinella Bat. & Peres (1960), anamorphic *Pezizomycotina*, Cpt.1eP.?. 1, Brazil. See Batista & Peres (*Publções Inst. Micol. Recife* **221**: 8, 1960).

Wardomyces F.T. Brooks & Hansf. (1923), anamorphic *Microascus*, Hso.0eP.13. 3, Europe; India. See Dickinson (*TBMS* **47**: 321, 1964), Gams (*Beih. Sydowia* **10**: 67, 1995), Issakainen *et al.* (*Medical Mycology* **41**: 31, 2003; phylogeny).

Wardomycopsis Udagawa & Furuya (1978), anamorphic *Microascus*, Hso.0eP.19. 4, Thailand. See Udagawa & Furuya (*Mycotaxon* **7**: 92, 1978).

Warkallisporonites Ramanujam & Rao (1979), Fossil Fungi. 1 (Miocene), India.

wart disease, of potatoes (*Synchytrium endobioticum*) See Curtis (*Phil. Trans. roy. Soc.* **B210**: 409, 1921; life history).

Wasson (R. Gordon; 1898-1986; USA). International banker, amateur mycologist and author. Wasson pioneered ethnomycology (q.v.), the study of traditions and traditional uses of fungi. He particularly investigated use of psychotropic fungi in religious rites in Greece, India and Mexico. *Publs.* (with Guercken) *Mushrooms, Russia and History* (1957); *Soma, Divine Mushroom of Immortality* (1968); *Maria Sabina and her Mazatec Mushroom Velada* (1976); (with Hoffmann & Ruck) *The Road to Eleusis, Unveiling the Secrets of the Mysteries* (1978). *Biogs, obits etc.* Anon. (*Mycologist* **1**: 135, 1987).

Waste utilization, see Biodegradation, Bioremediation, Biotechnology, Industrial mycology.

water activity, some fungi can grow at extremely reduced levels of water activity, e.g. *Xeromyces bisporus* at 0.75 a_w; see Hocking (*Microbiol. Sci.* **5**: 280, 1988; review).

water moulds, see Aquatic fungi.

Wawea Henssen & Kantvilas (1985), Arctomiaceae (L). 1, New Zealand; Tasmania. See Henssen & Kantvilas (*Lichenologist* **17**: 86, 1985), Miądlikowska & Lutzoni (*Am. J. Bot.* **91**: 449, 2004; phylogeny), Lumbsch *et al.* (*Lichenologist* **37**: 291, 2005; phylogeny).

Wawelia Namysl. (1908), Xylariaceae. 3 (coprophilous), Europe. *Chaenocarpus* is a possible earlier name. See Minter & Webster (*TBMS* **80**: 370, 1983), Lundqvist (*Persoonia* **14**: 417, 1992), Læssøe (*SA* **13**: 43, 1994; ? synonym of *Xylaria*), Webster *et al.* (*MR* **103**: 1604, 1999).

Waydora B. Sutton (1976), anamorphic *Pezizomycotina*, St.0eH.15. 1, widespread. See Sutton (*TBMS* **67**: 248, 1976).

Waynea Moberg (1990), Ramalinaceae (L). 3, widespread. See Moberg (*Lichenologist* **22**: 249, 1990), Roux & Clerc (*Bull. Soc. linn. Provence* **42**: 123, 1991; Europe), Roux *et al.* (*Biblthca Lichenol.* **58**: 383, 1995), Tretiach (*Nordic Jl Bot.* **18**: 721, 1998; Siberia), Ekman (*MR* **105**: 783, 2001; phylogeny), Llop (*Lichenologist* **38**: 519, 2006; Mediterranean).

WDCM. World Data Center on Microorganisms (National Institute of Genetics, Nishima, Japan; formerly University of Queensland), administered jointly by the World Federation of Culture Collections and MIRCEN, this centre maintains records of over 475 culture collections (including fungal collections) in over 60 countries; website (www.wfcc.info/datacenter.html), see also Genetic resource collections, MIRCEN.

Weathering. Physical, chemical and biological processes can bring about the weathering of rocks. All these factors ultimately result in fragmentation of the rock into smaller particles which can then contribute to the mineral fraction of soils. The biological weathering of rocks and their constituent minerals involves biogeochemical and biogeophysical weathering processes, the former bringing about a change in the chemical stability and composition of rocks and their minerals through the action of living organisms and their metabolic activities and the latter resulting in a mechanical disruption of the rocks and their constituent minerals (Silverman, *in* Trudinger & Swaine, *Studies in environmental science* **3**: 445, 1979).

One of the few detailed studies of life on rock surfaces was by Webley *et al.* (*J. Soil Sci.* **14**: 102, 1963) who recorded the numbers of fungi (as well as of bacteria, actinomycetes and other microorganisms) in the interior of porous weathered stones (none were found in unweathered stones). Fungi, notably *Trichoderma* sp. were found growing directly on sandstone (Williams & Rudolph, *Mycol.* **66**: 648, 1974). *Alternaria* sp., *Acremonium* sp. and various species of *Penicillium* were also noted by these authors. *Penicillium simplicissimum* has been isolated from the surface of weathered basalt (Silverman & Munoz, *Science* **169**: 985, 1970). The biodegradation of basalt by *P. simplicissimum* was studied by Mehta *et al.* (*Biotechnol. & Bioengin.* **21**: 875, 1979) who showed that organic acids were responsible for the decomposition processes although the organism was not isolated from the rock substratum. A range of fungi have been found from the interior of rock, mostly in deserts (including Antarctic dry deserts) but also from exposed rocks in other ecosystems, and even rocks of buildings and statues. These species, mostly black yeasts (see Yeasts: black) have been described as cryptoendolithic, and their biology has been extensively studied, particularly by Friedmann (q.v.), partly because of their importance in biodegradation but also, in the case of Antarctic fungi, as a model for possible life on Mars (Selbmann *et al.*, *Studies in Mycology* **51**: 1, 2005).

A high proportion of the organisms isolated by Webley *et al.* (1963) rendered silicates soluble when tested in pure culture and the authors (Henderson & Duff, *J. Soil Sci.* **14**: 236, 1963) concluded that fungi which produced citric and/or oxalic acid proved effective in decomposing certain natural silicates while an oxalic acid-producing strain also released metallic ions and silica from rocks and soils. Among the fungi studied were *Penicillium* spp. and species of *Trichoderma*, *Mucor* and *Spicaria*. Weathering of the edges of biotite flakes by *Aspergillus niger* was noted by Boyle & Voigt (*Science* **169**: 193, 1967) in pure culture studies and the activity was attributed to oxalic and citric acids. Cromack *et al.* (*Soil Biol. Biochem.* **11**: 463, 1979) have presented evidence for the accumulation of calcium oxalate in the fungal mats of the ectomycorrhizal fungus *Hysterangium crassum* associated with Douglas fir (*Pseudotsuga menziesii*) roots. This precipitation was brought about by oxalic acid secreted by the fungus and this was considered to be the reason for weathering of the igneous rock, andesite, in the root vicinity.

The weathering effects of crustose lichens on rock surfaces can manifest themselves as etch patterns on the rock minerals which can involve the transformation of certain minerals to siliceous relics and the formation of crystalline organic salts such as oxalates in the lichen thallus; the chemical composition of these oxalates is directly related to the substrate (Jones, *in* Galun (Ed.), *Handbook of lichenology* **3**: 109, 1988; Wilson & Jones, *in* Wilson (Ed.), *Residual deposits*, Sp. Publ. **11**: 5, 1983). See also Biodeterioration.

Websdanea Vánky (1997), Websdaneaceae. 1 (on *Restoniaceae*), Australia. See Vánky (*Mycol. Balcanica* **3**: 19, 2006).

Websdaneaceae Vánky (2001), Ustilaginales. 2 gen., 22 spp.

Lit.: Vánky (*Mycotaxon* **65**: 183, 1997), Piepenbring *et al.* (*Mycol.* **91**: 485, 1999), Vánky (*Fungal Diversity* **6**: 131, 2001), Vánky (*Aust. Syst. Bot.* **14**: 385, 2001), Vánky (*Mycol. Balcanica* **1**: 175, 2004).

Websteromyces W.A. Baker & Partridge (2000), anamorphic *Pezizomycotina*. 1, S. Africa. See Partridge *et al.* (*Mycotaxon* **74**: 488, 2000), Crous *et al.* (*MR* **110**: 264, 2006; key).

Weddellomyces D. Hawksw. (1986), Dacampiaceae. 8 (on lichens), Europe; Greenland. See Alstrup & Hawksworth (*Meddr Grønland* Biosc. **31**, 1990), Navarro-Rosiñes & Roux (*Mycotaxon* **53**: 161, 1995), Calatayud & Navarro-Rosinés (*Mycotaxon* **69**: 503, 1998), Navarro-Rosinés *et al.* (*C. r. Acad. Sc. Paris* sér. III, Sci. Vie **324**: 443, 2001; ultrastr.), Halici *et al.* (*Mycotaxon* **94**: 249, 2005; Turkey).

Weesea Höhn. (1920) nom. dub., Pezizomycotina. 1, Brazil. Type material is lost. See Rossman *et al.* (*Stud. Mycol.* **42**: 248 pp., 1999).

Wegea Aptroot & Tibell (1997), Arthoniales (L). 1, New Guinea. See Aptroot & Tibell (*Mycotaxon* **65**: 340, 1997), Grube (*Lichenologist* **33**: 387, 2001).

Wegelina Berl. (1900), Calosphaeriales. 1 (rotten wood and other fungi), widespread (temperate). See Barr (*Cryptog. Bryol.-Lichénol.* **19**: 169, 1998), Réblová *et al.* (*Stud. Mycol.* **50**: 533, 2004), Kummer *et al.* (*Z. Mykol.* **71**: 227, 2005), Mostert *et al.* (*Stud. Mycol.* **54**: 115 pp., 2006; key).

Wehmeyera J. Reid & C. Booth (1989), Diaporthales. 1, USA. See Barr (*Mycotaxon* **41**: 287, 1991).

Weinmannioscyphus Svrček (1977), Helotiaceae. 1, Europe. See Svrček (*Sb. nár. Mus. Praze* **40B**: 129, 1985).

Weinmannodora Fr. (1849) nom. dub., Fungi. See Petrak & Sydow (*Beih. Rep. spec. nov. regn. veg.* **42**, 1927).

Weinzettlia Velen. (1921) = Cortinarius fide Kuyper (*in litt.*).

Weiseria Doby & Saguez (1964), Microsporidia. 3.

Weissia Bat. & M.P. Herrera (1964), anamorphic *Pezizomycotina*, Cpt.1eH.?. 1, Brazil. See Batista & Herrera (*Anais Congr. Soc. Bot. Brasil* **13**: 469, 1962).

Weitenwebera Körb. (1863) = Chromatochlamys fide Mayrhofer & Poelt (*Herzogia* **7**: 13, 1985), Hawksworth *et al.* (*Dictionary of the Fungi* edn 8, 1995).

Weitenwebera Opiz (1857) [non *Weitenwebera* Opiz 1839, *Campanulaceae*] = Mycobilimbia fide Hafellner (*Beih. Nova Hedwigia* **79**: 241, 1984), Mayrhofer & Poelt (*Herzogia* **7**: 13, 1985).

Wentiomyces Koord. (1907), Pseudoperisporiaceae. *c.* 50, widespread (tropical). See Hansford (*Mycol. Pap.* **15**, 1946; sub *Dimeriella*), Bose & Müller (*Indian Phytopath.* **17**: 3, 1964), Farr (*Taxon* **14**: 18, 1965; nom. dub.), Roux *et al.* (*Mycotaxon* **50**: 459, 1994), Kondratyuk (*Muelleria* **9**: 93, 1996; Australasia).

Wenyingia Zheng Wang & Pfister (2001), Pyronemataceae. 1, Sichuan. See Wang & Pfister (*Mycotaxon* **79**: 397, 2001).

Weraroa Singer (1958), Strophariaceae. 8, New Zealand; S. America. Basidioma gasteroid. See Singer & Smith (*Bull. Torrey bot. Club* **85**: 324, 1958; key), Bougher & Lebel (*Aust. Syst. Bot.* **14**: 439, 2001).

Wernera Zschacke ex Werner (1934) = Thrombium fide Hawksworth *et al.* (*Dictionary of the Fungi* edn 8, 1995).

Wernerella Nav.-Ros., Cl. Roux & Giralt (1998), Dothideomycetes. 1, Morocco. See Navarro-Rosinés *et al.* (*Bull. Soc. linn. Provence* **47**: 167, 1996), Navarro-Rosinés *et al.* (*Bull. Soc. linn. Provence* **49**: 138, 1998).

Werneromyces Werner (1976) nom. inval. ≡ Wernera.

Westea H.J. Swart (1988), Dothideomycetes. 1, Australia. See Swart (*TBMS* **91**: 456, 1988).

Westerdijk (Johanna; 1883-1961; Netherlands). Student, University of Amsterdam, University of München and doctorate, University of Zürich (1906); Director, Phytopathological Laboratory 'Willie Commelin Scholten', Amsterdam (1906-1913); Director, Bureau pour la Distribution de Moisissures de l'Association Internationale des Botanistes (1907-1953) [this collection of 80 fungal species, formed in 1904, was the origin of and became the CBS (q.v.)]; collecting and travelling, Indonesia, Japan and the USA (1913-1915); Professor of Plant Pathology, Universities of Utrecht (1917-1930); Professor of Plant Pathology, University of Amsterdam (1930-1953). *Biogs, obits etc.* Beverwijk (*Mycopathologia et Mycologia Applicata* **17**: 359, 1962) [portrait]; Lohnis (*Johanna Westerdijk een Markante Persoonlijkheid*, 1963); Stafleu & Cowan (*TL-2* **7**: 210, 1988); Ten Houten (*Journal of General Microbiology* **32**: 1, 1963) [bibliography, portrait].

Westerdykella Stolk (1955), Sporormiaceae. 2 (from soil), Mozambique. See also *Pycnidiophora*. See Cejp & Milko (*Česká Mykol.* **18**: 82, 1964; key), von Arx & Storm (*Persoonia* **4**: 407, 1967), Ito & Nakagiri (*Mycoscience* **36**: 361, 1995), Lindemuth *et al.* (*MR* **105**: 1176, 2001), Kruys *et al.* (*MR* **110**: 527, 2006; phylogeny).

Weston (William Henry; 1890-1978; USA). Professor of Cryptogamic Botany, Harvard University (1928-1960). First President of the Mycological Society of America, and an influential teacher. *Biogs, obits etc.* Stafleu & Cowan (*TL-2* **7**: 217, 1988); Wilson (*Mycol.* **71**: 1103, 1980) [portrait].

wet bubble, a mushroom disease (*Mycogone perniciosa*); white mould. See under Mushroom cultivation.

Wettsteiniella Kuntze (1891) ≡ Arthrobotryum Ces.

Wettsteinina Höhn. (1907), Pleosporales. 29, Europe; N. America. See Barr (*Contr. Univ. Mich. Herb.* **9**: 523, 1972), Shoemaker & Babcock (*CJB* **65**: 373, 1987; key), Scheuer (*Mycotaxon* **54**: 173, 1995; nomencl.), Kodsueb *et al.* (*Mycol.* **98**: 571, 2006; phylogeny), Schoch *et al.* (*Mycol.* **98**: 1041, 2006; phylogeny).

Weufia Bhat & B. Sutton (1985), anamorphic *Pezizomycotina*, Hso.0bP.27. 1, Ethiopia. See Bhat & Sutton (*TBMS* **85**: 107, 1985).

WFCC, see Societies and organizations.

Whalleya J.D. Rogers, Y.M. Ju & F. San Martín (1997), Xylariaceae. 2, widespread (esp. tropical). See Rogers *et al.* (*Mycotaxon* **64**: 48, 1997), Triebel *et al.* (*Nova Hedwigia* **80**: 25, 2005; phylogeny), Ju *et al.* (*Mycol.* **99**: 612, 2007; phylogeny).

Whetstonia Lloyd (1906) = Dictyocephalos fide Long & Plunkett (*Mycol.* **32**: 637, 1940).

Whetzel (Herbert Hice; 1877-1944; USA). Student, Wabash College, Indiana (1897-1899); Postgraduate student (1902-1904) then Instructor (1904-1906); Assistant Professor of Botany (1906-1907) then Assistant Professor (1907) and later Professor of Plant Pathology, New York State College of Agriculture, Cornell University, New York. An influential plant pathologist and mycologist (especially expert in the *Sclerotiniaceae*), who in 1907 founded the first department of plant pathology in the USA. Contributed significantly to mycological explorations of the West Indies, with Chardón (q.v.), Seaver (q.v.) and others. *Publs. An Outline of the History of Plant Pathology* (1918); (with Hesler, Gregory & Rankin) *Laboratory Outlines in Plant Pathology* (1925) [revised edn]; also more than 200 other papers and books. *Biogs, obits etc.* Fitzpatrick (*Mycol.* **37**: 393, 1945, portraits); Barrus & Stakman (*Phytopathology* **35**: 659, 1945; bibliography); Stafleu & Cowan (*TL-2* **7**: 245, 1988).

Whetzelia Chardón & Toro (1934) = Vestergrenia Rehm fide von Arx & Müller (*Beitr. Kryptfl. Schweiz* **11** no. 1, 1954).

Whetzelia Zundel (1945) ≡ Ustacystis.

Whetzelinia Korf & Dumont (1972) nom. rej. prop. = Sclerotinia fide Kohn (*Mycotaxon* **9**: 365, 1979).

Whetzeliomyces Viégas (1945) = Anhellia fide von Arx (*Persoonia* **2**: 421, 1963).

whisky (**whiskey**, Ir, US), see spirits.

white blister (white 'rust'), a disease of plants caused by *Albugo*, esp. *A. candida* on crucifers.

white jelly fungus, basidioma of the edible *Tremella fuciformis*.

white piedraia, infection of hair shaft by *Trichophyton*.

whiteheads, a cereal disease (*Gaeumannomyces graminis*).

Whitfordia Murrill (1908) [non *Whitfordia* Elmer 1910, *Leguminosae*] = Amauroderma Murrill fide Pegler (*The polypores [Bull. BMS Suppl.]*, 1973).

Wickerhamia Soneda (1960), Saccharomycetales. 1, Japan. See Phaff & Miller *in* Kurtzman & Fell (Eds) (*Yeasts, a taxonomic study* 4th edn: 409, 1998), Suh *et al.* (*Mycol.* **98**: 1006, 2006).

Wickerhamiella Van der Walt (1973), Trichomonascaceae. 5, widespread. Perhaps allied to the *Dipodascaceae*. See Lachance *et al.* (*Can. J. Microbiol.* **46**: 1145, 2000; biogeography), Suh *et al.* (*Mycol.* **98**: 1006, 2006; phylogeny), Kurtzman & Robnett

(*FEMS Yeast Res.* **7**: 141, 2007).

Wielandomyces Raithelh. (1988), Bolbitiaceae. 1, Europe. See Raithelhuber (*Metrodiana* **16**: 48, 1988).

Wiesnerina Höhn. (1907) = Actiniceps fide Boedijn (*Persoonia* **1**: 11, 1959).

Wiesneriomyces Koord. (1907), anamorphic *Pezizomycotina*, Hsp.≡ eH.10. 2, widespread. See Maniotis & Strain (*Mycol.* **60**: 203, 1968), Kirk (*TBMS* **82**: 748, 1984; nomencl.).

Wilcoxina Chin S. Yang & Korf (1985), Pyronemataceae. Anamorph *Complexipes*. 3, Europe; N. America. See Yang & Korf (*Mycotaxon* **24**: 467, 1985; anamorph), Egger *et al.* (*MR* **95**: 866, 1991; as E-strain mycorrhizas), Egger *et al.* (*CJB* **74**: 773, 1996; DNA), Landvik *et al.* (*Nordic Jl Bot.* **17**: 403, 1997; DNA), Fujimura *et al.* (*Mycorrhiza* **15**: 79, 2005; after fire), Hansen & Pfister (*Mycol.* **98**: 1029, 2006; phylogeny), Tedersoo *et al.* (*New Phytol.* **170**: 581, 2006; mycorrhizas), Perry *et al.* (*MR* **111**: 549, 2007; phylogeny).

wild type (1) the naturally occuring species or taxon as opposed to morphological variants resulting from culture *in vitro* or biochemical mutants obtained therefrom; prototroph; (2) arbitrary designation for one or more strains chosen as genetic standards.

Willeya Müll. Arg. (1883) = Staurothele fide Zahlbruckner (*Catalogus Lichenum Universalis* **1**, 1921).

Willia E.C. Hansen (1904) [non *Willia* Müll. Hal. 1890, *Musci*] ≡ Hansenula.

Williopsis Zender (1925), Saccharomycetaceae. 11, widespread. See Naumov *et al.* (*Mikrobiologiya* **54**: 239, 1985; key 6 spp.), Naumov *et al.* (*Molek. Genet., Mikrobiol. Virusol.* **2**: 3, 1987), Liu & Kurtzman (*Antonie van Leeuwenhoek* **60**: 21, 1991; rRNA, key 5 spp.), Yamada *et al.* (*Biosc., Biotechn., Biochem.* **58**: 1236, 1994), James *et al.* (*Int. J. Syst. Bacteriol.* **48**: 591, 1998; phylogeny), Kurtzman *in* Kurtzman & Fell (Eds) (*Yeasts, a taxonomic study* 4th edn: 413, 1998), Naumova *et al.* (*Syst. Appl. Microbiol.* **27**: 192, 2004).

Wilmia Dianese, Inácio & Dorn.-Silva (2001), Phaeosphaeriaceae. 1, Brazil. See Dianese *et al.* (*Mycol.* **93**: 1014, 2001).

Wilmsia Körb. (1865) = Placynthium fide Henssen (*Symb. bot. upsal.* **18** no. 1, 1963).

Wilsonia Cheel & Dughi (1944) [non *Wilsonia* R. Br. 1810, *Convolvulaceae*] nom. inval. = Polychidium fide Hawksworth *et al.* (*Dictionary of the Fungi* edn 8, 1995).

Wilsoniellanom. dub.

Wilsonomyces Adask., J.M. Ogawa & E.E. Butler (1990) = Thyrostroma fide Sutton (*in litt.*), Adaskaveg (*CJB* **73**: 432, 1995), Sutton (*Arnoldia* **14**: 33, 1997).

wilt, a plant disease (esp. *Verticillium*, *Fusarium*) characterized by loss of turgidity and collapse of leaves. See Mace *et al.* (Eds) (*Fungal wilt diseases of plants*, 1981). Cf. hadromycosis.

Wiltshirea Bat. & Peres (1962) = Phaeosaccardinula fide von Arx & Müller (*Stud. Mycol.* **9**, 1975).

Wine making. E.C. Hansen (q.v.), later a director of the Carlsberg Physiological Laboratory, first showed that yeasts were invariably present in the soils of vineyards. Hansen's use of pure culture techniques in brewing were introduced into the wine industry between 1880 and 1890. The first distribution of pure yeast cultures for wine making was made from Geisenheimam Rhein in 1890. The term 'pure yeast culture' as used in wine making is not precise, since grape **must** (the juice produced after the grapes are crushed) is not sterile before the addition of the yeast starter. Sulphur dioxide and large amounts of yeast starter overwhelm the other organisms naturally present and result in a fermentation by *Saccharomyces cerevisiae*. Suitable strains must have: a high ethanol-producing ability (18-20% vol/vol), be cold resistant (fermenting at 4°C), be resistant to sulphite, ethanol (have an ability to start a new fermentation at 8-12% ethanol), tannin, and high concentrations of sugar (able to start fermentation above 30% wt/vol of sugar), be heat resistant (ability to ferment at 30-32°C), and be able to produce low volatile acidity (mostly as acetic acid). In wine making sulphur dioxide is added to the grape must at the rate of 50-200 ppm for 2 h prior to adding the yeast. The temperature of the fermentation (in vats) is controlled and usually below 30°C; high temperatures can destroy the wine. In general, red wines ferment for 3-6 days at 20-30°C, and white for 1-2 weeks at 10-21°C. The fermented juice is then drawn off from the residues and stored in a second vat for 7-11 days at 20-30°C. The wine is then racked, stored and aged to allow for clearing and flavour and colour development. Another form of wine depends on the 'noble rot' (*Botrytis cinerea*); grapes left on the vine until they are overripe become mouldy with *B. cinerea*, the grapes crack open, and the fungus lives on the juice and the sugar content increases; the mouldy and shrivelled grapes are hand-picked, and made into Sauterne (France), Tokai (Hungary), and Trockenbeerenauslesen (Germany).

Lit.: Ainsworth (1976), Amerine & Kunkee (*Ann. Rev. Microbiol.* **22**: 323, 1968), Benda (*in* Read, *Prescott & Dunn's industrial microbiology*, edn 4: 293, 1982), Beuchart (Ed.) (*Food and beverage mycology*, 1987), Kendrick (*The fifth kingdom*, edn 2, 1992), Rose (Ed.) (*Alcoholic beverages*, 1977), Varnam (*in* Jones, *Exploitation of micro-organisms*: 297, 1993) See also Brewing, Fermented foods and drinks, Food and beverage mycology, Food spoilage, Mycotoxicoses, Yeast.

Winfrenatia T.N. Taylor, Hass & Kerp (1997), Fossil Fungi. 1, Great Britain. See Taylor *et al.* (*Am. J. Bot.* **84**: 993, 1997).

Wingea Van der Walt (1971) = Debaryomyces Lodder & Kreger ex Kreger fide Kurtzman & Robnett (*Antonie van Leeuwenhoek* **66**: 337, 1994).

Winter (Heinrich Georg; 1848-1887; Germany). Student, Leipzig (1870-1872), then Munich (1872-1873), then Halle (1873); Lecturer in Botany, Zürich Polytechnic (1876-1883); Editor for Fungi, *Rabenhorst's Kryptogamen-Flora* edn 2, Leipzig (1883-1887). Editor of *Hedwigia* (1879-1887). *Publs*. Pilze in *Rabenhorst's Kryptogamen-Flora* edn 2 (1881-1887). *Biogs, obits etc*. Grummann (1974: 55); Pazschke (*Hedwigia* **26**: 185, 1887) [portrait]; Stafleu & Cowan (*TL-2* **7**: 380, 1988).

winter mushroom, see enokitake.

winter spore, a resting spore for overwintering, e.g. a teliospore of *Puccinia graminis*. See teliospore.

Winterella (Sacc.) Kuntze (1891) nom. dub. ? = Montagnula fide Reid & Booth (*CJB* **65**: 1320, 1987), Holm (*SA* **11**: 29, 1992), Castlebury *et al.* (*Mycol.* **94**: 1017, 2002), Mejía *et al.* (*MR* **112**: 23, 2008; phylogeny).

Winterella Berl. (1893) [non *Winterella* (Sacc.) Kuntze

1891] = Nitschkia fide Hawksworth *et al.* (*Dictionary of the Fungi* edn 8, 1995).

Winteria (Rehm) Sacc. (1883) ≡ Mycowinteria.

Winteria Sacc. (1878) ≡ Selinia fide Rossman *et al.* (*Stud. Mycol.* **42**: 248 pp., 1999).

Winterina Sacc. (1891) nom. ambig., Fungi.

Winterina Sacc. (1899) ≡ Calyculosphaeria.

Winteromyces Speg. (1912) = Gibbera fide Müller & von Arx (*Beitr. Kryptfl. Schweiz* **11** no. 2, 1962).

witches' brooms, massed outgrowths (proliferations) of the branches of woody plants caused by mites, viruses, etc., and fungi, esp. rusts (e.g. *Pucciniastrum goeppertianum* (*Vaccinium*), *Gymnosporangium ellisii* (*Chamaecyparis*), *Melampsorella cerastii* (*Abies* and *Picea*)) and *Taphrina* (e.g. *T. betulina* (birch), *T. cerasi* (cherry), *T. insititiae* (plum)); also the 'krulloten' of Cacao caused by *Crinipellis perniciosa*.

witches' butter, basidioma of *Exidia glandulosa*.

Wittmannia R. Czaker (1997), Microsporidia. 1. See Czaker (*J. Eukary. Microbiol.* **44**: 445, 1997).

Witwateromyces Hallbauer, Jahns & Van Warmelo (1977), Fossil Fungi, Ascomycota (?L). 1 (Precambrian), S. Africa.

Woessia D. Hawksw. & Poelt (1986) nom. rej., anamorphic *Lecanoraceae*, Hso.0fH.15 (L). 11, widespread. See Sérusiaux (*Biblthca Lichenol.* **58**: 411, 1995).

Wojnowicia Sacc. (1899), anamorphic *Pezizomycotina*, Cpd.≡ eP.15. 4, widespread. *W. hirta* (cereal footrot). See Sutton (*Česká Mykol.* **29**: 97, 1975), Farr & Bills (*Mycol.* **87**: 518, 1995; key).

Woldmaria W.B. Cooke (1961), Niaceae. 1, Europe; N. America. See Cooke (*Beih. Sydowia* **4**: 29, 1961), Bodensteiner *et al.* (*Mol. Phylogen. Evol.* **33**: 501, 2004; phylogeny).

Wolfina Seaver ex Eckblad (1968), Chorioactidaceae. 3, USA; China. See Pfister & Kurogi (*Mycotaxon* **89**: 277, 2004), Hansen & Pfister (*Mycol.* **98**: 1029, 2006; phylogeny), Pfister *et al.* (*MR* **112**: 513, 2008).

Wolfiporia Ryvarden & Gilb. (1984), Polyporaceae. 6, widespread (north temperate). See Ginns (*Mycotaxon* **21**: 332, 1984).

wolf's-moss, *Letharia vulpina*, from Scandinavian belief that it was poisonous to wolves.

Wolkia Ramsb. (1915), Pezizomycotina. 1, Java.

Wood-attacking fungi. Wood is not an easily degraded substratum. It is usually low in nitrogen and phosphorus, two elements essential for hyphal growth; hardwoods contain tannings, and softwoods various phenolic compounds, and these are generally toxic to fungi; wood also contains few easily assimilable sugars and starches, most of the organic material being in the form of cellulose, hemicellulose and lignin. Lignin in particular is resistant to enzymes, and covers other cellular material in the wood protecting it from decay. As a result, only certain specialized fungi are associated with wood decay. These, with the exception of a few ascomycetes (e.g. *Chaetomium*, *Ustulina*, *Xylaria*) are basidiomycetes, the following genera being specially important: *Armillaria*, *Coniophora*, *Collybia*, *Coriolus*, *Daedalea*, *Echinodontium*, *Fistulina*, *Fomes*, *Ganoderma*, *Heterobasidion*, *Hydnum*, *Hymenochaete*, *Lentinus*, *Lenzites*, *Merulius*, *Peniophora*, *Phellinus*, *Pholiota*, *Piptoporus*, *Polyporus*, *Poria*, *Serpula*, *Schizophyllum*, *Stereum*, *Trametes*.

There are three generally-recognized types of timber decay, based partly on symptoms: (1) **Brown rots** - cellulose and hemicellulose are decayed, but lignin remains more or less intact; symptoms are brick-shaped cracks, the wood becoming brown and crumbling when handled; production of the enzyme glucose oxidase is associated with this type of rot and may be a diagnostic feature; only basidiomycetes are associated with brown rots, typical examples being *Piptoporus betulinus* and *Serpula lacrymans*; (2) **White rots** - all components of the wood (cellulose, hemicellulose and lignin) are degraded more or less simultaneously, and the wood becomes progressively more fragile while remaining pale; *Armillaria* and *Heterobasidion* are two basidiomycete genera causing economically important white rots in standing timber, and white rots are also caused by some ascomycetes, e.g. species of *Xylaria*. (3) **Soft rots** - similar to brown rots (in that cellulose and hemicellulose are decayed, with lignin remaining more or less intact), but occurring in wood with a high moisture content and with nitrogen available; soft rot fungi are found in rotting window frames, decaying floor boards, jetties, cooling tower slats and fence posts (where, typically, nitrogen is obtained from the surrounding soil).

Rots of standing timber, due to colonization by fungi via wounds, broken branches and dead roots are sometimes classified by rot position, e.g. **top rot** (branches), **core** or **heart rot** (trunk), **butt rot** (base). Two other terms used to describe wood rots should be noted: (1) **Dry rot** - a brown rot caused by *Serpula lacrymans*. This is a major decay fungus of buildings in temperate areas. The name is misleading as stable damp conditions are needed for initiation of growth. As the fungus develops, it conducts water to drier wood through hyphal strands which penetrate masonry and plaster though not feeding on those materials. Eradication can be difficult and costly. See Jennings & Bravery (*Serpula lacrymans: fundamental biology and control strategies*, 1991). (2) **Wet rot** - caused by many fungal species, including brown, white and soft rot types. A common example is *Coniophora puteana*, the 'cellar fungus'.

Certain fungi are a cause of staining in wood. Stored wood may be stained by the growth of surface moulds (*Alternaria*, *Aspergillus*, *Mucor*, *Penicillium*, *Rhizopus*, etc.) or by fungi in the wood (e.g. *Ophiostoma*, a cause of 'blue stain'), sometimes of living trees. Brown oak (*Quercus*) wood stained by *Fistulina hepatica*; Cartwright, *TBMS* **21**: 68, 1937) is valued, and wood coloured green by *Chlorociboria aeruginascens* is used in making 'Tunbridge Ware'. Wood having 'zone lines' (q.v.) is sometimes used for ornaments (see spalted wood). Some wood staining fungi also damage wood pulp; see Melin & Nannfeldt (*Svenska Skogvs Fören. Tidskr.* **314**: 397, 1934).

Forest hygiene and good forestry practice are the chief methods by which decay in living trees is controlled. In cut trees, in stored wood, and wood in buildings, steps may be taken to maintain conditions in which fungal growth does not take place: water content of the wood and the air are kept low, with good ventilation, and fungicides used. A range of preservatives is available for preservation of wood, but subject to various mational regulations in their use. In the UK, for example, only products cleared for use under the Control of Pesticides Regulations (www.hse.gov.uk/biocides/copr/index.htm) should be used. Active ingredients may include coal tar creosote (a coal tar oil), copper and zinc naphthenates, organotin compounds, phenolic compounds, and copper, chromium and arsenic salts.

Lit.: Boyce (*Forest pathology*, edn 3, 1961), Bravery *et al.* (*Recognizing wood rot and insect damage in buildings*, 1987), Carroll & Wicklow (Eds) (*The fungal community*, edn 2, 1992), Cartwright & Findlay (*Decay of timber and its prevention*, edn 2, 1958), Cockroft (Ed.) (*Some wood-destroying basidiomycetes*, 1981), Deacon (*Fungal biology*, 2005), Dix & Webster (*The ecology of fungi*, 1995), Easton & Hale (*Wood. Decay, pests and protection*, 1993), Fergus (*Illustrated genera of wood decay fungi*, 1960), Findlay (*Dry rot and other timber troubles*, 1952), Findlay (Ed.) (*Preservation of timber in the tropics*, 1985), Gilbertson (*Mycol.* **72**: 1, 1980; wood-rotting fungi of N. Am., 119 refs), Hunt & Garratt (*Wood preservation*, 1938), Nobles (*CJB* **43**: 1097, 1965; identification of cultures; keys 149 spp.), Rayner & Boddy (*Fungal decomposition of wood*, 1988), Stalpers (*Stud. Mycol.* **16**, 1978; keys 550 spp. polypores in culture), Wang & Zabel (Eds) (*Identification manual for fungi from utility poles in Eastern United States*, 1990), Zainal (*Micro-morphological studies on soft rot fungi in wood* [*Bibl. Mycol.* **70**], 1981).

Woodiella Sacc. & P. Syd. (1899), Helotiales. 1, S. Africa.

Wood's light, Ultra-violet light filtered through nickel oxide-containing soda glass. *Microsporum*- but not *Trichophyton*-infected hairs show a bright greenish fluorescence in Wood's light, an effect made use of in diagnosis (see Davison & Gregory, *Can. J. Res.* **7**: 378, 1932). For examination of agarics by Wood's light see Deysson (*BSMF* **74**: 207, 1958).

Wormald (Hugh; 1879-1955; England). Plant pathologist, Wye College, Kent (1911-1922); Head of the Plant Pathology Section, East Malling Research Station (1923-1939). Noted for his work on diseases of fruit trees, especially bacterial diseases and brown rot. *Publs*. Brown rot diseases of fruit trees. *Bulletin of the Ministry of Agriculture* **88** (1935); *Technical Bulletin of the Ministry of Agriculture* **3** (1954); *Diseases of Fruits and Hops* (1939) [edn 3, 1955]. *Biogs, obits etc.* Barnes (*TBMS* **39**: 289, 1956) [portrait]; Harris (*Report East Malling Research Station* **1955**: 15, 1956) [portrait].

Woronichina Naumov (1951) [non *Woronichina* Elenkin 1933, *Algae*] = Stigmatula fide Hawksworth *et al.* (*Dictionary of the Fungi* edn 8, 1995).

Woronin (Michael Stepanovitch; 1838-1903; Russia). Student, University of St Petersburg (1854-1858), then Heidelberg (1858-1859, then Freiberg, with de Bary (q.v.) (1859-1860); visited Antibes (1860); masters degree, St Petersburg (1861); doctorate, Odesa (1874); private scientist, mostly St Petersburg (1861-1903); Member of the Imperial Academy of Sciences, St Petersburg (1898-1903). A pioneer plant pathologist in Russia, who elucidated the cause of club-root disease of *Brassica* crops and made successful recommendations for its control; gave his name to 'Woronin bodies' (q.v.); also celebrated for his discovery of nitrogen-fixing bacteria in nodules on legumes. *Publs.*Untersuchungen über die Entwicklung des Rostpilzes (*Puccinia helianthi*), welcher die Krankheit der Sonnenblumen verursacht. *Botanische Zeitung* (1872); *Plasmodiophora brassicae*, Urheber der Kohlpflanzenhernie. *Pringsheim's Jahrbücher für Wissenschaftliche Botanik* (1877) [English translation, *Phytopathological Classics* **4**, 1934]; Beitrag zur Kenntnis der *Monoblepharideen*. *Mémoires de l'Académie Impériale des Sciences* St Péterbourg (1904); Ueber *Sclerotinia cinerea* und *Sclerotinia fructigena*. *Mémoires de l'Académie Impériale des Sciences* St Péterbourg (1900); (with de Bary, q.v.) *Beiträge zur Morphologie und Physiologie der Pilze* (1864-1881) [a series published in *Abhandlungen von der Senckenbergischen Naturforschenden Gesellschaft*]. *Biogs, obits etc.* Grummann (1974: 568); Khokhriakov ([*Mikologiya i Fitopatologiya*] **11**: 451, 1977 [in Russian]); Smith (*Phytopathology* **2**: 1, 1912); Stafleu & Cowan (*TL-2* **7**: 455, 1988).

Woronin bodies. Rounded or elongated-oval highly refractive bodies in the cells of certain ascomycetes, particularly in association with septa (Buller, **5**: 127, 1933); in *Erysiphe graminis*, see McKeen (*Can. J. Microbiol.* **17**: 1557, 1971); Woronin bodies seal septal pores in response to damage of adjacent cells and, in *Magnaporthe grisea*, have been shown to provide a defence against the antagonistic and nutrient-limiting environment encountered within the host plant (Soundararajan *et al.*, *The plant cell* **16**: 1564, 2004); the genetic control and formation of Woronin bodies have been investigated (Wei Kiat Tey *et al.*, *Molecular biology of the cell* **16**: 2651, 2005; Liu *et al.*, *Journal of cell biology* **180**: 325, 2008). **Woronin's hypha** (of ascomycetes), a coiled hypha probably homologous with an archicarp (de Bary, 1887); a loosely coiled hypha of large diam., at the centre of a young perithecium, which later develops ascogenous hyphae (Miller, 1928); scolecite.

Woroninella Racib. (1898) = Synchytrium fide Gäumann (*Annls mycol.* **25**: 167, 1927).

Woroninula Mekht. (1979) = Arthrobotrys fide Scholler *et al.* (*Sydowia* **51**: 89, 1999).

wort, see Brewing.

wortmannin, an antibiotic from *Talaromyces wortmannii* (anamorph); antifungal esp. against *Botrytis*, *Cladosporium* and *Rhizopus* (Brian *et al.*, *TBMS* **40**: 365, 1957).

Wright (Jorge Eduardo; 1922-2005; Argentina). Mycologist, Laboratorio de Fitopatología, Ministerio de Agricultura y Ganadería (1945-1953); student, University of Michigan (1953-1955); Professor of Systematic Botany (1960-1988) then Professor emeritus (1988-2000), Universidad de Buenos Aires. A leading figure in Latin-American mycology, establishing a unique fungal culture and dried reference collection in Buenos Aires (**BACF**), and first President of the Asociación de Micología Carlos Spegazzini, Argentina's learned society for non-medical mycologists. *Publs. The genus Tulostoma (Gasteromycetes) - A World Monograph* (1987); (with Deschamps) *Patología Forestal del Cono Sur de América. vol. I* (1997); (with Alberto) *Guia de los Hongos de la Región Pampeana* 2 vols (2002). *Biogs, obits etc.* López (*Darwiniana* **43**: 300, 2005) [portrait]; López, Cabral & Cafaro (*Mycol.* **98**: 518, 2006).

Wrightiella Speg. (1923) [non *Wrightiella* F. Schmitz 1893, *Algae*] ≡ Leptodothis.

Wrightoporia Pouzar (1966), Bondarzewiaceae. 23, widespread. Polyphyletic. See David & Rajchenberg (*CJB* **65**: 202, 1987; key), Stalpers (*Stud. Mycol.* **40**: 129, 1996; key), Cui & Dai (*Nova Hedwigia* **83**: 159,

2006; China).

Wrightoporiaceae Jülich (1982), Agaricomycetes (inc. sed.). 1 gen., 31 spp. See Jülich (*Biblthca Mycol.* **85**: 396, 1981).

Wuestneia Auersw. ex Fuckel (1863), Diaporthales. Anamorph *Harknessia*. 15, widespread (temperate). Polyphyletic, most species are not congeneric with the type. See Holm (*Taxon* **24**: 475, 1975; nomencl.), Ananthapadmanaban (*TBMS* **91**: 517, 1988), Reid & Booth (*CJB* **67**: 909, 1989; key), Nag Raj (*Coelomycetous Anamorphs with Appendage-bearing Conidia*, 1993), Yuan & Mohammed (*MR* **101**: 195, 1997), Crous & Rogers (*Sydowia* **53**: 74, 2001), Castlebury *et al.* (*Mycol.* **94**: 1017, 2002; phylogeny), Lee *et al.* (*Stud. Mycol.* **50**: 235, 2004), Rossman *et al.* (*Mycoscience* **48**: 142, 2007), Rossman *et al.* (*Mycoscience* **48**: 135, 2007; review).

Wuestneiopsis J. Reid & Dowsett (1990), Diaporthales. Anamorph *Mastigosporella*. 1, N. America. See Reid & Dowsett (*CJB* **68**: 2398, 1990), Nag Raj (*Coelomycetous Anamorphs with Appendage-bearing Conidia*, 1993).

WWW, see Internet.

wyerone, a phytoalexin (q.v.) from broad bean (*Vicia faba*).

Wynnea Berk. & M.A. Curtis (1867), Sarcoscyphaceae. 4, widespread. See Pfister (*Mycol.* **71**: 144, 1979), Kaushal (*Res. Bull. Punjab Univ.* **34**: 29, 1983), Liu *et al.* (*Mycotaxon* **30**: 465, 1987; key 6 spp.), Li & Kimbrough (*CJB* **74**: 10, 1996; spore ontogeny), Harrington *et al.* (*Mycol.* **91**: 41, 1999; phylogeny), Zhuang (*Mycotaxon* **87**: 131, 2003; Asia), Zhuang (*Nova Hedwigia* **79**: 519, 2004; Asia), Padovan *et al.* (*J. Mol. Evol.* **60**: 726, 2005; phylogeny).

Wynnella Boud. (1885), Helvellaceae. 1, widespread. See Harmaja (*Karstenia* **14**: 102, 1974), Kimbrough (*MR* **95**: 421, 1991), O'Donnell *et al.* (*Mycol.* **89**: 48, 1997), Harrington *et al.* (*Mycol.* **91**: 41, 1999), Landvik *et al.* (*Mycol.* **91**: 278, 1999), Zhuang (*Mycotaxon* **90**: 35, 2004; China), Hansen & Pfister (*Mycol.* **98**: 1029, 2006; phylogeny).

Xanthagaricus (Heinem.) Little Flower, Hosag. & T.K. Abraham (1997), Agaricaceae. 12, widespread. See Little Flower & Hosagoudar (*New Botanist* **24**: 93, 1997).

Xanthidium.

Xanthoanaptychia S.Y. Kondr. & Kärnefelt (2003) = Teloschistes fide Kondratyuk & Kärnefelt (*Ukrainian Jour. Bot.* **60**: 427, 2003), Gaya *et al.* (*MR* **112**: 528, 2008).

Xanthocarpia A. Massal. & De Not. (1853) nom. rej. = Caloplaca fide Hawksworth *et al.* (*Dictionary of the Fungi* edn 8, 1995).

Xanthochroales. An order proposed for poroid basidiomycetes having uninflated hyphae lacking clamp connexions (Corner, *Clavaria*: 22, 1950).

xanthochroic (of a hymenomycete basidioma), having a reddish-brown or yellowish-brown context which darkens on treatment with KOH.

Xanthochrous Pat. (1897) ≡ Polystictus. or *Coltricia* (Hymenochaet.) fide Donk (1974).

Xanthoconium Singer (1944), Boletaceae. 7, cosmopolitan. See Wolfe (*CJB* **65**: 2142, 1987), Gori (*Boll. Gruppo Micol. 'G. Bresadola'* **40**: 239, 1997).

Xanthodactylon P.A. Duvign. (1941), Teloschistaceae (L). 1, S.W. Africa. See Kärnefelt (*Cryptog. Bot.* **1**: 147, 1989).

Xanthoglossum (Sacc.) Kuntze (1891) = Microglossum Gillet fide Hawksworth *et al.* (*Dictionary of the Fungi* edn 8, 1995).

Xanthomaculina Hale (1985), Parmeliaceae (L). 3, S. Africa. See Büdel & Wessels (*Dinteria* **18**: 3, 1986; vagrant *X. huena*), Henssen (*Cryptog. Bryol.-Lichénol.* **19**: 267, 1998; ontogeny, thallus), Blanco *et al.* (*Taxon* **53**: 959, 2004; phylogeny), Thell *et al.* (*Mycol. Progr.* **3**: 297, 2004; phylogeny), Thell *et al.* (*J. Hattori bot. Lab.* **100**: 797, 2006).

Xanthomendoza S.Y. Kondr. & Kärnefelt (1997), Teloschistaceae (L). 17, Americas. See Kondratyuk & Kärnefelt (*Progr. Probl. Lichenol. Nineties. Proc. Third Symp. Intern. Assoc. Lichenol. [Biblthca Lichenol.* **68**]: 26, 1997), Søchting *et al.* (*Mitteilungen aus dem Institut für Allgemeine Botanik Hamburg* **30-32**: 225, 2002), Lindblom (*Bryologist* **109**: 1, 2006), Miądlikowska *et al.* (*Mycol.* **98**: 1088, 2006; phylogeny), Gaya *et al.* (*MR* **112**: 528, 2008; phylogeny).

Xanthoparmelia (Vain.) Hale (1974), Parmeliaceae (L). *c.* 500, widespread (esp. southern hemisphere). See Elix *et al.* (*Bull. Br. Mus. nat. hist.* Bot. **15**: 163, 1986; key 117 spp. Australasia), Hale (*Smithson. Contr. Bot.* **74**, 1990; monogr.), Thomson (*Bryologist* **96**: 342, 1993; key 55 spp. N. Am.), Nash *et al.* (*Biblthca Lichenol.* **56**, 1995; key 77 spp. S.Am.), Calvelo (*Lichenology in Latin America. History, Current Knowledge and Applications* [Proceedings of GLAL-3, Terceiro Encontro do Grupo Latino-Americano de Liquenólogos, São Paulo, Brazil, 24-28 September, 1997]: 117, 1998; key, S America), Crespo & Cubero (*Lichenologist* **30**: 369, 1998; phylogeny), Mattsson & Wedin (*Lichenologist* **30**: 463, 1998; phylogeny), Elix (*Mycotaxon* **73**: 51, 1999; S Africa), Elix & Kantvilas (*Mycotaxon* **73**: 441, 1999; Australia), Elix (*Biblthca Lichenol.* **80**, 2001), Divakar & Upreti (*Nova Hedwigia* **75**: 507, 2002; India), Eliasaro & Adler (*Mitteilungen aus dem Institut für Allgemeine Botanik Hamburg* **30-32**: 25, 2002; Brazil), Elix (*Lichenologist* **34**: 283, 2002; Africa), Giordani *et al.* (*Lichenologist* **34**: 189, 2002; Italy), Hawksworth & Crespo (*Taxon* **51**: 807, 2002; nomencl.), Elix (*Mycotaxon* **87**: 395, 2003; generic limits), Elix (*Lichenologist* **35**: 291, 2003; Australia), Blanco *et al.* (*Taxon* **53**: 959, 2004; phylogeny), Elix (*Lichenologist* **36**: 277, 2004; Australia), Nash *et al.* (*Symb. bot. upsal.* **34** no. 1: 289, 2004; Mexico), Blanco *et al.* (*Mol. Phylogen. Evol.* **39**: 52, 2006; phylogeny), Elix (*J. Hattori bot. Lab.* **100**: 635, 2006; Australia), Miądlikowska *et al.* (*Mycol.* **98**: 1088, 2006; phylogeny), Thell *et al.* (*J. Hattori bot. Lab.* **100**: 797, 2006; phylogeny), Del Prado *et al.* (*MR* **111**: 685, 2007; ascospore anatomy), Honegger & Zippler (*MR* **111**: 424, 2007; mating systems), Pérez-Ortega & Elix (*Lichenologist* **39**: 297, 2007; Spain, key).

Xanthopeltis R. Sant. (1949), Teloschistaceae (L). 1, Chile. See Follmann (*Revta Universitaria (Chile)* **47**: 33, 1962), Kärnefelt (*Cryptog. bot.* **1**: 147, 1989).

Xanthophyllomyces Golubev (1995), Cystofilobasidiaceae. Anamorph *Phaffia*. 1, widespread. See Golubev (*Yeast* Chichester **11**: 101, 1995; taxonomy), Kucsera *et al.* (*Antonie van Leeuwenhoek* **73**: 163, 1998; life cycle), Medwid (*J. Industr. Microbiol. Biotechnol.* **21**: 228, 1998; ploidy), Fell *et al.* (*Int. J. Syst. Evol. Microbiol.* **50**: 1351, 2000; mol. phylogeny), Weiss *et al.* (*Frontiers in Basidiomycote Mycology*: 7, 2004).

Xanthoporia Murrill (1916) = Inonotus fide Pegler (*TBMS* **47**: 175, 1964).

Xanthoporina C.W. Dodge (1948) = Collemopsidium fide Aptroot (*in litt.*).

Xanthoporus Audet (2007), Agaricomycetes. 1.

Xanthopsis Acloque (1893) [non *Xanthopsis* (DC.) Koch 1851, *Compositae*] = Catolechia fide Hawksworth *et al.* (*Dictionary of the Fungi* edn 8, 1995).

Xanthopsora Gotth. Schneid. & W.A. Weber (1980) [non *Xanthopsora* Speg. 1922] ≡ Xanthopsorella.

Xanthopsora Speg. (1922) = Linochora fide Petrak & Sydow (*Annls mycol.* **33**: 176, 1935).

Xanthopsorella Kalb & Hafellner (1984), Catillariaceae (L). 1, Mexico; USA. See Kalb & Hafellner (*Nova Hedwigia* Beih. **79**: 384, 1984), Timdal (*Mycotaxon* **31**: 101, 1988).

Xanthopsorellaceae Hafellner (1984) = Catillariaceae.

Xanthopyrenia Bachm. (1919) = Collemopsidium fide Aptroot (*in litt.*).

Xanthopyreniaceae Zahlbr. (1926), Pyrenulales (L). 5 gen. (+ 4 syn.), 35 spp.

Lit.: Hawksworth (*J. Linn. Soc.* Bot. **96**: 3, 1988), Santesson (*Lichenologist* **24**: 7, 1992), Fałtynowicz & Sągin (*Acta Mycologica* Warszawa **30**: 147, 1995), Harada (*Bryologist* **102**: 50, 1999), McCarthy & Kantvilas (*Lichenologist* **31**: 227, 1999), Schultz *et al.* (*Pl. Biol.* **2**: 482, 2000), Mohr *et al.* (*MR* **108**: 515, 2004), Calatayud *et al.* (*Lichenologist* **39**: 129, 2007).

Xanthoria (Fr.) Th. Fr. (1860) nom. cons., Teloschistaceae (L). *c.* 56, widespread. See Thomson (*Am. midl. Nat.* **41**: 706, 1949; N. Am.), Almborn (*Bot. Notiser* **16**: 161, 1963; S. Afr.), Steiner & Poelt (*Pl. Syst. Evol.* **140**: 151, 1982; sect. *Xanthoriella*), Awasthi (*Proc. Indian Acad. Sci.* Pl. Sci. **96**: 227, 1985; Indian spp.), Janex-Favre & Ghaleb (*Cryptog. Bryol.-Lichénol.* **7**: 457, 1986; ontogeny), Poelt & Petutschnig (*Nova Hedwigia* **54**: 1, 1992; key *X. candelaria* group), Arnold & Poelt (*Biblthca Lichenol.* **57**: 49, 1995; chemistry), Castello (*Cryptog. Bryol.-Lichénol.* **16**: 79, 1995; Antarctica), Lindblom (*J. Hattori bot. Lab.* **83**: 75, 1997; N. Am.), Franc & Kärnefelt (*Graphis Scripta* **9**: 49, 1998; phylogeny), Kärnefelt *et al.* (*Lichenologist* **34**: 333, 2002; S Africa), Murtagh *et al.* (*MR* **106**: 1277, 2002; infraspecific variation), Gaya *et al.* (*Am. J. Bot.* **90**: 1095, 2003; phylogeny), Scherrer & Honegger (*New Phytol.* **158**: 375, 2003; hydrophobin genes), Søchting & Lutzoni (*MR* **107**: 1266, 2003; reln with *Caloplaca*), Wang *et al.* (*Mycosystema* **22**: 536, 2003; chemistry), Honegger *et al.* (*Lichenologist* **36**: 381, 2004; genetic diversity), Kondratyuk *et al.* (*Biblthca Lichenol.* **88**: 349, 2004; S Africa), Lindblom & Ekman (*MR* **109**: 187, 2005), Lindblom *et al.* (*Graphis Scripta* **17**: 12, 2005; Scandinavia), Scherrer *et al.* (*Fungal Genetics Biol.* **42**: 976, 2005; mating types), Sano *et al.* (*Jap. J. Med. Mycol.* **47**: 113, 2006; phylogeny), Lindblom & Ekman (*Lichenologist* **39**: 259, 2007; population biology), Gaya *et al.* (*MR* **112**: 528, 2008; phylogeny).

Xanthoriicola D. Hawksw. (1973), anamorphic *Pezizomycotina*, Hsp.0eP.15. 1 (on lichens, *Xanthoria*), Europe. See Hawksworth (*Bull. Br. Mus. nat. hist.* Bot. **6**: 183, 1979; SEM).

Xanthoriomyces E.A. Thomas ex Cif. & Tomas. (1953) ≡ Xanthoria.

Xanthothecium Arx & Samson (1973), Onygenaceae. 1, widespread. See Currah (*Mycotaxon* **24**: 1, 1985; posn).

Xeilaria, see *Cheilaria*.

Xenasma Donk (1957), Xenasmataceae. 9, Europe; N. America. See Hjortstam *et al.* (*Cortic. N. Europ.* **8**, 1988; key Eur. spp.) Sensu lato 48 spp., Liberta (*Mycol.* **52**: 884, 1962; key).

Xenasmataceae Oberw. (1966), Polyporales. 3 gen., 27 spp.

Lit.: Hallenberg (*Mycotaxon* **27**: 361, 1986), Hjortstam & Larsson (*Mycotaxon* **29**: 315, 1987), Lin & Chen (*Taiwania* **35**: 69, 1990), Nakasone (*Mycol. Mem.* **15**: 412 pp., 1990), Maekawa (*Rep. Tottori Mycol. Inst.* **31**: 1, 1993), Ginns (*Mycol.* **90**: 19, 1998), Greslebin & Rajchenberg (*N.Z. Jl Bot.* **41**: 437, 2003), Larsson *et al.* (*MR* **108**: 983, 2004).

Xenasmatella Oberw. (1966), Xenasmataceae. 14, widespread. See Hjortstam *et al.* (*Cortic. N. Europ.* **8**, 1988; sub *Phlebiella*).

Xenidiocercus Nag Raj (1993), anamorphic *Pezizomycotina*, Cpd.0eH.19. 3, India; W. Africa. See Nag Raj (*Coelomycetous Anamorphs with Appendage-bearing Conidia*: 975, 1993), Wu & Sutton (*Mycoscience* **36**: 271, 1995).

xenobiotic (1) a chemical not normally synthesized or metabolized by living organisms, e.g. a manufactured drug; (2) chemical waste or other pollutant toxic to a living organism.

Xenobotrytis R.F. Castañeda & W.B. Kendr. (1990) = Paracostantinella fide Kendrick (*in litt.*).

Xenocalonectria Crous & C.L. Schoch (2000), Nectriaceae. Anamorph *Xenocylindrocladium*. 1, Ecuador. See Schoch *et al.* (*Stud. Mycol.* **45**: 45, 2000), Crous (*Taxonomy and Pathology of Cylindrocladium (Calonectria) and Allied Genera*: 278 pp., 2002).

Xenochalara M.J. Wingf. & Crous (2000), anamorphic *Pezizomycotina*. 1, Netherlands. See Coetsee *et al.* (*S. Afr. J. Bot.* **66**: 101, 2000).

Xenochora Petr. (1948), anamorphic *Pezizomycotina*, Cpd.0eH/1eP.?. 1, Ecuador.

Xenocylindrocladium Decock, Hennebert & Crous (1997), anamorphic *Xenocalonectria*, Hso.?.?. 3, Ecuador. See Decock *et al.* (*MR* **101**: 788, 1997), Schoch *et al.* (*Stud. Mycol.* **45**: 45, 2000), Crous *et al.* (*Mycoscience* **42**: 559, 2001), Crous (*Taxonomy and Pathology of Cylindrocladium (Calonectria) and Allied Genera*: 278 pp., 2002).

Xenodiella Syd. (1935), anamorphic *Xenodium*, Hsp.0eH.?. 1, S. America.

Xenodimerium Petr. (1947) ? = Eudarluca fide Müller & von Arx (*Beitr. Kryptfl. Schweiz* **11** no. 2, 1962).

Xenodiscella Petr. (1954) = Rhagadolobium fide Müller & von Arx (*Beitr. Kryptfl. Schweiz* **11** no. 2, 1962).

Xenodium Syd. (1935), Elsinoaceae. Anamorph *Xenodiella*. 1, S. America.

Xenodochus Schltdl. (1826), Phragmidiaceae. 2 (on *Sanguisorba* (*Rosaceae*)), widespread (north temperate). See Sato & Sato (*TMSJ* **21**: 411, 1980; life cycle).

Xenodomus Petr. (1922), anamorphic *Pezizomycotina*, Cpd.0eH.?. 1, N. America.

Xenogliocladiopsis Crous & W.B. Kendr. (1994), anamorphic *Arnaudiella*, Hso.0eH.15. 1, S. Africa. See Crous & Kendrick (*CJB* **72**: 63, 1994).

Xenogloea Syd. & P. Syd. (1919) ≡ Kriegeria Bres.

Xenoheteroconium Bhat, W.B. Kendr. & Nag Raj (1993), anamorphic *Pezizomycotina*, Hso.≡ eP.1. 1, India. See Bhat & Kendrick (*Mycotaxon* **49**: 80, 1993).

Xenokylindria DiCosmo, S.M. Berch & W.B. Kendr. (1983), anamorphic *Pezizomycotina*, Hso.1eH.19. 1, Japan. See DiCosmo *et al.* (*Mycol.* **75**: 971, 1983).
Xenolachne D.P. Rogers (1947), Tremellales. 2, USA; Europe. See Grauwinkel (*Rheinland-Pfälzisches Pilzjournal* **3**: 3, 1993).
Xenolecia Hertel (1984), Porpidiaceae (L). 1, Chile. See Hertel (*Nova Hedwigia* Beih. **79**: 439, 1984).
Xenolophium Syd. (1925), Melanommataceae. 4, widespread (tropical). See Huhndorf (*Mycol.* **85**: 490, 1993).
Xenomeris Syd. (1924), Capnodiales. Anamorphs *Hormonema*-like, *Sclerophoma*. *c.* 11, widespread. *X. abietis* (dieback of conifers). See Funk & Shoemaker (*Mycol.* **63**: 567, 1971), Samuels *et al.* (*Brittonia* **40**: 392, 1988), Barr (*Sydowia* **41**: 25, 1989), Winton *et al.* (*Mycol.* **99**: 240, 2007; phylogeny).
Xenomyces Ces. (1879) = Sclerocystis fide Zycha *et al.* (*Mucorales*, 1969).
Xenomyxa Syd. (1939), Pezizomycotina. 1, Ecuador.
Xenonectria Höhn. (1920) = Byssosphaeria fide Barr (*Mycotaxon* **20**, 1984).
Xenonectriella Weese (1919), Nectriaceae. 4 (lichenicolous), widespread. See Rossman *et al.* (*Stud. Mycol.* **42**: 248 pp., 1999).
Xenopeltis Syd. & P. Syd. (1919), anamorphic *Pezizomycotina*, Cpt.0eH.?. 1, Philippines.
Xenoplaca Petr. (1949), anamorphic *Pezizomycotina*, Hso.0eP.?. 1, S. America. See von Arx (*Persoonia* **11**: 388, 1981; redescr.).
Xenopus Penz. & Sacc. (1901) nom. dub., Agaricomycetes. 'basidiomycetes'. See Hughes (*CJB* **36**: 747, 1958).
Xenosoma Syd. & P. Syd. (1921) nom. dub., Agaricomycetes. 'gasteromycetes' or ? *Trichosphaeriales*.
Xenosperma Oberw. (1965), Xenasmataceae. 4, widespread. See Oberwinkler (*Sydowia* **19**: 45, 1965).
Xenosphaeria Trevis. (1860) = Dacampia fide Alstrup & Hawksworth (*Meddr Grønland* Biosc. **31**, 1990).
xenospore, a spore dispersed from its place of origin (Gregory, *in* Madelin, 1966). Cf. memnospore.
Xenosporella Höhn. (1923) = Xenosporium fide Pirozynski (*Mycol. Pap.* **105**, 1966).
Xenosporium Penz. & Sacc. (1901), anamorphic *Thaxteriellopsis*, Hso.#eP.1. 13, widespread (esp. tropical). See Pirozynski (*Mycol. Pap.* **105**, 1966; key), Goos (*Mycol.* **82**: 742, 1990; key), Bussaban *et al.* (*Fungal Diversity* **14**: 61, 2003; Thailand), Zhao *et al.* (*Nova Hedwigia* **82**: 127, 2006).
Xenostele Syd. & P. Syd. (1921), Pucciniaceae. 4 (on *Lauraceae*), Asia. See Cummins (*Illustr. Gen. Rust Fungi*, 1959; as *Puccinia*), Zhuang (*Mycosystema* **3**: 29, 1990; species on *Lauraceae* (as *Puccinia*)).
Xenostigme Syd. (1930), ? Meliolaceae. 1 (from leaves), S. America.
Xenostigmella Petr. (1950) ? = Balladynopsis fide Sivanesan (*Mycol. Pap.* **146**, 1981).
Xenostigmina Crous (1998) ? = Pseudocercospora fide Crous (*Mycol. Mem.* **21**: 154, 1998), Crous & Braun (*CBS Diversity Ser.* **1**: 571 pp., 2003), Batzer *et al.* (*Mycol.* **97**: 1268, 2005).
Xenostilbum Petr. (1959) = Calostilbella fide von Arx (*Persoonia* **11**: 391, 1981).
Xenostomella Syd. (1930), Microthyriaceae. 2, S. America.
Xenostroma Höhn. (1915), anamorphic *Pezizomycotina*, St.0eH/1eP.?. 1, Europe.
Xenothecium Höhn. (1919), ? Hyponectriaceae. 1, S. America.
Xenotypa Petr. (1955), Valsaceae. 1, Europe; N. America. See Pirozynski (*CJB* **52**: 2129, 1974), Barr (*Mycol. Mem.* **7**, 1978).
Xenus Kohlm. & Volkm.-Kohlm. (1992), Pyrenulales. 1 (on coralline alga), Belize. See Eriksson & Hawksworth (*SA* **11**: 189, 1993).
Xepicula Nag Raj (1993), anamorphic *Pezizomycotina*, Ccu.0eH/1eP.15. 2, widespread. See Nag Raj (*Coelomycetous Anamorphs with Appendage-bearing Conidia*: 979, 1993), Seifert *et al.* (*Mycotaxon* **87**: 317, 2003).
Xepiculopsis Nag Raj (1993), anamorphic *Pezizomycotina*, Ccu.0eH/1eP.15. 2, widespread. See Nag Raj (*Coelomycetous Anamorphs with Appendage-bearing Conidia*: 9823, 1993), Seifert *et al.* (*Mycotaxon* **87**: 317, 2003).
xero- (prefix), dry; drought.
Xerocarpus P. Karst. (1881) [non *Xerocarpus* Guill., Perr. & A. Rich. 1832, *Papilionaceae*] = Phanerochaete fide Donk (*Taxon* **6**: 17, 1957).
Xerocomaceae Pegler & T.W.K. Young (1981) = Boletaceae.
Xerocomellus Šutara (2008) = Boletus Fr. fide Kuyper (*in litt.*).
Xerocomopsis Reichert (1940) ≡ Boletus Fr.
Xerocomus Quél. (1887) nom. cons. = Boletus Fr. fide Kuyper (*in litt.*).
Xeroconium D. Hawksw. (1981), anamorphic *Pezizomycotina*, Cpd.0eP.19. 1, Finland. See Hawksworth (*Bull. Br. Mus. nat. Hist.* Bot. **9**: 33, 1981).
Xerocoprinus Maire (1907), Agaricaceae. 1, Africa.
Xerodiscus Petr. (1943) = Arthonia fide Müller & von Arx (*Beitr. Kryptfl. Schweiz* **11** no. 2, 1962).
Xeromedulla Korf & W.Y. Zhuang (1987), Helotiaceae. 3, China; Philippines. See Zhuang & Korf (*Mycotaxon* **30**: 189, 1987).
Xeromphalia, see *Xeromphalina*.
Xeromphalina Kühner & Maire (1934), ? Mycenaceae. *c.* 30, widespread. See Smith (*Pap. Mich. Acad. Sci.* **38**: 53, 1953; N. Am.), Singer (*Boln Soc. argent. Bot.* **10**: 302, 1965), Miller (*Mycol.* **60**: 156, 1968; key 12 N. Am. spp.), Horak (*Sydowia* **32**: 131, 1979; Indomalaya & Austral. spp.), Klán (*Česká Mykol.* **38**: 205, 1984; Eur. spp.), Antonín (*Mykol. Listy* **82**: 1, 2002; key Centr. Eur. spp.), Antonín & Noordeloos (*A monograph of the genera Hemimycena,* , 2004), Clémençon (*Persoonia* **18**: 449, 2005; basidiome developm).
Xeromyces L.R. Fraser (1954) = Monascus fide Hawksworth & Pitt (*Aust. J. Bot.* **31**: 51, 1983; posn), Pitt & Hocking (*CSIRO Food Res. Q.* **42**: 1, 1983), Udagawa *et al.* (*TMSJ* **27**: 303, 1986), Kuraishi *et al.* (*Antonie van Leeuwenhoek* **77**: 179, 2000; ubiquinones), Park & Jong (*Mycoscience* **44**: 25, 2003; phylogeny), Stchigel *et al.* (*Stud. Mycol.* **50**: 299, 2004; phylogeny).
xerophilic, favouring habitats in which water is not available; either living in desert conditions or where water is not generally available because of the physiological status of cells (c.f. water activity).
xerophyte, a plant of dry habitats; sometimes incorrectly applied to fungi.
Xerosporae, hyphomycetes and coelomycetes with dry spores. Cf. *Gloiosporae* See Wakefield & Bisby (*TBMS* **25**: 49, 1942).
Xerotinus Rchb. (1828) ≡ Xerotus.
xerotolerant, able to grow under dry conditions (Pitt *in*

Duckworth (Ed.), *Water relations of foods*, 1975).

Xerotrema Sherwood & Coppins (1980), Odontotremataceae. 1, British Isles. See Sherwood & Coppins (*Notes R. bot. Gdn Edinb.* **38**: 368, 1980), Lumbsch *et al.* (*Mol. Phylogen. Evol.* **31**: 822, 2004; phylogeny).

Xerotus Fr. (1828), Polyporaceae. 4, Africa. See Pegler (*The polypores [Bull. BMS Suppl.]*, 1973; sub *Xerotinus*), Ryvarden (*Syn. Fung.* **5**, 1991).

Xerula Maire (1933), Physalacriaceae. *c.* 15, widespread (north temperate). See Dörfelt (*Mycotaxon* **15**: 62, 1982; nomencl., bibliogr.), Petersen & Nagasawa (*Rep. Tottori Mycol. Inst.* **43**, 2006).

Xerulaceae Jülich (1982) = Physalacriaceae.

Lit.: Redhead (*CJB* **66**: 479, 1988) however, they fit well in the *Physalacriaceae* clade; see, Moncalvo *et al.* (*Syst. Biol.* **49**: 278, 2000), Matheny *et al.* (*Mycol.* **98**: 982, 2006).

Xerulina Singer (1962) = Cyptotrama fide Kuyper (*in litt.*).

Xiambola Minter & Hol.-Jech. (1981), anamorphic *Pezizomycotina*, Ccu.1eH.38. 1, Czech Republic. See Minter & Holubová-Jechová (*Folia geobot. phytotax.* **16**: 213, 1981).

Xiphomyces Syd. & P. Syd. (1916), anamorphic *Pezizomycotina*, Hsp.0eP.?. 2, Philippines; Tristan da Cunha.

Xylaria Hill ex Grev. (1823) nom. rej. prop. ≡ Cordyceps.

Xylaria Hill ex Schrank (1789) nom. cons., Xylariaceae. Anamorphs *Moelleroclavus*, *Geniculosporium*-like, *Xylocoremium*. *c.* 300, widespread (esp. tropical). *X. digitata* (root rot of hardwoods); *X. hypoxylon*, the Candle-snuff fungus (black root rot of apple); *X. vaporaria* (an invader of mushroom beds). See Dennis (*Kew Bull.* **1956**: 401, 1957; trop. Am.), Dennis (*Revta Biol.* Lisb. **1**: 175, 1958; trop. Afr.), Dennis (*Bull. Jard. bot. Brux.* **31**: 111, 1961; Congo), Joly (*Revue Mycol.* Paris **33**: 155, 1966; Vietnam), Dennis *in* Parmasto (*Zhivaya priroda Dal'nego Vostoka*: 42, 1972; E. former USSR), Rogers (*Mycol.* **75**: 457, 1983), Bertault (*BSMF* **100**: 139, 1984; key 45 spp., Eur., N. Afr.), Rogers (*Mycol.* **76**: 23, 1984), Silveira & Rodrigues (*Acta Amazon.* Supl. **15**: 7, 1985; key 11 spp. Brazil), Rogers (*Mycotaxon* **26**: 85, 1986; key 30 spp. USA), Rogers (*Sydowia* **38**: 255, 1986; sectional classif., anamorphs), Rogers & Callan (*Mycol.* **78**: 391, 1986; *X. polymorpha*-group, USA), Rogers & Samuels (*N.Z. Jl Bot.* **24**: 615, 1986; key 19 NZ spp.), Rogers *et al.* (*Mycotaxon* **31**: 103, 1988; key 41 spp. Venezuela), San Martín & Rogers (*Mycotaxon* **34**: 283, 1989; key 63 spp. Mexico), Callan & Rogers (*Mycotaxon* **46**: 141, 1993; cultural features, 23 US spp.), Læssøe (*SA* **13**: 43, 1994; ? synonym of *Xylaria*), van der Gucht (*Mycotaxon* **60**: 327, 1996; New Guinea), Rogers *et al.* (*MR* **101**: 345, 1997; *Moelleroclavus* anam.), Ju & Rogers (*Mycotaxon* **73**: 343, 1999; Taiwan), Læssøe (*Kew Bull.* **54**: 605, 1999; *X. comosa* group), Lee *et al.* (*FEMS Microbiol. Lett.* **187**: 89, 2000; phylogeny), Ju & Rogers (*MR* **105**: 1134, 2001; anamorph), San Martín *et al.* (*Mycotaxon* **79**: 337, 2001; Mexico), Rogers *et al.* (*Sydowia* **54**: 91, 2002; fructicolous spp.), Davis *et al.* (*Am. J. Bot.* **90**: 1661, 2003; endophytes), Medardi (*Riv. Micol.* **46**: 25, 2003; Italy), Rogers & Ju (*Mycol. Progr.* **3**: 37, 2004; Costa Rica), Hladki & Romero (*Lilloa* **42**: 47, 2005; Argentina), Rogers *et al.* (*Mycol.* **97**: 914, 2005; on termite nests), Zhang *et al.* (*Mycol.* **98**: 1076, 2006; phylogeny), Ju & Hsieh (*Mycol.* **99**: 936, 2007; on termite nests, Taiwan), Stadler *et al.* (*Mycol. Progr.* **7**: 53, 2008; chemistry, entonaemoid spp.).

Xylariaceae Tul. & C. Tul. (1861), Xylariales. 85 gen. (+ 71 syn.), 1343 spp.

Lit.: Rogers (*Mycol.* **71**: 1, 1979; review), Petrini & Müller (*Mycol. Helv.* **1**: 501, 1986; keys 33 spp., culture), Petrini & Petrini (*Sydowia* **38**: 216, 1985; keys 22 spp.; culture), Whalley (*Sydowia* **38**: 369, 1986; ecology, distrib.), Dargan (*Beih. Nova Hedwigia* **44**: 489, 1987), Rogers *et al.* (*Mycotaxon* **29**: 113, 1987; keys 60 spp. Indonesia), Whalley & Edwards *in* Rayner *et al.* (Eds) (*Evolutionary biology of the fungi*: 423, 1987), Romero & Minter (*TBMS* **90**: 457, 1988; asci), González & Rogers (*Mycotaxon* **34**: 283, 1989), Barr (*Mycotaxon* **39**: 43, 1990; circumscr.), Silveira & Rodrigues (*Acta Amazon.* Supl. **20**: 8, 1990), Callan & Rogers (*Mycotaxon* **46**: 141, 1993), Chapela *et al.* (*MR* **97**: 157, 1993), Petrini & Petrini (*Biblthca Mycol.* **150**: 193, 1993; arctic-alp. spp.), Læssøe (*SA* **13**: 43, 1994), Læssøe & Spooner (*Kew Bull.* **49**: 1, 1993), Rappaz (*Mycol. Helv.* **7**: 99, 1995; on hardwood, Eur., N. Am.), Whalley & Edwards (*CJB* **73** Suppl. 1: S802, 1995; secondary metabolites), Ju & Rogers (*Mycol. Mem.* **20**: 365 pp., 1996), Whalley (*MR* **100**: 897, 1996; review), Ju *et al.* (*Mycotaxon* **61**: 243, 1997), Ju *et al.* (*Mycotaxon* **66**: 1, 1998), Rogers & Ju (*Mycotaxon* **68**: 345, 1998), Rogers *et al.* (*Mycotaxon* **67**: 61, 1998), Ju & Rogers (*Mycotaxon* **73**: 343, 1999; Taiwan), Johannesson *et al.* (*MR* **104**: 275, 2000), Lee *et al.* (*FEMS Microbiol. Lett.* **187**: 89, 2000), Rogers (*MR* **104**: 1412, 2000; review), Sánchez-Ballesteros *et al.* (*Mycol.* **92**: 964, 2000; molecular phylogeny), Collado *et al.* (*Mycol.* **93**: 875, 2001), Platas *et al.* (*MR* **108**: 71, 2004), Stadler *et al.* (*MR* **108**: 239, 2004), Hsieh *et al.* (*Mycol.* **97**: 844, 2005), Triebel *et al.* (*Nova Hedwigia* **80**: 25, 2005), Bitzer *et al.* (*MR* **112**: 251, 2008; phylogeny, chemistry).

Xylariales Nannf. (1932). Xylariomycetidae. 9 fam., 209 gen., 2487 spp. Stromata usually well-developed, mostly consisting only of fungal tissue. Ascomata perithecial, rarely cleistothecial, ± globose, superficial or immersed in the stroma, usually black- and thick-walled, the ostiole usually papillate, periphysate. Interascal tissue well-developed, of narrow paraphyses. Asci cylindrical, persistent, relatively thick-walled but without separable layers, with an often complex J+ apical ring, usually 8-spored, ± spherical in some cleistocarpous taxa and without apical apparatus. Ascospores usually pigmented, sometimes transversely septate, with germ pores or slits, sometimes with a mucous sheath or mucous appendages. Anamorphs varied, usually hyphomycetous, some pycnidial. Saprobes and plant parasites, mainly on bark or wood, some associated with termitaria, cosmop.

Probably polyphyletic; the *Amphisphaeriaceae* have been given their own order by some authors and other fams (esp. *Hyponectriaceae*) are still poorly understood. Fams:

(1) **Amphisphaeriaceae**
(2) **Cainiaceae**
(3) **Clypeosphaeriaceae**
(4) **Diatrypaceae**
(5) **Graphostromataceae**

(6) **Hyponectriaceae**
(7) **Iodosphaeriaceae**
(8) **Myelospermataceae**
(9) **Xylariaceae**

Lit.: von Arx (*Sydowia* **34**: 13, 1982; fams.), von Arx & Müller (*Beitr. Kryptogfl. Schweiz* **11**(1), 1954), Dennis (*British Ascomycetes*, 1978), Müller & von Arx (*Beitr. Kryptogfl. Schweiz* **11**(2), 1962; in Ainsworth *et al.*, *The Fungi* **4A**, 1973; keys gen.), Barr (*Mycotaxon* **37**: 43, 1990).

Xylariodiscus Henn. (1899) = Xylaria Hill ex Schrank fide Höhnel (*Sber. Akad. Wiss. Wien* Math.-naturw. Kl., Abt. 1 **119**: 928, 1910), Dennis (*Bull. Jard. bot. Brux.* **34**: 231, 1961), Læssøe (*SA* **13**: 43, 1994; ? synonym of *Xylaria*).

Xylariomycetidae O.E. Erikss. & Winka (1997), Sordariomycetes. Ord.:
Xylarinales
For *Lit.* see fam.

Xylariopsis F.L. Tai (1934) = Konradia fide Boedijn (*Annls mycol.* **33**: 229, 1935).

Xylasclerotes Stach & Pickh. (1957), Fossil Fungi. 1 (Carboniferous), Germany. See Stach & Pickhardt (*Palaeontologische Zeitschrift* **31**: 140, 1957).

Xylastra A. Massal. (1855) = Opegrapha Ach. fide Zahlbruckner (*Catalogus Lichenum Universalis* **2**, 1923).

Xyleborus R.C. Harris & Ladd (2007), Stereocaulaceae (L). 1, USA. See Harris & Ladd (*Opuscula Philolichenum* **4**: 57, 2007).

Xylissus Raf. (1808) nom. nud., Fungi. See Merrill (*Index Rafinesq.*, 1949).

Xylobolus P. Karst. (1881), Stereaceae. 3, widespread. See Chamuris (*Mycol. Mem.* **14**, 1988; key N. Am. spp.), Solheim & Hofton (*Sopp og Nyttevekster* **1**: 54, 2005; *Xylobolus frustulatus*).

Xylobotryum Pat. (1895), Dothideomycetes. 2 or 3, widespread (tropical). See Rossman (*Mycotaxon* **4**: 179, 1976), Barr (*Prodr. Cl. Loculoasc.*, 1987), Eriksson & Hawksworth (*SA* **7**: 95, 1988), Ju & Rogers (*Cryptog. Bot.* **4**: 346, 1994; ontogeny, posn).

Xyloceras A.L. Sm. (1901) = Xylobotryum fide Müller & von Arx *in* Ainsworth *et al.* (Eds) (*The Fungi* **4A**: 87, 1973).

Xylochia B. Sutton (1983), anamorphic *Pezizomycotina*, Hso.#eH-P.1/10. 1, India. See Sutton (*TBMS* **80**: 255, 1983).

Xylochoeras Fr. (1849) = Sclerotium fide Saccardo (*Syll. fung.* **14**: 1, 1899).

Xylochora Arx & E. Müll. (1954), Xylariales. 2 (from wood), Europe. See Kang *et al.* (*Fungal Diversity* **2**: 135, 1999).

Xylocladium P. Syd. ex Lindau (1900), anamorphic *Camillea*, Hsy.0eH.8. 10, widespread (esp. tropical). See Læssøe *et al.* (*MR* **93**: 152, 1989), Rogers *et al.* (*Nova Hedwigia* **71**: 431, 2000; Venezuela).

Xylocoremium J.D. Rogers (1984), anamorphic *Xylaria*, Hsy.0eH.14. 1, USA. See Rogers (*Mycol.* **76**: 913, 1984).

Xylocoryneum Sacc. [not traced] nom. dub., anamorphic *Pezizomycotina*.

Xylocrea Möller (1901) ? = Sarcoxylon fide Læssøe (*SA* **13**: 43, 1994).

Xylodon (Pers.) Gray (1821) nom. rej., Schizoporaceae. See Donk (*Taxon* **12**: 113, 1963).

xylogenous, living on wood.

Xyloglossum Pers. (1818) = Acrospermum fide Bonorden (*Handb. Mycol.*, 1851).

Xyloglyphis Clem. (1909), anamorphic *Pezizomycotina*, St.0eH.15. 1, Europe. See Sherwood (*Mycotaxon* **5**: 87, 1977).

Xylogone Arx & T. Nilsson (1969), Pezizomycotina. 1, Sweden. See Currah (*Mycotaxon* **21**: 1, 1985; posn).

Xylogramma Wallr. (1833), Helotiaceae. Anamorph *Cystotricha*. 18, widespread (north temperate). See Sherwood (*Mycotaxon* **5**: 1, 1977).

Xylographa (Fr.) Fr. (1836), Agyriaceae (±L). 8, Europe; N. America. See Brodo (*Graphis Scripta* **4**: 61, 1992; Europe), Lumbsch (*J. Hattori bot. Lab.* **83**: 1, 1997), Randlane (*Folia cryptog. Estonica* **42**: 105, 2006; Estonia), Lumbsch *et al.* (*MR* **111**: 1133, 2007).

Xylographaceae Tuck. (1888) = Trapeliaceae.

Xylographomyces Cif. & Tomas. (1953) ≡ Xylographa.

Xylohypha (Fr.) E.W. Mason ex Deighton (1960), anamorphic *Pezizomycotina*, Hso.0eP.3. 7, widespread. See Hughes & Sugiyama (*N.Z. Jl Bot.* **10**: 447, 1972), Padhye *et al.* (*J. Clin. Microbiol.* **26**: 702, 1988; from human), Sekhon *et al.* (*European Journal of Epidemiology* **8**: 387, 1992; phaeohyphomycosis, review), Hoog *et al.* (*J. Med. Vet. Mycol.* **33**: 339, 1995; physiology), Masclaux *et al.* (*J. Med. Vet. Mycol.* **33**: 327, 1995; phylogeny).

Xylohyphites Kalgutkar & Sigler (1995), Fossil Fungi, anamorphic *Pezizomycotina*. 1 (Tertiary), India. See Kalgutkar & Sigler (*MR* **99**: 514, 1995).

Xylohyphopsis W.A. Baker & Partridge (2000), anamorphic *Pezizomycotina*. 2, S. Africa. See Partridge *et al.* (*Mycotaxon* **74**: 486, 2000).

xyloma (of *Dothideales*), a sclerotium-like body producing sporogenous structures inside (de Bary) (obsol.).

Xyloma Pers. (1794) = Rhytisma fide Nannfeldt (*Nova Acta R. Soc. Scient. upsal.*, 1932).

Xyloma Raf. (1837) nom. dub., Fungi (L). No spp. included.

Xylomataceae Fr. (1821) = Rhytismataceae.

Xylomelasma Réblová (2006), Sordariomycetidae. 2, France; New Zealand. See Réblová (*Mycol.* **98**: 87, 2006).

Xylometron Paulet (1808) nom. rej. = Pycnoporus fide Donk (*Verh. K. ned. Akad. Wet.* tweede sect. **62**: 1, 1974).

Xylomides Schimp. (1869) ≡ Xylomites.

Xylomites Unger (1841), Fossil Fungi. 57 (Tertiary), Argentina; Europe.

Xylomyces Goos, R.D. Brooks & Lamore (1977), anamorphic *Aliquandostipitaceae*, Hso.≡ eP.3. 8, USA. See Goos *et al.* (*Mycol.* **69**: 282, 1977), Goh *et al.* (*MR* **101**: 1323, 1997), Kohlmeyer & Volkmann-Kohlmeyer (*Fungal Diversity* **1**: 159, 1998), Campbell *et al.* (*CJB* **85**: 873, 2007; phylogeny).

Xylomyzon Pers. (1825) = Serpula fide Cooke (*Spec. Publ. Div. Myc. Dis. Surv. US Department of Agriculture* **3**, 1953).

Xylopezia Höhn. (1917) = Exarmidium fide Sherwood (*Mycotaxon* **5**: 1, 1977), Sherwood-Pike & Boise (*Brittonia* **38**: 35, 1986), Aptroot (*Nova Hedwigia* **66**: 89, 1998).

Xylophagaceae Murrill (1903) = Coniophoraceae.

Xylophagus Link (1809) ≡ Serpula.

Xylophallus (Schltdl.) E. Fisch. (1933) = Mutinus fide Dring *in* Ainsworth *et al.* (Eds) (*The Fungi* **4B**: 458, 1973).

Xylopilus P. Karst. (1882) = Fomes fide Donk (*Per-*

soonia **1**: 173, 1960).

Xylopodium Mont. (1845) = Phellorinia fide Stalpers (*in litt.*).

Xyloschistes Vain. ex Zahlbr. (1903), ? Ostropales (L). 1, Lapland. See Redinger (*Feddes Repert.* **43**: 49, 1938), Etayo (*Monogr. Inst. Pirenaico Ecol.* **5**: 43, 1990).

Xyloschistomyces Cif. & Tomas. (1953) ≡ Xyloschistes.

Xyloschizon Syd. (1922), Rhytismataceae. 2, N. America.

Xylosorium Zundel (1939) = Pericladium fide Mundkur (*Mycol.* **36**: 287, 1944).

Xylosphaera Dumort. (1822) nom. rej. prop. = Xylaria Hill ex Schrank fide Hawksworth *et al.* (*Dictionary of the Fungi* edn 8, 1995).

Xylosphaeria Cooke (1879) nom. dub., Pezizomycotina. 1, British Isles. See Hawksworth *et al.* (*Dictionary of the Fungi* edn 8, 1995).

Xylosphaeria G.H. Otth (1869) ≡ Mycothyridium Petr.

Xylostroma Tode (1790) = Daedalea fide Donk (*Verh. K. ned. Akad. Wet.* tweede sect. **62**: 1, 1974; based on mycelium).

xylostromata, sheets of mycelium as in *Xylostroma*.

Xylotumulus J.D. Rogers, Y.M. Ju & Hemmes (2006), Xylariaceae. 1, Hawaii. See Rogers *et al.* (*Sydowia* **58**: 290, 2006).

Xynophila Malloch & Cain (1971) = Aphanoascus fide Cano & Guarro (*MR* **94**, 1990).

Xyphasma Rebent. (1844) nom. dub. ≡ Hyphasma.

Xystozukalia Theiss. (1916) ? = Scorias fide Luttrell *in* Ainsworth *et al.* (Eds) (*The Fungi* **4A**: 135, 1973).

Yalodendron Capr. (1962) nom. nud., anamorphic *Pezizomycotina*. Used for a yeast from fish. See Capriotti (*Arch. Mikrobiol.* **42**: 407, 1962).

Yalomyces Nag Raj (1993), anamorphic *Pezizomycotina*, Cpd.#eH.15/19. 1, India.

Yamadazyma Billon-Grand (1989), Saccharomycetales. See Fiol & Claisse (*J. gen. appl. Microbiol.* Tokyo **37**: 309, 1991; cytochromes), Yamada *et al.* (*Biosc., Biotechn., Biochem.* **59**: 445, 1995; rDNA), Yamada *et al.* (*Biosc., Biotechn., Biochem.* **59**: 1172, 1995; rDNA), Kurtzman *in* Kurtzman & Fell (Eds) (*Yeasts, a taxonomic study* 4th edn: 273, 1998), Suh *et al.* (*Mycol.* **98**: 1006, 2006; phylogeny).

Yamamotoa Bat. (1960), Asterinaceae. Anamorph *Peltasterella*. 2 or 3, Brazil. See Batista (*Publções Inst. Micol. Recife* **291**: 11, 1960).

Yarrowia Van der Walt & Arx (1981), Saccharomycetales. 1, USA. Perhaps allied to the *Dipodascaceae*. See Deák *et al.* (*Appl. Environm. Microbiol.* **66**: 4340, 2000; from poultry), Suzzi *et al.* (*Int. J. Food Microbiol.* **69**: 69, 2001; from cheese), Kurtzman (*Antonie van Leeuwenhoek* **88**: 121, 2005; phylogeny), Suh *et al.* (*Mycol.* **98**: 1006, 2006; phylogeny).

Yatesula Syd. & P. Syd. (1917), Chaetothyriaceae. 3 or 2, widespread (tropical). See Magnes (*Biblthca Mycol.* **165**, 1997).

Ybotromyces Rulamort (1986), anamorphic *Pezizomycotina*, Hso.0eP.38. 1 (from humans), Spain. See Benoldi *et al.* (*J. Med. Vet. Mycol.* **29**: 9, 1991; cutaneous phaeohyphomycosis, as *Botryomyces*).

yeasticidal, able to kill yeasts (q.v.).

Yeasts. Unicellular, budding fungi. The yeasts are not a formal taxonomic unit but a growth form exhibited by a range of unrelated fungi, and one exhibited in some cases by primarily filamentous forms as a part of the life-cycle or under particular environmental conditions. The yeast growth form is typically unicellular, with asexual reproduction occurring by budding (an at least some cases this seems to involve the same processes as conidiogenesis but in the context of organisms existing as separate individual cells); occasionally yeast cells may be seen in short branched or unbranched chains, and in some species hyphae may also be encountered. Different genera are included depending on the authority and definition used; Kurtzman & Fell (Eds) (1998) accept 100 gen. and ›700 spp., and Barnett *et al.* (*Yeasts: characteristics and identification*, 2000) 93 gen. and 678 spp. The best known fungi with a yeast form of growth are those used in baking and fermentation of alcoholic drinks, particularly *Saccharomyces cerevisiae*. The yeast form of growth also occurs in fungi in a wide range of other habitats, including leaf surfaces, soil (Sláviková & Vadkertiová, *J. Basic Microbiol.* **43**: 430, 2003) and medical and veterinary environments. They have also, more recently, been noted from insect guts (Suh *et al*, *MR* **103**: 261, 2005) and the deep ocean (Bass *et al.*, *Proc. Roy. Soc.* B. Biol. Sci. **274**: 3069, 2007), and probably many of these fungi remain undiscovered. The yeast form of growth makes these organisms ideal for genetic and other research (e.g. Ostergaard *et al.*, *Microbiol. Mol. Biol. Rev.* **64**: 34, 2000; d'Enfert & Hube (*Candida, comparative and functional genomics*, 2008), and the yeast *Saccharomyces cerevisiae* was the first eukaryote to have its genome fully sequenced (Williams, *Science* **272**: 481, 1996).

Sporogenous yeasts have teleomorphs in either the ascomycetes (esp. *Saccharomycetales* and *Schizosaccharomycetales*) or the basidiomycetes (*Sporidiales*, *Tremellales*); **asporogenous yeasts** are conidial fungi; **apiculate -**, yeasts (e.g. *Saccharomycodes*, *Nadsonia*, *Hanseniaspora*, *Kloeckera*) having minute polar projections which are multiple scars (annellides) (Streiblova *et al.*, *J. Bact.* **88**: 1104, 1964); **baker's, brewer's** or **beer -**, *Saccharomyces cerevisiae*; **black -**, yeast-like states of *Aureobasidium*, *Cladosporium*, *Moniliella*, etc., and esp. anamorphs of *Herpotrichielleae* (q.v.) incl. *Exophiala*, *Ramichloridium* and *Rhinocladiella* (de Hoog & Hermanides-Nijhof, *Stud. Mycol.* **15**, 1977; de Hoog (Ed.), *Stud. Mycol.* **19**, 1979; *Ann. Rep. Res. Center Path. Fungi, Chiba* **7**: 50, 1993, surveys); **bottom -**, one settling out at the bottom of a fermented liquid (the wort), e.g. *S. uvarum*; **top -**, one accumulating at the surface of the fermented wort, e.g. *S. cerevisiae*; **Chinese -**, *Amylomyces rouxii* and other fungi (Ellis *et al.*, *Mycol.* **68**: 131, 1976); **'flor' -**, one to which special qualities of wines (e.g. bouquet, taste) are due; **food -**, dry *Candida utilis* (q.v.) and other yeasts; **petite -**, a respiratory deficient mutant (Bulder, *Ant. v. Leeuwenhoek* **30**: 1, 1964); **scum -**, one (e.g. *Trichosporon cutaneum*) forming a surface scum or slime layer; **shadow (mirror) -**, *Bullera*, *Sporobolomyces* etc., producing ballistospores; **springer -**, the Institut Pasteur, Paris, strain of *S. cerevisiae*; **toddy -**, a mixture of yeasts which ferment the juice of the palmyra palm (*Borassus flabellifer*) (Ahmad *et al.*, *Mycol.* **46**: 708, 1954); **wine -**, races of *S. cerevisiae*.

Lit. (covering anamorphs and teleomorphs): **Identification**: Arx *et al.* (*Stud. Mycol.* **14**, 1977; genera), Barnett (*J. gen. Microbiol.* **99**: 183, 1977; nutritional tests), Barnett & Pankhurst (*A new key to the yeasts,*

1974), Barnett *et al.* (*A guide to identifying and classifying yeasts*, 1979; *Yeasts: characteristics and identification*, 1983; edn 2, 1990; edn 3, 2000), Belin (*Can. J. Microbiol.* **27**: 1235, 1981; spp. described since 1973), Kudrjanzev (*Sistematika drozheĭ*, 1954 [Russ., Germ., transl. *Die Systematik der Hefen*, 1960]), Kurtzman & Fell (Eds) (*The yeasts*, edn 4, 1998), Lodder (Ed.) (*The Yeasts*, edn 3, 1984; edn 2, 1970; edn 1, 1952), Lodder & Kreger-van Rij (*The yeasts*, 1952), Moore (*Bot. mar.* **23**: 361, 1980; basidiomycetous yeasts), Nakase *et al.* (*Jap. J. Med. Myc.* **32**: 21, 1991; systematics of basidiomycetous yeasts), Payne *et al.* (*J. gen. Microbiol.* **128**: 1265, 1982; computer generated keys gen.), Yarrow & Nakase (*Ant. v. Leeuwenhoek* **41**: 81, 1975; DNA base composition).

General: Fragner (*Česká Myk.* **39**: 234, 1985; spp. on humans), Fukazawa, *et al.* (*Handb. Appl. Mycol.: Humans, Animals & Insects* **2**: 425, 1991; serology and immunology of medically important yeasts), Guthrir & Fink (Ed.) (*Guide to yeast genetics and molecular biology*, 1991), Herskowitz (*Nature* **357**: 190, 1992; regulation hyphal growth), Jong *et al.* (*Mycotaxon* **31**: 207, 1988; coding strain features), Kirsop & Kertzman (Eds) (*Yeasts*, 1988; guide to sources), Odds (*J. med. vet. mycol.* **29**: 413, 1991; preservation in distilled water), Kurtzman (*Int. J. Syst. Bact.* **42**: 1, 1992; review), Prescott (Ed.) (*Methods in cell biology*, **11, 12**, *Yeast cells*, 1975), Rose & Péter (Eds) (Biodiversity and ecophysiology of yeasts. *The yeast handbook* **1**, 2006), Rose & Harrison (Eds) (*The yeasts*, **1**, *Biology of yeasts*, 1969, edn 2, 1987; **2**, *Physiology and biochemistry of yeasts*, edn 2, 1986; **3**, *Yeast technology*, 1970; **4**, *Yeast organelles*, edn 2, 1991), Rose *et al.* (*The Yeasts*, **6**, *Yeast genetics*, edn 2, 1995), Seehaus *et al.* (*Current Genetics* **10**: 103, 1985; gene probes), Sherman *et al.* (Eds) (*Methods in yeast genetics*, 1987), Skinner *et al.* (Eds) (*Biology and activities of yeasts*, 1980).

See also Boekhout *et al.* (*Yeasts of the world, morphology, physiology, sequences and identification*, 2002; compact disk).

See also *Blastomycota*, Anamorphic fungi, *Saccharomycetales*, *Schizosaccharomycetales*, *Ustomycetes*.

yellow rice, rice discoloured by *Penicillium islandicum* and rendered carcinogenic for rodents and possibly humans; see Mycotoxicoses.

yellows, of cabbage (*Fusarium oxysporum* f.sp. *conglutinans*).

Yelsemia J. Walker (2001), Melanotaeniaceae. 4 (on *Anthericaceae*, *Byblidaceae*, *Campanulaceae*, *Droseraceae*), Australia; N. America; S.E. Asia. See Walker (*Australasian Mycologist* **20**: 61, 2001; Australia), Shivas & Vánky (*Fungal Diversity* **13**: 131, 2003; Australia).

Yenia Liou (1949) = Ustilago fide Mordue (*Mycopathologia* **116**: 227, 1991).

Yinmingella Goh, K.M. Tsui & K.D. Hyde (1999), anamorphic *Pezizomycotina*, Hso.?.?. 1, Hong Kong. See Goh *et al.* (*CJB* **76**: 1693, 1998).

Yoshinagaia Henn. (1904), ? Dothioraceae. Anamorph *Japonia*. 1, Japan. See Eriksson & Hawksworth (*SA* **5**: 161, 1986; nomencl.), Sivanesan & Hsieh (*MR* **99**: 1295, 1995).

Yoshinagamyces Hara (1912) = Japonia fide Sivanesan & Hsieh (*MR* **99**: 1295, 1995).

Yoshinagella Höhn. (1913), Dothideales. 3, Japan; Hawaii.

Youngiomyces Y.J. Yao (1995), Endogonaceae. 4, N. America; Australasia. See Yao *et al.* (*Genera of Endogonales*: 229 pp., 1996).

Ypsilina J. Webster, Descals & Marvanová (1999), anamorphic *Pezizomycotina*, Hso.?.?. 1, British Isles. See Descals *et al.* (*CJB* **76**: 1658, 1998).

Ypsilon.

Ypsilonia Lév. (1846), anamorphic *Acanthotheciella*, Ccu.0bP.1. 5, widespread. See Nag Raj (*CJB* **55**: 1599, 1977).

Ypsilonidium Donk (1972) = Thanatephorus fide Langer (*Biblthca Mycol.* **158**, 1994).

Ypsilospora Cummins (1941), Raveneliaceae. 3 (on *Baphia*, *Inga* (*Leguminosae*)), W. Africa; America (tropical). See Ono & Hennen (*TBMS* **73**: 229, 1979), Eboh (*TBMS* **85**: 39, 1985; syn. of *Chaconia*), Hernández & Hennen (*Mycol.* **95**: 728, 2003).

YPSS, see Media.

Yuccamyces Gour, Dyko & B. Sutton (1979), anamorphic *Pezizomycotina*, Hsp.0fH.3. 5, India; Cuba. See Gour *et al.* (*TBMS* **72**: 413, 1979).

Yuea O.E. Erikss. (2003), Xylariales. 1, Chile. See Eriksson (*Mycotaxon* **85**: 314, 2003).

Yukonia R. Sprague (1962) = Buergenerula fide Barr (*Mycol.* **68**: 611, 1976).

Yunnania H.Z. Kong (1998), anamorphic *Pezizomycotina*, Hso.?.?. 1, China. See Kong (*Mycotaxon* **69**: 320, 1998).

Yurrajia.

Zaghouania Pat. (1901), Pucciniaceae. 2 (on *Oleaceae*), mediterranean; India; Japan; Taiwan. See Thirumalachar (*Bot. Gaz.* **107**: 74, 1945; morphology, life cycle [as *Cytopsora*]), Bahçecioglu & Gjaerum (*Mycotaxon* **90**: 55, 2004; Turkey).

Zaghouaniaceae Syd. & P. Syd. (1915) = Pucciniaceae.

Zahlbruckner (Alexander; 1860-1938; Slovakia, later Austria). Born, St Georgen [then in Hungary], near Pressburg [now Bratislava]; PhD, Vienna University (1883); member of staff (1883-1899), then Keeper (1899-1918) then Director (1918-), Naturhistorischen Museum, Vienna. Published important regional works on lichen-forming fungi of central Africa, China, Chile (Easter Island and Juan Fernández Island), Dalmatia, Formosa, Hawaii (partly with Magnusson, q.v.), Japan, Java, Samoa and South America; was the first author to provide a comprehensive account of genera and families of lichen-forming fungi. planned the sections for lichen-forming fungi in *Rabenhorst's Kryptogamenflora* (**9**, 1930-1960); published exsiccata (e.g. *Lichenes Rariores Exsiccati*). Collections in Padua (**PAD**) and Vienna (**W**). *Publs*. Flechten. *Engler & Prantl, Die Naturlichen Pflanzenfamilien* (1903-1907); *Catalogus Lichenum Universalis* 10 vols (1921-1940) [a major catalogue of lichen-forming fungi, publications in which their names have been used, and details of synonymy; it remains the basic nomenclatural reference work on lichen-forming fungi]. *Biogs, obits etc.* Dickinson (*Kew Bulletin* **1938**: 304, 1938); Grummann (1974: 444); Keissler (*Revue de Mycologie* **4**: 3, 1939) [portrait]; Lackovičova (*Dr Alexander Zahlbruckner (1860-1938) Osonosť a Dielo*, 1988) [bibliography, biography]; Redinger (*Annales de Cryptogamie Exotique* **6**: 85, 1933) [bibliography, portrait]; Stafleu & Cowan (*TL-2* **7**: 500, 1988).

Zahlbrucknera Herre (1910) [non *Zahlbrucknera* Rchb. 1832, *Saxifragaceae*] ≡ Zahlbrucknerella.

Zahlbrucknerella Herre (1912), Lichinaceae (L). 8, widespread. See Henssen (*Lichenologist* **9**: 17, 1977; key), Schultz & Büdel (*Lichenologist* **34**: 39, 2002; key), Henssen (*Biblthca Lichenol.* **88**: 195, 2004).

Zakatoshia B. Sutton (1973), anamorphic *Pezizomycotina*, Hso.0eH.15. 2 (on fungi), Canada; Austria. See Gams (*Windahlia* **16**: 59, 1986).

Zalerion R.T. Moore & Meyers (1962), anamorphic *Lulwoana*, Hsy.0-≡ hP.1. 9 (mainly marine), N. America; Europe. See Buczacki (*TBMS* **59**: 159, 1972), Fisher & Webster (*Nova Hedwigia* **54**: 77, 1992), Campbell *et al.* (*MR* **109**: 556, 2005; phylogeny, teleomorph), Baschien *et al.* (*Nova Hedwigia* **83**: 311, 2006).

Zamenhofia Clauzade & Cl. Roux (1985) = Porina Müll. Arg. fide Roux (*SA* **6**: 156, 1987), McCarthy (*Biblthca Lichenol.* **52**, 1993).

Zanchia Rick (1958) nom. dub., Basidiomycota. See Donk (*Taxon* **12**: 167, 1963).

Zanclospora S. Hughes & W.B. Kendr. (1965), anamorphic *Pezizomycotina*, Hso.0eH.15. 6, southern hemisphere. See Hughes & Kendrick (*N.Z. Jl Bot.* **3**: 151, 1965).

Zasmidium Fr. (1849) nom. dub., anamorphic *Mycosphaerellaceae*. fide Hawksworth & Riedl (*Taxon* **26**: 208, 1977), Arzanlou *et al.* (*Stud. Mycol.* **58**: 75, 2007).

zearalenone, a toxin of *Fusarium graminearum* (teleomorph *Gibberella zeae*); the cause of vulvovaginitis and infertility in cattle and pigs.

Zebrospora McKenzie (1991), anamorphic *Pezizomycotina*, Hso.≡ eP.10. 1, Australasia; Pacific Islands. See McKenzie (*Mycotaxon* **41**: 189, 1991).

Zelandiocoela Nag Raj (1993), anamorphic *Pezizomycotina*, St.0eH.15. 1, New Zealand. See Nag Raj (*Coelomycetous Anamorphs with Appendage-bearing Conidia*: 990, 1993).

Zelleromyces Singer & A.H. Sm. (1960) = Lactarius. Gasteroid forms are polyphyletic and have formerly been morphologically recognised as a separate genus. fide Miller *et al.* (*Mycol.* **93**: 344, 2001), Lebel (*Australasian Mycol.* **21**: 4, 2002; Australian spp.), Nuytinck *et al.* (*Belgian Jour. Bot.* **136**: 145, 2004; European spp.), Miller *et al.* (*Mycol.* **98**: 960, 2006).

Zeloasperisporium R.F. Castañeda (1996), anamorphic *Venturiaceae*, Hso.?.?. 2, Cuba. See Castañeda Ruíz *et al.* (*Mycotaxon* **60**: 284, 1996), Crous *et al.* (*Stud. Mycol.* **58**: 185, 2007; phylogeny).

Zelopelta B. Sutton & R.D. Gaur (1984), anamorphic *Pezizomycotina*, Cpt.1bH.10. 1, India. See Sutton & Gaur (*TBMS* **82**: 558, 1984).

Zelosatchmopsis Nag Raj (1991), anamorphic *Pezizomycotina*, Ccu.0eH.15. 1, Cuba. See Nag Raj (*CJB* **69**: 633, 1991).

Zelotriadelphia R.F. Castañeda, Saikawa, M. Stadler & Iturr. (2005), anamorphic *Pezizomycotina*. 1, Cuba. See Castañeda Ruiz *et al.* (*Mycotaxon* **91**: 340, 2005).

Zendera Redhead & Malloch (1977) = Dipodascus fide von Arx (*Persoonia* **9**: 393, 1977).

Zeora Fr. (1825) = Lecanora fide Hawksworth *et al.* (*Dictionary of the Fungi* edn 8, 1995).

zeorine (of apothecia), like those of *Zeora*.

Zephirea Velen. (1947) = Mycena fide Horak (*Synopsis generum Agaricalium*, 1968).

Zercosporidium Thor (1930) nom. dub., ? Fungi. 1, Svalbard.

Zernya Petr. (1947), anamorphic *Pezizomycotina*, Cpd.1eH.?. 1, Brazil.

Zerovaemyces Gorovij (1977) = Rhacophyllus fide Kuyper (*in litt.*).

Zerovaemycetaceae Gorovij (1977) = Agaricaceae.

Zeta Bat. & R. Garnier (1961) nom. dub., Pseudoperisporiaceae. 1, Brazil. See Rossman *et al.* (*Stud. Mycol.* **42**: 248 pp., 1999).

Zetesimomyces Nag Raj (1988), anamorphic *Pezizomycotina*, Ccu.1eH.15. 1, Cuba. See Nag Raj (*CJB* **66**: 2143, 1988).

Zetiasplozna Nag Raj (1993), anamorphic *Amphisphaeriaceae*, Cpd.≡ eP.19. 5, widespread. See Nag Raj (*Coelomycetous Anamorphs with Appendage-bearing Conidia*: 996, 1993), Jeewon *et al.* (*Mol. Phylogen. Evol.* **25**: 378, 2002).

Zeuctomorpha Sivan., P.M. Kirk & Govindu (1984), Pleosporaceae. 1, India. See Sivanesan *et al.* (*Bitunicate Ascomycetes and their Anamorphs*: 572, 1984).

Zeugandromyces Thaxt. (1912), Laboulbeniaceae. 3, widespread. See Nannfeldt (*Svensk bot. Tidskr.* **43**: 468, 1949).

zeugite, the organ in which fertilization is completed and the dikaryophase ends; e.g. an ascus or a basidium.

Zeus Minter & Diam. (1987), Rhytismataceae. 1, Greece. See Minter *et al.* (*TBMS* **88**: 55, 1987).

Zevadia J.C. David & D. Hawksw. (1995), anamorphic *Pezizomycotina*, Hso.?.?. 1, Ireland. See David & Hawksworth (*Biblthca Lichenol.* **58**: 64, 1995).

Zignoëlla Sacc. (1878) = Chaetosphaeria. A highly confused genus in need of further revision. fide Müller (*SA* **6**: 156, 1987), Cannon (*SA* **15**: 121, 1997), Réblová & Gams (*Czech Mycol.* **51**: 1, 1999), Fernández *et al.* (*Mycol.* **98**: 121, 2006).

Zignoina Cooke (1885) ? = Wallrothiella fide Eriksson & Hawksworth (*SA* **6**: 253, 1987), Réblová & Seifert (*Mycol.* **96**: 343, 2004).

Zilingia Petr. (1934), anamorphic *Pezizomycotina*, St.0eH.?. 1, Siberia.

Zimmermanniella Henn. (1902), Phyllachoraceae. 1 (from living leaves), S.E. Asia. See Petrak (*Hedwigia* **68**, 1928), Cannon (*IMI Descr. Fungi Bact.* **140**: no. 1140, 1992).

Zinzipegasa Nag Raj (1993), anamorphic *Pezizomycotina*, Cac.≡ eH.15/19. 1, Argentina. See Nag Raj (*Coelomycetous Anamorphs with Appendage-bearing Conidia*: 1005, 1993).

Zobelia Opiz (1855) = Choiromyces fide Trappe (*TBMS* **65**: 496, 1975).

Zodiomyces Thaxt. (1891), Laboulbeniaceae. 2, widespread. See Weir & Blackwell (*MR* **105**: 1182, 2001; phylogeny).

Zodiomycetaceae Nann. (1934) = Laboulbeniaceae.

Zoellneria Velen. (1934), Sclerotiniaceae. Anamorph *Amerosporium*. 1, Europe. See Spooner (*Biblthca Mycol.* **116**, 1987), Johnston & Gamundí (*N.Z. Jl Bot.* **38**: 493, 2000).

Zoggium Lar.N. Vassiljeva (2001), Mytilinidiaceae. 1, Europe. See Vasil'eva (*Mikol. Fitopatol.* **35**: 17, 2001).

Zografia Bogoyavl. (1922) = Coelomomyces fide Keilin (*Parasitol.* **19**: 365, 1927).

zonate, having concentric lines often forming alternating pale and darker zones near the margins; used of crustose lichen thalli, polypore surfaces, etc.

zonation (of cultures), regular concentric variation of

texture, pigmentation or sporulation frequently associated with fluctuations (esp. diurnal) in light, temperature, or other factors; 'Liesegang' phenomenon. See Bisby (*Mycol.* **17**: 89, 1925), Hein (*Am. J. Bot.* **17**: 143, 1930), Kafi & Tarr (*TBMS* **46**: 549, 1964). **ecological -**, see Ecology.

zone lines, narrow, dark brown, or black, lines (pseudosclerotia) or plates (pseudosclerotial plates) in decayed wood (esp. hardwoods) generally caused by fungi (Lopez *et al.*, *TBMS* **64**: 465, 1975); see also spalted (wood).

Zonilia Raf. (1815) nom. dub., Fungi. No spp. included.

Zonosporis Clem. (1931) ≡ Schwanniomyces.

zoogametes, a motile gamete; planogametes.

Zoogloea Eberth (1873) = Trichosporon fide Kirk (*in litt.*).

zoogloea (of bacteria), a colony embedded in a slimy substance.

zoogonidium (1) ? = zoospore (q.v.), (2) an aplanospore of a photobiont within the thallus of a lichen (obsol.).

Zoopagaceae Drechsler (1938), Zoopagales. 5 gen., 67 spp.

Lit.: Drechsler (*Mycol.* **26**: 135, 1934) *et seq.*; family proposed by, Drechsler (*Mycol.* **30**: 152, 1938), Duddington *in* Ainsworth *et al.* (Eds) (*The Fungi* **4B**: 231, 1973; emend. & keys), Dyal (*Sydowia* **27**: 293, 1973), Dayal (*Sydowia* **27**: 293, 1976; keys) Predacious fungi, Tanabe *et al.* (*Mol. Phylogen. Evol.* **16**: 253, 2000), Barron (*Biodiversity of Fungi* Inventory and Monitoring Methods: 435, 2004).

Zoopagales Bessey ex R.K. Benj. (1979). Zoopagomycotina. 5 fam., 22 gen., 190 spp. Asexual reproduction by conidia or merosporangia, sexual reproduction by zygospores; cosmop. parasites of fungi (mycoparasites), nematodes, amoebae, and other small terrestrial animals. Fams:

(1) **Cochlonemataceae**
(2) **Helicocephalidaceae**
(3) **Piptocephalidaceae**
(4) **Sigmoideomycetaceae**
(5) **Zoopagaceae**

Lit.: Duddington (*in* Ainsworth *et al.*, *The Fungi* **4B**: 231, 1973, *Biol. Rev.* **31**: 152, 1956), Drechsler (*Biol. Rev.* **16**: 265, 1941; review), Dyal (*Sydowia* **27**: 293, 1973; key spp. on amoebae & nematodes), Patil & Patil (*Indian Phytopath.* **47**: 217, 1994), Tanabe *et al.* (*Mol. Phylogen. Evol.* **16**: 253, 2000; phylogeny), White *et al.* (*Mycol.* **98**: 860, 2006; molecular phylogeny), Hibbett *et al.* (*MR* **111**: 109, 2007), and see under Families.

Zoopage Drechsler (1935), Zoopagaceae. 11, N. America. See Drechsler (*Mycol.* **27**: 30, 1935), Drechsler (*Mycol.* **28**: 363, 1936), Drechsler (*Mycol.* **29**: 229, 1937), Drechsler (*Mycol.* **39**: 379, 1947), Jones (*TBMS* **45**: 348, 1962), Dyal (*Sydowia* **27**: 293, 1976; keys parasites of nematodes and amoebae, bibliogr.).

Zoopagomycotina Benny (2007), Zygomycota. 22 gen., 190 spp., Ord.:

Zoopagales

Lit.: James *et al.* (*Mycol.* **98**: 860, 2006; molecular phylogeny), Hibbett *et al.* (*MR* **111**: 109, 2007), Hoffmann *et al.* (in: Gherbawy (Ed.) *Current Advances in Molecular Mycology*): in press, 2008).

Zoophagus Sommerst. (1911), Zoopagaceae. 5 (on *Algae*), Europe; N. America. See Dick *in* Ainsworth *et al.* (Eds) (*The Fungi* **4A**: 145, 1973), Coffell *et al.* (*Mycol.* **82**: 326, 1990; status; based on a misidentified *Cephaliophora*), Dick (*MR* **94**: 347, 1990; key, emend. of genus), Dick *et al.* (*Mycol.* **82**: 316, 1990; status), Powell *et al.* (*Mycol.* **82**: 460, 1990; status; based on a misidentified *Cephaliophora*), Morikawa *et al.* (*MR* **97**: 421, 1993; status), Glocking (*MR* **101**: 1179, 1997), Saikawa (*Mycol.* **89**: 268, 1997; ultrastr.), Liu *et al.* (*Mycosystema* **17**: 105, 1998), Tanabe *et al.* (*Mycol.* **91**: 830, 1999; phylogeny), White *et al.* (*Mycol.* **98**: 872, 2006; phylogeny).

zoophilic (of dermatophytes, etc.), preferentially pathogenic for animals; cf. anthrophilic.

Zoophthora A. Batko (1964), Entomophthoraceae. 32, widespread. See Remaudière & Hennebert (*Mycotaxon* **11**: 269, 1980), Ben-Ze'ev & Kenneth (*Mycotaxon* **14**: 456, 1982), Glare *et al.* (*Aust. J. Bot.* **35**: 49, 1987), Humber (*Mycotaxon* **34**: 441, 1989), Keller (*Sydowia* **43**: 39, 1991; key), Bałazy (*Flora Polska* **24**: 1, 1993), Keller & Petrini (*Sydowia* **57**: 23, 2005; key), Keller (*Sydowia* **58**: 38, 2006; Switzerland), Keller (*Sydowia* **59**: 75, 2007; Switzerland, comb.nov.).

zoosporangium (**zoosporange**), a sporangium producing zoospores.

zoospore, a motile sporangiospore, i.e. one having flagella; swarm spore; swarmer; simblospore; planospore; planont; cf. swarm-cell; see Waterhouse (*TBMS* **45**: 1, 1962), Fuller (*Mycol.* **69**: 1, 1977), Lange & Olson (*Dansk bot. Arkiv.* **33**, 1979; uniflagellate zoospores).

Zopf (Wilhelm; 1846-1909; Germany). Student under O. Brefeld (q.v.), Berlin (1874-1877); PhD, Halle (1878); Assistant Professor (1883) then Professor (1887), Halle University. Made major contributions to knowledge of an exceptionally wide range of fungi and lichens, including accounts of *Chaetomium*, lichenicolous fungi, and lichen chemistry. *Publs. Die pilze in Morphologischer, Physiologischer, Biologischer und Systematischer Beziehung* (1890); *Untersuchungen über die durch Parasitische Pilze Hervorgerufenen Krankheiten der Flechten* (1897-1898); *Die Flechtenstoffe* (1907). *Biogs, obits etc.* Grummann (1974: 55); Huneck *et al.* (*Willdenowia* **7**: 31, 1973) [application of Zopf's chemical names]; Stafleu & Cowan (*TL-2* **7**: 553, 1988); Tobler (*Bericht der Deutschen Botanischen Gesellschaft* **27**: (58), 1910) [portrait].

Zopfia Rabenh. (1874), Zopfiaceae. 5, widespread. See Hawksworth (*CJB* **57**: 91, 1979), Kruys *et al.* (*MR* **110**: 527, 2006; phylogeny).

Zopfiaceae G. Arnaud ex D. Hawksw. (1992), ? Pleosporales. 8 gen. (+ 2 syn.), 18 spp. Polyphyletic as currently circumscribed.

Lit.: Hawksworth (*CJB* **57**: 91, 1979), Hawksworth (*SA* **6**: 153, 1987), Hyde (*Trans. Mycol. Soc. Japan* **30**: 333, 1989), LoBuglio *et al.* (*Mol. Phylogen. Evol.* **6**: 287, 1996), Ranghoo & Hyde (*MR* **103**: 938, 1999), Kruys *et al.* (*MR* **110**: 527, 2006).

Zopfiella G. Winter (1884), Lasiosphaeriaceae. 24, widespread. See Malloch & Cain (*CJB* **49**: 869, 1971), Guarro *et al.* (*SA* **10**: 79, 1991; key), Huhndorf *et al.* (*Mycol.* **96**: 368, 2004; phylogeny), Miller & Huhndorf (*Mol. Phylogen. Evol.* **35**: 60, 2005; phylogeny), Cai *et al.* (*MR* **110**: 359, 2006; phylogeny).

Zopfinula Kirschst. (1939) = Keissleriella fide Bose (*Phytopath. Z.* **41**, 1961).

Zopfiofoveola D. Hawksw. (1979), Zopfiaceae. 1,

Sweden. See Hawksworth (*CJB* **57**: 91, 1979).

Zopheromyces B. Sutton & Hodges (1977), anamorphic *Pezizomycotina*, Hso.1eP.6. 1, Brazil. See Sutton & Hodges (*Nova Hedwigia* **28**: 493, 1977).

Zosterodiscus Hertel (1984) = Lecidea fide Hertel (*Mitt. bot. StSamml., München* **23**: 321, 1987).

Zschackea M. Choisy & Werner (1932) = Verrucaria Schrad. fide Hawksworth *et al.* (*Dictionary of the Fungi* edn 8, 1995).

Zugazaea Korf, Iturr. & Lizoň (1998), Helotiales. 1, Canary Islands. See Itturriaga *et al.* (*Mycol.* **90**: 697, 1998).

Zukalia Sacc. (1891) = Chaetothyrium fide von Arx & Müller (*Stud. Mycol.* **9**, 1975).

Zukalina Kuntze (1891) ? = Ascozonus fide Velenovský (*Monogr. Dicom. Boehm.*, 1934), van Brummelen (*Persoonia* **16**: 425, 1998).

Zukaliopsis Henn. (1904) = Molleriella fide von Arx (*Persoonia* **2**: 421, 1963).

Zundeliomyces Vánky (1987), Microbotryaceae. 1 (on *Polygonum*), Kazakhstan. See Vánky (*TBMS* **89**: 477, 1987), Piepenbring (*Frontiers in Basidiomycote Mycology*: 117, 2004; gall morphology).

Zundelula Thirum. & Naras. (1952) = Dermatosorus fide Langdon (*TBMS* **68**: 447, 1977).

Zunura Nag Raj (1993), anamorphic *Pezizomycotina*, Cac.0eH.19. 1, India. See Nag Raj (*Coelomycetous Anamorphs with Appendage-bearing Conidia*: 1006, 1993).

Zwackhia Körb. (1855) = Opegrapha Ach. fide Hawksworth *et al.* (*Dictionary of the Fungi* edn 8, 1995).

Zwackhiomyces Grube & Hafellner (1990), Xanthopyreniaceae. 9 (lichenicolous), widespread. See Grube & Hafellner (*Nova Hedwigia* **51**: 283, 1990; monogr.), Harris (*More Florida Lichens*, 1995), Aptroot (*Lichenologist* **30**: 501, 1998), Hoffmann & Hafellner (*Biblthca Lichenol.* **77**: 1, 2000), Calatayud *et al.* (*Lichenologist* **39**: 129, 2007; Spain, Iran, key).

Zychaea Benny & R.K. Benj. (1975), Syncephalastraceae. 1, Mexico. See Benny & Benjamin (*Aliso* **8**: 334, 1975), Voigt & Wöstemeyer (*Gene* **270**: 113, 2001; phylogeny).

Zygaenobia Weiser (1951), Entomophthorales. 1, former Czechoslovakia. See Weiser (*Ent. Listy* **14**: 134, 1951).

zygangium, gametangium of a zygomycete.

Zygnemomyces K. Miura (1973), Meristacraceae. 2, Japan; Australia. See Miura (*Rep. Tottori Mycol. Inst.* **10**: 520, 1973), McCulloch (*TBMS* **68**: 173, 1977), Tucker (*Mycotaxon* **13**: 481, 1981; key), Saikawa *et al.* (*CJB* **75**: 762, 1997; ultrastr.).

Zygoascus M.T. Sm. (1986), Trichomonascaceae. Anamorph *Candida*. 1, widespread. See von Arx & van der Walt (*Stud. Mycol.* **30**: 167, 1987), Smith *in* Kurtzman & Fell (Eds) (*Yeasts, a taxonomic study* 4th edn: 422, 1998), Brandt *et al.* (*J. Clin. Microbiol.* **42**: 3363, 2004; from human), Smith *et al.* (*Int. J. Syst. Evol. Microbiol.* **55**: 1353, 2005; phylogeny), Suh *et al.* (*Mycol.* **98**: 1006, 2006; phylogeny), Kurtzman & Robnett (*FEMS Yeast Res.* **7**: 141, 2007).

Zygochytrium Sorokīn (1874), ? Chytridiomycetes. 1 (on dead insect), Europe.

zygoconidium, Asexual propagule formed from the fusion or conjoining of two conidia generated simultaneously from adjacent conidiogenous loci. Kendrick & Watling (*The Whole Fungus*: 543, 1979) applied the term isthmospore (q.v.) to this structure and although similar in appearance, it is distinct in the nature of its origin. Zygoconidia are known in a number of genera, mainly basidiomycete anamorphs (e.g. *Anastomyces*, *Christiansenia* and *Zygogloea*). The term was first introduced by Boidin (*Bull. Soc. Linn. Lyon* **39**: 132, 1970) but more recently taken up by others following Oberwinkler & Bandoni (*Norw. J. Bot.* **2**: 501, 1982).

Zygodesmella Gonz. Frag. (1917) nom. dub., Fungi. See Donk (*Taxon* **11**: 103, 1962).

Zygodesmus Corda (1837) nom. dub., Pezizomycotina. See Rogers (*Mycol.* **40**: 633, 1948).

Zygofabospora Kudrjanzev (1960) = Kluyveromyces fide Naumov (*Mikrobiologiya* **57**: 114, 1988; key), Kurtzman *et al.* (*Taxon* **50**: 907, 2001; nomencl.), Naumov & Naumova (*FEMS Yeast Res.* **2**: 39, 2002), Naumova *et al.* (*FEMS Yeast Res.* **5**: 263, 2004; phylogeny).

Zygogloea P. Roberts (1994), Pucciniomycotina. 1, British Isles. See Roberts (*Mycotaxon* **52**: 241, 1994).

Zygohansenula Lodder (1932) = Pichia fide Hawksworth *et al.* (*Dictionary of the Fungi* edn 8, 1995).

Zygolipomyces Krassiln., Babeva & Meavahd (1967) = Lipomyces fide Lodder (*Yeasts, a taxonomic study* 2nd edn, 1970).

Zygomycetes G. Winter (1880), Zygomycota. Saprobes or parasites (esp. of arthropods). A polyphyletic taxon as traditionally circumscibed at this rank; see subphyla of the *Zygomycota*.

zygomycosis, a mycosis caused by a member of the Zygomycetes. Cf. Mucormycosis, phycomycosis.

Zygomycota Moreau (1954), Fungi. 10 ord., 27 fam., 168 gen., 1065 spp. A polyphyletic (or paraphyletic) taxon including four subphyla:

(1) **Entomophthoromycotina**
(2) **Kickxellomycotina**
(3) **Mucoromycotina**
(4) **Zoopagomycotina**

Lit.: Benjamin (*in* Kendrick (Ed.), *The whole fungus*: 579, 1979), Benny (*in* Parker, 1982, **1**: 184), O'Donnell (*Zygomycetes in culture*, 1979), Jeffries (*Bot. J. Linn. Soc.* **91**: 135, 1985; mycoparasitism), Schipper (*in* Rayner *et al.* (Ed.), *Evolutionary biology of the fungi*: 261, 1987), Morton & Benny (*Mycotaxon* **37**: 471, 1990; *Endogonales*, *Glomales*), Chien (*in* Peng & Chou (Eds), *Biodiversity & terrestrial ecosystems* (*Acad. Sinica Monogr.* **14**) p. 215, 1994), Weitzman & Della-Latta (*Clin. Microbiol. Newsletter* **19**: 81, 1997; zygomycosis), Cavalier-Smith (*Biol. Rev.* **73**: 203, 1998), Benny *et al.* (*in* McLaughlin *et al.* (Eds), *The Mycota* **7A**, 113, 2000), Tanabe *et al.* (*Mol. Phylogen. Evol.* **16**: 253, 2000), Strauss *et al.* (*South African Journal of Science* **96**: 597, 2000; medium for selective isolation of *Kickxellales*, *Mortierellales*, *Mucorales*, and *Piptocephalidaceae* [*Zoopagales*]), Shalchian-Tabrizi *et al.* (*PLoS ONE* **3**(3): e2098, 2008; phylogeny), Benny (*Aliso* **26**: 37, 2008; methods of isolation, culture, observation and preservation), and under Subphyla and Orders.

Zygomycotina Caval.-Sm. (1998), see *Zygomycota*.

Zygophiala E.W. Mason (1945), anamorphic *Schizothyrium*, Hso.1eP.10. 1, Jamaica. See Nasu & Kunoh (*TMSJ* **27**: 225, 1986; on grapes), Williamson & Sutton (*Pl. Dis.* **84**: 714, 2000; on apple), Batzer *et al.* (*Mycol.* **97**: 1268, 2005; phylogeny).

zygophore (of *Mucorales*), a special hyphal branch producing copulation branches.

Zygopichia (Klöcker) Kudrjanzev (1960) = Pichia fide

Batra *in* Subramanian (Ed.) (*Taxonomy of fungi* **1**: 187, 1978), Wu *et al.* (*FEMS Yeast Res.* **6**: 305, 2006).

Zygopichia E.K. Novák & Zsolt (1961) = Pichia fide von Arx (*Gen. Fungi Sporul. Cult.* Edn 3, 1981).

Zygopleurage Boedijn (1962), Lasiosphaeriaceae. 1 (coprophilous), widespread (tropical). See Lundqvist (*Symb. bot. upsal.* **20** no. 1, 1972; key), Huhndorf *et al.* (*Mycol.* **96**: 368, 2004), Miller & Huhndorf (*Mol. Phylogen. Evol.* **35**: 60, 2005), Cai *et al.* (*MR* **110**: 359, 2006; phylogeny).

Zygopolaris S.T. Moss, Lichtw. & Manier (1975), Legeriomycetaceae. 2 (in *Ephemeroptera*), Mexico; USA. See Moss & Lichtwardt (*CJB* **55**: 3099, 1977; ultrastr.), Lichtwardt (*The Trichomycetes. Fungal associates of arthropods*, 1986; key), White (*MR* **110**: 1011, 2006; phylogeny), Valle *et al.* (*Mycol.* **100**: 149, 2008; Mexico).

Zygorenospora Krassiln. (1954) = Zygofabospora fide Naumov (*Mikol. Fitopatol.* **21**: 131, 1987).

Zygorhizidium Löwenthal (1904), Chytridiaceae. *c.* 11, Europe; N. America. See Sparrow (*Aquatic Phycomycetes* Edn 2: 548, 1960; key), Doggett & Porter (*Mycol.* **88**: 720, 1996; sexual reproduction).

Zygorhynchus Vuill. (1903), Mucoraceae. 8 (esp. in soil), widespread. See Hesseltine *et al.* (*Mycol.* **51**: 173, 1959; key), Schipper *et al.* (*Persoonia* **8**: 321, 1975; zygospore ornamentation), O'Donnell *et al.* (*CJB* **56**: 1061, 1978; ontogeny), Heath & Rethoret (*Eur. J. Cell Biol.* **28**: 180, 1982; mitosis, ultrastr.), Schipper (*Persoonia* **13**: 97, 1986; key), Brown (*J. Phytopath.* **120**: 298, 1987), Taiwo *et al.* (*Microbios* **51**: 23, 1987), Taiwo *et al.* (*Microbios* **52**: 183, 1987; physiology), Brown (*J. Phytopath.* **123**: 222, 1988; biocontrol), Brown & Surgeoner (*Ann. appl. Biol.* **118**: 39, 1991; plant growth enhancement), Edelmann & Klomparens (*Mycol.* **87**: 304, 1995; ultrastr. zygospore dev.), Voigt & Wöstemeyer (*Gene* **270**: 113, 2001; phylogeny), Zheng (*Mycotaxon* **84**: 367, 2002; China).

Zygorrhynchus, see *Zygorhynchus*.

Zygosaccharis Clem. & Shear (1931) ≡ Zygosaccharomyces.

Zygosaccharomyces B.T.P. Barker (1901), Saccharomycetaceae. 11 (osmotolerant), widespread. See von Arx *et al.* (*Stud. Mycol.* **14**: 1, 1977), Kreger-van Rij (Ed.) (*Yeasts, a taxonomic study* 3rd edn, 1984; key), Kurtzman (*Yeast* Chichester **6**: 213, 1990; 9 spp. by DNA complementarity), James *et al.* (*Yeast* Chichester **10**: 871, 1994; molec. syst.), James *et al.* (*Int. J. Syst. Bacteriol.* **46**: 189, 1996; ITS), Kurtzman *in* Kurtzman & Fell (Eds) (*Yeasts, a taxonomic study* 4th edn: 424, 1998), Steels *et al.* (*Int. J. Syst. Bacteriol.* **49**: 319, 1999), Steels *et al.* (*FEMS Yeast Res.* **2**: 113, 2002; physiology), Esteve-Zarzoso *et al.* (*Syst. Appl. Microbiol.* **26**: 404, 2003; phylogeny), James & Stratford (*Yeasts in Food* Beneficial and Detrimental Aspects: 171, 2003; spoilage-causing spp.), Kurtzman (*FEMS Yeast Res.* **4**: 233, 2003), Duarte *et al.* (*Syst. Appl. Microbiol.* **27**: 436, 2004; isozymes), Rawsthorne & Phister (*Int. J. Food Microbiol.* **112**: 1, 2006; molecular detection), Suh *et al.* (*Mycol.* **98**: 1006, 2006; phylogeny).

Zygosaccharomycodes Nishiw. (1929) = Saccharomyces fide Batra *in* Subramanian (Ed.) (*Taxonomy of fungi* **1**: 187, 1978).

Zygospermella Cain (1935), Lasiosphaeriaceae. 2 (coprophilous), N. America; Europe. See Lundqvist (*Symb. bot. upsal.* **20** no. 1, 1972; key), Huhndorf *et al.* (*Mycol.* **96**: 368, 2004).

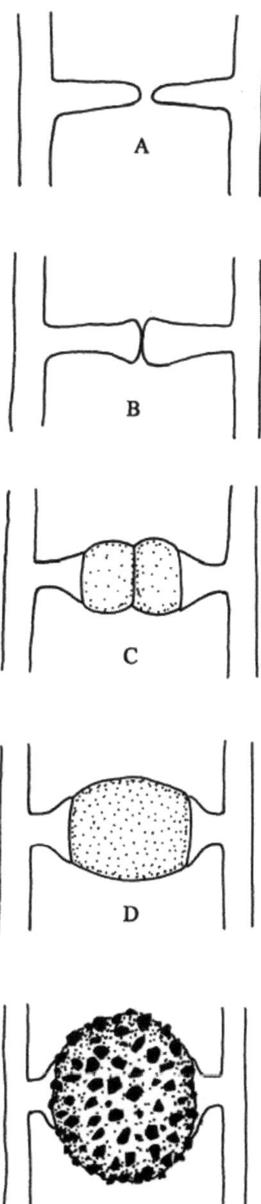

Fig. 27. Stages in zygospore formation in *Mucor mucedo*. A,B, young progametangia; C, gametangia and suspensors; D, young zygospore; E, mature zygospore. Not to scale.

Zygospermum Cain (1934) [non *Zygospermum* Thwaites ex Baill. 1858, *Euphorbiaceae*] ≡ Zygospermella.

zygospore, the resting spore resulting from the conjugation of isogametes or (in *Zygomycetes*), from the fusion of like gametangia (Fig. 27).

Zygosporites Will. (1880), Fossil Fungi, Zygomycetes. 1 (Carboniferous), British Isles. See Pia *in* Hirnier (*Handb. Paläobot.* **1**, 1927).

Zygosporium Mont. (1842), anamorphic *Pezizomycotina*, Hso.0eP.1. 11, widespread. See Hughes (*Mycol. Pap.* **44**, 1951), Ichinoe (*Bull. natn Inst. Hyg. Sci.* **89**: 135, 1971), Heredia Abarca & Mercado Sierra (*Mycotaxon* **68**: 137, 1998), Whitton *et al.* (*Fungal Diversity* **12**: 207, 2003; review).

zygote, the result of fusion of two gametes; a cell in which two nuclei of opposite sex have undergone fusion (Buller, 1941).

Zygothrix Reinsch ex Rabenh. (1866) nom. dub., Fungi. Based on sterile hyphae.

Zygotorulaspora Kurtzman (2003), Saccharomycetaceae. 2. See Kurtzman (*FEMS Yeast Res.* **4**: 243, 2003), Suh *et al.* (*Mycol.* **98**: 1006, 2006; phylogeny).

Zygowillia (Klöcker) Kudrjanzev (1960) = Pichia fide Lodder (*Yeasts, a taxonomic study* 2nd edn, 1970).

Zygowilliopsis Kudrjanzev (1960) = Williopsis fide Yamada *et al.* (*Biosc., Biotechn., Biochem.* **58**: 1236, 1994; status), Yamada *et al.* (*Biosc., Biotechn., Biochem.* **63**: 827, 1999), Naumova *et al.* (*Syst. Appl. Microbiol.* **27**: 192, 2004).

Zygozyma Van der Walt & Arx (1987) = Lipomyces fide van der Walt *et al.* (*Syst. Appl. Microbiol.* **12**: 288, 1989; key 3 spp.) but see van der Walt *et al.* (*Antonie van Leeuwenhoek* **59**: 77, 1991; ultrastr.), Doublés & McLaughlin (*Mycol.* **84**: 671, 1992), Smith *in* Kurtzman & Fell (Eds) (*Yeasts, a taxonomic study* 4th edn: 433, 1998), Suh *et al.* (*Mycol.* **98**: 1006, 2006; phylogeny), Kurtzman *et al.* (*FEMS Yeast Res.* **7**: 1027, 2007; phylogeny).

zymurgy, the practice of fermentation as in brewing and wine-making.

Zythia Fr. (1849), anamorphic *Pezizomycotina*, Cpd.0-1eH.?. 25, widespread.

ZythiaceaeClem. (1909). See *Nectrioidaceae*.

Zythiostroma Höhn. ex Falck (1923), anamorphic *Nectria*, St.0eH.15. 2 or 3, Europe; Java.

Zyxiphora B. Sutton (1981), anamorphic *Pezizomycotina*, Hsy.0eP.1. 1, India. See Sutton (*Kavaka* **8**: 55, 1980).

Achlya Nees (1823), Saprolegniaceae. *c.* 50, widespread (north temperate). See Johnson (*The genus Achlya*, 1956), Riethmüller *et al.* (*Mycol.* **94**: 834, 2002; phylogeny), Spencer *et al.* (*MR* **106**: 549, 2002; revision), Markovskaja (*Botanica Lithuanica* **10**: 141, 2004; Lithuania), Johnson *et al.* (*Mycotaxon* **92**: 11, 2005; systematics).

Achlyopsis De Wild. (1896) nom. dub., Pythiales. ? = Pythiogeton (Pythiogeton.). See Voglmayr *et al.* (*MR* **103**: 591, 1999).

Actinobotrys H. Hoffm. (1856) = Bremia fide Saccardo (*Syll. fung.* **7**: 1, 1888).

Adlerocystis Feldm.-Muhs. & Havivi (1963), ? Chromista. 2 (symbiotic on *Ornithodorus*), Israel. See Feldman-Muhsam & Havivi (*Parasitology* **53**: 183, 1963; related to *Blastocystis*).

Albuginaceae J. Schröt. (1893), Albuginales. 3 gen. (+ 1 syn.), 53 spp.

Albuginales Thines (2005). Albuginomycetidae. 1 fam., 3 gen., 53 spp. Fam.:
Albuginaceae
Lit.: Thines & Spring (*Mycotaxon* **92**: 443, 2005). See also *Lit.* under *Peronosporales*.

Albuginomycetidae Thines (2005), Oomycetes. Ord.:
Albuginales
For *Lit.* see fam.

Albugo (Pers.) Roussel (1806), Albuginaceae. *c.* 45, widespread. The 'White Blisters' or 'White Rusts'; *A. candida* (white blister of crucifers). See Biga (*Sydowia* **9**: 339, 1955; key), Pound & Williams (*Phytopathology* **53**: 1146, 1963; physiologic races *A. candida*), Choi & Priest (*Mycotaxon* **53**: 261, 1995; key), Priest (*Mycotaxon* **58**: 69, 1996; spp. on *Gentianaceae*), Legon *et al.* (*Mycologist* **16**: 114, 2002; *A. tragopogonis* var. *inulae* in UK), Thines & Spring (*Mycotaxon* **92**: 443, 2005; split into three genera), Choi *et al.* (*Molecular and Phylogenetic Evolution* **40**: 400, 2006), Constantinescu & Thines (*Sydowia* **58**: 178, 2006; sporangial dimorphism), Voglmayr & Riethmüller (*MR* **110**: 75, 2006; phylogeny), Choi *et al.* (*Fungal Diversity* **27**: 11, 2007; genetic diversity in *A. candida*).

Althornia E.B.G. Jones & Alderman (1972), Thraustochytriaceae. 1, British Isles. See Jones & Alderman (*Nova Hedwigia* **21**: 381, 1971), Raghukumar (*European Journal of Protistology* **38**: 127, 2002; ecology).

Anisolpidiaceae Karling (1943), Anisolpidiales. 1 gen., 3 or 6 spp.

Anisolpidiales M.W. Dick (2001). Chromista. 1 fam., 1 gen., 6 spp. Fam.:
Anisolpidiaceae
For *Lit.* see under fam.

Anisolpidium Karling (1943), Anisolpidiaceae. 6 (in marine algae), widespread. See Dick (*Straminipilous Fungi*: 670 pp., 2001).

Aphanodictyon Huneycutt ex M.W. Dick (1971), Leptolegniellaceae. 1, USA; Australia. See Dick (*TBMS* **57**: 422, 1971), Dick (*Straminipilous Fungi*: 670 pp., 2001).

Aphanomyces de Bary (1860), Leptolegniaceae. *c.* 15 (aquatic), widespread. *A. euteiches* (pea root rot); *A. astaci* (pathogenic for European crayfish, *Astacus*); *A. cochlioides* (sugar beet black rot); *A. invadans* [Epizootic Ulcerative Syndrome in fish]. See Scott (*Tech. Bull. Va agric. Exp. Stn* **151**, 1961; keys), Dick (*Straminipilous Fungi*: 670 pp., 2001), Levenfors & Fatehi (*MR* **108**: 682, 2004; molecular characterization of *Aphanomyces* species on legumes).

Aphanomycopsis Scherff. (1925), Leptolegniellaceae. 8, Europe; N. America. See Dick (*Straminipilous Fungi*: 670 pp., 2001).

Aplanes de Bary (1888) = Achlya fide Johnson (*The genus Achlya*, 1956).

Aplanochytrium Bahnweg & Sparrow (1972), Thraustochytriaceae. 8 (marine), Kerguelen. See Bahnweg & Sparrow (*Arch. Mikrobiol.* **81**: 46, 1972), Leander & Porter (*Mycotaxon* **76**: 439, 2000; monogr.), Moro *et al.* (*Protist* **154**: 331, 2003; sp. nov. from Ross Sea, Antarctica).

Aplanopsis Höhnk (1952), Saprolegniaceae. 2, Europe. See Höhnk (*Veröff. Inst. Meeresf. Bremerhaven* Sonderband **1**: 127, 1952), Spencer *et al.* (*MR* **106**: 549, 2002; revision).

Apodachlya Pringsh. (1883), Leptomitaceae. 3, widespread. See Jacobs (*Antonie van Leeuwenhoek* **48**: 389, 1982; key), Dick (*Straminipilous Fungi*: 670 pp., 2001).

Apodachlyella Indoh (1939), Apodachlyellaceae. 1, widespread. See Longcore *et al.* (*Mycol.* **79**: 621, 1987), Dick (*Straminipilous Fungi*: 670 pp., 2001).

Apodachlyellaceae M.W. Dick (1986), Leptomitales. 2 gen., 2 spp.

Apodya Cornu (1872) = Leptomitus fide Sparrow (*Aquatic Phycomycetes* Edn 2: 1187 pp., 1960).

Aqualinderella R. Emers. & W. Weston (1967), Rhipidiaceae. 1 (anaerobic), widespread. See Emerson (*Mycol.* **62**: 359, 1970; oogonium production), Held (*Mycol.* **62**: 339, 1970; nutrition), Dick (*Straminipilous Fungi*: 670 pp., 2001).

Araiospora Thaxt. (1896), Rhipidiaceae. 4, America; Europe. See Dick (*Straminipilous Fungi*: 670 pp., 2001).

Archilegnia Apinis (1935) nom. conf., Saprolegniaceae.

Artotrogus Mont. (1845) nom. rej. = Pythium Pringsh. fide Middleton (*Mem. Torrey bot. Club* **20**, 1943).

Atkinsiella Vishniac (1958), Saprolegniomycetidae. 1 (on *Crustacea*), Europe; N. America. See Sparrow & Gotelli (*Mycol.* **61**: 199, 1969), Martin (*Am. J. Bot.* **64**: 760, 1977; posn), Kitancharoen *et al.* (*Mycoscience* **35**: 265, 1994; disease of abalone), Dick (*MR* **102**: 1062, 1998; revision of taxa), Dick (*Straminipilous Fungi*: 670 pp., 2001), Hudspeth *et al* (*Fungal Diversity*: 577, 2004; phylogeny).

Basidiophora Roze & Cornu (1869), Peronosporaceae. 1, widespread. See Barreto & Dick (*J. Linn. Soc.* Bot. **107**: 313, 1991; key), Constantinescu (*Nova Hedwigia* **66**: 251, 1998; revision), Voglmayr *et al.* (*MR* **108**: 1011, 2004; phylogeny).

Benua Constant. (1998), Peronosporaceae. 1, USA. See Constantinescu (*Nova Hedwigia* **66**: 258, 1998), Thines *et al.* (*MR* **111**: 1377, 2007; phylogeny).

Bicilium H.E. Petersen (1910) = Eurychasma fide Dick (*Straminipilous Fungi*: 670 pp., 2001).

Blastocystis Alexeev (1911), Chromista. 3 (in alimentary tract of humans and animals), widespread. See Lavier (*Annls Parasit. hum. comp.* **27**: 339, 1952; synonymy), Lee (*TBMS* **54**: 313, 1970; ultrastr.), Nakamura *et al.* (*Mol. Biochem. Parasit.* **77**: 241, 1996; phylogeny), Silberman *et al.* (*Nature* **380**: 398, 1996; phylogeny), Stenzel & Boreham (*Clin. Microbiol. Rev.* **9**: 563, 1996; review), Yoshikawa *et al.* (*J. Eukary. Microbiol.* **43**: 127, 1996; DNA polymorphism), Konig & Müller (*Z. Bakteriol.* **286**: 435,

1997; incidence), Yoshikawa & Oishi (*Protoplasma* **200**: 31, 1997; biodiversity), Hanbold *et al.* (*Medical Mycol.* **36**: 263, 1998).

Blastulidiopsis Sigot (1931) = Blastulidium fide Manier (*Protistologia* **12**: 225, 1976).

Blastulidium Pérez (1903), Leptomitales. 1, France. See Manier (*Protistologia* **12**: 225, 1976), Dick (*Straminipilous Fungi*: 670 pp., 2001; asexual genus *inc. sed.*).

Blepharospora Petri (1917) = Phytophthora fide Buisman (*Mededelingen uit het Phytopathologisch Laboratorium 'Willie Commelin Scholten'* **11**: 1, 1927).

Branchiomyces Plehn (1912), ? Saprolegniales. 2 (on fish), widespread. See Peduzzi (*Mem. Ist. ital. Idrobiol.* **30**: 81, 1973), Neish & Hughes (*Diseases of Fish* **6**: 50, 1980).

Bremia Regel (1843), Peronosporaceae. 1, widespread (north temperate). *B. lactucae* (lettuce downy mildew). See Ling & Tai (*TBMS* **28**: 16, 1945), Marlatt (*Bull. agric. Exp. Stns Univ. Fla* **764**: 1, 1974), Michelmore & Ingram (*TBMS* **75**: 47, 1980; heterothallism), Crute & Dixon *in* Spencer (Ed.) (*The Downy Mildews (London)*: 421, 1981), Morgan (*CMI Descr.*: 682, 1981), Riethmüller *et al.* (*Mycol.* **94**: 834, 2002; phylogeny), Göker *et al.* (*CJB* **81**: 672, 2003; phylogeny), Voglmayr *et al.* (*MR* **108**: 1011, 2004; phylogeny), Thines *et al.* (*MR* **110**: 646, 2006; phylogeny; exclusion of *Bremia graminicola*), Choi *et al.* (*Mycopathologia* **163**: 91, 2007; ITS characteristics), Voglmayr & Constantinescu (*MR* **112**: 487, 2008; phylogeny).

Bremiella G.W. Wilson (1914) = Plasmopara fide Riethmüller *et al.* (*Mycol.* **94**: 834, 2002), Voglmayr *et al.* (*Mycotaxon* **100**: 11, 2007).

Brevilegnia Coker & Couch (1927), Saprolegniaceae. *c.* 15, widespread. See Johnson (*Mycol.* **69**: 287, 1977; 6 Scandinavian spp.), Inaba & Tokumasu (*Mycoscience* **43**: 59, 2002; Japan), Steciow (*Mycol.* **95**: 934, 2003; Argentina).

Brevilegniella M.W. Dick (1961), Leptolegniellaceae. 1 (from soil), USA; British Isles. See Dick (*Pap. Mich. Acad. Sci.* **46**: 195, 1961), Dick (*Straminipilous Fungi*: 670 pp., 2001).

Calyptralegnia Coker (1927), Saprolegniaceae. 3, N. America; Europe; Iraq. See Muhsin (*Polskie Archivum Hydrobiologii* **41**: 415, 1994; Iraq), Dick (*Straminipilous Fungi*: 670 pp., 2001).

Canteriomyces Sparrow (1960), Hyphochytriaceae. 1 (on freshwater algae), Europe; N. America. See Sparrow (*University of Michigan Studies, Science Series* **15**: 750, 1960), Dick (*Straminipilous Fungi*: 670 pp., 2001).

Chlamydomyxa W.A. Archer (1875) = Labyrinthula fide Olive (*The mycetozoans*: 215, 1975).

Chlamydomyzium M.W. Dick (1997), Myzocytiopsidaceae. 6 (parasitic in *Aschelminthes*), widespread. See Glockling & Dick (*MR* **101**: 883, 1997), Dick (*Straminipilous Fungi*: 670 pp., 2001), Glockling & Beakes (*MR* **110**: 1119, 2006; structure, development).

Chlorospora Speg. (1891) = Harzia fide Constantinescu (*Sydowia* **54**: 137, 2002).

CHROMISTA. 3 class., 16 ord., 29 fam., 126 gen. (+ 46 syn.), 1036 spp. (incl. Heterokonta, Pseudofungi, Pseudomycotina, stramenopiles). Kingdom of *Eukaryota*. Predominantly unicellular, filamentous, or colonial primarily phototrophic organisms; cell walls not of chitin and β-glucan (often cellulosic); chloroplasts (when present) located in the lumen of a generally rough endoplasmic reticulum, lack starch and phycobilisomes and have a two-membraned envelope inside a periplastid membrance, chlorophylls when present *a* and *c*; mitochonidia generally with tubular cristae; Golgi bodies and peroxisomes always present; flagella when present, with at least one with rigid, tubular, and usually tripartite flagellar hairs or mastigonemes (except haptophytes); mostly free-living; many microscopic in size (with major exceptions, e.g. brown algae).

Formerly often included in *Protoctista* (syn. *Protista*; e.g. Margulis *et al.*, 1990), but now recognized as a kingdom separate from the *Protozoa*. Includes 10 phyla (Corliss, 1994) or 5 phyla and 7 subphyla (Cavalier-Smith, 1998); encompasses a wide range of golden and brown algae, diatoms, chrysophytes and cryptomonads (but not chlorophyte and red algae; in *Plantae*). Dick (2001) formally proposes the kingdom *Straminipila* (q.v.) with a narrower delimitation than Cavalier-Smith (1998) but also includes the *Plasmodiophoromycetes*.

The 'fungal' phyla are interpreted as having lost chloroplasts secondarily, and are part of subkingdom *Heterokonta*. The classification within the kingdom is unsettled; 2 of the phyla Corliss (1994) accepted comprise organisms studied by mycologists: *Labyrinthomorpha* (incl. *Labyrinthulea* and *Thraustochytriaceae*) and *Pseudofungi* (incl. *Oomycetes* and *Hyphochytriomycetes*); Barr (1992) included the same fungi but used different ranks. Cavalier-Smith (1997, 1998) places the fungal groups in two separate phyla: *Sagenista* (*Labyrinthista* or *Labyrinthulomycetes*) and *Bigyra* (*Pseudofungi* or *Oomycetes* and *Hyphochytriomycetes*). Dick (2001) regards all these groups as belonging to the phylum *Heterokonta*.

Various arrangements and ranks have been proposed for the main groupings of 'fungi' within the *Chromista*; selected ones are compared in Table 3 (Fungi, q.v.). Three fungal phyla belonging to the *Chromista* are accepted in this edition of the *Dictionary*:

(1) **Hyphochytriomycota.**
(2) **Labyrinthulomycota.**
(3) **Oomycota.**

Lit.: Barr (*Mycol.* **84**: 1, 1992), Cantino (*Ann. Rev. Microbiol.* **13**: 103, 1959), Cavalier-Smith (*Prog. Phycol. Res.* **4**: 309, 1986; *in* Rayner *et al.* (Eds), *Evolutionary biology of the fungi*: 339, 1987; *in* Osawa & Honjo (Eds), *Evolution of life*: 271, 1991; *Microbiol. Rev.* **57**: 953, 1993; *Arch. Protistenkd.* **148**: 253, 1997; *Biol. Rev.* **73**: 203, 1998), Corliss (*Acta Protozool.* **33**: 1, 1994), Buczacki (Ed.) (*Zoosporic plant pathogens*, 1983), Cordá-Olmedo & Lipson (Eds) (*Phycomycetes*, 1987), Dick (*Straminipilous Fungi*, 2001), Fitzpatrick (1930), Fuller & Jaworski (Eds) (*Zoosporic fungi in teaching and research*, 1986), Margulis *et al.* (Eds) (*Handbook of Protoctista*, 1990) Sparrow (*Aquatic Phycomycetes*, edn 2, 1960; *in* Ainsworth *et al.* (Eds), *The Fungi* **4B**: 61, 1973; keys classes, orders, etc.); Yoon *et al.* (*proc. Natn Acad. Sci USA* **99**: 15507, 2002); Keeling (*Protist* **155**: 3, 2004).

See Kingdoms of fungi, Phylogeny, Protozoa, Straminipila.

Ciliatomyces I. Foissner & W. Foissner (1995), Lagenaceae. 1 (on cysts of *Kahliella simplex*), Austria. See Foissner & Foissner (*Phyton* Horn **35**: 116, 1995),

Dick (*Straminipilous Fungi*: 670 pp., 2001).

Ciliomyces I. Foissner & W. Foissner (1986) ≡ Ciliatomyces.

Cladolegnia Johannes (1955) = Saprolegnia fide Dick *in* Ainsworth *et al.* (Eds) (*The Fungi* **4A**: 145, 1973).

Cornumyces M.W. Dick (2001), Leptomitales. 8 (saprotrophic or parasitic in other fungi), widespread. See Dick (*Straminipilous Fungi*: 327, 2001).

Couchia W.W. Martin (1981), Saprolegniaceae. 3, USA. See Martin (*Mycol.* **73**: 1143, 1981), Martin (*Mycol.* **92**: 1149, 2000; sp. n. parasitic in midge eggs).

Crypticola Humber, Frances & A.W. Sweeney (1989), Crypticolaceae. 2, Australia. See Humber *et al.* (*J. Invert. Path.* **45**: 107, 1989), Dick (*Straminipilous Fungi*: 670 pp., 2001).

Crypticolaceae M.W. Dick (1998), Myzocytiopsidales. 1 gen. (+ 1 syn.), 2 spp.

Cymbanche Pfitzer (1869) ? = Ectrogella fide Dick (*in litt.*).

Cystochytrium Ivimey Cook (1932), Hyphochytriaceae. 1 (on *Veronica* roots), Europe. See Ivimey Cook (*TBMS* **16**: 251, 1932).

Cystopodaceae Bonord. (1884) = Albuginaceae.

Cystopus Lév. (1847) = Albugo fide Wilson (*Bull. Torrey Bot. Club* **34**: 61, 1907).

Cystosiphon Roze & Cornu (1869), Pythiaceae. 5, widespread. See Dick (*Straminipilous Fungi*: 670 pp., 2001).

Detonisia Gonz. Frag. (1925) nom. dub., ? Oomycota.

Developayella Tong (1995), Chromista. 1. Heterotrophic flagellate consistently clustering with the chromistan fungi in molecular phylogenies, frequently as a sister group to the hyphochytrids, labyrinthulids and thraustochytrids. See Tong (*Eur. J. Protist.* **31**: 24, 1995), Leipe *et al.* (*Eur. J. Protist.* **32**: 449, 1996; 16S-like rDNA), Cavalier-Smith *in* Coombs *et al.* (Eds) (*Evolutionary Relationships Among Protozoa*: 375, 1998; posn. phylum *Bigyra*: subphylum *Bigyromonada* as sister group to *Pseudofungi*).

Diasporangium Höhnk (1936), Pythiaceae. 1 (in soil), Europe; USA. See Voglmayr *et al.* (*MR* **103**: 591, 1999; posn), Dick (*Straminipilous Fungi*: 670 pp., 2001).

Dicksonomyces Thirum., P.N. Rao & M.A. Salam (1957) nom. conf., Peronosporales. based on a mixture of *Olpitrichum tenellum* and oospores of *Peronosclerospora sorghi*. See Waterhouse (*Mycol.* **60**: 977, 1968).

Dictypleiosporus Gandhe (2006), Saprolegniaceae. 1, India. See Gandhe (*Indian Phytopath.* **59**: 94, 2006).

Dictyuchus Leitg. (1868), Saprolegniaceae. 7, widespread (north temperate). See Blackwood & Powell (*Mycotaxon* **73**: 247, 1, 1999; gen. concept), Riethmüller *et al.* (*Mycol.* **94**: 834, 2002; phylogeny).

Diplanes Leitg. (1868) = Saprolegnia fide Coker (*The Saprolegniaceae with Notes on Other Water Molds*, 1923).

Diplophrys J.S.F. Barker (1868), Thraustochytriales. 2, widespread (north temperate). See Olive (*The mycetozoans*: 227, 1975), Dykstra & Porter (*Mycol.* **76**: 626, 1984; n.sp.), Leander & Porter (*Mycol.* **93**: 459, 2001; phylogenetic position in the *Labyrinthulomycota*).

Diplophysa J. Schröt. (1886) = Olpidiopsis fide Sparrow (*Aquatic Phycomycetes* Edn 2: 1187 pp., 1960).

Ducellieria Teiling (1957), Ducellieriaceae. 1 (aquatic), British Isles. See Kusel-Fetzmann & Novak (*Pl. Syst. Evol.* **138**: 199, 1981; posn), Hesse *et al.* (*Pl. Syst. Evol.* **165**: 1, 1989; life cycle, ultrastr.).

Ducellieriaceae M.W. Dick (2001), Leptomitales. 1 gen., 1 spp.

Ectrogella Zopf (1884), Ectrogellaceae. 5, widespread (north temperate). See Raghu Kumar (*CJB* **58**: 2557, 1980; ultrastr.), Dick (*Straminipilous Fungi*: 670 pp., 2001).

Ectrogellaceae Cejp (1959), Oomycota (inc. sed.). 1 gen. (+ 1 syn.), 5 spp.

Elina N.J. Artemczuk (1972), Thraustochytriaceae. 2, former USSR. See Karling (*Mycol.* **71**: 829, 1979), Dick (*Straminipilous Fungi*: 670 pp., 2001; systematic posn) But see.

Endosphaerium D'Eliscu (1977), Pythiales. 1 (on fish), USA. See D'Eliscu (*J. Invert. Path.* **30**: 420, 1977), Dick (*Straminipilous Fungi*: 670 pp., 2001).

Eupythium Nieuwl. (1916) ≡ Pythium Pringsh.

Eurychasma Magnus (1905), Eurychasmataceae. 1 (on *Ectocarpus*), Europe; Greenland. See Dick (*Straminipilous Fungi*: 670 pp., 2001), Cavalier-Smith (*J. Mol. Evol.* **62**: 388, 2006; phylogeny).

Eurychasmataceae H.E. Petersen (1905), Oomycota (inc. sed.). 2 gen., 4 spp.

Eurychasmidium Sparrow (1936), Eurychasmataceae. 3 (on marine *Rhodophyceae*), N. America; Europe. See Dick (*Straminipilous Fungi*: 670 pp., 2001).

Eurychasmopsis Canter & M.W. Dick (1994), Apodachlyellaceae. 1, British Isles. See Canter & Dick (*MR* **98**: 106, 1994), Dick (*Straminipilous Fungi*: 670 pp., 2001).

Geolegnia Coker (1925), Saprolegniaceae. 3, N. America; Europe. See Dick (*Straminipilous Fungi*: 670 pp., 2001).

Gilletia Sacc. & Penz. (1882) = Basidiophora fide Saccardo (*Syll. fung.* **12**: 267, 1897).

Gilletiaceae Nann. (1934) = Peronosporaceae.

Gonimochaete Drechsler (1946), Myzocytiopsidaceae. 4, widespread. See Drechsler (*Bull. Torrey bot. Club* **73**: 1, 1946), Barron (*CJB* **51**: 2451, 1973), Newell *et al.* (*Bull. Mar. Sci.* **27**: 189, 1977), Barron (*Mycol.* **77**: 17, 1985), Saikawa & Anazawa (*CJB* **63**: 2326, 1985; ultrastr.), Dick (*Straminipilous Fungi*: 670 pp., 2001).

Gracea M.W. Dick (1997), Olpidiopsidales. 2, British Isles. See Dick (*MR* **101**: 882, 1997), Dick (*Straminipilous Fungi*: 670 pp., 2001).

Graminivora Thines & Göker (2006), Peronosporaceae. 1 (on *Poaceae*), widespread. See Thines *et al.* (*MR* **110**: 651, 2006).

Haliphthoraceae Vishniac (1958), Oomycota (inc. sed.).

Lit.: Vishniac (*Mycol.* **50**: 75, 1958), Dick (*Pap. Mich. Acad. Sci.* **46**: 195, 1962).

Haliphthoros Vishniac (1958), Oomycota. 3 (on eggs and larvae of Crustacea and Mollusca), widespread (tropical). See Tharp & Bland (*CJB* **55**: 2936, 1977; biol. and host range), Dick (*Straminipilous Fungi*: 670 pp., 2001), Hudspeth *et al* (*Fungal Diversity*: 577, 2004; phylogeny).

Halocrusticida K. Nakam. & Hatai (1995), Oomycota. 6, widespread. See Nakamura & Hatai (*Mycoscience* **36**: 431, 1995), Hudspeth *et al.* (*Fungal Diversity* **13**: 47, 2003; phylogeny), Sekimoto *et al.* (*Mycoscience* **48**: 212, 2007; phylogeny).

Halodaphnea M.W. Dick (1998), Oomycota. 1. See Dick (*MR* **102**: 1065, 1998).

Halophytophthora H.H. Ho & S.C. Jong (1990), Peronosporaceae. 16, widespread. See Nakagiri (*TMSJ* **34**: 87, 1993; reproduction), Nakagiri (*Fungal Diversity* Res. Ser. **7**: 1, 2002; species from tropical and subtropical mangroves), Ho *et al.* (*Mycotaxon* **85**: 417, 2003; sp.n. Taiwan).

Hamidia Chaudhuri (1942) nom. dub., Saprolegniaceae.

Haptoglossa Drechsler (1940), Haptoglossaceae. 11, N. America; Europe. See Davidson & Barron (*CJB* **51**: 1317, 1973), Robb & Lee (*CJB* **64**: 1935, 1986; ultrastr. gun cell), Glockling & Beakes (*MR* **104**: 100, 2000), Dick (*Straminipilous Fungi*: 670 pp., 2001), Glockling & Beakes (*J. Linn. Soc.* Bot. **136**: 329, 2001).

Haptoglossaceae M.W. Dick (2001), Haptoglossales. 1 gen., 11 spp.

Haptoglossales M.W. Dick (2001). Oomycota. 1 fam., 1 gen., 11 spp. Fam.:
Haptoglossaceae
For *Lit.* see under fam.

Heterokonta. See *Chromista*.

Hyaloperonospora Constant. (2002), Peronosporaceae. 24, widespread. pathogens of *Brassicaceae*. See Constantinescu & Fatehi (*Nova Hedwigia* **74**: 310, 2002), Göker *et al.* (*CJB* **81**: 672, 2003; phylogeny), Göker *et al.* (*MR* **107**: 1314, 2003; phylogeny), Göker *et al.* (*Mycol. Progr.* **3**: 83, 2004; phylogeny).

Hydatinophagus Valkanov (1931), Saprolegniaceae. 2, Europe; USA. See Dick (*Straminipilous Fungi*: 670 pp., 2001).

Hydronema Carus ex Rchb. (1828) ≡ Achlya.

Hyphochytriaceae A. Fisch. (1892), Hyphochytriales. 3 gen. (+ 1 syn.), 8 spp.

Hyphochytriales E.A. Bessey ex P.M. Kirk, P.F. Cannon & J.C. David (2001). Hyphochytriomycetes. 2 fam., 6 gen., 24 spp. Fams:
(1) **Hyphochytriaceae**
(2) **Rhizidiomycetaceae**
For *Lit.* see under fam.

Hyphochytriomycetes Sparrow ex M.W. Dick (1983), Hyphochytriomycota. 1 ord., 2 fam., 6 gen., 24 spp. Ord.:
Hyphochytriales
For *Lit.* see ord. and fam.

Hyphochytriomycota Whittaker (1969), Chromista. 1 class., 1 ord., 2 fam., 6 gen., 24 spp. The whole (holocarpic) or part (eucarpic) of the thallus converting into a reproductive structure; holocarpic spp. often with branched rhizoids; zoospores with one anterior flagellum with mastigonemes; lacing protoplasmic and nucleus-associated microtubules. On algae and fungi in freshwater and soil as parasites or saprobes, also saprobic on plant and insect debris. One class: **Hyphochytriomycetes** and Ord.:
Hyphochytriales

Cavalier-Smith (*Arch. Protistenk.* **148**: 253, 1997) and Dick (2001, *in press*) assign the hyphochytrids to class rank within their subphyla *Pseudofungi* and *Peronosporomycotina* respectively. Dick (l.c.) introduces the order *Anisolpidiales* which he considers to be *incertae sedis* within the *Straminipila* (= *Chromista*).

Lit.: Fuller (*in* Margulis *et al.* (Eds), *Handbook of Protoctista*: 380, 1990), Paterson (*in* Parker, 1982, **1**: 179), Sparrow (*in* Ainsworth *et al.* (Eds), *The Fungi* **4B**: 61, 1973).

Hyphochytrium Zopf (1884), Hyphochytriaceae. 6, widespread (north temperate). See Fuller *in* Margulis *et al.* (Eds) (*Handbook of Protoctista*: 380, 1990), Dick (*Straminipilous Fungi*: 670 pp., 2001), Fuller *in* McLaughlin *et al.* (Eds) (*The Mycota* A Comprehensive Treatise on Fungi as Experimental Systems for Basic and Applied Research **7A**: 73, 2001).

Hyphophagus Minden (1911) ≡ Hyphochytrium.

Isoachlya Kauffman (1921), Saprolegniaceae. *c.* 10, widespread. See Seymour (*Nova Hedwigia* **19**: 1, 1970), Dick *in* Ainsworth *et al.* (Eds) (*The Fungi* **4A**: 145, 1973), Dick (*Straminipilous Fungi*: 670 pp., 2001).

Japonochytrium Kobayasi & M. Ôkubo (1953), Thraustochytriaceae. 1 (on *Gracilaria*), Japan. See Kobayasi & Ôkubo (*Bull. natn. Sci. Mus.* Tokyo, B **33**: 57, 1953), Cavalier-Smith & Chao (*J. Mol. Evol.* **62**: 388, 2006; phylogeny).

Jaraia Němec (1912) nom. dub., ? Saprolegniaceae.

Kawakamia Miyabe (1903) = Phytophthora fide Fitzpatrick (*The lower fungi. Phycomycetes*, 1930).

Labyrinthista, Chromista. 1 class., 2 ord., 2 fam., 12 gen., 53 spp. (Labyrinthomorpha, Labyrinthulomycota); labyrinthids and thraustochytrids. Trophic stage with an ectoplasmic network, and spindle-shaped or sphaerical cells that move by gliding within the network; uniquely contain bothrosomes; zoospores when present with two flagellae, one with mastigonemes; sexual reproduction known in some spp.

The spp. occur in salt and freshwater, often associated with plants and algal chromists, some as pathogens; culturable. Class:
Labyrinthulea (≡ *Labyrinthulomycota*)

Data from recent molecular investigations conflict with current morphological based classifications and suggest that new lineages need to be recognized.

Lit.: Barr & Allen (*CJB* **63**: 138, 1985; flagella), Corliss (1994), Honda *et al.* (*J. Eukaryot. Microbiol.* **46**: 637, 1999), Karling (*Predominantly holocarpic and eucarpic simple biflagellate phycomycetes*, 1981), Leander & Porter (*Mycol.* **93**: 459, 2001; phylogeny), Moss (*Bot. J. Linn. Soc.* **91**: 329, 1985), Olive (1975, 1982), Porter (*in* Margulis *et al.*, *Handbook of the Protoctista*: 388, 1990), Sparrow (*in* Ainsworth *et al.*, *The Fungi* **4B**: 61, 1973). See also *Mycetozoa*.

Labyrinthodictyon Valkanov (1969) = Labyrinthula fide Hawksworth *et al.* (*Dictionary of the Fungi* edn 8, 1995).

Labyrinthomorpha. See *Labyrinthista*.

Labyrinthomyxa Duboscq (1921) nom. conf., Labyrinthulaceae. See Olive (*The mycetozoans*: 215, 1975).

Labyrinthula Cienk. (1867), Labyrinthulaceae. *c/* 12 (in salt and fresh water), widespread (north temperate). *L. macrocystis* (wasting disease of eel-grass, *Zostera* spp.). See Porter (*Protoplasma* **61**: 1, 1969; ultrastr.), Porter *in* Margulis *et al.* (Eds) (*Handbook of Protoctista*: 388, 1990), Muehlstein *et al.* (*Mycol.* **82**: 180, 1991), Leander & Porter (*Mycol.* **93**: 459, 2001; phylogeny), Leander & Leander (*European Journal of Protistology* **40**: 317, 2004; phylogeny).

Labyrinthulaceae Cienk. (1867), Labyrinthulales. 1 gen. (+ 3 syn.), 12 spp.

Lit.: Dangeard (*Le Botaniste* **24**: 217, 1932).

Labyrinthulales E.A. Bessey ex P.M. Kirk, P.F. Cannon & J.C. David (2001). Labyrinthulea. 1 fam., 3 gen., 14 spp. Fam.:

Labyrinthulaceae
For *Lit.* see under fam.

Labyrinthulea, Labyrinthista. 2 ord., 2 fam., 12 gen., 53 spp. Ords:
(1) **Labyrinthulales**
(2) **Thraustochytriales**
For *Lit.* see ord. and fam.

Labyrinthuloides F.O. Perkins (1973) = Aplanochytrium fide Leander & Porter (*Mycotaxon* **76**: 439, 2000).

Labyrinthulomycetes. See *Labyrinthulea*.

Labyrinthulomycota. See *Labyrinthista*.

Lagena Vanterp. & Ledingham (1930), Lagenaceae. 1 (on wheat roots), N. America. The correct name for this genus when the *Chromista* are included in the *Protozoa* is *Lagenocystis*. See Dick (*Straminipilous Fungi*: 670 pp., 2001).

Lagenaceae M.W. Dick (2001), Oomycota (inc. sed.). 3 gen. (+ 2 syn.), 4 spp.

Lagenidiaceae J. Schröt. (1893) = Pythiaceae.

Lagenidiales Karling (1939) = Pythiales.

Lagenidicopsis N.J. Artemczuk (1972) nom. dub., Chromista.

Lagenidiopsis De Wild. (1896) = Cystosiphon fide Dick (*in litt.*).

Lagenidium Schenk (1857), Pythiaceae. 1, widespread. See Correll *et al.* (*Phytopathology* **83**: 1199, 1993; RAPD analysis of *C. orbiculare*), Dick (*Straminipilous Fungi*: 670 pp., 2001).

Lagenisma Drebes (1968), Lagenismataceae. 1 (marine on *Coscinodiscus*), widespread (northern hemisphere). See Schnepf & Drebes (*Helgol. wiss. Meersunters.* **29**: 291, 1977; development), Dick (*Straminipilous Fungi*: 670 pp., 2001).

Lagenismataceae M.W. Dick (2001), Lagenismatales. 1 gen., 1 spp.

Lagenismatales M.W. Dick (2001). Oomycetes. 1 fam., 1 gen., 1 spp. Thallus holocarpic, lacking a cell wall, pseudomycelial, coralloid or allantoid, nonseptate, walled at maturity, becoming transformed into a zoosporangium or a gametangium; asexual reproduction by zoosporangia; zoosporogenesis intrasporangial; zoospores diplanetic but not dimorphic; primary cysts with hollow spines; flagellar insertion lateral, heterokont and isokont; sexual reproduction oogamous by contact between gametic cysts; gametes present, flagellate, isogametic; resting spore formed in receptor cyst, aplerotic. Parasitic in marine diatoms. Fam.:
Lagenismataceae
Lit.: Dick (2001).

Lagenocystis H.F. Copel. (1956) ≡ Lagena.

Latrostium Zopf (1894), Rhizidiomycetaceae. 1 (on *Vaucheria*), Europe. See Dick (*Straminipilous Fungi*: 670 pp., 2001).

Leptolegnia de Bary (1888), Leptolegniaceae. 5, widespread (north temperate). See Dick *in* Ainsworth *et al.* (Eds) (*The Fungi* **4A**: 145, 1973), López Lasta *et al.* (*Revta Iberoamer. Micol.* **16**: 143, 1999), Johnson *et al.* (*Mycotaxon* **92**: 1, 2005).

Leptolegniaceae M.W. Dick (1999), Saprolegniales. 4 gen., 25 spp.

Leptolegniella Huneycutt (1952), Leptolegniellaceae. 5, widespread. See Huneycutt (*J. Elisha Mitchell scient. Soc.* **68**: 109, 1952), Dick (*Straminipilous Fungi*: 670 pp., 2001), Rocha & Pires Zottarelli (*Acta Bot. Brasilica* **16**: 287, 2002; Brazil).

Leptolegniellaceae M.W. Dick (1971), Leptomitales. 5 gen., 16 spp.
Lit.: Dick (*TBMS* **57**: 417, 1971).

Leptomitaceae Kütz. (1843), Leptomitales. 3 gen. (+ 1 syn.), 5 spp.

Leptomitales Kanouse (1927). Oomycetes. 4 fam., 13 gen., 33 spp. Mycelium of regularly constricted hyphae, cytoplasmic streaming obvious; asexual reproduction by zoosporangia, zoosporogenesis intrasporangial, release by dehiscence, discharge vesicles not formed, zoosporangia terminal; sexual reproduction by thin-walled, smooth oogonia, oospores single, plerotic, no periplasm. Freshwater saprobes and animal parasites, N. temp. Fams:
(1) **Apodachlyellaceae**
(2) **Ducellieriaceae**
(3) **Leptolegniellaceae**
(4) **Leptomitaceae**
Lit.: Dick (2001, *in* Margulis *et al.* (Eds), 661, 1990, *in* Ainsworth *et al.* (Eds), *The Fungi* **4B**: 145, 1973).

Leptomitus C. Agardh (1824), Leptomitaceae. 1, Europe; N. America. See Dick (*Straminipilous Fungi*: 670 pp., 2001).

Lucidium Lohde (1874) nom. dub., Pythiaceae.

Medusoides H. Voglmayr (1999), Pythiogetonaceae. 1 (on submerged leaves and twigs), USA. See Voglmayr *et al.* (*MR* **103**: 603, 1999), Dick (*Straminipilous Fungi*: 670 pp., 2001).

Mindeniella Kanouse (1927), Rhipidiaceae. 2, N. America. See Dick (*Straminipilous Fungi*: 670 pp., 2001).

Mindeniellaceae Cejp (1959) = Rhipidiaceae.

Mycelophagus L. Mangin (1903) = Phytophthora fide Clements & Shear (*Gen. Fung.*, 1931).

Myzocytiopsidaceae M.W. Dick (1997), Myzocytiopsidales. 4 gen. (+ 1 syn.), 33 spp.
Lit.: Dick (*MR* **101**: 878, 1997), Dick (*Mycol.*, 2001).

Myzocytiopsidales M.W. Dick (2001). Oomycetes. 2 fam., 9 gen., 56 spp. Thallus holocarpic over a period of time, pseudomycelial, rarely mycelial, coralloid, septate, without rhizoids; thalloid segments transformed into reproductive structures; asexual reproduction by sporangia, zoosporogenesis intrasporangial, spores zoosporic or aplanosporic, zoospores with lateral flagellar insertion; sexual reproduction homothallic and oogamous or automictic; oospores aplerotic; oospore walls smooth or ornamented. Obligate endoparasites of *Aschelmithes*, *Arthropoda* or algae. Fams:
(1) **Crypticolaceae**
(2) **Myzocytiopsidaceae**
Lit.: Dick (1995), Dick (2001), Glockling & Beakes (*Fungal Diversity* **4**: 1, 2000; nematophagous taxa).

Myzocytiopsis M.W. Dick (1997), Myzocytiopsidaceae. 15 (endobiotic in *Aschelminthes*), widespread. See Dick (*MR* **101**: 883, 1997), Dick (*Straminipilous Fungi*: 670 pp., 2001), Glockling & Beakes (*Mycol.* **98**: 1, 2006; ultrastr.).

Myzocytium Schenk (1858), Pythiaceae. 4, widespread. See Dick (*Straminipilous Fungi*: 670 pp., 2001), Kadlubowska (*Nova Hedwigia* **37**: 29, 2002; Poland), Pereira & Vélez (*Nova Hedwigia* **78**: 469, 2004; *Myzocytium megastonum* on *Rhizoclonium*).

Naegelia Reinsch (1878) ≡ Sapromyces fide Coker & Mathews (*N. Amer. Fl.* **2**, 1937).

Naegeliella J. Schröt. (1893) [non *Naegeliella* Correns

1892, *Algae*] ≡ Sapromyces fide Coker & Mathews (*N. Amer. Fl.* **2**, 1937).

Nellymyces A. Batko (1972), Rhipidiaceae. 1 (on immersed *Alnus glutinosa*), Poland. See Batko (*Acta Mycologica* Warszawa **7**: 256, 1971), Dick (*Straminipilous Fungi*: 670 pp., 2001).

Nematophthora Kerry & D.H. Crump (1980), Leptolegniellaceae. 1 (on nematodes), British Isles. See Kerry & Crump (*TBMS* **74**: 119, 1980), Dick (*Straminipilous Fungi*: 670 pp., 2001).

Nematosporangium (A. Fisch.) J. Schröt. (1893) = Pythium Pringsh. fide Fitzpatrick (*The lower fungi. Phycomycetes*, 1930).

Newbya M.W. Dick & Mark A. Spencer (2002), Saprolegniaceae. 11, widespread. See Spencer *et al.* (*MR* **106**: 558, 2002), Markovskaja (*Botanica Lithuanica* **10**: 141, 2004; Lithuania).

Novotelnova Voglmayr & Constant. (2008), Peronosporaceae. 1, Europe. See Voglmayr & Constantinescu (*MR* **112**: 487, 2008).

Nozemia Pethybr. (1913) = Phytophthora fide Lafferty & Pethybridge (*Sci. Proc. Roy. Dublin Soc.* N.S. **17**: 29, 1922).

Olpidiomorpha Scherff. (1926), Rozellopsidales. 1 (on *Pseudospora*), Europe. See Dick (*Straminipilous Fungi*: 670 pp., 2001).

Olpidiopsidaceae Sparrow ex Cejp (1959), Olpidiopsidales. 2 gen. (+ 5 syn.), 20 spp.

Olpidiopsidales M.W. Dick (2001). Oomycetes. 1 fam., 3 gen., 22 spp. Thallus holocarpic, initially plasmodial, olpidioid, subspherical or allantoid, nonseptate, without rhizoids; each thallus becoming transformed into a reproductive structure; asexual reproduction by zoosporangia, zoosporogenesis intrasporangial, zoospores heterokont with subapical flagellar insertion; sexual reproduction heterothallic, larger thalli functioning as oogonia, smaller contiguous thalli as antheridia (companion cells); oospores plerotic. Obligate parasites of flagellate fungi and green algae in freshwater and damp soil. Fam.: Olpidiopsidaceae

Lit.: Dick (*in* Margulis *et al.* (Eds), 1990), Dick (1995), Dick (2001).

Olpidiopsis Cornu (1872), Olpidiopsidaceae. 16 (on *Oomycetes*), widespread. See Dick (*Straminipilous Fungi*: 670 pp., 2001), Elíades *et al.* (*Darwiniana* **40**: 39, 2002; Argentina).

Oomycetes G. Winter (1880), Oomycota. 8 ord., 19 fam., 95 gen., 911 spp. Ords:

(1) **Albuginales**
(2) **Leptomitales**
(3) **Myzocytiopsidales**
(4) **Olpidiopsidales**
(5) **Peronosporales**
(6) **Pythiales**
(7) **Rhipidiales**
(8) **Saprolegniales**

For *Lit.* see ord. and fam.

Oomycota Arx (1967), Chromista. 1 class., 13 ord., 25 fam., 106 gen., 956 spp. Aquatic or terrestrial, freshwater or marine, saprobic or parasitic (some economically important on higher plants); thallus unicellular to mycelial (hyphae coenocytic), mainly aseptate; assimilative phase diploid (as in plants); zoospores with unequal (anisokont, heterokont) flagella, a tinsel one diverted with 2 rows of mastigonemes formed, and a whiplash one smooth or with fine flexous hairs backwards; with protoplasmic and nucleus-associated microtubules; the cell walls are a glucan-cellulose, rarely with minor amounts of chitin. Cosmop. and widespr.

Dick (*CJB* **73**(Suppl.): S712, 1995) and (2001) has proposed that the oomycetes should be recognized as class *Peronosporomycetes* and divided into three subclasses: *Peronosporomycetidae*; *Rhipidiomycetidae*; *Saprolegniomycetidae*. The sequence data in Riethmüller *et al.* (*CJB* **77**: 1790, 1999) support these subclasses.

Lit.: Buczaki (Ed.) (*Zoosporic plant pathogens: a modern perspective*, 1983), Cavalier-Smith (1987, *in* Green *et al.* (Eds), *The chromophyte algae*: 379, 1989, 1993), Cejp (*Oomycetes* 1, 1959 [Flora ČSR]), Corliss (1994), Dick (*New Phytol.* **71**: 1151, 1972; morphology and taxonomy; *in* Subramanian (Ed.), *Taxonomy of fungi*: 82, 1978, *in* Parker, 1982, **1**: 179), Dick & Win-Tin (*Biol. Rev.* **48**: 133, 1973; cytology), Fitzpatrick (1930), Fuller (*in* Margulis *et al.* (Eds), 1990: 380), Fuller & Jaworski (Eds) (*Zoosporic fungi in teaching and research*, 1986), Sparrow (*Aquatic phycomycetes* edn 2, 1960; *in* Ainsworth *et al.* (Eds), *The Fungi* **4B**: 61, 1973; ordinal classification, Dick *Bot. J. Linn. Soc.* **89**: 171, 1984, *in* Margulis *et al.* (Eds), 1990: 661, *CJB* **73**(suppl.): S712, 1995, *Straminipilous Fungi*, 2001). See also *Lit.* under Orders.

Pachymetra B.J. Croft & M.W. Dick (1989), Leptolegniaceae. 1 (on *Saccharum officinarum*), Australia. See Croft & Dick (*J. Linn. Soc.* Bot. **99**: 100, 1989), Riethmüller *et al.* (*Mycol.* **94**: 834, 2002; phylogeny; relationship to *Aphanomyces*).

Paraperonospora Constant. (1989), Peronosporaceae. 9, widespread. See Constantinescu (*Sydowia* **41**: 84, 1989), Riethmüller *et al.* (*Mycol.* **94**: 834, 2002; phylogeny), Göker *et al.* (*CJB* **81**: 672, 2003; phylogeny), Voglmayr *et al.* (*MR* **108**: 1011, 2004; phylogeny).

Perofascia Constant. (2002), Peronosporaceae. 1. See Constantinescu & Fatehi (*Nova Hedwigia* **74**: 324, 2002), Choi *et al.* (*MR* **107**: 1314, 2003; phylogeny), Voglmayr (*MR* **107**: 1132, 2003; phylogeny), Göker *et al.* (*Mycol. Progr.* **3**: 83, 2004; phylogeny), Scott *et al.* (*MR* **108**: 198, 2004; phylogeny).

Peronium Cohn (1854) ? = Olpidiopsis fide Dick (*in litt.*).

Peronophythora C.C. Chen ex W.H. Ko, H.S. Chang, H.J. Su, C.C. Chen & L.S. Leu (1978), Pythiaceae. 1, China; New Guinea; Taiwan; Vietnam. Arguments for both considering it a distinct genus or to merge with *Phytophthora* have been advanced. See Huang *et al.* (*Acta Mycol. Sin.* **2**: 201, 1983), Hall (*Mycopathologia* **106**: 189, 1989), Hudspeth *et al.* (*Mycol.* **92**: 674, 2000), Petersen & Rosendahl (*MR* **104**: 1295, 2000), Cook *et al.* (*Nova Hedwigia* Beiheft **122**: 231, 2001), Riethmüller *et al.* (*Mycol.* **94**: 834, 2002; phylogeny), Voglmayr (*MR* **107**: 1132, 2003; phylogeny), Zhang *et al.* (*Botanical Studies* **48**: 79, 2007).

Peronophythoraceae W.H. Ko, H.S. Chang, H.J. Su, C.C. Chen & L.S. Leu (1978) = Pythiaceae.

Peronoplasmopara (Berl.) G.P. Clinton (1905) ≡ Pseudoperonospora.

Peronosclerospora (S. Ito) Hara (1927), Peronosporaceae. 9, widespread. Sorghum downy mildew (*P. sorghi*). See Shaw (*Mycol.* **70**: 594, 1978), Williams (*Adv. Pl. Pathol.* **2**: 1, 1984), Bock *et al.* (*MR* **104**: 61, 2000), Ryley & Langdon (*Mycotaxon* **79**: 87, 2001; Australia), Perumal *et al.* (*MR* **110**: 471, 2006;

characterization, genetic distance analysis *Peronosclerospora sorghi*), Thines *et al.* (*MR* **112**: 345, 2008; phylogeny).

Peronosclerospora C.G. Shaw (1978) = Peronosclerospora (S. Ito) Hara fide Shaw & Waterhouse (*Mycol.* 72: 425, 1980).

Peronospora Corda (1837), Peronosporaceae. *c.* 75, widespread. Downy mildews, e.g. of beet (*Beta*) (*P. farinosa* f.sp. *betae*), clover (*P. trifoliorum*), crucifers (*P. parasitica*), onion (*Allium*) (*P. destructor*), spinach (*Spinacia*) (*P. farinosa* f.sp. *spinaciae*), tobacco (*P. hyoscyami* f.sp. *tabacina*). See Gäumann (*Beitr. Kryptfl. Schweiz* **5**, 1923; 260 spp. recognized), Lindquist (*Physis* **15**: 13, 1939; Argentina), Gustavsson (*Op. bot. Soc. bot. Lund* 3, 1959; Scandinavia), Yerkes & Shaw (*Phytopathology* **49**: 502, 1959; spp. on *Cruciferae*, *P. parasitica*, and *Chenopodiaceae*, *P. farinosa*), Ramfsjell (*Nytt Mag. Bot.* **8**: 147, 1960; Norway), Rao (*Nova Hedwigia* **16**: 269, 1968; India), Constantinescu (*Thunbergia* **15**, 1991; annotated list names), Petersen & Rosendahl (*MR* **104**: 1295, 2000; phylogeny), Dick (*Straminipilous Fungi*: 670 pp., 2001), Cooke *et al.* (*Advances in Downy Mildew Research*: 161, 2002; rel. between *Peronospora* and *Phytophthora*), Riethmüller *et al.* (*Mycol.* **94**: 834, 2002; phylogeny), Göker *et al.* (*CJB* **81**: 672, 2003; phylogeny), Voglmayr (*MR* **107**: 1132, 2003; phylogeny), Belbahri *et al.* (*MR* **109**: 1276, 2005; real time PCR detection), Cunnington (*Nova Hedwigia* **82**: 23, 2006; phylogeny, spp. on *Vicia*), Choi *et al.* (*MR* **111**: 381, 2007), Choi *et al.* (*Mycopathol.* **165**: 155, 2008; phylogeny, spp. on *Chenopodiaceae*), García-Blázquez *et al.* (*MR* **112**: 502, 2008; phylogeny, spp. on *Fabaceae*).

Peronosporaceae de Bary (1863), Peronosporales. 20 gen. (+ 15 syn.), 365 spp.

Peronosporales E. Fisch. (1892). Oomycetes. 1 fam., 20 gen., 365 spp. Mycelium intercellular with intracellular haustoria; dispersal by sporangia (zoosporangia) or conidia (conidiosporangia); sexual reproduction by oogamy: oospheres fertilized by nuclei from an antheridium, resulting in an oospore with thick and smooth or ornamented wall; cosmop., obligate parasites, almost totally confined to dicots, some economically important plant pathogens. Fam:
Peronosporaceae

Lit.: Lit. Note that some of publications listed do not match modern phylogenetic scheme.

(General): Fitzpatrick (*The Lower Fungi. Phycomycetes*: 85-233, 1930). Gäumann (*Die Pilze*: 49-65, 1949). Waterhouse (in Ainsworth *et al. The Fungi* **IVB**: 165-183, 1968). Spencer (Ed.) (*The Downy mildews*, 1981). Dick (*Straminipilous Fungi*, 2001). Spencer-Phillips *et al.* (Eds) (*Advances in Downy Mildew Research* **1**, 2002). Spencer-Phillips & Jeger (Eds) (*Advances in Downy Mildew Research* 2, 2004). Lebeda *et al.* (Eds) (*Advances in Downy Mildew Research* **3**, 2007).

(Monographs & lists): Jaczewski (*Mikologicheskaya Flora evropejskoj i aziatskoj Rossii. 1 Peronosporvye* 1901; Russia). Jaczewski & Jaczewski (*Opredelitel' Gribov* **I** *Fikomitsety* 3rd edn, 1931; world wide) Ito (*Mycological Flora of Japan* **I** *Phycomycetes* 1936; Japan). Lindtner (*Bull. Mus. Hist. Nat, Pays Serbe* Ser. B **9**: 1-153, 1957; former Yugoslavia). Ciferri (*Riv. Patol. Veg.* **1**: 333, 1961; Italy). Jörstad (*Nytt Mag. Bot.* **11**: 47. 1964; Norway). Palti & Chorin (*Phytopath. Medit.* **3**: 50, 1964; Israel). Ul'yanishchev (*Mikoflora Azerbajdjhana* **IV** *Peronosporovye Griby* 1967; Azerbaijan). Ubrizsy (*Acta Phytopath. Acad. Sci. Hung.* **2**: 153, 1967; Hungary). Osipyan (*Mikoflora Armyanskoj SSR* **I** *Peronosporovye Gariby*, 1968; Armenia). Kochman & Majewski (*Grzyby (Mycota)* **IV** *Phycomycetes, Peronosporales* 1970; Poland). Gaponenko (*Semejstvo Peronosporaceae Srednej Azii i Yujnogo Kazakhstana* 1972; Kazakstan). Benua & Karpova-Benua (*Parazitnye griby Yakutii* 1973; Russia, Yakutia). Vasyagina *et al.* (*Flora Sporovykh Rastenij Kazakhstana* **X** *Phycomycetes & Myxomycetes* 1977; Kazakhstan). Constantinescu & Negrean (*Mycotaxon* **16**: 537, 1983; Romania). Stanyaviciene (*Peronosporovye Griby Pribaltiki* 1984; Estonia, Latvia, Lithuania). Kirgizbaeva *et al.* (*Flora Gribov Uzbekistana* **2** *Nizshchie griby* 1985; Uzbekistan). Novotel'nova & Pystina (*Flora Sporovykh Rastenij SSSR* **XI** *Griby (3). Peronosporales* 1985; former USSR). Ul'yanishchev (*Opredelitel' Gribov Zakavkaz'ya. Peronosporovye Griby* 1985; Armenia, Azerbaijan, Georgia). Vanev *et al.* (*Gybite v Bylgariya* **2** *Razred Peronosporales* 1993, Bulgaria). Mazelaitis & Stanevičienėne (*Lietuvos Grybai* **I** *Myxomycota & Peronosporales* 1995; Lithuania). Yu (Ed.) (*Flora Fungorum Sinicorum* **6** *Peronosporales* 1998; China). Preece (*A Checklist of the Downy Mildews* (Peronosporaceae) *of the British Isles* 2002; British Isles). Shin & Choi (*Mycotaxon* **86**: 249. 2003; Korea). Brandenburger & Hagedorn (*Mitt. Biol. Bundesanstalt f. Land- u. Forstw. Berlin-Dahlem* **405**, 2006; Germany). García-Blázquez *et al.* (*Mycotaxon* **98**: 185. 2007; Andorra, Portugal, Spain & Balearic Islands).

Peronosporomycetes Locq. (1974), see *Oomycetes*.

Peronosporomycetidae M.W. Dick (2001), Oomycetes. See *Oomycetes*.

Peronosporomycotina, see *Oomycota*.

Petersenia Sparrow (1934), Pontismataceae. 3 (on marine *Rhodophyceae*), widespread. See Miller (*Mycol.* **54**: 422, 1963), Pueschel & van der Meer (*CJB* **63**: 409, 1985; ultrastr.), van der Meer & Pueschel (*CJB* **63**: 404, 1985; n.sp.), Molina *in* Moss (Ed.) (*The Biology of Marine Fungi*: 165, 1986), Dick (*Straminipilous Fungi*: 670 pp., 2001).

Phloeophthora Kleb. (1905) = Phytophthora fide Klebahn (*Krankh. Flied.*: 75 pp., 1909).

Phragmosporangium Seymour (2005), Saprolegniaceae. 1, Liberia. See Johnson *et al.* (*Mycotaxon* **92**: 6, 2005).

Physotheca, see *Rhysotheca*.

Phytophthora de Bary (1876), Peronosporaceae. *c.* 90, widespread. Includes important pathogens of: apple and pear (*P. cactorum* and *P. syringae*, fruit rot); cacao etc. (*P. palmivora*, pod rot and canker); chestnut (*P. cambivora*, ink disease); citrus (*P. citrophthora*, gummosis, foot and brown fruit rots); lima bean (*P. phaseoli*, downy mildew, *P. erythroseptica* and *P. megasperma*, pink rot); tomato and potato (*P. infestans*, blight; *P. cryptogea*, foot rot; *P. nicotianae*, foot and buck-eye rots in tobacco, tomato and many other plants). See Tucker (*Res. Bull. Mo agric. Exp. Stn* **153**, 1931; taxonomy), Tucker (*Res. Bull. Mo agric. Exp. Stn* **184**, 1933; distribution), Blackwell (*Mycol. Pap.* **30**, 1949; terminology), Waterhouse & Blackwell (*Mycol. Pap.* **57**, 1954; key Br. spp.), Hickman (*TBMS* **41**: 1, 1958; gen. review), Rangaswami (*Pythiaceous fungi*, 1962;

P. infestans), Waterhouse (*Mycol. Pap.* **92**, 1963; key), Chee (*RAM* **48**: 337, 1969; *P. palmivora*), Waterhouse (*Mycol. Pap.* **122**, 1970), Newhook & Podger (*Ann. Rev. Phytopath.* **10**: 299, 1972; hosts of *P. cinnamomi* in Australia, NZ), Gregory (Ed.) (*Phytophthora disease of cocoa*, 1974; pod rot and canker *P. palmivora*), Ribeiro (*A source book of the genus Phytophthora*, 1978), Brasier & Griffin (*TBMS* **72**: 111, 1979; taxonomy *P. palmivora*), Zentmyer (*P. cinnamomi and the diseases it causes* [*Am. phytopath. Soc. Monogr.*] **10**, 1980; *P. cinnamomi*), Stamps *et al.* (*Mycol. Pap.* **162**, 1990; tabular key), Ingram & Williams (*Adv. Pl. Path.* **7**, 1991; *P. infestans*), Lucas *et al.* (Eds) (*Phytophthora*, 1991; *P. infestans*), Erwin & Ribeiro (*Phytophthora Diseases Worldwide*, 1996), Goodwin (*Phytopathology* **87**: 462, 1997; popln genetics), Göker *et al.* (*CJB* **81**: 672, 2003; phylogeny), Martin & Tooley (*Mycol.* **95**: 269, 2003; phylogeny), Drenth & Guest (*ACIAR Monograph* **114**: 30, 2004; tropical spp.), Grünwald & Flier (*Annual Review of Phytopathology* **43**: 171, 2005; *Phytophthora infestans* at its center of origin), Drenth *et al.* (*Australas. Pl. Path.* **35**: 147, 2006; DNA-based detection and identification method), Schena *et al.* (*Molecular Plant Pathology* **7**: 365, 2006; detection and quantification with multiplex real-time PCR), Villa *et al.* (*Mycol.* **98**: 410, 2006; phylogeny).

Plasmopara J. Schröt. (1886), Peronosporaceae. *c.* 110, widespread. *P. viticola* (vine downy mildew), *P. halstedii* (downy mildew of sunflowers). See Constantinescu (*Mycotaxon* **43**: 471, 1992; spp. on *Umbelliferae*), Dick (*Advances in Downy Mildew Research*: 1, 2002; evolution), Riethmüller *et al.* (*Mycol.* **94**: 834, 2002; phylogeny), Göker *et al.* (*CJB* **81**: 672, 2003; phylogeny), Voglmayr *et al.* (*MR* **108**: 1011, 2004; phylogeny), Constantinescu *et al.* (*Taxon* **54**: 813, 2005; typification), Voglmayr *et al.* (*MR* **110**: 633, 2006; spp. on *Geraniaceae*).

Plasmoparopsis De Wild. (1896) = Pythiogeton fide Dick (*in litt.*).

Plasmoverna Constant., Voglmayr, Fatehi & Thines (2005), Peronosporaceae. 7, widespread. See Constantinescu *et al.* (*Taxon* **54**: 818, 2005).

Plectospira Drechsler (1927), Leptolegniaceae. 4, N. America; Japan. See Watanabe (*Mycol.* **79**: 77, 1987).

Pleocystidium C. Fisch (1884), Olpidiopsidaceae. 4 (parasitic in freshwater algae), widespread. See Dick (*Straminipilous Fungi*: 670 pp., 2001).

Pleotrachelus Zopf (1884), Rozellopsidales. 4 (parasitic on fungi), Europe; Canada. See Karling (*Chytriomyc. Iconogr.*, 1977; = *Olpidium*), Dick (*Straminipilous Fungi*, 2001).

Plerogone M.W. Dick (1986), Leptomitaceae. 1, USA; British Isles. See Dick (*J. Linn. Soc.* Bot. **93**: 228, 1986), Dick (*Straminipilous Fungi*: 670 pp., 2001).

Poakatesthia Thines & Göker (2007), Peronosporaceae. 1 (on *Pennisetum*), Ethiopia. See Thines *et al.* (*MR* **111**: 1380, 2007).

Pontisma H.E. Petersen (1905), Pontismataceae. 6, widespread. See Dick (*Straminipilous Fungi*: 670 pp., 2001).

Pontismataceae H.E. Petersen ex P.M. Kirk, P.F. Cannon & J.C. David (2001), Oomycota (inc. sed.). 2 gen., 9 spp.

Pringsheimina Kuntze (1891) ≡ Achlya.

Protascus P.A. Dang. (1903) nom. conf., Myzocytiopsidaceae. See Dick (*Straminipilous Fungi*: 670 pp., 2001).

Protoachlya Coker (1923), Saprolegniaceae. 6, N. America; Europe. See Dick (*Straminipilous Fungi*: 670 pp., 2001), Padgett & Johnson (*Mycol.* **96**: 205, 2004; zoosporangial discharge), Johnson *et al.* (*Mycotaxon* **92**: 11, 2005; systematics).

Protobremia Voglmayr, Riethm., Göker, Weiss & Oberw. (2004), Peronosporaceae. 1. See Voglmayr *et al.* (*MR* **108**: 1023, 2004).

Pseudolpidiaceae H.E. Petersen (1909) = Olpidiopsidaceae.

Pseudolpidiella Cejp (1959) = Sirolpidium fide Dick (*in litt.*).

Pseudolpidiopsis Minden (1911) = Pleocystidium fide Dick (*in litt.*).

Pseudolpidium A. Fisch. (1892) = Olpidiopsis fide Shanor (*J. Elisha Mitchell scient. Soc.* **55**: 179, 1939).

Pseudoperonospora Rostovzev (1903), Peronosporaceae. 6, widespread. *P. cubensis* (cucurbit downy mildew); *P. humuli* (hop (*Humulus*) downy mildew). See Waterhouse & Brothers (*Mycol. Pap.* **148**, 1981; monogr., keys), Constantinescu (*Crypt. Mycol.* **21**: 93, 2000; sporangium ultrastr.), Göker *et al.* (*CJB* **81**: 672, 2003; phylogeny), Voglmayr (*MR* **107**: 1132, 2003; phylogeny), Choi *et al.* (*MR* **109**: 841, 2005; *P. humuli* as syn. of *P. cubensis*).

Pseudoplasmodium Molisch (1925) = Labyrinthula fide Olive (*The mycetozoans*, 1975).

Pseudoplasmopara Sawada (1922) = Plasmopara fide Fitzpatrick (*The lower fungi. Phycomycetes*, 1930).

Pseudosphaerita P.A. Dang. (1895), Pseudosphaeritaceae. 4 (on *Algae*), France; Mexico. See Karling (*Bull. Torrey bot. Club* **99**: 223, 1973), Dick (*Straminipilous Fungi*: 670 pp., 2001).

Pseudosphaeritaceae M.W. Dick (2001), Rozellopsidales. 2 gen., 14 spp.

Pustula Thines (2005), Albuginaceae. 4. See Thines & Spring (*Mycotaxon* **92**: 454, 2005), Constantinescu & Thines (*Sydowia* **58**: 178, 2006; sporangial dimorphism).

Pyrrhosorus Juel (1901), ? Labyrinthulales. 1 (on red algae), Europe. See Dick (*Straminipilous Fungi*: 670 pp., 2001; systematic posn).

Pythiaceae J. Schröt. (1893), Pythiales. 7 gen. (+ 5 syn.), 163 spp.

Lit.: See Middleton (*Tijdschr. PlZiekt.* **58**: 226, 1952; generic concepts), Rangaswami (*Pythiaceous fungi*, 1962; host list), Dick *in* Margulis *et al.* (Eds) (*Handbook of Protoctista*: 661, 1990), Barr *et al.* (*CJB* **70**: 2163, 1992; flagellum ultrastr.).

Pythiacystis R.E. Sm. & E.H. Sm. (1906) = Phytophthora fide Leonian (*Am. J. Bot.* **12**: 444, 1925).

Pythiales M.W. Dick (2001). Oomycetes. 2 fam., 10 gen., 174 spp. Thallus mycelial or pseudomycelial. Asexual reproduction by sporangia or conidiosporangia; zoosporogenesis intrasporangial or extrasporangial in a vesicle; zoosporangia terminal, or sequential (percurrent or by internal or sympodial proliferation); sporangiophores rarely differentiated. Sexual reproduction by thin-walled oogonia; oospores usually single, aplerotic or nearly plerotic, periplasm absent or minimal and not persistent; oospore wall almost always smooth, ooplast solid and translucent. Cosmop., in plants, fungi and animals, or saprotrophic. Fams:

(1) **Pythiaceae**

(2) **Pythiogetonaceae**
Lit.: Dick *et al.* (1984), Dick (1995), Dick (2001).

Pythiella Couch (1935), Lagenaceae. 2 (in *Pythium*), Cuba; N. America. See Dick (*Straminipilous Fungi*: 670 pp., 2001).

Pythiogeton Minden (1916), Pythiogetonaceae. 9, widespread (north temperate). See Drechsler (*J. Wash. Acad. Sci.* **22**: 421, 1932), Batko (*Acta Mycologica* Warszawa **7**: 241, 1971; key), Voglmayr *et al.* (*MR* **103**: 591, 1999), Dick (*Straminipilous Fungi*: 670 pp., 2001), Ann *et al.* (*Mycol.* **98**: 116, 2006; sp.n. Taiwan).

Pythiogetonaceae M.W. Dick (1999), Pythiales. 2 gen. (+ 1 syn.), 10 spp.

Pythiomorpha H.E. Petersen (1909) = Phytophthora fide Blackwell *et al.* (*TBMS* **25**: 148, 1941), Waterhouse (*TBMS* **41**: 196, 1958).

Pythiopsis de Bary (1888), Saprolegniaceae. 5, widespread. See Spencer *et al.* (*MR* **106**: 549, 2002; revision), Johnson *et al.* (*Mycotaxon* **92**: 1, 2005; systematics).

Pythium Nees (1823) nom. rej., Saprolegniales.

Pythium Pringsh. (1858) nom. cons., Pythiaceae. *c.* 150, widespread. Causing damping-off, root diseases, and mycoparasitic; a few marine (*P. thalassium*; on marine pea-crab, *Pinnotheres*). See Middleton (*Mem. Torrey bot. Club* **20** no. 1, 1943), Atkins (*TBMS* **38**: 31, 1955; *P. thalassium*), Rangaswami (*Pythiaceous fungi*, 1962; host list), Waterhouse (*Mycol. Pap.* **110**, 1968; diagnoses, descriptions, figs. from original papers), Hendrix & Campbell (*Mycol.* **66**: 681, 1974), van der Plaats-Niterink (*Stud. Mycol.* **21**, 1981; key), Dick (*Keys to Pythium*, 1990; key), Bailey *et al.* (*FEMS Microbiol. Lett.* **207**: 153, 2002; identification of plant pathogenic spp.), Schurko *et al.* (*MR* **107**: 537, 2003; *Pythium insidiosum*, molecular phylogeny), Kong *et al.* (*FEMS Microbiol. Lett.* **240**: 229, 2004; species identification), Vanittanakom *et al.* (*J. Clin. Microbiol.* **42**: 3970, 2004; *Pythium insidiosum*, emerging human pathogen), Tambong *et al.* (*Appl. Environm. Microbiol.* **72**: 2691, 2006; identification and detection), Villa *et al.* (*Mycol.* **98**: 410, 2006; phylogeny).

Reessia C. Fisch (1883), Rhizidiomycetaceae. 2 (on *Lemna*), Europe.

Resticularia P.A. Dang. (1890) = Syzygangia fide Dick (*in litt.*).

Rheosporangium Edson (1915) = Pythium Pringsh. fide Fitzpatrick (*Mycol.* **15**: 166, 1923).

Rhipidiaceae Sparrow ex Cejp (1959), Rhipidiales. 6 gen. (+ 2 syn.), 15 spp.

Rhipidiales M.W. Dick (2001). Oomycetes. 1 fam., 6 gen., 15 spp. Thallus eucarpic, monocentric, saccate or pseudomycelial, with rhizoids. Asexual reproduction by zoosporangia; zoosporogenesis intrasporangial sometimes with a plasmamembranic vesicle; zoosporangia terminal; sporangiophores absent. Sexual reproduction by oogonia; oogonia more or less thick-walled; oospores usually single, aplerotic, periplasm persistent, sometimes permanent; oospore wall with a more or less rugose exospore. Saprotrophs in fresh or stagnant water; facultatively or obligately fermentative. Fam.:
Rhipidiaceae
Lit.: Dick *et al.* (1984), Dick (1995), Dick (2001).

Rhipidiomycetidae, Oomycetes. See *Oomycota*.

Rhipidium Cornu (1871) nom. cons., Rhipidiaceae. 4, widespread. See Dick (*Straminipilous Fungi*: 670 pp., 2001).

Rhizidiomyces Zopf (1884), Rhizidiomycetaceae. 13, widespread. See Sparrow (*Aquatic Phycomycetes* Edn 2: 751, 1960; key), Fuller (*Am. J. Bot.* **49**: 64, 1962; culture), Dick (*Straminipilous Fungi*: 670 pp., 2001), Fuller *in* McLaughlin *et al.* (Eds) (*The Mycota* A Comprehensive Treatise on Fungi as Experimental Systems for Basic and Applied Research **7A**: 73, 2001), Letcher *et al.* (*Australasian Mycologist* **22**: 99, 2004; Australia).

Rhizidiomycetaceae Karling ex P.M. Kirk, P.F. Cannon & J.C. David (2001), Hyphochytriales. 3 gen. (+ 1 syn.), 16 spp.

Rhizidiomycopsis Sparrow (1960) = Rhizidiomyces fide Longcore (*in litt.*).

Rhizoblepharis P.A. Dang. (1900) nom. dub., Chromista. 'phycomycetes'.

Rhysotheca G.W. Wilson (1907) = Plasmopara fide Fitzpatrick (*The lower fungi. Phycomycetes*, 1930).

Rozellopsidaceae M.W. Dick (2001), Rozellopsidales. 1 gen. (+ 1 syn.), 4 spp.

Rozellopsidales M.W. Dick (2001). Oomycota. 2 fam., 5 gen., 23 spp. Thallus holocarpic, plasmodial following infection and during the assimilative phase, plasmodium sometimes capable of vegetative division, transformed into a single sporangium or into discrete walled segments or sporangia, or into a single resting spore; asexual reproduction by zoospores, zoosporogenesis intrasporangial; zoospores motile within the zoosporangium prior to discharge; zoospores often with posterior refractive granules, and one or two flagella, biflagellate zoospores anisokont; sexual reproduction not established. Obligate endoparasites of flagellate fungi, algae or protoctists in freshwater. Fams:
(1) **Pseudosphaeritaceae**
(2) **Rozellopsidaceae**
Lit.: Dick (2001; systematics).

Rozellopsis Karling ex Cejp (1959), Rozellopsidaceae. 4, Europe; Japan. See Karling ex Cejp (*Fl. ČSR* **B(2), Oomycetes**: 342, 454, 1959), Dick (*Straminipilous Fungi*: 670 pp., 2001).

Salilagenidiaceae M.W. Dick (2001), Salilagenidiales. 1 gen., 6 spp.

Salilagenidiales M.W. Dick (2001). Saprolegniomycetidae. 1 fam., 1 gen., 6 spp. Fam.:
Salilagenidiaceae
For *Lit.* see under fam.

Salilagenidium M.W. Dick (2001), Salilagenidiaceae. 6, widespread. See Dick (*Straminipilous Fungi*: 314, 2001).

Saprolegnia Nees (1823), Saprolegniaceae. 15 (in freshwater), widespread (esp. north temperate). *S. parasitica* and other spp. pathogenic to fish and other aquatic vertebrates. See Tiffney (*Mycol.* **31**: 310, 1939; of fish), Scott *et al.* (*Virg. J. Sci.* **14**: 42, 1963), Seymour (*Nova Hedwigia* **19**: 1, 1970; gen. account, key 19 spp.), Mil'ko (*Mikol. Fitopatol.* **13**: 288, 1979), Neish & Hughes (*Fungal diseases of fishes* **6**, 1980), Willoughby *in* Roberts (Ed.) (*Microbial diseases of fish*, 1982; of salmon), Bruno & Wood *in* Wood & Bruno (Ed.) (*Fish Diseases & Disorders* **3**: 599, 1999), Steciow (*FEMS Microbiol. Lett.* **219**: 253, 2003; Argentina), Markovskaja (*Botanica Lithuanica* **12**: 97, 2006; Lithuania).

Saprolegniaceae Kütz. ex Warm. (1884), Saprolegniales. 19 gen. (+ 4 syn.), 145 spp.

Saprolegniales E. Fisch. (1892). Saprolegniomycetidae. 2 fam., 24 gen., 172 spp. 'water moulds'. Mycelium frequently stout and able to increase in diam., obvious cytoplasmic streaming. Asexual reproduction by sporangia; zoosporogenesis intrasporangial; zoospores released by papillate or operculate dehiscence or disintegration of sporangium, discharge vesicles not formed; zoosporangial formation sequential, percurrent or sympodial. Sexual reproduction by thick-walled, papillate or pitted oogonia, oosporogenesis centrifugal, periplasm absent or minimal and not persistent; oospores single or numerous, oospore wall almost always smooth. Freshwater, rarely marine, saprobic or parasitic; widespr. Fams:

(1) **Saprolegniaceae**

(2) **Leptolegniaceae**

Lit.: Coker (*The Saprolegniaceae with notes on other water molds*, 1923 [reprint 1968]), Fitzpatrick (1930), Cejp (1969), Seymour (1970), Dick (1973; *in* Margulis *et al.* (Eds), 1990: 661), Dick (2001).

Saprolegniomycetidae M.W. Dick (2001), see *Oomycota*.

Sapromyces Fritsch (1893), Rhipidiaceae. 3, widespread. See Dick (*Straminipilous Fungi*: 670 pp., 2001).

Schizochytrium S. Goldst. & Belsky (1964), Thraustochytriaceae. 2 (marine), widespread. See Raghu-Kumar (*TBMS* **90**: 627, 1988; key), Leander *et al.* (*European Journal of Protistology* **40**: 317, 2004; comparative morphology and molecular phylogeny), Cavalier-Smith & Chao (*J. Mol. Evol.* **62**: 388, 2006; phylogeny).

Sclerophthora Thirum., C.G. Shaw & Naras. (1953), Peronosporaceae. 7, widespread. *S. macrospora* (crazy top downy mildew). See Williams (*Adv. Pl. Pathol.* **2**: 1, 1984).

Sclerospora J. Schröt. (1879), Peronosporaceae. 3 (pathogens of *Poaceae*), widespread. See Narayanan (*Mycopathologia* **20**: 315, 1963; India), Waterhouse (*Misc. Publ. CMI* **17**, 1964; key), Williams (*Adv. Pl. Pathol.* **2**: 1, 1984), Dick (*Straminipilous Fungi*: 670 pp., 2001), Geetha *et al.* (*Journal of Mycology and Plant Pathology* **32**: 345, 2002; cellular fatty acid components in five pathotypes of *Sclerospora graminicola*), Göker *et al.* (*CJB* **81**: 672, 2003; phylogeny).

Sclerosporaceae M.W. Dick (1984) = Peronosporaceae.

Sclerosporales M.W. Dick (1984) = Peronosporales.

Scoliolegnia M.W. Dick (1969), Saprolegniaceae. 5, Europe. See Dick (*J. Linn. Soc.* Bot. **62**: 255, 1969), Dick (*Straminipilous Fungi*: 670 pp., 2001), Steciow *et al.* (*Mycotaxon* **91**: 381, 2005; status, key).

Siphopodium Reinsch (1875) nom. dub., Chromista. 'phycomycetes'.

Sirolpidiaceae Sparrow ex Cejp (1959), Oomycota (inc. sed.). 1 gen. (+ 1 syn.), 7 spp.

Sirolpidium H.E. Petersen (1905), Sirolpidiaceae. 7 (marine), widespread. *S. zoophthorum* on clam and oyster larvae. See Dick (*Straminipilous Fungi*: 670 pp., 2001), Rand (*Biodiversity of Fungi* Inventory and Monitoring Methods: 577, 2004).

Skirgiellopsis A. Batko (1978) = Rozellopsis fide Longcore (*in litt.*).

Sommerstorffia Arnautov (1923), Saprolegniaceae. 1 (on rotifers), Europe; N. America. See Saikawa & Hoshino (*Mycol.* **78**: 554, 1986; EM study), Dick (*Straminipilous Fungi*: 670 pp., 2001).

Sorodiplophrys L.S. Olive & Dykstra (1975), ? Labyrinthista. 1, Russia; USA. See Dykstra & Porter (*Mycol.* **76**: 626, 1984; affinity to *Diplophrys*).

Sphaerita P.A. Dang. (1886), Pseudosphaeritaceae. 10, widespread. See Sparrow (*Bull. Torrey bot. Club* **99**: 223, 1972), Dick (*Straminipilous Fungi*: 670 pp., 2001).

Sphaerosporangium Sparrow (1931) = Phytophthora fide Dick (*in litt.*).

Straminipila M.W. Dick (2001). Kingdom proposed by Dick to accommodate most organisms previously generally referred to the Kingdom *Chromista* (q.v.) and primarily characterized by the presence of tripartite tubular hairs on flagella or cysts. If this classification becomes generally accepted, the *Hyphochytriomycota* and *Oomycota* will need to be referred to it.

Synchaetophagus Apstein (1911) nom. dub., Chromista. See Vishniac (*Mycol.* **50**: 74, 1958) See, Karling (*Holocarpic Biflagellate Phycomycetes*: 206, 1981), Dick (*Straminipilous Fungi*: 670 pp., 2001; position uncertain).

Syzygangia M.W. Dick (1997), Myzocytiopsidaceae. 8 (intracellular parasites of freshwater algae), widespread. See Dick (*MR* **101**: 880, 1997), Dick (*Straminipilous Fungi*: 670 pp., 2001).

Tetradium Schltdl. (1852) [non *Tetradium* Lour. 1790, *Rutaceae*] = Bremia fide Fischer (*Rabenh. Krypt.-Fl.* **4** 2.1: 25, 1892).

Thanatostrea A. Franc & Arvy (1969) nom. inval., ? Labyrinthulales. 1, Portugal.

Thraustochytriaceae Sparrow ex Cejp (1959), Thraustochytriales. 7 gen., 36 spp.

Lit.: Sparrow *in* Ainsworth *et al.* (Eds) (*The Fungi* **4A**: 69, 1973).

Thraustochytriales Sparrow (1973). Labyrinthulea. 1 fam., 8 gen., 38 spp. Fam.:

Thraustochytriaceae

For *Lit.* see under fam.

Thraustochytrium Sparrow (1936), Thraustochytriaceae. 16 (on marine algae), N. America. See Gaertner (*Encycl. Cinematogr.*, 1971; life cycle), Porter & Jennings (*MR* **92**: 470, 1989), Ulken (*Biblthca Mycol.* **137**, 1990), Dick (*Straminipilous Fungi*: 670 pp., 2001), Chen & Chien (*Taiwania* **47**: 106, 2002; Taiwan), Leander *et al.* (*European Journal of Protistology* **40**: 317, 2004; comparative morphology and molecular phylogeny), Cavalier-Smith & Chao (*J. Mol. Evol.* **62**: 388, 2006; phylogeny).

Thraustotheca Humphrey (1893), Saprolegniaceae. 4, widespread. See Salvin (*Mycol.* **34**: 48, 1942), Blackwell & Powell (*Inoculum* **43**: 26, 1992; taxonomy), Blackwell & Powell (*Mycotaxon* **47**: 183, 1993), Steciow & Elíades (*Nova Hedwigia* **75**: 227, 2002; Argentina).

Trachysphaera Tabor & Bunting (1923), Pythiaceae. 1 (*T. fructigena*, mealy pod of cacao), Africa. See Dick (*Straminipilous Fungi*: 670 pp., 2001).

Trichothrauma Germ. (1850) nom. conf., ? Saprolegniales.

Ulkenia A. Gaertn. (1977), Thraustochytriaceae. 6, Europe. See Gaertner (*Veröffentlichungen des Instituts für Meeresforschung in Bremerhaven* **16**: 141, 1977), Dick (*Straminipilous Fungi*: 670 pp., 2001), Cavalier-Smith & Chao (*J. Mol. Evol.* **62**: 388, 2006; phylogeny).

Urophlyctidaceae Hadar (1982) = Physodermataceae.

Verrucalvaceae M.W. Dick (1984) = Saprolegniaceae.

Verrucalvus P. Wong & M.W. Dick (1985), Saprolegniaceae. 1, Australia. See Wong & Dick (*J. Linn.*

Soc. Bot. **89**: 174, 1984), Belkhiri *et al.* (*Molec. Phylogeny & Evol.* **9**: 1089, 1992), Dick (*Straminipilous Fungi*: 670 pp., 2001).

Viennotia Göker, Voglmayr, Riethm., Weiss & Oberw. (2003), Peronosporaceae. 1. See Göker *et al.* (*CJB* **81**: 682, 2003), Thines *et al.* (*MR* **110**: 646, 2006; phylogeny; rel. of *Graminivora graminicola*), Thines *et al.* (*MR* **112**: 345, 2008; phylogeny).

This page intentionally left blank

Acrasiaceae Poche (1913), Acrasida (inc. sed.). 1 gen., 2 spp. [*Acrasidae* (orth. zool.)].

Acrasiales Tiegh. ex J. Schröt. (1885). See *Acrasida*.

Acrasida J. Schröt. (1886) [*Acrasiales* (orth. bot.)]. Percolozoa. 4 fam., 6 gen., 14 spp. The only 'fungal' order of *Percolozoa* (q.v.) comprising fams:

(1) **Acrasiaceae**

(2) **Copromyxaceae**

(3) **Fonticulaceae**

(4) **Guttulinaceae**

For *Lit.* see under fam.

Acrasiomycetes. See *Heterolobosea*.

Acrasiomycota. See *Percolozoa*.

Acrasis Tiegh. (1880), Acrasiaceae. 2 (on beer yeast), widespread. See Dykstra & Keller *in* Lee, Leedale & Bradbury (Eds) (*An Illustrated Guide to the Protozoa, Organism Traditionally Referred to as Protozoa, or Newly Discovered Groups* **2**: 952, 2000), Spiegel *et al.* (*Biodiversity of Fungi* Inventory and Monitoring Methods: 547, 2004).

Acrocystis Ellis & Halst. ex Halst. (1890) = Cystospora fide Tabenhaus (*J. agric. Res.* **13**: 437, 1918).

Acytosteliaceae Raper ex Raper & Quinlan (1958), Dictyostelida (inc. sed.). 1 gen., 16 spp. [*Acytosteliidae* (orth. zool.)].

Acytostelium Raper (1956), Acytosteliaceae. 16, widespread. See Cavender & Vadell (*Mycol.* **92**: 992, 2000; monogr.), Cavender *et al.* (*Mycol.* **97**: 493, 2005; USA).

Aethaliaceae Fr. (1825) = Physaraceae.

Aethaliopsis Zopf (1885) = Fuligo fide Martin & Alexopoulos (*Myxomycetes*: 263, 1969).

Aethalium Link (1809) = Fuligo fide Martin & Alexopoulos (*Myxomycetes*: 263, 1969).

Alacrinella Manier & Ormières ex Manier (1968), Eccrinaceae. 2 (in *Isopoda*), France; USA. See Hibbets (*Syesis* **11**: 213, 1978; morphology, development), Lichtwardt (*The Trichomycetes. Fungal associates of arthropods*, 1986; key), Cafaro (*Mol. Phylogen. Evol.* **35**: 21, 2005; phylogeny).

Alwisia Berk. & Broome (1873) = Tubifera fide Martin & Alexopoulos (*Myxomycetes*, 1969).

Amallocystis Fage (1936) = Thalassomyces fide Kane (*N.Z. Jl Sci.* **7**: 289, 1964).

Amaurochaetaceae Rostaf. ex Cooke (1877) = Stemonitidaceae.

Amaurochaete Rostaf. (1873), Stemonitidaceae. 4, widespread. See Ing (*The Myxomycetes of Britain and Ireland, An Identification Handbook*, 1999; British Isles), Eliasson (*Karstenia* **40**: 31, 2000), Lado *et al.* (*Taxon* **54**: 543, 2005; nomencl.).

Amoebidiaceae J.L. Licht. ex P.M. Kirk, P.F. Cannon & J.C. David (2001), Amoebidiales. 2 gen., 15 spp.

Amoebidiales Léger & Duboscq (1929). Protozoa. 1 fam., 2 gen., 15 spp. Fam.:

Amoebidiaceae

Lit.: Manier (*Ann. Sci. nat., Bot.* sér. 12, **10**: 565, 1969; taxonomy), Trotter & Whisler (*CJB* **43**: 869, 1965; wall structure), Whisler (*J. Protozool.* **13**: 183, 1966; *Devel. Biol.* **17**: 562, 1968; culture and development), Moss (*in* Batra (Ed.), *Insect-fungus symbiosis*: 175, 1979), Lichtwardt (1986).

Amoebidium Cienk. (1861), Amoebidiaceae. 5 (on *Crustacea* and *Insecta*), widespread. See Whisler (*Am. J. Bot.* **49**: 193, 1962; life cycle, culture & nutrition), Trotter & Whisler (*CJB* **43**: 869, 1965; wall chemistry), Whisler (*Devel. Biol.* **17**: 562, 1968; development), Whisler & Fuller (*Mycol.* **60**: 1068, 1968; ultrastr.), Lichtwardt (*The Trichomycetes. Fungal associates of arthropods*, 1986; taxonomy, biology), Lichtwardt & Williams (*Mycol.* **84**: 376, 1992), Cavalier-Smith *in* Coombs *et al.* (Eds) (*Evolutionary Relationships Among Protozoa*: 375, 1998), Benny *in* McLaughlin *et al.* (Eds) (*The Mycota* A Comprehensive Treatise on Fungi as Experimental Systems for Basic and Applied Research **7A**, 2000), Benny & O'Donnell (*Mycol.* **92**: 1133, 2000) Cf. *Paramoebidium*, Tanabe *et al.* (*J. gen. appl. Microbiol.* Tokyo **51**: 267, 2005; phylogeny), White *et al.* (*Mycol.* **98**: 333, 2006; n.sp.).

Amoeboaphelidium Scherff. (1925), Protozoa. See monads.

Amoebosporus Ivimey Cook (1933), Protozoa. 2, West Indies.

Amoebozoa, Protozoa. 3 class., 7 ord., 20 fam., 82 gen., 1019 spp. A phylum of Protozoa (q.v.) containing the 'fungal' slime moulds [*Myxomycota* (orth. bot.)] and their allies.

Lit.: see under *Mycetozoa*.

Amphisporium Link (1815) = Didymium fide Martin & Alexopoulos (*Myxomycetes*: 383, 1969).

Amylotrogus Roze (1896) nom. dub., Myxogastria.

Ancyrophorus Raunk. (1888) = Enerthenema fide Torrend (*Brotéria* sér. bot. **7**: 72, 1908).

Andohaheloa Manier (1955) = Enterobryus fide Manier & Lichtwardt (*Annls Sci. Nat.* Bot., sér. 12 **9**: 519, 1968).

Angioridium Grev. (1827) = Physarum fide Saccardo (*Syll. Fung.* **7**: 347).

Anisomyxa Němec (1913) = Ligniera fide Kirk (*in litt.*).

Antonigeppia Kuntze (1898) = Lycogala fide Martin & Alexopoulos (*Myxomycetes*: 62, 1969).

Arcyodes O.F. Cook (1902), Trichiaceae. 1, N. America; Europe. See Ing (*The Myxomycetes of Britain and Ireland, An Identification Handbook*, 1999; British Isles).

Arcyrella (Rostaf.) Racib. (1884) = Arcyria fide Pando (*in litt.*).

Arcyria Hill ex F.H. Wigg. (1780), Arcyriaceae. *c.* 50, widespread. See Robbrecht (*Bull. Jard. bot. nat. Belg.* **44**: 303, 1974; key 13 spp.), Dörfelt *et al.* (*MR* **107**: 123, 2003; fossil), Yamamoto & Hagiwara (*Bull. natn. Sci. Mus.* Tokyo, N.S. **31**: 79, 2005; Japan).

Arcyriaceae Rostaf. ex Cooke (1877), Trichiida (inc. sed.). 2 gen. (+ 4 syn.), 51 spp.

Arcyriatella Hochg. & Gottsb. (1989), Trichiaceae. 1, Brazil. See Hochgesand & Gottsberger (*Nova Hedwigia* **48**: 485, 1989).

Arscyria, see *Arcyria*.

Arundinella L. Léger & Duboscq (1905) [non *Arundinella* Raddi 1823, *Gramineae*] = Arundinula fide Lichtwardt (*in litt.*).

Arundinula L. Léger & Duboscq (1906), Eccrinaceae. 6 (in *Decapoda*), widespread. See Lichtwardt (*Mycol.* **54**: 440, 1962; morphology), Hibbets (*Syesis* **11**: 213, 1978; development), Lichtwardt (*The Trichomycetes. Fungal associates of arthropods*, 1986; key), van Dover & Lichtwardt (*Biol. Bull.* **171**: 461, 1986), White *et al.* (*Mycol.* **98**: 333, 2006).

Astereptonema, see *Astreptonema*.

Astreptonema Hauptfl. (1895), Eccrinaceae. 5 (in *Amphipoda*), Europe; USA. See Moss (*TBMS* **65**: 115, 1975; ultrastr.), Hibbets (*Syesis* **11**: 213, 1978; development), Lichtwardt (*The Trichomycetes. Fun-

gal associates of arthropods, 1986; key), Cafaro (*Mol. Phylogen. Evol.* **35**: 21, 2005; phylogeny).

Bactridiopsis Henn. (1904) = Coccospora Wallr. fide Damon & Downing (*Mycol.* **46**: 209, 1954).

Bactriexta Preuss (1852) nom. dub., ? Protozoa.

Badhamia Berk. (1853), Physaraceae. *c.* 30, widespread. See Sekhon (*J. Indian bot. Soc.* **58**: 56, 1979; key 8 Indian spp.), Ing (*The Myxomycetes of Britain and Ireland, An Identification Handbook*, 1999; British Isles).

Badhamiopsis T.E. Brooks & H.W. Keller (1976), Physaraceae. 3, Australia; Hawaii; Japan; British Isles. See Brooks & Keller (*Mycol.* **68**: 835, 1976), Eliasson (*MR* **95**: 257, 1991; Hawaii), Ing (*The Myxomycetes of Britain and Ireland, An Identification Handbook*, 1999; British Isles), McHugh *et al.* (*N.Z. Jl Bot.* **41**: 487, 2003; Australia).

Barbeyella Meyl. (1914), Clastodermataceae. 1 (montane), widespread. See Schnittler *et al.* (*Karstenia* **40**: 159, 2000; ultrastr.), Schnittler *et al.* (*MR* **104**: 1518, 2000; ecol.).

Biomyxa Leidy (1875), Protozoa. 2 or. Amoeboid protist. See Anderson & Hoefler (*J. Ultrastr. Res.* **66**: 276, 1979; ultrastr.), Patterson (*Amer. Nat.* **154**: S96, 1999; systematic posn).

Brefeldia Rostaf. (1873), Stemonitidaceae. 1, widespread. See Ing (*The Myxomycetes of Britain and Ireland, An Identification Handbook*, 1999; British Isles), Crespo & Lugo (*Mycotaxon* **87**: 91, 2003; Argentina).

Brefeldiaceae Rostaf. ex Cooke (1877) = Stemonitidaceae.

Bursulla Sorokīn (1876), Echinosteliopsidaceae. 1 (coprophilous), former USSR. See Olive (*The mycetozoans*: 111, 1975).

Bursullaceae Zopf ex Berl. (1888) = Echinosteliopsidaceae.

Calomyxa Nieuwl. (1916), Dianemataceae. 2, widespread. See Ing (*The Myxomycetes of Britain and Ireland, An Identification Handbook*, 1999; British Isles), Mitchell (*Systematics and Geography of Plants* **74**: 261, 2004; key).

Calonema Morgan (1893), Trichiaceae. 4, widespread. See Rammeloo (*Icones Mycologicae*: [1], 1984; *Calonema aureum*), Estrada Torres *et al.* (*Mycol.* **95**: 354, 2003; Mexico sp.nov.), Mitchell (*Systematics and Geography of Plants* **74**: 261, 2004; key).

Calospeira G. Arnaud (1949) nom. inval., ? Dictyostelida.

Calyssosporium Corda (1831) nom. dub., ? Myxogastria.

Capillus Granata (1908) = Enterobryus fide Manier & Lichtwardt (*Annls Sci. Nat.* Bot., sér. 12 **9**: 519, 1968).

Carcerina Fr. (1849) = Diderma fide Pando (*in litt.*).

Cavosteliaceae L.S. Olive (1964), Protostelida (inc. sed.). 4 gen., 8 spp.

Lit.: Spiegel (*Handbook of Protoctista*: 484, 1990).

Cavostelium L.S. Olive (1965), Cavosteliaceae. 2, widespread (tropical). See Olive & Stoianovitch (*Am. J. Bot.* **56**: 979, 1969), Spiegel & Feldman (*Protoplasma* **163**: 189, 1991).

Ceratiomyxa J. Schröt. (1889), Ceratiomyxaceae. 4, widespread. See Ing (*The Myxomycetes of Britain and Ireland, An Identification Handbook*, 1999; British Isles), Stephenson (*Fungal Diversity Res. Ser.* **11**, 2003; New Zealand), Lado *et al.* (*Taxon* **54**: 543, 2005; nomencl.).

Ceratiomyxaceae J. Schröt. (1889), Protostelida (inc. sed.). 1 gen. (+ 3 syn.), 4 spp.

Lit.: Spiegel (*Handbook of Protoctista*: 484, 1990).

Ceratiomyxales G.W. Martin ex M.L. Farr & Alexop. (1977). See *Protostelida*.

Ceratiomyxella L.S. Olive & Stoian. (1971), Cavosteliaceae. 1, Brazil; Tahiti. See Olive & Stoianovitch (*Am. J. Bot.* **58**: 32, 1971), Glushchenko & Leontiev (*Mikol. Fitopatol.* **36**: 7, 2002; Ukraine).

Ceratiomyxomycetes. See *Protostelia*.

Ceratiopsis De Wild. (1896) = Ceratiomyxa fide Saccardo (*Syll. fung.* **14**, 1899).

Ceratium Alb. & Schwein. (1805) [non *Ceratium* Schrank 1793, *Algae*] ≡ Ceratiomyxa.

Cercozoa, Protozoa. 1 class., 1 ord., 2 fam., 15 gen., 50 spp. A phylum of Protozoa (q.v.) containing the 'fungal' plasmodiophorids [*Plasmodiophoromycota* (orth. bot.)] and their allies; included in the class *Phytomyxea*. Trophic phase intracellular in algal, fungal or plant host cells; plasmodia multinucleate, unwalled, cell division mitotic and cruciform (division producing a cross-like structure at metaphase), developing into either sporangia (forms zoospores or cytosori; resting structures); zoospores with two anteriorly directed whiplash flagella; flagellae smooth, lacking mastigonemes, equal in length. Obligate symbionts in soil or freshwater habitats; require coculturing with their hosts needed for laboratory studies. A single Ord.

Plasmodiophorida [*Plasmodiophorales* (orth. bot.)]

The placement of these fungi has varied. Dylewski (1990) kept them with the mastigomycete fungi (e.g. *Chromista*); they are retained in the *Protozoa* following Barr (1992) and Corliss (1994). Olive (1975) treated them as a separate class, while Corliss (1994) placed them as one of several orders in the phylum *Opalozoa*.

Lit.: Corliss (1994), Dylewski (*in* Margulis *et al.* (Eds), *Handbook of the Protoctista*: 399, 1990), Karling (*The Plasmodiophorales*, 1943; edn 2, 1968), Olive (1975), Waterhouse (*in* Ainsworth *et al.*, *The Fungi* **4B**: 75, 1973; key gen.), and see under subordinate taxa.

Cestodella Tuzet, Manier & Jolivet (1957) = Enterobryus fide Manier & Lichtwardt (*Annls Sci. Nat.* Bot., sér. 12 **9**: 519, 1968).

Chondrioderma Rostaf. (1873) = Diderma fide Martin & Alexopoulos (*Myxomycetes*: 347, 1969).

Cienkowskia Rostaf. (1873) [non *Cienkowskia* Regel & Rach 1859, *Boraginaceae*] ≡ Willkommlangea.

Cienkowskiaceae Rostaf. ex Cooke (1877) = Physaraceae.

Cionium Link (1809) ≡ Didymium.

Clastoderma A. Blytt (1880), Clastodermataceae. 2, widespread. See Eliasson & Keller (*MR* **100**: 610, 1996).

Clastodermataceae Alexop. & T.E. Brooks (1971), Echinostelida (inc. sed.). 2 gen. (+ 2 syn.), 3 spp.

Clastostelium L.S. Olive & Stoian. (1977), Protosteliaceae. 1, Guam. See Olive & Stoianovitch (*TBMS* **69**: 83, 1977), Spiegel & Feldman (*Mycol.* **80**: 525, 1988; trophic cells).

Clathrodastrum, see *Clathroidastrum P. Micheli ex Adans.*

Clathroidastrum Kuntze (1891) ≡ Stemonitis Gled.

Clathroidastrum P. Micheli ex Adans. (1763) ≡ Stemonitis Gled.

Clathroides P. Micheli (1729) = Arcyria fide Pando (*in* *litt.*).

Clathroptychiaceae Rostaf. ex Cooke (1877) = Reticulariaceae.

Clathroptychium Rostaf. (1875) = Dictydiaethalium fide Martin & Alexopoulos (*Myxomycetes*: 59, 1969).

Clathrosorus Ferd. & Winge (1920) = Spongospora fide Kirk (*in litt.*).

Clatroidastron, see *Clathroidastrum P. Micheli ex Adans.*

Claustria Fr. (1849) = Physarum fide Martin & Alexopoulos (*Myxomycetes*: 274, 1969).

Coccospora Wallr. (1833) nom. dub., Myxogastria. See Carris (*Sydowia* **45**: 92, 1993).

Coenonia Tiegh. (1884), Dictyosteliaceae. 1, Europe. See Dykstra & Keller *in* Lee, Leedale & Bradbury (Eds) (*An Illustrated Guide to the Protozoa, Organism Traditionally Referred to as Protozoa, or Newly Discovered Groups* **2**: 952, 2000).

Collaria Nann.-Bremek. (1975) = Comatricha fide Martin & Alexopoulos (*Myxomycetes*: 222, 1969).

Colloderma G. Lister (1910), Stemonitidaceae. 4, widespread. See Ing (*The Myxomycetes of Britain and Ireland, An Identification Handbook*, 1999; British Isles), Stephenson (*Fungal Diversity Res. Ser.* **11**, 2003; New Zealand), McHugh (*Mycotaxon* **92**: 107, 2005; Ecuador), Yamamoto & Hagiwara (*Bull. natn. Sci. Mus.* Tokyo, N.S. **31**: 79, 2005; Japan).

Collodermataceae Lister (1925) = Stemonitidaceae.

Comatricha Preuss (1851), Stemonitidaceae. *c.* 50, widespread. See Ing (*The Myxomycetes of Britain and Ireland, An Identification Handbook*, 1999; British Isles), Camino *et al.* (*Mycotaxon* **88**: 315, 2003; Cuba), Mitchell (*Systematics and Geography of Plants* **74**: 261, 2004; key gen.).

Comatrichoides Hertel (1956) = Comatricha fide Martin & Alexopoulos (*Myxomycetes*: 222, 1969).

Coniocephalum Brond. (1828) nom. dub., Myxogastria. See Martin (*in litt.*).

Copromyxa Zopf (1884), Copromyxaceae. 2, Europe; N. America. See Nesom & Olive (*Mycol.* **64**: 1359, 1972), Spiegel & Olive (*Mycol.* **70**: 843, 1978), Raper (*The dictyostelids*, 1984), Dykstra & Keller *in* Lee, Leedale & Bradbury (Eds) (*An Illustrated Guide to the Protozoa, Organism Traditionally Referred to as Protozoa, or Newly Discovered Groups* **2**: 952, 2000).

Copromyxaceae L.S. Olive & Stoian. (1975), Acrasida (inc. sed.). 2 gen., 6 spp. [*Copromyxidae* (orth. zool.)].

Copromyxella Raper, Worley & Kurzynski (1978), Copromyxaceae. 4, USA; Costa Rica. See Raper *et al.* (*Am. J. Bot.* **65**: 1013, 1978), Dykstra & Keller *in* Lee, Leedale & Bradbury (Eds) (*An Illustrated Guide to the Protozoa, Organism Traditionally Referred to as Protozoa, or Newly Discovered Groups* **2**: 952, 2000).

Cornuvia Rostaf. (1873), Arcyriaceae. 1, widespread. See Ing (*The Myxomycetes of Britain and Ireland, An Identification Handbook*, 1999; British Isles).

Coscinium Endl. [not traced] = Lamproderma fide Mussat (*Syll. fung.* **15**: 397, 1901).

Crateriachea Rostaf. (1873) = Physarum fide Martin & Alexopoulos (*Myxomycetes*: 274, 1969).

Craterium Trentep. (1797), Physaraceae. 16, widespread. See Ing (*The Myxomycetes of Britain and Ireland, An Identification Handbook*, 1999; British Isles), Stojanowska & Panek (*Acta Mycologica* Warszawa **38**: 65, 2003; Poland).

Cribraria Pers. (1794) nom. cons., Cribrariaceae. *c.* 40, widespread. See Novozhilov (*Mikol. Fitopatol.* **27**: 39, 1993), Lado *et al.* (*Mycol.* **91**: 157, 1999; *Cribraria zonatispora*), Li (*Mycoscience* **43**: 247, 2002; China), Lado *et al.* (*Taxon* **54**: 829, 2005; nomencl.).

Cribraria Schrad. ex J.F. Gmel. (1792) nom. rej., Myxogastria.

Cribrariaceae Corda (1838), Liceida (inc. sed.). 2 gen. (+ 2 syn.), 41 spp.

Cupularia Link (1833) = Craterium fide Pando (*in litt.*).

Cylichnium Wallr. (1833) = Licea fide Martin & Alexopoulos (*Myxomycetes*: 47, 1969).

Cystobacter J. Schröt. (1886) = Polyangium fide Buchanan *et al.* (*Index Bergeyana*, 1966).

Cystospora J.E. Elliot (1916) nom. dub., Plasmodiophoraceae. See Karling (*Plasmodiophorales*, 1968).

Cytidium Morgan (1896) = Physarum fide Martin & Alexopoulos (*Myxomycetes*: 292, 1969).

Daloala Tuzet, Manier & Vog.-Zuber (1952) = Enterobryus fide Manier & Lichtwardt (*Annls Sci. Nat.* Bot., sér. 12 **9**: 519, 1968).

Demordium Link (1809) nom. dub., Myxogastria. 1, Europe.

Dermocystidium Pérez (1908), Protozoa. *D. cochliopodii* on *Cochliopodium bilimbosum* (see Valkanov, *Nova Hedw.* **12**: 393, 1967). See Dykova & Lom (*J. appl. Ichthyology* **8**: 180, 1992; evidence of fungal nature) Now assigned to the DRIPs clade, q.v. *D. marinum* now belongs in the genus *Perkinsus*.

Dermocystis Pérez (1907) ≡ Dermocystidium.

Dermodium Link (1809) nom. conf., Myxogastria.

Dermodium Rostaf. (1875) = Lycogala fide Martin & Alexopoulos (*Myxomycetes*: 61, 1969).

Dermosporidium Carini (1940) = Rhinosporidium fide Azevedo (*Rev. Cien. Biol.* **1**: 97, 1963).

Diachaeella Höhn. (1909) = Diachea fide Martin & Alexopoulos (*Myxomycetes*: 177, 1969).

Diachea Fr. (1825), Stemonitidaceae. 10, widespread. See Ing (*The Myxomycetes of Britain and Ireland, An Identification Handbook*, 1999; British Isles), Stephenson (*Fungal Diversity Res. Ser.* **11**, 2003; New Zealand), Gaither & Keller (*Systematics and Geography of Plants* **74**: 217, 2004; SEM).

Diacheopsis Meyl. (1930), Stemonitidaceae. 16, widespread. See Kowalski (*Mycol.* **67**: 616, 1975; key), Ing (*The Myxomycetes of Britain and Ireland, An Identification Handbook*, 1999; British Isles), Mitchell (*Systematics and Geography of Plants* **74**: 261, 2004; key gen.), Moreno *et al.* (*Mycol. Progr.* **5**: 129, 2006; *Diacheopsis metallica* and *Diacheopsis kowalskii*).

Dianema Rex (1891), Dianemataceae. 9, widespread. See Kowalski (*Mycol.* **59**: 1080, 1968), Mitchell (*Systematics and Geography of Plants* **74**: 261, 2004; key to corticolous spp.).

Dianemataceae T. Macbr. (1899), Trichiida (inc. sed.). 2 gen. (+ 3 syn.), 11 spp.

Dianemina A.R. Loebl. & Tappan (1961) ≡ Dianema.

Dichosporium Nees (1816) nom. dub., Physaraceae.

Dictydiaethaliaceae Nann.-Bremek. ex H. Neubert, Nowotny & K. Baumann (1993), Liceida (inc. sed.). 1 gen. (+ 2 syn.), 2 spp.

Dictydiaethalium Rostaf. (1873), Dictydiaethaliaceae. 2, widespread. See Stephenson (*Fungal Diversity Res. Ser.* **11**, 2003; NZ), Mitchell (*Systematics and Geography of Plants* **74**: 261, 2004; key).

Dictydium Schrad. (1797) = Cribraria Pers. fide Lado & Pando (*Fl. Mycol. Iberica* **2**, 1997).

Dictyostelia, Mycetozoa. 1 ord., 2 fam., 4 gen., 93 spp. [*Dictyosteliomycetes* (orth. bot.)]. Dictyostelid slime moulds, dictyostelids. Ord.:
Dictyosteliales
For *Lit.* see ord. and fam.

Dictyosteliaceae Rostaf. ex Cooke (1877), Dictyostelida (inc. sed.). 3 gen. (+ 1 syn.), 77 spp. [*Dictyosteliidae* (orth. zool.)].

Dictyosteliales L.S. Olive ex P.M. Kirk, P.F. Cannon & J.C. David (2001). See *Dictyostelida.*

Dictyostelida [*Dictyosteliales* (orth. bot.)]. Dictyostelia. 2 fam., 4 gen., 93 spp. An order of 'fungal' slime moulds of *Mycetozoa* (q.v.) comprising Fams:
(1) **Actyosteliaceae**
(2) **Dictyosteliaceae**
See Cavender (1990; key 15 spp. Ohio).
For *Lit.* see under fam.

Dictyosteliomycetes. See *Dictyostelea.*

Dictyosteliomycota = Mycetozoa. See *Amoebozoa.*

Dictyostelium Bref. (1870), Dictyosteliaceae. *c.* 60, widespread. See Raper (*Quart. Rev. Biol.* **26**: 169, 1951; culture), Lee (*TMSJ* **12**: 142, 1971; Japanese spp.), Loomis (*Dictyostelium discoideum: a developmental system*, 1975), Olive (*The mycetozoans*: 55, 1975; key), Waddell (*Nature* **298**: 464, 1983; *D. caveatum* predacious on *Dictyostelium* spp.), Raper (*The dictyostelids*, 1984), Cavender *et al.* (*N.Z. Jl Bot.* **40**: 235, 2002; New Zealand), Glöckner *et al.* (*Nature* London **417**: 79, 2002; Sequence and analysis of chromosome 2 of *Dicyostelium discoideum*), Yeh (*Mycotaxon* **86**: 103, 2003; Taiwan), Cavender *et al.* (*Mycol.* **97**: 493, 2005; USA), Hagiwara (*Bull. natn. Sci. Mus.* Tokyo, N.S. **32**: 47, 2006), Hagiwara & Hosono (*Bull. natn. Sci. Mus.* Tokyo, N.S. **32**: 1, 2006; Japan).

Diderma Pers. (1794), Didymiaceae. *c.* 75, widespread. See Ing (*The Myxomycetes of Britain and Ireland, An Identification Handbook*, 1999; British Isles), Lado (*Cuadernos de Trabajo de Flora Micológica Ibérica* **16**, 2001; nomenclature), Stephenson (*Fungal Diversity Res. Ser.* **11**, 2003; New Zealand).

Didymiaceae Rostaf. ex Cooke (1877), Physarida (inc. sed.). 7 gen. (+ 12 syn.), 166 spp.

Didymium Schrad. (1797), Didymiaceae. *c.* 75, widespread. See Ing (*The Myxomycetes of Britain and Ireland, An Identification Handbook*, 1999; British Isles), Stephenson (*Fungal Diversity Res. Ser.* **11**, 2003; New Zealand).

Diphtherium Ehrenb. (1818) = Lycogala fide Martin & Alexopoulos (*Myxomycetes*: 64, 1969).

Diplophysalis Zopf (1884), Protozoa. (Algal parasite). Heterotrophic amoeboflagellate protist formerly included in the monads or proteomyxids. See Karling (*Am. J. Bot.* **17**: 928, 1930).

Disporium Léman (1819) = Didymium fide Martin & Alexopoulos (*Myxomycetes*: 383, 1969).

Eccrina Leidy (1852) = Enterobryus fide Manier & Lichtwardt (*Annls Sci. Nat.* Bot., sér. 12 **9**: 519, 1968).

Eccrinaceae L. Léger & Duboscq ex P.M. Kirk, P.F. Cannon & J.C. David (2001), Eccrinales. 14 gen. (+ 14 syn.), 53 spp.

Eccrinales Léger & Duboscq (1929). Protozoa. 3 fam., 17 gen., 63 spp. Thallus unbranched or branched only at the base, vegetatively coenocytic, with basal, secreted holdfast; asexual reproduction by endogenous sporangiospores produced in basipetal succession from the thallus tip, (a) multinucleate, thin-walled sporangiospores germinating within the digestive tract of the same host, (b) usually uninucleate, thick-walled sporangiospores which in aquatic forms are frequently appendaged and on release function as re-infestive spores; microthallus present in some genera; sexual reproduction known only in *Enteropogon*; endocommensals of aquatic and terrestrial *Crustacea*, *Diplopoda* and *Insecta*. Fams:
(1) Eccrinaceae
(2) Palavasciaceae
(3) Parataeniellaceae
Lit.: Lichtwardt (*Mycol.* **46**: 564, 1954, **52**: 410, 1960; taxonomy, 1986), Manier (1969, taxonomy), Hibbits (*Syesis* **11**: 213, 1978; development, sexual repr.), Moss (1979).

Eccrinella L. Léger & Duboscq (1933) = Astreptonema fide Manier (*Annls Sci. Nat.* Bot., sér. 12 **5**: 767, 1964).

Eccrinidus Manier (1970), Eccrinaceae. 1 (in *Diplopoda*), France. See Manier & Grizel (*C.R. Acad. Sci. Paris* **274**: 1159, 1972; ultrastr.), Lichtwardt (*The Trichomycetes. Fungal associates of arthropods*, 1986; key), Cafaro (*Mol. Phylogen. Evol.* **35**: 21, 2005; phylogeny).

Eccrinoides L. Léger & Duboscq (1929), Eccrinaceae. 4 (in *Isopoda* and *Diplopoda*), Europe. See Lichtwardt (*The Trichomycetes. Fungal associates of arthropods*, 1986).

Eccrinopsis L. Léger & Duboscq (1916) = Enterobryus fide Manier & Lichtwardt (*Annls Sci. Nat.* Bot., sér. 12 **9**: 519, 1968).

Echinosteliaceae Rostaf. ex Cooke (1877), Echinostelida (inc. sed.). 1 gen. (+ 1 syn.), 15 spp.

Echinosteliales G.W. Martin (1961). See *Echinostelida.*

Echinostelida [*Echinosteliales* (orth. bot.)]. Myxogastria. 2 fam., 3 gen., 18 spp. Spore mass white, pale pink, or yellow, columella present, peridium fugaceous. Fams:
(1) **Clastodermataceae**
(2) **Echinosteliaceae**
Lit.: Alexopoulos & Brooks (*Mycol.* **63**: 925, 1971).

Echinosteliopsidaceae L.S. Olive (1970), Protostelida (inc. sed.). 2 gen., 2 spp.

Echinosteliopsidales L.S. Olive (1970). See *Protostelida.*

Echinosteliopsis D.J. Reinh. & L.S. Olive (1967), Echinosteliopsidaceae. 1, widespread. See Reinhardt & Olive (*Mycol.* **58**: 967, 1967), Olive (*The mycetozoans*: 110, 1975).

Echinostelium de Bary (1873), Echinosteliaceae. 15, widespread. *E. lunatum*, the smallest known myxomycete. See Olive & Stoianovitch (*Mycol.* **63**: 1051, 1971), Whitney (*Mycol.* **72**: 950, 1980; key), Lado & Pando (*Flora Mycologica Iberica* **2**, 1997; Spain, Portugal), Mitchell (*Systematics and Geography of Plants* **74**: 261, 2004; key to corticolous spp.), Adamonytė (*Acta Mycologica* Warszawa **41**: 169, 2006; Lithuania).

Elaeomyxa Hagelst. (1942), Elaeomyxaceae. 2, widespread (north temperate). See Ing (*The Myxomycetes of Britain and Ireland, An Identification Handbook*, 1999; British Isles).

Elaeomyxaceae Hagelst. ex M.L. Farr & H.W. Keller (1982), Physarida (inc. sed.). 1 gen., 2 spp.

Ellobiocystis Coutière (1911) nom. dub., Protozoa. See Whisler *in* Margulis *et al.* (Eds) (*Handbook of Protoctista*: 715, 1990; systematic position).

Ellobiopsis Caullery (1910), Protozoa. 1 (parasite of marine copepods). See Whisler *in* Margulis *et al.* (Eds) (*Handbook of Protoctista*: 715, 1990; placed in the class *Ellobiopsida*).

Embolus Haller (1768) nom. dub., Mycetozoa.

Endemosarca L.S. Olive & Erdos (1971), Endemosarcaceae. 3, widespread. See Erdos (*Mycol.* **64**: 423, 1972; nuclear cycle), Olive (*The mycetozoans*: 212, 1975), Dick (*Straminipilous Fungi*: 670 pp., 2001).

Endemosarcaceae L.S. Olive & Erdos (1971), ? Plasmodiophorida (inc. sed.). 1 gen., 3 spp.

Endodromia Berk. (1841) nom. dub. ? = Echinostelium fide Höhnel (*Sitzungsber. Kaiserl. Akad. Wiss.* Math.-Naturwiss. Cl. Abt. 1 **123**: 97, 1914).

Endostelium L.S. Olive, W.E. Benn. & Deasey (1984), Protosteliaceae. 1, Papua New Guinea. See Olive *et al.* (*Mycol.* **76**: 884, 1984).

Enerthenema Bowman (1830), Stemonitidaceae. 3, widespread (north temperate). See Ing (*The Myxomycetes of Britain and Ireland, An Identification Handbook*, 1999; British Isles).

Enerthenemataceae Rostaf. ex Cooke (1877) = Stemonitidaceae.

Enigma G. Arnaud (1949) nom. inval., ? Protozoa.

Enteridiaceae M.L. Farr (1982) = Tubiferaceae.

Enteridium Ehrenb. (1819) nom. rej. = Reticularia Bull. fide Lado *et al.* (*Taxon* **47**: 109, 1998; nomencl.).

Enterobryus Leidy (1850), Eccrinaceae. 24 (in *Coleoptera*, *Diplopoda* and *Decapoda*), widespread. See Lichtwardt (*Mycol.* **46**: 564, 1954), Lichtwardt (*Mycol.* **49**: 463, 1957), Lichtwardt (*Mycol.* **49**: 734, 1957), Lichtwardt (*Mycol.* **50**: 550, 1958), Lichtwardt (*Mycol.* **52**: 248, 1960), Lichtwardt (*Mycol.* **52**: 410, 1960), Lichtwardt (*Mycol.* **52**: 743, 1960), Tuzet & Manier (*Protistologia* **3**: 413, 1967; ultrastr.), Manier *et al.* (*Biol. Gabon.* **3-4**: 305, 1972), Manier *et al.* (*Bull. Inst. Fund. Afrique Noire* **36**: 614, 1974), Wright (*Proc. Helminthol. Soc. Wash.* **46**: 213, 1979; ultrastr.), Lichtwardt (*The Trichomycetes. Fungal associates of arthropods*, 1986; key), Gorter (*Bothalia* **23**: 85, 1990), Cafaro (*Mol. Phylogen. Evol.* **35**: 21, 2005; phylogeny), White *et al.* (*Mycol.* **98**: 333, 2006).

Enteromyces Lichtw. (1961), Eccrinaceae. 1 (in *Decapoda*), widespread. See McCloskey & Caldwell (*J. Elisha Mitchell scient. Soc.* **81**: 114, 1965), Hibbets (*Syesis* **11**: 213, 1978), Cafaro (*Mol. Phylogen. Evol.* **35**: 21, 2005; phylogeny).

Enteromyxa Cienk. (1884), Protozoa. Amoeboflagellate protist formerly included in the monads or proteomyxids.

Enteropogon Hibbits (1979), Eccrinaceae. 2 (in *Decapoda*), USA. See Lichtwardt (*The Trichomycetes. Fungal associates of arthropods*, 1986), Chien & Hsieh (*Trichomycetes and Other Fungal Groups, Robert W. Lichwardt Commemoration Volume*: 55, 2001; Taiwan), Cafaro (*Mol. Phylogen. Evol.* **35**: 21, 2005; phylogeny).

Erionema Penz. (1898) = Fuligo fide Mitchell (*Systematics and Geography of Plants* **74**: 261, 2004).

Famintzinia Hazsl. (1877) nom. rej. = Ceratiomyxa fide Martin & Alexopoulos (*The myxomycetes*, 1969).

Fictoderma Preuss (1852), ? Protozoa.

Fonticula Worley, Raper & M. Hohl (1979), Fonticulaceae. 1, USA. See Worley *et al.* (*Mycol.* **71**: 746, 1979), Blanton *in* Margulis *et al.* (Eds) (*Handbook of Protoctista*: 75, 1990), Cavalier-Smith *in* Coombs *et al.* (Eds) (*Evolutionary Relationships Among Protozoa*: 375, 1998; placed in *Choanozoa*), Dykstra & Keller *in* Lee, Leedale & Bradbury (Eds) (*An Illustrated Guide to the Protozoa, Organism Traditionally Referred to as Protozoa, or Newly Discovered Groups* **2**: 952, 2000).

Fonticulaceae Worley, Raper & M. Hohl (1979), Acrasida (inc. sed.). 1 gen., 1 spp.

Frankiella Maire & A. Tison (1909) ? = Plasmodiophora fide Hawker & Fraymouth (*J. gen. Microbiol.* **5**: 369, 1951).

Fuliginaceae Chevall. (1826) = Physaraceae.

Fuligo Haller (1768), Physaraceae. 7, widespread. See Ing (*The Myxomycetes of Britain and Ireland, An Identification Handbook*, 1999; British Isles), Stephenson (*Fungal Diversity Res. Ser.* **11**, 2003; New Zealand).

Galeperdon Weber ex F.H. Wigg. (1780) ≡ Lycogala.

Galoperdon, see *Galeperdon*.

Guttulina Cienk. (1874) [non *Guttulina* Orb. 1839, *Invertebrata*] = Pocheina fide Ing (*in litt.*).

Guttulinaceae Zopf ex Berl. (1888), Acrasida (inc. sed.). 2 gen. (+ 1 syn.), 5 spp.

Guttulinopsidaceae L.S. Olive (1970) = Guttulinaceae.

Guttulinopsis L.S. Olive (1901), Guttulinaceae. 2, N. America. See Raper *et al.* (*Mycol.* **69**: 1016, 1977), Dykstra & Keller *in* Lee, Leedale & Bradbury (Eds) (*An Illustrated Guide to the Protozoa, Organism Traditionally Referred to as Protozoa, or Newly Discovered Groups* **2**: 952, 2000).

Gymnococcus Zopf (1884), Protozoa. 2 (Algal parasite). Amoeboflagellate protist formerly included in the monads or proteomyxids.

Gymnomycota. See *Mycetozoa*.

Gymnomyxa. See *Mycetozoa*.

Halterophora Endl. (1836) nom. dub., ? Myxogastria.

Heimerlia Höhn. (1903) nom. dub., ? Echinosteliaceae. See Martin *et al.* (*The genera of Myxomycetes*, 1983).

Heimerliaceae G. Arnaud (1949) = Echinosteliaceae.

Heliomycopsis Arnaud (1949) nom. inval., ? Dictyostelida.

Hemiarcyria Rostaf. (1875) ≡ Hemitrichia.

Hemitrichia Rostaf. (1873), Trichiaceae. 26, widespread. See Ing (*The Myxomycetes of Britain and Ireland, An Identification Handbook*, 1999; British Isles), Kuhnt (*Z. Mykol.* **71**: 165, 2005), Lado *et al.* (*Taxon* **54**: 543, 2005; nomencl.), Liu *et al.* (*Collection and Research* **19**: 27, 2006; Taiwan).

Heterodictyon Rostaf. (1873) = Cribraria Pers. fide Martin & Alexopoulos (*Myxomycetes*: 92, 1969).

Heterolobosea, Protozoa. 1 ord., 4 fam., 6 gen., 14 spp. [*Acrasiomycetes* (orth. bot.)]. An order of *Percolozoa* (q.v.) comprising the 'fungal' ord.:

Acrasida (syn. *Acrasiales*)

For *Lit.* see fam.

Heterotrichia Massee (1892) = Arcyria fide Martin & Alexopoulos (*Myxomycetes*: 121, 1969).

Hyalostilbum Oudem. (1885) = Dictyostelium fide Saccardo (*Syll. fung.* **6**: 1, 1888).

Hymenobolina Zukal (1893) = Licea fide Martin & Alexopoulos (*Myxomycetes*: 39, 1969).

Hymenobolus Zukal (1893) ≡ Hymenobolina.

Hyporhamma Corda (1854) nom. rej. = Hemitrichia fide Lado *et al.* (*Taxon* **54**: 829, 2005; nomencl.).

Hystricapsa Preuss (1851) nom. dub., ? Myxogastria. See Saccardo (*Syll. fung.* **18**, 1906).

Ichthyophonus Plehn & Mulsow (1911), Protozoa. 2 or 3 (in fish). See Sprague (*Syst. zool.* **14**: 110, 1965), Neish & Hughes (*Fungal diseases of fishes* **6**, 1980), Rand & Whitney (*Mycol. Soc. Amer. Newsl.*: 43 [abstract], 1987; ultrastr.), Grabda (*Marine fish parasitology. An outline*: 43, 1991), Spanggaard *et al.* (*J. Fish Dis.* **18**: 567, 1995; morphology), Ragan *et al.* (*Proc. natn Acad. Sci. U.S.A.* **93**: 11907, 1996; phylogeny), Baker *et al.* (*Microbiology* **145**: 1777, 1999; phylogeny), McVicar *in* Woo & Bruno (Eds) (*Fish Diseases & Disorders* **3**: 261, 1999; biology, treatment).

Ichthyosporea, Protozoa. Formal name for the organisms included in the DRIPs clade (q.v.) introduced at class level. See Cavalier-Smith (In: Coombs *et al.*, *Evolutionary Relationships of the Protozoa*: 375, 1998).

Iocraterium E. Jahn (1904) = Craterium fide Martin & Alexopoulos (*Myxomycetes*: 269, 1969).

Jundzillia Racib. ex L.F. Čelak. (1893) = Amaurochaete fide Martin & Alexopoulos (*Myxomycetes*: 174, 1969).

Kelleromyxa Eliasson (1991), ? Physarida. 1, Europe; N. America. See Eliasson (*MR* **95**: 1205, 1991).

Kleistobolus C. Lippert (1894) = Licea fide Martin & Alexopoulos (*Myxomycetes*: 39, 1969).

Labyrinthorhiza Chadef. (1956) nom. dub., Protozoa. See Olive (*The mycetozoans*: 215, 1975).

Lachnobolus (Fr.) Fr. (1849) = Arcyodes fide Pando (*in litt.*).

Lachnobolus Fr. (1825) nom. rej. = Amaurochaete fide Pando (*in litt.*).

Lactella Maessen (1955) = Enterobryus fide Manier & Lichtwardt (*Annls Sci. Nat.* Bot., sér. 12 **9**: 519, 1968).

Lajassiella Tuzet & Manier ex Manier (1968), Parataeniellaceae. 1 (in *Coleoptera*), France. See Lichtwardt (*The Trichomycetes. Fungal associates of arthropods*, 1986).

Lamproderma Rostaf. (1873), Stemonitidaceae. *c.* 45, widespread. See Kowalski (*Mycol.* **62**: 623, 1970; key), Ing (*The Myxomycetes of Britain and Ireland, An Identification Handbook*, 1999; British Isles), Singer *et al.* (*Öst. Z. Pilzk.* **10**: 25, 2001; nivicolous spp. Austria), Clark & Haskins (*Nova Hedwigia* **75**: 237, 2002; reproductive systems), Poulain *et al.* (*BSMF* **119**: 267, 2003; type studies).

Lamprodermataceae T. Macbr. (1899) = Stemonitidaceae.

Lamprodermopsis Meyl. (1910) = Dianema fide Martin & Alexopoulos (*Myxomycetes*: 104, 1969), Kowalski (*Mycol.* **67**: 456, 1975).

Leangium Link (1809) = Diderma fide Martin & Alexopoulos (*Myxomycetes*: 347, 1969).

Leidyomyces Lichtw., M.M. White, Cafaro & Misra (1999), Eccrinaceae. 1, USA. See Lichtwardt *et al.* (*Mycol.* **91**: 695, 1999).

Leocarpus Link (1809), Physaraceae. 1, widespread. See Ing (*The Myxomycetes of Britain and Ireland, An Identification Handbook*, 1999; British Isles), Stephenson (*Fungal Diversity Res. Ser.* **11**, 2003; New Zealand).

Lepidoderma de Bary (1873), Didymiaceae. 10 (nivicolous), widespread. See Kowalski (*Mycol.* **63**: 492, 1971; key), Ing (*The Myxomycetes of Britain and Ireland, An Identification Handbook*, 1999; British Isles), Stephenson (*Fungal Diversity Res. Ser.* **11**, 2003; New Zealand).

Lepidodermopsis Höhn. (1909) = Didymium fide Martin & Alexopoulos (*Myxomycetes*: 376, 1969) Used by, Lakhanpal (*Norw. Jl Bot.* **25**: 195, 1978).

Lepidodermopsis Wilczek & Meyl. (1934) nom. illegit. = Lepidoderma fide Martin & Alexopoulos
63: 492, 1971 (*Myxomycetes*: 403, 1969), Kowalski (*Mycol.* **63**: 492, 1971).

Leptoderma G. Lister (1913), Stemonitidaceae. 2, Europe; N. America. See Martin *et al.* (*The genera of Myxomycetes*, 1983), Ing (*The Myxomycetes of Britain and Ireland, An Identification Handbook*, 1999; British Isles).

Licaethalium Rostaf. (1873) = Reticularia Bull. fide Martin & Alexopoulos (*Myxomycetes*: 66, 1969).

Licea Schrad. (1797), Liceaceae. *c.* 66, widespread. See Keller & Brooks (*Mycol.* **69**: 669, 1977; key), Giler (*Nordic Jl Bot.* **16**: 515, 1996; ultrastr.), Flatau (*Stapfia* **73**: 63, 2000), Mitchell & McHugh (*Karstenia* **40**: 103, 2000; UK), Li *et al.* (*Mycotaxon* **90**: 437, 2004; China).

Liceaceae Chevall. (1826), Liceida (inc. sed.). 1 gen. (+ 9 syn.), 66 spp.

Liceales E. Jahn (1928). See *Liceida*.

Liceida [*Liceales* (orth. bot.)]. Myxogastria. 5 fam., 9 gen., 140 spp. Spore mass dingy or colourless, capillitium absent. Fams:

(1) **Cribrariaceae**
(2) **Dictydiaethaliaceae**
(3) **Liceaceae**
(4) **Listerellaceae**
(5) **Tubiferaceae**

Lit.: Martin (*Brittonia* **13**: 109, 1961; fam. key); Lado & Pando (*Flora Mycologica Iberica* **2**, 1997; keys Spain, Portugal).

Liceopsis Torrend (1908) = Reticularia Bull. fide Martin & Alexopoulos (*Myxomycetes*: 66, 1969).

Lignidium, see *Lignydium*.

Ligniera Maire & A. Tison (1911), Plasmodiophoraceae. 7, widespread (north temperate). See Karling (*The Plasmodiophorales* edn 2, 1968; key), Braselton *in* McLaughlin *et al.* (Eds) (*The Mycota* A Comprehensive Treatise on Fungi as Experimental Systems for Basic and Applied Research **7A**: 81, 2001).

Lignydium Link (1809) = Fuligo fide Martin (*Univ. Iowa Stud. nat. Hist.* **20**, 1966).

Lignyota Fr. (1849) nom. dub., Didymiaceae. See Martin & Alexopoulos (*Myxomycetes*: 416, 1969).

Lindbladia Fr. (1849), Cribrariaceae. 1, widespread (north temperate; Argentina; Sri Lanka). See Rex (*Bot. Gazette.* **17**: 201, 1982), Hatano (*Mycol.* **88**: 316, 1996).

Listerella E. Jahn (1906), Listerellaceae. 1 (on *Cladonia*), widespread. See Kowalski (*Mycol.* **59**: 1078, 1968), Eliasson & Gilert (*Nordic Jl Bot.* **2**: 249, 1982; SEM).

Listerellaceae E. Jahn ex H. Neubert, Nowotny & K. Baumann (1993) = Liceaceae.

Lycogala Pers. (1794), Tubiferaceae. *c.* 6, widespread. See Martin (*Mycol.* **59**: 158, 1967; key), Ing (*The Myxomycetes of Britain and Ireland, An Identification Handbook*, 1999; British Isles).

Lycogalaceae Corda (1828) nom. inval. = Tubiferaceae.

Lycoperdon Tourn. ex L. (1753) ≡ Lycogala.

Macbrideola H.C. Gilbert (1934), Stemonitidaceae. 13,

widespread. See Alexopoulos (*Mycol.* **59**: 103, 1967; key), Ing (*The Myxomycetes of Britain and Ireland, An Identification Handbook*, 1999; British Isles), Stephenson (*Fungal Diversity Res. Ser.* **11**, 2003; New Zealand).

Margarita Lister (1894) [non *Margarita* Gaudin 1829, *Compositae*] ≡ Calomyxa.

Margaritaceae Lister (1894) = Dianemataceae.

Matruchotia Skup. (1924) nom. illegit. = Amaurochaete fide Martin & Alexopoulos (*Myxomycetes*: 171, 1969).

Matruchotiella Skup. ex G. Lister (1925) nom. illegit. = Amaurochaete fide Martin & Alexopoulos (*Myxomycetes*: 171, 1969).

Maullinia I. Maier, E.R. Parodi, R. Westermeier & D.G. Müll. (2000), Plasmodiophoraceae. 1, France. See Maier *et al.* (*Protist* **151**: 235, 2000).

Membranosorus Ostenf. & H.E. Petersen (1930), Plasmodiophoraceae. 1, USA. See Ostenfeld & Petersen (*Zeitschrift für Botanik* **23**: 13, 1930), Dick (*Straminipilous Fungi*: 670 pp., 2001).

Mesenterica Tode (1790) nom. dub., Protozoa. Used for plasmodia fide Fries (1829).

Metatrichia Ing (1964), Trichiaceae. 7, Europe; tropical. See Lakhanpal & Mukerji (*Proc. Indian natn Sci. Acad.* Part B. Biol. Sci. **42**: 125, 1977), Ing (*The Myxomycetes of Britain and Ireland, An Identification Handbook*, 1999; British Isles).

Microcarpon Schrad. ex J.F. Gmel. (1792) nom. dub., ? Myxogastria.

Microglomus L.S. Olive & Stoian. (1977), Protosteliaceae. 1, Hawaii. See Olive *et al.* (*TBMS* **81**: 449, 1983), Spiegel (*Handbook of Protoctista*: 484, 1990).

Minakatella G. Lister (1921), Trichiaceae. 1, Japan; N. America. See Ing (*The Myxomycetes of Britain and Ireland, An Identification Handbook*, 1999; British Isles), Mitchell (*Systematics and Geography of Plants* **74**: 261, 2004).

Minakatellaceae Nann.-Bremek. ex H. Neubert, Nowotny & K. Baumann (1993) = Trichiaceae.

Minutularia P.A. Dang. (1891) [not traced] nom. dub., Protozoa. Based on a protozoan fide Dangeard [not traced].

Molliardia Maire & A. Tison (1911) = Tetramyxa fide Kirk (*in litt.*).

Mucilago Battarra (1755), Didymiaceae. 1, widespread. See Lado (*Cuad. Trab. Flora Micol. Ibér.* **7**: 305 pp., 1993; Spain, Portugal), Ing (*The Myxomycetes of Britain and Ireland, An Identification Handbook*, 1999; British Isles), Stephenson (*Fungal Diversity Res. Ser.* **11**, 2003; New Zealand).

Mycetozoa, Protozoa (Myxomycota, Myxostelida, Myxobionta, Gymnomycota, Gymnomyxa); plasmodial slime moulds; acellular slime moulds. 3 classes. Free-living, unicellular or plasmodial, non-flagellate in single- or multi-celled phagotrophic stages; mitochondrial cristae tubular; with uni- or multicellular sporocarps with single to many spores; spore walls cellulosic or chitinous; spores germinating to produce 1- or 2-flagellate cells.

The circumscription and ranking of the higher taxa of plasmodial slime moulds varies according to different authors. No formal name at all was used in Margulis *et al.* (1990), while Corliss (1994) used *Mycetozoa* inclusive of the dictyostelids (see *Dictyosteliomycota*). Three classes are recognized in this edition of the *Dictionary*:

(1) **Dictyostelea**

(2) **Myxogastrea**

(3) **Protostelea**

Corliss (1994) suggested that the taxon could be expanded to include the *Promycetozoida*, certain enigmatic marine protists with reticulopodia such as *Corallomyxa*, *Megamoebomyxa* and *Thalassomyxa* (see Grell, *Protistologica* **21**: 215, 1985; *Arch. Protistenkde* **140**: 303, 1991); these taxa have not been studied by mycologists and are not treated elsewhere in this *Dictionary*.

Lit.: Corliss (*Acta Protozool.* **33**: 1, 1994), Keller (*in* Parker (Ed.), **1**: 165, 1982), Margulis *et al.* (1990), Teixeira (*Rickia* suppl. **3**, 1970; gen. except Myxomycetes), Olive (*The mycetozoans*, 1975; *in* Parker, *Synopsis and classification of living organisms*: 521, 1982), Raper (*The dictyostelids*, 1994). See also *Protozoa*.

Myrosporium Corda (1831) nom. dub., ? Trichiida.

Myxogastres. See *Myxogastria*.

Myxogastria, Mycetozoa. 5 ord., 14 fam., 62 gen., 888 spp. [*Myxomycetes* (orth. bot.)]. True slime moulds. Trophic phase a free-living, multinucleate, coenocytic, saprobic plasmodium with a shuttle-movement of the protoplasm; plasmodium sometimes becoming a resting body or sclerodium under poor conditions, and the swarm spores of myxamoebae microcysts; sporangia sessile or stalked, often bright coloured; spores developed after meiosis, produced in masses with a persistent or evanescent peridium; swarm cells usually with two-anterior flagellae and no cell wall, forming myxamoebae directly or after loss of the flagella, sometimes undergoing division before capulation; reproduction usual, the resultant zygote becoming the plasmodium. Esp. on old wood or other plant material undergoing decomposition (see Madelin, *TBMS* **83**: 1, 1984; ecological significance).

The chief diagnostic characters are the type of sporocarp, the structure of the peridium and capillitium of the sporangium, calcium carbonate ('lime'), if present, and the size, colour and ornamentation of the spores. Ords:

(1) **Echinosteliales**

(2) **Liceales**

(3) **Physarales**

(4) **Stemonitales**

(5) **Trichiales**

Frederick (1990) and Newbert *et al.* (1993) include the *Ceratiomyxaceae* in this class, but see under *Protostelea*.

Lit.: *General*: Lister & Lister (*Mycetozoa*, edn 3, 1925), Jahn (*Nat. PflFam.* **2** (Aufl. 2), 1928), MacBride & Martin (*Myxomycetes*, 1934), Gray & Alexopoulos (*Biology of myxomycetes*, 1968), Martin (*Univ. Iowa Studies nat. Hist.* **20**(8), 1966; gen. list) Martin & Alexopoulos (*The myxomycetes*, 1969). Martin *et al.* (*The genera of Myxomycetes*, 1983), Olive (*Bot. Rev.* **36**: 59, 1970; keys, *The mycetozoans*, 1975), Teixeira (*Rickia* suppl. **4**, 1971; gen.), Alexopoulos (*in* Subramanian (Ed.), *Taxonomy of the fungi* **1**: 1, 1978; evolution), Mitchell (*Bull. BMS* **12**: 18, 90, **13**: 42, 1978-79; key corticolous spp.), Collins (*Bot. Rev.* **45**: 145, 1979; biosystematics), Eliasson & Lundqvist (*Bot. Notiser* **132**: 551, 1979; fimicolous spp.), Chassain (*Myxomycetes* **1**, 1979), Martin *et al.* (*The genera of Myxomycetes*, 1983), Frederick (*in* Margulis *et al.* (Eds) 1990: 467), Corliss (1994), Stephenson & Stempen (*Myxomycetes, A handbook of slime molds*, 1994).

Regional: **Argentina**, Deschamps (*Physis* Sect. C **34**: 159, **35**: 319, 1975-77). **Australia**, Mitchell (*Nova Hedw.* **60**: 269, 1995; list). **British Isles**, Ing (*A census catalogue of British myxomycetes*, 1968; revised, *Bull. BMS* **14**: 97, **16**: 26, 1980-82; **19**: 109, 1985). **Central & S. America**, Farr (*Rickia* **3**: 45, 1968; key, *Flora neotropica* **16**, 1976). **Costa Rica**, Alexopoulos & Sáenz (*Mycotaxon* **2**: 223, 1975). **Denmark**, Bjornekaer & Klinge (*Friesia* **7**: 149, 1964). **Germany**, etc. Neubert *et al.* (*Die Myxomyceten* **1**, 1993; keys, col. pls), Schinz (*Rabenh. Krypt.-Fl.* **1** (10), 1920). **Hawaii**, Eliasson (*MR* **95**: 257, 1991). **India**, Thind (*The myxomycetes of India*, 1977), Lakhanpal & Mukerji (*Indian myxomycetes*, 1981). **Italy**, Pirola & Credaro (*Giorn. bot. Ital.* **105**: 157, 1971). **Japan**, Emoto (*The myxomycetes of Japan*, 1977). **Korea**, Nakagawa (*Jl Chosen Nat. Hist. Soc.* **17**: 17, 1934). **Mexico**, Braun & Keller (*Mycotaxon* **3**: 297, 1976). **Netherlands**, Nannenga-Bremekamp (*De Nederlandse Myxomyceten*, 1975; *A guide to temperate myxomycetes* [Engl. transl.], 1992). **Nigeria**, Ing (*TBMS* **47**: 45, 1964). **N. America**, Hagelstein (*The Mycetozoa of North America*, 1944), Martin (*N. Am. Fl.* **1**(1), 1949). **Sierra Leone**, Ing (*TBMS* **50**: 549, 1967). **Spain**, Lado & Pando (*Flora Mycologica Iberica* **2**, 1997). **Sweden**, Santesson (*Svensk bot. Tidskr.* **58**: 113, 1964; list). **Tierra del Fuego**, Arambarri (*Fl. cript. Tierra del Fuego* **2**, 1975). **Uruguay**, Garcia-Zorron (*Mixomycetos del Uruguay*, 1967). See also *Mycetozoa*.

Myxomycetes. See *Myxogastria*.

Myxomycota. See *Mycetozoa*.

Myxothallophyta. See *Myxogastria*.

Nassula Fr. (1849) = Arcyria fide Martin & Alexopoulos (*Myxomycetes*: 121, 1969).

Nematostelium L.S. Olive & Stoian. (1970), Protosteliaceae. 2, widespread. See Olive & Stoianovitch (*Bot. Rev.* **36**: 68, 1970).

Nidularia With. (1787) nom. rej. prop. ≡ Craterium fide Martin & Alexopoulos (*Myxomycetes*: 172, 1969).

Nodocrinella D. Scheer (1977) nom. dub., Parataeniellaceae.

Octomyxa Couch, J. Leitn. & Whiffen (1939), Plasmodiophoraceae. 2 (on *Achlya* and other *Saprolegniales*), USA. See Sherwood (*J. Elisha Mitchell scient. Soc.* **84**: 52, 1968), Dick (*Straminipilous Fungi*: 670 pp., 2001).

Oligonema Rostaf. (1875), Trichiaceae. 7, widespread. See Ing (*The Myxomycetes of Britain and Ireland, An Identification Handbook*, 1999; British Isles), Haan *et al.* (*Systematics and Geography of Plants* **74**: 251, 2004; Belgium).

Ophiotheca Curr. (1854) = Perichaena fide Martin & Alexopoulos (*Myxomycetes*: 109, 1969).

Ophiuridium Hazsl. (1877) = Dictydiaethalium fide Martin & Alexopoulos (*Myxomycetes*: 59, 1969).

Orcadella Wingate (1889) = Licea fide Martin & Alexopoulos (*Myxomycetes*: 39, 1969).

Orcadellaceae T. Macbr. (1899) = Liceaceae.

Orthotricha Wingate (1886) = Clastoderma fide Martin & Alexopoulos (*Myxomycetes*: 204, 1969).

Ostenfeldiella Ferd. & Winge (1914) = Plasmodiophora fide Kirk (*in litt.*).

Ostracococcum Wallr. (1833) nom. dub., ? Myxogastria.

Palavascia Tuzet & Manier ex Lichtw. (1964), Palavasciaceae. 3 (in *Isopoda*), widespread. See Lichtwardt (*Mycol.* **56**: 318, 1964), Lichtwardt (*The Trichomycetes. Fungal associates of arthropods*: 256, 1986; key), Cafaro (*Mycol.* **92**: 361, 2000; n.sp. Patagonia), Cafaro (*Mol. Phylogen. Evol.* **35**: 21, 2005; phylogeny).

Paradiachea Hertel (1956), Stemonitidaceae. 5, Europe, N. America, Australasia. See Nannenga-Bremekamp (*Temperate myxomycetes*, 1991; key, icon. desc.).

Paradiacheopsis Hertel (1954) = Macbrideola fide Pando (*in litt.*).

Parallobiopsis Collin (1913), Protozoa. 1 (on invertebrate), Europe. See Whisler *in* Margulis *et al.* (Eds) (*Handbook of Protoctista*: 715, 1990; systematic position).

Paramacrinella Manier & Grizel (1971), Eccrinaceae. 1 (in *Amphipoda*), France. See Manier *et al.* (*Annls Sci. Nat.* Bot., sér. 12 **12**: 1, 1971), Lichtwardt (*The Trichomycetes. Fungal associates of arthropods*, 1986).

Paramoebidium L. Léger & Duboscq (1929), Amoebidiaceae. 10 (in *Diptera*, *Ephemeroptera*, *Plecoptera*), widespread. See Dang & Lichtwardt (*Am. J. Bot.* **66**: 1093, 1979; ultrastr.), Lichtwardt (*The Trichomycetes. Fungal associates of arthropods*, 1986; taxonomy), Williams & Lichtwardt (*CJB* **68**: 1045, 1990), Lichtwardt *et al.* (*Mycol.* **83**: 389, 1991), Lichtwardt & Williams (*Mycol.* **84**: 376, 1992), Lichtwardt & Arenas (*Mycol.* **88**: 844, 1996), Lichtwardt *et al.* (*Mycol.* **91**: 1060, 1999), Benny *in* McLaughlin *et al.* (Eds) (*The Mycota* A Comprehensive Treatise on Fungi as Experimental Systems for Basic and Applied Research **7A**, 2000), Tanabe *et al.* (*J. gen. appl. Microbiol.* Tokyo **51**: 267, 2005; phylogeny), Strongman & White (*CJB* **84**: 1478, 2006; n.sp. Canada).

Parataeniella R.A. Poiss. (1929), Parataeniellaceae. 6 (in *Isopoda*), widespread. See Lichtwardt & Chen (*Mycol.* **56**: 163, 1964), Scheer (*Arch. Protistenk.* **118**: 202, 1976), Lichtwardt (*The Trichomycetes. Fungal associates of arthropods*, 1986; key), Lichtwardt & Williams (*CJB* **68**: 1057, 1990), Chien & Hsieh (*Trichomycetes and Other Fungal Groups, Robert W. Lichwardt Commemoration Volume*: 55, 2001; Taiwan).

Paratrichella Manier (1947) = Enterobryus fide Manier & Lichtwardt (*Annls Sci. Nat.* Bot., sér. 12 **9**: 519, 1968).

Passalomyces Lichtw., M.M. White, Cafaro & Misra (1999), Eccrinaceae. 1, Dominica. See Lichtwardt *et al.* (*Mycol.* **91**: 695, 1999).

Pecila Lepell. (1822) nom. dub., Myxogastria.

Peltomyces L. Léger (1909), ? Plasmodiophoraceae. 3, Europe.

Percolozoa, Protozoa. 1 class., 1 ord., 4 fam., 6 gen., 14 spp. An order of *Protozoa* (q.v.) comprising the acrasid cellular slime moulds; acrasids. Trophic phase amoeboid, pseudopodia lobose; aggregating without streaming; nuclei with a compact centrally placed nucleolus; sporocarp sessile, independent and dividing when vegetative, some with simple supportive stalks; multispored, in chains or deliminted sori; flagellate cells usually absent; sexual reproduction unknown. On dung and isolated from a wide range of decaying plant materials and macromycetes, and also soil.

Raper (1973) included the dictyostelids and protostelids (see *Mycetozoa*) in the class *Acrasiomycetes*, later excluding the protostelids as a separate class (Raper, 1984). Molecular data show these three groups to be allied only at the phylum level.

Lit.: Olive (1975), Raper (*in* Ainsworth *et al.* (Eds), *The fungi* **4B**: 9, 1973; keys gen.; *The dictyostelids*, 1984), and under Families.

Perichaena Fr. (1817), Trichiaceae. 17, widespread. See Cavalcanti (*Rickia* **6**: 99, 1974; key), Keller & Eliasson (*MR* **96**: 1095, 1992; key), Ing (*The Myxomycetes of Britain and Ireland, An Identification Handbook*, 1999; British Isles), Moreno *et al.* (*Rivista di Micologia* **43**: 5, 2000; sp. nov.), Stephenson (*Fungal Diversity Res. Ser.* **11**, 2003; New Zealand).

Perichaenaceae Rostaf. ex Cooke (1877) = Trichiaceae.

Perkinsus Levine (1978), Protozoa. 1 (parasitic on oysters (*Crassostrea virginica*)), Gulf of Mexico. *P. marinus*, formerly *Dermocystidium marinum*, recognized in its own class (Levine, *J. Parasitol.* **64**: 549, 1978) in the *Apicomplexa* but see Siddall *&* *al.* (*Parasitology* **115**: 165, 1997) where it considered to be closer to the dinoflagellates. See Levine (*J. Shellfish Res.* **15**: 67, 1996; taxonomy, phylogeny), Siddall *et al.* (*Parasitology* **115**: 165, 1997) where it considered to be closer to the dinoflagellates.

Phagomyxa Karling (1944), Plasmodiophorida. 3 (algal parasite), USA. See Cavalier-Smith & Chao (*Arch. Protistenk.* **147**: 227, 1997; order *Phagomyxida* proposed), Schnepf *et al.* (*Helgoland Mar. Res.* **54**: 237, 2000; new spp.), Bulman *et al.* (*Protist* **152**: 43, 2001; systematic position).

Phlebomorpha Pers. (1822) nom. dub., Mycetozoa. Based on myxomycete plasmodia fide Martin (1966).

Physaraceae Chevall. (1826), Physarida (inc. sed.). 9 gen. (+ 18 syn.), 195 spp.

Physarales T. Macbr. (1922). See *Physarida*.

Physarella Peck (1882), Physaraceae. 1, widespread. See Shi & Li (*Mycosystema* **23**: 381, 2004; life cycle).

Physarida [*Physarales* (orth. bot.)]. Myxogastria. 3 fam., 18 gen., 364 spp. Spore mass dark-coloured, peridium or capillitium calcareous. Fams:
(1) **Didymiaceae**
(2) **Elaeomyxaceae**
(3) **Physaraceae**
For *Lit.* see under fam.

Physarina Höhn. (1909), Didymiaceae. 3, pantropical. See Höhnel (*Sber. Akad. Wiss. Wien* Math.-naturw. Kl., Abt. 1 **118**: 431, 1909; nomencl.).

Physarum Pers. (1794), Physaraceae. *c.* 135, widespread. See Anon. *in* Hutterman (Ed.) (*Physarum polycephalum*, 1973; use in cell-biology), Anon. *in* Aldrich & Daniel (Eds) (*Cell biology of Physarum and Didymium*, 1982), Chen *et al.* (*Mycosystema* **18**: 77, 1999; China), Ing (*The Myxomycetes of Britain and Ireland, An Identification Handbook*, 1999; British Isles), Clark & Stephenson (*Mycotaxon* **85**: 85, 2003), Stephenson (*Fungal Diversity Res. Ser.* **11**, 2003; New Zealand).

Phytoceratiomyxa Sawada (1929) nom. dub., ? Myxogastria.

Phytomyxa J. Schröt. (1886) nom. dub., ? Plasmodiophoraceae.

Phytomyxaceae J. Schröt. (1886) = Plasmodiophoraceae.

Phytomyxea, Cercozoa. 1 ord., 2 fam., 15 gen., 50 spp. See *Cercozoa*.

Phytosarcodina. See *Myxogastria*.

Pistillaria Jeekel, Tuzet, Manier & Jolivet (1959) = Enterobryus fide Manier & Lichtwardt (*Annls Sci. Nat.* Bot., sér. 12 **9**: 519, 1968).

Pittocarpium Link (1815) ? = Fuligo fide Martin (*Univ. Iowa Stud. nat. Hist.* **20**, 1966).

Planoprotostelium L.S. Olive & Stoian. (1971), Cavosteliaceae. 1, Brazil. See Olive & Stoianovitch (*J. Elisha Mitchell scient. Soc.* **87**: 115, 1971), Fiore-Donno & Berney (*J. Eukary. Microbiol.* **52**: 201, 2005; phylogeny).

Plasmodiophora Woronin (1877), Plasmodiophoraceae. 6 (obligate endoparasites of plants), widespread. *P. brassicae* (club root of crucifers). See Colhoun (*Phytopath. Pap.* **3**, 1958), Karling (*The Plasmodiophorales* edn 2, 1968; key), Ingram & Tommerup (*Proc. R. Soc. Lond.* B. Biol. Sci. **180**: 103, 1972; life history), Buczaki *et al.* (*TBMS* **65**: 295, 1975; specialization), Jousson *et al.* (*Acta Agric. Scand.* **25**: 261, 1975; bibliogr., 612 refs), Braselton *in* McLaughlin *et al.* (Eds) (*The Mycota* A Comprehensive Treatise on Fungi as Experimental Systems for Basic and Applied Research **7A**: 81, 2001), Bulman *et al.* (*Protist* **152**: 43, 2001; phylogeny), Braselton *in* Lee, Leedale & Bradbury (Eds) (*An Illustrated Guide to the Protozoa, Organism Traditionally Referred to as Protozoa, or Newly Discovered Groups* **2**: 1342, 2000).

Plasmodiophoraceae Zopf ex Berl. (1888), Plasmodiophorida (inc. sed.). 13 gen. (+ 7 syn.), 44 spp.

Plasmodiophorales F. Stevens (1919). See *Plasmodiophorida*.

Plasmodiophorida [*Plasmodiophorida* (orth. zool.)]. Phytomyxea. 2 fam., 15 gen., 50 spp. Obligate endoparasites of flowering plants (*Plasmodiophora*, *Spongospora*), algae (*Woronina* in *Vancheria*; *Sorodiscus* in *Chara*), and fungi (*Woronina* in *Saprolegniales*); frequently inducing hypertrophy of the infected cells. Fams:
(1) **Endemosarcaceae**
(2) **Plasmodiophoraceae**
For *Lit.* see under fam.

Plasmodiophoromycetes. See *Phytomyxea*.

Plasmodiophoromycota Whittaker (1969), see *Cercozoa*.

Pleiomorpha (Nann.-Bremek.) Dhillon (1978) = Licea fide Pando (*in litt.*).

Pocheina A.R. Loebl. & Tappan (1961), Guttulinaceae. 3, widespread. See Mitchell (*Systematics and Geography of Plants* **74**: 261, 2004; key).

Polymyxa Ledingham (1933), Plasmodiophoraceae. 2, Europe; N. America. See Karling (*The Plasmodiophorales* edn 2, 1968), Legrève *et al.* (*MR* **106**: 138, 2002; phylogeny).

Polyschismium Corda (1842) = Diderma fide Martin & Alexopoulos (*Myxomycetes*: 419, 1969).

Polysphondylium Bref. (1884), Dictyosteliaceae. 16, widespread. See Harper (*Bull. Torrey bot. Club* **59**: 49, 1932), Traub & Hohl (*Am. J. Bot.* **63**: 664, 1976), Raper (*The dictyostelids*, 1984), Cavender *et al.* (*N.Z. Jl Bot.* **40**: 235, 2002; New Zealand), Kawakami & Hagiwara (*Mycoscience* **43**: 453, 2002; mating groups), Cavender *et al.* (*Systematics and Geography of Plants* **74**: 243, 2004; Caribbean).

Promycetozoida. See *Mycetozoa*.

Protochytrium Borzí (1884) nom. dub., ? Protozoa.

Protoderma Rostaf. (1874) [non *Protoderma* Kütz. 1854, *Algae*] ≡ Licea.

Protodermataceae Rostaf. ex Cooke (1877) =

Liceaceae.

Protodermium Rostaf. ex Berl. (1888) ≡ Licea.

Protodermodium Kuntze (1891) ≡ Licea.

Protomonas Cienk. (1865), Protozoa. 1. Amoeboflagellate protist formerly included in the monads or proteomyxids.

Protomyxa Haeckel (1870), Protozoa. Amoeboflagellate protist formerly included in the monads or proteomyxids.

Protophysaraceae A. Castillo, Illana & G. Moreno (1998) = Physaraceae.

Protophysarum M. Blackw. & Alexop. (1975), Physaraceae. 1, northern subtropical. See Blackwell & Alexopoulos (*Mycol.* **67**: 33, 1975), Castillo *et al.* (*MR* **102**: 838, 1998), Lizárraga *et al.* (*Mycotaxon* **88**: 409, 2003; Mexico), Mitchell (*Systematics and Geography of Plants* **74**: 261, 2004; key gen.).

Protosporangium L.S. Olive & Stoian. (1972), Cavosteliaceae. 4, USA. See Raper (*The dictyostelids*, 1984), Bennett (*Mycol.* **78**: 857, 1986).

Protostelia, Amoebozoa. 1 ord., 4 fam., 16 gen., 38 spp. [*Protosteliomycetes* (orth. bot.)]. Protostelid slime moulds, protostelids. Trophic phase of simple amoeboid cells with filose pseudopodia, plasmodia not with a shuttle movement of the protoplasm; ± flagellate cells (present in 'amoeboflagellates'); sporulation not proceeding by the long grex action of myxamoebae; sporocarps stalked, stalks simple and often very delicate; spores formed singly or several together in some spp. germinating to produce 8 haploid flagellate spores). Isolated mainly by moist-chamber culture of aerial dead or decaying plant parts, especially bark; also on dung. Ord.:

Protostelida

The *Ceratiomyxaceae* were recognized as the basis of a separate class (*Ceratiomyxomycetes*) in the ninth edition of this *Dictionary*, but Spiegel (1990) did not distinguish any rank above family.

Spiegel (1990) noted that the protostelid genera could be placed in five more natural groups, based on the number of kinetoids per cell, the presence/absence and nature of the nuclear attachments, and the type of cell coats in addition to colonial morphology and other macroscopic features; these were not given formal names.

Lit.: Corliss (1994), Olive (1975, 1982), Raper (1973), Spiegel (*in* Margulis *et al.* (Eds), *Handbook of the Protoctista*: 484, 1990).

Protosteliaceae L.S. Olive (1962), Protostelida (inc. sed.). 9 gen. (+ 1 syn.), 24 spp.

Lit.: Spiegel (*Handbook of Protoctista*: 484, 1990).

Protosteliales L.S. Olive & Stoian. ex P.M. Kirk, P.F. Cannon & J.C. David (2001). See *Protostelida*.

Protostelida [*Protosteliales* (orth. bot.)]. Protostelia. 4 fam., 16 gen., 38 spp. Fams:

(1) **Cavosteliaceae**
(2) **Ceratiomyxaceae**
(3) **Echinosteliopsidaceae**
(4) **Protosteliaceae**

Lit.: see *Protostelia*.

Protostelidaceae, see *Protosteliaceae*.

Protosteliomycetes. See *Protostelia*.

Protosteliopsis L.S. Olive & Stoian. (1966), Protosteliaceae. 1, widespread (subtropical). See Olive & Stoianovitch (*Mycol.* **58**: 452, 1966), Spiegel (*Handbook of Protoctista*: 484, 1990).

Protostelium L.S. Olive & Stoian. (1960), Protosteliaceae. 8, widespread. See Olive & Stoianovitch (*Bull. Torrey bot. Club* **87**: 12, 1960), Spiegel *et al.* (*Mycol.* **98**: 144, 2006; ballistosporous n.sp.).

Prototrichia Rostaf. (1876), Trichiaceae. 1, widespread (temperate). See Ing (*The Myxomycetes of Britain and Ireland, An Identification Handbook*, 1999; British Isles), Stephenson (*Fungal Diversity Res. Ser.* **11**, 2003; New Zealand).

Prototrichiaceae T. Macbr. (1899) = Trichiaceae.

PROTOZOA. 6 class., 11 ord., 30 fam., 122 gen. (+ 118 syn.), 1161 spp. Kingdom of *Eukaryota*. Predominantly unicellular, plasmodial, or colonial phagotropic, wall-less in the trophic state; ciliary hairs never rigid or tubular; chloroplasts, where present lacking starch and phycobilisomes, with stalked thylakoid and 3 membranes; multicellular species with minimal cell differentiation and lacking collagenous connective tissue sandwiched between dissimilar epithelia.

The *Protozoa* have been included with the *Chromista* in a broader kingdom *Protoctista* (syn. *Protista*) by some (e.g. Margulis *et al.*, 1990) but the classification has been rejected by recent authors on both molecular and non-molecular grounds (Cavalier-Smith, 1993; Corliss, 1994).

Cavalier-Smith (1993) acceps 18 phyla in the *Protozoa* and Corliss (1994) 14; however, only one has been traditionally studied by mycologists: *Mycetozoa* (incl. 3 classes *Protostelea*, *Myxogastrea* and *Dictyostelea*). Other protozoan phyla are not treated further in this account or elsewhere in this edition of the *Dictionary*. Olive (1975) arranged the slime moulds into four classes and three subclasses; that system was adapted (using fungal terminations) in the seventh edition of the *Dictionary* which accepted *Acrasiomycetes*, *Ceratiomyxomycetes*, *Dictyosteliomycetes*, *Myxomycetes* and *Protosteliomycetes*. The eighth edition of the *Dictionary* recognized four phyla : *Acrasiomycota*; *Dictyosteliomycota*; *Myxomycota* (incl. *Ceratiomyxomycetes*, *Myxomycetes*, *Protosteliomycetes*); *Plasmodiophoromycota*. The ninth edition of the *Dictionary* recognized three phyla : *Acrasiomycota*; *Myxomycota* (incl. *Ceratiomyxomycetes*, *Dictyosteliomycetes*, *Myxomycetes*, *Protosteliomycetes*); *Plasmodiophoromycota*. More recent molecular and ultrastructural evidence has shown that the slime moulds and similar organisms are polyphyletic and appear to belong in five separate lineages:

(1) **Ramicristates** A group comprising the naked lobose amoebas, the filose and lobose testate ameobas, defined by the presence of branched mitochondrial cristae. Included are three slime mould classes: *Protostelea* (Protosteliomycetes), *Myxogastrea* (Myxomycetes) and *Dictylostelea* (Dictyosteliomycetes).

(2) **Heterolobosea** A group defined by flattened mitochondrial cristae and the mitochondria being associated with the endoplasmic reticulum. This group includes the schizopyrenid amoebae and the cellular slime moulds, now treated at the level of order, *Acrasida* (Acrasiales).

(3) **Copromyxida** *Copromyxa* was formerly included in the *Acrasiales* but the trophic phase lacks flagella and the mitochondrial cristae are tubular. For these reasons they are now regarded as distinct.

(4) **Fonticulida** *Fonticula*, also formerly included in the *Acrasiales*, has plate-like mitochondrial cristae, no flagella and the pseudopodia have subpseudopodia and are thus also distinct.

(5) **Plasmodiophorids** A group uniquely defined by the cruciate profile of the chromosomes and nucleolus at mitosis. One class included, *Plasmodiophorea* (Plasmodiophoromycetes).

The Heterolobosea are sometimes grouped with the Euglenozoa and some other smaller groups as the Excavates whilst the remaining fourare grouped as the Amoebozoa. There are many further protozoan organisms that have been studied or mentioned by mycologists and are listed in the *Dictionary*, some of which still await invesitgation using modern techniques. Larger groups include the Ellobiopsids, Nucleariids and Vampyrellids.

Lit.: Bonner (*The cellular slime molds*, 1959; edn 2, 1967), Cavalier-Smith (*Microbiol. Rev.* **57**: 953, 1993; *Biol. Rev.* **73**: 203, 1998), Coombs *et al.* (eds) [*Evolutionary Relationships Among Protozoa.* Systematics Association Special Volume 56, 1998], Corliss (*Acta Protozool.* **33**: 1, 1994), Heywood & Rothschild (*Biol. J. Linn. Soc.* **30**: 91, 1987; nomencl. higher taxa), Olive (*The mycetozoans*, 1975), Patterson, *Amer. Nat.* **154**: S96, 1999; Teixeira (*Rickia, suppl.* **3**, 1970; gen. except *Myxomycetes*).

See Classification, Kingdoms of fungi, Phylogeny.

Pseudospora Cienk. (1865), Protozoa. Amoeboflagellate protist formerly included in the monads or proteomyxids and listed by Saccardo (*Syll. Fung.* **7**: 453). See Canter & Lund (*Proc. Linn. Soc. Lond.* **179**: 203, 1968; biol.), Swale (*Arch. Mikrobiol.* **67**: 71, 1969; ultrastr.), Patterson (*Amer. Nat.* **154**: S91, 1999; current knowl.).

Pseudosporopsis Scherff. (1925), ? Protozoa.

Pygmomyces Arnaud (1949) nom. inval., Dictyostelida.

Pyxidium Gray (1821) = Perichaena fide Martin & Alexopoulos (*Myxomycetes*: 109, 1969).

Raciborskia Berl. (1888) = Comatricha fide Martin & Alexopoulos (*Myxomycetes*: 222, 1969).

Raciborskiaceae Berl. (1888) = Stemonitidaceae.

Ramacrinella Manier & Ormières (1962), Eccrinaceae. 1 (in *Amphipoda*), France. See Manier *et al.* (*Annls Sci. Nat.* Bot., sér. 12 **2**: 625, 1961), Lichtwardt (*The Trichomycetes. Fungal associates of arthropods*, 1986).

Recticoma D. Scheer (1935) = Enterobryus fide Manier & Lichtwardt (*Annls Sci. Nat.* Bot., sér. 12 **9**: 519, 1968).

Reticularia Bull. (1788) nom. cons., Tubiferaceae. 10, widespread. See Lado & Pando (*Fl. Mycol. Iberica* **2**, 1997; Spain (as *Enteridium*)), Lado & Pando (*Taxon* **47**: 453, 1998; nomencl.), Lado *et al.* (*Taxon* **47**: 109, 1998; nomencl.), Ing (*The Myxomycetes of Britain and Ireland, An Identification Handbook*, 1999; British Isles (as *Enteridium*)).

Reticulariaceae Chevall. (1826) = Tubiferaceae.

Reticulomyxa Nauss (1949), Protozoa. 1 (aquatic), USA. Amoeboid protist. See Patterson (*Amer. Nat.*.**154**: S96, 1999, systematic position).

Rhabdocystis Arnaud (1949) nom. inval., ? Dictyostelida.

Rhinosporidium Minchin & Fantham (1905), Protozoa. 1 (on humans and other animals causing rhinosporidiosis, q.v.), widespread. See Ashworth (*Trans. R. Soc. Edinb.* **53**: 301, 1923), Stoddart *et al.* (*J. Med. Microbiol.* **20**: 'x', 1985), Herr *et al.* (*J. Clin. Microbiol.* **37**: 2750, 1999; phylogeny).

Rhizellobiopsis Hovasse (1926), Protozoa. See Whisler *in* Margulis *et al.* (Eds) (*Handbook of Protoctista*: 715, 1990; systematic position).

Rhizomyxa Borzí (1884) ? = Ligniera fide Karling (*The Plasmodiophorales* edn 2, 1968).

Rostafinskia Racib. (1884) ≡ Raciborskia.

Schizoplasmodiopsis L.S. Olive (1967), Protosteliaceae. 6, widespread. See Olive & Stoianovitch (*Mycol.* **67**: 1088, 1975; key), Olive & Stoianovitch (*Am. J. Bot.* **63**: 1385, 1976), Spiegel & Feldman (*Mycol.* **85**: 894, 1993; ultrastr. *S. vulgare*), Glushchenko & Leontiev (*Mikol. Fitopatol.* **36**: 7, 2002; Ukraine).

Schizoplasmodium L.S. Olive & Stoian. (1966), Protosteliaceae. 3, widespread. See Olive & Stoianovich (*Am. J. Bot.* **63**: 1385, 1976; ballistosporic), Whitney (*Mycol.* **77**: 848, 1985; ultrastr.).

Scoriomyces Ellis & Sacc. (1885) nom. dub., Mycetozoa. Based on myxomycete sclerotia fide Carris (*Sydowia* **45**: 92, 1993).

Scyphium Rostaf. (1874) = Craterium fide Martin & Alexopoulos (*Myxomycetes*: 257, 1969).

Semimorula E.F. Haskins, McGuinn. & C.S. Berry (1983), Myxogastria. 1, USA. See Haskins *et al.* (*Mycol.* **75**: 153, 1983), Clark *et al.* (*Mycol.* **96**: 36, 2004; culture and reproductive systems).

Siphoptychium Rostaf. (1876) = Tubifera fide Martin & Alexopoulos (*Myxomycetes*: 54, 1969).

Soliformovum Spiegel (1994) nom. inval. = Protostelium fide Hawksworth *et al.* (*Dictionary of the Fungi* edn 8, 1995).

Sorodiscus Lagerh. & Winge (1913), Plasmodiophoraceae. 5 (in *Chara*), widespread. See Karling (*The Plasmodiophorales* edn 2, 1968; key), Dick (*Straminipilous Fungi*: 670 pp., 2001).

Sorolpidium Němec (1911) = Ligniera fide Karling (*The Plasmodiophorales* edn 2, 1968).

Sorophorae. See *Acrasiomycota*.

Sorosphaera J. Schröt. (1886), Plasmodiophoraceae. 3, Europe; N. America. See Dick (*Straminipilous Fungi*: 670 pp., 2001), Preece (*Mycologist* **16**: 27, 2002; *Sorosphaera veronicae* in UK), Kirchmair *et al.* (*Sydowia* **57**: 223, 2005; *Sorosphaera viticola*).

Sphaerocarpa Schumach. (1803) = Craterium fide Fries (*Syst. mycol.* **3**, 1829).

Sphaerocarpus Bull. (1791) [non *Sphaerocarpus* Adans. 1763, *Hepaticae*] = Physarum fide Martin (*Univ. Iowa Stud. nat. Hist.* **20**, 1966).

Spongospora Brunch. (1887), Plasmodiophoraceae. 3 (obligate endoparasites of plants), widespread. *S. subterranea* (potato powdery scab), *S. nasturtii* (crook root of watercress). See Tomlinson (*TBMS* **41**: 491, 1958; *S. nasturtii*, crook root of watercress), Karling (*The Plasmodiophorales* edn 2, 1968), Dick (*Straminipilous Fungi*: 670 pp., 2001), Down *et al.* (*MR* **106**: 1060, 2002; phylogeny).

Sporigastrum Link [not traced] nom. dub., ? Mycetozoa.

Sporomyxa L. Léger (1908), ? Plasmodiophoraceae. 2, Europe; Africa. See Karling (*The Plasmodiophorales* edn 2: 101, 1968).

Spumaria Pers. (1792) = Mucilago fide Martin & Alexopoulos (*Myxomycetes*: 374, 1969).

Spumariaceae Rostaf. ex Cooke (1877) = Didymiaceae.

Squamuloderma Kowalski (1973) = Didymium fide Martin *et al.* (*The genera of Myxomycetes*, 1983).

Staphylocystis Coutière (1911) = Thalassomyces fide Kane (*N.Z. Jl Sci.* **7**: 289, 1964).

Stegasma Corda (1843) = Perichaena fide Martin & Alexopoulos (*Myxomycetes*: 109, 1969).

Stemonaria Nann.-Bremek., R. Sharma & Y. Yamam. (1984), Stemonitidaceae. 12, widespread. See Nannenga-Bremekamp *et al.* (*Proc. K. Ned. Akad. Wet.* Ser. C, Biol. Med. Sci. **87**: 450, 1984), Ing (*The Myxomycetes of Britain and Ireland, An Identification Handbook*, 1999; British Isles).

Stemonitidales T. Macbr. (1922). See *Stemonitida*.

Stemonitida [*Stemonitales* (orth. bot.)]. Myxogastria. 1 fam., 15 gen., 201 spp. Spore mass dark-coloured, peridium and capillitium non-calcareous. Fam.: Stemonitidiaceae

For *Lit.* see under fam.

Stemonitidaceae Fr. (1829), Stemonitida (inc. sed.). 15 gen. (+ 16 syn.), 201 spp.

Stemonitis Gled. (1753), Stemonitidaceae. 16, widespread. See Ing (*The Myxomycetes of Britain and Ireland, An Identification Handbook*, 1999; British Isles), Mitchell (*Systematics and Geography of Plants* **74**: 261, 2004; key gen.).

Stemonitis Roth (1787) nom. illegit. = Stemonitis Gled. fide Farr & Alexopoulos (*Taxon* **30**: 357, 1981), Gams (*Taxon* **41**: 100, 1992).

Stemonitopsis (Nann.-Bremek.) Nann.-Bremek. (1975), Stemonitidaceae. 11, widespread. See Nannenga-Bremekamp (*Nederlandse Myxomyceten*: 203, 1974), Ing (*The Myxomycetes of Britain and Ireland, An Identification Handbook*, 1999; British Isles).

Strongylium Ditmar (1809) = Reticularia Bull. fide Martin & Alexopoulos (*Myxomycetes*: 172, 1969).

Stylonites Fr. (1848) nom. dub., Physarida. See Martin & Alexopoulos (*Myxomycetes*: 420, 1969).

Symphytocarpus Ing & Nann.-Bremek. (1967), Stemonitidaceae. 9, widespread (north temperate). See Ing & Nannenga-Bremekamp (*Proc. K. Ned. Akad. Wet.* Ser. C, Biol. Med. Sci. **70**: 218, 1967), Ing (*The Myxomycetes of Britain and Ireland, An Identification Handbook*, 1999; British Isles), Adamonyte (*Botanica Lithuanica* **9**: 55, 2003; Lithuania).

Taeniella L. Léger & Duboscq (1911), Eccrinaceae. 1 (in *Decapoda*), widespread. See Hibbets (*Syesis* **11**: 213, 1978), Lichtwardt (*The Trichomycetes. Fungal associates of arthropods*, 1986), Cafaro (*Mol. Phylogen. Evol.* **35**: 21, 2005; phylogeny).

Taeniellopsis R.A. Poiss. (1927), Eccrinaceae. 3 (in *Amphipoda*), Europe. See Manier (*Annls Sci. Nat.* Bot., sér. 12 **10**: 565, 1969), Lichtwardt (*The Trichomycetes. Fungal associates of arthropods*, 1986; key).

Tetramyxa K.I. Goebel (1884), Plasmodiophoraceae. 5, widespread (north temperate). See Karling (*The Plasmodiophorales* edn 2, 1968; key), Dick (*Straminipilous Fungi*: 670 pp., 2001).

Thalassomyces Niezab. (1913), Protozoa. See Kane (*N.Z. Jl Sci.* **7**: 289, 1964), Karling (*Holocarpic Biflagellate Phycomycetes*: 204, 1981), Whisler *in* Margulis *et al.* (Eds) (*Handbook of Protoctista*: 715, 1990; placed in class *Ellobiopsida*).

Tilmadoche Fr. (1849) = Physarum fide Martin & Alexopoulos (*Myxomycetes*: 274, 1969).

Tipularia Chevall. (1822) [non *Tipularia* Nutt. 1818, *Orchidaceae*] ≡ Halterophora.

Trabrooksia H.W. Keller (1980), Didymiaceae. 1, USA; Ecuador. See Keller (*Mycol.* **72**: 396, 1980), Ing (*The Myxomycetes of Britain and Ireland, An Identification Handbook*, 1999; British Isles), McHugh (*Mycotaxon* **92**: 107, 2005; Ecuador).

Trichamphora Jungh. (1838) = Physarum fide Pando (*in litt.*).

Trichella Léger & Duboscq (1929) = Enterobryus fide Manier & Lichtwardt (*Annls Sci. Nat.* Bot., sér. 12 **9**: 519, 1968).

Trichellopsis Maessen (1955) = Enterobryus fide Manier & Lichtwardt (*Annls Sci. Nat.* Bot., sér. 12 **9**: 519, 1968).

Trichia Haller (1768), Trichiaceae. *c.* 35, widespread. See Ing (*The Myxomycetes of Britain and Ireland, An Identification Handbook*, 1999; British Isles), Adamonyte (*Mycotaxon* **87**: 379, 2003; coprophilous sp.), Stephenson (*Fungal Diversity Res. Ser.* **11**, 2003; New Zealand).

Trichiaceae Chevall. (1826), Trichiida (inc. sed.). 10 gen. (+ 8 syn.), 100 spp.

Trichiales T. Macbr. (1922). See *Trichiida*.

Trichidium Raf. (1815) ≡ Trichia.

Trichiida [*Trichiales* (orth. bot.)]. Myxogastria. 3 fam., 14 gen., 162 spp. Spore mass bright coloured, columella absent, peridium persistent. Fams:
(1) **Arcyriaceae**
(2) **Dianemataceae**
(3) **Trichiaceae**

Lit.: Blackwell & Busard (*Mycotaxon* **7**: 61, 1978; pigments in spp. separation).

Trichodermataceae Fr. (1825) = Hypocreaceae.

Trichulius Schmidel ex Corda (1842) ? = Trichia fide Mussat (*Syll. fung.* **15**: 425, 1901).

Tripotrichia Corda (1837) = Leocarpus fide Martin & Alexopoulos (*Myxomycetes*: 245, 1969).

Tubifera J.F. Gmel. (1792) nom. cons. prop., Tubiferaceae. 7, widespread. See Nelson *et al.* (*Mycol.* **74**: 541, 1982; key).

Tubiferaceae T. Macbr. (1899), Liceida (inc. sed.). 4 gen. (+ 14 syn.), 30 spp.

Tubulifera O.F. Müll. ex Jacq. (1778) = Tubifera fide Kirk (*in litt.*).

Tubulina Pers. (1794) = Tubifera fide Pando (*in litt.*).

Tubulinaceae Lister (1894) = Reticulariaceae.

Tychosporium Spiegel (1995), Protosteliaceae. 1, USA. See Spiegel *et al.* (*Mycol.* **87**: 265, 1995).

Vampyrella Cienk. (1865), Protozoa. 21 or. Filose amoeboid protist formerly treated as a proteomyxid or monad. See Patterson (*Amer. Nat.* **154**: S96, 1999, systematic position).

Vampyrellidium Zopf (1885), Protozoa. 1. Amoeboid protist formerly treated as a proteomyxid or monad. See Patterson (*Amer. Nat.* **154**: S96, 1999, systematic position].

Vampyrelloides Schepotieff (1911), Protozoa. Filose amoeboid protist formerly treated as a proteomyxid or monad. See Patterson (*Amer. Nat.* **154**: S96, 1999, systematic position).

Verrucosia Teng (1932) = Lycogala fide Martin & Alexopoulos (*Myxomycetes*: 62, 1969).

Wilczekia Meyl. (1925) = Diderma fide Kowalski (*Mycol.* **67**: 448, 1975).

Willkommlangea Kuntze (1891), Physaraceae. 1, widespread (north temperate). See McHugh & Reed (*MR* **94**: 710, 1990; plasmocarp formation), Ing (*The Myxomycetes of Britain and Ireland, An Identification Handbook*, 1999; British Isles), Stephenson (*Fungal Diversity Res. Ser.* **11**, 2003; New Zealand).

Wingina Kuntze (1891) ≡ Orthotricha.

Woronina Cornu (1872), Plasmodiophoraceae. 4 (in *Vaucheria*, *Saprolegniales*), widespread (north temperate). See Karling (*The Plasmodiophorales* edn 2, 1968; key), Dick (*Straminipilous Fungi*: 670 pp., 2001).

Woroninaceae H.E. Petersen (1909) = Plasmodiophoraceae.
Xyloidium Czern. (1845) nom. dub., ? Myxogastria. See Streinz (*Nom. fung.*, 1862).
Xyloon, see *Xyloidium*.